MODERN PHYSICS

MODERN PHYSICS

Second Edition

S.L. Kakani
M.Sc., Ph.D.

Shubhra Kakani
M.Sc., Ph.D.

New Delhi | Mumbai | Chennai | Kolkata | Bengaluru | Hyderabad | Kochi | Guwahati

Publisher's note

Every possible effort has been made to ensure that the information contained in thisbook is accurate at the time of going to press, and the publisher and authors cannot accept responsibily for any errors or omissions, however caused. No responsibility for loss or damage occasioned to ay person acting, or refraining from action, as a result of the material in this publication can be acceptd by the editor, the publisher or the authors.

Every effort has been made to trace the owners of copyright material used in this bok. The authors and the publisher will be grateful for any omission brought to their notice for acknowlegement in the future editions of the book.

First Published 2006
Reprinted 2009, 2011
Second Edition 2013

Viva Books Private Limited

- 4737/23, Ansari Road, Daryaganj, New Delhi 110 002
  Tel. 42242200, 23258325, 23283121, Email: vivadelhi@vivagroupindia.net
- 76, Service Industries, Shirvane, Sector 1, Nerul, Navi Mumbai 400 706
  Tel. 27721273, 27721274, Email: vivamumbai@vivagroupindia.net
- Jamals Fazal Chambers, 26 Greams Road, Chennai 600 006
  Tel. 28294241, 28290304, Email: vivachennai@vivagroupindia.net
- B-103, Jindal Towers, 21/1A/3 Darga Road, Kolkata 700 017
  Tel. 22816713, Email: vivakolkata@vivagroupindia.net
- 7, GF, Sovereign Park Aptts., 56-58, K. R. Road, Basavanagudi, Bengaluru 560 004
  Tel. 26607409, Email: vivabangalore@vivagroupindia.net
- 101-102, Moghal Marc Apartments, 3-4-637 to 641, Narayanguda,
  Hyderabad 500 029, Tel. 27564481, Email: vivahyderabad@vivagroupindia.net
- First Floor, Beevi Towers, SRM Road, Kaloor, Kochi 682 018
  Tel. 0484-2403055, 2403056, Email: vivakochi@vivagroupindia.net
- 232, GNB Road, Beside UCO Bank, Silpukhuri, Guwahati 781 003
  Tel. 0361-2666386, Email: vivaguwahati@vivagroupindia.net

www.vivagroupindia.com

ISBN: 978-81-309-2307-9

Published by Vinod Vasishtha for Viva Books Private Limited, 4737/23 Ansari Road, Daryaganj, New Delhi 110 002.

Printed and bound by Raj Press, R-3, Inderpuri, New Delhi 110 012.

G0000/G1400

CONTENTS

PREFACE TO THE SECOND EDITION

In presenting the second edition of "Modern Physics", We acknowledge with a deep sense of gratitude, the overwhelming response received from students and fellow teachers. We are thankful to all those who have given us feedback and suggestions on the book.

In this second edition we have added chapters on *Nanomaterials, Atomic Nucleus, Elementary Particles, General Theory of Relativity* and *Cosmology*. The book has been thoroughly revised, updated, corrected and good number of new problems have been added. Several chapters have been updated. Summary and Objective type questions have been added. The Summary provided at the end of each chapter will definitely help students in reviewing the entire chapter in a short time.

We will be happy to receive any suggestions for the further improvement of the book.

Bhilwara
28.10.2011

S.L. KAKANI
SHUBHRA KAKANI

PART A

Atomic and Basics of Quantum Physics

1

Special Theory of Relativity

1.1 INTRODUCTION

Probably no physical theory in 20th century has been the subject of more discussion amongst philosophers and scientists, and at the same time caught the imagination of the (intelligent) layman, than the Einstein's theory of relativity discovered in vintage year 1905. The year 2005, marking the centenary of that remarkable year, has been declared the International year of physics by organisations such as the UN and UNESCO. This is essentially due to the fact that the concepts underlying the theory of relativity are not only radically new but also provide a framework which embraces practically all the branches of the physical sciences. Indeed, Einstein's revolutionary theory revolutionized our understanding of the physical universe in major ways. We shall devote this chapter to a discussion of the special theory of relativity.

1.2 INERTIAL FRAMES OF REFERENCE

Classical mechanics is based on the conception that the free space is empty and the position of a point in this free space can be given in terms of three real numbers (x, y, z), which are called the coordinates of a point since the position of a point or body can be stated only relative to some other points or bodies, the description of these observations requires a coordinate system or a frame of reference. A frame of reference is a technical term for the combination of a set of spatial coordinate axes and a time variable.

Galileo and other scientists realised that the form of the laws of nature depends on the choice of the frame of reference. Among all the possible frames of reference, there exists a class called the **inertial frame of reference**, in which the laws of nature take a simple form. Inertial frames of reference are those frames in which a particle or a body that is not acted upon by external forces, moves with constant velocity with respect to each other, and one of them is inertial frame then the other frame will also be an inertial frame Fig. (1.1). The laws of mechanics remains invariant in all inertial frames.

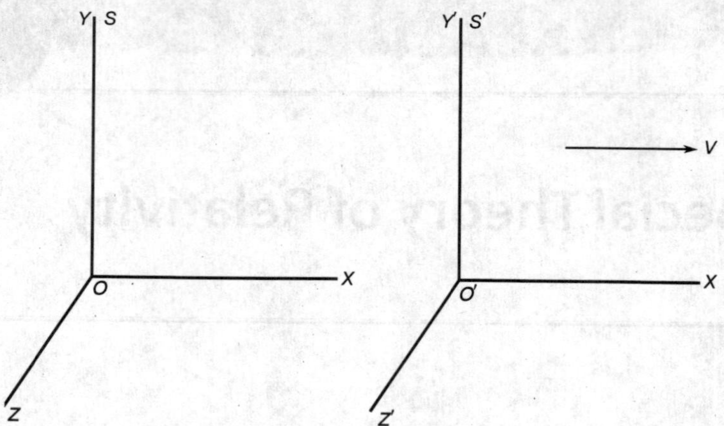

Fig. 1.1 Inertial frames of reference

No frame connected with any material object in the universe can serve the purpose of perfect inertial frame of reference for the earth and other planets as they revolve round the sun which itself is moving. Stars are also not fixed. Relation between the two coordinate systems is determined by the functional relation between their respective coordinates.

1.3 GALILEAN TRANSFORMATIONS

Consider two inertial frames of reference S and S', such that their coordinate axes coincide at time $t = 0$, and S' moves with velocity v along the x-axis w.r.t. S frame. Obviously, the relations between the coordinates, in the two frames can be expressed by the following equations

$$x' = x - vt$$
$$y' = y$$
$$z' = z$$
$$t' = t \tag{1.1}$$

These equations are called **Galilean transformations.** In obtaining Galilean transformation equations, it is assumed that (i) it is possible to define a time t which is the same for all inertial frames of reference, and (ii) the distance between two points is independent of frames of reference.

It is easy to show from Galilean transformations that the velocities and accelerations in the two frames of references are related by the equations:

$$\boldsymbol{u}' = \boldsymbol{u} - \mathbf{v} \tag{1.2}$$

where $\boldsymbol{v}$ is the relative velocity between two frames along the x-direction, and

$$\boldsymbol{a}' = \boldsymbol{a} \tag{1.3}$$

respectively. It is worth while to mention that the Galilean transformations are not consistent with Maxwell's electromagnetic theory.

1.4 VELOCITY OF LIGHT

It follows from Maxwell's electromagnetic theory that electromagnetic waves travel in vacuum with a speed equal to the ratio of electromagnetic unit to the electrostatic unit of charge. This ratio is essentially equal to the speed of light and hence light is considered as a form of electromagnetic radiation.

Now, the question arises, how does the velocity of light transform from one inertial frame to another? Galilean transformations reveal that the velocities are different in different frames of reference and are related by Eq. (1.2). However, Maxwell's equations for electromagnetic fields have no reference to the velocity of the inertial frame and hence imply that the speed of light is independent of the velocity of inertial frame. Michelson-Morley experiment also suggests that the speed of light is independent of the velocity of light.

1.5 THE SEARCH FOR THE ETHER

In spite of the great success of Maxwell's equations in describing the behaviour of electromagnetic phenomena in space and time, it was quite disturbing to find that they are not **covariant** under a Galilean transformation. Obviously, the Galilean transformation which permits a covariant description of mechanical forces, does not hold for electromagnetic forces.

Another serious difficulty arose from Maxwell's electromagnetic theory in connection with a medium, which was postulated to sustain wave character of light. We know that every wave motion has something that 'waves'. Sound waves have air and water waves have water. Surely it was argued that light waves must involve the waving of something even in free space. No one knew what it was, but it was given the name **luminiferous ether**.

Maxwell's equations and experimental observations, particularly on polarized light, show that light is a transverse wave motion. This implies that the ether is a solid. Transverse waves involve shear force and can occur only in solids, which can support shear. Obviously, the ether must be a rigid solid. The propagation velocity of mechanical waves in various materials depends on the elastic constants of the material. These are much greater for steel than for air. The very great velocity of light thus implies that they must have very large shear modulus. It is rather hard on the imagination to suppose that all space is filled with this rigid solid and the material objects move through this solid with resistance, yet it was supposed to exist. It was then natural to assume that the ether was in a state of absolute rest, the so-called 'stationary ether'. Accordingly, the ether becomes an **'absolute reference'**, through which the material bodies move without resistance.

If the ether is assumed to be at rest, then the interesting question is; how fast are we moving through the ether? Since all speculations about the ether stem from its properties as a medium for carrying light, an optical experiment is indicated. It is not very difficult to know how sensitive the apparatus must be required in order to measure the ether drift. Assuming, for the sake of simplicity, that the Sun has no ether drift, the velocity of the earth through the ether must be its orbital velocity. If the Sun has an ether drift, then the drift of the earth will be even greater than its orbital velocity for the same reasons.

Knowing that the radius of the earth's orbit is about 93 million miles, we find the orbital velocity to be about 18.5 miles/s (≈30 km/s). By performing the experiment at the best season of the year, we know that, we should be able to find an ether drift of at least 30 km/s. The velocity of light is 3×10^8 m/s. The orbital velocity of the earth is about 10^{-4} times the velocity of light. Obviously, we need a very sensitive experiment to detect the ether drift. A number of experiments were devised to measure the earth's motion through the ether, i.e., to detect an ether wind with respect to the earth. The most famous of these experiments was the **Michelson-Morley experiment**, which will be described below.

1.6 MICHELSON-MORLEY EXPERIMENT

Michelson, who became the First American to win a Nobel Prize in Physics in 1907, invented a new instrument of unprecedented sensitivity to determine the velocity of the earth through the ether, i.e., a medium for it to move in, then light should be dragged along by this ether as the earth moves along through space. In principle, the ether drift test consists simply of observing whether there is any shift of the interference fringes of light in Michelson interferometer when the entire instrument is turned through an angle of 90°.

The experimental arrangement is shown in Fig. (1.2). A monochromatic beam of light falls on a semi-silvered glass plate P placed at 45° to the beam and is partly reflected and partly transmitted. The reflected portion travels in a direction at right angles to that of the incident beam, falls normally at A on the plane mirror M_1, and is reflected back to P. The transmitted ray, travelling along the direction of the incident ray, falls normally at B on the plane mirror M_2 and is reflected back to P. The two beams, thus turned to P, interfere on their final journey towards the telescope T so that an interference pattern can be observed and studied. The beam coming from S and entering into T after being reflected from M_1 traverses the plate P thrice while that reflected from M_2 traverses the plate P only once. Thus the optical paths of two interfering beams are not equal. They are made equal by introducing a compensating plate (not shown in Fig. (1.2)) of the same thickness and material as P between P and M_2. Whole of the apparatus is floated on mercury, contained in a large vessel so that the interferometer may be rotated in a desired direction.

If the whole apparatus were at rest in ether the two rays would take the same time to return to P (distance PA and PB are kept equal). But in the actual experiment the whole apparatus is moving with the earth. Let us assume that the direction of motion of the earth coincides with the direction of the incident beam of light (in the short time of an experiment, this velocity can be considered as uniform and linear, since the earth makes one complete rotation round the Sun in one year). On account of motion at the apparatus along with the earth in the direction PM_2 with velocity v relative to the ether the paths of the two rays and the positions of their reflections from the mirrors will be as shown in Fig. (1.3). After reflection at P the ray proceeds towards the mirror M_1. In the time the ray reaches the mirror, it moves to the position M_1' so that the reflection occurs at this position

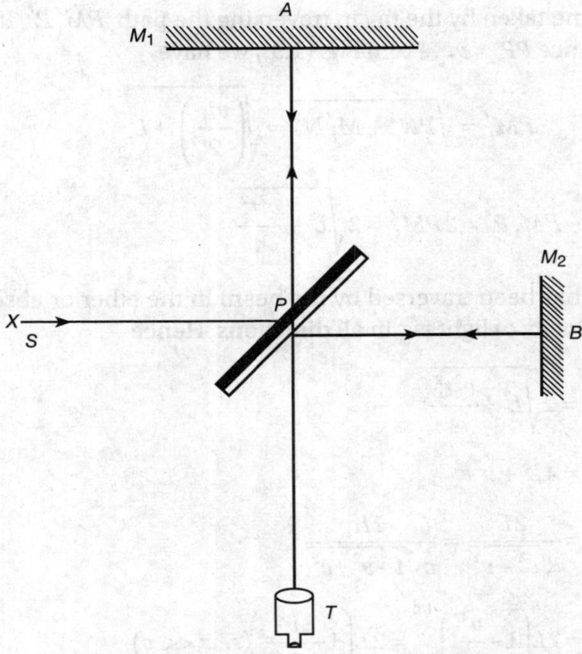

Fig. 1.2 Schematic diagram of Michelson-Morely experiment

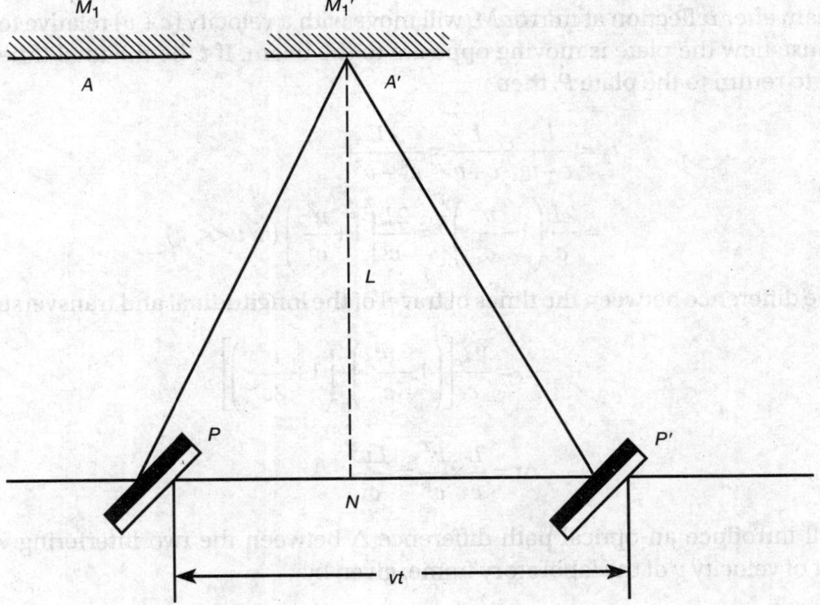

Fig. 1.3

of mirror M_1. If, t_1 is time taken by the ray in traversing the path $PM_1'P_1'$, in the meantime plate P will move distance $PP' = vt_1$. From Fig. (1.3), we have

$$PM_l' = \sqrt{PN^2 + M_l'N^2} = \sqrt{\left(\frac{vt_1}{2}\right)^2 + L^2}$$

$$\therefore \quad \text{Distance} \quad PM_1'P_1' = 2PM_1' = 2\sqrt{L^2 + \frac{v^2 t_1^2}{4}}$$

This distance $PM_1'P_1'$ has been traversed by the beam in the ether or absolute or inertial frame, in which the velocity of light is c in all directions. Hence

$$ct_1 = 2\sqrt{L^2 + \frac{v^2 t_1^2}{4}}$$

or

$$c^2 t_1^2 = 4L^2 + v^2 t_1^2$$

or

$$t_1 = \frac{2L}{\sqrt{c^2 - v^2}} = \frac{2L}{c\sqrt{1 - v^2/c^2}}$$

$$= 2L\left(1 - \frac{v^2}{c^2}\right)^{-1/2} = 2L\left(1 + \frac{v^2}{2c^2}\right) (\therefore v << c) \tag{1.4}$$

The beam transmitted through P and moving longitudinally towards M_2 has velocity $(c - v)$ relative to the mirror M_2 because the mirror is moving with the velocity of the earth (v). This beam after reflection at mirror M_2 will move with a velocity $(c + v)$ relative to the plate P, because now the plate is moving opposite to the beam. If t_2 be the total time taken by this ray to return to the plate P, then

$$t_2 = \frac{L}{c - v} + \frac{L}{c + v} = \frac{2L}{c^2 - v^2}$$

$$= \frac{2L}{c}\left(1 - \frac{v^2}{c^2}\right)^{-1} = \frac{2L}{c}\left(1 + \frac{v^2}{c^2}\right) (\because v << c) \tag{1.5}$$

Thus the difference between the times of travel of the longitudinal and transverse beams is

$$t_2 - t_1 = \frac{2L}{c}\left[\left(1 - \frac{v^2}{c^2}\right) - \left(1 + \frac{v^2}{2c^2}\right)\right]$$

or

$$\Delta t = \frac{L}{c} \times \frac{v^2}{c^2} = \frac{Lv^2}{c^3} \tag{1.6}$$

This will introduce an optical path difference Δ between the two interfering waves, on account of velocity v of the laboratory frame, given by

$$\Delta = c(t_2 - t_1) = c\Delta t = L\frac{v^2}{c^2} \tag{1.7}$$

The interference pattern produced by the two components when they recombine will thus shift from the position it would have occupied if v were equal to zero. This is the principle of Michelson-Morley experiment. If Δ corresponds to the shifting of n fringes, then

$$\Delta = n\lambda \tag{1.8}$$

where λ is the wavelength of the light used. Equating Eqs. (1.7) and (1.8) for Δ, one obtains

$$n = \frac{c\Delta t}{\lambda} = \frac{Lv^2}{\lambda c^2} \tag{1.9}$$

The path difference A is made large by making the total path traversed by the component beams long by multiple reflections. This way of elongating the paths is more convenient from the point of view of temperature control and mechanical rigidity than a direct increase in the length of the arms of the interferometer. To minimise the error due to vibrations, the whole apparatus is mounted on a block of stone resting on a special wooden ring which floats on mercury.

Now, if the whole apparatus is turned through 90° so that the other arm PM_1 becomes coincident with v. This causes the difference of path in the opposite direction and hence the displacement of fringes should be $2Lv^2/c^2$ corresponding to a path difference 2Δ. This procedure is adopted in experiments, since the different estimation of Δ from Eq. (1.7) involves the determination of the shift of the fringe system from its position under the condition $v = 0$ which cannot be realised. In actual experiment, distance L was taken nearly 11 m, the wavelength of light used was about 6000 Å. Taking $v = 3 \times 10^4$ m/s and $c = 3 \times 10^8$ m/s, one obtains,

$$2\Delta = \frac{2\times 11\times(3\times10^4)^2}{6\times10^{-7}\times(3\times10^8)^2}$$

$$= \frac{11}{3}\times\frac{1}{10} = 0.37 \text{ of a fringe} \simeq 0.4 \text{ fringe}$$

Thus a shift of about 0.4 fringe is expected in the fringe pattern on turning the apparatus by 90°. Michelson-Morley experiment was sensitive enough to detect a fringe shift of the order of 0.01 fringe, but in spite of taking into consideration all experimental errors no fringe shift was observed. After Michelson and Morley the experiment has been repeated with different wavelengths of light, using laser beam, at different places on the earth, at different altitudes, below the surface of the earth, etc., but the same negative result has been obtained every time. The negative results of Michelson-Morley experiment has following two consequences:

(i) The speed of light is the same in all directions.
(ii) The assumption of the **luminiferous ether** is not true as motion relative to it is undetectable.

Michelson tried to explain the negative result by assuming that the ether is dragged by the moving bodies, so that velocity of light is same in all directions. This idea was in contradiction to the original assumption of the ether as an all- pervasive, frictionless medium. Aberration experiment also contradicts this hypothesis.

1.6.1 Lorentz and Fitzgerald Hypothesis

Lorentz and Fitzgerald independently put forward another explanation to preserve the concept of ether frame. According to this hypothesis, along the direction of motion through the ether the size of all material bodies contracts (Lorentz-Fitzgerald contraction). Thus if we suppose that the distance in the direction of the motion of earth is shortened to l', such that

$$l' = l\sqrt{1-\frac{v^2}{c^2}} \tag{1.10}$$

Thus
$$t_1 = \frac{2L\left(1-\frac{v^2}{c^2}\right)^{1/2}}{c\left(1-\frac{v^2}{c^2}\right)} = \frac{2L}{c\sqrt{\left(1-\frac{v^2}{c^2}\right)}} \tag{1.11}$$

Thus t_1 and t_2 will have same value or

$$\Delta t = t_1 - t_2 = 0$$

No positive experimental confirmation could be provided for Lorentz-Fitzgerald hypothesis. The idea of such a contraction was later found to be correct, although the assumption of the presence of lumineferous ether was not found correct.

Example 1.1: In actual Michelson-Morley experiment, the total distance from the partially silvered mirror to each of the two mirrors was 10 meters. The wavelength of the light used was 5000 A°. If the orbital velocity of the earth is taken as 30 km/s, calculate the expected total fringe shift when the apparatus is rotated through 90°.

Solution Expected fringe shift $(\delta) = \frac{2Lv^2}{\lambda c^2}$

$$(\delta) = \frac{2\times10\times(30\times10^3)^2}{5\times10^{-7}\times(3\times10^8)^2} \qquad \begin{array}{|l} L = 10\,\text{m} \\ \lambda = 5000\text{Å} = 5000\times10^{-10}\,\text{m} \\ c = 3\times10^8\,\text{m/s} \\ v = 30\times10^3\,\text{m/s} \end{array}$$

$$= \frac{2}{5} = 0.4\,\text{fringe.}$$

Example 1.2: What will be expected fringe shift in Michelson Morley experiment, if the effective length of each part is 6 meters and wave length of light used is 6000 Å? Velocity of earth is 3×10^4 m/s.

Solution Expected fringe shift $\delta = \frac{2Lv^2}{c^2\lambda}$

$$\delta = \frac{2\times 6\times(3\times10^4)^2}{(3\times10^8)^2\times 6\times10^{-7}} \quad \left| \begin{array}{l} L = 6\text{m} \\ v = 3\times10^4\,\text{m/s} \\ \lambda = 6000\times10^{-10}\,\text{m} \\ c = 3\times10^8\,\text{m/s} \end{array} \right.$$

$$= \frac{1}{5}\text{ of a fringe}$$

Example 1.3: **In Michelson-Morley experiment the length of the paths of the two beams is 11 meters each. The wavelength of the light used is 6000 Å. The expected fringe shift is 0.4 fringe, calculate the velocity of earth relative to ether.**

Solution Expected fringe shift $\delta = \frac{2Lv^2}{c^2\lambda}$

$$\therefore \quad 0.4 = \frac{2\times 11\times v^2}{\left(3\times10^8\right)^2\times 6\times10^{-7}} \quad \left| \begin{array}{l} v = 3\times10^4\,\text{m/s} \\ \delta = 0.4\,\text{fringe} \\ L = 11\text{m} \\ c = 3\times10^8\,\text{m/s} \\ \lambda = 6000\times10^{-10}\,\text{m} \end{array} \right.$$

or $$v^2 = \frac{0.4\times(30\times10^8)^2\times 6\times10^{-7}}{2\times 11}$$

or $$v = \left[\frac{0.4\times(30\times10^8)^2\times 6\times10^{-7}}{2\times 11}\right]^{1/2} \simeq 1.2\times10^5\,\text{m/s}$$

1.7 EINSTEIN'S SPECIAL THEORY OF RELATIVITY

The failure of such efforts as the Michelson-Morley and others to discover a preferred frame for Maxwell's electromagnetic equations suggested that the latter must confirm to a principle of relativity. However, the fact that the Galilean principle of relativity-which was known to be valid for classical mechanics- failed for Maxwell's equations was a considerable source of frustration to physicists at the turn of the century. After a critical examination of the concepts of space, time and simultaneity, Einstein discarded the Galilean principle of relativity and postulated instead a principle of relativity for all physical laws. His postulates may be stated as follows:

(i) **The laws of physics have the same form in all inertial frames of reference moving with a constant velocity with respect-to one another.**

Obviously, all physical phenomena (mechanical, electromagnetic, etc.) proceed in exactly the same way, under the same conditions, in all inertial frames of reference, or, in other words, it is impossible to ascertain, by means of any experiments whatsoever, conducted in a closed system of bodies, whether the system is at rest or is travelling at uniform velocity in a straight line with respect to some inertial frame of reference.

(ii) **The speed of light in vacuum is the same for all observers who are in uniform, rectilinear, relative motion and is independent of the motion of the source. Its free space value is the universal constant given by Maxwell's equations.**

This postulate follows directly from the results of the Michelson-Morley experiment and many others.

Of these two postulates, the second represents an experimental fact, where as the first is a generalization from a wide range of physical experience. The first postulate is in no way self-evident; it is a hypothesis to be tested by experiment.

The above mentioned postulates are applicable to only inertial frames and due to this speciality the principle of relativity as applied to inertial frames is called special relativity. Einstein later generalized special relativity to include non-inertial frames also and the generalized theory is called **General Relativity**.

1.8 LORENTZ TRANSFORMATIONS

Einstein reexamined the concepts of space and time on the basis of two postulates. An observer at rest at the origin of a coordinate system *S* can specify any event by giving the coordinates (x, y, z) in *S* of the place where the event occurs and the time *t* in S′ when the event occurs. The same event can be specified by another observer at rest at the origin of a coordinate system *S′* which is in uniform rectilinear motion relative to *S*, by giving the coordinates (x', y', z') and the time t' of the event in *S′*. The algebraic equations connecting the two sets of coordinates and time are called the **transformation equations.**

Let us assume for our convenience that origin *O* and *O′* of the two inertial frames *S* and *S′* coincide at $t = t' = 0$. Now, consider a light signal emitted in free space at the instant $t = 0$ from the coincident origins of the systems *S* and *S′*. When the light signal reaches the point, *P*, let the position and times measured by the observers at *O* and *O′* be (x, y, z, t) and (x', y', z', t') respectively Fig. (1.4). If the velocity of light signal in vacuum is *c*, then the time required by the light signal to traverse the distance *OP* in frame *S* is

$$t = \frac{OP}{c} = \frac{\sqrt{x^2 + y^2 + z^2}}{c}$$

or
$$x^2 + y^2 + z^2 - c^2t^2 = 0 \tag{1.12}$$

The time required by the same light signal in trawelling the distance *O′P* in *S′* frame, as seen by an observer at rest at the origin of *S′* is

$$t' = \frac{O'P}{c} = \frac{\sqrt{x'^2 + y'^2 + z'^2}}{c}$$

or
$$x'^2 + y'^2 + z'^2 - c^2t'^2 = 0 \tag{1.13}$$

The velocity of light (*c*) is same in both the inertial frames of references.

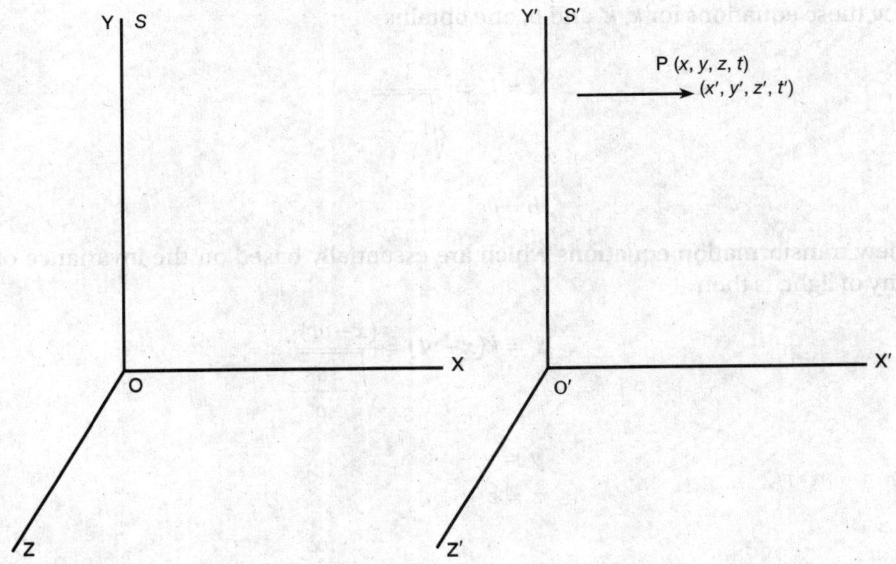

Fig. 1.4

The Galilean transformation connects measurement in the two frames according to Eq. (1.1). Using Eq. (1.1), Eq. (1.13) becomes $x^2 - 2xvt + vt^2 + y^2 + z^2 = c^2t^2$ and this is not in agreement with Eq. (1.13) in any way. This clearly indicates that the Galilean transformation fails. Therefore, we want new transformation equations which transform Eq. (1.13) to Eq. (1.12). The transformation must be linear in x and t, because we want to get a sphere which expands at a uniform rate. Let us try the relations:

$$\begin{aligned} x' &= k(x - vt) \\ y' &= y \\ z' &= z \text{ and } t' = k'(t - bx) \end{aligned} \tag{1.14}$$

Here k, k' and b are constants. Making these substitutions in Eq. (1.13), one obtains

$$k^2(x^2 - 2xvt + v^2t^2) + y^2 + z^2 = c^2k'^2(xt^2 - 2bxt + b^2x^2)$$

or

$$(k^2 - k'^2c^2b^2)x^2 - 2(k^2v + k'^2c^2b)xt + y^2 + z^2 - \left(k'^2 - \frac{k^2v^2}{c^2}\right)c^2t^2 = 0$$

This must be identical to Eq. (1.12). Equating the above with Eq. (1.12), one obtains

$$k^2 - k'^2b^2c^2 = 1$$

$$k^2v^2 - k'^2bc^2 = 0$$

$$k'^2 - \frac{k^2v^2}{c^2} = 1$$

Solving these equations for k, k' and b, one obtains

$$k = k' = \frac{1}{\sqrt{1-\frac{v^2}{c^2}}}$$

and
$$b = vc^2$$

The new transformation equations which are essentially based on the invariance of the velocity of light, is then

$$x' = k(x-vt) = \frac{(x-vt)}{\sqrt{1-\frac{v^2}{c^2}}}$$

$$y' = y$$

$$z' = z$$

$$t' = k(t-bx) = \frac{t-\frac{vx}{c^2}}{\sqrt{1-\frac{v^2}{c^2}}} \tag{1.15}$$

This set of transformation equations is known as the **Lorentz transformations**. These are symmetrical and retain their form in transformation from S' to S except that the sign of v changes. Thus

$$x = \frac{x'+vt'}{\sqrt{1-\frac{v^2}{c^2}}}$$

$$y = y'$$

$$z = z'$$

$$t = \frac{\left(t'+\frac{vx}{c^2}\right)}{\sqrt{1-\frac{v^2}{c^2}}} \tag{1.16}$$

These are inverse transformations. The Lorentz transformations are linear and at low velocities $\left(\frac{v}{c} << 1\right)$, they become Galilean transformations.

From Lorentz transformation equations, one can easily confirm the statement that events which happen at the same place at different times, as viewed from one frame, may be seen from another frame to happen at a different place as well. Similarly, a difference in spatial position with respect to one frame may correspond to a difference in both space and time with respect to another. Thus a space difference can be converted partly into a time difference or vice versa, merely by changing the frame of reference that is being used. Thus Lorentz transformation represent a most profound conceptual change specially with regard to space and time. Writing, $v/c = b$ and

and $$\left(1-\frac{v^2}{c^2}\right)^{-1/2} = \gamma \tag{1.17}$$

Lorentz transformation equations can be expressed as

s to s'	s' to s
$x' = \gamma(x - vt)$	$x = \gamma(x' + vt')$
$y' = y$	$y = y'$
$z' = z$	$z = z'$
$t' = \gamma(t - vx/c^2)$	$t = \gamma(t' + vx'/c^2)$
	where $\gamma = 1/\sqrt{1-v^2/c^2}$

(1.18)

To understand the consequences of Lorentz transformations we define an event, which is an occurrence taking place at a point (x, y, z) and at a time t. Thus a point in the space-time complex, i.e., (x, y, z, t) is an event. If we consider two events in S defined by (x_1, y_1, z_1, t_1) and (x_2, y_2, z_2, t_2), then the two events are called **simultaneous** when $\Delta t = t_2 - t_1 = 0$, they are called **colocal** when $\Delta x = x_2 - x_1 = 0$, $\Delta y = y_2 - y_1 = 0$ and $\Delta z = z_2 - z_1 = 0$ and when the two events take place at the same position ($\Delta x = \Delta y = \Delta z = 0$) and at the same time, i.e., $\Delta t = 0$, they are called **coincident.** The Lorentz transformations reveal that if two events are colocal in one inertial frame they may not be so in another frame (relativity of colocality) and if two events are simultaneous in one inertial frame they may not be observed at the same time in another inertial frame. This is called the **relativity of simultaneity.** However, if two events are coincident in one inertial frame, they will be coincident in all inertial frames. One can easily check the above results with the help of following relations. These have been obtained by differentiating Eq. (1.15).

$$\Delta x' = \frac{\Delta x - v\Delta t}{\sqrt{1-v^2/c^2}}$$

$$\Delta y' = \Delta y$$

$$\Delta z' = \Delta z$$

and $$\Delta t' = \frac{\Delta t - \dfrac{v\Delta x}{c^2}}{\sqrt{1-\dfrac{v^2}{c^2}}} \tag{1.19}$$

Obviously, if $\Delta x = \Delta y = \Delta z = 0$ but $\Delta t \neq 0$, then $\Delta x'$, $\Delta y'$ and $\Delta z'$ are not zero; if $\Delta t = 0$ but Δx, Δy, $\Delta z \neq 0$, then $\Delta t' = 0$; if $\Delta x = \Delta y = \Delta z = 0$ and also $\Delta t = 0$ then $\Delta x' = \Delta y' = \Delta z' = 0$ and $\Delta t' = 0$.

1.9 CONSEQUENCES OF LORENTZ TRANSFORMATIONS

1.9.1 Length Contraction

Consider a rod lying along the x-axis of a inertial frame S. Let the coordinates of the ends of the rod be x_1 and x_2 Fig. (1.5). Then the length L_0 of the rod in the inertial frame S in which it is at rest is given by

$$L_0 = x_2 - x_1 \tag{1.20}$$

Let S' be another inertial frame with its x-axis along that of S and y and z-axes parallel to the corresponding axes of S. Let S' be moving with a velocity v relative to S along x-axis. Let x_1' and x_2' be the coordinates of the ends of the rod as measured in S'. Then by Lorentz transformation equations, one obtains

$$x_1 = \frac{x_1' + vt'}{\sqrt{1-\frac{v^2}{c^2}}}$$

and

$$x_2 = \frac{x_2' + vt'}{\sqrt{1-\frac{v^2}{c^2}}} \tag{1.21}$$

Subtracting one from the other, we obtain

$$x_2 - x_1 = L_0 = x_1 = \frac{x_2' - x_1'}{\sqrt{1-\frac{v^2}{c^2}}} = \frac{L}{\sqrt{1-\frac{v^2}{c^2}}} \tag{1.22}$$

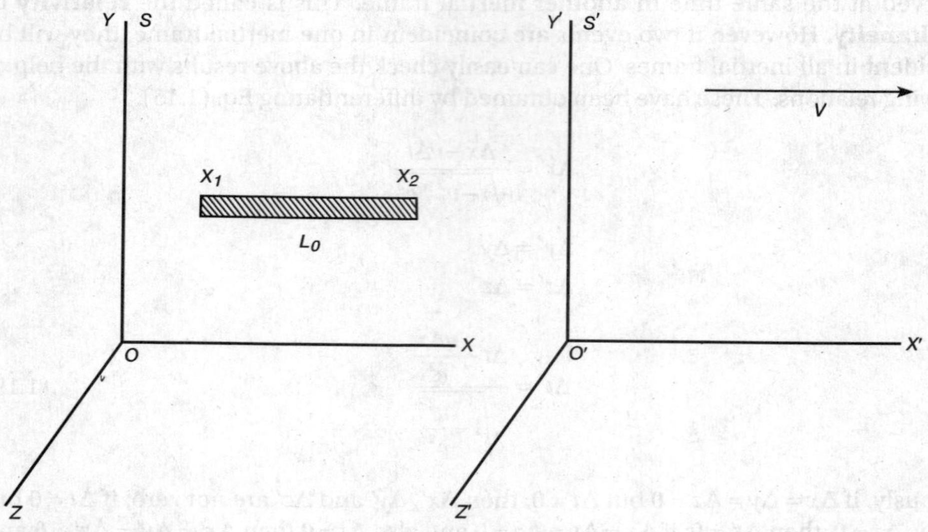

Fig. 1.5

where L is the length of the rod as measured in S'.

$$L = L_0\sqrt{1-\frac{v^2}{c^2}} \tag{1.23}$$

Obviously, $L < L_0$. Thus for an observer in motion with respect to an object the object appears to be shorter in the direction of motion than when he is at rest. It is important to note that Lorentz-Fitzgerald contraction effect is mutually reciprocal between different observers. The Lorentz contraction maybe observed, but, not 'seen' (since the eye and the instantaneous cameras record pictures formed by photons that arrive together).

1.9.2 Time Dilation: A Moving Clock Ticks More Slowly than a Clock at Rest

Consider a clock at a point x' in an inertial frame S'. Let S' be moving with a velocity v with respect to another inertial frame S along the common x-axis. Let a clock be situated in frame S at x. Let clock in S frame give out signals at two instants of time t_1 and t_2 as measured by an observer in S. Let an observer in frame S' measure these instants of time as t_1' and t_2' with his own clock. Then from Lorentz transformation equations, we obtain

$$t_1' = \frac{t_1 - \dfrac{vx}{c^2}}{\sqrt{1-\dfrac{v^2}{c^2}}}$$

$$t_2' = \frac{t_2 - \dfrac{vx}{c^2}}{\sqrt{1-\dfrac{v^2}{c^2}}} \tag{1.24}$$

Subtracting one from the other, we obtain

$$t_2' - t_1' = \frac{t_2 - t_1}{\sqrt{1-\dfrac{v^2}{c^2}}} \tag{1.25}$$

Letting $t_2' - t_1' = \Delta t$ and $t_2 - t_1 = \Delta t_0$, we have

$$\Delta t = \frac{\Delta t_0}{\sqrt{1-\dfrac{v^2}{c^2}}} \tag{1.26}$$

or

$$\Delta t_0 = \Delta t\sqrt{1-\frac{v^2}{c^2}} \tag{1.27}$$

For the observer in S frame the time interval between the two signals is $\Delta t = t_2 - t_1$ and for the observer in S' the time interval between the same two signals is $\Delta t' = t_2' - t_1'$. Eq. (1.26) reveals that Δt is larger than $\Delta t'$. Obviously, for the moving observer the time interval appears to be **elongated** or **dilated**. This phenomenon is called **time dilation**.

1.10 PROPER FRAME, PROPER LENGTH AND PROPER TIME

The inertial frame of reference in which the observed body is at rest is called the proper frame. The length of a rod as measured in the inertial frame in which it is at rest is called the **proper length**. The relation between the proper length (L_0) and the apparent or non-proper length (L) is as follows:

$$L = L_0\sqrt{1-\frac{v^2}{c^2}} \tag{1.28}$$

The time interval recorded by a clock fixed with respect to the observed event is called the **proper time interval.** The relation between the **proper time** (t_0) and apparent or non-proper time (t) is as follows:

$$t = t_0\bigg/\sqrt{1-\frac{v^2}{c^2}} \tag{1.29}$$

Obviously, **proper time** noted by a moving observer is always less than the corresponding interval of time in a stationary frame. A stationary observer finds that a moving clock runs slower than a stationary one. An event which repeats itself with a certain period in S' will appear to have a longer period when observed from frame S. It is important to note that this effect is mutually reciprocal between two observers.

The relation (1.29) suggests that any physical process occurring in S' appears, when viewed from S, to have slowed down compared with an identical process occurring in S. Thus spectrum of radiation emitted by atoms moving relative to a spectroscope will appear shifted towards the longer wavelength or the red side of the corresponding spectrum emitted by atoms which are at rest relative to the spectroscope. This effect due to time dilation has to be considered while considering **Doppler-shift in radiation.**

1.11 EXPERIMENTAL VERIFICATION OF TIME DILATION

μ-meson Decay: Time dilation has been verified in experiments on a nuclear particle, called μ-meson. Fast moving μ-mesons, are created in the cosmic rays at a height of about 10 kilometers from the surface of the earth and reach the earth in large numbers. These μ-mesons have a typical speed of 2.994×10^8 m/s, which is 0.998 of the speed of light c. A μ-meson is found to have an average life-time of 2×10^{-6} s after which it decays into an electron. Obviously, a μ-meson in its life-time can travel a distance of only $2.994 \times 10^8 \text{ m/s} \times 2 \times 10^{-6} \text{ s} \simeq 600$ m or 0.6 km. Then the question arises, how do μ-mesons travel a distance of 10 km to reach the earth? Rossi and Hall in 1941 attributed this result to the time-dilation effect. The μ-meson has a life time t_0 ($\approx 2 \times 10^{-6}$ s) in its own frame of reference. In the observer's frame of reference on the earth, however, the life time is lengthened owing to the relative motion, to the value t given by

$$t = \frac{t_0}{\sqrt{1-\frac{v^2}{c^2}}} = \frac{2\times10^{-6}}{\sqrt{1-(0.998)^2}}$$

$$= \frac{2\times10^{-6}}{0.063} = 3.17\times10^{-5}\,\text{s}$$

In 3.17×10^{-5} s, a meson whose speed is 0.998C(= 2.994×10^{8} m/s) can travel a distance

$$2.994 \times 10^{8} \times 3.17 \times 10^{-5} \simeq 9500\text{ m} \simeq 9.5\text{ km}$$

Hence, despite their brief life-time it is possible for the μ-mesons to reach the ground from the large altitudes at which they are actually formed. More recently, the dilation caused by the thermal vibration of the nuclei in certain crystals has also been verified.

A similar experiment was done with pions by Ayres in 1971. The proper life time, measured for point at rest, is known to be 26 μs. In one experiment, pions were observed in flight at $u/c = 0.913$, and their life-time was measured to be 63.7 μs (points decay to muons, so one can follow the exponential radioactive decay of the pions by observing the muons emitted as a result of decay). For pion moving at this speed, the expected dilated lifetime is in exact agreement with the measured value, once again confirming the time dilation effect.

Example 1.4: What will be the apparent length of a meter stick measured by an observer at rest, when the stick is moving along its velocity equal to $\frac{\sqrt{3}}{2}c$.

Solution We have

$$L = L_0\sqrt{1-\frac{v^2}{c^2}} \qquad \left|\; \begin{aligned} v &= \frac{\sqrt{3}}{2}c \\ L_0 &= 1\text{m} \end{aligned}\right.$$

$$\therefore \qquad L = 1\sqrt{1-\frac{3}{4}} = \sqrt{\frac{1}{4}} = 0.5\,\text{m}$$

Example 1.5: A light pulse is emitted at the origin of a frame of reference S' at time $t' = 0$. Its distance x' from the origin after a time t' is given by $x'^2 = c^2t'^2$. Use the Lorentz transformation to transform this equation to an equation in x and t and show that this is $x^2 = c^2t^2$. Discuss the implication of this result.

Solution $x'^2 = c^2t'^2$

From Lorentz transformation, we have

$$x' = \frac{x - vt}{\sqrt{1-\frac{v^2}{c^2}}} \quad \text{and} \quad t' = \frac{t - \frac{xv}{c^2}}{\sqrt{1-\frac{v^2}{c^2}}}$$

$$\therefore \qquad \frac{(x-vt)^2}{1-\frac{v^2}{c^2}} = \frac{c^2\left\{t - \frac{xv}{c^2}\right\}^2}{1-\frac{v^2}{c^2}}$$

or
$$(x-vt)^2 = c^2\left(t^2 + \frac{x^2v^2}{c^4} + \frac{2xtv}{c^2}\right)$$

or
$$x^2 + v^2t^2 - 2xtv = c^2t^2 + \frac{x^2v^2}{c^2} - 2xtv$$

or
$$x^2 - c^2t^2 - v^2t^2 - \frac{x^2v^2}{c^2} = 0$$

or
$$(x^2 - c^2t^2) - \frac{v^2}{c^2}(x^2 - c^2t^2) = 0$$

or
$$(x^2 - c^2t^2)\left(1 - \frac{v^2}{c^2}\right) = 0$$

Since
$$v \neq c \text{ and } \left(1 - \frac{v^2}{c^2}\right) \neq 0$$

$\therefore$
$$x^2 - c^2t^2 = 0$$
$$x^2 = c^2t^2$$

Obviously, the above result shows that the velocity of light is an absolute constant and independent of the frame of reference.

Example 1.6: **Calculate the length and the orientation of a rod of length 5 m in a frame of reference which is moving with a velocity equal to 0.6 c, in a direction making an angle of 30° with the rod.**

Solution **The proper length of the rod along the direction of motion of the frame is $L_{x_0} = 5\cos 30^\circ$ meter, and that perpendicular to this direction is $L_{y_0} = 5\sin 30^\circ$ meter.**

The apparent length along the moving frame is given by

$$L_x = L_{x_0}\sqrt{1 - \frac{v^2}{c^2}} \qquad \begin{aligned} v &= 0.6c \\ L_{x_0} &= 5\cos 30^\circ \\ &= \frac{5\sqrt{3}}{2} \end{aligned}$$

$$= \frac{5\sqrt{3}}{2}\sqrt{1-(.6)^2}$$

$$= \frac{5\sqrt{3}}{2} \times 0.8 = 2\sqrt{3}\text{ m}$$

The length perpendicular to the direction of motion remains the same,

i.e.,
$$L_y = L_{y_0} = 5\sin 30^\circ = \frac{5}{2} = 2.5\text{ m}$$

∴ Length observed in the moving frame

$$L = \sqrt{(L_{x^2} + L_{y^2})} = \sqrt{\left(12 + \frac{25}{4}\right)} = 4.27\,\text{m}$$

If the rod appears to make an angle θ with the direction of motion, then

$$\tan\theta = \frac{L_y}{L_x} = \frac{\frac{5}{2}}{2\sqrt{3}} = \frac{5}{4\sqrt{3}} = 0.72$$

Example 1.7: **Calculate the mean life of a burst of π^+ mesons travelling with $v = 0.3\,c$, if the proper mean life time is 2.5×10^{-7}s?**

Solution We have

$$\tau = \frac{\tau_0}{\sqrt{1 - \frac{v^2}{c^2}}} = \frac{2.5 \times 10^{-7}}{\sqrt{1 - (0.73)^2}} \quad \left| \begin{array}{l} v = 0.73c \\ \tau_0 = 2.5 \times 10^{-7}\,\text{s} \end{array} \right.$$

$$= \frac{2.5 \times 10^{-7}}{\sqrt{0.4671}} = 3.6 \times 10^{-8}\,\text{s}$$

Example 1.8: **Prove that $x^2 + y^2 + z^2 = c^2t^2$ is invariant under Lorentz transformation.**

Solution By Lorentz transformation, we have

$$x = \frac{x' + vt'}{\sqrt{1 - \frac{v^2}{c^2}}}, \quad y = y', \quad z = z'$$

$$t = \frac{t' + \frac{x'v}{c^2}}{\sqrt{1 - \frac{v^2}{c^2}}}$$

On making these substitution in the given expression, $x^2 + y^2 + z^2 = c^2t^2$, we obtain

$$\frac{(x' + vt')^2}{1 - \frac{v^2}{c^2}} + y'^2 + z'^2 = \frac{c^2\left[t' + \frac{x'v}{c^2}\right]^2}{1 - \frac{v^2}{c^2}}$$

or

$$\frac{x'^2 + v^2t'^2 + 2x't'v}{1 - \frac{v^2}{c^2}} + y'^2 + z'^2 = \frac{c^2t'^2 + \frac{x'^2v^2}{c^2} + 2x'vt'}{1 - \frac{v^2}{c^2}}$$

or $$y'^2+z'^2=\frac{1}{1-\frac{v^2}{c^2}}\left[\left(c^2t'^2-x'^2\right)\left(1-\frac{v^2}{c^2}\right)\right]=c^2t'^2-x'^2$$

∴ $$x'^2+y'^2+z'^2=c^2t'^2$$

Thus $$x^2+y^2+z^2-c^2t^2=x'^2+y'^2+z'^2-c^2t'^2$$

Clearly $x^2+y^2+z^2-c^2t^2=0$ is invariant under Lorentz transformation.

Example 1.9: **The length of a rocket is measured as 100 m before launching. During the flight apparent length is found to be 96 m when measured from the launching station. Calculate its speed.**

Solution

Proper length of the rocket $L_0=100$ m

Apparent length $L=96$ m

∴ $$L=L_0=\sqrt{1-\frac{v^2}{c^2}}$$

∴ $$96=100\sqrt{1-\frac{v^2}{c^2}}$$

or $$1-\frac{v^2}{c^2}=\left(\frac{96}{100}\right)^2=\left(\frac{24}{25}\right)^2=\frac{576}{625}$$

or $$\frac{v}{c}=\sqrt{1-\frac{576}{625}}=\sqrt{\frac{49}{625}}=\frac{7}{25}$$

∴ $$v=\frac{7}{25}c=\frac{7}{25}\times 3\times 10^8=8.4\times 10^7\text{ m/s}$$

Example 1.10: **μ-mesons with a proper life of 2.2 μs formed at the top of the earth's atmosphere have a velocity of 0.998c. What is the vertical distance they travel in their mean life time before decaying as observed by a person on the surface of the earth?**

Solution Apparent life time as observed by the observer on the earth's surface

$$t=\frac{t_0}{\sqrt{1-\frac{v^2}{c^2}}}=\frac{2.2\mu s}{\sqrt{1-(0.998)^2}}$$

∴ $$\text{Vertical distance travelled}=\text{speed}\times\text{time}=\frac{0.998\times 3\times 10^8\times 2.2\times 10^{-8}}{\sqrt{1-(0.998)^2}}$$

$$=164670\text{ m}=164.67\text{ km}$$

Example 1.11: **The average life time of a neutron as a free particle at rest is 15 minutes. It disintegrates spontaneously into an electron, proton and neutron. What is the average**

minimum velocity with which a neutron must leave the Sun in order to reach the earth before breaking up? Take the distance of earth from Sun as 11×10^{10} m.

Solution Let v be the velocity of the neutron. The average life time t of the moving neutron, as measured by an observer on earth would be

$$t = \frac{t_0}{\sqrt{1-\frac{v^2}{c^2}}}$$

where t_0 is the proper life time of the neutron at rest. The time t is also the time for reaching the neutron from Sun to earth before decay, Therefore

$$t = \frac{d}{v} = \frac{11\times10^{10}}{v}$$

$$\therefore \quad \frac{11\times10^{10}}{v^2} = \frac{15\times60}{\sqrt{1-\frac{v^2}{c^2}}}$$

Squaring, we obtain

$$v^2 = \frac{121\times10^{20}\times9}{850\times10^4}$$

or $$v = 1.13\times10^8\,\text{m/s}$$

1.12 RELATIVISTIC VELOCITY TRANSFORMATION EQUATIONS

An observer in the stationary system would define velocity in his system as $U_x = \left(\frac{x_2 - x_1}{t_2 - t_1}\right)$, i.e. an interval of distance divided by an interval of time. Similarly, an observer in the moving inertial frame would define the velocity as $U'_x = \left(\frac{x'_2 - x'_1}{t'_2 - t'_1}\right)$. If an observer in S frame is to observe the phenomenon in S' and wishes to express U' in terms of this position and time system, he would have to use the velocity transformation equations. Let us derive these relativistic velocity transformation equations. We have by definition

$$U_x = \frac{dx}{dt}, \quad U_y = \frac{dy}{dt}, \quad U_z = \frac{dz}{dt}$$

and $$U'_2 = \frac{dx'}{dt'}, \quad U'_y = \frac{dy'}{dt'}, \quad U'_z = \frac{dz'}{dt'} \tag{1.30}$$

From the Lorentz transformation equations, we have $dy' = dy$, $dz' = dz$

$$dx' = \frac{dx - vdt}{\sqrt{1-\frac{v^2}{c^2}}} = \gamma(dt - vdt)$$

and
$$dt' = \frac{dt - vdt}{\sqrt{1-\frac{v^2}{c^2}}} = \gamma(dt - \frac{vdx}{c^2})$$

$\therefore$
$$U'_x = \frac{dx'}{dt'} = \frac{\gamma(dx - vdt)}{\gamma\left(dt - \frac{vdx}{c^2}\right)}$$

or
$$U'_x = \frac{\frac{dx}{dt} - v}{1 - \frac{v}{c^2}\frac{dx}{dt}} = \left(\frac{U_x - v}{1 - \frac{v}{c^2}U_x}\right) \tag{1.31}$$

Similarly, one obtains
$$U'_y = \frac{U_y}{\gamma\left(1 - \frac{vU_x}{c^2}\right)} \tag{1.32}$$

and
$$U'_z = \frac{U_z}{\gamma\left(1 - \frac{vU_x}{c^2}\right)} \tag{1.33}$$

Eqs. (1.31) to (1.33) constitute the relativistic velocity transformation equations. It is important to note that the velocity components which are transverse to the relative motion of the reference frame, depend upon the relative motion but transverse distances do not.

From Eq. (1.31), it is clear that the denominator is less than 1. Obviously, the velocity as measured in the stationary system is less than $U'_x + v$, as we expect from classical theory. This reveals that observed from rest, moving objects in a moving system are relatively slowed.

From Eq. (1.31) if $U_x = c$, we obtain $U'_x + c$. Obviously, the velocity of light represents maximum observable velocity for a physical particle.

By interchanging the unprimed quantities and writing $-v$ in place inverse velocity transformation equations. Thus, we have
$$U_x = \frac{(U'_x + v)}{\left(1 + \frac{vU_x}{c^2}\right)} \tag{1.34}$$

$$U_y = \frac{U'_y\sqrt{1-\frac{v^2}{c^2}}}{1 + \frac{vU'_x}{c^2}} \tag{1.35}$$

and
$$U_z = \frac{U'_z\sqrt{1-\frac{v^2}{c^2}}}{1 + \frac{vU'_x}{c^2}} \tag{1.36}$$

The relativistic velocity transformation equation reduced to the classical laws of velocity transformations as $c \to \infty$, i.e. $U_x = U_x'$, $U_y = U_y'$ and $U_z = U_z'$

Example 1.12: A spaceship moving away from the earth with velocity 0.5*c* fires a rocket whose velocity relative to spaceship is 0.8*c* (a) away from the earth (b) towards the earth? What will be the velocity of the rocket as observed from the earth in the two cases.

Solution Let us regard the earth as *S* frame and spaceship as *S'* frame. The rocket is the object whose velocity in the S frame is to be determined.

We have
$$U = \frac{U' + v}{1 + \frac{U'v}{c^2}} \qquad \left| \begin{array}{l} U' = 0.8c \\ v = 0.5c \end{array} \right.$$

$$= \frac{0.8c + 0.5c}{1 + 0.8 \times 0.5}$$

In the second case,
$$U' = -0.8c$$

$$U = \frac{U' + v}{1 + \frac{U'v}{c^2}}$$

$$= \frac{-0.8c + 0.5c}{1 + \{(-0.8) \times (0.5)\}}$$

$$= \frac{-0.3c}{1 - 0.4} = -0.5c$$

Example 1.13: In the laboratory frame two particles are observed to travel in opposite directions with speed 2.8 × 10⁸ m/s. Deduce the relative speed of the particles.

Solution We have

$$U_x' = \frac{U_x + v}{1 - \frac{U_x v}{c^2}} \qquad \left| \begin{array}{l} U_x = 2.8 \times 10^8 \text{ m/s} \\ v = -2.8 \times 10^8 \text{ m/s} \\ c = 3 \times 10^8 \text{ m/s} \end{array} \right.$$

$$U_x' = \frac{0.8 \times 10^8 - (-2.8 \times 10^8)}{1 - \frac{(2.8 \times 10^8)(-2.8 \times 10^8)}{(3 \times 10^8)^2}}$$

Obviously, the velocity of the first particle relative to the second is 2.99 × 10⁸ m/s.

Example 1.14: A radioactive nucleus while moving with a velocity 0.2*c* in the lab-frame, emits a β-particle moves with a speed 0.6*c* relative to the nucleus. What is the velocity and direction of the β-particle if it is emitted in a direction (a) parallel (b) perpendicular to the direction of motion of the nucleus in the lab-frame?

Solution Let us consider that the frame of reference fixed on nucleus be *S'* and the lab-frame be *S*. The direction of motion of the nucleus be the *x*-direction.

(i) $U_x = \dfrac{U'_x + v}{1 + \dfrac{U'_x v}{c^2}}$ $\quad\Bigg|\quad U_{x'} = 0.6c,\ V = 0.2c,\ U'_y = U'_z = 0$

$$= \frac{0.6c + 0.2c}{1 + 0.12}$$

$$= \frac{0.8c}{1.12} = 0.714c$$

β-particle is emitted in x-direction

$$U_y = \frac{U'_y\sqrt{1 - \dfrac{v^2}{c^2}}}{1 + \dfrac{U'_x v}{c^2}} = 0$$

Similarly $\qquad U_z = 0$

Hence in the laboratory-frame the β-particle will move in x-direction with a velocity $0.714c$.

(ii) When β-particle is emitted in y-direction

$$U'_x = 0,\ U'_y = 0.6c,\ U'_z = 0;\ v = 0.2c$$

$$\therefore \qquad U_x = \frac{U'_x + v}{1 + \dfrac{U'_x v}{c^2}} = \frac{0 + 0.2c}{1 + 0} = 0.2c$$

$$U_y = \frac{0.6c\sqrt{1 - (0.2)^2}}{1 + 0} = 0.6c\sqrt{0.9c} = 0.588c$$

$$U_z = 0$$

$\therefore$ resultant velocity $\qquad U = \sqrt{U_x^2 + U_y^2}$

$$= c\sqrt{0.04 + 0.36 \times 0.96} = 0.621c$$

The direction of U in the $x - y$ plane will be

$$\alpha = \tan^{-1}\frac{U_y}{U_x} = \tan^{-1}\frac{0.588c}{0.2c} = \tan^{-1}(2.94) = 72.1^\circ$$

1.13 RELATIVITY OF MASS

In classical physics, the mass of a particle is taken to be same at all velocities, i.e. independent of the velocity of the particle, i.e., when masses are involved we say that a constant force produces a constant acceleration. In relativistic mechanics, however, a constant force does not produce a linear increase in velocity. As the velocity increases, the factor $\gamma = \sqrt{1 - \dfrac{v^2}{c^2}}$

becomes significantly different from 1 and the same force produces a smaller increase in velocity, it is because the mass has increased. In fact, one of the consequences of the relativity is that mass increases with increase in velocity. We know that the principle of conservation of momentum is a fundamental law of physics and it must hold good in all inertial frames of references, under Lorentz transformations. We find that mass is not absolute but varies with velocity in such a manner that it makes the principle of conservation of momentum hold good in all inertial frames. An expression for the variation in mass with velocity can be derived as follows:

Consider a collision between two bodies of equal masses in an inertial frame S'. For simplicity, let us assume that the bodies are moving with velocities U' and $-U'$ along the x-axis and after collision they coalesce into one body. Let the mass of each body as observed in S' frame be m'. After collision the coalesced body will have mass $2m'$. From the principle of conservation of momentum it will be at rest in S'.

Let S be another inertial frame such that S' is moving with a uniform relative velocity v in the x-direction with respect to S. The speeds of two bodies will not appear to be equal for an observer in frame S. If the velocities of the two bodies are observed as U_1 and U_2 in the frame S then by the relativistic law of addition of velocities

$$U_1 = \frac{U'+v}{1+\dfrac{U'v}{c^2}} \quad \text{and} \quad U_2 = \frac{-U'+v}{1-\dfrac{U'v}{c^2}} \tag{1.37}$$

If we now assume that mass is variable, the mass of the two bodies, in frame S, will appear to be different. Let m_1 be the mass of the body moving with a velocity U_1 and m_2 be the mass of the body moving with a velocity U_2. After collision as the coalesced body is at rest in frame S', it will appear to be moving with a velocity v in frame S and its mass will be (m_1+m_2). Fig. (1.6) depicts the situation as observed in the two frames of references. From the principle of conservation of momentum, we have

$$m_1U_1 + m_2U_2 = (m_1+m_2)V \tag{1.38}$$

or

$$m_1(U_1 - v) = m_2(v - U_2) \tag{1.39}$$

∴

$$\frac{m_1}{m_2} = \frac{v-U_2}{U_1-v} \tag{1.40}$$

Substituting the values of U_1 and U_2 from Eq. (1.37), we have

$$\frac{m_1}{m_2} = \frac{v - \dfrac{-U'+v}{1-\dfrac{U'v}{c^2}}}{\dfrac{U'+v}{1+\dfrac{U'v}{c^2}} - v} = \frac{1+\dfrac{U'v}{c^2}}{1-\dfrac{U'v}{c^2}} \tag{1.41}$$

Further from Eq. (1.37), we have

$$c^2 - U_1^2 = c^2 - \frac{U'+v}{1+\dfrac{U'v}{c^2}}$$

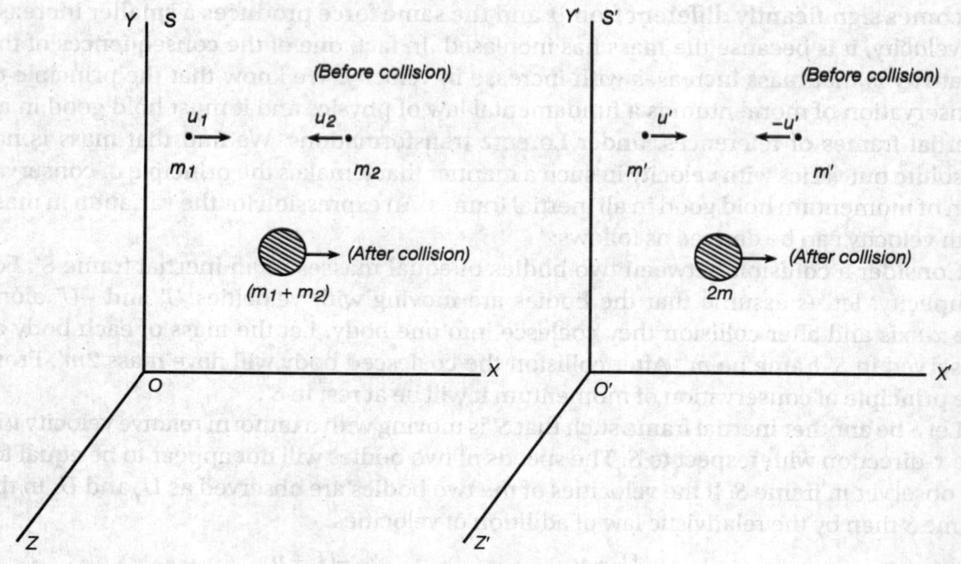

Fig. 1.6

or
$$c^2 - U_1^2 = c^2 - \frac{U'^2 + v^2 + 2U'v}{1 + \frac{U'^2 v^2}{c^4} + \frac{2U'v}{c^2}} \tag{1.42}$$

$$= \frac{\left(c^2 - U'^2\right)\left(1 - \frac{v^2}{c^2}\right)}{\left(1 + \frac{U'v}{c^2}\right)^2} \tag{1.43}$$

Similarly, we obtain (by replacing U' with $-U'$)

$$c^2 - U_2^2 = \frac{\left(c^2 - U'^2\right)\left(1 - \frac{v^2}{c^2}\right)}{\left(1 - \frac{U'v}{c^2}\right)^2} \tag{1.44}$$

Dividing Eq. (1.44) by Eq. (1.43), we obtain

$$\frac{1 + \frac{U'v}{c^2}}{1 - \frac{U'v}{c^2}} = \left(\frac{c^2 - U_2^2}{c^2 - U_1^2}\right)^{\frac{1}{2}} \tag{1.45}$$

or
$$\frac{m_1}{m_2} = \frac{\left(1-\frac{U_1^2}{c^2}\right)^{\frac{1}{2}}}{\left(1-\frac{U_2^2}{c^2}\right)^{\frac{1}{2}}} = \frac{1+\frac{U'v}{c^2}}{1-\frac{U'v}{c^2}} \tag{1.46}$$

or
$$m_1\sqrt{1-\frac{U_1^2}{c^2}} = m_2\sqrt{1-\frac{U_2^2}{c^2}} \tag{1.47}$$

Generalising, we obtain

$$m\sqrt{1-\frac{v^2}{c^2}} = \text{constant} \tag{1.48}$$

For a particle at rest, i.e., $v = 0$, if its mass is m_0 then

$$m\sqrt{1-\frac{v^2}{c^2}} = m_0(1-0) = m_0$$

or
$$m = \frac{m_0}{\sqrt{1-\frac{v^2}{c^2}}} \tag{1.49}$$

Obviously, the mass of a body moving with a speed v is always greater then its rest mass m_0. As v approaches c, the effective mass increases without limit Fig. (1.7). No body with finite rest mass m_0 can quite achieve the speed of light, when $\frac{v}{c} << 1$, the difference between m and m_0 becomes insignificant.

As v approaches c, the observed mass approaches infinity (∞). The idea of infinite mass seems ridiculous. It actually means that the velocity of the material body (whose rest mass m_0 is not zero) can never become equal to the velocity of light in vacuum, it has always to be less than it.

In 1909, Bucherer determined the ratio e/m as a function of velocity for fast electrons emitted from radioactive nuclei and showed that the mass of the electron varies with velocity according to Eq. (1.48).

1.13.1 Mass and Potential Energy

We know that mass is associated with potential as well as kinetic energy. Suppose that two masses are moving with equal and opposite velocities along the x-axis as seen by the observer in S frame collide with each other, a spring acting as a buffer between them. Just as they come to rest, let a lock snap shut and hold them combined into a single mass. Due to collision, the initial kinetic energy of the masses is converted into potential energy of the spring. Now, we examine this collision from a second frame S' that is moving

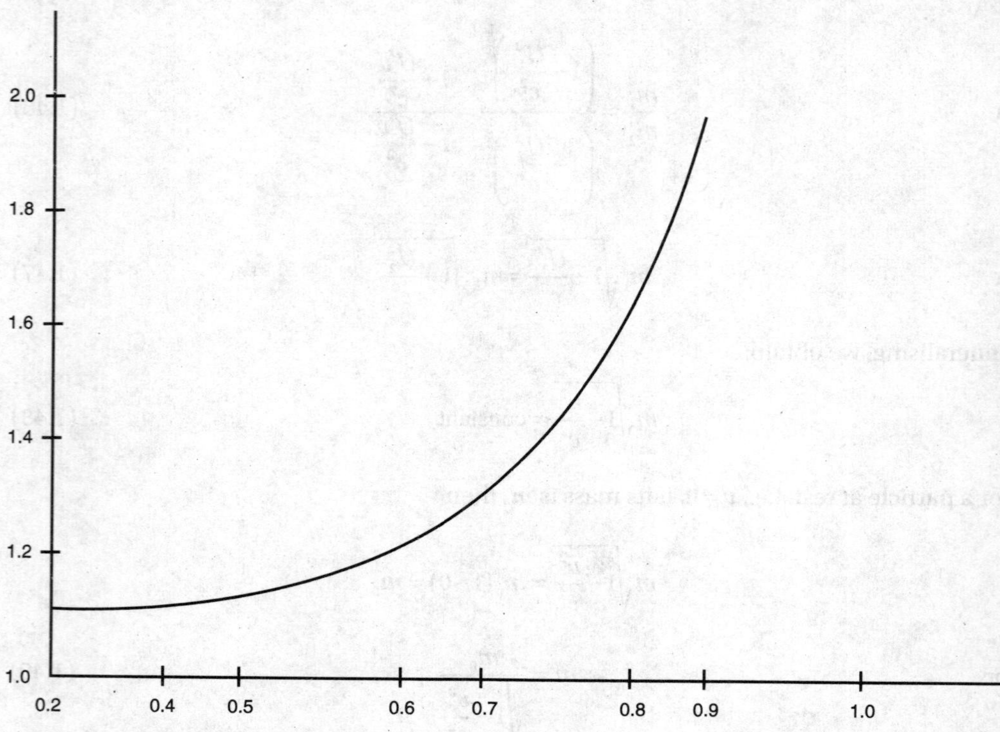

Fig. 1.7 Variation of mass with velocity

with velocity V parallel to x. Now, if the velocities of the two bodies prior to collision, as measured in S are $v_1 = v_{x_1} = v,\ v_2 = v_{x_2} = -v$, the same velocities as measured in S' are obtained as (using velocity transformation equations)

$$v_1' = v_{x_1}' = \frac{v - V}{1 - \frac{V^2}{c^2}} = -V + \left(1 - \frac{V^2}{c^2}\right)\frac{v}{1 - \frac{Vv}{c^2}}$$

$$v_2' = v_{x_2}' = \frac{-v - V}{1 - \frac{V^2}{c^2}} = -V - \left(1 - \frac{V^2}{c^2}\right)\frac{v}{1 + \frac{Vv}{c^2}}$$

$$\therefore \quad \frac{v_1'}{\left(1 - \frac{v_1'^2}{c^2}\right)^{\frac{1}{2}}} = \frac{V}{\left(1 - \frac{v_1'^2}{c^2}\right)^{\frac{1}{2}}} + \left(1 - \frac{V^2}{c^2}\right)\frac{v}{[1 - (V^2 + v^2)/c^2 + V^2/c^4]}$$

Therefore, the total momentum prior to collision is

$$\frac{m_0 v_1'}{\sqrt{1-\frac{v_1'^2}{c^2}}}+\frac{m_0 v_2'}{\sqrt{1-\frac{v_2'^2}{c^2}}}=-m_0 V\frac{1}{\sqrt{1-\frac{v_1'^2}{c^2}}}+\frac{1}{\sqrt{1-\frac{v_2'^2}{c^2}}} \tag{1.50}$$

Now, if the rest mass of the combined body after collision were merely the sum of the rest masses of the separate bodies, or $2m_0$. The total momentum after collision would be $-\frac{2m_0 V}{\sqrt{1-\frac{V^2}{c^2}}}$, since the velocity of the combined body is then $-V$ we can see that this not equal to the momentum prior to collision as given by Eq. (1.50). One can easily check this when v_1' and v_2' are both either greater or less than V.

We find that the conservation of momentum fails if only the rest masses are taken into account. Now, suppose we include in the mass of the combined body a mass associated with the potential energy of the spring. Then the total mass is proportional of the total energy and is constant. Clearly, the total mass after the collision is, therefore, the same as it was before the collisions, or

$$\frac{m_0}{\sqrt{1-\frac{v_1'^2}{c^2}}}+\frac{m_0}{\sqrt{1-\frac{v_2'^2}{c^2}}}$$

Now, if we multiply this value of the total mass by the common velocity of the bodies after the collision, which is $-V$, one obtains for the total momentum after collision exactly the same expression as that given by Eq. (1.50) for the total momentum prior to collision. Concluding, we can say that the principle of conservation of momentum thus holds here if and only if we assume that the potential energy in the spring makes its contribution to the mass and momentum of the system.

Example 1.15: With what velocity should a particle move so that the increase in its mass may be 25% of its rest mass?

Solution We have

$$m=\frac{m_0}{\sqrt{1-\frac{v^2}{c^2}}}$$

If m_0 is the rest mass of the particle, then its mass at the required velocity v is

$$m=m_0+\frac{25}{100}m_0=\frac{5}{4}m_0$$

$$\therefore \quad \frac{5}{4}m_0=\frac{m_0}{\sqrt{1-\frac{v^2}{c^2}}}$$

or $$1-\frac{v^2}{c^2}=\frac{16}{25}$$

or $$\frac{v^2}{c^2}=\frac{9}{25}$$

or $$v=\frac{3}{5}c=0.6c$$

Example 1.16: **Deduce the velocity at which the mass of a particle becomes 10 times its rest mass.**

Solution

$$m=\frac{m_0}{\sqrt{1-\frac{v^2}{c^2}}} \quad \left|\frac{m}{m_0}=10\right.$$

We have $$\frac{1}{\sqrt{1-\frac{v^2}{c^2}}}=10$$

or $$1-\frac{v^2}{c^2}=\frac{1}{100}$$

or $$\frac{v^2}{c^2}=1-\frac{1}{100}=\frac{99}{100}=0.99$$

or $$v=\sqrt{0.99}c=0.995\times3\times10^8\text{ m/s}=2.98\times10^8\text{ m/s}$$

1.14 RELATIVISTIC MOMENTUM

In classical mechanics, linear momentum of a particle is $\mathbf{P}=m\mathbf{v}$. This is conserved in a system of particles not acted upon by outside forces. Moreover, when an event such as a collision or an explosion occurs inside an insolated system, the vector sum of the momenta of its particles before the event is equal to their vector sum afterward. Now, the question arises whether $\mathbf{P}=m\mathbf{v}$ is valid as the definition of momentum in inertial frames in relative motion, and if not, what should be the relativistically correct definition? We now try to answer it.

Let $\mathbf{P}$ be conserved in a collision for all observers in relative motion at constant velocity, we may note that $\mathbf{P}=m\mathbf{v}$ holds in classical mechanics, i.e., for $v << c$. Obviously, whatever the relativistically correct $\mathbf{P}$ is, then it reduce to $\mathbf{P}=m\mathbf{v}$ for $\mathbf{v} << c$.

Let us consider a elastic collision between two particles A and B and two observers are observing it in S and S' inertial frames, where S' is moving in $+x$-direction with respect to S at the velocity $\boldsymbol{v}$. The origin of two frames coincide at $t=0$. The properties of A and B are identical when determined in reference frames in which they are at rest. The two frames S and S' are oriented as shown in Fig. [1.8(a)] with S' moving in the $+x$-direction w.r.t. x at the velocity $\boldsymbol{v}$.

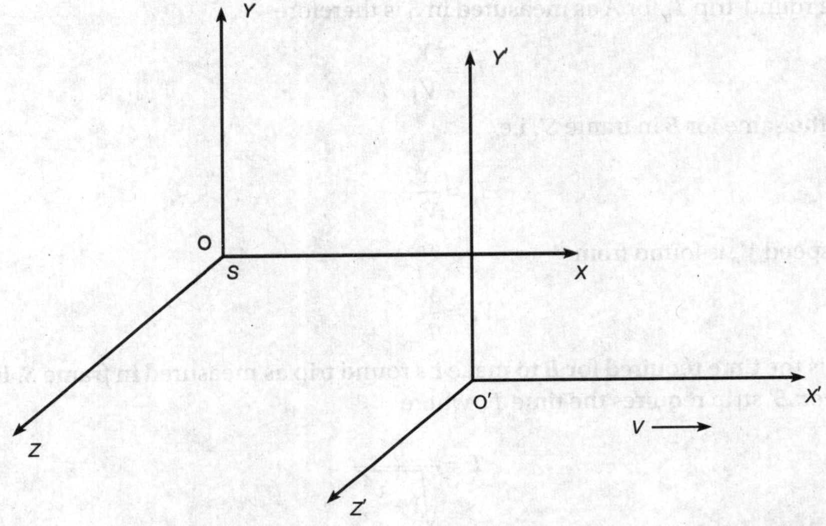

Fig. 1.8 (a) Two inertial frames in relative motion. An electric collision is observed by observers in these frames

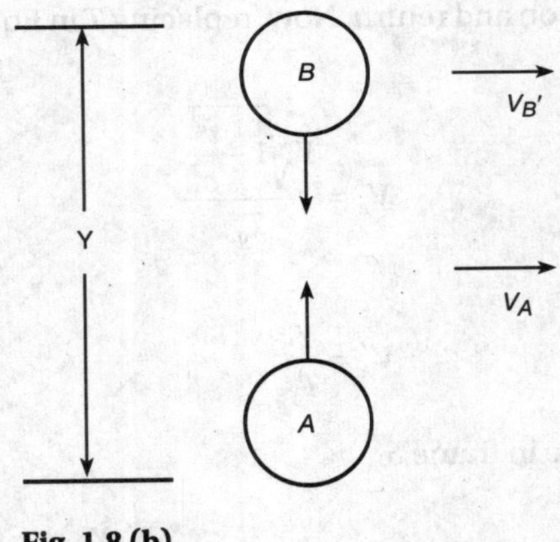

Fig. 1.8 (b)

Before the collision, particle A had been at rest in S and particle B in S'. Then, at the same instant, A was thrown in the $+y$-direction at the speed V_A while B was thrown in the $-y'$-direction at the speed V'_B, where

$$V_A = V'_B \tag{1.51}$$

Clearly, the behaviour of A as seen from S is exactly the same as the behaviour of B as seen from frame S'.

When the two particles A and B collide, A rebounds in the $-y$-direction at the speed V_A, while B rebounds in the $+y'$-direction at the speed V'_B. If the particles are thrown from positions y apart, an observer in frame S finds that the collision occurs at $y = \frac{1}{2}Y$ and the observer in S' finds that it occurs at $y' = y = \frac{1}{2}Y$.

Now, the round-trip T_0 for A as measured in S is therefore

$$T_0 = \frac{Y}{V_A} \tag{1.52}$$

and it is the same for B in frame S', i.e.,

$$T_0 = \frac{Y}{V_B} \tag{1.53}$$

In S the speed V_B is found from

$$V_B = \frac{Y}{T} \tag{1.54}$$

where T is the time required for B to make its round trip as measured in frame S. In frame S' however, B' strip requires the time T_0, where

$$T = \frac{T_0}{\sqrt{1-\frac{v^2}{c^2}}} \tag{1.55}$$

according to our previous results. Although observers in both frames see the same event, however, they disagree about the length of time the particle thrown from the other frame requires to make the collision and return. Now, replacing T in Eq. (1.54) with its equivalent in terms of T_0, one finds

$$V_B = \frac{Y\sqrt{1-\frac{v^2}{c^2}}}{T_0}$$

From Eq. (1.53),

$$V_A' = \frac{Y}{T_0}$$

Now, using $\mathbf{P} = m\mathbf{v}$, we have in frame S

$$P_A = m_A V_A = m_A \left(\frac{Y}{T_0}\right)$$

and

$$P_B = m_B V_B = m_B \sqrt{1-\frac{v^2}{c^2}}\left(\frac{Y}{T_0}\right)$$

This means that, in this frame, momentum will not be conserved if $m_A = m_B$, where m_A and m_B are the masses measured in S. However, if

$$m_B = \frac{m_A}{\sqrt{1-\frac{v^2}{c^2}}} \tag{1.56}$$

then momentum will be conserved.

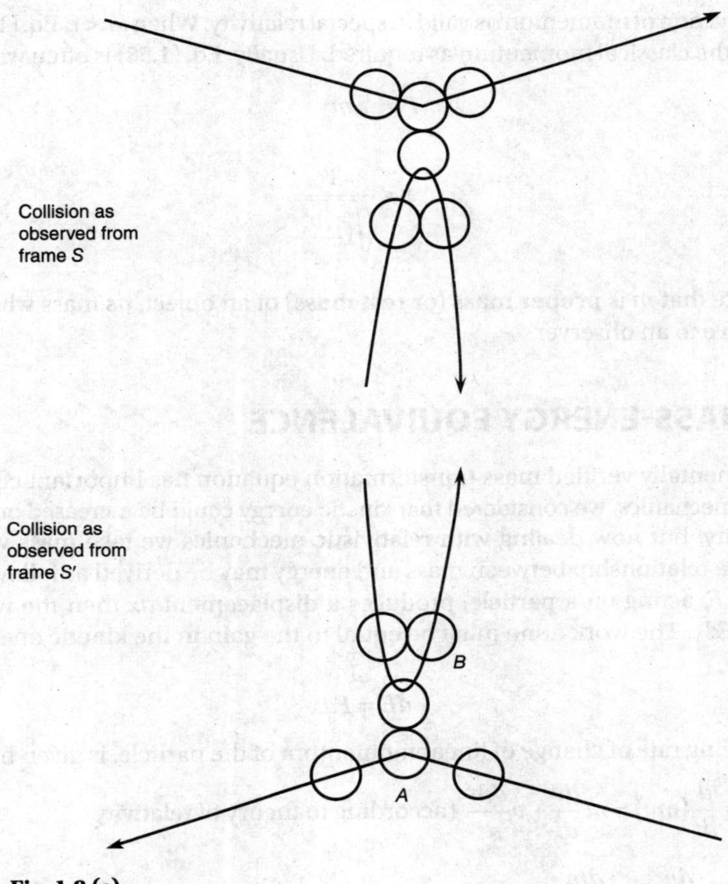

Fig. 1.8 (c)

In the collision of Fig. [1.8(c)] both *A* and *B* are moving in both *S* and *S′* frames. Suppose now that V_A and V_B' are very small compared with $\boldsymbol{v}$, the relative velocity of two frames *S* and *S′*. In this case an observer in *S* will see *B* approach *A* with the velocity $\boldsymbol{v}$, make a glancing collision (since $V_B' << v$) and then continue on. In the limit of $V_A = 0$, if *m* is the mass in *S* of *A* when *A* is at rest, then $m_A = m$. In the limit of $V_B' = 0$, if $m(\boldsymbol{v})$ is the mass in *S* of *B*, which is moving at velocity $\boldsymbol{v}$, then $m_B = m(\boldsymbol{v})$. Hence Eq. (1.56) takes the form

$$m(v) = \frac{m}{\sqrt{1-\frac{v^2}{c^2}}} \tag{1.57}$$

One can see that if linear momentum is defined as

$$\boldsymbol{P}(\text{Relativistic momentum}) = \frac{mv}{\sqrt{1-\frac{v^2}{c^2}}} \tag{1.58}$$

then conservation of momentum is valid in special relativity. When $v << c$, Eq. (1.58) becomes just $\boldsymbol{P} = m\boldsymbol{v}$ the classical momentum as required. Usually, Eq. (1.58) is often written as

$$\boldsymbol{P} = \gamma m \mathbf{v} \tag{1.59}$$

Where

$$\gamma = \frac{1}{\sqrt{1 - \frac{v^2}{c^2}}} \tag{1.60}$$

We may note that m is **proper mass** (or **rest mass**) of an object, its mass when measured at rest relative to an observer.

1.15 MASS-ENERGY EQUIVALENCE

The experimentally verified mass transformation equation has important consequences. In classical mechanics, we considered that kinetic energy could be increased only increasing by its velocity, but now dealing with relativistic mechanics we take mass variation into account. The relationship between mass and energy may be derived as follows:

If a force F, acting on a particle, produces a displacement dx then the work done by the force = Fdx. The work done must be equal to the gain in the kinetic energy dE of the particle, i.e.

$$dE = Fdx$$

But force, being rate of change of linear momentum of the particle, is given by

$$F = \frac{d}{dt}(mv) = m\frac{dv}{dt} + v\frac{dm}{dt} \quad \text{(according to theory of relativity)}$$

$$\therefore \quad dE = m\frac{dv}{dt}dx + \frac{dm}{dt}dx$$

$$= mvdv + v^2 dm \quad \because \left(\frac{dx}{dt} = v\right) \tag{1.61}$$

Now, from relativistic mass transformation relation

$$m = \frac{m_0}{\sqrt{1 - \frac{v^2}{c^2}}}$$

or

$$m^2 = \frac{m_0^2 c^2}{(c^2 - v^2)}$$

or

$$m^2c^2 - m^2v^2 = m_0^2 c^2$$

Differentiating, we obtain

$$c^2 2mdm - (m^2 2vdv + v^2 2mdm) = 0$$

Dividing throughout by $2m$, one obtains

$$c^2 dm-(mvdv+v^2 dm)=0$$

or
$$mvdv+v^2 dm=c^2 dm \tag{1.62}$$

Comparing Eqs. (1.61) and (1.62), one obtains

$$dE=c^2 dm \tag{1.63}$$

Eq. (1.63) shows that in relativity a change in kinetic energy can be expressed in terms of mass as variable. This equation is valid for a change in any form whatever, although it was derived here for change of kinetic energy. Since K.E. is zero, when $v=0$, then it is also zero when $m=m_0$. Therefore, on integrating Eq. (1.63), one obtains.

$$E=c^2\int_{m_0}^{m} dm =c^2(m-m_0)$$

or
$$E-mc^2-m_0c^2 \tag{1.64}$$

Eq. (1.64) is the relativistic expression for kinetic energy. For $v \ll c$, it reduces to the classical expression. Putting $m=\dfrac{m_0}{\sqrt{1-\dfrac{v^2}{c^2}}}$ in Eq. (1.64), one obtains

$$E=\frac{m_0c^2}{\sqrt{1-\dfrac{v^2}{c^2}}}-m_0c^2$$

For $v \ll c$, we have

$$\left(1-\frac{v^2}{c^2}\right)^{-\frac{1}{2}}=\left(1-\frac{v^2}{2c^2}+\frac{3\,v^4}{8\,c^4}+\cdots\right)$$

$$\therefore \quad E=m_0c^2\left(1-\frac{v^2}{2c^2}+\frac{3\,v^4}{8\,c^4}+\cdots-1\right)=\frac{1}{2}m_0v^2+\frac{3\,v^4}{8\,c^2} \tag{1.65}$$

The first term of Eq. (1.65) is the classical expression. Obviously, it is the only significant term at low velocities.

The equivalence between energy and mass proves to be universal. The best examples of conversion of mass into energy are the nuclear fission, nuclear fusion and nuclear reaction processes. Another most familiar example of mass energy equivalence is the phenomenon of pair production.

Example 1.17: What is the annual loss in the mass of the Sun, if the earth receives heat energy approximately 2 cal/cm²/min. The earth-Sun distance is about 150×10^6 km.

Solution Rate of energy radiated

$$=2\times 4.2\times 10^7 \text{ erg/cm}^2\text{/min}$$

Total energy radiated per minute

$$= 4\pi \times (150\times10^{11})^2 \times 2\times 4.2\times 10^7$$

Energy radiated per year

$$= 4\pi \times (150\times10^{11})^2 \times 2\times 4.2\times 10^7 \times 5.3\times 10^5$$

Annual loss in mass is $\frac{E}{c^2}$

$$= \frac{4\pi \times (150\times10^{11})^2 \times 2\times 4.2\times 10^7 \times 5.3\times 10^5}{9\times 10^{20}}$$

$$= 1.4\times 10^{14} \quad \text{tons per year}$$

Example 1.18: **A nucleus of mass m emits a gamma ray of frequency v_0. Show that the loss of internal energy by the nucleus is not hv_0 but it is $hv_0\left[1+\frac{hv_0}{2mc^2}\right]$.**

Solution The momentum of γ-ray photon is $p=\frac{hv_0}{c}$. According to the law of conservation of momentum, the nucleus having mass m will recoil with the momentum $\frac{hv_0}{c}$ in the backward direction. Therefore, the loss of energy of the nucleus in the recoiling is

$$E = \frac{1}{2}mv^2 = \frac{m^2v^2}{2m} = \frac{p^2}{2m} = \frac{(hv_0)^2}{2mc^2}$$

Energy of the emitted photon from the nucleus $= hv_0$

$\therefore$ Clearly, the total loss of internal energy of the nucleus

$$= hv_0 \frac{(hv_0)}{2m} = hv_0\left(1+\frac{hv_0}{2m}\right)$$

Example 1.19: **A certain accelerator produces a beam of neutral K-mesons or kaons ($m_Kc^2 = 498$ MeV) with kinetic energy 325 MeV. Consider a kaon that decays in flight into two pi-mesons or pions ($m_\pi c^2 = 140$MeV). Show that the kinetic energy of each pion in the special case in which the pions travel parallel or antiparallel to the direction of the kaon beam or 543 MeV and 0.6 MeV.**

Solution The initial relativistic total energy

$$E_K = K + m_Kc^2 = 325\text{MeV} + 498\text{MeV} = 823\text{MeV}$$

Total initial momentum

$$P_Kc = \sqrt{E_K^2 - (m_Kc^2)^2} = \sqrt{(823)^2 - (498)^2} = 655\text{MeV}$$

Now, the total energy of the final system consisting of the two pions is

$$E = E_1 + E_2 = \sqrt{(p_1c)^2 + (m_\pi c^2)^2} + \sqrt{(p_2c)^2 + (m_\pi c^2)^2} = 823\text{MeV} \qquad \text{(i)}$$

which, by conservation of total relativistic energy, we have equated to the initial total energy of 823 MeV. Obviously, we have one equation in the two unknowns p_1 and p_2, the momenta of two pions.

In order to find a second equation in the two unknowns, we apply conservation of momentum. The final momentum of the two pion system along the beam direction is $P_1 + P_2$, and setting this equal to the initial momentum P_K, one obtains

$$P_1c + P_2c = P_Kc = 655\text{MeV} \tag{ii}$$

We now have two equations in the two unknowns P_1 and P_2. Solving Eq. (ii) for P_2c and substituting this into Eq. (i), one obtains a quadratic equation for P_1c, which gives

$$P_1c = 668\text{MeV} \quad \text{or} \quad -13\text{MeV} \tag{iii}$$

Since the labels 1 and 2 of the two pions are arbitrary, one finds that one pion travelling parallel to the beam with momentum $P_1 = 668$ MeV/c, while the other pion travels in the opposite direction with momentum $P_2 = -13$ MeV/c. One obtains the corresponding kinetic energies from

$$K = \sqrt{(Pc)^2 + (m_0c^2)^2} - m_0c^2$$

$$K_1 = \sqrt{(668)^2 + (140)^2} - 140 = 543\text{MeV}$$

and

$$K_2 = \sqrt{(-13)^2 + (140)^2} - 140 = 0.6\text{MeV}$$

1.16 RELATION BETWEEN MOMENTUM AND ENERGY

Relativistic momentum of a particle moving with a velocity v is given by

$$\mathbf{P} = m\mathbf{v} \tag{1.66}$$

where $m = \dfrac{m_0c^2}{(1-\frac{v^2}{c^2})}$, m_0 being the rest mass of the particle.

From relativity, we have

$$E = mc^2 \tag{1.67}$$

From Eqs. (1.66) and (1.67), we have

$$E^2 - c^2p^2 = m^2c^4 - c^2m^2v^2$$

$$= \frac{m_0^2c^4}{1-\frac{v^2}{c^2}} - \frac{m_0^2}{1-\frac{v^2}{c^2}}c^2v^2 = m_0^2c^4\left[\frac{1-\frac{v^2}{c^2}}{1-\frac{v^2}{c^2}}\right]$$

$$= m_0^2c^4$$

or

$$E^2 = c^2p^2 + m_0^2c^4 \tag{1.68}$$

1.17 PARTICLES WITH ZERO REST MASS

Photon and Graviton are the familiar examples of particles with zero rest mass. A particles with zero rest mass always moves with the speed of light in vacuum. According to Eq. (1.68), if $m_0 = 0$, we have

$$E = Pc \tag{1.69}$$

But relativistic momentum is given by

$$P = mv = v\frac{E}{c^2} \qquad (\text{as } E = mc^2) \tag{1.70}$$

Hence from Eqs. (1.69) and (1.70), we have

$$P = v\frac{E}{c^2} = v\frac{Pc}{c^2}$$

or

$$v = c \tag{1.71}$$

i.e., a particle with zero mass rest mass always moves with the speed of light in vacuum. The velocity of the particle observed in some other inertial frame S', is

$$U' = \frac{U - v}{1 - \frac{Uv}{c^2}}$$

where v is the velocity of the frame S' with respect to the frame S in which the velocity of the particle is U. Hence if $U = c$, we have

$$U' = \frac{c - v}{1 - \frac{cv}{c^2}} = c$$

Clearly, the particle will have the same speed c and zero rest mass for all observers in inertial frames.

1.18 SPEED LIMIT FOR MATERIAL PARTICLES

Eqs. (1.23) and (1.26) show that for $v = c$, a body's length contracts to zero and any time interval between events on it would dilate into eternity as measured by an observer in a frame relatively at rest. The mass increases according to relation (48) tells us that the mass of a body whose velocity equals c would become infinite and so would its energy. These consequences are clearly absurd and leads to the conclusion that the velocity of light in vacuum is the maximum possible velocity and that no material body can acquire this velocity. However, when light traverse matter its velocity reduces to c/μ where μ is the refractive index of the material and is always greater than one. P. A. Cerenkov found that

in transparent materials it is possible for particles like electrons to travel faster than light through them. This effect is called *Cerenkov effect.*

The possibility of a particle velocity exceeding the free space velocity of light is not considered as this makes particle mass imaginary. In 1962, Bilaniuk and Sudershan proposed that particles of imaginary mass might exist if it is assumed that they always travel at a speed faster than that of light. Feinberg in 1967 named these particles as **tachyons**. So far these particles have not been detected.

1.18.1 The Cerenkov Effect

We have seen that the relativity of mass places a limit on the speed of a body that can have relative to an observer. As the body's speed v approaches the speed of light c the ratio v^2/c^2 approaches 1 and the quantity $\sqrt{1-v^2/c^2}$ approaches zero. Clearly, the mass m of the body given by

$$m = \frac{m_0}{\sqrt{1-v^2/c^2}}$$

approaches infinity as v approaches c. This means the relative speed v therefore must be less than c. In other words we can say that an infinite force would be needed to accelerate a body by a speed at which its mass is infinite, but there are neither infinite force nor infinite masses in the universe.

We know that velocity of light in vacuum is 3×10^8 m/s. In all material media, e.g. glass, water, air, etc. light travels more slowly than this, and atomic particles are capable of moving with higher speeds in such media than the speed of light in them, though never faster than the speed of light in vacuum. When an electrically charged particle moves though a substance at a speed exceedingly that of light, a cone of light waves is emitted in a process roughly similar to that in which a ship produces a bow wave as it moves through the water at a speed greater than that of water waves. These light waves are known as **cerenkov radiation**.

Fig. [1.8(d)] shows the production of cerenkov radiation. The medium through which the particle is moving has an index of refraction n. The speed of light in the medium is obviously c/n, which is less than the particle's speed v. As the particle travels through the medium, it creates disturbances along its path from which light waves spread out at the speed c/n. The envelope of these waves is a cone with the particle at the apex, where, from the Fig [1.8(d)], the angle θ between the direction of propagation of the radiation and the particles direction is specified by

$$\cos\theta = \frac{(c/n)t}{vt} = \frac{c}{nv}$$

When energetic atomic particles move through such media as transparent plastics, glass, etc. the cone of cerenkov radiation can be detected and the angle θ measured, making it possible to detect the speed of the particles. When an intense beam of such particles is involved, the cerenkov radiation is visible as a bluish glow.

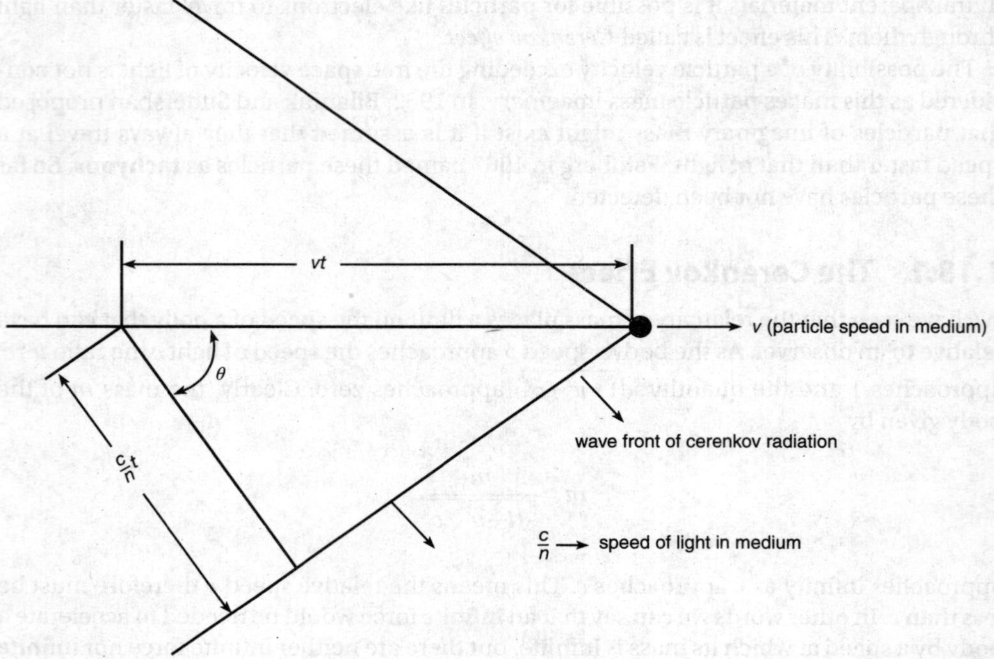

Fig. 1.8 (d)

1.19 SPACE AND TIME IN RELATIVITY

Space and time are no longer separate concepts in relativity but are interwoven in such a way that makes the introduction of four-dimensional space-time continuum very useful. In such a four dimensional space many of the laws of classical physics can be expressed in a particularly elegant form which greatly simplifies the shift from one inertial system to another.

Intervals: Four dimensional space is an imaginary concept of a space having four dimensions on whose axes the three coordinates *x*, *y* and *z* and the time *t* are plotted. Any event is represented by a point in four dimensional space, called **world point**. The motion of a certain particle in space and time is represented by a line in four-dimensional space, then the quantity

$$S_{12} = \sqrt{c^2(t_2 - t_1)^2 - (x_2 - x_1)^2 - (y_2 - y_1)^2 - (z_2 - z_1)^2} \tag{1.72}$$

is called the **interval** between the two events.

The interval between two infinitsimally close events is

$$ds = \sqrt{c^2 dt^2 - dx^2 - dy^2 - dz^2} = \sqrt{c^2 dt^2 - dl^2} \tag{1.73}$$

where $$dl^2 = dx^2 + dy^2 + dz^2$$

If the variable $\tau = ict$ (where $i = \sqrt{-1}$) is plotted instead of t along the time axis, the value $-ds^2 = dx^2 + dy^2 + dz^2 + d\tau^2$ can be regarded as square of an element in four-dimensional space.

The interval between two events is the same in all inertial frames of reference (invariance of a four dimensional interval). The invariance of an interval is a mathematical expression of the invariance of the velocity of light.

If x_1, y_1, z_1, t_1 and x_2, y_2, z_2, t_2 are the world points of two events in certain frame of reference Σ, then under the condition that $S_{12}^2 > 0$ (realness of the interval), there exists a frame of reference Σ' in which both events take place at a single point in space ($x_1^1 = x_2^1, y_1^1 = y_2^1$ and $z_1^1 = z_2^1$). Real intervals are said to be **time-like**. The time $t_{12} = t_2^1 - t_1^1$ that elapses between the two events in frame Σ' is equal to $t_{12}^1 = \frac{S_{12}}{c}$. The condition

$$S_{12}^2 = c^2 t_{12}^2 = l_{12}^2 = const > 0$$

where $t_{12} = t_2 - t_1$ and $l_{12}^2 = (x_2 - x_1)^2 + (y_2 - y_1)^2 + (z_2 - z_1)^2$, can be represented graphically as hyperbolas if $l = l_{12}$ and $ct = ct_{12}$ are plotted along the coordinate axes Fig. (1.9), where l_{12} and t_{12} correspond to the two given events in an arbitrary inertial frame of reference. Points A and A_1 correspond to event 2 which occurs at the same point in space as event 1 (point 0), but either later (point A) or earlier (point A_1).

For a **time like** interval it is always possible to find an inertial frame in which the two events occur at the same place. However, it is impossible to find an inertial frame in which the two events occurs simultaneously since in such a frame S_{12}^2 would be negative. There are events between which the interval, or separation is imaginary Thus,

$$S_{12}^2 = c^2 t_{12}^2 - l_{12}^2 < 0$$

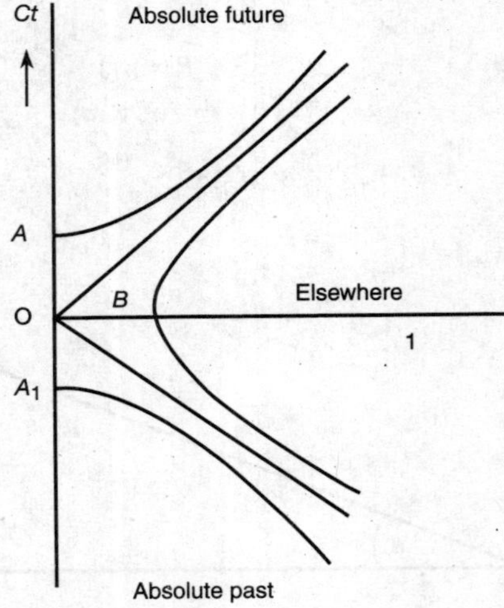

Fig. 1.9

Imaginary separations are said to be **space-like**. For any space like interval it is always possible to find some inertial frame in which the two events are **simultaneous**, but it is impossible to find a frame in which the events occur at the same place. Clearly the **space-like**, **time like** character of an interval is independent of the inertial frame of reference.

The 'locus' of zero intervals in four dimensional space (x, y, z, ict) is called the **'light cone'** for the given world point 0.

1.20 FOUR-VECTOR NOTATION

In ordinary three-space the components of a vector A in two inertial frame S and S' with a common origin are related by:

$$A_1^1 = A_1 \cos(x_1', x_1) + A_2 \cos(x_1', x_2) + A_3(x_1', x_3)$$

$$A_2^1 = A_1 \cos(x_2', x_1) + A_2 \cos(x_2', x_2) + A_3(x_2', x_3)$$

$$A_3^1 = A_1 \cos(x_3', x_1) + A_2 \cos(x_3', x_2) + A_3(x_3', x_3) \quad (1.74)$$

One can readily show that for a simple rotation through an angle θ in two dimensions this reduces to the form Fig. (1.10),

$$x' = x\cos\theta + y\sin\theta$$

$$y' = -x\sin\theta + y\cos\theta$$

$$z' = z \quad (1.75)$$

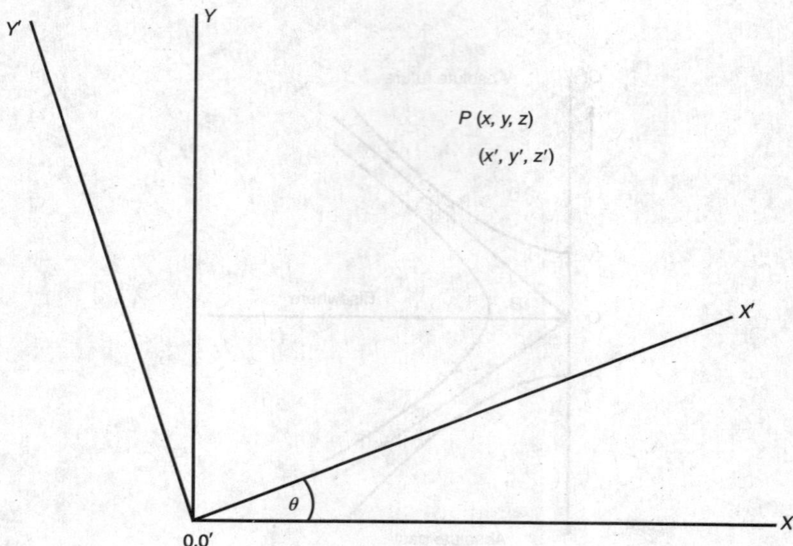

Fig. 1.10

Thus we see that the projections of x and y on x-axis form the vector x' and similarly the projections on y'-axis form the vector y'. Obviously, one can look at the Lorentz transformations also in the same way, where x' and t' depend on x and t in a way similar to the dependence of x' and y' on x and y. The Lorentz transformations

$$x' = \frac{x - vt}{\sqrt{1 - \frac{v^2}{c^2}}}, y' = y, z' = z$$

$$t' = \frac{t - \frac{vx}{c^2}}{\sqrt{1 - \frac{v^2}{c^2}}} \tag{1.76}$$

can also be put as

$$x' = \frac{x + \frac{iv}{c}(ict)}{\sqrt{1 + \left(\frac{iv}{c}\right)^2}}, \; y' = y, z' = z$$

and
$$ict' = \frac{ict - \left(\frac{iv}{c}\right)x}{\sqrt{1 + \left(\frac{iv}{c}\right)^2}} \tag{1.77}$$

We have
$$x' = x\left(\frac{1}{\sqrt{1 + \left(\frac{iv}{c}\right)^2}}\right) + ict\frac{\left(\frac{iv}{c}\right)}{\sqrt{1 + \left(\frac{iv}{c}\right)^2}}$$

and
$$ict' = ict\left(\frac{1}{\sqrt{1 + \left(\frac{iv}{c}\right)^2}}\right) - x\frac{\left(\frac{iv}{c}\right)}{\sqrt{1 + \left(\frac{iv}{c}\right)^2}} \tag{1.78}$$

Writing $\frac{1}{\sqrt{1 + \left(\frac{iv}{c}\right)^2}} = \cos\alpha$ and $\frac{\left(\frac{iv}{c}\right)}{\sqrt{1 + \left(\frac{iv}{c}\right)^2}} = \sin\alpha$, one obtains

$$x' = x\cos\alpha + (ict)\sin\alpha$$

or
$$ict' = (ict)\cos t\alpha - x\sin\alpha \tag{1.79}$$

Thus if we treat (ict) also as a vector like x, y and z, the Lorentz transformations results from the rotation of the coordinate frame by an angle θ in the $[x, (ict)]$ plane Fig. (1.11). The angle of rotation α is given by

$$\alpha = \tan^{-1}\left(\frac{iv}{c}\right) \tag{1.80}$$

Let us consider the above problem in another way also. In section 18 we have seen that a given point (x, y, z, t) in the four-dimensional space represents an event, the four coordinates tell us where it happens and when it happens. Let us introduce the notation

$$x_1 = x, x_2 = y, x_3 = z \quad \text{and} \quad x_4 = ict \tag{1.81}$$

the relation

$$x^2 + y^2 + z^2 - (ict)^2 = x'^2 + y'^2 + z'^2 - c^2t'^2$$

becomes

$$x_1^2 + x_2^2 + x_3^2 + x_4^2 = x_1'^2 + x_2'^2 + x_3'^2 + x_4'^2$$

or
$$\sum_{i=1}^{4} x_i^2 = \text{constant} \tag{1.82}$$

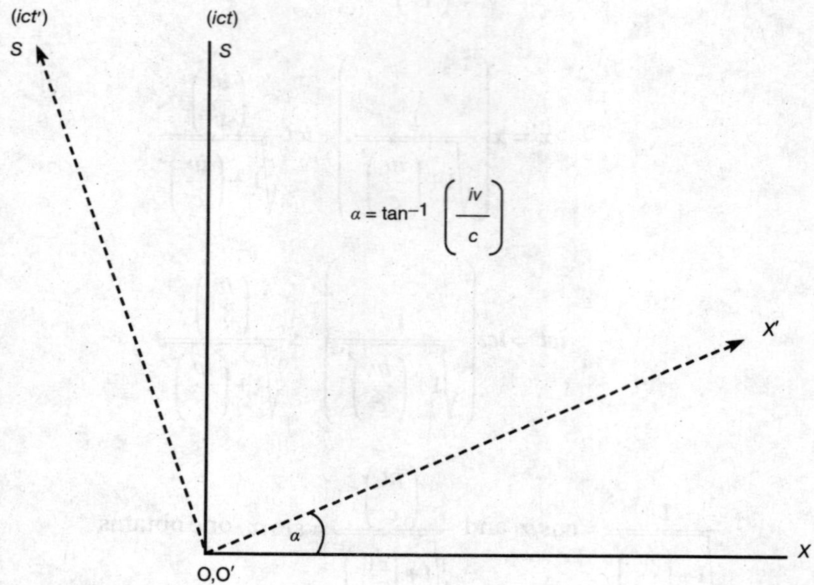

Fig. 1.11

The Lorentz transformation relations Eq. (1.10) can be written as

$$\begin{aligned} x_1' &= \gamma x_1 + i\beta\gamma x_4 \\ x_2' &= x_2 \\ x_3' &= x_3 \\ x_4' &= -i\beta\gamma x_1 + \gamma x_4 \end{aligned} \tag{1.83}$$

Eq. (1.83) in Matrix notation become

$$\begin{vmatrix} x_1' \\ x_2' \\ x_3' \\ x_4' \end{vmatrix} = \begin{vmatrix} \gamma & 0 & 0 & i\beta\gamma \\ 0 & 1 & 0 & 0 \\ 0 & 0 & 1 & 0 \\ -i\beta\gamma & 0 & 0 & \gamma \end{vmatrix} \begin{vmatrix} x_1 \\ x_2 \\ x_3 \\ x_4 \end{vmatrix} \tag{1.84}$$

Where

$$\gamma = \frac{1}{\sqrt{1+\left(\frac{iv}{c}\right)^2}} = \frac{1}{\sqrt{1+\frac{v^2}{c^2}}}$$

and

$$\beta = \frac{v}{c} \tag{1.85}$$

1.21 THE VELOCITY FOUR VECTOR

The components of the four-velocity are obtained by differentiating the position four-vector components with respect to the proper time as

$$v_i = \frac{dx_i}{d\tau}$$

where $i = 1, 2, 3$ and 4. Thus

$$\begin{aligned} v_1 &= \frac{dx_1}{d\tau} = \frac{dx_1}{dt}\frac{dt}{d\tau} = \gamma\frac{dx_1}{dt} = \gamma\dot{x} \\ v_2 &= \frac{dx_2}{d\tau} = \frac{dx_2}{dt}\frac{dt}{d\tau} = \gamma\frac{dx_2}{dt} = \gamma\dot{y} \\ v_3 &= \frac{dx_3}{d\tau} = \frac{dx_3}{dt}\frac{dt}{d\tau} = \gamma\frac{dx_3}{dt} = \gamma\dot{z} \\ v_4 &= \frac{dx_4}{d\tau} = \frac{dx_4}{dt}\frac{dt}{d\tau} = \gamma\frac{dx_4}{dt} = \gamma\frac{d(ict)}{dt} = ic\gamma \end{aligned} \tag{1.86}$$

$$= -c^2 \text{(invariant)} \tag{1.87}$$

Obviously, velocity transformations will be given by

$$\begin{vmatrix} v_1' \\ v_2' \\ v_3' \\ v_4' \end{vmatrix} = \begin{vmatrix} \gamma & 0 & 0 & i\beta\gamma \\ 0 & 1 & 0 & 0 \\ 0 & 0 & 1 & 0 \\ -i\beta\gamma & 0 & 0 & \gamma \end{vmatrix} \begin{vmatrix} v_1 \\ v_2 \\ v_3 \\ v_4 \end{vmatrix} \tag{1.88}$$

or

$$\begin{aligned} v_1' &= \gamma v_1 + i\beta\gamma v_4 \\ v_2' &= v_2 \\ v_3' &= v_3 \\ v_4' &= -i\beta\gamma v_1 + \gamma v_4 \end{aligned} \tag{1.89}$$

Let us see whether the relations (1.89) are the same as Eqs. (1.31), (1.32) and (1.33) obtained using Eq. (1.86), one obtains.

$$v_1' = \gamma' x' = \frac{v_x'}{\sqrt{1-\frac{v'^2}{c^2}}}$$

$$v_2' = \gamma' y' = \frac{v_y'}{\sqrt{1-\frac{v'^2}{c^2}}}$$

$$v_3' = \gamma' z' = \frac{v_z'}{\sqrt{1-\frac{v'^2}{c^2}}}$$

$$v_4' = ic\gamma' = \frac{ic}{\sqrt{1-\frac{v'^2}{c^2}}} \tag{1.90}$$

Hence from Eqs. (1.86), (1.89) and (1.90), one obtains

$$v_1' = \frac{U_x'}{\sqrt{1-\frac{v'^2}{c^2}}} = \frac{U_x}{\sqrt{1-\frac{v^2}{c^2}}} - \frac{v}{\left(1-\frac{v^2}{c^2}\right)} = \frac{U_x - v}{1-\frac{v^2}{c^2}} \tag{1.91}$$

$$v_2' = \frac{U_y'}{\sqrt{1-\frac{v'^2}{c^2}}} = \frac{U_y}{\sqrt{1-\frac{v^2}{c^2}}} \tag{1.92}$$

$$v'_3 = \frac{U'_z}{\sqrt{1-\frac{v'^2}{c^2}}} = \frac{U_z}{\sqrt{1-\frac{v^2}{c^2}}} \tag{1.93}$$

$$v'_4 = \frac{ic}{\sqrt{1-\frac{v'^2}{c^2}}} = \frac{-i\frac{v}{c}v_x}{\left(1-\frac{v^2}{c^2}\right)} = \frac{ic}{\left(1-\frac{v^2}{c^2}\right)}$$

$$= \frac{ic\left(1-\frac{vu_x}{c^2}\right)}{\left(1-\frac{v^2}{c^2}\right)} \tag{1.94}$$

From Eq. (1.90), we have

$$\frac{1}{\sqrt{1-\frac{v'^2}{c^2}}} = \frac{1-\frac{vu_x}{c^2}}{\left(1-\frac{v'^2}{c^2}\right)} \tag{1.95}$$

Substituting this value in Eq. (1.91), one obtains

$$v'_x = \frac{u_x - v}{1-\frac{vu_x}{c^2}}$$

$$v'_y = \frac{u_y\sqrt{1-\frac{v^2}{c^2}}}{\left(1-\frac{vu_x}{c^2}\right)}$$

$$v'_z = \frac{u_z\sqrt{1-\frac{v^2}{c^2}}}{\left(1-\frac{vu_x}{c^2}\right)} \tag{1.96}$$

We observe that the relations (1.96) are the same as the relations (1.31), (1.32) and (1.33). This proves the utility of the concept of four vector space.

1.22 THE MOMENTUM FOUR-VECTOR

We know that in Newtonian mechanics, momentum is defined as the product of mass and velocity. The four-momentum or the relativistic momentum in four vector space is given by the product of rest mass and the derivative of the position vector w.r.t. proper time, i.e., the four-momentum is formed simply by multiplying the four-velocity by the rest mass. Thus

$$P_i = m_0 v_i = m_0 \frac{dx_i}{d\tau} \tag{1.97}$$

Thus the four components of momentum vector are

$$\begin{aligned}
P_1 &= m_0 \frac{dx_1}{d\tau} = m_0 \gamma \dot{x} = m\dot{x} = P_x \\
P_2 &= m_0 \frac{dx_2}{d\tau} = m_0 \gamma \dot{y} = m\dot{y} = P_y \\
P_3 &= m_0 \frac{dx_3}{d\tau} = m_0 \gamma \dot{z} = m\dot{z} = P_z \\
P_4 &= m_0 \frac{dx_4}{d\tau} = m_0 (ic\gamma) = \frac{ic^2 m}{c} = \frac{iE}{c}
\end{aligned} \tag{1.98}$$

where $E = mc^2$ is the total relativistic energy. In the matrix from Eq. (1.98) can be expressed as

$$\rho_1 = \begin{pmatrix} \gamma m_0 u_x \\ \gamma m_0 u_y \\ \gamma m_0 u_z \\ \gamma m_0 ic \end{pmatrix} = \begin{pmatrix} mu_x \\ mu_y \\ mu_z \\ imc \end{pmatrix} = \begin{pmatrix} \rho_x \\ \rho_y \\ \rho_z \\ i\frac{E}{c} \end{pmatrix} \tag{1.99}$$

From Eqs. (1.98) and (1.99) it is clear that the spatial components P_1, P_2 and P_3 are the same as the components in ordinary three dimensional space, taking the mass as relativistic mass and squaring and adding the four components, one obtains

$$\sum_{i=4}^{4} P_i^2 = P_1^2 + P_2^2 + P_3^2 + P_4^2$$

$$= m_0^2 \gamma^2 (x^2 + y^2 + z^2 - c^2)$$

$$= m_0^2 \gamma^2 (v^2 - c^2)$$

$$= m_0^2 \frac{(v^2 - c^2)}{\left(1 - \frac{v^2}{c^2}\right)} - m_0^2 c^2 \tag{1.100}$$

Obviously, $\sum_{i=1}^{4} P_i^2$ is invariant. Writing the first three components in terms of P_x, P_y, P_z respectively and the fourth in terms of E and using Eq. (1.100), one obtains

$$\sum P_i^2 = \left(P_x^2 + P_y^2 + P_z^2 - \frac{E^2}{c^2} \right) = -m_0^2 c^2$$

or
$$P^2 - \frac{E^2}{c^2} = -m_0^2 c^2$$

or
$$E^2 = P^2 c^2 + m_0^2 c^4 \tag{1.101}$$

Which is same as relation Eq. (1.68), derived earlier.

1.23 THE FOUR-FORCE VECTOR

The components of the four-force are obtained by differentiating the four-momentum w.r.t. the proper time. For Example,

$$F_1 = \gamma \frac{d}{dt}, P_1 = \gamma \frac{dP_x}{dt} = \gamma F_x$$

Similarly
$$F_2 = \gamma F_y$$

$$F_3 = \gamma F_z$$

$$F_4 = \gamma \frac{dP_4}{dt} = \gamma \frac{i}{c} \frac{dE}{dt} = \gamma \frac{i}{c} \mathbf{F.u} \tag{1.102}$$

Thus, the first three components of the four-force are the three vector force components multiplied by γ.

$$F_i = \begin{pmatrix} \gamma F_x \\ \gamma F_y \\ \gamma F_z \\ \gamma \frac{i}{c} \boldsymbol{F.u} \end{pmatrix} \tag{1.103}$$

1.24 RELATIVITY AND ELECTROMAGNETISM

It is interesting to note that contrary to the situation in mechanics, review of the laws of electromagnetic field reveals them to be in harmony with relativity as they stand. One expects this in view of the fact that the theory of relativity actually developed out of experiments in that part of the field of electromagnetism which is called optics. However, there are many aspects of electromagnetic theory to which it is useful to apply theory of relativity. Use of four-space notation is found to be of great advantage.

We now postulate that charge of a particle is a scalar invariant. Let us consider a uniform charge distribution at rest in frames. The charge density in a small cube $\Delta x\, \Delta y\, \Delta z$ is $\rho_0 = \Sigma q_i/(\Delta x\, \Delta y\, \Delta z)$, where the sum is carried out over all charges within the cube. An observer O' in a moving frame S' agrees on the charge enclosed in a cube but in a accordance with the equation.

$$L = L_0\sqrt{\frac{1-v^2}{c^2}}$$

we have $\Delta x' = \Delta x\sqrt{1-v^2/c^2}$, although $\Delta y' = \Delta y$ and $\Delta z' = \Delta z$. Thus OO' finds the charge density to be $\rho' = \rho_0/\sqrt{1-\frac{v^2}{c^2}} = \gamma_T \rho_0$, where $\gamma_T = 1/\sqrt{1-v^2/c^2}$. Obviously, because of the Lorentz-Fitzgerald Contraction of the volume elements, ρ' exceeds ρ?.

Now, we consider a uniform charge distribution of density moving with velocity $\boldsymbol{v}$ relative to frame S. The current density is then $\boldsymbol{J} = \Sigma q_i \mathbf{v}/(\Delta x\, \Delta y\, \Delta z) = \rho \mathbf{v} = \gamma \rho_0 \mathbf{v}$, where $\gamma = 1/\sqrt{1-v^2/c^2}$. This shows that current density J' in S' is different, both because v' differs from v and also due to space contraction.

We may note that neither ρ nor $\boldsymbol{J}$ is invariant under Lorentz transformations. We introduce a **current density four-vector** as follows:

Let $\underline{u}$ be the four-velocity of the charges in the S-frame and ρ_0 be the **proper charge density**, viz, that in a frame in which the charges are at rest. Then the four vector $\underline{J} = \rho_0 \underline{u}$ is,

$$\begin{aligned} J &= 1_x \gamma \rho_0 v_x + 1_y \gamma \rho_0 v_y + 1_z \gamma \rho_0 v_z + 1_\omega i \gamma c \rho_0 \\ &= 1_{x'} J_x + 1_{y'} J_y + 1_{z'} J_z + 1_{\omega'} J_\omega \end{aligned} \tag{1.104}$$

Where $\gamma = 1/\sqrt{1-v^2/c^2}$

We may note that the spatial components of J correspond to the components of the current density, while the fourth component J_ω is $ic\gamma\, \rho_0 = ic\rho$.

We have the equation of continuity in electromagnetism, which follows from the conservation of charge

$$\nabla.\mathbf{J} + \frac{\partial \rho}{\partial t} = 0 \tag{1.105}$$

In four vector operator (□)

$$\Box = 1_x \frac{\partial}{\partial x} + 1_y \frac{\partial}{\partial y} + 1_z \frac{\partial}{\partial z} + 1_\omega \frac{\partial}{\partial \omega} \quad (1.106)$$

We can express the equation of continuity as

$$\Box \cdot J = 0 \quad (1.107)$$

1.25 MAXWELL'S EQUATIONS AND FOUR POTENTIAL

The ordinary force $\boldsymbol{F}$ acting on a charge q moving with velocity $\boldsymbol{v}$ is given by **Lorentz relation**.

$$\boldsymbol{F} = q[\boldsymbol{E} + (\mathbf{v} \times \boldsymbol{B})] \quad (1.108)$$

Where the electric field intensity $\boldsymbol{E}$ is the force per unit charge when the charge q is at rest relative to the reference frame. The difference $\boldsymbol{F} - q\boldsymbol{E} = q(\mathbf{v} \times \boldsymbol{B})$ is the force due to the magnetic induction $\boldsymbol{B}$. Obviously, the electric intensity $\boldsymbol{E}$ and the magnetic intensity $\boldsymbol{B}$ are thus defined for a particular reference frame; the distinction between electric and magnetic fields is in part a relative one, depending on the frame of reference, e.g. charges at rest relative to the S frame produce only an electrostatic field, but to an observer in the S' frame, these same charges constitute current elements with an associated magnetic field.

The electric displacement vector $\boldsymbol{D}$ is defined by the relation $\boldsymbol{D} = \varepsilon_0 \boldsymbol{E} + \boldsymbol{P}$, where $\boldsymbol{P}$ is the polarization (the electric dipole moment per unit volume) and E_0 is the permittivity of free space. Similarly, one can define the magnetizing field $\boldsymbol{H}$ by $\boldsymbol{H} = \boldsymbol{B}/\mu_0 + \boldsymbol{M}$, where $\boldsymbol{M}$ is the magnetization (magnetic moment per unit volume). If $\boldsymbol{J}$ represents current density and ρ charge density, these quantities are related through famous Maxwell's electromagnetic equations. We may note that though the theory of electromagnetic fields was formulated by Maxwell prior to the advent of the special theory of relativity, it is **form-invariant** under Lorentz transformations. In particular, it is consistent with the speed of propagation of electromagnetic radiation being the same in all inertial frames of references. Maxwell's equations are

$$\nabla \times \boldsymbol{E} = -\frac{\partial B}{\partial t} \quad (1.109)$$

$$\nabla \times \boldsymbol{D} = \rho \quad (1.110)$$

$$\nabla \times \boldsymbol{H} = \boldsymbol{J} + \frac{\partial \boldsymbol{D}}{\partial t} \quad (1.111)$$

$$\nabla \times \boldsymbol{B} = 0 \quad (1.112)$$

These four equations form the basis of all electromagnetic theory. It is most convenient to describe the electromagnetic fields in terms of the vector potential $\boldsymbol{A}$ and scalar potential ϕ. In view of Eq. (1.110), $\boldsymbol{B}$ can be expressed as the curl of a vector potential $\boldsymbol{A}$;

$$\boldsymbol{B} = \nabla \times \boldsymbol{A} \quad (1.113)$$

Substituting Eq. (1.113) in Eq. (1.109) to find **E**, one obtains

$$E=-\frac{\partial A}{\partial t}-\nabla\phi \tag{1.114}$$

where the constant of integeration is written as minus the gradient of a scalar electrostatic potential ϕ, determined by the distribution of charges in space. In order to reduce complications, one may confine this potential at any point *P* by making this point the origin of an *S* frame, i.e. S-coordinate system and evaluation of the integral, i.e.

$$\phi(t)=\frac{1}{4\pi\varepsilon_0}\int_{space}\frac{\rho(r-t-r/c)}{r}dV \tag{1.115}$$

Where dV is the volume element. In order to evaluate ϕ at time t, one must evaluate ρ at time $t-r/c$.

Similarly, one can see that the sources of the vector potential **A** are current, and **A** is given by

$$A(t)=\frac{\mu_0}{4\pi}\int_{space} J\frac{\rho(\mathbf{r},t-r/c)}{r}dV \tag{1.116}$$

Since in free space $c^2=1/\mu_0\varepsilon_0$, Eq. (1.115) can be expressed as

$$\frac{i\phi}{c}=\frac{\mu_0}{4\pi}\int_{space}\frac{ic\rho(\mathbf{r},t-r/c)}{r}dV$$

$$=\frac{\mu_0}{4\pi}\int_{space}\frac{J_\omega(\mathbf{r},t-r/c)}{r}dV$$

which represents the time component A_ω of a four-potential:

$$A=1_xA_x+1_yA_y+1_zA_z+1_\omega Å_\omega += A+1_\omega\frac{i\phi}{c} \tag{1.117}$$

In free space, polarization **P** and magnetization **M** are both zero, and so $D=\varepsilon_0 E$, and $B=\mu_0 H$. Now, taking the divergence of Eq. (1.114), which with the aid of Eq. (1.110) yields.

$$-\nabla\cdot E=\nabla^2\phi+\nabla\cdot\frac{\partial A}{\partial t}=-\frac{\rho}{\varepsilon_0} \tag{1.118}$$

While Eqs. (1.108) to (1.114) can be combined to give

$$\nabla^2 A-\nabla(\nabla\cdot A)-\mu_0\varepsilon_0\frac{\partial^2 A}{\partial t^2}-\mu_0\varepsilon_0\frac{\partial\nabla\phi}{\partial t}=-\mu_0 J \tag{1.119}$$

We may note that **A** is not completely defined by Eq. (1.113), since an infinity of vectors may have the same curl. We are further free to require that its divergence have some specified value; in this case one can follow Lorentz in imposing the condition.

$$\nabla A+\mu_0\varepsilon_0\frac{\partial\phi}{\partial t}=0 \tag{1.120}$$

Since A and ϕ are both continuous functions of x, y, z and t, one may exchange the order of differentiation. This together with Lorentz condition and $c^2 = 1/\mu_0\varepsilon_0$ permits one to write Eqs. (1.119) and (1.118) in the form

$$\nabla^2 \boldsymbol{A} - \frac{1}{c^2}\frac{\partial^2 \boldsymbol{A}}{\partial t^2} = -\mu_0 \boldsymbol{J} \tag{1.121}$$

$$\nabla^2 \varphi - \frac{1}{c^2}\frac{\partial^2 \phi}{\partial t^2} = -\frac{\rho}{\varepsilon_0} = -\mu_0 c^2 \rho \tag{1.122}$$

With $$\square^2 = \frac{\partial^2}{\partial x^2} + \frac{\partial^2}{\partial y^2} + \frac{\partial^2}{\partial z^2} + \frac{\partial^2}{\partial \omega^2} = \nabla^2 - \frac{1}{c^2}\frac{\partial^2}{\partial t^2}$$

These equations take the form

$$\square^2 \boldsymbol{A} = -\mu_0 \boldsymbol{J} \quad \square^2 \phi = -\rho/\varepsilon_0 = -\mu_0 c^2 \rho$$

or $$\square^2 \boldsymbol{A} = -\mu_0 \boldsymbol{J} \tag{1.123}$$

in four-vector notation. In the above form the Lorentz transformations of Maxwell's equations are most transparent. In terms of four potential $\boldsymbol{A}$ the Lorentz condition is simple $\square \cdot \times \underline{\boldsymbol{A}} = 0$. We may note that $c\varphi$ also transformers as a four vector.

1.25.1 The Current Density 'Four-Vector'

If ρ_0 is the charge density in a frame where the charges are at rest, then a current density four-vector may be defined as the product of ρ_0 and the four-velocity, i.e.

$$J_1 = P_0 U_i = \begin{pmatrix} \gamma\rho_0 u_x \\ \gamma\rho_0 u_y \\ \gamma\rho_0 u_z \\ \gamma\rho_0 ic \end{pmatrix} = \begin{pmatrix} \rho u_x \\ \rho u_y \\ \rho u_z \\ \rho ic \end{pmatrix} = \begin{pmatrix} J_x \\ J_y \\ J_z \\ \rho ic \end{pmatrix} \tag{1.124}$$

Here we have defined $\rho = \gamma\rho_0$, illustrating the increase in charge density due to the change in volume element. Then the product of local charge density and local velocity gives the ordinary three-space current density (J_x, J_y, J_z).

1.25.2 The Field Tensor

When one knows the components of the four potential in one frame, one can make a Lorentz transformation to a moving frame and, from the potentials, one can find the electric and magnetic intensities. Usually, one knows the field in one reference frame and like to find the fields in another frame without going through the potentials. One can achieve this through the field tensor, f_{ik}, given by

$$f_{ik} = \Box \times \mathbf{A} = \begin{bmatrix} 0 & \frac{\partial A_y}{\partial x} - \frac{\partial A_x}{\partial y} & \frac{\partial A_z}{\partial x} - \frac{\partial A_x}{\partial z} & \frac{\partial A_\omega}{\partial x} - \frac{\partial A_x}{\partial \omega} \\ \frac{\partial A_x}{\partial z} - \frac{\partial A_z}{\partial x} & 0 & \frac{\partial A_z}{\partial y} - \frac{\partial A_y}{\partial z} & \frac{\partial A_\omega}{\partial y} - \frac{\partial A_y}{\partial \omega} \\ \frac{\partial A_x}{\partial z} - \frac{\partial A_z}{\partial x} & \frac{\partial A_y}{\partial z} - \frac{\partial A_z}{\partial y} & 0 & \frac{\partial A_\omega}{\partial z} - \frac{\partial A_z}{\partial \omega} \\ \frac{\partial A_x}{\partial \omega} - \frac{\partial A_\omega}{\partial x} & \frac{\partial A_y}{\partial \omega} - \frac{\partial A_\omega}{\partial y} & \frac{\partial A_z}{\partial \omega} - \frac{\partial A_\omega}{\partial z} & 0 \end{bmatrix}$$

$k \rightarrow$

$$= \mathbf{i}\downarrow \begin{bmatrix} 0 & B_x & -B_y & \frac{iE_x}{c} \\ -B_z & 0 & B_x & \frac{iE_y}{c} \\ B_y & -B_x & 0 & \frac{iE_x}{c} \\ \frac{iE_x}{c} & \frac{iE_y}{c} & \frac{iE_z}{c} & 0 \end{bmatrix} \tag{1.125}$$

We may note that the field tensor is an antisymmetric second-rank tensor. In order to transform such a tensor from frame S to frame S' is some what more involved than to transform a vector. A four vector for example, transforms to

$$V_i' = \sum_{k=1}^{4} C_{ik} V_k \tag{1.126}$$

Where C_{ik} is the direction cosine between the i-axis in the S' frame and the k-axis in the S frame and V_k is respectively the x, y, z or ω component of V for $k = 1, 2, 3$ or 4. We may see that the terms of second-rank tensor have two indices and transform according to the following relation

$$T_{ik}^{j} = \sum_{i=1}^{4} \sum_{m=1}^{4} C_{ij} C_{km} T_{jm} \tag{1.127}$$

In matrix notation, one can express it as

$$\begin{bmatrix} T'_{11} & T'_{12} & T'_{13} & T'_{14} \\ T'_{21} & T'_{22} & T'_{23} & T'_{24} \\ T'_{31} & T'_{32} & T'_{33} & T'_{34} \\ T'_{41} & T'_{42} & T'_{43} & T'_{44} \end{bmatrix} = \begin{bmatrix} C_{11} & C_{12} & C_{13} & C_{14} \\ C_{21} & C_{22} & C_{23} & C_{24} \\ C_{31} & C_{32} & C_{33} & C_{34} \\ C_{41} & C_{42} & C_{43} & C_{44} \end{bmatrix} \begin{bmatrix} T_{11} & T_{12} & T_{13} & T_{14} \\ T_{21} & T_{22} & T_{23} & T_{24} \\ T_{31} & T_{32} & T_{33} & T_{34} \\ T_{41} & T_{42} & T_{43} & T_{44} \end{bmatrix} \begin{bmatrix} C_{11} & C_{12} & C_{13} & C_{14} \\ C_{12} & C_{22} & C_{32} & C_{42} \\ C_{13} & C_{23} & C_{33} & C_{43} \\ C_{14} & C_{24} & C_{34} & C_{44} \end{bmatrix} \tag{1.128}$$

One can easily verify the above by multiplying out the matrices and comparing the result with the expansion of Eq. (1.127). We may note that the third matrix on the right hand side of Eq. (1.128) is the transpose of the first, i.e., the matrix formed by interchanging rows and columns so that the elements in the ith row and k^{th} column of the original matrix appears in the kth row and i^{th} column of the transpose matrix.

The field tensor in the S' frame is therefore obtained as,

$$f_{ik}' = \begin{bmatrix} \gamma_T & 0 & 0 & i\beta_T\gamma_T \\ 0 & 1 & 0 & 0 \\ 0 & 0 & 1 & 0 \\ -i\beta_T\gamma_T & 0 & 0 & \gamma_T \end{bmatrix} \begin{bmatrix} 0 & B_x & -B_y & -\frac{iE_x}{c} \\ -B_x & 0 & B_z & -\frac{iE_y}{c} \\ B_y & -B_z & 0 & -\frac{iE_z}{c} \\ \frac{iE_x}{c} & \frac{iE_y}{c} & \frac{iE_z}{c} & 0 \end{bmatrix} \begin{bmatrix} \gamma_T & 0 & 0 & -i\beta_T\gamma_T \\ 0 & 1 & 0 & 0 \\ 0 & 0 & 1 & 0 \\ i\beta_T\gamma_T & 0 & 0 & \gamma_T \end{bmatrix} \tag{1.129}$$

from which we obtain

$$B_x' = B_x,\ B_y' = \gamma_T\left(B_y + \frac{VE_z}{c^2}\right) \quad \text{and} \quad B_z' = \gamma_T\left(B_z + \frac{VE_y}{c^2}\right) \tag{1.130}$$

$$E_x' = E_x,\ E_y' = \gamma_T(E_y - VB_z) \quad \text{and} \quad E_z' = \gamma_T(E_z - VB_y) \tag{1.131}$$

One can use these equations to ascertain the effect of uniform motion upon the field of a point charge. Let us consider that charge q is moving with speed v. One can take x-axis through the charge and in the direction of its motion. Now, we consider any point P on the x-axis and let this point and $\boldsymbol{v}$ establish the xy as shown in Fig. [1.12(a)].

One can choose an S' frame in which q is at rest; in this frame the field is purely electrostatic, and its components are

$$E_x' = \frac{qx'}{4\pi\varepsilon_0 r'^3},\quad E_y' = \frac{qy'}{4\pi\varepsilon_0 r'^3} \quad \text{and} \quad E_z' = 0 \tag{1.132}$$

Since the frame S is moving with a velocity $\mathbf{V} = -\mathbf{v}$ relative to frame S', the transformation equations to the frame S in accordance with Eq. (1.131) are

$$B_y = B_x' \quad B_y = \gamma\left(B_y' - \frac{vE_z'}{c^2}\right) \quad B_z = \gamma\left(B_z' + \frac{vE_y'}{c^2}\right)$$

$$E_x = E_x' \quad E_y = \gamma\left(E_y' - vE_z'\right) \quad B_z = \gamma\left(E_z' + vB_z'\right) \tag{1.133}$$

Now, in S' frame, $B' = 0$ and so $B_x' = 0 = B_y' = B_z'$. We have $x' = \gamma(x - vt)$ and $y' = y$. More over, $x - xt = r\cos\theta$, $y = r\sin\theta$, and

$$r'^2 = x'^2 + y'^2 = \gamma^2 (x - vt)^2 + y^2$$
$$= \gamma^2 r^2 \cos^2\theta + r^2 \sin^2\theta = \gamma^2 r^2[\cos^2\theta + (\sin^2\theta)/\gamma^2 = \gamma^2 r^2[1 - v^2(\sin^2\theta)/c^2]$$

Substituting these in Eqs. (1.133), one obtains

$$E_x = \frac{1}{4\pi\varepsilon_0 r^2}\frac{q\left(1-\frac{v^2}{c^2}\right)\cos\theta}{[1-v^2(\sin^2\theta)/c^2]^{1/2}}$$

$$E_y = \frac{q}{4\pi\varepsilon_0 r^2}\frac{(1-v^2/c^2)\sin\theta}{[1-v^2(\sin^2\theta)/c^2]^{1/2}}$$

$$E_z = 0 = B_x = B_y,\ B_z = \frac{qv\left(1-\frac{v^2}{c^2}\right)\sin\theta}{4\pi\varepsilon_0 r^2[1-v^2(\sin^2\theta)/c^2]^{1/2}} \tag{1.134}$$

Since $y/E_z = \tan\theta$, we have the electric field points radially outward from the instantaneously location of q. In spherical polar coordinates E is in the r-direction and $\boldsymbol{B}$ in the ϕ-direction in any inertial frame with its origin instantaneously at q. However the field of a moving charge is concentrated the equatorial plane Fig. [1.12(b)].

One can draw many other conclusions by transforming fields and potentials from one frame to another reference frame, e.g. one can easily show that **B.H – D.E, A.J–ρϕ** and **E.B** are invariants having the same values in both S and S' frames of references. For plane waves in free space $\mu_0 H^2 = \varepsilon_0 E^2$. Thus **B.H – D.E** = 0 in any inertial frame. This means to plane waves in another frame, although both direction of propagation and frequency are altered.

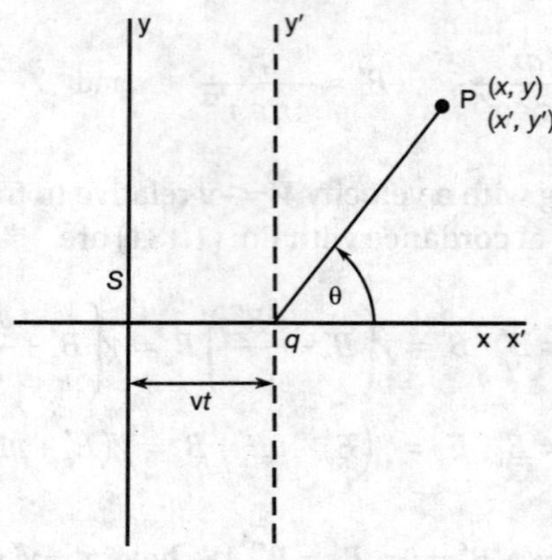

Fig. 1.12 (a) In frame S' charge q is at rest. S' is moving with speed v along the x-axis of the frame S

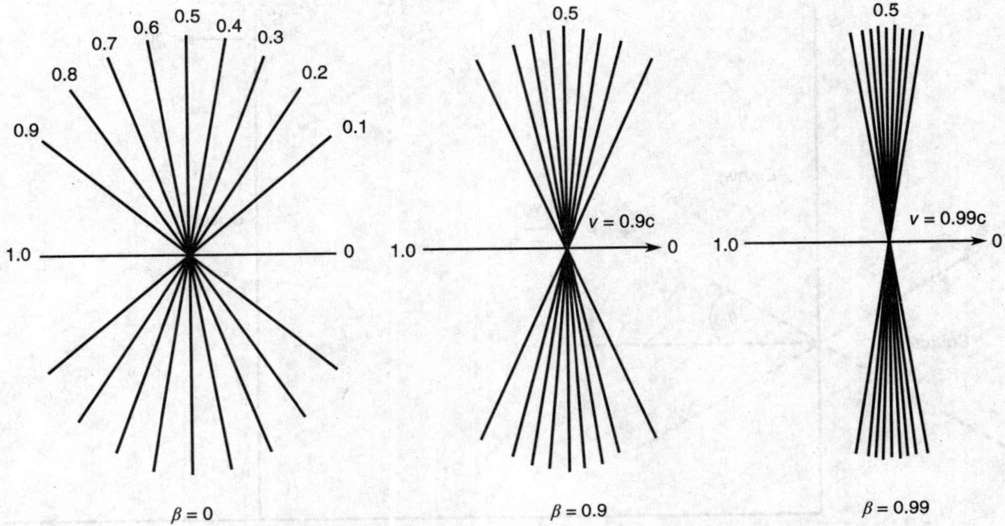

Fig. 1.12 (b) The concentration of the electric flux of a moving charge into equatorial plane, the decimals give the fraction of the total flux lying within a cone with axis v. We may note that inverse square law applies at all speeds

1.26 THE RELATIVISTIC DOPPLER EFFECT

The four-vector formalism can be used in a straightforward manner to obtain the Doppler frequency shift. One need only to incorporate the Planck expression for the energy of a photon. $E = h\nu$, and the relativistic expression for the momentum, $p = \frac{h\nu}{c}$, where c is the frequency. Consider a monochromatic source at rest in the S frame Fig. (1.13).

An observer moving to the right at velocity βc will detect a frequency which is lower than the proper frequency. For the case of the moving observer we have the transformation equation.

$$P_i' = \tau P_i$$

where τ is the following matrix

$$\tau = \begin{vmatrix} \gamma & 0 & 0 & i\beta\gamma \\ 0 & 1 & 0 & 0 \\ 0 & 0 & 1 & 0 \\ -i\beta\gamma & 0 & 0 & \gamma \end{vmatrix}$$

The equation for the fourth component becomes

$$(P_4)' = \gamma(P_4 - i\beta P_1)$$

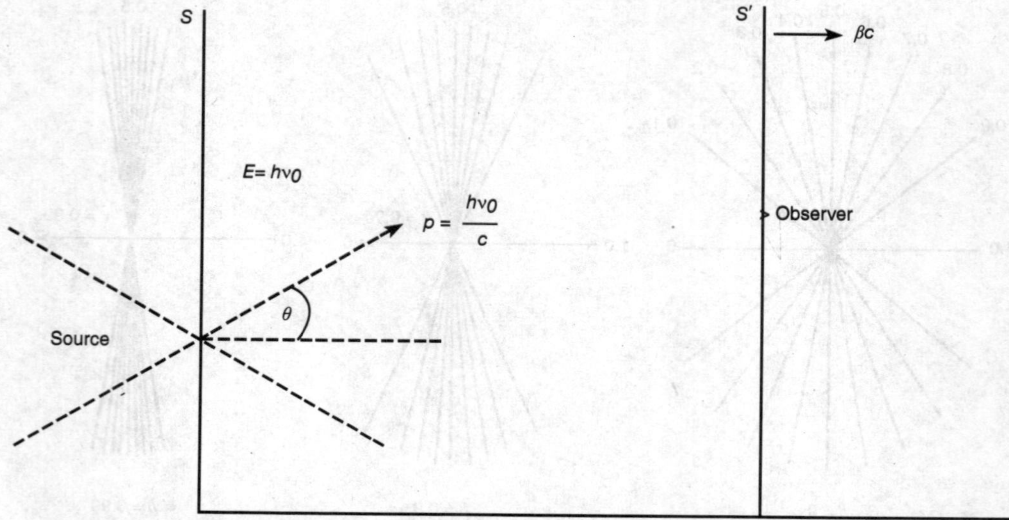

Fig. 1.13 Relative motion of observer and light source

or
$$\frac{ih\nu'}{c} = \gamma\left(\frac{ih\nu}{c} - i\beta\frac{h\nu_0}{c}\cos\theta\right)$$

and
$$\nu' = \frac{\nu_0(1-\beta\cos\theta)}{\sqrt{1-\beta^2}} \tag{1.135}$$

As the observer in the S' frame moves toward the source from the left, he sees light emitted at the angle $\theta = \pi$ and he measures the frequency,

$$\nu' = \frac{\nu_0\sqrt{1+\beta}}{\sqrt{1-\beta}} \quad \text{(approaching observer)} \tag{1.136}$$

As he passes the source he sees light emitted at the angle $\theta = \frac{\pi}{2}$ and he measures,

$$\nu' = \frac{\nu_0}{\sqrt{1-\beta^2}} \quad \text{(\textbf{transverse} Doppler effect)} \tag{1.137}$$

As he recedes to the right, the observer in S' measures light emitted at the angle $\theta = 0$ and he obtains

$$\nu' = \frac{\nu_0\sqrt{1-\beta}}{\sqrt{1+\beta}} \quad \text{(receding observer)} \tag{1.138}$$

Now, we consider the inverse transformation in which the observer is regarded as stationary in the *S* frame and source is moving. In matrix form, one can express it as

$$P_i = \tau^{-1} P_i' \tag{1.139}$$

The equation for the fourth component is

$$P_4 = \gamma(P_4' + i\beta P_1')$$

or

$$\upsilon = \frac{\upsilon_0 (1 + \beta \cos\theta')}{\sqrt{1-\beta^2}} \tag{1.140}$$

Again we have the three situations:

(i) Approaching source: $\theta' = 0$, and $\upsilon = \dfrac{\upsilon_0 \sqrt{1+\beta}}{\sqrt{1-\beta}}$

(ii) Receding source: $\theta' = \pi$, and $\upsilon = \dfrac{\upsilon_0 \sqrt{1-\beta}}{\sqrt{1+\beta}}$

(iii) Transverse motion: $\theta' = \dfrac{\pi}{2}$, and $\upsilon = \dfrac{\upsilon_0}{\sqrt{1-\beta^2}}$

The above results are consistent with the principle of relativity in that it is not possible to distinguish between motion of the source or motion of the observer. A relative velocity of separation decreases the observed frequency and a relative velocity of approach increase the observed frequency. It is worthwhile to mention that the transverse effect is simply an example of the phenomenon of time dilation, i.e., that moving clocks run slowly. These results can be summarized by the vector equation

$$\upsilon' = \frac{\upsilon_0 (1 - \beta \hat{n}.\hat{u})}{\sqrt{1-\beta^2}}$$

where $\beta \left(= \dfrac{u}{c} \right)$ is always positive, $\hat{n}$ is the unit vector, and $\hat{u}$ is the unit vector in the direction of the velocity of the primed frame.

Experimental confirmation of the relativistic Doppler effect first came from experiments done in 1938 by Ives and Stilwell. They sent a beam of hydrogen atoms, generated in a gas discharge, down a tube at a speed *u* Fig. (1.14). They could simultaneously observe light emitted by the atoms in a direction parallel to $\hat{u}$ (atom 1) and opposite to $\hat{u}$ (atom 2, reflected from the mirror).

They used a spectrograph to photograph the characterstic spectral lines from these atoms and also, on the same photographic plate, from atoms at rest. If the classical Doppler relation were valid, the wave lengths of the lines from atoms 1 and 2 would be placed at symmetric intervals $\Delta\lambda_1 = \pm\lambda_0(u/c)$ on either side of the line from the atoms at rest (wave length λ_0). The relativistic Doppler relation on the other hand, gives a small additional

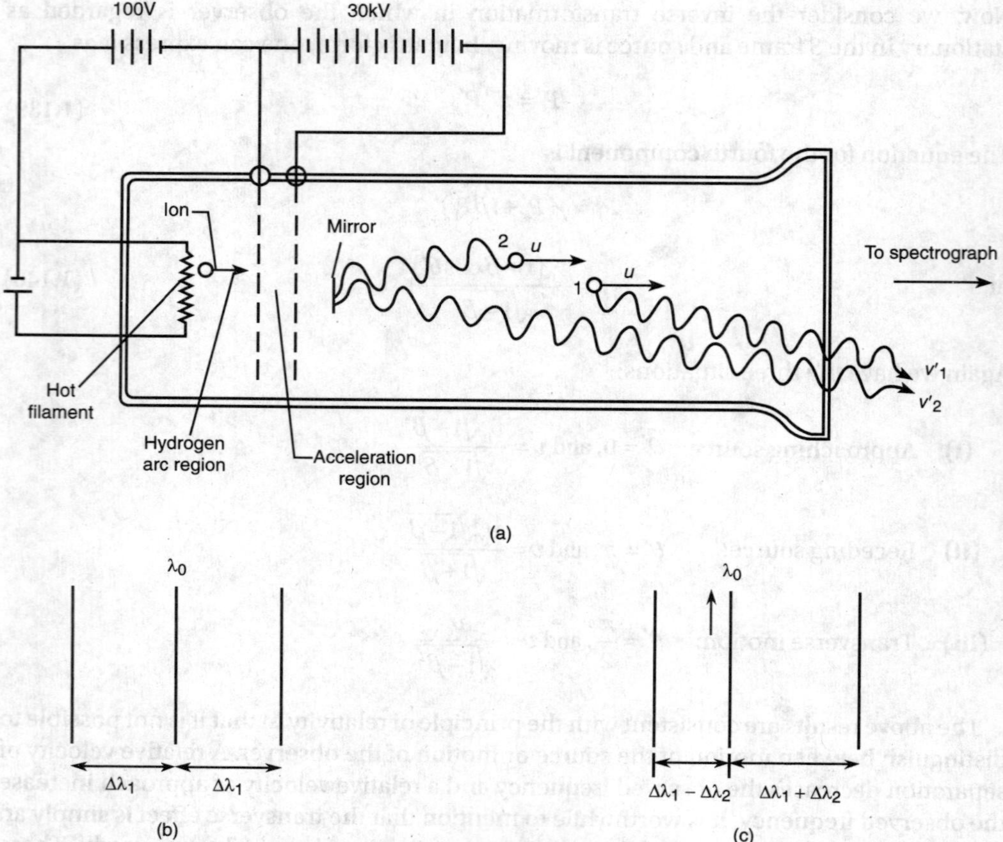

Fig. 1.14 Apparatus used in the Ives-Stilwell experiment (a) line spectrum effect (b) from classical Doppler effect (c) line spectrum expected from relativistic Doppler effect

symmetric shift $\Delta\lambda_2 = \frac{1}{2}\lambda_0(u/c)^2$ (computed for $u << c$ and hence higher order terms in u/c neglected). Fig. (1.15) shows the results of Ives and for one of hydrogen lines (the blue line of the Balmer series at $\lambda_0 = 486$ mm). The agreement between the experimental values and those predicted by relativistic theory is quite impressive.

Recent experiments with lasers have verified the relativistic formula at greater accuracy. These experiments are based on the absorption of laser light by an atom, when the radiation is absorbed, the atom changes from its lowest energy state (the ground state) to one of its excited states. The experiment consists essentially of comparing the laser wave length needed to excite atoms at rest with that needed for atoms in motion. One experiment used a beam of hydrogen atoms with kinetic energy 800 MeV (corresponding to $u/c = 0.84$) produced in a high energy proton accelerator. An ultraviolet laser was used to excite the atoms. This experiment verified the relativistic effect to an accuracy of about 3×10^{-4}. In another experiment MC Gowan, et al. in 1993

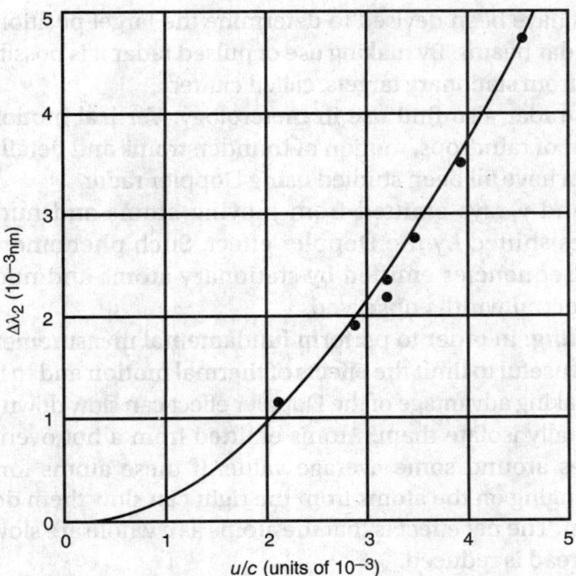

Fig. 1.15 Results of Ives and Stilwell experiment. According to classical Doppler theory, $\Delta\lambda_2 = 0$, while according to special relativity $\Delta\lambda_2$ depends on $(u/c)^2$

used a beam of neon atoms moving with a speed $u = 0.0036\,c$ was irradiated with light from a tunable laser. This experiment verified the relativistic Doppler effect shift to a precision of 2×10^{-6}.

1.27 APPLICATIONS OF THE DOPPLER EFFECT

Relativistic Doppler effect has many practical applications, three of which are discussed here.

(i) ***Astronomy:*** The Doppler shifts of known atomic transition frequencies determine the relative velocities of astronomical objects with respect to us. Such measurements are used to find the distances of unusual objects as *quasars* (objects having incredibly large masses that produce tremendous amounts of radiation). The Doppler effect has also been used to discover other mundane effects in astronomy, e.g. the rate of rotation of venus and the fact that venus rotates in the opposite direction of Earth – the Sun rises in the west on venus. This was determined by observing light reflected from both sides of Venus – on one side it is blue shifted and on the other side it is red shifted. The same technique has also been applied to determine the rate of rotation of stars.

(ii) ***Radar:*** when an electromagnetic radar signal reflects off a moving target, the so called *echo* signal will be shifted in frequency by the Doppler effect. Very small frequency shift can be determined by examining the beat frequency of the echo signal with a reference signal. The shift in frequency is proportional to the radial component of the target's is velocity. Navigation radar is quite complex, and ingenious techniques have been devised to determine the target position and velocity using multiple radar beams. By making use of pulsed radar it is possible to separate moving targets from stationary targets, called clutter.

Doppler radar also find use in meterology. Vertical motion of airdrafts, sizes and motion of raindrops, motion of thunderstroms and detailed patterns of wind distribution have all been studied using Doppler radar.

X-rays and γ-rays emitted from moving atoms and nuclei will have their frequencies shifted by the Doppler effect. Such phenomena help to broaden radiation frequencies emitted by stationary atoms and nuclei and add to the natural spectral widths observed.

(iii) ***Laser Cooling:*** In order to perform fundamental measurements in atomic physics, we find useful to limit the effects of thermal motion and to isolate single atoms. A method taking advantage of the Doppler effect can slow down even natural atoms and eventually isolate them. Atoms emitted from a hot oven will have a spread of velocities around some average value. If these atoms form a beam, a laser beam impinging on the atoms from the right can slow them down by transferring momentum. The net effect is that the atoms as a whole are slowed down and their velocity spread is reduced.

As the atoms slow down, they see that the Doppler shifted frequencies of the laser change, and the atoms no longer absorb the laser radiation. They continue with the same lower velocity and velocity spread. There are two methods to make the deceleration process continue: (i) The frequency of the laser beam is increased to keep the radiation consistent with the Doppler – shifted frequency needed to excite the atoms, and (ii) the laser frequency is kept constant, but carefully shaped magnetic fields are used to vary the frequencies needed to excite the atoms by changing the excited atomic energy levels. Using six intersecting laser beams coming in at different angles, an "*optical molasses*" has been created in which the atoms are essentially trapped and the average velocity is zero. In a remarkable series of experiments by researchers improving on the basic technique, atoms have been cooled to temperatures of 0.5 nK (5×10^{-10} K).

SUMMARY

1. The two postulates of the special theory of relativity are (i) the physical laws of nature are the same in all inertial frames of reference and (ii) the speed of light is the same in all inertial frames of reference. When physicists understood the implications of these two postulates, the concepts of Newtonian mechanics, although they had been broadly successful, had to give way to relativistic mechanics.
2. The Lorentz transformations are

$$x' = \frac{x - vt}{\sqrt{1-\beta^2}}$$

$$y' = y$$

$$z' = z$$

$$t' = \frac{t - vx/c^2}{\sqrt{1-\beta^2}}$$

The Lorentz transformation follows directly from the two postulates of the special relativity, any physical law which retains the same mathematical form under a Lorentz transformation automatically satisfies the principle of relativity. Such a law is said to be **Lorentz covariant law.**

3. Two important kinematical effects which derive from the theory of special relativity are the contraction of length parallel to the relative motion and the dilation of time.

$$L = L_0\sqrt{1-\frac{v^2}{c^2}},\ \Delta t = \Delta t_0\sqrt{1-\frac{v^2}{c^2}}$$

4. An important dynamical effect of the special theory of relativity is the increase of mass with velocity. It leads to the statement of the equivalence of mass and energy, $E = mc^2$. The conservation of total energy, which includes rest mass energy, now replaces the separate conservation laws for mass and energy in classical physics.
Energy and momentum are related by

$$E^2 = p^2c^2 + E_o^{\ 2}$$

where $E_o = m_oc^2$, m_o is the rest mass. In the case of massless particle, $E_o = 0$, so $E = pc$. Obviously, massless particles must travel at the speed of light.

5. The concept of 'four-vector' is developed by treating time as a fourth dimension on an equal footing with the three dimensions of Euclidean geometry. The invariance of the 'length' of a four-vector is a useful property in solving physical properties.
6. Maxwell's equations are invariant under transformations between any inertial reference frames. What appears as electric and magnetic magnetic fields is relative to the reference frame of the observer.

REVIEW QUESTIONS

1. Describe Michelson-Morley experiment and show how the negative results obtained there are interpreted?
2. Discuss the Michelson-Morley experiment to detect the motion of earth through ether. Give the difference in interpretation for the negative result as given by Einstein and Lorentz-Fitz-Gerald.

3. Show that the Lorentz-FitzGerald contraction can account for the null result of the Michelson-Morley experiment.
4. State postulates of special theory of relativity and deduce them from the Lorentz transformations.
5. Describe Michelson-Morley experiment and discuss its results.
6. Show that two simultaneous events at different positions in a frame of reference are not in general simultaneous in another inertial frame moving relative to this frame with velocity v.
7. Show how Lorentz transformations account for contraction of bodies in motion and the retardation of moving clocks.
8. Derive Lorentz transformation equations for two frames in uniform translation relative to each other.
9. What is time dilation in special relativity? Explain it by Lorentz transformations. Describe briefly one experiment in support of time dilation.
10. Derive the formula giving the variation of mass with velocity.
11. Write a note on Einstein mass-energy relation.
12. How does the law of conservation of momentum lead to the relativistic relation of variation of mass with velocity?
13. What do you understand by space-time continuum? Show that the Lorentz transformations result from the rotation of the four-vector coordinate frame.
14. Write the components of momentum in four-vector space and hence prove the invariance of $(E^2 - p^2c^2)$.
15. Define four-vector. Show that the magnitude of momentum four-vector is im_0 c.
16. Show that the Lorentz transformation is equivalent to a rotation in four-space.

PROBLEMS

1. Calculate the fringe shift in Michelson-Morley experiment. Given effective length of each path is 10 m, velocity of earth 3×10^4 m/s and wavelength of light used is 6000 Å. **[Ans.** 0.33 fringe]
2. In a Michelson-Morley experiment the optical path of each beam was 11 m and the wavelength of the light used was 5500 Å. Assuming v/c to be 10^{-4}, calculate the fringe shift that would have to be obtained if the ether existed. **[Ans.** 0.4 fringe]
3. In an experiment similar to the Michelson-Morley experiment the distance d of either mirror from the beam splitter is 25.9 m. The wavelength of the light used is 5890 Å and the apparatus is capable of detecting a time difference equal to one-hundredth of the period of this light. If no effect is observed when the apparatus is rotated by 90°, what is the maximum possible value of velocity of apparatus relative to the ether? **[Ans.** 3.25×10^3 m/s]
4. A meter stick lies at rest in the xy-plane at an angle of 45° with the x-axis. What are the length and orientation of the stick to an observer moving at relative velocity v in the x-direction?
5. In the laboratory the lifetime of a particle moving with speed 2.8×10^8 m/s is found to be 2.5 × m/s. Calculate the proper life time of the particle. **[Ans.** 8.9×10^8 s]

6. A man leaves the earth in a rocketship that makes a round trip to the nearest star which is 4 light years away at a speed of 0.8*c*. How much younger will he be on his return than his twin brother who preferred to stay behind? **[Ans.** 3.5 years]
7. If the galaxy has a diameter of 10^5 light-years, what will be the diameter appear to be to a cosmic ray particle travelling at the relative speed $v/c = 0.99$? How long will the trip take as measured by a clock riding with the particle?
8. Two particles are moving in opposite directions with speeds 0.9*c* as observed in laboratory frame. What is the velocity of one particle relative to the other? **[Ans.** 0.994*c*]
9. L_0^3 is the rest volume of a cube. It is viewed from a reference frame moving with a uniform velocity v parallel to an edge of the cube, show that the apparent volume will be $L_0^3\sqrt{1-\frac{v^2}{c^2}}$.
10. A cosmic ray μ-meson has a half life 2.2×10^{-6}s in a reference frame in which the meson is at rest. What will be its half life as measured by an observer on the earth, if it approaches the earth at a speed of 0.999*c*. **[Ans.** 45.5×10^{-6} s]
11. A radioactive atom is moving with respect to the laboratory at a speed of 0.3*c* in the *x*-direction. If it emits an electron having a speed of 0.8*c* in the rest frame of the atom, find the velocity of the electron with respect to a laboratory observer when: (a) the electron is ejected in the *x*-direction, (b) it is ejected in the negative *x*-direction and (c) it is ejected in the *y*-direction.
12. A beam of $10^4\pi^+$ mesons travels in a circular path of radius 20 m at a speed of 0.99*c*. Proper mean life time of the π^+ meson is 2.5×10^{-8} s. How many survive when the beam returns to the starting point after one revolution? **[Ans.** 925]
13. Find the velocity and the momentum of an electron whose kinetic energy equals its rest mass energy.
14. The total energy of a particle is exactly twice its rest energy. Find its speed? **[Ans.** 0.866*c*]
15. It is assumed that Sun gets energy by fusion of four hydrogen atoms into a helium atom. The rest mass of hydrogen and helium atoms are 1.0081 and 4.0039 a.m.u. respectively. Calculate energy released in each fusion process. [I amu = 1.67×10^{-27} kg] **[Ans.** 26.7 MeV]
16. What is the speed of an electron whose kinetic energy is equal to its rest energy? Does it depend upon the mass of the electron? **[Ans.** 0.866*c*, No]
17. Write the components of momentum four-vector, space and hence prove, the invariance of $(E^2 - p^2c^2)$.
18. Define four-vector. Show that the magnitude of momentum four-vector is (im_0c), where i is $\sqrt{-1}$, m_0 is rest mass and *c* is velocity of light.
19. Show that the Lorentz transformation from the primed to the unprimed system may be expressed by a matrix equation of the form $x = \Gamma^{-1}x'$, where Γ^{-1} is the matrix inverse of Γ given by

$$\Gamma^{-1} = \begin{vmatrix} \gamma & 0 & 0 & i\beta\gamma \\ 0 & 0 & 0 & 0 \\ 0 & 0 & 1 & 0 \\ i\beta\gamma & 0 & 0 & \gamma \end{vmatrix}$$

20. Find the scalar invariants associated with the four-velocity and the four-momentum.

21. Show that the Lorentz force per unit volume $q(\boldsymbol{E} + \mathbf{v} \times \boldsymbol{B})$ is the space part of a four vector. Determine the fourth component and show that it is a measure of the power expended by the electric field per unit volume.

22. A frame S_1 moves with velocity $v_1 = \beta_1 c$ with respect to a frame S. A second frame S_2 moves with velocity $v_2 = \beta_2 c$ relative to S_1. Find the velocity of S_2 relative to S and show that

$$\beta = \frac{\beta_1 + \beta_2}{1 + \beta_1 \beta_2}$$

23. Show that if a photon of energy $h\nu$ is absorbed by a stationary nucleus (or other particle) of rest mass M_0, the excited particle has a mass $M_0 + \frac{h\nu}{c^2}$ and a speed $\frac{h\nu c}{(h\nu + m_0 c^2)}$

24. Show that the four-velocity has the dimensions of speed. Prove that it transforms as a four vector by showing its magnitude is a constant independent of the inertial frame in which it is measured.

25. A particle of rest mass m_0 moves with a speed $\frac{c}{\sqrt{2}}$. Calculate its momentum, kinetic energy and total energy; **[Ans.** $m_0 c$, $m_0 c^2$ **and** $2\, m_0 c^2$**]**

26. Derive the following equation

$$\beta = v / c = \sqrt{1 - \left(\frac{E_o}{E_o + k}\right)^2}$$

where $E_o = m_o c^2$ and k is Kinetic energy.

SHORT ANSWER-QUESTIONS

1. How Lorentz-transformation follows directly from the postulates of the special theory of relativity?

Ans: Any physical law which retains the same mathematical form under a Lorentz transformation automatically satisfies the principle of relativity.

2. What are the two important kinematical effects which derive from the special theory of relativity?

Ans: (i) The contraction of length parallel to the relative motion and (ii) dilation of time.

3. What do you understand by Cerenkov radiation?

Ans: Very energetic particles are capable of moving faster in a material medium (water, glass etc.) than the velocity of light in that medium, though never faster than the velocity of light in free space. This phenomenon gives rise to the light waves known as Cerenkov radiation.

4. What do you understand by Tachyon?

Ans: Tachyon is a particle theoretically predicted and always travels with a speed greater than that of light. However, the existence of Tachyon has not been established so far.

5. What is the important dynamical effect of special theory of relativity?
Ans: An important dynamical effect of special theory of relativity is the increase of mass with increase in velocity, which leads to the statement of the equivalence of mass and energy, i.e., $E = mc^2$.
6. What is Proper time interval?
Ans: The proper time interval is the time interval measured by a clock attached to the observed body. Proper time interval (dt') noted by a moving observer is always less than the corresponding interval of time in a stationary frame.

$$dt' = dt\sqrt{1-\frac{v^2}{c^2}}$$

7. What do you understand by invariance of an interval?
Ans: The invariance of an interval is a mathematical expression of the invariance of the velocity of light. The interval between two events is the same in all inertial frames of reference (invariance of a four-dimensional interval).
8. What is *twin paradox*?
Ans: A paradox resulting from the special theory of relativity; if one of a pair of twins remains on earth while the other twin makes a journey to a distance star at close to the speed of light and subsequently returns to earth, the twins will have aged differently. The twin remaining on earth will have aged considerably more than the twin who travelled to a star.
9. *What is space - time continuum?*
Ans: A geometry that includes the three dimensions and a fourth dimension of time. In Newtonian physics, space and time are considered as separate entities and whether or not events are simultaneous is a matter that is regarded as obvious to any competent observer. In Einstein concept of the physical universe, based on a system of geometry devised by Minkowski, space and time are regarded as entwined, so that the two observers in relative motion could disagree regarding the simultaneity of distant events. In Minkowski's geometry, an event is identified by a *world point* in a four-dimensional continuum.
10. What is relativity?
Ans: The Laws of physics are same for all observers and don't depend on one's state of motion. That's the *principle of relativity*. Special relativity is restricted to inertial frames of reference. General relativity removes that restriction and in so doing becomes a theory of gravity both relativity theories radically alter our commonsense notions of space and time. In relativity measures of space and time—but not the laws of physics—become dependent on one's reference frame.
11. What are *Tachyons*?
Ans: Tachyons are postulated particles that travel faster than the speed of light. We have

$$E = \gamma mc^2 = \frac{mc^2}{\sqrt{1-\frac{v^2}{c^2}}}$$

Because $v > c$, the energy must be imaginary if the mass is real, or conversely, if one insist that energy be real, we must have an imaginary mass. Let us assume that energy is real and tachyon mass is imaginary. We must remember that ordinary matter must always travel at speed less than c and tachyons must always have speed greater than c. In order to slow down a tachyon, we must provide it more energy (see above equation). We may note that the energy must become infinite if we want to slow down a tachyon to speed c. If tachyon's energy is reduced, it speeds up!

As tachyons travel faster than c, we face a problem with casuality.

It has been proposed that tachyons might be created in high-energy particle collisions or in cosmic rays. No confirming evidence has been found. Tachyons, if charged could be detected from *Cerenkov radiation.*

OBJECTIVE QUESTIONS

1. The relativistic expression for the mass is

(a) $m = m_0\sqrt{1-\frac{v^2}{c^2}}$ (b) $m = m_0 / \sqrt{1-\frac{v^2}{c^2}}$

(c) $m = m_0\left(1-\frac{v^2}{c^2}\right)$ (d) $m_0 / m = \left(1-\frac{v^2}{c^2}\right)$

2. The relativistic relation between momentum and energy is

(a) $E^2 = c^2p^2 + m_0^2c^4$ (b) $E^2 = c^2p^2$

(c) $E^2 = m_0^2c^4$ (d) $E = mc^2$

3. The expression for the relativistic equivalence of mass and energy is

(a) $E = m^2c^2$ (b) $E = mc^2$

(c) $E = mc$ (d) $E = m/c^2$

4. An electron has a velocity 0.99c. Its energy will be

(a) 3.1 MeV (b) 0.31 MeV

(c) 31 MeV (d) 0.031 MeV

5. A metre rod is moving parallel to its length. When its apparent mass becomes twice the rest mass, its apparent length will be

(a) 1 m (b) 0.5 m

(c) 1.5m (d) 0.1m

6. A photon of energy $h\nu$ is absorbed by a stationary particle of rest mass M_0. The excited particle has a speed

(a) $\dfrac{h\nu c}{M_0c^2}$ (b) $\dfrac{h\nu c}{h\nu + M_0c^2}$

(c) $\dfrac{h\nu}{M_0c^2}$ (d) $h\nu + M_0c^2$

7. **An isolated excited nucleus of mass M emits a γ-ray in a transition to the ground state. If the excited state lies an energy ΔE above the ground state, the energy of the photon is**

(a) $\Delta E\left(1 - \dfrac{\Delta E}{2Mc^2}\right)$ (b) $\dfrac{\Delta E}{2Mc^2}$

(c) $\Delta E - Mc^2$ (d) $\Delta E + Mc^2$

ANSWERS

Question numbers	1	2	3	4	5	6	7
Answers	(b)	(a)	(b)	(a)	(b)	(b)	(a)

FILL IN THE BLANKS

1. Space time diagrams are useful to represent events _______. Time may be considered to be a fourth dimension for some purposes. **[geometrically]**
2. The space time interval for two events defined by _______ is an invariant between inertial systems. **[$\Delta s^2 = \Delta x^2 + \Delta y^2 + \Delta z^2 - c^2\Delta t^2$]**
3. The time interval between two events occurring at the same position in a system as measured by a clock at rest is called the _______. **[proper time]**
4. Massless particles must travel at the speed of _______. **[light]**
5. Maxwell's equations are _______ under transformations between any inertial frames. What appears as electric and magnetic fields is relative to the reference frame of the observer. **[invariant]**
6. Speed of light is the same in all _______ reference frames. **[inertial]**
7. Time dilation and length contraction are specific instances of the _______. **[Lorentz transformations]**
8. The laws of the conservation of mass and of energy are combined into one conservation law: _______. **[conservation of mass-energy]**
9. Energy and momentum are related by _______. **[$E^2 = p^2c^2 + E_0^2$]**
10. The Lorentz transformation has been tested for a hundred years, and no _______ has yet been detected. **[violation]**
11. Michelson and Morley proved in 1887 that either the elusive ether does not _______ or there must be significant problems with our understanding of nature. **[exist]**

12. In the case of massless particles, energy and momentum are related by______.
[E = p² c²]

13. The relativistic Doppler effect for light frequency f is given by______. where ($\beta = v/c$) is positive, when source and receiver are approaching one another and negative when they are receding

$$\left[f = \frac{\sqrt{1+\beta}}{\sqrt{1-\beta}} f_o \right]$$

14. $L = L_o\sqrt{1-\beta^2}$, where $\beta = v/c$ is known as ______, because objects are contracted in the direction of their motion. **[length or space contraction]**

15. Maxwell's equations are ______ under transformation between any inertial reference frames. **[invariant]**

2

Atomic Structure of Matter

2.1 INTRODUCTION

Dalton's atomic theory (1803) of the matter was the first scientific approach towards the understanding of the structure of the matter. According to this theory all matter was composed of very small particles called atoms. The atoms retained their individuality in all kinds of chemical combinations. Dalton further believed that the atoms of the same substance were all alike and differed from those of all other substances and further, that the atoms could not be subdivided, destroyed or created.

Prout in 1815, suggested that all the elements are made up of atoms of hydrogen. This suggestion was based on the fact that atomic weights of most of the elements were integral multiples of atomic weight of hydrogen atom, e.g., He = 4, C = 12, N = 14 etc. Later on it was found that atomic weights in most of the cases were not whole numbers, e.g., Cl = 35.5, Ne = 20.2. Therefore, Prout's hypothesis was abandoned.

2.2 THOMSON'S ATOM MODEL

In 1898, Sir J.J. Thomson discovered the electron and it was established that electron is a common constituent of all matter. He then proposed a physical model of the atom known as *'Plum pudding atom'*. The atom, as he pictured it, was a positive plum pudding in which were embedded negative electron raisins distributed in such a manner as to make the entire thing neutral Fig. (2.1).

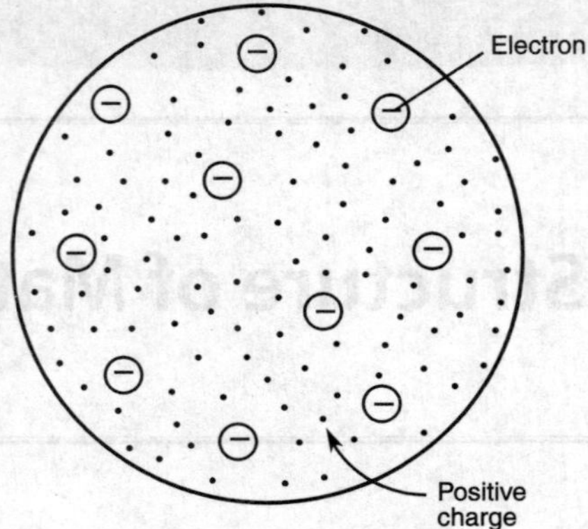

Fig. 2.1 Thomson's atom model

2.3 RUTHERFORD'S NUCLEAR ATOM MODEL

In 1911 Rutherford proposed a nuclear atom model on the basis of α-scattering experiment. He suggested that

(i) Atom is made up of a central core called nucleus of diameter ~ 10^{-15} m in which the entire positive charge and almost all mass of the atom is concentrated.

(ii) The nucleus is surrounded by as many electrons as there are positive charges on the nucleus. Since the dimensions of the atom are of the order of 10^{-10} m, most of the space within the atom is empty.

If the electrons were at rest they will be attracted and neutralized by the nucleus. To over come this difficulty, he suggested that the electrons are revolving round the nucleus in circular orbits Fig. (2.2 (a)).

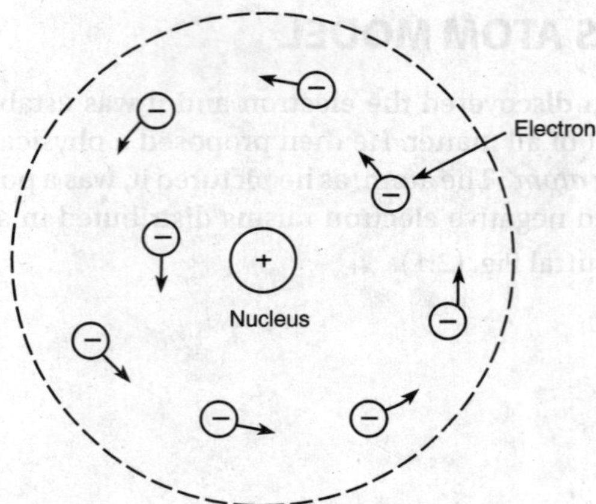

Fig. 2.2 (a) Rutherford's nuclear atom model

If r be the radius of the electronic orbit, v the velocity of the electron in its orbit and Ze be the charge on the nucleus, Z being the atomic number and e the charge on a proton or electron, then

$$\frac{mv^2}{r} = k\frac{Ze.e}{r^2} \tag{2.1}$$

Here $$k = \frac{1}{4\pi\varepsilon_0} = 9\times10^9 \text{ N-m}^2\text{-C}^2, \text{ and}$$

$$\varepsilon_0 = 8.85\times10^{-12}\,\text{F/m}$$

According to Eq. (2.1) it is possible to have an infinite number of orbits in which electrons can revolve round the nucleus.

2.3.1 Rutherford's Theory of Alpha Particle Scattering

Alpha particles (α) are highly energetic charged particles, emitted during the radioactive disintegration of certain heavy elements e.g. uranium, radium. The amount of positive charge on α-particle is equal to two units of electronic charge and their mass is equal to that of a helium atom. When a beam of α-particles is passed through thin foils of metallic elements of high atomic mass, it is observed that, while the majority of α-particles are scattered through small angles, a significant number is deviated through large angles (greater than 90°) (Figure 2.2(b)).

Geiger and Marsden observed that when a foil of gold of thickness 0.4×10^{-4} cm is used one α-particle in 20000 is deviated through an angle of $\frac{\pi}{2}$ or greater than $\frac{\pi}{2}$ and few are turned back into the direction from which they come. Such large angle scattering cannot be explained on the basis of Thomson's model of the atom. Thomson model consists of a uniform distribution of positive and negative charge, sometimes called the plum-pudding or chemical model.

In traversing the homogeneous atoms of metal according to the Thomson model (Figure 2.2(c)), the α-particles (helium ions) might be expected to be slightly deviated,

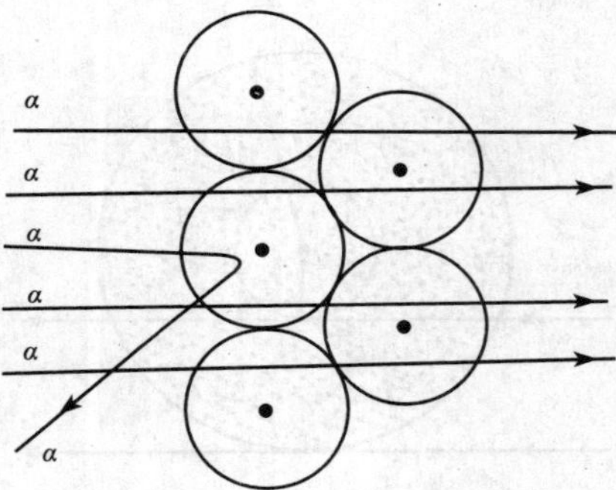

Fig. 2.2(b) Large and Small angle scattering of α-particle from an array of atoms

according to the quantity of the positive charge traversed, the number of electrons encountered, and so forth. These effects would be small, and the alpha particles would be expected to suffer only minor deviations.

The experimental results were astonishingly different. In the first place, most of the α-particles were not deviated at all (Figure 2.2(b)). Those that were deviated were turned through very large angles-some being even deflected back towards the source as shown in Figure 2.2(b).

To explain the observed large angle scattering of the α-particles, Rutherford in 1911 proposed a new model of the atom. He suggested that in a given atom the positive charge and the bulk of the mass is tightly concentrated in a "nucleus" of extremely small dimensions. The negatively charge, consisting of atom's electrons is distributed in the rest of the atom's volume which is thus largely empty space. Rutherford had discovered the nucleus which is a landmark in the history of the development of modern physics. Rutherford showed that the effect of the electrons in the atom outside the nucleus is negligible for the deflection of α-particles through angles greater than 1°. When an α-particle passes by very close to the nucleus, it experiences a strong electrostatic repulsive force due to the entire positive charge of the atom and hence scattered at a relatively large angle. At the suggestion of Rutherford experiments were conducted in 1911 by Geiger and Marsden to probe into the atoms and to find out what was inside the atom.

A schematic diagram of Rutherford and his colleagues experimental arrangement is shown in Figure 2.2(d). The apparatus obviously has to be enclosed in vacuum.

During the experiment a narrow collimated beam of mono-energetic α-particles is allowed to fall on a very thin foil of gold ($Z = 79$) or silver ($Z = 47$) which can be rolled or beaten to thickness of the order of 10^{-7} m or less. The α-particles do not lose any appreciable energy in passing through such thin metal foils. The scattered α-particles were detected and counted by a fluorescent screen-microscope arrangement. Rutherford and his colleagues observed that many α-particles went straight through the foil or were deflected by a very small angle. However, a few α-particles were deflected through very large angles; some of them even got back scattered. This was contrary of the Thomson's atom model, i.e. the positive charge is uniformly smeared over the entire volume of the atom.

To explain the above observations of the experiment, Rutherford assumed that the positive massive part of the atom was concentrated in a very small volume at the centre of

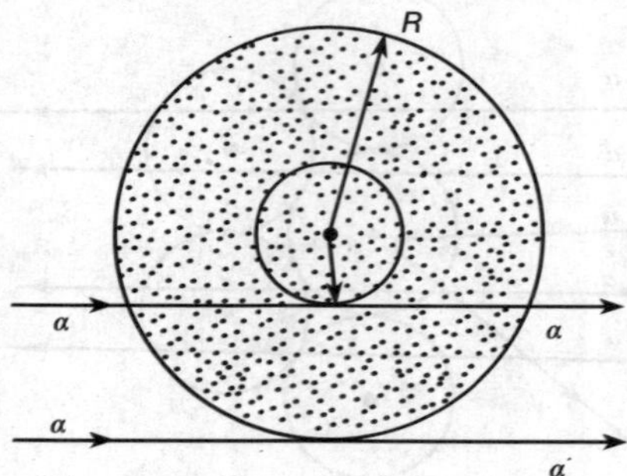

Fig. 2.2(c) Scattering of α-particle based on Thomson's model of the atom

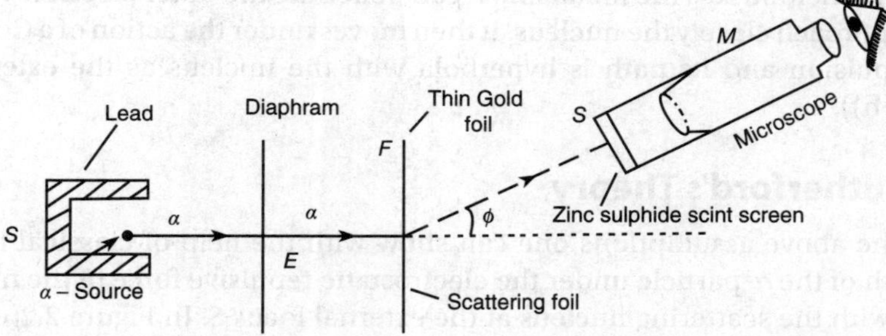

Fig. 2.2(d) A schematic diagram of the Rutherford scattering experiment. The system up to the scintillating screen, must be enclosed in a vaccuum

the atom. This core, now termed the *nucleus*, is surrounded by a cloud of electrons, which makes the entire atom electrically neutral. This nuclear model of the atom accounts for the large angle scattering of the α-particles in the following way. An α-particle approaching the centre of the atom experiences an increasingly large Coulomb repulsion. Since the atom is mostly empty space, most of the α-particles do not go sufficiently close to the nucleus to get sufficiently deflected and hence they pass the thin metal foil with practically no deviation. However, an α-particle that passes close to the nucleus is subjected to a very large Coulomb force exerted by massive positive core nucleus of charge $+Ze$ and is deflected at a large angle in a single encounter. The assumptions made by Rutherford during the explanation of the large angle α scattering are as follows:

(i) The positive charge of the atom is concentrated in an extremely small volume, known as 'nucleus' in the atom. The magnitude of this charge is an integer Z times the charge of an electron, where Z is an integer representing the atomic number of the element (Z is exactly equal to the number of orbital electrons in an atom). The negative charges (electrons) revolve around the nucleus in planetary orbits.

(ii) The whole mass of the atom (excluding the mass of the orbital electrons) is concentrated in the nucleus. Obviously, more than 99.99 % of the mass is concentrated within the small volume of the nucleus. Since the mass of the electron is negligible in comparison to that of the nucleon it follows that nuclei are very much denser. Obviously, the density of the nuclear matter is very high about 10^{15} times that of the density of the ordinary matter.

(iii) The scattering of α-particles takes place due to an encounter between the α-particle and the massive nucleus of the atom. The electrons outside the nucleus, owing to their very small mass, cannot appreciably deflect the far more massive α-particles.

(iv) Both the α-particle and the nucleus are small enough to be considered as point mass and point (positive) charge and the electrostatic repulsive force between them is governed by Coulomb's inverse square law.

(v) The nucleus of the atom (of the thin metal foil used) is so massive compared with the α-particle that it remains at rest during the encounter.

when an α-particle strikes the metal foil, it can penetrate the outer electron-cloud of an atom and approach closely the nucleus. It then moves under the action of a Coulombian force of repulsion and its path is hyperbola with the nucleus as the external focus (Figure 2.2(b)).

2.3.2 Rutherford's Theory

Based on the above assumptions one can show with the help of classical mechanics that the path of the α-particle under the electrostatic repulsive force of the nucleus is a hyperbola with the scattering nucleus at the external focus *S*. In Figure 2.2(e) the path of the α-particle far off from the target nucleus is along *AO*. At α-particle comes nearer *S*, the repulsive force on it varies as $\frac{1}{r^2}$ where *r* is the instantaneous separation between the α-particle and the nucleus. The α-particle gets deflected and finally moves along *OB*, tracing the hyperbolic curve *ADB*. Let the angle of deflection (or scattering angle) of the α-particle be ϕ which is also the angle between the asymptotic direction of approach of the α-particle and the asymptotic direction along which α-particle recedes. If there were no force acting on the α-particle, it would pass by the nucleus at the minimum distance equal to *p* which is the length of the perpendicular from *S* on the asymptote *AO*. This perpendicular distance is called *impact parameter*.

Let us take the centre o of the hyperbola as the origin. Obviously, the vertex *D* of the hyperbola is at $x = -a$. Similarly the vertex *E* of the second branch of the hyperbola is at $x = a$. Let (c, o) be the coordinates of the target nucleus and *d* the distance of vertex *D* from the nucleus which is evidently the distance of the nearest approach of the α-particle to the nucleus. Further, let θ be the angle which the asymptotic direction *AO* of the α-particle makes with the *x*-axis. From Figure 2.2(e), the distance of nearest approach

$$d = c + a = c\left(1 + \frac{a}{c}\right)$$

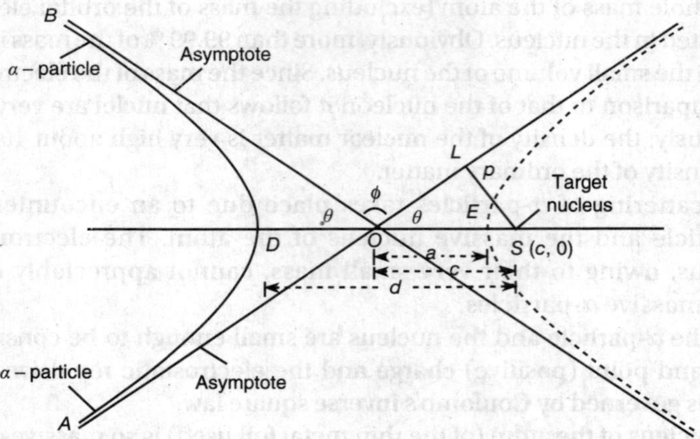

Fig. 2.2(e) Path of the α-particle during the scattering by a atomic nuclear

But $$\frac{a}{c} = \cos\theta^{*}$$

$\therefore$ $$d = c + c\cos\theta = c(1+\cos\theta)$$

Also $$\frac{p}{c} = \sin\theta \text{ or } c = \frac{p}{\sin\theta}$$

$\therefore$ $$d = \frac{p(1+\cos\theta)}{\sin\theta} = \frac{p\,2\cos^2\frac{\theta}{2}}{2\sin\frac{\theta}{2}\cos\frac{\theta}{2}} \tag{2.1(a)}$$

or $$d = p\cot\frac{\theta}{2} \tag{2.1(b)}$$

Let us now derive the relation between the impact parameter p and the angle of deflection ϕ. During the interaction between the α-particle and the nucleus, the laws of conservation of kinetic energy and the angular momentum must hold good. When the α-particle is at a large distance from the nucleus, its energy is only kinetic because the potential energy, being inversely proportional to distance, can be taken as zero. If v is the velocity of the α-particle at this distance, the total energy is $\frac{1}{2}mv^2$. Let the velocity of the α-particle when it reaches the vertex D is v_0. Its kinetic energy at D is $\frac{1}{2}mv_0^2$ and potential energy is $\frac{(Ze)2e}{4\pi\varepsilon_0 d} = \frac{Ze^2}{2\pi\varepsilon_0 d}$, where Ze is the charge on the nucleus ($Z \rightarrow$ atomic number and $2e \rightarrow$ charge on the α-particle). According to law of conservation of energy

$$\frac{1}{2}mv^2 = \frac{1}{2}mv_0^2 + \frac{Ze^2}{2\pi\varepsilon_0 d}$$

---

*The equation of hyperbola $\frac{x^2}{a^2} - \frac{y^2}{c^2-a^2} = 1$, where c is the x-coordinate of the focus S and is equal to $a\,e$ (e being the eccentricity of the hyperbola). The equation of the asymptote AO of the hyperbola is $y = \sqrt{c^2-a^2}\,\frac{x}{a}$ or $\sqrt{(c^2-a^2)}\,x - ay = 0$. The general formula for the length of the perpendicular from a point (x_1, y_1) on the line $ax + by + c = 0$ is $\frac{ax_1 + by_1 + c}{\sqrt{a^2+b^2}}$. Obviously, the length of the perpendicular from the focus (c, o) to the asymptote AO is

$$p = \frac{\left(\sqrt{c^2-a^2}\right)c - a(0)}{\sqrt{(c^2-a^2)+a^2}} = \sqrt{c^2-a^2}$$

$$OL = \sqrt{c^2-p^2} = \sqrt{c^2-(c^2-a^2)} = a$$

$$\frac{OL}{OS} = \frac{a}{c} = \cos\theta$$

Dividing throughout by $\frac{1}{2}mv^2$, one obtains

$$1 = \frac{v_0^2}{v^2} + \frac{Ze^2}{\pi\varepsilon_0 m v^2 d}$$

$$= \frac{v_0^2}{v^2} + \frac{b}{d} \quad \left(b = \frac{Ze^2}{\pi\varepsilon_0 m v^2}\right)$$

or $$\frac{v_0^2}{v^2} = 1 - \frac{b}{d}$$

Substituting the value of d from Equation (2.1(a)), one obtains

$$\frac{v_0^2}{v^2} = 1 - \frac{b\sin\theta}{b(1+\cos\theta)} \tag{2.1(c)}$$

From the law of conservation of angular momentum

$$mvp = mv_0 d$$

or $$\frac{v_0}{v} = \frac{p}{d} = \frac{\sin\theta}{1+\cos\theta}$$ (using Equation. (2.1(a))

Squaring, one obtains

$$\frac{v_0^2}{v^2} = \frac{\sin^2\theta}{(1+\cos\theta)^2} = \frac{1-\cos^2\theta}{(1+\cos\theta)^2} = \frac{1-\cos\theta}{1+\cos\theta} \tag{2.1(d)}$$

Comparing (2.1(c)) and (2.1(d)), one obtains

$$\frac{1-\cos\theta}{1+\cos\theta} = 1 - \frac{b\sin\theta}{p(1+\cos\theta)}$$

or $$\frac{b\sin\theta}{p(1+\cos\theta)} = 1 - \frac{1-\cos\theta}{1+\cos\theta} = \frac{2\cos\theta}{1+\cos\theta}$$

$\therefore$ $$\frac{b}{p} = 2\cot\theta$$

But $$\phi = \pi - 2\theta \quad \text{or} \quad \theta = \frac{\pi}{2} - \frac{\phi}{2}$$

$\therefore$ $$\frac{b}{p} = 2\cot\left(\frac{\pi}{2} - \frac{\phi}{2}\right) = 2\tan\frac{\phi}{2}$$

or $$p = \frac{b}{2}\cot\frac{\phi}{2}$$

$$= \frac{Ze^2}{2\pi\varepsilon_0 m v^2}\cot\frac{\phi}{2} \tag{2.1(e)}$$

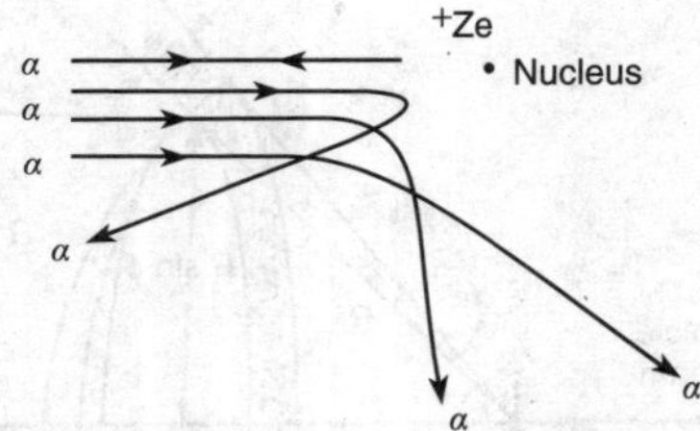

Fig. 2.2(f) α-particle scattering at different impact parameters

$$= \frac{(Ze)(2e)}{4\pi\varepsilon_0 mv^2}\cot\frac{\phi}{2} \tag{2.1(f)}$$

Equation (2.1(f)) gives the relation between the impact parameter and the angle of scattering. When the impact parameter p becomes smaller, the angle of scattering ϕ increases (Figure 2.2(f)). It further follows that an α-particle approaching a target nucleus with an impact parameter ranging from o to p will be scattered through an angle of ϕ or more. Obviously, an α-particle which is directed any where within an area πp^2 around the nucleus will be scattered through ϕ or more. The area πp^2 is called the *cross-section of interaction* and denoted by $\sigma\,(=\pi p^2)$.

Let us now consider beam of α-particles incident upon a thin foil of thickness t, containing n atoms per unit volume. Evidently, the number of target nuclei per unit area is nt so that an α-particle beam incident upon an area A faces ntA target nuclei. The total cross-section of interaction of the nuclei for a scattering angle of ϕ or more is equal to the number of target nuclei $nt\,A$ multiplied by the cross-section σ per nucleus for such scattering or equal to $n\,t\,A\,\sigma = nt\,A\,\pi p^2$. The fraction q of the α-particles scattered through an angle of ϕ or more is the ratio of the total cross-section of interaction and total target area A. Obviously,

$$q = \frac{\text{Number of } \alpha\text{-particles scattered through } \phi \text{ or more}}{\text{Number of incident } \alpha\text{-particles}}$$

$$= \frac{\text{Total cross section of interaction}}{\text{Total target area}}$$

$$= \frac{nt\,A\,\pi p^2}{A} = \pi nt\pi^2$$

Substituting p from (2.1(e)), one obtains

$$q = \pi nt\frac{b^2}{4}\cot^2\frac{\phi}{2} \tag{2.1(g)}$$

The fraction of the incident α-particles which gets scattered through an angle ϕ and $(\phi + d\phi)$ can be obtained by differentiating (2.1(g)) w · r · t · q. Hence

$$dq = -\pi\,nt\,\frac{b^2}{4}\cot\frac{\phi}{2}\,\text{cosec}^2\frac{\phi}{2}d\phi \tag{2.1(h)}$$

The negative sign indicates that dq decreases as ϕ increases.

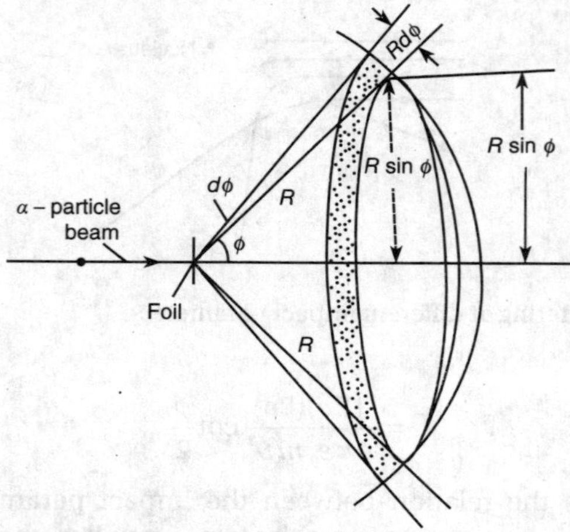

Fig. 2.2(g)

α-particles are detected by the scintillation which they produce on a fluorescent screen. Let a screen be placed at a distance R from a foil and suitably adjusted so as to receive the particles scattered through the angle between ϕ and $\phi + d\phi$ inside an annular ring of radius $R \sin\phi$ and width $Rd\phi$ (Figure 2.2(g)). The area of this ring is $2\pi R \sin\phi$.

$$R\,d\phi = 2\pi^2 \sin\phi\, d\phi = 4\pi R^2 \sin\frac{\phi}{2}\cos\frac{\phi}{2}\, d\phi$$

Thus the fraction of α-particles scattered between ϕ and $(\phi + d\phi)$ striking a unit area of the screen at a distance R from the scattering foil is given by

$$\frac{dq}{dA} = \frac{\pi n t b^2 \cot\frac{\phi}{2}\operatorname{cosec}^2\frac{\phi}{2}\, d\phi}{4\times 4\pi R^2 \sin\frac{\phi}{2}\cos\frac{\phi}{2}\, d\phi}$$

$$= \frac{nt\, b^2 \operatorname{cosec}^4\frac{\phi}{2}}{16R^2}$$

If N_0 is the number of α-particles actually incident normally on the foil per second, the number scattered to a unit area of the fluorescent screen at a distance R from the foil and at an angle ϕ with the original direction is given by

$$N = \frac{N_0\, nt\, b^2 \operatorname{cosec}^4\frac{\phi}{2}}{16\, R^2}$$

Substituting for b, we have

$$N = N_0 \frac{nt(Ze)^2 e^2 \operatorname{cosec}^4 \frac{\phi}{2}}{16\pi^2 \varepsilon_0^2 m^2 v^4 R^2} \quad (2.1(i))$$

This is *Rutherford's scattering formula.*

2.3.3 Distance of Closest Approach

The distance d of the closest approach of the α-particle can also be calculated in terms of scattering and other known parameters. From Equation (2.1(a)), we have

$$d = \frac{p(1+\cos\theta)}{\sin\theta}$$

But
$$\phi = \pi - 2\theta$$

$$\therefore \quad d = \frac{p\left[1+\cos\left(\frac{\pi}{2}-\frac{\phi}{2}\right)\right]}{\sin\left(\frac{\pi}{2}-\frac{\phi}{2}\right)} = \frac{p\left(1+\sin\frac{\phi}{2}\right)}{\cos\frac{\phi}{2}}$$

$$= p\left[\frac{1}{\cos\frac{\phi}{2}} + \tan\frac{\phi}{2}\right]$$

Substituting p from (2.1(e)), we have

$$d = \frac{b}{2}\cot\frac{\phi}{2}\left[\frac{1}{\cos\frac{\phi}{2}} + \tan\frac{\phi}{2}\right]$$

$$= \frac{b}{2}\left[\operatorname{cosec}\frac{\phi}{2} + 1\right]$$

$$= \frac{Ze^2}{2\pi\varepsilon_0 m v^2}\left(1+\operatorname{cosec}\frac{\phi}{2}\right) \quad (2.1(j))$$

2.3.4 Experimental Verification of Rutherford's Scattering Formula

From Equation (2.1(i)) we see that the number of particles falling on a unit area of zinc sulphide screen at a given distance R from the screening foil should be proportional to

(i) Cosec$^4 \frac{\phi}{2}$, where ϕ is the angle of scattering.

(ii) t, thickness of scattering foil

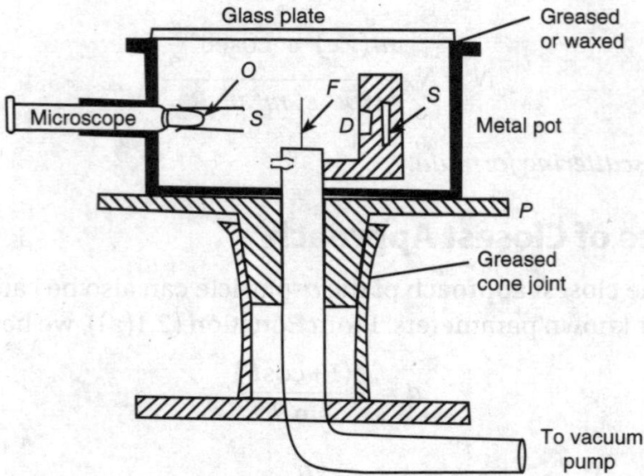

Fig. 2.2(h) Investigation of the scattering of α-particles by a thin metal foil

(iii) $\frac{1}{E^2}$ or $\frac{1}{v^4}$, i.e. reciprocal of the fourth power of velocity of α-particles ($E = \frac{1}{2}mv^2$ is the energy of α-particles), and

(iv) $(Ze)^2$, i.e. the square of the nuclear charge of the scattering material.

The above conclusions of Rutherford's α-particle scattering theory were tested by a series of experiments perpormed by N. Geiger and E. Marsdon in 1913, two of Rutherford's close associates. The experimental arrangement is shown in Figure 2.2(h).

Drawbacks of Rutherford's model

Inspite of strong experimental support, Rutherford's model has the following drawbacks:

(i) According to classical (electromagnetic) theory, a revolving electron should radiate energy continuously. If it is so, the electron will rapidly spiral inwards and fall into the nucleus and the atom would collapse. Calculations reveal that only about 10^{-16} s would be required for a stable hydrogen atom to collapse. However, atoms do not collapse, i.e., atoms are quite stable.

(ii) According to Rutherford's model the electrons can revolve in orbits of all possible radii and hence they should emit continuous radiation of all frequencies. This is not in agreement with experimental observations because atoms, like hydrogen, emit line spectrum of only certain fixed frequencies.

2.4 HYDROGEN SPECTRUM

The spectrum of hydrogen atom studied by several scientists and revealed towards the end of 19th century that it consists of a number of lines which fall into definite sets which are called '*series*'. The lines in each series are such that their separation and intensity decrease regularly towards shorter wavelengths converging to a limit called the '*series limit*'. An element may display several series. Each series can be represented

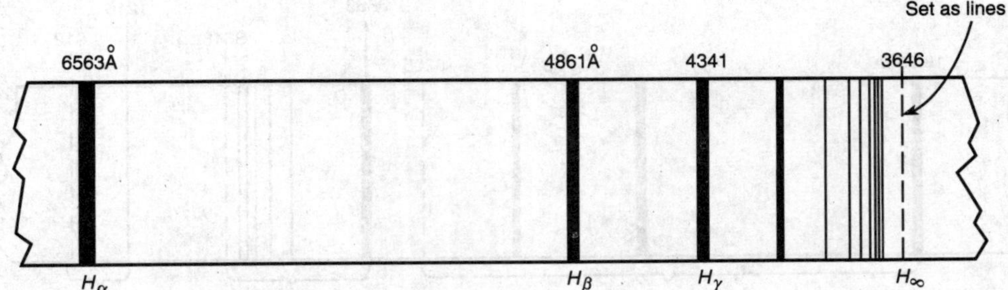

Fig. 2.3 The Balmer Series of hydrogen spectrum

by an empirical formula which has a similarity for all the series of a given element. The first such spectral series was reported by J.J.Balmer in 1885 and is called the *Balmer series of hydrogen*. The first line with the longest wavelength (6563Å) is named H_α, the next H_β and so on Fig. (2.3).

The series limit lies at 3646Å, beyond which is a faint continuous spectrum. Balmer's empirical formula for the wavelengths of the series is

$$\bar{\nu} = \frac{1}{\lambda} = R\left(\frac{1}{2^2} - \frac{1}{n^2}\right), \quad n = 3, 4, 5\ldots \tag{2.2}$$

The reciprocal of wavelength is called the *wave number* and is designated by $\bar{\nu}$. Obviously, the wave number is the number of waves in unit length. R is a constant called the *'Rydberg constant'* and has the value

$$R = 1.097\times10^7\,\text{m}^{-1}$$

One can obtain the wavelength of H_α line by writing $n = 3$, that of the H_β line by writing $n = 4$ and so on in Eq. (2.2). By writing $n = \infty$, one obtains the series limit at $\bar{\nu} = R/4$.

The Balmer series contains only those spectral lines which fall in the visible part of the hydrogen spectrum and this is also the reason why this series was the first to be observed. An examination of the ultraviolet and infrared regions of the hydrogen spectrum revealed the existence of other series. The lines in falling in the ultraviolet region form the *'Lyman series'* whose wave numbers and wavelengths are given by

$$\bar{\nu} = \frac{1}{\lambda} = R\left(\frac{1}{1^2} - \frac{1}{n^2}\right), \quad n = 2, 3, 4\ldots \ (\textit{Lyman series}) \tag{2.3}$$

In the infra-red region three spectral series have been observed whose wave numbers and wavelengths are given by

$$\bar{\nu} = \frac{1}{\lambda} = R\left(\frac{1}{3^2} - \frac{1}{n^2}\right), \quad (n = 4, 5, 6\ldots) \ (\textit{Paschen series})$$

$$\bar{\nu} = \frac{1}{\lambda} = R\left(\frac{1}{4^2} - \frac{1}{n^2}\right), \quad (n = 5, 6, 7\ldots) \ (\textit{Brackett series})$$

$$\bar{\nu} = \frac{1}{\lambda} = R\left(\frac{1}{5^2} - \frac{1}{n^2}\right), \quad (n = 6, 7, 8\ldots) \ (\textit{Pfund series})$$

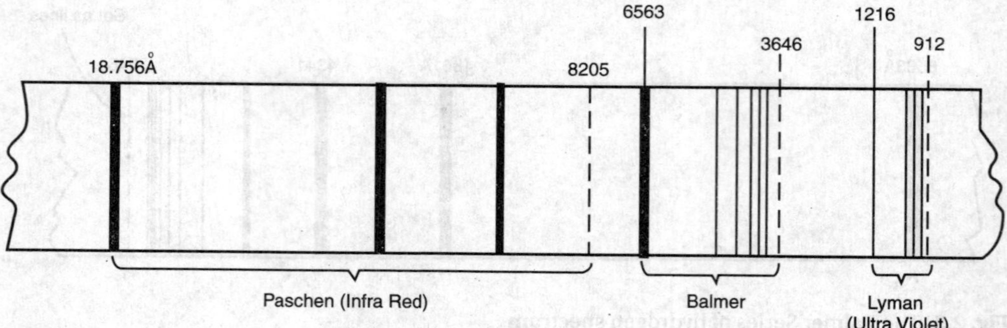

Fig. 2.4

We must note that the value of Rydberg constant R is the same for all the series. Some of the intense lines of Paschen, Balmer and Lyman series plotted in terms of wavelength are shown in Fig. (2.4).

Any successful model of the atom must be able to account for the remarkable regularity of spectral lines in the series of hydrogen spectrum and the spectrum of other elements.

2.5 BOHR'S THEORY OF ATOMIC STRUCTURE

Neils Bohr, a Danish physicist, in 1913, trying to find a way out of the difficulties associated with the Rutherford Model of the atom and to find the cause of the emission of spectral lines by hydrogen, proposed the following postulates:

(i) The electrons revolve around the nucleus in various stationary (or non-radiating) circular orbits. The necessary centripetal force to keep the electron in the closed circular orbit is provided by the electrostatic attraction between the electron and the nucleus.

If m is the mass of electron, v its velocity and r the radius of the circular orbit, then necessary centripetal force is given by $\frac{mv^2}{r}$. The electrostatic force of attraction between the nucleus having charge Ze and electron of charge e is KZe^2/r^2, where $K = 1/4\pi\varepsilon_0 = 9 \times 10^9$ Nm²/(Coulomb)².

Thus according to Bohr's first postulate

$$\frac{mv^2}{r} = K\frac{Ze^2}{r^2} \qquad [K = 1/4\pi\varepsilon_0 = 1 \text{ in C.G.S system}] \tag{2.4}$$

(ii) The only orbits that are allowed are those in which the angular momentum of the revolving electron is an integral multiple of $h/2\pi = \hbar$ ($= 1.05 \times 10^{-34}$ J-s). The angular momentum of the only permissible orbits is given by

$$L = mvr = \frac{nh}{2\pi}, \quad n = 1, 2, 3\ldots \tag{2.5}$$

$n = 1, 2, 3$ etc. is called the *principal quantum* number and this condition is usually called the *Bohr's quantum condition.*

(iii) When an electron is in an allowed orbit, the atom does not radiate energy (classical electromagnetic theory predicts that any accelerated charge will radiate electromagnetic waves).

(iv) If the electron jumps from an initial stationary orbit of energy E_i to a final stationary orbit of energy $E_f (E_i > E_f)$, a photon of frequency is emitted.

$$\nu = \frac{(E_i - E_f)}{h} \tag{2.6}$$

2.6 BOHR'S THEORY OF THE HYDROGEN ATOM

The greatest success of Bohr's was its ability to explain the structure and spectrum of hydrogen. According to Bohr's atom model, the structure of the hydrogen atom can be represented as shown in Fig. (2.5). For hydrogen Z (atomic number) = 1 and hence the charge on the nucleus $Ze = e$.

Let m be the mass of the electron and v its velocity along a circular orbit of radius r. As the centripetal force is provided by the electrostatic attraction, we have

$$\frac{mv^2}{r} = \frac{Ke^2}{r^2}$$

or

$$mv^2 = \frac{Ke^2}{r} \tag{2.7}$$

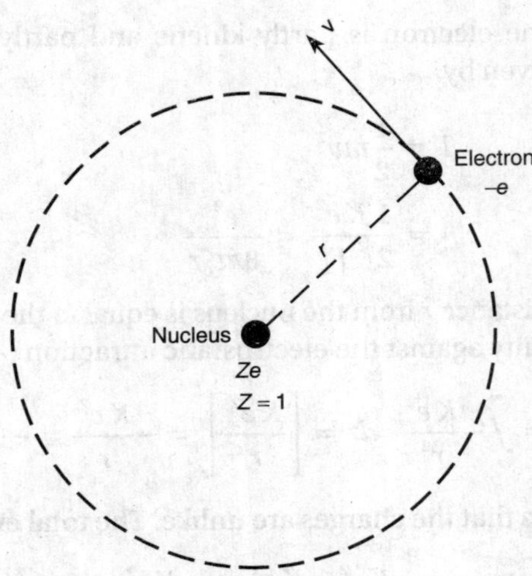

Fig. 2.5 Hydrogen atom

From Bohr's second postulate, the angular momentum of the electron in a permitted orbit is given by

$$mvr = nh/2\pi, n = 1, 2, 3\ldots$$

$$\therefore \quad v = \frac{nh}{2\pi mr} \text{ (velocity of electron)} \tag{2.8}$$

Squaring Eq. (2.8) and substituting in Eq. (2.7), one obtains

$$m\frac{n^2h^2}{4\pi^2m^2r^2} = K\frac{e^2}{r}$$

or

$$\frac{n^2h^2}{4\pi^2mr} = Ke^2$$

$$\therefore \quad r = \frac{n^2h^2}{4\pi^2mKe^2} = \frac{\varepsilon_0 n^2h^2}{\pi me^2} \tag{2.9}$$

This is the expression of the *radius of the permitted orbit.* Obviously, the radii of the permitted orbits are proportional to n^2 and increase in the ratio of 1, 4, 9, 16... from one orbit to the next. Substituting the values of *m, e, h,* ε_0, the radius of the first hydrogen orbit ($n = 1$) is given by

$$r = 5.29\times10^{-11}\,\text{m} = 0.529\,\text{Å}$$

The above value is found to be in good agreement with the value of radius of the orbit obtained by other methods.

Substituting the value of r from Eq. (2.9) in Eq. (2.8), one obtains

$$v = \frac{2\pi Ke^2}{nh} = \frac{e^2}{2\varepsilon_0 nh} \tag{2.10}$$

Energy: The energy of the electron is partly kinetic and partly potential. The kinetic energy of the electron is given by

$$T = \frac{1}{2}mv^2$$

$$= \frac{1}{2}\frac{Ke^2}{r} = \frac{e^2}{8\pi\varepsilon_0 r} \tag{2.11}$$

The potential energy at a distance r from the nucleus is equal to the work done in removing the electron from r to infinity against the electrostatic attraction ($-e^2/r$), and is given by

$$U = \int_r^\infty -\frac{Ke^2}{r^2}\,dr = \left[\frac{Ke^2}{r}\right]_r^\infty = -\frac{Ke^2}{r} = -\frac{e^2}{4\pi\varepsilon_0 r} \tag{2.12}$$

The negative sign indicates that the charges are unlike. The total energy of the electron is

$$E = K + U = \frac{Ke^2}{2r} - \frac{Ke^2}{r} = \frac{-Ke^2}{2r} = \frac{-e^2}{8\pi\varepsilon_0 r} \tag{2.13}$$

Substituting for *r* from Eq. (2.9), one obtains

$$E = E_n = -\frac{e^2}{8\pi\varepsilon_0} \times \frac{\pi m e^2}{\varepsilon_0 n^2 h^2} = -\frac{me^4}{8\varepsilon_0^2 h^2}\left(\frac{1}{n^2}\right) \tag{2.14}$$

This gives the energy of the electron revolving in the nth orbit. The negative sign indicates that the electron cannot escape from the orbit, i.e., the electron is bound to the nucleus and some work is required to be done to pull it away. Eq. (2.14) also reveals that the energy of the electron is *quantized.* It further follows from Eq. (2.14) that as n increases E becomes less negative and hence its algebraic value increases. Obviously, the electron has minimum energy when it is in its innermost orbit or ground state. An electron or atom must be imparted some energy in order that the electron may jump to an outer orbit. Similarly, energy is emitted by an electron when it jumps from an outer to an inner orbit. When an electron is forced to revolve in an outer orbit, the state of the atom is called an *excited state.* An atom in an excited state is not stable and in due course of time it must jump to an inner orbit and emit the difference of energy in the form of a quantum of radiation.

Let E_i and E_f be the energies corresponding to the initial (higher) and final (lower) orbits of an excited atom (Hydrogen $Z = 1$), we have

$$E_i = -\frac{me^4}{8\varepsilon_0^2 h^2}\frac{1}{n_i^2}$$

and

$$E_f = -\frac{me^4}{8\varepsilon_0^2 h^2}\frac{1}{n_f^2}$$

Where n_i and n_f are the corresponding quantum numbers. Then according to Bohr's fourth postulate, the frequency f of the emitted radiation is given by

$$h\nu = E_i - E_f = -\frac{me^4}{8\varepsilon_0^2 h^2}\frac{1}{n_i^2} + \frac{me^4}{8\varepsilon_0^2 h^2}\frac{1}{n_f^2}$$

$$= \frac{me^4}{8\varepsilon_0^2 h^3}\left(\frac{1}{n_f^2} - \frac{1}{n_i^2}\right)$$

or

$$\nu = \frac{me^4}{8\varepsilon_0^2 h^3}\left(\frac{1}{n_f^2} - \frac{1}{n_i^2}\right) \tag{2.15}$$

The corresponding wavelength λ is given by

$$\frac{1}{\lambda} = \frac{\nu}{c} = \frac{me^4}{8\varepsilon_0^2 c h^3}\left(\frac{1}{n_f^2} - \frac{1}{n_i^2}\right) \tag{2.16}$$

Eq. (2.16) indicates that since n_i and n_f can take only integral values, the radiation emitted by excited hydrogen atoms should contain certain discrete wavelengths only. In terms of wave number, Eq. (2.16) can be expressed as

$$\bar{\nu} = R\left(\frac{1}{n_f^2} - \frac{1}{n_i^2}\right) \tag{2.17}$$

where $R = \dfrac{me^4}{8\varepsilon_0^2 ch^3}$ is a constant and its value comes out to be the same as the Rydberg constant R in the Balmer's formula derived empirically. The value of Rydberg constant R comes out to be 1.096×10^7 m^{-1}. This value is in good agreement with the value obtained from the spectroscopic data of the Balmer series.

Spectral series of Hydrogen atom: The various series observed in the spectrum of hydrogen atom can be easily explained from Eq. (2.17) as follows:

(i) **Lyman Series:** Putting $n_f = 1$ and $n_i = 2, 3, 4...$ in Eq. (2.17), one obtains the wave numbers of spectral lines of *Lyman series* in *ultra-violet region.* Thus for this series

$$\bar{\nu} = R\left(\frac{1}{1^2} - \frac{1}{n^2}\right) \qquad n_i = 2, 3, 4... \tag{2.18}$$

(ii) **Balmer series:** If we put $n_f = 2$ and $n_i = 3, 4, 5...$, in Eq. (2.17), one obtains the wave numbers of spectral lines of Balmer series in the *visible region.* Thus for this series

$$\bar{\nu} = R\left(\frac{1}{2^2} - \frac{1}{n_i^2}\right) \tag{2.19}$$

(iii) **Paschen series:** Putting $n_f = 3$ and $n_i = 4, 5, 6...$, one obtains the wave numbers of spectral lines of Paschen series in the *infra-red region.* Thus for this series

$$\bar{\nu} = R\left(\frac{1}{3^2} - \frac{1}{n_i^2}\right) \qquad n_i = 4, 5, 6... \tag{2.20}$$

(iv) **Brackett Series:** Putting $n_f = 4$ and $n_i = 5, 6, 7...$, one obtains the wave numbers of spectral lines of Brackett series in the *infra-red region.* Thus for this series, we have

$$\bar{\nu} = R\left(\frac{1}{4^2} - \frac{1}{n_i^2}\right) \qquad n_i = 5, 6, 7... \tag{2.21}$$

(v) **Pfund series:** Putting $n_f = 5$ and $n_i = 6, 7, 8...$, one obtains the wave numbers of spectral lines of Pfund series in the *infra-red region.* Thus for this series,

$$\bar{\nu} = R\left(\frac{1}{5^2} - \frac{1}{n_i^2}\right) \qquad n_i = 6, 7, 8... \tag{2.22}$$

Note: Transitions between states with negative energy will give rise to line spectra, whereas transitions between states with positive energy $E > 0$ and states $E < 0$ will result in a continuous spectrum.

The transitions for all above series are shown in Fig. (2.6).

Atomic Energy levels: From Eq. (2.14) we have

$$E_n = -\frac{me^4}{8\pi^2\varepsilon_0^2 h^2}\cdot\frac{1}{n^2}$$

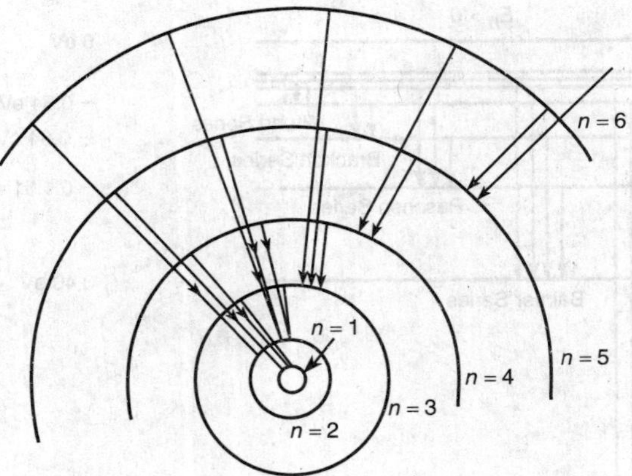

Fig. 2.6 Orbital diagram

Where n takes the values of n = 1, 2, 3 ..., By using the values of $m = 9.11 \times 10^{-31}$ kg, $e = 1.6 \times 10^{-19}$ C for the mass and charge of the electron, we obtain

$$E_n = -\frac{13.6}{n^2}\text{eV} \quad \text{for} \quad n = 1, 2, 3... \tag{2.23}$$

$$1\text{eV} = 1.6\times10^{-19}\,\text{C}\times1\,\text{Volt}$$

$$= 1.6\times10^{-19}\,\text{J}$$

The state of lowest energy or ground state corresponds to $n = 1$ and its energy is $E_i = -13.6$ eV.

Fig. (2.7) is an energy level diagram representing the allowed energies for the hydrogen atom. Note that all the states from $n = 1$ to $n = \infty$ are bound states, since they have negative energies. When n increases and approaches $n = \infty$, the energy states get closer and closer together until the energy difference between two consecutive states becomes so small that the distribution gives a practically continuous spectrum. Above the line given by $n = \infty$, the energy states have positive energy, $E > 0$, and the spectrum of the states is continuous. The system is then unbound, meaning that the electron is free.

Shortcomings of Bohr's Atomic Model

Bohr's theory, although very successful in explaining the spectrum of hydrogen atom and giving valuable information about atomic structure, has the following drawbacks:

(i) The 'fine structure', i.e., individual line of hydrogen spectrum accompanied by a number of faint lines, cannot be explained by Bohr's theory as such. The fine structure of spectral lines can only be explained when (a) the relativistic variation in the mass of the electron and (b) the electron 'spin' are taken into account.

(ii) Bohr's model fails to explain the variation in intensity of the spectral lines of an element. The intensity of the spectral lines can be explained by quantum mechanics.

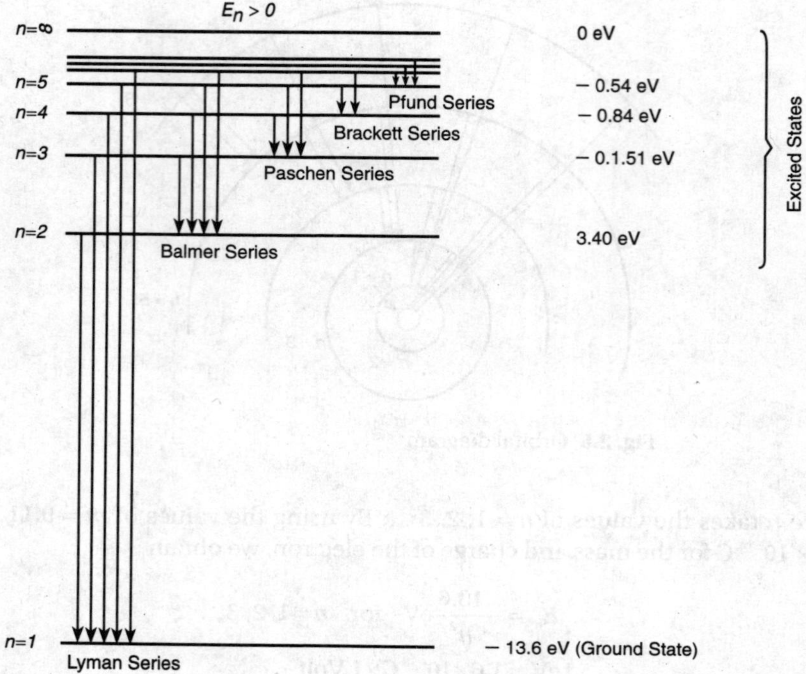

Fig. 2.7

(iii) Bohr's model fails to explain the spectra of complex atoms. It is only applicable to one-electron atoms such as hydrogen, hydrogen isotopes, ionised helium etc.

(iv) Bohr's theory fails to explain satisfactorily the distribution of electrons in atoms.

(v) The success of Bohr's theory in explaining the effect of magnetic field on spectral lines is only partial, i.e., it cannot explain the 'anomalous' Zeeman effect.

We will study about these in subsequent chapters.

2.7 SPECTRUM OF IONISED HELIUM

Helium ($^4_2$He) has an atomic number (Z) 2 and atomic mass number (A) 4. Obviously, it has 2 protons and 2 neutrons in its nucleus and 2 electrons in the extra-nuclear system. If we consider He$^+$, i.e., one of the outer electrons is removed by ionisation and the remaining electron may be assumed to go round the nucleus in a circular orbit. He$^+$ has a nuclear charge +2e so that $Z = 2$ for it. One can write the equation for its term values as

$$E_n = -\frac{mz^2e^4}{32\pi^2\varepsilon_0^2n^2\hbar^2} = -\frac{me^4}{8\pi^2\varepsilon_0^2n^2\hbar^2} \tag{2.24}$$

Obviously, the term values of He$^+$ are four times larger than the corresponding hydrogen terms. If there is a transition from an initial energy level n_i to a final level n_f, then the emitted radiation has the wave number and wavelength

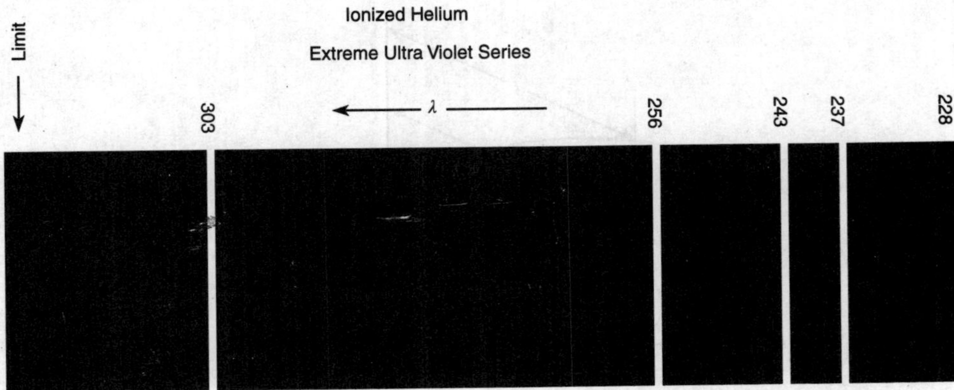

Fig. 2.8 Ionised helium spectrum

$$\bar{\nu} = \frac{1}{\lambda} = \frac{me^4}{16\pi^3 \varepsilon_0^2 c\hbar^3}\left(\frac{1}{n_f^2} - \frac{1}{n_i^2}\right)$$

$$= 4R_H\left(\frac{1}{n_f^2} - \frac{1}{n_i^2}\right) \qquad (2.25)$$

Where R_H is the Rydberg constant for hydrogen. Considering transitions from $n_i = 5, 6, 7, 8, \ldots$ levels of He^+ to the final level with $n_f = 4$, the series is called *Pickering* series Fig. (2.8).

One can easily see that the lines of the Pickering series should be given by

$$\bar{\nu} = 4R_H\left(\frac{1}{4^2} - \frac{1}{n_i^2}\right) = R_H\left(\frac{1}{2^2} - \frac{4}{n_i^2}\right) \qquad (2.26)$$

Obviously, the alternate lines of the Pickering series with n_i even ($n_i = 6, 8, 10\ldots$) of He^+ should have the same wavelengths as the spectral lines of the Balmer series of hydrogen. This means that the alternate lines of the Pickering series of ionised helium should coincide with the Balmer series of hydrogen. Careful measurements have shown that this coincidence predicted by the simple Bohr theory is however not quite correct. The lines of the Pickering series of the He^+ have slightly higher wave numbers than the corresponding lines of Balmer series.

The reason for the above difference is due to the finite masses of nuclei of hydrogen and helium.

2.8 FINITE MASS CORRECTION

In the original Bohr's theory, it is assumed that the electron revolves round a nucleus to be at rest. This assumption would be correct if the nucleus is infinitely heavy. However a nucleus of finite mass cannot be at rest. Hence the combined electron-nucleus system should be considered to revolve round the common centre of mass (Fig. 2.9) with the same angular velocity.

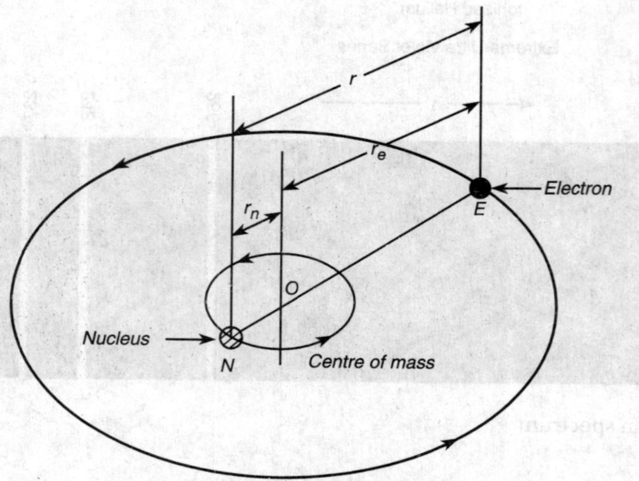

Fig. 2.9 Motion of the electron and nucleus around the centre of mass

In Fig. (2.9), let N represent the nucleus and E the electron revolving round the common centre of mass 0. The mass of hydrogen nucleus is 1836 times heavier than the electron mass and the mass of helium nucleus is four times heavier than the mass of the hydrogen nucleus. Even though these masses are much higher than the electronic mass (m), one cannot consider them to be infinitely heavy.

If r_e and r_n are the radii of the orbits of the electron and the nucleus and M_H and m are the masses of hydrogen nucleus and electron respectively. From the definition of centre of mass E, O and N are always in a straight line, then one can write

$$M_H r_n = m r_e$$

$$\therefore \quad \frac{r_n}{r_e} = \frac{m}{M_H} \quad \text{or} \quad \frac{r_n}{r_e + r_n} = \frac{m}{m + M_H}$$

If the electron-nucleus distance be r, then

$$r = r_e + r_n$$

Then

$$r_n = \frac{mr}{m + M_H} \tag{2.27}$$

and

$$r_e = \frac{M_H r}{m + M_H} \tag{2.28}$$

Let ω be the angular velocity of the system about the centre of mass. The combined angular momentum of the electron and the nucleus is

$$m\omega r_e^2 + M_H \omega r_n^2 = L = \frac{nh}{2\pi} \tag{2.29}$$

Using Eqs. (2.27) and (2.28), one obtains from Eq. (2.29)

$$\left(\frac{mM_{\rm H}}{m+M_{\rm H}}\right)\omega r^2 = \frac{nh}{2\pi} \tag{2.30}$$

Suppose
$$\mu = \frac{mM_{\rm H}}{m+M_{\rm H}} = \frac{m}{1+\dfrac{m}{M_{\rm H}}} \tag{2.31}$$

μ is known as *reduced mass* of the electron.
Then one obtains

$$\mu\omega r^2 = \frac{nh}{2\pi} = n\hbar \tag{2.32}$$

Comparing with the simple Bohr theory result $mvr = m\omega r^2 = n\hbar$ we note that the result of taking into account the finite mass of the nucleus is to replace the mass m of the electron by $\mu\left(=\dfrac{mM_{\rm H}}{m+M_{\rm H}}\right)$. The Rydberg constant will now be

$$R_{\rm H} = \frac{\mu e^4}{8\varepsilon_0^2 c h^3} = \frac{\mu e^4}{64\pi^3\varepsilon_0^2 c\hbar^3} \tag{2.33}$$

or
$$R_{\rm H} = \frac{R_\infty}{1+\dfrac{m}{M_{\rm H}}} \tag{2.34}$$

where R_∞ is the Rydberg constant for an infinitely heavy nucleus given by

$$R_\infty = \frac{me^4}{64\pi^3\varepsilon_0^2 c\hbar^3} = 1.09737\times 10^7\ \text{m}^{-1} \tag{2.35}$$

The transition between two energy levels results in the emission of electromagnetic radiation. The wave number is obtained from

$$\bar{\nu} = \frac{1}{\lambda} = R_{\rm H}Z^2\left(\frac{1}{n_f^2}-\frac{1}{n_i^2}\right) = \frac{R_\infty Z^2}{1+\dfrac{m}{M_{\rm H}}}\left(\frac{1}{n_f^2}-\frac{1}{n_i^2}\right) \tag{2.36}$$

One can see that the Rydberg constant depends on the nuclear mass M. Since $M >> m$, the difference between R_M and R_∞ is small. We obtain for hydrogen and helium as

$$R_{\rm H} = 1.09677\times 10^7\ \text{m}^{-1} \text{ and } R_{\rm He} = 1.09722\times 10^7\ \text{m}^{-1}$$

The value of $R_{\rm H}$ agrees very well with the experimental value derived from the study of the hydrogen spectrum ($= 1.09678\times 10^7\text{m}^{-1}$). Similarly in the case of helium too, the agreement between the experimental value and the theoretical value is very good.

We note that the reduced mass $\mu < m$, and it increases with increasing nuclear mass M.μ becomes equal to the electronic mass m as $M \to \infty$. Thus $R_{\rm H}$ increases as $M_{\rm H}$ increases. When $M_{\rm H} \to \infty$, $R_{\rm H} \to R_\infty$. Since $R_{\rm He} > R_{\rm H}$ and hence the spectral lines of the Pickering

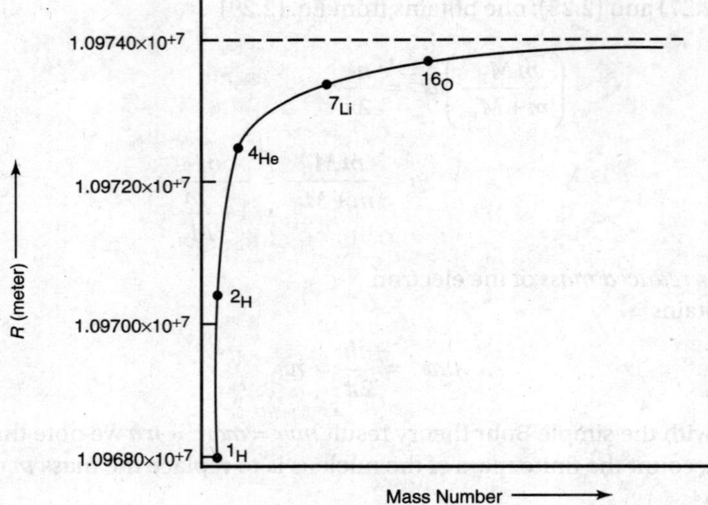

Fig. 2.10

series of He^+ should have slightly higher wave numbers or shorter wavelengths than the corresponding lines of the Balmer series of hydrogen. The variation of the Rydberg constant from element to element is shown in Fig. (2.10). One can see that the large change occurs between ^1H and ^2H and with increasing mass number the Rydberg constant approaches more and more closely to R_∞.

2.9 DISCOVERY OF HEAVY HYDROGEN (DEUTRON)

The variation of Rydberg constant with the mass of the nucleus resulted in the discovery of *deutrium* (*heavy hydrogen*), an isotope of naturally occurring hydrogen. Its nucleus is known as *deutron.* ^1H and ^2H have the same chemical properties. The atoms of both types of hydrogen i.e., ^1H and ^2H have only one electron revolving in their orbits. Their nuclei have the same charge but different masses, i.e. mass of the deutron nucleus $M_D = 2M_H$. So the Rydberg constant for the deuterium is slightly greater than that for hydrogen

$$R_H = \frac{R_\infty}{1+m/M_H},\ R_D = \frac{R_\infty}{1+m/M_D}$$

Calculations yield $R_D = 1.09707 \times 10^{+7}\ m^{-1}$. Obviously, the lines of the deutrium spectra will have slightly higher wave numbers than the corresponding lines of the hydrogen spectra. However, the general nature of the two spectra is exactly same.

Urey and coworkers in 1931, photographed hydrogen spectrum by means of a large concave grating. They observed that each hydrogen line was accompanied by a very faint line slightly on the short wavelength side. They measured the wavelength difference and attributed the faint line to some isotope of hydrogen. This isotope of naturally occurring hydrogen is now known as 'heavy hydrogen' or 'deutrium.' From the faintness of the lines, it is obvious that its abundance in naturally occuring hydrogen is quite low. Later, improved methods of measurements have yielded a relative abundance of only about 1 part in 7000 ($^1_1\text{H}-99.985\%$, $^2_1\text{H(D)}-0.015\%$) for D.

One can determine the wavelength difference between the corresponding spectral lines of ^{1}H and ^{2}H as follows:

$$\lambda_{\text{H}} = \frac{1}{\bar{\nu}_{\text{H}}} = \frac{1}{R_{\text{H}}}\left(\frac{1}{n_2^2} - \frac{1}{n_1^2}\right)^{-1} = \frac{1+\dfrac{m}{M_{\text{H}}}}{R_\infty}\left(\frac{1}{n_2^2} - \frac{1}{n_1^2}\right)^{-1}$$

$$\lambda_{\text{D}} = \frac{1}{\bar{\nu}_{\text{D}}} = \frac{1}{R_{\text{D}}}\left(\frac{1}{n_2^2} - \frac{1}{n_1^2}\right)^{-1} = \frac{1+\dfrac{m}{M_{\text{D}}}}{R_\infty}\left(\frac{1}{n_2^2} - \frac{1}{n_1^2}\right)^{-1}$$

$$\therefore \Delta\lambda = \lambda_{\text{H}} - \lambda_{\text{D}} = \frac{\dfrac{m}{M_{\text{H}}} - \dfrac{m}{M_{\text{D}}}}{R_\infty\left(\dfrac{1}{n_2^2} - \dfrac{1}{n_1^2}\right)}$$

m/M_{H} and m/M_{D} are very small numbers and hence one can express the denominator in the above expression to be approximately equal to $1/\lambda_{\text{H}}$. Also $M_{\text{D}} \approx 2M_{\text{H}}$ and therefore one obtains finally

$$\Delta\lambda = \lambda_{\text{H}}\frac{m}{2M_{\text{H}}} = \frac{\lambda_{\text{H}}}{3672}$$

We have $n_1 = 4 \rightarrow n_1 = 2$, a H_β line of the Balmer series of hydrogen and it has wavelength $\lambda_{\text{H}} = 4681$Å. Making use of this, one obtains from the above expression $\Delta\lambda = 1.28$Å for the wavelength difference of the H_β lines of hydrogen and deutrium.

2.10 BOHR'S CORRESPONDENCE PRINCIPLE

This principle establishes a correlation between the results obtained from the classical and quantum theories. The laws of classical mechanics are applicable to explain the motion of macroscopic bodies, starting from the very large heavenly bodies to the very small objects like the atoms and molecules. Kinetic theory of matter is based on classical mechanics. Planck's quantum theory of radiation, Einstein's light quantum hypothesis to explain photoelectric effect and de Broglie's hypothesis concerning the wave-like behaviour of particles of matter provided a strong ground work for the formulation of quantum physics. The principles of quantum physics, so different from those of classical physics in the microworld that lies beyond the reach of our senses, must nevertheless yield results identical with those of classical physics in the domain where experiment indicates the latter to be valid.

Bohr pointed out that the new laws of quantum mechanics required to describe the motion of the sub atomic particles (like the electron inside the atoms) have to be developed in such a way that these laws must go over to the laws of classical mechanics in the limit of larger macroscopic bodies. This is known as *'Bohr's Correspondence Principle'.* This forms an important basis in formulating the laws of quantum mechanics.

The frequency f_n of revolution of an electron in the hydrogen atom rotating in its orbit of principal quantum number n is given by

$$f_n = \frac{\omega_n}{2\pi} = \frac{V}{2\pi r} = \frac{1}{2\pi}\left(\frac{1}{4\pi\varepsilon_0}\frac{Ze^2}{mr_n^3}\right)^{1/2} = \frac{me^4}{32\pi^3\varepsilon_0^2 n^3\hbar^3}$$

Now the frequency of the radiation emitted due to transition from the $(n+1)^{\text{th}}$ to the nth orbit of the hydrogen atom is

$$\nu = \frac{\omega}{2\pi} = \frac{me^4}{64\pi^3\varepsilon_0^2\hbar^3}\left[\frac{1}{n^2} - \frac{1}{(n+1)^2}\right]$$

$$= \frac{me^4}{64\pi^3\varepsilon_0^2\hbar^3}\frac{2n+1}{n^2(n+1)^2}$$

If $n >> 1$, then one obtains

$$\nu = \frac{me^4}{64\pi^3\varepsilon_0^2\hbar^3}\frac{2n}{n^4} = \frac{me^4}{32\pi^3\varepsilon_0^2 n^3\hbar^3} = f_n$$

Obviously, in the limit of very large quantum numbers, the results of Bohr atom model are the same as those obtained from classical electromagnetic theory, as required by Bohr's correspondence principle.

In the above discussion, it is assumed that the transition takes place from the $(n+1)^{\text{th}}$ to the nth orbit and the change of quantum number is by 1 unit. However, when the change of quantum number is 2, 3... units, i.e., when the transition is from $(n+2)^{\text{th}}$, $(n+3)^{\text{th}}$, $(n+4)^{\text{th}}$ etc., orbit to n^{th} orbit, then for large n, one finds the frequencies of the emitted radiation to be twice, thrice, four times etc. of the frequency of radiation. The classical electromagnetic theory predicts the emission of radiation of these frequencies also.

2.11 HYDROGEN LIKE SPECTRA

The spectra of hydrogen and helium are the simplest in appearance. Other hydrogen like spectra have been observed for doubly ionised Lithium $Li^{++}(Z=3)$, triply ionised beryllium, $Be^{+++}(Z=4)$ and so on up to multiply ionised oxygen. The spectra of most other elements show an extremely large number of lines and one finds it extremely difficult at first sight to classify them into series. However, with some difficulty, one can seperate the lines of such a complicated spectrum into series. We must note that the spectra given out by the elements in any particular column of periodic table are similar to each other, e.g., the spectra of the alkali elements and the alkali earth metals bear a resemblance to the hydrogen spectrum and therefore they are called *hydrogen-like spectra.* If we examine minutely the Balmer series, then we find that each line in this series is really a group of very close lines. Obviously, the lines are said to exhibit a *fine structure.* One can easily see that the explanation of the complex series found in all hydrogen-like spectra is closely related to the explanation of the fine structure of the Balmer series. Sommerfeld provided the explanation by postulating that the orbit of the electron is elliptic and not circular as assumed by Bohr.

2.12 ALKALI ATOMIC SPECTRA

From the Bohr's model of the atom we have seen that the wave numbers of various spectral lines are expressed as

$$\bar{\nu} = R\left(\frac{1}{m^2} - \frac{1}{n^2}\right) \tag{2.37}$$

where m is fixed for a particular series and is termed as the fixed term and n takes running values greater than m and is termed as the *running term.*

The atomic spectra of all the elements consisting of atoms having one electron in their outermost orbit (like hydrogen) are similar and are called alkali atomic spectra. The alkali metals are lithium, sodium, potassium, rubidium and cesium. The other examples exhibiting alkali like spectra are simply ionised atoms of beryllium, magnesium, calcium and doubly ionised atoms of boron and aluminium etc.

Rydberg, using all the experimental data then available and Hartley's law, was able to classify the alkali atomic spectra of alkali elements into three series, namely, the *sharp*, the *diffuse* and the *principal* series to which was added later the *fundamental* or the *Bergmann* series. One can put his formula for the wave numbers $\bar{\nu}$ of the members of any of these series in the form

$$\bar{\nu} = R\left[\frac{1}{(m-\alpha)^2} - \frac{1}{(n-\beta)^2}\right] \tag{2.38}$$

where α and β are characteristic constants called *quantum defects.* The various series found in the alkali atomic spectra are the following:

(i) *Principal series:* $\bar{\nu}_P = R\left[\frac{1}{(1-\alpha)^2} - \frac{1}{(n-\beta_P)^2}\right] \quad n \geq 2$

(ii) *Sharp series:* $\bar{\nu}_S = R\left[\frac{1}{(2-\alpha)^2} - \frac{1}{(n-\beta_S)^2}\right] \quad n \geq 2$

(iii) *Diffuse series:* $\bar{\nu}_D = R\left[\frac{1}{(2-\alpha)^2} - \frac{1}{(n-\beta_D)^2}\right] \quad n \geq 3$

(iv) *Fundamental or Bergmann series:* $\bar{\nu}_F = R\left[\frac{1}{(2-\alpha)^2} - \frac{1}{(n-\beta_F)^2}\right] \quad n \geq 4$

Where $\beta_P, \beta_S, \beta_D, \beta_F$ are the characterstic constants of principal, sharp, diffuse and fundamental series respectively.

In particular, a spectral series in the case of lithium atom can be given by

$$\bar{\nu} = R\left[\frac{1}{(2-0.040)^2} - \frac{1}{(n-0.405)^2}\right] \qquad n = 3, 4, 5\ldots$$

One can write the other series by *Ritz combination principle*, which states that any one of the running terms in one series can be used as the fixed term in another to obtain a new spectral line. Thus one can write another series for lithium atom as

$$\bar{\nu} = R\left[\frac{1}{(2-0.405)^2} - \frac{1}{(n-0.040)^2}\right] \qquad n = 3, 4, 5\ldots$$

and

$$\bar{\nu} = R\left[\frac{1}{(3-0.405)^2} - \frac{1}{(n-0.040)^2}\right] \qquad n = 4, 5, 6\ldots$$

2.13 SOMMERFELD'S EXTENSION OF BOHR THEORY OF ATOM

Within a short span after the announcement of Bohr theory of atom, German physicist Arnold Sommerfeld in 1916 extended the theory to include elliptic orbits as well. In Bohr model the motion of the electron is essentially one-dimensional. Obviously, one quantum number (n) was sufficient to specify the state of the atom. Now an electron in an elliptic orbit has two degrees of freedom. Hence an electron moving in an elliptic orbit Fig. (2.11) requires two quantum numbers. An electron in an elliptic orbit has two degrees of freedom, namely the radial distance r and the azimuthal angle θ. Sommerfeld and Wilson independently showed that Bohr's quantisation rule is a particular case of a more generalised quantum condition. If p is a momentum and q the corresponding position coordinate the Sommerfeld and Wilson's general condition of quantisation can be expressed as

$$\int p.dq = nh \qquad (2.39)$$

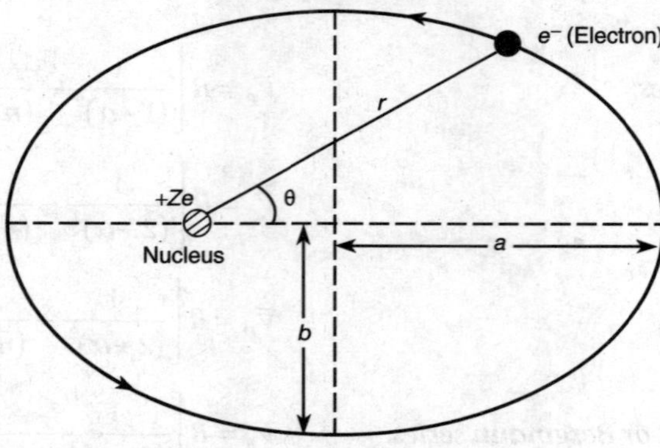

Fig. 2.11 Elliptic orbit of the electron

If p_θ and p_r be the angular momenta of the electron, then according to Eq. (2.39), we must have

$$\int_0^{2\pi} p_\theta . d\theta = kh, \quad k = 1, 2, 3... \tag{2.40}$$

and

$$\oint p_r . dr = n_r h, \quad n_r = 0, 1, 2... \tag{2.41}$$

Here K and n_r are integers, called respectively *'azimuthal'* and *'radial'* quantum numbers. Making use of polar equation of the ellipse and integrating over a complete revolution, one can easily show that

$$\frac{k}{k+n_r} = \frac{b}{a} \tag{2.42}$$

where a is semi-major axis and b is semi-minor axis of the ellipse. Since both k and n_r are both integers, one can put

$$k + n_r = n \ (n = 1, 2, 3...) \tag{2.43}$$

where n is called the *'total* or *principal quantum number'* of the electron. Then

$$\frac{k}{n} = \frac{b}{a} \tag{2.44}$$

This is the condition of quantisation for the orbits. Thus only *those elliptic orbits are permitted for the electron for which the ratio of the major to the minor axis is the ratio of two integers.*

Sommerfeld calculated the total energy E of an electron in a quantized elliptic orbit. The total energy is the sum of the kinetic energy K and potential energy U.

Thus

$$E = K + U$$

$$= \frac{1}{2} m \, (\dot{r}^2 + r^2 \dot{\theta}^2) - \frac{1}{4\pi\varepsilon_0} \frac{Ze^2}{r} \tag{2.45}$$

Since $p_r = m\dot{r}$ and $p_\theta = mr^2\dot{\theta}$ and hence Eq. (2.45) can be rewritten as

$$E = \frac{1}{2m}\left\{p_r^2 + \frac{p_\theta^2}{r^2}\right\} - \frac{1}{4\pi\varepsilon_0}\frac{Ze^2}{r} \tag{2.46}$$

Now,

$$p_r = m\dot{r} = m\frac{dr}{dt} = m\frac{dr}{d\theta}\frac{d\theta}{dt} = m\frac{dr}{d\theta}\dot{\theta}$$

$$= (mr^2\dot{\theta})\frac{1}{r^2}\frac{dr}{d\theta} = \frac{p_\theta}{r^2}\frac{dr}{d\theta} \tag{2.47}$$

Using Eqs. (2.47) and (2.46) becomes

$$E = \frac{p_\theta^2}{2mr^2}\left[\left(\frac{1}{r}\frac{dr}{d\theta}\right)^2 + 1\right] - \frac{1}{4\pi\varepsilon_0}\frac{Ze^2}{r} \tag{2.48}$$

Now, the polar equation of the ellipse is

$$\frac{1}{r} = \frac{1}{a}\left(\frac{1-\varepsilon\cos\theta}{1-\varepsilon^2}\right) \tag{2.49}$$

where ε is the eccentricity of the ellipse and $1-\varepsilon^2 = \frac{b^2}{a^2}$. Putting the values of r and $\frac{dr}{d\theta}$ from Eq. (2.49) in Eq. (2.48), one obtains

$$E = -\frac{mZ^2e^4(1-\varepsilon^2)}{(4\pi\varepsilon_0)^2 2p_\theta^2} \tag{2.50}$$

For an isolated system, the angular momentum p_θ is constant. Then from Eq. (2.40), one obtains

$$p_\theta = \frac{kh}{2\pi}$$

Also $$1-\varepsilon^2 = \frac{b^2}{a^2} = \frac{k^2}{n^2}$$

Making these substitutions in Eq. (2.50), one obtains

$$E = -\frac{mZ^2e^4\left(\frac{k^2}{n^2}\right)}{(4\pi\varepsilon_0)^2 2\left(\frac{kh}{2\pi}\right)^2}$$

$$= -\frac{mZ^2e^4}{8\varepsilon_0^2h^2}\left(\frac{1}{n^2}\right) \tag{2.51}$$

$$= -\frac{RZ^2hc}{n^2} \tag{2.52}$$

Obviously, the total energy does not depend on the azimuthal quantum number k, it depends on the principal quantum number n only. The value of E as given above is exactly the same as found from Bohr's theory. This means the principal quantum number n in the Sommerfeld theory is the principal quantum number of the Bohr theory.

Now, for a given value of $n(= k + n_r)k$ can take n possible different values ($k = 1, 2, 3 \ldots n$). Obviously, for a given n, there are n orbits of different eccentricities ($\because \frac{b}{a} = \frac{k}{n}$), which can be occupied by the electron. However, the energy E for all these orbits is the same, because E is independent of k and depends upon n only. This means that *the mere introduction of elliptic orbits adds no new energy levels.*

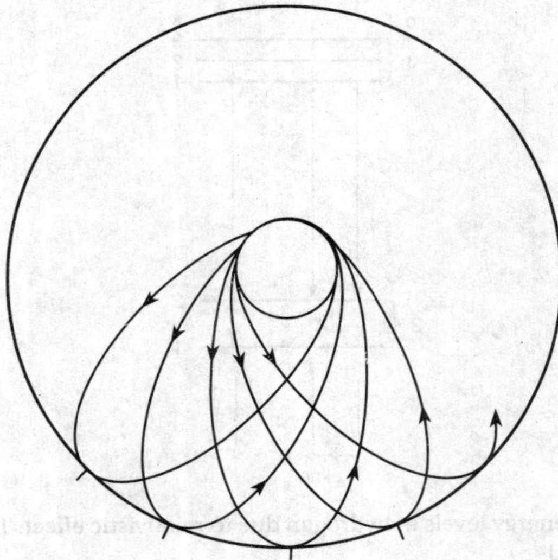

Fig. 2.12

Relativistic Correction

Sommerfeld pointed out that the origin of the fine structure of the spectral lines of the hydrogen-like atoms was due to the relativistic variation of mass of the electron revolving in an elliptic orbit. In an elliptic orbit the velocity of the electron varies, becoming very large. When the electron passes close to the nucleus. According to the special theory of relativity, the mass m of a particle moving with velocity v is related to its rest mass m_0 by the relation $m = m_0 / \sqrt{1 - v^2 / c^2}$. Relativistic effect makes the energy of the electron in a more elliptic (i.e. of greater eccentricity) orbit than that in a non-relativistic elliptic orbit. One can show that describing the path of the electron in an elliptic orbit in the non-relativistic case is to be replaced by the following equation, when the relativistic variation of mass is taken into account

$$r = \frac{a(1-\varepsilon^2)}{1+\varepsilon\cos\gamma\theta} \tag{2.53}$$

where

$$\gamma^2 = 1 - \frac{Z^2 e^4}{(4\pi\varepsilon_0)^2 p^2 c^2} \tag{2.54}$$

Here we have written $p_\theta = p$ and this is a constant. For the non-relativistic case (c $\rightarrow \infty$), $\gamma = 1$ and one finds that r return to the same value when θ changes by 2π so that it has the same periodicity as θ. As a result of the variation of the electron mass due to differences in its velocities at different points on its elliptic orbit there is slow precession of the major axis in the plane of the ellipse about an axis through one of the foci. The trajectory of the electron in this case is a *rosette* as shown in Fig. (2.12).

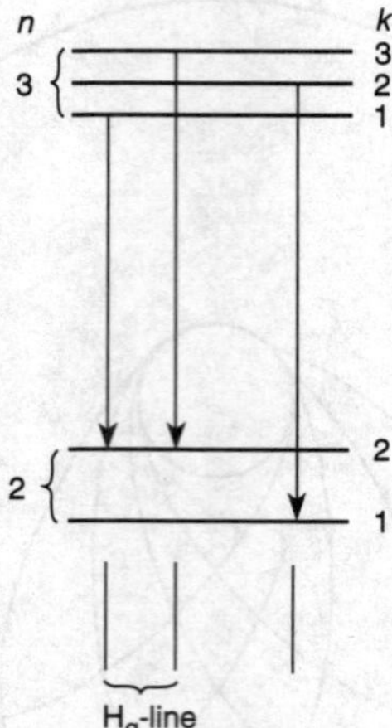

Fig. 2.13 Splitting of the energy levels of hydrogen due to relativistic effect. The splitting shown are highly magnified

Sommerfeld have shown that the different k-orbits with a given n have slightly different energies and calculated the total energy to the electron to be

$$E_{nk} = -\frac{\mu Z^2 e^4}{32\pi^2 \varepsilon_0^2 n^2 \hbar^2}\left[1+\frac{Z^2\alpha^2}{n^2}\left(\frac{n}{k}-\frac{3}{4}\right)\right] \tag{2.55}$$

where $\alpha << 1$ ($\approx 1/137$) is a pure number, the difference in energy from Bohr's theory as given by the second term with the bracket is quite small. α is called the *fine structure constant.*

One can now explain the fine structure of H_α line which is emitted when the electron jumps from the $n = 3$ to $n = 2$ level. The $n = 3$ level consists of 3 sub-levels corresponding to $k = 1,2,3...$ and $n = 2$ level consists of 2 sublevels corresponding to $k = 1, 2$ (Fig. 2.13).

Only those electron-jumps from the upper sub-levels to the lower sub-levels are allowed for which the k changes by unity, i.e., follows the *selection* rule:

$$\Delta k = k_f - k_i = \pm 1 \tag{2.56}$$

Obviously, the possible transitions consistent with Eq. (2.56) are

$$k_i = 1 \text{ and } k_f = 2 \qquad (\Delta k = +1)$$
$$k_i = 2 \text{ and } k_f = 1 \qquad (\Delta k = -1)$$
$$k_i = 3 \text{ and } k_f = 2 \qquad (\Delta k = -1)$$

Thus there are three possible jumps, showing that the H_α-line actually consists of three components.

We must note that the azimuthal quantum number k of the Bohr Sommerfeld theory has been replaced by the new azimuthal quantum number $l = (k-1)$ in quantum mechanics. We shall see later that for a given value of n, the possible values of l are 0, 1, 2, 3... $(n - 1)$.

Table 2.1

Orbital quantum number (l)	0	1	2	3	4	5	6
Optical electron designation	s	p	d	f	g	h	i
State of atom	S	P	D	F	G	H	I

The orbital angular momentum of the electron has a value $p_l = \sqrt{l(l+1)}\ \hbar$. The value of n defines a shell and each value of l determines one subshell. Obviously, a shell corresponding to principal quantum number n has n subshells. The optical electron can occupy any of the subshells. When an electron is in a particular subshell, it is designated by a small letter (*s, p, d, f*...), which specifies the subshell. In the case of atoms which have a single optical electron and when this electron is in a particular subshell the atom is said to be in a particular state designated by the corresponding capital letter (S, P, D, F...). These designations are shown in Table 2.1.

While designating an electron the letter designating it is written after the number that indicates the total quantum number n, e.g., 3*p* electron, here 3 is total quantum number and orbital quantum number corresponding to *p* is 1. We must note that for a given value of n the highest possible l value is $(n - 1)$. Obviously, one can have 4*d*, 5*f*, 2*p* and 2*s* electrons, where as 1*p*, 2*d*, 3*f* electrons do not exist.

2.14 EXCITATION AND IONISATION POTENTIALS

When an electron in an atom is raised from its ground state to a higher energy state the atom is said to be in the *excited* state. An atom can remain in an excited state only for a very short interval of time ($\sim 10^{-8}$s). The atom returns to its ground state either by a single quantum jump or by a series of such jumps between intermediate steps,. The minimum energy required to raise an atom from one energy level to another is called an *excitation energy*. When excitation energy is expressed in electron volt (eV) it is termed as *excitation or resonance or radiation potential*. In some cases e.g., inert gases and mercury, the excited atom can remain in the excited state for a period longer than 10^{-8}s, say from 10^{-6} to 2 s, such excited states are termed *metastable states*. The potential required to raise an atom to a metastable state is termed as a metastable potential. If an electron is removed completely free from the influence of the nucleus of that atom then the atom is said to be *ionised*. The energy required to ionise an atom is termed as *ionisation energy*. When expressed in electron-volt, the ionisation energy is called ionisation potential. Ionisation potential for hydrogen atom is 13.6 eV. The energy required to raise the hydrogen atom from the ground state to the first excited state is given by 13.6 – 3.4 = 10.2 eV; that required to rise it to the second excited state is given by (13.6 – 1.51) = 12.09 eV. Obviously, the discrete amounts of energy are needed to excite the hydrogen atom to different energy states. It thus follows that 10.9 eV and 12.09 eV are excitation potentials. The name *critical potential* is applied collectively to excitation, metastable and ionisation potentials.

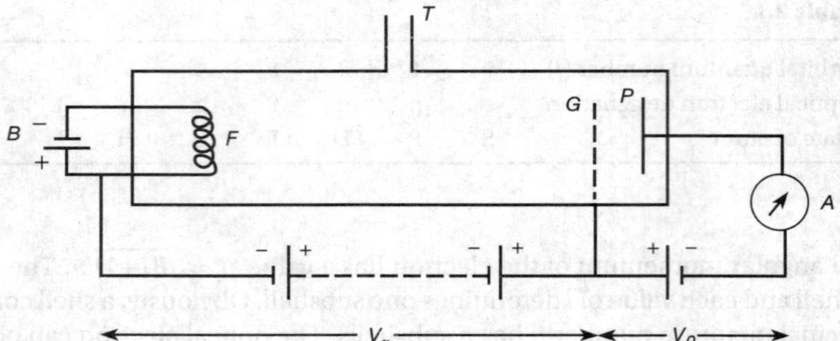

Fig. 2.14 Experimental set-up for the Frank-Hertz experiment

2.15 EXPERIMENTAL EVIDENCE FOR QUANTISATION AND DETERMINATION OF CRITICAL POTENTIAL

Direct evidence for the existence of discrete energy states within an atom was provided by Frank and Hertz in 1914 and Devis and Goucher and others.

Frank and Hertz Experiment

The apparatus used by Frank and Hertz is shown in Fig. 2.14. It consists of a glass tube in which are mounted a thermionic filament *F*, a grid *G* and a plate *P*. The glass tube can be evacuated by connecting the side tube *T* to a pump. The atoms of the element to be studied are introduced in the tube in the form of a gas (e.g., mercury vapour) at low pressure of about 1mm of mercury. The distance between *F* and *G* is kept considerably greater than the mean free path of the electrons. The distance between the grid and the plate is kept less than the mean free path. The filament *F* is heated by a battery *B*. An accelerating potential V_α is applied between the filament (*F*) and the grid (*G*) and a small fixed retarding potential V_0 (−0.5 V) between grid (*G*) and plate (*P*).

The electrons emitted from the filament (*F*) are accelerated between *F* and *G* by the potential V_α and retarded between *G* and *P* by the potential V_0. Obviously only those electrons will reach *P* which have energies greater than eV_0 at *G*. The ammeter *A* measures the plate current.

Fig. (2.15) shows current vs accelerating potentialy potentialy V_α plot. The curve shows a series of regularly spaced peaks.

As V_α is increased, the plate current increases as in any electronic tube, except that a significant decrease in the plate current occurs each time the accelerating potential is increased by approximately 5 V. Some of these electrons with energies slightly greater than 4.88 eV will experience inelastic collisions and will be left with such small energy that they will not be able to reach the plate, because of the presence of the retarding voltage. If V_α is increased by an additional 5 V, some of the electrons that were left with almost no kinetic energy will experience another inelastic collision and will not reach the plate. This explains the second dip at a potential of about 5 V greater than the first dip. Hence, this

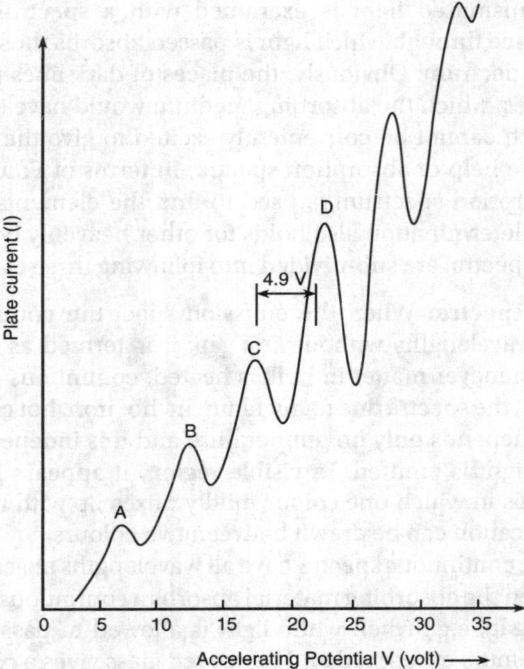

Fig. 2.15 Plate current vs. accelerating potential in Frank-Hertz experiment. The separation between any two consecutive peaks is about 4.9 V

second dip corresponds to those electrons that have experienced two inelastic collisions; the third dip corresponds to three inelastic collisions, and so on. Each time there is an inelastic collision, the mercury atoms will be excited and return to the ground state by the emission of photons. Spectroscopic measurements reveal that the wavelength of the radiation coming from the tube is 2536 Å. This corresponds to transitions from the first excited state of mercury to the ground. This result, together with the fact that the energy difference between two consecutive dips is about 4.9 V, shows in a very convincing way the existence of discrete energy level in the mercury atom. It is also possible by using different voltages and improved resolution, to measure the excitation of other atomic energy levels. The potential difference between any two peaks of the curve gives a critical potential. The curve for mercury shows a first excitation potential or *resonance potential* of 4.9 V. Frank and Hertz were awarded the Nobel prize in Physics in 1925 for this work.

2.16 TYPES OF SPECTRA

The spectra may be divided into two types (i) *Emission spectra* and (ii) *Absorption spectra*. The emission spectra are obtained when light coming directly from the source is examined with the help of a spectroscope. The absorption spectra are obtained when light from a source which gives continuous emission spectrum is passed through an absorbing material and the transmitted light is examined with a spectroscope. In absorption spectrum, the substance through which light is passed absorbs those wavelengths, which it emits in emission spectrum. Obviously, the places of dark lines in absorption spectra correspond to the lines, which the absorbing medium would have emitted on excitation. The substances, which cannot be conveniently excited to give the emission spectra are easily studied with the help of absorption spectra. In terms of Fraunhoffer dark lines in solar spectrum, absorption spectrum is used to find the elements present in the solar atmosphere. Similar determination also holds for other heavenly bodies. Both, the emission and absorption spectra, are sub-divided into following three classes:

(i) Continuous spectra: When the emission spectrum contains a wide range of continuous wavelengths without any gap, it is termed as continuous emission spectrum. Whenever matter in bulk is heated, continuous emission spectrum is obtained, e.g., the spectra due to hot filament, hot iron, hot charcoal, etc. This type of spectrum depends only on temperature and it is independent of the material from which light is emitted. In visible region, it appears in the form of multi-coloured strips in which one colour mildly mixes up with its neighbours and no line of demarcation can be drawn between two colours.

Absorption continuous spectra have all wavelengths absent. Such a spectrum is obtained when the absorbing material absorbs a continuous range of wavelengths without any gap, e.g., when white light is allowed to pass through red glass, it absorbs all colours except red and hence red glass gives a continuous absorption spectrum.

(ii) Line spectra: Line emission spectra consists of a series of coloured sharp lines separated by dark spaces in between two lines. Line emission spectrum is emitted whenever matter in atomic state is excited. Hence line spectra are also called *atomic spectra*. It is different for different types of atoms.

The line absorption spectra are obtained when white light is passed through absorbing material in atomic state, e.g., a gas or a vapour. The characteristic dark lines appear in the absorption at the places where the coloured lines would have been present when the absorbent is excited in its atomic state. A line absorption spectrum is also called atomic absorption spectrum. A familiar example of line absorption spectrum is the solar spectrum.

(iii) Band spectra: Whenever matter in molecular state is excited, ***band emission spectrum*** is emitted. It has coloured bands separated by the dark spaces on two sides. In such a spectrum a band of wavelengths is separated by a gap. Such spectra are also called as molecular spectra. It is different for different types of molecules.

Band absorption spectra are obtained when white light is passed through absorbing material in molecular state. At certain places the dark bands appear with coloured strips on their either sides. The aqueous solution $KMnO_4$ gives five bands in green region. The positions of dark bands depend on the nature of the absorbing materials, whereas the width and number of dark bands depend upon the thickness of the absorbing material.

Out of the spectra discussed above, line spectra is important as it has precision and measurements can be made over it.

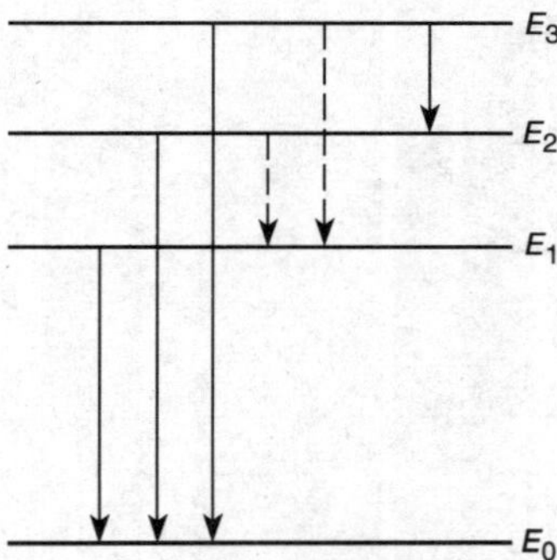

Fig. 2.16 Emission spectra in the form of energy level diagram

2.17 EMISSION AND ABSORPTION LINE (ATOMIC) SPECTRA

2.17.1 Emission Atomic spectra

Such spectra are obtained whenever the emitting substance in atomic state is excited.

Let mercury vapour be excited in atomic state by bombarding with electrons of variable known energy. When the energy of the electrons is bombarded is low, no light is observed. When the bombarding energy of the electrons is 4.86 eV, i.e., the first excitation energy of mercury vapour, the light is emitted suddenly. A sharp line of wavelength 2537Å is observed, when it is examined with an ultraviolet spectroscope. Similarly, when the energy of bombarding electrons reaches other excitation energies of mercury (i.e., 6.67 eV, 8.84 eV, 10.4 eV) other lines are observed. This reveals that the mercury atoms can emit light of discrete wavelengths.

A mercury atom absorbs 4.86 eV energy and is excited to the first excited state. This does not remain in excited state for long and returns back to the ground state emitting a photon of energy 4.86 eV which corresponds to a line of wavelength 2537Å in the emission spectrum in the ultraviolet region. Similarly, when mercury atom absorbs energy of 6.67 eV, 8.84 eV, 10.4 eV, it reaches its second, third, fourth excited states and so the photons of energy 6.67 eV, 8.84 eV, 10.4 eV are emitted when atom returns to its ground state and hence the lines of different wavelengths are observed in the emission spectrum of mercury.

Let E_0, E_1, E_2, E_3 (Fig. 2.16) represent the energies of an atom in the ground state, first excited state, second excited and third excited states respectively. When an atom returns from its first excited state to ground state, a photon of energy $(E_1 - E_0)$ is emitted, whose wavelength is given by

$$\frac{hc}{\lambda} = E_1 - E_0 \text{ or } \lambda = \frac{hc}{(E_1 - E_0)}$$

Similarly, one can calculate the wavelength of other lines in the emission spectra.

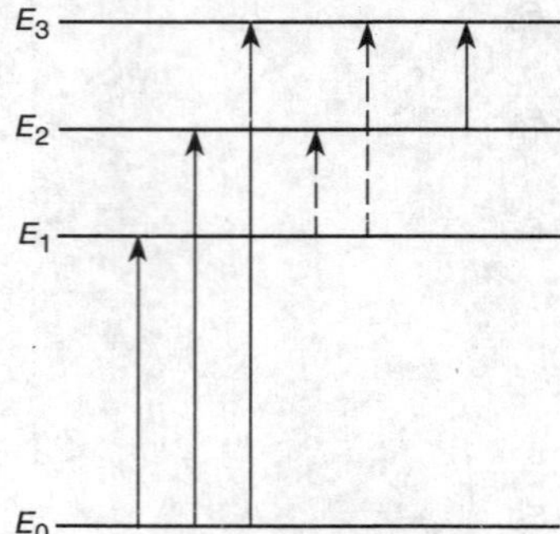

Fig. 2.17 Absorption spectrum in the form of energy level diagram

2.17.2 Absorption Atomic spectra

It is obtained when the absorbing substance is in the atomic state. The phenomenon of absorption is quite opposite to the phenomenon of emission. The gas or vapour atoms absorb the photons of allowed energy so that the absorbing atoms reach up to one of the allowed energy states. The absence of these photons correspond to the dark (missing) lines in the spectrum (Fig. 2.17). Obviously, if white light is passed through mercury vapours, those few lines will be missing which are observed in the emission spectrum of the mercury.

It is worthwhile to mention that all the lines in the emission spectrum of an element are not observed as dark lines in the absorption spectrum, but all the lines of absorption spectrum are found in the emission spectrum of the same element. This difference in the emission and absorption spectra of an element is due to the different physical conditions of the substance at the time of emission and absorption.

2.18 FLUORESCENCE AND PHOSPHORESCENCE

We have read that when atom falls from an excited state to the ground state or lower energy state, the energy in the form of electromagnetic radiations is emitted. Emission of light by an atom or substance due to causes other than high temperature is called as *luminescence.* The transition to the ground level of atoms of a substance which are excited to a higher energy level due to irradiation with visible light, X-rays, or γ-rays, is called *photoluminescence.* Photoluminescence phenomenon is of two kinds:

2.18.1 Fluorescence

There are several chemical compounds which absorb radiations of one wavelength and then immediately re-emit radiations of longer wavelengths, e.g., they may absorb violet or ultraviolet radiations and then emit radiations of blue or yellow colour. This phenomenon of absorbing light of one wavelength by a substance and the re-emission of light of greater wavelength is called 'fluorescence'. The emission of light ceases with the shutting off of incident light. The phenomenon was first observed by fluorspar, i.e., calcium fluoride. Fluorspar, Fluorescein, Quinine sulphate, uranium oxide, etc., exhibit the phenomenon of fluorescence. Barium platinocyanide fluoresces with bright greenish yellow colour when X-rays fall on it. The effect is also strongly marked in uranium glass, calcite and some fossils.

Fluorescence is not excited if the incident radiation has a longer wavelength than the fluorescent radiation. Atoms of the fluorescent substances are excited by incident photons. Major part of their energy is spent for excitation, whereas the remainder being taken up by various non-optical processes. Obviously, due to the presence of non-optical processes the energy of fluorescent radiation is always less than that of the incident radiation, i.e., the wavelength of fluorescent light is always greater than that of the incident light. This is known as *Stoke's law*.

Fluorescent light is emitted only as long as the excitation of emission is maintained. When the incident light is cut off, the emission decays exponentially and ceases within a short time. The time interval between excitation and the emission being less than 10^{-8}s.

The phenomenon of fluorescence is put to large practical uses, e.g.,

(i) Fluorescent light emitted by a substance is characteristic of that substance. One can identify a substance by the fluorescent light it emits. Very minute quantities of a substance can be detected by fluorescence analysis. This method is widely used in industry, medicine, biology etc.

(ii) Gems such as diamonds are often recognized by the characteristic glow given off by them in the presence of ultraviolet light.

(iii) A live tooth at once fluoresces in the presence of ultraviolet light and one can easily distinguish it from a dead one.

(iv) One can detect the presence of some costly mineral ores by their fluorescence caused by ultraviolet radiations.

(v) Fluorescent materials have been widely used in the manufacture of fluorescent lamps. These lamps are used for general illumination.

(vi) Several luminescent paints have been developed in recent years, which fluoresce when subjected to some suitable radiations. Such paints containing a very small quantity of some radium compounds are used to paint the dials of watches and clocks and the buttons of light switches. They emit α-rays, which excite the luminescent paint and cause it to emit greenish yellow light.

(vii) Certain fluorescent compounds have been developed which when added to washing powder produce bluish fluorescence, which neutralises the yellow brown tint of the washed fabrics. This gives a white appearance to the fabrics. Similar fluorescent compounds are also used to increase the whiteness of paper.

Fluorescent Lamp: The fluorescent lamps are thinly coated inside with a fluorescent material e.g., magnesium tungstate, zinc silicate, cadmium borate, zinc beyllium silicate, etc. the strong ultraviolet radiations from mercury vapour are absorbed and transformed by this material into visible radiations, which increases the total luminous flux. A glass plate coated with barium platinocyanide which fluoresces when X-rays fall on it, is used as a screen for the X-ray examination of the internal organs of the body.

2.18.2 Phosphorescence

We have read that the phenomenon of fluorescence last so long as the stimulating radiations are present. But there are also certain substances which continue to emit visible radiations even after the incident light is cut off. The phenomenon is known as *phosphorescence* and the substances possessing this property are called phosphorescent substances. The time interval between excitation of the atom and emission of radiation in these substances is greater than 10^{-8} s, ranging from 10^{-4} s to several minutes, or hours or even days.

Phosphorescence is the characteristic of solids. The centers of phosphorescence are atoms, ions or groups of ions located near a lattice imperfection. Lattice imperfection may be produced by introducing an impurity atom or activator.

Cadmium sulphide (Balmain's luminous paint) is strongly phosphorescent. When cadmium sulphide is illuminated with ultraviolet light, it emits a pale blue glow for about an hour. Barium sulphide, Strontium sulphide, sulphides of uranium and some other salts of uranium are also phosphorescent. Inorganic non-fluorescent basic substances, e.g., sulphides of calcium, barium, Strontium, Zinc, etc., activated by a trace of a metal or some other inorganic compounds form a class of substances known as *complex phophors*. These complex phosphors exhibit strong phosphorescence and are used as luminous paints. Phosphorescence of ruby and emerald is due to the presence of the trace of chromium.

Example 2.1: The wavelength of the first member of the Balmer series in the hydrogen spectrum is 6563 Å. Calculate the wavelength of the first member of the Lyman series.

Solution We have for the first member of the Balmer series

$$\frac{1}{\lambda}=R\left[\frac{1}{2^2}-\frac{1}{3^2}\right]=\frac{5R}{36} \qquad \text{(i)}$$

For the first member of the Lymann series

$$\frac{1}{\lambda'}=R\left[\frac{1}{1^2}-\frac{1}{2^2}\right]=\frac{3R}{4} \qquad \text{(ii)}$$

Dividing (i) by (ii), one obtains

$$\frac{\lambda'}{\lambda}=\frac{5\times 4}{36\times 3}$$

$$=\frac{5}{27}$$

$$\therefore \qquad \lambda'=\frac{5}{27}\lambda$$

$$=\frac{5}{27}\times 6563$$

$$=1215\text{Å}.$$

Example 2.2: The Rydberg constant for hydrogen is 10967700 m^{-1}. Calculate the short and long wavelength limits of Lyman series.

Solution For Lyman series, the wave number is given by

$$\bar{\nu} = \frac{1}{\lambda} = R_{\text{H}}\left(\frac{1}{1^2} - \frac{1}{n^2}\right)$$

For the short wavelength limit ($\lambda = \lambda_s$), $n = \infty$

or
$$\bar{\nu}_s = \frac{1}{\lambda_s}$$

$$= R_{\text{H}}\left(\frac{1}{1^2} - \frac{1}{\infty^2}\right) = R_{\text{H}}$$

∴
$$\lambda_s = \frac{1}{R_{\text{H}}} = \frac{1}{10967700}\text{m}$$

For long wavelength limit ($\lambda = \lambda_L$), $n = 2$

∴
$$\bar{\nu}_L = \frac{1}{\lambda_L}$$

$$= R_{\text{H}}\left(\frac{1}{1^2} - \frac{1}{2^2}\right)$$

$$= \frac{3}{4}R_{\text{H}}$$

∴
$$\lambda_L = \frac{4}{3R_{\text{H}}}$$

$$= \frac{4}{3} \times 991.6\text{Å}. = 1215\text{Å}$$

Example 2.3: Hydrogen atom in its ground state is excited by means of a monochromatic radiation of wavelength 970.6Å. How many different wavelengths are possible in the resulting emission spectrum? Find the longest wavelength amongst these.

Solution We have

$$E = h\nu = \frac{hc}{\lambda}$$

$$= \frac{6.6 \times 10^{-34} \times 3 \times 10^{8}}{970.6 \times 10^{-10} \times 1.6 \times 10^{-19}}$$

$$= 12.75\text{eV}.$$

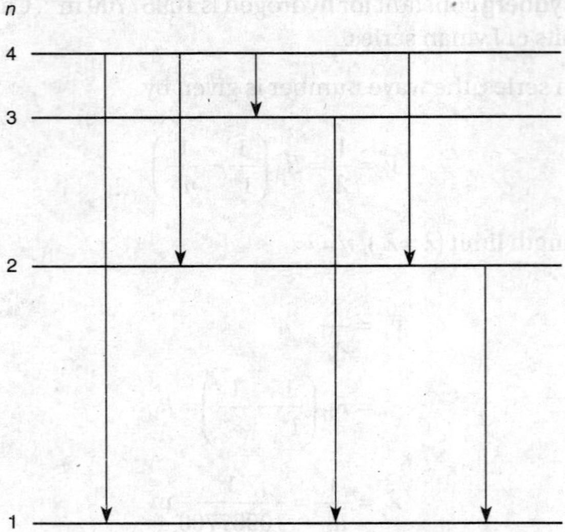

Fig. 2.18

$\therefore$ Energy of the excited state

$$E_n = -13.6 + 12.75 = -0.85\text{eV}$$

We know that

$$E_n = \frac{-13.6}{n^2} \quad \text{or} \quad n^2 = \frac{-13.6}{E_n}$$

$$= \frac{-13.6}{-0.85}$$

$$= 16$$

$\therefore$ $$n = 4$$

The number of possible transitions in going to the ground state and hence the number of different wavelength in the spectrum will be six (Fig. 2.18).

The longest wavelength corresponds to minimum energy difference, i.e. for the transition from $n = 4$ to $n = 3$. We have

$$E_3 = \frac{-13.6}{3^2} = -1.51\text{eV}$$

$$\frac{h_c}{\lambda_{max}} = E_4 - E_3$$

or $$\lambda_{max} = \frac{6.6\times10^{-34}\times3\times10^{8}}{(1.51-0.85)\times1.6\times10^{-19}}$$

$$= 18.75\times10^{-7}\,\text{m} = 18750\text{Å}$$

Example 2.4: **In a hydrogen like atom the ionisation energy equals 4 times Rydberg's constant for hydrogen is 2.2 × 10⁻¹⁸ J. What is the wavelength of radiation emitted when a jump takes place from the first excited state to the ground state? Find the radius of first Bohr's orbit.**

Solution The ionisation energy of hydrogen like Bohr atom of atomic number Z is given by

$$E = -RZ^2$$

$$= -\frac{2\pi k^2 Z^2 me^4}{n^2 h^2}$$

where Rydberg constant,

$$R = \frac{2\pi^2 k\, me^4}{n^2 h^2} = 2.2\times10^{-18}\ J$$

As ionisation energy $E = 4 \times$ Rydberg constant $= 4R$, one obtains $4R = RZ^2 \therefore Z = 2$.

(i) Energy of radiation emitted E when the electron jumps from the first excited state to the ground state is given by

$$E = RZ^2\left(\frac{1}{1^2} - \frac{1}{2^2}\right)$$

$$= 4R\left(1 - \frac{1}{4}\right)$$

$$= 3R$$

$$= 3\times2.2\times10^{-18}$$

$$= 6.6\times10^{-18}\ \text{J}$$

$$\therefore \quad \lambda = \frac{hc}{E}$$

$$= \frac{6.6\times10^{-34}\times3\times10^{8}}{6.6\times10^{-18}}$$

$$= 3\times10^{-8}\ \text{m}$$

(ii) Radius of the first Bohr orbit $= \dfrac{\text{Bohr radius of hydrogen atom}}{Z}$

$$= \frac{5\times10^{-11}}{2}\ \text{m}$$

$$= 2.5\times10^{-11}\ \text{m}$$

Example 2.5: **A doubly ionised lithium atom is hydrogen-like with atomic number $(Z) = 3$;**

(i) **Find the wavelength of the radiation required to excite the electron in Li^{++} from the first to the third Bohr orbit. (Take ionisation energy of the hydrogen atom = 13.6 eV).**

(ii) **How many spectral lines are observed in the emission spectrum of the above excited system?**

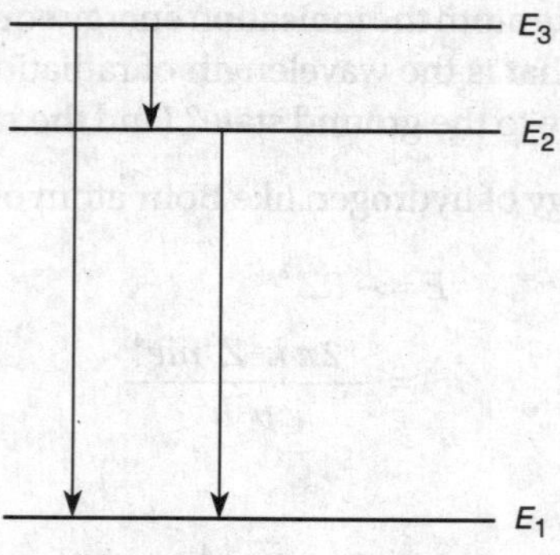

Fig. 2.19

Solution **(i) The energy difference of electron in Li⁺⁺ between the first and third orbit**

$$E_3 - E_1 = 13.6 \times Z^2 \left(\frac{1}{n_1^2} - \frac{1}{n_2^2} \right)$$

$$= 13.6 \times (3)^2 \left(\frac{1}{1^2} - \frac{1}{3^2} \right)$$

$$= 13.6 \times 9 \times \frac{8}{9} \times 1.6 \times 10^{-19}\ \text{J}$$

Now $$E_3 - E_1 = \frac{hc}{\lambda}$$

$\therefore$ $$\lambda = \frac{hc}{E_3 - E_1}$$

$$= \frac{6.6 \times 10^{-34} \times 3 \times 10^8}{13.6 \times 8 \times 1.6 \times 10^{-19}}$$

$$= 113.7\text{Å}.$$

(ii) **The following three spectral lines are observed due to the following transitions Fig. (2.19).**
3rd to 1st orbit
3rd to 2nd orbit
2nd to 1st orbit

Example 2.6: **A particle of charge equal to that of an electron and mass 208 times the mass of the electron called (μ-meson) moves in a circular orbit around a nucleus of charge +3*e*. Take mass of the nucleus to be infinite. Assuming the Bohr model of the atom to be applicable to this system.**

(i) **Derive an expression for the radius of the nth Bohr orbit,**

(ii) **Find the value of *n* for which the radius of the orbit is approximately the same as that of the first Bohr orbit for the hydrogen atom, and**

(iii) **Find the wavelength of the radiation emitted when the μ-meson jumps from the third orbit to the first orbit. (Rydberg constant R = $1.097 \times 10^7\ \text{m}^{-1}$).**

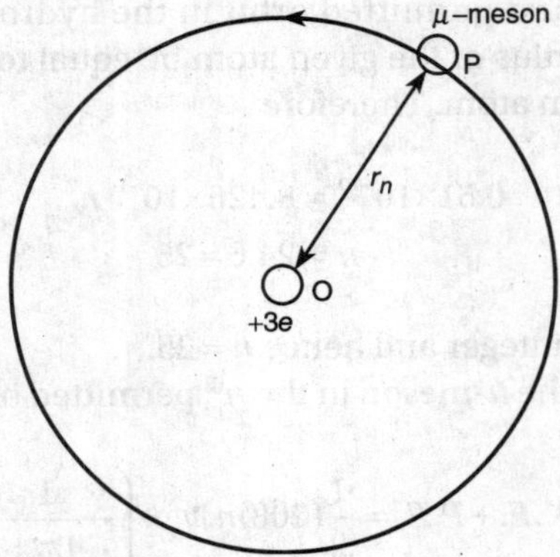

Fig. 2.20

Solution

(i) As shown in Fig. (2.20), P is μ-meson revolving round the nucleus O having a charge +3e in the nth circular orbit of radius r_n. Mass of the μ-meson = 208 m, where m is the mass of the electron. For the first orbit, the electrostatic force of attraction of nucleus on the particle P, provides the centripetal force.

$$\therefore \qquad \frac{1}{4\pi\varepsilon_0}\frac{(3e)e}{r_n^2} = \frac{(208m)v_n^2}{r_n}$$

$$\therefore \qquad r_n^{-1} = 4\pi\varepsilon_0\left(\frac{208mv_n^2}{3e^2}\right) \qquad \text{(i)}$$

Applying Bohr's quantum condition, we have

$$(208m)v_n r_n = \frac{nh}{2\pi} \qquad \text{(ii)}$$

or

$$v_n = \frac{nh}{2\pi(208m)r_n} \qquad \text{(iii)}$$

Eliminating v_n from Eq. (i) and (iii), one obtains

$$\frac{1}{r_n} = \frac{208m}{3e^2}4\pi\varepsilon_o\left(\frac{nh}{2\pi(208m)r_n}\right)^2$$

or

$$r_n = \frac{4\pi\varepsilon_0 n^2 h^2}{12\pi^2 e^2 \times (208m)} \qquad \text{(iv)}$$

$$= \frac{n^2 \times (6.625\times10^{-34})^2}{9\times10^9 \times 12\times(3.142)^2 \times (1.6\times10^{-19})^2 \times 208\times9.1\times10^{-31}}$$

$$= 8.426\times10^{-14} n^2 \qquad \text{(v)}$$

(ii) The radius of the first permitted orbit in the hydrogen atom is 0.51×10^{-10}m. In order that the radius of the given atom be equal to the radius of the first Bohr orbit of the hydrogen atom. Therefore

$$0.51\times10^{-10} = 8.426\times10^{-14}\,n^2$$

or

$$n = 24.6 \approx 25$$

Since n must be an integer and hence $n = 25$.

(iii) The total energy of the μ-meson in the nth permitted orbit is

$$E_n = K.E. + P.E. = \frac{1}{2}(208m)v_n^2 + \left(-\frac{1}{4\pi\varepsilon_0}\frac{3e^2}{r_n}\right)$$

$$= -\frac{1}{2}\left(\frac{1}{4\pi\varepsilon_0}\frac{3e^2}{r_n}\right) \qquad \text{(vi)}$$

Substituting r_n from Eq. (iv) in Eq. (vi), one obtains

$$\therefore \quad E_n = -\frac{1}{2}\left[\frac{1\times3e^2\times12\pi^2e^2(208m)}{(4\pi\varepsilon_0)\times(4\pi\varepsilon_0)n^2h^2}\right]$$

$$= -\frac{3744\pi^2me^4}{(4\pi\varepsilon_o)^2n^2h^2} \qquad \text{(vii)}$$

We have,

$$hv = \frac{hc}{\lambda} = E_{n_1} - E_{n_2}$$

$$\therefore \quad \frac{1}{\lambda} = 1872\left[\frac{2\pi^2me^4}{(4\pi\varepsilon_0)^2ch^3}\right]\left(\frac{1}{n_2^2} - \frac{1}{n_1^2}\right)$$

$$R = 2\pi^2me^4/(4\pi\varepsilon_o)^2ch^3 \qquad \text{(viii)}$$

$$\therefore \quad \frac{1}{\lambda} = 1872R\left(\frac{1}{n_2^2} - \frac{1}{n_1^2}\right)$$

$$= 1872\times1.097\times10^7\left(\frac{1}{n_2^2} - \frac{1}{n_1^2}\right)$$

$$= 1872\times1.097\times10^7\left(\frac{1}{1^2} - \frac{1}{3^2}\right)$$

$$= 1825.408\times10^7\,\text{m}^{-1}$$

$$\therefore \quad \lambda = 5.478\times10^{-11}\,\text{m}$$

Example 2.7: The wavelength of the first Balmer line (H_α) of hydrogen is 6562.8Å. Calculate the wavelength of the same line for tritium ($_1H^3$)

Solution The wavelengths of H_α line ($n = 3 \rightarrow n = 2$) of hydrogen (^1H) and tritium (^3H) are given by

$$\frac{1}{\lambda_H} = R_H\left(\frac{1}{2^2} - \frac{1}{3^2}\right) = \frac{5}{36}R_H$$

and
$$\frac{1}{\lambda_T} = R_T\left(\frac{1}{2^2} - \frac{1}{3^2}\right) = \frac{5}{36}R_T$$

Where R_H and R_T are Rydberg constants for hydrogen and tritium respectively ($Z = 1$ for both).
From this, one obtains

$$\frac{\lambda_T}{\lambda_H} = \frac{R_H}{R_T} \qquad \text{(i)}$$

We have the Rydberg constant for an atom of mass M is given by

$$R_m = \frac{R_\infty}{1 + \dfrac{m}{M}}$$

Where R_∞ is the Rydberg constant for any atom of infinitely heavy nucleus. One obtains from Eq. (i)

$$\frac{\lambda_T}{\lambda_H} = \frac{1 + \dfrac{m}{M_T}}{1 + \dfrac{m}{M_H}}$$

Now, $M_T = 3M_H$ and $M_H = 1836$ m. Thus

$$\frac{\lambda_T}{\lambda_H} = \frac{1 + \dfrac{1}{3 \times 1836}}{1 + \dfrac{1}{1836}} = \frac{5509}{5511}$$

$$\therefore \qquad \lambda_T = \frac{5509}{5511} \times \lambda_H = \frac{5509}{5511} \times 6562.8 = 6560.4\text{Å}$$

Example 2.8: The first two excitation potentials of atomic hydrogen in Franck-Hertz experiment are 10.2 and 12.09 V. Draw an energy level diagram and show all possible transitions for emission and absorption along their wavelengths.

Solution The energy level diagram and three possible transitions marked by (a), (b) and (c) are shown in Fig. (2.21).

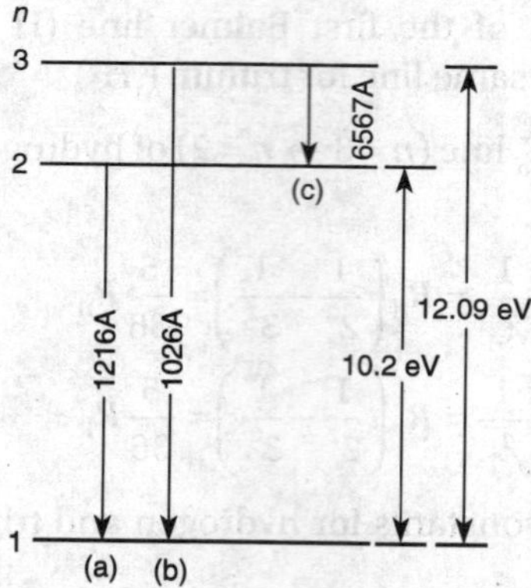

Fig. 2.21

The frequency of radiation resulting from the transitions (a) is given by

$$\nu_a = \frac{c}{\lambda_a} = \frac{E_2 - E_1}{h}$$

or
$$\lambda_a = \frac{hc}{E_2 - E_1} \qquad \text{(i)}$$

or
$$\lambda_a = \frac{6.63\times10^{-34}\times3\times10^{8}}{10.2\times1.6\times10^{-19}} = 1216\times10^{-10}\,\text{m} = 1216\text{Å}$$

Similarly,
$$\lambda_b = \frac{hc}{E_2 - E_1}$$

$$= \frac{6.63\times10^{-34}\times3\times10^{8}}{12.09\times1.6\times10^{-19}} = 1.026\times10^{-7}\,\text{m} = 1026\text{Å}$$

and
$$\lambda_c = \frac{hc}{E_3 - E_2} = \frac{hc}{(E_3 - E_1)-(E_2 - E_1)}$$

$$= \frac{hc}{(12.09-10.2)eV} = \frac{hc}{1.89\text{eV}}$$

$$= \frac{6.63\times10^{-34}\times3\times10^{8}}{1.89\times1.6\times10^{-19}} = 6.567\times10^{-7}\,\text{m} = 6567\text{Å}$$

We must note that in absorption, only the transitions starting from $n = 1$ shall be observed, which correspond to 1216Å and 1026Å.

Example 2.9 A 5 meV α-particle approaches a gold (Z = 79) nucleus with an impact parameter of 2.6×10^{-13} m. Through what angle will it be scattered? ($e = 1.6 \times 10^{-19}$ C, 1 eV $= 1.6 \times 10^{-19}$ J).

Solution We have $\cot\frac{\phi}{2} = \frac{2p}{b} = \frac{(4\pi\varepsilon_0)pE_k}{Ze^2}$

$$\therefore \quad \cot\frac{\phi}{2} = \frac{2.6\times10^{-13}\times8\times10^{-13}}{9\times10^9\times79\times(1.6\times10^{-19})^2}$$

$$= 11.42$$

$$\therefore \quad \frac{\phi}{2} = \cot^{-1}(11.42) = 5^\circ$$

or $\quad \phi = 10^\circ$

$E_K = 5$ meV

$= 8\times10^{-13}$ J

$p = 2.6\times10^{-13}$ J

$Z = 79$

$e = 1.6\times10^{-19}$ C

$\frac{1}{4\pi\varepsilon_0} = 9\times10^9\,\text{N}\cdot\text{m}^2/\text{C}^2$

Example 2.10 **In Rutherford scattering experiment the number of particles observed at an angle of 10° is one million per minute. How many particles per minute will be observed at 90° and 180°?**

Solution $N(\phi) \propto \frac{1}{\sin^4\left(\frac{\phi}{2}\right)} = \frac{K}{\left(\sin^4\frac{\phi}{2}\right)}$

Here K is constant. Now $N(10°) = 10°$/min.

$$\therefore \quad 10^6 = K\frac{1}{\sin^4\left(\frac{10}{2}\right)^\circ} = \frac{K}{(0.0872)^4}$$

$$\therefore \quad K = (0.872)^4\times10^6$$

Now
$$N(90^\circ) = K\frac{1}{\sin^4\left(\frac{90^\circ}{2}\right)} = (0.0872)^4\times10^6\times\frac{1}{\left(\frac{1}{\sqrt{2}}\right)^4}$$

$$= 232$$

Similarly, one obtains $N(180°) = 58$

Example 2.11 **A beam of α-particles of energy 5.5 MeV is incident normally on a gold ($Z = 79$) foil at a rate of 10^4 particles per sec. An α-particles counter with an aperture 4 sq. cm is placed at a distance 0.25 m from the foil. Calculate the counts at an angle of 60°, if the thickness of the foil is 10^{-5} cm. The atomic weight of gold is 197 and its density is 19.3 gm/cm³. (Agvogadro number is 6.023×10^{23}).**

Solution We have

$N_o = 10^4$/sec.

n (number of atoms in the foil per unit volume) $= \dfrac{\text{Avogadro number}}{\text{Atomic weight}} \times \text{density}$

$$= \frac{6.023\times10^{23}}{197}\times 19.3$$

$$= 5.9 \times 10^{22} \text{ per cm}^3 = 5.9 \times 10^{28}/\text{m}^3$$

$$t = 10^{-5}\text{ cm} = 10^{-7}\text{ m}$$

r (distance of the counter from the foil) = 0.25 cm

$$Z = 79$$

$$E_k = 5.5 \times 1.6 \times 10^{-13}\text{ J}$$

$$\phi = 60^\circ \therefore \sin\frac{60^\circ}{2} = \frac{1}{2}$$

$$N(\phi) = \frac{N_0\, nt\, Z^2 e^4}{4r^2 (4\pi\varepsilon_0)^2 E_k^2}\frac{1}{\sin^4\left(\dfrac{\phi}{2}\right)}$$

$$\therefore \quad N(\phi) = \frac{10^4\times5.9\times10^{28}\times10^{-7}\times(79)^2\times\left(1.6\times10^{-19}\right)^4}{4\times(0.25)^2\times\dfrac{1}{\left(9\times10^9\right)^2}\times\left(5.5\times1.6\times10^{-13}\right)^2}\times\frac{1}{\left(\dfrac{1}{2}\right)^4}$$

$$= 1.6$$

The area of the aperature of the counter is 4 sq. cm = 4×10^{-4} m^2

∴ The number of α-particles entering the counter $= 1.6 \times 4 \times 10^{-4} = 6.4 \times 10^{-4}$/sec

Example 2.12 A thin (1 mg/cm^2) target of ^{48}Ca is bombarded with a 10 nA beam of α-particles. A detector subtending a solid angle of 2×10^{-3} steradians, records 15 protons per second. If the angular distribution is measured to be isotropic, determine the total cross-section (in mb) for the ^{48}Ca (α, p) reaction. Take the atomic mass of ^{48}Ca = 48 u.

Solution Since $\sigma(\theta)$ is isotropic, the total number of protons emitted into the full 4π solid angle is $\dfrac{15\times4\pi}{\left(2\times10^{-3}\right)} = 1$. Using

$$l = \frac{l_0 \sigma m N_A}{A}$$

We have $$\sigma = \frac{l_A}{(l_0\ m\ N_A)}$$

where m is the mass of the target in units of mass per unit area. Substituting values, one obtains $\sigma = 241$ mb.

SUMMARY

1. Rutherford proposed the nuclear model of atom, in which positive charge is concentrated at the central nucleus, which is heavier.
2. Geiger and Marsden scattering experiments with energetic α-particles showed that the number of backward scattered α-particles could be accounted for only if the Rutherford's model of atom is correct.
3. Rutherford atomic model could not pass the test of stability and also failed to explain emission and absorption spectra of elements.
4. Bohr proposed the model for hydrogen atom. He assumed that orbits and energies of atom are quantized, i.e. restricted orbits and energies of atom are permitted, like Planck's quantization of harmonic oscillators. This means that the laws of classical Physics are inapplicable to atomic world and one must use laws of quantum physics.

REVIEW QUESTIONS

1. State the postulates of Bohr's theory and deduce an expression for the energy of the nth orbit of hydrogen atom. What interpretation do you give to the negative sign of the energy value?
2. Find an expression for the energy required to ionise hydrogen atom in the lowest energy level.
3. Discuss early views on atomic structure. How did Bohr modify Rutherford's model to make it dynamically stable? In what respects Bohr's model of the atom found to be lacking? How these defects of Bohr's model have been removed in Sommerfield's model?
4. State Bohr's postulates. Derive an expression for the frequencies if radiation emitted in hydrogen atom. Hence explain the various spectral series of this atom.
5. Why in Bohr's theory of hydrogen atom the quantum number n, cannot take the value zero?
6. Lyman series can be observed in absorption spectra while the Balmer series cannot be observed. Explain.
7. Can a hydrogen atom absorb a photon of energy greater than the binding energy of the atom?

8. Explain the effect of nuclear mass on the Rydberg constant and how it led to the discovery of heavy hydrogen.
9. Discuss the effect of nuclear motion ('finite' nuclear mass) on the spectra of hydrogen like atoms.
10. Explain, How Bohr's theory can be applied to a singly-ionised helium atom.
11. What relationship do you expect between the hydrogen spectrum and the ionised helium spectrum? Show that the ratio of the mass of electron to the mass of proton can be determined from the above two spectra.
12. Discuss relativistic Sommerfeld theory of hydrogen atom. How does it explain the fine structure of H_α line.
13. Give a qualitative discussion of the improvements made by Sommerfeld in Bohr's atomic model. Show how fine structure of Balmer lines was explained on this.
14. Discuss Bohr's correspondence principle.
15. Show that for electron transitions in large adjacent quantum energy levels, Bohr's atom model corresponds asymptotically to the classical theory and that the orbital frequency of the electron is the frequency of the emitted spectral lines.
16. What do you understand by excitation and ionisation potentials of an atom? Describe Franck and Hertz experiment and show that it provides a direct evidence of the existence of discrete energy levels in an atom. State the limitations of the experiment.

PROBLEMS

1. The wavelength of first member of Balmer series in hydrogen spectrum is 6563Å. Calculate the wavelength of the first member of the Lyman series in the same spectrum. **[Ans.** 1215Å]
2. Show that the velocity of electron in the first hydrogen atom orbit is $c/137$ where c is the velocity of light.
3. Calculate the time taken by the electron to traverse the first orbit in the hydrogen atom. Electron mass and charge are 9.1×10^{-31} Kg and 1.6×10^{-19} C and $h = 6.63 \times 10^{-34}$ J.s. **[Ans.** 1.5×10^{-16} s]
4. The first member of Balmer series of hydrogen has a wavelength 6563Å. Calculate the wavelength of second member. **[Ans.** 4861.5Å]
5. Calculate the radius of the second Bohr orbit of hydrogen atom and the velocity of electron in this orbit. [$1/4\pi\varepsilon_0 = 9 \times 10^9$N-m^2/c^2, m $= 9.1 \times 10^{-31}$Kg, $e = 1.6 \times 10^{-19}$ C and $= 6.63 \times 10^{-34}$J.s.] **[Ans.** 2.1Å, 1.1×10^6m/s]
6. The Rydberg constant for hydrogen is 10967700 m^{-1}. Calculate the short and long wavelength limits of Lyman series. **[Ans.** (i) $\lambda_s = 911.6$Å, (ii) $\lambda_L = 1215$Å]
7. How many times does the electron go round the first Bohr orbit of hydrogen in 1 s? **[Ans.** 6.57×10^{15} Hz]
8. Electrons of energy 12.2 eV are fired at the hydrogen atoms in a gas discharge tube. Determine the wavelength of the lines that can be examined by hydrogen. **[Ans.** $\lambda_1 = 6563$Å, $\lambda_2 = 1215$Å, $\lambda_3 = 1026$Å corresponding to $n = 3 \rightarrow 2$, $n = 2 \rightarrow 1$ and $n = 3 \rightarrow 1$ transitions respectively]

9. Calculate the wavelength of the photon emitted when the hydrogen atom goes from $n = 10$ state to the ground state. ($R_H = 1.097 \times 10^{-3}$Å). **[Ans.** 921Å]
10. Calculate the wavelength of the eighth line of the Balmer series of the hydrogen atom. Rydberg constant for hydrogen atom is $1.097 \times 10^7 m^{-1}$.
[Hint: For eighth line, $n = 10$] **[Ans.** 3798Å]
11. The series limit of Balmer series is 3646 Å. Calculate the wavelength of the first member of this series. **[Ans.** 6563Å]
12. An orange photon of wavelength 600 nm is emitted from an atom. Find the difference in energy in the two atomic states involved. Find the same result for the red 6563 Å line of hydrogen. **[Ans.** 2.07 eV, 1.9 eV]
13. The wavelength of the first line of Balmer series of hydrogen is 6562.8 Å. Calculate (i) the ionisation potential and (ii) the first excitation potential of the hydrogen atom. **[Ans.** (i) 13.6 eV, (ii) 10.2 eV]
14. With Franck-Hertz type of experiment on sodium, the first spectral line to appear is the D-line, $\lambda = 5890$ Å. What is the first excitation potential of the sodium? **[Ans.** 2.1V]
15. In a Franck-Hertz experiment the first dip in the current-voltage graph for hydrogen was observed at 10.2 V. Calculate the wavelength of light emitted by hydrogen atom when excited to the first excitation level. **[Ans.** 1217Å]

SHORT ANSWER-QUESTIONS

1. In a hydrogen atom, there is only one electron, but its emission spectrum shows many lines, why?
Ans: The electron in a hydrogen atom revolves round the nucleus in stationary orbits. The emission spectral line is produced due to the transition of the orbital electron from a higher orbit to a lower one. When electron jumps from $n = 2, 3, 4...$ to $n = 1$ orbit we obtain Lyman series and so on. Thus we obtain various series and each series contains several lines.
2. In Bohr's model of the atom the mass of the electron is taken as constant. Is it correct?
Ans: It is not correct to take the mass of the electron constant in Bohr's atom model. In a stationary orbit the velocity of a revolving electron is of the order of velocity of light and hence relativistic correction must be applied, i.e., $m = m_0 / \sqrt{1 - v^2 / c^2}$.
3. In what respect did classical physics fail to account for the structure of the atom?
Ans: Classical physics failed to account for the stability of an atom. According to classical physics an electron revolving round the nucleus should emit radiation (electromagnetic waves). Thus a revolving electron loses energy and moves into a smaller orbit and ultimately it should fall into the nucleus.
4. Can a hydrogen atom absorb a photon of energy greater than the binding energy (B.E.) of the atom?
Ans: The B.E. of the hydrogen atom is numerically equal to the energy of the lowest state, i.e., –13.6 eV. This is the energy, which must be supplied to it in order to remove its electron from the lowest state(level) to a state of zero total energy.

We know that the highest energy level of the atom corresponds to $n = \infty$ and hence it is $E = 0$. Above this level there are energy states of the system consisting of unbound electron plus the ionised atom. The total energy of an unbound electron is not quantized and is positive. Obviously, any energy $E > 0$ is possible for the electron, and the energy states form a continuum. When a hydrogen atom receives a photon of any energy which is greater than its B.E. of 13.6 eV, the atom absorbs the photon and the electron of the atom passes from its discrete ground state to a continuum state.

5. **A hydrogen atom is in the ground state. What is the quantum number of state to which it will be excited absorbing a photon of energy 12.09 eV?**
   Ans: 3.
6. **The energy of an electron revolving in the nth orbit of the hydrogen atom is given by $E_n = -Rch/n^2$, where R is Rydberg's constant. What are the dimensions of R?**

   Ans: $R = \left[\dfrac{E_n n^2}{ch}\right] = \left[\dfrac{ML^2T^{-2}}{LT^{-1}ML^2T^{-2}T}\right] = [L^{-1}]$

7. **What are the dimensions of the expression $\left(\dfrac{2\pi ke^2}{ch}\right)$, where $k = 1/4\pi\varepsilon_0$? What is this expression called?**

   Ans: $\dfrac{2\pi ke^2}{\text{ch}} = \dfrac{(\text{Nm}^2\text{C}^{-2}(\text{c}^2)}{(\text{ms}^{-1})(\text{Js})} = \dfrac{[\text{ML T}^{-2}\text{L}^2(\text{AS})^{-2}][(\text{As})^2]}{[(\text{LT}^{-1})][(\text{ML}^2\text{T}^{-2}\text{T})]}$

   $$= \frac{[\text{ML}^3\text{ T}^{-2}]}{[\text{ML}^3\text{T}^{-3}]} = \text{M}^0\text{L}^0\text{T}^0\text{Å}$$

8. **An electron is in the third excited state. How many different photon wavelengths are possible?**
   Ans: 6. $n = 4 \rightarrow n = 3$, $n = 4 \rightarrow n = 2$, $n = 4 \rightarrow n = 1$, $n = 3 \rightarrow n = 2$, $n = 3 \rightarrow n = 1$, $n = 2 \rightarrow n = 1$.
9. **What is the ground state energy of electron in case of Li⁺?**
   Ans: We have

   $$E_n = -\frac{13.6}{n^2}(Z^2)\text{eV}$$

   In case of Li atom, $Z = 3$ and $n = 2$. Therefore

   $$E_1 = \frac{-13.6\times 9}{1^2}\text{eV} = -122.4\text{eV}.$$

10. **Does introduction of elliptic orbits adds to new energy levels?**
    Ans: No.
11. **What are critical potentials of the atom?**
    Ans: The excitation potentials and ionisation potential together are called the critical potentials of the atom.

12. What are the limitations of the Franck-Hertz experiment?
Ans: Atoms have more than one excitation potentials and also an ionisation potential. Therefore in an actual experiment the curve obtained is quite complicated and one cannot distinguish between excitation potentials and ionisation potential.

13. "In the limit of large quantum numbers, classical and quantum physics provide identical results." Explain this

14. For a harmonic oscillator with momentum p and position q, write the Wilson - Sommerfeld quantum condition.
Ans: $\oint pdq = nh$, n is principal quantum number

15. What element has a hydroden like spectrum whose lines have wavelength four times shorter than those of atomic hydrogen?
[**Hint:** $\frac{\lambda_1}{\lambda_1} = \frac{Z_2^{\,2}}{Z_1^{\,2}}$]
Ans: $Z_2 = 2$ (Helium)

16. What minimum energy must an electron have for all lines of all the series of the hydrogen spectrum to appear when the hydrogen atoms are excited by impacts of the electrons?
Ans: 13.58 eV

OBJECTIVE QUESTIONS

1. The shortest wavelength in the Lyman's series is 911.6Å. Then the longest wavelength in the Lyman series is
(a) 2430Å (b) 1215Å
(c) 600Å (d) ∞

2. Which of the following transitions in a hydrogen atom emits the photon of lowest frequency?
(a) $n = 4 \rightarrow n = 3$ (b) $n = 3 \rightarrow n = 1$
(c) $n = 4 \rightarrow n = 2$ (d) $n = 2 \rightarrow n = 1$

3. According to Bohr's theory of the hydrogen atom, the radii r_n of stationary electron orbits are related to principal quantum number n as
(a) $r_n \alpha n^2$ (b) $r_n \alpha 1/n^2$
(c) $r_n \alpha 1/n^3$ (d) $r_n \alpha 1/n$

4. The velocity of electron in the first Bohr's orbit of the hydrogen atom is about ($c \rightarrow$ velocity of light)
(a) $\frac{c}{173}$ (b) $\frac{c}{137}$
(c) $\frac{c}{371}$ (d) $\frac{c}{731}$

5. Hydrogen atom does not emit X-rays because

(a) Its energy levels arte too far apart (b) Its energy levels are too close to each other

(c) It has a very small mass (d) It has a single electron

6. For a given set of values of n and l, the possible number of electron in a closed shell can be

(a) $2n$ (b) $2(2l+1)$

(c) $2n(2l+1)$ (d) nl

7. According to Bohr's theory the ratio of the times taken by the electron in a hydrogen atom to complete one revolution in orbits corresponding to ground and first excited states is

(a) 1:8 (b) 8:1

(c) 1:4 (d) 4:1

8. An electron jumps from the fourth to the second orbit of hydrogen atom. Given the Rydberg constant $R = 10^7\ \text{m}^{-1}$, the frequency of the emitted radiation, in Hz, is

(a) $(9/4)\ 10^{15}$ (b) $(9/16)\ 10^{15}$

(c) $(3/16)\ 10^{15}$ (d) $(3/4)\ 10^{15}$

9. The state of triply ionised beryllium, which has the same orbital radius as the ground state of hydrogen has the principal quantum number

(a) 6 (b) 5

(c) 4 (d) 2

10. The hydrogen-like atom has one electron revolving around a stationary nucleus. The energy required to excite the electron from the second orbit to the third orbit is 47.2 eV. The atomic number of the atom is

(a) 6 (b) 5

(c) 4 (d) 3

ANSWERS

Question numbers	1	2	3	4	5	6	7	8	9	10
Answers	(b)	(a)	(a)	(b)	(b)	(b)	(a)	(b)	(d)	(b)

1. (b) For the Lyman series we have $\dfrac{\lambda_L}{\lambda_S} = \dfrac{4}{3}$ $\quad \therefore \lambda_L = \dfrac{4}{3} \times 911.6 = 1215\text{Å}$

2. (a) As n increases, the energy levels get progressively closer. One can see that from the given sets, the closest to each other, i,e., having minimum energy difference are $n = 4 \rightarrow n = 3$.

FILL IN THE BLANKS

1. The negative total energy $\left(E=-\frac{1}{2}\frac{1}{4\pi\varepsilon_0}\frac{Ze^2}{\gamma}\right)$ of electron moving around the nucleus in circular orbit means that the electron in the atom is ______ to the ______. **[bound, nucleus]**
2. P fund series is found in far ______ region. **[infrared]**
3. Bohr asserted that only those orbits are allowed in which the ______ of electron is integral multiple of $\hbar\,(=h/2\pi)$. **[angular momentum]**
4. The electron in the first orbit ($n = 1$) of hydrogen atom ($Z = 1$) has energy equal to ______. **[−13.6 eV]**
5. First ionisation energy of hydrogen atom is ______. **[13.6 eV]**
6. The energy level for $n = 1$ is called the ______ and has energy equal to ______. **[ground state, −13.6 eV]**
7. $\mu=\frac{mM}{m+M}$, where m and M are masses of the electron and nucleus respectively, μ is called ______. **[reduced mass]**
8. An atom which has charge equal to that of electron and mass equal to 207 times the electronic mass is called ______ atom. **[muonic (mesic)]**
9. The Rydberg constant for an atom depends on nuclear mass hence it will have ______ values for different isotopes. **[different]**
10. Franck and Hertz experiment provides strong and conclusive evidence in support of the evidence of ______ energy states of atoms. **[discrete]**
11. Bohr pointed out that ______ and ______ theories yield indentical result in the region of high quantum numbers. **[quantum, classical]**
12. In Wilson-Sommerfeld quantum condition $\oint pdq = nh$, the integral on the left hand side represents the area of ______ trajectory and is equal to ______. **[phase, πab]**

3

Alkali Spectra, Space Quantization and Periodicity

3.1 INTRODUCTION

The vector atom model is the consequence as an empirical extension of Bohr-Sommerfeld atomic model when scientists noticed that the detailed motion of electrons is too complex. Bohr, Sommerfeld, Uhlenbeck, Goudsmit, Pauling, Lande, Stern, Gerlach and other physicists mainly contributed to the development of vector atom model. The aim of these physicists was to cover new research fields and at the same time to overcome the deficiencies of the early atomic models. Interestingly, the empirical rules used in vector atom model to explain the behaviour of the atom have been substantiated later by quantum mechanics. Vector atom model enables one to remember and use the results of quantum mechanics in a very convenient way. In this chapter, we shall delineate the main features of the vector atom model and explain some of the phenomena in terms of it. Prior to introduction of vector atom model it would be proper to give an account of the alkali spectra.

3.2 ALKALI SPECTRA

The alkali atoms Li, Na, K, ... etc have similar properties and readily give up one electron to form positive ions. However, the energy required to remove one electron from alkali atoms is small, e.g., 5.1 eV in case of Na, but that required to remove a second electron is much larger (47.3 eV). This clearly suggests that out of all the electrons in an alkali atom, one electron is loosely bound to the atom. This electron is called 'optical' or valence electron and the spectral lines of an alkali atom arise due to the transitions of this electron only. This is why the alkali atom spectrum is called 'one electron' spectrum. We may note that alkali atom spectrum has a striking similarity to hydrogen spectra. Fig. (3.1) shows energy level diagram of lithium and sodium.

The corresponding energies of hydrogen are included for comparison. Energy level diagram for Na exhibit the fine structure splitting of the $3p$ state makes each transition involving that state in to a closely spaced doublet. We may note that fine structure splitting increases with Z and hence it is negligible in Li.

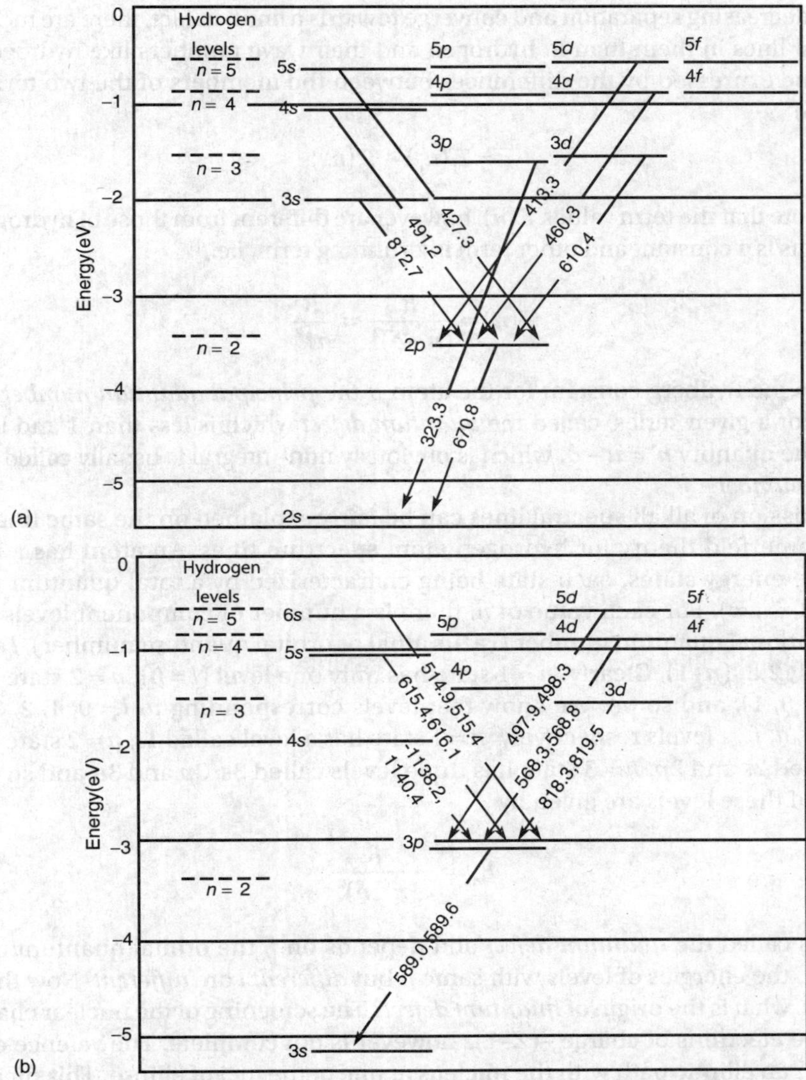

Fig. 3.1 (a) Energy level diagram of Li, showing some transitions (wavelength values marked in nanometers) in the optical region. For comparison, we have included the corresponding energies of hydrogen. (b) Energy level diagram of sodium. The fine structure splitting of the $3p$ state makes each transition involving that state into a closely spaced doublet we may note that fine structure splitting increases with Z, it is negligible in Li

After hydrogen, alkali atom spectrum is the most simple. One can classify alkali spectrum in four series: principal series, sharp series, diffuse series and fundamental series. The principal series in the alkali spectrum is the most prominent and can be observed in emission as well as in absorption spectrum. The other series are observed in emission spectrum only. The lines in a particular series in alkali spectra, like those of hydrogen have a regular decreasing separation and converge towards a limit. In fact, there are more series of spectral lines in them than in hydrogen and their wave numbers like hydrogen lines, can also be expressed by the differences between the members of the two term series of the type

$$\bar{\nu} = T_1(n_1) - T_2(n_2) \tag{3.1}$$

We may note that the term values $T(n)$, however are different from those of hydrogen. One of the terms is a constant and other term is a running term, i.e.,

$$T(n) = \frac{R}{(n-\delta)^2} = \frac{R}{n^{*2}} \tag{3.2}$$

where R is the Rydberg constant for the atom n the *principal quantum number* and δ a constant for a given series, called the *quantum defect* which is less than 1 and is not an integer. The quantity $n^* = n - \delta$, which is obviously non-integral is usually called *effective quantum number.*

The emission of alkali spectral lines can be fairly explained on the same lines as the Bohr-Sommerfeld theory for hydrogen atom spectrum lines. An atom has a number of discrete energy states, each state being characterized by a total quantum number $n(= 1, 2, 3, ..., \infty)$. For each value of n, there is a number of component levels labeled by an additional quantum number l (azimuthal or orbital quantum number), l can take values 0, 1, 2,... $(n-1)$. Clearly, $n = 1$ state has only one level $(l = 0)$; $n = 2$ state has two levels $(l = 0, 1)$, and so on. We know that levels corresponding to $l = 0, 1, 2, 3, 4$, are called *s, p, d, f*, ... levels respectively. $n = 1$ state has a level called 1*s*, $n = 2$ state has two levels called 2*s* and 2*p*, $n = 3$ state has three levels called 3*s*, 3*p* and 3*d* and so on. The energies of these levels are given by

$$E_{n,l} = -\frac{R}{(n-\delta)^2} \tag{3.3}$$

where δ is called the *quantum defect* and depends on l, the orbital quantum number. Obviously, the energies of levels with same n but *different l* are *different*. Now the question arises, what is the origin of *quantum defect*? The screening of the nuclear charge $+Ze$ by the core electrons of charge $-(Z-1)e$ however is not complete. The valence electron revolves in an elliptic path with the nucleus at one of the foci of ellipse. This means that the valence electron is at varying distances from the nucleus and may thus penetrate the *inner core* Fig. [3.2(b)]. This penetration causes the screening less effective and *mean effective nuclear charge* $+Z''e$ greater than the effective nuclear charge $+Z'e$ outside the electron core. Due to this its energy would be reduced, i.e., more negative and become which is the form of energy term required.

$$E_{n,l} = -R\left(\frac{Z''}{n}\right)^2 = -R\left(\frac{Z'}{n-\delta}\right)^2 = -\frac{R}{(n-\delta)^2} \tag{3.4}$$

The value of δ depends on l, the orbital quantum number for the lowering of energy would be greatest for orbits of maximum eccentricity or smallest l, since for these the average

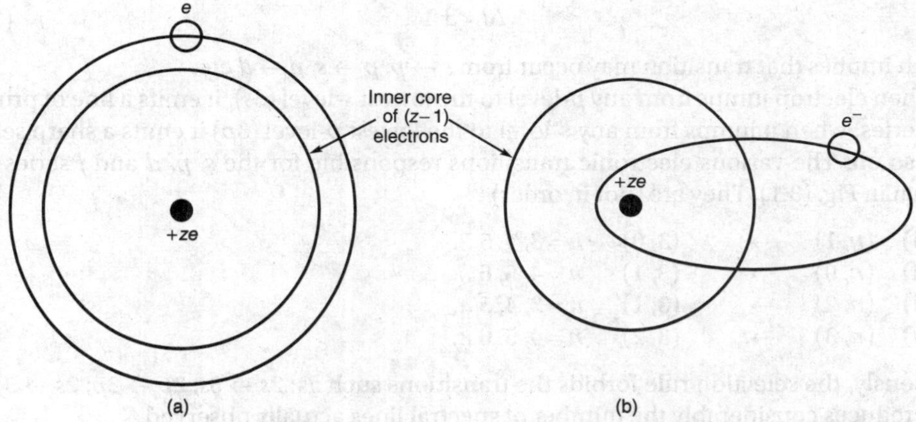

Fig. 3.2 (a) Similarity between hydrogen atom and alkali atoms. Both have one valence electron. (b) Origin of quantum defect

effective nuclear charge $+Z''e$ will be a maximum. One finds that the lowest energy state will be the one having the greatest quantum defect δ so that for states

$$\begin{array}{lccccccccc} & & & S & & P & & D & & F \\ \text{with} & \delta & = & \delta_0 & > & \delta_1 & > & \delta_2 & > & \delta_3 \\ & l & = & 0 & & 1 & & 2 & & 3 \end{array}$$

the orbital quntum number l determines the energies of the sub-levels of equal principal quantum number n.

We may note that spectroscopists designate term values of alkali atoms of different l by special symbols; e.g., $l = 0$ terms are S-terms, $l = 1$ are P terms, $l = 2$ terms are D-terms $l = 3$ terms are F-terms etc. Here S stand for sharp, P for principal, D for diffuse and F for fundamental.

Orbits corresponding to $l = 0, 1, 2, \ldots$ are designated as $s, p, d, f, \ldots$ and electrons in these orbits are called $s, p, d, f, \ldots$ electrons. In order of increasing energy, electrons with different values of principal quantum number n, orbital quantum number l are described as follows:

n	1	2	2	3	3	3	etc.
$l(\geq n-1)$	0	0	1	0	1	2	etc.
electrons are designated as	$1s$	$2s$	$2p$	$3s$	$3p$	$3d$	etc.

However, $1p, 2d, 3f, \ldots$ orbits which require that $n = l$ do not exist.

Let us consider an alkali atom, say Na. Measurement of spectral wavelengths and the ionization potential shows that the ground state of the sodium atom is $3s$, i.e., the valence or optical electron occupies the $3s$ level in the normal atom. However, when the atom is excited by any external means the electron leaves the $3s$ level and goes to any of the higher energy levels $3p, 3d, 4s, 4p, 4d, 4f, \ldots$ etc. From the higher energy levels electron jumps back to the lower energy levels controlled by the *selection rule* Fig. [3.1(b)].

$$\Delta l = \pm 1 \tag{3.5}$$

which implies that transition may occur from $s \to p, p \to s, p \to d$ etc.

When electron jumps from any p-level to the lowest s-level ($3s$), it emits a line of principal series; when it jumps from any s-level to the lowest p-level ($3p$) it emits a sharp series and so on. The various electronic transitions responsible for the s, p, d and f series are shown in Fig. (3.1). They are (not in order)

(i)	$(n, 1)$	$\to$	$(3, 0)$	$n = 3, 4, 5 \ldots$
(ii)	$(n, 0)$	$\to$	$(3, 1)$	$n = 4, 5, 6 \ldots$
(iii)	$(n, 2)$	$\to$	$(3, 1)$	$n = 3, 4, 5 \ldots$
(iv)	$(n, 3)$	$\to$	$(3, 2)$	$n = 4, 5, 6 \ldots$

Obviously, the selection rule forbids the transitions such as: $2s \to 3s$; $2p \to 3p$; $2s \to 3d$;... This reduces considerably the number of spectral lines actually observed.

From the above, we may note that (i) series (ii) above cannot start at $n = 3$ because the (3, 0) level is below the (3, 1) level. Experimentally it is reported that there is no level with the quantum number (n, n) and thus series (iv) must also start at $n = 4$.

We may note the departure of n^* from integral values for *Li*-terms Fig. [3.1(b)]. The integers as given on the left side correspond to hydrogen terms. We may further note that as l increases progressively, the departure of *Li*-terms from *H*-terms decreases. It is maximum for *S*-terms and least for *F*-terms.

The well-known *D*-lines of Na originates from transitions $3p \to 3s$, the so called *principal* series Fig. (3.3). From the foregoing discussion, each spectral line of any one of the alkali spectral lines should be *singlet*, but in fact spectral lines of alkali atoms are *doublets*. Sodium *D*-line is also found to consist of two closely lying components, known as D_1 and D_2 lines having wavelengths 5896 Å respectively Fig. (3.4). One can easily resolve these components by means of an ordinary spectrometer. We may note that such doublet structure of spectral lines is a characteristic feature of the spectra of all alkali atoms. The origin of doublet structure can however, be explained on the basis of the concept of *electron spin*.

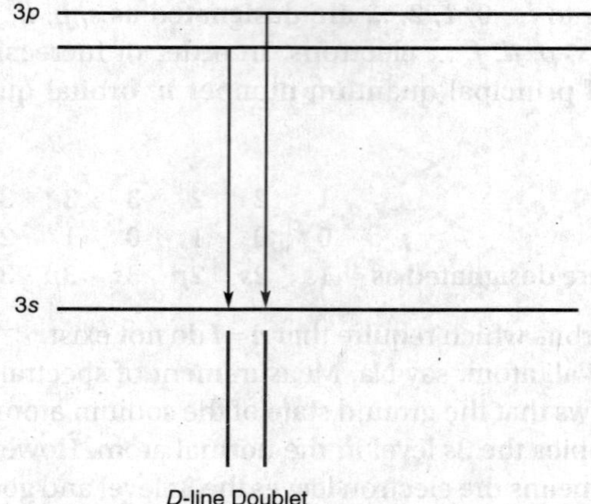

Fig. 3.3

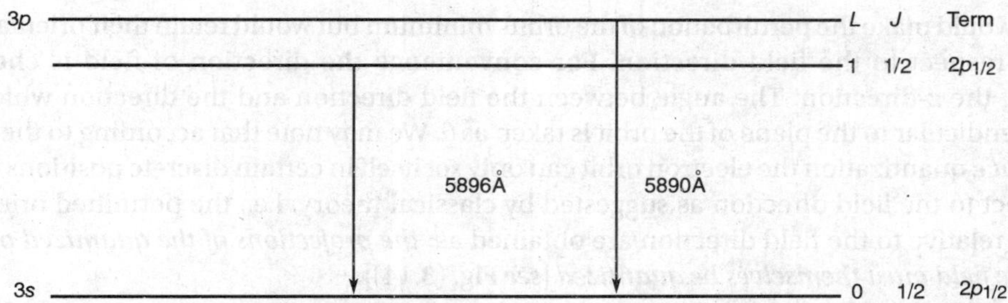

Fig. 3.4 Doublet structure of the energy levels of sodium (origin of D_1 and D_2 lines). The wavelengths are given in angstrom units

3.3 CHARACTERISTICS OF VECTOR ATOM MODEL

The two essential elements that characterize the vector atom model are: (i) *Space quantization* and (ii) concept of *electron spin*. We will now discuss these characteristics in more detail.

3.3.1 Space Quantization

Sommerfeld introduced the concept of space quantization in order to interpret the magnetic splitting of spectral lines (Zeeman effect). According to Bohr-Sommerfeld theory, two quantum numbers are required to explain the motion of the electron in an atom, i.e., principal quantum number n and the azimuthal quantum number k. In quantum mechanics k is replaced by quantum number $l = k - 1$. Description of electron motion in terms of n and k essentially implies that the motion of electron is confined to the orbital plane, there being only two degrees of freedom r and θ. The periodicity of the radial motion is described in terms of the radial quantum $n_r = (n-k)$ which is thus not an independent quantum number. Thus the two quantum numbers, one corresponding to radius vector n_r and other corresponding to angles n_φ are sufficient to describe the electron orbit. In general, the motion of the electron in an atom is three-dimensional and therefore possesses three degrees of freedom. Thus an additional quantum condition is required to describe the actual state of affairs. According to classical mechanics, when an electron moves in three-dimensional orbits it may have all orientations in three-dimensions with respect to a fixed orbit. According to Sommerfeld theory the continuous manifold positions of orbits in space, permitted classically are not permitted. In vector atom model, which is based on quantum theory, the orbits are to be quantized no only in *magnitude* (i.e., *size* and *form*) but also in *direction* or in spatial orientation. Such space quantization makes the orbit vector quantities. Thus the third condition quantises the orientation of elliptical orbit in three dimensions and does not alter the original Sommerfeld orbits in regard to their size and shape.

To quantized spatially, a certain favoured direction, as reference is necessary with respect to which the orbits may receive their orientation. An external field might suit the purpose, i.e., such a direction, but the difficulty is that the external field would perturb the orbits and thus deform them. In order to avoid this situation, one may choose the external field so small, i.e., the limiting case when the field tends to zero, it does not deform the orbits.

This would make the perturbation of the orbits minimum but would retain their orientation with respect to the field direction. For convenience the direction of field is chosen along the z-direction. The angle between the field direction and the direction which is perpendicular to the plane of the orbit is taken as θ. We may note that according to the rule of space quantization the electron orbit can only set itself in certain discrete positions with respect to the field direction as suggested by classical theory, i.e., the permitted orientations relative to the field direction are obtained as: *the projections of the quantized orbits on the field must themselves be quantized* [*see* Fig. (3.14)].

3.3.2 Spinning Electron

Uhlenbeck and Goudsmith in 1925 first introduced the idea of spinning electron. In order to explain the *multiple structure* (the fine structure) of spectral lines, they found it necessary to assume that each electron had an additional angular momentum equal to $\frac{1}{2}h/2\pi (= \frac{1}{2}\hbar)$. This angular momentum is usually referred to as the *spin angular momentum* or *spin* of the electron.

According to these hypotheses the electron revolves not only in an orbit round the nucleus but also spin about its own axis. Clearly, this has two types of motion viz., *orbital motion* and *spin motion*. If again, the rotating body carries an electric charge, the charge itself would rotate giving rise to a circuital current and the consequent magnetic moment. We may note that the picture of an electron a point particle, spinning about its axis should not be taken too literally. According to quantum mechanics, the spin motion like orbital motion should be quantized. Hence a new quantum number, known as spin quantum number s such that for a single electron $s = \frac{1}{2}$ in angular momentum unit of $h/2\pi$. Obviously, the orbital and spin motions are quantized not only in magnitude but also in direction according to the concept of spatial quantisation. The two are considered as quantized vectors and hence the model of atom is referred as *vector atom model*.

In the older model of the atom, the electron gas was endowed with only the orbital angular momentum and the orbital magnetic moment. In the light of spin, electron has now, in addition, yet another angular momentum, the *spin angular momentum* and another magnetic moment, the spin magnetic moment as shown in Fig. (3.5). The total angular momentum of the atom is obviously the resultant of the orbital and the spin

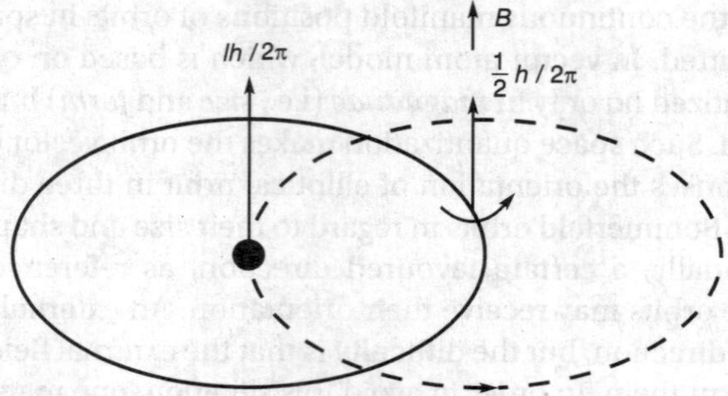

Fig. 3.5 Spinning electron

angular momenta. Similarly, the total magnetic moment is the sum of the orbital and the spin magnetic moments. Concluding, one can say that the atomic magnet is the resultant of *two* magnets, i.e., one arising from the *orbital motion* and the other from the *spin motion*.

The two fundamental concepts introduced about the vector atom model have been experimentally verified by Stern and Gerlach while studying the behaviour of atoms in non-homogeneous magnetic fields.

3.4 ANGULAR MOMENTA AND MAGNETIC MOMENTA

We now examine the connection between the angular momentum and the associated magnetic moment.

The orbital angular momentum p_l of an electron in an orbit is the cross-product of the radius vector r and the linear momentum $p_l(=\boldsymbol{mv})$, i.e.,

$$\boldsymbol{pl}=\boldsymbol{r}\times\boldsymbol{p}=\boldsymbol{r}\times\boldsymbol{mv} \tag{3.6}$$

If the orbit is circular, $\vec{\boldsymbol{p}}_l$ is perpendicular to the plane of the orbit and is mrv in magnitude. However, according to Bohr's quantum condition, the above magnitude is quantized in units of $h/2\pi$, i.e.,

$$p_l = l\frac{h}{2\pi} = l\hbar, \quad \text{where } l = 0, 1, 2, \ldots\text{etc.} \tag{3.7}$$

(p_l = magnitude of $\boldsymbol{p}_l = mr\omega = mr^2\omega$, where ω is the circular frequency in its orbit: $\omega = 2\pi\nu$, ν being the frequency of revolution).

We know that a circulating electron is equivalent to a current in the loop. It would thus give rise to a *magnetic dipole* of moment $\mu_l = iA$, where I is the current strength and A the area commanded by the loop. Now, for a circular orbit $A = \pi r^2$ and $i = e\nu = e\omega/2\pi$. Thus one obtains

$$\begin{aligned}\mu_l &= \text{magnetic moment associated with orbital motion}\\ &= (e\omega/2\pi)\times\pi r^2 = e\omega r^2/2\end{aligned} \tag{3.8}$$

The magnetic moment has a direction at right angles to the plane of the orbit as shown in Fig. (3.6). We may note that μ_l is opposite to that of $\boldsymbol{p}_l$ of the electron, since the electron has a negative charge.

Now, the ratio of the magnetic moment to the angular momentum is therefore

$$\frac{\mu_l}{p_l} = \frac{e\omega^2 r/2}{mr^2\omega} = \frac{e}{2m} \tag{3.9}$$

Obviously, the above ratio involves the charge and mass of the electron and this ratio is known as the *gyromagnetic ratio*. One obtains the same expression for the gyromagnetic ratio even if one assumes that electron rotate in an elliptic orbit instead of a circular orbit.

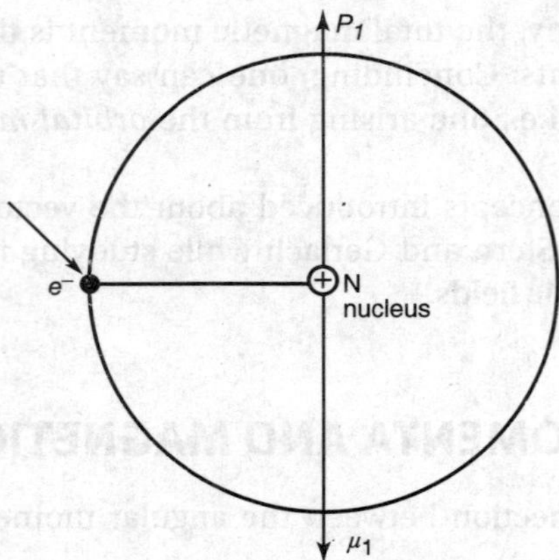

Fig. 3.6 The angular momentum and magnetic moment due to orbital motion of the electron

But $$p_l = \frac{lh}{2\pi}, \quad \text{where} \quad l = 0, 1, 2, \ldots$$

$$\mu_l = \frac{e}{2m} p_l = l\frac{eh}{4\pi m} = l\mu_B = \frac{e}{2m}\left(\frac{lh}{2\pi}\right) \tag{3.10}$$

Where $$\mu_b = \frac{eh}{4\pi m} = 9.2741 \times 10^{-24}\ \text{J/Tesla}\left(= \frac{J}{(\text{Weber} \times m^2)}\right) \tag{3.11}$$

$$= 5.788 \times 10^{-5}\ \text{eV/T}$$

Expression Eq. (3.10) is the expression for the magnetic moment of the atom *due to the orbital motion of an electron in it.*

Bohr Magneton (μ_B): $\mu_B = \dfrac{eh}{4\pi m}$ is known as the Bohr magneton and it is the basic unit of the atomic magnetic moment. This is the minimum magnetic moment that an electron would possess as a result of orbital motion.

We may note that there is also a magnetic moment due to the spin motion of the electron. This is found to have a value of 1 μ_B.

Spin Magnetic moment (μ_s): Similar to the orbital angular momentum p_l, it is necessary to introduce a spin angular momentum p_s, and an associated magnetic moment μ_s. However, the ratio μ_s/p_s is different from the gyromagnetic ratio for the orbital motion. We have

$$\frac{\mu_s}{p_s} = g_s\left(\frac{e}{2m}\right)$$

experimentally, $$g_s = 2$$

$\therefore$ $$\mu_s = 2\frac{e}{2m} p_s = \frac{e}{m}\left(s\frac{h}{2\pi}\right)$$

since p_s, like p_l is also quantized and is given by

$$p_s = s\frac{h}{2\pi}$$

where s is known as the *spin quantum number* and $s = \frac{1}{2}$ always.

$$\therefore \quad \mu_s = \frac{e}{m}\frac{1}{2}\frac{h}{2\pi} = \frac{eh}{4\pi m} = \mu_B$$

Clearly, the *spin magnetic moment* (μ_s) of the electron is equal to the *Bohr magneton* μ_B.

We may note that μ_s and p_s may also be obtained from the Dirac's relativistic quantum mechanical theory of electron. Electron is a point particle and its spin cannot be visualized classically. This is why spin magnetic moment is termed *intrinsic magnetic moment*. Although Uhlenbeck and Goudsmit introduced the spin as an *adhoc* hypothesis but spin appears naturally by solving Dirac relativistic equation for electron.

3.5 ORBITAL ANGULAR MOMENTUM

Bohr's theory of hydrogen atom assume that the electron moves in circular orbit with an angular velocity ω. Due to this angular velocity the electron has a definite angular momentum. According to Bohr's theory, the angular momentum is an integral multiple of $h/2\pi$ and represented by $nh/2\pi$. However, in Sommerfeld theory, the angular momentum is represented by $n_\varphi h/2\pi$, where n_φ is the azimuthal quantum number. In vector atom model of the atom, the azimuthal quantum number n_φ is replaced by l which is known as *orbital quantum number* l is an integer and for a given n (principal quantum number) l can take up values 1, 2, 3, ... $(n-1)$. It arises from quantisation of the magnitude of orbital angular momentum p_l of the electron, and is given by

$$p_l = l\frac{h}{2\pi}, \qquad \text{where} \qquad l = 0, 1, 2, \ldots \text{etc.} \tag{3.12}$$

We may note that orbital angular mometum (p_l) is a vector quantity and is represented by a straight line which is parallel to the axis of rotation (Fig. (3.6)). Its positive sense is taken by the direction of advance of a right hand screw rotated with the motion.

According to quantum mechanics, p_l is not equal to $l\,(h/2\pi)$, i.e., $p_l \neq l\,(h/2\pi)$ but this is given by

$$p_l = \sqrt{l(l+1)}\frac{h}{2\pi} \tag{3.13}$$

Purely, by convention the electron is called as *s, p, d, f*, etc. electron according as

Orbital angular momentum (l)	0	1	2	3	4	5 ...
Electron designation	*s*	*p*	*d*	*f*	*g*	*h* ...

Here the word *s, p, d, f*, ... etc. stand for *sharp, principal, diffuse, fundamental* etc.

When $n = 3$ $l = 0$ the electron is said to be 3*s*
$n = 3$ $l = 1$ the electron is said to be 3*p*
$n = 3$ $l = 2$ the electron is said to be 3*d*

The different subshells can be tabulated as shown in Table 3.1.

The maximum number of electrons allowed in a subshell is given by $2(2l + 1)$. The distribution of electrons in different subshells is shown in the following Table 3.2.

We may note that *l*-value determines the shape of the orbit where as the principal quantum number (*n*) determines the size of the electron-orbit and arises from quantization of the energy of the electron.

In the case of more complex atoms the total orbital angular momentum is denoted by $\vec{L}$. $\vec{L}$ is a vector sum of all orbital angular momenta of individual electrons. However, the vector sum $\vec{L}$ is restricted by quantum conditions to integral values only. Let us consider an example of two electrons having orbital angular momentum $l = 1$ and $l = 2$. The total angular momentum (L) for these two situations may have values from $L = 2 + 1 = 3$ to $L = 2–1 = 1$ and all integral values in between the two values of l, i.e., l may have values 3, 2 and 1. Fig. (3.7) illustrates the vector addition of these vectors.

Table 3.1 Different subshells

Shells	***n***	***Subshell***				
		***l* = 0**	**1**	**2**	**3**	**4**
K	1	1*s*				
L	2	2*s*	2*p*			
M	3	3*s*	3*p*	3*d*		
N	4	4*s*	4*p*	4*d*	4*f*	
O	5	5*s*	5*p*	5*d*	5*f*	5g

Table 3.2 Distribution of electrons in different subshells

Shells	***Principal quantum number n***	***Orbital quantum number n***	***Possible electrons***	***Maximum number of electrons 2(2l + 1)***	***Total electrons 2n²***
K	1	0	1*s*	2	2
L	2	0	2*s*	2	8
		1	2*p*	6	
M	3	0	3*s*	2	
		1	3*p*	6	18
		2	3*d*	10	
N	4	0	4*s*	2	
		1	4*p*	6	32
		2	4*d*	10	
		3	4*f*	14	

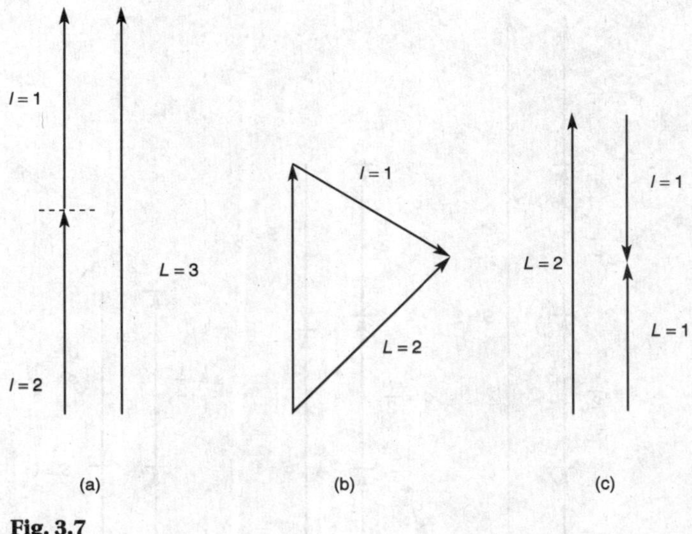

Fig. 3.7

3.6 ELECTRON SPIN AND SPIN QUANTUM NUMBER

Spin quantum number is symbolized by *s*. *s* always takes the value $\frac{1}{2}$ for electron so that it is fixed and does not depend on the orbit. It arises from quantization of the magnitude of the spin angular momentum p_s, given by

$$p_s = s\frac{h}{2\pi} = \frac{1}{2}\frac{h}{2\pi}, \qquad \text{where} \quad s = \frac{1}{2} \tag{3.14}$$

We may note that the half-integer value contradicts the integral multiple rule of quantization. But this was adjusted for the agreement with the experimental results. However, the quantum mechanical treatment of spinning electron by Dirac automatically led to this half-integral property.

According to wave mechanics, the spin angular momentum p_s of the electron is given by

$$p_s = \sqrt{s(s+1)}\frac{h}{2\pi} \tag{3.15}$$

In the absence of any external field, the motion of an electron in an atom can be completely described in terms of *n*, *l* and *s*.

One can obtain the spin angular momenta of several electrons as follows:

Let us consider the case of four electrons. All the electrons may have the spins in one direction, i.e., total $\vec{S}$ may be 2. Out of these four electrons, the spins of three electrons may have one direction and fourth electron may have spin in opposite direction. Obviously, the vector sum $\vec{S}$ in this case will be 1. There is also a possibility that the two electrons have their spins in one direction and rest two electrons in opposite direction. Clearly, the vector sum in this case will be zero. Thus in one of four electrons the spins can have values 0, 1 or 2. Fig. (3.8) depicts this situation. Obviously, in case of three electrons the spin values may be 1/2 or 3/2.

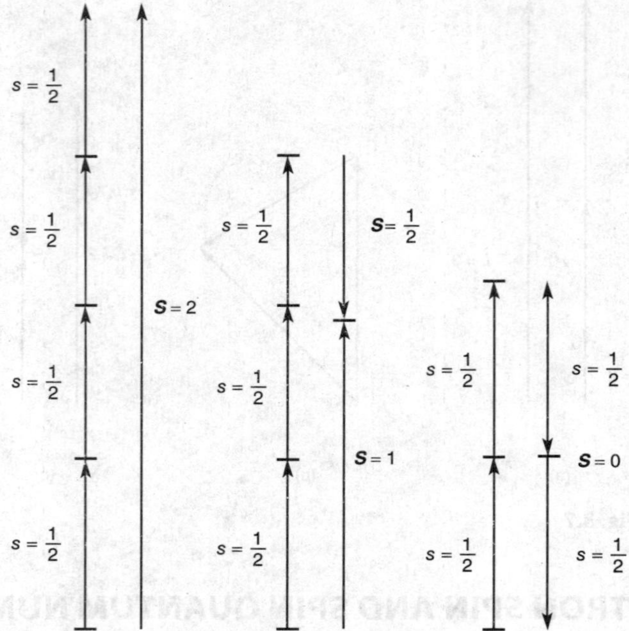

Fig. 3.8

We may note that for an odd number of electrons S must be an odd multiple of 1/2, while for even number of electrons, S must be an even multiple of 1/2.

3.7 TOTAL ANGULAR MOMENTUM OF ELECTRON

We have read that an electron has orbital angular momentum p_l and spin angular momentum p_s. The total angular momentum j of a single electron is the vector sum of the orbital and spin angular momenta of the single electron. The magnitude of total angular momentum J is given by

$$J = j\frac{h}{2\pi} \tag{3.16}$$

According to quantum mechanics, however, $J = \sqrt{j(j+1)}h/2$. j is sometimes termed as the *inner quantum number*. The values of total angular momentum j are $J = l + s$ and $J = l - s$ with the restriction that j must be positive. We may note that here positive sign is taken when s is parallel to l and minus sign when s is anti-parallel to l. This is depicted in Fig. (3.9). For example,

for $l = 2$ and $s = \frac{1}{2}$, J can have values $2 + \frac{1}{2} (= 5/2)$ and $2 - 1/2 (= 3/2)$.

In accordance with the wave mechanics p_s and p_l couple vectorially as shown in Fig. (3.10).

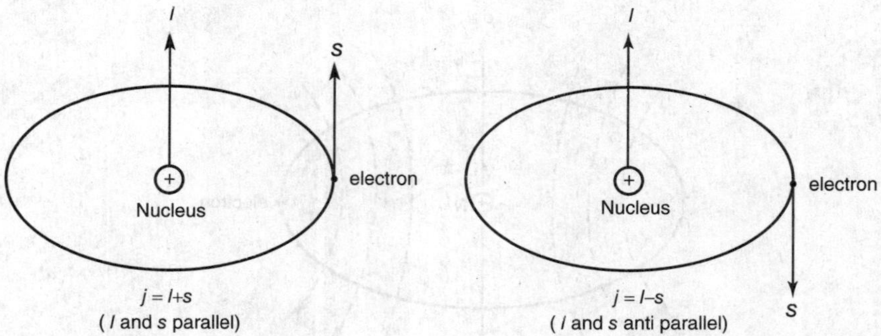

Fig. 3.9

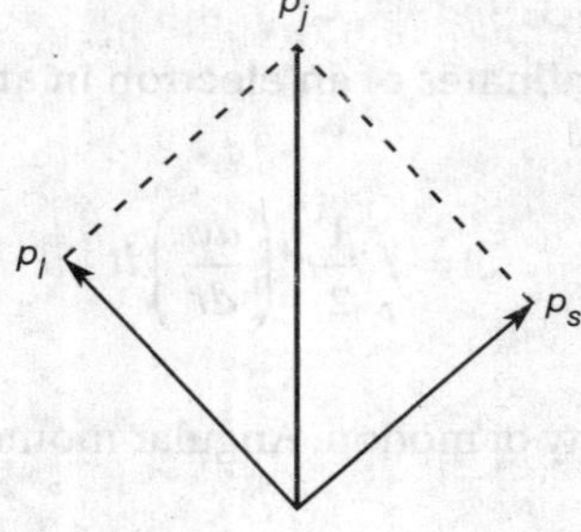

Fig. 3.10

3.8 MAGNETIC MOMENT OF AN ORBITAL ELECTRON

There are two magnetic momenta associated with the electron, one due to its orbital motion and other due to spin motion of electron. The orbital motion of the electron is just like the flow of a current since the electron is a negatively charged particle although the current is not confined in a wire. Due to the flow of the current a magnetic field is set up as shown in Fig. (3.11).

Let us consider the simple case of an electron which is revolving in a central orbit like an ellipse. The current i due to the motion of the electron round the nucleus is given by

$$i = \frac{e}{T} \quad \text{(MKS unit)} \tag{3.17}$$

where i is current in amperes, e is charge on electron in coulomb and T is the periodic time in seconds. This current i gives rise to a magnetic moment μ_l as follows:

$$\mu_l = iA = \frac{eA}{T} \tag{3.18}$$

where A is the area enclosed by the orbit.

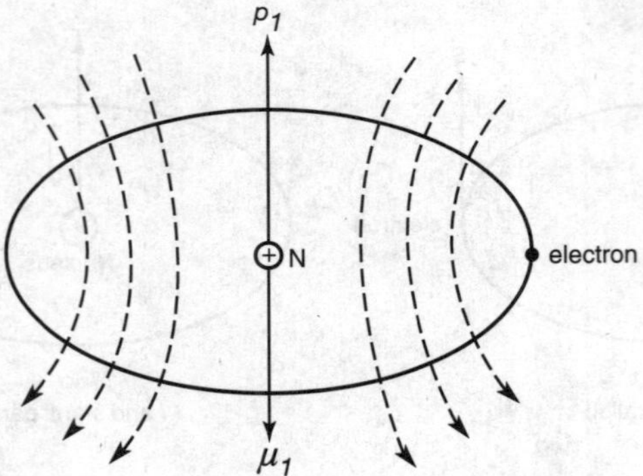

Fig. 3.11

Let r and ϕ be the polar coordinates of an electron in an elliptical orbit, then the area A is given by

$$A=\int_0^T \frac{1}{2}r^2\left(\frac{d\phi}{dt}\right)dt \tag{3.19}$$

where $\frac{1}{2}r^2\left(\frac{d\phi}{dt}\right)$ is a real velocity of motion. Angular momentum p_l is given by

$$p_l = mr^2\left(\frac{d\phi}{dt}\right)$$

or

$$\frac{p_l}{2m}=\frac{1}{2}r^2\left(\frac{d\phi}{dt}\right) \tag{3.20}$$

Substituting the value of $\frac{1}{2}r^2\left(\frac{d\phi}{dt}\right)$ from Eq. (3.20) in Eq. (3.19), one obtains

$$A=\int_0^T \frac{p_l}{2m}dt$$

$$=\frac{p_l}{2m}[t]_0^T=\frac{p_l}{2m}T \tag{3.21}$$

(Since p_l and m are constants.)
Substituting the value of A from Eq. (3.21) in Eq. (3.18) one obtains

$$\mu_l=\frac{e}{T}\frac{p_l}{2m}T=\frac{e}{2m}p_l \tag{3.22}$$

We have

$$p_l = l\frac{h}{2\pi}$$

$\therefore$

$$\mu_l=\frac{e}{2m}l\frac{h}{2\pi}=\frac{eh}{4\pi m}l \tag{3.23}$$

The expression Eq. (3.23) shows that orbital magnetic moment μ_l is directly proportional to l. Now, if $l=0, \mu_l=0$ and when $l=1$, $\mu_l = \frac{eh}{4\pi m}$. We have seen that $eh/4\pi m$ is the smallest constant unit, known as *Bohr magneton* ($\mu_B = \frac{eh}{4\pi m}$). Obviously,

$$\mu_l = \mu_B l = 9.264\times10^{-24}\, l$$

Where $\mu_B = 9.274\times10^{-24}$ Ampere-m^2 or J/T

$$= 5.788\times10^{-5}\,\text{eV/T} \tag{3.24}$$

3.8.1 Gyromagnetic Ratio

The ratio μ_l/p_l is known as gyromagnetic ratio and represented by G. We have

$$G = \frac{\mu_l}{p_l} = \frac{l(eh/4\pi m)}{l(h/2\pi)} = e/2m \tag{3.25}$$

one can also express this ratio as

$$G = g\frac{e}{2m} \tag{3.26}$$

where g is called *Lande splitting factor*. We may note that Lande splitting factor is 1 for orbital motion of electron.

3.9 MAGNETIC MOMENT DUE TO ELECTRON SPIN

The spinning electron behaves like a spinning tinny magnet. Electron has spin magnetic moment due to its spin. Fig. (3.12) shows the magnetic field generated by the spinning electron and its spinning magnetic moment μ_s. The spin magnetic moment of the electrons is given by

$$\mu_s = 2\frac{e}{2m}p_s \tag{3.27}$$

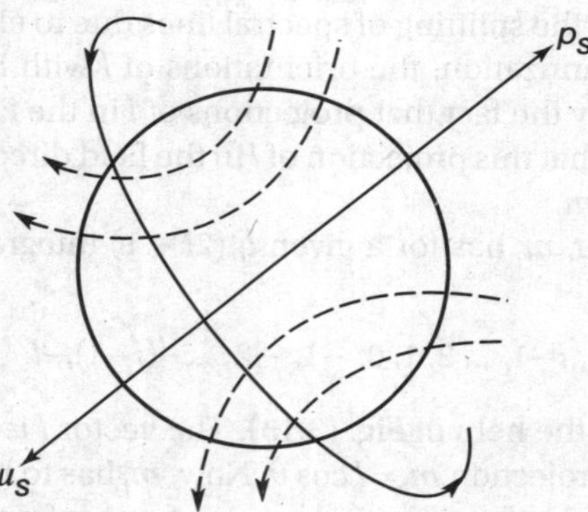

Fig. 3.12 Magnetic moment due to electron spin

we have the spin angular momentum p_s,

$$p_s = s\frac{h}{2\pi} \tag{3.28}$$

From Eqs. (3.28) and (3.27), one obtains

$$\mu_s = 2\frac{e}{2m}s\frac{h}{2\pi} = 2s\frac{eh}{4\pi m}$$

But $$s = \frac{1}{2}$$

$\therefore$ $$\mu_s = \frac{eh}{4\pi m} = \mu_B \text{(One Bohr magneton)}$$

$$= 9.274\times10^{-24}\ \text{J/T} = 5.788\times10^{-5}\ \text{eV/T}$$

Obviously, the magnetic moment due to electron spin is equal to one *Bohr magneton.*

We may note that the doubling of energy levels in atoms having one electron is due to the interaction between the two fields, i.e., the magnetic field of electron orbit and magnetic field of electron spin.

3.10 MAGNETIC QUANTUM NUMBERS (m_l, m_s)

We have seen that a magnetic field has a directive influence on the orbital and spin motion of the electron. The orbital and spin quantum numbers would have some preferential values in the magnetic field and the electron would thus have two more quantum numbers: m_l and m_s.

3.10.1 Magnetic Orbital Quantum Number (m_l)

When an atom is placed in a strong magnetic field, the electrons with same values of principal quantum number n and orbital quantum number l, may still differ in their behviour. In order to account for one more, quantum number, m_l known as magnetic orbital quantum number has been introduced. This is supported by *Zeeman effect* and *Stark effect.* We will see that Zeeman effect refers to the splitting of spectral lines due to magnetic field, while the Stark effect refers to the splitting of spectral lines due to electric field. In accordance to the rule of space quantization, the orientations of l with H (magnetic field) are limited and are determined by the fact that projections of l in the field direction assumes integral values. We may note that this projection of l in the field direction is known as magnetic field quantum number, m_l.

Magnetic quantum number, m_l has for a given l, $(2l + 1)$ integral values from $-l$ to $+l$ through O.

$\therefore$ $$m_l = l,\ l-1,\ \ldots 2,\ 1,\ 0,\ -1,\ -2,\ \ldots -(l-1), -l$$

We can understand this with the help of Fig. (3.13). The vector l is inclined at an angle θ with the field direction. The projection $m_l = l\cos\theta$. Now, m_l has to be an integer and $\cos\theta$ never exceed unity, and therefore the permitted values of m_l are from $-l$ to l at unit interval including 0, i.e., $(2l + 1)$ possible values. Clearly, from quantization of the direction of the

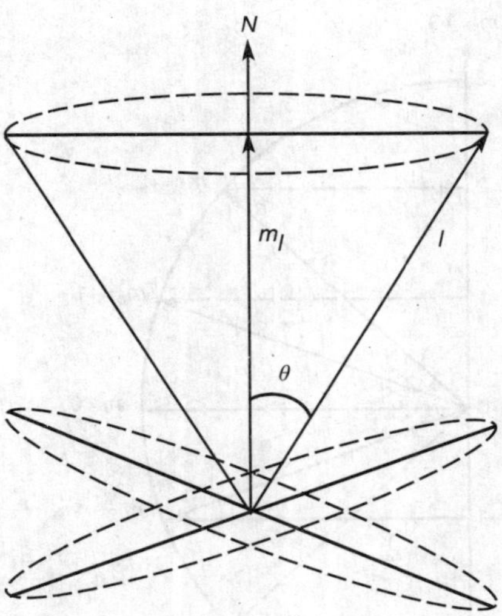

Fig. 3.13 Magnetic moment due to electron spin

orbital angular momentum p_l and determines the orientation of the orbit in space. As stated above the directions are given by

$$\cos\theta = m_l / l$$

quantum mechanically $$\cos\theta = m_l \sqrt{l(l+1)}$$

For $l = 0$ $m_l = 0$, and for each value of the principal quantum number n, there is only one orientation, i.e., s orbital. For $l = 1$ i.e., p orbital, $m_l = +1, 0, -1$, i.e., three permitted orientations are possible. In absence of external magnetic field these orbitals are equivalent in energy and are said to be three fold degenerate. Now, for $l = 2$ (d orbital), $m_l = +2, +1, 0, -1, -2$, i.e., five permitted orientations. In the absence of external magnetic field, five permitted orbitals are equivalent in energy and they are said to be the five degenerate. For f orbital, i.e., for $l = 3$, $m_l = +3, +2, 0, -1, -2, -3$, i.e., there are seven permitted orientations. Space quantization is shown in Fig. (3.14). We may note that in the absence of the magnetic field, these seven orbitals are said to be seven fold degenerate.

3.10.2 Magnetic Spin Quantum Number (m_s)

Just as the vector representing orbital angular momentum processes in the presence of magnetic field, so does the vector representing the spin angular momentum, m_s. The numerical value of magnetic spin quantum number m_s is the projection of the spin vector s on the field direction. By analogy with the orbital vector, magnetic spin quantum number m_s can have any of the $(2s + 1)$ values from $-s$ to $+s$ excluding zero, i.e., $-1/2$ and $+1/2$ only (since s is only $\frac{1}{2}$ and never zero). It arises from quantisation of the spin angular momentum p_s and determines the orientation of the spin either up or down as shown in Fig. (3.15).

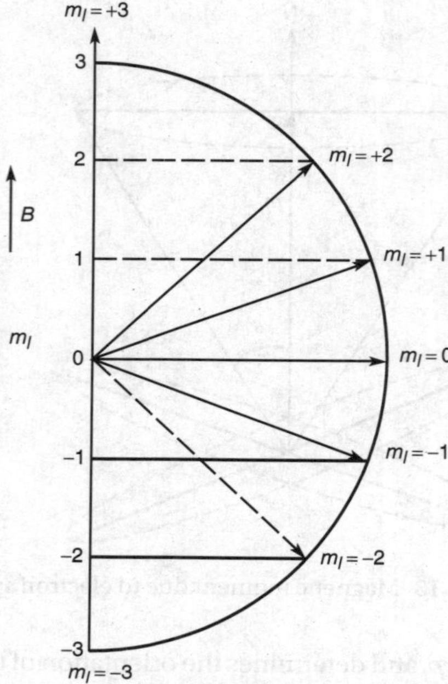

Fig. 3.14 Space quantisation of orbital angular momentum

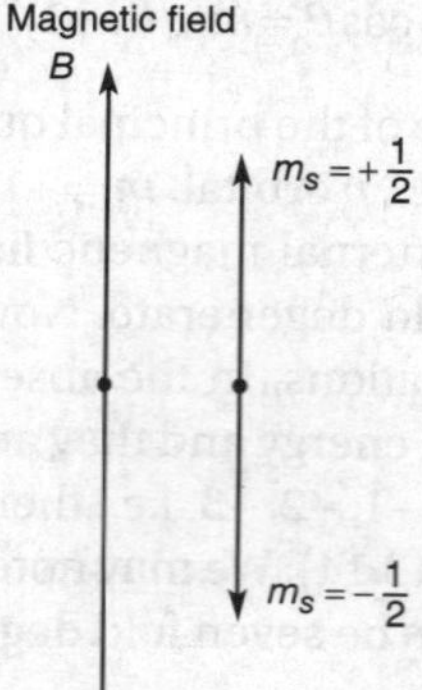

Fig. 3.15 Orientation of spin angular momentum of electron with respect to the magnetic field

3.10.3 Magnetic Total Angular Momentum Quantum Number (m_J)

It is the vector sum of the quantum numbers m_l and m_s. Its numerical value is the projection of the total angular momentum vector J on the field direction. Since J can have half integral values, m_J can also assume half integral values. The possible values of m_J are $(-J)$, $(-J-1), \ldots -1, 0, +1, \ldots (J-1), J$, excluding zero for a single electron.

We may note that the permitted orientations of J are $(2J+1)$ and hence m_J can have only $(2J+1)$ values form $-J$ to $+J$ excluding zero.

This reveals that the precessional motions of the electron orbit in strong magnetic field contribute energy to the electron and each electron level is split into $(2J + 1)$ sublevels. Obviously, this explains the splitting of spectral lines in a strong magnetic field. If n, l, m_l and m_s are known, J and m_J could be found and the electron in an atom is completely defined.

Obviously, the four quantum numbers n, l, m_l and m_s are considered as *labels* that completely specify the state of an electron in an atom model and thus determine its energy and angular momentum.

We may note that in electronic transitions, some or all of these quantum numbers (and also the associated properties) change. However, the *charge, rest mass* and *spin* are the intrinsic properties of the electron and do not change in transitions.

3.11 SPIN-ORBIT INTERACTION OR SPLITTING

We have read that all electrons have a spin quantum number s, which has the value 1/2 and adds vectorially to the orbital quantum number l. Since the magnetic field arising from orbital motion of the electron is parallel to $lh/2\pi(l\hbar)$, the magnetic moment and the associated angular momentum $sh/2\pi(=s\hbar)$ must orient either parallel or anti-parallel to it. Obviously, l and s behave as vectors and one may represent them as shown in Fig. (3.16). The resultant of l and s vector addition is J which obviously has the value $J = l \pm 1/2$, subject however to the restriction that J must always be positive.

Doublet structure of sodium lines

With the help of *l-s* coupling, one can easily explain the doublet structure of sodium D-lines and more complicated multiplet structure of other transitions not involving a level with $l = 0$. Electrons having the same value of n and l but different J have different energies

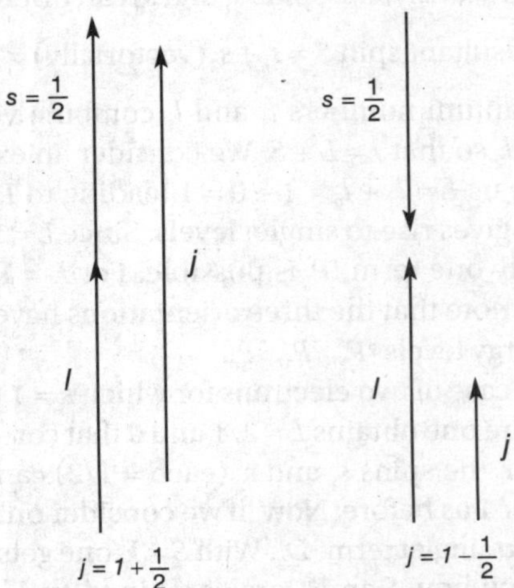

Fig. 3.16 *l-s* coupling for a single electron forming J

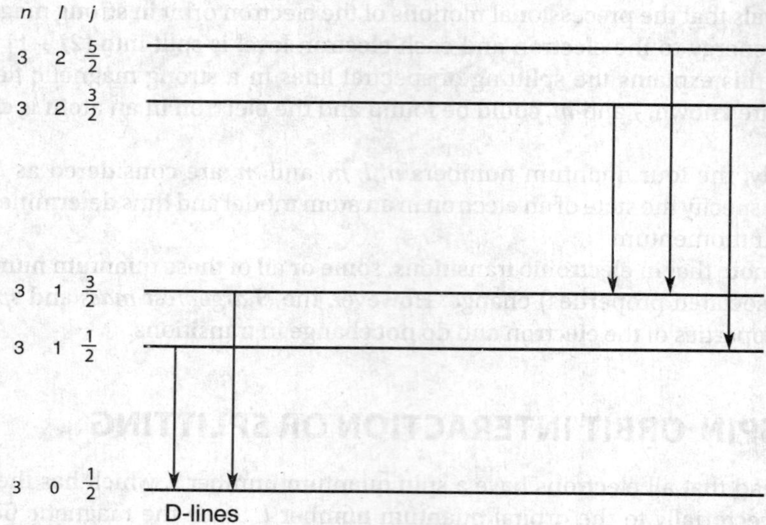

Fig. 3.17 Spin-orbit splitting doublet structure of sodium D-lines

as illustrated in Fig. (3.17). This is termed as spin-orbit splitting. On careful analysis, it is found that this spin-orbit splitting between the levels of the alkali spectra decreases with n for a given Z, but increases with Z for a given n.

Explanation of complex spectra involving active electron > 1

Interstingly, the spin-orbit interaction not only explains the structure of alkali spectra, but also explains the complexity of the spectra containing more than one active electron, e.g., in Ca, Zn etc. In these elements there are two active electrons each having a spin 1/2 so that $s_1 = 1/2$ and $s_2 = 1/2$. Obviously, the spins s_1 and s_2 can orient parallel or anti-parallel to each other and the total resultant spin $S = s_1 + s_2$ (vectorially) $= \frac{1}{2} \pm \frac{1}{2} = 0$ or 1. Further, the corresponding orbital quantum numbers l_1 and l_2 combine vectorially to form the total quantum number $L = l_1 + l_2$ so that $J = L + S$. We consider an example; an atom with $l_1 = 1$ and $l_2 = 0$, this would give us $L = l_1 + l_2 = 1 + 0 = 1$ leading to P levels. The total resultant spin $S = 0$ or 1. Now, $S = 0$, gives rise to singlet levels. Since $L = 1$, $S = 0$, one finds the value of $J = 0 + 1 = 1$ so that only one term ${}^1P_1$ is possible. For $L = 1$, $S = 1$, J has values 2, 1, 0 [See Fig. (3.18a)]. We may note that the three orientations have slightly different energies giving rise to separate energy levels ${}^3P_2$, ${}^3P_1$, ${}^3P_0$.

Let us now consider the case of two electrons for which $l_1 = 1$ and $l_2 = 1$. Now, combining l_1 and l_2 vectorially, as before one obtains $L = 2$, 1 and 0 that correspond to D, P and S terms (see Fig. (3.18b)). However, the spins s_1 and s_2 (each = 1/2) can set themselves parallel of anti-parallel so that $S = 0$ or 1 as before. Now, if we consider only d-term for $S = 0$, we have $J = L + S = 2 + 0 = 2$ leading to singlet term ${}^1D_2$. With $S = 1$, one gets a triplet term. The integral values of J depend solely on how S and L are combined and [see Fig. 3.18(c)] one finds $J = 3, 2, 1$ leading to ${}^3D_3$, ${}^3D_2$ and ${}^3D_1$ terms.

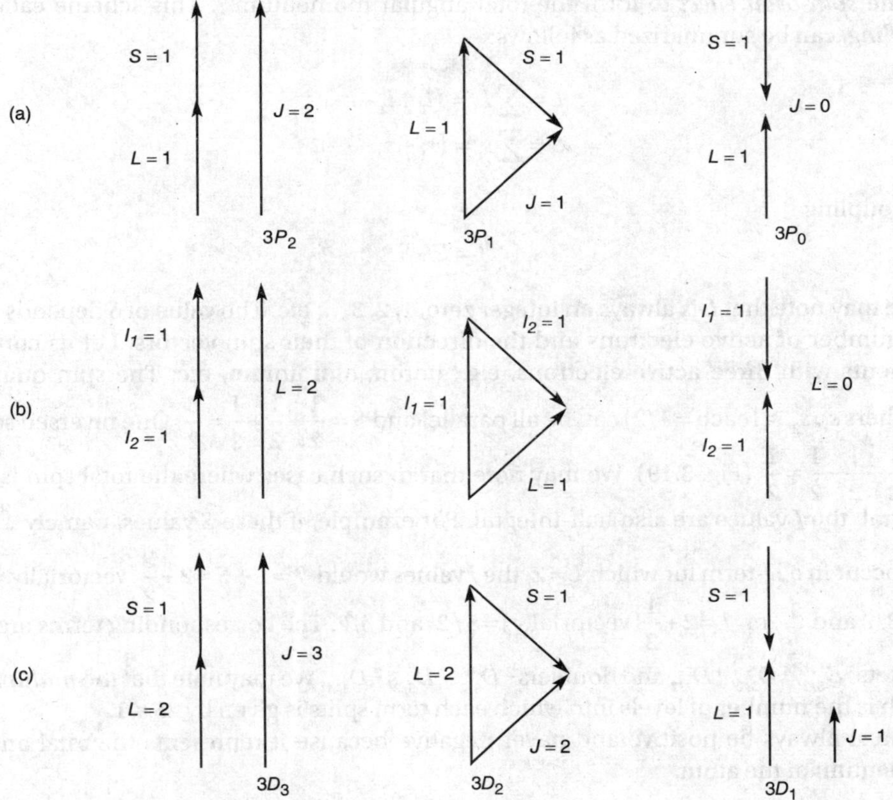

Fig. 3.18 Vector addition of L and S gives vector J

3.12 COUPLING SCHEMES

When more than one electron contributes orbital and spin angular momenta to the total angular momentum J of an atom J is still the vector sum of these individual momenta. Generally two types of coupling known as *Russell-Saunders* or *L-S coupling* and *J-J* coupling occurs. We now describe these coupling schemes:

3.12.1 *L-S* Coupling or Russell-Saunders Coupling

This is also called as normal coupling as this occurs most frequently. When more than one electron contributes orbital and spin angular momenta to the total angular momentum J of an atom, J is still the vector sum of these individual momenta. In this scheme, the usual pattern for all but the heaviest atoms is that the orbital angular momenta L of the various electrons are coupled together into a single resultant L. The spin angular mometa S_i are also coupled together into another single resultant S. The momenta L and S then interact via the *spin-orbit effect* to form the total angular momentum J. This scheme called *LS coupling*, can be summarized as follows:

$$L = \sum L_i = (l_1 + l_2 + \cdots)$$

and

$$S = \sum s_i = (s_1 + s_2 + \cdots)$$

L-S coupling

$$J = L + S$$

We may note that L is always an integer zero, 1, 2, 3, ... etc. The value of S depends upon the number of active electrons and the direction of their spin vectors. Let us consider elements with three active electrons, e.g., boron, aluminium, etc. The spin quantum numbers s_1, s_2, s_3 (each = 1/2) can be all parallel and $S = \frac{1}{2} + \frac{1}{2} + \frac{1}{2} = \frac{3}{2}$; One reversed so that $S = \frac{1}{2} + \frac{1}{2} - \frac{1}{2} = \frac{1}{2}$ (Fig. 3.19). We may note that in such cases where the total spin is half-integral, the J values are also half-integral. For example, if these S values, namely $\frac{3}{2}$ and 1/2, occur in a D-term for which $L = 2$, the J values would $J = L + S = 2 + \frac{3}{2}$ vectorially = 7/2, 5/2, 3/2 and $\frac{1}{2}$, or $J = 2 + \frac{1}{2}$ (vectorially) $= 5/2$ and 3/2. The corresponding terms are thus quartets ${}^4D_{5/2}$, ${}^4D_{3/2}$, ${}^4D_{1/2}$ and doublets ${}^2D_{5/2}$, ${}^2D_{5/2}$, ${}^2D_{3/2}$. We may note that the *multiplicity*, which is the number of levels into which each term splits is given by $2S + 1$.

J must always be positive and never negative because it represents the total angular momentum of the atom.

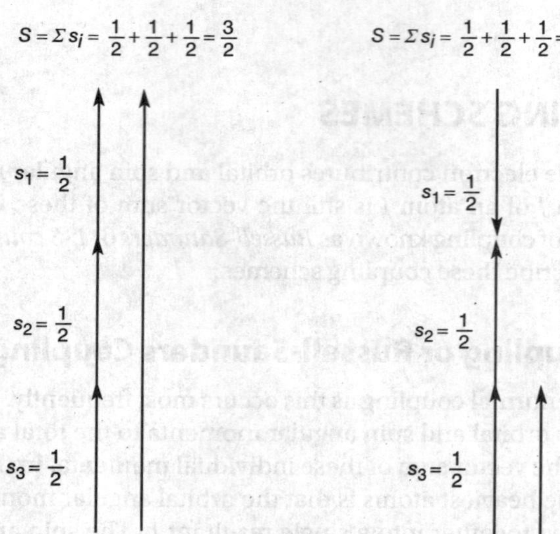

Fig. 3.19 For three active electrons the composition of three vectors

The angular momentum magnitudes L, S, J and their Z components L_z, S_Z and J_z are all quantized in the usual ways, with the respective quantum numbers L, S, JM_L, M_S and M_J. Hence

$$
\begin{aligned}
L &= \sqrt{L(L+1)}\,\hbar \\
L_z &= M_L\hbar \\
S &= \sqrt{S(S+1)}\,\hbar \\
S_z &= M_s\hbar \\
J &= \sqrt{J(J+1)}\,\hbar \\
J_z &= M_J\hbar
\end{aligned} \tag{3.29}
$$

Both L and M_L are always integers or 0, while the other quantum numbers are half integer if an odd number of electrons is involved and integral or 0 if an even number of electrons is involved. When $L > S$, J can have $2S + 1$ values; When $L < S$, J can have $2L + 1$ values.

Interestingly, atomic nuclei also have intrinsic momenta and magnetic moments, and these contribute to the total atomic angular momenta and magnetic moments. Such contributions are small because nuclear magnetic moments are $\sim 10^{-3}$ the magnitude of electronic moments. They lead to the *hyperfine structure* of spectral lines with typical spacings between components of $\sim 10^{-3}$ nm as compared with typical fine structure spacings a hundred times greater.

L-S coupling applies to a large number of elements including all the lighter ones. This coupling scheme is the basis of the usual spectroscopic notation as it applies to a large number of elements.

***J-J* coupling**

In case of heavier elements, particularly in excited states, the coupling between the various *s*-vectors and *l* vectors of individual electrons become weak and a tendency to form a *J*-vector by combining the *l* and *s* for each separate electron is markedly developed. In the extreme case, *l* and *s* for each electron combine to form a separate *J* and all the *J*-vectors are then vectorially added to form *J*. Thus for each active electron $J_1 = l_1 + s_1$; $J_2 = l_2 + s_2$, $J_3 = l_3 + s_3$, ... etc. and $J = J_1 + J_2 + J_3 + \ldots = \sum J_i$. This coupling scheme is known as *J-J* coupling which is distinctly different from *L-S* coupling scheme.

Pure *J-J* coupling is seldom found and in most of the known cases *L-S* coupling is effective. However, there are many heavy elements that have spectra that require for their interpretation a coupling intermediate between the *L-S* and *J-J*.

3.13 SELECTION RULES

There are number of energy states in an atom. It is observed that all the transitions between different energy states as spectral lines do not appear. The spectral lines in the spectrum are governed by certain principles known as *shape selection rules*. There are three selection rules as follows:

(i) **Selection rule for L:** Only those lines observed for which the value of L changes by ± 1, i.e., $\Delta L = \pm 1$.

(ii) **Selection rule for *J*:** Only those spectral lines are observed when the transitions take place between two energy states for which $\Delta J = \pm 1$ or 0. We may note that $0 \rightarrow 0$ transition is not permitted.

(iii) **Selection rule for *S*:** Only those spectral lines are observed for which the value of S changes by 0, i.e., $\Delta S = 0$.

In the presence of magnetic field, it is observed that the orbital magnetic quantum number m_l and spin magnetic quantum number m_s play very important role in the transition. The selection rules for m_l and m_s are:

$$\Delta m_l = 0 \text{ or } \pm 1 \text{ and } \Delta m_s = 0$$

and in consequence,

$$\Delta m_J = 0 \text{ or } \pm 1$$

3.14 SPECTRAL INTENSITY RULES

The spectral intensity rules are as follows:

(i) The intensity of transition (i.e., a spectral line) is strong for which L and J change in the same sense.

(ii) The intensity of the transition (i.e., a spectral line) is weak for which L and J change in opposite senese.

(iii) The intensity of a spectral line is strong when the transition is in the decreasing sense, i.e., from $L \rightarrow (L-1)$.

(iv) The intensity of a spectral line is weak when the transition is in the increasing sense, i.e., from $L \rightarrow (L+1)$.

(v) The oppositely directed spectral transitions do not occur.

One can summarize the above rules as follows:

$\Delta L = -1,$	$\Delta J = -1$	most intense line
$\Delta L = -1,$	$\Delta J = 0$	less intense line
$\Delta L = +1,$	$\Delta J = 1$	weaker line
$\Delta L = +1,$	$\Delta J = 0$	weakest line
$\Delta L = -1,$	$\Delta J = +1$	no spectral transition
$\Delta L = +1,$	$\Delta J = -1$	no spectral transition

3.15 PAULI EXCLUSION PRINCIPLE

Wolfgang Pauli in 1925, on empirical considerations, enunciated from a study of the experimental data on the atomic spectra of helium and other similar atoms, his famous exclusion principle that governs the way the electrons and atoms are distributed among the various shells and subshells and determine the maximum number of electrons in a given energy level. It reads as follows:

No two electrons in a single atom can have the same set of quantum numbers n, l, m_l, m_s, i.e., *the same quantum state*. Alternatively, *no more than one electron can have a given set of quantum numbers*.

Pauli principle is the most important rule governing the structure of atoms and no study of the properties of atoms can be attempted without a thorough understanding of Pauli principle.

Let us consider the example of helium ($Z = 2$). The first electron in helium in the 1*s* ground state, has quantum numbers $n = 1$, $l = 0$, $m_l = 0$, $m_s = +1/2$ or $-1/2$. The second electron can have the same *n*, *l* and m_l but it cannot have the same m_s, since then Pauli exclusion principle would be violated. Thus if the first 1*s* electron must have $m_s = -1/2$. Now let us consider the case of lithium ($Z = 3$). Just as with helium, the first two electrons will have quantum numbers $(n, l, m_l, m_s) = (1, 0, 0, +1/2)$ and $(1, 0, 0, -1/2)$. Since the third electron, according to the Pauli principle, cannot have the same set of quantum numbers as the first two, it cannot go into the $n = 1$ level, since there are only two different set of quantum numbers available in the $n = 1$ level, and both of those sets have already been used. The third electron must therefore go into one of the $n = 2$ levels; experiments indicate that the next level available of the two $n = 2$ levels (2*s* or 2*p*) is the 2*s* level so that the third electron might have quantum numbers $(n, l, m_l, m_s) = (2, 0, 0, +1/2)$ or $(2, 0, 0, -1/2)$. The fourth electron, in the case of beryllium ($Z = 4$), would have the same *n*, *l*, and m_l, but the opposite m_s value as the third electron. When one reaches to boron, with $Z = 5$, the fifth electron cannot go into the 2*s* state, since we have already assigned both possible sets of quantum numbers in that level; the fifth electron then goes into one of the 2*p* levels. One might therefore expect that the properties of boron, with a 2*p* electron, would be different from the properties of lithium or beryllium, which have only 2*s* electrons.

It is this process of first using up all of the possible quantum numbers for one level, and then placing electrons in the next level, which accounts for the variations in the chemical and physical properties of the elements.

We may note that Pauli principle enunciates the *indistinguishability* or the equivaence of electrons. This is why, sometimes it is also called as the *equivalence principle*. Pauli's principle is applicable to normal as well as excited states of the atom and has been extremely useful in explaining the atomic and spectral phenomena, particularly the periodicity of the elements in the periodic table of elements. One can say that Pauli exclusion principle is one of the basic pillars of atomic and quantum physics.

3.16 ELECTRON CONFIGURATION OF ATOMS

We have read that a nucleus has a positive charge of *Z* units and around the nucleus *Z*-electrons are so arranged that the atom as a whole is neutral. A system of particles is said to be stable when its total energy is a minimum. This means that the various electrons in a complex atom should occupy the lowest quantum state. However, this is prevented by the Pauli's exclusion principle, which guides the arrangement of the extra-nuclear electrons in an atom. In the light of these two basic rules: (i) principle of minimum energy and (ii) Pauli's exclusion principle, we shall determine the electron configuration. On the basis of these principles one finds:

(i) The total number of electrons distributed about the nucleus on an electrically neutral atom is equal to the atomic number of atom.

(ii) The total number of electrons that can be accommodated with principal quantum number n is $2n^2$. Electrons having the same total quantum number n are said to belong to the same shell. They are on the average, roughly at the same distance from the atomic nucleus and therefore interact virtually with the same electric field and have similar energies.

(iii) A main energy level is subdivided into subshells. For a main energy level corresponding to principal quantum number n, there are n subshells. Notations for sub-shells are given in Table 3.3.

(iv) Each subshell can have a maximum capacity of $2(2l + 1)$ electrons, i.e., the maximum number of electrons in *s, p, d, f, g*, ... sub-shells are 2, 6, 10, 14, 18, ... respectively (*see* Table 3.3). The energy of an electron in a particular shell depends to some extent on l-value although the dependence is not so great as on n-value. We may note that electrons occupy the lowest energy states available. The electrons in each shell increase, in general, in energy with increasing l. However, in a subshell the electrons have almost identical energies, since the dependence of electron-energy on m_l and m_s is minor.

The filling of various subshells is done in the following ways:

(i) Electrons are added one by one to various shells as one moves from one atom to the next in the order of increasing atomic numbers.

(ii) Based on *Aufbau principle*, the electron enter the various subshells in the order of increasing energy. In accordance with this principle, the subshell of lowest energy is filled up first while the subshell of higher energy is filled up later. Obviously, *the system of particles is stable corresponding to the condition that the total energy of the system is minimum.* In increasing order of filling of various subshells in increasing order of energy is shown in Fig. (3.20) and the sequence of quantum states in an atom is shown in Fig. (3.21).

(iii) In accordance with Pauli's exclusion principle, an orbital cannot accommodate more than two electrons.

(iv) In general, the electrons in a subshell remain unpaired, i.e., have parallel spins, whenever possible. Table 3.4. In other words, according to Hund's rule no electron

Table 3.3 Notation for subshells

• Principal (Total) Quantum number (n)	1	2	3	4	5	---
• Orbital Quantum number (l)	0	1	2	3	4	---
• Notation for subshell	*s*	*p*	*d*	*f*	*g*	---
• Number of electrons in a sub-shell, i.e., capacity: $2(2l + 1)$	2	6	10	14	18	---

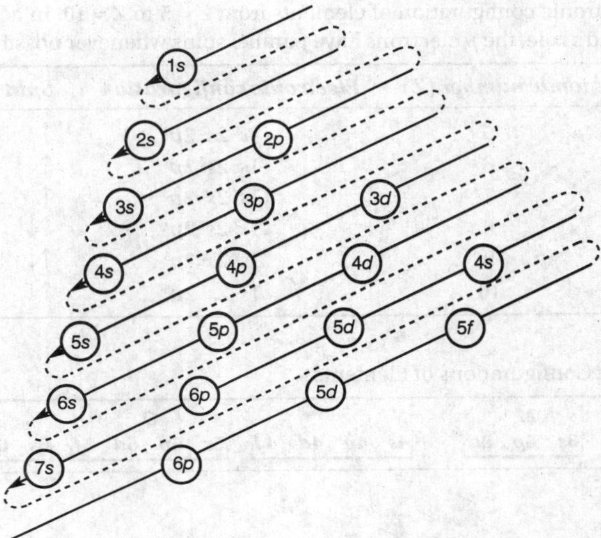

Fig. 3.20 Order of filling of various subshells

Fig. 3.21 The sequence of quantum states in an atom not to scale

pairing takes place in *p, d* and *f* orbitals until all the orbitals of the given subshells contain one electron each parallel spins.

(v) **Fully filled and half filled orbitals:** These orbitals have extra stability, i.e., these orbitals have lower energy and hence electrons prefer to enter those subshells, which are fully filled or half filled. This means orbitals with p^3, p^6, d^5, d^{10}, f^7 and f^{14} configuration are more stable.

Electronic configuration is shown in Table (3.5). We may note from the table that the hydrogen atom has only one electron and this electron goes in the orbital with lowest energy,

Table 3.4 Electronic configuration of elements from $Z = 5$ to $Z = 10$. In accordance with Hund's rule, the p electrons have parallel spins whenever possible

Elements	*Atomic number (Z)*	*Electronic configuration*	*Spin of p electrons*		
Boron	5	$1s^2\,2s^2\,2p^1$	↑		
Carbon	6	$1s^2\,2s^2\,2p^2$	↑	↑	
Nitrogen	7	$1s^2\,2s^2\,2p^3$	↑	↑	↑
Oxygen	8	$1s^2\,2s^2\,2p^4$	↑↓	↑	↑
Fluorine	9	$1s^2\,2s^2\,2p^5$	↑↓	↑↓	↑
Neon	10	$1s^2\,2s^2\,2p^6$	↑↓	↑↓	↑↓

Table 3.5 Electronic Configurations of Elements

	K	*L*		*M*			*N*				*O*				*P*			*Q*	
	1s	**2s**	**2p**	**3s**	**3p**	**3d**	**4s**	**4p**	**4d**	**4f**	**5s**	**5p**	**5d**	**5f**	**6s**	**6p**	**6d**	**7s**	
1 H	1																		
2 He	2																		← Inert gas
3 Li	2	1																	← Alkali
4 Be	2	2																	metal
5 B	2	2	1																
6 C	2	2	2																
7 N	2	2	3																
8 O	2	2	4																
9 F	2	2	5																← Halogen
10 Ne	2	2	6																← Inert gas
11 Na	2	2	6	1															← Alkali metal
12 Mg	2	2	6	2															
13 Al	2	2	6	2	1														
14 Si	2	2	6	2	2														
15 P	2	2	6	2	3														
16 S	2	2	6	2	4														
17 Cl	2	2	6	2	5														← Halogen
18 Ar	2	2	6	2	6														← Inert gas
19 K	2	2	6	2	6		1												← Alkali
20 Ca	2	2	6	2	6		2												metal
21 Sc	2	2	6	2	6	1	2												
22 Ti	2	2	6	2	6	2	2												
23 V	2	2	6	2	6	3	2												
24 Cr	2	2	6	2	6	4	1												
25 Mn	2	2	6	2	6	5	2												Transition
26 Fe	2	2	6	2	6	6	2												elements
27 Co	2	2	6	2	6	7	2												
28 Ni	2	2	6	2	6	8	2												
29 Cu	2	2	6	2	6	9	1												
30 Zn	2	2	6	2	6	10	2												

(Continued)

Table 3.5 (*Continued*)

31 Ga	2	2	6	2	6	10	2	1							
32 Ge	2	2	6	2	6	10	2	2							
33 As	2	2	6	2	6	10	2	3							
34 Se	2	2	6	2	6	10	2	4							
35 Br	2	2	6	2	6	10	2	5 ← Halogen							
36 Kr	2	2	6	2	6	10	2	6 ← Inert gas							
37 Rb	2	2	6	2	6	10	2	6			1 ← Alkali metal				
38 Sr	2	2	6	2	6	10	2	6			2				Transition elements (38 Sr – 48 Cd)
39 Y	2	2	6	2	6	10	2	6	1		2				
40 Zr	2	2	6	2	6	10	2	6	2		2				
41 Nb	2	2	6	2	6	10	2	6	4		1				
42 Mo	2	2	6	2	6	10	2	6	5		1				
43 Tc	2	2	6	2	6	10	2	6	6		2				
44 Ru	2	2	6	2	6	10	2	6	7		1				
45 Rh	2	2	6	2	6	10	2	6	8		1				
46 Pd	2	2	6	2	6	10	2	6	10						
47 Ag	2	2	6	2	6	10	2	6	10		1				
48 Cd	2	2	6	2	6	10	2	6	10		2				
49 In	2	2	6	2	6	10	2	6	10		2	1			
50 Sn	2	2	6	2	6	10	2	6	10		2	2			
51 Sb	2	2	6	2	6	10	2	6	10		2	3			
52 Te	2	2	6	2	6	10	2	6	10		2	4			
53 I	2	2	6	2	6	10	2	6	10		2	5 ← Halogen			
54 Xe	2	2	6	2	6	10	2	6	10		2	6 ← Inert gas			
55 Cs	2	2	6	2	6	10	2	6	10		2	6		1 ← Alkali metal	
56 Ba	2	2	6	2	6	10	2	6	10		2	6		2	
57 La	2	2	6	2	6	10	2	6	10		2	6	1	2	
58 Ce	2	2	6	2	6	10	2	6	10	2	2	6		2	
59 Pr	2	2	6	2	6	10	2	6	10	3	2	6		2	Lanthanides (59 Pr – 70 Yb); Transition elements (59 Pr – 74 W)
60 Nd	2	2	6	2	6	10	2	6	10	4	2	6		2	
61 Pm	2	2	6	2	6	10	2	6	10	5	2	6		2	
62 Sm	2	2	6	2	6	10	2	6	10	6	2	6		2	
63 Eu	2	2	6	2	6	10	2	6	10	7	2	6		2	
64 Gd	2	2	6	2	6	10	2	6	10	8	2	6	1	2	
65 Tb	2	2	6	2	6	10	2	6	10	9	2	6		2	
66Dy	2	2	6	2	6	10	2	6	10	10	2	6		2	
67 Ho	2	2	6	2	6	10	2	6	10	11	2	6		2	
68 Er	2	2	6	2	6	10	2	6	10	12	2	6		2	
69 Tm	2	2	6	2	6	10	2	6	10	13	2	6		2	
70 Yb	2	2	6	2	6	10	2	6	10	14	2	6		2	
71 Lu	2	2	6	2	6	10	2	6	10	14	2	6	1	2	
72 Hf	2	2	6	2	6	10	2	6	10	14	2	6	2	2	
73 Ta	2	2	6	2	6	10	2	6	10	14	2	6	3	2	
74 W	2	2	6	2	6	10	2	6	10	14	2	6	4	2	

(*Continued*)

Table 3.5 (*Continued*)

	K	*L*		*M*			*N*				*O*				*P*			*Q*	
	1*s*	2*s*	2*p*	3*s*	3*p*	3*d*	4*s*	4*p*	4*d*	4*f*	5*s*	5*p*	5*d*	5*f*	6*s*	6*p*	6*d*	7*s*	
75 Re	2	2	6	2	6	10	2	6	10	14	2	6	5			2			
76 Os	2	2	6	2	6	10	2	6	10	14	2	6	6			2			
77 Ir	2	2	6	2	6	10	2	6	10	14	2	6	7			2			
78 Pt	2	2	6	2	6	10	2	6	10	14	2	6	9			1			
79 Au	2	2	6	2	6	10	2	6	10	14	2	6	10			1			
80 Hg	2	2	6	2	6	10	2	6	10	14	2	6	10			2			
81 Tl	2	2	6	2	6	10	2	6	10	14	2	6	10		2	1			
82 Pb	2	2	6	2	6	10	2	6	10	14	2	6	10		2	2			
83 Bi	2	2	6	2	6	10	2	6	10	14	2	6	10		2	3			
84 Po	2	2	6	2	6	10	2	6	10	14	2	6	10		2	4			
85 At	2	2	6	2	6	10	2	6	10	14	2	6	10		2	5 ← Halogen			
86 Rn	2	2	6	2	6	10	2	6	10	14	2	6	10		2	6 ← Inert gas			
87 Fr	2	2	6	2	6	10	2	6	10	14	2	6	10		2	6		1	← Alkali
88 Ra	2	2	6	2	6	10	2	6	10	14	2	6	10		2	6		2	metal
89 Ac	2	2	6	2	6	10	2	6	10	14	2	6	10		2	6	1	2	
90 Th	2	2	6	2	6	10	2	6	10	14	2	6	10		2	6	2	2	Actinides
91 Pa	2	2	6	2	6	10	2	6	10	14	2	6	10	2	2	6	1	2	
92 U	2	2	6	2	6	10	2	6	10	14	2	6	10	3	2	6	1	2	
93 Np	2	2	6	2	6	10	2	6	10	14	2	6	10	4	2	6	1	2	
94 Pu	2	2	6	2	6	10	2	6	10	14	2	6	10	5	2	6	1	2	
95 Am	2	2	6	2	6	10	2	6	10	14	2	6	10	6	2	6	1	2	
96 Cm	2	2	6	2	6	10	2	6	10	14	2	6	10	7	2	6	1	2	
97 Bk	2	2	6	2	6	10	2	6	10	14	2	6	10	8	2	6	1	2	
98 Cf	2	2	6	2	6	10	2	6	10	14	2	6	10	10	2	6		2	
99 Es	2	2	6	2	6	10	2	6	10	14	2	6	10	11	2	6		2	
100 Fm	2	2	6	2	6	10	2	6	10	14	2	6	10	12	2	6		2	
101 Md	2	2	6	2	6	10	2	6	10	14	2	6	10	13	2	6		2	
102 No	2	2	6	2	6	10	2	6	10	14	2	6	10	14	2	6		2	
103 Lr	2	2	6	2	6	10	2	6	10	14	2	6	10	14	2	6	1	2	

i.e., 1*s*. Hence the electronic configuration of hydrogen is $1s^1$. Helum has two electrons and second electron also occupies the 1*s* orbital. Its configuration is $1s^2$. The next element in the table is lithium and it has three elecrons. Since the third electron cannot be allowed in 1*s* orbital because of Pauli exclusion principle and hence this electron takes the next available choice, i.e., 2*s* orbital. Now, the electronic configuration of lithium is $1s^2 2s^1$. The 2*s* orbital can accomodate one more electron and hence the electronic configuration of beryllium –4($Z = 4$) is $1s^2 2s^2$.

The next added electron has the choice to occupy the 2*p* orbital. 2*p* orbital can accommodate 6 electrons and hence the next six elements in the Table 3.5, i.e., B, C, N, O, F and Ne have the electronic configurations $1s^2\ 2s^2\ 2p^1$, $1s^2\ 2s^2\ 2p^2$, $1s^2\ 2s^2\ 2p^3$, $1s^2\ 2s^2\ 2p^4$, $1s^2\ 2s^2\ 2p^5$, $1s^2\ 2s^2\ 2p^6$ respectively. We may note that the electronic configurations of the elements from sodium to argon follow exactly the same pattern as elements from lithium to neon. Obviously, with this the 3*p* orbitals are filled.

Now, the next added electron should occupy the $3d$ orbitals but as the energy of $4s$ orbital is less than the energy of $3d$ orbital and therefore the electron occupies the $4s$ orbital. Thus, we can write the electronic configurations of potassium and calcium as $1s^2 2s^2 2p^6 3s^2 3p^6 4s^1$ and $1s^2 2s^2 2p^6 3s^2 3p^6 4s^2$ respectively. Now the next added electron goes to $3d$ orbital. We may note that from Scandium (Sc) to Zinc(Zn), the $3d$ orbitals are progressively occupied. Now, comes the puzzling problem of electronic configurations of chromium (Cr) and copper (Cu), which have five and ten electrons in $3d$ orbitals rather than four and nine. This is due to the fact that fully filled and half filled orbitals have extra stability. This is why, Cr and Cu adopt the d^5 and d^{10} configuration in preference rather than d^4 and d^9 electronic configurations.

Similarly, one can consider the electronic configuration of other elements.

3.17 PERIODIC TABLE

Mendeleev in 1869 formulated the periodic law whose statement reads as *when the elements are listed in order of atomic weights, elements with similar chemical and physical properties recur at regular intervals.*

A *periodic table* is an arrangement of the elements according to atomic number (Z) in a series of rows such that elements with similar properties form vertical columns. Nowadays commonly used long form of the periodic table is shown in Table (3.6).

Elements with similar properties form the *groups* shown as vertical columns in Table (3.6). Group 1 consists of hydrogen ($Z = 1$) plus the alkali metals, which are all soft elements, have low melting points and are very active chemically, e.g., Li, Na and K. Although, hydrogen is physically a non-metal behaves chemically much like an active metal. Group 7 consists of halogens, volatile in the gaseous state. Like the alkali metals, the halogens are chemically active elements, but as oxidizing agents rather than as reducing agents, e.g., fluorine, chlorine, bromine and iodine; fluorine is so active that it can corrode platinum. Group 8 consists of inert gases, of which, helium, neon and argon are examples. Inert gases are inactive chemically and virtually they do not form compounds with other elements and moreover their atoms do not join together into molecules.

In Table 3.6, the horizontal rows are called *periods*. We may note that the first three periods are broken in order to keep their members aligned with the most closely related elements of the long periods below. Most of the elements are metals. Across the period is a more or less steady transition from an active metal through less active metals and weakly active metals and weakly active nonmetals to highly active in the nonmetals and finally to an inert gas. Within each column there are also regular changes in properties, but they are far less conspicuous than those in each period, e.g., increasing atomic number in the alkali metals is accompanied by greater chemical activity, while the reverse is true in the halogens.

A series of *transition elements* appears in each period after the third between the Group 2 and Group 3 elements of the Periodic Table. The transition elements are metals and in general these are hard and brittle with high melting points, that have similar chemical behaviour. Fifteen of the transition elements in Period 6 are virtually indistinguishable in their properties and are known as *lanthanide* or *rare earths* elements. One finds another group of closely related metals in Period 7, known as *actinide elements*. One finds the following more or less steady transition along each period:

Table 3.6 The Periodic table of the elements

The number above the symbol of each element is its atomic number, and the number below its name is its average atomic mass. The elements whose atomic masses are given in parantheses do not occur in nature but have been created in nuclear reactions. The atomic mass in such a case is the mass number of the most long-lived radioisotope of the element. Elements with atomic numbers 110, 111, 112, 114, and 116 have also been created but not yet named

Group	1	2											3	4	5	6	7	8
Period								Transition metals										
1	1 H Hydrogen 1.008																	2 He Helium 4.003
2	3 Li Lithium 6.941	4 Be Beryllium 9.012											5 B Boron 10.81	5 C Carbon 12.01	7 N Nitrogen 14.01	8 O Oxygen 16.00	9 F Fluorine 19.00	10 Ne Neon 20.18
3	11 Na Sodium 22.99	12 Mg Magnesium 24.31											13 Al Aluminium 26.98	14 Si Silicon 28.90	15 P Phosphorus 30.97	16 S Sulfur 32.07	17 Cl Chlorine 35.45	18 Ar Argon 39.95
4	19 K Potassium 39.10	20 Ca Calcium 40.08	21 Sc Scandium 44.96	22 Ti Titanium 47.88	23 V Vanadium 50.94	24 Cr Chromium 52.00	25 Mn Manganese 54.94	26 Fe Iron 55.8	27 Co Cobalt 58.93	28 Ni Nickel 58.69	29 Cu Copper 63.55	30 Zn Zinc 65.39	31 Ga Gallium 69.72	32 Ge Germanium 72.59	33 As Arsenic 74.92	34 Se Selenium 78.96	35 Br Bromine 79.90	36 Kr Krypton 83.80
5	37 Rb Rubidium 85.47	38 Sr Strontium 87.62	39 Y Yttrium 88.91	40 Zr Zicronium 91.22	41 Nb Niobium 92.91	42 Mo Molybdenum 54.94	43 Tc Technetium (98)	44 Ru Ruthenium 101.1	45 Rh Rhodium 102.9	46 Pd Palladium 106.4	47 Ag Silver 107.9	48 Cd Cadmium 112.4	49 In Iridium 114.8	50 Sn Tin 118.7	51 Sb Antimony 121.9	52 Te Tellurium 127.6	53 I Iodine 126.9	54 Xe Xenon 131.8
6	55 Cs Cesium 132.9	56 Ba Barium 137.3		72 Hf Hafnium 178.5	73 Ta Tantalum 180.9	74 W Tungsten 183.9	75 Re Rhenium 186.2	76 Os Osmium 190.2	77 Ir Iridium 192.9	78 Pt Platinum 195.1	79 Au Gold 197.0	80 Hg Mercury 200.6	81 Tl Thallium 204.4	82 Pb Lead 207.2	83 Bi Bismuth 209.0	84 Po Polonium (209)	85 At Astatine (210)	86 Rn Radon (222)
7	87 Fr Francium (223)	88 Ra Radium 226.0		104 Rf Rutherfordium (261)	105 Db Dubnium (262)	106 Sg Seaborgium (263)	107 Ns Nielsbohrium (262)	108 Hs Hassium (264)	109 Mt Meitnerium (266)									
	Alkali metals																Halogens	Inert gases

Lanthanides (rare earths)

57 La Lanthanum 138.9	58 Ce Cerium 140.1	59 Pr Praseodymium 140.9	60 Nd Neodymium 144.2	61 Pm Promethium (145)	62 Sm Samarium 150.4	63 Eu Europium 152.0	64 Gd Gadolinium 157.3	65 Tb Terbium 158.9	66 Dy Dysprosium 162.5	67 Ho Holmium 184.9	68 Er Erbium 167.3	69 Tm Thulium 168.9	70 Yb Ytterbium 173.0	71 Lu Lutetium 175.0
89 Ac Actinium (227)	90 Th Thorium 232.0	91 Pa Protactinium 231.0	92 U Uranium 238.0	93 Np Neptounium (237)	94 Pu Plutonium (244)	95 Am Americium (243)	96 Cm Curium (247)	97 Bk Berkolium (247)	98 Cf Californium (251)	99 Es Einsteinium (252)	100 Fm Fermium (257)	101 Md Mendelevium (260)	102 No Nobelium (259)	103 Lr Lawrencium (262)

Actinides

Active metal → less active metal → weakly active non-metal → highly active non-metal → an inert gas.

Although there are regular changes in properties within each group, they are not so prominent as within each period. Thus, the alkali metals show greater chemical activity with increasing atomic number. The reverse is the case with halogen.

In each period after the third, there appears, between Group 2 and Group 3, a series of elements called as *transition elements*. Transition elements are metals having similar chemical properties, but they hardly show any resemblance with the elements in the major groups of periodic table. For example, there are 15 virtually indistinguishable transition elements in Period 6, called the lanthanides or rare earths and a similar group of 15 elements beyond uranium in Period 7, called the *actinides*.

For over a century the periodic law has been indispensable to chemists because it provides a framework for organizing their knowledge of elements. We may say that it is one of the triumphs of the quantum theory of atom that it enables one to account in a natural way for the periodic law without invoking any new assumptions.

One finds that the number of stable or nearly stable elements are into $Z = 102$. Elements beyond 102 are unstable. Nuclear theory predicts that some super heavy nuclei can be more nearly stable.

3.18 PROPERTIES OF THE ELEMENTS

Our study about elements reveals:

(i) Filled subshells are normally very stable configurations. An atom with one electron beyond a closed shell will readily give up that electron to another atom to form a chemical bond. Similarly, an atom lacking one electron from a closed shell will readily accept an additional electron from another atom in forming a chemical bond.

(ii) Filled subshells do not normally contribute to the chemical or physical properties of an atom. Only the electrons in the unfilled subshells need be considered. However, X-ray energies are an exception to this rule.

We now consider a number of different physical properties of the elements:

(i) **Atomic radii:** The radii of an atom is not a precisely defined quantity, since the electron probability density determines the "size" of an atom. The radii are similarly hard to define experimentally, and in fact different kinds of experiments may give different values for the radii. One way of defining the radius is by means of the spacing between the atoms in a crystal containing that element. The variation of typical atomic radii with Z is shown in Fig. (3.22).

(ii) **Ionization energy:** The minimum energy needed to remove an electron from an atom is called its *ionization energy*. Hydrogen has an ionization energy of 13.6 eV. Helium has an ionization energy of 24.6 eV for the first electron and 54.4 eV for the second electron. Variation of ionization energy for elements with atomic number Z is shown in Fig. (3.23).

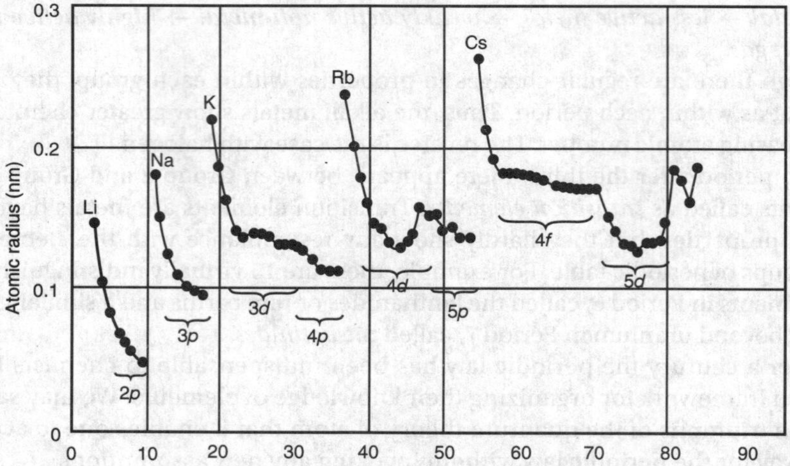

Fig. 3.22 Atomic radii of elements, determined from ionic crystal atomic separations. These radii are different from the mean radii of the *electron cloud* for free atoms

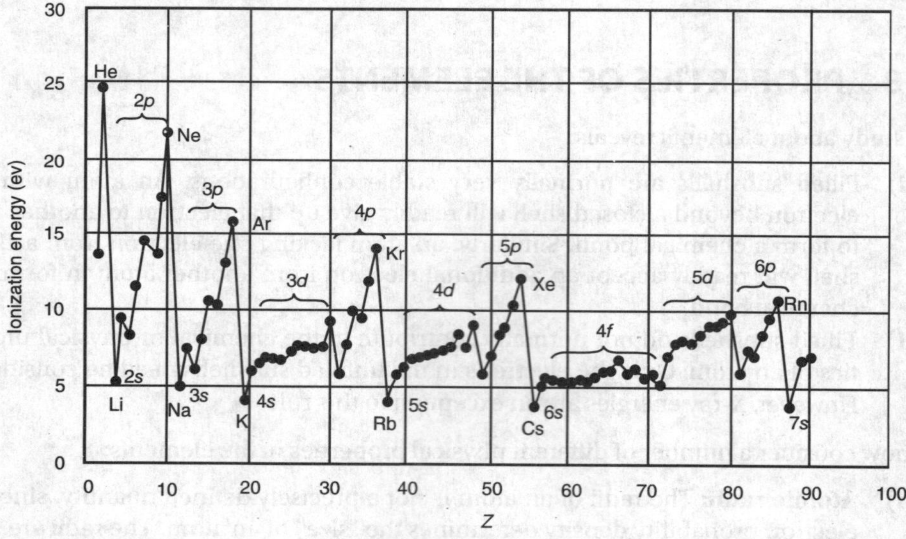

Fig. 3.23 Variation of ionization energies of neutral atoms of the elements with atomic number Z

(iii) Electrical resistivity (ρ): An electric current flows in bulk materials, when a potential difference (voltage) is applied across the material. The current I and voltage V are related as $V = IR$, where R is the electrical resistance of the material. If the material is uniform with length L and cross-sectional area A, then the resistance (R) is

$$R = \rho \frac{L}{A}$$

We may note that the resistivity ρ is characteristic of the kind of material and is measured in units of ohm-m. A good electrical conductor has a small resistivity, e.g., $\rho = 1.7 \times 10^{-8}$ Ω-m for copper. A poor conductor has a large resistivity, e.g., $\rho = 2 \times 10^{15}$ Ω-m for sulphur. From the atomic point of view, current depends on the movement of relatively loosely bound electrons, which can be removed from their atoms be applied electric potential difference, and also on the ability of the electrons to travel from one atom to another. Obviously, elements with *s* electrons, which are the least tightly bound and which also travel farthest from the nucleus, are expected to have small resitivities. On the basis of resistivity, materials are classified as *conductors, semiconductors and insulators.*

(iv) Magnetic susceptibility: When a material is placed in a magnetic field of intensity *B*, the material becomes "magnetized" and acquires a magnetization *M*, which for materials in proportional to *B*:

$$\mu_0 M = \chi B$$

where χ is a dimensionless constant called the *magnetic susceptibility*. Materials for which $\chi > 0$ are called *paramagnetic* and those for which $\chi < 0$ are called *diamagnetic.* Materials that remain permanently magnetized even when *B* is removed are known as *ferromagnetic* and χ is undefined for such materials.

From the atomic point of view, the magnetism of atoms depends on the *L* and *S* of the electrons in unfilled subshells, since the atomic magnetic moment μ_L and μ_s are proportional to *L* and *S*.

3.19 SPECTROSCOPIC NOTATION

We two often encounter with notations like 2 $D_{5/2}$, 2 $D_{7/2}$, 2 $D_{3/2}$, etc., in spectroscopy (atomic spectra). These are spectroscopic notations.

A spectroscopic notation, using the capital letter for the corresponding orbital quantum number *l*, i.e.,

$$\begin{array}{ccccccc} l & = & 0 & 1 & 2 & 3 & 4 \quad 5 \\ & & S & P & D & F & G \quad H \end{array}$$

specify the energy state of each electron. The concise short-hand notation of the form

$$n^{2n+1}D_J$$

designate the following:

n is the principal quantum number, 2*s* + 1 represents the multiplicity of the state, *D* means that $l = 2$ and $j = l \pm s$ is the total angular momentum quantum number.

Obviously, in spectroscopic notation, one can designate a 3*d*-electron either by

$$3^2 D_{5/2} \quad \text{or by} \quad 3^2 D_{3/2}$$

where $n = 3$, $l = 2$, $2s + 1 = 2\left(\frac{1}{2}\right) + 1 = 2$ and $j = 2 \pm \frac{1}{2} = 5/2, 3/2$.

In these spectroscopic notations, therefore the two most prominent lines of sodium (yellow lines or D-lines) represent the transitions:

$$3^2P_{3/2} \rightarrow 3^2S_{1/2} \quad (\lambda = 5889.9\text{Å})$$
$$3^2P_{1/2} \rightarrow 3^2S_{1/2} \quad (\lambda = 5895.9\text{Å})$$

Multiplicity (r): A superscript number before the letter (^2P, for instance) is used to indicate the *multiplicity* of the state, which is the number of different possible orientations of L and S and hence the number of different possible values of J. The multiplicity is equal to $2S + 1$ in the usual situation where $L > S$, since J ranges from $L + S$ to $L - S$. Thus when $S = 0$ the multiplicity is 1 (a singlet state) and $J = L$; When $S = 1/2$, the multiplicity is 2(doublet state) and $J = L \pm 1/2$; when $S = 1$, the multiplicity is 3 (a triplet state) and $J = L + 1$, L or $L - 1$; and so on. In a configuration in which $S > L$, the multiplicity is given by $2L + 1$. The total angular momentum quantum number J is used as a subscript after the letter, so that a $^2P_{3/2}$ state (read as doublet "P three halves") refers to an electronic configuration in which $S = \frac{1}{2}$, $L = 1$ and $J = 3/2$. These designations are called *term symbols.*

When the angular momentum of the atom arises from a single outer electron, the principal quantum number n of this electron is used as a prefix. Thus the ground state of the sodium atom is described by $3^2S_{1/2}$, since its electronic configuration has an electron with $n = 3$, $l = 0$ and $s = 1/2$ (and hence $j = 1/2$) outside closed $n = 1$ and $n = 2$ shells. For consistency it is conventional to denote the above state by $3^2P_{1/2}$ with the subscript 2 indicating a doublet, even though there is only a single possibility for J since $L = 0$.

Ground state of hydrogen and alkali atoms

The ground state configuration of valence electron is written as ns^1. Since the valence shell has only one electron with $l = 0$ and hence

$$L = l = 0, \quad S = s = 1/2$$

$\therefore$
$$J = L + S = 0 = \frac{1}{2} = \frac{1}{2}$$

Multiplicity
$$r = 2l + 1 = 1$$

Excited states of alkali atoms

Let the valence electron in an alkali atom in promoted s to p state then

$$L = l_1 = 1, \quad S = s_1 = 1/2$$

Multiplicity
$$r = s_1 = 1/2$$

or
$$r = 2S + 1 = 2$$

$$J = 1 \pm \frac{1}{2} = \frac{3}{2}, \frac{1}{2}$$

Spectroscopic symbol $\quad 2P_{1/2}, \; ^2P_{3/2}$

Now, if the valence electron is promoted to *d* state then

$$L = 2, S = 1/2, J = 2 \pm \frac{1}{2} = \frac{5}{2}, \frac{3}{2}$$

Multiplicity $r = 2S + 1 = 2$

Spectroscopic symbol ${}^2D_{3/2}, {}^2D_{/2}$

Obviously, each term (except ground state) split into two sets called multiplets. In the present case it is *doublet.*

3.20 HUND'S RULES (DETERMINATION OF GROUND STATE)

The rules for finding the ground state quantum number are known as Hund's rules.

(i) Of the terms belonging to a given electronic configuration, the term with the greatest possible *S* and greatest possible value of *L* at this *S* will have the lowest energy.

(ii) The multiplets formed by equivalent electrons are normal, i.e., the energy of the state grows with the increase in the value of *J* if not more than half of the subshell is filled.

(iii) The multiplets are inverted, i.e., the energy diminishes with an increase in *J* if more than half of the subshell is filled.

Obviously, when not more than half of a subshell is filled the component of the multiplet with $J = L - S$ has the lowest energy.

In the case of carbon the two 2*p* electrons can be combined to give $L = 0$, 1 or 2 and $S = 0$ or 1. The ground state of the carbon will be identified by only one particular choice of *L* or *S*. In accordance with Hund's rules, the maximum value M_s for carbon is +1, obtained when the two valence electrons both have $m_s = +1/2$. Thus $S = 1$ with only two electrons in the 2*p* shell, the Pauli principle places no restriction. The maximum value of m_l for the first electron is +1 and for the second electron in accordance with Pauli principle will be 0. Thus maximum value of $M_L = +1$ and $L = 1$. Clearly, ground state of carbon is characterized by $S = 1$ and $L = 1$.

In case of nitrogen (electronic configration $1s^2 2s^2 2p^3$), there are three electrons in the *p* subshell. With $m_s = +\frac{1}{2}$, we get $M_s = +3/2$ and therefore $S = 3/2$. Each of three electrons has quantum numbers $(2, 1, m_l, +\frac{1}{2})$. To maximize M_L, we assign the first electron the maximum value of m_l, namely +1. The maximum value of m_l left for the second electron is 0, and the third electron must therefore have $m_l = -1$. The total M_L is $1 + 0 + (-1) = 0$, so $L = 0$. Thus $L = 0$, $S = 3/2$ are the ground state quantum number for nitrogen.

3.21 APPLICATIONS OF VECTOR ATOM MODEL

3.21.1 Fine Structure of H_α Line

From the hydrogen spectrum, it is clear that H_α line results from the transition of electron from the energy level corresponding to $n = 3$ to $n = 2$. The entire state of the H-atom is determined by its singlet valence electron.

There are two s-sublevel ($l = 0$) and p-sub-level ($l = 1$) corresponding to $n = 2$. When electron is in s-sublevel ($l = 0$)

$$L = l = 0, S = s = 1/2 \quad \therefore \quad J = L \pm S = 1 \pm \frac{1}{2} = 1/2$$

This state is $^2S_{1/2}$.

Now, when the electron is in p-sublevel ($l = 1$), we have

$$L = l - 1, S = s = 1/2 \quad \therefore \quad J = L \pm S = 1 \pm \frac{1}{2} = \frac{1}{3}, \frac{1}{2}$$

This state is represented by $^2P_{3/2}$, $^2P_{1/2}$.

Corresponding to $n = 3$, there are three sublevels. When the electron is in s-sublevel ($l = 0$)

$$L = l = 0, S = s = 1/2 \quad \therefore J = L \pm S = 1 \pm \frac{1}{2} = \frac{3}{2}, \frac{1}{2}$$

The corresponding states are $^2P_{3/2}$ and $^2P_{1/2}$.

Fig. (3.24) shows the energy levels corresponding to $n = 3$ to $n = 2$. One can show that a state with lower value of J has smaller energy that a state with higher value of J. In all fifteen transitions are possible but selection rules forbid some of these transitions. Only those transitions are permitted in which L changes by ± 1 or J changes by 0 or ± 1, i.e.,

$$\Delta L = \pm 1, \Delta J = 0, \pm 1$$

Transitions permitted by the selection rules are only seven and they are as follows:

$$^2D_{5/2} \to {}^2P_{3/2}, \quad {}^2D_{3/2} \to {}^2P_{1/2}, \quad {}^2P_{1/2} \to {}^2S_{3/2}, \quad {}^2D_{3/2} \to {}^2P_{3/2}, \quad {}^2P_{3/2} \to {}^2S_{1/2}.$$

$$^2S_{1/2} \to {}^2P_{3/2}, \quad {}^2S_{1/2} \to {}^2P_{1/2}.$$

We may note that the states for which J values are the same but L values differ by unity are coincident. In the upper the coincident pair of levels are ($^2D_{3/2}$, $^2P_{3/2}$) and ($^2P_{1/2}$, $^2S_{1/2}$). The pair having the same energy in the lower state are ($^2P_{1/2}$, $^2S_{1/2}$). Obviously, the transitions 4, 4 and 5, 5 are identical. Thus the number of transitions now reduces to five only. Only doublets are observed experimentally. This may be due to *Doppler broadening*. The Doppler broadening arises due to thermal motion of atoms and five components merge together and therefore it becomes difficult to record all the five lines and only doublets are observed.

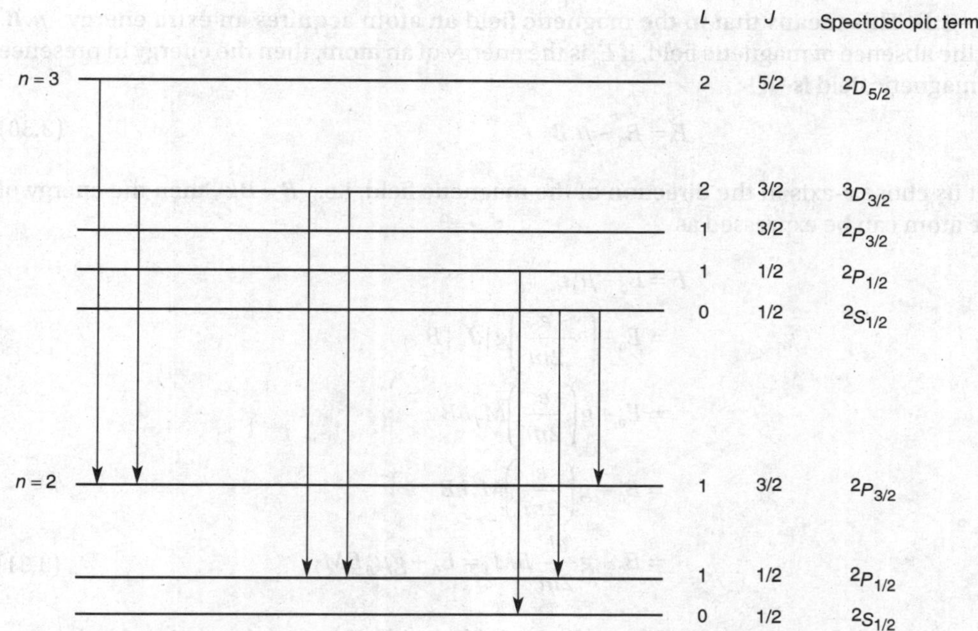

Fig. 3.24 Fine structure of H_α line

3.21.2 Fine Structure of Sodium *D* lines

The two *D*-lines of sodium in the spectrum result from the transitions of electron from 3*p* to 3*s* level. In the ground state of sodium atom the valence electron lies in the 3*s* energy level and in this state,

$$L = l = 0, S = s = 1/2 \quad \text{and} \quad J = L \pm S = 0 \pm \frac{1}{2} = 1/2$$

Obviously, the ground state is $^2S_{1/2}$. When the valence electron is excited to 3*p* level, we have

$$L = l = 1, S = s = 1/2 \quad \text{and} \quad J = L \pm S = 1 \pm \frac{1}{2} = \frac{3}{2}, \frac{1}{2}$$

This state is denoted by $^2P_{3/2}$, $^2P_{1/2}$. The energy level diagram of 3*s* and 3*p* levels are as shown in Fig. (3.4). Three transitions from the upper level to lower level are possible. However, the selection rules $\Delta L = \pm 1$, $\Delta J = 0, \pm 1$ allow only two transitions, i.e., one transition is forbidden. The D_1 line corresponding to $\lambda = 5896$Å originates from the transition $^2P_{1/2} \rightarrow {}^2S_{1/2}$ and D_2 line ($\lambda = 5890$Å) from the transition $^2P_{3/2} \rightarrow {}^2S_{1/2}$.

3.22 ENERGY OF AN ATOM IN MAGNETIC FIELD

When an atom possessing magnetic moment is placed in a uniform magnetic field, atom experiences a torque equal to $\mu \times \boldsymbol{B}$. Referring to a zero of potential energy when μ and $\boldsymbol{B}$ are perpendicular to each other, the potential energy in an arbitrary orientation is given by $-\mu.\boldsymbol{B}$. This means that in the magnetic field an atom acquires an extra energy $-\mu.\boldsymbol{B}$. In the absence of magnetic field, if E_0 is the energy of an atom, then the energy in presence of magnetic field is

$$E = E_0 - \mu.\boldsymbol{B} \tag{3.30}$$

Let us chose z-axis in the direction of the magnetic field, i.e., $\boldsymbol{B} = B\hat{k}$, then the energy of the atom can be expressed as

$$\begin{aligned} E &= E_0 - \mu_z B \\ &= E_0 - \left(-\frac{e}{2m}\right) g\,|J_z|\,B \\ &= E_0 + g\left(\frac{e}{2m}\right) M_J \hbar B \\ &= E_0 + g\left(\frac{e}{2m}\right) M_J \hbar B \\ &= E_0 + g\frac{e\hbar}{2m} BM_J = E_0 + g\mu_B BM_J \end{aligned} \tag{3.31}$$

where $M_J = J, J+1, \ldots 0, -1, -2, \ldots -(J-1), J$. Obviously, M_J can take on $2J+1$ values, an atomic energy level splits into $2J+1$ equally spaced sublevel Fig. (3.25). We may note that the splitting of an energy level results from the interaction of magnetic field with the magnetic moment of the atom. We have already read that an atomic level with $g = 0$ does not split at all, e.g., the state $^4D_{5/2}$ has $L = 0$, $S = 3/2$, $J = 1/2$ and $g = 0$. Similarly a state with $J = 0$ is called singlet as no splitting occurs, e.g., ^1S state.

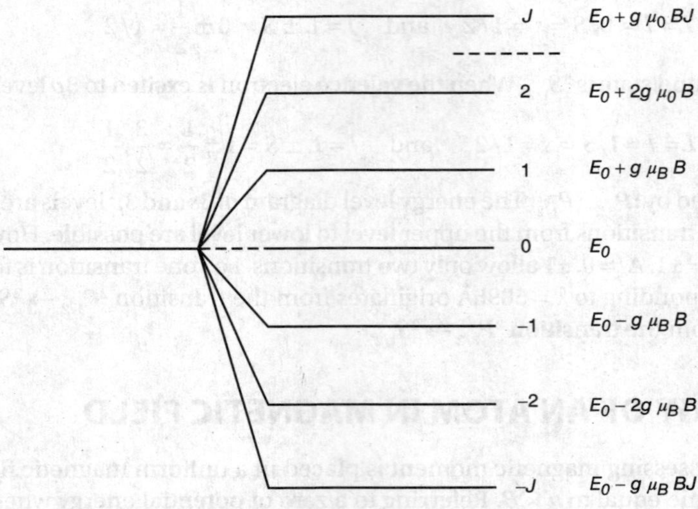

Fig. 3.25 Splitting of energy level of an atom when placed in a magnetic field

In magnetic field the separation between the two adjacent level is $g\mu_0 B$ and the total splitting is given by

$$E = 2g\mu_B BJ$$

Zeeman effect

Zeeman in 1896 discovered that when a source of light *S* emitting a line spectrum is placed in a strong magnetic field of flux density *B*, each spectral line undergoes the following striking modifications Fig. (3.26):

(i) When we examine the spectral line of frequency ν parallel to the field (longitudional view), we observe that each line appears to split into two, on the opposite sides of and equivalent from the original line. The light having frequency $\nu \pm \Delta\nu$ is *circularly polarized* in opposite sense, the higher frequency line is along the direction of the magnetising current.

(ii) When we observe *perpendicular* to the field (transverse view) *three* lines or *triplet* appear; the previous two of frequencies $\nu \pm \Delta\nu$ are now found to be accompanied by a line at the position of the original line of frequency ν. The side lines are found to be *plane polarised,* the vibrations being perpendicular to the magnetic field. The middle line is also *plane polarised* but with vibrations parallel to the magnetic field. This phenomenon is known as Zeeman effect. Lorentz also predicted the Zeeman effect from his electron theory. He showed that

$$\Delta\nu' = \frac{eB}{4\pi m}$$

where *e* and *m* are respectively the charge and mass of the electron. Soon after this prediction, Zeeman triplets were reported in the singlet of Cd and Zn. We may note that historically the first determination of *e/m* for electron was made

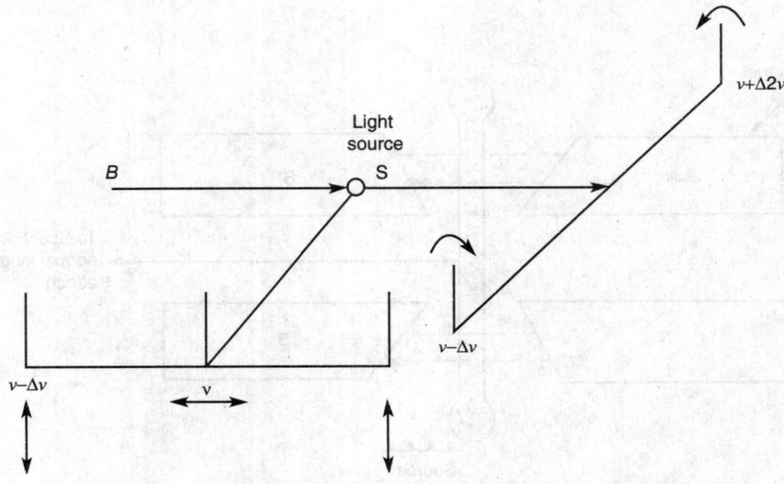

Fig. 3.26 Normal Zeeman effect: Longitudional and transverse view

from a knowledge of the Zeeman shift $\Delta\nu$ and the magnetic field B value of e/m so determined agreed well with the value published soon after by J. J. Thomson.

However, it is found that very very few spectral lines show normal triplets in accordance with Lorentz theory. When the magnetic field is comparatively weak, each line splits into more than three components, e.g., four, six, or even more and have shifts different from $\Delta\nu$ as calculated from Lorentz electron theory. This is known as *anomalous Zeeman* effect. The explanation of anomalous Zeeman effect had to wait till the quantum theory was developed and the idea of spinning electron was introduced. At this stage, we can easily say the anomalous Zeeman effect arises because the spin magnetic moment of the electron is $\frac{h}{2\pi}\left(\frac{e}{2m}\right)$ instead of $\frac{1}{2}\frac{h}{2\pi}\left(\frac{e}{2m}\right)$ and the normal Zeeman effect occurs when the net electron spin is zero.

Experimental arrangement

The experimental arrangement for observing normal Zeeman effect is as shown in Fig. (3.27). The source of light radiation such as sodium flame or a mercury arc is placed between the pole pieces of a powerful electromagnet. The electromagnet has conical pole-pieces and the holes are drilled along the length so that light radiation can pass through it and spectrum can be examined by a spectroscope of high resolving power along the direction of magnetic field. One may also observe spectral lines in a direction perpendicular to the magnetic field.

In the normal Zeeman effect, when the light radiation is viewed (through spectroscope) perpendicular to the direction of magnetic field (normal transverse Zeeman effect) three components are observed. One in the same position as the original line having original frequency ν_0 and the two components being on either side of the central line and equallyseparted from it having frequencies $\nu_1 = (\nu + \Delta\nu)\left(\text{where}\,\Delta\nu = \frac{eB}{4\pi m}\right)$ and $\nu_2 = (\nu - \Delta\nu)\left(\text{where}\,\Delta\nu = \frac{eB}{4\pi m}\right)$. The central component is linearly polarized having

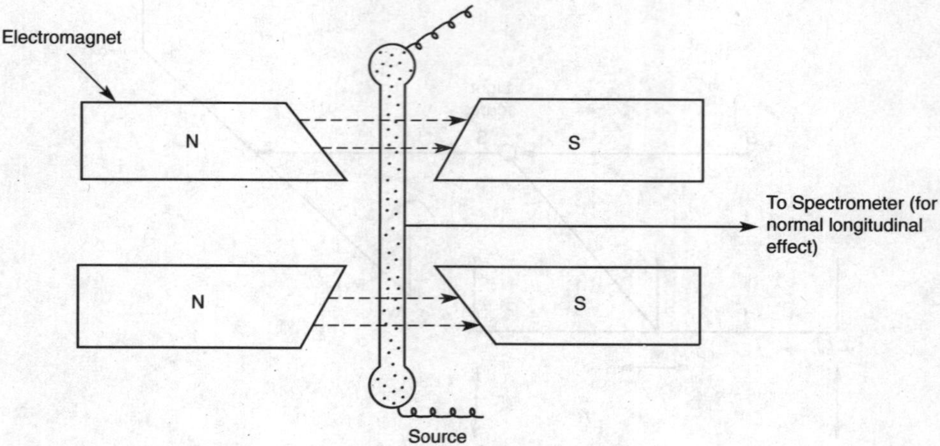

Fig. 3.27 Experimental arrangement for Zeeman effect

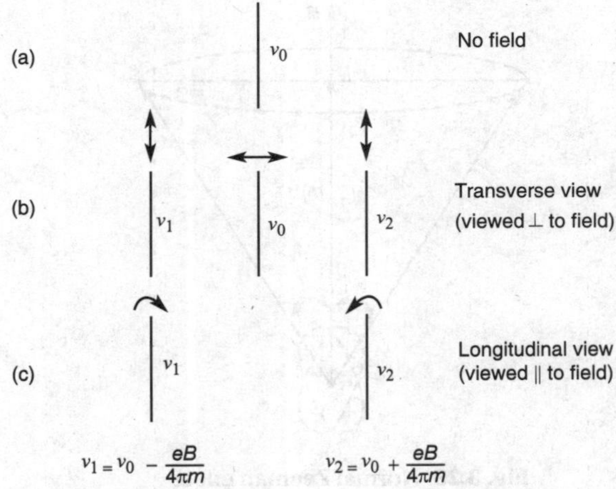

Fig. 3.28 Normal Zeeman effect

vibrations parallel to the magnetic field, while the two other component are linearly polarized with vibration at right angles to the magnetic field Fig. (3.28). This splitting is known as *normal transverse Zeeman effect.*

When the light is viewed (through the hole) in a direction parallel to the magnetic field, only the two outer components are seen and there is no central component. Fig. [3.28(c)]. These components are circularly polarized in opposite sents, i.e., one component being clockwise and the other anticlockwise. This is known as *normal longituditional zeeman effect.* We may note that for making longituditional observation possible, one of the pole pieces carries a hole drilled parallel to the magnetic field direction, to allow the light to pass through and fall on the spectrograph for photographing the Zeeman lines.

Theory of Normal Zeeman effect

One can explain the normal Zeeman effect on the basis of vector atom model. The space quantization of orbits provides a satisfactory explanation of this effect.

Let us consider an atom having an electron in its elliptical orbit. Here one can ignore the spin of the electron. The angular momentum of the electron p_l is given by $lh/2\pi$, where l takes integral values. The corresponding magnetic moment μ_l of the electron is given by $leh/4\pi m = (e/2m)\ p_l$. The vectors l and μ_l act along the same line, i.e., perpendicular to the orbital, but in opposite direction because the electron is negatively charged.

When the atom (which is a tiny magnet) is placed in a uniform external magnetic field B, there is no net translational force on the atom, but there is a torque which causes the electron orbit to rotate, or precess, about the direction of the magnetic field B. The precession results in the precession of elliptical orbit about an axis through the nucleus parallel to the magnetic field (Larmar theorem). This precession is represented in the Fig. (3.29) by the precession of angular momentum vector about B. The angular velocity ω_p of this precession is given by

$$\omega_p = \frac{eB}{2m} \tag{3.33}$$

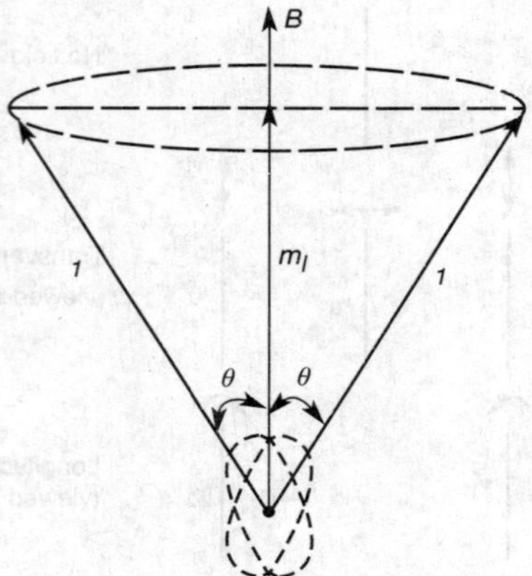

Fig. 3.29 Normal Zeeman effect

Due to the precession of elliptical orbit with angular velocity ω_p, 1 also precesses with same angular velocity ω_p. Due to this precession, the kinetic energy of the electron is increased, of course, the potential energy remains unaffected. Let ΔE be the increased energy. Now, we compute ΔE.

The work done in rotating a magnet of magnetic moment μ in a magnetic field B through an angle θ is $\mu\ B \cos\theta$. Therefore, the change in the energy of the atom due to its precession about the field is

$$\Delta E = -\,\mu.B = -\,\mu B\cos\theta \tag{3.34}$$

where θ is the orientation of the electron-orbit in the magnetic field. We know that this orientation is quantized (space quantisation) and θ can take only certain discrete values given by

$$\cos\theta = \frac{m_l}{l} \tag{3.35}$$

where m_l is the magnetic orbital quantum number and take only the values from $-l$ to l, i.e.,

$$m_l = -\,l, (-l+1), (-l+2), \ldots 0, 1, 2, \ldots, (l-1), l,$$

a total of $(2l + 1)$ possible values. Obviously, the change in energy is also quantized and is given by

$$\Delta E = -\mu B\cos\theta = -\mu B\frac{m_l}{l} \tag{3.36}$$

When the electron is moving in elliptical orbit, the magnetic moment μ of the electron is given by

$$\mu = l\frac{eh}{4\pi m} \tag{3.37}$$

Substituting this value of μ from Eq. (3.37) in Eq. (3.36), one obtains

$$\Delta E = -l\left(\frac{eh}{4\pi m}\right)B\frac{m_l}{l}$$

$$= -m_l \frac{eh}{4\pi m}B \quad (3.38)$$

Since m_l has $(2l+1)$ possible values, therefore the original energy level E splits into $(2l+1)$ energy levels, called *Zeeman levels*. Therefore the number of possible transitions between energy levels, and hence the number of spectral lines is increased.

Theory shows that only those transitions are allowed for which the quantum number m_l changes by 0 or ±1, i.e.,

$$\Delta m_l = 0, \pm 1$$

Therefore, from Eq. (3.38) the energy change due to the Zeeman splitting of levels is

$$\Delta E = 0, \pm\frac{eh}{4\pi m}B$$

Hence the change in the frequency of the emitted spectral line is given by

$$\Delta \nu = \frac{\Delta E}{h} = 0 \pm \frac{eB}{4\pi m} \quad (3.39)$$

Thus the original line of frequency ν_0 is splitted into three components of frequencies $\nu \pm \Delta\nu$, given by $\nu_0, \nu_1 = \nu_0 + \frac{eB}{4\pi m}$ and $\nu_2, \nu_1 = \nu_0 - \frac{eB}{4\pi m}$, i.e., one undisplaced component $\nu_0(\Delta m_l = 0)$, the two equally displaced components ν_1 $(\Delta m_l = 1)$ and $\nu_2(\Delta m_l = -1)$ on either side of the ν_0 component. We thus get the *normal zeeman effect* (Triplet). There are only three components because of the selection rule. The way by which a single line becomes a triplet on application of a magnetic field is shown in Fig. (3.30).

By measuring the separation $\Delta\nu$, the value of e/m can be obtained by expression Eq. (3.39). The value of e/m so obtained is found to be in very good agreement with that obtained by other methods.

When we consider the spin of the electron, the energy change of the electron due to the magnetic field is given by

$$\Delta E = -g\mu_B m_j B \quad (3.40)$$

where g is called *Lande splitting factor*, which depends on the values of l, s and j.

For $l = 0$, $g = g_s = 2$. The m_j is the Z-component of the total angular momentum having $2j + 1$ values from $-j$ to $+j$ in steps of 1. Obviously, in a fairly *weak* magnetic field, the lines should have a structure — *anomalous Zeeman effect* and an examination of the optical spectra emitted by atoms in a weak magnetic field shows that they do have a structure. We may note that this is consistent with the splitting of the energy levels in accordance with the above equation.

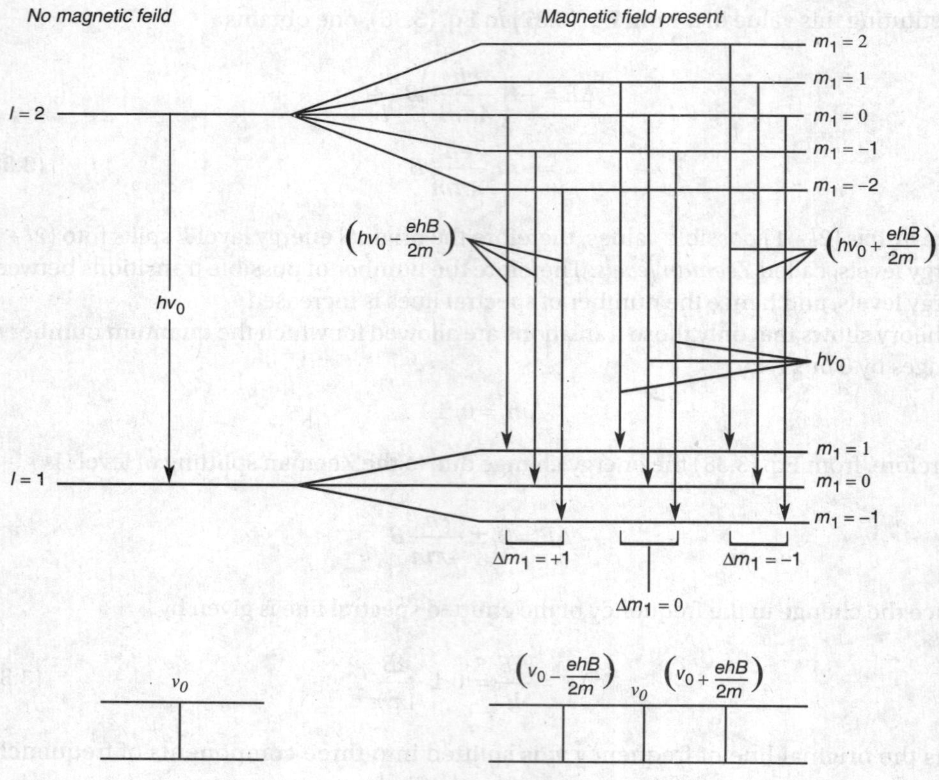

Fig. 3.30 Normal Zeeman effect. The splitting of a spectral line into three components when the radiating atoms are in a magnetic field of magnitude B

Measurements of observed Zeeman splitting yield values of the g-factor and therefrom it can also be deduced that the magnitude of J is indeed $\sqrt{j(j+1)}h/2\pi$ and similarly for L and S vectors.

3.23 DETERMINATION OF e/m FROM ZEEMAN EFFECT

The change in the frequency of a spectral line, when the light source is placed in a magnetic field B, is given by

$$\Delta\nu = \frac{eB}{4\pi m}$$

but $\quad \lambda = c/\nu \quad$ or $\quad \nu = c/\lambda$

$$\therefore \quad \Delta\nu = -\frac{c}{\lambda^2}\Delta\lambda$$

$$\therefore \quad \Delta\lambda = \frac{\lambda^2 \Delta\nu}{c} = \frac{\lambda^2}{c}\frac{eB}{4\pi m} = \frac{eB\lambda^2}{4\pi mc}$$

or
$$e/m = \frac{4\pi c}{B}\frac{d\lambda}{\lambda^2} \tag{3.41}$$

Obviously, by measuring the wavelength change e/m can be evaluated. The value of e/m obtained from Zeeman effect is in very good agreement with the value obtained by other methods.

Importance: Zeeman effect have made it an important phenomenon. This effect (i) furnishes a spectroscopic method for the determination of e/m of electron, (ii) presents the fine structure of spectral lines, (iii) vindicates the magnetic character of atom and confirms the electromagnetic theory of light.

3.24 ANOMALOUS ZEEMAN EFFECT (LANDGE'S G FACTOR)

When we place a source of radiation, giving line spectrum is placed in a magnetic field, the spectral lines are split into a number of component lines, symmetrically distributed about the original line. Doublets, triplets and even more complex systems are observed. This effect is known as *Zeeman effect.* When the magnetic field is *comparatively weak,* each line splits into more than three components. Fig. (3.31) shows the normal and anomalous Zeeman effects in various spectral lines.

One can explain the anomalous Zeeman effect on the basis of vector atom model by considering the *electron spin* in addition to the *orbital motion* of the electron. The total angular momentum $\boldsymbol{J}$ of the atom is now the sum of orbital and spin angular momentum $\boldsymbol{L}$ and $\boldsymbol{S}$, i.e.,

$$J = L + S \tag{3.42}$$

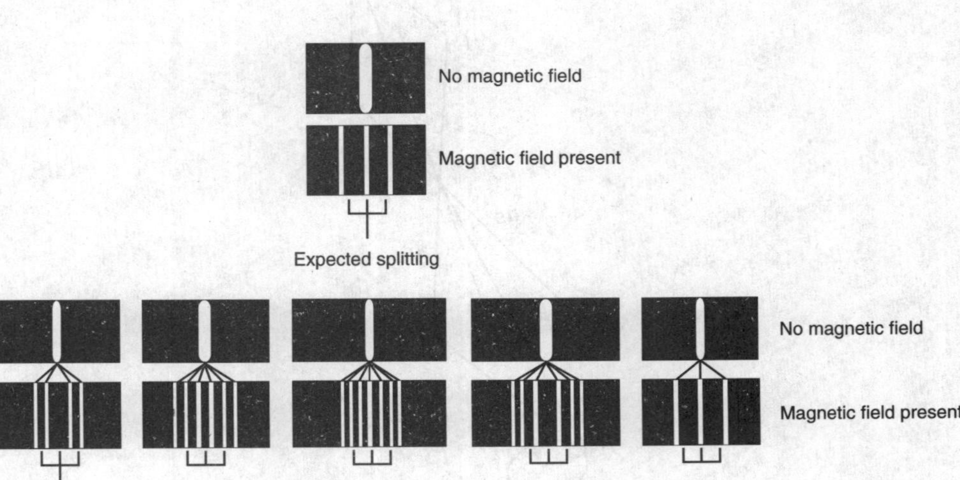

Fig. 3.31 The normal and anomalous Zeeman effect in various spectral lines

The magnetic moment due to the orbital motion μ_L is given by

$$\mu_L = L\frac{eh}{4\pi m} = L\frac{h}{2\pi}\frac{e}{2m} = p_l\frac{e}{2m} \tag{3.43}$$

Due to the negative of electron, μ_l is directed opposite to the direction of L.

The magnetic moment due to spin of electron, μ_s is given by

$$\mu_S = S\frac{eh}{4\pi m} = S\frac{h}{2\pi}\frac{e}{2m} = p_s\frac{e}{2m} \tag{3.44}$$

Due to the negative charge of the electron, μ_s is also directed to the direction of S.

Now, the total magnetic moment μ of the atom is the vector sum of μ_L and μ_s, i.e.,

$$\mu = \mu_L + \mu_S \tag{3.45}$$

From Eqs. (3.45) and (3.44) it is clear that the ratios of μ_l/p_l and $\frac{\mu_s}{p_s}$ are not equal. This means that the direction of μ does not coincide with the direction of J. Fig. (3.32) shows the relationships between magnetic moments and angular moments. Clearly, any vector not in the direction of J will precess around it and therefore vector μ will also precess about the same direction. Let us now resolve vector μ in two components, one parallel to J, i.e., $\mu_{||}J$, i.e., $\mu_{||}$ and other perpendicular to it i.e., $\mu_\perp$. If the precession of μ about J occurs sufficiently rapidly, then the value of perpendicular component $\mu_\perp$ (which would be constantly changing in direction), averaged over the period of the motion will be zero, i.e., $\mu_\perp = 0$. But the component of μ_J of μ along J, i.e., μ_J is equal to the sum of the components of μ_L and μ_s along J. Thus we have

$$\mu_J = \mu_L\cos(L,J) + \mu_S\cos(S,J) \tag{3.46}$$

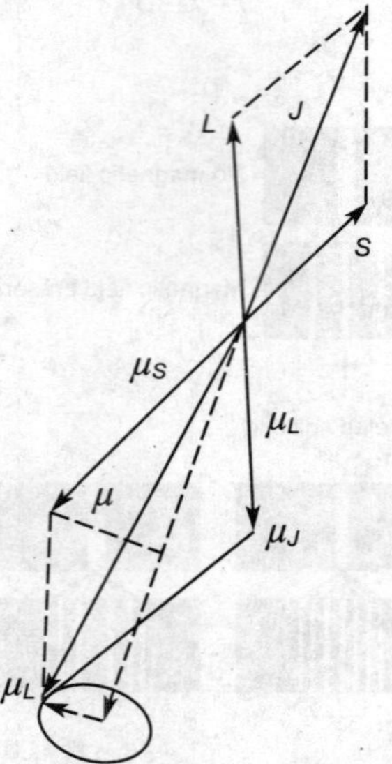

Fig. 3.32 Relationship between magnetic moments and angular moments and origin of Landge's G-factor

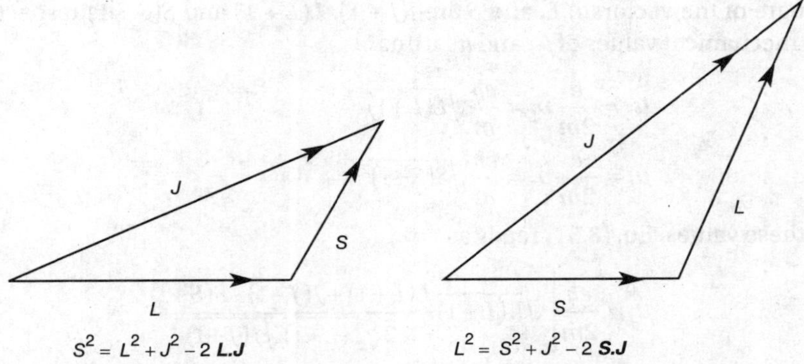

Fig. 3.33 Orientations of vectors ***L*** and ***S*** with respect to *J*

where cos(*L*, *J*) and cos(*S*, *J*) represent the cosine of the angle between *L* and *J* and cosine of angle between *S* and *J* respectively. Substituting the values of μ_L and μ_S from Eqs. (3.43) and (3.44) in Eq. (3.46), one obtains,

$$\mu_j = \frac{eh}{4\pi m}[L\cos(L,J) + 2S\cos(S,J)] \tag{3.47}$$

From Fig. (3.33), we have

$$S^2 = L^2 + J^2 - 2L.J,\ L^2 = S^2 + J^2 - 2S.J \tag{3.48}$$

$$\therefore \quad \cos(L,J) = \frac{L^2+J^2-S^2}{2LJ} = \frac{L(L+1)+J(J+1)-S(S+1)}{2\sqrt{L(L+1)}\sqrt{J(J+1)}} \tag{3.49}$$

$$\cos(S,J) = \frac{S^2+J^2-L^2}{2SJ} = \frac{S(S+1)+J(J+1)-L(L+1)}{2\sqrt{S(S+1)}\sqrt{J(J+1)}} \tag{3.50}$$

Since the square of the vectors *J*, *L* and *S* are *J*(*J* + 1), *L*(*L* + 1) and *S*(*S* + 1).

Substituting Eqs. (3.50) and (3.49) in Eq. (3.47), one obtains

$$\mu_J = \frac{eh}{4\pi m}\left[\frac{L^2+J^2-S^2}{2J} + \frac{S^2+J^2-L^2}{J}\right]$$

$$= \frac{eh}{4\pi m}\left[\frac{3J^2+S^2-L^2}{2J}\right] = \frac{ehJ}{4\pi m}\left[1+\frac{J^2+S^2-L^2}{2J^2}\right] \tag{3.51}$$

Eq. (3.51) shows the *inner precession*, i.e., it reveals what happens in the interior of the atom in the absence of any external magnetic field. When we apply an external magnetic field of flux density *B* to an atom, it precesses about the direction of applied field. This precession is called as *outer precession*. If the external magnetic field is small compared with the internal field, the coupling between *L* and *S* remains intact. In the present case we combine *L* and *S* to form *J* an then we consider the precession of *J* about *B*. On the other hand, if the external field is large compared with the internal, then the coupling between *L* and *S* is loosened or even broken. In the present case, we consider the precession of *L* and *S* about *B* independently.

The square of the vectors *J*, *L*, and *S* are *J*(*J* + 1), *L*(*L* + 1) and *S*(*S* + 1) respectively; the quantum mechanical values of μ_L and μ_s are as

$$\mu_L = \frac{e}{2m} p_L = \frac{e\hbar}{m}\sqrt{L(L+1)}$$

$$\mu_S = \frac{e}{2m} p_S = \frac{e\hbar}{m}\sqrt{S(S+1)}$$

Using all these values, Eq. (3.51) reads as

$$\mu_j = \frac{e\hbar}{2m}\left[\sqrt{L(L+1)}\frac{L(L+1)+J(J+1)-S(S+1)}{2\sqrt{L(L+1)}\sqrt{J(J+1)}} + 2\sqrt{S(S+1)}\frac{S(S+1)+J(J+1)-L(L+1)}{2\sqrt{S(S+1)}\sqrt{J(J+1)}}\right]$$

On simplification, one obtains

$$\mu_J = \frac{e\hbar}{2m}\sqrt{J(J+1)}\left[1+\frac{J(J+1)+S(S+1)-L(L+1)}{2J(J+1)}\right] \tag{3.52}$$

∴
$$\mu_J = \frac{e\hbar}{2m} g\sqrt{J(J+1)} \tag{3.53}$$

where
$$g = 1+\frac{J(J+1)+S(S+1)-L(L+1)}{2J(J+1)} \tag{3.54}$$

is well known *Lande g-factor* or *splitting factor*.

∴
$$\mu_J = \sqrt{J(J+1)}g\mu_B \tag{3.55}$$

Lande-g factor determines the splitting of energy levels in the presence of a weak external field. It is obvious from Eq. (3.54) that the splitting depends upon the values of *L*, *S* and *J*. μ_B in Eq. (3.55) is Bohr magneton.

The splitting of atomic energy levels in a magnetic field of flux density *B* is given by

$$\begin{aligned}\Delta E &= \mu_J B\cos(J,B)\\ &= \frac{e\hbar}{2m}\sqrt{J(J+1)}gB\cos(J,B)\\ &= g\mu_B B\sqrt{J(J+1)}\cos(J,B)\\ &= g\mu_B BM_J\end{aligned} \tag{3.56}$$

Since $\sqrt{J(J+1)}\cos(J,B)$ is the projection of vector *J* on the direction of magnetic field, i.e., component of *J* along magnetic field and hence this is denoted by M_J, which is magnetic quantum number.

Now, the value of *g* depends on the values of *L*, *S* and *J* and is thus different for different energy levels, which therefore would split differently in a magnetic field and M_J has (2*J* + 1) possible values. Obviously, there will be (2*J* + 1) different terms for a given value of *J*.

We have read that the transitions between the magnetically split levels are controlled by the following selection rules:

$$\Delta M_J = 0, \pm 1$$

There will be thus more lines in the hyper-fine structure of *anomalous Zeeman effect*.

Anomalous Zeeman patterns of sodium lines

We now consider the anomalous magnetic splitting with *D*-lines of sodium. Sodium belongs to one electron system. In the spectrum of sodium we have two line: D_1 line (λ = 5890Å) and D_2-line (λ = 5896Å). D_1 line arises due to the transition $3^2\,P_{1/2} \rightarrow 3^2\,S_{1/2}$, while D_2 line arises due to the transition $3^2\,P_{3/2} \rightarrow 3^2\,S_{1/2}$. When we apply a magnetic field of the order of 30,000 Oersteds and the spectrum is viewed through a high resolving power spectroscope, each line exhibits the anomalous Zeeman effect. D_1 line splits into four-components and D_2 line into 6 as shown in Fig. (3.34).
The splitting of energy levels are given by the relation:

$$\Delta E = \mu_B g\, B M_J$$

The quantity, $\mu_B\, B$ is known as *Lorentz unit* and it is customary to express ΔE in this unit. Thus we have

$$\Delta E = M_J g \text{ Lorentz unit}$$

Now, we shall express the $M_J\,g$ values for three levels $^2P_{3/2}$, $^2P_{1/2}$ and $^2S_{1/2}$. For example, for the level term $^2P_{3/2}$, we have $L = 1$, $S = 1/2$ and $J = 3/2$. The value of g is given by

$$g = 1 + \frac{J(J+1) + S(S+1) - L(L+1)}{2J(J+1)}$$

$$= 1 + \frac{(3/2)(5/2) + (1/2)(3/2) - (1)(2)}{2(3/2)(5/2)} = \frac{4}{3}$$

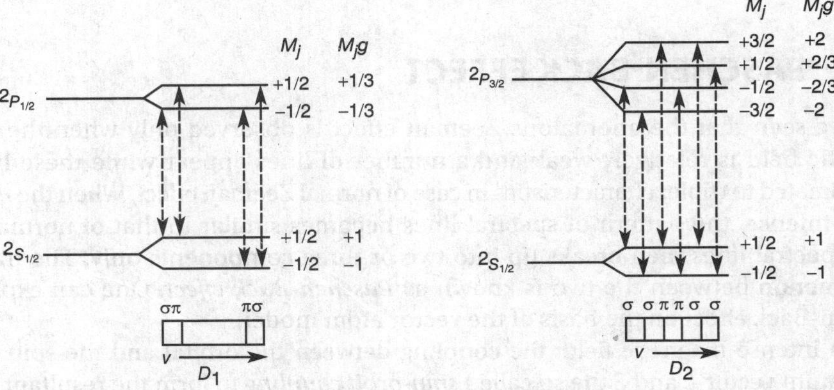

Fig. 3.34 Origin of anomalous Zeeman effect for sodium *D*-lines

Table 3.7

Term	*L*	*S*	*J*	*g*	M_J	$M_J g$
$^2P_{3/2}$	1	1/2	3/2	4/3	$\frac{3}{2},\frac{1}{2},\frac{-3}{2},-1/2$	$2,\frac{3}{2},-2/3,-2$
$^2P_{1/2}$	1	1/2	1/2	2/3	$\frac{1}{2},-1/2$	$\frac{1}{3},\frac{-1}{3}$
$^2S_{1/2}$	0	1/2	1/2	2	1/2, −1/2	1. −1

Now, M_J can have $(2J+1)$ values at unit intervals. Since $J = 3/2$, M_J has four values: 3/2, 1/2, −1/2 and −3/2. Similarly, one can obtain the values of g and M_J for other terms, e.g., $^2P_{1/2}$ and $^2S_{1/2}$. These values are shown in Table (3.7). The splitting of levels and the allowed transitions are as shown in Fig. (3.34). We may note that these are in accord with the observed splitting of D_1 and D_2 lines, as mentioned earlier.

Sodium D_1 line arises from the transition $^2P_{1/2} \rightarrow {}^2S_{1/2}$. $P_{1/2}$ level splits into two levels and $S_{1/2}$ also splits into two levels. Thus there will be four levels, i.e., four transitions possible. Now, applying the selection rules $\Delta M_J = 0$ or ± 1 all the four transitions are possible. Clearly, D_1 line splits into four components, on either side of the original line at distances +4/3, +2/3, −2/3, −4/3 in Lorentz units as shown in Fig. (3.34).

Sodium D_2 line arises from $^2P_{3/2} \rightarrow {}^2S_{1/2}$. $^2P_{3/2}$ level splits into two levels. Obviously, there will be six levels i.e., eight transitions are possible. Out of these possible eight transitions, two transitions are forbidden by the selection rules $\Delta M_J = 0$ or ± 1. Obviously, only six transitions are allowed. Clearly, D_2 line splits into six components symmetrically about the D_2 line at distance +5/3, +3/3, +1/3, −1/3, −3/3 and −5/3 as shown in Fig. (3.34).

We may note that the anomalous Zeeman effect is observed when the magnetic field is not very strong and in most cases, a spectral line splits into more than three components. However, for elements for which the ground states are $^1S_0$, the spectral line splits even in relatively weak magnetic field into three components only (Normal zeeman effect), for these cases $S = 0$ and $J = L$ so that lande g factor = 1 implying that the upper and the lower states have the same splittings, independent of L and J. Few examples are Mg, Ca, Sr, Cd, etc.

3.25 PASCHEN-BACK EFFECT

We have seen that the anomalous Zeeman effect is observed only when the applied magnetic field is relatively weak and a number of lines appear while these lines are approximated to triplet characteristics in case of normal Zeeman effect. When the *magnetic field* is intense, the pattern of spectral lines becomes similar to that of normal effect, i.e., a spectral lines then breaks up into two or three components only. The transition phenomenon between the two is known as *Paschen-Back effect.* One can explain the Paschen-Back effect on the basis of the vector atom model.

In an intense magnetic field, the coupling between the orbital and the spin angular momentum vector L and S the so called *spin-orbit coupling* to form the resultant angular momentum vector J gets broken or loosen. In this situation, one cannot replace the

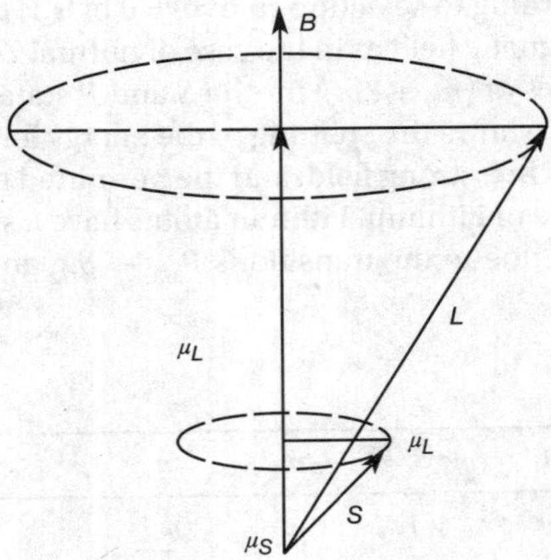

Fig. 3.35

magnetic moment μ of the atom by the parallel component μ_J because $\mu_\perp$ when averaged over a period will not be equal to zero. This is the region of Paschen-Back effect. If we increase the magnetic field further, the coupling between L and S becomes practically annulled and the orbital and spin vectors L and S independently precess about the magnetic field direction with different Larmor angular speeds due to the difference in the corresponding gyromagnetic ratio. These precessions will be faster than the precession of L and S round J so that angle (L, S) is no longer invariant. Now, there are $(2L + 1)$ possible components of L in the magnetic field direction and these are given by

$$m_L = L, L-1, L-2, \ldots, 1, 0, -1, \ldots -(L-1), -L$$

Similarly we have $(2S + 1)$ possible components of S in the magnetic field direction, given by

$$m_S = S, S-1, S-2, \ldots, 1, 0, -1, \ldots, -S$$

In the presence of magnetic field, the energy change ΔE will be sum of two parts, one arising from precession of L about B and the other from the precession of S about B. Hence we have

$$\begin{aligned}\Delta E &= (\Delta E)_L + (\Delta E)_S \\ &= \mu_L B\cos(L,B) + \mu_B B\cos(S,B) \\ &= \frac{eh}{4\pi m}B[L\cos(L,B) + 2S\cos(S,B)] \\ &= \frac{eh}{4\pi m}B[m_L + 2m_s] \\ &= \mu_B B[m_L + 2m_s] \qquad (3.57)\end{aligned}$$

where μ_B is the Bohr magneton and the factor of 2 in the second term is the value of g_s (gyromagnetic ratio) for the spin moment. The quantity $(m_L + 2m_s)$ is known as *strong field* quantum number. We know that m_L can take only integer values while m_s can have either integer or half integer values. Obviously, $(m_L + 2m_s)$ can have only integer values. Since S or m_s cannot change during the transition, the only change is due to m_L in the quantity $(m_L + 2m_s)$. According to selection rule $m_L = 0$ or ± 1, the given line split in three components in strong magnetic field as in the case of normal Zeeman effect.

Table 3.8 gives the values of $(m_L + 2m_s)$ for the S and P states of an atom with a single optical electron. Fig. (3.36) shows the splitting of the energy lervels $3P$ and $3S$, associated with the D-lines of sodium in a strong field, and the permitted transitions.

Let us consider an example of Lithium. Lithium atoms have a single optical electron. The fine structure is a doublet due to the transitions $P_{3/2} \rightarrow S_{1/2}$ and $P_{1/2}$ and $S_{1/2}$. In a strong

Table 3.8

State	***L***	***S***	m_L	m_s	$(m_L + 2m_s)$
S State	0	$\frac{1}{2}$	0	$+\frac{1}{2}$	+1
				$-\frac{1}{2}$	−1
P State ($P_{3/2}$, $P_{1/2}$)	1	$\frac{1}{2}$	+1	+1/2	+2
				−1/2	0
			0	+1/2	+1
				−1/2	−1
			−1	+1/2	0
				−1/2	−2

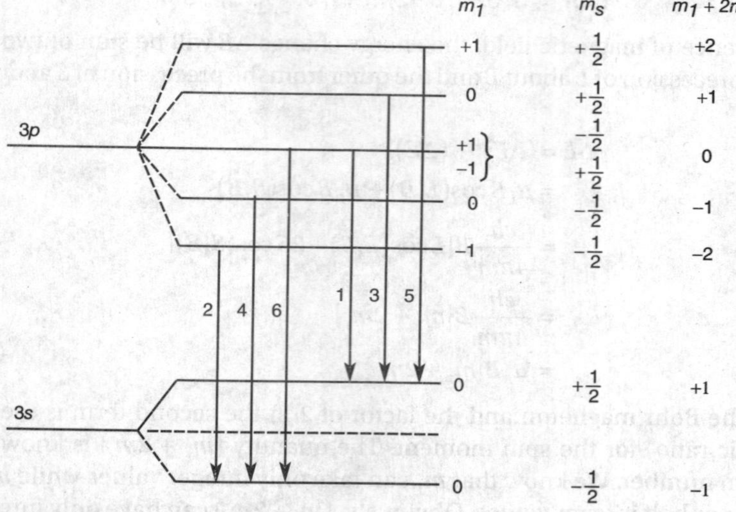

Fig. 3.36 Paschen-Back effect

magnetic field, the coupling between L and S is broken hence the two sub-states $P_{3/2}$ and $P_{1/2}$ have no definite meaning. According to strong field quantum number $(m_L + 2m_s)$, there will be five components of the upper P state and two components of the lower S state. Theoretically, ten transitions are possible between five components of P state and two components of S state. Selection rule $\Delta(m_L + 2m_s) = 0$ or ± 1 permits only six transitions Fig. (3.36). Out of these six permitted transitions (1, 2), (3, 4) and (5, 6) coincide because the values of $\Delta(m_L + 2m_s)$ for them are +1, 0, −1 respectively.

Summarising, we can say that the spinning electron with $s = 1/2$ and $g_s = 2$ thus enables one to explain the anomalous Zeeman effect and Paschen-Back effect.

We may note that there exists a *residual interaction* between the vectors L and S even in strong magnetic field and we have neglected it here. This residual interaction makes the splitting of lines more complicated.

3.26 STARK EFFECT

This is a similar effect to the Zeeman effect discovered by J.Stark in 1913. This effect occurs due to the influence of electric field, i.e., by the action of an electric field a spectral line gets split into a number of components.

When positive rays produced behind the perforated cathode in a discharge tube allowed to pass between two metal plates across which a strong electric field is maintained, emitted spectral lines that were split into *several* components, e.g., in the case of Balmer series of hydrogen each line is split into components whose *number* increases with frequency. Moreover, the *frequency shift*, i.e., the difference of frequency from that of the undistributed line, is *proportional to the electric field*. As in Zeeman effect, the component lines exhibit various degrees of polarization. We may note that the light emitted at right angles to the field, as in Zeeman effect, is plane polarised. The pattern of splitting in an intense field and that in a relatively weak field are quite different from each other and we can see that this bears a resemblance with the Zeeman effect and Paschen Back effect.

3.27 STERN-GERLACH EXPERIMENT (SPACE QUANTIZATION)

A direct and confirming evidence of the space quantization of angular momentum and hence of magnetic moment associated with electron spin and the way in which it orientates in a magnetic field came from the celebrated experiment conducted by Stern and Gerlach in 1922, which was originally devised to measure the magnetic moment of individual silver atoms.

Theory: When a magnetic dipole is placed in a *homogeneous* magnetic field, it will not be acted upon by any *resultant* force as the magnetic forces acting on the two poles of a magnet are equal and oppositely directed. Clearly, dipole will merely experience a torque that would tend it to align with the magnetic field direction about which it will process.

However, the situation changes in a *non-homogeneous* field, i.e., B-field is not homogeneous. Obviously, potential energy is no longer constant. In this situation, each pole of the dipole is subjected to a force of different magnitude. Dipole then experiences a net

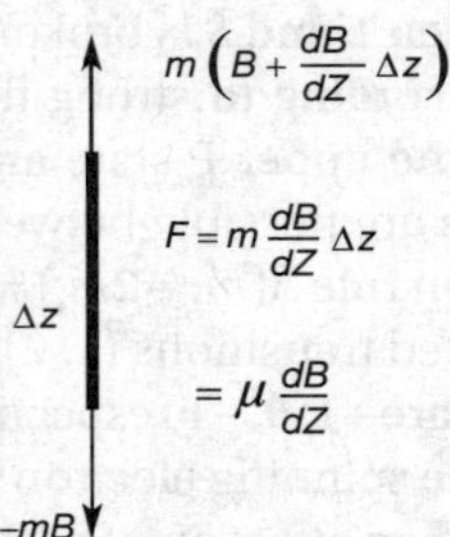

Fig. 3.37 Force on an atomic magnet in a non-homogeneous magnetic field

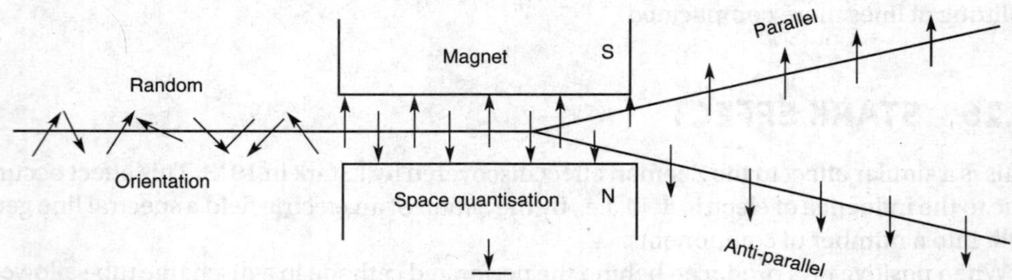

Fig. 3.38 Motion of atomic magnetic dipoles in a non-homogeneous field

resultant *translatory force* that varies with its orientation relative to the magnetic field. The components of the force will be $F_x = -\partial U/\partial x$, $F_y = -\partial U/\partial y$, $F_z = -\partial U/\partial z$. This causes dipole to move bodily with an acceleration in the direction of magnetic field (parallel) or opposite to it (anti-parallel). Fig. (3.37) and Fig. (3.38) illustrates this.

The magnet depicted in Fig. (3.39) gives rise to a magnetic field that has its largest component, the z-component of the field. Because of the slight bending of the field lines shown in Fig. (3.39), the z-component has a small z-dependence. Thus there is a force in the z-direction given by

$$F_z = -\frac{\partial U}{\partial z} = -\mu_z\frac{\partial B_z}{\partial z} \tag{3.58}$$

where B_z is the z-component of the field and μ_z the component of the magnetic moment μ in the Z-direction.

Classically, one finds that $\mu_z = \mu\cos\theta$, where θ is the angle between the vector μ and the Z-axis. Thus F_Z can have all values between $+(\partial B_z/\partial z)\,\mu$ and $-(\partial B_z/\partial z)\,\mu$. If then a magnetic dipole flies through a non-homogeneous magnetic field, it would be deflected being a function of the degree of inhomogeneity of the field. This reveals that classically a continuous range of deflection would be observed if a beam of dipoles i.e., a beam of atoms pass through the magnetic field, resulting in the broadening of the beam.

Now, if *quantization* is a reality, all setting of the magnetic moment relative to the field are not possible, but only a certain discrete component beams are permitted. What Stern and Gerlach found was that the beam was split into *discrete components*, each of which made a separate trace on the as shown in plate as shown in Fig. [3.40(c)].

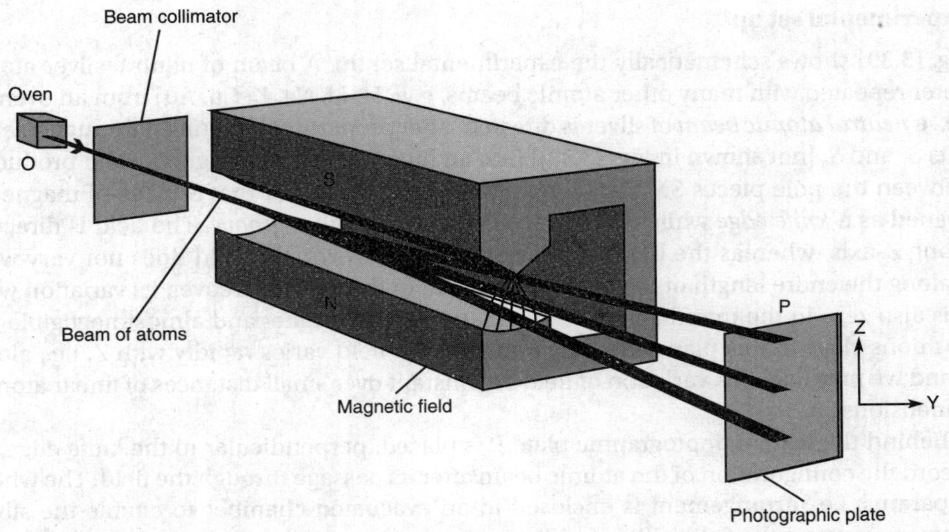

Fig. 3.39 Stern-Gerlach apparatus. The pole pieces of the magnet are shaped to give a magnetic field that is primarily oriented in the Z-direction, but varies for different values of Z

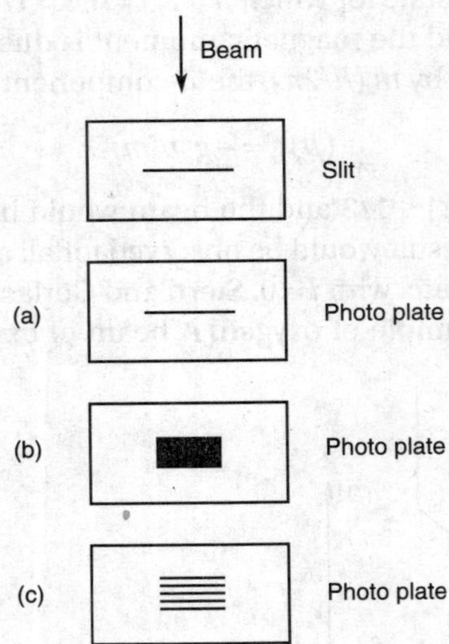

Fig. 3.40 A beam of atoms in the form of flat ray, broad in the X-direction, but narrow in the Z-direction passes through a Stern-Gerlach experimental set up. The inhomogeneity in the magnetic field deflects a magnetic dipole in the Z-direction by an amount proportional to L_z. (a) in the absence of magnetic field, the beam passes straight through without deflection, (b) With a magnetic field present the beam would be smeared in the Z-direction if the angular momentum of the atoms took on classically (i.e., arbitrary) values, (c) Since the possible values of L_z are quantized and therefore we picture five allowed values of L_z

Experimental set up

Fig. (3.39) shows schematically the experimental set up. A beam of neutral silver atoms (later repeated with many other atomic beams, e.g., H, Li, Na, K, Cu, Au) from an oven O, i.e., a *neutral atomic beam* of silver is directed, after properly collimating it through a set of slits S_1 and S_2 [not shown in Fig. (3.39)] into an inhomogeneous magnetic field produced between the pole pieces SN Fig. (3.41). From figure it is clear that S-piece of magnet is shaped as a *knife edge* while the N-piece is provided with a *groove*. The field is directed along Z-axis, whereas the beam passes along Y. However, the field does not vary with Y along the entire length of the path of the beam in the field. Moreover, its variation with X is also zero in the meridian plane (YZ plane) of the magnet and almost negligible for positions close to this plane. We may note that the field varies rapidly with Z, i.e., along B and we may take this variation of field as constant over small distances of linear atomic dimensions.

Behind the field, a photographic plate P is placed, perpendicular to the knife edge, to record the configuration of the atomic beam after its passage through the field. The whole apparatus, i.e., arrangement is enclosed in an evacuated chamber to enable the silver atoms to traverse the field without collision.

When there is no magnetic field, the atomic beam travels undeflected towards the plate P. However, in the non-homogeneous field, each silver atom in the beam is acted upon by translational force, as discussed earlier, at right angles to the path of the beam. The atoms are not incident on plate P at a point off the axis. The normal states of the atoms are $^2S_{1/2}$ and if they are in the ground state for which $n = 2$, $l = 0$, $s = 1/2$ then since $l = 0$, the angular magnetic moment is zero and the magnetic moment is due to spin only. Now, denoting the spin angular momentum by $m_s(h/2\pi)$, the Z-component of the magnetic moment is

$$(\mu_s)_z = -g_s \mu_s m_s$$

Since $s = 1/2$, one expects $|m_s| \leq 1/2$ and the beam would be *split into two* components. We may note that the same result would be observed for all atoms having a single valence electron, e.g., Li, Na, Cu, Au, etc. with $l = 0$. Stern and Gerlach exactly observed the same. We consider the another example of oxygen. A beam of oxygen atom was split into *five*

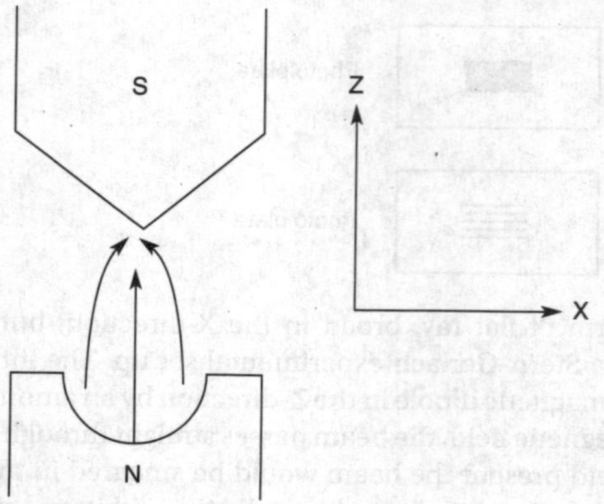

Fig. 3.41 Pole pieces of a magnet used in Stern-Gerlach experiment

components, which suggest that atoms had angular momentum $l = 2$, with the five different angles of deflection corresponding to $m = 2, 1, 0, -1$ and -2. The principle of quantization of spin magnetic moment was thus verified beyond doubt. When the splitting of the beam into spatially distinct components was first observed people called the phenomenon *space quantization*, a meaningless name that still appears in the literature.

We have seen that in the case of one electron atoms, the splitting gives two short lines on the photographic plate and it is possible, from the distance between the lines, to determine the magnetic moment of silver atoms and hence the magnetic moment.

3.28 HYPERFINE STRUCTURE OF SPECTRAL LINES

There are many contributions to the energy spectrum of hydrogen beyond those we have already discussed; the clarification of some of them has led to the theory of *quantum electrodynamics*.

There are two types of nuclear effects that produce hyperfine structure of spectral lines, when examined under instruments of highest possible resolution. We may note that hyperfine structure of spectral lines is caused by properties of the atomic nucleus. The first nuclear effect is due to the presence of different *isotopic species* in the element under investigation. The energy levels of isotopes are slightly displaced relative to each other, producing spectral transitions at slightly different wave numbers. In light atoms the *isotope shift* may arise from simple differences in the nuclear motion. The simplest example is that of hydrogen, in which the isotope shift led Urey et al., to the discovery of heavy hydrogen or deuterium. Different isotopes of an element have the same number and arrangement of extra nuclear electrons but differ from one another in mass. Since the nuclear mass enters into the Rydberg constant for an atom (through the reduced mass of the atom), different isotopes have slightly different value of Rydberg constant. The Rydberg constants for the two kinds of hydrogen atom can be found by substituting $M = M_1$ and $M = 2M_1$ successively in

$$m_r = \frac{m}{1 + \dfrac{m}{M}}$$

and making use of $R = mre^4 / 8\varepsilon_0^2 h^3 c$, one obtains

$$R_1 = \frac{M_1 R_\infty}{M_1 + m} \text{ and } R_2 = \frac{2M_1 R_\infty}{2M_1 + m}$$

M_1 being the mass of a proton and m the mass of an electron. The difference in wavelength for a given line of wavelength λ is then

$$\Delta\lambda = \frac{\lambda(R_2 - R_1)}{R_\infty} \approx -\frac{\lambda m}{2M_1} \tag{3.59}$$

Since $M_1/m = 1, 836$, it follows that the μ_B line for deuterium should lie $4861/3672 = 1.32$ Å on the violet side of that for ordinary hydrogen. This line was observed to be faintly visible, in the expected position in the spectrum from a sample of common hydrogen; it increased in strength as the relative concentration of deuterium was increased.

In the spectrum of heavy atom, isotope shifts are found, in general, to be proportional to the differences in atomic mass. For e.g., Tungsten consists of chiefly of four isotopes, three having mass numbers 182, 184 and 186 about equally abundant, and 183 about half as abundant as the others. The three observed lines are ascribed to the more abundant isotopes, any effect of 183 being assumed to be masked by the others.

In such heavy atoms, the isotope shift cannot be explained as a simple difference in the effects of nuclear motion, which are far too small. The cause lies in a departure from simple coulomb interaction between nucleus and electrons associated with the finite size and shape of the nucleus, the effect varying from isotope to isotope. Many nuclei have significant electric quadrupole and magnetic dipole moments.

The second type of effect arises because the charged nucleus can possess spin angular momentum and the associated magnetic moment. The latter interacts with the magnetic moment. The electron gives rise to splittings called *hyperfine structure*. The hyperfine interaction is small, but it has important applications in both radio astronomy and medical technology. As nuclear magnetic moments are smaller that electronic magnetic moments by ~10^{-3}, the hyperfine splitting is smaller that the spin-orbit splitting by the same factor.

The ordinary fine structure due to spin orbit interaction does not exhaust the possibilities of fine details in spectral lines. After the discovery of isotopes, it was thought that each component of a hyperfine pattern of lines was emitted by a different isotope. Later hyperfine structure was discovered in the spectra of some elements, which consist of only one isotope, e.g., Bismuth, the line λ = 3596Å contains 6 hyperfine components spread over a range of 0.3Å or 2.3cm^{-1}. Pauli in 1924 suggested that the nucleus has an angular momentum and an associated magnetic moment. If the atomic nucleus possesses an intrinsic angular momentum I with which is associated a magnetic moment μ_I, i.e.,

$$\mu_I = +g_I \frac{e}{2m_p} I \tag{3.60}$$

where e and m_p are respectively the charge and mass of the proton. The quantity g is called the *nuclear g factor* or *gyromagnetic constant* and it is measured value for proton is 5.56. The reversal in sign compared with electron comes from the positive charge on the proton. The magnitude of magnetic moment is

$$\begin{aligned}\mu_I &= g_I\left(\frac{e}{2m_p}\right)\sqrt{I(I+1)}\frac{h}{2\pi}\\ &= g_I\sqrt{I(I+1)}\frac{eh}{4\pi m_p}\end{aligned} \tag{3.61}$$

The quantity $\dfrac{eh}{4\pi m_p}$ forms a natural unit for the measurement of nuclear magnetic moment and is called the *nuclear magneton*, μ_N. It is 1/1836 times the Bohr magneton (because mass of proton is 1836 times the mass of electron. Thus

$$\mu_I = g_I\sqrt{I(I+1)}\ \mu_N \tag{3.62}$$

We can easily see that only electrons in the $l = 0$ state couple to the nuclear spin, the lone remaining part of the electron's total angular momentum is the electron spin **S** itself. Thus the coupling between the proton spin and the electron angular momentum is a coupling between the proton spin and the electron spin. This spin-spin interaction contains the factor ***I.S***. Let total spin of electron-proton is $\boldsymbol{S}_{tot}$. We can easily show that the amount of the hyperfine energy splittings depends on the total spin. The result of a complete calculation is that for $l = 0$ the total spin-1 (triplet) state lies above the total spin 0 (state) by an amount $\Delta E = h\nu$ where the frequency ν is

$$\nu = 1{,}420 \text{ MHz}$$

The wavelength corresponding to this frequency is 21 cm. Light actually a radio signal of this wavelength is emitted when there is a transition between the two levels.

This transition is of great importance in astrophysics. Its importance lies in the fact that such radio waves are not absorbed by interstellar dust, so that we can map hydrogen in space using the presence or absence of the 21 cm radiation. Excitation to the higher total spin-1 state occurs through collisions involving atleast one hydrogen atom. The 21 cm radio waves can be detected and from the intensity the distribution of atomic hydrogen in the plane of the galaxy can be trapped.

Example 3.1: Show that for a given principal quantum number n there are n^2 possible each of which can accommodate a maximum of 2 electrons.

Solution The orbital quantum number l corresponding to n has the following values,

$$l = 0, 1, 2, \ldots (n-1)$$

For each value of l, the magnetic quantum number m_l has the following values:

$$m_l = +l, +(l-1), +(l-2), \ldots 0, -1, -2, \ldots -(l-2), -(l-1), -l$$

Now, the total number of possible states is given by

$$\sum_{l=0}^{n-1}(2l+1) = 1 + 3 + 5 + \ldots + (2n-1)$$

R.H.S. is an A.P. series whose sum is given by

$$= n\left(\frac{a+d}{2}\right) = n\left[\frac{1+(2n-1)}{2}\right] = n^2$$

Obviously, the total number of states is n^2 corresponding to two different values of spin, each state can accommodate a maximum of 2 electrons.

Example 3.2: On the basis of vector atom model show that the possible values of the total angular momentum of a f electron are

$$\frac{3\sqrt{7}}{2}\frac{h}{2\pi} \text{ and } \frac{\sqrt{35}}{2}\frac{h}{2\pi}.$$ [Meerut]

Solution We know that an f-electron has the orbital quantum number $l = 3$ and its spin quantum number $s = \pm 1/2$. Thus, the total angular momentum quantum number is

$$j = l \pm s = 3 \pm \frac{1}{2} = \frac{7}{2}, \frac{5}{2}$$

We have the total angular momentum quantum number,

$$J = \sqrt{j(j+1)}\frac{h}{2\pi} = \frac{3\sqrt{7}}{2}\frac{h}{2\pi}; \frac{\sqrt{35}}{2}\frac{h}{2\pi}$$

Example 3.3: Show that the wavelength separation of the fine structure resulting from the spin-orbit interaction within the hydrogen atom at $\lambda = 4000$Å is 13.27 Wb/m². [Udaipur]

Solution In accordance with the Biot-Savart law, we have the magnetic field due to orbital motion of the electron around the proton (i.e., nucleus of hydrogen atom) is given by

$$B = \frac{\mu_0}{4\pi}\frac{2\pi I}{r} = \frac{\mu_0}{4\pi}\frac{2\pi e f}{r}$$

$$\therefore \quad B = 10^{-7} \times \frac{2 \times 3.14 \times (1.60 \times 10^{-19}) \times 0.75 \times 10^{15}}{53 \times 10^{-12}} \qquad \left| \begin{array}{l} r = 0.53\text{Å} \\ \\ f = 0.75 \times 10^{15}\,\text{Hz} \end{array} \right.$$

$$= 13.27\ \text{Wb/m}^2$$

Example 3.4: An electron beam enters a uniform field of flux density 1.2 Wb/m². Show that the difference in energy in eV between electrons having spins parallel and antiparallel to the field is 13.95×10^{-5} eV. [Kerala, Lucknow]

Solution We have the energy of a magnet in the field

$$= -\mu_B B \cos\theta$$

When spin is parallel to field, i.e., $\theta = 0$, we have the energy

$$E_1 = -\mu_B B$$

When spin is antiparallel to field, we have $\theta = 180°$ and

$$E_2 = \mu_B B$$

$$\therefore \quad \text{difference in energy} = 2\mu_B B$$

$$= 2 \times 9.3 \times 10^{-24} \times 1.2\text{J}$$

$$= 13.95 \times 10^{-5}\text{eV}.$$

Example 3.5: Indicate the possible (n, l, j) values of electrons in the first three shells of an atom. The symbols have their usual meanings. [Mysore]

Solution (i) $n = 1$(First shell): In this case we have $l = 0$ and $j = l \pm s = 0 \pm \frac{1}{2}$ or $j = 1/2$ because negative value of j is not permitted. Obviously, the possible values of (n, l, j) are (1, 0, 1/2).

(ii) $n = 2$ (second shell): In this case $l = 0$ and 1. Now, $j = \left(0 \pm \frac{1}{2}\right)$ or 1/2 and (1 ± 1/2), i.e.,

$\frac{3}{2}$ and $\frac{1}{2}$ $\therefore j = \frac{1}{2}$, 3/2 and 1/2

This means three states are possible. These states are (2, 0, 1/2); (2, 1, 3/2) and (2, 1, 1/2). One can show these possible states as follows

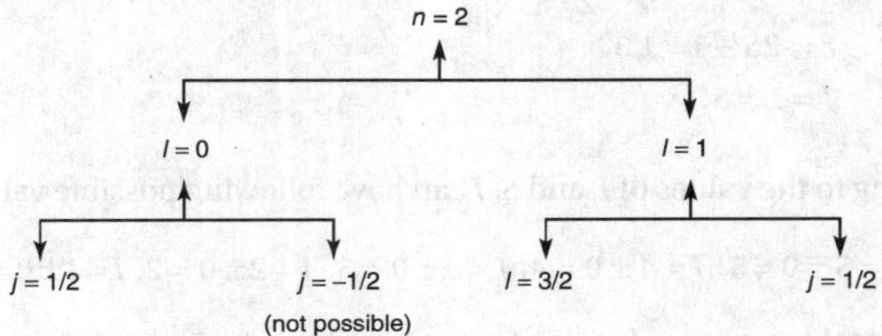

(iii) $n = 3$ (third shell): For this case $l = 0$, 1, 2 and

$$j = \left(0 \pm \frac{1}{2}\right); (1 \pm 1/2); (2 \pm 1/2) = 1/2; (3/2, 1/2); (5/2, 3/2)$$

Obviously, five states as follows are possible

$$(3, 0, 1/2), (3, 1, 3/2), (3, 1, 1/2), (3, 2, 5/2) \text{ and } (3, 2, 3/2)$$

The possible states are shown below.

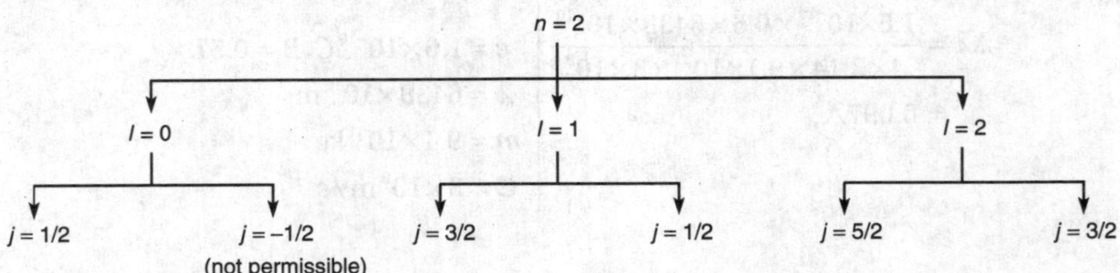

Example 3.6: Show that the precessional frequency of an electron orbit when placed in a magnetic field of 6Wb/m² is about 8.4×10^{10}Hz. [MDS]

Solution Larmor precessional frequency is given by

$$\Delta\nu = \frac{eB}{4\pi m}$$

$$\therefore \quad \Delta\nu = \frac{1.6 \times 10^{-19} \times 6}{4 \times 3.14 \times 9.1 \times 10^{-31}} = 8.39 \times 10^{10}\,\text{Hz} \cong 8.4 \times 10^{10}\,\text{Hz}$$

$B = 6\,\text{Wb/m}^2$
$e = 1.6 \times 10^{-19}\,\text{C}$
$m = 9.1 \times 10^{-31}\,\text{kg}$

Example 3.7: How many different types of terms can a two electron system consisting of *d* and *f* electrons possess? [Meerut, Agra]

Solution For *d electrons* we have $l_1 = 2$ and *f* electrons $l_2 = 3$

$$\therefore \quad L = l_1 + l_2 = 2 \pm 3. \text{ Possible values} = 5, 4, 3, 2, 1$$

$$S = S_1 \pm S_2 = \frac{1}{2} \pm \frac{1}{2} = 0,1$$

$$\therefore \quad r = 2S + 1 = 1,3$$

now, $$J = L \pm S$$

$\therefore$ corresponding to the values of *L* and *S*, *J* can have following possible values

(i) $J=5\pm 0=5,\ J=4\pm 0=4, J=3\pm 0=3,\ J=2\pm 0=2, J=1\pm 0=1,$

We may note that these are *single terms* for $S = 1$, we have $J = 5 \pm 1 = 6, 5, 4; J = 4 \pm 1 = 5, 4, 3;$ $3 \pm 1 = 4, 3, 2; J = 2 \pm 1 = 3, 2, 1;$ and $J = 1 \pm 1 = 2, 1, 0;$
All these are triplet states.

Example 3.8: Show that the normal Zeeman splitting of the line 6438Å in a magnetic field of 0.5T is 0.097Å. [Mysore]

Solution we have Zeeman shift

$$d\lambda = \frac{eB\lambda^2}{4\pi mc}$$

$$\therefore \quad \Delta\lambda = \frac{1.6\times 10^{-19}\times 0.5\times 6438\times 10^{-10}}{4\times 3.14\times 9.1\times 10^{31}\times 3\times 10^{8}}$$

$$= 0.097\text{Å}$$

$e = 1.6\times 10^{-19}C, B = 0.5T$

$\lambda = 6438\times 10^{10}\,\text{m}$

$m = 9.1\times 10^{31}\,\text{kg}$

$C = 3\times 10^{8}\,\text{m/s}.$

Example 3.9: A sample of certain element is placed in a 0.3 T magnetic field and suddenly excited. How far apart are the Zeeman components of the 450 nm spectral line of the element? [Jodhpur]

Solution The separation of the Zeeman components is given by

$$\Delta v = \frac{eB}{4\pi m} \quad \Big| \quad v = c/\lambda,\ dv = -c\frac{d\lambda}{\lambda^2}$$

we obtain $$\Delta\lambda = \frac{\lambda^2 \Delta v}{c} = \frac{eB\lambda^2}{4\pi mc}$$ Disregarding the minus sign.

Substituting the proper values one obtains,

$$\Delta\lambda = \frac{(1.6\times10^{-19})\times0.3T\times(4.5\times10^{-7})^2}{4\times3.14\times9.1\times10^{31}\times3\times10^8}$$

$$= 2.83\times10^{-12}\,\text{m} = 0.00283\,\text{nm}.$$

Example 3.10: Determine the number of subshells in which the following terms split in a weak magnetic field: (i) $^3P_0$ (ii) $^2F_{5/2}$ (iii) $^4D_{1/2}$.

Solution (i) $^3P_0$ state: For this state we have $J=0, M_J=0$. Clearly, the energy level does not split. (ii) $^3F_{5/2}$ state: We have $g \neq 0$, $J = 5/2$, M_J has $2J + 1 = 6$ values. Clearly this splits into 6 subshells. (iii) $^4D_{1/2}$ state: For this state $g = 0$ and hence it does not split at all.

Example 3.11: A spectrometer can resolve spectral lines in the region $\lambda \simeq 6000$Å when separated by $\Delta\lambda = 0.1$Å. Show that the external field required to confirm the normal Zeeman triplet is 0.59 W/m².

Solution From Example 3.9, we have

$$\Delta v = \frac{c\Delta\lambda}{\lambda^2}$$

but $\quad \Delta E = \mu_B B$ and $\Delta v = \mu_B B/h \qquad (\because \Delta E = h\Delta v)$

$$\therefore \qquad B = \frac{hc\Delta\lambda}{\mu_B\lambda^2}$$

Substituting the proper values, one obtains

$$= \frac{6.62\times10^{-34}\times3\times10^8\times0.1\times10^{-10}}{9.3\times10^{-24}\times(6\times10^{-7})^2}$$

$$= 0.59\,\text{Wb/m}^2.$$

Example 3.12: In Stern-Gerlach type of experiment, the magnetic field varies with distance in the z-direction according to $\frac{dB_Z}{dz} = 1.4\text{T/mm}$. The silver atoms travel a distance $x = 3.5$ cm through the magnet. The most probable speed of the atoms emerging from the oven is $v = 750$m/s. Show that the separation of the two beams as they leave the magnet is 0.16 mm. Given: the mass of a silver atom is 1.8×10^{-25} kg, and its magnetic moment is about 1 Bohr magneton. [Punjab]

Solution The potential energy of the magnetic moments in the magnetic field is

$$U = -\mu.B = -\mu_z B_z \qquad \text{(i)}$$

because the field along the central axis of the magnet has only a z-component. The force on the atom

$$F_z = -\frac{dU}{dz} = \mu_z\frac{dB_z}{dz} \qquad \text{(ii)}$$

The acceleration of a silver atom as it passes through the magnet is

$$a = \frac{F_z}{m} = \frac{\mu_z\left(\frac{dB_z}{dz}\right)}{m} \quad \text{(iii)}$$

One can find the vertical deflection ΔZ of either beam from $\Delta Z = \frac{1}{2}at^2$, where t the time to traverse the magnet, equals x/v. Since each beam is deflected by this amount, the net separation d is $2\Delta z$, or

$$d = \frac{\mu_z (dB_z / dz)x^2}{mv^2}$$

$$= \frac{9.27\times10^{-24}\times1.4\times10^3\times3.5\times10^{-2}}{1.8\times10^{-25}\times750}$$

$$= 1.6\times10^{-4} = 0.16\,\text{mm}$$

One finds that the above result is consistent with experimental observation.

SUMMARY

1. The restrictions on the quantum numbers are as follows:

   (i) *Principal quantum number* (n) = 1, 2, 3, ... This determines the energy levels

   (ii) *Orbital quantum number* (l) determines the angular momentum:

   $$|l| = \sqrt{l(l+1)}\,\hbar$$

   where l ranges from 0 to $n - 1$, i.e.

   $$l = 0, 1, 2, ..., (l - 1)$$

   All states with the same principal quantum number n form a *shell*, identified by the letters *K, L, M*, ... (corresponding to $n = 1, 2, 3, ...$). All states with the same values of both n and l form a *subshell*, designated by *s, p, d, f*, ... (corresponding to $l = 0, 1, 2, 3, ...$).

   (iii) *Orbital magnetic quantum number* (m_l) determines the component of the angular momentum along any given axis:

   $$L_z = m_l \hbar$$

   This is *space quantization*, where m_l ranges from $-l$ to l, i.e.

   $$m_l = -l, -l+1, ..., l-1, l$$

   (iv) *Spin magnetic quantum number* (m_s) Electrons are spin 1/2 particles called fermions. The component of their spin angular momentum on a given axis is $\pm\frac{1}{2}\hbar$, i.e. spin magnetic quantum number can have only two values $\pm\frac{1}{2}$. Thus, this additional quantum number doubles the number of allowed states specified by quantum numbers n, l, and m_l.

The electron has an intrinsic angular momentum called *spin angular momentum* (S). Its magnitude is

$$S = \frac{\sqrt{3}}{2}\hbar$$

Electronic spin can be described by a quantum number $s = 1/2$.

2. The quantity $e\hbar/2m_e$ is called the *Bohr magneton* (μ_B) and has the numerical value = 9.274×10^{-24} J/T.
3. *Pauli exclusion principle* states that no two electrons in an atom can have the same set of quantum numbers n, l, m_l, and m_s. Using this, one can determine the electronic configuration of the elements. This procedure forms a basis for understanding atomic structure and the chemical properties of the elements.
4. The allowed electronic transitions between any two states in an atom are governed by the *selection rules*

$$\Delta l = \pm 1, \text{ and } \Delta m_l = 0, \text{ or } \pm 1.$$

5. *Spin-orbit coupling* results in *fine-structure* splitting of atomic energy levels.

REVIEW QUESTIONS

1. Give an account of different quantum numbers necessary to specify the state of an electron in an atom. State Pauli's exclusion principle and explain its importance.
2. Describe two essential features that characterise the vector atom model of atom. Discuss the various quantum numbers associated with this model.
3. Enunciate Pauli's exclusion principle and show that the maximum number of electrons in a shell with principal quantum number n given by $2n^2$.
4. Explain Pauli's exclusion principle as applied to electrons in atoms. Elucidate how this principle assists in the interpretation of the periodic system of elements. **[Rohtak]**
5. Explain Russel-Sanders (L-S) and (J-J) couplings of orbital and spin angular momenta. How are the atomic states, i.e., term symbols written making use of L-S coupling? **[Mysore]**
6. Indicate the possible (n, l, m_l, s) values of electrons in the first three shells of an atom. The symbols have their usual meanings. **[Mysore]**
7. Using vector atom model of atom explain the fine structure of H_α line of hydrogen spectrum and sodium D-lines. **[Kerala]**
8. Explain space quantization? Describe Stern and Gerlach experiment and explain the results of this experiment. **[Kerala]**
9. Describe briefly with theory, the Stern-Gerlach experiment. Discuss critically the results in the case of some typical atoms on the basis of quantum theory. **[Mysore]**
10. What is Zeeman effect? Obtain an expression for normal Zeeman shift on the basis of vector atom model.

11. Explain anomalous Zeeman effect. Taking example of splitting of sodium D-lines, explain this effect. **[Meerut]**
12. Using vector atom model, explain the fine structure of H_α line of hydrogen spectrum and sodium D-lines.
13. What is normal Zeeman effect? Give the theory of normal Zeeman effect and show how you can determine the value of the specific charge (e/m) of electron with its help? **[Meerut]**
14. Show that the magnetic moment of an atom can be expressed as $\mu_J = -\mu_B g\sqrt{j(j+1)}$. Also show that its projection along any direction as $(\mu_j)_z = -\mu_B g M_j$, where the symbols have usual meanings.
15. State and explain Pauli's exclusion principle and its use in determining the electronic shell structure of atoms. How does it account for the periodic classification of elements?
16. Give a brief account of the various quantum numbers of an electron related in the vector atom model of an atom. State Pauli's exclusion principle and deduce from it the maximum number of electrons, which can occupy a given shell. Hence explain building up of the electronic configurations of elements with examples of H, He, Li, Ne and Na. **[Meerut]**
17. Write short notes on the following:
    (a) Vector atom model
    (b) Quantum numbers
    (c) Pauli's exclusion principle
    (d) Gyromagnetic ratio
    (e) Zeeman effect
    (f) Paschen-Back effect
    (g) Stark effect
    (h) Space quantization
    (i) Stern-Gerlach experiment

PROBLEMS

1. Express the quantum numbers l, j, m_j for $n = 1, 2$ and 3.

Ans. $n=1$: $\left(1, 0, +\frac{1}{2}, +\frac{1}{2}\right)$ and $\left(1, 0, +\frac{1}{2}, -\frac{1}{2}\right)$

$n=2$: n	l	j	m_j
2	1	3/2	−3/2
2	1	3/2	−1/2
2	1	3/2	$+\frac{1}{2}$
2	1	3/2	+3/2
2	1	1/2	−1/2
2	1	1/2	+1/2

$n = 3$:	n	l	f	m_j
	3	2	5/2	−5/2
	3	2	5/2	−3/2
	3	2	5/2	−1/2
	3	2	5/2	+1/2
	3	2	5/2	+3/2
	3	2	5/2	+5/2
	3	2	3/2	−3/2
	3	2	3/2	−1/2
	3	2	3/2	+1/2
	3	2	2	+3/2

2. Show that the maximum potential energy associated with the magnetic dipole moment of $4p$ electron placed in an external magnetic field of 0.25Wb/m^2 is 2.05×10^{-5} eV. **[Kerala]**

3. An unknown magnetic field produces a set of lines for a transition from a state with $l = 3$ to one with $l = 2$ between which there is a maximum energy difference of 6.1×10^{-4} eV. Show that there are only 15 lines if $\Delta m = -1, 0, 1$ is allowed. Further show that field strength is 11 Tesla.

4. Given that the neutron and proton each have spin 1/2, show that no combination of spins and orbital angular momenta can make the nucleus of ^{3}He, which consists of $2p$ and $1n$ have integral total angular momentum. **[Mysore]**

5. Given that neutron (n) and proton (p) each have spin 1/2, show that no combinations of spins and orbital angular momenta can make the nucleus of ^{3}He, which consists of two protons and a neutron, have integer total angular momentum. **[Delhi]**

6. Given that neutron (n) and proton (p) each have spin 1/2, show that no combinations of spins and orbital angular momenta can make the nucleus of ^{4}He, which consists of two protons and a neutron, have integer total angular momentum. **[Meerut]**

7. What angular momentum states result when angular momentum $l_1 = 3$ is added to angular momentum $l_2 = 2$? Check that the degeneracies $2l + 1$ for these states add up to the total number of states $(2l_1 + 1)(2l_2 + 1)$. [$j = 5, 4, 3, 2, 1$] **[Kerala]**

8. Show that the value of external magnetic field for which the energy splittings due to the Zeeman effect are comparable in magnitude to spin-orbit effects and is of the order of 6 tesla. What is the relevance of this problem?

[**Ans.** The relevance of this problem is that for fields much stronger than $6T$, the spin-orbit splitting can be neglected in comparison with the splittings induced by the external field.]

9. Compute the change in wavelength of the $(2p - 1s)$ photon when a hydrogen atom is placed in a magnetic field of $2T$.

[Hint: the energy of the photon is

$$E = -13.6\text{eV}\left(\frac{1}{2^2} - \frac{1}{1^2}\right) = 10.2 \text{ eV}$$

$$\therefore \qquad \lambda = \frac{hc}{E} = \frac{1240\text{eV.nm}}{10.2\text{eV}} = 122 \text{ nm}$$

the change in energy ΔE of the level is

$$\Delta E = \mu_B B = (9.27\times 10^{-24}\ J/T)(2T)$$
$$= 18.5\times 10^{-24}\ J = 11.6\times 10^{-5}\ \text{eV}$$
$$\Delta\lambda = \frac{\lambda^2}{ch}\Delta E = \frac{(122\ \text{nm})^2}{1240\ \text{eV-nm}}\times 11.6\times 10^{-5}\ \text{eV}$$
$$= 0.00139\ \text{nm}$$

we note that for such a relatively large magnetic field as $2T$, the change in wavelength is very small, but easily measurable with the help of optical spectrometer].

10. Explain why each of the following sets of quantum numbers (n, l, m_l, m_s) is not permitted for hydrogen: **[Hyderabad]**

(a) (2, 2, –1, +1/2) (b) (3, 1, +2, –1/2)
(c) (4, 1, +1, –1/2) (d) (2, –1, +1, +1/2)

11. For each l value, the number of possible states is $2(2l + 1)$. Show explicitly that the total number of states for each principal quantum number is

$$\sum_{i=0}^{n-1} 2(2l+1) = 2n^2$$

Dose this gives the degeneracy of each energy level?

12. The magnetic moment associated with the orbital motion of an atomic electron has magnitude 3.213×10^{-23} A-m². Show that the total angular momentum quantum number, l of the electron's orbit is 3.

13. In a Frank Hertz experiment the first dip in the current voltage graph for hydrogen was observed at 10.2V. Show that the wavelength of the light emitted for hydrogen when excited to the first excited level is 1215Å. **[Kerala]**

14. A hydrogen atom is placed in an intense magnetic field of 0.8Wb/m². Assuming that the magnetic field produces normal Zeeman triplet, show that the separation between two consecutive states when $\lambda = 5000$Å is 0.0941Å **[Mysore]**

15. A sample of certain element is placed in a magnetic field of flux density 0.3Wb/m². Show that the Zeeman components of a spectral line of wavelength are separated by 0.0283Å. **[Punjab]**

16. An electron beam enters a uniform magnetic field of flux density 1.2 Wb/m². Show that the energy difference between electrons whose spins are parallel and antiparallel to the field is 1.39×10^{-4}eV. **[Punjab]**

SHORT ANSWER-QUESTIONS

1. A D term consists of five components. What are the multiplicities of this term?
Ans. 5, 6, 7, ...

2. An electron in a state with non-vanishing angular momentum forms a current loop, or equivalently, a magnetic moment that can be measured by looking at shifts in atomic energy levels in an external field. The degenerate energy levels are split and the magnitude of the component of the quantized angular momentum along the direction of the magnetic field can be measured. What is this effect?
Ans. Zeeman effect.

3. What is the importance of Zeeman experiment?
Ans. Zeeman spectroscopy shows that even in a state with zero angular momentum the electron has a magnetic moment. This quantity is attributed to an intrinsic angular momentum of the electron, called its spin.
4. What are the peculiar properties of spin.
Ans. (i) Spin is characterized by a quantum number that, in contrast to the integer values of the orbital angular momentum, has the value 1/2.
(ii) The magnetic moment is related in a non-classical way to the angular momentum.
5. What is spin-orbit interaction?
Ans. An electron interacts with the magnetic field it experiences because it moves in a coulomb field. This interaction is called spin-orbit interaction.
6. What is the basis of magnetic resonance imaging?
Ans. Constitutents of the nucleus, i.e., neutron and proton also have spin and can also interact with the magnetic field associated with the moving electron. This effect is the basis of magnetic resonance imaging.
7. What is the minimum possible value of total angular momentum of lithium atom whose valence electron is in a state $n = 3$. What is the corresponding state?
Ans. $\frac{\sqrt{35}}{2}\hbar^2, D_{5/2}$
8. What was the first conclusive evidence of space quantization?
Ans. Separated images observed on the screen in Stern-Gerlach experiment.
9. Is intrinsic spin is a characteristic of every fundamental particle?
Ans. Yes, e.g., the proton and neutron have a spin quantum number of 1/2. The photon has a spin quantum number of 1, while the pi-meson (pion) has $s = 0$.
10. What is the magnetic moment of hydrogen atom in the ground state.
Ans. $\sqrt{3}\mu_B$ since $L = 0$, $S = \frac{1}{2}$, $J = \frac{1}{2}$ and $g = 2$
11. Write the spectral term of an atom with $S = 1/2$, $J = 5/2$ and $g = 6/7$.
Ans. $^2F_{5/2}$
12. A carbon atom with electronic configuration $1s^2 2s^2 2p3d$ has maximum possible total angular momentum at such a configuration. What is the magnetic moment of the atom in this.
Ans. $\mu + \frac{5\sqrt{5}}{2}\mu_B$. We have $L = 1 \pm 2 = 3, 2, 1$, $S_{max} = 1$ and $L_{max} = 3 \therefore J_{max} = 4$

OBJECTIVE QUESTIONS

1. Corresponding to principal quantum number $n = 1$, we have
(a) $l = 0, m_l = 0, m_s = \pm 1/2$ (b) $l = 0, m_l = 1, m_s = \pm 1/2$
(c) $l = 0, m_l = 0, m_s = 0$ (d) $l = 0, m_l = 1, m_s = 0$
2. The K-shell contains only
(a) one sub-shell denoted by *s* (b) one sub-shell denoted by *p*
(c) two sub shells denoted by *s* and *p* (d) none of the above

3. The ionization potential of sodium is

(a) 1.5 eV (b) 5.1 eV

(c) 3.5 eV (d) 10.1 eV

4. In terms of quantum defect (Δ), the energy (E) of the valence electron in alkali metals is written as

(a) $E = -\dfrac{Rch}{n-\Delta}$ (b) $E = -\dfrac{Rch}{(n-\Delta)^{1/2}}$

(c) $E = -\dfrac{Rch}{(n-\Delta)^2}$ (d) $E = -\dfrac{Rch}{(n-\Delta)^3}$

5. The effective quantum number for the ground state of rubidium is 1.805. If $R = 109737$ cm^{-1}, the ionization potential of the atom is

(a) 1.176 eV (b) 2.176 eV

(c) 3.176 eV (d) 4.176 eV

6. The first member of principal series of sodium has a wave length of 5890 A°. The first excited S state of sodium lies 3.18 eV above the ground state. The wave length of first member of sharp series is

(a) 3481 A° (b) 5481 A°

(c) 7481 A° (d) 11481 A°

7. The principal and sharp series for sodium atom converge to continum at 41450 and 24477 cm^{-1} respectively. The ionization potential of sodium atom is

(a) 1.5 eV (b) 3.5 eV

(c) 7.5 eV (d) 5.139 eV

8. The doublet splitting of the first excited state 2 ^2P state in an atom with $Z = 2$ is 5.84 cm^{-1}. The corresponding separation in hydrogen atom is

(a) 0.365 cm^{-1} (b) 0.165 cm^{-1}

(c) 0.565 cm^{-1} (d) 0.765 cm^{-1}

ANSWERS

Question numbers	1	2	3	4	5	6	7	8
Answers	(a)	(a)	(b)	(c)	(d)	(d)	(d)	(a)

FILL IN THE BLANKS

1. ^1S term will split in ________ components in magnetic field. **[3]**

2. The Russel-Saunders terms for the configuration 2s 2p are ________. **[$^3P_2$, $^3P_1$, $^3P_0$, $^1P_1$]**

3. The D-lines of sodium originate from ______. **[$3P \rightarrow 3S$]**

4. $^2D_{5/2}$ term will split into ______ components in magnetic field. **[6]**

5. μ and the allowed values of the projection of an atom in the state $^2D_{3/2}$ are ______.

$$\left[L=2, S=1/2, J=3/2, g=4/5, \mu=2\sqrt{\frac{3}{5}}\mu_B\right]$$

6. The spectral term of an atom with $S=1, L=2, g=4/3$ is ______. **[$^3D_3$]**

7. Magnetic moment (μ_J), of an atom can be expressed as ______. **[$\mu_B = -\mu_B g\sqrt{J(J+1)}$]**

8. A D term consists of five components. The multiplicities of this term are ______. **[5, 6, 7, ...]**

9. The possible terms of atoms with 2*p*3*s* configurations are ______. **[$^1P_1^3P_2, ^3P_1, ^3P_0$]**

10. The spectral term of an atom with $S=1/2, J=5/2, g=6/7$ is______. **[$^2F_{5/2}$]**

11. Wave function ψ of an electron is characterized by ______, ______, ______, and ______ quantum numbers.

[principal quantum number *n*, orbital quantum number *l*, magnetic quantum number m_l, spin quantum number m_s]

12. Principal quantum number determines the ______ of electron in the atom and the ______ of electron from nucleus. **[total energy, average distance]**

13. Principal quantum number *n* can take ______ 1, 2, 3, 4, ... **[integer values]**

14. Orbital quantum number determines the orbital ______ of electron and its magnitude is given by ______. **[angular momentum, $|l| = \sqrt{l(l+1)}\hbar$]**

15. Relativistic quantum mechanics shows that electron possesses an intrinsic ______ *s* whose magnitude is given by ______. **[angular momentum, $|s| = \sqrt{s(s+1)}\hbar$]**

16. Due to spin-orbit interaction each energy level in an atom characterized by a *L* value splits into a group of sub-levels called ______ levels. **[fine-structure]**

17. The collection of fine-structure levels belonging to a given *L* value is called a ______. **[multiplet]**

18. In light element containing more than one valence electron, the electrostatic interaction between electrons is ______ than the spin-orbit interaction. **[greater]**

4 Molecular Spectra

4.1 INTRODUCTION

We have read that the *line spectrum* originates due to the atoms of different elements and can be adequately explained with the help of Bohr-sommerfeld model of the atoms, and also by quantum mechanics.

In some special cases, the spectrum of light emitted from discharge tubes has the appearance of the *band* of the light, instead of discrete lines. This type of spectrum has its origin in the molecules of different elements and compounds. Such a spectrum is known as the *band spectrum*, i.e. such spectrum has the appearance of a band of light, instead of discrete lines and is usually studied in absorption. However, the spectra of diatomic molecules may also be conveniently studied in emission and are found to supplement the absorption data.

Usually the study of band spectrum is done by expereiments with the *absorption spectrum*. However, the spectra of many diatomic molecules are studied more conveniently in emission and are found to supplement the absorption data.

On closer examination of the band spectrum, one finds that each band in a band spectrum extends upto a certrain definite limit at one end and the intensity slowly fades away at the opposite end as shown in Fig. (4.1). The former end is known as the *band head*. The intensity of the band gradually fades away towards the other end.

In a high resulotion spectroscope, however a band is found to be made up of a very large number of closely spaced spectral lines in a regular pattern. These lines are found to be arranged in a regular manner or pattern. The lines spread from a fixed point in the band, called *band origin*, towards both the ends but the *crowding* of the lines being *more dense* towards the *band head*. Generally a number of regularly spaced bands is observed for a given substance and they are termed as *band groups* or a *groups of bands*. A number of such groups for the *given substance constitutes a (band system)*. There may be more than one band system in a given molecular spectrum of a substance.

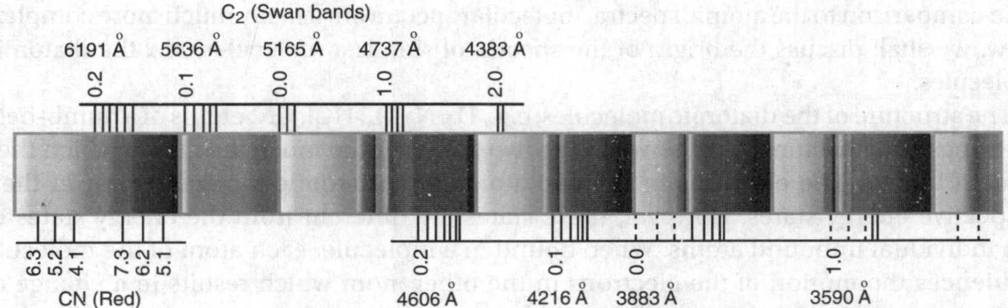

Fig. 4.1 Band spectrum of diatomic molecule: CN and C_2. Band heads on the left of the band are to be noted

The band spectrum may be divided into three distinct classes:

(i) **Pure rotation band**: The wavelengths of pure rotation band extend from 150μ to 30μ ($1\mu = 1$ micron $= 10^{-4}$cm) in the *far infra-red*. We may note that the wavelength is much longer than that of visible light. Each spectrum is composed of a series of nearly equidistant lines. Like rotation-vibration spectra, the pure rotational spectra are observed only for the hetronuclear diatomic molecules.

(ii) **Rotation-vibration band**: The wavelength of these bands extend from about 5μ to 1μ in *near infra red*. These bands, like electronic bands have a fine structure. They are observed only for the heteronuclear molecules. The homonuclear molecules like H_2, N_2, O_2 do not produce vibrational-rotational bands.

(iii) **Electronic bands**: The wavelengths of these bands normally extend from about 1000 to 7000Å in the visible or ultraviolet-region and both in emission and absorption. The electronic spectra are observed of the heteronuclear as well as for the homonuclear diatomic molecules. Electronic spectra consists of quite a large number of bands. Each individual band has a sharp edge, called band head, where the intensity falls suddenly to zero and from this edge the intensity falls off gradually to the other of the band. Infact a band is composed side of a series of lines which draw further apart as the distance from the band-head increases. The band is said to be 'degraded' toward the side opposite to the band-head.

4.2 ORIGIN OF THE BAND SPECTRUM

The origin of the band spectra lie, in the molecules of substance. When a substance giving rise to band spectrum is heated then the nature of the spectrum is changed. At high temperature, the band disapper and discrete spectral lines characteristic of the constitutent elements of the molecules of the substance are observed, e.g., when an asbestos fibre intergrated with the solution of common salt is held in the flame of bunsen burner, then yellow D-lines of sodium appear in the spectrum. We may note that at the high temperature of the flame, NaCl molecule dissociates into its constitutents, i.e., sodium and chlorine atoms. As a result the characteristic lines of sodium are observed.

In comparison to the atomic spectra, molecular spectra are usually much more complex. Now, we shall discuss the origin of the spectra of simplest molecules, viz. the diatomic molecules.

The structure of the diatomic molecules, e.g., H_2, N_2, O_2, HCl, CN, etc., is of a dumb-bell. A strong attractive force acts between the two atoms of the molecule due to which they remain bound. The electrons in the two atoms of a diatomic molecule exist in their respective energy states. However, these states are different from the energy states of the individual unbound atoms. When bound in a molecule, each atom of the molecule influences the motion of the electrons in the other atom which results in a change of their energy states. We have read that in the case of the atomic energy levels, the electronic states of the molecules have energies (E_e) of the order of a few electron-volts. Transitions between such energy states result in the emission of light in the visible or ultraviolet regions.

We may note that the energies of the molecular energy levels depend not only on the motion of the orbital electrons, but also on the motion of the atoms as a whole inside the molecule. The molecular energy states may arise from:

(i) **Rotation of the molecule as a whole**: Due to rotation, it will have some rotational energy, E_r. This also affects the energy of the molecular levels, The rotation may be about an axis joining the two atoms or about an axis perpendicular to the former. We may note that the rotational energy depends on the *moment of inertia* (I) of the molecule. We know that moment of inertia of a body depends upon the mass of the body and the distances of the different parts of the body from the axis of the rotation. Atomic mass is almost totally confined within its nucleus having dimensions of the order of 10^{-15} m which is small compared to atomic dimensions ($\sim 10^{-10}$ m). Moreover, the distances of the different parts of the nucleus from the line joining them is negligibly small so that the moment of inertia about this line is negligible. This means that the contribution of the rotation about this axis to the total rotational energy is negligibly small. However, the rotation about the axes perpendicular to the line joining the two atoms have to be considered.

(ii) **Vibration due to constitutent atoms:** The energies of the molecular energy levels of a molecule also depend on the motion of the atoms as a whole inside the molecule. In a diatomic molecule, the two atoms usually vibrate along the line joining them, just like a linear harmonic oscillator, due to which the molecule possesses some *vibration energy*, E_v, which affects the molecular energy levels.

(iii) **Changes involving electronic configuration of molecules:** The energies of rotation and vibration in a molecule are due to the motion of its atomic nuclei, which contain virtually all the molecule's mass. The molecule's electrons also can be excited to higher energy levels than those corresponding to its ground state. However, the spacing of these levels is much greater than the spacing of rotational or vibrational levels.

Electronic transistions involve radiation in the visible or ultraviolet parts of the spectrum. Each transition appears as a series of closely spaced lines, called a band, due

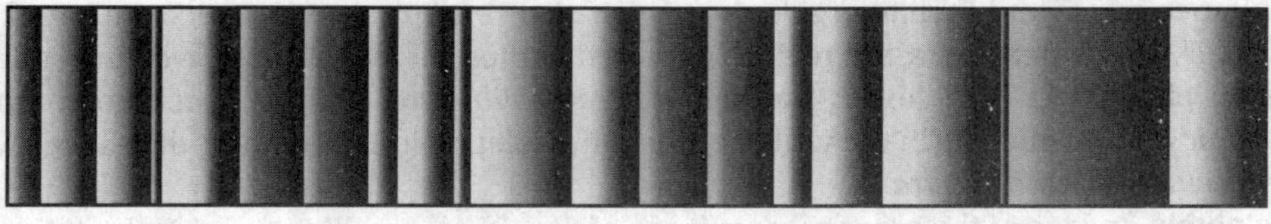

Fig. 4.2 A portion of the band spectrum of PN

to the presence of difference rotational and vibrational states in each electronic state (Fig. (4.2)). All molecules exhibit electronic spectra, since a dipole moment change always accompanies a change in the electronic configuration of a molecule. Therefore homonuclear molecules, such as H_2 and N_2, which have neither rotational nor vibrational spectra because they lack permanent dipole moments, nevertheless have electronic spectra whose rotational and vibrational fine structures enable moments of inertia and bond force constants to be found.

Electronic excitation in a polyatomic molecule often leads to a change in the molecule's shape which can be determined from the rotational fine structure in its band spectrum. The origin of such changes lies in the different characters of wave functions of electrons in different states, which lead to correspondingly different band geometries. e.g., the molecule beryllium hydride, Be H_2, is linear (H—Be—H) in one state and bent (H—Be) in another.

H

In a quantum mechanical theory of the molecular structure, it is necessary not only to consider the above mention three types of motion, but we have to consider the mutual coupling of these three different types of motion. However, it is difficult to obtain an exact solution of the Schrodinger wave equation when all these are taken into consideration. It is however possible to obtain an approximate solution by neglecting the coupling between the different types of motions as suggested by Born and Oppenheimer. In the light of this approximation, the energy of the molecular energy level may be written as

$$E = E_e + E_v + E_r \tag{4.1}$$

Where E_e, E_v and E_r are respectively the energies imparted to the molecule due to changes in electronic configuration, vibration and rotation and each one of them is quantised. The electronic energy E_e is of the order of a few electron-volts, where as vibrational energy E_v is much smaller than E_e and the rotational energy E_r is small compared to E_v. Thus, we have

$$E_e >> E_v >> E_r \tag{4.2}$$

Thus with each electronic state, several vibrational states are associated and with each vibrational state, a number of rotational states are involved.

The spectra arising from transitions involving change in E_r alone produce *pure rotation spectra* which usually lie in the farinfra-red region, as stated above. Transitions in which simultaneously E_v and E_r change give rise to *vibration-rotation spectra,* which are found to lie usually in the near infra-red region. As stated earlier, the transitions involving simultaneous changes in E_e, E_v and E_r produce *electronic-vibration-rotation* spectra or simply *electronic bands* which lie in the visible or the ultraviolet region. Generally, they are called as *molecular or band spectra.*

To obtain an idea of the order of magnitude of the electronic, vibrational and rotational energies of a diatomic molecule, one finds that for a sepration *a* between the nuclei of the two atoms in the molecule, the uncertainity principle gives the momentum of the valence electrons to be $p_e \approx \Delta p = \hbar / a$ so that the kinetic energies of the electrons should be of the order of

$$E_e = p_e^2 / 2m = \hbar^2 / 2m_e a^2 \tag{4.3}$$

Since a ~1 Å, one obtains E_e ~ 3.80 eV. Obviously, the electronic energy in the molecule is typically of the same order of magnitude as that for the valence electrons in the atom.

In order to get an estimate of the *vibrational energy* of the nuclei, one may use the classical argument. If *F* be the force which binds the electrons to the molecule, the nuclei must be bound by an equal and opposite force. Now, if we assume that the force is simple harmonic with a force constant *k*, then we can write $F = -kx$ so that the angular frequency of electronic motion is $\omega_e \sim (k/m_e)^{1/2}$ correspondly, one finds the angular frequency of the vibrational motion of the nuclei as $\omega_N \sim (k/m)^{1/2}$, where *M* is the nuclear mass. Clearly, the ratio of the energies of the vibrational motion of the nuclei to that of the electronic motion is typically

$$E_v / E_e = \hbar\omega_N / \hbar\omega_e = \sqrt{\frac{m_e}{M}} \tag{4.4}$$

The ratio m_e/M is of the order of 10^{-3} to 10^{-5} and hence the ratio E_v/E_e is of the order of 10^{-2}. Since the electronic energy is ~1eV, producing lines in the visible or ultra-violet regions the vibrational energy should be in the infra-red region.

Let us now estimate the rotational energy, E_r. We take a diatomic molecule with two nuclei of mass *M* each separted by a distance *a*, the rotational energy is

$$E_r \sim \hbar / 2I = \hbar / Ma^2 \tag{4.5}$$

Where $I = Ma^2/2$ is the moment of inertia. Now using Eq. (4.3), one obtains

$$E_r = (m_e / M) E_e \tag{4.6}$$

Obviously, the rotational energy of nuclear motion is lower than the electronic energy by a factor ~(m_e/M) and by a factor ~$(m_e/M)^{1/2}$ compared to the vibrational energy. This means that they are typically ~10^{-3}eV or less while the vibrational energy is typically 10^{-1} to 10^{-2}eV.

We may note that the above estimates justify the Bohr-Oppenheimer approximation regarding the uncoupling of the electronic, vibrational and rotational motions just mentioned above.

Now, if E_1 and E_2 be the energies of two levels in a molecule, then the energy of the radiation resulting from a transition between the two levels is

$$E_2 - E_1 = hv \tag{4.7}$$

one can explain the origin of the *pure-rotational or of the rotation-vibration* spectra of molecules with the help of Eqs. (4.1) and (4.3).

Classically, the rotational energy is given by

$$E_r = \frac{L^2}{2I} \tag{4.8}$$

Where L is the angular momentum of the system about the axis of rotation. Quantization of the magnitude of angular momentum gives

$$L^2 = r(r+1)\hbar^2$$

with *rotational quantum number* $r = 0, 1, 2, \ldots$, so that

$$E_r = \frac{\hbar^2}{2I} r(r+1) \tag{4.9}$$

Successive rotational levels will be separated in energy by

$$\Delta E_r = E_r - E_{r-1} = \frac{\hbar^2}{2I}[r(r+1)-(r-1)r] = \frac{\hbar^2}{I} r \tag{4.10}$$

The quantity $\frac{\hbar^2}{I}$ for the typical molecule has a value of about 10^{-4}eV to 10^{-3}eV, so little energy is needed to raise a molecule to an excited rotational state. At room temperature, for example, the translational thermal energy of molecules is 2.5×10^{-2} eV, so that ordinary collisions can transfer the necessary energy of excitation. At any given temperature the rotational state population obey the Boltzmann distribution, since they are spread over many states so each population is small.

4.2.1 Pure Rotational Spectrum of a Diatomic Molecule

Let us suppose that the electronic energy E_e of a molecule remains unchanged. In addition, let us assume that the vibrational energy E_v of the molecule also remains unchanged. This is possible if we assume that the two atoms of the diatomic molecule are as if joined together by a *rigid rod* as shown in Fig. (4.3), so that the distance between them (interatomic distance) remains unchanged as the molecule rotates as a whole. We call such a model of the molecule as a *rigid rotator model* as mentioned above, only the rotation about an axis perpendicular to the line joining the two atoms has to be considered.

Let the masses of the two atoms separated by a distance r be m_1 and m_2 and their distances from the centre of mass be r_1 and r_2 then, one can write

$$m_1 r_1 = m_2 r_2 \text{ and } r = r_1 + r_2 \tag{4.11}$$

The moment of inertia of the molecule (I) about an axis passing through CM and perpendicular to the bond is

$$\begin{aligned} I &= m_1 r_1^2 + m_2 r_2^2 \\ &= m_2 r_2 r_1 + m_1 m_2 r_1 \qquad \text{using Eq. (4.11)} \\ &= (m_1 + m_2) r_1 r_2 \end{aligned}$$

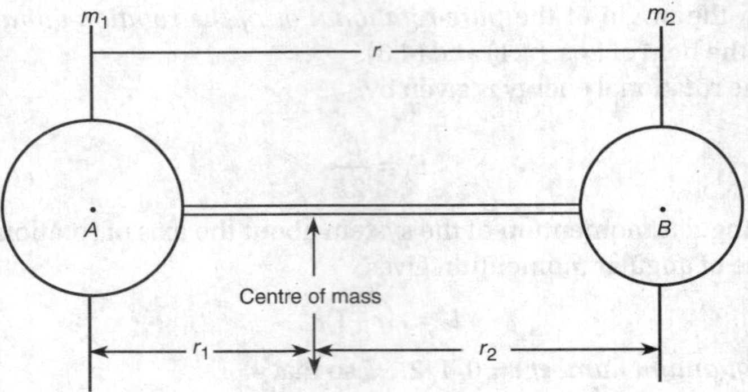

Fig. 4.3 (a) Diatomic molecule: rigid rotator

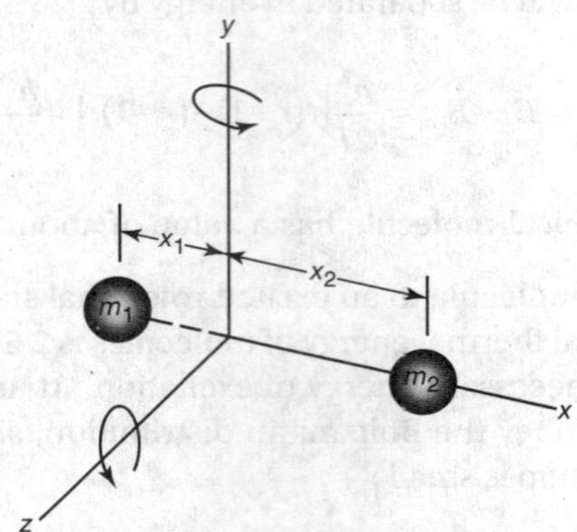

Fig. 4.3 (b) A rotating diatomic molecule

but

$$m_1 r_1 = m_2 r_2 = m_2 (r - r_1)$$

$$\therefore \qquad r_1 = \frac{m_2 r}{m_1 + m_2} \quad \text{and} \quad r_2 = \frac{m_1}{m_1 + m_2} r \tag{4.12}$$

Using Eq. (4.12) we obtain from Eq. (4.11) the moment of inertia of the molecule as

$$I = \frac{m_1 m_2}{m_1 + m_2} r^2 = \mu r^2 \tag{4.13}$$

where
$$\mu = \frac{m_1 m_2}{m_1 + m_2} \tag{4.14}$$

is called the *reduced mass* of the system, i.e., the molecule equivalent to that of a single point-mass μ about an axis distant r from it.

If ω be the angular velocity of rotation then the angular momentum L and the rotational kinetic energy E_r of the molecule are

$$L = I\omega,\ E_r = \frac{1}{2}I\omega^2$$

This gives

$$E_v = \frac{L^2}{2I} \tag{4.15}$$

For a rigid rotator, there is no variation of r so that the Hamiltonian becomes (See Section 4.2)

$$H = -\frac{\hbar^2}{2I}\left\{\frac{1}{\sin\theta}\frac{\partial}{\partial\theta}\sin\theta\frac{\partial}{\partial\theta} + \frac{1}{\sin^2\theta}\frac{\partial^2}{\partial\phi^2}\right\} \tag{4.16}$$

We may note that the potential energy term in the Hamiltonian is absent since there is no vibration. One can solve the Schrodinger wave equation $H\Psi = E\Psi$ by the separation of variables (See Section 4.2). The two equations in θ and ϕ can be solved to yield the eigen-values of the square of the angular momentum operator L^2 as $J(J+1)\hbar$. Thus we have

$$L_r = \sqrt{J(J+1)}\,\hbar \qquad (\hbar = h/2\pi) \tag{4.17}$$

where J is the *rotational quantum number* and the allowed values of J are given by $J = 0$, 1, 2, ... Clearly, the molecule can thus have a discrete set of rotational energy levels with different values of J.

Now the energy of a rotating molecule is $\frac{1}{2}I\omega^2$, so the rotational energy states are

$$E_r = \frac{1}{2}I\omega^2 = \frac{L_r^2}{2I}$$

$$= \frac{\hbar^2}{2I}J(J+1)$$

$$= \frac{h^2}{8\pi I}J(J+1) \tag{4.18}$$

The molecule can thus have a discrete set of rotational energy levels, each corresponding to a given J-value. These constitute the pure rotational levels as shown in Fig. (4.4).

Like atomic spectra, the energies of the different molecular levels are expressed by their *term values* obtained by dividing the energy by $2\pi\hbar c$. We may note that the unit of the term value is m^{-1}. The difference between the term values of two levels gives the *wave number* of the emitted (or absorbed) spectral line due to transition between the two levels. The reciprocal of this gives us the wavelength.

Expressing the energy in *wave number*, one obtains

$$\frac{E_r}{ch} = \frac{h}{8\pi^2 Ic}J(J+1) = BJ(J+1)\ F(J)$$

where

$$B = \frac{\hbar}{4\pi Ic} \text{ and } F(J) = J(J+1)$$

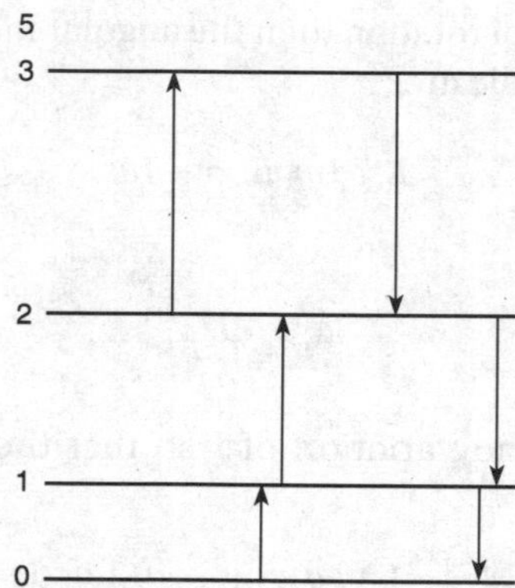

Fig. 4.4 Discrete set of pure rotational energy levels of a molecule

For a given diatomic molecule, B is a constant and is known as the *rotational constant*. $F(J)$ is referred as the *rotational term value*, having unit m^{-1}.

According to electromagnetic theory of light, rotational motion of only the *polar molecules* e.g., HCl, HF, CO, etc., lead to the emission of radiation. In such polar molecules, the centres of positive and negative charges are permanently separated so that they posses permanent electric dipole moments. In non-polar coincident, so that they do not posses permanent electric dipole moments. In non-polar molecules, e.g., *homonuclear* molecules like H_2, O_2, N_2 etc., the centers of positive and negative charges are *coincident so that do not posses* any permanent dipole moments. In accordance with the e.m. theory, the radiation of a dipole can lead to the emission of radiation, there will be emission of radiation due to pure rotation of only polar molecules. Applying Bohr's correspondence principle, one can conclude that even in the quantum mechanical case, pure rotational spectrum can arise only in the case of polar molecules. The allowed transitions between the rotational states of a diatomic molecule are however restricted by the selection rule

$$\Delta J = \pm 1 \tag{4.19}$$

i.e., the transitions leading to a change in J by one unit are only allowed, the rest being forbidden.

The change in rotational energy in allowed transition is

$$\Delta E = E_{J+1} - E_J = \frac{2h^2}{8\pi^2 I}(J+1) = \frac{\hbar^2}{I}(J+1)$$

but

$$\Delta E = h\nu$$

The photon involved in the transition has the frequency υ, given by

$$\nu = \frac{\Delta E}{H} = \frac{2h}{8\pi^2 I}(J+1) = \frac{\hbar}{2\pi I}(J+1)$$

∴ The *wave number* of the rotational spectral line is given by

$$\bar{\nu} = \frac{\nu}{c} = \frac{2h}{8\pi^2 Ic}(J+1) = \frac{\hbar}{2\pi Ic}(J+1) = 2B(J+1) \tag{4.20}$$

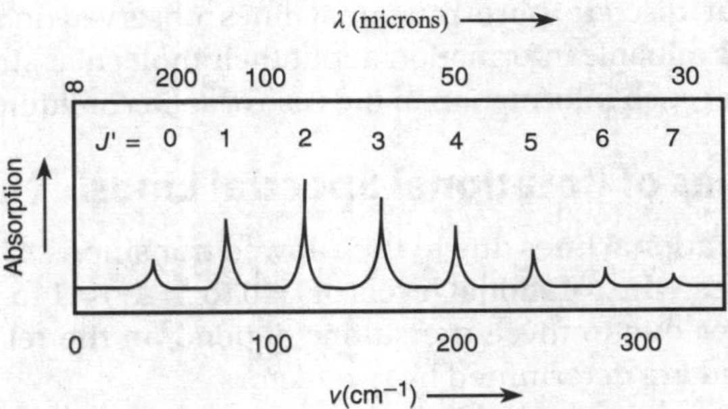

Fig. 4.5 Equispaced pure rotational spectral lines in absorption spectrum of HF molecule

Obviously, the spacing between the consecutive rotational lines is thus independent of J, given by

$$\Delta\bar{\nu} = 2B = \hbar / 2\pi Ic$$

Thus the pure rotational spectrum of a diatomic molecule of the type under consideration consists (Fig. 4.4) of a series of *equally spaced lines*; the *spacing between the consecutive lines* in wave number is $2B$ (the first line occurring at $2B$), independent of J. Fig. (4.5) shows the pure rotational absorption spectrum of HF molecules. For an emission line, J refers to the initial state and for an absorption line to the final state for molecular transitions. The wave number difference between the consecutive rotational lines is found to be $\Delta\bar{\nu} = 4050\,\text{m}^{-1}$ in this case.

From Fig. (4.5), the wave numbers $\bar{\nu}$ of the spectral lines in this case are of the order of a few hundred cm^{-1}. Writing $\bar{\nu} = 200\,\text{cm}^{-1} = 20{,}000\,\text{m}^{-1}$ for a typical rotational absorption line, we obtain its wavelength as $\lambda = 1/\bar{\nu} = 50\times10^{-6}\,\text{m} = 50\mu$, which is in the far infra-red. The pure rotational lines lie in the extreme infra-red and microwave region: 10^{-4} to 10^{-1} m in wavelength.

One can estimate the moment of inertia and hence the distance between the two atoms (bond-length) of a diatomic molecule from the measured value of the rotational constant B. For the HF molecule, one obtains

$$B = \frac{\Delta\bar{\nu}}{2} = \frac{4050}{2} = 2025\,\text{m}^{-1}$$

$$\therefore \qquad B = \frac{\hbar}{4\pi Ic} = 2025$$

$$\therefore \qquad I = \frac{\hbar}{4\pi Bc} = \frac{1.05\times10^{-34}}{4\pi\times2025\times3\times10^{8}} = 1.38\times10^{-47}\,\text{kg-m}^2$$

The reduced mass of HF molecule is

$$\mu = \frac{m_1 m_2}{m_1 + m_2} = \frac{1\times19}{1+19}\times1.66\times10^{-27}$$

Since $I = \mu r^2$, we get the interatomic distance $r = \sqrt{\dfrac{I}{\mu}} = \left(\dfrac{1.38\times10^{-47}}{1.58\times10^{-27}}\right)^{1/2}$

$$= 0.935\times10^{-10}\,\text{m} = 0.935\text{Å}$$

We may note that discrete pure rotational lines observed in the case of gaseous substances can yield valuable information about their molecular structure. However, it is not an easy task to get such information in the case of solids or liquids.

4.2.2 Intensities of Rotational Spectral Lines

The intensities of rotational lines due to the allowed transitions $\Delta J = \pm 1$, i.e., transitions will take place between the rotational levels of $j = 0$ to $j = 1$, $j = 1$ to $j = 2$, $j = 2$ to $j = 3$ and so on. The intensities due to these transitions depend on the relative probabilities of these transitions, and are determined by two factors:

(i) The relative number of molecules present initially at each level- the Boltzmann factor and

(ii) The *degeneracy* or the *statistical weights* of these states.

According to Maxwell-Boltzmann (MB) statistics, the relative population in the levels is given by

$$\frac{N_J}{N_o} = \exp(-E_J / kT) \qquad (4.21)$$

$$= \exp[-BhcJ\,(J+1)/kT] \qquad (4.22)$$

where B is the rotational constant, $k_B = 1.38 \times 10^{-23}$ J/K is the Boltzmann constant.

At room temperature, say $T = 300$ K and $B = 2\text{ cm}^{-1}$, the relative population in the state $J = 1$ is

$$N_1 = N_0 \exp\left[\frac{-200 \times 6.63 \times 10^{-34} \times 3 \times 10^{10} \times 1 \times 2}{1.38 \times 10^{-28} \times 300}\right]$$

$$= N_0 \exp(-0.019) \simeq 0.98 N_0$$

The above estimate clearly reveals that at room temperature, the number of molecules in the $J = 1$ state is almost equal to the number of molecules in the lowest ($J = 0$) state.

The *degeneracy* or the *statistical weight* of the state of a given J is $(2J + 1)$, i.e., $(2J + 1)$ fold degeneracy. The total population of the level J is determined by the factor $\beta = (2J + 1)$ $\exp(E_J/k_B T)$. When plotted against J, this shows a maximum at a J-value determined by B and T and then diminishes.

Thus the intensities of the rotational lines for transitions $J \rightarrow J + 1$ will be maximum for

$$J = \sqrt{\frac{kT}{2Bhc}} - \frac{1}{2} \qquad (4.23)$$

This is obtained readily by differentiating β and equating $d\beta/dJ$ to zero.

4.3 ROTATION-VIBRATION SPECTRA

We can say that the rigid rotator model of a diatomic molecule is an idealized model. A diatomic molecule, when sufficiently excited, can vibrate as well as rotate. However, the vibrational energy, being greater than the rotational energy, the vibration is always accompanied by rotation.

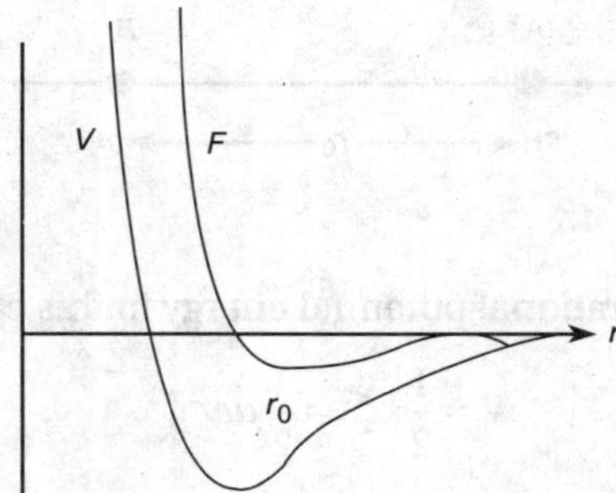

Fig. 4.6 Plot of potential energy V and interatomic force F against the distance r between two atoms of a molecule

Actually the two atoms vibrate along the line joining them, as if they are connected by a massless spring. The minimum distance between the two atoms is the equilibrium distance r_0 about which the molecule vibrates along the bond length. Thus the distance r between the atoms changes in a periodic manner about a mean value r_0. If the amplitude of vibration Δr is small in comparison to r_0, then the vibration may by regarded as simple harmonic. Otherwise it will not be simple harmonic.

Any attempt to distort the bond between the atoms requires an input of energy. The force F between the two atoms in the molecule is a function of the distance r between them corresponding to force F, the molecule has a potential energy $V = V(r)$ such that $F = -\partial V/\partial r$. We may plot the potential energy V against the internuclear distance r (Fig. 4.6). We assume $V = 0$ for very large r. Near the minimum of the curve, marked V which corresponds to the normal configuration of the molecule, we note that the shape of the curve is a close approximation to a parabola. It is evident from the figure that at $r = r_0$, V has the minimum value $-V_0$ and $F = 0$ at this point.

For $r > r_0$, the force F is negative, i.e., attractive, while for $r < r_0$ it is positive, i.e., repulsive. Under the influence of such a force the two atoms would be at rest at the equilibrium distance r_0. If however, atoms are disturb even slightly they will vibrate about this point. If the system is given a kinetic energy greater than $-V_0$ then the atoms will fly apart and the molecule will dissociate. Near the minimum of the curve, marked V which corresponds to the normal configuration of the system, i.e., molecule, the shape of the curve (Fig. 4.6) is a close approximation to a parabola. In this region, one finds

$$V = V_0 + \frac{1}{2}k(r - r_0)^2 \tag{4.24}$$

where V represents the potential energy and r_0 the equilibrium separation of the atoms. The interatomic force F giving rise to this potential energy is given by the slope of the potential energy (V) curve, i.e.,

$$F = -\frac{dV}{dr} = -k(r - r_0) \tag{4.25}$$

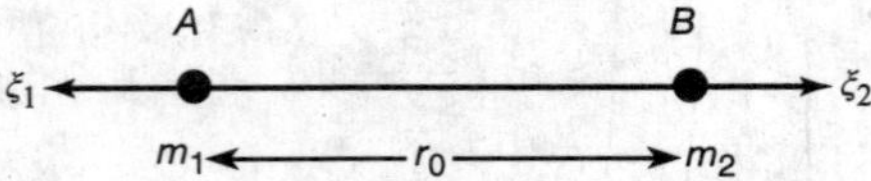

Fig. 4.7

Writing $r - r_0 = \Delta r = \xi$, the vibrational potential energy in this case can be written as

$$V = \frac{1}{2}k\xi^2 = \frac{1}{2}\mu\omega^2\xi^2 \tag{4.26}$$

where $k = \mu\omega_0^2$ is the *force constant* and $\mu = \dfrac{m_1 m_2}{m_1 + m_2}$ is the reduced mass. m_1 and m_2 are the masses of two atoms Fig. (4.7). Let ξ_1 and ξ_2 represent the instantaneous distances of the two atoms of masses m_1 and m_2 from their equilibrium positions A and B, then one can write

$$\xi = \xi_1 + \xi_2 \tag{4.27}$$

$$m_1\xi_1 = m_2\xi_2 \tag{4.28}$$

From Eqs. (4.27) and (4.28), one obtains

$$\xi_1 = \frac{m_2\xi}{m_1 + m_2}, \xi_2 = \frac{m_1\xi}{m_1 + m_2} \tag{4.29}$$

Now, the vibrational kinetic energy of the atoms is

$$E_k = \frac{1}{2}m_1\dot{\xi}_1^2 + \frac{1}{2}m_2\dot{\xi}_2^2 = \frac{1}{2}\mu\dot{\xi}^2 \tag{4.30}$$

The total vibrational energy is

$$E_v = E_k + V = \frac{1}{2}\mu\dot{\xi}^2 + \frac{1}{2}k\xi^2 = p^2/2\mu \tag{4.31}$$

One can write the Hamiltonian for the vibrational motion as

$$H = -\frac{p^2}{2\mu} + V = -\frac{\hbar}{2\mu}\frac{\partial^2}{\partial\xi^2} + \frac{1}{2}\mu\omega_0^2\xi^2 \tag{4.32}$$

Schrodinger equation can be written as

$$H\psi = -\frac{\hbar^2}{2\mu}\frac{\partial^2\psi}{\partial\xi^2} + \frac{1}{2}\mu\omega_0^2\xi^2\psi = E_v\psi \tag{4.33}$$

The solution of Eq. (4.33) is

$$\psi = \left(\frac{\sqrt{\alpha}}{2^v v!\sqrt{\pi}}\right)^{1/2} \exp(-\alpha\xi^2/2)H_v(\sqrt{\alpha}\xi) \tag{4.34}$$

where $\alpha = m\omega_0/\hbar . H_v(\sqrt{\alpha}\xi)$ is the Hermite polynomial of order v. Now, writing $\sqrt{\alpha}\xi = q$, one gets

$$H_v(q) = (-1)^v \exp(q^2)\left(\frac{d}{dq}\right)^v \exp(-q^2) \tag{4.35}$$

The energy eigen values are obtained as

$$E_v = (v+\frac{1}{2})\hbar\omega_0 = (v+\frac{1}{2})\hbar\nu_0 \tag{4.36}$$

where $v = 0, 1, 2, 3, \ldots$ Obviously, the frequencies are thus quantized and v is known as the *vibrational quantum number*. Expressing the energy in wave number the vibrational term becomes

$$G_v = \frac{E_v}{hc} = (v+\frac{1}{2})\bar{\nu}_0 \tag{4.37}$$

where

$$\bar{\nu}_0 = \nu_0/c = \omega_0/2\pi c$$

Quantum-mechanically, therefore the energy can have only certain *equally spaced discrete values* given by relation in Eq. (4.36). For $v = 0$, we obtain the *zero-point energy*, $E_0 = \hbar\omega/2 = \frac{h\nu_0}{2}$, where v_0 is called proper frequency. The selection rule for the allowed transitions for linear harmonic oscillator is $\Delta v = \pm 1$ (rotational motion is not taken into account and only pure vibrational levels are governed by this selection rule). Obviously, the transition between its vibrational states always results in radiation of a single wave number $\bar{\nu}_0$, no matter what are the vibrational quantum numbers of the combining states. The spacing between such idealized pure vibrational levels are given by

$$\Delta G(v) = \bar{\nu}_0 \tag{4.38}$$

When electromagnetic radiation is incident on a diatomic molecule, its electric vector may cause a transition if the electric dipole moment of the molecule varies with the vibrational coordinate ξ. We may note that *homonuclear* diatomic molecules e.g., O_2, N_2, etc., have no permanent electric dipole moments for any value of ξ, there will be no such transition for them. Interestingly homonuclear molecules do not have any vibrational infrared spectrum. Such molecules also do not have pure rotational infrared spectrum either.

For a *heteronuclear* diatomic molecule the dipole moment is found to be a function of the internuclear separation. If we assume linear variation of the dipole moment with the distance between the nuclei, one can easily show from the time dependent perturbation theory that transitions between the different vibrational levels in such molecules are governed by the selection rule: $\Delta v = \pm 1$. The energy level spectrum corresponds to that of a linear harmonic oscillator so that the energy levels are equispaced (Fig. 4.8). The selection rules $\Delta v = \pm 1$ show that the allowed transitions between the vibrational levels give rise to a single *absorption band* with the wave number given by

$$\bar{\nu}_v = G(v+1) - G(v) = \bar{\nu}_0 \tag{4.39}$$

We may note that this agrees approximately with the experimental observation. The infrared spectra of heteronuclear diatomic molecules, e.g., HCl or CO show only one strong absorption band. One can determine the force constant for the concerned molecule from a measurement of the wave number as

$$\bar{\nu}_v = \bar{\nu}_0 = \frac{\nu_0}{c} = \frac{\omega_0}{2\pi c} = \frac{1}{2\pi c}\sqrt{\frac{k}{\mu}}$$

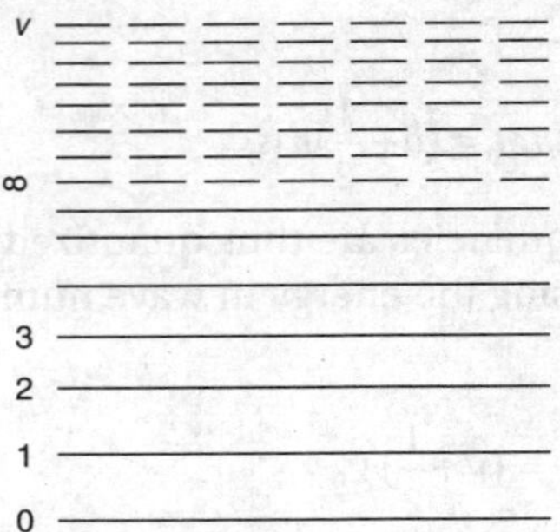

Fig. 4.8 Converging energy levels of a diatomic molecule

The force constant, k is given by

$$k = 4\pi^2 c^2 \mu \bar{\nu}_0^2 \tag{4.40}$$

Experimental values of $\bar{\nu}_0$ for HCl and CO molecules are $2.886 \times 10^5\ \text{m}^{-1}$ and $2.143 \times 10^5\ \text{m}^{-1}$ respectively. From these values, the force constants are obtained as

$$k(\text{HCl}) = 480.7\ \text{N/m and k(CO)} = 1855\ \text{N/m}$$

The above results clearly reveals that the CO bond strength is much stronger than the HCl bond strength. The amplitude ξ_0 of the classical turning point for the round state with $\nu = 0$ is given by

$$E_0 = \frac{1}{2}\mu\omega_0^2\xi_0^2 = \frac{\hbar\omega_0}{2}$$

$$\therefore \quad \xi_0 = \left(\frac{\hbar}{\mu\omega_0}\right)^{1/2} = \left(\frac{h}{4\pi^2 \mu \bar{\nu}_0 c}\right)^{1/2} \tag{4.41}$$

For the values of $\bar{\nu}_\nu = \bar{\nu}_0$ given above, one obtains $\xi_0(\text{HCl}) = 0.11$Å and $\xi_0(\text{CO}) = 0.05$Å. The corresponding equilibrium bond lengths of these molecules are obtained as 1.275Å and 1.128Å respectively. Obviously, even in the ground state, the amplitudes of vibration of the atoms in the diatomic molecules are 5% to 10% of the bond lengths.

According to Eq. (4.36), the pure vibrational levels should be equispaced. However, for large vibrational quantum numbers, the vibrational amplitude ξ becomes progressively larger and hence the vibration is no longer simple harmonic. Consequently, for large ν the energy levels are equispaced but converge slowly as shown in Fig. (4.8). As $\nu \to \infty$, the internuclear separation r becomes very large and the energy levels extend upto a *convergence limit.* At this point the molecule dissociates into its constituent atoms.

4.3.1 Anharmonic Oscillator

One finds that the selection rules, $\Delta\nu = \pm 1$ are not strictly obeyed even for lower values of ν. This is due to the fact that even for $\nu = 0$, ξ is a significant fraction of bond length. It is found that the permanent dipole moments of heteronuclear molecules are strictly linear in ξ only for small values of ξ in which case the selection rules, $\Delta\nu = \pm 1$ are strictly obeyed.

However, due to relatively large values of ξ, anharmonic vibrational bands appear with $\Delta v > 1$, e.g., in the case of HCl, anharmonic band for 2–0 ($v' = 0$ to $v'' = 2$) and 3–0 transitions appear at $\bar{v} = 5.668 \times 10^5 \text{ m}^{-1}$ and $\bar{v} = 8.347 \times 10^5 \text{ m}^{-1}$. However, these are slightly smaller than twice and three times the wave number $2.866 \times 10^5 \text{ m}^{-1}$ of the fundamental 1–0 band mentioned above. This reveals that even for such low values of v, these energy levels are not strictly equispaced.

One can represent the anharmonicity of the vibrational levels by a relation of the type

$$G_v = \bar{v}_e(v + \frac{1}{2}) - \alpha\bar{v}_e(v + \frac{1}{2})^2 \tag{4.42}$$

where α is the anharmonicity parameter which is small and positive so that the energy levels crowd together with increasing v.

We can see that relation Eq. (4.42) replaces the relation Eq. (4.36) for the harmonic oscillator. $\bar{v}_e$ depends on the force constant k for infinitesimal vibrations. For HCl, one obtains the values of $\bar{v}_e$ and $\alpha\bar{v}_e$ as $2.990 \times 10^5 \text{ m}^{-1}$ and $5.2 \times 10^3 \text{ m}^{-1}$ respectively. Obviously, $\bar{v}_e$ differs slightly from $\bar{v}_0$ in Eq. (4.34). The latter is $2.886 \times 10^5 \text{ m}^{-1}$ for HCl which determines the mean force constant for the amplitude of $v = 1$ level.

Sometimes a small cubical term is to be added on the R.H.S. of Eq. (4.42) for more accurate representation of the anharmonicity of the levels.

P.M. Morse has suggested an empirical potential of the form to explain the gradual convergence of the vibrational levels for higher values of v.

$$V(r) = D[1 - \exp\{-\beta(r - r_o)\}]^2 \tag{4.43}$$

where D and β are constants. Morse potential is a fairly correct representation for larger r. However, Morse potential does not go to infinity as $r \to \infty$ but, remains constant (Fig. 4.6). If one substitutes it in the Schrodinger equation in place of the harmonic oscillator potential, then the equation can be solved exactly. The term values are then given exactly by Eq. (4.39). However the eigen functions no longer can be classified according to parity as in the case of the linear harmonic oscillator. We may note that they have greater amplitudes for $\xi > 0$ than for $\xi < 0$ (anharmoic about $\xi = 0$) which is responsible for the violation of the selection rules.

We can write the selection rules for the anharmonic oscillators as $\Delta v = \pm 1, \pm 2, \pm 3, \ldots$. However, in practice the transitions with the vibrational level spacings are of the order of 10^5 m^{-1}, the ratio of the populations between the $v = 1$ and $v = 0$ levels is (at room temperature).

$$\begin{aligned} \frac{N_1}{N_0} &= \exp(-E_1 / k_B T) = \exp(\Delta G_0 hc / k_B T) \\ &= \exp[\frac{10^5 \times 6.63 \times 10^{-34} \times 3 \times 10^8}{1.38 \times 10^{-23} \times 300}] \\ &= \exp(-4.8) = 0.008 \end{aligned} \tag{4.44}$$

Obviously, the population of the $v = 1$ state is less than 1% of the population in the $v = 0$ state. Thus, one can take all absorptive type of vibrational transitions to originate from the $v = 0$ level. The three transitions of importance are as follows:

(i) $v = 0$ to $v' = 1$ ($\Delta v = 1$);

$$\Delta G_v = G_1 - G_0 = 2\bar{v}_e(1-2\alpha) \tag{4.45}$$

This has a large intensity.

(ii) $v = 0$ to $v' = 2$ ($\Delta v = 2$);

$$\Delta G_v = G_2 - G_0 = 2\bar{v}_e(1-3\alpha) \quad \text{and} \tag{4.46}$$

(iii) $v = 0$ to $v' = 3$($\Delta v = 3$);

$$\Delta G_v = G_3 - G_0 = 3\bar{v}_e(1-4\alpha) \tag{4.47}$$

We may note that the last two bands are of diminishing intensities, much weaker than the first.

The value of α is small (~0.001) and therefore the three lines lie close to $\bar{v}_e$, $2\bar{v}_e$, and $3\bar{v}_e$. The first of these is termed as the *fundamental absorption band* and the other two bands are the first and second *overtones*. The above conclusions agree with the observed wave numbers in the case of the HCl bands.

It is found that at higher temperatures the population of the $v = 1$ level may become appreciable so that the transition $v = 1$ to $v' = 2$ may have appreciable probability. At $T =$ 600 K, the population of the $v = 1$ level is about 9% of that of the $v = 0$ level and the 1–2 band has an intensity of about 10% of that of the 0–1 band.

4.3.2 Molecular Energy Levels due to both Rotation and Vibration

The molecular energy due to both rotation and vibration can be expressed by

$$E = E_v + E_r = (v + \frac{1}{2})\hbar\omega_0 + \frac{\hbar^2}{2I}l(l+1) \tag{4.48}$$

In general, the quantity $\hbar^2/2I$ is much smaller (~10^{-4}eV) than $\hbar\omega_0$(~10^{-1} eV), and one may say that to each vibrational level there correspond several rotational levels as shown in Fig. 4.9. When the selection rules $\Delta v = \pm 1$ and $\Delta l = \pm 1$ are taken into account for a transition between two rotational levels belonging to two adjacent vibrational levels, the rotation-vibration spectrum results. This consists of lines whose frequencies are given by

$$\nu_{v,r} = \nu_0 \pm 2Bc\,(l+1)$$
$$\tilde{\nu}_{v,r} = \tilde{\nu}_0 \pm 2B\,(l+1), \qquad l = 0, 1, 2, \ldots \tag{4.49}$$

where l refers to the lower rotational level. The frequencies are equally spaced by an amount $2Bc$ on both sides of the central frequency ν_0. The line of frequency ν_0 is missing from the spectrum because it requires a transition with $\Delta l = 0$, which is forbidden. Fig. (4.10) illustrates the possible rotational transitions for the rotational transitions $v = 1 \rightarrow v = 0$. Those lines with $\Delta l = +1$ are called the R-branch of the spectrum; the name P-branch is used to designate those lines with $\Delta l = -1$. The intensities of different transitions for emission and absorption depend on the population of the initial energy level, which in turn depends on the temperature of the substance.

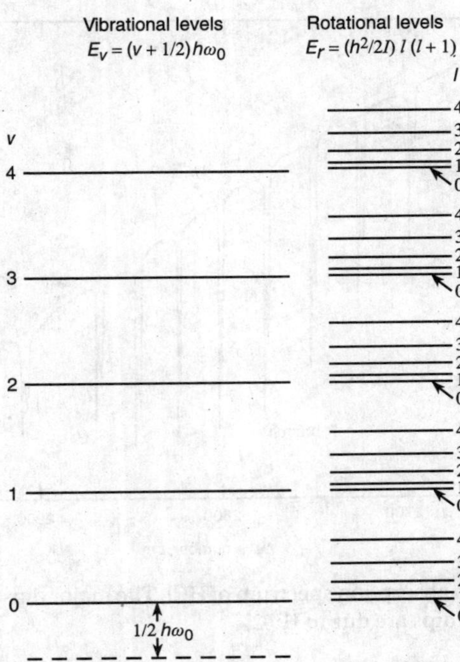

Fig. 4.9 Vibrational and rotational energy levels of a diatomic molecule

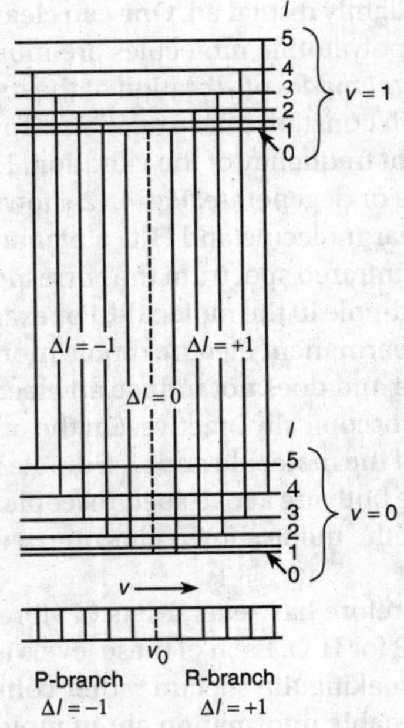

Fig. 4.10 Vibration-rotation transitions in a diatomic molecule

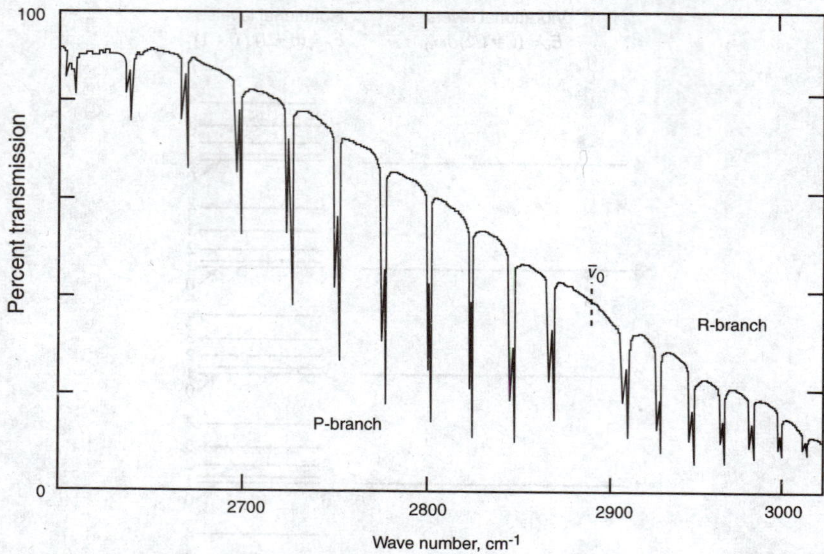

Fig. 4.11 Vibration-rotation absorption spectrum of HCl. The major dips in the spectrum are due to $H^{35}Cl$ and smaller dips are due to $H^{37}Cl$

The rotation-vibration spectrum of HCl is shown in Fig. 4.11. Because both ν_0 and I depend on the mass of the molecules, the spectra of molecules having different isotopes, such as $H^{35}Cl$ and $H^{37}Cl$, are slightly displaced. One can clearly recognize this in the figure.

The vibrational spectra of polyatomic molecules are more complex. These spectra are analyzed in terms of the *normal modes* of vibration of the molecules, which are vibrations in which all nuclei vibrate with constant phase relations. To each normal mode there corresponds in general, a different frequency of the vibration. However, symmetry may make some of the frequencies equal or degenerate. Fig. 4.12 shows the normal vibrations of two simple molecules: CO_2, a linear molecule and H_2O a planar molecule. For a frequency to be active in the vibrational or infrared spectrum the corresponding normal vibration must induce an oscillating electric dipole in the molecule. For example, consider the symmetric CO_2 molecule, which has no permanent electric dipole moment. Oscillation ω_1 preserves the symmetry of the molecule and does not induce an electric dipole moment; therefore the vibrational mode is spectroscopically inactive. On the other hand, ω_3 induces a dipole molecule parallel to the axis of the molecule and ω_2 a dipole moment perpendicular to the axis of the molecule; therefore both are active spectroscopically. Incidentally, ω_2 is doubly degenerate, since it consists if the mode shown plus one in which the atoms oscillate perpendicular to the page.

A polyatomic molecule therefore has several sets of vibrational energy levels. Some of them are illustrated in Fig. 4.13 for H_2O. Each of these levels is associated with corresponding rotational levels, thereby making the spectra rather complex. An analysis of rotation-vibration spectra provides valuable information about molecular dimensions, geometry, dissociation energy and so on.

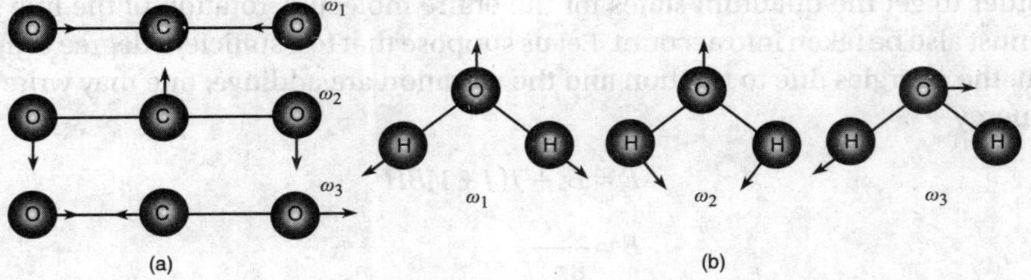

Fig. 4.12 Normal vibrations of polyatomic molecules:

(a) $CO_2 = \omega_1 = 1337 cm^{-1}, \omega_2 = 667 cm^{-1}\ \omega_3 = 2349 cm^{-1}$

(b) $H_2O = \omega_1 = 3657 cm^{-1}, \omega_2 = 1595 cm^{-1}\ \omega_3 = 3756 cm^{-1}$

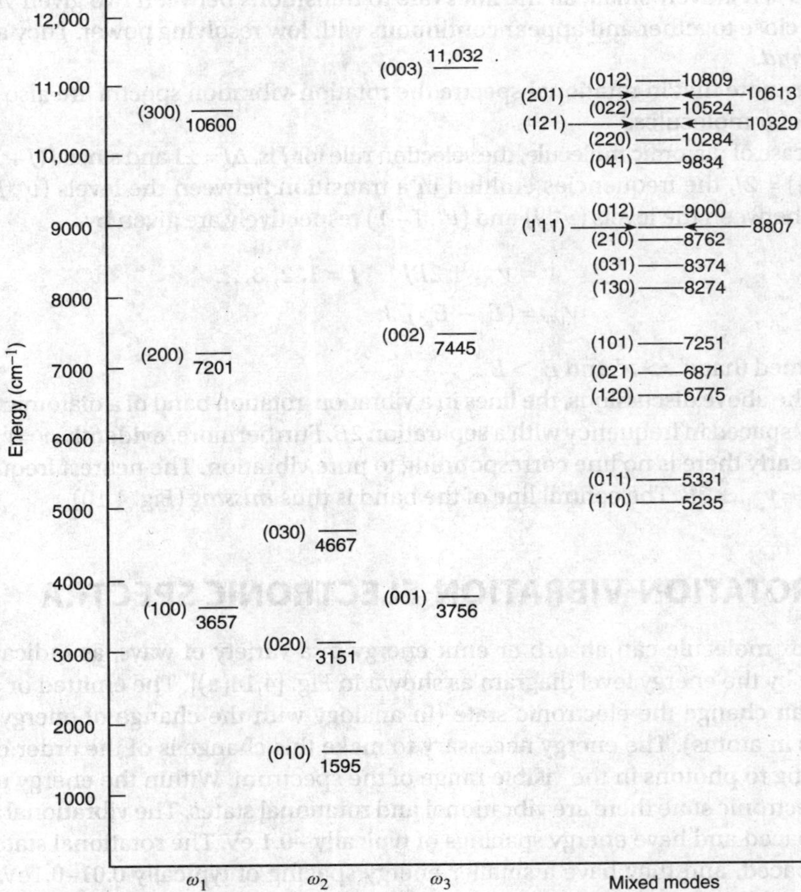

Fig. 4.13 Vibrational levels of H_2O in the gaseous phase. The energies are expressed in cm^{-1}. The numbers in parentheses correspond to the vibrational mode (n_1, n_2, n_3)

In order to get the quantum states for the entire molecule, rotation of the line joining them must also be taken into account. Let us suppose that to a sufficient degree of approximation, the energies due to rotation and the vibration are additive, one may write for the total energy

$$E = E_v + J(J+1)BH \tag{4.50}$$

where $$B = \frac{h}{8\pi^2 I}$$

The different E_v-values are called *vibrational levels* of the molecule and different E values, the *vibration-rotation levels*. We may note that Bh is usuallv very small, compared to the difference between successive values of E_v, the rotational levels corresponding to a given vibrational level form a closely spaced group. Fig. (4.13) shows the general arrangement of the rotation-vibration levels.

Since B is relatively small, all the lines due to transitions between two given vibrational levels lie close together and appear continuous with low resolving power. They are said to form a *band.*

We may note that in rotational spectra the rotation-vibration spectra are also observed only in *polar* molecules.

In the case of diatomic molecule, the selection rule for J is, $\Delta J = \pm 1$ and since $J(J+1) - (J-1)(J-1+1) = 2J$, the frequencies emitted in a transition between the levels $(v', J-1)$ and (v'', J) or between the levels (v', J) and $(v'', J-1)$ respectively are given by

$$\nu = \nu_{v'v''} \pm 2BJ, \quad J = 1, 2, 3, \ldots$$

$$\therefore \quad \nu_{v'v''} = (E_{v'} - E_{v''})/h$$

It is assumed that $v' >> v''$ and $E_{v'} > E_{v''}$.

From the above discussions, the lines in a vibration-rotation band of a diatomic molecule is equally spaced in frequency with a separation $2B$. Furthermore, evidently, $\nu = \nu_{v'v''}$ cannot occur. Clearly there is no line corresponding to pure vibration. The nearest frequencies to this are $\nu = \nu_{v'v''} \pm 2B$. The central line of the band is thus *missing* (Fig. 4.10).

4.4 ROTATION-VIBRATION-ELECTRONIC SPECTRA

A complex molecule can absorb or emit energy in a variety of ways, as indicated schematically by the energy level diagram as shown in Fig. [4.14(a)]. The emitted or absorbed energy can change the electronic state (in analogy with the change of energy levels of electrons in atoms). The energy necessary to make this change is of the order of eV, corresponding to photons in the visible range of the spectrum. Within the energy minimum of any electronic state there are vibrational and rotational states. The vibrational states are evenly spaced and have energy spacings of typically –0.1 eV. The rotational states are not evenly spaced, and they have a smaller energy spacing of typically 0.01–0.1eV. Because the rotational spacing is much smaller than the vibrational spacings, it is convenient to consider each vibrational state as providing the basis on which a sequence of rotational states is built.

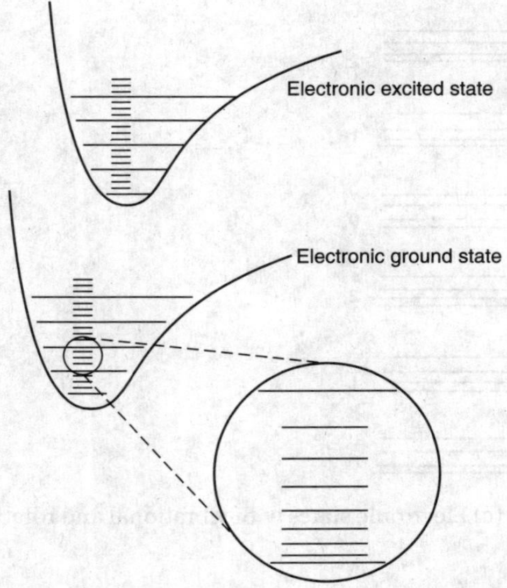

Fig. 4.14 (a) Molecular electronic, vibrational and relational energy levels

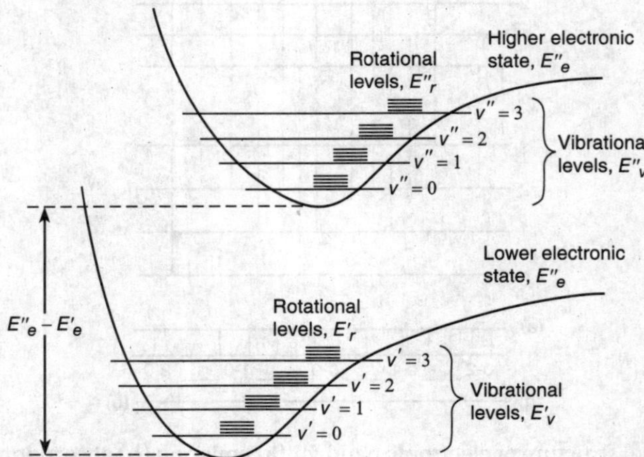

Fig. 4.14 (b) Vibrational and rotational energy levels associated with two electronic states

When enough energy is available in the molecules, molecular electronic states may as well be excited along with those of vibration and rotation. This leads to the most general type of transitions and the general molecular spectra Fig. [4.14(b)] and [4.14(c)].

The transitions between the levels in the two electronic states give rise to the *electronic band spectrum*. The transitions are governed by the selection rules

$$\Delta v = \pm 1, \pm 2, \pm 3, \ldots$$
$$\Delta J = \pm 1 \tag{4.51}$$

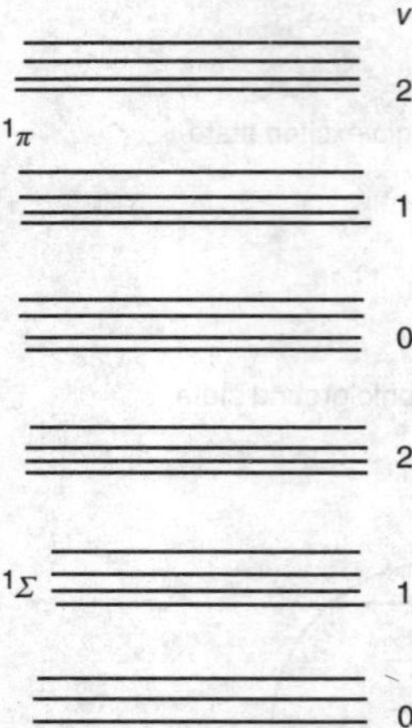

Fig. 4.14 (c) Electronic states with vibrational and rotational level

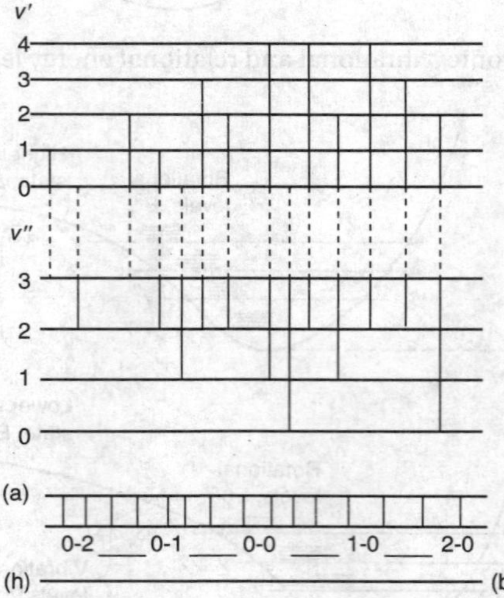

Fig. 4.15 Vibrational structure of electronic band for (a) emission (b) absorption

We may note that the electronic bands have the same character as the rotation-vibration bands. Only difference is that they, i.e., in the visible region, which makes their study easier than the rotation-vibration bands. One can study the electronic bands both in emission and absorption. At the room temperatures, most of the molecules are in the $v'' = 0$ vibrational level of the lowest electronic state. When we irradiate with visible radiation, they are raised to the different vibrational levels v' of the upper electronic state. These absorption bands constitute a *progression*. Fig. (4.15) shows the absorption bands 0–0, 1–0, 2–0 etc. Fig. (4.15) shows the progression of the vibrational levels of the upper electronic state. From the study of these bonds one can assign the correct vibrational quantum numbers to these levels from a study of these bands. On the other hand, the emission spectra are usually much more complex. At the higher temperatures necessary to excite these spectra, a large number of molecules are raised to the different vibrational levels of the upper electronic states. The transitions from these to the different vibrational levels of the lower electronic states give rise to a large number of emission bands. Obviously, it is much more difficult to assign correct vibrational quantum numbers to the associated levels from a study of the emission bonds. However, one important advantage of the emission spectra is that it is possible to study the high vibrational levels of the ground state.

As stated above, when enough energy k is available in the molecules, their electronic states may as well be excited along with those of vibration and rotation. This leads to the most general type of transitions and general molecular spectra. These transitions would produce photons of energy. Since different energies can be added,

we have $$E = E_e + E_v + E_r$$

Where E_e refers to the electronic energy at the minimum of the potential energy curve. In an electronic transition all three energy may change. The frequency ω_0 and the moment of inertia I are, in general, different for both electronic states.

or $$h\nu = h\nu_e + (v + \frac{1}{2})\, h\nu_0 + J(J+1)\frac{h^2}{8\pi^2 I} \tag{4.52}$$

The electrons which are involved in the process are the outermost electrons of the atoms constituting the molecules. The transitions are energetic enough to produce electronic bands in the visible and ultraviolet regions. As stated above for every electronic state, there are a number of vibration states and for each of the vibration states there are a good number of associated rotational states. In transitions involving a given pair of electronic sates, and characterised by given values of v' and v'' of v, various changes of J may occur. The resulting lines form a *single band.* All bands due to transition between a given pair of elecronic states for all possible v' and v'', constitute a *band system*. We may note that various electronic jumps are possible and therefore the band spectra consist of many band systems. The selection rules for lines in a given electronic band are

$$\Delta J = 0 \quad \text{or} \quad \pm 1$$

Lines for which ΔJ has sign opposite to Δv are said to form the P branch of the band and those for $\Delta J = 0$ constitute the Q branch; those for which ΔJ is in the same branch as ΔV form the R branch. However, for special reasons the Q-Branch is frequently missing. Fig. (4.16) shows graphically the variation of wave numbers of the lines of the P, Q and R branches. These graphs are called *Fortat diagrams.*

4.5 FRANK-CONDON PRINCIPLE: INTENSITY DISTRIBUTION

This principle is used in determining the intensity of band spectra. The time required for molecular vibrations (~10^{-12} s) is much longer than that required for an electronic transition (~10^{-15} s), so that the electronic transition in a molecule takes place so rapidly in comparison to vibrational motion that immediately afterwards the nuclei still have very much the same relative position and velocity as before the transition.

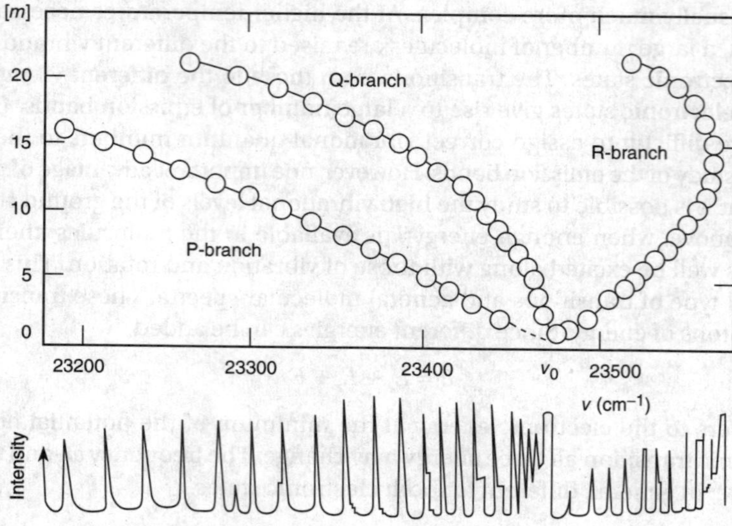

Fig. 4.16 Fortrat diagram (AIM band for $\lambda = 4291$Å)

During an electronic transition, there is no restriction imposed by quantum mechanics on the change of vibrational quantum number. However, it is observed that during electronic transition, the vibrational lines in a progression are not of the same intensity but vary in a complicated manner from one molecule to another. The intensity distribution yields valuable information about the vibrational structure, particularly when rotational analysis is not possible.

The intensity distribution can be explained with the help of Frank-Condon principle. It was formulated first in classical terms by James Frank and reformulated according to quantum mechanics by E.U. Condon about 1920. It states that *an electronic transition in a molecule occurs so rapidly, compared to vibrational motion of atoms, that during the transition a vibrating molecule does not change its internuclear distance appreciably. The nuclei of the atoms are essentially at the same relative positions and have the same relative velocity as those immediately before the transition.* In other words, the most probable transitions occur when, at the time of transition, the nuclei are at one end of their oscillation and at the same internuclear distance for both the initial and final states.

Classically, the vibrational velocity of an atom in a diatomic molecule is zero at the two mechanics, we would say that the most probable transitions are those for which the vibrational wave function suffers the least distortion at the time of transition. Since the oscillatory wave functions are very similar at the classical ends of the oscillation, we reach the same conclusion with the quantum mechanical model as we do with the classical model. Obviously, the vibrating atom would thus spend most of its time at the turning point of its motion implying that the probability of transition is highest if one of the turning points of a level in the other electronic state-the so called *vertical transitions.*

Quantum mechanically, we can say that the classical results agrees for high values of vibrational quantum number except for $v = 0$. For $v = 0$ the atom is most likely to be found at the centre of motion, i.e., at equilibrium internuclear distance. For $v = 1, 2, 3, \ldots$ etc.,

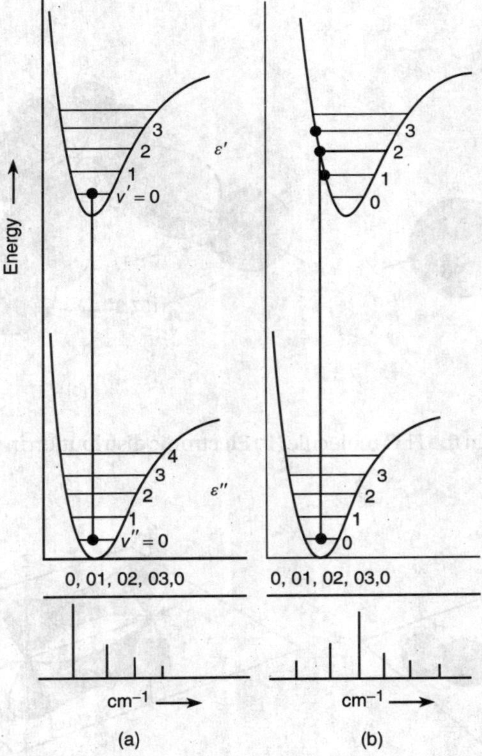

Fig. 4.17 Frank-Condon principle. Intensity distribution of lines

the most probable positions increasingly approach the extremities and classical as well as quantum pictures merge for high values of vibrational quantum number v.

Fig. [4.17(a)] shows the vibrational potential energy curves in the two electronic states. We may see that figure shows no relative displacement along r for these two states. We can see that the vertical transition being the most probable transition, the strongest spectral line would be $(0 \rightarrow 0)$. We may note that the chances of other transitions being much less, he $(1 \rightarrow 0)$, $(2 \rightarrow 0)$ etc., lines rapidly diminish in intensity.

Fig. [4.17(b)] shows the two potential curves in the two electronic states with appreciable relative displacement along r. Vertical transition indicates that transitions from $v'' = 0$ level to $v' = 2$ is most likely to occur, transitions to lower and higher v' states being less likely. Clearly the intensity of $(0 \rightarrow 2)$ line would be the strongest.

4.6 POLYATOMIC MOLECULES

For molecules with more than two atoms, an important new element enters into the analysis of electronic motion: the geometrical arrangement of the electrons and the nuclei (or in other words, the molecular symmetry). We may note from the diatomic case that bonding is favoured when the bonding electrons can fit within the region joining the two atoms.

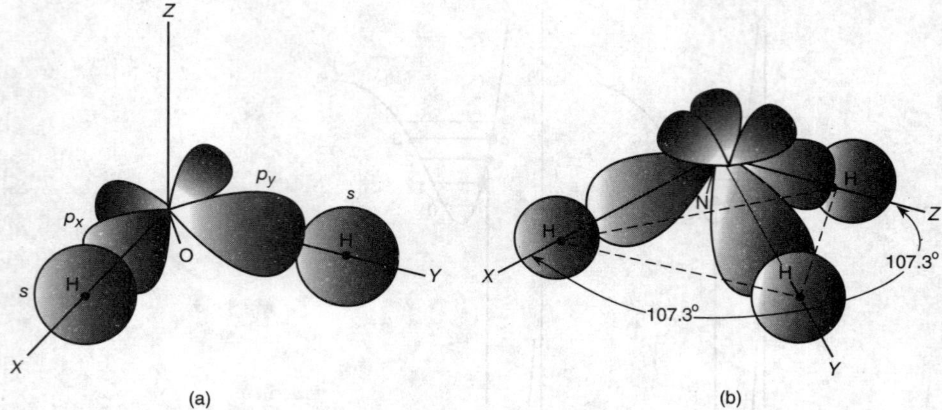

Fig. 4.18 (a) Electronic distribution in the H_2O molecule (b) Electronic distribution in the NH_3 molecule

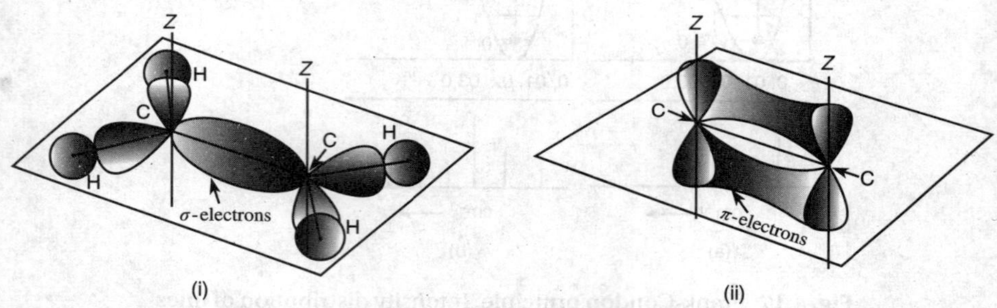

Fig. 4.18 (c) Electronic distribution in ethylene sp^2 hybridization (i) σ-bond, (ii) π-bond

This requires that the atomic wave functions which pertain to the two atoms and which make up the molecular orbital extend or "overlap" as much as possible along the line joining the two nuclei, favouring the condition in which a pair of electrons (one from each atom) are concentrated between two atoms. One may state this requirement of "maximum overlapping" of atomic wave functions as follows:

A bond between two atoms occurs in the direction in which the respective atomic wave functions making up the molecular orbital are concentrated or overlap; the strength of the bond depends on the degree of over lapping of the atomic wave functions.

The electronic distribution in the H_2O, NH_3, ethylene, butadiene molecules are shown in Fig. (4.18). Butadiene(C_4H_6)is an example of *conjugated* molecules.

Experimental as well as theoretical investigation of electronic spectra of polyatomic molecules is much more complex than that of diatomic molecules. Predissociation is much more common in polyatomic molecules, so that in only a few classes of molecules it is possible to study spectra in emission, even to observe the rotational structure in absorption. We may note that most of the polyatomic molecules are sufficiently involatile and can be studied in solution or in the crystalline state where the rotational structure is lost. However, there are also molecules for which a spectrum has been observed in the gaseous phase.

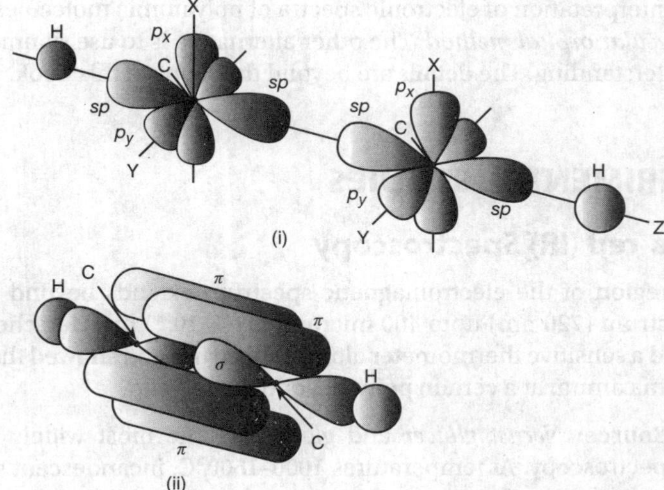

Fig. 4.18 (d) Electronic distribution in acetylene using *sp* hybridization (i) wavefunctions with atoms separated (ii) σ and π-bonds

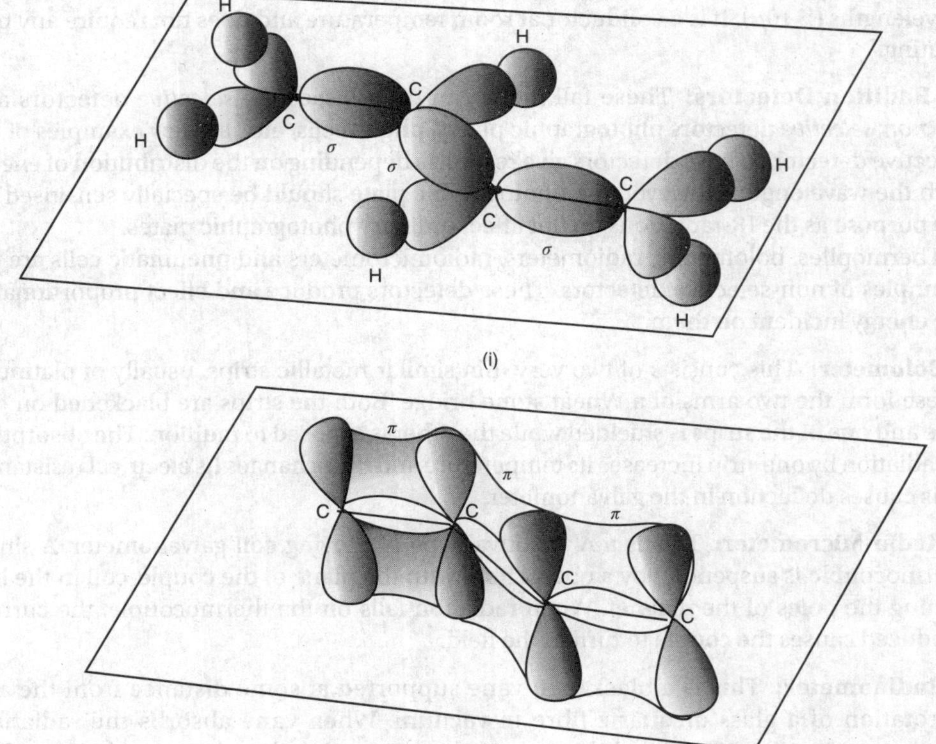

Fig. 4.18 (e) Electronic distribution in butdiene: (i) Localized σ-bonds; (ii) unlocalized π-bonds

Theoretical interpretation of electronic spectra of polyatomic molecules is best carried out by the *molecular orbital method.* The other alternative is to use symmetry considerations for its understanding. The details are beyond the scope of this book.

4.7 EXPERIMENTAL STUDIES

4.7.1 Infra-red (IR) Spectroscopy

The infra-red region of the electromagnetic spectrum extends beyond the red end of the visible spectrum (720 nm) upto 400 microns. ($1\mu = 10^{-4}$ cm). Herschel discovered it. Herschel moved a sensitive thermometer along a spectrum and showed that the temperature reached a maximum at a certain point beyond the red end.

IR-Radiation Source: *Nernst glower* and *globar* are the most widely used radiation sources in IR spectroscopy. At temperatures 1000–1500°C, incandescent solids also emit radiations throughout IR region.

Glower: This consists of a filament composed of a mixture of rare earth oxides particularly Zn, Ce, etc. It is non-conducting at room temperature but becomes conducting when heated. The globar is a rod of silicon carbide called as carborundum. It is useful at longer wavelengths ($> 10\mu$). It is a conductor at room temperature and does not require any preheating.

IR-Radition Detectors: These fall mainly in two classes: (i) *selective* detectors and (ii) *non-selective* detectors photographic plates, photo-cells, etc., are few examples of the selective detectors. These detectors give response depending on the distribution of energy with the wavelength. However, the photographic plate should be specially sensitised for the purpose as the IR-radiations do not affect ordinary photographic plates.

Thermopiles, bolometers, radiometers, radiomicrometers and pneumatic cells are the examples of non-selective detectors. These detectors produce and effect proportional to the energy incident on them.

A Bolometer: This consists of two very thin similar metallic strips, usually of platinum. These form the two arms of a Wheat stone bridge. Both the strips are blackened on one side and one of the strips is shielded while the other is exposed to radition. The absorption of radiation by one strip increases its temperature and thus changes its electrical resistance. This causes deflection in the galvanometer.

A Radio-Micrometer: This is a Δ' arsonval type of moving coil galvanometer. A single thermocouple is suspended by a quartz fibre with the plane of the couple-coil in the line joining the poles of the magnet. When radiation falls on the thermocouple, the current produced causes the couple to turn in the field.

A Radiaometer: This is a blackened vane supported at some distance from the axis of rotation of a glass or quartz fibre in vacuum. When vane absorbs the radiation, the supporting fibre is twisted through angles proportional to the amount if incident radiation.

Production of Spectrum: By the method of *focal isolation* the radiations from the source are first sorted out so that the infra-red spectrum of definite range may be produced. Next the spectrum is produced either by a *prism* or a *grating-type* radio-spectrometer.

The material of the prism in the prism-method should be so selected so as to suit the region of the spectrum to be studied. We may note that no one material is transparent to the entire region. The list of the suitable materials for ranges are shown against the materials in the following table.

Prism Spectrometer: Two widely used arrangements of prism spectrometer are: (i) *Wadsworth mounting* and (ii) *Littrow mounting.*

Wadsworth Mounting: Fig. (4.19) shows the Wadsworth mounting of prism spectrometer. 60° prism *P* and the plane mirror *M* are rotated together about the vertical axis through the centre of the prism base. The angular deviation is exactly nullified by reflection from the plane mirror, but the lateral shift that results in 60° prism *P* is independent of the wavelength. Obviously, if the system is so adjusted that the radiation of a given wavelength λ which traverses the prism at minimum deviation is focussed at slit S_2. The entire spectral region may be scanned at minimum deviation simply by the rotation of

Table 4.1 Ranges of IR-Radiation for Various Materials

Materials	*Range*
Glass	visible–2.5μ
Quartz	visible–4.5μ
Fluorspar	4μ–9.0μ
Rocksalt (Nacl)	8μ–16μ
Sylvine (KCl)	15μ–20μ
Potassium Bromide (KBr)	19μ–28 μ
Thallium Bromide	25μ–40μ

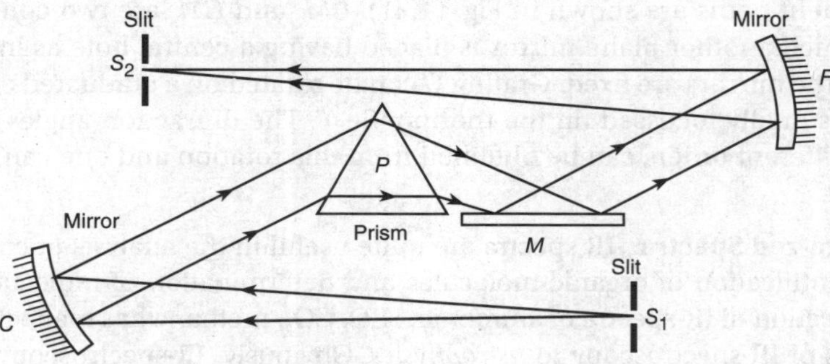

Fig. 4.19 Wadsworth mounting of prism spectrometer

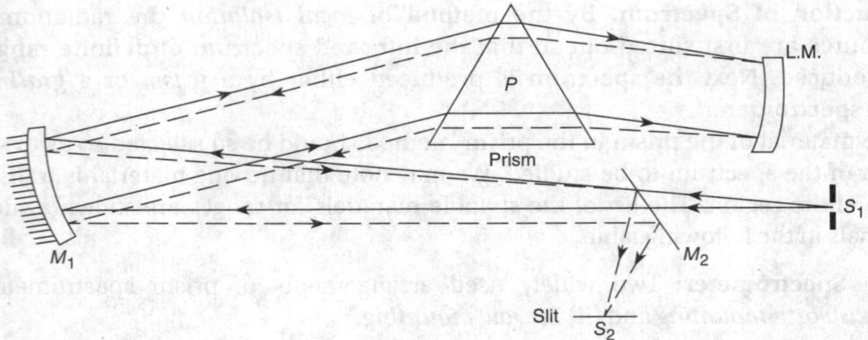

Fig. 4.20 Littrow mounting of prism spectrometer

prism-mirror combination. The radiation enters the system through the slit S_1 and falls on prism P after being rflected from the concave mirror C, and before being focussed at S_2 experiences a reflection at concave mirror F.

The spectrum is scanned by rotating the Littrow mirror (L. M) about the vertical axis at A in the *Littrow mounting* (Fig. 4.20). The beam emerging from the slit S_2 is focussed on a thermal detector [not shown in Fig. (4.20)] by an elliptic mirror. One can evaluate the wavelength from the calibration curves by studying the absorption spectrum of known molecules like water, CO_2, H_2O, methanol etc.

Grating Spectrometer: According to the range desired, one can use either a wire grating, an echellete grating, a concave grating or a laminar grating in the grating spectrometer. Using this arrangement one can determine the wavelengths of the lines absolutely from the knowledge of the grating space.

A grating with 60,000 lines/m permits one to measure in infra-red spectra upto 30,000Å. However, for investigation of longer wavelength coarser gratings are to be used. Echellete gratings can be used for almost any desired wavelength as these gratings concentrate the greater part of infra-red energy in the first order spectrum.

Properly sorted out infra-red radiations are focussed on a small hole at the centre of plane metallic mirror M_1. The hole is at the focus of the concave mirror C_1M_1 and the subsequent paths are shown in Fig. (4.21). CM_2 and CM_3 are two concave mirrors between which another plane mirror is placed having a central hole as in M_1. G is the grating. All the mirrors are fixed. Grating G can be rotated on a graduated circular scale. The beam is finally focussed on the thermopile T. The diffraction angles corresponding to the different orders can be obtained from this rotation and one can evaluate the wavelengths.

Uses of Infra-red Spectra: IR spectra are quite useful in the analysis of chemical compounds, identification of organic molecules and determination of molecular structure. The identification of IR-spectra of ammonia. H_2O, CO_2, methane has made it possible the applications of IR-spectroscopy to *astrophysics*. Obviously, IR-spectroscopy has opened up a new gateway to our knowledge of *astral bodies*.

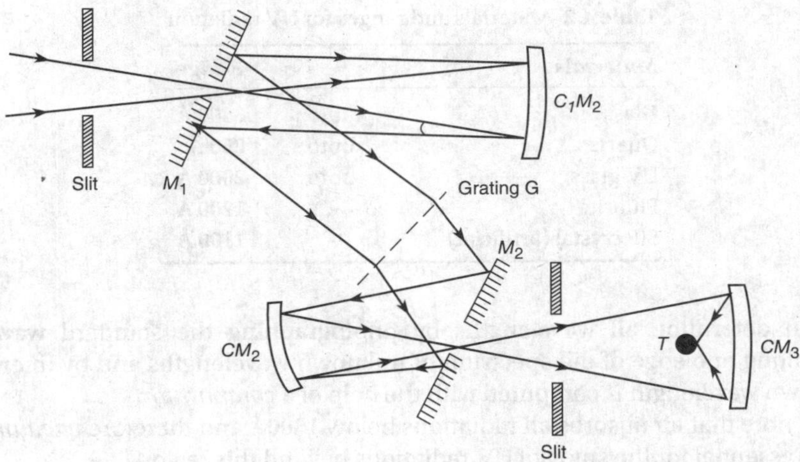

Fig. 4.21 Grating spectrometer (used for studying IR-spectrometer)

4.7.2 Ultraviolet (UV) Spectroscopy

The UV region of electomagnetic spectrum extends from 3900Å to about 100Å. Ritter found that the blackening of AgCl-the photochemical change-increases more and more towards the violet end of the spectrum and tends to maximum at some distance beyond the violet. Thus the ultraviolet spectroscopy was discovered.

UV-radiation Sources: Usual laboratory sources for UV-rdiation are iron or copper arcs and the mercury are suitably encased in a tube that absorbs UV-radiations only negligibly. The arcs should be preferably vacuum-arcs as UV-radiations are absorbed by the commonly used transparent materials and also by air. One must use vacuum spark discharge for extrenely short waves.

Production of UV-spectrum: UV-spectrum is in general produced either by a prism or by a grating.

Prism method: In this method the prism should be such that it does not absorb the radiations to be studied. Materials suitable for ranges shown against each are summarized in the following Table (4.2):

We may note that the associated lenses should also be of the same material. This factor puts a limit to the region that could be investigated by this method.

Grating method: In this method the range is extended without limit with the use of the reflection gratings, concave or plane. However, in this case the radiation is not to pass through the material of the grating. Below 1100Å, grating is the *only method* for the production of UV spectrum.

Recording of Spectrum: The photographic method using special plates such as Ilford Q-plates is generally used to record the UV-spectra.

Table 4.2 Materials and ranges for UV radiation

Materials		***Range***
Glass	upto	3300 Å
Quartz	upto	850 Å
UV-glass	upto	2000 Å
Fluorite	---	1200 Å
SiF crystal (artificial)	---	1100 Å

One can determine all wavelengths by photographing the standard wavelengths touching along one edge of the spectrum of unknown wavelengths and by interpolation the unknown wavelength is computed with the help of a *comparator*.

We may note that air absorbs all radiations below 1800Å and therefore *vacuum specto-graphs* are essential for the study of UV-radiations beyond this region.

Ultraviolet(UV)-spectrographs: There are several types of UV-spectrographs which are in used for different regions of the spectrum:

Quartz region: It consists of near and short ultraviolet regions. A quartz spectrograph is used here. The quartz polarises the radiation it transmits, produces double images due to birefringence and has the property of rotating the plane of polarisation. This is why fused quartz is preferred in constructing prisms and lenses of this spectrograph. One can eliminate the birefrigence effect by cutting the quartz lenses and prisms such that light propagates along the optic axis. By taking left-handed and right handed quartz crystals in equal parts, the rotation doubling is compensated.

Obviously, a simple quartz spectrograph consists of a collimating lens and a camera lens of opposite variety of quartz and a 60° prism, being a combination of two 30°-prisms of opposite types of quartz. All these are encased in a vacuum chamber.

One can obtain the same effect with much economy of material by a *Littrow-spectrograph* with a single quartz lens L and a total prism T and a 30° quartz prism (quartz half prism) backed by a mirror H as shown in Fig (4.22). The rotational doubling in the forward passage of light through lens and prism is nullified by the return journey along the same path on being reflected from the mirror or the metallic coating on the back of the prism.

Schumann region (1850–1100Å): Air absorbs UV beyond 1700Å and quartz does not transmit UV beyond 1850Å. This is why the vacuum spectrographs with CaF_2 (Fluorite) prisms and lenses are used for Schumann region. The whole arrangement is similar to the arrangement for previous region except that they must be housed in an evacuated chamber B.

Lyman region (1100Å -X-rays): This region is extreme UV region. Generally, concave grating vacuum spectrographs with Eagle's mounting are used upto 300Å. Fig. (4.23) shows a concave grating vacuum spectrograph for studying Lymann region. V is the vacuum chamber, S_1 the slit, S_2 the source, G the concave grating and PP the photographic plate. We can see that P.P. and G are on the Rowland circle.

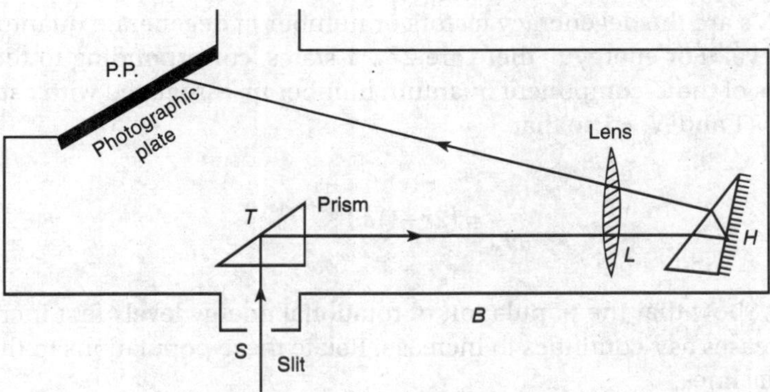

Fig. 4.22 Ultraviolet Littrow spectrograph

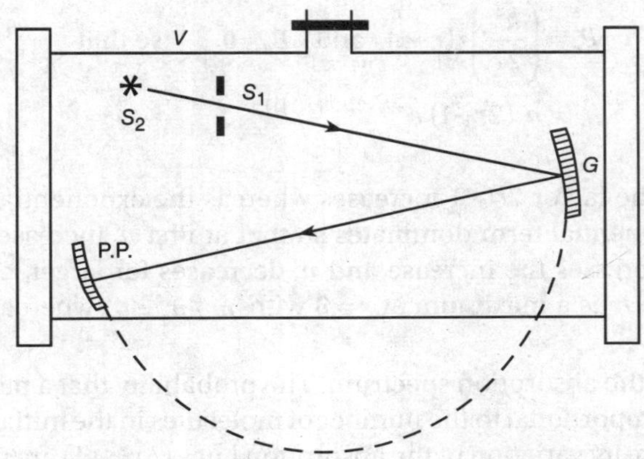

Fig. 4.23 Concave grating vacuum spectrograph

Millikan used a concave reflection grating at grazing incidence below 300Å. He went upto 136Å. Sieghbham succeeded in getting the X-ray spectrum of the same wavelength bridging the gap between X-ray and ultraviolet.

Uses of UV-spectra: Scientists have used the UV-absorption measurements for the determination of the structure of vitamin A and standardising fish liver oil concentrates. The photochemical and photobiological effects of UV-radiation have been utilised by scientists and doctors for the treatment of rickets, skin infection, diseases of bones etc. Using UV-spectroscopy the structure of several organic compounds have been revealed. UV-spectra are also used in high power microscopy. There are many more applications.

Example 4.1: Find the ratio of n_r, the number of molecules in rotational level r, to n_0 the number in the $r = 0$ level, in sample in equilibrium at temperature T.
Solution We have this from the Boltzmann factor,

$$\frac{n_r}{n_o} = \frac{N_r}{N_o} e^{-(E_r - E_0)/kT}$$

in which the N's are the degeneracy factors or number of degenerate quantum states for each energy level. For energy E_r there are $2r + 1$ states, corresponding to the number of possible values of the Z-component quantum number m_r associated with each value of r. Hence $N_r = 2r + 1$ and $N_0 = 1$ so that

$$\frac{N_r}{N_0} = (2r+1)\, e^{-(E_r - E_0)/kT}$$

Example 4.2: Show that the population of rotational energy levels first increases with r and then decreases as r continues to increase. Relate these populations to the intensities of the rotational lines.

Solution We have

$$E_r = \left(\frac{\hbar^2}{2I}\right) r(r+1) \quad \text{and} \quad E_0 = 0, \qquad \text{so that}$$

$$n_r = n_0 (2r+1)\, e^{-(\hbar^2/2IKT)\, r(r+1)}$$

Now as r increases the factor $2r + 1$ increases whereas the exponential factor decreases. For larger r the exponential term dominates so that at first n_r increases with r, but soon the exponential suppresses the increase and n_r decreases for larger r. For example HBr at room temperature n_r is a maximum at $r = 3$ with $n_r / n_0 \simeq 4$, whereas by $r = 9$ we have $n_9 / n_0 \simeq 1/2$.

Now, we consider the absorption spectrum. The probability that a particular frequency will be absorbed is proportional to the number of molecules in the initial rotational energy level. Hence the intensity variation of the absorption lines ($\Delta r = +1$) are proportional to the populations of the initial rotational energy level. Fig. (4.23). One can similarly argue for emission spectrum.

Example 4.3:

(a) Given that the equivalent force constant C of a vibrating HCl molecule is about 470N/m, estimate the energy difference between the lowest and the first vibrational state of HCl.

(b) Given that the rotational inertia of HCl has the value $I = 2.66 \times 10^{-47}$ kg-m^2, estimate the energy difference between the lowest and first excited rotational state of HCl.

(c) At room temperature, collisions of HCl molecules in a gas can transfer sufficient kinetic energy to internal energy to excite many *rotational* states. At what temperature would the number of molecules in the first excited *vibrational* state be equal to $1/e$ (about 37%) of the number in the ground vibrational state?

Solution

(a) We have for HCl

$$\mu = \frac{35}{36} m_H \quad \text{and} \quad C = 470\text{N/m}$$

and also
$$m_H = \frac{1}{6.02\times10^{23}}\text{g} = \frac{1}{6.02\times10^{26}}\text{kg}$$

We have $\Delta E = h\,\nu_0$, where $\nu_0 = \frac{1}{2\pi}\sqrt{c/\mu}$. Hence, using the given data, one obtains the energy difference as $h\nu_0 = \frac{h}{2\pi}\sqrt{c/\mu} = 0.59\times10^{-19}\text{J} = 0.37\,\text{eV}$

(b) Since $E_r = \frac{\hbar^2}{2I} r(r+1)$, the lowest rotational state has an energy $E_0 = 0$ and the first excited rotational state has an energy $E_1 = \left(\frac{\hbar^2}{2I}\right)^2 = \hbar^2 / I$. The required energy difference then is $\Delta E = \hbar^2 / I$. Hence,

$$\frac{\hbar^2}{I} = \frac{(6.63\times10^{-34}\,\text{J}-\text{s})}{(2\pi)^2\times2.26\times10^{-47}\,\text{kg}-\text{m}^2} = 4.2\times10^{-22}\,\text{J} = 2.6\times10^{-3}\,\text{eV}$$

Thus the energy difference between the two lowest vibrational levels is greater by a factor 142 (i.e., 0.37/2.6 × 10^{-3}) than that between the two lowest rotational levels in HCL.

(c) We have

$$\frac{n_1}{n_0} = \frac{N_1}{N_0} e^{-(E_1 - E_0)/kT}$$

where the subscripts refer to $v = 0$ to $v = 1$. The vibrational states are not degenerate so that $N_1 = 1 = N_0$. Also $(E_1 - E_0) = h\nu_0$, so that

$$\frac{n_1}{n_0} = e^{-h\nu_0/kT}$$

and if
$$kT = h\nu_0$$
$$n_1 = n_0 e^{-1}$$

Hence
$$T = \frac{h\nu_0}{k} = \frac{0.59\times10^{-19}\,J}{1.38\times10^{-23}\,J/^0K} \simeq 4300°\text{K}$$

is the temperature at which the number of HCl molecules in the first excited vibrational state is about 37% of the number in the ground state. Clearly the number of HCl molecules in the $v = 1$ state at room temperature is negligible compared to the number in the ground state.

Example 4.4: Calculate the energies and wavelengths of the three lowest radiations emitted by molecular H_2.

Solution The photon energy equals the energy difference between the levels which is given by

$$\Delta E = (L+1)\hbar^2 / mR_{eq}^2 \qquad \text{(i)}$$

Where R_{eq} is the equilibrium separation and l is angular momentum. To evaluate this energy we must first determine the rotational quantity $\hbar^2 / mR_{eq}^2$. For H_2 the reduced mass m is half the mass of a hydrogen atom. Then

$$\frac{\hbar^2}{mR_{eq}} = \frac{\hbar^2 c^2}{mc^2 R_{eq}^2} = \frac{(hc)^2}{4\pi^2 mc^2 R_{eq}^2}$$

$$= \frac{(1240\text{eV-nm})^2}{4\pi^2(0.5\times 1.008\,\text{amu}\times 931.5\frac{\text{MeV}}{\text{amu}})\times(0.074\,\text{nm})^2}$$

Eq. (i) now gives the energies directly and the corresponding wavelengths are found from $\lambda = hc/\Delta E$.

$$L=1 \text{ to } L=0 \quad \Delta E = 0.0152\text{eV} \quad \lambda = 81.6\mu\text{m}$$
$$L=2 \text{ to } L=1 \quad \Delta E = 0.0304\text{eV} \quad \lambda = 40.8\mu\text{m}$$
$$L=3 \text{ to } L=2 \quad \Delta E = 0.0456\text{eV} \quad \lambda = 27.2\mu\text{m}$$

The emitted energies from a sequence ΔE, $2\Delta E$, $3\Delta E$ ... and the emitted wavelength are λ, $\lambda/2$, $\lambda/3$ ...

We may note that rotational transitions are (like the vibrational transitions) in the infrared region, but the wavelengths are 1–2 orders of magnitude larger than those of the vibrational transitions. This region of the spectrum corresponds to the far infrared and microwave radiations.

Example 4.5: Show that the bond length of HF molecule is about 0.93Å. Given that the rotational constant for this molecule is 2025 m^{-1}. [Calcutta]

Solution We have the rotational constant for a diatomic molecule is given by

$$B = \frac{\hbar}{4\pi IC} \quad \text{or} \quad I = \frac{\hbar}{4\pi Bc}$$

$$\therefore \qquad I = \frac{1.05\times 10^{-34}}{4\times 3.14\times 2025\times 3\times 10^8} = 1.38\times 10^{-47}\,\text{kg-m}^2$$

The reduced mass of HF molecule,

$$\mu = \frac{m_1 m_2}{m_1 + m_2} = \frac{1\times19}{1+19}u$$

$$= \frac{19}{20}\times1.66\times10^{-27}\text{ kg}$$

$$= 1.58\times10^{-27}\text{kg}$$

Now, $$I = \mu r^2 \quad \therefore r = \sqrt{I/\mu} = \sqrt{\frac{1.38\times10^{-47}}{1.58\times10^{-27}}} \simeq 0.93\times10^{-10}\text{ m}$$

$$\simeq 0.93\text{Å}$$

Example 4.6: Calculate the moment of inertia (I) of NO molecule. Given the nuclear distance = 1.15Å. [Kerala]

Solution We have, $I = \mu r^2$, where r is the nuclear distance and μ is reduced mass. Now,

$$\mu = \frac{m_1 m_2}{m_1 + m_2} = \frac{14\times16}{14+16}u = \frac{224}{30}\times1.66\times10^{-27}\ kg$$

$$\therefore \quad I = \frac{224}{30}\times1.66\times10^{-27}\times(1.15\times10^{-10})^2 \quad \Big| \; r = 1.5\times10^{-10}\text{ m}$$

$$= 1.64\times10^{-46}\text{ kg-m}^2$$

Example 4.7: The moment of inertia of the CO molecule is 1.46×10^{-46} kg-m². Calculate the energy in eV and the angular velocity in the lowest rotational energy level of the CO-molecule. [Meerut]

Solution The quantum mechanical energy of the rotating molecule is given by

$$E = \frac{\hbar^2}{2I}J(J+1)$$

where I is the moment of inertia of the molecule about the axis of rotation. We know that the lowest rotational energy level corresponding to $J = 1$ and for this level in CO, we have

$$E_{J=1} = \frac{h^2}{4\pi^2 I} = \frac{(6.62\times10^{-34})^2}{4\times(3.14)^2\times(1.46\times10^{-46})}$$

$$= 7.6\times10^{-23}\text{ J}$$

$$= \frac{7.6\times10^{-23}}{1.6\times10^{-19}} = 4.75\times10^{-4}\text{ eV}$$

The angular velocity of the CO molecule when $J = 1$ is

$$\omega = \sqrt{\frac{2E}{I}} = \sqrt{\frac{2\times7.6\times10^{-23}}{1.46\times10^{-46}}} = 10.2\times10^{11}\ \frac{\text{rad}}{\text{s}}$$

Example 4.8: When CO is dissolved in liquid carbon tetrachloride, infrared radiation of frequency 6.42 × 10^{13} Hz is absorbed. Carbon tetrachloride by itself is transparent at this frequency, so the absorption must be due to the CO. Find the force constant of the bond in the CO molecule. Also determine the spacing between its vibrational energy levels.

Solution The reduced mass of the CO molecule

$$\mu = \frac{m_1 m_2}{m_1 + m_2} = \frac{12 \times 16}{12 + 16} = 1.14 \times 10^{-26}\,\text{kg}$$

Now the frequency of oscillation of such a two body oscillator

$$\nu_0 = \frac{1}{2\pi}\sqrt{\frac{k}{\mu}}$$

or force constant, $k = 4\pi^2 \nu_0^2 \mu$

$$= 4 \times (3.14)^2 \times (6.42 \times 10^{13}\,\text{Hz})(1.4 \times 10^{-26}\,\text{kg})$$
$$= 1.86 \times 10^3\,\text{N/m}$$

The separation ΔE between the vibrational levels in CO is

$$\Delta E = E_{\nu+1} - E_\nu = h\nu_0 = (6.63 \times 10^{-34}\,\text{J-s})\,(6.42 \times 10^{13}\,\text{Hz})$$
$$= 4.26 \times 10^{-20}\,\text{J} = 0.266\text{eV}$$

We may note that this is considerably more than the spacing between its rotational energy levels. Because $\Delta E > kT$ for vibrational states in a sample at room temperature, most of the molecules in such a sample exist in the $\nu = 0$ state with only their zero point energies. We may note that this situation is very different from that characteristic of rotational states, where the much smaller energies mean that the majority of the molecules in a room temperature sample are excited to higher states.

Example 4.9: When benzene is irradiated with the mercury line 4358Å a Raman line is observed at the same position as the argon line of $\lambda = 4201$Å. Show that the wave number of the corresponding absorption line of benzene is 86000m^{-1}. [Bombay]

Solution

$$\bar{\nu} = \frac{1}{10^{-10}}\left(\frac{1}{4201} - \frac{1}{4358}\right) = 86000\,\text{m}^{-1}$$
$$= 8600\,\text{m}^{-1}$$

4358Å = 4358×10^{-10} m
4201Å = 4201×10^{-10} m

Example 4.10: With exciting line 2536Å a Raman line for a sample is observed at 2612Å. Show that the Raman shift is 1147 cm^{-1}.

Solution We have $\bar{\nu}\ (= 1/\lambda)$ of exciting line

$$= \frac{1}{2536 \times 10^{-8}\,\text{cm}} = 39432\,\text{cm}^{-1}$$

$$\bar{\nu} \text{ of Raman line} = \frac{1}{2612 \times 10^{-8}\,\text{cm}} = 38285\,\text{cm}^{-1}$$

$\therefore$ Raman shift, $\Delta\bar{\nu} = (39432 - 38285)\,\text{cm}^{-1}$

$$= 1147\,\text{cm}^{-1}$$

Example 4.11: **The exciting line in an experiment is 5460Å and the Stokes' line is at 5520Å. Show that the wavelength of the anti-Stokes line is 5401Å.** **[Kerala]**

Solution

$$\bar{\nu} \text{ of the exciting line} = \frac{1}{5460 \times 10^{-8}\,\text{cm}} = 18315\,\text{cm}^{-1}$$

$$\bar{\nu}_s \text{ of the exiting line} = \frac{1}{5520 \times 10^{-8}\,\text{cm}} = 18116\,\text{cm}^{-1}$$

$\therefore$ $\Delta\bar{\nu}$ = wave number shift = $(18315 - 18116)\,\text{cm}^{-1} = 199\,\text{cm}^{-1}$

Anti-Stokes' line will have an equal shift on the other side of the exciting line.
$\therefore$ wave number of anti-Stokes' line

$$\bar{\nu}_{A,S} = \bar{\nu} + \Delta\bar{\nu} = (18315 + 199)\,\text{cm}^{-1} = 18514\,\text{cm}^{-1}$$

or

$$\lambda_{A,S} = \frac{1}{18514\,\text{cm}^{-1}} = 5.401 \times 10^{-5}\,\text{cm} = 5.401 \times 10^{-7}\,\text{m}$$

Example 4.12: **Show that the vibrational frequency and photon energy for H_2 are 1.3×10^{14} Hz and 0.54eV respectively.**

Solution We have

$$\nu = \frac{1}{2\pi}\sqrt{\frac{k}{m}}$$

To estimate k, we consider the molecule to behave like a simple harmonic oscillator in the vicinity of equilibrium separation R_{eq}, and one can fit the parabolic oscillator potential energy $U = \frac{1}{2}kx^2$ to the molecular energy in the region. Fig. (4.24) shows the region of the energy minimum and a parabola that approximates the curve near the minimum.
The equation of the parabola

$$E - E_{min} = \frac{1}{2}k(R - R_{eq})^2 \qquad \text{(i)}$$

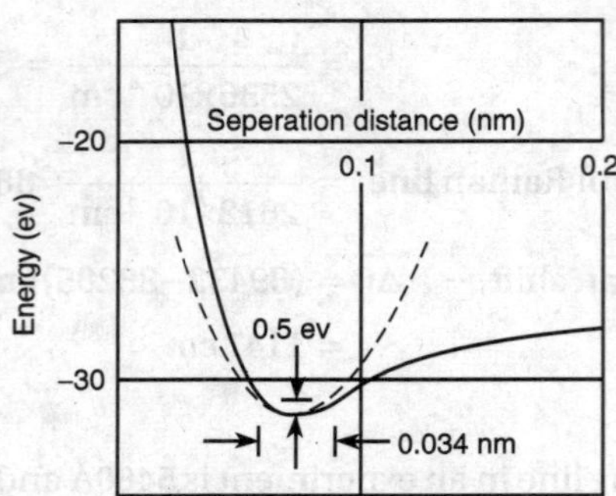

Fig. 4.24 Fitting a parabola (dashed line) to the energy minimum of H_2

Now, by finding the value of $R - R_{eq}$ from the graph one can find k. From Fig. (4.24) when $E - E_{min} = 0.50$ eV, one obtains $(R - R_{eq}) = \frac{1}{2} \times 0.034\,\text{nm} = 0.017\,\text{nm}$. Thus,

$$k = \frac{2(E - E_{min})}{(R - R_{eq})^2} = \frac{2 \times 0.50\text{eV}}{(0.017\text{nm})^2} = 36 \times 10^{20}\ \text{eV/m}^2$$

The reduced mass of molecular hydrogen is half the mass of a hydrogen atom. Now,

$$\nu = \frac{1}{2\pi}\sqrt{\frac{k}{m}} = \frac{1}{2\pi}\sqrt{\frac{k}{mc^2}}$$

$$= \frac{1}{2\pi}\sqrt{\frac{(36 \times 10^{20}\ \text{eV/m})^2 (9.0 \times 10^{16}\ \text{m}^2/\text{s}^2)}{0.5\,(1.008\ \text{amu})(931.5\text{V/amu})}}$$

$$= 1.3 \times 10^{14}\ \text{Hz}$$

and

$$E = h\nu = (4.14 \times 10^{-15}\ \text{eV.s})(1.3 \times 10^{14}\ \text{Hz})$$

$$= 0.54\text{eV}$$

The wavelength of this radiation is 2.3 μm, which is in the infrared region of the spectrum. We may note that molecular vibrations typically give photons in the infrared.

We may note that the parabola gives a reasonable approximation to the actual energy curve only for energies upto about 1eV above the minimum, which corresponds to the excited state with $n = 2$. Now, if we were to excite the H_2 molecule to energies greater than 1eV above its ground state, one would expect some deviations from the behaviour predicted by the simple harmonic oscillator. In particular, all the transitions would no longer have the same energy, and changes other than $\Delta N = \pm 1$ may occur. We may note that in other molecules, the harmonic oscillator approximation may remain valid for larger vibrational quantum numbers, but eventually it will fail in all cases at sufficiently large n.

Example 4.13: Fig. (4.25) shows a portion of rotational absorption spectrum of a molecule. Show that the rotational inertia of the molecule is 0.016 u-nm^2

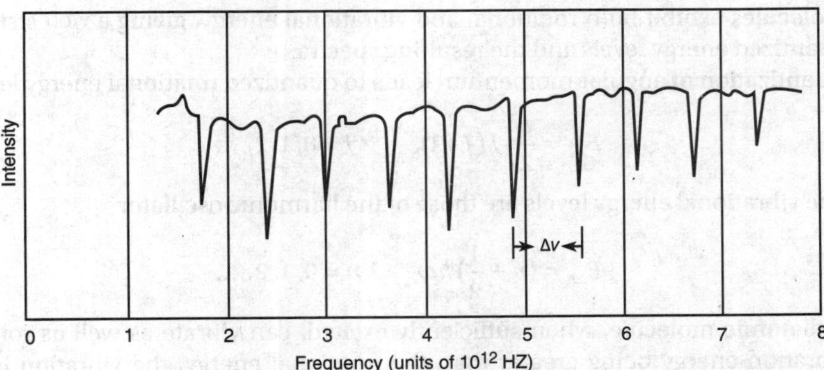

Fig. 4.25 A molecular absorption spectrum

Solution We may note that each peak in the absorption spectrum corresponds to the molecule moving from one level of Fig. (4.25) to the next higher level after absorbing a photon of proper energy or frequency. We have

$$\nu = \frac{\Delta E}{h} = (L+1)\frac{\hbar}{2\pi I}$$

where $$I = mR_{eq}^2$$

where R_{eq} is the equilibrium separation between the molecules and I is the rotational inertia. However, relation (1) gives the frequency of each peak in figure but does not allow us to find I since we don't know the L value for the peaks. Now, we calculate the difference in frequency $\Delta\nu$ between the adjacent peaks L and $L + 1$:

We have $$\Delta\nu = (L+2)\frac{\hbar}{2\pi I} - (L+1)\frac{\hbar}{2\pi I} = \frac{\hbar}{2\pi I}$$

From Fig. (4.28) $$\Delta\nu = 6.2\times10^{11}\,\text{Hz}$$

$$\therefore \quad I = \frac{\hbar}{2\pi\Delta\nu} = \frac{1.05\times10^{-34}\,\text{J–s}}{2\pi\times6.2\times10^{11}\,\text{Hz}}$$

$$= 2.7\times10^{-47}\,Kg-m^2 = 0.0164\,\text{nm}^2$$

SUMMARY

1. The binding of atoms into molecules involves both electric forces and quantum-mechanical effects with the Pauli exclusion principle. Although individual atoms are electrically neutral, the distribution of charge within them gives rise to attractive or repulsive forces.
2. Common types of molecular bonding include ionic, covalent, van der Waals, and metallic bonding.

3. Molecules exhibit both rotational and vibrational energy, giving a rich structure of quantized energy levels and the resulting spectra.
4. Quantization of angular momentum leads to quantized rotational energy levels

$$E_{rot} = \frac{\hbar^2}{2I} J(J+1), \qquad J = 0, 1, 2, \ldots$$

5. The vibrational energy levels are those of the harmonic oscillator

$$E_{vib} = (n + \frac{1}{2})\hbar\omega \qquad n = 0, 1, 2, \ldots$$

6. A diatomic molecule, when sufficiently excited, can vibrate as well as rotate. The vibration energy being greater than the rotational energy, the vibration is always accompanied by rotation.
7. When enough energy is available in the molecules, their electronic states may as well be excited along with those of vibration and rotation. This would lead to the most general type of transitions and the general molecular spectra. These transitions would produce photons of energy (assuming that the different energy could be added)

$$h\nu = h\nu_e + (v + \frac{1}{2})h\nu_o + J(J+1)\frac{\hbar^2}{2I}\nu$$

REVIEW QUESTIONS

1. What is the band spectrum? Explain its origin. What are the three distinct classes of band spectrum?
2. What do you understand by molecular spectra? How the molecular energy states arise?
3. What are the emission or absorption spectra?
4. What is pure rotational spectra? How this spectra arises? What is a rigid rotator model?
5. What are heteronuclear and homonuclear molecules? Give few examples of each type. Can pure rotational spectrum arise in both types? Explain the importance of pure rotational spectrum.
6. Show that the intensities of the rotational lines for transitions will be maximum for

$$J = \sqrt{\frac{kT}{2Bhc}} - \frac{1}{2}$$

7. What is vibration rotation spectrum? Show that the lines in a vibration rotation band of a diatomic molecule is equally spaced in frequency with a separation $2B$.
8. What is rotation-vibration-electronic spectra? What are Fortrat diagrams?
9. Show that for pure rotational spectrum of a diatomic molecule the wave number difference between lines $\Delta\bar{v}_r = \hbar/2\pi cI$. **[Kerala]**
10. Show that the lines of the Q branch of the electronic band of a diatomic molecule can be represented by the formula:

$$Q(J) = \bar{v}_0 + (B' - B'')J(J+1)$$

where B' and B'' are the rotational constants for the two electronic states and $\bar{v}_0$ is the wave number of the band-origin. **[Mysore]**

11. Show that the interatomic distance in a diatomic molecule increases by ΔR due to rotation of molecule is given by

$$\Delta R = \frac{\hbar^2}{\mu k} \frac{J(J+1)}{R_0^3}$$

where μ is the reduced mass, k is the force constant, J is the rotational quantum number and R_0 is the equilibrium distance between the nuclei when there is no rotation. **[Kerala]**

12. Describe any suitable apparatus for the ultraviolet spectroscopy. Discuss the pecularities of the different components of the apparatus used for the purpose. To what wavelength could this be used? How would you go beyond? **[Calcutta]**

13. What are emission and absorption spectra? Can you use an ordinary glass prism spectrograph to study ultraviolet spectra? Explain with reasons. **[Meerut]**

14. Discuss the pure rotation spectra of a heteronuclear diatomic molecule and show that the pure rotation spectrum of such a molecule consists of a series of equally spaced lines separated by a constant wave number difference, $2B$. **[Ujjain]**

15. What is Raman effect? Discuss the characteristics of Raman lines and explain the importance of Raman effect. Why Stokes' lines are more intense than antistokes' lines.

16. Explain Raman effect and compare the phenomenon with the ordinary scattering of light. Giving an elementary theory for the Raman effect, describe an experiment for its study in liquids.

17. Compare Raman effect and Compton effect. What are the practical uses of Raman spectra? Compare Raman spectra with infrared spectra.

18. Write short-notes on the following
    (i) Raman effect
    (ii) Study of Raman effect using laser beam
    (iii) Stoke and antistokes Raman lines
    (iv) Frank Condon principle
    (v) Infrared spectroscopy

PROBLEMS

1. The interatomic distance of the CO molecule is 1.13Å. What is the energy difference between the $j = 0$ and $j = 1$ rotational states of this molecule? **[Ans.** 0.000479 eV**]**

2. Show that the energy in calories per mole required to raise NO from its lowest rotational level to the level with $J = 4$ is 97.42. **[Ujjain]**

3. If $\Delta \bar{v}_r = 4050\text{m}^{-1}$ for the HF molecule, show that the interatomic distance of this molecule is 0.937Å.

4. The atoms of CsI are separated by 3.31Å. Show that its reduced mass and moment of inertia are 64.96 amu, 1.18×10^{-44} kg-m^2 respectively.

5. The most intense vibrational bands of CO and HCl molecules have the wave numbers $2.143 \times 10^5\text{m}^{-1}$ and $2.886 \times 10^5\text{m}^{-1}$ respectively. Show that the force constants of these molecules are 1855N/m and 480.7N/m respectively. **[Punjab]**

6. Calculate the stokes and anti-stokes Raman frequencies in units of $h/8\pi^2 I$ from rotational transitions into a final state of quantum number. $J = 4$ **[14, –22]**

7. The term value of the first rotational level of the molecule CO is 384m^{-1}. Show that the ratio of the number of molecules in this level to that in the ground state at 300 K is 0.98. **[Kerala]**

8. Show that the ratio of the intensities of the violet and red secondary lines in the Raman scattering of CCl_4 at 300K is 0.35, 0.22, 0.11 and 0.024, when the frequency of the normal vibrations of CCl_4 expressed in Hz are 217, 315, 457 and 774. **[Agra]**

9. The dissociation energy of H_2 molecule is 4.74 eV. While its fundamental vibrational wave number is 4395 cm^{-1}. Assuming the depth of the potential well and the force constant for the deuterium molecule D_2 to be equal to those for H_2 molecule, show that the dissociation energy of the D_2 molecule is 4.56eV.

10. The stokes radiation in the Raman spectrum of a diatomic molecule excited by $\lambda_0 = 4350$Å appears at $\lambda = 4735$Å. One of the atoms constituting the molecule is ^{16}O. If the force constant of the molecule has the value 1540Nm^{-1}, atomic mass of ^{16}O is 15.995 amu, show that the other atom is *N*.

SHORT ANSWER-QUESTIONS

1. What is rotational spectra?
Ans. Rotational spectra arise from transitions between rotational energy states. Only molecules that have electric dipole moments can absorb or emit electromagnetic photons in such transitions.

2. Does non-polar diatomic molecules such as H_2 and symmetric polyatomic molecules such as CO_2 and CH_4 exhibit rotational spectra?
Ans. No

3. Write the selection rule for rotational transitions in the case of rigid diatomic molecule.
Ans. $\Delta J = \pm 1$

4. How can you specify the vibrational energy level of a diatomic molecule?
Ans. $E_v = (v + \frac{1}{2})\hbar\sqrt{k/\mu}$

where v, the *vibrational quantum number* and μ the *reduced mass*. v may have the values $v = 0, 1, 2, 3, ...$

5. Write the selection rule for transitions between vibrational states.
Ans. $\Delta v = \pm 1$ in the harmonic oscillator approximation.

6. How many modes of vibration can have a complex molecule?
Ans. A complex molecule may have many different modes of vibration. Some of these modes involve the entire molecule, but others ("local modes") involve only groups of atoms whose vibrations occur more or less independently of the rest of the molecule.

7. What are vibration-rotation spectra?
Ans. Pure rotational spectra are observed only in liquids where interactions between adjacent molecules inhibit rotation. Because the excitation energies involved in molecular rotation are much smaller than those involved in vibration, the freely moving molecules in a gas or vapour nearly always are rotating, regardless of their vibrational state.

8. Does electronic transitions involve radiation in the visible or ultraviolet parts of the spectrum?
Ans. Yes.

9. Does electronic excitation in a polyatomic molecule often leads to a change in the molecule's shape?
Ans. Yes. The change in shape can be determined from the rotational fine structure in its band spectrum.

10. The bond between the hydrogen and chlorine atoms in a HCl molecule has a force constant of 470N/m. It is likely that a HCl molecule will be vibrating in its lowest vibrational state at room temperature?
Ans. No.

OBJECTIVE QUESTIONS

1. A simplest diatomic molecule (e.g. H_2, O_2, CO, HCl) may be considered as two point masses m_1 and m_2 connected by a spring like force. Such a molecule can vibrate along the line joining the atoms and may be considered as a two-body oscillator. The frequency of vibration of this oscillator is given by

(a) $\nu = \frac{1}{2\pi}\sqrt{\frac{k}{\mu}}$ (b) $\nu = \frac{1}{2\pi}\frac{\sqrt{k}}{\mu}$

(c) $\nu = \frac{1}{2\pi}\left(\frac{k}{\mu}\right)^{3/2}$ (d) $\nu = \frac{1}{2\pi}\left(\frac{k}{\mu}\right)^{2}$

2. The rotational energy of a molecular arises due to its
(a) rotation about all three axes
(b) rotation about one of its axes
(c) rotation and vibration about one of its axes
(d) none of the above

3. The rotational kinetic energy (E) of a molecule of moment of inertia I is given by

(a) $E = I\omega$ (b) $E = \frac{1}{2}I\omega^2$

(c) $E = \sqrt{I\omega}$ (d) $E = \sqrt{\frac{1}{2}I\omega^2}$

4. Quantization of angular momentum leads to quantized rotational energy levels given by

(a) $E_{rot} = \frac{\hbar^2}{2I}l(l+1), \quad l = 0, 1, 2, \ldots$ (b) $E_{rot} = \frac{\hbar^2}{2I}l, \quad l = 0, 1, 2, \ldots$

(c) $E_{rot} = \frac{\hbar^2}{2I}\sqrt{l(l+1)}, \quad l = 0, 1, 2, \ldots$ (d) $E_{rot} = \frac{\hbar^2}{2I}l^2, \quad l = 0, 1, 2, \ldots$

5. The vibrational energy levels of a molecule are those of the harmonic oscillator given by

(a) $E_{vib} = h\nu$ (b) $E_{vib} = n\hbar w, \quad n = 0, 1, 2, \ldots$

(c) $E_{vib} = (n + \frac{1}{2})\hbar\omega, \quad n = 0, 1, 2, \ldots$ (d) $E_{vib} = \frac{1}{2}I\omega^2$

6. A molecule with vibrational quantum number n and rotational quantum number l can undergo transitions obeying selection rules:
(a) $\Delta n = 0$, and $\Delta l = \pm 1$
(b) $\Delta n = \pm 1$ and $\Delta l = 0$
(c) $\Delta n = 0$ and $\Delta l = 0$
(d) $\Delta n = \pm 1$ and $\Delta l = \pm 1$

7. For HF molecule, $\Delta \bar{\nu} = 4050\ m^{-1}$, rotational constant, $B = 1/2$, and M.I., $I = 1.38 \times 10^{-47}$ kg–m² the internuclear distance r is
(**Hint:** $r = \sqrt{I/\mu}$)
(a) 0.49 A°
(b) 0.19 A°
(c) 0.94 A°
(d) 1.94 A°

ANSWERS

Question numbers	1	2	3	4	5	6	7
Answers	(a)	(b)	(b)	(a)	(c)	(d)	(c)

FILL IN THE BLANKS

1. Just as atomic spectroscopy gives us a way to unidentify atoms from their charactertistic emission or absorption spectrum, molecular sepectroscopy enables us to identify molecules by the radiation they ______ or ______. **[absorb, emit]**
2. The absorption spectra of our atmosphere can be used to identify trace amounts of various ______, and thus molecular spectroscopy helps to measure the purity of our ______. **[pollutants, air]**
3. The rotational states are not evenly spaced and they have a smaller energy spacing of typically ______. **[0.01–0.1eV]**
4. The frequency-differences of Raman lines depend on the nature of the scatterer but is ______ of the frequency of the exciting radiation. **[independent]**
5. The intensities of Raman lines are quite small compared to the parent line, being only about ______ of the intensity of the ______ line. **[1%, unmodified]**
6. The total energy (E) of a molecule is made up of three parts ______, ______, and ______. **[electronic energy (E_e), vibrational energy (E_v), rotational energy (E_r)]**
7. The radiation corresponding to wave number $4.03 \times 10^6\ m^{-1}$ lies in ______ of the electromagnetic spectrum. **[ultraviolet part]**
8. The vibrational energy of a molecule is due to ______ of its constitutent atoms. **[vibration]**
9. The rotational energy of a molecule arises due to its ______ about one of its axes. **[rotation]**
10. The radiation corresponding to wave number $\bar{\nu} = 2.98 \times 10^2\ m^{-1}$ lies in ______. **[microwave region]**
11. The isotopic exchange of atoms in a diatomic molecule alters the moment of inertia but not the ______. **[inter nuclear distance]**
12. The region in which vibration-rotation bands are found extends approximately from ______ A° to ______ A°. **[8000, 50,000]**

5

X-Rays

"X-rays provide us with a kind of supermicroscope, by means of which one can 'see' not only atoms and their arrangement in crystals but also even the interior of the atoms itself".

5.1 INTRODUCTION

In 1985 X-rays were discovered by Rontgen while studying the phenomenon of discharge of electricity in rarefied gases. An induction coil was connected to an evacuated tube in which at the ends of the tube there was a cathode and an anode. The tube was covered with thin black paper. During the experiment he observed that a paper screen washed with barium-platino-cyanide continued to fluorescent whether treated side or other he put towards the discharge tube. The fluorescence was observable 2 meters away from the apparatus. Rontgen soon convinced that the agency, causing fluorescence, originate in the discharge tube at the point where cathode-stream struck the tube. As the nature of these new rays was unknown, Rontgen called them "X-rays".

With the discovery of X-rays, Rontgen proceeded to study these rays. These rays are not deflected by magnetic or electric field. Hence these do not possess any charge. These cause many other substances to fluoresce e.g., calcium compounds, Uranium glass, rock salt etc., and also affect photographic plate. These travel in straight lines, causing shadow of the objects placed in their path. He produced a radiograph of a human hand showing clearly the bones against the flesh and asserted that X-rays would prove invaluable in diagnosis in medical practice.

5.2 PRODUCTION OF X-RAYS: COOLIDGE TUBE

The X-rays are produced by bombarding high energy electrons on a matter. The tube, used for the purpose of producing X-rays, is called an X-ray tube. Rontgen designed X-ray tube, but it has been improved soon. The basic principle of the production of X-rays demands the following parts in the tube: (i) A source of electron (ii) A target of high atomic weight and (iii) High tension to accelerate electrons.

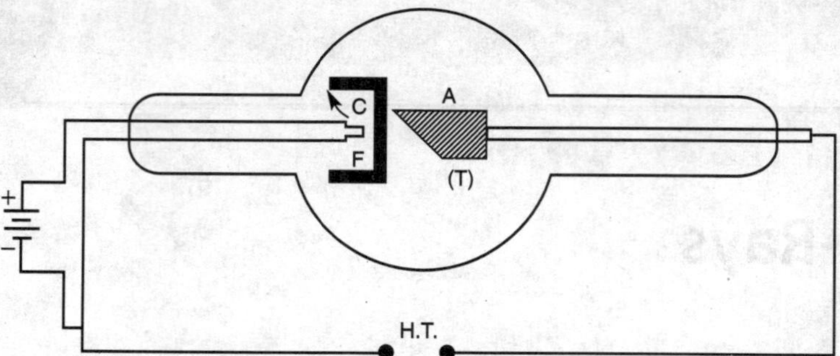

Fig. 5.1 Coolidge X-ray tube

Nowadays Coolidge tube (Fig. 5.1) is used for the production of X-rays. This tube was designed by an American physicist Coolidge in 1913. In this tube the intensity and quality of X-rays can be controlled separately. It consists of a cathode, a tungsten filament F, from which electrons are emitted. The filament is placed inside a metal cup C through which a narrow beam of electrons moves towards the target A, placed at an angle of 45° with the path of the beam, made of tungsten or molybdenum. The target A, is fixed to a copper rod projecting outside and is kept cold by circulating cold water since enormous amount of heat is generated during the working of the tube. Filament and target both are placed inside a vacuum glass tube evacuated to a pressure of about 10^{-8}m of mercury. The filament is heated by a current 3–5 A, which emits electrons. When a high voltage of the order of 60 kV to 1000 kV is applied to the target (also called anti cathode), then this electron beam is accelerated towards the target, and hit it to produce X-rays. By adjusting the amount of current in the filament, the number of emitting electrons and hence intensity of emitted X-rays can be controlled. The penetrating power of the X-rays depends upon the energy of incident electrons. Hence by adjusting proper voltage of cathode and anti cathode, the energy of the electrons and hence the penetrating power of X-rays can be controlled. By using accelerated electrons from a betatron by which electrons can be accelerated up to energies of the order of 10^8 eV, highly energetic and penetrating X-rays are produced. These X-rays have been used to investigate the structure of atomic nuclei.

5.3 PROPERTIES OF X-RAYS

The important properties of X-rays are following:

(i) They are highly penetrating rays and can penetrate through most substances. However, their penetrability is different in different substances. X-rays are able to penetrate more in low density substances whereas less in high density substances, e.g., ordinary glass is quite transparent to X-rays, but lead glass is almost completely opaque to them.

(ii) X-rays ionise the gas through which they pass.

(iii) Like light, they affect photographic plate and the effect is more intense.

(iv) They are not deflected by electric and magnetic fields which shows that unlike cathode rays or positive rays they are not beams of charged particles.

(v) They cause fluorescence in several materials, e.g., zinc sulphide, a plate coated with barium platinocyanide, etc., becomes luminous when exposed to X-rays.

(vi) They exhibit the phenomena of reflection, refraction, interference, diffraction and polarisation.

(vii) They travel in straight lines with the speed of light.

(viii) They liberate photoelectrons from the metals. When strike metals, they produce secondary X-rays.

(ix) X-rays have a harmful effect upon human body. A long exposure of any part of human body to X-rays kills the body tissues and causes incurable sores.

From the above characteristics it is obvious that X-rays are electromagnetic radiation of wavelengths much shorter than those of visible or ultraviolet radiation. One can measure these wavelengths by diffraction experiments. Wavelength of X-rays are ~ 10^{-2} to 100Å.

5.4 MEASUREMENT OF THE INTENSITY OF X-RAYS

One can measure the intensity of X-rays by photographic and ionisation methods. The determination of the intensity with ionisation method is more accurate. For such measurements, ionisation chamber is usually employed as shown in Fig. (5.2).

C is cylindrical chamber made of brass. There are two open ends of which are closed by two thin sheets *W*, *W* of some low *Z*-metal, e.g., aluminium, These are called windows of the chamber. A beam of X-rays can enter the chamber through one of the windows and come out through the other window. The length of the chamber is usually 20 to 100 cm. The chamber is filled with some gas or vapour e.g., methyl bromide, air etc.

While passing through the chamber the X-ray beam ionises the gas within it. This causes the production of positive ions and negative electrons in the gas. Inside the chamber a metal rod *A* insulated from the wall and parallel to the axis of the cylinder is placed.

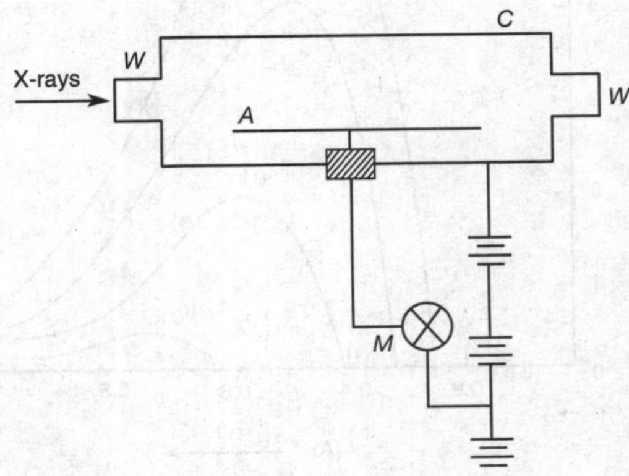

Fig. 5.2 Ionisation chamber

A potential difference of few hundred volts is maintained between A and C. Due to the movement of positive ions toward cathode and negative ions toward anode ionisation current is produced in the chamber. The metal rod A is connected to an electrometer M, which can measure the very small ionisation current produced in the chamber by the X-rays. It is found that the ionisation current is linearly proportional to the intensity of X-rays passing through the chamber. Obviously, again from Fig. (5.3) it is evident that λ_m decreases with increasing potential difference (p.d.). For a given wavelength, the intensity is higher when the p.d. is higher. Following conclusions can be made from Fig. (5.3).

(i) At small wavelength side there are no radiations emitted for a given (p.d.). Thus each continuous spectrum shows a sharp limit on shorter wavelength side.

(ii) The intensity of the spectrum increases for all wavelengths, at all potentials of the tube. After attaining certain maxima, it starts decreasing for higher wavelengths.

(iii) The total energy emitted increases with the applied voltage. Furthermore, the maxima of intensity shifts towards the shorter wavelength, as applied voltage is increased.

(iv) The total intensity of continuous X-rays for given tube potential V is proportional to atomic number Z of the target. For a given target total intensity depends upon V^2 and electron beam current.

Further, it has been found that minimum wavelength limit of continuous X-rays does not depend upon the element of the target.

When an electron passes through the electric field of the atomic nucleus, it loses energy by radiation. This energy appears as a *continuous X-ray spectrum* and is called

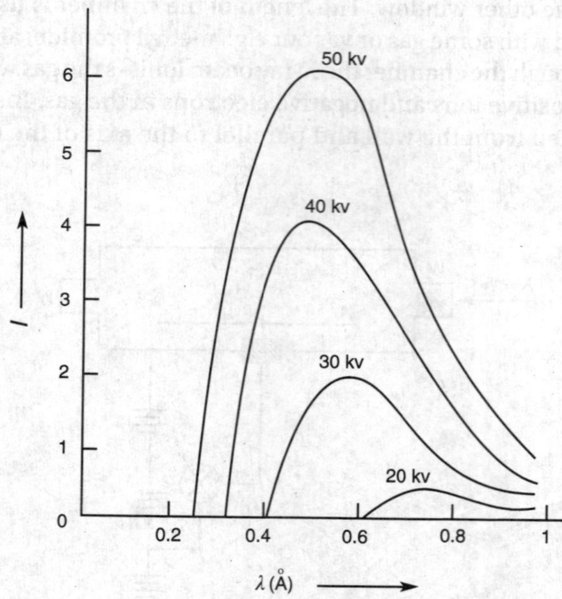

Fig. 5.3 Intensity vs wavelength emitted from tungsten at different voltages. Continuous X-ray spectrum

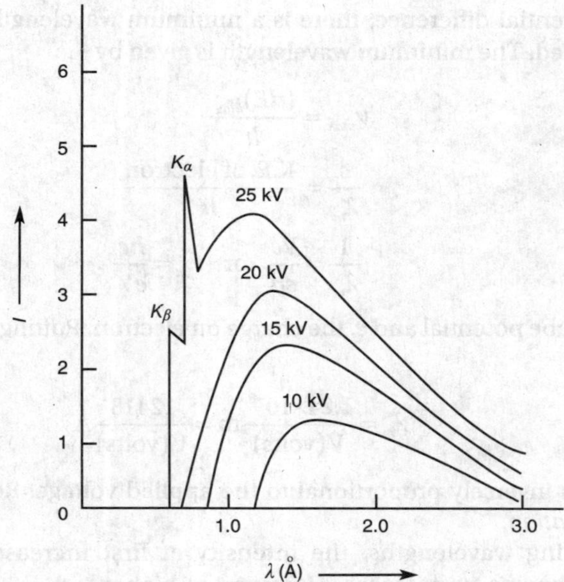

Fig. 5.4 Intensity vs wavelength current for X-rays emitted from molybdenum target ($Z = 42$) for different voltages. There appears peaks at higher voltages

bremsstrahlung or breaking radiation. This means continuous radiations are actually produced by the 'braking' or slowing down of incident charged particles. The continuous distribution of wavelengths emitted is analogous to the continuous distribution of wavelengths in visible white light. Obviously, this type of X-radiation is also sometime called *white radiation*.

When the target is of low Z, e.g., molybdenum ($Z = 42$), the variation of I with λ has the appearance as shown in Fig. (5.4). In this case also, one finds a continuous variation of I with λ for $V < 20{,}000$ Volts, the measurement of the ionisation current gives a measure of the X-ray beam.

5.5 VARIATION OF THE X-RAY INTENSITY WITH WAVELENGTH

5.5.1 Continuous and Characteristic X-rays

It is reported that X-rays of different wavelengths are emitted from the X-ray tube. If one measures the intensity I of the X-rays as function of λ (wavelength) and variation of I with λ is plotted graphically, then one obtains the graphs as shown in Figs. (5.3) and (5.4). When the anode (i.e., the target) is made of some high Z-metal, e.g., tungsten ($Z = 74$), it is found that I (intensity of the emitted X-rays) varies continuously with λ up to potential differences of the order of 50,000 volts between the anode and the cathode Fig. (5.3).

For a given potential difference, there is a minimum wavelength (λ_m) below which no X-rays are emitted. The minimum wavelength is given by

$$\nu_{max} = \frac{(dE)_{max}}{h}$$

or

$$\frac{c}{\lambda_m} = \frac{\text{K.E. of electron}}{h} \tag{5.1}$$

$\therefore$

$$\frac{1}{\lambda_m} = \frac{Ve}{ch} \quad \text{or} \quad \lambda_m = \frac{hc}{eV}$$

where V is the tube potential and e, the charge on electron. Putting the values of h, c and e, one obtains

$$\lambda_m = \frac{1.24\times10^{-6}}{V(\text{volts})}\text{m} = \frac{12413}{V(\text{volts})}\text{Å}$$

Obviously, λ_m is inversely proportional to the applied voltage. Relation (1) is known as *Duane-Hunt relation.*

With increasing wavelengths, the intensity at first increases and after attaining a maximum it begins to decrease. However, at higher p.d. ~25,000V or even higher, several discrete peaks appear, superimposed upon the continuous background at certain definite wavelengths. Experimentally, it is observed that for targets with still lower Z, e.g., copper ($Z = 29$), such peaks begin to appear at still low p.d. ($V \gtrsim 8000$ volts).

We must note that the wavelengths at which the peaks appear are different for the copper target from those for the molybdenum target. This means the characteristics of the target material and are different for different target elements.

The characteristics X-ray spectrum for molybdenum target (Fig. (5.4)) exhibit two sharp peaks of high intensity occuring at wavelengths 0.71 A and 0.63 A. Barkla called these as K lines. Barkla also showed the presence of L-lines, which are less penetrating (Fig. (5.5)). Afterwards M-series has also been discovered. Mosley called these K-lines as K_α and K_β lines. These lines generally occur in the form of small groups. The group in the short wavelength region is termed as K-series, the next group in the direction of increasing

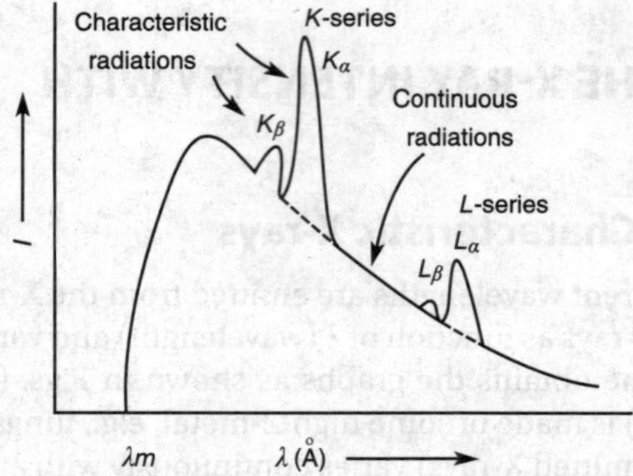

Fig. 5.5 The variation of I with λ at sufficiently high voltage

wavelength as *L*-series, then *M*-series and so on. Further, there is a definite short wavelength limit to the continuous X-ray spectrum.

5.6 ORIGIN OF CONTINUOUS SPECTRUM

According to electromagnetic theory of light, the accelerated or decelerated light charged particles, e.g., electrons or protons emit electromagnetic radiation of different frequencies. A part of the kinetic energies of these particles is transformed into the energy of the emitted radiation. When the electrons accelerated within the X-ray tube hit the anode (i.e., target), their motion becomes decelerated, i.e., their velocities decrease. During the process, the electrons experiences strong coulomb's force of attraction due to the nucleus of the target atom and is suddenly slowed down, at the same time suffering a deflection in its path. As a result electrons emit electromagnetic radiation with a continuous distribution of wavelengths starting from a minimum.

Let an incident electron moving with velocity v be slowed down to a velocity v', then by the conservation of energy principle, the frequency of the emitted photon is given by

$$\frac{1}{2}mv^2 = -\frac{1}{2}mv'^2 = h\nu = \frac{hc}{\lambda} \tag{5.2}$$

where $h\nu$ is the energy of the photon of frequency ν (i.e., wavelength $\lambda \because \nu = c/\lambda$). If the electron is completely stopped during a collision, $v' = 0$, so that the maximum energy of the emitted photon will be equal to

$$E_m = h\nu_m = hc/\lambda_m = eV \tag{5.3}$$

where V is potential difference between the cathode and the anode within the X-ray tube. Obviously, the electrons emitted from the cathode acquire an energy eV when they hit the target (anode). Obviously, the wavelength of the emitted radiation depends on the fraction of the incident electron energy (eV) that is transformed into X-radiation. We must remember that many of the impinging electrons penetrate several atomic layers in the target before they are brought to rest. Thus not all the impinging electrons give rise to X-ray photons of the wavelength in accordance with Eq. (5.3). As a result a continuous spectrum is obtained with the minimum wavelength limit set by Eq. (5.3). The increase in applied p.d. decreases the minimum wavelength (λ_{min}).

Using Eq. (5.3), one can estimate the maximum ν_m and λ_m as follows:

$$\lambda_m = \frac{hc}{eV} = \frac{12413}{V(\text{volts})} \text{Å}$$

For $V = 10{,}000$ volts, one obtains $\lambda_m = 1.24$Å and for $V = 50{,}000$ volts, $\lambda_m = 0.248$Å.

It is worthwhile to mention that sometimes X-rays are classified according to their penetrating power. The most penetrating radiation is termed as *hard radiation* whereas the radiation with relatively less penetrating power is called *medium* radiation. The radiation with very low penetrating power is termed as *soft radiation.*

One can understand the nature of the spectral distribution of the continuous X-rays on the basis of the theory of bremsstrahlung based on electromagnetic theory.

Due to deceleration of the electron beam in the direction of its motion within the target inside the X-ray tube, we have to take the spectral energy distribution of the bremsstrahlung X-ray for electrons of different energies and assign different weights to these curves. These are based on the absorption of X-rays within the X-ray tube in air and in window of the ionisation chamber, the variation of reflecting power of the crystal grating for different wavelengths and the different ionising powers of the beams of different wavelengths.

One finds that the spectral energy distribution of the bremsstrahlung radiation is independent of the frequency and can be expressed as

$$I_\lambda = I_\nu(\nu^2/c^2) \propto 1/\lambda^2 \tag{5.4}$$

Eq. (5.4) explain the observed spectral distribution of the continuous X-rays.

5.7 ORIGIN OF THE CHARACTERISTIC X-RAY SPECTRUM (X-RAY PEAKS)

When an impinging electron has sufficiently large energy to enable it to penetrate into the interior of a target atom, it may knock off an electron in the *K*-shell of the target atom Fig. [5.6 (a)]. Obviously, the atom is excited. The atom can remain in the excited state only for a very short interval of time (~10^{-8}s). An electron from the *L* shell may jump into the vacant space in the *K*-shell as shown in Fig. [5.6 (b)]. The difference in energy between *K* and *L* shell is emitted in the form of X-ray photon. The resulting line is the first line of the *K*-series. If an electron from the *M*-shell jumps into the vacant space in the *K*-shell, one obtains the second line of the *K*-series. Obviously, the jumps from the various outer shell of the atoms to the *K*-shell constitute the group of lines in the *K*-series. These radiations have the maximum energy and are the most penetrating. Similarly, the spectral lines in the *L*-series arise from the transitions of the electrons from the outer shells to the *L*-shell and so on. Clearly, the frequencies of *L*-series are smaller than those of the *K*-series. As the energies of various levels in target atom depends on the element of which the target is made the wavelengths (or the frequencies) of the lines given out are different in the case of different elements.

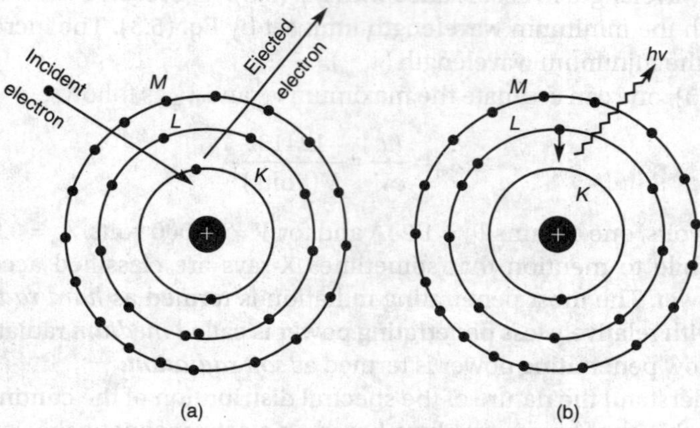

Fig. 5.6

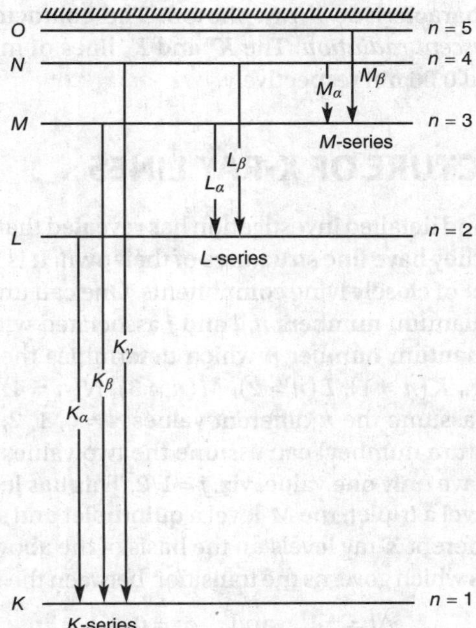

Fig. 5.7 X-ray energy level diagram

This is the reason that these spectral lines are called characteristic lines of the X-ray spectrum. One can better understand the production of X-ray spectral lines from the energy level diagram shown in Fig. (5.7). Here *K, L, M* etc., show the various energy levels of the orbits. A jump from *L* shell to *K* shell gives rise to K_α line and so on. The X-ray characteristic lines originating from the transitions between the different X-ray levels are usually designated by the symbols α, β, γ etc., e.g.,
K-level to *L*-level transition $\rightarrow K_\alpha$ line
L-level to *M*-level transition $\rightarrow L_\alpha$ line
L-level to *N*-level transition $\rightarrow L_\beta$ line and so on. The frequencies of these lines are given by

$$\nu(K_\alpha) = \frac{E_K - E_L}{h}$$

$$\nu(L_\alpha) = \frac{E_L - E_M}{h} \tag{5.5}$$

and so on. One can also produce the characteristic radiation by allowing the continuous X-rays generated by the bremsstrahlung process in the X-ray tube to be scattered from different elements outside the tube. The bremsstrahlung process can be represented symbolically by the equation; *electron* $\rightarrow$ *electron* + *photon*. These X-ray photons can eject electrons from the inner *K, L, M, N* etc., orbits of the target atom of the scatterer if their energy *hν* is greater than the binding energies of the electrons in these orbits. Obviously, the vacancy thus created in an inner orbit is subsequently filled up by the transition of an electron from an outer orbit. The surplus energy due to such transition in an atom is carried away by a characteristic X-ray photon. The characteristic radiation is also sometimes called *fluorescent radiation.* The K_α and K_β lines of molybdenum have wavelengths of about 0.07 and 0.06 nm respectively.

5.8 FINE STRUCTURE OF X-RAY LINES

Careful measurements and detailed investigation has revealed that the characteristic X-ray lines are not single, but they have fine structures of their own. It is found that each of them is composed of a number of closely lying components. One can understand their origin by considering the three quantum numbers n, l and j associated with each electronic level. For a given principal quantum number n which determines the positions of the different electronic shells, e.g., $K(n = 1)$, $L(n = 2)$, $M(n = 3)$, $N(n = 4)$,.... etc., l(the azimuthal quantum number) can assume the n different values: $l = 0, 1, 2, 3, ... (n - 1)$. Again for a given l, j (the total quantum number) can assume the two values: $j = l + 1/2$ or $j = l - 1/2$, whereas for $l = 0$, j can have only one value, viz. $j = 1/2$. This has led to the conclusion that K-level is single, the L-level a triplet, the M-level a quintuplet and so on. Table 5.1 provides a classification of the different X-ray levels on the basis of the above quantum numbers.

The two selection rules which governs the transition between the different X-ray levels are:

$$\Delta l = \pm 1 \quad \text{and} \quad \Delta j = 0, \pm 1 \tag{5.6}$$

Fig. (5.8) shows the fine structure of the X-ray levels and the possible transitions between them. From the figure, it is evident that K_α line is composed of the two components $K_{\alpha 1}$ and $K_{\alpha 2}$; K_β line is composed of two components $K_{\beta 1}$ and $K_{\beta 2}$. We can see from Fig. (5.8) that $K_{\alpha 1}$ originates due to transition from the K-level to the L_{III} level for which $n = 2$, $l = 1$ and $j = 3/2$. We note that $K_{\alpha 2}$ component originates from the transition between K-level to the L_{II} level for which $n = 2$, $l = 1$ and $j = 1/2$. One finds that the splitting of L, M, N etc. lines are more complex.

We know that the binding energy of an electron in the K-shell is much higher than that of an electron in the L-shell and hence the energies and obviously, the frequencies of the

Table 5.1 Classification of different X-ray levels on the basis of their quantum numbers n, l and j

Shell	***n***	***l***	***J***	***Energy level***	***Multiplicity of energy level***
K	1	0	1/2	K_1	1
L	2	0	1/2	L_1	
		1	1/2, 3/2	L_{II}, L_{III}	3
M	3	0	1/2	M_1	
		1	1/2, 3/2	M_{II}, M_{III}	5
		2	3/2, 5/2	M_{IV}, M_V	
N	4	0	1/2	N_1	
		1	1/2, 3/2	N_{II}, N_{III}	7
		2	3/2, 5/2	N_{IV}, N_V	
		3	5/2, 7/2	N_{IV}, N_{VII}	

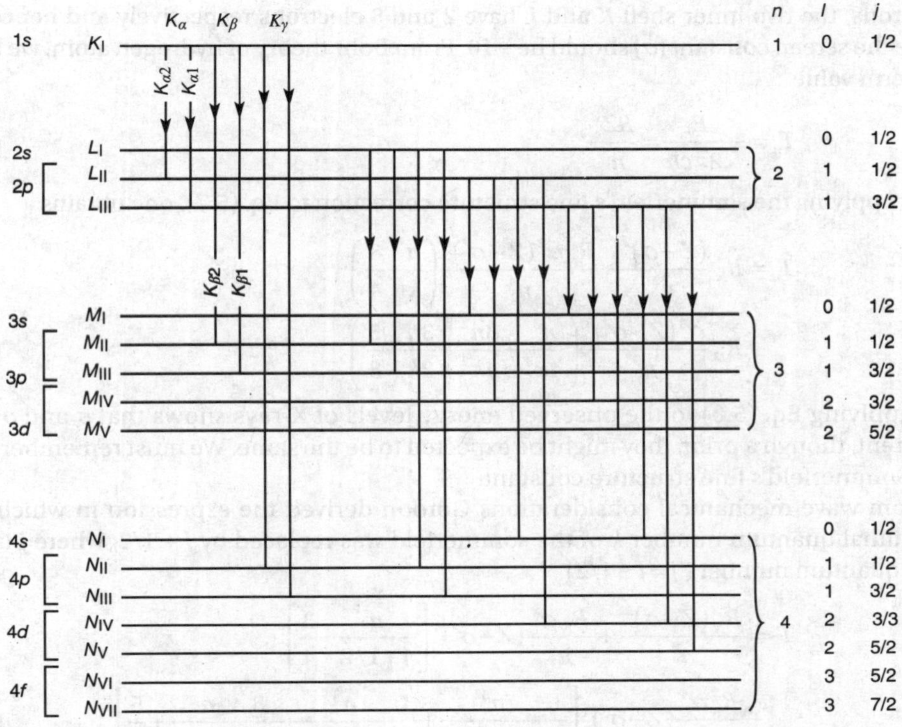

Fig. 5.8 Fine structure of X-ray lines and the possible transitions between them

K-lines are much higher than those of the *L*-lines. So that the wavelength of *K*-lines are much shorter than those of the *L*-lines. Similarly, one finds that the wavelengths of *L*-lines are much shorter than those of the *M*-lines and so on, i.e.,

$$\nu_k > \nu_L > \nu_M > \nu_N \ldots$$

or

$$\lambda_k > \lambda_L > \lambda_M > \lambda_N \ldots$$

5.9 THEORETICAL EXPLANATION OF THE FINE STRUCTURE

While studying hydrogen like atoms, we have seen that the radii of the orbits of the hydrogen-like atoms are inversely proportional to *Z*. This means that, these radii for heavy atoms of the target materials of the X-ray tubes will have quite small values. Thus one regards that the electrons in the *L*-shell of such a heavy atom may be in the coulomb field of the nuclear charge reduced by two units due to the screening of the former by the two *K*-electrons. This is termed as *internal screening effect*. There will also be much smaller *external screening effect* due to the electrons in the orbits outside the *L*-orbit. Both these effects combine to reduce the effective charge of the nucleus from *Ze* to $(Z-\sigma)e$, where σ is called the combined screening effect and its value is about 2 for *L*-electrons. For the *M*-shell electrons, the two inner shell K and L have 2 and 8 electrons respectively and hence for these the screen constant (σ) should be ~ 10. From Bohr theory of hydrogen atom, we have the term value

$$T_n = -\frac{E_n}{2\pi c\hbar} = \frac{RZ_e^2}{n^2} \tag{5.7}$$

After applying the sommerfeld's fine structure correction to Eq. (5.7), one obtains

$$T_n = R_N \frac{(Z-\sigma)^2}{n^2} + \frac{R_N \alpha^2 (Z-\sigma')^4}{n^4}\left(\frac{n}{k} - \frac{3}{4}\right) + R_N \frac{\alpha^4 (Z-\sigma')^6}{n^6}\left(\frac{n^3}{4k^3} + \frac{3n^2}{4k^2} - \frac{3n}{2k} + \frac{5}{8}\right) \tag{5.8}$$

On applying Eq. (5.8) to the observed energy levels of X-rays shows that σ and σ' are different, though a prior, they might be expected to be the same. We must remember that α is Sommerfeld's fine structure constant.

From wave mechanical considerations Gordon derived the expression in which the azimuthal quantum number k of the sommerfeld was replaced by $j + 1/2$, where j is the total quantum number ($j = l \pm 1/2$).

$$T_n = \frac{R_N (Z-\sigma)^2}{n^2} + \frac{R_N \alpha^2}{n^4}(Z-\sigma')^4\left[\frac{n}{j+1/2} - \frac{3}{4}\right] + \frac{R_N \alpha^4}{n^6}(Z-\sigma')^6\left[\frac{1}{4}\frac{n^3}{(j+1/2)^3} + \frac{3}{4}\frac{n^3}{(j+1/2)^2} - \frac{3}{2}\frac{n}{j+1/2} + \frac{5}{8}\right] \tag{5.9}$$

The correction terms involving j in Eq. (5.9) results from the spin-orbit interaction as well as from relativistic effects and are termed to as the *spin-relativistic correction*. One can use Eq. (5.9) to calculate the wave number difference between $L_{II} - L_{III}$ and $M_{II} - M_{III}$ levels for which n, s and l are the same but j is different (Fig. (5.8)). For $L_{II} - L_{III}$, we have $n = 2$, $s = 1/2$, $l = 1$ and $j = 1/2, 3/2$. Thus one obtains the wave number separation as

$$\Delta \upsilon_L = \Delta \bar{\nu}(L_{II} - L_{III}) = \frac{R_N \alpha^2}{16}(Z-\sigma_L')^4\left[1 + \frac{5}{8}\alpha^2 (Z-\sigma_L')^2\right] \tag{5.10}$$

Similarly, we have for $M_{II} - M_{III}$, $n = 3$, $s = 1/2$, $l = 1$ and $j = 1/2, 3/2$. Thus one obtains

$$\Delta \nu_m = \Delta \bar{\nu}(M_{II} - M_{III}) = \frac{R_N \alpha^2}{81}(Z-\sigma_M')^4\left[\frac{3}{2} + \frac{31}{32}\alpha^2 (Z-\sigma_M')^2\right] \tag{5.11}$$

From experimental data for these separations, one obtains $\sigma_L' \approx 3.5$ and $\sigma_m' \approx 8.5$ independent of the value of Z, i.e., these screening constants have the above values for all elements. These values agree with the values obtained by Pauling from purely theoretical considerations. Further, to a first approximation, one expect from Eq. (5.11) the spin-relativity doublet wave number separations to be proportional to $(Z-\sigma')^4$; this is termed as *regular doublet law* and is confirmed by experiments.

If we retain only the first term on the R.H.S. of Eq. (5.9), one finds that the difference in term values for the doublets such $L_I - L_{II}$, $M_I - M_{II}$, $M_{III} - M_{IV}$ etc., which have the same n, s

and j but different L (Fig. (5.8)) values, is proportional to the difference between the values of the screening constants σ of the sub-shells (of a given shell) under consideration and is independent of Z. This is known as the Hertz *irregular doublet law*. It is found to be supported, in a general way, be experiments.

5.10 AUGER TRANSITIONS

We have seen that the characteristic X-radiations result when a transition takes place with singly ionized states. They are referred to as *first-order* or *diagram radiations*. The improved experimental techniques have shown that these intense first-order radiations are accompanied on the short wavelength side by faint radiations. These *satellite radiations* are referred to as *nondiagram* or *second-order radiations*.

When a K-electron is ejected from an atom either by electron impact or by the absorption of an X-ray photon, the resulting vacancy in the K-shell is filled by the transition of an electron from an outer shell, e.g., the L-shell. This corresponds to a transition from K-level to the L-level of the atom and this is usually associated with the emission of the characteristic X-ray photon of energy $h\nu_{KL} = E_K - E_L$. As stated above that in some cases, the surplus energy in such transition may be directly absorbed by another L-electron, which is there by emitted. Obviously, there are now two vacancies in the L-shell, the resulting energy level of the atom will be different from the L-level and can be referred as LL-level, its energy being given by E_{LL}. One can easily that the kinetic energy of the ejected second electron will then be

$$\frac{1}{2}mu^2 = E_k - E_{LL} \tag{5.12}$$

Such a non-radiative transitions resulting in the emission of electrons from the same atom first observed by Auger in 1925 are called *Auger transitions* and the effect is known as *Auger effect*.

We must note that the surplus energy in the transition of the first L-electron to the vacant K-shell may also be absorbed by an M-electron instead of a second L-electron as stated above. Obviously, in this case the second electron would be an M-electron and its energy being slightly different from that in the previous case. One may expect similar ejection from other outer shells also.

Radiationless transitions are also reported in the transitions of electrons from the outermost shell of atoms, as also in the case of nuclear transitions giving rise to the emission of internal conversion electrons.

5.11 MOSELEY'S LAW

Moseley made a systematic study of the characteristic spectra of a large number of metallic elements. The relationship between the frequency of emitting X-ray line and the special atomic number of the target element is known as *Moseley's law*. The law states that the *the frequency of a given spectral line of the characteristic X-ray increases with atomic number of the element and is proportional to the square of the atomic number Z of the element.* Moseley found that the spectra of different elements are very similar, and with increasing atomic number Z, the spectral lines merely shift towards shorter wavelengths or higher frequencies (Fig. (5.9)). Moseley plotted a graph of the K-series between the square root of frequency ($\sqrt{\nu}$) and atomic number (Z) and found it very approximately to be a straight line (Fig. (5.10)). One can represent these straight-line graphs by an equation of the form (Moseley's law),

$$\sqrt{\nu} = c_1(Z-b) \tag{5.13}$$

where C_1 and b are constant for a given series. The constant b is called the *screening factor.* The X-ray spectrographs of all elements of the periodic table prove the validity of Moseley's law. The law is applicable to other series also but with different values of C_1 and b. Rewriting Eq. (5.13) as

$$\nu = c(Z-b)^2 \tag{5.14}$$

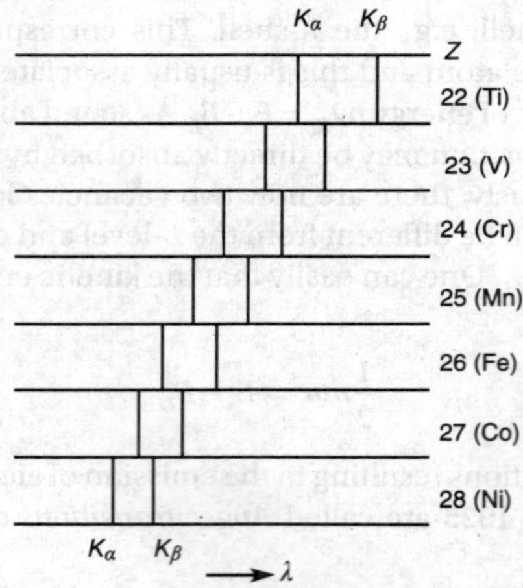

Fig. 5.9 Fine structure of X-ray lines and the possible transitions between them

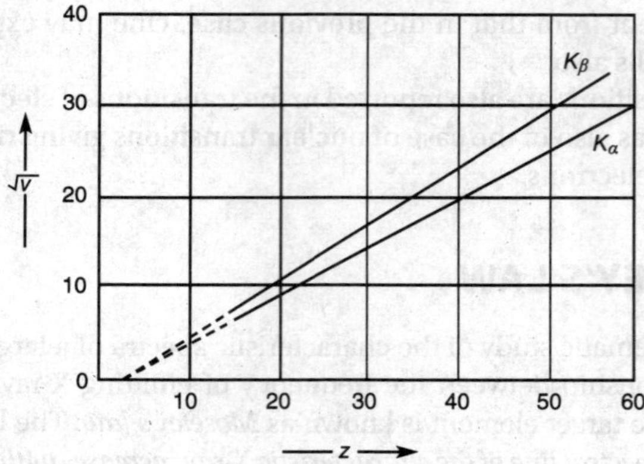

Fig. 5.10 Moseley diagram

C is another constant. For K_α radiation, it is reported that $C = \frac{3}{4}Rc$, where R is the Rydberg constant and c is the velocity of light in vacuum. For K_α radiation, $b = 1$ and hence one can write

$$\nu_{K\alpha} = \frac{3}{4}Rc(Z-1)^2 = Rc(Z-1)^2\left[\frac{1}{1^2} - \frac{1}{2^2}\right] \tag{5.15}$$

One can see that Eq. (5.15) is similar to the expression for the first line of the Lyman series of hydrogen, the only difference being the presence of the factor $(Z–1)^2$ in the present case in place of $Z^2 = 1$ for the case of hydrogen atom.

We have seen earlier that Mendeleev had arranged the elements in the periodic table in the order of the increasing atomic weight. However, Mendeleev's table had a few anomalies, e.g., the properties of Argon, Cobalt and Tellurium were not in agreement with the properties of the other elements of their group.

Moseley's work on X-ray spectra clearly indicated that an element is distinguished by its atomic number Z (number of positive charges on the nucleus of the atom). Rutherford in the same year also confirmed that the atomic number was the fundamental constant, which decided the chemical properties of the atom. Moseley then pointed out that the elements in the periodic table of elements must be arranged in the order of increasing atomic number (instead of atomic weight). In the light of this Mendeleev changed the positions of certain elements in the Mendeleev's periodic table. For example, Mendeleev found that the atomic number of cobalt $\left({}^{58.9}_{27}\text{Co}\right)$ is less than that of nickel $\left({}^{58.7}_{28}\text{Ni}\right)$ even though its atomic weight is greater. Keeping this in view he arranged them in the order Co, Ni, instead of Ni, Co. Similarly Mendeleev placed Argon $\left({}^{40}_{18}\text{A}\right)$ before potassium $\left({}^{39}_{19}\text{K}\right)$, and tellurium $\left({}^{127.6}_{52}\text{Te}\right)$ before iodine $\left({}^{127}_{53}\text{I}\right)$. When he made these changes, the anomalies of Mendeleev's periodic table disappeared.

Moseley had to left certain gaps while arranging the elements in the order of increasing atomic number, e.g., at $Z = 43$ and 72. He predicted that the corresponding rare earth elements could exist but had not then been discovered. These elements were subsequently discovered and are named as technetium and hafnium respectively.

5.12 ABSORPTION OF X-RAYS

When a parallel beam of X-rays is incident over a slab of matter, a portion of it is being absorbed by the slab and the reamaining part is transmitted. The absorbed portion of X-rays depends upon its intensity and thickness of the slab. Let a beam of X-rays of wavelength λ and intensity I be incident over the slab of the material of uniform thickness dx. If corresponding diminution in intensity is dI, we have

$$dI \propto -Idx$$

or

$$dI = -Idx \tag{5.16}$$

where the constant of proportionality μ is called the linear coefficient of absorption of the material and is equal to the fractional decrease in the intensity of the X-rays per unit thickness of the absorber.

Arranging Eq. (5.5) and integrating, one obtains

$$\int \frac{dI}{I} = -\int \mu dx$$

or $$\log_e I = -\mu x + c$$

where c is integration constant, which can be determined from initial conditions. Thus at $x = 0$, the intensity $I = I_0$. Substituting, we have

$$\log_e I_0 = c$$

If after traversing a distance x in the slab, the corresponding intensity is I, we have

$$\log_e I = -\mu x + \log_e I_0$$

or $$\log_e (I/I_0) = -\mu x$$

or $$I = I_0 \exp(-\mu x) \tag{5.17}$$

Using Eq. (5.17) and knowing initial intensity and intensity after traversing x in the material, μ can be determined. μ has the dimension of m^{-1}. The plot of $\ln_e I$ against x is a straight line of slope μ (Fig. (5.12)).

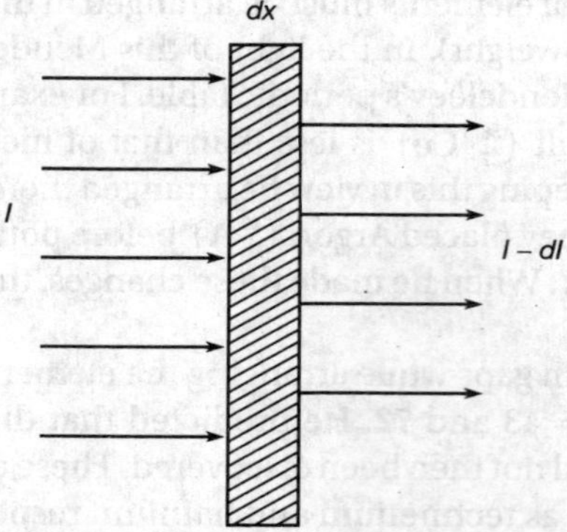

Fig. 5.11 Absorption of X-rays

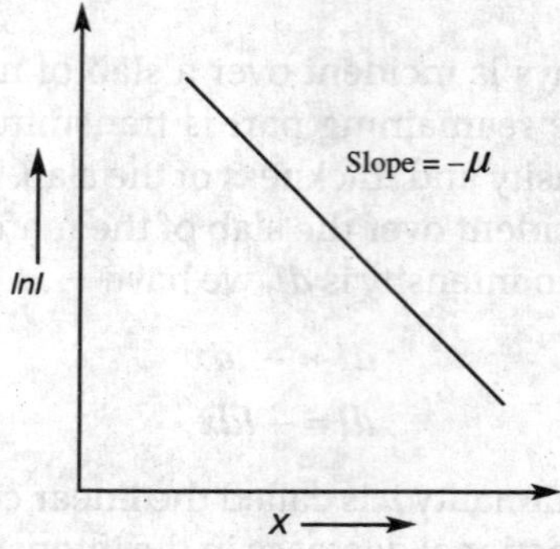

Fig. 5.12 Plot of $\log_e I$ vs x

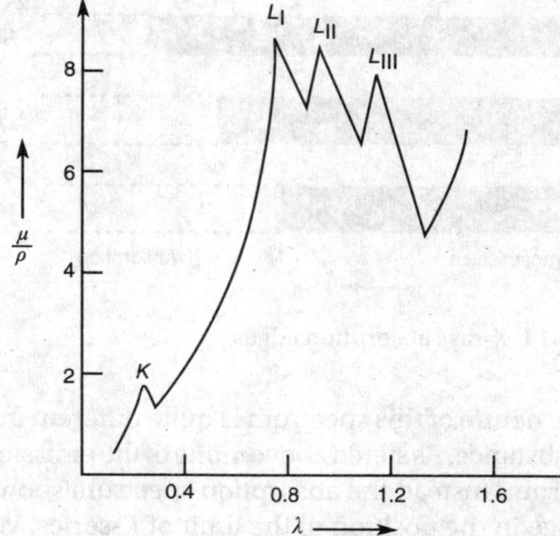

Fig. 5.13 Typical observation curve for lead. Variation of $\mu_m(\mu/\rho)$ with λ

The intensity of X-rays depend upon the energy transmitted through unit area placed perpendicular to the direction of propagation, whereas its absorption depend upon the number of atoms of the absorbing material. Hence linear coefficient of absorption also depends upon the density of the material. If the density of material is ρ, the quantity μ/e is called the *mass absorption coefficient* (μ_m) of the material. μ_m of the material depends upon the nature of the material (element) and represents the amount of energy absorbed by unit mass of it when the beam of X-rays is transmitted through unit area. The absorption by specific atom is called the *atomic absorption coefficient* (μ_A) and is given by

$$\mu_A = \frac{\mu / e}{N / M} = \frac{\mu M}{N / M} = \frac{\mu M}{\rho N} \tag{5.18}$$

where M and N are atomic weight of the material and Avagadro number respectively.

In Fig. (5.13) the variation of mass absorption coefficient for lead and corresponding wavelength of X-rays is shown. The mass absorption coefficient depends upon the cube of the wavelength (λ^3). At points K, L_I, L_{II} and L_{III}, the mass absorption coefficient drops and then again increases. These are known as limits of absorption. Whenever the wavelength of X-rays exceeds these limits, its absorption increases in the material. We must note that the material does not absorb the characteristic X-rays. Similarly corresponding to orbits M and N also for certain elements, there are 5 and 7 absorption limits.

5.13 ABSORPTION EDGES

We should note that unlike in the visible region, no discrete lines are observed in the X-ray absorption spectrum. A absorption spectrum is obtained when a beam of continuous X-rays is passed through the film of a substance and the transmitted beam is sent into a

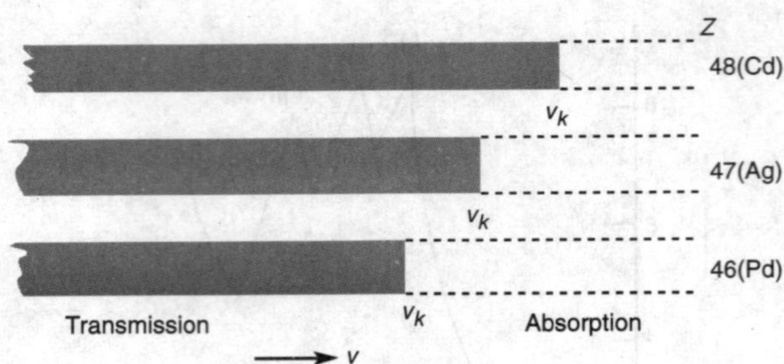

Fig. 5.14 X-rays absorption edges

X-ray spectrograph. The nature of this spectrum is quite different from the X-ray emission spectrum of the same substance. As stated above, none of the emission line K_α, K_β ··· appears in the absorption spectrum. Instead the absorption spectrum shows a continuous region bounded by a sharp edge in the position of the limit of *K*-series. When one uses successive elements in the periodic table as absorbers, the edge shifts toward higher frequency. (Fig. (5.14)). Obviously, these edges are characteristic of the absorber element and are referred as *absorption edges.*

Since the energies of the *K, L, M* etc., absorption edges are equal to the binding energies E_K, E_L, E_M etc., of the electrons in the corresponding shells and hence one can obtain the energy differences between the different absorption edges. This method is used in measuring the energies of the levels in X-ray emissions.

5.14 SCATTERING OF X-RAYS

When a beam of X-rays is allowed to fall on a plate of some chosen element, then along with the absorption of these rays, which transformed into heat etc., a portion is transmitted through the plate. The rays going out from the plate consist of a part of primary X-rays and a part of secondary radiation. The secondary radiation contains four important types of radiations:

(i) **Scattered X-rays:** These are similar to incident X-rays which have only changed their direction of propagation by passing through the absorbing material. Like primary rays these are heterogeneous and independent of the nature of the absorbing material.

(ii) **Characteristic X-rays:** These are typical of the nature of the absorber material. These are homogeneous and don't depend on the quality of the primary rays.

(iii) **Scattered β-ray:** These are nothing but fast moving electrons (β-particles) produced by photoelectric process. Like scattered X-rays, these are independent of the nature of the absorber material but depend on the quality of the primary or incident rays.

(iv) **Characteristic β-rays:** These are similar to scattered β-rays with the exception of being characteristic of the absorber material.

Fig. (5.15) depicts the phenomenon of interaction of X-rays with matter.

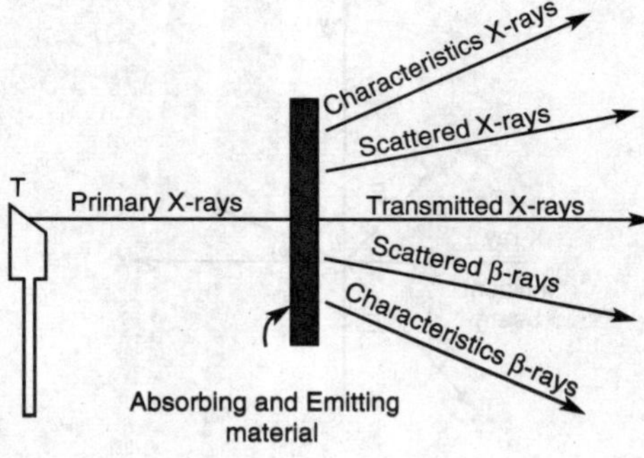

Fig. 5.15

5.15 THOMSON'S THEORY OF SCATTERING OF X-RAYS

The theory of scattering of X-rays at a target was developed by Thomson in 1900, using classical electromagnetic theory. The theory is based on the following assumptions:

(i) The wavelength of X-ray is small as compared to the dimension of the atom and hence the scattered X-rays from different electrons of atoms would have no phase relations. Thus interaction of any electron with X-rays is independent of the interactions of the other electrons.

(ii) The energy of the quantum of X-rays is very large as compared to the binding energy of the electron in the atom. Obviously, the electron can be considered to be free.

(iii) The energy of the quantum of X-rays is small as compared to rest mass energy of an electron. Thus one can ignore the quantum aspect of X-rays.

Let us consider that the incident X-ray beam travelling in the direction of x-axis strike an electron of scatterer at origin of coordinates at O. If E is the instantaneous strength of the electric field in the direction as shown in Fig. (5.16) at the location of electron, then the electron would experience a force F given by

$$F = eE \tag{5.19}$$

So the acceleration of the electron

$$a = \frac{F}{m} = \frac{eE}{m} \tag{5.20}$$

E, F and a all oscillate sinusoidally with the same frequency ν as that of incident electromagnetic waves. Under the influence of the acceleration, the electron will emit electromagnetic waves of the same frequency and in phase with incident waves. In this way electrons absorb energy of the incident beam and scatter in all directions.

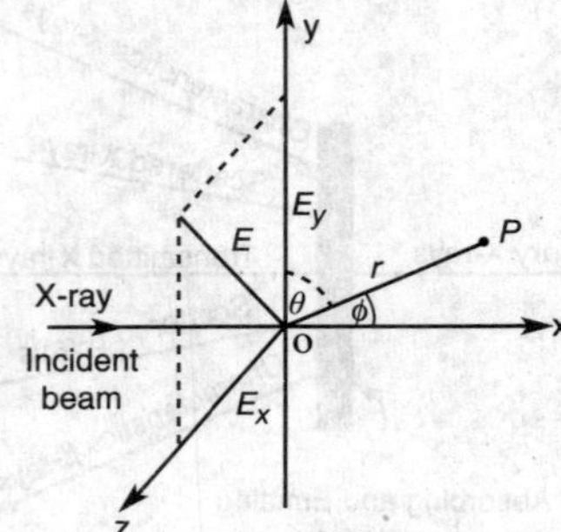

Fig. 5.16 X-rays scattering. Resolution of the electric vector $(\vec{E})$ in the incident beam

The electric field of this wave at a point *P*, distant *r* from origin making an angle θ with the direction of acceleration is given by

$$E_\theta = \frac{qa\sin\theta}{4\pi\varepsilon_0 c^2 r} = \frac{(-e)\left(-\dfrac{Ee}{m}\right)\sin\theta}{4\pi\varepsilon_0 c^2 r}$$

$$= \frac{Ee^2 \sin\theta}{4\pi\varepsilon_0 mc^2 r} \tag{5.21}$$

According to Poynting's theorem, the intensity of the incident beam is

$$I = \varepsilon_0 cE^2 \tag{5.22}$$

Similarly the intensity of the scattered beam is

$$I_\theta = \varepsilon_0 cE_\theta^2 = \frac{\varepsilon_0 cE^2 e^4 \sin^2\theta}{16\pi^2\varepsilon_0^2 m^2 c^4 r^2}$$

$$= \frac{e^4 E^2}{16\pi^2 m^2 c^3 \varepsilon_0} \frac{\sin^2\theta}{r^2} \tag{5.23}$$

Thus one obtains

$$\frac{I_\theta}{I} = \left(\frac{E_\theta}{E}\right)^2 = \left(\frac{e^2\sin\theta}{4\pi\varepsilon_0 mc^2 r}\right)^2 \tag{5.24}$$

Relation (Eq. (5.24)) gives the intensity of the scattered beam in terms of the incident intensity in a direction θ w.r.t. the direction of the *y*-axis. Actually the incident radiation is unpolarized. The electric vector $\vec{E}$ can be resolved into components E_y and E_z parallel to *y* and *z*-axis respectively, such that

$$E^2 = E_y^2 + E_z^2 \tag{5.25}$$

For an interval long compared to the period of X-rays the average value of E_y^2 and E_z^2 would be same for an unpolarized beam. Thus

$$E_y^2 = E_z^2 = \frac{E^2}{2} \tag{5.26}$$

Hence corresponding intensities are given by

$$I_y = I_z = I/2 \tag{5.27}$$

The intensity I_y of the scattered beam at point P due y component is given by

$$I_{\theta y} = \frac{I_y e^4 \sin^2\theta}{16\pi^2\varepsilon_0^2 m^2 c^4 r^2} = \frac{I_y e^4 \cos^2\phi}{16\pi^2\varepsilon_0^2 m^2 c^4 r^2} \qquad (\because \phi = 90° - \theta)$$

Putting for $I_y = I/2$, one obtains

$$I_{\theta y} = \frac{Ie^4\cos^2\phi}{32\pi^2\varepsilon_0^2 m^2 c^4 r^2} \tag{5.28}$$

Similarly, one obtains the intensity $I_{\theta z}$ of scattered beam due to Z-component as

$$I_{\theta z} = \frac{Ie^4 \sin^2\theta_z}{m^2c^4r^2}$$

But $I_z = I/2$ and $\theta_z = \pi/2$ as OP is perpendicular to Z-axis. Hence,

$$I_{\theta z} = \frac{Ie^4}{32\pi^2\varepsilon_0^2 m^2 c^4 r^2} \tag{5.29}$$

Now, $$I = I_{\theta y} + I_{\theta z} = \frac{Ie^4}{32\pi^2\varepsilon_0^2 c^4 r^2}(1+\cos^2\phi)$$

However, it is not possible to conduct experiment on a single electron. Hence if there are n_e electrons per unit volume of the scattering material, we have total intensity I_s due to n electrons as

$$I_s = \frac{n_e Ie^4}{32\pi^2\varepsilon_0^2 m^2 c^4 r^2}(1+\cos^2\phi) \tag{5.30}$$

While deriving Eq. (3.30) we have assumed that the electrons are completely free; i.e., the binding of the electrons in the atoms has been neglected. This equation gives angular distribution of the scattered X-ray radiation. The term $(1 + \cos^2\phi)$ is known as *polarization factor*.

Since I_s represents the scattered energy radiated per unit area in one sec. of wave surface of radius r, from unit volume of the scatterer, the total energy per unit volume can be obtained by integrating I_s over whole surface. Consider an elementary ring of radius $r \sin\phi$ and width $rd\phi$ as shown in Fig. (3.17), we have scattered energy in this ring as

$$dW_s = \frac{n_e Ie^4}{32\pi^2\varepsilon_0^2 m^2 c^4 r^2}(1+\cos^2\phi).2\pi r\sin\phi.rd\phi$$

$$= \frac{2\pi n_e Ie^4 r^2}{32\pi^2\varepsilon_0^2 m^2 c^4 r^2}(1+\cos^2\phi)\sin\phi\, d\phi$$

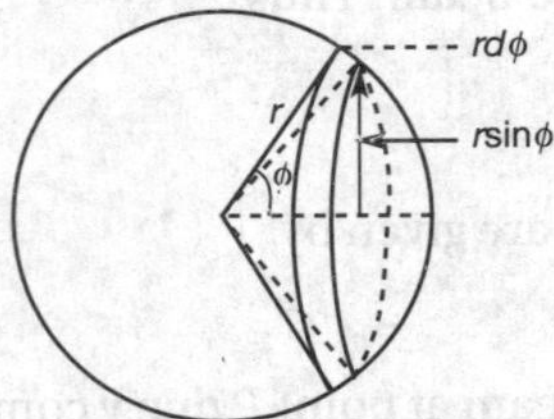

Fig. 5.17

Hence the total energy of the scattered beam is

$$W_s = \int dW_s = \frac{\pi n_e I e^4}{32\pi^2 \varepsilon_0 m^2 c^4} \int_0^{\pi} (1+\cos^2\phi)\sin\phi \, d\phi$$

$$= \frac{8}{3}\pi n_e I \left(\frac{e^2}{4\pi\varepsilon_0 mc^2}\right)^2 \quad (5.31)$$

The *classical scattering coefficient* σ is defined as the ratio of power radiated to the primary intensity, we have

$$\sigma_e = \frac{W}{I} = \frac{8}{3}\pi \left(\frac{e^2}{4\pi\varepsilon_0 mc^2}\right)^2 = \frac{\sigma}{n_e} \quad (5.32)$$

σ is the scattering coefficient and also referred as *Thomson cross-section* and this type of scattering is known as *Thomson scattering.* The term $(e^2/4\pi\,\varepsilon_0 mc^2)$ is referred as classical electron radius (r_0). There is no change of wavelength associated with this type of scattering and hence it is sometimes called *unmodified scattering.* The classical scattering cross-section for a free electron

$$\sigma_e = \frac{\sigma}{n_e} = \frac{8\pi}{3} r_0^2 = 6.63 \times 10^{-29}\,\text{m}^2 \quad (5.33)$$

where
$$r_o = \frac{e^2}{4\pi\varepsilon_0 mc^2} = 2.81 \times 10^{-15}\,\text{m} \quad (5.34)$$

Both σ_e and r_0 are universal constants, depending only on the three fundamental constants *e, m* and *c*. σ has the dimension of an area and r_0 the dimension of a length.

If a collimated beam of radiation of intensity *I* is incident on a surface placed normal to the beam, then the energy falling in a unit area of the surface per second is *I*. The fraction of this energy which is scattered by a single electron is equal to $I\sigma$. Obviously, one can regard σ as the effective area of cross-section presented by the electron to the incident beam. This means the terms classical cross-section of scattering by an electron for σ and r_0 find justification. The mass scattering coefficient σ_m of the scatterer is defined as the ratio of the amount of radiation per sec. by all the electrons of unit mass of scatterer to that the intensity of incident beam. This is given by

$$\sigma_m = \frac{N}{A} Z\sigma \quad (5.35)$$

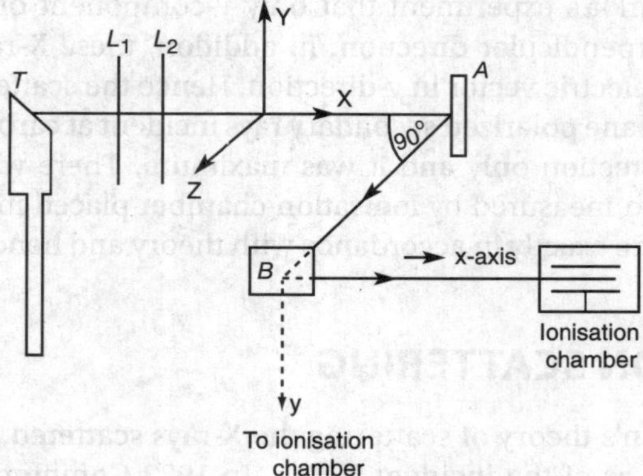

Fig. 5.18 Polarisation of scattered X-rays. Proof of Thomson scattering

Where N is the Avagadro's number, $Z \rightarrow$ the atomic number, A be the kilogram-atomic weight of the scatterer. Obviously, knowing σ_m, one can calculate the number of electrons per atom. σ_m for carbon is 0.02 m²/kg.

Barkla etal. performed experiments and proved Thomson scattering and the assumption that X-rays are a form of electromagnetic radiations. His schematic experimental arrangement is shown in Fig. (5.18). A beam of X-rays from X-ray tube T, passing through aligned slits were allowed to incident over a block A if carbon along X-axis. The secondary X-rays from A along Z-axis were incident upon another block B of carbon. The intensity of tertiary X-rays along x and y directions were measured by two ionisation chambers. According to Thomson's theory, here $\phi = \pi/2$

$$\therefore \qquad I_s = \frac{n_e I e^4}{32\pi^2 \varepsilon_0^2 m^2 c^4 r^2}$$

Barkla determined Z for the carbon atom. Experimental determination provided, for the mass scattering coefficient of carbon, the value $\sigma/\rho = 0.02\text{m}^2/\text{kg}$. Here σ is expressed in m⁻¹ and ρ in Kg/m³. From Eq. (3.33), one obtains

$$n_e = \frac{3\sigma}{8\pi r_0^2} = \frac{\sigma}{6.63\times10^{-29}} = \frac{0.02\rho}{6.63\times10^{-29}}$$

One obtains the number of carbon atoms per unit volume as

$$n_a = \frac{N_0\rho}{12} = \frac{6.025\times10^{23}\,\rho\times10^3}{12} = 5.02\times10^{23}\,\rho$$

Obviously, the number of electrons per carbon atom is

$$Z = \frac{n_e}{n_a} = \frac{3\times10^{26}}{5.02\times10^{25}} \approx 6$$

Obviously, the atomic number of carbon is equal to the number of electrons in the carbon atom. This conclusion was also found to be true for other elements. This means that the nuclear charge of an atom is equal to the total negative charge of the electrons in the atom.

It is found from Barkla's experiment that only y-component of secondary scattered X-rays exist in the perpendicular direction. In addition, these X-rays were found to be plane polarized, with electric vector in y-direction. Hence the scatter A acted as polariser for X-rays. When the plane polarized secondary rays incident at carbon block B, the intensity scatter along x-direction only and it was maximum. There was no intensity along y-direction as has been measured by ionisation chamber placed in this position, acts as analyzer. These facts are exactly in accordance with theory and hence verify the same.

5.16 COMPTON SCATTERING

According to Thomson's theory of scattering the X-rays scattered from the matter have the same wavelength as of the incident X-rays. In 1922 Compton made the measurement of the wavelengths of the scattered X-rays from light elements and found that in the scattered beam not only the wavelength of incident X-rays but also X-rays of slightly greater wavelength were present. This modified radiation is called *Compton radiation* and the effect is known as *Compton effect.* A detailed description of this effect is provided in chapter 6.

Example 5.1: Calculate the shortest wavelength of X-rays produced in a tube when the applied voltage is 12.4 kV.

Solution We have

$$eV = h\nu_{max} = \frac{hc}{\lambda_{min}}$$

$$\therefore \quad \lambda_{min} = \frac{hc}{eV} \qquad h = 6.63\times10^{34}\,\text{J-s}$$

$$= \frac{6.6\times10^{-34}\times3\times10^{8}}{1.6\times10^{-19}\times12400} \qquad c = 3\times10^{8}\,\text{m/s}$$

$$= 1\times10^{-10}\,\text{m} = 1\,\text{Å} \qquad e = 1.6\times10^{-19}\,\text{C}$$

Example 5.2: What should be the kinetic energy of an electron so that its de Broglie wave length is same as that of X-rays produced in an X-ray tube operating at 30 kV?

Solution We have

$$\lambda_{min} = \frac{hc}{eV} \tag{i}$$

If p be the momentum and K be the kinetic energy of the electron of mass m, we have

$$K = \frac{P^2}{2m} \tag{ii}$$

But $P = \dfrac{h}{\lambda}$, where λ is the de Broglie wavelength of the electron

$$\therefore \quad K = \frac{h^2}{2m\lambda^2} \qquad \begin{array}{|l} h = 6.63\times 10^{-34} \text{ J-s} \\ m = 9.1\times 10^{-31} \text{kg} \\ e = 1.6\times 10^{-19}\text{C} \\ V = 30000 \text{ Volt} \end{array}$$

Using (i), one obtains

$$K = \frac{e^2V^2}{2mc^2}$$

$$\therefore \quad K = \frac{(1.6\times 10^{-19})^2 \times (30{,}000)^2}{2\times(9.1\times 10^{-31})\times(3\times 10^8)^2} = 1.4\times 10^{-16}\text{ J}$$

Example 5.3: The series limit of the Balmer series for hydrogen is 364.6 nm. Find the atomic number of the element, which gives X-ray wavelengths down to 0.1 nm.

Solution The Balmer series of hydrogen is given by

$$\frac{1}{\lambda} = R\left(\frac{1}{2^2} - \frac{1}{n^2}\right); \qquad n = 3, 4, 5\ldots, \infty$$

Now, for the series limit ($n = \infty$), one gets

$$\frac{1}{\lambda_\infty} = \frac{R}{4}$$

or

$$R = \frac{4}{\lambda_\infty} = \frac{4}{364.6\text{nm}}$$

We have the X-ray wavelengths of the K-series are given by

$$\frac{1}{\lambda} = R(Z-1)^2\left(\frac{1}{1^2} - \frac{1}{n^2}\right)$$

The minimum wavelength of this series occurs when $n = \infty$ and thus

$$\frac{1}{\lambda} = R(Z-1)^2 \qquad \begin{array}{|l} \lambda = 0.1\text{nm} \\ R = \dfrac{4}{364.6}\text{nm} \end{array}$$

or

$$(Z-1)^2 = \frac{1}{\lambda R} = \frac{364.6\text{mn}}{0.1\text{nm}\times 4} = 911.5$$

or

$$Z - 1 = 30.2 \;\therefore\; Z = 31\,(\text{Gallium})$$

Example 5.4: Calculate the quantum of energy emitted when an L-electron (i.e., an electron from the $n = 2$ state) drops into the empty K-state in copper ($Z = 29$)? Given $R = 109737 \times 10^2\text{cm}^{-1}$; $1\text{m}^{-1} = 1.239 \times 10^{-2}\text{eV}$.

Solution The electron drops from $n = 2$ to $n = 1$ state, we have

$$\frac{1}{\lambda} = R(Z-b)^2\left[\frac{1}{1^2} - \frac{1}{2^2}\right]$$

For the K-line, $b = 1$. Thus

$$\frac{1}{\lambda} = \frac{3}{4}R(Z-1)^2$$

$$= \frac{3}{4}\times(109737\times10^2\,\mathrm{m}^{-1})(29-1)^2$$

$$= 6.4525\times10^9\,\mathrm{m}^{-1} = 7994.6\mathrm{eV}$$

Example 5.5: The wavelength of L_α X-ray lines of silver and platinum are 4.1538Å and 1.3216Å respectively. An unknown substance emits L_α X-ray with a wavelength of 0.966Å. If the atomic numbers of silver and platinum be 47 and 78 respectively, determine atomic number of the unknown substance.

Solution We have Moseley's Law

$$\sqrt{\nu} = k(Z-b) = \sqrt{\frac{c}{\lambda}}$$

$\therefore$
$$\sqrt{\frac{3\times10^8}{4.1538\times10^{-10}}} = k(47-b) \quad \text{(i)}$$

and
$$\sqrt{\frac{3\times10^8}{1.3216\times10^{-10}}} = k(78-b) \quad \text{(ii)}$$

For unknown substance

$$\sqrt{\frac{3\times10^8}{0.9666\times10^{-8}}} = k(Z-b) \quad \text{(iii)}$$

Dividing (i) (ii) one obtains,

$$\sqrt{\frac{1.3216}{4.1538}} = \frac{47-b}{78-b} \quad \therefore b \approx 7$$

Dividing (iii) by (i), obtains

$$\sqrt{\frac{1.3216}{0.966}} = \frac{Z-7}{78-7}$$

or
$$1.142(78-7) = Z-7 \quad \therefore Z = 90.$$

5.17 X-RAY DIFFRACTION AND CRYSTAL AS THREE-DIMENSIONAL GRATING

Soon after the discovery of X-rays in 1895, the nature of X-rays became a matter of concern. Results of early experiments indicated that if X-rays consisted of waves, their wavelengths were of the order of 10^{-8} to 10^{-9} cm. Laue suggested that the ordered arrangement of atoms in a crystal must make it to act as a three-dimensional grating. Since the spacing between the layers of these atoms in a crystal is of the order of 10^{-8} cm. (which is of the order of wavelength of X-rays), so a crystal would be suitable for the diffraction of X-rays.

Laue Experiment

The experimental arrangement of Laue and his co-workers is shown in Fig. (5.19). A thin pencil of X-rays after passing through aligned slit was allowed to pass through a thin plate of crystal of zinc blende. The transmitted beam is received on a photographic plate. After the exposure of several hours, when the plate was developed, it has been found that in addition to the central spot, it consists of other fainter spots arranged regularly (Fig. (5.19)). These spots are known as *Laue spots*.

Each spot in the Laue pattern arises due to the constructive interference between the waves reflected from one of the various sets of parallel plane in the crystal. Fig. (5.20) shows two such sets of plane by means of continuous and dotted lines. The corresponding planes will be $\perp^r$ to the plane of the paper. A simple interpretation was provided by W.L. Bragg.

Bragg's Law

Consider a set of parallel planes of the crystal, separated by spacing d. The crystal acts as a series of parallel reflecting planes. Let a beam of monochromatic X-rays strike the crystal.

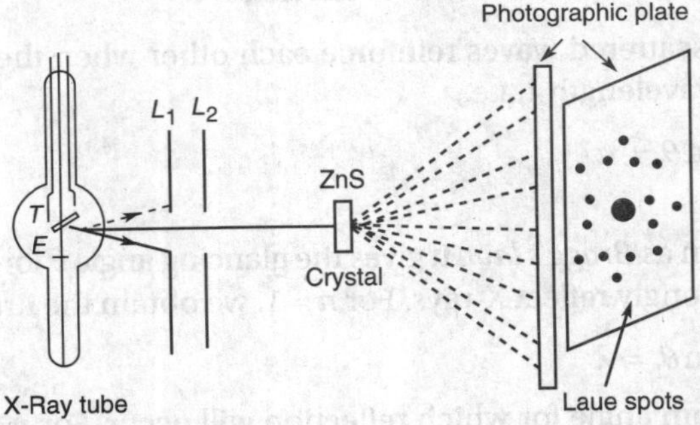

Fig. 5.19

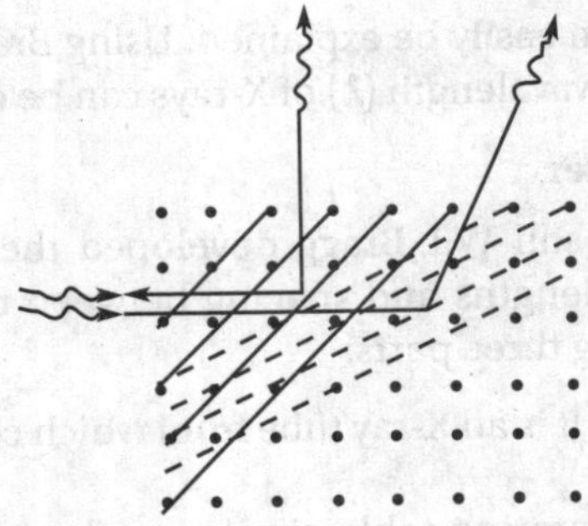

Fig. 5.20

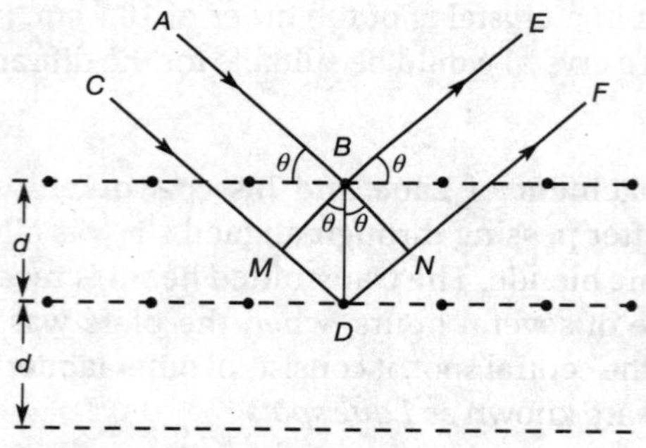

Fig. 5.21

Since X-rays can penetrate the crystal there would be partial reflection from every plane till it would be absorbed completely. The rays reflected from various planes will interfere constructively if they are in same phase. Consider two parallel rays *AB* and *CD* incident over the plane at glancing angle θ. Corresponding reflected rays from atoms *B* and *D* are *BE* and *DF*. The path difference between these rays is $MD + DN$, where *M* and *N* are feet of $\perp$ from *B* on *CD* and *DF* respectively. Then

$$MD = DN = d\sin\theta$$

$$\therefore \quad MD + DN = 2d\sin\theta \tag{5.36}$$

Now the two sets of scattered waves reinforce each other when the path difference is an integral multiple of wavelength λ, i.e.,

$$2d\sin\theta = n\lambda \tag{5.37}$$

Here n is an integer.

(Eq. (5.37)) is known as *Bragg's law*. It gives the glancing angle θ for which a set of parallel atomic planes will strongly reflect X-rays. For $n = 1$, we obtain the first order reflection as

$$2d\sin\theta_1 = \lambda \tag{5.38}$$

Here θ_1 is the minimum angle for which reflection will occur. For $n = 2$, i.e., second order reflection, we have

$$2d\sin\theta_2 = 2\lambda \tag{5.39}$$

Obviously, the Laue spots can easily be explained. Using Bragg's relation (Eq. (5.37)) and knowing lattice space d, the wavelength (λ) of X-rays can be determined.

The Bragg X-ray spectrometer

In 1913 W.H.Bragg and his son W.L.Bragg developed the X-ray spectrometer for the measurement of X-ray wavelengths and spacing between the lattice planes of crystals. It consists mainly of following three parts:

(i) **Source of X-rays(*T*):** It is an X-ray tube from which collimated fine pencil of X-ray is obtained.

(ii) **Turn table:** It is a rotating table similar to the ordinary spectrometer. Over this turntable is mounted the crystal (C). A fine pencil of X-ray is allowed to

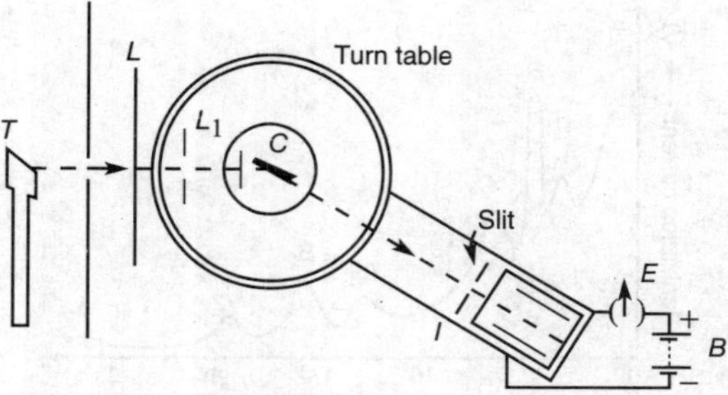

Fig. 5.22 Bragg's Spectrometer

fall over the crystal at glancing angle. The angular position of the table and the angle of incidence can be measured with the help of verniers attached to the table.

(iii) Detecting device: The reflected beam from the crystal (C) is detected by an ionisation chamber (I). The beam, through a slit, is allowed to enter the chamber filled with ethyl bromide. The ionisation current is measured by the electrometer (E). The arm carrying ionisation chamber can be rotated and be recorded by the help of vernier. For recording the intensity of reflected beam, in place of ionisation chamber, a photographic plate may also be used. Fig. (5.22) shows the essential parts of the spectrometer. The ionisation current produced by the X-rays is a measure of the intensity of X-rays diffracted by an angle 2θ.

At each particular setting of the crystal, i.e., a certain value of θ only the particular wavelength λ which satisfies the Bragg's relation

$$2d \sin\theta = n\lambda$$

is reflected to the photographic plate or ionisation chamber. Thus if the crystal spacing d is known then corresponding to the known value of θ the wavelength of X-rays can be determined. Alternatively, if the X-ray wavelength is known the crystal lattice spacing can be determined.

To obtain a spectrum of X-ray wavelengths produced by a certain source, the ionisation current (I) is measured as a function of the glancing angle (θ). Fig. (5.23) shows a typical plot obtained with a platinum target. The ionisation current is seen to increase suddenly at some sharply defined angles; the three peaks A_1, B_1 and C_1 represent first order X-ray spectrum lines. Second order of these lines appear at A_2, B_2 and C_2 at glancing angles whose sines are twice those of A_1, B_1 and C_1. The relative intensities of the second order spectral lines are also in the same ratio of those of the corresponding lines in the first order. This line spectrum is superimposed on continuous spectrum.

The X-ray spectral lines are characteristic of the nature of the material used as target. The choice of the crystal to be used is governed be the range of wavelengths to be measured, the case with which a good crystal surface can be obtained and the reflecting power of the crystal. Commonly used crystals are calcite, rock salt and quartz.

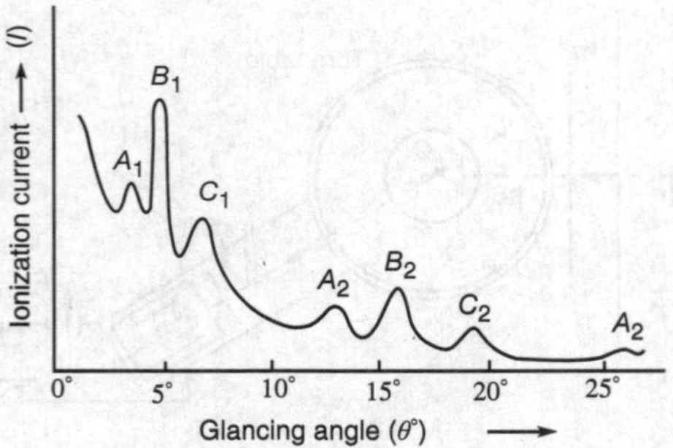

Fig. 5.23

Determination of wavelength

To determine wavelength of X-rays the grating spacing d must be known accurately. As an example consider the case of NaCl crystal (Fig. (5.7)). X-ray studies of this crystal show that the sodium and chlorine atoms occupy alternate corners of the cube lattice. Hence, on the average the volume occupied by every atom is d^3 with mass ρd^3 where ρ is the density. Thus

$$\rho d^3 = \frac{M}{2N}$$

Here $M \rightarrow$ Molecular weight of sodium chloride = 58.454

$\therefore$ Mass of one mole of NaCl = 58.454 g

$N_0 \rightarrow$ Avagadro number $= 6.03 \times 10^{23}$

Since each mole of any substance consists of $N_0 = 6.03 \times 10^{23}$ molecules, the mass of each NaCl molecule

$$= \frac{58.454}{6.03 \times 10^{23}} = 9.71 \times 10^{-23} \text{ g/mole}$$

or mass of half a molecule or that of an elementary cube of NaCl

$$= \frac{9.71 \times 10^{-23}}{2} = 4.855 \times 10^{23} \text{ g}$$

The density of NaCl $= 2.16 \text{ g/cm}^3$

$\therefore$ Volume of the elementary cube of NaCl $= \dfrac{\text{Mass}}{\text{density}}$

$$= \frac{4.855 \times 10^{23}}{2.16} \text{ cm}^3$$

But volume of elementary cube = d^3

$$\therefore \quad d^3 = \frac{4.855\times10^{23}}{2.16}$$

$$\therefore \quad d = \left(\frac{4.855\times10^{23}}{2.16}\right)^{1/3}$$

$$= 2.82\times10^{-8}\,\text{cm} = 2.82\text{Å}$$

Example 5.6: X-rays wavelength 0.36Å diffracted in a Bragg spectrometer at an angle of 4°48'. Find the effective value of atomic spacing.

Solution We have from Bragg's law

$$\lambda = 2d\sin\theta \quad \Big| \quad \lambda = 0.36\times10^{-8}\,\text{cm}$$

$$\therefore \quad d = \frac{\lambda}{2\sin\theta} \quad \Big| \quad \theta = 4°48', \quad \sin 4°48' = 0.0837$$

$$\therefore \quad d = \frac{0.36\times10^{-8}}{2\times0.0837} = 2.15\times10^{-8}\,\text{cm} = 2.15\text{Å}$$

Example 5.7: The glancing angle for the first order reflection from a nickel crystal is 9.5°. If the wavelength of X-rays is 0.58Å, what is the spacing between the atomic planes of the crystal?

Solution

$$2d\sin\theta = n\lambda \quad \Big| \quad n = 1$$

$$\therefore \quad d = \frac{n\lambda}{2\sin\theta} \quad \Big| \quad \lambda = 0.58\times10^{-10}\,\text{m}, \quad \theta = 9.5°, \quad \sin 9.5° = 0.165$$

$$= \frac{0.58\times10^{-10}}{2\times0.165} = 1.76\times10^{-10}\,\text{m} = 1.76\text{Å}$$

Example 5.8: (a) The Bragg angle for reflection from the (110) plane in bcc Iron is 22° for an X-ray wavelength of λ = 1.54Å. Compute the cube edge for iron (b) What is the Bragg angle for the reflection from the (111) planes? (c) Calculate the density of bcc iron. The atomic weight of Fe is 55.8.

Solution (a) If the spacing between the (110) planes is d then we have,

$$2d\sin\theta = \lambda \text{ for } 1^{\text{st}} \text{ order}$$

or

$$d = \frac{\lambda}{2\sin\theta} = \frac{1.54}{2\times\sin 22} = 2.055\text{Å}$$

In cubic crystals

$$d_{hkl} = \frac{a}{(h^2+k^2+l^2)^{1/2}}$$

$$\therefore \qquad d_{110} = \frac{a}{\sqrt{2}} = 2.055$$

$$\therefore \qquad \text{Cubelength } a = 2.055 \times \sqrt{2}$$

(b) For (111) planes the separation is

$$d_{111} = \frac{a}{\sqrt{3}} = \frac{2.9}{\sqrt{3}}$$

(c) If θ is Bragg angle, then

$$2d_{111} \sin\theta = \lambda$$

or

$$\sin\theta = \frac{1.54 \times \sqrt{3}}{2 \times 2.9} = 0.459$$

Example 5.9: Calculate the glancing angle on the cube face (1 0 0) of rock salt crystal ($a = 2.814$Å) corresponding to second order reflection of X-rays of wavelength $\lambda = 0.710$Å.

Solution For {1 0 0} plans the spacing d is equal to a and using Bragg's condition for direction

$$2d \sin\theta = n\lambda$$

We have for second order $n = 2$

$$2 \times 2.184 \sin\theta = 2 \times 0.710$$

or

$$\sin\theta = \frac{2 \times 0.710}{2 \times 2.814}$$

$$= 0.25$$

or

$$\theta = \sin^{-1}(0.25) = 14.6°$$

Example 5.10: Monochromatic X-rays of $\lambda = 0.814$Å are allowed to fall on a crystal. Two of the lines are observed at angles 158° and 159° from the direct beam. If these are third order maxima due to two sets of lattice planes, deduce the difference of their spacing.

Solution If the spacing between the planes in the tow orientations is d_1 and d_2 then we have

$$2d_1 \sin\theta_1 = n\lambda$$

and

$$2d_2 \sin\theta_2 = n\lambda$$

when n is 3 (order of spectrum) and

$$\theta_1 = \frac{158}{2} = 79° \quad \text{and} \quad \theta_2 = \frac{159}{2} = 79.5°$$

$$\therefore \qquad d_1 = \frac{n\lambda}{2\sin\theta_1} = \frac{3 \times 0.814}{2 \times \sin 79} = 1.248\text{Å}$$

$$d_2 = \frac{n\lambda}{2\sin\theta_1} = \frac{3 \times 0.814}{2 \times \sin 79.5} = 1.246\text{Å}$$

and

$$\theta = 27.37°$$

(c) Density is given by

$$\rho = \frac{nA}{NV}$$

Here structure is BCC, $n = 2$ and volume of unit cell $= a^3$

$$\therefore \qquad \rho = \frac{2 \times 55.8}{6.02 \times 10^{26} \times (2.9 \times 10^{-10})^3}$$

$$= 7.6 \times 10^3 \, \text{kg}^{-3}$$

5.18 LAUE EQUATIONS

Now, we take a more general approach in which X-rays scattered from different individual atoms can re-combine and obey the so-called Laue equations. This will also provide a test for validity of Bragg picture, i.e., from Laue equations, Bragg's condition (Eq. (5.40)) will be derived.

Let us assume that the only effect of an atom on the X-rays is to change its direction i.e., we shall restrict to elastic scattering only. Let us examine the radiation scattered by two identical scattering centres in the crystal separated by a distance r. Let $\hat{S}_0$ and $\hat{S}$ be unit vectors in the direction of the incident and reflected waves respectively (Fig. (5.24)). The incident radiation is assumed to be a parallel beam, and the scattered beam is assumed to be detected at a very distant observation point. The path difference between the waves scattered from the point O and A is

$$\Delta = OB - CA = \Delta_2 - \Delta_1$$

But

$$\Delta_2 = \vec{r}\,\hat{S} \text{ and } \Delta_l = \vec{r}\,\hat{S}_0$$

$$\Delta = \vec{r}\cdot\hat{S} - \vec{r}\cdot\hat{S}_0 = \vec{r}\cdot(\hat{S} - \hat{S}_0) = \vec{r}\cdot\vec{s} \tag{5.40}$$

where

$$\vec{S} = \hat{S} - \hat{S}_0$$

The vector $\vec{S}$ happens to be the normal to a plane from which the reflection takes place. This plane is called the reflecting plane or it is the plane of the reflecting atoms for the Bragg derivation. If the angle between $\hat{S}_0$ and $\hat{S}$ be 2θ then the magnitude of $S = 2\sin\theta$. Since $\hat{S}_0$ and $\hat{S}$ are unit vectors.

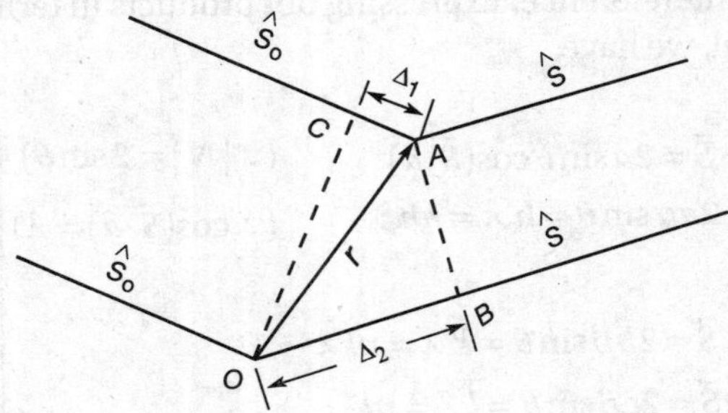

Fig. 5.24

The phase difference ϕ between the radiation scattered at the two points is

$$\begin{aligned}\phi &= \frac{2\pi}{\lambda}(\text{path difference}) \\ &= \frac{2\pi}{\lambda}(\vec{r}\cdot\vec{S}) \qquad (5.41)\end{aligned}$$

In order that the direction $\hat{S}$ be a diffraction maxima the scattering contribution from every atom in the crystal in that direction must differ in phase by an integral multiple of 2π. In order for this to hold good, it is only necessary for the radiation from atoms separated by the primitive lattice vectors $(\vec{a},\vec{b},\vec{c})$ to add in phase for them the contribution from other atoms separated from the origin by integral combinations of these vectors will certainly add in phase. If scattering contribution from neighbouring atoms were to differ in phase in any other manner, it would then always be possible to found an atom some where in the crystal which would contribute radiation. Just π out of phase with the contribution from a given atom; these contribution should then cause atom by atom, giving no diffracted beam.

Now, let us relate this to the crystal structure. The significant distances (r) in a crystal are the primitive translation vectors because they are the nearest-neighbour distances, so let us express the phase difference in terms of adjacent atoms, which are separated by the primitive translation distance. Thus we have

$$\left.\begin{aligned}\phi_a &= \frac{2\pi}{\lambda}(\vec{a}\cdot\vec{S}) = 2\pi h' = 2\pi nh \\ \phi_b &= \frac{2\pi}{\lambda}(\vec{b}\cdot\vec{S}) = 2\pi k' = 2\pi nk \\ \phi_c &= \frac{2\pi}{\lambda}(\vec{c}\cdot\vec{S}) = 2\pi l' = 2\pi nl\end{aligned}\right\} \qquad (5.42)$$

When h', k', l' are any integers with n as common factor among them. h, k, l are also integers with no common factor greater than 1.

This brings us from a consideration of any pair of atoms to consideration of adjacent atoms in the crystal. All the conditions out-lined in Eq. (5.42) must be satisfied simultaneously for constructive intereference. Expressing dot products in terms of direction cosines α, β and γ of the normal, we have,

$$\left.\begin{aligned}\vec{a}\cdot\vec{S} &= 2a\sin\theta\cos(\vec{S}\cdot\vec{a}) && (\because |N| = 2\sin\theta) \\ &= 2a\alpha\sin\theta = h'\lambda = nh\lambda && (\because \cos(\vec{S}\cdot\vec{a}) = \lambda) \\ \text{Similarly} \quad & \\ \vec{b}\cdot\vec{S} &= 2b\beta\sin\theta = k'\lambda = nk\lambda \\ \vec{c}\cdot\vec{S} &= 2c\beta\sin\theta = l'\lambda = nl\lambda\end{aligned}\right\} \qquad (5.43)$$

These are the Laue equations, which determine the direction of constructive interference for a particular wavelength. These equations can also be written as

$$\left.\begin{aligned}\frac{2\alpha a}{h}\sin\theta = n\lambda\\ \frac{2\beta b}{k}\sin\theta = n\lambda\\ \frac{2\gamma c}{l}\sin\theta = n\lambda\end{aligned}\right\} \tag{5.44}$$

Expressing Eq. (5.44) in the form

$$\left.\begin{aligned}\frac{2\alpha\sin\theta}{n\lambda} = \frac{h}{a}\\ \frac{2\beta\sin\theta}{n\lambda} = \frac{k}{b}\\ \frac{2\gamma\sin\theta}{n\lambda} = \frac{l}{c}\end{aligned}\right\} \tag{5.45}$$

Where θ is the angle between the incident or reflected direction and the reflecting plane, we observe that in a direction of constructive interference the direction cosines are proportional to $\frac{h}{a}, \frac{k}{b}$ and $\frac{1}{c}$. The direction cosines are taken with respect to the primitive translation vectors. These direction cosines are the direction cosines of the scattering normal $\vec{S}$ can be seen from Eq. (5.43). to be proportional to $\frac{h}{a}, \frac{k}{b}$ and $\frac{1}{c}$. Therefore the lattice planes (h, k, l) are parallel to the "reflecting plane" of X-ray diffraction. This is because their normals having the same direction cosines, must be parallel. In other words, we can get the diffraction maxima a by reflecting in lattice plane. This is why Bragg's law works.

The Bragg's law can be obtained from Laue equation. Squaring and adding Eq. (5.45), we have,

$$\frac{4^2\sin^2\theta}{n^2\lambda^2}(\alpha^2+\beta^2+\gamma^2) = \frac{h^2}{a^2}+\frac{k^2}{b^2}+\frac{l^2}{c^2} \tag{5.46}$$

But $\alpha^2+\beta^2+\gamma^2 = 1$ being the sum of the squares of direction cosines and

$$\frac{h^2}{a^2}+\frac{k^2}{b^2}+\frac{l^2}{c^2} = \frac{1}{d^2}$$

Here d is the distance between adjacent planes with indices (h, k, l)

$$\therefore \quad \frac{4^2\sin^2\theta}{n^2\lambda^2} = \frac{1}{d^2} \tag{5.47}$$

or

$$2d\sin\theta = n\lambda$$

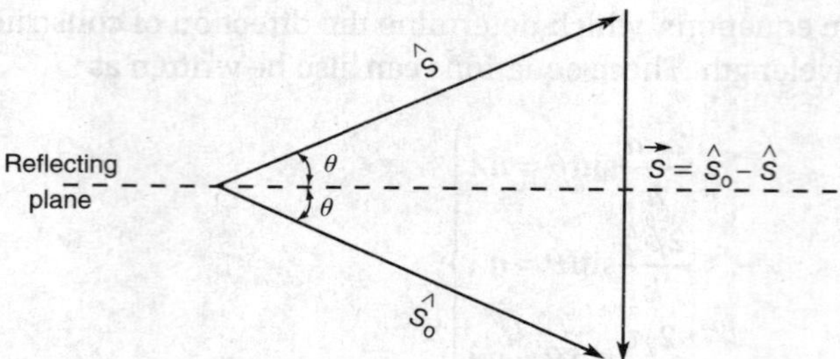

Fig. 5.25

In the derivation of Laue equations, we began with the scattering of X-rays by atoms. If there were no atoms, there would be no scattering; so those planes containing no atoms will not reflect. From the observed directions of reflection one can infer which of the lattice planes are rich in atoms and there by obtain the crystal structure.

In order to reflect, the planes must be full of atoms in equivalent sites. This is necessary for constructive interference. An arbitrary plane would have equivalent atoms only at large separation.

The basic idea underlying the use of Bragg's law in studying crystal structure is apparent from Eq. (5.47). Since λ can be determined independently, and θ can be measured directly from reflection experiments (it is half the angle between incident and diffracted beams shown in Fig. (5.25) and using Eq. (5.47) the interplanar distance d can easily be calculated. Thus the cell constant can be determined using different orientations of the crystal.

SUMMARY

1. x-rays are produced when electrons accelerated by a high voltage V strike the metal target in the x-ray tube. The spectrum of wavelengths emitted by the tube consists of two parts: a *continuous spectrum* and a *series of peaks superimposed* on it. The continuous spectrum depends on the voltage across the tube whereas the peaks are characterstics of the material of the target (*characteristic spectrum*).
2. The emission of x-ray from declerating electron is called *Bremsstrahlung process*. This process is sometimes referred as *inverse photoelectric effect.*
3. Moseley established that square root of the frequencies of the spectral lines, for e.g. K_α line, were proportional to the atomic number Z of the element
$$\sqrt{\nu} = a(Z - b)$$
where a and b are constants for a given series.
4. When a collimated beam of x-rays is made to pass through a medium, the intensity of beam decreases in accordance with
$$I = I_o e^{-\mu x}$$
where μ is a constant called the linear absorption coefficient, I_o be the intensity of incident beam, I the intensity after traversing a distance x.
5. Bragg's Law is
$$2d \sin\theta = n\lambda, \quad n = 1, 2, 3, \ldots$$

REVIEW QUESTIONS

1. Give an account of the production and properties of X-rays. How can it be established that these are electromagnetic radiations?
2. What are continuous and characteristic X-rays? Discuss their salient features.
3. How are continuous X-rays produced? What is the significance of the short wavelength limit and how it is related with the voltage applied across the X-ray tube?
4. What is characteristic X-ray spectrum? Explain its origin.
5. Discuss Moseley's law and explain it on the basis of Bohr atom model. What do you understand by Bremsstrahlung?
6. Derive Moseley's law on the basis of Bohr's theory and discuss its importance.
7. Define scattering and mass scattering coefficients. How Thomson's Theory can be verified experimentally?
8. The optical spectra of elements show periodic characteristics with increasing Z, while X-ray characteristic spectra show continuous order through all Z-values. Explain.
9. Mention the applications of Moseley's law.
10. How Moseley's law helped in rearranging various elements on a rational basis?
11. How can you account for the origin of characteristic lines in different K and L series of X-ray spectra.
12. What factors govern intensity and penetrating power of X-rays? What changes will take place in the character of X-ray spectrum with the increase of tube potential.
13. Give a brief account of experiments to establish (i) the transverse character and (ii) the particle character of X-rays.
14. Discuss the diffraction of X-rays by a three-dimensional lattice with special reference to Laue spots. Interpret them.

PROBLEMS

1. An X-ray tube works at a potential difference of 100,000V. Only 0.1% of the energy of cathode rays is converted into X-ray radiation and heat is generated in the target at the rate of 120 calorie per second. What current does the tube pass and what is the energy and velocity of an electron when it reaches the target.
[**Hint:** Operating potential = 10^5V, electron energy = 10^5eV = $.6\times10^{-14}$J. Velocity of the electron

$$V=\left(\frac{2\times1.6\times10^{-14}}{9.1\times10^{-31}}\right)^{1/2}=1.875\times10^{8}\,\text{m/s}$$

Since the velocity is quite high, one has to apply relativistic correction. We have for K.E.,

$$E=c^2(m-m_0)=c^2m_0(\beta-1),\ \therefore\ \beta=1/\sqrt{1-(v^2/c^2)}$$

$$\therefore\quad 1-\beta=E/m_0c^2=\frac{1.6\times10^{-14}}{9.1\times10^{-31}\times9\times10^{16}}=\frac{16}{81.9}=\frac{1}{5.12}$$

$$\therefore\quad \beta=1.1953\text{ and thus }1-\frac{v^2}{c^2}=\frac{1}{\beta^2}\cong0.7\text{ or }v=1.643\times10^{8}\text{m/s}$$

If n electrons reach the target in one second, the amount of n electrons converted into heat

$$= n \times 159.84 \times 10^{-14} J \therefore \quad n \times 159.84 \times 10^{-14} = 120 \times 4.2$$

$$\therefore \quad n = 3.16 \times 10^{14} s$$

$$\text{current} = 3.16 \times 10^{14} \times 1.6 \times 10^{-19} c/s = 50.6 \mu A$$

2. What are the most energetic X-rays emitted in each case when 40 keV electrons bombard (i) Mo target ($Z = 42$) (ii) Cu target ($Z = 29$)? [**Ans.** 40 keV, whichever be the target]

3. Electrons are accelerated in television tubes through potential differences of about 10 kV. Find the highest frequency of the electromagnetic waves emitted when these electrons strike the screen of the tube. What kind of waves are these? [**Ans.** 2.41 × 10^18 Hz, X-rays]

4. An X-ray tube operating at 66 kV is producing continuous X-rays. Calculate the shortest wavelength limit of X-rays. [**Ans.** 0.19Å]

5. The minimum wavelength recorded in the continuous X-ray spectrum of a 100-kV tube is 12.35 × 10^-12 m. Find the value of Planck's constant. [**Ans.** $h = 6.59 \times 10^{-34}$ J-s]

6. An X-ray tube with copper target is operated at 25kV. The glancing angle for a NaCl crystal for the Cu K_α line is 15.8°. Find the wavelength of this line. Also find the glancing angle for photons at the short wavelength limit. (d for NaCl = 2.82Å). [**Ans.** $\lambda_{min} = 0.496$Å, $\theta = 5.1°$]

7. The K_α line from the anticathode of an X-ray tube is reflected in the first order at an angle of 6° by a crystal of NaCl mounted on a Bragg spectrometer. Calculate the wavelength of K_α line. Given density of NaCl = 2.17 × 10^3 kg/m^3, $N = 6.023 \times 10^{26}$/kg-mole; molecular weight of NaCl = 58.5. [**Ans.** 0.59Å]

8. The K_α line from molybdenum ($Z = 42$) has a wavelength of 0.7078Å. Calculate the wavelength of K_α line of Zinc ($Z = 30$). [**Ans.** 1.4148Å]

9. If the short series limit of the Balmer series for hydrogen is 3646Å. Calculate the atomic number of the element, which gives X-ray wavelengths down to 1Å. Identify the element. [**Ans.** ($Z = 31$), Gallium]

10. If the K, L, M energy levels of platinum involved in the emission of the K-series of X-rays of this element are 78000, 12000 and 30000 eV respectively, calculate the wavelength of the K_α and K_β lines of platinum. [**Ans.** $\lambda_{k\alpha} = 0.188$Å, $\lambda_{k\beta} = 0.165$Å]

11. A material whose K-absorption edge is 0.15Å is irradiated with 0.13Å X-rays. Calculate the maximum kinetic energy of photoelectrons that are emitted from the K-shell? [**Ans.** 12.68 keV]

12. For a certain metal the K-absorption edge is at 0.172Å. The wavelength of K_α, K_β and K_γ lines of K-series are 0.210Å, 0.192Å and 0.180Å respectively. Calculate the energies of K, L, M and N shells for the metal.

[**Ans.** $E_{k_\alpha} = 58.93$ keV, $E_{k_\beta} = 64.45$ keV, $E_{k_\gamma} = 68.75$ keV $\therefore$ Energy of K-shell = –71.95keV, Energy of L-shell = –13.02 KeV, Energy of M-shell = –7.50 KeV and energy of N-shell = –3.20 keV]

13. The following K_α-lines have been measured; Magnesium: 9.87Å; Sulphur: 5.36Å Calcium: 3.35Å; Chromium: 2.29Å Cobalt: 1.79Å; Copper: 1.54Å; Rubidium: 0.93Å; Tungsten: 0.21Å
Plot the square root of the K_α-frequency against the atomic number of the element and verify Moseley relation $\sqrt{\nu}=a(Z-b)$. Determine a and b.

14. Calculate the glancing angle on the cube face [100] of a rock salt crystal (a = 5.63Å) corresponding to second order diffraction maximum for the X-rays of wavelength 0.71Å.

15. Monochromatic X-rays of wavelength 0.124Å are scattered from a block of Carbon. Calculate the maximum energy of recoil electron. **[Ans.** 28.13 keV]

16. The glancing angle of the first order spectrum in Bragg's spectrometer is found to be 8°. Calculate the wavelength of X-rays if the spacing between the planes be 2.83×10^{-10}m.

SHORT ANSWER-QUESTIONS

1. What is the order of wavelength of X-rays?
Ans: 10^{-10}m.

2. X-ray photon collides with an electron and bounces off. Comment on its new frequency.
Ans: The new frequency of the photon, which bounces off is either equal or less than the frequency of the incident photon.

3. The anode in the Coolidge tube has high atomic weight; why?
Ans: (i) To produce hard X-rays and (ii) high melting point to withstand the high temperature.

4. Of which wavelength will the energy of X-rays be higher (i) 6Å (ii) 36Å
Ans: The energy of 6Å X-rays will be higher and it will be 6 times that of 36Å.

5. On what factor the shortest wavelength of X-rays emitted from an X-ray tube depend?
Ans: The shortest wavelength of the X-rays emitted depends on the energy of the electrons incident on the target, which in turn depends on the potential they have fallen. We have $\lambda_{min} = hc/e\text{V}$ ($\because h,c,e$ are constants).

6. Why X-rays cannot be diffracted by means of an ordinary grating?
Ans: The short wavelength of the X-rays (compared to the grating constant of optical grating) makes it difficult to observe X-ray diffraction with ordinary gratings.

7. Is there exist a sharp limit on the short wavelength side for each continuous X-ray spectrum.
Ans: Yes, the sharp limit on the short wavelength side is dependent on the voltage applied to the incident electrons and is given by, $\lambda_{min} = hc/e\text{V}$.

OBJECTIVE QUESTIONS

1. Molybdenum is used as a target element for production of x-rays because it is

(a) A heavy element and can easily absorb high velocity electrons.

(b) A heavy element with a high melting point.

(c) An element having high thermal conductivity.

(d) Heavy and can easily detect electrons.

2. The characteristic X-ray radiation is emitted when

(a) The electrons are accelerated to a fixed energy.

(b) The source of electrons emits a monoenergetic beam.

(c) The bombarding electrons knock out electrons from the inner shell of the target atoms and one of the outer electrons falls into this vacancy.

(d) The valence electrons in the target atoms are removed as a result of the collision.

3. Hydrogen atom does not emit X-rays because

(a) Its energy levels are close to each other.

(b) Its energy levels are too far apart.

(c) It is too small in size.

(d) It has a single electron.

4. Bragg's equation will have no solution, if

(a) $\lambda > 2d$ (b) $\lambda < 2d$

(c) $\lambda < d$ (d) $\lambda = d$

5. The energy of X-ray photon is 3.3×10^{-16} J. Its frequency per second would be

(a) 5×10^{17} (b) 5×10^{-18}

(c) 6.62×10^{18} (d) 2×10^{-18}

6. In producing X-rays a beam of electrons accelerated by a potential difference V is made to strike a metal target. For what value of V the X-rays have the lowest wavelength, $\lambda_m = 3094$Å

(a) 10 kV (b) 20 kV

(c) 30 kV (d) 40 kV

7. In radio-therapy X-rays are used to

(a) Detect bone fracture. (b) Treat cancer by control.

(c) Detect heart disease. (d) Detect fault in radio receiving circuits.

8. Moseley's Law relates

(a) Frequency and applied voltage. (b) Frequency and atomic number.

(c) Wavelength and intensity of X-rays. (d) Wavelength and angle of scattering.

9. The frequency of K_α-line of a source of atomic number Z is proportional to

(a) Z^2 (b) $(Z-1)^2$

(c) $1/Z$ (d) Z

10. The wavelength of K_α-line from an element of atomic number 41 is λ. Then the wavelength of K_α-line of an element of atomic number 21 is

(a) $\lambda/4$ (b) 4λ

(c) 3.08λ (d) 0.26λ

[**Hint:** $\frac{\nu_{41}}{\nu_{21}} = \frac{41-1}{21-1} = 4 \therefore \lambda_{41}:\lambda_{21} = 1:4$]

ANSWERS

Question numbers	1	2	3	4	5	6	7	8	9	10
Answers	(b)	(c)	(d)	(a)	(a)	(d)	(b)	(b)	(b)	(a)

FILL IN THE BLANKS

1. The emission of x-ray from decelerating electron is called ______. **[Bremsstrahlung process]**
2. Bremsstrahlung process is sometimes called ______. **[inverse photoelectric effect]**
3. The x-ray photon of shortest wavelength (highest frequency) is emitted when the incident electron loses all of its ______ in a single encounter. **[kinetic energy]**
4. In x-ray spectroscopy a neutral atom in the ground state is assigned ______ energy and that having vacancy (hole) in the inner level is assigned ______ energy. **[zero, positive]**
5. When a hole makes transition from K to L, M, N, ... levels etc., the x-ray lines emitted are said to constitute ______ series. Similarly, when a hole moves from L to M, N, ... etc., level, the lines of ______ series originate. The allowed transitions obey the selection rules: ______, ______. **[$K, L, \Delta l = \pm 1, \Delta j = 0, \pm 1$]**
6. The intensity of x-rays falls ______ with the distance x traversed in the absorbing material in accordance with ______. **[exponential, $I = I_o e^{-\mu x}$]**
7. The primary interaction processes responsible for the absorption of electromagnetic radiation are ______, ______, and ______. **[photoelectric effect, Compton scattering, pair production]**
8. The wavelength of K_α line of an element is 1.54 A°. The atomic number of target element is ______. **[29]**

   [**Hint:** $\frac{1}{\lambda} = \frac{3}{\nu} R(z-1)^2, \quad z = 1 + \sqrt{\frac{4}{3\lambda R}}$]

9. The elements in a row between those whose wavelengths of K_α line are equal to 250 pm and 179 pm are $z =$ ______. **[24, 25, 26]**

10. The binding energy of a K-electron in $z = 23$ vanadium atom whose L absorption edge has a wavelength $\lambda_L = 2.4$ nm is ______. **[516 *K*eV]**

[**Hint:** $E_L = E_L = \dfrac{ch}{\lambda_L}$]

11. For tungsten the K absorption edge is 0.18 A°. It is irradiated with x-rays of wavelength 0.10 A°. The maximum kinetic energy of photoelectrons emitted from K shell is ______. **[55.11 eV]**

[**Hint**: Energy of incident photon

$$E = \frac{ch}{\lambda} = 124\ KeV$$

Binding energy of K-electron

$$E_K = \frac{ch}{\lambda_K} = 68.89\ KeV$$

∴ maximum K.E. of ejected electron $= 124 - 68.89 = 55.11$ eV]

6

Origin of Quantum Theory, Wave Mechanic Concepts and Uncertainty Principle

6.1 INADEQUACY OF CLASSICAL MECHANICS

By the end of 19th century experimental evidence began to accumulate which indicated that the interaction of radiation with matter was not entirely in accordance with the Maxwell's theory of electromagnetism. At the same time the theory of atomic structure of matter was developing, mainly as a result of the discovery of the electron by J.J. Thomson, and the confirmation of nuclear model of the atom by Rutherford. Still another series of experiments forced the physicists to review his concepts of the motion of subatomic particles, since they apparently did not move precisely in accordance with the laws of classical mechanics. To explain the new observations, a sequence of new ideas, introduced in a more or less ad hoc fashion, were incorporated by several physicists. With the passage of time, and by the efforts of many great physicists and mathematicians these ideas evolved until they became what is now known as **Quantum theory**. In this chapter we shall review the more important experimental bases of quantum mechanics.

6.1.1 Electromagnetic Waves

Maxwell in 1864 made the remarkable suggestion that accelerated electric charged generate linked electric and magnetic disturbances that can travel indefinitely through the space. If the electric charges oscillate periodically, the disturbances are waves whose electric and magnetic components are perpendicular to each other and to the direction of propagation Fig. [6.1(a)].

From the earlier work of Faraday, Maxwell knew that a changing magnetic field can induce a current in a wire loop. Thus a changing magnetic field is equivalent in its effects to an electric field. Maxwell proposed the converse, i.e., a changing electric field has a magnetic field associated with it. Maxwell established that optics is a branch of electromagnetism.

Maxwell developed a complete **electromagnetic theory**. There are four fundamental equations of Maxwell. These equations relate the electric and magnetic field vectors to the electric charge density ρ and the current density $\boldsymbol{J}$ producing these fields.

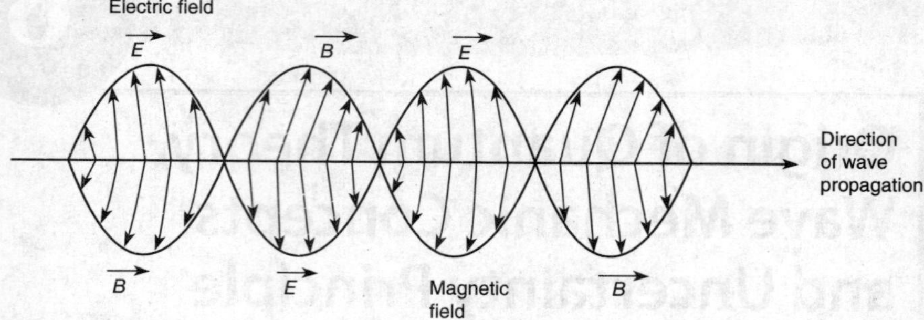

Fig. 6.1(a) The electric and magnetic fields in an electromagnetic wave are perpendicular to each other and to the direction of propagation of the wave and vary together

In S.I. system of units, the **electromagnetic field equations in integral and differential forms** are as follows:

Integral form:

$$\oint_s D.ds = \int_V \rho dv \tag{6.1}$$

$$\oint B.ds = 0 \tag{6.2}$$

$$\oint E.dl = \int_s \left(-\frac{\partial B}{\partial t}\right).ds \tag{6.3}$$

$$\oint H.dl = \int_s \left(J + \frac{\partial D}{\partial t}\right).ds \tag{6.4}$$

Differential form:

$$\nabla.\boldsymbol{D} = \rho \tag{6.5}$$

$$\nabla.\boldsymbol{B} = O \tag{6.6}$$

$$\nabla \times E = -\frac{\partial B}{\partial t} \tag{6.7}$$

$$\nabla \times H = J + \frac{\partial D}{\partial t} \tag{6.8}$$

where $\boldsymbol{D}$ is electric displacement, $\boldsymbol{E}$ and $\boldsymbol{H}$ are electric and magnetic field vectors respectively. Eqs. (6.5) and (6.6) are simply differential form of **Gauss' s law** in electrostatics and magnetostatics. Eq. (6.6) also signifies the non-existence of magnetic monopole. Eq. (6.7) represents the differential form of Faraday's law of electromagnetic induction. Eq. (6.8) is the modified form of Ampere's circuital law. Clearly, Maxwell field equations represent generations of fundamental experimental laws of electricity and magnetism.

A solution of Maxwell's field equations in a non-conducting isotropic dielectric medium yields the result that the electric and magnetic fields can travel through space in the form of waves. Such waves are called electromagnetic waves. The relevant wave equations for the **electric (*E*)** and **magnetic (*B*)** field vectors are:

$$\nabla^2 \boldsymbol{E} = \varepsilon\mu \frac{\partial^2 \boldsymbol{E}}{\partial t^2} \tag{6.9}$$

and
$$\nabla^2 \boldsymbol{H} = \varepsilon\mu \frac{\partial^2 \boldsymbol{H}}{\partial t^2} \tag{6.10}$$

where $\nabla^2 = \left(\frac{\partial^2}{\partial x^2} + \frac{\partial^2}{\partial y^2} + \frac{\partial^2}{\partial z^2}\right)$ is the **Laplacian operator.**

We may note that there are many solutions of Eqs. (6.9) and (6.10) corresponding to different kinds of electromagnetic waves, i.e., **plane, spherical and cylindrical.** We may consider the simplest case of plane electromagnetic waves where, ***E*** and ***H*** are functions of say, *y* and *t* only (taking the direction of propagation along the *y*-axis). Now, the Eqs. (6.9) and (6.10) reduce to:

$$\frac{\partial^2 \boldsymbol{E}}{\partial y^2} = \varepsilon\mu \frac{\partial^2 \boldsymbol{E}}{\partial t^2} \tag{6.11}$$

and
$$\frac{\partial^2 \boldsymbol{H}}{\partial y^2} = \varepsilon\mu \frac{\partial^2 \boldsymbol{H}}{\partial t^2} \tag{6.12}$$

One can easily show that each of the Eqs. (6.11) and (6.12) describe a **transverse, plane, harmonically** varying wave moving in the positive *y*-direction with phase velocity given by

$$c = \frac{1}{\sqrt{\varepsilon\mu}} \tag{6.13}$$

In free space or vacuum $\varepsilon = \varepsilon_0$ and $\mu = \mu_0$, so that the velocity of electromagnetic propagation in free space is

$$v = \frac{1}{\sqrt{\varepsilon_0 \mu_0}} \tag{6.14}$$

where ε_0 is the electric permittivity of free space and μ_0 is its magnetic permeability. Substituting the values of ε_0 and μ_0 in Eq. (6.14), one obtains

$$v = 2.998 \times 10^8 \text{ m/s}$$

which is the velocity of light in free space, as measured experimentally. From this, Maxwell concluded that light consists of electromagnetic waves.

For a plane electromagnetic wave travelling is vacuum, ***E*** and ***H*** are mutually perpendicular at all times and they are also perpendicular to the direction of propagation of the wave. The vectors ***E, H*** and the propagation vector ***k*** form a right-handed orthogonal set.

Light is not the only example of an e.m. wave. The wavelength range from the vary short (nuclear gamma rays) to the very long (radio waves). Although all such waves have the same fundamental nature, many features of their interaction with matter depend upon their frequencies. A characteristic property of all waves is that they obey **principle of superposition**.

When two or more trains of light waves meet in a region, they **interfere** to produce a new wave whose instantaneous amplitude is the sum of those of the original waves. The interference of light waves was first demonstrated in 1801 by Young, who used a pair of slits illuminated by monochromatic light from single source. From each slit secondary waves spread out as though originating at the slit; this is an example of **diffraction,** which, like interference, is a characteristic wave phenomenon.

Interference and diffraction are found only in waves, the particles we are familiar with do not behave in those ways. If light consisted of stream of classical particles, the entire screen in Young's double slit experiment would be dark. Clearly, Young's double slit experiment is proof that light consists of waves. Maxwell's electromagnetic theory further tells us that light waves are electromagnetic. However, some experiments with light and other electromagnetic radiation clearly indicate that the wave interpretation is not a complete description of the properties of electromagnetic radiation.

6.1.2 Black-body Radiation

December, 14, 1940 must be considered the birthday of quantum mechanics. It was on this day that Max Planck expounded his famous formula for the spectral distribution of **black-body radiation**. By a black body, we mean a body, which absorbs all the radiations it receives. For practical purposes an isothermal cavity with a small aperture through which radiation from outside may be admitted is considered a black body. The cavity always contains radiations emitted by the walls. For such a cavity, a typical plot of the spectral distribution of energy for several absolute temperatures is shown in Fig. [6.1(b)].

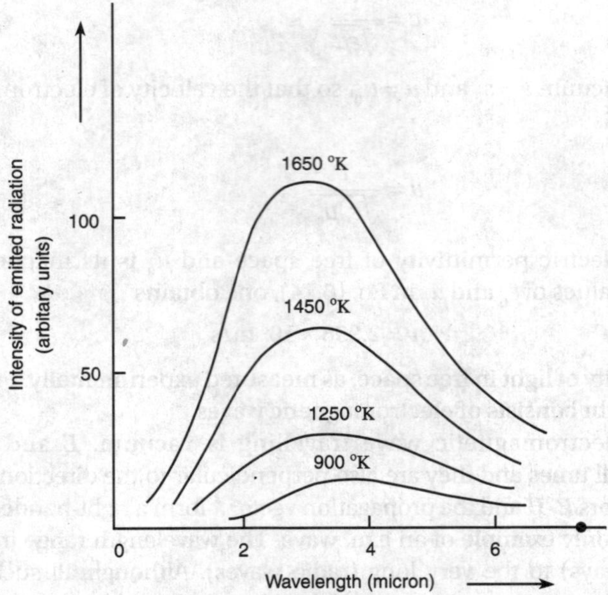

Fig. 6.1(b) Spectral distribution of black-body radiation for several different temperatures

The curves shown were obtained by Lummer and Pringsheim in 1900, although their general characteristics were known earlier. A study of these curves reveal the following important facts:

(i) At a given temperature the energy is not uniformly distributed in the spectrum of black-body radiation.

(ii) At a given temperature, the intensity of radiations increases with increase in wavelength and becomes maximum at a particular wavelength (λ_m), which is the characteristic of the temperature of the radiating body. With further increase in wavelength, the intensity of radiation decreases.

(iii) Raising the temperature increases the energy of all spectral components. The short wavelength cutoff advances towards the origin as the temperature increases and the peak of the curve shifts to shorter wavelengths.

The shift of the peak of the curve was found to obey the following empirical relationship, commonly called **Wien's displacement law.**

$$\lambda_p T = \text{constant} \tag{6.15}$$

where λ_p refers to the wavelength corresponding to the peak of the curve. With the help of classical thermodynamics, Wien has further shown that the amount of energy contained in the spectral region included within the wavelength λ and $\lambda + d\lambda$ emitted by a black body is given by

$$E_\lambda d\lambda = \frac{A}{\lambda^5} e^{-B/\lambda T} d\lambda \tag{6.16}$$

The quantities A and B are not derived but the simply curve-fitting parameters. Although Wien's radiation law is consistent with both the Stefan-Boltzmann law [$E(T) = \sigma T^4$, where $\sigma = 5.6699 \times 10^{-8}$ Watts-m^{-2} deg^{-4}] and the displacement law, it is quite unsatisfactory as a theory since it is not derivable from a physical model. That is, there was no attempt to relate the emitted radiation to physical processes taking place within the radiating body.

Wien's law works well only for short wavelengths. There are considerable deviation from Wien's law at long wavelengths and high temperatures. From Eq. (6.16), one obtains finite energy even for $T = \infty$ (infinite). Rayleigh argued that it is unlikely that energy (E) should be finite for $T = \infty$.

Rayleigh and Jeans, by considering that the radiations in black body have degrees of freedom and using **classical law of equipartition of energy,** showed that

$$E_\lambda d\lambda = \frac{A}{\lambda^4} T d\lambda \tag{6.17}$$

where A is a constant ($= 8\pi k$). Here k is the Boltzmann constant and it should not be confused with the wave vector $\boldsymbol{K}$.

A difficulty with Eq. (6.17) immediately appears when we consider very small wavelengths. From Eq. (6.17), it is clear that the energy radiated in a given wavelength range $d\lambda$, increases rapidly as λ decreases and approaches infinity for very short wavelengths, which is not true. Obviously, Rayleigh-Jean's law describes the experimental curve quite well in the long wavelength region, it diverges as the wavelengths approach zero. This failure was such a crushing blow to classical physics that it is historically referred to as **"the ultraviolet catastrophe"**. The failure of Rayleigh-Jeans law to explain the spectral energy curves of black-body radiation indicated that the fundamental assumptions of classical physics required suitable modification.

Being well aware of the shortcomings of both the Rayleigh-Jeans and Wien radiation laws, Planck examined the mathematical statistics of these two laws to ascertain what changes, if any, might result in a reasonable description of the experimental radiation curve. As a result of this work, he was led to certain conclusions about the nature of the electromagnetic oscillators which are in equilibrium with the energy density within the black-body cavity. These revolutionary postulates, which have become the foundation of the quantum theory of radiation, are as follows:

(i) The amount of energy emitted or absorbed by an oscillator is proportional to its frequency. Calling the constant of proportionality h, we then write for the change in oscillator energy,

$$\Delta\varepsilon = h\nu$$

(ii) An oscillator cannot have an arbitrary energy but must occupy one of a discrete set of energy states given by $\varepsilon_n = nh\nu$

where n is an integer or zero. It was assumed that the ground state corresponds to the zero energy states. The value of Planck's constant h is 6.626×10^{-27} erg/s ($= 6.626196 \times 10^{-34}$ Js).[1] Planck's constant is a **universal constant** which plays an important role in all quantum phenomena.

The previous picture of a continuum of oscillator states is now replaced by a discrete set of "quantized" states. Furthermore, the amount of energy emitted or absorbed is also quantized, since each quantum must correspond to the energy difference between two states of a given oscillator. Each quantum of electromagnetic energy is called a **photon**. Using Planck's quantum hypothesis, we now derive the famous Planck's radiation formula.

Let N be the total number of Planck's oscillators and E is their total energy, then the average energy per oscillator, $\bar{\varepsilon}$, is given by $\bar{\varepsilon} = \dfrac{E}{N}$.

Let N_0, N_1, N_2,.......N_n.........*etc*... be the number of oscillators having energies, 0, ε, 2ε, ...$n\varepsilon$,...... respectively, then

$$N = N_0 + N_1 + N_2 + \ldots\ldots\ldots + N_n + \ldots\ldots = \sum_{n=0}^{\infty} N_n \tag{6.18}$$

and

$$E = 0 + \varepsilon N_1 + 2\varepsilon N_2 + \ldots + n\varepsilon N_n + \ldots\ldots = \sum_{n=0}^{\infty} n\varepsilon N_n \tag{6.19}$$

[1]The constant $\hbar = \dfrac{h}{2\pi}$, which is also called Planck's constant, is equal to $\hbar = 1.055 \times 10^{-27}$ erg-sec (1.055×10^{-34} J-sec).

According to Maxwell-Boltzmann distribution function, the number of resonators or oscillators having energy $n\varepsilon$ will be

$$N_n = N_0 e^{-n\varepsilon/kT} \tag{6.20}$$

Obviously, the higher energy states are thus less likely to be populated, and as the energy increases indefinitely the number of such oscillators becomes vanishingly small. Using Eqs. (6.19) and (6.21), Eq. (6.20) becomes

$$\begin{aligned} N &= N_0 + N_0 e^{-\varepsilon/kT} + N_0 e^{-2\varepsilon/kT} + \ldots + N_n e^{-n\varepsilon/kT} \\ &= N_0[1 + e^{-\varepsilon/kT} + e^{-2\varepsilon/kT} + \ldots] \end{aligned} \tag{6.22}$$

and

$$\begin{aligned} E &= \varepsilon[N_0 e^{-\varepsilon/kT} + N_0 e^{-2\varepsilon/kT} + \ldots] \\ &= N_0 \varepsilon[e^{-\varepsilon/kT} + e^{-2\varepsilon/kT} + \ldots e^{-n\varepsilon/kT} + \ldots] \end{aligned} \tag{6.23}$$

$\therefore$

$$\bar{\varepsilon} = \frac{E}{N} = \frac{\sum_{n=0}^{\infty} n\varepsilon e^{-n\varepsilon/kT}}{\sum_{n=0}^{\infty} e^{-n\varepsilon/kT}}$$

Putting $e^{-\varepsilon/kT} = x$, one obtains $\sum_{n=0}^{\infty} e^{-n\varepsilon/kT}$

$$= 1 + x + x^2 + \ldots = \frac{1}{1-x} = \frac{1}{1 - e^{-\varepsilon/kT}}$$

and

$$\begin{aligned} \sum_{n=0}^{\infty} n\varepsilon e^{-n\varepsilon/kT} &= \varepsilon(x + 2x^2 + 3x^2 + 4x^3 + \ldots) \\ &= x\varepsilon \frac{d}{dx}(1 + x + x^2 + x^3 + \ldots) = x\varepsilon \frac{d}{dx}\left(\frac{1}{1-x}\right) \\ &= \frac{x\varepsilon}{(1-x)^2} = \frac{xe^{-\varepsilon/kT}}{(1 - e^{-\varepsilon/kT})^2} \end{aligned}$$

$\therefore$

$$\bar{\varepsilon} = \frac{\varepsilon e^{-\varepsilon/kT}}{1 - e^{-\varepsilon/kT}} = \frac{\varepsilon}{e^{\varepsilon/kT} - 1} = \frac{h\nu}{e^{h\nu/kT} - 1} \tag{6.24}$$

Obviously, the average energy of an oscillator is not kT, but given by Eq. (6.24). The number of resonators per unit volume in frequency range ν and $\nu + d\nu$ is given by

$$N = \frac{8\pi\nu^2}{c^3} d\nu \tag{6.25}$$

The energy density belonging to range $d\nu$ can be easily obtained by multiplying the average energy of an oscillator by the number of resonators per unit volume in the frequency range ν and $\nu + d\nu$ as

$$E_\nu\, d\nu = \frac{8\pi h\,\nu^3}{c^3} d\nu \frac{1}{e^{h\nu/kT}-1} = \frac{8\pi h\nu^3}{c^3}\frac{1}{e^{h\nu/kT}-1} d\nu \tag{6.26}$$

This is famous Planck's radiation law. One can express Eq. (6.26) in terms of wavelength by using the relation $\nu = \frac{c}{\lambda}$ and hence $|d\nu| = \left|\frac{c}{\lambda^2} d\lambda\right|$ as

$$E_\lambda d\lambda = \frac{8\pi hc}{\lambda^5}\frac{1}{e^{-ch/\lambda kT}-1} d\lambda \tag{6.27}$$

For small temperatures and short wavelength, λT is small and hence $e^{ch/\lambda kT} >> 1$. Obviously, 1 in the denominator of Eq. (6.27) can be neglected and hence Eq. (6.27) reduces to

$$E_\lambda d\lambda = \frac{8\pi hc}{\lambda^5} e^{-ch/\lambda kT} d\lambda \tag{6.28}$$

This is **Wien's law** with $A = 8\pi ch$ and $B = -\frac{hc}{k}$.

For higher temperatures and large wavelength $e^{hc/\lambda/kT} \approx 1 + \frac{hc}{\lambda kT}$ ($\because$ T is large). Using it in Eq. (6.27), one obtains

$$E_\lambda d\lambda = \frac{8\pi kT}{\lambda^4} d\lambda \tag{6.29}$$

This is **Rayleigh-Jean's law** with $A = 8\pi k$.

Obviously, the Planck's theory of radiation incorporates all that is valid from the older theories as special cases. It thus serves as an excellent example of a conceptual advance, which opened exciting new frontiers while still preserving much of the older physics.

Planck's theory tell that radiant energy is absorbed or emitted as quanta was so revolutionary that many scientists were slow to realize its significance. Even Planck did not like the result of the theory!

But Einstein and Niels Bohr applied the quantum theory to the problems of **photoelectric emission** and to the **structure of the atom,** and with its use they could explain a good deal that 19th century physics could not. The importance of Planck's quantum theory was solidly established and he won a Nobel Prize in 1918 for his work.

Let us now reflect for a moment on the significance of Planck's postulate. It states that the energy of a harmonic oscillator, expressed in terms of its frequency of vibration, may on the average occur in units no smaller than $h\nu$, ($E = nh\nu$ where $n = 0, 1, 2, 3, \ldots$). It does not say what is the energy of the oscillator at any instant, nor does it say how the value of h is to be calculated. From this we may conclude as:

(i) Planck's postulate is empirical, that is, it contains a parameter h whose value cannot be calculated except through the comparison of some expression in which it appears with the appropriate experimental results. There is no theory of Planck's constant. The only physical reason we can give for its existence is that it is invaluable for the elucidation of microscopic phenomena.

(ii) The postulate is statistical in that it refers only to the time averaged behaviour of the microscopic oscillator. Nothing is said about energy except that it varies directly with the frequency if evaluated over period of time long compared with ν^{-1}. Bose and Einstein developed statistics of photons. Particles obeying **BE statistics** are called **'bosons'**. According to BE statistics the density and temperature of a gas are statistical concepts and in a similar manner, intensity and temperature of radiation are statistical concepts. Though the phenomenon of BE-Condensation (BEC) was predicted by Einstein in 1925 (based on the new statistics of Bose), it was observed only in 1995. Nobel prize in physics for 2001 has been jointly awarded to Eric A. Cornell and Wolfgang Ketterle for the achievement of BEC in dilute gases of alkali atoms, and for the early fundamental studies of the properties of the condensates.

(iii) The postulate does not in itself suggest that a flaw exists in the Maxwell's theory of electromagnetism. It states only what energy an oscillator must have on the average if it is to remain in thermodynamic equilibrium with a radiation field. We should not conclude from this that the latter must necessarily exhibit discrete energies when interacting with matter-we have verified the quantum hypothesis only for the emission of radiation by a heated solid!

Planck Constant h: It denotes a quantity of action or an elementary quantum. It is this constant h which is responsible for the discrete individuality of the photon. This constant makes radiation to behave like a particle. The dimensions of h are that of angular momentum.

$$h = \frac{[\text{Energy}]}{[\text{Frequency}]} = \frac{ML^2T^{-2}}{T^{-1}} = [ML^2T^{-2}] = \text{Angular momentum.}$$

Obviously, 'h' is defined as the smallest quantum of angular momentum of a particle.

We know that momentum is always associated with motion, and hence h always represents motion. It was Niels Bohr who for the first time postulated that electrons of an atom could revolve round in such orbits whose angular momentum is an integral multiple of $h/2\pi(=\hbar)$.

Mass and Momentum of Photon

The energy of a photon is $h\nu$. Photons move with velocity of light c. Obviously, a mass can be ascribed to photon, in accordance with the theory of relativity, i.e.,

$$\text{Mass of photon}(m) = h\nu/c^2$$

$$\therefore \quad \text{Momentum of photon} = \text{mass} \times \text{velocity}, \ \frac{h\nu}{c^2} \times c = h\nu / c$$

$$\therefore \quad \text{Energy of photon} = h\nu = mc^2 = \frac{m_0}{\sqrt{1 - v^2/c^2}} \times c^2$$

$$\therefore \quad \text{Rest mass, } m_0 = \frac{h\nu}{c^2}\sqrt{1 - \frac{v^2}{c^2}}$$

We know that photon always travel with the velocity of light c while its energy content $h\nu$ is always finite. This means that the rest mass m_0 of the photon approaches zero in such a way that as ν approaches c, photon's energy always approaches the value $h\nu$. This means photon is not a particle in complete sense because a material particle must have a rest mass. Thus a photon always has wave structure and sometimes behave in a compact form.

6.1.3 Photoelectric Effect

The importance of the Planck's hypothesis of energy quantization was quickly realised by Albert Einstein. Einstein extended this idea to all kinds of electromagnetic radiations and explained the important phenomenon of photoelectric effect.

When ultraviolet light or X-rays strike a piece of metal it frees electrons from its surface and it is experimentally possible to count these electrons and to measure their energy. A quantitative investigation of the effect shows that the light intensity determines the number of electrons thus freed but has no influence on their energy. This is contrary to what we would expect from Maxwell's electromagnetic theory. Since the intensity of a light wave is proportional to the square of the amplitude of its electric field vector (E), the energy of the electrons is, surprisingly, determined only by the colour of the light, i.e., its frequency.

The following facts must be explained by any satisfactory theory of the photoelectric effect:

(i) If the incident light has no frequency component above a certain threshold frequency, no photoelectrons will be emitted.

(ii) The threshold frequency appears to depend upon the properties of the metal.

(iii) Provided that the threshold frequency is exceeded, photoelectrons will be emitted instantly, regardless of how low the intensity of light source is.

(iv) The photoelectric current (number of electrons emitted per second) increases with the intensity of light but is independent of the frequency.

(v) The maximum kinetic energy of the photoelectrons is independent of the intensity of light and depends only upon its frequency.

Attempts to explain the above phenomena by means of classical radiation theory met with defeat. Classically, an electron would be ejected as soon as it could accumulate enough energy to overcome the force which binds it to the metal. This energy is called the **work function** of the metal. Since an electromagnetic wave has an energy density associated with it, one would expect that the metal should accumulate enough energy, ultimately, to eject an electron, regardless of the frequency or the intensity of the source. However, the classical picture was completely at a loss to explain all of the above experimental facts except number four. From the classical point of view it was reasonable to expect the photoemission to increase when the energy density of the incident radiation is increases.

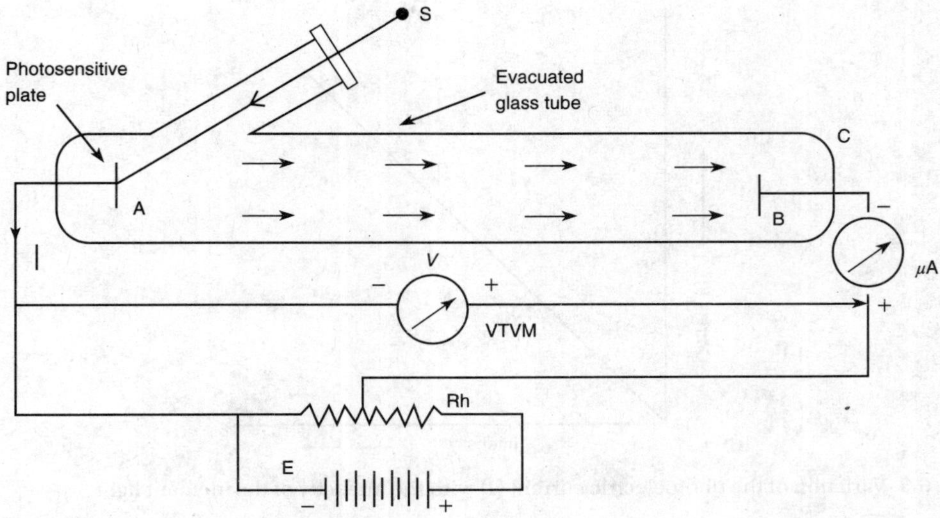

Fig. 6.2 Experimental arrangement to study the photoelectric effect

Einstein used Planck's idea of energy quantization to explain the above mentioned observations. He proposed that electromagnetic radiation (light) could be considered to have properties of the both **wave** and **particles**. He called the particles of light **'photons'**. Each photon of light possessed an energy equal to $h\nu$ (quantum of light). Before discussing the Einstein's photoelectric effect, we could like to provide details of the experimental study of the photoelectric effect.

Photoelectric effect: When a radiation of high frequency, e.g., ultraviolet light or X-rays is incident on a clean metal surface, electrons are emitted from the surface of the metal. The emitted electrons are called **photoelectrons** and the phenomenon is called the **photoelectric effect**. The entire range of electromagnetic radiation from gamma and X-rays to the ultraviolet, the visible and the infra-red produce this effect. The photoelectric effect is also observed in solids, liquids and gases.

The credit for the discovery of photoelectric effect goes to Heinrich Hertz (1887). Hallwachs undertaken further experimental study in 1888. Lenard in 1899 showed that the carriers of electricity emitted from metal surface, under the action of high frequency, were electrons.

Experimental arrangement to study the photoelectric effect: A simple experimental arrangement for the study of photoelectric effect is shown in (Fig. 6.2).

The emitting surface, i.e., photosensitive plate A is mounted opposite a metal plate B in a highly evacuated glass tube C. The two plates A and B form two electrodes, to which a variable potential difference can be applied. The evacuated glass tube is fitted with quartz window D, through which ultraviolet light from the source *S* is transmitted and allowed to fall on the plate A.

When the applied potential difference between the plates A and B is such that A is at negative potential with respect to B, the photoelectrons emitted from A are accelerated towards plate B.

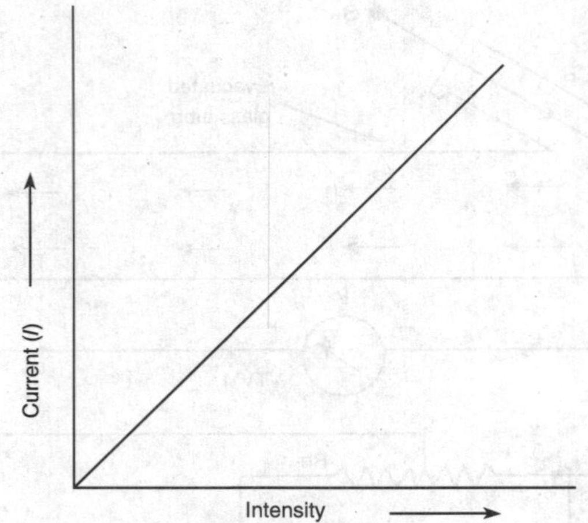

Fig. 6.3 Variation of the photoelectric current (I) with the intensity of the incident light

The resulting photoelectric current I flowing in the circuit is measured by the μA (micro ammeter) and the accelerating potential difference V is measured by vacuum tube voltmeter. If the potential differcnce is reversed so that plate B is now at negative potential with respect to A the photoelectrons arc repelled towards A by the retarding potential. Consequently, the photoelectric current is reduced. It is found that the photoelectric current depend on the following factors: (i) the frequency of the incident radiation (ii) the intensity of incident radiation (iii) the potential difference between the electrodes and (iv) the nature of the emitting surface.

The important experimental observations are as follows:

(i) **Effect of Frequency on the Photoelectric Current:** The collector plate B is made sufficiently positive w.r.t. the emitter photosensitive plate A. The surface of plate A is illuminated with monochromatic light of different frequencies. It is observed that the photoelectric current is produced only when the frequency υ of the incident light is greater than a certain minimum value υ_0. This minimum value of the frequency is called the threshold frequency for the given surface.

(ii) **Effect of the intensity of the incident Light:** To study the effect of the intensity of the incident light, the collector plate B is made sufficiently positive with respect to the emitter plate A. By keeping the frequency of the incident light and the potential difference V constant, the photoelectric current I is measured for the various intensities of the incident light. (Fig. 6.3) shows the variation of photoelectric current with the intensity. Obviously, the photoelectric current is proportional to the intensity of incident light.

(iii) **Effect of Potential Difference:** Keeping the frequency and the intensity X, of the incident light constant the potential difference between the electrodes is varied. Starting with a small positive potential, it is reduced to zero and then made negative with the help of a reversing key. The variation of photoelectric current with the potential difference is as shown in (Fig. 6.4). When the collector plate B is positive, the photo-electric current remains constant. When the collector plate

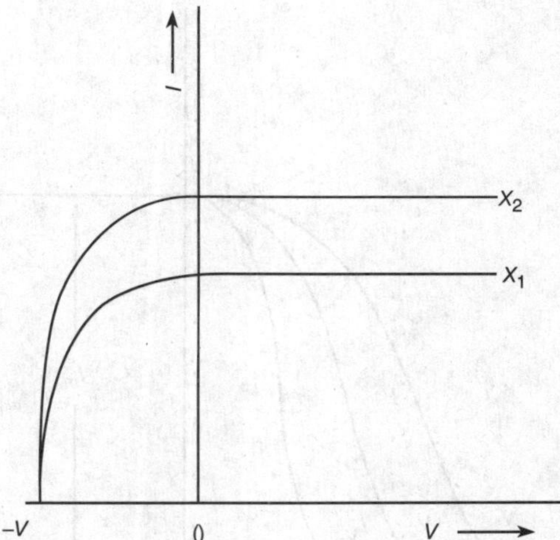

Fig. 6.4 Variation of photoelectric current with the potential difference

is made more and more negative, the photoelectric current goes on decreasing until it is stopped entirely. The retarding potential that stops the photo current is called the slopping potential. If the. intensity of the incident light is increased to X_2 then initial photo current for potential V_0 is found to be the same for the light of the same frequency.

The above experimental observation is explained by assuming that when light of a certain frequency ν ($\nu > \nu_0$) is incident on a emitting surface, photoelectrons with kinetic energies ranging from zero to a certain maximum value are emitted from the surface. The photo-electrons which are emitted with the maximum kinetic energy are stopped by the stopping potential V_0 volts. Obviously, the work done by the retarding potential on the faster electrons must be equal to their kinetic energy, i.e.,

$$eV_0 = \frac{1}{2}mv_{max}^2 \tag{6.30}$$

where m is the mass of the electron, e is the electronic charge and v_{max} is the maximum velocity of emission of the electrons.

(iv) Effect of frequency on the Stopping Potential: To study the effect of the frequency of the incident light on the stopping potential, the intensities of light of different frequencies are adjusted to produce the same maximum (I_m) value of the photoelectric current when collector B is positive. The potential of collector B is then reduced in steps and made zero; it is then made more and more negative with the help of the reversing key. For the given photo-cathode, the graphs obtained are as shown in (Fig. 6.5). From (Fig. 6.5), it is evident that the stopping potential increases as the frequency of the incident light is increased. Thus if the frequencies are in the increasing order, i.e., $\nu_1 > \nu_2 > \nu_3$ then the corresponding stopping potentials are also in the increasing order, i.e., $V_{01} > V_{02} > V_{03}$.

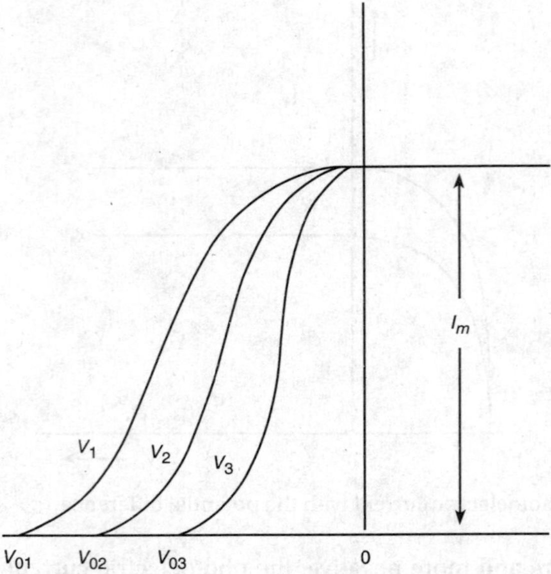

Fig. 6.5

(v) Effect of frequency on the maximum kinetic energy: The magnitudes of the stopping potential at different frequencies are converted into the maximum kinetic energies. A plot of maximum kinetic energy against the frequency is as shown in (Fig. 6.6). A straight line is obtained. The straight line a meets the frequency axis at a certain frequency ν_0. This reveals that ν_0 is the minimum frequency for which the maximum kinetic energy is zero. This minimum value of frequency, is called the **threshold frequency** for the emitting surface. Thus the threshold frequency for a given emitting surface defined as minimum frequency of incident light, below which no photoelectrons are emitted, however intense the incident light may be. From (Fig. 6.6) it is evident for another emitting surface, a straight line b parallel to a is obtained. On the basis of above experimental observations, we can summarize the following important characteristics of the photoelectric effect:

(i) For given photosensitive surface, there is a minimum frequency ν_0 of incident radiation, below which there is no emission of photoelectrons. This frequency (ν_0) is called the threshold frequency for the surface and its value depends on the material and the nature of the emitting surface.

(ii) For a given frequency ν ($\nu > \nu_0$) the number of photoelectrons emitted per second from a given surface, under a constant accelerating potential difference is directly proportional to the intensity of the incident radiation.

(iii) When radiation of a given frequency ν ($\nu > \nu_0$) is incident on a given surface, photoelectrons with kinetic energies ranging from zero to a certain maximum value, are emitted from the surface. The stopping potential for the photo-electrons and hence the maximum kinetic energy is independent of the intensity of the radiation.

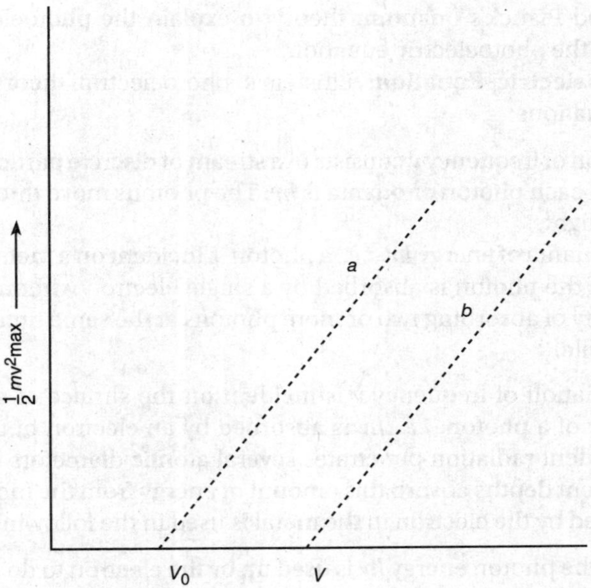

Fig. 6.6

(iv) **The maximum kinetic energy $\left(\frac{1}{2}mv_{\max}^2 = eV_0\right)$ of the photoelectrons emitted from a given surface increases linearly with the frequency of the incident radiation. The straight line so obtained for the given surface meets the frequency axis at the threshold frequency ν_0, for the surface. The slope of the straight line is the same for all emitting surfaces and is equal to Planck's constant h.**

(v) **The emission of photoelectrons from photosensitive surfaces is an instantaneous process. As soon as radiation of frequency greater than the threshold frequency is incident on a given photosensitive surface, there is emission of photoelectrons.**

Failure of Classical Theory: The classical theory which assume that light is propagated as electromagnetic waves, fails to explain the above experimental observations on electromagnetic waves. The drawbacks of classical theory are:

(i) **According to the classical theory, if the intensity of illumination of incident radiation were increased, electrons would be emitted with greater energies. This is contrary to the experimental observation, i.e., on increasing the intensity of incident radiation the photoelectrons are ejected with the same kinetic energy.**

(ii) **The existence of threshold frequency for a given material cannot be explained on the basis of wave theory of radiation.**

(iii) **The instantaneous ejection of electrons from the metal surface also cannot be explained from classical theory. If the illumination is very faint, the classical theory indicates that time of several minutes would have to elapse before a single electron is released and this is contrary to the above experimental observation,**

Einstein applied Planck's quantum theory to explain the photoelectric effect and in 1905 he obtained the photoelectric equation.

Einstein's Photoelectric Equation: Einstein's photoelectric theory is based on the following two equations:

(i) A radiation of frequency ν consists of a stream of discrete particles, called photons. Energy of each photon or quanta is $h\nu$. The photons move through space with the speed of light.

(ii) When a quanta of energy $h\nu$, *i.e.*, a photon is incident on a metal surface, the entire energy of the photon is absorbed by a single electron without any time lag. The probability of absorbing two or more photons at the same time by a metal surface is negligible.

Suppose that radiation of frequency ν is incident on the surface of a metal in vacuum. The entire energy of a photon, *i.e.*, $h\nu$ is absorbed by an electron of the metal. We must note that the incident radiation penetrates several atomic diameters into the metal, and electrons at different depths absorb this amount of energy from the incident photons. The energy $h\nu$ absorbed by the electron in the metal is used in the following two ways:

(i) A part of the photon energy $h\nu$ is used up by the electron to do a certain minimum amount of work to overcome attractive forces of positive ions of the metal. This minimum amount of energy is called the **photoelectric work function** and usually denoted by W of the metal. The work function (W) depends on nature of the emitting surface.

(ii) The remaining energy ($h\nu - W$) appears as the kinetic energy of the photoelectron. If the electron with this kinetic energy comes from the surface, it does not suffer any collision with another electron. In this situation, it will be emitted from the surface into vacuum with this energy as maximum kinetic energy $\frac{1}{2}mv_{max}^2$, where m is the mass of the electron and v_{max} its maximum speed for the given frequency. Thus for such an electron, we have

$$\frac{1}{2}mv_{max}^2 = h\nu - W \tag{6.31}$$

The above equation is called **Einstein's photoelectric equation**. Suppose the frequency of the incident radiation is reduced to a certain value ν_0 such that an electron is just emitted from the surface with zero kinetic energy. Then we have when $\nu = \nu_0$, $\frac{1}{2}mv_{max}^2 = 0$. Then from Eq. (6.31), we have

$$0 = h\nu_0 - W \quad \text{or} \quad W = h\nu_0 \tag{6.32}$$

Substituting for $W = h\nu_0$ from Eq. (6.32) in Eq. (6.31) we have

$$\frac{1}{2}mv_{max}^2 = h\nu - h\nu_0 = h(\nu - \nu_0) \tag{6.33}$$

Eq. (6.33) is another form of Einstein's photoelectric equation. The frequency ν_0 is called the *threshold frequency*. On the basis of Einstein's theory, one can define the threshold frequency as follows:

It is the frequency of incident photon which has just sufficient energy to liberate an electron with zero kinetic energy from the emitting metal surface.

It is worthwhile to mention that the energy of light photon incident on the surface of the metal is not divided, but the whole energy of a photon is used up by the same electron in ejecting from the parent atom and imparting kinetic energy.

Explanation of the characteristics of the photoelectric effect from Einstein's equation

(i) Eq. (6.33) shows that the graph of $\frac{1}{2}mv^2_{max}$ against ν is a straight line for an emitting surface. For all emitting surfaces the straight lines will have the same slope equal to Planck's constant h.

(ii) When the intensity of incident radiation is increased, then the increased number of photons in the incident radiation will cause the emission of more photoelectrons per second. Obviously, the photoelectric current is proportional to the intensity of the incident radiation. The increase of the intensity does not affect the maximum kinetic energy.

(iii) From Eqs. (6.32) and (6.33), we have

(a) For $\nu < \nu_0$, no photoelectrons can be emitted.

(b) For $\nu = \nu_0$, photoelectrons are just emitted with zero kinetic energy.

(c) For $\nu > \nu_0$, photoelectrons are emitted with kinetic energies ranging from zero to a maximum value.

(d) For $\nu > \nu_0$, and the intensity of incident light is increased, the number of photoelectrons emitted per second will increase.

Some recent important observations about the photoelectric effect are:

(i) Only a small fraction (~5%) of the incident photon succeeds in ejecting photoelectrons from photo-metal while rest of the photons are absorbed by the system as a whole and generate thermal energy.

(ii) Photoelectric effect is also observed for isolated atoms in the form of a gas. e.g., Na, K vapour and this process is called **photo-ionisation.**

(iii) The energy required for ejecting the electrons may also be provided by heating the metal, which results in the **thermionic emission** of the electrons.

(iv) The advent of lasers capable of emitting coherent radiation at high power levels, two or multi photon photo-emission has now been observed. Theory predicts that the **double quantum photo current** should be proportional to the square of the power of the incident radiation as opposed to the nearly linear relationship that holds for the single quantum photo effect. In a **multi photon photoelectric effect,** the kinetic energy of the photo-electron is given by

$$\frac{1}{2}mv^2_{max} = Nh\nu - eV_s \tag{6.34}$$

and the critical frequency is $\frac{eV_s}{Nh}$ which is smaller than the corresponding frequency for single photon process by a factor of $1/N$.

In the analysis of the photoelectric effect, we have considered a photon as a wave packet of energy, while not making any statements for the quantization of momentum. In fact, since a significant amount of momentum was carried away by the metal, conservation of momentum could not be used for the photoelectron system. A photon as a particle must have both a quantum of energy as well as a quantum of momentum. A. H. Compton in 1921 demonstrated in the scattering of X-rays by electrons that a photon have both a quantum of energy as well a quantum of momentum. We will discuss it later on.

Millikan's Verification of Einstein's Equation: From Eqs. (6.15) and (6.18), we have

$$eV_0 = h\nu - h\nu_0 \quad \therefore \quad V_0 = \frac{h}{e}\nu - \frac{h}{e}\nu_0 \tag{6.35}$$

The above equation shows that the stopping potential V_0 for photoelectrons from a given surface varies linearly with the frequency ν of the incident light. This equation was used by Millikan in 1916 for determining Planck's constant.

The experimental arrangement used by Millikan is shown in (Fig. 6.7). Three different alkali metal surfaces are mounted on a drum D suitably mounted inside an evacuated chamber. The drum D can be rotated from outside so that the surface of any one metal is first scraped clean by sharp knife edge K and then brought before the quartz window W. A beam of monochromatic light passing through the window W falls on cleaned surface C.

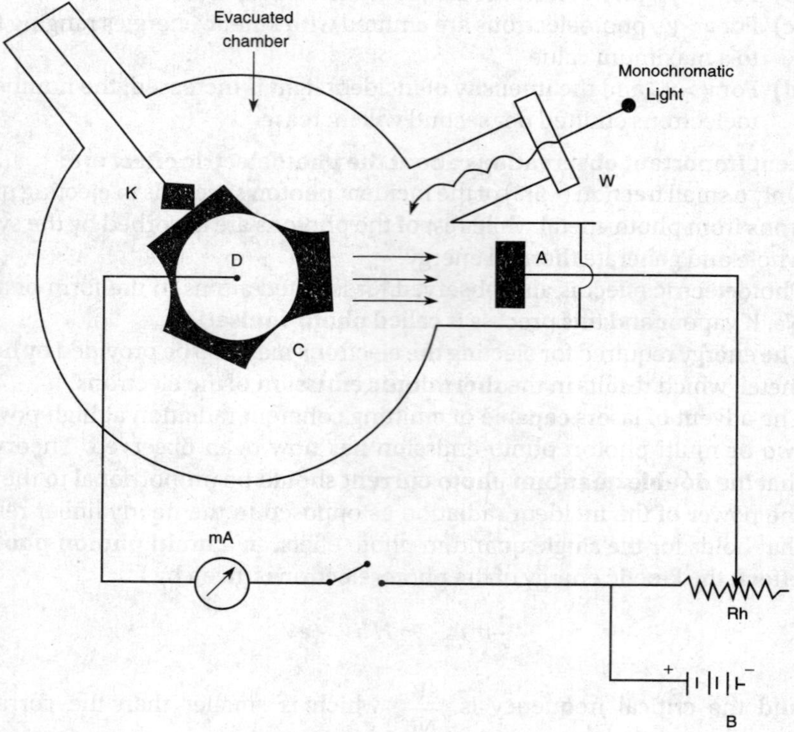

Fig. 6.7

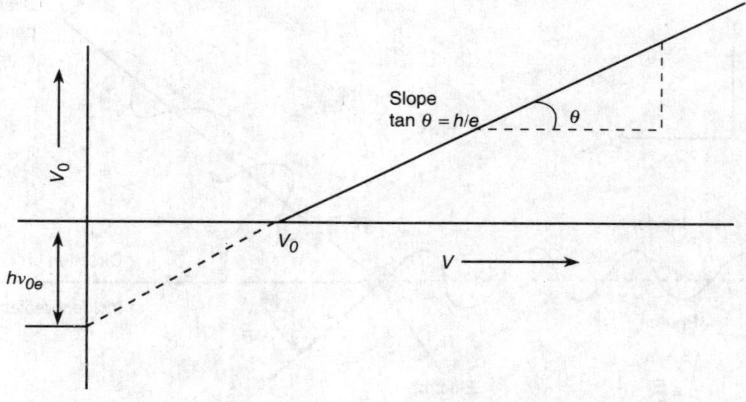

Fig. 6.8

The emitted photo electrons are collected by an electrode A which may be maintained at positive or negative potential w.r.t. C (or drum D) by means of a potential divider potential. The millimeter (mA) measures this photoelectric current.

Working: The anode *A* is first given a negative potential w.r.t. C. The electrons emitted from C are retarded. If the potential is small, only the slower electrons are pulled back and the current falls. With the increase in negative potential, more and more faster electrons are pulled back. Ultimately, at a certain potential V_0 called the stopping potential, even the fastest elections fail to reach A. At this stage the photoelectric current is zero. Obviously, the kinetic energy of fastest electrons just reduces to zero when they move across the retarding potential V_0. Thus one obtains

$$\frac{1}{2}mv_{\text{max}}^2 = eV_0 \tag{6.36}$$

Obviously, the stopping potential multiplied by electronic charge gives the maximum kinetic energy of the photoelectrons.

Millikan plotted a graph between V_0 and the frequency of incident light over a wide range of frequencies and obtained a straight line as shown in (Fig. 6.8). Parallel lines were obtained by Millikan for other photo-metals.

6.1.4 Compton Effect

A.H. Compton in 1923, while studying the scattering of X-rays by a block of paraffin, observed that the wavelength of the scattered radiation was greater than the wavelength of incident radiation and the shift in wavelength was dependent on the angle of the scattering. The wavelength of the scattered radiation is found maximum at right angles to the incident beam. This effect is called Compton effect. Essentially, it is a demonstration of the scattering of a photon by an electron as a particle-particle scattering.

Theory: The theory of Compton effect was given by Compton and also by Debye at about the same time. Compton considered the radiation to be quantized and assumed the electron of the atom of scattering material to be loosely bound (almost free) and at rest before collision

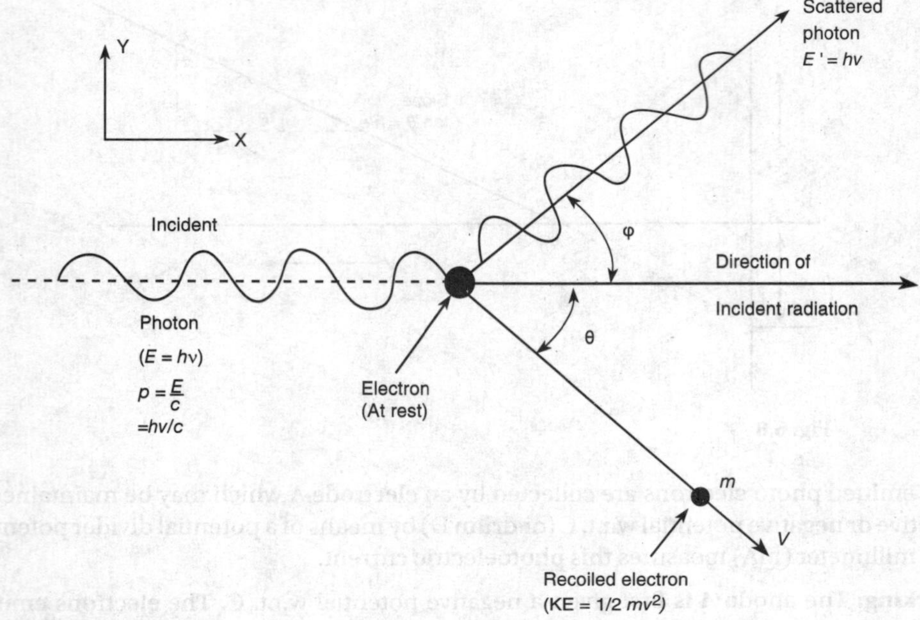

Fig. 6.9 Compton scattering

with the photon. Let λ and ν be the wavelength and frequency of the incident radiation. According to the quantum theory of radiation, the energy of each photon associated with this radiation is $h\nu$. When this photon of energy $h\nu$ strikes a loosely bound (almost free) electron at rest, then as a result of collision the electron gains momentum and recoils in a direction making an angle θ, with the direction of incident radiation (Fig. 6.9). At the same time, the photon is scattered at an angle ϕ with the direction of the incident radiation with lesser energy $h\nu$ ($\nu' < \nu$). It is assumed that laws of conservation of energy and momentum holds good.

Let m_0 be the rest mass of the electron and m be the mass of the electron when it is recoiled with a velocity v, then from the special theory of relativity, we have

$$m\frac{m_0}{\sqrt{1-v^2/c^2}} \tag{6.37}$$

The momentum of the photon of energy $h\nu$ is given by

$$p=\frac{E}{c}=\frac{h\nu}{c} \tag{6.38}$$

The energy of the electron (at rest) before collision will be m_0c^2 and after collision it will be mc^2. From the law of conservation of energy, we have

$$h\nu + m_0c^2 = h\nu' + mc^2 \tag{6.39}$$

From the law of conservation of momentum, in the x-direction

$$\frac{h\nu}{c}=\frac{h\nu'}{c}\cos\phi + mv\cos\theta \tag{6.40}$$

and in y-direction $$0 = \frac{h\nu'}{c} = \sin\phi + mv\sin\theta \tag{6.41}$$

From Eqs. (6.40) and (6.31), one obtains

$$mvc\cos\theta = h(\nu - \nu'\cos\phi) \tag{6.42}$$

$$mvc\sin\theta = h\nu'\sin\phi \tag{6.43}$$

Squaring and adding Eqs. (6.42) and (6.43), one obtains

$$m^2v^2c^2 = h^2[(\nu - \nu'\cos\phi)^2 + (\nu'\sin\phi)^2] = h^2(\nu^2 + \nu'^2 - 2\nu\nu'\cos\phi) \tag{6.44}$$

Squaring Eq. (6.39), one obtains

$$m^2c^4 = [h(\nu - \nu') + m_0c^2]^2 = h^2(\nu^2 - 2\nu\nu' + \nu'^2) + m_0^2c^4 + 2hm_0c^2(\nu - \nu') \tag{6.45}$$

Subtracting Eq. (6.44) from Eq. (6.45), one obtains

$$m^2c^4\left(1 - \frac{v^2}{c^2}\right) = 2h^2\nu\nu'(\cos\phi - 1) + m_0^2c^4 + 2hm_0c^2(\nu - \nu') \tag{6.46}$$

But $$mc^2 = \frac{m_0c^2}{\sqrt{1 - v^2/c^2}} \quad \text{or} \quad m^2c^2\left(1 - \frac{v^2}{c^2}\right) = m_0^2c^4 \tag{6.47}$$

Using Eq. (6.47), Eq. (6.46) becomes

$$2h^2\,\nu\nu'(1 - \cos\phi) = 2hm_0c^2(\nu - \nu') \tag{6.48}$$

$$h(1 - \cos\phi) = m_0c^2\left(\frac{1}{\nu'} - \frac{1}{\nu}\right) = m_0c(\lambda' - \lambda) \tag{6.49}$$

Obviously, the shift in wavelength of the scattered photon as compared to that of the incident photon is

$$\lambda' - \lambda = \Delta\lambda = \frac{h}{m_0c}(1 - \cos\phi) \tag{6.50}$$

$$\lambda' - \lambda = \Delta\lambda = \frac{h}{m_0c}\sin^2\phi/2 \tag{6.51}$$

This is Compton expression for the shift in the wavelength of the scattered X-rays. h/m_0c is denoted by λ_c and is called Compton wavelength. λ_c has the dimensions of length and it has a value

$$\frac{h}{m_0c} = \frac{6.6\times10^{-34}}{9.1\times10^{-31}\times3\times10^{8}} = 0.0242\times10^{-10}\quad m = 0.0242\text{Å}$$

Thus $$\Delta\lambda = \lambda_c(1 - \cos\phi) \tag{6.52}$$

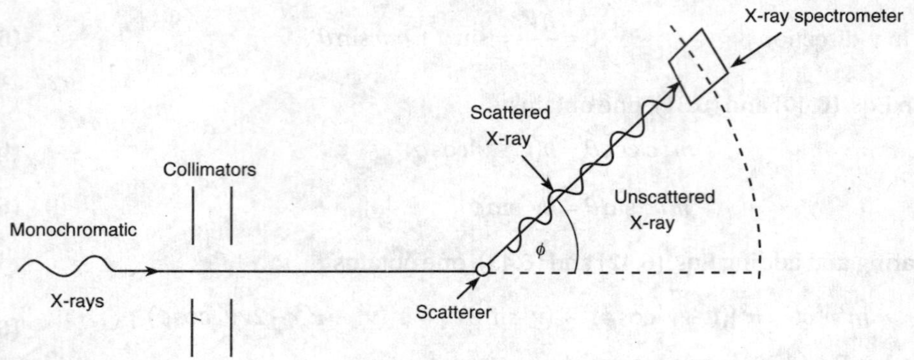

Fig. 6.10

From Eq. (6.50) it is evident that λ' is always greater than λ as $\cos\phi$ is always less than 1. Furthermore $\Delta\lambda$ is also dependent on $\cos\phi$. Thus maximum value of $(1-\cos\phi)=[1-(-1)]=2$, at $\phi = 180°$ hence the maximum value of the shift in wavelength can be $(\Delta\lambda)_{max} = 0.0484$Å.

Thus to have an appreciable relative change in wavelength i.e., $(\Delta\lambda/\lambda)$, the wavelength of the incident X-ray must be quite small. Obviously, the Compton effect is easily observable for smaller wavelengths.

Experimental Verification of Compton Effect

The experimental arrangement to demonstrate the Compton effect is shown in (Fig. 6.10). A beam of monochromatic X-rays of known wavelength (λ) is made incident on a graphite scatterer. The distribution of intensity with wavelength of the monochromatic X-rays scattered at various angles is measured by X-ray spectro-meter. The intensity distribution versus wavelength curves for various scattering angles are shown in (Fig. 6.11). Obviously, the curves have two peaks, one corresponding to modified X-ray, and the other corresponding to unmodified radiation. The difference between two peaks on wavelength axis gives the Compton effect or shift.

From the various curves it is obvious that greater is the scattering angle, greater is the Compton shift in accordance with Eq. (6.27). At $\phi = 90°$, $\Delta\lambda = 0.024$ Å is in good agreement with the value obtained from Compton formula given by Eq. (6.27).

Compton's work provided rather convincing evidence that a photon can undergo particle-like collision with both atoms and unbound electrons. Later studies of the recoil electrons and their energies added further confirmation to the prediction of the theory. Some additional salient features of Compton effect are:

(i) Bethe and Geiger in 1925 by observing the scattered electron confirmed that in Compton experiment the shift in the wavelength of the radiation is indeed due to the scattering of the radiation by the electron.

(ii) A spread in the wavelength of the Compton shifted X-rays is observed. It is due to the fact that the target electron is in general not stationary but has a momentum spread even inside the atom. One can take into account the correction due to the binding of the electron in terms of its momentum distribution.

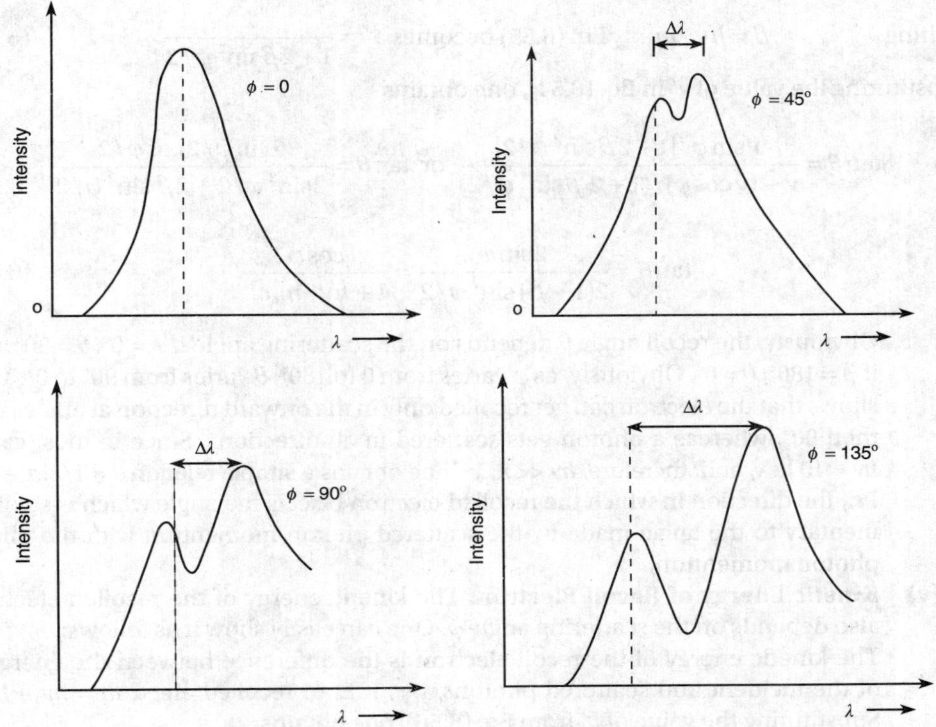

Fig. 6.11 Intensity distribution versus wavelength curves for various scattering angles

(iii) **The Compton shift $(\lambda - \lambda_0)$ is independent of λ_0 but the intensity of scattering depends on λ_0. Compton shift $(\lambda - \lambda_0)$ actually increases as $\lambda_0 \to 0$ and hence the effect is more easily observable for X-rays than for lower frequency radiation.**

(iv) **The direction of recoil electron (θ) and scattering angle are given by**

$$\tan\theta = \frac{\cot\varphi/2}{1 + h\nu/m_0c^2} \tag{6.53}$$

One can derive the above relation as follows: Dividing Eq. (6.43) by Eq. (6.42), one obtains

$$\tan\theta = \frac{h\nu'\sin\varphi/2}{h(\nu - \nu'\cos\varphi)} = \frac{\nu'\sin\varphi}{(\nu - \nu'\cos\varphi)} \tag{6.54}$$

Using Eq. (6.34), one obtains

or $$\frac{1}{\nu'} = \frac{1}{\nu} + \frac{h}{m_0c^2}(1 - \cos\varphi) = \frac{1}{\nu} + \frac{h}{(m_0c^2)}2\sin^2\varphi/2$$

or $$1/\nu' = \frac{1 + \left(\dfrac{h\nu}{m_0c^2}\right)2\sin^2\varphi/2}{\nu} \text{ or } 1/\nu' = \frac{\nu}{1 + \left(\dfrac{h\nu}{m_0c^2}\right)2\sin^2\varphi/2} \tag{6.55}$$

Writing $\beta = h\nu / m_0 c^2$, Eq. (6.55) becomes $\nu' = \dfrac{\nu}{1 + 2\beta \sin^2 \varphi/2}$ (6.56)

Substituting the value of ν' in Eq. (6.54), one obtains

$$\tan\theta = \frac{\nu \sin\varphi / [1 + 2\beta \sin^2 \varphi/2]}{\nu - (\nu \cos\varphi)/(1 + 2\beta \sin^2 \varphi/2)} \text{ or } \tan\theta = \frac{2\sin\varphi/2 \cos\varphi/2}{2\sin^2\varphi/2 + 2\beta \sin^2\varphi/2}$$

or

$$\tan\theta = \frac{2\sin\varphi}{2(1+\beta)\sin^2\varphi/2} = \frac{\cos\varphi}{1 + h\nu/m_0c^2} \tag{6.57}$$

Obviously, the recoil angle θ depends on the scattering angle if $\varphi = 0°$; $\theta = 90°$ and if $\varphi = 180°$; $\theta = 0°$. Obviously, as φ varies from 0 to 180°, θ varies from 90° to 0°. This shows that the electron can get recoiled only in the onward direction at angles less then 90°, whereas a photon gets scattered in all directions. Since in most cases $h\nu \approx 10$ keV, and, therefore, $h\nu << m_0c^2$, one obtains a simple relation $\theta = 1/2(\pi - \varphi)$, *i.e.*, the direction in which the recoiled electron bisects the angle which is supplementary to the angle made by the scattered photon momentum with the initial photon momentum.

(v) **Kinetic Energy of Recoil Electron:** The kinetic energy of the recoiled electron also depends on the scattering angle ϕ. One can easily show it as follows:
The kinetic energy of the recoil electron is the difference between the energies of the incident and scattered photons, i.e., K.E. of recoiled electron = $h\nu - h\nu'$. Substituting the value of υ' from Eq. (6.56), one obtains

$$\text{K.E.} = h\nu - h\left[\frac{\nu}{1 + 2\beta \sin^2(\phi/2)}\right] = h\nu\left[\frac{2\beta \sin^2\phi/2}{1 + 2\beta\sin^2(\phi/2)}\right] \tag{6.58}$$

Comparison of Compton Effect and Photoelectric Effect

(i) In Compton effect a photon cannot transfer all of its energy to a free electron, where as in photoelectric effect, an electron absorbs the incident photon completely before it leaves the photometal surface. To understand it clearly, let us consider a Compton collision between a photon and free electron (initially at rest). The incident photon has an energy $h\nu$ and a momentum $h\nu/c$. Let us suppose that this photon transfer its total energy (and momentum) to the electron who acquires a velocity ν. From the laws of conservation of momentum and energy, we have

$$h\nu = \frac{1}{2}m\upsilon^2 \quad \text{and} \quad h\nu/c = m\upsilon$$

Eliminating $h\nu$ between the above two equations, one obtains

$$\upsilon = 2c$$

But υ cannot exceed c (velocity of light) and hence a free electron cannot absorb all the energy of a photon. In photoelectric effect, an electron absorbs the incident photon completely before it leaves the photo-metal surface as photoelectron. Thus the electron in the photometal surface cannot be free, i.e., it must be bound to an atom otherwise it could not absorb the incident photon completely.

(ii) Photoelectric effect can be explained by mere quantum considerations, where as one has to use relativistic considerations in applying laws of conservation of momentum and energy in case of explaining Compton effect with quantum theory. Obviously, we can say that Compton effect is **relativistic photoelectric effect.**

6.1.5 Specific Heat of Solids

We know from the law of equipartition of energy that the average kinetic energy per atom in each degree of freedom is $\frac{1}{2}kT$. A solid may be supposed to consist of individual atoms which are bound to their neighbours, but they can vibrate freely when heated. As the atom has 3 degrees of freedom in oscillatory motion, its total energy is $\frac{3}{2}kT$. Thus the kinetic energy of oscillation of all the atoms in 1 gm-atom of a solid is given by

$$\text{K.E.} = N \times \frac{3}{2}kT = \frac{3}{2}RT$$

where N is Avogadro's number and R is the molar gas constant.

Now, in simple oscillatory motion the average kinetic energy is equal to the average potential energy. Obviously, the average potential energy per atom in each degree of freedom will also be $\frac{1}{2}kT$. Therefore, the potential energy of each atomic oscillator is $\frac{3}{2}kT$. Thus, the potential energy for 1 gm-atom is given by

$$\text{P.E.} = N \times \frac{3}{2}kT = \frac{3}{2}RT$$

Therefore the total energy of 1 gm-atom of the solid is given by

$$U = \frac{3}{2}RT + \frac{3}{2}RT = 3RT \tag{6.59}$$

The specific heat of a solid at constant volume is defined as

$$C_v = \left(\frac{dU}{dT}\right)_v = \frac{d}{dT}(3RT) = 3R \tag{6.60}$$

Since $R \simeq 2$ cal/(g-atom-°K) $\quad\therefore\quad C_v \simeq 6$ cal/(g-atom-°K)

This is *Dulong and Petit law*. According to the *Dulong and Petit law*, the specific heat of all solids must be about $3R$ and it should not vary with temperature. However, the specific heat for a number of substances was found to be much smaller than $3R$. Moreover, it was found that the specific heat of solids was found to vary with temperature, tending to zero at absolute zero Fig. [6.12(a)]. Thus the classical theory of specific heat of solids fails lo explain the behaviour of the specific heat of solids with temperature.

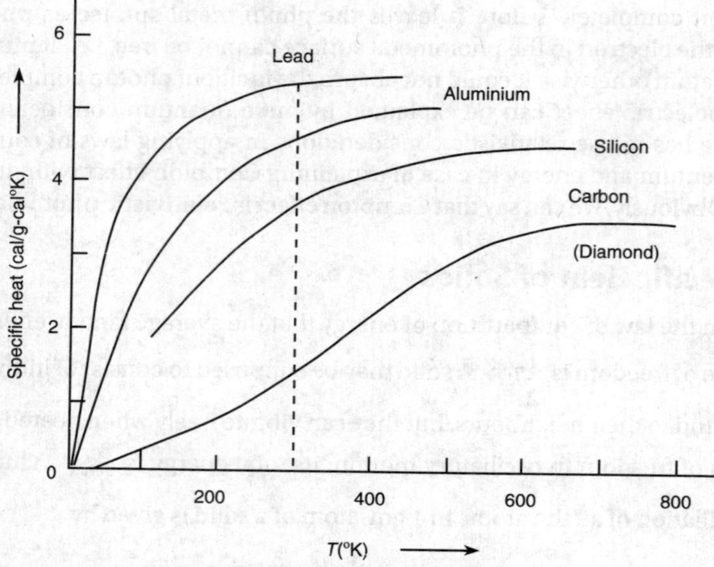

Fig. 6.12(a) Specific heat of solids

Einstein's Quantum Theory of Specific Heat: A long step toward the explanation of the temperature variation of sp. heat of solids was taken by Einstein in 1907. Einstein treated each atom as a three dimensional harmonic oscillator of single frequency ν behaving independently of its neighbours. He replaced the classical value kT for the mean energy of an oscillator by the value derived by Planck for the mean energy of an oscillator given by

$$\bar{\varepsilon} = \frac{h\nu}{(e^{h\nu/kT} - 1)} \tag{6.61}$$

Therefore, the total energy of a gm-atom of the solid consisting of N atoms is

$$U = \frac{h\nu}{e^{h\nu/kT} - 1}$$

Therefore, the specific heat is given by

$$C_V = \left(\frac{dU}{dT}\right)_v = 3Nh\nu \frac{e^{h\nu/KT}\left(\frac{h\nu}{kT^2}\right)}{(e^{h\nu/kT} - 1)^2}$$

$$= 3Nk\left(\frac{h\nu}{kT}\right)^2 \frac{e^{h\nu/kT}}{(e^{h\nu/kT} - 1)^2} = 3R\left(\frac{h\nu}{kT}\right)^2 \frac{e^{h\nu/kT}}{(e^{h\nu/kT} - 1)^2} \tag{6.62}$$

At sufficient high temperatures,

$$\frac{h\nu}{kT} \to 0 \text{ and } (e^{h\nu/kT} - 1)^2 \approx \left(\frac{h\nu}{kT}\right)^2$$

and hence Eq. (6.60) reduces to $3R$. This is in agreement with Dulong and Petit's law.

At very low temperatures $T << \frac{h\nu}{k}$, we have $\frac{h\nu}{kT} >> 1$, so that $e^{h\nu/kT} >> 1$, and Eq. (6.62) gives

$$C_V = 3R\left(\frac{h\nu}{kT}\right)^2 e^{-h\nu/kT} \tag{6.63}$$

This shows that as $T \to 0$, C_V also approaches zero. This is in general agreement with experimental results. At very low temperatures careful measurements show that C_V varies as T^3 near 0 °K, thereby indicating the need of improvement even in Einstein's theory. However, Einstein's theory clearly displays that harmonic oscillators can take only discrete energy values and thereby confirms the inadequacy of classical theory.

Debye's Theory: Debye considered solid as a continuous elastic body, in which the vibrations of the atoms generate stationary waves. The internal energy of the solid rests in these elastic stationary waves.

Let us consider that elastic waves are set up in a unit volume of a continuous solids. The number of possible waves of any kind with frequencies ν and $\nu + d\nu$ is given by $\frac{4\pi}{V^3}\nu^2 d\nu$, where v is the speed of the wave. The elastic weaves that can occur in a solid are of two kinds (i) longitudinal waves, which travel with the velocity $V_l\sqrt{\left\{(K+\frac{4}{3}\eta)/\rho\right\}}$, where η is the coefficient of rigidity, K the bulk modules and ρ the density of the solid and (ii) transverse waves which travel with the velocity $V_t = \sqrt{\frac{\eta}{\rho}}$.

Since there are two perpendicular directions of polarisation for a transverse wave, and hence the total number of possible elastic stationary waves with frequencies between ν and $\nu + d\nu$ is

$$4\pi\nu^2\left(\frac{1}{V_l^3}+\frac{2}{V_t^3}\right)d\nu$$

If V is the volume of a g-atom of a particular solid, the number of modes of vibration with frequencies between ν and $\nu + d\nu$ in it will be

$$4\pi V\left(\frac{1}{V_l^3}+\frac{2}{V_t^3}\right)\nu^2\, d\nu$$

Hence the total number of modes comprising all possible frequencies is given by

$$4\pi V\left(\frac{1}{V_l^3}+\frac{2}{V_t^3}\right)\int_0^{\nu_m}\nu^2 d\nu = 3N$$

or

$$\frac{4}{3}\pi V\left(\frac{1}{V_l^3}+\frac{2}{V_t^3}\right)\nu_m^3 = 3N$$

$$\therefore \qquad \nu_m^3 = \frac{3N}{\left[\frac{4}{3}\pi V\left(\frac{1}{v_l^3} + \frac{2}{v_t^3}\right)\right]} \qquad (6.64)$$

Obviously, the elastic spectrum is cut at upper limit ν_m which has a definite value for each substance. The total energy of the solid is, therefore, given by

$$U = 4\pi V\left(\frac{1}{v_l^3} + \frac{2}{v_t^3}\right)\int_0^{\nu_m}\frac{h\nu}{e^{h\nu/kT}-1}\nu^2\,d\nu = \frac{9N}{\nu_m^3}\int_0^{\nu_m}\frac{h\nu^3}{e^{h\nu/kT}-1} \quad \text{(from Eq. (6.64))}$$

Let $\frac{h\nu}{kT} = x$, so that $\nu = \frac{kT}{h}x$ and $d\nu \frac{kT}{h}dx$, then

$$U = 9RT\left(\frac{T}{T_c}\right)^3\int_0^{T_c/T}\frac{x^3}{e^x-1}dx$$

$$\therefore \qquad C_v = \left(\frac{dU}{dT}\right)_v = 9R\left[4\left(\frac{T}{T_c}\right)^3\int_0^{T_c/T}\left[\frac{x^3}{e^x-1}dx - \frac{T_c}{T}\times\frac{1}{e^{T_c/T}-1}\right]\right] \qquad (6.65)$$

At high temperatures x and T_c/T are very small and hence $e^x \lambda\ 1 + x$ and $e^{T_c/T} \cong 1 + T_c/T$. Therefore

$$\therefore \qquad C_v = 9R\left[4\left(\frac{T}{T_c}\right)^3\int_0^{T_c/T}x^2dx - 1\right] = 9R\left[\frac{4}{3} - 1\right] = 3R \qquad (6.66)$$

which is the Dulong-Petit value in agreement with experiment at result at high temperatures.

At very low temperature $x \to \infty$, and $\frac{T_c}{T} \to \infty$, so that second term in the bracket of Eq. (6.65) becomes negligible, and one can replace the upper limit of integration in the first term by ∞. Thus, one obtains

$$C = 9R\left[4\left(\frac{T}{T_c}\right)^3\right]\int_0^{\infty}\frac{x^3}{e^x-1}dx = 36R\left(\frac{T}{T_c}\right)^3\left(\frac{\pi^4}{15}\right) = 234\left(\frac{T}{T_c}\right)^3 \qquad (6.67)$$

Since T_c is a constant for a solid, one obtains

$$C_v\, \alpha T^3 \qquad (6.68)$$

This is **Debye T³ Law**. At intermediate temperatures the integral appearing in Eq. (6.65) is evaluated numerically. Debye's theory explains quite successfully the observed variation of specific heat of solids.

Although Compton scattering and the photoelectric effect provided the earliest experimental evidence in support of the quantization (particle like behaviour) of

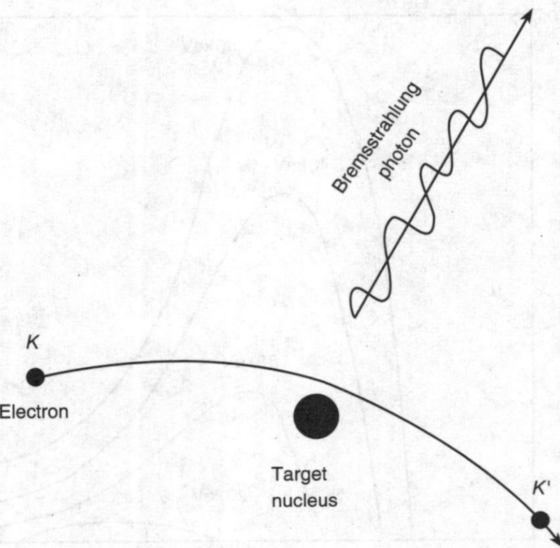

Fig. 6.12 (b) The Bremsstrahlung process responsible for the production of X-rays in the continuous spectrum

electromagnetic radiation, there are numerous other experiments that can also be interpreted correctly only if we assume the existence of photons as discrete quanta of electromagnetic radiation, e.g., **Bremsstrahlung** and **X-ray production, pair production, Raman effect,** etc. Now, we discuss some of these processes, which can not be understood if we consider only the wave nature of electromagnetic radiation.

6.1.6 Bremsstrahlung and X-ray Production

The quantum nature of radiation is further confirmed by the existence of a short wavelength limit of the Bremsstrahlung X-ray spectrum Fig. [6.12(b)].

The continuous X-radiation (Fig. 6.13) is often called Bremsstrahlung, from the Greman breams (= braking, i.e., decelerating) + strahlung (= radiation).

The Bremsstrahlung process occurs not only in X-ray tubes but wherever fast electrons collide with matter, as in cosmic rays, in the van Allen radiation belts which surround the earth and in stopping of electron emerging from accelerators or radioactive nuclei. The Bremsstrahlung process can be considered as an **inverse photoelectric effect**. In the photoelectric effect, a photon is absorbed, its energy and momentum going to an electron and a recoiling nucleus. In the Bremsstrahlung process, a photon is created, its energy and momentum coming from a colliding electron and nucleus. We may note that we deal with the creation of photons in the Bremsstrahlung process, rather than with their absorption or scattering by matter. Symbolically we can write the Bremsstrahlung process as

electron $\rightarrow$ electron + photon

whereas in the photoelectric effect as

electron + photon $\rightarrow$ electron

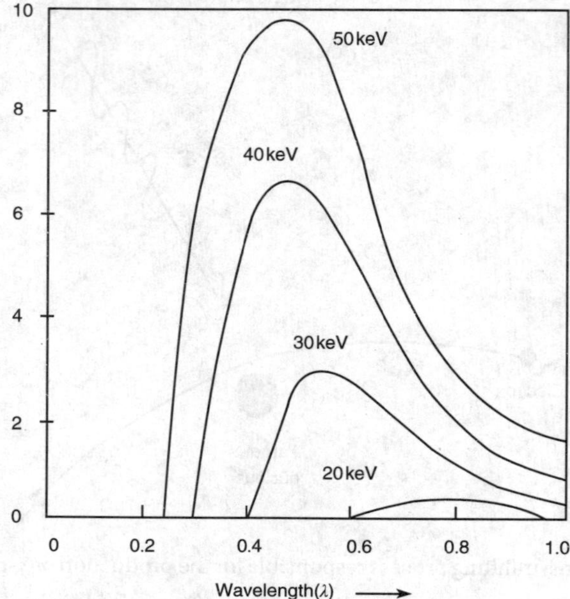

Fig. 6.13 The continuous X-ray spectrum emitted from a tungsten target for four different values of eV, the incident electron energy

However neither process occurs for free electrons. In both cases there must be a heavy atom in the neighbourhood to take care of the recoil momentum.

6.1.7 Pair Production and Pair Annihilation

In addition to the photoelectric and compton effect there is another process whereby photons lose all their energy in interactions with matter, namely the process of pair production: in the process two particles are created: an electron and a position. Here we have an example of creation of rest energy, i.e., radiant energy converted into rest mass energy as well as into kinetic energy. In this process (Fig. 6.14), a high energy photon loses all of its energy $h\nu$ (photon) in an encounter with a nucleus, creating an pair (an electron and an positron) and endowing them with kinetic energies. In this process pair production energy taken by the recoil of the nucleus is negligible because it is so massive and thus the balance of total relativistic energy in the process is simply.

$h\nu = E_- + E_+ = (m_0c^2 + K_-) + (m_0c^2 + K_+) = K_- + K_+ + 2m_0c^2$, where E_- and E_+ are the total relativistic energies, K_- and K_+ are the kinetic energies of the electron and positron respectively. The photon must have an energy of at least $2\ m_0c^2 = 1.02$ MeV in order for this process to occur, such high energy photons are in the region of **nuclear gamma rays**. Symbolically,

$$\text{Photon} \rightarrow \text{electron} + \text{positron}$$

This process, like brehmmstrahlung, will not occur unless there is an atom nearby to supply the necessary recoil momentum. The reverse process

$$\text{electron} + \text{positron} \rightarrow \text{photon}$$

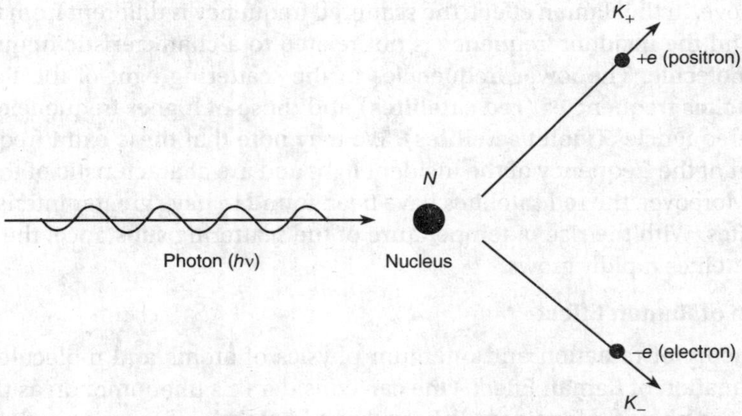

Fig. 6.14 Pair production process

also occurs; this process is known as **electron-position annihilation** and occur for free electrons and positrons as long as at least two photons are created. In this process the electrons and positron disppear and are replaced by two photons. Conservation of energy requires that, if E_1 and E_2 are the photon energies, then

$$(m_0c^2 + K_+) + (m_0c^2 + K_-) = E_1 + E_2$$

Usually K_+ and K_- are negligibly small and we can assume the positron and electron to be essentially at rest. Momentum conservations then demands the two photons to have equal energies of m_0c^2 (= 0.511 MeV) and to move in exactly opposite directions.

Electron-positron pairs are produced in nature by cosmic-ray photons and in the laboratory by Bremsstrahlung photons from particle accelerators. Other particle pairs, such as proton and antiproton, can be produced as well if the initiating photon has sufficient energy. Because the electron and position have the smallest rest mass of known particles, the threshold energy of their production is the smallest. Experiment verifies the quantum picture of pair production process. There is no satisfactory explanation whatever of this phenomenon in classical physics.

6.1.8 Raman Effect

An interesting effect which gives much information about molecular quantum states was discovered experimentally by C.V. Raman in 1928, while investigating the scattering of light by molecules of transparent liquids, gases and solids. Raman found that in addition to the frequency of incident light, the scattered radiation had a number of other frequencies higher and lower than the incident one. This phenomenon is called **Raman effect**. We are familiar with other light scattering processes. In ordinary **Rayliegh scattering** by molecules, the scattered frequency is the same as the incident frequency. In the **fluorescence process**, the frequency of the incident light coincides with an absorption frequency of the scattering gas molecules; this is a **resonance** phenomenon in which the molecule is raised to an excited state and after a short life time there, reemits light at a different frequency.

As stated above, in the Raman effect, the scattered frequency is different from the incident frequency and the incident frequency is not related to a characteristic frequency of the scattering molecule. The lower frequencies in the scattering light of the Raman effect are called **stokes** frequencies **(red satellites)** and those of higher frequencies are called **antistokes** frequencies **(violet satellites)**. We may note that these extra frequencies are independent of the frequency of the incident light and are characteristic of the scattering substance. Moreover, the red satellites have been found to have greater intensity than the violet satellites. With the rise of temperature of the scattering substance, the intensity of the violet sattelites rapidly grows.

Explanation of Raman Effect

Quantum theory of radiation and quantum physics of atoms and molecules provide a simple explanation of Raman Effect. One can consider this phenomenon as the **inelastic collision** of the photon of incident light with the molecule of the scatterer. Basically, in the Raman effect it is the frequency shifts in the scattered spectrum that are determined by the nature of the scatterer rather than the frequencies themselves. Let us consider the collision of a photon of frequency ω_0 with a molecule, which is initially in the energy state E_i. During the process of the collision of photon with the molecule, the photon either gives up its energy to the molecule or receives energy from the molecule of the scatterer and the molecule goes to the energy state E_f. Now, if the scattered photon has the frequency ω then the law of conservation of energy demands that

$$\hbar\omega_0 + E_i = \hbar\omega + E_f \tag{6.69}$$

or

$$\omega = \omega_0 + \frac{E_i - E_f}{\hbar} \tag{6.70}$$

or

$$\omega = \omega_0 + \Delta\omega \tag{6.71}$$

we may note that in terms of such ideas the Raman effect had been predicted at a much earlier date by Smekal.

In the Eq. (6.66) the term $\Delta\omega$ can be interpreted as a frequency that the atom might conceivably emit or absorb in the usual way, in jumking from the first state, E_i to the final state, E_f. **The difference between the frequency of each Raman line and the frequency of the incident light is thus equal to the frequency of some conceivable emission or absorption line of the scattering atom or molecule.** Now, two cases of interest arise:

(i) $E_i < E_f$: In this case $\Delta\omega$ is negative and the frequency of the scattered photon is given by

$$\omega = \omega_0 - |\Delta\omega| \quad \text{(Stokes' lines)} \tag{6.72}$$

This situation arises when the molecule is initially in the ground state and the photon transfers its energy to the molecule. Obviously, the photon energy and hence the frequency diminishes and the molecule jumps to the excited state. This explains the origin of Stokes' lines in Raman spectra, Fig. [6.15(a)]. We may note that since there are large number of excited states, the transitions of molecules from the ground state to the excited states are accompanied by many spectral lines.

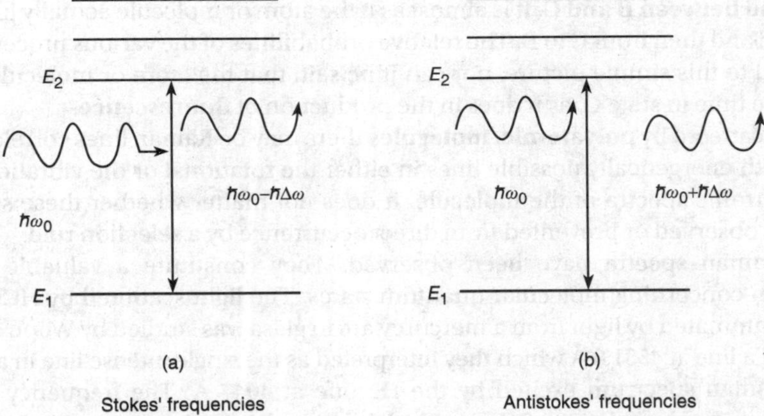

Fig. 6.15 Origin of Stokes' and AntiStokes' Raman lines

We know that at ordinary temperature, the number of molecules in the ground state exceeds those in the excited state, the probability of upward transitions is greater than the downward transitions. Obviously, the intensities of red satellites are greater than those of violet satellites.

(ii) $E_i > E_f$: In this case $\Delta\omega$ is positive and the frequency of the scattered photon is obtained as

$$\omega = \omega_0 + |\Delta\omega| \quad \text{(antistokes' lines)} \tag{6.73}$$

This situation arises when the molecule is initially in one of its excited state Fig. [6.15(b)]. The incident photon cause the excited molecule to jump to the ground state. During this process the photon gets energy from the molecule and thereby the scattering photon has higher energy, i.e., frequency. **The violet satellites** appear in this process.

The number of molecules at ordinary temperature is smaller in excited states than those in the ground state and therefore the number of downward transitions from the excited states to the ground state is smaller. This is why the violet satellites are weaker in intensity. On raising the temperature of the scattering substance, the number of molecules in the excited states increases and thereby the downward transitions. Clearly, with the increase in temperature of the scattering substance an increase in intensity of the violet companions increases.

The Raman shift can be expressed in terms of wave numbers as ($\bar{\nu} = \lambda^{-1} = \omega/2\pi c$):

$$\text{Raman shift} = \bar{\nu}_0 \pm \Delta\bar{\nu}$$

where positive sign stands for **antistokes'** lines and negative sign for **Stokes'** lines.

The intensity of a Raman line has nothing to do with the intensity of the emission or absorption line that is thus correlated with it. The selection for the two are quite different; transitions that are forbidden in ordinary spectra occur freely in the Raman effect. This is one reason for its great theoretical interest. According to wave mechanics, **a Raman jump is possible between two atomic or molecular levels A and B only when there exists at least one third level C, such that ordinary relative transitions are allowed between**

A and C and between B and C. It is almost as if the atom or molecule actually jumped first from A to C and then from C to B. The relative probabilities of the various processes do not correspond to this simple picture, nor can it be said that the atom or molecule exists for any definite time in state C, as it does in the production of **fluorescence.**

In light scattered by **polyatomic, molecules** there may be Raman lines correlated in this manner with energetically possible lines in either the rotational or the vibration-rotation or the electronic spectra of the molecule. It does not matter whether these spectra can actually be observed or prevented from direct occurrence by a selection rule.

Many Raman spectra have been observed. They constitute a valuable source of information concerning molecular quantum states. The light scattered by HCl gas when strongly illuminated by light from a mercurcy arc in glass was studied by Wood and Dieke. They found a line at 4581.8 Å which they interpreted as the single intense line in a vibration-rotation Raman spectrum excited by the Hg line at 4047 Å. The frequency difference between these two lines is 2886.0 Cm^{-1}, which is in excellent agreement with the frequency of the missing central line in the 3.5 μ band from HCl, viz, 2885.4 Cm^{-1}. Much closer to the exciting line and on both sides of it there were also a number of lines interpreted as a rotational Raman spectrum. In the corresponding spectrum adjacent to the mercury line at 4358 Å, the measured spacing between the lines lay mostly between 41 and 42 Cm^{-1}, in good agreement with double the spacing in the ordinary rotation spectrum of HCl.

Raman spectroscopy is quite useful even for homonuclear molecules (two identical nuclei) such as O_2 and N_2. We cannot directly observe rotational spectra or vibration-rotation spectra for such molecules because they have no electric dipole moment. We can, however obtain a spectrum, corresponding to vibration and rotation of such molecules in the Raman scattering. It is as though the incident radiation polarizes the molecule, thereby inducing an electric dipole moment; this permits absorption and emission of radiation corresponding to rotational and vibrational motions of the molecule. Of course, in an electronic transition in O_2 or N_2 the fine structure of the spectrum reveals the vibrational and rotational structure but such a spectrum lines in the ultraviolet and the fine structure is very difficult to resolve.

With the advent of lasers, **laser Raman Spectroscopy** have been developed. This has opened a new era of research and applications of Raman effect.

Experimental Arrangement for Studying Raman Effect

A Schematical experimental arrangement for studying Raman effect is shown in Fig. 6.16. The figure shows tube whose one end is flat and the other end extends into a horn shape. The scattering substance is taken in this tube. The horn shape part of tube is blackened and filled with mercury to provide dark background. In front of the flat face of the tube a spectrograph is mounted. To keep the temperature of the liquid constant, the tube is surrounded by a jacket in which water is circulated. To illuminate the experimental liquid a monochromatic source with suitable filter is used. After suffering scattering through liquid, the light enters spectrograph through slits S_1 and S_2. The Raman lines are recorded on a photographic plate.

We may note that Raman effect is characteristic of molecules and hence it is widely used in the study of structure of various molecules. Raman spectra constitute a valuable source of information concerning molecular quantum states. Even homo-nuclear molecules such as O_2 and N_2 give Raman spectra corresponding to vibration-rotation and rotational spectra, although the latter cannot be observed disectly. Obviously, Raman effect provides a proof in favour of quantum nature of radiation.

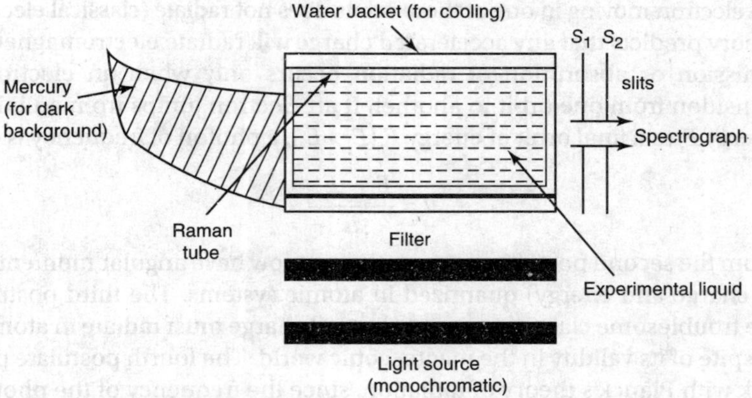

Fig. 6.16 Experimental arrangement for observing Raman Spectra

6.2 ATOMIC STRUCTURE AND ATOMIC SPECTRA

The assumption that electrons are constituents of all atoms was reasonable from such experiments as the photoelectric effect in metals, the ionisation of gases in discharge tubes. X-ray bombardment, and so forth. A difficulty arose, however, in attempting to design a stable atom which could contain both electrons and the positive charges necessary to make the atom electrically neutral. In 1911, Rutherford from his unexpected large-angle scattering of α-particles experiment concluded that the atom consists of very small but massive nucleus (size of the order of 10^{-14} m) carrying a charge $+Ze$. Around this central region are located the Z electrons of the neutral atom. The diameter of an atom is about 10^{-10} m or 10,000 times greater than the size of the nucleus. We consider this model to be a dynamic model. If a static model is assumed, all the electrons surrounding the nucleus would be attracted towards it because of the Coulombic force between the nucleus and the electrons, and the atom would soon collapse. In the dynamic planetary model the massive nucleus is essentially at rest. with the electrons revolving about it in circular and elliptical paths. However, in spite of initial achievements physicists found that Rutherford's atomic model had to be abandoned because, according to classical electrodynamics, (i) an accelerated charge must continuously radiate classical energy and (ii) the frequency of the emitted radiation must be equal to the frequency of revolution.

In 1913, Bohr proposed a theory which was successful in explaining the radiation spectra of one-electron atoms, although it is in direct disagreement with the classical theory of radiation. Bohr's postulates may he summarized as follows:

(i) The Coulombic force on a planetary electron provides the centripetal acceleration required for a dynamical stable circular orbit.

(ii) The only permissible orbits are those in the discrete set for which the angular momentum of the electron equals an integral times $\hbar = h/2\pi$. The angular momentum of the only permissible orbits are given by

$$mvr = \frac{nh}{2\pi} = n\hbar \;\; (n = 1, 2, 3\ldots) \tag{6.74}$$

(iii) An electron moving in one of these orbits does not radiate (classical electromagnetic theory predicts that any accelerated charge will radiate electromagnetic theory).

(iv) Emission or absorption of radiation occurs only when an electron makes a transition from one orbit to another. If an electron jumps from an initial orbit of energy E_i to a final orbit of energy $E_f(E_i > E_f)$, a photon of frequency is emitted.

$$\nu = \frac{E_i - E_f}{h} \tag{6.75}$$

From the second postulate we note that we now have angular momentum (as well as charge and energy) quantized in atomic systems. The third postulate rejects the troublesome claim that an accelerated charge must radiate in atomic systems, in spite of its validity in the macroscopic world. The fourth postulate provides the link with Planck's theory of radiation, since the frequency of the photon emitted or absorbed is given by the energy difference of the two states divided by h.

6.2.1 Bohr's Theory and Hydrogen Spectrum

Bohr applied the quantization rule to the Rutherford's atom model (one-electron atom) and successfully explained the observed spectrum of hydrogen atom.

Let us consider that an electron of charge e is rotating in an orbit of radius r. The centripetal force acting on it will be $\frac{mv^2}{r}$. If Ze is the charge of nucleus, the electrostatic force on the electron will be Ze^2h^2. Then from the first postulate,

$$\frac{Ze^2}{r^2} = \frac{mv^2}{r} \tag{6.76}$$

From the second postulate, we have

$$mvr = n\hbar = \frac{nh}{2\pi} \tag{6.77}$$

or

$$2\pi m\, vr = nh \tag{6.78}$$

By combining Eqs. (6.76) and (6.78), one obtains the expression for the radius of nth orbit,

$$r_n = \frac{n^2 h^2}{4\pi^2 Ze^2 m} \tag{6.79}$$

The quantity $a_0 = \frac{h^2}{4\pi^2 m_0 e^2} \sim 0.53\text{Å}$ is the radius of the first orbit of hydrogen calculated for a fixed nucleus. It is often simply called the **Bohr radius**. From Eq. (6.79) it is evident that the radii of the allowed orbits are proportional to the squares of the intergers. Obviously, the spacing between adjacent orbits increases progressively.

From Eqs. (6.76) and (6.77), one obtains the expression for the velocity of electron in an orbit as

$$v = \frac{2\pi Ze^2}{nh} \tag{6.80}$$

Using Eq. (6.76), we may write the kinetic energy of the electron as

$$T = \frac{1}{2}mv^2 = \frac{Ze^2}{2r} \tag{6.81}$$

Since the potential energy of the electron in the Coulomb field of the nucleus is

$$U = \frac{Ze^2}{r} \tag{6.82}$$

then the total energy is

$$E_n = T + U = +\frac{Ze^2}{2r} - \frac{Ze^2}{r} = -\frac{Ze^2}{2r} = -\frac{2\pi^2 Z^2 e^4 m}{n^2 h^2} \quad (n = 1, 2 \ldots) = \frac{Z^2 me^4}{2n^2 \hbar^2} \tag{6.83}$$

The state of the lowest energy, i.e., the ground state corresponds to $n = 1$. The energies of the other excited states increases with n, tending to zero as $n \to \infty$ for the hydrogen. We obtain the magnitude of the ground state energy for a fixed nucleus as

$$E_0 = \frac{2\pi^2 me^4}{h^2} \cong 13.6\,\text{eV}$$

Eq. (6.82) indicates that the quantization of energy has risen as a result of assuming that the angular momentum is quantized.

When the correction is made for the motion of the nucleus around the centre of mass of the hydrogen atom, the expression for the energy of the system becomes.

$$E_n = \frac{2\pi^2 Ze^4 \mu}{n^2 h^2} \tag{6.84}$$

where $\mu = \dfrac{mM}{M + m}$ is the reduced mass of the nucleus and electron system and M is the mass of the nucleus.

The frequency of a photon emitted by a transition from the nth level to the kth level, where $n > k$, is given by

$$h\nu = hc\left(\frac{1}{\lambda}\right) = hc\overline{\nu} = E_n - E_k = \frac{m}{2}\frac{(Ze^2)^2}{\hbar^2}\left[\frac{1}{k^2} - \frac{1}{n^2}\right]$$

Expressed in wave numbers, this becomes

$$\overline{\nu} = \frac{2\pi^2 m^2 Ze^4}{h^3 c}\left[\frac{1}{k^2} - \frac{1}{n^2}\right] = R_y\left[\frac{1}{k^2} - \frac{1}{n^2}\right] \tag{6.85}$$

where R_y is the Rydberg constant calculated from the reduced mass. For hydrogen $R_y = 109681$ cm^{-1} whereas the experimental value from optical spectroscopy is $R_y = 109677.576 \pm 0.012$ cm^{-1}. The Rydberg constant calculated for a fixed nucleus (i.e., for $\mu = m$) is $R_y = 109737.309$ cm^{-1}. Furthermore, the experimentally established spectral series for hydrogen are all explained by Eq. (6.85) by setting the value of k and letting n run through the integers such that $n > k$. For example, $k = 1$, $n = 2, 3, 4, \ldots$ corresponds

Table 6.1 Hydrogen Spectral Series

Series	*Quantum Number*		*Region*
	Lower state (m)	*Upper state (n)*	
Lyman series	$m = 1$	$n = 2, 3, 4, \ldots$	Ultraviolet
Balmer series	$m = 2$	$n = 3, 4, 5, \ldots$	Visible
Paschen series	$m = 3$	$n = 4, 5, 6, \ldots$	Infrared
Brackett series	$m = 4$	$n = 5, 6, 7, \ldots$	Infrared
P fund series	$m = 5$	$n = 6, 7, 8, \ldots$	Infrared

to the **Lyman series;** $k = 2$, $n = 3, 4, 5 \ldots$ corresponds to the **Balmer series**: $k = 3, 4, 5 \ldots$ corresponds to the **Paschen, Brackett** and **Pfund** series, respectively Eq. (6.24) can also be used to calculate the ionisation potential of hydrogen by letting $n \to \infty$. Then we obtain

$$V = \frac{chR_y}{1.6 \times 10^{-12}} \sim 13.6V$$

The excellent agreement between the spectroscopic values for the Rydberg constant, the ionisation potential, and the emission spectrum of hydrogen was a great triumph for the Bohr theory. However, it was immediately apparent that the Bohr theory was good only for a one electron atom, since it failed badly in the cases of helium and lithium, the next simplest neutral atoms.

6.2.2 Generalization of Bohr's Quantum Condition by Sommerfeld and Wilson (Old Quantum Theory)

In 1916 Sommerfeld and Wilson showed independently that Bohr's condition of quantization of angular momentum of hydrogen atom was a particular example of a general law of periodic system with N degrees of freedom described by (q_i, p_i), where q_i are coordinates and p_i are canonically conjugate momenta. In a three dimensional coordinate

$$q_i = x, q_2 = y, q_3 = z$$

system and the corresponding momenta will be p_x, p_y and p_z. Wilson and Sommerfeld quantum condition is

$$\oint p_i dq_i = nh$$

where $$n = 0, 1, 2, 3, \ldots \quad (6.86)$$

This condition, is known as the **Wilson-Sommerfeld quantization rule**. For his elliptical orbits Sommerfeld therefore took.

$$\oint p_r dr = n'h \quad (6.87)$$

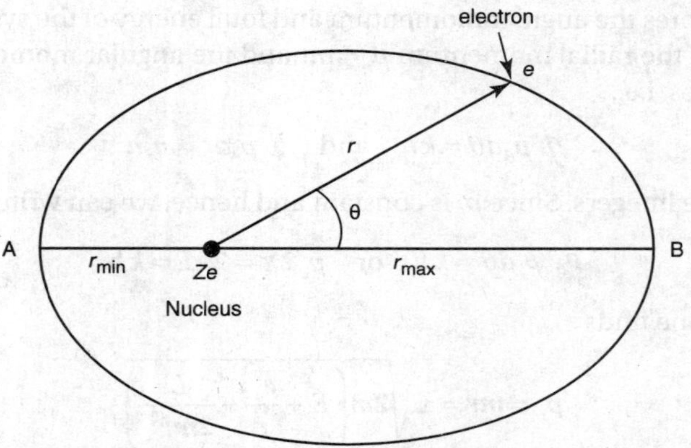

Fig. 6.17(a) An ellipse with parameters

$$\oint p_\theta d\theta = kh \tag{6.88}$$

where the integer n' and k are known as **the radial and azimuthal quantum numbers** respectively.

Sommerfeld assumed that n' could be 0, 1, 2, 3 ... while the azimuthal quantum number k took on values 1, 2, 3 ... with 0 excluded since that would correspond to harmonic oscillation of the electron directly through the nucleus.

Sommerfeld then showed, however, that if allowance is made for the known variation of electronic mass with speed, the energy of the elliptical motion is slightly different from that of the circular motion, and in this way he arrived at a splitting, or **fine structure,** of the levels of one-electron atoms, which appeared to be in quantitative agreement with observation.

Elliptic Orbits of Hydrogen

We know that the simplest force is associated with point particles is the mutual force acting along the line joining the two. This force is usually termed as central force. Let us consider the two particles of hydrogen atom with the nucleus of charge Ze fixed at the origin and an electron of mass m and charge e moving relative to the nucleus as shown in Fig. [6.17(a)]. In accordance with the classical mechanics, the first integrals of the equations of motion of the system are

$$mr^2\theta = L \qquad \text{(a constant)} \tag{6.89}$$

and $$\frac{1}{2}mv^2 + \frac{1}{2}\frac{L^2}{mr^2} + V(r) = E \qquad \text{(a constant)} \tag{6.90}$$

with $$V(r) = -\frac{e^2}{r} \tag{6.91}$$

Here L and E denotes the angular momentum and total energy of the system respectively. We may note that the radial momentum $p_r = mr$ and the angular momentum $L = p_\theta$ obey the quantized rules, i.e.,

$$\oint p_\theta d\theta = kh \quad \text{and} \quad \oint p_r dr = n_r h \tag{6.92}$$

where k and n_r are integers. Since p_θ is constant and hence, we can write

$$p_\theta \oint d\theta = kh \quad \text{or} \quad p_\theta 2\pi = 2\pi L = kh \tag{6.93}$$

From Eq. (6.90), one finds

$$p_r = mr = \pm\sqrt{2m\left(E + \frac{e^2}{r} - \frac{L}{2mr^2}\right)} \tag{6.94}$$

Here the positive valve of p_r corresponds to the increase of r from $r_{\min}$ and the negative value to the other half of the elliptical path. Thus, we have $\oint p_r dr = 2 \int_{r_{\min}}^{r_{\max}} \sqrt{2m\left(E + \frac{e^2}{r} - \frac{L^2}{2mr^2}\right)} dr = 2\pi\left(e^2\sqrt{\frac{m}{-2E}} - L\right)$. Making use of Eqs. (6.92) and (6.93), the above result yields,

$$2\pi e^2\sqrt{\frac{m}{-2E}} = (k + n_r)h = nh \tag{6.95}$$

where, $n = n_r + k$, is the **principal quantum number**. Solving for E, one obtains

$$E = -\frac{me^4}{2\hbar^2 n^2} = -\frac{me^4}{(4\pi E_0)^2 2\hbar^2 n^2}, \; n = 1, 2, 3, \ldots \tag{6.96}$$

which is same as the Bohr's Eq. (6.84). We may note that no restriction is imposed to circular orbits while deriving Eq. (6.96).

At points A and B of ellipse, $dr/dt = 0$ and this happens for $r = r_{\min}$ and $r = r_{\max}$. When $dr/dt = 0$, from Eq. (6.96), one obtains

$$r^2 + \frac{e^2 r}{E} - \frac{L^2}{2mE} = 0 \tag{6.97}$$

The sum of the roots of Eq. (6.92) is given by

$$r_{\min} + r_{\max} = -\frac{e^2}{E}$$

Since $r_{\min} + r_{\max} = 2a$, the semi major axis of the ellipse is

$$a = -\frac{e^2}{2E} \tag{6.98}$$

On substitution of the value of E, one obtains

$$a_n = \frac{n^2\hbar^2}{me^2} = n^2 a_0 \tag{6.99}$$

where a_0 is the Bohr radius.

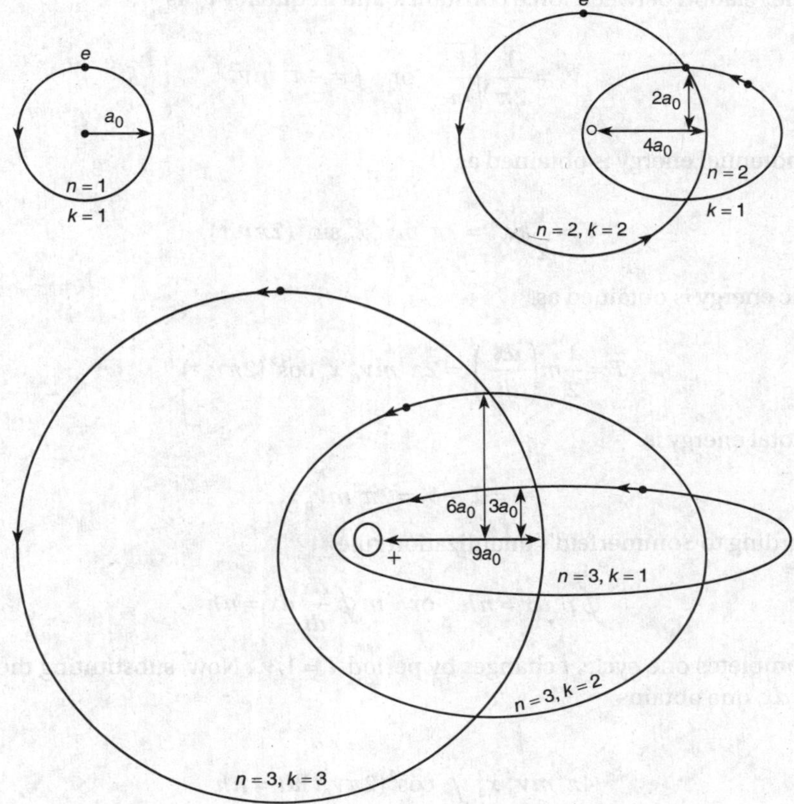

Fig. 6.17(b) Sommerfeld's elliptical orbits of electrons for $n = 1, 2, 3$

Sommerfeld first solved elliptical orbits of hydrogen atom as an example of the general quantization rule. The energy is dependent on $n(=k+n_r)$, the principal quantum number, therefore, all orbits for which n has the same value, will have the same energy. States of motion belonging to the same energy are said to be **degenerate**. Among all orbits having the same value of n, one orbit will be circular one for which $n_r = 0$ and $k = n$. This clearly reveals that why the energy expression is the same as one gets by Bohr's theory for circular orbits. However, sommerfeld ruled out the case $k = 0$, as it corresponds to motion of the electron along a straight line through the nucleus. Fig. 6.17(b) show electron orbits for $n = 1, 2, 3$.

Applications of Sommerfeld Theory

(i) Harmonic oscillator

One can represent the displacement x with t of a harmonic oscillator of frequency ν_0 by

$$x = x_0 \sin(2\pi\nu_0 t) \tag{6.100}$$

we have the relation between force constant k and frequency ν_0 as

$$\nu_0 = \frac{1}{2\pi}\sqrt{\frac{k}{m}} \quad \text{or} \quad k = 4\pi^2 m\nu_0^2 \tag{6.101}$$

Thus the potential energy is obtained as

$$V = \frac{1}{2}kx^2 = 2\pi^2 m\nu_0^2 x_0^2 \sin^2(2\pi\nu_0 t)$$

The kinetic energy is obtained as

$$T = \frac{1}{2}m\left(\frac{dx}{dt}\right)^2 = 2\pi^2 m\nu_0^2\, x_0^2 \cos^2(2\pi\nu_0 t)$$

Thus the total energy is

$$E = T + V = 2\pi^2 m\nu_0^2 x_0^2 \tag{6.102}$$

Now, according to Sommerfeld's quantization rule

$$\oint p_x dx = nh \quad \text{or} \quad m\oint \frac{dx}{dt}dx = nh \tag{6.103}$$

When x completes one cycle, t changes by period $T = 1/\nu_0$. Now, substituting the value of dx/dt and dx, one obtains

$$4\pi^2 m\nu_0^2\, x_0^2 \int_0^{1/\nu_0} \cos^2(2\pi\nu_0 t)dt = nh$$

On solving, one obtains

$$2\pi^2 m\nu_0 x^2 = nh \quad \text{or} \quad x_0 = \sqrt{\frac{nh}{2\pi^2 m\nu_0}} \tag{6.104}$$

on substituting the value of x_0 in Eq. (6.102), one obtains

$$E_n = nh\nu_0 = n\hbar\omega, \qquad n = 0, 1, 2, \ldots \tag{6.105}$$

Thus, according to old quantum theory, the energy of linear harmonic oscillator are integeral multiples of $\hbar\omega\ (= h\nu_0)$.

(ii) Rigid Rotator

A rigid rotator consists of two mass points which are connected by a massless rod. A rigid rotator restricted to move in a plane is described by an angle coordinate θ and the momentum congugate to the angle coordinate $p_\theta = I\dot{\theta} = I\omega$, where I is a constant of motion and called the moment of inertia. Now, the quantization rule for this case reduces to

$$\int_0^{2\pi} p_\theta d\theta = Jh \quad \text{or} \quad p_\theta = \frac{Jh}{2\pi}, J = 0, 1, 2, \ldots \tag{6.106}$$

Obviously, the angular momentum p_θ is an integral multiple of $h/2\pi$. With $p_\theta = I\omega$, the energy of a classical rotator $E = I\omega^2/2$ reduces to

$$E_J = \frac{J^2 h^2}{8\pi^2 I}, \; J = 0, 1, 2, \ldots \tag{6.107}$$

we know that to describe a rigid rotator in space, one requires two coordinate θ and ϕ of the polar coordinates. On applying the quantizaion rule, one obtains the same expression as the one in Eq. (6.100) for the total angular momentum and

$$p_\phi = m\hbar \; m = -J, -J+1, \ldots, J-1, J$$

for the component of angular momentum along the z-axis. We see that energy is independent of the quantum number m and is given by Eq. (6.101). We further note that each energy level is $(2J + 1)$ fold degenerate.

(iii) Particle in a Box

Let us consider a particle of mass m moving inside a rectangular box of sides a, b and c without any influence of force. We further assume that the collision with the walls of the box is perfectly elastic. We take the cartesian axes along the edges of the box. The linear momentum p_x, p_y and p_z are constants of motion and change sign during collision with the walls. Now, applying quantization rule, one finds

$$\oint p_x dx = 2ap_x = n_x h$$

We obtain

$$p_x = \frac{n_x h}{2a}, \quad n_x = 0, 1, 2, \ldots$$

Similarly,

$$p_y = \frac{n_y h}{2b}, \quad n_y = 0, 1, 2, \ldots$$

and

$$p_z = \frac{n_z h}{2c}, \quad n_z = 0, 1, 2, \ldots$$

Now, the total energy

$$E_{n_x n_y n_z} = \frac{1}{2m}(p_x^2 + p_y^2 + p_z^2) = \frac{h^2}{2m}\left(\frac{n_x^2}{a^2} + \frac{n_y^2}{b^2} + \frac{n_z^2}{c^2}\right) \tag{6.108}$$

for a cubical box of side a, i.e., $a = b = c$, we have

$$E_{n_x n_y n_z} = \frac{h^2}{8ma^2}\left(n_x^2 + n_y^2 + n_z^2\right) \tag{6.109}$$

Obviously the energy of the system is quantized.

6.3 THE FRANCK-HERTZ EXPERIMENT

A direct and striking demonstration of the existence of the discrete stationary states postulated by the Bohr theory of the atom was first provided by the experiment designed by J. Franck and G. Hertz. In this experiment mercury vapour was bombarded with electrons of the known kinetic energy. When this kinetic energy is less than the energy of the first excited state of atomic mercury, the only energy which the colliding electron loses is the small amount of kinetic energy (about 1 part in 10^5) that it can transfer to the massive mercury atom by an elastic collision. However, when the kinetic energy of the incident electron just exceeds the energy of the first excited state of mercury, then the electron gives up virtually all of its kinetic energy to the mercury atom in an elastic collision. Obviously, by comparing the kinetic energy of the electrons before and after the collision one can determine how much energy is transferred to the target atoms.

Franck and Hertz used a simple apparatus consisting of a mercury tube containing a cathode, a plate, the accelerating grid which was located physically near the plate (Fig. 6.18). Both the plate and the grid were maintained at positive potentials w.r.t. the cathode, but the plate potential was slightly lower than the grid potential. This small retarding potential on the plate had the effect of preventing contributions to the plate current from electrons having negligible kinetic energy. Hence, if electrons were to lose most of their kinetic energy in inelastic collisions with mercury atoms, the retarding potential would prevent them from reaching the plate and a drop in plate current would result. Such a drop in plate current was found to occur at 4.9 V Fig. (6.19). Furthermore, Franck and Hertz noted that at this voltage, the 2536 Å spectral line of mercury appeared in the emission spectrum of the[2] vapour. A simple calculation

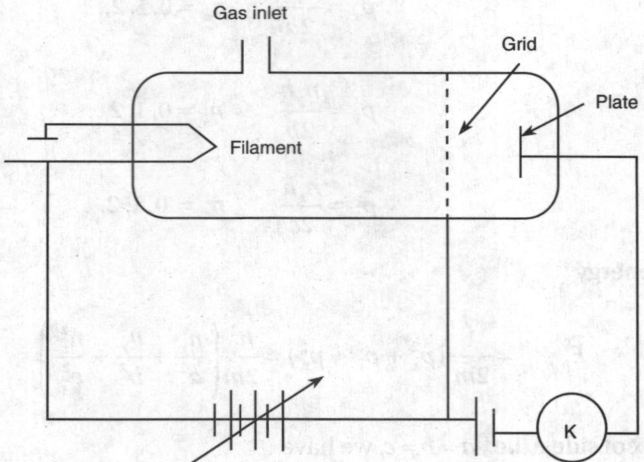

Fig. 6.18 Franck-Hertz experiment

[2]**Note:** It is now known that Wilson-Sommerfeld quantization condition $\oint p_q d_q = n_q h$ is not a generally valid condition. The idea of sharp orbits it also not in accord with modern quantum mechanics.

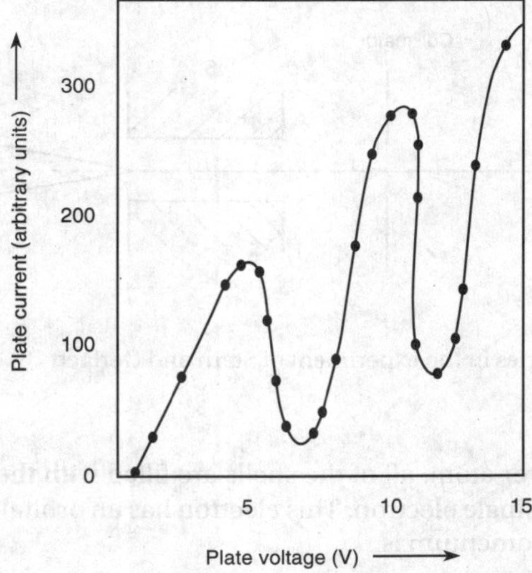

Fig. 6.19 Current vs. Voltage in the Franck-Hertz experiment

reveals that the photon energy of the 2536 Å line corresponds to 4.86 eV. At slightly higher voltages a large drop in the plate current occurs and new lines appear in the emission spectrum of the mercury vapour. This behaviour is repeated at multiples of 4.9 V as shown in (Fig. 6.14). We must conclude from this that the Bohr concept of discrete energy states is qualitatively correct.

6.4 SPATIAL (SPACE) QUANTIZATION: STERN AND GERLACH EXPERIMENT

So far we have seen that there are ample evidences, which demonstrate that the energy is quantized. A different kind of quantization was revealed by an atomic beam experiment performed by Stern and Gerlach in 1922, in which it was demonstrated that the vector magnetic moment of an atom placed in an external magnetic field can be in only one or another set of discrete directions.

In the Stern-Gerlach experiment, silver atoms were evaporated in an oven and collimated into a narrow beam, which was passed through an inhomogeneous magnetic field as shown in (Fig. 6.20) between the shaped poles of a magnet and fall on to a photographic plate. Classical theory predicts that a single smeared pattern should be produced on the photographic plate. Instead, the stream of silver atoms splits into two distinct lines after passing through the inhomogeneous magnetic field.

As the shells of electrons about the nucleus are filled, the electrons are alternately aligned such that each new pair consists of one electron with spin up and one electron with spin-down. As a result, each completed shell has a resultant spin angular momentum

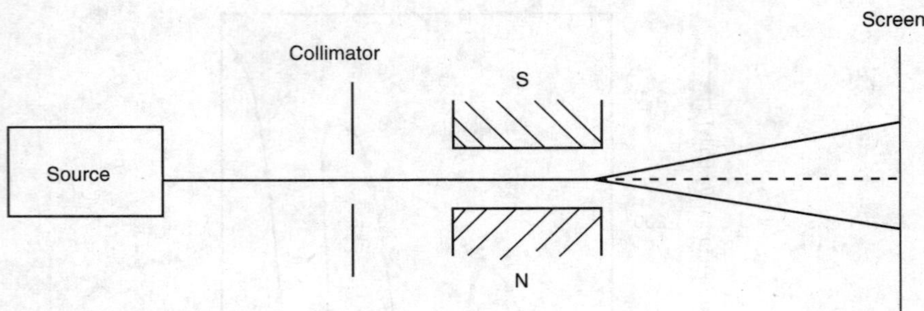

Fig. 6.20 Particle trajectories in the experiment of Stern and Gerlach

of zero. In a normal silver atom, all of the shells are filled with the exception of the outer shell, which contains a single electron. This electron has an orbital quantum number $l = 0$, and the total angular momentum is

$$J = \sqrt{J(J+1)}\,\hbar \text{ or } J = \sqrt{\frac{1}{2}\left(\frac{1}{2}+1\right)}\hbar = \frac{1}{2}\sqrt{3}\,\hbar\text{, since } j = \frac{1}{2} \text{ When } l > 0\text{, when } J = l \pm s$$

The spin angular momentum is quantized according to $L_s = \sqrt{s(s+1)}\,\hbar$, and w.r.t. the z-direction determined by the magnetic field it has only components

$$L_{sz} = L_s \cos\theta = m_s \hbar \tag{6.110}$$

A spinning electron creates a magnetic dipole moment μ_s, which under the action of the magnetic field B experiences a torque

$$\tau = \mu_s \times \boldsymbol{B} \tag{6.111}$$

that lends to align the dipole with the applied field B. The energy associated with this magnetic dipole is

$$\Delta E = -\mu_s.\boldsymbol{B} = -\mu_s B \cos\alpha \tag{6.112}$$

In addition to the rotational effect given by the torque, there is a net translational force

$$F = -\frac{d}{dz}(\Delta E) \tag{6.113}$$

which pushes the dipoles that are somewhat aligned with the field in the direction of the most intense part of the field; or for these antialigned, they are pushed away from the intense portion of the field. The energy associated with the interaction of $\boldsymbol{L}_s$ and the external field $\boldsymbol{B}$ is

$$\Delta E = m_s 2\left(\frac{e\hbar}{2m}\right)B = 2m_s\mu_B \boldsymbol{B} \tag{6.114}$$

where $\mu_B = e\hbar/2m$ is the Bohr magneton. The net force acting on the atoms is

$$F = -\frac{d}{dz}(\Delta E) = -m_s 2\left(\frac{e\bar{h}}{2m}\right)\frac{dB}{dZ} \tag{6.115}$$

Those atoms with $m_s = +1/2$ are displaced downward in (Fig. 6.15), since from Eq. (6.94), $F < 0$ because $\frac{dB}{dZ} > 0$ and $\boldsymbol{B}$ increases in + z-direction. For atoms with $m_s = -1/2$, $F > 0$, and the atoms are displaced upward.

In general, as the total angular momentum is $J = \sqrt{J(J+1)}\,\hbar$, the number of lines in the Stern-Gerlach pattern is having multiplicity = $2J + 1$. Obviously, Stern and Gerlach's experiment provides a direct proof of the idea of space quantization. Since $J = J = \frac{1}{2}$ in the present case, there are $(2J + 1)$ or two possible orientation of J in the magnetic field. Hence the beam splits into two parts in the magnetic field. This was interpreted that an electron has an intrinsic (built in) magnetic dipole moment μ_s due to the fact that it has an intrinsic angular momentum S called its spin.

Thus, in addition to silver atoms, any atom with a single outer electron in some state $l = 0$ should show a pattern of two distinct lines.

Obviously, the conclusion from the Stern-Gerlach experiment is in contradiction with classical physics, according to which the vector moment could be in any direction. Thus. we are forced to conclude that the electronic orbits in the atom are not capable of taking on every orientation and dimension conceivable, but that the structure of the atom is ordered, that is to say, spatially discrete. This again confirms the inadequacy of classical physics.

What is a Photon?

We have established the firm experimental basis for the particle like nature of light. One can now describe photons by giving a few of their properties:

(i) Like an electromagnetic wave, photons move with the speed of light.
(ii) Photons have zero mass and rest energy.
(iii) Photons carry energy and momentum which are related to the frequency and wavelength of the electromagnetic wave by $E = h\nu$ and $p = h/\lambda$.
(iv) Photons can be created or destroyed when radiations is emitted or absorbed.
(v) Photons can have particle like collisions with other particles such as electrons.

Other properties of photons energy from more detailed studies, e.g., the role of photons in transmitting the electromagnetic force according to the modern field theory of electromagnetism, and the deflection of photons by a gravitational field.

6.5 LIMITATIONS OF OLD QUANTUM THEORY

Although the old quantum theory achieved many successes in atomic and molecular spectroscopy, it was an incomplete theory in the sense that none of its receipts for quantization were derived from first principles. Since it could not be applied to periodic systems, most collision and scattering problems were beyond its pale. Furthermore, it contained errors, contradictions and ambiguities. It did have some virtues, however. It predicted a large body of experimental results from a few simple rules, and it set the stage for the new quantum mechanics which soon replaced it. We will now discuss the wave mechanics of de-Broglie and Schrodinger.

6.6 THE WAVE NATURE OF PARTICLES: MATTER WAVES

Another important revolution in physical concepts took place at the end of the first quarter of 20th century and radically changed our approach to the description of motion of a particle.

Louis de-Broglie in 1924 made the bold suggestion that if radiation had a particle like nature, then it should follow from basic concepts of symmetry that particles ought to possess a wave-like nature of some kind. De-Broglie's work was motivated by the mystery of the Bohr orbits, which he attempted to explain by fitting a standing wave around the circumference of each orbit. Thus de-Broglie required that $n\lambda = 2\pi r$, where λ is the wavelength associated with the nth orbit and r is its radius. Combining this with Bohr's condition

$$mvr = n\hbar$$

We immediately obtain

$$\lambda = \frac{h}{mv} = \frac{h}{P} \tag{6.116}$$

Assuming the existence of a natural symmetry in the properties of matter and energy, he proposed that a material particle of total energy E and momentum P must be accompanied by a phase wave, analogous to that associated to the photon. A photon of light of frequency v has the momentum $P = \frac{hv}{c}$, which can be expressed in terms of wavelength λ as $P = \frac{h}{\lambda}$ $\quad (\because \quad \lambda v = c)$.

The wavelength of a photon is therefore specified by its momentum according to the relation $\lambda = \frac{h}{P}$. De-Broglie suggested that this formula is a completely general one that applies to material particles as well as to photons. The momentum of a particle of mass m and velocity v is $P = mv$, and its de-Broglie wavelength is accordingly

$$\lambda = \frac{h}{mv} \tag{6.117}$$

The greater the particle's momentum, the shorter its wavelength. In Eq. (6.117) m is the relativistic mass

$$m = \frac{m_0}{\sqrt{1 - v^2 / c^2}} \tag{6.118}$$

de-Broglie made this suggestion in his Ph. D. thesis at the University of Paris and nearly failed for putting forward such an unlikely idea. Einstein was passing through Paris that weekend and the thesis was referred to him for comment. It was only because Einstein suggested there might conceivably be something in the proposition that de-Broglie was awarded his doctorate for what later turned to be one of the brilliant concepts of the century.

Relation Eq. (6.116) can also be expressed in terms of kinetic energy (K) as

$$K = \frac{1}{2} m v^2 \quad \therefore \quad mv = \sqrt{2mK}$$

$$\lambda = \frac{h}{\sqrt{2mK}} \tag{6.119}$$

If a charge particle carrying a charge q is accelerated through a potential difference V Volts, then Kinetic energy

$$K = qV$$

and de-Broglie wavelength

$$\lambda = \frac{h}{\sqrt{2mqV}} \tag{6.120}$$

The expression for the de-Broglie wavelength associated with an electron in the non-relativistic case will be

$$\lambda = \frac{h}{\sqrt{2m_0 eV}} \tag{6.121}$$

For electron $m_0 = 9.1 \times 10^{-31}$ kg and $e = 1.6 \times 10^{-19}$C, we obtain for voltage $V = 100$ V

$$\lambda = \frac{6.63 \times 10^{-34}}{\sqrt{2 \times 9.1 \times 10^{-31} \times 100 \times 1.6 \times 10^{-19}}} = 1.226 \times 10^{-10} \text{ m} = 1.226\text{Å}$$

Such wavelengths are therefore of the same order of magnitude as those of X-rays. It would therefore appear that the regular array of atoms in a crystal, which have a spacing of several Å between them, might well provide a conventional diffraction grating for electron waves as they had previously for X-rays. Experiments on the scattering of electron beam by the atoms at the surface of the crystal, which were being undertaken by Davisson and Germer were in fact the first experimental proof of the existence of matter waves. Introducing the wave number $K = 2\pi/\lambda$, and the angular frequency $\omega = 2\pi v$, one may write the de-Broglie relation in the more symmetric form

$$\lambda = \frac{h}{P} \quad \text{or} \quad \frac{2\pi}{K} = \frac{h}{P} \quad \text{or} \quad P = \frac{h}{2\pi}K = \hbar K \tag{6.122}$$

and

$$E = \frac{h}{2\pi}\omega = \hbar\omega \tag{6.123}$$

Certain of de-Broglie's ideas were developed by the German mathematician E. Schrodinger into a precise mathematical theory. Developed by Born, Heisenberg, Dirac and other great physicists of this century, this theory has now become the highly successful quantum physics of today.

6.7 EXPERIMENTAL CONFIRMATION OF PARTICLE WAVES

In 1925 a series of experiments were begun on secondary electron emission under C. J. Davisson and C.H. Kunsman and later Davisson and L.H. Germer. They have demonstrated that a beam of electrons does suffer diffraction.

Davisson and Germer Experiment: The experimental arrangement used by Davisson and Germer is shown in (Fig. 6.21). A beam of electrons is obtained from a hot tungsten Filament *F* by thermionic emission. These electrons are accelerated to desired velocity by

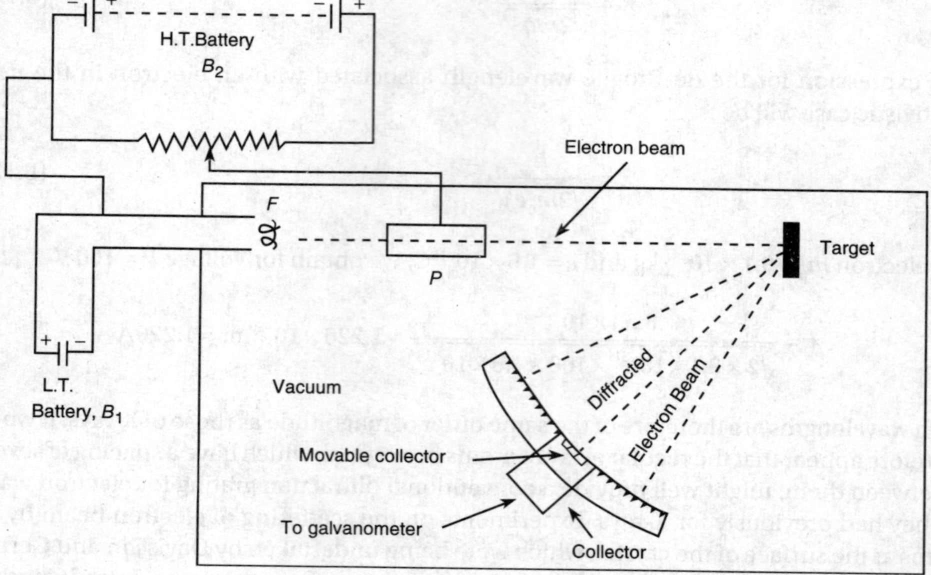

Fig. 6.21 Davisson and Germer experiment

applying suitable potential from a high-tension battery B_2. The accelerated electrons are collimated o a fine beam by allowing them to pass through a system of pin holes provided in the cylinder P.

The collimated beam of electrons strikes the nickel crystal N at normal incidence. The crystal is an ordered array of atoms and each atom acts as a scatterer of electrons.

The nickel crystal is capable of rotating about an axis $\perp^r$ to the plane of the paper. The electrons are scattered in all directions by the atoms of the nickel crystal. The intensity of the electron beam scattered by the crystal in a given direction is measured by allowing it to enter an electron collector which can be rotated about the same axis as the target. The collector is connected to a sensitive galvanometer whose deflection is proportional to the square of the electron beam entering the collector. The whole apparatus is evacuated in a highly evacuated chamber. The deflections of the galvanometer for the various positions give the distribution at different azimuths the crystal is rotated about the axis OX.

In one set of experiments, the (111) face of the nickel crystal was presented to the incident beam of electrons. The plots of intensity of the scattered beam at different angles of scattering for seven accelerating voltages are shown. These curves are plotted in polar coordinates, i.e., the length of the radius vector corresponding to a certain angle was drawn directly proportional to the collector current. It is observed that well defined maximum appears in the direction $\theta = 50°$ when the accelerating potential is 54 V. Beyond 54 V, the bump again decreased to become insignificant at about 68 V (Fig. 6.22). If the intensity were the same at all scattering angles the curves would be circles centred on the point of scattering.

The bump in its most prominent state ($\theta = 50°$ and $V = 54$ V) verifies the existence of electron waves. For, according to de-Broglie, the wavelength associated with an electron accelerated through ($V = 54$ V) is

$$\lambda = \frac{12.3}{\sqrt{V}} = \frac{12.3}{\sqrt{54}} = 1.67\text{Å}$$

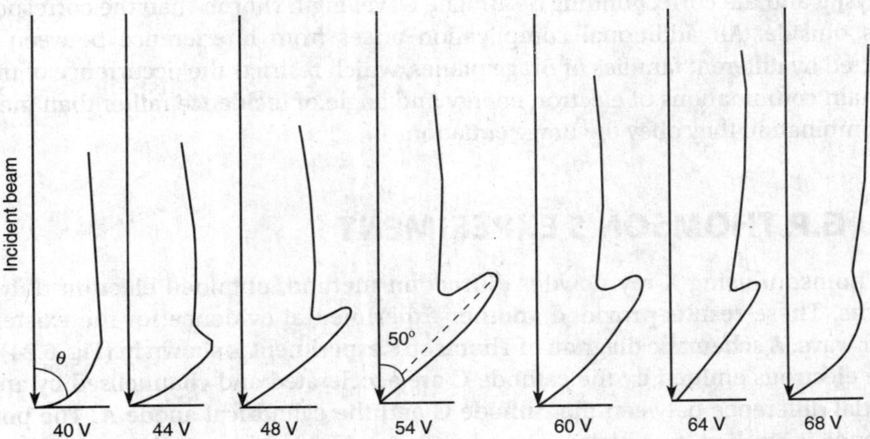

Fig. 6.22

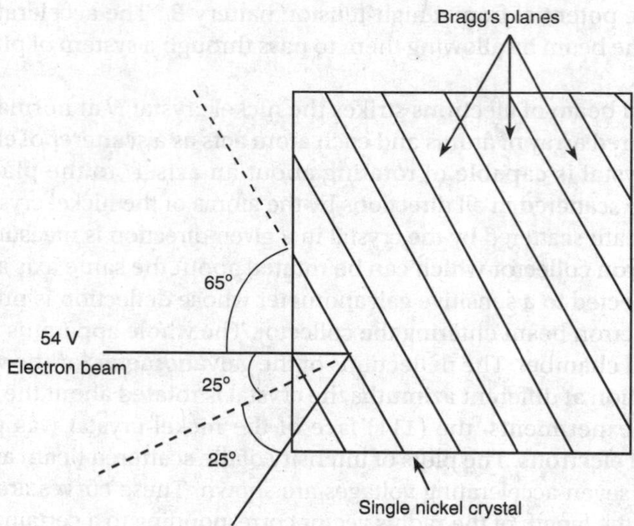

Fig. 6.23

Now, on the basis of a wave diffracted from the Bragg planes within the crystal, a first order diffraction ($n = 1$) from the nickel crystal with spacing $d = 0.91$ Å will give a wavelength (Fig. 6.23)

$$\lambda = 2d\sin\theta = 2\times 0.91\times \sin 65^\circ = 1.65\text{Å}$$

which is the wavelength of a diffracted wave. This comparison is good and gives strong evidence that; indeed, the electrons do have a wave character as particle character.

The analysis of the Davisson Germer experiment is actually less straightforward since the energy of an electron increases when it enters a crystal by an amount equal to the work function of the surface. Hence the electron speeds in the experiment were greater inside the crystal and the corresponding de-Broglie wavelength shorter than the corresponding values outside. An additional complication arises from interference between waves diffracted by different families of Bragg planes, which restricts the occurrence of maxima to certain combinations of electron energy and angle of incidence rather than merely to any combination they obey the Bragg equation.

6.8 G.P. THOMSON'S EXPERIMENT

G.P. Thomson, using X-ray powder diffraction method, obtained electron diffraction patterns. These results provided another experimental evidence for the existence of matter wave. A schematic diagram of Thomson's experiment is shown in (Fig. 6.24).

The electrons emitted by the cathode C are accelerated and channelised by applying potential difference between the cathode C and the cylindrical anode A. The potential difference is applied by a high tension battery and varies from 10,000 to 50,000 V. The accelerated fine beam of electrons is then made to bombard against a thin film of metals

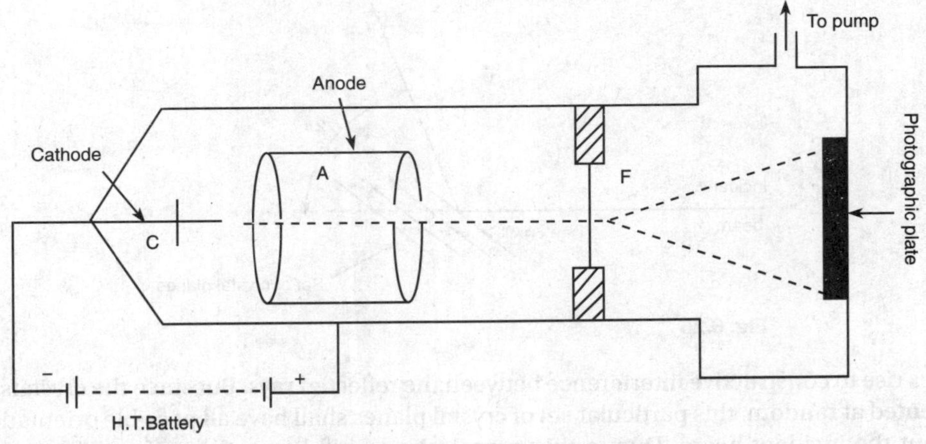

Fig. 6.24 G.P. Thomson's experiment for electron diffraction

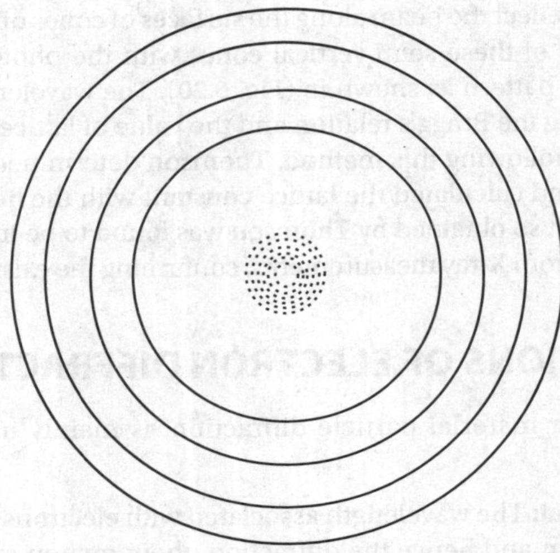

Fig. 6.25 Electron diffraction pattern

like gold, silver, aluminium, etc. After passing through the film, the electrons are received upon a photographic plate. The developed photographic plate shows a pattern consisting of a series of well defined concentric rings about a central spot (Fig. 6.25). It is worthwhile to mention that the diffraction pattern can be disturbed by a magnetic field, showing that the pattern was produced by electrons and not by any radiation.

In Thomson's experiment, one can imagine the metallic film to be composed of a very large number of tiny crystal oriented at random. Due to the random orientation of these millions of micro-crystals, a given electron beam is intercepted by a particular set of crystal planes at all possible angles. In a particular micro crystal, this set of crystal planes is having suitable angle of orientation so as to satisfy Bragg's law ($2d \sin\theta = n\lambda$) and this

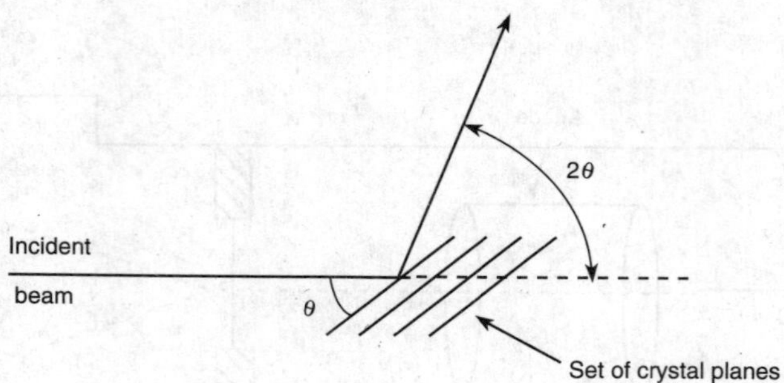

Fig. 6.26

gives rise to constructive interference between the reflected rays. But since the crystals are oriented at random, this particular set of crystal planes shall have all possible orientations about the incident beam. Due to this crystal planes will lie on the surface of a cone of semi vertical angle 2θ, where θ is the angle which the beam makes with this set of crystal planes as shown in (Fig. 6.26). Other sets of crystal planes suitably oriented so as to satisfy Braggs condition will reflect the beam along the surfaces of cones of different semi-vertical angles. The intercepts of these semi-vertical cones with the photographic film or plate give rise to diffraction pattern as shown in (Fig. 6.20). The wavelength of electron waves can be calculated using the Bragg's relation and the value of lattice constant known from X-ray data. Instead of following this method, Thomson determined the wavelength from de-Broglie's relation and calculated the lattice constant with the help of Bragg's law. The value of lattice constant so obtained by Thomson was found to be in good agreement with the value determined from X-ray measurements, confirming the existence of matter waves.

6.9 APPLICATIONS OF ELECTRON DIFFRACTION

Electron diffraction or material particle diffraction, is mainly used for the following purposes:

(i) **Study of crystal:** The wavelength associated with electrons and neutrons is of the order of 10^{-10} m and hence the diffraction phenomenon exhibited by these particles is used to study the crystal structure like X-rays. The electron beams can be conveniently produced and controlled as compared to X-rays. Electron diffraction method is superior to X-ray diffraction method to study the surface structure as the electrons interact with the surface atoms more intimately. Moreover, electrons do not penetrate the matter like X-rays.

(ii) **Electron Microscope:** it is based upon the principle that particle such as electron exhibit wave properties like light rays but have much shorter wavelength and electron can be focused by suitable electric and magnetic field just as light rays can be focused with the help of glass lenses. The resolving power of electron microscope is very high about 10^3 times greater than that of optical microscope.

A schematic diagram of electron microscope is shown in (Fig. 6.27). The main parts of the electron microscopes are (i) **electron gun** and (ii) **magnetic or electrostatic lenses.**

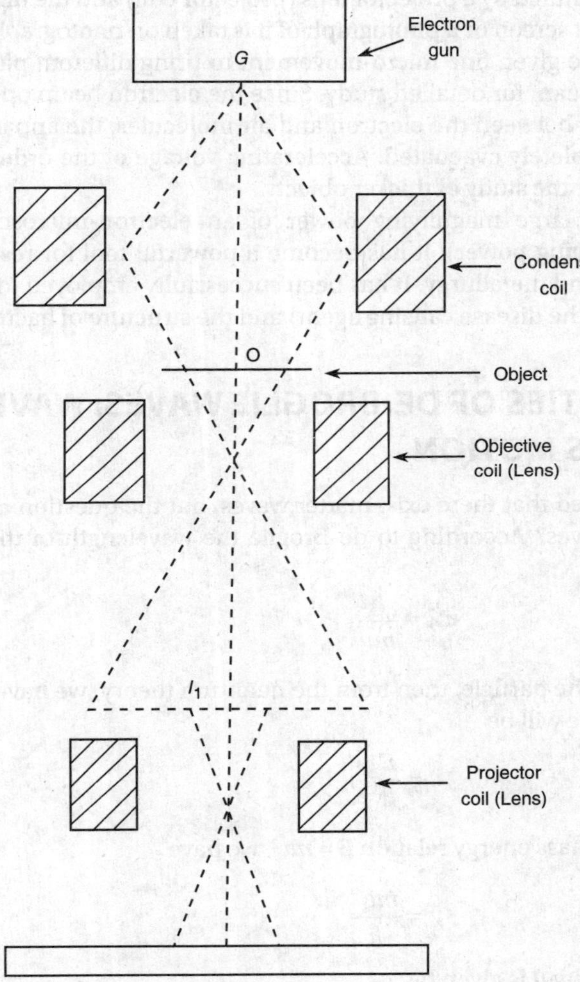

Fig. 6.27 Electron microscope

Working: The working of an electron microscope is similar to that of an optical microscope. The electron beam whose energy is of the order of tens to the hundred of keV, is produced by an electron gun G. This electron beam is concentrated on the object by a condenser lens coil for magnetic focusing. The object lies outside the focus of a magnetic lens which serves as the objective. The object must be very thin so that electrons may pass through it. The electron beam passes through the object and electrons are scattered by the more dense parts of the object which after passing through the objective form the first magnified image I_1. This serves the purpose of an objective lens in an optical microscope. The first magnified image I_1 lies just above the plane of the magnetic lens. A part of this image is further magnified by a projector lens (projector coil) and the final image is either seen on a fluorescent screen or a photograph of it is taken on photographic plate.

The objects can be given fine micro-movement to bring different parts of the objects under the incident beam for detailed study. Since the electron beam operates in vacuum to prevent collisions between the electron and air molecules, the apparatus is mounted in a glass tube completely evacuated. Accelerating voltage of the order of 3,00,000 V is usually employed for the study of thicker objects.

Due to the very large magnifying power of an electron-microscope (~10^5 limes with adequate resolving power), it has become a powerful tool for research in physics, chemistry, biology and metallurgy. It has been successfully employed to study very small objects like viruses (the disease causing agent) and the structure of bacteria.

6.10 VELOCITIES OF DE-BROGLIE WAVES: WAVE PACKET AND ITS MOTION

So far we have studied that there exist matter waves, but the question arises: what is the velocity of these waves? According to de-Broglie the wavelength of the matter wave is given by

$$\lambda = \frac{h}{mv} \tag{6.124}$$

If E is the energy of the particle, then from the quantum theory, we have $E = h\nu$. Thus the frequency of the wave will be

$$\nu = \frac{E}{h} \tag{6.125}$$

But from Einstein's mass energy relation $E = mc^2$, we have

$$\nu = \frac{mc^2}{h} \tag{6.126}$$

The de-Broglie velocity u is given by

$$u = \nu\lambda = \frac{mc^2}{h} \cdot \frac{h}{mv} = \frac{c^2}{v} \tag{6.127}$$

For a material particle, the above relation has following disturbing features.

(i) The velocity u of the de-Broglie wave is not the same as the velocity of the moving particle.

(ii) Since the velocity of the material particle is always less than the velocity of light c, the velocity u of the de-Broglie wave come out to greater than c. This violates the fundamental postulate of the special theory of relativity.

To overcome the above difficulties Schrodinger postulated that a moving material particle is equivalent to a **wave-packet** rather than a single wave. A **wave-packet** comprises a group of waves, each with slightly different velocity and wavelength, with phases and

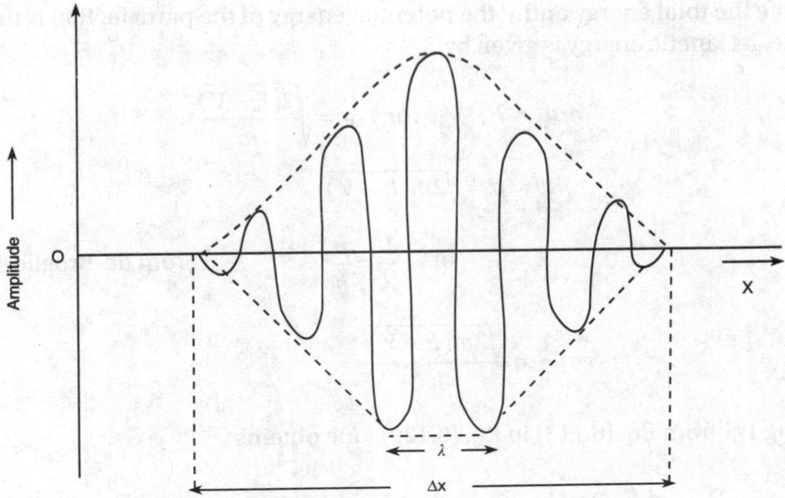

Fig. 6.28 A wave packet in one-dimension

amplitudes so chosen that they interfere constructively over only a small region of space where the article can be located, outside of which they produce destructive interference so that the amplitude reduces to zero rapidly. Obviously, a wave-packet is formed as a result of the superposition of many plane waves of different frequencies grouped around some central frequency. Since the envelope of the packet must have a finite spatial extent, the sum of the amplitudes of all of its plane wave components must be zero everywhere except where the packet is localized.

Form of a typical wave-packet in one dimension is shown in (Fig. 6.28). The velocity with which the wave packet moves is called the **group velocity**. The energy carried by the packet as its group velocity is thus analogous to the kinetic energy of a free particle of the same velocity. The individual waves forming the wave packet possess an average velocity, u called the **phase velocity**. One can easily show that the velocity of the material particle v is same as group velocity of wave packet.

Let us suppose that v is the velocity of the moving particle, g is the group velocity. u is the velocity of an individual wave train comprising the group or wave-packet and λ is the wave-length of the individual wave. The group velocity g of a wave-packet is found to be given by

$$g = u - \lambda \frac{du}{d\lambda} \tag{6.128}$$

Writing Eq. (6.128) as $g = \lambda^2 \left(\frac{u}{\lambda^2} - \frac{1}{\lambda} \frac{du}{d\lambda} \right) = -\lambda^2 \left(\frac{1}{\lambda} \frac{du}{d\lambda} - \frac{u}{\lambda^2} \right) = \lambda^2 \frac{d}{d\lambda} \left(\frac{u}{\lambda} \right)$

$$= -\lambda^2 \left(\frac{d\nu}{d\lambda} \right) \qquad (\because \nu = u/\lambda) \text{ (Here } \nu \text{ is frequency)}$$

$$\frac{1}{g} = -\frac{1}{\lambda^2} \left(\frac{d\lambda}{d\nu} \right) = \frac{d}{d\nu} \left(\frac{1}{\lambda} \right) \tag{6.129}$$

Let E denote the total energy and V the potential energy of the particle. If m is the mass of the particle, its kinetic energy is given by

$$\frac{1}{2}mv^2 = E - V \quad \text{or} \quad v = \sqrt{\frac{2(E-V)}{m}}$$

or

$$mv = p = \sqrt{2m(E-V)} \tag{6.130}$$

But

$$\lambda = \frac{h}{p} \quad \text{or} \quad \frac{1}{\lambda} = \frac{p}{h} \qquad \text{(From de-Broglie's relation)}$$

or

$$\frac{1}{\lambda} = \frac{\sqrt{2m(E-V)}}{h} \tag{6.131}$$

Substituting $1/\lambda$ from Eq. (6.113) in Eq. (6.129), one obtains

$$\frac{1}{g} = \frac{d}{d\nu}\left(\frac{\sqrt{2m(E-V)}}{h}\right) = \frac{1}{h}\frac{d}{d\nu}\sqrt{2m(h\nu - V)} \quad (\because \quad E = h\nu) \tag{6.132}$$

$$= \frac{1}{h}\frac{1}{2}[2m(h\nu - V)^{-1/2}]2mh = \sqrt{\frac{m}{2(h\nu - V)}} = \frac{1}{v} \qquad \text{(from Eq. (6.132))}$$

$$g = v \tag{6.133}$$

Obviously, the group velocity g of de-Broglie wave group or wave-packet associated with a moving particle is the same as the classical velocity v of the particle. However, it is important to note that the motion of a particle and motion of a wave-packet are equivalent with only one difference, i.e., the wave packet has a spatial dimension whereas a particle (a mass-point) has not. If the wave packet can be made small enough, the distinction between the two becomes insignificant.

6.11 PRINCIPLE OF COMPLEMENTARITY

Niels Bohr was in a habit of saying: **the waves and corpuscular views are complementary.** By this he means: if we prove the corpuscular character of an experiment, then it is impossible at the same time to prove its wave character, and conversely. To illustrate it, let us consider Young's interference experiment with the two slits.

In the Young's interference experiment with the double slits we have on the screen a system of interference fringes. By replacing the screen by a photoelectric cell, we can demonstrate the corpuscular character of light even in the fringes. It therefore appears as if we had here an experiment in which waves and particles are demonstrated simultaneously. Really, however, it is not so, for to speak of a particle means nothing unless at least two points of its path can be specified experimentally, and similarly with a wave, unless at least two maxima (interference) are observed. If then we propose to carry out the **"demonstration of a corpuscle"**, we must settle the question whether its path has gone through the upper or the lower of the two slits to the receiver. We therefore repeat the experiment, not only setting up a photoelectrically sensitive instrument as receiver, but also providing some contrivance which shows whether the light has passed through the upper slit (say a thin photographic film or the like). This contrivance in the slit, however, necessarily throws the light quantum out of its undisturbed path; the probability of getting it in the receiver (the screen) is therefore not the same as it was originally, i.e., the preliminary calculation by wave theory of the interference phenomenon is illusory. Thus if pure interference is to be observed, we are necessarily precluded from making an observation of any point of the path of the light quantum before it strikes the screen. Summarizing we can say **that in no single experiment does a photon show both wave and particle properties, nor does a particle simultaneously show both particle and wave properties**. In other words, an attempt to determine the particle properties exactly eliminates any possibility of simultaneously observing any aspect of wave-like properties and vice versa. This suggests that the wave and particle pictures give complementary descriptions of the same system.

The question arises, what is the ultimate connection between the wave behaviour and the particle behaviour. We know that the intensity of the wave is proportional to the square of the amplitude, we then have

$$(\text{Probability to observe photons})^2 \propto |\text{electric field amplitude}|^2$$

This expression provides the ultimate connection between the wave behaviour and the particle behaviour, we can also see that a similar expression connects the wave and the particle aspects of those objects, e.g., electrons, etc.

6.12 THE UNCERTAINTY PRINCIPLE

The description of material particles in terms of waves, and the setting up of a connection between energy and momentum, on one side, and frequency and wavelength, on the other, lead to an indefiniteness that had not been previously suspected in the mechanics of particles. This indefiniteness has become famous as the **'uncertainty principle'** or **'indeterminacy principle'** put forward by Heisenberg in 1927. This principle can be shown to follow from the experiments already discussed, or it can be shown to follow from the wave-packet concept that in turn follows from these experiments. We shall first state this principle and then show how it is related to experiments and wave packets.

The uncertainty principle is concerned with the simultaneous measurement of certain pairs of variables. It is possible to state it very generally, but such a statement would necessarily be rather abstract. Instead, we shall state two aspects of the principle that involve by far the most important pairs of variables and that illuminate the physical meaning of the principle.

The first aspect involves the simultaneous measurement of the momentum and position of a particle (e.g., an electron or a photon). The principle states that the experiment cannot fix these to an unlimited precision, but that the momentum P_x is determinable only to within a range Δp_x and the position (x) within a range Δx, where

$$\Delta px \, \Delta x \geq \hbar \qquad (6.134)$$

(An expression of this type also holds for other components of linear momentum and for angular momentum and angular position). This result was first enunciated by Heisenberg in 1927. He called it the principle of **Unbeslimmlheil**. This term has been variously translated as **indefiniteness, indeterminacy** or **uncertainty**. We could devise an experiment that would give much poorer determinations of p_x and x than $\Delta p_x \Delta x = \hbar$. The uncertainty principle states that we cannot do better than this. It is important to mention here that there is no restriction on Δx or on Δp_x, but only on their product. Therefore, we could, for example, devise an experiment to measure x very accurately, but only at the sacrifice of the knowledge of p_x. Note also that there is no restriction on product like $\Delta p_x \Delta y$.

The second aspect of the uncertainty principle is a similar limitation on the simultaneous measurement of the energy E and time t. E might be the energy of a photon and t the time it was emitted. The principle states that

$$\Delta E\, \Delta t \geq \hbar \tag{6.135}$$

Here again there are no restrictions on the accuracy with which E or t can be measured, but only on the product $\Delta E\, \Delta t$.

We have
$$\Delta E = \Delta\left(\frac{p^2}{2m}\right) = \frac{2p\Delta p}{2m} = \frac{p\Delta p}{m}$$

$$\Delta E\, \Delta t = \Delta p \frac{p}{m} \Delta t = \Delta p \frac{mv}{m} \Delta t = \Delta p.\Delta x$$

By Heisenberg's principle

$$\Delta E\, \Delta t \geq \hbar$$

This is known as **alternate form of Heisenberg's principle.**

The uncertainty principle is quite foreign to classical mechanics, which recognises no fundamental limitations on measurements of any kind. Note that in common with all quantum theories and phenomena, Planck's constant (h) appears here.

Uncertainty relation for angular momentum and angle variables. The Heisenberg's uncertainty principle discussed in the previous section applies to any pair of variables which are canonically conjugate in the Hamiltonian formulation of mechanics. The uncertainty relationship for angle variables often expressed in the form

$$\Delta\theta\, \Delta A_\theta \geq \frac{\hbar}{2} \tag{6.136}$$

where ΔA_θ is the uncertainty associated with angular momentum A_θ and $\Delta\theta$ is the indefiniteness in the corresponding angle. However, θ has a finite range of extent 2π and as such $\Delta\theta$ must be finite.

6.13 UNCERTAINTY AND MEASUREMENTS

The limits on the possibility of simultaneous determination of position and momentum implied by the Heisenberg's uncertainty principle are merely logical derivations from our assumptions about matter waves and their probability interpretation. Before we can be sure of the validity of these limits, it is necessary to construct a quantum theory of the processes of measurement, and to show that this theory leads to the same result. In other words, we must see in detail what happens in actual measuring processes that prevent us from making measurements that would permit us to determine momentum and position or energy and time simultaneously with unlimited accuracy.

In any measurement, it is necessary to have some system that we may regard as the measuring apparatus and from whose state we can draw inferences about the systems we are observing. In order that this be possible, it is necessary that the measuring apparatus interacts with what is observed in a known and calculable fashion. Suppose that we wish to measure the position and momentum of something at a certain moment. To accomplish this, we must provide it with something else that is to carry the desired information back to us; that is, we have to touch it with our finger, illuminate with light, or interact with it in some other way. To make it more clear now we discuss the few experiments.

6.13.1 Localization Experiment-γ-ray Microscope

Let us consider an experiment where by a particle (say electron) might be observed in a microscope. Light of frequency v, falls upon the particle and is scattered. If it is to be observed at all the scattered photon must enter the objective of the microscope as shown in (Fig. 6.29). There is, therefore, an uncertainty in its momentum given by $\Delta p = p \sin \alpha$ $\left(\text{from (Fig. 6.24) } \tan \alpha = \frac{\Delta p}{p}\right)$ where α is small, we can write $\tan \alpha = \sin \alpha = \frac{\Delta p}{p}$ since it can go anywhere within a cone of semi-vertical angle α. The particle at p, recoils with an equal and opposite amount of momentum in accordance with the Compton effect.

The position of the particle is defined by the resolving power of the microscope such that

$$\Delta x = \frac{\lambda}{\sin \alpha} \tag{6.137}$$

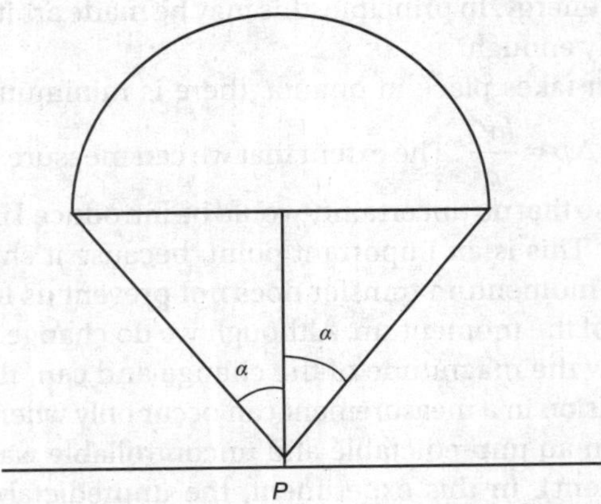

Fig. 6.29 Measurement of position and momentum of a particle in a hypothetical experiment by means of a γ-ray microscope

Again, we have $\sin\alpha = \frac{\lambda}{\Delta x} = \frac{\Delta p}{p}$ and therefore $\Delta p.\Delta x = \lambda p = \hbar$.

To improve the resolving power we can increase the aperture α of the microscope, but in so doing Δp is also increased. Decreasing the wavelength λ would also increase the resolving power, but a correspondingly large amount of momentum would be transferred to the particle from the photon due to a larger Compton effect. Similar relationships exist for the momenta along each of the cartesian axes, so that we have

$$\Delta p_x \Delta x \geq \hbar,\ \Delta p_y \Delta y \geq \hbar,\ \Delta p_z \Delta z \geq \hbar$$

6.13.2 Momentum Determination Experiment

Let us consider the method of measuring momentum by measuring the velocity with the aid of the Doppler shift of the light radiated by the particle. This is actually done in measuring the velocity of radiating atoms[3]. The

$$\nu' \equiv \nu\left(1 - \frac{v}{c}\right)$$

or

$$\frac{v}{c} = \frac{\nu - \nu'}{\nu} = \frac{\Delta\nu}{\nu}$$

where ν = frequency radiated by the atom when at rest, ν' = frequency radiated by the atom when in motion. Note that this is a non-relativistic approximation.

Now, the position at $t = 0$ can, in principle, be fixed arbitrarily with high accuracy. This could be done, for example, by having the particle come through a very fine slit at this time of course, we are trying to measure. The uncertainty in the velocity depends on the accuracy with which we can measure ν', and it is well known that the uncertainty in ν' is $\Delta\nu' = \frac{1}{\tau}$, where τ is the time duration of the wave train of radiated light. We, therefore, desire a long wave train. Now, the length of the wave train is determined by the time the atom takes to radiate its energy. In principle, this may be made arbitrarily long by choosing atoms that radiate slowly enough.

Because the radiation takes place in quanta, there is minimum possible momentum transfer to the electron, $\Delta p = \frac{h\nu'}{c}$. The extent that we can measure ν', we can calculate the transfer of momentum, so that no uncertainty would be introduced if we made a completely accurate measurement. This is an important point, because it shows that the existence of a minimum possible momentum transfer does not prevent us from making arbitrarily precise measurements of the momentum. Although we do change the momentum in this measurement. We know the magnitude of the change and can, therefore, correct for it. An inherent lack of precision in a measurement can occur only when some crucial property of the system changes in an unpredictable and uncontrollable way during the course of experiment (measurement). In this experiment, the unpredictable and uncontrollable

[3]The velocity of a star, for example, is frequently obtained by measuring the red shift.

quantity is the time of quantum. All we know is that the emission of the quantum was emitted at sometime between 0 and τ. We are not allowed to measure this time more precisely, for its measurement would reduce the accuracy with which we can measure the frequency of quantum.

But when the quantum is emitted, there will be an abrupt change in the electronic velocity given by $\Delta v = \dfrac{\Delta p}{m} = \dfrac{h\nu'}{mc}$. Hence, there will be an interval of time, anywhere, between 0 and τ during which it may have been travelling at a velocity different from the one that we measure. This will result in an uncertainty in the distance the particle covers, equal to

$$\Delta x \cong \tau \Delta v \cong \frac{h\nu'\tau}{mc}$$

Thus, we get
$$\Delta x \Delta p = m \Delta x \Delta v = \frac{h\nu'\tau\Delta\nu}{c}$$

or
$$\Delta \nu \cong \frac{c\Delta\nu}{\nu} = \frac{c}{\nu\tau}$$

For the non-relativistic case,
$$\frac{\nu'}{\nu} \cong 1$$

Thus we obtain
$$\Delta x \Delta p \cong \hbar$$

Here a rather interesting effect occurred as a result of the incompletely predictable and controllable time of emission of the quantum. Although the position before the measurement (at $t = 0$) was fairly well-defined, the deterministic relation between this position and the position reached after the transfer of quantum is destroyed while the measurement takes place. Thus, although the momentum is made more definite in the course of the measurement, the position is made less definite.

6.13.3 The Double Slit Experiment

To crystallize the ideas we have been discussing about the wave-particle duality and to emphasize the uncertainties introduced in the process of measurement, we take the double-slit diffraction experiment as one more example. A schematic diagram of the experiment is shown in (Fig. 6.30). The source emits a beam, which is incident on two slits separated by a distance D. The beam could be electromagnetic, that is, a stream of photons or a stream of electrons, that is a beam of **cathode rays. The following analysis applies to either**, but for convenience we shall consider the stream to be photons. At a distance from the slits, which is large, compared to D, we have a detector.

The detector will show the characteristic interference pattern indicated by the curve labelled intensity on the figure. The positions of the maxima and minima are found by considering the superposition of the wave from one slit with the other; that is the interference pattern is a prediction of wave model. The angular separation of the central maxima and the adjacent minima can be determined from $D \sin\theta = \lambda/2 = h/2p_x$.

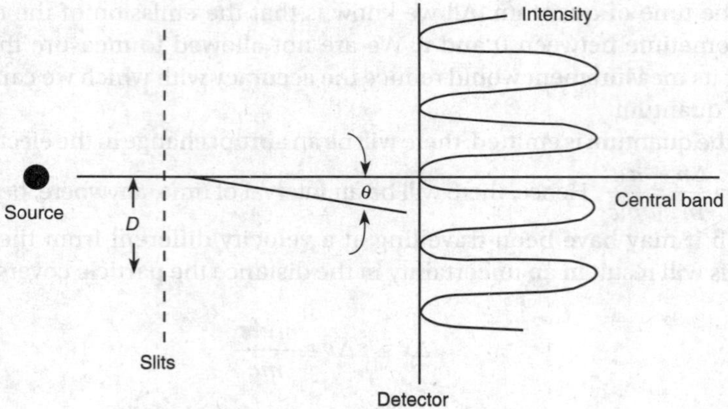

Fig. 6.30 Schematic diagram of the double slit experiment

When θ is small, it become $\theta = \dfrac{h}{2p_x D}$, where θ is the angle between any adjacent maxima and minima.

If one of the slits is covered during the experiment, the resulting single slit pattern will differ from the preceding one. If a cover is placed over the bottom slit for the first half of the experiment and then the cover is placed over the top slit for the second half, the total exposure at the detector is different than when both slits are uncovered. The positions on the screen, when the photon will most likely strike, change. If we consider the light beam to be a stream of photons, a single photon must pass through either one slit or the other. The fact that the pattern is different for two slits than one slit means that the motion of the photon is influenced by the other slit. It appears that the photon, as it passes through one slit, 'knows' whether the other slit is open. Thus the particle model seems to lead to an absurd result.

An analysis of the double slit diffraction experiment shows that no information was obtained experimentally about the passage of a photon through either slit. Consequently, a statement about such passage is meaningless. This emphasizes that physics deals with predictions for experimental results. The definition of every physical quantity must contain, explicity or implicitly, the operation necessary to measure the physical quantity. It makes no sense to talk about results which cannot be observed. If we want to know through which slit the photon passes, we must revise the experiment so that we can determine this. To detect the passage of a photon we place many small particles first to the right of the slits. After passing through a slit, the photon will collide with one of these particles. Observation of the recoil of the small particles will collide with one of these particles. Observation of the recoil of the small particles will allow us to determine the slit used by the photon. In this ideal experiment the only uncertainties introduced will be those predicted by the uncertainty principle; we will assume that the precision achieved in constructing the apparatus was so high that there are no uncertainties in dimensions, such as the separation of the slits.

To be certain of the slit from which the small particle recoiled, the uncertainty in its y-position must be much smaller than the separation of the slits; we must have

$$\Delta y << D$$

There will be an exchange of momentum during the collision, but there is an uncertainty in the amount since we don't know the details of the collision. The uncertainty in the y-component of the momentum of the photon, Δp_y must not be great enough to cause it to deviate from the interference pattern. This means we must have

$$\Delta p_y / p_x << \theta = \frac{\hbar}{2p_x D} \quad \text{or} \quad \Delta p_y << \frac{\hbar}{2D}$$

Since momentum is conserved, Δp_y is also the uncertainty in the y-component of the momentum of the small particle. If we multiply the uncertainty in the y-position of small particle with the uncertainty in its y-component of momentum, we get

$$\Delta y \, \Delta p_y << \frac{D\hbar}{2D} = \frac{\hbar}{2} \tag{6.138}$$

If we compare this with the prediction of the uncertainty principle

$$\Delta y \Delta p_y \simeq h,$$

we see that the small uncertainties necessary in Eq. (6.138) are not possible. If Δy is small enough that we can determine through which slit the photon passes, Δp_y is so large that we do not have the interference pattern at the detector. **We cannot detect both the particle and wave nature,** as indicated by the collision with the small particles, and the wave nature, as indicated by the interference pattern. This example also illustrates how the uncertainty principle is used to resolve apparent conflicts between the wave model and particle model.

One of the immediate practical applications of uncertainty principle is to predict **a 'natural line width'** for any spectral line, which will exist simply because of the limited time the atom spends in its excited state.

6.14 ELEMENTARY PROOF OF UNCERTAINTY PRINCIPLE

(i) $\Delta x \Delta p_x \geq \hbar$

Consider a particle moving along the x-axis. The de-Broglie relation between the wavelength λ of the associated wave and momentum p_x of the particle along x-axis is given by

$$\lambda = \frac{h}{p_x} \qquad \therefore p_x = \frac{h}{\lambda} = \frac{h}{2\pi}\frac{2\pi}{\lambda} = \hbar k \tag{6.139}$$

The particle in motion is represented by a wave-packet formed by the superposition of a number of plane waves of wavelengths different from λ. In the simplest way one can consider the wave-packet as the superposition of two simple harmonic plane waves of propagation constant k and $k + \Delta k$. The length of such a wave-packet is given by

$$\Delta x = \frac{2\pi}{\Delta k} \tag{6.140}$$

Since the particle must be somewhere within the wave-packet, Δx is the uncertainty in the position. The uncertainty in the propagation constant of the associated wave is Δk. Obviously the uncertainty Δp_x, in the momentum will be from Eq. (6.139)

$$\Delta p_x = \hbar \Delta k \tag{6.141}$$

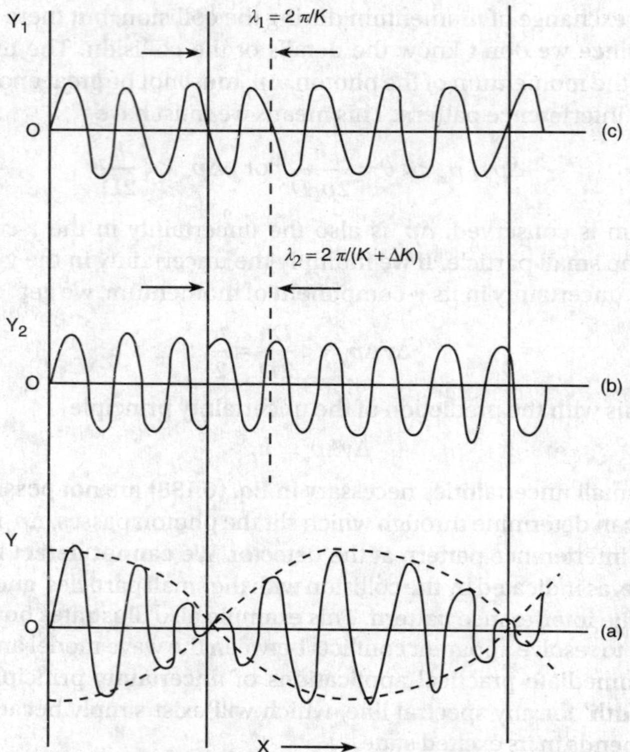

Fig. 6.31 Wave packet

Multiplying Eqs. (6.140) and (6.141), one obtains

$$\Delta x \Delta p_x = \left(\frac{2\pi}{\Delta k}\right)\hbar \Delta k = 2\pi\hbar$$

If the wave packets have shapes different from that shown in Fig. [6.31 (a)] then the sign of equality can be replaced by sign ≥.

$$\therefore \qquad \Delta x \Delta p_x \geq \hbar \qquad (6.142)$$

In obtaining the uncertainty relation between Δx and Δp_x we have considered the wave packet formed by the superposition of two plane waves. But superposition of two plane waves gives rise to a series of wave packets instead of a single wave packet. A single wave packet is formed by superposition of an infinite number of plane waves of propagation constant slightly different from each other. Using Fourier analysis of a single wave-packet in one dimension, one can easily show that the width Δx of the wave packet and the range Δk of the propagation are given by

$$\Delta x \Delta k \approx 1 \qquad \therefore \Delta x \approx \frac{\pi}{\Delta k} \qquad (6.143)$$

Multiplying Eqs. (6.141) and (6.143) together, one obtains

$$\Delta x \, \Delta p_x = \hbar$$

Since this is the minimum limit of the product of the uncertainties, one obtains

$$\Delta x \Delta p_x \geq \hbar \tag{6.144}$$

Significance: One can draw following conclusions from Eq. (6.144):

(a) If the position coordinate x of a particle in motion is accurately determined at some instant, so that $\Delta x = 0$, then at the same instant the uncertainty Δp_x in the determination of the momentum of the particle becomes infinite.

(b) If the momentum p_x of a particle is accurately determined at some instant, so that $\Delta p_x = 0$, then at the same instant the uncertainty Δx in the determination of the position coordinate becomes infinite. Obviously, if an experiment is designed to measure x or p_x accurately, the other quantity will become completely uncertain. One can measure both quantities by means of an experiment, but only within certain limits of accuracy specified by the uncertainty relation.

(c) For a particle of mass m moving with velocity v, the product of uncertainty Δx and Δv is given by

$$\Delta x \Delta v \geq \frac{\hbar}{m}$$

For a heavy particle $\hbar/m$ is very small, and, therefore, the product $\Delta x \ \Delta v$ of the two uncertainties becomes very small. For such particles both x and v can be determined accurately.

For very heavy particles, if m is such that $\hbar/m \approx 0$, the uncertainties vanish and one can measure x and v with perfect accuracy. Obviously, this is the limiting case of classical mechanics. This means, **classical mechanics is true for heavy bodies and uncertainties are a characteristic of atomic and subnuclear particles.**

(ii) $\Delta E \Delta t \geq \hbar$

The uncertainty relation between energy and time is obtained by considering the motion of a wave-packet. A particle in motion is represented by a wave-packet. The group velocity of the wave-packet is equal to the particle velocity. Let Δx be the width of the wave-packet moving along the x-axis. Let v_g be the group velocity of the wave-packet and v_x the particle velocity along x-axis. Suppose the wave packet moves through Δx in time Δt.

Since Δx is the uncertainty in the x-coordinates of the particles and Δt is the uncertainty in time t and it is given by

$$\Delta t = \frac{\Delta x}{v_g} \tag{6.145}$$

The energy E of a particle is a function of its momentum p_x, we have

$$\Delta E = \frac{\partial E}{\partial p_x} \Delta p_x$$

But
$$\frac{\partial E}{\partial p_x} = v_x = v_g$$

$$\Delta E = v_g \Delta p_x \tag{6.146}$$

If Δp_x, is the uncertainty in the momentum p_x at time t, then Eq. (6.146) gives the uncertainty ΔE in the energy of the particle at the same time. Multiplying Eqs. (6.147) and (6.146), one obtains

$$\Delta E \, \Delta t = \Delta x \, \Delta p_x \tag{6.147}$$

According to the uncertainty relation between position and momentum, we have

$$\Delta x \, \Delta p_x \geq \hbar \tag{6.148}$$

From Eq. (6.147), we have $\Delta E \, \Delta t \geq \hbar$

which is **energy time uncertainty relation.**

Significance

The physical significance of the energy-time uncertainty relation is quite different from that of the position momentum uncertainly relation. If ΔE is the maximum uncertainty in the determination of the energy of a system in a particular state, then according to Eq. (6.143) the minimum time interval for which the system remains in that state is given by and, if a system remains in a

$$\Delta t = \frac{\hbar}{\Delta E}$$

particular state for a maximum time interval Δt, then the minimum uncertainty in the energy of the system in that states is given by

$$\Delta E = \frac{\hbar}{\Delta t}$$

6.15 APPLICATIONS OF UNCERTAINTY PRINCIPLE

According to uncertainty principle it is not possible to measure simultaneously the position and momentum of the particle with the desired accuracy. On the other hand, according to classical mechanics, it is possible to know the position and momentum of any particle with desired accuracy provided its initial position and momentum are known. Further, quantum mechanics introduces the laws of probability and therefore probability takes the place of certainty in classical mechanics. Further, the quantum mechanics introduces a new concept of particle according to which the term particle is something composed of a very deep and essential unison of both the corpuscle and the wave resulting in a certain unavoidable uncertainty. The principle of uncertainty explains a large number of facts which could not be explained by classical mechanics. A few of them are discussed below:

6.15.1 Non-existence of Free Electrons in the Nucleus

We know that (i) the maximum possible kinetic energy of an electron emitted by a radioactive nuclei is about 4 MeV, (ii) the rest mass of an electron is $m_e = 9.11 \times 10^{-31}$ kg and (iii) the diameter of the nucleus is $\approx 2 \times 10^{-14}$ m.

If an electron exists inside the nucleus, it can be anywhere within the diameter of the nucleus. Obviously, the maximum uncertainty Δx in the position of the electron is the same as the diameter of the nucleus, i.e.,

$$\Delta x = 2\times 10^{-14} \tag{6.149}$$

According to Heisenberg uncertainty relation, the product of the uncertainty Δx in the position of the electron and the uncertainty Δp_x in the x-component of its momentum is given by $\Delta x.\Delta p_x \geq \hbar$. Therefore, the uncertainty in the momentum Δp_x is

$$\Delta p_x \geq \frac{\hbar}{\Delta x}$$

$\therefore$ Minimum uncertainty in the momentum is given by

$$\Delta p_x = \frac{\hbar}{\Delta x} = \frac{6.63\times 10^{-34}}{2\pi\times 2\times 10^{-14}} = \frac{6.63\times 10^{-20}}{4\pi} = 5.28\times 10^{-21} \text{ kg-m/s} \tag{6.150}$$

This means that if the electron exists inside the nucleus, then its minimum momentum should be $P_{min} = 5.28 \times 10^{-21}$ kg-m/s.

According to theory of relativity, the total relativistic energy E of the particle is given by

$$E^2 = p^2c^2 + m_0^2c^4 \tag{6.151}$$

For the electron having minimum momentum, the minimum energy is given by

$$\begin{aligned} E_{min}^2 &= p_{min}^2c^2 + m_0^2c^4 \\ &= (5.28\times 10^{-21}\times 3\times 10^8)^2 + (9.11\times 10^{-31})^2\times (3\times 10^8)^4 \qquad (6.152) \\ &= 2.5\times 10^{-24} + 6.27\times 10^{-27} \text{ J} \end{aligned}$$

Obviously, the second term is much smaller than the first and hence it can be neglected.

Then $$E_{min} = \sqrt{2.5\times 10^{-24}} = 1.58\times 10^{-12} \text{ J} = \frac{1.58\times 10^{-12}}{1.6\times 10^{-19}} \text{eV}$$

$$= 0.9875\times 10^7 \text{ eV} \tag{6.153}$$

$$= 9.875\times 10^6 \text{ eV} = 9.875 \text{ MeV} \tag{6.154}$$

Obviously, if free electron exists in the nucleus, it must have a minimum energy of about 9 MeV.

As stated in the beginning that the maximum kinetic energy which a β-particle (an electron), emitted from a radioactive nuclei can have is 4 MeV. This clearly indicates that free electrons cannot be present inside the nuclei.

Strength of Nuclear Force: Let us consider that a nucleon is confined within the sphere of radius $r_0(= 1.2 \times 10^{-14}$ m). Then the momentum must be of the order of $p = \hbar/r_0$. Obviously, kinetic energy of the nucleon (proton or neutron) of mass M will be of the order

$$\text{K.E.}\ (T) \approx p^2/2M \approx \frac{1}{2M}(\hbar^2/r_0^2) \approx 10 \text{ MeV}$$

Now if the nucleon is bound in the nucleus, then the binding energy must be greater than kinetic energy and opposite in sign. We must remember that the nuclear binding energy is of the order of 10 MeV.

6.15.2 Ground State Energy of H-atom and Radius of the Bohr's First Orbit

Suppose the electron in the hydrogen atom moves around the nucleus, i.e., proton in a circular orbit of radius r. Then the maximum uncertainty in the determination of its position w.r.t. the proton can be taken equal to r, i.e., $\Delta r = r$.

From the uncertainty relation, the minimum uncertainty Δp in the simultaneous determination of its momentum p in the plane of the orbit is given by $p = \dfrac{\hbar}{\Delta r} = \dfrac{\hbar}{r}$.

Since the momentum p cannot be less than the uncertainty Δp, and hence the minimum possible momentum is

$$p = \frac{\hbar}{r}$$

The kinetic energy K of the electron is given by

$$K = \frac{p^2}{2m} = \frac{\hbar^2}{2mr^2} \tag{6.155}$$

And the electrostatic potential energy is given by

$$U = \frac{e^2}{4\pi\varepsilon_0 r} \tag{6.156}$$

Therefore, the total energy of the electron in the hydrogen atom is

$$E = \frac{\hbar^2}{2mr^2} - \frac{e^2}{4\pi\varepsilon_0 r} \tag{6.157}$$

The ground state energy must be a minimum value of E. Suppose r_1 is the distance of the electron from the proton in the ground state. For $E = E_{\text{min}}$, we have

$$\left(\frac{dE}{dr}\right)_r = r_1 = 0$$

$\therefore$

$$-\frac{\hbar^2}{mr_1^3} + \frac{e^2}{4\pi\varepsilon_0 r_1^2} = 0$$

or

$$\frac{e^2}{4\pi\varepsilon_0} = \frac{\hbar}{mr_1}$$

$$\therefore \qquad r_1 = \frac{4\pi\varepsilon_0\hbar^2}{me^2} = \frac{4\pi\varepsilon_0 h^2}{me^2\times 4\pi^2} = \frac{\varepsilon_0 h^2}{\pi me^2} \tag{6.158}$$

Eq. (6.158) is the well known expression for the radii of the Bohr's first orbit. Using the value: $\hbar = 6.63\times10^{-31}$ J.s, $m = 9.11\times10^{-31}$ kg, $e = 1.6\times10^{-19}$ C, $\varepsilon_0 = 8.85\times10^{-12}$ C²/N-m², one obtains: $r_1 = 5.31\times10^{-11}$ m. This is the radius of the hydrogen atom in the ground state. The expression for the energy E_1 in the ground state is obtained by substituting

$$r = r_1 = \frac{4\pi\varepsilon_0\hbar^2}{me^2}$$

From Eq. (6.157), we have $E_1 = \dfrac{\hbar^2}{2mr_1^2} - \dfrac{e^2}{4\pi\varepsilon_0 r_1}$

$$= \frac{\hbar^2}{2m}\left(\frac{-me^2}{4\pi\varepsilon_0\hbar^2}\right)^2 - \frac{e^2}{4\pi\varepsilon_0}\left(\frac{me^2}{4\pi\varepsilon_0\hbar^2}\right)$$

$$= \frac{me^4}{32\pi^2\varepsilon_0^2\hbar^2} - \frac{me^4}{16\pi^2\varepsilon_0^2\hbar^2} = -\frac{me^4}{32\pi^2\varepsilon_0^2\hbar^2}$$

$$= \frac{me^4\times4\pi^2}{32\pi^2\varepsilon_0^2\hbar^2} = -\frac{me^4}{8\varepsilon_0^2\hbar^2} \tag{6.159}$$

Substituting the numerical values of various constants, appearing in Eq. (6.159) and using the fact that 1 eV = 1.6×10^{-19} J, one obtains, $E_1 = -13.6$ eV. This is the ground state energy of the hydrogen atom.

6.15.3 Ground State Energy of Harmonic Oscillator

The Hamiltonian of a linear harmonic oscillator is given by

$$\hat{H} = \hat{H}(x,p) = \frac{\hat{p}^2}{2m} + \frac{1}{2}k\hat{x}^2 \tag{6.160}$$

where $\hat{x}$ and $\hat{p}$ are Hermitian operators corresponding to x and p respectively.

Let ψ represent eigen vector[4] of $\hat{H}$ belonging to eigen value E. Also, let $\hat{A}$ and $\hat{B}$ be, respectively, the operators corresponding to position and momentum, one can write

$$\Delta x = \{<\overline{x}^2>\}^{1/2} \text{ and } \Delta p = \{<p^2>\}^{1/2} \tag{6.161}$$

where angular brackets < > represent the average. Since both x^2 and p^2 are positive for the oscillator, we have

$$\Delta x > 0;\ \Delta p > 0 \text{ so that } \Delta x\Delta p > 0$$

According to Heisenberg uncertainty principle

$$\Delta x\,\Delta p \ge \hbar/2 \tag{6.162}$$

[4]**Note:** About eigen vectors and eigen values etc, we will study later.

Now, $E = <\psi|\hat{H}|\psi> = <\hat{H}>$

$$= \frac{1}{2m} <\hat{p}^2> + \frac{1}{2} k <x^2> = \frac{1}{2m} (\Delta p)^2 + \frac{k}{2} (\Delta x)^2 = \frac{1}{2m} (\Delta p)^2 + \frac{\hbar^2 k}{8} \frac{1}{(\Delta p)^2} \quad (6.163)$$

The minimum value of E is given by the minimum of the expression,

$$\frac{1}{2m} (\Delta p)^2 + \frac{\hbar^2 k}{8} \frac{1}{(\Delta p)^2}$$

i.e.,

$$E_{\min} = E_0 = \left[\frac{(\Delta p)^2}{2m} + \frac{\hbar^2 k}{8(\Delta p)^2} \right]_{\min} = \frac{(\Delta p_0)^2}{2m} + \frac{\hbar^2 k}{8(\Delta p_0)^2} \quad (6.164)$$

where

$$(\Delta p_0)^2 = \frac{1}{2} \sqrt{\hbar^2 km} = \frac{m}{2} \hbar \omega_0 \quad (6.165)$$

ω_0 being the classifical frequency of the oscillator, given by $\omega_0 = \sqrt{\frac{k}{m}}$ (6.166)

Thus

$$E_0 = \frac{1}{2} \hbar \omega_0 \quad (6.167)$$

The above energy, i.e., $E_0 = \frac{1}{2} \hbar \omega_0$ is called the **lowest energy** or **'Zero-point energy'**.

6.15.4 Natural Width of Spectral Line

According to Heisenberg's uncertainty principle of energy and time

$$\Delta E \Delta t \geq \hbar$$

Since the lifetime of electron in an excited orbit is finite ($\approx 10^{-8}$sec), therefore the energy levels of the atom given by $\left(\Delta E = \frac{\hbar}{\Delta t} \right)$ must have a finite width. This means that the excited energy levels have a finite energy spread thereby indicating that the radiation emitted when an electron jumps truly monochromatic. In other words the spectral lines can never be infinitely sharp but must have a natural finite width.

6.15.5 Stability of an Atom

We have read that the ground state energy is the lowest possible energy with which the atom can exist. This energy is the sum of kinetic and potential energies. We must remember that the sign of kinetic and potential energies are opposite. If the electron is confined in a small region, the kinetic energy becomes large because the wave will be carrying large momentum. In other words, if the wave is spread in a large region, the potential energy will be small. This means that the ground state refers to the best possible compromise and this provides the explanation for the stability of the atom against collapse.

The uncertainty in position measurement is of the order of r_0 and this is true for all directions. This reveals that there occurs no significance of circular orbit of radius r_0.

6.15.6 Light Quanta

Let us consider that an electromagnetic wave packet is made by opening a shutter for a time interval Δt. Obviously, one obtains a pulse of radiation which is passing any specific point in time Δt. Further the electric field will be large during time interval Δt.

Let us suppose that the pulse of radiation has one quantum only. Let us consider that this pulse is incident on a target having several atoms, but only one of the atoms of the target will absorb the quantum. Obviously, the probability of the absorption will be proportional to $|E|^2$, where E denotes the strength of the electric field. This means that the momentum will be absorbed within time interval Δt, because during this time, there is maximum electric field in the region having the target. This means that Δt may be considered as uncertainty in the time during which the absorption of quantum takes place. However, the pulse of radiation is having a range of angular frequencies given as follows: $\Delta\omega \geq 1/\Delta t$

$\therefore$ The range of energies is obtained as

$$\therefore \qquad \Delta E = h\Delta\nu = \frac{h\Delta(2\pi\nu)}{2\pi} = \hbar\Delta\omega \geq \hbar/\Delta t$$

One obtains $\Delta E \Delta t \geq \hbar$. Obviously, in any process in which there occurs a transfer of quantum from radiation to the matter (or **vice versa**), the product of uncertanties in the time of transfer and in the quantity of energy transferred is never less than $\hbar$, i.e., $\Delta E \Delta t \approx \hbar$.

We must note that in the above discussion, we have carefully avoided the quantum nature of light concept.

6.16 SOME REMARKS ABOUT UNCERTAINTY PRINCIPLE

The discussion in the preceding sections proves that according to the uncertainty principle, we can never obtain precise knowledge of a physical system, regardless of improvements in experimental techniques. We can only obtain probable values, and this is a characteristic of nature.

This principle provided a tremendous support for the methods of wave mechanics. It shows that there was an uncertainty inherent in nature, and this was precisely what wave mechanics predicted. The theory has been immensely successful in providing correct solution to many scientific problems, and is one of the corner-stones of contemporary physics.

A few scientists, however, are not completely satisfied. Some, like David Bohm, consider that we may yet discover "hidden variables" at subatomic levels, the result of whose behaviour is the uncertainty law. They use the analogy of water molecules: these impart to bulk water and ice such characteristics as capillarity and crystal structure which the molecules do not themselves possess. In the same way, the "hidden variables" may not themselves follow the uncertainty principle, and the behaviour of particles would once again be completely determined. At the moment there is little to favour this view.

Summarizing we can say that the Heisenberg's uncertainty principle is by no means a statement indicating fundamental limitations to our knowledge of the microcosm. It only reflects the limited applicability of the concepts of classical physics to the region of microcosm.

6.17 BOHR'S CORRESPONDENCE PRINCIPLE

Now, we study a principle which was a great heuristic aid in the early development of quantum mechanics before the logical foundations were well established. This principle is called as Bohr's correspondence principle and it continued to be a useful test for some theories. The Bohr's correspondence principle states that the predictions of quantum mechanics and classical physics agree in the limit of large sizes and energies, as the quantum number n approaches infinity. The quantum mechanical solutions of problems approach classical mechanical solutions as the legnth involved becomes much larger than the de-Broglie wavelength of the particle involved. This principle merely states a result that could be proved for each separate problem directly from the Schrodinger equation; it is not an additional assumption of quantum mechanics. The correspondence principle implies that quantum mechanics may be successfully extrapolated to the laboratory scale of sizes. The converse (extrapolation of classical mechanics to atomic sizes) is certainly not successful. It can be seen that this requirement is satisfied by the Planck's quantum theory of radiation and the wave theory of matter, and also by Bohr's theory of atom.

The correspondence principle is of most interest in the problem of the absorption or radiation of energy. Let us consider an electrified particle, vibrating sinusoidally according to classical mechanics, with a frequency ν, will emit radiation of this frequency ν, gradually losing energy until it comes to rest. In the quantum theory, however, we know that an oscillator has an infinite set of energy levels, given by

$$E_n = nh\nu$$

where n is the quantum number. If we start with a system in a highly excited state, with a large value for n, it will find lower levels whose energy is less by $h\nu$, $2h\nu$, $3h\nu$, etc. It could, then make transitions to the states whose energy was lower by each of these values. The emitted photons would then have energies $h\nu$, $2h\nu$, $3h\nu$, etc., so that the frequencies of the emitted lines would be ν, 2ν, 3ν etc.

This, however, would be contrary to the classical behaviour of the oscillator, in which only the frequency ν is emitted. Bohr assumed, therefore, that there must be a selection principle operating in this case, which would make the probability of transition equal to zero for any transition in which the quantum number n changes by more than one unit ($\Delta n = \pm 1$). This statement is called a selection rule, and the quantum-mechanical theory should produce this selection rule for the harmonic oscillator. That it does will be dmonstrated later on.

6.18 INADEQUACY OF QUANTUM THEORY

The quantum theory developed by Planck, Einstein and Bohr was e.g., black body radiation, photoelectric effect, spectral lines of hydrogen atom, harmonic oscillator, rigid rotator, etc. However, this theory was found inadequate to explain number of cases. Few of them are as follows:

(i) The general quantization rule is found to be valid only for periodic systems, i.e., it is not valid applicable to non-periodic cases.

(ii) Application of this quantum theory to hydrogen molecule, helium atom etc. led to results contrary to experimental results.

(iii) In the case of hydrogen spectrum, this theory could explain only broad features, i.e., failed to explain fine structure.

(iv) This theory failed to account for the concept of **spin**, i.e., failed to account the concept of half odd integer quantum numbers which are necessary for zeeman effect and fine structure of spectral lines.

(v) This theory was found inadequate for evaluating transition probabilities and intensities of spectral lines.

(vi) This theory was unsuccessful in explaining the dispersion of light.

These shortcomings clearly indicate that the quantum theory developed by Planck, Bohr, Einstein, etc. is not very general quantum theory and short comings, defects, etc. of this theory cannot be removed by extensions and simple modifications. Obviously, one has to look for a more general and basic theory, which may account for all basic thenomena.

Most physicists, but not all, view quantum mechanics a fundamental theory of nature. However, Albert Einstein, of the founders of quantum mechanics, never accepted it as fundamental; he always considered the probabilistic nature of quantum phenomena as the outcome of a deeper, as yet undiscovered physics. Einstein stated, "Quantum physics is certainly imposing. But an inner voice tells me it is not yet the real thing. The quantum theory says a lot, but does not really bring us closer to the secret of the 'old one'.

Example 6.1: The photoelectric threshold for a certain metal is 3600 Å. Determine the maximum energy of the ejected photoelectrons by a radiation of 2000 Å.

Solution Work function $\phi = h\nu_0 = \dfrac{hc}{\lambda_0}$

$$\phi = h\nu_0 = \frac{hc}{\lambda_0}$$

$$= \frac{6.63\times10^{-34}\times3\times10^{8}}{3600\times10^{-10}} = 0.5521\times10^{-18}\text{ J}$$

$$= \frac{0.5521\times10^{-18}}{1.6\times10^{-19}}\text{eV} = 3.45\text{ eV}$$

$h = 6.63\times10^{-34}$ Js

$c = 3\times10^{8}$ m/s.

$\lambda_0 = \text{Å} = 3600\times10^{-10}$ m

Now, for incident light of wavelength $\lambda'_0 = 2000\text{Å}$

$$\phi = \frac{hc}{\lambda'_0} = \frac{6.63\times10^{-34}\times3\times10^{8}}{2000\times10^{-10}\times1.6\times10^{-19}}\text{eV} = 6.21\text{eV}$$

From Einstein's photoelectric equation

$$h\nu = \varphi + \frac{1}{2}mv^2$$

$\therefore$ Kinetic energy of ejected electron: $\dfrac{1}{2}mv^2 = h\nu - \phi = 6.21 - 3.45 = 2.76\text{eV}$

Example 6.2: Using Planck's law

$$E_\lambda d\lambda = \frac{8\pi hc}{\lambda^3}\frac{1}{\left[\exp\dfrac{hc}{\lambda kT} - 1\right]}$$

show that (i) it reduces to Rayleigh-Jeans law in the limit as λ gets large and (ii) reduces to Wien's law in the short wavelength limit. Comment on the result.

Solution (i) For large wavelength region, we have

$$e^{hc/\lambda kT} = 1 + \frac{hc}{\lambda kT} + \cdots$$

Using this expansion, Planck's law reduces to

$$E_\lambda d\lambda = \frac{8\pi kT}{\lambda^4} d\lambda$$

This is Rayleigh-Jean's law with $A = 8\pi k$.

(ii) For short wavelength limit λT is small and hence $e^{ch/\lambda kT} = 1$, therefore, in the denominator of Planck law expression, 1 can be neglected in comparison to $e^{ch/\lambda kT}$, hence one obtains

$$E_\lambda d\lambda = \frac{8\pi ch}{\lambda^5} e^{-ch/\lambda kT} d\lambda$$

This is Wien's law with $A = 8\pi ch$ and $B = -ch/k$.

This problem clearly illustrates that Planck's theory of radiation incorporates all that is valid from the older theories as special cases. It thus serves as an excellent example of a conceptual advance, which opened new frontiers while still preserving much of the classical physics.

Example 6.3: Show that a photon cannot transfer all of its energy to a free electron. What is the maximum recoil kinetic energy that can be given to an electron by a photon of energy? E_0?

Solution We are all familiar with the 'pair production' phenomenon in which a positron-electron pair is produced by the annihilation of a gamma ray quantum. Since the electron and positron each require 0.51 MeV of rest mass energy, it is evident that the threshold photon energy for this process is slightly over I MeV. Pair production occurs only in the presence of a nucleus of some other particle-which can be given some of the initial momentum of the photon. This clearly illustrates that a photon cannot transfer all of its energy to a free electron.

The recoil energy of the electron

$$E = E_0 - h\nu'$$

where $E_0 = h\nu$ is the energy of the incident photon and $h\nu'$ is the energy of the scattered photon.

$$\therefore \qquad E = h\nu - h\nu' = h\nu\left(1 - \frac{\nu'}{\nu}\right) = h\nu\left(\frac{\lambda' - \lambda}{\lambda'}\right)$$

or

$$E = \frac{h\nu \dfrac{h}{\lambda m_0 c}(1-\cos\phi)}{1 + \dfrac{h}{\lambda m_0 c}(1-\cos\phi)}$$

or

$$\frac{E}{h\nu} = \frac{\alpha(1-\cos\phi)}{1+\phi(1-\cos\phi)}$$

where $\alpha = \dfrac{h}{\lambda m_0 c^2}$

Clearly, recoil energy of the electron depends only on the angle of scattering ϕ. As angle of scattering increase $\cos\phi$ decreases and when $\phi = \pi/2$, $\cos\phi = 0$.

Example 6.4: A photon of wavelength 3Å suffers Compton scattering by a free electron originally at rest. If the angle of scattering is 90°, what is the kinetic energy of recoil electron?

Solution

$$\Delta\lambda = \frac{\lambda}{m_0 c}(1-\cos\phi)$$

Here

$$\phi = 90°, \cos 90° = 0$$

$$\Delta\lambda = \frac{\lambda}{m_0 c} = 0.024\,\text{Å}$$

From the law of conservation of energy, we have

$$h\nu + m_0 c^2 = h\nu' + mc^2$$

or

$$h(\nu - \nu') + m_0 c^2 = mc^2$$

∴ Recoil energy of electron

$$\frac{1}{2}m_0 v^2 = h(\nu - \nu') = hc\left(\frac{1}{\lambda} - \frac{1}{\lambda'}\right) = hc\frac{\Delta\lambda}{\lambda(\lambda + \Delta\lambda)}$$

$$= \frac{6.6\times 10^{-34} \times 3\times 10^{8} \times 0.024\times 10^{-10}}{3\times 10^{-10} \times 3.024\times 10^{-10}} = 33.3\text{ eV}$$

Example 6.5: Calculate the maximum percentage change in wavelength due to Compton scattering for incident photon of wavelength 1Å and 10Å. What conclusion do you draw from these results?

Solution

$$\Delta\lambda = \frac{h}{m_0 c}(1-\cos\phi)$$

For $\Delta\lambda$ maximum, $\phi = 180°$

$$\therefore \quad (\Delta\lambda)_{max} = \frac{2h}{m_0 c} = \frac{2\times 6.6\times 10^{-34}}{9.1\times 10^{-31} \times 3\times 10^{8}} m = 0.048\,\text{Å}$$

Maximum change of percentage in wavelength

(i) $\lambda = 1\,\text{Å}$

$$\frac{(\Delta\lambda)_{max}}{\lambda}\times 100 = \frac{0.048}{10}\times 100 = 4.8\%$$

(ii) $\lambda = 10\,\text{Å}$

$$\frac{(\Delta\lambda)_{max}}{\lambda}\times 100 = \frac{0.048}{10}\times 100 = 0.48\%$$

Results indicate that $\dfrac{(\Delta\lambda)_{max}}{\lambda}$ for higher wavelength is small and hence its measurement is difficult.

Example 6.6: A particle of rest mass m_0 has a kinetic energy T. Show that its de-Broglie wavelength is given by

$$\lambda = \frac{hc}{\sqrt{T(T + m_0c^2)}}$$

Find the value of λ in the limiting case (a) $T << m_0c^2$ (b) $T >> m_0c^2$

Solution We have from the theory of relativity

$$E^2 = p^2c^2 + m^2c^4 \tag{i}$$

where 'p' is the momentum and m_0 is the rest mass of the particle. If T be the particle's kinetic energy then

$$E = T + m_0c^2 \tag{ii}$$

From (i) and (ii), one obtains

$$p^2c^2 = (T + m_0c^2)m_0^2c^4$$

Corresponding de-Broglie wavelength

$$\lambda = \frac{h}{p} = \frac{hc}{\sqrt{T(T + 2m_0c^2)}}$$

(i) $T << m_0c^2$, we have

$$\lambda = \frac{hc}{\sqrt{2Tm_0c^2)}}\left[1 + \frac{T}{2m_0c^2}\right]^{-1/2} = \frac{hc}{\sqrt{2m_0c^2T}}\left(1 - \frac{T}{4m_0c^2}\right)$$

Substituting the known values, we obtain

$$\lambda = \frac{12.27}{\sqrt{V}}\left(1 - \frac{\text{eV}}{4m_0c^2}\right)$$

(ii) $T >> m_0c^2$, we obtain

$$\lambda = \frac{ch}{T}\left(1 + \frac{2m_0c^2}{T}\right)^{-1/2} = \frac{ch}{T}\left(1 - \frac{m_0c^2}{T}\ldots\ldots\right)$$

Example 6.7: Prove that the de-Broglie wavelength for a material particle at rest mass m_0 and charge q, accelerated from rest through a potential difference of V volts relativistically is given by

$$\lambda = \frac{h}{\left[2m_0qv\left(1 + \frac{qV}{2m_0c^2}\right)\right]^{1/2}}$$

Solution From relativistic formulae have

$$E^2 = p^2c^2 + m_0^2c^4 \tag{i}$$

and

$$E = T + m_0c^2 = qV + m_0c^2 \tag{ii}$$

$$p^2c^2 = E^2 - m_0^2c^4 = (qV + m_0c^2)^2 - m_0^2c^4$$

$$= q^2V^2 + 2m_0c^2qV$$

$$p^2 = 2m_0qV\left(1+\frac{qV}{2m_0c^2}\right)$$

or
$$p = \sqrt{2m_0qV\left[1+\frac{qV}{2m_0c^2}\right]}$$

$\therefore$
$$\lambda = \frac{h}{p} = \frac{h}{\sqrt{2m_0qV\left[1+\frac{qV}{2m_0c^2}\right]}}$$

Example 6.8: According to classical theory, a charge *e*, subjected to an acceleration $\frac{d^2r}{dt^2}$, radiates energy at the rate $\frac{dE}{dt} = \frac{2}{3}\frac{e^2}{c^3}\left(\frac{d^2r}{dr^2}\right)^2$. Considering classical model of hydrogen atom, within an electron revolving in a circular orbit round the nucleus, calculate the time during which the radius of the atom will shrink to half. Initial radius is $r_0 = 0.5$Å.

Solution If *r* be the radius of the hydrogen atom, the total energy of the hydrogen atom can be expressed as

$$E = -\frac{e^2}{2r} \tag{i}$$

Acceleration
$$\frac{d^2r}{dr^2} = \frac{v^2}{r} = \frac{e^2}{mr^2} \tag{ii}$$

The equation for radiation rate may be written as

$$\frac{dE}{dt} = \frac{e^2}{2r^2}\frac{dr}{dt} = \frac{-2e^2}{3c^2}\left(\frac{e^2}{mr^2}\right)^2$$

$\therefore$
$$dt = -\frac{3}{4}\frac{m^2c^2}{e^4}r^2 \tag{iii}$$

The time *t* for the radius to half, i.e., from r_0 to $r_0/2$, then

$$t = -\frac{3}{4}\frac{m^2c^2}{e^2}\int_0^{r_0/2} r^2 dr = \frac{7}{32}\frac{m^2c^2r_0^2}{e^4}$$

If we substitute proper values of *m*, *c*, r_0 and *e*, we obtain, $t = 1.4 \times 10^{-11}$ s.
It is worthwhile to mention that there is no quantum analogue of this problem.

Example 6.9: Using uncertainty principle, estimate the ground state of the helium atom.

Solution Let us suppose that two electrons of Helium atom can be localized in the regions of dimensions Δx_1 and Δx_2 of the space. Then the spread in momenta of the electrons can be expressed as

$$\Delta p_1 > \hbar/\Delta x_1 \text{ and } \Delta p_2 > \hbar/\Delta x_2$$

Therefore, the spread of K.E. can be expressed as

$$\Delta E > \frac{\hbar^2}{2m}\left[\frac{1}{(\Delta x_1)^2} + \frac{1}{(\Delta x_2)^2}\right]$$

The spread of interaction energy between electrons

$$\Delta U_1 = \frac{e^2}{\Delta x_1 + \Delta x_2}$$

and the spread of the potential energy of the interaction of the electrons with the nucleus ($Z = 2$) is

$$\Delta U_2 = -2e^2\left(\frac{1}{\Delta x_1} + \frac{1}{\Delta x_2}\right)$$

$\therefore$ Total energy, $$\Delta E \geq \frac{\hbar^2}{2m}\left[\frac{1}{(\Delta x_1)^2} + \frac{1}{(\Delta x_2)^2}\right] - 2e^2\left[\frac{1}{\Delta x_1} + \frac{1}{\Delta x_2}\right]\frac{e^2}{\Delta x_1 + \Delta x_2}$$

In the ground state, ΔE must be minimum, i.e.,

$$\frac{d(\Delta E)}{d(\Delta x_1)} = 0 \quad \text{and} \quad \frac{d(\Delta E)}{d(\Delta x_2)} = 0$$

$\therefore$ $$\frac{d(\Delta E)}{d(\Delta x_1)} = -\frac{\hbar^2}{m(\Delta x_1)^3} + \frac{2e^2}{(\Delta x_1)^2} - \frac{e^2}{(\Delta x_1 - \Delta x_2)^2} = 0$$

and $$\frac{d(\Delta E)}{d(\Delta x_2)} = -\frac{\hbar^2}{m(\Delta x_2)^2} - \frac{2e^2}{(\Delta x_2)^2} - \frac{e^2}{(\Delta x_1 + \Delta x_2)^2} = 0$$

Solving last two equations, one obtains

$$\Delta x_1 = \Delta x_2 = \frac{4\hbar^2}{7me^2}$$

$\therefore$ Total energy in the ground state

$$\Delta E = -\frac{49}{16}\frac{m^2e^4}{\hbar^2}$$

On substituting proper values for m, e, $\hbar$ and then solving, we obtain $\Delta E = -2.1$eV.

Example 6.10: For the study of the structure of atomic nuclei, several laboratories have constructed electron accelerators with energy up to 6 GeV. What is the electron de-Broglie wavelength in that case, and what is the need for such high energies?

Solution de-Broglie wavelength

$$\lambda = \frac{ch}{E}$$

Here $c = 3 \times 10^8$ m/s

$$= \frac{3\times 10^8 \times 6.6\times 10^{-34}}{6\times 10^9 \times 1.6\times 10^{-19}} = 2.07\times 10^{-16}\,\text{m}$$

$h = 6.6 \times 10^{-34}$ Js

$E = 6$ GeV $= 6 \times 10^9$ eV

The high energy particles with wavelength much smaller than 10^{-14} m are needed to study the nuclear structure which has features of dimensions of the order of 10^{-14} m.

Example 6.11: Bearing in mind the wave properties of particles, find the limits of the applicability of the classical concepts for an electron of 10 eV energy and a proton with 1 MeV energy.

Solution It is necessary that the dimensions of the regions or "obstacle" limiting the motion of the particle be appreciably larger than their de-Broglie wavelength (compare the situation in wave optics). We find thus that these dimensions must be much greater than the de-Broglie wavelength for the electron and de-Broglie wavelength for proton i.e., $>> 4 \times 10^{-10}$ m for the electron (de-Broglie wavelength for electron) and $>> 3 \times 10^{-10}$ m for the proton (de-Broglie wavelength for the proton).

Example 6.12: An electron is inside a spherical metallic particle of volume 10^{-12} m³ and has a kinetic energy of about 10 eV. Use the uncertainty relation to find the relative uncertainty in the electron velocity.

Solution The volume of metallic sphere is

$$V = \frac{4}{3}\pi r^3$$

$$\therefore \quad r = \left(\frac{3V}{4\pi}\right)^{1/3} = \left(\frac{3\times 10^{-12}}{4\times 3.12}\right)^{1/3} = 6\times 10^{-3}\ \text{cm} = 6\times 10^{-5}\ \text{m}$$

Thus the uncertainty in the momentum must be at least of the order of

$$p \sim \frac{\hbar}{2}$$

After substituting proper values one gets the uncertainity in velocity. [**Ans:** 6×10^{-7}]

Example 6.13: The position of the centre mass of a globule of mass 2 mg can be determined with an accuracy of 2μ. In this case is there, any sense applying the uncertainity relations when determining the velocity of globule?

Solution $\Delta v = \dfrac{\hbar}{2\mu} = \dfrac{1.05\times 10^{-27}}{2\times 10^{-6}} \sim 5.3\times 10^{-21}\ \text{cm/s} \sim 5.3\times 10^{-23}\ \text{m/s}$

Clearly, Heisenberg's uncertainty relations do not play a role here.

Example 6.14: To see whether an electron belongs to a given atom, one must determine its position with an accuracy of 1 Å. Compare the resulting uncertainty in the electron velocity with the magnitude of the velocity itself. Assume that the electron is in the first Bohr orbit in the hydrogen atom.

Solution Velocity of electron in the first Bohr orbit

$$v_1 = \frac{2\pi Ze^2}{nh} = \frac{2\times 3.14\times 1\times (4.8\times 10^{-10})^2}{1\times 6.6\times 10^{-27}} \quad \text{(C.G.S. system)}$$

$$= 2.2\times 10^8\ \text{cm/s} = 10^6\ \text{m/s}$$

Given $\quad \Delta v_1 = 10^8$ cm/s

Example 6.15: The trajectory of a particle in a Wilson chamber is a chain of droplets of linear dimensions of about 1μ. Could one observe deviations from the classical motion for an electron with an energy 1 keV?

Solution The ratio of the momentum component perpendicular to the trajectory to the total momentum, i.e., $\frac{\Delta p}{p} \sim 6\times10^{-7}$, which is very small. Under such conditions the concept of a trajectory is completely reasonable and one should not expect any deviations from the laws of classical mechanics.

Example 6.16: Discuss how the Heisenberg relations can be applied to estimate a value for the mass μ_π, for the π-meson, given that the range of nuclear forces is about 1.4 fermi.

Solution We can get an estimate for μ_π by using the Heisenberg relation $\Delta E.\Delta t \geq \hbar$. As $\Delta t \sim \frac{a}{c}$, where a is the range of the nuclear forces, we find

$$\Delta E \sim \frac{\hbar c}{a}$$

and

$$\mu_\pi \sim \frac{\Delta E}{c^2} \sim \frac{\hbar}{ac}$$

Substituting the values

$$\mu_\pi \sim \frac{1.05\times10^{-34}\ \text{J}-\text{s}}{14\times10^{-15}\ \text{m}\times3\times10^{8}\ \text{cm}} \sim 270m_e$$

Example 6.17: For what value of the relative uncertainty in the angular momentum of the electron in the first Bohr orbit does its angular coordinate become completely undetermined?

Solution From the Heisenberg's uncertainty principle for angle variables

$$\Delta\theta\Delta A_\theta \geq \hbar$$

we find that θ becomes completely undetermined ($\Delta\theta \geq \pi$) if

$$\frac{\Delta A_\theta}{A_\theta} \leq 32\%$$

Example 6.18: The diameter of an atomic nucleus is of the order of 10^{-13} cm. You want to obtain information about the size and shape of nuclei and you have decided to do this by bombarding them with fast protons. What is the approximate energy to which you have to accelerate the protons?

Solution In a light microscope the wavelength of the illuminating light has to be short compared to dimensions of the object to be observed. Similarly, if we 'illuminate' a nucleus with a proton, we must require that the de-Broglie wavelength of the proton is short compared to the nuclear dimensions if we are to observe any detail. In optical microscopy we settle for nothing less than an image of the object.

Nuclear physicists are less demanding and are willing to calculate the size of their object from diffraction patterns. Thus we assume that useful information can be obtained even if the de-Broglie wavelength is equal to the nuclear diameter. Hence

$$\lambda = \frac{h}{p} = 10^{-15}\, m \quad \text{and} \quad P = \sqrt{2mE}$$

$$E = \frac{h^2.10^{26}}{2\text{m}} = \frac{(6.626)^2 \times 10^{-68} \times 10^{30}}{2 \times 1.67 \times 10^{-19}} = 820\,\text{MeV}$$

Using relativistic connection between energy and momentum, value of E comes out to be 617 MeV.

Example 6.19: Show that the de-Broglie wavelength associated with an electron of energy V electron volts approximately $12.27/\sqrt{V}$ Å.

Solution We have $\lambda = \dfrac{h}{\sqrt{2meV}}$

Substituting the value of h, m and e, one obtains

$$\lambda = \frac{6.6 \times 10^{-34}}{\sqrt{2 \times 9.1 \times 10^{-34} \times 1.6 \times 10^{-19}\, V}} = \frac{12.27 \times 10^{-10}}{\sqrt{V}} = \frac{12.27}{\sqrt{V}}\,\text{Å}$$

Example 6.20: Find de-Broglie wavelength of a neutron of energy 12.8 MeV. Given mass of neutron = 1.675×10^{-17} kg, $h = 6.62 \times 10^{-34}$ J-s, $c = 3 \times 10^8$ m/sec and 1 eV = 1.6×10^{-19} J.

Solution The rest mass energy of neutron

$$= m_0c^2 = 1.675 \times 10^{-27} \times (3 \times 10^8)^2 = 15.075 \times 10^{-11}\,\text{J}$$

$$= \frac{15.075 \times 10^{-11}}{1.6 \times 10^{-19}} = 9.42 \times 10^8 = 942\text{ MeV}$$

Obviously, the given energy for neutron 12.8 MeV is very small compared with the rest mass energy and so relativistic effect can be ignored. Thus the de-Broglie wavelength associated with the neutron of energy 12.8 MeV is

$$\therefore \qquad \lambda = \frac{h}{\sqrt{2m_0K}} = 12.8\text{ MeV} = 12.8 \times 10^6 \times 1.6 \times 10^{-19}\,\text{J}$$

$$= 12.8 \times 1.6 \times 10^{-13}\,\text{J}, \quad \lambda = \frac{6.62 \times 10^{-34}}{\sqrt{2 \times 1.675 \times 10^{-27} \times 12.8 \times 1.6 \times 10^{-13}}}$$

$$= 0.8 \times 10^{-14}\text{ m} = 0.8 \times 10^{-4}\,\text{Å}$$

Example 6.21: Bohr model of atom was proposed ten years before quantum mechanics and uncertainty relation. How good the Bohr model is?

Solution Following table shows the correct and wrong aspects of Bohr model of course we can consider modification of relativistic effect.

Bohr Model

	Qualitative	*Quantitative*	Uncertainty
Orbit	Wrong		
$r_0 = a_0 n^2$		Wrong	$\left\langle \frac{1}{r} \right\rangle = 1/(n^2 a_0)$
			$< r >_{nl} = \frac{1}{2}[3n^2 - l(l+1)]a_0$
			$r_{mp} = (l = l_{\max} = n-1) = n^2 a_0$
E_n	Correct	Correct	$E_n \sim \left\langle \frac{1}{r} \right\rangle_n$
L	Correct	Wrong	$L < n\hbar$

Example 6.22: **Assume the uncertainty in position of a hydrogen molecule of mass about 2×10^{-27} kg is of the order of the diameter about 10^{-10} m. Determine the uncertainty in momentum.**

Solution The uncertainty in momentum is given by

$$\Delta p_x > \frac{h}{\Delta x} \geq \frac{1.054 \times 10^{-34}}{10^{-16}} \geq 1.054 \times 10^{-24} \text{ kg-m/s}$$

Example 6.23: **The average period that elapses between the excitation of an atom and the time it emits radiation is 10^{-8} sec. Find the uncertainty in the energy emitted and the uncertainty in the frequency of light emitted. Given**

$$\hbar = 1.054 \times 10^{-34} \text{ Js and } h = 6.63 \times 10^{-34} \text{ Js}$$

Solution The uncertainty in the energy is given by

$$\Delta E = \frac{\hbar}{\Delta t} = \frac{1.054 \times 10^{-34}}{10^{-8}} = 1.054 \times 10^{-26} \text{ J}$$

The uncertainty in the frequency of light emitted is given by

$$\Delta \nu = \frac{\Delta E}{h} = \frac{1.054 \times 10^{-26}}{6.63 \times 10^{-34}} = 1.59 \times 10^{7} \text{ Hz}$$

Example 6.24: **A proton is confined to a space of radius 10^{-14} m. Calculate the minimum uncertainty in its momentum. Also calculate the minimum kinetic energy the proton should have mass of the proton is 1.672×10^{-27} kg and $h = 6.63 \times 10^{-34}$ Js.**

Solution The diameter of the space in which the proton is confined is 2×10^{-14} m. The minimum uncertainty in the position of proton in this space will be $\Delta x = 2 \times 10^{-14}$. The corresponding minimum uncertainty in its momentum is given by

$$\Delta p = \frac{\hbar}{\Delta x} = \frac{1.054 \times 10^{-34}}{2 \times 10^{-14}} = 0.53 \times 10^{-20} \text{ kg-m/s}$$

The corresponding (minimum) kinetic energy of the proton is

$$\text{K.E.} = \frac{p^2}{2m} = \frac{(0.53\times10^{-20})^2}{2\times1.672\times10^{-27}} = 8.4\times10^{-15}\text{ J} = 5.25\times10^4\text{ eV} = 0.0525\text{MeV}$$

Example 6.25: Calculate the smallest possible uncertainty in the position of an electron moving with velocity 3×10^7 m/s.

Solution The maximum uncertainty in the momentum is given by

$$\Delta p_x = mv = \frac{m_0 v}{\sqrt{1 - v^2/c^2}}$$

The minimum uncertainty in the position is given by

$$\Delta x = \frac{\hbar}{\Delta p_x} = \frac{\hbar}{m_0 v/\sqrt{1 - v^2/c^2}} = \frac{\hbar\sqrt{1 - v^2/c^2}}{m_0 v}$$

But $\quad \dfrac{v}{c} = \dfrac{3\times10^7}{3\times10^8} = \dfrac{1}{10}$

$$\therefore \quad \Delta x = \frac{h}{m_0 v}\sqrt{1 - \frac{1}{100}} = \frac{1.05\times10^{-34}\sqrt{99}}{9.11\times10^{-31}\times3\times10^7\times10} = 0.38\times10^{-11}\text{ m} = 0.038\,\text{Å}$$

Example 6.26: **A nucleon is confined to a nucleus of radius 5×10^{-15} m. Calculate the minimum uncertainty in the momentum of the nucleon. Also calculate the minimum kinetic energy of the nucleon. Given, mass of nucleon $m = 1.67 \times 10^{-27}$ kg and Planck's constant $h = 6.6 \times 10^{-34}$ J-s.**

Solution We have $\quad (\Delta p)_{\min} (\Delta x)_{\max} = h$

Given $\quad (\Delta x)_{\max} = 2 \times 5 \times 10^{-15} m = 10^{-14} m$

$$(\Delta p)_{\min} = \frac{h}{(\Delta x)_{\max}} = \frac{6.6\times10^{-34}}{10^{-14}} = 6.6\times10^{-20}\text{ kg-m/s}$$

Since p cannot be less than $(\Delta p)_{\min}$, so we have

$$p_{\min} = (\Delta p)_{\min}$$

$$\therefore \quad E_{\min} = \frac{p_{\min}^2}{2\text{m}} = \frac{(6.6\times10^{-20})^2}{2\times1.67\times10^{-27}} = 1.32\times10^{-12}\text{ J} = 8.25\times10^6\text{ eV} = 8.25\text{MeV}$$

Example 6.27: **Show that the ratio of uncertainties in the velocities of an electron and a proton confined to a 10^{-18} m box is about 5.5×10^{-4}. Given mass of the proton = 1.66×10^{-27} kg.**

Solution **We know that the uncertainty in the measurement of position for both the electron and the proton in same. Obviously, the uncertainties in the measurement of momentum must also be the same for both particles.**

$$\therefore \quad \frac{\text{uncertainty in velocity of electron}}{\text{uncertainty in velocity of proton}} = \frac{9.1\times10^{-31}}{1.66\times10^{-27}} = 5.48\times10^{-4} = 5.5\times10^{-4}$$

Example 6.28: An electron has a speed of 600 m/s with an accuracy of 0.005%. Show that the certainty with which the position of the electron can be located is about 10^{-30}.

Solution We have electron speed = 600 m/s. This is quite small compared to that of light and hence there is no need to apply relativity correction of mass. Hence the accuracy in the measurement of momentum is also 0.005% or

$$\Delta p = \frac{0.005}{100} = 5\times10^{-5}$$

$\therefore$ Accuracy in the measurement of position

$$\Delta x = \frac{h}{2\pi\Delta p} = \frac{6.63\times10^{-34}}{2\times3.14\times5\times10^{-5}} = 0.528\times10^{-30} \simeq 10^{-30}$$

Example 6.29: The speed of a bullet ($m = 50$ g) and a speed of electron ($m = 9.1\times10^{-28}$ g) are measured to be same, namely 300 m/s, with an uncertainty of 0.01%. With what fundamental accuracy could we have located the position of each, if the position is measured simultaneously with the speed in the same experiment?

Solution For the electron

$$p = mv = 9.1\times10^{-31}\text{ kg}\times300\text{ m/s} = 2.7\times10^{-28}\text{ kg-m/s}$$

and $$\Delta p = m\Delta v = 0.0001\times2.7\times10^{-28}\text{ kg-m-s} = 2.7\times10^{-32}\text{ kg-m/s}$$

so that $$\Delta x \geq \frac{h}{4\pi\Delta p} = \frac{6.6\times10^{-34}\text{ J-s}}{4\pi\times2.7\times10^{-32}\text{ kg-m/s}} = 2\times10^{-3}\text{ m} = 0.2\text{m}$$

For the bullet

$$p = mv = 0.05\text{ kg}\times300\text{ m/s} = 15\text{ kg-m/s}$$

$$\therefore \quad \Delta x \geq \frac{h}{4\pi\Delta p} = \frac{6.6\times10^{-34}\text{ J-s}}{4\pi\times1.5\times10^{-3}\text{ kg-m/s}} = 3\times10^{-3}\text{ m}$$

Hence, for macroscopic objects such as bullets the uncertainty principle sets no practical limit to our measuring procedure. Δx in this case being about 10^{-17} times the diameter of a nucleus but, for microscopic objects such as electrons, there are practical limits, Δx in this case being about 10^{7} times the diameter of an atom.

Example 6.30: The signal from a television station contains pulses of full-width $\Delta t \sim 10^{6}s$. Explain why it is not feasible to transmit television in the AM broadcasting band.

Solution The full-width range of frequencies in the signal is, from $\Delta t\Delta\nu > 14\pi$, $\Delta\nu \sim 1/10^{-6}\text{s} = 10^{6}$ $\text{s}^{-1} \simeq 10^{6}$ Hz. Thus the entire broadcast band ($\nu \simeq 0.5\times10^{6}$ Hz to $\nu \simeq 1.5\times10^{6}$ Hz) would be able to accommodate only a single television "channel". There could also be serious difficulties in building transmitters and receivers with such a very large fractional band pass.

At frequencies used in television transmission ($\nu \approx 10^8$ Hz) many channels fit into a reasonable portion of the spectrum, and the band pass requirements are nominal.

Example 6.31: Consider a microscopic particle moving freely along the x-axis. Assume that at the instant $t = 0$ the position of the particle is measured and is uncertain by the amount Δx_0. Calculate the uncertainty in the measured position of the particle at some later time t.

Solution The uncertainty in the momentum of the particle at $t = 0$ is at least

$$\Delta p_x = \hbar / 2\Delta x_0$$

$\therefore$ The velocity of the particle at that instant is uncertain by at least

$$\Delta v_x = \Delta p_x / m = \hbar / 2m\Delta x_0$$

and the distance x travelled by the particle in the time t cannot be known more accurately than within

$$\Delta x = t\Delta v_x = \hbar t / 2m\Delta x_0$$

If by a measurement at $t = 0$ we have localized the particle within the range Δx_0, then in a measurement of its position at time t the particle could be found anywhere within a range of at least as large as Δx.

We must note that Δx is inversely proportional to Δx_0, so that the more carefully we localize the panicle at the initial instant, the less we shall know about its final position. Also, the uncertainty Δx increases linearly with time t. This corresponds to a spreading out, as time goes on, of the group of waves associated with the motion of the particle.

Example 6.32: In a measurement of water waves, 10 wave crests are counted in a distance of 200 cm. What is the minimum uncertainty in the wavelength that might be obtained from this experiment?

Solution We have $k = \dfrac{2\pi}{\lambda}$

or $$dk = -\frac{2\pi}{\lambda^2} d\lambda$$

Now, replacing the differential with small intervals and taking magnitudes only, one obtains:

$$\Delta k = \frac{2\pi}{\lambda^2} \Delta\lambda$$

we have $$\Delta x \Delta k \sim 1$$

so $$\Delta x \left(\frac{2\pi}{\lambda^2} \Delta\lambda \right) \sim 1$$

or $$\Delta x \Delta\lambda \sim \lambda / 2\pi$$

or $$\Delta\lambda \sim \frac{1}{\Delta x} \frac{\lambda^2}{2\pi} = \frac{1}{200} \times \frac{(20)^2}{2\pi} \sim 0.3 \text{ cm}$$

Example 6.33: Certain ocean waves travel with a phase velocity $V_{\text{phase}} = \sqrt{g\lambda/2\pi}$, where g is the acceleration due to gravity. What is the group velocity of a "wave-packet" of these waves? Express your result in terms of phase velocity.

Solution We have $K = 2\pi/\lambda$,

$$v_{\text{phase}} = \sqrt{g/k}$$

But $$v_{\text{phase}} = \omega/k \qquad \therefore\ \omega/k = \sqrt{g/k}$$

or $$\omega = \sqrt{gk}$$

and $$d\omega = \sqrt{g}\left(\frac{1}{2}k^{-1/2}\right)dk$$

$$\therefore \quad \frac{d\omega}{dk} = \frac{1}{2}\sqrt{g/k}$$

$$\therefore \quad v_{\text{group}} = \frac{1}{2}v_{\text{phase}}$$

Example 6.34: An electron moves in the x-direction with the speed of 3.6×10^6 m/s. We can measure its speed to a precision of 1% (i) With what precision can we simultaneously measure its position? (ii) What you can say about its motion in y-direction?

Solution (i) We have, electron momentum, $p_x = mv_x$

$$= 9.1 \times 10^{-31} \times 3.6 \times 10^6$$
$$= 3.3 \times 10^{-24} \text{ kg-m/s}$$

The uncertainty Δp_x is 1% of this value. Thus

$$\Delta p_x = 3.3 \times 10^{-26} \text{kg-m/s}$$

Now, the uncertainity in position $\Delta x \sim \dfrac{\hbar}{\Delta p_x} = \dfrac{1.05 \times 10^{-34}}{3.3 \times 10^{-26}} = 3.2 \text{ nm}$, which is roughly 10 atomic diameters.

(ii) If the electron is moving in the x-direction, then we know its speed in the y-direction precisely; thus $\Delta p_y = 0$. The uncertainty relationship $\Delta y\, \Delta p_y \sim \hbar$, then it demands that Δy be infinite. This means that the we know nothing at all about the electron's y-coordinate.

SUMMARY

1. The characteristic of *blackbody radiation* cannot be explained by classical concepts. Planck argued that the atomic oscillators responsible for this radiation exist only in discrete *quantum states.*
2. In the *photoelectric effect,* electrons (photoelectrons) are ejected from a metallic surface when light is incident on that surface. Einstein successfully explained this effect

by extending Planck's quantum theory to the electromagnetic field. Einstein viewed light as a stream of particles called *photons*, each with energy $E = h\nu$, where ν is frequency and h is Planck's constant. The maximum kinetic energy of the ejected photoelectron from the metallic surface is given by

$$K_{max} = h\nu - \phi$$

where ϕ is the *work function* of the metal.

3. x-rays striking a target are scattered at various angles by electrons in a target. A shift in wavelength is observed for the scattered x-rays, and the phenomenon is called *Compton effect.* Classical physics fails to explain this effect. If we treat x-ray as a photon, conservation of energy and momentum applied to the isolated system of photon and the electron yield the following expression for compton shift

$$\lambda' - \lambda = \frac{h}{mc}(1 - \cos\theta)$$

where m is mass of the electron and θ is the scattering angle.

4. In 1924, Louis de Broglie proposed that photons have wave and particle characteristics and therefore (perhaps) every object of mass m and momentum p has wavelike properties, with a *de Broglie wavelength* given by

$$\lambda = \frac{h}{p} = \frac{h}{mv}$$

5. The *wave-particle duality* is the basis of quantum mechanics. It can be interpreted by imagining a particle to be made of combination of large number of waves. These waves interfere constructively in a small region of space called a *wave packet*.

6. Heisenberg in 1927 stated that according to fundamental equations of quantum mechanics, in principle one cannot measure the position and momentum (or any other canonically conjugate coordinates, such as angle and angular momentum) simultaneously with precision. This statement is known as *Heisenberg's uncertainty principle.* If Δx and Δp_x are the uncertainties in position and corresponding momentum of a particle, then

$$\Delta x \, \Delta p_x \geq \frac{\hbar}{2}$$

or if ΔE and Δt are precise energy at a precise time, then $\Delta E \, \Delta t \geq \hbar/2$. The uncertainty principle is a natural out growth of the wave packet model and the said limit on measurement is independent of measurement procedure or instrument. Heisenberg uncertainty principle are consistent with the complimentarity principle and statistical interpretation.

7. *Correspondence principle*: Quantum effects become important and measurable only on the submicroscopic level of atoms and molecules. Quantum results must blend smoothly with classical results when quantum number becomes large.

REVIEW QUESTIONS

1. Explain the distribution of energy in a black body spectrum. How far the classical theories explain the distribution? Write a note on Planck's law of distribution.

2. Discuss important features of the spectral distribution of energy in a black body radiation. How far the classical theories explain it? Write Planck's radiation law and using it explain the spectral distribution of black body radiation.

3. State Wien's displacement law and Rayleigh Jean's law of radiation and indicate how far two laws could explain the curves.

4. Discuss Planck's hypothesis for a black body radiation. Show that the average energy of a Planck's oscillator of frequency v in thermal equilibrium with a heat reservoir of temperature T is given by $\overline{\varepsilon} = hv/[\exp(hv/kT) - 1]$

5. Discuss the distribution of energy in black body radiation. Discuss the success and limitations of the classical theory in explaining it. Derive Planck's radiation formula. Explain how the short comings of classical theory were overcome.

6. How Einstein's photoelectric equation explains the phenomenon of photoelectric effect? What information is derived for the nature of light?

7. What is photoelectric effect? Give an account of the photoelectric emission of the electrons. Give Einstein's interpretation of the same

8. Write laws of photoelectric effect. How has photoelectric effect been explained by Einstein?

9. What is Compton effect? Derive an expression for the change in wavelength of a photon when it is scattered by an electron. Calculate the maximum change in wave-length in a Compton scattering experiment.

10. Describe Compton effect. Derive the expression for the frequency of the scattered photon in terms of the frequency of the incident radiation and scattering angle.

11. **(i)** Explain why Compton effect is not observed experimentally for visible rays?
(ii) Both photoelectric effect and Compton effect arise due to action of photons on electrons but the two effects are not the same. Explain.
(iii) Explain the presence of unmodified radiation in Compton scattering.

12. What is de-Broglie hypothesis? Show that the wavelength λ associated with a particle of mass m and kinetic energy E is given by $\lambda = h\sqrt{2mE}$, where h is Planck's constant.

13. Describe Davisson and Germer's experiment and mention the important conclusions.

14. **(a)** Draw the schematic diagram of Davisson and Germer experimental arrangement to study the electron diffraction. Explain how the experiment was carried out.
(b) Show that the results of this experiment are closely in agreement with the electron wavelength calculated by de-Broglie relation.

15. State Heisenberg's uncertainty principle. Is this principle the outcome of the wave description of a particle? Describe at least one experiment to prove its validity.

16. State the energy time uncertainty principle. Explain the natural width of spectral lines with its help.

17. What is correct statement of uncertainty principle? Discuss its significance and importance. From the uncertainty principle, show that the electrons cannot reside in the nucleus of the atom. Given that nuclear radius = 10^{-14} m and the maximum possible kinetic energy of an electron in an atom = 4 MeV.

18. State and explain the significance of the time energy uncertainty relation.
19. Define the uncertainties Δx and Δp_x and prove the Heisenberg's uncertainty relation $\Delta x \Delta p_x \geq \hbar/2$.
20. State and explain the uncertainty principle. Describe the experiment based on gamma-ray microscope or any other experiment which supports the uncertainty principle. Discuss the non existence of the electron in the nucleus using uncertainty priniciple.
21. Write short notes on:
    (i) Franck-Hertz experiment (ii) Space Quantization
    (iii) Matter waves (iv) Principle of complementarity
    (v) Rigid Rotaor (vi) Raman effect
    (vii) Pair production (viii) Bremsstrahlung
    (ix) Sommerfeld quantum theory (x) Bremsstrahlung and X-ray production

PROBLEMS

1. Show that the ratio of the de-Broglie wavelength to the Compton wavelength for the same particle is equal to $\sqrt{\left(\frac{c}{v}\right)^2 - 1}$
2. What is the de-Broglie wavelength of an (α particle (He nucleus)) with a kinetic energy of 7.7 MeV? In the Rutherford's scattering experiments, distances of the order of 10^{-13} m were involved yet the analysis of the experiment did not include the wave properties of the a particle. Was this justified? **[Ans.** 5.18×10^{-15} m]
3. Consider a pendulum bob of mass 0.10 kg moving at 3 m/s. Suppose that the momentum p_x need not be known more accurately then $\Delta p_x = 10^{-6} p_x$. What limitation does the uncertainty principle impose on the simultaneous measurement of x?
4. The lifetime of an excited state of an atom is about 10^{-5} s, an atom can radiate at any time from $t = 0$ to $t = \infty$ after it is excited, but the average time is ~10^{-6} s. Using this as Δt for the emission of a photon, compute the minimum $\Delta\nu$ permitted by the uncertainty principle. What fraction of ν is this if the wavelength of spectral line involved is 5000 Å (This calculation gives the limiting sharpness of a spectral line if no other processes, such as Doppler effect, broaden the line). **[Ans.** 10^8 s^{-1}, 1.67×10^{-7}]
5. The velocity of a proton in x-direction is measured to an accuracy of 10^{-7} m-s^{-1}. Determine the limit of accuracy with which the proton can be located simultaneously (a) along the x-axis, (*b*) along the y-axis. Repeat for a case in which the particle is an electron.
   [**Ans.** (a) for the proton: 3.9 m, for an electron: 7.3×10^3 m (b) with unlimited precision for both cases].
6. If the velocity of (i) an electron and (ii) a droplet of water of radius 10^{-2} mm are known to an accuracy of 10^{-3} ms^{-1}, *m* calculate the maximum accuracy with which their position coordinates can be known.

7. Given that the radius of a typical nucleus is 5×10^{-15} m, and that the average binding energy for nucleons within the nucleus is 8 MeV. Show from the uncertainty principle that it is not possible for an electron to exist within the nucleus.
8. The position and momentum of a 1 KeV electron are simultaneously determined. If the position is located to within 1 Å, what is the percentage of uncertainty in its momentum? **[Ans.** 6.2%]
9. Compare the uncertainties in the velocities of an electron and a proton confined in 10 Å box. **[Ans.** 1.16×10^5 m/s; 63 m/s]
10. The atoms in a solid possess a certain minimum zero-point energy even at 0° K, while no such restrictions hold for the molecules in an ideal gas. Use uncertainty principle to explain these statements.
[**Hint:** Each atom in a solid is limited to a certain definite region of space otherwise the assembly of atoms would not be a solid. The uncertainty in the position of each atom is therefore finite, and its momentum and hence energy cannot be zero. There is no restriction on the position of a molecule in an ideal gas, and the uncertainty in its position is effectively infinite and its momentum and hence energy can be zero.]
11. Verify that the uncertainty principle can be expressed in the form $J > h$, where $\Delta J \Delta\theta \geq \hbar$, is the uncertainty principle and ΔJ is the uncertainty in the angular momentum of a body and $\Delta\theta$ is the uncertainty in its angular position.
[**Hint:** Consider a particle moving in a circle]
12. A high-speed shutter is placed between a monochromatic light source (a laser) and a high resolution spectrometer. First the shutter is held open for a long time while a measurement of photon energy is made. During a later experiment the shutter is opened for only 10^{-9} s while the photon energy is measured. How does the result of the second measurement compare with that of the first? (a) Qualitatively give reasons, (b) Quantitatively, assuming the spectrometer has definite resolution.
13. The threshold wavelength for photoelectric emission in tungsten is 2300 Å. What wavelength of light must be used in order for electrons with a maximum energy of 1.5 eV to be ejected?
14. What do you understand by the wave nature of a particle? Compare the wavelength of 10 KeV electrons with that of 10 KeV photon, $c = 3 \times 10^8$ m/s, $m_0 = 9.1 \times 10^{-31}$ kg.
15. Calculate the percentage change in wavelength in a 0.015 nm photon which undergoes a 120° Compton scattering from an electron. **[Ans.** 24.3%]
16. Assuming the nuclear force between two protons to have a range = 10^{-15} m due to π-meson exchange, with the help of uncertainty relation $\Delta E \Delta t \geq \hbar$, determine the approximate mass of π-meson in MeV. **[Ans.** = 12.4×10^2 MeV]
17. Excited nuclear states may have life time as short as 10^{-15}s. Estimate the natural line width of the states. **[Ans.** 10^{15}/s]
18. Show that when the energy of a particle is much larger than its rest energy then its de-Broglie wavelength is nearly the same as that of a photon having the same energy.
19. The life time of the energy level in a hydrogen atom is of the order of 10 ns. Show that the ratio of the uncertainty in the energy of the predicted state with the energy of the state is $\sim 3.2 \times 10^{-9}$.

20. An excited atom has an average life-time of 10^{-8} s. During this period it emits a photon and returns to the ground state. Show that the minimum uncertainty in the frequency of this photon is 8×10^6s.

21. Using uncertainty principle, calculate ground state energy of a two electron atom of nuclear charge q. [**Ans.** $E = -\frac{me^2}{\hbar^2}\left(q - \frac{e}{4}\right)^2$]

22. Show that the de-Broglie wavelength of a particle, of charge e, rest mass m_0, moving at relativistic speeds is given as a function of the accelerating potential V as

$$\lambda = \frac{h}{\sqrt{2m_0eV}}\left(1 + \frac{eV}{2m_0c^2}\right)^{-1/2}$$

Show how this agrees with $\lambda = h/p$ in the non-relativistic limit.

23. Derive a relation between kinetic energy K of the recoil electron and the energy E of the incident photon in the Compton effect and show that

$$\frac{K}{E} = \frac{\left(\frac{2h\nu}{m_0c^2}\right)\sin^2(\theta/2)}{\left[1 + \left(\frac{2h\nu}{m_0c^2}\right)\sin^2(\theta/2)\right]}$$

SHORT ANSWER-QUESTIONS

1. Even though the photon mass is zero light of a suitable wavelength is able to eject electron from a metal surface. How this is possible?
2. In the domain of everyday experience quantum effects are imperceptible. Explain clearly and give an example.
3. Name the various types of emissions of electrons from metals.
   Ans: (i) thermionic emission (ii) photoelectron emission (iii) high field emission (iv) secondary emission.
4. What are photoelectrons? Are they in any way different from ordinary electrons?
   Ans: The electrons emitted during photoelectric effect are called photoelectrons. They are in no way different from ordinary electrons.
5. Is photoelectric emission a spontaneous process?
   Ans: No, it is not a spontaneous process.
6. Our bodies are continuously bombarded by photons of radiowaves whose frequencies range from 1 to 1,000 MHz. How is it that they are not dangerous?
7. Can photoelectric effect occur with free electrons? Explain.
8. What are photons?
9. Why a hole in the wall of a cavity radiator is termed as a black body?
   Ans: A black body is one that completely absorbs all the radiations, whatever be the wavelength, falling on it and reflects none. When any radiation from outside incident on the fine hole made in the walls of a cavity radiator is completely absorbed by successive reflections at the inner walls and none is reflected back to the eye. This is why a hole behaves as a black body.

10. In the photoelectric effect a photon moving in one direction effects an electron in some other direction. Does this mean that the law of conservation of moementum is not obeyed in photo electric effect? Explain.

11. What happens to the velocity of emitted electrons when the wavelength of incident light is decreased?

Ans: The velocity of emitted electrons will increase.

12. Increasing the intensity of the incident light does not alter the maximum kinetic energy of the ejected photoelectron. What happens to the extra energy in the incident beam?

13. Can you observe Compton effect if a proton is substituted in place of electron?

Ans: A proton is much heavier than electron and therefore the Compton shift will be too small to be observed.

14. Explain why Compton effect is not observed experimentally for visible light rays?

Ans: Compton effect can be detected only for those radiations whose wavelength is not greater than a few Å, e.g., for $\lambda = 5$ Å, there is a maximum wavelength change of 1%. while for $\lambda = 1$Å there is a 5% change. For visible light ($\lambda \approx 5000$Å) the maximum wavelength change (0.05 Å) is only about 0.0001% of the initial wavelength which is undetectable.

15. Can matter waves travel faster than light? Justify your answer.

16. Explain why the wave nature of matter is not apparent in our daily life?

17. Show that if the uncertainty in the location of the particle is equal to the de-Broglie wavelength, the uncertainty in its velocity is equal to its velocity divided by 4π.

18. Does the concept of Bohr orbits violate the Heisenberg's uncertainty principle?

19. In accordance with the uncertainty principle, the electron cannot exist inside the nucleus of an atom. Whether the same argument apply to a proton? Explain.

20. In the photoelectric experiments, the current (number of electrons emitted per unit time) is proportional to the intensity of light. Can this result alone be used to distinguish between classical and quantum theories?

21. Why it is that even for incident radiation that is monochromatic, photoelectrons are emitted with a spread of velocities?

22. The existence of a cut off frequency in the photoelectric effect is often regarded as the most potent objection to a wave theory, explain.

23. Why are photoelectric measurements very sensitive to the nature of photoelectric surface?

24. Do the results of photoelectric experiments invalidate Young's interference experiment?

25. Can you use the device of letting $h \to 0$ to obtain classical results from quantum results in the case of the photoelectric effect? Explain.

26. Why, in Compton scattering, would you expect $\Delta\lambda$ to be independent of the materials of which the scatterer is composed?

27. Do you observe a Compton effect with visible light? Why?

28. Can you think of experimental evidence contradicting the idea that vacuum is a sea of electrons in negative energy states?

29. If, in the de-Broglie formula, we let $m \to \infty$, do we get the classical result for macroscopic particles?

30. Can we measure the frequency ν for de-Broglie waves? If so, how?

31. What is photoelectric effect?

Ans: This involves the ejection of electrons from a metal surface illuminated with electromagnetic waves. Explanation of the effect led Einstein to propose photons as the quanta of electromagnetic wave energy.

32. What is the importance of Planck's constant?

Ans: Planck's constant $h = 6.63 \times 10^{-34}$ Js sets the fundamental scale of quantization. It is also expressed as $\hbar = h/2\pi$.

33. What inference you can draw from Compton effect?

Ans: Compton effect shows that photons interact with free electrons exactly like colliding particles, losing energy and emerging with longer wavelength.

34. A beam of aluminium atoms is used to dope a semiconductor chip to set its electrical properties. If the atom's velocity is known to within 0.2 m/s, how accurately can they be positioned?

Ans: $\Delta x = \hbar / \Delta p$ (i)

$$\Delta p = m\Delta v = (29.684)(1.66 \times 10^{-27} kg)(0.20 m/s)$$

$$= 9 \times 10^{-27} kg\text{-}m/s$$

$$\therefore \quad \Delta x = \frac{6.63 \times 10^{-34}}{2\pi \times 9 \times 10^{-27}} = 12\ nm$$

35. How are wave-particle duality and the uncertainity principle related?

36. Why uncertainities in the particle's position and momentum arises?

Ans: The uncertainties arise from the quantum structure of matter.

37. How can you Judge that application of quantum mechanics is relevant or not for a particular system?

Ans: In general, application of quantum mechanics to a particular system or particle is relevant when the de Broglic wavelength of the system or particle in question (h/p) is greater than the characteristic size of the system (d) or particle.

OBJECTIVE QUESTIONS

1. Wien's displacement law is

(a) $\lambda_m T$ = constant (b) λ_m/T = constant

(c) λ_m/T^2 = constant (d) $\lambda_m T^2$ = constant

2. The average energy of Planck's oscillator of frequency ν is

(a) $h\nu[e^{h}\nu/kT - 1]$ (b) $h\nu/[e^{h\nu/kT} - 1]$

(c) $h\nu$ (d) $h\nu/kT$

3. Planck's radiation formula in terms of wavelength is

(a) $\rho(\lambda)d\lambda = \frac{8\pi hc}{\lambda^5}d\lambda$ (b) $\rho(\lambda)d\lambda = \exp[-ch/\lambda kT]d\lambda$

(c) $\rho(\lambda)d\lambda = \frac{8\pi hc}{\lambda^5}\exp.[-ch/\lambda cT]d\lambda$ (d) $\rho(\lambda)d\lambda = \left[\frac{8\pi hc}{\lambda^5} - 1\right]d\lambda$

4. Einstein's photoelectric equation is

(a) $h\nu = \left(\frac{1}{2}mv^2\right)_{max}$ (b) $h\nu = h\nu_0 + \left(\frac{1}{2}mv^2\right)_{max}$

(c) $h\nu = 2h\nu_0$ (d) $h\nu = h\nu_0/2$

5. The shift in the wavelength in Compton effect depends

(a) only on the energy of the incident photon

(b) only on the scattering angle

(c) scattering angle as well as energy of the incident photon

(d) none of the above

6. The change in wavelength of an X-ray photon when it is scattered through an angle of 90° by a free electron is of the order of

(a) 10-8 m (b) 10-10 m

(c) 10-12 m (d) 10-15 m

[**Hint:** $\Delta\lambda = \frac{h}{m_0 c}(1-\cos\phi) = \frac{6.6\times10^{-34}}{9.1\times10^{-31}\times3\times10^{8}}(1-\cos 90°) = 2.4\times10^{-12}$ m]

7. For the study of Compton effect, the best suited rays are:

(a) gamma rays (b) visible rays

(c) X-rays (d) ultraviolet rays

8. In Compton scattering for a given photon of energy E, the maximum kinetic energy gained by scattered electron is

(a) $2E^2/(2E + m_0c^2)$ (b) $2E^2/m_0c^2$

(c) $E^2/2m_0c^2$ (d) $4E^2/(E + 2m_0c^2)$

9. The de-Broglie wave group associated with a body in motion travels with the velocity

(a) with which the body is moving

(b) half of the de-Broglie wave group velocity

(c) one fourth of the velocity with which the body is moving

(d) none of the above

10. Bohr's quantum condition is

(a) $mvr = h/2\pi$ (b) $mvr = (n+1)h/2\pi$

(c) $mvr = nh/2\pi$ (d) $mvr = n^2h/2\pi$

11. The uncertainty in the location of a particle is equal to its de-Broglie wavelength. The uncertainty in its velocity is

(a) $v/4\pi$ (b) $v/2$

(c) $v/8\pi$ (d) $v/16\pi$

[**Hint:** $\Delta x \Delta p = h/4\pi$, we have $\Delta x = h/p$ $\frac{h}{p}\Delta p = \frac{h}{4\pi}$ or $\Delta p = p/4\pi$ or $\Delta(mv) = mv/4\pi$ $\therefore \Delta v = v/4\pi$]

12. Assume we have two small opaque bodies a large distance from one another supported by fine threads in a large evacuated enclosure whose walls are opaque and kept at a constant temperature. In such a case the bodies and walls can exchange energy only by means of radiation. Let e represent the rate of emission of radiation energy by a body and let a represent the rate of absorption of radiation energy by a body. At equilibrium,

(a) $\frac{e_1}{a_1} = \frac{e_2}{a_2} = 0$ (b) $\frac{e_1}{a_1} = \frac{e_2}{a_2} = 1$

(c) $\frac{e_1}{a_1^2} = \frac{e_2}{a_2^2} = 1$ (d) $\frac{e_1^2}{a_1} = \frac{e_2^2}{e_2} = 1$

13. An atom can radiate at any time after it is excited. It is found that in a typical case the average excited atom has a lifetime of about 10^{-8} s, i.e., during this period it emits a photon and is deexcited. The minimum uncertainty Δv in the frequency of the photon is

(a) $8 \times 10^6\ s^{-1}$ (b) $10^4\ s^{-1}$

(c) $5 \times 10^8\ s^{-1}$ (d) $8 \times 10^9\ s^{-1}$

ANSWERS

Question numbers	1	2	3	4	5	6	7	8	9	10	11	12	13
Answers	(a)	(b)	(c)	(b)	(c)	(c)	(c)	(a)	(a)	(c)	(a)	(b)	(a)

FILL IN THE BLANKS

1. Quantum effect become important and measurable only on _______ level of atoms and molecules. **[sub microscopic]**
2. Quantum results must blend smoothly with classical results when the quantum number becomes large. This statement is known as the _______. **[correspondence principle]**
3. The wave-particle duality is the basis of _______. **[quantum particle model]**

4. x-rays striking a target are scattered at various angles by electrons in the target. A shift in wavelength is observed for the scattered x-rays, and the phenomenon is known as the ______. **[compton effect]**

5. A proton is moving with the speed of 1.8×10^7 m/s. Its de Broglie wavelength is ______. **[2.2×10^{-14} m]**

6. An electron is accelerated through a potential difference of 150 V from rest. Its de Broglie wavelength is ______. **[1×10^{-10} m]**

7. The ______ nature of moving electrons is the principle of an electron microscope. **[wave]**

8. The combination of wave and particle properties is called ______ duality. **[wave-particle]**

9. Louis de Broglie proposed that photons have wave and particle characteristics and therefore (perhaps) all forms of matter have wave as well as ______ characteristics. **[particle]**

10. The de Broglie wavelength of an electron is 0.21 nm. Its kinetic energy is ______. **[5.47×10^{-18} J]**

11. The de Broglie wavelength of an electron is 0.7 A°. Its velocity is ______. **[10.4×10^6 m/s]**

12. The minimum energy of a photon required to produce an electron-positron pair must be equal to ______. **[1.02 MeV]**

13. The energy of the electrons that we need to use in an electron microscope to resolve a separation of 0.27nm is ______. **[20.6 MeV]**

[**Hint:** $E = \dfrac{2\pi^2\hbar^2}{m_e\lambda^2} = \dfrac{2\pi(\hbar c)^2}{(m_e c^2)\lambda^2}$

$$= \frac{2\pi(197.33\times10^{-15}\ \mathrm{MeV}m)^2}{(0.511\,\mathrm{MeV})(0.27\times10^{-9}\ m^2)} = 20.6\,\mathrm{MeV}]$$

14. The maximum kinetic energy transferred to a proton when hit by a photon of energy $h\nu$ is ______.

$$\left[\frac{h\nu}{1+m_p c^2/(2h\nu)}\right]$$

7

Schrodinger Wave Equation and Principles of Quantum Mechanics

7.1 NEED FOR A WAVE EQUATION AND JUSTIFICATION

The experiments described in Chapter 1 clearly reveal that classical physics is inadequate for atomic scale phenomena and for interaction between electrons and electromagnetic waves. We could proceed directly from these experiments to an explanation of the structure of atoms, molecules, and solids, but such a procedure would have to be very vague. The ideas expressed in $E = h\nu$ and $\lambda = h/mv$ are basic to all these problems, but it is not possible to build a complete understanding of atomic phenomena directly upon these ideas.

We succeeded in obtaining a mathematical substitute for the classical mechanics of a particle. Instead of solving Newtonian equations of motion, we may solve the wave equation and obtain a wave-packet which moves in space in exactly the same way as a particle moves according to classical mechanics. However, there is one difference, i.e., the wave-packet has a dimension whereas the classical particle is assumed to be dimensionless. But for all practical purposes the wave-packet may be treated as a point and classical mechanics as a special case of wave or quantum mechanics. But our position at this point is similar to the situation in classical mechanics before Newton's laws were developed. There existed a variety of experimental informations, such as the motion of pendulums, falling bodies, and bodies on inclined planes; there was also some sort of explanation of each experiment. But there was no unifying theory and no way of predicting the results of problems, where experiments could not be performed.

What we require is a theory that is based on the experiments mentioned in Chapter 1 and that enables us to predict and explain more complicated atomic phenomena. The theory that accomplishes this is the **wave mechanics or quantum mechanics** and the heart of this theory is the **Schrodinger equation**. Schrodinger, in 1925 using de-Broglie's ideas of matter waves set up a wave equation to describe the new mechanics of the particles, and successfully applied it to various atomic problems. His methods were rapidly adopted by other investigators, and applied with such success that there is hardly a field of physics and chemistry that has remained untouched by Schrodinger's work.

7.2 WAVE EQUATION

In classical mechanics, a wave equation is a second order differential equation in space and time. Solution of this wave equation represents wave disturbances in a medium. Obviously, a wave equation is the usual basis of mathematical theory of wave motion. As an example of wave equation we can consider the equation of an electromagnetic wave, travelling in the x-direction, i.e.,

$$\frac{\partial^2 E_y}{\partial x^2} = \frac{1}{c^2}\frac{\partial^2 E_y}{\partial t^2}$$

where E_y is the y-component of the electric intensity.

We must note that a differential equation for the wave associated with a particle in motion cannot be derived from first principles. One can develop the equation by any one of the following procedures:

(i) The classical equations of motion are transformed into a wave equation in accordance with wave properties of matter based on de-Broglie's hypothesis.

(ii) A complex variable quantity, called the **wave function,** is assumed to represent a plane simple harmonic wave associated with a free particle, and the classical expression for the total energy is used.

(iii) A particle at a given position and at given time is represented by **a wave-packet** which is obtained by superposition of a group of plane waves of nearly the same wavelength, which interfere destructively everywhere except at the wave-packet, and the classical expression for the total energy is used.

(iv) In the classical expression for the total energy of a particle, the dynamical quantities are replaced by their corresponding operators. These operators are allowed to operate upon the wave function.

Any one of the above procedure can be followed to develop the Schrodinger equation.

7.2.1 Time-Independent Schrodinger's Wave Equation

If a particle has wave properties, as de-Broglie proposed, it is expected that there should be some sort of wave equation, which describes the behaviour of the wave function. The usual basis of the mathematical theory of the wave motion is furnished by a differential equation. In accordance with the Maxwell's relations the electromagnetic equation is expressed as follows:

$$\frac{\partial^2 E}{\partial x^2} + \frac{\partial^2 E}{\partial y^2} + \frac{\partial^2 E}{\partial z^2} - \frac{1}{c^2}\frac{\partial^2 E}{\partial t^2} = 0 \tag{7.1}$$

where E is the wave amplitude and c is the velocity of electromagnetic wave.

We cannot follow a logical development from first principle to find the differential equations for matter waves, but we can guess a wave equation in harmony with the properties, which we find the waves to have and then test our guess by comparison of experimental data with predictions deduced from this wave equation.

Let us consider a system of stationary waves to be associated with the particle. Let $\psi(x, y, z, t)$ be the wave function for de-Broglie waves at any time t. Then the differential equation for the motion of particle in waves in accordance with Eq. (7.1) can be written as follows:

$$\frac{\partial^2\psi}{\partial x^2}+\frac{\partial^2\psi}{\partial y^2}+\frac{\partial^2\psi}{\partial z^2}-\frac{1}{u^2}\frac{\partial^2\psi}{\partial t^2}=0 \tag{7.2}$$

where u is the wave velocity of particle waves.

The solution of differential Eq. (7.2) gives $\psi(x, y, z, t)$ as a periodic displacement in terms of time as below:

$$\psi(x,y,z,t) = \psi_0(x,y,z)e^{-i\omega t} \tag{7.3}$$

where $\psi_0 \rightarrow$ amplitude of particle waves at the point considered. It is the function of position (x, y, z) only and independent of time (t). If the position vector is $\boldsymbol{r}$, then $\boldsymbol{r} = \hat{i}x+\hat{j}y+\hat{k}z$.

Therefore Eq. (7.3) can be written as $\psi(r,t) = \psi_0(r)e^{-i\omega t}$ (7.4)

Let us differentiate Eq. (7.4) twice with respect to t, then

$$\frac{\partial^2\psi}{\partial t^2}=-\omega^2\psi_0 e^{-i\omega t} \quad \text{or} \quad \frac{\partial^2\psi}{\partial t^2}=-\omega^2\psi \tag{7.5}$$

From Eqs. (7.2) and (7.5), we obtain

$$\frac{\partial^2\psi}{\partial x^2}+\frac{\partial^2\psi}{a y^2}+\frac{\partial^2\psi}{\partial z^2}+\frac{\omega^2}{u^4}\psi=0 \tag{7.6}$$

But $\omega = 2\pi\nu = 2\pi\frac{u}{\lambda} \quad \therefore \quad \frac{\omega}{u}=\frac{2\pi}{\lambda}$ (7.7)

From Eqs. (7.6) and (7.7), we obtain

$$\frac{\partial^2\psi}{\partial x^2}+\frac{\partial^2\psi}{\partial y^2}+\frac{\partial^2\psi}{\partial z^2}+\frac{4\pi^2}{\lambda^2}\psi=0$$

Writing $\nabla^2 = \frac{\partial^2}{\partial x^2}+\frac{\partial^2}{\partial y^2}+\frac{\partial^2}{\partial z^2}$, we get $\nabla^2\psi+\frac{4\pi^2}{\lambda^2}\psi=0$ (7.8)

Now, from de-Broglie's relation $\lambda = \frac{h}{mv}$

we get $\nabla^2\psi+\frac{4\pi^2 m^2 v^2}{h^2}\psi = 0 \quad \text{or} \quad \nabla^2\psi+\frac{p^2}{\hbar^2}\psi = 0$ (7.9)

Since in mechanics $p(= mv)$ is a function of time, this equation for space function $\psi(x, y, z)$ would contain p and therefore, also time, and this, of course, would be obvious nonsense.

However, by means of the principle of conservation of energy which in non-relativistic mechanics is of the form

$$= E \text{ or } \frac{p^2}{2m} + V(x,y,z) = E$$

We can express p as a function of space coordinates: $p^2 = 2m(E - V)$. Then from Eq. (7.9), we have time independent Schrodinger wave equation

$$\nabla^2\psi + \frac{2m}{\hbar^2}(E - V)\psi = 0 \tag{7.10}$$

Here, of course $V(x, y, z)$ is the potential energy.

7.2.2 Schrodinger's Wave Equation for a Free Particle

We have seen that in classical mechanics a free particle may be either at rest or moving with constant momentum p. In either case its total energy E is a constant.

To find the behaviour predicted by quantum mechanics for a free particle, we solve the time independent Schrodinger Eq. (7.10), setting $V(x, y, z) = 0$, we get Schrodinger's equation for a free particle as below:

$$\nabla^2\psi + \frac{2mE}{\hbar^2}\psi = 0 \tag{7.11}$$

For one-dimensional motion, the equation is

$$-\frac{\hbar^2}{2m}\frac{d^2\psi(x)}{dx^2} = E\psi(x) \tag{7.12}$$

The solutions are the eigen function $\psi(x)$, and the wave function $\psi(x, t)$ are

$$\psi(x,t) = \psi(x)\exp[-i\,Et/\hbar] \tag{7.13}$$

The eigen value E are equal to the total energy of the particle. An acceptable solution of the time-independent Schrodinger equation for this non-binding potential should exist for any value $E \geq 0$.

The form of the free particle wave function is

$$\psi(x,t) = \cos(kx - \omega t) + i\sin(kx - \omega t) = \exp(ikx - \omega t) = \exp(ikx)\exp(-i\omega t)$$

where $$k = \frac{\sqrt{2mE}}{\hbar} \quad \text{and} \quad \omega = E/\hbar \tag{7.14}$$

Comparing Eq. (7.14) with the general form of the wave function $\psi(x,t) = \psi(x)\exp[-i\,Et/\hbar]$, one finds that

$$\psi(x) = \exp(ikx) \tag{7.15}$$

Obviously, the complex exponential of Eq. (7.15) gives the form of a free particle eigen function corresponding to the eigen value E. We must remember that it is a travelling wave free particle eigen function because the corresponding wave function, $\psi(x, t) = \exp[i(kx - \omega t)]$, represents a travelling wave.

7.2.3 Time Dependent Schrodinger's Wave Equation

To obtain the time dependent Schrodinger's wave equation, we differentiate Eq. (7.4) with respect to t and obtain

$$\frac{\partial \psi}{\partial t} = -i\omega\psi_0 e^{-i\omega t} = -i2\pi\nu\psi_0 e^{-i\omega t} = -i\frac{E}{\hbar}\psi$$

or
$$E\psi = i\hbar\frac{\partial \psi}{\partial t} \tag{7.16}$$

From Eq. (7.16), substituting the value of $E\psi$ in Eq. (7.10), we obtain

$$\nabla^2\psi + \frac{2m}{\hbar^2}\left(i\hbar\frac{\partial \psi}{\partial t} - V\psi\right) = 0$$

or
$$-\frac{\hbar^2}{2m}\nabla^2\psi + V\psi = \hbar\frac{\partial \psi}{\partial t} \text{ or } -\frac{\hbar^2}{2m}\nabla^2\psi + V\psi + V\psi = -\frac{\hbar}{i}\frac{\partial \psi}{\partial t} \tag{7.17}$$

Eq. (7.17) is Schrodinger's famous wave equation containing the time. The agreement of results deduced from this equation justifies the belief that the equation is valid for the matter waves associated with a particle of mass m as long as relativistic effects can be neglected. The equation is remarkable among the differential equations of mathematical physics in that it contains $i = \sqrt{-1}$. But for this factor it would resemble the equation for the flow of heat in a solid body, which likewise contains the first derivative with respect to time. The motion of matter waves combines, in fact, some of the features of heat flow in a solid with other features resembling the progression of mechanical disturbances, such as sound waves. Eq. (7.17) can also be expressed as

$$\left(-\frac{\hbar^2}{2m}\nabla^2 + V\right)\psi = i\hbar\frac{\partial \psi}{\partial t} \text{ or } \left(-\frac{\hbar^2}{2m}\nabla^2 + V\right)\psi = -\frac{\hbar}{i}\frac{\partial \psi}{\partial t} \tag{7.18}$$

The operator $\left(-\frac{\hbar^2}{2m}\nabla^2 + V\right)$ is called **Hamiltonian** and is written as H; while operator $-\frac{\hbar}{i}\frac{\partial}{\partial t}$, operated on ψ, gives E [see Eq. (7.16)]. Therefore, Eq. (7.18) can be written as follows

$$H\psi = E\psi \tag{7.19}$$

This is known as the **Hamiltonian form** of the Schrodinger equation where H is the Hamiltonian coinciding with the total energy operator. Although, H and E both are energy operators but H depends on the space coordinates and derivatives whereas E depends upon the time derivative. As stated earlier all the three forms of Schrodinger's equation describe the motion of a non-relativistic particle.

Alternative Derivation of Time-Dependent Schrodinger Wave Equation

Consider a free particle of momentum p and energy E, moving in one-dimension with velocity v. Being a free particle, its total energy E is entirely kinetic, i.e.,

$$E = \frac{1}{2}mv^2 = \frac{1}{2m}(mv)^2 = \frac{p^2}{2m} \tag{7.20}$$

We have read that a moving particle is associated with de-Broglie waves which have wave length λ and frequency v. We have

$$p = \frac{\hbar}{\lambda} = \hbar k \quad \text{(de-Brogile Relation)} \tag{7.21}$$

and
$$E = hv = \hbar\omega \quad \text{(Einstein Relation)} \tag{7.22}$$

In Eq. (7.21), $k = 2\pi/\lambda$ is propagation constant and $\omega = 2\pi v$ is angular frequency of waves. Combining Eqs. (7.20), (7.21) and (7.22), one obtains

$$\hbar\omega = \frac{\hbar^2 k^2}{2m} \tag{7.23}$$

Now, we want a wave equation which describes these travelling waves involving Eq. (7.23). We must ensure that the function should be a harmonic one because such functions can be superimposed, i.e., yielding a wave-packet which will represent the particle. Obviously, this means that it must be sine, cosine or exponential function of $(kx - \omega t)$. Moreover, such a function be the exponential function, $\exp[i(kx - \omega t)]$ provided the differential wave equation be of the form:

$$i\hbar\frac{\partial\psi}{\partial t} = -\frac{\hbar^2}{2m}\frac{\partial^2\psi}{\partial x^2} \quad (\text{with } \psi = \text{constant}, \exp[i(kx - \omega t)]) \tag{7.24}$$

$$\therefore \quad i\hbar\frac{\partial}{\partial t}[\exp[i(kx-\omega t)]] = i\hbar(-i\omega)\exp[i(kx-\omega t)] = \hbar\omega$$

and
$$-\frac{\hbar^2}{2m}\frac{\partial^2}{\partial x^2}\exp[i(kx-\omega t)] = \frac{\hbar^2}{2m}(ik)^2\exp[i(kx-\omega t)] = \frac{\hbar k}{2m}\psi.$$ Thus one obtains

$$\hbar\omega\psi = \frac{\hbar^2 k^2}{2m}\psi \tag{7.25}$$

Eq. (7.24) is the Schrodinger wave equation for a free particle. We note that this equation is linear in ψ and its derivatives. Obviously, one can represent the wave-packet by Fourier superposition of individual waves

$$\psi = \sum_k A(k)\exp[i(kx-\omega t)] = \int A(k)\exp[i(kx-\omega t)]dk \tag{7.26}$$

Extension of Wave Equation to Three-Dimensions

Proceeding exactly in a manner described for one-dimension, we can write

$$E = \frac{|p|^2}{2m} = \frac{\hbar^2}{2m}|k|^2 = \frac{\hbar^2}{2m}(k_x^2 + k_y^2 + k_z^2)$$

and $$\psi = \text{const}\exp[i(k_x x + k_y y + k_z z - \omega t)] = A\exp[i(kr - \omega t)] \tag{7.27}$$

where $\boldsymbol{r}$ is position vector for the particle. Obviously, such a wave function $\psi(\boldsymbol{r}, \boldsymbol{t})$ will satisfy the following differential equation

$$i\hbar\frac{\partial \psi}{\partial t} = -\frac{\hbar^2}{2m}\nabla^2\psi \tag{7.28}$$

We note that there exists striking resemblance between Schrodinger wave equation and the classical momentum equation for a free particle. Thus, we have

$$E(\psi) = \frac{p^2}{2m}\psi$$

or $$i\hbar\frac{\partial}{\partial t}\psi = \frac{\hbar^2}{2m}\nabla^2\psi = \left(-\frac{i\hbar^2}{2m}\nabla\right)^2\psi \tag{7.29}$$

We note that $i\hbar\frac{\partial}{\partial t}$ and $-i\hbar\nabla$ operators are operating on the wave function ψ. This means that Schrodinger equation is the quantum mechanical analogue of classical equation. One can extend this procedure for a bound particle, i.e., one can include the effects of forces acting on the particle. Let V represent the potential energy of the particle when force $\boldsymbol{F}$ is acting on it. We have

$$\boldsymbol{F} = -\nabla V(\boldsymbol{r}, t) \tag{7.30}$$

Thus the total energy E of the particle,

$$E = T + V = \frac{p^2}{2m} + V \tag{7.31}$$

Replacing E by $\hbar\frac{\partial}{\partial t}$ and $\boldsymbol{p}$ by $-i\hbar\nabla$, one can write the Schrodinger equation for a bound particle under the influence of forces, i.e., $E\psi = \frac{p^2}{2m}\psi + V\psi$ as

$$i\hbar\frac{\partial \psi}{\partial t} = -\frac{\hbar^2}{2m}\nabla^2\psi + V\psi \tag{7.32}$$

This is **Three-Dimensional Time-Dependent Schrodinger Equation.**

Time-independent or Stationary State Schrodinger Equation

When the field, in which the particle is moving does not depend on time, i.e., it will only be a function of position $\boldsymbol{r}$, then the corresponding wave equation is called the **stationary**

state Schrodinger wave equation or time-independent Schrodinger wave equation. We have

$$\psi(r,t)=\psi(r)f(t) \tag{7.33}$$

Now, we can put time-independent wave equation as follows:

$$E\psi=\frac{p^2}{2m}\psi+V\psi \quad\Rightarrow\quad i\hbar\frac{\partial\psi}{\partial t}=\frac{\hbar^2}{2m}\nabla^2\psi+V\psi \tag{7.34}$$

Substituting Eq. (7.34) in Eq. (7.33), one obtains

$$i\hbar\psi\frac{\partial f}{\partial t}=-\frac{\hbar^2}{2m}f\nabla^2\psi+Vf\psi$$

Dividing the above equation throughout by $f\psi$, one obtains

$$\frac{1}{f}i\hbar\frac{\partial f}{\partial t}=-\frac{\hbar^2}{2m}\frac{1}{\psi}\nabla^2\psi+V(r) \tag{7.35}$$

We note that LHS of Eq. (7.35) depends on t only, whereas RHS only on $\mathbf{r}$. Obviously, this equality implies that both are to be independent of r and t and have to be equal to constant E. Now, one can put the equation for f as

$$i\hbar\frac{\partial f}{\partial t}=Ef \tag{7.36}$$

and for ψ as follows:

$$\left[-\frac{\hbar^2}{2m}\nabla^2\psi+V(r)\right]\psi=E\psi \quad\text{or}\quad H\psi=E\psi \tag{7.37}$$

where $H=-\frac{\hbar^2}{2m}\nabla^2\psi+V(r)$ is the Hamiltonian operator. Corresponding to the total energy, we have $\frac{p^2}{2m}+V=T+V$. Solving Eq. (7.33) for f, one obtains

$$f=c\exp[(-i\hbar)Et]=c\exp(-i\omega t) \tag{7.38}$$

where $E=\hbar\omega$ and c is an arbitrary constant. We note that f depends on t harmonically with frequency ω. The arbitrary constant is having the dimensions of energy and is called the separation constant. Now, one can put the normalised solution of the time-dependent Eq. (7.34) as

$$\psi(\mathbf{r},t)=\psi(\mathbf{r})\exp(-i\omega t)=\psi(\mathbf{r})\exp\left(-\frac{i}{\hbar}Et\right) \tag{7.39}$$

We can easily see that the probability density $\psi^*\psi$ is independent of time. We have

$$\psi^*\psi=\psi^*(\mathbf{r})\exp\left(\frac{i}{\hbar}Et\right)\psi(\mathbf{r})\exp\left(-\frac{i}{\hbar}Et\right)=\psi^*(\mathbf{r})\psi(\mathbf{r})=\text{independent of time.}$$

We note that $\psi^*\psi$ is constant in time. This means that wave Eq. (7.39) represents a stationary state of the particle.

7.2.4 Justification of the Wave Equation

Consider again a broad wave-packet representing a rather sharply defined momentum and moving from a region of space in which the particle is free forward a region in which a force is present. For simplicity we shall restrict to a one-dimensional description. We also suppose that the forces are conservative, so that classically the particle has, and maintains throughout its motion, a constant

$$E = \frac{p^2}{2m} + V \tag{7.40}$$

Accordingly, the momentum is a function of position. Fourier decomposition of incident wave-packet is

$$\psi(x,0) = \frac{1}{\sqrt{2\pi}} \int_0^\infty a(k) e^{ikx} dk \tag{7.41}$$

By omitting the range of integration from $-\infty$ to 0, we have included the assumption that the wave-packet moves toward positive x, and contains no negative k components. Using the substitution

$$k = \sqrt{\frac{2mE}{\hbar^2}} \tag{7.42}$$

we can transform Eq. (7.41) into an integral over the energy of the free particle:

$$\psi(x,0) = \int_0^\infty f(E) e^{ik(E_x)} dE_x \tag{7.43}$$

where $f(E)$ is assumed to be a smoothly varying positive function of E with a single peak. Thus Eq. (7.43) is a rather specialized wave-packet. When this wave-packet travels in force-free space, it spreads in time as follows:

$$\psi(x,t) = \int f(E) e^{\frac{i}{\hbar}(px-Et)} dE \tag{7.44}$$

Now, question arises: how this wave-packet will move when it enters the region where a force present?

Let us assume that the frequency remains **unaltered** when the wave-packet approaches the region of changing potential energy. This is basic assumption and recommends itself by analogy with other types of wave motion. This is further supported by principle of conservation of energy, i.e., constant energy implies constant frequency according to the equation $E = h\nu = \hbar\omega$.

In the spirit of these remarks we may compare the motion of a particle through a force field with the propagation of a wave through a strange medium whose detailed microscopic properties we do not propose to analyse here, but which can somehow be set into "sympathetic" vibrations of the same frequency and which thus interacts with the incident wave to modify its properties of propagation, i.e., its velocity of propagation and its wavelength. Thus the form of wave-packet moving through a region of force may be represented by

$$\psi(x,t)=\int f(E)e^{\frac{i}{\hbar}[a(x,E)-Et]}dE \tag{7.45}$$

This is a generalization of Eq. (7.44) in which a (x, E) is yet undetermined function. This function must be chosen so that Eq. (7.45) represents a wave-packet classically. Since $f(E)$ is assumed to be a smooth positive function of E, the peak of the amplitude of Eq. (7.45) occurs if x and t are chosen so as to make the phase $a(x, E) - Et$ stationary when $f(E)$ is near its peak. This ensures that the Fourier components near the peak of $f(E)$ are as much in phase as possible and are therefore added constructively. Mathematically, the phase is stationary if

$$\frac{\partial}{\partial E}[a(x,E)-Et]=0 \tag{7.46}$$

After differentiation, E in this equation is to be evaluated at the energy which corresponds to the peak of $f(E)$. Eq. (7.46) leads to the condition

$$t=\frac{\partial a}{\partial E} \tag{7.47}$$

On the other hand, classically for a conservative motion

$$t=\int_x \frac{dx'}{V(x')}=\int_x \frac{\partial p(x,E)}{\partial E}dx' \tag{7.48}$$

where $p(x, E)$ is the momentum as defined by Eq. (7.40). Comparison of Eq. (7.47) with Eq. (7.48) gives the possible choice

$$a(x,E)=\int_x p(x',E)dx' \tag{7.49}$$

Thus we obtain

$$\psi(x,t)=\int f(E)e^{\frac{i}{\hbar}\left(\int^x p(x',E)\,dx'-Et\right)}dE \tag{7.50}$$

If the particle is free and moves with constant momentum p corresponding to a wavelength, $\lambda=\frac{h}{p}$, Eq. (7.50) is an exact equation, instead of an approximation. It is reasonable to assume that Eq. (7.50) is better approximation than the more nearly $\lambda(x) = h|p(x)|$ is independent of x:

$$\left|\frac{d\lambda}{dx}\right| << 1 \tag{7.51}$$

If this is true, the wave is quasi-harmonic and $\lambda(x)$ can be considered as it is slowly varying wavelength. Eq. (7.51) can also be written as

$$\lambda\left|\frac{dp}{dx}\right| << p^2 \tag{7.52}$$

or

$$\lambda(x)\left|\frac{dV}{dx}\right| << \frac{[p(x)]^2}{m} \tag{7.53}$$

This condition implies that the potential energy changes vary little over the distance of a wavelength. The particle can then be considered as nearly free over a distance of many wavelengths. For this reason $\hbar\left|\frac{dp}{dx}\right|$ can be neglected compared to p^2 in expressions derived from Eq. (7.50).

One obstacle in the way of applying Eq. (7.50) extensively is that it may be quite inaccurate in regions where the kinetic energy varies rapidly with x, particularly near the classical turning points of the motion, where $p(x, E) = 0$ and where Eq. (7.52) can never be satisfied. Hence, Eq. (7.50) does not provide us with an adequate formula for the description of the particle motion, where it is least classical and most interesting.

It is at this point that we ask ourselves whether a more generally satisfactory wave function could not be found if we abandoned the explicit construction of wave-packets, which represent quite special semi-classical wave forms. A wave-packet must always be given in terms of a particular initial wave $\psi(x, 0)$. Let us ask instead for some general relation which $\psi(x, t)$ must satisfy, no matter what the particular initial condition may be. Our experience with classical physics suggests that such a relation is most likely going to be a **differential equation** for $\psi(x, t)$. In general, we expect that ψ waves are complex quantities and that $\psi(x, 0)$ alone, without knowledge of time derivative, determines the future of the wave. Hence, we shall attempt to obtain a first-order differential equation in t. Following this idea, let us differentiate Eq. (7.50) w.r.t. time:

$$\frac{\partial\psi}{\partial t} = \int f(E)\frac{i}{\hbar}Ee^{\left[\frac{i}{\hbar}\left(\int^x pdx' - Et\right)\right]}dE \tag{7.54}$$

We would like to eliminate in the initial condition $f(E)$ from R.H.S. of Eq. (7.54). This can almost be accomplished if we note

$$\frac{\partial\psi}{\partial t} \simeq \frac{i}{\hbar}\int f(E)p(x,E)e^{\left[\frac{i}{\hbar}\left(\int^x pdx' - Et\right)\right]}dE$$

and

$$\frac{\partial^2\psi}{\partial t^2} \simeq \frac{i}{\hbar^2}\int f(E)p^2(x,E)e^{\left[\frac{i}{\hbar}\left(\int^x pdx' - Et\right)\right]}dE$$

$$+\frac{i}{\hbar}\int f(E)\frac{\partial p}{\partial x}e^{\left[\frac{i}{\hbar}\left(\int^x pdx' - Et\right)\right]}dE \tag{7.55}$$

But according to Eq. (7.52) the last term is negligible in comparison to first. Using Eqs. (7.51), (7.54) and (7.55), we obtain at least as an approximation

$$i\hbar\frac{\partial\psi}{\partial t} = -\frac{\hbar^2}{2m}\frac{\partial^2\psi}{\partial t^2} + V(x)\psi \tag{7.56}$$

Eq. (7.56) is valid even when Eq. (7.50) is not. In fact Eq. (7.56) is the quantum mechanical wave equation and thus justifies the Schrodinger's wave equation.

7.3 WAVE FUNCTION AND ITS INTERPRETATION

At this stage it would be interesting as well as befitting to define or interpret the significance of the wave function ψ in terms of the observable properties associated with the particle or the system. It is, however, an essential element of the quantum theory that the wave function associated with a physical system contains all relevant information about the behaviour of the system and thus describes it completely. In other words, any meaningful question about the result of an experiment performed upon the system can be answered if the wave function is known. Thus far, the only interpretative guide available to us is that the wave function be large where the particle is likely to be and small elsewhere. It is apparent, therefore, that this has to be supplemented with more detailed statements or postulates that enable us to get out of $\psi(\boldsymbol{r}, t)$ the maximum amount of information permitted by nature.

It is assumed that the wave associated with a particle in motion is represented by a complex variable quantity $\psi(x, y, z, t)$ or $\psi(\boldsymbol{r}, t)$, called the wave function. This complex wave function $\psi(\boldsymbol{r}, t)$ is defined to describe the entire physical state of a particle or a system of particles such as an atom or molecule. Let us assume that the wave function is specified in the x-direction by

$$\psi = A\exp\left[-i\omega\left(t-\frac{x}{v}\right)\right] \tag{7.57}$$

where A is constant and $i=\sqrt{-1}$. The phase velocity v_p is given by

$$v_p = \nu\lambda = 2\pi\nu\frac{\lambda}{2\pi} = \frac{\omega}{k} \tag{7.58}$$

$$\therefore \qquad \frac{\omega}{v_p} = k \tag{7.59}$$

Using Eqs. (7.58), Eq. (7.59) can be expressed as

$$\psi(x,t) = A\exp[i(\omega t - kx)] \tag{7.60}$$

Here the propagation constant $K = 2\pi/\lambda$ and the momentum in x-direction $p_x = \frac{h}{\lambda} = \frac{h}{2\pi}\frac{2\pi}{\lambda} = \hbar k$ and energy $E = h\nu = \frac{h}{2\pi}2\pi\nu = \hbar\omega$. Using these results, one obtains from Eq. (7.60),

$$\psi(x,t) = A\exp\left[-i\left(\frac{Et}{\hbar} - p_x\frac{x}{\hbar}\right)\right] = A\exp\left[-\frac{i}{\hbar}(Et - p_x x)\right] \tag{7.61}$$

$$= A\exp\left[-\frac{i}{\hbar}(p_x x - Et)\right] \tag{7.62}$$

Eq. (7.61) or Eq. (7.62) represents a plane continuous simple harmonic wave associated with a particle of accurately known momentum p_x and total energy E moving in the positive x-direction. For a plane wave travelling in an arbitrary direction, the wave function is written as

$$\psi(\boldsymbol{r},t)=A\exp\left[\frac{i}{\hbar}(pr-Et)\right] \tag{7.63}$$

where the vector $\boldsymbol{p}$ is the momentum and the vector $\boldsymbol{r}$ is the position vector of the particle. Using the above form of the wave function we will derive Schrodinger equation.

We have assumed that a wave function is a complex quantity and therefore it cannot have a direct physical meaning. It may be expressed in the form

$$\psi(x,y,z,t)=a+ib \tag{7.64}$$

where a and b are real functions of the variables (x, y, z, t). One can obtain the complex conjugate of ψ which is denoted by ψ^* by changing i to $-i$, i.e.,

$$\psi^*(x,y,z,t)=a-ib \tag{7.65}$$

Multiplying Eq. (7.64) by Eq. (7.65), one obtains

$$\psi(x,y,z,t)\psi^*(x,y,z,t)=a^2+b^2$$

On the L.H.S. of Eq. (7.65) the product of $\psi\psi^*$ is denoted by P, i.e.,

$$P=\psi(x,y,z,t)\psi^*(x,y,z,t)=|\psi(x,y,z,t)|^2 \tag{7.66}$$

$$P=|\psi(x,y,z,t)|^2=a^2+b^2 \tag{7.67}$$

Obviously, the product of ψ and ψ^* is real and positive if $\psi \neq 0$. Its positive square root is denoted by $|\psi(x, y, z, t)|$ and is called modulus of ψ. The quantity $|\psi(x, y, z, t)|^2$ is called the **probability density**. Maxborn has used the result Eq. (7.67) and interpreted $\psi(x, y, z, t)$ as follows:

For the motion of a particle, the quantity

$$PdV=\psi(x,y,z,t)\psi^*(x,y,z,t)dV=|\psi(x,y,z,t)|^2\,dV \tag{7.68}$$

is the probability that the particle will be found in a volume element dV surrounding the point at position (x, y, z) at time t. Thus the total probability for finding the particle anywhere in space in question is proportional to the integral of $|\psi(x, t)|^2$ over all the space. For the motion of a particle in one dimension the quantity.

$$Pdx=\psi(x,t)\psi^*(x,t)=|\psi(x,t)|^2\,dx \tag{7.69}$$

In this case $|\psi(x, t)|^2$ is called the **probability per unit distance**. Thus the total probability for the one-dimensional space along the x-axis is

$$P_x=\int_{-\infty}^{\infty}|\psi(x,t)|^2\,dx \tag{7.70}$$

For three-dimensional space, we have

$$P = \int_{-\infty}^{\infty} |\psi(x,y,z,t)|^2 \, dV \tag{7.71}$$

The above integrals must be finite in order to represent a real particle. The above discussion clearly reveals that the quantum laws and results of their measurements have probabilistic interpretation. The wave function $\psi(\boldsymbol{r}, t)$ is sometimes called probability **amplitude** for a position of the particle. Since probability is a dimensionless quantity it follows that the dimensions of the wave function must be $(\text{length})^{3/2}$.

It is convenient to define the probability density for the particle as

$$\rho(x,t) = \frac{|\psi(x,t)|^2}{\int_{-\infty}^{\infty} |\psi(x,t)|^2 \, dx} \tag{7.72}$$

where

$$\int_{-\infty}^{\infty} \rho(x,t) dt = 1 \tag{7.73}$$

For physical acceptable wave functions it is always possible to introduce an appropriate factor in the wave function such that

$$\int_{-\infty}^{\infty} |\psi(x,t)|^2 \, dx \int_{-\infty}^{\infty} \psi^*(x,t)\psi(x,t) dx = 1 \tag{7.74}$$

When Eq. (7.74) is true, wave functions ψ is said to be **normalised**. A wave function which satisfies the condition Eq. (7.74) is said to be normalised (to unity). The normalising condition for the wave function for the motion of a particle in three dimension is

$$\int_{-\infty}^{\infty} |\psi(x,y,z)|^2 \, dV = 1 \tag{7.75}$$

where ψ is a function of the space coordinates (x, y, z).

If a wave function does not satisfy the normalisation condition Eq. (7.75), then it must be multiplied by a constant factor, called the **normalisation factor**. For example

$$\int \varphi^2 dx = N (N \neq 1)$$

Then the normalisation factor will be $\frac{1}{\sqrt{N}}$ and the normalized wave function will be $\frac{1}{\sqrt{N}}\varphi$. One can easily see that $\frac{1}{\sqrt{N}}\varphi$ will be as good a wave function as ψ itself.

From Eq. (7.72), it is obvious that when a wave function ψ is normalised the probability density is simply the square of its amplitude. To arrive at a result consistent with physical observation, several additional requirements are imposed on the wave function ψ:

(i) ψ must be **finite** for all values of x, y, z.
(ii) ψ must be **well behaved,** i.e., **single valued and continuous everywhere,** obviously, for each set of values of x, y, z, ψ must have one value only.

(iii) ψ **must be continuous in all regions except in those regions where the potential energy** $V(x, y, z) = \infty$.

(iv) **If a particle can be present in the states described by the wave functions** $\psi_1, \psi_2, \psi_3 ......, \psi_n$, **it can also be in all states described by the wave function constructed from** $\psi_1, \psi_2, \psi_3 ......, \psi_n$, **by the linear transformation**

$$\psi = a_1\psi_1 + a_2\psi_2 + ... + a_n\psi_n \tag{7.76}$$

where $a_1, a_2, a_3 ......, a_n$ are arbitrary complex numbers. This is known as **Principle of Superposition.**

(v) The wave function ψ must approach zero as x tends to $\pm\infty$.

(vi) A wave function satisfying the following condition is said to be an **orthogonal wave function**

$$\int_{-\infty}^{\infty} \psi_i^*(x)\psi_j(x)dx = 0 \quad (i \neq j) \tag{7.77}$$

The orthogonality condition guarantee the non-interference of the wave function representing different states.

(vii) **Degeneracy:** There may, however, be situations when two wave functions ψ_j and ψ_k correspond to the same energy E. This is a case of 'degeneracy'; the system in such an energy state is said to be in a degenerate state. In such a situation, the degenerate wave function ψ_j and ψ_K will not be necessary be 'orthogonal.'

(viii) **Orthonormal set:** The normalisation and orthogonality conditions may be combined as follows:

$$\int \psi_i \psi_j dV = 1 \text{ if } i \neq j$$

$$\int \psi_i \psi_j dV = 0 \text{ if } i \neq j \tag{7.78}$$

Wave functions satisfying Eq. (7.78) are said to form **"Orthonormal"** set of wave functions.

7.3.1 Statistical Interpretation of Wave Function ψ by Max Born

The fundamental postulate relating to the physical interpretation is due to Max Born which is contained in the statement known as Bern's first assumption. According to this assumption, a particle such as an electron is still a point indivisible and without dimension. But its position may not always be exactly known. In general only its **probable** positions may be known. Bern's first assumption states that:

The quantity $\psi^*\psi = |\psi|^2$ is to be interpreted as a **probability density or probability per unit volume** for a particle in the state ψ. More precisely, if a particle is described by a wave function ψ, then, in a measurement of the position of the particle, the probability $P(\mathbf{r})d\mathbf{r}$ of finding it in a volume element $d\tau = dxdydz$ at the point $\mathbf{r}$ is proportional to $|\psi(\mathbf{r})|^2 d\mathbf{r}$. Thus

$$P(\mathbf{r},t) = |\psi|(\mathbf{r},t)|^2 \tag{7.79}$$

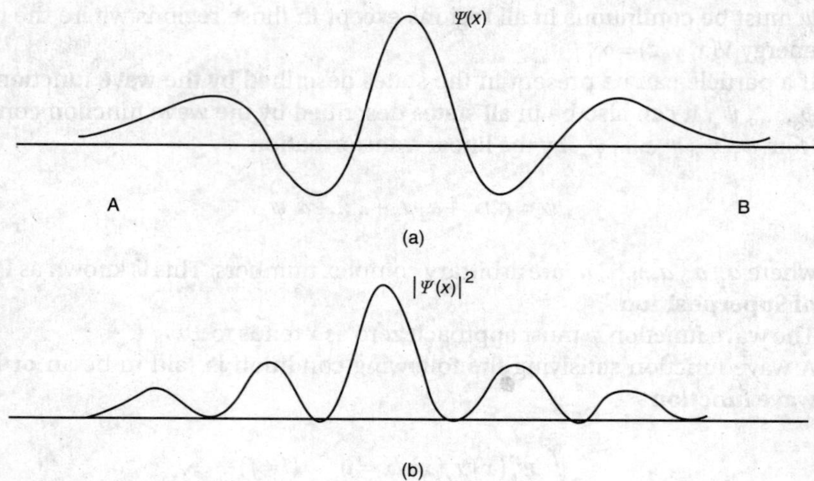

Fig. 7.1 (a) Wave function of a particle moving between A and B
(b) Probability distribution corresponding to the wave function shown in (a)

This means that $P(\boldsymbol{r}, t)\, dx\,dy\,dz$ is to be the possibility of finding the particle in its volume element $dx\,dy\,dz$ about its point $\boldsymbol{r}$ at the time t, when a large number of precise position measurements are made on independent particles each of which is described by the one-particle wave function $\psi(\boldsymbol{r}, t)$. The wave function $\psi(\boldsymbol{r}, t)$ can be either real or complex. The only quantity having a physical meaning is the probability density $P(\boldsymbol{r}, t)$. The probability of finding a particle in a volume element $\tau = dx\,dy\,dz$ at the point $\boldsymbol{r}$ at time t is proportional to $\psi^*\psi d\tau$, i.e.,

$$P(r)d\tau \propto |\psi(r,t)|^2\, d\tau \tag{7.80}$$

This postulate suggested by Max Born and later generalized by Bohr, Dirac, Heisenberg and others shows that the quantum laws and results of their measurements have probabilistic interpretation. The wave function $\psi(\boldsymbol{r}, t)$ is sometimes called a **probability amplitude** for a position of the particle.

7.3.2 Normalization of the Wave Function

In physical situations, waves are frequently encountered where the particle is bound by forces to a limited region or space. The atomic electron, the particle in a box with impenetrable walls are examples of this kind. In such cases, the particle is certainly to be found if the entire space is investigated, i.e., the probability of finding the particle in this space[1] is unity, we require

$$\int P(r,t)d\tau = 1 \tag{7.81}$$

[1]When the particle is bound to a limited space, the probability of finding the particle at infinite distance is zero, i.e., $\psi\psi^* = 0$ (at $x = \infty$).

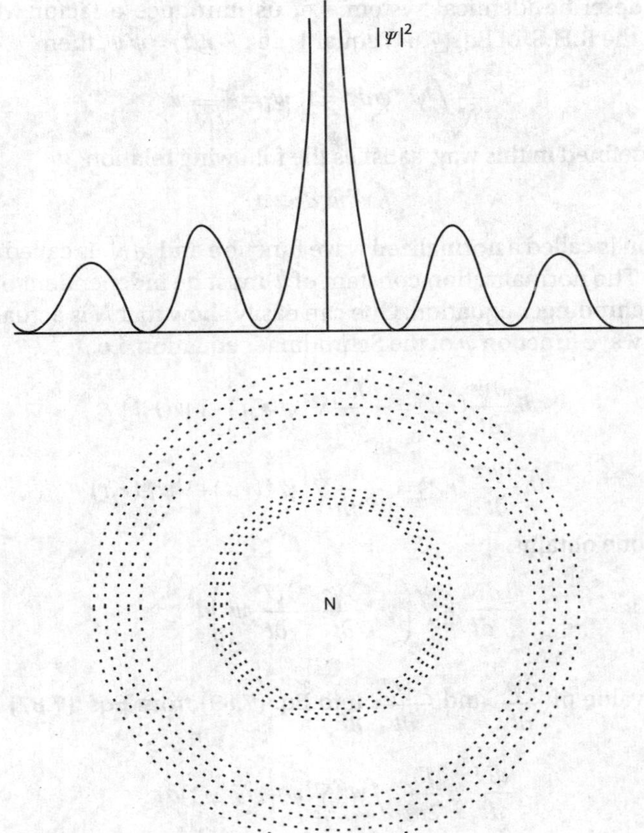

Fig. 7.2 Probability distribution for an electron in an atom

The integration extends over all space. The proportionality between $P(\boldsymbol{r}, t)$ and $\psi^*\psi$ in Eq. (7.80) introduces a constant which can be easily determined from Eq. (7.81). From Eq. (7.79), one obtains

$$1=\int P(\boldsymbol{r},t)d\tau=C\int \psi^*\psi d\tau \tag{7.82}$$

where C is the constant of proportionality and is given by

$$C=1/\int \psi^*\psi d\tau \tag{7.83}$$

Obviously, the probability, is given by

$$P(\boldsymbol{r},t)=\frac{\psi^*\psi}{\int \psi^*\psi d\tau} \tag{7.84}$$

From Eq. (7.84) it is obvious that $P(\boldsymbol{r}, t)$ remains unchanged if $\psi(\boldsymbol{r}, t)$ is multiplied by any constant factor. This does not change the fact that ψ is a solution of the Schrodinger equation (superposition principle). Obviously, two wave functions differing only by a constant factor describe identical system. Let us introduce a factor which makes the denominator on the R.H.S. of Eq. (7.84) equal 1, i.e., $P(\mathbf{r}, t) = \psi^*\psi$, then

$$\frac{1}{N}\int \psi^*\psi d\tau = 1,\ \psi_1 = \frac{1}{\sqrt{N}}\psi \tag{7.85}$$

The function ψ_1 defined in this way, satisfies the following relation

$$\int \psi_1^*\psi_1 d\tau = 1 \tag{7.86}$$

The wave function is called a normalized wave function and $\sqrt{N}$ is called **the normalization coefficient**. The normalization constant of ψ must be independent of time so that ψ_1 may satisfy the Schrodinger equation. One can easily show that N is actually independent of time for every wave function ψ of the Schrodinger equation, i.e.,

$$i\hbar\frac{\partial\psi}{\partial t}(\mathbf{r},t) = -\frac{\hbar^2}{2m}\nabla^2\psi(\mathbf{r},t) + V\psi(\mathbf{r},t) \tag{7.87}$$

and

$$-i\hbar\frac{\partial\psi^*}{\partial t}(\mathbf{r},t) = -\frac{\hbar^2}{2m}\nabla^2\psi^*(\mathbf{r},t) + V\psi^*(\mathbf{r},t) \tag{7.88}$$

From Eq. (7.85), one obtains,

$$\frac{dN}{dt} = \int\left(\psi^*\frac{\partial\psi}{\partial t} + \frac{\partial\psi^*}{\partial t}\psi\right)d^3\tau \tag{7.89}$$

Substituting the value of $\frac{\partial\psi}{\partial t}$ and $\frac{\partial\psi^*}{\partial t}$ into Eq. (7.89) from Eqs. (7.87) and (7.88), one obtains

$$\frac{dN}{dt} = \frac{i\hbar}{2m}\int \psi^*\nabla^2\psi - \psi\nabla^2\psi^*)d\tau \tag{7.90}$$

Transforming the above integral into a surface integral by the use of Green's theorem, one obtains

$$\frac{dN}{dt} = \frac{i\hbar}{2m}\int_s (\psi^*\nabla\psi - \psi\nabla\psi^*).\hat{n}dS \tag{7.91}$$

Here S is the surface enclosing the volume of integration and $\hat{n}$ is the unit vector normal to the surface element dS. Since the wave function ψ should vanish at large distances in order to define the position of the particle, and hence the surface integral in Eq. (7.91) approaches to zero as the surface of integration recedes to infinity. Thus, we have $\frac{dN}{dt} = 0$, i.e., the normalization constant is independent of time. Two wave functions are said to be orthogonal when

$$\int \psi_i^*\psi_j d\tau = 0 \quad i \neq j \tag{7.92}$$

the integration being over all space. If the wave functions are orthogonal as well as normalized, then we shall call them as **orthonormal wave functions**. In general, the orthonormal wave functions satisfy the condition

$$\int \psi_i^*\psi_j d\tau = \delta_{ij} \quad i \neq j \tag{7.93}$$

where δ_{ij} is the Kronecker delta function defined as

$$\delta_{ij} = \begin{cases} 1\,f \text{ or } & i = j \\ 0\,f \text{ or } & i \neq j \end{cases}$$

7.3.3 Boundary Condition at Large Distance

In the preceding section we have seen that any wave function, which has the property

$$\int |\psi(r,t)|^2 \, d\tau = 1$$

is said to be normalized. Evidently, $\psi_1 = \frac{1}{\sqrt{N}}\psi$ through which the normalized wave function is defined, makes sense only if the normalization coefficient, $\sqrt{N}$ is finite. The finiteness implies that $|\psi(\boldsymbol{r}, t)|^2$ vanishes at infinity and hence

$$\psi(r,t) \to 0 \quad \text{as} \quad r \to \infty \tag{7.94}$$

Eq. (7.94) is called a boundary condition, which holds for all those wave functions, which can be normalized.

7.3.4 Acceptable or Well-Behaved Wave Function

The interpretation given to ψ or ψ^2 imposes certain restrictions on the acceptability of wave function:

(i) The wave function ψ must always remain finite; and infinite value of ψ at any point would mean infinite probability of finding the particle at that point, which is assured, because no real wave can have infinite amplitude. (ii) ψ must be single valued at all point, i.e., it must have one and only one value at a particular point. This restriction on wave function arises.

Because the probability at any given point must be an unambiguous quantity. (iii) ψ and functions $\partial\psi/dx$, $\partial\psi/\partial y$ and $\partial\psi/\partial z$ must be continuous, i.e., their values must not change abruptly at any point. This is because of the fact that in the Schrodinger wave equation itself $H\psi$ i.e., the second derivative of wave function ψ must be defined at every point in space. Let for example, the probability at x were 80% and that at the neighbouring point $x + dx$ only 20%, it would be a highly ambiguous situation. Thus any function, which is finite, single valued and continuous can be taken as an acceptable or well behaved function.

7.3.5 Symmetric and Antisymmetric Wave Functions

One can express the wave function $\psi(1, 2, 3, \ldots n)$ of a system of non interacting particles as the product of the wave functions $\psi(1), \psi(2), \psi(3), \ldots \psi(n)$ of the individual particles i.e.,

$$\psi(1,2,3,\ldots N) = \psi(1)\psi(2)\psi(3)\ldots\psi(n) \tag{7.95}$$

Let us use Eq. (7.95) to look into the kinds of wave functions that can be used to describe a system of two particles. Let us suppose that one of the particles is in quantum state a and the other in quantum state b. Because the particles are identical, it should make no difference in the probability density $|\psi|^2$ of the system if the particles are exchanged, with the one in state *a* replacing the one in state *b*, and vice versa. Symbolically, one require that

$$|\psi(1,2)|^2 = |\psi(2,1)|^2 \tag{7.96}$$

The wave function $\psi(2, 1)$ that represents the exchanged particles can be either

$$\textbf{Symmetric} \quad \psi(2,1) = \psi(1,2) \tag{7.97}$$

or

$$\textbf{Antisymmetric} \quad \psi(2,1) = -\psi(1,2) \tag{7.98}$$

and still fulfill Eq. (7.96). We may note that the wave function of the system is not itself a measurable quantity, and so it can be in sign by the exchange of the particle. Wave functions that are unaffected by an exchange of particles are said to be **symmetric**, while those that reverse sign upon such exchange are said to be **antisymmetric**.

If particle 1 is in state *a* and particle 2 is in state *b*, the wave function of the system is, according to Eq. (7.95)

$$\psi_I = \psi_a(1)\psi_b(2) \tag{7.99}$$

If particle 2 is in state *a* and the particle 1 is state *b*, the wave function is

$$\psi_{II} = \psi_a(2)\psi_b(1) \tag{7.100}$$

Because the two particles are indistinguishable, one have no way to know at any moment whether ψ_I is correct at any moment is the same as the likelihood that ψ_{II} is correct.

Equivalently, one can say that the system spends half the time in the configuration whose wave function is ψ_I and the other half in the configuration whose wave function is ψ_{II}. Obviously, a linear combination of ψ_I and ψ_{II} is the proper description of the system. There are two such combinations, symmetric and antisymmetric possible, i.e.,

$$\textbf{Symmetric:} \quad \psi_s = \frac{1}{\sqrt{2}}[\psi_a(1)\psi_b(2)+\psi_a(2)\psi_b(1)] \tag{7.101}$$

$$\textbf{Antisymmetric:} \quad \psi_A = \frac{1}{\sqrt{2}}[\psi_a(1)\psi_b(2)-\psi_a(2)\psi_b(1)] \tag{7.102}$$

There are a number of important distinctions between the behaviour of particles in systems whose wave functions are symmetric and that of particles in systems whose wave functions are antisymmetric. In the symmetric case, both particles 1 and 2 can simultaneously exist in the same state, with $a = b$. In the antisymmetric case, if we set $a = b$, one finds that

$$\psi_A = \frac{1}{\sqrt{2}}[\psi_a(1)\psi_b(2)-\psi_a(2)\psi_b(1)] = 0$$

Clearly, the two particles cannot be in the same quantum state. Pauli found that no two electrons in an atom can be in the same quantum state, so one can conclude that systems of electrons are described by wave functions that reverse sign upon the exchange of any pair of them.

All the particles which have odd half-integral spins $\left(\frac{1}{2},\frac{3}{2},\frac{5}{2},\ldots\right)$ have wave functions that are antisymmetric to an exchange of any pair of them. Such particles, which include protons and neutrons as well as electrons, obey the Pauli exclusion principle when they are in the same system. That is, when they move in a common force field, each member of the system must be in a different quantum state. Particles of odd half-integral spin are often referred to as **fermions**.

Particles whose spins are 0 or an integer (1, 2, 3, ...) have wave functions that are symmetric to an exchange of any pair of them. These particles, which include photons, alpha particles and helium atoms, do not obey the Pauli exclusion principle. Particles of 0 or integer spin are often referred to as **bosons**.

We may note that there are other consequences of the symmetry or antisymmetry of particle wave function besides that expressed in the Pauli exclusion principle. It is these consequences that make it useful to classify particles according to the nature of their wave functions rather than merely according to whether or not they obey the Pauli exclusion principle.

7.4 CONTINUITY EQUATION AND CONSERVATION OF PROBABILITY OR PROBABILITY CURRENT DENSITY

Immediately doubts arise over the consistency of the probabilistic interpretation of the wave function as discussed in section 7.3. If the probability of finding the particle in some bounded region of space decreases as time goes on, then the probability of finding it outside this region must increase by the same amount. The probability interpretation of the ψ waves can be made consistently only if this **conservation of probability** is guaranteed. This requirement is fulfilled, owing to Gauss's integral theorem, if it is possible to define a **probability current density** $\boldsymbol{J}$ which together with the probability density $\rho = \psi^*\psi$ satisfies a continuity equation

$$\frac{\partial \rho}{\partial t} + \nabla . \boldsymbol{J} = 0 \tag{7.103}$$

exactly as in the case of the conservation of matter in hydrodynamics, or conservation of charge in electrodynamics. A relation of the form Eq. (7.103) can easily be deduced from the wave equation. As stated earlier that the probability equation

$$P(r)dr = \int_V \psi^* \psi d\tau \tag{7.104}$$

gives the probability of finding the particle in a region of space of volume V bounded by the surface A. In order to determine the probability flow of the particle, we must know the changes in the probability with time within the region. Therefore differentiating Eq. (7.104) with time, we get

$$\frac{dP}{dt} = \frac{d}{dt}\int_V \psi^* \psi d\tau = \int_V \left(\frac{d\psi^*}{dt}\psi + \psi^* \frac{d\psi}{dt} \right) d\tau \tag{7.105}$$

Time independent Schrondinger's equation and its complex conjugate are given by

$$-\frac{\hbar^2}{2m}\nabla^2 \psi + V\psi = i\hbar \frac{\partial \psi}{\partial t} \tag{7.106}$$

$$-\frac{\hbar^2}{2m}\nabla^2 \psi^* + V\psi^* = -i\hbar \frac{\partial \psi^*}{\partial t} \tag{7.107}$$

Multiply Eq. (7.106) by ψ^* and Eq. (7.107) by ψ and then subtracting Eq. (7.106) from Eq. (7.107), we get

$$-\frac{\hbar^2}{2m}\psi\nabla^2\psi^*+\frac{\hbar^2}{2m}\psi^*\nabla^2\psi=-i\hbar\frac{\partial\psi^*}{\partial t}-i\hbar\psi^*\frac{\partial\psi}{\partial t}$$

or

$$\psi\frac{\partial\psi^*}{\partial t}+\psi^*\frac{\partial\psi}{\partial t}=-\frac{\hbar}{2im}(\psi^*\nabla^2\psi-\psi\nabla^2\psi^*) \tag{7.108}$$

From Eqs. (7.105) and (7.108), we obtain

$$\frac{dP}{dt}=-\frac{\hbar}{2im}\int_V(\psi^*\nabla^2\psi-\psi\nabla^2\psi^*)d\tau \tag{7.109}$$

This integral can be transformed into integral over the surface A by Green's theorem and we have

$$\frac{dP}{dt}=-\frac{\hbar}{2im}\int_A(\psi^*\nabla\psi-\psi\nabla\psi^*).dA \tag{7.110}$$

or

$$\frac{dP}{dt}+\nabla.J=0 \tag{7.111}$$

where $J=\frac{\hbar}{2im}(\psi^*\nabla\psi-\psi\nabla\psi^*)$. J is called the **probability current density.**

Clearly Eq. (7.111) is analogous to Eq. (7.103). Therefore it is quite reasonable to interpret the vector J as a **probability current density**. If Eq. (7.111) is multiplied by charge e of the particle then eP give the charge density and eJ give the **current density of charge in quantum mechanical description**. Eq. (7.111) can also be expressed as

$$\frac{dP}{dt}+(-\nabla).J=-\int_A JdA \tag{7.112}$$

This shows that the rate of change of probability of the particle inside a surface is equal to the negative of probability density current through the surface A. Relation is generally useful for the determination of flux of particles in the scattering problems. The idea of probability density current is generally useful for the description of unbounded systems.

It is important to note that the operator $-i\hbar\nabla$ represents the momentum, even when a force is present. Then the velocity will be represented by the operator, $-\frac{i\hbar}{m}\nabla$ or $\frac{\hbar}{im}\nabla$ and it is apparent that

$$J=\text{real part of}\left(\psi^*\frac{h}{im}\nabla\psi\right)\text{or real part of}\left(\psi\frac{h}{im}\nabla\psi^*\right) \tag{7.113}$$

While this interpretation of J is suggestive, it must be realised that J is not susceptible to direct measurement in the sense in which P is.

7.5 SOLUTION OF TIME-DEPENDENT SCHRODINGER'S EQUATION

The time dependent Schrodinger's equation is expressed as

$$\left(-\frac{\hbar^2}{2m}\nabla^2+V\right)\psi=-\frac{\hbar}{i}\frac{\partial\psi}{\partial t} \tag{7.114}$$

As stated earlier that Eq. (7.114) is a partial differential equation in four variables (x, y, z, t). In the cases where V is independent of time and function of position coordinates only. Schrodinger suggested seeking solution of Eq. (7.114) which represents the standing waves. If ψ represents standing waves, it may be expressed as

$$\psi=f(x,y,z)\phi(t) \tag{7.115}$$

where $f(x, y, z)$ is independent of time t and $\phi(t)$ is independent of position coordinates (x, y, z). From Eqs. (7.114) and (7.115), we have

$$\left[-\frac{\hbar^2}{2m}\nabla^2+V(r)\right]f(r)\phi(t)=-\frac{\hbar}{i}\frac{\partial}{\partial t}[f(r)\phi(t)] \tag{7.116}$$

Dividing Eq. (7.116) by $f(\boldsymbol{r})\phi(t)$, we obtain

$$\frac{1}{f(r)}\left[-\frac{\hbar}{2m}\nabla^2+V(r)\right]f(r)=-\frac{\hbar}{i}\frac{1}{\phi(t)}\frac{\partial}{\partial t}\phi(t) \tag{7.117}$$

In Eq. (7.117) we note that R.H.S. is independent of position coordinates (x, y, z or say $\boldsymbol{r}$) while L.H.S. is independent of time. Therefore, if Eq. (7.117) is to be satisfied, each side of this Eq. (7.117) must be equal to a constant say, E. We have

$$\frac{1}{f(r)}\left[-\frac{\hbar^2}{2m}\nabla^2+V(\boldsymbol{r})\right]f(\boldsymbol{r})=E \tag{7.118}$$

or

$$\nabla^2 f+\frac{2m}{\hbar^2}(E-V)f=0$$

and

$$-\frac{\hbar}{i}\frac{1}{\phi(t)}\frac{\partial}{\partial t}\phi(t)=E \quad \text{or} \quad -\frac{\hbar}{i}\frac{1}{\phi(t)}\frac{\partial\phi(t)}{\partial t}=E \tag{7.119}$$

The solution of Eq. (7.119) is given by

$$\phi(t)=e^{Et/i\hbar} \tag{7.120}$$

On comparing Eq. (7.120) with the equation of plane wave: $\psi=ae^{\frac{2\pi i}{h}(Et-p.r)}$ we notice that E must have the dimensions of energy. Therefore, E appearing in Eq. (7.119) designates the total energy of the particle. Substituting the value of $\phi(t)$ from Eq. (7.120) in Eq. (7.115), we obtain

$$\psi=f(x,y,z)e^{-Et/i\hbar}=f(r)e^{-\frac{iEt}{\hbar}} \tag{7.121}$$

The solutions to the Eq. (7.119) called **eigen functions**[2], exist only for definite values of E, called **eigen values.** The whole set of eigen values of E is called the energy spectrum of the particle or system of particles. The most important aim of quantum mechanics is to determine the eigen values and eigen functions of particles or system of particles. Eigen values are also termed as **characteristic or proper** values whereas eigen functions are termed as characteristic functions or wave functions. However, Schrodinger's equation is only one of a large number of possible eigen value equations, and other eigen value equations must frequently be solved. The eigen value is not only the average value of the quantity obtained from a large number of measurement but each individual measurement gives the same value. It is, therefore, also called the **sharp value**. The fact of a physical quantity leaving a sharp value in a given experimental situation is the classical equivalent of conservation of physical quantity.

Even values have the remarkable property of being invariant under linear transformations of the coordinates, as will be clear as we proceed along.

Usually, one vector is associated with each value which satisfies the eigen value equation. It is referred to as the **eigen vector** of the operator involved. Further, the different eigen values yield a set of co-directional eigen vectors or an **eigen line**. In cases of degeneracy, however, the eigen lines are partially indeterminate.

The wave functions of quantum mechanics which are of interest are **well behaved** functions and satisfy the continuity conditions and boundary conditions. Let $\psi(x)$ be such a function of a state of the system then by result of operation with an operator A_{op}, we get the same function as

$$A_{op}\psi(x) = a\psi(x) \tag{7.122}$$

where a is complex number, $\psi(x)$ is called the eigen function of the operator A_{op} except for multiplication by the number a, $\psi(x)$ is not changed by the operation A_{op} and in this it is sense invariant. The number a is called the eigen value of A_{op} associated with the eigen function ψ. As an example consider a well behaved function sin $2x$. On operated by an operator $-\dfrac{d^2}{dx^2}$ we get

$$-\frac{d^2}{dx^2}(\sin 2x) = 4(\sin 2x) \tag{7.123}$$

From Eqs. (7.122) and (7.123), we get that 4 is the eigen value of the operator $-\dfrac{d^2}{dx^2}$ associated with eigen function sin $2x$.

7.5.1 Boundary and Continuity Conditions on the Wave Function

Schrodinger steady slate equation in three-dimension is

$$\frac{\partial^2\psi}{\partial x^2} + \frac{\partial^2\psi}{\partial y^2} + \frac{\partial^2\psi}{\partial z^2} + \frac{2m}{\hbar^2}(E-V)\psi = 0 \tag{7.124}$$

[2]A quantum state represented by an eigen function is said to be an eigen state.

Schrodinger equation (steady state) in one-dimension is

$$\frac{\partial^2\psi(x)}{\partial x^2}+\frac{2m}{\hbar^2}(E-V)\psi(x)=0 \tag{7.125}$$

The time independent or steady state Schrodinger wave equation is a second order linear differential equation in x (in general in $\boldsymbol{r}(x, y, z)$). In general, Schrodinger's steady state equation can be solved only for certain values of the energy E. What is meant by this statement has nothing to do with any mathematical difficulties that may be present, but is something much more fundamental. To "solve" Schrodinger's equation for a given system means to obtain a wave function ψ that not only obeys the equation and whatever boundary conditions there are but also fulfills the requirements for an acceptable wave function-namely that it and its derivatives, i.e., ψ and $\frac{\partial\psi}{\partial x}$ **be continuous, finite** and **single valued**. If there is no such wave function, the system cannot exist in a steady state. Hence, while obtaining the solutions of steady state Schrodinger equation it will be required that the wave function and its gradient are continuous, finite and single valued at every point in space.

Thus energy quantization appears in wave mechanics as a natural element of the theory and energy quantization in the physical world is revealed as a universal phenomenon characteristic of all stable systems.

7.6 FREE PARTICLE SYSTEM

A free particle is one which is not subjected to any force or potential barrier and is free to move in limitless space. For a linear motion of the particle, the Hamiltonian H is given as,

$$H=-\frac{\hbar^2}{2m}\frac{d^2}{dx^2}+V \tag{7.126}$$

and the wave function ψ (which is also eigen function of H here) is $\psi(x)$.

Since the particle is not subjected to any external force, its potential energy V is constant, which for convenience, may be assumed to be zero. Thus, the Schrodinger steady state equation for the particle may be written explicitly as

$$\frac{d^2\psi}{dx^2}+\frac{2m}{\hbar^2}E\psi=0 \tag{7.127}$$

The general solution of Eq. (7.127) is

$$\psi=Ae^{ikx}+Be^{-ikx} \tag{7.128}$$

where
$$k=\sqrt{\frac{2m}{\hbar^2}E} \tag{7.129}$$

and A and B are constants.

Eq. (7.128) gives the steady state or time independent part of the wave function. The complete wave function for particle is given by

$$\psi(x,t) = \psi(x)e^{-Et/\hbar} = (Ae^{ikx} + Be^{-ikx})e^{-iEt/\hbar} \tag{7.130}$$

$$= Ae^{ikx}e^{i\omega t} + Be^{-ikx}e^{-i\omega t} \qquad (\because \omega = \frac{E}{\hbar})$$

$$= Ae^{-i(\omega t - kx)} + Be^{-i(\omega t + kx)} \tag{7.131}$$

Eq. (7.131) represents the sum of two stationary solutions in principle of superposition, i.e., a continuous plane simple harmonic wave. The first term on the R.H.S. of Eq. (7.131) represents the wave travelling in the positive x-direction and the second term represents the wave travelling in the negative x-direction. The wavelength of each travelling plane wave is $\lambda = 2\pi/k$. The phase velocity of these waves is

$$v = \nu\lambda = \frac{E}{h}\frac{2\pi}{k} = \frac{E}{\hbar k}$$

According to de-Broglie's hypothesis these waves correspond to a particle of momentum

$$p = \frac{h}{\lambda} = \frac{hk}{2\pi} = \hbar k \text{ and energy } E = \left(= \frac{p^2}{2m} = \frac{\hbar^2 k^2}{2m} = \frac{\hbar^2 2mE}{2m\hbar^2} = E \right)$$

Therefore, for the motion of the particle in the positive x-direction, we have

$$\psi(x) = Ae^{ikx} \tag{7.132}$$

or

$$\psi(x,t) = Ae^{-i(\omega t - kx)} \tag{7.133}$$

The momentum operator, $\frac{\hbar}{i}\frac{\partial}{\partial x}$, operating on the wave function $\psi(x, t)$, gives

$$\frac{\hbar}{i}\frac{\partial}{\partial x} = \frac{\hbar}{i}[A(ik)e^{-i(\omega t - kx)}] = \hbar k A e^{-i(\omega t - kx)} = \hbar k \psi = p_x \psi \tag{7.134}$$

This equation shows that the wave function $\psi(x, t)$ for the particle is an eigen function of the linear momentum operator, and the momentum p_x is the eigen value of the operator. Hence the momentum remains sharp with the value p_x.

Now, as the constant k and hence energy E are completely arbitrary; i.e., no constraint has been applied on these, the particle can have any energy between 0 and $+\infty$, i.e., the particle has a continuous energy spectrum. The two sets of waves represented by Eq. (7.131) are travelling in opposite directions in $+x$ and $-x$-directions. Thus corresponding to the same eigen value E, there are two eigen functions as with the energy operator $i\hbar\frac{\partial}{\partial t}$, one obtains

$$i\hbar\frac{\partial}{\partial t}(Ae^{ikx}e^{-iEt/\hbar}) = E(Ae^{ikx}e^{-iEt/\hbar}) \tag{7.135}$$

and

$$i\hbar\frac{\partial}{\partial t}(Ae^{-ikx}e^{-iEt/\hbar}) = E(Ae^{-ikx}e^{-iEt/\hbar}) \tag{7.136}$$

i.e., in the two cases, we have

$$i\hbar\frac{\partial\psi}{\partial t} = E\psi \tag{7.137}$$

Obviously, ψ has two different, values but the eigen value E remains the same. Such eigen functions are termed as **"degenerate functions"** and in the present case of a free particle the degeneracy is **two fold.**

The probability of finding the particle between x and $x + dx$ is given by

$$pdx = \psi(x,t)\psi^*(x,t)dx = A^2 dx \tag{7.138}$$

Therefore, the probability density P for the position of the particle with the definite value of momentum is constant over the x-axis, i.e., all positions of the particle are equally probable. This conclusion is also obtained from the principle of uncertainty.

According to the interpretation of the wave function the probability of finding the particle somewhere in space must be equal to 1, i.e.,

$$\int_{-\infty}^{\infty} \psi(x,t)\psi^*(x,t)dx = 1 \tag{7.139}$$

In this case $\psi(x, t)\psi^*(x, t)dx$ = constant = A^2. Hence the wave function for the particle cannot be normalized and A must remain arbitrary. The difficulty arises because we are dealing with an arbitrary case. In practice, we cannot have an absolutely free particle. The particle will always be confined within an enclosure in the laboratory, and hence its position can be determined. This means that its momentum cannot be determined with absolute accuracy.

7.7 EHRENFEST'S THEOREM

We now proceed to show that the basic law of classical mechanics, i.e., Newton's second law of motion, may be derived, with certain approximations from the basic law of quantum mechanics, namely, the time dependent Schrodinger equation. This is referred to as Ehrenfest's theorem. This theorem can also be stated in another form as follows:

For a particle for which the uncertainty in position and momentum can be neglected, the quantum mechanical prediction of the motion agrees with the classical prediction or the average motion of a wave-packet agrees with the motion of the corresponding classical particle.

(i) Ehrenfest's Theorem (Newton's Second Law of Motion as a Special Case): The theorem states that quantum mechanics gives the same results as classical mechanics for a particle for which the average or expectation values of dynamical quantities are involved.

We prove the Ehrenfest's theorem for one-dimensional motion of a particle by showing that

$$\text{(i)}\ \frac{d\langle x\rangle}{dt} = \frac{\langle p_x\rangle}{m} \quad \text{and} \quad \text{(ii)}\ \frac{d\langle p_x\rangle}{dt} = \langle F_x\rangle$$

Proof: (i) $$\frac{d<x>}{dt} = \frac{<p_x>}{m}$$

Let x be the position coordinate of a particle of mass m at time t. The expectation value of x is given by

$$<x> = \int_{-\infty}^{\infty} \psi^*(x,t)x\psi(x,t)dx \tag{7.140}$$

Differentiating Eq. (7.140) w.r.t. t, we have

$$\frac{d<x>}{dt} = \int_{-\infty}^{\infty} x\frac{\partial(\psi\psi^*)}{\partial t}dx \tag{7.141}$$

Now, we have

$$\frac{\partial(\psi\psi^*)}{\partial t} = \frac{i\hbar}{2m}\frac{\partial}{\partial x}\left[\psi^*\frac{\partial\psi}{dx} - \psi\frac{\partial\psi^*}{dx}\right] \tag{7.142}$$

Substituting for $\frac{\partial(\psi\psi^*)}{\partial t}$ from Eq. (7.142) in Eq. (7.141), we have

$$\frac{d<x>}{dt} = \frac{i\hbar}{2m}x\left[\psi^*\frac{\partial\psi}{dx} - \psi\frac{\partial\psi^*}{dx}\right]_{-\infty}^{\infty} - \frac{i\hbar}{2m}\int_{-\infty}^{\infty}\left(\psi^*\frac{\partial\psi}{dx} - \psi\frac{\partial\psi^*}{dx}\right)dx$$

As x approach either ∞ or $-\infty$, ψ and $\frac{\partial\psi}{\partial x}$ approaches zero, and therefore the first term becomes zero. Hence, one obtains

$$\frac{d<x>}{dt} = -\frac{i\hbar}{2m}\int\left(\psi^*\frac{\partial\psi}{dx} - \psi\frac{\partial\psi^*}{dx}\right)dx \tag{7.143}$$

The expectation value of p_x is given by $<p_x> = \int_{-\infty}^{\infty}\psi^*\frac{\hbar}{i}\frac{\partial\psi}{\partial x}dx$

$$\int_{-\infty}^{\infty}\psi^*\frac{\partial\psi}{\partial x}dx = \frac{i}{\hbar}<p_x> \tag{7.144}$$

Similarly,

$$\int_{-\infty}^{\infty}\psi\frac{\partial\psi^*}{\partial x}dx = -\frac{i}{\hbar}<p_x> \tag{7.145}$$

Substituting the value of integrals from Eqs. (7.144) and (7.145) in Eq. (7.143), one obtains

$$\frac{d<x>}{dt} = -\frac{i\hbar}{2m}\left[\frac{i}{\hbar}<p_x> + \frac{i}{\hbar}<p_x>\right] = \frac{<p_x>}{m} \tag{7.146}$$

and $$\frac{d<p_x>}{dt} = <F_x>$$

The expectation value of the momentum p_x is given by

$$<p_x> = \int_{-\infty}^{\infty} \psi^*(x,t)\frac{\hbar}{i}\frac{\partial}{\partial x}\psi(x,t)dx \tag{7.147}$$

$$= \frac{\hbar}{i}\int_{-\infty}^{\infty} \psi^*\frac{\partial \psi}{\partial x}dx \tag{7.148}$$

Differentiating Eq. (7.148) w.r.t. t, one obtains

$$\frac{d<p_x>}{dt} = \frac{\hbar}{i}\left[\frac{\partial \psi^*}{\partial t}\frac{\partial \psi}{\partial x} + \psi^*\frac{\partial^2 \psi}{\partial x \partial t}\right]dx \tag{7.149}$$

The time-independent Schrodinger equations for ψ and ψ^* are

$$i\hbar\frac{\partial \psi}{\partial t} = -\frac{\hbar^2}{2m}\frac{\partial^2 \psi}{\partial x^2} + V\psi \tag{7.150}$$

$$i\hbar\frac{\partial \psi}{\partial t} = -\frac{\hbar^2}{2m}\frac{\partial^2 \psi^*}{\partial x^2} + V\psi^* \tag{7.151}$$

Differentiating Eq. (7.150) w.r.t. x, one obtains

$$i\hbar\frac{\partial^2 \psi}{\partial x \partial t} = -\frac{\hbar^2}{2m}\frac{\partial^2 \psi^*}{\partial x^2} + \frac{\partial (V\psi)}{\partial x} \tag{7.152}$$

We write Eq. (7.149) in the form

$$\frac{d<p_x>}{dt} = \int_{-\infty}^{\infty}\left(-i\hbar\frac{\partial \psi^*}{\partial t}\frac{\partial \psi}{\partial x} - \psi^* i\hbar\frac{\partial^2 \psi}{\partial x \partial t}\right)dx$$

Substituting the expression for $-i\hbar\frac{\partial \psi^*}{\partial t}$ and $i\hbar\frac{\partial^2 \psi}{\partial t}$ in these equations, one obtains

$$\frac{d}{dt}<p_x> = \int_{-\infty}^{\infty}\left[\left(\frac{-\hbar^2}{2m}\frac{\partial^2 \psi^*}{\partial x^2} + V\psi^*\right)\frac{\partial \psi}{\partial x} - \psi^*\left(\frac{-\hbar^2}{2m}\frac{\partial^2 \psi}{\partial x^2} + \frac{\partial (V\psi)}{\partial x}\right)\right]dx$$

$$= \frac{-\hbar^2}{2m}\int_{-\infty}^{\infty}\left[\frac{\partial^2 \psi^*}{\partial x^2}\frac{\partial \psi}{\partial x} - \frac{\partial^2 \psi}{\partial x^2}\right]dx + \int_{-\infty}^{\infty}\left[V\psi^*\frac{\partial \psi}{\partial x} - \psi^*\frac{\partial V(\psi)}{\partial x}\right]dx$$

$$= \frac{-\hbar^2}{2m}\int_{-\infty}^{\infty}\left[\frac{\partial}{\partial x}\left(\frac{\partial \psi}{\partial x}\frac{\partial \psi^*}{\partial x} - \psi^*\frac{\partial^2 \psi}{\partial x^2}\right)\right]dx + \int_{-\infty}^{\infty} V\psi^*\frac{\partial \psi}{\partial x} - \psi^*\left(\psi\frac{\partial V}{\partial x} + V\frac{\partial \psi}{\partial x}\right)dx$$

$$= \frac{\hbar^2}{2m}\left[\frac{\partial \psi}{\partial x}\frac{\partial \psi^*}{\partial x} - \psi^*\frac{\partial^2 \psi}{\partial x^2}\right]_{-\infty}^{\infty} - \int_{-\infty}^{\infty}\psi^*\frac{\partial \psi}{\partial x}\psi^* d\,x \tag{7.153}$$

As x approaches either ∞ or $-\infty$, ψ and $\frac{\partial \psi}{\partial x}$ approaches zero. Therefore, the first term on the R.H.S. of Eq. (7.153) is zero. The second term represents the expectation value of the differential coefficient of the potential V w.r.t. x, i.e.,

$$<\frac{\partial V}{\partial x}> = \int_{-\infty}^{\infty} \psi^* \frac{\partial V}{\partial x} \psi dx$$

$$\therefore \qquad \frac{d<p_x>}{dt} = -<\frac{\partial V}{\partial x}>$$

But $-\frac{\partial V}{\partial x}$ is the classical force F_x and hence $\frac{d<p_x>}{dt} = -<F_x>$ (7.154)

Hence a wave-packet will move like a classical particle provided the expectation value yields a good representation of the classical variable. One can easily generalize the above results for three dimensional case also.

(ii) Generalization to Three Dimensions: Considering the origin of Schrodinger equation, Ehrenfest's theorem is an expected result. The derivation will not give as anything new physically, Nevertheless, it is instructive to see how the same result may be obtained in a different manner.

For a particle described by a wave-packet $\psi(x, y, z)$, we define a point $(\bar{x},\bar{y},\bar{z})$ called the centre of gravity of the wave-packet, by forgoing equations:

$$\bar{x} = \int_{-\infty}^{\infty} x\psi^* \psi d\tau,^3 \quad \bar{y} = \int_{-\infty}^{\infty} y\psi^* \psi d\tau, \quad \bar{z} = \int_{-\infty}^{\infty} z\psi^* \psi d\tau \tag{7.155}$$

$\bar{x},\bar{y},\bar{z}$ may be considered as the average values of the three coordinates of the particle. When the size of the wave-packet is made smaller and smaller, the whole packet reduces to a point identical with $(\bar{x},\bar{y},\bar{z})$. Thus the point $(\bar{x},\bar{y},\bar{z})$ is the classical limits of a wave-packet. The position of this point changes in time because ψ^* and ψ are time-dependent. The behaviour of ψ^* and ψ in the course of time is determined by the time-dependent Schrodinger's equation. We shall see that the motion of the point $(\bar{x},\bar{y},\bar{z})$ is identical with the motion of the particle determined by Newton's second law of motion. Time dependent Schrodinger's equation is

$$\nabla^2\psi + \frac{2m}{\hbar^2} V\psi = -\frac{2mi}{\hbar}\frac{\partial \psi}{\partial t} \tag{7.156}$$

Complex conjugate of this equation is

$$\nabla^2\psi^* + \frac{2m}{\hbar^2} V\psi^* = \frac{2mi}{\hbar}\frac{\partial \psi^*}{\partial t} \tag{7.157}$$

---

[3] $d\tau = dx\,dy\,dz$

Eqs. (7.156) and (7.157) are completely equivalent, one following the other automatically. The velocity of the point $(\bar{x},\bar{y},\bar{z})$ may be obtained by differentiation w.r.t. time. As the dependence of $\bar{x},\bar{y},\bar{z}$ on time comes in through the time dependence of ψ^* and ψ only x, y, z will be held as constants in differentiation w.r.t. time.

$$\frac{d\bar{x}}{dt} = \int_{-\infty}^{\infty} \frac{d\psi^*}{dt} x\psi d\tau + \int_{-\infty}^{\infty} \psi^* x \frac{d\psi}{dt} d\tau = \int_{-\infty}^{\infty} \frac{\hbar}{2mi}\left(\nabla^2\psi^* - \frac{2m}{\hbar^2}V\psi^*\right)x\psi d\tau$$

$$-\int_{-\infty}^{\infty} \psi^* p \frac{\hbar}{2mi}\left(\nabla^2\psi - \frac{2m}{\hbar^2}V\psi^*\right)d\tau = \frac{\hbar}{2mi}\int_{-\infty}^{\infty} [(\nabla^2\psi^*)x\psi - \psi^* x(\nabla^2\psi)]d\tau \quad (7.158)$$

Transforming the first term of the last expression by the well-known procedure of integration by parts in three-dimensions

$$\int_{-\infty}^{\infty} [(\nabla^2\psi^*)(x\psi)d\tau = \int_{-\infty}^{\infty} \{\nabla.[(\nabla\psi^*)(x\psi)] - (\nabla\psi^*).\nabla(x\psi)\}d\tau \quad (7.159)$$

The first term may be transformed to an integral over a surface at infinity. Since the $\psi(x, y, z, t)$ of a wave-packet, by definition, vanishes at infinity, the surface integral drops out. The second term of Eq. (7.166) is symmetric w.r.t. ψ^* and $x\psi$. Thus we have

$$\int_{-\infty}^{\infty} (\nabla^2\psi^*)(x\psi)d\tau = \int_{-\infty}^{\infty} \psi^*\nabla^2(x\psi)d\tau \quad (7.160)$$

It follows that

$$\frac{d\bar{x}}{dt} = \frac{\hbar}{2mi}\int_{-\infty}^{\infty} \psi^*[\nabla^2(x\psi) - x\nabla^2\psi]d\tau = \frac{\hbar}{mi}\int_{-\infty}^{\infty} \psi^*\frac{d\psi}{dx}d\tau \quad (7.161)$$

Differentiating once again and multiplying by m, we have

$$m\frac{d^2\bar{x}}{dt^2} = \frac{\hbar}{i}\int_{-\infty}^{\infty} \frac{\partial\psi^*}{\partial t}\frac{\partial\psi}{\partial t}d\tau + \frac{\hbar}{i}\int_{-\infty}^{\infty} \psi^*\frac{\partial}{\partial x}\left(\frac{\partial\psi}{\partial t}\right)d\tau$$

$$= \frac{\hbar}{i}\int_{-\infty}^{\infty} \frac{\hbar}{2mi}\left(\nabla^2\psi^* - \frac{2m}{\hbar^2}V\psi^*\right)\frac{\partial\psi}{\partial x}d\tau - \frac{\hbar}{i}\int_{-\infty}^{\infty} \psi^*\frac{\hbar}{2mi}\left[\frac{\partial}{\partial x}\nabla^2\psi - \frac{2m}{\hbar}\frac{\partial}{\partial x}(V\psi)\right]d\tau$$

$$= \frac{\hbar}{2m}\int_{-\infty}^{\infty} (\nabla^2\psi^*)\frac{\partial\psi}{\partial x} - \psi^*\nabla^2\left(\frac{\partial\psi}{\partial x}\right) + \psi^*\frac{2m}{\hbar^2}\frac{\partial V}{\partial x}\psi d\tau \quad (7.162)$$

The first two terms cancel out after we perform the same kind of integration by parts as we have just done. Thus we have

$$m\frac{d^2\bar{x}}{dt^2} = \int_{-\infty}^{\infty} -\left(\frac{\partial V}{\partial x}\right)\psi^*\psi d\tau \quad (7.163)$$

The R.H.S. is just the average value for force $-\dfrac{\partial V}{\partial x}$ over the wave-packet. When the dimension of wave-packet is reduced to a point, the above relation states that mass times acceleration is equal to the force acting on the particle, which is Newton's second law of motion.

7.8 STATIONARY STATE SOLUTIONS

We now again turn our attention to the wave equation and its stationary state solutions. The time-independent Schrodinger's equation is

$$-\frac{\hbar^2}{2m}\nabla^2\psi(\boldsymbol{r}) + V\psi(\boldsymbol{r}) = E\psi(\boldsymbol{r})$$

The great importance of the Schrodinger time-independent equation derives from the fact that the **separation of variables yield** not just some particular solution of wave equation, but generally yields all solutions of physical interest.

Under very mild conditions imposed on the wave function we can prove that the separation constant E is a real number. The proof can be based on the continuity equation,

$$\frac{dp}{dt} + \nabla.\boldsymbol{J} = 0$$

we obtain from $\psi(\boldsymbol{r}, t) + f(t)\psi(\boldsymbol{r})$

$$p = \psi^*(r,t)\psi(r,t) = e^{\left[-\frac{i}{\hbar}(E-E^*)t\right]}\psi^*(r)\psi(r)$$

Upon substituting in continuity equation and integrating over space, we obtain

$$(E - E^*)\int\psi^*(r,t)\psi(r,t)d\tau = \frac{\hbar}{i}\int\nabla.d\tau = \frac{\hbar}{i}\int\boldsymbol{J}.dS$$

If the total probability flux over the boundary surface vanishes, we obtain $E = E^*$ and E must be real. If E is real, the time-dependent factor $f(t) = e^{-iEt/\hbar}$ is purely oscillatory. Hence in a separable state $\psi(\boldsymbol{r}, t) = f(t)\psi(\boldsymbol{r})$ and the function $\psi(\boldsymbol{r}, t)$ and $\psi(\boldsymbol{r})$ differ only by a time-dependent phase factor of constant amplitude. This justifies the designation of spatial factor $\psi(\boldsymbol{r})$ as a time-independent wave function.

For such a state the expectation value of any physical quantity which does not depend on time explicitly is constant and may be calculated from the time-independent wave function alone.

$$<A> = \int\psi^*(r,t)A\psi(r,t)d\tau = \int\psi^*(r)A\psi(r)d\tau$$

We thus see that the particular solutions

$$\psi(r,t) = e^{-iEt/\hbar}\psi E(r) \tag{7.164}$$

has intriguing properties, $|\psi(\boldsymbol{r}, t)|^2$ and all probabilities are constant in time. For this reason wave functions of the form Eq. (7.164) are said to represent **stationary states**. For such a state $\nabla\boldsymbol{J} = 0$. It is important to note that although the state is stationary, the particle so described is in motion and not stationary at all. Thus we can define the stationary state as that **particular state for which the probability distribution function $\psi\psi^*$ is independent of time**.

Let $\psi\psi^*$ represent the probability distribution function for a system in the state represented by the wave function

$$\psi(r,t) = \sum_{m=1}^{\infty}A_n\phi_n(r)e^{-iE_nt/\hbar} \tag{7.165}$$

The complex conjugate of Eq. (7.165) is given by

$$\psi^*(r,t) = \sum_{m=1}^{\infty} A_m^* \phi_m^*(r) e^{-iE_m t/\hbar} \qquad (7.166)$$

So that the probability distribution function

$$\psi\psi^* = \left[\sum_{n=1}^{\infty} A_n \phi_n(r) e^{-iE_n t/\hbar}\right]\left[\sum_{n=1}^{\infty} A_m^* \phi_m^*(r) e^{-iE_m t/\hbar}\right]$$

$$= \sum A_n A_n^* \phi_n(r)\phi_n^*(r) + \sum_m \sum_n{}' A_n A_m^* \phi_n(r)\phi_m^*(r) e^{\frac{i(E_m - E_n)t}{\hbar}} \qquad (7.167)$$

In relation Eq. (7.167) the prime on the double summation shows that the terms $m = n$ are not included. This clearly indicates that time enters in the probability distribution function $\psi\psi^*$ and hence ψ is not a stationary state solutions Eq. (7.165). If A_n' s are zero for all values except for one value of E_n, then $\psi\psi^*$ will be independent of time t. In such a case one finds,

$$\psi_n(r,t) = \phi_n(r) e^{-\frac{iE_n t}{\hbar}} \qquad (7.168)$$

The function ϕ_n, appearing in Eq. (7.168) is a particular solution of time-dependent Schrodinger's equation but the solution represented by this equation is stationary state solution[4] and this is independent of time.

A knowledge of stationary states is of fundamental importance in understanding chemical problems. If the state of a system corresponds to one of these stationary states, every physical property of the system e.g., energy, dipole moment, charge density etc. will be independent of time. We must note that atomic and molecular spectra arise due to transitions between phase stationary states brought about by the action of radiation.

7.9 FUNDAMENTAL POSTULATES OF QUANTUM MECHANICS

Postulate 1: The dynamical states of a particle or a quantum mechanical system can be described by a wave function which contains all that can be known about the particle. The wave function is a single valued function of the coordinates of the particles and of the time. In general it is a complex function and may he multiplied by an arbitrary complex number without changing its physical significance. When wave function ψ is properly normalized, then

$$\int_\tau \psi^* \psi \, d\tau = 1$$

Two non-zero wave functions are said to be orthogonal if their scalar product is zero, i.e., $\int_\tau \psi_i^* \psi_j \, d\tau = 0$. Wave functions that are both normalized and orthogonal satisfy the combined orthonormality conditions $\int_\tau \psi_i^* \psi_j \, d\tau = \delta_{ij}$, where $\delta_{ij} = 1$ for $i = j$ and $i \neq j$.

[4]Since $\psi\psi^* = \phi_n \phi_n^*$

Postulate 2: The superposition principle is valid for functions representing physical states expressed as

$$\psi = \sum_i c_i \psi_i = c_1\psi_1 + c_2\psi_2 + ...$$

where the $c_i^{'}$ are the expansion coefficients and the ψ_i are the known discrete states.
Postulate 3: The Schrodinger wave equation describes the behaviour of a wave function in space and time. If a wave function $\psi(\boldsymbol{r}, t)$ is known at some particular time, then its development is given by the Schrodinger wave equation

$$H\psi(\boldsymbol{r},t) = i\hbar\frac{\partial}{\partial t}\psi(\boldsymbol{r},t) \tag{7.169}$$

Here H is the Hamiltonian operator for the system. In the particular case of a time independent potential, Eq. (7.169) is separable in space and time and $\psi(\boldsymbol{r}, t)$ may be written as the product of a spatial function and a time function. The spatial part is an energy eigen function, i.e., it is one of the solutions of the energy eigen equation, i.e.,

$$H\psi_s(\boldsymbol{r}) = E_n\psi(\boldsymbol{r}) \tag{7.170}$$

Here the eigen value n^{th} is the energy of the E_n state. Since the energy E_n remains constant in time, the solutions are known as stationary states. The nth particular solution of Eq. (7.170) is

$$\psi_n(r,t) = \psi_s(r)\exp[-iE_n t/\hbar] \tag{7.171}$$

Using the superposition principle, a general solution of Eq. (7.171) may then be expanded in terms of solution just obtained, i.e.,

$$\psi(r,t) = \Sigma C_n \psi E_n(r)\exp[-iE_n t/\hbar] \tag{7.172}$$

The sufficient condition for Eq. (7.172) to be a valid expansion for a general wave function ψ is that the functions ψ_n form a complete, orthonormal set of functions.
Postulate 4: Observables are represented by linear operators which multiply, differentiate, or otherwise act on the wave function ψ to produce a new function.
The operator Q is linear if it commutes with constants and if it obeys the distributive law, i.e.,

$$Q(C\psi) = CQ\psi$$

and if
$$Q(\psi_1 + \psi_2) = Q(\psi_1) + Q(\psi_2)$$

The operator Q is **Hermitian** if it satisfies the following equality,

$$\int_\tau (Q\psi)^*\psi d\tau = \int_\tau \psi^*(Q\psi)d\tau \tag{7.173}$$

Postulate 5: When the linear operator Q operates on a particular wave function, ϕ_s say, it may produce a function which differs from ϕ_s, only by constant multiplying factor q_s i.e.,

$$Q\phi_s = q_s\phi_s \tag{7.174}$$

If so, then a measurement of Q is certain to yield the numerical value q_s. The particle is said to be in an **eigen state** of Q belonging to the eigen value q_s, and the wave function is called an eigen function of Q belonging to the eigen value q_s. In general,

$$\psi = \sum_{s=1}^{p} C_s \psi_s \tag{7.175}$$

Postulate 6: When a given system is in a state ψ, the expected mean or *expectation value* of a sequence of measurements on the observable represented by the operator Q is given by

$$q = \frac{\int \psi^* Q\psi d\tau}{\int \psi^* \psi d\tau} \tag{7.176}$$

If ψ is normalised, then $q = \int \psi^* Q\psi d\tau$. If ψ is an eigen function of Q such that $Q\psi = q\psi$, then ψ is said to be a pure state, and the expectation value of Q is simply the eigen value q each time the measurement is made.

7.9.1 Important Theorems Relating to Basic Postulates

(i) If two operators A and B commute, they have the same set of eigen functions. Let ψ_A be the eigen function of a, with an eigen value a. Thus $A\psi_A = a\psi_A$. Given $AB = BA$, i.e., $A(B\psi_A) = BA\psi_A = Ba\psi_A = a(B\psi_A)$. This means $(B\psi_A)$ is also an eigen function of A with the eigen value a, which is possible only if $B\psi_A$ is just a multiple of ψ, i.e., $B\psi_A = C\psi_A$ or ψ_A is also an eigen function of B.

(ii) If two operators commute, then it is possible to find out a set of functions that are simultaneously the eigen function of both the operators.
Suppose ψ is an eigen functions of operator A with eigen value λ and B is another operator that commutes with A, one can write $A(B\psi) = BA\psi = B\lambda\psi = \lambda(B\psi)$ because λ is a constant. Obviously, the above equation reveals that $(B\psi)$ is an eigen function of A with the same eigen value λ. This is possible only if the function $B\psi$ is obtained by multiplying ψ with a factor μ say, i.e., $B\psi = \mu\lambda$, i.e., μ is also eigen function of B with eigen value λ.

(iii) A physical quantity (e.g., E, p_x, etc.) of a system can be represented in quantum mechanics by its Hermitian operators. If two Hermitian operators commute, it is possible to determine accurately both corresponding quantities simultaneously.

7.10 OPERATORS

The formal treatment of quantum mechanics begins with the concept and properties of operators, eigen values, eigen functions and a set of basic postulates, also known as quantum mechanical laws. Since an understanding of the properties of operators is necessary for following quantum mechanical postulates. We start with an introduction to quantum mechanical operators.

Operators: An operator is a symbol for a rule for transforming a given mathematical function into another function. It has no physical meaning if written alone. For example, A is an operator which in itself does not mean anything, but if a quantity or a number is put with it, it transforms that quantity into its square root, another quantity similarly, $\frac{d}{dx}$ is an operator which transforms a function into its first derivative w.r.t. x; e.g., $\frac{d}{dx}$ transforms the function $\sin x$ into the function $\cos x$. When $\frac{d}{dx}$ operates on a function $f(x) = x^n$, then $\frac{df}{dx} = nx^{n-1}$.

In general, if A denotes an operator which transforms the function $f(x)$ into the function $g(x)$, then one can write $Af(x) = g(x)$. To illustrate it, we consider following examples:

(i) Let A be $\frac{d}{dx}$ and $f(x) = ax^3$; then $Af(x) = \frac{d}{dx}(ax^3) = 3ax^2$, i.e., $g(x) = 3ax^2$.

(ii) Let A be a (i.e., multiplication by a) and $f(x) = x^3 + x^2 + c$, then $Af(x) = a(x^3 + x^2 + c)$ $= ax^3 + ax^2 + ac$, i.e., $g(x) = ax^3 + ax^2 + ac$.

Algebra of Operators: Although operators do not have any physical meaning, they can be added, subtracted, multiplied and have some other properties.

Addition and Subtraction: The addition or subtraction of operators yields new operators, the sum or the difference of operators being defined by

$$(A \pm B)f(x) = Af(x) \pm Bf(x)$$

For example A be $\log_e$ and B be $\frac{d}{dx}$ and $f(x) = x^3$, then

$$(A \pm B)f(x) = \left(\log_e + \frac{d}{dx}\right)x^3 = \log_e x^3 + \frac{d}{dx}(x^3) = \log_e x^3 + 3x^2 = Af(x) \pm Bf(x)$$

Multiplication: Multiplication of two operators means operations by the two operators one after the other, the order of operation being from right to left, for e.g., $ABf(x)$ means that the function $f(x)$ is first operated on by B to yield a new function $g(x)$ which is then operated on by A to yield the final function $h(x)$,

$$AB\ f(x) = A[B\,f(x)] = A\,g(x) = h(x)$$

For example, let A be $4x^2$, B be $\frac{d}{dx}$ and $f(x) = ax^3$, then

$$AB\ f(x) = 4x^2\frac{d}{dx}(ax^3) = 4x^2(3ax^2) = 12ax^4$$

The square of an operator means that the same operator is applied successively twice, i.e.,

$$A^2 f(x) = AAf(x).\text{e.g., let } A = \frac{d}{dx} \text{ and } f(x) = \sin x, \text{then}$$

$$A^2(\sin x) = \frac{d}{dx}\left[\frac{d}{dx}(\sin x)\right] = \frac{d}{dx}(\cos x) = -\sin x$$

Commutative Property: When a series of operations are performed on a function successively the result depends on the sequence in which the operations are performed, in other words, in operator algebra it is not necessary that $AB\,f(x) = BA\,f(x)$. For e.g., let A denote $\frac{d}{dx}$, B stand for $3x^2$, and the function $f(x)$ be sin x; then

$$AB\ f(x) = \frac{d}{dx}[3x^2(\sin x)] = \frac{d}{dx}(3x^2 \sin x) = 6x \sin x + 3x^2 \cos x$$

and

$$BA\ f(x) = 3x^2 \frac{d}{dx}(\sin x) = 3x^2 \cos x$$

If two operators are such that the result of their successive applications is the same irrespective of the order of operations then the two operators are said to be commutative. In the above example, the two operators are non cummutative, let A stand for 3+, B for 4+ and $f(x)$ be ax; then

$$AB\,f(x) = 3 + 4 + (ax) = 3 + (4 + ax) = 7 + ax$$
$$AB\,f(x) = 4 + 3 + (ax) = 4 + (3 + ax) = 7 + ax$$

Obviously, A and B commute.

Linear Operator: An operator is said to be linear if its application on the sum of two functions separately, i.e., if

$$A[f(x) + g(x)] = A\,f(x) + A\,g(x)$$

or

$$AC\,f(x) = C.A.\,f(x) \text{ where } C \text{ is constant}$$

This is illustrated through the following examples:

(i) $\frac{d}{dx}$ is a linear operator because $\frac{d}{dx}(ax^m + bx^n) = \frac{d}{dx}(ax^m) + \frac{d}{dx}(bx^n)$

(ii) $\surd$ is not a linear operator because $\sqrt{f(x) + g(x)} \neq \sqrt{f(x)} + \sqrt{g(x)}$

Two linear operators of fundamental importance are:

Null or Zero Operator: It is defined as $0\psi = 0$, i.e., it destroys the function to which it is applied.

Unit or Identity Operator: It is defined as $1\psi = \psi$, i.e., unit operator makes no change in the function on which it operates.

Algebra of Linear Operators: If the operates A and B are such that for every wave function ψ, we have $A\psi = B\psi$, then the operators A and B are defined to be equal to each other. If $A\psi \neq B\psi$ then A and B are termed **as unusual operators.**

(i) **Sum of Two Linear Operators:** If $C\psi = A\psi + B\psi$, then $C = A + B$, i.e., the operator C is defined to be the sum of operators A and B.

(ii) **Product of Two Linear Operators:** If $C\psi = A\psi(B\psi)$ then C is the product of operators A and B. It should be noted that AB and BA may not be equal, i.e., the order of the operators in the product is very important and cannot be changed in general.

Commutator Operator: For any two operators A and B, the difference $AB - BA$, which is simply denoted by $AB - BA$ or $[A, B]$ is called **"Commutative operator"**. If A and B commute then $[A, B] = 0$, where 0 is called the zero operator which means multiplying a function with zero. Let us consider a example, where $A\frac{d}{dx}$, $B = 3x^2$, and $f(x) = \sin x$, the commutator is obtained as follows:

$$[A, B]f(x) = [AB - BA]f(x) = (6x\sin x + 3x^2\cos x) - 3x^2\cos x = 6x\sin x$$
$$= 6x\,f(x), \text{ i.e., } [A, B] = 6x$$

Non-Commutating Operators: If $AB \neq BA$, i.e., $[A, B] \neq 0$, then the operators are said to be non-commutating operators.

Reciprocal Operator: If $[AB - BA] = 1$, then A is said to be reciprocal of B and vice versa, i.e., $A^{-1} = B$ and $B^{-1} = A$.

Square of Operator: The square of the operator A is defined as $A^2 = A.A$. Similarly, one can define the higher powers $A^3 = A^2 . A = A.A^2$.

The Operator ∇ and ∇^2: So far we have restricted to the example of simple one dimensional operators, viz, $\frac{d}{dx}$, $2x$, etc., which operate on functions of a single variable like $f(x)$ or $\{f(x) + g(x)\}$. But there may be two or thee dimensional operators which operate on functions of more than one variable, i.e., $f(x, y, z)$. Thus the operator $\left(\frac{\partial}{\partial x} + \frac{\partial}{\partial y} + \frac{\partial}{\partial z}\right)$ when applied to a function f, where f stands for $f(x, y, z)$ gives the result,

$$\left(\frac{\partial}{\partial x} + \frac{\partial}{\partial y} + \frac{\partial}{\partial z}\right)f = \left(\frac{\partial f}{\partial x} + \frac{\partial f}{\partial y} + \frac{\partial f}{\partial z}\right)$$

A very important differential operator, known as "del" or ∇ operator, or a vector operator, is defined as,

$$\nabla = \left(\hat{i}\frac{\partial}{\partial x} + \hat{j}\frac{\partial}{\partial y} + \hat{k}\frac{\partial}{\partial z}\right) \tag{7.177}$$

where $\hat{i}$, $\hat{j}$ and $\hat{k}$ are unit vectors along the x, y and z-axis. This is a vector operator in the sense that when applied to a scalar function it transforms it into its gradient which is a vector,

$$\nabla = \hat{i}\frac{\partial f}{\partial x} + \hat{j}\frac{\partial f}{\partial y} + \hat{k}\frac{\partial f}{\partial z} \tag{7.178}$$

The square of the ∇ operator ∇^2, also called Laplacian operator, is defined as,

$$\nabla^2 = \frac{\partial^2}{\partial x^2} + \frac{\partial^2}{\partial y^2} + \frac{\partial^2}{\partial z^2} \tag{7.179}$$

7.11 EIGEN FUNCTIONS AND EIGEN VALUES OF OPERATORS

Let the result of the operation of an operator A upon wave function ψ is $A\psi = \psi'$, where ψ' is a function of the same variables as function ψ. In general function ψ' will be quite different form function ψ, but in some cases, one may obtain

$$A\psi = \lambda\psi \tag{7.180}$$

where λ is some constant. In all such cases function ψ is called an **eigen function** of the operator A and λ is called the **eigen value** of A associated with wave function ψ. Eq. (7.180) is called the eigen value equation for the operator.

Operators in quantum mechanics play very important role. Each dynamical variable of the motion of a system can be represented by a linear operator and the eigen values of the operator give the result of a precise measurement of the dynamical variable represented by that operator.

In the preceding sections we will see that $-\hbar\nabla$ and $i\hbar\frac{\partial}{\partial t}$ are operators in the quantum mechanics corresponding to linear momentum (P) and the total energy (E) of a system. When $-\hbar\nabla$ and $i\hbar\frac{\partial}{\partial t}$ are operated on their respective eigen functions, we obtain the momentum and the energy of the system respectively, i.e.,

$$(-i\hbar\nabla)\psi = p\psi; \qquad \psi = (2\pi\hbar)^{-3/2}\, e^{-ip.r/\hbar} \tag{7.181}$$

and

$$\left[i\hbar\frac{\partial}{\partial t}\right]\psi = E\psi; \qquad \psi = \phi(r)\exp(-i^{Et/\hbar}) \tag{7.182}$$

In Eqs. (7.181) and (7.182), p and E are the eigen values of the momentum and energy operators respectively and gives us the momentum and total energy of the system respectively. Hamiltonian operators (H) is

$$H = -\frac{\hbar^2}{2m}\nabla^2 + V(r) \tag{7.183}$$

Obviously, H represents the total energy of a time independent system. There are some more operators which correspond to the dynamical variables of a system.

Example 7.1: **Show that $\sin 2x$ is not an eigen function of the operator $\frac{d}{dx}$ but of $\frac{d^2}{dx^2}$. What is the eigen value?**

Solution $\frac{d}{dx}(\sin 2x) = 2\cos 2x$. Obviously, $\sin 2x$ is not an eigen function of the operator $\frac{d}{dx}$.

Now, $\frac{d^2}{dx^2}(\sin 2x) = -4(\sin 2x)$ obviously, the eigen value is –4.

Example 7.2: Show that if ψ is an eigen function of linear operator A with an eigen value λ, then $C\lambda$, where C is an arbitrary constant, is also an eigen function of A with same eigen value.

Solution $A\psi = \lambda\psi$

Since A is linear operator, we have

$$A(C\psi) = \lambda(C\psi) = C(\lambda\psi) = \lambda(C\psi)$$

Since C and λ are just numbers.

Operator for Momentum: The wave function for a free particle moving along the positive x-direction is

$$\psi(x,t) = Ae^{\frac{i}{\hbar}(p_x x - Et)} \tag{7.184}$$

Differentiating Eq. (7.184) w.r.t. x, one obtains

$$\frac{\partial\psi}{\partial x} = A\left(\frac{i}{\hbar}\right)p_x e^{\frac{i}{\hbar}(p_x x - Et)} = \frac{i}{\hbar}p_x\psi \text{ or } \frac{\hbar}{i}\frac{\partial\psi}{\partial x} = p_x\psi \tag{7.185}$$

This is the eigen value equation for the x-component of the momentum. The equation shows that (i) $\frac{\hbar}{i}\frac{\partial}{\partial x}$ is the operator, (ii) $\psi(x, t)$ is the eigen function of the operator (iii) p_x is the eigen value of the operator. Obviously, $\frac{\hbar}{i}\frac{\partial}{\partial x}$ is the momentum operator for the x-component of the momentum. The operator is denoted by p_x. Thus

$$p_x = \frac{\hbar}{i}\frac{\partial}{\partial x} \tag{7.186}$$

Similarly, for the y and z components of the momentum, the operators are

$$p_y = \frac{\hbar}{i}\frac{\partial}{\partial y} \tag{7.187}$$

$$p_z = \frac{\hbar}{i}\frac{\partial}{\partial z} \tag{7.188}$$

The momentum operator is defined as the operator which, operating on the wave function, reproduces the wave function multiplied by the momentum. In three dimensions, the operator for the momentum p is

$$p = \frac{\hbar}{i}\nabla \tag{7.189}$$

Operator for Kinetic Energy: From the momentum operator, we have

$$\frac{\hbar}{i}\frac{\partial\psi}{\partial x} = p_x\psi \tag{7.190}$$

Differentiating the above equation w.r.t. x, one obtains

$$\frac{\hbar}{i}\frac{\partial^2}{\partial x^2} = p_x \frac{\partial \psi}{\partial x}$$

But $$\frac{\partial \psi}{\partial x} = \frac{i}{\hbar} p_x \psi$$

$$\frac{\hbar}{i}\frac{\partial^2 \psi}{\partial x^2} = p_x \frac{i}{\hbar} p_x \psi$$

or $$-\hbar^2 \frac{\partial^2 \psi}{\partial x^2} = p_x^2 \psi$$

$\therefore$ $$-\frac{\hbar^2}{2m}\frac{\partial^2 \psi}{\partial x^2} = \frac{p_x^2}{2m}\psi \quad (7.191)$$

or $$-\frac{\hbar^2}{2m}\frac{\partial^2 \psi}{\partial x^2} = K\psi \quad (7.192)$$

where $K = \frac{p_x^2}{2m}$ is the kinetic energy of the particle moving along the x-direction. The equation reveals that (i) $-\frac{\hbar}{2m}\frac{\partial^2}{\partial x^2}$ is the operator (ii) $\psi(x, t)$ is the eigen function of the operator (iii) $K = \frac{p_x^2}{2m}$ is the eigen value of the operator. Hence $-\frac{\hbar}{2m}\frac{\partial^2}{\partial x^2}$ is the kinetic energy operator for motion of the particle along the x-direction. Thus the operator for the kinetic energy is

$$K = -\frac{\hbar^2}{2m}\frac{\partial}{\partial x^2} \quad (7.193)$$

In three dimensions the operator for K is

$$K = \frac{\hbar^2}{2m}\nabla^2 \quad (7.194)$$

Operator for Total Energy: The total energy of the particle moving along the x-axis is given by

$$E = \frac{p_x^2}{2m} + V(x)$$

where $V(x)$ is the potential energy. Substituting the value of $p_x^2/2m$ form Eq. (7.191), one obtains

$$-\frac{\hbar^2}{2m}\frac{\partial^2 \psi}{\partial x^2} = (E-V)\psi \quad \text{or} \quad -\frac{\hbar^2}{2m}\frac{\partial^2 \psi}{\partial x^2} + V\psi = E\psi$$

or $$\left[-\frac{\hbar^2}{2m}\frac{\partial^2}{\partial x^2} + V\right]\psi = E\psi \quad (7.195)$$

This is the eigen value equation for the total energy E of the particle moving along the x-direction. The above equation reveals that (i) $\left(-\frac{\hbar^2}{2m}\frac{\partial^2}{\partial x^2} + V\right)$ is the operator, (ii) $\psi(x, t)$ is the eigen function of the operator, and (iii) E is the eigen value of the operator.

Hence $\left(-\frac{\hbar^2}{2m}\frac{\partial^2}{\partial x^2} + V\right)$ is the total energy operator for the motion of the particle along the x-direction. This operator is denoted by H_x and is called the Hamiltonian operator for the one-dimensional motion. Thus $H_x = \frac{\hbar^2}{2m}\frac{\partial^2}{\partial x^2} + V$.

For the three dimensional motion, the total energy operator will be

$$-\frac{\hbar^2}{2m}\left(\frac{\partial^2}{\partial x^2} + \frac{\partial^2}{\partial y^2} + \frac{\partial^2}{\partial z^2}\right) + V = -\frac{\hbar^2}{2m}\nabla^2 + V$$

The total energy operator is denoted by H, i.e.,

$$H = -\frac{\hbar^2}{2m}\nabla^2 + V \tag{7.196}$$

Obviously, the Hamiltonian operator H is the differential operator, which operating on the wave function, reduces the same function multiplied by the total energy. We must note that we have used H for the total energy operator and not E because the total energy is assumed to be expressed in terms of the momentum, similar to the Hamiltonian function H, which stands for the total energy in terms of the position and momentum coordinates.

Importance of Hamiltonian Operator: Let $u(x, y, z)$ be the wave function associated with a particle in motion. The Schrodinger equation is

$$-\frac{\hbar^2}{2m}\nabla^2 u + Vu = Eu$$

or

$$\left[-\frac{\hbar^2}{2m}\nabla^2 + V\right]u = Eu \tag{7.197}$$

or

$$Hu = Eu \tag{7.198}$$

In Eq. (7.119), we note that (i) on the L.H.S., the operator is the Hamiltonian operator which operates on the wave function $u(x, y, z)$. (ii) on the R.H.S., Eu is the product of the total energy E and the wave function u. Obviously, the Hamiltonian operator operating on the wave function reproduces the same function multiplied by the total energy E. This means that the solution of the time-independent Schrodinger equation are the eigen functions of the Hamiltonian operator, and the eigen values E of the operator are the only possible values of the total energy that a quantum mechanical system may possess.

Operator for the total energy in terms of the differential with respect to time: The wave function for a free particle moving along the positive x-direction is

$$\psi(x,t) = Ae^{\frac{i}{\hbar}(p_x x - Et)}$$

Differentiating with respect to t, one obtains

$$\frac{\partial \psi}{\partial t} = A\left(\frac{i}{\hbar}\right)(-E)e^{\frac{i}{\hbar}(p_x x - Et)} = -\frac{i}{\hbar}E\psi = \frac{1}{i\hbar}E\psi$$

$$\therefore \qquad i\hbar\frac{\partial \psi}{\partial t} = E\psi \tag{7.199}$$

Obviously, the total energy operator E is $i\hbar\frac{\partial}{\partial t}$, i.e.,

$$E = i\hbar\frac{\partial}{\partial t} \tag{7.200}$$

The total energy operator is the operator, which, operating on the wave function reproduces the wave function multiplied by the total energy.

Operator for Angular Momentum: In classical mechanics the angular momentum $\boldsymbol{L}$ of a particle is defined by

$$\boldsymbol{L} = \boldsymbol{r} \times \boldsymbol{p} \tag{7.201}$$

where $\boldsymbol{r}$ is the position vector of the particle relative to some arbitrary origin and $\boldsymbol{p}$ is its linear momentum. From Eq. (7.201) the components L_x, L_y and L_z of $\boldsymbol{L}$ along x, y and z axes, are respectively given by

$$L_x = yp_z - zp_y,\; L_y = zp_x - xp_z,\; L_z = xp_y - zp_x \tag{7.202}$$

Now, using the expressions for the linear momentum operators, the operators for L_x, L_y and L_z are given by

$$L_x = \frac{\hbar}{i}\left[y\frac{\partial}{\partial z} - z\frac{\partial}{\partial y}\right], \quad L_y = \frac{\hbar}{i}\left[z\frac{\partial}{\partial z} - x\frac{\partial}{\partial z}\right]$$

and

$$L_z = \frac{\hbar}{i}\left[x\frac{\partial}{\partial y} - y\frac{\partial}{\partial x}\right] \tag{7.203}$$

Transforming these operators into spherical polar coordinates, one obtains

$$L_x = \frac{\hbar}{i}\left[-\sin\varphi\frac{\partial}{\partial\theta} - \cot\theta\cos\phi\frac{\partial}{\partial\varphi}\right]$$

$$L_y = \frac{\hbar}{i}\left[\cos\varphi\frac{\partial}{\partial\theta} - \cot\theta\sin\varphi\frac{\partial}{\partial\varphi}\right]$$

$$L_z = \frac{\hbar}{i}\frac{\partial}{\partial\theta} \tag{7.204}$$

Now $$L^2 = L_x^2 + L_y^2 + L_z^2 = -\hbar^2\left[\frac{1}{\sin\theta}\frac{\partial}{\partial\theta}\left(\sin\theta\frac{\partial}{\partial\theta}\right) + \frac{1}{\sin^2\theta}\frac{\partial^2}{\partial\varphi^2}\right] \tag{7.205}$$

Quantum mechanical operators to various physical quantities are listed in Table 7.1.

Rules for Setting up a Quantum Mechanical Operators:

(i) Write down the expression for the physical quantity in classical terms, i.e., in terms of cartesian coordinates of position (x, y, z) and momenta (p_x, p_y, p_z).

(ii) Replace these coordinates and momenta by their corresponding operators using Table 7.1.

(iii) Operator for a coordinate (say x) is multiplication by that variable x itself.

(iv) Operator for a coordinate of momentum (say p_x) is $\frac{\hbar}{i}\frac{\partial}{\partial x}$.

Example 7.3: Express total energy operator in quantum mechanical operator from, when a single particle is moving in x-direction.

Solution Total energy operator

$$H = \frac{p_x^2}{2m} + V(x)$$

$$\text{Now, K.E. operator } T_x = \frac{p_x^2}{2m} = \frac{1}{2m}\left(\frac{\hbar}{i}\frac{\partial}{\partial x}\right)^2 = -\frac{\hbar}{2m}\frac{\partial^2}{\partial x^2}$$

$$\text{P.E. operator } V(x) = V(x)$$

$$\therefore \quad \text{Total energy operator } H = -\frac{\hbar}{2m}\frac{\partial^2}{\partial x^2} + V(x)$$

Table 7.1 Quantum Mechanical Operators Corresponding to Various Physical Quantities

Physical Quantity of Dynamical Variable	*Symbol*	*Quantum Mechanical Operator*
Position	x	x
Position	r	r
Momentum	P_x	$\frac{\hbar}{i}\frac{\partial}{\partial x}$
	P_y	$\frac{\hbar}{i}\frac{\partial}{\partial y}$
	P_z	$\frac{\hbar}{i}\frac{\partial}{\partial z}$
	P	$\frac{\hbar}{i}\nabla$
Kinetic energy	T	$-\frac{\hbar^2}{2m}\nabla^2$
x Component of Kinetic energy	T_x	$-\frac{\hbar^2}{2m}\frac{\partial^2}{\partial x^2}$
Potential energy	V	V
Total energy operator	E	$i\hbar\frac{\partial}{\partial t}$
Total energy	H	$-\frac{\hbar^2}{2m}\nabla^2 + V$

Table 7.2

Variable	*Coordinate representation*	*Momentum representation*
x	x	$i\hbar\partial / \partial p_x$
x^2	x^2	$-\hbar^2\partial^2 / p_x^2$
p_x	$-i\hbar\partial / \partial x$	p_x
p_x^2	$-\hbar^2\partial^2 / \partial x^2$	p_x^2
E	$i\hbar\partial / \partial t$	$i\hbar\partial / \partial t$

The coordinate representation and momentum representation of some commonly used operators in quantum mechanics are given in Table 7.2.

We must note that the momentum and energy eigen values take on a continuous set of values. The set of eigen values and eigen functions may also be discrete. In general, for a bound state, the sets of eigen values and eigen functions are discrete sets and vice versa for a free state.

Let A is the operator to some dynamical variable and suppose we make a measurement of that variable for a system represented by a state function ψ, then the result of measurement 'λ' is given by

$$A\psi = \lambda\psi$$

Further, the expectation value of A in a series of measurements performed on an ensemble of systems in identical states is given by

$$<A> = \int \psi^* A\psi d\tau \Big/ \int \psi^* \psi d\tau \tag{7.206}$$

$$= \int \psi * \lambda\psi d\tau \Big/ \int \psi * \psi d\tau = \lambda \tag{7.207}$$

Obviously, the expectation value of an operator A in a state represented by an eigen function of A is the corresponding eigen value. Since the physical quantity corresponding to the operator A is an observable, it must be real. Clearly, the eigen values of the operators corresponding to the dynamical variables are real.

Let $\psi_1, \psi_2, \psi_3, \ldots \psi_m, \ldots$ etc., be eigen functions corresponding to discrete eigen values. If for any two eigen functions, we have

$$\int \psi_m^* \psi_n d\tau = 0: \qquad m \neq n \text{ and } m, n = 1, 2 \text{ etc.}$$

then the function ψ_m and ψ_n are called as orthogonal to each other and the set of eigen functions is called an **orthogonal set of eigen functions**.

It the two functions ψ_m and ψ_n satisfy the following condition

$$\int \psi_m^* \psi_n d\tau = 1 \text{ for } m = n = 1, 2, 3\ldots \text{ etc.}$$

then the set of eigen functions is called a **normalized set of eigen functions**.

If the set of eigen functions is orthogonal as well as normalized, then it is called as an **orthonormal set of eigen functions and it satisfies the following condition:**

$$\int \psi_m^* \psi_n d\tau = \delta_{mn} \begin{cases} 1 & \text{for} \quad m = n \\ 0 & \text{for} \quad m \neq n \end{cases}$$

7.12 COMPLETE SET OF EIGEN FUNCTIONS

Consider a linear combination of the members of an orthonormal set of eigen functions of some operator as

$$\sum_{i=1}^{\infty} c_i \psi_i \tag{7.208}$$

Here $c_i's$ are the complex numbers. If the series represented by Eq. (7.208) converges, then it defines a function

$$f = \sum_{i=1}^{\infty} c_i \psi_i \tag{7.209}$$

One can express, large number of functions as a sum of Eq. (7.207). The set of functions ψ is called a complete set w.r.t. the members of such a class of functions.

We have read that any normalized arbitrary wave function 'ψ' can be expressed as

$$\psi = \sum_{i=1}^{\infty} c_i \psi_i \tag{7.210}$$

Here $\psi_i's$ are the eigen functions of an operator corresponding to the dynamical observable to the system. Obviously, the set of the functions $\psi_i's$ is complete w.r.t. all the wave functions $\psi_i's$. To obtain the coefficients $c_i's$, one may make use of orthonormality of wave functions. We have

$$\int \psi_j^* \psi_i d\tau = \sum_{i=1}^{n} c_i \int \psi_j^* \psi_i d\tau = \sum_{i=1}^{n} c_i \delta_{ji} = c_i$$

Obviously,
$$c_i = \int \psi_i^* \psi_i d\tau \tag{7.211}$$

Concluding, we can say that every physical quantity can be represented by an operator with eigen functions $\{\psi_1, \psi_2, \ldots, \psi_i \ldots\}$ which forms a complete set of orthonormal functions w.r.t. every physical state ψ expressed as Eq. (7.210), where the coefficients are given by Eq. (7.211).

7.13 COMPLETENESS RELATION

If $\{\psi_1, \psi_2, \ldots \psi_i \ldots\}$ be a complete set of eigen functions of some operator corresponding to a dynamical observable of some system, then an arbitrary state ψ can be expressed as

$$\psi = \sum_{i=1}^{n} c_i \psi_i; \qquad c_i = \int \psi_i^* \psi d\tau$$

$$\therefore \quad \int \psi^* \psi d\tau = \int \sum_{i} \sum_{j=1}^{\infty} c_i^* \psi_i^* c_j \psi_j d\tau = \sum_{i,j=1}^{\infty} c_i^* c_j \int \psi_i^* \psi_i d\tau$$

$$= \sum_{i,j=1}^{\infty} c_i^* c_i \delta_{ij} = \sum_{i=1}^{\infty} c_i^* c_i = \sum_{i=1}^{\infty} |c_i|^2 \qquad (7.212)$$

Eq. (7.212) is called the **completeness relation** for the given set. Eq. (7.212) is the **necessary** as well as sufficient conditions for a set of functions to be complete. If Eq. (7.212) is satisfied, one will have to show that $\sum_{i=1}^{\infty} c_i \psi_i$ indeed converges to ψ.

7.14 HERMITIAN OPERATORS

A linear operator is Hermitian if it satisfies the following equality

$$\int_{\tau} (A\psi)^* \psi d\tau = \int_{\tau} \psi^* (Q\psi) d\tau \qquad (7.213)$$

If $A = A^+$, then A is called self adjoint or **Hermitian** if $A = -A^+$, then A is called **anti-Hermitian.** The expectation value of the operator, which is a real number is given by $< A > = \int \psi^* A\psi d\tau$.

More generally, a Hermitian operator satisfies the following condition

$$\int (A\psi)^* \phi d\tau = \int \psi^* (A\phi) d\tau \qquad (7.214)$$

where ψ and ϕ are any two functions.

Properties of Hermitian operators: Hermitian operators have the following important properties:

(i) The Sum of two Hermitian Operators is Hermitian: Let A and B be Hermitian operators, then for any two arbitrary wave functions ψ and ϕ, we have

$$\int \psi^*(A\phi) d\tau = \int (A\psi)^* \phi d\tau \quad \text{and} \quad \int \psi^*(B\phi) d\tau = \int (B\psi)^* \phi d\tau$$

Adding the above two equations, one obtains

$$\int \psi^*(A\phi) d\tau + \int \psi^*(B\phi) d\tau = \int A(\psi)^* \phi d\tau + \int (B\psi)^* \phi d\tau$$

or

$$\int \psi^*(A + B)\phi d\tau = \int \{(A + B)\psi\}^* \phi d\tau$$

obviously, $(A + B)$ is a Hermitain operator.

(ii) The Product of Commuting Hermitian Operator A and B is Hermitian: Let A and B be two commuting operators, i.e., $AB = BA$. Then for any two arbitrary states ψ and ϕ, we consider the integral

$$\int \psi^* AB \psi d\tau \qquad (7.215)$$

For the Hermitian operator A and the state ψ and $B\phi$, one obtains the condition

$$\int \psi^* A(B\psi) d\tau = \int (A\psi)^* B\phi d\tau \qquad (7.216)$$

Similarly, for the Hermitian operator B and the states $A\psi$ and ϕ, we obtain

$$\int (A\psi)^* B d\tau = \int [B(A\psi)]^* \phi d\tau \qquad (7.217)$$

Using Eqs. (7.216), (7.217) and (7.213) becomes

$$\int \psi^* AB\psi d\tau = \int \psi^* A(B\phi)\phi d\tau = \int [B(A\psi)]^* \phi d\tau$$

or

$$\int \psi^* AB\phi d\tau = \phi(AB\psi)^* d\tau \qquad (AB = BA)$$

obviously AB is Hermitian.

(iii) The eigen values of a Hermitian operator are real: Let ψ be any eigen function of a Hermitian operator A with eigen value λ, then we have

$$\int \psi^* A\psi d\tau = \int (A\psi)^* \psi d\tau$$

But

$$A\psi = \lambda\psi$$

$$\int \psi^* \lambda\psi d\tau = \int (\lambda\psi)^* \psi d\tau$$

or

$$\lambda \int \psi^* \psi d\tau = \lambda^* \int \psi^* \psi d\tau$$

Since

$$\psi = 0, \text{ and hence } \int \psi^* \psi d\tau \neq 0$$

Obviously, $\lambda = \lambda^*$, i.e., λ is real.

(iv) The Eigen function of Hermitian Operator are orthogonal if they correspond to distinct eigen values: Let A be a Hermitian Operator, and ψ_1, ψ_2 be its two eigen functions with distinct eigen values λ_1 and λ_2, we have

$$A\psi_1 = \lambda_1\psi_1 \text{ and } A\psi_2 = \lambda_2\psi_2 (\lambda_1 \neq \lambda_2)$$

Using the property of Hermitian operator, we have

$$\int \psi_1^* A\psi_2 d\tau = \int (A\psi_1)^* \psi_2 d\tau$$

or

$$\int \psi_1^* \lambda_2 \psi_2 d\tau = \int (\lambda_1\psi_1)^* \psi_2 d\tau$$

or

$$\lambda_2 \int \psi_1^* \psi_2 d\tau = \lambda_1^* \int \psi_1^* \psi_2 d\tau = \lambda_1 \int \psi_1^* \psi_2 d\tau$$

($\because \lambda_1$ must be real being the eigen value of a Hermitian operator)

$\therefore$

$$(\lambda_1 - \lambda_2) \int \psi_1^* \psi_2 d\tau = 0$$

i.e., $$\int \psi_1^* \psi_2 d\tau = 0 \text{ since } \lambda_1 \neq \lambda_2$$

This shows that ψ_1 and ψ_2 are orthogonal.

(v) Every linear combination of two linearly independent eigen functions of A belonging to the same eigen value will also be an eigen function with the same eigen value: Let ψ_1 and ψ_2 are two linearly independent eigen functions of A and both have the eigen value λ, i.e.,

$$A\psi_1 = \lambda\psi_1 \quad \text{and} \quad A\psi_2 = \lambda\psi_2$$

Let $c_1\psi_1 + c_2\psi_2$ be any linear combination of ψ_1 and ψ_2, then

$$A(c_1\psi_1 + c_2\psi_2) = c_1 A\psi_1 + c_2 A\psi_2 = \lambda(c_1\psi_1 + c_2\psi_2)$$

where c_1 and c_2 are arbitrary constants. One can always construct orthogonal eigen functions by selecting these arbitrary constants.

(vi) Schmidt Orthogonalisation Process: One can always construct pair of ϕ_1 and ϕ_2 by taking suitable linear combinations of ψ_1 and ψ_2.

Let us consider that ψ_1 and ψ_2 be two linearly independent degenerate functions which are normalised but not orthogonal. Let $\phi_1 = \psi_1$ and $\phi_2 = a_1\psi_1 + a_2\psi_2$. Now let us consider $\int \psi_1 \psi_2 d\tau = S$. Obviously we have to determine a_1 and a_2 such that

$$\int \phi_1\phi_2 d\tau = 0 \quad \text{and} \quad \int \phi_2\phi_2 d\tau = 1$$

or $$\int \psi_1(a_1\psi_1 + a_2\psi_2)d\tau = 0$$

or $$a_1 \int \psi_1^2 d\tau + a_2 \int \psi_1\psi_2 d\tau = 0$$

or $$a_1 + a_2 S = 0$$

or $$a_1 = -a_2 S \tag{7.218}$$

or $$\int (a_1\psi_1 + a_2\psi_2)^2 d\tau = 1$$

or $$a_1^2 \int \psi_1^2 d\tau + a_2^2 \int \psi_2^2 d\tau + 2a_1 a_2 \int \psi_1\psi_2 d\tau = 1$$

or $$a_1^2 + a_2^2 S + 2a_1 a_2 S = 1 \tag{7.219}$$

Substituting for a_1 from Eq. (7.218) in Eq. (7.219), one obtains

$$a_2^2 S^2 + a^2 - 2a_2^2 S^2 = 1$$

or $$a_2 = (1 - S^2)^{-1/2} \tag{7.220}$$

$$a_1 = -S(1 - S^2)^{-1/2} \tag{7.221}$$

Thus the desired orthogonal pair will be

$$\phi_1 = \psi_1$$

and
$$\phi_2 = a_1\psi_1 + a_2\psi_2 = -S(1-S^2)^{-1/2}\psi_1 + (1-S^2)^{-1/2}\psi_2$$

$$= (1-S^2)^{-1/2}(\psi_2 - \psi_1 S) \tag{7.222}$$

This is called **"Schmidt orthogonalisation"** process. We note that eigen functions of a Hermitian operator form a complete orthonormal set.

One can easily conclude that a series of function $\psi_1, \psi_2, \ldots$, which are subjected to certain restrictions (e.g., single valuedness, finiteness, continuity, boundary conditions), is said to form a complete set if any arbitrary function F having the same restrictions and can be expressed as a linear combination of these functions, i.e.,

$$F = a_1\psi_1 + a_2\psi_2 + \cdots = \sum_i a_i\psi_i \tag{7.223}$$

7.15 ADJOINT OF AN OPERATOR

If A is an arbitrary operator, then A^+ (A-dagger) is called its adjoint.

If $A = A^+$ then A is called Hermitain and its and if $A = -A^+$, then A^+ is called anti-Hermitian. Obviously, Hermitian is a **synonym for self-adjoint**. One can easily prove it as follows:

If A is a Hermitian operator, then

$$\int \psi^* A\phi d\tau = \int (A\psi)^* \phi d\tau \tag{7.224}$$

Adjoint is defined as

$$\int \psi^* A^+\phi d\tau = \int (A\psi)^* \phi d\tau \tag{7.225}$$

where ψ and ϕ are two arbitrary wave functions. Comparing Eqs. (7.224) and (7.225), one obtains,

$A^+ = A$, i.e., A is Hermitian, if it is self adjoint. The converse of this is also true, i.e., if A is Hermitian, then $A = A^+$. The product of two operators AB will be Hermitain if and only if A and B commute. One can prove it as follows:

$$\int \psi(AB)^+ \phi d\tau = \int (AB\psi)^* \phi d\tau = \int (B\psi)^* A^+\phi d\tau = \int \psi B^+ A^+ \phi d\tau$$

If A and B are both Hermitian, then $(AB)^+ = B^+ A^+ = BA$.

7.16 SIMULTANEOUS MEASUREMENT AND COMMUTING OPERATORS

Consider two operators A and B associated with the observables p and q, respectively. We wish to know what to expect if a measurement of p is followed by a measurement of q and, conservely, if a measurement of q is followed by a measurement of p. These two expectation values are expressed mathematically as

$$<QP> = \int \psi^* QP\psi d\tau \quad \text{and} \quad <PQ> = \int \psi^* PQ\psi d\tau \tag{7.226}$$

provided that ψ is normalized. Now, if ψ is simultaneously an eigen function of each operator, then we have the relations

$$A\psi = P_1\psi;\ B\psi = q_1\psi$$

and the two expectation values become:

$$<BA> = p_1 \int \psi^* Q\psi d\tau = p_1 q_1 \tag{7.227}$$

and

$$<AB> = q_1 \int \psi^* P\psi d\tau = q_1 p_1 \tag{7.228}$$

Since p_1 and q_1 are simply numbers, they commute and the order of the measurements is immaterial. Physically, this means that a system that is in simultaneous eigen states of two operators will be undisturbed by any sequence of measurement of the observables associated with these two operators. Mathematically, we see that for this to occur the operators must commute, i.e., $[B, A] = BA - AB = 0$.

7.16.1 Commutator Algebra

If A and B are two operators which do not commute, the physical quantities A and B cannot both be sharply defined simultaneously. The degree to which an inevitable lack of precision in A and B has to be admitted is measured by their commutator.

$$AB - BA = [A, B]$$

Since expressions of the type $(AB - BA)$ occur frequently in quantum mechanics, it has been given special name and special symbol, $[A, B]$ which is called the **commutator** of operators A and B. If $[A, B] = 0$, then operators are said to commute. It satisfied the following identities:

$$\left.\begin{aligned} [A,B] &\equiv -[B,A] \\ [A,BC]. &\equiv [A,B]C + B[A,C] \\ [AB,C] &\equiv [A,C]B + A[B,C] \\ [A[B,C]] &\equiv [B,[C,A]] + [C,[A,B]] \equiv 0 \end{aligned}\right\} \tag{7.229}$$

The commutator of two operators A and B will in most general case be another operator, for example D.

7.16.2 The Fundamental Commutation Relations

We have seen that the fundamental quantities in terms of which all the dynamical observables of classical motion can be expressed are the position coordinate x_c and the linear momentum p_{xc} along the direction of motion (e.g., x-axis). The connection between the classical and quantum mechanical theories is given by the following correspondence:

$$x_c \to x;\ p_{xc} \to p_x = i\hbar\frac{\partial}{\partial x} \tag{7.230}$$

Here x and $p_x = i\hbar\frac{\partial}{\partial x}$ are the quantum mechanical operators, corresponding to the classical quantities x_c and p_{xc} respectively. The commutation relation for x and p_x is called the fundamental **commutation relation** and it is expressed as

$$[x, p_x] = i\hbar \tag{7.231}$$

One can easily prove this relation. We have for any wave function ψ

$$xp_x\psi = -i\hbar\frac{\partial}{\partial x}\psi$$

and
$$p_x x\psi = -i\hbar\frac{\partial}{\partial x}(x\psi) = -i\hbar x\frac{\partial \psi}{\partial x} - i\hbar\psi$$

∴
$$xp_x\psi - p_x x\psi$$

or
$$[x, p_x] = i\hbar\psi$$

Similarly, one can show that

$$[y, p_y] = [z, p_z] = i\hbar \tag{7.232}$$

Eqs. (7.231) and (7.232) are called as the **fundamental commutation relations.**

7.17 EQUATION OF MOTION FOR OPERATORS

The expectation value of an operator A in the state ψ is represented as

$$<A> = \int \psi^* A\psi d\tau$$

∴
$$\frac{d}{dt}<A> = \int \frac{\partial \psi^*}{\partial t} A\psi d\psi + \int \psi^* A\frac{\partial A}{\partial t}\psi d\tau + \int \psi^* A\frac{\partial \psi}{\partial t} d\tau \tag{7.233}$$

We know that according to the correspondence between the classical dynamical observables and the quantum mechanical operators, A will be function of x and p_x. Since the time does not appear explicitly in it and hence $\frac{\partial A}{\partial t} = 0$. Thus, one obtains

$$\frac{d}{dt} < A > = \int \frac{\partial \phi^*}{\partial t} A \psi d\tau + \int \psi^* A \frac{\partial A}{\partial t} d\tau$$

Time dependent Schrodinger's equation gives us

$$\frac{\partial \psi}{\partial t} = \frac{1}{i\hbar} H\psi \text{ and } \frac{\partial \psi^*}{\partial t} = -\frac{1}{i\hbar} H\psi^*$$

Using these results, Eq. (7.233) becomes

$$\begin{aligned}
\frac{d < A >}{dt} &= \frac{1}{i\hbar} \int H\psi^* A \psi d\tau + \frac{1}{i\hbar} \int \psi^* A H \psi d\tau \\
&= \frac{1}{i\hbar} \int \psi^* (AH - HA) \psi d\tau \qquad (\because H \text{ is a Hermitian operator}) \\
&= (i\hbar)^{-1} < (AH - HA) = (i\hbar)^{-1} < [A, H] >
\end{aligned}$$

Thus
$$\frac{dA}{dt} = \frac{1}{i\hbar} [A, H] \tag{7.234}$$

Obviously, the time derivative of an operator A is equal to the commutator of A and H, multiplied by $(i\hbar)^{-1}$. Eq. (7.234) is usually called as the **Heisenberg's equation of motion or equation of motion for an operator** A. Eq. (7.234) reveals that if some operator commutes with the Harmiltonian, its time rate of change will be zero, i.e., the operator will be constant in time and hence it represents a constant of motion. Obviously, if the operator, corresponding to any quantity, commutes with H, then the quantity will be a constant of motion.

7.18 UNITARY OPERATORS

A linear operator U, whose inverse (U^{-1}) and adjoint (U^+) are identical is called unitary operator. Thus

$$U^{-1} = U^+ \text{ or } UU^+ = U^+U = 1 \tag{7.235}$$

If A is a Hermitian operator, then a unitary operator may be constructed from it as follows:

$$U = \frac{1 + iA}{1 - iA}, \qquad U^+ = \frac{1 - iA}{1 + iA}, UU^+ = 1 \tag{7.236}$$

Unitary operator can also be constructed from a Hermitian operator, as follows:

$$U = \frac{U + U^+}{2} + i\frac{U - U^+}{2i} = A + iB \text{ (say)}$$

$$A = \frac{U + U^+}{2} \text{ and } B = \frac{U - U^+}{2i}$$

Now, $$A^+ = \left(\frac{U + U^+}{2}\right)^+ = \frac{U^+ + U}{2} = \frac{U + U^+}{2} = A$$

and $$B^+ = \left(\frac{U - U^+}{2i}\right) = \frac{U^+ - U}{-2i} = \frac{U - U^+}{2i} = B$$

Obviously, A and B are Hermitian operators, we have

$$AB = \frac{U + U^+}{2}\frac{U - U^+}{2i} = \frac{UU - UU^+ + U^+U - U^+U^+}{4i} = \frac{UU - U^+U^+}{4i}$$

and $$BA = \frac{U - U^+}{2i}\frac{U + U^+}{2} = \frac{UU + U^+U^+ + UU^+ - U^+U}{4i} = \frac{UU - U^+U^+}{4i}$$

Thus $AB = BA$, i.e., A and B commute with one another. Thus one can find a simultaneous eigen function ψ for A and B.

$$A\psi = A'\psi \text{ and } B\psi = B'\psi$$

Obviously, A' and B' are real numbers, because A and B are Hermitian operators.

$\therefore$ $$U\psi = (A + iB)\psi = (A' + iB')\psi + U'\psi \quad \text{(say)}$$

Obviously, the eigen value of U is $U' = A' + iB'$

Now, $$A^2 + B^2 = \frac{1}{4}(U + U^+)^2 - \frac{1}{4}(U - U^+)^2 = U^+U = 1$$

$\therefore$ $$(A^2 + B^2)\ \psi(A'^2 + B'^2)\ \psi = 1\psi$$

Thus $$A'^2 + B'^2 = 1, \text{ i.e., } A' = |A' + iB'|$$

and hence $$\sqrt{A'^2 + B'^2} = 1$$

Obviously, the eigen value of any arbitrary unitary operator U has the absolute value unity. One can, therefore express it as $U' = \exp(ic')$, where c' is real number. Let us define a Hermitain operator C, such that $C\psi = C'\psi$, where C' is the eigen value of C.

$$\because \quad \exp[iC]\psi = \left(1 + iC + C\frac{(iC)^2}{2!} + \ldots\right)\psi = \left(1 + iC' + \frac{(iC)^2}{2!} + \ldots\right)\psi$$

$$= \exp(iC')\psi = U'\psi = U\psi$$

Obviously, $U = \exp(iC)$, i.e., C is a Hermitian operator.

7.19 DEGENERACY

Form Schrodinger equation, $H\psi = E\psi$, we have already proved that for every wave function ψ, there must be a corresponding energy eigen value E, i.e., for every eigen functions there will be one eigen value. It means that each energy state must have a wave function which will be characteristic of the system. There may, however, be situations when two wave functions ψ_j and ψ_k correspond to the same energy E. This is a 'case' of 'degeneracy'; the system in such an energy state is said to be in a degenerate state. In such a situation, the degenerate wave function ψ_j and ψ_k will not necessarily be orthogonal.

7.20 FORBIDDEN TRANSITIONS

When certain particular eigen functions are multiplied together, they give zero product, i.e., $\psi_p \psi_q^* = 0$ (we should not confuse this situation with orthogonality). In such a situation, the simultaneous execution of the two corresponding vibrations by the electron will not cause the emission of radiation. These transitions are termed as **forbidden transitions**. Due to forbidden transitions, certain lines will be missing in spectral lines. In the presence of a strong electric and magnetic field these lines may actually appear. In such a situation, the eigen functions are distorted by the strong field, and hence the product $\psi_p \psi_q^*$ is no longer zero.

Example 7.4: Show that the average value of the square of a Hermitian operator is positive.

Solution In the formula

$$\overline{L^2} = \frac{\int \psi^* L^2 \psi d\tau}{\int \psi^* \psi d\tau}$$

where L is a Hermitian operator. The denominator is, of course positive so that we must prove that numerator is positive. Since L is Hermitian, we get

$$\int \psi^* L L^* \psi d\tau = \int (L\psi)(L^* \psi^*) d\tau = \int |L\psi|^2 d\tau > 0$$

Example 7.5: Does taking the complex conjugate correspond to: (a) a linear operator; (b) a Hermitian operator; (c) an operator which is its own complex conjugate?

Solution

(a) It is not linear. Indeed denoting the operator of complex conjugation by a, we have $a(c_1\psi_1 + c_2\psi_2) = c_1^* a\psi_1 = c_2^* a\psi_2 \neq c_1 a^*\psi_1 + ca\psi_2$

(b) It is not Hermitian. Any Hermitian operator satisfies the condition

$$\int \psi^* L\phi d\tau = \int \phi L^* \psi^* d\tau \tag{i}$$

In the present case we have $\int \psi^* a\phi d\tau = \int \psi^* \phi^* d\tau$ and $\int \phi a^* \psi^* d\tau = \int \phi\psi d\tau$ so that condition Eq. (i) is not satisfied.

(c) The operator a is its own complex conjugate, that is, it is real. Indeed $a\psi = \psi^*$ so that $a^*\psi^* = \psi$. Denoting ψ^* by ϕ we have $a^*\psi = \phi^* = a\psi$ or $a^* = a$. (The operator of complex conjugation changes i to $-i$, and its complex conjugate changes $-i$ to i, which is the same).

Example 7.6: Prove that the function ψ, which at $t = 0$ describes a state with a well defined value of the physical quantity A, will at $t > 0$ be an eigen function of the operator $A(-t)$ corresponding to the same eigen value. The quantity $A(\tau)$ is the operator corresponding to the quantity A in the Heisenberg representation at time τ.

Solution From the Schrodinger equation

$$i\hbar\frac{\partial\psi}{\partial t} = H\psi$$

It follows that any state changes with time according to

$$\psi(\xi,t) = \exp(-iHt/\hbar)\psi(\xi,0) \tag{i}$$

where the set of independent variables on which the wave function of the system depends is denoted by ζ. The operator A in the Heisenberg representation is of the form

$$A(t) = e^{\left(\frac{iHt}{\hbar}\right)} A e^{\left(-\frac{iHt}{\hbar}\right)} \tag{ii}$$

where A is operator corresponding to the physical quantity A in the Schrodinger representation. For our problem, we have

$$A\psi(\zeta,0) = A\psi(\xi,0) \tag{iii}$$

Substituting here $\psi(\xi,0) = \exp(iHt/\hbar)\psi(\xi,t)$ and operating on it from the left with $\exp(-iHt/\hbar)$, one finds

$$e^{-\frac{iHt}{\hbar}} A e^{\frac{iHt}{\hbar}} \psi(\xi,t) = A\psi(\xi,t) \tag{iv}$$

Comparing the operator on the left-hand side of this equation with Eq. (ii), we find $A(-t)\psi(\zeta, t) = A\psi(\xi, t)$. Out proof is valid for the case where the Hamiltonian H of the system does not depend explicitly on the time.

Example 7.7: If A and B are two operators such that both these operator commute with their commutator $[A, B]$, prove that

$$\text{(i)} \quad [A, B^n] = nB^{n-1}[A, B] \quad \text{and} \quad \text{(ii)} \quad [A^n, B] = nA^{n-1}[A, B]$$

Solution These two identifies can be proved by the method of induction

$$\text{L.H.S.} = [A, B^n] = [A, B^{n-1}B] = B^{n-1}[A, B] + [A, B^{n-1}]B$$

Since A and B commute with their commutator $[A, B]$. We have

$$[B,[A, B]] = 0 \text{ or } B[A, B] - [A, B]B = 0$$

$$B[A, B] = [A, B]B$$

$$\therefore \quad \text{L.H.S.} = B^{n-1}[A, B] + [A, B^{n-1}]B = B^{n-1}[A, B] + [A, B^{n-2}B]B \quad \text{(i)}$$

$$= B^{n-1}[A, B] + [A, B^{n-2}]BB + B^{n-2}[A, B]B$$

$$= B^{n-1}[A, B] + [A, B^{n-2}]B^2 + B^{n-2}B[A, B]$$

$$= B^{n-1}[A, B] + [A, B^{n-2}]B^2 + B^{n-1}[A, B]$$

$$= 2B^{n-1}[A, B] + [A, B^{n-2}]B^2 \quad \text{(ii)}$$

$$\therefore \quad \text{R.H.S.} = 2B^{n-1}[A, B] + [A, B^{n-3}B]B^2$$

$$= 2B^{n-1}[A, B] + B^{n-3}[A, B]B^2 + [A, B^{n-3}]B.B^2$$

$$= 2B^{n-1}[A, B] + B^{n-3}[A, B]BB + [A, B^{n-3}]B^2$$

$$= 2B^{n-1}[A, B] + B^{n-3}B[A, B]B + [A, B^{n-3}]B^3$$

$$= 2B^{n-1}[A, B] + B^{n-3}BB[A, B]B + [A, B^{n-3}]B^3$$

$$= 2B^{n-1}[A, B] + B^{n-1}[A, B] + [A, B^{n-3}]B^3$$

$$= 3B^{n-1}[A, B] + [A, B^{n-3}]B^3 \quad \text{(iii)}$$

Let $n = 2$;

$$[A,B^2] = B[A, B] + [A, B]B = B[A, B] + [A, B] = 2B[A, B]$$

Let $n = 3$;

$$[A,B^2] = 2B^2[A, B] + [A, B]BB = 2B^2[A, B] + B[A, B]B$$

$$= 2B^2[A, B] + BB[A, B]$$

$$= 2B^2[A, B] + B^2[A, B] = 3B^2[A, B]$$

Let $n = 4$;

$$[A,B^4] = 3B^3[A,B] + [A,B]BBB = 3B^3[A,B] + B[A,B]B$$
$$= 3B^3[A,B] + BB[A,B]B = 3B^3[A,B] + BBB[A,B]$$
$$= 3B^3[A,B] + B^3[A,B] = 4B^3[A,B]$$

Thus we see that the identity is true for $n = 2$, 3 and 4. In this way we can show that the identity is true for $n = 5, 6, 7$ ... etc. Let us assume that it is true for $n - 1$, i.e.,

$$[A, B^{n-1}] = (n-1)B^{n-2}[A,B]$$

Then
$$[A,B^n] = [A,B^{n-1}B] = B^{n-1}[A,B] = [A,B^{n-1}]B$$
$$= B^{n-1}[A,B] + (n-1)B^{n-2}[A,B]B$$
$$= B^{n-1}[A,B] + (n-1)B^{n-2}[A,B]$$
$$= B^{n-1}[A,B] + (n-1)B^{n-1}[A,B] = nB^{n-1}[A,B]$$

Thus we see that if the identity is true for $(n-1)$, it is true for n also; since it is true for $n = 2, 3, 4, 5$ ... etc., it is true for all positive integral values of n.

Proceeding in a similar way, we can prove the second identity.

Example 7.8: If A and B are two operators which commute with their commutator $[A, B]$, prove that $e^A e^B = e^{A+B+\frac{1}{2}[A,B]}$

Solution We shall follow the proof given by Glauber. Let us assume a function $f(\lambda)$ defined by $f(\lambda) = e^{\lambda A}e^{\lambda B}$, then

$$\frac{df}{d\lambda} = Ae^{\lambda A}e^{\lambda B} + e^{\lambda A}Be^{-\lambda B}$$
$$= A^{\lambda A}e^{\lambda B} + e^{\lambda A}Be^{-\lambda A}e^{\lambda A}e^{\lambda B}$$
$$= (A + e^{\lambda A}Be^{-\lambda A})e^{\lambda A}e^{\lambda B}$$
$$= (A + e^{\lambda A}Be^{-\lambda A})f(\lambda)$$

Since $[B, A]$ commutes with A, $[B, A^n] = nA^{n-1}[B, A]$.

$$[B, e^{-\lambda A}] = \sum_n (-1)^n \frac{\lambda^n}{n!}[B, A^n]$$
$$= \sum_n (-1)^n \frac{\lambda^n}{n(n-1)!} nA^{n-1}[B, A]$$
$$= \sum_n (-1)^n \frac{\lambda^n}{(n-1)!} A^{n-1}[B, A]$$

or $$Be^{-\lambda A} - e^{-\lambda A}B = \sum_n (-1)^{n-1}\frac{\lambda^{n-1}}{(n-1)!}(-\lambda)A^{n-1}[B, A]$$

$$= -\lambda e^{-\lambda A}[B, A]$$

or $$Be^{-\lambda A} = e^{-\lambda A}B - \lambda e^{-\lambda A}[B, A]$$

or $$Be^{-\lambda A} = e^{-\lambda A} + B + \lambda e^{-\lambda A}[A, B]$$

or $$e^{\lambda A}Be^{-\lambda A} = e^{\lambda A}e^{-\lambda A}B + \lambda e^{\lambda A}e^{-\lambda A}[A, B]$$

or $$e^{\lambda A}Be^{-\lambda A} = B + \lambda[A, B]$$

$$\frac{df}{d\lambda} = (A + B + \lambda)[A, B])f(\lambda)$$

Integrating the above equation, we obtain

$$\int \frac{df}{f} = \int (A + B + \lambda)[A, B])d\lambda$$

or $$\log_e f = (A + B)\lambda + [A, B]\lambda^2/2$$

or $$f = \exp(A\lambda + B\lambda) + [A, B]\lambda^2/2$$

or $$e^{\lambda A}e^{\lambda B} = \exp(A\lambda + B\lambda) + [A, B]\lambda^2/2$$

Putting $\lambda = 1$ in the above equation, we get

$$e^A e^B = \exp(A + B + [A, B] \times \frac{1}{2})$$

Example 7.9: Prove the identity

$$e^A Be^{-A} = B + [A, B] + \frac{1}{2!}[A,(A, B)] + \frac{1}{3!}[A,[A,[A,[A.B]]]+\cdots$$ where $e^{\lambda A}$ is defined by

$$e^A = 1 + A + \frac{AA}{2!} + \frac{AAA}{3!} + \frac{AAAA}{4!} + \cdots$$

Solution Let us consider a function $f(\lambda)$ given by $f(\lambda) = e^{\lambda A}Be^{-\lambda A}$

(i) $f(0) = e^{(0)A}Be^{-(0)A} = B$

(ii) $$\frac{df}{d\lambda} = Ae^{\lambda A}Be^{-\lambda A} + e^{\lambda A}B(-A)e^{-\lambda A}$$

$$= Ae^{\lambda A}Be^{-\lambda A} - e^{\lambda A}Be^{-\lambda A}A$$

$$= A(f) - f(\lambda)A$$

$$= [A, f(\lambda)]$$

$$\left.\frac{df}{d\lambda}\right|_{\lambda=0} = [A, f(0)]$$

$$= [A, B]$$

(iii) $$\frac{d^2 f}{d\lambda^2} = \frac{d}{d\lambda}\left(\frac{df}{d\lambda}\right) = \left[A, \frac{df}{d\lambda}\right] = [A, [A, f(\lambda)]]$$

$$\therefore \quad \left.\frac{d^2 f}{d\lambda^2}\right|_{\lambda=0} = [A, [A, f(0)]] = [A, [A, B]]$$

Expanding $f(\lambda)$ as a power series in λ using Taylor series expansion, we have

$$f(\lambda) = f(0) + \frac{\lambda}{1!}\left(\frac{df}{d\lambda}\right)_{\lambda=0} + \frac{\lambda^2}{2!}\left(\frac{d^2 f}{d\lambda^2}\right)_{\lambda=0} + \frac{\lambda^3}{3!}\left(\frac{d^3 f}{d\lambda^3}\right)_{\lambda=0} + \cdots$$

or $$e^{\lambda A}Be^{-\lambda A} = B + \frac{\lambda}{1!}[A, B] + \frac{\lambda^2}{2!}[A[A, B]] + \frac{\lambda^3}{3!}[A, [A, [A, B]]] + \cdots$$

$$[A,[A,[A,B]]] + \cdots$$

Letting $\lambda = 1$, in the above expansion, we get the required result

$$e^A Be^{-A} = B + [A, B] + \frac{1}{2!}[A, [A, B]] + \frac{1}{3!}[A, [A, B]]] + \cdots$$

Example 7.10: If $[A, B] = rB$ where r is a number, the result obtained in example 7.6 reduces to

$$e^A Be^{-A} = e^r B$$

Solution

Given $\quad [A, B] = rB$

Then $\quad [A, [A, B]] = [A, rB] = r[A, B] = rrB = r^2 B$

and $\quad [A, [A, [A, B]]] = [A, [A, rB]] = r[A, [A, B]] = r[A, rB] = rr[A, B] = r^3 B$

$$\text{L.H.S.} = e^A Be^{-A} = B + \frac{1}{1!}[A, B] + \frac{1}{2!}[A,[A, B]] + \frac{1}{3!}[A,[A,[A, B]]] + \cdots$$

$$= B + rB + r^2 / \sqrt{2}B + \cdots = (1 + r + r^2 + 2! + \ldots.)B = e^r B$$

Example 7.11: **If $\{\phi_1(x), \phi_2(x)....\}$ is a complete set of orthonormal functions, then show that the following closure relation satisfied:**

$$\sum_{n=1}^{\infty} \phi_n^*(x')\phi_n(x) = \delta(x - x')$$

Solution Let us consider any arbitrary function $\psi(x)$. We can express this function in terms of the given complete set of the orthonormal function as

$$\psi(x) = \sum_{n=1}^{\infty} C_n \phi_n(x)$$

Here $$C_n = \int \phi_n^*(x)\psi(x)dx$$

$$\psi(x) = \sum_{n=1}^{\infty} \int \phi_n^*(x')\psi(x')dx'\phi_n(x)$$

$$= \int \psi(x')\left[\sum_{-\infty}^{\infty} \phi_n^*(x')\phi_n(x)\right]dx'$$

$$= \int \psi(x')\delta(x - x')dx'$$

∴ Obviously, $$\sum_{-\infty}^{\infty} \phi_n^*(x')\phi_n(x) = \delta(x - x')$$

Example 7.12: Show that the momentum operator $P = -i\hbar\nabla$ is a Hermitian operator.

Solution If P is a Hermitian operator it must satisfy the relation

$$\int \psi^*(-\hbar\nabla\phi)d\tau = \int (-i\hbar\nabla\psi)^*\phi d\tau \qquad \text{(i)}$$

or $$-i\hbar \int \psi^*(\nabla\psi)d\tau = -i\hbar \int (\nabla\psi)^*\psi d\tau$$

we have $$\nabla(\psi^*\psi) = \psi^*(\nabla\psi) + (\nabla\psi^*)\psi$$

or $$\psi^*(\nabla\psi) = \nabla(\psi^*\psi) - (\nabla\psi^*)\psi$$

Integrating, one obtains

$$\int \psi^*(\nabla\psi)d\tau = \int (\nabla\psi^*\psi)d\tau - \int (\nabla\psi^*)\psi d\tau$$

$$= \nabla \int \psi^*\psi d\tau - \int (\nabla\psi^*)\psi d\tau = -\int (\nabla\psi^*)\psi d\tau \qquad \text{(ii)}$$

$$\left[\because \int \psi^*\psi d\tau \text{ is a constant}\right]$$

Multiplying Eq. (ii) on both sides by $-i\hbar$, we obtain Eq. (i). This shows that momentum operator is a Hermitian operator.

Example 7.13: If $H = \frac{1}{2}mx^2 + V(x)$, then show that $i\hbar\dot{x} = (x,H)$, is satisfied if $xp_x - p_x x = i\hbar$ where $V(x)$ is a potential function.

Solution

Here $$H = \frac{1}{2}mx^2 + V(x) \quad \text{(i)}$$

$$p_x = m\dot{x} \quad \therefore \quad \dot{x} = p_x/m \quad \text{(ii)}$$

Thus, $$H = \frac{1}{2}mp_x^2/m^2 + V(x) = \frac{p_x^2}{2m} + V(x) \quad \text{(iii)}$$

Now, $$[x,H] = \left[x, \frac{p_x^2}{2m} + V(x)\right] = 1/2m[x,p_x^2] + 1/2m[x.V(x)] \quad \text{(iv)}$$

Second term in Eq. (iv) will vanish as $V(x)$ is a function of x and it will commute with x.

$$\therefore \quad [x,H] = \frac{1}{2}[mp_x^2] = \frac{1}{2m}[x,p_x]p_x + \frac{1}{2m}p_x[x,p_x]$$

$$= \frac{1}{2m}i\hbar p_x + \frac{1}{2m}p_x i\hbar = i\hbar m\dot{x}/m = i\hbar\dot{x}$$

Example 7.14: Show that the momentum p of a free particle is a constant of motion.

Solution The free particle Hamiltonian is given by

$$H = \frac{p_x^2}{2m} = -\frac{\hbar^2}{2m}\frac{\partial^2}{\partial x^2} \quad (V = 0 \text{ for a free particle, } p_x = -i\hbar\partial/\partial x)$$

$$\therefore \quad p_x H\psi(x) = -i\hbar\frac{\partial}{\partial x}\left[-\frac{\hbar^2}{2m}\frac{\partial^2}{\partial x^2}\psi(x)\right] = -\frac{i\hbar^3}{2m}\frac{\partial^3}{\partial x^3}\psi(x) \quad \text{(i)}$$

and $$Hp_x\psi(x) = -\frac{\hbar^2}{2m}\frac{\partial^2}{\partial x^2}\left[-i\hbar\frac{\partial}{\partial x}\psi(x)\right] = \frac{i\hbar^3}{2m}\frac{\partial^3}{\partial x^3}\psi(x) \quad \text{(ii)}$$

$$\therefore \quad [p_x, H] = p_x H - Hp_x = 0$$

Obviously, the momentum of a free particle commutes with the free particle Hamiltonian and hence it is called as a constant of motion.

Example 7.15: If $H = p^2/2\mu + \frac{1}{2}\mu\omega^2 x^2$, then show that

(a) $[x,H] = i\hbar p/\mu$ and (b) $[p,H] = -i\hbar\mu\omega^2 x$

Solution

(a) $$[x,H] = \left[x,\frac{p^2}{2\mu} + \frac{1}{2}\mu\omega^2 x^2\right] = 1/2\mu[x,p^2] + \frac{1}{2}\mu\omega^2[x,x^2]$$

The second commutator vanishe as x is commuting with itself.

$$\therefore \quad [x,H] = xH - Hx = 1/2\mu[x,p^2] = (1/2\mu)2i\hbar p = i\hbar p/\mu$$

(b) $$[p,H] = [p,p^2/2\mu + \frac{1}{2}\mu\omega^2 x^2] = 1/2\mu\underbrace{[p,p^2]}_{0} + \frac{1}{2}\mu\omega^2[p,x^2]$$

$$= \frac{1}{2}\mu\omega^2(p,x^2) = \frac{1}{2}\mu\omega^2(-2i\hbar x) = -i\hbar\omega^2 x \quad (\because [x^2,p]=2i\hbar x)$$

Example 7.16: If A and B are constants of motion and H is the Hamiltonian, then show that $[A, B]$ is also a constant of motion.

Solution We know that the constant of motion commutes with the Hamiltonian and hence

$$[A,H] = 0 \text{ and } [B,H] = 0$$

$$[(A,B),H] = [AB - BA,H] = [AB,H] - [BA,H]$$

$$= A[B,H] + [A,H]B - B[A,H] - [B,H]A = 0$$

This shows that $[A, B]$ commutes with the Hamiltonian and hence $[A, B]$ is a constant of motion.

Example 7.17: Show that an operator cannot have more than one reciprocal.

Solution Let an operator A has two reciprocals B_1 and B_2, then

$$AB_1 = B_1A = I \quad \text{and} \quad AB_2 = B_2A = I$$

$$AB_1 = AB_2 \tag{i}$$

Multiplying both sides of Eq. (i) from left by A^{-1}, one obtains

$$A^{-1}AB_1 = A^{-1}AB_2$$

or $$B_1 = B_2 \; [A^{-1}A = 1]$$

Obviously, an operator cannot have more than one reciprocal.

Example 7.18: **If $H = p^2/2m + V(x)$, then find the value of $[x.(x, H)]$.**

Solution $$[x,H] = [x,p^2/2m + V(x)] = [x,p^2/2m] + \underset{0}{\underset{\|}{[x,V(x)]}}$$

$$\therefore \quad [x,H] = [x,p^2/2m] = \frac{1}{2m}[x,p^2] = \frac{1}{2m}2i\hbar p = i\hbar p/mx$$

$$\therefore \quad [x,(x,H)] = [x,i\hbar p/m] = i\hbar/m[x,p] = i\hbar/mi\hbar = -\hbar^2/m$$

Example 7.19: Show the difference of the operators $\left(x^2p_x^2 + p_x^2x^2\right)$ and $\frac{1}{2}(xp_x + p_xx)^2$ is equal to $-3/4\hbar^2$.

Solution

$$(x^2p_x^2 - p_x^2x^2) - \frac{1}{2}(xp_x + p_xx)^2$$

$$= \frac{1}{2}(x^2p_x^2 + x^2p_x^2 + p_x^2x^2 + p_x^2x^2) - \frac{1}{2}(xp_xxp_x + xp_x^2x + p_xx^2p_x + p_xxp_xx)$$

$$= \frac{1}{2}\{(x^2p_x^2 + xp_x^2x) + (p_x^2x^2 + p_xx^2p_x) + (x^2p_x^2 - xp_xxp_x) + (p_x^2x^2 - p_xxp_xx)\}$$

[Using $[x,p_x] = i\hbar$ and $[p_x,x] = -i\hbar$, one obtains]

$$[x,p_x^2] = [x,p_x] + p_x[x,p_x] = i\hbar p_x - p_xi\hbar = 2i\hbar p_x$$

and $$[px,x^2] = x[p_x,x] + [p_x,x]x = -xi\hbar + (-i\hbar x) = -2i\hbar x$$

$$= \frac{1}{2}(2i\hbar xp_x - 2i\hbar p_xx + i\hbar xp_x - i\hbar p_xx)$$

$$= \frac{1}{2}(2i\hbar[x,p_x] + i\hbar[x,p_x]) = \frac{3}{4}i\hbar[x,p_x] = \frac{3}{1}i\hbar.i\hbar = -\frac{3}{4}\hbar^2$$

Example 7.20: By using the following results, show that (i) $[f(x),p] = i\hbar\partial f/\partial x$ and (ii) $[x,f(p)] = i\hbar\partial f/\partial p$ where $f(x)$ and $f(p)$ are polynomials in x and p respectively.

(i) $[x^n,p] = i\hbar nx^{n-1}$ (ii) $[x,p^n] = i\hbar np^{n-1}$

Solution Any polynomial $f(x)$ can be expressed as $f(x)\sum_{n=0}^{k} na_nx^n$ where n is the degree of the polynomial

$$\therefore \quad \frac{\partial f(x)}{\partial x} = \sum_{n=0}^{k} na_nx^{n-1}$$

$$\therefore \quad [f(x),s] = \left[\sum_{n=0}^{k} a_nx^n,p\right] = \sum_{n=0}^{k} a_n[x^n,p] \text{ or } \sum_{n=0}^{k} a_nnx^{n-1} = i\hbar\frac{\partial f(x)}{\partial x}$$

Similarly, one can show that $[x, f(p)] = i\hbar\partial f / \partial p$.

Example 7.21: Show that $(xp_x + p_x x)$ is a self adjoint operator, where xp_x is not.

Solution Since position operator x and momentum operator p_x are both self adjoint and hence their product xp_x can be self adjoint if they commute with each other. But we have $[x, p_x] = i\hbar \neq 0$, i.e., x and p_x do not commute and hence xp_x cannot be self adjoint.

$$(xp_x + p_x x)^* = (xp_x)^* + (p_x x)^+$$
$$= p_x^+ x^+ + x^+ p_x^+ = p_x x + xp_x = (xp_x + p_x x)$$

i.e., adjoint of $(xp_x + p_x x)$ is equal to $(xp_x + p_x x)$. Thus $(xp_x + p_x x)$ is self adjoint.

Example 7.22: Show that $i[A, B]$ will be Hermitian, if A and B are Hermitian operators.

Solution

Let $$D = I[A,B] = i(AB - BA)$$

$\therefore$ $$D^+ = -i\{(AB)^+ - (BA)^+\}$$
$$= -i\{B^+A^+ - A^+B^+\} = -i(BA - AB) = i[A,B]$$

(A, B are Hermitian operators).

Example 7.23: The Hamiltonian of a particle is given by $H = p^2/2m + \frac{1}{2}m\omega^2x^2 + \alpha x^3$ where p and x are the momentum and the position operators for the particle. Find p and x.

Solution We have $[x, p] = i\hbar$

Form the Heisenberg's equation for an operator, we have (i)

$$\dot{x} = \frac{\partial x}{\partial t} = \frac{1}{i\hbar}[x, H] = \frac{1}{i\hbar}\left[x, \frac{p^2}{2m} + \frac{1}{2}m\omega^2x^2 + \alpha x^3\right]$$

$$= \frac{1}{2mi\hbar}[x, p^2] + \frac{1}{2i\hbar}m\omega^2\underset{\substack{\parallel \\ 0}}{[x, x^2]} + \frac{\alpha}{i\hbar}\underset{\substack{\parallel \\ 0}}{[x, x^3]}$$

$$= \frac{1}{2mi\hbar}[x, p^2] = \frac{1}{2i\hbar m}\{[x,p]p + p[x,p]\} = \frac{1}{2i\hbar m}\{i\hbar p + pi\hbar\} = p/m$$

Similarly, $$\dot{p} = \frac{dp}{dt} = \frac{1}{i\hbar}[p, H] = \frac{1}{i\hbar}[p.p^2/2m + \frac{1}{2}m\omega^2x^2 + \alpha x^3] = -m\omega^2x - 3\alpha x^2$$

Example 7.24: Show that the parity operator (π) is a Hermitian and linear operator. Find its eigen value. If the potential energy in the Hamiltonian $H = p^2/2m + V(x)$ is an even function, then prove that π commutes with H.

Solution Operator π is defined as, $\pi\psi(x) = \psi(-x)$

(i) Let $\psi_1(x)$ and $\psi_2(x)$ be two functions. Then from the property of π, we have

$$\pi[\psi_1(x) + \psi_2(x)] = [\psi_1(-x) + \psi_2(-x)] = \pi\psi_1(x) + \pi\psi_2(x)$$

and $\pi(\alpha\psi_1(x)) = \alpha\psi_1(-x) = \alpha\pi\psi_1(x)$

Here α is any arbitrary constant.

(ii) If π is Hermitian operator, then

$$\int \psi^*(x)\pi\phi(x)d^3x = \int (\pi\psi(x))^*\phi(x)d^3x \text{ or } \int \psi^*(x)\phi(-x)d^3x$$

$$= \int \psi^*(-x)\phi(x)d^3x$$

Writing $x' = -x$, one obtains

$$\text{L.H.S.} = \int \psi^*(-x')\phi(x')d^3x'$$

$$= \int \psi^*(-x)\phi(x)d^3x = \text{R.H.S.}$$

Obviously, π is a Hermitian operator.

(iii) Let $\pi\psi = \lambda\psi$, λ is the eigen value of operator π.

or $\pi\pi\psi = \pi\lambda\psi = \lambda^2\psi$ or $\psi = \lambda^2\psi$

$$\lambda^2 = 1 \text{ or } \lambda = \pm 1$$

Obviously, the eigen functions belonging $\lambda = \pm 1$ of π are the even functions ψ_e, i.e.,

$$\psi_e(x) = \psi_e(-x)$$

and eigen functions belong to $\lambda = -1$ or π are odd functions ψ_0, i.e.,

$$\psi_0(x) = -\psi_0(-x)$$

Thus, we can express $\psi(x)$ as

$$\psi(x) = \frac{1}{2}[\psi(x) + \psi(-x)] + \frac{1}{2}[\psi(x) - \psi(-x)]$$

Obviously, eigen functions of π are a complete set w.r.t. the class of function x.

(iv) $H = p^2/2m + V(x)$

$$H\psi(x) = -\psi^2 m\frac{d^2}{dx^2}\psi(x) + V(x)\psi(x) \left[\because p^2 = -i\hbar\frac{d}{dx}\right]$$

$$\therefore \quad \pi H\psi(x) = -\frac{\hbar^2}{2m}\pi\frac{d^2}{dx^2}\psi(x) + \pi V(x)\psi(x)$$

$$= -\frac{\psi^2}{2m}\frac{d^2}{dx^2}\psi(-x) + V(-x)\psi(-x)$$

$$= -\frac{\hbar^2}{2m}\frac{d^2}{dx^2}\psi(-x) + V(x)\psi(-x)$$

$\because$ $V(x)$ is even function

$\therefore$ $V(x) = V(-x)$

$$\therefore \quad \pi H\psi(x) = \left[-\frac{\hbar^2}{2m}\frac{d^2}{dx^2}+V(x)\right]\psi(-x)$$

$$\pi H\psi(x) = H\psi(-x)=H\pi\psi(x)$$

or $[H,\pi] = 0$ i.e., π commutes with H.

Example 7.25: Derive the operator equation $\frac{d}{dx}x^n = nx^{n-1} + x^n\frac{d}{dx}$ and then show that $\left[\frac{d}{dx},x^n\right] = x^{n-1}$

Solution Let us consider wave function $\psi(x)$. Now operating $\frac{d}{dx}x^n$ on $\psi(x)$, one obtains

$$\frac{d}{dx}[x^n,\psi(x)] = x^n\frac{d}{dx}\psi(x) + nx^{n-1}\psi(x) \text{ or } \frac{d}{dx}x^n = x^n + nx^{n-1}$$

Also, $$\left[\frac{d}{dx},x^n\right] = \frac{d}{dx}x^n - x^n\frac{d}{dx} = nx^{n-1} + x^n\frac{d}{dx} = nx^{n-1}$$

Example 7.26: The Hamiltonian for a charged particle in an electromagnetic field is given by

$$H = \frac{1}{2m}\left\{p-\frac{e}{c}A(x)\right\}^2 + e\phi(x), \text{ determine } x$$

Solution

We have $\dot{x} = (i\hbar)^{-1}[x,H]$

$$\therefore \quad [x,H] = \left[x,\frac{1}{2m}\left\{p - \frac{e}{c}A(x)\right\}^2 + e\phi(x)\right] = \left[x,\frac{1}{2m}\left\{p - \frac{e}{c}A(x)\right\}\right] + \left[\underset{\substack{\parallel\\0}}{x,e\phi(x)}\right]$$

$\therefore$ $[x,e\phi(x)] = 0$ and hence H commutes with x only.

$$[x,H] = \frac{1}{2m}\left[x,\left\{p-\frac{e}{c}A(x)\right\}^2\right] = \frac{1}{2m}\left(p - \frac{e}{c}A(x)\right)\left[x,\left\{p - \frac{e}{c}A(x)\right\}\right]$$

$$+\frac{1}{2m}\left[x\left\{p-\frac{e}{c}A(x)\right\}\right]\left\{p-\frac{e}{c}A(x)\right\}$$

But $\left[x,p-\frac{e}{c}A(x)\right] = [x,p]-\frac{e}{c}\underset{0}{\underset{\|}{[x,A(x)]}} = i\hbar$

$$\therefore \qquad [x,H]=\frac{1}{2m}\left(p-\frac{e}{c}A(x)i\hbar\right)+\frac{1}{2m}i\hbar\left(p-\frac{e}{c}A(x)\right)=i\hbar/m\left(p-\frac{e}{c}A(x)\right)$$

Equation of motion $\dot{x} = \frac{l}{i\hbar}\frac{i\hbar}{m}\left(p-\frac{e}{c}A(x)\right)=\frac{1}{m}\left(p-\frac{e}{c}A(x)\right)$

It is worthwhile to mention that if one treats x and p as the classical dynamical variable representing the position and momentum of a particle in the presence of e.m. field, then one finds that the above expression is exactly the classical relation between velocity and momentum.

Example 7.27: That operator A is defined as

$$A\psi(x)=\int_{-\infty}^{\infty}\theta(x-x')\psi(x')dx'$$

Here $\theta(x)$ is the step function defined by $\theta(x)=\begin{cases}0, & x<0\\ 1, & x>0\end{cases}$

If ψ and $d\psi/dx$ are integrable for $1x1\to\infty$; show that A is inverse of d/dx.

Solution

We have to show that $\qquad A=(d/dx)^{-1}$

or $\qquad A\frac{d}{dx}=\frac{d}{dx}A=1$

Given $\qquad \frac{d}{dx}A\psi(x)=\frac{d}{dx}\left[\int_{-\infty}^{\infty}\theta(x-x')\psi(x')dx'\right]$

$$=\frac{d}{dx}\left[\int_{-\infty}^{x}\theta(x-x)\psi(x')dx'+\int_{x}^{\infty}\theta(x-x')\psi(x')dx'\right]$$

$$=\frac{d}{dx}\left[\int_{-\infty}^{x}\psi(x')dx'+0\right]=\psi(x)$$

Also we have

$$A\frac{d}{dx}\psi(x)=\int_{-\infty}^{\infty}\theta(x-x')\frac{d}{dx}\psi(x')dx'=\int_{-\infty}^{x}\theta(x-x')\frac{d}{dx'}\psi(x')dx'$$

$$+\int_{x}^{\infty}\theta(x-x')\frac{d}{dx'}\psi(x')dx' = \int_{-\infty}^{x}\frac{d}{dx'}\psi(x')dx = \psi(x)$$

Obviously, $\frac{d}{dx}A\psi(x) = A\frac{d}{dx}\psi(x)$ or $\frac{d}{dx}A = A\frac{d}{dx}$, i.e., $A = \left(\frac{d}{dx}\right)^{-1}$

Example 7.28: Show that every operator can be expressed as the combination of the two operators, each of them is Hermitian operator.

Solution Let A be an operator whose adjoint is A^+. We can express A as

$$A = \frac{A + A^+}{2} + i\frac{A - A^+}{2i}$$

Let $B = \frac{A + A^+}{2}, C = \frac{A - A^+}{2i} \quad (\therefore\ A = B + iC)$

Then $B^+ = \frac{A^+ + (A^+)^+}{2} = \frac{A^+ + A}{2} = B$

and $C^+ = \frac{A^+ - (A^+)^+}{2(-i)} = \frac{A^+ - A}{-2i} = \frac{A - A^+}{2i} = C$

Obviously, B and C are both Hermitian and A can be expressed as $B + iC$.

Example 7.29: Prove that $\left[x_i, p_j\right] = i\hbar\delta_{ij}$

Solution $x_i p_j \psi = i\hbar x_i \frac{\partial\psi}{\partial x_i}$

$$p_j x_i \psi = i\hbar\frac{\partial}{\partial x_i}(x_j\psi) = -i\hbar\left(\delta_{ij}\psi + x_i\frac{\partial\psi}{\partial x_j}\right)$$

$\because$ $[x,p] = i\hbar$ and $\frac{\partial x_i}{\partial x_j} = \delta_{ij}$

$\therefore$ $(x_i p_j - p_j x_i)\psi = i\hbar\delta_{ij}\psi$

Since this is true for any wave function ψ and hence one can conclude that $[x_i, p_j] = i\hbar\delta_{ij}$.

Example 7.30: If Δx and Δp_x represent the uncertainties in position and momentum operators x and p_x respectively, then prove that they satisfy the uncertainty relation: $\Delta x \Delta p_x \geq \hbar/2$.

Solution We have $[A, B] = iC; C \neq 0$.
Here A and B are operators. Also we have $\Delta A\,\Delta B \geq \langle C \rangle/2$.
For the operators x and p_x we have

$$[x, p_x] = i\hbar \quad \text{and} \quad \Delta x \Delta p_x \geq \langle \hbar/2 \rangle \geq \hbar/2$$

Example 7.31: **Show that the complex conjugation is not a linear operator.**

Solution **Complex conjugation can be defined by the process: $A\psi = \psi^*$**

But ψ can be expressed as a linear combination of function, i.e.,

$$\psi = C_1\psi_1 + C_2\psi_2$$

$$\therefore \quad A\psi = A(C_1\psi_1 + C_2\psi_2)$$

$$= (C_1\psi_1 + C_2\psi_2)^*$$

$$= C_1^*\psi_1^* + C_2^*\psi_2^*$$

Also $$A(C_1\psi_1 + C_2\psi_2) = C_1^*(A\psi_1) + C_2^*(A\psi_2)$$

Obviously, this operation is not a linear operation. Such type of operators in quantum mechanics are termed as antilinear operators.

Example 7.32: **A particle is in motion along a line between $x = 0$ and $x = a$ with zero potential energy, and at point for which $x < 0$ and $x > a$ the potential energy infinite. The wave function for the particle in the nth state is given by $\psi_n = A\sin\dfrac{n\pi x}{a}$. Find the expression for the normalized wave function.**

Solution **For the nth state probability that the particle will be between x and $x + dx$ is**

$$pdx = |\psi_n(x)|^2\,dx = A^2\sin\frac{2n\pi x}{a}dx$$

There is only one particle in the region from $x = 0$ to $x = a$ at all times. It is somewhere in this region. Therefore

$$\int pdx = \int_0^a |\psi_n(x)|^2\,dx = 1 \quad \text{or} \quad \int_0^a A^2\sin^2\frac{n\pi x}{a}dx = 1$$

$$\text{or} \quad A^2\int_0^a \frac{1}{2}\left[1 - \cos\frac{2n\pi x}{a}\right]dx = 1 \quad \text{or} \quad \frac{A^2}{2}\left[x - \frac{a}{2\pi n}\sin\frac{2n\pi x}{a}\right]_0^a = 1$$

The second term of the integrated expression is zero at both the limits. Therefore,

$$\frac{A^2}{a} = a \quad \text{or} \quad A = \sqrt{2/a}$$

the normalized wave function is $\psi_n(x) = \sqrt{2/a}\,\sin\dfrac{n\pi x}{a}$.

Example 7.33: **Two trains of travelling waves may be specified by the function:**

(i) **$\psi_1 = \sin(ax + bt)$ and $\psi_2 = \sin(ax - bt)$. For each of these trains find the speed and direction of propagation and the wavelength.**

(ii) **Show that $\psi_+ = \psi_1 + \psi_2$ and $\psi_- = \psi_1 - \psi_2$ represent standing waves.**

Solution

(i) $\psi_1 = 0$ when $ax + bt = \pi n$, where $n = 0, n = \pm 1, n = \pm 2, \ldots..$

$$\therefore \qquad x = \frac{\pi n - bt}{a}$$

Each such nodal point is moving along the x-axis in the negative direction with velocity $\frac{dx}{dt} = -\frac{b}{a}$. Similarly for ψ_2, $\frac{dx}{dt} = b/a$. This wave propagates along the x-axis in the positive direction. The wavelength in each case will be twice the distance between adjacent nodes, i.e.,

$$\lambda = 2(x_{n+1} - x_n) = \frac{2\pi}{a}$$

(ii) $\psi_+ = \sin(ax + bt) + \sin(ax - bt)$

$$= \sin ax \cos bt + \cos ax \sin bt + \sin ax \cos bt - \cos ax \sin bt$$

$$= 2 \sin ax \cos bt$$

This function will have a node wherever $x = \frac{\pi n}{a}$. These nodes do not move with time and the function therefore represents a standing wave. Similarly $\psi = 2 \cos ax$ sin bt. This is node a wherever $x = \frac{\pi(n + 1/2)}{a}$.

Example 7.34: What are the expressions for the following operators:

(a) $\left(\frac{d}{dx} + x\right)^2$, (b) $\left(x\frac{d}{dx}\right)^2$ and (c) $\left(\frac{d}{dx}x\right)^2$

Solution

(a)
$$\left(\frac{d}{dx} + x\right)^2 \psi = \left(\frac{d}{dx} + x\right)\left(\frac{d}{dx} + x\right)\psi$$

$$= \left(\frac{d}{dx} + x\right)\left(\frac{d\psi}{dx} + x\psi\right)$$

$$= \left(\frac{d^2\psi}{dx^2} + 2x\frac{d\psi}{dx} + x^2\psi + \psi\right)$$

$$= \left(\frac{d^2}{dx^2} + 2x\frac{d}{dx} + x^2 + 1\right)\psi$$

or
$$\left(\frac{d}{dx} + x\right)^2 = \left(\frac{d^2}{dx^2} + 2x\frac{d}{dx} + x^2 + 1\right)$$

(b)
$$\left(x\frac{d}{dx}\right)^2\psi = \left(x\frac{d}{dx}\right)\left(x\frac{d}{dx}\right)\psi$$
$$= \left(x\frac{d}{dx}\right)\left(x\frac{d\psi}{dx}\right)$$
$$= x\left(x\frac{d^2\psi}{dx^2} + \frac{d\psi}{dx}\right)$$
$$= \left(x^2\frac{d^2}{dx^2} + x\frac{d}{dx}\right)\psi$$

or
$$\left(x\frac{d}{dx}\right)^2 = x^2\frac{d^2}{dx^2} + x\frac{d}{dx}$$

(c)
$$\left(\frac{d}{dx}x\right)^2\psi = \left(\frac{d}{dx}x\right)\left(\frac{d}{dx}x\right)\psi$$
$$= \left(\frac{d}{dx}x\right)\frac{d}{dx}(x\psi)$$
$$= \left(\frac{d}{dx}x\right)\left(x\frac{d\psi}{dx}+\psi\right)$$
$$= \frac{d}{dx}\left(x^2\frac{d\psi}{dx} + x\psi\right)$$
$$= x^2\frac{d^2\psi}{dx^2} + 2x\frac{d\psi}{dx} + x\frac{d\psi}{dx} + \psi$$
$$= \left(x^2\frac{d^2}{dx^2} + 2x\frac{d}{dx} + x\frac{d}{dx} + 1\right)\psi$$

or
$$\left(\frac{d}{dx} + x\right)^2 = x^2\frac{d^2}{dx^2} + 3x\frac{d}{dx} + 1$$

Example 7.35: Which of the functions sin 3x, 6 cos 4x, 5x^3, 1/x, 3e^{-5e}, 1$_n$ 2x are eigen functions of operator $\frac{d^2}{dx^2}$? For each eigen function find the value.

Solution

(a) $\frac{d^2}{dx}(\sin 3x) = -9(\sin 3x)$

Obviously, sin 3x is the eigen function of $\frac{d^2}{dx^2}$ with eigen value –9.

(b) $\frac{d^2}{dx^2}(5x^3) = 30(x)$. Obviously, 5$x^3$ is the eigen function of $\frac{d^2}{dx^2}$

(c) $\frac{d^2}{dx^2}(1/x) = 2\frac{1}{x^3}$. Obviously, $1/x$ is not the eigen function of $\frac{d^2}{dx^2}$

(d) $\frac{d^2}{dx^2}(3e^{-5x}) = 25(3e^{-5x})$. Obviously, $3e^{-5x}$ is the eigen function of $\frac{d^2}{dx^2}$ with eigen value 25.

(e) $\frac{d^2}{dx^2}.(ln 2x) = \left(-\frac{1}{x^2}\right)$. Obviously, $(l_n 2x)$ is not the eigen function of $\frac{d^2}{dx^2}$

Example 7.36: If φ_1 and φ_2 are the two degenerate eigen functions of the linear operator H, show that $\psi = C_1\varphi_1 + C_2\varphi_2$ is an eigen function of H with the same eigen value as that for φ_1 and φ_2.

Solution

Since
$$H\varphi_1 = \lambda\varphi_1 \text{ and } H\varphi_2 = \lambda\varphi_2 \quad \text{(i)}$$

$$H(C_1\varphi_1 + C_2\varphi_2) = H(C_1\varphi_1) + H(C_2\varphi_2) \quad \text{(ii)}$$
$$= C_1 H\varphi_1 + C_2 H\varphi_2$$
$$= C_1\lambda\varphi_1 + C_2\lambda\varphi_2 \text{ [from (i) and (ii)]}$$
$$= \lambda(C_1\varphi_1 + C_2\varphi_2)$$

Example 7.37: Consider the operator $R = -\frac{d^2}{dx^2}$ and the eigen value equation $R\psi = \lambda\psi$. Write out the possible eigen functions and discuss the condition under which they are well behaved and when λ is an eigen value.

Solution We have $\frac{d^2\psi}{dx^2} = \lambda\psi$. General solution of the above differential equation is

$$\psi = Ae^{\sqrt{-\lambda x}} + Be^{-\sqrt{\lambda x}} \text{ for } \lambda \neq 0 \text{ and } \psi = Cx + D \text{ for } \lambda = 0$$

where A, B, C and D are constants. Three cases of interest are

(i) $\lambda < 0$; ψ is not well behaved as it does not remain finite as $|x| \to \infty$ so λ is not an eigen value.

(ii) $\lambda < 0$; ψ is well behaved only if $C = 0$.

(iii) $\lambda < 0$; $\psi = (A + B)\cos\sqrt{\lambda x} + i(A - B)\sin\sqrt{\lambda x}$, which is well behaved for all values of x. Hence λ represents an eigen value.

Example 7.38: Examine the solutions of $R\psi = \pm C\psi$, where $R\frac{d^2}{dx^2}$, and the eigen functions for the space $-\infty < x < \infty$. (C is a positive constant).

Solution Let us first consider the positive sign. The solution of $\frac{d^2\psi}{dx^2} = C\psi$ is $\psi = Ae^{\sqrt{Cx}} \pm Be^{\sqrt{Cx}}$. The above function diverges as $x \to \pm\infty$, so it is not suitable eigen function. However, if A were zero, the function $\psi = Be^{-\sqrt{Cx}}$ would be well behaved in the positive half space ($x > 0$). For $B = 0$, the function $\psi = Ae^{\sqrt{Cx}}$ would be well behaved in the negative half space ($x > 0$). For $B = 0$ the function $\psi = Ae^{\sqrt{Cx}}$ would be well behaved in the negative half space ($x > 0$). To obtain well behaved eigen functions for all space, we must consider eigen value $-C$. Then

$$\frac{d^2\psi}{dx^2} = -C\psi$$

and the solution of the above equation is: $\psi = D\sin\sqrt{C}x + E\cos\sqrt{C}x$.

The above function is well-behaved for all x, and an eigen function of R exists for all eigen values $-C$. For this situation, one obtains a continuum of possible eigen values, whereas in many physical problems the eigen values form a continuum of possible eigen values, whereas in many physical problems the eigen values form a discrete set of numbers.

Example 7.39: Find the eigen functions and eigen values of linear momentum operator p_x. Show that in the absence of any restriction there is a continuous series of eigen values of p_x.

Solution

We have
$$p_x = \lambda\psi$$

or
$$\frac{h}{2\pi i}\frac{d\psi}{dx} = \lambda\psi$$

or
$$\frac{d\psi}{dx} = ik\psi \qquad \left(\text{Here } k = \frac{2\pi\lambda}{h}\right)$$

or
$$\psi = A\exp(ikx) \qquad \text{(Eigen function)}$$

we have
$$\lambda = \frac{kh}{2\pi} \qquad \text{(eigen values)}$$

we note that λ depends on k; if there is no restriction on k there will be no restriction on λ provided ψ is finite, single valued and continuous for all values of x.

Example 7.40: Show that the function $\psi = \exp[i(k_x x + k_y y + k_z z)]$ is simultaneously an eigen function of both operators $p = \frac{h}{2\pi i}\nabla$ and $-\hbar^2\nabla^2$. Find the eigen values.

Solution

(i)
$$p\psi = \frac{h}{2\pi i}\left[\hat{i}\frac{\partial}{\partial x} + \hat{j}\frac{\partial}{\partial y} + \hat{k}\frac{\partial}{\partial z}\right][\exp i(k_x x + k_y y + k_z z)]$$

$$= \frac{h}{2\pi}[\hat{i}k_x + \hat{j}k_y + \hat{k}k_z][\exp i(k_x x + k_y y + k_z z)]$$

$\therefore$ Eigen value
$$= \frac{h}{2\pi}(\hat{i}k_x + \hat{j}k_y + \hat{k}k_z)$$

(ii) $$-\hbar^2\nabla^2\psi = -\hbar^2\nabla^2[\exp i(k_x x + k_y y + k_z z)]$$
$$= -\hbar^2[-(k_x^2 + k_y^2 + k_z^2)\{\exp i(k_x x + k_y y + k_z z)\}]$$

$\therefore$ Eigen value is $= -\hbar^2(k_x^2 + k_y^2 + k_z^2)$

Example 7.41: An operator $\dfrac{d}{dr}$ which operates on a set of eigen functions exp(–*ar*), *a* being real and positive and real eigen values is given. Is the operator necessarily Hermitian?

Solution We have $\dfrac{d}{dr}[\exp(-ar)] = -a\exp(-ar)$ eigen value = –*a*. Let us take two of the eigen functions be with $a_1 = 1$ and $a_2 = 2$.

$\psi_1 = \exp(-r)$, eigen value = –1 and $\psi_2 = \exp(-2r)$, eigen value = –2

we have $$\int_0^\infty \exp(-r)\frac{d}{dr}[\exp(-2r)]r^2 dr = -2\int_0^\infty r^2\exp(-3r)dr = -\frac{4}{27}$$

and $$\int_0^\infty \exp(-2r)\frac{d}{dr}[\exp(-2r)]r^2 dr = -2\int_0^\infty r^2\exp(-4r)dr = -\frac{1}{16}$$

(we have made use of Table of integrals for evaluating integrals).

Obviously, $\dfrac{d}{dr}$ is not Hermitian unless $a_1 = a_2$.

Example 7.42: Show that if a normalized wave function is expanded in terms of an orthonormal set of functions the sum of the squares of the coefficients is unity.

Solution

Let $$\psi = c_1\phi_1 + c_2\phi_2 + c_3\phi_3 + \cdots$$

$$\int \psi^2 dt = \left[c_1^2\int\phi_1^2 d\tau + c_2^2\int\phi_2^2 d\tau + c_3^2 + c_3^2\int\phi_2^2 d\tau + \cdots\right]$$
$$+ 2c_1c_2\int\phi_1\phi_2 d\tau + 2c_1c_3 + 2c_2c_3\int\phi_2\phi_3 d\tau + \cdots = 1$$

or $$c_1^2 + c_2^2 + c_3^2 + \cdots = 1$$

($\phi_1, \phi_2, \phi_3, \ldots$ are normalized and orthogonal wave function).

Example 7.43: ψ and ϕ are two non-orthogonal eigen functions of an operator. Find the normalization constant for a wave function $\psi' = \phi - S\psi$.

Solution We have $\psi' = \phi - S\psi$, where $S = \int\psi^*\phi d\tau$

Let *N* be the normalization constant for ψ'. We have

$$N^2\left[\int\phi^2 d\tau + \int S^2\psi^2 d\tau - \int 2S\phi\psi d\tau\right] = 1$$

or $$N^2[1 + S^2 - 2S^2] = 1 \text{ or } N = (1 - S^2)^{-1/2}$$

Example 7.44: Two normalized but nonorthogonal eigen functions are $\psi = [\exp(-r)]/\sqrt{\pi}$ and $\phi = [r\exp(-r)]/\sqrt{3\pi}$. Construct a new function ψ' which is orthogonal to ψ and is normalised.

Solution

We have $$S = \int \psi\phi d\tau = \int_0^\infty \frac{e^{-r}}{\sqrt{\pi}} \frac{re^{-r}}{\sqrt{3\pi}} dr \int_0^\pi \sin\theta d\theta \int_0^{2\pi} d\phi$$

or $$S = \frac{4}{\sqrt{3}} \int_0^\infty r^3 e^{-2r} dr = \frac{\sqrt{3}}{2} \qquad \left(\because \int_0^\pi \sin\theta d\theta \int_0^{2\pi} d\phi = 4\pi \right)$$

(with the help of table of integrals)

Since the new function ψ' is ortogonal to $\psi = [\exp(-r)/\sqrt{\pi}]$ and hence

$$\psi' = \phi - S\psi = \frac{1}{\sqrt{3\pi}} re^{-r} - \frac{\sqrt{3}}{2} \frac{1}{\sqrt{\pi}} \exp(-r) = \frac{1}{\sqrt{3\pi}} \left(r - \frac{3}{2} \right) e^{-r}$$

Let N be the normalization factor, then

$$\int (N\psi')d\tau = N^2 \int_0^\infty \left[\frac{1}{\sqrt{3\pi}} \left(r - \frac{3}{2} \right) e^{-r} \right]^2 r^2 dr \int_0^\pi \sin\theta d\theta \int_0^{2\pi} d\theta$$

or $$\frac{N^2 4\pi}{3\pi} \int_0^\infty \left[\left(r^2 - 3r + \frac{9}{4} \right) r^2 e^{-2r} \right] dr = 1$$

or $$\frac{4N^2}{3} \left[\frac{3}{4} - \frac{9}{8} + \frac{9}{16} \right] = 1 \text{ or } N^2 = 4 \therefore N = 2$$

Obviously, the normalised wave function is $\psi' = \dfrac{2}{\sqrt{3\pi}} \left(r - \dfrac{3}{2} \right) e^{-r}$

SUMMARY

1. *Quantum mechanics* is the theory of the microworld based on wave functions and probabilities develoved by Heiseinberg and Schrodinger.
2. In quantum mechanics particles are represented by a *wave function* $\psi(x,y,z)$. The *probability density* that a particle will be found at a point is $|\psi|^2$. If the particle is confined to moving along the x-axis, the probability that it will be located in an interval dx is given by $|\psi|^2 dx$. Moreover, the wave function must be *normalized*.

$$\int_{-\infty}^{\infty} |\psi|^2 \, dx = 1$$

3. The measured position x of the particle, averaged over several trials, is termed the *expectation value* of x and is defined as

$$<x> = \int_{-\infty}^{\infty} \psi^* x \psi \, dx$$

4. A fundamental equation of quantum mechanics, which relates probability wave amplitudes to the forces acting or the system is called *Schrodinger's wave equation.* *Schrodinger's time dependent* wave equation is

$$i\hbar \frac{\partial \psi(x,t)}{\partial t} = -\frac{\hbar^2}{2m} \frac{\partial^2 \psi(x,t)}{\partial x^2} + v\psi(x,t)$$

Schrodinger's time independent form for the spatial dependence (in one dimension) of $\psi(x)$, where $\psi(x,t) = \psi(x)\exp(-iEt/\hbar)$, is

$$-\frac{\hbar^2}{2m} \frac{\partial^2 \psi(x)}{\partial x^2} + v(x)\psi(x) = E\,\psi(x)$$

Schrodinger's equation is as basic to quantum mechanics as Newton's laws of motion are to classical mechanics.

5. Certain properties of ψ and $\frac{\partial \psi}{\partial x}$ lead to quantized behaviour. The wave function $\psi(r, t)$ must be finite, single valued and continuous; $\frac{\partial \psi}{\partial x}$ must be continuous. The wave function must be normalised for use in determining probabilities.

REVIEW QUESTIONS AND PROBLEMS

1. Obtain Schrodinger time-dependent wave equation and separate it into space and time dependent parts. Give the probability interpretation of the wave functions and show that probability density P and probability current density J satisfy the continuity equation $\frac{dP}{dt} + \nabla . \mathbf{J} = 0$. Explain the physical significance of this equation.

2. Show that one dimensional wave equation (Schrodinger) has a solution of the form $\psi(x)\,\phi\,(t)$ if the potential energy V of the particle associated with the wave function $\psi(x)$ is independent of time t. Show that the most general solution of Schrodinger equation for such a particle can be written as: $\psi = \sum_{n=1}^{\infty} A_n \psi_n(x) \exp[-iEt/\hbar]$.

3. Ehrenfest's theorem states that for a particle for which the uncertainty in position and momentum can be neglected, the quantum mechanical prediction of the motion agrees with the classical prediction. Show that this theorem is valid for a one-dimensional motion for which $P = P(x)$ by showing that

(a) $\frac{d<x>}{dt} = \int \psi^* \frac{i}{\hbar}(Hx - xH)\psi dx = \frac{<p_x>}{m}$ and (b) that $\frac{d(p_x)}{dt} = <-\frac{dP}{dx}>$

4. Prove that the operator $\alpha = c$ is only one for which any arbitrary function is an eigen function corresponding to the eigen value c.
(**Hint:** Note that $\alpha = \beta$ if $\alpha\psi = \beta\psi$ for every permissible function ψ).
5. A particle is represented by the wave function $\psi(x) = e^{-(x-x_0)^2/2\alpha}\sin kx$. Plot the wave function $\psi(x)$ and the probability distribution $|\psi(x)|^2$. Estimate the uncertainty in the position and momentum of the particle.
6. Given that $\psi(x,t) = c_1\psi_1(x)\exp[-iEt/\hbar] + c_2\psi_2(x)\exp[-iE_2t/\hbar]$, where the proper functions $\psi_1(x)$ and $\psi_2(x)$ are orthogonal and normalized, show that

$$c_1 = \int \psi_1^*(x)\psi(x,0)dx \quad \text{and} \quad c_2 = \int \psi_2^*(x)\psi(x,0)dx$$

This shows that the wave function at any time can be found in terms of the initial wave function $\psi(x, 0)$. Extend the result to the case in which: $\psi(x,t) = \sum_n c_n\psi_n(x)e^{-iE_nt/\hbar}$
7. Show that momentum operator $-i\hbar\dfrac{d}{dx}$ is Hermitian.
8. If $\psi_1(x, t)$ and $\psi_2(x, t)$ are both solutions of Schrodinger's equation for a given potential $V(x)$, show that the linear combination $\psi = a_1\psi_1 + a_2\psi_2$ in which a_1 and a_2 are arbitrary constants is also a solution.
9. The possible values obtained from measurement of a discrete variable, called x are 1, 2, 3 and 4. (a) Given that the respective probabilities are $\frac{1}{4}, \frac{1}{4}, \frac{1}{4}$ and $\frac{1}{4}$, calculate the expectation values of x and x^2. (b) Given that the respective probabilities are $\frac{1}{12}, \frac{5}{12}, \frac{5}{12}$ and $\frac{1}{12}$, calculate the expectation values x and x^2. (c) Compare the results for parts, (a) and (b). **[Ans.** (a) 2.5, 7.5; (b) 2.6, 6.83]
10. If $\psi(x,t) = A\sin\dfrac{\pi x}{a}\exp[-iE_0t/\hbar]$ represent a normalized function in the range $0 \le x \le a$; calculate the value of the following: (a) constant A, (b) expectation values of x and x^2.
11. (a) State basic postulates of quantum mechanics,
(b) if A is a non-Hermitian operator show that $(A + A^*)$ and $i(A - A^*)$ are both Hermitian.
(c) By mathematical induction on n, show that
$[x^n, p] = i\hbar n x^{n-1}$ and $[x, p^n] = i\hbar n p^{n-1}$, where n is any positive integer. Hence prove that
$[f(x), p] = i\hbar\dfrac{\partial f}{\partial x}$ and $[x, f(p)] = i\hbar\dfrac{\partial f}{\partial p}$, where $f(x)$ is any polynomial in x and $f(p)$ is a polynomial in p.
12. (a) Show that if two observables commute they have a complete set of simultaneous eigen vectors, (b) Show that all the eigen values of a Hermitian operator are real and eigen kets belonging to different eigen values are orthogonal.

13. (a) Prove that the chances for the existence of simultaneous eigen states for two commuting observables are most favourable. (b) If the coordinates x and the momenta p are the dynamical variables of a free particle, show that

$$<p'|x^2|p''> = \left((\hbar/i)\frac{\partial}{\partial p}\right)^2 \delta(p' - p'')$$

14. Show that $<x'|p^2|x''> = \left(\frac{\hbar}{i}\frac{\partial}{\partial x'}\right)^2 \delta(x' - x'')$

15. Show that if A and B are two non-commuting Hermitian operators, the corresponding observations cannot both be defined sharply. Apply the results to show that $\Delta x \Delta p_x \geq \hbar/2$, the symbols having the usual meaning.

16. Show that the product of two commuting Hermitian operators is Hermitian.

17. Prove the commutation relation $[P_x, f(x)] = (\hbar/i)\frac{dt}{dx}$ where $f(x)$ is a function that can be expanded in a power series.

18. Show that the eigen values of a Hermitian operator are real.

19. An operator like $U = 1 + i \in F$, where $\in$ is real infinitesimal quantity is said to produce an infinitesimal transformation. Prove that if the above operator is unitary F ought to be Hermitian. Also prove that in such an infinitesimal transformation, the transformed operator

$$\hat{A} = A + i\in[F, A]$$

20. If A and B are two operators such that both commute with their commutator $[A, B]$, then prove that $[A, B^n] = nB^{n-1}[A, B]$.

21. F and G are two Hermitian operators. Under what conditions the product FG is also a Hermitian operator?

22. Show that if the Hamiltonian of a system is invariant under some coordinate transformation, the operator of the transformation will commute with the Hamiltonian.

23. Show that the average value of the square of a Hamiltonian operator is positive.

24. (a) Show explicitly that the operator p_x is not Hermitian (p_x is the x component of momentum).

(b) Show that if two observables are simultaneously measurable, they must commute.

25. Prove that the commutation relation $[p_x f(x)] = \frac{\hbar}{i}\frac{df}{dx}$, where $f(x)$ is a function that can be expanded in a power series.

26. Determine which pairs among the following pairs commute (a) $\partial/\partial x$ (b) $\partial/\partial t$ (c) x (d) t.

27. If A and B are two operators having the commutation relation $AB - BA = l$, determine $A^nB - BA^n$.

28. Prove that a Hermitian operator can have only real eigen values.

29. Show that two operators having simultaneous eigen functions commute.

30. Explain the concept of parity. Show that the eigen values of parity operator are +1 and −1.

31. In quantum mechanics commutator of two operators A and B is denoted by $[A, B]$ and defined as $[A, B] = AB - BA$. Show that (i) $[x, P_x] = i\hbar = [A, BC] = B[A, C] + [A, B]C$.

32. (a) Formulate the time-dependent Schrodinger equation. Compare it with classical wave equation $\nabla^2\psi = \dfrac{1}{v^2}\dfrac{\partial^2\psi}{\partial t^2}$. Give physical significance of the wave function.

(b) Starting from the time-dependent Schrodinger equation, obtain the equation of continuity. Mention the assumptions made and discuss the physical significance of the equation of continuity. Using the expression for probability current density, prove that it is always real.

33. Derive one-dimensional Schrodinger equation from a wave-packet and hence show that the energy operator and momentum operator can be represented as follows:

$$E \rightarrow i\hbar\partial/\partial t \text{ and } p \rightarrow i\hbar\partial/\partial x$$

34. Starting from the time independent Schrodinger equation of the type $\nabla.\boldsymbol{J} + \dfrac{\partial\rho}{\partial t} = 0$. Discuss the above equation physically. What will be $\boldsymbol{J}$ for $\psi = \exp[ik, r]$? Interpret the result.

35. State Ehrenfest's theorem. Prove that $\dfrac{d}{dt}<x> = \dfrac{<p_x>}{m}$, where symbols have usual meanings.

36. (i) Starting with the time-dependent Schrodinger's equations, show that the probability is conserved when the probability density is defined as

$$S = \frac{i\hbar}{2m}\left(\psi^*\frac{\partial\psi}{\partial x} - \psi\frac{\partial\psi^*}{\partial x}\right)$$

(ii) What will be $\boldsymbol{S}$ for $\psi = A\exp[-i/\hbar(Et - px)]$. Why $\boldsymbol{S}$ so obtained is not true in general? State the condition under which it will be a valid approximation.

37. (a) Why should any wave function be finite, continuous and differentiable at every point?

(b) When and why do we call a solution of the Schrodinger's equation a stationary one? Show that they occur only if the potential energy does not depend explicitly on the time.

38. Explain the significance of the properties of completeness and orthogonality of the wave functions and also of unitary transformation in quantum mechanics.

39. If F is a non-Hermitian operator, show that $(F + F^+)$ and $i(F - F^+)$ are Hermitian operators.

40. (a) Discuss the statistical interpretation of the wave function. Obtain an expression for the probability current density.

(b) Explain the general rules for forming the Schrodinger's equation by correspondence principle.

41. The uncertainty in measurement of a physical quantity A is defined as $\Delta A = [\hat{A}^2 - (\hat{A})^2]^{1/2}$. Show that inspite of the uncertainty of time (Δt), one can exactly determine the total ground state energy (E_0) of a system.

[**Hint:** We have $\quad <E> = \int\psi_0 H^2\psi_0 d\tau = \int\psi_0 E_0\psi_0 d\tau = E_0\int\psi_0^2 d\tau = E_0$

(if ψ_0 is normalized)

Now $$<E^2> = \int \psi_0 H^2 \psi_0 d\tau = \int \psi_0 H(H\psi_0) d\tau = \int \psi_0 H = (E_0\psi_0) d\tau$$

$$= E_0 \int \psi_0 H d\tau = E_0 \int \psi_0 E_0 \psi_0 d\tau = E_0^2$$

$$\therefore \quad \Delta E = [<E_0^2> - (<E_0>)^2 = 0]$$

42. The state of a system is described by the function $\psi(r, t)$, which is a function of both position and time, but the expected value of any physical quantity is independent of time. Explain it.

[**Hint:** We have $\psi(r, t) = \psi(x)\,\phi\,(t) = \psi(x) a \exp(2\pi i E t/h)$

$$<\bar{A}> = \int \psi^*(x) a \exp\left(\frac{2\pi i E t}{h}\right) A\psi(x) a \exp\left(-\frac{2\pi i E t}{h}\right) = a^2 \int \psi^+(x) A\psi(x) dx$$

43. Show that the operator p_x for linear momentum is Hermitian. Explain its physical significance.

$$\left[\textbf{Hint}: \int_{-\infty}^{\infty} \psi^* \left(\frac{\hbar}{i}\frac{d}{dx}\right) \phi d\tau = \int_{-\infty}^{\infty} \phi \left(\frac{\hbar}{i}\frac{d}{dx}\right)^* \psi^* dx\right.$$

Integrating by parts, one obtains L.H.S. as

$$\left.\frac{\hbar}{i}[\psi^*\phi]_{-\infty}^{\infty} - \frac{\hbar}{i}\int_{-\infty}^{\infty} \phi \frac{d\psi^*}{dx} dx = 0 + \int_{-\infty}^{\infty} \phi \left(-\frac{\hbar}{i}\frac{d}{dx}\right) \psi dx = \int_{-\infty}^{\infty} \left(\frac{\hbar}{i}\frac{d}{dx}\right)^* \psi^* dx\right]$$

44. A system is described by the Hamiltonian operator, $H = -\dfrac{d^2}{dx^2} + x^2$ (i) Show that $\psi = Ax\exp(-x^2/2)$ is eigen function of H. Determine eigen value (ii) For the function to the normalised determine A and (iii) Determine $<x>$ for the state $\psi = Ax\exp\left(-\dfrac{x^2}{2}\right)$.

[**Hint:** One can easily see that ψ is a well behaved function.

(i) $$-\frac{d^2}{dx^2}[Ax\exp(-x^2/2)] = -A\frac{d}{dx}[1-x^2]\exp(-x^2/2) = A(3-x^2)x\exp(-x^2/2)$$

$$\therefore \quad H\psi = \left(-\frac{d^2}{dx^2} + x^2\right)\left[A\exp\left(-\frac{x^2}{2}\right)\right] = 3\left[Ax\exp\left(-\frac{x^2}{2}\right)\right]$$

Obviously, eigen value is 3.

(ii) $$\int_{-\infty}^{\infty} \psi^2 dx = \int_{-\infty}^{\infty} A^2 x^2 \exp(-x^2) dx = A^2\sqrt{\pi/4} = 1 \text{ (Normalization condition)}$$

$$A = (4/\pi)^{1/8}$$

(iii) The time-independent state function is $\psi = Ax = \exp(-x^2/2)$

$$<x> = \int_{-\infty}^{\infty}\left[Ax\exp\left(-\frac{x^2}{2}\right)\right]x\left[Ax\exp\left(-\frac{x^2}{2}\right)\right]dx$$

$$= \int_{-\infty}^{\infty}\left[A^2x^3\exp(-x^2)dx + \int_{0}^{\infty}A^2x^3\exp(-x^2)dx = 0\right]$$

45. (i) A particle in a ring is in a state with $\psi = \frac{1}{\sqrt{\pi}}\cos 2\phi$, calculate the expectation value of angular momentum $< p_\phi >$, where $p_\phi = \frac{\hbar}{i}\frac{d}{d\phi}$. (ii) Express ψ as a linear combination of exponential and evaluate the expectation value of the angular momentum using the relation $< p_0 > = \Sigma|c_n|^2\, p_{\phi n}$ is the eigen value for the nth exponential and c_n is the appropriate coefficient.

[**Hint:** (i) $$< p_{\phi n} > = \int_0^{2\pi}\frac{1}{\sqrt{\pi}}\cos 2\phi\frac{\hbar}{i}\frac{d}{d\phi}\left(\frac{1}{\sqrt{\pi}}\cos 2\phi\right)d\phi$$

$$= -\frac{h}{2\pi^2 i}\int_0^{2\pi}2\cos 2\phi d\phi = -\frac{\hbar}{2\pi^2 i}\int_0^{2\pi}\sin 2\phi d\phi = 0$$

(ii) $< p_\phi > \left(=\frac{\hbar}{i}\frac{d}{d\phi}\right)$. Normalized eigen functions of this are $\frac{1}{\sqrt{2\pi}}\exp(\pm im\phi)$ with eigen value $\pm mh/2\pi$. $\therefore\ \psi = \frac{1}{\sqrt{2}}\left[\frac{1}{\sqrt{2\pi}}\exp(im\phi) + \frac{1}{\sqrt{2\pi}}\exp(im\phi)\right]$

where $\frac{1}{\sqrt{2}}$ is normalized factor

and $$< p_\phi > = \int_0^{2\pi}\psi^* p_\phi \psi d\phi = \left(\frac{1}{\sqrt{2}}\right)^2\frac{mh}{2\pi} + \left(\frac{1}{\sqrt{2}}\right)^2\left(-\frac{mh}{2\pi}\right) = 0.$$

SHORT ANSWER-QUESTIONS

1. Set up the Schrodinger equation in the time dependent form for a single particle.
2. Derive the time dependent Schrodinger equation.
3. What is the physical, significance of a wave function?
4. Explain the terms eigen values and eigen functions.
5. Explain the term state function.
6. Describe the situation which led to the development of wave mechanics.
7. What do you understand by quantum mechanical operators?
8. Write postulates of wave mechanics.

9. What are Hermitian operators? Write their properties.

10. Show that eigen values of a Hermitian operator are real.

11. Show that eigen functions of a Hermitian operator corresponding to different eigen values are orthogonal.

12. What is Schmidt orthogonalisation process?

13. Write Schrodinger equation as an eigen value equation. [**Ans.** $H\psi = E\psi$]

14. What are stationary states?

15. Why are there difficulties in applying, the de Brogile postulate $\lambda = h/p$ to a particle whose linear momentum is of changing magnitude?

16. How does the de-Broglie postulate enter into the Shrodinger theory?

17. What would be the effect on the Schrodinger theory of changing the definition of total energy in the relation $v = E/h$ by adding the constant rest mass energy of the particle.

18. Explain, why the Schrodinger equation is written in terms of the potential energy and not in terms of the force.

19. Why it is so important for the Schrodinger equation to be linear in the wave function?

20. The mass m of the particle appears explicitly in Schrodinger's equation, but its charge e does not, even though both may affect its motion, why?

21. Why does the Schrodinger's equation contain a first time derivative?

22. What is the basic connection between the properties of a wave function and the behaviour of associated particle?

23. Explain briefly the meaning of a well-behaved eigen function in the context of Schrodinger quantum mechanics.

24. Are there other examples in science, engineering, or mathematics in which differential operators are related to physical quantities?

25. State in words the meaning of expectation value of x.

26. Why does Schrodinger quantum mechanics provide only statistical information? In your opinion, does this reflect a failing of the theory, or a property of nature?

27. Explain briefly how the quantization of energy is related to the well behaved character of acceptable eigen functions.

28. Why is an ψ acceptable eigen functions if $V(x) < E$?

29. Why does ψ tend to go to infinity if $V(x) > E$?

30. If a particle is not bound in a potential, its total energy is not quantized. Does this mean the potential has no effect on the behaviour of the particle? What effect would you expect it have? [**Ans.** Particle is a free particle]

31. Can there be another bridge, i.e. different from de Broglie concept, between particles and waves?

Ans. Let us think about electromagnetic radiation from the particle point of view. The probability per unit volume of finding a photon in a given region of space at an instant of time is proportional to the number of photons(N) per unit volume at that time

$$\frac{\text{Probability}(P)}{V} \propto \frac{N}{V} \qquad \text{(i)}$$

Now, the number of photons per unit volume (N/V) is proportional to the intensity of the radiation:

$$\frac{N}{V} \propto I \tag{ii}$$

Since the intensity of electromagnetic radiation is proportional to the square of the electric field amplitude for the electromagnetic wave, we have

$$I \propto E^2 \tag{iii}$$

Equating (i) and (iii), one obtains

$$\frac{\text{Probability}(P)}{V} \propto E^2 \tag{iv}$$

Obviously, for electromagnetic radiation, the probability per unit volume of finding a particle associated with this radiation (the photon) is proportional to the square of the amplitude of the wave associated with the particle.

Recognizing the wave-particle duality of both electromagnetic radiation and matter, one may suspect a parallel proportionality for a material particle, i.e. the probability per unit volume of finding the particle is proportional to the square of the amplitude of a wave representing the particle. There is a de Broglie wave associated with every particle. The amplitude of the de Broglie wave associated with a particle is not a measurable quantity (as the wave function representing a particle is generally a complex quantity). In contrast the electric field is a measurable quantity for an electromagnetic wave. The matter analog to (iv) relates the square of the wave's amplitude to the probability per unit volume of finding the particle. As a result, one can call the amplitude of the wave associated with the particle the *probability amplitude*, or the *wave function* (ψ). This was first suggested by Max Born in 1913.

OBJECTIVE QUESTIONS

1. The Laplacian operator is

(a) ∇ (b) ∇^2

(c) ∇^3 (d) ∇^4

2. Eigen values of a Hermitian operator are

(a) imaginary (b) real

(c) can be real or imaginary (d) zero

3. Eigen functions of a Hermitian operator corresponding to different eigen values are

(a) orthogonal (b) orthonormal

(c) orthogonal as well as orthonormal (d) none of the above

4. Quantum mechanical operator of momentum p is

(a) $\frac{\hbar}{i}\nabla$ (b) $-\frac{\hbar^2}{2m}\nabla^2$

(c) $\frac{\hbar}{i}\nabla^2$ (d) $\nabla^2/8\pi$

5. Quantum mechanical operator of kinetic energy T is

(a) $\frac{\hbar}{i}\nabla$ (b) $-\frac{\hbar^2}{2m}\nabla^2$

(c) $\hbar\nabla^2$ (d) $\hbar\nabla^2/8\pi$

6. Quantum mechanical operator for the x component of angular momentum L_x is

(a) $\frac{h}{2\pi i}\left(x\frac{\partial}{\partial y}-y\frac{\partial}{\partial x}\right)$ (b) $\frac{h}{2\pi i}\left(z\frac{\partial}{\partial x}-x\frac{\partial}{\partial z}\right)$

(c) $\frac{h}{2\pi i}\left(y\frac{\partial}{\partial z}-z\frac{\partial}{\partial y}\right)$ (d) none of the above

7. If A is the operator, operating on a state function ψ, then the possible values of any physical quantity of a system (e.g., energy, angular momentum, etc.) in the operator equation $A\psi = a\psi$ are given by

(a) eigen functions ψ

(b) eigen values a

(c) eigen value a as well as eigen functions ψ

(d) none of the above is correct

8. $\psi(x) = A\exp(\pm 2\pi i\, p_x x/h)$ is a well behaved function with A as constant and p_x is the x-component of momentum. If $k = 2\pi\, p_x/h$, then the eigen function of momentum operator in $p_x\psi(x) = p_x\psi(x)$ is

(a) $\psi = A\exp(\pm ikx)$ (b) $\psi = A\exp(ikx)$

(c) $\psi = A\exp(-ikx)$ (d) none of the above

9. The eigen values of p_x in Q. 8 are

(a) $\frac{kh}{2\pi}$ (b) $k/2\pi$

(c) $\frac{k^2}{h^2}$ (d) zero

10. If $A = 3x^2$ and $B = \frac{d}{dx}$, then

(a) $AB = BA$ (b) $AB \neq BA$

(c) $AB = BA/2$ (d) $AB = 2BA$

11. If two operators A and B are Hermitian, then their product (AB) is also Hermitian if and only if A and B

(a) commute (b) do not commute

(c) are non-zero (d) none of the above

12. If the wave function is well behaved, then the Hamiltonian operator for the total energy of a system is

(a) zero (b) Hermitian

(c) Non-Hermitian (d) non of the above

13. A certain system is described by the Hamiltonian operator $H = \frac{d^2}{dx^2} + x^2$, then $\psi = Ax \exp(-x^2/2)$ is

(a) eigen value of H

(b) eigen function of H

(c) eigen value as well as eigen function of H

(d) none of the above

14. The expectation value of x for the state described by $\psi = Ax \exp(-x^2/2)$ is

(a) A (b) $A^2/2$

(c) $A/4$ (d) 0

15. If $a(k) = \begin{cases} 1/\sqrt{\alpha} & -\alpha/2 \le k \le \alpha/2 \\ 0 & |k| > \alpha/2 \end{cases}$

where α is some constant and $\hbar k = p$, the momentum of the particle. If wave function is $\psi(q) = \frac{1}{\sqrt{2\pi}} \int_{-\infty}^{\infty} a(k)\exp(ikq)dk$, then

(a) $(\Delta k)(\Delta q) \sim 0$ (b) $(\Delta k)(\Delta q) \sim 1$

(c) $(\Delta k)(\Delta q) \sim \infty$ (d) none of the above

[**Hint:** $\psi(q) = \frac{1}{\sqrt{\pi}} \int_{-\infty}^{\infty} a(k)\exp(ikq)dk$

$$= \frac{1}{\sqrt{2\pi}} \int_{-\alpha/2}^{\alpha/2} \frac{1}{\sqrt{\alpha}} \exp(ikq)dk = \frac{1}{\sqrt{2\pi}} 2\left[\sin\left(\frac{q\alpha}{2}\right)\right]/q$$

we see that the first two zeros of $\psi(q)$ near the origin are given by $q\alpha/2 = \pm\pi$ or $q = \pm 2\pi/\alpha$. Obviously, $\Delta k \sim \alpha$ and $\Delta q \sim 1/\alpha$ or $(\Delta k)(\Delta q) \sim 1$].

16. The Hamiltonian operator for a charged particle in the electromagnetic field is

(a) Unitary (b) Hermitian

(c) Orthogonal (d) Orthonormal

17. The linear momentum operator is

(a) Unitary (b) Orthogonal

(c) Orthonormal (d) Hermitian

18. Let the wave function for a system of two particles is written as $\psi(r_1, r_2) = \psi(r_1)\psi_2(r_2)$. H_1 and H_2 are Hamiltonians corresponding to particles 1 and 2 respectively, where H_1 acts on the functions of r_1 and H_2 acts on functions of r_2 then the Hamiltonian of the system may be put of the form

(a) $H = H_1 H_2$
(b) $H = H_1 + H_2$
(c) $H = H_1/H_2$
(d) $H = H_2/H_1$

19. Which of the following functions are acceptable wave functions?

(a) $\psi = x$
(b) $\psi = x^2$
(c) $\psi = \sin x$
(d) $\psi = e^x$
(e) $\psi = e^{-x}$
(f) $\psi = e^{-x^2}$
(g) $\psi = \tan x$

Ans.
(a) Not acceptable, as $\psi \to \infty$ as $x \to \infty$
(b) Not acceptable
(c) Acceptable
(d) Not acceptable, $\psi \to \infty$ as $x \to \infty$
(e) Not acceptable, $\psi \to \infty$ as $x \to -\infty$
(f) Acceptable, $\psi \to 0$ as $x \to \pm\infty$
(g) Not acceptable, as $\psi \to \infty$ as $x = \pi/2\infty$

20. $\psi = \sin x$ is a wave function, (i) It is acceptable (ii) Is it normalized?
[**Hint:** (i) Yes (ii) $\int \sin^2 x dx \neq 1$]

ANSWERS

Question numbers	1	2	3	4	5	6	7	8	9	10
Answers	(b)	(b)	(a)	(a)	(b)	(c)	(b)	(a)	(a)	(b)
	11	12	13	14	15	16	17	18		
	(a)	(b)	(b)	(d)	(b)	(b)	(d)	(b)		

FILL IN THE BLANKS

1. A wave function obeying ______ condition is said be normalized.

$$\left[\int_{-\infty}^{\infty} |\psi|^2 \, dV = 1 \right]$$

2. A wave function can be normalized by ______ it with an appropriate constant. **[multiplying]**

3. Schrodinger equation cannot be derived from other basic principles of physics. It is a ______ in itself. **[basic principle]**

4. The values of energy E_n for which Schrodinger's time-dependent equation can be solved are called ______ and the corresponding wave functions ψ_n are called ______. **[eigen values, eigen functions]**
5. The probability of finding the particle in the arbitarily sized interval $a \le x \le b$ is ______.

$$[P_{ab} = \int_a^b |\psi|^2\, dx]$$

6. The Schrodinger equation provides a description of physical reality in remarkable ______ with ______. **[agreement, experiments]**
7. Schrodinger description agrees with Newtonian mechanics for ______ where quantum effects are small. **[macroscopic systems]**
8. Time-independent Schrodinger equation is developed by merging de Broglie's matter wave hypothesis, $\lambda = h/p$ with the Newtonian relation $K = p^2/2m$; we expect the equation to hold only for ______ particles. **[non-relativistic]**
9. The Schrodinger equation gives the wave for non-relativistic particles and leads to energy ______ for confined particles. **[quantization]**
10. The square of the wave function is the probability density, and in dimension the probability of finding the particle in some small interval dx at position x is ______.

$$[p(x) = \psi^2(x)dx]$$

8

Simple Applications of Quantum Mechanics and One-Dimensional Energy Eigen Value Problems

8.1 INTRODUCTION

As we go from one region to another region such that potential changes, then there is a change in wavelength of the waves also. Since the equation of motion is a second order differential equation in *x*, it is necessary that ψ and $\frac{d\psi}{dx}$ be continuous at the boundaries of discontinuity. It is an outcome of the finiteness of ψ, *E* and *V* (everywhere). ψ must be finite, if we want to retain its physical interpretation in terms of probability density. Also *E* and *V* are finite, as systems possessing infinite energies do not occur in nature.

From Schrodinger's time-independent equation of motion, we see that $\frac{d^2\psi}{dx^2}$ is finite (not necessarily continuous) everywhere which is possible only if $\frac{d\psi}{dx}$ is finite and continuous. This gives us one boundary condition. In order that $\frac{d\psi}{dx}$ exists everywhere and be finite (as is implied by the mere use of a differential equation) it is also necessary that $\frac{d\psi}{dx}$ be continuous everywhere. This gives us second boundary condition.

If ψ and $\frac{d\psi}{dx}$ are not continuous everywhere, then sources or sinks of probability density $(=\psi^*\psi d\tau)$ will exist and it will not be possible to normalize ψ independent of time. This may be summarised as follows: **The wave function ψ must be single valued and finite everywhere further so long as $V(x)$ is finite (whether continuous or not) ψ and $\frac{d\psi}{dx}$ must be continuous.**

If $V(x)$ is infinite anywhere the appropriate boundary condition is established by a limiting process that starts from a finite value of $V(x)$ and the above boundary conditions. The result is that the surface at which there is an infinite potential jump, the wave function is zero and the first derivative of ψ is not determined.

8.2 THE STEP POTENTIAL (ENERGY LESS THAN STEP HEIGHT)

Consider a particle oving in the potential Fig. [8.1(a)] which represents particles colliding with a thick wall

$$V(x)=\begin{cases}0 & x<0\\ V_0 & x>0\end{cases}$$

We may think $V(x)$ as an approximate representation of the potential energy function for a charged particle moving along the axis of a system of two electrodes, separated by a very narrow gap, which are held at different voltages. The upper half of Fig. [8.1(b)] illustrates this system, and the lower half illustrates the corresponding potential energy function. As the gas decreases, the potential function approaches the idealization illustrated in Fig. [8.1(a)].

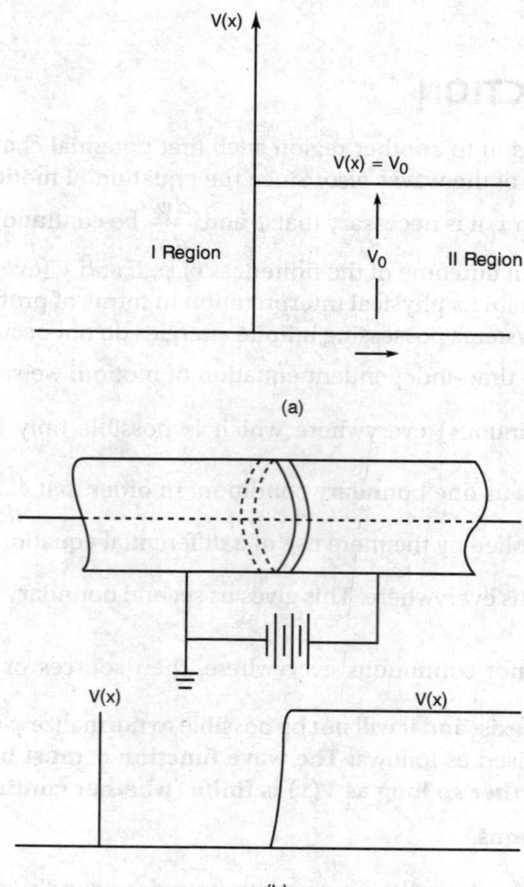

Fig. 8.1 (a) Potential step **(b)** Illustrating a physical system with a potential energy function that can be approximated by a step potential. A charged particle moves along the axis of two cylindrical electrodes held at different voltages. Its potential energy is constant when it is inside either electrode, but it changes very rapidly when passing from one electrode to the other

To the left of $x = 0$, the problem is same as that of a free particle. Classically, the particles would be reflected without loss of speed $E < V_0$, they would be transmitted passing over the potential step with reduced speed

$$v_{\text{red}} = \sqrt{\frac{2(E - V_0)}{\mu}} \quad \text{if} \quad E > V_0$$

Quantum mechanically, all information about the beam can be obtained from a wave function that satisfies Schrodinger's time independent equation of motion:

$$\frac{d^2\psi}{dx^2} + \frac{2\mu E}{\hbar^2}\psi_I = 0 \quad \text{for} \quad x < 0 \tag{8.1}$$

$$\frac{d^2\psi_{II}}{dx^2} + \frac{2\mu(E - V_0)}{\hbar^2}\psi_{II} = 0 \quad x < 0 \tag{8.2}$$

or

$$\frac{d^2\psi_I}{dx^2} + k_1^2\psi_I = 0 \quad x < 0 \tag{8.3}$$

$$\frac{d^2\psi_{II}}{dx^2} + k_2^2\psi_{II} = 0 \quad x > 0 \tag{8.4}$$

where

$$k_1 = \sqrt{\frac{2\mu E}{\hbar^2}} \quad \text{and} \quad k_2 = \sqrt{\frac{2\mu(E - V_0)}{\hbar^2}} \tag{8.5}$$

The general solution of Eqs. (8.3) and (8.4) is given by

$$\psi_I = Ae^{ik_1^x} + Be^{-ik_1^x} \tag{8.6}$$

and

$$\psi_{II} = Ce^{ik_2x} + De^{-ik_2x} \tag{8.7}$$

where A, B, C and D are constants of integration.

The functions e^{ik_1x} and e^{-ik_1x} $\left[\text{when multiplied by the factor } \exp\left(\frac{-Et}{\hbar}\right)\right]$ represents wave travelling towards positive and negative directions of A respectively Fig. (8.2). Because of the discontinuity in V, form of Schrodinger equation changes discontinuously at $x = 0$; the constants of integration are to be chosen in such a way as to make ψ and $\frac{d\psi}{dx}$ continuous at $x = 0$.

Therefore, boundary conditions are

$$[\psi_I]_{x=0} = [\psi_{II}]_{x=0} \tag{8.8}$$

and

$$\left[\frac{d\psi_I}{dx}\right]_{x=0} = \left[\frac{d\psi_{II}}{dx}\right]_{x=0} \tag{8.9}$$

From Eqs. (8.8) and (8.9), we get

$$A + B = C + D$$

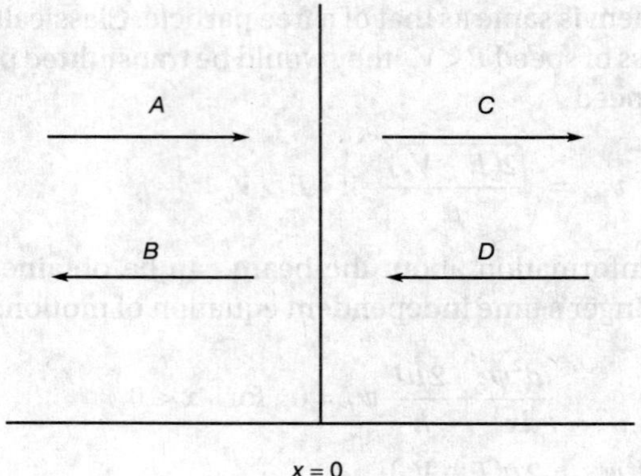

Fig. 8.2 The wave function as a superposition of travelling waves

or
$$k_1(A+B)=k_1(C+D)$$

and
$$k_1(A-B)=k_2(C-D)$$

Adding these two equations, we get

$$2k_1A=(k_1+k_2)C+(k_1-k_2)D$$

or
$$C=\left[\frac{2k_1}{k_1+k_2}\right]A-\left[\frac{k_1-k_2}{k_1+k_2}\right]D \tag{8.10}$$

$$B=C+D-A$$

$\therefore$
$$B=\frac{2k_1}{k_1+k_2}A-\frac{k_1-k_2}{k_1+k_2}D+D-A=\frac{k_1-k_2}{k_1+k_2}A-\frac{2k_1}{k_1+k_2}D \tag{8.11}$$

$\therefore$
$$\psi_1=Ae^{ik_1x}+\left[\frac{k_1-k_2}{k_1+k_2}\right]Ae^{ik_1x}+\left[\frac{2k_2}{k_1+k_2}\right]De^{-ik_1x}$$

and
$$\psi_{II}=\frac{2k_1}{k_1+k_2}Ae^{ik_2x}-\left[\frac{k_1-k_2}{k_1+k_2}\right]De^{ik_2x}+De^{-ik_2x}$$

The general solution ψ may be written as a linear combination of two solutions

$$\psi=A\psi_1+D\psi_2$$

where
$$\psi_1=\begin{cases} e^{ik_1x}\dfrac{k_1-k_2}{k_1+k_2}e^{-ik_1x} & \text{for } x<0 \quad (8.12a)\\[2mm] \dfrac{2k_1}{k_1+k_2}e^{ik_2x} & \text{for } x>0 \quad (8.12b)\end{cases}$$

and
$$\psi_2 = \begin{cases} \left(\dfrac{2k_2}{k_1 + k_2}\right) e^{-ik_1 x} & \text{for } x < 0 \quad (8.13a) \\ e^{-ik_2 x} - \left(\dfrac{k_1 - k_2}{k_1 + k_2}\right) e^{ik_2 x} & \text{for } x > 0 \quad (8.13b) \end{cases}$$

The continuity conditions at $x = 0$ are satisfied by ψ_1 and ψ_2 both and hence by their linear combination also. However, another condition to be imposed upon the wave functions is that it must remain finite as $|x|$ becomes infinite. The behaviour of the functions ψ_1 and ψ_2 for large values of $|x|$ is governed by the relative magnitude of E and V_0. There are two important cases worth consideration:

8.2.1 $E > V_0$

The numbers k_1 and k_2 are real and may be assumed to be positive. Under these conditions ψ_1 and ψ_2 are finite everywhere and hence admissible as solutions of the problem. The general solution ψ is a linear combination of two linearly independent solutions ψ_1 and ψ_2 and therefore ψ is doubly degenerate.

Interpretation of ψ_1: The function ψ_1 represents a state of the system in which a wave is incident from the left, i.e., the direction of the negative $x(e^{ik_1 x})$. Second term in Eq. [8.12(a)] represents a wave reflected from the step and travelling towards x-direction, while Eq. [8.12(b)] represents a wave transmitted beyond the step towards right. A plot of $|\psi(x)|^2$ with x is shown in Fig. (8.3).

$$\psi_1 = e^{-ik_1 x} + Re^{-ik_1 x} \qquad x < 0$$

and

$$\psi_1 = Te^{ik_2 x} \qquad x > 0$$

where

$$R = \frac{k_1 - k_2}{k_1 + k_2} \quad \text{and} \quad T = \frac{2k_1}{k_1 + k_2}$$

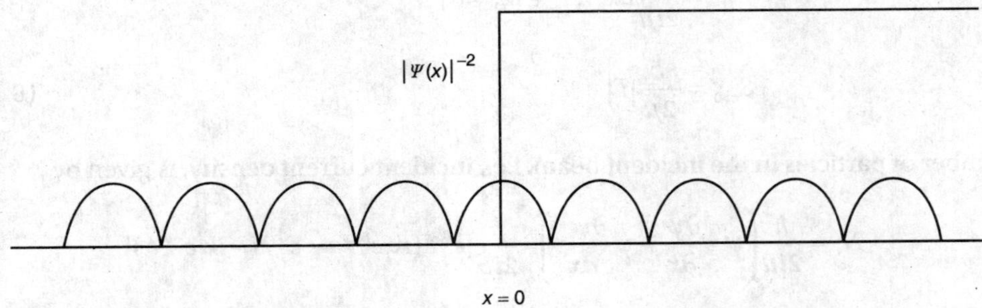

Fig. 8.3 The absolute square of the real part of the wave function representing particles approaching a potential step from the left with energy higher than the height of the step

At this stage, we define reflectance (R) and transmittance (T) as follows.

Reflectance, Reflectivity or Reflection coefficient (R): It is defined as the ratio of the magnitude of reflected current to the magnitude of incident current, i.e.,

$$R = \frac{\text{Magnitude of reflected current}}{\text{Magnitude of incident current}}$$

Transmittance (T): At the potential discontinuity the transmittance is defined as the ratio of the magnitude of transmitted current to the magnitude of incident current, i.e.,

$$T = \frac{\text{Magnitude of transmitted current}}{\text{Magnitude of incident current}}$$

R and T are amplitudes of the reflected and transmitted waves relative to the amplitude of the incident wave.

Probability current $S = \frac{\hbar}{2i\mu}\left[\psi^*\left(\frac{\partial\psi}{\partial x}\right) - \left(\frac{\partial\psi^*}{\partial x}\right)\psi\right]$

For $x < 0$, $S_{x<0}$

$$= \frac{\hbar}{2i\mu}[(e^{-ik_1x} + R^*e^{ik_1x})(ik_1e^{ik_1x} - ik_1Re^{-ik_1x}) - (-ik_1e^{-ik_1x} + ik_1R^*e^{ik_1x})(e^{ik_1x} + Re^{ik_1x})]$$

$$= \frac{\hbar}{2i\mu}[(ik_1 + ik_1R^*e^{2ik_1x} - ik_1Re^{-2ik_1x} - ik_1|R|^2 + ik_1 - ik_1R^*e^{2ik_1x} + ik_1Re^{-2ik_1x} - ik_1|R|^2)]$$

$$= \frac{\hbar}{2i\mu}[2ik_1 - 2ik_1|R|^2]$$

or $$S_{x<0} = \frac{\hbar k_1}{\mu}[1 - |R|^2] \tag{8.14}$$

For $x > 0$; $$S_{x>0} = \frac{\hbar}{2i\mu}[(T^*e^{-ik_3x})(ik_2Te^{ik_2x}) - (-ik_2T^*e^{-ik_2x})(Te^{ik_2x})]$$

$$= \frac{\hbar}{2i\mu}[ik_2|T|^2 + ik_2|T|^2]$$

or $$S_{x>0} = \frac{\hbar k_2}{2\mu}|T|^2 \tag{8.15}$$

Number of particles in the incident beam, i.e., incident current density, is given by

$$N_i = \frac{\hbar}{2i\mu}\left(\psi^*\frac{\partial\psi}{\partial x} - \psi\frac{\partial\psi^*}{dx}\right) = \frac{\hbar}{2i\mu}[e^{ik_1x}(ik_1e^{ik_1x}) - e^{ik_1x}(-ik_1e^{-ik_1x})]$$

or $$N_i = \frac{\hbar}{2i\mu}(ik_1 + ik_1) = \frac{\hbar k_1}{\mu}$$

Number of particles in the transmitted beam i.e., transmitted current density, is given as

$$N_T = \frac{\hbar}{2i\mu}[(T^* e^{-ik_2 x})(ik_2 T e^{ik_1 x}) - (-ik_2 T^* e^{-ik_1 x})(T e^{ik_1 x})]$$

or
$$N_T = -\frac{\hbar}{2i\mu}[ik_2 |T|^2 + ik_2 |T|^2] = \frac{\hbar k_2}{\mu}|T|^2$$

Number of particles in the reflected beam, i.e., reflected current density, is given to be

$$N_R = \frac{\hbar}{2i\mu}(-)[(R^* e^{ik_1 x})(-ik_1 R e^{-ik_1 x}) - (ik_1 R^* e^{-ik_1 x})(R e^{-ik_1 x})]$$

or
$$N_R = \frac{\hbar}{2i\mu}[-ik_1 |R|^2 - ik_1 |R|^2]$$

or
$$N_R = \frac{\hbar k_1}{\mu}|R|^2$$

Transmissivity (Probability of Transmission)

$$P_T = \frac{N_T}{N_i} = \frac{k_2}{k_1}|T|^2 = \frac{4k_1 k_2}{(k_1 + k_2)^2} \tag{8.16}$$

Reflectivity (Probability of Reflection)

$$P_R = \frac{N_R}{N_i} = |R|^2 = \frac{(k_1 - k_2)^2}{(k_1 + k_2)^2} \tag{8.17}$$

∴
$$P_T + P_R = 1 \tag{8.18}$$

and
$$S_{x<0} = S_{x>0} \tag{8.19}$$

Thus we see that the reflectivity of a particle from a potential step (whose height is less than the energy of the particle) is finite in contradiction to the classical result, which is zero.

In terms of the refractive index (μ) of the *I* medium with respect to *II*,

$$2\mu_1 = \frac{k_2}{k_1} = \frac{\sqrt{\frac{2\mu}{\hbar^2}(E - V_0)}}{\sqrt{\frac{2\mu}{\hbar^2}E}} = \sqrt{\frac{E - V_0}{E}} = \sqrt{\frac{1 - V_0}{E}}$$

Reflectivity $$P_R = \frac{(k_1 - k_2)^2}{(k_1 + k_2)} = \frac{(1-\mu)^2}{(1+\mu)^2} = |R|^2$$

Transmissivity $$P_T = \frac{4k_1 k_2}{(k_1 k_2)^2} = \frac{4\mu}{(1+\mu)^2} = \mu|T|^2$$

A plot of transmissivity and reflectivity with respect to μ is shown in the Fig. (8.4).

We note that reflectivity approaches zero as μ approaches unity (i.e., $k_2 \to k_1$) and it tends to one as μ approaches zero (i.e., $k_2 \to 0$). Also reflectivity becomes large when V_0 is comparable with the energy of the particle. There is always a finite probability of reflection

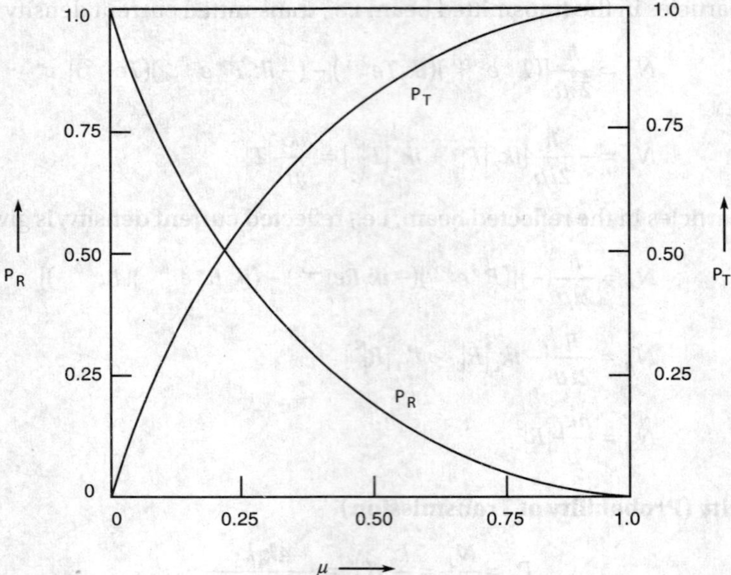

Fig. 8.4 Variation of P_R and P_T with μ

(may be very small), no matter how small V_0 is. A similar interpretation can be made for the other admissible solution ψ_2.

8.2.2 $E < V_0$

In this case propagation vector k_1 is real, but k_2 is purely imaginary.

$$k_2 = \sqrt{\frac{2\mu}{\hbar^2}(E - V_0)} = i\sqrt{\frac{2\mu}{\hbar^2}(V_0 - E)} = ik$$

The solution of Schrodinger's equation may now be written as

$$\psi_1 = Ae^{ik_1x} + Be^{-ik_1x} \qquad x < 0 \tag{8.20}$$

$$\psi_{11} = Ce^{-kx} + De^{kx} \qquad x > 0 \tag{8.21}$$

In order that the solution ψ_{II} is well behaved, i.e., it remains finite as x becomes very large, we should retain only the negative exponential term (positive exponential term must be equal to zero);

$$\therefore \qquad D = 0 \text{ and } \psi_{II} = Ce^{-kx}$$

The continuity conditions now give

$$A + B = C \text{ and } ik_1(A - B) = -kC \tag{8.22}$$

or

$$A - B = -\frac{k}{ik_1}C = \frac{ik}{k_1}C \tag{8.23}$$

From Eqs. (8.22) and (8.23), we obtain

$$2A = C + \frac{ik}{k_1}C = \frac{k_1 + ik}{k_1}C$$

or
$$C = \frac{2k_1}{k_1 + ik}A$$

and
$$B = C - A = \frac{2k_1}{k_1 + ik}A - A = \frac{k_1 - ik}{k_1 + ik}A$$

The solution in the two regions is now obtained as

$$\psi_1 = Ae^{ik_1x} + \frac{k_1 - ik}{k_1 + ik}Ae^{-ik_1x} \text{ and } \psi_{II} = \frac{2k_1}{k_1 + ik}Ae^{-kx}$$

The problem can now be considered in two different ways:

(i) Continuous stream model

In this case we consider a continuous stream of particles impinging on the barrier from left by taking a single monochromatic wave.

The flux of particles incident on the barrier from the left $= \frac{\hbar k_1}{\mu}|A|^2$.

Flux of particles reflected from the barrier (moving towards left in the region $x < 0$)

$$= |B|^2 \frac{\hbar k_1}{\mu} = |A|^2 \left|\frac{k_1 - ik}{k_1 + ik}\right|^2 \frac{\hbar k_1}{\mu} = \frac{\hbar k_1}{\mu}|A|^2$$

Thus, we see that all the particles are reflected by the barrier in the region $x < 0$. However, there is a finite probability of finding the particle in the region $x > 0$; even though this region is forbidden, classically. The phase of the reflected wave is not the same as that of the incident wave. In the region to the left of $x = 0$, we may write

$$\psi_1 = A\left[e^{ik_1x} + \frac{k_1 - ik}{k_1 + ik}e^{-ik_1x}\right]$$

or
$$\psi_1 = A\left[e^{ik_1x} + \frac{(k_1 - ik)^2}{k_1^2 + k^2}e^{-ik_1x}\right]$$

Let
$$k_1 - ik = \sqrt{k_1^2 + k^2 e^{-i\theta}} \quad \text{where} \quad \theta = \tan^{-1}\left(\frac{k}{k_1}\right)$$

Then
$$\psi_1 = A\left(e^{ik_1x} + e^{-2i\theta}e^{-ik_1x}\right)$$

or
$$\psi_1 = Ae^{-i\theta}[e^{i(k_1k+\theta)} + e^{-i(k_1x+\theta)}]$$

or
$$\psi_1 = 2Ae^{-i\theta}\cos(k_1x + \theta) \tag{8.24}$$

A plot of $|\psi(x)|^2$ vs x is shown in Fig. (8.5).

(ii) Wave-packet approach (Motion of a single particle is followed)

In order to follow the motion of a single particle, we must construct a wave-packet from incident waves that are continuous solution of the Schrodinger equation from

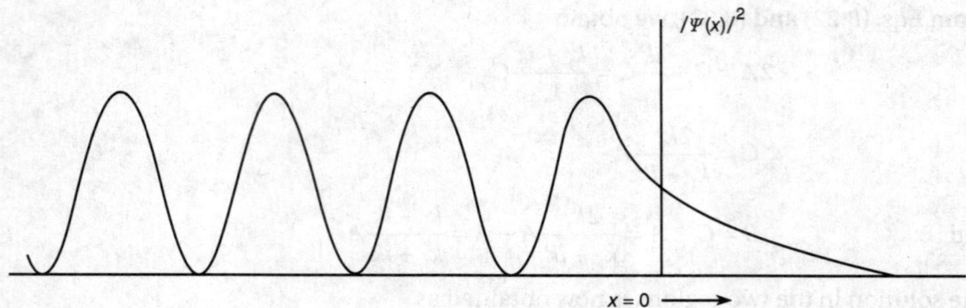

Fig. 8.5 The absolute square of the real part of the wave function representing particles impinging on the step from left

$x = -\infty$ to $+\infty$. These waves have the simple form $e^{i(k'x-\omega t)}$ for negative x. The packet has the form

$$\psi_{inc}(x,t) = \int_{-\infty}^{+\infty} \phi(k'-k_1)e^{i(k'x-\omega t)}dk' \quad \text{for} \quad x<0 \tag{8.25}$$

The centre of the packet is given by the position where phase is extremum; therefore we get $x = v_g t$ for the centre of the packet. Starting at negative t, the centre of the packet moves with constant velocity until it reaches $x = 0$ at $t = 0$. Each incident wave $\{\exp(ik'x - i\omega t)\}$ gives rise to a reflected wave $\left(\dfrac{k'-ik}{k'+ik}\right)e^{-i(k'x+\omega t)}$

Therefore, shape of the reflected wave packet is given by

$$\psi_{ref.}(x,t) = \int_{-\infty}^{+\infty} \phi(k'-k_1)\left\{\left[\left(\frac{k'-ik}{k'+ik}\right)e^{-i(k'x+\omega t)}\right\}dk'\right.$$

$$= \int_{-\infty}^{+\infty} \phi(k'-k_1)e^{-i\{(k'x+\omega t+2\tan^{-1}(k/k')\}}dk' \tag{8.26}$$

The centre of the packet follows a path where the phase is extremum. Therefore,

$$\frac{\partial}{\partial k'}\left[k'x + \omega t + 2\tan^{-1}\left(\frac{k}{k'}\right)\right] = 0$$

or
$$x + \frac{\partial \omega}{\partial k'}t - \frac{2k}{k^2+k'^2} = 0$$

or
$$x = -\left(\frac{\partial \omega}{\partial k'}\right)t + \frac{2k}{k^2+k'^2} = -v_g t + \frac{2k}{k^2+k'^2}$$

or
$$x = -v_g\left[t - \frac{2k}{v_g(k^2+k'^2)}\right] \tag{8.27}$$

Thus the reflected packet moves to the left of the step with constant-velocity v_g but there is a time delay in the appearance of the reflected packet.

$$\text{Time delay} = \Delta t = \frac{2k}{v_g(k^2 + k'^2)}$$

This time delay is caused by the sinking of the incident packet into the step, before it reappears as a reflected wave-packet.

8.2.3 Infinite Potential Jump

An infinite potential jump is an idealization of a rigid boundary. For the purpose of examining the modification of the continuity condition on the wave function by the presence of rigid boundaries, we treat infinite potential jump as the limiting case of a finite potential jump as V_0 tends to infinity.

Keeping E fixed (i.e., finite) we see that k_1 remains unchanged while $k_2 = ik \to \infty$ as V_0 approaches infinity.

$$A = \frac{1}{2}\left(1 + \frac{ik}{k_1}\right)C = \frac{1}{2}\left(1 + \frac{k_2}{k_1}\right)C \text{ and } B = \frac{1}{2}\left(1 - \frac{ik}{k_1}\right)C = \frac{1}{2}\left(1 - \frac{k_2}{k_1}\right)C$$

Thus A and B ceases to be finite unless C tends to zero. By putting

$$C = \frac{k_1 C_1}{ik_2} = \frac{-k_1 C_1}{k}, \text{ we obtain}$$

$$A = \frac{C}{2} + \frac{k_2 C}{2k_1} \to \frac{C_1}{2i} \quad \text{and} \quad B = \frac{C}{2} - \frac{k_2 C}{2k_1} \to -\frac{C_1}{2i}$$

$C \to 0$ as V_0 approaches infinity.

Then $\quad \psi_1 = Ae^{ik_1 x} + Be^{-ik_1 x}$

or $\quad \psi_1 = \frac{C_1}{2i}e^{ik_1 x} - \frac{C_1}{2i}e^{-ik_1 x}$

$$\left.\begin{aligned} &\text{or} \quad \psi_1 = C_1 \sin k_1 x \quad x < 0 \\ &\text{and} \quad \psi_{II} = Ce^{-kx} = 0 \quad x > 0 \end{aligned}\right\} \tag{8.28}$$

Thus we see that wave function remains continuous everywhere, but its derivative $\frac{d\psi}{dx}$ is continuous at the boundary of the step. The condition that $\frac{d\psi}{dx}$ be continuous is replaced by the condition that $\psi = 0$ in regions were $V_0 = \infty$.

8.3 PARTICLE IN A POTENTIAL WELL (ONE-DIMENSIONAL BOX)

Any particle that is bound by external forces to a limited region of space can be described approximately as moving under the influence of a box potential. Any such potential as shown in Fig. [8.6(a)] can be represented mathematically as

$$V(x) = \begin{cases} \infty & x \le 0 \text{ and } x \ge L \\ 0 & 0 < x < L \end{cases} \tag{8.29}$$

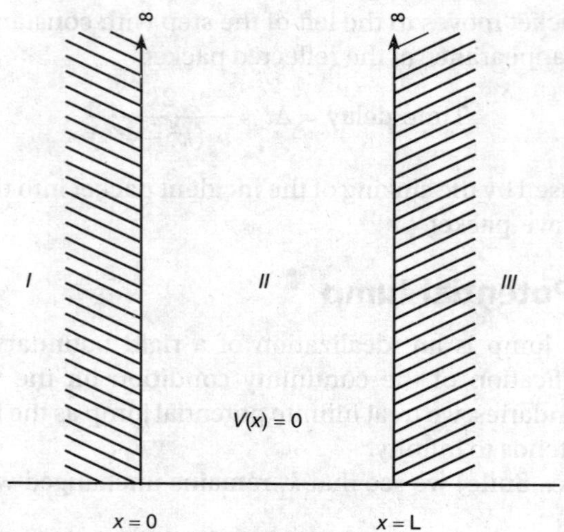

Fig. 8.6 (a) Potential for a particle in a box

From the results of previous section, we see that the solution will vanish in the region $x \leq 0$ and $x > L$. The Schrodinger equation in region II is

$$\frac{d^2\psi}{dx^2} + \frac{2\mu}{\hbar^2}(E)\psi = 0$$

or
$$\frac{d^2\psi}{dx^2} + k^2\psi = 0$$

with
$$k = \sqrt{\frac{2\mu E}{\hbar^2}}$$

$\therefore$
$$\psi = Ae^{ikx} + Be^{-ikx} \tag{8.30}$$

Boundary conditions are $\psi(x) = 0$ for $x = 0$ and $x = L$

$\therefore$ $A + B = 0$ or $B = -A$

and $Ae^{ikL} + Be^{-ikL} = 0$

or $Ae^{ikL} - Ae^{-ikL} = 0$

or $2iA \therefore \sin kL = 0$

since $A \neq 0$ $\sin kL = 0$

or
$$kL = n\pi \tag{8.31}$$

where n is an integer.

$$\psi(x) = Ae^{ikx} + Be^{-ikx} = Ae^{ikx} - Ae^{-ikx}$$

$$= A(e^{ikx} - e^{-ikx}) = 2iA\sin kx$$

or
$$\psi(x) = A'\sin kx = A'\sin\frac{n\pi x}{L} \tag{8.32}$$

Allowed values of energy or eigen values of energy of the particle are given by

$$E_n = \frac{\hbar^2 k_n^2}{2\mu} = \frac{\hbar^2 \pi^2 \hbar^2}{2\mu L^2} = \frac{\hbar^2 h^2}{8\mu L^2} \tag{8.33}$$

Eq. (8.33) represents the eigen values of energy. These values are called the energy levels of the particle. One can draw the following conclusions:

(i) The lowest energy of the particle is obtained by putting $n = 1$ in Eq. (8.99). One obtains for $n = 1$ as

$$E_1 = \frac{h^2}{8\mu L^2} \tag{8.34}$$

It is called the **ground state energy level** of the particle. The value of E_n in terms are expressed as

$$E_n = n^2 E_1 \tag{8.35}$$

Here n are called quantum numbers while the corresponding E values are called energy levels. The quantum states with higher values of $n(= 2, 3, 4...\text{etc.})$ are termed as excited states. According to quantum mechanics (but not in accordance to classical mechanics) the kinetic energy is not zero in ground state.

(ii) The possible values of energy of the particle in the potential box are discrete corresponding to $n = 1, 2, 3, ...$ The classical mechanics does not require any such discrete energy levels for bound systems. The quantization of energy conclusion is universal, i.e., whenever a particle is confined to move in a limited space, its energy is quantized, e.g., energies of electrons bound to atoms.

(iii) **The Energy levels:** The quantisation of energy is shown graphically in Fig. [8.6(b)]. The allowed energy levels are shown as horizontal lines. That these line are straight emphasizes the fact that energy (E) is the same irrespective of the position (x) of the particle. Since for a particular value of n, the energy (E) does not change with x, the latter is often termed as "constant of motion", further, since the H or E is independent of time (t) the system under consideration is a good and simple example of **"Conservative system"**.

The lowest value of E is not zero but $\frac{\pi^2 \hbar^2}{2mL^2}$ (for $n = 1$). Evidently, the particle in a box will continue to move to and fro in the region O to L even at the temperature of absolute zero; for this region it is called the **"zero-point energy"**. This conclusion is universal and valid for any system undergoing to and fro motion.

The dependence of E on n^2 results in the increased spacing between energy levels as n increases

$$\Delta E = E_{N+1} - E_n = [(n+1)^2 - n^2]E_1 = (2n+1)E_1 \tag{8.36}$$

Since the energy E_1 is proportional to $1/a^2$, the smaller the box the greater are the energy spacings (ΔE), i.e., the quantisation is more pronounced. Alternatively, as the box is made wider ΔE decreases and in the limited case of $L \to \infty$ (free particle) the quantization vanishes.

The wave functions ψ_n corresponding to E_n are called Eigen functions.

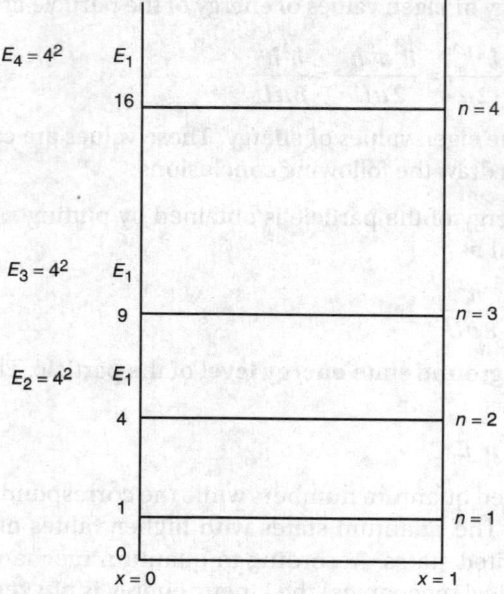

Fig. 8.6 (b)

Normalization of the Wave Functions

The wave functions for the motion of the particle are

$$\psi_n(x) = A\sin\frac{n\pi x}{L} \tag{8.37}$$

in the region $0 < x < L$ and $\psi_n = 0$, in the region $x < 0$; $x > L$. This wave function is still incomplete in as much as the constant A is not determined. A can be determined by making the use of the fact that the total probability that the particle is somewhere in the box must be unity, i.e.,

$$\int_0^a \left|\psi_n(x)^2\right| dx = 1 \tag{8.38}$$

or $$\int_0^a A^2 \sin^2\frac{n\pi x}{L} dx = 1$$

or $$A^2 \int_0^a \frac{1}{2}\left(1 - \cos\frac{2n\pi x}{L}\right) dx = 1$$

or $$\frac{A^2}{2}\left[x - \frac{L}{2\pi n}\sin\frac{2n\pi x}{L}\right]_0^L = 1$$

The second term of the integrated expression becomes zero at both limits. Therefore

$$\frac{A^2 L}{2} = 1 \text{ or } A = \sqrt{\frac{2}{L}} \tag{8.39}$$

The normalized wave function is thus

$$\psi_n = \sqrt{\frac{2}{L}} \sin \frac{n\pi x}{L} \tag{8.40}$$

Note: Actually $\psi = \pm\sqrt{\frac{2}{L}} \sin\frac{n\pi x}{L}$. However, since probability is given by $|\psi^2|$ and not by ψ, it is immaterial whether positive or negative A is chosen. Customarily, we select the positive series of normalised ψ functions.

The notion of trapping particles in potential wells is used in the burgeoning field of *nanotechnology*, which refers to the design and application of devices having dimensions ranging from 1 to 100 nm. The fabrication of these devices often involves manipulating single atoms or small groups of atoms to form structures such as the *quantum corral.*

One area of nanotechnology of interest to researchers is the *quantum dot.* The quantum dot, a small region that is grown in a silicon crystal, acts as a potential well. This region can trap electrons into states with quantized energies. The wave functions for a particle in a quantum dot look similar to those in Fig. 8.7(a) if L is on the order of nanometers. The storage of binary information using quantum dots is an active field of research. A simple

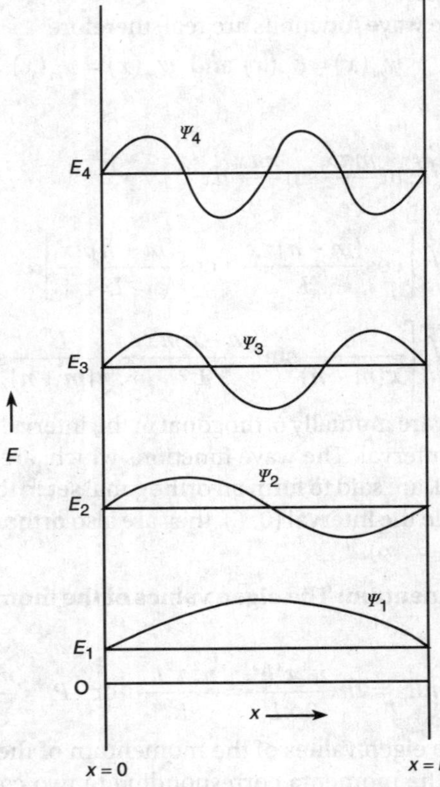

Fig. 8.7 (a) Energy and eigen functions of first few states of a particle in a box

binary scheme would involve associating a one with a quantum dot containing an electron and a zero with an empty dot. Other schemes involve cells of multiple dots such that arrangement of electrons among the dots correspond to ones and zeros.

Forms of the Wave Functions

The expression for the real form of the wave function can be used conveniently to draw its graphical representatives. The wave functions for the first three values of n are shown in Fig. [8.7(a)]. It is evident that the wave function ψ_1 has two nodes at $x = 0$ and at $x = L$. The wave function ψ_2 has three nodes at $x = 0$, $x = L/2$ and at $x = L$. The wave function ψ_3 has four nodes at $x = 0$, $x = L/3$, $x = 2L/3$ and at $x = L$. Thus the wave function ψ_n will have $(n + 1)$ nodes. Obviously the increasing number nodes is consistent with increasing energy of the system.

Orthogonality of the Wave function

Let $\psi_n(x)$ and $\psi_m(x)$ be the normalized wave functions of the particle in the interval $(0, a)$ corresponding to the different energy levels E_n and E_m respectively. These wave functions are

$$\psi_n(x) = \sqrt{\frac{2}{L}}\sin\frac{n\pi x}{L} \quad \text{and} \quad \psi_m(x) = \sqrt{\frac{2}{L}}\sin\frac{m\pi x}{L}$$

where m and n are integers.

In the present case the wave functions are real, therefore

$$\psi_m^*(x) = \psi_n(x) \text{ and } \psi_m^*(x) = \psi_m(x)$$

Now we have, when $m \neq n$

$$\int_0^a \psi_m^*\psi_n dx = \frac{2}{L}\int_0^L \sin\frac{m\pi x}{L}\sin\frac{n\pi x}{L}dx$$

$$= \frac{1}{L}\int_0^L \left[\cos\frac{(m-n)\pi x}{L} - \cos\frac{(m+n)\pi x}{L}\right]dx$$

$$= \frac{1}{L}\int_0^L \left[\frac{L}{\pi(m-n)}\sin\frac{(m-n)\pi x}{L} - \frac{L}{\pi(m+n)}\sin\frac{(m+n)\pi x}{a}\right]_0^L = 0$$

Obviously, the functions are mutually orthogonal in the interval $(0, L)$. These functions are also normalised in this interval. The wave functions which, are normalised and mutually orthogonal in an interval, are said to form an orthogonal set in this interval. Since the wave functions are zero outside the interval $(0, L)$, they are also orthonormal in the whole range of x, i.e., in the interval $(-\infty, \infty)$.

Eigen Values of the Momentum: The eigen values of the momentum P_n along the x-axis are given by

$$P_n^2 = 2mE_n = 2m\frac{n^2\pi^2\hbar^2}{2mL^2} = \frac{n^2\pi^2\hbar^2}{L^2} \quad \text{or} \quad P_n = \frac{n\pi\hbar}{L} = \frac{nh}{2L} \tag{8.41}$$

Eq. (8.41) shows that the eigen values of the momentum of the particle are discrete, and the difference between the momenta corresponding to two consecutive energy levels is constant equal to $h/2L$.

Probability of location of the particle over a small range between x and $x + dx$

According to quantum mechanics, the probability $P(x)\,dx$ that the particle be found over a small distance dx at position x is given by

$$P(x)dx = \int |\psi_n|^2\, dx$$

$$= \frac{2}{a}\sin^2\left(\frac{n\pi x}{L}\right)dx \tag{8.42}$$

therefore, the probability density for the one dimensional motion is

$$P(x) = \frac{2}{a}\ \sin^2\left(\frac{n\pi x}{L}\right) \tag{8.43}$$

The probability density is maximum when

$$\frac{n\pi x}{a} = \frac{\pi}{2}, \frac{3\pi}{2}, \frac{5\pi}{2}$$

or

$$x = \frac{L}{2n}, \frac{3L}{2n}, \frac{5L}{2n} \tag{8.44}$$

Obviously, for the state defined by $n = 1$, the most probable position of the particle is at $x = L/2$. For $n = 2$, the most probable positions are at $x = L/4$ and $x = 3L/4$. For $n = 3$, the most probable positions are at $x = L/6$, $x = 3L/6$, $x = 5L/6$. The variation of the probability densities with x for first three values of n are shown in Fig. [8.7(b)].

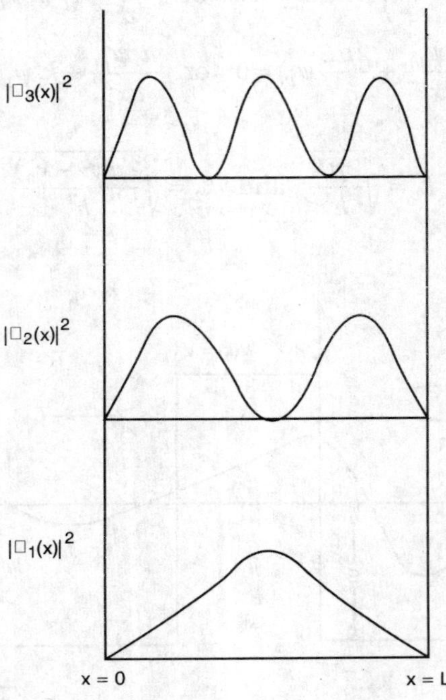

Fig. 8.7 (b)

Thus the quantum mechanical result is quite different from the classical one. According to classical mechanics a particle in such a potential box would travel with a uniform velocity from one wall to the other, and at the walls it would be perfectly reflected. Therefore, the probability of finding the particle within a small distance dx any where in the one-dimensional box is the same and is equal to dx/L. The probability $1/L$ is represented by a straight line at height $1/L$ above the x-axis.

The problem of particle in a one dimensional box is not merely an exercise in mathematics; it can serve as a good model for the calculation of approximate energies in atomic and molecular systems; besides accounting for the effect of quantization.

8.4 POTENTIAL BARRIER

A potential fucntion of the form shown in Fig. (8.8) is known as rectangular potential barrier and can be represented mathematically by the following equation:

8.4.1 Case I ($E > V_0$)

Schrodinger equation for such a particle in three regions of space may be written as;

$$\left.\begin{aligned}
x<0 \qquad & \frac{d^2\psi_I}{dx^2}+\frac{2\mu E}{\hbar^2}\psi_I=0 \quad \text{or} \quad \frac{d^2\psi_I}{dx^2}+k_1^2\psi_I=0\\
0<x<a \qquad & \frac{d^2\psi_{II}}{dx^2}+\frac{2\mu(E-V_0)}{\hbar^2}\psi_{II}=0 \quad \text{or} \quad \frac{d^2\psi_{II}}{dx^2}+k_2^2\psi_{II}=0\\
0>a \qquad & \frac{d^2\psi_{III}}{dx^2}+\frac{2\mu E}{\hbar^2}\psi_{III}=0 \quad \text{or} \quad \frac{d^2\psi_{III}}{dx^2}+k_1^2\psi_{III}=0
\end{aligned}\right\} \tag{8.45}$$

where

$$k_1=\sqrt{\frac{2\mu E}{\hbar^2}} \quad \text{and} \quad k_2=\sqrt{\frac{2\mu(E-V_0)}{\hbar^2}} \tag{8.46}$$

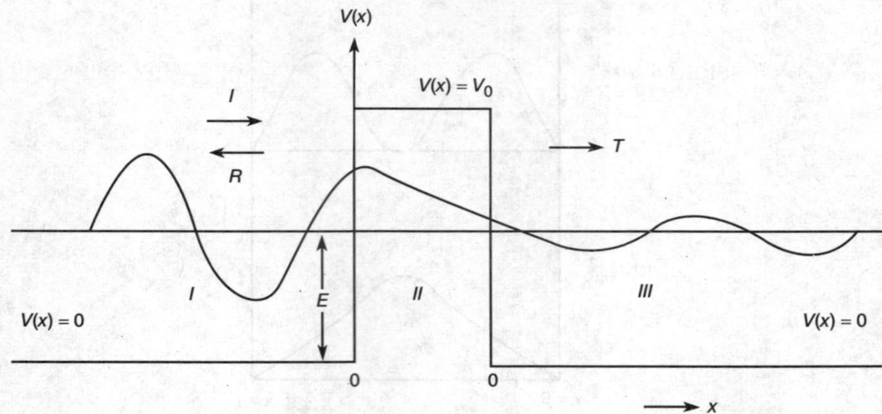

Fig. 8.8 A rectangular potential barrier of width a and height V_0

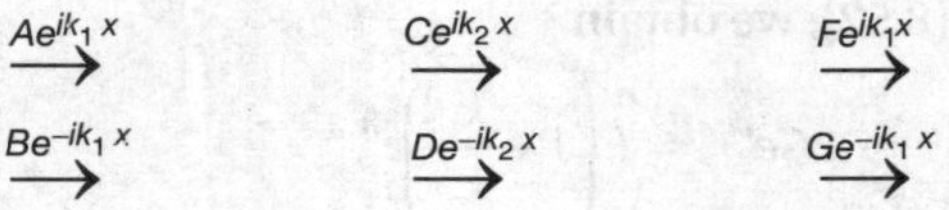

$\xrightarrow{Ae^{ik_1x}}$ $\xrightarrow{Ce^{ik_2x}}$ $\xrightarrow{Fe^{ik_1x}}$

$\xrightarrow{Be^{-ik_1x}}$ $\xrightarrow{De^{-ik_2x}}$ $\xrightarrow{Ge^{-ik_1x}}$

Fig. 8.9

Wave functions in *I*, *II* and *III* regions can evidently be written as

$$\left.\begin{aligned} \psi_I &= Ae^{ik_1x} + Be^{-ik_1x} \\ \psi_{II} &= Ae^{ik_2x} + De^{-ik_2x} \\ \psi_{III} &= Ee^{ik_1x} + Ge^{-ik_1x} \end{aligned}\right\} \tag{8.47}$$

On multiplying them (ψ_I, ψ_{II} and ψ_{III}) by exp. $(-i\omega t)$, they represent a superposition of waves travelling in different directions as shown in Fig. (8.9).
Continuity of wave functions and its derivative at $x = 0$ and $x = a$ demands

$$\psi_I\big|_{x=0} = \psi_{II}\big|_{x=0}$$

$$\left.\frac{d\psi_I}{dx}\right|_{x=0} = \left.\frac{d\psi_{II}}{dx}\right|_{x=0}$$

$$\psi_{II}\big|_{x=a} = \psi_{III}\big|_{x=a}$$

$$\left.\frac{d\psi_{II}}{dx}\right|_{x=a} = \left.\frac{d\psi_{III}}{dx}\right|_{x=a}$$

giving us four equations connecting the unknown coefficients *A*, *B*, *C*, *D*, *F* and *G*.

$$\left.\begin{aligned} A + B &= C + D \\ ik_1(A - B) &= ik_2(C - D) \\ Ce^{ik_2a} + De^{-ik_2a} &= Fe^{ik_1a} + Ge^{-ik_1a} \\ ik_2Ce^{ik_2a} - ik_2De^{-ik_2a} &= ik_1Fe^{ik_1a} - ik_1aGe^{-ik_1a} \end{aligned}\right\} \tag{8.48}$$

Eq. (8.47) represents a wave (Ae^{ik_1x}) incident on the barrier from the left and another wave (Ge^{-ik_1x}) incident on the barrier from the right. Now if we consider the incidence of waves on the barrier only from left, then $G = 0$ and set of Eq. (8.48) now reduces to

$$A + B = C + D \tag{8.49}$$

$$ik_1(A - B) = ik_2(C - D)$$

or
$$A - B = \frac{k_2}{k_1}(C - D) \tag{8.50}$$

$$Ce^{ik_2a} - De^{-ik_2a} = Fe^{ik_1a} \tag{8.51}$$

$$ik_2Ce^{ik_2a} + ik_2De^{-ik_2a} = ik_1Fe^{ik_1a}$$

or
$$Ce^{ik_2a} - De^{-ik_2a} = \frac{k_1}{k_2}Fe^{-ik_1a} \tag{8.52}$$

Solving Eqs. (8.51) and (8.52), we obtain

$$2Ce^{ik_2a} = F\left(1+\frac{k_1}{k_2}\right)e^{ik_1a}$$

or
$$C = F\left(\frac{k_1+k_2}{2k_2}\right)e^{i(k_1-k_2)a} \tag{8.53}$$

$$2De^{-ik_2a} = F\left(1-\frac{k_1}{k_2}\right)e^{ik_1a}$$

or
$$D = F\left(\frac{k_2+k_1}{2k_2}\right)e^{i(k_1-k_2)a} \tag{8.54}$$

Solving Eqs. (8.49) and (8.50), we obtain

$$2A = \left(1+\frac{k_2}{k_1}\right)C + \left(1-\frac{k_2}{k_1}\right)D$$

or
$$A = \left(\frac{k_1+k_2}{2k_1}\right)C + \left(\frac{k_1-k_2}{2k_1}\right)D$$

or
$$A = F\left[\left(\frac{k_1+k_2}{2k_1}\right)\left(\frac{k_1+k_2}{2k_1}\right)e^{-i(k_1-k_2)} + \left(\frac{k_1-k_2}{2k_1}\right)\left(\frac{k_2-k_2}{2k_1}\right)e^{i(k_1-k_2)a}\right]$$

$$= \frac{Fe^{ik_1a}}{4k_1k_2}[(k_1^2+k_2^2+2k_1k_2)e^{-ik_1a} + (k_1^2-k_2^2+2k_1k_2)e^{ik_2a}]$$

$$= \frac{Fe^{ik_1a}}{4k_1k_2}[2k_1k_2(e^{ik_2a}+e^{-ik_2a}) - (k_1^2+k_2^2)(e^{ik_2a}-e^{-ik_2a})]$$

$$= \frac{Fe^{ik_1a}}{4k_1k_2}[2k_1k_2\cos k_2a - 2i(k_1^2+k_2^2)\sin k_2a]$$

$\therefore$
$$\frac{F}{A} = \frac{4k_1k_2e^{-ik_1a}}{4k_1k_2\cos k_2a - 2i(k_1^2+k_2^2)\sin k_2a}$$

or
$$\frac{F}{A} = \frac{2k_1k_2e^{-ik_1a}}{2k_1k_2\cos k_2a - i(k_1^2+k_2^2)\sin k_2a} \tag{8.55}$$

and
$$B = \left(1+\frac{k_2}{k_1}\right)\frac{D}{2} + \left(1-\frac{k_2}{k_1}\right)\frac{C}{2}$$

$$= F\left[\left(\frac{k_1-k_2}{2k_1}\right)\left(\frac{k_1+k_2}{2k_2}\right)e^{(k_1-k_2)a} + \left(\frac{k_1+k_2}{2k_1}\right)\left(\frac{k_2-k_1}{2k_2}\right)e^{i(k_1-k_2)a}\right]$$

$$= -F\frac{(k_1-k_2)(k_1+k_2)}{4k_1k_2}e^{ik_1a}\left[e^{ik_2a} - e^{-ik_2a}\right]$$

$$= -A\left[\frac{4k_1k_2e^{-ik_1a}}{4k_1k_2\cos k_2a - 2i(k_1^2+k_2^2)\sin k_2a}\right]\left[\frac{(k_1^2-k_2^2)}{4k_1k_2}e^{ik_1a}2i\sin k_2a\right]$$

$$= \frac{B}{A}\left[\frac{-2i\sin k_2 a}{4k_1k_2\cos k_2 a\cos k_2 a - i(k_1^2+k_2^2)\sin k_2 a}\right]$$

$$= \frac{B}{A}\left[\frac{-i\sin k_2 a}{2k_1k_2\cos k_2 a - i(k_1^2+k_2^2)\sin k_2 a}\right] \tag{8.56}$$

Transmissivity

$$P_T = \left|\frac{F}{A}\right|^2 = \frac{4k_1^2k_2^2}{4k_1^2k_2^2\cos^2 k_2 a + (k_1^2+k_2^2)^2\sin^2 k_2 a}$$

or
$$P_T = \frac{4k_1^2k_2^2}{4k_1^2k_2^2 + \{(k_1^2+k_2^2)^2 - 4k_1^2k_2^2\}\sin^2 k_2 a}$$

or
$$P_T = \frac{4k_1^2k_2^2}{4k_1^2k_2^2 + (k_1^2+k_2^2)^2 - \sin^2 k_2 a} \tag{8.57}$$

or
$$P_T = \left\{1 + \frac{(k_1^2+k_2^2)^2}{4k_1^2k_2^2}\sin^2 k_2 a\right\}^{-1}$$

or
$$P_T = \left[1 + \frac{k_1^2a^2}{4}\left(1 - \frac{k_2^2}{k_1^2}\right)^2\left(\frac{\sin k_2 a}{k_2 a}\right)^2\right]^{-1}$$

In the limiting case when $E \to V_0$ and $k_2 \to 0$, we obtain

$$P_T = \left(1 + \frac{k_1 a^2}{4}\right)^{-1} \tag{8.58}$$

The transmissivity becomes unity, whenever $k_2 a = n\pi$

or
$$\left\{\sqrt{\frac{2\mu(E-V_0)}{\hbar^2}}\right\}a = n\pi$$

or
$$\frac{2\mu a^2}{\hbar^2}(E - V_0) = n^2\pi^2$$

or
$$E = V_0 + \frac{\hbar^2 n^2\pi^2}{2\mu a^2} = V_0\left[1 + \frac{\hbar^2 n^2\pi^2}{8\mu V_0 a^2}\right] \tag{8.59}$$

Transmissivity is minimum when $k_2 a = (2n+1)\frac{\pi}{2}$

or
$$\left\{\sqrt{\frac{2\mu(E-V_0)}{\hbar^2}}\right\}a = (2n+1)\frac{\pi}{2}$$

or
$$\frac{2\mu a^2(E-V_0)}{\hbar} = (2n+1)^2\frac{\pi^2}{4}$$

or
$$E - V_0 + \frac{\hbar^2(2n+1)^2\pi^2}{8\mu a^2} = V_0\left[1+\frac{(2n+1)^2h^2}{32\mu V_0 a^2}\right] \tag{8.60}$$

Minimum value of transmissivity

$$(P_T)_{\min} = \frac{4k_1^2k_2^2}{(k_1^2+k_2^2)^2} = \frac{4E(E-V_0)}{(2E-V_0)^2}$$

or
$$(P_T)_{\min} = \frac{E(E-V_0)}{(E-V_0/2)} \tag{8.61}$$

Thus we see that transmissivity increased from $\left(\frac{1}{1+k_1^2a^2/4}\right)$, with the increase of energy of the incident particle and becomes maximum (equal to unity) for the first time when energy of particle becomes equal to $V_0\left(1+\frac{h^2}{8\mu V_0 a^2}\right)$; then it oscillates between its maximum value (unity) and minimum value given by Eq. (8.61).

Reflectivity
$$P_R = \left|\frac{B}{A}\right|^2 = \frac{(k_1^2-k_2^2)\sin^2 k_2a}{4k_1^2k_2^2\cos^2 k_2a + (k_1^2+k_2^2)\sin^2 k_2a}$$

or
$$P_R = \frac{(k_1^2-k_2^2)^2\sin^2 k_2a}{4k_1^2k_2^2 + (k_1^2-k_2^2)\sin^2 k_2a} \tag{8.62}$$

Also
$$P_T + P_R = 1 \tag{8.63}$$

A plot of real part of the wave function representing particles incident on the barrier from left with energy greater than the height of the barrier is shown in Fig. (8.10). Variation of transmissivity with the energy of incident particle is shown in Fig. (8.11).

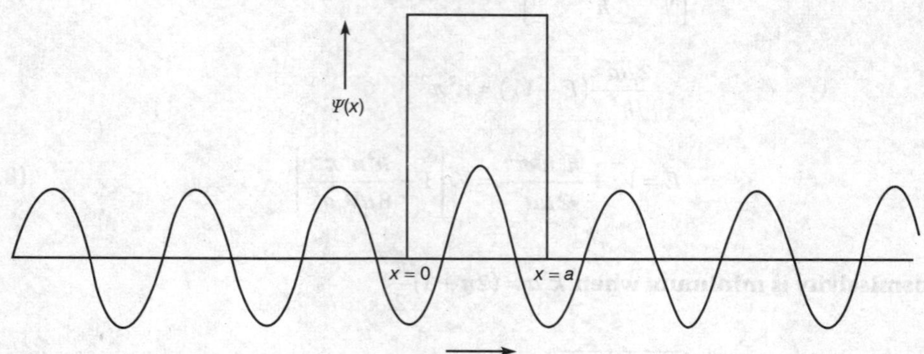

Fig. 8.10

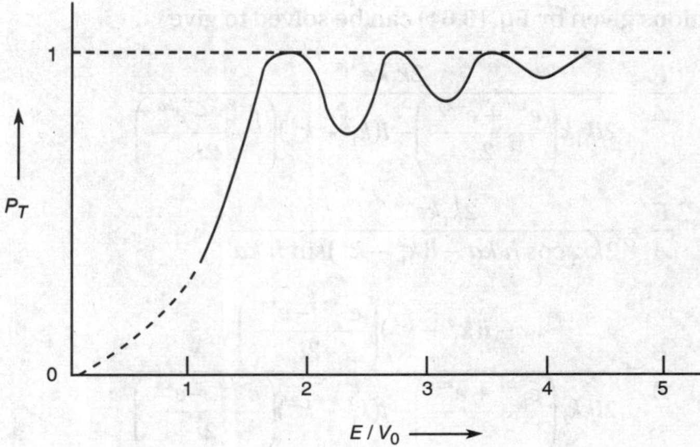

Fig. 8.11

8.4.2 Class II ($E < V_0$): Quantum Mechanical Tunnelling

For a particle or wave incident on the barrier from the left, Schrodinger equation in three regions may be written as:

$$x < 0 \qquad \frac{d^2\psi_1}{dx^2} + \frac{2\mu E}{\hbar^2}\psi_1 = 0 \quad \text{or} \quad \frac{d^2\psi_1}{dx^2} + k_1^2\psi_1 = 0$$

$$0 < x < a \qquad \frac{d^2\psi_{II}}{dx^2} + \frac{2\mu(E - V_0)}{\hbar^2}\psi_{II} = 0 \quad \text{or} \quad \frac{d^2\psi_{II}}{dx^2} - \frac{2\mu(V_0 - E)}{\hbar^2}\psi_{II} = 0$$

$$\text{or} \qquad \frac{d^2\psi_{II}}{dx^2} + k^2\psi_{II} = 0$$

$$x > a \qquad \frac{d^2\psi_{III}}{dx^2} + \frac{2\mu E}{\hbar^2}\psi_{III} = 0 \quad \text{or} \quad \frac{d^2\psi_{III}}{dx^2} + k_1^2\psi_{III} = 0$$

Solutions of these equations may be written as

$$\psi_I = Ae^{ik_1x} + Be^{-ik_1x}$$
$$\psi_{II} = Ce^{-kx} + De^{+kx}$$
$$\psi_{III} = Fe^{ik_1x}$$

Boundary conditions at $x = 0$ and $x = a$ gives

$$\left.\begin{aligned} A + B &= C + D \\ ik_1(A - B) &= -k(C - D) \\ Ce^{-ka} + De^{ka} &= Fe^{ik_1a} \\ -kCe^{-ka} + kDe^{ka} &= ik_1Fe^{ik_1a} \end{aligned}\right] \tag{8.64}$$

The set of equations given by Eq. (8.64) can be solved to give

$$\frac{F}{A} = \frac{2k_1 k e^{ik_1 a}}{2ik_1 k\left(\frac{e^{ka} + e^{-ka}}{2}\right) - i(k_1^2 - k^2)\left(\frac{e^{-ka} - e^{ka}}{2i}\right)}$$

or

$$\frac{F}{A} = \frac{2k_1 k e^{-ik_1 a}}{2kk_1 \cos h\, ka - i(k_1^2 - k^2) \sin h\, ka}$$

and

$$\frac{B}{A} = \frac{-i(k_1^2 - k^2)\left(\frac{e^{-ke} - e^{ka}}{2i}\right)}{2ikk_1\left(\frac{e^{-ka} + e^{ka}}{2}\right) - i(k_1^2 - k^2)\left(\frac{e^{-ka} - e^{ka}}{2i}\right)}$$

or

$$\frac{B}{A} = \frac{(k_1^2 + k^2)\sin h\, ka}{2ikk_1 \cos h\, ka + (k_1^2 - k^2)\sin h\, ka}$$

Transmissivity: $P_T = \left|\frac{F}{A}\right|^2 = \frac{4k_1^2 k^2}{4k_1^2 k^2 \cos h^2\, ka + (k_1^2 - k^2)^2 \sin h^2\, ka}$

or

$$P_T = \frac{4k_1^2 k^2}{4k_1^2 k^2 (1 + \sin h^2\, ka) + (k_1^2 - k^2)^2 \sin h^2\, ka}$$

or

$$P_T = \frac{4k_1^2 k^2}{4k_1^2 k^2 + (k_1^2 - k^2)^2 \sin h^2\, ka} \quad (8.65)$$

Reflectivity: $P_R = \left|\frac{B}{A}\right|^2 = \frac{(k_1^2 + k^2)^2 \sin h^2\, ka}{4k_1^2 k^2 \cos h^2 ka + (k_1^2 - k^2)^2 \sin h^2\, ka}$

or

$$P_R = \frac{(k_1^2 + k^2)^2 \sin h^2\, ka}{4k_1^2 k^2 + (k_1^2 + k^2)^2 \sin h^2\, ka} \quad (8.66)$$

Reflectivity and transmissivity (for a barrier) are plotted in Fig. (8.12) as functions of its width for fixed values E and V_0 (i.e., fixed value of k_1 and k). The wave function is depicted by Fig. (8.13). Dashed portion in Fig. (8.11) gives transmission probability when the energy of the particle is less than the height of the barrier. The transmissivity decreases monotonically from the value given by Eq. (8.58) as E decreases below V_0. This barrier is often referred to as tunnelling through the barrier. When $ka >> 1$, the transmission coefficient becomes very small and is given approximately by

$$P_T = \frac{4k_1^2 k^2}{(k_1^2 + k^2)^2 \left(\frac{e^{ka} - e^{-ka}}{2}\right)^2}$$

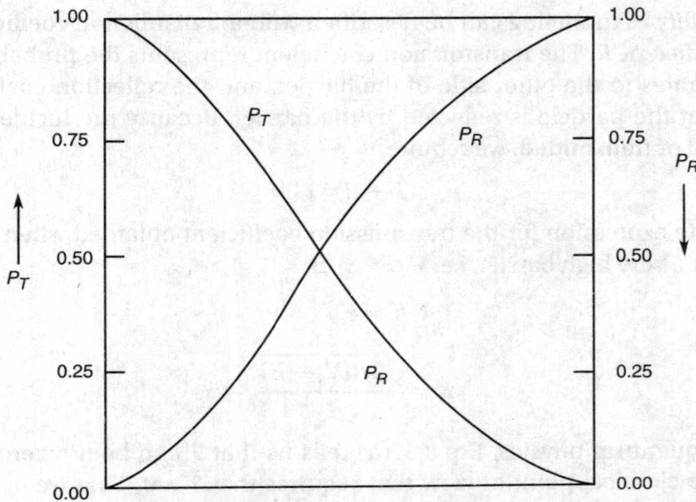

Fig. 8.12

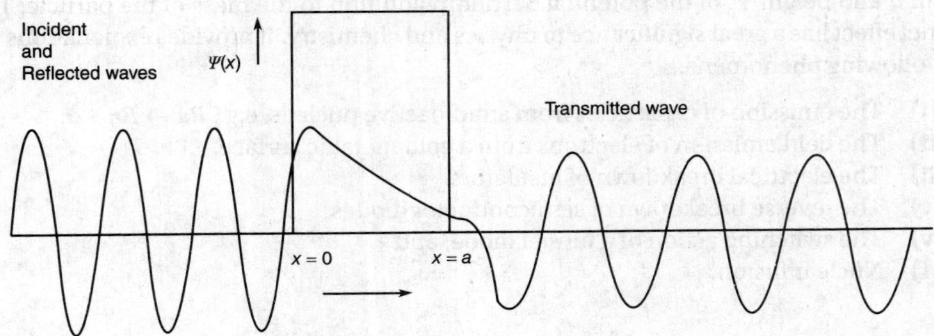

Fig. 8.13

or
$$P_T = \frac{\dfrac{4(2\mu E)}{\hbar^2}\left\{\dfrac{2\mu(V_0 - E)}{\hbar^2}\right\}}{\left\{\dfrac{2\mu E}{\hbar^2} + \dfrac{2\mu(V_0 - E)}{\hbar^2}\right\}^2} 4e^{-2ka}$$

or
$$P_T = \frac{16E(V_0 - E)}{V_0^2} e^{-2ka} \tag{8.67}$$

This explains the emission of α-particles from nuclei, flow of electric current between two metals in contact in spite of an oxide layer between them and flow of electrons from metallic electrode to solution in which they are placed.

The probability of tunneling can be described with a transmission coefficient T and a reflection coefficient R. The transmission coefficient represents the probability that the particle penetrates to the other side of the barrier, and the reflection coefficient is the probability that the particle is reflected by the barrier. Because the incident particle is either reflected or transmitted, we require

$$T + R = 1 \tag{8.67(a)}$$

An approximate expression for the transmission coefficient obtained when $T = 1$ (a very wide barrier or a very high barrier, i.e. $V_0 >> E$), is

$$T \approx e^{-2KL} \tag{8.67(b)}$$

where

$$K = \frac{\sqrt{2\mu(V_0 - E)}}{\hbar} \tag{8.67(c)}$$

According to quantum physics, Eq. 8.67(a) tells us that T can be nonzero, which is in contrast to the classical point of view that requires that $T = 0$. That we experimentally observe the phenomenon of tunneling provides further confidence in the principle of quantum mechanics.

The quantum mechanical tunnel effect or the transmission coefficient depends on the width a and height V_0 of the potential barrier in addition to the mass of the particle. The tunnel effect has a great significance in physics and chemistry. It provides explanations for the following phenomena.

(i) The emission of α-particles from a radioactive nucleus, e.g., $Ra \rightarrow Rn + \alpha$.
(ii) The field emission of electrons from a cold metallic surface.
(iii) The electrical breakdown of insulators.
(iv) The reverse breakdown of semiconductor diodes.
(v) The switching action of a tunnel diode, and
(vi) Nuclear fusion.

8.5 EMISSION OF α-PARTICLES FROM A RADIOACTIVE ELEMENT

A direct experimental evidence of quantum mechanical tunnelling phenomenon comes from emission of α-rays from a radioactive nucleus. For example, in the radioactive process; $Ra \rightarrow Rn + \alpha$, emitting α-particle from the radium nucleus has to encounter a strong potential barrier due to Coulombic repulsion by the radon nucleus having a large nuclear charge. The energy (E) of the α-particle is not sufficient to cross over the barrier. Yet, the particle is able to come out of the radium nucleus by quantum mechanical tunnelling through the barrier. We discuss it in detail as follows:

Potential Energy Curve

Elements with atomic number $Z \leq 81$ are usually unstable and consequently undergo spontaneous disintegration with the emission of α- and β-particles and γ-rays. An α-particle is helium nucleus ($_2He^4$). This phenomenon of spontaneous disintegration is called **natural radioactivity**. There is a narrow range of kinetic energies of α-particles emitted by radioactive elements. An α-particle emitted by thorium ($_{90}Th^{232}$) atom has lowest energy (4.007 MeV) and that emitted by thorium C′ ($_{84}Po^{212}$) has highest energy (8.95 MeV). Since an α-particle carries a positive charge 2e, when it has been emitted from a nucleus of atomic number Z, its potential energy at a large distance from the residual nucleus of atomic number (Z – 2) is obtained according to relation

$$V(r) = K\frac{(Z-2)e(2e)}{r} = \frac{2(Z-2)e^2}{4\pi\varepsilon_0 r} \tag{8.68}$$

where Z is the charge number of parent nucleus, r is the distance of α-particle from the centre of the nucleus and $k = \frac{1}{4\pi\varepsilon_0} = 9\times10^9$ MKS units. If the distance r is decreased, the potential energy increases till it becomes maximum equal to V_0 at a position $r = r_0$ just outside the nucleus. The maximum potential energy is equal to the work which must be done against the repulsive electrostatic force to move the α-particle from $r = \infty$ to $r = r_0$. When α-particle is inside the nucleus, it is acted upon by short range and extremely powerful attractive forces due to the nucleus. The nuclear forces are effective over a short distance r_0 from the centre of the nucleus. This distance is called the effective radius of the nucleus. Due to the attractive forces the potential energy of alpha (α) particle inside the nucleus is negative. The potential energy curve for an α-particle, which is in the form of a potential well is shown in Fig. (8.14).

The maximum positive potential energy of an α-particle just outside the nucleus is given by

$$V_0 = \frac{2(Z-2)e}{4\pi\varepsilon_0 r_0} \tag{8.69}$$

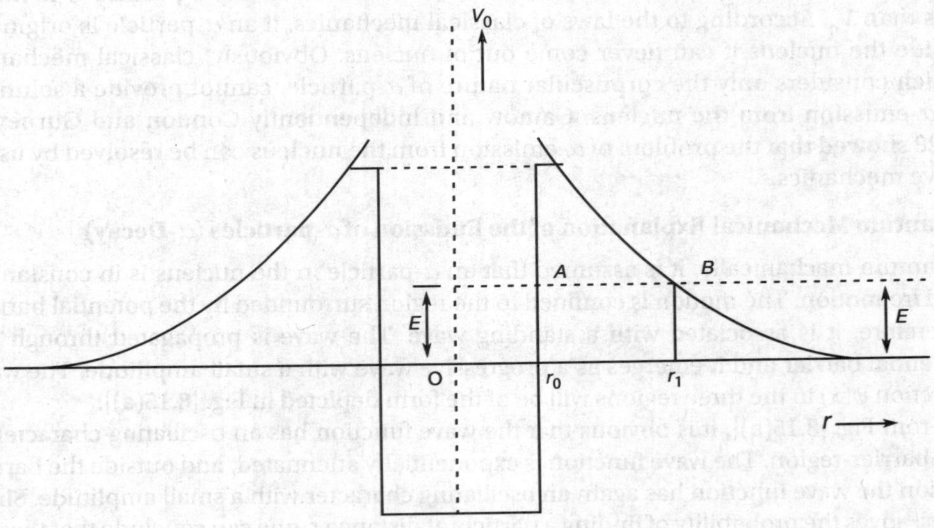

Fig. 8.14

For $_{84}Po^{210}$, which emits an α-particle with kinetic energy 5.3 MeV, we have $Z = 84$, $e = 1.6 \times 10^{-19}$ C, $\frac{1}{4\pi\varepsilon_0} = 9\times10^9$ MKS units and $r_0 = 1.5 \times 10^{-15} A^{1/3} = 1.5 \times 10^{-15} \times (210)^{1/3}$ m. Substituting these values in the above relation, one obtains, $V_0 = 27$MeV (Here, we have estimated the nuclear radius R by an empirical relation, $R = R_0A^{1/3}$ where A is the mass number of the nucleus and R_0 is a constant having a value of the order of 1.5×10^{-15} m).

One can easily see that the kinetic energy of an α-particle emitted from a radio-active nucleus is much less than the maximum positive potential energy just outside the nucleus.

Failure of Classical Mechanics to explain the Emission of α-particles from the nucleus

When an α-particle is emitted from a radioactive nucleus with kinetic energy E, this energy must be equal to the total energy (kinetic + potential) of the α-particle before the ejection from the nucleus. This energy is shown by the dashed line AB in potential energy curve in Fig. (8.14). The energy E is equal to the potential energy of the α-particle at $r = r_1$.

$$E = \frac{2(Z-2)e^2}{4\pi\varepsilon_0 r_1} \tag{8.70}$$

In the region from $r = r_0$ to $r = r_1$ the kinetic energy K of the α-particle is given by

$$K = \frac{2(Z-2)e^2}{4\pi\varepsilon_0 r} - \frac{2(Z-2)e^2}{4\pi\varepsilon_0 r_1}, \text{ where } r_0 < r < r_1 \text{ or } K = -\frac{2(Z-2)e^2}{4\pi\varepsilon_0}\left(\frac{1}{r} - \frac{1}{r_1}\right) \tag{8.71}$$

Since $r < r_1$, this region is the region of negative kinetic energy for the α-particle. This region of electrostatic field around the nucleus is called **potential barrier region** for the α-particle. The distance $AB = r_1 - r_0$ is the width of the barrier for an α-particle of kinetic energy E, and the maximum potential energy V_0 of the α-particle just outside the nucleus is the height of the potential barrier, since for an α-particle E is much less than V_0. According to the laws of classical mechanics, if an α-particle is originally inside the nucleus it can never come out of nucleus. Obviously, classical mechanics which considers only the corpuscular nature of α-particle, cannot provide a solution, of α-emission from the nucleus. Gamow and independently Condon and Gurney in 1928 showed that the problem of α-emission from the nucleus can be resolved by using wave mechanics.

Quantum Mechanical Explanation of the Emission of α-particles (α-Decay)

Quantum mechanically, it is assumed that an α-particle in the nucleus is in constant to and fro motion. The motion is confined to the region surrounded by the potential barrier. Therefore, it is associated with a standing wave. The wave is propagated through the potential barrier and it emerges as a progressive wave with a small amplitude. The wave function $\psi(x)$ in the three regions will be of the form depicted in Fig. [8.15(a)].

From Fig. [8.15(a)], it is obvious that the wave function has an oscillating character in the barrier-region. The wave function is exponentially attenuated, and outside the barrier region the wave function has again an oscillating character with a small amplitude. Since $|\psi(r)|^2$ gives the probability of finding a particle at distance r, one can conclude that there is a finite probability of transmission of an α-particle through the potential barrier.

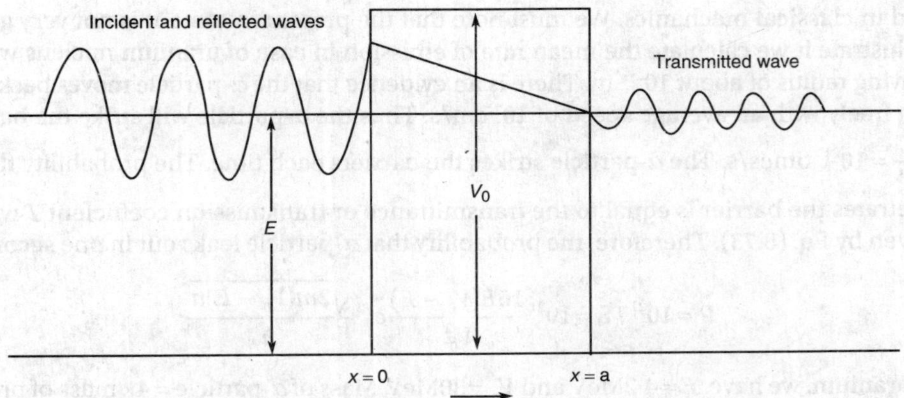

Fig. 8.15 (a)

The quantum mechanical theory of the emission of α-particles was first given by G. Camow, and independently by R.W. Gurney and E.U. Condon in 1928. According to the theory, an approximate expression for the transmission coefficient is given by

$$T = T_0 e\left[\frac{-2\sqrt{2m}}{\hbar}\int_{r_0}^{r_1}(V(r)-E)^{1/2}\,dr\right] \tag{8.72}$$

If v_0 be the velocity of the α-particle inside the potential well, it will hit the barrier roughly v_0/r times in one second. Then the transmission coefficient T is the probability that the α-particle may leak out at each collision with the barrier. Hence the probability (P) of emission of an α-particle per second is

$$P = T_0\left(\frac{V_0}{R}\right)e^{\frac{-2\sqrt{2m}}{\hbar}\int_{r}^{r_1}[V(r)-E)]^{1/2}\,dr} \tag{8.73}$$

where T_0 is usually of the order of unity, m is the mass of the α-particle and V is the potential energy. The probability of emission of an α-particle per second, is nothing but decay constant, λ. Therefore,

$$\lambda = P = \frac{V_0}{R}e^{\frac{-2\sqrt{2m}}{\hbar}\int_{r}^{r_1}[V(r)-E)]^{1/2}\,dr} \tag{8.74}$$

One can easily show that the above reduces to the **Geiger-Nuttal law.**

$$\log_e \lambda = a - \frac{b(Z-2)}{V_0} \tag{8.75}$$

The above theory removes the difficulties of classical mechanics. By assigning the wave aspect to the particle, there always exists certain probability of the particle to penetrate through the barrier. Since, in quantum mechanics, we are only concerned with the probability of doing something; the probability may become reality in a few years or after many hundred or thousand years. Thus no question of casualty exists as faced in classical mechanics. We must note that the present treatment is not very good. To illustrate it we calculate the mean rate of emission in case of uranium nucleus which is having radius of about 10^{-14} m. There is an evidence that the α-particle moves back and forth freely with an average speed of 10^7 cm/s. Thus the α-particle will strike the barrier $\frac{10^7}{10^{-14}} = 10^{21}$ times/s. The α-particle strikes the barriers each time. The probability that it penetrates the barrier is equal to the transmittance or transmission coefficient T which is given by Eq. (8.73). Therefore, the probability that α-particle leaks out in one second is

$$P = 10^{21} TS = 10^{21} \frac{16E(V_0 - E)}{V_0^2} e^{-2} \frac{\sqrt{2m(V_0 - E)a}}{\hbar}$$

For uranium, we have, $E = 4.2$MeV and $V_0 = 30$MeV. Mass of α-particle = 4× mass of proton = $4 \times 1.6 \times 10^{-24}$ gm, $\hbar = 1.05 \times 10^{-27}$ erg-s, $a = 2 \times 10^{-12}$ cm

$$\therefore \quad P = 10^{21} \times 16 \times \frac{(4.2\text{MeV}) \times (30 - 4.2)\text{MeV}}{(30\text{MeV})^2} \exp\left[-2\sqrt{2m(V_0 - E)}\frac{a}{\hbar}\right]$$

$$= 10^{21} \times \frac{16 \times 4.2 \times 25.8}{30 \times 30}$$

$$\exp\left[\frac{-2\sqrt{(2 \times 4 \times 1.6 \times 10^{-24} \times 25.8 \times 10^4 \times 1.6 \times 10^{-12})} \times 2 \times 10^{-12}}{1.05 \times 10^{-27}}\right]$$

$$= 1.9 \times 10^{23} \quad e^{(=-90)} = 10^{-18} \text{ /s} = 10^{-11} \text{ /year}$$

∴ The average lifetime of the nucleus is equal to the reciprocal of P, i.e.,

$$\tau = \frac{1}{10^{-11}} \text{ years} \quad = 10^{11} \text{ years}$$

However, experimentally τ has been found to be 10^{19} years. The discrepancy between theory and experimental results has been ascribed to the fact that the barrier is not rectangular.

Barrier penetration of atoms takes place in the periodic inversion of the ammonia molecule, NH_3. Fig. [8.15(b)] illustrates schematically the structure of this molecule. It consists of three H-atoms arranged in a plane, and equidistant from the N-atom. There are two completely equivalent equilibrium positions for the N-atom, one on either side of the plane containing the H-atoms. Fig. [8.15(c)] indicates the potential energy acting on the N-atom, as a function of its distance x from the plane. The potential function $V(x)$ has two minima, corresponding to the two equilibrium positions, which are symmetrically disposed about a low maximum located at $x = 0$. This maximum, which constitutes a barrier separating the two binding regions, arises the repulsive Coulomb forces that act on the N-atom if it penetrates the plane of the H-atoms. The forces are strong enough that in classical mechanics the N-atom is not able to cross the barrier, if the molecule is in one of its low-lying energy states, i.e., the lower allowed energies of this binding potential are below the top of the barrier, as shown in the figure. But penetration of the classically excluded region allows the N-atom to tunnel through the barrier. If it is initially on one side,

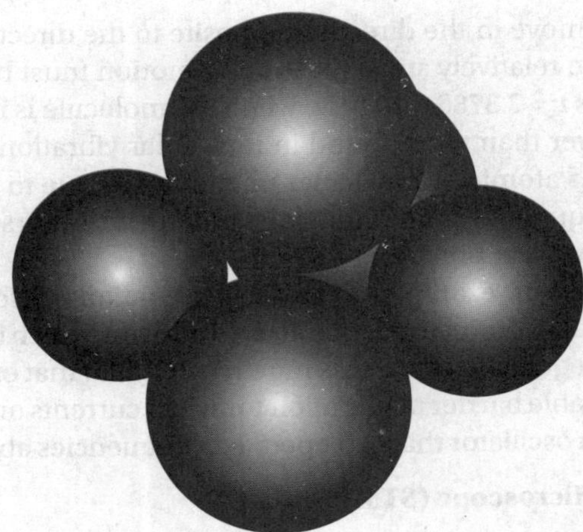

Fig. 8.15 (b) A schematic illustration of the NH_3 molecule. The light spheres represent the three H-atoms arranged in a plane. The dark spheres represent two equivalent equilibrium-positions of the single N-atom

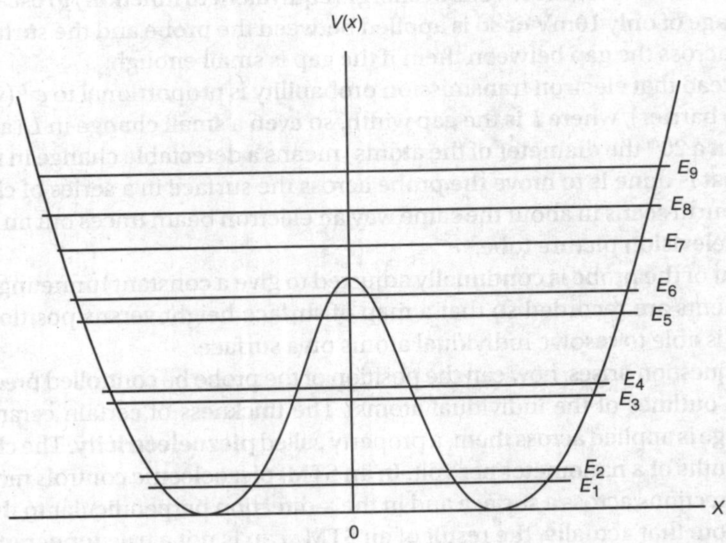

Fig. 8.15 (c) The potential energy of the N-atom in the NH_3 molecule, as a function of all distances from the plane containing the three H-atoms, which lies at $x = 0$. In its lower energy states, the total energy of the molecule lies below the top of the barrier separating the two minima, as indicated by the eigen values of the potential shown in the figure

it will tunnel through and eventually appear on the other side. Then it will do it again in the opposite direction. The position of the N-atom with respect to the plane containing the H-atoms actually oscillates slowly back and forth across the plane (Since the molecule's centre of mass remains fixed in an inertial reference frame, in such a reference frame the

H-atoms must always move in the direction opposite to the direction of the N-atom and since the H-atoms have relatively small mass, their motion must be relatively large). The oscillation frequency is $\upsilon = 2.3786 \times 10^{10}$ Hz, when the molecule is in its ground state. This frequency is much lower than those found in molecular vibrations not involving barrier penetration, or in other atomic or molecular phenomena. Due to the resulting technical simplifications the frequency was used as a standard in the first atomic clocks which measure time with maximum precision.

A recent, and very useful, example of barrier penetration of electrons is found in the tunnel diode. This is a semiconductor device, like a transistor, which is used in fast electronic circuits since its high frequency response is much better than that of any transistor. Tunnel diode employs controllable barrier penetration to switch currents on or off so rapidly that it can be used to make an oscillator that can operate at frequencies about 10^{11} Hz.

Scanning Tunneling Microscope (STM)

The ability of electrons to tunnel through a potential barrier is used in an ingenious way in STM to study surfaces on an atomic scale of size. In an STM a metal probe with a point so fine that its tip is a single atom is brought close to the surface of conducting or semiconducting material. Normally even the most loosely bound electrons in an atom on a surface need several electron-volts of energy (equivalent to function) to escape. However, when a voltage of only 10mV or so is applied between the probe and the surface, electrons can tunnel across the gap between them if the gap is small enough.

We have read that electron transmission probability is proportional to e^{-L} (where L is the width of the barrier), where L is the gap width, so even a small change in L (as little as 0.01 nm, less than a 20th the diameter of the atoms) means a detectable change in the tunneling current. What is done is to move the probe across the surface in a series of closely spaced back-and-fourth scans in about the same way an electron beam traces out an image on the screen of a television picture tube.

The height of the probe is continually adjusted to give a constant tunneling current, and the adjustments are recorded so that a map of surface height versus position is built up. Such a map is able to resolve individual atoms on a surface.

Now, the question arises, how can the position of the probe be controlled precisely enough to travel the outlines of the individual atoms? The thickness of certain ceramics changes when a voltage is applied across them, a property called **piezoelectricity**. The changes might be several tenths of a nanometer per volt. In an STM, piezoelectric controls move the probe in x and y-directions across a surface and in the z-direction perpendicular to the surface.

We may note that actually, the result of an STM scan is not a true topographical map of surface height but a contour map of constant electron density on the surface. Obviously, this means that atoms of different elements appear differently, which greatly increases the value of the STM as a research tool.

There are many biological materials which conduct electricity, but they also by the flow of ions rather than of electrons and so cannot be studied with STMs, very recently developed, **atomic force microscope** (AFM) can be used on any surface, although with somewhat less resolution than an STM. In an AFM, the sharp tip of a fractured diamond passes gently against the atoms on a surface. A spring keeps the pressure of the tip constant, and a record is made of the directions of the tip as it moves across the surface. The result is a map showing contours of the constant repulsive force between the electrons of the probe and the electrons of the surface atoms. Even relatively soft biological materials can be examined with an AFM and changes in them monitored, for e.g., the linking together of molecules of the blood protein fibrin, which occurs when blood clots, has been watched with an AFM.

Nuclear Fusion

The basic reactions that powers the sun and, indirectly, almost everything else in the solar system is *fusion*. In one step of the process that occurs at the core of the sun, protons must approach each other to within a small distance that they to form a deuterium nucleus. According to classical physics, these protons cannot overcome and penetrate the barrier caused by their mutual electrical repulsion. Quantum mechanically, however, the protons are able to tunnel through the barrier and fuse together.

Example 8.1: Show that near a resonance, transmissivity is given by

$$P_T = 1 - \frac{\mu V_0^2 a^2}{2E\hbar^2}\left\{1 - \frac{n\pi\hbar}{a\sqrt{2\mu(E-V_0)}}\right\}^2$$

Solution Transmissivity from a rectangular barrier of height V_0 and width a [for a particle of energy $E(E > V_0)$] is given by Eq. (8.57).

$$P_T = \frac{4k_1^2k_2^2}{4k_1^2k_2^2 + (k_1^2 - k_2^2)^2 \sin^2 k_2 a}$$

or
$$P_T = \left[1 + \frac{(k_1^2 - k_2^2)^2 \sin^2 k_2 a}{4k_1^2k_2^2}\right]^{-1}$$

At the resonance, we have seen that $k_2 a = n\pi$. So we can expect $k_2 a$ to be nearly equal to $n\pi$ when we are just near a resonance; sin $k_2 a$ may be replaced by $(k_2 a \pm n\pi)$.

$\therefore$
$$P_T = \left[1 - \frac{(k_1^2 - k_2^2)^2 (k_2 a \pm n\pi)^2}{4k_1^2k_2^2}\right]$$

Near a resonance we expect $P_T \approx 1$. In agreement with this result, we retain only negative sign with $n\pi$. Then $P_T = \left[1 - \frac{(k_1^2 - k_2^2)^2 (k_2 a - n\pi)^2}{4k_1^2k_2^2}\right]$

or
$$P_T = 1 - \frac{(k_1^2 - k_2^2)^2 k_2^2 a^2}{4k_1^2k_2^2}\left(1 - \frac{n\pi}{k_2 a}\right)^2$$

or
$$P_T = 1 - \frac{(k_1^2 - k_2^2)^2 a^2}{4k_1^2}\left(1 - \frac{n\pi}{k_2 a}\right)^2$$

$$k_1^2 - k_2^2 = \frac{2\mu E}{\hbar^2} - \frac{2\mu(E - V_0)}{\hbar^2} = \frac{2\mu}{\hbar^2}(E - E + V_0) = \frac{2\mu V_0}{\hbar^2}$$

$\therefore$
$$P_T = 1 - \left(\frac{2\mu V_0}{\hbar^2}\right)^2 \frac{a^2}{4\left(\frac{2\mu E}{\hbar^2}\right)}\left[1 - \frac{n\pi\hbar}{a\sqrt{(2\mu(E - V_0)}}\right]^2$$

or $$P_T = 1 - \frac{V_0^2 a^2 \mu}{2E\hbar^2}\left[1 - \frac{n\pi\hbar}{a\sqrt{(2\mu(E - V_0)}}\right]^2 \qquad \text{(i)}$$

8.6 THE SQUARE WELL

A one-dimensional potential well shown in Fig. (8.16) is defined by the equation

$$V(x) = \begin{cases} 0 & x < 0 \text{ and } x > a \\ -V_0 & 0 < x < a \end{cases} \qquad (8.76)$$

where V_0 is a positive quantity. There are three important cases:

(i) Total energy of the particle is negative, it is confined to the well and forms bound slates ($E < 0$);

(ii) Total energy of the particle is positive; it is unconfined and is scattered by the potential ($E > 0$);

(iii) The energy of the particle is negative but well depth becomes infinitely large ($E < 0$ and $V_0 \to \infty$).

8.6.1 Case I ($-V_o < E < 0$): Discrete Energy Levels

The potential energy of the particle is negative and kinetic energy is positive but the sum of these two (i.e., total energy) is negative. Schrodinger equation for such a particle in three regions of space may be written as:

$$\left.\begin{aligned} x < 0 \qquad & \frac{d^2\psi_1}{dx^2} + \frac{2\mu E}{\hbar^2}\psi_1 = 0 \text{ or } \frac{d^2\psi_1}{dx^2} - k_3^2\psi_1 = 0 \\ 0 < x < 0 \qquad & \frac{d^2\psi_{II}}{dx^2} + \frac{2\mu(E + V_0)}{\hbar^2}\psi_{II} = 0 \text{ or } \frac{d^2\psi_{II}}{dx^2} + k_4^2\psi_{II} = 0 \\ x > a \qquad & \frac{d^2\psi_{II}}{dx^2} + \frac{2\mu E}{\hbar^2}\psi_{II} = 0 \text{ or } \frac{d^2\psi_{II}}{dx^2} - k_3^2\psi_{III} = 0 \end{aligned}\right\} \qquad (8.77)$$

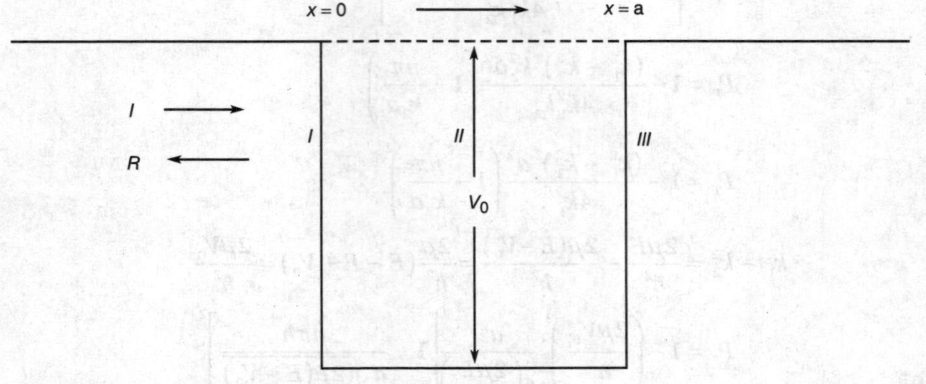

Fig. 8.16 A square well of depth V_0 and width a

where $$k_3 = \sqrt{\frac{-2\mu E}{\hbar^2}} \quad \text{and} \quad k_4 = \sqrt{\frac{2\mu(E+V_0)}{\hbar^2}} \tag{8.78}$$

Possible solutions of Eq. (8.77) in three regions may be written as

$$\psi_I = Ae^{k_3 x} + Be^{-k_3 x} \tag{8.79}$$

$$\psi_{II} = Ce^{ik_4 x} + De^{-ik_4 x} \tag{8.80}$$

$$\psi_{III} = Fe^{k_3 x} + Ge^{k_4 x} \tag{8.81}$$

Eq. (8.79) represents the solution of Schrodinger equation in the I region (i.e., when x is negative). Finiteness of the solution ψ_I as x approaches minus infinity requires that $B = 0$.

Similarly Eq. (8.81) represents the solution of Schrodinger equation in III region (i.e., when x is greater than a). Finiteness of the solution ψ_{III} as x approaches minus infinity requires that $G = 0$.

Thus the solution is an exponentially decreasing solution in I and III region whereas it is oscillatory in II region.

$$\psi_I = Ae^{k_3 x}$$

$$\psi_{II} = Ce^{ik_4 x} + De^{-ik_4 x}$$

$$\psi_{III} = Fe^{-k_3 x}$$

Continuity conditions at $x = 0$ and $x = a$ requires that

$$[\psi_1]_{x=0} = [\psi_{II}]_{x=0}$$

$$\left[\frac{d\psi_I}{dx}\right]_{x=0} = \left[\frac{d\psi_{II}}{dx}\right]_{x=0}$$

$$[\psi_{II}]_{x=a} = [\psi_{III}]_{x=a}$$

$$\left[\frac{d\psi_{II}}{dx}\right]_{x=a} = \left[\frac{d\psi_{III}}{dx}\right]_{x=a}$$

giving us four equations connecting A, C, D and F as

$$\left.\begin{aligned} A &= C + D \\ k_3 A &= ik_4 C - ik_4 D \\ Ce^{ik_4 a} + De^{-ik_4 a} &= Fe^{-ik_3 a} \\ ik_4 Ce^{ik_4 a} - ik_4 De^{-ik_4 a} &= -k_3 Fe^{-ik_3 a} \end{aligned}\right] \tag{8.82}$$

Eq. (8.82) represents a set of four homogeneous equations, among four unknown quantities A, C, D and F. This will give a unique solution, only if the determinant formed by the coefficients of A, C, D and F vanishes.

$$\text{The determinant} = \begin{vmatrix} 1 & -1 & 1 & 0 \\ k_3 & -ik_4 & ik_4 & 0 \\ 0 & e^{ik_4 a} & e^{-ik_4 a} & -e^{-ik_3 a} \\ 0 & ik_4 e^{ik_4 a} & -ik_4 e^{-ik_4 a} & k_3 e^{-k_3 a} \end{vmatrix}$$

$$\text{or} \qquad \text{The determinant} = \begin{vmatrix} 1 & 0 & 0 & 0 \\ k_3 & k_3 - ik_4 & k_3 + ik_4 & 0 \\ 0 & e^{ik_4a} & e^{-ik_4a} & -e^{-k_3a} \\ 0 & ik_4e^{ik_4a} & -ik_4e^{-ik_4a} & k_3e^{-k_3a} \end{vmatrix}$$

$$= \begin{vmatrix} k_3 - ik_4 & k_3 + ik_4 & 0 \\ e^{ik_4a} & e^{-ik_4a} & -e^{-ik_3a} \\ ik_4e^{ik_4a} & -ik_4e^{-ik_4a} & k_3e^{-k_3a} \end{vmatrix}$$

$$= (k_3 - ik_4)(k_3 - ik_4)e^{-(k_3+ik_4)a} - (k_3 + ik_4)(k_3 + ik_4)e^{-(k_3-ik_4)a}$$

$$= [-(k_3^2 - k_4^2)(e^{ik_4a} - e^{-ik_4a}) - 2ik_3k_4(e^{ik_4a} + e^{-ik_4a})e^{-k_3a}]$$

$$= -2ie^{ik_3a}\left[(k_3^2 - k_4^2)\sin k_4a + 2k_3k_4\cos k_4a\right]$$

The necessary condition that a unique solution may exist is given by

$$(k_2^2 - k_4^2)\sin k_4a + 2k_2k_4\cos k_4a = 0$$

or
$$\tan k_4a = -\frac{2k_3k_4}{k_3^2 - k_4^2} \tag{8.83}$$

This is a transcendental equation in k_4 and can be solved either by geometrical or by numerical methods. To solve Eq. (8.83) by geometrical method, we assume

$$y_1 = \tan k_4a \quad \text{and} \quad y_2 = \frac{2k_3k_4}{k_3^2 - k_4^2}$$

For different values of k_4a, we plot y_1 vs. k_4a (solid curve) and y_2 vs. k_4a (dotted curve) Fig. (8.17). The points of intersection of these two curves give the values k_4a for which Eq. (8.83) is satisfied. k_4a varies from 0 to a maximum value $\beta = \sqrt{\frac{2\mu V_0a^2}{\hbar^2}}$.

If the value of β is such that $(n-1)\pi < \beta \le n\pi$ (where n is an integer). Then there are n points of intersection and hence n allowed values of energy of the particle. The number of nodes (points where wave function vanishes except for $|x| = \infty$) in this case is equal to $(n-1)$. The value of n is given by

$$\left(\frac{\beta}{\pi} + 1\right) > n \ge \frac{\beta}{\pi} \tag{8.84}$$

If the value of β is less than or equal to π, then there is only one allowed value of energy and wave function is approximately given by Fig. (8.18).

With the increase in depth of the well (or β), new energy levels appear in the well. According to classical mechanics, energy of the particle can vary continuously from $-V_0$ to 0 while according to quantum mechanics only a few discrete values of energy are permissible; the number of these levels depend on the depth and width of the well and mass of the particle. Another important difference between the two descriptions is that according to quantum description there is a finite probability of finding the particle outside the well (classically forbidden region).

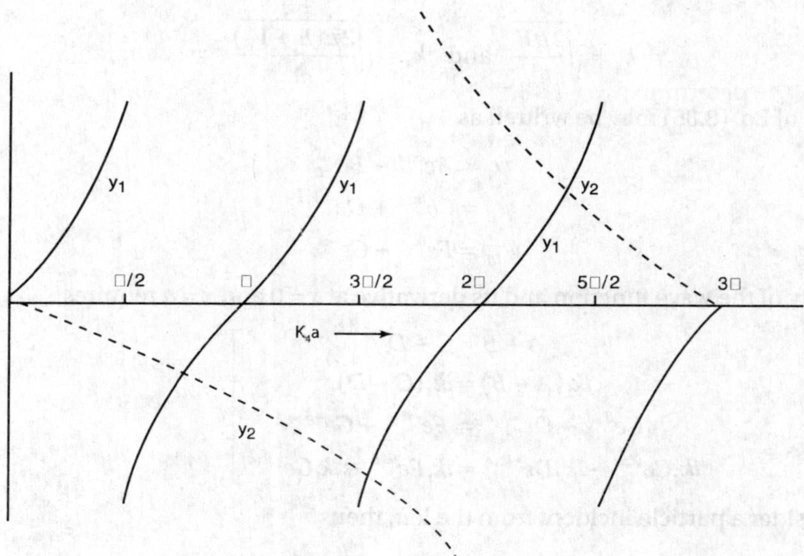

Fig. 8.17 Graphical determination of energy levels in a square well with $\beta = 9$

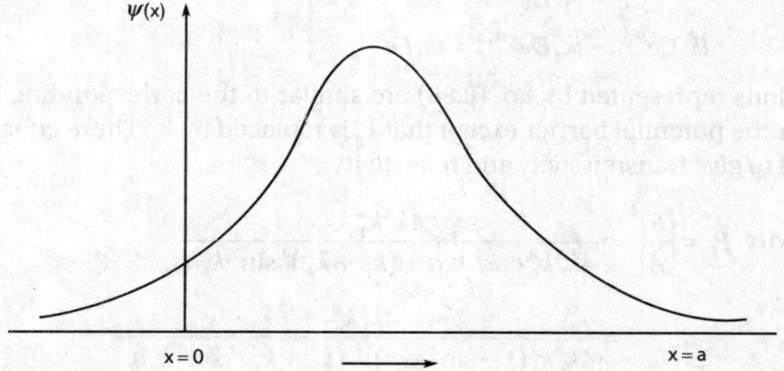

Fig. 8.18 Real part of $\psi(x)$ vs x

8.6.2 Case II ($E > 0$): Scattering

Let us discuss briefly what happens to a particle incident from a great distance from the left on a square well, when it is scattered. Schrodinger equation in three regions may be written as

$$\left.\begin{array}{lll} x < 0 & \dfrac{d^2\psi_1}{dx^2} + \dfrac{2\mu E}{\hbar^2}\psi_1 = 0 \quad \text{or} \quad \dfrac{d^2\psi_1}{dx^2} + k_1^2\psi_1 = 0 \\ 0 < x < a & \dfrac{d^2\psi_{II}}{dx^2} + \dfrac{2\mu(E + V_0)}{\hbar^2}\psi_{II} = 0 \quad \text{or} \quad \dfrac{d^2\psi_{II}}{dx^2} + k_4^2\psi_{II} = 0 \\ x > a & \dfrac{d^2\psi_{III}}{dx^2} + \dfrac{2\mu E}{\hbar^2}\psi_{III} = 0 \quad \text{or} \quad \dfrac{d^2\psi_{III}}{dx^2} + k_1^2\psi_{III} = 0 \end{array}\right] \tag{8.85}$$

where $$k_3 = \sqrt{\frac{2\mu E}{\hbar^2}} \quad \text{and} \quad k_4 = \sqrt{\frac{2\mu(E+V_0)}{\hbar^2}} \tag{8.86}$$

Solutions of Eq. (8.85) may be written as

$$\psi_I = Ae^{ik_1x} + Be^{-ik_1x}$$
$$\psi_{II} = Ce^{ik_4x} + De^{-ik_4x}$$
$$\psi_{III} = Fe^{ik_1x} + Ge^{-ik_1x}$$

Continuity of the wave function and its derivative at $x = 0$ and $x = a$ requires

$$\left.\begin{aligned} A + B &= C + D \\ ik_1(A - B) &= ik_4(C - D) \\ Ce^{ik_4a} + De^{-ik_4a} &= Fe^{-ik_1a} + Ge^{-ik_1a} \\ ik_4Ce^{ik_4a} - ik_4De^{-ik_4a} &= ik_1Fe^{ik_1a} - ik_1Ge^{-ik_1a} \end{aligned}\right] \tag{8.87}$$

If we consider a particle incident from the left, then

$$\left.\begin{aligned} G = 0 \quad \text{and} \quad A + B &= C + D \\ ik_1(A - B) &= ik_4(C - D) \\ Ce^{ik_4a} + De^{-ik_4a} &= Fe^{ik_1a} \\ ik_4Ce^{ik_4a} - ik_4De^{-ik_4a} &= ik_1Fe^{ik_1a} \end{aligned}\right] \tag{8.88}$$

Set of equations represented by Eq. (8.88) are similar to the corresponding Eq. (8.49) to Eq. (8.52) for the potential barrier except that k_2 is replaced by k_4. These equations can be easily solved to give transmissivity and reflectivity.

Transmissivity $$P_T = \left|\frac{F}{A}\right|^2 = \frac{4k_1^2k_4^2}{4k_1^2k_4^2\cos^2 k_4a + (k_1^2 + k_4^2)^2\sin^2 k_4a}$$

$$= \frac{4k_1^2k_4^2}{4k_1^2k_4^2(1 - \sin^2 k_4a) + (k_1^2 + k_4^2)^2\sin^2 k_4a}$$

$$= \frac{4k_1^2k_4^2}{4k_1^2k_4^2 + \{(k_1^2 + k_4^2)^2 - 4k_1^2k_4^2\}\sin^2 k_4a}$$

or $$P_T = \frac{4k_1^2k_4^2}{4k_1^2k_4^2 + (k_1^2 - k_4^2)^2\sin^2 k_4a} \tag{8.89}$$

Probability of Reflection of Reflectivity

$$P_R = \left|\frac{B}{A}\right|^2 = \frac{(k_1^2 - k_4^2)^2\sin^2 k_4a}{4k_1^2k_4^2\cos^2 k_4a + (k_1^2 + k_4^2)^2\sin^2 k_4a}$$

or $$P_R = \frac{(k_1^2 - k_4^2)^2\sin^2 k_4a}{4k_1^2k_4^2 + (k_1^2 - k_4^2)^2\sin^2 k_4a} \tag{8.90}$$

From Eq. (8.89) we see that maximum value of transmissivity is equal to 1, whenever k_4a is n integral multiple of π; the minimum value is obtained when k_4a is an odd integral multiple of $\pi/2$.

$$(P_T)_{\min} = \frac{4k_1^2k_4^2}{4k_1^2k_4^2 + (k_1^2 - k_4^2)^2} = \frac{4k_1^2k_4^2}{(k_1^2 + k_4^2)}$$

This minimum value of transmissivity [i.e., $(P_T)_{\min}$] increases with the increases in energy of the incident particle. The above result can be interpreted as follows. The wave which reflects from the surface xa arrives back at $x = 0$ with a phase shift of $n\pi$ then it interferes constructively with the next wave coming in and consequently the transmitted wave reinforced. Obviously, for certain wave lengths the transmittance is unity. One can represent the transmittance as a wave number as shown in Fig. (8.19).

The transmission coefficient vs E/V_0 has been plotted for a square well with $\beta = 9$ Fig. (8.20). The spikes on the left are at the positions of the discrete-energy levels.

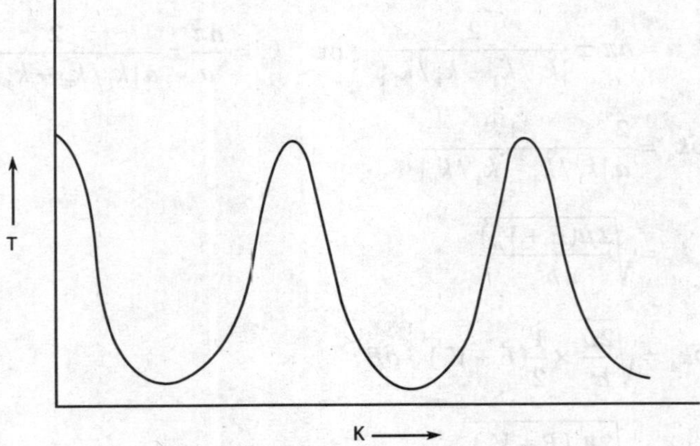

Fig. 8.19 Real part of $\psi(x)$ vs x

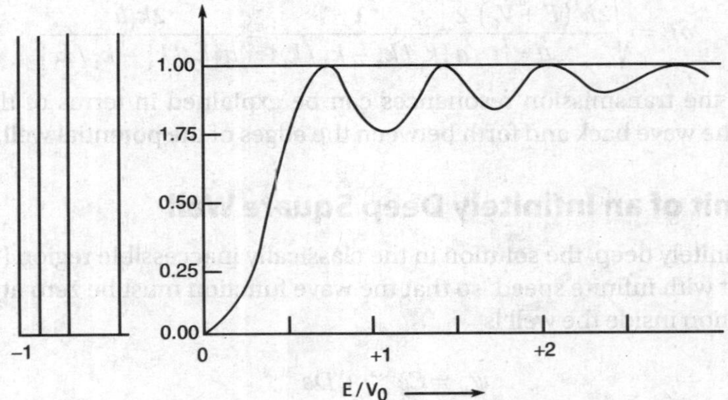

Fig. 8.20 Transmissivity vs E/V_0 for a square well with $\beta = 9$. The spikes on the left are at the positions of discrete-energy levels

Width of Transmission Resonance Peaks: To compute the width of a resonance peak, we assume that $\frac{k_4}{k_1}$ is large, so that the peaks are sharp. Position of the peaks is given by $k_4 a = n\pi$. The width of a peak is given by the energy difference between two energy values (on the same peak) where transmission coefficient is equal to 1/2.

$$\frac{4k_1^2k_4^2}{4k_1^2k_4^2+(k_1^2-k_4^2)^2\sin^2 k_4a}=\frac{1}{2}$$

or $$(k_1^2-k_4^2)^2\sin^2 k_4a = 4k_1^2k_4^2$$

or $$\sin^2 k_4a = \frac{2}{(k_1/k_4-k_4/k_1)^2} \quad \text{or} \quad \sin k_4a = \pm\frac{2}{|k_1/k_4-k_4/k_1|} \tag{8.91}$$

If the denominator on the right of Eq. (8.91) is large, then k_4a will differ only slightly from $n\pi$ at this point, and we can write

$$k_4a \simeq n\pi \pm \frac{2}{|k_1/k_4-k_4/k_1|} \quad \text{or} \quad k_4 \simeq \frac{n\pi}{a} \pm \frac{2}{a|k_1/k_4-k_4/k_1|}$$

or $$\delta k_4 \simeq \frac{2}{a}\frac{1}{|k_1/k_4-k_4/k_1|}$$

Also $$k_4 = \sqrt{\frac{2\mu(E+V_0)}{\hbar^2}}$$

$$\therefore \quad \delta k_4 = \sqrt{\frac{2\mu}{\hbar^2}}\times\frac{1}{2}(E+V_0)^{-\frac{1}{2}}\delta E$$

$$\delta E = \sqrt{\frac{2\hbar^2(E+V_0)}{\mu}}\delta k_4$$

or $$\delta E = \sqrt{\frac{2\hbar^2(E+V_0)}{\mu}}\frac{2}{a|k_1/k_4-k_4/k_1|} = \frac{2k_4\hbar}{a|k_1/k_4-k_4/k_1|} \tag{8.92}$$

The width of the transmission resonances can be explained in terms of the process of reflection of the wave back and forth between the edges of the potential well.

8.6.3 Limit of an Infinitely Deep Square Well

If a well is infinitely deep, the solution in the classically inaccessible region (i.e., $x<0$ and $x>a$) dies out with infinite speed, so that the wave function must be zero at each edge of the well. Solution inside the well is

$$\psi_{II} = Ce^{ik_4x}+De^{-ik_4x}$$

Boundary conditions are $\quad [\psi_{II}]_{x=a}=0$ and $[\psi_{III}]_{x=a}=0$

I Condition gives, $C + D = 0$ i.e., $D = -C$

and $\therefore$ $\psi_{II} = Ce^{ik_4x} + (-C)e^{ik_4x} = 2iC\sin k_4x$

or $\psi_{II} = C'\sin k_4x$

II Condition requires $\sin k_4 a = 0$

or $k_4 a = n\pi$ where n is an integer $\therefore\ k_4 = \dfrac{n\pi}{a}$ (8.93)

and $\psi_{III} = C'\sin k_4 x = C'\sin\left(\dfrac{n\pi x}{a}\right)$ (8.94)

From Eq. (8.92), we get an expression for energy levels $\left\{\sqrt{\dfrac{2\mu}{\hbar^2}(V_0 + E)}\right\}a = n\pi$

or $$\frac{2\mu}{\hbar^2}(V_0 + E) = \frac{n^2\pi^2}{a^2} \quad \text{or} \quad (V_0 + E) = \frac{n^2\pi^2\hbar^2}{2\mu a^2} \tag{8.95}$$

or $$V_0 - |E| = \frac{n^2\pi^2\hbar^2}{2\mu a^2}$$

Since total energy E of the particle is negative in this case, $E + V_0$, the distance in energy from the bottom of the well is just the kinetic energy of the particle in the well. (Since $k_4 \to \infty =$ as $V_0 \to -\infty$ and $t = -\infty$ wave function itself vanishes at $x = 0$ and $x = a$. Taking into account a shift V_0 of the zero of energy and making the identification $a = L$, the energy levels given by Eq. (8.95) coincides with the energy levels given by Eq. (8.33).

8.7 PERIODIC POTENTIAL (KRONIG-PENNEY MODEL OF SOLIDS)

In a periodic potential, we assume that potential hills and valleys follow each other in periodic succession indefinitely in both directions. Mathematically the potential shown in Fig. (8.21) may be expressed by the equation

$$V(x) = \begin{cases} 0 & nl < x < nl + a \\ V_0 & nl + a < x < (n+1)l \end{cases} \tag{8.96}$$
$$n = 0, \pm 1, \pm 2, \ldots$$

where b is the width of the barriers, a is the width of the valleys and $l = n + b$ is the period of the potential. Solutions can be obtained by solving one-dimensional Schrodinger equation

$$\frac{d^2\psi}{dx^2} + \frac{2\mu}{\hbar^2}[E - V(x)]\psi = 0$$

In the valleys $V(x) = 0$, and writing $\sqrt{\dfrac{2\mu E}{\hbar}} = k_1$, we can write

$$\psi_I(x) = Ae^{ik_1x} + Be^{-ik_1x}$$

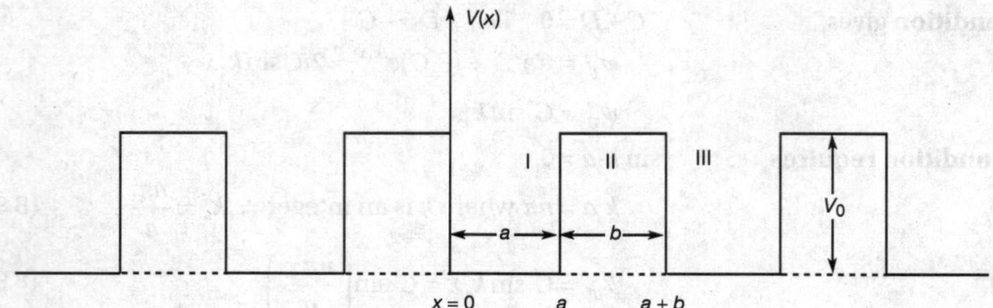

Fig. 8.21

In the barrier region $V(x) = V_0$ and writing $\sqrt{\frac{2\mu(V_0 - E)}{\hbar^2}}k$, one obtains

$$\psi_{II}(x) = Ce^{kx} + De^{-ikx}$$

Since the points x and $x + nl$ are symmetrically equivalent (where n is an integer), we expect that

$$|\psi(x)|^2 = |\psi(x + nl)|^2 \tag{8.97}$$

Therefore the most general form for $\psi(x)$ may be written as

$$\psi(x) = \exp(i\alpha x)U(x)$$

where α is a real number and $U(x)$ is any periodic function such that

$$U(x + l) = U(x)$$

In terms of U the solution may be written as:

$$U_I = Ae^{i(k_1 - \alpha)x} + Be^{-i(k_1 + \alpha)x}$$
$$U_{II} = Ce^{(k - i\alpha)x} + De^{-(k - i\alpha)x} \tag{8.98}$$

The requirement of the continuity of the wave function and its first derivative at $x = 0$ and $x = -b$ demands that the wave function $U(x)$ must also satisfy these conditions, since $e^{i\alpha x}$ is a well-behaved solution. Applying these boundary conditions and recalling that $U(x)$ has the periodicity of the lattice, i.e., $U(a) = U(-b)$, we obtain

$$[U_I]_{x=0} = [U_{II}]_{x=0}$$
$$\left[\frac{dU_I}{dx}\right]_{x=0} = \left[\frac{dU_{II}}{dx}\right]_{x=0}$$
$$[U_I]_{x=a} = [U_{II}]_{x=-b}$$
$$\left[\frac{dU_I}{dx}\right]_{x=a} = \left[\frac{dU_{II}}{dx}\right]_{x=-b}$$

giving
$$\left.\begin{aligned} &A+B=C+D\\ &i(k_1-\alpha)A-i(k_1+\alpha)B=(k-i\alpha)C-(k+i\alpha)D\\ &Ae^{i(k_1-\alpha)a}+Be^{-(k_1\alpha)a}=Ce^{-(k-i\alpha)b}+De^{(k+i\alpha)b}\\ &i(k_1-\alpha)Ae^{i(k_1-\alpha)a}-i(k_1+\alpha)Be^{-(k_1+\alpha)a}\\ &=(k-i\alpha)Ce^{-(k+\alpha)b}-(k+i\alpha)De^{(k+i\alpha)b}\end{aligned}\right] \tag{8.99}$$

The coefficient A, B, C and D can be determined from the equations represented by Eq. (8.99). Eq. (8.99) is a set of four linear, homogeneous equations among four unknowns A, B, C, D and will possess a unique solution only if the determinant formed by the coefficients of A, B, C, D vanishes.

$$\text{Determinant} = \begin{vmatrix} 1 & 1 & -1 & -1 \\ i(k_1-\alpha) & -i(k_1+\alpha) & -(k-i\alpha) & (k+i\alpha) \\ e^{i(k_1-\alpha)a} & e^{-i(k_1-\alpha)a} & -e^{-i(k-i\alpha)b} & -e^{(k+i\alpha)b} \\ i(k_1-\alpha)e^{i(k_1-\alpha)a} & -i(k_1+\alpha)e^{-i(k_1+\alpha)a} & -(k-i\alpha)e^{i(k-i\alpha)b} & (k-i\alpha)e^{i(k+i\alpha)b} \end{vmatrix} = 0$$

After a little algebra, we get a relationship

$$\left(\frac{k^2-k_1^2}{2kk_1}\right)\sin hkb\sin k_1a+\cos hkb\cos k_1a=\cos\alpha(a+b) \tag{8.100}$$

The permissible values of E are restricted to those for which Eq. (8.100) holds. A necessary condition for E is that the left hand side $f(E)$ which by virtue of k and k_1 is a function of E, be restricted to the range $-1 \le f(E) \le +1$ since $\cos\alpha(a+b)=\cos\alpha l$ can have its values between -1 and $+1$, regardless of the value of αl. We have

$f(E)=\cos k_1a\cos hkb+\left(\dfrac{k_2^2-k_1^2}{2kk_1}\right)\sin hkb\sin k_1a$ $f(E)$ vs E/V_0 is plotted in Fig. (8.22).

The darkened bands of energy are the range of which $|f(E)| \le 1$. The only permissible values of energy lies in these ranges.

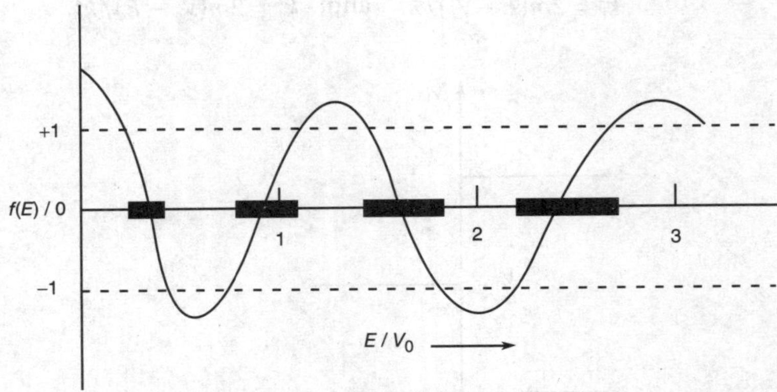

Fig. 8.22 Solid line showing allowed values of energy

In the limiting case, when $V_0 \to \alpha$, the wells (or valleys) become well separated by an infinite barrier and

$$k = \sqrt{\frac{2\mu(V_0 - E)}{\hbar^2}} \to \sqrt{\frac{2\mu V_0}{\hbar^2}} \quad \text{and} \quad k >> k_1$$

$\cos hkb = e^{kb}/2$ and $\sin hkb \simeq e^{kb}/2$

$$f(E) = \left[(\cos k_1 a)\frac{e^{kb}}{2}\right] + \left[\frac{k}{2k_1}\frac{e^{kb}}{2}\right]\sin k_1 a = \cos\alpha(a+b)$$

or
$$\left(\cos k_1 a + \frac{k}{2k_1}\sin k_1 a\right)\frac{\exp(kb)}{2} = \cos\alpha(a+b)$$

since exp (kb) is very large, the equation can be satisfied only if the quantity within the brackets is nearly equal to zero. In this limiting case

$$\cos k_1 a + \frac{k}{2k_1}\sin k_1 a \approx 0 \quad \text{or} \quad \cot k_1 a \approx \frac{-k}{2k_1} \approx -\infty \quad \text{or} \quad k_1 a = n\pi$$

Thus we obtain the same condition as that of a particle confined in a box.

Example 8.2: Consider a particle incident on a potential step as shown in Fig. (8.23). Obtain the solution of Schrodinger equation for (i) $V_2 > E > V_1$ (ii) $E > V_2$.

Solution
$$\frac{d^2\psi}{dx^2} + \frac{2m}{\hbar^2}(E - V_1)\psi = 0 \quad x > 0$$

$$\frac{d^2\psi}{dx^2} + \frac{2m}{\hbar^2}(E - V_2)\psi = 0 \quad x < 0$$

For $V_2 > E > V_1$, we have

$$\psi(x) = \begin{cases} A\exp(ik_1 x) + B\exp(-ik_1 x) & x > 0 \\ C\exp(k_2 x) + D\exp(-k_2 x) & x < 0 \end{cases} \qquad \text{(i)}$$

where
$$k_1^2 = 2m(E - V_1)/\hbar^2 \quad \text{and} \quad k_2^2 = 2m(V_2 - E)/\hbar^2$$

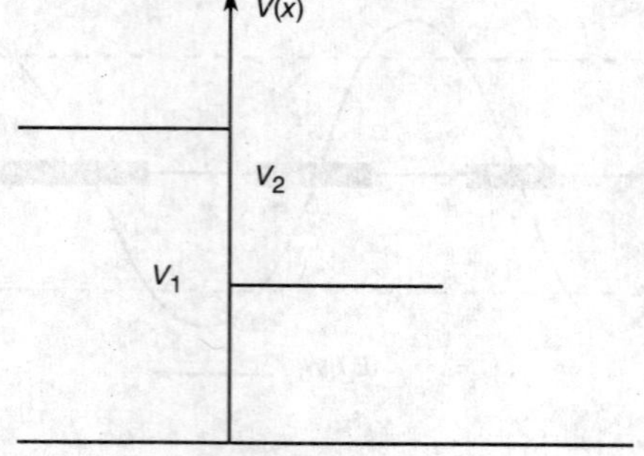

Fig. 8.23

From boundary conditions, at $x = \pm\infty$, $D = 0$ and from the continuity of ψ and $\partial\psi/\partial x$ at $x = 0$, one obtains

$$A/B = (ik_1 + k_2)/(ik_1 - k_2) = \exp(i\alpha)$$

where α is real and $\tan\alpha = 2k_1k_2/(k_2^2 - k_1^2)$ and $C/B = 1 + \exp(i\alpha)$

Now, $$\psi(x) = \begin{cases} B\exp[i(k_1x + \alpha) + A\exp(-ik_1x) & x > 0 \\ B(1 + \exp i\alpha)\exp(k_2x) & x < 0 \end{cases} \quad \text{(ii)}$$

We can see that the functions in the two regions can be joined smoothly at $x = 0$ for all values of E between V_1 and V_2 and obviously the energy spectrum is continuous. The solution given by Eq. (ii) represents a plane wave of amplitude B incident from the right and a reflected wave with an amplitude A such that $|A|^2 = |B|^2$, i.e., the reflection is total. The transmitted wave is rapidly attenuated and we note that there is no permanent penetration. From Eq. (ii), one finds the current density $S = 0$ for all x. Now, for $x > 0$, the incident and reflected currents cancel while for $x < 0$ the current due to exp (k_2x) is 0. For $E > V_2$, let us assume that the particle to be incident from the right. We have

$$\psi(x) = \begin{cases} A\exp(ik_1x) + B\exp(-ik_1x) & x > 0 \\ D\exp(-k_2x) & x < 0 \end{cases}$$

where $k_2^2 = 2m(E - V_2)/\hbar^2$. Applying the continuity of ψ and $\partial\psi/\partial x$ at $x = 0$, one obtains

$$A/B = (k_1 - k_2)/(k_1 + k_2) \quad \text{and} \quad D/B = 2k_1/(k_1 + k_2)$$

One can easily find the current density as

$$S = \begin{cases} \hbar k_1(|A|^2 - |B|^2/m & x > 0 \\ -\hbar k_2|D|^2/m & x < 0 \end{cases}$$

Since dS/dx has to be zero and hence $|A|^2/|B|^2 = 1 - k_2\,|D|^2k_1\,|B|^2$ and now $R = |A|^2/|B|^2$ and $T = k_2\,|D|^2/k_1 + |B|^2$ with $R + T = 1$. Thus we obtain $R = (k_1 - k_2)^2/(k_1^2 + k_2^2)$ and $T = \dfrac{4k_1k_2}{(k_1 + k_2)^2}$.

Example 8.3: What is the energy spectrum of a particle in a periodically changing potential field. The field is given by the relations

$$V(x) = 0 \text{ for } nl \le x \le nl + a(n = 0, \pm 1, \pm 2, \ldots) = V_0 \text{ for } nl - b \le x \le nl$$

The quantity $l = a + b$ is the period of the potential.

Solution Let us first link the solutions $\psi(x)$ and $\psi(x + l)$. The Schrodinger equations for these cases are

$$\frac{-\hbar^2}{2m}\frac{d^2\psi(x)}{dx^2} + V(x)\psi(x) = E\psi(x) \quad \text{(i)}$$

$$\frac{-\hbar^2}{2m}\frac{d^2\psi(x+l)}{d(x+l)^2}+V(x+l)\psi(x+l)=E\psi(x+l) \qquad \text{(ii)}$$

We see that because $V(x + l) = V(x)$ and $\dfrac{d^2}{d(x+l)^2}=\dfrac{d^2}{dx^2}$, both functions correspond to one energy level, E. Considering E a simple eigen value, we find that these functions differ only in a constant factor, i.e., $\psi(x+l)=\rho\psi(x)$. In the general case, we can write

$$\psi(x+nl)=\rho^n\psi(x) \qquad (n=0,\pm 1,\pm 2,\ldots) \qquad \text{(iii)}$$

It follows then from the requirement of the finiteness of $\psi(x)$ that $|\rho| = 1$, i.e., $\rho = \exp(ikl)$, where k is any real number. Thus, we have

$$\psi(x+l)=\exp(ikl)\psi(x) \qquad \text{(iv)}$$

If we solve the Schrodinger equation for $-b \le x \le 0$ (region I), for $0 \le x \le a$ (region *II*), and for $a \le x \le b$ (region *III*), we obviously will have a solution for all x's.

First region: $V = V_0$. The equation takes the form

$$\frac{-\hbar^2}{2m}\frac{d^2\psi_1}{dx^2}+V_0\psi_1=E\psi_1$$

Writing $\lambda^2=\dfrac{2m(V_0-E)}{\hbar^2}$, we can write $\psi_1 = C_1\exp(\lambda x)+C_2\exp(-\lambda x)$

Second region: $V = 0$. Here the equation is

$$\frac{-\hbar^2}{2m}\frac{d^2\psi_{II}}{dx^2}=E\psi_{II}$$

and the solution is $\psi_{II} = C_3\exp(\mathrm{i}\alpha x)+C_4\exp(-\mathrm{i}\alpha x)$ where $\alpha^2 = 2mE/\hbar^2$.

Third region: $V = V_0$. Obviously, the equation is same as in the first region, i.e.,

$$-\frac{\hbar^2}{2m}\frac{d\psi_{III}}{dx^2}+V_0\psi_{III}=E\psi_{III}$$

and its solution is $\psi_{III} = C_5\exp(\lambda x)+C_6\exp(-\lambda x)$

If x lies in the I region, $x + l$ lies in the III region, and according to Eq. (iv), the solutions are linked by the condition

$$\psi_{III}(x+l)=\exp(ikl)\psi_I(x)$$

Hence $\qquad C_5 = C_1\exp(ikl-\lambda l)$ and $C_6 = C_2\exp(ikl+\lambda l)$

i.e., $\qquad \psi_{III}\ \exp(ikl)[C_1\exp\{\lambda(x-l)+C_2\exp\{-\lambda(x-l)\}$

We must remember that solutions ψ_I, ψ_{II} and ψ_{III} must be continuous with their first derivatives in passing from the first region to the second ($x = 0$) and from the second to the third ($x = a$). One obtains the following conditions:

$$\psi_I(0) = \psi_{II}(0) \qquad \text{and } C_1 + C_2 = C_3 + C_4$$

$$\left.\frac{d\psi_I}{dx}\right|_{x=0} = \left.\frac{d\psi_{II}}{dx}\right|_{x=0} \qquad \text{and } \lambda(C_1 - C_2) = i\alpha(C_3 - C_4)$$

$$\psi_{II}(a) = \psi_{III}(a) \qquad \text{and } C_3 \exp(i\alpha a) + C_4 \exp(-i\alpha a)$$

$$= \exp(ikl)[C_1 \exp(-\lambda b + C_2 \exp(\lambda b)]$$

$$\left.\frac{d\psi_{II}}{dx}\right|_{x=0} = \left.\frac{d\psi_{III}}{dx}\right|_{x=0} \qquad \text{and } i\alpha[C_3 \exp(i\alpha a) - C_4 \exp(-i\alpha a)]$$

$$= \lambda \exp(ikl)[C_1 \exp(-\lambda b) - C_2 \exp(\lambda b)]$$

We have a system of four homogeneous linear equations for the unknowns C_1, C_2, C_3, and C_4. For there to be a nontrivial solution, the system determinant, as we know, must be zero:

$$\begin{vmatrix} 1 & 1 & -1 & -1 \\ \lambda & -\lambda & -i\alpha & i\alpha \\ -\exp(ikl - \lambda b) & -\exp(ikl - \lambda b) & \exp(i\alpha a) & \exp(-i\alpha a) \\ -\lambda \exp(ikl - \lambda b) & \lambda \exp(ikl - \lambda b) & i\alpha \exp(i\alpha a) & -i\alpha \exp(-i\alpha a) \end{vmatrix} = 0$$

This brings us to an equation that determines the energy of the particle in the periodic field:

$$f(E) = \cosh \lambda b \cos \alpha a + \frac{\lambda^2 - \alpha^2}{2\lambda\alpha} \sinh \lambda b \sin \alpha a = \cos kl \qquad \text{(v)}$$

On the LHS, the energy enters through α and l. It is obvious that for this equation to be valid, $|f(E)|$ must not exceed unity. We can see that at $\alpha a = n\pi$, this is not so, since then $f(E) = \pm \cosh \lambda b$ and $|f(E)| > 1$. Hence the energies $E = \dfrac{\hbar^2 n^2 \pi^2}{2ma^2}$ prove to be forbidden for a particle in a periodic field, e.g., an electron in a crystal.

Example 8.4: A free particle is in one-dimensional motion, subject to boundary conditions called **Born-Von-Karman** conditions

$$\psi(a) = \psi(b); \left.\frac{d\psi}{dx}\right|_{x=a} = \left.\frac{d\psi}{dx}\right|_{x=b}$$

Determine (i) stationary states of the system (ii) $< p >$ when the particle is in the nth eigen state and (iii) for a stationary state, show that $\dfrac{d}{dx} < x > = 0$

Solution From the given boundary conditions: $\exp(ika) = \exp(ikb)$

$$k(b - a) = 2\pi n (n = 0, 1, 2, \ldots)$$

Therefore, the eigen functions satisfying the boundary conditions are

(i) $\psi_n(x) = A \exp\{(-i2\pi nx)/(a - b)\}$

If it is required that $\int_a^b |\psi_n(x)|^2 = 1$, then $A = 1/\sqrt{(b-a)}$

(ii) $$<p_x> = \int_n^b \psi_n(x)\frac{\hbar}{i}\frac{\partial\psi_n}{dx}dx = \frac{nb}{b-a}$$

Obviously, with states for $n \neq 0$, $<p_x> \neq 0$

(iii) $$\frac{d}{dt}<x> = \frac{1}{i\hbar}<[x,H]> 0$$

From Ehrenfest's theorem $\frac{d}{dt}<x> = <p_x> = m$

Obviously, we note that there is a contradiction between (ii) and (iii).

Example 8.5: **A beam of monoenergetic electrons strikes the surface of a metal at normal incidence. Calcualte the reflection probability of these electrons if $E = 0.1$ eV and $V_0 = 8$ eV.**

Solution Since the electrons with energy 0.1 eV encounters a potential drop −8 eV at the metal surface, they will all enter the metal according to classical mechanics, and, because of the law of conservation of energy they will acquire a final momentum $p_1 = \frac{\sqrt{2m(E+V_0)}}{\hbar}$ after doing so. From potential step expressions, we have

$$R = \frac{(p_2 - p_1)^2}{(p_2 + p_1)^2} \quad \text{and} \quad T = \frac{4p_1p_2}{(p_1 + p_2)^2}$$

Thus for $E = 0.1$ eV, $V_0 = 8$ eV, we have

$$R = \left[\frac{\sqrt{E} - \sqrt{(E+V_0)}}{\sqrt{E} + \sqrt{(E+V_0)}}\right]^2 = \left[\frac{1 - \sqrt{(1+V_0/E)}}{1 + \sqrt{(1+V_0/E)}}\right]^2 = \left(\frac{8}{10}\right)^2 = 0.64.$$

Example 8.6: **Find the height of the potential barrier for α-particles emitted from Radon ($_{86}Rn^{222}$) assuming that the effective nuclear radius is given by $r_0 = 1.5 \times 10^{-15} A^{1/3}$ m.**

Solution $$V_0 = \frac{2(Z-2)e^2}{4\pi\varepsilon_0 r_0} = \frac{2(86-2)(1.6\times10^{-19})^2}{4\times3.14\times8.85\times10^{-12}\times1.5\times10^{-15}\times(222)^{1/3}}\text{J}$$

$$= 0.426\times10^{-11}\ \text{J} = 26.62\,\text{MeV}.$$

Example 8.7: **Find the width of the potential barrier for an α-particle emitted with kinetic energy 5.5 MeV from $_{86}Rn^{222}$.**

Solution The width of the potential barrier due to the nucleus is given by, $a = r_1 - r_0$, where r_0 is the effective nuclear radius, and r_1 is the radius from the centre of the nucleus, at which the potential energy of the α-particle is equal to the kinetic energy of the emitted α-particle. Here $r_0 = 1.5 \times 10^{-15} A^{1/3} = 1.5 \times 10^{-15} \times (222)^{1/3} = 9.08 \times 10^{-15}$ m.

r_1 is given by $$E = \frac{2(Z-2)e^2}{4\pi\varepsilon_0 r_1}$$

$$\therefore \qquad r_1 = \frac{2(Z-2)e^2}{4\pi\varepsilon_0 E} = \frac{2\times(86-2)\times(1.6\times10^{-19})^2}{4\pi\times8.85\times10^{-12}\times5.5\times1.6\times10^{-13}} = \frac{2\times84\times(1.6)^2\times10^{-13}}{4\pi\times8.85\times5.5\times1.6}$$

$$= 34.97\times10^{-15}\text{ m}$$

Now $\qquad a = r_1 - r_0 = (43.97\times10^{-15} - 9.08\times10^{-15})\text{m} = 43.98\times10^{-15}\text{ m.}$

Example 8.8: **A stream of electrons, each, energy $E = 3$ eV is incident on a potential barrier of height $V_0 = 4$ eV. The width of the barrier is 20Å. Calculate the percentage transmission of the beam through this barrier.**

Solution The probability of transmission through a potential barrier is given by

$$T = \frac{16E}{V_0}\left(1-\frac{E}{V_0}\right)e^{-2\beta a}$$

where $\qquad 2\beta a = 2a\sqrt{\frac{2m(V_0-E)}{\hbar}} = \frac{2a}{\hbar}\sqrt{2m(V_0-E)}$

Given $\qquad E = 3\text{eV} = 3\times1.6\times10^{-19}\text{ J},\ V_0 = 4\text{eV} = 4\times1.6\times10^{-19}\text{ J}$

$$a = 20A = 20\times10^{-10} = 2\times10^{-19}\text{ m}$$

$$\therefore \qquad 2\beta a = \frac{2a}{\hbar}\sqrt{2m(V_0-E)} = \frac{2\times2\times10^{-9}}{1.054\times10^{-34}}\sqrt{2\times9.11\times10^{-31}\times1.6\times10^{-19}}$$

$$= 20.49$$

$$T = \frac{16E}{V_0}\left(1-\frac{E}{V_0}\right)e^{-2\beta a} = \frac{16\times3}{4}\left(1-\frac{3}{4}\right)e^{-20.49} = 12\times\frac{1}{4}\times\frac{1}{(2.718)^{20.49}}$$

$$= 3.797\times10^{-9}$$

$\therefore \qquad$ Percentage transmission $= 3.797\times10^{-7}\%$

Example 8.9: **An α-particle having energy 10 MeV approaches a potential barrier of height equal to 30 MeV. Determine the width of the potential barrier if the transmission coefficient is 2×10^{-3}. Given mass of an α-particle $= 4\times1.67\times10^{-27}$ kg and $\hbar = 1.054\times10^{-34}$ J-s.**

Solution $\qquad T = \frac{16E}{V_0}\left(1-\frac{E}{V_0}\right)e^{-2\beta a}$ and $e^{2\beta a} = \frac{16E}{TV_0}\left(1-\frac{E}{V_0}\right)$

$$\therefore \qquad 2\beta a = 2.303\ \log_{10}\frac{16E}{TV_0}\left(1-\frac{E}{V_0}\right)$$

or $$a = \frac{1}{2\beta} \times 2.303 \log_{10} \frac{16E}{TV_0}\left(1 - \frac{E}{V_0}\right) \text{ and } \beta = \frac{1}{\hbar}\sqrt{2m(E - V_0)}$$

$$= \frac{1}{1.054 \times 10^{-34}} \times \sqrt{2 \times 4 \times 1.67 \times 10^{-27} \times (30 - 10) \times 10^6 \times 1.6 \times 10^{-19}}$$

$$= \frac{10^{14}}{1.054 \times 10^{-34}}\sqrt{2 \times 4 \times 1.67 \times 20 \times 1.6} = \frac{20.68 \times 10^{14}}{1.054} = 19.62 \times 10^{14}$$

Hence $$a = \frac{10^{-14}}{2 \times 19.62} \times 2.303 \log_{10}\left[\frac{16 \times 10}{2 \times 10^{-3} \times 30}\left(1 - \frac{10}{30}\right)\right]$$

$$= \frac{10^{-14}}{2 \times 19.62} 2.303 \log_{10} \frac{16}{9} \times 10^3 = \frac{10^{-14} \times 2.303 \times 3.25}{39.24} = 1.91 \times 10^{-15} \text{m.}$$

Example 8.10: **A particle of total energy E enters the barrier $V = V(x)$ as shown in Fig. (8.24) at the point $x - x_1$ and leaves it at the point $x = x_2$. Assuming the potential energy curve $V(x)$ to be sufficiently smooth, calculate the approximate value of the transmission coefficient T.**

Solution Let us divide the interval (x_1, x_2) into intervals of length Δx_i large compared with the relative penetration depth $d_i = \sqrt{\frac{\hbar^2}{8m(V(x_i) - E)}}$, of a particle in the rectangular harrier so obtained. Now, the transmission coefficient T_i for the ith rectangular barrier is given by

$$T_i = \exp\left[-2\sqrt{\frac{2m(V(x_i) - E)}{\hbar^2}}\Delta x_i\right]$$

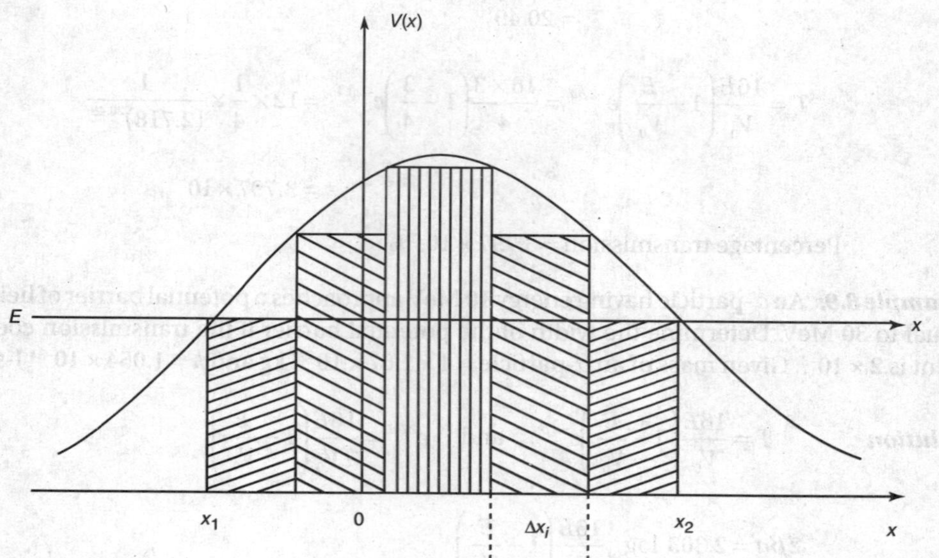

Fig. 8.24

Hence the transmission coefficient through the whole barrier will be given by the product

$$T = TT_iT_i = \exp\left[-2\sum_i \sqrt{\frac{2m(V(x_i)-E)}{\hbar^2}}\Delta x_i\right]$$

Increasing the subdivision of the interval x_1, x_2 as far as is allowed by the condition $d_i << \Delta x$, the approximate formula,

$$T = \exp\left[-2\int_{x_1}^{x_2} \sqrt{\frac{2m(V(x)-E)}{\hbar^2}}dx\right]$$

is obtained. The above expression is called as **"Gamow's penetrability factor"**.

Example 8.11: Find the decay constant λ if four α-decay. The transmission of the potential barrier D and λ are related $\lambda = nD$ and if the potential is of the form $V = -V_0$ for $r < r_0$ and for $r \geq r_0$ the α-particle interact with the nucleus via the Coulomb law. The electric charge of the nucleus is Ze, and $r_0 << \frac{2Ze^2}{E}$. The factor $n\alpha v_i/r_0$, where v_i is the particle's velocity inside the nucleus and r_0 is the radius of the nucleus, is proportional to the number of collisions with the "wall" of the nucleus per unit time.

Solution The α-particle in a nucleus lies in a deep potential well. Assuming approximately that $V = -V_0$ for $r \leq r_0$; r_0 characterizes the range of nuclear forces. For $r \geq r_0$ (outside the nucleus), $V = \frac{2Ze^2}{r}$. The transmittance of the particle through a barrier is limited by a straight line at $r = r_0$ and by a hyperbola $V = \frac{2Ze^2}{r}$, for $r > r_0$ is determined as

$$D = D_0 \exp\left(-\frac{2\sqrt{2m}}{\hbar}\int_{r_0}^{r_2}\sqrt{\frac{2Ze^2}{r} - E}.dr\right),$$

where E is the energy of the particle falling on the barrier and r_0 and r_2 are the points of retrogression, at which $V = E$, i.e., $r_2 = 2Ze^2/E$.

To evaluate the integral in (i), let us introduce $\cos^2 u = r/r_2$. Obviously, at $r = r_2$, we have $u = 0$. Denoting $\cos^2 u_0 = r_0/r_2$, one can write

$$I = \int_0^{u_0}\sqrt{\frac{1}{\cos^2 u} - 1}\;\; 2\sin u \cos u \, du\sqrt{E}\frac{2Ze^2}{E} = \frac{2Ze^2}{\sqrt{E}}\left[u_0 - \frac{\sin 2u_0}{2}\right]$$

Assuming $\sqrt{\frac{r_0}{r_2}} << 1$, one can expand u_0 in a series: $u_0 = \text{arc}\cos\sqrt{\frac{r_0}{r_2}} = \frac{\pi}{2} - \sqrt{\frac{r_0}{r_2}}$

Then $\sin 2u_0 = 2\sqrt{\frac{r_0}{r_2}}$

Hence $I \approx \frac{2Ze^2}{\sqrt{E}}\left[\frac{\pi}{2} - 2\sqrt{\frac{r_0}{r_2}}\right]$ and $D \approx D_0 \exp\left[-\frac{4}{\hbar}\left(\frac{\pi Ze^2}{v_\infty} + \sqrt{2mZe^2 r_0}\right)\right]$

where $v_\infty = \sqrt{2E/m}$ is the velocity of the emitted α-particle measured far from the nucleus (where $v = 0$), and $\lambda = nD_0 \exp\left[-\frac{4}{\hbar}\left(\frac{\lambda Ze^2}{v_\infty} + \sqrt{2mZe^2 r_0}\right)\right]$

Example 8.12: An electron is trapped in a one-dimensional region of length 1×10^{-10} m (a typical atomic diameter). (a) How much energy must be supplied to excite the electron from the ground state to the first excited state? (b) In the ground state, what is the probability of finding the electron in the region from $x = 0.09 \times 10^{-10}$ m to 0.110×10^{-10} m? (c) In the first excited state, what is the probability of finding the electron between $x = 0$ and $x = 0.25 \times 10^{-10}$ m?

Solution

(a) $$E_1 = \frac{\hbar^2\pi^2}{2mL^2} = \frac{(1.05\times10^{-34})^2\times(3.14)^2}{2\times(9.1\times10^{-31})(1\times10^{-10})^2} = 6\times10^{-18}\,\text{J} = 37\,\text{eV}$$

with $E_2 = 4E_1$, the energy difference ΔE is

$$\Delta E = E_2 - E_1 = 4E_1 - E_1 = 3E_1 = 3\times 37 = 111\,\text{eV}$$

(b) $$\text{Probability} = \int_{x_1}^{x_2} |\psi|^2\,dx = \frac{2}{L}\int_{x_1}^{x_2} \sin^2\frac{\pi x}{L}\,dx = \left(\frac{x}{L} - \frac{1}{2\pi}\sin\frac{2\pi x}{L}\right)\Bigg|_{x_1}^{x_2}$$

For $x_1 = 0.09\times10^{-10}$ m and $x_2 = 0.11\times10^{-10}$ m, one obtains, Probability $= 0.0038 = 0.38\%$

(c) $$\text{Probability} = \int_{x_1}^{x_2}\left(\frac{2}{L}\right)\sin^2\frac{2\pi x}{L}\,dx = \left(\frac{x}{L} - \frac{1}{4\pi}\sin\frac{4\pi x}{L}\right)\Bigg|_{x_1}^{x_2} = 0.25$$

Example 8.13: Show that the energy and wave function of a particle in a square well of finite depth V_0 reduces to the energy and wave function of a square well with rigid walls in the limit $V_0 \to \infty$.

Solution For a well of finite depth, we obtain $ka \tan ka = \alpha a$ or $\tan(\alpha a) = \alpha/k$

where $$k^2 = \frac{2mE}{\hbar^2} \quad\text{and}\quad \alpha = \frac{2m}{\hbar^2}(V_0 - E)$$

$$\therefore \qquad \tan ka = \sqrt{\frac{V_0 - E}{E}}$$

when $V_0 \to \infty$, $\tan(ka) \to ka$. Then, one finds $ka = \frac{n\pi}{2}$ or $k^2a^2 = \frac{n^2\pi^2}{4}$

or $$E_n = \frac{\pi^2\hbar^2 n^2}{8ma^2}$$

The wave function for square well potential with finite walls, when $V_0 \to \infty$, $\alpha \to \infty$ reduces to wave function of a particle in a square well with rigid walls.

Example 8.14: **A 30 eV electron is incident on a square barrier of height 40 eV.**

(i) What is the probability that the electron will tunnel through if the barrier width is 1.0 nm?

(ii) What is the probability that the electron will tunnel through it if the barrier width is 0.10 nm?

Solution

(i) We assume that the probability of transmission is low so that one can use the relation

$$T \approx e^{-2KL} \quad \text{(i)}$$

where

$$K = \frac{\sqrt{2\mu(V_o - E)}}{\hbar} \quad \text{(ii)}$$

For the given barrier height and electron energy, the quantity ($V_o - E$) has the value

$$V_o - E = (40\,\text{eV} - 30\,\text{eV}) = 10\,\text{eV}$$
$$= 1.6 \times 10^{-18}\,\text{J}$$

Using (ii),

$$2KL = 2\frac{\sqrt{2 \times (9.11 \times 10^{-31}\,\text{kg}) \times (1.6 \times 10^{-18}\,\text{J})}}{1.054 \times 10^{-34}\,\text{J.s}} \times (1.0 \times 10^{-9}\,\text{m})$$
$$= 32.4$$

Therefore, the probability of tunneling through the barrier is

$$T \approx e^{-2KL} = e^{-32.4} = 8.5 \times 10^{-15}$$

Clearly, the electron has only about 1 chance in 10^{14} to tunnel through the 1.0 nm wide barrier.

(ii) $L = 0.10$ nm, we have $2kL = 3.24$ and $T \approx e^{2KL} = e^{-3.24} = 0.039$. Obviously, electron has relatively high probability, about 4% of penetrating the 0.10 nm barrier.

Example 8.15: **A particle of mass 1.673×10^{-27} kg is in a sphere of diameter 50 fm. Calculate its second energy level using particle in a box model.**

Solution **We have**

$$E_n = \frac{h^2 n^2}{8mL^2}$$

Here $h = 6.6 \times 10^{-34}$ J.s, $n = 2$, $m = 1.673 \times 10^{-27}$ kg, $L = 50 \times 10^{-15}$ m

$$\therefore \quad E_n = \frac{(6.626 \times 10^{-34})^2 (2)^2}{8 \times 1.673 \times 10^{-27} \times (50 \times 10^{-15})^2 \times 1.6 \times 10^{-19}}$$
$$= 0.328\ \text{MeV}$$

SUMMARY

1. If a particle of mass m is confined in a one dimensional box of width L whose walls are perfectly rigid, the allowed wave functions for the particle are:

$$\psi(x) = A\sin\left(\frac{n\pi x}{L}\right)$$

where n is an integer quantum number starting at 1. The particle has well defined wavelength λ whose values are such that the width L of the box is equal to an integer number and half wave lengths, i.e., $L = n\lambda/2$. The energies of a particle in a box are quantized and are given by

$$E_n = \left(\frac{h^2}{8mL^2}\right)n^2, n = 1, 2, 3, \ldots$$

2. The infinite square well potential is a particularly simple application of Schrödinger wave equation, and it leads to quantized energy levels and quantum numbers. The three-dimensional infinite square-well potential leads to the concept of degenerate states, different physical states with the same energy.
3. Quantum systems generally involve a particle under some constraints due to environment, which forms the basis of quantum particle system under *boundary conditions*. The wave function for quantum-particle system is found and application of boundary conditions on the system allows a determination of the allowed energies and unknown constants in the wave function.
4. When a particle of energy E strikes a potential barrier of height V, where $E < V$ (negative kinetic energy), the particle has a finite probability of penetrating the barrier. (Classically, the entry of the particle in this region is forbidden). This quantum process is called tunneling and is studied by considering various potential barrier shapes. The quantum tunneling is the basic mechanism that explains the operation of the scanning tunneling microscopes, tunnel diodes and the phenomenon of α-decay in some radioactive nuclei. Quantum tunneling process is consistent with the uncertainly principle and occurs only for short distances.
5. The values of energy E_n for which Schrodinger equation can be solved are called *eigen values* and the corresponding wave function ψ_n are called *eigen functions*.

REVIEW QUESTIONS AND PROBLEMS

1. Show that for a particle in a one dimensional box, the mean square deviation of the particle from the centre of the box is

$$<x^2> - <x>^2 = \frac{L^2}{12}\left(1 - \frac{6}{n^2\pi^2}\right)$$

2. A particle of mass μ moves in a potential $V(x)$ defined by

$$V(x) = \begin{cases} 0 & x \le 0 \\ -V_0 & 0 < x < a \\ 0 & x \ge a \end{cases}$$

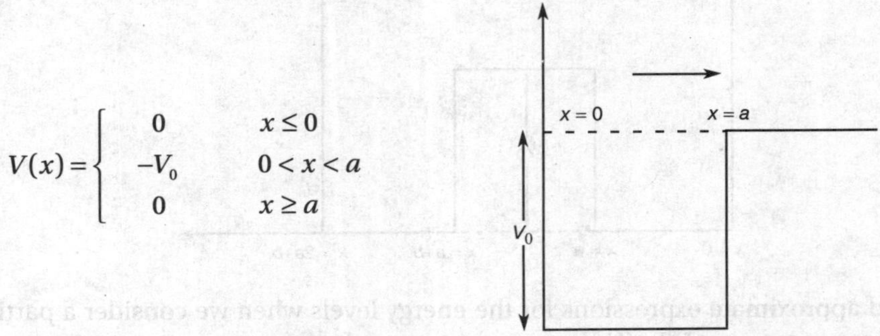

The energy E of the particle satisfies the relation $-V_0 < E < 0$. Solve the Schrodinger equation using appropriate boundary conditions and find the allowed values of energy of particle.

3. Consider a potential step as shown in figure.

$$V(x) = \begin{cases} 0 & x < 0 \\ V_0 & 0 < x < d \\ 2V_0 & x < a \end{cases}$$

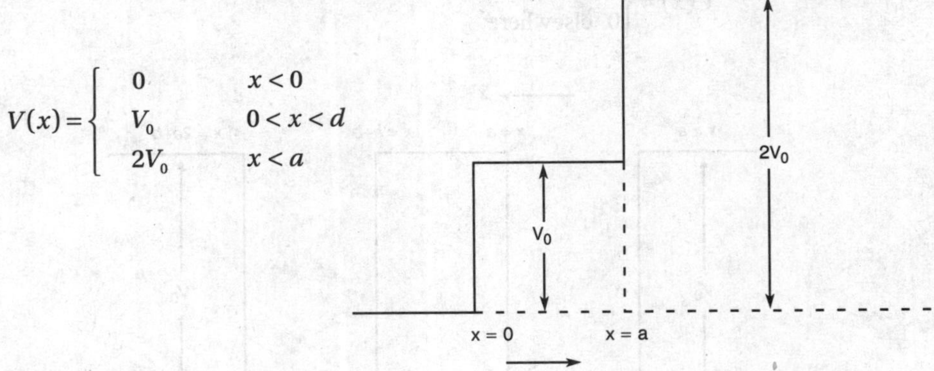

A particle of mass μ having energy $E = 3V_0$ is incident on the barrier. Calculate the probability of transmission; give a plot of transmissivity vs a and calculate the values of a that give maximum and minimum transmission.

4. Consider a potential of the form shown in figure.

$$V(x)=\begin{cases} V_0 & a<x<a+b \\ 0 & 0<x<a \quad \text{and} \quad a+b<x<2a+b \\ \infty & x<0 \quad \text{and} \quad x\geq 2a+b \end{cases}$$

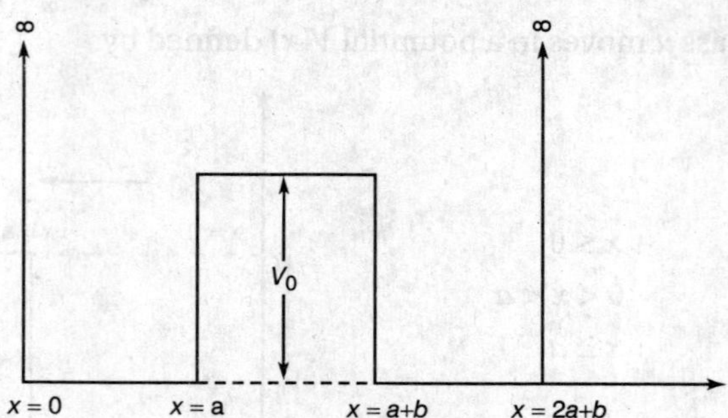

Find approximate expressions for the energy levels when we consider a particle of mass μ and energy $E(E<V_0)$.

5. Two identical potential wells are separated by a distance b, comparable with the width of the well a. Find the allowed values of bound state energy levels.

$$V(x)=\begin{cases} -V_0 & \text{for} \quad 0<x<a, (a+b)<x<(2a+b) \\ 0 & \text{elsewhere} \end{cases}$$

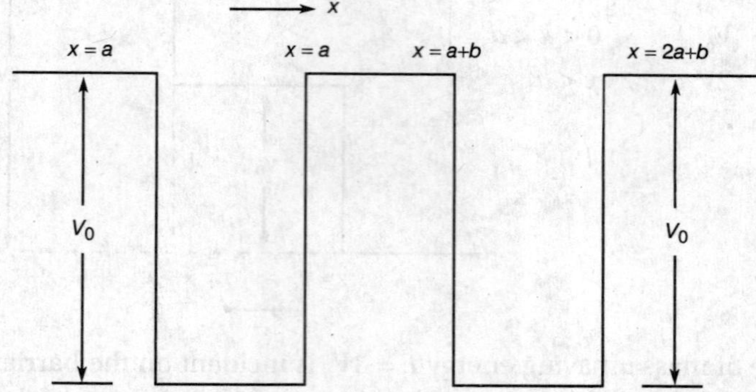

6. Determine the energy levels of a particle in a potential well shown in figure.

$$V(x) = \begin{cases} -V_0 & 0 < x < a \\ +V_0 & 0 > a \\ 0 & x < a \end{cases}$$

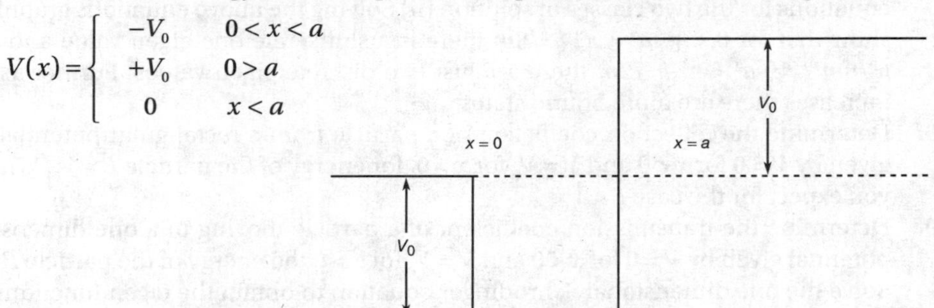

7. (a) A proton or (b) an electron are trapped in a one dimensional box of 10^{-10} m length. What is the minimum energy these particles can have?
8. A particle is trapped in a one dimensional box whose length is L. Its energy is $E = \pi^2 \hbar^2/2mL^2$, wherein the box the particle is most likely to be found?
9. An electron whose kinetic energy is 10 eV is trapped in a potential well whose walls are 10.01 V high. How thin must the walls of the well be in order for the electron to have an appreciable chance to escape?
10. A junction between two superconducting metals exhibits the a.c. Josephson effect. A voltage of 1.0 mV is applied across the junction. What is the frequency of the electromagnetic radiations emitted by the junction.
11. Explain the problem of the looking of a particle through rectangular potential barrier and explain the theory of alpha decay.
12. Calculate the transmission coefficient of electrons through one dimensional rectangular potential barrier.
13. A beam of electrons of energy 100 eV falls on one dimensional rectangular potential barrier of height 110 eV and width 1 mm. Calculate the percentage of electrons transmitted.
14. A particle travelling with energy E along x-axis has in its path a rectangular potential barrier of height $V > E$ and width a. Calculate the transmission coefficient of the particle and discuss briefly its applications to the phenomenon of alpha decay in nuclei.
15. Formulate the problem of square well potential by writing the appropriate potential and Schrodinger equations. Obtain the solutions and an expression for the transmissivity (transmission coefficient). Discuss the expression with special emphasis on resonance.
16. A particle travelling with energy E along x-axis has in its path a rectangular potential barrier of height $V > E$ and width a. Calculate the transmission coefficient of the particle and discuss briefly its application to the observed phenomenon of α-decay in nuclei.
17. Determine the transmission coefficient for a particle of energy $E < V_0$ for a rectangular one-dimensional potential barrier given by $V = 0$ for $x < a$ and $x > a$ and $V = V_0$ for $-a < x < a$. Discuss briefly its application to the observed phenomenon of α-decay.
18. Consider the one dimensional finite square-well potential. Assume the potential energy to be of the form $V(x) = 0$ for $-a < x < a$ and $V(x) = V_0$ for $x < -a$ and $x > a$.

(i) Solve the one dimensional Schrodinger's equation for $E < V_0$ to determine the equations for the two classes of solution (ii) Solving the above equations graphically show that for $0 < V_0 a^2 < \pi^2 \hbar^2/8m$ there is just discrete one eigen value and or $\pi^2 \hbar^2/8m < V_0 a^2 < \pi^2 \hbar^2/2m$ there are just two discrete eigen values. Further as $V_0 a^2$ increases there are more bound states.

19. Determine the reflection coefficient of a particle from a rectangular potential well given by $V = 0$ for $x < 0$ and $V = V_0$ for $x > 0$, for energy of the particle $E > V_0$. What do you expect for the case $E < V_0$?

20. Determine the transmission coefficient of a particle moving in a one dimensional potential given by $V = 0$ for $x < 0$ and $V = V_0$ for $x > 0$ the energy of the particle $E > V_0$.

21. Solve the one dimensional Schrodinger equation to obtain the eigen functions and eigen values for a particle moving in a square potential of the type $V(x) = \infty$ for $x = 0$ and $x < L$ and $V(x) = 0$ for $0 < x < L$.

22. Solve the Schrodinger equation for a particle moving in one dimensional potential well of depth V_0 and width 'a' for energy $E < 0$. Explain the graphical method to calculate its bound state energy levels.

23. Describe the tunnelling phenomena in quantum mechanics. Obtain the transmission coefficient for *a* particle of energy E_0 crossing a barrier defined by $V(x) = 0$ for $x < 0$, $V(x) = V_0$ for $0 < x < a$ and $V(x) = 0$ for $x > a$. Comment on the solution when the thickness a is a multiple of the de-Broglie wavelength.

24. (a) Determine the energy levels and normalised eigen functions of a particle in a potential box. Potential energy of a particle is $V = \infty$ for $x \le 0$ and $x \ge 0$ and $V = 0$ for $0 < x < a$ (b) By using the results for reflection and transmission coefficients, show that the particles are completely transmitted (even it $E < V_0$) if the width of the barrier is integral multiple of half the de Broglie wavelength.

[**Hint:** you may use the relation:

$$T = \left[1 + \frac{\sinh^2(K_2 a)}{4E/V_0(1 - E/V_0)}\right]^{-1} \quad \text{and} \quad K_2 = \sqrt{\frac{2m}{\hbar^2}(V_0 - E)}\Bigg]$$

25. A particle of energy E is travelling in a potential barrier $V(x) = V_0$ for $0 < x < a$ and $V(x) = 0$ for $x \le 0$ and $a \le x$. Can the particle penetrate through the barrier if $E < V_0$? Explain classically and quantum mechanically. Derive the necessary formulae.

26. A particle encounters a one dimensional barrier, $V(x) = 0$ for $x < 0$ and $x > a$ and $V(x) = V_0$ for $0 < x < a$. Write down the Schrodinger wave equation and solve for scattering coefficient for $E > V_0$ when E is the energy of the particle.

27. Set up the problem of a one-dimensional square potential barrier $V(x) = V_0$ for $|x| < a$ and $V(x) = 0$ for $|x| > a$ and obtain the expressions for reflection and transmission coefficients.

28. A beam of particles each of mass m, is incident on a step potential barrier given by $V(x) = 0$ for $x < 0$ and $V(x) = V_0$ for $x > 0$. The particles approach the barrier with velocity V (in x-direction) such that $\frac{1}{2}mv^2 > V_0$. Find the current densities for reflected and transmitted particles assuming unit incident current density and show that if $\frac{1}{2}mv^2 > V_0$, all particles will be reflected. What is the nature of the wave for $x > 0$ in this case?

29. Write the Schrodinger equation for a rectangular potential problem. Solve it to obtain expression for the transmission coefficient for a particle having energy lesser than the barrier height. What do you mean by tunnel effect?
30. What is meant by quantum mechanical tunnelling? Deduce the conditions for a particle to tunnel through a potential barrier.
31. A beam of electrons strikes an energy barrier of height $V = 0.03$ eV. What is the percentage of electrons reflected and transmitted if the energy of the electrons is (i) 00.025 eV (ii) 0.04 eV.
[**Hint**. For $E < V$, $\psi_1 = A\exp(ik_1x) + B\exp(ik_1x)$ and $\psi_{II} = C\exp(-k_1x) + D\exp(k_2x)$. One obtains from the continuity conditions
$$[A+B=C \quad \text{and} \quad A-B=-ck_2/ik_1]$$
32. For an electron in the Bloch state, what are the possible momentum values? What are their probabilities?
33. Sketch the following potentials: (i) a single potential step (barrier) (ii) a potential well (two barriers) with origin at one end, (iii) a potential well with origin at the middle, and (iv) a potential well (single barrier of a certain width). Write Schrodinger equation for each potential and find the eigen functions.
34. Sketch ψ and ψ^2 for the states $n = 4$ and 5 of a particle in a one-dimensional box.
35. Treating the nucleus (size ~ 10^{-14} in) as a one-dimensional box, show why electron does not exist inside the nucleus.
36. Show that the probability of finding the particle in the vicinity of the centre of a one-dimensional box of length 1 nm is less than the probability of finding it in the range $0 < x < 0.2$ nm corresponding to the state $n = 1$.

SHORT QUESTION-ANSWERS

1. Under what circumstances is a discontinuous potential function a reasonable approximation to an actual system?
2. If a potential function has a discontinuity at a certain point, do its eigen functions have discontinuities at that point? if not, why?
3. What exactly is meant by the statement that the reflection coefficient is one for a particle incident on a potential step with total energy less than the step height? What is meant by the statement that the reflection coefficient is less than one if the total energy is greater than the step height? Can reflection coefficient ever be greater than one?
4. A real exponential is a non oscillatory function. Why is a complex exponential an oscillatory function?
5. What is the fallacy in the following statement? "Since a particle cannot he detected while tunnelling through a barrier, it is senseless to say that the process actually happens".
6. A particle is incident on a potential barrier, with total energy less than the barrier height, and it is reflected. Does the reflection involve only the potential discontinuity facing its direction of incidence? If the other discontinuity were removed, so that the barrier were changed into a step, is the reflection coefficient changes?

7. What would a standing wave eigen function for an unbound eigen value of a finite square well look like?
8. Why do the lowest eigen values and eigen functions of an infinite square well provide the best approximation to the corresponding eigen values and eigen functions of a finite square well?
9. Are there any measurable consequences of the penetration of a classically excluded region which is of finite length? Consider a bound particle in a finite square well potential.
10. In the sun, two nuclei of low mass in violent thermal motion can collide by penetrating the Coulomb barrier which separates them. The mass of the single nucleus formed is less than the sun of the masses of the two nuclei, so energy is liberated. This fusion process is responsible for the heat output of the sun. What would be the consequences to the life on earth if it could not happen because barriers were impenetrable?
11. Show from a qualitative argument that a one-dimensional finite square well potential always has one bound eigen value, no matter how shallow the binding region. What would the eigen function look like by the binding region were very shallow?

OBJECTIVE QUESTIONS

1. For a particle encountering a potential barrier, the sum of the reflection and transmission coefficients is always

   (a) 0 (b) infinite
   (c) 1 (d) 0.5

2. The transmission coefficient of a particle moving along the x-axis through a delta function potential barrier a $\delta(x-x_0)$ is

   (a) $T = 1/\left[1+\dfrac{m^2a^2}{2mE\hbar^2}\right]$ (b) $T = \dfrac{m^2a^2}{\hbar^2k^2}\dfrac{1}{\left(1+\dfrac{m^2a^2}{\hbar^4k^2}\right)}$
   (c) 1 (d) 0

3. An electron of mass 9.1×10^{31} kg is moving back and forth between potential barriers 10^9 m apart. The energy E_0 for the electron is about

   (a) 6×10^{-20} J (b) 6×10^{-13} J
   (c) 6×10^{-7} J (d) 6×10^{-3} J

   [**Hint:** $E_0 = \dfrac{h^2}{8ma^2} = \dfrac{(6.63\times10^{-34})^2}{(8\times9.1\times10^{-31})\times(10^{-9})^2} = 6\times10^{-20}$ J]

4. The tunnelling through a potential barrier increases if

   (a) height and width of the barrier are increased
   (b) height and width of the barrier are reduced
   (c) height is increased and width reduced
   (d) height is reduced and width increased

5. The energy of a particle of mass n in a one-dimensional box of length $2L$ with ends at $\pm L$, i.e., the origin at the middle point is

(a) $E_n = \dfrac{n^2h^2}{16\mu L^2}$ (b) $E_n = \dfrac{h^2}{\mu L^2}$

(c) $E_n = \dfrac{n^2h^2}{\mu L^2}$ (d) $E_0 = 0$

6. An integral number of half wavelengths of a single particle of mass, can be fitted between the walls of a one-dimensional box of length L (origin is taken at one end of the box). The expression for the quantized energy levels is

(a) $E_n = \dfrac{n^2h^2}{4\mu L^2}$ (b) $E_n = \dfrac{n^2h^2}{8\mu L^2}$

(c) $E_n = \dfrac{n^2h^2}{16\mu L^2}$ (d) $E_n = \dfrac{n^2h^2}{32\mu L^2}$

7. The general solution for a particle in a one-dimensional box having infinite potential wells is

(a) $\psi = A\sin kx + B\cos kx$ (b) $\psi = \dfrac{A\sin kx}{B\cos kx}$

(c) $\psi = A\cos kx/B\sin kx$ (d) none of the above

8. The length L of a one-dimensional box for which the difference between the lowest energy levels of a molecule of mass m becomes comparable to its kinetic energy at a given temperature T is

(a) $L = \dfrac{h}{2}\sqrt{\dfrac{n}{mkT}}$ (b) $L = \dfrac{h}{2}\sqrt{\dfrac{2n}{mkT}}$

(c) $L = \dfrac{h}{2}\left(\dfrac{2n+1}{mkT}\right)^{1/2}$ (d) $L = \dfrac{h}{2}\sqrt{\dfrac{n}{16mkT}}$

$$\left[\textbf{Hint: } E_{n+1} - E = (2n+1)\frac{h^2}{8mL^2} = \frac{1}{2}kT \text{ or } L = \frac{h}{2}\left(\frac{2n+1}{mkT}\right)^{1/2}\right]$$

9. For a particle on the state $n = 3$ of a one dimensional box of length L, the probability that the particle is in the region $0 \le x < L/4$ is

(a) 0.25 (b) 0.09

(c) 0.3031 (d) 0.909

$$\left[\textbf{Hint: } P = \frac{2}{L}\int_0^{L/4} \sin^2\frac{n\pi x}{L}dx = \frac{1}{4} - \frac{\sin n\pi/2}{2n\pi} \text{ for } n = 3, P = 0.3031\right]$$

10. The probability of finding the particle in a one-dimensional box in the region $L/4$ and $3L/4$ for n even is

(a) 0.1 (b) 0.3

(c) 0.5 (d) 1

$$\left[\textbf{Hint: } \int_{L/4}^{3L/4} \psi_n^2 dx = \frac{2}{L}\int_{L/4}^{3L/4} \sin^2\frac{n\pi x}{L}dx = \frac{1}{2} + \frac{1}{n\pi}\left(\sin\frac{n\pi}{2} - \sin\frac{3n\pi}{2}\right) = \frac{1}{2} \text{ for even } n\right]$$

11. In question 10, the probability for n odd is

(a) 0 (b) 1

(c) 4 (d) $\frac{1}{2} + \frac{(-1)^k}{n\pi}$

12. The average value of P_x^2 for the particle in a box is

(a) mE (b) $6m\,E$

(c) $2m\,E$ (d) $4m\,E$

$$\left[\textbf{Hint: } P_x^2 = \frac{2}{L}\int_0^L \sin\frac{n\pi x}{L}\left[-\frac{h^2}{4\pi^2}\frac{d^2}{dx^2}\left(\sin\frac{n\pi x}{L}\right)\right]dx = \frac{2h^2}{4\pi^2 L}\left(\frac{n\pi}{L}\right)^2 \int_0^L \sin^2\frac{n\pi x}{L}dx = \frac{n^2h^2}{4L^2} = 2mE\right]$$

13. For a particle in a box, the fractional difference in the energy between adjacent eigen values is:

(a) $\frac{\Delta E_n}{E_n} = \frac{1}{n^2}$ (b) $\frac{\Delta E_n}{E_n} = \frac{2n+1}{n^2}$

(c) $\frac{\Delta E_n}{E_n} = 0$ (d) $\frac{\Delta E_n}{E_n} = \frac{n-1}{n+1}$

14. Which one is not the example of potential barrier penetration:

(a) emission of α-particles in the decay of radioactive nuclei.

(b) periodic inversion of ammonia molecule.

(c) the tunnel diode used as a switching unit in fast electronic circuits.

(d) free particle motion.

ANSWERS

Question numbers	1	2	3	4	5	6	7	8	9	10
Answers	(c)	(a)	(a)	(b)	(a)	(b)	(a)	(c)	(c)	(c)
	11	12	13	14						
	(d)	(c)	(b)	(d)						

FILL IN THE BLANKS

1. The energy of a particle confined to a one dimensional box treated quantum mechanically is _______. **[quantized]**
2. An electron is confined between two impenetrable walls 0.2 nm apart. The allowed energies of the particle for the quantum states described by $n = 1, 3$ are _______, and _______ respectively. **[9.42 eV, 84.8 eV]**
3. A 1 mg object is confined between two rigid walls separated by 1 cm. Its minimum speed is _______ m/s. **[3.3×10^{-26}]**
4. The allowed values of a particle in a box are given by E_n = _______, n=1, 2, 3,.... **[$\frac{h^2}{8mL^2}n^2$]**
5. According to quantum mechanics, the particle in a box can never be at _______. **[rest]**
6. The quantum dot, a small region that is grown in a silicon crystal, acts as a _______. **[potential well]**
7. A particle of mass m is confined to a one-dimensional box between $x = 0$ and $x = L$. The expectation value of the position x of the particle for a state with quantum number n is _______. **[$L/2$]**
8. If a particle of mass m is confined to moving in a one-dimensional box of width L whose walls are perfectly rigid, the allowed wave functions for the particle are _______, where n is an integer quantum number starting at 1. **[$\psi_n = A \sin\frac{n\pi x}{L}$]**
9. Quantum systems generally involve a particle under some constraints imposed by the environment, which is the basis of _______ particle under boundary conditions model. **[quantum]**
10. When a particle of energy E meets a barrier of height V, where $E < V$, the particle has a finite _______ of penetrating the barrier. **[probability]**
11. The process of penetrating a barrier of height V by a particle of energy E ($E < V$) is called _______. **[tunneling]**
12. Quantum tunneling is the basis of the mechanism that explains the operation of _______ and the phenomenon of _______ in some radioactive nuclei. **[scanning tunneling microscope, α-decay]**

9

Central Potential Problems, Hydrogen Atom and Rigid Rotator

9.1 PARTICLE IN A RECTANGULAR THREE-DIMENSIONAL BOX

We have seen that a particle moving freely in a one-dimensional box (P.E; $V = 0$) serves as a very convenient model for several types of atomic and molecular systems. Calculations, though approximate, agree fairly well with observed results. Electronic motions in atoms and molecules are, however three-dimensional and a three dimensional box model should be more appropriate. Though electronic motions in atoms and molecules are complicated due to some other factors, let us see how far the results of quantum mechanical treatment of a single particle moving in a three-dimensional rectangular box are of interest.

Let a particle of mass m be in motion in a rectangular potential box as shown in Fig. [9.1(a)] with sides of lengths a, b, and c parallel to the x, y and z axes respectively. Suppose there is no force acting on the particle inside the box, so that in the region $0 < x < a$, $0 < y < b$ and $0 < z < c$ the potential energy $V(x, y, z) = 0$ and outside the box $V(x, y, z) = \infty$.

9.1.1 Wave Equation for the Particle

For the motion of the particle inside the box, the Schrodinger time-independent equation reduces to

$$\nabla^2\psi + \frac{2m}{\hbar^2}E\psi = 0 \tag{9.1}$$

or

$$\frac{\partial^2\psi}{\partial x^2} + \frac{\partial^2\psi}{\partial y^2} + \frac{\partial^2\psi}{\partial z^2} + \frac{2m}{\hbar^2}E\psi = 0 \tag{9.2}$$

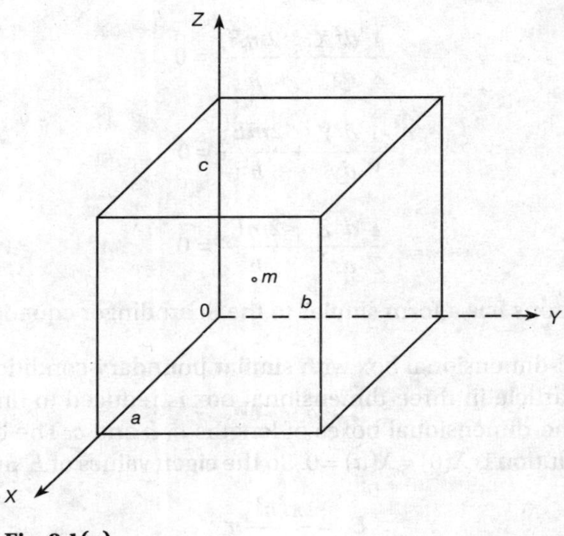

Fig. 9.1(a)

This is partial differential equation where the wave function ψ is a function of x, y and z. The three dimension Eq. (9.2) can be reduced to three ordinary one-dimensional equations by the method of separation of variables. We assume that the wave function $\psi(x, y, z)$ is equal to the product of three function x, y, and z each of which is a function of one variable only. Thus we have

$$\psi(x,y,z) = X(x)Y(y)Z(z) \tag{9.3}$$

Substituting Eq. (9.3) in Eq. (9.2), one obtains

$$YZ\frac{d^2X}{dx^2} + ZX\frac{d^2Y}{dy^2} + XY\frac{d^2Z}{dz^2} + \frac{2mE}{\hbar^2}XYZ = 0 \tag{9.4}$$

We have used ordinary derivatives instead of partial derivatives because each of the function x, y and z is a function of one variable only. Dividing Eq. (9.4) by XYZ, one obtains

$$\frac{1}{X}\frac{d^2X}{dx^2} + \frac{1}{Y}\frac{d^2Y}{dy^2} + \frac{1}{Z}\frac{d^2Z}{dz^2} + \frac{2m}{\hbar^2}E = 0 \tag{9.5}$$

In Eq. (9.5), $\frac{2mE}{\hbar^2}$ is a constant for a particular value of kinetic energy. Since the velocity of the particle being a vector quantity, can be resolved into three components along the coordinate axes, the kinetic energy E can be expressed as the sum of the corresponding terms E_x, E_y and E_z. Hence

$$E = E_x + E_y + E_z \tag{9.6}$$

From Eqs. (9.5) and (9.6), we obtain the following three ordinary differential equations, each comprising a single variable.

$$\frac{1}{X}\frac{d^2X}{dx^2}+\frac{2mE_x}{\hbar^2}=0 \tag{9.7}$$

$$\frac{1}{Y}\frac{d^2Y}{dy^2}+\frac{2mE_y}{\hbar^2}=0 \tag{9.8}$$

$$\frac{1}{Z}\frac{d^2Z}{dz^2}+\frac{2mE_z}{\hbar^2}=0 \tag{9.9}$$

Each of these equations has a form similar to the Schrodinger equation, $\frac{d^2\psi}{dx^2}+K^2\psi=0$ for a particle in one-dimensional box with similar boundary conditions. In other words, the problem of a particle in three-dimensional box is reduced to one of three particles in three separate one-dimensional boxes of lengths *a*, *b* and *c*. The boundary condition applicable to the solution is $X(0)=X(a)=0$. So the eigen values of E_x are given by

$$E_x=\frac{\pi^2\hbar^2}{2ma^2}n_x^2 \tag{9.10}$$

where $n = 1, 2, 3 \ldots$ and the corresponding normalized eigen functions are given by

$$X(x)=\sqrt{\frac{2}{a}}\sin\frac{n_x\pi x}{a} \tag{9.11}$$

The solutions for *y* and *z* are of the same form, one obtains

$$Y(y)=\sqrt{\frac{2}{a}}\sin\frac{n_y\pi y}{b} \tag{9.12}$$

$$E_y=\frac{\pi^2\hbar^2}{2mb^2}n_y^2 \tag{9.13}$$

$$Z(z)=\sqrt{\frac{2}{c}}\sin\frac{n_z\pi z}{c} \tag{9.14}$$

and

$$E_z=\frac{\pi^2\hbar^2}{2mc^2}n_z^2 \tag{9.15}$$

Where n_x, n_y and n_z are now the quantum numbers for the three separate one-dimensional boxes.

Eigen values of energy

Substituting the expressions of E_x, E_y and E_z in Eq. (9.6), one obtains

$$E_{n_x,n_y,n_z}=\frac{\pi^2\hbar^2}{2m}\left[\frac{n_x^2}{a^2}+\frac{n_y^2}{b^2}+\frac{n_z^2}{c^2}\right] \tag{9.16}$$

where $n_x = 1, 2, 3 \ldots$, $n_y = 1, 2, 3 \ldots$ and $n_z = 1, 2, 3 \ldots$

Eq. (9.16) gives the eigen values of the energy of the particle. These values are called the energy levels of the particles. The lowest (ground state) energy E_1 corresponds to $n_x = n_y = n_z = 1$.

Wave Functions: The total normalized wave functions inside the box for the stationary states are given by

$$\psi_{n_x, n_y, n_z}(x, y, z) = X(x)Y(y)Z(z) = \sqrt{\frac{8}{abc}} \sin\frac{n_x \pi x}{a} \sin\frac{n_y \pi y}{b} \sin\frac{n_z \pi z}{c} \tag{9.17}$$

where n_x, n_y and n_z are integers. The wave functions are zero outside the box. It may easily be proved that the wave function is normalized, because

$$\frac{8}{abc}\int_0^a \left(\sin\frac{n_x \pi x}{a}\right)^2 dx \int_0^b \left(\sin\frac{n_y \pi y}{b}\right)^2 dy \int_0^c \left(\sin\frac{n_z \pi z}{a}\right)^2 dz = 1$$

From Eqs. (9.16) and (9.17) one can draw following conclusions

(i) Three integers n_x, n_y and n_z which are called quantum numbers are required to describe each state. It we change the sign of the quantum numbers, there is no change in the energy and in the wave function except that the minus sign appears on R.H.S. of Eq. (9.17). Obviously, all the stationary states are given by the positive integral values of n_x, n_y and n_z. No quantum number can be zero, because if any one of them is taken zero, then $\psi(x, y, z) = 0$ which would mean that the particle does not exist inside the box.

(ii) The lowest possible energy, i.e., the ground state occur when $n_x = n_y = n_z = 1$ and it depends on the values of *a*, *b* and *c*.

(iii) If the particle is confined in a cubical box in which $a = b = c = L$ (say), the eigen values of energy are given by

$$E_{n_x, n_y, n_z} = \frac{\pi^2 \hbar^2}{2mL^2}(n_x^2 + n_y^2 + n_z^2) \tag{9.18}$$

In this case the energy of the particle in the ground state is given by

$$E_{111} = \frac{3\pi^2 \hbar^2}{2mL^2} \tag{9.19}$$

No other state will have this energy, and this state has only one wave function. Therefore, the ground state and the energy level is said to be **non-degenerate.** We note that E_{111} is three times that for a particle in one-dimensional box of length *L*. The corresponding wave function is

$$\psi_{111} = \sqrt{\frac{8}{L^2}} \sin\frac{\pi x}{L} \sin\frac{\pi y}{L} \sin\frac{\pi z}{L} \tag{9.20}$$

(The subscripts in *E* and ψ denote the quantum numbers.)

(iv) **Degeneracy:** An interesting situation arises in a cubical box. In cubical box, the energy depends on the sum of the squares of the quantum numbers. Consequently the particle having the same energy in a excited state will have several different stationary states or different wave functions. Such states and energy levels are said to be degenerate. For Example, we consider the first excited sates when only one of the quantum number is 2, while the others are 1 each, i.e.,

n_x	n_y	n_z
2	1	1
1	2	1
1	1	2

Obviously there are three independent states corresponding to the following three combinations of n_x, n_y and n_z: (2, 1, 1), (1, 2, 1) and (1, 1, 2). Each state has the same energy.

$$E_{211} = E_{121} = E_{112} = \frac{6\pi^2\hbar^2}{2mL^2} \tag{9.21}$$

but different wave functions

$$\left.\begin{aligned} \psi_{211} &= \sqrt{\frac{8}{L^3}}\sin\frac{2\pi x}{L}\sin\frac{\pi y}{L}\sin\frac{\pi z}{L} \\ \psi_{121} &= \sqrt{\frac{8}{L^3}}\sin\frac{\pi x}{L}\sin\frac{2\pi y}{L}\sin\frac{\pi z}{L} \\ \psi_{112} &= \sqrt{\frac{8}{L^3}}\sin\frac{\pi x}{L}\sin\frac{\pi y}{L}\sin\frac{2\pi z}{L} \end{aligned}\right\} \tag{9.22}$$

Obviously, the first excited state is **triply degenerate or threefold degenerate.** The first three energy levels of a particle in a cubical box are shown in Fig. [9.1(b)]. The degeneracy arises whenever there is an integral relation among *a*, *b* and *c*. But if $n_x = n_y = n_z$, the state will be non-degenerate. A slight distortion of the system may lead to the breakdown of degeneracy.

E

$E_{222} = \frac{3\pi^2h^2}{2mL^2}$ — 222 — Non-degenerate

$E_{221} = E_{212} = E_{111} = \frac{9\pi^2h^2}{2mL^2}$ — 221 212 122 — Triply-degenerate

$E_{221} = E_{121} = E_{122} = \frac{6\pi^2h^2}{2mL^2}$ — 211 121 112 — Triply-degenerate

$E_{111} = \frac{12\pi^2h^2}{2mL^2}$ — 111 — Non-degenerate

Fig. 9.1(b)

Table 9.1

Energy levels	*Quantum numbers*	*Degree of degeneracy*
$\frac{3\pi^2\hbar^2}{2mL^2}$	(111)	Non-degenerate
$\frac{6\pi^2\hbar^2}{2mL^2}$	(211), (121), (112)	Threefold degenerate
$\frac{9\pi^2\hbar^2}{2mL^2}$	(221), (212), (122)	Threefold degenerate
$\frac{11\pi^2\hbar^2}{2mL^2}$	(311), (131), (113)	Threefold degenerate
$\frac{12\pi^2\hbar^2}{2mL^2}$	(222)	Non-degenerate
$\frac{14\pi^2\hbar^2}{2mL^2}$	(123), (132), (213), (231), (312), (321)	Sixfold degenerate
$\frac{17\pi^2\hbar^2}{2mL^2}$	(322), (232), (223)	Threefold degenerate
$\frac{18\pi^2\hbar^2}{2mL^2}$	(411), (141), (114)	Threefold degenerate
$\frac{19\pi^2\hbar^2}{2mL^2}$	(331), (313), (133)	Threefold degenerate
$\frac{21\pi^2\hbar^2}{2mL^2}$	(421), (412), (241), (214), (124), (142)	Sixfold degenerate

Degree of Degeneracy: The number of independent wave functions for the stationary states of an energy level is called the degeneracy of the level.

For a particle in a cubical box, number of energy levels with the corresponding quantum numbers and the degree of degeneracy are given in the Table 9.1.

9.1.2 Physical Representation of the Wave Function

Because the function ψ involves three variables x, y and z, it cannot be represented graphically. A qualitative understanding of the wave function can be achieved, however, from the nature of the expression Eq. (9.17). The value of the function in the ground state ($n_x = n_y = n_z = 1$) is maximum when $x = a/2$, $y = b/2$ and $z = c/2$ (i.e., at the box centre) and is positive everywhere. In the excited states, the function changes sign and exhibits modes in some particular regions. For example, for the state ψ_{211} the function vanishes at $x = a/2$ irrespective of the values of y and z, i.e., along the YZ plane which is, therefore, the **"nodal plane"**. Similarly, the XZ and YZ planes passing through $b/2$ and $c/2$ are **nodal planes** for the state ψ_{112} respectively.

One can arrive at a better picture by considering $|\psi|^2$ function, i.e., probability. For the state ψ_{111} in a cubical box, the probability of finding the particle is maximum at the centre of the box and decreases uniformly toward each side. The probability distribution may

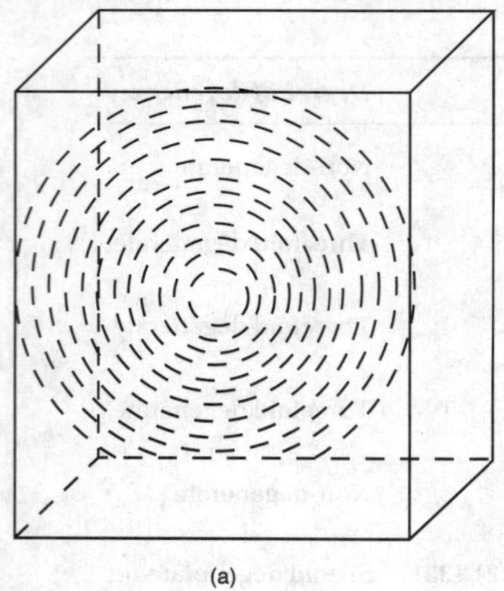
(a)

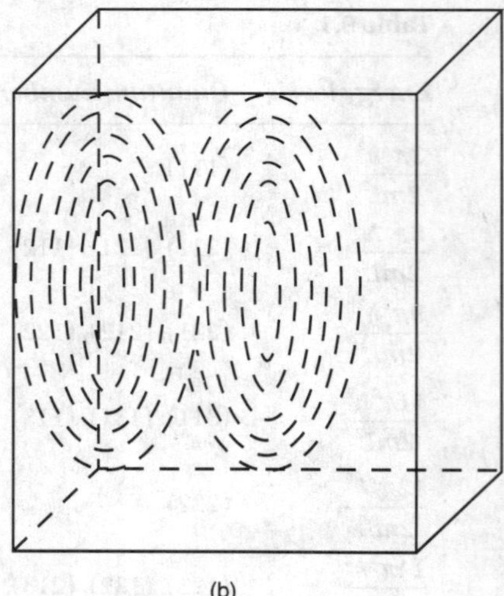
(b)

Fig. 9.2 Probability distribution in a cube

be shown pictorially as in Fig. (9.2). In the next higher state, the probability distribution exhibits two equal halves and the probability is zero along the vertical planes passing through the box centre (nodal planes) as described above.

The state (2, 2, 2) has three vertical nodal planes along each side, i.e., $YZ = XZ$ and XY. It is apparent that the energy level increases the probability distribution and is spread out more and more inside the box and in the limiting case tends to correspond to the classical prediction.

9.2 THE RIGID ROTATOR

Two point masses m_1 and m_2 joined together by a rigid massless rod and rotating like a dumbbell about an axis through their centre of mass C normal to the line jointing the two masses is called the rigid rotator Fig. (9.3). In practice as a first approximation, a rotating diatomic molecule whose nuclei are supposed to be separated by a define mean distance may be treated as a rigid rotator with free axis of rotation.

In a diatomic molecule the atoms are not fixed but vibrating w.r.t. each other and so the distance between atoms will not always be constant but the distance apart in the equilibrium position will be constant.

In Fig. (9.3), two point masses m_1 and m_2 are joined by a rigid bar whose length r is given by

$$r = r_1 + r_2 \tag{9.23}$$

where r_1 and r_2 are the distances of point masses m_1 and m_2 from the centre of mass C of the system about which the system rotates end over end as shown in Fig. (9.3). In this case the moment of interia about C is defined by

$$I = m_1 r_1^2 + m_2 r_2^2 \tag{9.24}$$

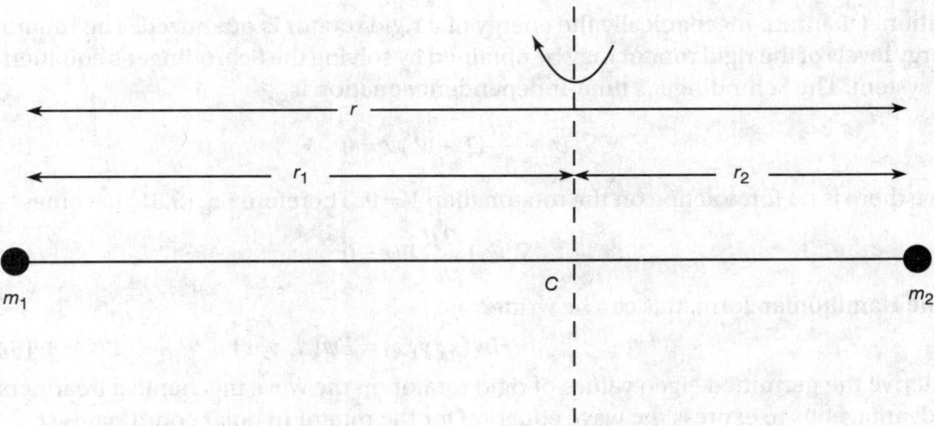

Fig. 9.3 Rigid rotator

As the system is balanced about its centre of mass *C*, one may write

$$m_1 r_1 = m_2 r_2 \tag{9.25}$$

Substituting (9.25) in (9.24), one obtains

$$I = m_2 r_2 r_1 + m_1 r_1 r_2 = r_1 r_2 (m_1 + m_2) \tag{9.26}$$

From Eqs. (9.23) and (9.25), one obtains

$$m_1 r_1 = m_2 r_2 = m_2 (r - r_1) \tag{9.27}$$

One solving Eqs. (9.26) and (9.27), one obtains.

$$r_1 = \frac{m_2}{m_1 + m_2} \text{ and } r_2 = \frac{m_1}{m_1 + m_2} \tag{9.28}$$

On putting these values of r_1 and r_2 in Eq. (9.175), one obtains

$$I = \frac{m_1 m_2^2}{(m_1 + m_2)^2} r^2 + \frac{m_1^2 m_2}{(m_1 + m_2)^2} r^2 = \frac{m_1 m_2}{m_1 + m_2} r^2 = \mu r^2 \tag{9.29}$$

Where μ is the **reduced mass** of the rigid rotator and its value is

$$\mu = \frac{m_1 m_2}{m_1 + m_2} \tag{9.30}$$

Obviously, we have reduced the rotating rigid rotator by a single mass μ at a distance r from the axis of rotation. Thus a rigid rotator can also be defined as particle of mass $\mu = \left(\frac{m_1 m_2}{m_1 + m_2} \right)$ rotating in space at a constant distance from fixed centre.

Classically the regular momentum and energy of this simple rigid rotator are given respectively by $P = I\omega$, and $E = \frac{1}{2} I\omega^2$, where ω is the uniform angular speed of rotation; we have taken the potential energy of rigid rotator as zero, since the two point masses are rigidly joined together, with the distance between them remaining unaffected due to the rotation. Quantum mechanically the energy of a rigid rotator is quantized. The rotational energy levels of the rigid rotator may be obtained by solving the Schrodinger's equation for the system. The Schrodinger's time-independent equation is

$$\nabla^2\psi + \frac{2\mu}{\hbar^2}(E - V)\psi = 0 \tag{9.31}$$

Since there is no force acting on the rotator, then $V = 0$. Therefore Eq. (9.31) becomes

$$\nabla^2\psi + \frac{2\mu}{\hbar^2}E\psi = 0 \tag{9.32}$$

In the Hamiltonian form, this can be written as

$$H\psi(x, y, z) = E\psi(x, y, z) \tag{9.33}$$

To derive the permitted eigen values of rigid rotator on the wave mechanical treatment, it is advantageous to express the wave equation for the rotator in polar coordinates (r, θ, φ), i.e.,

$$H\psi(r, \theta, \varphi) = E\psi(r, \theta, \varphi) \tag{9.34}$$

where

$$H = \frac{\hbar^2}{2\mu}\nabla^2 \tag{9.35}$$

∇^2 can, be expressed in polar coordinates as

$$\nabla^2 = \frac{1}{r^2}\frac{\partial}{\partial r}\left(r^2\frac{\partial}{\partial r}\right) + \frac{1}{r^2\sin\theta}\frac{\partial}{\partial\theta}\left(\sin\theta\frac{\partial}{\partial\theta}\right) + \frac{1}{r^2\sin^2\theta}\frac{\partial^2}{\partial\varphi^2} \tag{9.36}$$

Using Eqs. (9.36), (9.34) can be written as

$$\left\{\frac{1}{r^2}\frac{\partial}{\partial}\left(r^2\frac{\partial}{\partial r}\right) + \frac{1}{r^2\sin\theta}\frac{\partial}{\partial\theta}\left(\sin\theta\frac{\partial}{\partial\theta}\right) + \frac{1}{r^2\sin^2\theta}\frac{\partial^2}{\partial\varphi^2}\right\}\psi(r,\theta,\varphi) + \frac{2\mu E}{\hbar^2}\psi(r, \theta, \varphi) = 0 \tag{9.37}$$

But r is constant. Therefore, first factor becomes,

$$\frac{1}{r^2}\frac{\partial}{\partial r}\left(r^2\frac{\partial\psi}{\partial r}(r,\theta,\varphi)\right) = 0 \tag{9.38}$$

Now, Schrodinger equation of rigid rotator in space may be put as follows:

$$\left[\frac{1}{\sin\theta}\frac{\partial}{\partial\theta}\left(\sin\theta\frac{\partial}{\partial\theta}\right) + \frac{1}{\sin^2\theta}\frac{\partial^2}{\partial\varphi^2}\right]\psi(\theta,\varphi) + \lambda\psi(\theta,\varphi) = 0 \tag{9.39}$$

where

$$\lambda = \frac{2\mu ER^2}{\hbar^2} \tag{9.40}$$

To obtain the solution of Eq. (9.38), one can use the method of separation of variable by assuming

$$\psi(\theta,\varphi) = Q(\theta)S(\varphi) = QS \tag{9.41}$$

where Q is a function of θ only while S is a function of φ only. Substituting Eq. (9.41) in Eq. (9.38) and dividing both sides by $QS/\sin^2\theta$, one obtains

$$\frac{\sin}{Q}\frac{d}{d\theta}\left(\sin\theta\frac{dQ}{d\theta}\right) + \lambda\sin^2\theta = \frac{1}{S}\frac{d^2S}{d\varphi^2} \tag{9.42}$$

Let a quantity m^2 be defined by

$$\frac{1}{S}\frac{d^2S}{d\varphi^2} = -m^2 \quad \text{or} \quad \frac{d^2S}{d\varphi^2} + m^2 S = 0 \tag{9.43}$$

and
$$\frac{\sin\theta}{Q}\frac{d}{d\theta}\left(\sin\theta\frac{dQ}{d\theta}\right) + \lambda\sin^2\theta = m^2$$

or
$$\frac{1}{\sin\theta}\frac{d}{d\theta}\left(\sin\theta\frac{dQ}{d\theta}\right) + \left(\lambda - \frac{m^2}{\sin^2\theta}\right)Q = 0 \tag{9.44}$$

The solution of the Eq. (9.43) may be put as

$$S = Ae^{\pm im\varphi} \tag{9.45}$$

Where m is a positive integer.
It is possible to obtain the value of the constant A by applying the normalization condition, i.e.,

$$\int SS^* d\tau = 1$$

$$\therefore \quad A^2 \int_0^{2\pi} e^{im\varphi}e^{-im\varphi}d\varphi = 1 \quad \text{or} \quad A^2 2\pi = 1 \qquad \therefore \quad A = \frac{1}{\sqrt{2\pi}} \tag{9.46}$$

Hence solution of Eq. (9.43) may be put as follows

$$S = \frac{1}{\sqrt{2\pi}}e^{im\varphi} \tag{9.47}$$

Now, we consider Eq. (9.44):

$$\frac{1}{\sin}\frac{d}{d\theta}\left(\sin\theta\frac{dQ}{d\theta}\right) + \left(\lambda - \frac{m^2}{\sin^2\theta}\right)Q = 0$$

To solve this equation, we make the following substitutions

$$\cos\theta = x, \textit{ then } -\sin\theta d\theta = +dx \text{ and } \lambda = \frac{2\pi E}{\hbar^2} = l(l+1)$$

$$\therefore \quad \sin\theta\frac{dQ}{d\theta} = \frac{\sin^2\theta}{\sin\theta}\frac{dQ}{d\theta} = -(1-x^2)\frac{dQ}{dx} \text{ and } \frac{1}{\sin\theta}\frac{d}{d\theta} = -\frac{d}{dx}$$

Using these results, the above equations transforms as

$$(1-x^2)\frac{d^2Q}{dx^2} - 2x\frac{dQ}{dx} + \left(\lambda - \frac{m^2}{1-x^2}\right)Q = 0 \tag{9.48}$$

Eq. (9.48) is **an associate Legendre equation** having solutions as $P_l^m(x)$,

where $\quad P_l^m = (1-x^2)^{m/2}\frac{d^m}{dx^m}P_l(x)$

Hence the general solution of Eq. (9.44) may be put as follows

$$Q_l^m(\theta)B\,P_l^m(\cos\theta) \tag{9.49}$$

The value of normalized constant B may be given as follows:

$$B=\left\{\left(\frac{2l+1}{2}\right)\frac{(l-m!)!}{(1+m!)!}\right\}^{\sqrt{2}} \tag{9.50}$$

$$\therefore \quad Q_l^m(\theta)=\left\{\frac{2l+1}{2}\frac{(l-m!)!}{(l+m!)!}\right\}^{\sqrt{2}} P_l^m(\cos\theta) \tag{9.51}$$

Hence the total wave function may be put as follows.

$$\psi_{l,m}(\theta,\phi)=\left\{\frac{2l+1}{2}\frac{(l-m!)!}{(l+m!)!}\right\}^{\sqrt{2}}\frac{1}{\sqrt{2\pi}}P_l^m(\cos\theta)e^{+im\phi} \tag{9.52}$$

The corresponding eigen value of ψ_{lm} may be put as follows:

$$\frac{2\mu E_l R^2}{\hbar^2}=l(l+1) \tag{9.53}$$

or

$$E_l=\frac{\hbar^2}{2l}l(l+1) \tag{9.54}$$

[where $l=0, 1, 2, 3, ...$ and $I=\mu R^2$ is M.I. of rigid rotator]
when

$$l=0, \quad E_0=0$$

$$l=1, \quad E_1=\frac{\hbar}{2I}2=2B(\text{say})$$

$$l=2, \quad E_2=\frac{\hbar}{2I}6=6B$$

$$l=3, \quad E_3=\frac{\hbar}{2I}12=12B$$

The energy level diagram of a rigid rotator is depicted in Fig. (9.4). Thus we have a series of discrete energy levels whose energy increases quadratically with increasing l.

If we compare Eq. (9.54) with the classical formula for the system, in the quantum state l i.e.,

$$P=\frac{h}{2\pi}\sqrt{l(l+1)}=\frac{h}{2\pi}L \tag{9.55}$$

We note that this agrees with the general result for the magnitude of angular momentum vectors in quantum mechanics, i.e.,

$$L=\sqrt{l(l+1)}\hbar$$

The results of rigid rotator are widely used in the study of spectra of diatomic molecules.

Since only certain discrete energy values and angular momenta of the rigid rotator are possible, it follows that only **certain rotational frequencies** are possible. Since classically the angular momentum $P=I\omega$, one obtains for the angular velocity

$$\omega=\frac{\hbar}{I}\sqrt{l(l-1)}=\frac{\hbar}{I}L$$

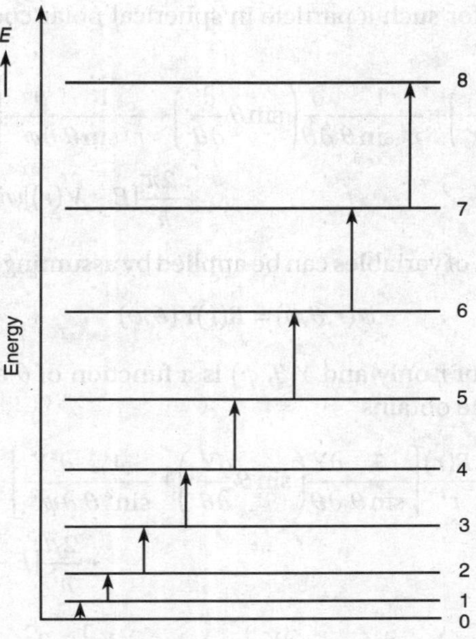

Fig. 9.4 Energy level diagram of a rigid rotator

Therefore, the rotational frequency is

$$\nu_{rot} = \frac{\hbar}{2\pi I}\sqrt{l(l+1)} = \frac{\hbar}{2\pi I}L \tag{9.56}$$

i.e., the rotational frequency increases approximately linearly with L.

9.2.1 Eigen Function

The Schrodinger equation of the rigid rotator is closely related to that of the hydrogen atom. The only difference being the absence of the potential energy term. Since this term for the hydrogen atom affects only the radial dependence of the eigen functions it is clear that the eigen functions ψ_r of the rigid rotator are identical with those obtained from the hydrogen eigen functions by dropping the factor depending on r, i.e., they are the so called **surface harmonics.**

9.3 SPHERICALLY SYMMETRIC POTENTIALS

When a particle of mass μ, move under the influence of a potential which depends only upon the distance of the particle from a fixed point (say, origin) then the potential energy of the system can be represented by a function of r only (independent of θ and ϕ). Spherical polar coordinates are most suitable for such a system, as separation of variables is possible.

Schrodinger equation for such a particle in spherical polar coordinate system may be written as.

$$\left[\frac{1}{r^2}\frac{\partial}{\partial r}\left(r^2\frac{\partial}{\partial r}\right)+\frac{1}{r^2\sin\theta}\frac{\partial}{\partial\theta}\left(\sin\theta\frac{\partial}{\partial\theta}\right)+\frac{1}{r^2\sin\theta}\frac{\partial^2}{\partial\varphi^2}\right]\psi(r,\theta,\phi)$$
$$+\frac{2\pi}{\hbar^2}[E-V(r)]\psi(r,\theta,\phi)=0 \tag{9.57}$$

The method of separation of variables can be applied by assuming the solution in the form.

$$\psi(r,\theta,\phi)=R(r)Y(\theta,\phi) \tag{9.58}$$

where $R(r)$ is a function of r only and $Y(\theta, \phi)$ is a function of θ and ϕ only. Substituting Eq. (9.58) in Eq. (9.57), one obtains

$$Y(\theta,\phi)\frac{1}{r^2}\frac{\partial}{\partial r}\left(r^2\frac{\partial R}{\partial r}\right)+\frac{R(r)}{r^2}\left[\frac{1}{\sin\theta}\frac{\partial Y}{\partial\theta}\left(\sin\theta\frac{\partial Y}{\partial\theta}\right)+\frac{1}{\sin^2\theta}\frac{\partial^2 Y}{\partial\psi^2}\right]$$
$$+\frac{2\mu}{\hbar^2}[E-V(r)]R(r)Y(\theta,\phi)=0$$

or $$\frac{1}{R}\frac{\partial}{\partial r}\left(r^2\frac{\partial R}{\partial r}\right)+\frac{1}{Y}\left[\frac{1}{\sin\theta}\frac{\partial}{\partial\theta}\left(\sin\theta\frac{\partial Y}{\partial\theta}\right)+\frac{1}{\sin^2\theta}\frac{\partial^2 Y}{\partial\phi^2}\right]+\frac{2\mu}{\hbar^2}\times[E-V(r)]r^2=0$$

or $$\frac{1}{R}\frac{\partial}{\partial r}\left(r^2\frac{\partial R}{\partial r}\right)+\frac{2\mu}{\hbar^2}[E-V(r)]r^2=-\frac{1}{Y}\left[\frac{1}{\sin\theta}\frac{\partial}{\partial\theta}\left(\sin\theta\frac{\partial Y}{\partial\theta}\right)+\frac{1}{\sin^2\theta}\frac{\partial^2 Y}{\partial\phi^2}\right] \tag{9.59}$$

The L.H.S. of Eq. (9.59) depends only on r, and the R.H.S. only on θ and ϕ. Since the functions of independent variables must be equal regardless of the values of the variables, each side must be equal to a constant called the separation constants, denoting this by λ, we get separate equations of radial and angular parts. For radial part, we get

$$\frac{I}{R}\frac{\partial}{\partial r}\left(r^2\frac{\partial R}{\partial r}\right)+\frac{2\mu}{\hbar^2}[E-V(r)]r^2=\beta$$

or $$\frac{I}{R}\frac{\partial}{\partial r}\left(r^2\frac{\partial R}{\partial r}\right)+\left[\frac{2\mu}{\hbar^2}\times\{E-V(r)\}\}-\frac{\beta}{r^2}\right]R=0 \tag{9.60}$$

Equation satisfied by angular part is

$$-\frac{1}{Y}\left[\frac{1}{\sin\theta}\frac{\partial}{\partial\theta}\left(\sin\theta\frac{\partial Y}{\partial\theta}\right)+\frac{1}{\sin^2\theta}\frac{\partial^2 Y}{\partial\phi^2}\right]=\beta$$

or $$\frac{1}{\sin\theta}\frac{\partial}{\partial\theta}\left(\sin\theta\frac{\partial Y}{\partial\theta}\right)+\frac{1}{\sin^2\theta}\frac{\partial^2 Y}{\partial\phi^2}+\beta Y=0 \tag{9.61}$$

This equation contains no expression for total potential energy and is therefore, completely general. So that, solving it once means solving it for all cases, where V does not depend on A or ϕ.

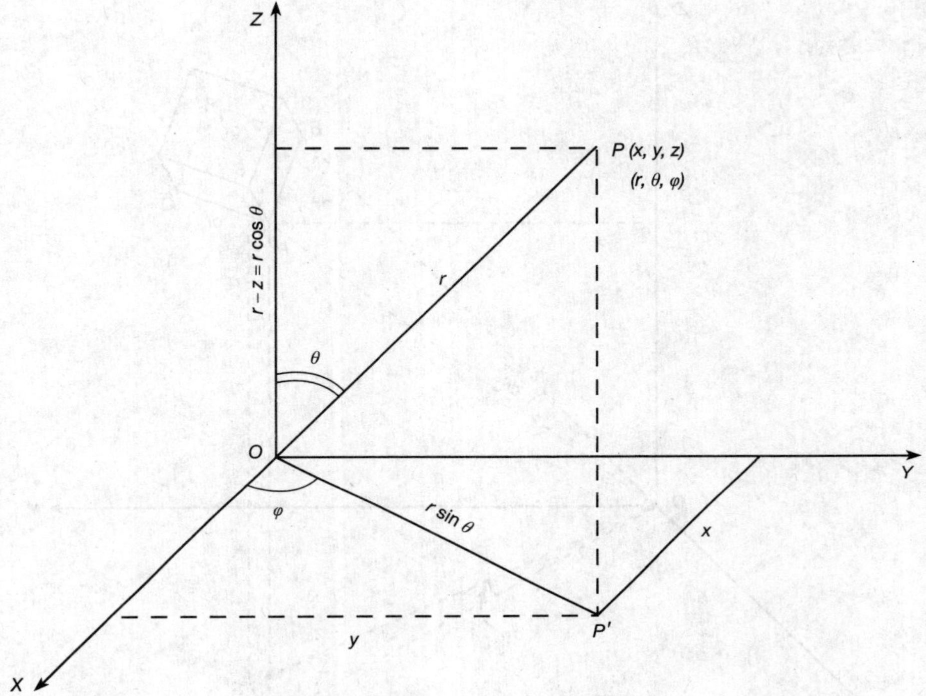

Fig. 9.5(a) Spherical polar coordinates of a point $P(r, \theta, \phi)$

Now Eq. (9.61) can be further separated into two parts by substituting

$$Y(\theta,\phi) = \Phi(\theta), \Phi(\phi)$$

$$\Phi(\phi)\frac{1}{\sin\theta}\frac{\partial}{\partial\theta}\left(\sin\theta\frac{\partial\Phi}{\partial\theta}\right)\frac{\Phi(\theta)}{\sin^2\theta}\frac{\partial^2\Phi}{\partial\phi^2} + \beta\Phi(\theta) = 0$$

$$\frac{1}{\Phi}\sin\theta\frac{\partial}{\partial\theta}\left(\sin\theta - \frac{\partial\Phi}{\partial\theta}\right) + \lambda\sin^2\theta = -\Phi(\phi)\frac{\partial^2\Phi}{\partial\phi^2} \qquad (9.62)$$

The L.H.S. of this equation depends only on θ, and the R.H.S. on Φ; therefore each side must be equal to a constant, denoting it by m^2, one obtains

$$\frac{1}{\Phi}\sin\theta\frac{\partial}{\partial\theta}\left(\sin\theta - \frac{\partial\Phi}{\partial\theta}\right) + \beta\sin^2\theta = m^2$$

or
$$\frac{1}{\sin\theta}\frac{\partial}{\partial\theta}\left(\sin\theta\frac{\partial\Phi}{\partial\theta}\right) + \left(\beta - \frac{m^2}{\sin^2\theta}\right)\theta = 0 \qquad (9.63)$$

and
$$-\Phi(\varphi)\frac{\partial^2\Phi}{\partial\varphi^2} = m^2 \quad \text{or} \quad \frac{\partial^2\Phi}{\partial\varphi^2} + m^2\Phi = 0 \qquad (9.64)$$

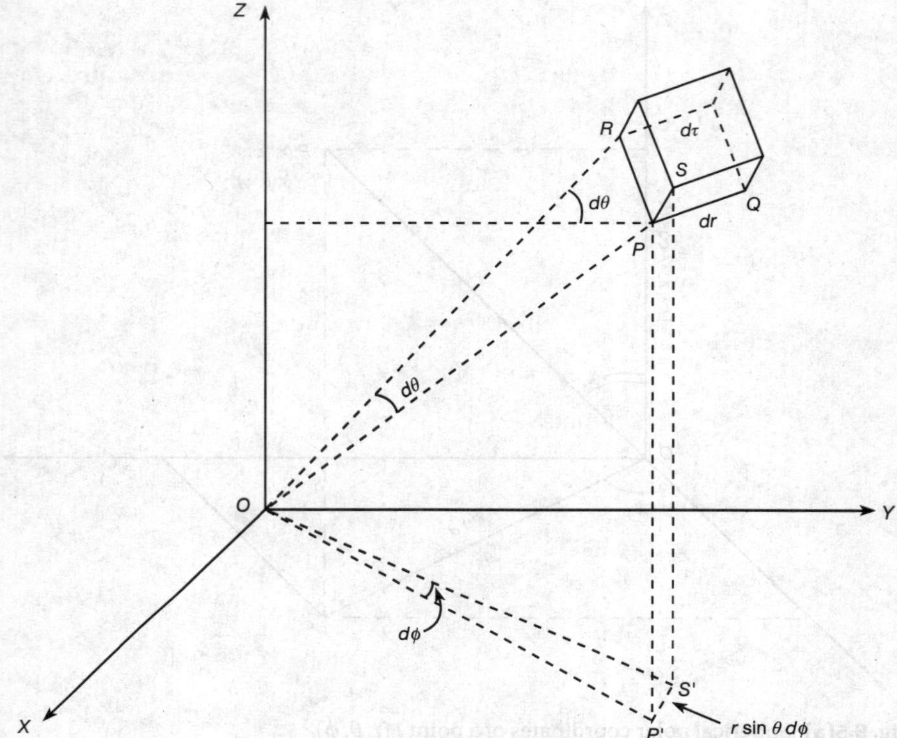

Fig. 9.5(b) The volume element in spherical coordinates

Eq. (9.64) can be solved at once; its general solution can be written as

$$\Phi = ae^{im\varphi} + be^{-im\varphi} \qquad m \neq 0$$
$$\Phi = a + b\varphi \qquad m = 0$$

where a and b are constants to be determined from the boundary conditions. The azimuthal angles ϕ and $\phi + 2\pi$ represent the same physical point, therefore $\Phi(\varphi + 2\pi) = \Phi(\varphi)$. This requires m to be a real integer including zero.

$$\Phi_m = ae^{im\varphi} \qquad m = 0, \pm 1, \pm 2, \pm 3, \ldots$$

Any linear combination of these solutions will also be a solution of Eq. (9.64) the value of a can be determined from the normalization of Φ, which requires

$$\int_0^{2\pi} \Phi_m^* \Phi_m d\varphi = 1 \text{ or } \int_0^{2\pi} a^2 d\varphi = 1$$

or
$$a^2 2\pi = 1 \text{ or } a = (2\pi)^{-1/2}$$

The normalized eigen functions are

$$\Phi_m = (2\pi)^{-1/2} e^{im\varphi} \tag{9.65}$$

Were $m = 0, \pm 1, \pm 2, \pm 3, \ldots$, implies that positive and negative values of m are corresponding to the distinct solution while for $m = 0$, there is only one solution. m **is usually termed as the magnetic quantum number which is having obviously the integral nature.** One can obtain real solutions of ϕ equation by taking linear combinations of complex solutions for ϕ as follows

$$\phi = \frac{1}{\sqrt{\pi}}\sin m\phi \quad \text{or} \quad \frac{1}{\sqrt{\pi}}\cos m\phi \quad \text{for} \quad m = \pm 1, \pm 2, \ldots.$$

and

$$\phi = \frac{1}{\sqrt{2\pi}} \text{ for } m = 0$$

Solution of Θ equation

The equation satisfied by Θ is

$$\frac{1}{\sin\theta}\frac{\partial}{\partial\theta}\left(\sin\theta\frac{\partial\Theta}{\partial\theta}\right) + \left(\beta - \frac{m^2}{\sin^2\theta}\right)\Theta = 0$$

In order to solve the above equation, we introduce a new parameter x defined by the equation, $x = \cos\theta$, x varies between the limits −1 to + 1; corresponding to the limits $\theta = 0$ to π.

$$\frac{d\Theta}{d\Theta} = \frac{d\Theta}{dx}\cdot\frac{dx}{d\theta} = -\sin\theta\frac{d\Theta}{dx}$$

$$\therefore \quad \frac{d}{d\theta}\left(\sin\theta\frac{\partial\Theta}{\partial\theta}\right) = \frac{d}{dx}\left[-\sin^2\theta\frac{d\Theta}{dx}\right]\frac{dx}{d\theta} = \sin\theta\frac{d}{dx}\left[(1-x^2)\frac{d\Theta}{dx}\right]$$

$$\therefore \quad \frac{1}{\sin\theta}\frac{d}{d\Theta}\left(\sin\theta\frac{\partial\Theta}{\partial\theta}\right) = \frac{d}{dx}\left[(1-x^2)\frac{d\Theta}{dx}\right]$$

Equation satisfied by Θ may now be written as

$$\frac{d}{dx}\left[(1-x^2)\frac{d\Theta}{dx}\right] + \left[\beta - \frac{m^2}{1-x^2}\right]\Theta = 0$$

or

$$(1-x^2)\frac{d^2\Theta}{dx^2} - 2x\frac{d\Theta}{dx} + \left(\beta - \frac{m^2}{1-x^2}\right)\Theta = 0 \tag{9.66}$$

The solution of Eq. (9.66) for Θ depends on the value of mth in the Θ_m function, i.e., different solution [for Θ] are obtained for different values of m, therefore Θ is labelled as θ^m. Eq. (9.66) possesses a singularity at $x = \pm 1$. To remove this singularity we try a solution of the from

$$\Theta^m(x) = (-x^2)^\alpha P^m(x) \tag{9.67}$$

where α is a constant and the superscript m denotes, m solution.

$$\frac{d\Theta^m(x)}{dx} = (1-x^2)^\alpha\frac{dP^m(x)}{dx} - 2\alpha(1-x^2)^{\alpha-1}xP^m(x)$$

and

$$\frac{d^2\Theta^m(x)}{dx^2} = (1-x^2)^\alpha\frac{d^2P^m(x)}{dx^2} - 4\alpha(1-x^2)^{\alpha-1}x^2\frac{dP^m(x)}{dx}$$

$$-2\alpha\left[(1-x^2)^{\alpha-1}-2(\alpha-1)(1-x^3)^{\alpha-2}x^2\right]P_m(x)$$

$$\therefore \quad (1-x^2)^{\alpha+1}\frac{d^2P^m(x)}{dx^2}-2(2\alpha+1)(1-x^2)^{\alpha}x\frac{dP^m(x)}{dx}$$

$$-(1-x^2)^{\alpha-1}\left[2\alpha(1-x^2)-4\alpha(\alpha-1)x^2-4\alpha x^2-\beta(1-x^2)+m^2\right]P_m(x)=0$$

or
$$(1-x^2)^{\alpha+1}\frac{d^2P^m(x)}{dx^2}-2(2\alpha+1)x(1-x)^{2\alpha}\frac{dP^m(x)}{dx}$$

$$-(1-x^2)^{\alpha-1}[2\alpha-\beta+m^2-(4\alpha^4+2\alpha-\beta)x^2]P^m(x)=0 \qquad (9.68)$$

The coefficient of $P^m(x)$ in Eq. (9.68) is given by

$$(1-x^2)^{\alpha-1}[(-2\alpha+\beta)(1-x^2)+4\alpha^2x^2-m^2]$$

If we choose value of α such that this coefficient is equal to zero at $x=\pm1$, then the coefficient of $P^m(x)$ will not be singular at $x=\pm1$. Thus α may be chosen to satisfy the relationship

$$4x^2(\pm1)^2=m^2 \quad \text{or} \quad \alpha=\pm m/2$$

But α cannot be negative, as becomes infinite at $x=\pm1$. Therefore, only the choice

$$\alpha=\frac{|m|}{2} \qquad (9.69)$$

leads to a physically acceptable solution.

Thus Eq. (9.66) in having a singularity at $x=\pm1$, has been transformed into Eq. (9.68) which does not have a singularity if $\alpha=\frac{|m|}{2}$.

Substituting the value of α in Eq. (9.68), we get an equation satisfied by $P^m(x)$;

$$(1-x^2)^{\frac{|m|}{2}+1}\frac{d^2P_m(x)}{dx^2}-2(|m|+1)x(1-x^2)^{\frac{|m|}{2}}\frac{dP^m(x)}{dx}$$

$$-(1-x^2)^{\frac{|m|}{2}-1}[|m|-\beta+m^2-m^2x^2-|m|x^2+\beta x^2]P^m(x)=0$$

or
$$-(1-x^2)^{\frac{|m|}{2}+1}\frac{d^2P_m(x)}{dx^2}-2(|m|+1)x(1-x^2)^{\frac{|m|}{2}}\frac{dP^m(x)}{dx}+(1-x^2)^{\frac{|m|}{2}-1}$$

$$\{\beta-(|m|)(|m|+1)\}(1-x^2)P^m(x)=0$$

or
$$(1-x^2)^{\frac{|m|}{2}}\left[(1-x^2\frac{d^2P_m(x)}{dx^2}-2(|m|+1)x\frac{dP^m(x)}{dx}+\{\beta-(|m|)(|m|)+1)\}P^m(x)\right]=0$$

or
$$(1-x^2)\frac{d^2P_m(x)}{dx^2}-2(|m|+1)x\frac{dP^m(x)}{dx}+\{\beta-(|m|)(|m|)+1)\}P^m(x)=0 \qquad (9.70)$$

To find a solution of Eq. (9.70), we express $P^m(x)$ by an infinite series in x;

$$P^m(x)=\sum_{v=0}^{\infty}a_vx^v$$

Then $$\frac{dP^m(x)}{dx} = \sum_\nu a_\nu x^{\nu-1} \quad \text{and} \quad \frac{d^2P^m(x)}{dx^2} = \sum_\nu a_\nu \nu(\nu-1)x^{\nu-2}$$

$$\therefore \quad (1-x^2)\sum_\nu \nu(\nu-1)a_\nu x^{\nu-2} - 2(|m|+1)\sum_\nu \nu a_\nu x^\nu + \{\beta - |m|(|m|+1)\}\sum_\nu a^\nu x^\nu = 0 \quad (9.71)$$

Eq. (9.71) is an identity in x and hence the coefficients of individual powers of x must vanish

$$x^0 \to 2x1a_2 + \{\beta - (|m|)(|m|+1)\}a_0 = 0$$

$$x^1 \to 3x2a_2 + 2(|m|+1)a_1 + \{\beta - (|m|)(|m|+1)\}a_1 = 0$$

$$x^2 \to 4x4a_4 - 2x1a_2 - 4(|m|+1)a_2 + \{\beta - (|m|)(|m|+1)\}a_2 = 0$$

or in general, for the coefficient of x^ν;

$$(\nu+2)(\nu+1)a_{\nu+2} - \nu(\nu-1)a_\nu - 2(|m|+1)\nu a_\nu + \{\beta - (|m|)(|m|+)\}a_\nu = 0$$

or $$a_{\nu+2} = \frac{\nu^2 - \nu + 2\nu|m| + 2\nu - \beta + (|m|)(|m|+1)}{(\nu+1)(\nu+2)} a_\nu$$

or $$a_{\nu+2} = \frac{\nu^2 + \nu|m| + \nu(|m|+1) + (|m|)(|m|+1) - \beta}{(\nu+1)(\nu+2)} a_\nu$$

or $$a_{\nu+2} = \frac{(\nu+|m|)(\nu+|m|+1) - \beta}{(\nu+1)(\nu+2)} a_\nu \quad (9.72)$$

Thus for any particular value of a_υ Eq. (9.72) permits us to calculate $a_{\nu+2}$, $a_{\nu+4}$, ... and so forth. Let us examine the behaviour of the ratio $a_{\nu+2}/a_\nu$ as $n \to \infty$, to determine the

behaviour of the series as $x \to \pm 1$. For large values of ν, $\dfrac{a_\nu + 2}{a_\nu} \cong \dfrac{\nu^2}{\nu^2} = 1$.

Also $$(1-x^2)^{-1} = 1 + x^2 + x^4 + x^6 + \cdots + x^\nu + x^{\nu+2} + \cdots = \sum_{\nu=0}^{\infty} x^{2\nu} \quad (9.73)$$

$$\frac{\text{Coefficient of } x^{\nu+2}}{\text{Coefficient of } x^\nu} = 1$$

Thus we conclude that for values of $|x|$ approaching unity, the function $P^m(x)$ behaves identically to the series represented by Eq. (9.73). This series goes to **infinity** as $x \to \pm 1$ and therefore so does the infinite series $P^m(x)$ with coefficients given by Eq. (9.72) [however the series converges for $-1 < x < 1$ (for any value of $|m|$ and β)] and in the consequence does not correspond to an acceptable wave function. In order to be satisfactory, then our series for $P^m(x)$ must be truncated after a certain term in μ, i.e., it must contain only a finite numbers of terms. Then either the even or the odd series can be broken off at the term $x_{\nu'}$ by putting

$$\beta = (\nu' + |m|)(\nu' + 1|m| + 1) \quad (9.74)$$

where ν' can be 0, 1, 2, 3, ... etc. The other series can be made to vanish by equating either a_1 or a_0 to zero. Then function $P^m(x)$ contains only odd or even powers or x as ν' is odd or even.

It is convenient to introduce another quantum number l instead of ν', defined by the equation

$$l = \nu' + |m| \tag{9.75}$$

The allowed values of l are $(|m|)$, $(|m| + 1)$, $(|m| + 2)$, ... Alternatively, we may say that for any particular value of the Θ equations are

$$P_i^{|m|}(x) = \Theta = (1 - x^2)^{\frac{|m|}{2}} P^{|m|}(x) \tag{9.76}$$

where $P^{|m|}(x)$ is defined by the recurrence relations Eqs. (9.72) and (9.75) and $P_l^m(x)$ is known as **Associated Legendre Polynomial** corresponding to quantum numbers l and m. After extensive calculations these solutions can be shown to have the from (normalised)

$$P_l^{|m|}(x) = (-1)^m \frac{(1 - x^2)^{|m|/2}}{2^l l!} \left(\frac{d}{dx} \right)^{l+|m|} (x^2 - 1)^l \tag{9.77}$$

The angular part $y_{lm}(\theta, \phi)$ of the complete wave function, which is a solution of Eq. (9.61) with $\beta = l(l + 1)$ is known as a **spherical harmonic.**

$$Y_{lm}(\theta,\phi) = N_{lm} P_l^m(\cos\theta)\phi_m(\varphi) \tag{9.78}$$

N_{lm} is the normalization constant for the associated Legendre function. It can easily seen that the eigen functions corresponding to different even values of λ or l are orthogonal. For a given value of l there exist $(2l + 1)$ linearly independent solutions [corresponding to all integral values of m between $+ l$ and $-l$] and hence solutions are $(2l + 1)$ fold degenerate. The choice of function $\phi_m(\varphi)$ given by Eq. (9.65) makes these degenerate eigen functions orthogonal.

The integral $\int_{\theta=0}^{\pi} \int_{\varphi=0}^{2\pi} Y_{lm}^{*}(\theta,\varphi) Y_{l',m.}(\theta,\varphi) \sin\theta d\theta d\phi$ vanishes unless $l = l'$ and $m = m'$.

The ϕ part of the integral vanishes when $m \neq m'$ without regard for the l values: the θ part of the integral vanishes only when $l \neq l'$ and $|m| \neq |m'|$, since for $m \neq m'$ the orthogonality is taken care of by the integration over ϕ. The integral

$$\int_{\theta=0}^{\pi} P_l^{|m|}(\cos\theta) P_{l'}^{|m|}(\cos\theta) \sin\theta d\theta = \int_{-1}^{+1} P_l^{|m|}(x) P_{l'}^{|m|}(x) dx \tag{9.79}$$

can be evaluated by using the generating functions for the associated legendre polynomial. The integral (9.79) vanishes unless $l = l'$; for $l = l'$ its values is equal to $\left\{ \frac{2}{(2l+1)} \times \frac{(l+|m|)!}{(l-|m|)!} \right\}$; thus N_{lm} is reciprocal of the square root of this quantity, except for an arbitrary complex phase factor of unit magnitude.

$$\therefore \quad Y_{lm}(\theta,\phi) = (-1)^m \frac{1}{\sqrt{2\pi}} e^{im\phi} \left[\left(\frac{2l+1}{2} \right) \frac{(l-|m|)!}{(l+|m|)!} \right]^{1/2} P_l^{|m|}(\cos\theta) \tag{9.80}$$

or

$$Y_{lm}(\theta,\phi) = (-1)^m \left[\left(\frac{2l+1}{4\pi} \right) \frac{(l-|m|)!}{(l+|m|)!} \right]^{1/2} P_l^{|m|}(\cos\theta) e^{im\phi} \tag{9.81}$$

The first few spherical harmonics are

$$\left.\begin{aligned}
Y_{0,0} &= \frac{1}{(4\pi)^{1/2}} \\
Y_{1,0} &= \left(\frac{3}{4\pi}\right)^{1/2} \sin\theta \\
Y_{1,\pm1} &= \mp\left(\frac{3}{8\pi}\right)^{1/2} \sin\theta e^{\pm\phi} \\
Y_{2,0} &= \left(\frac{5}{16\pi}\right)^{1/2} (3\cos^2\theta - 1) \\
Y_{2,\pm1} &= \mp\left(\frac{15}{8\pi}\right)^{1/2} \sin\theta\cos\theta e^{\pm i\phi} \\
Y_{2,\pm2} &= \left(\frac{15}{32\pi}\right)^{1/2} \sin^2\theta e^{\pm 2i\varphi} \\
Y_{3,0} &= \mp\left(\frac{63}{16\pi}\right)^{1/2} \sin\left(\frac{5}{3}\cos^3\theta - \cos\theta\right) \\
Y_{3,\pm1} &= \mp\left(\frac{21}{64\pi}\right)^{1/2} \sin\theta(5\cos^2\theta - 1)e^{\pm\varphi} \\
Y_{3,\pm2} &= \left(\frac{105}{32\pi}\right)^{1/2} \sin^2\theta\cos\theta e^{\pm\varphi};\ Y_{3,\pm3} = \mp\left(\frac{35}{64\pi}\right)^{1/2} \sin^2\theta e^{\pm 3i\varphi}
\end{aligned}\right] \tag{9.82}$$

The normalised polar wave functions for $l = 0, 1, 2$ and $m = 0, \pm 1, \pm 2$ are given in Table 9.2.

Parity

The concept of parity introduced earlier can now be extended to spherically symmetric potential problems. When the position coordinate $\boldsymbol{r}$ is reflected through the origin, then $\boldsymbol{r}$ is replaced by $-r$, in cartesian coordinates. This is equivalent to a replacement of x by $-x$, y by $-y$ and z by $-z$ or in spherical polar coordinates this is equivalent to a replacement of θ by $\pi - \theta$, ϕ by $\pi + \varphi$ with r unchanged. Thus

$$\psi(r,\theta,\varphi) \to \psi(r, \pi - \theta, \pi + \phi) \tag{9.83}$$

A wave function has a –ve or +ve (or an odd or even) parity according as it changes or does not change sign on reflection through the origin. Thus $\psi(-r) = -\psi(r)$, **negative parity** and $\psi(-r) = \psi(r)$, **positive parity.**

The energy levels for a spherically symmetric potential are degenerate at least with respect to the quantum number m, for $l > 0$. When $\boldsymbol{r}$ is reflected through the origin, the radial part $R(r)$ of the solution is unchanged. For φ part

$$\Theta(\varphi) = ae^{im\varphi} \to ae^{im(\pi+\varphi)} = ae^{im\pi}e^{ima} = (-1)^m eа^{im\varphi}$$

Table 9.2 Normalised polar and radial wave functions

n	*l*	*m*	*Qm* (θ)	$R_{nl}(r)$
1	0	0	$\frac{1}{\sqrt{2}}$	$2\left(\frac{1}{a_0}\right)^{3/2} e^{-r/2a}$
2	0	0	$\frac{1}{\sqrt{2}}$	$\frac{1}{2\sqrt{2}}\left(\frac{1}{a_0}\right)^{3/2}\left(2-\frac{r}{a_0}\right)e^{-r/2a}$
2	1	0	$\frac{\sqrt{6}}{2}\cos\theta$	$\frac{1}{2\sqrt{6}}\left(\frac{1}{a_0}\right)^{3/2}\frac{r}{a_0}e^{-r/2a}$
1	1	±1	$\frac{\sqrt{3}}{2}\sin\theta$	$\frac{1}{2\sqrt{6}}\left(\frac{1}{a_0}\right)^{3/2}\frac{r}{a_0}e^{-r/2a}$
3	0	0	$\frac{1}{\sqrt{2}}$	$\frac{2}{81\sqrt{3}}\left(\frac{1}{a_0}\right)^{3/2}\left(27-18\frac{r}{a_0}+2\frac{r^2}{a_0^2}\right)e^{-r/3a_0}$
3	1	0	$\frac{\sqrt{6}}{2}\cos\theta$	$\frac{4}{81\sqrt{6}}\left(\frac{1}{a_0}\right)^{3/2}\left(6-\frac{r}{a_0}\right)\frac{r}{a_0}e^{-r/3a_0}$
3	0	±1	$\frac{\sqrt{3}}{2}\sin\theta$	$\frac{4}{81\sqrt{6}}\left(\frac{1}{a_0}\right)^{3/2}\left(6-\frac{r}{a_0}\right)\frac{r}{a_0}e^{-r/3a_0}$
3	2	0	$\frac{\sqrt{10}}{4}(3\cos^2\theta-1)$	$\frac{4}{81\sqrt{30}}\left(\frac{1}{a_0}\right)^{3/2}\frac{r^2}{a_0^2}e^{-r/3a_0}$
3	2	±1	$\frac{\sqrt{15}}{2}\sin\theta\cos\theta$	$\frac{4}{81\sqrt{30}}\left(\frac{1}{a_0}\right)^{3/2}\frac{r^2}{a^2}e^{-r/3a_0}$
3	2	±2	$\frac{\sqrt{15}}{4}\sin^2\theta$	$\frac{4}{81\sqrt{30}}\left(\frac{1}{a_0}\right)^{3/2}\frac{r^2}{a_0^2}e^{-r/3a_0}$

or

$$\Theta(\varphi)\to(-1)^m\,\Theta(\varphi) \tag{9.84}$$

For θ part

$$x=\cos\theta\to\cos(\pi-\theta)=-\cos\theta=-x \qquad \therefore\ x\to -x \tag{9.85}$$

and

$$P_l^{|m|}(x)\to P_l^{|m|}(-x)=(-1)^{l-|m|}P_l^{|m|}(x)$$

$$\therefore\quad Y_l^m(\theta,\phi)\to Y_l^m(\pi-\theta,\pi+\theta)=(-1)^{l-|m|}(-1)^m Y_l^m(\theta,\phi)=(-1)^l Y_l^m(\theta,\phi)$$

Thus we see that all degenerate eigen functions have the same parity, which is even (+1) or odd (–1) according as l is even or odd.

It will be seen from the above that in spherical coordinates, the parity of a wave function is determined only by the angular part of the function. When this is a spherical harmonic function, the value of l uniquely determines the parity given by

Radial part parity = $(-1)^l$.

Let us now investigate the radial part of the equation [substituting $\beta = l(l+1)$ in Eq. (9.60)]

$$\frac{1}{r^2}\frac{\partial}{\partial r}\left(r^2\frac{\partial R_l}{\partial r}\right)+\frac{2\mu}{\hbar^2}\left[E-V(r)-\frac{l(l+1)}{r^2}\frac{\hbar^2}{2\mu}\right]R_l = 0 \tag{9.86}$$

Here we attached a subscript l to $R(r)$ corresponding to the l th solution, where $\beta = l(l+1)$. Since the possible values of E are determined by Eq. (9.86) and this equation does not contain m, the energy levels must be independent of quantum number m. Let us make the substitution

$$R_l(r)=\frac{U_l(r)}{r},\ \frac{dR_l}{dr}=\frac{1}{r}\frac{dU_l}{dr}-\frac{1}{r^2}U_l$$

$$\therefore \qquad r^2\frac{dR_l}{dr}=r\frac{dU_l}{dr}=-U_l$$

$$\frac{d}{dr}\left(r^2\frac{dR_l}{dr}\right)=r\frac{d^2U_l}{dr^2}\quad \therefore\quad \frac{1}{r^2}\frac{d}{dr}=\left(r^2\frac{dR_l}{dr}\right)=\frac{1}{r}\frac{d^2U_l}{dr^2}$$

Eq. (9.87) reduces to

$$\frac{1}{r}\frac{d^2U_l}{dr^2}+\frac{2\mu}{\hbar^2}\left[E-V(r)-\frac{l(l+1)}{r^2}\frac{\hbar^2}{2\mu}\right]\frac{U_l}{r}=0$$

or

$$\frac{d^2U_l}{dr^2}+\frac{2\mu}{\hbar^2}\left[E-V(r)-\frac{l(l+1)}{r^2}\frac{\hbar^2}{2\mu}\right]U_l=0 \tag{9.87}$$

Eq. (9.87) is called the radial wave equation and is equivalent to one-dimensional Schrodinger equation for a particle of mass μ moving under the influence of a potential

$$V'=V(r)+\frac{l(l+1)\hbar^2}{2\mu r^2} \tag{9.88}$$

The additional "potential energy" term is known **centifugal potential barrier** and is related to the angular momentum. A classical particle having an angular momentum L about an axis through the origin and perpendicular to the plane of the orbit has an angular velocity $\omega = L/\mu r^2$, when its radial distance from the origin is r. An inward force $\mu\omega^2 r = L/\mu r^2$ is required to keep the particle in this path; this **"centripetal force"** is supplied by the potential energy and hence adds to $V(r)$ that appears in the radial motion as an additional centrifugal potential energy equal to $\dfrac{L^2}{2\mu r^2}$.

9.4 A PARTICLE IN A SPHERICAL BOX

Let us consider a particle moving in a spherical box potential defined by the equation.

$$V(r)=\begin{cases}0 & \text{for } r<a\\ \infty & \text{for } r\geq a\end{cases} \tag{9.89}$$

in analogy to the problem of a particle moving under the influence of a one-dimensional box potential, $\psi = 0$ for $r \geq$a.

Inside the box, we use spherical polar coordinates to discuss the motion of the particle. Since the potential is spherically symmetric, we can have separation of variables, as is done in section (9.1).

$$\psi(r,\theta,\phi) = \left[\frac{U_l(r)}{r}\right] Y_{lm}(\theta,\phi)$$

where $Y_l^m(\theta,\phi)$ is given by Eq. (9.80) and $U_l(r)$ is a solution of Eq. (9.87).

Inside the box $V(r) = 0$; so Eq. (9.31) reduces to the from

$$\frac{d^2U_l(r)}{dr^2} + \frac{2\mu}{\hbar^2}\left[E - \frac{\hbar^2 l(l+1)}{2\mu r^2}\right]U_l(r) = 0 \qquad r < a \tag{9.90}$$

Boundary conditions to be satisfied by U_l are: $U_l(a) = 0$ since $V(a) = \infty$ and $U_l(0) = 0$, otherwise $\psi(r) = \left[\frac{U_l(r)}{r}\right] Y_{lm}(\theta,\phi)$ tends to infinity in the limit r tending to zero. Eq. (9.78) is a one-dimensional Schrodinger equation for a particle moving under the influence of an effective potential

$$V(r) = \frac{l(l+1)\hbar^2}{2\mu r^2}$$

This effective potential, known as **centrifugal potential,** tries to keep away the particles of higher angular momenta from origin. If a particle comes very close to the nucleus. It should be speeded up in order to conserve angular momentum and in this process conservation of energy is violated. However, if we look at the problem in the light of quantum mechanics, we see that the probability of finding a particle with high-angular momentum (l) near the origin is greatly reduced (though not zero) just like exponentially decaying wave function in classically forbidden region in one-dimensional potential problems.

(i) Zero angular momentum: ($l = 0$). In the simplest case of zero angular momentum, Eq. (9.90) reduces to

$$\frac{d^2U_0(r)}{dr^2} + \frac{2\mu E}{\hbar^2}U_0(r) = 0$$

or

$$\frac{d^2U_0(r)}{dr^2} + k^2U_0(r) = 0 \tag{9.91}$$

where $k = \sqrt{\frac{2\mu E}{\hbar^2}}$. Solution of Eq. (9.92) may be written as

$$U_0(r) = A\sin kr + B\cos kr \tag{9.92}$$

where A and B are constants, whose values can be determined from the boundary conditions. Since $U_0(0) = 0$, therefore, B must be chosen to be equal to zero. Also $U_0(a) = 0$.

A sin $ka = 0$ or sin $ka = 0$ (since A cannot be zero) or $ka = n\pi$

Where n is integer. Thus the energy levels are quantized just as for a particle in a one-dimensional box. Energy levels are given by

$$k^2a^2 = \frac{2\mu E_n a^2}{\hbar^2} = n^2\pi^2 \quad \text{or} \quad E_n = \frac{n^{\pi^2}\hbar^2}{2\mu a} = \frac{n^2\hbar^2}{8\mu a^2} \tag{9.93}$$

The total wave function may be written in the from,

$$\psi_{n00}(r,\theta,\phi)=\frac{A}{r}\sin\left(\frac{n\pi r}{a}\right)Y_{00}(\theta,\phi)$$

where (n, l, m) refer to the three quantum numbers used, nth radial solution corresponding to any value of l is denoted by $U_{nl}(r)$. The normalization factor A can be determined by using the relation

$$\int_0^a\left[\frac{A\sin\left(\frac{n\pi r}{a}\right)}{r}\right]^2 r^2 dr=1$$

This gives
$$A^2\cdot\frac{a}{2}=1 \text{ or } A=\sqrt{\frac{a}{2}}$$

$$\psi_{n00}(r,\theta,\phi)=\sqrt{\frac{2}{a}}\left[\frac{\sin\left(\frac{n\pi r}{a}\right)}{r}\right]_{Y_{00}(\theta,\phi)} \tag{9.94}$$

(ii) Solution for orbitrary value of l: In general, when $l \neq 0$, let us try a solution of the from $U_l(kr)=r^{1/2}f_1(kr)$

or
$$\frac{dU_l}{dr}=r^{1/2}\frac{df_l}{dr}+\frac{1}{2}r^{-1/2}f_l \text{ and } \frac{d^2U_l}{dr}=r^{1/2}\frac{d^2f_l}{dr^2}+r^{-1/2}\frac{df_l}{dr}-\frac{1}{4}r^{-3/2}f_l$$

$\therefore$
$$r^{1/2}\frac{d^2f_l}{dr^2}\times r^{-1/2}\frac{df_l}{dr}-\frac{1}{4}r^{-3/2}f_l+\left\{\frac{2\mu E}{\hbar^2}+\frac{l(l+1)}{r^2}\right\}(f_l)r^{1/2}=0$$

or
$$r^2\frac{d^2f_l}{dr^2}+r\frac{df_l}{dr}\left[k^2r^2-\left(l^2+l+\frac{1}{4}\right)\right]f_l=0$$

or
$$r^2\frac{d^2f_l}{dr^2}+r\frac{df_l}{dr}\left[k^2r^2-\left(l+\frac{1}{2}\right)^2\right]f_l=0 \tag{9.95}$$

or
$$(kr)^2\frac{d^2f_l}{d(kr)^2}+(kr)\frac{df_l}{d(kr)}+\left[(kr)^2-\left(l+\frac{1}{2}\right)^2\right]f_l=0 \tag{9.96}$$

This is **Bessel's differential equation of order** $\pm\left(l+\frac{1}{2}\right)$ having argument kr. Solution of Eq. (9.96) may be written as

$$f_l(r)=AJ_{l+1/2}(kr)+BJ_{(l+1/2)}(kr) \tag{9.97}$$

where A and B are arbitrary constant.
Since $J_{-(l+1/2)}(0)\neq 0$ $\therefore f_l(0)=0$, requires that $B=0$.
Boundary condition at $r=a$, gives possible values of the energy of the particle

$$J_{l+1/2}(k_na)=0 \tag{9.98}$$

Spherical Bessel function $J_l(kr)$ is defined by the equation

$$J_l(kr) = \sqrt{\frac{\pi}{2kr}} J_{l+1/2}(kr)$$

The constant has been arbitrarily chosen to give normalised functions.

$$\therefore \quad R_l(k_n r) = \frac{U_l(k_n r)}{r} = \frac{f_l(k_n r)}{r^{1/2}} = \frac{AJ_{l+1/2}(k_n r)}{r^{1/2}} J_l(k_n r) \quad (9.99)$$

Explicit expression for first few spherical Bessel functions is given by

$$\left.\begin{aligned} J_0(kr) &= \frac{\sin kr}{kr}, \; J_1(kr) = \frac{\sin kr}{(kr)^2} - \frac{\cos kr}{kr} \\ J_2(kr) &= \left[\frac{3}{(kr)^3} - \frac{1}{kr}\right]\sin kr - \frac{3}{(kr)^2}\cos kr \\ J_3(kr) &= \left(\frac{15}{k^4 r^4} - \frac{6}{kr^2}\right)\sin kr - \left(\frac{15}{k^3 r^2} - \frac{1}{kr}\right)\cos kr \end{aligned}\right] \quad (9.100)$$

Let us examine the behaviour of these solutions, near origin: $U_l(k_n r) = r^\alpha V_l(k_n r)$ where $V_l(kr)$ is a power series in r. We are interested in the behaviour of the wave function near $r = 0$, therefore we consider only the lowest of r in $U_l(k_n r)$.

Also $\dfrac{dU_l}{dr} = \alpha r^{\alpha-1} V_l + \cdots$ and $\dfrac{d^2 U_l}{dr^2} = \alpha(\alpha-1) r^{\alpha-2} V_l + \cdots$

On substituting these values in Eq. (9.90) (neglecting high powers of r), one obtains,

$$\alpha(\alpha-1) r^{\alpha-2} T_l + \alpha r^{\alpha-1} V_l + \cdots + \left\{k^2 - \frac{l(l+1)}{r^2}\right\} r^\alpha V_l = 0 \quad (9.101)$$

Near the origin, k^2 may be neglected in comparison to $\dfrac{l(l+1)}{r^2}$, so Eq. (9.101) becomes

$$[\alpha(\alpha-1) - l(l+1)] r^{\alpha-2} T_l \cong 0 \text{ or } \alpha^2 - \alpha - l^2 - l = 0$$

or

$$(\alpha - l - 1)(\alpha + l) = 0$$

or

$$\alpha = l+1, -l \quad (9.102)$$

Thus there are two possible values of $\alpha : l+1$ or $-l$. The latter value is not physically significant as this value of α gives a singularity at the origin. Therefore, α must be chosen as $(l+1)$ and

$$U_l(k_n r) = r^{l+1} V_l(k_n r) \quad \text{and} \quad R_l(k_n r) = \frac{U_l(k_n r)}{r} = r^l V_l(k_n r)$$

For small r, we can set $V_l(k_n r) \cong A$ constant, so that

$$R_l(k_n r) \sim r^l \quad (9.103)$$

Thus if R_l vs r is plotted near the origin for various values of l as in the Fig. (9.6), it follows that with the increase in the value of l, the wave function falls off more rapidly close to the origin.

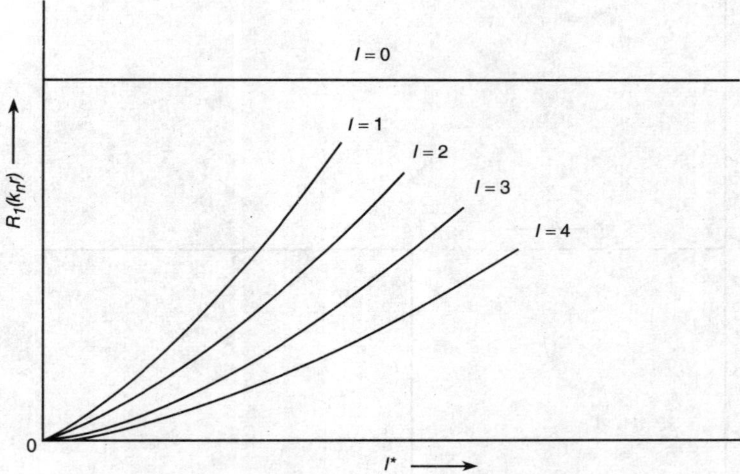

Fig. 9.6 Radial wave functions near the origin for various values of l

The energy levels of a particle in a spherical box are given by the solutions of Eq. (9.98). For each level E_{nl}, the quantum number m can be 0, ± 1, ± 2, ... ± l. The wave function for each state (n, l, m) is given by

$\psi_{n,l,m}(r, \theta, \varphi) = R_l(k_n r) y_{l,m}(\theta, \varphi)$. The probability of finding the particle between distance r and $r + dr$ from the origin is given by $|R_l|^2\, 4\pi r^2 dr$.

They approach zero at $r = a$ and with the exception of $l = 0$, they go to zero at $r = 0$.

9.5 THREE-DIMENSIONAL SQUARE WELL POTENTIAL

Let us consider a square well potential of finite depth [as shown in Fig. (9.7)] defined by the equation:

$$V(r) = \begin{cases} V_0 & r < a \\ 0 & r > a \end{cases} \tag{9.104}$$

where V_0 is positive.

A spherical region of this type in which the potential is less than that of the surroundings serves to attract a particle. For bound states, *i.e*, negative energies of the particle, Schrodinger equation in the two regions may be written as

$$r < a \left[\frac{1}{r^2}\frac{\partial}{\partial r}\left(r^2\frac{\partial}{\partial r}\right) + \frac{1}{r^2 \sin\theta}\frac{\partial}{d\theta}\left(\sin\theta\frac{\partial}{\partial\theta}\right) + \frac{1}{r^2\sin^2\theta}\frac{\partial^2}{\partial\varphi^2}\right]\psi_I(r,\theta,\varphi) + \frac{2\mu}{\hbar^2}(E - V_0)\psi_I(r,\theta,\varphi) = 0$$

or

$$\left[\frac{1}{r^2}\frac{\partial}{\partial r}\left(r^2\frac{\partial}{\partial r}\right) + \frac{1}{r^2 \sin\theta}\frac{\partial}{d\theta}\left(\sin\theta\frac{\partial}{\partial\theta}\right) + \frac{1}{r^2\sin^2\theta}\frac{\partial^2}{\partial\varphi^2}\right]\psi_I(r,\theta,\varphi) + k_4^2\psi_I(r,\theta,\varphi) = 0 \tag{9.105}$$

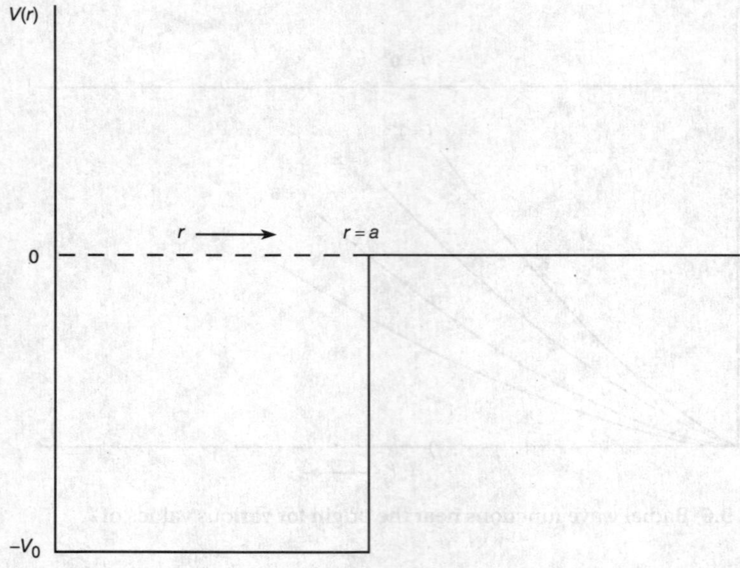

Fig. 9.7 Spherically symmetric square well potential of depth V_0 and radius a

$$r < a \left[\frac{1}{r^2}\frac{\partial}{\partial r}\left(r^2\frac{\partial}{\partial r}\right)+\frac{1}{r^2\sin\theta}\frac{\partial}{d\theta}\left(\sin\theta\frac{\partial}{\partial\theta}\right)+\frac{1}{r^2\sin^2\theta}\frac{\partial^2}{\partial\varphi^2}\right]\psi_{II}(r,\theta,\varphi)$$

$$+\frac{2\mu}{\hbar^2}\psi_{II}(r,\theta,\varphi)=0$$

or

$$\left[\frac{1}{r^2}\frac{\partial}{\partial r}\left(r^2\frac{\partial}{\partial r}\right)+\frac{1}{r^2\sin\theta}\frac{\partial}{d\theta}\left(\sin\theta\frac{\partial}{\partial\theta}\right)+\frac{1}{r^2\sin^2\theta}\frac{\partial^2}{\partial\varphi^2}\right]\psi_{II}(r,\theta,\varphi)$$

$$-k_3^2\psi_{II}(r,\theta,\varphi)=0 \tag{9.106}$$

where $k_3=\sqrt{\dfrac{-2\mu E}{\hbar^2}}$ and $k_4=\sqrt{\dfrac{2\mu(V_0+E)}{\hbar^2}}$

Since the potential is spherically symmetric, separation of variables is possible, writing

$$\psi(r,\theta,\varphi)=\frac{U_I(r)}{r}y_{lm}(\theta,\phi).$$

We get radial equation in I and II region as

$$\frac{d^2U_{II}}{dr^2}+\left[k_4^2-\frac{l(l+1)}{r^2}\right]U_{II}=0 \qquad r<a \tag{9.107}$$

$$\frac{d^2U_{III}}{dr^2}+\left[-k_3^2-\frac{l(l+1)}{r^2}\right]U_{III}=0 \qquad r<a \tag{9.108}$$

Wave function and its derivative must be continuous at the boundary of discontinuity, i.e., at $r = a$, ψ and $\frac{\partial \psi}{\partial r}$ must be continuous. Eqs. (9.107) and (9.108) are analogous to the one-dimensional Schrodinger equations for a particle in classically accessible and classically forbidden region. Zero angular momentum ($l = 0$): Eqs. (9.107) and (9.108) now reduce to a simple from,

$$\frac{d^2U_{0I}}{dr^2} + k_4^2 U_{0I} = 0 \qquad r < a \tag{9.109}$$

and

$$\frac{d^2U_{0II}}{dr^2} - k_3^2 U_{0II} = 0 \qquad r > a$$

The solution of Eq. (9.109) is the same as that obtained for the potential well of finite depth except for the following points:

(i) The domain of r is 0 to ∞ in place of the domain $-\infty$ to $+\infty$ for x;

(ii) The boundary condition that wave functions should not become infinite (ψ should be zero as $|x| \to \infty$) at $x - \infty$ is now replaced by the condition that it should not become infinite at $r = 0$.

The solutions are

$$U_{0I} = A \sin k_4 r + B \cos k_4 r, \; U_{0II} = C \exp(-k_3 r) \tag{9.110}$$

We are interested in the bound state energy levels for which $E < 0$. The requirement that $R(r)$ be finite at $r = 0$; demands $B = 0$. Thus the solution has the from of the odd parity solution of the one-dimensional potential well.

The energy levels are obtained by demanding the continuity of logarithmic derivative at $r = a$;

$$\left[\frac{1}{U_{0I}} \frac{dU_{0I}}{dr}\right]_{r=a} = \left[\frac{1}{U_{0II}} \frac{dU_{0II}}{dr}\right]_{r=a}$$

$$k_4 \cot k_4 a = -k_3 \quad \text{or} \quad \cot k_4 a = -k_3 / k_4 \tag{9.111}$$

or

$$\cot \sqrt{\frac{2\mu a^2 (V_x 0 + E)}{\hbar^2}} = -\sqrt{\frac{-E}{V_0 + E}} \tag{9.112}$$

This is a transcendental equation and can be solved by graphical method.

Let

$$\beta = \sqrt{\frac{2\mu V_0 a^2}{\hbar^2}}$$

$\therefore$

$$\beta - k_3^2 a^2 = \frac{2\mu a^2 V_0}{\hbar^2} + \frac{2\mu a^2 E}{\hbar^2} = \frac{2\mu a^2 (E + V_0)}{\hbar^2} = k_4^2 a^2$$

or

$$\beta^2 - k_4^2 a^2 = k_3^2 a^2$$

$\therefore$

$$k_4 a \cot k_4 a = -k_3 \;\; a = -\sqrt{\beta^2 + k_4^2 a^2}$$

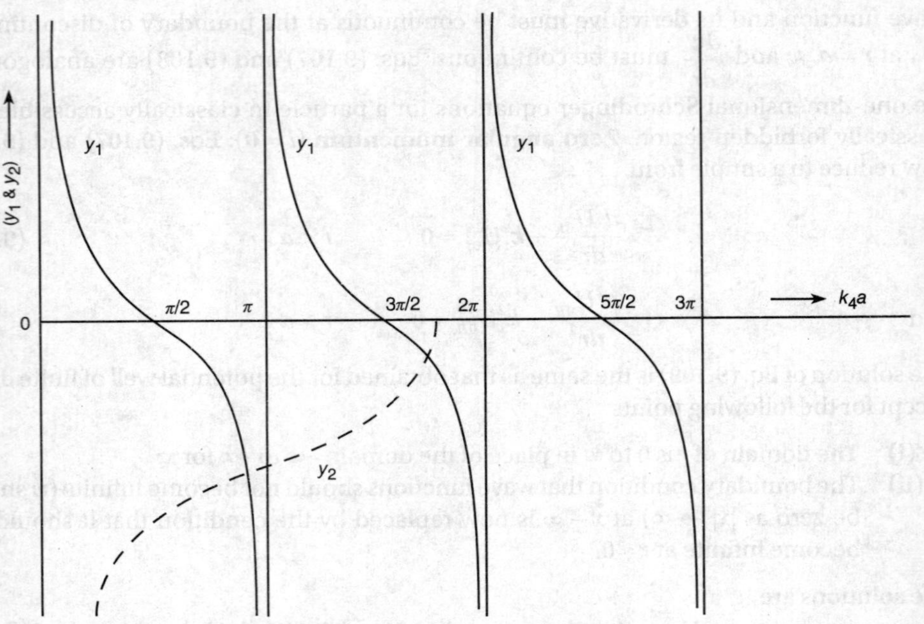

Fig. 9.8 Graphical determination of energy levels in a square well with $\beta = 5.5$

or $$\operatorname{Cot} k_4 a = \sqrt{\frac{\beta^2 - k_4^2 a^2}{k_4^2 a^2}} \tag{9.113}$$

$$y_l = \cot k_4 a \quad \text{and} \quad y_2 = \frac{\sqrt{\beta^2 - k_4^2 a^2}}{k_4^2 a^2} \tag{9.114}$$

y_l vs $k_4 a$ (solid curve) and y_2 vs $K_3 a$ (broken curve) for $\beta = 5.5$ are plotted on the same graph. Point of intersection of these two curves give the allowed values of energies. From Fig. (9.4), we see that there is no level unless

$$(E + V_0) > \frac{\hbar \pi^2}{8 \mu a^4} \text{ or } V_0 > \frac{\hbar^2 \pi^2}{8 \mu a^2} \quad \text{(since } E \text{ is negative)} \tag{9.115}$$

$$\text{There is one bound state if } \frac{\hbar \pi^2}{8 \mu a^4} < V_0 \le \frac{9 \hbar^2 \pi^2}{8 \mu a^2} \tag{9.116}$$

Solution for arbitrary l: To solve the radial equation for an arbitrary value of l (non-zero); we make the substitution

$$U_l = r^{1/2} f_l$$

$$\frac{dU_l}{dr} = r^{1/2} \frac{df_l}{dr} + \frac{1}{2} r^{-1/2} f_l \quad \text{and} \quad \frac{d^2 U_l}{dr^2} = r^{1/2} \frac{d^2 f_l}{dr^2} + r^{-1/2} \frac{df_l}{dr} - \frac{1}{4} r^{-3/2} f_l$$

Eq. (9.107) for $r < a$, now becomes

$$r^{1/2}\frac{d^2 f_{II}}{dr^2} + r^{-1/2}\frac{df_{II}}{dr} + \left[k_4^2 - \frac{l(l+1)}{r^2}\right]r^{1/2} f_{II} - \frac{1}{4}r^{-3/2} f_{II} = 0$$

or

$$r^2\frac{d^2 f_{II}}{dr^2} + r\frac{df_{II}}{dr} + \left[(k_4 r)^2 - \left(l+\frac{1}{2}\right)^2\right]f_{II} = 0$$

or

$$(k_4 r)^2\frac{d^2 f_{II}}{d(k_4 r)^2} + (k_4 r)\frac{df_{II}}{d(k_4 r)} + \left[(k_4 r)^2 - \left(l+\frac{1}{2}\right)^2\right]f_{II} = 0$$

or

This is Bessel's differential equation of order $\pm\left(l+\frac{1}{2}\right)$ having argument $k_4 r$.

$$\therefore \quad f_{II}(r) = A' J_{l+1/2}(k_4 r) + B' J_{-(R+1/2)}(k_4 r)$$

where A' and B' are arbitrary constants. Since $J_{-(l+1/2)}(0) \neq (0)$, therefore $f_{I1}(\theta) = 0$, demands that $B' = 0$. We define spherical Bessel function by the equation

$$J_l(kr) = \sqrt{\frac{\pi}{2kr}} J_{l+1/2}(kr)$$

$$\because \quad R_{II}(k_4 r) = \frac{V_{II}(k_4 r)}{r} = \frac{f_{II}(k_4 r)}{r^{1/2}} = \frac{A' J_{(l+1/2)}(k_4 r)}{r^{1/2}} = A J_l^{(k_4 r)} \qquad (9.117)$$

The leading terms for small $k_4 r$ are

$$J_l(k_4 r) \to \frac{(k_4 r)^l}{(2l+l)!} \qquad (9.118)$$

where $(2l+1)! = 1 \times 3 \times 5 \times \ldots \times (2l+1)$

Spherical Bessel functions of order 0, 1, 2, ... etc. have been defined earlier. Radial Eq. (9.108) for $r > a$, becomes

$$r^{1/2}\frac{d^2 f_{I11}}{dr^2} + r^{-1/2}\frac{df_{I11}}{dr} + \left[-k_3^2 - \frac{l(l+1)}{r^2}\right]r^{1/2} f_{I11} - \frac{1}{4}r^{-3/2} f_{I11} = 0$$

or

$$r^2\frac{d^2 f_{I11}}{dr^2} + r\frac{df_{I11}}{dr}\left[-k_3^2 r^2 - \left(\frac{l(l+1)}{r^2}\right)^2\right]f_{I11} = 0$$

or

$$(ik_3 r)^2\frac{d^2 f_{III}}{d(ik_3 r)^2} + (ik_3 r)\frac{df_{III}}{d(ik_3 r)}\left[(ik_3 r)^2 - \left(l+\frac{1}{2}\right)^2\right]f_l = 0$$

This is Bessel's differential equation of order $\pm\left(l+\frac{1}{2}\right)$ and argument $(ik_3 r)$. Solutions of this equation may be written as

$$f_{III} = A'' J_{l+1/2}(ik_3 r) + B'' J_{-(l+1/2)}(ik_3 r)$$

We define spherical Neumann functions, for half integral (negative) Bessel function by equation

$$n_l(kr) = (-1)^{l+1}\left(\frac{\pi}{2kr}\right)^{1/2} J_{-(l+1/2)}(kr) \tag{9.119}$$

In this region of space, both spherical Bessel function and spherical Neumann function are allowed solutions and hence their linear combination. Now we define

$$h_l^{(1)}(kr) = J_l(kr) + i\eta_l(kr)$$
$$h_l^{(2)}(kr) = J_l(kr) + i\eta_l(kr)$$

$h_l^{(1)}$ and $h_l^{(2)}$ are known as **spherical Hankel function** of the first and second kind respectively. Exponential forms of spherical Bessel, Neumann and Hankel functions are given by

$$J_l(kr)_{\overline{(kr\to\infty)}} \frac{\sin\left(kr - l\frac{\pi}{2}\right)}{kr} = \frac{\cos\left\{kr - \frac{(l+1)\pi}{2}\right\}}{kr}$$

$$n_l'(kr)_{\overline{(kr\to\infty)}} - \frac{\cos\left(kr - \frac{l\pi}{2}\right)}{kr} = \frac{\sin\left\{kr - \frac{(l+1)\pi}{2}\right\}}{kr}$$

$$h_l^{(1)}(kr)_{\overline{(kr\to\infty)}} \frac{\cos\left\{kr - \frac{(l+1)\pi}{2}\right\}}{kr} + i\frac{\sin\left\{kr - \frac{(l+1)\pi}{2}\right\}}{kr} = \frac{e^{i}\left\{\rho - \frac{(l+1)\pi}{2}\right\}}{kr}$$

$$h_l^{(2)}(kr)_{\overline{(kr\to\infty)}} \frac{\cos\left\{kr - \frac{(l+1)\pi}{2}\right\}}{kr} + i\frac{\sin\left\{kr - \frac{(l+1)\pi}{2}\right\}}{kr} = \frac{e^{-i}\left\{\rho - \frac{(l+1)\pi}{2}\right\}}{kr}$$

Boundary conditions in this case demands that the wave function must approach zero in the limit $r \to \infty$; only these solutions are allowed which decrease exponentially for large values of r. Hence spherical Hankel functions of the second kind are not acceptable, as they increase exponentially. Therefore allowed solution for $r > a$may be written as

$$R_{III} = Bh_l^{(1)}(ik_3r) - B[J_l(ik_3r) + ir_{il}(ik_3r)] \tag{9.120}$$

Explicit expressions for spherical Neumann and spherical Hankel functions for $l = 0, 1, 2, \ldots$ are given below:
Spherical Neumann functions

$$\eta_0(kr) = -\frac{\cos kr}{kr}$$

$$\eta_1(kr) = -\frac{\cos kr}{k^2r^2} - \frac{\sin kr}{kr}$$

$$\eta_2(kr) = -\left(\frac{3}{k^3r^3} - \frac{1}{kr}\right)\cos kr - \frac{3}{k^2r^2}\sin kr$$

$$\eta_3(kr) = \left(\frac{6}{k^2r^2} - \frac{15}{k^4r^4}\right)\cos kr + \left(\frac{1}{kr} - \frac{15}{k^3r^3}\right)\sin kr$$

Spherical Hankel functions of the first kind

$$h_0^{(1)}(ikr) = \frac{e^{-kr}}{kr}$$

$$h_1^{(1)}(ikr) = i\left(\frac{1}{kr} + \frac{1}{k^2r^2}\right)e^{-kr}$$

$$h_2^{(1)}(ikr) = \left(\frac{1}{kr} + \frac{3}{k^2r^2} + \frac{3}{k^3r^3}\right)e^{-kr}$$

$$h_3^{(1)}(ikr) = -i\left(\frac{1}{kr} + \frac{6}{k^2r^2} + \frac{15}{k^3r^3} + \frac{15}{k^4r^4}\right)e^{-kr}$$

Spherical Hankel functions of the second kind are:

$$h_0^{(2)}(ikr) = e^{ikr}\,kr$$

$$h_1^{(2)}(ikr) = \left(-\frac{1}{kr} - \frac{1}{k^2r^2}\right)e^{kr}$$

$$h_2^{(2)}(ikr) = \left(\frac{1}{kr} + \frac{3}{k^2r^2} - \frac{3}{k^3r^3}\right)e^{kr}$$

$$h_3^{(2)}(ikr) = i\left(-\frac{1}{kr} + \frac{6}{k^2r^2} - \frac{15}{k^3r^3} + \frac{15}{k^4r^4}\right)e^{kr}$$

Also $\quad \eta_l(kr)_{\overrightarrow{(kr\to 0)}} - \dfrac{(2l-1)!}{(kr)^{l+1}}$

where $(2l-1)! = 1\times 3\times 5\times \cdots \times (2l-1)$

A few properties of J' s of η' s are

(i) $\int J_0^2(\rho)\rho^2 d\rho = \frac{1}{2}\rho^3[J_0^2(\rho) + \eta_0(\rho)J_1(\rho)]$

(ii) $\int \eta_0^2(\rho)\rho^2 d\rho = \frac{1}{2}\rho^3[\eta_0^2(\rho) - J_0(\rho)\eta_1(\rho)]$

(iii) $\eta_{l-1}(\rho)J_l(\rho) - \eta_l(\rho)J_{l-1}(\rho) = \dfrac{1}{\rho^2}$ for $l > 0$

(iv) $J_l(\rho)\dfrac{d\eta_l(\rho)}{d\rho} - \eta_l(\rho)\dfrac{dJ_l(\rho)}{d\rho} = \dfrac{1}{\rho^2}$

(v) $J_{l-1}(\rho) + J_{l+1}(\rho) = \dfrac{2l+1}{\rho}J_l(\rho)$ for $l > 0$

(vi) $\eta_{l-1}(\rho) + \eta_{l+1}(\rho) = \dfrac{2l+1}{\rho}\eta_l(\rho)$ for $l > 0$

(vii) $\dfrac{dJ_l(\rho)}{d\rho} = \dfrac{1}{2l+1}[lJ_{l-1}(\rho) - (l+1)J_{l+1}(\rho)] = J_{l-1}(\rho) - \dfrac{l+1}{\rho}J_l(\rho)$

(viii) $\dfrac{d\eta_l(\rho)}{d\rho} = \dfrac{1}{2l+1}[\eta_{l-1}(l+1)_{\eta+1}(\rho)] = \eta_{l-1}(\rho) - \dfrac{l+1}{\rho}\eta_l(\rho)$

(ix) $\dfrac{d}{d\rho} = [\rho^{l+1}J_l(\rho)] = \rho^{l+1}J_{l-1}(\rho)$ for $l > 0$

(x) $\frac{d}{d\rho}=[\rho^{l+1}\eta_l(\rho)]=\rho^{l+1}\eta_{l-1}(\rho)$ for $l>0$

(xi) $\frac{d}{d\rho}=[\rho^{-l}J_l(\rho)]=\rho^{-l}J_{l+1}(\rho)$

(xii) $\frac{d}{d\rho}=[\rho^{-l}\eta_l(\rho)]=\rho^{-l}\eta_{l+1}(\rho)$

(xiii) $\int J_l^2(\rho)\rho^2 d\rho=\frac{1}{2}\rho^3[J_l^2(\rho)-J_{l-1}(\rho)J_{l+1}(\rho)]$ for $l>0$

(xiv) $\int \eta_l^2(\rho)\rho^2 d\rho=\frac{1}{2}\rho^3[\eta_l^2(\rho)-\eta_{l-1}(\rho)\eta_{l+1}(\rho)]$ for $l>0$

Energy levels: The energy levels are obtained by demanding the continuity of logarithmic derivative of $R(r)$ at $r=a$. This gives

$$\left[\frac{1}{AJ_l(k_4r)}\times\frac{d}{dr}\{AJ_l^1(k_4r)\}\right]_{r=a}=\left[\frac{1}{Bh_l^{(1)}(ik_3r)}\frac{d}{dr}\{Bh_l^{(1)}(ik_3r)\}\right]_{r=a}$$

or
$$\frac{k_4}{J_l(k_4a)}\left[\frac{dJ_l(k_4a)}{d(k_4a)}\right]_{r=a}=\frac{ik_3}{h_l^{(1)}(ik_3a)}\left[\frac{dh_l^{(1)}(ik_3r)}{d(ik_3r)}\right]_{r=a}$$

or
$$\frac{k_4}{J_l(k_4a)}\left[J_{l-1}(k_4a)-\frac{l+1}{k_4a}\right]=\frac{k_3}{h_l^{(1)}(ik_3a)}\left[h_{l-1}^{(1)}(ik_4)-\frac{l+1}{ik_3a}h_l^{(1)}(ik_3a)\right]$$

or
$$k_4\frac{J_{l-1}(k_4a)}{J_l(k_4a)}-\frac{l+1}{a}=ik_3\frac{h_{l-1}^{(1)}(ik_3a)}{h_l^{(1)}(ik_3a)}-\frac{l+1}{a}$$

or
$$k_4\frac{J_{l-1}(k_4a)}{J_l(k_4a)}=ik_3\left[\frac{h_{l-1}(ik_3a)+i\eta_{l-1}(ik_3a)}{J_l(k_3a)+i\eta_l(ik_3a)}\right] \tag{9.121}$$

Eq. (9.121) can be solved to give allowed values of energy for any value of angular momentum l. For $l=1$, Eq. (9.121) takes a simple from

$$k_4\left[\frac{J_0(k_4a)}{J_1(k_4a)}\right]=ik_3\left[\frac{J_0(ik_3a)+i\eta_0(ik_3a)}{J_1(ik_3a)+i\eta_1(ik_3a)}\right]=Ik_3\left[\frac{h_0^{(1)}(ik_3a)}{h_1^{(1)}(ik_3a)}\right]$$

or
$$k_4\left\{\left(\frac{\sin k_4a}{k_4a}\right)\frac{1}{\left(\frac{\sin k_4a-k_4a\cos k_4a}{k_4^2a^2}\right)}\right\}=ik_3\left[\left\{-\frac{e^{-ik_3a}}{k_3a}\right\}\frac{1}{\left\{\frac{i(1+k_3a)e^{-ik_3a}}{k_3^2a^2}\right\}}\right]$$

or
$$k_4^2a\left[\frac{\sin k_4a}{\sin k_4a-k_4a\cos k_4a}\right]=-\frac{k_3^2a}{1+k_3a}$$

or
$$\left[\frac{\sin k_4a-k_4a\cos k_4a}{\sin k_4a}\right]\left(\frac{1}{k_4^2}\right)=-\left(\frac{1+k_3a}{k_3^2}\right)$$

or $$1 - k_4 a \cot k_4 a = -\frac{k_4^2}{k_3^2}(1 + k_3 a)$$

or $$\frac{\cot k_4 a}{k_4 a} - \frac{1}{k_4^2 a^2} = \frac{1 - k_3 a}{k_3^2 a^2} = \frac{1}{k_3 a} + \frac{1}{k_3^2 a^2} \tag{9.122}$$

and $$k_3^2 a^2 + k_4^2 a^2 = -\frac{2\mu a^2 V}{\hbar^2} + \frac{2\mu a^2 (V + V_0)}{\hbar^2} = \frac{2\mu V_0 a^2}{\hbar^2} = \beta^2 \tag{9.123}$$

Eq. (9.122) can be solved graphically (or numerically) in a manner similar to the case $l = 0$. A new level appears whenever $k_3 a$ is zero or cot $k_4 a$ is infinite. This occurs whenever $k_4 a$ is an integral multiple of π; thus bound state solutions are obtained whenever $k_4 a$ is equal to π, 2π, 3π ...., etc. Thus there is no energy level with $l = 1$, if $V_0 \le \frac{\pi^2 \hbar^2}{2\mu a^2}$. There will be one bound state with $l - 1$ if

$$\frac{\pi^2 \hbar^2}{2\mu a^2} < V_0 \le \frac{2\pi^2 \hbar^2}{\mu a^2}$$

The smallest value of $V_0 a^2$ for which a bound state with $l = 1$ exists higher than the corresponding value of $V_0 a^2$ for $l = 0$. This can be justified as follows: the presence of a centrifugal term in potential energy for a particle possessing angular momentum requires a stronger potential to bind it than a particle with no angular momentum.

On solving Eq. (9.121), we find that the minimum square well potential strength "$V_0 a^2$" required to bind a particle of angular momentum quantum number l increases in the value l.

9.6 FREE PARTICLE IN THREE DIMENSIONS

Schrodinger equation for particle of a mass μ is

$$\frac{-\hbar}{2\mu} \nabla^2 \psi = E\psi \tag{9.124}$$

Here $V = 0$ everywhere, using cartesian coordinates, we can separate the variables getting one equation for each variable x, y and z. We obtain

$$\psi(x, y, z) = X(x)Y(y)Z(z)$$

Substituting this value of ψ in Eq. (9.124), one obtains

$$\frac{d^2 X}{dx^2} + \frac{2\mu E_x}{\hbar^2} X = \frac{d^2 X}{dx^2} + k_x^2 X = 0 \tag{9.125}$$

$$\frac{d^2 Y}{dy^2} + \frac{2\mu E_y}{\hbar^2} Y = \frac{d^2 Y}{dy^2} + k_x^2 Y = 0 \tag{9.126}$$

$$\frac{d^2 Z}{dz^2} + \frac{2\mu E_z}{\hbar^2} Z = \frac{d^2 Z}{dz^2} + k_x^2 Z = 0 \tag{9.127}$$

where $$k^2 = k_x^2 + k_y^2 + k_z^2 = \frac{2\mu}{\hbar^2}(E_x + E_y + E_z) = \frac{2\mu E}{\hbar^2} \tag{9.128}$$

Solution of Eq. (9.125) may be written as

$$X(x) = Ae^{ik_x x} + Be^{-ik_x x}, Y(y) = A_{1}e^{ik_y y} + B_{1}e^{-ik_y y}, Z(z) = A_{2}e^{ik_z z} + B_{2}e^{-ik_z Z}$$

$$\frac{\hbar k_x}{\mu}\{1|A|^2 - |B|^2\}, \frac{\hbar k_y}{\mu}\{|A_1|^2 - |B_1|^2\} \text{ and } \frac{\hbar k_z}{\mu}\{|A_z|^2 - |B_z|^2\}$$

represents net flux of particles (No. of particles incident per unit area per unit time) along $+x$, $+y$ and $+z$-directions respectively. The wave function that describes a flux of particles in some directions of space (say along propagation vector ($\boldsymbol{k}$)) moving with velocity $\frac{\hbar|k|}{\mu}$ is given by

$$\psi(r) = Ae^{ik.r} \tag{9.129}$$

where r is the position vector of a point from an arbitrarily fixed origin. Solution Eq. (9.129) is known as **plane wave solution** since it represents an infinite wavefront propagating though space. A' is a constant which can be determined by Dirac-δ function normalization. Here $\boldsymbol{k}$ must be real, if the eigen functions are to remain finite at large distances. Since every positive value of E is allowed, the spectrum is continuous and degeneracy is infinite for each positive value of energy; Eq. (9.128) restricts the magnitude of $\boldsymbol{k}$, but not its direction. Plane wave solutions Eq. (9.129) with energy E are also functions of linear momentum operator $-i\hbar\nabla$ with eigen value $\hbar\boldsymbol{k}$.$\boldsymbol{k}$-normalization for free particle eigen functions is given as

$$\int \psi_k^*(r)\psi_{r'}(r)d\tau = \delta(\boldsymbol{k} - \boldsymbol{k'}) = \delta(k_x - k_x')\delta(k_y - k_y')\delta(k_z - k_z') \tag{9.130}$$

where $d\tau = dxdydz$ is the volume element.

$\therefore$
$$|A|^2 \int e^{-ik.r} e^{-ik'.r} d\tau = \delta(\boldsymbol{k} - \boldsymbol{k'})$$

Since
$$\int e^{-(k'-r).r} d\tau = (2\pi)^3 \delta(\boldsymbol{k} - \boldsymbol{k'})$$

$\therefore$
$$|A|^2 = (2\pi)^{-3} \text{ or } A = (2\pi)^{-3/2}$$

within an arbitrary phase factor, which we choose zero.

$\therefore$ Normalised plane wave solutions are

$$\psi(\boldsymbol{r}) = (2\pi)^{-3/2} e^{ik.r} \tag{9.131}$$

The closure relation applied to the eigen functions is

$$\int \psi_k^*(\boldsymbol{r})\psi_k(\boldsymbol{r'})d^3k = \delta(\boldsymbol{r} - \boldsymbol{r'})$$

or
$$\frac{1}{(2\pi)^3}\int e^{ik.(r'-r)} d^3k = \delta(\boldsymbol{r} - \boldsymbol{r'}) \tag{9.132}$$

in agreement with the definition of **Dirac delta function.**

Periodic Boundary Conditions: If we desire to have discrete eigen values, we impose periodic boundary conditions by demanding that upon translation of L in the direction of (any one of three) the coordinate axes all eigen functions of Schodinger equation will remain unchanged.

The eigen functions are of the same form as given by Eq. (9.129) but k_x, k_y, k_z now satisfy the conditions

$$k_x^2 + k_y^2 + k_z^2 = \frac{2\mu E}{\hbar^2}$$

and
$$k_x = \frac{2\pi}{L}n_x,\ k_y = \frac{2\pi}{L}n_y \text{ and } k_z = \frac{2\pi}{L}n_z \tag{9.133}$$

where n_x, n_y ane n_z are integers (positive, negative or zero).

The wave functions is now normalized in a cube of side length L, thus giving

$$\int_0^L \int_0^L \int_0^L |\psi_k(r)|^2\, dx\, dy\, dz = 1 \text{ or } \int_0^L \int_0^L \int_0^L dx\, dy\, dz = 1$$

or
$$A^2 L^3 = 1 \ \therefore\ A = L^{-3/2}$$

Hence normalised eigen functions are $\psi k(\mathbf{r}) = L^{-3/2} e^{-ik.r}$

Density of states: The number of states whose x component of $\mathbf{k}$ lies between k_x and $k_x + dk_x$ is simply the number of integers lying between $\frac{Lk_x}{2\pi}$ and $\frac{L}{2\pi}(k_x + dk_x)$. This number is approximately equal to $\frac{L}{2\pi}dk_x$. Thus there are $\left(\frac{L}{2\pi}\right)^3 dk_x dk_y dk_z$ states in the range $dk_x dk_y dk_z$ of $\mathbf{k}$. If $g(k, r)\, dkdr$ denotes the number of states whose $|\mathbf{k}|$ lies between k and $k + dk$ within the solid angle $d\Omega$, then

$$g(k, r)\, dk d\Omega = \left(\frac{L}{2\pi}\right)^3 k^2\, dk d\Omega = \frac{V k^2\, dk \sin\theta d\theta d\varphi}{8\pi^3}$$

Since
$$E = \frac{\hbar^2 k^2}{\mu} \ \therefore\ dE = \frac{\hbar^2 k}{\mu} dk$$

$$g(E, \Omega)\, dE d\Omega = \frac{V}{8\pi^3}\left(\frac{2\mu E}{\hbar^2}\right)\frac{\mu\, dE}{\hbar^2}\left(\frac{\hbar^2}{2E\mu}\right)^{1/2} d\Omega$$

or
$$g(E, r)\, dE\, d\Omega = \frac{\mu V}{8\pi^3 \hbar^2}\left(\frac{2\mu E}{\hbar^2}\right)^{1/2} dE d\Omega$$

Denoting $\rho(E) = g(E, \Omega) d\Omega$, we obtain

$$\rho(E) = \frac{\mu V}{3\pi^3 \hbar^2}\left(\frac{2\mu E}{\hbar^2}\right)^{1/2} d\Omega = \frac{\mu V}{8\pi^3 \hbar^2} k d\Omega$$

Total number of states between E and $E + dE$ is obtained by integrating over the solid angle

$$g(E) dE = \frac{\mu V}{8\pi^3 \hbar^2} \frac{(2\mu)^{1/2} E^{1/2}}{\hbar} dE 4\pi = \frac{V(2\mu)^{3/2}}{4\pi^2 \hbar^2} E^{1/2} dE \tag{9.134}$$

The free particle problem can also be regarded as a central force problem since $V = 0$ is a special case of a potential that depends only on r. Eigen functions are separable in spherical polar coordinates and angular momentum must be constant of motion.

Schrodinger equation for $V = 0$ may be written as

$$\left[\frac{1}{r^2}\frac{\partial}{\partial r}\left(r^2\frac{\partial}{\partial r}\right)+\frac{1}{r^2\sin\theta}\frac{\partial}{\partial\theta}\left(\sin\theta\frac{\partial}{\partial\theta}\right)+\frac{1}{r^2\sin\theta}\frac{\partial^2}{\partial\varphi^2}\right]\psi(r,\theta,\varphi)+\frac{2\mu E}{\hbar^2}\psi(r,\theta,\varphi)=0$$

Writing $\psi(r, \theta, \phi) = R_l(r)Y_{lm}(\theta, \phi)$, we can separate the above equation into angular and radial parts. Equation satisfied by radial function $R_l(r)$ may be written as

$$\frac{1}{r^2}\frac{d}{dr}\left(r^2\frac{d}{dr}\right)R_l(r)+\left[\frac{2\mu E}{\hbar^2}-\frac{l(l+1)}{r^2}\right]R_l(r)=0$$

or
$$\frac{1}{r^2}\frac{d}{dr}\left(r^2\frac{d}{dr}\right)R_l(r)+\left[k^2-\frac{l(l+1)}{r^2}\right]R_l(r)=0 \tag{9.135}$$

The solutions of Eq. (9.135) have been discussed in previous sections; the general solution may be written as

$$R_l(r)=\frac{U_l(r)}{r}AJ_l(kr)+B\eta_l(kr)=0 \tag{9.136}$$

where A and B are arbitrary constants. Since $\eta(0) \neq 0$, the coefficient B must be chosen to be equal to zero in order that $\psi(0)$ be finite. An equivalent representation of the solution is

$$R_l(r)=\frac{U_l(r)}{r}A'h_l^{(1)}(kr)-B'h_l^{(2)}(kr) \tag{9.137}$$

These functions correspond to imaginary exponential solutions in one dimension. Thus there are two representations of free particles:

(i) A collimated beam of particles having precisely known energy E and momentum $P = \hbar k$ and direction $\hat{k}$ but completely unknown position represented by plane wave solutions.

(ii) A flux of particles is emitted from a source with known values of energy E, angular momentum l and z-component of angular momentum represented by spherical waves.

$$\psi_{klm}(r,t)=A''hl^{(1)}(kr)Y_{lm}(\theta,\varphi)e^{-i\omega t} \tag{9.138}$$

Let us consider a scattering problem in which a collimated beam of particles impinges on a scattering centre of forcewhich can be represented by a spherically symmetric potential for any particle in the incident beam. Incident beam can be represented by plane wave, but it is convenient to use spherical wave to study the interaction at the scattering centre. After scattering of the incident beam, it is convenient to use plane wave description. Thus within the range of the potentials (k, l, m) description is preferred while outside the range of the potential (k_x, k_y, k_z) description is preferred. Let us therefore express ψ_{k_x,k_y,k_z} as an expansion in the functions ψ_{klm}

$$\psi_{k_x,k_y,k_z}(r,t)=\sum_k\sum_l\sum_m a_{klm}(k,\theta_k,\phi_k)\psi_{klm}(r,t) \tag{9.139}$$

where k, θ_k, ϕ_k are the spherical polar coordinates of the vector $\boldsymbol{k}$. In order that the two sides of Eq. (9.175) may have the same time dependence, we must choose $k=\sqrt{k_x^2+k_y^2+k_z^2}$.

Therefore, only a single term from the k-summation contributes in Eq. (9.139). The direction of x-axis for the specification of position of a particle is chosen as $\mathbf{k}$-direction. Thus if time dependence and k summation is omitted in Eq. (9.139), we get

$$\frac{1}{(2\pi)^{3/2}} e^{ikr\cos\theta} = \sum_l \sum_m a_{lm}(\theta_k, \phi_k) J_l(kr) Y_{lm}(\theta, \phi) \tag{9.140}$$

Since the L.H.S. of Eq. (9.140) does not depend on ϕ, neither can the R.H.S. Only Y_{lm} with $m=0$ are independent of φ, therefore only such terms can contribute to the summation and

$$\frac{1}{(2\pi)^{3/2}} e^{ikr\cos\theta} = \sum_l a_{l0}(\theta_k, \phi) J_l(kr) \frac{1}{\sqrt{2\pi}} P_l(\cos\theta)$$

Using $\cos\theta = x$ and multiplying by $P_l'(x)$ and integrating over x, we obtain

$$\frac{1}{(2\pi)^{3/2}} \int_{-1}^{+1} e^{ikrx} P_l'(x) dx = \sum_l a_{l0} J_l(kr) \frac{1}{\sqrt{2\pi}}) \int_{-1}^{+1} P_l'(x) P_l(x) dx$$

or
$$\frac{1}{(2\pi)^{3/2}} [2i^{l'} J_l'(kr)] = \sum_l a_{l0} J_l(kr) \frac{1}{\sqrt{2\pi}} \left(\frac{2}{2l+1}\right) \delta_{ll'}$$

or
$$\frac{1}{(2\pi)^{3/2}} [2i^{l'} J_l'(kr)] = \frac{2}{(2l+1)\sqrt{2\pi}} a_l' J_l'(kr)$$

where the integral on the L.H.S. has been evaluated by parts. We have

$$\therefore \qquad a_{l0}' = \frac{(2l'+1) i^{l'}}{2\pi}$$

$$\therefore \qquad \frac{1}{(2\pi)^{3/2}} e^{ikr\cos\theta} = \frac{1}{(2\pi)^{3/2}} \sum i^l (2l+1) J_l(kr)(\cos\theta) J\, P_l$$

The reverse transformation from ψ_{klm} to ψ_{k_x,k_y,k_z} can also be derived. To do this we must integrate over all possible orientation of ψ_{k_x,k_y,k_z}. Thus

$$\int \psi_{klm} = (r,t) = \int\int \psi_{k_x,k_y,k_z}(r,t) b_{lm}(\theta_k, \phi_k) \sin\theta_k, d\theta_k d\phi_k$$

It can be shown that $b_{lm} = a_{lm}^*$ (9.141)

9.7 HYDROGEN ATOM

A system consisting of two particles, a proton and electron is known as hydrogen atom. Hamiltonian for this system may be written as

$$H = \frac{P_1^2}{2m_l} + \frac{P_2^2}{2m_2} + V(|r_1 - r_2|) \tag{9.142}$$

here the subscripts 1 and 2 denote the electron and proton respectively. We change the variables from $\boldsymbol{r}_1$ and $\boldsymbol{r}_2$ to $\boldsymbol{R}$ (coordinate of centre of mass) and $\boldsymbol{r}$ (relative separation between the two particles) defined by the following equations:

$$\boldsymbol{R}(x, y, z) = \frac{m_1 r_1 + m_2 r_2}{m_1 + m_2} \tag{9.143}$$

and $$\boldsymbol{r}(x, y, z) = r_1 - r_2 \tag{9.144}$$

$$\therefore \quad \boldsymbol{R} = \frac{m_1 r_1 + m_2(r_1 - r)}{m_1 + m_2} + \frac{(m_1 + m_2)r_1 - m_2 r}{m_1 + m_2} = r_1 - \frac{m_2}{m_1 + m_2} r = r_1 - \frac{\mu r}{m_1}$$

$$\therefore \quad r_1 = \boldsymbol{R} + \frac{\mu}{m_1} r \text{ and } r_2 = r_1 - r = \boldsymbol{R} + \frac{m_2}{m_1 + m_2} r - r = \boldsymbol{R} - \frac{\mu r}{m_2}$$

where $\mu = \dfrac{m_1 m_2}{m_1 + m_2}$ is known as reduced mass of the system. For partial derivatives, we have

$$\frac{\partial X}{\partial x_1} = \frac{m_1}{m_1 + m_2}, \frac{\partial X}{\partial x_2} = \frac{m_2}{m_1 + m_2}, \frac{\partial x}{\partial x_1} = 1, \frac{\partial x}{\partial x_2} = -1$$

and similar equations for y and z.

$$\frac{\partial \psi}{\partial x_1} = \frac{\partial \psi}{\partial X}.\frac{\partial X}{\partial x_1} + \frac{\partial \psi}{\partial x}.\frac{\partial x}{\partial x_1} = \frac{m_1}{m_1 + m_2}\frac{\partial \psi}{\partial X} + (1)\frac{\partial \psi}{\partial x} = \frac{\mu}{m_2}.\frac{\partial \psi}{\partial X} + \frac{\partial \psi}{\partial x}$$

$$\frac{\partial \psi}{\partial x_2} = \frac{\partial \psi}{\partial X}.\frac{\partial X}{\partial x_2} + \frac{\partial \psi}{\partial x}.\frac{\partial x}{\partial x_2} = \frac{m_2}{m_1 + m_2}.\frac{\partial \psi}{\partial X} + (-1)\frac{\partial \psi}{\partial x} = \frac{\mu}{m_1}.\frac{\partial \psi}{\partial X} - \frac{\partial \psi}{\partial x}$$

$$\therefore \quad \nabla_1 = \frac{\mu}{m_1}\nabla_R + \nabla_r$$

and $$\nabla_2 = \frac{\mu}{m_2}\nabla_R - \nabla_r$$

where $$\nabla_R = \left(\frac{\partial}{\partial X}, \frac{\partial}{\partial Y}, \frac{\partial}{\partial X}\right) \text{ and } \nabla_r = \left(-\frac{\partial}{\partial x}, \frac{\partial}{\partial y}, \frac{\partial}{\partial z}\right)$$

$$\nabla_1^2 \psi = \left(\frac{\mu}{m_2}\nabla_R + \nabla_r\right)\left(\frac{\mu}{m_2}\nabla_R \psi + \nabla_r \psi\right) = \frac{\mu^2}{m_2^2}\nabla_R^2 \psi + \frac{\mu}{m^2}(\nabla_R \nabla_r + \nabla_r \nabla_R)\psi + \nabla_r^2 \psi$$

and $$\nabla_2^2 \psi = \left(\frac{\mu}{m_1}\nabla_R - \nabla_r\right)\left(\frac{\mu}{m_1}\nabla_R \psi - \nabla_r \psi\right) \frac{\mu^2}{m_1^2}\nabla_R^2 \psi - \frac{\mu}{m_1}(\nabla_R \nabla_r + \nabla_r \nabla_R)\psi + \nabla_r^2 \psi$$

The wave function $\psi(\boldsymbol{r}_1 - \boldsymbol{r}_2)$ is a function of six coordinates and is a solution of the Schrodinger equation

$$H\psi = \left[-\frac{\hbar^2}{2m_1}\nabla_1^2 - \frac{\hbar^2}{2m_2}\nabla_2^2 + V(r)\right]\psi = E_t \psi$$

or $$\left[-\frac{\hbar^2}{2m_1}\left\{\frac{\mu^2}{m_2^2}\nabla_R^2+\frac{\mu}{m_2}(\nabla_R\nabla_r+\nabla_r\nabla_R)+\nabla_r^2\right\}\right.$$

$$\left.-\frac{\hbar^2}{2m_2}\left\{\frac{\mu^2}{m_1^2}\nabla_R^2-\frac{\mu}{m_1}(\nabla_R\nabla_r+\nabla_r\nabla_R)+\nabla_r^2\right\}+V(r)\right]\psi=E_t\psi$$

or $$\left\{-\frac{\hbar^2\mu^2}{2m_1m_2^2}-\frac{\hbar^2\mu^2}{2m_1^2m_2}\right\}\nabla_R^2\psi+\left\{-\frac{\hbar^2}{2m_1}-\frac{\hbar^2}{2m_2}\right\}\nabla_r^2\psi$$

$$-\frac{\hbar^2\mu}{2m_1m_2}(\nabla_R\nabla_r+\nabla_r\nabla_R)\psi+\frac{\hbar^2\mu}{2m_1m_2}\{\nabla_R\nabla_r+\nabla_r\nabla_R\}\psi+V(r)\psi=E_t\psi$$

or $$-\frac{\hbar^2\mu^2}{2m_1^2m_2^2}(m_1+m_2)\nabla_R^2\psi-\frac{\hbar^2}{2m_1m_2}(m_2+m_1)\nabla_r^2\psi+V(r)\psi=E_t\psi$$

or $$-\frac{\hbar^2}{2(m_1+m_2)}\nabla_R^2\psi-\frac{\hbar^2}{2\mu}\nabla_r^2\psi+V(r)\psi=E_t\psi \tag{9.145}$$

Eq. (9.145) can now be separated into two parts, by writing ψ ($\boldsymbol{R}$, $\boldsymbol{r}$) as the product of two functions $\Phi(\boldsymbol{R})$ and $\psi(\boldsymbol{r})$; $\psi(\boldsymbol{R}, r)=\Phi(\boldsymbol{R})\psi(\boldsymbol{r})$. By usual procedure we find that Φ and ψ satisfy equations.

$$-\frac{\hbar^2}{2M}\nabla_R^2\Phi(R)=E_e\Phi(R) \tag{9.146}$$

$$-\frac{\hbar^2}{2\mu}\nabla_r^2\psi(r)+V(r)\psi(r)=E\psi(r) \tag{9.147}$$

Where $M=m_1+m_2$ is total mass of the system and $E_t=E_e+E$ is total energy of the system. Eq. (9.146) has a solution of the form

$$\Phi(R)=(\text{Constant})\times\exp\left(\frac{i\boldsymbol{P}.\boldsymbol{R}}{\hbar}\right) \tag{9.148}$$

Where $\boldsymbol{P}$ is the momentum of the centre of mass, having magnitude $\sqrt{2ME_e}$ and arbitrary direction. The function $\Phi(R)$ given by Eq. (9.148) represents the motion of the centre of mass of the system, as that of a single particle of mass $M=m_1+m_2$ and energy $E_e=\dfrac{P^2}{2M}$. It is plane wave solution indicating that the centre of mass moves as if it were a free particle. This corresponds to the classical result that the centre of mass moves in a straight line with uniform speed.

The potential energy for hydrogen and hydrogen like atoms is given by $V(r)=-\dfrac{Ze^2}{r}$. Eq. (9.147) becomes (in spherical polar coordinates) $\left[\nabla_r^2+\dfrac{2\mu}{\hbar^2}\left(E+\dfrac{Ze^2}{r}\right)\right]\psi=0$

or $$\left[\frac{1}{r^2}\frac{\partial}{\partial r}\left(r^2\frac{\partial}{\partial r}\right)+\frac{1}{r^2\sin\theta}\frac{\partial}{\partial\theta}\left(\sin\theta\frac{\partial}{\partial\theta}\right)+\frac{1}{r^2\sin\theta}\frac{\partial^2}{\partial\varphi^2}\right]\psi+\frac{2\mu}{\hbar^2}\left(E+\frac{Ze^2}{r}\right)\psi=0 \tag{9.149}$$

Since the potential is spherically symmetric, we can carry out the separation of variables by assuming $\psi(r) = R(r)Y(\theta, \varphi)$. Equations satisfied by angular and radial parts may be written as

$$\frac{1}{\sin\theta}\frac{\partial}{\partial\theta}\left(\sin\theta\frac{\partial Y}{\partial\theta}\right)+\frac{1}{\sin^2\theta}\frac{\partial^2 Y}{\partial\varphi^2}+\beta Y=0 \tag{9.150}$$

$$\frac{1}{r^2}\frac{\partial}{\partial r}\left(r^2\frac{\partial R}{\partial r}\right)+\left[\frac{2\mu}{\hbar^2}\left(E+\frac{Ze^2}{r}\right)-\frac{\beta}{r^2}\right]R=0 \tag{9.151}$$

Solutions of Eq. (9.150) have already been discussed in section (9.3). These solutions are

$$Y_{lm}(\theta,\varphi)=\in\left[\left(\frac{2l+1}{4\pi}\right)\frac{|1-|m||}{|l+|m||}\right]^{1/2}P_l^{|m|}(\cos\theta)e^{im\phi} \tag{9.152}$$

where
$$\varepsilon=\begin{cases}(-1)^m & \text{for} \quad m>0\\ 1 & \text{for} \quad m<0\end{cases} \tag{9.153}$$

Radial Equation

$$\frac{1}{r^2}\frac{\partial}{\partial r}\left(r^2\frac{\partial R}{\partial r}\right)+\left[\frac{2\mu}{\hbar^2}\left(E+\frac{Ze^2}{r}\right)-\frac{l(l+1)}{r^2}\right]R=0 \tag{9.154}$$

Case I: $E > 0$: When the energy E of the particle is positive it is found that acceptable solution exist for all values of energy. Since these are of the little importance for us, we don't discuss them here. A unified treatment can be given for the continuous as well as discrete spectrum. In the next section, we shall discuss the case in which total energy of the particle is positive.

Case II: $E < 0$: Let us now consider the situation in which total energy (E) is negative, i.e., the energy is insufficient to ionise the atom. We introduce the symbols.

$$k_3=\sqrt{\frac{2-\mu E}{\hbar^2}} \tag{9.155}$$

$$\lambda=\frac{\mu Ze^2}{\hbar^2 k_3} \tag{9.156}$$

$$\rho=2k_3 r \tag{9.157}$$

and
$$R(r)=S(\rho)$$

$$\frac{1}{r^2}\frac{d}{dr}\left(r^2\frac{dR}{dr}\right)=\left(\frac{4k_3^2}{\rho^2}\right)\frac{d}{d\rho}\left(\frac{\rho^2}{4k_3^2}2k_3\frac{dS}{d\rho}\right)\frac{d\rho}{dr}=\frac{2k_2}{\rho^2}\frac{d}{d\rho}\left(\rho^2\frac{dS}{d\rho}\right)2k_3=\frac{4k_3^2}{\rho^2}\frac{d}{d\rho}\left(\rho^2\frac{dS}{d\rho}\right)$$

Eq. (9.154) now becomes

$$\frac{4k_3^2}{\rho^2}\frac{d}{d\rho}\left(\rho^2\frac{dS}{d\rho}\right)+\left[\frac{2\mu E}{\hbar^2}+\frac{2\mu Ze^2}{\hbar^2 k_3}\frac{k_3}{r}-\frac{l(l+1)}{r^2}\right]S=0$$

or
$$\frac{4k_3^2}{\rho^2}\frac{d}{d\rho}\left(\rho^2\frac{dS}{d\rho}\right)+\left[-k_3^2+2\lambda\frac{2k_3^2}{2k_3 r}-\frac{4k_3^2(l+1)}{4k_3^2 r^2}\right]S=0$$

or $$\frac{4k_3^2}{\rho^2}\frac{d}{d\rho}\left(\rho^2\frac{dS}{d\rho}\right)+\left[-k_3^2+\frac{4k_3^2\lambda}{\rho}-\frac{4k_3^2(l+1)}{\rho^2}\right]S=0$$

$$\therefore \quad \frac{1}{\rho^2}\frac{d}{d\rho}\left(\rho^2\frac{dS}{d\rho}\right)+\left[-\frac{1}{4}+\frac{\lambda}{\rho}-\frac{l(l+1)}{\rho^2}\right]S=0 \tag{9.158}$$

Asymptotic solutions: For large values of ρ, Eq. (9.158) takes a particularly simple form

$$\frac{d^2S}{d\rho^2}=\frac{1}{4}S \tag{9.159}$$

Solutions of this equation are $S=e^{\rho/2}$ and $S=e^{-\rho/2}$

Only the second solution vanishes in the limit $\rho\to\infty$ and is therefore satisfactory as a wave function. The complete solution for Eq. (9.158) may now be assumed to have the form

$$S(\rho)=e^{-\rho/2}F(\rho) \tag{9.160}$$

$$\frac{dS}{d\rho}=-\frac{1}{2}e^{-\rho/2}F(\rho)+e^{-\rho/2}\frac{dF(\rho)}{d\rho}$$

$$\frac{d^2S}{d\rho^2}=e^{-\rho/2}\frac{d^2F(\rho)}{d\rho^2}-e^{-\rho/2}\frac{dF(\rho)}{d\rho}+\frac{1}{4}e^{-\rho/2}F(\rho)$$

Substituting these values of S, $\frac{dS}{d\rho}$ and $\frac{d^2S}{d\rho^2}$ in Eq. (9.158) we get an equation for $F(\rho)$;

$$e^{-\rho/2}\frac{d^2F}{d\rho^2}-e^{-\rho/2}\frac{dF}{d\rho}+\frac{1}{4}e^{-\rho/2}F+\frac{2}{\rho}e^{-\rho/2}\frac{dF}{d\rho}-\frac{1}{\rho}e^{-\rho/2}F+\left[\frac{\lambda}{\rho}-\frac{1}{4}-\frac{l(l+1)}{\rho^2}\right]e^{-\rho/2}F=0$$

or $$\frac{d^2F}{d\rho^2}+\left(\frac{2}{\rho}-1\right)\frac{dF}{d\rho}+\left[\frac{\lambda-1}{\rho}-\frac{l(l+1)}{\rho^2}\right]F=0 \tag{9.161}$$

The coefficient of $\frac{dF}{d\rho}$ as well as F posses a singularity at the origin, which is a regular point.
To remove this singularity we makes the substitution

$$F(\rho)=\rho^S L(\rho)=\sum_v a_v\rho^{v+S} \tag{9.162}$$

Here $L(\rho)$ is a power series in ρ beginning with a non-vanishing constant term.

$$\frac{dF}{d\rho}=\rho^S\frac{dL}{d\rho}+S\rho^{S-1}L$$

$$\frac{d^2F}{d\rho^2}=\rho^S\frac{d^2L}{d\rho^2}+2S\rho^{S-1}\frac{dL}{d\rho}+S(S-1)\rho^{S-2}L$$

$$\therefore \quad \frac{d^2F}{d^2\rho}+2S\frac{dL}{d\rho}\rho^{S-1}+S(S-1)\rho^{S-2}L+\left(\frac{2}{\rho}-1\right)\left(\rho^S\frac{dL}{d\rho}+S\rho^{S-1}L\right)$$
$$+\left[\frac{\lambda-1}{\rho}-\frac{l(l-1)}{\rho^2}\right]\rho^S L=0$$

or $$\frac{d^2L}{d\rho^2}\rho^s + [(2S+2)\rho^{s-1} - \rho^s]\frac{dL}{d\rho} + [(\lambda - 1 - S)\rho^{s-1} + \{S(S-1) + 2S - l(l+1)\}\rho^{s-2}]L = 0$$

or $$\frac{d^2L}{d\rho^2}\rho^s + [2(S+1)\rho^{s-1} - \rho^s]\frac{dL}{d\rho} + [(\lambda - S - S)\rho^{s-1} + \{S(S+1) - l(l+1)\}\rho^{s-2}]L = 0 \tag{9.163}$$

Since L begins with the term a_0, the coefficient of the term having lowest power ρ, i.e., ρ^{S-2} is $[S(S+1) - l(l+1)]a_0$; since a_0 does not vanish, we require that

$$S(S+1) - l(l+1) = 0 \quad \text{or} \quad S^2 + S - l = 0$$

or $$(S-l)(S+l+1) = 0 \quad \text{or} \quad S - l = 0 \quad \text{and} \quad -(l+1)$$

of these two values, the solution with $S = -(l+1)$ does not lead to an acceptable wave function (it has a singularity at the origin) therefore we require $S = l$ in order to remove singularity at the origin.

$$F(\rho) = \rho^l L(\rho) \tag{9.164}$$

and $$\rho^l\frac{d^2L}{d\rho^2} + \frac{dL}{d\rho}[2(l+1)\rho^{l-1} - \rho^l](\lambda - l - 1)\rho^{l-1}L = 0$$

or $$\rho\frac{d^2L}{d\rho^2} + [2(l+1) - \rho]\frac{dL}{d\rho} + (\lambda - l - 1)L = 0 \tag{9.165}$$

Substituting

$$L = \sum_v a_v\rho^v, \frac{dL}{d\rho} = \sum_v va_v\rho^{v-1}, \frac{d^2L}{d\rho^2} = \sum_v v(v-1)a_v\rho^{v-2}$$

$$\therefore \quad \sum_v v(v-1)a^v\rho^{v-1} + 2(l+1)\sum_v va_v\rho^{v-1} - \sum_v va_v\rho^v + (\lambda - l - 1)\sum_v a_v\rho^v = 0$$

This is an identity in powers of ρ, therefore, coefficients of individual powers ρ must vanish. Equating to zero, the coefficients of various powers of ρ, we get following equations:

Coeff. of ρ^0 $\quad 2(l+1)a_1 + (\lambda - l - 1)a_0 = 0$
Coeff. of ρ^1 $\quad 2\times 1a_2 + 4(l+1)a_2 - a_1 + (\lambda - l - 1)a_1 = 0$
Coeff. of ρ^2 $\quad 3\times 2a_3 + 6(l+1)a_3 - 2a_2 + (\lambda - l - 1)a_2 = 0$
- - - - - - - - -
- - - - - - - - -
- - - - - - - - -

Coeff. of $\quad \rho^v\, v(v+1)a_{v+1} + 2(l+1)(v+1)a_{v+1} - va_v + (\lambda - l - 1)a_v = 0$

or $$a_{v+1} = \frac{v - \lambda + l + 1}{v(v+1) + 2(l+1)(v+1)}a_v = \frac{v - \lambda + l + 1}{(v+1)(v+2l+2)}a_v \tag{9.166}$$

For any value of λ and l; if the series (9.162) does not terminate, its dominant asymptotic behaviour can be inferred from the coefficients of the terms having very large power;

$$\frac{a_{\nu+1}}{a_\nu} \underset{(\nu\to\infty)}{} \frac{\nu}{\nu^2} = \frac{1}{\nu}$$

Also $$e^\rho = 1 + \rho + \frac{\rho^2}{2!} + \frac{\rho^2}{3!} + \cdots \frac{\rho^\nu}{\nu!} + \frac{\rho^{\nu+1}}{\nu+1!} + \cdots$$

$$\frac{\text{Coefficient of } \rho^{\nu+1}}{\text{Coefficient of } \rho^\nu} + \frac{\nu!}{(\nu+1)!} = \frac{\nu!}{(\nu+1)\nu!} = \frac{1}{(\nu+1)} \to \frac{1}{\nu}$$

Thus we conclude that for large values of ρ, the function $L(\rho)$ consisting of an infinite number of terms behaves as e^ρ. The product $S(\rho) = e^{-\rho/2}\rho^l L(\rho)$ will be- have as $e^{\rho/2}$ for large values of ρ. Since this function increases to infinity at $\rho = \infty$, we conclude that the boundary condition [$\psi(\rho) = 0$ and $\rho = \infty$] is not satisfied and these mathematical solutions are therefore not acceptable as solutions for the problem on physical ground. The only way this catastrophe can be avoided is if λ happens to have certain particular values such that

$$\lambda = n' + l + 1 = n \tag{9.167}$$

where n' is an integer. n' is known as **radial quantum number** while n the **total quantum number.** In this case, the numerator of Eq. (9.166) becomes zero when ν takes the value n' and $a'_{n+1} = a'_{n+2} = a'_{n+3} = \cdots = 0$.

For such values of λ, the infinite series expansion for $L(\rho)$ is truncated and becomes a polynomial, known as **associated Laguerre polynomials.** The product $e^{-\rho/2}\rho^l L(\rho)$ approaches zero as $\rho \to \infty$, since $e^{-\rho/2}$ goes to zero at a faster rate than the function $F(\rho)$ goes to infinity.

Energy Levels

Introducing for λ, the value given by Eq. (9.167) in Eq. (9.158), we obtain

$$n = \frac{\mu Z e^2}{\hbar^2 k_3} \quad \text{Also } k_3 = \frac{\mu Z e^2}{n\hbar^2} = \frac{Z}{n a_0}$$

or $$n^2 = \frac{\mu^2 Z^2 e^4}{\hbar^4 k_3^2} = \frac{\mu^2 Z^2 e^4}{\hbar^4}\left(-\frac{\hbar^2}{2\mu E}\right)$$

or $$E = \frac{\mu Z^2 e^4}{2n^4\hbar^2} \quad \text{or} \quad E = \frac{2\pi^2 \mu Z^2 e^4}{h^2 n^2} = \frac{RchZ^2}{n^2} \tag{9.168}$$

where $R = \dfrac{2\pi^2 \mu e^4}{ch^2} \equiv$ Rydberg's constant and $a_0 = \dfrac{\hbar^2}{\mu e^2} \equiv$ Bohr radius

It is seen that the energy levels of a hydrogen like atom in a state represented by the quantum number n', l, m does not depend on the individual values, but only on the value of the total quantum number $n.n'$ as well as l can assume all positive integral values including zero and hence the allowed values of n are 1, 2, 3, ... (excluding zero). The energy levels of hydrogen atom (for which $Z = 1$) are shown in Fig. (9.9). The levels pack very close together

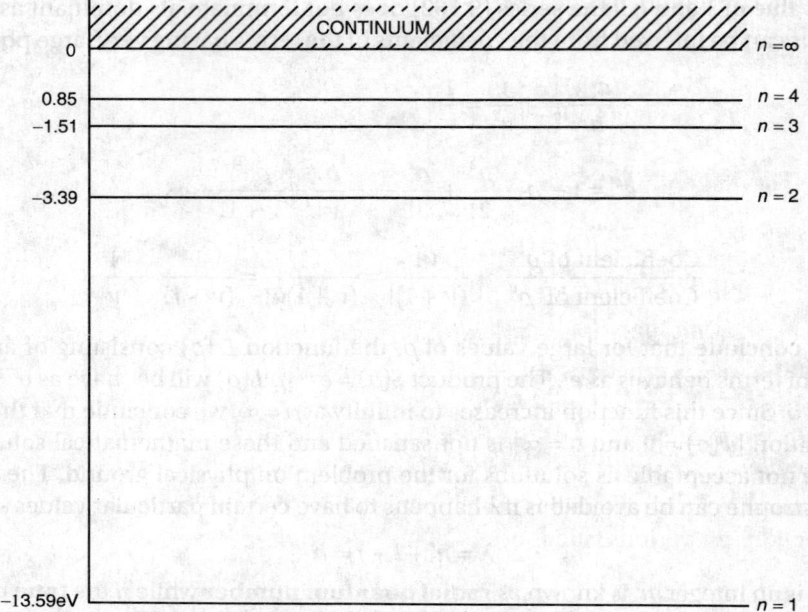

Fig. 9.9 Energy levels of Hydrogen atom vs principal quantum number

as n increases, reaching 13.60 eV above the ground state in the limit $n \to \infty$, which corresponds to the electron at infinite separation from the nucleus or atom ionised. The states with $l = 0, 1, 2, 3, \ldots$ are denoted as *s, p, d, f*, ... states.

Degeneracy: Except for $n = 1$, each energy level is degenerate; being represented by more than one independent solution of the wave equation. If we introduce the quantum numbers *n, l, m* as subscripts, the wave functions $\psi(r, \theta, \phi)$ may be written as;

$$\psi_{nlm}(r,\theta,\varphi) = R_{nl}(r)Y_{lm}(\theta,\phi) = R_{nl}(r)\theta_{\ln}(\theta)\Phi_{m}(\varphi)$$

The allowed values of the quantum numbers are

$$m = 0, \pm 1, \pm 2, \ldots, l = |m|, |m|+1, |m|+2, \ldots\ n = l+1,\ l+2,\ l+3, \ldots$$

This may be rewritten in the form: *Total quantum number* $n = 1, 2, 3, \ldots$, *Azimuthal quantum number* $l = 0, 1, 2, 3, \ldots.\ n - 1$ and *Magnetic quantum number* $m = l, -l+1, -l+2, \ldots -1, 0, +1, \ldots 1, l-1, l$. Thus there are $(2l + l)$ independent wave functions for a given value of n and l and n^2 independent wave functions with a given value of n as can be seen below:

$$\sum_{l=0}^{n-1}(2l+1) = 2\sum_{l=0}^{n-1} l + \sum_{l=0}^{n-1}(1) = 2\left[\frac{(n-1)n}{2}\right] + n = n^2 - n + n = n^2$$

$(2l + l)$ wave functions having same value of n and l are said to form a **completed, subgroup** and the n^2 wave functions having same n forms a **completed group.** Radial wave function may be written as

$$R_{nl}(r) = Ae^{-k_3 r}(2k_3 r)^l L_{n+1}^{2l+1}(2k_3 r) = A^{-\rho/2}\rho^l L_{n+l}^{2l+1}(\rho)$$

where A is a constant; its value can be determined from normalization condition, which requires that

$$\int_0^\infty [R_{nl}(r)]^2 r^2 dr = 1$$

$$\therefore \quad A^2 \int_0^\infty e^{-\rho}\rho^{2l}[L_{n+l}^{(2l+1)}(\rho)]^2 \left(\frac{\rho^2}{4k_3^2}\right)\frac{d\rho}{2k_3} = 1 \text{ or } \frac{A^2}{8k_3^3}\int_0^\infty \rho^{2l+2}e^{-\rho}[L_{n+l}^{2l+1}(\rho)]^2 = 1$$

or

$$\frac{A^2}{\left(\frac{2Z}{na_0}\right)^3}\frac{2n|(n+l)^3}{|n-l-1} = 1 \text{ or } A = -\left(\frac{2Z}{na_0}\right)^{3/2}\left[\frac{|(n-l-1)}{2n|(n+l)^3}\right]^{1/2}$$

$\therefore$ Normalised radial functions

$$R(r) = -\left[\left(\frac{2Z}{na_0}\right)^3 \frac{|(n-l-1)}{2n|(n+l)^3}\right]^{1/2} e^{-\rho/2}\rho^l L_{n+1}^{2l+1}(\rho)$$

Here minus sign is introduced to make the function positive for small value of r. Radial wave functions for various values of n and l may be written as

$$n=1, l=0 \qquad R_{10} = \left(\frac{Z}{a_0}\right)^{3/2} 2e^{-\rho/2}$$

$$n=2, l=0 \qquad R_{20} = \left(\frac{Z}{a_0}\right)^{3/2} \frac{1}{2\sqrt{2}}(2-\rho)e^{-\rho/2}$$

$$n=2, l=1 \qquad R_{21} = \left(\frac{Z}{a_0}\right)^{3/2} \frac{\rho}{2\sqrt{6}}e^{-\rho/2}$$

$$n=3, l=0 \qquad R_{30} = \left(\frac{Z}{a_0}\right)^{3/2} \frac{(6-6\rho+\rho^2)}{9\sqrt{3}}e^{-\rho/2}$$

$$n=3, l=1 \qquad R_{31} = \left(\frac{Z}{a_0}\right)^{3/2} \frac{(4-\rho)\rho}{9\sqrt{6}}e^{-\rho/2}$$

$$n=3, l=2 \qquad R_{32} = \left(\frac{Z}{a_0}\right)^{3/2} \frac{\rho^2}{9\sqrt{30}}e^{-\rho/2}$$

$$n=4, l=0 \qquad R_{40} = \left(\frac{Z}{a_0}\right)^{3/2} \left(\frac{24-36\rho+12\rho^2-\rho^3}{96}\right)e^{-\rho/2}$$

$$n=4, l=1 \qquad R_{41} = \left(\frac{Z}{a_0}\right)^{3/2} \frac{(20-10\rho+\rho^2)\rho}{32\sqrt{15}}e^{-\rho/2}$$

$$n=4, l=2 \qquad R_{42} = \left(\frac{Z}{a_0}\right)^{3/2} \frac{(6-\rho)\rho^2}{96\sqrt{5}}e^{-\rho/2}$$

$$n=4, l=3 \qquad R_{43} = \left(\frac{Z}{a_0}\right)^{3/2} \frac{\rho^3}{96\sqrt{35}}e^{-\rho/2}$$

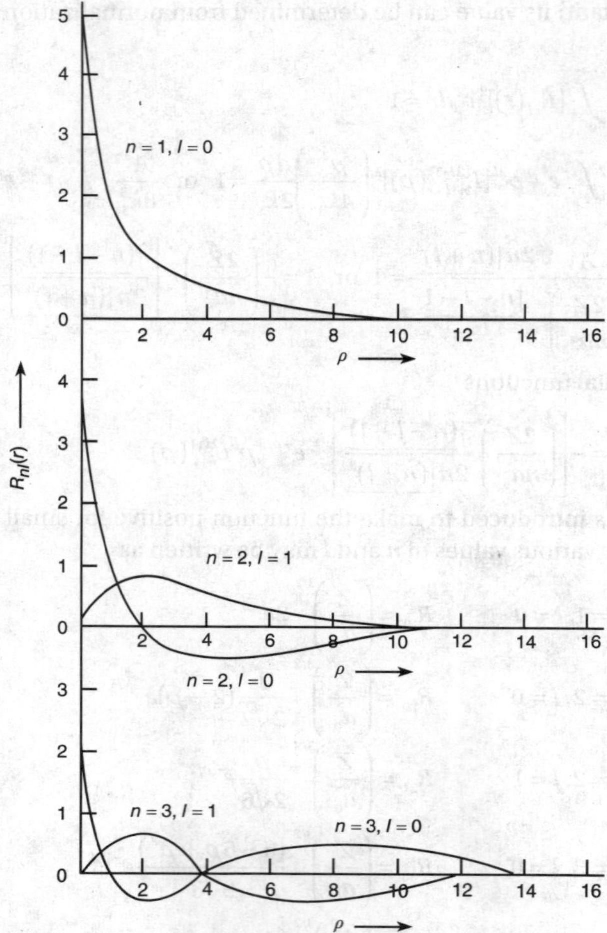

Fig. 9.10 Radial functions for hydrogen atom. $R_{nl}(r)$ for $n = 1, 2, 3$ and $l = 0, 1$

A plot of these functions is shown in Fig. (9.10); near the origin $R_{nl}(r)$varies as $r^l l = 0$ corresponds to zero angular momentum and we may think of electron as moving back and forth through the nucleus and therefore $R_{nl}(0) \neq 0$ in this case. The radial function lies farther out from the origin for states with higher $l(l \neq 0)$. The wave function crosses r-axis $(n - l - 1)$ times in the region between $r = 0$ and $r = \infty$. The radial distribution function $D_{nl}(r)$ is given by $D_{nl}(r) = 4\pi r^2 R_{nl^2}(r)$.

$D_{nl}(r)$ is proportional to the probability of finding the electron in a spherical shell of radius r. A plot of $D_{nl}(r)$ *vs* ρ is given in Fig. (9.11). If we define

$$<r^{\pm p}> = \int_0^\infty r^{\pm p}[R_{nl}(r)^2 r^2]\,dr$$

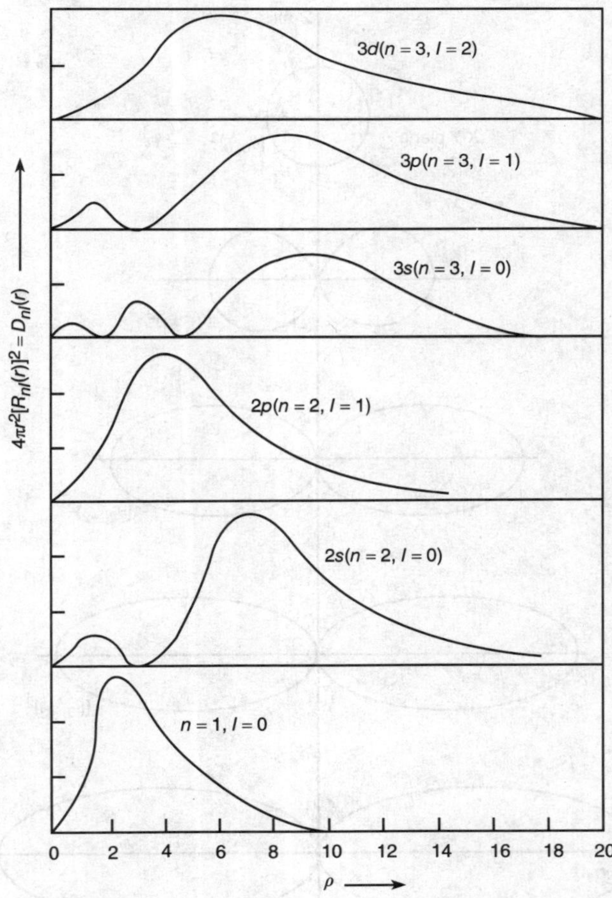

Fig. 9.11 Radial probability distribution function for Hydrogen atom in various states

Then $$<\frac{1}{r}>=\frac{1}{n^2}\left(\frac{Z}{a_0}\right),\ <\frac{1}{r^2}>=\frac{1}{n^3(l+1/2)}\left(\frac{Z}{a_0}\right)^2$$

$$<\frac{1}{r^3}>=\frac{1}{n^3 l(l+1/2)(l+1)}\left(\frac{Z}{a_0}\right)^3$$

$$<\frac{1}{r^4}>=\frac{[n^2-l(l+1)]}{2n^5 l(l-1/2)(l+1/2)(l+1)(l+3/2)}\left(\frac{Z}{a_0}\right)^4$$

$$<r>=n^2\left[1+\frac{1}{2}\left\{1-\frac{l(l+1)}{n^2}\right\}\right]\left(\frac{a_0}{Z}\right)$$

$$<r^2>=\frac{1}{2}[5n^2+1-3l(l+1)]n^2\left(\frac{a_0}{Z}\right)^2$$

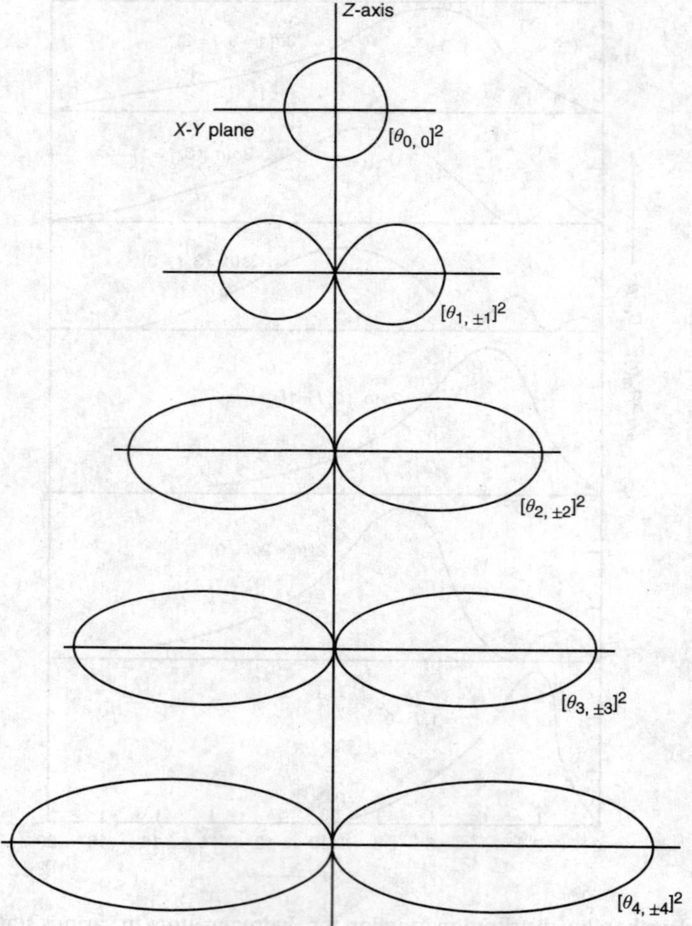

Fig. 9.12

In general

$$\frac{s+1}{n^2}<r^s>-(2s+1)\frac{a_0}{Z}<r^{s-1}>+\frac{s}{4}[(2l+1)^3-s^2]\left(\frac{a_0}{Z}\right)^2<r^{s-2}>=0$$

With the condition $s > 2l - 3$.

The angular part of the wave function is given by the function $Y_{lm}(\theta, \phi)\ |Y_{lm}(\theta, \varphi)|^2$ for $m = \pm l$ has been plotted for various l values in Fig. (9.12). It is seen that with the increase in the value of orbital angular momentum quantum number, the probability distribution function becomes more and more concentrated about the *XY* plane.

The behaviour of the distribution function for other values of m is similarly shown in Fig. (9.13). It is seen that the function tend to be concentrated in direction corresponding to the plane of the oriented Bohr orbit.

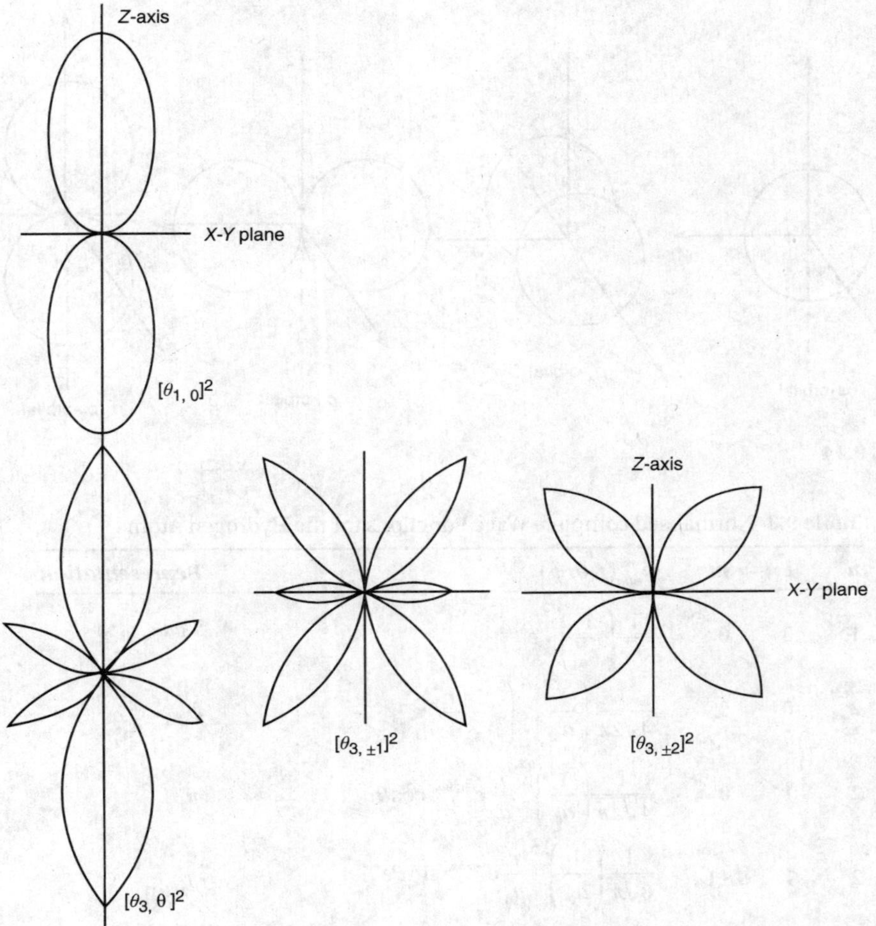

Fig. 9.13

The angular dependence of the probability distribution function for s and p orbitals (real part) is shown in Fig. (9.14). The function s is spherically symmetric and the functions p_x, p_y and p_z are equivalent except for orientation.

Complete Wave Functions

The complete wave functions of hydrogen atom are given by

$$\psi_{nlm}(r,\theta,\varphi) = R_{nl}(r)Q_{ml}(\theta)F_m(\varphi) = \left[\left(\frac{2}{na_0}\right)^3 \frac{(n-l-1)!}{2n\{(n+1)!\}^3}\right] e^{-r/na_0}\left(\frac{2r}{na_0}\right)^l$$

$$L_{n+l}^{2l+1}\left(\frac{2r}{na_0}\right)\left[\frac{2l+1}{2}\frac{(l-m!)}{(l+m!)}\right]P_l^m(\cos\theta)\frac{1}{\sqrt{2\pi}}e^{im\varphi} \tag{9.169}$$

where $n = 1, 2, 3, 4, \ldots.,$ $l = 0, 1, 2, 3, \ldots\ldots (n-1)$, and $m = 0, \pm1, \pm2, +3, \ldots. \pm l$. The complete wave functions for the first three values of n are given in Table 9.3.

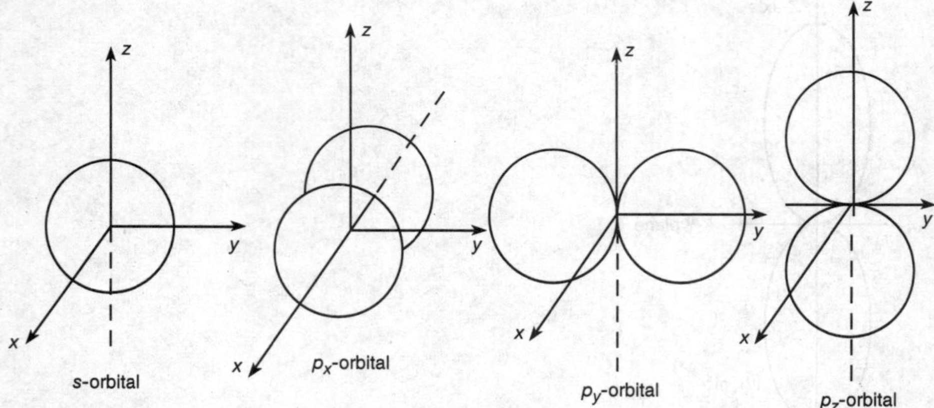

Fig. 9.14

Table 9.3 Normalised complete Wave Functions for the Hydrogen Atom

n	*l*	*m*	$\psi_{nlm}(r, \theta, \varphi)$	*Representation*
1	0	0	$\frac{1}{\sqrt{\pi}}\left(\frac{1}{a_0}\right)^{3/2} e^{-r/a_0}$	ψ_{100}
2	0	0	$\frac{1}{4\sqrt{2\pi}}\left(\frac{1}{a_0}\right)^{3/2}\left(2-\frac{r}{a_0}\right)e^{-r/2a_0}$	ψ_{200}
2	1	0	$\frac{1}{4\sqrt{2\pi}}\left(\frac{1}{a_0}\right)^{3/2}\frac{r}{a_0}e^{-r/2a_0}\cos\theta$	ψ_{210}
2	1	±1	$\frac{1}{8\sqrt{\pi}}\left(\frac{1}{a_0}\right)^{3/2}\frac{r}{a_0}e^{-r/2a_0}\sin\theta e^{+\varphi}$	$\psi_{21,(\pm 1)}$
3	0	0	$\frac{1}{81\sqrt{3\pi}}\left(\frac{1}{a_0}\right)^{3/2}\left(27-18\frac{r}{a_0}+2\frac{r}{a_0^2}\right)e^{-r/3a_0}$	ψ_{300}
3	1	0	$\frac{\sqrt{2}}{8\pi\sqrt{\pi}}\left(\frac{1}{a_0}\right)^{3/2}\left(6-\frac{r}{a_0}\right)\frac{r}{a_0}e^{-r/3a_0}\cos\theta$	ψ_{310}
3	1	±1	$\frac{1}{81\sqrt{\pi}}\left(\frac{1}{a_0}\right)^{3/2}\left(6-\frac{r}{a_0}\right)\frac{r}{a_0}e^{-r/3a_0}\sin\theta e^{\pm i\varphi}$	$\psi_{31,(\pm 1)}$
3	2	0	$\frac{1}{81\sqrt{6\pi}}\left(\frac{1}{a_0}\right)^{3/2}\frac{r^2}{a_0^2}e^{-r/3a_0}(3\cos^2\theta-1)$	ψ_{320}
3	2	±1	$\frac{1}{81\sqrt{\pi}}\left(\frac{1}{a_0}\right)^{3/2}\frac{r^2}{a_0^2}e^{-r/3a_0}\sin\theta\cos\theta e^{\pm l\varphi}$	$\psi_{32,(\pm 1)}$
3	2	±2	$\frac{1}{162\sqrt{\pi}}\left(\frac{1}{a_0}\right)^{3/2}\frac{r^2}{a_0^2}e^{-r/3a_0}\sin^2\theta e^{\pm 2l\varphi}$	$\psi_{32,(\pm 2)}$

Degeneracy: Except for $n = 1$, each energy level is degenerate, being represented by more than one independent solution of the wave equations. The alalowed values of quantum numbers are: Total quantum number $n = 1, 2, 3, ...$, Azimuthal quantum number $l = 0, 1, 2, 3, ... (n-1)$, and Magnetic quantum number $m = -l, -l+1, -l+2 ..., -1, 0, +1, ... l-1, l$.

Thus there are $(2l+1)$ independent wave functions for a given value of n and l and n^2 independent wave functions with a given value of n as can be seen below

$$\sum_{l=0}^{n-1}(2l+1) = 2\sum_{l=0}^{n-1} l + \sum_{l=0}^{n-1}(l) = 2\frac{(n-1)n}{2} + n = n^2 - n + n = n^2$$

The degeneracy with respect to m is the characteristic of central force field, i.e., the force field depending only on the radial distance r, while the degeneracy with respect to l is characteristic of Coulomb field. In presence of some external field such as a magnetic field $(2l+1)$ fold m degeneracy disappears and the level is split up into n^2 different energy levels.

$$a_0 = \frac{4\pi\varepsilon_0\hbar^2}{m_0 e^2} = 0.53\text{Å} = \text{radius of the first Bohr orbit.}$$

When $n = 2$, $l = 0$, $m = 0$,$[l = 1, m = 0]$ and $l = 1$, $m = \pm 1$, therefore yielding four wave functions or quantum states in all. For $n = 3$, there are nine states and for $n = 4$, there are sixteen states and so on. In Fig. (9.15) different eigen states in case of hydrogen like atom have been depicted.

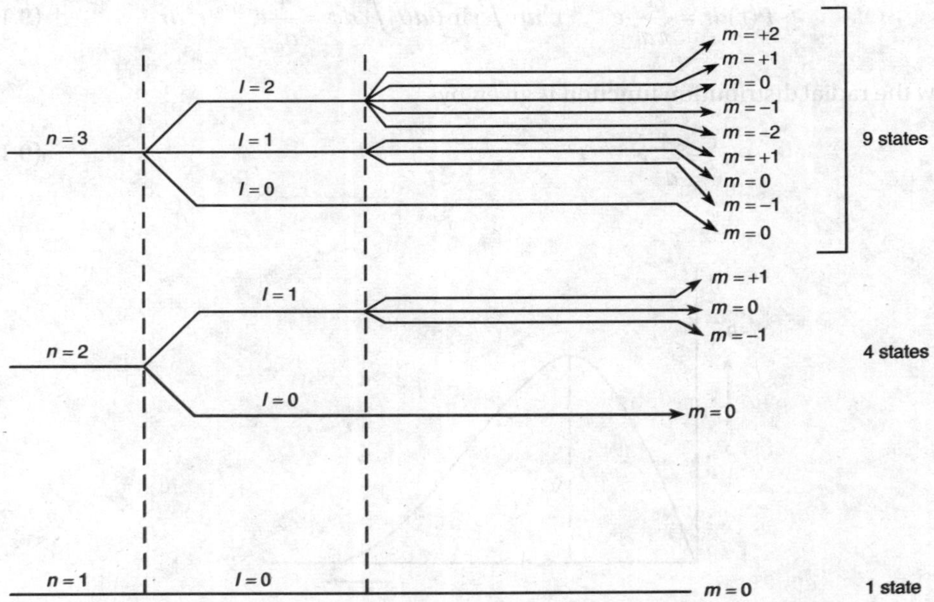

Fig. 9.15 Different Eigen states in hydrogen like atom

Ground State of Hydrogen Atom

The potential energy is spherically symmetrical. Therefore, there will be a set of wave functions which are spherically symmetrical. Such wave functions are functions of r only, i.e.,

$$\psi = R(r) \tag{9.170}$$

In the ground state $n = 1$, $l = 0$ and $m = 0$. The wave functions is given by

$$\psi_{100} = \frac{1}{\sqrt{\{\pi a_0^3\}}} e^{-r/a_0} \tag{9.171}$$

The probability distribution function of the electron relative to nucleus is given by

$$\psi_{100}^{*}\psi_{100} = \frac{1}{\pi a_0} e^{-2r/a_0} \tag{9.172}$$

Eq. (9.172) is independent of θ and φ, hence the normal hydrogen atom is spherically symmetric. The probability that the electron will lie in a small volume $dV = r^2 \sin\theta dr\, d\theta d\varphi$ given by

$$\psi_{100}^{*}\psi_{100} r^2 dr \sin\theta d\Phi = \frac{1}{\pi a_0^3} e^{-2r/a_0} r^2 \sin\theta d\theta d\varphi \tag{9.173}$$

The probability that the electron will lie between the distance r and $r + dr$ from the nucleus irrespective of its angular distribution is given by

$$P(r)dr = \frac{1}{\pi a_0^3} e^{-2r/a_0} r^2 dr \int_0^{\pi} \sin\theta d\theta \int_0^{2\pi} d\phi = \frac{4}{a_0^3} e^{-2r/a_0} r^2 dr \tag{9.174}$$

Now the radial distribution function is given by

$$P(r) = \frac{4}{a_0^3} e^{-2r/a_0} r^2 \tag{9.175}$$

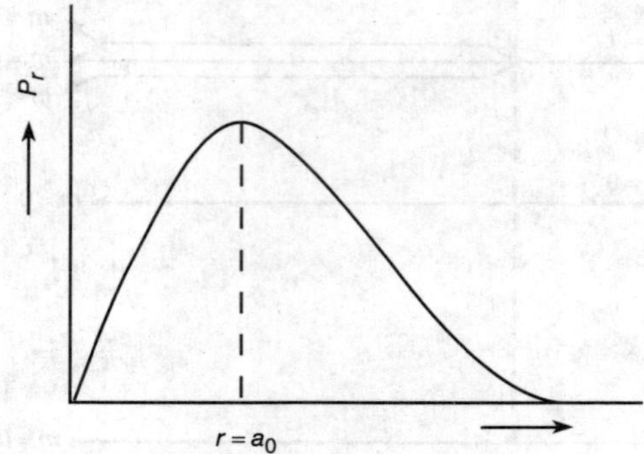

Fig. 9.16 Most probable position of the electron

For the most probable distance r, $P(r)$ should be maximum, i.e.,

$$\frac{d}{dr}P(r)=0$$

$$\therefore \qquad \frac{d}{dr}\left[\frac{4}{a_0^3}e^{-2r/a_0}r^2\right]=0$$

This gives $r = a_0$ = radius of the first Bohr's orbit. Obviously, the most probable distance of the electron from the nucleus in the normal state of hydrogen atom is equal to Bohr's radius. The wave function $R(r) = e^{-r/a_0}$ therefore, represents an s-state with a spherical orbit.

Average Value of the radius of H-atom

One is often concerned with the average distance r of the electron from the nucleus as some atomic properties are related to it. For hydrogen in the $1s$ state, this is given by

$$\overline{r_{ls}} = \int \psi_{ls}\hat{r}\psi_{ls}dV = \int_0^{\pi} \exp(-r)\hat{r}\frac{1}{\sqrt{\pi}}\exp(r)r^2dr \int_0^{\pi}\sin\theta d\theta \int_0^{\pi} d\varphi$$

$$= \frac{1}{\pi}\int r^2\exp(-2r)(2)(2\pi)dr = 4\frac{3!}{2^4}(\text{using table of integrals}) = \frac{3}{2}\text{Å} = 1.5\text{Å}$$

Note: Instead of ψ one could use $R_{10}(r)$, then $r = \int_0^{\infty} R_{10}(r)\hat{r}R_{10}(r)r^2dr$

9.8 UNSOLD'S THEOREM

This theorem states that the sum of the probability distribution functions for a given value of l and all permissible values of m is constant, i.e.,

$$\sum_{m=-1}^{+1}\theta^{m*}(\theta)\Phi^*_m(\varphi)\theta_l^m(\theta)\Phi_m(\varphi) = \text{constant}$$

This theorem means that every closed subshell atom or ion has a spherically symmetric distribution of electric charge.

9.9 VIRIAL THEOREM

This theorem states for a system of charged particles having columbic forces $\left(\text{i.e., of the form } \frac{Z_1Z_2e^2}{r_{12}}\right)$ as inter particle force, expectation value of potential energy is twice the expectation value of kinetic energy with negative sign; mathematically $<V> = -2<T>$.

Example 9.1: The Hamiltonian for hydrogen atom and the $1s$ wave function in atomic units are, $H = -\frac{1}{2}\nabla^2 - \frac{1}{r}$ and $\psi_{ls} = \frac{1}{\sqrt{\pi}}e^{-r}$ respectively. Calculate the ground state energy of hydrogen atoms

Solution

$$\overline{E}_{1s} = \int \psi^*_{1s} H \psi_{1s} dV$$

$$= \frac{1}{\pi}\int_0^\infty e^{-r}\left[-\frac{1}{2r^2}\frac{d}{dr}\left(r^2\frac{d}{dr}\right)-\frac{1}{r}\right]e^{-r}r^2dr\int_0^\pi \sin\theta d\theta\int_0^{2\pi} d\varphi = -\frac{4\pi}{2\pi}\int_0^\infty r^2e^{-2r}dr$$

$$= \frac{1}{2}\text{atomic mass unit} = -2.18\times10^{-18}\,\text{J}-\text{s} = -13.6\,\text{eV}.$$

Example 9.2: Derive the atomic orbital for $l = 1$, $m = 0$

Solution $Y_{10}(\theta,\varphi) = \left(\frac{3}{2}\right)^{1/2}\cos\theta\frac{1}{\sqrt{2\pi}}\exp(i0\varphi) = \frac{\sqrt{3}}{2\pi}\cos\theta = \frac{Z}{r}$

Obviously, the orbital is p_z.

Example 9.3: A hydrogen-like wave function is given below with r in atomic units

$$\psi = \frac{\sqrt{2}}{81\sqrt{\pi}}Z^{3/2}(6-Zr)Zr\exp\left(-\frac{Zr}{3}\right)\cos\theta$$

Determine the quantum numbers n, l and m by inspection.

Solution The exponential factor $\left(-\frac{Zr}{3}\right)$ shows that $n = 3$. The radial factor $r^l = r$ shows that $l = 1$. The angular factor $\cos\theta$ shows that $m = 0$.

Example 9.4: Show that the probability of finding the electron in a hydrogen like atom is independent of the angle φ.

Solution $\psi^*\psi = \left[R(r)P(\theta)\frac{1}{\sqrt{2\pi}}e^{-im\varphi}\right]\left[R(r)P(\theta)\frac{1}{\sqrt{2\pi}}e^{im\varphi}\right] = [R(r)P(\theta)]^2$

Example 9.5: Show that ψ_{1s} and ψ_{2s}, are orthonormal.

Solution $\psi_{1s} = \frac{1}{\sqrt{\pi}}\left(\frac{Z}{a_0}\right)^{3/2}\exp(-Zr/a_0)$ and $d\tau = r^2 dr\sin\theta d\theta d\varphi$

$$\int \psi_{1s}^2 d\tau = \frac{Z^3}{\pi a_0^3}\int_0^\infty \exp\left(-\frac{-2Zr}{a_0}\right)r^2dr\int_0^\pi \sin\theta d\theta\int_0^{2\pi} d\varphi = \frac{4\pi Z^3}{\pi a_0^3}Z\frac{2!}{(2/a_0)^3} = 1$$

$$\psi_{2s} = \frac{1}{4\sqrt{2\pi}}\left(\frac{Z}{a_0}\right)^{3/2}\left(2-\frac{Zr}{a_0}\right)\exp\left(-\frac{Zr}{2a_0}\right)$$

$$\int \psi_{2s}^2 d\tau = \frac{1}{32\pi}\left(\frac{Z}{a_0}\right)\int_0^\infty\left[\left(2-\frac{Zr}{a_0}\right)^2\exp\left(-\frac{Zr}{a_0}\right)\right]r^2dr\int_0^\pi \sin\theta d\theta\int_0^{2\pi} d\varphi$$

$$= \frac{4\pi}{32\pi}\left(\frac{Z}{a_0}\right)^3\left[\int_0^\infty 4r^3\exp(-Zr/a_0)dr - \int_0^\infty \frac{4Zr^8}{a_0}\exp\left(-\frac{Zr}{a_0}\right)dr + \int_0^\infty \frac{Z^2r^4}{a_0^2}\exp\left(-\frac{Zr}{a_0}\right)dr\right]$$

$$= \frac{1}{8}\left(\frac{Z}{a_0}\right)^3 \left[\frac{4\times 2!}{(Z/a_0)^3} - \frac{4(Z/a_0)3!}{(Z/a_0)^4} + \frac{\frac{Z^2}{a_0^2}\times 4!}{(Z/a_0)^5}\right]$$

$$\therefore \quad \int \psi_{1s}\psi_{2s}d\tau = \int_0^\infty \sqrt{\frac{Z^3}{\pi a_0^3}}\exp\left(-\frac{Zr}{a_0}\right)\frac{1}{\sqrt{32\pi}}\left(\frac{Z}{a_0}\right)^{3/2}\left(2-\frac{Zr}{a_0}\right)\exp\left(-\frac{Zr}{2a_0}\right)r^2 dr$$

$$\int_0^\infty \sin\theta d\theta \int_0^{2\pi} d\varphi$$

$$= \frac{4\pi}{\sqrt{32\pi}}\left(\frac{Z}{a_0}\right)^3 \int_0^\infty \left(2-\frac{Zr}{a_0}\right)\exp(-3Zr/2a_0)r^2 dr$$

$$= \frac{1}{\sqrt{2}}\left(\frac{Z}{a_0}\right)^3\left[\int_0^\infty 2r^2\exp(-3Zr/2a_0 - \int_0^\infty \frac{Zr^3}{a_0}\exp(-3Zr/2a_0)dr\right]$$

$$= \frac{1}{\sqrt{2}}\left(\frac{Z}{a_0}\right)^3\left[\frac{2\times 2!}{(3Z/2a_0)^3} - \frac{(Z/a_0)3!}{(3Z/2a_0)^4}\right] = 0$$

Example 9.6: **The ψ_{2s} atomic orbitals for hydrogen-like atoms are given as:**

$$\psi_{2s} = \frac{1}{4\sqrt{2\pi}}\left(\frac{Z}{a_0}\right)^{3/2}\left(2-\frac{Zr}{a_0}\right)\exp\left(\frac{Zr}{2a_0}\right)$$

Calculate the radii of the nodal surface of H.

Solution $\psi_{2s} = 0$ at $r = \dfrac{2a_0}{Z} = 1.058\times 10^{-10}$ m.

Example 9.7: **Show that the total probability density for any value of *n* is spherically symmetric irrespective of the value of *l*.**

Solution $\psi^2_{1s} = \dfrac{1}{\pi}e^{-2r}$ (in atomic unit)

$$\psi^2_{2S} = \frac{1}{32\pi}(2-r)^2 \text{ (in a.u.)}$$

$$\psi^2_{2p} = \psi^2_{2p_x} + \psi^2_{2p_y} + \psi^2_{2p_z} = \frac{1}{32\pi}r^2e^{-r}(\sin^2\theta\cos^2\varphi + \sin^2\varphi + \cos^2\theta)$$

$$= \frac{1}{32\pi}r^2e^{-r}(\sin^2\theta + \cos^2\theta) = \frac{1}{32\pi}r^2e^{-r} \text{ (in a.u.) obviously, a function of } r \text{ only}$$

$$\psi^2_{3d} = \psi^2_{3d_{xy}} + \psi^2_{3d_{yz}} + \psi^2_{3d_{xz}} + \psi^2_{3d_{x^2-y^2}} + \psi_{3d_{x^2}} = \text{a function of } r \text{ only.}$$

Example 9.8: Calculate the average distance of the electron from the nucleus of H-atom in the 2*s* state.

Solution $\bar{r}_{2s} = \dfrac{1}{32\pi a_0^3} \int_0^\infty \left(2 - \dfrac{r}{a}\right)^2 \exp\left(-\dfrac{r}{a}\right) r r^2 dr \int_0^\pi \sin\theta d\theta \int_0^{2\pi} d\varphi$

$$\frac{1}{8a_0^3} \int_0^\infty \left(4r^3 - \frac{4r^4}{a_0} + \frac{r^5}{a_0^2}\right) \exp(-r/a_0) dr = \frac{1}{8a_0^3}\left[\frac{4\times 3!}{\left(\frac{1}{a_0}\right)^4} - \frac{4\times 4!}{a_0\left(\frac{1}{a_0}\right)^5} + \frac{5!}{a_0^2\left(\frac{1}{a_0}\right)^6}\right] = 6a_0$$

Example 9.9: Write the Hamiltonian for hydrogen atom and reduce the two body hydrogen problem in one body problem.

Solution In Hydrogen atom, there is one proton at the nucleus and one electron revolves round it. There is a coulomb force which acts between proton (+ ve charged) and electron (–ve charged). The potential is spherical symmetric. The total energy of proton and neutron is

$$E = \frac{p_e^2}{2m} + \frac{p_p^2}{2M} - \frac{e^2}{|r_e - r_p|} \quad \text{(i)}$$

where p_e and p_p are momentum of electron and proton respectively, m and M are masses of electron and proton respectively and $|r_e - r_p|$ is distance between electron and proton. Now we can write the Hamiltonian for hydrogen atom is

$$H = \frac{p_e^2}{2m} + \frac{p_p^2}{2m} + V(|r_e - r_p|) \quad \text{(ii)}$$

$$= -\frac{\hbar^2}{2m}\nabla_r^2 - \frac{\hbar^2}{2M}\nabla_p^2 + V(|r_e - r_p|) \quad \text{(iii)}$$

Here, we have used the momentum operator $p \to i\hbar\nabla$. Using $H\psi = E\psi$, we obtain

$$\left[-\frac{\hbar^2}{2m}\nabla_e^2 - \frac{\hbar^2}{2M}\nabla_p^2 - \frac{e^2}{|r_e - r_p|}\right]\psi = E\psi \quad \text{(iv)}$$

Here ψ is the total wave function whose value depends on the coordinates of electron and proton, i.e., $\psi \to \psi(r_e, r_p)$

If we take $\psi = \psi(r_e)\,\psi(r_p)$, still it is not possible to separate E.q. (iv) in two separate equations. However Eq. (iv) can be separated into two equations: (i) Which controls the centre of mass motion (ii) Which describes the relative motion of both the particles. For this purpose, we can define the coordinate vector for centre of mass as

$$R = \frac{mr_e + Mr_p}{m + M} \quad \text{(v)}$$

The relative coordinate w.r.t. both particles is

$$r = r_e - r_p \tag{vi}$$

One can express the Hamiltonian in terms of these coordinates as

$$H = \frac{\hbar^2}{2(m+M)}\nabla_R^2 - \frac{\hbar^2}{2\mu}\nabla_r^2 - \frac{e^2}{r} \tag{vii}$$

where $\mu = \frac{mM}{m+M}$ is the reduced mass of electron-proton system. Now, Schrodinger equation becomes

$$\left(\frac{\hbar^2}{2(m+M)}\nabla_R^2 + \frac{\hbar^2}{2\mu}\nabla_r^2\right)\psi(r,R) + (E-V)\psi(r,R) = 0 \tag{viii}$$

Now, we write
$$\psi(r,R) = \varphi(R)Q(r) \tag{ix}$$

Using Eqs. (ix), (viii) becomes

$$\frac{\hbar^2}{2(m+M)\varphi(R)}\nabla^2\varphi + \frac{\hbar^2}{2\mu Q}\nabla_r^2 Q(r) + (E-V)\phi(R)\varphi(R) = 0 \tag{x}$$

Obviously, Eq. (x) is a combination of two equations, out of which one depends on r and other on R only. Clearly, the value of each constant must be equal to the same constant, i.e.,

$$\frac{\hbar^2}{2(m+M)\varphi}\nabla_r^2\varphi = \frac{\hbar^2}{2\mu Q}\nabla_r^2 Q - (E-V)\phi(R)\varphi(r) \tag{xi}$$

or
$$\frac{\hbar^2}{2(m+M)\varphi}\nabla_r^2\varphi = -E_R\phi(R) \quad \text{and} \quad \frac{\hbar^2}{2\mu Q}\nabla_r^2 Q(r) - (E-V)\varphi(r) = -E_R\varphi(r)$$

or
$$\frac{\hbar^2}{2(m+\mu)}\nabla_r^2\varphi(R) + E_R\phi(R) = 0 \tag{xii}$$

and
$$\frac{\hbar^2}{2\mu}\nabla_r^2(Q)r + (E-E_R)Q(r) = 0 \tag{xiii}$$

Putting $E_r = E - E_R$, one obtains from Eq. (xiii)

$$\frac{\hbar^2}{2\mu}\nabla_r^2 Q(r) + \{E_r - V(r)\}Q(r) = 0 \tag{xiv}$$

Here E_R is associated with centre of mass and E_r is the value of relative energy.

From Eq. (xii) it is clear that it is free from V and the mass of free particle is $(m+M)$. Obviously, its solution will be plane waves. Therefore, this corresponds to entire motion of hydrogen atom. This is why, we are not interested in this equation.

Eq. (xiv) describes the motion of a particle of mass μ in a potential field $V(r)$, where r corresponds to respective coordinates. Since $\mu = \frac{mM}{m+M}$ and $M \approx 1850$ m and therefore we can consider $\mu = M$. Therefore $E_R = \frac{\hbar^2 p^2}{2(m+M)}$ being very small, one can put $E = E_r$.

Thus we have reduced the problem of two body Schrodinger's equation to one body problem equation using reduced mass and relative coordinates. However, the hydrogen problem can be solved by considering the proton at rest and by putting m in place of μ.

Example 9.10: Verify that

$$R_{1,0}(r)=2\left(\frac{Z}{a_0}\right)^{3/2}\exp(-Zr/a_0),R_{2,0}(r)=\frac{1}{2\sqrt{2}}\left(\frac{Z}{a_0}\right)^{3/2}\left(2-\frac{Zr}{a_0}\right)\exp(-Zr/2a_0)$$

and $R_{2,1}(r)=\dfrac{1}{2\sqrt{6}}\left(\dfrac{Z}{a_0}\right)^{3/2} r\exp(-Zr/2a_0)$

are normalised radial eigen functions of hydrogen like atoms.

Solution We have the radial operator: $-\dfrac{\hbar}{2m}\dfrac{1}{r^2}\dfrac{d}{dr}\left(r^2\dfrac{d}{dr}\right)-\dfrac{Ze^2}{r}$

$$\therefore \qquad -\frac{\hbar^2}{2m}\frac{1}{r^2}\left(\frac{d}{dr}\right)\left(r^2\frac{d}{dr}\right)-\frac{Ze^2}{r}\Bigg]\left\{2\left(\frac{Z}{a_0}\right)^{3/2}\exp(-Zr/a_0)\right\}$$

$$=-\frac{e^2}{2a_0}\left[2\left(\frac{Z}{a_0}\right)^{3/2}\exp(-Zr/a_0)\right]$$

Obviously, eigen value $= -\dfrac{e^2}{2a_0}$

Now $$\int_0^\infty\left[2\left(\frac{Z}{a_0}\right)^{3/2}\exp(-Zr/a_0)\right]^2 r^2dr=4\left(\frac{Z}{a_0}\right)^3\frac{L^2}{(Z/a_0)^3}=1$$

Obviously, this is a normalised radial eigen function. Similarly, one can proceed for other functions.

Example 9.11: Show that the radial equation

$$\frac{1}{r^2}\frac{d}{dr}\left(r^2\frac{dR}{dr}\right)+\left\{\frac{2m}{\hbar^2}[E-V(r)]-\frac{\lambda}{r^2}\right\}R=0$$

can be put in the form $-\dfrac{\hbar^2}{2m}\dfrac{d^2x}{dr^2}+\left[V(r)+\dfrac{l(l+1)\hbar^2}{2mr^2}\right]\chi=E\chi$, where $\chi=R(r)r$.

Solution Substituting

$$\frac{dR}{dr}=\frac{1}{r}\frac{d\chi}{dr}-\frac{\chi(r)}{r^2}\text{ and }\frac{d^2R}{dr^2}=\frac{1}{r}\frac{d^2\chi}{dr^2}-\frac{2}{r^2}\frac{d\chi}{dr}+\frac{2\chi}{r}$$

in the radial equation, one obtains

$$\frac{1}{r^2}\frac{d}{dr}\left(r^2\frac{dR}{dr}\right)+\left\{\frac{2m}{\hbar}[E-V(r)]-\frac{l(l+1)}{r^2}\right\}R=0 \qquad \text{or}$$

$$-\frac{\hbar^2}{2m}\frac{d^2\chi}{dr^2}+\left[V(r)+\frac{l(l+1)\hbar^2}{2mr^2}\right]\chi=E\chi$$

We note that the radial motion is similar to the one dimensional motion of a particle in a potential $V(r)+l\frac{(l+1)\hbar^2}{2mr^2}$.

Example 9.12: Find the most probable radial distance for the electron for a hydrogen atom in its ground state.

Solution We have the radial probability function, i.e., the probability that the electron is between r and $r + dr$ from the nucleus is.

$$Q(r)dr=\left(\frac{4}{a_0^3}\right)\exp\left(-\frac{2r}{a_0}\right)r^2dr$$

One obtains the value of r at which $Q(r)$ is maximum by

$$\frac{dQ}{dr}=0, \text{ i.e., } \frac{4}{a_0^3}\left[-\frac{2r^2}{a_0}+2r\right]\exp\left(-\frac{2r}{a_0}\right)=0 \quad \therefore \quad r=a_0 \quad \text{or} \quad r=0$$

Obviously, $\frac{d^2Q}{dr^2}$ is negative at $r=a_0$ and positive at $r=0$. This means that $Q(r)$ is maximum at $r=a_0$.

Example 9.13: Show that the most probable distance r_m of an electron in the $2p$ state of H-atom is 4Å.

Solution $P(r)=r^2[R_{21}(r)]^2=\frac{1}{24}r^4e^{-r}\text{ Å}.$

Now, setting $\frac{dP}{dr}=0$, we have $(4r^3-r^4)e^{-r}=0$ or $r=4$ Å (or $4a_0$)

Obviously 4 Å is the most probable distance.

Example 9.14: Show that the average distance of the electron from the nucleus of hydrogen atom in the $2s$ state is $6a_0$.

Solution

$$\overline{r_{2s}}=\frac{1}{32\pi a_0^3}\int_0^\infty\left(2-\frac{r}{a_0}\right)^2\exp\left(-\frac{r}{a_0}\right)rr^2dr\int_0^\pi\sin\theta d\theta\int_0^{2\pi}d\varphi$$

$$=\frac{1}{8\pi a_0^3}\int_0^\infty\left(4r^3-\frac{4r^4}{a_0^2}+\frac{r^5}{a_0^2}\right)\exp\left(-\frac{r}{a_0}\right)dr$$

$$=\frac{1}{8a_0^3}\left[\frac{4\times 3!}{\left(\frac{1}{a_0}\right)^4}-\frac{4\times 4!}{a_0\left(\frac{1}{a_0}\right)^5}+\frac{5!}{a_0^2\left(\frac{1}{a_0}\right)^6}\right]=6a_0$$

Example 9.15: Show that the average distance of the electron in the $2p_z$ state from the nucleus of an H-atom is $5a_0$ or 5 Å.

Solution $\bar{r} = \frac{1}{32\pi a_0^5} \int_0^\infty r \exp(-r/2a_0)^2 rr^2 dr \int_0^\pi \sin\theta d\theta \int_0^{2\pi} d\phi$

$$= \frac{4\pi}{32\pi a_0^5} \int_0^\infty r^5 \exp\left(-\frac{r}{a_0}\right) dr = \frac{5!}{24a_0^5\left(\frac{1}{a_0}\right)^6} = 5a_0 \text{ i.e., } 5\text{Å}$$

Examle 9.16: **Write the analytical expressions for the ground and first excited state wave functions of H-atom. Sketch the function.**

Solution $\psi_{1s}^H = \frac{1}{\sqrt{\pi}}\exp(-r); \psi_{2s}^H = \frac{1}{4\sqrt{2\pi}}(2-r)\exp\left(-\frac{r}{2}\right)$

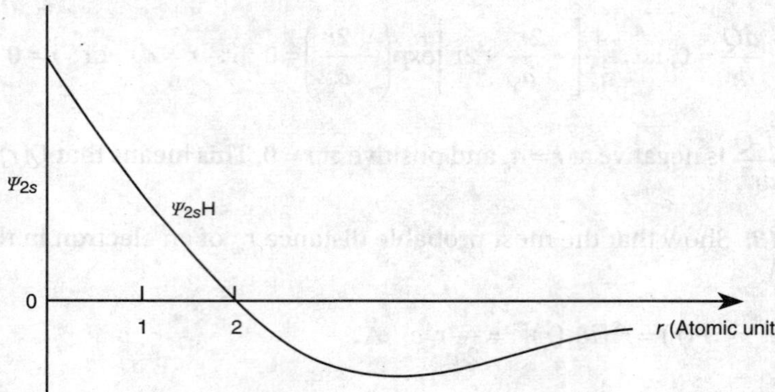

SUMMARY

1. Hydrogen atom is the first system treated by Schrodinger quantum mechanically. The results agree with those predicted by Bohr and observed experimentally. This is of great importance as it forms the foundation for the treatment of multielectron atoms and molecules. The system has two particles which make the solution of Schrödinger equation much more difficult.
2. The emergence of three quantum numbers n, l, and m is a consequence of the fact that the time-independent Schrödinger equation for hydrogen contains three independent variables.
3. The wave function $\psi_{nlm}(r,\theta,\phi)$ which describes the motion of an electron in a hydrogen atom is called an *atomic orbital.*
4. A diatomic molecule consists of two masses m_1 and m_2 connected by a bond of length r. If bond is rigid, it is called *rigid rotator.* The line joining the two masses is the axis of the rotator.
5. The allowed energy eigen functions of a rigid rotator are

$$E_n = \frac{\hbar^2 n(n+1)}{2I}, \text{ for } n = 0,1,2,\ldots$$

where $I = \mu r^2$ is the moment of inertia.

REVIEW QUESTIONS AND PROBLEMS

1. Find the bound state wave functions and corresponding enerey levels for a particle of mass μ (with zero angular momentum) moving under the influence of a three-dimensional potential of the type

$$V(r) = V_0 \exp\left(-\frac{r}{a}\right)$$

2. Show that the energy levels of a particle moving in a spherically symmetric potential $V(r) = \frac{\alpha}{r^2} + \beta r^2$ (with α and β, both positive) are given by

$$E_n = \sqrt{\frac{\beta\hbar^2}{2\mu}}\left[4n + 2\left\{(2l+1)^2 \frac{8\mu\alpha}{\hbar^2}\right\}^{1/2}\right], \text{ where } n = 0, 1, 2, 3, \ldots \text{ etc.}$$

3. A particle of mass μ under the influence of a spherically symmetric potential given by $V(r) = \frac{\alpha}{r^2} - \beta/r$ (α and β both positive). Show that for total negative energy of the particle, permissible values of energy are given by

$$E_n = \frac{2\mu\beta^2}{\hbar^2}\left[(2n+1) + \left\{(2l+1)^2 + \frac{8\mu\alpha}{\hbar^2}\right\}^{1/2}\right]^{-2}$$

4. A particle of mass μ moves in a cubical box of length L placed in the first octet with one of its corner at the origin.

$$V(x,y,z) = \begin{cases} = 0 & \text{for} \begin{cases} 0 < x < L \\ 0 < y < L \\ 0 < z < L \end{cases} \\ = 0 & \text{for all other points} \end{cases}$$

5. Determine discrete energy levels and corresponding eigen functions. Show that the density of states is given by $g(E) = \frac{\mu V}{\pi^2\hbar^2}\left(\frac{1}{2}\mu E\right)^{1/2}$, where $V = L^3$

(a) The deutron nucleus consists of a neutron and proton. Assuming hard core model for the potential

$$V(r) = \begin{cases} \infty & \text{for } r > a \\ -V_0 & \text{for } a < r < a + b \\ 0 & \text{for } r > a + b \end{cases}$$

Calculate the allowed values of energy in the zero angular-momentum state.

(b) Assuming $V_0 = 40$ MeV and Binding energy of deutron = 2.228 MeV, show that the minimum value of b (deutron radius) is 1.895 fermi.

6. Pure rotalional lines in the absorption spectrum of HCl are given by wave numbers 21.18, 42.36, 63.54, 84.72 and 105.91 cm^{-1}. Calculate the value of internuclear separation; given that

$$M_D = 35.5 M_H \quad \text{and} \quad = M_H = 1.67 \times 10^{-24} \text{ g.}$$

7. Find the probability that the electron in a hydrogen atom will be found at a distance greater than the classically permissible value. Assume the system to be in its ground state.
8. If the interatomic distances of HF, HCl, HBr and HI are all assumed to be constant and equal to one another, what would be the difference among the energy levels of the four molecules?
9. Find the energy levels for a mesic atom (negative μ-meson and a proton) or an alkali atom. The potential being given by the expression $V(r) = -\frac{e^2}{r}\left(1+\frac{c}{r}\right)$ where c is a constant. The factor $\left(1+\frac{c}{r}\right)$ accounts for the increase of effective nuclear charge as r decreases.
10. Verify, by application of appropriate operators, that the quantum mechanics of the $2p$ state of the hydrogen atom are indeed $l = 1$ and $m = 0, \pm 1$.
11. Calculate the expectation value of the kinetic and potential energy of the hydrogen ground state.
12. Show that a hydrogen atom in the $3d$ state is spherically symmetric as long as no external perturbation distinguishes a particular coordinates system.
13. A rigid diatomic molecule is free to rotate in a fixed plane. Set up the Schrodinger equation to describe its motion and show that the rotational energy eigen values are given by $\frac{\hbar^2 m^2}{8\pi^2 I}$, where I is the moment of inertia of the molecule and m takes the values of 0, ± 1, ±2, ±3 etc. How are the results modify if the axis is free?
14. Calculate the rotational energy levels of a molecule consisting of two atoms with a fixed distance between them. How can the theoretical deductions be checked experimentally?
15. Solve the radical equation for the hydrogen atom. Calculate the most probable distance of the electron from the nucleus. Discuss the degeneracy of the energy levels and mention, briefy, the conditions under which the degeneracy will be removed.
16. The radial wave function for the ground state ($n = 1, l = 0$) for the hydrogen atom is $R_{1,0}(r) = \left(\frac{1}{a_0}\right)^{\frac{3}{2}} 2e^{-\frac{r}{a_0}}$ where $a_0 = \frac{\hbar^2}{me^2}$, the Bohr radius. If the probability for finding the electron between r and $r + dr$ is $P(r)\, dr$, draw a rough sketch to show how $P(r)$ varies with r for the ground state. Discuss briefly how the result compares with the prediction of Bohr's simple model of the hydrogen atom.
17. A one-dimensional "hydrogen" atom is one in which an electron confined to the x-axis is acted upon by a force inversely proportional to the square of its distance from the origin. Find the energy eigen values and the eigen functions of this system.
18. Show directly that the degree of degeneracy of the nth hydrogen atom energy eigen value is equal to n by considering the solution of the Schrodinger's equation in parabolic coordinates.
19. Solve the radial part of Schrodinger equation for hydrogen atom and obtain its energy eigen values.
20. For a particle moving in a spherically symmetric potential $V(r)$, the eigen functions are $V = R(r) Y_{l,m}(\theta, \phi)$. Which quantum number decides the parity? How?

21. The radial wave function of an electron in the ground state of a hydrogen atom is proportional to exp $(Zr - r/a_0)$ where Z and a_0 are constants. Normalize this function and obtain the normalized radial function.
22. Write the Schrodinger equation in spherical polar coordinates for a hydrogen-like atom. Obtain the three equations from this one equation by the method of separation of variables. State the acceptable solutions of these equations (derivation not required). Mention the conditions under which lie acceptable solutions are possible. Hence find an expression for energy eigen value. Discuss the degeneracy.
23. (a) Write the Schrodinger equation for a particle moving in a Coulomb field and hence obtain radial Schrodinger equation, (b) Solve the radial equation to obtain the energy eign values of the hydrogen atom. What is the degeneracy for any value of l?
24. Starting from the angular part of the wave equation in a spherically symmetric potential, show that the spherical harmonics are given by

$$Y_{l,m}(\theta,\phi) = \left[\frac{2l+1}{4\pi}\frac{(l-|m|)!}{(l+|m|)!}\right]^{1/2} P_l^m(\cos\theta)e^{im\phi}$$

where P_l^m $(\cos\theta)$ is the associated Legendre function.
25. (a) The radial equation for a hydrogen-like atom is

$$\frac{1}{r^2}\frac{d}{dr}\left[r^2\frac{dR}{dr}\right]+\frac{2\mu}{\hbar^2}\left[E+\frac{Ae^2}{r}-\frac{\hbar^2 l(l+1)}{2\mu r^2}\right]R=0$$

Show that the energy levels are given by $E_n = -\dfrac{\mu Z^2 e^4}{2\hbar^2 n^2}$

(b) What is the degeneracy (considering spin also) of a slate of total quantum number n? Discuss briefly the degeneracy for an alkali atom like Na.
26. (a) The wave function of a hydrogen atom consists of a radial part and an angular part. Explain the physical significance of each. (b) The radial wave function for the ground state ($n=1, l=0$) for the hydrogen atom is.

$$R_{1.0}(r) = a_0^{-3/2}\exp(-r/a_0)$$

where $a_0 = \dfrac{\hbar^2(4\pi\varepsilon_0)}{me^2}$, the Bohr radius. What is the probability $P(r')\,dr$ for finding electron between r and $r + dr$? Draw a rough sketch to show how $P(r)$ varies with r for the ground state.
27. A rigid diatomic molecule is free to rotate in fixed plane. Set up the Schrodinger equation to describe its motion and show that the rotational energy eigen values are given by $\hbar^2 m^2/2I$, where I is the M.I. of the molecule and l takes the values 0, ± 1, ± 2, ± 3, etc., How are the results modified if the axis is free.
28. From the definitions of the centre of mass and moment of inertia I, show that for a diatomic molecule $I = \mu r^2$, where μ is reduced mass.
29. Show that on the rigid rotator (rotor) model the energy difference between the adjacent lines in a rotation spectrum of a diatomic molecule is constant.

30. What is a rigid rotator? Write Schrodinger's equation for it and obtain eigen values and eigen functions.

31. Calculate the classical frequency of rotation of a rigid body whose energy is given by $E_J = \frac{J(J+1)\hbar^2}{2l}$, where symbols have their usual meanings. For states $J = J$ and $J = J + 1$ show that the frquency of the spectral line associated with a transition between these states is intermediate between the rotational frequencies of the states.

SHORT ANSWER-QUESTIONS

1. Using atomic units, express the complete Hamiltonian for electronic motion in hydrogen atom in spherical coordinates. Write the radial and angular parts.

$$\left[\textbf{Hint. } H = -\frac{1}{2}\nabla^2 - \frac{1}{r} = -\frac{1}{2r^2}\left[\frac{\partial}{\partial r}\left(r^2\frac{\partial}{\partial r}\right) + \frac{1}{\sin\theta}\frac{\partial}{\partial\theta}\left(\sin\theta\frac{\partial}{\partial\theta}\right) + \frac{1}{\sin^2\theta}\frac{\partial^2}{\partial\phi^2}\right] - \frac{1}{r}\right]$$

Obviously, the radial part is $= -\frac{1}{2r^2}\left[\frac{\partial}{\partial r}\left(r^2\frac{\partial}{\partial r}\right)\right] - \frac{1}{r}$. We can see that the second and third terms within the square brackets together constitute the operator for the square of angular momentum, i.e., L^2. We note that the operator L^2 does not contain r and commutes with H.

Obviously, both H and L^2 can have the same set of eigen functions $\psi(r, \theta, \phi)$ and that $\psi(r, \theta, \phi)$ can be expressed as $\psi(r, \theta, \phi) = R(r)A(\theta, \phi)\, \psi(r, \theta, \phi)$ and $A(\theta, \phi)$ will be the eigen functions of L^2. Thus the angular momentum operator L^2 is

$$L^2 = \frac{1}{\sin\theta}\frac{\partial}{\partial\theta}\left(\sin\theta\frac{\partial}{\partial\phi}\right) + \frac{1}{\sin^2\theta}\frac{\partial^2}{\partial\phi^2}\Bigg]$$

2. Show that the functions ψ_{2s} and ψ_{2p_z} of $H\,(z = 1)$ are orthogonal.

$$\left[\textbf{Hint. } \int \psi_{2s}\psi_{2p_z}\, d\tau = \int_0^{\infty}\int_0^{\pi}\int_0^{2\pi} \frac{1}{\sqrt{32\pi}}(2-r)\exp(-r/2)\frac{1}{\sqrt{32\pi}} re^{-r/2}\cos\theta r^2\, dr \sin\theta d\theta d\phi\right.$$

$$\left.\text{we have} \int_0^{\pi} \cos\theta\sin\theta d\theta = \frac{1}{2}\int_0^{\infty} \sin 2\theta d\theta = \frac{1}{4}[\cos 2\theta]_0^{\pi} = 0\right]$$

3. Determine the quantum numbers n, l, m by inspection of the following function

$$\psi = \frac{\sqrt{2}}{81\sqrt{\pi}} Z^{3/2}(6 - Zr)Zr\exp(-Zr/3)\cos\theta$$

[**Hint.** From the exponential factor exp(–Zr/3), we note that $n = 3$. The radial factor $r^l = r$ shows that $l = 1$. From the angular factor $\cos\theta$, we note that $m = 0$]

4. Show that the probability of finding the electron in a hydrogen-like atom is independent of the angle ϕ.

[**Hint.** We have $\psi^*\psi = \left[R(r)P(\theta)\frac{1}{\sqrt{2\pi}}e^{-im\phi}\right]\left[R(r)P(\theta)\frac{1}{2\pi}e^{im\phi}\right] = \{R(r)P(\theta)\}^2$ clearly, probability is independent of ϕ]

5. Show that the ψ_{2p_z} is anti-symmetrical with respect to the xy-plane but its probability distrubution is symmetrical.

[**Hint.** $\psi_{2p_z} = \frac{1}{4\sqrt{2\pi}} z^{5/2} re^{-zr/2} \cos\theta$. We note that $\cos\theta$ changes sign at $\theta = \pi/2$ but $\cos^2\theta$ does not and hence probability of distribution involving $\psi^*_{2p_z}\psi_{2p_z}$ is symmetrical.]

6. Find the probability density for $\theta = \pi/2$ and $\theta = \pi/4$ at a distance of 1 Å from the nucleus for an electron in $2p_z$ orbital of an h-atom.

[**Hint.** (i) $\psi^2 = \frac{1}{32\pi} r^2 \exp(-r)\cos^2\theta$, (i) $\theta = \pi/2$, we have $\psi^2 = 0$

(ii) $\theta = \pi/4$, $\psi^2 = \frac{1}{64\pi} r^2 \exp(-r) = \frac{1}{64\pi}\exp(-1)$ for $r = 1$]

7. Determine the probability of an electron being found at a distance of 0, 0.5 Å from the nucleus of h-atom in $1s$ state?

[**Hint.** $P(r) = r^2 R^2(r) = \frac{1}{\pi} r^2 \exp(-2r)$. Putting $r = 0, 0.5$, one obtains $P(r) = 0, \frac{1}{4\pi}\exp(-1)$]

8. Show that the maximum electron density in H-atom in the $2s$ state is a_0^3, where $a_0 = 0.529 \times 10^{-10} m$

[**Hint.** $\psi_{2s}^2 = \frac{1}{4\pi a_0^3}\left(2 - \frac{r}{a_0}\right)^2 \exp(-r/a_0)$

This will be maximum when exp $(-r/a_0) = 1$ or $r = 0$. Thus we get the value as $1/4\pi\, a_0^3$]

9. Consider that interaction between two particles depends only on the relative distance between them, then the two-body problem can be treated as two one-body problems. Explain it by taking the example of hydrogen atom.

[**Hint.** Express the Hamiltonian in terms of the centre of mass coordinates $R = \left(\frac{m_1 r_1 + m_2 r_2}{m_1 + m_2}\right)$ and relative coordinates $r = r_1 - r_2$ where r_1 and r_2 are position vectors of the proton of mass m_1 and the electron of mass m_2]

OBJECTIVE QUESTIONS

1. The wave functions of the hydrogen atom for $1s$ state is

(a) $\psi_{1s} = \frac{1}{\sqrt{\pi}} \frac{1}{(a_0)^{3/2}} \exp(-r/a_0)$

(b) $\psi_{1s} = \frac{1}{\sqrt{\pi}} e^{-r}$

(c) $\psi_{1s} = \frac{1}{4\sqrt{2\pi}}(2 - r)e^{-r/2}$

(d) $\psi_{1s} = \frac{1}{4\sqrt{2\pi}}\left(\frac{1}{a_0}\right)^{3/2}\left(2 - \frac{r}{a_0}\right)\exp\left(-\frac{r}{2a_0}\right)$

2. The potential energy (V) of the electron having charge e in the hydrogen atom of r is

(a) $V = \frac{e^2}{r}$ (b) $V = -\frac{e^2}{r}$

(c) $V = \frac{4e^2}{r}$ (d) $V = -\frac{4e^2}{r}$

3. If the wave function for 1s orbital of hydrogen atom is $\psi_{100} = \frac{1}{\sqrt{\pi a_0^3}} \exp(-r/a_0)$, then the average distance of 1s electron from the nucleus is

(a) a_0 (b) $\frac{1}{2}a_0$

(c) $\frac{3}{2}a_0$ (d) $\frac{7}{2}a_0$

4. A rigid rotator (diatomic molecule) is free to rotate in fixed plane. The rotational energy eigen values are given by if I is the moment of inertia of the molecule and in takes the values 0, ±1, ±2, ±3, etc.

(a) $\frac{\hbar^2 m^2}{2I}$ (b) $\frac{\hbar^2 m^2}{8I}$

(c) $\frac{\hbar^2 m^2}{16I}$ (d) $\frac{\hbar^2 m^2}{32I}$

5. The radial function $R_{1,\theta} = 2\left(\frac{Z}{a_0}\right)^{3/2} \exp(-Zr/a_0)$ is the radial eigen functions of

(a) Helium-like atoms (b) Sodium-like atoms

(c) Chlorine-like atoms (d) Hydrogen-like atoms

6. $\psi_{1s} = \frac{1}{\sqrt{\pi}}\left(\frac{Z}{a_0}\right)^{3/2} \exp(-Zr/a_0)$ and $\psi_{2s} = \frac{1}{4\sqrt{\pi}}\left(\frac{Z}{a_0}\right)^{3/2}\left(2 - \frac{Zr}{a_0}\right)\exp(-Zr/2a_0)$ are

(a) orthogonal functions

(b) orthonormal functions

(c) orthogonal as well as orthonormal functions

(d) none of the above

7. Functions ψ_{2s} and ψ_{2p_z} of hydrogen atom are

(a) orthogonal (b) orthonormal

(c) orthogonal as well as orthonormal (d) none of the above

8. The probability of finding the electron in a hydrogen-like atom is independent of

(a) r (b) θ

(c) ϕ (d) r as well as θ

9. The average value of $1/r$ for $1s$ orbital of an H-atom in atomic unit is

(a) 2 (b) 0.5

(c) 2 (d) 1

10. The average potential energy of $1s$ orbital of an H-atom in atomic unit is

(a) −1 (b) −2

(c) −4 (d) −8

11. The average kinetic energy of $1s$ orbital of an H-atom in atomic unit is

(a) 1 (b) 2

(c) 0.5 (d) 3

12. In the case of the 3-dimensional, isotropic harmonic oscillator, the energy levels are given by

(a) $E_n = \left(n + \frac{1}{2}\right)\hbar\omega$ (b) $E_n = \left(n + \frac{3}{2}\right)\hbar\omega$

(c) $E_n = \left(n + \frac{5}{2}\right)\hbar\omega$ (d) $E_n = n\hbar\omega$

13. The most probable value of r for a $2p$ electron in a hydrogen atom is $4a_0$. This is the same as the radius of the

(a) $n = 1$ Bohr orbit (b) $n = 2$ Bohr orbit

(c) $n = 3$ Bohr orbit (d) $n = 4$ Bohr orbit

14. The wave function for a hydrogen atom in a $2p$ state varies with direction as well as with

(a) distance from the nucleus (b) n^2

(c) n^4 (d) none of the above

15. The probability of finding an atomic electron whose radial wave function is $R(r)$ outside a sphere of radius r_0 centred on the nucleus is

(a) $\int_{r_0}^{\infty} |R(r)|^2 r^2 dr$ (b) $\int_{0}^{r_0} |R(r)|^2 r^2 dr$

(c) $\int_{0}^{\infty} |R(r)|^2 r^2 dr$ (d) $\int_{r_0}^{\infty} |R(r)|^2 dr$

16. The location at which the radial probability density is a maximum tor the $n = 2$, $l = 1$ state of the hydrogen atom is

(a) $4a_0$ (b) $2a_0$

(c) a_0 (d) $8a_0$

17. The expectation value of the radial coordinate in $n = 2$, $l = 1$ state is

(a) $4a_0$ (b) $5a_0$

(c) $3a_0$ (d) $8a_0$

18. If $d(R_{nl})^2/dr=0$, the most probable values of r for the $l=n-1$ states in hydrogen atom are

(a) $\bar{r}=n^2a_0$ (b) $\bar{r}=(n-1)^2a_0$

(c) $\bar{r}=a_0$ (d) $\bar{r}=(n-3)^2a_0$

19. For a cylindrical box of radius a and height b the wave equation has the Bessel equation of the from

(a) $\left[\frac{1}{x^2}+1-\frac{m_2}{x^2}\right]J_m(x)=0$ (b) $\left[\frac{1}{x^2}\left(x\frac{d}{dx}\right)^2+1\right]J_m(x)=0$

(c) $\left[\frac{1}{x^2}\left(x\frac{d}{dx}\right)^2+1-\frac{m^2}{x^2}\right]J_m(x)=0$ (d) $\left(1-\frac{m^2}{x^2}\right)J_m(x)=0$

20. For a cylindrical box of radius a and height b, $(p_e^2=\hbar^2\frac{\partial}{\partial\rho}(\rho\partial/\partial\rho))$

(a) $H=L_z^2\rho^{-2}/2\mu$ (b) $H=(p_r^2+p_z^2+p_z^2)\rho^{-2}/2\mu$

(c) $H=L^2/2\mu$ (d) $H=\rho^2L^2/2\mu$

ANSWERS

Question numbers	1	2	3	4	5	6	7	8	9	10
Answers	(a)	(b)	(c)	(a)	(d)	(b)	(a)	(c)	(d)	(a)
	11	12	13	14	15	16	17	18	19	20
	(c)	(b)	(b)	(a)	(a)	(a)	(b)	(a)	(c)	(b)

FILL IN THE BLANKS

1. The simplest wave function for a hydrogen atom describing 1s state and designated $\psi_{1s}(r)$ is ______, where a_0 is the Bohr radius and wave function is normalised. $\left[\frac{1}{\sqrt{\pi a_0^3}}e^{-r/a_0}\right]$

2. The probability of finding the electron in a volume element dV is ______. **[$|\psi|^2$ dV]**

3. The radial probability density function P(r) is defined as the ______ of finding the electron in a spherical shell of radius r and thickness dr. **[probability per unit radial distance]**

4. The probability that the electron in the ground state of hydrogen will be found outside the Bohr radius is ______. **[67.7%]**

5. The methods of quantum mechanics can be applied to the hydrogen atom using the appropriate potential energy function. The solution to this equation yields the wave function for the allowed states and allowed energies. The allowed energy depends only on the ______ n. **[principal quantum number]**

10

Harmonic Oscillator (Vibrational Motion of a Particle)

10.1 INTRODUCTION

We are all familiar with simple harmonic motion (vibrational or oscillation) a characteristic motion often observed in a physical system. SHM is a back and forth motion along the same path, in which the displacement forms an equilibrium (rest) position varies periodically with time. The vibration is said to be simple harmonic if it has a fixed time period (T) and amplitude a. A swinging pendulum through a very small angle, atoms in a molecule executing vibratory motion along a bond with bond length and bond angles changing periodically are few examples constituting harmonic oscillators. The simple harmonic oscillator is of tremendous importance in physics, and all fields based on physics, because it is the prototype for any system involving oscillations. For instance, it is used in the study of: the vibration of atoms in diatomic molecules, the acoustic and thermal properties of solids which arise from atomic vibrations, magnetic properties of solids that involve vibrations in the orientation of nuclei, and electrodynamics of quantum systems in which electromagnetic waves are vibrating. Generally speaking, the simple harmonic oscillator can be used to describe almost any system in which an entity is executing small vibrations about a point of stable equilibrium.

In all such harmonic oscillators when a particle is displaced from its equilibrium position, a force starts acting on it in the opposite direction in order to restore it to its equilibrium position. If x be the displacement, the restoring force is,

$$F = -kx \tag{10.1}$$

where k (proportionality constant) is called **force constant**. The minus sign appearing in the relation indicate that the force acts in the opposite direction to the direction of displacement. The above relation is known as **Hooke's law**.

Classical treatment of harmonic oscillator gives different result form those obtained from quantum mechanical treatment. However, classical treatment is restricted to macro particles (bodies) and the later to micro particles (atoms, molecules, nucleii, etc.,).

10.1.1 One-Dimensional Harmonic Oscillator: Classical Mechanical Treatment

According to Newton's second law of motion

$$F = m\frac{d^2x}{dt^2} \tag{10.2}$$

Combining Eqs. (10.1) and (10.2), one obtains

$$m\frac{d^2x}{dt^2} = -kx \quad \text{or} \quad \frac{d^2x}{dt^2} = -\left(\frac{k}{m}\right)x = -\omega^2 x \tag{10.3}$$

where $\omega = \sqrt{k/m}$

The general solution of Eq. (10.3) is

$$x = A\sin\omega t + B\cos\omega t \tag{10.4}$$

where A and B are constants of integration. Eq. (10.4) can also be expressed as

$$x = a\sin(\omega t + \theta) \tag{10.5}$$

If we put $A = a\cos\theta$ and $B = a\sin\theta$, where a and θ are two other constants. These constants can be determined from the initial conditions. Eq. (10.5) shows that particle performs simple harmonic oscillations with frequency

$$v = \frac{\omega}{2\pi} = \frac{1}{2\pi}\sqrt{\frac{k}{m}} \tag{10.6}$$

where $\omega = 2\pi v$ is called the angular frequency of oscillations. The potential energy $V(x)$ at the displacement x is given by

$$V = -\int_0^V dv = -\int_0^x \frac{dv}{dx}dx$$

But
$$\frac{dv}{dx} = -F = -(-kx) = kx$$

$\therefore$
$$V = \int_0^x kx\,dx = \frac{1}{2}kx^2 = \frac{1}{2}m\omega^2x^2 \tag{10.7}$$

and the kinetic energy T at the displacement x is given by

$$T = \frac{1}{2}m\left(\frac{dx}{dt}\right)^2 = \frac{1}{2}ma^2\omega^2\cos^2(\omega t + \theta)$$

$$= \frac{1}{2}m\omega^2a^2\,[1 - \sin^2(\omega t + \theta)] = \frac{1}{2}m\omega^2(a^2 - x^2) \tag{10.8}$$

Therefore, according to classical mechanics the total energy E of the oscillator is given by

$$E = T + V = \frac{1}{2}m\omega^2(a^2 - x^2) + \frac{1}{2}m\omega^2x^2 = \frac{1}{2}m\omega^2a^2 = \frac{1}{2}ka^2 \tag{10.9}$$

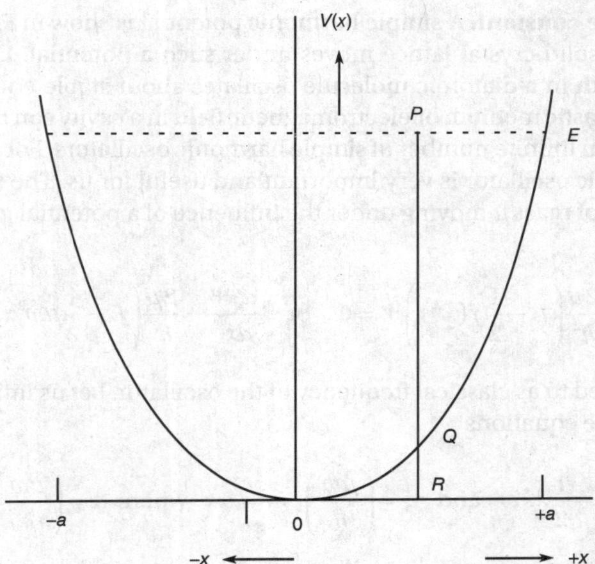

Fig. 10.1 The simple harmonic oscillator potential

Eq. (10.9) shows that the oscillator can have any positive value of *E*. Eq. (10.9) also shows that the potential energy has a parabolic form Fig. (10.1). It means that the particle moves in a potential well of parabolic form Fig. (10.1). Eq. (10.1) further shows that the total energy of a particle oscillating with a constant frequency $\nu = \omega/2\pi$ is proportional to the square of the amplitude. According to classical mechanics the particle can oscillate with any amplitude, continuously with increase in the amplitude.

At the extreme point (x = ± a) of oscillation, all the energy is potential energy $\left(E = V = \frac{1}{2}ka^2\right)$, while at the equilibrium point ($x = 0$), all the energy is kinetic energy. Fig. (10.1) shows that at any point within the extremes, the vertical line *PQ* represents kinetic energy and *QR* the potential energy; the line *PR* representing the total energy *E*.

10.1.2 One-Dimensional Harmonic Oscillator: Quantum Mechanical Treatment

In most of the problems concerning us, we have a particle in stable equilibrium condition; on displacing it slightly from the position of equilibrium, a restoring force acts on it. Using Taylor series expansion near the equilibrium position (say $x = a$) we get an expression for potential energy

$$V(x) = V(a) + V'(a)(x-a) + \frac{(x-a)^2}{L^2}V''(a) + \cdots$$

$V(x)$ has a minimum at $x = a$, i.e., $V'(a) = 0$ and $V''(a) > 0$. If a is origin of coordinate system and $V(a)$ as zero reference for energy measurement, then $V(x)$ may be given by the expression

$$V(x) = V''(0)\frac{x^2}{L^2} = \frac{1}{2}f_x^2 \tag{10.10}$$

f is known as **force constant**. A simple harmonic potential is show in Fig. (10.1). An atom or molecule in a solid crystal lattice moves under such a potential. Due to vibration of atoms, bond length in a diatomic molecule oscillates about stable equilibrium position. Vibrations of an elastic medium or electromagnetic field in a cavity can be described by the superposition of an infinite number of simple harmonic oscillators. For this reason a study of simple harmonic oscillator is very important and useful for us. The Schrodinger equation for a particle of mass μ moving under the influence of a potential given by Eq. (10.10) may be written as

$$\frac{d^2\Psi}{dx^2} + \frac{2\mu}{\hbar^2}\left(E - \frac{1}{2}f(x^2)\right)\Psi = 0 \quad \text{or} \quad \frac{d^2\Psi}{dx^2} + \frac{2\mu}{\hbar^2}\left(E - \frac{1}{2}\mu\omega^2x^2\right)\Psi = 0 \qquad (10.11)$$

ω is usually referred to as classical frequency of the oscillator. Let us introduce the parameters λ and ξ by the equations

$$E = \frac{1}{2}\lambda\hbar\omega \text{ and } \xi = \left(\frac{\mu\omega}{\hbar}\right)^{1/2} x = \alpha x \text{ where } \alpha = \left(\frac{\mu\omega}{\hbar}\right)^{1/2}$$

$$\frac{d\Psi}{dx} = \frac{d\Psi}{d\xi}.\frac{d\xi}{dx} = \left(\frac{\mu\omega}{\hbar}\right)^{1/2}\frac{d\Psi}{d\xi}$$

and
$$\frac{d^2\Psi}{dx^2} = \frac{d}{dx}\left(\frac{d\Psi}{dx}\right) = \frac{d}{d\xi}\left[\left(\frac{\mu\omega}{\hbar}\right)^{1/2}\frac{d\Psi}{d\xi}\right].\frac{d\xi}{dx} = \left(\frac{\mu\omega}{\hbar}\right)\frac{d^2\Psi}{d\xi^2}$$

$$\therefore \quad \left(\frac{\mu\omega}{\hbar}\right)\frac{d^2\Psi}{d\xi^2} + \frac{2\mu}{\hbar^2}\left[\frac{1}{2}\lambda\hbar\omega - \frac{1}{2}\mu\omega^2.\left(\frac{\hbar}{\mu\omega}\right)\xi^2\right]\Psi = 0$$

or
$$\frac{d^2\Psi}{d\xi^2} + (\lambda - \xi^2)\Psi = 0 \qquad (10.12)$$

the solution of Eq. (10.12) is facilitated by first examining the dominating behaviour of Ψ in the asymptotic region $\xi \to \pm\infty$. The potential $[V(x)$ or $V(\xi) \to \infty$ as $\to \pm\infty]$, the only possible solution of Eq. (10.12) for $\xi \to \pm\infty$ must be $\Psi \equiv 0$. This is because a region of infinitely large potential must have zero probability of containing the particle. Therefore, we will study only those solutions which satisfy the boundary condition $\Psi(\xi) = 0$ at $\xi = \pm\infty$. For large values of ξ, Eq. (10.12) may be written as

$$\frac{d\Psi}{d\xi^2} - \xi^2\Psi = 0 \qquad (10.13)$$

where we have used the assumption that $\xi^2 >> \lambda$. Let us now consider the function $\exp(\pm\xi^2/2)$. On substituting this solution in Eq. (10.13), we get

$$\frac{d^2}{d\xi^2}[\exp(\pm\xi^2/2)] = (\xi^2 \pm 1)\exp(\pm\xi^2/2) \qquad (10.14)$$

since $\xi^2 >> 1$, the term '±1' in Eq. (10.14) can be neglected in comparison to ξ^2. Thus we see that $e^{\pm\xi^2/2}$ is a solution of Eq. (10.13). $\Psi(\xi)$ must approach zero at $\xi = \pm\infty$, therefore only the negative exponent is acceptable as a solution on physical grounds even though both positive and negative exponents are mathematically admissible. Therefore, a general solution valid for all values of ξ may be written as

$$\Psi(\xi) = e^{-\xi^2/2}\Phi(\xi) \tag{10.15}$$

where $\Phi(\xi)$ is some function of ξ. We have

$$\frac{d\Psi}{d\xi} = e^{-\xi^2/2}\frac{d^2\Phi}{d\xi} - \xi e^{-\xi^{2/2}}\Phi$$

$$\frac{d^2\Psi}{d\xi^2} = e^{-\xi^{2/2}/2}\frac{d^2\Phi}{d\xi^2} - 2\xi e^{-\xi^2/2}\frac{d\Phi}{d\xi} - e^{-\xi^2/2}(1-\xi^2)\Phi$$

Equation satisfied by $\Phi(\xi)$ may be written as

$$e^{-\xi^2/2}\left[\frac{d^2\Phi}{d\xi^2} - 2\xi\frac{d\Phi}{d\xi} - (1-\xi^2)\Phi + (\lambda-\xi^2)\Phi\right] = 0$$

or

$$\frac{d^2\Phi}{d\xi^2} - 2\xi\frac{d\Phi}{d\xi} + (\lambda-1)\Phi = 0 \tag{10.16}$$

We try a power-series expansion in ξ for $\Phi(\xi)$,

$$\Phi(\xi) = \xi^s(a_1\xi + a_2\xi^2 + a_3\xi^3 + \cdots)$$

or

$$\Phi(\xi) = \sum_{\nu=0}^{\infty} a_\nu \xi^{\nu+S} \tag{10.17}$$

where a' s are expansion coefficients to be determined.

$$\frac{d\Phi}{d\xi} = \sum_\nu (\nu+S)a_\nu\xi^{\nu+S-1}$$

$$\frac{d^2\Phi}{d\xi^2} = \sum_\nu (\nu+S)(\nu+S-1)a_\nu\xi^{\nu+S-2}$$

Substituting the value of Φ, $\frac{d\Phi}{d\xi}$ and $\frac{d^2\Phi}{d\xi^2}$ in Eq. (10.16), we obtain

$$\sum_\nu(\nu+S)(\nu+S-1)a_\nu\xi^{\nu+S-2} - 2\sum_\nu a_\nu(\nu+S)\xi^{\nu+S} + \sum_\nu(\lambda-1)a_\nu\xi^{\nu+S} = 0 \tag{10.18}$$

since the left hand side of Eq. (10.18) should be zero for all value of ξ, the coefficient of each power of ξ must be identically equal to zero. Equating the coefficients of various powers of ξ to zero, we get

$$\text{coeff. of } \xi^{S-2} \rightarrow S(S-1)a_0 = 0 \tag{10.19}$$

$$\text{coeff.of } \xi^{S-1} \rightarrow (S+1)Sa_1 = 0 \quad (10.20)$$

$$\text{coeff.of } \xi^{S} \rightarrow (S+2)(S+1)a_2 - (2S+1-\lambda)a_0 = 0 \quad (10.21)$$

$$\text{coeff.of } \xi^{S+1} \rightarrow (S+3)(S+2)a_3 - \{2(S+1)+1-\lambda\}a_1 = 0 \quad (10.22)$$

$$\text{coeff.of } \xi^{S+1} \rightarrow (S+\nu+2)(S+\nu+1)a_{\nu+2} - \{2(\nu+S)+1-\lambda\}a_\nu = 0 \quad (10.23)$$

where ν is an integer, Eq. (10.10) tells us that $S = 0$ or $S = 1$. From Eq. (10.11) we can conclude that either $S = 0$ or $a_1 = 0$ or both Eq. (10.14) gives us a **recursion formula** connecting even (odd) coefficients to even (odd)

$$a_{\nu+2} = \frac{2\nu + 2S + 1 - \lambda}{(S+\nu+2)(S+\nu+1)} a_\nu \quad (10.24)$$

Oscillator wave function can be chosen to be either even or odd in ξ. Then we should choose a_1 (and all other odd subscripts) to be zero or a_0 (and all other even subscripts) to be zero. If we choose a_1 to be zero then function is even or odd according as $S = 1$ or 0. Similarly, if we choose a_0 to be zero, then Ψ or Φ is even or odd according as $S = 1$ or 0. It follows from Eq. (10.24) that the presence of a finite or an infinite number of terms in Eq. (10.8) will be determined form the choice of S and λ. If the series does not terminate, its dominant asymptotic behaviour can be inferred from the coefficients of the terms having very large power:

$$\frac{a_\nu + 2}{a_\nu} \underset{\nu\to\infty}{\longrightarrow} \frac{2}{\nu}$$

Also $$\xi^2 = 1 + \xi^2 + \frac{\xi^4}{\underline{|2}} + \frac{\xi^6}{\underline{|3}} + \frac{\xi^8}{\underline{|4}} + \cdots + \frac{\xi^\nu}{\underline{|\nu/2}} + \frac{\xi^{\nu+2}}{\underline{|\nu+2/2}} + \cdots$$

$$\frac{\text{coefficients of } \xi^{\nu+2}}{\text{coefficients of } \xi^\nu} = \frac{(\nu/2)!}{(\nu/2+2)!} = \frac{(\nu/2)!}{(\nu/2+1)(\nu/2)!} = \frac{2}{\nu+2} \rightarrow \frac{2}{\nu}$$

Thus we conclude that for large $\pm\xi$, the function $\Phi(\xi)$ consisting of an infinite number of terms behaves as $e^{\xi^2/2}$. The product $\Psi(\xi) = e^{-\xi^2/2}\Phi(\xi)$ will then behave as $e^{-\xi^2/2}$ for large value of ξ. Since this function increases to infinity at $\xi = \pm\infty$, we conclude that the boundary conditions [$\Psi(\xi) = 0$ at $\xi = \pm\infty$] are not satisfied and these mathematical solutions are therefore not acceptable as solutions for the problem on physical ground. The only way this catastrophe can be avoided is if λ happens to have certain particular values such that $\lambda = (2n + 1)$, where n is a positive integer including zero.

In this case, the numerator of Eq. (10.24) becomes zero when ν takes the value n, and $a_{n+2} = a_{n+4} = a_{n+6} = a_{n+8}$. For such value of λ, the infinite series expression for $\Phi(\xi)$ is truncated and becomes a polynomial. The product $e^{-\xi^2/2}\Phi(\xi)$ is truncated and becomes a polynomial. The product Ψ_0^2 approaches zero as $\xi \rightarrow \pm\infty$, since $e^{-\xi^2/2}$ goes to zero at a faster rate than the polynomial in ξ goes to infinity. The only mathematical solutions of Eq. (10.22) that are physically acceptable are those for which $\lambda = (2n + 1)$ where n can be 0, 1, 2, 3, ... etc. The general solution $\Psi(\xi)$ can be written as

$$\Psi(\xi) = e^{-\xi^2/2}\left[\sum_{\nu=0,2,4\ldots} a_\nu \xi^\nu + \sum_{\nu=1,2,3\ldots} a_\nu \xi^\nu\right] \quad (10.25)$$

Each summation contains one arbitrary coefficient a_0 or a_1. All other coefficients can be defined in terms of these coefficients. If n is even, then only the even series is truncated, therefore we must choose $a_1 = 0$ in order to make all odd coefficients zero. Similarly, when n is odd, then we must choose $a_0 = 0$. Thus, only physically significant solutions are those for which $\Phi(\xi)$ is a finite polynomial consisting of either even powers of ξ or of odd powers of ξ only.

The restriction on possible value of λ is equivalent to a restriction of possible values of E:

$$E_n = \frac{1}{2}(2n+1)\hbar\omega = \left(n + \frac{1}{2}\right)\hbar\omega \tag{10.26}$$

Thus the energy levels of a particle in a harmonic oscillator potential are quantized into equispaced energies $\frac{1}{2}\hbar\omega$, $\frac{3}{2}\hbar\omega$, $\frac{5}{2}\hbar\omega$, ... For each value of n, energy is given by Eq. (10.26).

10.1.3 Zero-point Energy

The finite (non-zero) value of ground state energy level $\frac{1}{2}\hbar\omega$ is known as zero-point energy. This result is a characteristic of quantum mechanics and is related to the uncertainty principle. The total energy is of the order of $\frac{(\Delta p)^2}{2\mu} + \frac{1}{2} f(\Delta x)^2$, where Δp and Δx are measures of the spreads in momentum and position. If this is minimized, taking account of the uncertainty relation, then the minimum Δp is of the order of $(f\mu\hbar^2)^{1/4}$, so that minimum total energy or zero-point energy is to the order of $\frac{\hbar}{2}\left(\frac{f}{\mu}\right)^{1/2}$ or $\frac{\hbar\omega}{2}$.

Eigen Values of Energy one obtains the eigen-values of the total energy E_n:

$$\frac{2E_n}{\hbar\omega} = (2n+1)$$

or
$$E_n = \frac{1}{2}(2n+1)\hbar\omega = \left(n + \frac{1}{2}\right)\hbar\omega = \left(n + \frac{1}{2}\right)h\nu \tag{10.27}$$

where $n = 0, 1, 2, \ldots$ ω is the angular frequency and ν is the frequency of the classical oscillator given by

$$\nu = \frac{\omega}{2\pi} = \frac{1}{2\pi}\sqrt{\frac{k}{m}}$$

From Eq. (10.26), one can easily draw the following conclusions:

1. The energy of the harmonic oscillator is quantized, contrary to the classical view, and this result emerges from the solution of the Schrodinger's equation.
2. The energies are determined by the parameter n which is called '**vibrational quantum number**', of the oscillator.
3. The lowest energy of the oscillator is obtained by putting $n = 0$ in Eq. (10.26) and it is $E_0 = \frac{1}{2}h\nu = \frac{1}{2}\hbar\omega$.

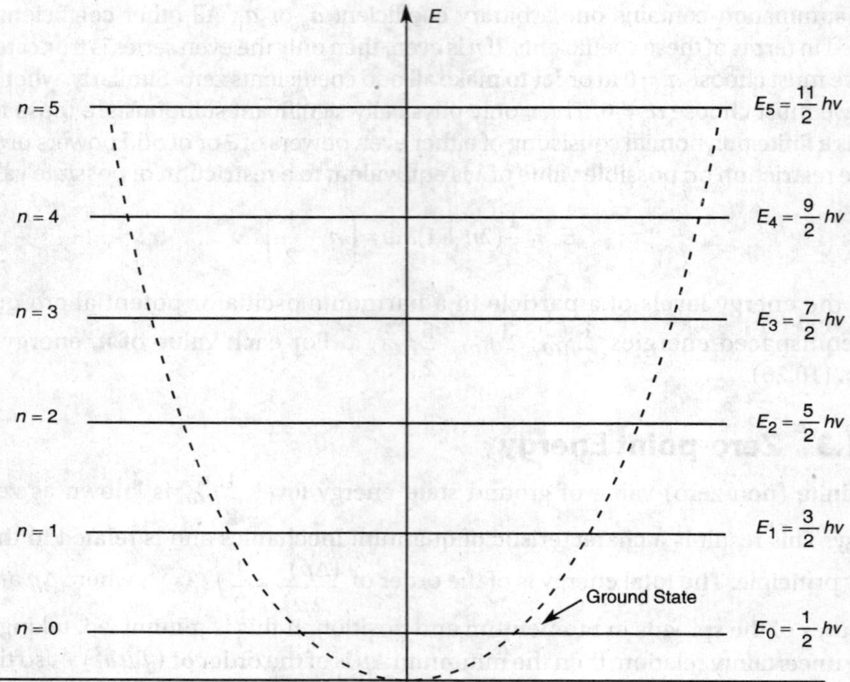

Fig. 10.1(a) Energy level diagram for the harmonic oscillator

This is called the **ground state energy** or the **zero-point vibrational energy** of the harmonic oscillator. The zero-point energy is the characteristic result of quantum mechanics. The value of $E_n = (2n + 1)\, E_0$, where $n = 0, 1, 2, 3, \ldots$. Classically the lowest values of the energy, i.e., the ground state energy of the harmonic oscillator is zero.

4. The eigen values of the total energy E_n depend only on one quantum number, i.e., vibrational quantum number n. Therefore all the energy levels of the harmonic oscillator are non-degenerate.
5. As n increases by unity, the energy increases by $h\nu$ so that the energy levels are equally spaced as shown in Fig. [10.1(a)].

Eq. (10.16) may now be written as

$$\frac{d^2\Phi}{d\xi^2} - 2\xi\frac{d\Phi}{d\xi} + 2n\Phi(\xi) = 0$$

For each integral value of n, solution of Eq. (17) are known as **Hermite polynomials** $H_n(\xi) = H_n(\alpha x)$ of nth degree.

10.1.4 Hermite Polynomials

The Hermite polynomial $H_n(\xi)$ is a solution of the differential equation

$$H_n'' - 2\xi H_n' + 2nH_n = 0 \tag{10.28}$$

On differentiating this equation with respect to ξ, we obtain

$$(H_n'') - 2\xi(H_n')' - 2H_n' + 2nH_n' = 0$$

or

$$(H_n')'' - 2\xi(H_n')' + 2(n-1)(H_n') = 0$$

H_n' satisfies Eq. (10.28) except that n is replaced by $n - 1$. We therefore conclude that apart from a constant factor, (H_n') is the same as H_{n-1} and

$$H_n' = CH_{n-1}$$

By comparison of the terms in x^{n-1}, C can be evaluated;

$$2^n(nx^{n-1}) = C2^{n-1}x^{n-1} \qquad \therefore\ C = 2n$$

and

$$H_n' = 2nH_{n-1} \tag{10.29}$$

Also

$$H_n'' = (H_n')' = (2nH_{n-1})' 2nH'_{n-1} = 2n \times 2(n-1)H_{n-2}$$

$$= 4n(n-1)H_{n-2} \tag{10.30}$$

Let us consider the equation for H_{n+1}

$$H_{n+1}'' - 2\xi H'_{n+1} + 2(n+1)\, H_{n+1} = 0$$

On substituting the values of H_{n+1}' and H''_{n+1} from Eqs. (10.29) and Eq. (10.30), we have

$$4(n+1)nH_{n-1} - 2\xi \times 2(n+1)\, H_n + 2(n+1)\, H_{n+1} = 0$$

or

$$2nH_{n-1} - 2\xi H_n + H_{n+1} = 0$$

or

$$H_{n+1} - 2\xi H_n + 2nH_{n-1} = 0 \tag{10.31}$$

Also

$$H_{n+1} + H_n' = H_{n+1} + 2nH_{n-1} = 2\xi H_n$$

or

$$H_{n+1} - 2\xi H_n + H_n' = 0 \tag{10.32}$$

The generating function $g(\xi, h)$ of Hermite polynomials is defined by

$$g(\xi, h) = \sum_{n=0}^{\infty} H_n(\xi)\frac{h^n}{n!} \tag{10.33}$$

$$\frac{\partial g}{\partial h} = \sum_{n=0}^{\infty} H_n(\xi)\frac{nh^{n-1}}{n!} = \sum_{n=1}^{\infty} H_n(\xi)\frac{h^{n-1}}{(n-1)!} = \sum_{n=0}^{\infty} H_{n+1}(\xi)\frac{h^n}{n!}$$

or

$$\frac{\partial g}{\partial h} = \sum_{n=0}^{\infty}(2\xi H_n - 2nH_{n-1})\frac{h^n}{n!} = \sum_{n=0}^{\infty} 2\xi H_n\frac{h^n}{n!} - 2\sum_{n=0}^{\infty} nH_{n-1}\frac{h^n}{n!}$$

or $$\frac{\partial g}{\partial h} = 2g\xi - 2\sum_{n=1}^{\infty} H_{n-1}\frac{h^{n-1}}{(n-1)}h = 2g\xi - 2h\sum_{n=0}^{\infty} H_n\frac{h^n}{n!} = 2g\xi - 2hg$$

or $$\frac{\partial g}{\partial h} = 2(\xi - h)g$$

This partial differential equation in h can be solved to give the result

$$g = C(\xi)e^{2\xi h - h^2} \tag{10.34}$$

The constant of integration $c(\xi)$ may possibly depend upon ξ. Substituting $h = 0$ in Eq. (10.34) and (10.26) we get $g(\xi, 0) = C(\xi) \therefore C(\xi) = 1$

and $$g(\xi, h) = e^{2\xi h - h^2} \tag{10.35}$$

Using Taylor series expansion, one obtains

$$H_n(\xi) = \left[\frac{\partial^n}{\partial h^n} g(\xi,h)\right]_{h=0} = \left[\frac{\partial^n (e^{2\xi h - h^2})}{\partial h^n}\right]_{h=0}$$

or $$H_n(\xi) = e^{\xi^2}\left[\frac{\partial^n \left\{e^{-(\xi-h)^2}\right\}}{\partial h^n}\right]_{h=0}$$

or $$H_n(\xi) = e^{\xi^2}(-1)^n\left[\frac{\partial^n}{\partial \xi^n}\left\{e^{-(\xi-h)^2}\right\}\right]_{h=0}$$

$$\left[\text{since for any function } g(t-y), \frac{\partial g}{\partial t} = -\frac{\partial g}{\partial y}\right]$$

$\therefore$ $$H_n(\xi) = e^{\xi^2}(-1)^n\left[\frac{\partial^n}{\partial \xi^n}\left(e^{-\xi^2}\right)\right] \tag{10.36}$$

This is an explicit formula for $H_n(\xi)$; although the direct calculation of $P_n(\xi)$ by Eq. (10.36) is a little tedious.

10.1.5 Orthogonality of Hermite Polynomials

The generating function for hermite polynomials can be used to find the orthogonality relations, for hermite polynomials.

Let us consider an integral containing the product of two generating functions:

$$\int_{-\infty}^{+\infty} e^{-S^2+2S\xi} e^{-r^2+2R\xi} e^{-\xi^2} d\xi = \int_{-\infty}^{+\infty} \left(\sum_n \frac{S^n H_n(\xi)}{n!}\right)\left(\sum_m \frac{t^m H_m(\xi)}{m!}\right) e^{-\xi^2} d\xi$$

$$= \sum_n \sum_m \frac{S^n t^m}{n!m!} \int_{-\infty}^{+\infty} H_n(\xi) H_m(\xi) e^{-\xi^2} d\xi$$

$$\int_{-\infty}^{+\infty} e^{S^2-2S\xi-t^2+2r\xi-\xi^2} k\xi = \int_{-\infty}^{+\infty} e^{-\{\xi^2-2\xi(S+t)+(S+t)^2\}} e^{St} d\xi$$

$$= e^{2St} \int_{-\infty}^{+\infty} e^{-(\xi-S-t)^2} d\xi = e^{2St} \pi^{\frac{1}{2}} \sum_{n=0}^{\infty} \frac{(2St)^n}{n!}$$

$$\therefore \quad \sum_n \sum_m \frac{S^n t^m}{n!m!} \int_{-\infty}^{+\infty} H_n(\xi) H_m(\xi) e^{-\xi^2} d\xi = \pi^{1/2} \sum_{n=0}^{\infty} \frac{2^n S^n t^n}{n!}$$

If equal powers of S and t are equated on both sides, we get

$$\int_{-\infty}^{+\infty} H_n(\xi) H_m(\xi) e^{-\xi^2} d\xi = 2^n n! \pi^{1/2} \tag{10.37}$$

$$\int H_n(\xi) H_m(\xi) e^{-\xi^2} d\xi = 0 \quad n \neq m \tag{10.38}$$

10.1.6 Harmonic Oscillator Wave Functions

The normalized Harmonic oscillator wave functions are:

$$\Psi_n(\xi)E = N_n \exp\left(\frac{-\xi^2}{2}\right) H_n(\xi) = N_n \exp\left(-\frac{1}{2}\alpha^2 d^2\right) H_n(\alpha x)$$

Normalization constant N_n can be calculated as follows:

$$\int_{-\infty}^{+\infty} |\Psi_n(x)|^2 dx = 1$$

or

$$|N_n|^2 \int_{-\infty}^{+\infty} H_n^2(\alpha x) \exp(-\alpha^2 x^2) dx = 1$$

or

$$\frac{|N_n|^2}{\alpha} \int_{-\infty}^{+\infty} H_n^2(\xi) \exp(-\xi^2) d\xi = 1$$

or

$$\frac{|N_n|^2}{\alpha} 2^n n! \pi^{1/2} = 1$$

$$N_n = \left[\frac{\alpha}{2^n n! \pi^{1/2}}\right]^{1/2}$$

$$\therefore \qquad \Psi_n(x) = \left[\frac{\alpha}{2^n n! \pi^{1/2}}\right]^{1/2} \exp\left(\frac{-\alpha^2 x^2}{2}\right) H_n(ax) \qquad (10.39)$$

First few Hermite polynomials may be written as

$$H_0(\xi) = 1$$
$$H_1(\xi) = 2\xi$$
$$H_2(\xi) = 4\xi^2 - 2$$
$$H_3(\xi) = 8\xi^3 - 12\xi$$
$$H_4(\xi) = 16\xi^4 - 48\xi^2 + 12$$
$$H_5(\xi) = 32\xi^5 - 160\xi^3 + 120\xi$$
$$H_6(\xi) = 64\xi^6 - 480\xi^4 + 720\xi^2 - 120$$
$$H_7(\xi) = 128\xi^7 - 1344\xi^5 + 3360\xi^3 - 1680\xi$$

10.1.7 Ground State

Wave function representing ground state of the harmonic oscillator is

$$\Psi_0(x) = \left(\frac{\alpha^2}{\pi}\right)^{1/4} \exp\left(\frac{-\alpha^2 x^2}{2}\right) \qquad (10.40)$$

In Fig. [10.(2a)], we have plotted $\Psi_0(\xi)$ as a function of ξ, while in Fig. [10.2(b)] we have plotted $|\Psi_0(\xi)|^2$ vs $\xi |\Psi_0(x)|^2 dx$ gives the probability of finding the particle between

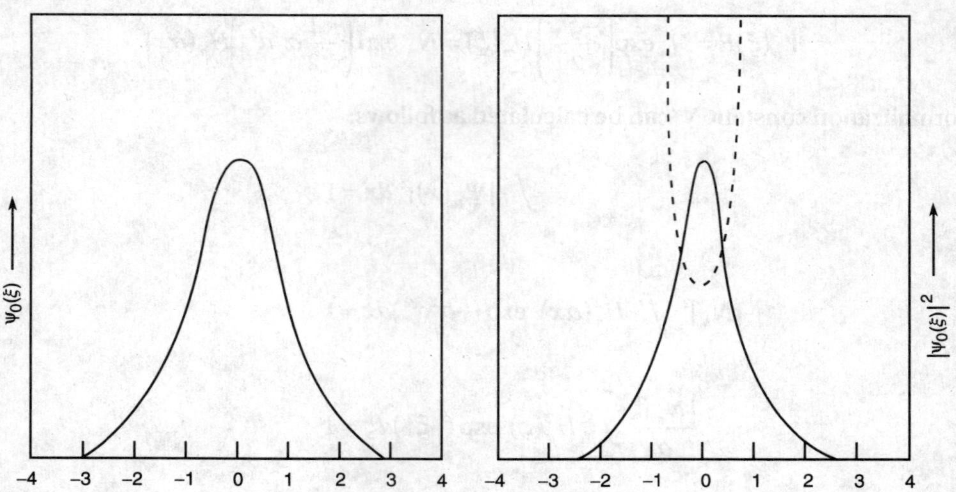

Fig. 10.2 (a), (b) The wave function $\Psi_0(\xi)$ for the ground state of the Harmonic oscillator and the corresponding probability distribution function $|\Psi_0(\xi)|^2$. The classical distribution for an oscillator with the same value of energy is shown by dotted curve

x and $x + dx$. We see from the Fig. [10.2(b)] that the result of quantum mechanics is much different from the probability function which is calculated from classical considerations for harmonic oscillator with the same energy. Classically the particle is most likely to be found at the ends of its motion, which are well defined points (classical turning points), whereas Ψ_0^2 has its maximum at the origin. Furthermore, quantum-mechanically there is a finite probability of finding the particle outside the region allowed classically. This result that particle can penetrate into a region in which its total energy is less then its potential energy is connected with Heisenberg's uncertainty principle, which leads to the conclusion that it is not possible to measure exactly both, the position and the velocity of the particle at the same time.

10.1.8 Probability of Finding the Particle in Classical Limits (In Ground State)

Energy of an oscillating particle in the ground state is equal to $\frac{1}{2}\hbar\omega^2$. Classically the total energy of the particle is $\frac{1}{2}\mu\omega^2 a^2$, where a is the amplitude.

$$\therefore \quad \frac{1}{2}\mu\omega^2 a^2 = \frac{1}{2}\hbar\omega \quad \text{or} \quad a^2 = \frac{\hbar}{\mu\omega}$$

$$\text{or} \quad a = \sqrt{\frac{\hbar}{\mu\omega}} \tag{10.41}$$

$$\text{since} \quad \xi = \sqrt{\frac{\hbar}{\mu\omega}}; \quad x = \pm a, \text{ corresponds to } \xi = \pm 1. \tag{10.42}$$

Probability of finding the particle in classical region

$$= \int_{-a}^{+a} \left(\frac{\alpha}{\pi^{1/2}}\right) e^{-\alpha^2 x^2} dx = \int_{-1}^{+1} \frac{1}{\sqrt{\pi}} e^{-\xi^2} d\xi = \frac{2}{\sqrt{\pi}} \int_0^1 \exp(-\xi^2) d\xi$$

$$= \frac{2}{\sqrt{\pi}} \int_0^1 \left(1 - \xi^2 + \frac{\xi^4}{2!} - \frac{\xi^6}{3!} + \frac{\xi^8}{4!} - \cdots\right) d\xi = \frac{2}{\sqrt{\pi}} \left[\xi - \frac{\xi^3}{3} + \frac{\xi^5}{10} - \frac{\xi^7}{42} + \frac{\xi^9}{216} - \cdots\right]_0^1$$

$$= \frac{2}{1.772}\left[1 - \frac{1}{3} + \frac{1}{10} - \frac{1}{42} + \frac{1}{216} - \cdots\right] = 1.129[1 - 0.333 + 0.10 - 0.024 + 0.004 - \cdots]$$

$$= 0.843 \tag{10.43}$$

Thus there is only 84.3% probability of finding the particle in classical region.

Fig. (10.3) shows the form of ψ_n for larger values of n. Fig. (10.4) shows the variation of $\psi^2(x)$ for the state $n = 10$, compared with the classical probability curve for the same value of energy $\left(\frac{21\hbar\omega}{2}\right)$. It is observed that, apart from the rapid fluctuation in the curve obtained from quantum mechanics, there is a general agreement between the two functions, as is expected (Correspondence-Principle).

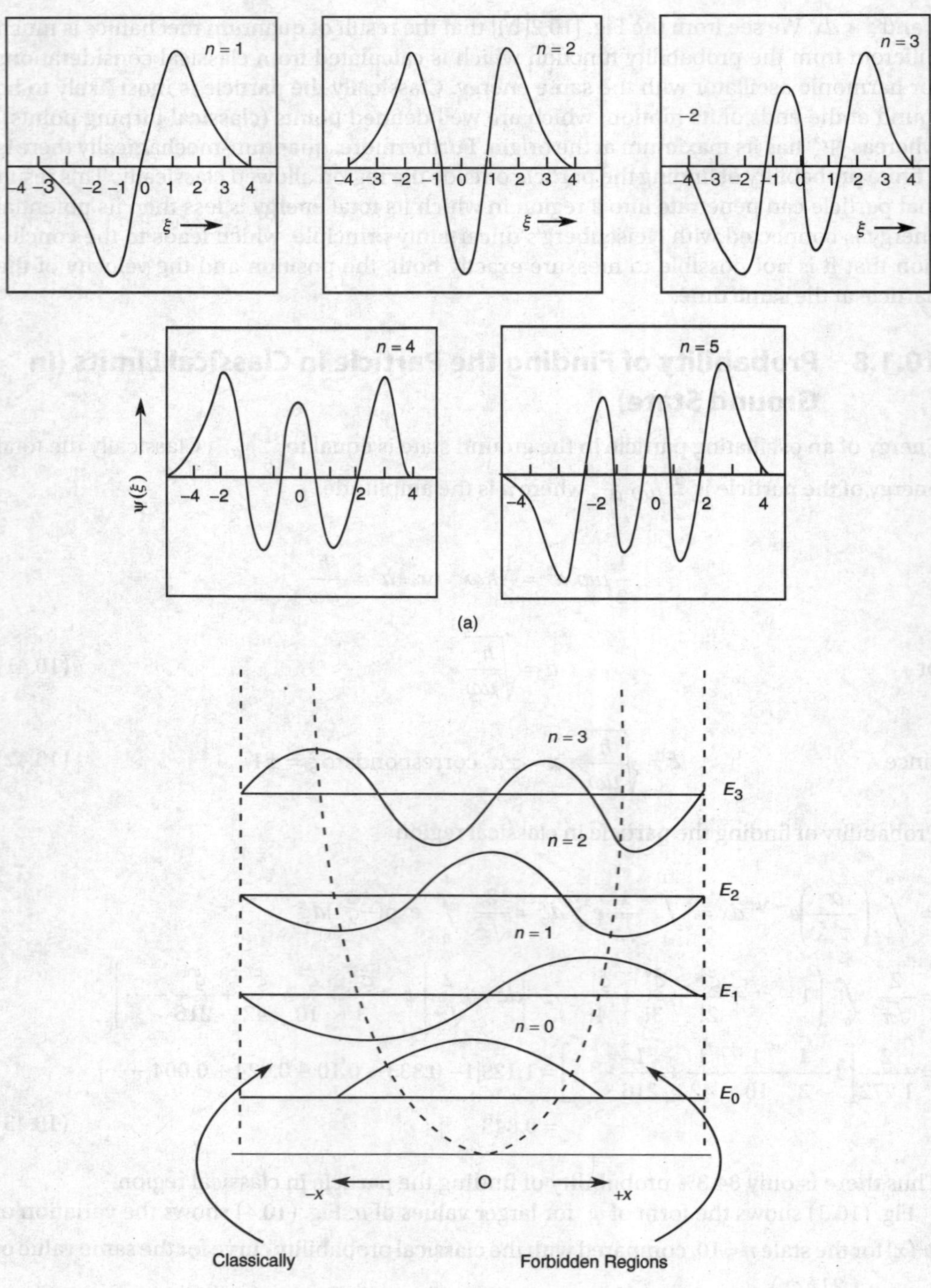

Fig. 10.3(a) The wave functions ψ_1, ψ_2, ψ_3, ψ_4, ψ_5 for the harmonic oscillator. **(b)** Classically Forbidden regions

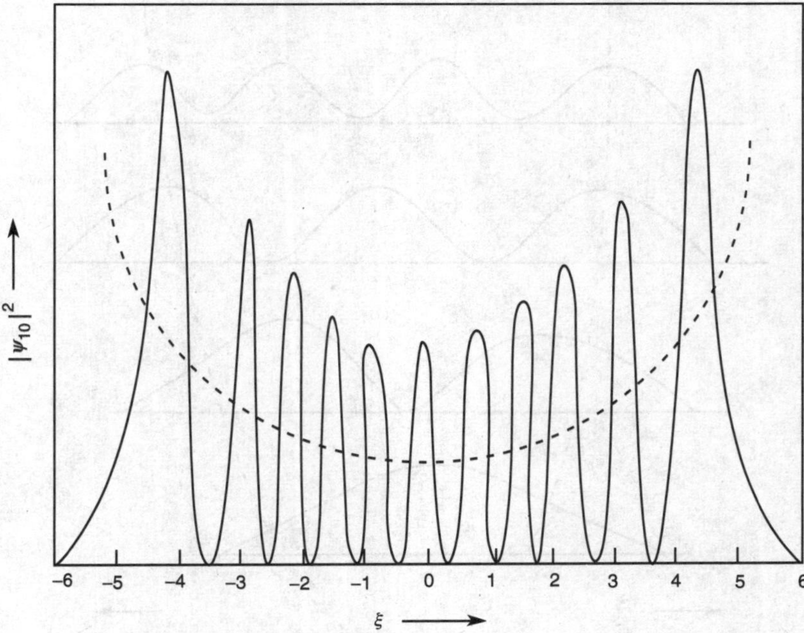

Fig. 10.4 The probability density for the state n = 10 of an harmonic oscillator (solid curve) and dotted curve shows the classical probability curve for the same value of energy

The number of zero's (i.e., points where the probability is zero except for the end points) in the nth excited state of the oscillator are equal to n and all of them are contained within the classical interval (i.e., from $-a$ to $+a$).

The plot of probability ($|\psi^2|$) against x is shown in Fig. (10.5). The plots of ψ and $|\psi^2|$ reveal several interesting features of quantum mechanical harmonic oscillator.

(i) The wave functions are either symmetric (even) or antisymmetric (odd) with respect to $x = 0$. We have seen that the exponential part of the wave function, i.e., exp $(-\alpha x^2/2)$ is necessarily symmetric because of the quantity x^2 the polynomial part i.e., $H_n(x\sqrt{\alpha})$ will be metric for $x^0, x^2, x^4, \ldots$ and antisymmetric for $x, x^3, x^5, \ldots$

(ii) The functions $|\psi^2|$ are never negative and hence, are necessarily symmetric.

(iii) In the ground state ($n = 0$) there are no nodes; and each successive state has one additional node.

(iv) The presence of nodes in the probability plots may appear curious-there is zero probability in the middle and finite probability on either side.

(v) Still more curious is the fact the ψ and ψ^2 are non zero even for large values of x, i.e., even beyond the classical turning points, i.e., the particle can **'penetrate'** the potential barrier.

(vi) The ψ^2 plot reveals the fact that the particle in the ground state has the maximum probability of being found at the central position ($x = 0$), while in the excited states the probability increases towards the classical turning points (expected from the classical results).

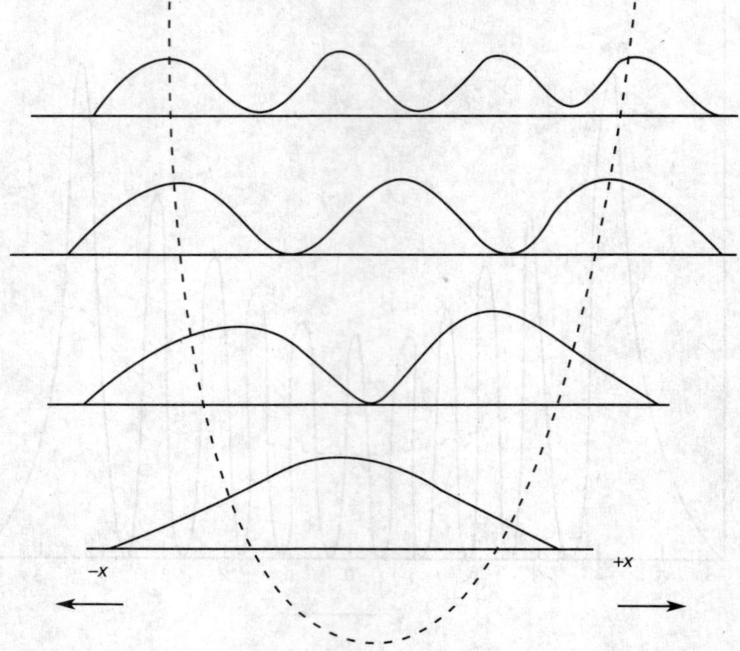

Fig. 10.5

Comparison of the Classical Probability Density and Quantum Mechanical Probability Density:

According to classical mechanics the probability-density of a linear simple harmonic oscillator having amplitude A and energy $E_n = \left(n + \frac{1}{2}\right)\hbar\omega$, is given by

$$P(x) = \frac{1}{\pi\sqrt{A^2 - x^2}} = \frac{1}{\pi\sqrt{\frac{2E_n}{m\omega^2} - x^2}}$$

The above equation shows that the probability of finding the particle is maximum at $x = \pm$ A, i.e., at the extremities of the path, and it is minimum at $x = 0$, i.e., at the equlibrium position. The classical probability-density for $n = 0$ and $n = 10$ is shown by dotted curves in Figs. [10.6(a)] and Fig. [10.6(b)].

According to quantum mechanics the probability density of the particle in the nth energy state is given by $P(x) = \psi_n(x)\psi_n^*(x)$. The wave function for the linear harmonic oscillator is

$$\psi_n(x) = \left(\frac{\alpha}{\pi}\right)^{1/4} (2^n n!)^{-1/2} e^{-\alpha x^2/2} H_n(x\sqrt{\alpha})$$

In the ground state, i.e., $n = 0$, the above equation becomes

$$\psi_0(x) = \left(\frac{\alpha}{\pi}\right)^{1/4} 1e^{-\alpha x^2/2}$$

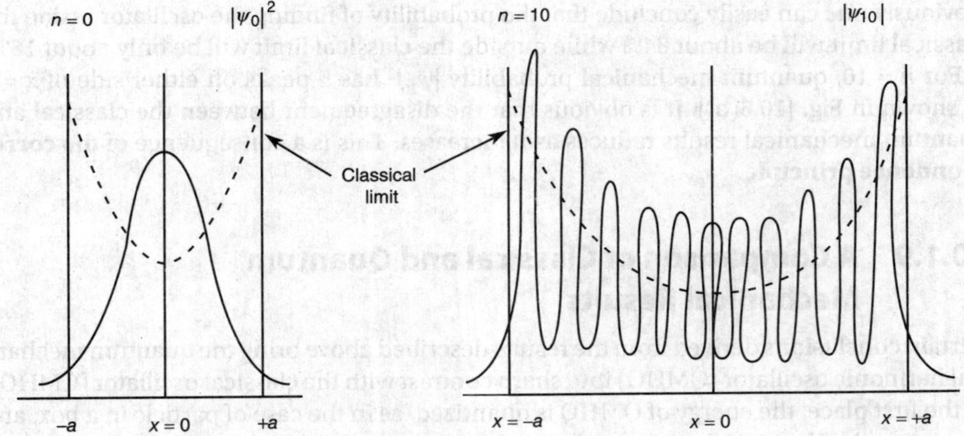

Fig. 10.6 (a), (b) Plots of $|\psi|^2$ for a harmonic oscillator

The classical probability becomes

$$P_c = \int_{-a}^{a} |\psi|^2\, dx = 2\int_{-a}^{a}\left(\frac{\alpha}{\pi}\right)^{-1/2} e^{-\alpha x^2}\, dx = 2\int_{-a}^{a}\left(\frac{\alpha}{\pi}\right)^{-1/2} e^{-\alpha x^2}\, dx$$

(The amplitude varies from $-a$ to a)

According to quantum mechanics, we have

$$E = \frac{1}{2}\hbar\omega_0 = \frac{1}{2}\hbar\sqrt{\frac{k}{m}} = \frac{1}{2}kx^2 \quad \text{or} \quad x^2 = \frac{\hbar}{\sqrt{mk}} = \frac{y^2}{\alpha} = \frac{1}{\beta^2}$$

i.e., quantum mechanically x tends to vary from $-\frac{1}{\beta}$ to $+\frac{1}{\beta}$. The Classical and quantum mechanical limits for $n = 0$ have been depicted in Fig. [10.6(b)]. Hence quantum mechanical probability is given by

$$P_q = 2\int_0^{1/\beta} \frac{\beta}{\sqrt{\pi}} \exp(-\beta^2 x^2)dx$$

If we put $\beta x = t$, one obtains,

$$P_q = \frac{2}{\sqrt{x}}\int_0^1 e^{-t^2}\, dt$$

$$P_q = \frac{2}{\sqrt{\pi}}\int_0^1\left[1 - \frac{t^2}{1!} + \frac{t^4}{2!} - \frac{t^6}{3!} + \frac{t^8}{4!}\cdots\right]dt$$

$$= \frac{2}{\sqrt{\pi}}\left[t - \frac{t^3}{3(1)!} + \frac{t^5}{5(2)!} - \frac{t^7}{7(3)!} + \frac{t^9}{9(4)!} - \cdots\right]_0^1$$

$$= \frac{2}{\sqrt{\pi}}\left[1 - \frac{1}{3(1)!} + \frac{1}{5(2)!} - \frac{1}{7(3)!} + \frac{1}{9(4)!} - \cdots\right] = 0.83 = 83\%$$

Obviously one can easily conclude that the probability of finding the oscillator inside the classical limit will be about 84% while outside the classical limit will be only about 16%.

For $n = 10$, quantum mechanical probability $|\psi_{10}|^2$ has 5 peaks on either side of $x = 0$ as shown in Fig. [10.6(b)]. It is obvious that the disagreement between the classical and quantum mechanical results reduces as n increases. This is a consequence of the **correspondence principle**.

10.1.9 A Comparison of Classical and Quantum Mechanical Results

Certain conclusions derived from the results described above bring the quantum mechanical harmonic oscillator (QMHO) into sharp contrast with the classical oscillator (CMHO). In the first place, the energy of QMHO is quantized, as in the case of particle in a box, and the energy levels are given by a vibrational quantum number. The energy of CMHO, on the other hand, is continous. The occurrence of the vibrational energy levels in QMHO is responsible for the vibrational spectra which the classical treatment could not explain.

The lowest energy or zero point energy (E_0) or a QMHO is $\frac{1}{2}h\upsilon$ or $\frac{1}{2}\hbar\omega$, while that of a CMHO is zero (no zero-point energy). This behaviour is again similar to that of a particle in a box having infinite walls. In fact, the existence of a zero-point energy is characteristic of any quantum mechanical system executing to and fro, i.e., harmonic oscillatory motion.

Wave functions (ψ) and probability functions ($|\psi|^2$) reveal still more interesting features. Probability calculations show that, in the ground state, the most probable position of a QMHO is at the equilibrium point (no displacement) while for a CMHO it lies at the maximum displacement (turning) points. This distinction gradually disappears as n increases, i.e., in the excited states when the points of maximum probability even for the former shift more and more towards the classical turning points, reaching these in the limiting case. This again is similar to the case of particle in a box.

The highly interesting feature of the QMHO is its penetration into classical forbidden regions, i.e., regions of $E < V$. But then it should not be surprising if one remembers that the potential barrier of the HO is not infinite but $\frac{1}{2}kx^2$ and the barrier penetration is characteristic of a quantum mechanical system, e.g., particle in a potential well with finite height.

Applications to Molecular Vibrations: An obvious application of the harmonic oscillator solutions is to the vibrational spectra of linear diatomic molecules.

10.1.10 Parity

We see that harmonic oscillator functions are even functions of x if n is even, and odd functions of x if n is odd. This property arises because the Hamiltonian itself is an even function of x, i.e., $H(-x) = H(x)$.

The Schrodinger equation can be written as

$$H(x)\psi(x) = E\psi(x) \tag{10.44}$$

On changing $x \to -x$ everywhere, we get

$$H(-x)\psi(-x) \equiv E\psi(-x) \qquad \text{or} \qquad H(x)\psi(-x) = E\psi(-x) \tag{10.45}$$

Comparing Eqs. (10.44) and (10.45), we see that since the eigen functions are non-degenerate (i.e., for each value of energy, there is only one eigen function), $\psi(-x)$ must be the same as $\psi(x)$ up to a constant factor

$\therefore$
$$\psi(-x) = \lambda\psi(x)$$

and
$$\psi(x) = \lambda\psi(-x) = \lambda \times \lambda\psi(x) = \lambda^2\psi(x)$$

so that
$$\lambda^2 = 1 \quad \text{or} \quad \lambda = \pm 1$$

The eigen functions which belong to the class $\lambda = \pm 1$ are said to have **even parity** and those with $\lambda = -1$ are said to have **odd parity**. If the system under study is invariant to the transformation $x \to -x$, then conclusions as to energy levels, etc., cannot be influenced by the choice of which the two directions along x is to be positive. In other words, if $H(x)$ is an even function of x, the eigen functions of non-degenerate states have definite parity. For states of definite parity, it can be easily seen (from parity considerations) that expectation value of any odd operator is always zero without any reference to the explicit form of ψ.

10.1.11 Recurrence Relations

From the recurrence relations for Hermite polynomials we can obtain recurrence relations for oscillator wave functions. Starting from Eq. (10.31), we obtain

$$H_{n+1} - 2\xi H_n + 2nH_{n-1} = 0$$

Multiplying above equation by $\left[\dfrac{\alpha}{2^n n! \pi^{1/2}}\right]^{1/2} \exp\left(-\dfrac{\alpha^2 x^2}{2}\right)$

we get
$$\left[\frac{\alpha}{2^n n! \pi^{1/2}}\right]^{1/2} \exp\left(\frac{-\alpha^2 x^2}{2}\right) H_{n+1}$$

$$-2\xi\left[\frac{\alpha}{2^n n! \pi^{1/2}}\right]^{1/2} \exp\left(\frac{-\alpha^2 x^2}{2}\right) H_n + 2n\left[\frac{\alpha}{2^n n! (\pi)^{1/2}}\right]^{1/2} \exp\left(\frac{-\alpha^2 x^2}{2}\right) H_{n-1} = 0$$

or
$$\sqrt{2(n+1)}\left[\frac{\alpha}{2^{n+1}(n+1)! \pi^{1/2}}\right]^{1/2} \exp\left(-\frac{\alpha^2 x^2}{2}\right) H_{n+1} - 2\alpha x\left[\frac{\alpha}{2^n n! \pi^{1/2}}\right]^{1/2}$$

$$\exp\left(\frac{-\alpha^2 x^2}{2}\right) H_{n-1} + \sqrt{2n}\left[\frac{\alpha}{2^{n-1}(n-1)! \pi^{1/2}}\right]^{1/2} \exp\left(-\frac{\alpha^2 x^2}{2}\right) H_{n-1} = 0$$

or
$$\sqrt{2(n+1)}\,\psi_{n+1}(x) - 2\alpha x\psi(x) + \sqrt{2n}\,\psi_{n-1}(x) = 0$$

or $$x\psi_n(x) = \frac{1}{\alpha}\left[\sqrt{\frac{n+1}{2}}\psi_{n+1}(x) + \sqrt{\frac{n}{2}}\psi_{n+1}(x)\right] \tag{10.46}$$

or $$x\psi_n(x) = \left(\frac{\hbar}{\mu\omega}\right)^{1/2}\left[\sqrt{\frac{n+1}{2}}\psi_{n+1}(x) + \sqrt{\frac{n}{2}}\psi_{n-1}(x)\right] \tag{10.47}$$

or $$\psi_n(x) = \left[\frac{\alpha}{2^n n!\pi^{1/2}}\right]^{1/2}\exp\left(-\frac{\alpha^2 x^2}{2}\right)H_n(\alpha x)$$

$\therefore$ $$\frac{d\psi_n(x)}{dx} = \left[\frac{\alpha}{2^n n!\pi^{1/2}}\right]^{1/2}\exp\left[(-\alpha^2 x)\exp\left(-\frac{\alpha^2 x^2}{2}\right)H_n(\alpha x) + \exp\left(-\frac{\alpha^2 x^2}{2}\right)\frac{dH_n(\alpha x)}{dx}\right]$$

or $$\frac{d\psi_n(x)}{dx} = -(\alpha^2 x)\left[\frac{\alpha}{2^n n!\pi^{1/2}}\right]^{1/2}\exp\left(-\frac{\alpha^2 x^2}{2}\right)H_n(\alpha x) + \left(\frac{\alpha}{2^n n!\pi^{1/2}}\right)^{1/2}\exp\left(\frac{\alpha^2 x^2}{2}\right)\frac{dH_n(\alpha x)}{d(\alpha x)}\alpha$$

or $$\frac{d\psi_n(x)}{dx} = -\alpha^2 x\psi_n(x) + \left[\frac{\alpha}{2^n n!\pi^{1/2}}\right]^{1/2}\exp\left(-\frac{\alpha^2 x^2}{2}\right)[\alpha 2n H_{n-1}(\alpha x)]$$

or $$\frac{d\psi_n(x)}{dx} = -\alpha^2 \times \frac{1}{\alpha}\left[\sqrt{\frac{n+1}{2}}\psi_{n+1}(x) + \sqrt{\frac{n}{2}}\psi_{n-1}(x)\right] + 2\alpha n \times \frac{1}{\sqrt{2n}}\left[\frac{\alpha}{2^{n-1}(n-1)!\pi^{1/2}}\right]^{1/2}\exp\left(-\frac{\alpha^2 x^2}{2}\right)\times H_{n-1}(\alpha x)$$

or $$\frac{d\psi_n(x)}{dx} = -\alpha\left[\sqrt{\frac{n+1}{2}}\psi_{n+1}(x) + \sqrt{\frac{n}{2}}\psi_{n-1}(x)\right] + 2\alpha\sqrt{\frac{n}{2}}\psi_{n-1}(x)$$

or $$\frac{d\psi_n(x)}{dx} = \alpha\left[\sqrt{\frac{n}{2}}\psi_{n-1}(x) - \sqrt{\frac{n+1}{2}}\psi_{n+1}(x)\right] \tag{10.48}$$

10.1.12 Uncertainty Relations

Uncertainty in x can be calculated if we know the expectation value of x and x^2; similarly uncertainty in p can be calculated by a knowledge of the expectation values of p and p^2.

$$<x> = \int_{-\infty}^{+\infty} \psi_n^* x \psi_n dx = \int_{-\infty}^{+\infty} \psi_n^* \alpha \left\{ \sqrt{\frac{n}{2}} \psi_{n-1} + \sqrt{\frac{n+1}{2}} \psi_{n+1} \right\} dx$$

$$= \alpha \sqrt{\frac{n}{2}} \int_{-\infty}^{+\infty} \psi_n^* \psi_{n-1} dx + \alpha \sqrt{\frac{n+1}{2}} \int_{-\infty}^{+\infty} \psi_n^* \psi_{n+1} dx$$

Each of these integrals is equal to zero; therefore expectation value of x is equal to zero for any state of the harmonic oscillator, i.e.,

$$<x> = 0 \tag{10.49}$$

$$< x > = \int_{-\infty}^{+\infty} \psi_n^* x \psi_n dx = \int_{-\infty}^{+\infty} \psi_n^* x \left(\frac{1}{\alpha}\right) \left\{ \sqrt{\frac{n}{2}} \psi_{n-1} + \sqrt{\frac{n+1}{2}} \psi_{n+1} \right\} dx$$

$$= \frac{1}{\alpha} \sqrt{\frac{n}{2}} \int_{-\infty}^{+\infty} \psi_n^* x \psi_{n-1} dx + \frac{1}{\alpha} \sqrt{\frac{n+1}{2}} \int_{-\infty}^{+\infty} \psi_n^* x \psi_{n+1} dx$$

$$= \left(\frac{1}{\alpha}\right) \sqrt{\frac{n}{2}} \int_{-\infty}^{+\infty} \psi_n^* \left(\frac{1}{\alpha}\right) \left(\sqrt{\frac{n+1}{2}} \psi_{n-2} + \sqrt{\frac{n}{2}} \psi_n \right) dx$$

$$+ \left(\frac{1}{\alpha}\right) \sqrt{\frac{n+1}{2}} \int_{-\infty}^{+\infty} \psi_n^* \left(\frac{1}{\alpha}\right) \left(\sqrt{\frac{n+1}{2}} \psi_n + \sqrt{\frac{n+2}{2}} \psi_{n+2} \right) dx$$

$$= \left(\frac{1}{\alpha^2}\right) \sqrt{\frac{n(n-1)}{4}} \int_{-\infty}^{+\infty} \psi_n^* \psi_{n-2} dx + \left(\frac{1}{\alpha^2}\right) \left(\frac{n}{2}\right) \int_{-\infty}^{+\infty} \psi_n^* \psi_n dx + \left(\frac{1}{\alpha^2}\right) \left(\frac{n+1}{2}\right)$$

$$\int_{-\infty}^{+\infty} \psi_n^* \psi_n dx + \left(\frac{1}{\alpha^2}\right) \sqrt{\frac{(n+1)(n+2)}{4}} \int_{-\infty}^{+\infty} \psi_n^* \psi_{n+2} dx = \left(\frac{1}{\alpha^2}\right) \left(\frac{2n+1}{2}\right)$$

(Using orthogonality condition for the oscillator wave functions)

or
$$< x^2 > = \frac{\hbar}{\mu\omega} \left(n + \frac{1}{2} \right) \tag{10.50}$$

∴
$$(\Delta x)^2 = < x^2 > - < x >^2 = < x^2 > = \frac{\hbar}{\mu\omega} \left(n + \frac{1}{2} \right)$$

or
$$\Delta x = \sqrt{\frac{\hbar}{\mu\omega}} \left(n + \frac{1}{2} \right)^{1/2} \tag{10.51}$$

Now, $$<P>\int_{-\infty}^{+\infty}\psi_n^* = \psi_n dx = \int_{-\infty}^{+\infty}\psi_n^*\left(-i\hbar\frac{d\psi_n}{dx}\right)dx$$

or $$<p> = -i\hbar\int_{-\infty}^{+\infty}\psi_n^*\alpha\left\{\sqrt{\frac{n}{2}}\psi_{n-1} - \sqrt{\frac{n+1}{2}}\psi_{n+1}\right\}dx$$

or $$<p> = i\hbar\alpha\sqrt{\frac{n}{2}}\int_{-\infty}^{+\infty}\psi_n^*\psi_{n-1}dx + i\hbar\alpha\sqrt{\frac{n+1}{2}}\int_{-\infty}^{+\infty}\psi_n^*\psi_{n+1}dx$$

or $$<p> = 0 \tag{10.52}$$

$$<p^2> = \int_{-\infty}^{+\infty}\psi_n^* p^2\psi_n dx = \int_{-\infty}^{+\infty}\psi_n^*\left(-\hbar^2\frac{d^2}{dx^2}\psi_n\right)dx = -\hbar^2\int_{-\infty}^{+\infty}\psi_n^*\frac{d}{dx}\left(\frac{d\psi_n}{dx}\right)dx$$

$$= -\hbar^2\int_{-\infty}^{+\infty}\psi_n^*\frac{d}{dx}\left[\alpha\left\{\sqrt{\frac{n}{2}}\psi_{n-1} - \sqrt{\frac{n+1}{2}}\psi_{n+1}\right\}\right]dx$$

$$= -\hbar^2\alpha\sqrt{\frac{n}{2}}\int_{-\infty}^{+\infty}\psi_n^*\frac{d\psi_{n-1}}{dx} + \hbar^2\alpha\sqrt{\frac{n+1}{2}}\int_{-\infty}^{+\infty}\psi_n^*\frac{d\psi_{n+1}}{dx}dx$$

$$= -\hbar^2\alpha\sqrt{\frac{n}{2}}\int_{-\infty}^{+\infty}\psi_n^*\alpha\left\{\sqrt{\frac{n-1}{2}}\psi_{n-2} - \sqrt{\frac{n}{2}}\psi_n\right\}dx$$

$$+ \hbar^2\alpha\sqrt{\frac{n+1}{2}}\int_{-\infty}^{+\infty}\psi_n^*\alpha\left(\sqrt{\frac{n+1}{2}}\psi_n - \sqrt{\frac{n+1}{2}}\psi_{n+1}\right)dx$$

$$-\hbar^2\alpha^2\sqrt{\frac{n(n-1)}{2}}\int_{-\infty}^{+\infty}\psi_n^*\psi_{n-2}dx + \hbar^2\alpha^2\left(\frac{n}{2}\right)\int_{-\infty}^{+\infty}\psi_n^*\psi_n dx$$

$$+ \hbar^2\alpha^2\left(\frac{n+1}{2}\right)\int_{-\infty}^{+\infty}\psi_n^*\psi_n dx - \hbar^2\alpha^2\sqrt{\frac{(n+1)(n+2)}{4}}\int_{-\infty}^{+\infty}\psi_n^*\psi_{n+2}dx$$

$$= \hbar^2\alpha^2\frac{n}{2} + \hbar^2\alpha^3\left(\frac{n+1}{2}\right) = \hbar^2\alpha^2\left(\frac{2n+1}{2}\right)$$

$$= \hbar^2\left(\frac{\mu\omega}{\hbar}\right)\left(\frac{2n+1}{2}\right) = \hbar\mu\omega\left(n+\frac{1}{2}\right) \tag{10.53}$$

$\therefore$ $$(\Delta p)^2 = <p^2> - <p^2> = <p^2> = \hbar\mu\omega\left(n+\frac{1}{2}\right)$$

$\therefore$ $$\Delta p = \sqrt{\hbar\mu\omega}\left(n+\frac{1}{2}\right)^{1/2} \tag{10.54}$$

$$\therefore \qquad (\Delta x)(\Delta p) = \sqrt{\frac{\hbar}{\mu\omega}}\left(n+\frac{1}{2}\right)^{1/2}\sqrt{\hbar\mu\omega}\left(n+\frac{1}{2}\right)^{1/2} = \hbar\left(n+\frac{1}{2}\right)$$

Thus the minimum possible value is $\frac{\hbar}{2}$ for the ground state wave function,

$$\psi_0(\alpha) = \left(\frac{\alpha^2}{\pi}\right)^{1/4}\exp\left(-\frac{1}{2}\alpha^2 x^2\right)$$

which would be expected since the from the wave packet is Gaussian as discussed earlier.

10.1.13 Momentum Representation for Oscillator

In the momenum, p-representation

$$\left(p_x \to p_x, x \to i\hbar\frac{\partial}{\partial p_x}\right), \quad \text{we have}$$

$$H = \frac{p_x^2}{2m} + \frac{1}{2}k_x x^2 = \frac{p_x^2}{2m} - \frac{1}{2}k_x\hbar^2\frac{\partial^2}{\partial p_x^2} \tag{10.55}$$

Now
$$H\phi = \left(\frac{p_x^2}{2m} - \frac{1}{2}k_x\hbar^2\frac{\partial^2}{\partial p_x^2}\right) = E\phi \tag{10.56}$$

Using,
$$p_x = p_x/\beta,\ E = \frac{1}{2}\hbar\omega\lambda,\ \beta = (k_x m\hbar^2)^{1/4}, \omega = (k_x/m)^{1/2} \tag{10.57}$$

Eq. (10.56) becomes

$$\left(\frac{\partial^2}{\partial p_x^2} + \lambda - p_x^2\right)\phi = 0 \tag{10.58}$$

10.2 THE THREE-DIMENSIONAL HARMONIC OSCILLATOR (CARTESIAN COORDINATES)

A particle attracted towards a fixed point by a force which is proportional to the distance from the fixed point is represented by a potential energy of the form:

$$V(r) = \frac{1}{2}fr^2 = \frac{1}{2}f(x^2 + y^2 + z^2) \tag{10.59}$$

The potential is spherically symmetric. The Schrodinger equation for such a particle is

$$\nabla^2\psi + \frac{2\mu}{\hbar^2}\left(E - \frac{1}{2}fr^2\right)\psi = 0 \quad \text{or} \quad \nabla^2\psi + \left(\frac{2\mu E}{\hbar^2} - \frac{\mu f}{\hbar^2}r^2\right)\psi = 0$$

or $$\nabla^2\psi + \left[\frac{2\mu E}{\hbar^2} - \frac{\mu^2\omega^2}{\hbar^2}r^2\right]\psi = 0$$

Substituting $E = \dfrac{\lambda\hbar\omega}{2}$ and $\xi = \left(\dfrac{\mu\omega}{\hbar}\right)^{1/2} r = \alpha r$ and $\rho = \left(\dfrac{\mu\omega}{\hbar^2}\right)^{1/3} x$, then

we have $$\xi^2 = \rho^2 + \eta^2 + \sigma^2 = \left(\frac{\mu\omega}{\hbar^2}\right)(x^2 + y^2 + z^2) = \frac{\mu\omega}{\hbar^2}r^2,$$ where

$$\eta = \left(\frac{\mu\omega}{\hbar^2}\right)^{1/2} y \quad \text{and} \quad \sigma = \left(\frac{\mu\omega}{\hbar}\right)^{1/2} z$$

In the above equation, we obtain

$$\left(\frac{d^2}{d\rho^2} + \frac{d^2}{d\eta^2} + \frac{d^2}{d\sigma^2}\right)\psi + [\lambda + (\rho^2 + \eta^2 + \sigma^2)]\psi = 0 \tag{10.60}$$

In rectangular coordinates, Eq. (10.60) is separable, we make the substitution

$$\psi = X(\rho)Y(\eta)Z(\sigma) = X(\alpha x)Y(\alpha y)Z(\alpha z)$$

Then $$\left(YZ\frac{d^2X}{d\rho^2} + XZ\frac{d^2Y}{d\eta^2} + XY\frac{d^2Z}{d\sigma^2}\right) + [\lambda - (\rho^2 + \eta^2 + \sigma^2)]XYZ = 0$$

or $$\left(\frac{1}{X}\frac{d^2X}{d\rho^2} - \rho^2\right) + \left(\frac{1}{Y}\frac{d^2Y}{d\eta^2} - \eta^2\right)\left\{\frac{1}{Z}\left(\frac{d^2Z}{d\sigma^2}\right) - \sigma^2\right\} = -\lambda$$

This gives us three equations:

$$\frac{d^2X}{d\rho^2} + (\lambda_x - \rho^2)X = 0 \tag{10.61}$$

$$\frac{d^2Y}{d\eta^2} + (\lambda_y - \eta^2)Y = 0 \tag{10.62}$$

$$\frac{d^2Z}{d\sigma^2} + (\lambda_z - \sigma^2)Z = 0 \tag{10.63}$$

where the constants λ_x, λ_y and λ_z are related by

$$\lambda_x + \lambda_y + \lambda_z = \lambda \tag{10.64}$$

Each of Eqs. (10.61), (10.62) and (10.63) is the differential equation for a one-dimensional harmonic oscillator, for which the normalized wave functions are given by Eq. (10.39). Thus finite solutions are possible only if:

$$\lambda_x = 2n_x + 1, \qquad \lambda_y = 2n_y + 1, \qquad \lambda_z = 2n_z + 1$$

where n_x, n_y and n_z are non negative integers. Obviously, there are three quantum numbers (n_1, n_2 and n_3) to characterize the energy, instead of one. The corresponding normalized wave functions is

$$\psi_{n_x n_y n_z} = \left[\frac{\alpha^3}{\pi^{3/2} 2^2 n_x n_y ! n_z !}\right]^{1/2} e^{-\alpha^2 r^2/2} Hn_x(\alpha x) Hn_y(\alpha y) Hn_z(\alpha z) \tag{10.65}$$

where
$$n = n_x + n_y + n_z \tag{10.66}$$

According to Eq. (10.64), the energy is

$$E = \left(n_x + \frac{1}{2}\right)\hbar\omega + \left(n_y + \frac{1}{2}\right)\hbar\omega + \left(n_z + \frac{1}{2}\right)\hbar\omega$$

or
$$E = \left(n_x + n_y + n_z + \frac{3}{2}\right)\hbar\omega \text{ or } E = \left(n + \frac{3}{2}\right)\hbar\omega \tag{10.67}$$

For given value of n, the wave functions (10.59) are a set of linearly independent functions, each of which corresponds to one of the sets of integers n_x, n_y and n_z which satisfy Eq. (10.66). The states of energy E are therefore degenerate. To calculate the degree of degeneracy of a level with a given n, we note that if n_1 is also fixed, the number of combinations of n_2 and n_3 are $(0, n - n_1)$; $(1, n - n_1 - 1)$; $(2, n - n_1 - 2)$; ....; $(n - n_1, 0)$, thus therefore $(n - n_1 + 1)$ states. Now n_1 can also vary from 0 to n; thus degree of degeneracy is

$$(n+1) + n + (n-1) + \cdots + (1) = \frac{(1+n)(n+2)}{2}$$

$$\Downarrow \quad \Downarrow \quad \Downarrow \quad \Downarrow$$

$$n_1 = 0 \quad n_1 = 1 \quad n_1 = 2 \quad n_1 = n$$

We can similarly calculate the eigen values and eigen functions of a two-dimensional oscillator.

10.2.1 Spherical Polar Coordinates

For a three a three-dimensional isotropic harmonic oscillator, potential energy has the form

$$V(r) = \frac{1}{2}\mu\omega^2 r^2$$

The potential is spherically symmetric and therefore the Schrodinger equation can be solved in spherical polar coordinates. The angular part of the wave function will be the spherical harmonics $Y_{l,m}(\theta, \phi)$ and the wave function can be written in the form

$$\psi(r,\theta,\phi) = R(r) Y_{l,m}(\theta,\phi)$$

Radial part of the wave function, $R(r)$ satisfies the following equation:

$$\frac{1}{r^2}\frac{d}{dr}\left(r^2\frac{dR}{dr}\right) + \frac{2\mu}{\hbar^2}\left[E - \frac{\hbar^2}{2\mu}\frac{l(l+1)}{r^2} - \frac{1}{2}\mu\omega^2 r^2\right]R(r) = 0$$

By substituting the variable ξ, given by $\xi = \left[\frac{\mu\omega}{\hbar}\right]^{1/2} r$ and $E = \frac{\lambda\hbar\omega}{2}$, one obtains

$$\frac{1}{\xi^2}\frac{d}{d\xi}\left(\xi^2\frac{dR}{d\xi}\right) + \left[\lambda - \frac{l(l+1)}{\xi^2} - \xi^2\right]R = 0$$

or
$$\frac{d^2R}{d\xi^2} + \frac{2}{\xi}\frac{dR}{d\xi} + \left[\lambda - \frac{l(l+1)}{\xi^2} - \xi^2\right]R = 0 \tag{10.68}$$

The asymptotic form of the solution (i.e., for $\xi \to \infty$) is obviously

$$R(\xi) - \exp\left(\pm\frac{\xi^2}{2}\right)$$

$R(\xi)$ must approach zero at $\xi = \infty$, therefore only the negative exponent is acceptable as a solution on physical grounds, although both positive and negative exponents are mathematically possible.

$$\therefore \quad R(\xi) = e^{-\xi^2/2}\phi(\xi)$$

$$\frac{dR}{d\xi} = e^{-\xi^2/2}\frac{d\phi}{d\xi} - \xi e^{-\xi^2/2}\phi$$

$$\frac{d^2R}{d\xi^2} = e^{-\xi^2/2}\frac{d^2\phi}{d\xi^2} - 2\xi e^{-\xi^2/2}\frac{d\phi}{d\xi} - (1-\xi^2)e^{-\xi^2/2}\phi$$

$$\therefore \quad e^{-\xi^2/2}\frac{d^2\phi}{d\xi^2} + \left(-2\xi e^{-\xi^2/2} + \frac{2}{\xi}e^{-\xi^2/2}\right)\frac{d\phi}{d\xi} + \left[\xi^2 - 1 - 2 + \lambda - \frac{l(l+1)}{\xi^2} - \xi^2\right]e^{-\xi^2/2}\phi = 0$$

or
$$\frac{d^2\phi}{d\xi^2} + \left(\frac{2}{\xi} - 2\xi\right)\frac{d\phi}{d\xi} + \left[\lambda - 3 - \frac{l(l+1)}{\xi^2}\right]\phi = 0 \tag{10.69}$$

In analogy with the earlier treatment of Eq. (10.47), this equation can also be solved by the series method. We assume

$$\phi(\xi) = \sum_{v=0}^{\infty} a_v \xi^{v+s} \tag{10.70}$$

$$\frac{d\phi}{d\xi} = \sum_v a_v (v+s)\xi^{v+s-1} \tag{10.71}$$

$$\frac{d^2\phi}{d\xi^2} = \sum_v a_v (v+s)(v+s-1)\xi^{v+s-2} \tag{10.72}$$

Substituting these values in Eq. (10.69), we obtain

$$\sum_{v} a_v (v+s)(v+s-1)\xi^{v+s-2} + 2\sum_{v} a_v (v+s)\left(\frac{1}{\xi}-\xi\right)\xi^{v+s-1}$$
$$+\sum_{v} a_v \left[\lambda - 3 - \frac{l(l+1)}{\xi^2}\right]\xi^{v+s} = 0$$

Comparing the coefficient of ξ^{s-1} on both sides of the above equation, we obtain

$$a_0 s(s-1) + 2a_0 s - l(l+1)a_0 = 0$$

or $$a_0[s(s+1) - l(l+1)] = 0 \quad \text{Since } a_0 \neq 0$$

$\therefore$ $$s = (s+1) - l(l+1) = 0$$

or $$s = l_1 - (l+1) \tag{10.73}$$

The value $s = -(l+1)$ leads to a series beginning with $\xi^{-(l+1)}$ which is finite at $r = 0$. Consequently, only $s = l$ is allowed, and the series can be written as

$$\phi(\xi) = \xi^l \sum_{v=0}^{\infty} a_v \xi^v$$

Comparing the coefficient of ξ^{l-1} on both sides, we obtain

$$s(s+1)a_1 + 2(s+1)\,a_1 = 0 \quad \text{or} \quad (s+1)(s+2)a_1 = 0$$

Since $$(s+1)(s+2) \neq 0 \;\therefore\; a_1 = 0$$

A recurrence relation between various coefficients can be obtained by comparing the coefficient of ξ^{v+s}.

$$a_{v+2}(v+s+2)(v+s+1) + 2a_{v+2}(v+s+2) - 2a_v(v+s)$$
$$+ (\lambda - 3)a_v - l(l+1)a_{v+2} = 0$$

or $$a_{v+2}[(v+s+2)(v+s+1) + 2(v+s+2) - l(l+1)] = [3 + 2(v+s) - \lambda]a_v$$

or $$a_{v+2} = \left[\frac{3 + 2(v+l) - \lambda}{(v+l+2)(v+l+1) + 2(v+l+2) - l(l+1)}\right]a_v$$

or $$a_{v+2} = \left[\frac{3 + 2(v+l) - \lambda}{(v+2)(v+2l+4)}\right]a_v \tag{10.74}$$

$$a_1 = a_3 = a_5 = a_7 = \cdots = 0 \tag{10.75}$$

If the series does not terminate, its dominant asymptotic behaviour can be inferred from the coefficients of the terms having very large power:

$$\frac{a_{v+2}}{a_v} \underset{v\to\infty}{\longrightarrow} \frac{2}{v}$$

Also $$e^{\xi^2} = 1+\xi^2+\frac{\xi^4}{2!}+\frac{\xi^6}{3!}+\cdots+\frac{\xi^v}{(v/2)!}+\frac{\xi^{v+2}}{(v/2)+1!}+\cdots$$

$$\frac{\text{coefficient of } \xi^{v+2}}{\text{coefficient of } \xi^v} = \frac{(v/2)!}{(v/2+1)!} = \frac{v/2!}{(v/2+1)(v/2)!} = \frac{2}{v+2} \to \frac{2}{v}$$

Thus we conclude that for large values of ξ, the infinite series behaves like e^{ξ^2} and the function R behaves very much like $\exp(\xi^2/2)$. Since this function increases to infinity at $\xi = \infty$, we conclude that the boundary conditions $[R(\xi) = 0$ at $\xi = +\infty]$ are not satisfied and these mathematical solutions are therefore not acceptable as solution for the problem on physical grounds. The only way this catastrophe can be avoided is if λ happens to have certain particular values such that

$$\lambda = 2(n'+l)+3 = 2n+3 \tag{10.76}$$

where $n = n' + l \equiv$ (l plus an even integer).

When this condition holds, the infinite series becomes a polynomial.

Thus we have the quantum condition

$$E = \frac{1}{2}[2(n'+l)+3]\,\hbar\omega = \left(n'+l+\frac{3}{2}\right)\hbar\omega = \left(n+\frac{3}{2}\right)\hbar\omega \tag{10.77}$$

The eigen functions may be written as

$$\psi = e^{-\xi/2}\phi_{n,l}(\xi)Y_{lm}(\theta,\phi)\ (n \ge l) \tag{10.78}$$

$\phi_{nl}(\xi)$ is a polynomial in ξ of degree n and the lowest term in ξ is ξ^l. These eigen functions can be classified according to n and l as shown in Fig. (10.7). The entries in the figure are the numbers ($2l$+1) of linearly independent states belonging to the corresponding values of l. There is one state for $n = 0$, there are 3 states for $n = 1$, $1 + 5 = 6$ states for $n = 2$, $3 + 7 = 10$ states for $n = 3$. In general there are $\frac{1}{2}(n+1)(n+2)$ states for each value of n in agreement with the conclusion reached earlier in section (10.2). The functions represented by Eqs. (10.65) and (10.78) are merely different representations of the degenerate states belonging to a given value of energy and the total number of such states must be independent of the system of coordinates in terms of which they are expressed.

Each of the functions represented by Eq. (10.65) is a linear combination of the $\left[\frac{(n+1)(n+2)}{2}\right]$ functions expressed by Eq. (10.78) for a given value of n.

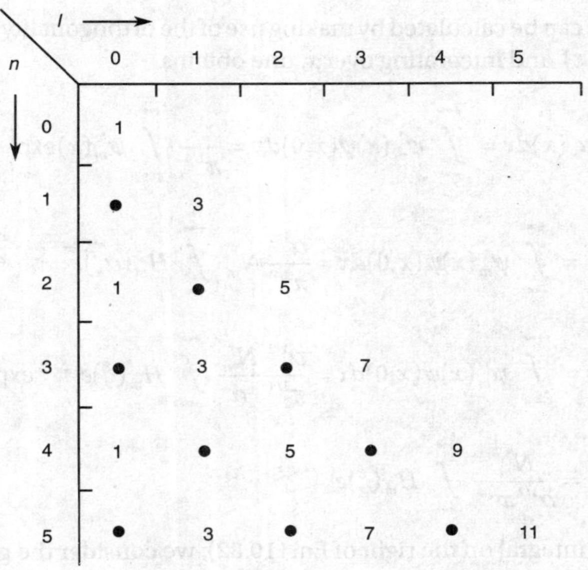

Fig. 10.7

10.3 MOTION OF A WAVE PACKET

The general solution of the Schrodinger's time-dependent equation for the harmonic oscillator is given by

$$i\hbar\frac{\partial\psi(x,t)}{\partial t}=\left[-\frac{\hbar^2}{2\mu}\frac{\partial^2}{\partial x^2}-\frac{1}{2}fx^2\right]\psi(x,t) \tag{10.79}$$

$\psi(x, t)$ can be expanded in terms of stationary wave functions:

$$\psi(x,t)=\sum_{n=0}^{\infty}C_n\psi_n(x)e^{-\frac{iE_n}{\hbar}}=\sum_{n=0}^{\infty}C_n\psi_n(x)e^{-i\omega t(n+\frac{1}{2})}$$

or

$$\psi(x,t)=\left(e^{-\frac{i\omega t}{2}}\right)=\sum_{n=0}^{\infty}C_n\psi_n(x)\exp(-i\omega_n t) \tag{10.80}$$

where the C_n^{*}s are the arbitrary expansion coefficients. Thus apart from the phase factor, $\psi(x, t)$ is a periodic function of t with the period of classical harmonic oscillator, i.e., $2\pi/\omega$. This suggests the possibility of finding a solution in the form of a wave packet, whose centre oscillates with the period of classical motion.

To investigate this possibility, we assume that the form of ψ at $t=0$ is given by a Gaussian function (properly normalized) with its centre at x_0, i.e.,

$$\psi(x,0)=\sum_{n=0}^{\infty}C_n\psi_n(x)=\frac{\alpha^{1/2}}{\pi^{1/4}}\exp\left[-\frac{\alpha^2(x-x_0)^2}{2}\right] \tag{10.81}$$

The coefficients C_n can be calculated by making use of the orthogonality of ψ_n's. Multiplying Eq. (10.81) by $\psi_m^*(x)$ and integrating over x, one obtains

$$\sum_n C_n \int_{-\infty}^{+\infty} \psi_m^*(x)\psi_n(x)dx = \int_{-\infty}^{+\infty} \psi_m^*(x)\psi(x,0)dx = \frac{\alpha^{1/2}}{\pi^{1/4}} \int_{-\infty}^{+\infty} \psi_m^*(x)\exp\left[-\frac{\alpha^2}{2}(x-x_0)^2\right]dx$$

or $$\sum_n C_n\delta_{mn} = \int_{-\infty}^{+\infty} \psi_m^*(x)\psi(x,0)dx = \frac{\alpha^{1/2}}{\pi^{1/2}}N_m^* \int_{-\infty}^{+\infty} H_m(\alpha_n)e^{-\frac{a^2x^2}{2}-\frac{a^2}{2}(x-x_0)^2}dx$$

or $$C_m = \int_{-\infty}^{+\infty} \psi_m^*(x)\psi(x,0)dx = \frac{\alpha^{1/2}}{\pi^{1/4}}\frac{N_m^*}{\alpha} \int_{-\infty}^{+\infty} H_m(\xi)e^{-\xi^2/2}\exp[-(\xi-\xi_0)^2/2]dx$$

or $$C_m = \frac{N_m^*}{\alpha^{1/2}\pi^{1/4}} \int_{-\infty}^{+\infty} H_m(\xi)e^{(-\xi^2+\xi\xi_0-\xi^2/2)} \tag{10.82}$$

For evaluating the integral on the right of Eq. (10.82); we consider the generating function for Hermit polynomials, then

$$\int_{-\infty}^{+\infty} e^{(-S^2+2S\xi)-(\xi^2+\xi\xi_0-\xi_0^2)/2}d\xi = \int_{-\infty}^{+\infty}\left\{\sum_{n=0}^{\infty}\frac{S^n}{n!}H_n(\xi)\right\}e^{-\xi^2+\xi\xi_0-\xi_0^2/2}d\xi$$

or $$\int_{-\infty}^{+\infty} \exp\left[-\left\{\xi^2-2\xi\left(S+\frac{\xi_n}{2}\right)+\left(S+\frac{\xi_0}{2}\right)^2\right\}-\frac{\xi_0^2}{4}+S\xi_0\right]d\xi$$

$$= \sum_{n=0}^{\infty}\frac{S^n}{n!}\int_{-\infty}^{+\infty} H_n(\xi)e^{(-\xi^2+\xi\xi_0-\xi_0^2/2)}d\xi$$

or $$\int_{-\infty}^{+\infty} \exp\left[-\left\{\xi-\left(S+\frac{\xi_n}{2}\right)\right\}^2\right]d\xi\times e^{-\xi_0^{2/4}+S\xi_0} = \sum_{n=0}^{\infty}\frac{S^n}{n!}\int_{-\infty}^{+\infty} H_n(\xi)e^{(-\xi^2+\xi\xi_0-\xi_0^2/2)}d\xi$$

or $$\pi^{1/2}\exp[-\xi_0^{2/4}]\sum_{n=0}^{\infty}\frac{(S\xi_0)^2}{n!} = \sum_{n=0}^{\infty}\frac{S^n}{n!} = \int_{-\infty}^{+\infty} H_n(\xi)e^{(-\xi^2+\xi\xi_0-\xi_0^2/2)}d\xi$$

∴ $$\int_{-\infty}^{+\infty} H_n(\xi)e^{(-\xi^2+\xi\xi_0-\xi_0^2)/2}d\xi = \pi^{1/2}\exp[-\xi_0^{2/4}]\xi_0^n$$

and $$C_m = \frac{N_m^*}{\alpha^{1/2}\pi^{1/4}}\pi^{1/2}\exp(-\xi_0^{2/4})\xi_0^m = \left[\frac{\alpha}{2^m m!\pi^{1/2}}\right]^{1/2}\cdot\frac{\pi^{1/2}}{\alpha^{1/2}\pi^{1/4}}e^{-\xi_0^{2/4}}\xi_0^m$$

or $$C_m = \frac{\xi_0^m e^{-\xi_0^{2/4}}}{(2^m m!)^{1/2}} \tag{10.83}$$

Substituting the value of C_m in Eq. (10.80), we obtain

$$\psi(x,t) = \left(e^{-i\omega t/2}\right)\sum_{n=0}^{\infty}\left[\frac{\xi_0^n e^{-\xi_0^2/2}}{2^n n^{1/2}!}\right]\left[\frac{\alpha}{2^n n! \pi^{1/2}}\right]^{1/2} H_n(\xi) e^{-\xi_0^2/2} e^{in\omega t}$$

or $$\psi(x,t) = \frac{a^{1/2}}{\pi^{1/4}}\exp\left(-\frac{\xi^2}{2}-\frac{\xi_0^2}{4}-\frac{i\omega t}{2}\right)\sum_{n=0}^{\infty}\frac{H_n(\xi)}{n!}\left(\frac{\xi_0 e^{-i\omega t}}{2}\right)^n$$

or $$\psi(x,t) = \frac{a^{1/2}}{\pi^{1/4}}\exp\left(-\frac{\xi^2}{2}-\frac{\xi_0^2}{4}-\frac{i\omega t}{2}\right)\left\{\exp\left[-\left(\frac{\xi_0 e^{-i\omega t}}{2}\right)^2+2\left(\frac{\xi_0 e^{-i\omega t}}{2}\right)\xi\right]\right\}$$

$$= \frac{\alpha^{1/2}}{\pi^{1/4}}\exp\left[-\frac{\xi^2}{2}-\frac{\xi_0^2}{4}-\frac{i\omega t}{2}-\frac{\xi_0^2\exp(-2i\omega t)}{4}+\xi\xi_0\exp(-i\omega t)\right]$$

$$= \frac{\alpha^{1/2}}{\pi^{1/4}}\exp\left[-\frac{\xi^2}{2}-\frac{\xi_0^2}{4}-\frac{i\omega t}{2}-\frac{\xi_0^2}{4}(\cos 2\omega t - i\sin 2\omega t)+\xi\xi_0(\cos\omega t - i\sin\omega t)\right]$$

$$= \frac{\alpha^{1/2}}{\pi^{1/4}}\exp\left[-\frac{\xi^2}{2}-\frac{\xi_0^2}{4}(1+\cos 2\omega t)+\xi\xi_0\cos\omega t-\frac{i\omega t}{2}+\frac{i\xi_0^2}{4}\sin 2\omega t - i\xi\xi_0 i\omega t\right]$$

$$= \frac{\alpha^{1/2}}{\pi^{1/4}}\exp\left(-\frac{\xi^2}{2}-\frac{\xi_0^2}{4}\times 2\cos^2\omega t+\xi\xi_0\cos\omega t\right)$$

$$\exp\left[-i\left(\frac{\omega t}{2}-\frac{\xi_0^2}{4}\sin 2\omega t+\xi\xi_0\sin\omega t\right)\right]$$

$$= \frac{\alpha^{1/2}}{\pi^{1/4}}\exp\left[-\frac{1}{2}(\xi-\xi_0\cos\omega t)^2\right]\exp\left[-i\left(\frac{\omega t}{2}-\frac{\xi_0^2}{4}\sin 2\omega t+\xi\xi_0\sin\omega t\right)\right]$$

The absolute square of this wave function gives a position probability density.

$$|\psi(x,t)|^2 = \frac{\alpha}{\pi^{1/2}}\exp\{-(\xi-\xi_0\cos\omega t)^2\}$$

or $$|\psi(x,t)|^2 = \frac{\alpha}{\pi^{1/2}}e^{-\alpha^2(x-x_0\cos\omega t)^2} \tag{10.84}$$

This shows that ψ represents a **wave-packet** oscillating about the point $x=0$ with an amplitude x_0 and frequency ω. The shape of the wave-packet however remains unchanged.

As $x_0 \to x$, ψ approaches the lowest energy eigen function ψ_0. With the increase in the value of x_0, the involvement of higher excited states (having a large value of n) increases and the value of n_0 for which C_n has a maximum also increases.

The energy level $E_{n_0} = \left(n_0+\frac{1}{2}\right)\hbar\omega$, from whose neighbourhood most of the contribution of ψ comes, is approximately equal to $\frac{1}{2}fx^2$ for the classical oscillator that has the same amplitude.

Example 10.1: **The energy eigen values of a molecule indicate that the molecule is a one-dimensional harmonic oscillator. In going from the first excited state to the ground state the molecule emits a photon of energy $h\nu = 0.1$ eV. Assuming that the oscillating part of the molecule is a proton, calculate the probability that the proton is at a distance from the origin that would be forbidden to it by classical mechanics.**

Solution If $E = h\nu = 0.1$ eV, we have $E_0 = \frac{\hbar\omega}{2} = 0.05$, Force constant $C = m\omega^2$.

A classical harmonic oscillator with this force constant and a total energy E_0 has a amplitude of $A = \frac{2E_0}{C} = \frac{\hbar}{\sqrt{2mE_0}}$. The ground state wave function is $U_0 = B_0 e^{-x^2/2} = \frac{1}{\sqrt{\pi}} e^{-x^2/2}$.

Using the normalization, the probability of finding the particle outside the classically defined limit A is then

$$P(A) = 1 - \frac{1}{\sqrt{\pi}} \int_{-A}^{A} e^{-x^2} dx = 1 - \frac{2}{\sqrt{\pi}} \int_{0}^{A} e^{-x^2} dx$$

Example 10.2: **Verify that the following functions are eigen functions of the simple Harmonic oscillator, (*a*) $\psi = \exp\left(-\frac{1}{2}\beta x^2\right)$ (*b*) $\psi = x\exp\left(-\frac{1}{2}\beta^2 x\right)$, where $\beta = \frac{2\pi(mk)^{1/2}}{h}$, $k \rightarrow$ force constant. What is the eigen value in each case?**

Solution

$$\text{(a) } H\psi = \left[\frac{-\hbar^2}{8\pi^2 m}\frac{d^2}{dx^2} + \frac{1}{2}kx^2\right]\exp\left(-\frac{1}{2}\beta x^2\right)$$

$$= \left[-\frac{\hbar^2}{8\pi^2 m}(\beta^2 x^2 - \beta) + \frac{1}{2}kx^2\right]\exp\left(-\frac{1}{2}\beta x^2\right)$$

Now $\quad \nu = \frac{1}{2\pi}\sqrt{\frac{k}{m}}$ or $k = 4\pi^2\nu^2 m \qquad \therefore \quad \beta = \frac{2\pi\sqrt{mk}}{m} = \frac{4\pi^2 \nu m}{h}$

Substituting for β in the above equation, one obtains

$$H\psi = \left[-\frac{h^2}{8\pi^2 m}\frac{16\pi^4\nu^2 m^2}{h^2}x^2 + \frac{h^2}{8\pi^2 m}\frac{4\pi^2\nu m}{h} + \frac{1}{2}kx^2\right]\exp\left(-\frac{1}{2}\beta x^2\right)$$

$$= \left[-\frac{1}{2}kx^2 + \frac{1}{2}h\nu + \frac{1}{2}kx^2\right]\exp\left(-\frac{1}{2}\beta x^2\right) = \left(\frac{1}{2}h\nu\right)\exp\left(-\frac{1}{2}\beta x^2\right) = \left(\frac{1}{2}h\nu\right)\psi$$

$\therefore$ Eigen value is $\frac{1}{2}h\nu$.

(b) One can easily show that eigen value in this case will be $\frac{3}{4}h\nu$.

Example 10.3: **Calculate the average potential and kinetic energies of harmonic oscillator in the ground state. Compare the results obtained with the classical results.**

Solution

$$<V> = \int_{-\infty}^{\infty} \psi_0^* V \psi_0 dx = \left(\frac{\beta}{\pi}\right)^{1/2} \left(-\frac{1}{2}k\right) \int_{-\infty}^{\infty} x^2 \exp(-\beta x^2) dx$$

$$= \left(\frac{\beta}{\pi}\right)^{1/2} \frac{1}{2} k \frac{1}{2} \left(\frac{\pi}{\beta^3}\right)^{1/2} \text{(From table of integrals)} = \frac{k}{4\beta} = \frac{k}{4} \frac{h}{2\pi\sqrt{mk}} = \frac{h}{8\pi}\sqrt{\frac{k}{m}}$$

The energy $E = \frac{h}{4\pi}\sqrt{\frac{k}{m}}, <V> = \frac{1}{2}E$ = average K.E. (because E = P.E. + K.E.). This is the same as the result obtained from classical consideration.

Example 10.4:

(a) Write the value of the harmonic oscillator for the second excited state.

(b) The eigen function corresponding to this eigen value may be written as

$$\psi_2 = N(x^2 + kx + c)\exp\left(\frac{-\beta x^2}{2}\right)$$

where N is the normalization factor and $\beta = 2\pi / h\sqrt{mk}$ knowing that this function must be even (symmetric) and that it is orthogonal to the ground state eigen function, determine the value of b and c.

(c) Determine the normalization factor N.

(d) Show that the resulting function is an eigen function of H and calculate the eigen value.

Solution (a) The function is even, therefore $B = 0$. We have

$$\int_{-\infty}^{\infty} \psi_0 \psi_2 dx = 0$$

or $$\int_{-\infty}^{\infty} \left(\frac{\beta}{\pi}\right)^{1/4} \exp\left(-\frac{1}{2}\beta x^2\right) N(x^2 + c) \exp\left(\frac{-\beta x^2}{2}\right) dx = 0$$

or $$N\left(\frac{\beta}{\pi}\right)^{1/4} \int_{-\infty}^{\infty} (x^2 + c) \exp(-\beta x^2) dx = 0$$

or $$N\left(\frac{\beta}{\pi}\right)^{1/4} \left[\frac{1}{2\beta}\left(\frac{\pi}{\beta}\right)^{1/2} + c\left(\frac{\pi}{\beta}\right)^{1/2}\right] = 0$$

or $$c = -\frac{1}{2\beta}$$

(b) $\therefore$ $$\psi_2 = N\left(x^2 - \frac{1}{2\beta}\right)\exp\left(-\beta\frac{x^2}{2}\right)$$

(c) $$\int_{-\infty}^{\infty}\left[N\left(x^2-\frac{1}{2\beta}\right)\exp\left(-\beta\frac{x^2}{2}\right)\right]^2 dx = 1$$

or $$N^2\left[\int_{-\infty}^{\infty} x^4\exp(-\beta x^2)dx - \int_{-\infty}^{\infty}\frac{1}{\beta}x^2\exp(-\beta x^2)dx + \int_{-\infty}^{\infty}\frac{1}{4\beta^2}\exp(-\beta x^2)dx\right] = 1$$

Using table of integrals, we have

$$\int_0^{\infty} x^{2n}\exp(-\beta x^2)dx = \frac{1\times3\times5\ldots.(2n-1)}{2^{n+1}\beta^n}\left(\frac{\pi}{\beta}\right)^{1/2}$$

Here $n = 2$

$$\int_{-\infty}^{\infty} x^4\exp(-\beta x^2)dx = \frac{1\times3}{8\beta^2}\left(\frac{\pi}{\beta}\right)^{1/2}.\text{ Therefore }\int x^4\exp(-\beta x^2)dx = \frac{2\times3}{8\beta^2}\left(\frac{\pi}{\beta}\right)^{1/2}$$

$$\int_{-\infty}^{\infty} x^2\exp(-\beta x^2)dx = \frac{1}{2\beta}\left(\frac{\pi}{\beta}\right)^{1/2}\quad\text{and}\quad\int_{-\infty}^{\infty}\exp(-\beta x^2)dx = \left(\frac{\pi}{\beta}\right)^{1/2} = 1$$

Therefore, $$N^2\left[\frac{3}{4\beta^2}-\frac{1}{2\beta^2}+\frac{1}{4\beta^2}\right]\left(\frac{\pi}{\beta}\right)^{1/2} = 1$$

or $$N^2\frac{1}{2\beta^2}\left(\frac{\pi}{\beta}\right)^{1/2} = 1 \text{ or } N\left[2\beta^2\left(\frac{\pi}{\beta}\right)^{1/2}\right]^{1/2}$$

(d) $$\psi_2 = \left[2\beta^2\left(\frac{\beta}{\pi}\right)^{1/2}\left(x^2-\frac{1}{2\beta}\right)\exp\left(-\frac{\beta x^2}{2}\right)\right]$$

$$= \frac{1}{\sqrt{2}}\left(\frac{\beta}{\pi}\right)^{1/4}(2\beta x^2-1)\exp\left(\frac{-\beta x^2}{2}\right)$$

$$H\psi_2 = \left[-\frac{h^2}{8\pi^2 m}\frac{d^2}{dx^2}+\frac{1}{2}kx^2\right]$$

$$\therefore\quad \left[\frac{1}{\sqrt{2}}\left(\frac{\beta}{\pi}\right)^{1/4}(2\beta x^2-1)\exp\left(-\frac{\beta x^2}{2}\right)\right] = \frac{5h}{4\pi}\sqrt{\frac{k}{m}}\psi_2.$$

Example 10.5: **The potential energy of a simple harmonic oscillator of mass m, oscillating with angular frequency ω is $V(x) = \frac{1}{2}m\omega^2 x^2$. (a) Write the time-independent Schrodinger equation for a simple harmonic oscillator, (b) Given the eigen function of the Hamiltonian operator for the ground state as $\psi_0 = \left(\frac{\alpha}{\pi}\right)^{1/4} e^{-\alpha x^2/2}$, where $\alpha = \frac{m\omega}{\hbar}$. Calculate the energy eigen value in the ground state.**

Solution The Hamiltonian operator for the harmonic oscillator is

$$H = -\frac{\hbar^2}{2m}\frac{d^2}{dx^2} + \frac{1}{2}m\omega^2 x^2$$

Therefore, the time-independent Hamiltonian form of the Schrodinger equation is

$$\left[-\frac{\hbar^2}{2m}\frac{d^2}{dx^2} + \frac{1}{2}m\omega^2 x^2\right]\psi_0 = E_0\psi_0 \qquad \text{(i)}$$

Now

$$\psi_0 = \left(\frac{\alpha}{\pi}\right)^{1/4} e^{-\alpha x^2/2}$$

∴

$$\frac{d\psi_0}{dx} = \left(\frac{\alpha}{\pi}\right)^{1/4} e^{-\alpha x^2/2}(-\alpha x)$$

$$= -\alpha x\left(\frac{\alpha}{\pi}\right)^{1/4} e^{-\alpha x^2/2} = -\alpha x\psi_0 \qquad \text{(ii)}$$

and

$$\frac{d^2\psi_0}{dx^2} = -\alpha x\frac{d\psi_0}{dx} - \alpha\psi_0 = -\alpha x(-\alpha x\psi_0) - \alpha\psi_0 = \alpha^2 x^2\psi_0 - \alpha\psi_0$$

Multiplying the above equation throughout by $-\hbar^2/2m$, one obtains

$$\frac{-\hbar^2}{2m}\frac{d^2\psi_0}{dx^2} = -\frac{\hbar^2}{2m}\alpha^2 x^2\psi_0 + \frac{\hbar}{2m}\alpha\psi_0$$

Substituting $\alpha = \dfrac{m\omega}{\hbar}$ and simplifying, one obtains

$$\left[-\frac{\hbar^2}{2m}\frac{d^2}{dx^2} + \frac{1}{2}m\omega^2 x^2\right]\psi_0 = \frac{1}{2}\hbar\omega\psi_0 \qquad \text{(iii)}$$

Comparing Eqs. (i) and (iii), one obtains $E_0 = \dfrac{1}{2}\hbar\omega_0$.

Example 10.6: Prove that if the power series $H(y) = A_0 + A_1 y + A_2 y^2 + A_3 y^3 + \cdots = \sum_{n=0}^{\infty} A_0 y^n$ whose coefficients are connected by $A_{n+2} = \dfrac{2n+1-\lambda}{(n+1)(n-2)}A_n$, does not terminate for some value of n, then the wave function $\psi = e^{-\alpha x^2/2}\,H(y)$ will be infinite as y approaches infinity.

Solution In n is a large number, then in the recursion formula one can neglect all constant numbers which appear in the sum or difference with n. So, for large values of n the recursion formula becomes

$$A_{n+2} = \frac{2n}{n^2}A_n = \frac{2}{n}A_n \qquad \text{(i)}$$

Now, we have exponential series

$$e^{x^2/2} = 1 + y^2 + \frac{y^4}{2!} + \dots + \frac{y^n}{\left(\frac{n}{2}\right)!} + \frac{y^{n+2}}{\left(\frac{n+2}{2}\right)!} + \cdots = \sum_{n=0}^{\infty} y^n \text{, where } n \text{ is even.}$$

It is evident that the coefficients B_n and B_{n+2} of y^n and y^{n+2} respectively are

$$B_n = \frac{1}{(n/2)!} \quad \text{and} \quad B_{n+2} = \frac{1}{\left(\frac{n+2}{2}\right)!}$$

Therefore the recursion formula for the exponential series is

$$\frac{B_{n+2}}{B_n} = \frac{(n/2)!}{\left(\frac{n+2}{2}\right)!} = \frac{(n/2)!}{\left(\frac{n}{2}+1\right)!} = \frac{1}{\left(\frac{n}{2}+1\right)}$$

∴ For large value of n, $$B_{n+2} = \frac{2}{n} B_n \qquad \text{(ii)}$$

Dividing (i) by (ii), one obtains $\frac{A_{n+2}}{B_{n+2}} = \frac{A_n}{B_n} = C$

where C is a constant

$$\therefore \qquad A_n = CB_n$$

Thus for large values of n, we have

$$\sum A_n y^n = \sum C B_n y^n = C \sum B_n y^n = Ce^{y^2}$$

Hence $$H(y) = f(y) = Ce^{y^2}$$

where $f(y)$ is that part of the series which does not contain high powers of y. The complete wave function is then:

$$\psi(y) = e^{-y^2/2} H(y) = e^{-y^2/2}\left[f(y) + Ce^{y^2}\right] = e^{-y^2/2} f(y) + Ce^{y^2/2} \qquad \text{(iii)}$$

Eq. (iii) shows that as y approaches infinity, the last term on the right hand side becomes infinite. Hence the condition that $\psi(y)$ must be finite for all values of y will be satisfied, if the powers $H(y)$ terminates after a finite number of terms so that only the first term $e^{-y^2/2} f(y)$ remains in (iii).

Example 10.7: Calculate the probability of finding a simple harmonic oscillator with the classical limits if the oscillator is in its normal state.

Solution The wave function of the harmonic oscillator in the ground state is

$$\psi_0(x) = \left(\frac{m\omega}{\hbar\pi}\right)^{1/4} e^{-y^2/2} \quad \text{where } y = \left(\frac{m\omega}{\hbar\pi}\right)^{1/2} x$$

Classical limit of Harmonic oscillator. Let harmonic oscillator exists between the limits A and $+A$. In the ground state, its energy is $E_0 = \frac{1}{2}\hbar\omega$ and its total energy is $E = \frac{1}{2}mA^2\omega^2$.

Therefore
$$\frac{1}{2}mA^2\omega^2 = \frac{1}{2}\hbar\omega$$

If $x = \pm A$, then $y = \pm 1$. Therefore the probability of finding the particle within the classical limits is

$$P = \int_{-A}^{A} |\psi_0|^2\, dx = \int_{-A}^{A}\left(\frac{m\omega}{k\pi}\right)^{1/2} e^{-y^2/2}e^{-y^2/2}dx$$

$$= \int_{-1}^{1}\frac{1}{\sqrt{\pi}}e^{-y^2}dy = \frac{2}{\sqrt{\pi}}\int_0^1 e^{-y^2}dy \left[\because dy = \left(\frac{m\omega}{k\pi}\right)^{1/2}dx\right]$$

$$= \int_{-1}^{1}\frac{1}{\sqrt{\pi}}e^{-y^2}dy = \frac{2}{\sqrt{\pi}}\int_0^1 e^{-y^2}dy \left[\because dy = \left(\frac{m\omega}{k\pi}\right)^{1/2}dx\right]$$

$$= \frac{2}{\pi}\left[\int_0^2\left(1 - \frac{y^2}{1!} + \frac{y^4}{2!} - \frac{y^6}{3!} + \cdots\right)dy\right]$$

$$= \frac{2}{\sqrt{\pi}}\left[y - \frac{y^3}{3(1)!} + \frac{y^3}{5(2)!} - \frac{y^5}{7(3)!} + \cdots\right]_0^1 = \frac{2}{\sqrt{\pi}}\left[1 - \frac{1}{3} + \frac{1}{10} - \frac{1}{42} + \cdots\right]$$

$$= \frac{2}{1.77}[1 - 0.33 + 0.1 - 0.024 + \cdots] = 1.33[1.1 - 0.354] = 0.84 = 84\%.$$

Obviously, the probability of finding a simple harmonic oscillator (particle) within the classical limits is approximately 84% and outside the classical limits is 16%.

Example 10.8: Normalize the ground state wave function for the one-dimensional oscillator so that the total probability of finding the particle is unity.

Solution $\int_0^{\infty} |\psi_0|^2\, dx = 1$

$\therefore$
$$\int_{-\infty}^{\infty} |A|^2 \exp\left(-\frac{m\omega}{\hbar}x^2\right)dx = 1$$

or
$$2|A|^2\int_{-\infty}^{\infty}\exp\left(-\frac{m\omega}{\hbar}x^2\right)dx = 1 \text{ or } |A|^2\frac{2\sqrt{\pi\hbar}}{m\omega} = 1 \text{ or } A = \left(\frac{m\omega}{\pi\hbar}\right)^{1/4}$$

Thus the normalized wave function is

$$\psi_0 = \left(\frac{m\omega}{\pi\hbar}\right)^{1/4}\exp\left(\frac{-m\omega x^2}{2\hbar}\right)$$

Example 10.9: The Hamiltonian for a one-dimensional harmonic oscillator of classical frequency ω is $-\frac{\hbar^2}{2m}\frac{d^2}{dx^2} + \frac{1}{2}m\omega^2x^2$. Obtain the Schrodinger equation in terms of the dimensionless variable $q = \alpha x$ and a dimensionless Eigen value ε.

Solution $\dfrac{d^2\psi}{dx^2} = a^2 \dfrac{d^2\psi}{dq^2}$

$$\therefore \quad \frac{d^2\psi(q)}{dq^2} + \left(\frac{2mE}{\hbar^2\alpha^2} - \frac{m^2\omega^2}{\hbar^2\alpha^2}q^2\right)\psi(q) = 0 \quad \text{or} \quad \frac{d^2\psi}{dq^2} + (\varepsilon - q^2)\psi = 0$$

where
$$\varepsilon \frac{2mE}{\hbar^2\alpha^2} \quad \text{and} \quad \alpha^2 = \frac{m^2\omega^2}{\hbar^2}$$

Example 10.10: **Write down the Schrodinger equation for an oscillator in the momentum representation. Find the wave function in the momentum space.**

Solution Hamiltonian for linear oscillator in the coordinate space is

$$H = \frac{p^2}{2m} + \frac{m\omega^2 x^2}{2}$$

The above Hamiltonian in the momentum space is

$$H = \frac{p^2}{2m} - \frac{\hbar^2 m\omega^2}{2}\frac{d^2}{dk^2}$$

Thus the Schrodinger equation in the momentum representation is

$$\left(\frac{p^2}{2m} - \frac{\hbar^2}{2}m\omega^2\frac{d^2}{dk^2}\right)a(p) = E_a(p)$$

The method of solving this equation is similar to that of the equation in the coordinate representation $a_n(p) = \dfrac{1}{2^n[n\sqrt{\pi m\omega h}]}\exp(-p^2/2m\omega\hbar)H_n\left(\dfrac{P}{\sqrt{m\omega\hbar}}\right)$

Example 10.11: **Show that for the quantum number *n* large, the mechanical probability of finding the harmonic oscillator between *x* and *x* + *dx* is same as that obtain from classical mechanics.**

Solution Classical equation describing the motion of a harmonic oscillator is $x = a\sin\omega t$ velocity $v = \dfrac{dx}{dt} = a\omega\cos\omega t = \omega\sqrt{a^2 - x^2}$. Time spent by the oscillator during the interval dx is

$$\frac{2dt}{T} = \frac{2}{T}\frac{dx}{v} = \frac{2}{T}\frac{dx}{\omega\sqrt{a^2 - x^2}} = \frac{2}{T}\frac{T}{2\pi}\frac{2x}{\sqrt{a^2 - x^2}} = \frac{1}{\pi}\frac{dx}{\omega\sqrt{a^2 - x^2}}$$

Classical probability of finding the harmonic oscillator in the interval x and $x + dx$ is

$$P(x)dx = \frac{1}{\pi}\frac{dx}{\sqrt{a^2 - x^2}} \quad \text{or} \quad P(x) = \frac{1}{\pi\sqrt{a^2 - x^2}}$$

Quantum mechanical probability of finding the harmonic oscillator in the interval x and $x + dx$ is

$$P(x) = |\psi_n|^2 = \left(\frac{m\omega}{\pi\hbar}\right)^{1/2} e^{-y^2} H_n^2(y)\frac{1}{2^n n!} \qquad \text{for } n \text{ large}$$

$$P(x) = |\psi_n|^2 = \frac{\sqrt{m\omega}}{\hbar}\frac{1}{\pi\sqrt{2n - y^2}}\cos^2\left[\left(2n + \frac{1}{2}\right)\frac{y}{\sqrt{2}} - \frac{n\pi}{2}\right]$$

For n large, we can take $\cos^2\left[\left(2n + \frac{1}{2}\right)\frac{y.}{\sqrt{2}} - \frac{n\pi}{2}\right] = 1/2$

$$\therefore \qquad P(x) = |\psi_n|^2 = \frac{1}{\pi}\sqrt{\frac{m\omega}{\hbar}}\frac{1}{\sqrt{2n - y^2}}$$

Now $2n - y^2 = 2n - \frac{\omega}{\hbar}y^2$. But classical energy of harmonic oscillator $E = \frac{1}{2}ma^2\omega^2$ and quantum mechanically the energy of nth level is:

$$E_n = \left(n + \frac{1}{2}\right)\hbar\omega = (2n + 1)\frac{\hbar\omega}{2}$$

$$\frac{1}{2}ma^2\omega^2 = \frac{\hbar\omega}{2}(2n + 1) = \frac{\hbar\omega}{2}2n \text{ (for } n \text{ large)} = \hbar\omega n \quad \text{or} \quad 2n = \frac{ma^2\omega}{\hbar}$$

$$2n - y^2 = \frac{ma^2\omega}{\hbar} - \frac{m\omega}{\hbar}x^2 = \frac{m\omega}{\hbar}(a^2 - x^2)$$

On substituting the value of $(2n - y^2)$ for n large, one obtains

$$P(x) = |\psi_n|^2 = \sqrt{\frac{m\omega}{\hbar}}\frac{1}{\pi}\sqrt{\frac{\hbar}{m\omega}}\frac{1}{\sqrt{a^2 - x^2}} = \frac{1}{\pi\sqrt{a^2 - x^2}}$$

Example 10.12: Find the expectation value of the energy when the state of the harmonic oscillator is described by the following wave function

$$\psi(x, t) = \frac{1}{\sqrt{2}} = [\psi_0(x, t) + \psi_1(x, t)]$$

where $\psi_0(x, t)$ and $\psi_1(x, t)$ are the wave functions for the ground state and the first excited states respectively.

Solution The expression for expectation value of energy E is

$$< E > = \int \psi^*(x, t)E\psi(x, t)dx$$

But $\qquad \psi(x, t) = \frac{1}{\sqrt{2}}[\psi_0(x, t) + \psi_1(x, t)]$ and its complex conjugate

$$\psi^{*}(x,t)=\frac{1}{\sqrt{2}}[\psi_0^{*}(x,t)+\psi_1^{*}(x,t)]$$

The value of $\psi_0(x,t)=\left[\frac{\alpha}{\sqrt{\pi}}\right]^{1/2} e^{-\alpha^2}x^2/2$ and $\psi_1(x,t)=\left[\frac{\alpha}{2\sqrt{\pi}}\right]^{1/2} e^{-\frac{\alpha^2x^2}{2}2\alpha x}$

Therefore, $\psi_0^{*}=\left[\frac{\alpha}{\sqrt{\pi}}\right]^{1/2} e^{-\frac{\alpha^2x^2}{2}}=\psi_0(x,t)$

$$\psi_1^{*}=\left[\frac{\alpha}{2\sqrt{\pi}}\right]^{1/2} e^{-\frac{\alpha^2x^2}{2}}=\psi_1(x,t)$$

Therefore the value of

$$\psi^{*}(x,t)\psi(x,t)=\frac{1}{\sqrt{2}}[\psi_0^{*}(x,t)+\psi_1^{*}(x,t)]\frac{1}{\sqrt{2}}[\psi_0(x,t)+\psi_1(x,t)]$$

Substituting the values, we get $\frac{1}{2}[\psi_0(x,t)+\psi_1(x,t)]^2$

Expectation value of the energy

$$<E>=\frac{1}{2}\left[\int_{-\infty}^{-\infty}\psi_0^2(x,t)E_0dx+\int_{-\infty}^{\infty}\psi_1(x,t)E_1dx+2\int_{-\infty}^{-\infty}\psi_0(x,t)\,\psi_1(x,t)Edx\right.$$

But $$\int_{-\infty}^{\infty}\psi_0(x,t)\psi_1(x,t)Edx=0$$

$\therefore$ $$<E>=\frac{1}{2}\int_{-\infty}^{-\infty}\psi_0^2(x,t)E_0dx+\int_{-\infty}^{\infty}\psi_1^2(x,t)$$

$$<E>=\frac{1}{2}\int_{-\infty}^{-\infty}\frac{1}{2}\hbar\omega\psi_0^2(x,t)dx+\frac{1}{2}\int_{-\infty}^{\infty}\hbar\omega\psi_1^2(x,t)dx\left(\text{as } E_1=\frac{3}{2}\hbar\omega \text{ and } E_2=\frac{1}{2}\hbar\omega\right)$$

Now putting the values of $\psi_0^2(x,t)$ and $\psi_1^2(x,t)$, we get

$$<E>=\frac{1}{4}\hbar\omega\int_{-\infty}^{\infty}\frac{\alpha}{\sqrt{\pi}}e^{-\alpha^2x^2}dx+\frac{3}{4}\hbar\omega\int_{-\infty}^{\infty}\frac{\alpha}{2\sqrt{\pi}}e^{-\alpha^2x^2}\,4\alpha^2x^2dx=\frac{1}{4}\hbar\omega+\frac{3}{4}\hbar\omega=\hbar\omega.$$

Example 10.13: You are given the task of finding out whether a certain element is a mixture of several isotopes or not? For evidence would you turn to atomic or molecular spectra? Why?

Solution We know that the mass M of the nucleus enters the energy level of an atom only through the reduced mass $\mu = \dfrac{m_e M}{m_e + M}$. Since $m_e << M$ the small difference in M between different isotopes has a negligible influence on m. The atomic spectra of the various isotopes of an atom differ, thus very little. A molecule in its vibrational modes approaches harmonic oscillator.

The energy values of the harmonic oscillator $E_n = \dfrac{(2n+1)\hbar\omega}{2} = \dfrac{(2n+1)\hbar}{2}\sqrt{\dfrac{C}{m}}$ are inversely proportional to the square root of the mass m of the vibrating particle.

Assuming that Δm the mass difference between two isotopes and that $\Delta m << M$. In this case the ratio of the molecular energy values for the two isotopes will be

$$R = \sqrt{\frac{M + \Delta m}{M}} = 1 + \frac{1}{2}\frac{\Delta M}{M}$$

For the same ratio we get in the case of an atomic spectra

$$\left[\frac{m_e M + \Delta M}{m_e + M + \Delta M}\right]\left[\frac{m_e + M}{m_e M}\right] \approx 1 + \frac{m_e}{M}\frac{\Delta M}{M}$$

Example 10.14: Find the energy levels and the wave functions of a one-dimensional harmonic oscillator that is located in a constant electric field *E*. The electric charge of the oscillator is *e*.

Solution For this case the Schrodinger equation is

$$-\frac{\hbar^2}{2m}\frac{d^2\psi}{dx^2} + \left(\frac{m\omega^2}{2}x^2 - e|E|x\right)\psi = E\psi$$

can be reduced to the problem of the harmonic oscillator by completing squares in the expression for the potential energy. Introducing new variables, one obtains

$$x_1 = x - \frac{e|E|}{m\omega^2}, \xi = \sqrt{\frac{\hbar}{m\omega}}x_1, \quad \text{and} \quad E_1 = E + \frac{e^2|E|^2}{2m\omega^2}, \text{ one obtains}$$

$$-\frac{\hbar^2}{2m}\frac{d^2\psi}{dx_1^2} + \frac{m\omega^2}{2}x_1^2\psi = E_1\psi$$

Solving this, one obtains the eigen functions

$$\psi_n = C_n \exp(-\xi^2/2)H_n(\xi)$$

and eigen values

$$E_n = \left(n + \frac{1}{2}\right)\hbar\omega - \frac{e^2|E|^2}{2m\omega^2}$$

Example 10.15: Find the average kinetic energy of a one-dimensional harmonic oscillator whose energy is $\frac{7}{2}\hbar\omega$.

Solution For the one-dimensional case the kinetic energy is $T = -\frac{\hbar^2}{2m}\frac{d^2}{dx^2}$, it follows that

$$<T> = -\frac{\hbar^2}{2m}\int_{-\infty}^{\infty}\psi^*\frac{d^2\psi}{dx^2}dx = \frac{\hbar^2}{2m}\int_{-\infty}^{\infty}\left|\frac{d\psi}{dx}\right|^2 dx$$

In the given case $n = 3$ and $\psi = \psi_3 = C_3 \exp\left(\frac{-\xi^2}{2}\right)H_3(\xi)$, where $H_3 = 8\xi^3 - 12\xi$. Changing variable x to ξ, we obatin: $x = \xi(m\omega/\hbar)^{-1/2}$ and substituting $C_3^2 = \sqrt{\frac{m\omega}{\hbar\pi}}\frac{1}{2^3 3!}$, one obtains

$$<T>_3 = \frac{\hbar\omega}{2\sqrt{\pi}}\frac{1}{2^3 3!}\int_{-\infty}^{\infty}\left\{\frac{d}{d\xi}[\exp(-\xi^2/2)(8\xi^3 - 12\xi)]\right\}^2 d\xi$$

$$= \frac{\hbar\omega}{3!\sqrt{\pi}}\int_{-\infty}^{\infty}\exp(-\xi^2)(4\xi^8 - 36\xi^6 + 93\xi^4 - 54\xi^2 + 9)d\xi$$

One can easily evaluate the integrals if we differentiate w.r.t. a parameter.

Since
$$\int_{-\infty}^{\infty} e^{-a\xi^2}d\xi = \sqrt{\pi/\alpha}$$

or
$$\int_{-\infty}^{\infty}\xi^{2n}e^{-a\xi^2}d\xi = -\left(\frac{\partial}{\partial\alpha}\right)^n\int_{-\infty}^{\infty}e^{-a\xi^2}d\xi = \frac{1\times 3\times\cdots\times(2n-1)}{2^n}\sqrt{\frac{\pi}{\alpha^{2n+1}}}$$

$\therefore$
$$<T>_3 = \frac{\hbar\omega}{3!\sqrt{\pi}}\times\frac{21}{2}\times\sqrt{\pi} = \frac{1}{2}\hbar\omega\left(3 + \frac{1}{2}\right)$$

SUMMARY

1. Whenever a system of some kind vibrates about an equilibrium position, harmonic motion occurs.
2. Many systems exist for which harmonic oscillators provide very good approximations, e.g. at low temperatures atoms vibrating about their mean positions in molecules or crystal lattices can be regarded as good approximations to harmonic oscillators in quantum mechanics.
3. The simple harmonic oscillator, where the potential is $V(x) = kx^2/2$, is an important application of Schrodinger's wave equation because it approximates many complex systems in nature but is exactly soluble. The energy levels of simple harmonic

oscillator are $E_n = (n+\frac{1}{2})\hbar\omega$, where $n=0$ represents the ground state energy $E_0 = \hbar w/2$. Obviously, the minimum energy of harmonic oscillator is not zero, i.e., the oscillator exhibits zero-point energy – is a consequence of the uncertainity principle.

REVIEW QUESTIONS AND PROBLEMS

1. A simple pendulum may be regarded as a linear harmonic oscillator to the first approximation. Discuss the motion of the simple pendulum consisting of mass of 0.001 kg attached to a string of length 9.8 cm, quantum mechanically. Find the quantum mechanical solution corresponding to classical motion of amplitude 0.01 m.
2. Find the momentum distribution of energy eigen function ψ_n of a linear harmonic oscillator.
3. Solve the anisotropic, three-dimensional harmonic oscillator problem with a potential $V(x,y,z) = \frac{1}{2}\left(k_x x^2 - k_y^2 y^2 + k_z z^2\right)$. Correlate the energy levels with those of section (10.2) by letting $k_x \to k$, $k_y \to k$ and $k_z \to k$.
4. An electron is trapped in a one dimensional harmonic oscillator potential. The frequency difference between the spectral lines emitted by the electron is $\Delta\nu = 10^{15}$ cycles/s. The electron is in its ground state. What is the probability of finding the electron at a distance of $(10^{-7} \pm 1 \times 10^{-9})$ cm from the origin? What is the probability of finding the electron at the centre of the potential well in a interval of $\pm 10^{-9}$ cm?
5. Give an outline of quantum mechanical description of a simple harmonic oscillator. In what way is this description different from classical description? Discuss the properties of the oscillator in the lowest three energy states.
6. Establish Schrodinger's equation for a linear harmonic oscillator and solve it to obtain its eigen value and eigen functions. Discuss the significance of zero-point energy.
7. Determine the energy levels of a linear harmonic oscillator on the basis of Schrodinger's equation.
8. A linear harmonic oscillator is defined by the point mass μ and a restoring force kq, where k is the positive constant and q the coordinate of the point. Show that the wave function $\phi = e^{-\frac{x^2}{2}}(4x^2+2)$ is an eigen function for Schrodinger's equation of this oscillator, where x is defined by the substitution $q^2 = x^2\,\hbar/\sqrt{k\mu}$. Find the value of the energy belonging to given eigen function.
9. Calculate the probability of finding a simple harmonic oscillator within the classical limit if the oscillator is its normal state.
10. Find the Schrodinger's equation for a linear harmonic oscillator and obtain the energy eigen values. Find the expectation value of the energy when the state of the oscillator is described by the following wave function.

$$\psi(x,t) = \frac{1}{\sqrt{2}}[\psi_0(x,t) + \psi_1(x,t)]$$

where $\psi_0(x, t)$ and $\psi_1(x, t)$ represent the wave functions for the ground state and the first excited state respectively.

11. Solve the Schrodinger's equation for the linear harmonic oscillator and determine the normalized wave function and the energy levels of the oscillator.

12. Use Schrodinger's equation to calculate the normalized wave functions of a linear harmonic oscillator. Show that, if the oscillator is in its normal state, then the probability of finding the particle outside the classical limits in approximately 16%.

13. Solve the Schrodinger equation for the linear harmonic oscillator and obtain the energy eigen values. In what way is this quantum mechanical description different from classical description?

14. The normalized eigen function of one-dimensional harmonic oscillator is

$$\psi_n(x) = \left[\frac{\alpha}{2^n\, n!\sqrt{\pi}}\right]^{1/2} \exp\left[-\alpha^2 \frac{x^2}{2} H_n(\alpha x)\right]$$

where $\alpha = \sqrt{\dfrac{m\omega}{\hbar}}$ and $H_n(\alpha x)$ is a Hermite polynomial of degree n.

(a) Find the expectation value of the energy when the state of the oscillator is described by the following wave function: $\psi(x,t) = \dfrac{1}{\sqrt{2}}[\psi_0(x,t) + \psi_1(x,t)]$.

(b) Explain parity and illustrate your answer by considering the parity of the oscillator wave functions.

15. The eigen functions of the one-dimensional harmonic oscillator are the Hermite functions $\psi_n(\xi) = N_n \exp(-\xi^2/2)H_n(\xi)$, where $\xi = \sqrt{\dfrac{m\omega}{\hbar}}x$ and $H_n(\xi)$ is a Hermite Polynomial : (i) Find the normalization constant (ii) Calculate $<x^2>$ and $<p^2>$ for the nth state, (iii) Calculate the value of the energy for the nth state, (iv) Explain the significance of the zero-point energy of the oscillator, (v) draw rough sketches of the energy eigen functions $\psi_n(\xi)$ for $n = 0, 1, 2$.

16. Show that the average potential energy of a linear harmonic oscillator for the nth state is given by $<v> = 1/2(n + 1/2)\, \hbar\omega$. The orthogonality relation $\int_{-\infty}^{\infty} e^{-\xi^2} H_m(\xi)H_n(\xi)d\xi = 2^n n!\sqrt{\pi}\delta_{mn}$ may be used. The oscillator's wave functions is $\psi_n = N_n \exp(-\xi^2/2)H_n(\xi)$.

where $\sqrt{\dfrac{m\omega}{\hbar}}x$ and $N_n = \left(\dfrac{m\omega}{\pi\hbar}\right)^{1/4}\left(\dfrac{1}{2^n\, n!}\right)^{1/2}$.

(b) Explain the significance of the zero-point energy of the oscillator.

17. (a) Solve the Schrodinger equation for linear harmonic oscillator and obtain the energy eigen values.

(b) Draw a rough sketch of the eigen functions of the oscillator and comment on their parity.

SHORT ANSWER-QUESTIONS

1. Assuming that the following functions are eigen functions of the simple harmonic oscillator, find the eigen value in each case :

(i) $\psi = \exp\left(-\frac{1}{2}\beta x^2\right)$ (ii) $\psi = x\exp\left(-\frac{1}{2}\beta x^2\right)$

where $\beta = \sqrt{mk/\hbar}$, k is force constant.

[**Ans:** (i) $\frac{1}{2}h\nu$ (ii) $\frac{3}{4}h\nu$]

2. Show that the average potential energy of a one-dimensional harmonic oscillator in its ground state is $\hbar\omega/4$.

3. A linear harmonic oscillator is defined by the point mass μ and a restoring force $-kq$, where k is positive constant and q is the coordinate of the point, Show that the following wave function $\psi = [\exp(-x^2/2)][4x^2 + 2]$ is an eigen function for Schrodigner's wave equation of this oscillator, where x is given by $q = \sqrt{\frac{\hbar x^2}{\sqrt{k\mu}}}$

4. Consider a one-dimensional harmonic oscillator that is on its nth energy level. Find $< x^2 >$ and the expectation value of the potential energy for such a case.

$$\left[\textbf{Ans:} < x^2 >_n = C_n^2\left(\frac{\hbar}{m\omega}\right)^{3/2}\int_{-\infty}^{\infty} e^{-\xi^2}\frac{d_n(H n\xi^2)}{d\xi^n}d\xi = \frac{\hbar}{m\omega}(n+1/2) \text{ and} < V >_n = \frac{E_n}{2}\right]$$

5. Show that, if the linear harmonic oscillator is in its normal state then the probability of finding the particle outside the classical limit is approximately 16%.

6. A simple pendulum of mass 100 gm and length 1m has an amplitude of 1 cm. Show that the energy separation between the adjacent allowed states, $E_{n+1} - E_n = 3.3 \times 10^{-34}$ J.

7. Explain in simplest terms the relation between the zero-point energy and uncertainty principle.

8. Would you expect the zero-point energy to have much effect on the heat capacity of matter at low temperature? Justify your answer.

9. If the eigen functions of a potential have definite parities, the one of lowest energy has even parity. Explain why?

10. Are there analogies in classical physics to the quantum mechanical concept of parity?

11. Are there unbound states for a simple harmonic oscillator potential? How many bound states are there? How realistic is the potential?

12. What is the probability of finding the harmonic oscillator within the classical limits?
Ans: 83%

13. What is the ground state probability of finding the oscillator outside the classical limits?
Ans: 17%

14. What is the zero-point energy of harmonic oscillator?
Ans: 1/2 $\hbar\omega$.

OBJECTIVE QUESTIONS

1. In a harmonic oscillator, the particle varies in position from $-A$ to $+A$ and in momentum $A/\sqrt{2}$ and $\Delta p = p_0/\sqrt{2}$. The minimum energy of this harmonic oscillator is

(a) $\frac{7}{2}h\nu$ (b) $\frac{5}{2}h\nu$

(c) $\frac{3}{2}h\nu$ (d) $\frac{1}{2}h\nu$

2. First three harmonic-oscillator wave functions are normalized solutions of

(a) Uncertainty relation (b) Schrodinger's wave equation

(c) Ehrentest's theorem (d) Bohr's quantum condition

3. If the oscillator is in its normal state, then the probability of finding the particle outside the classical limits is approximately.

(a) 84% (b) 36%

(c) 16% (d) 0

4. The zero-point energy of a linear harmonic oscillator of frequency 50 Hz is

(a) 1.66×10^{-32} J (b) 1.66×10^{-39} J

(c) 1.66×10^{-25} J (d) 1.66×10^{-13} J

[**Hint** : Zero-point energy $= \frac{h\nu}{2} = \frac{6.63\times10^{-34}\times50}{2} = 1.66\times10^{-32}$ J]

5. The period of a linear harmonic oscillator is 1*s*. Its zero point energy is

(a) 3.3×10^{-27} J (b) 3.3×10^{-34} J

(c) 3.3×10^{-41} J (d) 3.3×10^{-13} J

[**Hint** : $\frac{h\nu}{2} = \frac{6.63\times10^{-34}\times1}{2} = 3.3\times10^{-34}$ J]

6. The energy of a linear harmonic oscillator in its third excited state is 0.1 eV. Its frequency of vibration is

(a) 6.89×10^{12} Hz (b) 6.89×10^{3} Hz

(c) 6.89×10^{6} Hz (d) 6.89×10^{9} Hz

$$\left[\textbf{Hint}: \nu = \frac{E_n}{\left(n+\frac{1}{2}\right)h} = \frac{0.1\times1.6\times10^{-19}\times1}{\left(3+\frac{1}{2}\right)\times6.63\times10^{34}} = 6.89\times10^{12}\,Hz\right]$$

7. The frequency of a linear harmonic oscillator is 6×10^{12}*Hz*. The energy of the oscillator in the fourth excited state is

(a) 1.66 eV (b) 16.6 eV

(c) 166 eV (d) 0.166 eV

8. The normalized oscillator wave function for $n = 5$ is

(a) $\psi(q) = \frac{1}{2}\left(\frac{1}{15\sqrt{\pi}}\right)^{1/2} (4q^5 - 20q^3 + 15q)\exp(-q^2/2)$

(b) $\psi(q) = \frac{1}{2}\left(\frac{1}{6\sqrt{\pi}}\right)^{1/2} (4q^4 - 12q^2 + 3)\exp(-q^2/2)$

(c) $\psi(q) = \left(\frac{1}{3\sqrt{\pi}}\right)^{1/2} (2q^2 - 3q)\exp(-q^2/2)$

(d) $\psi(q) = \left(\frac{1}{2\sqrt{\pi}}\right)^{1/2} (2q^2 - 1)\exp(-q^2/2)$

9. The classical probability of finding the linear harmonic oscillator in dx at x is given by

(a) $p(x)dx = \frac{dx}{\pi a}$ (b) $p(x)dx = \frac{dx}{\pi\sqrt{a^2 - x^2}}$

(c) $p(x)dx = \frac{dx}{\pi x}$ (d) $p(x)dx = \frac{dx}{\pi(a - x)}$

10. The probability density of the linear harmonic oscillator at the classical limits is

(a) $1/e^{th}$ of that at the origin (b) $1/e^2$ of that at the origin

(c) $1/2e^2$ of that at the origin (d) $1/8\pi e^2$ of that at the origin

11. The normalized wave function for the harmonic oscillator with x as the variable and H_n Hermite function is

(a) $\psi_n = \left\{\frac{\sqrt{\alpha}}{2^n n!\sqrt{\pi}}\right\}^{1/2} \exp\left(\frac{-\alpha x^2}{2}\right) H_n(\sqrt{\alpha}x)$

(b) $\psi_n = \left\{\frac{\sqrt{\alpha}}{2^n n!\sqrt{\pi}}\right\}^{1/2} H_n(\sqrt{\alpha}x)$

(c) $\psi_n = \left\{\frac{\sqrt{\alpha}}{2^n n!\sqrt{\pi}}\right\}^{1/2} \exp\left(\frac{\alpha x^2}{2}\right)$

(d) $\psi_n = \frac{\alpha}{\pi\sqrt{\alpha^2 - x^2}} \exp\left(-\frac{\alpha x^2}{2}\right)$

12. The zero-point energy first observed in the

(a) Scattering of α-particles (b) Scattering of x-rays by crystals

(c) Scattering of neutrons (d) Scattering of protons

13. In the state n of the oscillator, which one of the following relation is correct (T → K.E. and V → P.E.)

(a) $<T>=<V>$ (b) $<T>\neq<V>$

(c) $<T>=<V^2>$ (d) $<T>=0$

14. The energy levels of the linear oscillator are

(a) degenerate (b) non-degenerate

15. A charged particle oscillator will

(a) radiate energy in an (stationary) eigen state

(b) not radiate energy in an (stationary) eigen state

ANSWERS

Questions Numbers	1	2	3	4	5	6	7	8	9	10
Answers	(d)	(b)	(c)	(a)	(b)	(a)	(d)	(a)	(b)	(a)
	11	12	13	14	15					
	(a)	(b)	(a)	(b)	(b)					

FILL IN THE BLANKS

1. The energy value (E_n) of an linear harmonic oscillator based on quantum therory is E_n = ______, where n = 0, 1, 2, ... **[$n\hbar\omega$]**
2. The zero-point energy of linear harmonic oscillator is ______. **[$\frac{1}{2}\hbar\omega$]**
3. $E = 0$ for a linear harmonic oscillator mean that both position and momentum are well defined which is a violation of ______. **[uncertainty principle]**
4. Classically, the probability of finding the harmonic oscillator at a given point is inversely proportional to its ______ at that point. In the quantum case, for $n = 0$, the probability is maximum at ______ and as the quantum number increases the maximum probability moves towards the extreme positions. **[velocity, $x = 0$]**
5. A harmonic oscillator is in the ground state. Its probability density is maximum at ______ given by $P(o) = \left(\frac{m\omega}{\pi\hbar}\right)^{1/2}$. **[$x = 0$]**
6. The probability density of the linear harmonic oscillator in an arbitrary superposition state is ______ with the period equal to the period of the ______. **[periodic, oscillator]**
7. For the harmonic oscillator, $<x^2>$ = ______. **[$\frac{\hbar}{2m\omega}(2n+1)$]**
8. For the harmonic oscillator, $<p^2>$ = ______. **[$(n + 1/2)\, m\hbar\omega$]**
9. For the harmonic oscillator, the expectation value of kinetic energy, $<T>$ = ______. **[$\frac{1}{2}(n+1/2)\hbar\omega$]**

11

Orbital Angular Momentum and Spin

11.1 INTRODUCTION

Angular momentum $\boldsymbol{L}$ is an important and very interesting property of physical systems in classical as well as quantum mechanics. Classically the angular momentum of a particle is represented by the vector $\boldsymbol{L} = \boldsymbol{r} \times \boldsymbol{p}$, where $\boldsymbol{r}$ is the position vector that locates the particle and $\boldsymbol{p}$ is its linear momentum. The direction of $\boldsymbol{L}$ is perpendicular to the plane of the orbit. Quantum theory gives a relationship between the length of the angular momentum vector and angular momentum quantum number l: $|\boldsymbol{L}| = \sqrt{l(l+1)}\hbar$. We may note that the length of the vector $|\boldsymbol{L}|$ is always greater than $l\hbar$ since $\sqrt{l(l+1)}$ is always greater than l. Moreover these values of $|\boldsymbol{L}|$, which we can interpret as the "magnitude" of the electron's angular momentum, are totally different from those found in the Bohr model.

In this chapter, we shall consider the operators representing angular momentum components and their eigen values, eigen vectors and electron spin.

11.2 ORBITAL ANGULAR MOMENTUM OPERATORS

Orbital momentum $\boldsymbol{L}$ of a particle is defined classically as the product of its position vector $\boldsymbol{r}$ and the linear momentum $\boldsymbol{p}$:

$$\boldsymbol{L} = \boldsymbol{r} \times \boldsymbol{p} = (xp_z - zp_y)\hat{i} + (zp_x - xp_z)\hat{j} + (xp_y - yp_x)\hat{k} \tag{11.1}$$

The components of classical orbital angular momentum $\boldsymbol{L}$ are given by

$$\begin{aligned} L_x &= (xp_z - zp_y), \\ L_y &= (zp_x - xp_z), \\ L_z &= (xp_y - yp_x) \end{aligned} \tag{11.2}$$

Now replacing the momentum components p_q by $-i\hbar\dfrac{\partial}{\partial q}$, where q is the coordinate. We can express operators of L_x, L_y, L_z as

$$L_x = -i\hbar\left(y\frac{\partial}{\partial z} - z\frac{\partial}{\partial y}\right) \tag{11.3}$$

$$L_y = -i\hbar\left(z\frac{\partial}{\partial x} - x\frac{\partial}{\partial z}\right) \tag{11.4}$$

$$L_z = -i\hbar\left(x\frac{\partial}{\partial y} - y\frac{\partial}{\partial x}\right) \tag{11.5}$$

Let us transform the above equations to a system of spherical polar coordinates with the same origin using the relations

$$\frac{\partial}{\partial x} = \frac{\partial}{\partial r}\frac{\partial r}{\partial x} + \frac{\partial}{\partial \theta}\frac{\partial \theta}{\partial x} + \frac{\partial}{\partial \phi}\frac{\partial \phi}{\partial x}$$

$$\frac{\partial}{\partial y} = \frac{\partial}{\partial r}\frac{\partial r}{\partial y} + \frac{\partial}{\partial \theta}\frac{\partial \theta}{\partial y} + \frac{\partial}{\partial \phi}\frac{\partial \phi}{\partial y}$$

$$\frac{\partial}{\partial z} = \frac{\partial}{\partial r}\frac{\partial r}{\partial z} + \frac{\partial}{\partial \theta}\frac{\partial \theta}{\partial z} + \frac{\partial}{\partial \phi}\frac{\partial \phi}{\partial z} \tag{11.6}$$

The spherical polar coordinates of the point P (Fig. (11.1)) are related to cartesian coordinates by

$$x = r\sin\theta\cos\theta$$

$$y = r\sin\theta\sin\theta$$

$$z = r\cos\theta \tag{11.7}$$

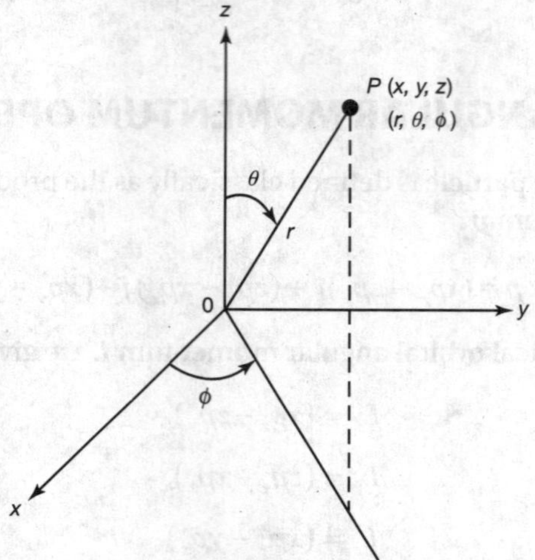

Fig. 11.1 Spherical polar coordinates r, θ, ϕ of an arbitrary point P in space

Therefore

$$r = (x^2 + y^2 + z^2)^{1/2} \tag{11.8}$$

$$\cos\theta = \frac{z}{r} = \frac{z}{(x^2 + y^2 + z^2)^{1/2}} \tag{11.9}$$

$$\tan\phi = \frac{y}{x} \tag{11.10}$$

on differentiating Eq. (11.8) and making use of Eq. (11.7), one obtains

$$\frac{\partial r}{\partial x} = \frac{1}{2}(x^2 + y^2 + z^2)^{-1/2} 2x = \frac{x}{r} = \sin\theta\cos\phi \tag{11.11}$$

Similarly on differentiating Eq. (11.9) and then making use of Eqs. (11.8) and (11.5), one obtains

$$-\sin\theta\frac{\partial\theta}{\partial x} = z\left(-\frac{1}{2}\right)(x^2 + y^2 + z^2)^{-3/2} 2x$$

or

$$\frac{\partial\theta}{\partial x} = z/(x^2 + y^2 + z^2)^{1/2} x/(x^2 + y^2 + z^2)\sin\theta$$

$$= \frac{z}{r}\frac{r\sin\theta\cos\theta}{r^2\sin\theta} = \cos\theta\cos\phi/r \tag{11.12}$$

Similarly differentiating Eq. (11.11) and then making use of Eq. (11.12), one obtains

$$\sec^2\phi\frac{\partial\phi}{\partial x} = -y/x^2$$

or

$$\partial\phi/\partial x = -\frac{r\sin\theta\sin\phi}{r^2\sin^2\theta\cos^2\phi\sec^2\phi} = -\frac{\sin\phi}{r\sin\theta} \tag{11.13}$$

Similarly, proceeding in the similar way, one can easily show that

$$\frac{\partial r}{\partial y} = \sin\theta\sin\phi, \frac{\partial\theta}{\partial y} = \cos\theta\frac{\sin\phi}{r}, \frac{\partial\phi}{\partial y} = \frac{\cos\phi}{r\sin\theta} \tag{11.14}$$

and

$$\frac{\partial r}{\partial z} = \cos\theta, \frac{\partial\theta}{\partial z} = -\frac{\sin\theta}{r}\frac{\partial\phi}{\partial z} = 0 \tag{11.15}$$

Substituting Eqs. (11.6) and (11.7) into Eqs. [(11.3, 11.4, 11.5)] and then substituting the above relations into the resulting equations, one obtains

$$L_x = -i\hbar\left[r\sin\theta\sin\phi\left\{\cos\theta\frac{\partial}{\partial r} - \frac{\sin\theta}{r}\frac{\partial}{\partial\theta} + 0\right\}\right.$$

$$\left. - r\cos\theta\left\{\sin\theta\sin\phi\frac{\partial}{\partial r} + \cos\theta\frac{\sin\phi}{r}\frac{\partial}{\partial\theta} + \frac{\cos\phi}{r\sin\theta}\frac{\partial}{\partial\phi}\right\}\right]$$

$$= i\hbar\left[\sin^2\theta\sin\phi\frac{\partial}{\partial\theta} + \cos^2\theta\sin\phi\frac{\partial}{\partial\theta} + \cot\theta\cos\phi\frac{\partial}{\partial\phi}\right]$$

$$= i\hbar\left[\sin\phi\frac{\partial}{\partial\theta} + \cot\theta\cos\phi\frac{\partial}{\partial\phi}\right] \quad (11.16)$$

$$L_y = -i\hbar\left[r\cos\theta\left\{\sin\theta\cos\phi\frac{\partial}{\partial r} + \cos\theta\frac{\cos\phi}{r}\frac{\partial}{\partial\theta} - \frac{\sin\phi}{r\sin\theta}\frac{\partial}{\partial\phi}\right\}\right.$$

$$\left.- r\sin\theta\cos\phi\left\{\cos\theta\frac{\partial}{\partial r} - \frac{\sin\theta}{r}\frac{\partial}{\partial\theta} + 0\right\}\right]$$

$$= -i\hbar\left(\cos\phi\frac{\partial}{\partial\theta} - \cot\theta\sin\phi\frac{\partial}{\partial\theta}\right) \quad (11.17)$$

$$L_z = -i\hbar\frac{\partial}{\partial\phi} \quad (11.18)$$

We can define

$$L^2 = L_x^2 + L_y^2 + L_z^2 = L_xL_x + L_yL_y + L_zL_z \quad (11.19)$$

Substituting for L_x, L_y and L_z from Eqs (11.16) and (11.18) into the above one obtains

$$L^2 = -\hbar^2\left[\frac{1}{\sin\theta}\frac{\partial}{\partial\theta}(\sin\theta\frac{\partial}{\partial\theta}) + \frac{1}{\sin^2\theta}\frac{\partial^2}{\partial\phi^2}\right] \quad (11.20)$$

11.3 ANGULAR MOMENTUM COMMUTATION RELATIONS

One can obtain the commutation relations of the components of $\boldsymbol{L}$ in cartesian coordinates as

$$[L_x, L_y] = [(yp_z - zp_y), (zp_x - xp_z)]$$

$$= [yp_z - zp_x] - [yp_z, xp_z] - [zp_y - zp_x] + [zp_y, xp_z] \quad (11.21)$$

We may remember that in these commutators, p_q stands for $-i\hbar\frac{\partial}{\partial q}$. We may also note that in the second and third terms on the R.H.S of Eq. (11.21), all the variables involved commute with each other. Hence both of them vanish. Since y and p_x commute with z and p_z, we have

$$[zp_z, zp_x] = yp_x[p_z, z] = -i\hbar\, yp_x \quad (11.22)$$

Similarly, one obtains

$$[zp_y, xp_z] = p_y x[z, p_z] = -i\hbar xp_y \quad (11.23)$$

Hence,

$$[L_x, L_y] = i\hbar(xp_y - yp_x) = i\,\hbar L_z$$

Similarly, one can obtain the other commutators. We obtain

$$[L_x, L_y] = i\hbar L_z, \quad [L_y, L_z] = i\hbar L_x, \quad [L_z, L_x] = i\hbar L_y \quad (11.24)$$

We may note that the components of angular momentum do not commute with one another and therefore they are not measurable simultaneously. In other words, we can say that if the system is in an eigenstate of one angular momentum component, it will not be simultaneously in an eigenstate of either of the others. Moreover, these commutation relations hold for the components of total angular momentum $\boldsymbol{L} = \Sigma\, L_i$ of a system of particles also. One can write the commutation relations in Eq. (11.24) in a compact form as

$$\boldsymbol{L} \times \boldsymbol{L} = i\,\hbar \boldsymbol{L} \tag{11.25}$$

We may note that in a usual sense the vector product of a vector with itself is zero. Clearly, we have to consider $\boldsymbol{L}$ as a vector operator and not as a usual vector. The L.H.S of Eq. (11.25) has to be considered as a determinant and has to be expanded the term by term comparison with the right hand side.

Now we consider the commutator of L^2 with the components of $\boldsymbol{L}$

$$\begin{aligned}[L^2, L_x] &= [L_x^2, L_x] + [L_y^2, L_x] + [L_z^2, L_x] \\ &= 0 + L_y[L_y, L_x] + [L_y, L_x]L_y + L_z[L_z, L_x] + [L_z, L_x]L_z \\ &= L_y(-i\hbar L_z) + (-i\hbar L_z)L_y + i\hbar L_z L_y + i\hbar L_y L_z = 0\end{aligned}$$

Thus, we can conclude

$$[L^2, L_x] = [L^2, L_y] = [L^2, L_z] = 0 \tag{11.26}$$

Clearly, the square of angular momentum commutes with its components, i.e. one can measure total angular momentum simultaneously with any other components. As the components among themselves are non-commuting, we cannot measure L^2, L_x, L_y, L_z simultaneously. This means, we cannot have a representation in which all the four are diagonal. The eigen value problem is therefore approached in such a way that L^2 and one of the components, say L_z, have simultaneous eigen vectors. We may note that the selection of L_z is completely arbitrary. The commutation relations in Eqs. (11.24) and (11.26) form the basis and foundation for the theory of angular momentum. It is found more convenient to work with the operators L_+ and L_- defined as

$$L_+ = L_x + iL_y \text{ and } L_- = L_x - L_y \tag{11.27}$$

the operator L_+ is called the *raising operator* and L_-, the *lowering operator*. These operators commute with L^2, we have

$$[L^2, L_+] = 0, [L^2, L_-] = 0 \tag{11.28}$$

and

$$\begin{aligned}[L_z, L_+] &= [L_z, L_x] + i[L_z, L_y] \\ &= i\hbar L_y + \hbar L_x = \hbar L_+\end{aligned} \tag{11.29}$$

In the same way, one can also evaluate the other commutators. One obtains the following relations

$$\left.\begin{aligned}[L_z, L_-] &= -\hbar L_- \\ [L_x, L_+] &= -\hbar L_z \\ [L_x, L_-] &= \hbar L_z \\ [L_y, L_+] &= -i\hbar L_z \\ [L_y, L_-] &= i\hbar L_z\end{aligned}\right] \tag{11.30}$$

one also obtains

$$[L_+, L_-] = 2\hbar L_z \tag{11.31}$$

and

$$L_+L_- = L^2 - L_z^2 + \hbar L_z, \quad L_-L_+ = L^2 - L_z^2 - \hbar L_z \tag{11.32}$$

11.4 EIGEN VALUES AND EIGEN FUNCTION OF L^2, AND L_z

Eq. (11.24) tells us that we cannot find a complete set of simultaneous eigen functions of the three components of **L**, because the three components do not commute. One can seek a complete set of simultaneous eigen functions of L^2 and one component of **L**, because L^2 with $L\pm$.

The eigen value equation for L^2 in spherical polar coordinates can be written as

$$-\hbar^2\left[\frac{1}{\sin\theta}\frac{\partial}{\partial\theta}\left(\sin\theta\frac{\partial}{\partial\theta}\right)+\frac{1}{\sin^2\theta}\frac{\partial^2}{\partial\phi^2}\right]Y = \lambda\hbar^2 Y \tag{11.33}$$

Eq. (11.33) is same as the angular part of the Schrodinger equation of a system moving in a potential $V(r)$. The solution gives $\lambda = l(l+1)$ with eigen functions

$$Y = Y_{lm}(\theta,\phi) = \varepsilon\left[\frac{2l+1}{4\pi}\frac{(l-|m|)!}{(l+|m|)!}\right]^{1/2} p_l^{|m|}\cos\theta e^{|m|\phi} \tag{11.34}$$

where $l = 0, 1, 2, 3,$ and $m = 0, \pm1, \pm2, \ldots \pm l$ and $\varepsilon = (-1)^m$ for $m > 0$ and $\varepsilon = 1$ for $m \le 0$. Now, the form of L_z, in Eq. (11.34) immediately gives

$$L_z Y_{lm} = -i\hbar\frac{\partial}{\partial\phi}Y_{lm}(\theta,\phi) = m\hbar Y_{lm}(\theta\phi) \tag{11.35}$$

Obviously, the spherical harmonics $Y_{lm}(\theta, \phi)$ are eigen functions of L^2 and L_z with eigen values $l(l+1)\,\hbar^2$ and m^{th} respectively. Accordingly l is given the name *orbital quantum number*.

We may note that L^2 and L_z commute and hence they have simulataneous eigen functions. Since L_x, L_y and L_z do note commute and hence spherical harmonics Y_{lm} are note the eigen functions of L_x and L_y. We may further note that the eign functions of L_z are much simpler than those of L_x and L_y is the result of the fact that Z-axis was chosen as the axis of the polar coordinate system as shown in Fig. (11.1).

Eq. (11.35) shows that the eigen values of L_z are $m\hbar$. Thus the measurement of L_z can yield as a result only the values $0, \pm h, \pm 2\hbar$ .... Since the Z-axis points in an arbitrary chosen direction it must be true that the angular momentum about any axis is quantized (called space quantization of angular momentum) and can upon measurment only reveal one of these discrete values.

The integral m_l is given the name *magnetic quantum number* because of the part played by this number in describing the effect of a uniform magnetic field **B** on a charged paraticle moving in a central force field in which the potential energy depends on the distance r of the moving particle from the fixed point, usually the coordinate origin.

The possible value for m_l for a given value of l range from $-l$ to $+l$ through 0, so that the number of possible orientations of $\boldsymbol{L}$ in a magnetic field is $(2l+1)$. When $l=0$, that is, $m_l=0$, L_z can have only single value 0 which means $\boldsymbol{L}$ is $\perp^r$ to $\boldsymbol{B}$. When $l=1$, L_z may be $-\hbar$, 0, $\hbar$, when $l=2$, L_z may be $-2\hbar$, $-\hbar$, 0, $+2\hbar$; and so on.

Now, the question arises, why is only one component of $\boldsymbol{L}$, is quantized? The answer is simple. If $\boldsymbol{L}$, is fixed in space, say in z-direction, then the electron would be confined to a definite plane xy at all times. Therfore its z-coordinate will be zero. That amounts to infinite uncertainty in p_z ($\Delta z\, \Delta p_z = \hbar/2$) which is impossible if electron is part of an atom. Hence $\boldsymbol{L}$, precesses about Z-axis (or direction of $\boldsymbol{L}$, is constantly changing) making a cone such that its protection $L_z = m_l\hbar$ always and the average values of L_x and L_y are zero. Therefore only z-component of $\boldsymbol{L}$ will be quantized.

The above calculations show that the allowed eigenvalues of simultaneous eigen functions of the angular momentum operators L^2 and L_z are

$$L^2 : \hbar^2\, l(l+1),\ l = 0, 1/2, 1, 3/2, \ldots,$$

$$L_z : m\hbar \text{ or } m_l\,\hbar,\ m \text{ or } m_l = -l, -l+1, \cdots, l-1, l \tag{11.36}$$

11.5 CONSERVATION OF ANGULAR MOMENTUM

The symemetry of the Hamiltonian under rotations implies conservation of angular momentum. Let us consider the single-particle Hamiltonian

$$H = \frac{p^2}{2m} + V(r) \tag{11.37}$$

The potential, $V(r)$ is a function only of distance r from the origin, so it is unaffected by a rotation. Let us check that this means H commutes with the components of $\boldsymbol{L}$.

We have $\quad [L_z, p_x] = [xp_y - yp_x, p_x] = i\,\hbar p_y$

$$[L_z, p_y] = -i\hbar\, p_x$$

$$[L_z, p_y] = 0 \tag{11.38}$$

This yields $\quad [l_z, p^2] = 0 \qquad (11.39)$

Since $\boldsymbol{L}$ generate rotations about $\boldsymbol{r} = 0$ and $V(r)$ is rotationally symmetric, V must commute with $\boldsymbol{L}$:

$$[\boldsymbol{L}, V(r)] = 0 \tag{11.40}$$

Therefore, the components of $\boldsymbol{L}$ commute with the Hamiltonian

$$[H, \boldsymbol{L}] = 0 \tag{11.41}$$

11.6 GENERAL ANGULAR MOMENTUM

Number of experimental results e.g., spectra of alkali metals, anomalous Zeeman effect, stern-Gerlach experiment, etc., could only be explained by invoking the concept that the electron in an atom possesses an additional intrinsic angular momentum involving half integral quantum numbers. However, the definition of angular momentum based on classical physics led to the $(2l + 1)$ integral values of $m\hbar$, $m = 0, \pm1, \pm2, \pm3 \ldots \pm l$ for the z-component of the angular momentum. Obviously, the definition of angular momentum $\boldsymbol{L} = \boldsymbol{r} \times \boldsymbol{p}$ is not general enough to include half integral quantum numbers. Thus one may take the definition of general angular momentum as the commutation relations given by

$$[J_x, J_y] = i\hbar J_z, [J_y, J_z] = i\hbar J_x \text{ and } [J_z, J_x] = i\hbar J_y \tag{11.42}$$

we may note that $\boldsymbol{J}$ denotes the *general angular momentum*. The components of $\boldsymbol{J}$ and the operators J^2, $J_+ = J_x + iJ_y$ and $J_- = J_x - iJ_y$, obey all the commutation relations holds good for operators $\boldsymbol{L}$, L_+ and L_-.

The square of the general angular momentum $\boldsymbol{J}$ commutes with its components J_x, J_y and J_z. However, the components among themselves are non-commuting. The details are beyond the scope of this book.

11.7 ELECTRON SPIN AND PAULI THEORY

So far we have considered the electron to be a mass point or point particle. This particle possesses three degrees of freedom associated with its three coordinates x, y and z. The quantum theory that we have developed is based on the assumption that the state of the particle (or say an electron state) at a given time is characterized by a wave function $\phi(x, y, z)$. This wave function depends only on the spatial coordinates x, y and z. However, based on this theory, it is not possible to explain or interpret several experimental observations, few of them are:

(i) In Stern and Gerlach experiment, a beam of neutral silver atoms passed through a inhomogeneous (or non-uniform) magnetic field is split into two components with equal and opposite deflection from the undeviated beam. The outermost single electron in these silver atoms is in the ground state. We may note that silver atom has 47 electrons and its electronic configuration is $4d^{10}5s^1$. Obviously, this has zero angular momentum, i.e., the electron is in the s-state which corresponds to $l = 0$. This means $m_l = 0$ as m_l can take all the values from $-l$ to $+l$ through 0. This means the magnetic quantum number must be zero and there should be no deflection at all (as it is proportional to the magnetic moment μ_z or L_z and $L_z = m_l \hbar = 0$). However, the experimental observed deviation corresponds to a magnetic moment of one Bohr magneton ($\mu_B = e\hbar/2m$).

(ii) When an atom is subjected to uniform magnetic field it is observed that each of its spectral line spilts into a certain odd number of equidistant lines. The interval between these lines is found to be proportional to the magnetic field. This is called *normal Zeeman effect*. One can explain this experimental observation qualitatively with the quantum theory developed so far. However, it is observed that in the case of atoms with odd atomic number (in particular for the H-atom), the energy levels are found to be divided into an even number of Zeeman levels. This effect is called *anomalous Zeeman effect.* One cannot explain this observation on the quantum theory developed so far as this number should always be odd, being equal to $2l+1$ with l being in integer (0, 1, 2, ...).

In order to resolve the said difficulties Uhlenbeck and Goudsmit proposed the electron spin hypothesis as follows:

(i) an electron has an intrinsic angular momentum (not associated with the orbital motion) called *spin* or *spin angular momentum*

$$\boldsymbol{S} = \frac{1}{2}\hbar \quad \text{and}$$

(ii) an intrinsic magnetic momentum is associated with this angular momentum,

$$\mu_s = \frac{e}{m}\boldsymbol{s}\left(= \frac{e}{m} \times \frac{1}{2}\hbar = \frac{e\hbar}{2m} = \mu_B \right)$$

Moreover, this magnetic moment can have only two values (projections).

We may note that the above is more precisely stated by Pauli. We may note that the coefficient of proportionality between the angular momentum and the magnetic moment is twice as large as $\mu_l = -\frac{e}{2m}\boldsymbol{L}$. Obviously, this means that *spin gyromagnetic ratio* is twice the *orbital gyromagnetic ratio.*

Dirac while developing *relativistic quantum mechanics,* has shown that associated with the electron is a spin angular momentum $\boldsymbol{S} = \frac{1}{2}\hbar$. Clearly, spin has no classical analogue. We may note that spin does not depend upon $\boldsymbol{r}$ and $\boldsymbol{p}$ variables.

Now, the spin-(1/2) system (electron) has four degrees of freedom, three position coordinates (*x, y, z*) and the fourth observable pertaining to spin. Now taking the *Z*-component S_z as the fourth observable, the electron wave function can be written as $\psi(\boldsymbol{r}, S_z)$ or $\psi(\boldsymbol{r}, m_s)$, the coordinates m_s takes the values +1/2 or −1/2. When the interaction between the spin and space parts is negligible, one can write the wave function as

$$\Psi(\boldsymbol{r}, m_s) = \phi(\boldsymbol{r})\,\chi\,(m_s) \tag{11.43}$$

where $\phi(\boldsymbol{r})$ represents the part that depends on space coordinates and $\chi(m_s)$ the part that depends on the spin coordinates.

One can easily obtain the eigenvectors of the spin materices S_x, S_y and S_z by writing the eigen value equations. Since the spin matrices are 2 × 2 the eigenvectors must be coulmn vectors with two components. We have the eigen value equation for S_z with eigen value $\hbar/2$ as

$$\frac{1}{2}\hbar\begin{pmatrix} 1 & 0 \\ 0 & -1 \end{pmatrix}\begin{pmatrix} a_1 \\ a_2 \end{pmatrix} = \frac{1}{2}\hbar\begin{pmatrix} a_1 \\ a_2 \end{pmatrix}$$

or

$$\begin{pmatrix} a_1 \\ -a_2 \end{pmatrix} = \begin{pmatrix} a_1 \\ a_2 \end{pmatrix}$$

This means $a_2 = 0$. The normalization condition yields

$$|a_1|^2 = 1 \text{ or } a_1 = 1$$

The eigen vector of the matrix S_z corresponding to the eigen value $\hbar/2$ is clearly $\begin{pmatrix}1\\0\end{pmatrix}$.

Similarly, one can easily show that the eigen vector for the eigen value $\hbar/2$ is $\begin{pmatrix}0\\1\end{pmatrix}$.

We denote these eigen vectors by α and β which are usually called the *spin-up* and *spin-down* states respectively, i.e.,

$$\propto = \begin{pmatrix}1\\0\end{pmatrix} \text{ and } \beta = \begin{pmatrix}0\\1\end{pmatrix}$$

We may note that the two component eigen vectors of spin-(1/2) particles are also called as *spinors*. One can find the eigen vectors of S_x and S_y in the same way. Table 11.1 lists the spin matrices of a spin-(1/2) system along with eigen values and eigen vectors.

Example 11.1: Show that $L^2 = \frac{1}{2}(L_+ L_- + L_- L_+) + L_z^2$

where $L_+ = L_x + iL_y$ and $L_- = L_x - iL_y$

Solution We have

$$L_+ L_- = (L_x + iL_y)(L_x - iL_y)$$

$$= L_x^2 + L_y^2 + iL_y L_x - iL_x L_y$$

and

$$L_- L_+ = (L_x - iL_y)(L_x + iL_y)$$

$$= L_x^2 + L_y^2 - iL_y L_x + iL_x L_y$$

or

$$L^2 = L_x^2 + L_y^2 + L_z^2 = \frac{1}{2}(L_+ L_- + L_- L_+) + L_z^2$$

Example 11.2: Show that

$$L^2 = L_+ L_- + L_z^2 - \hbar L_z$$

Table 11.1 Spin matrices S_x, S_y and S_z of a spin $-1/2$ system. Their eigen values and eigen-vectors are also given

Spin component	*Spin matrix*	*Eigen value*	*Eigen vectors*
S_x	$\frac{\hbar}{2}\begin{pmatrix}0 & 1\\1 & 0\end{pmatrix}$	$\hbar/2$, $-\hbar/2$	$\frac{1}{\sqrt{2}}\begin{pmatrix}1\\1\end{pmatrix}$, $\frac{1}{\sqrt{2}}\begin{pmatrix}1\\-1\end{pmatrix}$
S_y	$\hbar\begin{pmatrix}0 & -i\\i & 0\end{pmatrix}$	$\hbar/2$, $-\hbar/2$	$\frac{1}{\sqrt{2}}\begin{pmatrix}1\\i\end{pmatrix}$, $\frac{1}{\sqrt{2}}\begin{pmatrix}1\\-i\end{pmatrix}$
S_z	$\frac{\hbar}{2}\begin{pmatrix}1 & 0\\0 & -1\end{pmatrix}$	$\hbar/2$, $-\hbar/2$	$\begin{pmatrix}1\\0\end{pmatrix}$, $\begin{pmatrix}1\\0\end{pmatrix}$

Solution We have

$$L_+L_- = L_x^2 + L_y^2 - i(L_x L_y - L_y L_x)$$
$$= L_x^2 + L_y^2 + \hbar L_z$$

Hence
$$L_+L_- + L_z^2 = L_x^2 + L_y^2 + L_z^2 + \hbar L_z$$
$$= L^2 + \hbar l_z$$

or
$$L^2 = L_+L_- + L_z^2 - \hbar L_z$$

Example 11.3: Show that

$$[L^2, L_\pm] = 0$$

Solution We have

$$L^2 = \frac{1}{2}(L_+L_- + L_-L_+) + L_z^2$$

Since the components of L commute with L^2, i.e., $[L^2, L_x] = 0$, $[L^2, L_y] = 0$, $[L^2, L_z] = 0$ and hence L_+ and L_- commute with L^2, i.e., $[L^2, L_\pm] = 0$

Example 11.4: For Pauli's matrices prove that:

(i) $[\sigma_x, \sigma_y] = 2i\sigma_z$
(ii) $\sigma_x\sigma_y\sigma_z = i$

Solution We have

$$S = \frac{1}{2}\hbar\sigma \quad \text{and} \quad [S_x, S_y] = i\hbar S_z$$

(i) Now substituting the values of S_x, S_y, S_z one obtains

$$\left[\frac{1}{2}\hbar\sigma_x, \frac{1}{2}\hbar\sigma_y\right] = i\hbar\frac{1}{2}\hbar\sigma_z \quad \text{or} \quad [\sigma_x, \sigma_y] = 2i\,\sigma_z$$

(ii) We have

$$\sigma_x\sigma_y\sigma_z = \begin{pmatrix} 0 & 1 \\ 1 & 0 \end{pmatrix}\begin{pmatrix} 0 & -i \\ i & 0 \end{pmatrix}\begin{pmatrix} 1 & 0 \\ 0 & -1 \end{pmatrix}$$
$$= \begin{pmatrix} i & 0 \\ 0 & -i \end{pmatrix}\begin{pmatrix} 1 & 0 \\ 0 & -1 \end{pmatrix}$$
$$= \begin{pmatrix} i & 0 \\ 0 & i \end{pmatrix} = i$$

SUMMARY

1. In quantum mechanics, an angular momentum operator $\boldsymbol{J}$ is defined by the following commutation relations. These relations are satisfied by its components: $[J_x, J_y] = i\hbar\, J_z$, (cyclic)
2. $[J^2, J] = 0$, i.e. J^2 commutes with angular momentum J. So J^2 and J have simultaneous eigen functions and are simultaneously measurable.

3. An electron has an intrinsic angular momentum (not associated with the orbital motion) called spin or spin angular momentum

$$\boldsymbol{S} = \frac{1}{2}\hbar$$

and an intrinsic magnetic moment $\mu_S = \frac{e}{m}\boldsymbol{S} = \mu_B$ is assovciated with this angular momentum. Further, this magnetic moment can have two values, $\pm\mu_B$.

REVIEW QUESTIONS

1. State the commutation relations obeyed by the components of angular momentum.
2. The definition of angular momentum given by $\boldsymbol{L} = \boldsymbol{r} \times \boldsymbol{p}$ is not a general one. Explain why? Define a general angular momentum.
3. What do you understand by "spin" of an electron?
4. Evaluate the following commutators:
   (i) $[L_x, [L_y, L_z]]$ (ii) $[L_x^2, L_y^2]$ (iii) $[L_y^2, L_x]$
5. Prove that spin matrices S_x and S_y have $\pm\hbar/2$ eigen values.
6. Show that
   (i) $[L_+, L_-] = 2\hbar L_z$ (ii) $[L_z, L_-] = -i\hbar L_-$
7. Compute the length of the angular momentum vectors and represent the orbital motion of an electron in a state with $l = 1$ and in another state with $l = 2$.
   [**Hint**: $|L| = \sqrt{l(l+1)}\hbar = \sqrt{2}\,\hbar$ for $l = 1$ and $|L| = \sqrt{l(l+1)}\,\hbar = \sqrt{6}\,\hbar$ for $l = 2$]
8. What are the possible z-components of the vector $\boldsymbol{L}$ which represents the orbital angular momentum of a state with $l = 2$?
   [**Hint:** The possible m_l values for $l = 2$ are +2, +1, 0, −1, −2 and so the $\boldsymbol{L}$ vector can have any five possible z-components: $L_z = 2\hbar, 1\hbar, 0, -1\,\hbar$ or $-2\hbar$. The length of the vector $\boldsymbol{L}$ is $\sqrt{6}\,h$]
9. What does it mean that $\boldsymbol{L}$ precesses about the z-axis? Can we observe the precession?

PROBLEMS

1. Find the value of the uncertainty product $(\Delta L_x)(\Delta L_y)$ in a representation in which L^2 and L_z have simultaneous eigen functions. $[\geq \frac{m\hbar^2}{2}]$
2. Show that the commutator of L_x with L_y is, $[L_x, L_y] = i\hbar\, L_z$

FILL IN THE BLANKS

1. In quantum mechanics the linear momentum ($\boldsymbol{p}$) of the particle is replaced by the differential operator, _______. **$[\boldsymbol{p} = -i\hbar\nabla]$**
2. Eigen values of L^2 are _______. **$[L^2\phi = l(l+1)\,\hbar^2\phi]$**
3. The magnetic moment of a current loop is given by $\mu_l = -\frac{e}{2m}\boldsymbol{L}$, the quantity $-e/2m$ is called the _______. **[gyromagnetic ratio]**
4. Spin angular momentum of a electron is _______. **$[\boldsymbol{S} = \frac{1}{2}\hbar]$**

12

Statistical Physics

12.1 INTRODUCTION

Statistical mechanics is the branch of physics which attempts to relate the macroscopic properties of an assembly of particles to the microscopic properties of particles themselves. This branch of physics is not concerned with the actual motions or interactions of individual particles, but investigates instead their *most probable* behaviour. Although statistical mechanics cannot help us to determine the life history of a particular particle but it is able to inform us of the likelihood that a particle (exactly which one particle we cannot know in advance) has certain position and momentum at a certain instant. Since so many phenomena in the physical world involve assemblies of particles, the value of a statistical rather than deterministic approach is clear. One can apply statistical mechanics with equal facility to classical problems, e.g., molecules of a gas and quantum mechanical problems, eg., free electrons in a metal or photons in a box. Obviously, statistical mechanics is one of the most powerful tools of theoretical physics.

One can classify the study of statistical mechanics under two categories: (i) Classical statistics or *Maxwell-Boltzmann (MB) statistics* and (ii) *Quantum statistics*. The former utilizes the classical results or Maxwell's law of molecular velocities and Boltzmann theorem relating entropy and probability. *MB statistics is used in case of identical particles of any spin that are sufficiently widely separated to be distinguished*, for e.g., the molecules of gas. *MB* statistics explains successfully the phenomena like temperature, pressure, energy etc., of a gas but fails to explain experimentally observed facts, e.g., blackbody radiation, photoelectric effect, specific heat at low temperature etc. Bose and Einstein (*BE*), Fermi and Dirac (FD) developed *quantum statistics*. Bose-Einstein statistics is applicable to particles which have zero or integer spin, whereas Fermi-Dirac statistics is applicable to all fundamental particles with spin one-half. The particles obeying BE statistics are known as *bosons* and they do not obey the Pauli exclusion principle. Examples of bosons are photons (spin 1), neutral helium atoms in the ground state (spin 0) and α-particles (spin 0). The fundamental particles obeying Fermi-Dirac statistics are known as *fermions* and these particles obey the Pauli exclusion principle. Electrons, protons, neutrons etc., are the examples of fermions.

12.2 PHASE SPACE

One can completely specify the state of a system of particles classically at a particular instant if the position and momentum of each of its constitutent particles are known. Since position and momentum are vectors with three components a piece, one must know six quantities, x, y, z, p_x, p_y, p_z for each particle.

We know that position of a particle is a point having the coordinates x, y, z in ordinary three-dimensional space. It is more convenient to generalize this conception by imagining a six-dimensional space in which a point has six coordinates x, y, z, p_x, p_y, p_z. Obviously a point in such a space represents position as well as momentum and this combined position and momentum space is called *phase space*. We have introduced the notion of phase space to enable one to develop statistical mechanics in a geometrical frame work there by permitting a simpler and more straight forward method of analysis than an equivalent one wholly abstract in character. Phase of the motion of a particle, i.e., position and momentum of the particle can be represented by a point in the phase space. We may note that a point in phase space corresponds to a particular position and momentum, while a point in ordinary space corresponds to a particular position only. Fig. (12.1) illustrates various situations.

Obviously, every particle is completely specified by a point in phase space, and the state of a system of particles corresponds to a certain distribution of points in phase space.

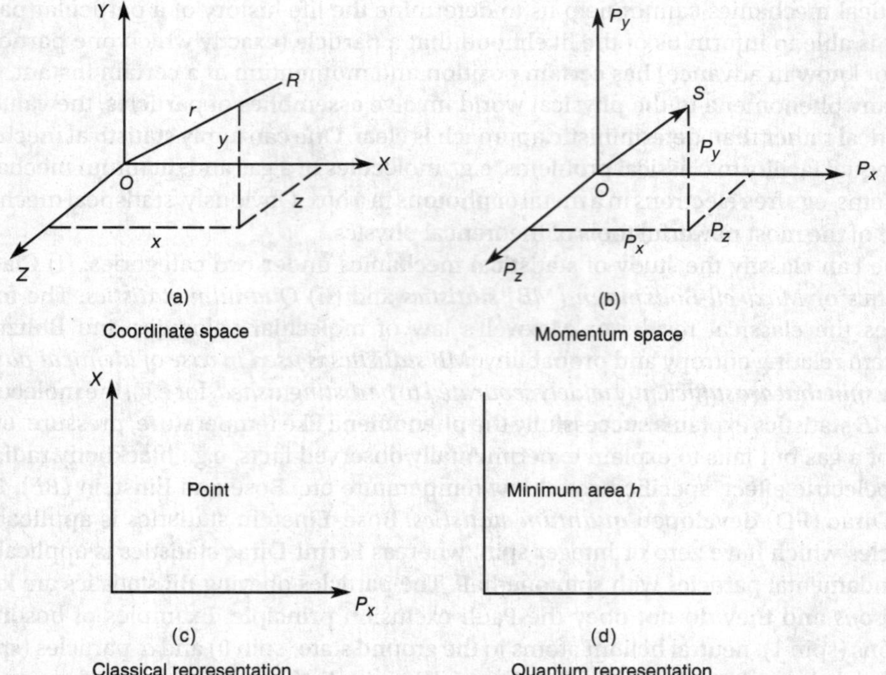

Fig. 12.1 (a) Coordinate space **(b)** Momentum space **(c)** classical representation **(d)** quantum representation

The uncertainty principle complies one to elaborate what we mean by a "point" in phase space. Let us divide phase space into tiny six-dimensional cells whose sides are dx, dy, dz, dp_x, dp_y, dp_z. As one reduces the size of the cells, one approaches more or more closely to the limit of a point in phase space. However, the volume of each of these cells is

$$\tau = dx\,dy\,dz\,dp_x\,dp_y\,dp_z$$

and in accordance with uncertainty principle, we have

$$dx\,dp_x \geq \hbar$$
$$dy\,dp_y \geq \hbar$$
$$dz\,dp_z \geq \hbar$$

clearly we have $\tau \geq \hbar^3$.

Thus a "point" in phase space is actually a cell whose minimum volume is of the order of $\hbar^3$. One can think of a particle in phase space as being located somewhere in such a cell centered at some location x, y, z, p_x, p_y, p_z instead of being precisely at the point itself.

A detailed analysis reveals that each cell in phase space actually has the volume h^3. This does not contradict the uncertainty principle argument $h^3 > \hbar^3$. In general, one can say that each cell in a phase space consising of k-coordinates and k-momenta occupies a volume of h^k. One can readily verify this statement for a particle in a one-dimensional box with infinite walls, where $k = 1$.

The task of the statistical mechanics is to determine the state of the system by investigating that how the particles are ditributed among the cells in phase space.

12.3 PROBABILITY OF A DISTRIBUTION

Let us first understand the meaning of probability? We consider the example of tossing of a coin. The probability or the chance that the head or the tail falling upwards is 1/2, i.e., 50%. If the coin is tossed for a large number of times, the chances of getting the head or tail upward are equal, i.e., 50% i.e., head would be up half equal number of times and for another half, the tail will be up. Thus the *probability (P) of an event can be defined as the ratio of number of favourable cases of observing an event to the total number of events*, i.e.,

$$P = \frac{\text{Number of cases of observing an event}}{\text{Total number of events}}$$

We must understand clearly that if a coin is tossed 1000 times it does mean that head or the tail will fall upwards exactly 500 times but its estimated value will be 500 times.

Let us now consider the tossing of two distinct coins say one made up of copper and another of silver. The probability (P) that heads or tails of both the coins would be up is only $\frac{1}{2} \times \frac{1}{2} = \frac{1}{4}$, i.e., product of probability for individual coin.

Probability distribution: The probability of distribution (W) mainly depends upon the two factors: (i) A *priori probability distribution (G)*, and (ii) *Thermodynamic probability distribution (M)*. Now we discuss these in detail.

A priori probability distribution (G)

Let us consider the case of a large box which is divided into k cells having areas a_1, a_2, a_3, ... a_k. Now suppose that without favouring any portion of the said box, we throw N balls in the box in a completely haphazard manner. We assume that all the balls are identical and distiguishable. Now, we are interested in observing that how many balls fall into each cell. We repeat the procedure a number of times. From observations, we note that a particular distribution of balls in the cells is most probable. However, this most probable distribution may or may not be found. The most probable distribution of balls in the cells is one in which number of balls in each cell is proportional to the area of the cell. Obviously, larger is the area of the cell, greater is the number of balls in it. Let A be total area of the cells, i.e.,

$$A = a_1 + a_2 + a_3 + \cdots a_k \tag{12.1}$$

then a priori probability g_i that a ball falls into the ith cell is the ratio between the area a_i of the cell and the total area A of the entire box. That is

$$g_i = \frac{a_i}{A} \tag{12.2}$$

where a_i is the area of the ith cell.

The sum of the priori probabilities for all of the cells must be 1, since the balls by hypothesis falls into the box somewhere, i.e.,

$$\sum g_i = g_1 + g_2 + g_3 + \cdots g_k = 1 \tag{12.3}$$

The probability that two balls fall in the ith cell is g_i^2, the product of the probabilities g_i that each one separately fall in. Similarly, the priori probabilities for n_i ball to fall in the ith cell is $(g_i)^{n_i}$. Now, a priori probability G of any particular *distribution* of the N balls among the k cells is therefore the product of k probabilities of the form $(g_i)^{n_i}$. Thus we have

$$G = (g_1)^{n_1}(g_2)^{n_2}(g_3)^{n_3}\cdots(g_k)^{n_k} \tag{12.4}$$

subject to the condition that the total number of balls equal to N i.e.,

$$\sum n_i = n_1 + n_2 + n_3 + \cdots n_k = N \tag{12.5}$$

We may note that when the cells are equal in size they all have the same a priori probability g and

$$G = g^N = g^{\Sigma(n_1 + n_2 + n_3 + \cdots n_k)} \tag{12.6}$$

The thermodynamic distribution (M)

In case of a priori probability, we have assumed that n_1 balls are in the first cell, n_2 balls are in the second cell, and so on through n_k balls in the cell k, there are $\lfloor n_1$, $\lfloor n_2$, $\lfloor n_3$, ... $\lfloor n_k$ irrelevant permutations. There may be number of ways in which one can achieve this configuration. This is why, it is essential to introduce the idea of thermodynamical probability which concerns the number of distinct arrangements of balls in the cells which correspond to a particular configuration. One can define the thermodynamic probability M of the distribution as the total number of irrelevant permutation, or

$$M = \frac{\lfloor N}{\lfloor n_1, \lfloor n_2, \lfloor n_3, \cdots \lfloor n_k} \tag{12.7}$$

Consider the case of n_1 balls out of total N balls that fall in the first cell. One can achieve this in the following way

$$^{N}C_{n_1} = \frac{\lfloor N}{\lfloor n_1 \lfloor N - n_1} \tag{12.8}$$

Now, any n_2 balls out of the remaining $(N - n_1)$ balls may fall in the second cell. One can achieve this in the following way

$$^{(N-n_1)}C_{n_2} = \frac{\lfloor N - n_1}{\lfloor n_2 \lfloor N - n_1 - n_2} \tag{12.9}$$

Obviously, the total number of ways of choosing n_1 and n_2 balls for first and second cells will be

$$\frac{\lfloor N}{\lfloor n_1 \lfloor (N - n_1)} \times \frac{\lfloor N - n_1}{\lfloor n_2 \lfloor (N - n_1 - n_2)} = \frac{\lfloor N}{\lfloor n_1 \lfloor n_2 \lfloor (N - n_1 - n_2)} \tag{12.10}$$

Similarly one can find the total number of ways choosing the balls for k^{th} cell. We have

$$\frac{\lfloor N}{\lfloor n_1 \lfloor (N - n_1)} \times \frac{\lfloor (N - n_1)}{\lfloor n_2 \lfloor N - n_1 - n_2)} \times \cdots \times \frac{\lfloor (n - n_1 - n_2 \cdots - n_{k-1})}{\lfloor n_k \lfloor (N - n_1 - n_2 \cdots - n_k)}$$

$$= \frac{\lfloor N}{\lfloor n_1 \lfloor n_2 \cdots \lfloor n_k} \quad (\because n_1 + n_2 + n_3 + \cdots n_k = N) \tag{12.11}$$

Now, the total probability W of the distribution is the product GM of a priori probability [Eq. (12.4)] and the thermodynamic probability Eq. (12.7), we have

$$W = \frac{\lfloor N}{\lfloor n_1 \lfloor n_2 \cdots \lfloor n_k} (g_1)^{n_1} (g_2)^{n_2} (g_3)^{n_3} \cdots (g_k)^{n_k} \tag{12.12}$$

To verify that Eq. (12.12) is correct, we add together the probabilities for all possible distributions and obtain

$$\sum W = \sum \frac{\lfloor N}{\lfloor n_1 \lfloor n_2 \cdots \lfloor n_k} (g_1)^{n_1} (g_2)^{n_2} (g_3)^{n_3} \cdots (g_k)^{n_k}$$

$$= (g_1 + g_2 + g_3 + \cdots g_k)^N$$

[With the help of *Multinomial theorem* of algebra (a generalization of binomial theorem)]. Since from Eq. (12.3), the sum of the priori probabilities g_i is 1 and $1^N = 1$, the sum of the probabilities for all possible distributions is 1. The meaning of

$$\sum W = 1$$

is that all N balls are certain to fall in the box some where which agrees with our initial hypothesis and also confirms the correctness of Eq. (12.12).

12.4 MAXWELL-BOLTZMANN DISTRIBUTION LAW

Let us consider the case of a vessel containing non-interacting gas molecules at a temperature T (say). The molecules of gas have same mass and are moving freely within the vessel in all possible directions. The molecules of gas acquire all possible velocities and energies because they collide with each other. We may note all the molecules of gas are individually identified. At a certain temperature, the entire gas has a certain value of energy but the different molecules of the gas have different energies. Maxwell in 1859 considered that how a fixed total amount of energy is distributed among the various members of an assembly. The expression covering this idea is called as Maxwell-Boltzmann distribution law.

Let us suppose that there are N molecules in the gas. Out of these N molecules let n_1 have energy ε_1, n_2 have energy ε_2, n_3 have energy ε_3 and so on. In this way one may divide the entire assembly of molecules into different energy states with energies $\varepsilon_1, \varepsilon_2, \varepsilon_3, \ldots \varepsilon_k$ and having $n_1, n_2, n_3, \ldots n_k$ molecules respectively. However the total number of molecules N and the total energy of gas molecules is constant and therefore, one finds

(i) For conservation of total numbers of particles in the gas

$$n_1 + n_2 + n_3 + \cdots + n_k = \sum_i n_i = N \tag{12.13}$$

(ii) For conservation of energy of gas molecules

$$n_1\varepsilon_1 + n_2\varepsilon_2 + n_3\varepsilon_3 + \cdots + n_k\varepsilon_k = \sum_i n_i\varepsilon_i = E \tag{12.14}$$

The probability W for any distribution is the product of a priori probability and the thermodynamic probability. We have

$$W = \frac{\lfloor N}{\lfloor n_1 \lfloor n_2 \cdots \lfloor n_k}(g_1)^{n_1}(g_2)^{n_2}\cdots(g_k)^{n_k} \tag{12.15}$$

or

$$W = \lfloor N \sum \frac{(g_i)^{n_i}}{\lfloor n_i} \tag{12.16}$$

where g_i is the priori probability for a molecule to have energy ε_i.

Let us now consider, which distribution yields the largest value of W. Taking natural logarithm of Eq. (12.16), one obtains

$$\log W = \log\lfloor N + \sum_i \log\left(\frac{g_i n_i}{\lfloor n_i}\right)$$

or

$$\log W = \log\lfloor N + \sum_i n_i \log g_i - \sum_i \log\lfloor n_i \tag{12.17}$$

we have *Stirling's formula*

$$\log\lfloor N = N\log N - N \qquad N >> 1 \tag{12.18}$$

Using Eq. (12.18), one obtains Eq. (12.17) as

$$\log W = N\log N - N + \sum_i n_i \log g_i - \left[\sum_i n_i \log n_i - n_i\right]$$

or $$\log W = N\log N - \sum_i n_i \log g_i - \sum_i n_i \log n_i \quad \left[\because \sum_i n_i = N\right] \tag{12.19}$$

We have an equation for log W rather than for W itself, this is no handicap since

$$(\log W)_{\max} = \log W_{\max}$$

The condition for a distribution to be most probable one is that small changes δn_i in any of the n_i's should not affect the value of W. We may note that if the n_i's were continuous variables instead of being restricted to integral values, we would express this condition in the usual way as $\partial W/\partial x_i = 0$ for maximum probability. However if changes in $\log W$ corresponding to a change in n_i of δn_i is $\delta \log W$, the condition for the maximum probability would be

$$\delta \log W = 0 \tag{12.20}$$

Using Eq. (12.20), Eq. (12.19) becomes

$$\delta \log W_{\max} = \delta(N\log N) + \sum_i \delta(n_i \log g_i) - \sum_i \delta(n_i \log n_i) = 0$$

or $$\delta \log W_{\max} = \log g_i.\delta n_i - \sum_i \log n_i \delta n_i - \sum_i n_i \delta \log n_i = 0$$

or $$\delta \log\, W_{\max} = \sum_i \log\, g_i \delta n_i - \sum \log n_i \delta n_i \tag{12.21}$$

or $$\sum_i n_i \delta \log n_i = 0^1$$

One can write Eqs. (12.13) and (12.14) as

$$\delta n_1 + \delta n_2 + \delta n_3 + \cdots + \delta n_k = \sum \delta n_i = dN = 0 \tag{12.22}$$

and $$\varepsilon_1 \delta n_1 + \varepsilon_2 \delta n_2 + \varepsilon_3 \delta n_3 + \cdots + \varepsilon_k \delta n_k = \sum_i \varepsilon_i \delta n_i = dE = 0 \tag{12.23}$$

In order to incorporate Eqs. (12.22) and (12.23) on various δn_i in Eq. (12.21), one can make use of *Lagrange's method of undetermined multipliers.* In this method one multiplies

---

[1] $\sum_i n_i \delta \log n_i = \sum_i n_i \frac{1}{n_i} \delta n_i = \sum_i \delta n_i$

Now, $\sum_i \delta n_i = \delta n_1 + \delta n_2 + \delta n_3 + \cdots + \delta n_n$

Since total number of molecules is constant and hence

$\sum_i \delta n_i = dN = 0$

Eq. (12.22) by $-\alpha$ and Eq. (12.23) by $-\beta$, where α and β are quantities independent of n_i's and add them to Eq. (12.21), one obtains

$$\sum_i -\alpha\delta n_i = -\alpha dN = 0 \tag{12.24}$$

$$\sum_i -\beta\varepsilon_i\delta n_i = -\beta dE = 0 \tag{12.25}$$

$$\delta\log W_{\max} - \alpha dN - \beta dE = 0$$

and
$$\sum(\log g_i - \log n_i - \alpha - \beta\varepsilon_i)\delta n_i = 0 \tag{12.26}$$

Eq. (12.26) will be satisfied, in general, only when each term in summation is separately zero; i.e.,

$$\log g_i - \log n_i - \alpha - \beta\varepsilon_i = 0$$

or
$$\log\left(\frac{n_i}{g_i}\right) = -\alpha - \beta\varepsilon_i$$

or
$$n_i = g_i \exp[-(\alpha + \beta\varepsilon_i)] \tag{12.27}$$

Eq. (12.27) gives the distribution of molecules over the various energy states for the most probable configuration. This is known as *Maxwell-Boltzmann* distribution.

Evaluation of constants α and β

For evaluation of α and β, one can consider the continuous distribution of energies rather than the discrete set $\varepsilon_1, \varepsilon_2, \varepsilon_3, \ldots \varepsilon_k$ considered so far. If $n(E)dE$ be the number of molecules whose energies lie between ε and $\varepsilon + d\varepsilon$, then Eq. (12.27) can be expressed as

$$n(\varepsilon)d\varepsilon = g(\varepsilon) - e^{(\alpha+\beta\varepsilon)d\varepsilon} \tag{12.28}$$

One can express the Eq. (12.28) in momentum form as

$$n(p)dp = g(p)e^{-(\alpha+\beta p^2/2m)dp} \quad (\because \varepsilon = p^2/2m) \tag{12.29}$$

where $g(p)$ is a priori probability that a molecule has momentum between p and $p + dp$. This is equal to the number of cells in phase space within which such a molecule may exist. If each cell has an infinitesimal volume h^3, then

$$g(p)dp = \frac{\int\int\int\int\int\int dx\,dy\,dz\,dp_x\,dp_y\,dp_z}{h^3}$$

where the numerator is the phase-space volume occupied by particles with the specified momenta in the range p and $p + dp$. Here

$$\int\int\int dxdydz = V$$

where V is the volume occupied by the gas in ordinary position space, and

$$\int\int\int dp_x dp_y dp_z = 4\pi^2 dp$$

where $4\pi p^2 dp$ is the volume of a spherical shell of radius p and thickness dp in momentum space. We have

$$g(p)dp = \frac{4\pi p^2 V\,dp}{h^3} \tag{12.30}$$

Making use of Eq. (12.30), Eq. (12.29) becomes

$$n(p)dp = \frac{4\pi p^2 V}{h^3}\exp[-(\alpha + \beta p^2/2m)]dp \tag{12.31}$$

Now, integrating Eq. (12.31) between the limits 0 and ∞, one obtains

$$\int_0^\infty n(p)dp = \frac{4\pi V e^{-\alpha}}{h^3}\int_0^\infty p^2 e^{-\beta p^2/2m}dp \tag{12.32}$$

We are now able to find $e^{-\alpha}$, since

$$\int_0^\infty n(p)dp = N$$

Now,

$$N = \frac{4\pi V e^{-\alpha}}{h^3} I \tag{12.33}$$

where

$$I = \int_0^\infty p^2 e^{-\beta p^2/2m}dp$$

Using the standard definite integral

$$\int_0^\infty x^2 e^{-ax^2}dx = \frac{1}{4}\left(\frac{\pi}{a^3}\right)^{1/2}$$

We obtain

$$I = \frac{1}{4}\left[\frac{\pi}{(\beta/2m)^3}\right]^{1/2} \tag{12.34}$$

Using Eq. (12.34), Eq. (12.33) becomes

$$N = \frac{4\pi V e^{-\alpha}}{h^3}\frac{1}{4}\left[\frac{\pi}{(\beta/2m)^3}\right]^{1/2}$$

$$= \frac{e^{-\alpha}V}{h^3}\left(\frac{2\pi m}{\beta}\right)^{3/2}$$

Hence

$$e^{-\alpha} = \frac{Nh^3}{V}\left(\frac{\beta}{2\pi m}\right)^{3/2} \tag{12.35}$$

Now, Eq. (12.31) takes the form

$$n(p)dp = 4\pi N\left(\frac{\beta}{2\pi m}\right)^{3/2} p^2 e^{-\beta p^2/2m}dp \tag{12.36}$$

In order to evaluate β, we compute the total energy E of the assembly of the molecules. Since

$$p^2 = 2m\varepsilon \quad \text{and} \quad dp = \frac{m d\varepsilon}{\sqrt{2m\varepsilon}}$$

One can write Eq. (12.36) in the form

$$n(\varepsilon)d\varepsilon = 4\pi N\left(\frac{\beta}{2\pi m}\right)^{3/2} 2m\varepsilon\, e^{-\beta\varepsilon}\frac{m d\varepsilon}{\sqrt{2m\varepsilon}}$$

$$= \frac{2N\beta^{3/2}}{\sqrt{\pi}}\sqrt{\varepsilon}e^{-\beta\varepsilon}d\varepsilon \tag{12.37}$$

The total energy E is

$$E = \int_0^\infty \varepsilon n(\varepsilon)d\varepsilon$$

$$= \frac{2N\beta^{3/2}}{\sqrt{\pi}}\int_0^\infty \varepsilon^{3/2}e^{-\beta\varepsilon}d\varepsilon$$

$$= \frac{2N\beta^{3/2}}{\sqrt{\pi}}\frac{3}{4\beta^2}\sqrt{\frac{\pi}{\beta}} = \frac{3}{2}\frac{N}{\beta} \tag{12.38}$$

Where we have made use of the definite integral

$$\int_0^\infty x^{3/2}e^{-\alpha x}dx = \frac{3}{4a^2}\left(\frac{\pi}{a}\right)^{1/2}$$

According to kinetic theory of gases, the total energy E of N molecules of an ideal gas at the absolute temperature T is

$$E = \frac{3}{2}NkT \tag{12.39}$$

where $k = 1.380 \times 10^{-23}$ J/mol-deg is the Boltzmann constant.

Comparing Eqs. (12.38) and (12.39), one obtains

$$\beta = 1/kT \tag{12.40}$$

Using Eq. (12.40) in Eq. (12.37), one can write the Maxwell-Botzmann distribution law in the final form as

$$n(\varepsilon)d\varepsilon = \frac{2\pi N}{(\pi kT)^{3/2}}\sqrt{\varepsilon}e^{-\varepsilon/kT}d\varepsilon \tag{12.41}$$

Eq. (12.41) represents the number of molecules with energies between ε and $\varepsilon + d\varepsilon$ in a sample of an ideal gas that contains a total of N molecules and whose absolute temperture is T.

Maxwell-Boltzmann Distribution for molecular speeds

One can obtain the distribution for molecular speeds from Eq. (12.41) by substituting

$$\varepsilon = \frac{p^2}{2m} = \frac{1}{2}mv^2 \quad \text{and} \quad d\varepsilon = mvdv$$

$$n(p)dp = \frac{2\pi N}{(\pi kT)^{3/2}} \frac{p}{\sqrt{2m}} e^{-p^2/2mkT} \frac{p}{m} dp$$

$$= \frac{\sqrt{2}\pi N}{(\pi kT)^{3/2}} p^2 e^{-p^2/2mkT} dp \qquad (12.42)$$

Eq. (12.42) gives the number of molecules having momenta between p and $p + dp$.

Again substituting ε in terms of velocity in Eq. (12.41), one obtains the Botzmann distribution of speeds as

$$n(v)dv = \frac{\sqrt{2}\pi Nm^{3/2}}{(\pi kT)^{3/2}} v^2 e^{-mv^2/2kT} dv \qquad (12.43)$$

The above equation gives the number of molecules having speeds between v and $v + dv$.

12.5 MEAN, ROOT MEAN SQUARE AND MOST PROBABLE VELOCITIES

At a given temperature, the molecules of a gas are in constant haphazard motion. The *mean velocity* of gas molecules must be zero, but their mean speed will have a certain value. We will call the mean speed as mean velocity.

Let us consider that $f(v)dv$ represents the probability of gas molecules having its total velocity in the range v and $v + dv$. By definition we have,

$$f(v)dv = \frac{n(v)dv}{N} \qquad (12.44)$$

Now, substituting the value of $n(v)dv$ from Eq. (12.43), one obtains

$$f(v)dv = \frac{\sqrt{2}\pi m^{3/2}}{(\pi kN)^{3/2}} v^2 e^{-mv^2/2kT} dv \qquad (12.45)$$

The mean velocity of the molecule is now given as

$$\bar{v} = \int_0^\infty v f(v) dv \qquad (12.46)$$

Since theoretically the gas molecules can take all possible velocities and hence we have taken the limits of integration from 0 to ∞.

Now, substituting the value of $f(v)dv$ from Eq. (12.45) in Eq. (12.46), one obtains

$$\bar{v} = \int_0^\infty 4\pi \left(\frac{m}{2\pi kT} \right)^{3/2} e^{-mv^2/2kT} v^2 dv$$

or

$$\bar{v} = 4\pi \left(\frac{m}{2\pi kT} \right)^{3/2} \int_0^\infty e^{-mv^2/2kT} v^2 dv$$

or $$\bar{v} = 4\pi\left(\frac{m}{2\pi kT}\right)^{3/2}\frac{1}{2}\left(\frac{2kT}{m}\right)^2 \quad \left[\because \int_0^\infty e^{-ax^2}x^2 dx = \frac{1}{2a^2}\right]$$

or $$v = \sqrt{\frac{8kT}{\pi m}} = 1.59\sqrt{\frac{kT}{m}} \tag{12.47}$$

We may note that Eq.(12.47) represents the *average velocity* of a gas molecule. $\bar{v}$ is independent of the molecule's mass. The value of $\bar{v}$ at room temperature is about 0.04eV. One can obtain the *mean square velocity* ($\bar{v}^2$) in a similar way. Thus

$$\bar{v}^2 = \int_0^\infty v^2 f(v)dv$$

$$= \int_0^\infty 4\pi\left(\frac{m}{2\pi kT}\right)^{3/2} e^{-mv^2/2kT}v^4 dv$$

$$= 4\pi\left(\frac{m}{2\pi kT}\right)^{3/2}\frac{3}{8}\sqrt{\pi}\left(\frac{2kT}{m}\right)^{5/2} = \frac{3kT}{m} \tag{12.48}$$

$$\left[\because \int_0^\infty e^{-ax^2}x^4 dx = \frac{3}{8}\frac{\sqrt{\pi}}{(a)^{5/2}}\right]$$

Now, the *root mean square velocity* (rms) of molecules will be

$$v_{\text{rms}} = \sqrt{v^2}\sqrt{\frac{3kT}{m}} = 1.73\sqrt{\frac{kT}{m}} \approx 1{,}09\bar{v} \tag{12.49}$$

so that r.m.s speed is about 9% greater than mean velocity. The *most probable velocity* (v_p) is the velocity for which $f(v)$ is maximum, i.e.,

$$\frac{d}{dv}f(v) = 0$$

or $$\frac{d}{dv}\left[4\pi\left(\frac{m}{2\pi kT}\right)^{3/2} v^2 e^{-mv^2/2kT}\right] = 0$$

or $$4\pi\left(\frac{m}{2\pi kT}\right)^{3/2}\left[2ve^{-mv^2/kT} - \frac{mv^2}{kT}e^{-mv^2/kT}\right] = 0$$

Since $4\pi\left(\frac{m}{2\pi kT}\right)^{3/2} \neq 0$, hence one finds

$$2ve^{-mv^2/kT} - \frac{mv^2}{kT}e^{-mv^2/kT} = 0$$

or $$2v_p - \frac{mv_p^2}{kT} = 0$$ (Now we have replaced v, by v_p)

or $$v_p = \sqrt{\frac{2kT}{m}} = 1.41\sqrt{\frac{kT}{m}} \tag{12.50}$$

Fig. (12.2) shows *Maxwell-Boltzmann velocity distribution*.

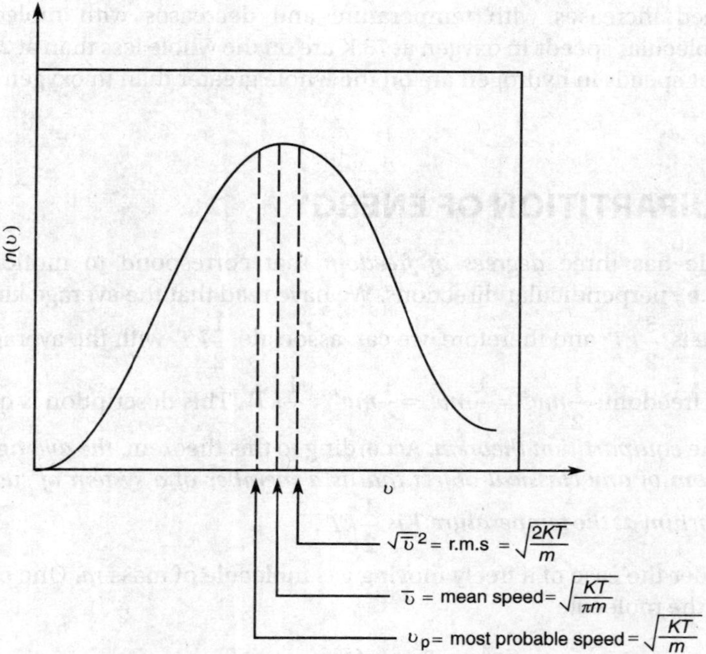

Fig. 12.2 *Maxwell-Boltzmann velocity distribution*

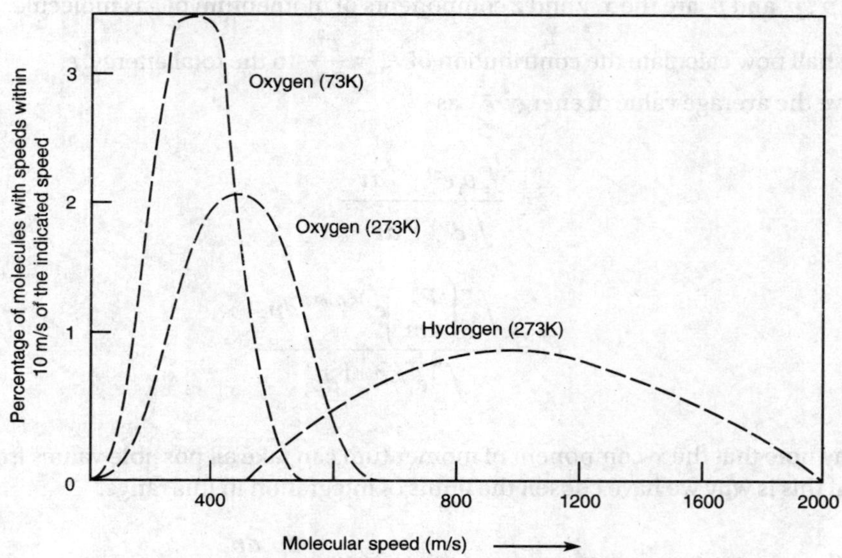

Fig. 12.3 Distribution of molecular speeds in oxygen at three different temperatures

We may note that molecular speeds in a gas vary considerably on either side of v_p. The distribution of molecular speeds in oxygen at 73 K (–200 °C), in oxygen at 273 K (0 C) and in hydrogen at 273 K is shown in Fig. (12.3). We can see that the most probable molecular speed increases with temperature and decreases with molecular mass. Accordingly molecular speeds in oxygen at 73 K are on the whole less than at 273 K, and at 273 K molecular speeds in hydrogen are on the whole greater than in oxygen at the same temperature.

12.6 EQUIPARTITION OF ENERGY

A gas molecule has three *degrees of freedom* that correspond to motions in three independent, i.e., perpendicular directions. We have read that the average kinetic energy of the molecule is $\frac{3}{2}kT$ and therefore, we can associate $\frac{1}{2}kT$ with the average energy of each degree of freedom: $\frac{1}{2}mv_x^2 = \frac{1}{2}mv_y^2 = \frac{1}{2}mv_z^2 = \frac{1}{2}kT$. This description is quite general and is called the *equipartition theorem.* According to this theorem, *the average energy per degree of freedom of any classical object that is a member of a system of such objects in thermal equilibrium at the temperature T is* $\frac{1}{2}kT$.

Let us consider the case of a freely moving gas molecule of mass *m.* One can write the total energy of the molecule

$$\varepsilon = \frac{p_x^2}{2m} + \frac{p_y^2}{2m} + \frac{p_z^2}{2m} = \varepsilon_x + \varepsilon_y + \varepsilon_z \tag{12.51}$$

where p_x, p_y and p_z are the *x, y* and *z* components of momentum of gas molecule.

We shall now calculate the contribution of $\varepsilon_x = \frac{p_x^2}{2m}$ to the total energy, ε.
We have the average value of energy $\overline{\varepsilon_x}$ as

$$\overline{\varepsilon_x} = \frac{\int_\tau \varepsilon_x e^{-\varepsilon_x/kT} d\tau}{\int_\tau e^{-\varepsilon_x/kT} d\tau} \tag{12.52}$$

or

$$\overline{\varepsilon_x} = \frac{\int_{-\infty}^{\infty} \left(\frac{p_x^2}{2m}\right) e^{-p_x^2/2mkT} dp_x}{\int_{-\infty}^{\infty} e^{-p_x^2/2mkT} dp_x} \tag{12.53}$$

We may note that the *x*-component of momentum can take all possible values from $-\infty$ to $+\infty$ and this is why we have chosen the limits of integration in this range.

Putting
$$x^2 = \frac{p_x^2}{2mkT} \quad \text{or} \quad 2xdx = \frac{2p_x dp_x}{2mkT}$$

or
$$dp_x = 2mkTx / p_x dx = \sqrt{2mkT}\, dx$$

in the integral Eq. (12.53), we obtain

$$\overline{\varepsilon_x} = \frac{\int_{-\infty}^{\infty}(kTx^2)e^{-x^2}\sqrt{2mkT}\,dx}{\int_{-\infty}^{\infty}e^{-x^2}\sqrt{2mkT}\,dx}$$

$$= kT\frac{\int_{-\infty}^{\infty}x^2e^{-x^2}\,dx}{\int_{-\infty}^{\infty}e^{-x^2}\,dx}$$

Integrating the integral in the numerator by parts, we have

$$\int_{-\infty}^{\infty}x^2e^{-x^2}\,dx = \frac{1}{2}\int_{-\infty}^{\infty}e^{-x^2}\,dx$$

$$\therefore \quad \overline{\varepsilon_x} = \frac{kT\frac{1}{2}\int_{-\infty}^{\infty}e^{-x^2}\,dx}{\int_{-\infty}^{\infty}e^{-x^2}\,dx} = \frac{1}{2}kT \tag{12.54}$$

Similarly, one can show that

$$\overline{\varepsilon_y} = \frac{1}{2}kT \quad \text{and} \quad \overline{\varepsilon_z} = \frac{1}{2}kT$$

We know that total average energy of a molecule is $\frac{3}{2}kT$. Obviously, energy is equally divided among its degrees of freedom.

We may note that degrees of freedom are not limited to linear velocity components and each variable that appeared square in the formula for the energy of a particular object represents a degree of freedom. Thus each component ω_l of angular velocity (provided it involves a moment of inertia I_l), is a degree of freedom, so that $\frac{1}{2}I_l\omega_l^2 = \frac{1}{2}kT$. Obviously, a diatomic molecule has five degrees of freedom, one each for motion in the x, y and z directions and two for rotations about axis perpendicular to its symmetry axis.

A degree of freedom is similarly associated with each component Δs_l of the displacement of an object that gives rise to a potential energy proportional to $(\Delta s_l)^2$, e.g., a one-dimensional harmonic oscillator has two degrees of freedom, one that corresponds to its kinetic energy $\frac{1}{2}mv_x^2$ and other to its potential energy $\frac{1}{2}k(\Delta x)^2$, where k is the force constant. Each oscillator in a system is in thermal equilibrium accordingly has a total average energy $(2\frac{1}{2}kT) = kT$ provided, one can disregard quantization. To a first approximation we can say that the constitutent particles (atoms, ions or molecules) of a solid behave thermally like a system of classical harmonic oscillators.

We may note that the equipartition theorem also applies to nonmechanical systems, e.g., to thermal fluctuations("noise") in electrical circuits.

12.7 QUANTUM STATISTICS

The *Maxwell-Boltzmann (MB)* distribution function holds for systems of identical particles that can be distinguished one from another, which means paricles whose wave functions do not overlap very much, i.e., identical particles which can be distinguished from one another in some way. Molecules in a gas fit this description and obey *Maxwell-Boltzmann Statistics (MBS)*. Now, if the wave functions do overlap appreciably, the situation changes because the particles cannot now be distinguished, although they can still be counted. This means that the distribution functions for the indistinguishable particles of quantum physics are different from those of classical physics. In accordance with the quantum-mechanical consequences of indistinguishability, the systems of particles with overlapping wave functions fall into two categories:

(i) Bosons: These are the particles with *zero*(0) or *integral* spins. These particles do not obey the Pauli exclusion principle, and the wave function of a system of bosons is not affected by the exchange of any pair of them. A wave function of this kind is called *symmetric*. We may note that any number of bosons can exist in the statistics dealing with bosons is called *Bose-Einstein Statistics* (*BES*). *BES* governs identical particles which cannot be distinguished apart, though they can be counted. In *BES* all quantum states are assumed to have equal a priori probabilities. A distribution function for bosons is known as the *Bose-Einstein distribution* function and has the form

$$f_{BE}(\varepsilon) = \frac{1}{Ae^{-\varepsilon/kT} - 1} \tag{12.55}$$

(ii) Fermions: These are the particles with odd half-integral spins $(\frac{1}{2}, \frac{3}{2}, \frac{5}{2}, \ldots)$. These particles obey the Pauli exclusion principle, and wave function of a system of fermions changes sign upon the exchange of any part of them. A wave function of this kind is called *antisymmetric*. Only one fermion can exist in a particular quantum state of the system. The statistics dealing with these particles is called *Fermi-Dirac Statistics (FDS)*. Obviously, *FDS* is applicable to indistinguishable particles which are governed by the Pauli exclusion principle. The distribution function for fermions is known as *Fermi-Dirac distribution* and has the form

$$f_{FD}(\varepsilon) = \frac{1}{Ae^{\varepsilon/kT} + 1} \tag{12.56}$$

Now, we shall derive distribution laws.

12.8 BOSE-EINSTEIN DISTRIBUTION LAW

Let us consider a system consisting of N indistinguishable objects. Let these objets be divided into quantum states such that there are n_1 objects in state 1 with energy ε_1, n_2 objects in state n_2 with energy ε_2, ... n_i elements in state n_2 with energy ε_i respectively. Let g_i be the degeneracy of the ith state, i.e., g_i represents the number of cells or states

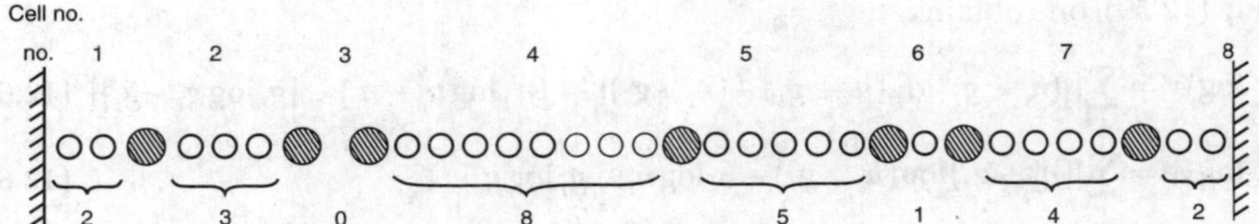

Fig. 12.4 A series of 25 indistinguishable particles separated by 7 partitions into 8 cells. Obviously, $n_i = 25$, $g_i - 1 = 7$ and $g_i = 8$

that have the same energy ε_i. Let us now consider the number of ways in which n_i indistinguishable elements can be distributed in g_i cells. The required number of different ways is just equivalent to that in which n_i indistinguishable particles are to be distinguished in a box of g_i components, without any restrictions as to the number of particles in any of the compartments.

In order to carry out the required distribution, let us consider a series of n_i white balls and $g_i - 1$ black balls placed in a line as shown in Fig. (12.4). We note that $g_i - 1$ of the balls can be regarded as partitions separating a total of g_i intervals, with the entire series therefore representing n_i balls arranged in g_i cells. In the figure $n_i = 25$ indistinguishable particles distributed into $g_i = 8$ cells. The number of partitions between cells will be $g_i - 1 = 8 - 1 = 7$(black balls). The first cell contains two particles, the second three particles third none, and so on as shown in Fig. (12.4).

There are $\lfloor(n_i + g_i - 1)$ possible permutations among $(n_i + g_i - 1)$ objects, but of these the $\lfloor n_i$ permutations of n_i particles among themselves and $\lfloor(g_i - 1)$ permutations of the $g_i - 1$ particles among themself do not affect. Clearly, there are $\dfrac{\lfloor(n_i + g_i - 1)}{\lfloor n_i \lfloor(g_i - 1)}$ possible distinguishably different arrangements of the n_i indistinguishable particles among the g_i cells.

The number of ways W, in which particles can be distributed among different states, i.e., probability W of the entire distribution of N particles is the product

$$W = \sum_i \frac{\lfloor(n_i + g_i - 1)}{\lfloor n_i \lfloor(g_i - 1)} \tag{12.57}$$

We now assume that $(n_i + g_i) >> 1$, so that one can replace $(n_i + g_i - 1)$ by $(n_i + g_i)$ and $(g_i - 1)$ by g_i. Hence,

$$W = \sum_i \frac{\lfloor(n_i + g_i)}{\lfloor n_i \lfloor g_i} \tag{12.58}$$

Taking the natural logarithm of both sides of Eq. (12.58), one obtains

$$\log W = \sum_i (\log\lfloor(n_i + g_i) - \log\lfloor n_i - \log\lfloor g_i) \tag{12.59}$$

Making use of Stirling's formula

$$\log\lfloor n = n\log n - n$$

in Eq. (12.59), one obtains

$$\log W = \sum_i [\{(n_i + g_i)\log(n_i + g_i) - (n_i + g_i)\} - \{n_i \log n_i - n_i\} - \{g_i \log g_i - g_i\}] \quad (12.60)$$

$$\text{or} \quad \log W = \sum_i [(n_i + g_i)\log(n_i + g_i) - n_i \log n_i - g_i \log g_i] \quad (12.61)$$

The condition that a particular configuration is most probable is that small changes δn_i in any individual n_i's do not affect the value of probability W. Thus if a change in log W of δ log W occurs when n_i changes by δn_i, one can express the above condition as

$$\delta \log W = 0 \quad (12.62)$$

From Eq. (12.61), one obtains

$$\begin{aligned} \delta \log W &= \sum_i [\delta\{(n_i + g_i)\log(n_i + g_i)\} - \delta\{n_i \log n_i\} - \delta g_i \log g_i\}] \\ &= \sum_i [\log(n_i + g_i)\delta n_i + \delta n_i - \log n_i \delta n_i - \delta n_i] \quad (\because \delta g_i = 0) \\ &= \sum_i [\log(n_i + g_i) - \log n_i]\delta n_i = 0 \end{aligned} \quad (12.63)$$

Using Eq. (12.62), one obtains

$$\sum_i [\log(n_i + g_i) - \log n_i]\delta n_i = 0 \quad (12.64)$$

We incorporate the conservation of particles expressed in the form

$$\sum \delta n_i = 0 \quad (12.65)$$

and the conservation of energy, expressed in the form

$$\delta \sum_i n_i \varepsilon_i = \sum \varepsilon_i \delta n_i = 0 \quad (12.66)$$

Now, we are in a position to incorporate the conditions of conservations of number of particles and energy in Eq. (12.64). For this purpose, one can use the method of undetermined multipliers. We multiply Eq. (12.65) by $-\alpha$ and Eq. (12.66) by $-\beta$ and adding these two equations in Eq. (12.64), one obtains

$$\sum_i [\log(n_i + g_i) - \log n_i - \alpha - \beta\varepsilon_i]\delta n_i = 0 \quad (12.67)$$

Since the δn_i's are independent, the quantity in bracket must vanish for each value of i. Thus

$$[\log(n_i + g_i) - \log n_i - \alpha - \beta\varepsilon_i] = 0 \quad (12.68)$$

$$\text{or} \quad \log\frac{(n_i + g_i)}{n_i} - \alpha - \beta\varepsilon_i = 0 \quad (12.69)$$

$$1 + \frac{g_i}{n_i} = \exp(\alpha + \beta\varepsilon_i)$$

$$\text{or} \quad \frac{g_i}{n_i} = \exp(\alpha + \beta\varepsilon_i) - 1$$

or $$n_i = \frac{g_i}{e^{(\alpha+\beta\varepsilon_i)} - 1} \tag{12.70}$$

$$= \frac{g_i}{e^{(\alpha+\varepsilon_i/kT)} - 1} \quad (\because \beta = 1/kT) \tag{12.71}$$

$$= \frac{g_i}{e^{\alpha} e^{\varepsilon_i/kT} - 1} \tag{12.72}$$

$$= \frac{g_i}{Ae^{\varepsilon_i/kT} - 1} \tag{12.73}$$

where $A = e^{\alpha}$. Eqs. (12.70) to (12.73) represents the *most probable* distribution of particles among various energy levels for a system of particles obeying Bose-Einstein statistics.

12.8.1 Planck's Radiation Law

Let us consider the case of a black body, i.e., a cavity in some opaque material. In the case of Maxwell-Boltzmann statistics, we treat the cavity as a box filled with electromagnetic radiation, but now we consider the radiation to be in the form of photons. We assume the box to be filled with a 'gas' of photons. Photons have spin 1/2, so they are *bosons* and obey BE statistics. Accordingly, we have

$$n_i = \frac{g_i}{e^{(\alpha+\varepsilon_i/kT)} - 1} \tag{12.74}$$

In the derivation of expression Eq. (12.74), we assumed $\sum_i \delta n_i = 0$ or $\alpha \sum_i \delta n_i = 0$ because the number of particles was to be considered constant. Because photons are continuously created and destroyed as radiation is emitted or absorbed by the walls of the cavity, the total number of particles is not constant, and the parameter $A = e^{\alpha}$ loses its significance, i.e.,

$$\sum_i \delta n_i \neq 0$$

when $$\sum_i \delta n_i \neq 0 \text{ means } \alpha = 0$$

Eq. (12.74) reduces to

$$n_i = \frac{g_i}{e^{\varepsilon_i/kT} - 1} \tag{12.75}$$

Let us consider the distribution of photons in a small interval of frequency lying between ν and $\nu + d\nu$, in a cavity at absolute temperature T, we have

$$n(\nu)d\nu = \frac{n(\nu)d\nu}{e^{\varepsilon_i/kT} - 1} \tag{12.76}$$

Now, we calculate $g(\nu)d\nu$.

The number of states $g(p)$ in which a photon can have momentum p and $p + dp$ is given by

$$g(p)dp = 2 \times \frac{4\pi p^2 V dp}{h^3} \tag{12.77}$$

We have multiplied by 2 because the photons can have two different directions of polarisation, i.e., circularly clockwise and circularly counter clockwise.

For photons $p = h\nu/c$ and $dp = hd\nu/c$

$$\therefore \qquad p^2 dp = \left(\frac{h\nu}{c}\right)^2 \times \frac{hd\nu}{c} = \frac{h^3\nu^2 d\nu}{c^2}$$

Substituting this value in Eq. (12.77) and expressing $g(p)dp$ in terms of $g(\nu)d\nu$, one obtains

$$g(\nu)d\nu = \frac{8\pi V}{h^3} \times \frac{h^3\nu^2 d\nu}{c^3} = \frac{8\pi V}{c^3}\nu^2 d\nu \tag{12.78}$$

Substituting $g(\nu)d\nu$ from Eq. (12.78) in Eq. (12.76), one obtains

$$n(\nu)d\nu = \frac{8hV}{c^3}\frac{\nu^2 d\nu}{e^{h\nu/kT} - 1} \qquad (\because \varepsilon_i = h\nu) \tag{12.79}$$

The corresponding spectral density is given be

$$\varepsilon(\nu)d\nu = h\nu\frac{n(\nu)d\nu}{V}$$

$$= h\nu\left(\frac{8\pi V}{c^3 V}\right)\left(\frac{\nu^2 d\nu}{e^{h\nu/kT} - 1}\right)$$

$$= \frac{8\pi h\nu^3 d\nu}{c^3}\frac{1}{e^{h\nu/kT} - 1} \tag{12.80}$$

which is *Planck's law of radiation* and agrees with the experimental findings.

Although Planck got the right formula of radiation, his derivation is from quantum mechanics point of view seriously flawed. We now know from quantum mechanics that the harmonic oscillators in the cavity walls have the energy $\varepsilon_n = (n + \frac{1}{2})h\nu$ and not $nh\nu$. Including the zero-point energy if $\frac{1}{2}h\nu$ does not lead to average energy, then

$$\bar{\varepsilon} = \frac{h\nu}{e^{h\nu/kT} - 1} \tag{12.81}$$

does not lead to average energy of Eq. (12.81) when *MB* statistics are used. The proper procedure is to consider the electromagnetic waves in a cavity as photon gas subject to *Bose-Einstein statistics (BES)*, since the spin of photon is 1. The average number of photons $f(\nu)$ in each state of energy $E = h\nu$ is therefore given by the *BE* distribution function of Eq. (12.74).

The value of α in Eq. (12.72) depends on the number of particles in the system being considered. However, the number of photons in a cavity need not be considered; unlike gas molecules or electrons, photons are created and destroyed all the time. Although the total radiant energy in a cavity at a given temperature remains constant, the number of photons that incorporate this energy can change. The non-conservation of photon means that $\alpha = 0$. Obviously, the *BE* distribution for photon is

$$f_{BE}(\nu) = \frac{1}{e^{h\nu/kT} - 1} \tag{12.82}$$

Clearly, the Planck theory of black body radiation, which was so successful in accounting for experimental results, follows directly from the *BE* distribution of photons.

12.8.2 Liquid Helium

Liquid helium is one of the most remarkable substances one can study in the laboratory. Some of the important properties of liquid helium are:

(i) Helium gas is the most inert gas of the all inert gases. Under normal conditions, helium gas forms no compounds, and it has the lowest boiling point, 4.18K, of any material.

(ii) Helium behaves much like an ordinary liquid below 4.18K. As the helium gas boils, the escaping gas forms bubbles, like a boiling pot of water. On further cooling the liquid, a sudden transition occurs at a temperature 2.18K and the violent boiling stops and the liquid becomes absolutely still. However, the evaporation continues, but only from the surface.

(iii) On further cooling the liquid below 2.18K the specific heat and the thermal conductivity both increases suddenly and discontinuously. The transition point at 2.28K is known as *Lambda* (λ) *point* transition (Fig. (12.5)). The thermal conductivity rises at the λ-point transition by a factor of perhaps 10^6.

(iv) A liquid, when flows through a capillary tube causes some resistance to the flow due to its viscosity. When liquid helium flows through a capillary tube, its viscosity drops at the lambda transition point by a factor ~10^6. Below λ transition point liquid

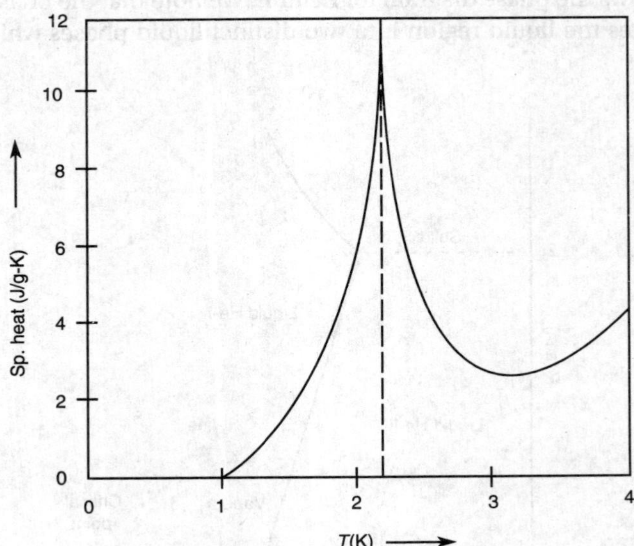

Fig. 12.5 The specific heat of liquid helium exhibit discontinunity at 2.18K, known as λ-point transition

helium flows easily through narrow capillary tube (or some narrow constrictions), through which it would not flow above the λ-point transition.

(v) Below the λ-point transition, liquid helium has the power to seemingly defy gravity, flowing up and over the walls of container. The helium forms a thin film which lines the walls of the container, the remaining liquid is then drawn up by the film like a siphon, and one can see the helium dripping from the bottom of the container.

All these strange properties of liquid helium occur because liquid helium obeys *Bose-Einstein statistics*. Ordinary helium ($_2He^4$) has two electrons filling the 1*s* shell, so the total angular momentum of electrons is zero. It happens also that the helium nucleus (α-particle) also has a spin zero. Obviously, the total spin of the atom (electron spin + nuclear spin) is zero. This means helium atom behaves like a boson. At λ-point, transition i.e., 2.1K, a change of phase takes place in liquid helium. Above λ-point helium behaves like an ordinary helium and this phase is called the He–I. Below λ-point the phase of liquid helium is called the He–II phase and its special properties are due to its *superfluid* component. When we decrease the temperature from the λ-point toward absolute zero, the relative concentration of the normal fluid decreases and that of the super fluid increases. We may note that the unusual properties of liquid helium are all caused by the superfluid component below λ-point, which is also known as *quantum liquid*. Because the helium atoms obey B-E statistics and therefore Pauli exclusion principle does not prevent all the atoms from being in the same quantum state. This begins to happen at λ-point and this effect is usually referred as *Bose-condensation*. One can think of the superfluid as being a singly quantum state made up of a very large number of atoms; the atoms behave in a cooperative way, giving the super fluid its unusual properties.

Fig. (12.6) shows the phase diagram for helium. We note that the phase diagram shows that λ-line divides the liquid region into two distinct liquid phases which are called the

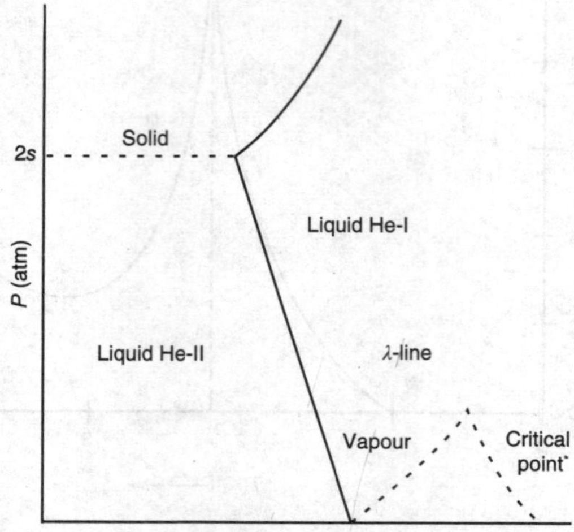

Fig. 12.6 Phase diagram

He-I and He-II. As stated above the liquid phase He-II is termed as a superfluid with following properites.

(i) Liquid He-II is a very good conductor of heat.

(ii) Heat flow in He-II is propagated in the form of temperature waves which is also known as the *second sound*. At λ-point the velocity of *second sound* is zero.

(iii) In He-II the ordinary sound velocity rises to a steady value of about 240m/s at and below 1K upto absolute zero.

(iv) The viscosity of He-II is about million times less than that of the He-I and it dispappears completely if the flow velocity is less than 10cm/s.

Recently, researchers have been interested in studying the vortex motion of superfluid helium (^{3}He). Upon rotation the superfluid tends to break into separate vortices - small regions of rotation with rotation pattern distinct from those in neighbouring regions. We may note that the vortex patterns are similar to those observed in superconductors. This discovery which connects superconductivity and superfluidity in ^{3}He is of prime importance. Superfluidity in ^{3}He was discovered in 1970s. This makes it a relatively recent discovery, compared with superfluidity in ^{4}He.

^{3}He has three distinct superfluid phases. Fig. 12.5 (a) shows two phases. A third phase can be produced only in the presence of a magnetic field (Fig. 12.5(b)). We may note that this raises an interesting point about ^{3}He. With three spin 1/2 particles in ^{3}He nucleus, it has net magnetic moment. Interestingly, the magnetic behaviour of ^{3}He, along with the fermion/boson distinction is largely responsible for the different behaviour of two isotopes. We may further note that it also helps one to distinguish between the fermion

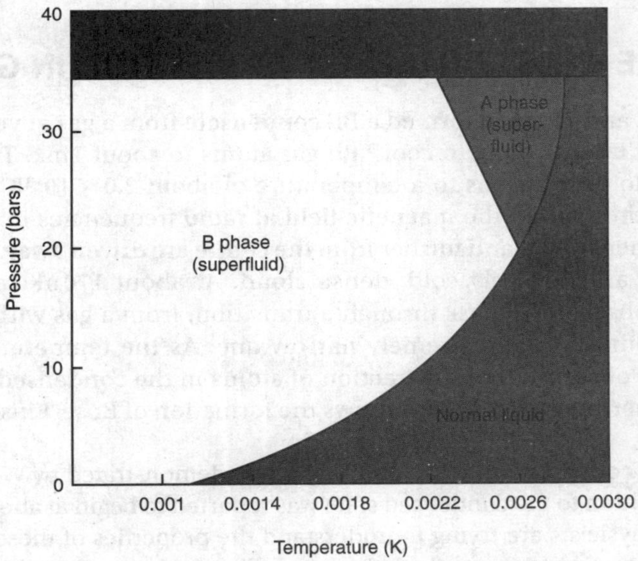

Fig. 12.5(a) Phase diagram for ^{3}He (zero applied magnetic field). There are two superfluid phases A and B. Phase A exists only at very high pressures (1 bar = 10^5 Pa $\approx$ 1 atm)

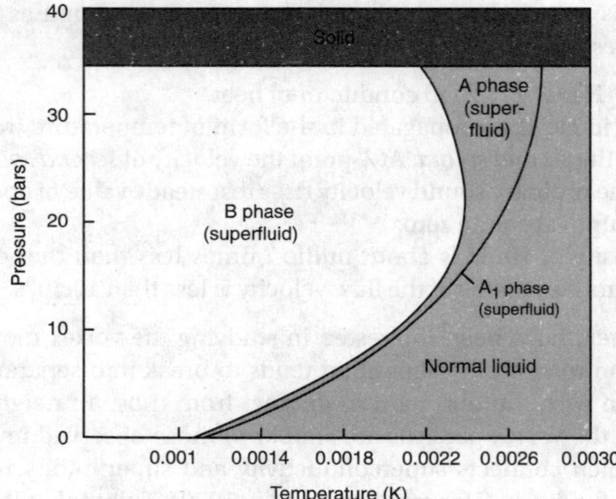

Fig. 12.5(b) Phase diagram for liquid ^{3}He in the presence of strong magnetic field

pairing mechanism in superfluid ^{3}He and fermion pairing in superconductors. In case of ^{3}He, pairing of atoms is formed in such a way as to have a net spin (the magnetic moments of two atoms tend to align with each other) and net angular momentum (atoms also tend to revolve around one another), but Cooper pairs in a superconductor have zero spin and zero angular momentum.

12.9 BOSE–EINSTEIN (BE) CONDENSATION IN GASES

In 1995 Cornell and Wieman formed a BE condensate from a gas at very temperature. First, they used laser cooling to cool ^{87}Rb gas atoms to about 1mk. Then they used a magnetic trap to cool the gas to a temperature of about 2.0×10^{-8}K (20nK). In their magnetic trap they varied the magnetic field at radio frequencies in such a way that atoms with higher speeds and further from the centre are driven away. What remained ultimately was an extremely cold, dense cloud. At about 170nK temperature, the Rb cloud was observed to pass through a transition, from a gas with normal, broad velocity distribution to an extremely narrow one. As the temperature was further lowered, it was observed that the fraction of atoms in the condensed state increased (just as in a superfluid). Fig. 12.5(c) shows the formation of Bose-Einstein condensate as gas is cooled.

In 1995, Bose condensation with sodium gas was demonstrated by W. Ketterle at MIT. For Na the transition to the condensed state was reported to begin at about 2μK.

Nowadays, physicists are trying to understand the properties of these and other Bose condensates and trying to reconcile their observations with quantum theory.

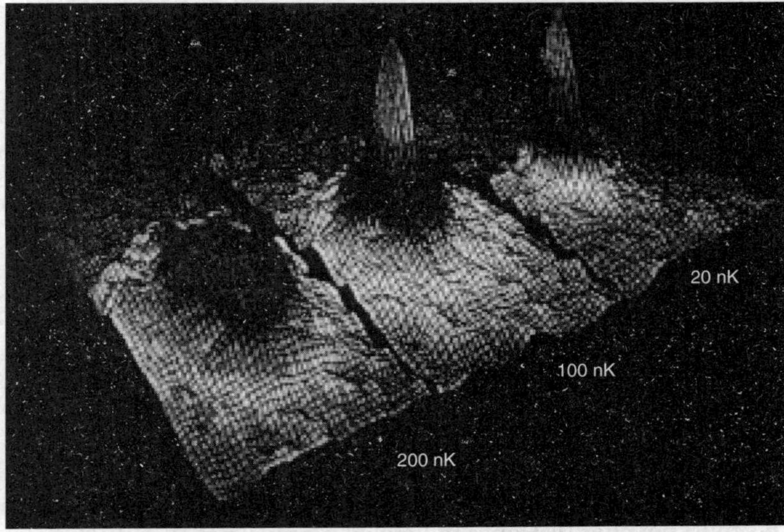

Fig. 12.5(c) Formation of Bose-Einstein condensate as a sample of atoms is cooled. Two horizontal axes represent velocity components, whereas the vertical axis represents the number of atoms having those velocities. Obviously, a higher and sharper peak around the center of the graph (at zero velocity) shows the condensation of atoms into a small space

12.10 FERMI-DIRAC (FD) DISTRIBUTION LAW

Particles of half integral spin $(\frac{1}{2}, \frac{3}{2}, \ldots)$ which obey the Pauli exclusion principle, such as nucleons or electrons are known a *fermions. Fermi-Dirac statistics* apply to these indistinguishable particles. Our derivation of the FD distribution law will therefore be parallel that of the BE distribution law except that now each cell (i.e., quantum state) can be occupied by at most one particle.

Let us consider that the total number of ways in which n_i particles can be distributed in g_i cells[2] having same energy e_i. We may note that $(g_i - n_i)$ cells are vacant. The g_i cells can be rearranged in $\lfloor g_i$ different ways, but the $\lfloor n_i$ permutations of filled cells among themselves are irrelevant here as the particles are indistinguishable. Further, $\lfloor (g_i - n_i)$ permutations of the vacant cells among themselves are also irrelevant since the cells are not occupied. Obviously, the number of distinguishable arrangements of the particles among the cells is therefore

$$\frac{\lfloor g_i}{\lfloor n_i \lfloor (g_i - n_i)}$$

[2](g_i must be greater than or equal to n_i because there must at least one cell be available for every particle).

The probability W of the entire distribution of the particles is given by

$$W = \frac{\lfloor g_i}{\sum_i \lfloor n_i \lfloor (g_i - n_i)} \tag{12.83}$$

Taking the natural logarithm of both sides of Eq. (12.83), one obtains

$$\log W = \sum_i [\lfloor \log g_i - \lfloor \log n_i - \lfloor \log (g_i - n_i)] \tag{12.84}$$

Making use of Stirling's formula,

$$\lfloor \log n = n \log n - n$$

Eq. (12.84) becomes

$$\log W = \sum_i g_i \log g_i - g_i - n_i \log n_i + n_i - (g_i - n_i) \log (g_i - n_i) + (g_i + n_i)$$

$$= \sum_i [g_i \log g_i - n_i \log n_i - (g_i - n_i) \log (g_i - n_i)] \tag{12.85}$$

For this distribution to represent maximum probability, small changes δn_i in any of the individual n_i's must not alter W. Hence individual n_i's must not alter W. Hence

$$\delta \log W_{\max} = \sum_i \delta [g_i \log g_i - n_i \log g_i - (g_i - n_i) \log (g_i - n_i)] = 0$$

or

$$\sum_i [0 - \delta n_i - \log n_i \delta n_i + \delta n_i + \log (g_i - n_i) \delta n_i] = 0$$

or

$$\sum_i [- \log n_i + \log (g_i - n_i)] \delta n_i = 0 \tag{12.86}$$

Taking into account the conservation of particles and of energy by adding

$$\sum_i \delta n_i = 0 \tag{12.87}$$

$$\sum_i \varepsilon_i \delta n_i = 0 \tag{12.88}$$

Multiplying Eq. (12.87) by $-\alpha$ and Eq. (12.88) by $-\beta$ and then adding to Eq. (12.86), one obtains

$$\sum_i [- \log n_i + \log (g_i + n_i) - \alpha - \beta \varepsilon_i] \delta n_i = 0 \tag{12.89}$$

Since the δn_i's are independent, the quantity in brackets must vanish for each value of i, and so

$$- \log n_i + \log (g_i - n_i) - \alpha - \beta \varepsilon_i = 0$$

or

$$\log \left(\frac{g_i - n_i}{n_i} \right) = (\alpha + \beta \varepsilon_i)$$

or

$$\frac{g_i}{n_i} - 1 = e^{(\alpha + \beta \varepsilon_i)}$$

or

$$n_i = \frac{g_i}{1 + e^{(\alpha + \beta \varepsilon_i)}} = \frac{g_i}{1 + e^{(\alpha + \varepsilon_i / kT)} - 1} \quad (\because \beta = 1/kT) \tag{12.90}$$

Eq. (12.90) represents the Fermi-Dirac distibution law.

12.10.1 Application of Fermi-Dirac Statistics: Fermi Energy of Free Electrons

The most important application of the FD distribution law is in the free electron theory of metals. In a metal, the valence electrons are not very strongly bound to individual atoms, and consequently travel rather freely throughout the volume of the metal. One can treat these electrons of a "gas" that obeys FD distribution, i.e., in thermal equilibrium at absolute zero, the energy levels from zero to a certain value of E_F are occupied by the particles. Above E_F, the energy levels of others are left vacant. The energy E_F is known as *Fermi energy*.

According to Fermi-Dirac distribution law, the number of particles having their energy lying between ε and $\varepsilon + d\varepsilon$ is given by

$$n(\varepsilon)d\varepsilon = \frac{g(\varepsilon)d\varepsilon}{e^{(\alpha+\varepsilon_i/kT)}+1} = g(\varepsilon)d\varepsilon f(\varepsilon) \tag{12.91}$$

where $f(\varepsilon) = \dfrac{1}{e^{(\alpha+\varepsilon_i/kT+1)}}$ is FD distribution function.

For a system of free electrons enclosed in a volume V, one finds

$$g(\varepsilon)d\varepsilon = 2\times\frac{4\pi p^2 dpV}{h^3} \tag{12.92}$$

We multiplied this expression by two, in order to take account for the two spin states of an electron.

Using Eq. (12.92), Eq. (12.91) takes the form

$$n(\varepsilon)d\varepsilon = \frac{8\pi p^2 dpV}{h^3} f(\varepsilon) \tag{12.93}$$

The total number of electons in the system is given by

$$N = \int_{\varepsilon=0}^{\infty} n(\varepsilon)d\varepsilon = \int_{E=0}^{\infty} \frac{8\pi p^2 dpV}{h^3} f(\varepsilon) \tag{12.94}$$

We have at absolute zero, Fermi function $F(E) = 1$ for $\varepsilon < \varepsilon_F$ and $f(E) = 0$ for $\varepsilon > \varepsilon_F$

$$N = \int_{\varepsilon=0}^{\varepsilon_F} \frac{8\pi p^2 dpV}{h^3} = \frac{8\pi V}{h^3}\left[\frac{p^3}{3}\right]_{\varepsilon=0}^{\varepsilon_F} \tag{12.95}$$

Now, $p^2 = 2m\varepsilon$, where m is the mass of the electron. When $\varepsilon = 0$, $p = 0$ and when $\varepsilon = \varepsilon_F$, $p = \sqrt{2m\varepsilon_F}$. We have

$$N = \frac{8\pi V}{h^3}\left[\frac{(\sqrt{2m\varepsilon_F})^3}{3}\right]$$

$$= \frac{8\pi V}{h^3}\frac{(2m\varepsilon_F)^{3/2}}{3}$$

$$3h^3 N = 8\pi V(2m)^{3/2}\varepsilon_F^{3/2}$$

or

$$\varepsilon_F = \left[\frac{3h^3 N}{8\pi V(2m)^{3/2}}\right]^{2/3}$$

or Fermi energy at $T = 0$,

$$\varepsilon_F = \frac{h^2}{2m}\left(\frac{3n}{8\pi}\right)^{2/3} \tag{12.96}$$

where $n = N/V$ is the density of electrons in the assembly.

One can find the mean energy of the electrons

$$\varepsilon_m = \frac{1}{N}\int_0^\infty \varepsilon p(\varepsilon)d\varepsilon \tag{12.97}$$

After simplification, one obtains

$$\varepsilon_m = \frac{3}{5}\varepsilon_F \tag{12.98}$$

12.11 COMPARISON OF THREE DISTRIBUTION FUNCTIONS

A comparison of three statistical distribution functions is shown in Fig. (12.7).

The properties of three statistical distribution functions are summarized in Table (12.1).

In the limiting case when $\frac{g_i}{n_i} >> 1$, one can neglect 1 in the denominators of *BE* and *FD* distribution laws, because

$$\frac{g_i}{n_i} \cong \frac{g_i}{n_i} + 1 \cong \frac{g_i}{n_i} - 1$$

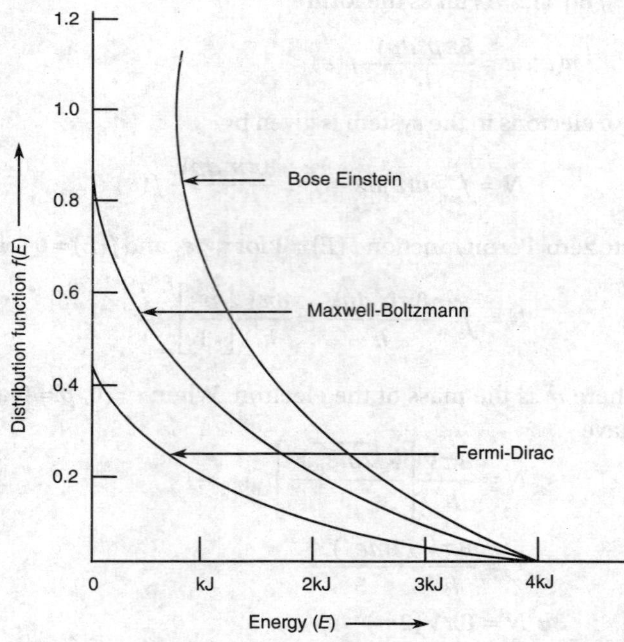

Fig. 12.7 A comparison of three statistical distribution functions for the same value of α. B.E. function is always higher than the MB function, which is a pure exponential, and the FD function is always lower. The functions give the probability of occupancy of a state of energy ε at absolute temperature T

Table 12.1 Three statistical distribution functions

Properties	***Maxwell-Boltzmann (MB)***	***Bose-Einstein (BE)***	***Fermi-Dirac (FD)***
Distribution function (number of particles in each state ε at the temperature T)	$f_{MB}(\varepsilon) = Ae^{-\varepsilon/kT}$	$f_{BE}(\varepsilon) = \dfrac{1}{e^{A}e^{\varepsilon/kT} - 1}$	$f_{FD}(\varepsilon) = \dfrac{1}{e^{(\varepsilon - \varepsilon_F)/kT} + 1}$
Systems to whom applied	Identical distinguishable particles	Identical, indistinguishable particles that do not obey Pauli exclusion principle	Identical, indistinguishable particles that obey Pauli exclusion principle
Categories of particles	Classical	Bosons	Fermions
Properties of particles	All particles with any spin, particles far enough apart so wave functions do not overlap	Spin 0, 1, 2, ... particles wave functions are symmetrical to intercharge of particles labels	Spin $\frac{1}{2}, \frac{3}{2}, \frac{5}{2}, \ldots$ wave functions are anti-symmetric to interchange of particle labels
Examples	Molecules of a gas	Photons in a cavity, phonons in a solid, liquid; helium at low temperatures	Free electrons in a metal electrons in a star whose atoms have collapsed (white dwarf stars)
Properties of distribution	No limit to number of particles per state	No limit to number of particles per state; more particles per state than $f_{MB}(\varepsilon)$ at low energies approaches $f_{MB}(\varepsilon)$ at high energies	Never greater than 1 particle per state; fewer particles per state than $f_{MB}(\varepsilon)$ at low energies; approaches $f_{MB}(\varepsilon)$ at high energies

Obviously, all the three distribution laws become equivalent to that of the *MB* distribution law. The limiting condition $\dfrac{g_i}{n_i} >> 1$ can be considered to be fulfilled for temperature not too low and pressure not too high. There are however, certain exceptional cases like liquid He-II, electron gas inside metals, distribution in radiation, etc., where the classical approximation fails.

Example 12.1: Assuming that each atom of copper contributes one free electron, calculate the Fermi energy. Given: Atomic weight of copper = 63.54 kg/K-mole, Density of copper = 8.93×10^3 kg/m^3, Avagadro number = 6.02×10^{23} kg, $h = 6.63 \times 10^{-34}$ J-s.

Solution We have

Fermi energy, $$\varepsilon_F = \frac{h^2}{2m}\left(\frac{3n}{8\pi}\right)^{3/2} \qquad \text{(i)}$$

Number of free electrons per unit volume

$$n = \frac{N}{V} = \frac{(\text{atoms/mole}) \times (\text{mass/volume})}{\text{mass/mole}} = \frac{N_o \rho}{M} \quad \text{(ii)}$$

Using Eq. (ii), Eq. (i) takes the form

$$\varepsilon_F = \frac{h^2}{2m}\left(\frac{2N_0\rho}{M \times 8\pi}\right) \quad \text{(iii)}$$

where $N_0 \rightarrow$ number of atoms per mole, i.e., Avagadro's number when one atom contributes free electrons.

Substituting the given values, one obtains

$$\varepsilon_F = \frac{(6.63\times10^{-34})^2}{3\times9.1\times10^{-31}} \times \left(\frac{3\times6.02\times10^{26}\times8.93\times10^3}{63.53\times8\times3.14}\right)$$

$$= 6.95\times1.6\times10^{-19}\,\text{J} = 6.95\,\text{eV}$$

Example 12.2: Compute the Fermi energy for Sodium.

Solution Each sodium atom contributes one valence electron to the metal, and so the number of electrons per unit volume, N/V, is equal to the number of sodium atoms per unit volume. One can find this from the density of sodium and the mass of a sodium atom:

$$\frac{N}{V} = \text{atoms per unit volume} = \frac{\rho N_A}{M}$$

$$= \frac{(0.971\times10^3\,\text{kg/m}^3)\times(6.02\times10^{26}\,\text{atoms/kilomole})}{23\times10^{-3}\,\text{kg/mole}}$$

$$= 2.54\times10^{28}\,\text{m}^{-3}$$

$$\varepsilon_F = \frac{h^2}{2m}\left(\frac{3}{8\pi}\frac{N}{V}\right)^{2/3} = \frac{(hc)^2}{2mc^2}\left(\frac{3}{8\pi}\times2.54\times10^{28}\,\text{m}^{-3}\right)^{2/3}$$

$$= \frac{(1240\text{eV} - \text{nm})^2}{2(0.511\times10^6\,\text{eV})}(2.09\times10^{18}\,\text{m}^{-2})\left(\frac{10^{-9}\,\text{m}}{\text{nm}}\right)^2$$

$$= 3.15\,\text{eV}$$

The average energy of the valence electron is $\frac{3}{5}\varepsilon_F = 1.89\text{eV}$. Obviously, even at the absolute zero of temperature, the electrons still have quite a large average energy.

Example 12.3: Show that the average energy of a free electron in a metal at absolute zero of temperature, $\bar{\varepsilon} = \frac{3}{5}\varepsilon_F$.

Solution We have for a system of free electrons in a thermal equilibrium, we have

$$n(\varepsilon)d\varepsilon = \frac{8\pi p^2 dpV}{h^3} f(\varepsilon) \quad \text{(i)}$$

where $n(\varepsilon)d\varepsilon$ represents the number of electrons having energies lying between ε and $\varepsilon + d\varepsilon$. $f(\varepsilon)$ is Fermi distribution function.

The total number of electrons is given by

$$N = \int_{\varepsilon=0}^{\infty} n(\varepsilon)d\varepsilon = \int_{\varepsilon=0}^{\infty} \frac{8\pi p^2}{h^3} f(\varepsilon) \qquad \text{(ii)}$$

Now, the total energy ε_T of electrons is given by

$$\varepsilon_T = \int_{\varepsilon=0}^{\infty} n(\varepsilon)d\varepsilon = \int_{\varepsilon=0}^{\infty} \frac{p^2}{2m}\frac{8\pi p^2 dp}{h^3} V f(\varepsilon) \qquad (\because \varepsilon = p^2/2m) \qquad \text{(iii)}$$

We have at absolute zero of temperature, $f(\varepsilon) = 1$ for $\varepsilon = \varepsilon_F$ and $f(\varepsilon) = 0$ for $\varepsilon > \varepsilon_F$. Hence

$$\varepsilon_T = \int_{\varepsilon=0}^{\varepsilon_F} \frac{8\pi V}{2mh^3} p^4 dp = \frac{8\pi V}{2mh^3}\left[\frac{p^5}{5}\right]_{\varepsilon=0}^{\varepsilon_F}$$

but $$p = \sqrt{2m\varepsilon} = \sqrt{2m\varepsilon_F} \qquad (\because \varepsilon = \varepsilon_F) \qquad \text{(iv)}$$

$\therefore$ $$\varepsilon_T = \frac{8\pi V}{2mh^3}\left(\sqrt{2m\varepsilon_F}\right)^5 \qquad \text{(v)}$$

From Eq. (ii), we have

$$N = \int_{\varepsilon=0}^{\varepsilon_F} \frac{8\pi V}{h^3} p^2 dp = \frac{8\pi V}{h^3}\left[\frac{p^3}{3}\right]_{\varepsilon=0}^{\varepsilon_F}$$

or $$N = \frac{8\pi V}{2mh^3}\left(\sqrt{2m\varepsilon_F}\right)^3 \qquad \text{(vi)}$$

At 0K, the average energy of the electron can be obtained from Eqs. (v) and (vi), i.e.,

$$\bar{\varepsilon} = \frac{\varepsilon_T}{N} = \frac{3}{10m}\left(\sqrt{2m\varepsilon_F}\right)^3 = \frac{3}{10m}2m\varepsilon_F = \frac{3}{5}\varepsilon_F$$

Example 12.4: Determine the internal energy of electron gas per unit volume at 0K if the zero degree Fermi-energy is 8.8 × 10^{-19}J. The density and atomic weight of silver is 10.3 kg/m^3 and 108 respectively.

Solution $\varepsilon = \frac{3}{5}\left(\frac{N}{V}\right)\varepsilon_{F(0)}$

The required internal energy is

$$\varepsilon = \frac{3}{5}\left(\frac{N}{V}\right)\varepsilon_{F(0)} = \frac{3}{5}\times 5.9\times 10^{28}\times 8.8\times 10^{-19}\,\text{J/m}^3$$

$$= 3.1\times 10^{10}\,\text{J/m}^3$$

Example 12.5: Evaluate the temperature at which there is one percent probability that a state with an energy 0.5eV above the Fermi energy, will be occupied by an electron.

Solution We have the Fermi-Distribution function

$$f(E) = \frac{1}{\exp(\varepsilon - \varepsilon_F)/\text{kT} + 1}$$

where ε_g is the Fermi-energy we have $\varepsilon = (\varepsilon_F + 0.5)$eV

Thus $$f(E) = \frac{1}{100} = \frac{1}{1 + \exp(0.5/kT)}$$

or $$0.01 = \frac{1}{1+e^x} \quad \text{where} \quad x = 0.5/kT$$

or $$0.01 + 0.01e^x = 1$$

or $$e^x = \frac{0.99}{0.01}$$

or $$x = 2.303\log\left(\frac{99}{100}\right) = \frac{0.5}{kT}$$

$\therefore$ $$T = \frac{0.109 \times 1.6 \times 10^{-19}}{1.38 \times 10^{-23}} = 1264^\circ \text{K}$$

Example 12.6: The fermi energy of silver is 5.51 eV. What is the average energy of the free electrons in silver at 0K? What temperature is necessary for the average molecular energy in an ideal gas to have this value? [MDS]

Solution The average electron energy at 0K

$$\overline{\varepsilon_0} = \frac{3}{5}\varepsilon_F = \frac{3}{5} \times 5.51 = 3.306\,\text{eV}$$

The average molecular energy in an ideal gas at temperature, TK is

$$\frac{3}{2}kT = \overline{\varepsilon_0}$$

or $$T = \frac{2}{3}(\overline{\varepsilon}/k) = \frac{2 \times 3.306 \times 1.6 \times 10^{-19}}{3 \times 1.38 \times 10^{-23}}$$
$$= 2.56 \times 10^4\,\text{K}$$

SUMMARY

1. *Maxwell Boltzmann* (*M.B.*) statistics is applicable to a system of identical distinguishable particles.
2. *M. B.* statistics can be applied to dilute gases at room and higher temperature.
3. *Bose-Einstein* (*B.E.*) statistics is applicable to the system of identical, indistinguishable particles, which have integral spin (0, 1, 2, 3, ...). Particles with integral spins are called *bosons*. Photons (spin 1), phonons (quantum of acoustic vibration), pions, α-particle, helium atom, etc. are few examples of boson.
4. *Fermi-Dirac (F.D) statistics* is applicable to the system of identical, indistinguishable particles, which have odd half integral spin (1/2, 3/2, 5/2...). Particles with odd-half-integral spin are called fermions and they obey Pauli exclusion principle. Electrons, positrons, protons, neutrons, μ-mesons, etc., are examples of fermions.

5. The statistical factors associated with *MB* (classical) and quantum (*BE* and *FD*) distributions are given by

$$F_{MB} = A\exp(-\beta E)$$

$$F_{FD} = \frac{1}{B_1 \exp(\beta E) + 1}$$

$$F_{BE} = \frac{1}{B_2 \exp(\beta E) - 1}$$

where A, B_1, B_2 are normalization factors. In each case the distribution function $n(E)$, can be expressed as the product of the density of states $g(E)$ and the appropriate statistical factor. Density of states $g(E)$ is defined as the number of states available per unit energy range.

6. *FD* statistics predict the correct behaviour of conduction electrons in a metal. *FD* statistics can be used to find the electrical conductivity (and the correct temperature dependence thereof) as well as the electronic contribution to the specific heat of a metal.
7. *BE* statistics can be used to derive the Planck's law for blackbody radiation.
8. *BE* statistics help us to understand some properties of liquid helium, which undergoes a transition to the superfluid state at the temperature $T_c = 2.17$ K. The extraordinary low viscosity of a superfluid is due to the fact that the molecules in a superfluid state obey *BE* statistics. *BE* condensation has been reported in a variety of new materials in the past few years.

REVIEW QUESTIONS

1. Explain:
   (a) a priori probability
   (b) thermodynamic probability
   (c) phase space
   What is the significance of a point in phase space. **[MDS]**
2. Give a free description of phase space in generalised coordinates.
3. Derive Maxwell-Boltzmann distribution law of energy. Find the relation betwen r.m.s and average value of velocity of molecules. **[Udaipur]**
4. Show that at absolute temperature *T* the quantity *kT* is roughly equal to the mean energy per degree of freedom.
5. Assuming Maxwell-Boltzmann distribution law of energy, derive Maxwell's law of distribution of velocities for the molecules of a perfect gas. **[Raj]**
6. Obtain an expession for Maxwell-Boltzmann distribution function from statistical considerations applied to a dilute gas.
7. On the basis of Maxwell's law of distribution of speed, calculate the average speed and r.m.s velocity of molecule of a perfect gas.
8. State and derive the law of equipartition of energy.

9. Give an account of *BE* statistics. What are the basic changes made by Bose in classical *MB* statistics? Derive *BE* distribution law. **[Udaipur]**

10. On the basic of *BE* statistics, derive Planck's law of black body radiation.

11. Derive an expression for Fermi-Dirac distribution law. How it differs from *MB* and *BE* distribution laws?

12. Explain the distinguishing features of *MB, BE* and *FD* statitstics? Give a comparison of these three statistics.

13. Define Fermi energy. Write an expression for Fermi-Dirac distribution law. Obtain an expression for Fermi energy of a system of free electrons. **[Punjab]**

PROBLEMS

1. In a system of particles, 20% of the particles are moving with a speed of 2m/s, 30% with a speed of 4m/s and 50% with a speed of 3m/s. Calculate the average speed, r.m.s speed and the most probable speed. If the mass of each particle is 2×10^{-3} kg, show that mean kinetic energy of the particle is 10.1 mJ.

[**Ans.** $V_{av} = 3.1$m/s, $V_{rms} = 3.18$ m/s, $V_{mp} = 3$m/s]

2. If the Fermi energy of a metal is 10eV, calculate:
(a) the average energy of free electrons in the metal at 0K.
(b) corresponding classical temperature and
(c) the speed of the electron with average energy calculated above.

[**Ans.** (i) 6eV (ii) 4.6×10^4K (iii) 1.4×10^6m/s]

3. The Fermi energy in silver is 5.51eV. Calculate (i) average energy of free electrons in silver at 0K. (ii) temepratur e necessary for the average molecular energy in an ideal gas with above energy. (iii) Speed of electron with above energy.

[**Ans.** (i) 3.3eV (ii) 2.56×10^4K (iii) 1.08×10^6m/s]

4. Find the r.m.s speed of N_2 molecule in the air at 27°C. [**Ans.** $V_{rms} = 517$m/s]

5. Show that the most probable speed of a molecule of an ideal gas is equal to $\sqrt{2kT/m}$.

6. Show that the average speed of a molecule of an ideal gas is equal to $\sqrt{8kT/\pi m}$.

SHORT ANSWER-QUESTIONS

1. Explain the consequences of finite *h*.
Ans: The finite value of h ($= 6.6 \times 10^{-34}$J-s) ensures the occupancy factor for the phase cells of microstates a definite value for each situation.

2. Is Maxwellian velocity distribution law applicable to non-ideal gases?
Ans: Maxwellian velocity distribution law is not applicable to non-ideal gases as potenetial energy due to inter-molecular interactions is present. We may note that if the mean potential energy per molecule is a small fraction of the mean translational energy per molecule, then the interaction is said to be weak and velocity distribution deviates very little from the Maxwellian distribution.

3. Is *MB* statistics an approximation of the *BE* or *FD* statistics?
Ans: Yes.

OBJECTIVE QUESTIONS

1. In quantum statistics, the volume of phase cells is
   (a) zero (b) h
   (c) h^3 (d) $1/h^3$
2. The ratio of the number of cases in which the event occurs to the total number of cases is called the probability of an event.
   (a) True (b) False
3. A collection of large number of identical independent system is called an ensemble.
   (a) True (b) False
4. *MB* statistics is an approximation of *BE* or *FD* statistics under appropriate approximations.
   (a) True (b) False
5. *BES* is applicable to the identical ______ particles of zero and integral spin. These particles are called ______. **[indistinguishable, bosons]**
6. In classical statistics each particle has a recognizable ______. **[individuality]**
7. *FD* statistics is applicable to the identical, indistinguishables of ______ spin. These particles are called ______. **[half, Fermion]**
8. Quantum statistics was first formulated by ______. **[Bose]**
9. Depending on the nature of the particles, three kinds of statistics or distribution laws are used to describe the properties of the system. These are: ______, ______ and ______. **[Maxwell-Boltzmann, Bose-Einstein, Fermi-Dirac]**
10. *M.B.* statistics is applicable to the system of ______, ______ particles. **[identical, distinguishable]**
11. *B.E.* statistics is applicable to the system of ______, ______ particles, which have integral spin (0, 1, 2, ...). **[identical, indistinguishable]**
12. *F.D.* statistics is applicable to the system of ______, ______ particles, which have odd-half integral spin (1/2, 3/2, 5/2, ...). **[identical, industinguishable]**
13. Statistical mechanics is that branch of physics which studies macroscopic systems from a ______ or ______ point of view. **[microscopic, molecular]**
14. Two particles are to be distributed in an energy level, which is 3-fold degenerate. If particles are distinguishable, the possible microstates are ______. **[6]**
15. Two particles are to be distributed in an energy level, which is 3-fold degenerate. If particles are indistinguishable, the number of possible microstates are ______. **[3]**
16. The ensemble average is the ______ at a fixed time over all the elements in an ensemble. **[average]**

ANSWERS

Question numbers	1	2	3	4
Answers	(c)	(a)	(a)	(a)

PART B

Basics of Solid State

13

Crystal Structure and Defects

13.1 INTRODUCTION

Solids may be classified by structural organization into **crystalline, polycrystalline**, and **amorphous** types. An amorphous solid does not have a well-defined structure, in fact, it is distinguished by formlessness Fig. [13.1(a)]. In a poly-crystalline solid Fig. [13.1(b)], there are many small regions, each having a well-organized structure but differing from its neighbouring regions. Such a material can be produced inexpensively and is used extensively in micro electronics. Polycrystalline materials are separated by well defined boundaries called as grain boundaries. Polycrystalline materials are stronger than ordinary materials because crystals in this type of materials have different orientations with respect to each other and grain boundaries obstruct the movement of dislocations. Such crystals are called **isotropic** because they exhibit same properties in every plane and direction, where as single crystal is called **aniostropic**. In crystalline solid, atoms are arranged in orderly array. These are solids which have a regular periodic arrangement in their component particles, bounded by flat faces, orderly arranged in reference to one another, which converge at the edges and vertices Fig. [13.1(c)]. It is possible to specify a unit cell which, repeating itself, produces the crystalline solid. A crystal is symmetrical about its certain elements like points, lines or planes and if it is rotated about these elements, it is not possible to distinguish its new position from the original position. This symmetry is an important characteristic based on internal structure of crystal. Symmetry helps one to classify crystals and describing their behaviour. At temperatures below that of crystallization, the crystallization state is stable for all solids.

13.1.1 Single Crystal

Most of the materials exist in polycrystalline form, but there are some materials, which exist in the form of single crystals, e.g., sugar, sodium chloride (common salt), diamond, etc.

Single crystals represent a material in its ideal condition and are produced artificially from their vapour or liquid state. These crystals help us in studying behaviour and defects of the materials in ideal conditions.

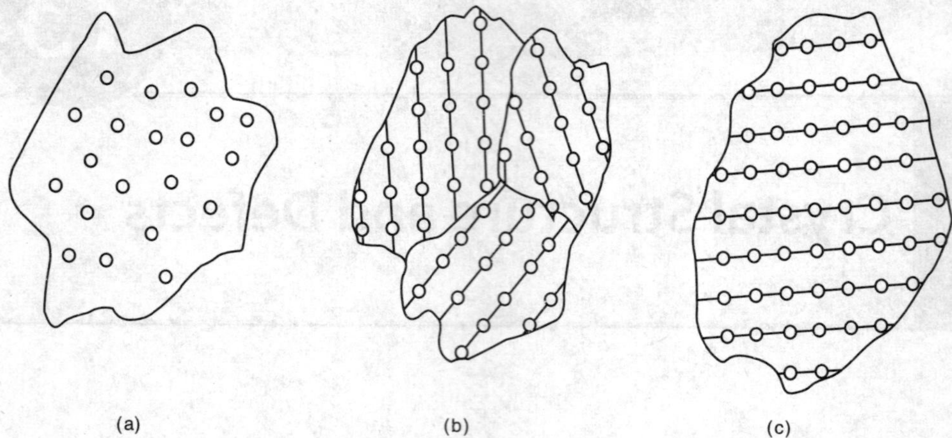

Fig. 13.1 Classification of solids based on structural organization (a) Amorphous (b) Polycrystalline (c) Crystalline

13.1.2 Whiskers

These are very thin filaments, hair like single crystals of about 13 mm length and perhaps 10^{-4} cm diameter. These are produced as dislocations of free crystals and are without any structural defect. This is why whiskers are far stronger than polycrystals of same material. These are used as reinforcement in materials to increase strength by embedding fibres of one material in a matrix of another. The properties of these fibre or whisker-reinforced composites can often be tailored for a specific application. The increase in diameter of the whiskers decreases its strength and increases its ductility. The cost of whiskers and the expensive fabrication is the major disadvantage of the method.

Whiskers are the most defect-free crystalline solids available today. The best known composite are probably, fibre glass, which consists of glass-reinforcing fibres in a matrix of either an epoxy polymer or polyester whiskers can bear considerably high stresses both at low and relatively elevated temperatures. Single crystals of SiC, Al_2O_3, S-glass, graphite, boron, iron, silver, copper and tin can be produced by means of special techniques.

Whiskers of a wide variety of substances, e.g., mercury, graphite, sodium and potassium chlorides, copper, iron, and aluminium oxide have been grown from super satured media. Whiskers grown in this way are usually a few micrometers in diameter and up to few inches long. Some are exceptionally strong, both in bend tests and in tension tests.

In addition to exceptional strength, whiskers often have unique electrical, magnetic, or surface properties.

13.2 SPACE LATTICES AND CRYSTAL STRUCTURE

Solids could be divided into classes according to type of binding between the atoms constituting the solid. We now wish to discover how the atoms-arrange themselves around each other in a crystal. We want to find out whether there are any laws governing

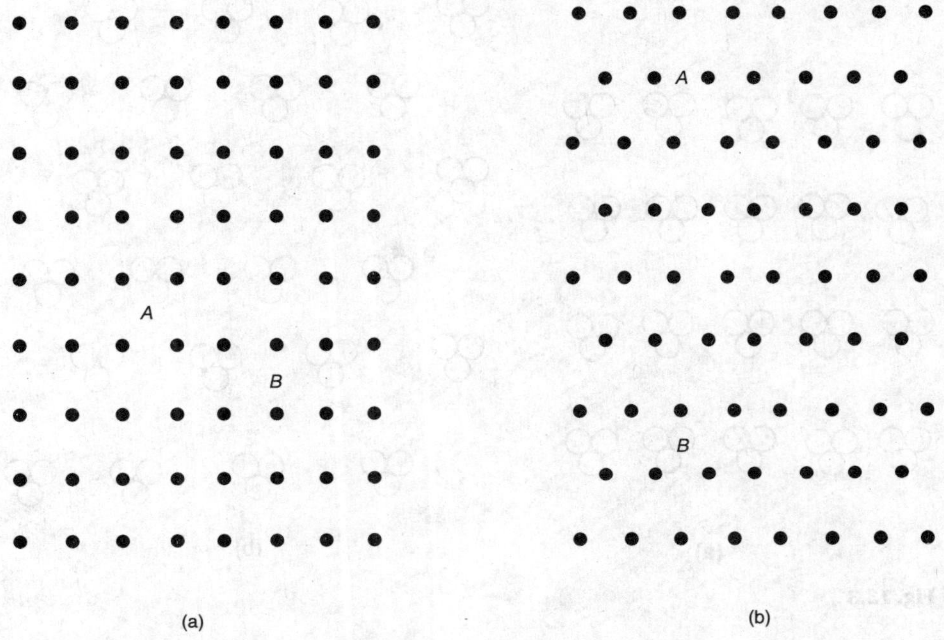

Fig. 13.2

the construction of crystals and whether there is any limit to the possible number of symmetrical arrangements available.

Crystals are made up of regular and periodic three dimensional patterns of atoms in space, called the crystal structure. Crystal structures are described in terms of an idealised geometrical concept called a **space lattice.** A **space lattice is an array of points in space such that the environment about each point is the same**. Fig [13.2(a)] represents a two dimensional lattice where as Fig [13.2(b)] is not a lattice though both of them contain an array of points in two dimensions. This is because the environments about any two points *A* and *B* in Fig. [13.2(b)] are not identical, where as the environments about any two points *A* and *B* in Fig. [13.3(a)] are identical. The same argument can be extended to the three dimensional space. These lattice points may or may not be atom sites. But we associate every lattice point with one or more atoms called the **basis** or the **pattern,** which when repeated with correct periodicity in all directions gives the actual crystal structure. The crystal structure is real, where as the lattice is imaginary. We can imagine the crystal structure to be built by a lattice array of the basis or the pattern. Fig (13.3) shows a possible basis or pattern representing each lattice point. We find that basis in this case consists of three different atoms. It must be noted that all the bases into which the crystal structure is divided contain the same composition of atoms, same argument and the same orientation. The lattice picture of the crystal is found to be very convenient for the study of the symmetry properties of the crystal. Again there are 14 lattices to describe the infinite crystal structures. Thus the lattice concept simplifies the crystal structure analysis. That is why we use this imaginary concept of lattice. Thus it is clear that every crystal structure

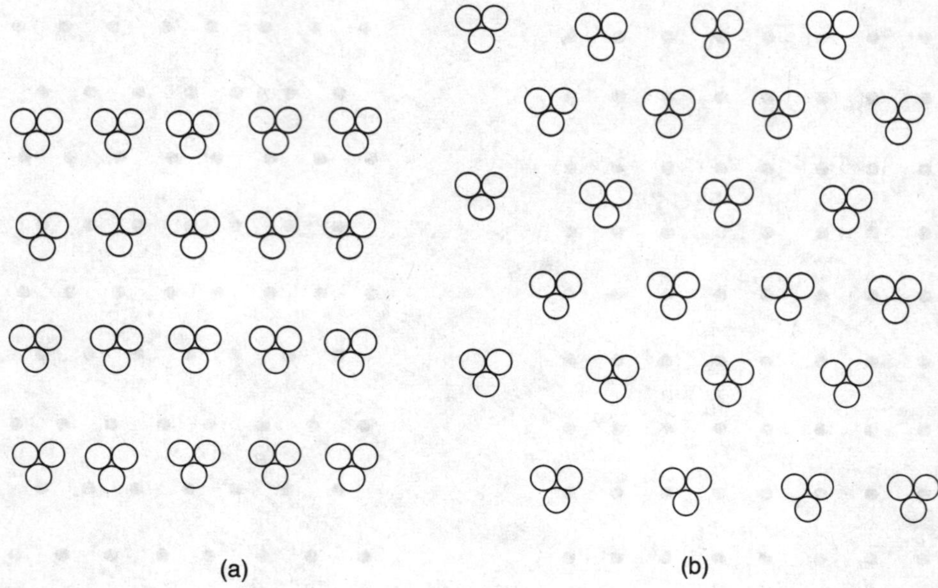

Fig. 13.3

can be reduced to its corresponding lattice and knowing the basis of the lattice points, it is possible to get back to the crystal structure.

13.3 BRAVIS LATTICE

Bravis showed that the lattice points can be arranged only in 14 different ways in space, i.e., there are only 14 ways of arranging points in space so that the environment looks the same from each point. These lattices are called the Bravis lattices named after the discoverer.

In order to represent the lattice points in space we require a coordinate system. All the lattice points in space can be generated by translating three non-coplanar fundamental vectors. The three non-coplanar directions constitute a coordinate system. If ***a, b*** and ***c*** are fundamental vectors in any three non-caplanar directions, which need not be mutually orthogonal, then the position vector ***r*** of any lattice point can be given by

$$\boldsymbol{r} = n_1\boldsymbol{a} + n_2\boldsymbol{b} + n_3\boldsymbol{c} \tag{13.1}$$

where n_1, n_2 and n_3 are integers. Any vector of this form is called a lattice vector. In fact one coordinate system or crystal system is enough to represent all the lattice points. In such a coordinate system the three non-coplanar vectors and the three angles between the three non-coplanar directions may in general satisfy the inequality

$$a \neq b \neq c \quad \text{and} \quad \alpha \neq \beta \neq \gamma \neq 90^\circ$$

Table 13.1

System	*No. of lattices*	*Lattice symbol*	*Nature of unit cell*	*Examples*
Triclinic	1	P (Simple)	$a \neq b \neq c$ $\alpha \neq \beta \neq \gamma$	$FeSO_4 . 5H_2O$ $CaS_2O_36H_2O$
Monoclinic	2	P (Simple) C (Base centred)	$a \neq b \neq c$ $\alpha = \gamma = 90° \neq \beta$	$Li_2SO_4H_2O$ Borax
Orthorhombic	4	P (Simple) C (Base centred) I (Body centred) F (Face centred)	$a \neq b \neq c$ $\alpha = \beta = \gamma = 90°$	$PbSO_4$ $C_6H_2(NO_2)_4OH$
Tetragonal	2	P (Simple) I (Body centred)	$a = b \neq c$ $\alpha = \beta = \gamma = 90°$	In, MgCl,Sn
Cubic	3	P (Simple) I (Body centred) F (Face centred)	$a = b = c$ $\alpha = \beta = \gamma = 90°$	Au, Cu, Ag, Fe, NaCl, (Most metals and their alloys)
Trigonal (Rhombohedral)	1	R (Simple)	$a = b = c$ $\alpha = \beta = \gamma < 120° \neq 90°$	$CaCO_3$ $MgSO_4.6H_2O$
Hexagonal	1	P (Simple)	$a = b \neq c$ $\alpha = \beta = 90°, \gamma = 120°$	$Be_3Al_2(Si_6O_{18})$ Be, Mg, Zn, Cd

But the difficulty in representing all the lattice points in the above coordinate system is that we cannot visualise the crystal with all its symmetric features. To bring out the essential symmetry of the lattice, the coordinate axes have to be chosen such that they represent the symmetry axes of the crystal. It is found that seven coordinate systems or crystal systems are required to specify the 14 lattices bringing out completely the essential symmetry.

The characteristics of the seven crystal systems used to represent the 14 Bravis lattices are summarised in the Table 13.1.

The 14 Bravis lattices are shown in Fig. (13.4).

The choice of the seven coordinate systems is not unique, but there are only seven judicious selections out of an infinite choices. In order to understand this let us consider two dimensional lattice shown in Fig (13.5). We see that the same lattice can be represented by coordinate axes *OA* and *OB* or *OP* and *OA* or *OT* and *OA*. Now the question arises as to which pair of directions one must chose to describe the lattice? As we have already stated that it is found most convenient to chose those coordinate axes which best represents the symmetry of the lattice. Again the question arises why there are 14 lattices when we have selected only seven crystal systems. Are not the seven lattices enough? Since lattice array is merely a periodice repetition of a fundamental unit it is not necessary to consider the entire array. It is enough if we consider a representative fundamental unit which is being repeated. Let us see how the fundamental unit is chosen. To understand this here again we will consider two dimensional case and extend the argument to the three dimensional

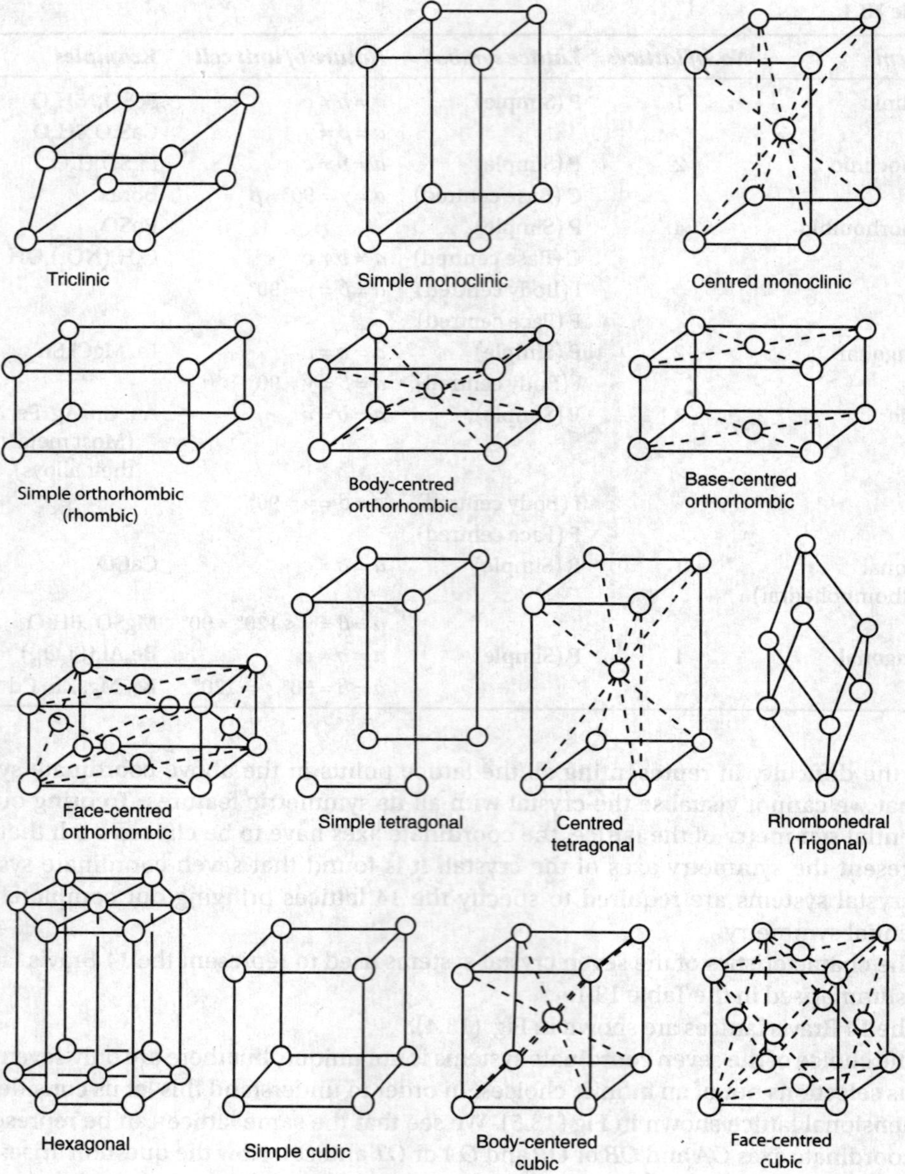

Fig. 13.4 14 Bravis lattices

case. In Fig. (13.6) the lattice points are joined by two inter-secting sets of parallel lines parallel to the coordinate axes suitable for that lattice are shown. It is found that these intersecting lines divide the entire lattice into a set of equal sized parallelograms packed together, as shows in Fig. (13.7). We can consider each elementary parallelogram (or a parallelopiped in three dimension) as the fundamental unit or the basic cell. It is clear that by

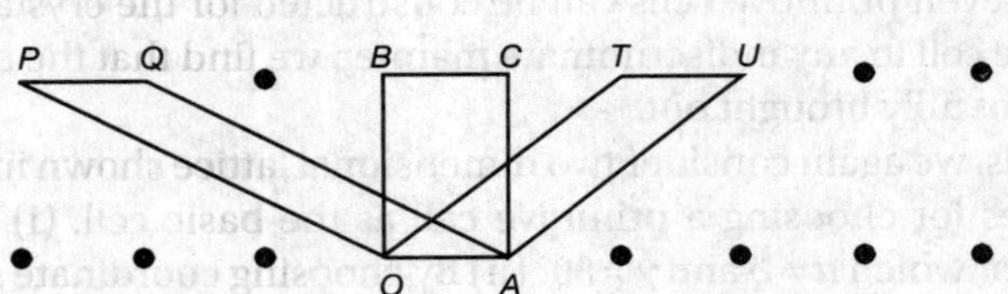

Fig. 13.5 Two-dimensional lattice

Fig. 13.6

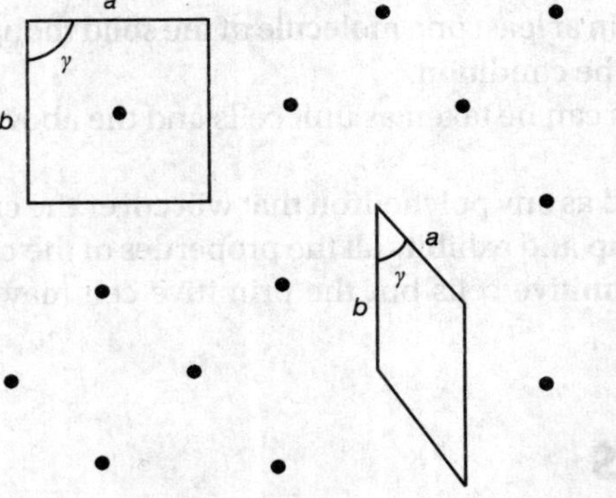

Fig. 13.7

translating the basic cell along its edges the entire lattice can be generated. Parallelopipeds formed by lattice points only at their corners are called primitive or simple cells denoted by the symbol (*P*). Seven primitive cells can be constructed for the crystal systems. When we chose a primitive cell in any indiscriminate manner, we find that the essential symmetry of the lattice is not fully brought out.

To understand this, we again consider two dimensional lattice shown in Fig. (13.7). Here we have two choices for choosing a primitive cell as the basic cell. (i) By choosing the coordinate system for which $a \neq b$ and $\gamma \neq 90°$ (ii) By choosing coordinate system for which $a = b$ and $\gamma = 90°$. In first case the lattice points lie only at the corners, i.e., the basic cell is primitive. In second case the basic cell in addition to the points at the corner has a lattice point at its centre also. This cell is no longer primitive but is called multiple primitive. Which coordinate system has to be chosen in the above case? Since the choice of the rectangular coordinate-system brings out the lattice symmetry much better, it has to be chosen. Thus a rectangular coordinate system in two dimensions can have two possible lattices, one being primitive and the other being centred. In three dimensions the lattice points can be at the body centre of the cell or at the face centre of the cell. This is the reason for possible 14 Bravis lattices.

13.4 UNIT CELL

Apart from the basic cell, there is another choice for the cell called unit cell. If the solid is made up of identical atoms as is the case with metallic elements, where every lattice point is associated with one atom, the fundamental cell or the basic cell can be taken as the unit cell. In ionic solids like NaCI, KCl, etc., the basic cell cannot be taken as the unit cell, since it does not contain one complete molecule of the solid. This does not mean that the unit cell must contain only one molecule. For example in Ferrite the volume of space lattice of one molecule contains fraction number of basic cells. The choice of unit cell is made consistent with the following considerations:

(a) The unit cell must be integer times the basic cell.

(b) The unit cell must contain at least one molecule of the solid though it may contain more in order to satisfy the condition.

(c) In fact much bigger cells can be taken as unit cells and the above choice is one of convention.

A unit cell is thus defined as any polyhedron that will cover the crystal by application of translational group and exhibits all the properties of the crystal as a whole. All the crystals have primitive cells but the primitive cell may not be the most convenient unit cell.

13.5 ATOMIC RADIUS

This is defined as half the distance between the nearest neighbours in the crystal structure of a pure element. Atomic radius is denoted by *r* and expressed in terms of the cube edge element *a*. One can calculate the atomic radius by assuming that atoms are spheres in contact in a crystal. Calculation of atomic radius in various crystal structures is illustrated as below.

(i) Simple Cubic (SC) Structure: In SC Structure, atoms touch each other along the lattice Fig. [(13.8(a))].

We have

$$a = 2r$$

$$\therefore \quad r = a/2$$

$$\text{Area} \quad a^2 = 4r^2$$

(ii) BCC Structure: The atoms touch each other along the diagonal of the cube as shown in Fig. [(13.8(b)], Obviously, the diagonal in this case is $4r$. Also,

$$AC^2 = AB^2 + BC^2 = a^2 + a^2 = 2a^2$$

$$\therefore \quad AD^2 = AC^2 + CD^2 = 2a^2 + a^2 = 3a^2$$

$$(4r)^2 = 3a^2$$

$$\text{or} \quad a = \frac{4r}{\sqrt{3}} \quad \text{and} \quad 2r = \frac{a\sqrt{3}}{2}$$

$$\therefore \quad r = \frac{a\sqrt{3}}{4}$$

$$\text{area} \quad a^2 = \frac{16r^2}{3}$$

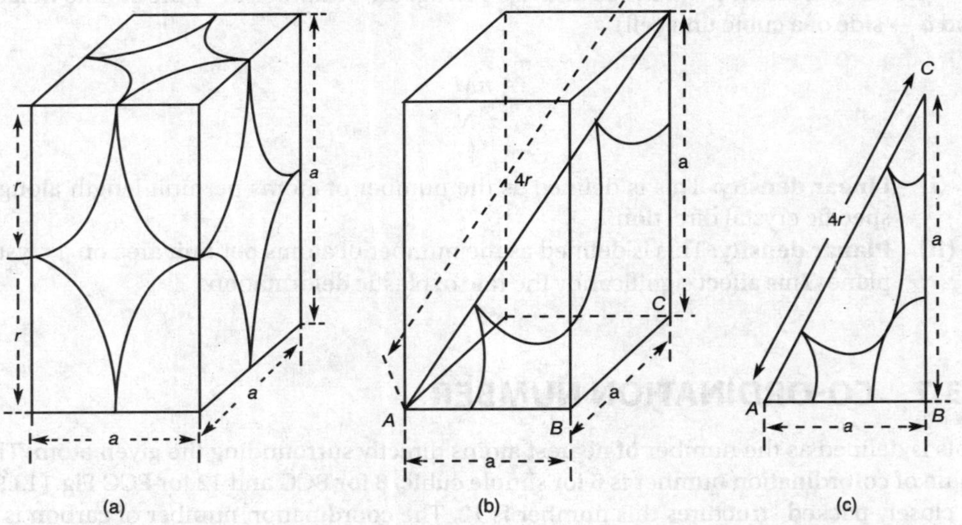

Fig. 13.8 Atomic radii for three unit cells (a) Simple Cubic (SC) (b) Body-Centred Cubic (BCC) and (c) Face-Centred Cubic (FCC)

(iii) FCC Structure: Atoms within this structure touch along the diagonal of any face of the cube Fig. (13.8). The diagonal has a length of $4r$.

$$AC^2 = AB^2 + BC^2$$

or $$(4r)^2 = a^2 + a^2$$

$\therefore$ $$r^2 = \frac{2a^2}{16}$$

or $$r = \frac{\sqrt{2}a}{4} = \frac{a}{2\sqrt{2}}$$

and $$\text{area} = a^2 = 8r^2$$

13.6 DENSITY OF CRYSTAL

$$\text{Density of the crystal } (\rho) = \frac{\text{Mass of unit cell}}{\text{Volume of unit cell}}$$

$$\text{Mass of unit cell} = \frac{\text{Atomic mass}}{\text{Avogadro number}} \times \text{Number of atoms per unit cell}$$

$$= \frac{M}{N_A} \times n$$

($n \rightarrow$ number of atoms per unit cell and $N_A \rightarrow$ Avogadro's number, $M \rightarrow$ the atomic weight, and $a \rightarrow$ side of a cubic unit cell)

$$= \frac{nM}{a^3 N_A}$$

(i) Linear density: This is defined as the number of atoms per unit length along a specific crystal direction.

(ii) Planar density: This is defined as the number of atoms per unit area on a crystal plane. This affect significantly the rate of plastic deformation.

13.7 CO-ORDINATION NUMBER

This is defined as the number of nearest atoms directly surrounding the given atom. The value of co-ordination number is 6 for simple cubic, 8 for BCC and 12 for FCC Fig. (13.9). In closely packed structures this number is 12. The coordination number of carbon is 4, i.e., number of nearest neighbours of carbon atom is 4. This low coordination, results in a relatively inefficient packing of the carbon atoms in the crystal.

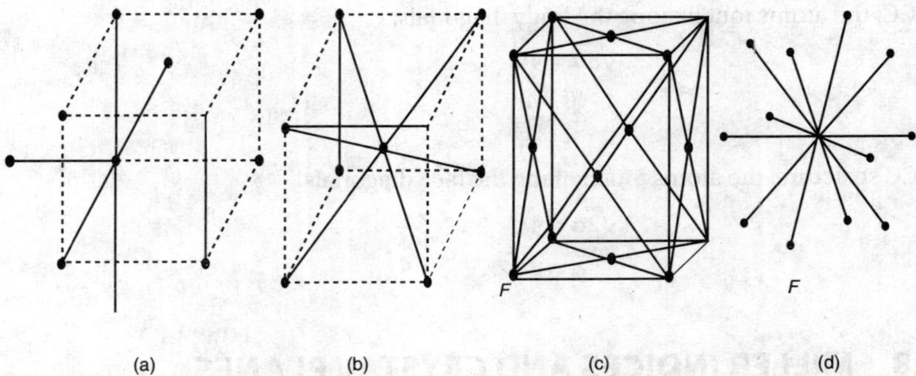

Fig. 13.9 The coordination number in three cubic lattices

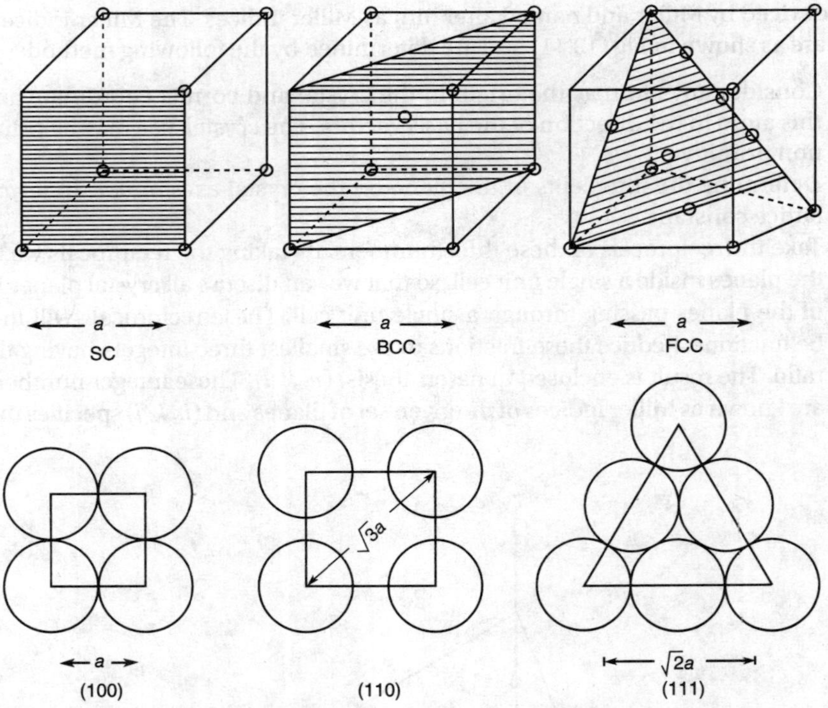

Fig. 13.10 Monoatomic, SC, BCC and FCC structures along with their (1 0 0), (1 1 0) and (1 1 1) type planes respectively

Example 13.1: Determine the relationship between the lattice parameter *a* and the atomic radius *r* for monoatomic, SC, BCC and FCC structures shown in Fig. (13.10).

Solution In simple cubic (SC) structure, the atom touch one another along the cube edges,

$$\therefore \qquad a = 2r$$

In BCC, the atoms touch along the body diagonals,

$$\therefore \qquad \sqrt{3}a = 4r$$

$$\text{or} \qquad a = \frac{4r}{\sqrt{3}}$$

In FCC structure, the atoms touch along the face diagonals,

$$\therefore \qquad \sqrt{2}a = 4r$$

$$\text{or} \qquad a = 2\sqrt{2}r.$$

13.8 MILLER INDICES AND CRYSTAL PLANES

In order to study the properties of crystals it is useful to have a means of describing the different planes and different directions in a crystal. This is done by a system of crystallographic indices deviced by Miller and named after him as Miller indices. The Miller indices for the plane *P* are as shown in Fig (13.11) and are determined by the following method:

(i) Consider any atom as the origin in the crystal and correct coordinate axes from this atom in the direction of the basis vectors. The crystal axis may be primitive or non primitive.

(ii) Determine the intercepts of the plane on the crystal axes *a, b, c,* in terms of the lattice constant.

(iii) Take the reciprocals of these three numbers. By taking the reciprocals we bring all the planes inside a single unit cell, so that we can discuss all crystal planes in terms of the planes passing through a single unit cell. These reciprocals will in general be fractions. Reduce these fractions to the smallest three integers having the same ratio. The result is enclosed in paren-thesis: (*h, k, l*). These integer numbers, *h, k, l* are known as Miller indices of the given set of planes and (*h, k, l*) specifies the plane.

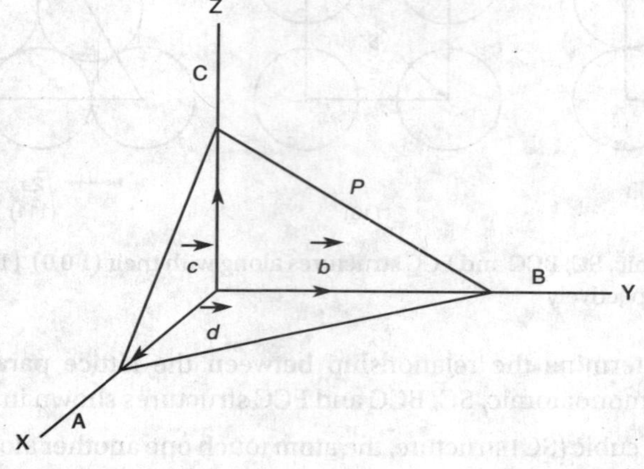

Fig. 13.11

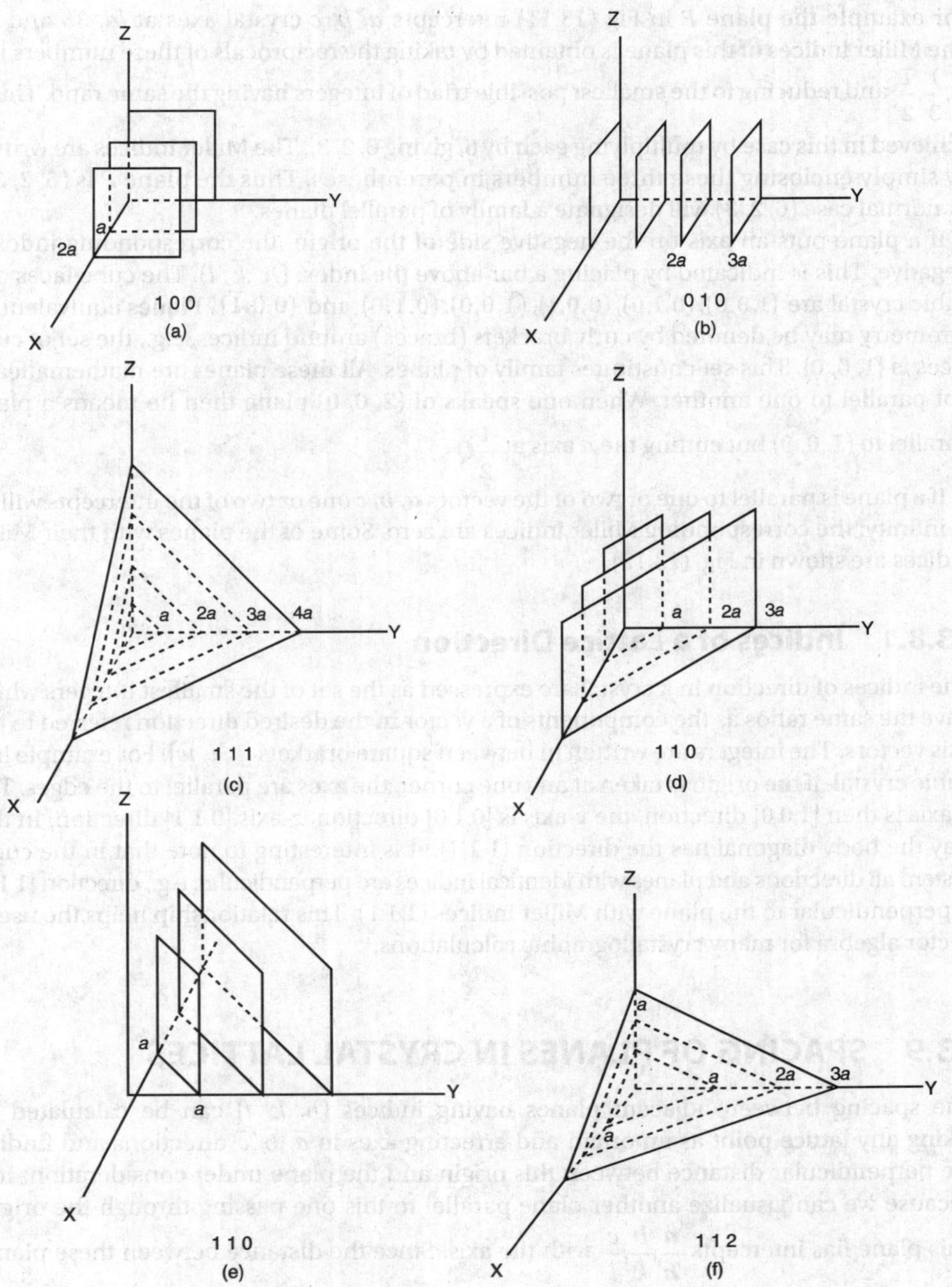

Z
Y
X
a
2a
1 0 0
(a)
Z
Y
X
2a
3a
0 1 0
(b)
Z
Y
X
a
2a
3a
4a
1 1 1
(c)
Z
Y
X
a
2a
3a
1 1 0
(d)
Z
Y
X
a
a
1 1 0
(e)
Z
Y
X
a
a
2a
3a
a
1 1 2
(f)

Fig. 13.12

For example the plane P in Fig (13.11) intercepts a, b, c crystal axes at la, $3b$ and $2c$. The Miller indices of this plane is obtained by taking the reciprocals of these numbers i.e., $\frac{1}{1}, \frac{1}{3}, \frac{1}{2}$ and reducing to the smallest possible triad of integers having the same ratio. This is achieved in this case by multiplying each by 6, giving 6, 2, 3,. The Miller indices are written by simply enclosing these three numbers in parentheses. Thus the plane P is (6, 2, 3,). In normal case (6, 2, 3) will designate a family of parallel planes.

If a plane puts an axis on the negative side of the origin, the corresponding index is negative. This is indicated by placing a bar above the index: $(h, \bar{k}, l)$. The cube faces of a cubic crystal are $(1,0,0), (0,1,0), (0,0,1), (\bar{1},0,0), (0,\bar{1},0)$ and $(0,0,\bar{1})$. Planes equivalent by symmetry may be denoted by curly brackets (braces) around indices, e.g., the set of cube faces is {1, 0, 0}. This set constitutes family of planes. All these planes are mathematically not parallel to one another. When one speaks of (2, 0, 0) plane then he means a plane parallel to (1, 0, 0) but cutting the a axis at $\frac{1}{2}Q$.

If a plane is parallel to one or two of the vectors ***a, b, c*** one or two of the intercepts will be at infinity: the corresponding Miller indices are zero. Some of the planes with their Miller indices are shown in Fig. (13.12).

13.8.1 Indices of a Lattice Direction

The indices of direction in a crystal are expressed as the set of the smallest integers which have the same ratios as the components of a vector in the desired direction referred to the axis vectors. The integers are written in between square brarkets $[u, v, w]$. For example in a cubic crystal, if the origin is taken at any one corner, the axes are parallel to the edges. The x-axis is then [1 0 0] direction, the y-axis is [0 1 0] direction, z-axis [0 1 1] direction. In this way the body diagonal has the direction [1 1 1]. It is interesting to note that in the cubic system all directions and planes with identical indices are perpendicular, e.g., direction [1 1 1] is perpendicular to the plane with Miller indices (1 1 1). This relationship helps the use of vector algebra for many crystallographic calculations.

13.9 SPACING OF PLANES IN CRYSTAL LATTICES

The spacing between adjacent planes having indices (h, k, l) can be calculated by taking any lattice point as an origin and errecting axes in $\bar{a}$, $\bar{b}$, $\bar{c}$ directions and finding the perpendicular distance between this origin and the plane under consideration. It is because we can visualize another plane parallel to this one passing through the origin. This plane has intercepts $\frac{a}{h}, \frac{b}{k}, \frac{c}{l}$ with the axis. Since the distance between these planes

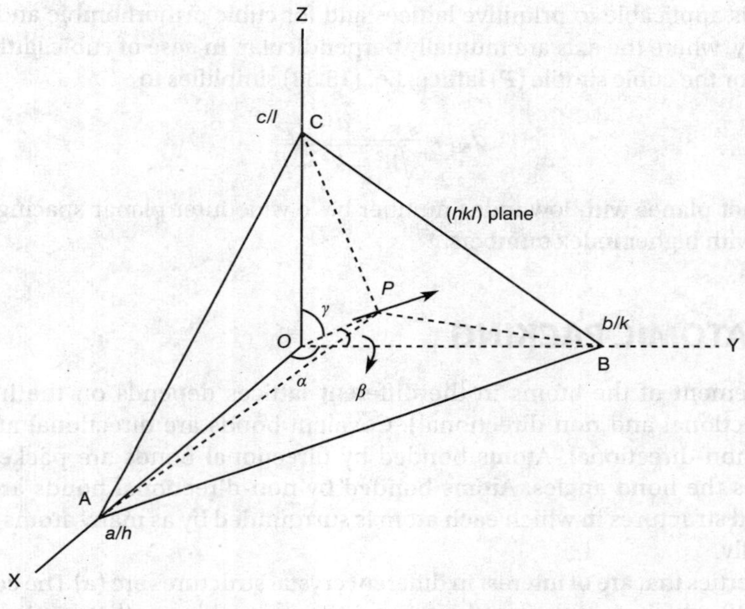

Fig. 13.13

(h, k, l) is simply the length of normal line drawn from origin to the plane shown, then it is evident from the Fig. (13.13) that this distance d_{hkl} is given by

$$\left.\begin{aligned} d_{hkl} &= \frac{a}{h}\cos\alpha \\ &= \frac{b}{k}\cos\beta \\ &= \frac{c}{l}\cos\gamma \end{aligned}\right] \tag{13.2}$$

Here α, β, γ are the angles between the normal to the plane and $\bar{a}, \bar{b}, \bar{c}$ axes respectively or $\cos\alpha$, $\cos\beta$ and $\cos\gamma$ are the direction cosines of the normal with respect of the axis. Then from Eq. (13.2), we have

$$\cos\alpha = \frac{h}{a}d_{hkl}, \cos\beta = \frac{k}{b}d_{hkl} \text{ and } \cos\gamma = \frac{l}{c}d_{hkl}$$

But
$$\cos^2\alpha + \cos^2\beta + \cos^2\gamma = 1$$

∴
$$\frac{h^2}{a^2}d_{hkl}^2 + \frac{k^2}{b^2}d_{hkl}^2 + \frac{l^2}{c^2}d_{hkl}^2 = 1$$

or
$$d_{hkl} = \frac{1}{\sqrt{\frac{h^2}{a^2} + \frac{k^2}{b^2} + \frac{l^2}{c^2}}} \tag{13.3}$$

Eq. (13.38) is applicable to primitive lattices and for cubic orthorhombic and tetragonal systems only, where the axis are mutually perpendicular. In case of cubic lattice $a = b = c$ and hence for the cubic simple (P) lattice, Eq. (13.38) simplifies to

$$d_{hkl} = \frac{a}{\sqrt{h^2 + k^2 + l^2}} \tag{13.4}$$

It is clear that planes with low index number have wide inter planar spacing compared with those with higher index numbers.

13.10 ATOMIC PACKING

The arrangement of the atoms in the different lattices depends on the inter-atomic bonds (directional and non-directional). Covalent bonds are directional and metallic bonds are non-directional. Atoms bonded by directional bonds are packed in a way that satisfies the bond angles. Atoms bonded by non-directional bonds are generally close packed structures in which each atom is surrounded by as many atoms as possible geometrically.

The properties that are of interest in different crystal structures are (a) The coordination number (CW), which gives the number of equidistant neighbours that an atom has in the given structure, (b) The radius of the atom which is defined as half the distance between the nearest neighbours which are supposed to be in contact with each other along the line of the bonding directions, (c) The relative packing density or simply the packing factor defined by the fraction of the volume occupied by spherical atoms to the total volume of the unit cell relating to that structure.

13.10.1 Atomic Packing (Two-dimensional)

Consider the packing arrangement of atomic spheres in two dimensions as shown in Fig. [13.14(a)]. Here each atom is surrounded by 4 nearest neighbours in contact. We say the coordination number of each atom is 4. In Fig. [13.14 (b)] each atom has

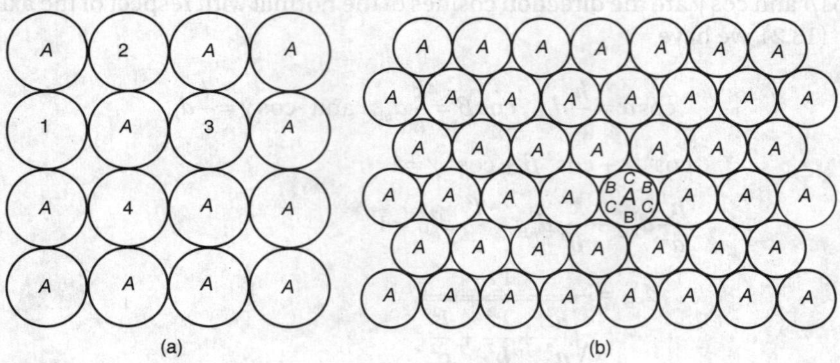

Fig. 13.14

6 nearest neighbours in contact or the coordination number of each atom is 6. Greater the coordination number greater is the closeness of packing. The arrangement in Fig. [13.14(b)] represents the closest packing of identical spheres in two dimensions. Geometrically it is not possible to place a greater number of indentical spheres in contact with any one sphere. Since crystals are three dimensional structures the crystal structure is obtained by stacking the different atomic planes one above the other. Different methods of stacking the different atomic planes gives rise to different structures. Seventy percent of metallic structures crystallise in face centred cubic or base centred cubic or hexagonal closed packed structures. The remaining metallic elements are either hexagonal rhombohedral or orthorhombic in structure. Simple cubic is rare in metals. There is also a certain periodicity in the occurrence of the different structure. The alkali metals Li, Na, K, Rb, Cs etc., crystallise in bee structures. The elements of the second group crystallise in hexagonal (hcp) structures. The sub group metals Cu, Ag and generally at least one allotropic form of the transition elements and rare earths crystallize in fcc (Face centred) structures.

13.11 SIMPLE CUBIC

In Fig. [13.14(a)] the atom sites are marked by A. Suppose the next layers and the other successive layers are stacked exactly above the A sites we get the simple cubic (SC) structure. In the SC structure the atoms are situated at the corners of the cube touching each other along the edges. Each atom is surrounded by 6 nearest neighbours. So the coordination number is 6. In Fig. [13.15(a)] one unit cell of the simple cubic structure in shown and in Fig. [13.15(b)] the atoms are shown as hard spheres arranged in the SC structure.

The atomic radius r is $\frac{a}{2}$, where a is the cube edge. Since each atom is common to the orders of 8 cubes the contribution of each atom to the unit cell is 1/8. Since there are eight

(a) Simple cubic (SC)

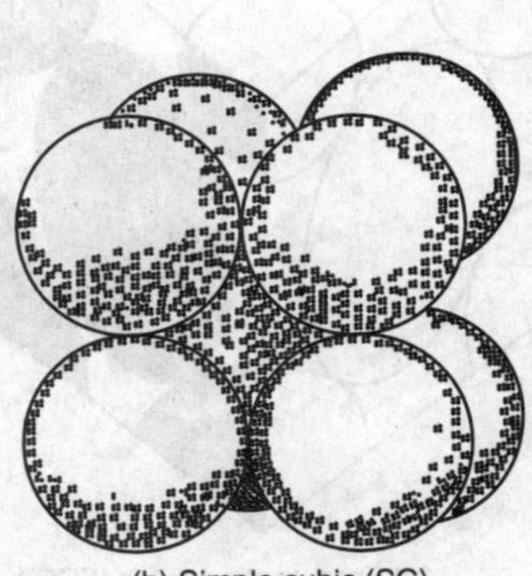

(b) Simple cubic (SC)

Fig. 13.15 (a) Simple Cubic (SC) (b) Simple Cubic (SC)

corners in each unit cell the number of atoms per unit cell is 1. Volume of unit cell is a^3. Volume of the atoms in the unit cell is $\frac{4}{3}\pi r^3$. The relative packing density or the packing factor is defined as the ratio of the volume of the atoms occupying the unit cell to the volume of the unit cell. Thus in the SC structure the packing factor is

$$\text{Packing factor} = \frac{\text{Volume of one atom (assummed hardsphere)}}{\text{Volume of unit cell}}$$

$$= \frac{\frac{4}{3}\pi r^3}{a^3} = \frac{4}{3}\pi\left(\frac{a}{2}\right)^3\frac{1}{a^3} = \frac{\pi}{6} \quad \because \quad \left(r = \frac{a}{2}\right)$$

This structure is exhibited by P_o at a certain temperature.

13.12 FACE CENTRED CUBIC LATTICE STRUCTURE: (FCC)

In Fig. [13.14(b)] the atom sites are marked *A* and the depressions are marked either *B* or *C*. The second layer of atoms can be arranged either on *A* or *B* or *C* sites. The corresponding arrangements are termed *AA* or *AB* or *AC*. If the repetition sequence is *ABC ABC ABC A* the structure is fcc. Here in Fig [13.16.(a)] one unit cell of the fcc structure is shown and in Fig. [13.16(b)] the atoms are shown as hard spheres arranged in the FCC structure.

In this case there are atoms at the corners of the unit cell and at the centre of each face. This structure is symmetrical with respect to rotation by 90° about any one of the cube edges. The face centred cubic structure is also closed-packed. Each unit cell

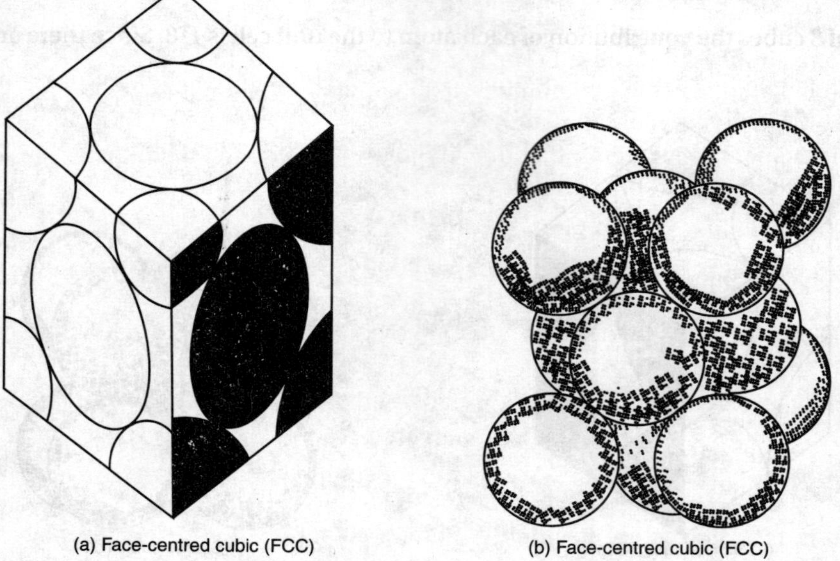

(a) Face-centred cubic (FCC) (b) Face-centred cubic (FCC)

Fig. 13.16 (a) Face-Centered Cubic (FCC) (b) Face-Centered Cubic (FCC)

of volume a^3 contain 4 atoms and each atom has twelve nearest neighbours. Thus the coordination number is 12. In the close packing, the hard sphere radius is determined using the condition that face diagonal is equal to $4r$ (r is radius of the sphere) because the atoms are in contact along diagonal of faces. Thus

$$(4r)^2 = 2a^2, \qquad \text{where } a \text{ is cube edge}$$

or
$$r = \frac{a\sqrt{2}}{4} = \frac{a}{2\sqrt{2}}$$

The elements having face-centred cubic structure are Ag, Al, Au, Cu, Pb, Ca, Pt, etc. The fraction of the total volume filled, i.e., packing factor is

$$\text{Packing factor} = \frac{\text{Volume of atoms (assumed hard spheres in unit cell)}}{\text{Volume of unit cell}}$$

$$= \frac{4\times\frac{4}{3}\pi r^3}{a^3} = 4\times\frac{4}{3}\pi\left(\frac{a}{2\sqrt{2}}\right)^3 \cdot \frac{1}{a^3} = \frac{\pi}{3\sqrt{2}} = 0.74$$

13.13 BODY CENTRED CUBIC STRUCTURE (BCC)

Another cubic structure, only slightly less closely packed than fcc is the body centred cubic structure (BCC). The repetition sequence of this is *AB AC AB AC* Fig. [13.14(b)]. The frame work for the atoms is a cubic array, but in addition there is an atom at the centre of the cube Fig. [13.17(a)] and [13.17(b)]. There are two atoms per unit cell of volume a^3. Since a corner atom is shared among 8 cells, the contribution of each atom to unit cell is 1/8. Since there are eight corners in each unit cell thus the contribution of these atoms towards unit cell is 1. The body centred atom belongs exclusively to the cell under consideration. Thus the number of atoms per unit cell is 2. The cell is therefore non-primitive.

In a closed packing, the atoms assumed as hard spheres are in contact along the body diagonal. Thus the hard sphere radius is determined by the condition

$$(4r)^2 = a^2 + \left(\sqrt{2}a\right)^2 = 3a^2 \qquad \text{or} \qquad r = \frac{\sqrt{3}}{4}a$$

$$\text{Packing factor} = \frac{\text{Volume of all atoms (assumed hardspheres in unit cell)}}{\text{Volume of unit cell}}$$

$$= \frac{2\times\frac{4}{3}\pi r^3}{a^3} = 2\times\frac{4}{3}\pi\left(\frac{\sqrt{3}}{4}a\right)^3 \cdot \frac{1}{a^3} = \frac{\sqrt{3}\pi}{8} = 0.68$$

Each atom in this structure has only 8 nearest neighbours and thus the coordination number is 8. This structure is characteristic of many monoatomic crystals such as metals. The alkali and many of the alkaline-earth metals crystallize in the bcc structure.

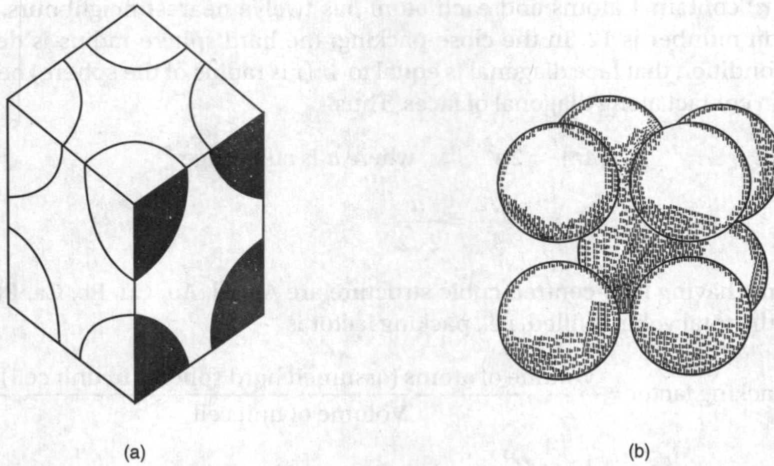

(a) (b)

Fig. 13.17 (a) Body-Centred Cubic (BCC) (b) Body-Centred Cubic (BCC)

13.14 HEXAGONAL CLOSED-PACKED STRUCTURE (HCP)

If the stacking sequence in Fig. [13.14(b)] is *AB AB ABA*, the structure obtained is called hexagonal close packed structure. In this case the frame work for atoms is a hexagonal frame work in the basal plane. Corresponding to close packing of hard spheres-like balls on a table. The next plane of atoms has the same arrangement but is shifted in such a way that the atoms fit between the gaps in the first plane. The third plane again has the same arrangement and each atom lies directly above corresponding atom in the first plane. The fourth plane is identical to the second and so on Fig. [13.18(a)] and Fig. [13.18(b)].

The unit cell Fig. [13.18(b)] is infact a primitive one, although it has a basis of two atoms at (0, 0, 0) and $\left(\frac{2}{3},\frac{1}{3},\frac{1}{2}\right)$ because, the atomic position do not form a Bravis lattice. Each atom has 6 neighbours at a distance a that lie in the basal plane through the atom at 0. There are 6 others at a distance three lying above and three below the basal plane. The ideal case of the hexagonal closed

$$a_1^1 = \left[\frac{a^2}{3}+\frac{c^2}{4}\right]^{\frac{1}{2}}$$

packing occurs when all 12 neighbours are at the same distance ($a = a_1^1$), i.e.,

$$a^2 = \frac{a^2}{3}+\frac{c^2}{4}\ {}^1$$

or
$$\frac{c}{a} = \sqrt{\frac{8}{3}} = 1.633$$

[1]The sections of hexagonal packing are shown in Fig. (13.19) in basal plane distance $OC = A/\sqrt{3}$ and $a_1^2 = (a/\sqrt{3})^2 + (c/2)^2$ or $a_1^2 = a^2/3 + c^2/4$ and $a_1 = a = 2r$ where r is atomic radius.

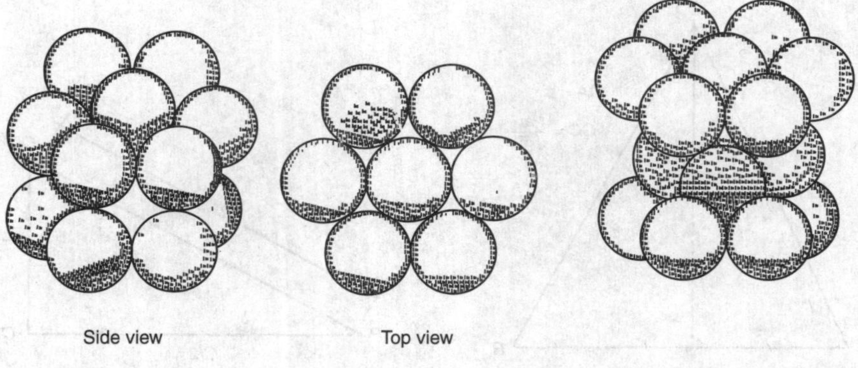

Fig. 13.18 (a)

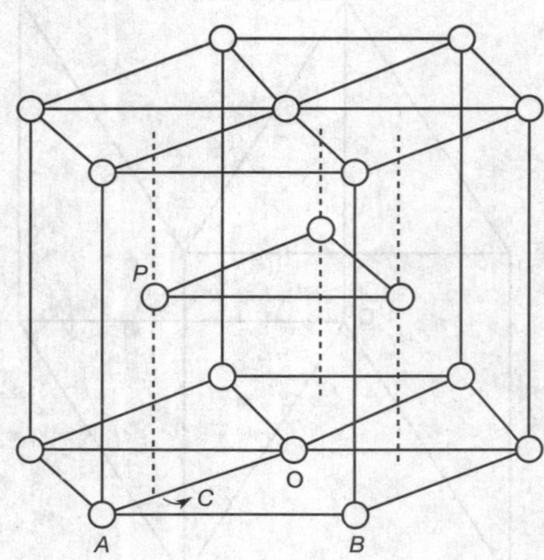

Fig. 13.18 (b)

In real crystals the values of $\frac{c}{a}$ deviates only slightly from the ideal value

$$\text{Packing fraction} = \frac{\text{Volume of the atoms in unit cell}}{\text{Volume of unit cell}}$$

In case of ideal hcp structure, the radius of hard spheres is given by $r = \frac{a}{2}$
The number of atoms per unit cell are 2

and volume of unit cell is $= 2 \times \frac{1}{2} \times \frac{\sqrt{3}}{2} a.ac = \frac{\sqrt{3}}{2} a^2 c$

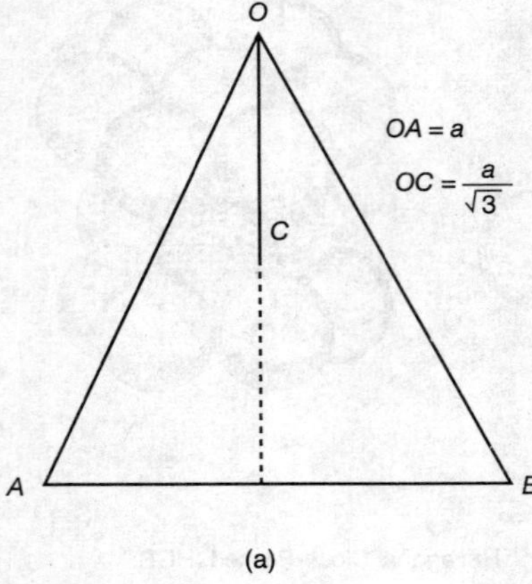

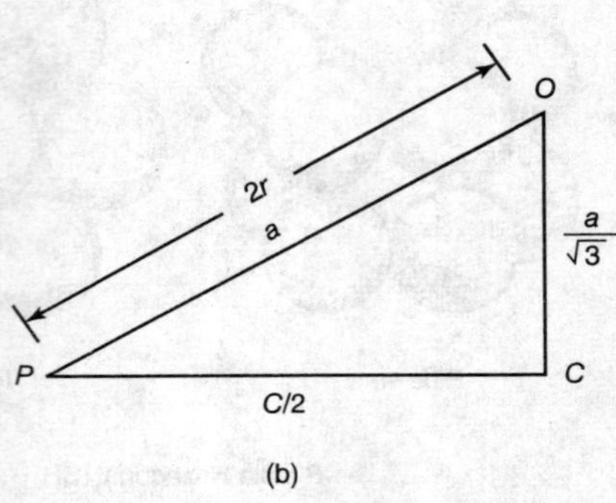

Fig. 13.19

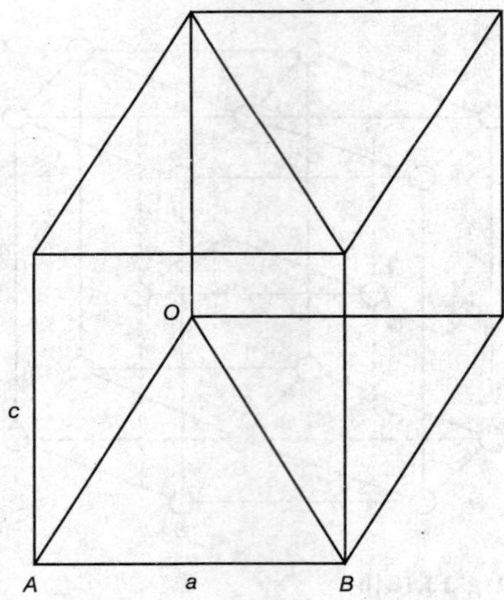

Fig. 13.20

Thus

$$\text{Packing fraction} = \frac{2 \times \frac{4}{3}\pi r^3}{\frac{\sqrt{3}}{2}a^2 c} = \frac{2 \times \frac{4}{3}\pi \left(\frac{a}{2}\right)r^3}{\frac{\sqrt{3}}{2}a^2 c}$$

$$= \frac{2\pi}{3\sqrt{3}}\frac{a}{c} = \frac{2\pi}{3\sqrt{3}}\sqrt{\frac{3}{8}} \qquad \because \frac{c}{a} = \sqrt{\frac{8}{3}}$$

$$= \frac{\pi}{3\sqrt{2}} = 0.74$$

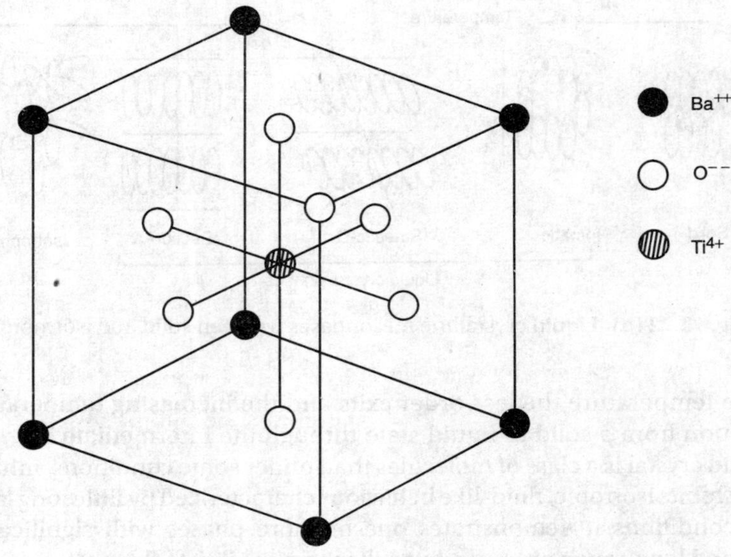

Fig. 13.21

13.15 PERVOSKITE STRUCTURE

Pervoskite structure is characteristics of the barium titanate (Ba TiO_3) class of ferro electrics in the unpolarised phase. The crystal is cubic with Ba^{++} ions at the cube corners. O^{--} ions at the centres of the cube faces and Ti^{4+} ions at the cube centre Fig. (13.21).

The first transition is to a tetragonal structure, the positive ions being displaced relative to the negative ones, along a [1 0 0] direction. The Pervoskite structure is an example of a cubic crystal in which every ion is not at a point of full cubic symmetry (the Ba^{++} and Ti^{4+} are but the O^{--} ions are not). Therefore the local field acting on the oxygen ions is more complicated than that given by the simple Lorentz formula. This is important in understanding the mechanism for the Ferro electricity.

13.16 LIQUID CRYSTALS

It is observed that many long chain organic molecules when heated undergo one or more phase transitions in passing from the solid to liquid phase, indicating the existence of one or more intermediate (meso) phases. The molecular ordering in these mesophases lie between that of a solid and an isotropic liquid. These mesophases are *'ordered fluid phases'.* It is this simultaneous possession of liquid-like and solid-like molecular order in a single phase with anistropic molecules which give rises rise to a unique, fascinating, and potential technologically relevant behaviour and exhibits a wide range of interesting properties.

The difference between above three states can be attributed to the temperature of the substance. Temperature is a measure of randomness of the molecules and therefore the

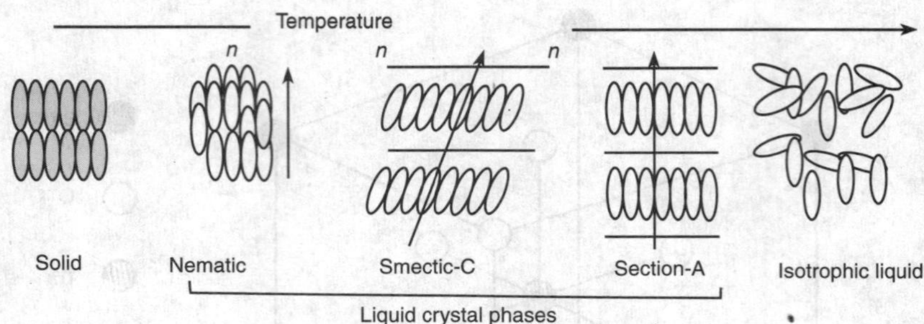

Fig. 13.22 (a) Liquid crystalline mesophases between solid and isotropic liquid

higher the temperature the less order exits and the increasing temperature will cause the transition from a solid to liquid state through the intermediate *liquid crystal state.* Thus, liquid crystal is a class of molecules that, under some conditions, inhibits a phase in which it exhibits isotropic, fluid-like behaviour characterized by little long-range ordering. In other conditions it demonstrates one or more phases with significant anisotropic structure and long-range ordering, but still with an ability to flow.

Fig. 13.22(a) shows the liquid crystalline mesophases between solid and isotropic phases.

The liquid crystals are composed of moderate and long size organic molecules with highly exotic shapes. Because of elongated shape, under appropriate conditions, the molecules can exhibit orientational order, such that all the axes line up in a particular direction. The average direction of the molecules is called the director, *n*.

Liquid crystal materials exhibit several common characteristics, e.g. rod-like molecular structure, rigidness of the long axis, and strong dipoles and/or easily polarisable.

Liquid crystals exhibit 'anistropic' physical (optical, mechanical, symmetrical, etc.,) properties and also exhibit birefringence and give interference patterns in polarised light. We may remember that true liquids are 'isotropic'.

The temperature at which an organic solid transforms to liquid crystal phase is called the *phase transition temperature* and the higher temperature at which the true isotropic liquid is obtained is referred to as the *melting temperature.*

Liquid crystals are classified into two classes:

(i) *Nematic liquid crystals*: These crystals have long molecules all alligned in the same direction, but otherwise randomly arranged. Most nematics are uniaxial but biaxial ones are also known.

(ii) *Cholesteric and Smectic liquid crystals:* These crystals also have aligned molecules, which are arranged in distinct layers. In cholesteric liquid crystals, the axes of the molecules are parallel to the plane of the layers, whereas in smectic crystals they are perpendicular. Smectic phase includes smectic A, smectic C and hexatic phases. Besides nematic phase, cholesteric phase, smectic phase, there also exist columnar phases.

Concepts from the theory of *phase transitions,* such as *order parameters* and broken symmetry, have proved useful in analysing the properties of liquid crystals.

Liquid crystals find wide use in *displays,* which rely on the change in optical properties of certain liquid crystalline molecules in the presence or absence of an electric field.

In the presence of an electric field, these molecules align with the electric field, altering the polarization of the light in a certain way. The ordering of liquid crystalline phases is extensive on the molecular scale, but does not extend to the macroscopic scale as might be found in classical crystalline solids. The ordering in a liquid crystal might extend along one dimension, but along another dimension might have significant disorder.

The application of liquid crystal display technology can be found in digital watches, calculators, flat TV displays, thermometers which show the body temperature in colour, and computers with fantastic colour graphics.

13.17 POLYMERS

A *polymer* is a large molecule built by repetition of small, simple unit called *monomer*. The repetiting chemical units are covalently linked to each other in a macromolecule. The properties of a polymer are entirely different from monomers, from which it is formed. The molar masses of polymers vary from 10^3 to 10^7.

A *monomer* is a simple molecule with two or more binding sites through which it forms covalent linkages with other monomer molecules to form the macromolecule.

Obviously, monomers are the building blocks of polymers. We may note that all simple molecules cannot behave as monomers but only those with two or more bonding sites can act as monomers. Clearly, the molecules like ammonia, water, ethanol, etc., are not monomers. Examples of monomers are: alkenes, vinyl chloride, adipic acid, glycol with two bonding sides.

Polymerization is defined as the chemical reaction in which a monomer is converted to the polymer under specific conditions. We may note that monomer alone cannot undergo polymerization, but requires the presence of a chemical called *initiator*.

$$\text{Monomer} + \xrightarrow[\text{Pressure}]{\text{Initiator required}} \text{Polymer}$$

$$n[CH_2 = \underset{\displaystyle |}{\overset{\displaystyle Cl}{C}}H \xrightarrow[\text{Initiator}]{\text{Peroxide}} (-CH_2 - \underset{\substack{\text{PVC} \\ \text{(polymer)}}}{\overset{\displaystyle Cl \\ |}{CH}} -)_n]$$

Few examples of monomers and polymeters are as follows:
Monomers: Vinyl chloride, Vinyl acetate, Ethylene, Styrene
Polymers: Polyvinyl chloride, Polyvinyl acetate, Polyethylene, Polystyrene.
Polymers can be classified under different methods:

(i) *Natural Polymers and Synthetic (manmade) Polymers:* Starch, cellulose, proteins, silk, jute, wool rubber falls under the category of natural polymers, whereas polythene, polypropene, polystyrene, PVC, baklite, nylon, polysters, etc., fall under the category of synthetic or manmade polymers.

(ii) *Addition* Polymers and *Condensation Polymers*: Addition polymers are formed by the addition of olifinic monomers without the elimination of by-products. Polythene, polypropylene, plexiglass, PVC etc., are few examples. Condensation polymers are formed by the intermolecular condensation of two different monomers with the elimination of byproducts like ammonia, water, HCl, etc.. Bakelite, nylon polyester are few examples.

(iii) According to the structure and end use, polymers are classified into four types:

(a) *Elastomers*: These undergo large elongation when stretched and retain their shape on releasing the force. Natural rubber, Buna-s rubber, etc., are few examples.

(b) *Fibres*: These are thin thread like polymers and do not undergo deformation. Wood, silk, cotton, nylon, terylene, polysters, etc., are few examples.

(c) *Resins*: These are polymers of low molar mass in liquid or semiliquid form. They may go further polymerization to give plastics (higher molar mass). Examples of resins are phenol - for maldehyde resin, urea formaldehyde resin, epoxy resin, etc.

(d) *Plastics*: These are polymers which can be moulded into desired articles by application of pressure and temperature. PVC and plexiglass are examples of plastics.

Polymerization is of two types:

(i) *Additional Polymerization*: When molecules just add to form the polymer, the process is called 'addition polymerisation'.

(ii) *Condensation Polymerisation*: When molecules do not just add on but also undergo some reaction in forming the polymer, the process is called condensation polymerisation.

Important applications of Polymers are given in Table 13.2.

Table 13.2 Applications of polymers

Group I	*Group II*
Electrostatic materials	Molecular electronics
Conducting adhesives	Electrical displays
Electromagnetic shielding	Chemical and biochemical sensors
Printed circuit board electrolytes	Rechargeable batteries and solid
Artificial nerves	Drug release systems
Antistatic clothing	Optical computers
Thermal sensors	Ion exchange membranes
Piezoceramics	Electromechanical actuators
Active electronics	Smart structures
Aircraft structures	

13.18 QUASI CRYSTALS

We have read that crystals in nature cannot have lattice periodicity different from 1, 2, 3, 4 and 6 fold symmetries (i.e., 5-fold symmetry is non-existent. This is due to the fact that for having a crystalline phase, an essential characteristic necessary for the formation of crystal is the existence of translational periodicity. In two dimensions, a solid with a

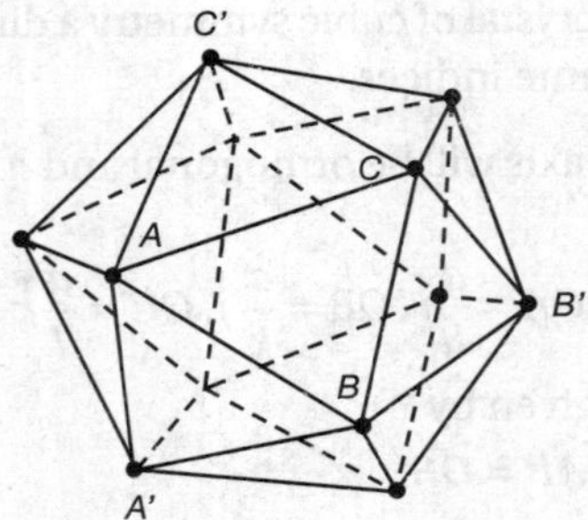

Fig. 13.22 (b) An icosahedron

five-fold symmetry of a pentagon is an example of a point symmetry, which is not allowed in crystallography, but for which quasicrystals exist. In three dimensions the symmetry of an icosahedron is not allowed in crystallography, but quasicrystals with this symmetry exist. *Basically, quasicrystal is a structure which cannot be indexed to any Bravais lattice and which has a symmetry intermediate between a crystal and a liquid.* It has a well-defined discrete point group symmetry, like a crystal, but one which is explicitly incompatible with periodic translational order (e.g. exhibiting five, eight, or twelve-fold symmetry axes). Instead, quasicrystals, possess a particular kind of translational order known as *quasi-periodicity* (a solid structure in which there is long-range incommensurate translational order). Diffraction patterns for quasicrystals have Bragg peaks, with the desity of Bragg peaks in each plane being higher than would be expected for a perfect periodic crystal. Few examples of quasicrystals are: $Al_{0.86}$ $Mn_{0.14}$, $Mg_{0.38}$ $Zn_{0.41}$, and $Cu_{0.2}$ $Al_{0.65}$ $Fe_{0.15}$.

It has been though that in their electronic band structure, there would be distribution of band gaps within the allowed energy base. This might lead to exciting new electronic properties.

The full point symmetry group of the experimentally observed diffraction patterns of these five-fold symmetry crystals is found to correspond to a platonic solid called an *icosahedron* (Fig. 13.22b).

13.19 NANOMATERIALS

Nanomaterials are characterized by atleast one dimension in the nanometer (1 nm = 10^{-9} m) range. Nanostructures constitute a bridge between molecules and infinite bulk systems. Individual nanostructures include clusters, quantum dots, nanocrystals, nanowires, and nanotubes, while collections of nanostructures involve arrays, assemblies, and superlattices of the individual nanostructures. The physical and chemical properties of nanomaterials can differ significantly from those of the atomic-molecular or the bulk materials of same composition. These assembled materials, of single or multi-phase systems, with "building blocks" in the nanometre range are called *nanocrystalline, nanostructured* or *nanophase* materials. The uniqueness of the structural characteristics, energetics, response, dynamics, and chemistry of nanostructures constitutes the basis of nanoscience. Suitable control of properties and response of nanostructures can lead to new devices and technologies. Chapter 24 is devoted to this field.

Example 13.1: Prove that in a crystal of cubic symmetry a direction (*k, k, l*) is perpendicular to the plane (*h, k, l*) with the same indices.

Solution For cubic symmetry axis will be orthogonal and $a = b = c$. If plane *ABC* has Miller indices (*h, k, l*), then

$$OA = \frac{a}{h}\hat{i},\ OB = \frac{a}{k}\hat{j},\ OC = \frac{a}{l}\hat{k}$$

Vector *AB* in the plane *ABC* is given by

$$\boldsymbol{OA} + \boldsymbol{AB} = \boldsymbol{OB}$$

or

$$\boldsymbol{AB} = \boldsymbol{OB} - OA = \frac{a}{k}\hat{j} - \frac{a}{h}\hat{i} \tag{13.5}$$

and the vector along the direction [*h, k, l*] is given the

$$\boldsymbol{R} = h\hat{i} + k\hat{j} + l\hat{k} \tag{13.6}$$

Taking dot product of *AB* and ***R*** we have

$$AB.R = \left(\frac{a}{k}\hat{j} - \frac{a}{h}\hat{i}\right)\left(h\hat{j}\hat{i} + k\hat{j} + l\hat{k}\right) = 0$$

This shows that the plane [*h, k, l*] is perpendicular to the direction [*h, k, l*]. Here $\hat{i}, \hat{j}, \hat{k}$ are the unit vectors along the three orthogonal axis.

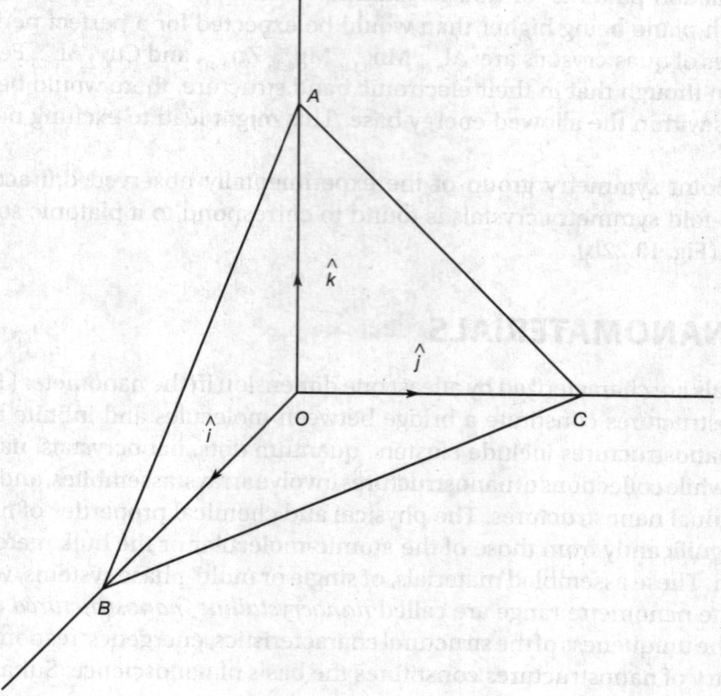

Fig. 13.23

Example 13.2: Determine the primitive vectors and primitive cell volume for a BCC lattice.

Solution Fig. (13.24) shows primitive translation vectors ***a, b*** and ***c*** of body centred cubic lattice. These vectors connect the lattice point at the origin to lattice points at the body centre. The primitive cell is obtained on completing the rhombohedron. In terms of the cube edge a primitive translation vectors are

$$\boldsymbol{a}' = \frac{a}{2}\left(\hat{i}+\hat{j}+\hat{k}\right),\ \boldsymbol{b}' = \frac{a}{2}\left(-\hat{i}+\hat{j}+\hat{k}\right) \text{ and } \boldsymbol{c}' = \frac{a}{2}\left(\hat{i}+\hat{j}+\hat{k}\right)$$

$$\therefore \quad |\boldsymbol{a}'| = \frac{\sqrt{3}a}{2} = |\boldsymbol{b}'| = |\boldsymbol{c}'|$$

$$\cos\gamma = \frac{\boldsymbol{a}'\boldsymbol{b}'}{|\boldsymbol{a}'||\boldsymbol{b}'|} = \frac{-\frac{1}{4}+\frac{1}{4}-\frac{1}{4}}{\frac{3}{4}} = -\frac{1}{3}$$

or
$$\gamma = \cos^{-1}\left(-\frac{1}{3}\right) = 109^{\circ}28'$$

where γ is the angle between primitive vectors $\boldsymbol{a}'$ and $\boldsymbol{b}'$. Thus the angle between vectors $\boldsymbol{b}'$ and $\boldsymbol{c}'$ and between $\boldsymbol{c}'$ and $\boldsymbol{a}'$ is also $\gamma[= \cos^{-1}(-1/3)]$. Volume of primitive cell $= \boldsymbol{a}'.\,\boldsymbol{b}'\times\boldsymbol{c}' = \frac{1}{2}a^3$

Example 13.3: NaCl crystallises as cubic. Taking the molecular weight of NaCl as 58.46 and density at room temperature as 2.167×10^3 kg/m^3. Calculate lattice constant.

$$\left(\text{Avogadro number} = 6.023\times10^{26} \text{ per kg-mol}\right)$$

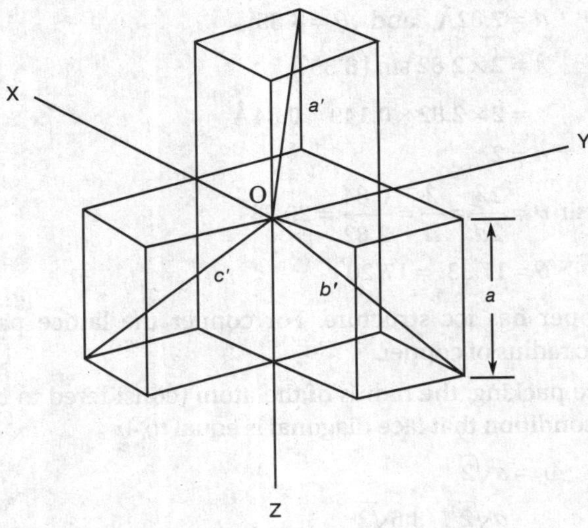

Fig. 13.24

Solution Mass of NaCl molecule is

$$= \frac{58.46}{6.02 \times 10^{26}} = 9.7 \times 10^{-26}\ \text{kg}$$

and the number of NaCl molecules per unit volume

$$= \frac{2.167 \times 10^3}{9.7 \times 10^{-26}} = 2.23 \times 10^{28}\ \text{molecules/m}^3$$

Since NaCl is diatomic, the number of atoms per unit volume

$$= 2 \times 2.23 \times 10^{28} = 4.46 \times 10^{28}\ \text{atoms/m}^3$$

Let a be the lattice constant, i.e., the distance measured along the edge of the cube between adjacent atoms in the crystal and n be number of atoms along the edge of a 1 meter cube, then the length of an edge is na and the volume of unit cube is n^3a^3 since n^3 is the number of atoms in 1m^3, hence

$$n^3 a^3 = 4.46 \times 10^{28} \times a^3 = 1$$

or
$$a^3 = \frac{1}{4.46 \times 10^{28}}$$

or
$$a = 2.82 \times 10^{-10}\ \text{m} = 2.82\ \text{Å}.$$

Example 13.4: The spacing between successive (100) planes in NaCl is 2.82 Å. X-rays incident upon the surface of this crystal is found to give rise to first order Bragg reflection at a grazing angle of 8° 35′. Calculate the wave length of the X-ray and the angle at which the second order Bragg reflection would occur.

Solution We have for Bragg reflectionv

$$2d \sin\theta = n\lambda$$

For
$$n = 1, 2d \sin\theta = \lambda$$

here
$$d = 2.82\text{Å} \quad \text{and} \quad \theta = 8^\circ 35'$$

∴
$$\lambda = 2 \times 2.82 \sin\left(8^\circ 35'\right)$$

$$= 2 \times 2.82 \times 0.149 = 0.84\,\text{Å}$$

For
$$n = 2$$

$$\sin\theta = \frac{2\lambda}{2d} = \frac{\lambda}{d} = \frac{0.84}{2.82} = 29.78$$

∴
$$\theta = 17.33^\circ = 17^\circ 20'$$

Example 13.5: Copper has fcc structure. For copper the lattice parameter a = 3.6Å. Calculate the atomic radius of copper.

Solution In fcc close packing, the radius of the atom (considered to be hard sphere) r is determined by the condition that face diagonal is equal to $4r$

∴
$$4r = a\sqrt{2}$$

or
$$r = \frac{a\sqrt{2}}{4} = \frac{3.6\sqrt{2}}{4} = 1.273\text{Å}.$$

Example 13.6: **Show that the density of lattice points in lattice planes is $\frac{nd}{V}$ were V is the volume of unit cell, d the spacing between neighboring planes and n is number of lattice points per unit cell.**

Solution Consider N successive lattice planes with spacing d and area of cross section A. The volume of this part of the lattice is NAd.

If the density of lattice points in these planes is δ (i.e. the number of lattice points per unit area is δ), then the total number of lattice points in the same part of the lattice is $A\delta N$. Further if the volume of unit cell is V, there are n lattice points per unit cell, we get

$$A\delta N = \frac{NA}{V}n$$

or

$$\delta = \frac{nd}{V}$$

For primitive lattice, there is one lattice point per unit cell ($n = 1$), for such cases,

$$\delta = \frac{d}{V}$$

Example 13.7: Copper has fcc structure and its atomic radius is 1.273 Å. Obtain (i) the bond length (ii) the density of copper. (atomic weight of copper = 63.5 and Avogadro number = 6.02×10^{23} molecules per gm mole).

Solution (i) In the fcc system of crystals, atomic radius $r = \frac{a}{2\sqrt{2}}$, where a is lattice parameter or bond length. So that bond length $a = 2\sqrt{2}(r) = 2\sqrt{2} \times 1.273 = 3.6$Å

(ii) We know that density is given by $\rho = \frac{nA}{Na^3}$

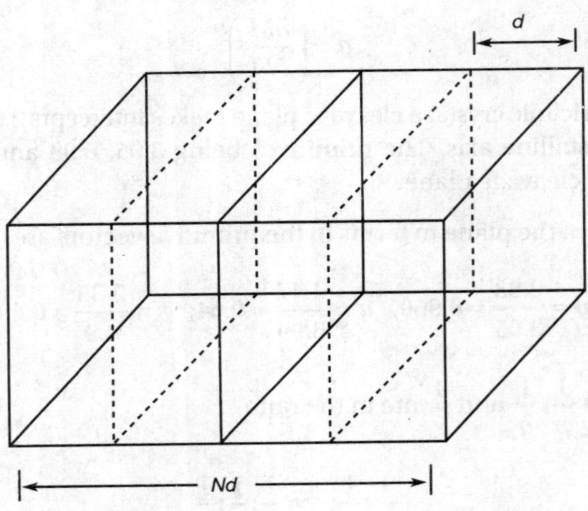

Fig. 13.25

Where n is number of atoms per unit cell, A is atomic weight N is Avogadro number and a lattice parameter. Since copper has fcc structure and hence the number of atoms per unit cell, $n = 4$.

$$\therefore \qquad \rho = \frac{4 \times 63.5}{6.02 \times 10^{26} \times \left(3.6 \times 10^{-10}\right)^3}$$

$$N = 6.02 \times 10^{26} \text{ molecules per kg mole}$$

$$\therefore \qquad \rho = 0.9043 \times 10^4 = 9.043 \times 10^3 \text{ Kg/m}^3.$$

Example 13.8: Show that for cubic lattice the parameter a is given by $a = \left(\frac{nM}{Nd}\right)^{\frac{1}{3}}$, where M is molecular weight, d is density, N is Avogadro's number and n is number of molecules per unit cell.

Solution If M is the molecular weight of the material and N the Avogadro number, we have mass of one molecule $= \frac{M}{N}$. If n be the number of atoms in unit cell, we have mass of all the molecules in unit cell $m = \frac{n.M}{N}$. If volume of the unit cell is V, then density

$$\rho = \frac{m}{V} = \frac{nM}{VN}$$

In cubic lattice if a is lattice parameter then $V = a^3$

$$\therefore \qquad \rho = \frac{nM}{Na^3}$$

or

$$a^3 = \frac{nM}{N\rho}$$

or

$$a = \left(\frac{nM}{Nd}\right)^{\frac{1}{3}}$$

Example 13.9: In a triclinic crystal a cleavage plane makes intercepts 2.93, 4.47 and 2.35 mm along the three crystalline axis. The primitives being 3.05, 6.99 and 4.99Å. Reduce the Miller indices of the cleavage plane.

Solution Intercepts of the plane in terms of the primitive vectors are

$$p = \frac{2.93}{3.05} = 0.960, \quad q = \frac{4.47}{6.99} = 0.64, \quad r = \frac{2.35}{4.9} = 0.48$$

Thus the reciprocals $\frac{1}{p}, \frac{1}{q}$ and $\frac{1}{r}$ are in the ratio

$$\frac{1}{p} : \frac{1}{q} : \frac{1}{r} = \frac{1}{6} : \frac{1}{4} : \frac{1}{3}$$

and the smallest integers having the ratio $\frac{1}{6}:\frac{1}{4}:\frac{1}{3}$ are (2: 3: 4). Thus the Miller indices of the plane are (2 3 4).

Example 13.10: In an orthorhombic crystal whose primitives are 1.21, 1.84 and 1.97 Å. A cleavage plane whose Miller indices are (231) cuts intercepts 1.68 mm along the *x*-axis. Deduce the lengths of intercepts along *y* and *z*-axis. Also deduce the lattice spacing between the planes.

Solution The Miller indices of the family of planes are 2, 3, 1. Their reciprocals are in the ratio

$$\frac{1}{h}:\frac{1}{k}:\frac{1}{l}=\frac{1}{2}:\frac{1}{3}:\left(-\frac{1}{1}\right)=3:2:-6$$

If l_1, l_2 and l_3 are the intercepts and axis, then

$$\begin{aligned} l_1:l_2:l_3 &= 3a:2b:-6c \\ &= 3(1.21):2(1.84):-6(1.97) \\ &= 3.63:3.68:(-11.82) \end{aligned}$$

But $$l_1 = 1.98 \text{ mm}$$

Then $$l_2 = \frac{3.68}{3.63}\times 1.68 = 1.70 \text{ mm}$$

and $$l_3 = -\frac{11.82}{3.63}\times 1.68 = -5.47 \text{ mm}$$

The spacing between the planes is given by

$$\begin{aligned} d_{hkl} &= \frac{1}{\left[\frac{h^2}{a^2}+\frac{k^2}{b^2}+\frac{l^2}{c^2}\right]^{1/2}} \\ &= \frac{1}{\left[\left(\frac{2}{1.21}\right)^2+\left(\frac{3}{1.84}\right)^2+\left(\frac{-1}{1.97}\right)^2\right]^{1/2}} \\ &= \frac{1}{[2.73+2.66+0.28]^{1/2}} = \frac{1}{[5.67]^{1/2}} = 0.42\text{Å}. \end{aligned}$$

Example 13.11: An atomic plane in a crystal lattice makes intercept of 3*a*, 4*b* and 6*c* with the crystallographic axis where *a*, *b* and *c* are the dimensions of the unit cell. Show that the Miller indices of the atomic plane are (4 3 2).

Solution In terms of the lattice constants, the intercepts are 3,4 and 6. Their reciprocals are 1/3, 1/4 1/6. On reducing to a common denominator, they become 4/12, 3/12 and 2/12. Obviously, the Miller indices of the plane are (4 3 2).

Example 13.12: In a single cubic crystal find the ratio of the intercepts on the three axis by (1 2 3) plane.

Solution The reciprocals of Miller indices are 1/1, 1/2 and 1/3. On reducing to a common denominator they become 6, 3 and 2. Intercepts on the three axis are $6a$, $3b$ and $2c$, where a, b and c are the lattice constants along the three axis.

Example 13.13: Draw (1 0 1) and (1 1 1) planes in a unit cubic cell. Find the Miller indices of the directions which are common to both the planes.

Solution Intercepts of the plane (1 0 1) with the three axis are 1/1, 1/0 and 1/1, i.e., 1, ∞ and 1. Similarly, the intercepts of the (1 1 1) with the three axis are 1/1, 1/1 and 1/1 i.e., 1, 1 and 1.

Now, taking the point 0 as origin and lines *OA*, *OB* and *OC* as the axis *a*, *b* and *c* respectively, the plane with the intercepts 1, ∞ and 1 is the plane *ADGC* and that with intercepts 1, 1 and 1 is plane *ABC* Fig. (13.26). From figure it is obvious that the line common to both the planes is the line *AC*. This line corresponds to two directions, i.e., *AC* and *CA*.

Projections of the direction *AC* on the axis are –1, 0 and 1. Projection of direction *CA* on the axis are 1, 0 and –1.

Thus the required indices are and $[\bar{1}\ 0\ 1]$ and $[1\ 0\ \bar{1}]$.

Example 13.14: Draw the following planes and directions in the case of a FCC structure: (1 1 2), (0 0 1) and (1 0 1).

Solution

(i) **Plane and direction (1 1 2):** In this case, we have $h = 1$, $k = 1$ and $l = 2$. The reciprocals of h, k and l are 1/1, 1/1, 1/2, i.e., 1, 1, 0.5.

Now, we can sketch the plane with intercepts 1, 1, 0.5 along $x - x$, $y - y$ and $z - z$ axis respectively Fig. [13.27(a)].

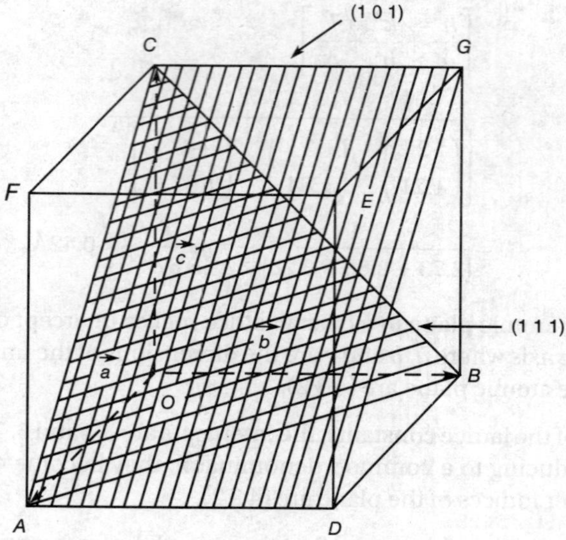

Fig. 13.26

(ii) **Plane of (0 0 1):** Here $h = 0$, $k = 0$, $l = 1$. The reciprocals are 1/0, 1/0, 1/1, i.e., ∞, ∞ 1. We obtain the sketch of the plane as shown in Fig. [13.27(b)].

(iii) **Plane of (1 0 1):** Here $h = 1$, $k = 0$ and $l = 1$. The reciprocals are 1/1, 1/0 and 1/1 i.e., 1, ∞ and 1. We obtain the sketch as shown in Fig. [13.27(c)].

Example 13.15: A plane makes intercepts of 1, 2 and 0.5 Å on the crystallographic axis of an orthorhombic crystals with $a:b:c = 3:2:1$. Find the Miller indices of this plane.

Solution Taking the lengths of the axis *OA, OB* and *OC* as 3, 2 and 1 Å respectively, we obtain the plane with intercepts of 1, 2 and 0.5 Å on the axis is the plane *DBC* Fig. (13.28). The intercepts of this plane relative to full lengths of the axis are: 1/3, 2/2 and 0.5/1 or 1/3, 1 and 1/2. Reciprocals are: 3, 1 and 2. Obviously, the Miller indices of the plane *DBE* are (312).

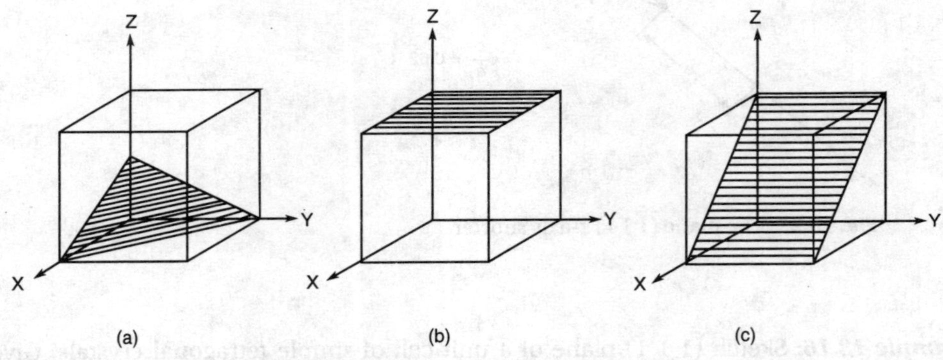

Fig. 13.27 (a) Plane (1 1 2) (b) Plane (0 0 1) (c) Plane (1 0 1)

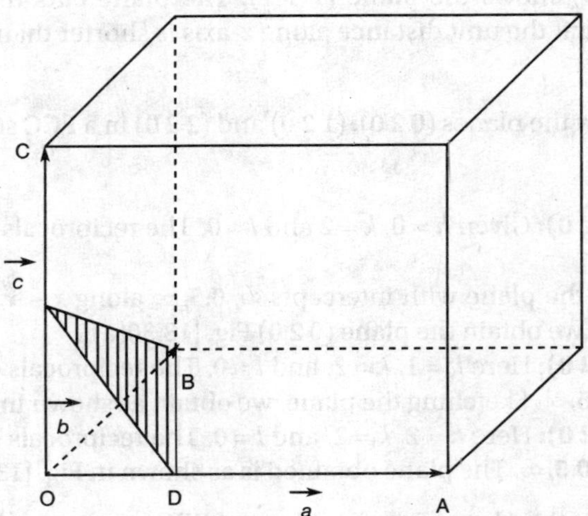

Fig. 13.28

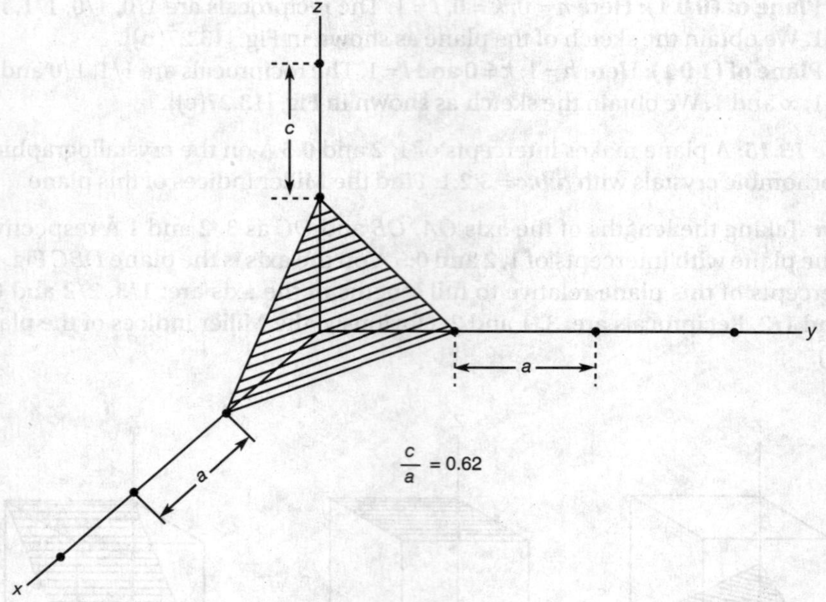

Fig. 13.29 The plane (1 1 1) *z*-axis shorter

Example 13.16: Sketch (1 1 1) plane of a unit cell of simple tetragonal crystals. Given $c/a = 0.62$.

Solution Fig. (13.29) shows the plane (1 1 1). The plane cuts the three axis at unit distances. We note that the unit distance along *z*-axis is shorter than the unit distance of other two axis.

Example 13.17: Draw the planes (0 2 0), (1 2 0) and (2 2 0) in a FCC structure.

Solution

(i) **Plane of (0 2 0):** Given $h = 0$, $k = 2$ and $l = 0$. The reciprocals are 1/0, 1/2, 1/0 i.e., ∞, 0.5, ∞.

Sketching the plane with intercepts ∞, 0.5, ∞ along $x - x$, $y - y$ and $z - z$ axis respectively, we obtain the plane (0 2 0) Fig. [13.30(a)].

(ii) **Plane of (1 2 0):** Here $h = 1$, $k = 2$, and $l = 0$. The reciprocals of *h, k, l* are 1/1, 1/2, 1/0, i.e., 1, 0.5, ∞. Sketching the plane, we obtain as shown in Fig. [13.30(b)].

(iii) **Plane of (2 2 0):** Here $h = 2$, $k = 2$, and $l = 0$. The reciprocals of *h, k, l* are 1/2, 1/2, 1/0, i.e., 0.5, 0.5, ∞. The plane obtained is as shown in Fig. [13.30(c)].

Example 13.18: Draw the planes and directions of FCC structures (3 2 1), (1 0 2), (2 0 1) and (1 1 1).

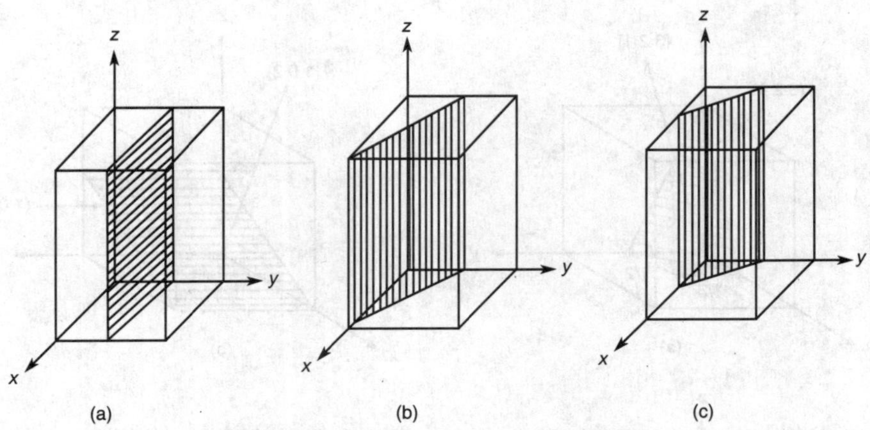

Fig. 13.30 (a) Plane (0 2 0) (b) Plane (1 2 0) (c) Plane (2 2 0)

Solution

(i) **Plane and Direction of (3 2 1):** Here $h = 3$, $k = 2$, and $l = 1$. The reciprocals are 1/3, 1/2, 1/1, i.e., 0.3, 0.5, 1. The sketch of the plane with intercepts 0.3, 0.5 and 1 is as shown in Fig. [13.31(a)]. A line drawn normal to this sketched plane and passing through the origin gives the required direction.

(ii) **Plane and Direction of (1 0 2):** Here $h = 1$, $k = 0$ and $l = 2$. The reciprocals of these are: 1/1, 1/0 and 1/2, i.e., 1, ∞, 0.5. The sketch of the plane with these intercepts is as shown in Fig. [13.31(b)]. A line drawn normal to this sketched plane and passing through the origin gives the required direction.

(iii) **Plane and Direction of (0 1 1):** Here $h = 0$, $k = 1$ and $l = 1$. The reciprocals of h, k and l are: 1/0, 1/1, 1/1 i.e., ∞, 1,1. The sketch of the plane with these intercepts is shown in Fig. [13.31(c)]. A line drawn normal to this sketched plane and passing through the origin gives the required direction.

(iv) **Plane and Direction of (0 0 1):** Here $h = 0$, $k = 0$ and $l = 1$. The reciprocals of h, k and l are: 1/0, 1/1, i.e., ∞,∞ 1. The sketch of the plane with these intercepts is shown in Fig. [13.31(d)]. A line drawn normal to this sketched plane and passing through the origin gives the required direction.

Example 13.19: In a cubic unit cell, find the angle between normals to the plane (1 1 1) and (1 2 1).

Solution Since the given crystal is cubic, the normals to the planes (1 1 1) and (1 2 1) are the directions [1 1 1] and [1 2 1] respectively. If θ be the angle between the normals, then

$$\cos\theta = \frac{h_1h_2 + k_1k_2 + l_1l_2}{\left(h_1^2 + k_1^2 + l_1^2\right)^{1/2}\left(h_2^2 + k_2^2 + l_2^2\right)^{1/2}}$$

or $$\cos\theta = 0.9428$$

or $$\theta = 19.47^\circ \text{ or } 19^\circ 28'.$$

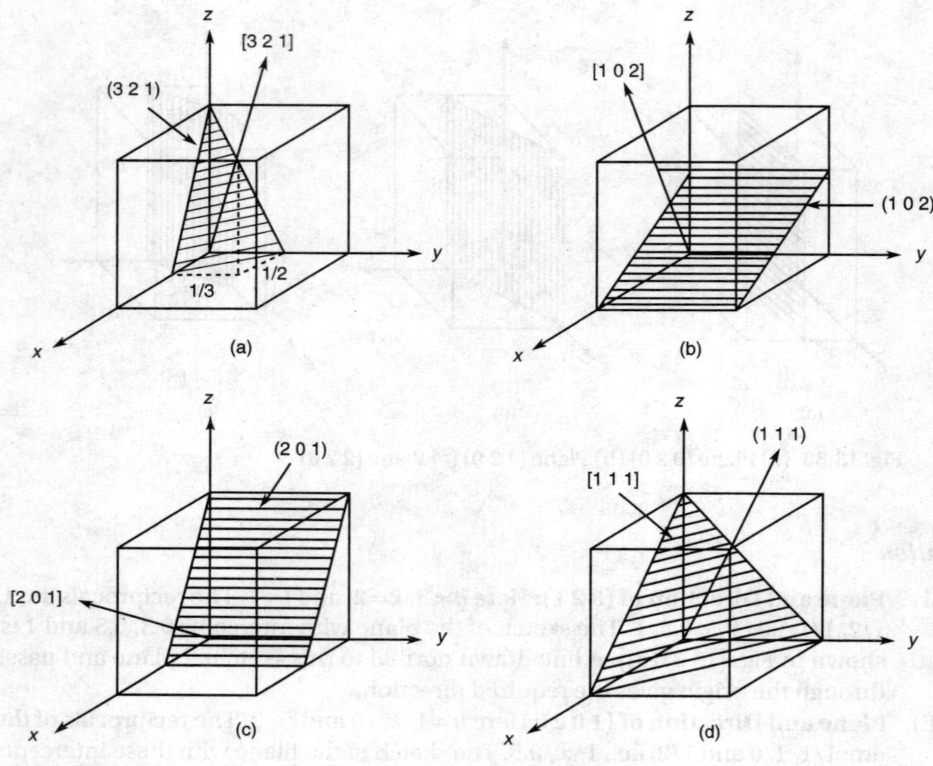

Fig. 13.31

Example 13.20: **Determine the packing efficiency and density of sodium chloride from the following data: (i) radius of the sodium ion = 0.98 Å, (ii) radius of chlorine ion = 1.81 Å (iii) atomic mass of sodium = 22.99 amu and atomic mass of chlorine = 35.45 amu.**

Solution The unit cell structure of NaCl is shown in Fig. (13.32). We can see that the Na^+ and Cl^- ions touch along the cube edges.

Lattice Parameter, $a = 2(\text{radius of Na}^+ + \text{radius of Cl}^-)$

$$= 2(0.98 + 1.81) = 5.58\text{Å}$$

$$\text{Atomic Packing Fraction} = \frac{\text{Volume of ions present in the unit cell}}{\text{Volume of the unit cell}}$$

$$= \frac{4(4/3)\pi r_{Na^+}^3 + 4\left(\frac{4}{3}\right)\pi r_{Cl^-}^3}{a^3}$$

$$= \frac{16\pi}{3}\left[\frac{(0.98)^3 + (1.81)^3}{(5.58)^3}\right] = 0.663 \text{ or } 66.3\%$$

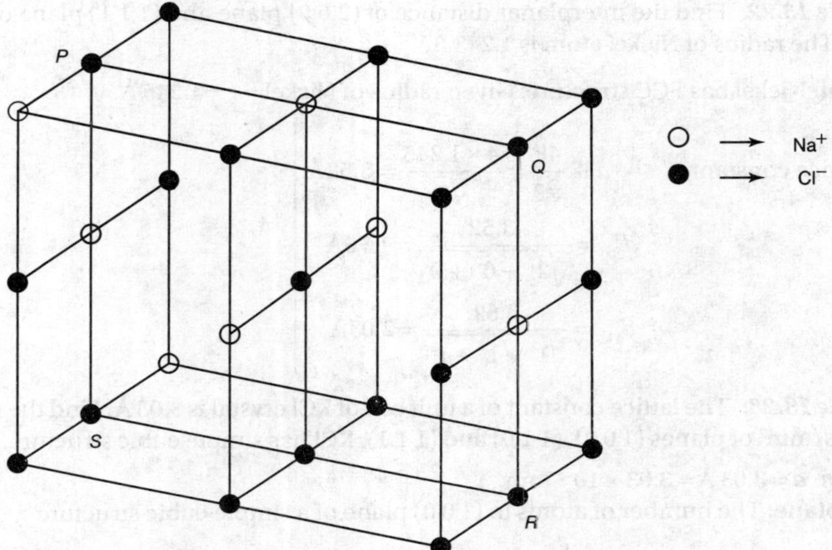

Fig. 13.32 Sodium chloride structure

$$\text{Density} = \frac{\text{Mass of the unit cell}}{\text{Volume of the unit cell}}$$

$$= \frac{4(22.99 + 35.45) \times 1.66 \times 10^{-27}}{(5.58 \times 10^{-10})^3} \text{kg/m}^3$$

$$= 2234 \text{ kg/m}^3 \text{ or } 2.33 \text{ g/cm}^3.$$

Example 13.21: Aluminium has FCC structure. Its density is 2700 kg/m³. Find the unit cell dimensions and atomic diameter. Given at weight of Al = 26.98.

Solution

$$\text{Density} = \frac{nm}{a^3} N_A$$

$$= 2700 \text{ kg/m}^3 = 2.7 \text{ g/cm}^3$$

$$= \frac{4 \times 26.98}{a^3 \times 6.023 \times 10^{23}}$$

$$\therefore \quad a^3 = \frac{4 \times 26.98}{2.7 \times 6.023 \times 10^{23}} = 6.6 \times 10^{-23} \text{ /cm}^3$$

$$\therefore \quad a = 4.048 \times 10^{-10} \text{ m} = 4.048 \text{Å}$$

For FCP structure,

$$r = \frac{a}{2 \times 1.414} = \frac{4.048}{2 \times 1.414} = 1.43 \text{Å}$$

$$\text{Diameter} = 2r = 2.86 \text{Å}.$$

Example 13.22: Find the interplanar distance of (2 0 0) plane and (1 1 1) plane of Nickel crystal. The radius of Nickel atom is 1.245 Å.

Solution Nickel has FCC structure. Given radius of Nickel = r = 1.245 Å.

$$\therefore \quad \text{Lattice constant,} \quad \left.\begin{aligned} a &= \frac{4r}{\sqrt{2}} = \frac{4\times 1.245}{\sqrt{2}} = 3.52\text{Å} \\ d_{200} &= \frac{3.52}{\sqrt{2^2+0^2+0^2}} = 1.76\text{Å} \\ d_{111} &= \frac{3.52}{\sqrt{1^2+1^2+1^2}} = 2.03\text{Å} \end{aligned}\right\}$$

Example 13.23: The lattice constant of a unit cell of KCl crystal is 3.03 Å. Find the number of atoms/mm² of planes (1 0 0), (1 1 0) and (1 1 1). KCl has simple cubic structure.

Solution a = 3.03 Å = 3.03 × 10⁻⁷ mm.

(1 0 0) plane: The number of atoms in (1 0 0) plane of a simple cubic structure

$$= \frac{1}{a^2} = \frac{1}{\left(3.03\times 10^{-7}\right)^2} = 10.9\times 10^{12}$$

(1 1 0) plane: The number of atoms in (1 1 0) plane of simple cubic structure

$$= \frac{0.707}{a^2} = \frac{0.707}{\left(3.03\times 10^{-7}\right)^2} = 7.7\times 10^{12}$$

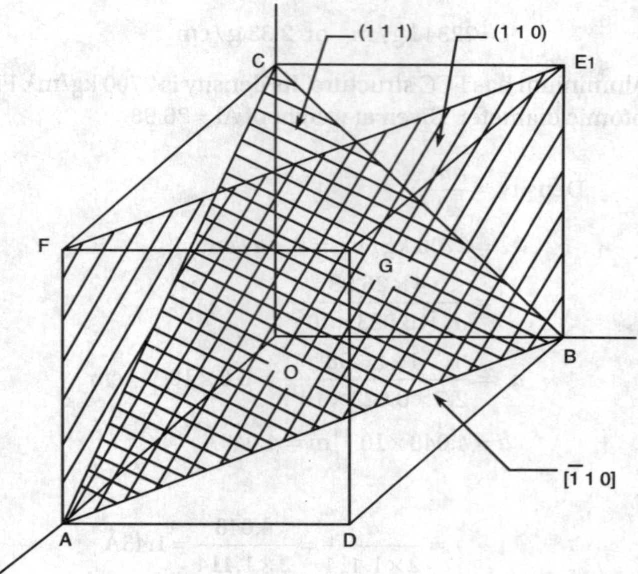

Fig. 13.33 Plane in a unit cell of Nickel crystal

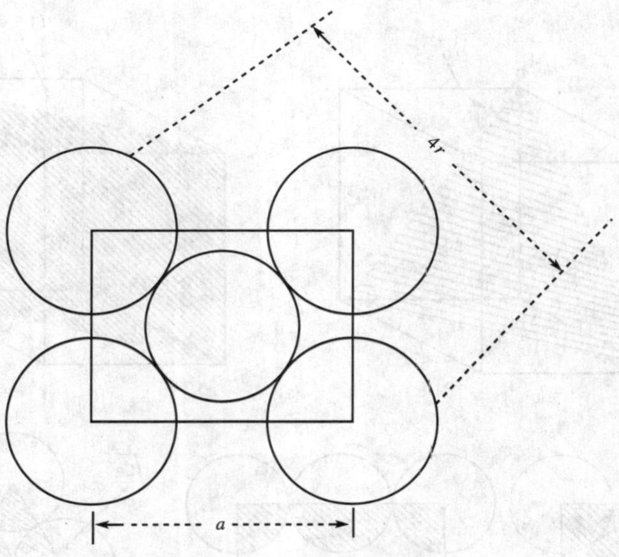

Fig. 13.34 Interplanar distances in Ni crystal

(1 1 1) plane: The number of atoms in (1 1 1) plane of a simple cubic structure

$$= \frac{0.58}{a^2} = \frac{0.58}{\left(3.03 \times 10^{-7}\right)^2} = 6.3 \times 10^{12}.$$

Example 13.24: Determine the planar density of Ni (FCC structure) in (1 0 0) plane. Given, the radius of Ni atom = 1.245 Å.

Solution From Fig. (13.34), we have number of atoms in (1 0 0) plane

$$= 1 + \frac{1}{4} \times 4 = 2$$

$$\therefore \qquad \text{Area of plane} = a^2,$$

$$a = \frac{4r}{\sqrt{2}} = 3.52\text{Å}$$

$$\therefore \qquad \text{Planar density} = \frac{2}{\left(3.52 \times 10^{-7}\text{mm}^2\right)} = 16.1 \times 10^{12}\ \text{atoms/mm}^2$$

Example 13.25: Calculate the planar atomic densities of planes (1 0 0), (1 1 0) and (1 1 0) in FCC unit cell and apply your result for lead (FCC form).

Solution

(i) **Plane (1 0 0):** Fig. [13.35(a)] shows an FCC unit cell with planes (1 1 1) and (1 1 1). Fig. [13.35 (b)] shows the plane (1 0 0) with atoms on it. Similarly Fig. [13.35(c)] and [13.35(d)] shows the plane (110) and (1 1 1) with atoms contained in them respectively.

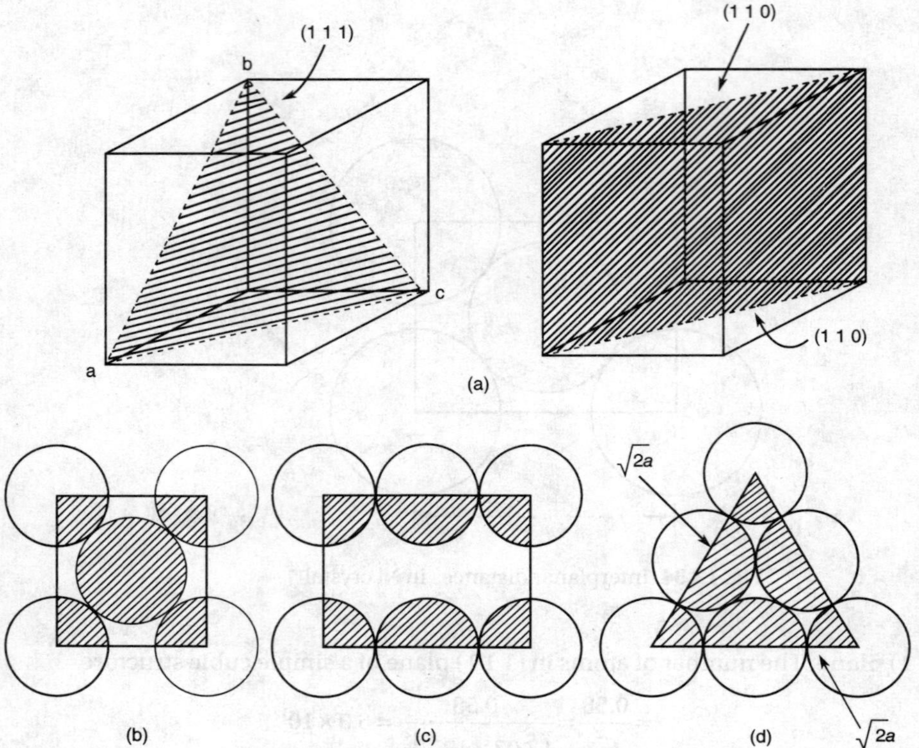

Fig. 13.35 Distribution of atoms in planes (1 0 0), (1 1 0) and (1 1 1) in FCC unit cell

Number of atoms contained in (1 0 0) plane is $4\times\frac{1}{4}+1=2$

Let a be the edge of the unit cell and r the radius of the atom, then $a=2\sqrt{2}r$

$$\therefore \qquad \text{Planar density of plane (100)} = \frac{2}{4\times 2r^2} = \frac{0.25}{r^2}$$

The radius of lead atom is 1.75 Å. The planar density of (1 0 0) plane of lead

$$= \frac{0.25}{\left(1.75\times10^{-7}\right)^2} = 8.2\times10^{12}\ \text{atoms/mm}^2$$

$$= 8.2\times10^{18}\ \ \text{atoms/m}^2$$

(ii) **Plane (1 1 0):** From Fig. [13.35(c)], we have the number of atoms contained in plane (1 1 0)

$$=4\times\frac{1}{4}+2\times\frac{1}{2}=2$$

The top edge of the plane (1 1 0) is 4*r*, whereas the vertical edge = $a = 2\sqrt{2}r$. Thus the planar density of (1 1 0)

$$= \frac{2}{8\sqrt{2r^2}} = 0.177/r^2$$

In case of (1 1 0) plane in lead, we have planar density $= \frac{0.177}{(1.75\times10^{-2})^2}$.

(iii) From Fig. [13.35(d)], we have the number of atoms contained in the plane (1 1 1)

$$= 3\times\frac{1}{6}+\frac{3}{2} = 2$$

$$\text{Area of (111) plane} = \frac{1}{2}\sqrt{\frac{3}{2}}a\sqrt{2}a = 4\sqrt{3}r^2$$

$$\therefore \quad \text{Planar density of (111)} = \frac{2}{4\sqrt{3}r^2} = \frac{0.29}{r^2}$$

For lead crystal, we obtain the value 9.5 × 10^{10} atoms / mm^2.

Example 13.26: Determine the linear atomic density in the (1 1 0) and (1 1 1) directions of copper crystal lattice, Lattice constant of copper (FCC) is 3.61 × 10^{-10}m.

Solution From Fig. (13.36) and earlier discussions, we note that the face diagonal along (1 1 0) direction intersects two half diameters and one full diameters. Thus the number of diameters of atom along (1 1 0) direction

$$= \frac{1}{2}+1+\frac{1}{2} = 2$$

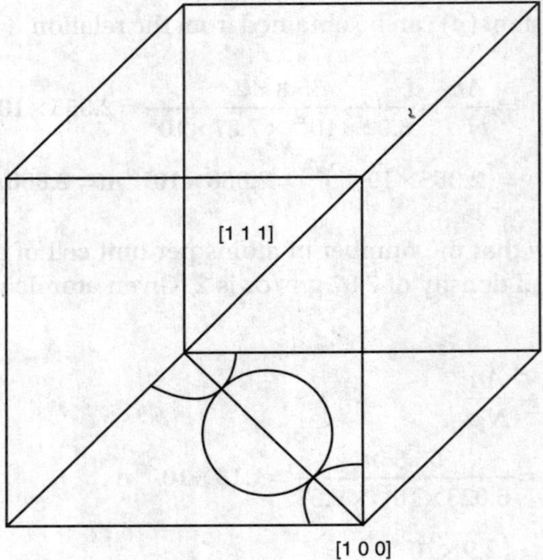

Fig. 13.36

Let a be the lattice constant, then the length of the face diagonal = $\sqrt{2}a$

∴ Linear density of (1 1 0)within unit cell

$$= \frac{2}{\sqrt{2}a} = \frac{\sqrt{2}}{a}$$

Since $$a = 3.61 \times 10^{-10}\,\text{m} = 3.61 \times 10^{-7}\,\text{mm}$$

∴ $$\rho_{110} = \frac{\sqrt{2}}{3.61 \times 10^{-7}} = 3.91 \times 10^{6}\ \text{atoms/mm}$$

The direction (1 1 1) is along the body diagonal. From Fig. (13.36), the length of the diagonal along (1 1 1)

$$= \sqrt{\left(\sqrt{2}a\right)^2 + a^2} = \sqrt{3}a$$

The number of atomic diameters intersected by diagonal along (1 1 1) is $\frac{1}{2} + \frac{1}{2}$ at two ends. Thus the linear density along (1 1 1) within the crystal unit cell

$$= \frac{1}{\sqrt{3}a} = \frac{1}{\sqrt{3} \times 3.61 \times 10^{-7}} = 1.6 \times 10^{6}\ \text{atoms/mm}$$

Obviously, in FCC the linear density along (1 1 0) direction is greater than that along (1 1 1) direction.

Example 13.27: The density of α – Fe is 7.87 × 10³ kg/m³. **Atomic weight of Fe is 55.8. If α –Fe crystallizes in BCC space lattice, find lattice constant. Given Avogadro's number (N) = 6.02 × 10²⁶ /kg/mole.**

Solution Lattice constant (a) can be obtained from the relation

$$a^3 = \frac{An}{N\rho} = \frac{55.8 \times 2}{6.02 \times 10^{26} \times 7.87 \times 10^{3}} = 2.355 \times 10^{-29}$$

$$= \left(2.355 \times 10^{-29}\right)^{1/3} = 2.866 \times 10^{-10}\,\text{m} = 2.866\text{Å}.$$

Example 13.28: Show that the number of atoms per unit cell of a metal having a lattice parameter of 2.9 Å and density of 7.87 gm/cc is 2. Given atomic weight of metal = 55.85 and N = 6.023 × 10²³.

Solution $$a^3 = \frac{An}{N\rho}$$

or $$\left(2.9 \times 10^{-3}\right)^3 = \frac{55.85n}{6.023 \times 10^{23} \times 7.87} = 1.18 \times 10^{-23}\,\text{n}$$

∴ $$n = \frac{\left(2.9 \times 10^{-8}\right)^3}{1.18 \times 10^{-23}} = 2.$$

13.20 DEFECTS OR IMPERFECTIONS IN CRYSTALS

Up to now, we have described perfectly regular crystal structures, called **ideal crystals** and obtained by combining a basis with an infinite space lattice. In ideal crystals atoms were arranged in a regular way. However, the structure of real crystals differs from that of ideal ones. **Real crystals always have certain defects or imperfections**, and therefore, the arrangement of atoms in the volume of a crystals is far from being perfectly regular.

Natural crystals always contain defects, often in abundance, due to the uncontrolled conditions under which they were formed. The presence of defects which affect the colour can make these crystals valuable as gems, as in ruby (chromium replacing a small fraction of the aluminium in aluminium oxide: Al_2O_3). Crystal prepared in laboratory will also always contain defects, although considerable control may be exercised over their type, concentration, and distribution.

The importance of defects depends upon the material, type of defect, and properties which are being considered. Some properties, such as density and elastic constants, are proportional to the concentration of defects, and so a small defect concentration will have a very small effect on these. Other properties, e.g., the colour of an insulating crystal or the conductivity of a semiconductor crystal, may be much more sensitive to the presence of small number of defects. Indeed, while the term defect carries with it the connotation of undersirable qualities, defects are responsible for many of the important properties of materials and much of material science involves the study and engineering of defects so that solids will have desired properties. A defect free, i.e., ideal silicon crystal would be of little use in modern electronics, the use of silicon in electronic devices is dependent upon small concentrations of chemical impurities such as phosphorus and arsenic which give it desired properties. Some simple defects in a lattice are shown in Fig. (13.37).

There are some properties of materials such as stiffness, density and electrical conductivity which are termed structure-insensitive, are not affected by the presence of defects in crystals while there are many properties of greatest technical importance such

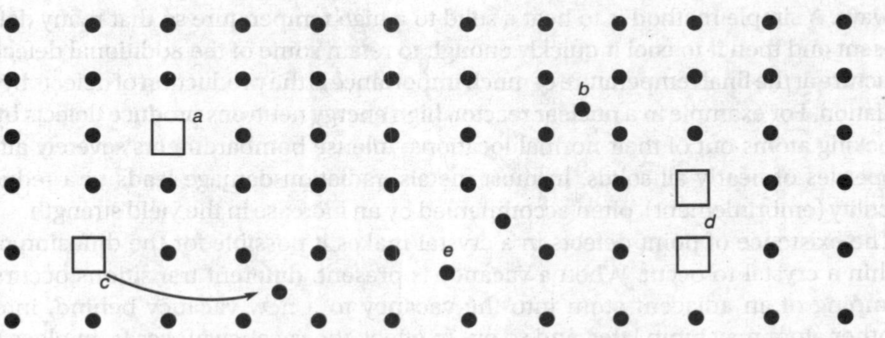

Fig. 13.37 Some simple defects in a lattice

as mechanical strength, ductility, crystal growth, magnetic hysteresis, dielectric strength, condition in semiconductors, which are

Key:

a = vacancy (Schottky defect)
b = interstitial
c = vacancy-interstitial pair (Frenkel defect)
d = divacancy
e = split interstitial
□ = vacant site

termed structure sensitive are greatly affected by the relatively minor changes in crystal structure caused by defects or imperfections. Crystalline defects can be classified on the basis of their geometry as follows:

(i) Point imperfections
(ii) Line imperfections
(iii) Surface and grain boundary imperfections
(iv) Volume imperfections

The dimensions of a point defect are close to those of an interatomic space. With linear defects, their length is several orders of magnitude greater than the width. Surface defects have a small depth, while their width and length may be several orders larger. Volume defects (pores and cracks) may have substantial dimensions in all measurements, i.e., at least a few tens of Å. Now, we will discuss these crystalline imperfections.

(i) Point Defects

Localized point defect is the simplest category of defect. Point defects in an ionic crystal must be such as to leave the crystal electrically neutral. A Schottky defect is the simultaneous occurrence of a positive ion vacancy and a negative ion vacancy Fig. (13.37). A Frankel defect is the simultaneous occurrence of a positive ion interstitial and a positive ion vacancy.

Frankel and Schottky both defects are present in all ionic solids. However, one or the other defect prominates in any given crystal because their energies of formation are usually different. This is why Schottky defects prevail in alkali halides and Frenkel defects in silver halides.

Point defects in excess of the number in thermal equilibrium may be created in a number of ways. A simple method is to heat a solid to a high temperature so that many defects are present and then it to cool it quickly enough to retain some of the additional defects in the structure at the final temperature of much importance is the production of defects by particle radiation. For example in a nuclear reactor, high energy neutrons produce defects by simply knocking atoms out of their normal locations. Intense bombardments severely affects the properties of nearly all solids. In most metals, radiation damage leads to a reduction in ductility (embrittlement), often accompanied by an increase in the yield strength.

The existence of point defects in a crystal makes it possible for the diffusion of atoms within a crystal to occur. When a vacancy is present, different transitions occurs by the slumping of an adjacent atom into the vacancy to a new vacancy behind, into which another atom may jump later, and so on. In effect, the vacancy moves from place to place through the crystal. When an interstitial is present, diffusion occurs as it migrates from one interstitial location to a neighbouring one. In both cases each change in position requires a certain amount of extra energy (~ 1 eV), in order for the moving atom to squeeze past the atoms that project in its path. We know that diffusion in a solid is strongly temperature dependent, increasing from a usually negligible rate at room temperature to one not far from that characteristic of the corresponding liquid near the melting point.

Diffusion in an ionic crystal enables it to conduct electric current, since the moving defects are either charged interstitials or vacancies that behave as though they carry charges opposite to those of the missing ions. We may note that the study of the electrical conductivity of ionic crystals have provided much detailed information on point defects, information that would have been difficult to obtain in any other way.

13.20.1 Dislocations

Line defects in a crystal are called dislocations. A dislocation is a line of atoms with different coordinations from those of other atoms in the crystal. Dislocations are of two types: **(i) edge dislocation (ii) screw dislocation**. Fig. (13.38) shows an edge dislocation. Edge dislocation can be visualized in terms of the removal of part of one layer of atoms and the subsequent accomodation of the array of the defect. In Fig. (13.38), the dislocation itself is indicated by the symbol $\perp$, and in its immediate neighborhood the crystal structure is severely distorted.

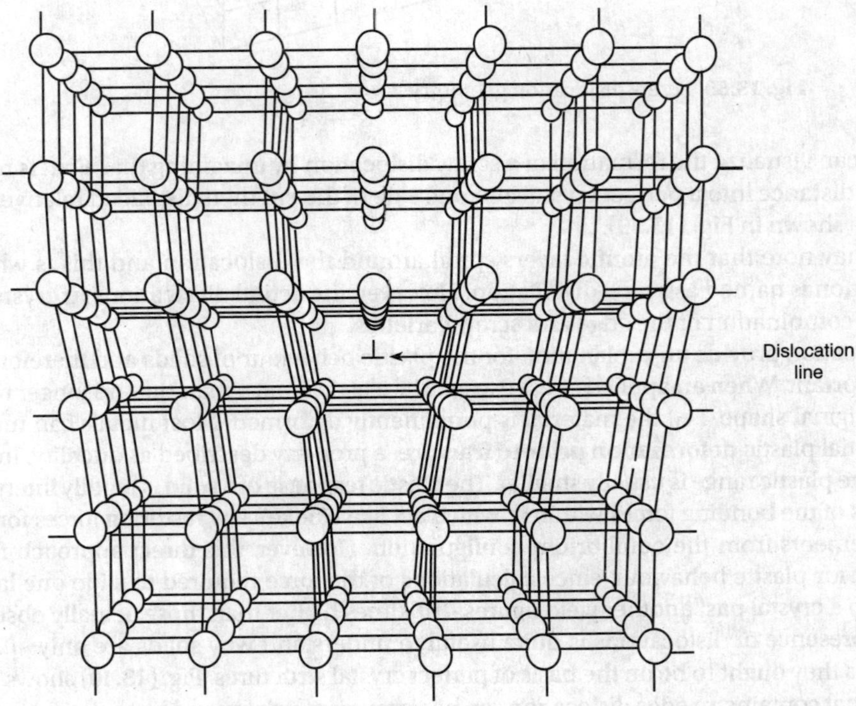

Fig. 13.38 Edge dislocation in a simple cubic crystal

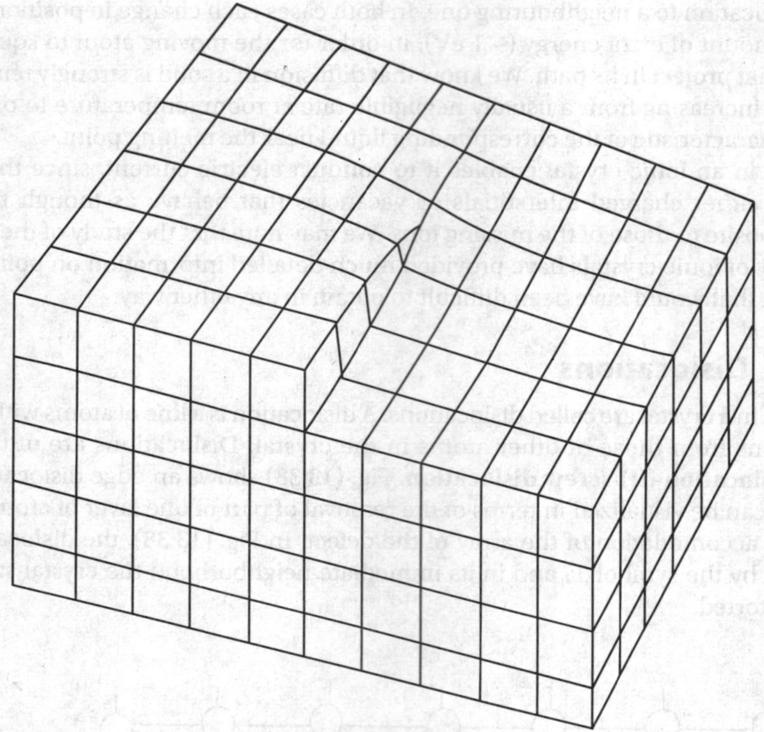

Fig. 13.39 Screw dislocation geometry

One can visualize the formation of a screw dislocation by imagining that a cut is made a certain distance into a perfect crystal and one side of the cut then displaced relative to the other as shown in Fig. (13.39).

We may note that the atomic layers spiral around the dislocation and this is why this dislocation is named as screw dislocation. However, the actual dislocations in crystals are usually combination of the edge and screw varieties.

Dislocation provide an explanation for the plastic behaviour of solids and therefore they are important. When an applied stress exceeds its elastic limit, a materials no longer returns to its original shape, i.e., the material is permanently deformed. Most metals can undergo substantial plastic **deformation prior to fracture**, a property described as ductility. In other solids the plastic range is usually smaller. The elastic response of a solid is readily interpreted in terms of the bonding forces within it, which act like Hooke's law restoring forces for small displacements from the equilibrium configuration. However, this direct approach fails to account for plastic behaviour, since calculations of the force required to slide one layer of atoms in a crystal past another yield figures ~10^3 times higher than those actually observed.

The presence of dislocations is quite useful to understand why solids are only ~0.1% as strong as they ought to be on the basis of perfect crystal structures. Fig. (13.40) shows how a crystal that contains an edge dislocation can be permanently deformed by a relatively modest shearing stress. When the shear is applied the line of atoms below and to the right of the

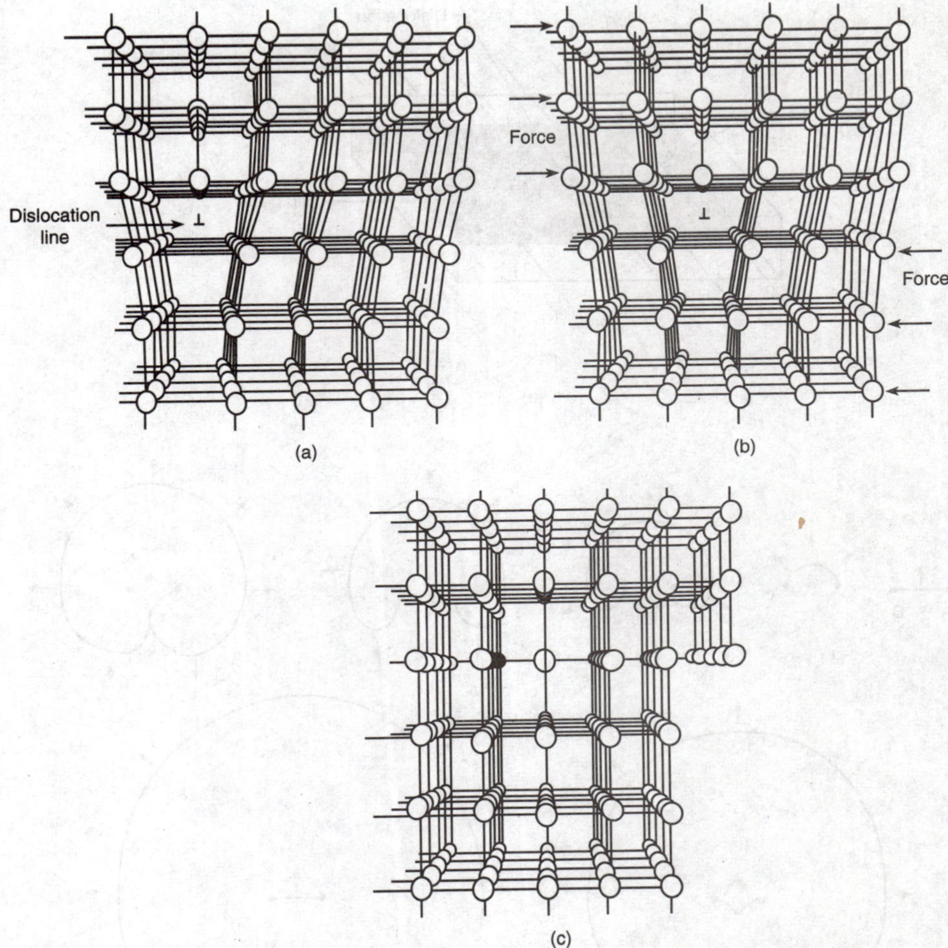

Fig. 13.40 The motion of a dislocation through a crystal under stress causes slip. (a) Initial configuration of a crystal (b) As the atoms in the layer under it successively shift their bonds with those of the layer under it successively shift their bonds with those of the upper layer one line at a time (c) A crystal after acquiring a permanent deformation

dislocation shifts its bonds to the line of atoms directly about it. This causes the dislocation to move one atom spacing to the right. The process is repeated until the dislocation arrives at the edge of the crystal, and the deformation is then permanent. The entire process is called **slip**. The plane along which dislocation moves is called slip plane. In slip, the atomic bonds holding one layer to the next are broken only line at a time, whereas in a perfect crystal all the bonds between two layers would have to be broken more or less simultaneously for plastic flow to take place. Obviously, this is a much more formidable proposition.

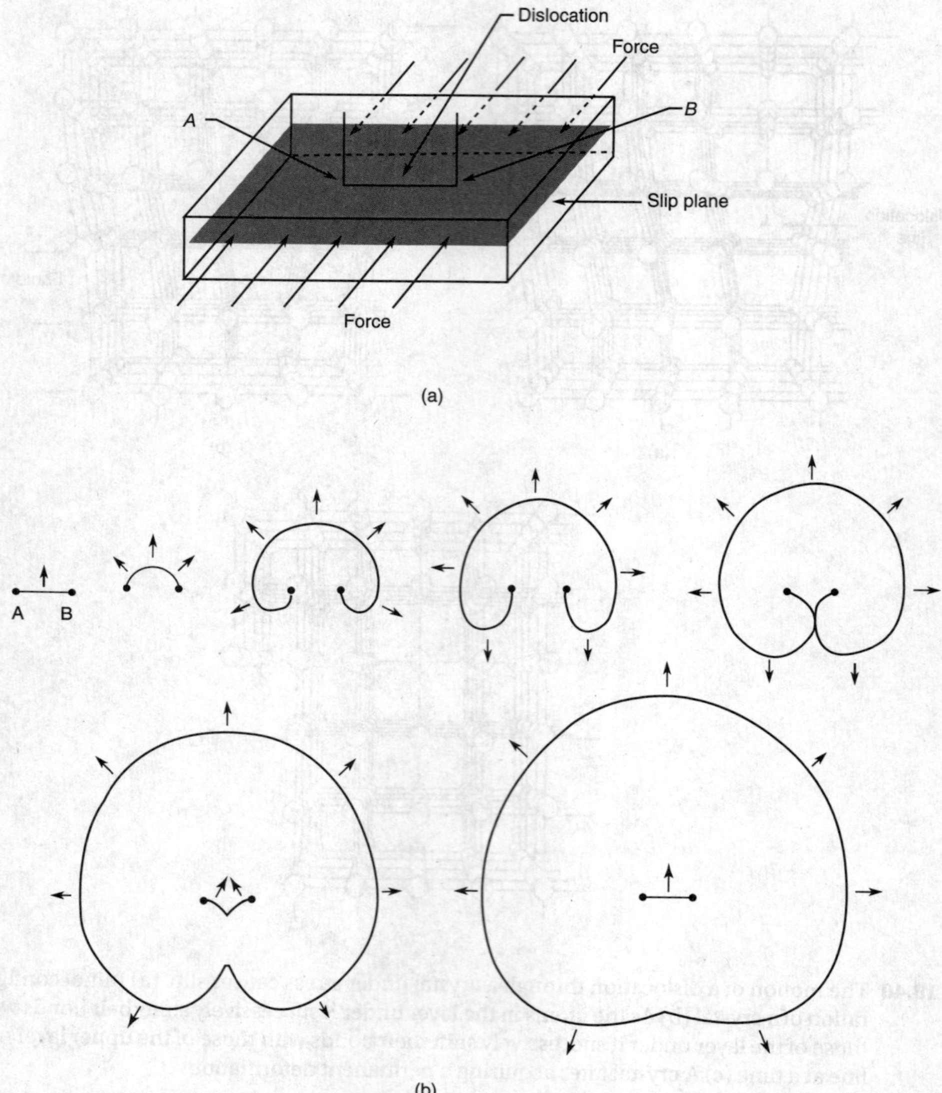

Fig. 13.41 Frank-Read source that multiplies dislocations in a crystals under stress (a) Initially, the dislocation segment *AB* in the slip plane. We may note that the ends of the segment are anchored at *A* and *B* since the dislocation changes the direction at these points and slip only occurs in the indicated plane. The applied shear causes the dislocation segment to move outward, and it eventually expands until the spiral atoms come together at the rear to form a complete dislocation loop. Under the influence of the shear, the remaining cusp then becomes a straight segment again and the process repeats itself

The available evidence indicates that dislocations come into being naturally in crystals during their formations. We can say that it is exceedingly difficult to produce crystals free of dislocations. We may note that such perfect crystals turn out to be phenomenally strong. Moreover, dislocations multiply during plastic flow and the contiued deformation of a solid increases its dislocation content. This means that the dislocations become so numerous and so tangled together that they impede one another's motion. This causes to decrease the plasticity of the materials. This phenomenon is named as work hardening and becomes prominent when there are about 10^8 dislocations/sq.cm. in a slip plane. When a work hardened crystal is heated its disordered atomic array tends to return to regularity. The process of heating a work hardened crystal to restore its ductility is called annealing.

Besides work hardening, another means of impeding slip in a metal is the delibrate introduction of foreign atoms to interfere with the motion of dislocations.

SUMMARY

1. A *crystal* is formed by the three-dimensional periodic arrangement of atoms, ions, or molecules.

2. The arrangement of atoms in a crystal is periodic. One can replace each atom by a geometrical point-located at the equilibrium position of that atom. The resulting pattern of these imaginary spatial points is called *crystal lattice* and the points are termed as *lattice* points.

3. The position of any lattice point w.r.t. the origin of coordinates at the lattice point O can be written as

$$\boldsymbol{T} = n_1\boldsymbol{a} + n_2\boldsymbol{b}$$

where n_1 and n_2 are arbitrary integers. **T** is called the lattice (or crystal) *translation vector.*

4. A space lattice can be constructed by using the basis vectors $\boldsymbol{a}$, $\boldsymbol{b}$ and $\boldsymbol{c}$ along the three directions x, y, and z of the crystallographic axes. With respect to any point as origin, the location of any other lattice point is

$$\boldsymbol{T} = n_1\boldsymbol{a} + n_2\boldsymbol{b} + n_3\boldsymbol{c}$$

5. $\underset{\text{(solid)}}{\text{Crystral structrure}} = \text{Lattice} + \text{Basis}.$

The arrangment of atoms/molecules in a crystal is termed as the *crystal structure.* One can obtain the entire structure of a crystal by repetition of a certain pattern unit called *unit cell.* The basic building block of the unit cell is known as the *basis* of the *unit cell.*

6. A *unit cell* is defined as a parallelopiped bound by the three translation vectors $\boldsymbol{a}$, $\boldsymbol{b}$, and $\boldsymbol{c}$ along the three crystallographic axes x, y, and z respectively. The volume of this unit cell is $(\boldsymbol{a} \times \boldsymbol{b}) \cdot \boldsymbol{c}$ and α, β and γ are angles between three three basis vectors $\boldsymbol{a}$, $\boldsymbol{b}$ and $\boldsymbol{c}$. If this volume in space contains only one lattice point, we call it as the *primitive unit cell.* A larger cell containing more than one lattice point as a unit cell is called a *non-primitive unit cell.*

7. A symmetry operation is one which leaves the crystal invariant.
8. There are 32 allowed crystal point groups and 230 space groups in three dimensions.
9. There are seven crystal systems corresponding to different possible shapes the unit cell have and are characterised by a set of symmetry elements associated with each point in Bravais lattice.
10. There are 14 Bravais lattices (of which one is general and 13 are special) which accomodate the 32 point groups or 230 space groups.
11. *Coordination number* is the number of equidistant nearest neighbours that an atom has in a given crystal structure.
12. The distance between the centres of two nearest neighbouring atom is called *nearest neighbour distance.*
13. For a crystal containing atoms of same size, the *atomic radius* is half the distance between nearest neighbours.
14. The ratio of volume of atoms occupying the unit cell to the volume of unit cell relating to that structure is called *atomic packing factor* (APF).
15. A crystal lattice may be considered as a set of parallel planes passing through the lattice points. The orientation of a crystal plane is represented by the three numbers h, k, l, called the *Miller indices.* If the lengths of the intercepts on the X, Y and Z axes by a crystal plane are l_1, l_2 and l_3, respectively, then
$$h:k:l=\frac{1}{l_1}:\frac{1}{l_2}:\frac{1}{l_3}$$
16. Most metals crystallize in one of the three structures: (i) face centred cubic, (ii) body centred cubic of cubic system, and (iii) hexagonal closed - packed structure, (a special example of the simple hexagonal lattice.
17. The set of actual symmetry elements of the crystalline lattice is called its *space group.*
18. We know that crystals in nature cannot have lattice periodicity different from 1, 2, 3, 4 and 6 - fold symmetries (i.e., 5-fold symmetry is non-existent). This is due to the fact that having a crystalline phase, an essential characteristic necessary for crystal formation is the existence of translational periodicity. A solid with a 5-fold symmetry will not have translational periodicity, and hence thought to be non-existent. However, there are physical systems with many properties of usual crystalline state but without three-dimensional lattice periodicity. Such quasiperiodic systems are termed as *quasicrystals.*
19. An arrangement of certain kinds of long molecules in a liquid with the result that the liquid does not suffer an entire loss of fluidity is termed as *liquid crystal.* These crystals have characteristic optical properties.
20. The concept of an ideal crystal with a perfect arrangement of atoms is strictly speaking, valid only at the absolute zero temperature because then there is no entropy contribution. At a finite temperature, a certain native configurational disorder is introduced into the structure and the solid becomes structurally imperfect. Almost, in all cases, a crystal cannot be grown without an initial imperfection.
21. There are three broad categories of imperfections in a crystal: (i) point defects, (ii) line defects, and (iii) surface defects.

REVIEW QUESTIONS

1. Explain the concepts of lattice, and crystal structure. How are they related?
2. Draw primitive cells corresponding to BCC and FCC unit cells.
3. Describe, how atoms are arranged in cubic crystal?
4. What is Bravis lattice? What is the maximum number of Bravis lattices possible? How will you account for the existence of thousands of structures from these lattice?
5. Write a short note on space lattice.
6. Draw net sketches of unit cells of simple cubic, BCC and FCC crystal structures. Calculate the number of atoms in each case.
7. The end-centred orthorhombic is one of the Bravis lattices but the end centred tetragonal is not. Give reasons.
8. What is the number of atoms per unit cell in the FCC structure?
9. Give a brief account of crystal directions and crystallographic planes.
10. Draw the following:
    (a) $[1\bar{1}1]$, $[1\bar{2}1]$ and $[1\bar{2}1]$ directions in cubic and tetragonal lattices.
    (b) $[\bar{1}11]$, $[1\bar{1}2]$ and $[\bar{2}10]$ planes is cubic and orthorhombic lattices.

3. Prove that in a cubic lattice the direction [*hkl*] is perpendicular to the plane (*hkl*).
4. Calculate the atomic packing factor of simple cubic, BCC and FCC crystals.
5. Draw a hexagonal closed packed unit cell. Calculate the number of atoms per unit cell. Calculate its atomic packing factor.
6. Draw a plan view of sodium chloride structure. In how many ways can this structure be interpreted?
7. Write the points of similarity and differences of the monoatomic SC, monoatomic BCC and CsCl structure.
8. What are Miller indices? How are they determined?
9. Draw atomic arrangement in planar surface imperfection with a boundary.
10. Discuss in brief the point defects. How these defects affect the properties of materials?
11. Explain Schottky and stacking faults.
12. Explain Miller Bravis system? How is this system used to find the index of different planes?
13. Write short notes on
    (a) Bravis Lattices
    (b) Coordination number
    (c) Crystal defects
    (d) Crystal growth
    (e) Crystal Whiskers
    (f) Grain boundaries
    (g) Crystallographic indices
    (h) Liquid crystals
    (i) Polymers
    (j) Quasi crystals

PROBLEMS

1. Draw a $(1\bar{1}0)$ plane in a cubic unit cell. Show all the <1 1 1> directions that lie on this plane and give the Miller indices of each direction. [**Ans.** ([1 1 1], $[11\bar{1}]$, $[\bar{1}\bar{1}1]$ and $[\bar{1}\bar{1}\bar{1}]$)]
2. A plane makes intercepts of 1,2 and 3 Å on the crystallographic axis of an orthorhombic crystal with $a : b : c = 3 : 2 : 1$. Determine the Miller indices of this plane. [**Ans.** (9 3 1)]
3. Aluminium has an FCC structure. Its density is 2700 kg/m³ (sp. gravity = 2.7). Find the unit cell dimension and the atomic diameter. [**Ans.** (4.05 Å, 2.86 Å)]
4. Show that the shortest centre to centre distance between any two atoms among the four atoms located inside the diamond-cubic unit cell is $a/\sqrt{2}$.
5. Determine the number of the nearest neighbours and the closest distance of approach in terms of lattice parameter of monoatomic SC, BCC and FCC structures. [**Ans.** (6,a; 8,$a\sqrt{3}/2$;12, $a/\sqrt{2}$)]
6. Prove that c/a ratio for an ideal HCP structure is 1.633.
7. Zinc (HCP) has lattice parameters a and c as 2.66 Å and 4.95 Å respectively. Calculate the packing fraction and density of zinc, given the atomic radius and the atomic mass of Zn as 1.31 Å and 65.37 amu respectively. [**Ans.** (62%, 7155 kg-m^{-3})]
8. Calculate the density of CsCl crystal. Given radii of Cs^+ = 1.65 Å and Cl^- = 1.81 Å. [**Ans.** (4380 kg–m^{-3})]
9. A crystal plane makes intercepts $4a$, $2b$ and $-2c$ on the crystallographic axis of a crystal whose unit cell has dimensions a, b and c. Find the Miller indices of the plane. [**Ans.** (1 2 2)]
10. Show the maximum occupied volumes by hard spheres arranged in the structures given below are

Simple cubic	$\pi/6$
FCC	$\pi 2^{1/2}/6$
Diamond	$\pi 3^{1/2}/16$

11. Copper (FCC) has density of 8960 kg-m^{-3}. Calculate the unit cell dimension and radius of Cu atom, given the atomic mass of Cu = 63.54 amu. [**Ans.** (3.61 Å, 1.28 Å)]
12. A crystal reflects monochromatic X-rays strongly when the Bragg's glancing angle for a first order reflection is 15°. Show that the glancing angles for the second and third order reflections of the same type are 31.2° and 50.9° respectively.
13. Copper (FCC) has a lattice parameter of 3.61 Å. The first order Bragg reflection from (1 1 1) planes appears at an angle 21.7°. Show that the wavelength of X-rays used is 1.54 Å.
14. An aluminium crystal has dislocation density of 10^{10} m^{-2}. The shear modulus of aluminium is 25.94 GN-m^{-2}. Show that the elastic energy of line imperfections stored in the crystal is 10.6 J-m^{-3}.
15. Calculate the spacing between dislocations in a tilt boundary in FCC nickel. Given, the angle of tilt is 2°.
16. Calculate the dislocation energy per m³ of copper (FCC) with a dislocation density of 10^{10} m^{-2}. Given, shear modulus of copper = 45 GN m^{-2} and lattice parameter = 3.61 Å. [**Ans.** (14.7 J-m^2)]

SHORT ANSWER-QUESTIONS

1. What is difference between crystalline and non crystalline solids?
Ans: Atoms in crystalline solids are positioned in an orderly and repeated pattern whereas in non crystalline or amorphous materials there is random and disordered atomic distribution.
2. How various crystal structures are specified?
Ans: The various crystal structures are specified in terms of parallelopiped unit cells, which are characterized by geometry and atom positions within.
3. What is a space lattice?
Ans: A space lattice is an infinite array of points, all with identical surroundings.
4. How a crystal structures is obtained?
Ans: A crystal structure is obtained by combining a space lattice with a basis. The basis must give the number of atoms per lattice point, their types, mutual orientations and distances of separation.
5. How many types of space lattices are there?
Ans: Fourteen. however, crystal structure run into thousands.
6. What are the three relatively simply crystal structures in which most common metals exist?
Ans: FCC, BCC and HCP.
7. Mention tow features of crystal structure.
Ans: (i) Coordination number (or number of nearest-neighbour atoms) and (ii) atomic packing (the fraction of solid sphere volume in the unit cell). Coordination number and atomic packing fraction are the same for both FCC and HCP crystal structure, each of which may be generated by the stacking of closed-packed planes of atoms.
8. How crystal directions and crystal planes are denoted?
Ans: These are denoted by Miller indices. A family of crystal directions or planes includes all possible combinations of the indices, both positive and negative. For a given crystal structure, planes having identical atomic packing yet different Miller indices belong to the same family.
9. What is a single crystal?
Ans: Single crystals are materials in which the atomic order extends uninterrupted over the entire specimen; under some circumstances, they may have flat faces and regular geometric shapes. We may note that vast majority of crystalline solids, however, are polycrystalline, being composed of many small crystals or grains having different crystallographic orientations.
10. How the space lattice and its dimensions can be determined from the powder method for X-ray diffraction?
Ans: From the position of lines in a powder pattern and from the extinction rules for different cubic crystals, the space lattice and its dimension can be determined.
11. What do you understand by crystal imperfections?
Ans: All solid materials contain large number of imperfections or deviations from crystalline perfection, and several types of them may be classified according to their geometry, and size.

12. What are point defects?
Ans: Points defects are those associated with one of two atomic positions, including vacancies (or vacant lattice sites), self-interstitials (host atoms that occupy interstitial sites), and impurity atoms. They are Frenkel and Schottky imperfections in ionic crystals.

13. What are the edge and screw dislocations?
Ans: These are the limiting types of lime imperfections. One can resolve any general dislocation into edge and screw components.

OBJECTIVE QUESTIONS

1. The number of lattice points in a primitive cell are
   (a) 3/2 (b) 1/2
   (c) 2 (d) 1
2. The number of lattice point in a rhombohedral unit cell is
   (a) 1 (b) 2
   (c) 4 (d) 16
3. The number of atoms present in the unit cell of HCP structure is
   (a) 2 (b) 4
   (c) 6 (d) 12
4. The minimum number of ions in the unit cell of an ionic crystal with FCC space lattice is
   (a) 16 (b) 12
   (c) 8 (d) 2
5. If r is the radius of the atom in a crystal crystallizing in the simple cubic structure, then the nearest neighbour distance is
   (a) $8r$ (b) $4r$
   (c) $3r$ (d) r
6. The space lattice with two parameters does not belong to the crystal system
   (a) triclinic (b) rhombohedral
   (c) hexagonal (d) tetragonal
7. The number of diad axis of symmetry elements that are present in a cubic crystal are
   (a) 6 (b) 3
   (c) 2 (d) 1
8. The nearest neighbour distance in BCC structure is
   (a) $a/\sqrt{2}$ (b) $2a/\sqrt{3}$
   (c) $a\sqrt{3}/2$ (d) $2a/\sqrt{3}$

9. The unit cell with three lattice parameter is
(a) monoclinic (b) tetragonal
(c) triclinic (d) orthorhombic

10. Which one of the following metals crystallizes in FCC structure
(a) zinc (b) sodium
(c) aluminium (d) cesium chloride

11. The atomic diameter of an FCC crystal (lattice parameters a) is
(a) $\frac{a}{2}$ (b) $a\sqrt{2}$
(c) $\frac{a\sqrt{2}}{2}$ (d) $\frac{a}{2\sqrt{2}}$

12. The number of Bravis lattices with two lattice point are
(a) 5 (b) 6
(c) 3 (d) 1

13. The packing factor of diamond cubic crystal structure is
(a) 90% (b) 45%
(c) 34% (d) none of these

14. The unit cell has $a = 5$ Å, $b = 8$ Å, $c = 3$ Å, $\alpha = 90°, \beta = 65°$ and $\gamma = 54°$. The space lattice for this unit cell is
(a) monoclinic (b) rhombohedral
(c) triclinic (d) orthorhombic

15. The faces in a tetragon are
(a) 24 (b) 12
(c) 6 (d) 2

16. Magnesium crystallizes in HCP structure. If the lattice constant is 0.32 nm, the nearest neighbour distance in magnesium is
(a) 0.64 nm (b) 0.32 nm
(c) 0.16 nm (d) 0.8 nm

17. Zinc crystallizes in HCP structure. If r is the radius of zinc atom, the height of the unit cell is
(a) $r\sqrt{3/8}$ (b) $2r\sqrt{8/3}$
(c) $r\sqrt{\frac{8}{3}}$ (d) r

18. The lattice constant of a BCC unit cell with atomic radius of 1.24 Å is
(a) 1.432 Å (b) 2.864 Å
(c) 1.754 Å (d) 1.432 Å

19. The correct order of coordination number in BCC, FCC and HCP unit cells is

(a) 12, 8, 6 (b) 8, 12, 12

(c) 6, 8, 12 (d) 12, 8, 24

20. The number of sulphide atoms in the unit cell of zinc sulphide crystal is

(a) 6 (b) 4

(c) 2 (d) 1

21. The Miller indices of the line of intersection of a $(1\bar{1}1)$ and a $(1\bar{1}0)$ planes are

(a) [1 1 0] (b) $[1\bar{1}1]$

(c) $[\bar{1}\bar{1}0]$ (d) $[1\bar{1}0]$

22. The Miller indices of the plane parallel to *y* and *z*-axis are

(a) (0 0 1) (b) (1 1 1)

(c) (0 1 0) (d) (1 0 0)

23. The Miller indices of a plane in an orthorhombic unit cell ($a: b: c = 1: 2: 3$) making intercepts of 3, 2, 1 Å on *a, b, c* axis respectively are

(a) (1 2 3) (b) (3 2 1)

(c) (9 3 1) (d) (1 3 9)

24. If (3 2 6) are the Miller indices of a plane, the intercepts made by the plane on the three crystallographic axis are

(a) (a, b, c) (b) $(2a, 3b, c)$

(c) $(a, 2b, 3c)$ (d) $(2a, b, 3c)$

25. In a simple cubic lattice $d_{100}: d_{110} : d_{111}$ is

(a) $\sqrt{6}:\sqrt{3}:\sqrt{2}$ (b) $\sqrt{6}:\sqrt{3}:2\sqrt{2}$

(c) 6: 3: 4 (d) $\sqrt{3}:\sqrt{2}$

26. The acute angle between [101] and $[\bar{1}01]$ directions in a tetragonal crystal with $c/a = 1.5$ is

(a) 67.38° (b) 45°

(c) 33.69° (d) 90°

27. The maximum radius of the interstitial sphere that can fit into the void between the body centred atom of BCC structure is

(a) $r\left[\sqrt{3}-1\right]$ (b) $r\left[\sqrt{2}-1\right]$

(c) $r\left[\left(2\sqrt{3}\right)-1\right]$ (d) $r/\sqrt{2}$

28. The interplanar spacing of the first reflecting plane (lowest θ) in FCC crystal is

(a) $a/\sqrt{3}$ (b) $2a/\sqrt{3}$

(c) a (d) $2a/\sqrt{5}$

29. A plane in a unit cell is described by its Miller indices (632). The plane intersects x,y,z respectively at points whose distances from origin are

(a) 6, 3 and 2 units (b) 1/3, 2/3 and 1 units
(c) 1/3, 1/2 and 1/1 units (d) 1/6, 1/3 and 1/2 units

30. If the first reflection from an FCC crystal has a Bragg angle $\theta = 21.5°$, the θ corresponding to second reflection is

(a) 13.5° (b) 18.5°
(c) 25° (d) 36.8°

31. The Miller indices of the fifth reflection in a FCC crystal is

(a) 111 (b) 222
(c) 311 (d) 331

32. The number of atoms per unit area of the plane (010) of a simple cubic crystal is

(a) $1/2r$ (b) $1/4r^2$
(c) $2r^2$ (d) $4r^2$

33. The highest linear density, i.e., the densest direction in FCC and BCC crystals are respectively

(a) $1/2r$<111> (b) <1 1 1> and <101>
(c) <100> and <110> (d) <1 1 1> and <110>

34. The two dimensional lattice with the highest rotational symmetry is

(a) honeycomb lattice (b) hexagonal lattice
(c) square lattice (d) triangular lattice

35. The Wigner-Seitz cell for any two dimensional Bravis lattice is

(a) a circle (b) always a triangle
(c) always a square (d) either a hexagonal or rectangle

36. The angle between any two of the lines joining a site of the diamond structure to its four nearest neighbours is

(a) 120° (b) 90°
(c) 30° (d) $\cos^{-1}(-1/3)$

ANSWERS

Question numbers	1	2	3	4	5	6	7	8	9	10	11	12
Answers	(d)	(a)	(c)	(c)	(c)	(a)	(a)	(c)	(c)	(c)	(c)	(a)
	13	14	15	16	17	18	19	20	21	22	23	24
	(d)	(c)	(b)	(b)	(b)	(b)	(b)	(b)	(a)	(d)	(d)	(b)
	25	26	27	28	29	30	31	32	33	34	35	36
	(a)	(a)	(c)	(b)	(c)	(c)	(b)	(b)	(a)	(d)	(d)	(d)

FILL IN THE BLANKS

1. Crystal lattice + basis → ______. **[crystal structure]**
2. The coordination number is equal to ______ for simple cubic, ______ for BCC and ______ for FCC lattice. **[6, 8, 12]**
3. In a FCC structure, the unit cell is a cubic with an ______ at each of the eight corners and an atom at each of six ______. **[atom, cube faces]**
4. Packing fraction for FCC structure is ______. **[0.74]**
5. The diamond structure is characterised by a ______ bond arrangement. **[tetrahedral]**
6. The Bravais lattice of KCl crystal is ______. **[FCC]**
7. Graphite is an ______ of carbon in which sp^2 hybridisation binds each carbon atom in a layer to ______ nearest neighbour so that strong covalent bonds exist in layers. **[allotrope, three]**
8. Fullerene consists of a hollow spherical cluster of ______ carbon atoms, a single molecule is denoted by ______. **[60, C_{60}]**
9. In a cubic crystal, the distance between the successive planes of indices (*hkl*) is given by d_{hkl} = ______. **[$\frac{a}{\sqrt{h^2+k^2+l^2}}$]**
10. In a tetragonal lattice $a = b = 2.5$ Å, $c = 1.8$ Å, the lattice spacing between (1 1 1) planes is ______. **[1.26 Å]**
11. In an ideal hexagonal close packed (HCP) structure c/a ratio is equal to ______. **[$(8/a)^{1/2}$]**
12. The number of atoms in SC, BCC and FCC units cells are ______, ______ and ______ respectively. **[1, 2, 4]**
13. The packing efficiences for SC, BCC and FCC structures are ______, ______ and ______ respectively. **[52%, 68% 74%**
14. If $|a| = 4.50$ Å, $|b| = 3.50$ Å and $|c| = 2.50$ Å the lattice spacing for (1 1 0) planes in an orthorhombic lattice is ______. **[0.36 Å]**
15. There are ______ different types of three-dimensional Bravais lattices. **[14]**
16. Miller indices are a set of ______ integers that specify the orientations of a crystal plane. **[three]**
17. The packing fraction of a unit cell is the ratio of the volume of ______ to the volume of the ______. **[atomic spheres, unit cell]**
18. Three broad categories of crystal imperfections are ______, ______ and ______. **[point defects, line defects, surface defects]**
19. The introduction of point defects into crystal increases its ______ energy as compared to that of a ______ crystal. **[internal, perfect]**
20. A self ______ defect in an ionic solid is known as Frenkel defect. **[interstitial]**

14

Crystal Bonding

14.1 INTRODUCTION

A solid is defined as a piece of matter in which the atoms and the molecules comprising it are very closely packed. The materials in solid form exhibit quite a lot of interesting properties. It is clear that the reason for the close packing of the atoms and molecules in a solid is the existence of some type of the binding forces between them. The natural question then arises, what is the origin of these binding forces which bind the different particles in different ways, giving rise to different properties in different solids. Thus the most convenient basis for classification of crystals is the character of the inter-atomic binding force(s) in various types of crystalline materials. In 1940 H. Seitz classified solids into five types according to the bonding of atoms. The theme adopted by Seitz is generally accepted for classification. We shall also use this classification and list the solid types starting from the solids with largest binding energy to the most weakly bound as follows:

The distinction in the said five categories is not sharp one, because some materials may belong to more then one class. Bearing this in mind, we discuss the various categories of crystals in turn in this chapter and illustrate how binding occurs in each category.

14.2 BONDING IN CRYSTALS

From the very existence of solids we may draw two general conclusions: (i) There must act an attractive force between the atoms or molecules in a solid which keep them together (ii) There must be repulsive forces acting between the atoms as well, since large external pressures are required to compress a solid to appreciable extent. In order to understand the importance of these two types of forces, let us consider the simplest system, e.g.,

Table 14.1

Solids	*Type of bond*	*Formation*	*Binding energy eV/atom*	*Typical examples*
Covalent	Covalent homopolar or pair bonds	Electron shared between two atoms	2–6	Carbon (Diamond), Germanium, Silicon etc.
Ionic	Ionic	Electron transfer and Columb interaction between cations and anions	0–2	Alkali Halides
Metallic	Metallic	Freely moving, electrons in an array of positive ions	1–5	All metals and alloys
Molecular (Vander Waals)	Molecules	Weak attractive forces due to dipole interaction between pairs	0.002–0.1	Noble gases
Hydrogen bond	Hydrogen atom attracted between two other atoms	electrostatic bond of H atom with an electronegative atom	0.5	Ice, organic compounds, biological materials

simple pair of atoms A and B which form a stable chemical compound. Let us assume that the potential energy of atom B due to the presence of atom A is given by

$$u(r) = -\frac{\alpha}{r^n} + \frac{\beta}{r^m} \tag{14.1}$$

Here r is the distance between the nuclei of the two atoms, α, β, m and n are constants characteristics for the molecule AB. The zero of the energy is chosen such that for the finite separation $u = 0$. In Eq. (14.1) the first term $\frac{\alpha}{r^n}$ which is negative, corresponds to the energy associated with the forces of attraction. The second term, $\frac{\beta}{r^m}$ which is positive corresponds to the forces of repulsion. The force between the two atoms A and B as a function of r is given by

$$F(r) = \frac{du}{dr} = -\frac{n\alpha}{r^{n+1}} + \frac{m\beta}{r^{m+1}} \tag{14.2}$$

The energy and the force between two atoms A and B which form a chemical compound are plotted in Fig. (14.1).

The stable configuration of the system corresponds to the minimum in $u(r)$ curve which occurs for the particular separation $r = r_0$. The corresponding energy $u(r_0)$ is called the binding energy and is negative, i.e., B.E. = $-u(r_0)$. This means that the energy in magnitude equals $u(r_0)$ is required to dissociate the molecule, i.e., to separate the two atoms. As such this energy is called the **dissociation energy**. Dissociation occurs at high temperatures

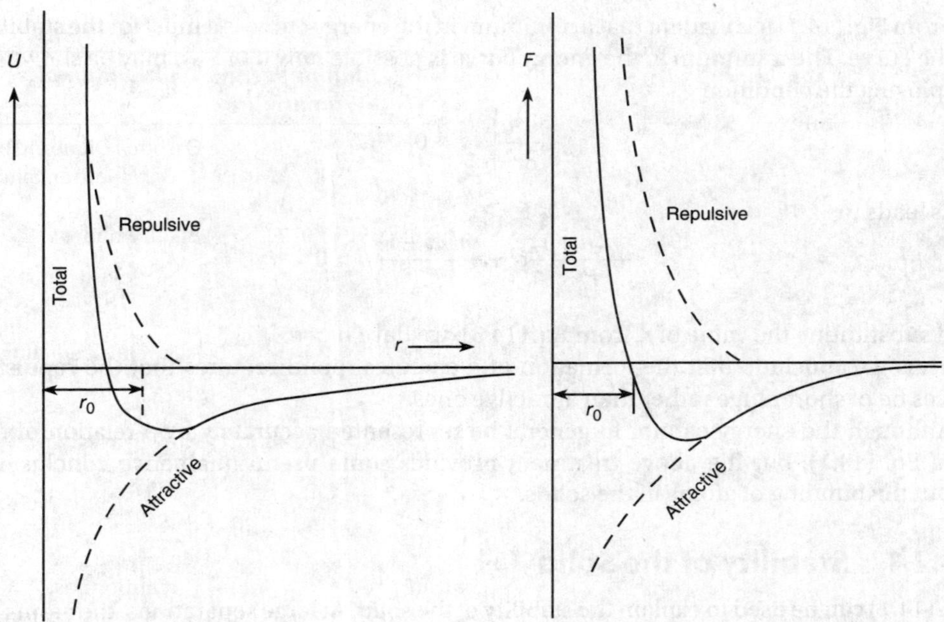

Fig. 14.1

or as a result of other processes in which the molecules can absorb sufficient energy. The dissociation energies are of the order of 1 eV to few eV.

At the equilibrium separation $r = r_0$ the potential energy must be minimum or the first derivative of Eq. (14.1) must vanish, i.e.,

$$\left.\frac{du}{dr}\right|_{r=r_0} = 0 = \frac{n\alpha}{r^{n+1}} - \frac{m\beta}{r^{m+1}}$$

From this it follows that

$$\beta = \alpha\frac{n}{m}[r_0]^{\{(m+1)-(n+1)\}}$$

or

$$[r_0]^{m-n} = \left(\frac{\beta m}{\alpha n}\right) \tag{14.3}$$

This means that at $r = r_0$ the attractive and repulsive forces balance each other, i.e., $F(r_0) = 0$. Thus equilibrium state energy is.

$$u(r_0) = -\frac{\alpha}{r_0^n} + \frac{\beta}{r_0^m} = -\frac{\alpha}{r_0^n}\left[1 - \frac{n}{m}\right] \tag{14.4}$$

It must be noted that though the attractive and repulsive forces are equal, the attractive and repulsive energies are not equal since $n \neq m$. If $m >> n$ the total binding energy is essentially the energy of attraction and is determined by $-\frac{\alpha}{r^n}$.

From Fig. (14.1) it is evident that a minimum in the energy curve is a must for the stability of the curve. The minimum in the energy curve is possible only if $m >> n$ may be shown by employing the condition

$$\left.\frac{\partial^2 u}{\partial r^2}\right|_{r=r_0} > 0$$

This leads to

$$-\frac{n(n+1)}{r_0^{n+2}}\alpha + \frac{m(m+1)}{r_0^{m+2}}\beta > 0$$

and substituting the value of r_0 from Eq. (14.3), we have $m > n$.

Thus we conclude that the formation of a chemical bond requires that the repulsive forces be of short range rather than attractive ones.

Although the energy cannot in general be represented accurately by a relation of the type Eq. (14.1), but the above treatment provides some useful qualitative conclusions about the bonding of atoms in the solids.

14.2.1 Stability of the Solid

Fig. (14.1) can be used to explain the stability of the solid. At large separations, the atoms do not interact with each other so that $u = 0$. As (the atoms approach each other, they experience attractive forces due to each other. Primarily because of the attraction of positive and negative charges in the atoms. As the inter atomic separation decreases to the order of one or two atomic diameters, the attractive forces start to become appreciable and also, the repulsive forces come into play. The repulsive forces arises as a consequence of the Pauli's exclusion-principle and has the effect of increasing the potential energy. Since this increase is not consistent with the stability of the system, the further reduction in the interatomic separation is checked. At still smaller distances the energy is dominantly repulsive. It is thus now quite clear that because of the repulsive forces the atoms cannot reduce their interatomic separation in an arbitrary fashion. This result explains the importance of the repulsive forces. In absence of the repulsive force the atoms would ultimately interpenetrate into each other because of the increasing attractive forces and would occupy the same site. The equilibrium is then established only when a compromise between the attractive and repulsive forces is reached so that F is equal to zero. It is clear from Fig. (14.1) that this occurs at $r = r_0$ and this situation corresponds to a minimum in the energy curve at this point.

14.3 IONIC BOND

Certain neutral atoms tend to loose their valence electron relatively easily because of the smallness of the ionisation potentials and thus acquire a net positive charge then they are called **ions**. These atoms are called **electropositive** and their ions are known as **cations**. Most of these atoms form **metallic crystals**. For examples Na-atom with the electronic configuration $1s^2 2s^2 2p^6 3s^1$ has the first ionisation potential of 5.13 eV. On removal of $3s^1$ electron the Na^+ cation assumes the inert gas configuration $1s^2 2s^2 2p^6$ and ion becomes stable with essentially spherical distribution of electron charge (due to 10 electrons) around the sodium nucleus (containing 11 protons and 12 neutrons) with 11 units of positive charge. The net charge of the ion is then +1 in units of electronic charge.

Atoms having non-metallic properties like O, N, F, Cl etc., are known as electronegative atoms as they have a tendency to accept and bind additional electrons and there by form negatively charged ions. These ions are known as anions. For example Cl with $Z = 17$ has the electronic configuration $1s^2\ 2s^2 2p^6 3s^2 3p^5$ and can form an inert gas (Argon structure just by accepting one extra electron and there by forming the ion Cl^- with a net negative charge of –1.

The positive (Na^+) and negative (Cl^-) ions then attract electrostatically and are bound together by an Ionic bond. In addition to this attractive force, there is also a repulsive force, which becomes operative, when the two negatively charged electron clouds of the ions try to overlap; when the two ions try to come near each other, the molecule is at the equilibrium distance when the two opposing forces balance each other. The binding or cohesive energy is of the order of 5 to 10 eV. The charge distribution of each ion is spherically symmetric. Consequently Ionic solids crystallize in closed packed structures of which the NaCl and CsCl structures Fig (14.2) are the commonest having cubic structures. The melting points are quite high of the order or 800°C for NaCl.

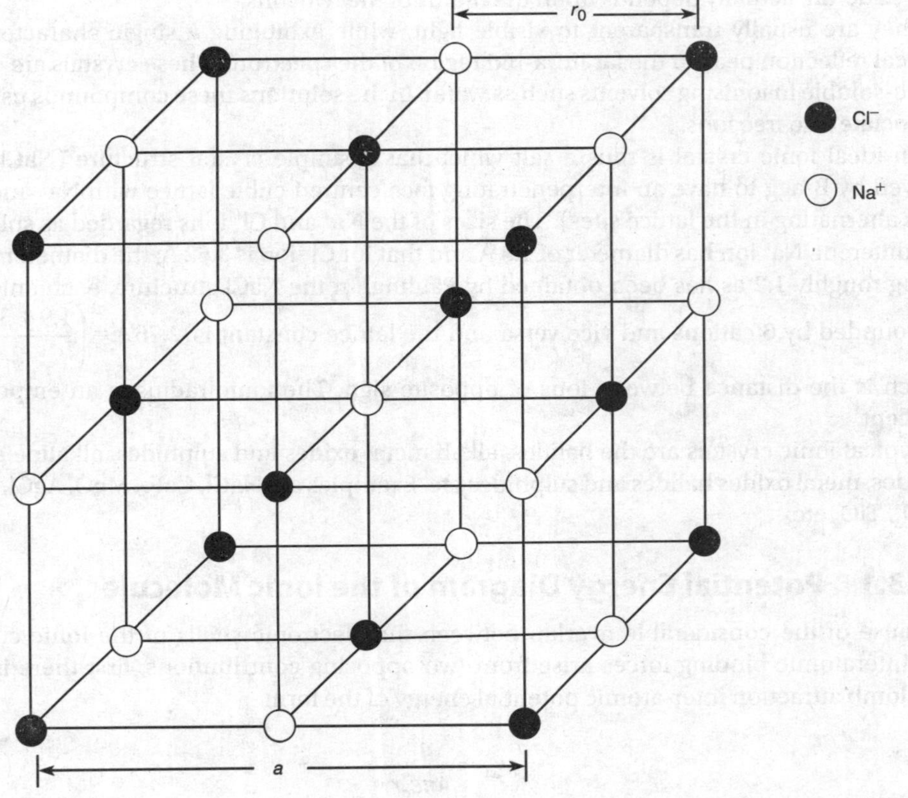

Fig. 14.2

The ionic bond is strong when compared with other bonds, a typical, value for the binding energy of a pair of atoms being about 5 eV. This strength is attributed due to the strength of the Coulomb force responsible for the bonding. Experimentally, this strength is characterised by the high melting temperatures associated with ionic crytals. Thus the melting temperatures for the ionic crystal NaCl is 801°C while the melting temperatures for Na and K metals are 97.8 °C and 63 °C respectively.

They are poor conductors of electricity at normal temperatures because the valence electrons are found quite tightly to the ionic nuclei. At elevated temperatures, however the ions themselves become mobile and ionic conductivity results.

The positive and negative ions have the same electrical structure as the nearest inert gas atoms, except that they are electrically charged. In case of NaCl, Na^+ have the neon electron structure and Cl^- have argon electron structure. The charge on the ion is, therefore, spherically distributed (with some distortion near the region of contact with neighbouring atoms of course) and an ion is thus like a small charged sphere and the packing of the ions is closest one. Because ions having like charges repel each other, a stable packing is obtained in which ion is surrounded with ions having an opposite charge. A periodic array results in which the environment of all similar atoms is the same and the sum of all positive and negative charges adds up to zero, so that ionic solid is electrically neutral. They usually crystallize in the relatively close packed NaCl and CsCl structure Fig. (14.2). The structure actually depends upon the radius of the two ions.

They are usually transparent to visible light, while exhibiting a single characteristic optical reflection peak in the far infra-red region of the spectrum. These crystals are often quite-soluble in ionising solvents such as water. In the solutions these compounds usually dissociate into free ions.

An ideal ionic crystal is thus a salt which has a simple crystal structure (NaCl was proved by Bragg to have an interpenetrating face centred cubic lattice with Na^+ and Cl^- ions alternating in the lattice sites). The sizes of the Na^+ and Cl^- ions regarded as spheres are different. Na^+ ion has diameter of 1.9 Å and that for Cl^- ion is 3.62 Å; the diameter ratio being roughly 1:2 as has been obtained by Pauling. In the NaCl structure, each anion is surrounded by 6 cations and vice versa and the lattice constant is $2.76\,\text{Å} = \left(\frac{1.9 + 3.62}{2}\right)$ which is the distance between ions of opposite sign. The ionic radius is an emperical concept.

Typical ionic crystals are the halides, alkali metal oxides and sulphides, alkaline earth halides, metal oxides halides and sulphides etc. Examples are NaCl, CaF_2, MgO, AgCl, ZnS, Al_2O_3, TiO_2 etc.

14.3.1 Potential Energy Diagram of the Ionic Molecule

Because of the considerable overlap between the electronic shells of the ionic crystal the interatomic binding forces arise from two opposing contributions, first there is the Coulomb attraction inter-atomic potential energy of the form

$$u_{\text{att}} = \frac{q^2}{4\pi\varepsilon_0 r} \tag{14.5}$$

where q is the charge of an ion and r is the distance of separation between them. Second here is the interaction due to the overlap of the electronic shells of neighbouring atoms. In general the repulsive energy is of the form

$$u_{\text{rep}} = \frac{\beta}{r^m} \quad (14.6)$$

More exact form of the repulsive energy in the ionic crystal is given by Born-Mayer as

$$u_{\text{rep}} = A \exp\left(-\frac{r}{b}\right) \quad (14.7)$$

where A and b are empirical constants. The resulting interaction energy of two ions is

$$u = A \exp\left(-\frac{r}{b}\right) - \frac{q^2}{4\pi\varepsilon_0 r} \quad (14.8)$$

This energy is shown in Fig. (14.3).

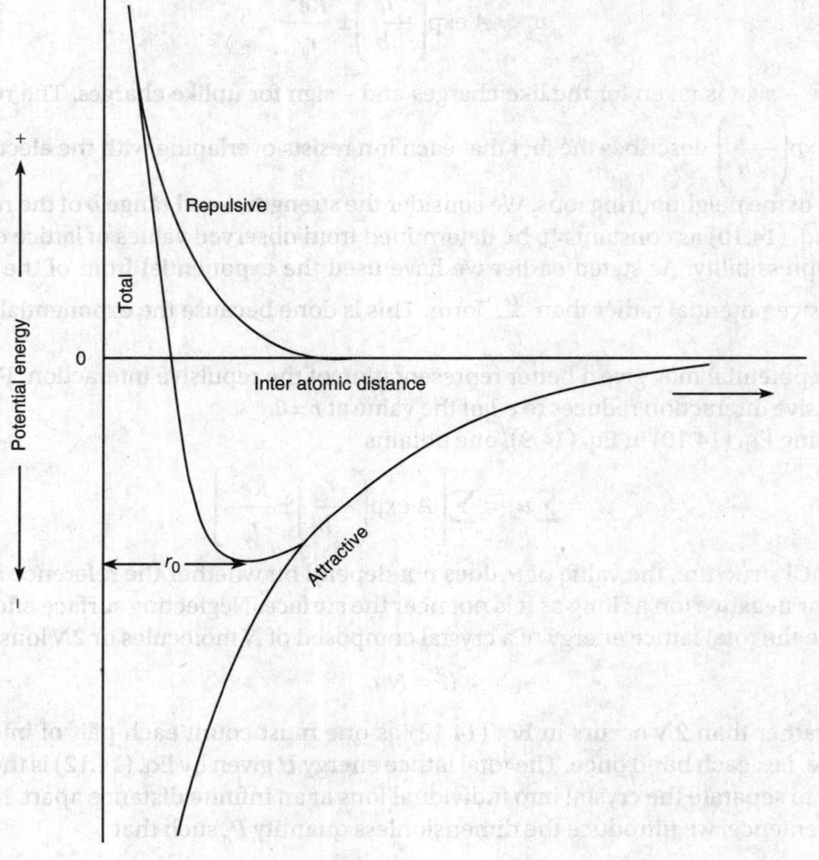

Fig. 14.3

14.4 BINDING ENERGY OF IONIC CRYSTAL

The most common crystal structure found for ionic crystal is NaCl. We shall consider it for the calculation of binding energy of ionic crystal. Let u_{ij} be the interaction energy between i^{th} and j^{th} ions of the crystal. We define a sum u_i which represents the total energy of interaction between ion i and all other ions of the crystal, i.e.,

$$u_i = \sum_{g} u_{ij} \tag{14.9}$$

where the summation includes all ions except $j = i$, because the binding between ion i and j itself is not possible. The interaction term u_{ij} in Eq. (14.9) may be written as the sum of two terms (i) Coulomb interaction energy energy $\pm \frac{Ke^2}{r_{ig}}\left(\text{here } K = \frac{1}{4\pi\varepsilon_0}\right)$ where r_{ij} is the distance between the ions i and j and (ii) energy resulting from a repulsive force which becomes effective only at very small inter-ionic separation. Thus the total energy of interaction between the two ions i and j becomes

$$u_{ij} = A \exp\left(-\frac{r_{ij}}{b}\right) \pm \frac{Ke^2}{r_{ij}} \tag{14.10}$$

where the + sign is taken for the like charges and – sign for unlike charges. The repulsive term $A \exp\left(-\frac{r_{ij}}{b}\right)$ describes the fact that each ion resists overlaping with the electron distribution of the neighbouring ions. We consider the strength A and range b of the repulsive term in Eq. (14.10) as constants to be determined from observed values of lattice constant and compressibility. As stated earlier we have used the exponential from of the empirical repulsive potential rather than $\frac{a}{r_{ij}^n}$ form. This is done because the exponential form of repulsive potential may give a better representation of the repulsive interaction. For $r_{ij} = b$ the repulsive interaction reduces to e^{-1} of the value at $r = 0$.

Substituting Eq. (14.10) in Eq. (14.9), one obtains

$$u_i = \sum_j u_{ij} = \sum_j \left[A \exp\left(-\frac{r_{ij}}{b}\right) \pm \frac{Ke^2}{r_{ij}} \right] \tag{14.11}$$

In the NaCl structure, the value of u_i does not depend on whether the reference ion i is a positive or negative ion as long as it is not near the surface. Neglecting surface effects, one may write the total lattice energy of a crystal composed of N molecules or $2N$ ions as

$$U = Nu_i \tag{14.12}$$

Here N rather than $2N$ occurs in Eq. (14.12) as one must count each pair of interaction only once, i.e., each bond once. The total lattice energy U given by Eq. (14.12) is the energy required to separate the crystal into individual ions at an infinite distance apart.

For convenience, we introduce the dimensionless quantity P_{ij} such that

$$r_{ij} = P_{ij} R \tag{14.13}$$

Here R is the nearest neighbours separation in the crystal. The Eq. (14.11) then becomes

$$u_i = \sum_j \left[A \exp\left(-\frac{P_{ij} R}{b} \right) \pm \frac{Ke^2}{P_{ij} R} \right] \tag{14.14}$$

If we include the repulsive interaction only among nearest neighbours, we have

$$u_i = \begin{cases} A \exp\left(-\dfrac{P_{ij} R}{b} \right) - \dfrac{Ke^2}{R} & \text{(nearest neighbours)} \\ \pm \dfrac{1}{P_{ij}} \dfrac{Ke^2}{R} & \text{(other wise)} \end{cases}$$

Thus

$$U = \sum_j u_{ij} = N u_i = N\left(ZAe^{\frac{-R}{b}} - \alpha \frac{Ke^2}{R} \right) \tag{14.15}$$

where Z is the number of nearest neighbours of any ion and

$$U = \sum_j \pm \frac{1}{p_{ij}} \tag{14.16}$$

is called **Madelung constant.** The sum should include the nearest neighbour contribution which is given by Z. In the theory of ionic crystals the value of Madelung constant is of prime importance. Clearly the value of Medelung constant depends on the lattice structure. Medelung constant is really nothing but a correction factor that tells how much of an error will be there if only nearest neighbours are considered.

14.4.1 Medelung Constant for a Linear Chain Ionic Solid

We have said that the value of the Madelung constant depends on the crystal structure and for the sake of simplicity we will first calculate the constant for a linear ionic solid. In such a linear chain the ions of alternate sign are arranged along an infinite line as shown in Fig. [14.4(a)]. Let us consider a negative ion as reference ion and let R be interatomic spacing. The reference atom has two positive ions as its nearest neighbours situated at a distance R on other side.

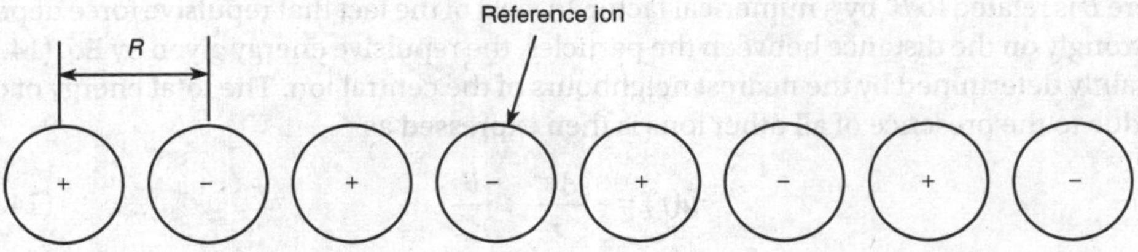

Fig. 14.4(a)

Attractive Coulomb energy due to the two nearest neighbours *is* $-\frac{2Ke^2}{R}$.

Repulsive Coulomb energy due to the two next nearest neighbours is $+\frac{2Ke^2}{2R}$.

Attractive Coulomb energy due to the two next nearest neighbours is $-\frac{2Ke^2}{3R}$.

Thus total Coulomb energy due to all the ions in the linear chain ionic solid is

$$U = -\frac{2Ke^2}{R} + \frac{2Ke^2}{2R} - \frac{2Ke^2}{3R} + \cdots$$

$$= -\frac{2Ke^2}{R}\left[1 - \frac{1}{2} + \frac{1}{3} - \frac{1}{4} + \cdots\right] \tag{14.17}$$

And comparision with Eq. (14.16) gives

$$\alpha = 2\left[1 - \frac{1}{2} + \frac{1}{3} - \frac{1}{4} + \cdots\right] = 2\log_e^2 \tag{14.18}$$

We see that the total Coulomb energy of a linear chain ionic lattice is directly proportional to $\frac{Ke^2}{R}$. The constant of proportionality is the Madelung constant per molecule for the ionic solid, if n is the number of molecules per unit volume of the solid, the Madelung constant for one dimensional solid comes out to $2n \log 2$.

14.5 DETERMINATION OF THE REPULSIVE EXPONENT

As already discussed, the repulsive forces become noticeable when the electron shells of neighbouring ions begin to overlap, and they increase strongly in this region with decreasing values of r. These forces can best be studied on the basis of wave mechanics because they are non classical in nature. Born in his early work made the simple assumption that the repulsive energy between two ions as function of their separation could be expressed by a power law of the type $\frac{B'}{r_{m'}}$, where B' and m are constants characteristics of the ions in solid under consideration. Here m is called the repulsive exponent. Thus we may write for the repulsive energy of an ion due to the presence of all other ions as

$$u_{\text{rep}} = \frac{B}{r^m} \tag{14.19}$$

where B is related to B' by a numerical factor. In view of the fact that repulsive force depend so strongly on the distance between the particles, the repulsive energy given by Eq. (14.19) is mainly determined by the nearest neighbours of the central ion. The total energy of one ion due to the presence of all other ions is then expressed as

$$u(r) = -\frac{Ae^2}{r} + \frac{B}{r^m} \tag{14.20}$$

Here r is interatomic distance and $A = K\alpha$.

The value of the repulsive exponent m can be obtained from the compressibility data. The compressibility (β) of a solid is given by the expression

$$\beta = -\frac{1}{V}\left(\frac{\partial V}{\partial P}\right) \tag{14.21}$$

The first law of thermodynamics gives

$$du = \delta Q - \delta W$$

where δQ and δW are the heat flow and work done by the system respectively. If we consider experiments carried at near absolute zero δQ, can be neglected because $\delta Q = C_v dT$ and C_v is vanishingly small near absolute zero. Therefore,

$$du = -\,\delta W = -\,PdV$$

or

$$P = -\frac{du}{dV}$$

and

$$\frac{dP}{dV} = -\frac{d^2u}{dV^2} \tag{14.22}$$

Substituting in Eq. (14.21), one obtains at absolute zero,

$$\frac{1}{\beta_0} = -\,V_0\left(\frac{\partial P}{\partial V}\right) = V_0\frac{d^2u}{dV^2} \tag{14.23}$$

where V_0 is the volume of the crystal corresponding to an inter-ionic distance r_0 (V corresponds to the variable r, the interatomic distance).

For NaCl structure, since r is the interionic distance and the lattice constant $a = 2r$ (Fig. (14.2), NaCl has face centred cubic structure. In a unit cell a corner Na^+ ion is shared by 8 unit cells and there are 8 corners so that there is $\frac{1}{8}\times 8 = 1$ ion of Na^+. Each Na^+ ion, the face centred is shared by two unit cells and there are 6 such ions yielding $6\times\frac{1}{2} = 3$ Na^+ ions. Thus making a total of 4 Na^+ ions in the unit cell. There are 12 Cl^- ions in the middle of the edges and one at the body centre of the unit cell. Since each edge is shared by 4 unit cells, the number of Cl^- ions in the unit cell is $12\times\frac{1}{4}+1 = 4$. Hence a unit cell of side a contains 4 molecules of NaCl. The volume per molecule is then $\frac{1}{4}a^3 = \frac{1}{4}(2r)^3 = 2r^3$. If we assume a mole of NaCl the number of molecules is N_0, the Avogadro number and so

$$V = 2N_0 r^3 \tag{14.24}$$

$$\therefore \qquad \frac{dr}{dV} = \frac{1}{6N_0 r^2}$$

and

$$\left(\frac{dr}{dV}\right)^2_{r=r_0} = \frac{1}{36N_0^2 r_0^4} \tag{14.25}$$

We have

$$\frac{du}{dV} = \frac{du}{dr}\frac{dr}{dV}$$

and

$$\frac{d^2u}{dV^2} = \frac{du}{dr}\frac{d^2r}{dV^2} + \frac{d^2u}{dr^2}\left(\frac{dr}{dV}\right)^2 \tag{14.26}$$

But at equilibrium distance, we have $\frac{du}{dr} = 0$ and from Eq. (14.20), we have

$$\frac{du}{dr} = \frac{Ae^2}{r^2} - \frac{mB}{r^{m+1}} = 0 \text{ for } r = r_0$$

$$\therefore \quad \frac{B}{r_0^m} = \frac{Ae^2}{mr_0}$$

Thus equilibrium **cohesive** energy is

$$u(r_0) = -\frac{Ae^2}{r_0} + \frac{B}{r_0^m} = -\frac{Ae^2}{r_0} + \frac{Ae^2}{mr_0}$$

$$= \frac{Ae^2}{r_0}\left[1 - \frac{1}{m}\right] \tag{14.27}$$

and equilibrium compressibility β_0 is given by

$$\frac{1}{\beta_0} = V_0\frac{d^2u}{dV^2} = V_0\left(\frac{d^2u}{dr^2}\right)_{r=r_0}\left(\frac{dr}{dV}\right)^2_{r=r_0} \tag{14.28}$$

But

$$\frac{d^2u}{dr^2} = -\frac{2Ae^2}{r^3} + \frac{m(m+1)B}{r^{m+2}}$$

$$\therefore \quad \left(\frac{d^2u}{dr^2}\right)_{r=r_0} = -\frac{2Ae^2}{r_0^3} + \frac{m(m+1)B}{r_0^{m+2}}$$

$$= -\frac{2Ae^2}{r^3} + \frac{(m+1)Ae^2}{r_0^3} = \frac{Ae^2}{r_0^3}(m-1) \tag{14.29}$$

and substituting the values of $\left(\frac{d^2u}{dr^2}\right)_{r=r_0}$, $\left(\frac{dr}{dV}\right)_{r=r_0}$ and V_0, We have

$$\frac{1}{\beta_0} = 2N_0r_0^3\frac{Ae^2}{r_0^3}(m-1)\frac{1}{36N_0^2r_0^4} = \frac{Ae^2(m-1)}{18N_0r_0^4} \tag{14.30}$$

From the experimental measurement of the bulk modulus as a function of temperature and by extra-polation to absolute zero, one obtains β_0 and hence from Eq. (14.30) the value of parameter *m*. For NaCl, $r_0 = 2.91$Å and one obtains $m \approx 9$. It is found that for alkali halides *m* varies from 6 to 10. The equilibrium cohesive energy is then obtained using Eq. (14.27) and Eq. (14.30).

Though there is marked variation in the value of m from one crystal to another, even then an appreciation error in m leads to a relatively small error in the lattice energy which is proportional to $\left(1-\frac{1}{m}\right)$. If we change m by unity $u(r_0)$ changes by only 1 or 2 percent. According to Eq. (14.27) and in view of the relatively large values of m, most of the lattice energy is due to the Coulomb interaction and the repulsion contributes only a relatively small fraction. On the other hand, the repulsive and attractive forces action on any one just balance $r = r_0$ and thus are equal in magnitude.

14.6 COVALENT BONDING

The characteristic feature of the covalent bond is that electrons are shared between rather than transfered from one to another. Covalent bonds are formed generally by the formation of shared electron pairs between the valance electrons in the incomplete outer shell of atoms.

The simple covalent bond occurs between atoms in the molecules H_2, CH_4, CCI_4. It is a strong bond and the bond between the two carbon atoms in the diamond has a cohesive (binding) energy of the order 7.3 eV with respect to separated neutral atoms. This is comparable with cohesive energy of ionic crystals. Because of this many authors choose not to make any sharp distinction between the ionic and covalent bonds.

Covalent crystals are quite hard but brittle, they have variable electrical resistance, i.e., quite sensitive to the presence of impurity atoms, the resistivity being low at high temperatures where conduction is almost as good as in some metals. The optical properties of the covalent crystals are characterised by high refractive index (μ) and high dielectric constant. The crystals being transparent to long-wave length radiations, but opaque to shorter wave lengths.

The classic example of covalently bonded crystals are the group IV semiconductors, C, Si, and Ge. The nature and origin of covalent bond have received considerable attention recently.

The most striking feature of the covalent bond is its strong directional properties. Thus C, Si and Ge have the diamond lattice structure in which each carbon atom, say, is connected to four neighbouring carbon atoms at tetrahedral angles to one another even through this arrangements gives a low filling of space.

The other important characteristic property of the covalent bond is its saturability. Saturability means that each atom can form covalent bonds only with limited number of neighbours. For example each hydrogen atom can form covalent bonds only with one of its neighbours. The electron pair constituting such a bond has antiparallel spins and occupies one quantum state. A third atom in this case instead of being attracted will be repelled.

The covalent bond is usually formed from two electrons, one from each atom participating in the bond as illustrated below.

$$:\ddot{\text{Cl}}\cdot + \cdot\ddot{\text{Cl}}: \rightarrow :\ddot{\text{Cl}}:\ddot{\text{Cl}}:$$

Unpaired electrons

The electrons forming the covalent bond tend to be partly localised in the region between the atoms joined by the bond. In accordance with the Pauli Principle, the spins of the two electrons in the covalent bond are anti-parallel.

The binding of molecular hydrogen is a simple example of a covalent bond. The strongest binding occurs when the spins of the two electrons are anti-parallel. The binding depends on the relative spin orientation not because there are strong magnetic dipole force between the spins but because the Pauli principle modifies the distribution of charge according to the spin orientation. A spin dependent Coulomb energy of the system is found. This is called the **exchange interaction**.

Covalent bonding also prevails in crystals of the from $A^N B^{8-N}$ composed of elements A and B with N and $8 - N$ valence electrons per atom of A and B, i.e., with 8 valence electrons per atom pair. The nature and origin of this bond have been excellently characterised by Philips.

14.7 METALLIC BONDING

Nearly 70% of the elements in the periodic table are metals. Metals are characterised by the availability of large number of valence electrons which can be easily removed from the parent atoms. A simple and rather crude picture of metallic bonding can be conceived as arising because of the fact that barring the repulsions which are also important, the energy of the valance electrons is lowered when they come close to more than one nucleus. These valance electrons called the electron gas are supposed to move about the crystal lattice freely more or less like gas molecules. The positive ion cores may be supposed to be held together by some sort of resultant electrostatic attraction between the fixed positive ions and the mobile negative electrons. The free electrons are mainly responsible for the high thermal and electrical conductivity, since they can absorb easily any energy from lattice vibrations or electromagnetic radiations and there by increase their energy and mobility. Metals are opaque since these free electrons absorb photons in the visible region and get excited to higher energies. High reflectivity of metals is due the fact that these electrons re-emit this energy-when they fall back to lower energy levels. The above theory due to Drude-Lorentz though successful in explaining the general nature of metals is found to be quite inadequate to account for several experimental results like temperature dependence of resistance para-magnetism of free electrons, electronic specific heat, etc., The dependence of metallic character of solids on the number of valence electrons is another feature which the above theory fails to explain. Elements like sodium, potasium, etc., which have only one electron in the outer most orbit are marked by metallic. When there are 2 or 3 electrons in the outer most orbit as in the case of Magnesium, Aluminium etc., the elements are just metallic. When the number of valence electrons reach 4 as in the case of germanium, Tin, Lead, etc., the metallic character becomes much less. When the number of valence electrons become 6 or 7, the elements definitely loose their metallic character.

With the advent of Quantum mechanics a better picture of the metallic bonding was evolved. It is convenient to regard the metallic bonding as an unsaturated covalent bond with possible hybridisation in some structure. Let us now consider the bonding between hydrogen atoms and lithium atoms both belonging to the 1st group. Hydrogen is monovalent and has one electron in its outer orbit. When two hydrogen atoms come nearer the valence electrons of the two hydrogen atoms share the same shell of the molecule; since it has two allowed quantum levels one for spin up and other for spin down. The shell becomes complete. This electron sharing between the two atoms results in the lowering of electron energies and hence a bonding of the two hydrogen atoms occurs with the consequent for motion of the hydrogen molecule. When the 3rd hydrogen atom comes nearer to the hydrogen molecule, it is no longer attracted since there is no room in the shell. It has to occupy only a higher energy level, which means that the force is repulsive, i.e., the bonding between that force is repulsive. We say that the bonding between the two hydrogen atoms is saturated. A hydrogen atom cannot bond with more than one hydrogen atom. It is a typical covalent bond, where the number of bonding is exactly equal to the number of valance electrons. Let us now consider the monovalent Lithium atoms whose electron configuration is $1s^2 2s^1$. When another lithium atom comes nearer, we must expect it to form a covalent bond with the original atom since there is a vacant quantum state in the $2s$ subshell. We must also expect the bonding to become saturated since the $2s$ subshell becomes complete. But in the Lithium atom there are 6 more quantum states belonging to $2p$ sub-shell, whose energies are only slightly greater than the energies of $2s$ levels. So when a third Lithium atom comes near the molecule, the third electron occupies a different quantum state without violating Pauli's exclusion principle. This means that the third electron can also form an electron pair bond with the electron of the first atom. It turns out that quite a large number of atoms can surround a single Lithium atom. Since the central atom has only unpaired electron, this electron must take turns to form electron pair bonds with its nearest neighbours. Since Lithium crystallises to become a body centred cubic crystal; in which each atom has 8 nearest neighbours. The central atom has to form covalent bonds with the 8 neighbours in turn. A covalent bond requires 2 electrons per bond. But lithium atom has to satisfy itself with only a quarter electron pair available per bond. The bonding is far from saturation. This situation is described as unsaturated covalent bond. This is the reason why as many atoms as possible from cluster around metallic atom in an effort to attain saturation. But geometric considerations limit the number of closest neighbours. In general metallic atoms have the largest number of closest neighbours and metals are in general close packed structures. The unsaturated nature of the bonding is the cause of weakness in metals. The binding energies of Na and Al metals are 1.12 eV and 3.23 eV respectively as compared to 7.4 eV in the case of covalent bonded diamond. Because of the large number of vacancies in other orbits, metals do not exhibit preferences in forming the bonds. This helps easy rearrangement of atoms without loss of strengths, i.e., the metals can be easily deformed. Because of the large number of vacancies the electron can wander from atom to atom without being attached to any particular atom as in ionic crystals or covalent crystals. Thus the valency electrons, behave in a manner similar to the molecules in a gas.

From quantum mechanics point of view we say that the bonding arises because of overlap of the orbitals which results in lowered energy states. In the case of metallic atoms the wave functions of the valence electrons are spread out more in space. This means that the probablity of occupation of the valance electrons, i.e., $|\psi^2|$ functions are also spread out. This results in the increase of the mean radius of the electron in the free atom. In addition the kinetic energy of the valence electron is lowered by the extended ψ function. The spatial extension of the wave function is a consequence of the increase in de-Broglie wave length of the valence electron. From the de-Broglie relation $\lambda = \frac{h}{P}$, we find that increased wave-length results in decreasing momentum and hence decreasing kinetic energy. Further potential energy is also lowered, because in the solid state, the valence electrons come closer to one or other nucleus than they are in a free atom. It is thus lowering of both the potential and the kinetic energies of the valence electrons that is responsible for the metallic bonding. In general, the smaller the number of valence electrons an atom has, the more loosely they are held and hence greater is the metallic bonding. As the number of valence electrons increases the tightness with which they are held to the nucleus increases and hence they become more localised in space, increasing the covalent nature of the bonding. The transition metals like Fe, Ni, W and Ti etc., have in-complete *d* sub-shells. This leads to the formation of **hybrid covalent bonds** due to the arrangement of electrons in the *s* and *d* levels. This accounts for their high bonding energies of the order of 4.5eV.

14.7.1 How to Predict Predominance of Metallic or Covalent Bond

We have seen that the lowering of the kinetic and the potential energy of the valence electrons results in the metallic bonding. The relative amount of the potential and kinetic energies lower varies from metal to metal. The potential energy effect is larger in monovalent metals, but the kinetic energy effect is larger in most other metals. We can say in general, metallic bonding is favoured if the number of valence electrons per atom is small, since in that case only the lowest kinetic energy states need to be occupied, due to the smaller number of valence electrons. Lower kinetic energy configuration means large extension of spatial wave functions and which again means that the valence electrons can have more like electron gas which is the special feature in metals. The elements on the right side of the peroiodic table are definitely non-metallic because the valency electrons became more lightly bound to the nucleus and the electron gas behaviour is completely absent. For elements in the middle of the periodic table, the competition between these two structures is very severe. Diamond exhibits almost pure covalent bonding where as silicon and germanium are more metallic. Tin exhibits in two forms: Metallic tin (white) is stable above 13°C, Covalent tin (grey) is stable below this temperature, lead is mostly metallic.

14.8 MOLECULAR OR VAN DER WAAL'S BONDING

Though the inert gases do not have any valence electrons because of their completed shell structures to form any type of bonding, they do become liquids and solids at sufficiently low temperatures and high pressures. This is explained by the fact that an atom surrounded by revolving electrons constitute fluctuating dipoles or rotating dipoles. When two such atoms come near, there is dipolar interaction between these fluctuating dipoles leading to attractive forces. Higher order polar interactions also become possible but they are extremely feeble. The bonding which arises from the fluctuating dipole nature of the atom is called the Vander Waal's bond. Vander Waal's forces arise out of induced dipoles, permanent dipoles and dispersion forces. For atoms or molecules which have no permanent dipoles the interaction energy due to induced dipoles is given by

$$u_i = -\frac{2\mu^2\alpha}{r^6} \tag{14.31}$$

where α is polarizability of the atom or the molecule, μ is the induced dipole moment and r is the interatomic distance.

When the atom or the molecule has permanent dipole moment the energy due to the dipolar interaction is given by

$$u_P = -\frac{2\mu^4}{3r^6 kT} \tag{14.32}$$

where μ is the permanent dipole moment, k is Boltzmann's constant and T-temperature in Kelvin.

Dispersion forces are understood better from quantum mechanical considerations. This is the interaction between the fluctuating dipolar atoms or molecules. The interaction energy u_d is given by

$$u_d = -\frac{3\alpha^2 h\upsilon_0}{4r^6} \tag{14.33}$$

υ_0 is the oscillation frequency of the atom or the molecule before the interaction. So the total Vander Waal's attraction energy is given by

$$u = u_i + u_P + u_d$$

or

$$u = -\frac{2\mu^2\alpha}{r^6} - \frac{2\mu^4}{3r^6 kT} - \frac{3\alpha^2 h\upsilon_0}{4r^6} \tag{14.34}$$

Eq. (14.34) predict values of right order of magnitude for many substances. Of course, Vander Waal's bonds are very weak as compared to covalent, metallic and ionic bonds. But they play an important role in determining the final arrangement of groups of atoms in solids specially in polymers.

Weak Vander Waal's forces occur in graphite. Graphite consists of carbon atoms in tetragonal arrays with each atom bonded to three atoms. Fig. [14.4(b)] shows graphite structure.

14.9 HYDROGEN BONDING

The Hydrongen bond is formed between an atom of hydrogen and an extremely electronegative atom, e.g., an atom of oxygen, fluorine, nitrogen, chlorine. Such an atom attracts the bonding electrons and becomes negatively charged. The hydrogen atom after loosing the bonding electron assumes a positive charge. The hydrogen bond is a result of electro-static attraction of these charges. Hydrogen bond in ice is illustrated in Fig. (14.5).

The hydrogen bond is essentially ionic in character and the cohesive energy may be discussed in terms of Coulomb interactions between charges and Leonard-Jones Potentials. The binding energy of ice or water is almost 0.5 eV.

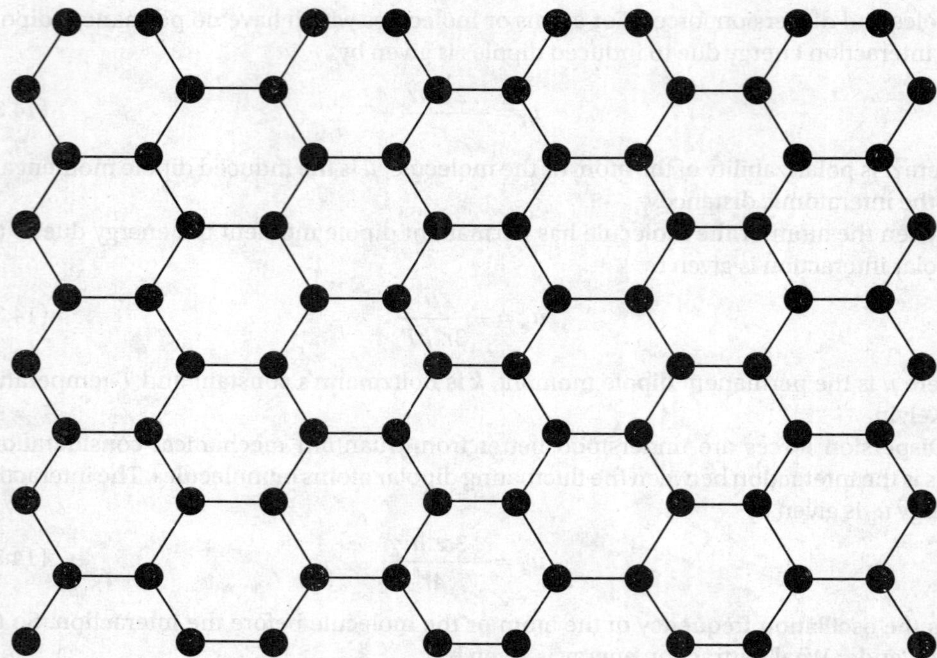

Fig. 14.4(b) Graphite structure. It consists of carbon atoms in tetragonal arrays, with each atom bonded to three others. The layers are bonded together by weak Vander Waal's forces

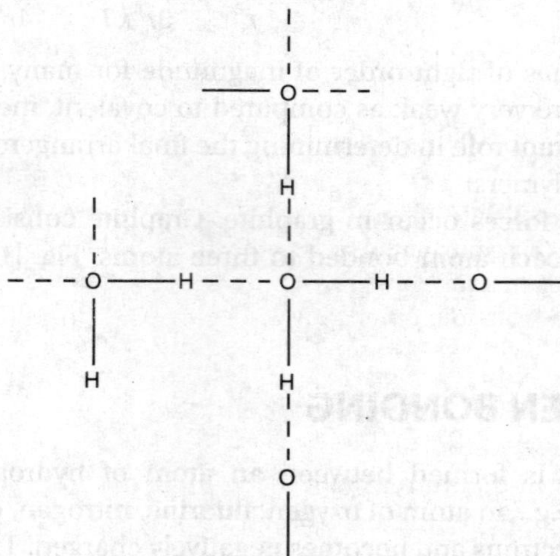

Fig. 14.5 Hydrogen bond

The hydrogen bond is the cause of association of molecules of liquid (such as H_2O, acids, spirits, etc.,) which results in greater viscosity, higher boiling point, abnormal thermal expansion etc., Bonding in H_2O molecule is shown in Fig. (14.6). Should there be no hydrogen bonds between the molecules of water, its boiling point at atmospheric

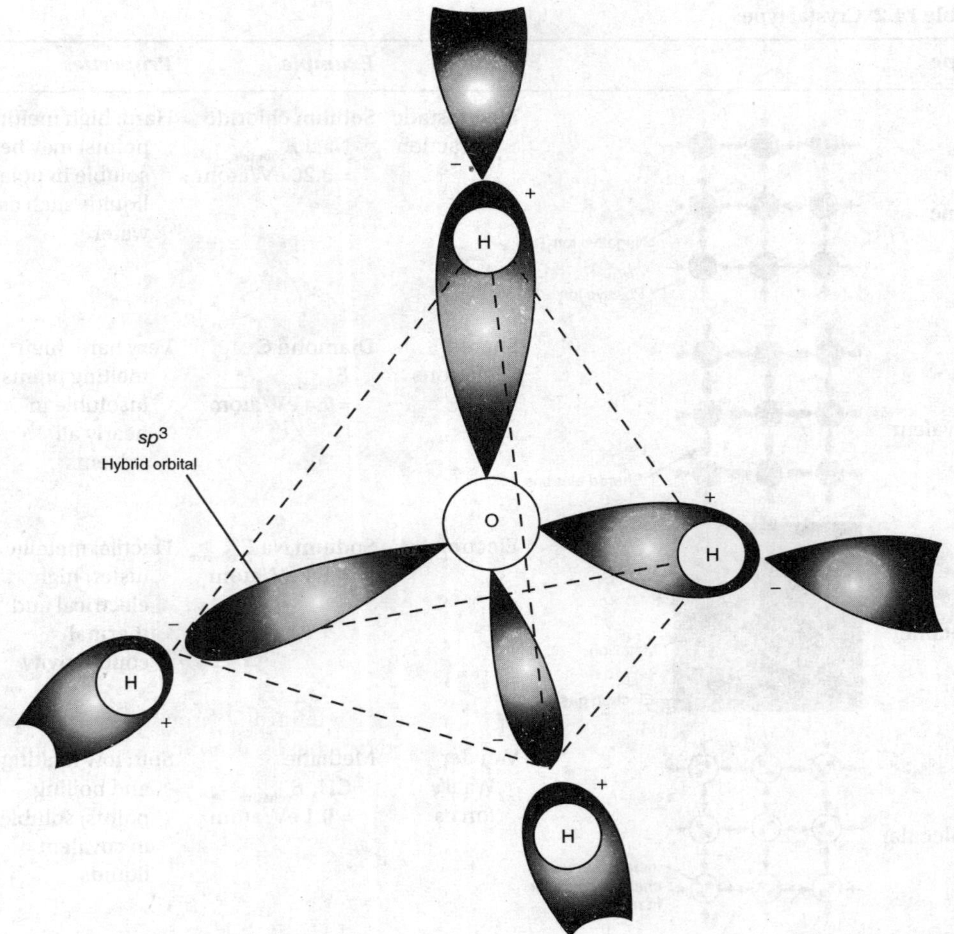

Fig. 14.6 Bonding in H_2O molecule. In an H_2O molecule, the four pairs of valence electrons around the oxygen atom (six contributed by the O atom and one each by the H atom) occupy four sp^3 hybrid orbitals that form a tetrahedral pattern. We may note that each H_2O molecule can form H bonds with four other H_2O molecules

pressure would not be +100°C but 80°C and its viscosity would be lower by almost an order of magnitude. When water is heated above 0°C hydrogen bond is broken.

It should be pointed out that in real solids not one type of bonds discussed above ever exist purely by itself. Practically there is always a superposition of two or more types of bonds, one of them as a rule plays a dominant part in determining the structure and the properties of the solid.

The various crystal types are summarized in Table 14.2.

Example 14.1: An atom *A* has an ionisation energy of 5 eV and another atom *b* has an electron affinity of 4 eV. What is the energy released or absorbed in placing the two atoms 5A apart?

Table 14.2 Crystal types

Type	*Bond*	*Example*	*Properties*
Ionic	Electrostatic attraction	Sodium chloride Nacl $E_{cohesive}$ = 3.28 eV/atom	Hard; high melting points; may be soluble in polar liquids such as water
Covalent	Shared eletrons	Diamond C $E_{cohesive}$ = 7.4 eV/atom	Very hard; high melting points; insoluble in nearly all solvents
Metallic	Electron gas	Sodium Na $E_{cohesive}$ = 1.1 eV/atom	Ductile; metallic luster; high electrical and thermal conductivity
Molecular	Van der Waal's forces	Methane CH_4 $E_{cohesive}$ = 0.1 eV/atom	Soft; low melting and boiling points; soluble in covalent liquids

Solution Since ionization energy of atom A is 5 eV that means the cost of removing one electron from atom A is a 5 eV, i.e., when 5 eV energy is supplied the atom A becomes positive ion A^+.

The electron affinity of *b* is 4 eV, means when this atom absorbs an electron, it releases an energy of 4 eV and becomes electro-negative.

These two will attract and when they are placed at a distance of 5A, the Coulombian potential energy then becomes

$$u_c = -\frac{Kq_1q_2}{r} \qquad q_1 = q_2 = e = 1.6\times10^{-19}\,\text{C}$$

$$= -\frac{9\times10^9\times1.6\times10^{-19}\times1.6\times10^{-19}}{5\times10^{-10}}\text{J} = -\frac{9\times10^9\times1.6\times10^{-19}}{5\times10^{-10}}\text{eV}$$

$$= -2.88\text{ eV}$$

Thus total energy absorbed = +5 – 4 + 2.88 = +3.88 eV.

Example 14.2: **If the potential energy function is expressed as**

$$u(r) = -\frac{\alpha}{r^6} + \frac{\beta}{r^{12}}$$

show that

(a) The intermolecular distance r_0 for which the potential energy is minimum is given by $\left(\frac{2\beta}{\alpha}\right)^{\frac{1}{6}}$

(b) The interatomic distance r_0' for which the potential energy is zero is given by $r_0\,(2)^{-\frac{1}{6}}$

(c) The minimum potential energy in given by $u_{min} = -\frac{\alpha^2}{4\beta}$

Solution

(a) The condition for the minimum potential energy is

$$\left(\frac{du}{dr}\right) = 0$$

$$\therefore \quad \frac{\partial}{\partial r}\left(-\frac{\alpha}{r^6} + \frac{\beta}{r^{12}}\right) = \frac{6\alpha}{r^7} - \frac{12\beta}{r^{13}} = 0$$

$$\text{or} \quad r^6 = \frac{2\beta}{\alpha} \quad \text{or} \quad r\left(\frac{2\beta}{\alpha}\right)^{\frac{1}{6}}$$

The inter atomic distance for minimum potential energy u_{min} is r_0

$$\therefore \quad r = r_0 = \left(\frac{2\beta}{\alpha}\right)^{\frac{1}{6}} \qquad \text{(i)}$$

(b) if $r = r_0'$ for which $u(r) = 0$, then

$$-\frac{\alpha}{r_0^6} + \frac{\beta}{r_0'^{12}} = 0$$

$$\text{or} \quad r_0'^6 = \frac{\beta}{\alpha} = \frac{r_0^6}{2} \qquad \text{from (i)}$$

$$\text{or} \quad r_0' = r_0\left(\frac{1}{2}\right)^{\frac{1}{6}} = r_0(14.2)^{-1/6} \qquad \text{(ii)}$$

(c) The minimum potential energy is

$$u_{min} = -\frac{\alpha}{r_0^6} + \frac{\beta}{r_0^{12}} = -\frac{\alpha}{r_0^6}\left[1 - \frac{\beta}{\alpha}\left(\frac{1}{r^0}\right)^6\right]$$

$$= -\frac{\alpha}{\left(\frac{2\beta}{\alpha}\right)}\left[1 - \frac{\beta}{\alpha}\cdot\frac{\alpha}{2\beta}\right]\left(\because r_0^6 = \frac{2\beta}{\alpha}\right)$$

$$= -\frac{\alpha^2}{2\beta}\cdot\frac{1}{2} = -\frac{\alpha^2}{4\beta}$$

Example 14.3: Sort out from the following substances (giving reasons) those having (i) ionic (ii) covalent (iii) metallic and (iv) Vander Waal's bonds. What sort of structure (compact or loose) do you expect for them.

Cu, Zn, Pb, Ni, Mg, Ne, Ar, NaCl, NaI, CuNi and SiC.

Solution

(i) NaCl and NaI must have ionic bonds, because the Na atoms have one single electron in its outer most shell and both Cl and I atoms have in their outer-most shell 7 electrons each, i.e., one short for completion of the shell. The former, therefore, easily transfers its lone outer shell electron to the outer shell of the latter and we thus have an N^+ ion and a Cl^- or I^- ion with the outer shells of both types of ions complete. The electrostatic attraction between the two oppositely charged ions then provides the ionic bond between them. Since the ionic bond is the strongest of all other bonds both NaCl and NaI must have a compact or close-packed structure.

(ii) SiC must have covalent bonds, since both Si and C have four electrons each in their outer most shells. which form covalent bonds. These bonds being next only in strength to ionic bonds (on account of their directional nature) the structure of SiC too must be quite compact.

(iii) Metals Cu, Zn, Pb, Ni and Mg, as also the metallic alloy CuNi must naturally have metallic bonds. These bonds though weaker than ionic and covalent bonds are however non-directional. These metals and the alloy (CuNi) must also, have a compact and close-packed structure.

(iv) Both Ne and Argon, being inert gases must have Vander Waal's bonds, which are the weakest of all. Since they are non-directional, the structure of Ne and Argon in the solid state must be expected to be fairly compact.

Example 14.4: The ionisation energy of lithium is 5.4 eV and the electron affinity of bromine is 3.5 eV (i) What is the net amount of energy required to form an Li^+ and Br^- ion pair. (ii) Considering them as a joint charge how close together must an Li^+ and Br^- ion be for the total energy of pair to be zero?

Solution

(i) Energy required to transfer an electron from an Li to a Br atom or to form an Li^+ and Br^- pair

$$= 5.4 \text{ eV} - 3.5 \text{ eV} = 1.9 \text{ eV}$$

(ii) Let Li^+ and Br^- be point charges with charge (+e) and (–e) and r be the distance between them. Then potential energy of charges is

$$-\frac{Kq_1q_2}{r} = \frac{Ke^2}{r} \quad \text{where } K = 9\times10^9 \frac{NM^2}{C^2}$$

$$\therefore \quad \text{potential energy} = \frac{Ke}{r}\text{eV}$$

For total energy of pair to be zero, we have

$$\frac{Ke}{r} = 1.9$$

or $$r = \frac{Ke}{1.9} = \frac{9 \times 10^9 \times 1.6 \times 10^{-19}}{1.9} = 7.58 \times 10^{-10}\,\text{m}$$

Example 14.5: The potential energy of a system of two atoms is given by the relation

$$U = A/r^2 + B/r^{10}$$

A stable molecule is formed with the release of 8 eV energy when the interatomic distance is 2.8 A. Find A and B and the force needed to dissociate this molecule into atoms and the interatomic distance at which the dissociation occurs.

Solution We have the potential energy of the system

$$U = A/r^2 + B/r^{10} \qquad \text{(i)}$$

$$\therefore \quad F = -\frac{dU}{dr} = -\frac{2A}{r^3} + \frac{10B}{r^{11}} \qquad \text{(ii)}$$

(i) At equilibrium distance r_0, $F = 0$. Therefore Eq. (ii) gives

$$A = \frac{5B}{r_0^8} = \frac{5B}{(2.8 \times 10^{-10})^8} = 1.32 \times 10^{17}\, B \qquad \text{(iii)}$$

∴ Bond energy or the dissociation energy, U_0 is obtained from Eq. (i) by putting $r = r_0$,

$$U_0 = -\frac{A}{r_0^2} + \frac{B}{r_0^{10}} = -\frac{A}{r_0^2}\left(1 - \frac{B}{Ar_0^8}\right)$$

Using Eq. (iii), one obtains

$$U_0 = -\left(\frac{A}{r_0^2}\right)\left(1 - \frac{1}{5}\right) = -\frac{4}{5}\frac{A}{r_0^2}$$

Negative sign signifies the release of energy.

$$\therefore \quad A = \left(\frac{5}{4}\right)(2.8 \times 10^{-10})^2 (8.0 \times 1.6 \times 10^{-19}) = 1.256 \times 10^{-37}\,\text{J-m}^2$$

and $$B = 9.52 \times 10^{-115}\,\text{J-m}^2$$

The critical interatomic distance r_c is obtained from the condition

$$\left.\frac{dF}{dr}\right|_{r=r_c} = 0$$

From Eq.(iii) $$\frac{6A}{r_0^4} - 110\frac{B}{r_c^{12}} = 0$$

It yields $$r_c = (110B/6A)^{1/8}$$

$$= \left(\frac{110 \times 9.52 \times 10^{-115}}{6 \times 1.256 \times 10^{-37}}\right)^{1/8} = 3.3 \times 10^{-10}\,\text{m}$$

Putting $r = r_c$ in Eq. (ii), the force required to dissociate the molecule is

$$F = -\left(-\frac{2}{r_c^3}\right)\left(A - \frac{5B}{r_c^8}\right)$$

Using Eq. (iv), it becomes

$$F = -\left(-\frac{2}{r_c^3}\right)\left[A - 5B\left(\frac{6A}{110B}\right)\right] = \left(\frac{2}{r_c^3}\right)\left(\frac{8A}{11}\right)$$

$$= -\frac{2}{(3.3\times10^{-10})^3}\left(\frac{8\times1.256\times10^{-37}}{11}\right) = -5.08\times10^{-9}\ \text{N}$$

The negative sign indicates that the force existing at $r = r_c$ is the force of attraction. The force needed to dissociate the molecule is 5.08×10^{-9} N.

Example 14.6: **The lattice energy of KCl crystal containing *N*-molecules of KCl is given by $U = -N(Mq^2/r - B/r^n)$. Find the repulsive exponent *n*. Given nearest neighbour equilibrium distance, R_0 = 3.14Å, compressibility of KCl, $K = 5.747 \times 10^{-11}$ m²/*N* Madelung constant $M = 1.748$.**

Solution $U = -N\left(\dfrac{Mq^2}{4\pi\varepsilon_0 R} - \dfrac{B}{R^n}\right)$

Differentiating, we obtain

$$\frac{dU}{dR} = \frac{NMq^2}{4\pi\varepsilon_0 R^2} - \frac{nNB}{R^{n+1}} \quad \text{(i)}$$

For $R = R_0$, $dU/dR = 0$, this yields

$$B = \frac{Mq^2}{4\pi\varepsilon_0 n} R_0^{n-1} \quad \text{(ii)}$$

The structure of KCl is identical to that of NaCl. The compressibility is reciprocal to the bulk modulus, and hence

$$\frac{1}{K} = \left(\frac{1}{18NR_0}\right)\left.\frac{d^2U}{dR^2}\right|_{R=R_0} \quad \text{(iii)}$$

From Eq. (i), we have

$$\left.\frac{d^2U}{dR^2}\right|_{R=R_0} = -\frac{2NMq^2}{4\pi\varepsilon_0 R_0^3} + \frac{n(n+1)NB}{R_0^{n+2}}$$

$$= -\frac{2NMq^2}{4\pi\varepsilon_0 R_0^3}(n-1) \quad \text{(Using Eq.(ii))}$$

Hence from Eq. (iii), we have

$$n = 1 + \frac{18R_0^4(4\pi\varepsilon_0)}{KMq^2}$$

$$= 1 + \frac{18 \times \left(3.147 \times 10^{-10}\right)^4}{5.747 \times 10^{-11} \times 1.748 \times \left(1.6 \times 10^{-19}\right)^2 \times 9 \times 10^9}$$

$$= 1 + 7.63 = 8.63$$

Example 14.7: **The force of attraction between ions of Na and Cl is 3.02×10^{-9} N when the two ions just touch each other. Given: ionic radius of Na^+ ion is 0.95 Å, $e = 1.6 \times 10^{-19}$ C, $\varepsilon_0 = 8.854 \times 10^{-12}$ C²/N m². Find the radius of Cl^- ion.**

Solution We have $F_1 = \dfrac{Z_1 Z_2 e^2}{4\pi\varepsilon_0 r^2}$ (force of attraction)

$$\therefore \quad F = 3.02 \times 10^{-19} N = \frac{(+1)(-1)(1.6)^2 \times \left(10^{-19}\right)^2}{4\pi \times 8.854 \times 10^{-12} r^2}$$

$$= \frac{0.23}{r^2} \times 10^{-26} N$$

$$\therefore \quad r^2 = \frac{0.023 \times 10^{-26}}{3.02 \times 10^{-9}} = 0.0076 \times 10^{-17}$$

$$\therefore \quad r = 0.276 \times 10^{-9} \text{ m} = r_{Na^+} + r_{Cl^-}$$

$$\therefore \quad r_{Cl^-} = 2.76 \times 10^{-10} - 0.95 \times 10^{-10} = 1.81 \times 10^{-10} \text{ m}$$

$$= 1.81 \text{Å}$$

$Z_1 = +1$ for Na^+

$Z_2 = -1$ for Cl^-

$e = 1.6 \times 10^{-19}$ C

$\varepsilon_0 = 8.854 \times 10^{-12}$

Example 14.8: **The ionic radii of Mg^{++} and S^{--} respectively are 0.65 Å. Calculate the force of attraction between these ions. Given: $e = 1.6 \times 10^{-19}$C, $\varepsilon_0 = 8.854 \times 10^{-12}$C²/N-m²**

Solution Force of attraction

$$F_1 = \frac{Z_1 Z_2 e^2}{4\pi\varepsilon_0 r^2}$$

$$= \frac{2 \times (-2) \times \left(1.60 \times 10^{-19}\right)^2}{4\pi \times 8.854 \times 10^{-12} \times \left(2.49 \times 10^{-10}\right)^2}$$

$$= \frac{4 \times 2.56 \times 10^{-38}}{111.27 \times 6.2 \times 10^{-32}} N = 0.015 \times 10^{-6} N$$

$$= 1.5 \times 10^{-8} N$$

$Z_1 = +2$ for Mg^{++}

$Z_2 = -2$ for S^{--}

r = radius of Mg^{++} ion + radius of S^-

$= (0.65 + 1.8\text{Å})$

$= 2.49 \times 10^{-10}$ m

Example 14.9: For NaCl, the Na atom has an ionisation energy 5.14 eV and the Cl atom has an electron affinity 3.61 eV. The equilibrium separation between the ion-pair is 0.282 nm. Show that the energy required to transfer an electron from Na to Cl is about-7.395 eV.

[Kerala]

Solution We have the Coulomb energy of NaCl as

$$E = \frac{1.75e}{4\pi\varepsilon_0 r_0}\text{eV}$$

$$= \frac{1.6\times10^{-19}\times1.75}{4\pi\times8.85\times10^{-12}\times0.282\times10^{-9}} = 8.925\text{eV}$$

Now, the energy required to transfer an electron from Na to Cl

$$=(5.14-3.61-8.925)\text{eV} = -7.395\text{eV}$$

SUMMARY

1. Atoms in molecules/solids are held together by chemical bonds. This holding together phenomenon is called as *bonding.*
2. In any solid, the mutual interatomic forces are basically *electrostatic* in nature, and the primary differences among different types of solids depend on the ways in which the valence electrons of constitutent elements are distributed, i.e., it depends on the bonding.
3. On the basis of bonding type, there are five types of solids: (i) *Ionic* solids (e.g. alkali halides, alkaline oxides), (ii) *Covalent* solids (diamond, silicon, etc.,), (iii) *van der Waal* bonded molecules (O_2, H_2, Kr, Xe, solid He), (iv) *Hydrogen-bonded* solids (i.e., some fluorides and compounds having water of crystallization), and (v) *Metallic* solids (various metals and alloys).
4. In a simple diatomic molecule, two atoms exert attractive forces between them which keep them together and repulsive forces between them since external energy is required to bind them together. At equilibrium, the two forces become equal. At this stage the system has minimum energy given by

$$V(r) = \frac{-A}{r^n} + \frac{B}{r^m}$$

   where r is the distance between two nuclei and A, B, m, and n are constants characterstic of the molecule.
5. The energy at the equilibrium position $V(r_0)$ is called the *binding energy* or the *cohesive energy* of the molecule.
6. Based on bond strength, atomic bonds are classified into *primary* and secondary. Primary bonds have energies in the range 0.1 to 10 eV. Ionic, covalent and Metallic bonds belong to this category. Secondary bonds have energies in the range 0.1 to 0.5 eV. Hydrogen and van der Waals bonds are examples of secondary bonds.

7. An *ionic bond* is formed when atoms that have low ionization energies, and hence can lose electrons readily, interact with other atoms that have high electron afinity. This bond is formed between an electropositive element (Group I or Group II elements like Na, K, Mg, etc.,) and an electronegative element (Group VI or Group VII elements like Cl, Br, O, etc.,).
8. A *Covalent bond* is formed when one or more pairs of electrons are shared between the incomplete outer shells of two atoms, e.g. H_2 molecule. The spins of two electrons in a shared-pair are antiparallel. The covalent bond is also known as a *homopolar* or *electron pair bond.*
9. The *metallic bond* is formed between electropositive elements having low ionisation energy. The binding energy of metallic bonds are usually in the range of 1–3 eV. The free electrons are responsible for the high electrical and thermal conductivity of metals.
10. A hydrogen bond is formed when a hydrogen atom makes a bond with an extremely electronagnetive atom (O, F, N, Cl, etc.,).
11. van der Waals bonds are usually observed in inert gases. The weak short range forces that are weaker than the atomic bonding forces, existing between atoms and molecules are called van der Waals forces. These bonds are formed due to dipole attraction.

REVIEW QUESTIONS

1. Describe the nature and origin of various forces existing between the atoms of a crystal. Explain the formation of a stable bond using the potential energy versus interatomic distance curve.
2. List the various types of bonds occurring in a crystals. Discuss any one of them in brief.
3. Explain electron cloud. What is the role of electron cloud in metallic bond?
4. List the different types of bonds occurring in a crystal. Describe the characteristics of metallic bonds.
5. Describe the binding of atoms in metals.
6. Distinguish between ionic and metallic bonds in solids, illustrate with example.
7. Why are close packed structures experienced mostly in metals and not in ionic and covalent bonds?
8. Explain the origin of metallic bonding. How does it differ from ionic bonding? Explain with suitable examples.
9. How are atoms held together in a metallic bond? Explain diagramatically.
10. What is hydrogen bond? How is it different from a dipole bond? Explain the role of hydrogen bond during the formation of ice?
11. Derive an expression for the binding energy of an ionic crystal and obtain the expression for Madelung constant. Evaluate Madelung constant for a linear ionic crystal.
12. What are various types of bonding in materials? Explain the different types of bonding and illustrate them with suitable examples.
13. How do the melting points of the ionic solids vary?

14. Why do inert gases get liquified and solidified at very low temperatures?
15. Define cohesive energy and determine its value for crystals of inert gases.
16. Using the concept of wave mechanics, explain why a covalent bond tends to be the strongest in the directions when $|\psi^2|$ is a maximum.
17. Explain why carbon atoms in diamond bond covalently, while lead atoms bond metallically, even though carbon and lead have four valence electrons each.
18. Explain with the help of suitable sketches the various types of bonding in crystals.
19. Describe the ionic bonding and giving example show that the bond energy of ionic bonding decreases as the size of electrons increases. Explain this behaviour.
20. What is a secondary bond? Explain the concept of dipole bonding. Why are some compounds non-polar even if they have dipoles?
21. Compare and contrast metallic, covalent and ionic bonds. Give examples.
22. What do you understand by percentage ionic factor of a covalent bonds?
23. Explain in which of the ionic and covalent bonds wave characteristics of electrons play important role?
24. Describe the essential features of the following with one example of each category: (a) metallic bond (b) ionic bond (c) covalent bond.
25. Explain briefly: (a) covalent bonding (b) ionic bonding.
26. It is observed that CCl_4 has no net dipole moment. What inference do you draw about C-Cl bond in this compound?
27. Describe compounds which show mixed ionic and covalent bonds?
28. The potential energy of a diatomic molecule is given in terms of the interatomic distance r by the expression: $U(r) = -\dfrac{a}{r^m} + \dfrac{b}{r^n}$, where symbols have usual meaning. Derive an expression for equal spacing of the atoms and hence obtain the dissociation energy.
29. Explain the origin of Vander Waal's force in molecular crystals. How does the binding due to Vander Waal's forces differ from the binding the valence crystals.

PROBLEMS

1. The potential energy of a system of two atoms is given by $U = -\alpha/r^4 + \beta/r^{12}$. Calculate the amount of energy released when the atoms form a stable bond. Also determine the bond length. $\left[\left(4\alpha^3/27\beta\right)^{1/2}, \left(3\beta/\alpha\right)^{1/8}\right]$
2. Assume that the energy of two particles in the field of each other is given by the following function: $U = -a/r + b/r^8$, where a and b are constants and r is the distance between the centres of the particles. Show that if particles are pulled apart, the molecules will break as soon as $r = (36b/a)^{1/7} = r_0(4.5)^{1/7}$.
3. The interaction energy of a system of two atoms is given by $U = -A/r^6 + B/r^{12}$. The atoms form a stable bond with bond length of 3A and bond energy of 1.8 eV. Calculate A and B. Compute the forces required to break the molecule and the critical interatomic distance for which it occurs. [4.19×10^{-76} J-m^6, 1.53×10^{-133} J-m^{12}, 2.56×10^{-9} N, 3.33×10^{-10} m].

4. Assume that the energy of two particles in the field of each other is given by $u = -\frac{2\mu^2\alpha}{r^6} - \frac{2\mu^4}{3r^6kT} - \frac{3\alpha^2 h\nu_0}{4r^6}$ where α, β are constants and r is the distance between centres of the particles. Show that the two particles form a stable compound for $r = r_0 = (8\beta/\alpha)^{1/7}$ and that in the stable configuration, the energy attraction is 8 times energy of repulsion.
5. Consider a line of alternate positive and negative ions each carrying a charge q. The repulsive potential energy between neighbours is given by A/r^n. Show that for a total of $2N$ ions, the equilibrium potential energy of the system is $U_0 = -(2Nq^2 ln2/r_0)(1 - 1/n)$, where r_0 is the equilibrium interionic distance.
6. The potential energy U of a system of two atoms varies as a function of their distance of separation r as $U = -A/r^n + B/r^m$. Show that in equilibrium: (i) $r = r_0 = (mB/nA)^{1/(m-n)}$ (ii) the energy of attraction is m/n times the energy of repulsion and (iii) the bond energy, $U_0 = \frac{A}{r_0^n}\left(\frac{m-n}{m}\right)$
7. The distance between the nearest positive and negative ions in KCl is 3.14 Å. The structure is similar to NaCl. Find the ionic bond energy in the solid in kilocalories per mole.

SHORT ANSWER-QUESTIONS

1. How atomic bonding in solids are considered?
   Ans: These are considered in terms of attractive and repulsive forces.
2. How many types of primary bonds are?
   Ans: Three: ionic, covalent and metallic.
3. Give names of secondary bonds.
   Ans: Vander Waals and hydrogen bonds. These are weak in comparison to primary bonds.
4. Mention the range of energies of primary and secondary bonds.
   Ans: Primary bond energies are in the range of 100–1000 kJ/mol, whereas that of secondary bonds are in the range of 1–50 kJ/mol.
5. How non directional ionic bond is developed?
   Ans: For ionic bonds, electrically charged ions are formed by the transference of valence electrons from one atom of another. The forces are Coulombic.
6. What is covalent bonding?
   Ans: In a covalent bonding there is sharing of valence electrons between adjacent atoms.
7. What is a metal?
   Ans: A metal is an array of positive ions which are held together in a cloud of free electrons.

8. What is metallic bonding?
Ans: With metallic bonding, the valence electrons form a 'sea of electrons' that is uniformly dispersed around the metal ion cores and acts as a form of glue for them. The metallic bond is directional and generally weaker than ionic and covalent bond.

9. What is the relation between melting and boiling points of materials and bond strength?
Ans: Melting and boiling points of materials increase with increasing bond strength. Strong and directional bonds result in hard and brittle solids.

10. How secondary bonds are formed?
Ans: Secondary bonds from attractive forces between electric dipoles, of which there are two types: induced and permanent.

11. What happens when hydrogen covalently bonds to a non metallic element such as fluorine?
Ans: Highly polar molecule is formed.

OBJECTIVE QUESTIONS

1. Sharing of electrons between neighbouring atoms results in
(a) metallic bond (b) ionic bond
(c) covalent bond (d) none of above

2. Primary bonds have energy range in kJ/mol.
(a) 10–100 (b) 10–1000
(c) 1–10 (d) 100–5000

3. The nature of binding for a crystal with alternate and evenly spaced positive and negative ions is
(a) dipole (b) ionic
(c) metallic (d) covalent

4. The atomic bond in NaCl is
(a) ionic (b) metallic
(c) covalent (d) van der Waal's

5. Thermal expansion of materials arises from
(a) thermal vibrations (b) weak bonds
(c) strong bonds (d) asymmetry of potential energy curve

6. To break one H–Cl bond, 4.4 eV of energy is required. This energy is equal to
(a) 420 kJ/mol (b) 420×10^3 kJ/mol
(c) 420 J/k-mol (d) 42 J/k-mol

7. The electron affinity of helium in kJ/mol is
(a) 4 (b) 16
(c) –16 (d) 0

8. The length of H–H bonds is

(a) 1 nm (b) 0.1 nm

(c) 0.037 nm (d) 0.074 nm

9. The Fe–Fe bond length is 2.48 Å. The radius of iron atom is

(a) 0.62 Å (b) 1.24 Å

(c) 2.48 Å (d) 3.96 Å

10. Which of the following elements is a covalently bonded crystal?

(a) germanium (b) aluminium

(c) lead (d) sodium chloride

11. Hydrogen bonds are stronger than

(a) ionic bonds (b) metallic bonds

(c) covalent bonds (d) van der Waal's bonds

12. Which of the following relation represent the partial energy of a diatomic molecule?

(a) $Ar^m + Br^n$ (b) $Ar^m - Br^n$

(c) $A/r^m - B/r^n$ (d) $-A/r^m + b/r^n$

13. The radius of anion is r_a and of cation is r_c, the bond length is

(a) $\sqrt{3}(r_c + r_a)$ (b) $r_a - r_c$

(c) $(r_c + r_a)$ (d) $(r_c + r_a)/2$

14. If 0.28 nm is the spacing between the nearest neighbor ions is NaCl crystal lattice, the unit cell parameter is

(a) 5.6 Å (b) 1.6 Å

(c) 0.7 Å (d) 1.4 Å

15. The bond energy of NaCl molecule is given by the relation:

(a) $V = e/4\pi\varepsilon_0 r_0^2$ (b) $V = -e/4\pi\varepsilon_0 r_0^2$

(c) $V = e^2/4\pi\varepsilon_0 r_0$ (d) $V = -e^2/4\pi\varepsilon_0 r_0$

16. Which of the following gives the lattice energy per k-mol of NaCl crystal? (Symbols have their usual meanings)

(a) $-\dfrac{AN_A e^2}{4\pi\varepsilon_0 r_0}\left(\dfrac{n-1}{n}\right)$ (b) $\dfrac{AN_A r^2}{4\pi\varepsilon_0 r_0}\left(\dfrac{n}{n-1}\right)$

(c) $-\dfrac{AN_A e^2}{4\pi\varepsilon_0 r_0^2}\left(\dfrac{n-1}{n}\right)$ (d) $\dfrac{AN_A r^2}{4\pi\varepsilon_0 e}\left(\dfrac{n}{n-1}\right)$

17. Metallic bond is not characterized by

(a) opacity (b) ductility

(c) high conductivity (d) directionality

18. Mixed ionic-covalent bonds are found in
(a) high strength materials (b) semiconductors
(c) heat insulators (d) none of the above

19. The bond energy of C–C covalent bond in kJ/mol is
(a) 370,800 (b) 640
(c) 1640 (d) zero

20. The sublimation energy of sodium is
(a) 1.08×10^3 kJ/k-mol (b) 10.8×10^3 kJ/k-mol
(c) 108×10^3 kJ/k-mol (d) 0.108×10^3 kJ/k-mol

21. Which of the following has hydrogen bonding?
(a) HF (b) C
(c) CH_4 (d) CsCl

22. A semiconductor has
(a) van der Waal's bonding (b) covalent bonding
(c) metallic bonding (d) ionic bonding

23. The directional interatomic bond is
(a) van der Waal's (b) metallic
(c) covalent (d) ionic

24. The energy of cohesive energy in case of van der Waal's bonding is
(a) 0.1–0.8 eV/atom (b) 0.002–0.1 eV/atom
(c) 5–8 eV/atom (d) 8–12 eV/atom

25. The solids which are always opaque to visible radiation:
(a) covalent (b) ionic
(c) metallic (d) van der waals

ANSWERS

Question numbers	1	2	3	4	5	6	7	8	9	10
Answers	(c)	(b)	(b)	(a)	(d)	(b)	(d)	(d)	(b)	(a)
	11	12	13	14	15	16	17	18	19	20
	(b)	(d)	(c)	(a)	(d)	(a)	(d)	(b)	(a)	(c)
	21	22	23	24	25					
	(b)	(b)	(c)	(b)	(c)					

FILL IN THE BLANKS

1. In a solid material, forces acting between constitutent atoms are: (i) ______ and (ii) ______. **[attractive forces, repulsive forces]**
2. An ionic bond is formed when atoms have low ______, and hence can lose electrons readily. **[ionisation energies]**
3. The lattic energy may be defined as the energy released when the constitutent ions are placed in their respective positions in the ______. **[crystal lattice]**
4. In sodilum chloride, the equilibrium distance between ions (r_0) = 2.31 A°, Madelung constant (α) = 1.748. The potential energy per ion pair for m = 9 is ______. **[7.49 eV]**
5. All ionic solids are rigid and crystalline in nature, the constitutent of the crystals are ______ and not ______. **[ions, atoms]**
6. A covalent bond is formed when one or more ______ of electrons are shared between the incomplete ______ shells of the two atoms. The spins of two electrons in a shared-pair are ______. **[pairs, outer, antiparallel]**
7. The hybridisation is the mixing of orbitals of valence shell of the same atom to form new orbitals with the same energy and ______. **[similar shape]**
8. A covalent bond is a bond between the atoms of the same ______, hence it is also called as ______ bond. **[polarity, homopolar bond]**
9. Covalent crystals are typically semiconductors (e.g. Si and Ge) which are electrical insulators at 0 K. The electrical conductivity (σ) rises exponentially with rise in temperature in accordance with $\sigma \propto$ ______, where E_g is constant having the dimensions of energy. **$[-\frac{E_g}{2k_BT}]$**
10. Covalent substances are insoluble in ______ solvents, but are soluble in ______ solvents. **[polar, non polar]**
11. The metallic bond is formed between ______ elements having low ______ energy. **[electropositive, ionisation]**
12. The metallic bond has the cohesive energy of ______ per atom. **[1–5 eV]**
13. Metals are opaque to light, since the light energy is ______ by free ______. **[absorbed, electrons]**
14. Metallic bond may be considered to be ______ between ionic and covalent bond. **[intermediate]**
15. Diamond is a very good ______, silicon and germanium are ______, while tin is a good ______. **[insulator, semiconductors, conductor]**

15

Band Theory of Solids

15.1 FORMATION OF BANDS

In order to study qualitatively the energy spectrum of an electron moving in a crystalline solid we begin the discussion by considering the spectrum of a free atom, and see how this spectrum is gradually modified as atoms are assembled to form the solid.

Let us consider lithium as a concrete example of a free atom. In free lithium atom, the electron moves in a potential well as shown in Fig. [15.1(a)]. On solving Schrodinger equation, we obtain a series of discrete energy levels, as shown. As in the case of the hydrogen atom, these level are denoted by 1*s*, 2*s*, 2*p*, etc., The lithium atom contains three electrons, two of which occupy the 1*s* shell (completely filled), and the third the 2*s* subshell.

Let us now consider the situation when two lithium atoms combine to form the lithium molecule Li_2. The potential as "seen" by the electron is now the double well as shown in Fig. [15.1(b)]. The energy spectrum now consists of a set of discrete doublets. Each of the atomic levels, i.e., the 1*s*, 2*s*, 2*p*, etc., has split into two closely spaced levels. These levels are also nomenclatured as the 1*s*, 2*s*, 2*p* etc., molecular level with the fact that each of these is in fact, composed of two sublevels.

It is easy to explain as to why the atomic levels split into two, and only two sublevels. When the two atoms are far apart, the influence of one atom on an electron on the other atoms is very small and may be treated as perturbation. In this approximation, the unperturbed levels 1*s*, 2*s*, etc., are each doubly degenerate, because an electron in a 1*s* level, for instant, may occupy that level in either atom; and since there are two atoms, the energy is thus doubly degenerate. This degeneracy is strictly valid only if the interaction between the atoms is neglected entirely. When this interaction is included, the double degeneracy is lifted and each level is split into its two sublevels. The molecular orbitals corresponding to these sublevels are usually taken to the symmetric and antisymmetric combinations of the corresponding atomic orbitals.

According to the Pauli's exclusion principle, each molecular level can accommodate almost two electrons of opposite spins, since Li_2 molecule has six electrons out of which four occupy the 1*s* molecular doublet and the other two the lower levels of the 2*s* doublet.

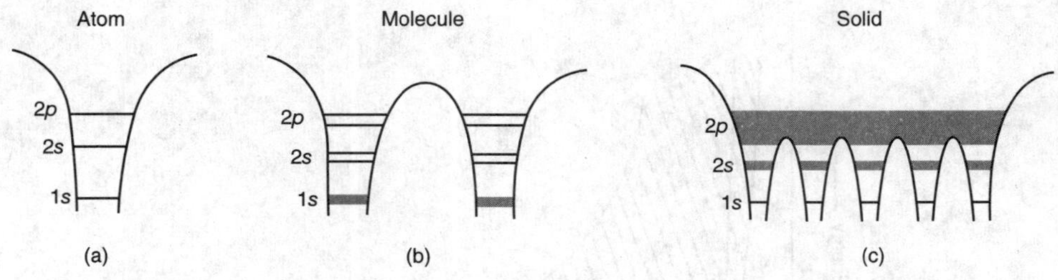

Fig. 15.1

The amount of the splitting between the sublevels depends strongly on the inter nuclear distance of the two atoms in the molecule. The closer the two nuclei, the stronger the perturbation and the larger the splitting. The splitting also depends on the atomic orbitals. The splitting of the $2p$ level is larger than that of the $2s$ level, which is larger still than that of the $1s$ level. The reason is that the radius of the $1s$ orbital, for instant, is very small, and the orbital is therefore tightly bound to its own nucleus. It is not greatly affected by the perturbation. The same is not true for the $2s$ and $2p$ orbitals, which have larger radii and are only loosely bound to their own nuclei. It thus follows that the higher the energy, the greater the splitting incurred.

The above discussion may be generalised for a polyatomic Li molecule containing an arbitrary number of atoms. Thus in a molecule with 3-atoms, each atomic level is split into a triplet, in a 4-atoms molecule into a quadruplet and so forth. The Lithium solid may then be viewed as the limiting case in which the number of atoms has become very larger, resulting in a gigantic lithium molecule. As a result each of the atomic levels is split into N closely spaced sub levels, where N is the number of atoms in the solid. But since N is very large ($\approx 10^{23}$), the sublevels are so extremely closed to each other that they appear to form an energy band. Thus the $1s$, $2s$, $2p$ levels give rise respectively to the $1s$, $2s$, and $2p$ bands as shown in Fig. [15.1(c)]. Here it is important to mention that the energy spectrum for solid in fact is discrete with energy interval between two adjacent levels of the order of (10^{-20} ~ 10^{-23}eV). Since this is an extremely small value, the individual sublevels are indistinguishable, so we can consider their distribution as a continuous energy band.

Clearly the energy spectrum in a solid consists of a set of energy bands with intervening regions separating these bands. The intervening regions are regions of forbidden energy called energy gaps. There are the regions which cannot be occupied by electrons. This situation is different from that of a free atom or molecule, in which the allowed energies form a set of discrete levels. This broadening of descrete levels into bands is one of the most fundamental properties of a solid.

The width of the band varies, but in general the higher the band the greater is its width because as we have seen in case of molecules a high energy state corresponds to a large atomic radius and hence a strong perturbation, which is the cause of the **level broadening** in the first place. On the other hand, the low energy states correspond to tightly bound orbitals, which are affected but slightly by the perturbation.

Fig. (15.2), shows $2s$ and $2p$ bands for metallic lithium plotted as functions of the lattice constant a. We may note that the band widths increase as a decrease as is expected.

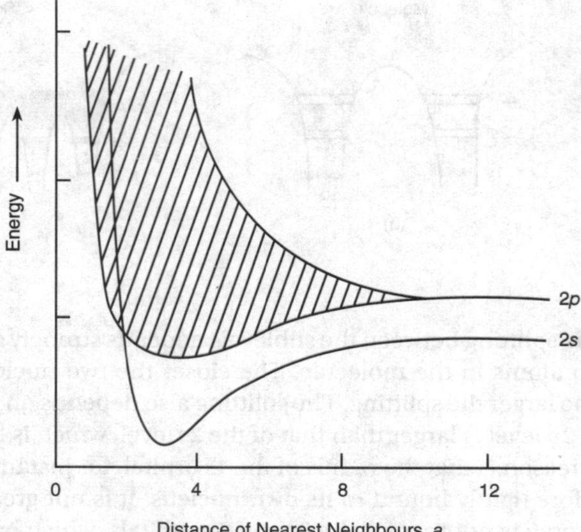

Fig. 15.2

Also that for $a < 6a_o$, the 2*s* and 2*p* bands broaden to the point at which they begin to overlap; and the gap between them vanishes entirely.

The crystal orbitals, i.e., the wave functions describing the electronic states in the bands-extend through-out the solid, unlike the atomic orbitals which are localised around particular atoms, and decay exponentially away from those atoms. In this sense, we refer to solid wave functions as delocalised orbitals. In fact these orbitals actually describe electron waves travelling in the solid, the concept of delocalisation is a basic one. It is responsible for all electronic transport phenomenon in solid, e.g., electrical conduction.

15.2 WAVE FUNCTIONS IN A PERIODIC LATTICE AND BLOCH THEOREM

The behaviour of an electron in a crystalline solid is determined by studying the appropriate Schrodinger equation with crystal potential as seen by the electron. A reasonable approximation of the potential experienced by an electron in passing through the crystal is one which is perfectly periodic with the period that of the lattice, i.e.,

$$V(x) = V(x + a) \tag{15.1}$$

where *a* is lattice vector. Such a potential in shown schematically in Fig. (15.3).

The periodic potential with period that of lattice may be thought of as arising from the periodic distribution of the positive charge associated with the ion cores situated on the lattice sites, plus the constant (average) contribution due to all other free electrons

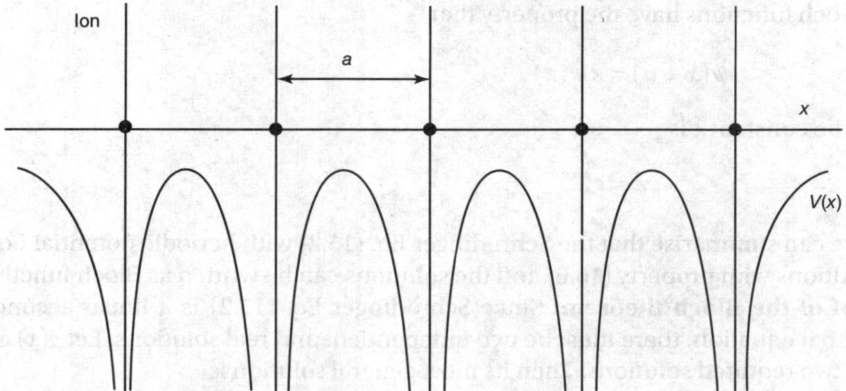

Fig. 15.3

of the crystal. The latter contribution to the potential accounts, in an average sense the interaction effects of the single electron with all others. The periodicity character of the potential extends to infinity in all directions inside the crystal. However at the surface of the crystal it is interrupted and potential there looks some what as shown at the right-hand edge of the Fig. (15.3). It is because, the lattice spacing is not quite uniform there. We shall, however, ignore this irregularity and assume that it is perfectly periodic right from the site of the surface. We now describe the behaviour of the electron in this potential using one electron approximation through-out. This is done by finding the electron wave function.

If $V(x)$ denotes the potential energy of an electron in a linear lattice of lattice constant a, then assumming it to be perfectly periodic satisfying Eq. (15.1). The wave functions of the electron are then obtained from Schrodinger equation.

$$\frac{\partial^2 \psi}{\partial x^2} + \left(\frac{2m}{\hbar^2}\right)[E - V(x)]\psi = 0 \tag{15.2}$$

With reference to the solutions of this equation, there is an important theorem known as **Bloch theorem:** which states that **the solutions are plane waves modulated by the function $u_k(x)$ which has the same periodicity as the lattice**. In the theory of differential equations it is known as **Floquet's theorem**. Mathematically the solutions are of the form

$$\psi(x) = e^{\pm ikx} u_k(x) \tag{15.3}$$

where $$u_k(x) = u_k(x + a) \tag{15.4}$$

The wave functions of the type Eq. (15.3) are called Bloch functions. We note that the Bloch functions Eq. (15.3) have the property

$$\psi(x + a) = e^{\pm ikx(x+a)} u_k(x + a) \ = e^{\pm ika}\psi(x) \tag{15.5}$$

Thus Bloch functions have the property that

$$\psi(x+a)=\lambda\psi(x) \tag{15.6}$$

where the constant λ is

$$\lambda=e^{\pm ika} \tag{15.7}$$

Thus we can summarise that the Schrodinger Eq. (15.2) with periodic potential Eq. (15.1) has solutions with property (15.6) and the solutions can be written as Bloch functions.

Proof of the Bloch theorem: Since Schrodinger Eq. (15.2) is a linear second order differential equation, there must be two independent and real solutions. Let $g(x)$ and $f(x)$ are the two required solutions. Then its most general solution is

$$\psi(x)=Ag(x)+Bf(x) \tag{15.8}$$

where A and B are arbitrary constants. Since $\psi(x+a)=\psi(x)$, not only $g(x)$ and $f(x)$ but also $g(x+a)$ and $f(x+a)$ satisfy Schrodinger equation (15.2). Now, since a differential equation of the second order has only two independent solutions, these must be expressible as linear combination of the independent ones. Hence we must have the relations

$$\left.\begin{aligned} g(x+a)&=\alpha_1 g(x)+a_2 f(x)\\ f(x+a)&=\beta_1 g(x)+\beta_2 f(x)\end{aligned}\right\} \tag{15.9}$$

where α_1 and $\beta_i (i=1,2)$ are constants and real functions of energy E, then

$$\begin{aligned}\psi(x+a)&=Ag(x+a)+Bf(x+a)\\ &=A[\alpha_1 g(x)+\alpha_2 f(x)]+B[\beta_1 g(x)+\beta_2 f(x)]\\ &=(A\alpha_1+B\beta_1)g(x)+(A\alpha_2+B\beta_2)f(x)\end{aligned} \tag{15.10}$$

Now $\psi(x+a)$ can always be expressed in the form

$$\psi(x+a)=\lambda\psi(x)=\lambda Ag(x)+\lambda Bf(x) \tag{15.11}$$

Comparing the co-efficents of $g(x)$ and $f(x)$ in Eqs. (15.10) and (15.11), we have

$$\left.\begin{aligned}(\alpha_1-\lambda)A+\beta_1 B&=0\\ \text{and}\qquad \alpha_2 A+(\beta_2-\lambda)B&=0\end{aligned}\right\} \tag{15.12}$$

The Eq. (15.12), however have non-vanishing solutions for A and B only if the determinant of their co-efficients vanishes, that is if

$$\begin{vmatrix}(\alpha_1-\lambda) & \beta_1\\ \alpha_2 & (\beta_2-\lambda)\end{vmatrix}=0$$

or

$$\lambda^2-(\alpha_1+\beta_2)\lambda+(\alpha_1\beta_2-\beta_1\lambda_2)=0 \tag{15.13}$$

Now one can easily prove [1] that $\alpha_1\beta_2 - \beta_1\alpha_2 = 1$. Thus we have

$$\lambda^2 - (\alpha_1 + \beta_2)\lambda + 1 = 0 \tag{15.14}$$

The solution of this quadratic equation in λ helps to determine the two possible values of λ for which Eq. (15.11) is true. Hence there are two functions $\psi_1(x)$ and $\psi_2(x)$ which exhibit the property (11). Thus, there are two cases of interest.

(i) Energy ranges for which $(\alpha_1 + \beta_2)^2 < 4$.

The two roots λ_1 and λ_2 of Eq. (15.14) will be complex and since $\lambda_1\lambda_2 = 1$, they will be conjugate. In these regions of energy, we way write.

$$\left.\begin{aligned} \lambda_1 &= e^{ika} \\ \text{and} \qquad \lambda_2 &= e^{-ika} \end{aligned}\right\} \tag{15.15}$$

The corresponding functions $\psi_1(x)$ and $\psi_2(x)$ then have the property

$$\left.\begin{aligned} \psi_1(x+a) &= e^{ikx}\psi_1(x) \\ \text{and} \qquad \psi_2(x+a) &= e^{-ikx}\psi_2(x) \end{aligned}\right\} \tag{15.16}$$

Such functions are called Bloch functions. The functions are written in the form, where $u_k(x)$

$$\psi_k(x) = e^{\pm ikx}u_k(x) \tag{15.17}$$

is a periodic function of period a. Thus the solutions of Schrodingers wave equation in this region can be expressed in the form of Bloch functions.

---

[1] Since $g(x)$ and $f(x)$ are real solutions of Eq. (15.2) then we have

$$\frac{d^2g(x)}{dx^2} = \frac{2m}{\hbar^2}[E - V(x)]g(x) = 0 \tag{i}$$

and

$$\frac{d^2}{dx^2}f(x) + \frac{2m}{\hbar^2}[E - V(x)]f(x) = 0 \tag{ii}$$

Multiplying (i) by $f(x)$ and (ii) by $g(x)$ and subtracting, one obtains

$$f(x)\frac{d^2}{dx^2}g(x) - g(x)\frac{d^2}{dx^2}f(x) = 0$$

or

$$\frac{d}{dx}\left[f(x)\frac{d}{dx}g(x) - g(x)\frac{d}{dx}f(x)\right] = 0$$

or

$$w(x) = f(x)\frac{d}{dx}g(x) - g(x)\frac{d}{dx}f(x) = \text{Constant}$$

This is known as **Green's identity**. R.H.S. of this equation is called Wronskian, $w(x)$ of the two solutions and is constant in this case. Further from Eq. (15.9)

$$f(x+a)\frac{d}{dx}(x+a) - g(x+a)\frac{d}{dx}f(x+a) = (\alpha_1\beta_2 - \alpha_2\beta_1)\left[f(x)\frac{dg(x)}{dx} - g(x)\frac{df(x)}{dx}\right]$$

or

$$w(x+a) = (\alpha_1\beta_2 - \alpha_2\beta_1)w(x) \quad \text{or} \quad \alpha_1\beta_2 - \alpha_2\beta_1 = 1$$

(ii) For energy range in which $(\alpha_1 + \beta_2)^2 > 4$: In this case the two roots λ_1 and λ_2 are real. The solutions of Schrodinger's equation are of the types,

$$\left.\begin{aligned} \psi_1(x) &= e^{\mu x}u(x) \\ \text{and} \qquad \psi_2(x) &= e^{-\mu x}u(x) \end{aligned}\right\} \tag{15.18}$$

where μ is a real quantity. Although such solutions are mathematically sound, they cannot in general be accepted as wave functions describing electrons, since they are not bounded. Thus there are no electronic states in the energy region corresponding to real roots λ_1 and λ_2. The above discussion thus leads also to the notion that the energy spectrum of an electron in a periodic potential consists of allowed and forbidden energy regions or bands.

The state function ψ_k of the form Eq. (15.17) known as Bloch function has several interesting properties:

(a) It has the form of a travelling plane wave, represented by the factor e^{ikx} which implies that the electron propagates through the crystal like a free particle. The effect of the function $u_k(x)$ is to modulate this wave so that the amplitude oscillates periodically from one cell to the next, as shown in Fig. (15.4) (Fig. 15.4 shows Bloch function or wave in which smooth curve represents the wave e^{ikx} which is modulated by atomic-like 'wiggly' function $u_k(x)$ but this does not affect the basic character of the state function, which is that of a travelling wave).

If the electrons were indeed entirely free, the state function $\psi_k(x)$ would be given by $\sqrt{\frac{2}{L}}\,e^{ikx}$, that is the function $u_k(x)$ is a constant. But the electron is not free, since it interacts with the lattice, and this determines the special character of the periodic function u_k.

(b) Because the electron behaves as a wave of vector $\boldsymbol{k}$, it has a de-Broglie wave length $\lambda = 2\pi/k$ and hence a momentum

$$P = \hbar k \tag{15.19}$$

according to the de-Broglie relation. We shall call the vector the crystal momentum of the electron.

(c) The Bloch function ψ_k is crystal orbital, as it is localized through-out the solid and not localized around any particular atom. Thus the electron is shared by the whole crystal. This is, of course consistent with property (a) above, in which we described the electron as a travelling wave. Note also that the function ψ_k is so chosen that the electron probability distribution $|\psi_k|^2$ is periodic in the crystal.

In the above discussion, we have stressed the analogy between a crystalline electron and a free one. This is very helpful in understanding the properties of electrons in crystals. One should not however, jump to the conclusion that the two are identical in their behaviour. The Bloch function electron exhibits many intriguing properties not exhibited by a free electron properties which result from the interaction of the electron with the lattice.

Energy Bands: So far we have centred our discussion on the state function and have said nothing about energy. The energy spectrum results from solving the Schrodinger's equation (15.2). Substituting for ψ_k as

$$\psi_k(x) = e^{ikx}u_x(x) \tag{15.20}$$

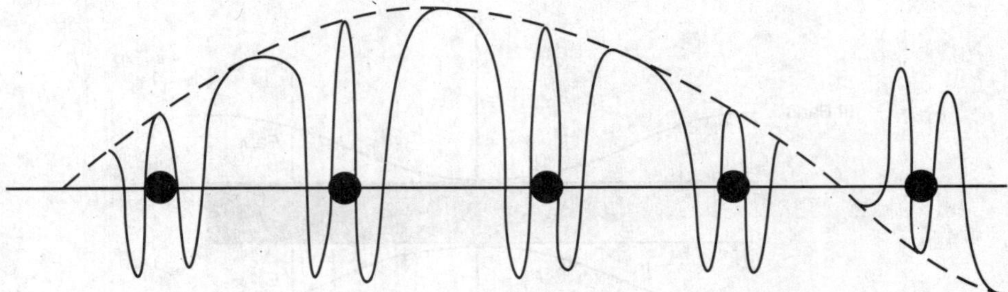

Fig. 15.4

from Bloch form Eq. (15.3), We have

$$\frac{\partial^2}{\partial x^2}\left[e^{ikx}u_k(x)\right]+\frac{2m}{\hbar^2}[E_k - V(x)]e^{ikx}u_k(x)=0 \tag{15.21}$$

Differentiating and rearranging the terms, one obtains

$$\left[-\frac{\hbar^2}{2m}\left(\frac{\partial}{\partial x}+ik\right)^2+V(x)\right]u_k(x)=E_k u_k(x) \tag{15.22}$$

and in three dimensions, we have

$$\left[-\frac{\hbar^2}{2m}(\nabla+i\boldsymbol{k})^2+V(r)\right]u_k(r)=E_k u_k(r) \tag{15.23}$$

Eq. (15.23) in fact is the wave equation for the periodic function $u_k(r)$. This is an eigen equation like Schrodinger equation and can therefore be solved in a similar manner. Note that the operator in the brackets is an explicit function of $\boldsymbol{k}$; and hence both the eigen functions and eigen values depend on $\boldsymbol{k}$; a fact we have already used explicitly by labeling them with the vector $\boldsymbol{k}$. An eigen value equation leads, however, not to one but to many solutions. For each value of $\boldsymbol{k}$, therefore we find a large number of solutions, giving a set of discrete energies $E_{1,k}$, $E_{2,k}$, ... as shown in Fig. (15.5). It is obvious that the energy is a multivalued function of $\boldsymbol{k}$. Since the energies depend on $\boldsymbol{k}$, they vary continuously as $\boldsymbol{k}$ varies over its range of values. Each level leads to an energy band as shown in Fig. (15.5). We shall hence forth write the energy eigen values as $E_n(k)$ and refer to the subscript n as the band index, for obvious reasons.

The number of bands is large usually infinite but only the lowest ones are occupied by electrons. Each band covers a certain energy range, extending from the lowest to the highest value it takes when plotted in $\boldsymbol{k}$-space. The energy intervals interspersed between the band constitute the energy gaps which are forbidden energies that cannot be occupied by electrons.

Further, since $\boldsymbol{k}$ is a vector quantity, a diagram such as Fig. (15.5), is a plot of the energy bands in only one particular direction in $\boldsymbol{k}$-space. If these bands are plotted in different $\boldsymbol{k}$-directions, their appearance would change, in general. A complete representation of the bands therefore requires one to specify the energy values throughout the $\boldsymbol{k}$-space. Often this is accomplished at least partially by drawing the energy contours in $\boldsymbol{k}$-space for the

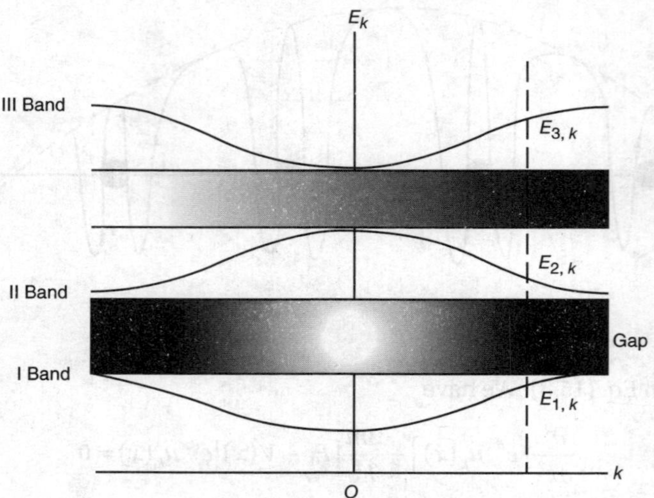

Fig. 15.5

various bands. The bands satisfy certain important symmetry relations that enables us to restrict our consideration to relatively small regions in ***k***-space.

The energy bands which have emerged from this analysis are the same as those discussed already and in fact we can establish a one-to-one correspondence between the energy bands and the atomic levels from which they arise. The particular significance of the present results is that here we can classify the electron state within the band according to their momentum as given by ***k***. Such a classification is found extremely useful and not evident from the previous discussion.

15.3 THE CRYSTAL POTENTIAL

The crystal potential $V(r)$, which acts on the electron is composed of two parts: (i) the interaction of the electron with the ion cores forming the lattice and (ii) its interaction with other Bloch electrons moving through the lattice. In metallic sodium, for example, an electron in the 3s band interacts with Na^+ ions forming the bcc structure, as well as with other electrons in this band. We therefore write $V(r)$ as the sum

$$V(\mathbf{r}) = V_i(\mathbf{r}) + V_e(\mathbf{r}) \tag{15.24}$$

where the first term on right represents the interaction with the ion cores and the second the interaction with the electrons. The ion part may be written as

$$V_i(\mathbf{r}) = \sum_j V_i(\mathbf{r} - \mathbf{R}_j) \tag{15.25}$$

where $V_i(\mathbf{r} - \mathbf{R}_j)$ is the potential of an ion located at the lattice vector $\mathbf{R}_j$ as shown in Fig. [15.6(a)] and summation is over all the ions. The potential $V_i(\mathbf{r})$ obviously has the same periodicity as that the lattice.

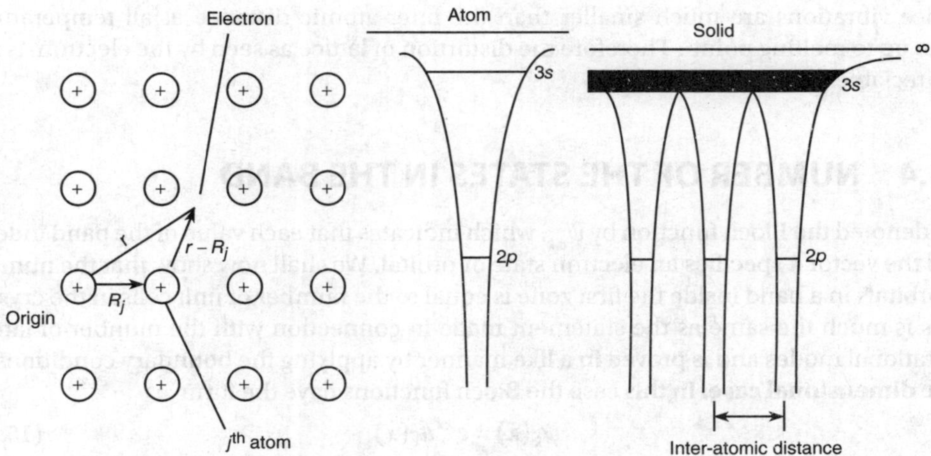

Fig. 15.6

The electronic potential $V_e(\boldsymbol{r})$, the so called electron-electron interaction presents several hundreds which make its treatment very difficult. Firstly, because we can evaluate this term only if we know the states for all other electrons, but these states are not given in advance. Infact, they are the states we are trying to find. Secondly the potential $V_e(\boldsymbol{r})$ is not strictly periodic, since the electrons are in constant motion throughout the lattice. In addition to this, for a proper treatment one should really consider the dynamics of all the electrons simultaneously and not one electron at a time as we have done above. This is a typical example of the many-body problems which are often encountered in solid-state physics.

Inspite of all these difficulties, it is a matter of consolation that electron-electron interaction is very weak and these difficulties are far less serious than they could arise otherwise. The major effect of this interaction is that the electrons distribute themselves primarily around the ions, so that they screen these ions from other electrons. This has the additional effect of making the electron-ion interaction weak even at long range.
So we can write an approximate expression for the potential as

$$V(\boldsymbol{r}) = \sum_j V_s(\boldsymbol{r} - \boldsymbol{R}_j) \tag{15.26}$$

where $V_s(\boldsymbol{r} - \boldsymbol{R}_j)$ is the potential of the second ion located at the lattice point $\boldsymbol{R}_j$, and precisely because this potential is once again periodic, it satisfies the requirements of the Bloch theorem. Fig. [15.6(b)] shows the crystal potential for Na.

In discussing the crystal potential, we have so far tacitly assumed that the atoms are at rest at their lattice sites. However they are not in fact stationary. They are in a constant state of oscillation as a result of their thermal excitation. As such, then our assumption of a stationary lattice is an approximation, and the question now arises is: How good is our approximation? One may answer this pragmatically by pointing out that band structures calculated on the basis of a stationary lattice are usually in good agreement with the experiment; except at temperatures close to the meeting point of the solid. The reason that the stationary lattice approximation seems to hold so well is that the amplitudes of lattice vibrations are much smaller than the inter-atomic distance at all temperatures even up to melting point[2]. Therefore the distortion of lattice as seen by the electron, is not appreciable.

15.4 NUMBER OF THE STATES IN THE BAND

We denoted the Bloch function by $\psi_{n,k}$, which indicates that each value of the band index n and the vector $\boldsymbol{k}$ specifies an electron state or orbital. We shall now show that the number of orbitals in a band inside the first zone is equal to the number of unit cells in the crystal. This is much the same as the statement made in connection with the number of lattice vibrational modes and is proved in a like manner by applying the boundary conditions.

One dimensional case: In this case the Bloch functions have the form

$$\psi_k(x) = e^{ikx}u_k(x) \tag{15.27}$$

If we impose the periodic boundary condition on this function it follows that the only allowed values of k are given by

$$k = \frac{2\pi n}{L} \tag{15.28}$$

where $n = 0, \pm1, \pm 2, \pm 3$, etc., (Note that $u_k(x)$ is intrinsically periodic, so the condition $u_k(x + L) = u_k(x)$ is automatically satisfied). The allowed values of k form a uniform mesh whose unit spacing is $\frac{2\pi}{L}$. The number of states inside the first zone, whose length is $\frac{2\pi}{a}$ is therefore equal to ,

$$\frac{2\pi}{a}\cdot\frac{L}{2\pi} = \frac{L}{a} = N \tag{15.29}$$

where N is the number of unit cells. This is in agreement with the assertion made earlier.

A similar argument may be used to establish the validity of the statement in two and three-dimensional lattices. Here the separation between the allowed $\boldsymbol{K}$ values in the $\boldsymbol{k}$-space, in each of the three orthogonal directions is $\frac{2\pi}{L}$. This means that each point in $\boldsymbol{k}$-space occupies volume $\frac{8\pi^3}{L^3}$. The volume of the first zone in the simple cubic lattice is $\frac{8\pi^3}{a^3}$. So the number of $\boldsymbol{K}$ values in each zone is given by

$$N = \frac{8\pi^3/a^3}{8\pi^3/L^3} = \frac{L^3}{a^3} \tag{15.30}$$

where N is the number of unit cells in volume L^3 of the crystals.

It has been shown that each band has N-states inside the first zone. Since each such state can accommodate at most two electrons, of opposite spins, in accordance with the Pauli exclusion principle, it follows that the maximum number of electrons that may occupy a single band is $2N$. This result is significant, as it will explain whether a solid is going to behave as a metal or insulator.

[2] The average amplitude of the atomic oscillation due to thermal excitation at the melting point is about 5% of the interatomic distance.

15.5 KRONIG PENNY MODEL

In previous sections we studied the general properties of the state functions, and also the energies of an electron moving in a crystalline solid. To obtain explicit results, however, we must solve Schrodinger equation for the actual potential $V(r)$ in the particular solid of interest. But the process of solving the schrodinger equation for any but the simplest potentials is a tedious and time consuming task. Although this is essential for obtaining results that may be compared with experiments, it is preferable to start the discussion of explicit solutions by using rather simplified potentials. The advantage is that we can solve Schrodinger equation with only minimal mathematical effort and thus concentrate on the new physical concepts involved.

The behaviour of electrons in a one dimensional periodic potential has been treated by Kronig and Penny by assuming a simple model of periodic potential as shown in Fig. (15.7). The period of potential in $(a + b)$ region such as $0 < x < a$, the potential energy is assumed to be zero and in region such as $-b < x < 0$ the potential energy is V_0.

For such an array of periodic square potentials the exact solution of schrodinger equation is possible. Although the potential represented by Fig. (15.7) is some what idealised periodic potential which is only a crude approximation to that found in the actual crystal. Even then it is nevertheless very useful because it serves to illustrate in a most explicit way many of the important characteristic features of the quantum behaviour of electrons in periodic lattice. The wave functions associated with this model may be calculated in the one-electron approximation by solving Schrodinger equation

$$\frac{\partial^2\psi(x)}{\partial x^2}+\frac{2m}{\hbar^2}[E-V(x)]\psi(x)=0 \tag{15.31}$$

Since the potential is periodic, the wave function must have the Block form, and we may expect.

$$\psi(x)=e^{ikx}u(x) \tag{15.32}$$

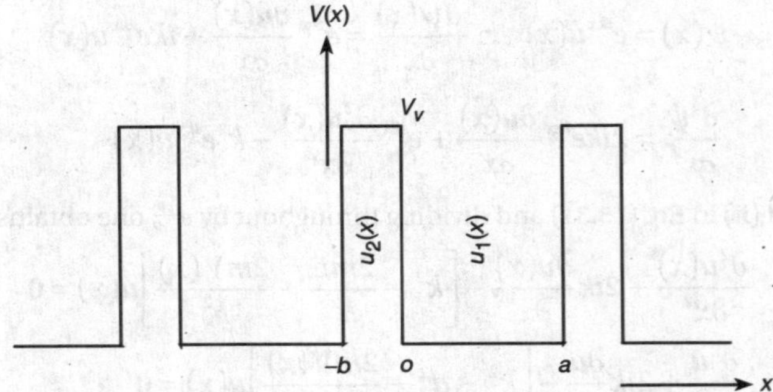

Fig. 15.7

Substituting Eq. (15.32) in Eq. (15.31) and putting $a = \left(\frac{2mE}{\hbar^2}\right)^{1/2}$ it is found that $u(x)$ must satisfy[3]

$$\frac{d^2u}{dx^2} + 2ik\frac{du}{dx} - \left[k^2 - a^2\frac{2mV(x)}{\hbar^2}\right]u(x) = 0 \tag{15.33}$$

If $u_1(x)$ and $u_2(x)$ represent the values of $u(x)$, in the two regions of potential function, i.e., in the intervals $(0 < x < a)$ and $(b < x < 0)$ respectively, we have,

$$\frac{d^2u}{dx^2} + 2ik\frac{du_1}{dx} - (k^2 - \alpha^2)u_1 = 0 \text{ for } 0 < x < a \because V(x) = 0 \tag{15.34}$$

and
$$\frac{d^2u_2}{dx^2} + 2ik\frac{du_2}{dx} - (k^2 - \beta^2)u_2 = 0 \text{ for } -b < x < 0 \tag{15.35}$$

where $\beta = \left[\frac{2m(E - V_0)}{\hbar^2}\right]^{1/2}$ and is purely imaginary quantity for $0 < E < V_0$.

The solutions of the differential Eqs. (15.34) and (15.35) are

$$u_1 = Ae^{i(\alpha - k)x} + Be^{-i(\alpha + k)x} \text{ for } (0 < x < a) \tag{15.36}$$

$$u_2 = Ce^{i(\beta - k)x} + De^{-i(\beta + k)x} \text{ for } (-b < x < 0) \tag{15.37}$$

where A, B, C and D are constants. Remembering that $u(x)$ has the periodicity of the lattice and the wave function $\psi(x)$ and its derivative at $x = a$ and $x = -b$ demands that wave function $u(x)$ must satisfy boundary conditions:

$$\left.\begin{aligned} &u_1(0) = u_2(0), u_1(a) = u_2(-b) \\ \text{and} \quad &\left(\frac{du_1}{dx}\right)_{x=0} = \left(\frac{du_2}{dx}\right)_{x=0}, \left(\frac{du_1}{dx}\right)_{x=0} = \left(\frac{du_2}{dx}\right)_{x=0} \end{aligned}\right\} \tag{15.38}$$

Using boundary conditions, one obtains

$$A + B = C + D \tag{15.39}$$

$$A(\alpha - k) - (\alpha + k)B = (\beta - k)C - (\beta + k)D \tag{15.40}$$

3

$$\psi(x) = e^{ikx}u(x) \quad \therefore \quad \frac{\partial\psi(x)}{\partial x} = e^{ikx}\frac{\partial u(x)}{\partial x} + ike^{ikx}u(x) \tag{i}$$

and
$$\frac{\partial^2\psi}{\partial x^2} = 2ike^{ikx}\frac{\partial u(x)}{\partial x} + e^{ikx}\frac{\partial^2 u(x)}{\partial x^2} - k^2e^{ikx}u(x) \tag{ii}$$

Substituting (i) and (ii) in Eq. (15.31) and dividing throughout by e^{ikx}, one obtains

$$\frac{\partial^2 u(x)}{\partial x^2} + 2ik\frac{\partial u(x)}{\partial x} - \left[k^2 - \frac{2mE}{\hbar^2} + \frac{2mV(x)}{\hbar^2}\right]u(x) = 0$$

or
$$\frac{\partial^2 u}{\partial x^2} + 2ik\frac{\partial u}{\partial x} - \left[k^2 - d^2 + \frac{2mV(x)}{\hbar^2}\right]u(x) = 0$$

$$Ae^{i(\alpha-k)^a} + Be^{-i(\alpha+k)a} = Ce^{-i(\beta-k)b} + De^{i(\beta+k)b} \tag{15.41}$$

$$(\alpha-k)Ae^{i(\alpha-k)a} - (\alpha+k)Be^{-i(\alpha+k)a} = (\beta-k)Ce^{-i(\beta-k)b} - (\beta+k)De^{i(\beta+k)b} \tag{15.42}$$

The co-efficients *A, B, C* and *D* can thus be determined and the wave function can be calculated. However, for our purpose we are more interested in determining the values of the energy for which satisfactory solutions are obtained. The Eq. (15.39) have non-vanishing solutions if and only if the determinant of the co-efficients vanishes. This requires that

$$\begin{vmatrix} 1 & 1 & 1 & 1 \\ \alpha-k & -(\alpha+k) & \beta-k & -(\beta+k) \\ e^{i(\alpha-k)a} & e^{i(\alpha+k)a} & e^{i(\beta-k)a} & e^{i(\beta+k)a} \\ (\alpha-k)e^{i(\alpha-k)a} & -(\alpha+k)e^{-i(\alpha-k)a} & (\beta-k)e^{-i(\beta-k)b} & -(\beta+k)e^{i(\beta+k)b} \end{vmatrix} = 0$$

Expanding the determinant, it can be shown that this leads to the following conditions:

$$-\frac{\alpha^2+\beta^2}{2a\beta}\sin\alpha a \sin\beta b + \cos\alpha a\cos\beta b + \cos k(a+b) \tag{15.43}$$

Since in the range $0 < E < V_0$, β as defined earlier is imaginary and for these values of energy it is most convenient to express

$$\beta = i\gamma$$

in this region, and noting that $\cos ix = \cos hx$ and $\sin ix = i\sin hx$, Eq. (15.43) can be written as,

$$\frac{\gamma^2-\alpha^2}{2a\gamma}\sinh\gamma b\sin\alpha a + \cosh\gamma b\cos\alpha a = \cos k(a+b) \tag{15.44}$$

where γ is a real quantity in the interval $0 < E < V_0$ just as is β in the interval $V_0 < E < \infty$. Thus we may use Eq. (15.43) more conveniently when $V_0 < E < \infty$ and Eq. (15.44) when $0 < E < V_0$.

The wave vector k is real and the right hand side of Eq. (15.44) lies only between –1 and +1. If we write the left hand side as $F(\alpha a, \gamma b)$, one sees that F cannot assume all values continuously but that for certain values of Energy E, i.e., (α, γ), $\cos k(a+b)$ would become greater or less than +1 and –1 respectively and hence will not correspond to physical reality. This means that there exist forbidden energy eigen values and so the allowed energy values then become divided into "energy band" separated by regions of forbidden energy.

In arriving this important conclusion clearly and without too much algebric labor, Kronig and Penny considered the simplification of the problem by assuming that $b \to 0$ and $V_0 \to \infty$, such that the product bV_0 is finite. This assumption is equivalent to the consideration of a setting of Dirac δ-functions type potential energies V_0 separated by a distance a the potential energy being zero δ-function spikes Fig. (15.8). We have

$$\gamma b = \left[\frac{2mb^2(V_0-E)}{\hbar^2}\right]^{1/2}$$

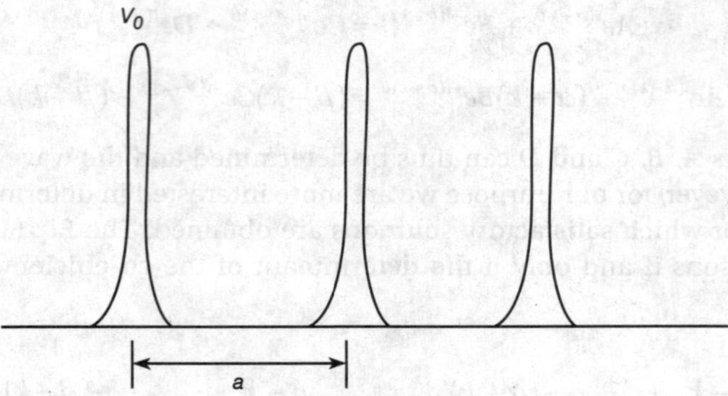

Fig. 15.8

and then taking V_0 to be large in comparison to E we have

$$\gamma b = \left[\frac{2mb^2V_0}{\hbar^2}\right]^{1/2} = \left(\frac{2m}{\hbar^2 V_0}\right)^{1/2}(bV_0) \to 0 \text{ as } V_0 \to \infty$$

and
$$\frac{\gamma^2 - \alpha^2}{2\alpha\gamma} = \frac{1}{\alpha a}\left[\frac{a}{2\gamma}(\gamma^2 - \alpha^2)\right] = \frac{1}{\alpha a}\cdot\frac{\gamma a}{2}$$

$$\sin h\,\gamma b \approx \gamma b$$

and so
$$\frac{\gamma^2 - \alpha^2}{2\alpha\gamma}\sin\gamma b \to \frac{1}{\alpha a}\cdot\frac{\gamma a}{2}\cdot\gamma b = \frac{1}{\alpha a}\left(\frac{1}{2}\gamma^2 ab\right) = \frac{P}{\alpha a}$$

where
$$P = \alpha t.\frac{1}{2}\gamma^2 ab \approx \alpha t.\left(\frac{mV_0 ab}{\hbar^2}\right)$$

The $E - k$, i.e., $(\alpha - k)$ relation Eq. (15.44) for the electron is of the form (in the limit $b \to 0$, $V_0 \to \infty$ and $V_0 b$ finite.)

$$p\frac{\sin\alpha a}{\alpha a} + \cos\alpha a = \cos ka \tag{15.45}$$

The left hand side of Eq. (15.45) which is some function of αa $[\phi\ (\alpha a)]$ can be plotted as a function of αa for a given value of p.

The general nature of graph is shown in Fig. (15.9). Eq. (15.45) gives the condition for the solutions of the wave equation to exist. We see that this is satisfied only for those values of αa for which its left-hand side lies between +1 and −1, this is because its right-hand side falls in this range. Such values of αa will-therefore, represent the wave like solutions Eq. (15.32) and are allowed. This can be further understood from Fig. (15.9), which represents the plot of $\phi(\alpha a) = p\frac{\sin\alpha a}{\alpha a} + \cos\alpha a$ versus αa for the value of $p = \frac{3\pi}{2}$. The part of the vertical axis lying between the horizontal lines represents the range, acceptable to the left-hand side

$$\phi(\alpha a) = p\frac{\sin\alpha a}{\alpha a} + \cos\alpha a$$

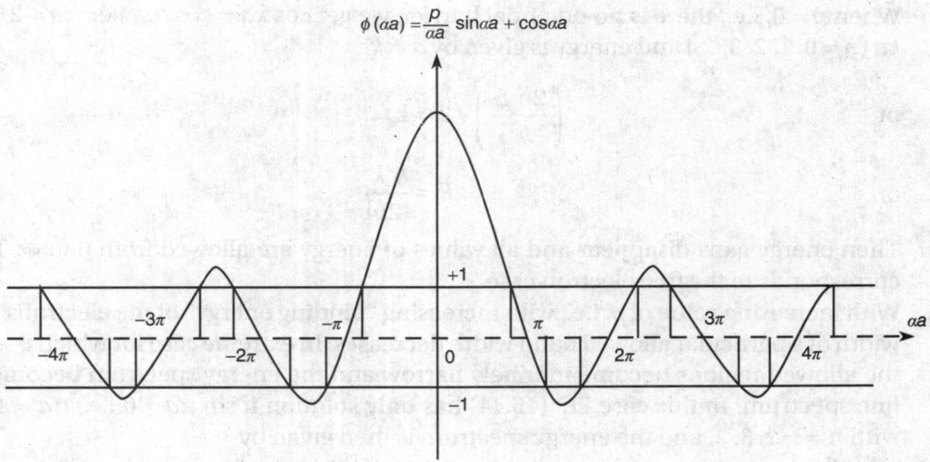

Fig. 15.9

Remembering that α^2 is proportional to the energy E, the abscissa will be a measure for the energy. The following conclusions can be drawn:

1. When $\alpha a = n\pi + \delta_n$, where δ_n is a small angle
$$\phi(\alpha a) = (-1)^2(1 + p\delta_n)$$
which is greater than +1 when n is even and less than −1 when n is odd. Hence forbidden energy ranges appear for all values of p. Thus the energy spectrum of the electron consists of a number of allowed energy bands separated by forbidden regions.
2. The edge of the energy band (i.e., maximum energy in band) is obtained when
$$\alpha a = n\pi \quad n = 1,2,3,\ldots$$
and the energy corresponding to this values of α is
$$E_n = \frac{\hbar^2}{2m}\left(\frac{\pi}{a}\right)^2 n^2$$
The corresponding values of k is
$$ka = \pm n\pi, \text{ i.e., } k_n = \pm\frac{n\pi}{a}$$
The energy gap therefore occurs at $k = \pm\frac{\pi}{\alpha}, \pm\frac{2\pi}{a}\ldots$ and the width of allowed energy bands increases with increasing value of αa, i.e., with increasing energy; this is a consequence of the fact that the first term of Eq. (15.45) decreases on an average with increasing αa.

3. When $p = 0$, i.e., there is no potential barrier, we get cos ka = cos αa, i.e., $\alpha a = 2n\pi \pm ka$ ($n = 0, 1, 2, 3, \ldots$) and energy is given by $\alpha = k$

or
$$\left(\frac{2mE}{\hbar^2}\right)^1 /2 = k$$

or
$$E = \frac{\hbar^2 k^2}{2m}$$

Then energy gaps disappear and all values of energy are allowed from 0 to ∞. This corresponds to the free electron case.

4. With increasing value of p, i.e., with increasing "binding energy" of the electrons, the width of a particular allowed band width decreases. In extreme case for which $p \to \infty$, the allowed regions become infinitely narrow and the energy spectrum becomes a line spectrum. In this case Eq. (15.44) has only solution if sin $\alpha a = 0$ i.e., $\alpha a = \pm n\pi$ with $n = 1, 2, 3, \ldots$ and the energy spectrum is then given by

$$E_n = \left(\frac{\pi^2 \hbar^2}{2na^2}\right) n^2 \quad \text{for } p \to \infty$$

which are the energy levels of a particle in a constant box of atomic dimensions. Physically this could be expected because for large p tunneling through the barriers becomes improbable.

These conclusions are evident from Fig. (15.10), where the energy spectrum is given as function of p for $p = 0$. We simply have the free electron model and the energy spectrum is (quasi) continuous. For $p = \infty$ a live spectrum results as discussed above. For a given value of p, the position and width of the allowed and forbidden bands are obtained by erecting a vertical line, the shaded areas correspond to allowed bands.

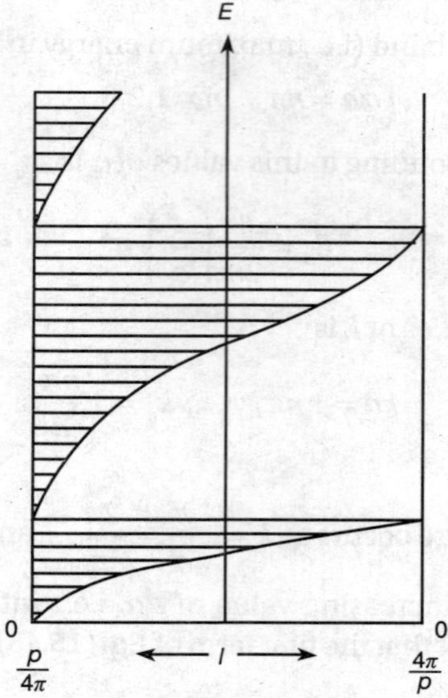

Fig. 15.10

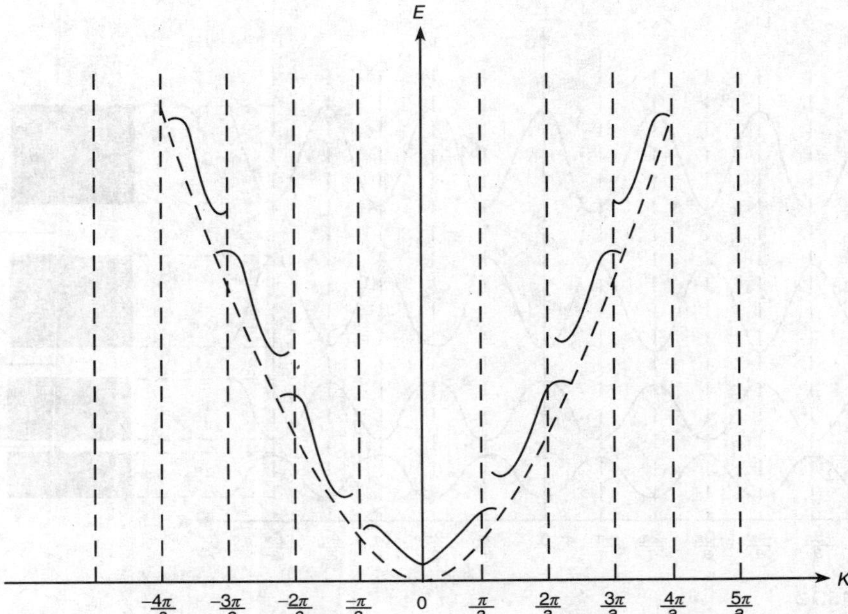

Fig. 15.11

5. The discontinuities in the E versus k curve occur for $k = \frac{n\pi}{a}$ with $n = 1, 2, 3$ ... These k values define the boundaries of first, second etc., Brillouin zones. Fig. (15.11) gives the E versus k (dispersion) curve. Here the first zone extends form $-\frac{\pi}{a}$ to $+\frac{\pi}{a}$. Similarly the second zone consists of two parts from $\frac{\pi}{a}$ to $\frac{2\pi}{a}$ and another from $-\frac{\pi}{a}$ to $-\frac{2\pi}{a}$. Showing that the discontinuities occur at zone boundaries. The significance of these discontinuities can be further seen by substituting $k = \frac{2\pi}{\lambda}$, we have

$$k = \frac{2\pi}{\lambda} = \frac{n\pi}{a} \quad \text{or} \quad n\lambda = 2a$$

This equation may be compared with the Bragg's law ($2d \sin\theta = n\lambda$) for reflection at normal incidence ($\theta = 90°$). We can conclude that this defines the wave length (and hence k) for which total reflection occurs.

6. It is clear from Eq. (15.45) that cos ka takes a specific value for each allowed energy value E. Since cos ka is an even periodic function of ka, the ka itself can accept either a positive value or a negative value or ($ka = \pm 2\pi n$) (n is any integer) for a given value of energy. Consequently, the energy E of the electrons is also an even periodic function of k with periodicity of π/a. This periodic repetition of the allowed energy values, corresponding to first few allowed bands in shown in Fig. (15.12). This representation

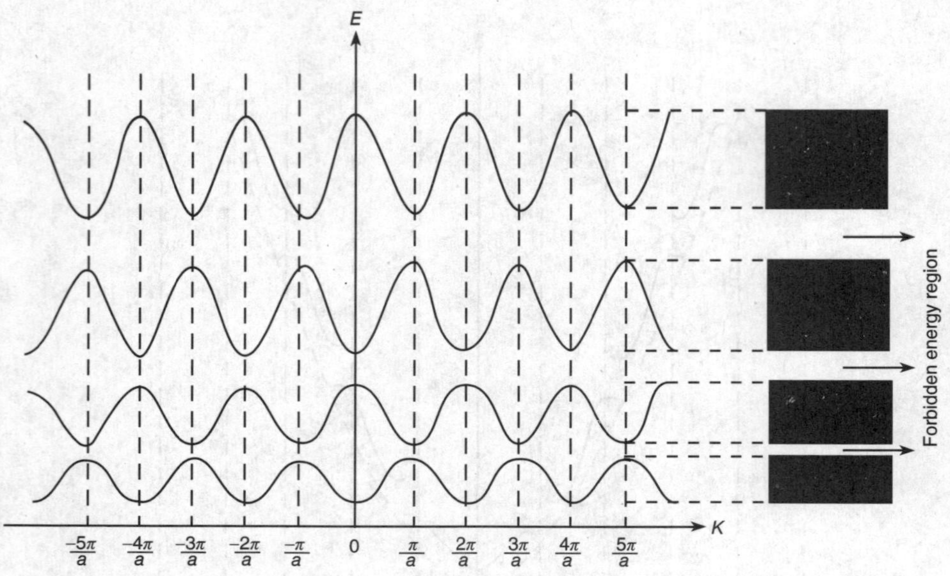

Fig. 15.12

can be considered as the periodic repetition of the region $-\frac{\pi}{a} < k < \frac{\pi}{a}$ (first Brillouin zone) through all of k-space. This representation is know to form the **periodic or repeated zone scheme.**

This scheme, however, gives a large number of k-values corresponding to a particular allowed energy value E. This can be easily seen by drawing a horizontal line corresponding to the energy E. Therefore, k is not uniquely determined here. It is therefore frequently convenient to introduce the "reduced wave vector" Which is limited to the region

$$-\frac{\pi}{a} < k < \frac{\pi}{a}$$

The energy versus reduced wave vector is represented in Fig. (15.13). It may be noted here that the fact that k is not uniquely determined also follows quite generally from the form of Bloch function given by Eq. (15.32). Consider the function $e^{ikx}u_x(x)$ and introduce a new wave vector $k' = k + \frac{2\pi n}{a}$; where n is an integer. One may write

$$\psi_{k'} = e^{ik'x}u_{k'}(x) = e^{i\left(k+\frac{2\pi x}{a}\right)x}u_{k'}(x) = e^{ikx}u_{k'}(x)e^{iGx}$$

where $G = \frac{2\pi x}{a}$ is a reciprocal lattice vector

Here e^{iGx} and $u_{k'}(x)$ are both periodic with the periodicity of the lattice. So $e^{iGx}u_{k'}(x)$ is also periodic with the same periodicity. Denoting it by $u_k(x)$, we have

$$\psi_{k'} = e^{ik'x}u_{k'}(x) = \psi_k \qquad (15.46)$$

Hence $\psi_{k'}$ is just as good a Bloch function as the initial function ψ_k.

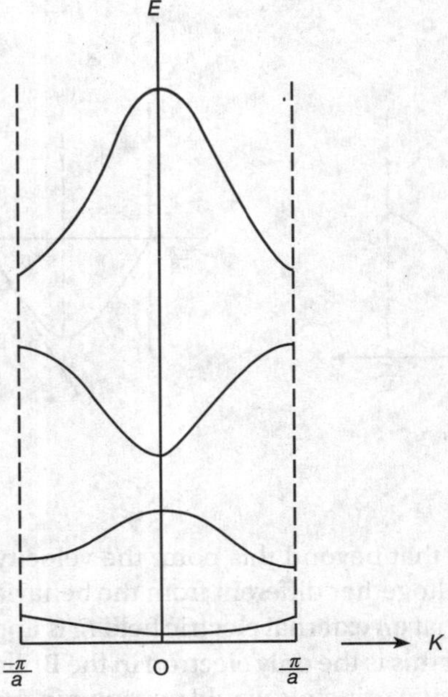

Fig. 15.13

15.6 VELOCITY OF THE BLOCH ELECTRONS AND THE DYNAMICAL EFFECTIVE MASS VELOCITY

For convenience in visualising the motion of the Bloch electrons in solid, we have to localise the wave function by superposing solutions having different values of k. If this is done, then the group velocity associated with electron wave packet is

$$v_g = \frac{d\omega}{dk} = \frac{1}{\hbar}\frac{dE}{dk} \tag{15.47}$$

Here a is the angular frequency of the electron waves and is related to the energy E of the particle by $E = \hbar\omega$. Now for free electrons $E = \frac{\hbar^2 k^2}{2m}$, then Eq. (15.47) leads us to write

$$v_g = \frac{\hbar k}{m} = \frac{p}{m} \tag{15.48}$$

giving linear variation of v_g with the k. In band theory however E is not in general proportional to k^2 as is evident from Fig. (15.13). Employing an $E(k)$ curve as represented in Fig. [15.14(a)], one obtains velocity as function of k using Eq. (15.47). This curve is given in Fig. [15.14.(b)] (Note that for free electrons v is proportional to k and is represented by dashed line in Fig. [15.14(b)]. At the top and bottom of the energy band $v = 0$ because from periodicity of the $E(k)$ curves it follows that $\frac{dE}{dk} = 0$, the absolute value of the velocity reaches a maximum for $k = k_0$, where k_0 corresponds to the inflexion point of the $E(k)$ curve.

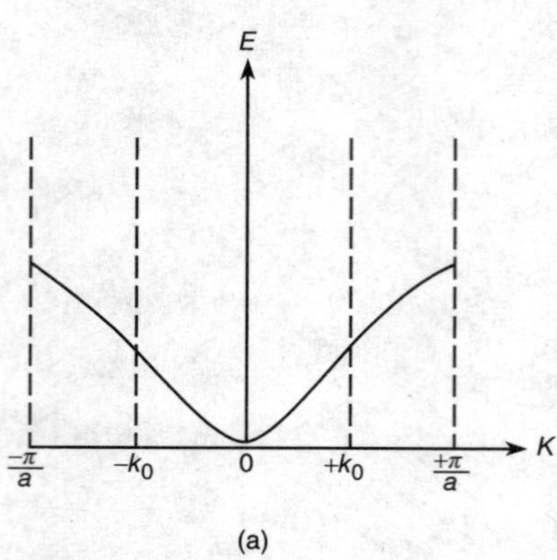

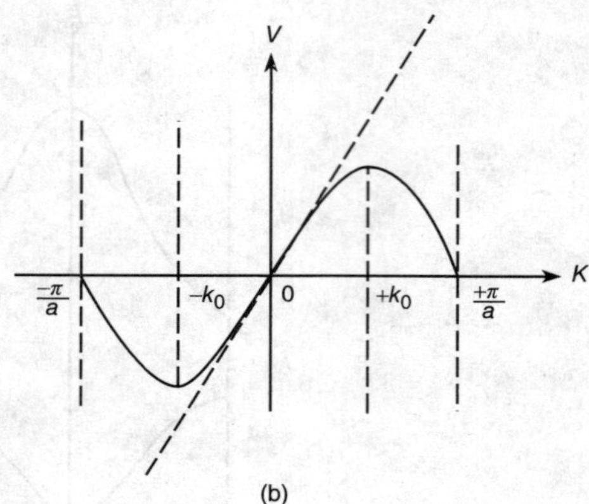

Fig. 15.14

It is of importance to note that beyond this point the velocity decreases with increasing energy, a feature which is altogether different from the behaviour of free electrons.

Effective mass: Consider that an external electric field E_0 is applied to an electron, initially in a state k. We assume that this is the only electron in the Brillouin zone under consideration, so that the Pauli exclusion principle would not enter in the discussion. When the field E_0 has acted on the electron for a small time t, suppose that it has gained a velocity v over a distance dx during this time and an energy dE given by,

$$dE = -eE_0 dx = -eE_0 v dt = -\frac{eE_0}{\hbar}\frac{dE}{dk}dt$$

here $$v = \frac{1}{\hbar}\frac{dE}{dt}$$

Hence we have

$$dk = -\frac{eE_0}{\hbar}dt$$

or $$\hbar\frac{dk}{dt} = \frac{dp}{dt} = -eE_0 = F \tag{15.49}$$

where p now denotes crystal momentum. Eq. (15.49) simply states that the rate of change of k is proportional to, and is in the same direction as the force on the electron. This relation is a very important one in the dynamics of Bloch electrons and is known as acceleration theorem.

Let us consider the consequences of this acceleration theorem. Considering one dimensional case and re-writing Eq. (15.49) as

$$\frac{dk}{dt} = \frac{F}{\hbar}$$

We observe that the wave vector k increases uniformly with time. Thus as t increases, the electron traverses the k-space at a uniform rate Fig. (15.15). The electron starting from $k = 0$ for example moves up the band until it reaches the top (point A) and then starts to

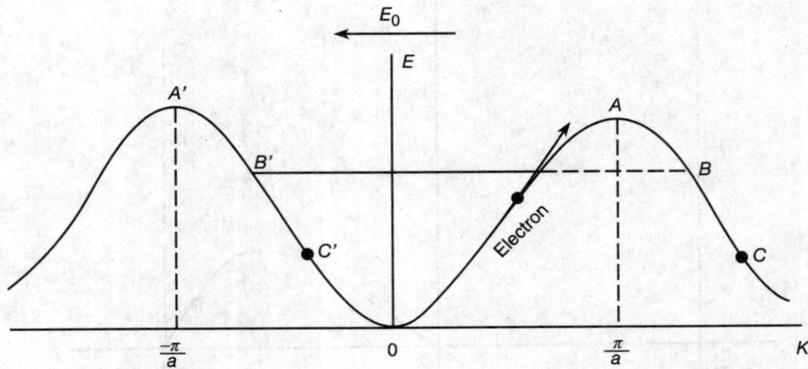

Fig. 15.15

descend along the path *BC*. If we use the reduced zone scheme, then once the electron passes the zone edge at *A*, it immediately reappears at the equivalent point *A*′, then continues to descend along the path *A*′, *B*′, *C*′. Remember, because of the translational symmetry property, the points *B*′, *C*′ are respectively equivalent to the points, *B*, *C*.

Note that, in the presence of electric field the electron is in constant motion in *k*-space, it is never at rest. Also note that the momentum in *k*-space is periodic in reduced zone scheme; since after traversing the zone once, the electron repeats the motion. The period of motion is readily found to be

$$T = \frac{2\pi\hbar}{Fa} = \frac{2\pi\hbar}{eE_0 a} \tag{15.50}$$

Fig. (15.16) shows the velocity of the electron as it traverses the *k*-axis. Starting at $k = 0$, as the time passes on, the velocity increases, reaches maximum, decreases and the vanishes at the zone edge. The electron then turns round and acquires a negative velocity and so forth. The velocity we are discussing is the velocity in real space, i.e., the usual physical velocity. It follows that a Bloch electron, in the presence of static electric field, executes an oscillatory periodic motion in real space very much unlike a free electron. This is one of the surprising conclusions of electron dynamics in crystal.

Now differentiating Eq. (15.47) with respect to time, we get for the acceleration of the electron

$$\frac{dv_g}{dt} = \frac{1}{\hbar}\frac{d}{dt}\left(\frac{dE}{dk}\right) = \frac{1}{\hbar}\frac{d^2E}{dk^2}.\frac{dE}{dt}$$

and using Eq. (15.49), we can write as

$$\frac{dv_g}{dt} = -\frac{eE_0}{\hbar}\frac{d^2E}{dk^2} \tag{15.51}$$

or

$$F = -eE_0 = \frac{\hbar^2}{\left(\frac{d^2E}{dk^2}\right)}\frac{dv_g}{dt} \tag{15.52}$$

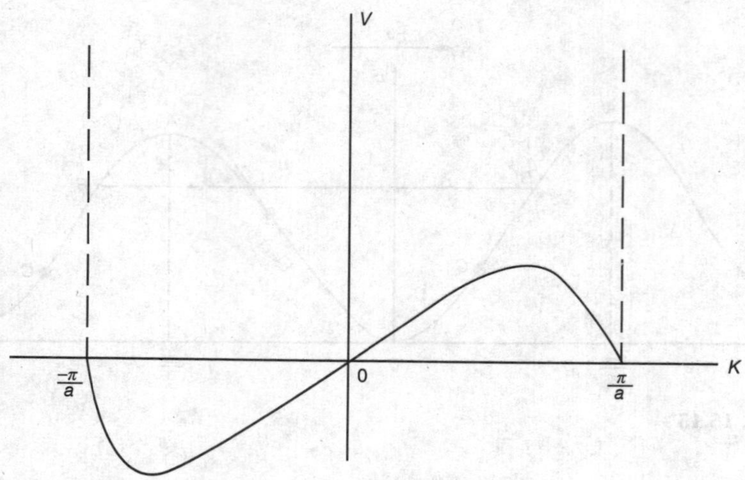

Fig. 15.16

This equation connects the force $F(=-eE_0)$ on the electron with the acceleration $\left(\frac{dv_g}{dt}\right)$ through the proportionality factor $\frac{\hbar^2}{\left(d^2E/dk^2\right)}$ and resembles with the Newton's force equation. Therefore the quantity $\frac{\hbar^2}{\left(d^2E/dk^2\right)}$ may be regarded as the mass-called the **effective mass** (m^*) of the electron. We thus have[4]

$$m^* = \frac{\hbar^2}{\left(d^2E/dk^2\right)} \tag{15.53}$$

as the effective mass of the electron moving in a crystal. For an electron moving in a uniform potential (i.e., for a free electron)

$$E = \frac{\hbar^2 k^2}{2m}$$

so
$$\frac{d^2E}{dk^2} = \frac{\hbar^2}{m} \quad \text{or} \quad m^* = m \quad \text{for all values of } k$$

[4] It may be remembered here that the effective mass given by Eq. (15.53) is applicable only to one dimensional crystal. For a general three dimensional crystal the effective mass must be represented by **tensor** $\frac{1}{m_{ij}^*} = \frac{1}{\hbar^2}\frac{\partial^2 E}{\partial k_i \partial k_j}$ with $i, j = x, y, z$, i.e., in three dimensions there may be nine components of effective mass. In the regions of spherical energy surfaces, it is however indentical with Eq. (15.53).

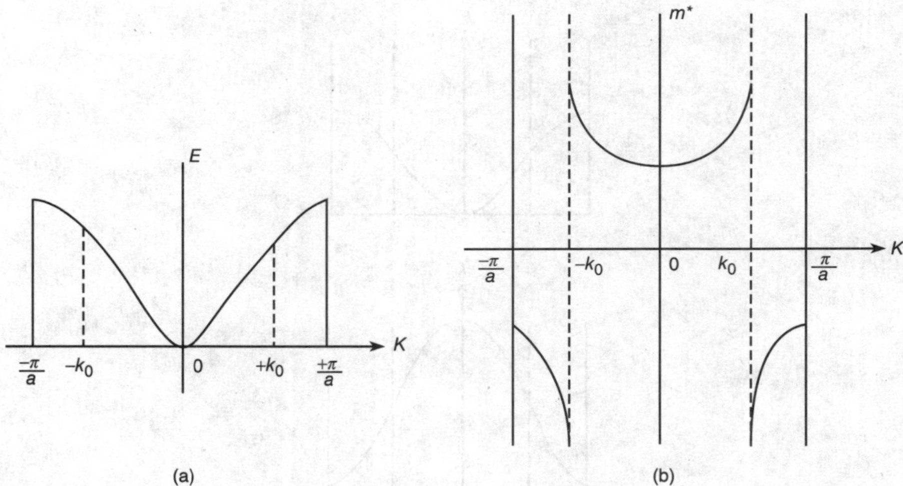

Fig. 15.17

But for an electron moving in a periodic potential, E does not vary with k in the above manner and so $m^* \neq m$. Neither m^* will be same for all energy values; rather it will vary because of the complex form of the $E - k$ curve. The Bloch electron in the crystal will thus behave dynamically just like a particle with variable effective mass and the whole effect of the periodic potential on the motion of the electron is to replace the free electron mass with proper effective mass. The significance of this is that the free electron theory of metals is justified to a great extent, provided the free electrons are considered to have the appropriate effective mass (determined by $E - k$ curve) instead of the true mass. One may thus conclude that all the results of the free electron theory are correct provided that m in each case is replaced by suitable m^*.

The effective mass m^* is inversely proportional to the curvature of the band. Where the curvature is large, i.e., $\frac{d^2E}{dk^2}$ is large, the mass is small, a small curvature implies a large mass. For several transition metals it is usually much higher and for semiconductors it is lower. The effective mass need not always be greater than m. More interesting, however is the fact that it can be negative even. Fig. [15.17(a)] gives the energy variation with k in a band and Eq. [15.17(b)] gives the variation of m^* with k near $k = 0$, the effective mass approaches m; As the value of k increases, m^* increases reaching its maximum value at the point of inflexion on the curve $E - k$ (for $k = k_o$). Above the point of inflexion ($k > k_0$), m^* is negative and as $k \to \pi/a$ it decreases to small negative value.

It is observed that beyond the point of inflexion the mass m^* becomes negative. It is because the region is now close to the top of the band and a negative mass is to be expected. The negative mass can be seen dynamically by noting that the velocity decreases for $k > k_o$. Thus acceleration is negative, i.e., opposite to the applied force, implying a negative mass. This means that in this region of k-space the lattice exerts such a large retarding force on

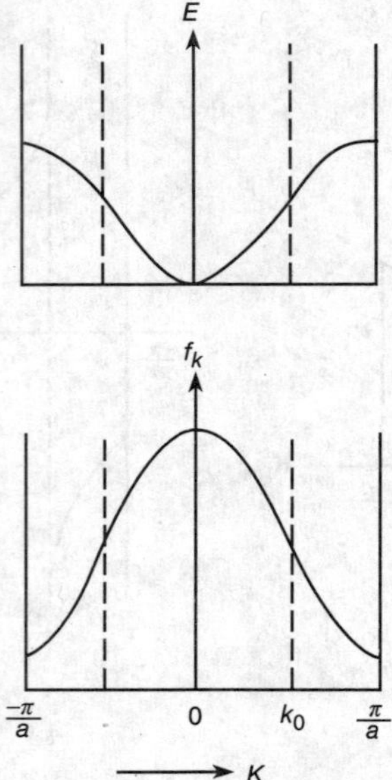

Fig. 15.18

the electron that it overcomes the applied force and produces negative acceleration. It is frequently convenient to introduce a factor

$$f_k = \frac{m}{m^*} = \frac{m}{\hbar^2}\frac{d^2E}{dk^2} \tag{15.54}$$

where f_k is a measure to the extent to which an electron in state k is free. If m^* is large, f_k is small i.e., the particle behaves as a **"heavy"** particle. When $f_k = 1$ the electron behaves as a free electron. Note that f_k is positive in the lower half of the band and negative in the upper half as shown in Fig. (15.18).

15.7 MOMENTUM, CRYSTAL MOMENTUM, AND PHYSICAL ORIGIN OF THE EFFECTIVE MASS

We have said that Bloch electron in the state ψ_k behaves as if it had a momentum $\hbar k$. Basically, there are three different reasons to support this statement

(i) Bloch functions has the form

$$\psi_k = e^{ik.r}u_k(r)$$

which, since u_k is periodic, appears essentially a plane wave with k, a constant of motion and that $\hbar k$ will have the dimensions of momentum and as the energy of the electron increases (the particle becomes thereby more nearly free) the value of k in general approximate those of the free-particle momentum divided by $\hbar$.

(ii) When an electric field is applied, the wave vector varies with time according to Eq. (15.52), i.e.,

$$\frac{d(\hbar k)}{dt} = F_{ext}$$

again indicating that $\hbar k$ acts like a momentum. Here F_{ext} refers to the external force applied to the crystal.

(iii) In collision process involving a Bloch electron, the electron contributes a momentum equal to $\hbar k$.

It can be easily seen that these conclusions must hold no matter what particular form periodic potential takes. It is therefore customary to refer as the crystal momentum and not the actual momentum of Bloch electrons. To make the distinction between the actual momentum of electron (p) and the crystal momentum p_c(=$\hbar k$), we evaluate the average momentum using quantum mechanical method

$$p = \langle \psi_k | -i\hbar\nabla | \psi_k \rangle \tag{15.55}$$

where $i\hbar\nabla$ is the momentum operator and ψ_k is the Bloch function. On solving this integral, using properties of the wave function ψ_k, we find that

$$p = m_0 v$$

where m_0 is the mass of free electron and v its velocity given by Eq. (15.47). Thus the true average momentum of electron is equal to the true mass m_0 times the actual velocity v. If $p_0 = \hbar k$ were the true momentum, then the force appearing on the right of Eq. (15.49) should have been the total force and not just the external force because there is a force exerted by the lattice, yet this force does not appear to influence p_c.

The above idea may now be used to give a physical interpretation of the effective mass. Since the vector $p = m_0 v$ is equal to the true momentum, one may write

$$m_0 \frac{dv}{dt} = F_{tot} = F_{ext} + F_L \tag{15.56}$$

where F_{tot} and F_L are respectively, the total force and the lattice force acting on the electron. By lattice force, we mean the force exerted by the lattice on the electron as a result of its interaction with crystal potential. Eq. (15.56) can be expressed in terms of the effective mass as

$$m_0 \frac{dv}{dt} = m_0 \frac{F_{ext}}{m^*}$$

where m^* is given by

$$m^* = \frac{m_0 F_{ext}}{F_{ext} + F_L} \tag{15.57}$$

Now we see the reason why m^* is different from m_0, the free mass. If F_L were to vanish the effective mass would become equal to true mass.

The effective mass may be smaller or larger than m_0 or even negative, depending on the lattice force. Suppose that the electron is "piled up" primarily near the top of the crystal

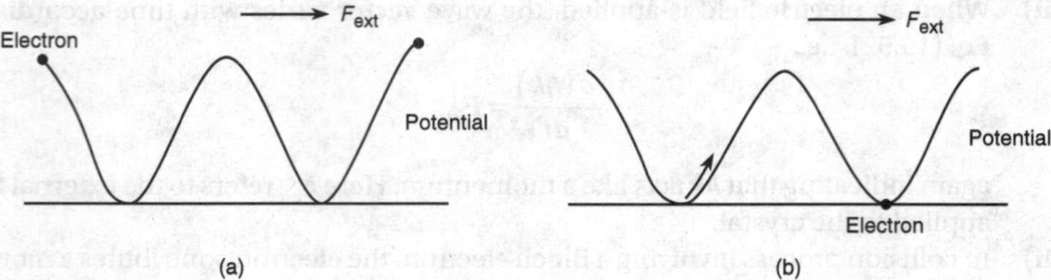

Fig. 15.19

potential as shown in Fig. [15.19(a)]. When an external force is applied, it causes the electron to "roll down hill" along the potential curve. As a result a positive force becomes operative and hence according to Eq. (15.57) $m^* < m_0$. This is what happens in alkali metals for instant in the conduction band in semi-conductors. Here m^* is less than m_0 because the lattice force assists the external force.

On the other hand when electron is piled mainly near the bottom of the potential curve Fig. [15.19(b)] then clearly the lattice force tends to oppose the external force resulting in $m^* > m_0$. This is the situation in alkali halides. If the potential curve is sufficiently steep, then F_L becomes larger than F_{ext} and m^* becomes negative.

How, the crystal momentum $p_c = \hbar k$ is still a very useful quantity? In problems of electrodynamics in crystal fields, crystal momentum is much more useful than true momentum. Since it is easier to follow motion in k-space than in real space. Therefore we shall continue to use p_c and refer to it as the momentum, when there is no ambiguity and even drop subscript.

In other words, the effective mass m^* and crystal momentum $\hbar k$ are artifices which allow us formally test or ignore the lattice force and concentrate on the external force only. This is very useful because lattice force is not known a prior nor is it easily found and manipulated as the external force.

15.8 NEGATIVE EFFECTIVE MASS AND HOLES

Let us now consider the significance of the negative effective mass m^*. It may be remembered that the negative m^* occurs for $k > k_0$ where k_0 corresponds to the point of inflexion of the $E - k$ curve. In other words we can say that it appears only near the top of the energy bands. Clearly for simple metals it is not of much interest because there-in the band it is only half filled. But for materials with nearly filled bands, there will be certain electrons populating the states at the top of the band and will thus have negative m^*. It is these electrons the consideration of which is significant.

For simplicity we consider a band with all states filled except one at the top. The effect of applying an electric field would be to shift a group of electrons up to this empty state and to produce an empty state some where in the interior of the band. The depth of this empty state would however depend on the magnitude of the field. In other words, the effect is equivalent to as if one electron has moved from some where in the filled portion of the band to the top and the empty state has moved from the top to the interior, although a group of electrons undoubtedly has participated to cause the effect. Thus the effect of the electric field can also be described by the motion of the empty state above. We call this empty state as a hole. As a matter of fact, in a nearly filled band. It is much to talk of these few holes than of the comparatively large electrons; their motion specifies the motion of the whole electron assembly.

15.8.1 Characteristics of the Holes

In order to determine the characteristics of the holes, we shall take up the same example of a single hole at the top of an otherwise filled band and consider the collective behaviour of the whole electron assembly when an electric field is applied. Denoting the charge of an electron by e and velocities by v we have for the current associated with all the electrons in the band.

$$I = -e\sum_i v_i = e\sum j(v_j - v_k) \tag{15.58}$$

Here $\sum_i v_i$ represents the total velocity of all occupied electron states, $\sum_j v_j$ represents the total velocity of all possible electron states in the band and v_k the velocity of the unfilled state (hole). The sum is $\sum_j v_j$, however zero because for every positive values of velocity there is always a negative value of equal magnitude as is evident from Fig. (15.14). Then Eq. (15.58) reduces to

$$I = -e\sum_i v_i = ev_k \tag{15.59}$$

where I is the current contributed by all the electrons in the band and u_k is the velocity of the hole state. Eq. (15.58) thus makes it possible to conclude that the current from the nearly filled band is equivalent to that produced from the motion of the unfilled electronic state or hole which behaves like a positive particle, rather than a very large number of electrons. Now consider the rate of change of current I due to the field. We have

$$\frac{dI}{dt} = e\left(\frac{dv_k}{dt}\right) = -\frac{e^2 E_0}{m_k^*} \tag{15.60}$$

Here we have made use of Eqs. (15.52) and (15.53). Thus we have

$$\frac{dv_k}{dt} = i\frac{eE_0}{m_k^*} \tag{15.61}$$

Since the hole tends to lie in the upper part of the nearly filled band so m_k^* is negative and right hand side of Eq. (15.61) positive. We thus have instead

$$\frac{dv_k}{dt} = i\frac{eE_0}{|m_k^*|} \tag{15.62}$$

It is clear that the sign of the charge must match with the sign of the mass. Since e is essentially positive, $|m_k^*|$ must also be positive. In other words, we may account for the behaviour of a nearly filled band by means of a relatively small number of positive particles of positive mass; which are referred to as holes. This concept of a nearly filled band is of great importance in the theory of semi-conductors. The physical basis associated with the idea of a hole comes from Hall effect, which measures the sign of the moving charges.

15.9 THE DISTINCTION BETWEEN METALS, INSULATORS, AND INTRINSIC SEMICONDUCTORS

Although a proper distinction between these three groups of materials is possible only by considering the results of a three dimensional periodic potential, it is instructive at this point to indicate how the band theory leads naturally to possibility of such a distinction. To see, this let us consider a particular energy band which we shall assume to be filled with electrons up to a certain value k as shown in Fig. (15.20).

As far as the effect of an external electric field is concerned, one would like to know with how many "free electrons" the N electrons in the band are equivalent. Presumably, once we knew the answer to this question, it would be possible to draw conclusions about the conductivity associated with this band. The effective number of free electrons in the band is in accordance with the preceeding section equal to

$$N_{\text{eff}} = \sum f_k \tag{15.63}$$

where the summation extends over the all occupied states in the band. Now using Eq. (15.28), the number of states in an interval dk (excluding the spin) for a one-dimensional lattice of length L is equal to $L\dfrac{dk^4}{2\pi}$ [5]. Because two electrons occupy each of these states in the shaded region of Fig. (15.20), one may write instead of Eq. (15.63),

$$N_{\text{eff}} = 2\frac{L}{2\pi}.\int_{-k_1}^{+k_1} f_k dk = \frac{L}{\pi}.\int_{-k_1}^{+k_1} \frac{m}{\hbar^2}.\frac{d^2E}{dk^2}.dk$$

$$= \frac{2L_m}{\pi\hbar^2}\int_0^{k_1} \frac{d^2E}{dk^2}dk \tag{15.64}$$

Here we have used Eq. (15.54). Thus the effective number of electrons in the band in

$$N_{\text{eff}} = \frac{2L_m}{\pi\hbar^2}\left(\frac{dE}{dk}\right)_{k=k_1} \tag{15.65}$$

From this result one can draw the following conclusions:

(i) The effective number of electrons in a completely filled band vanishes, because $\dfrac{dE}{dk}$ vanishes at the top of the band.

(ii) The effective number of electrons reach a maximum for a band filled to the inflexion point of the curve $E(k)$ because then $\dfrac{dE}{dk}$ is maximum.

---

[5] From Eq. (15.28) we have allowed states $n = \dfrac{Lk}{2\pi}$ hence allowed states in an interval is $dn = \dfrac{L}{2\pi}dk.$

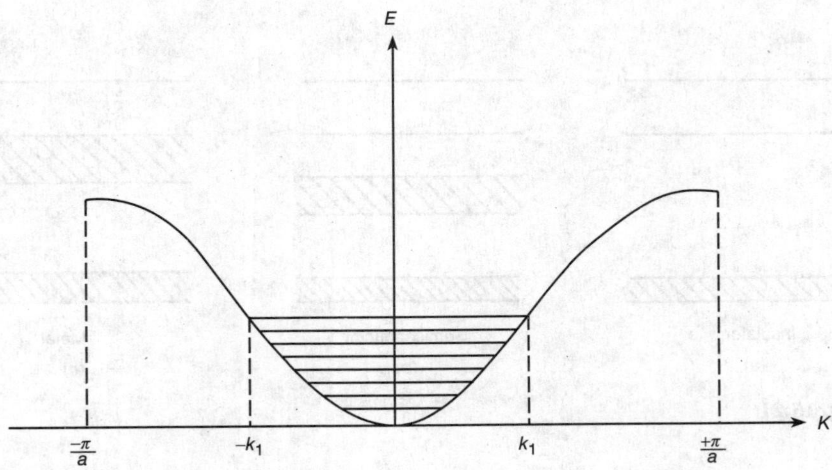

Fig. 15.20

From the above discussion it follows that a solid for which a certain number of energy bands are completely filled, the other bands being completely empty is an insulator Fig. [15.21(a)]. It is evident that the situation depicted by Fig. [15.21(a)] can occur actually only at absolute zero, when the crystal is in its lowest energy state. At temperatures different from zero, some electrons from the upper filled band will be excited into the next empty band ("conduction band") and conduction becomes possible. If the forbidden energy gap is of the order of several electron volts, however, the solid will remain an insulator for all practical purposes. An example is diamond, for which the forbidden gap in about 7 eV.

For a small gap width say about 1 eV; the number of thermally excited electrons may become appreciable and in this case one speaks of an intrinsic semiconductor. Examples are germanium and silicon. It is evident that the distinction between insulators and intrinsic semiconductors is only a quantitative one. In fact, all intrinsic semiconductors are insulators at $T = 0$, where as all insulators may be considered semiconductors at $T > 0$. It may be noted here that the conductivity of semiconductors in general increases with increasing temperature, where as the conductivity of metals decreases with increasing temperature.

We may note that three dimensional models allow the possibility of over lapping of bands, i.e., a solid which in one-dimensional model should be an insulator may turn out to be a metal: the divalent metals are examples of this.

Example 15.1: A certain cubic crystal has cube edge $a = 5.7$Å. Calculate the four lowest electron energies if wave vector k in the reduced zone scheme has magnitude $\dfrac{\pi}{2a}$ and is normal to cube face.

Solution Consider a cartesian coordinate system with its axis parallel to cube edges and take $k = \dfrac{2\pi}{a}\hat{i}$. Reciprocal lattice vectors have the from

$$G = h\left(\frac{2\pi}{a}\right)\hat{i} + k\left(\frac{2\pi}{a}\right)\hat{j} + l\left(\frac{2\pi}{a}\right)\hat{k}$$

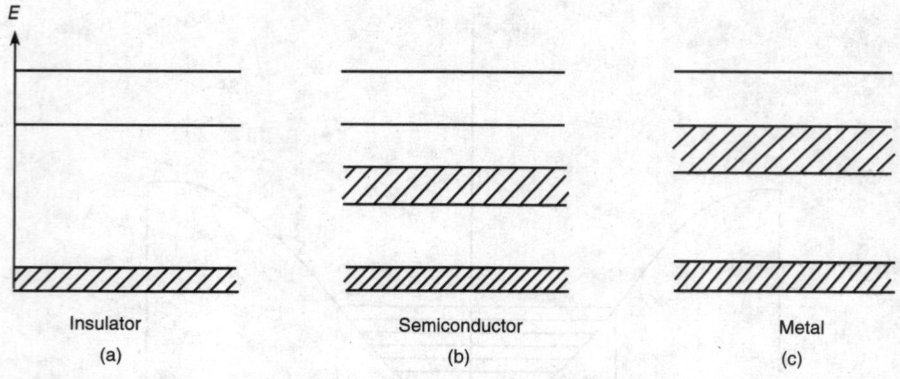

Fig. 15.21

where h, k, l are integers and $\hat{i}, \hat{j}, \hat{k}$ are unit vectors. For $u_0 = 0$ the energy levels are given by

$$E(k) = \frac{\hbar^2}{2m}\left[\left(\frac{\pi}{2a} + \frac{2\pi h}{a}\right)^2 + \left(\frac{2\pi k}{a}\right)^2 + \left(\frac{2\pi l}{a}\right)^2\right]$$

$$= \frac{2\hbar^2\pi^2}{m}\left[\left(\frac{1}{4} + h\right)^2 + k^2 + l^2\right] = 7.37\times10^{-20}\left[\left(\frac{1}{4} + h\right)^2 + k^2 + l^2\right] J$$

Selecting integer values for h,k,l, we obtain the lowest four values for

for (0 0 0), $E = \dfrac{7.37\times10^{-20}}{16} = 4.59\times10^{-20}\,\text{J} = 2.87\times10^{-1}\,\text{eV}$

for (1 0 0), $E = 7.37\times10^{-20}\left[\left(\frac{1}{4} - 1\right)^2\right] = 7.37\times10^{-20}\times\frac{9}{16}$

$= 4.14\times10^{-19}\,\text{J} = 2.59\,\text{eV}$

for (0 1 0), $E = 7.37\times10^{-20}\left[\left(\frac{1}{4}\right)^2 + 1\right] = 7.37\times10^{-20}\times\frac{17}{16}$

$= 7.81\times10^{-19}\,\text{J} = 4.88\,\text{eV}$

for (1 0 0), $E = 7.37\times10^{-20}\left[\left(\frac{1}{4}\right)^2 + 1\right] = 7.37\times10^{-20}\times\frac{25}{16}$

$= 1.15\times10^{-18}\,\text{J} = 7.18\,\text{eV}$

Example 15.2: Show that for each band there are $2N^2\tau/(2\pi)^3$ states per unit volume of reciprocal space, where τ is the volume of a direct lattice unit cell.

Solution A brillouin zone is a primitive unit cell of the reciprocal lattice, and its volume is

$$A^*\left(B^* \times C^*\right) = \frac{(2\pi)^3}{a.b\times c} = \frac{(2\pi)^3}{\tau}$$

There are N^3 propagation vectors in the Brillouin zone and two states in each band for each propagation vector; so for each band there are $\frac{2N^2\tau}{(2\pi)^3}$ states per unit volume of reciprocal space.

Example 15.3: **Show that if in the potential model used in connection with Kronig-Penny calculation, V_0 is allowed to become infinitely large and b is allowed to approach zero, such that the product $P = -\frac{\beta^2 ab}{2}$ remains fixed, then allowed energy values are given by the solutions of equation.**

$$P\frac{\sin \alpha a}{\alpha a} + \cos \alpha a = \cos ka$$

Also show that this leads to allowed and forbidden energy bands.

Solution While discussing Kronig-Penny model we have shown that for non-vanishing solutions of Schrodinger equation for periodic potential, the condition is

$$-\frac{\alpha^2 + \beta^2}{2\alpha\beta}\sin \alpha a \sin \beta b + \cos \alpha a \cos \beta b = \cos k(a+b)$$

where $\alpha = \left[\frac{2mE}{\hbar^2}\right]^{1/2}, \beta\left[\frac{2m(E - V_0)}{\hbar^2}\right]^{1/2}$

and potential energy is zero for $0 < x < a$ and V_0 for $-b < x < 0$, so that period of potential is $(a + b)$. It is apparent that with in energy range $0 < E < V_0$, β is imaginary and putting $\beta = i\gamma$. We have

$$\frac{\gamma^2 - \alpha^2}{2\alpha\gamma}\sinh \gamma b \sin \alpha a + \cos h\gamma b \cos \alpha a = \cos k(a+b)$$

Substituting the values of γ and α, we have

$$\gamma^2 - \alpha^2 = \frac{2mV_0}{\hbar^2}$$

and since $b \to 0$, $\sin h\gamma b \to \gamma$ band $\cosh \gamma\, b \to 1$, hence we have

$$\frac{2mV_0}{\hbar^2 2\alpha\gamma}\gamma b \sin \alpha a + \cos \alpha a = \cos ka$$

or
$$\frac{mV_0 b}{\alpha\hbar^2}\sin \alpha a + \cos \alpha a = \cos ka$$

Since $V_0 \to \infty$ and neglecting E in comparison to V_0 we have

$$P = -\frac{\beta^2 ab}{2} = \frac{mV_0 b}{\hbar^2}a$$

or
$$\frac{mV_0 b}{\hbar^2} = \frac{P}{a}$$

$\therefore$
$$\frac{P}{\alpha a}\sin \alpha a + \cos \alpha a = \cos ka$$

The quantity P is regarded as the scattering power of a single spike. The plot of $\left[\frac{P}{\alpha a}\sin \alpha a + \cos \alpha a\right]$ versus αa is shown in Fig. (15.9).

We know that α^2 is proportional to the energy, i.e., abscissa is a measure of the energy. Further more it is important to note that right hand side can accept only values between −1 and +1 as indicated by horizontal line. As such the allowed energy spectrum of the electron consists of a number of allowed energy bands separated by forbidden regions.

Example 15.4: The energy near a valence band edge is given by $E(k) = -1 \times 10^{-33}k^2$ Joules. An electron is removed from $k = 1 \times 10^9 \hat{k}_x \text{m}^{-1}$. The band is otherwise full:

(a) What is the sign and magnitude of the effective mass of the hole?
(b) What is the direction and magnitude of the wave vector of the hole?
(c) What is the crystal momentum of the hole?
(d) What is the velocity of the hole?
(e) The energy of the hole, referred to the valence band edge? and (*f*) the electric current carried by the hole?

Solution If the energy is a quadratic function of k, we may write $E = \frac{\hbar^2}{2m^*}k^2$, where m^* is the effective mass of electron. Comparing with given energy, we have

$$\frac{\hbar^2}{2m^*} = -1\times 10^{-33}$$

or

$$m^* = -\frac{\hbar^2}{2\times 10^{-33}}$$

(a) The effective mass of a hole is opposite to the effective mass of an electron at the same point of the energy band, we have

$$\therefore \quad m_h^* = -m_e^* = \frac{\hbar^2}{2\times 10^{-33}} = \frac{1.05\times 10^{-34}\times 1.05\times 10^{-34}}{2\times 10^{-33}}$$

$$= 0.55\times 10^{-35}\text{ kg}$$

(b) If the electron is missing from the state of wave-vector k_e then the wave vector of (the hole is)

$$k_h = -k_e$$

But

$$k_e = 1\times 10^9\,\hat{k}_x\text{m}^{-1}$$

$$k_h = -k_e = -1\times 10^9\,\hat{k}_x\text{m}^{-1}$$

(c) The crystal momentum is given by $\hbar k$. Crystal momentum of hole is

$$P_h = \hbar k_h = -1.05\times 10^{-34}\times 10^9\,\hat{k}_x = -1.05\times 10^{-25}\,\hat{k}_x\text{ kg ms}^{-1}$$

(d) If V_e is velocity of an electron in the state k_e, then the velocity to be assigned to the hole of wave vector $k_h = -k_e$ is

$$v_h = v_e = \hbar^{-1}\nabla(k_e)$$

where gradient is taken with respect to k_e.

$$\therefore \qquad v_h = \hbar^{-1}\frac{\partial}{\partial k}(E_k) = \hbar^{-1}\left(-2\times10^{-33}\right)\hat{k}$$

$$\text{But} \qquad \hat{k} = 1\times10^{9}\,\hat{k}_x$$

$$\therefore \qquad v_h = \frac{-2\times10^{-33}\times10^{9}}{1.05\times10^{-34}}\hat{k}_x = -1.9\times10^{8}\,\hat{k}_x \text{ ms}^{-1}$$

(e) The energy of the hole referred to zero for a filled band is positive and is

$$E_h(k_h) = -E(k_e) = 1\times10^{-33}\text{ J}$$

(f) If an electron is missing from the state E, the net electric current carried by the band is that of the unpaired electron in the state G is,

$$j = -ev(G) = e[-v(G)]$$

This current may be viewed as that of a positive charge e moving with velocity $-v(G)$. Now $-v(G)$ is equal to the velocity $v(E)$ of the orbital at E, the orbital from which the electron was removed. Thus the current may be written as

$$j = ev(E)$$

$$= 1.6\times10^{-19}\times1.9\times10^{8}\,\hat{k}_x$$

$$= 3.04\times10^{-11}\,\hat{k}_x \text{ amp}$$

Example 15.5: **A certain simple cubic structure has a cube edge of 4.85 Å. Take the zero of energy to be at the bottom of the lowest free electron band**

(a) Assume that the electrons are completely free and calculate the energy of the lowest energy state with propagation vector at the centre of a Brillouin Zone.

(b) Suppose $u(G_1)$ = 0.24 eV, where G_1 is the reciprocal lattice vector perpendicular to the Brillouin zone face of part (*a*). Calculate the energy of the two lowest free electron states with the propagation vector used in (*a*).

Solution

(a) At the zone face centre $k = \dfrac{\pi}{a}$ and energy $E = \dfrac{\hbar^2 k^2}{2m}$. Assuming totally free electron, we have

$$= \frac{\left(1.05\times10^{-34}\right)^2\left(\dfrac{\pi}{4.85\times10^{-10}}\right)^2}{2\times9.1\times10^{-31}} - 5 \quad \because\ a = 4.85\text{Å}$$

$$= 2.54\times10^{-19}\,J = 1.59\text{eV}$$

(b) Since energy $u(G_1)$ = 0.24 eV hence there are two distinct levels, one $|u(G_1)|$ below the free electron given by

$$1.59 - 0.24 = 1.35\text{eV}$$

and other above the free electron level given by

$$=1.59+0.24=1.83\,\text{eV}$$

Example 15.6: Show that the electron densities at which the free electron spherical Fermi surface first touches the zone boundary in a bcc metal in given by,

$$n = 2\sqrt{2}x\pi/3a^3$$ [Mysore]

Solution Let us take that k_m be the minimum distance in $\mathbf{k}$-space from the origin to the boundary of the first Brillouin zone. We have for bcc metal,

$$G = \frac{2\pi}{a}(1, 1, 0) = \frac{2\pi\sqrt{2}}{a}$$

$$k_m = \frac{G}{2} = \frac{\pi\sqrt{2}}{a}$$

Since the Fermi surface touches the zone boundary then

$$k_F = k_m$$

or $$(3\pi^2 n)^{1/3} = \sqrt{2}\pi / a$$

or $$n = \frac{1}{3\pi^2}\frac{2\sqrt{2}\pi^3}{a^3} = \frac{2\sqrt{2}}{3}\frac{\pi}{a^3}$$

Example 15.7: The valence band of a simple cubic metal has the form: $E = Ak^2 + B$, where

$A = 10^{-38}$ J-m^2. Show that $m^*/m = 0.6$ [Kerala]

Solution We have $E = Ak^2 + B$

$$\frac{\partial E}{\partial k} = 2Ak \quad \text{and} \quad \frac{\partial^2 E}{\partial k^2} = 2A$$

Now, $$m^* = \frac{\hbar^2}{\partial^2 E/\partial k^2} = \frac{(1.05\times10^{-34})^2}{(9.1\times10^{-31})2A}$$

$$= \frac{1.1\times10^{-68}}{2\times9.1\times10^{-31}\times10^{-38}} = 0.604$$

Example 15.8: Show that the Fermi radius of two-dimensional square lattice of parameter a is 0.399 $(2\pi/a)$. [MDS]

Solution We know that reciprocal lattice of a square lattice is again a square lattice of lattice dimension, $2\pi/a$. Thus the area of the first Brillouin zone = $(2\pi/a)^2 = 4\pi^2/a^2$. Now if the metal is monovalent, then the area occupied by electrons will be half the area of Brillouin zone, which can be occupied by two electrons, and which allows the two possible spin states for each electron. This means that in two-dimensions, Fermi sphere becomes a Fermi circle of radius k_F, we have

$$\pi k_F^2 = \frac{1}{2}(4\pi^2 / a^2)$$

or
$$k_F = \frac{1}{\sqrt{2\pi}}(2\pi / a) = 0.399\left(\frac{2\pi}{a}\right)$$

SUMMARY

1. The theory that treats electrons as freely moving within the body of a metal crystal or fluid is called *free electron* theory. In a free electron gas the particle concentration is ~10^{29} electrons/m^3.
2. The highest level filled at absolute zero is called the *Fermi level.*
3. Electrons are *fermions*. The probability that an energy state E at any temperature T is filled is given by *Fermi-Dirac distribution function*

$$f(E) = \frac{1}{\exp[(E - E_F)/kT] + 1}$$

where E_F is the Fermi energy.
4. Density of states $D(E)$ is a measure of the number of states within a given arbitrary narrow interval of energy that the electron may possess

$$D(E) = \frac{V}{2\pi^2}\left(\frac{2m}{\hbar^2}\right)^{3/2} E^{1/2}$$

where V is the volume of the crystal.
5. The maximum possible velocity of the electron is at the *Fermi level,* and is denoted by v_F. Hence, the surface obtained with v_F as radius is called *Fermi surface.* The equation for conductivity becomes

$$\sigma = \frac{ne^2 l}{m v_F} = \frac{ne^2 \tau_F}{m}$$

where τ_F is collision time.
6. Thermal conductivity

$$K = \frac{\pi^2 N_A k^2 T \tau_F}{3m}$$

7. The ratio between thermal and electrical conductivities is same for all metals at any particular temperature,

$$\frac{K}{\sigma T} = \frac{1}{3}\left(\frac{\pi k}{e}\right)^2$$

This ratio is called *Lorentz number* ($L = 2.45 \times 10^{-8}$ WΩ K^{-2} at reasonably high temperature). This is known as *Wiedmann-Franz law.*

8. According to *Kronig-Penny model*, the periodic field due to a one-dimensional lattice may be assumed in the form of potential wells of zero potential energy and width a_0 (i.e., the interionic separation) and separated by rectangular barriers of height U_b and width $(a - a_0)$, the ionic size.
9. The production of an e.m.f. within a conductor or semiconductor through which a current is flowing when there is a strong transverse magnetic field. The potential difference develops at right angles to both the current and the field. It is caused by the deflection of the charge carriers by the field. This effect is called *Hall effect*.
10. According to '*Band theory*', in crystals, electrons fall into allowed energy bands. In different solids these bands may remain separate or overlap. The regions separating these bands are the regions of *forbidden energy*. This region of forbidden energy is also referred to as *energy gap*.
11. Physically, effective mass (m^*) represents an exchange of momentum with the lattice, as the electron gets retarded near the Brillouin zone boundary. In general, effective mass is a tensor.

$$m^* = \hbar^2 \Big/ \left(\partial^2 E \big/ \partial \vec{k}^2 \right)$$

REVIEW QUESTIONS AND PROBLEMS

1. Investigate the motion of an electron in a periodic field. Discuss the electronic levels in crystals.
2. **(a)** Given the solution of Schrodinger equation for a one dimensional periodic lattice (with symbols having their usual meaning).

$$p\frac{\sin \alpha a}{\alpha a} + \cos \alpha a = \cos ka, \qquad \text{where } \alpha = \left(\frac{2mE}{\hbar^2}\right)^{1/2}$$

discuss the formation of energy band in a solid.
(b) Prove that the effective mass of an electron in an energy band is given by

$$\left(\frac{1}{m^*}\right)_{ij} = \frac{1}{\hbar^2}\frac{\partial^2 E}{\partial k_i \partial k_j}$$

3. Derive an expression for the effective mass m of the electron in crystal and explain its physical basis.
4. Discuss the essential features of the electron energy band structure of solid on the basis of Kronig-Penny model.
5. Illustrate the motion of an electron in one-dimensional weak periodic potential by drawing schematic dispersion curves for the first, second and third Brillouin zone in the reduced and extended zone scheme. How do you physically explain the existence of energy gaps in this problem? Does this result enables you to have a model for the energy of semiconductor. Explain your answer.
6. Explain the origin of the forbidden energy gap in crystalline solids and the concept of effective mass of an electron in a lattice.

7. State and prove Bloch theorem.
8. Discuss the origin of energy band gap. Explain by using Kronig-Penny model the variation of energy band structure of solid.
9. (a) Explain the concept of effective mass of electron in crystals by considering the motion of a wave packet in an applied electric field. What is the physical importance of this concept?
10. The periodic potential of one-dimensional lattice spacing can be approximated by a box-shaped wave form which has a value at each atom of $-V$ and which changes to zero at a distance 0.1 a on either side of each lattice point. Estimate the width of the first energy gap in the electron energy system. [(**Ans.** 0.3.7V)]
11. Discuss the formation of the bands (qualitatively) in solids.

SHORT ANSWER-QUESTIONS

1. What is effective mass?
Ans: When an external electromagnetic field acts upon electrons in a solid, the resultant motion is not what one expect, if the electrons were free. In fact, the electrons behave as though they were free but with a different mass, which is called the effective mass. One can obtain the effective mass from the dependence of electron energy on the wave vector.
2. What is a Kronig-Penney model?
Ans: Kronig-Penney model is an idealized, one-dimensional model of a crystal which exhibits many of the basic features of the electronic structure of real crystals.
3. What is Bloch theorem?
Ans: This theorem states that the wave function of an electron moving in a periodic potential (such as in a crystal lattice) has potential (such as in a crystal lattice) has the from of a plane wave, modulated by a function which has the same periodicity as the potential.
4. What is a Fermi surface?
Ans: Fermi surface is a surface of constant energy in a space defined by the components of the wave vectors of a system of half-integral spin particles. This is not a real surface, but as a geometrical description of the dynamical behaviour of conduction electrons in solids. Its name comes from the fact that half integral spin particles obey Fermi-Dirac statistics and at the zero of the temperature T fill energy levels up to a maximum energy, the Fermi energy, and no energy levels are occupied above this energy. The wave vectors of electrons having, the Fermi energy at $T = 0$ define the Fermi-surface.
5. What are energy bands?
Ans: These are the widened discrete energy levels due to the interaction between the atoms of the solids.
6. How will you specify first Brillouin zone?
Ans: The first Brillovin zone is defined as the region in ***k***-space lying in the range, $-\pi/a < k < \pi/a$.
7. Is the area of all Brillouin zones in two dimensions is same?
Ans: Yes.

8. What is the Bragg's diffraction condition for the calculation of Brillouin zones?
Ans: $2\,\mathbf{k}.G + G^2 = 0$ **which geometrically means that one can construct the Brillouin zones by taking any reciprocal lattice point as centre and then drawing perpendicular bisectors between that point and all surrounding reciprocal lattice points. The smallest volume enclosed by the right bisector is the first Brillouin zone. Likewise, one can define higher order zones.**

OBJECTIVE QUESTIONS

1. The valence band of a solid.
(a) is the highest occupied band
(b) can never be empty
(c) is always partially filled
(d) is less electron energy than conduction band

2. Each energy band of small piece of solid containing 50 atoms will have closely spaced energy levels numbering.
(a) 25 (b) 50
(c) 100 (d) 5

3. The effective mass of a localized electron is
(a) unity (b) infinite
(c) zero none of the above

4. The electron having a negative mass is located in the band
(a) in the energy gap
(b) anywhere in the energy
(c) near the bottom of the band
(d) near the zone boundary

5. In the Kronig-Penny model, if there exists no potential barrier, then
(a) there is a periodic dependence of energy E on k
(b) e is not continuous function of k
(c) there are no forbidden energy regions
(d) all values of energy E are not allowed

6. From the study of Fermi surface, one can study.
(a) effective mass of the electron
(b) crystal potential
(c) the density of conduction electrons
(d) all the above

7. The Fermi energy of Al is

(a) 5 eV (b) 9 eV

(c) 12 eV (d) 23 eV

8. The effective mass of an electron.

(a) can be positive, negative as well as infinite

(b) ban never be negative

(c) can never be infinite

(d) depends on the effective charge of the electron

ANSWERS

Question numbers	1	2	3	4	5	6	7	8
Answers	(a),(c),(d)	(b)	(b)	(d)	(c)	(d)	(c)	(a)

FILL IN THE BLANKS

1. The band theory of solids has great success in accounting the ______ properties of metals, insulators and semiconductors. **[conduction]**
2. According to band theory of solids, the energy spectrum of electron in solids consists of ______ allowed energy band and ______ energy bands. **[alternate, forbidden]**
3. The effect of lattice on the motion of electron can be taken into account by replacing the true mass m by its ______ in the equation of motion. $$\left[m^* \left(= \hbar^2 \Big/ \left(\frac{d^2E}{dk^2}\right)\right)\right]$$
4. $\psi(x) = e^{ikx} u_k(x)$, where $u_k(x)$ is a periodic function, i.e., $u_k(x+a) = u_k(x)$. This is known as ______. **[Bloch's theorem]**
5. Li, Na, K, Rb, Cs, Ag, Au are examples of ______ metals. **[monovalent]**
6. Ge, Si, InSb, GaAs, AlSb, InAs, InP are examples of ______. **[semiconductors]**
7. One dimensional attractive potential is given by $V(z) = -\lambda\delta(z)$, if $k = \frac{m\lambda}{\hbar^2}$, the lowest energy state occurs at ______. $$\left[\varepsilon_0 = -\frac{\hbar^2 K^2}{2m}\right]$$
8. The total number of states in a band is equal to the number of ______. **[unit cells]**
9. Assuming that silver is monovalent metal (with spherical Fermi surface) having density 10.5 g/cm^3 and atomic weight 107.87, its Fermi energy is ______. **[5.51 eV]**

16

Lattice Vibrations (Phonons) and Thermal Properties of Solids

16.1 INTRODUCTION

In classical thermodynamics we study the relationship between heat and work in a physical system or body by applying the first and second law of thermodynamics to the state of the system. But we are unable with these tools alone to determine certain thermal properties of the system, such as specific heat, thermal conductivity and thermal expansion, primarily because these quantities depend on the internal energy of the system and we do not have any means to determine expression for internal energy (u). As such we must either develop some kind of theory for internal energy (u).

We will now discuss the various models specially **Debye model** which can explain the behaviour of thermal heat capacity with temperature. We will also consider the effect of vibrating electrons on the thermal capacity. The features of more exact calculations for thermal conductivity of metals are also indicated.

16.2 LATTICE VIBRATIONS AND THERMAL ENERGY

As discussed earlier, the solids are regular aggregates of large numbers of atoms (or molecules) held in their positions of fixed equilibrium by strong cohesive forces resulting from their electrical interactions. The only individual motions of the atoms are small vibrations around their equilibrium positions. But the coupling between the atoms is so strong that you can not set one atom in vibration without disturbing the nearby atoms, and eventually the whole solid. Therefore we must consider collective vibrational excitations of the solid, called **lattice vibrations.**

These lattice vibrations set up standing waves in solid whose frequencies depend on the shape and size of the body and in a way are similar to standing waves in cavity. Although the possible frequencies have a discrete spectrum, their spacing is very small. If the body is relatively large compared with atomic dimensions and the spectrum may be assumed to be continuous. These standing waves have the same nature as elastic waves propagated through a solid, and therefore their propagation velocity is the same as the velocity of the elastic waves, i.e. the velocity of sound.

Thermal energy in a solid is contributed mostly by lattice vibrations and electron vibrations. The essential feature of the lattice vibrations is that their energies are quantized just like the electromagnetic radiations are quantized in radiation and each quantum of energy is given the name **photon** and in lattice vibrations each quantum of energy is given the name phonon. **Thus a phonon is a quantum of sound. This is the smallest unit of energy of a vibration corresponding to a sound wave.** The vibration spectrum of the phonon wave occupy a wide range of frequency from 10^4 to 10^{13} Hz. The high frequency part of the spectrum is in the infra-red range and the low frequency part is in the accoustic range. The energy of the elastic waves of the individual vibrations of frequency ν is taken $h\nu$ where h is Planck's constant. Each quantum of energy $h\nu$ is called the phonon. Since the elastic waves in a solid are called the sound waves, these phonons which are a sort of elastic waves in a solid travel with the same velocity as sound in the solid medium, just like phótons travels with the velocity of light in a medium.

The process of quantization of elastic energy is closely analogous to that used in quantizing the energy of an electromagnetic field. In case of elastic waves the particle like entity called the phonon carries the unit energy of the elastic field in a particular mode given by

$$\varepsilon = h\nu = \hbar\omega$$

Since the phonon also represents a travelling wave, it carries a momentum of its own. By analogy with photon, the momentum of the phonon is given by

$$P = \frac{h}{\lambda} = \hbar K$$

where λ is the wave length and $\boldsymbol{K}$ is the wave vector.

Just as we think of an electromagnetic wave as a stream of photons, we now view an elastic sound wave as a stream of phonons which carry the energy and momentum of the wave.

The number of phonon in a mode at thermal equilibrium can be determined using analogy with photons. Since the energy per phonon is equal to and the average energy of phonon in a mode is given by

$$<\varepsilon> = \frac{\hbar\omega}{e^{\hbar\omega/kT}} - 1 \quad \text{(neglecting zero point energy)}$$

It follows that the average number of phonons in a mode is given by

$$<n> = \frac{1}{e^{\hbar\omega/kT}} - 1$$

This number depends on the temperature: at $T = 0$, $< n > = 0$ but as T increases $< n >$ also increases, eventually reaching the value $kT/\hbar$ at high temperatures. Here we observe an interesting point. Phonons are created simply by raising the temperature and therefore their number in the system is not conserved. This is unlike the case of the more familiar elementary particles of physics, e.g., electrons or protons- in which the number is conserved.

The concept of the phonon is an extremely important in solid state physics. Their interaction with other forms of radiations such as x-rays, neutrons and light not only confirms for the energy and momentum of the phonon but also furnishes valuable information on the state of vibration of the solid.

Summarizing we can say that the energy of the phonons is the thermal energy of the solid. Who are then, the major contributors of phonons in solid? Lattice vibrations contribute a major share of the phonons in a solid. Electron vibrations contribute a small amount of phonons in a solid, in addition such processes as the distortion of long-range atomic arrangement in alloys (order-disorder), the randomization of electron spins in a ferro-magnetic materials and the change in distribution of electrons in superconductors also contribute phonons in some cases.

16.3 VIBRATIONS OF ONE DIMENSIONS MONOATOMIC LINEAR LATTICE

A lattice is an array of atoms connected by elastic spring. When any atom within the lattice is excited by external force, its motion is shared by all the atoms in the crystal. Obviously, the crystal as a whole vibrates. Let us suppose that we have one dimensional chain of indentical atoms connected by Hooke's law along the x-axis Fig. [16.1(a)]. Let the atoms are spaced at distance a from each other and each atom have a mass m. We are interested to investigate the vibrational motion of this lattice.

Whenever the vibrational motion is excited, all the atoms of the crystal execute simple harmonic motion about their equilibrium positions. We now consider that the displacement of the nth atom from its equilibrium positions is given by x_n. In order to set up a force equation, let us assume:

(i) The force to produce an atomic displacement is proportional to the displacement as the atoms are connected by elastic springs.

(ii) The significant forces between atoms in a crystal are due to direct nearest neighbouring interactions, i.e. we can consider the influence of atoms $n + 2$, $n + 3$, $n + 4$, $n + 5$,... etc., upon atom n negligible.

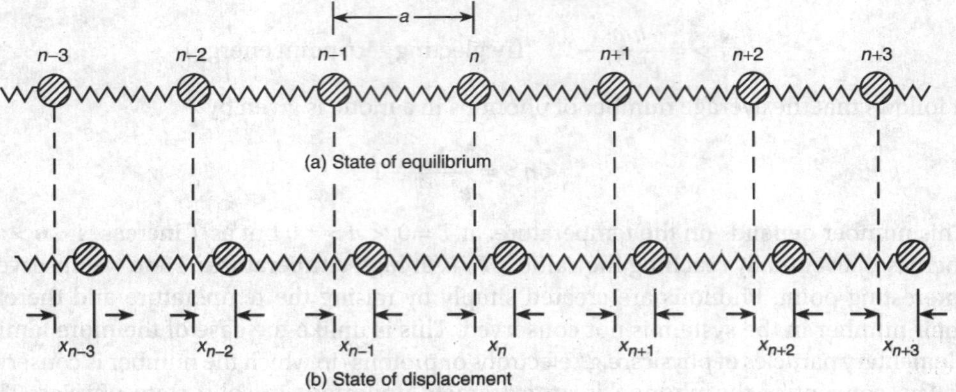

Fig. 16.1 One dimensional monoatomic linear lattice

On the basic of these two assumptions, one can write the net force acting upon the nth atom in terms of the extension of two springs which binds it to two atoms $(n+1)$ and $(n-1)$. Thus, one finds

$$F_n = \beta(x_{n+1} - x_n) - \beta(x_n - x_{n-1}) \tag{16.1}$$

where β is spring constant.

Applying Newton's law (second), one obtains the equation of motion as

$$m\frac{d^2 x_n}{dt^2} = \beta(x_{n+1} + x_{n-1} - 2x_n) \tag{16.2}$$

Obviously, Eq. (16.2) represents the motion of nth atom. One finds the solution of Eq. (16.2) of the form

$$x_n = A\exp[i(\omega t - kna)] \tag{16.3}$$

where $k = 2\pi/\lambda$ and the x-coordinate of the nth atom of the chain is na. Thus, we have

$$x_{n+1} = A\exp[i\{\omega t - k(n+1)a\}]$$

and

$$x_{n-1} = A\exp[i\{\omega t - k(n-1)a\}] \tag{16.4}$$

From Eq. (16.3), one obtains

$$\frac{dx_n}{dt} = Ai\omega \exp[i(\omega t - kna)]$$

and

$$\frac{d^2x_n}{dt^2} = A(i\omega)^2 \exp[i(\omega t - kna)] \tag{16.5}$$

Substituting these values in Eq. (16.2), one obtains

$$-A\omega^2 m \exp[i(\omega t - kna)] = \beta[A\exp\{i(\omega t - k(n+1)a)\}]$$
$$+ A\exp i[\omega t - k(n-1)a] - 2A\exp i[\omega t - kna]$$

or

$$-m\omega^2 = \beta[e^{-ika} + e^{ika} - 2] = 2\beta[\cos ka - 1]$$
$$= -4\beta\sin^2\frac{ka}{2}$$

or

$$\omega^2 = \frac{4\beta}{m}\sin^2\frac{ka}{2}$$

or

$$\omega = \pm\sqrt{\frac{4\beta}{m}}\sin\frac{ka}{2}$$

Here, the frequency is an essentially positive quantity, irrespective of whether k is positive or negative. Now, considering magnitude of right hand side of above equation, one finds

$$\omega = \sqrt{\frac{4\beta}{m}}\sin\frac{ka}{2} \tag{16.6}$$

One can draw the following significant conclusions:

(i) At Low frequencies ($k \to 0$, the long wavelength limit)

For low values of k which are much less than π/a or for long wavelength, λ, one finds that ω is approximately linear with k because

$$\sin\frac{ka}{2} \cong \frac{ka}{2}$$

Thus we have $$\omega \sim \sqrt{\frac{\beta}{m}}ka$$

In this case, one obtains the *phase velocity* as,

$$v_p = \frac{\omega}{k} = \sqrt{\beta/m}\,a \tag{16.7}$$

and *group velocity* $$v_g = \frac{d\omega}{dk} = \sqrt{\frac{\beta}{m}}a \tag{16.8}$$

We can see that phase and group velocities are equal and constant in this long wavelength limit. This is expected as the waves of long wavelength would not be sensitive to the discreteness of the structure, i.e. large number of atoms are participating in all displacements as on a homogeneous line. Fig. [16.2(a)], shows this situation. We may note that in this case the medium acts as elastic, continuous and homogeneous.

(ii) At higher frequencies

When k increases, we can see that ω is not a linear function of k and therefore group velocity and phase velocity are not equal. One obtains,

$$v_p = \frac{\omega}{k}\sqrt{\left(\frac{\beta}{m}\right)}\left|\frac{\sin\frac{ka}{2}}{\frac{ka}{2}}\right|$$

$$v_g = \frac{d\omega}{dk}\sqrt{\left(\frac{\beta}{m}\right)}a\left|\cos\frac{ka}{2}\right| \tag{16.9}$$

We may note that both v_p and v_g are functions of frequency, ω.

We find that this situation is a similar situation as light passes through a medium where the refractive index is a function of frequency. The phenomenon is known as **dispersion** and the medium is called as **dispersive medium.** Clearly, in this case the medium is a dispersion medium. We may note that for very long wavelengths, the dispersion effects are negligible and the same medium behaves like a homogeneous continuous medium.

(iii) At frequencies $\omega_{max} = \sqrt{4\beta/m}$

In this case, v_p is finite and $v_g = 0$, i.e. signal or energy is being propagated. Obviously, the wave is a standing wave. Fig. [16.2(b)] shows this situation.

We may note that only frequencies for which $\omega \le \left(\frac{4\beta}{m}\right)^{1/2}$ will be propagated down a linear lattice. Clearly, in this way the chain may be considered as a low

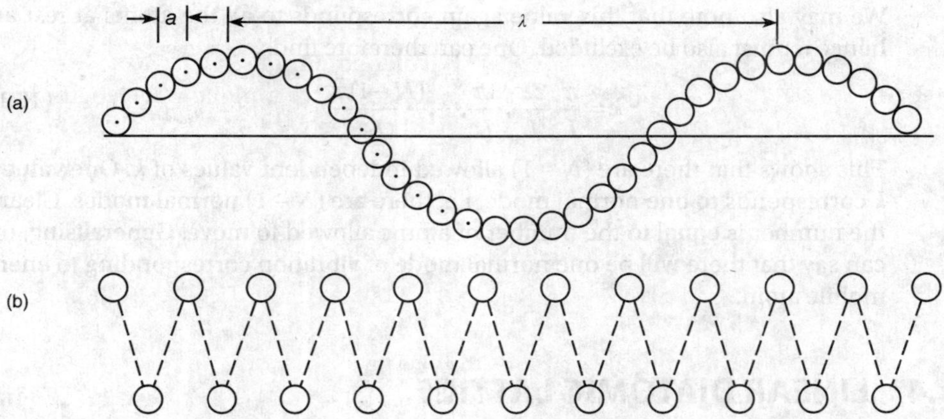

Fig. 16.2 (a) Motion appears as a homogeneous line (b) Motion is a standing wave

pass filter which transmits frequencies between zero and $\left(\dfrac{4\beta}{m}\right)^{1/2}$. However, in contrast with this, the continuous string has no such frequency limit. Further, in case of continuous medium, the velocity of propagation is independent of wavelength, while in case arrays of mass points, the velocity of propagation become smaller as the wavelength decreases.

Normal modes of vibration in a finite length of the lattice: Let us now determine the number of possible modes of vibration, i.e. independent solution of wave equation (16.2) which satisfy a given set of boundary conditions, associated with a lattice of length L. Further, we consider that there are $(N+1)$ atoms numbered from $n = 0$ to $n = N$ in the lattice at a separation a. Let us further suppose that the atoms at the two ends are fixed when a wave is moving in a lattice, it is reflected form the fixed end and as a result a standing wave is set up in the lattice. One can represent the standing wave as

$$x_n = x_0 \sin kna \sin \omega t \tag{16.10}$$

Eq. (16.10) will represent a normal mode provided it satisfies the boundary conditions. Obviously, according to the boundary condition, the displacement x_n at the two ends of the lattice must be zero, i.e. $x_n = x_0 = 0$. This requires that, $\sin kna = \sin kNa = 0$

or $$k = \frac{1\pi}{Na}, \frac{2\pi}{Na}, \frac{3\pi}{Na}, \ldots, \frac{N\pi}{Na}$$

or $$k = \frac{\pi}{L}, \frac{2\pi}{L}, \frac{3\pi}{L}, \ldots, \frac{N\pi}{L} \tag{16.11}$$

We may note that here $k = 0$ is excluded as it corresponds to all the atoms at rest. Clearly, in order that Eq. (16.10) represents the normal modes of vibration, the values of k must be restricted as given by Eq. (16.10). Now, for $k = 4\pi k^2 dk$, one finds

$$x_n = x_0 \sin n\pi \sin \omega t = 0 \tag{16.12}$$

We may also note that this value again corresponds to all the atoms at rest and hence it must also be excluded. One can therefore finds,

$$k = \frac{\pi}{L}, \frac{2\pi}{L}, \frac{3\pi}{L}, \ldots, \frac{(N-1)\pi}{L} \tag{16.13}$$

This shows that there are $(N - 1)$ allowed independent values of k. One value of k corresponds to one normal mode, i.e. there are $(N - 1)$ normal modes. Clearly, the number is equal to the number of atoms allowed to move. Generalising, one can say that there will be one normal mode of vibration corresponding to energy mobile atom.

16.4 LINEAR DIATOMIC LATTICE

Let us consider a linear lattice of two different kinds of atoms, arranged as shown in Fig. (16.3). We may note that this is diatomic lattice (lattice with two atoms per primitive call as in the case of NaCl crystal structure). Let us take the distance between nearest neighbours is a, i.e. atoms of the same nature in the lattice are at a distance of $2a$ apart. Let m and M be the masses of lighter and heavier toms respectively.

Let us assume that there is only nearest neighbour interaction. One can write the separate equations of motion for light and heavy atoms as follows:

$$F_{2n} = m\frac{d^2 x_{2n}}{dt^2} = \beta[x_{2n+1} + x_{2n-1} - 2x_{2n}] \tag{16.14}$$

$$F_{2n+1} = M\frac{d^2 x_{2n+1}}{dt^2} = \beta[x_{2n+2} + x_{2n} - 2x_{2n+1}] \tag{16.15}$$

The solution of Eqs. (16.14) and (16.15) are of the form

$$x_{2n} = B\exp[i(\omega t - 2kna)] \tag{16.16}$$

and

$$x_{2n+1} = B\exp[i(\omega t - k(2n+1)a] \tag{16.17}$$

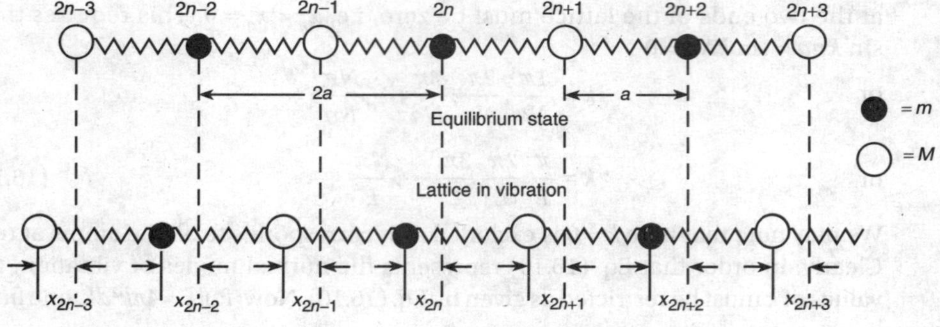

Fig. 16.3 A linear diatomic lattice

where k is the wave vector of a particular mode of vibration and the frequency of two types of atoms is being considered the same because both types of atoms participate in the single wave motion. The amplitudes of vibration are different due to the difference in the masses of both types of atoms. From Eqs. (16.16) and (16.17), one obtains

$$x_{2n+2} = A\exp[i\{\omega t - (2n+2)ka\}] = x_{2n}\exp\{-i2ka\} \tag{16.18}$$

$$x_{2n-2} = B\exp[i(\omega t - (2n-2)ka] = x_{2n+1}\exp(2ika) \tag{16.19}$$

On differentating Eqs. (16.16) and (16.17) twice w.r.t. t, one obtains

$$\frac{d^2x_{2n}}{dt^2} = -\omega^2 A\exp[i(\omega t - 2kna)] \tag{16.20}$$

$$\frac{d^2x_{2n+1}}{dt^2} = -\omega^2 B\exp[i(\omega t - (2n+1)ka)] \tag{16.21}$$

Substituting Eqs. (16.20) and (16.21) in Eqs. (16.14) and (16.15) respectively and simplifying, one obtains

$$-m\omega^2 A = \beta B(e^{ika} + e^{-ika}) - 2\beta A$$

or
$$(2\beta - m\omega^2)A - 2\beta B\cos ka = 0 \tag{16.22}$$

and
$$-M\omega^2 B = \beta A(e^{ika} + e^{-ika}) - 2\beta B \tag{16.23}$$

or
$$(-2\beta\cos ka)A + (2\beta - M\omega^2)B = 0$$

$$\begin{vmatrix} 2\beta - m\omega^2 & -2\beta\cos ka \\ -2\beta\cos ka & 2\beta - M\omega^2 \end{vmatrix} = 0 \tag{16.24}$$

or
$$(2\beta - M\omega^2)(2\beta - m\omega^2) - 4\beta^2\cos^2 ka = 0$$

Rearranging and collecting like power of ω, we have

$$\omega^4 - \frac{2\beta(m+M)}{mM}\omega^2 + \frac{4\beta^2\sin^2 ka}{mM} = 0 \tag{16.25}$$

We note that Eq. (16.25) is a quadratic equation in ω^2. This has two solutions. Let these two solutions are ω_+^2 and ω_-^2 according to positive and negative signs in quadratic formula

$$\omega_\pm^2 = \beta\left(\frac{1}{m} + \frac{1}{M}\right) \pm \beta\left(\frac{1}{m} - \frac{1}{M}\right)^2 - \left[\frac{4\sin^2 ka}{mM}\right]^{1/2} \tag{16.26}$$

Eq. (16.26) reveals that there are two angular frequencies ω_+ and ω_- corresponding to a single value of k. A plot of ω vs k is shown in Fig. (16.4).

From Fig. (16.4), one finds that there are two branches: (i) One corresponding to ω_- and (ii) other corresponding to ω_+. These branches are usually called as **acoustical and optical branches** respectively.

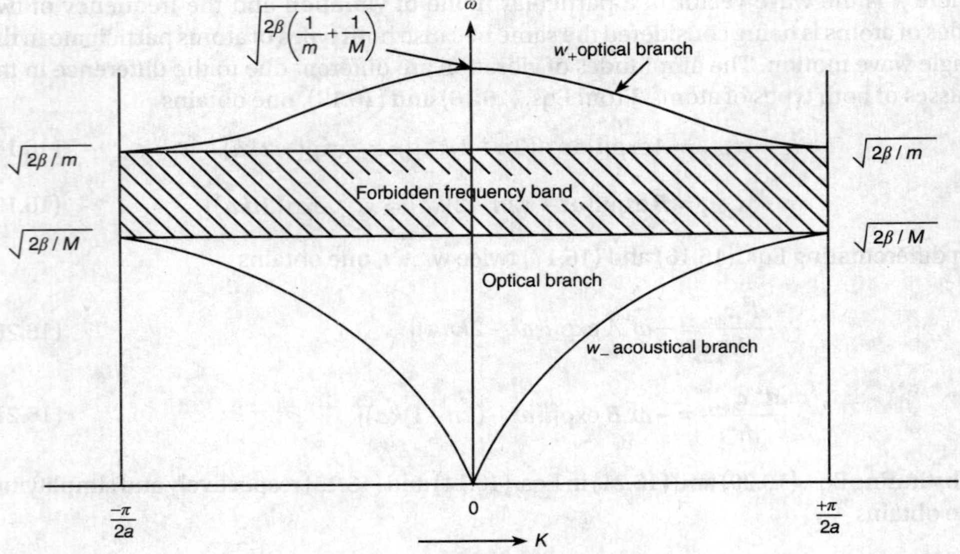

Fig. 16.4 Corresponding to diatomic lattice, optical and acoustical branches

For small k, i.e. $k \to 0$, one can neglect the sign term in Eq. (16.26). For positive root (ω_+), one finds

$$\omega_+^2 \cong 2\beta\left(\frac{1}{m}+\frac{1}{M}\right) \quad \text{or} \quad \omega_+ = \left\{2\beta\left(\frac{1}{m}+\frac{1}{M}\right)\right\}^{1/2} \tag{16.27}$$

Obviously, this corresponds to **optical branch.** When $k \to \pi/2a$, the sine term in Eq. (16.26) becomes $-4m/M$ and one finds

$$\omega_+ = \sqrt{\frac{2\beta}{m}} \tag{16.28}$$

Let us now consider the negative root, i.e. ω_-. Now for $k \to 0$, one can not neglect the sine term in this case as it is only non zero term. One finds its value as $4k^2a^2/mM$. Thus, we have

$$\omega_-^2 = \beta\left(\frac{1}{m}+\frac{1}{M}\right) - \beta\left[\left(\frac{1}{m}+\frac{1}{M}\right)^2 - \frac{4k^2a^2}{mM}\right]^{1/2}$$

$$= \beta\left(\frac{1}{m}+\frac{1}{M}\right) - \beta\left(\frac{1}{m}+\frac{1}{M}\right)\left[1 - \frac{mM}{(m+M)^2}4k^2a^2\right]^{1/2}$$

$$= \beta\left(\frac{1}{m}+\frac{1}{M}\right)\left[1 - \left\{1 - \frac{mM}{(m-M)^2}4k^2a^2\right\}^{1/2}\right]$$

$$= \beta\left(\frac{1}{m}+\frac{1}{M}\right)\left[1-1+\frac{mM}{(m+M)^2}2k^2a^2\right] = \frac{2\beta}{m+M}k^2a^2$$

$$\therefore \qquad \omega_- = ka\left(\frac{2\beta}{m+M}\right)^{1/2} \qquad (16.29)$$

This corresponds to **acoustic branch**. For $k \to \pi/2a$, one finds $\omega_- = (2\beta/m)^{1/2}$. Thus we have

(i) From Fig. (16.4), we note that the allowed frequency range of propagation split into two branches. The upper branch is called **optical branch**, whereas the lower branch is called acoustic branch. We may also note that the acoustical branch is similar to a monoatomic lattice but the optical branch represents a completely different type of wave motion.

(ii) As shown in Fig. (16.4), we observe that the characterstic feature of the elastic waves in diatomic lattices is that there is a bond of frequencies between the two branches for which wave like solution do not exist. The existence of such a band clearly reveals that it is not possible to excite vibrations of frequency lying within this band. This is why the band is called as **forbidden frequency band**. The width of this band depends upon the mass ratio, M/m Larger is this ratio, the wider is the frequency band. When $m = M$, i.e. two masses are equal, the two branches join together and the forbidden band disappears.

Displacement of particle in two branches

To study the physical difference between the vibrations represented by two branches, let us calculate the ratio of A / B for the two branches as $k \to 0$ and hence $\cos ka \to 1$. Thus from Eq. (16.23), one obtains

$$-\omega^2 mA = 2\beta B - 2\beta A$$

or
$$\frac{\omega^2}{2\beta} = -\left(\frac{B-A}{mA}\right)$$

and
$$-\omega^2 MB = 2\beta A - 2\beta B$$

or
$$\frac{\omega^2}{2\beta} = \left(\frac{A-B}{MB}\right)$$

$$\therefore \qquad \left(\frac{A-B}{MB}\right) = \frac{B-A}{mA}$$

or
$$\frac{A}{B} = -\frac{M}{m} \qquad (16.30)$$

Eq. (16.30) shows that the two masses move opposite to each other and their amplitudes are inversely proportional in the ratio of their masses in such a manner that their centres of mass remain at rest during the periodic i.e. motion as shown in Fig. [16.5(a)].

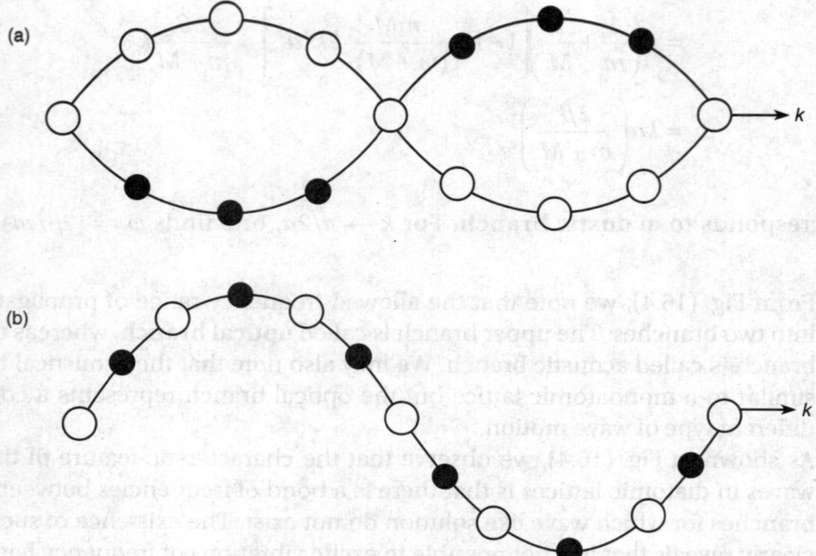

Fig. 16.5 Particle displacements in two branches (a) optical mode, and (b) acoustical mode

For acoustical branch

In this case, one finds

$$-m\omega^2 A = 2\beta B\left(1-\frac{k^2a^2}{2}\right)-2\beta A$$

and
$$-M\omega^2 B = 2\beta A\left(1-\frac{k^2a^2}{2}\right)-2\beta B$$

Adding the above two equations, one obtains

$$-\omega^2(mA+MB) = -k^2a^2\beta(A+B)$$

Substituting the value of ω^2 from Eq. (16.28) and solving, one obtains

$$A/B = 1 \tag{16.31}$$

Obviously, in this case, the two masses move in the same direction with the same amplitude see Fig. [16.5(b)].

One can excite the optical branch vibrations by a force that produces opposite effect on two atoms, e.g., ionic crystal. These crystals respond to this kind of excitation when they are irradiated by light beam, because these crystals have two types of oppositely charged ions. Clearly, the optical branch has been derived from this mode of excitation.

The acoustical branch vibrations can be excited by some kind of force that forces due to all the atoms in the crystal to go in the same direction. When a beam of sound waves is directed at the surface of the crystal, it produces the desired effect. This is why such vibrations are called acoustical vibrations and constitutes the acoustical branch in dispersion relation. We may note that a monoatomic lattice can respond only to this kind of excitation.

Number of possible normal modes of vibration in a band

Let us calculate the number of possible normal modes of vibration in a band of a crystal having finite length L. We have used a running wave picture and therefore it is convenient to use the boundary condition:

$$x_{2n}(x) = x_{2n}(x+L) \tag{16.32}$$

Using Eq. (16.16), one obtains

$$A\exp[i(\omega t - 2kna)] = A\exp[i(\omega t - 2k(na+L))]$$

or

$$\exp(-2ikL) = 1$$

$$\therefore \quad k = \pm\frac{\pi}{L}, \pm\frac{2\pi}{L}, \pm\frac{3\pi}{L}, \ldots \pm\frac{N\pi}{L} \tag{16.33}$$

where N is the number of unit cells in the crystal, thus $Na = L$. To keep it to $\pm\pi/2a$, the maximum value of k has been taken to be $\pm\frac{N\pi}{2L}$. Since each value of k represents a normal mode, one can conclude that the total number of possible normal modes in any band is equal to N, i.e. the number of unit cells.

Idea of phonon is a useful concept in describing the properties of solids where the idea of a gas of phonons can be used to understand the heat capacity. Phonons in a solid have a mean free path due to collisions among themselves via the phonon-phonon interaction and also due to collisions with impurity atoms. In insulators heat is transmitted by phonons and the magnitude of thermal conductivity is determined by the mean free paths. **The electron-Phonon interaction** is important in giving an attractive interaction between pairs of electrons in some metals and alloys that leads to superconductivity.

16.5 PHONONS, INTERNAL ENERGY AND SPECIFIC HEAT

The vibrational energy of the atomic and subatomic particles (the lattice and the electrons) constitute the thermodynamic internal energy of the solid. So the phonon energy is also the same as the internal energy of the solid. As the temperature of the solid increases, phonons increase, phonon energy increases and hence its internal energy also increases. The specific heat at constant volume (C_V) is defined as the rate of change of internal energy with temperature (T) at constant volume, i.e.,

$$C_V = \left(\frac{\partial u}{\partial T}\right)_V \tag{16.34}$$

Since the internal energy u comprises of the energy due to lattice vibrations and electron vibrations we can write

$$u = u_{\text{lattice}} + u_{\text{electron}} \tag{16.35}$$

and

$$C_V = (C_V)_{\text{lattice}} + (C_V)_{\text{electron}} \tag{16.36}$$

Since the electronic contribution to the internal energy and hence to the specific heat of the solid is very small at room temperature for most of the solids, we will first take into account the lattice contribution to the internal energy and to the specific heat of the solid. Our objective now is to find an expression for the internal energy due to lattice vibrations in a solid assuming a suitable model of a solid and hence calculate its lattice specific heat $(C_V)_{\text{lattice}}$. The model so assumed must yield such results as can explain the following experimental facts about the solids.

(i) In the room temperature range C_V is independent of temperature and equal to $3R$ or $3Nk$ or 25J mol^{-1} deg^{-1} for all solids. This is Dulong and Petit's law. It holds just below the melting point at which C_V starts to rise.

(ii) At low temperatures C_V begins to fall off and approaches zero as $T \to 0$.

(iii) The approach to zero near $T = 0$ follows T^3 law i.e.

$$C_V = (\text{constant})T^3 \tag{16.37}$$

in insulators and in metals it follow the law

$$C_V = (\text{constant})T \tag{16.38}$$

If the metal becomes a superconductor the drop is faster than T.

(iv) In magnetic solids there is a large contribution to the heat capacity over the temperature range in which the magnetic moments become ordered. Below 1 K the ordering of nuclear moments may give very large heat capacity.

Let us now evaluate C_V theoretically and compare the value so obtained with the experimental results. We shall throughout deal with a crystal composed of n atoms which are held together in a periodic manner. These atoms are assumed to be free to vibrate about their equilibrium positions subject to constraining forces which to a first approximation obey Hook's law.

16.5.1 The Various Theories of Lattice Specific Heat

As stated above the vibrational energy of the lattice constitute the thermodynamical internal energy of the solid, as such the vibrational energy of a linear chain of N atoms may be expressed as the energy of N harmonic oscillators. Extending the arguments to the three-dimensional case one can conclude that "The vibrational energy of a crystal containing N atoms is equivalent to the energy of a system of $3N$ harmonic oscillators". This feature is common to all theories of the specific heat and the distinction between the various theories is based on their differences in the proposed frequency spectrum of the oscillators. The central problem in the theory of the specific heat is therefore the calculation of the wave lengths and frequencies of the possible modes of vibration of the crystal under consideration. The different approaches to this problem are discussed below.

The following qualitative explanation will clarify the above statement with regard to the harmonic oscillator representation. Suppose if it is possible to fix the position of all the nuclei in a crystal and such that they are all in their equilibrium position. If one of the nuclei were now displaced over a distance small compared with the shortest interatomic distances, and then set free again, the displaced atom would carry out harmonic vibrations about its equilibrium position and its energy of vibration would be the same as that of three one-dimensional harmonic oscillators, one for each direction of motion. Applying the same reasoning to the other atom in the crystal, one arrives at a system of $3N$ harmonic oscillators representing the vibrations of the crystal as a whole.

16.5.2 Classical Theory of Specific Heat

Dulong and Petit in 1819, observed that the product of the atomic weight and the specific heat of each solid is a constant, 6.0 cal/deg-At wt. at room temperature and constant volume. This was called as Dulong and Petit law. The justification of this law was obtained from Boltzmann's consideration of the law of equipartition of energy. One can express the kinetic energy of a representative atom of mass m of a monoatomic solid as

$$\frac{1}{2}Mv^2 = \frac{1}{2}M\left(v_x^2 + v_y^2 + v_z^2\right) \tag{16.39}$$

where v is the root mean square velocity, we have from the law of equipartition of energy

$$\frac{1}{2}mv_x^2 = \frac{1}{2}mv_y^2 = \frac{1}{2}mv_z^2 = \frac{1}{2}kT \tag{16.40}$$

where k is Boltzmann's constant. This yields

$$C_V = \frac{3}{2}Nk = \frac{3}{2}A \tag{16.41}$$

where N is Avogadro's number and R is universal gas constant. We have read that in a solid, the atoms are bound and oscillate about their mean positions, the mean kinetic energy being one half the total energy. Obviously, if $3R/2$ is the contribution of specific heat associated with mean kinetic energy, then an equal contribution will be there from the potential energy part. Now, the total specific heat is

$$C_V = 3R = 5.96 \text{ cal/deg.-At.wt.} \tag{16.42}$$

Dulong and Petit found that this law is not valid in the case of diamond, boron and silicon, the value of sp. heat at 20°C found to be 1.92, 3.32 and 5.11 cal / deg-At. wt. respectively. This observed variation and also variations observed in some other substances could not be explained on the basis of kinetic theory of matter. Moreover, Nernst reported that sp. heat of a solid decreases with decrease in temperature and at absolute zero, the specific heat tends to zero. Further more, Nernst found that specific heat increases with increase in temperature and tends to maximum value. Clearly, classical theory of sp. heat was inadequate.

16.5.3 Break Down of the Classical Theory

The energy of a harmonic oscillator of natural angular frequency is given by

$$\varepsilon = \frac{P^2}{2m} + \frac{m\omega^2 q^2}{2} \tag{16.43}$$

where the first term on right represents the kinetic energy (P is momentum) and the second term represents the potential energy (q is the displacement from the equilibrium position).

According to classical statistical mechanics the average energy of harmonic oscillator is given by:

$$<\varepsilon> = \frac{\int_0^\infty \varepsilon e^{-\varepsilon/kT} d\varepsilon}{\int_0^\infty e^{-\varepsilon/kT} d\varepsilon} = kT \tag{16.44}$$

where k is Boltzmann's constant and T is the absolute temperature. It is important to note that the frequency does not enter in this result. Thus the vibrational energy of a crystal of N atoms is classically always equal to

$$E = 3NkT \tag{16.45}$$

In other words the vibrational energy of crystal is independent of the assumed frequency distribution of the oscillators used in the model. Now as long as the volume of the solid is kept constant, the Eq. (16.45) gives that the total vibrational energy of the crystal is only temperature dependent. Thus for a solid containing one type of atoms putting N equal to Avogadro number, one obtains molar specific heat as

$$C_V = 3Nk = 3R = 25 \text{ J mole}^{-1} \text{ deg}^{-1} \tag{16.46}$$

Here R is gas constant. Similarly, if the solid consists of N atoms of kind A and N atoms of kind B, the specific heat per mole would be $6R$, etc. The result obtained is in quantitative agreement with experiment at high temperatures only. In other words, it does not explain the variation of the specific heat at low temperatures as already mentioned. This discrepancy is essentially removed when quantum theory is used. It may be noted that the classical theory led to similar difficulties in the specific heat of molecules.

16.5.4 The Einstein's Model

The physical model employed by Einstein was over simplified but his results definitely indicate that quantum theory contained the answer to the difficulties encountered in classical theory. The salient features of Einstein's model are:

(i) All the atoms vibrate independently of each other.

(ii) The solid containing N atoms is considered to be equivalent to SN harmonic oscillators.

(iii) All the atomic oscillators have the same frequency (ν) because of their assumed identical surroundings.

(iv) The energy spectrum of oscillators is note continuous but discrete. The possible energy levels of a harmonic oscillator are given by

$$\varepsilon_n = \left(n + \frac{1}{2}\right) h\nu \tag{16.47}$$

(v) Any number of oscillators may be in the same quantum state of the system.

(vi) Since these atomic oscillators form an assembly of systems which are distinguishable or identifiable by virtue of their location at separate and distinct sites, hence the oscillators are regarded as having Maxwell-Boltzmann distribution of energies.

Eq. (16.47) refers to an isolated oscillator but the atomic oscillators in solid are not isolated. They are continually exchanging energy with the ambient thermal bath surrounding the solid. The energy of the oscillators is therefore continually changing but its average value at thermal equilibrium is given by

$$<\varepsilon> = \frac{\sum_{n=0}^{\infty} \varepsilon_n e^{-\varepsilon_n/kT}}{\sum_{n=0}^{\infty} e^{-\varepsilon_n/kT}}$$

$$= \frac{\sum_{n=0}^{\infty}\left(n+\frac{1}{2}\right)h\nu e^{-\left(n+\frac{1}{2}\right)h\nu/kt}}{\sum_{n=0}^{\infty} e^{-\left(n+\frac{1}{2}\right)h\nu/kT}}$$

$$= \frac{h\nu\sum_{n=0}^{\infty}\left(n+\frac{1}{2}\right)e^{-\left(n+\frac{1}{2}\right)x}}{\sum_{n=0}^{\infty} e^{-\left(n+\frac{1}{2}\right)x}} \tag{16.48}$$

where $$x = \frac{h\nu}{kT} \tag{16.49}$$

Eq. (16.48) can be written as

$$<\varepsilon> = h\nu\frac{\frac{1}{2}e^{x/2}+\frac{3}{2}e^{3x/2}+\frac{5}{2}e^{5x/2}+\cdots}{e^{x/2}+e^{3x/2}+e^{5x/2}+\cdots}$$

or $$<\varepsilon_n> = h\nu\frac{d}{dx}\log[e^{x/2}+e^{3x/2}+e^{5x/2}+\cdots]$$

$$= h\nu\frac{d}{dx}\log[e^{x/2}(I+e^{x}+e^{2x}+e^{3x}+\cdots)]$$

$$= h\nu\frac{d}{dx}\log\left[\frac{e^{x/2}}{I-e^{x}}\right] = h\nu\frac{d}{dx}\left[\frac{1}{2}x-\log(I-e^{x})\right]$$

$$= h\nu\left[\frac{1}{2}+\frac{e^{x}}{I-e^{x}}\right] = h\nu\left[1+\frac{1}{e^{-x}-1}\right]$$

$$= h\nu\left[\frac{1}{2}+\frac{I}{e^{h\nu/kT}-1}\right] \tag{16.50}$$

The internal lattice thermal energy is obtained by multiplying the average energy per oscillator by the number of oscillators (3N). This given us

$$E = 3N<\varepsilon_n> = \frac{3Nh\nu}{2}+\frac{3Nh\nu}{e^{h\nu/kT}-1} \tag{16.51}$$

$$C_V = \frac{\partial E}{\partial T} = 3NK\left(\frac{h\nu}{kT}\right)^2 \frac{e^{h\nu/kT}}{(e^{h\nu/kT}-1)^2} \tag{16.52}$$

Putting $$h\nu = \theta_E K \tag{16.53}$$

Here θ_E is called "Einstein's temperature", we have

$$C_V = 3Nk\left(\frac{\theta_E}{T}\right)^2 \frac{e^{\theta_E/T}}{(e^{\theta_E/T}-1)^2} \tag{16.54}$$

The purpose behind defining the Einstein's temperature is that the transition from the high temperature region in which the classical laws hold good to the low temperature region where the quantum laws take charge is determined by equating thermal energy kT to the quantum of energy $h\nu$. Thus θ_E is an adjustable parameter chosen to produce the best fit to the measured values over the whole temperature range.

The plot of C_V versus T using Eq. (16.54) is shown in Fig. (16.6), which, indicate that the theory is now in agreement with experiment at least qualitatively, over the entire temperature range. Note that in particular that $C_V \to 0$ as $T \to 0$ a new and important feature which was lacking in the classical theory.

Let is now examine the behaviour of C_V as given by Eq. (16.54) in extreme temperature limits. If $kT >> h\nu$, i.e. $T >> \theta_E$ high temperature limit, we can write

$$e^{h\nu/T} = 1 + \frac{h\nu}{kT}$$

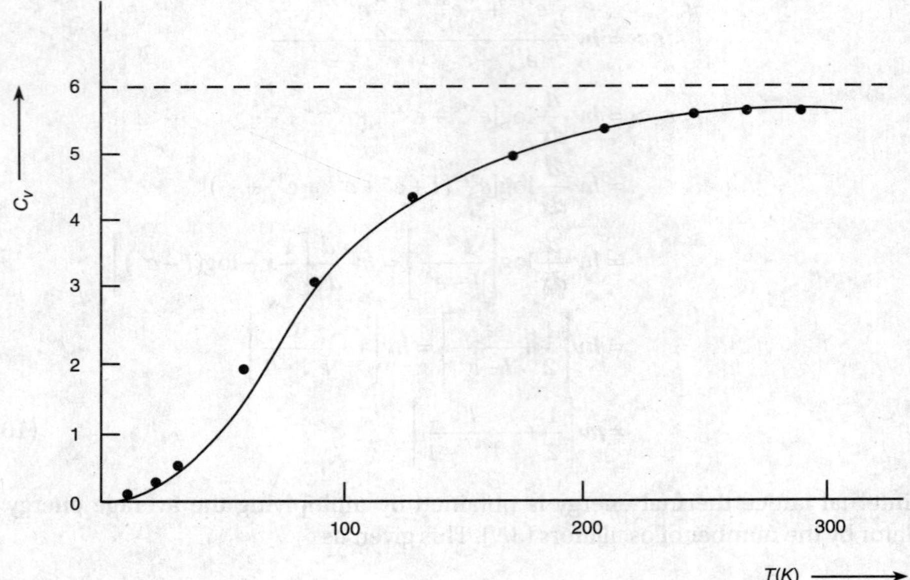

Fig. 16.6

and $$E = \frac{3Nh\nu}{2} + 3NkT$$

and $$C_V = \frac{\partial E}{\partial T} = 3Nk = 3R$$

Which is **Dulong and Petit law.** Thus the model is satisfactory at the high temperature limit because the quantum aspects become irrelevant. At low temperatures $T << \theta_E$

$$e^{h\nu/kT} >> 1$$

and $$C_V = (3Nk)\left(\frac{\theta_E}{T}\right)^2 e^{-(\theta_E/T)} \tag{16.55}$$

The specific heat approaches zero exponentially as temperature approaches zero. The reason that the specific heat becomes quit small at low temperatures can be understood by assuming that the crystal is placed in contact with an external heat bath consisting of an ideal monoatomic gas at some given temperature and allowed to absorb energy of the ideal gas. The average energy of the gas atoms is $\frac{3}{2}kT$ and if temperature of the system is high enough so that kT is of the order of or greater than the energy $h\nu$ required to excite one of the vibrating atoms of the crystal to a higher energy state, then such excitations will occur frequently when gas atoms from the heat bath collide with the crystal. These collisions will then be inelastic in mechanical sense and energy initially belonging to the gas atoms will readily be transfered to the crystal lattice as vibrational energy.

At low temperatures the mean thermal energy kT is very much less than the quantum of energy $h\nu$ and is insufficient to excite the thermal vibrations of oscillators-most of them are in the state of zero point energy of frozen energy state and execute zero point oscillations and not the thermal oscillation. Under these conditions an occassional gas atom having much higher energy than average will be capable of effecting such an excitation of a small fraction of oscillator. Hence the mean thermal energy of oscillator will be much less than kT. Thus at low temperatures the specific heat of quantum oscillator will be less than the specific heat of classical oscillator.

In many respects the Einstein's model has been a remarkable; its results are in good agreement with experiments over most of the temperature range. However the drawbacks of this model are:

(i) It explains the specific heat curves up to a certain temperature ($\approx 0.2\theta_E$) but fails completely at extremely low temperatures (below $0.2\theta_E$) where T^3 law holds true.

(ii) ν and θ_E are obtained emperically and can not be verified from any other independent physical data.

(iii) Einstein himself recognised that the assumption of monoatomic vibrations did not correspond to actual facts. The vibrations of a particular atom must be very complex indeed because it is under the field of force of thousands of other vibrating atoms. Nevertheless Einstein's model was of great step forward in the right direction.

16.6 VIBRATIONAL MODES OF CONTINUOUS MEDIUM (DISTRIBUTION FUNCTION)

In the preceding section it was pointed out that the discrepancy between the Einstein's theory and experimental results in the low temperature region was a consequence of the over simplified model employed by Einstein. Debye in 1912 tackled the problem from a different point of view with great success. Debye realised that it is possible to propagate waves through solids covering a wave length region extending from low frequencies (sound waves) up to short waves (infrared absorption). The essential difference between the Debye model and the Einstein model is that Debye considers the vibrational modes of a crystal as a whole where as Einstein's starting point was to consider the vibration of a single atom assuming the atomic vibrations to be independent of each other.

Let us first consider for simplicity the vibrational modes of one-dimensional line, of lattice constant a. Here we will use the method of periodic boundary conditions. In this method we shall consider the medium as unbounded, but require that the solutions to be periodic over a large distance $L = Na$. Thus, we have

$$S(na) = S(na + L) \tag{16.56}$$

where S represents the displacement of the particle. Then for progressive wave solution, where displacement for wave vector k and frequency ω is given by

$$S_n = Ae^{r(\omega t + kna)} \tag{16.57}$$

Using Eq. (16.56), we have

$$Ae^{r(\omega t+kna)} = Ae^{r[(\omega t+(na+L))k]}$$

or

$$e^{ikL} = 1 \tag{16.58}$$

Thus the allowed values of k which are confined to region $-\dfrac{\pi}{a} \le k \le +\dfrac{\pi}{a}$ are

$$kL = \pm\, 2\pi m \tag{16.59}$$

Where m is an integer i.e. k takes up values

$$K = 0, \pm\frac{2\pi}{L}, \pm\frac{4\pi}{L}, \pm\frac{6\pi}{L}, \dots \pm\frac{N\pi}{L} \tag{16.60}$$

The value $k = 0$, corresponds to all the particles at rest and this value must be omitted. Here we observe that the frequency spectrum is discrete, one frequency (mode of vibration) corresponds to each integer value of m. The interval between two successive valves of k is then $\dfrac{2\pi}{L}$.

Each k value of Eq. (16.59) represents a mode of vibration. Now suppose we choose an arbitrary interval dk in k-space and look for the number of modes whose k's be in this interval. We assume here that L is large so that the frequency spectrum is quasi continuous, this is true, of course for the macroscopic solids with which we are dealing. Thus the number modes is[1]

$$dm = \frac{L}{2\pi}dK \tag{16.61}$$

[1] $kL = 2\pi m \;\therefore\; dkL = 2\pi dm$

or dm the number of modes within range dk are $dm = \dfrac{1}{2\pi}dk.$

But k and frequency ω are interelated via the dispersion relation and we may well seek the number of modes in frequency range lying between ω and $\omega + d\omega$. The density of states $g(\omega)$ is defined such that $g(\omega)d\omega$ gives the number, comparing this definition with Eq. (16.61), one may write:

$$g(\omega)d\omega = \frac{L}{2\pi}dk$$

or
$$g(\omega) = \frac{L}{2\pi}\frac{dk}{d\omega} = \frac{L}{2\pi}\frac{1}{d\omega/dk} \tag{16.62}$$

We note from Fig. (16.7), however that in calculating $g(\omega)$ we must include the modes lying in the negative k-region as well as in the positive region. The former represent waves travelling to the left and the later waves travelling to the right. The effect is to multiply the Eq. (16.63) by a factor of two, that is

$$g(\omega) = \frac{L}{\pi}\frac{1}{d\omega/dk} \tag{16.63}$$

This is a general result for the one-dimensional case and we see that the density of states $g(\omega)$ is determined by the dispersion relation. For the linear relation $\dfrac{d\omega}{dk} = v_0$, where v_0 is the velocity of propagation of the wave and therefore

$$g(\omega) = \frac{L}{\pi}\cdot\frac{1}{v_0} \tag{16.64}$$

which is constant in and dependent of ω.

Density of modes in three dimensions

Let us consider a crystal with dimensions (L_x, L_y, L_z). The boundary conditions are

$$\left.\begin{aligned} S(L_x,Y,z,t) &= S(0,y,z,t) \\ S(x,L_y,z,t) &= S(x,0,z,t) \\ S(x,Y,L_z,t) &= S(x,y,0,t) \end{aligned}\right\} \tag{16.65}$$

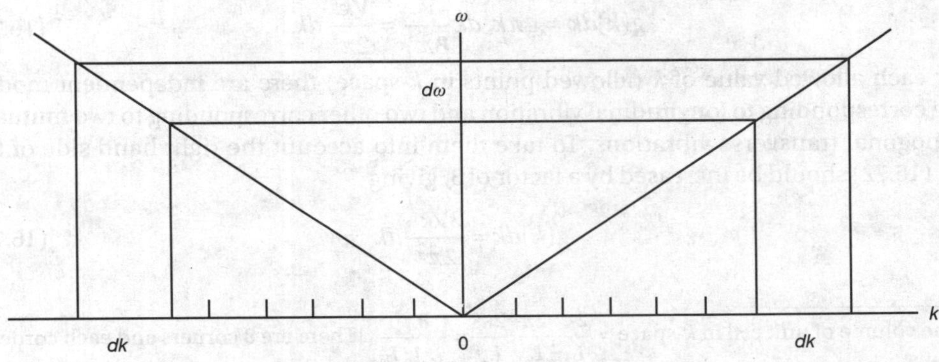

Fig. 16.7

The equations of motion then lead to vibrational solution of the form

$$S(x,y,z,t) = Ae^{i[k_x+k_y y+k_z z-\omega t]} = Ae^{i[k.r-\omega t]} \tag{16.66}$$

where

$$k^2 = k_x^2 + k_y^2 + k_z^2 \tag{16.67}$$

Using boundary conditions Eq. (16.65), we obtain

$$Ae^{i[k_y y+k_z z-\omega t]} = Ae^{i[k_x L_x+k_y y+k_z z-\omega t]}$$

or

$$e^{ik_x L_x} = 1$$

and

$$k_x = \frac{2\pi n_x}{L_x} \tag{16.68}$$

Similarly

$$k_y = \frac{2\pi n_y}{L_y} \tag{16.69}$$

and

$$k_z = \frac{2\pi n_z}{L_z} \tag{16.70}$$

where n_x, n_y, n_z are integers (positive or negative). A set of allowed values of k_x, k_y and k_z corresponds to a single normal mode of vibration satisfying the periodic boundary conditions. The various allowed normal modes can thus be represented as points in an orthogonal k-space shown in Fig. (16.8). Therefore there is one allowed value of k per $\frac{8\pi^2}{L_x, L_y, L_z}$ volume is k-space or the allowed values of k per unit volume of k-space[2] for each polarisation are

$$\frac{L_x L_y L_z}{8\pi^3} = \frac{V}{8\pi^3} \tag{16.71}$$

Here $L_x L_y L_z = V$ the volume of the specimen, The constant k value surfaces

$$k^2 = k_x^2 + k_y^2 + k_z^2$$

These are spheres in k-space and hence one can calculate the allowed number of modes $g(k)dk$ in an internal dk about the wave vector k by evaluating the volume of spherical surface of radius k and $k + dk$ of k-space. The required volume is $4\pi k^2 dk$. Hence the allowed modes are

$$g(k)dk = 4\pi k^2 dk \frac{V}{8\pi^3} = \frac{Vk^2}{2\pi^2} dk \tag{16.72}$$

For each allowed value of k (allowed points in k-space) there are independent modes, one corresponding to longitudinal vibration and two other corresponding to two mutually orthogonal transverse vibrations. To take them into account the right hand side of the Eq. (16.72) should be increased by a factor of 3, giving

$$g(k)dk = \frac{3Vk^2}{2\pi^2} dk \tag{16.73}$$

[2] The volume of unit cell in k-space $= \frac{2\pi}{L_x} . \frac{2\pi}{L_y} . \frac{2\pi}{L_z} = \frac{8\pi^3}{L_x L_y L_z}$. There are 8 corners and each corner is shared by 8 such unit cells. Hence there is one point per unit cell of volume $\frac{8\pi^3}{L_x L_y L_z}$.

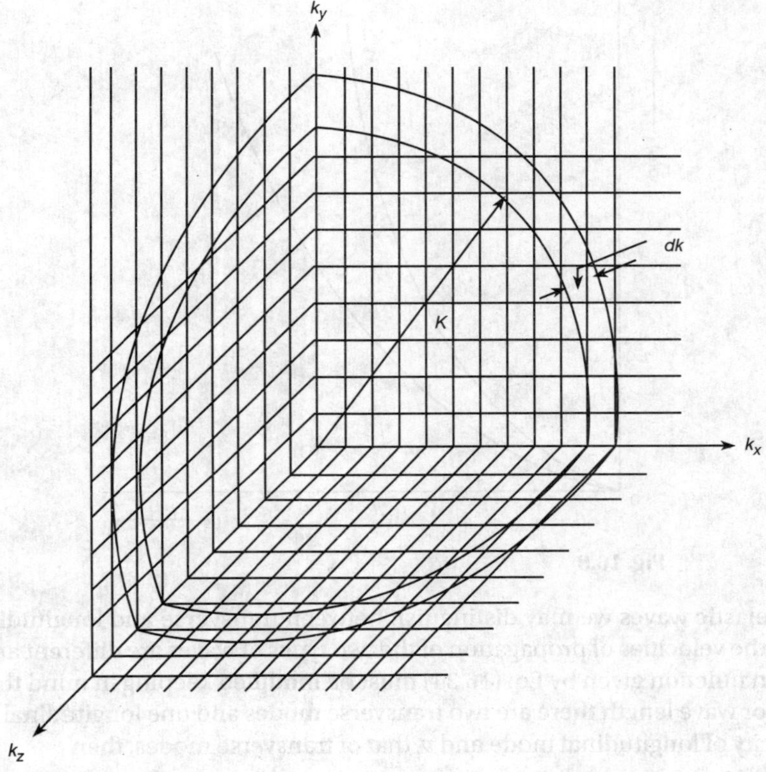

Fig. 16.8

The density of states $g(\omega)$ is defined as the number of modes whose frequencies lie in the interval ω and $\omega + d\omega$. This number can be obtained from Eq. (16.73) by making a change of variable k to ω, which may be accomplished by the use of dispersion relation. Unfortunately such a relation is not known in any convenient form. Debye simplified the problem by assuming the continuum approximation in which phase velocity v_0 is taken to be constant which he assumed to be equal to the velocity of long wave length elastic vibrations propagating through the crystal, i.e.,

$$\omega(k) = v_0 k$$

Substituting in Eq. (16.73) we have

$$g(\omega)d\omega = \frac{3V}{2\pi^2}\frac{\omega^2}{v_0^2}\frac{d\omega}{v_0} = \frac{3V}{2\pi^2}\frac{\omega^2}{v_0^3}d\omega \tag{16.74}$$

Thus the density of states is given by

$$g(\omega) = \frac{3V}{2\pi^2}\frac{\omega^2}{v_0^3} \tag{16.75}$$

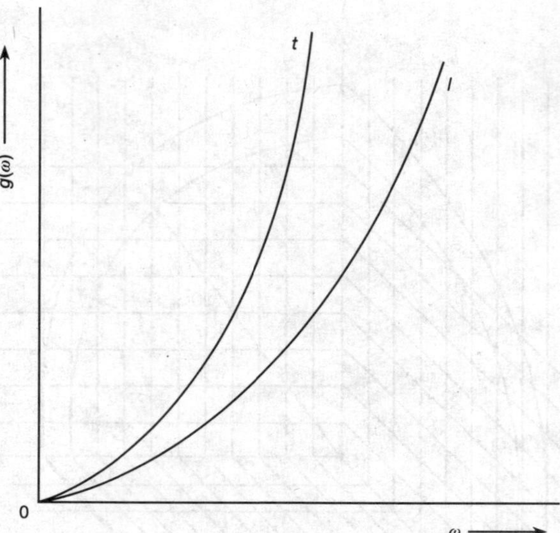

Fig. 16.9

In case of elastic waves we may distinguish between transverse and longitudinal waves. In general the velocities of propagation of the two types of waves are different and thus the distribution function given by Eq. (16.34) must be modified keeping in mind that for each frequency or wave length there are two transverse modes and one longitudinal mode. If v_l is the velocity of longitudinal mode and v_t that of transverse modes, then

$$g(\omega) = \frac{V}{2\pi^2}\cdot\left(\frac{1}{v_\ell^2} + \frac{2}{v_t^3}\right)\omega^2 \tag{16.76}$$

The function $g(\omega)$ is plotted versus in Fig (16.9). We see that $g(\omega)$ increases as ω^2, unlike the one-dimensional case in which $g(\omega)$ was constant. The increase in the present case is apparent of the fact that the volume of the spherical shell Fig (16.8) increases as k^2 and hence as ω^2 since ω is proportional to k.

It is worth while to mention here that, when the wave lengths of the modes are small compared with the dimensions of the sample the density of states function $g(\omega)$ is independent of the choice of boundary conditions. In using the periodic conditions we have made the choice which is mathematically most convenient for our purposes.

16.7 DEBYE MODEL

The main difference between Einstein's model and Debye's model is that, Debye considers the vibrational modes of a crystal as whole, i.e. the dispersion relation is taken for homogeneous solid. Thus

$$\omega = v_0 k \tag{16.77}$$

On the other hand Einstein's starting point was to consider the vibrations of a single atom, assuming the atomic vibrations to be independent of each other.

The Debye model is based on the following assumptions:

(i) A solid is treated as an isotropic elastic continuum.

(ii) The Debye approximation considers the thermal energy in accoustical mode, where as Einstein's approximation consider the thermal energy in optical mode.

(iii) The atoms of the solid have $3N$ frequencies ranging from zero to a maximum ω_m. The restriction to the maximum frequency is a consequence of the fact that the waves of the half wave length shorter than interatomic $\left(\frac{\lambda}{2} < a\right)$ distance cannot be propagated through the crystal.

(iv) With each frequency Debye associated a distribution function $g(\omega)$ such that $g(\omega)d\omega$ represents the number of modes of vibrations between the frequency interval ω and $\omega + d\omega$.

(v) These modes of vibration are distributed in energy according to the Maxwell-Boltzmann distribution law.

(vi) Since any vibrational motion of the system can be thought of as a superposition of independent normal-modes of vibrations, these normal modes are regarded as independent harmonic oscillators whose allowed energy levels are given by Eq. (16.47) and whose average energy $\langle E_n(\omega)\rangle$ is given by Eq. (16.50).

(vii) The contribution to the internal vibrational energy of the crystal from these modes of vibration is given by

$$dE = \langle E(\omega)\rangle g(\omega)d\omega \tag{16.78}$$

The total energy of the crystal is then obtained by integration. Thus

$$E = \int_0^{\omega_m} \langle \epsilon(\omega)\rangle g(\omega)d(\omega) \tag{16.79}$$

Before evaluating the internal energy E Eq. (16.79), we need to know lower and upper ends of the frequency spectrum. The lower limit is evidently $\omega = 0$. The upper cut off frequency was determined by Debye using (ii) assumption, i.e. the total number of frequencies is $3N$. Mathematically this means that

$$\int_0^{\omega_m} g(\omega)d\omega = 3N$$

or

$$\int_0^{\omega_m} \frac{V}{2\pi^2}\cdot\left(\frac{1}{v_\ell^3}+\frac{2}{v_t^3}\right)\frac{\omega_m^3}{3} = 3N$$

or

$$\omega_m^3 = 9N\frac{2\pi^2}{V}\left(\frac{1}{v_\ell^3}+\frac{2}{v_t^3}\right)^{-1} \tag{16.80}$$

Here Debye cut-off takes at Debye frequency ω_m in which it is assumed to be common to transverse and longitudinal modes.[3]

---

[3] In Born's modification, the cut off takes place at a common minimum wave length, corresponding to the maximum frequencies for the longitudinal and transverse modes respectively.

The Debye frequency is now found by substituting the values in Eq. (16.79), we have

$$E = \int_0^{\omega_m} \hbar\omega \left[\frac{1}{2} + \frac{1}{e^{\hbar\omega/kT} - 1}\right] \frac{9N}{\omega_m^3} \omega^2 d\omega \tag{16.81}$$

Here we have used $h\nu = \hbar\omega$ and

$$g(\omega) = \frac{V}{2\pi^2}\left(\frac{1}{v_\ell^3} + \frac{1}{v_t^3}\right)\omega^2 = \frac{9N}{\omega_m^3}\omega^2 \tag{16.82}$$

Re-writing Eq. (16.81), we have

$$E = \frac{9N\hbar}{2\omega_m^3}\int_0^{\omega_m} \omega^3 d\omega + \frac{9N\hbar}{\omega_m^3}\int_0^{\omega_m} \frac{\omega^3}{e^{\hbar\omega/kT} - 1}$$

$$= \frac{9}{8}N\hbar\omega_m + \frac{9N\hbar}{\omega_m^3}\int_0^{\omega_m} \frac{\omega^3}{e^{\hbar\omega/kT} - 1} \tag{16.83}$$

The first term in Eq. (16.83) is zero point contribution to the internal energy which, however, is independent of temperature and contribute nothing to specific heat. Let us now put

$$x = \frac{\hbar\omega}{kT} \quad \therefore \; dx = \frac{\hbar}{kT}d\omega$$

Then

$$x_m = \frac{\hbar\omega_m}{kT} \tag{16.84}$$

and

$$\theta_D = \frac{\hbar\omega_m}{T} \tag{16.85}$$

The parameter θ_D usually referred to as Debye temperature, is seen to have dimensions of temperature and plays the role of a characteristic temperature in much the same way as does the Einstein's temperature in Einstein's theory. It is of course independent of temperature, except for the slight temperature variation introduced by variation of volume and velocity with temperature.

Substituting Eqs. (16.84) and (16.86) in Eq. (16.83). We have[4]

$$E = \frac{9}{8}Nk\theta_D + 9NkT\left(\frac{T}{\theta_D}\right)\int_0^{3x_m} \frac{x^3}{e^x - 1}dx \tag{16.86}$$

---

[4] $$\frac{9N\hbar}{\omega_m^3}\int_0^{\omega_m} \frac{\omega^3}{e^{\hbar\nu/kT} - 1}\;\frac{9N\hbar}{\omega_m^3}\int_0^{x} \frac{\left(\frac{kT}{\hbar}\right)x^3}{e^x - 1}\frac{kT}{\hbar}dx$$

$$= 9NkT\left(\frac{T}{\hbar\omega_m}\right)^3 \int_0^{\omega_m} \frac{x^3}{e^x - 1}dx$$

$$= 9NkT\left(\frac{T}{\theta_B}\right)^3 \int_0^{\omega_m} \frac{x^3}{e^x - 1}dx$$

On solving the above integral we will get the internal thermal energy of the solid. Differentiating Eq. (16.81) with respect to temperature, one obtains

$$C_V = \frac{\partial E}{\partial T} = \frac{9N\hbar}{\omega_m^3}\int_0^{\omega_m} \frac{d}{dT}\left[\frac{\omega^3}{e^{\hbar\omega/kT}-1}d\omega\right]$$

Since integral is with respect to ω and differentiation is with respect to temperature which are independent of each other. Hence we can differentiate first and then integrate. Differentiation gives

$$C_V = \frac{9N\hbar}{\omega_m^3}\int_0^{\omega_m} \frac{\omega^3 e^{\hbar\omega/kT}}{[e^{\hbar\omega/kT}-1]^2}\left(\frac{\hbar\omega}{kT^2}\right)d\omega$$

$$= \frac{9N\hbar}{\omega_m^3}\int_0^{\omega_m} \frac{x^3 e^x}{(e^x-1)^2}\left(\frac{\hbar}{kT^2}\right)\left(\frac{kT}{\hbar}\right)^4\left(\frac{kT}{\hbar}\right)dx \quad \left(\because \omega = \frac{kT}{\hbar}x\right)$$

$$= 9Nk\left(\frac{kT}{\hbar\omega_m}\right)^3 \int_0^{x_m} \frac{x^4}{(e^x-1)^2}dx$$

$$= 9Nk\left(\frac{T}{\theta_D}\right)^3 \int_0^{\theta_D/T} \frac{x^4 e^x}{(e^x-1)^2}dx \tag{16.87}$$

Here $$x_m = \frac{\hbar\omega_m}{kT} = \frac{\theta_D}{T} \tag{16.88}$$

To compare Eq. (16.87) with experimental results one must know the Debye temperature θ_D. This is determined by choosing the value which when substituted into Eq. (16.87) yields the best fit over the whole temperature range. We see from Eq. (16.87) that if C_V is plotted versus a reduced temperature T/θ_D then the same curve should be obtained for all substances that is a universal curve for specific heat. This is found correct for different substances Fig. (16.10), where we see that it holds to a remarkable degree.

Let us try to evaluate the integral in Eq. (16.86) at the extreme values of temperature, i.e., $T \to 0$ and $T \to \infty$. As $T \to 0$, i.e., $T >> \theta_D$. Here we can write

$$e^x = 1 + x \quad \text{(neglecting other terms)}$$

$$\therefore \quad \int_0^{x_m} \frac{x^3 dx}{e^x - 1} \to \int_0^{x_m} x^2 = \frac{x_m^3}{3} = \frac{1}{3}\left(\frac{\theta_D}{T}\right)^3 \tag{16.89}$$

$$\therefore \quad E = \frac{9}{8}Nk\theta_D + 9NkT\left(\frac{T}{\theta_D}\right)^3 \frac{1}{3}\left(\frac{\theta_D}{T}\right)^3$$

$$= \frac{9}{8}Nk\theta_D + 3NkT \tag{16.90}$$

and $$C_V = \frac{\partial E}{\partial T} = 3Nk = 3R \tag{16.91}$$

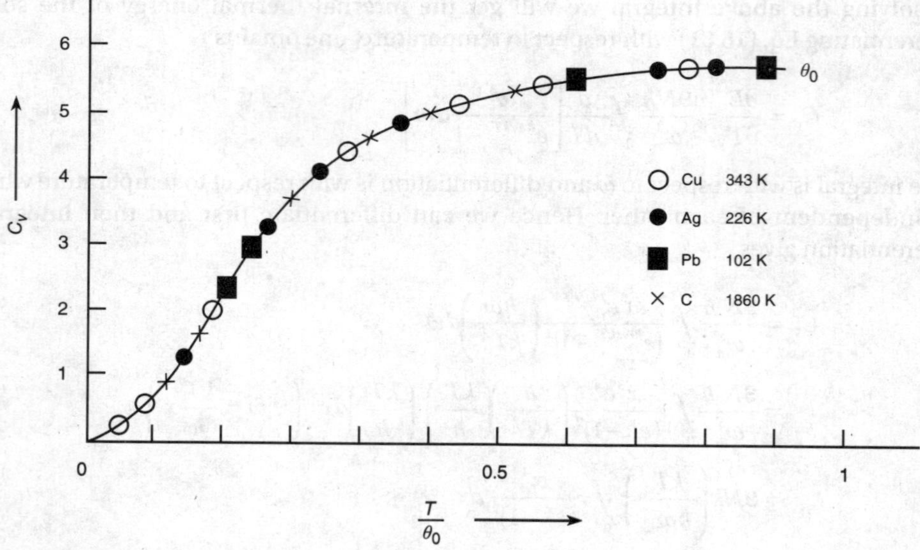

Fig. 16.10

The Dulong and petit law is in agreement with both the classical and Einstein's approximations. Physically this means that at high temperatures ($T \gg \theta_D$) there is sufficient thermal energy to excite all $3N$ modes regardless of their frequency so that they all are contributing to C_V. It is thus equivalent to the Einstein's approximation because the frequency is unimportant and within the classical approximation the quantum discontinuities are negligible.

At low temperatures, $T \to 0$ implies that $x \to \infty$ so that we can let the upper limit of the integration go to infinity instead of x_m. The resulting integral can be solved by contour integration method[5] and comes out to be equal to $\frac{x^4}{15}$, i.e.,

$$\int_0^\infty \frac{x^3}{e^x - 1} dx = \frac{x^4}{15}$$

giving

$$E = \frac{9}{8} Nk\theta_D + 9NkT\left(\frac{T}{\theta_D}\right)^3 \frac{x^4}{15}$$

$$= \frac{9}{8} Nk\theta_D + \frac{3}{5} x^4 NkT\left(\frac{T}{\theta_D}\right)^3 \qquad (16.92)$$

[5] Details of the method can be found in any standard text on complex variables, e.g., Mathematical Physics by Kakani and Hemrajani, CBS Pub., New Delhi.

and
$$C_V = \frac{\partial E}{\partial T} = \frac{12}{5}\pi^4 Nk\left(\frac{T}{\theta_D}\right)^3 \tag{16.93}$$

Thus the energy of vibration is proportional to T^4 at low temperatures which is analogous to the theory of black body radiation at any temperature because in case of black body radiation, the upper limit to the frequency does not exist. Equation (50) exhibits the Debye T^3 law. Debye's predictions for specific heat in the low temperature region are in good agreement with experimental data for many substances.

Criticism of Debye Model

(i) The continuum model is valid for long wave lengths, i.e. the model assumes that only low frequencies are active in the solid.

(ii) Debye assumed that the number of modes of vibrations are $3N$ but this is contrary to the assumption that solid is an elastic continum; which is supposed to possess infinite frequencies, i.e. there is little justification for the equation.

(iii) ω_m is taken to be the same for longitudinal and transverse waves. But this assumption cannot be justified on physical grounds because of the difference in nature of the two waves. Born however proposed to cut off the spectrum in such a way that the longitudinal and transverse modes have a common minimum wave length (λ_{min}), so that one obtains two Debye frequencies, one for longitudinal modes ($v_l = C_1/\lambda_{min}$) and other for transverse modes ($v_t = C_t/\lambda_{min}$). This is actually theoretically more sound and in line with the theory of lattice vibrations developed by Born and Von Karmann.

(iv) If the Debye model were strictly valid the values of θ_D so obtained should be independent of temperature. Instead, one finds that θ_D varies with temperature. The deviation reaches as much as 10% or even more in some cases.

(v) Actual crystalline nature of the solid has not been taken into account.

16.8 ELECTRONIC CONTRIBUTION TO THE INTERNAL ENERGY AND HENCE TO THE SPECIFIC HEAT OF METALS

In the free electron model proposed by Drude and Lorentz, the conduction electrons are treated as free particles which obey the classical laws of machanics, electro-magnetism and statistical mechanics. According to this model if we assume the valence electrons in a metal to behave like free particles just like classical gas, we must expect an internal energy contribution of $\frac{3}{2}kNTZ$.

Where Z in the number of valence electrons per atom. So the specific heat contribution works out to be $\frac{3}{2}RZ$, i.e. numerically for an atom with one valence electron the atomic specific heat (electronic) works out to $\frac{3}{2}R$. At temperatures large compared to Debye temperatures the total specific heat of the metals can be expressed as heat of the metals, i.e.

$$C_V = C_V\,(\text{lattice}) + C_V\,(\text{electronic})$$

$$= 3Nk + \frac{3}{2}Nk = \frac{9}{2}Nk = \frac{9}{2}R \tag{16.94}$$

Experiments on heat capacity in metals show, however that specific heat (C_V) is very nearly equal to $3R$ at high temperatures as in case of insulators. Accurate measurements in which the contributions of electrons to the total heat capacity are isolated show that (C_V) (electronic) is smaller than the classical value $\frac{3}{2}R$ by a factor of about 10^{-2}. To explain this discrepancy, we must once again turn to quantum concepts.

The energy of the electron in a metal is quantized according to quantum mechanics. The various quantum energy levels are shown in Fig 11(a). The electrons in the metals occupy these levels. In doing so, they follow a very important quantum principle, the Pauli exclusion principle, according to which an energy level can accomodate at the most two electrons, one with spin up and the other with spin down. Thus in filling the energy levels, two electrons occupy the lowest level, two more the next level and so forth, until all the electrons in the metal have been accommodated. The energy of the highest occupied level is called the Fermi energy, E_F.

The situation described above is obtainable at $T = 0\,\text{K}$ even at the lowest possible temperature, the electron system has a considerable amount of energy by virtue of the exclusion principle. If it were not for this principle all the electrons would fall into the lowest level, and the total energy of the system would be negligible. This corresponds to the assertion, usually made in classical mechanics that as $T \to 0\,\text{K}$ all motion ceases and the energy vanishes. This assertion does not apply to the conduction electrons.

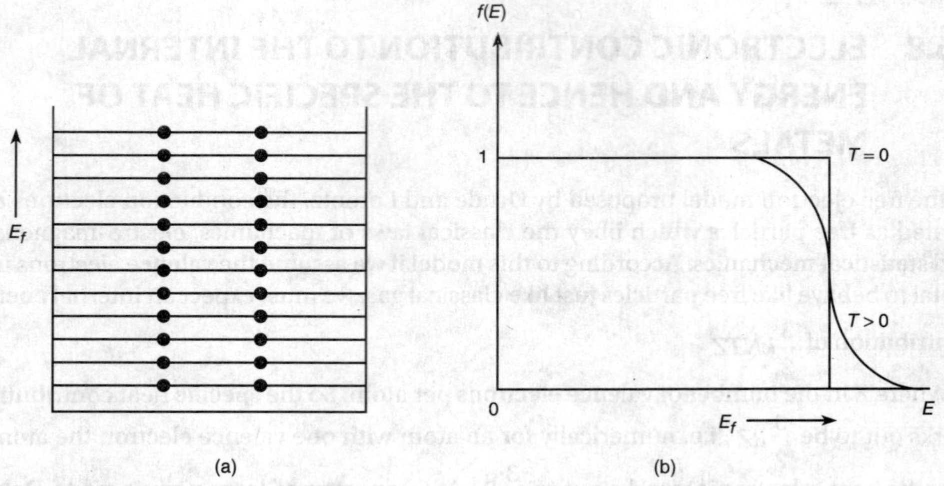

Fig. 16.11

The distribution of electrons among the levels is usually described by the distribution function $f(E)$: Which is defined as the probability that the level with energy E is occupied by an electron. Thus if the level is certainly empty, then $f(E) = 0$ which if it is certainly full them $f(E) = 1$. In general $f(E)$ has value between zero and unity.

It follows from the preceding discussion that the distribution function for electrons at $T = 0$ K has the from

$$f(E) = \begin{cases} 1 & \text{for} \quad E < E_F \\ 0 & \text{for} \quad E > E_F \end{cases} \tag{16.95}$$

That is all levels below E_F are completely filled, and all those above E_F are completely empty. This function is plotted in Fig. [16.11(b)], which shows the discontinuity at the Fermy energy.

So far we have restricted our treatment to the temperature at absolute zero. When the system is heated ($T > 0$ K) thermal energy excites the electrons. But this energy is not shared equally by all electrons, as would be the case in the classical treatment, because the electrons lying well below the Fermi level can not absorb energy. If they did so they would move to higher level which is already occupied and hence the exclusion principle would be violated.

Here we may mention that the energy available to electron for absorbtion is of the order of kT(= 0.025 eV at room temperature), which is much smaller than E_F which is of the order of 5 eV. Therefore only those electrons close to the Fermi level can be excited, because the levels above E_F are empty, and hence when those electrons move to a higher level there is no violation of the exclusion principle. Thus only those electrons which are a small fraction of the total number are capable of being thermally excited and this explains the low electronic specific heat.

The distribution function $f(E)$ at temperatures other than 0 K is given by

$$f(E) = \frac{1}{e^{(E-E_F)/kT} + 1} \tag{16.96}$$

This is known as the Fermi-Dirac distribution function. This function is also plotted in Fig. [16.11(b)] which shows that it is substantially the same as the distribution at $T = 0$ K except very close to Fermi level, where some of the electrons are excited from below E_F to above it. This is of course expected in view of the above discussion.

Thus we have seen that only the electrons near the Fermi level can gain energy when the metal is heated. An electron at the Fermi level may increase its energy from E_F to $(E_F + kT)$ at the most. Electrons having energies less than $(E_F - kT)$ do not gain any energy and so obviously cannot contribute any thing to the specific heat of the metal. Since all the states below $(E_F - kT)$ are occupied, so we see that a fraction kT/E_F of electrons contribute to the electronic specific heat and this works out to be much less than even 1%[6] of the valance electrons responsible for the electronic specific heat. Assuming that NkT/E_F electrons per mole contribute to the specific heat (Where N is Avogadro number) and each electron

[6] $kT = 0.025$ eV, $E_F = 5$ eV

⊠ $\frac{kT}{E_F} \times 100 = \frac{0.025}{5} \times 100 = 0.5\%$

absorbs energy kT on an average, it follows that the electronic contribution to the thermal energy per mole is them

$$E = \frac{NkT}{E_F} kT = \frac{N(kT)^2}{E_F}$$

and specific that

$$C_V = \frac{\partial E}{\partial T} = 2Nk\frac{kT}{E_F} = 2R\frac{kT}{E_F} \tag{16.97}$$

Since at room temperature kT/E_F is of the order of 0.01 for most metals, it is clear why the electronic contribution to the specific heat is very small about 1% of the lattice vibrations contribution. It is apparent from Eq. (16.97) that the electronic specific heat varies nearly with temperature quite in accord with experimental observations. This is unlike the lattice heat capacity which is constant at high temperatures and proportional to T^3 at low temperatures. We also notice that the electronic specific heat becomes important only at low temperatures when the lattice specific heat goes to zero.

Sometimes Fermi temperatures is used in this context, It is defined as $E_F = kT_F$ and the electronic specific heat in terms of Fermi temperatures is

$$C_V = 2R\frac{kT}{E_F} = 2R\frac{kT}{kT_F} = 2R\frac{T}{T_F} \tag{16.98}$$

A typical value for T_F corresponding to $E_F = 5$ eV is 60000 K. Thus in order for the specific heat of the electrons in a solid to reach its classical value, the solid must be heated to a temperature comparable to T_F. But this is not possible because the solid will melt and evaporate long before this temperature is reached. At all practical temperatures, therefore the specific heat of electrons is far below its classical value.

An exact evaluation of the electronic heat capacity yields the value.

$$C_V = \frac{\pi^2}{2} R\frac{kT}{E_F} = \frac{\pi^2 R}{2}\frac{T}{T_F} \tag{16.99}$$

which is clearly of the same order of magnitude as the approximate expression (16.99).

Now the total specific heat of a metal at low temperature may be given by the sum of lattice and electronic contribution, i.e.

$$C_V = AT^3 + \lambda T \tag{16.100}$$

Where γ and A are constants characteristics of the material and are given by Eqs. (16.93) and (16.99). The electronic term is linear in T and is dominant at sufficiently low temperatures. The measurements made in this limit determines primarily the electronic contributions and the direct comparison of the theory with experiment is possible. A plot of $\frac{C_V}{T}$ versus T^2 Fig. (16.12) gives a straight from which the constants A and λ can be found out. These values will be useful for checking the theoretically calculated values of specific heats of metals.

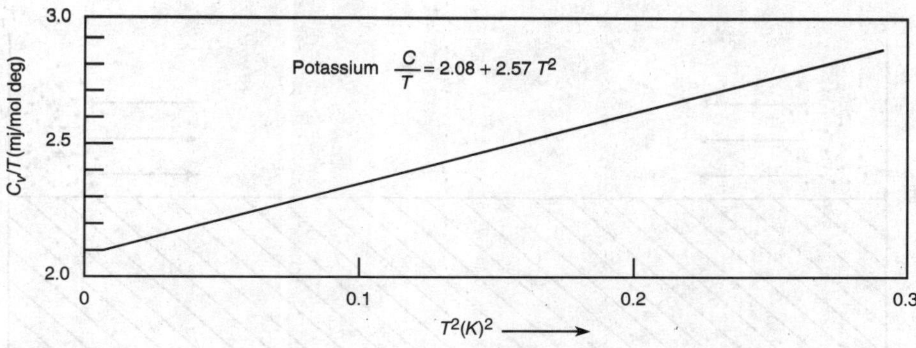

Fig. 16.12

16.9 THERMAL CONDUCTIVITY OF METALS

When the ends of a metallic wire are at different temperatures, i.e. a temperature gradient exists across the wire, the transport of the thermal energy occurs due to the influence of this gradient. The basic experimental fact is that the heat current Q, i.e. the amount of thermal energy crossing a unit area per unit time-is proportional to the temperature gradient,

$$Q = -k\frac{dT}{dx} \tag{16.101}$$

where k is thermal conductivity. In insulators, heat is carried entirely by phonons, but in metals heat many be transported by both electrons and phonons. The conductivity k is therefore equal to the sum the two contributions,

$$k = k_e + k_{ph} \tag{16.102}$$

where k_e and k_{ph} refers to the thermal conductivity contributions due to electrons and phonons respectively. In most metals the contribution of the electrons greatly exceeds that of the phonons because of the great concentration of electrons: typically $k_{ph} \approx 10^{-2}k_e$. This means that the conductivity of the phonons can be ignored while discussing the thermal conductivity of metals.

The physical process by which heat-conduction takes place through electrons is explained in Fig (16.13). The electrons at the hot end (left) travel in all directions but a certain fraction travel to the right and carry energy to the cold end (to the right). Similarly a certain fraction of the electrons at the cold end (on the right travel to the left and carry energy to the hot end, i.e. to the left). These oppositely travelling electron currents are equal but because those at the hot end are more energetic on the average than those on the right, a net energy is transferred to the right resulting in a heat current. Note that heat is transported almost entirely by those electrons which are near the Fermi levels, because those well below this level cancel each other's contributions. Once more it is observed that the electrons at the Fermi surface play a primary role in transport phenomena.

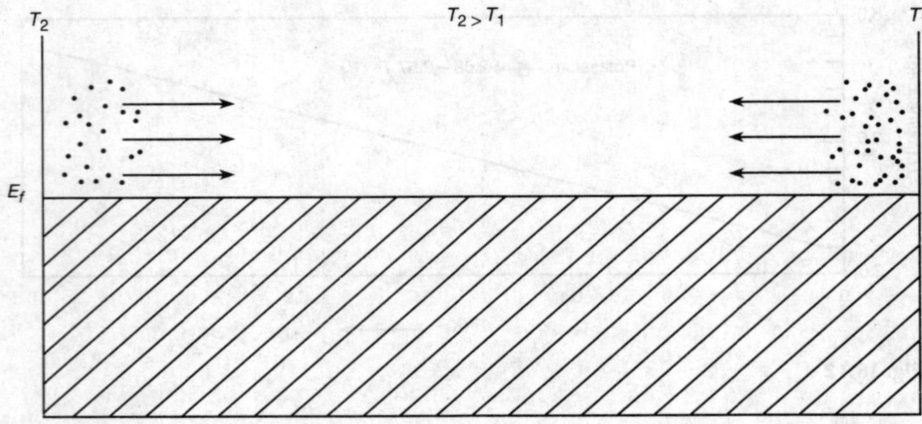

Fig. 16.13

In order to calculate thermal conductivity k qualitatively, we use the following formula[7]

$$k = \frac{1}{3} C_V v \lambda \tag{16.103}$$

Here C_V is the molecular specific heat at constant volume, v is mean spead of the molecules and λ mean free path.

In the present case where the electrons are involved C_V is the electronic specific heat. This is given by Eq. (16.99).

$$C_V = \frac{\pi^2}{2} R \frac{k_T}{E_F} = \frac{\pi^2 R}{2} \frac{T}{T_F} \tag{16.104}$$

where E_F is Fermi energy and T_F Fermi temperature. In above relation R is to be replaced by Nk, because here we deal with unit volume rather than a mole. In addition v and λ should be replaced by v_F and λ_F (corresponding values at Fermi level) because only electrons of Fermi level are effective. Thus

$$E_F = \frac{1}{2} m v_F^2$$

and we have

$$k = \frac{1}{3} \cdot \frac{x^2}{2} \frac{Nkk_T}{1/2 m v_F^2} \cdot v_F \lambda_F = \frac{x^2}{3} \frac{Nk^2 T}{m v_F} \lambda_F \tag{16.105}$$

If the relaxation time for electrons, i.e. time interval between two collisions at the Fermi level be taken as $\tau\,(E_F)$, then

$$\lambda_F = v_F \tau(E_F)$$

---

[7] This formula is proved using the kinetic theory of gases (transport phenomenon).

Then the expression for thermal conductivity becomes

$$k = \frac{x^2}{2} \cdot \frac{Nk^2T}{mv_F} \cdot v_F \tau(E_F) = \frac{\pi^2}{3} \frac{Nk^2T}{m} \tau(E_F) \qquad (16.106)$$

Relation Eq. (16.105) expresses thermal conductivity in terms of electronic properties of metal.

Variation of electronic thermal conductivity with temperature

At room temperature normal pure metals tend to have values of the thermal conductivity one or two orders of magnitude higher than for dielectric solids, so that under these conditions the electrons must carry almost all the heat current. In pure metals the electronic contribution is dominant at all temperatures. In impure metals or in disordered alloys the phonon contribution may be comparable with electronic contribution.

Since τ (E_F) varies as $1/T$ for most metals, k is nearly temperature independent. However it implies only to higher temperature. The theory for low temperature region is very complicated but it can be said that at very low temperature τ (E_F) becomes constant and k_{el} thus varies directly with temperature and becomes zero as $T \rightarrow 0$. The variation of thermal conductivity of copper with temperature is shown in Fig. (16.14).

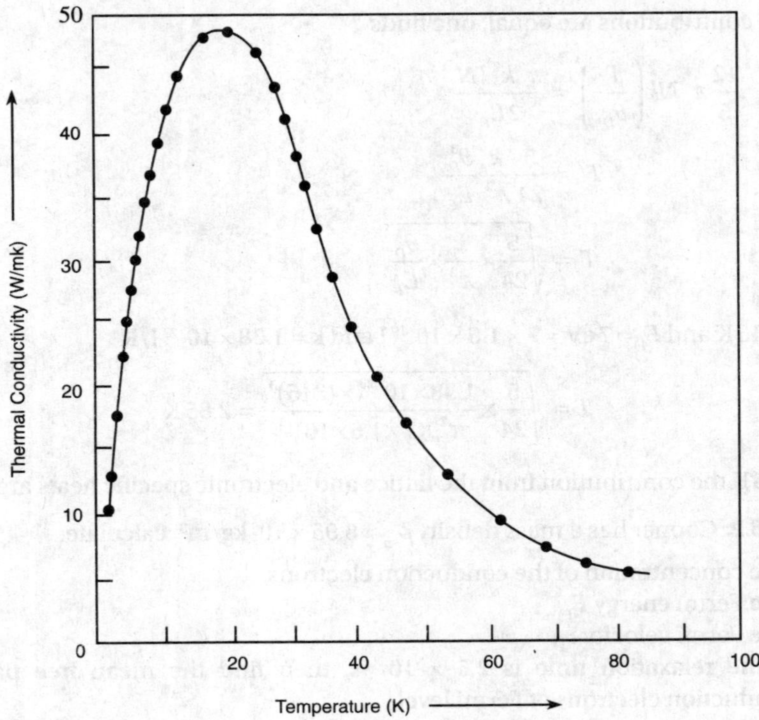

Fig. 16.14

It may also be mentioned that if the metal contains impurities, then besides the electron-phonon scattering there will also be an electron-impurity scattering. In such a case if the two types of scattering are independent, we may write for the total conductivity as

$$\frac{1}{k} = \frac{1}{k_l} + \frac{1}{k_i} \tag{16.107}$$

where k_ℓ is the contribution arising from electron lattice scattering and k_i is the contribution from electron impurity scattering. It is found that addition of impurities decreases the thermal conductivity.

Example 16.1: For copper the Debye temperature is 315 K, the Fermi energy is 7 eV and the number of valance electrons per atom is 1. At what temperature are the contributions from the lattice and the electronic specific heats equal?

Solution The lattice contribution is given by Debye law, i.e.,

$$C_V = \frac{12}{5}\pi^4 Nk\left(\frac{T}{\theta_D}\right)^3$$

and the electronic contribution assuming 1 free electron per atom is given by

$$C_V = \frac{\pi^2 k^2 TN}{2E_F}$$

When their contributions are equal, one finds

$$\frac{12}{5}\pi^4 Nk\left(\frac{T}{\theta_D}\right)^3 = \frac{\pi^2 k^2 TN}{2E_F}$$

or
$$T^2 = \frac{5}{24}\frac{k}{\pi^2}\frac{\theta_D^3}{E_F}$$

or
$$T = \sqrt{\frac{5}{24}\times\frac{k}{\pi^2}\times\frac{\theta_D^3}{E_F}}$$

Here $\theta_D = 315$ K and $F_F = 7$ eV $= 7 \times 1.6 \times 10^{-19}$ J and $k = 1.38 \times 10^{-23}$ J/K

∴
$$T = \sqrt{\frac{5}{24}\times\frac{1.38\times10^{-23}\times(315)^3}{\pi^2\times7\times1.6\times10^{-19}}} = 2.85\,\text{K}$$

Thus at 2.85 K the contribution from the lattice and electronic specific heats are equal.

Example 16.2: Copper has a mass density $\rho_m = 8.95 \times 10^3$ kg/m³, calculate:

(a) The concentration of the conduction electrons,
(b) The Fermi energy E_F,
(c) The Fermi velocity v_F,
(d) If the relaxation time is 2.5×10^{-14} s, then find the mean free path of the conduction electrons at Fermi level.

Assume 1 free electron per atom. Atomic weight of copper is 64.

Solution

(a) 1 gm. mol. of copper weights 64 gm. and its volume V is

$$V = \frac{64}{\rho_m}$$

and it contains 6.06 × 10²³ atoms (Avogadro number). Hence the concentration of conduction electrons assuming one free electron per atom.

$$\frac{N}{V} = \frac{1\times 6.06\times 10^{26}\times 8.95\times 10^{3}}{64}\left[\because \begin{array}{l}6.06\times 10^{23}\ \text{atom/g.mol}\\ = 6.06\times 10^{26}\ \text{atom/kg.mol}\end{array}\right]$$

$$= 8.47\times 10^{28}\ \text{m}^3$$

(b) Fermi Energy is given by

$$E_F = \frac{\hbar^2}{2m}\left[3\pi^2\frac{N}{V}\right]^{2/3} \quad \text{where } m \text{ is mass of the electron}$$

$$= \frac{\hbar^2}{2m}\left[\frac{3n}{8\pi}\right]^{2/3} \quad \text{where } n = \frac{N}{V}\ \text{no. of free electron per unit volume}$$

$$= \frac{(6.6\times 10^{-34})^2}{2\times 9.1\times 10^{-31}}\left[\frac{3\times 8.47\times 10^{28}}{8\pi}\right]^{2/3}$$

$$= 11.2\times 10^{-19}\ \text{Joule} = 7.0\ \text{eV}$$

(c) Fermi velocity v_F is given by

$$E_F = \frac{1}{2}mv_F^2$$

$$\text{or} \quad v_F = \sqrt{\frac{2E_F}{m}} = \sqrt{\frac{2\times 11.2\times 10^{-19}}{9.1\times 10^{-31}}} = 1.57\times 10^6\ \text{m/s}$$

(d) The mean free path

$\lambda_F = v_F \tau_F = 1.57 \times 10^6 \times 2.5 \times 10^{-14} = 3.925 \times 10^{-8}\text{m} = 392.5$ Å

Example 16.3: At low temperatures the phonon contribution to the heat capacity can be written as aNT^3, where a is constant. Take the Debye temperature for germanium to the 363K, and calculate the value of a. Then calculate the phonon contribution per unit cell to the heat capacity at 10, 20 and 30K, assuming the Debye approximation holds.

Solution According to Debye approximation, the phonon contribution to heat capacity is given by

$$C_V = \frac{12}{3}\pi^2 kN\frac{T^3}{\theta_D^3}$$

where T is the temperature and θ_D is Debye temperature. We have

$$C_V = aNT^3$$

Comparing the two, one obtains

$$a = \frac{12}{5}\frac{\pi^4 k}{\theta_D^3}(k = 1.38\times10^{-23}\,\text{J/K and } \theta_D = 363\,\text{K})$$

$$= \frac{12}{5}\frac{\pi^4 \times 1.38\times10^{-23}}{(363)^3} = 6.7\times10^{-29}\,\text{J/K}^4$$

(a) At 10 K

The phonon contribution to heat capacity per unit cell is

$$\frac{C_V}{N} = aT^3 = 6.7\times10^{-29}\times(10)^3 = 6.7\times10^{-26}\,\text{J/K}$$

(b) At 20 K

$$\frac{C_V}{N} = aT^3 = 6.7\times10^{-29}\times(20)^3 = 5.36\times10^{-25}\,\text{J/K}$$

(c) At 30 K

$$\frac{C_V}{N} = aT^3 = 6.7\times10^{-29}\times(30)^3 = 1.81\times10^{-24}\,\text{J/K}$$

Example 16.4: Calculate the density of available electronic states within energy range E and $E + dE$.

Solution In the ground state of system of N free electrons, the occupied orbitals may be represented as points inside a sphere is k-space. The energy at the surface of the sphere is the Fermi energy, the wave vector at the Fermi surface have a magnitude k_F such that

$$E_{f(o)} = \frac{\hbar^2 k_F^2}{2m} \qquad \text{(i)}$$

But we know that there is one allowed wave vector for the volume element $\left(\frac{2\pi}{L}\right)^3$ of k-space. Thus in the sphere called Fermi sphere of volume $\left(\frac{4\pi}{3}k_F^3\right)$, the total number of orbitals is given by

$$\frac{2\frac{4\pi}{3}k_F^3}{(2\pi/L)^3} = \frac{V}{3\pi^3}k_F^3 = N \qquad \text{(ii)}$$

where the factor 2 on the left come from the two allowed values of m_s, the spin quantum number of each allowed value of k. We have put number of orbitals equal to N, i.e., the number of electrons and $V = L^3$, the volume of the solid. Thus

$$k_f = \left(\frac{3\pi N}{V}\right)^{1/3} \qquad \text{(iii)}$$

This depends only on the particle concentration and not on the mass

$$E_{F(o)} = \frac{\hbar^2}{2m}\left(\frac{3\pi^2 N}{V}\right)^{1/3} \qquad \text{(iv)}$$

This relates the Fermi energy to the electron concentration $\frac{N}{V}$ and the mass m.

The density of states function $g(E)$ defined from the fact that all the energy states below $E_{F(o)}$ are occupied and this is equal to the total number of electrons, i.e.,

$$\int_0^{E_{F(o)}} g(E)dE = N \tag{v}$$

Substituting the value to N, one obtains

$$\int_0^{E_{F(o)}} g(E)dE = \frac{V}{3\pi^2}\left(\frac{2mE_{F(o)}}{\hbar^2}\right)^{3/2}$$

Expressing the integral as an indefinite integral we have

$$\int g(E)dE = \frac{V}{3\pi^2}\left(\frac{2mE}{\hbar^2}\right)^{3/2}$$

or

$$g(E) = \frac{V}{3\pi^2}\left(\frac{2mE}{\hbar^2}\right)^{3/2} E^{1/2} = CE^{1/2} \tag{vi}$$

where

$$C = \frac{V}{3\pi^2}\left(\frac{2m}{\hbar^2}\right)^{3/2}$$

Example 16.5: Use the Debye approximation and the free electron model to compare phonon and electron contributions to the constant volume heat capacity for potassium at 0.01 and 10K. Polassim has bcc structure with a cube edge of 5.225 Å and its Debye temperature 89K. Each atom contributes 1 electron to the lowest free electron bond. Take the effective mass to be the same as the mass of a free electron.

Solution We have Fermi energy for potassium as

$$E_F = \frac{\hbar^2}{2m}\left[\frac{3n}{8\pi}\right]^{2/3}$$

where n is the no. of free electrons per unit volume. Since potassium has bcc structure hence the volume of unit cell is $\frac{a^3}{2}$, where a is cube length. There are 2 atoms per unit cell and assuming one free electron per atom we have

$$n = \frac{2}{a^3/2} = \frac{4}{a^3}$$

and since $a = 5.225\,A = 5.225 \times 10^{-10}$ m

$$n = \frac{4}{(5.225\times10^{-10})^3} = 2.8\times10^{28}\,\text{m}^{-3}$$

and $$E_F = \frac{6.6\times10^{-34}\times6.6\times10^{-34}}{2\times9.1\times10^{-31}}\times\left(\frac{3\times2.8\times10^{28}}{8\pi}\right)^{2/3} \text{ J}$$

$$= \frac{6.6\times10^{-34}\times6.6\times10^{-34}}{2\times9.1\times10^{-31}\times1.6\times10^{-19}}\times\left(\frac{3\times2.8\times10^{28}}{8\pi}\right)^{2/3} \text{ eV}$$

Electron contribution to heat capacity is given by

$$C_V = \frac{\pi^2 k^2 NT}{2E_F}$$

and phonon contribution is given by

$$C_V = \frac{12}{5}\pi^4 Nk\left(\frac{T}{\theta_D}\right)^3$$

At 0.1 K temperature

$$(C_V)_e = \frac{\pi^2 k^2 NT}{2E_F} = \frac{\pi^2(1.38\times10^{-23})^2\times6.02\times10^{23}\times0.1}{2\times3.34\times1.6\times10^{-19}}$$

Here N is taken as Avogadro number for molar specific heat

$$\therefore \quad (C_V)_e = 1.0587\times10^{-4} = 0.10587 \text{ J/mol-K}$$

and phonon contribution

$$(C_V)_p = \frac{12}{5}\pi^2 Nk\left(\frac{T}{\theta_D}\right)^3$$

$$= \frac{12}{5}\pi^2\times6.02\times10^{23}\times1.38\times10^{-23}\left(\frac{0.1}{89}\right)^3$$

$$= 2.8\times10^{-7} \text{ J/mol-K}$$

At 10K temperature

electron contribution

$$(C_V)_e = \frac{\pi^2(1.38\times10^{-23})^2\times6.02\times10^{23}\times10}{2\times3.34\times1.6\times10^{-19}}$$

$$= 1.06\times10^{-2} \text{ J/mole-K}$$

Phonon contribution

$$(C_V)_p = \frac{12}{5}\pi^2\times6.02\times10^{23}\times1.38\times10^{-23}\times\left(\frac{10}{89}\right)^3$$

$$= 2.8\times10^{-1} = 0.28 \text{ J/mol-K}$$

We see that with the rise in temperature the electron contribution decreases rapidly where as phonon contribution increases.

Example 16.6: Calculate the lattice specific heat capacity of an array of N identical atoms on the basis of Debye approximation.

Solution In case of one dimensional array of N atoms we have the number of modes $g(\omega)$ $d\omega$ in the frequency range $d\omega$ at ω is given by

$$g(\omega)d\omega = \frac{L}{\pi}\frac{d\omega}{d\omega/dk}$$

where L is the length in which atoms are arranged.
Assuming the dispersion relation (Debye approximation)

$$\omega = v_0 k$$

We have density of states as

$$g(\omega) = \frac{L}{\pi v_0}$$

where v_0 is the phase velocity taken to be constant.

Assuming the **Maxwell-Boltzmann** distribution law, the average energy $<\varepsilon>$ of the mode can be taken as

$$<\varepsilon> = \frac{\hbar\omega}{e^{\hbar\omega/kT} - 1}$$

The total energy is then given by

$$u = \varepsilon t \int_0^{\omega_m} <\varepsilon> g(\omega)d\omega = \int_0^{\omega_m} \frac{L}{\pi v_0}\cdot\frac{\hbar\omega}{e^{\hbar\omega/kT} - 1}d\omega$$

$$= \frac{\hbar L}{\pi v_0}\int_0^{\omega_m} \frac{\omega}{e^{\hbar\omega/kT-1}}d\omega \qquad \text{(i)}$$

where ω_m is defined as

$$\int_0^{\omega_m} \frac{L}{\pi v_0} d\omega = N$$

or $$\omega_m = \frac{L\pi v_0}{L}$$

Since there are N identical atoms in any array, hence the total number of modes will be N.

Let $$\frac{\hbar\omega}{kT} = x \;\therefore\; x_m = \frac{\hbar\omega_m}{kT} = \frac{\hbar}{kT}\left[\frac{N\pi v_0}{L}\right] = \frac{\theta_D}{T}$$

where parameter θ_D is Debye temperature.
Eq. (i) then becomes

$$u = \frac{Lk^2T^2}{\pi v_0 \hbar}\int_0^{\theta_D/T} \frac{x}{e^x - 1}dx$$

Solving the integral under approximation $\frac{\theta_D}{T} \to \infty$, i.e., for low temperature using contour method, we have

$$\int_0^\infty \frac{x}{e^x - 1} dx = 1.51$$

$$u = \frac{Lk^2T^2}{\pi v_0 \hbar} = 1.51$$

and
$$C_V = \frac{3.02Lk^2}{\pi v_0 \hbar} T$$

Thus at low temperature the specific heat is proportional to temperature.
For **high temperature** $e^x \approx 1 + x$ leaving higher terms, we obtain

$$u = \frac{Lk^2T^2}{\pi v_0 \hbar} \int_0^{\theta_D/T} \frac{x}{x} dx = \frac{Lk^2T^2}{\pi v_0 \hbar} \cdot \frac{\theta_D}{T} = \frac{Lk^2T}{\pi v_0 \hbar}$$

Obviously, i.e., C_V is independent of temperature.

Example 16.7: In a three dimensional sample with one atom in the basis and approximating the dispersion relation by $\omega = vk$, where v is the speed of sound (which is same for all directions of k). Also assume that all the branches of the spectrum have angular frequency ω_m: (a) show that the total number of phonons is given by

$$n_T = \frac{3V}{2\pi^2 v^3} \left[\frac{k_B T}{\hbar}\right]^3 \int_0^{x_m} \frac{x^2 dx}{e^x - 1}$$

where $x_m = \frac{\hbar \omega_m}{k_B T}$ (b) show that for $k_B T \ll \hbar\omega$ the total number of phonons is proportional to T^3.

Solution If $< E >$ is the average energy of the modes and $\hbar\omega$ the energy of the mode, then number of phonons in the mode of frequency ω are

$$dn = \frac{< E >}{\hbar \omega} = \frac{< E >}{\hbar v k} \quad \text{phonons per mode}$$

But we have
$$<E> = \frac{\hbar \omega}{e^{\hbar\omega/kT} - 1}$$

∴
$$dn = \frac{1}{e^{\hbar\omega/kT} - 1} \quad \text{phonons per mode}$$

But we know that the number of modes within frequency range $d\omega$ at ω is

$$g(\omega) d\omega = \frac{3V\omega^2}{2\pi^2 v^3} d\omega$$

So that total number of phonons will be

$$n = \int_0^{\omega_m} (\text{Phonons per mode})(\text{no. of modes per frequency})$$

or
$$n = \int_0^{\omega_m} \frac{1}{e^{\hbar\omega/k_B T} - 1}\left(\frac{3V\omega^2}{2\pi^2 v^3}\right) d\omega$$

$$= \frac{3V}{2\pi^2 v^3}\int_0^{\omega_m} \frac{\omega^2 d\omega}{e^{\hbar\omega/k_B T} - 1}$$

Let us put $\frac{\hbar\omega}{k_B T}$ then $\omega = \frac{k_B T}{\hbar} x$ and $d\omega = \frac{k_B T}{\hbar} dx$ and $\omega_m = \frac{k_B T}{\hbar} x_m$

$\therefore$
$$n = \frac{3V}{2\pi^2 v^3}\int_0^{x_m}\left(\frac{k_B T}{\hbar}\right)^2 x^2 \left(\frac{k_B T}{\hbar}\right) dx . \frac{1}{e^x - 1}$$

$$= \frac{3V}{2\pi^2 v^3}\left(\frac{k_B T}{\hbar}\right)^3 \int_0^{x_m} \frac{x^2}{e^x - 1} dx$$

(b) For $k_B T << \hbar\omega$ the limit $x_m \to \infty$ and the integral becomes[8]

$$\int_0^{x_m} \frac{x^2}{e^x - 1} dx = \int_0^{\infty} \frac{x^2}{e^x - 1} dx = \sum_{n=1}^{\infty} \frac{1}{n^2}$$

But
$$\sum_{n=1}^{\infty} \frac{1}{n^2} = 1.51$$

$\therefore$
$$n = \frac{3V}{2\pi v^3}(1.51)\left[\frac{k_B T}{\hbar}\right]^3$$

Moreover
$$\omega_m^3 = v^3 \frac{6\pi N}{V}$$

$\therefore$
$$n = 9N\left[\frac{k_B T}{\hbar\omega_m}\right] + 1.51$$

$$= 13.59N(T/\theta_0)^3, \quad \text{Where } \theta_0 = \frac{\hbar\omega_m}{k_B} \text{ is Debye temperature}$$

i.e.
$$n \propto T^3$$

[8]Such type of integrals are solved by using Remain Zeta functions in which we have

$$\int_0^{\infty} \frac{x^n}{e^x - 1} dx = (n-1)\sum_{s=1}^{\varepsilon} \frac{1}{n^s} = \sqrt{n}\xi(n)$$

Example 16.8: A system consists of 10^{23} atoms acting as simple harmonic oscillators each having a frequency 10^{12} Hz. Ignoring the zero point energy show that the mean energy of the system at 20 K is 6.6 K.

Solution The mean energy of the system

$$<E> = \frac{n\hbar\omega}{e^{\hbar\omega}/k_BT}$$

$$n = 10^{23},\ \hbar = 1.05\times10^{-34},\ \omega = 2\pi\times10^{12},\ k_B = 1.38\times10^{-23}\ T = 20\text{K}$$

Substituting these values, one obtains

$$<E> = \frac{10^{23}\times1.05\times10^{-34}\times10^{12}\times2\pi}{[e^{2\pi\times1.05\times10^{-34}\times10^{12}/1.38\times10^{-23}\times20}-1]} = 6.6\text{ J}$$

Example 16.9: The Fermi velocity of the electron in a metal is 0.86×10^6 m/s. Determine the Fermi energy and the Fermi temperature [MDS, Ajmer]

Solution Fermi energy, $E_F = \frac{1}{2}mv_F^2$

$$= \frac{1}{2}\times\frac{9.1\times10^{-31}\times(0.86)^2\times10^{12}}{1.6\times10^{-19}}\text{eV} = 2.1\text{ eV}$$

Fermi temperature, $T_F = \frac{E_F}{k_B} = \frac{2.1\times1.6\times10^{-19}}{1.38\times10^{-23}} = 2.43\times10^4\text{K}$

Example 16.10: Taking Fermi energy as 7.04 eV determine the electronic specific heat of copper [Raj]

Solution

$$C_V^{el} = \frac{\pi^2k_B^2T}{2E_F}N_A$$

$$= \frac{(3.14)^2\times(1.38\times10^{-23})^2\times600\times6.02\times10^{26}}{2\times7.04\times1.6\times10^{-19}}$$

$$= 3\times10^2\text{ J k mol/K}$$

Example 16.11: Assuming the longitudional and the transverse velocities to be equal, using following data show that Debye temperature for diamond is 2780K. Atomic weight of diamond = 12, Young's modulus = 10^{12} N/m^2 and density = 3500 kg/m^3.

Solution Debye temperature,

$$\theta_D = \frac{h\nu}{k_B}\left[\frac{9N}{12\pi v}\right]^{1/3} = \frac{h}{k_B}\sqrt{\left[\frac{Y}{\rho}\right]}\left[\frac{3\rho}{4\pi m}\right]^{1/3}$$

$$= \frac{6.62\times10^{-34}}{1.38\times10^{-23}}\left[\frac{10^{12}}{3500}\right]^{1/2}\left[\frac{3\times3500}{4\times3.14\times1.67\times10^{-27}}\right] = 2780\text{ K}$$

SUMMARY

1. Phonon is a quantum of crystal lattice vibrational energy having a magnitude $h\nu$, where h is Planck's constant and ν is the frequency of the vibration.
2. Phonons are analogous to the quanta of light, i.e. photons.
3. The concept of phonons is useful in the treatment of the thermal conductivity of nonmetallic solids and, through consideration of electron-phonon interactions, the temperature dependence of the electrical conductivity of metals and understanding the phenomenon of superconductivity.
4. Phonons are indistinguishable particles and obey BE statistics.
5. There are two types of phonons: (i) acoustical phonons, (ii) optical phonons, which are associated with acoustical and optical modes of vibration respectively.
6. When a lattice vibration (phonon) is involved in an interaction, not only is its energy always an integral multiple of $\hbar\omega$, but also, at a time, only a single phonon is absorbed or excited. We may remember that photon interaction may be viewed as a scattering collision.
7. Dulong and Petit's law

$$C_V = \frac{dU}{dT} = 3R$$

This law is found to be obeyed by heavy elements (At. Wt. > 40) at high temperature. However, at low temperatures, C_v is found to decrease with temperature and finally reaches zero at 0 K.

8. In Einstein model, the atoms of a solid are considered as identical oscillators vibrating independently with the same frequency ν_E. The molar specific heat

$$C_V = 3R\left(\frac{\theta_E}{T}\right)^2 \frac{\exp(\theta_E/T)}{[\exp(\theta_E/T)-1]^2}$$

where $\theta_E = {h\nu_E}/{k}$ is *Einstein temperature* At higher temperature, $T >> \theta_E$ and $C_v = 3R$ which agrees with classical result. At low temperatures, $T << \theta_E$ and therefore

$$C_V = 3R\left(\frac{\theta_E}{T}\right)^2 \exp\left(-\frac{\theta_E}{T}\right)$$

i.e., C_v drops to zero exponentially as $T \to 0$. However, experiments reveal that C_v approaches zero as T^3. The decrease predicted by above equation is very rapid than that obtained by the experimental observations.

9. *Debye Model* is based on the assumption that the atoms do not vibrate independently with a constant frequency as suggested by Einstein. A solid of N atoms has $3N$ modes. All these modes have different frequencies up to a maximum frequency (ν_o) instead of a single frequency in Einstein's model. Debye obtained

$$C_V = \frac{12}{5}\pi^4 Nk\left({T}/{\theta_D}\right)^3$$

which is known as *Debye T^3 law*. This found in good agreement with experimental observations.

REVIEW QUESTIONS AND PROBLEMS

1. What is the Einstein's model for the phonon frequency distribution. Use this model to calculate the internal energy per mole for a solid. Show that the specific heat varies exponentially with $-1/T$ in this model.
2. What is the observed dependence of the lattice specific heat at low temperatures? Deduce the temperature dependence of the specific heat predicted by Einstein's model. Does the Einstein's model explain the observed behaviour of the specific heat at very low temperatures? What is the nature of agreement between Einstein's model and experimental results at high temperatures and why?
3. how does the Debye model differ from Einstein's model. Mention essential differences in the model and explain the consequences of these differences.
4. Derive an expression for the specific heat of solids following Einstein's model. How does specific heat depend on temperature and to what extent does this model agree with experimental results?
5. Derive an expression for the specific heat of a solid on the basis of the Einstein's theory. Clearly outline the assumptions of the theory. Discuss the variation of specific heat with temperature?
6. Show that the heat capacity of mono-atomic lattice in one dimension in Debye approximation is proportional to T/θ for the low temperatures ($T<<\theta$), where θ is the effective Debye temperature in one dimension. What is the corresponding dependence in three-dimensions?
7. Describe the vibrations of one dimensional monoatomic linear lattice.
8. Discuss the normal modes of vibration in a finite lengths of the lattice.
9. Describe the vibrations of linear diatomic lattice.
10. Explain particle displacement in two branches in case of Linear diatomic lattice.
11. Discuss number of possible normal modes of vibration in a bar in case of linear diatomic lattice.
12. Calculate specific heat of a free electron gas by using Fermi Dirac distribution. Explain the agreement of experimental results with theoretical results.
13. Discuss the reasons for the failure of Dulong and Petit's law to predict specific heats at low temperatures. Why should the law is valid at high temperatures?
14. Discuss the salient features of Debye's theory of specific heats and show how far it agrees with the experimental values.
15. Find an expression for the specific heat of a solid on the Einstein model and show that it converges to $3R$ at high temperatures.
16. Show that the electronic specific heat of a conductor is very small compared to the lattice specific heat at room temperature. Calculate the value of electronic specific heal C_V.
17. Calculate the ratio of electron to lattice heat capacity for C_V at $T = 0.3$K, 4K, 20K, 77K and 300K.
18. Define thermal conductivity. Derive an expression for the thermal conductivity in metals.
19. Discuss the variation of thermal conductivity of metals with temperature. Show how the thermal conductivity of metals differs from that of insulators.

20. Unlike the number of electrons, the number of phonons in a solid is not conserved as the temperatures changes. Consider a three-dimensional sample with one atom in the basis and approximate the dispersion relationship by $\omega = vk$, where v is the speed of sound. To simplify the calculations, assume the speed of sound is the same for all directions of k and assume all three branches of the spectrum have the same maximum angular frequency ω_m.

(a) Show that the total number of phonons given by

$$n_T = \frac{3V}{2\pi^2 v^3}\left[\frac{k_B T}{\hbar}\right]^3 \int_0^{x_m} \frac{x^2\,dx}{e^x - 1} \text{ where } x_m = \frac{\hbar\omega_m}{k_B T}$$

(b) Show that for $k_B T << \hbar\omega$ the total number of phonons is proportions to T^3.

21. Strontium has fcc structure with cube edge a = 6.08A. Each atom contributes 2 electrons to nearly free electron bonds. Assume the electrons to be absolutely free. Calculate (a) radius of the Fermi sphere (b) The fermi energy.

[**Ans.** $E_F = 109$ eV, $K_F = 5.38 \times 10^{10}\,\text{m}^{-1}$]

22. For potassium estimate the temperature at which the phonon and electron contributions to the constant heat capacity are equal. Debye temperature for potassium is 91K and Fermi energy is 3.34 eV. [**Ans.** 0.64K]

23. Find the expression for the mean energy of the atomic oscillator in the case of gas and in the case of solid. Hence find expressions for the specific heat of a gas and a solid.

SHORT ANSWER-QUESTIONS

1. What is a phonon?

Ans: Phonon is a quantum of sound, the smallest energy of a vibration corresponding to a sound wave. A phonon is analogous to a photon or light quantum. We can say that a phonon is a packet of sound waves which behaves like a mass less particle with energy $h\nu$ and momentum h/λ, where ν is the frequency of vibration and λ is the wavelength.

2. Is Dulong-Petit's law is quite accurate at room temperature?

Ans: Yes.

3. What is the basic assumption of Debye theory?

Ans: Debye assumed that a crystalline solid can be represented by an isotropic elastic continuum. Moreover, the atoms in a solid do not vibrate with same frequency, but have some dispersion. As there are finite number of atoms, however, not all values of the frequency are allowed. There is a maximum allowed frequency ω_D that any atom can have. In general, the total number of modes of vibration per gm atom is equal $3N$.

4. What is Sommerfeld's theory of metals?

Ans: Sommerfeld's theory shows that the conduction electrons can make an important contribution to the heat capacity. According to Sommerfeld, there must be a linear term in the temperature included in the expression for C_V in order to account for the electron contribution, i.e.,

$$C_V = \gamma T + (12\pi^4 R/S)(T/\Theta)^3$$

γ is sometimes called **Sommerfeld gamma.**

5. What is the classical value of molar lattice specific heat?
Ans: 3 R_H.
6. What do you understand by electronic specific heat?
Ans: This is the contribution to the total specific heat due to transitions of electrons to state of higher energy.
7. What is strange in classical, Einstein and Debye theories of specific heats of solids?
Ans: These theories apply with equal degrees of success to both metals and non-metals, which is strange because they ignore presence of free electrons in metals.

OBJECTIVE QUESTIONS

1. The classical expression for the mean energy of a one dimensional oscillators is

(a) $\left(\frac{3}{k_BT}+\frac{2}{k_BT}\right)$ (b) $\left(\frac{k_BT}{2}+\frac{k_BT}{2}\right)$

(c) $\left(\frac{3}{2}k_BT\right)$ (d) $\left(\frac{1}{2}k_BT\right)$

2. In a Phonon gas

(a) Phonon do not collide with each other
(b) there is no conservation of energy in the collisions of phonons
(c) number of phonons are conserved in collisions
(d) phonons collide with the surface of crystals and also with impurity atoms

3. The value of molar specific heat according to classical theory is

(a) 3*R* (b) *R*/2
(c) *R* (d) 3*R*

4. The total molar specific heat of a metal at temperatures above Debye temperature θ_D is

(a) $\left(3R+\frac{3}{2}R\right)$ (b) $\left(\frac{3}{2}R+\frac{3}{2}R\right)$

(c) $\left(\frac{1}{2}R+\frac{1}{2}R\right)$ (d) $\left(R+\frac{R}{2}\right)$

5. The zero point energy of a solid on the Debye model is

(a) $\frac{3}{5}R\theta_D$ (b) $\frac{1}{2}R\theta_D$

(c) $\frac{9}{8}R\theta_D$ (d) $\frac{3}{4}R\theta_D$

6. At low temperatures the lattice specific heat varies as

(a) T (b) T^{-1}

(c) T^2 (d) T^3

7. The discrete values of energy for the harmonic oscillator can have

(a) $n\hbar\omega$ (b) $n\hbar\omega/2$

(c) $2n\hbar\omega$ (d) $(n\hbar\omega)^2$

8. According to Einstein's model of a solid, the molar specific heat at constant volume is

(a) $C_V = \frac{3}{2}R\left(\frac{h\nu}{k_BT}\right)^2\left[\frac{e^{h\nu/k_BT}}{(e^{h\nu/k_BT}-1)^2}\right]$ (b) $C_V = 3R\left(\frac{h\nu}{k_BT}\right)^2\left[\frac{e^{h\nu/k_BT}}{2}\right]$

(c) $C_V = 3R\left(\frac{h\nu}{k_BT}\right)^2\left[\frac{e^{h\nu/k_BT}}{(e^{h\nu/k_BT}-1)^2}\right]$ (d) $C_V = \frac{3}{2}R\,(h\nu/k_BT)^2$

ANSWERS

Question Numbers	1	2	3	4	5	6	7	8
Answers	(b)	(d)	(c)	(a)	(c)	(a)	(a)	(c)

FILL IN THE BLANKS

1. Phonons are _______ particles and obey _______ statistics.
[indistinguishable, Base-Einstein]

2. There are two types of phonons _______ and _______. **[acoustical, optical]**

3. When a lattice vibration is involved in an interaction, not only is its energy always an integral multiple of _______, but also, at a time, only a single _______ is absorbed or excited. **[$\hbar\omega$, phonon]**

4. In analogy with a quantum of electro-magnetic wave, i.e. photon, an energy quantum of lattice vibration is called a _______. **[Phonon]**

5. It is practically observed that the lattice contribution to heat capacity at zero temperature is zero. This is only possible when lattice waves is _______. **[quantised]**

6. In general, the specific heat of a solid is expressed as : $C_{\text{solid}} =$ _______.
[$C_{\text{lattice}} + C_{\text{electronic}}$]

7. According to Dulong and Petit's law, the molar heat capacity of the solids is _______, i.e. independent of _______ and _______. **[constant, temperature, frequency]**

8. According to Einstein's theory of specific heat of solids, a crystal consists of atoms which may regarded as _______ and independent _______.
[identical, harmonic oscillators]

9. There is no quantitative agreement between Einstein's theory of specific heat of solids and experiment at _______. **[low temperatures]**

10. In Debye's theory of lattice heat capacity of solids, at very low temperatures, $C_v \propto T^3$. This is known as Debye's _______. According to this, $C_v \rightarrow 0$, as _______. **[T^3 law, $T \rightarrow 0$]**

11. Debye's model of lattice heat capacity is valid for _______ only. **[long wavelengths (low frequencies)]**

12. The Einstein frequency for copper $\nu_E = 4.8 \times 10^{12}$ Hz. Einstein temperature for copper is _______. **[229.6 K]**

$$\left(\textbf{Hint}: \theta_E = \frac{h\nu_E}{k_B}\right)$$

13. Debye's model of lattice heat capacity cannot be applied to crystals comprised of _______ than one type of _______. **[more, atoms]**

14. There is no quantitative agreement between Einstein's theory of specific heat of solids and experiment at _______ temperatures. The discrepancy is due to oversimplified assumption of the Einstein's theory in which the atomic oscillators have been considered to vibrate _______ at the same frequency. **[low, independently]**

15. In Debye's model of lattice specific heat capacity, it is assumed that the total number of vibrational modes to be $3N$. But, the solid is an elastic _______ and should possess _______ frequencies. **[continuum, infinite]**

17

Free Electron Theory of Metals

17.1 INTRODUCTION

Metals are one of the important class of solids and two-third of elements are solids. Obviously, the natural tendency of elements is to favour the metallic state. From the begining of twentieth century several attempts have been made to model a metal from the begining of the 20th century. The striking physical properties of metals make them so attractive that a successful theory should emerge as natural consequence. In the most simplified model, it is assumed that outermost electrons in a metal behave like free electrons, e.g., the electronic configuration of copper and magnesium are $1s^2\ 2s^2\ 2p^6\ 3s^2\ 3p^6\ 3d^{10}\ 4s^1$ and $1s^2\ 2s^2\ 2p^6\ 2s^2$ respectively and the outermost orbit electrons $4s^1$ and $2s^2$ respectively are assumed to behave as free electrons. We may note that in this free electron model the movement of the free electrons is considered to be independent of the details of the motion of the nuclei and the 'core' electrons in the metal. Now, we study the various theories based on free electron model of metals.

17.2 DRUDE-LORENTZ'S CLASSICAL THEORY (FREE ELECTRON GAS MODEL)

Drude in 1900, postulated a free electron gas model essentially based on the classical kinetic theory of gases. According to this model the metal consists of positive ion cores with the valence electrons moving freely among these cores. The electrons are, however, bound to move within the metal due to electrostatic attraction between the positive ion cores and electrons. Drude's theory is mainly based on the following assumptions:

(i) The electron-electron and electron-positive ion interactions are neglected. However, the electrons remain confined within the metal. This is only possible if the electron-positive ion attractive force is active.

(ii) Electrons move opposite to the external electric direction and make collisions with immobile and impenetrable positive ion cores. However, an électron bumps, (i.e., the collision involving physical contact unrealistic) from positive ion to positive ion and between successive collisions moves in a straight line in accordance with the Newton's laws of motion.

(iii) All free electrons in a metal move with *root mean square* (RMS) speed of Maxwell-Boltzmann distribution representing the random or thermal velocity of electrons. Immediately after the collision, the average electron velocity is zero.

Lorentz in 1909 postulated that free electrons in a metal obey Maxwell-Boltzmann statistics under equilibrium conditions. The combined models of Drude and Lorentz constitute the *Drude-Lorentz theory*. This is also called as classical theory as it is based on Maxwell-Boltzmann classical statistics.

17.2.1 DC Electrical Conductivity of Metals

Let us apply Drude's theory to derive an expression for dc conductivity. During the time interval between two successive collisions an external static electric field can affect electron velocity. Since the average velocity immediately after the collision is zero and hence the gain in velocity is destroyed each time a collision occurs. One finds that a large influence of electric field is reflected in a larger mean free time or relaxation time τ. $1/\tau$ is the probability of collision per unit time. Obviously, the probability that a collision occurs in a small time interval dt is simply dt/τ.

The acceleration of electron of mass m can be taken as eE/m. We obtain the mean drift velocity

$$v_d = \left(\frac{-eE}{m}\right)\tau \tag{17.1}$$

Let there be a total of n number of electrons per m^3 in the metal, all with constant drift velocity v_d. The relation for the net electric current density is

$$\begin{aligned} J &= -ne\,v_d = \left(\frac{ne^2\tau}{m}\right)E \\ &= \sigma E \end{aligned} \tag{17.2}$$

which is Ohm's law, where the positive scalar quantity σ is given by

$$\sigma = \frac{ne^2\tau}{m} \tag{17.3}$$

σ is defined as the electrical conductivity and it is reciprocal of electrical resistivity ρ. We may note that the form of the relation Eq. (17.3) remains the same in all models including those obtained from quantum mechanics. The difference in different modes lies only in the way n, m and τ are defined.

We may note that the electrical conductivity σ is not universally scalar since in some complicated situations $\boldsymbol{J}$ becomes non-linear with respect to $\boldsymbol{E}$ making σ to behave as a *tensor*. Usually the electrical conductivity is expressed in terms of the drift mobility of electron μ as

$$\sigma = ne\mu \tag{17.4}$$

Where
$$|\mu| = \frac{v_d}{E} \tag{17.5}$$

By exploring the ideas of kinetic theory such as to define τ in terms of thermal velocity v_{th}, Drude obtained the following relation.

$$\tau = \frac{\wedge}{v_{th}} \tag{17.6}$$

$$\frac{3}{2}k_B T = \frac{1}{2}mv_{rms}^2 = \frac{1}{2}mv_{th}^2 \tag{17.7}$$

where $\wedge$ is the electron mean free path. Using these relations, one can write Eq. (17.3) as

$$\sigma = \frac{ne^2 A}{(3k_B mT)^{1/2}} \tag{17.8}$$

We find that Eq. (17.8) is another achievement of Drude theory since it gives the right magnitude of electrical conductivity and moreover correctly describes that conductivity increases with decrease in temperature.

17.2.2 Thermal Conductivity of Metals

It is assumed in Drude's model of metals that the major contribution to thermal conductivity of metals comes from conduction electrons. Obviously, this assumption has basis in the general observation that metals are far better conductors of heat than insulators. However, in Drude's model phonon contribution is neglected on account of its relatively much small measure.

Drude obtained the following expression for thermal conductivity of electrons

$$k_{el} = \frac{3}{2}\left(\frac{nk_B^2\tau}{m}\right)T \tag{17.9}$$

Where
$$C_{el} = \frac{3}{2}nk_B$$

In 1835 Wiedemann and Franz were able to see that all good conductors of electricity are good conductors of heat too. They have shown that the ratio of thermal conductivity to electrical conductivity is proportional to the temperature for a large number of metals.

Later scientists found that the proportionality constant equalling $k/\sigma T$ has almost the same value for most of the common metals. This constant is usually called as the *Lorentz number* and the empirical law became famous as the *Wiedemann-Franz law.* From Eqs. (17.3) and (17.9), Drude obtained

$$\frac{K}{\sigma} = \frac{3}{2}\left(\frac{k_B}{e}\right)^2 T \tag{17.10}$$

The coefficient of T in Eq. (17.10) is a constant. Obviously, Drude was able to derive the Wiedmann-Franz law. This was acknowledged as the biggest success of the Drude model. The Lorentz number L is defined as

$$L = \frac{k}{\sigma T} = \frac{3}{2}\left(\frac{k_B}{e}\right)^2 \tag{17.11}$$

$$= 1.11 \times 10^{-8}\, \text{W}\Omega\text{K}^{-2}$$

$$= 1.11 \times 10^{-8}\, (\text{VK}^{-1})^2$$

But the value of L so obtained is about half the experimental value. Although, Drude's model describes correctly the Wiedmann-Franz law but this disagreement indicates that Drude's model is lacking in some respects.

17.2.3 Lorentz Modification of the Drude's Theory

Lorentz modified the oversimplified Drude's model on the basis of the following points:

(i) Lorentz abandoned the assumption that all electrons move with the same thermal velocity

(ii) By the presence of an electric or thermal gradient the classical Maxwell-Boltzmann law of distribution of velocity is perturbed. Both of these tend to displace the equilibrium velocity distribution and also distort its symmetry.

(iii) To describe the transport of charge and kinetic energy of electrons by a statistical distribution of mobile electrons constituting the electron gas, the approach of Boltzmann transport equation is followed.

On the basis to above modifications, Lorentz obtained the following expression for electrical conductivity:

$$\sigma_L = \left(\frac{8}{3\pi}\right)^{1/2} \frac{ne^2 \wedge}{(3mk_BT)^{1/2}} \tag{17.12}$$

If $\sigma_B = \dfrac{ne^2 \wedge}{(3mk_BT)^{1/2}}$ denote the Drude conductivity, then

$$\sigma_D = \left(\frac{3\pi}{8}\right)^{1/2} \sigma_L = 1.09\sigma_L \tag{17.13}$$

From Eq. (17.13) it is obvious that the Lorentz modifications provide neither any noteworthy change in the quantitative measure and nor any change in the temperature dependence of electrical conductivity. Lorentz also analyzed the effect of a uniform magnetic field on current carrying conductors. This effect is known as *Hall effect.* We know that the net force on an electron moving with velocity $\boldsymbol{v}$ under the action of static electric field $\boldsymbol{E}$ and a uniform magnetic field $\boldsymbol{B}$ is

$$\boldsymbol{F} = -e[\boldsymbol{E} + \boldsymbol{v} \times \boldsymbol{B}] \tag{17.14}$$

The force $\boldsymbol{F}$ is called Lorentz force.

Lorentz also solved the Boltzmann transport equation under the above conditions and for the Hall coefficient obtained the following expression.

$$R_H = -\left(\frac{3\pi}{8}\right)\frac{1}{ne} \tag{17.15}$$

By feeding the measured value of R_H, one can estimate the electron density n. The estimate is found consistent with the contribution of one conduction electron per atom in sodium, where as not so good for silver and there is no agreement for other metals. This clearly indicates that free electron approximation is best applicable to alkali metals with the lone loosely bound outermost $3s^1$ electron. We may note that subject is best dealt by the band theory of solids, which also accounts for the observed positive Hall coefficient in certain metals.

One finds other important consequences of the Lorentz solution of Boltzmann transport equation as:

(i) **Thomson effect:** An electromotive force is developed in conductor when a temperature gradient is maintained between its two ends.

(ii) **Magnetoresistance effect:** When a metal is placed in a magnetic field, the resistivity of metal increases. We may note that Drude model does not predict this.

Drude-Lorentz theory met with a number of failures. This theory correctly predicted the room temperature resistivity of various metals but the temperature dependence of resistivity could not be established correctly. This theory predicted that resistivity of metal varies as $T^{1/2}$ where as actually it varies linearly with T. Moreover, this theory also failed to explain the heat capacity and paramagnetic susceptibility of the conduction electrons. We may note that these properties are based on interaction of free electrons with external sources of energy which may be thermal or magnetic in nature. The application of MB statistics to Drude-Lorentz theory allows all the free electrons to gain energy which results in much higher values of heat capacity and paramagnetic susceptibility. However, the use of FD statistics or quantum statistics allows only a fraction of the total number of free electrons to gain energy and the values of heat capacity and paramagnetic susceptibility thus obtained agrees with the observed values. We may note that these are the principal achievements of *Sommerfeld's quantum theory*. Drude-Lorentz theory is also unable to account for the occurence of long mean free path (~10^8 to 10^9 interatomic spacings or greater than 1 cm) at low temperatures.

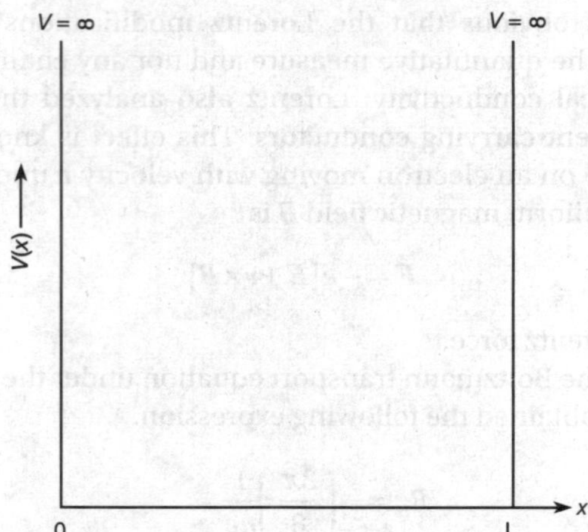

Fig. 17.1 One-dimensional potential box bounded by infinite potential ($V = \infty$) barriers

17.3 SOMMERFELD'S QUANTUM MODEL

Considering the free electron model of Drude-Lorentz classical theory, Sommerfeld treated the problem quantum mechanically using FD statistics in place of classical MB statistics. The possible electronic energy states in the potential energy box and the distribution of electrons in these energy states are then determined with the help of quantum statistics. One can study the one-dimensional and three-dimensional cases separately.

(i) Free Electron gas in one-dimensional case

Let us consider an electron of mass m bounded to move in a one-dimensional crystal of length L Fig. (17.1). This electron is prevented from leaving the crystal due to large potential energy barrier ($V = \infty$) at its surfaces. Although the potential energy barriers extend over a few atomic layers near the surface, we have taken them infinitely large for the sake of simplicity. We can see that this problem is identical to that of an electron moving in a one-dimensional potential box which is bounded by ($V = \infty$) barriers as shown in Fig. (17.1). Within the box or crystal, the crystal energy is assumed to be zero. The boundary conditions for the wave function are:

$$V(x) = 0 \quad \text{for} \quad 0 < x < L$$
$$V(x) = \infty \quad \text{for} \quad x \leq 0 \quad \text{and} \quad x \geq L \tag{17.16}$$

One can determine the wave function ψ_n of the electron occupying the nth state from the solution of the Schrodinger's equation, i.e.,

$$\frac{d^2\psi_n}{dx^2} + \frac{2m}{\hbar^2}(E_n - V)\psi_n = 0 \tag{17.17}$$

where E_n reprsents the kinetic energy of the electron in the nth state and V is the potential energy. Within the box, $V = 0$ and therefore Eq. (17.17) takes the form

$$\frac{d^2\psi_n}{dx^2} + \frac{2m}{\hbar^2}E_n\psi_n = 0 \tag{17.18}$$

The general solution of Eq. (17.18) is

$$\psi_n(x) = A\sin kx + B\cos kx \tag{17.19}$$

Where

$$k = \sqrt{\frac{2mE_n}{\hbar}} \tag{17.20}$$

and A and B are constants to be determined from boundary conditions. We have the boundary conditions

$$\psi_n(0) = 0 \quad \text{and} \quad \psi_n(L) = 0 \tag{17.21}$$

We may note that these boundary conditions are based on the fact that at $x = 0$ and L, $V \to \infty$ and $V(x)\psi_n(x)$ in Eq.(17.17) also approaches ∞. For wave function $\psi_n(x)$ may be continuous, the kinetic energy E_n must also become infinite which is not feasible. Obviously, $\psi_n(x)$ must vanish for $x = 0$ and L.
For $x = 0$, Eq. (17.19) yields $B = 0$ and Eq. (17.19) reduces

$$\psi_n(x) = A\sin kx \tag{17.22}$$

For $\psi_n(L) = 0$, Eq. (17.22) gives

$$\sin kL = 0$$

or

$$k = n\pi/L \tag{17.23}$$

where $n = 1, 2, 3$ .... For the allowed wave function, Eq. (17.22) reduces

$$\psi_n(x) = A\sin\left(\frac{n\pi}{L}x\right) \tag{17.24}$$

One can obtain the allowed energy values from Eqs. (17.20) and (17.23) as

$$E_n = \frac{\hbar^2}{2m}\left(\frac{n\pi}{L}\right)^2 = \frac{n^2\hbar^2}{8mL^2} \tag{17.25}$$

or

$$E_n \propto n^2 \tag{17.26}$$

From Eqs. (17.24) and (17.25) we can see that the allowed wave functions $\psi_n(x)$ and the allowed energy values E_n exist only for integral values of n. n is called the quantum number. This means that the energy spectrum consists of discrete energy levels where the spacing between the levels is determined by the values of n and L. Moreover it decreases with increase in L. We may note that if n is of the order of a few centimeters, the energy levels form almost a continuum. However, if L has atomic dimensions, the spacing between the levels becomes appreciable. Fig. (17.2) shows E_n vs n plot.

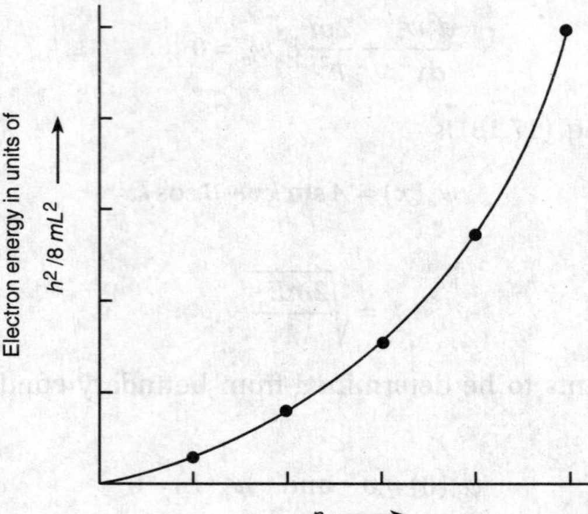

Fig. 17.2 For one-dimensional crystal, E_n vs n curve

One can determine the constant A in Eq. (17.24) by using the condition that the probability of finding an electron somewhere on the line is unity, i.e.,

$$\int_0^L \psi_n^*(x)\psi_n(x)\,dx = 1$$

Using Eq. (17.24), one obtains

$$A^2\int_0^L \sin^2\left(\frac{n\pi}{L}x\right)dx = 1$$

or
$$\frac{A^2}{2}\int_0^L\left[1-\cos\left(\frac{2n\pi}{L}x\right)\right]dx = 1$$

or
$$\frac{A^2}{2}\int_0^L dx = 1$$

or
$$A = \sqrt{\frac{2}{L}}$$

Substituting the above value of A in Eq. (17.24), one obtains

$$\psi_n(x) = \sqrt{\frac{2}{L}}\sin\left(\frac{n\pi}{L}x\right) \tag{17.27}$$

Fig. (17.3) shows the energy levels and the wave functions corresponding to $n = 1, 2, 3, 4, \ldots$.

(ii) Fermi energy

In accordance with the Pauli's exclusion principle which states that no two electrons can have all their quantum numbers identical, i.e., each orbital or quantum state can

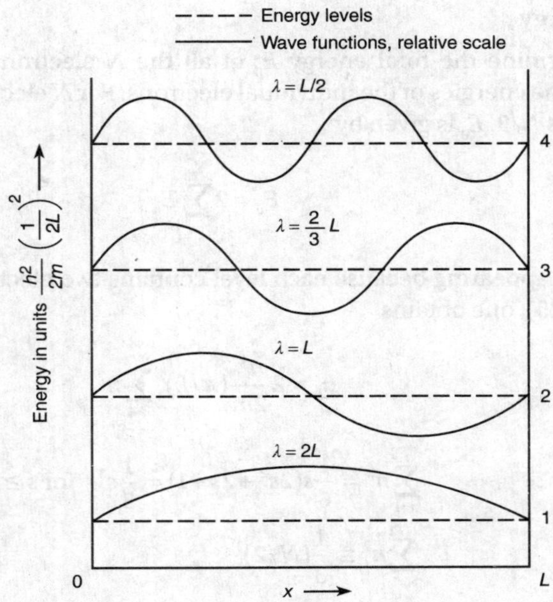

Fig. 17.3 Eigen values and eigen functions of first four states of a free electron confined in a potential box of length L

be occupied by at the most one elecrron, the electrons are distributed among the various possible energy levels. However, the different states may have different values (non-degenerate case) or the same value (degenerate case) of energy. We know that in a solid, an electron in a conduction electronic state has the quantum numbers n and m_s, where n is the principal quantum number. Each set of values of quantum numbers n and m_s is the magnetic spin quantum number and m_s define a quantum state and for each value of n, m_s can have two possible values +1/2 and −1/2. Obviously, each energy level defined by quantum number n can have two quantum states and thus accomodate a maximum of two electrons, one with spin up and the other with spin down, i.e., each energy level is doubly degenerate. The topmost filled energy level at 0K is known as the *Fermi level* and the energy corresponding to the Fermi level is denoted by E_F and is called *Fermi energy*.

If there are N total number of electrons to be accommodated on the line then, for even n, one can write

$$N = 2n_F \tag{17.28}$$

where n_F is the principal quantum number of the Fermi level. Using Eq. (17.28), Eq. (17.25) gives

$$E_F = \frac{\hbar^2}{2m}\left(\frac{n_F \pi}{L}\right)^2 = \frac{\hbar^2}{2m}\left(\frac{N\pi}{2L}\right)^2 \tag{17.29}$$

(iii) Total energy

One can determine the total energy E_0 of all the N electrons in the ground state by summing up the energies of the individual electrons. For N electrons the number of filled energy levels is $N/2$. E_0 is given by

$$E_0 = 2\sum_{n=1}^{N/2} E_n$$

The factor 2 is appearing because each level contains two electrons with equal energies. Using Eq. (17.25), one obtains

$$E_0 = 2\frac{\bar{h}^2}{2m}(\pi/L)^2 \sum_{n=1}^{N/2} n^2 \tag{17.30}$$

since $$\sum_{n=1}^{s} n^2 = \frac{1}{6}s(2s^2+2s+1) = \frac{1}{3}s^3 \quad \text{for } s \geq 1$$

$$\therefore \quad \sum_{n=1}^{N/2} n^2 \cong \frac{1}{3}(N/2)^3$$

Thus Eq. (17.30) takes the form

$$E_0 = 2\frac{\hbar^2}{2m}\left(\frac{\pi}{L}\right)^2 \frac{1}{3}\left(\frac{N}{2}\right)^2 = \frac{1}{3}\frac{\hbar^2}{2m}\left(\frac{N\pi}{2L}\right)^2 N$$

or $$E_0 = \frac{1}{3}NE_F \tag{17.31}$$

This shows that for one-dimensional crystal, the average kinetic energy in the ground state is one-third of the Fermi energy.

(iv) Density of states, D(E)

This is defined as the number of electronic states present in a unit energy range. Thus

$$D(E) = \frac{dn}{dE} \tag{17.32}$$

where dn represents the number of eletronic quantum states present in the energy interval E and $E + dE$. For a free electron gas, as stated earlier, each energy level contains two electronic states, one with spin up and other with spin down, the actual density of states is twice that of given by Eq. (17.32), thus

$$D(E) = 2\frac{dn}{dE} \tag{17.33}$$

From Eq. (17.25) we obtain

$$\frac{1}{n} = \left(\frac{h^2}{8\pi L^2 E}\right)^{1/2} \tag{17.34}$$

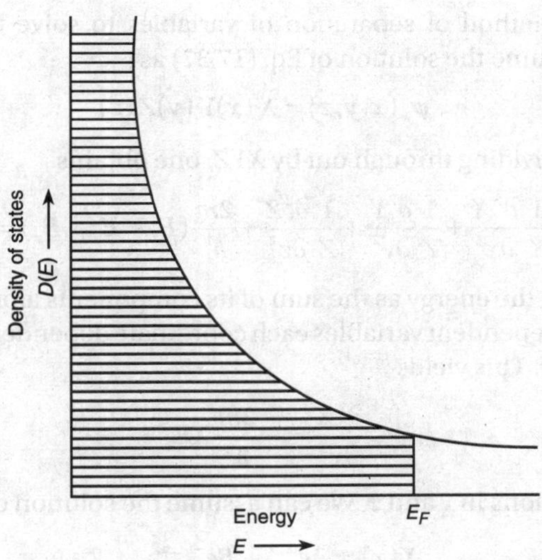

Fig. 17.4 For one-dimensional metallic crystal, variation of density of electronic states with energy. All the states upto Fermi level are filled at 0 K

Hence
$$D(E) = \left(\frac{8mL^2}{h^2E}\right)^{1/2} = \frac{4L}{h}\left(\frac{m}{2E}\right)^{1/2} \tag{17.35}$$

Fig. (17.4) shows the plot of $D(E)$ versus E. From Fig. (17.4) it is clear that all the energy levels present below the Fermi level are filled and all those present above it are empty.

We may note that this type of situation in fact exists at absolute zero. Clearly at 0K, the Fermi level divides the filled and unfilled levels in the metallic crystals.

17.3.1 Free Electron Gas in Three Dimensions

We now consider the behaviour of the free electron gas in a three-dimensional metallic crystal. The potential of the electrons inside the crystal is zero and we may take it zero, whereas potential has large value outside the crystal. For simplicity, we may regard the three-dimensional crystal as a cubical box with side L. The free particle Schrodinger equation in three dimensions is

$$\nabla^2\psi_k\left(r + \frac{2m}{\hbar^2}E_K\psi_k(r)\right) = 0 \tag{17.36}$$

In cartesian coordinates, the *Laplacian operator* ∇^2 can be expanded to give

$$\frac{\partial^2\psi_k(r)}{\partial x^2} + \frac{\partial^2\psi_k(r)}{\partial y^2} + \frac{\partial^2\psi_k(r)}{\partial z^2} + \frac{2m}{\hbar^2}E_K\psi_k(r) = 0 \tag{17.37}$$

where E_K is the total energy (kinetic energy in the present case) of the electron in the k-state.

One can use the method of separation of variables to solve this partial differential Eq. (17.37). Let us assume the solution of Eq. (17.37) as

$$\psi_k(x,y,z) = X(x)Y(y)Z(z) \tag{17.38}$$

On substituting and dividing through out by XYZ, one obtains

$$\frac{1}{X}\frac{\partial^2 X}{\partial x^2} + \frac{1}{Y}\frac{\partial^2 Y}{\partial y^2} + \frac{1}{Z}\frac{\partial^2 Z}{\partial z^2} + \frac{2m}{\hbar^2}(E_x + E_y + E_z) = 0 \tag{17.39}$$

where $E_k = E_x + E_y + E_z$, the energy as the sum of its components along x, y and z directions. Since x, y and z are independent variables each coordinate dependent term must separately be equal to a constant. This yields

$$\frac{\partial^2 X}{\partial x^2} = -\frac{2m}{\hbar^2} X E_x \tag{17.40}$$

and two similar equations in y and z. We can assume the solution of Eq. (17.40) as

$$X(x) = Ae^{ik_x x} + Be^{-ik_x x} \tag{17.41}$$

Substituting Eq. (17.41) in Eq. (17.40), one obtains

$$k_x^2 = \frac{2m}{\hbar^2} E_x \tag{17.42}$$

The boundary conditions $\psi_k = 0$ or $X(x) = 0$ at $x = 0$ and $x = L$, yields

$$A + B = 0 \tag{17.43}$$

$$Ae^{ik_x L} + Be^{-ik_x L} = 0 \tag{17.44}$$

These are satisfied, if we have

$$k_x = \frac{\pi n_x}{L}, \quad n_x = 1, 2, \ldots \tag{17.45}$$

Substituting this in Eq. (17.42), one obtains the energy E_x and the function $X(x)$ as

$$E_x = \frac{\hbar^2 k_x^2}{2m} = \frac{\hbar^2 \pi^2 n_x^2}{2mL^2}, \quad n_x = 1, 2, \ldots \tag{17.46}$$

$$X(x) = C\sin(xk_x) \tag{17.47}$$

where C is constant. Further normalization of the function $X(x)$ requires

$$C^2 \int_0^L \sin^2 \frac{\pi n_x x}{L} dx = 1 \tag{17.48}$$

or

$$C = \sqrt{2/L} \tag{17.49}$$

Using Eq.(17.49), one can wite the final solution of Eq. (17.41) as

$$X(x) = \sqrt{\frac{2}{L}} \sin \frac{\pi_x n_x}{L}, \quad n_x = 1, 2, \cdots \tag{17.50}$$

Corresponding to energy

$$E_x = \frac{\hbar^2 \pi^2 n_x^2}{2mL^2} \quad (17.51)$$

with n_x as the quantum number describing this state. Similarly, one can obtain the solutions for y and z. On combining all the three solutions for x, y and z, the electron wave function is obtained as

$$\psi_k(\mathbf{r}) = \left(\frac{2}{L}\right)^{3/2} \sin\frac{\pi x n_x}{L} \sin\frac{\pi y n_y}{L} \sin\frac{\pi z n_z}{L} \quad (17.52)$$

corresponding to the allowed energies given by

$$E_k = \frac{\hbar^2 \pi^2}{2mL^2}\left(n_x^2 + n_y^2 + n_z^2\right) = \frac{\hbar^2}{2m}|\mathbf{k}|^2$$

where $$\mathbf{k} = \hat{i}\frac{\pi n_x}{L} + \hat{j}\frac{\pi n_y}{L} + \hat{k}\frac{\pi n_z}{L} \quad (17.53)$$

For a wave function ψ_k as Eq. (17.52) to exist, all n_x, n_y and n_z have to be positive non-zero integers. Therefore the least values of n_x, n_y and n_z (i.e., 1, 1, 1) will define the ground state energy level as

$$E_{(111)} = \frac{3\hbar^2 \pi^2}{2mL^2} \quad (17.54)$$

The energy level, an excited state is for n_x, n_y and n_z having values 2, 1, 1. We obtain

$$E_{(211)} = \frac{6\hbar^2 \pi^2}{2mL^2} \quad (17.55)$$

We may note that this has degeneracy of 3 as 2 1 1, 1 2 1 and 1 1 2. These are all equivalent. Higher values of n_x, n_y and n_z will give higher allowed energy levels with their degeneracies depending on the values of n_x, n_y and n_z, e.g., $\frac{27\hbar^2 n^2}{2mL^2}$ is a four-fold degenerate state corresponding to n_x, n_y and n_z having values 3, 3, 3; 5, 1, 1; 1, 5, 1 and 1, 1, 5.

The above results clearly reveal that the energy spectrum consists of discrete energy levels. We may note that these levels are generally very close to each other (energy difference ~10^{-15} eV) and may be regarded as continuous for most of the purposes. Moreover, these levels are also said to be *quasi continuous*. Truly speaking, the spcing between the levels depends upon the dimensions of the box. For atomic dimensions, the energy levels are widely spaced and discrete.

One can show that the normalised wave function is

$$\psi_k(\mathbf{r}) = \left(\frac{1}{L^3}\right)^{1/2} e^{i\mathbf{k}.\mathbf{r}} = \left(\frac{1}{V}\right)^{1/2} e^{i(\mathbf{k}).(\mathbf{r})} \quad (17.56)$$

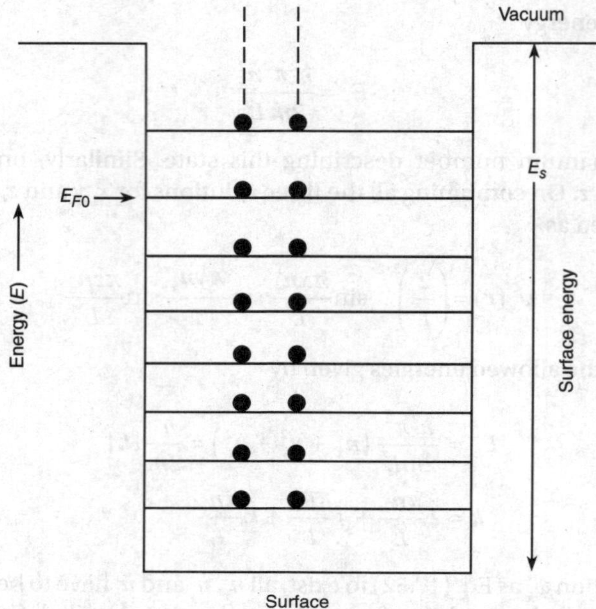

Fig. 17.5 Sommerfeld's free electron model at 0K. All the levels below Fermi level E_{Fo} are filled where as all levels those above E_{Fo} are empty

(i) Filling of energy levels

We have read that the distribution of electrons in the allowed energy levels follow the Pauli's exclusion principle. In the present case, there are a set of four quantum numbers k_x, k_y, k_z and m_s, which represents a state. Apparently, each state can be occupied by at the most one electron. From Eq. (17.53) it is obvious that a set of values of k_x, K_y and k_z defines an energy level. Now, since for each set of k_x, k_y and k_z the fourth quantum number can have two values ±1/2, it follows that each energy level contains two quantum states of orbitals and therefore can accomodate a maximum of two electrons, one with spin up and the other with spin down. Obviously, each energy level is doubly degenerate. This means that a total of N non-interacting electrons at 0K can be filled in $N/2$ energy levels. The topmost filled level is the $N/2^{th}$ level and all the levels lying above it are empty. This energy level is therefore, the *Fermi level*. Fermi level divides the filled and empty level at 0K. This Fermi energy or the energy of the Fermi level is denoted by E_{Fo}. Fig. (17.5) shows the distribution of electrons among the energy levels present inside the box at 0K. This reveals that unlike the classical theory, Sommerfeld's quantum theory does not allow the condensation of all the electrons into the state of zero energy even at absolute zero. We can say that these electrons are rather distribuited among the discrete energy levels having energies ranging from 0 to E_{Fo}.

(ii) Fermi-Dirac (F.D.) statistics and Fermi-Dirac distribution function

We have already stated that the Sommerfeld theory differs from the classical theories in the replacement of Maxwell-Boltzmann statistics by the Fermi-Dirac statistics. In order to

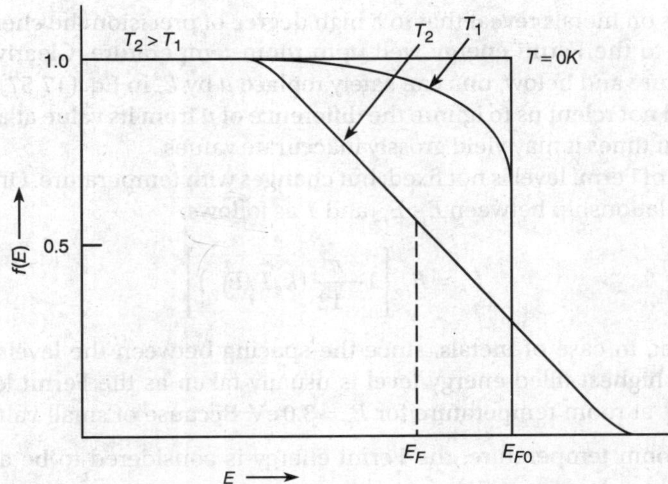

Fig. 17.6 Fermi distribution function for three different temperatures. Figure also shows the variation of the Fermi energy with temperature

study the distribution of electrons above absolute zero, we introduce the FD distribution of electrons above absolute zero i.e., we introduce the FD distribution function $f(E)$ given by

$$f(E) = \frac{1}{\exp[(E-\mu)/k_BT]+1} = \frac{1}{\exp[(E-E_F)/k_BT+1]} \tag{17.57}$$

where μ is *chemical potential* and is a function of temperature. The highest energy level populated at absolute zero is designated as E_F and is called the Fermi level. Its value is decided so as to get the correct number for the total number of particles (here, electrons) using the FD distribution function given by Eq. (17.57).

At absolute zero, for E_F the Eq. (17.57) gives $f(E) = 0$ and for $E < E_f$, $f(E) = 1$. This is what we have read in the last section that below Fermi level, all states are occupied and the probability of finding an electron on a level with energy greater that Fermi energy is zero. The temperature dependence of $f(E)$ may be summarized as follows:

$$\begin{aligned} \text{At } T = 0, \quad & f(E) = 0 \ \text{ for } \ E > E_F \\ & \quad\;\;\; = 1 \ \text{ for } \ E < E_F \end{aligned} \tag{17.58}$$

$$\text{At } \ T > 0, \qquad f(E) = \frac{1}{2} \ \text{ for } \ E = E_F \tag{17.59}$$

$$\begin{aligned} \text{At } \ T \to 0, \quad & \lim_{T\to 0} f(E) = 1 \ \text{ for } \ E < \mu \\ & \qquad\quad\;\; = 0 \ \text{ for } \ E > \mu \end{aligned} \tag{17.60}$$

For these conditions to be true, it is required that

$$\lim_{T\to 0} \mu = E_F \tag{17.61}$$

The plot of $f(E)$ versus E for different temperatures is shown in Fig. (17.6).

Observations on metals reveal that to a high degree of precision the chemical potential remains equal to the Fermi energy well upto room temperature. Clearly, for metals at room temperature and below, one can safely replace μ by E_F in Eq. (17.57). However, this practice should not relent us to ignore the difference of μ from its value at absolute zero in general, since at times it may yield grossly inaccurate values.

The position of Fermi level is not fixed, but changes with temperature. One can write the approximate relationship between E_F, E_{Fo} and T as follows:

$$E_F = E_{Fo}\left[1-\frac{\pi^2}{12}(k_B T/E_{Fo})\right] \tag{17.62}$$

As stated earlier, in case of metals, since the spacing between the levels is quite small (~10^{-19}eV), the highest filled energy level is usually taken as the Fermi level. The value of $k_B T/E_F \cong 0.01$ at room temperature for $E_F = 3.0$ eV. Because of small value of the factor $\left(k_B T/E_{F_o}\right)$ at room temperature, the Fermi energy is considered to be a constant with respect to temperature and therefore 'o' subscript is dropped from Eq. (17.62).

(iii) Density of available electron states, D(E)

The density of states, $D(E)$ is the total number of available electronic states (or orbitals) per unit energy range at energy E. Linear momentum operator ($\boldsymbol{p}$) in quantum mechanics is

$$\boldsymbol{p} = -i\hbar\nabla \tag{17.63}$$

when operated on the state $\psi_k(\boldsymbol{r})$, one obtains

$$p\psi_k(\boldsymbol{r}) = -i\hbar\nabla\,\psi_k(\boldsymbol{r}) = \hbar k\psi_k(\boldsymbol{r}) \tag{17.64}$$

Obviously, the plane wave function $\psi_k(\boldsymbol{r})$ is an eigen function of the linear momentum with the eigen value $\hbar k$. The particle velocity is given by

$$\boldsymbol{v} = \frac{\hbar \boldsymbol{k}}{m} \tag{17.65}$$

In a system of N-free electrons the occupied states or orbitals in the ground state may be represented by points inside a sphere in the $\boldsymbol{k}$-space. Clearly, the energy corresponding to the surface of the sphere then reperesents the Fermi energy. Let $\boldsymbol{k}_F$ be the wave vector from the origin of the $\boldsymbol{k}$-space to the surface of the sphere. (Fig. 17.7). Now, the Fermi energy is given by

$$E_F = \frac{\hbar^2}{2m}k_F^2 \tag{17.66}$$

We may note that there is one allowed wave vector or one distinct triplet of quantum numbers k_x, k_y and k_z which corresponds to the volume element $(2\pi/L)^3$ of ($\boldsymbol{k}$)-space. Obviously in the sphere called the *Fermi sphere of volume* $(4\pi/3)k_F^3$, the total number of electronic states or orbitals is

$$N = 2\frac{\frac{4\pi}{3}k_F^3}{(2\pi/L)^3} = \frac{L^3}{3\pi^2}k_F^3 = \frac{V}{3\pi^2}k_F^3 \tag{17.67}$$

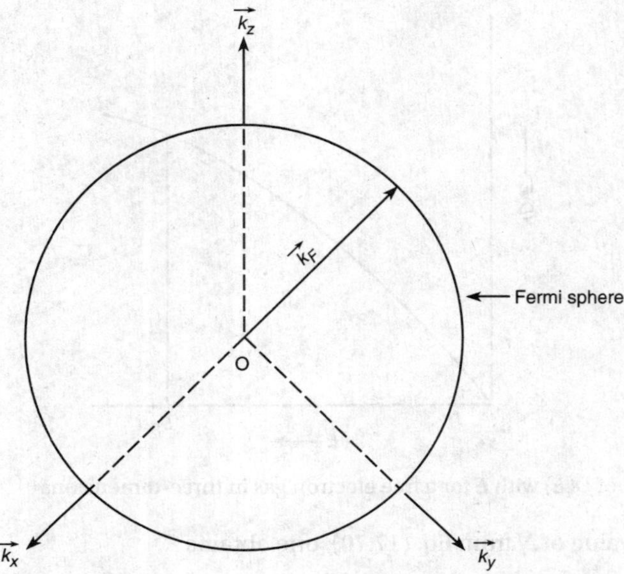

Fig. 17.7 A sphere in the ***k***-space having radius k_F sphere contains the points which represents the occupied states of a system of *N*-free particles (electrons) in the ground state

where the factor 2 takes account of spin up and down degeneracy. This means that there are two electronic states corresponding to each value of *k*. We have set the number of electrons, *N*. From Eq. (17.67), we obtain

$$k_F = \left(\frac{3\pi^2 N}{V}\right)^{1/3} \tag{17.68}$$

From Eq. (17.68), we note that k_F depends on the concentration of electrons, N/V, and is independent of the mass of the electron. Using Eq. (17.68) in Eq. (17.67), one obtains

$$E_F = \frac{\hbar^2}{2m}\left(\frac{3\pi^2 N}{V}\right)^{2/3} \tag{17.69}$$

This shows that E_F depends on both the electronic concentration and mass. Now, the total number of electrons is given by

$$N = \frac{V}{3\pi^2}\left(\frac{2mE_F}{\hbar^2}\right)^{3/2} \tag{17.70}$$

The electron velocity, V_F at the fermi surface is obtained from Eqs. (17.64) and (17.67) as

$$V_F = \frac{\hbar k_F}{m} \frac{\hbar}{m}\left(\frac{3\pi^2 N}{V}\right)^{1/3} \tag{17.71}$$

In the ground state of the system, all the energy states below E_F are occupied and the total number of states is equal to the total number of electrons. Using this fact, the density of states function $D(E)$ is obtained as

$$\int_0^L D(E)\, dE = N \tag{17.72}$$

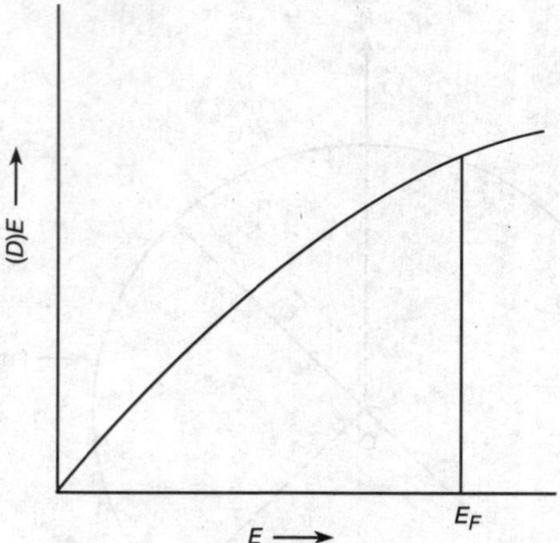

Fig. 17.8 Variation of $D(E)$ with E for a free electron gas in three-dimensions

Substituting the value of N from Eq. (17.70), one obtains

$$\int_0^L D(E)\,dE = \frac{V}{3\pi^2}\left(\frac{2mE_F}{\hbar^2}\right)^{3/2}$$

Now, expressing the integral in an indefinite form, one obtains

$$\int_0^L D(E)\,dE = \frac{V}{3\pi^2}\left(\frac{2mE}{\hbar^2}\right) = \int_0^L = \frac{V}{2\pi^2}\left(\frac{2m}{\hbar^2}\right)^{3/2} E^{1/2}dE$$

or
$$D(E) = \frac{V}{2\pi^2}\left(\frac{2m}{\hbar^2}\right)^{3/2} E^{1/2} \tag{17.73}$$

The variation of $D(E)$ with E is shown in Fig. (17.8). We may note that $D(E)$ also increases with increase in crystal volume, this is to accomodate the total number of electrons N which also increases with volume of the crystal. One can obtain the density of states at Fermi surface by putting $E = E_F$ in Eq. (17.73), i.e.,

$$D(E_F) = \frac{dN}{dE_F} = \frac{V}{2\pi^2}\left(\frac{2m}{\hbar^2}\right)E_F^{1/2} \tag{17.74}$$

One can obtain the density of filled electronic states, $N(E)$ at any temperature T by multiplying the density of states, $D(E)$ at 0 K by the probability of occupation, $f(E)$ of the quantum state E at that temperature, i.e.,

$$N(E) = D(E)\,f(E) \tag{17.75}$$

The function $f(E)$ is the Fermi function and is defined by Eq. (17.57). Now the number of filled electronic states in the energy range E and $E + dE$ at temperature T is given by

$$N(E)\,dE = D(E)\,f(E)\,dE \tag{17.76}$$

Now, we discuss some parameters of free electron gas at absolute zero temperature,

Fermi Energy, E_{Fo}: One can obtain the total number of electrons by integrating Eq. (17.76), i.e.,

$$\int_0^\infty N(E)\,dE = \int_0^\infty D(E)f(E)\,dE \tag{17.77}$$

Using Eq.(17.73), one obtains

$$N = \frac{V}{2\pi^2}\left(\frac{2m}{\hbar^2}\right)^{3/2}\int_0^\infty E^{1/2} f(E)\,dE$$

$$= \frac{V}{2\pi^2}\left(\frac{2m}{\hbar^2}\right)^{3/2}\left[\int_0^\infty E^{1/2} f(E)\,dE + \right]\int_{E_{Fo}}^\infty E^{1/2} f(E)\,dE$$

Using Eqs. (17.57), (17.58) and (17.59), one obtains

$$N = \frac{V}{2\pi^2}\left(\frac{2m}{\hbar^2}\right)^{3/2}\int_0^{E_{Fo}} E^{1/2}\,dE = \frac{V}{3\pi^2}\left(\frac{2m}{\hbar^2}\right)^{3/2} E_{Fo}^{3/2}$$

or

$$E_{Fo} = \frac{\hbar^2}{2m}\left(\frac{3\pi^2 N}{V}\right)^{2/3} = \frac{\hbar^2}{2m}(3\pi^2 n)^{2/3} \tag{17.78}$$

$$= \frac{\hbar^2}{8m}\left(\frac{3n}{\pi}\right)^{2/3} \quad (17.79)$$

where $n = N/V$ is the electron concentration. Eq. (17.78) clearly reveals that the Fermi energy at 0K can be calculated simply from $n\ (= N/V)$. One also finds that the electrons have some definite energy at absolute zero. We may note that this is entirely different from the classical theory according to which the electrons should have zero energy at absolute zero.

Average kinetic energy, $\overline{E}_0$: One can calculate the average kinetic energy of an electron at absolute zero as

$$\overline{E}_0 = \frac{1}{N}\int_0^\infty E D(E) f(E)\,dE$$

Using Eq. (17.73) for $D(E)$, one obtains

$$\overline{E}_0 = \frac{V}{2\pi^2}\left(\frac{2m}{\hbar^2}\right)^{3/2}\frac{1}{N}\left[\int_0^{E_{Fo}} E E^{1/2} f(E)\,dE + \int_0^{E_{Fo}} E E^{1/2} f(E)\,dE\right]$$

$$= \frac{V}{2\pi^2}\left(\frac{2m}{\hbar^2}\right)^{3/2}\frac{1}{N}\int_0^{E_{Fo}} E^{3/2}\,dE$$

$$= \frac{2}{5}\frac{V}{2\pi^2}\left(\frac{2m}{\hbar^2}\right)^{3/2}\frac{1}{N}E_{Fo}^{5/2} \tag{17.80}$$

or

$$\overline{E}_0 = \frac{3}{5}E_{Fo} \tag{17.81}$$

Clearly, Eq. (17.81) relates the average kinetic energy of an electron at absolute zero to the Fermi energy at the same temperature.

17.4 APPLICATONS OF FREE ELECTRON GAS MODEL

One can broadly divide the electronic properties of metals into two groups: (i) static properties and (ii) Transport properties.

The *static properties* of the metals are those properties which one can treat effectively by considering just the energy levels or the distribution of energy levels to which the electrons belong. Specifically, we need to consider the overall change in potential of the electrons without investigating deeply the detailed process producing the transitions. These properties of metals include various *electron emission* properties, the *magnetic properties, contact potential, heat capacity*, etc. We may note that these properties arise as a result of the excitations of the electrons by light, thermal energy, strong electric fields, etc.

Transport properties of metals include electrical and thermal conductivities, Hall effect, thermoelectricity etc. These properties can be treated by considering the detailed response of the electrons to an external field, i.e., by considering the accelerated properties of the electrons.

17.4.1 Electronic Specific Heat

Sommerfeld was able to resole the most glaring anomaly concerning the electron heat capacity of metals, as encountered in the Drude model with the use of quantum statistics. Sommerfeld was able to show that the specific heat of the electrons in a metal is very much small compared to the lattice specific heat. In order to calculate the electronic specific heat we need to know the average energy in the metal at temperature T. The average kinetic energy of a free electron as given by the classical statistical mechanics is

$$\overline{E}_0 = \frac{3}{2} k_B T \tag{17.82}$$

If the metal contains N free electrons, then the total kinetic energy becomes

$$E = N\overline{E}_0 = \frac{3}{2} N k_B T$$

Thus, the elecronic specific heat is therefore given by

$$C_V = \frac{3}{2} N k_B \tag{17.83}$$

The measurement of optical reflection coefficient of metals and other experimental observation clearly indicates that all the electrons might not contribute to the sp. heat, but, only a small fraction of the total number of electrons might contribute. We find that only the uppermost electrons equalling a fraction of roughly $(4K_B T/E_F)$ are affected by the presence of field gradient (thermal or electrical). Sommerfeld thus concluded that electrons occupying states only close to the Fermi level E_F contributes to most of the properties of metals. This view is in accordance with the quantum mechanics.

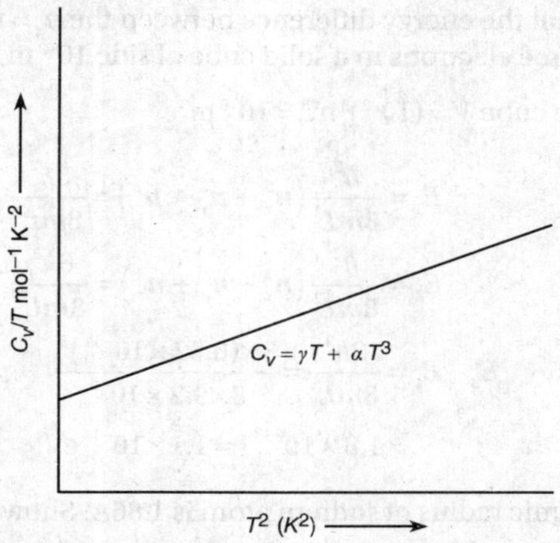

Fig. 17.9 Variation of C_V/T as a function of T^2 for alkali metals

According to quantum mechanics, only those electrons contribute to the specific heat which lie within an energy range k_BT below the Fermi level, E_F. This happens because when the electron gas is heated upto a temperature T, these electrons in a metal acquire energy of the order k_BT and jump to the empty higher energy states. Obviously, the deep lying electrons cannot do so because the unfilled energy states are not available to these electrons, as such, do not contribute to the specific heat of metal. One can easily show that the number of electrons which contribute to the specific heat is of the order $N\left(\frac{k_BT}{E_F}\right)$ or $N\,(T/T_F)$, where T_F is called the *Fermi temperature.* Fermi temperature is defined by the equation

$$E_F = k_BT_F$$

Now using the effective value of N, Eq. (17.83) takes the form

$$C_V = C_{el} = \frac{3}{2}Nk_B\left(\frac{T}{T_F}\right) \tag{17.84}$$

which as $T << T_F$ at normal temperatures, is much less than 3/2 k_B suggested by classical free electron model and agrees much better with experimental results. We may note that electronic specific heat varies linearly with T and at low temperatures (below the Debye temperature and Fermi temperature), lattice sp.heat follows Debye T^3 law and the heat capacity of a metal is expressed as

$$C_V = \gamma T + \alpha T^3 \tag{17.85}$$

where the first term represents the electron contribution and the second term is the usual Debye term contributed by *phonons.* γ and α are constants and characterstic of the material and can be determined from the conventional graph between C_V/T and T^2 using the measured values of C_V. This is the behaviour observed at low temperatures for alkali metals. A plot of $\frac{C_V}{T}$ vs T^2 yields a straight line with slope α and intercept γ a shown in Fig. (17.9).

Example 17.1: **Show that the energy difference between the $n_x = n_y = n_z = 1$ level and next higher energy level for free electrons in a solid cube of side 10^{-2} m is about 10^{-14} eV.**

Solution Volume of the cube $V = (10^{-2})^3 \text{ m}^3 = 10^{-6} \text{ m}^3$

$$E_1 = \frac{\hbar^2}{8mL^2}\left(n_x^2 + n_y^2 + n_z^2\right) = \frac{3\hbar^2}{8mL^2}$$

and
$$E_2 = \frac{\hbar^2}{8mL^2}\left(n_x^2 + n_y^2 + n_z^2\right) = \frac{6\hbar^2}{8mL^2}$$

$\therefore$
$$E_2 - E_1 = \frac{3\hbar^2}{8mL^2} = \frac{3\times(6.62\times10^{-34})^2}{8\times9.2\times10^{-6}}$$

$$= 1.8\times10^{-33}\text{ J} = 1.1\times10^{-14}\text{ eV.}$$

Example 17.2: **The atomic radius of sodium atom is 1.86Å. Show that the Fermi energy of sodium at absolute zero is 3.1 eV.**

Solution We have, Fermi energy at 0K given by

$$E_{Fo} = \frac{\hbar^2}{2m}\left(\frac{3\pi^2 N}{V}\right)^{2/3}$$

where N is the number of free electrons present in volume V of the metal. Let us consider a unit cell of sodium atom

Volume of unit cell,
$$V = a^3 = \left(\frac{4r}{\sqrt{3}}\right)^3$$

Given
$$r = 1.86\times10^{-10}\text{ m}$$

$\therefore$
$$V = \left(\frac{4\times1.86\times10^{-10}}{\sqrt{3}}\right)^3 = 7.93\times10^{-29}\text{ m}^3$$

An atom of sodium has only one valence electron. The number of electrons in a unit cell of sodium is 2. We have

$$\frac{N}{V} = \frac{2}{7.93\times10^{-29}} = 2.52\times10^{26}\ \frac{\text{electrons}}{\text{m}^3}$$

$\therefore$
$$E_{Fo} = \frac{(1.05\times10^{-34})^2}{2\times9.1\times10^{-31}}\times(3\pi^2)(2.52\times10^{28})^{2/3}$$

$$= 4.98\times10^{-19}\text{ J} = 3.1\text{eV.}$$

Example 17.3: **Show that the number of energy states available for the electrons in a cubical box of 0.01 m side lying below an energy of 4eV is 36.1 × 10[21].**

Solution The number of energy states below 4eV is

$$= \frac{V}{2\pi^2}\left(\frac{2m}{\hbar^2}\right)^{3/2}\int_0^4 E^{1/2}\, dE$$

$$= \frac{V}{2\pi^2}\left(\frac{2m}{\hbar^2}\right)^{3/2} \frac{2E}{3}\Bigg|_0^4$$

$$= \frac{(0.01)^3}{2\times(3.14)^2}\left(\frac{2\times 9.1\times 10^{-31}}{1.05\times 1.05\times 10^{-68}}\right)\times\frac{2}{3}\times 8\times 1.6\times 10^{-19}$$

$$= 36.1\times 10^{21}$$

Example 17.4: The Fermi velocity of the electron in a metal is 0.86×10^6 m/s. Determine the Fermi energy and Fermi temperature. [Banglore]

Solution

(i) Fermi energy,

$$E_F = \frac{1}{2}mv_F^2 = \frac{1}{2}\times\frac{9.1\times 10^{-31}\times(0.86)^2\times 10^{12}}{1.6\times 10^{-19}}\text{eV}$$

$$= 2.1\text{eV}$$

(ii) Fermi temperature,

$$T_F = \frac{E_F}{k_B} = \frac{2.1\times 1.6\times 10^{-19}}{1.38\times 10^{-23}}$$

$$= 2.4\times 10^4\,\text{K}.$$

Example 17.5: The Fermi energy of copper at 0K is 7.04eV. Show that the Fermi energy at 300K is 7.0399eV.

Solution We have

$$E_F = E_{Fo}\left[1 - \frac{\pi^2}{12}\left(\frac{k_B T}{E_{Fo}}\right)^2\right]$$

$$= 7.04\left[1 - \frac{\pi^2}{12}\left(\frac{1.38\times 10^{-23}\times 300}{7.04\times 1.6\times 10^{-19}}\right)^2\right]\text{eV}$$

$$= 7.0399\text{eV}$$

SUMMARY

1. **About two-third of elements are metals. Obviously, the natural tendency of elements is to favour the metallic state.**

2. **The theory of metals heavily relies on the idea that the valence electrons of metal atoms can be treated as free.**

3. **A metallic crystal may be envisaged as the superposition of a periodic array of positive ion cores and quasi-uniform negative charge density of the free electron gas.**

4. According to Drude the metal must have two types of particles: (i) free valence electrons, and (ii) the remainder of atom, a positive ion carrying the major portion of the atomic mass, i.e. particles as heavy and immobile.
5. Electrical conductivity from Drude's theory is given by

$$\sigma = \frac{ne^2 \wedge}{(3\, m\, k_B T)^{1/2}}$$

where $\wedge = \tau v_{\text{th}}$, τ is the relaxation time. Obviously, conductivity increases with decrease in temperature. However, dependence of conductivity on $T^{-1/2}$ is not in agreement with the observed T^{-1} dependence in most of common metals.
6. Drude model assumes that the major contribution to thermal conductivity of metals comes from conduction electrons.

$$K_{el} = \frac{3}{2}\left(\frac{n k_B^2 \tau}{m}\right) T$$

7. Drude obtained

$$\frac{K}{\sigma} = \frac{3}{2}\left(\frac{k_B}{e}\right)^2 T$$

where $L = \frac{3}{2}\left(\frac{K_B}{e}\right)^2$ is constant and known as Lorentz number (= $1.11 \times 10^{-8}\ W\Omega K^{-2}$ = $1.11 \times 10^{-8}\ (V K^{-1})^2$. However, this number is half the experimental value.
8. Lorentz modified Drude model and he abandoned the assumption that all electrons move with same thermal velocity. He applied Boltzmann transport equation approach in place of classical Maxwell Boltzmann velocity distribution. He obtained

$$\sigma_L = \left(\frac{8}{3\pi}\right)^{1/2} \frac{ne^2 \wedge}{(3 m k_B T)^{1/2}} = 1.09\sigma_D$$

where σ_D is Drude conductivity. Obviously, Lorentz modifications provide neither any noteworthy change in the quantitative measure nor any change in the temperature dependence of electrical conductivity.
9. Lorentz obtained the following expression for the Hall coefficient,

$$R_H = -\left(\frac{3\pi}{8}\right)\frac{1}{ne}$$

10. Sommerfeld described the cube of a metal crystal as a three dimensional potential box with an infinite barrier at the surfaces taking the potential energy as a constant within the box.
11. Sommerfeld succeeded in resolving the most glaring thermal anomaly concerning the electron heat capacity, as encountered in the Drude model with the use of quantum statistics.

REVIEW QUESTIONS

1. What is free electron gas model of metals? Explain, how this model help to understand the lattice heat capacity of metals?
2. Derive the expressions for the free energy, the total energy and the density of states on the basis of free electron gas model in one-dimension. How density of states varies with energy? **[Kerala]**
3. What is the Fermi level? What does it signify? How Fermi level change with temperature and concentration of electrons? **[Delhi]**
4. Write the expression for the Fermi-Dirac distribution function. How this function behaves with temperature?
5. Give a comparison of the equilibrium distribution functions representing Maxwell-Boltzmann and Fermi-Dirac statics. Explain why only former depends on the particle density? **[Mysore]**
6. Derive expressions for the Fermi energy and density of states for a free electron gas in three-dimensions. **[Punjab]**
7. Show that the fraction x of electrons within $k_B T$ of the Fermi level is given by

$$x = \frac{3}{2}\frac{k_B T}{E_F}$$

8. Show that the average kinetic energy of a three-dimensional free electron gas at 0K is

$$\overline{E}_0 = \frac{3}{5}E_{Fo}$$ **[Raj]**

PROBLEMS

1. Derive pressure versus volume relationship for a free electron gas at 0K. Hence obtain an expression for the bulk modulus in terms of the average kinetic energy.
2. Assuming Drude model, show that at any moment the time between the last and the next collision averaged over all electrons is 2τ. Keeping this result in view, comment on Drude's mistake of a factor of 1/2 in the calculation of electrical conductivity of metals.
3. Show that the wavelength associated with an electron having an energy equal to Fermi energy is given by

$$\lambda_F = 2(\pi/3n)^{1/2}$$

4. A particle of mass m is confined in a field free region between impenetrable walls at $x = 0$ and $x = a$. Show that the stationary energy levels of the particle are given by:

$$E_n = \frac{n^2 h^2}{8ma^2}$$

5. Silver (FCC) has an atomic radius of 1.44Å. Assuming silver to be monovalent metal show that the values of Fermi energy, Fermi temperature and Fermi velocity are 5.5 eV, 63850 K and 1.39×10^6 m/s respectively.

6. Evaluate the Fermi function for an energy $k_B T$ above the Fermi energy. [0.269]
7. Assuming the electrons to be free, calculate the total number of states below $E = 5\text{eV}$ in a volume 10^{-5}m^3. [5.1×10^{23}]

SHORT ANSWER-QUESTIONS

1. Write the expression for electrical conductivity of metals on the basis of Drude's model.

Ans: $\sigma = \dfrac{ne^2 \wedge}{(3mk_B T)^{1/2}}$, where $\wedge$ is the electron mean free path.

2. How experimentally the dc electrical conductivity of metal varies?

Ans: $\sigma \propto T^{-1}$

3. The Drude model correctly describes the Widemann-Franz law, but gives a wrong magnitude of the Lorentz number L. Comment on it.

Ans: The value of L obtained in the Drude theory is given by

$$L = \frac{k}{\sigma T} = \frac{3}{2}\left(\frac{k_B}{e}\right)^2$$

where k is the thermal conductivity. This value is about half the measured value.

4. What is the relation between the dc electrical conductivities in the Drude model and Lorentz model

Ans: $\sigma_D = \left(\dfrac{3\pi}{8}\right)^{1/2} \sigma_L \quad \begin{vmatrix} D \rightarrow \text{Drude} \\ L \rightarrow \text{Lorentz} \end{vmatrix}$

OBJECTIVE QUESTIONS

1. The density of allowed states between E and $E + dE$ is proportional to

(a) E (b) E^2

(c) $E^{3/2}$ (d) $E^{1/2}$

2. The classical expression for electrical resistivity exhibiting its temperature dependence is

(a) $\rho \propto 1/T^2$ (b) $\rho \propto T^{1/2}$

(c) $\rho \propto T^2$ (d) $\rho \propto 1/T$

3. Mobility of the electron is

(a) Reciprocal of conductivity

(b) Flow of electron per unit electric field.

(c) Average electron drift velocity per unit electric field.

(d) None of the above

4. At absolute zero temperature, all the allowed states of energy upto Fermi level will be
   (a) occupied (b) partly occupied and partly empty
   (c) empty (d) half empty and half occupied
5. Wiedmann-Franz law is
   (a) $\sigma/\sigma_T = LT$ (b) $\sigma/\sigma_T = T/L$
   (c) $\sigma_T/\sigma = L/T$ (d) $\sigma_T/\sigma = LT$
6. At temperature above zero, for $E << E_F$, the Fermi-Dirac function approaches
   (a) unity (b) zero
   (c) e^{-E/k_BT} (d) infinity
7. The Fermi energy of a given metal is 1.4eV. The Fermi temperature of the metal is
   (a) 1.6×10^5 K (b) 1.6×10^6 K
   (c) 1.6×10^3 K (d) 1.6×10^4 K
8. Most widely used conducting materials are
   (a) gold and silver (b) copper and aluminium
   (c) germanium and aluminium (d) tungusten and platinium

ANSWERS

Question numbers	1	2	3	4	5	6	7	8
Answers	(d)	(b)	(c)	(a)	(d)	(a)	(d)	(b)

FILL IN THE BLANKS

1. In the classical free-electron model, in a metal containing N ion cores per unit volume, each of effective scattering radius R, the mean free path of an electron would be $\lambda =$ ____. **[$1/\pi R^2N$]**
2. DC electrical conductivity (σ) of a metal is observed to vary with temperature as $\sigma \propto$ ____. This reveals that all the free electrons of a metal do not take part in ____, rather only the electrons near the Fermi level are involved in ____ and ____ conduction. **[$\sigma \propto T^{-1}$, conduction, thermal, electrical]**
3. The Boltzmann transport equation is concerned with the determination of the ____ distribution function of the system under influence of ____ fields. **[steady state, external]**
4. Sommerfeld gave the theory of electrical conduction in metals using ____ statistics. **[Fermi–Dirac (FD)]**
5. Sommerfeld's theory of electrical conduction gave results in confirmation with ____ law. **[Wiedemann-Franz]**

6. Thermal conductivity of electron gas can be defined as the ratio of the _____ to the _____ which give rise to the flow of energy.

[energy flux, temperature gradient]

7. Wiedemann and Franz in 1853 discovered that all good electrical conductors are also good _____ and at a particular temperature the ratio of thermal conductivity (K) to the electrical conductivity (σ) is the _____ for all metals, i.e. _____.

[thermal conductors, same, $\frac{K}{\sigma T}$ = constant]

8. The value of constant in Wiedemann-Franz law, $\frac{K}{\sigma T}$ = constant is _____ $W\Omega K^{-2}$.

[2.45×10^{-8}]

9. The constant $L = \pi^2 k^2/3e^2$ in Widemann-Franz law is known as _____.

[Lorentz number]

10. Sommerfeld free electron theory of metals treated the conduction electrons in a metal as a gas of _____ obeying _____ statistics and _____.

[free electrons, Fermi-Dirac, Pauli exclusion principle]

11. A function that gives the number of available electron states per unit volume in an energy interval between E and $E + dE$ is designated by $D(E)dE$, where $D(E)$ is called as the _____ and represents the _____.

[density of states, number of states per unit volume per electron volt]

12. The density of states $D(E)$ for a particle of mass m^* with energy between E and $E + dE$ is given by _____.

[$D(E)dE = \frac{4\pi}{\hbar^3}(2m^*)^{3/2} E^{1/2} dE$]

13. The Fermi energy of a given material is the energy of that quantum state which has the _____ equal to half of being occupied by the _____ electrons.

[probability, conduction]

14. The Fermi energy is an important part of an electron's energy band diagram of a material because many electronic properties, e.g. _____, _____, and _____ properties depend upon the location of Fermi level within a band.

[optical, electrical, magnetic]

15. If E_F^o is the Fermi energy at 0K, the Fermi energy at room temperature $T(E_F(T))$ is equal to _____.

[$E_F^o\left\{1-\frac{\pi^2}{12}\left(\frac{kT}{E_F^o}\right)^2\right\}$]

16. Using density of states at Fermi level, one finds the electronic specific heat C_e = _____.

[$\left(\frac{\pi^2 N k_B^2}{2E_F}\right)T$]

17. The wavelength associated with an electron having an energy equal to the Fermi energy = _____.

[$2\sqrt[3]{\pi/3n}$]

18

Electrical Conductivity

18.1 DEFINITIONS AND EXPERIMENTAL FACTS

Certain materials possess the property of conducting electric charge from point to point under the influence of an electric field, that is if an electric field E_0 is applied to them, an electric density J is set up in the field direction. The law of electrical conduction in metals: **ohm's law** is

$$I = \frac{V}{R} \tag{18.1}$$

Where I is the current, V is the potential difference, and R is the resistance of the conductor (wire). We want to express this law in a from which is independent of the length and cross-section of the conductor, since these factors are, after all irrelevant to the basic physics of conduction. Suppose that L and A are respectively the length and cross-section of the wire, then

$$R = \frac{LP}{A} \quad \text{and} \quad \sigma = \frac{I}{\rho}$$

where ρ is resistivity and σ is conductivity. The dimensions of ρ are ohm-m and that of σ are mhos-m^{-1}. Ohm's law then becomes

$$I = \frac{VA}{L\rho}$$

or

$$\frac{I}{A} = \frac{V}{L}\sigma$$

But $\frac{I}{A} = J$ the current density (current per unit area) the electric field and

$$E_0 = \frac{V}{L}$$

Then ohm's law can be written as

$$J = \sigma E_0 \tag{18.2}$$

Clearly the electrical conductivity (σ) measures the capability of the solid to conduct electric charge under the influence of an electric field. For most common materials it is a scalar quantity.

Since the current density is given by the product of the electrons flux nv_D and the electronic charge, where n is the number of electrons per unit volume and v_D the average drift velocity of the electrons in the field direction. Eq. (18.2) can also be expressed as

$$J = nev_D = \sigma E_0$$

or

$$\sigma = ne\frac{v_D}{E_0} \tag{18.3}$$

But $\frac{v_D}{E_0} = \mu$ and is called mobility and Eq. (18.3) assumes the form

$$\sigma = ne\mu \tag{18.4}$$

This is the form in which conductivity (σ) is usually expressed. Since here n and σ are the constants, σ is determined essentially by the mobility μ. It must obviously be intimately related to the resistance experienced by the electrons while flowing through the materials.

We presently deal with the electrical conductivity of metals. All the metals are characterized by a high value of the electrical conductivity, of the order of 10^7 mhos/m. Any theory of the electrical conductivity of metals must explain a number of pertinent experimental facts. Apart from deviation in special cases or under extreme conditions. The general experimental features of the electrical conductivity of metals are:

(i) The metallic conductors obey ohm's law, which states that the current density in the steady state is proportional to the electric field strength.

(ii) The specific resistivity of metals at room temperature is of the order of 10^{-7} ohm-m.

(iii) The resistivity of metals above the Debye temperature varies linearly with temperature.

(iv) At low temperatures, but above approximately 20 K the resistivity of many metals is proportional to T^5. At liquid helium temperature some metals exhibit a minimum in the resistivity versus temperature curve.

(v) The resistivity of an impure specimen can be given by the Mathiessen's rule.

$$\rho = \rho_0 + \rho(T) \tag{18.5}$$

where ρ is a constant for the given impure specimen (which increases with impurity concentration) and $\rho(T)$ is the temperature dependent resistivity of the pure specimen.

(vi) The resistivity of alloys which exhibit order-disorder transitions show pronounced minima corresponding to ordered phases.

(vii) For most metals the resistivity decreases with increasing pressure.

(viii) Above the Debye temperature, the ratio of thermal conductivity to electrical conductivity is proportional to temperature (T). The constant of proportionality being approximately the same for all metals. This is known as **Weidemann-Franz law.**

(ix) A number of metals exhibit the phenomenon of super-conductivity. For such metals the resistivity disappears in the temperature range $0 < T < T_C$. Here T_C is called critical temperature and for Hg, $T_C \approx 4.5\,\text{K}$.

(x) Conductivity varies in the presence of magnetic field, an effect known as **magneto-resistance.**

18.2 DRUDE-LORENTZ (FREE ELECTRON) THEORY OF ELECTRICAL CONDUCTIVITY

The earliest theory proposed in the beginning of 20th century to explain the mechanism of electrical conduction in metals was due to Drude-Lorentz. According to this theory, called the **Free Electron Theory,** a metal is supposed to consist of positive metal ion fixed in the lattice, whose valance electrons move about haphazardly in the positive sea, just like gas molecules in a vessel. The metal is supposed to be held together by the electrostatic forces between the positive charged ions and the free electrons. Further the classical gas laws are applicable to the free electrons.

Drude and Lorentz discussed the steady state conditions of this classical free electron gas under the influence of an applied electric field. It we suppose that there are n electrons per unit volume and imagine that the electrons are moving around in a random way, then in the absence of the field, the average velocity is defined by

$$\langle v_D \rangle = \frac{1}{n}\sum_{i=1}^{n} v_i \tag{18.6}$$

will be zero. This is because in equilibrium there are just as many electrons moving in one direction as in the opposite direction, (v_D) is called the average drift velocity.

Let us see what happens to the random motion of the electron gas when the metal is subjected to an electric field. The electron gets accelerated against the direction of the electric field. But this acceleration does not continue indefinitely since they suffer collisions with the massive atoms. Every collision of the electron with the atom will bring back the electron to its original state of random motion and electron will start a fresh on its journey without any "memory" of what happened before the collision. The mutual repulsions of the electrons themselves are neglected in this theory, since they do not result in the transfer of any momentum to the electrons in the direction of the applied field. The mean velocity acquired by the electron gas due to the application of the electric field is called the **drift velocity** and this is superimposed over its already existing random motion as required by the classical gas.

Let us now consider an electron of mass m starting with an initial zero drift velocity under the influence of an electro-static field E_0. In the presence of the field, however a free electron obtains an acceleration in the direction of the field of magnitude $\left(-\frac{eE_0}{m}\right)$ and thus its drift velocity v_D increases with time. We can now write

$$\left(\frac{\partial v_D}{\partial t}\right)_{\text{Field}} = -\frac{eE_0}{m} \tag{18.7}$$

We now consider the effect of electron-lattice interactions (collisions), as the accelerated electrons must collide with the lattice. Thus collision processes are considered by introducing the concept of the relaxation time τ. It is the time which is closely related to the mean time of flight between collisions. All details of collision processes occurring in the electron gas are summarised for the present purpose under this consideration. It is introduced into the theory by supposing that the probability for an electron to collide with the lattice during a small time interval dt is $\frac{dt}{d\tau}$. We now assume for the sake of simplicity that τ is constant (equal to an average value), independent of the energy of the electron and of the direction of motion. We assume also that in a collision with the lattice, the electron loses all the energy it has acquired from the external field and that its velocity after collision is independent of the direction before the collision. With this model for the electron-lattice collision we may argue to write

$$\left(\frac{dv_D}{dt}\right)_{\text{colli}} = -\frac{v_D}{\tau} \tag{18.8}$$

Since $1/\tau$ is the probability for a collision per second, and after the collision the velocities are random. In steady state of the electron gas in which case $\frac{dv_D}{dt}$ is zero. We thus have in the steady state,

$$\frac{dv_D}{dt} = 0 = \left(\frac{dv_D}{dt}\right)_{\text{Field}} + \left(\frac{dv_D}{dt}\right)_{\text{colli}}$$

or $$-\frac{eE_0}{m} - \frac{v_D}{\tau} = 0$$

$\therefore$ $$v_D = -\frac{eE_0\tau}{m} \tag{18.9}$$

The electric field (together with collisions) thus gives rise to a constant value of the drift velocity. Using now the definition Eq. (18.3), we have for the electrical conductivity:

$$\sigma = \frac{nev_D}{E_0} = \frac{ne^2\tau}{m} \tag{18.10}$$

We observe that the electrical conductivity is a linear function of the relaxation time. Eq. (18.10) is in accordance with ohm's law, in that the conductivity of the metal is independent of the electric field. If we define a mean free path λ for the electron, we can put $\tau = \frac{\lambda}{<v>}$ where $<v>$ is the average thermal velocity. But the classical expression for average thermal velocity based on kinetic theory is

$$\frac{1}{2}m<v^2> = \frac{3}{2}kT \quad \text{or} \quad <v> = \sqrt{\frac{3kT}{m}}$$

Here k is Boltzmann's constant and T the temperature. Thus we have for electrical conductivity,

$$\sigma = \frac{ne^2}{m} \cdot \frac{\lambda}{\sqrt{\frac{3kT}{m}}} = \frac{ne^2 \lambda}{\sqrt{3kTm}} \tag{18.11}$$

Assuming λ to be of the order of inter-atomic distance this formula predicts correctly the order of magnitude of the electrical conductivity of metals. This result indicates that

$$\sigma \propto T^{-1/2}$$

a variation with temperature which is not in agreement with the experimental observation. The variation

$$\sigma \propto T^{-1}$$

is closer to the experimental facts as indicated earlier. This difficulty is associated with almost all the metals. Apart from these serious discrepancies, it is also not correct to assume that the mean free path should be independent of temperature. Obviously the free electron theory is incomplete. The advent of the quantum theory gave further clue to tackle this problem. Sommerfeld in 1928 applied the quantum statistics developed by Fermi-Dirac and tried to rectify the short comings to the free electron theory and this is known as **quantum free electron theory.** The remarkable thing is that the new theory also gives a similar expression for the conductivity with the difference that it is interpreted in a different way.

Relaxation time: This is a new concept very useful in the study of solid state physics phenomenon. We have introduced this concept in connection with the electron-lattice interaction and we found that this relaxation time is closely related to the mean free time between two successive collisions of the electron with lattice. In the derivation for the electrical conductivity σ we have taken the mean free time as τ. We know that when an electron distribution is subjected to an external field the electron reaches a mean drift velocity (v_{D_0}) called the **steady state velocity** on the equilibrium velocity attained as a result of the two opposing influences the applied electric field on one hand and the opposing influence due to the electron lattice collisions on the other hand.

Suppose the electric field is suddenly switched off, the mean drift velocity of the electron is found to decay exponentially. This can be easily obtained by finding the solution of Eq. (18.6). This gives

$$v_D(t) = v_D(o)e^{-t/\tau} \tag{18.12}$$

where $v_D(t)$ is the drift velocity at time t and τ is relaxation time, i.e., τ is the characteristic time governing the establishment of equilibrium condition. In the light of this model the relaxation time may be interpreted as being associated with the governance of the establishment of equilibrium (through collisions) from an initially disturbed situation in which $v_D \neq 0$. For the particular type of collisions postulated above, τ also represents the mean free time between collision. It must be emphasised however, that the relaxation time, and the mean free time between collisions are identical only if the velocity after collision is random. For example, if the scattering is not isotropic and τ is the mean free time between collision, the relaxation time can easily be shown to be

$$\tau = \frac{\tau_c}{(1-\langle\cos\beta\rangle)} \tag{18.13}$$

where $\langle\cos\beta\rangle$ is the average of the cosine of the scattering angle. Thus when nearly all collisions involve small angles, the electron has a rather strong memory and it takes a relatively large number of collision to erase this memory, i.e., $\tau \gg \tau_c$ in that case.

18.3 BOLTZMANN TRANSPORT EQUATION

Boltzmann transport equation is an equation which specify the change of the distribution function with time when non-equilibrium state is established by allowing the electric current or thermal current to flow. This non-equilibrium distribution function will be some what different from the equilibrium distribution function. In the limit when the perturbation from the equilibrium condition vanishes, naturally, the non-equilibrium distribution function must approach the appropriate equilibrium distribution. The resulting equation known as Boltzmann transport equation will form the basis of transport-calculations.

To derive this equation, consider a region of phase space about the point (x, y, z, P_x, P_y, P_z). The number of particles entering this region in time dt is equal to the number which were in the region of phase space at $(x - v_x\,dt, y - v_y\,dt, z - v_z\,dt, P_x - F_x\,dt, P_y - F_y\,dt, P_z - F_z dt)$ at time dt earlier. Here F represents the force acting on the particles of the distribution at the point (x, y, z) at time t, and it follows from Newton's law that $F = \frac{dP}{dt}$. If $f\{x, y, z, P_x, P_y, P_z\}$ is the distribution function, which expresses the number of particles per quantum state in the region, then the change df which occurs during time dt due to the motion of the particles in coordinate space and due to the fact that force field acting on the particles tend to move them from one region to other in momentum space is $df = f(x - v_x\,dt, y - v_y\,dt, z - v_z\,dt, P_x - F_x\,dt, P_y - F_y\,dt, P_z - F_z\,dt) - f(x, y, z, P_x, P_y, P_z)$. Using Taylor's expansion and retaining only first order terms in the limit $dt \to 0$, one obtains

$$df = f(x,y,z,P_x,P_y,P_z) + \left(-v_x\frac{\partial f}{\partial x} - v_y\frac{\partial f}{\partial y} - v_z\frac{\partial f}{\partial z} - F_x\frac{\partial f}{\partial P_x} - F_y\frac{\partial f}{\partial P_y} - F_z\frac{\partial f}{\partial P_z}\right) - f(x,y,z,P_x,P_y,P_z)$$

or $$df = \left(-v_x\frac{\partial f}{\partial x} - v_y\frac{\partial f}{\partial y} - v_z\frac{\partial f}{\partial z} - F_x\frac{\partial f}{\partial P_x} - F_y\frac{\partial f}{\partial P_y} - F_z\frac{\partial f}{\partial P_z}\right)$$

or $$df = -v.\nabla f - F.\nabla_\rho f \tag{18.14}$$

where the symbol ∇_p refers to the gradient operator in momentum space (P_x, P_y, P_z). So far only the change in the distribution function due to motion of the particles in coordinate space and due the momentum changes arising from force field acting on the particles have been accounted for. Particles may also be transferred into or out of a given region of phase space by collision or scattering interactions involving other particles of the distribution or scattering centre external to the assembly of particles under consideration. If the rate of change of the distribution function due to collisions or scattering is denoted by $\left(\frac{\partial f}{\partial t}\right)_{\text{colli}}$ then the total rate of change of f may be obtained by simply adding this quantity to the right hand side of Eq. (18.14), giving

$$\frac{df}{dt} = -v.\nabla f - F.\nabla_p f + \left(\frac{\partial f}{\partial t}\right)_{\text{colli}} \tag{18.15}$$

This is the **Boltzmann's equation** giving the rate at which the distribution function at a given position and momentum changes and contains three contributors. The first is a drift term, the electrons in question are leaving that region of space with velocity v and if the distribution function varies in space, the number leaving the region will be different from the number entering. The second term similarly represents the change arising because the particles at the moment in question are being accelerated into different momentum states. If the distribution function varies with momentum, the number being accelerated out will be different from the number accelerated into the momentum region of interest. Finally there is a change because electrons may be scattered out at a different rate than they are being scattered in.

To discuss the properties of the system which is displaced from equilibrium by a small perturbing force such as an electric field or a temperature gradient, when the distribution function f_0 which is the equilibrium distribution function in the absence of the perturbation is known, a relaxation time approximation is made in transport calculations. This may be written mathematically in the form

$$\left(\frac{\partial f}{\partial t}\right)_{\text{colli}} = \frac{f - f_0}{\tau} \tag{18.16}$$

where τ is relaxation time. The relaxation time approximation is a very plausible one and agrees with a wide range of experiments; however it certainly cannot be true in details. If for example, the scattering events are primarily elastic, they will tend to cause a decay of any current to the equilibrium value of zero, but will not so effectively cause the decay of any isotropic deviation from equilibrium of the distribution as a function of energy. Thus it may be necessary to define different relaxation time and for different phenomenon being studied. In addition we must take care in what we write for equilibrium distribution function f_0 in $\left(\frac{\partial f}{\partial t}\right)_{\text{colli}}$. If, for example distribution function is inhomogeneous, i.e., different total densities of electrons at different points, then in the scattering term, we must write the equilibrium distribution function corresponding to the local density of particles $f_0\{n(r)\}$, otherwise we will be introducing scattering events that instantaneously transfer electrons from one position to another. However, when relaxation-time approximation is used carefully it leads to a good description of many properties. Using this form we may now write the Boltzmann equation in the collision time approximation.

$$\frac{df}{dt} + -v.\nabla f + F.\nabla_{\rho} f = -\frac{f - f_0}{\tau} \tag{18.17}$$

18.4 SOMMERFELD THEORY OF ELECTRICAL CONDUCTIVITY

A theory of metallic conductivity based on average velocities, was developed by Drude in 1900. Lorentz in 1905 reinvestigated the problem using Boltzmann's transport equation and a simplified model for the collisions between electrons and atoms in the lattice. However, the use of classical statistics led to serious difficulties. In 1928 Sommerfeld recalculated the conductivities along the lines of Lorentz theory but replacing classical statistics by Fermi Dirac statistics. Sommerfeld did not investigate the actual mechanism of interaction between the electrons and the lattice any further but assumed that a relaxation time can be defined which is a function of the energy of the electron only. As an application to the Boltzmann transport equation we shall discuss below **Sommerfeld's** theory of electrical conductivity based on the free electron approximation.

We shall consider first a uniform isotropic substance at constant temperature in a steady state under the influence of a constant applied electric field E_0. Since the system is in a steady state, $\frac{df}{dt} = 0$ and since the material is uniform and at constant temperature, the distribution must be the same in every part of the sample, where by $\nabla f = 0$. Under these conditions the force upon an electron is $-E_0 e$ and the Boltzmann equation in the relaxation time approximation may be written as

$$eE_0.\nabla_{\rho} f = \frac{f - f_0}{\tau} \tag{18.18}$$

If the co-ordinate system is chosen so that z-axis is in the direction of E_0 then E_0 has only a z-component and Eq. (18.18) can be written as

$$\frac{eE_0}{m} \cdot \frac{\partial f}{\partial v_z}$$ [1]

---

[1] $eE.\nabla_P f = eE_0 \hat{k}\left(\frac{\partial f}{\partial P_x}\hat{i} + \frac{\partial f}{\partial P_y}\hat{j} + \frac{\partial f}{\partial P_z}\hat{k}\right) = eE_0 \frac{\partial f}{\partial P_z}$

But $P_z = mv_z$

$$\therefore \quad eE_0 \nabla_P f = eE_0 \frac{\partial f}{\partial (mv_z)} = \frac{eE_0}{m}\frac{\partial f}{\partial v_z} = \frac{f - f_0}{\tau}$$

or
$$\frac{eE_0\tau}{m}\cdot\frac{\partial f}{\partial v_z} = f - f_0 \tag{18.19}$$

In Eq. (18.19) the difference $(f - f_0)$ is assumed to be quite small so that $\frac{\partial f}{\partial v_z}$ can be approximated by $\frac{\partial f_0}{\partial v_z}$. In addition since the electrons obey Pauli exclusion principle and hence we should in fact use the Fermi-Dirac distribution function. The equilibrium Fermi-Dirac distribution function is

$$f_0(E) = \frac{1}{\exp\left(\frac{E - E_f}{kT}\right) + 1}$$

hence
$$\frac{\partial f_0}{\partial v} = \frac{\partial f_0}{\partial E}\cdot\frac{\partial E}{\partial v_z} = -\frac{mv_z eP[-(E - E_f)/kT]}{kT[1 + \exp(E - E_f)/kT]^2}$$

$$= \frac{mv_z}{kT}\cdot f_0(1 - f_0) \tag{18.20}$$

Substituting this result into Eq. (18.19), one obtains

$$f = f_0 - \frac{eE_0\tau}{kT} f_0(1 - f_0)v\cos\theta \tag{18.21}$$

Since $v_z = v\cos\theta$ where θ is the angle between the polar z-axis and the direction v. The electrical current-density J_z is defined by

$$J_z = ne\langle v_z\rangle \tag{18.22}$$

Where n is the number of electrons per unit volume and $\langle v_z\rangle$ is expressed as

$$\langle v_z\rangle = \frac{\int v_z f(v) g(v) d^3V}{\int f(v) g(v) d^3V} \tag{18.23}$$

where $g(v)$ is the density of states per unit volume of velocity space and d^3V is an appropriate volume element in velocity space. If we consider spherical coordinates (r, θ, ϕ) in velocity space, we have volume elements d^3V as

$$d^3V = r^2 \sin\theta\, d\theta\, d\phi\, dV \tag{18.24}$$

and knowing that the density of states in velocity space is constant $\left(\frac{2m^3V}{h^3}\right)$, we can write for average value of v_z as

$$\langle v_z\rangle = \frac{\int_0^\infty\int_0^\pi \left[f_0 - \frac{eE_0}{kT} f_0(1 - f_0)\, v\cos\theta\right] v^3 \sin\theta\cos\theta\, d\theta\, dV}{\int_0^\infty\int_0^\pi \left[f_0 - \frac{eE_0\tau}{kT} f_0(1 - f_0)\, v\cos\theta\right] v^2 \sin\theta\, d\theta\, dV} \tag{18.25}$$

In Eq. (18.25) the integration is from 0 to 2π has already been performed. Since the integrands are both independent of ϕ and the factor 2π obtained in numerator and denominator cancel. Since $\int_0^{\pi} \sin\theta\cos\theta d\theta = 0$, hence first term in the numerator and second term in denominator vanish. Now integrating over θ from 0 to π, one obtains

$$\langle v_z \rangle = -\frac{eE_0 \int_0^{\infty} f_0(1-f_0)\tau v^4 dV}{3kT \int_0^{\infty} v^2 f_0 dV} \qquad (18.26)^2$$

Although it is impossible to evaluate these integrals over the Fermi-Dirac distribution function analytically in general case; it is quite easy to obtain an approximate result which is very nearly correct for $T << T_F$ (Fermi temperature)[3] which of course, is usually the case for most metallic conductors under conditions which are encountered in practice. To accomplish this, it is most convenient to express the integrals in Eq. (18.26) as integral over energy rather than velocity by putting $E = \frac{1}{2}mv^2 \left(\text{or } v = \sqrt{\frac{2E}{m}} \right)$ where by we obtain,

$$\langle v_z \rangle = -\frac{2eE_0}{3mkT} \frac{\int_0^{\infty} E^{3/2}\tau(E) f_0(E)(1-f_0(E))dE}{\int_0^{\infty} E^{1/2} f_0(E)dE} \qquad (18.27)$$

It must now be noted that for $T < T_F$ ($kT << E_f$) the Fermi function $F(E_0)$ is very small for E significantly in excess of E_f and the complimentary function $[1 - f_0(E)]$ is very small for values of E which are substantially smaller than E_f as shown in Fig. (18.1). It is easy to show analytically that $f_0(1 - f_0)$ is maximum at $E = E_f$ at which point $f_0(1-f_0) = \frac{1}{4}$ (since f_0 = 1/2 for $E = E_f$) and that the width of the region over which $f_0(1 - f_0)$ is not small compared to this maximum value is of order of kT. There will thus be no significant contribution to the integral in the numerator of Eq. (18.27) except where E takes on values with a few kT units E_f. If $E^{3/2}\tau$ (E) is a function which does not vary much over this energy range, we may regard $f_0(1 - f_0)$ as essentially **Dirac-Delta (δ) function** of $(E - E_f)$ and write

$$f_0(E)[1 - f_0(E)] \approx A\delta(E - E_f) \qquad (18.28)$$

---

[2] $\int_0^{\infty} \cos^2\theta \sin\theta d\theta = \frac{2}{3}$ and $\int_0^{\infty} \sin\theta d\theta = 2$

[3] For temperatures less than T_F the electron will occupy all the energy levels up to $E = E_F$. According to the Pauli exclusion principle and there by form a "degenerate gas" of average kinetic energy $3/5E_F$.

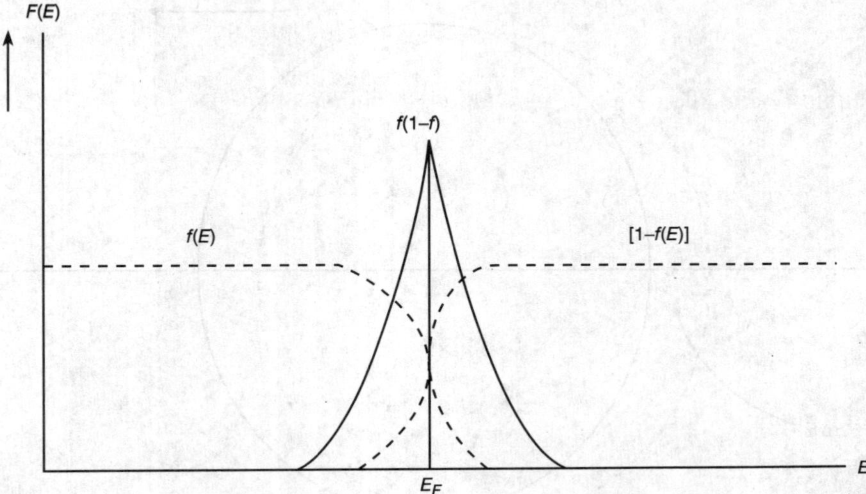

Fig. 18.1

Where A is a normalizing factor, which on actual integration[4] comes out to be kT. Substituting Eq. (18.28) in Eq. (18.27) and in denominator using $f_0(E) = 1$ for $0 < E < E_f$ and zero for $E < E_f$ which is good approximation for $kT << E_f$, we obtain

$$\langle v_z \rangle = -\frac{2eE_0}{3m} \frac{\int_0^\infty E^{3/2}\tau(E)\delta(E-E_f)dE}{\int_0^\infty E^{1/2} dE}$$

$$= -\frac{eE_0 \tau(E_f)}{m} \qquad (18.29)$$

Here we have used the property $\int f(x)\delta(x-a)dx = f(a)$ of the Dirac-δ-function.

The current density $J_z = -ne\langle v_z \rangle$ is obtained by substituting $\langle v_z \rangle$, thus

$$J_z = \frac{ne^2 E_0 \tau(E_f)}{m}$$

---

4 $\int_0^\infty f_0(E)[1 - f_0(E)]dE = \int_0^\infty a\delta(E - E_f)dE = A$

L.H.S. is integrated by putting $x = \dfrac{E - E_f}{kT}$ and remembering that $f_0(E) = \dfrac{1}{e^{(E-E_F)}/kT}$

we have $\int_0^\infty f_0(E)[1 - f_0(E)]dE = kT - E_F / kT \int_{-E_F/kT}^{\infty} \dfrac{e^x dx}{(1+e^x)^2} = \dfrac{kT}{1 + e^{E_f/kT}} \approx kT$

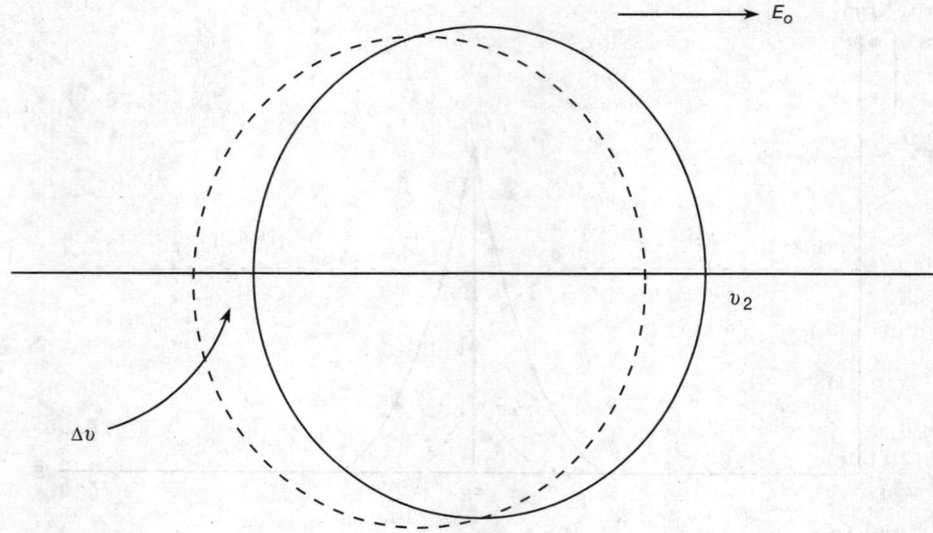

Fig. 18.2

and conductivity

$$\sigma = \frac{j_z}{E_0} = \frac{ne^2\tau(E_f)}{m} = ne\mu \tag{18.30}$$

Where μ is mobility and is defined as

$$\mu = \left|\frac{\langle v_z \rangle}{E_0}\right|$$

It is interesting to note that although all electrons take part in the conduction mechanism only the relaxation time of electrons at the Fermi surface occurs in the conductivity. The reason for this may be explained with reference to Fig. (18.2). The full circle represents the Fermi distribution for a two dimension case in the absence of an external field. In the presence of field along the z-direction, the velocity of all electrons is shifted by an amount Δv (the average drift velocity) leading to the dashed circle. It is evident that the distribution is changed only in the vicinity of the Fermi level, so that only the relaxation time of electrons near E_f is of importance.

The expression for electrical conductivity Eq. (18.30) obtained by Sommerfeld is exactly the same as derived by Drude-Lorentz Eq. (18.10) by the application of the classical free electron of v_D and λ (mean free path). Drude Lorentz incorrectly predicted that the temperature dependence was due to the variation of average energy or the velocity of the electrons and that λ was of the same order of magnitude as the interatomic distance. According to the quantum mechanical theory due to Sommerfeld, the average energy ($\langle E \rangle$) of conduction electron vary very little with temperature. It is therefore necessary to assume that the observed variation of resistivity is caused by the change in mean free path.

To understand the variation property, it is necessary to consider the wave nature of the electrons and to treat the collision between the electrons and the atoms by a scattering process similar to the diffraction of X-rays. When the atoms in a metallic crystals are essentially at rest, the electron waves are coherrently scattered by the perfectly periodic array and proceed in a forward direction without disturbance. It is similar to the passing of the light wave through a perfect crystal, so it follows that electrons can be passed through a perfect crystal without any resistance what so ever for its passage. If all the atoms were to be at rest the mean free path for the electron scattering would be infinite. What then is the mechanism which gives rise to resistance to the passage of electrons through a metallic crystal? Deviation from the periodicity of the lattice are the essential cause for the phenomenon of electrical resistivity. Only at absolute zero of temperature the atomic array may tend to be periodic. Quantum mechanics tells, that even at absolute zero of temperature the atoms make zero point vibrations which may not be perfectly **harmonic.** As temperature increases the periodicity of the atomic array is considerably disturbed by an-harmonic oscillations executed by the atomic oscillators. These atomic displacements cause an in-coherent scattering of the electron waves, since they are not in phase with each other. The periodicity of lattice is not only disturbed by the lattice vibrations but also by the lattice defects, such as vacancies interstitial and dislocations, presence of foreign impurity atoms and boundaries. The mean free path is inversely proportional to the square of the atomic displacements which however are directly proportional to the temperature. Thus

$$\lambda \propto \frac{1}{T} \quad \text{or} \quad \sigma \propto \frac{1}{T} \quad \text{and} \quad \rho \propto T \tag{18.31}$$

Actually this is correct only above a critical temperature called the **Debye temperature,** and below this critical temperature the temperature dependence is non-linear.

18.5 MATHIESSEN'S RULE

The resistivity of metals arises due to the scattering of the conduction electrons by lattice vibrations (or phonon scattering as it is called), by impurities and by structural imperfections. At low temperatures the impurity scattering is dominant and the resistivity due to it is found to be practically temperature independent. The total resistivity (ρ) of a metal can be given as the sum of three components, namely ρ_ℓ the resistivity due to lattice scattering which is temperature dependent, ρ_i the resistivity due to scattering by impurities and ρ_d the resistivity due to defects or dislocations in the structure, i.e.,

$$\rho = \rho_\ell + \rho_i + \rho_d \tag{18.32}$$

This relation is known as Mathiessen's Rule. The reason for the above behaviour of resistivity lies in the mechanism of scattering process. The probability of scattering P_s of the electrons per unit time is given by the reciprocal of the relaxation time τ. That is

$$\tau = \frac{1}{P_s} \tag{18.33}$$

This means that larger (the relaxation time, smaller is the probability of scattering and greater is the electrical conductivity as given by the expression $\sigma = \dfrac{ne^2\tau}{m}$.

We will define another quantity called the **'Scattering probability per unit length'** (Σ_s). If the electrons are assumed to move with a mean velocity v_f (the velocity corresponding to the Fermi energy), we can write the scattering probability as

$$\rho_s = \Sigma_s v_f$$

or

$$\tau = \frac{1}{P_s} = \frac{1}{\Sigma_s v_f} \tag{18.34}$$

The calculations of Σ_s is some what difficult so we will make an approximation by assuming that Σ_s is proportional to the number of scattering centres per unit volume (n_s) and so we can write

$$\Sigma_s = A_s n_s \tag{18.35}$$

where the constant A_s is called the scattering cross-section per centre so

$$\tau = \frac{1}{v_s A_s n_s} \tag{18.36}$$

Any variation in the value of τ is only due to variation in A_s since v_s and n_s are constants for a given metal. It can be shown by rigorous theoretical calculations that for lattice vibrations A_s is proportional to the square of the amplitude (A) of the lattice vibrations, that is

$$A_s \propto A^2 \tag{18.37}$$

Since the energy of the atomic oscillator is proportional to the square of the amplitude and again since the vibrational energy of oscillator is proportional to the absolute temperature (T), we can say

$$A_s \propto T \tag{18.38}$$

and thus

$$\tau \propto \frac{1}{T} \tag{18.39}$$

Hence the conductivity $\sigma \propto \frac{1}{T}$ and this relation is in accordance with experimental results. We can now make the reasonable assumption that the scattering probability per unit length (Σ_s) is

$$\Sigma_s = \Sigma_{s\ell} + \Sigma_{si} + \Sigma_{sd}$$

substituting this value of Σ_s in Eq. (18.34), one obtains

$$\frac{1}{\tau} = v_f(\Sigma_{s\ell} + \Sigma_{si} + \Sigma_{sd})$$

and resistivity $\rho = \frac{m}{ne^2\tau}$ and hence

$$\begin{aligned}\rho &= \frac{m}{ne^2} v_f(\Sigma_{s\ell} + \Sigma_{si} + \Sigma_{sd}) \\ &= \frac{mv_f}{ne^2}\Sigma_{s\ell} + \frac{mv_f}{ne^2}\Sigma_{si} + \frac{mv_f}{ne^2}\Sigma_{sd} \\ &= \rho_\ell + \rho_i + \rho_d\end{aligned} \tag{18.40}$$

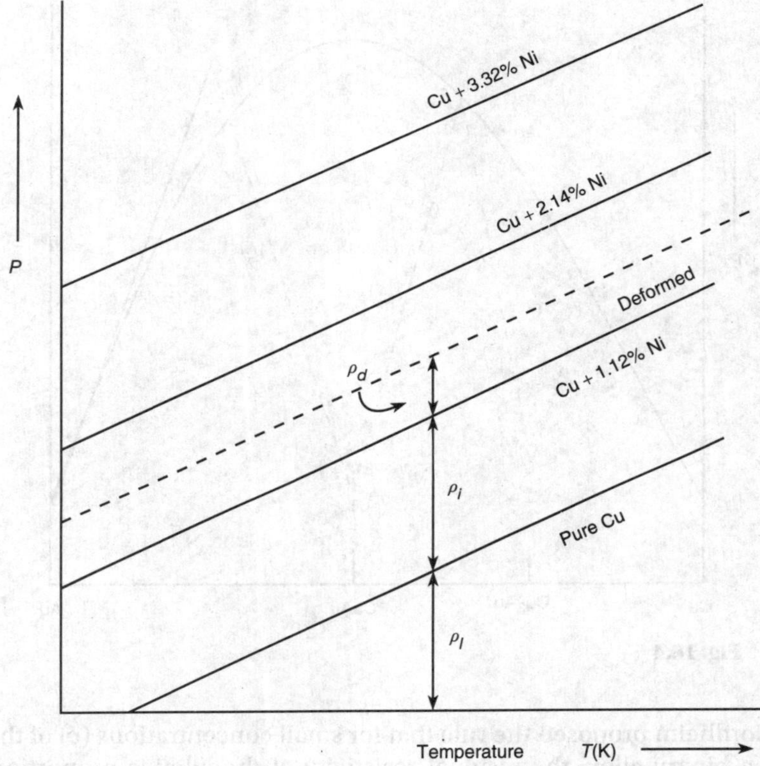

Fig. 18.3

This additive nature of the resistivity is illustrated in Fig. (18.3) in the case of Cu-Ni alloy. The residual resistivity or the extra polated resistivity at 0 K gives the value of the resistivity due to impurities and defects.

18.5.1 Electrical Resistivity of Alloys

A metallic alloy is formed when one metal is dissolved in another and remains dissolved in the solid state. The solution is then called a solid solution. Brass is an alloy of copper and zinc (in the ratio 6:40); steel is formed by dissolving small amount of carbon in iron. It is observed that two types of solid solutions occur: (i) the **substitutional solid solutions** of which brass is an example in which the atoms of the solute elements occupy some of the lattice point previosuly occupied by the solvent atoms with a random distribution. The atomic size of these atoms are the same within 10–15%. (ii) the **interstitial solid solution** in which the solute atoms (like C, H) have such a small size that they can be accommodated within the "empty" space between the solute atoms when the latter are packed to form a crystalline lattice. Since the solute atoms because of their atomic size differences produce a change of the Fourier co-efficient of the lattice potential they produce a change in the residual resistivity of solvent metal.

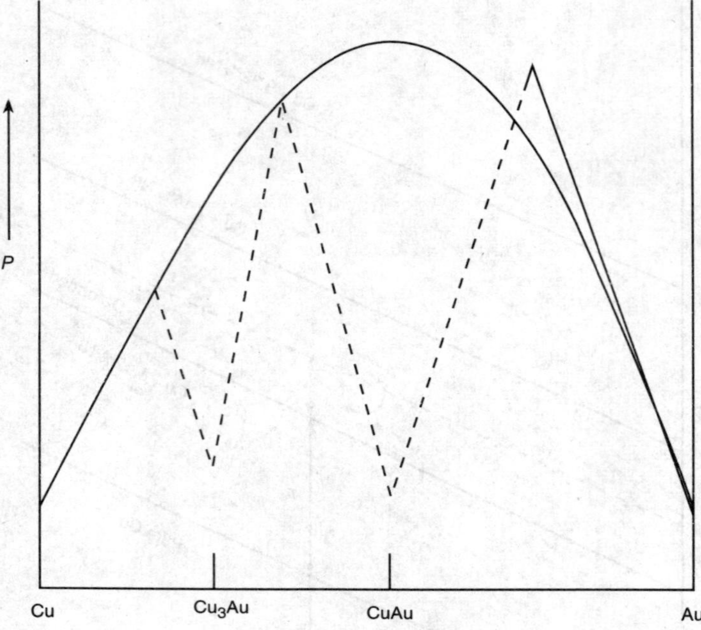

Fig. 18.4

In 1931 Nordheim proposed the rule that for small concentrations (c) of the alloying element in a binary alloy, the residual resistivity of the alloy is proportional to the product of the solute concentration c and the solvent concentration $(1 - c)$. Thus Nordheim rule states that

$$\rho(c) = Ac(1 - c) \tag{18.41}$$

where A is a constant. This rule holds well for alloys which do not contain transition elements in solution or do not under-go phase transitions.

Many alloys form super lattices or ordered substitutional solid solutions below a critical temperature by a first order transition. Examples are copper-gold alloys (25% gold or above 400°C) betabrass (5% Cu, 470°C), etc. X-ray investigations show that in the super-lattice, i.e., in the ordered state, alternate planes in the lattice of the alloys are composed entirely of copper or gold atoms, where as above the critical temperature the gold atoms are distributed at random over the lattice points. The electrical resistivity of super-lattice as a function of the heat treatment and composition show peculiarities due to long range ordering. For annealed alloys, the resistivity decreases sharply by a factor 2 or 3 at compositions for which the alloy is fully ordered Fig. (18.4). For disordered alloys (when the disorder existing at high temperatures is retained by quenching, i.e., cooling the alloy fast enough to low temperatures) the resistivity variation with composition is smooth and obeys Nordheim's rule.

18.6 THERMAL CONDUCTIVITY AND WIEDEMANN-FRANZ LAW

The high thermal conductivity of metals show that metallic electrons are good carriers of heat. Considering the electrons to form a perfect gas one can show from kinetic theory consideration that the thermal conductivity is given by:

$$k = \frac{1}{3} C_V \lambda C \tag{18.42}$$

Where C_V is heat capacity, λ mean free path given by $\lambda = c\tau$, where τ is relaxation time and c average thermal velocity (root mean square velocity). The thermal conductivity of free electron gas can be obtained by using the expression

$$C_V = \frac{\pi^2}{2} \frac{Nk_B T}{T_F}$$

for heat capacity. And since only electrons at Fermi level are effective, we have

$$E_f = \frac{1}{2} mv_F^2 = k_B T_F$$

Thus we have

$$K_{c1} = \frac{1}{3} \frac{\pi^2}{2} \frac{Nk_B T}{T_F} v_F \lambda$$

$$= \frac{1}{3} \frac{\pi^2}{2} . \frac{Nk_B Tk_B}{\frac{1}{2} mv_f^2} . v_F v_f \tau(E_f)$$

$$= \frac{\pi^2}{3} \frac{Nk_B^2 T}{m} \tau(E_f) \tag{18.43}$$

where $\lambda = v_f \tau(E_f)$, $\tau(E_f)$ is the collision time or relaxation time and N is electron concentration.

The Wiedemann-Franz law states that for metals at not too low temperature the ratio of the thermal conductivity to the electrical conductivity is directly proportional to the temperature, with the value of the constant of proportionality independent of the particular metal. This result was most important in the history of the theory of metals as it supported the picture of an electron gas. Using Eq. (18.30) for electrical conductivity and Eq. (18.43) for thermal conductivity, the Wiedemann Franz ratio can be written as

$$\frac{k}{\sigma} = \frac{\pi^2 k_B^2 TN\tau(E_f)}{3m} \times \frac{m}{Ne^2 \tau(E_f)} = \frac{\pi^2}{3} \left(\frac{k_B}{e} \right)^2 T \tag{18.44}$$

The Lorentz number L is defined as

$$L = \frac{k}{\sigma T} = \frac{1}{3} \left(\frac{\pi k}{e} \right)^2 \tag{18.45}$$

This remarkable result involves neither N nor m. It does not involve τ, if the relaxation times are identical for electrical and thermal processes. The experimental values of L at various temperatures are found to be in good agreement with relation Eq. (18.45).

At low temperatures ($T << \theta_0$) the value of Lorentz number tends to decrease. For pure copper near 15 K the observed value is attributed to a difference in the collision averages involved in the thermal and electrical conductivities, so that the thermal and electrical relaxation times are not identical in value.

18.7 THE HALL EFFECT

When a magnetic field is applied perpendicular to a conductor carrying current, a voltage is developed across the specimen in the direction perperdicular to both the current and the magnetic field. This phenomenon is known as the *Hall effect.*

The physical process under-lying the Hall effect is illustrated in Fig. (18.5). Suppose that an electric current I_x is flowing in a wire in the x-direction and a magnetic field B_z is applied normal to the wire in the z-direction. We shall show that this leads to an additional electric field, normal to both I_x and B_z, that is in the y-direction.

To see how this comes about, let us first consider the situation before the magnetic field is introduced. There is an electric current flowing in the positive x-direction, which means that the conduction electrons are drifting with a velocity v in the negative x-direction. When the magnetic field is introduced, the Lorentz force $\boldsymbol{F} = e(\boldsymbol{v} \times \boldsymbol{B})$ causes the electrons to bend down-wards as shown in Fig. (18.5). As a result, electrons accumulate on the lower surface, producing a net negative charge there. Simultaneously a net positive charge appears on the upper surface because of the deficiency of electrons there. This combination of positive and negative surface charges creates a down ward electric field; which is called the Hall field.

Let us evaluate this Hall field. The Lorentz force F_L which produces the charge accumulation in the first place is in the negative y-direction, and has the value

$$F_L = ev_x B$$

Now the field created by the surface charges produce a force which opposes this Lorentz force. The accumulation process continues untill the Hall force completely cancels the Lorentz force. This is the steady state $F_H = F_L$. If the Hall field be E_H, then Hall force

$$F_H = eE_H$$

in the positive y-direction and in steady state we, have

$$eE_H = ev_x B$$

or

$$F_H = v_x B \tag{18.46}$$

It is convenient to express this interms of measurable quantities and for this purpose the velocity is expresed in terms of the current density $I_x = N(-e)\,v_x$. This leads to

$$E_H = -\frac{1}{Ne} J_x B \tag{18.47}$$

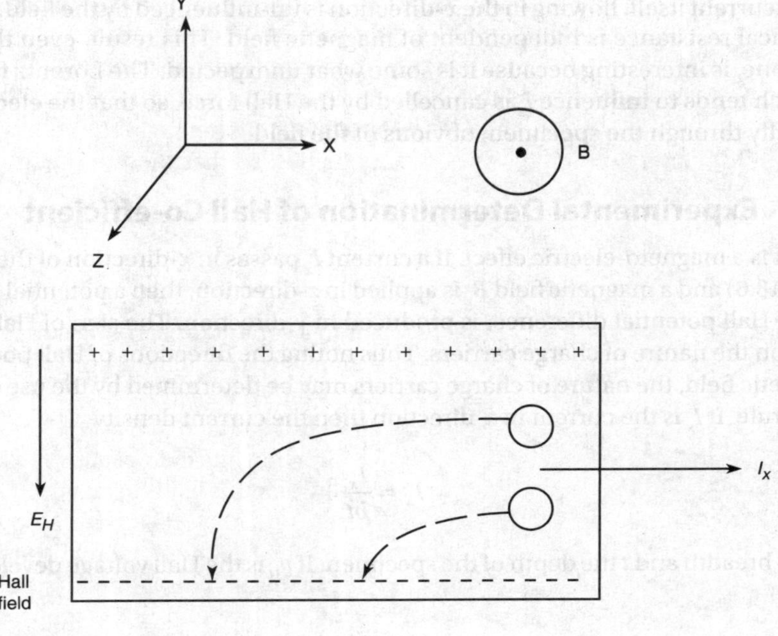

Fig. 18.5

The Hall field is thus proportional both to the current and to the magnetic field. The proportionality constant, i.e.,

$$R_H = \frac{E_H}{j_x B} = \frac{1}{Ne} \tag{18.48}$$

R_H is known as **Hall constant.** The result Eq. (18.48) is very useful one in practice. Since R_H is inversely proportional to the electron concentration N, it follows that we can determine it by measuring Hall field. Infact, this is the standard technique for determining electron concentration. The technique is particularly valuable because apart from N the only other quantity on which R_H depends is the charge on the electron ($-e$) which is a fundamental physical constant.

Another useful feature of the Hall constant is that its sign depends on the sign of the charge of the current carriers. Thus electrons, being negatively charged, lead to a negative Hall constant. By contrast the Hall co-efficient due to conduction by holes (which are positively charged) is positive[5]. Thus the sign of R_H indicates the sign of the carriers involved which is very valuable information, particularly in the case of semi-conductors. For example, the Hall constant for both Zn and Cd are positive indicating that the current in these substances is carried by holes. The above analysis shows another interesting aspect of the transport phenomenon in the presence of a magnetic

[5] These holes are different from Fermi holes as already discussed in this chapter. These concern with the semi-conductors (See chapter on semi-conductors).

field. The current itself, flowing in the x-direction is un-influenced by the field. Therefore the electrical resistance is independent of magnetic field. This result, even though it is negative one, is interesting because it is some what unexpected. The Lorentz force of the field, which tends to influence J_x is cancelled by the Hall force, so that the electrons flow horizontally through the specimen, obvious of the field.

18.7.1 Experimental Determination of Hall Co-efficient

Hall effect is a magneto-electric effect. If a current I_x passes in x-direction of the specimen slab Fig. (18.6) and a magnetic field B_z is applied in z-direction, then a potential difference, called the Hall potential difference, is produced in y-direction. The sign of Hall potential depends on the nature of charge carriers. Thus noting the directions of Hall potential and the magnetic field, the nature of charge carriers may be determined by the use of Fleming left hand rule. If I_x is the current in x-direction then the current density

$$J_x = \frac{I_x}{bt}$$

where b is breadth and t the depth of the specimen. If v_H is the Hall voltage developed, then Hall field

$$E_H = \frac{V_H}{b}$$

Then Hall Co-efficient

$$R_H = \frac{E_H}{J_x B_z} = \frac{V_H}{b.\frac{I_x}{bt}B_z} = \frac{V_H \cdot t}{I_x B_z}$$

Thus the measurement of R_H involves the measurement of v_H.

The specimen is placed in the magnetic field and the circuit is completed as shown in Fig. (18.7). B is a battery and Rh is a rheostat which acts as potential divider. v_x is the voltmeter to measure the applied potential difference across the specimen and A is the ammeter connected in series to read the current (I_x) passing through the specimen, mV is the sensitive digital milli voltmeter or a standardized vernier potentio-meter to measure the Hall potential difference developed in the specimen along y-direction.

The magnetic field is applied by switching on an electromagnet. Key K_1 is closed to allow a current I_x in the specimen. The current I_x in ammeter and potential difference v_x across the specimen length is noted. Key K_2 is also closed to switch on the milli-voltmeter (or potentiometer) circuit and Hall potential difference v_H is noted. The value of I_x is varied and the corresponding values of V_x are measured.

The magnetic field strength is measured with a flux meter or a precaliberated ballistic galvanometer. The actual value of the magnetic field inside the crystal is found by using the relation

$$B_z = \mu B$$

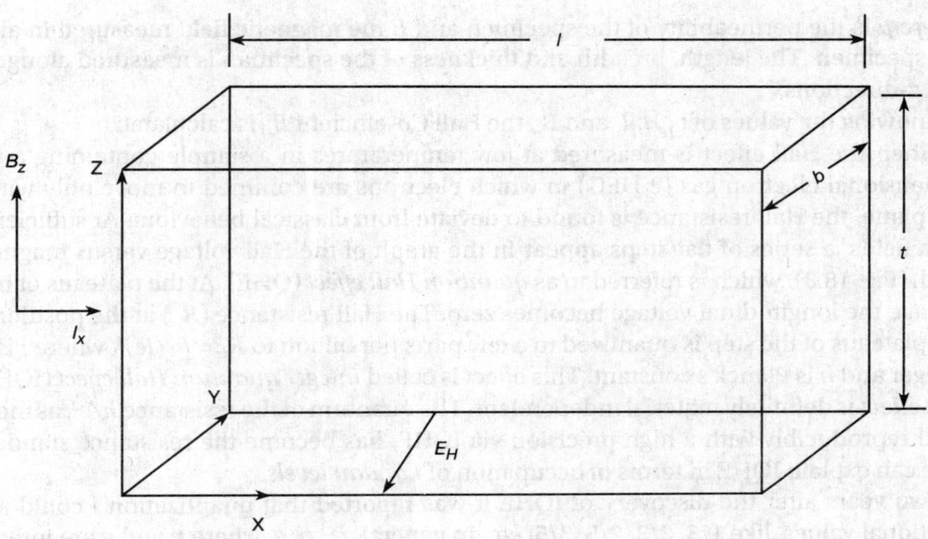

Fig. 18.6

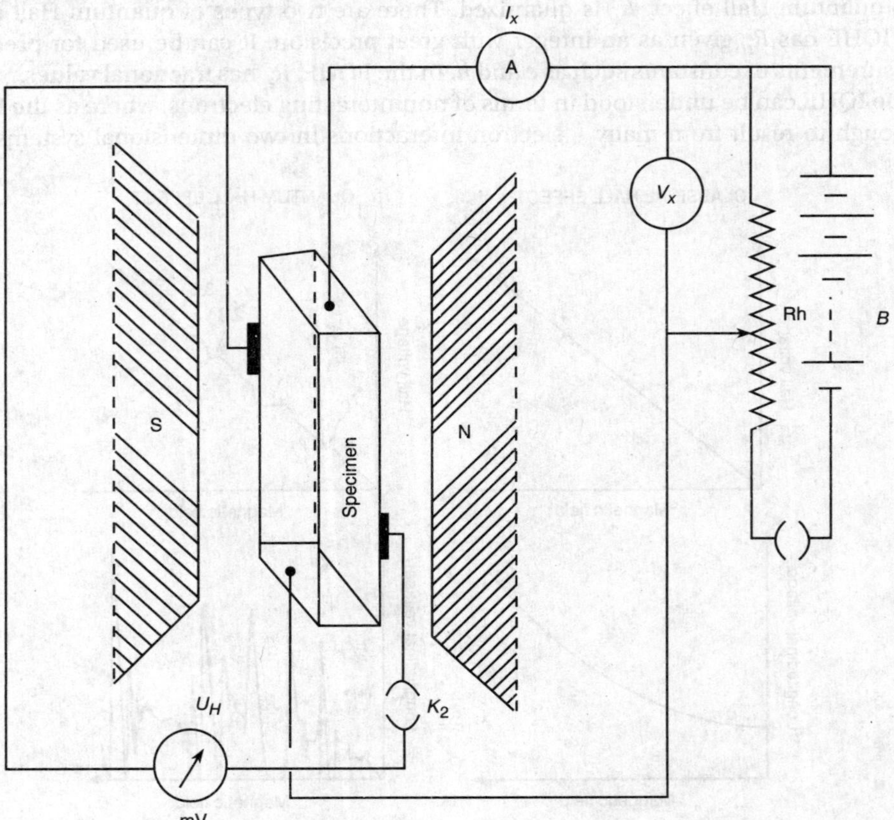

Fig. 18.7

where μ is the permeability of the specimen and B the magnetic field measured in air at the specimen. The length, breadth and thickness of the speciman is measured along x, y and z-directions.

Knowing the values of v_H, t, I_x and B_z, the Hall Co-efficient B_H is calculated.

When the Hall effect is measured at low temperatures in a sample containing a two dimensional electron gas (2 DEG) in which electrons are confined to move only within the plane, the Hall resistance is found to deviate from classical behaviour. At sufficiently high fields, a series of flat steps appear in the graph of the Hall voltage versus magnetic field. (Fig. 18.8), which is referred to as *quantum Hall effect* (QHE). At the plateaus of high voltage the longitudinal voltage becomes zero. The Hall resistance (R_H) at the position of the plateaus of the step is quantized to a few parts per billion to $R_H = h/(ie)^2$, where i is an integer and h is Planck's constant. This effect is called *integer quantum Hall effect* (IQHE). The effect is definitely material independent. The quantum of the resistance h/e^2, as measured reproducibly with a high precision via IQHE, has become the resistance standard. One can explain IQHE in terms of occupation of *Landau levels.*

Two years after the discovery of IQHE it was reported that quantization i could take fractional values, like 1/3, 2/3, 2/5, 3/5, etc. In general, $i = p/q$, where p and q are integers with q odd. The phenomenon was named as *fractional quantum Hall effect* (FQHE).

In quantum Hall effect, R_H is quantized. There are two types of quantum Hall effect. The IQHE has R_H given as an integer with great precision. It can be used for precision measurements of constants such as e and h. In the FQHE, R_H has fractional values.

The IQHE can be understood in terms of noninteracting electrons, where as the FQHE is though to result from many - electron interactions in two dimensional systems, and

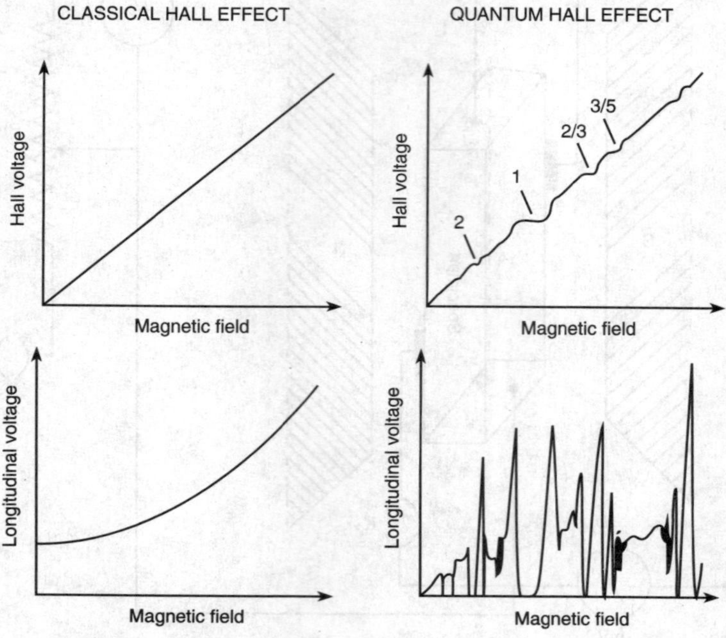

Fig. 18.8 Variation of the longitudinal voltage drop and the Hall voltage as a function of the magnetic field in the classic and quantum Hall effect

can be an example of *anyons* (these quasi particles have statistics intermediate between bosons and fermions).

We may note that analogous to the quantum Hall effect exist for spin.

Example 18.1: The electrical resistivity of a certain copper sample is 1.77 × 10^{-8} Ωm. Use the free electron approximation to estimate (a) the relaxation time and (b) the average speed of electrons in a field of 100 Vm^{-1}. Copper is FCC with a cube edge of 3.61 Å and each atom contributes one electron to a nearly free electron band.

Solution (a) Copper being FCC has 4 atoms per unit cell. Thus the concentration of electrons in the band is

$$n = \frac{4}{a^3} = \frac{4}{(3.61\times10^{-10})^3} = 8.5\times10^{28} \text{ electrons m}^{-3}$$

Conductivity σ is given by $$\sigma = \frac{ne^2\tau}{m}$$

or relaxation time $$\tau = \frac{\sigma m}{ne^2} = \frac{m}{ne^2\rho}$$

∵ Conductivity σ is the reciprocal of the resistivity ρ, and therefore

$$\therefore \quad \tau = \frac{9.1\times10^{-31}}{8.5\times10^{28}\times(1.6\times10^{-9})^2\times1.77\times10^{-8}} = 2.37\times10^{-14}\text{ s}$$

(b) The average speed is

$$\langle v\rangle = \frac{e\tau}{m}E_0 = \frac{1.6\times10^{-19}\times2.37\times10^{-14}}{9.1\times10^{-31}}\times100 = 0.416\text{ ms}^{-1}$$

This speed is much less than the speed of an electron at the Fermi surface which is about 1.6 × 10^6 m/sec for copper.

Example 18.2: In a certain copper sample the electron drift velocity is 2.16 ms^{-1} in an electric field of 500 V/m. Estimate (a) the electron mobility (b) the relaxation time.

Solution (a) We know that mobility is given by

$$\mu = \frac{\langle v\rangle}{E_0} = \frac{2.16}{500} = 4.32\times10^{-3}\text{ m}^2\text{/V-s}$$

(b) The relaxation time in terms of mobility is given by

$$\tau = \frac{m^*\mu}{e}$$

Taking m^* to be free electron mass then

$$\tau = \frac{9.1\times10^{-31}\times4.32\times10^{-3}}{1.6\times10^{-19}} = 2.46\times10^{-14}\,\text{s}$$

Example 18.3: Consider a slab of copper 0.1 mm thick, 1.0 mm wide and 10.0 mm long:

(a) If we drive a current 1 A down the length of the slab, what is the current density?

(b) If we then put the slab in the magnetic field of 1 T with the field perpendicular to 1 mm × 10 mm face, what Hall field will be produced, if the Hall coefficient is 0.55 × 10^{10} m^3/c.

(c) What Hall Voltage will be observed across the slab?

Solution

(a) Current density is given by

$$J_x = \frac{\text{Current}}{\text{Area of cross-section}} = \frac{1}{(\text{breadth})\times(\text{thickness})}$$

$$= \frac{1}{1\times10^{-3}\times0.1\times10^{-3}} = 10^7\,\text{A/m}^2$$

(b) Hall field is given by

$$E_H = R_H . J_x B$$

$$= -0.55\times10^{-10}\times10^7\times1 = 5.5\times10^{-4}\,\text{v/m in magnitude}$$

(c) Hall Voltage is given by

$$V_H = E_H \times \text{breadth} = 5.5\times10^{-4}\times1\times10^{-3} = 5.5\times10^{-7}\,\text{V} = 0.55\mu\text{V}$$

Example 18.4: Show that if the mean free path is independent of the velocity, the conductivity of Maxwell-Boltzmann free electron gas may be expressed as

$$\sigma = \frac{4}{3}\frac{ne^2\lambda}{\sqrt{2\pi mk_B T}}$$

Solution When λ and τ are independent of velocity, the Maxwell-Boltzmann distribution law give average value of τ as

$$\tau = \frac{\lambda<v>}{<v^2>}$$

But kinetic theory and law of equipartition of energy gives

$$\frac{1}{2}m\langle v^2\rangle = \frac{3}{2}k_BT$$

or

$$\langle v^2\rangle = 3k_BT/m$$

Hence

$$\tau = \frac{\lambda\langle v\rangle}{\dfrac{3k_BT}{m}}$$

But

$$\langle v\rangle = \sqrt{\frac{8k_BT}{\pi m}}$$

$\therefore$

$$\tau = \lambda\sqrt{\frac{8k_BT}{\pi m}}\frac{m}{3k_BT}$$

But the electrical conductivity is given by

$$\sigma = \frac{ne^2\tau}{m} = \frac{ne^2}{m}\lambda\sqrt{\frac{8k_BT}{\pi m}}\frac{m}{3k_BT}$$

$$= \frac{ne^2\lambda}{m}\sqrt{\frac{8}{\pi mk_BT}} = \frac{4}{3}\frac{ne^2\lambda}{\sqrt{2\pi mk_BT}}$$

Example 18.5: Estimate the collision time for free electrons in copper applying classical free electron theory. Find the corresponding mean free path and compare it with the experimental value

Given: conductivity $\sigma = 6.8 \times 10^7 \Omega^{-1} - \text{m}^{-1}$, $m_e = 9.1 \times 10^{-31}$ kg and $e = 1.6 \times 10^{-19}$C.

Solution $\sigma\dfrac{ne^2\tau}{m}$ or $\tau = \dfrac{\sigma m}{ne^2}$ and $n = \dfrac{N_A\rho_x}{M_A}$

Here $N_A = 6.02 \times 10^{26}$/k-mol, $\rho_x = 8.92 \times 10^3$ kg/m³ (density) and $M_A = 63.55$. Substituting these values, we obtain,

$$n = \left(\frac{6.02\times10^{26}\times8.92\times10^3}{63.55}\right) = 8.5\times10^{28}/\text{m}^3$$

Now,

$$\tau = \frac{\sigma m}{ne^2} = \frac{6.8\times10^7\times9.1\times10^{-31}}{8.5\times10^{28}\times(1.6\times10^{-19})^2} = 2.84\times10^{-14}\text{ s}$$

Example 18.6: The density and electrical conductivity of copper are 8.95×10^3 kg/m³ and $6.4 \times 10^7\ \Omega^{-1}$-m^{-1} respectively at room temperature. Determine the density of free electrons and relaxation time. Given $M_{\text{Cu}} = 63.54$ kg.

Solution Density of free electrons,

$$n = \frac{N\rho}{M} = \frac{6.02\times10^{26}\times8.95\times10^3}{63.54} = 8.5\times10^{28}/\text{m}^3$$

Relaxation time,

$$\tau = \frac{\sigma m}{ne^2} = \frac{6.4\times10^7\times9.1\times10^{-31}}{8.5\times10^{28}\times(1.6\times10^{-19})^2} = 2.67\times10^{-14}\text{ s}$$

Example 18.7: The electrical conductivity at copper of 300 K is 6.4 × 10^7/ohm/m. Assuming Wiedemann-Franz law to be valid, show that the electronic contribution to the thermal conductivity of copper is 470.4 W/m/K.

Solution We have $K = \sigma T \dfrac{\pi^2}{3} \dfrac{k_B^2}{e^2}$

$$= 6.4 \times 10^7 \times 300 \times 2.45 \times 10^{-8} = 470.4 \text{ W/m/K}$$

Example 18.8: An alloy of a metal is found to have a resistivity of 10^{-6}Ω at 300K. When the alloy is heated to a temperature of 1000 K, the resistivity increased by 7%. Using Mtthiessen's rule, show that the resistivity of the alloy is 0.97 × 10^{-10}Ω-m. [Vikram]

Solution

$$\rho = \rho_0 + \alpha T_1 \tag{i}$$

and
$$\rho + 0.07\rho = \rho_0 + \alpha T_2 \tag{ii}$$

Subtracting and simplifying, one obtains

$$\alpha = \frac{0.07\rho}{T_2 - T_1} = 0.07 \times \frac{10^{-6}}{700} = 1 \times 10^{-6}\,\Omega\text{-m/K}$$

Again, $\rho = \rho_0 + dT$

∴ $\rho_0 = \rho - dT_1 = 10^{-6} - 1 \times 10^{-10} \times 300 \;\; = 0.97 \times 10^{-10}\,\Omega\text{-m}$

SUMMARY

1. Ohm's law is the law of conduction in metals:

$$I = \frac{V}{R} \quad \text{or} \quad J = \sigma E_0$$

where σ is electrical conductivity, which measures the capability of the solid to conduct electric charge under the influence of an electric field.

$$\sigma = ne\mu$$

where μ is the mobility.

2. The resistivity of an impure specimen can be expressed by *Mathiessen's rule*

$$\rho = \rho_0 + \rho(T)$$

3. Conductivity varies in the presence of magnetic field. This effect is termed as *magneto resistance.*
4. The mean velocity acquired by the electron gas due to the application of the electric field is called the *drift velocity.*

5. Boltzmann transport equation specify the change of the distribution function with time when non-equilibrium state is established by allowing the electric current or thermal current to flow.
6. The resistivity of metals arises due to the scattering of conduction electrons by lattice vibrations (or phonon scattering) by impurities and by structural imperations.

$$\rho = \rho_l + \rho_i + \rho_a$$

REVIEW QUESTIONS AND PROBLEMS

1. Derive an expression for Hall constant by free electron theory. Describe an experimental set up to observe the Hall effect. What important informations can be obtained from measurement of this effect.
2. Derive an expression for the electrical conductivity of a free electron gas using the collision time concept. Does this result explain the experimental values of resistivity of a normal metal like sodium.
3. Write down Boltzmann transport equation for electrons gas under external electric field. Solve it to obtain an expression for electrical conductivity in solids.
4. On the basis of free electron theory derive an expression for the electrical and thermal conductivity of metal and hence Wiedemann Franz Law.
5. Consider standard geometry for Hall effect where a rod-shaped specimen is placed in a magnetic field along z-direction. An electric field is applied along the length of rod which is also taken as x-direction. Show that the resistance is independent of the magnetic field, provided we assume single relaxation time for all the electrons.
6. What is Wiedemann-Fanz Law? Show that the Lorentz number is equal to $\frac{\pi^2}{3}\left(\frac{k}{e}\right)^2$, where the symbols have their usual meanings.
7. Calculate the electric conductivity of free electron Fermi gas using Boltzmann equation and the relaxation time approximation.
8. Explain why pure metals show a constant resistivity at low temperature.
9. Derive an expression for the electronic thermal conductivity of a metal and deduce Wiedeman-Franz law. What is Lorentz number?
10. A copper wire of cross-sectional area 5×10^{-9} sq.cm. carries a steady current of 50 amperes. Assuming one free electron per atom, calculate:
(a) the density of free electrons (b) the average drift velocity (c) The relaxation time. Given the resistivity of copper as $1.7 \times 10^{-6}\ \Omega$-cm.
11. What variables determine relaxation time τ, and How?
12. Explain qualitatively why the Hall constant for a current of positive charges is positive?
13. The residual resistivity for 1 atomic percent of As impurities in Cu is $6.8 \times 10^{-8}\ \Omega$-m. Calculate the cross-section for the scattering of an electron by one As impurity in Cu.
14. The electrical resistivity of copper at room temperature is 1.55×10^{-8} ohm-m and density 8.95×10^{3} kg-m^{-3}. Calculate the relaxation time and mean free path at Fermi level. Assume one free electron per atom. Also calculate the average drift velocity of these electrons in a field of 1 Volt per cm and compare the result with the average velocity in the absence of a field.

15. Show that on the basis of the classical picture of electron scattering by rigid spheres (the atom) and on the assumption that the electrons obey Boltzmann statistics, the electrical conductivity should be proportional to $T^{-1/2}$. How does it compare with experiment?
16. Consider a group of similar particles which at the instant $t = 0$ all move in the x-direction with the same velocity v_{ox}. Suppose the particles are scattered by obstacles such that the average time between collision is τ_c. The scattering is not isotropic. Show that the average velocity of the group measured along the x-direction decreases exponentially to zero with a relaxation time $\tau = \frac{\tau_c}{[1-\langle \cos\beta \rangle]}$ where β is the scattering angle and $< \cos\beta >$ is the average of a $\cos\beta$.
17. The atomic radius of silver which crystallizes in FCC structure is 1.4. Calculate the Fermi energy of silver at 0 K. **[5.7 eV]**
18. Show that the wavelength associated with an electron having an energy equal to the Fermi energy is given by $\lambda_F = 2(2\pi/3n)^{1/3}$.
19. Estimate the fraction of electrons excited about the Fermi level at room temperature for Na and Cu. **[0.8%, 0.36%]**
20. A Copper wire has a resistivity 1.8×10^{-8} Ω-m at 300 K. Assuming copper is very pure, estimate the resistivity at 700°C and the percentage change in the resistivity from room temperature to 700°C. **[5.84 × 10^{-8} Ω-m and 22.4%]**

SHORT ANSWER-QUESTIONS

1. What are metallic conductors?

   Ans: Metallic conductors obey Ohm's law, which states that the current in the steady state is proportional to the electric field strength.
2. What are the characteristics of metallic materials?

   Ans: Electrical resistivity increases with temperature, impurity content and plastic deformation. The contribution of each to the total resistivity is additive. The resistivity of an impure specimen is given by Mathiessen's rule

$$\rho = \rho_0 + \rho(T)$$

   Where ρ_0 is a constant for the impure specimen and $\rho(T)$ is the temperature dependent resistivity of the pure specimen.
3. How the resistivity of a metal varies with temperature at low temperatures?

   Ans: At low temperatures, the resistivity (ρ) is proportional to the fifth power of absolute temperature, i.e., $\rho \propto T^5$.
4. What is magnetic resonance?

   Ans: The conductivity of a metal varies in the presence of magnetic field. This effect is known as magnetic resonance.
5. What do you know about electric and thermal conductivities of metals?

   Ans: Metals have high electric and thermal conductivities.

6. What is Wiedemann-Franz law?
   Ans: The ratio of electrical to thermal conductivities of a metal is directly proportional to the absolute temperature.
7. Where pure metals are preferred?
   Ans: Pure metals are chosen for high conductivity and large temperature coefficient of resistance.
8. How the resistivity is affected when the mobility of electrons in a metal increases?
   Ans: Resistivity decreases.
9. The collision time and root mean square velocity of the electron at room temperature are 2.5×10^{-14} s and 1×10^5 m/s respectively. What will be the classical value of the mean free path of the electron?
   Ans: 2.5 nm
10. How the electrical resistivity of a metal depends on temperature according to classical theory?
    Ans: $\rho \propto T^{1/2}$
11. If the Fermi energy of silver at 0 K is 5 eV, what will be the mean energy of electron in silver at 0 K?
    Ans: 3 eV
12. How electric field E, conductivity σ and current density J, are related in accordance to ohm's law?
    Ans: $J = E/\sigma$
13. Write the unit of Hall coefficient.
    Ans: $Vm^3\ A^{-1}\ wb^{-1}$
14. If e, μ and n respectively represent the charge, mobility and concentration of electrons respectively, what will be the electrical conductivity of metal?
    Ans: $ne\mu$

OBJECTIVE QUESTIONS

1. If the mobility of electrons in a metal increases, the resistivity
   (a) decreased (b) increases
   (c) first decreases and then increases (d) first increases and then decreases
2. Mobility of the electron is
   (a) reciprocal of electrical conductivity
   (b) flow of electron per unit electric field
   (c) average electron drift velocity per unit electric field
   (d) none of the above
3. Most widely used conducting materials are
   (a) gold and silver (b) copper and aluminium
   (c) tungsten and platinum (d) germanium and silk

4. The average drift velocity v of electrons in a metal is related to the electric field E and collision time τ as

(a) $v = \dfrac{eE\tau}{m}$ (b) $v = \dfrac{m}{\sqrt{eE\tau}}$

(c) $v = \sqrt{\dfrac{eE\tau}{m}}$ (d) $v = \left(\dfrac{eE\tau}{m}\right)^{2/3}$

5. The temperature dependence of the classical expression for electrical resistivity of a metal is

(a) $\rho \propto \sqrt{T}$ (b) $\rho \propto T^{-1}$

(c) $\rho \propto T^{-2}$ (d) $\rho \propto T^{2}$

6. If e, μ and n represent the charge, mobility and concentration of electrons respectively, then the electrical conductivity of the metal is given by

(a) ne (b) $ne\mu$

(c) $\dfrac{\mu e}{n}$ (d) $n/\mu e$

7. The Fermi energy of silver at 0K is 5 eV, the mean energy of electron in silver at 0 K is

(a) 3 eV (b) 1.5 eV

(c) 4.5 eV (d) 6 eV

8. If the Fermi energy of a metal is 1.4 eV, the Fermi temperature of the metal is approximately

(a) 1.6×10^{8} K (b) 1.6×10^{6} K

(c) 1.6×10^{4} K (d) 1.6×10^{2} K

9. Which of the following metals has the lowest temperature coefficient of resistance?

(a) Silver (b) Gold

(c) Copper (d) Aluminium

10. At very high temperatures, the mean free path and collision time of an electron in a metal are proportional to

(a) $1/T^{2}$ (b) $1/T$

(c) T^{2} (d) independent of temperature

11. If the Fermi energy of a metal is 1.4 eV, the Fermi temperature of the metal is approximately

(a) 1.6×10^4 K (b) 1.6×10^2 K

(c) 1.6×10^3 K (d) 1.6×10^8 K

12. If σ and E are the electric conductivity and applied field respectively on a current carring conductor, the heat developed per unit volume per second is

(a) σ/E (b) σ^2/E

(c) E^2/σ (d) σ/E^2

ANSWERS

Question Numbers	1	2	3	4	5	6	7	8	9	10	11	12
Answers	(a)	(c)	(b)	(a)	(a)	(b)	(a)	(c)	(b)	(b)	(a)	(c)

FILL IN THE BLANKS

1. When a magnetic field is applied perpendicular to a conductor carrying current, a voltage is developed across the specimen in the direction perpendicular to both the ______ and ______ field. The phenomenon is called the ______. **[current, magnetic, Hall effect]**
2. In a conducting strip placed in a magnetic field perpendicular to the current flow in a strip, the resistance across the current path depends linearly on the strength of the magnetic field. The effect is called ______. **[Hall effect]**
3. Because of its independence from the sample geometry, the classical Hall effect has become a standard technique for determination of ______, ______ and ______ of free carriers in metals and semiconductors. **[type, density, mobility]**
4. When the Hall effect is measured at low temperatures in a sample containing a two dimensional electron gas (2 DEG) in which electrons are confined to move only within the plane, the Hall resistance (R_H) is found to ______ from classical behaviour. At sufficiently high fields, a series of flat steps appear in the graph of Hall voltage (V_H) versus magnetic field. This is referred to as ______ Hall effect. **[deviate, quantum]**
5. The mobility (m) is defined as the velocity acquired by the ______ particles per unit ______. **[current carrying, electric field]**
6. Hall constant (R_H) is given by ______. **[$-1/Ne$]**

19

Magnetism

19.1 INTRODUCTION

The magnetic properties of matter have fascinated physicists, chemists and engineers for many years. In recent times, these properties have been the subject of special interest because of the information they yield about the constitution of matter and the interaction involved there in. This information, being truely interdisciplinary, is of interest not only to physicists but also to other scientists and engineers. Magnetic materials have wide ranging technical applications, from transformer cores in electrical machinery, to magnetic tapes in computers.

A knowledge of the magnetic behaviour of materials and its origin will be of great help in the proper utilisation of such materials. We know that out of all the different materials only a few like iron, cobalt and nickel are strongly magnetic and others do not get magnetised. We can trace the magnetism in materials to the magnetism exhibited by its constituents vis a vis atoms. We will be interested to know what are the contributing factors to the magnetism of atoms and why atoms of iron, cobalt and nickel alone exhibit pronounced magnetic properties and that of other atoms do not exhibit the same amount of magnetic behaviour.

19.2 CLASSIFICATION OF MAGNETIC MATERIALS

It is possible to classify the magnetism observed in solids into five-types as follows:

(a) Diamagnetism
(b) Paramagnetism
(c) Ferromagnetism
(d) Antiferromagnetism
(e) Ferrimagnetism

The objective now is to explore the origin of magnetism in substances and explain the magnetic behaviour of different substances from the atomic point of view and see how far this picture gives the clue to get ordered magnetism, i.e., magnetic substance with any desired characteristics.

Table 19.1 Summary of different types of the magnetic behaviour in terms of susceptibility

Type of Magnetism	*Magnetic susceptibility* (χ)	*Temperature dependence*	*Examples*
Diamagnetism	Small, negative intermediate, negative	Independent varies with magnetic field and temperature below 0° K	Organic materials, light elements, Alkali earths, Bismuth
	Large, negative	Exists only below a critical temperature	Superconducting metals
Paramagnetism	Small, positive	Independent	Alkali metals, transition metals
	Large, positive	$\chi = \dfrac{c}{T}$	Rare earths
Ferromagnetism	Very large, positive	For $T > \theta_c$ $\chi = \dfrac{c}{T-\theta_c}$ For $T < \theta_c$ $\chi \to$ complex θ_c- Curie temperature	Some transition metals and rare earth metals
Antiferromagnetism	Small, positive	For $T > T_N$ $\chi = \dfrac{c}{T-\theta_c}$ For $T < T_N$ $\chi \propto T$ T_N-Neel temperature	Salts of transition elements
Ferrimagnetism	Very large, positive	For $T > T_N$ $\chi = \dfrac{c}{T \pm \theta_c}$ For $T < T_N$ χ complex	Ferrites

19.3 REVIEW OF BASIC FORMULAE

Let us begin with the fundamental concept of the magnetic dipole moment. Consider two magnetic charges q_m and $-q_m$, which are equal and opposite. They form a magnetic dipole whose moment is given by

$$\mu_m = q_m \boldsymbol{d} \tag{19.1}$$

where d is the vector joining the negative to the positive charge Fig. (19.1). Note the similarity between this definition and that is used in connection with the moment of an electric dipole.

When our magnetic dipole is placed in a magnetic field whose induction is $\boldsymbol{B}$, then because of a charge q_m experience force

$$\boldsymbol{F} = q_m \boldsymbol{B} \tag{19.2}$$

The dipole itself (Composed of two opposite charges) experiences a couple whose torque is

$$\tau = \mu_m \times \beta \tag{19.3}$$

The effect of this torque is to turn the dipole and align it with the field in a manner similar to the way an electric field aligns an electric dipole. Because of the torque, dipole has an orientation potential energy given by

$$V = -\mu_m . \beta = -\mu_m B \tag{19.4}$$

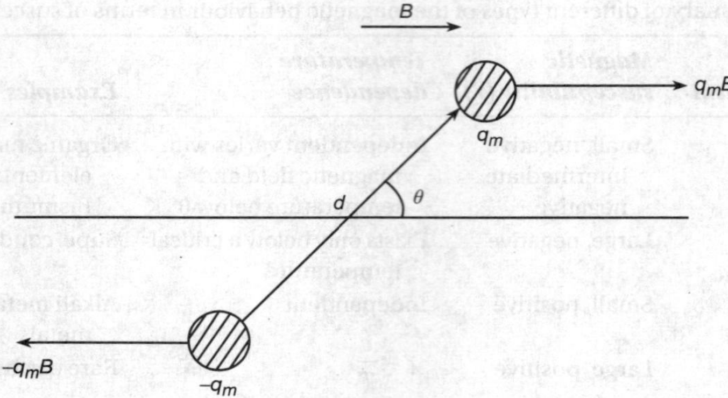

Fig. 19.1

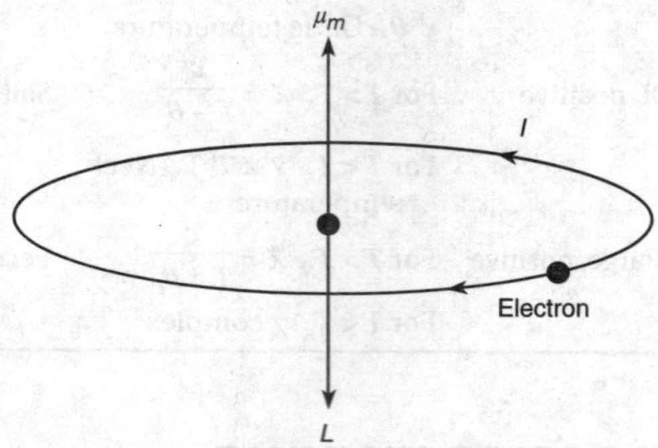

Fig. 19.2

where θ is the angle between the field and the dipole direction. The minimum energy occurs at $\theta = 0$, where the dipole lies along the field. The maximum energy is achieved at $\theta = \pi$, where the dipole is oriented opposite to the field.

We have defined the magnetic dipole in terms of magnetic charges, but such charges do not infact exist. All the known magnetic properties of matter are attributable in the rotation of electric charges. We recall from elementary physics that an electric current loop acts like a magnetic dipole of moment.

$$\boldsymbol{\mu}_m = IA \tag{19.5}$$

where I is the current and A the area of the loop. The direction of $\boldsymbol{\mu}_m$ which is a vector is normal to the plane of the loop and such that the current flows counter clockwise relative to an observer, standing along $\boldsymbol{\mu}_m$ Fig. (19.2).

19.4 ORIGIN OF ATOMIC MAGNETISM

In our classical model of the atom, we know that electrons revolve round the central positive nucleus in different orbits, which we take to be circular for the sake of simplicity. A moving electron constitutes an electric current. Thus the revolving electrons in an atom constitute circular currents and this gives us a possible explanation for the origin of atomic magnetism. Modern quantum theory tells us that in addition to electric currents produced by the revolving electrons, the spin of the electrons as well as the spin of the nucleus contribute to the atomic magnetism, though they are not strictly electric currents according to our classical notion. We will now consider their contribution one by one.

(a) Magnetic contribution due to the orbital motion of the electron

Consider an electron of charge *e* revolving round a nucleus with a time period *T* in a circular orbit of radius *r*. The magnitude of the circular current is given by $\frac{e}{T}$. But *T* is equal to $\frac{2\pi}{\omega_0}$, where ω_0 is the angular velocity of the election and so the current become

$$I = \frac{e\omega_0}{2\pi}$$

The magnetic dipole moment μ_m associated with a circular current is given by the product of the current strength and the area of the circular current and this acts perpendicular to the plane of the circular current. Thus, we find

$$\mu_m = -\frac{e\omega_0}{2\pi}\pi r^2 = -\frac{e\omega_0 r^2}{2} \tag{19.6}$$

The negative sign shows that the magnetic moment vector is opposite to the angular velocity vector. It is useful to express the magnetic moment of the electrons is terms of its angular momentum. If *m* is the mass of the electron, the orbital angular momentum *L* of the electron is $m\omega_0 r^2$. So the magnetic moment of the electron in terms of its angular momentum is given by

$$\mu_L = -\frac{e\omega_0 m r^2}{2m} = -\frac{e}{2m}L \tag{19.7}$$

The constant $\frac{e}{2m}$ is called the *gyro-magnetic ratio* and is repesented by the symbol γ. It is defined as the ratio of the magnitude of the magnetic moment to the magnitude of the angular momentum of any particle or system. We know from Bohr's theory that angular momentum of the orbital electron is determined by the orbital quantum number ℓ which can take only the restricted set of values.

$$\ell = 0, 1, 2, 3, \ldots (n-1) \tag{19.8}$$

where *n* is the principal quantum number of the electron. The angular momentum of the electron associated with a particular value of 'ℓ' is given by

$$L = \frac{\ell h}{2\pi} = \ell\hbar \tag{19.9}$$

Electrons associated with states $\ell = 0, 1, 2, 3$, etc., are called the *s, p, d, f, g*, etc., electrons respectively. Substituting the value of *L* in Eq. (19.7), we have

$$\mu_L = -\frac{e}{2m}\ell\hbar = -\frac{e\ell h}{4\pi m} \tag{19.10}$$

From Eq. (19.10) we see that the electron in an atom can take only certain specified values of magnetic moment depending on ℓ values. For S electron $\ell = 0$ and hence the magnetic moment contribution by the S electrons is Zero.

Again we know from quantum theory that not only the magnitudes of the ℓ vector are quantised but also their directions. This restriction is called space quantisation and is arrived at as follows. Suppose the atomic magnet is placed in an external field B. This magnetic field tend to rotate the orientation of the atomic magnet. The atomic magnetic moment vector cannot take all possible orientations but only those orientations for which the projections of the angular momentum vector along the field direction are integers specified by another set of quantum numbers called the magnetic orbital quantum number m_ℓ. The restricted set of value which m_ℓ can take are

$$m_\ell = n, (n-1), \cdots 0, \cdots -(n-1), -n \tag{19.11}$$

We observe that m_ℓ can take $(2\ell + 1)$ values. For example a d electron has the possible components of angular momentum along the direction of magnetic field $2\hbar$, $\hbar$, 0, $-\hbar$ and $-2\hbar$. Consequently, the possible magnetic moment components along the direction of the applied field are

$$\frac{2\hbar}{2m}, \frac{\hbar}{2m}, 0, -\frac{\hbar}{2m} \text{ and } -\frac{2\hbar}{2m} \tag{19.12}$$

Fig. (19.3) shows the 5 possible orientations of the angular momentum vector, specified by the orbital quantum number $\ell = 2$. We also observe that the projections

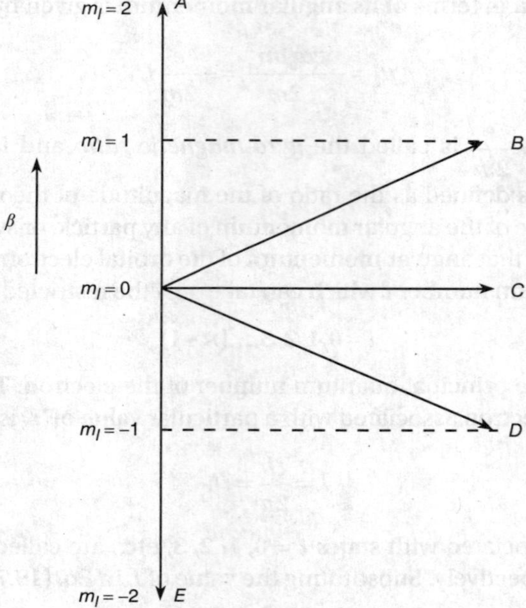

Fig. 19.3

of these angular momentum vectors ***OA, OB, OC, OD,*** and ***OE*** along the field direction B are 2, 1, 0, –1, –2 respectively. The quantity $\frac{e\hbar}{2m}(=\mu_B)$ is called Bohr magneton and its values is 0.92 × 10^{-23} J/T. In general, the components of the orbital magnetic moment along an external applied magnetic field is equal to m_ℓ Bohr magnetons. The total magnetic moment contributed by the orbital motion of all the electrons is an atom can be vectorially added. For example in the case of Hydrogen and Helium atoms there are only S electrons and so the contribution to the magnetic moment by the orbital motion is zero. We can show that a completely filled electronic shell also does not contribute any thing to the magnetic moment of the atom. Consider for example M shell corresponding to $n = 3$. The possible values ℓ are 0, 1 and 2. For $\ell = 0$ there is no magnetic moment ($\mu_\ell = 0$) contribution. For $\ell = 1$, m_ℓ can take three values 1, 0, –1 and if these states are occupied, the vector sum of their magnetic moment becomes

$$\frac{e\hbar}{2m}+0-\frac{e\hbar}{2m}=0 \tag{19.13}$$

for $\ell = 2$, m_ℓ can take 5 values (2, 1, 0, –1, –2) and if all the states are filled, we find that the magnetic moment contribution is zero. From a look at the electron configuration of many elements in the periodic table, we find they have incomplete electronic shells and so they have resultant magnetic moments. But their contribution to the magnetic properties of solids is negligible, since it is found that the orbital magnetic moments of these elements and their compounds are "frozen in", i.e., they behave like immobile magnetic dipoles which can not be aligned by external applied magnetic field.

(b) Magnetic contribution due to the Spin of the electron

Modern Quantum theory tells us that the electrons must have an intrinsic angular momentum due to what is called "Spin". Spin is a relativistic effect and we must not consider it in the same way as orbital motion or consider it analogous to the motion of a spinning body. Spin is a fundamental property of the electron. The spin values are also quantised in magnitude as well as in direction. The corresponding quantum numbers are S and m_s respectively. The spin quantum number can take only one value, i.e., 1/2. The magnetic spin quantum number can take either of the two values +1/2 or –1/2. The intrinsic angular momentum of the electron due to the spin becomes $\frac{1}{2}\hbar\left(\text{or}\ \frac{h}{4\pi}\right)$. Just as we have expressed the orbit magnetic moment in terms of the orbital angular momentum, we can also express the spin magnetic moment in term of the spin angular momentum s and it comes out to be

$$\mu_s = -\frac{2e}{2m}S = -\frac{e}{m}S \tag{19.14}$$

We find that this relation differs from Eq. (19.7) by a factor 2 on the right hand side, i.e., the spin gyromagnetic ratio (μ_s/S) is twice the value obtained for orbital motion (μ_m/L). And since spin angular momentum $s = \frac{1}{2}\hbar$, so the intrinsic magnetic moment of the electrons works out to be

$$\mu_s = \frac{e}{m}\frac{1}{2}\hbar = \frac{eh}{4\pi m} = \mu_B \text{ (Bohr magneton)} \tag{19.15}$$

When an electron spin magnetic system is placed in an external field, the spin magnetic moment along the field direction can take only of the two values,

$$+\frac{eh}{4\pi m} \quad \text{or} \quad -\frac{eh}{4\pi m} \tag{19.16}$$

i.e., either parallel to the field or anti parallel to the field, and the other orientations are forbidden as per spatial quantisation rule. We have seen that in expression connecting electron spin magnetic moment and spin angular momentum Eq. (19.14) a factor 2 appears, where as this factor become in the expression connecting the orbital magnetic moment and orbital angular momentum Eq. (19.7). So in general we can replace this factor by a symbol g and write the relation connecting magnetic moment and the angular momentum as

$$\mu = -g\frac{e}{2m}L \tag{19.17}$$

The symbol g is called the **Spectroscopic Splitting factor** or the **Lande's Splitting factor**. This was originally introduced by Lande to explain the anamalous zeeman effect and latter theories have justified this assumption theoretically.

(c) The nuclear magnetic moment

The angular momentum associated with the nuclear Spin is also expressed in units of $\hbar = \left(\frac{h}{2\pi}\right)$, then the nuclear magnetic moment is given by

$$\mu_n = g_n \frac{eh}{4\pi M} \tag{19.18}$$

where M is the mass of the proton and g_n nuclear g factor. The mass of the nucleus, however is of the order of 2000 times of the electron hence the magnetic moment associated with the nuclear spin, a quantity called the nuclear magneton, is of the order of 10^{-3} Bohr magneton.

(d) The total magnetic moment

The nuclear magnetic moment is very small and hence we can neglect it. Then the vector addition of the orbital angular momentum and the spin momentum for simple case, i.e., one electron is represented by the quantum number j. Where $j = (\ell + s)$ and is expressed as $\hbar j$. For an atom having z-electrons, the rules for adding the orbital and spin angular moments are to be studied in detail.

The addition can be performed in several ways but one very common and generally used in magnetism is discussed here.

(i) The resultant angular momentum L is determined by the vector sum of individual orbital angular momenta and is expressed as $\hbar L$. The maximum value of L is $\sum_1^2 \ell$ and minimum value is zero, i.e.,

$$L = 0, 1, 2, \ldots \sum_1^2 \ell \tag{19.19}$$

(ii) The resultant spin angular momentum S is also determined as vector addition of individual spin angular momenta and is expressed as $\hbar S$. The value of S varies from maximum of $\sum_1^z S = \frac{z}{2}$, by intervals of $\frac{1}{2}$ to a minimum of zero or $\frac{1}{2}$ depending on whether Z is even or odd. While determining S, care must be taken to ensure that Pauli exclusion principle is not violeted and further the determination of L is consistent with the determination of S. The later is a part of Hund's rule.

(iii) The total angular momentum (J) is obtained by the Vector sum of L and S. J can have possible values from $(L - S)$ to $(L + S)$ in steps of 1. The total possible values are $(2S + 1)$. This is called the multiplicity. The component of J, along any specified direction, say in the direction of the applied magnetic field along z-axis in expressed as $M_J\hbar$, where M_J varies from $-J$ to $+J$ in steps of 1. Thus M_J can have $(2J + 1)$ values, i.e., J can have $(2J + 1)$ orientations. This type of coupling in which first L and S are separately found and then resultant J is found by vector sum of L and S is known as (orbital and Spin motions of the electron produce magnetic field resulting in mutual perturbation) **"Russell-Saunder's Coupling"** of the Spin and orbital momenta (In short we write L-S coupling). If the factor were the same for spin and orbital momentum the resultant magnetic moment can be easily be obtained by algebric addition. The Lande g factor or spectroscopic splitting factor has the general value.

$$g = 1 + \frac{J(J+1) + S(S+1) - L(L+1)}{2J(J+1)} \tag{19.20}$$

So the total magnetic moment of the atomic system become

$$\mu_m = gJ\frac{e\hbar}{2m} = gJ\frac{eh}{4\pi m} = gJ\mu_B \tag{19.21}$$

g and J are known from spectroscopic data and so the total magnetic moment (μ) of a free atom can be calulated. If the atom or the ion is part of a solid element, (μ) is not the same as the magnetic moment for the free atom of ion, because of mutual interactions between the neighboring atoms or ions in the lattice. For example in the iron atom there are 6 electrons in complete $3d$ shell. The orbital magnetic moments are frozen in as we have said already. Out of the six spinning electrons 5 spins are pointing is one direction and one spin is pointing in the opposite direction. Quantum mechanics tells us that this is the minimum energy arrangement of the spins. The resultant spin is 4 and so the iron atom must have a magnetic moment of 4 Bohr magneton. But in metallic irons, the average number of Bohr magneton per atom is only 2. This is because of the mutual interaction between neighboring atoms in the lattice.

19.5 DYNAMICS OF CLASSICAL DIPOLE IN MAGNETIC FIELD

A magnetic dipole (μ_m) in magnetic field $\boldsymbol{B}$ experiences a torque τ given by

$$\tau = \mu_m \times \boldsymbol{B}$$

Then equation of motion is given by

$$\tau = \frac{d\boldsymbol{L}}{dt} = \mu_m \times \boldsymbol{B} \tag{19.22}$$

where $\boldsymbol{L}$ is angular momentum. Writing magnetic moment in terms of angular momentum we have

$$\mu_m = -g\left(\frac{e}{2m}\right)\boldsymbol{L}$$

and
$$\frac{d\boldsymbol{L}}{dt} = -\frac{2m}{ge}\cdot\frac{d\mu_m}{dt} \tag{19.23}$$

Subsitituting in Eq. (19.22), one obtains

$$\frac{d\mu_m}{dt} = -\frac{ge}{2m}(\mu_m \times \boldsymbol{B}) \tag{19.24}$$

This equation represents a processional motion Fig. (19.4) of frequency known as the Larmor frequency. The dipole simply processes around the direction of the field always maintaining the same angle. For an electron at $\boldsymbol{B} = 0.1\,\text{W/m}^2$, one obtains $\omega_L \approx 10\,\text{GHz}$, where

$$\omega_L = \frac{e\boldsymbol{B}}{2m} \tag{19.25}$$

The statement concerning the Larmor precession seems to cast some doubt on our earlier assumption that the dipole tends to align itself with the field. But in the precession, the dipole merely rotates around $\boldsymbol{B}$ without ever getting closer to the direction of the field. The point is well taken. In a pure Larmor precession no alignment takes place. In practical situation however this precession is usually accompanied by numerous collisions, during which dipole loses energy. As it does so, it gradually approaches the direction of $\boldsymbol{B}$ until eventually it lines up exactly with the field. This process of gradual magnetisation is referred to as relaxation.

The potential energy of the dipole, i.e., Eq. (19.4) can also be written as

$$E = -\mu_m \boldsymbol{B}\cos\theta = \frac{ge}{2m}L_z\boldsymbol{B} \tag{19.26}$$

where we have used Eq. (19.17). Here $\boldsymbol{B}$ is taken to be in z-direction and L_z is the z-component of angular momentum.

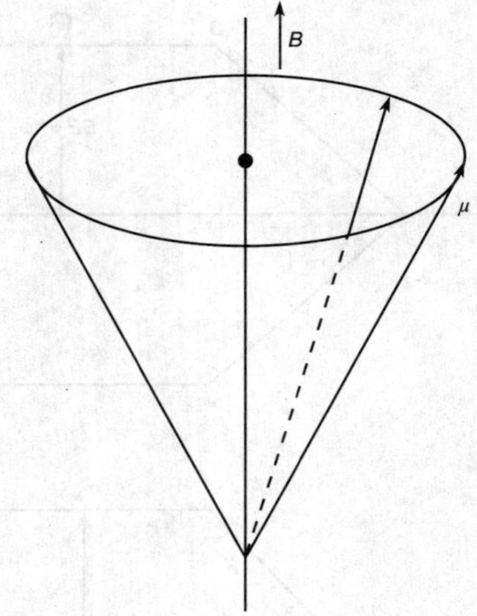

Fig. 19.4

For orbital motion $g = 1$ and dipole energy is

$$E = \frac{e}{2m} L_z \boldsymbol{B} \tag{19.27}$$

and according to quantum mechanics the component L_z is quantized by $L_z = m_\ell \hbar$ where m_ℓ is an integer which takes the values $-\ell$, $(-\ell + 1)...(\ell - 1)$, ℓ where ℓ is the orbital quantum number for angular momentum of the electron. This Eq. (19.27) may also be written as

$$E = \left(\frac{e\hbar}{2m}\right) \boldsymbol{B} m_\ell = \mu_B . \boldsymbol{B} m_\ell \tag{19.28}$$

As m_ℓ takes its various allowed values, the energy also takes its appropriate values in the presence of magnetic field for $\ell = 1$, m_ℓ takes -1, 0, and 1 and the corresponding three levels are illustrated in Fig. (19.5). The Splitting of an atomic level by the magnetic field in the manner just described is known as Zeeman splitting. The interval between any two adjacent levels is the same and is given by

$$\Delta E = \mu_B \boldsymbol{B} \tag{19.29}$$

Note that the lowest level $m_\ell = -1$, corresponds to the orientation in which $\boldsymbol{L}$ is opposite to B and hence $\boldsymbol{\mu}$ is parallel to $\boldsymbol{B}$ in agreement with the classical picture. Similarly, the highest level $m_t = 1$, corresponds to the orientation in which $\boldsymbol{\mu}$ is opposite to the field. In case of the spin, $g = 2$, the Zeeman energy is by

$$E = 2\mu_B \boldsymbol{B} m_s \tag{19.30}$$

Since the Spin quantum number $s = 1/2$, the allowed values of m_s are

$$m_s = +\frac{1}{2} \text{ and } -\frac{1}{2}$$

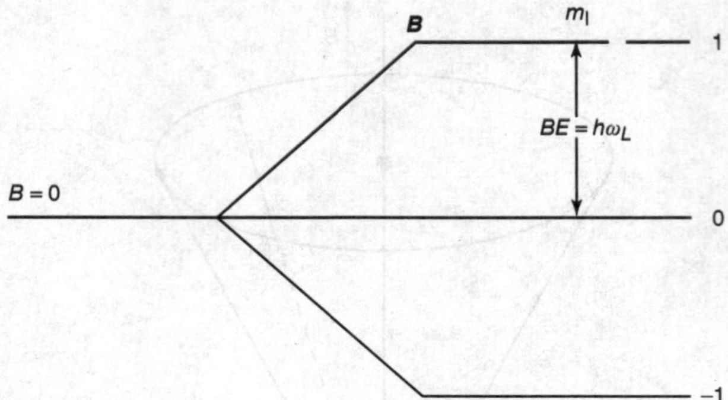

Fig. 19.5

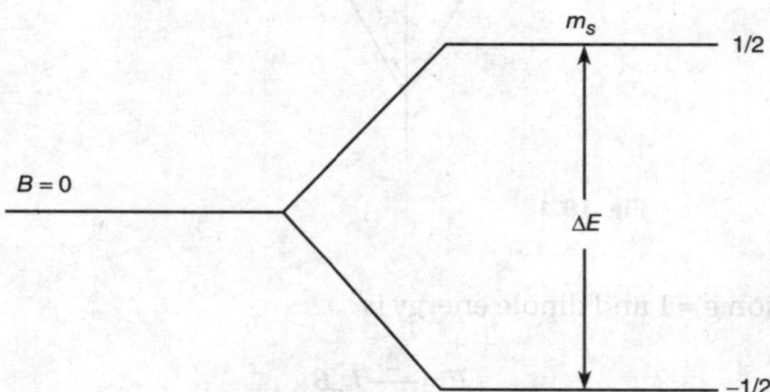

Fig. 19.6

The corresponding Zeeman splitting composed of only two lines as shown in Fig. (19.6). The difference in energy between two levels is

$$\Delta E = 2\mu_B \boldsymbol{B} \tag{19.31}$$

which is double the orbital separation. It is useful to establish a correspondence between the classical and quantum descriptions of the dipole motion in a magnetic field. Note that for the orbital case, the energy splitting $\Delta E = \mu_B \boldsymbol{B}$ is also equal to $\hbar\omega_L$, where ω_L is Larmor frequency given by Eq. (19.25). Thus the quantum frequency for the transition between the Zeeman spacing is twice as large.

19.6 MAGNETIC SUSCEPTIBILITY

A magnetic field can be described by either of two vectors, the magnetic induction $\boldsymbol{B}$ or the magnetic field intensity $\boldsymbol{H}$, which are related in vacuum by

$$\boldsymbol{B} = \mu_0 \boldsymbol{H} \tag{19.32}$$

where $\mu_0 = 4\pi \times 10^{-7}$ $\boldsymbol{H}$ m^{-1} is permability of free space.

When a material medium is placed in a magnetic field, the medium is magnetised. The magnetisation is described by the magnetisation vector $\boldsymbol{M}$ called the dipole moment per unit volume. The magnetic induction inside the medium is then given by the relation

$$\boldsymbol{B} = \mu_0 \boldsymbol{H} + \mu_0 \boldsymbol{M} \tag{19.33}$$

Eq. (19.33) comes from Gauss theorem in magnetism. The induction is composed of two parts. The part $\mu_0 \boldsymbol{H}$ generated by the external field and part $\mu_0 \boldsymbol{M}$ due to magnetisation of the medium.

Since the magnetisation is induced by the field, we may assume that $\boldsymbol{M}$ is proportional to $\boldsymbol{H}$, that is

$$\boldsymbol{M} = \chi \boldsymbol{H} \tag{19.34}$$

The proportionality constant χ being known as the magnetic susceptibility of the medium[1]. Substituting Eq. (19.34) in Eq. (19.33), we have

$$\boldsymbol{B} = \mu_0 \boldsymbol{H} + \chi \boldsymbol{H} \mu_0 = \mu_0 \boldsymbol{H}(1+\chi) = \mu_0 \mu_r \boldsymbol{H} \tag{19.35}$$

$\because$ $$\mu_r = 1 + \chi$$

or $$\boldsymbol{B} = \left(\frac{1+\chi}{\chi}\right)\mu_0 \boldsymbol{M}$$

Further we have

$$\boldsymbol{M} = x\left(\frac{\boldsymbol{B}}{\mu_0} - \boldsymbol{M}\right)$$

We may note that except ferromagnetic materials, $\chi << 1$ and hence the above relation can be written

$$\boldsymbol{B} = \frac{\mu_0 \boldsymbol{M}}{\chi}$$

Hence $$\chi = \frac{\mu_0 \boldsymbol{M}}{\boldsymbol{B}}$$

Thus the Vector $\boldsymbol{B}$ and $\boldsymbol{H}$ are proportional to each other

$$\boldsymbol{B} = \mu \boldsymbol{H} \tag{19.36}$$

Where μ is the proportionality constant

$$\mu = \mu_0(1+\chi) \tag{19.37}$$

is known as the **permeability of the medium**. It is often more convenient to use the relative permeability μ_r, which is defined as

$$\mu_r = \frac{\mu}{\mu_0}$$

Therefore

$$\mu_r = 1 + \chi \tag{19.38}$$

a relation connecting the permeability and susceptibility of the medium.

Our discussion assume that the medium is magnetically isotropic but real crystals are anisotropic and the susceptibility and permeability are represented by second-rank tensors. In order to avoid mathematical complications, however, we shall ignore an-isotropic effects in our treatment.

---

[1]The magnetic susceptibility χ bears no physical relationship to the electric susceptibility although the same symbol is used for both. No confusion should arise however, since electirc susceptibility will not appear in this chapter.

19.7 PHENOMENON OF DIAMAGNETISM

Diamagnetism is a universal phenomenon exhibited by all substances. If it is not explicity observed in many cases, it only means that other stronger magnetic effects like paramagnetism and ferromagnetism completely mask the diamagnetic effects. The atoms or molecules of any substance when paced in a magnetic field will develop a negative moment giving rise to a magnetic field opposing the applied field, this is the basic diamagnetic action. The diamagnetism arises because of the reaction produced in the electronic orbits of atoms and molecules placed in a magnetic field. When you seek to change the flux linked with the electron orbits in an atom or molecules by introducing external field, it is opposed in accordance with the Lenz's law of eletromagnetic induction. The currents induced in the electron orbits by introduction of the external magnetic field are in *a*-direction opposite to the direction of original electron currents and it is these currents induced in opposite direction that give rise to an opposing moment or negative moment to the atoms or the molecules. This negative magnetic moment is very small and so it can be observed only with difficulty, in cases where the atoms or the molecules initially had no net magnetic moments. If the atoms or the molecules had any net resultant moment, the external field will have the effect of aligning the magnetic dipoles giving rise to a stronger positive magnetic moment, i.e., a magnetic moment in the same direction as that of the external applied field. Thus the atoms and the molecules with zero resultant magnetic moments are called diamagnetics.

Langevin Theory of Diamagnetism

We will now establish the fact that the effect of a magnetic field on the orbital motion of an electron is such as to produce diamagnetic susceptibility. Consider an electron rotating about the nucleus in a particular orbit, and let a magnetic field be applied perpendicular to the plane of paper as shown in Fig. [19.7(a)]. Prior to the application of this yield, we have

$$F_0 = m\omega_0^2 r \tag{19.39}$$

where F_0 is the attractive Coulomb force between the nucleus and the electron, which provides the necessary centripetal force. ω_0 is the angular velocity. The magnetic moment of electron is

$$\mu_0 = IA = \frac{e}{2}\omega_0 r^2 \tag{19.40}$$

where r is the radius of the electron's orbit. This moment is perpendicular to the electron orbit, i.e., parallel to the field and sense of rotation shown in Fig. [19.7(a)] and [19.7(b)].

When the field is applied, an additional force starts to act on the electron; the Lorentz force $-e(\boldsymbol{V} \times \boldsymbol{B})$. For the geometry of Fig. [19.7(a)] the effect is to produce a radially outward force given by $eBr\omega$ and Eq. (19.14) therefore mended to

$$F_0 - e\boldsymbol{B}r\omega = m\omega^2 r \tag{19.41}$$

Thus the angular frequency is now different from ω_0 and its value may be determined from this relation.

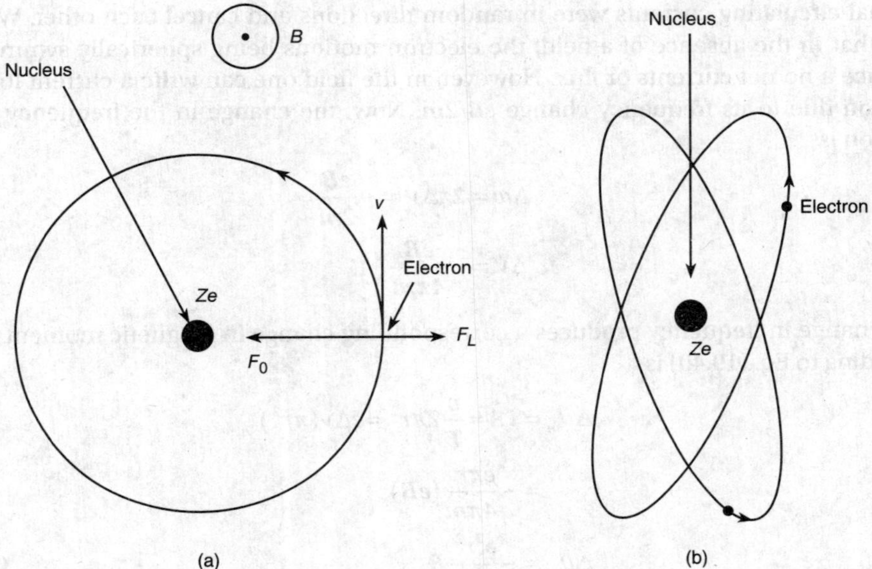

Fig. 19.7

Eq. (19.41) can be written as

$$\frac{Ze^2}{4\pi\epsilon_0 r^2} - eB\omega r = mr\omega^2$$

or
$$\omega^2 + \frac{eB\omega}{r} - \frac{Ze^2}{4\pi\epsilon_0 mr^3} = 0$$

or
$$\omega = \frac{\frac{-eB}{m} \pm \sqrt{\frac{e^2B^2}{m^2} + \frac{4ze^2}{4\pi\epsilon_0 mr^3}}}{2}$$

$$= \frac{-eB}{2m} \pm \sqrt{\frac{ze^2}{4\pi\epsilon_0 mr^3} + \frac{e^2B^2}{4m^2}}$$

or
$$\omega = \frac{-eB}{2m} \pm \sqrt{\omega_0^2 + \frac{e^2B^2}{4m^2}}$$

$$\omega = \omega_0 - \frac{eB}{2m} \quad \text{if} \quad \frac{eB}{2m} << \omega_0 \tag{19.42}$$

The ± sign on ω_0 reveals that those electrons whose orbital moments were parallel to the field are slowed down and those whose moments were anti-parallel are speeded ω_0 by an amount eB/m. This discussion results and conclusions are usually called as **Larmor theorem**. This change of frequency gives rise to a magnetic polarization or magnetization.

The reason for this is that the frequency change is equivalant to an additional current, and moreover this current component in every atom is in the same direction, where as original circulating currents were in random directions and cancel each other. We may note that in the absence of a field, the electron motions being spherically symmetrical produce a no net currents or flux. However, in the field one can write a current for each electron due to its frequency change $eB/2m$. Now, the change in the frequency of the electron is

$$\Delta\omega = 2\pi\Delta\nu = -\frac{eB}{2m}$$

or

$$\Delta\nu = -\frac{eB}{4\pi m}$$

This change in frequency produces a corresponding change in magnetic moment which according to Eq. (19.40) is

$$\Delta\mu_m = i\,A = \frac{e}{T}2\pi r^2 = e\Delta\nu(\pi r^2)$$

$$= -\frac{e\pi r^2}{4\pi m}(e\boldsymbol{B})$$

or

$$\Delta\mu_m = -\frac{e^2 r^2}{4m}\boldsymbol{B} \tag{19.43}$$

Since the moment parallel to the field has been reduced, the induced moment is opposite to the field, i.e., the response of the electron is diamagnetic.

It can be readily appreciated that if we initially chose an electron which was rotating counter clock wise, the initial moment would be opposite to the z-axis, i.e., negative. The effect of the field would then be to speed up the electron resulting in an even more negative moment. That is, the induced moment would again be negative diamagnetic and given by Eq. (19.43). Thus the diamagnetic response of an orbiting electron holds good in general and infact may be shown to follow directly from the familiar Lenz's Law.

When applied to atom Eq. (19.43) requires more modifications, because the electron orbits around a spherical surface rather than in a circle Fig. [19.7(b)]. However only the cross section normal to the field is effective in the diamagnetic response, and hence on the average we should replace r^2 in Eq. (19.43) by $\frac{2}{3}r^2$ (the mean value of all possible orbital radii and for spherical change distribution it is $\frac{2}{3}r^2$), the new r being the radius of sphere, which leads to

$$\Delta\mu_m = -\left(\frac{e^2 r^2}{6m}\right)\boldsymbol{B} \tag{19.44}$$

We can now readily evaluate magnetic susceptibility. Given that the atom has z-electrons and that there are n-atoms per unit volume the susceptibility.

$$\chi = \frac{\boldsymbol{M}}{\boldsymbol{H}} = \frac{\mu_0 N Z \Delta\mu_m}{\boldsymbol{B}}$$

or

$$\chi = -\frac{\mu_0 e^2}{6m}(N Z r^2) \tag{19.45}$$

Table 19.2

Element	*He*	*Ne*	*Ar*	*Kr*	*Xe*	F^-	Cl^-	Br^-	Li^-	Na^+	K^+	Mg^{++}	Ca^{++}	Sr^{++}
$\chi \times 10^{-6}$ Per gm ion	−1.9	−7.6	−19	−29	−44	−9.4	−24.2	−34.5	0.7	6.1	14.6	4.3	10.7	18.0

where r^2 is the average square radius of the electron. The averaging is done over all the occupied orbitals in the same order (~10^{-5}) of magnitude as those obtained by measurements.

Diamagnetic susceptibility is observed most clearly in those solids in which the atomic shells are completely filled. Examples of these are provided by rare-gas crystals, and also ionic crystals whose magnetic susceptibilities are given in Table 19.2.

In case of covalent crystals, which are also diamagnetic Eq. (19.45) can be applied only to the core electrons. The electrons forming the bond have orbits which are far from circular, and hence the derivation leading to Eq. (19.45) does not apply here. The susceptibility of covalent crystal may be written as

$$\chi = \chi_i + \lambda \tag{19.46}$$

where χ_i includes the effect of the core electrons (the ions) and λ the effect of the bonding electrons. One can determine the value of λ for a specific bond emperically in a given compound, and use this value in other compounds in which it occurs.

When some of the atomic shells in a solid are incompletely filled, the substance then has a paramagnetic contribution in addition to the diamagnetic contribution. The net susceptibility is the difference between the two contributions, but since the paramagnetic one is usually larger, it masks the diamagnetic contribution.

19.8 PARAMAGNETISM

Paramagnetism, associated with positive susceptibility, occurs because many atoms have a permanent magnetic moment which over takes the diamagnetic effect. It is the magnetic analogue of the electrical orientational polarizability. The direction of magnetisation is parallel to the applied magnetic field. The permanent moments of atoms result from the following contributions:

(i) Intrinsic spin, moment of the electrons,
(ii) the orbital motion of the electrons,
(iii) the nuclear spin magnetic moment.

Paramagnetism is found in

(i) Metals.
(ii) A few compounds with an even number of electrons, including molecular oxygen and organic bi-radicals.
(iii) Atoms, molecules, and lattice defects possessing an odd number of electrons, as here the total spin of the system cannot be zero for example, gaseous nitric oxide,

free sodium atoms, organic free radicals e.g., triphenyl methyl $C(C_6H_5)$ F centres in alkali halides.

(iv) Free atoms and ions with a partially filled inner shell, rare-earth and actinide elements, transition elements, ions iso-electronic with transition elements, for example Mn^{++}, Gd^{++}, U^{++++}. Paramagnetism is exhibited by many of these ions when incorporated into solids but not invariably.

19.9 LANGEVIN'S CLASSICAL THEORY OF PARAMAGNETISM

Classical theory of paramagnetism developed by Langevin, considers the paramagnetic solids in terms of a paramagnetic gas in which each particle is assumed to bear a permanent magnetic moment. When a mass of such a gas is placed in a magnetic field each particle tends to orient itself with its magnetic axis parallel to it. The thermal agitation to which each particle is always subjected at ordinary temperature, however resists this tendency of the field to orient the particles. In thermal equilibrium, the particles therefore set themselves oriented at certain angles with the applied magnetic field. This gives rise to the magnetisation of the medium. The classical theory calculates it by assuming that the only aligning force is the external field.

Let us consider a paramagnetic gas containing no particles, each bearing a permanent magnetic moment μ_m. When the magnetic field is in direction $\boldsymbol{B}$ is applied, the potential energy of particles whose magnetic axis makes an angle θ with the field is $-\mu_m B \cos\theta$ Fig. (19.8). The minimum dipole energy corresponds to $\theta = 0$. Therefore all the dipoles tend to orient themselves in the direction of the external field, which is hampered by thermal motion.

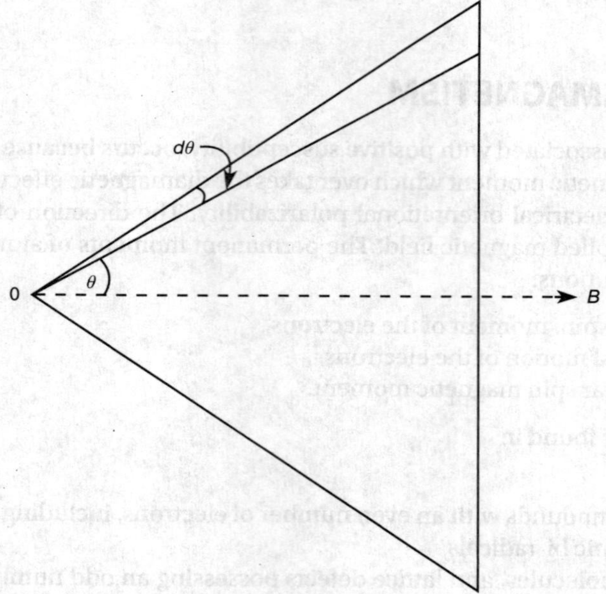

Fig. 19.8

The probability that a dipole is inclined at an angle θ to the direction of the field B in thermal equilibrium is proportional to the Boltzmann factor

$$\exp(\mu_m \boldsymbol{B} \cos\theta / k_B T)$$

This is in accordance with the classical Maxwell-Boltzmann distribution Law. Next let us imagine that all the particles of the gas have been collected and placed with their centres at a point 0, and their axes pointing in the some directions in which they actually pointing before they were collected Fig. (19.8). Then the number dn of particles whose axes fall within the sold angle $d\omega$ between two hollow cones of semi-angle θ and $\theta + d\theta$ is given by

$$dn = n_0 \exp(\mu_m \boldsymbol{B} \cos\theta / k_B T) d\omega \tag{19.47}$$

where solid angle $d\omega = 2\pi \sin\theta d\theta$. Now each of the particles contributes a component of magnetic moment $\mu_m \cos\theta$ parallel to the field, while by symmetry, the component perpendicular to the field annul each other, hence the mean magnetic moment $<\mu_m>$ (or $\bar{\mu}_m$) in the direction of the applied field is obtained by dividing the sum of the resolved components of the magnetic moments of all the dipoles in the direction of the field by the total number of dipoles, i.e.,

$$\bar{\mu}_m = \frac{\int \mu_m \cos\theta . dn}{\int dn} = \frac{\mu_m \int_0^\pi \cos\theta \exp(\mu_m B \cos\theta / k_B T) \sin\theta d\theta}{\int_0^\pi \exp(\mu_m B \cos\theta / k_B T) \sin\theta d\theta}$$

putting

$$\frac{\mu_m \boldsymbol{B}}{k_B T} = a$$

We obtain

$$\bar{\mu}_m = \frac{\mu_m \int_0^\pi \cos\theta \exp(a\cos\theta) \sin\theta d\theta}{\int_0^\pi \sin\theta \exp(a\cos\theta) d\theta}{}^2$$

$$= \mu_m \left(\coth a - \frac{1}{a} \right) = \mu_m L(a) \tag{19.48}$$

where $L(a) = \left(\coth a - \dfrac{1}{a} \right)$ is called the Langevin function. Fig. (19.9) shows the plot of the Langevin function $L(a)$ with α.

---

[2] Putting $\cos\theta = x$ the integral becomes

$$\frac{\int_{+1}^{-1} xe^{ax} dx}{\int_{+1}^{-1} e^{ax} dx} = \frac{\int_{-1}^{+1} xe^{ax} dx}{\int_{-1}^{+1} e^{ax} dx} = \frac{\left[\frac{xe^{ax}}{a} - \frac{e^{ax}}{a^2} \right]_{-1}^{+1}}{\left[\frac{e^{ax}}{a} \right]_{-1}^{+1}}$$

$$= \frac{\frac{e^a}{a} - \frac{e^{-a}}{a} - \left(-\frac{e^{-a}}{a} - \frac{e^{-a}}{a^2} \right)}{\frac{e^{-a}}{a} - \frac{e^{-a}}{a}} = \frac{\frac{1}{a}(e^a + e^{-a}) - \frac{1}{a^2}(e^a - e^{-a})}{\frac{1}{a}(e^a + e^{-a})}$$

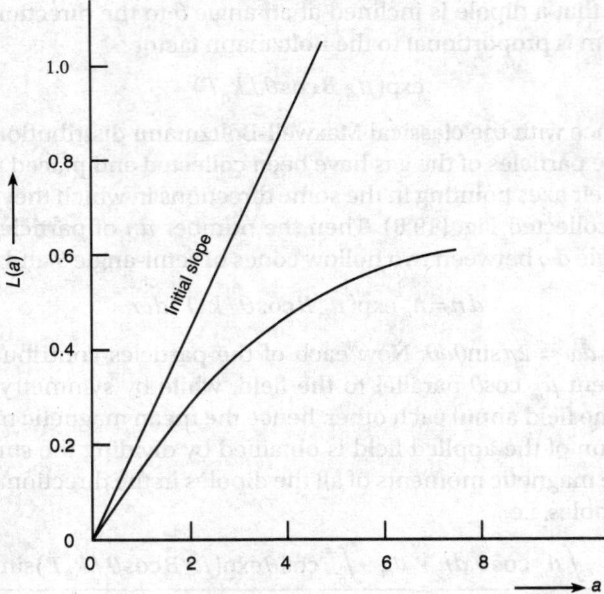

Fig. 19.9

Case 1: $\mu_B >> k_B T$

At low temperatures of large applied field, we observe that the Langevin function tends to unity. Hence magnetisation (magnetic moment per volume)

$$\boldsymbol{M} = n\mu_m \boldsymbol{M}_s \tag{19.49}$$

where n is the number of atoms per units volume and M the saturation magnetisation. Although saturation effects have been observed at low temperatures in Gadolium sulphate $Gd_2\,(SO_4)_3\,8H_2O$ but the results do not fit the theory.

Case 2: $\mu_B \boldsymbol{B} << k_B T$

For $\dfrac{\mu_m \boldsymbol{B}}{k_B T} << 1$, i.e., when T is large, we can expand coth a in a power series

$$\text{Coth}\, a = a^{-1} + \frac{a}{3} - \frac{a^2}{45} + \ldots$$

For $a << 1$, we can limit ourselves with the first two terms of the expansion. Then

$$\bar{\mu}_m = \frac{\mu_m a}{3} \tag{19.50}$$

and the magnetisation

$$M = \frac{\mu_m n a}{3} = \frac{\mu_m n}{3} \cdot \frac{\mu_m \boldsymbol{B}}{k_B T} = \frac{\mu_m^2 n \boldsymbol{B}}{3 k_B T} \tag{19.51}$$

The paramagnetic susceptibility

$$\chi = \frac{M}{H} = \frac{\mu_0 \mu_m^2 n}{3k_B T} = \frac{C}{T} \qquad (19.52)$$

where $C = \frac{\mu_m^2 \mu_0 n}{3k_B}$ is called curie constant. This is referred to as the Curie law. If one substitutes $n = 10^{26}\,m^{-3}$ and $T = 100\,K$ one finds that $x \approx 10^{-5}$ in agreement with observation.

The susceptibility given by Eq. (19.52) is also referred to as the Langevin paramagnetic susceptibility. Note in particular that χ is inversely proportional to temperature. This is in marked contrast to the diamagnetic susceptibility, which is essentially temperature independent.

Short comings of the theory: At low temperatures and large applied field the theory predicts the saturation effects which have been observed at low temperatures in gadolium-sulphate but the results do not fit the theory.

Besides this the derivation of the Curie Law is also open to a serious criticism which was pointed out by Min Van Leeuwan. It is not clear that if the permanent magnetic moment is to be associated with, and proportional to the angular momentum of moving electric charges, why should all the particles bear the same magnetic moment $1/m$. Infact, the magnetic moment of different particles should range from $-\infty$ to $+\infty$. But when the Max Well Boltzmann Statistics are applied to a system of such particles, the diamagnetic and paramagnetic contributions to the susceptibility cancel each other and susceptibility is thus zero. Recall that paramagnetism does not exist alone, diamagnetism also exists at the same time which is usually much smaller. Hence Langevin's theory must be treated as complete. A more satisfactory theory originates from quantum mechanics.

Quantum theory: The classical theory given by Langevin is incapable of providing a consistent explanation of magnetic phenomenon as a result of the motion of electronic charges. The assumption that all orientations of the magnetic moments w.r.t. ***B*** are possible, which is the basis of Langevin's theory is wrong. These difficulties have on the whole been overcome by the quantum theory of paramagnetism.

The magnetic moment of an atom or ion in free space is given by

$$\mu_J = \gamma \hbar \boldsymbol{J} = g \mu_B \boldsymbol{J} \qquad (19.53)$$

where the total angular momentum J is the vector sum of the orbital $\boldsymbol{L}$ and spin $\boldsymbol{S}$ angular momenta, γ is called the gyro magnetic ratio and μ_B is Bohr magnetion, g is Lande g factor and J is total angular momentum quantum number. For a free atom the g factor is given by Eq. (19.20)

In presence of a magnetic field B, the energy of the magnetic moment $\boldsymbol{\mu}_J$ is given by the vector product of theses quantities i.e.,

$$\mu = \boldsymbol{\mu} - \boldsymbol{B} = \boldsymbol{M}_J g \mu_B B \qquad (19.54)$$

where $\boldsymbol{M}_J = J(J-1), \ldots \{-(J-1)\}, -(-J)$ i.e., $\boldsymbol{M}_J$ is the magnetic quantum number associated with J and there are $(2J+1)$ ways in which it can align itself in a magnetic field. Thus an energy difference exists between spin magnetic dipole moments parallel and antiparallel

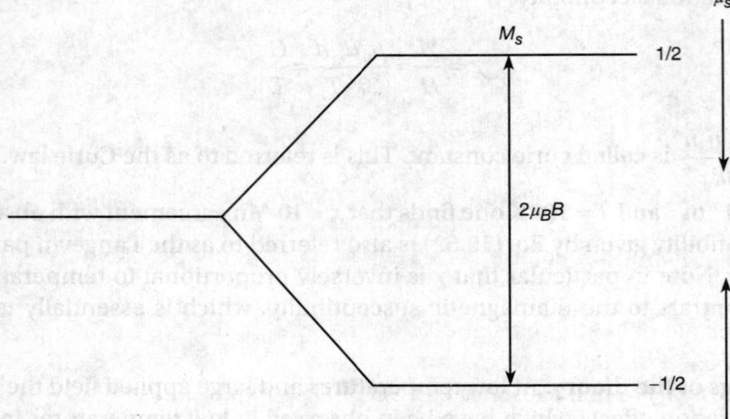

Fig. 19.10

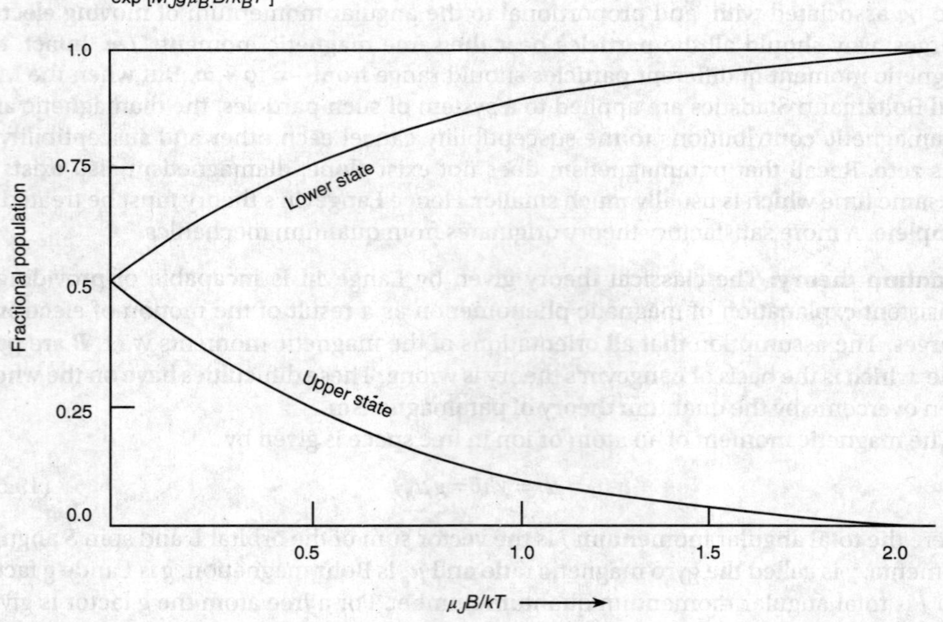

Fig. 19.11

to the field. The magnitude of the energy difference being $2\mu_B B$. This is shown in Fig. (19.11). Fig. (19.10) and upper levels are plotted.

The fractional population of lower level and using Maxwell Boltzmann distribution, the number of atoms with the particular value M_J will be proportional to

$$\exp[M_J g\mu_B.B / k_B T]$$

Consider now the unit volume containing n atoms placed in the magnetic field the magnetisation along the magnetic field will be $M = n \times$ statistical average of the magnetic moment component per atom along B.

$$\boldsymbol{M} = n\frac{\sum_{M_J=-J}^{+J} \boldsymbol{M}_g g\mu_B \exp[\boldsymbol{M}_J g\mu_B \boldsymbol{B}/k_B T]}{\sum_{M_J=-J}^{+J} \exp[\boldsymbol{M}_J g\mu_B \boldsymbol{B}/k_B T]} \tag{19.55}$$

Summation sign is used because $\boldsymbol{M}_J$ assumes only the quantised values. We may now study two cases:

Case 1: At normal field strengths and ordinary temperatures, the value of the exponent $[\boldsymbol{M}_J g\mu_B \boldsymbol{B}/k_B T]$ is very small. At $\boldsymbol{B} = 1T$ and $T = 300$ K with $g = 2$ and $J = 1/2$ this is less than 0.0002. It is possible to make a series expansion of exponentials and to the first order this can be written as[3]

$$\boldsymbol{M} = n\frac{g\mu_B \sum_{M_J=-J}^{+J} \boldsymbol{M}_J[1 + \boldsymbol{M}_J g\mu_B \boldsymbol{B}/k_B T]}{\sum_{M_J=-J}^{+J} [1 + \boldsymbol{M}J g\mu_B \boldsymbol{B}/k_B T]^3}$$

$$= n\frac{g\mu_B\left[0 + \frac{g\mu_B \boldsymbol{B}}{k_B T}\frac{J(J+1)(2J+1)}{3}\right]}{(2J+1)+0}$$

$$= \frac{ng^2\mu_B^2 \boldsymbol{B}}{3k_B T}J(J+1) \tag{19.56}$$

and the paramagnetic susceptibility

$$\chi = \frac{\boldsymbol{M}}{\boldsymbol{H}} = \frac{\mu_0 n g^2 \mu_B^2}{3k_B T}J(J+1) \qquad (\because \boldsymbol{B} = \mu_0 \boldsymbol{H})$$

$$= \frac{\mu_0 n \mu_{\text{eff}}^2}{3k_B T} = \frac{C}{T} \tag{19.57}$$

where $\mu_{\text{eff}} = P\mu_B$ and $P = g\sqrt{J(J+1)}$ (19.58)

---

[3] 1 Number is $\sum_{M_J=-J}^{+J} M_J[1 + M_J g\mu_B B/k_B T] = \sum_{M_1=-1}^{+J} M_J + \sum_{M_J=-1}^{+J} M_J^2 g\mu_B B/\mu_B T$

$$\sum_{M_J=-J}^{+J} M_J = J + (J-1) + (J-2) + \cdots + \{-(J-2)\} + \{-(J-1)\} + (-J) = 0$$

$$\sum_{M_J=-J}^{+J} M_J^2 = J^2 + (J-1)^2 + (J-2)^2 + \cdots + \{-(J-2)\}^2 + \{-(J-1)\}^2 + (-J)^2 = 0$$

$$= 2[1 + 1^2 + 3^2 + \cdots + J^2] = \frac{2J(J+1)(2J+1)}{6} = \frac{J(J+1)(2J+1)}{3}$$

Denominator is $\sum[1 + M_J g\mu_B B/k_B T] = \sum_{M_J=-J}^{+J} 1 + \frac{g\mu_B B}{k_B T}\sum M_J = (2J+1) + 0$

Here the number P is called the effective number of Bohr magnetons for the atom. The Eq. (19.58) is identical with the classical result Eq. (19.52) with

$$\mu_m = P\mu_B$$

and the Curie law can be deduced. It must be emphasized, however that in deducing this equation we have assumed that every particle is in the same ground state. This may be true only in the case of wide multiplets where so much energy is required to raise them to higher energy states, so that nearly all the atoms must be in the ground state. This situation is found to exist with many ions of the rare earth group.

In case of narrow multiplets, since a particle requires a small energy in order to excite to higher energy states, the atoms must exist in a variety of different energy states. In this case we may therefore reasonably assume that the coupling of L with S is small compared with that of either L or S with B. This is equivalent to the assumption that an atom behaves as if it is in strong field so that the orbital and spin moments are separately quantised. With

$$m_L = L, L-1, \cdots (-L) g_L = 1$$
$$M_s = s(s-1), \cdots (-s) g_s = 2$$

We then have for the magnetic moment of unit volume containing n atoms, the expression.

$$\boldsymbol{M} = n\mu B \frac{\sum_{M_L=-L}^{+L} M_L \exp[M_L \mu_B \boldsymbol{B}/k_B T]}{\sum_{M_L=-L}^{+L} \exp[M_L \mu_B \boldsymbol{B}/k_B T]} + n\mu B \frac{\sum_{M_s=-s}^{+s} 2M_s \exp[2M_s \mu_B \boldsymbol{B}/k_B T]}{\sum_{M_s=-s}^{+s} \exp[2M_s \mu_B \boldsymbol{B}/k_B T]}$$

Since exponentials are still small we have

$$\boldsymbol{M} = n\mu_B \left[\frac{1}{2} L(L+1) \frac{\mu_B \boldsymbol{B}}{k_B T} + \frac{4}{3} s(s+1) \frac{\mu_B \boldsymbol{B}}{k_B T} \right] \tag{19.59}$$

or

$$\chi = \frac{n\mu_0 \mu_{\text{eff}}^2}{3k_B T} = \frac{C}{T} \tag{19.60}$$

where

$$\mu_{\text{eff}} = P\mu_B \text{ and } P = \sqrt{L(L+1) + 4s(s+1)} \tag{19.61}$$

We see that curie law is again deduced, but the value of P is different from that in the case of wide multiplets. It is found that a modified from Eq. (19.61) represents the magnetic behaviour of ions of elements of the iron group.

When multiplets are comparable with $k_B T$ as in the case of Sn^{+3} and Eu^{3+} and it also becomes necessary to take account of the temperature independent paramagnetism contribution. This is called **Van Vleck Paramagnetism**. Curie Law is not obeyed is such cases.

Case 2: We now calculate the magnetisation under the condition of intense field and low temperatures. Under this condition

$$M_j g\mu_B \boldsymbol{B}/kBT >> 1$$

and it is not possible to make a series expansion of the exponentials of the Eq. (19.55). In this case we proceed as under:

Let

$$\chi = \frac{g\mu_B B}{k_B T}, \quad \text{then}$$

$$M = \frac{ng\mu_B \sum_{M_J=-J}^{+J} M_J e^{M_J x}}{\sum_{M_J=-J}^{+J} e^{M_J x}}$$

$$= ng\mu_B \frac{d}{dx}\left[\log_e \sum_{M_J=-J}^{+J} e^{M_J x}\right]$$

$$= ng\mu_B \frac{d}{dx}\left[\log_e \frac{e^{Jx}(1-e^{-(2J+1)x})^2}{1-e^{-x}}\right]$$

as $\sum_{M_J=-J}^{+J} e^{M_J x}$ is a geometrical progression with $(2J+1)$ terms and common ratio e^{-x} and first term is e^{Jx}.

Multiplying with $e^{x/2}$ and dividing with $e^{x/2}$, we have

$$M = ng\mu_B \frac{d}{dx}\left[\log_e \frac{e^{(j+1/2)}x}{e^{x/2}-e^{-x/2}}\right] = ng\mu_B \frac{d}{dx}\left[\log_e \frac{\sinh(j+1/2)x}{\sinh x/2}\right]$$

$$= ng\mu_B\left[\frac{2j+1/2}{2j}\coth\left(\frac{2j+1}{2j}\right)a - \frac{1}{2j}\coth\left(\frac{a}{2j}\right)\right]$$

where $a = jx$ or $x = a/j$

Now defining the function $B(a)$ as,

$$B(a) = \frac{2J+1}{2J}\coth\frac{(2J+1)}{2J}a - \frac{1}{2J}\coth\left(\frac{x}{2J}\right)$$

The magnetisation is given by

$$M = ng\,J\mu_B B(a) \tag{19.62}$$

$B(a)$ is called the **Brillouin function.** In the limit $J \to \infty$, it approaches the Langevin function $L(a)$. We should expect from the fact that $J = \infty$ means an infinite number of possible orientations are allowed. It is found that Eq. (19.62) expresses very satisfactorily the results obtained on the susceptibility of hydrated gadolium sulphate $Gd_2(SO_4)\,8\,H_2O$ at liquid helium temperature Fig. (19.12).

Here with very high field (5*T*) applied at very low temperatures (4K and lower), we get the phenomenon of paramagnetic saturation, a situation in which all the dipoles would be ultimately directed along *B*. The value of the saturation magnetic moment, however, is not $ng\sqrt{J(J+1)}\mu_B$ but $ngJ\mu_B$. This is because the greatest component of

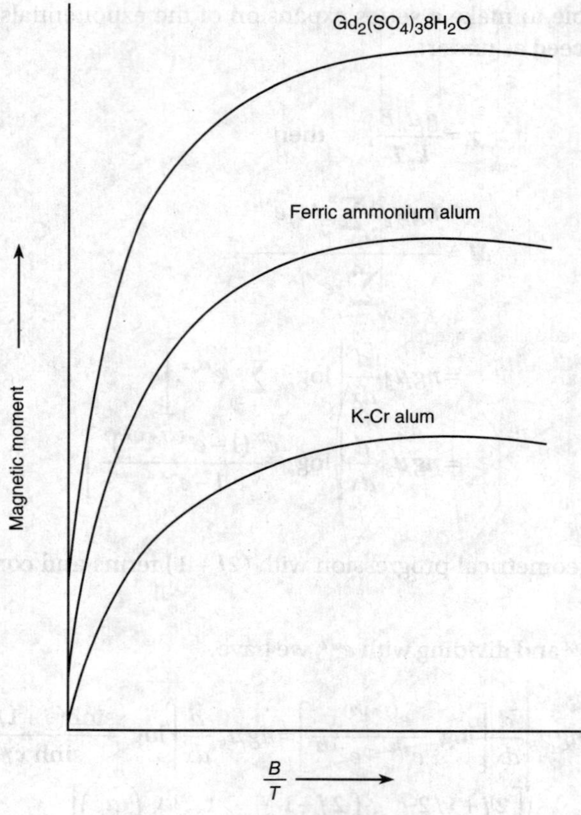

Fig. 19.12

each moment parallel to $\boldsymbol{B}$ in $Jg\mu_B$ and the actual magnetic moment always processes at a finite angle to the field.

19.10 FERROMAGNETISM

Ferromagnetism is the existence of a spontaneous magnetisation even in zero applied field. Pure Iron, Cobalt, Nickel as well as many transition metal alloys exhibit this property, i.e., there is found to be remanent magnetic polarization in them. A spontaneous moment suggests that electron spins and magnetic moment are arranged in a regular manner. The order need not be simple. Various spin arrangements are shown in Fig. (19.13).

Ferromagnetism is exactly opposite to that or pure paramagnetism, i.e., in a ferromagnetic crystal unlike paramagnetic crystal all the neighbours of a given magnetic ion a lattice are also magnetic which result in spontaneous magnetisation. Just as in ferroelectricity and super conductivity the ferromagnetic property disappear completely above a critical temperature termed the Curie temperature. Above this temperature, where C is curie

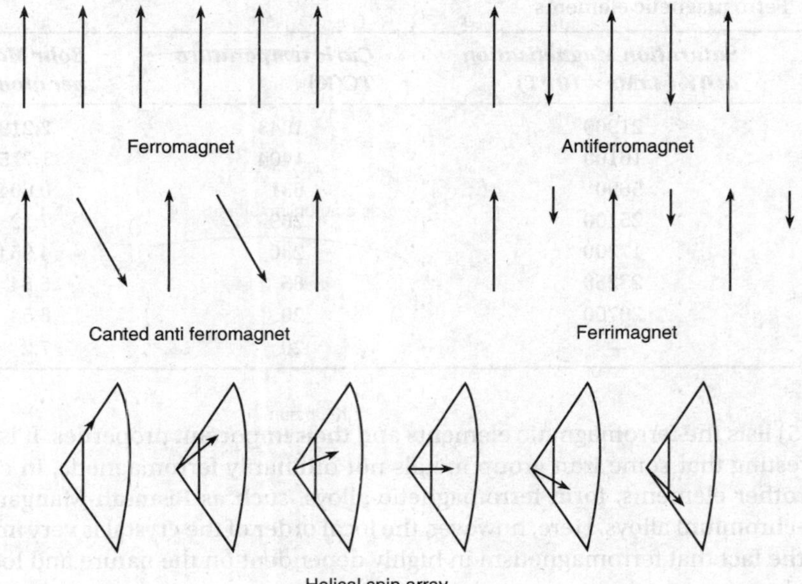

Fig. 19.13

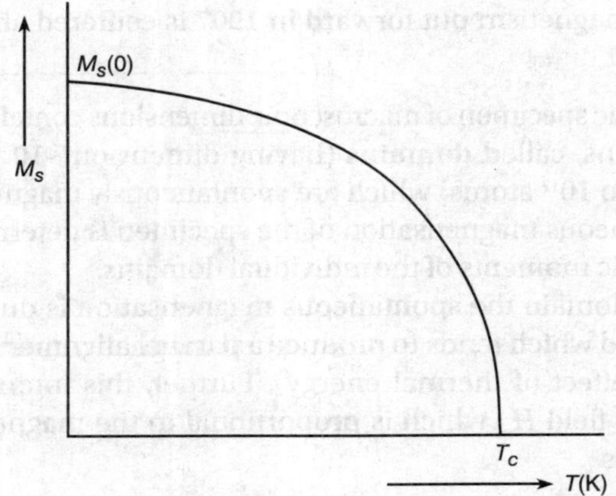

Fig. 19.14

constant these materials are only paramagnetic, ferromagnetic materials are characterised by the following properties.

(i) A temperature dependent spontaneous magnetisation varying from a maximum $M_s(0)$ at $T = 0$ K to 0 at T_C as shown in Fig. (19.14). M_s is usually called the **saturation magnetisation.**

(ii) Well above Curie temperature (T_C) the susceptibility follows the **Curie-Weiss law.**

$$\chi = \frac{C}{T - T_C} \tag{19.63}$$

(iii) The formation of domains which are spontaneously magnetised.

Table 19.3 Ferro magnetic elements

Element	*Saturation Magnetisation at 0 K ($4\pi Ms \times 10^{-4}T$)*	*Curie temperature TC(K)*	*Bohr Magnetons per atom (0 K)*
Fe	21900	1043	2.219
Co	16100	1404	1.715
N	5690	631	0.604
Gd	25100	289	7.12
Tb	17200	230	4.95 (or 9.1)
Dy	23250	85	6.84
Hc	28700	20	8.54
Er	—	20	7.2

Table (19.5) lists the ferromagnetic elements and their important properties. It is particularly interesting that some iron group metals not ordinarily ferromagnetic, in combination with other elements, form ferromagnetic alloys, such as Bismuth-Manganese and Platinum-chromium alloys. Here, however, the local order of the crystal is very important, stressing the fact that ferromagnetism in highly dependent on the nature and location of neighbouring atoms.

Weiss theory of Ferromagnetism and Weiss field

Weiss theory of ferromagnetism put forward in 1907 is centered about the following two hypothesis:

(i) A ferromagnetic specimen of macroscopic dimensions contains in general a number of small regions, called **domains** (having dimension ~10^{-4} m and containing as many as 10^{15} to 10^{17} atoms) which are spontaneously magnetised. The magnitude of the spontaneous magnetisation of the specimen is determined by a vector sum of the magnetic moments of the individual domains.

(ii) Within each domain the spontaneous magnetisation is due to the existence of a molecular field which tends to produce a parallel alignment of the atomic dipoles (despite the effect of thermal energy). Further, this internal field is equivalent to a magnetic field H_m which is proportional to the magnetisation M within the domain, that is

$$B_m = \lambda M \tag{19.64}$$

where λ is the Weiss field constant and is independent of temperature. λ is determined emperically. The effective field (B_{eff}) may be regarded as the vector sum of the external field B and B_m. Thus

$$B_{eff} = B + \mu_0 \lambda \tag{19.65}$$

Let us consider a ferromagnetic solid containing N atoms per unit volume, each with a total angular momentum quantum number J. One may then write the magnetisation.

$$\boldsymbol{M} = Ng\mu_B J\, \boldsymbol{B}(\alpha) \tag{19.66}$$

Where Brillouin function $\boldsymbol{B}(a)$ is given by

$$\boldsymbol{B}(a) = \frac{2J+1}{2J}\cot h\frac{(2J+1)}{2J}a - \frac{1}{2J}\coth\left(\frac{a}{2J}\right)$$

and
$$a = \frac{g\mu_B J}{k_B T}\boldsymbol{B}_{\text{eff}} = \frac{g\mu_B J}{k_B T}(\boldsymbol{B} + \mu_0 \lambda \boldsymbol{M})$$

For spontaneous magnetisation, the external field is not required. Hence in the absence of external field, i.e., when B is zero, one obtains

$$a = \frac{g\mu_B J}{k_B T}\lambda \boldsymbol{M}(T)\mu_0 \tag{19.67}$$

Note that we are using the equation derived for paramagnetic specimens. This is because we may consider ferromagnetic substance as the paramagnetic substances with an internal molecular field which is equivalent to the magnetic field λM, and which is in addition to the external field. Here we have also written $M(T)$ instead of M to emphasize temperature dependence. If M_s denotes the value of magnetic moment per unit volume corresponding to complete alignment of the magnetic moments parallel to the field, i.e., the saturation magnetic moment, then

$$M_s(0) = Ng\mu_B J \tag{19.68}$$

Then
$$\frac{\boldsymbol{M}(T)}{\boldsymbol{M}_s(0)} = \frac{k_B Ta}{\mu_0 N\lambda g^2 \mu_B^2 J^2} \tag{19.69}$$

and
$$\frac{\boldsymbol{M}(T)}{\boldsymbol{M}_s(0)} = B(a) \tag{19.70}$$

Before we attempt the simultaneous solution of last two Eqs. (19.69) and (19.70), note that $B(a) \rightarrow 1$ as $x \rightarrow \infty$, i.e., as $T \rightarrow 0$. Thus $M_s(0)$ is the maximum value of the magnetisation and would occur if the temperature were reduced to zero.

Now one can obtain the simultaneous solution of Eqs. (19.69) and (19.70) by plotting both functions for $\frac{\boldsymbol{M}(T)}{M_s(0)}$ against $\boldsymbol{a}$. These are shown in Fig. (19.15). Below a critical temperature $T = T_c$ K where the straight line is tangential to the Brillouin function at the origin, there are two intersections of the curves. The intersection at $a = 0$ is classically unstable non-equilibrium condition and the higher intersection corresponds to a spontaneous magnetisation.

Thus below a certain temperature there can exist in a ferromagnetic solid a magnetisation within the material even in the absence of magnetic field. At or above the critical temperature, called the Curie temperature the only intersection is at $M(T)/M_s(0)$ which means that the spontaneous magnetisation vanishes.

One observes that the temperature dependence of the saturation magnetisation is of the form

$$M_s(T) \propto (T - T_c)^{-1/2} \quad \text{for} \quad T < T_c$$

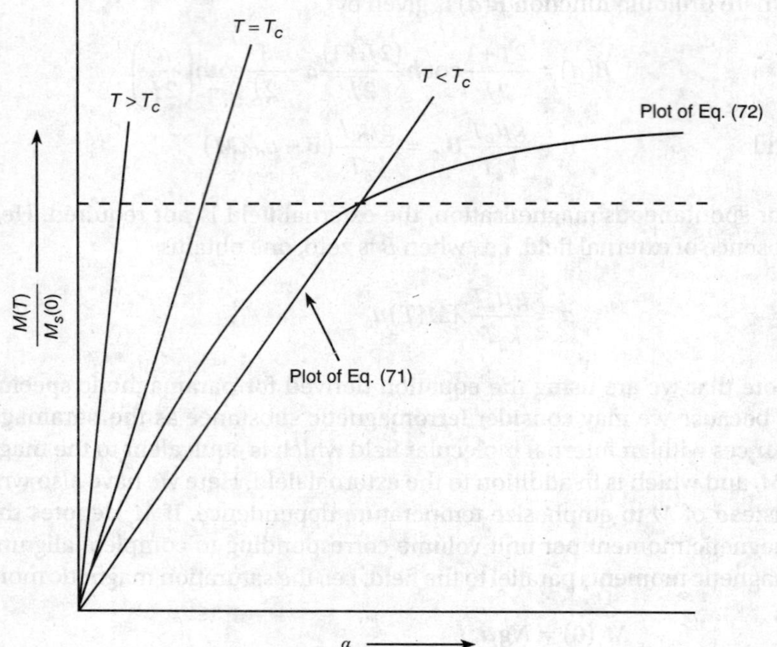

Fig. 19.15

Fig. (19.16) shows the spontaneous magnetisation versus the temperature variation. This behaviour of M_s enable one to classify the paramagnetic-ferromagnetic transition as a second order phase transition, i.e., a transition characterised by an order parameter (in this case M_s) that is non-zero only below T_c. We will discuss it in detail later. It will be evident that there must exist a relation between the Curie temperature (T_c) and molecular dimension less field constant λ. In fact one expects T_c to increase with λ because the tendancy for parallel alignment increases as λ becomes larger. In order to establish this relationship, one makes use of the fact that for a << 1 (near the origin Fig. (19.17) the Brillouin function $B(a)$ is approximated as

$$B(a)=\frac{J+1}{3J}a-\frac{J+1}{3J}\left(\frac{2J^2+2J+1}{30J^2}\right)a^3\approx\frac{J+1}{3J}a \qquad (19.71)$$

Hence the tangent of curve Eq. (19.70) at the origin has a slope equal to $\frac{J+1}{3J}$. Also the slope of the Eq. (19.70) is $\frac{k_BT}{\mu_0 N\lambda g^2\mu_B^2 J^2}$. Equating these two slopes, one obtains

$$\frac{k_BT}{\mu_0 N\lambda g^2\mu_B^2 J^2}=\frac{J+1}{3J}$$

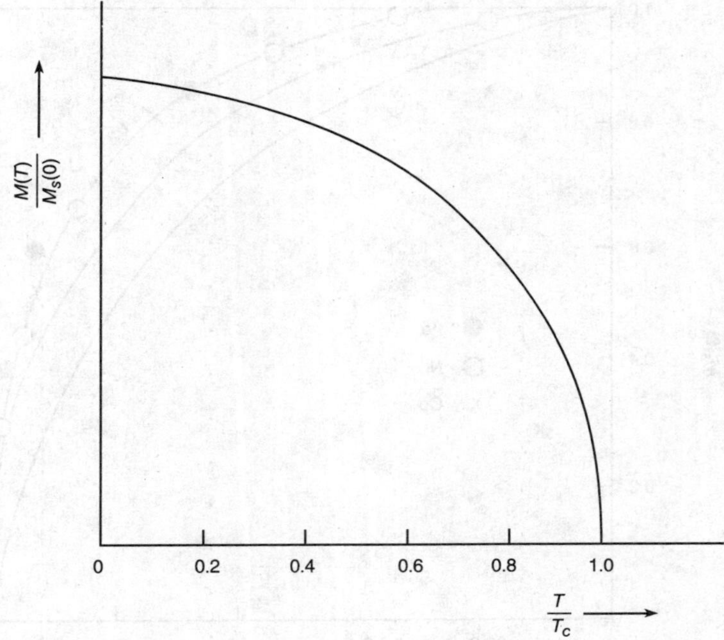

Fig. 19.16

or
$$T = \frac{\mu_0 N g^2 \mu_B^2 J(J+1)\lambda}{3k_B}$$

For
$$T = T_C$$

$$\frac{3k_B T_C}{\lambda} = \mu_0 N g^2 \mu_B^2 J(J+1) = N\mu_m^2 \mu_0 \tag{19.72}$$

Here $\mu_m = g\mu_B\sqrt{J(J+1)}$ is the **total magnetic moment per atom**. Eq. (19.72) reveals that T_c is proportional to molecular field constant.

19.11 TEMPERATURE DEPENDENCE OF SATURATION OR SPONTANEOUS MAGNETISATION

Combining Eqs. (19.69) and (19.72) we obtain

$$\frac{\boldsymbol{M}(T)}{M_s(0)} = \frac{J+1}{3J}\left(\frac{T}{T_C}\right)a \tag{19.73}$$

Here M_s (0) is the maximum value of the saturation magnetisation. $\boldsymbol{M}(T)$ is the value of spontaneous magnetisation at temperature T. $\frac{\boldsymbol{M}(T)}{M_s(0)}$ must satisfy both values given by Eqs. (19.70) and (19.73) hence it can be solved by intersection method indicated

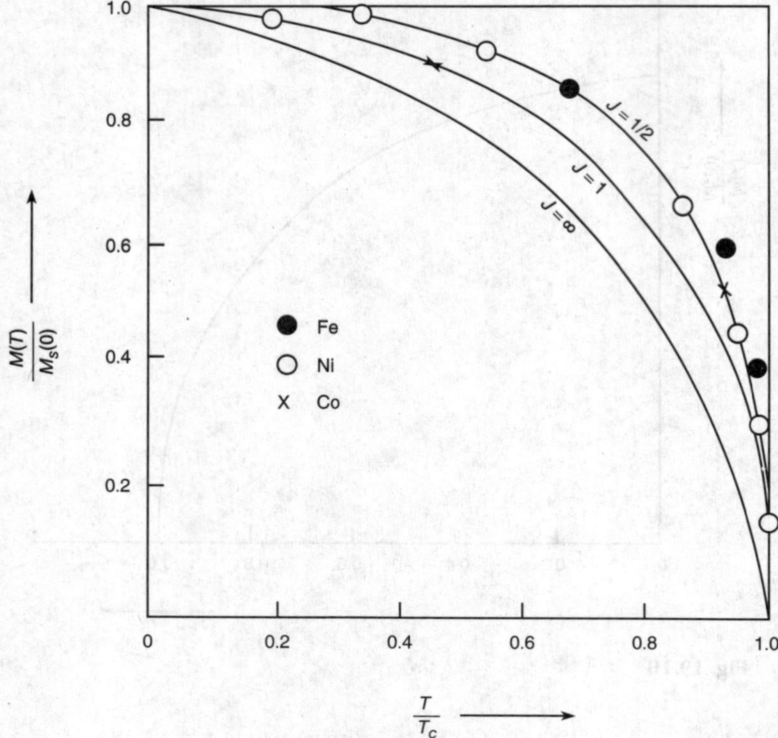

Fig. 19.17

in Fig. (19.17). As mentioned earlier and also evident from Eqs. (19.70) and (19.73) that for a given value of J, this procedure leads to universal curve. In Fig. (19.17), we have represented such curves for $J = 1/2$, $J = 1$ and $J = \infty$, the latter case corresponds to the results of classical Langevin theory for freely rotating dipoles. It is observed that the curve $J = 1/2$ fits the data best, indicating that the magnetisation comes from spin moments rather than from the orbital moments of electrons and that the appropriate value for $g = 2$. For the orbital motion the value of $g = 1$.

We note that the magnetisation decreases very slowly as the temperature is first raised above absolute zero and then drops more steeply at higher temperature until finally falls precipitously to zero at the Curie temperature. However the theory is by no means perfect. Infact, the more exact Block spin wave treatments is applicable in the low temperature range (this is beyond the scope of this book) according to which

$$\boldsymbol{M}(T) = M_s(0)\left[1 - A\left(\frac{k_B T}{J}\right)^{3/2}\right] \text{ for } T << T_c$$

is the relation between $\boldsymbol{M}(T)$ and $M_s(0)$. Here A is a numerical constant equal to 0.1174 for the simple cubic lattice. The result is known on Block $T^{3/2}$ Law.

19.12 THE PARAMAGNETIC REGION

Above the Curie temperature the spontaneous magnetisation is zero. One can see at small values of *a*, i.e., at high temperature, Brillouin function is

$$B(a) \approx \frac{J+1}{3J} a$$

and magnetisation is

$$\boldsymbol{M} = Ng\mu_B J\, B(a) = Ng\mu_B J \frac{(J+1)a}{3J} = \frac{1}{3} Ng\mu_B (J+1)$$

and in the presence of external field in given by

$$a = \frac{g\mu_B (B + \mu_0 \lambda M) J}{k_B T}$$

Elimination or *a* gives

$$\boldsymbol{M} = \frac{Ng^2 \mu_B^2 J(J+1)}{3k_B T}(B + \lambda \boldsymbol{M} \mu_0) \tag{19.74}$$

or

$$\boldsymbol{M} = \frac{C(B + \lambda \boldsymbol{M} \mu_0)}{\mu_0 T} \quad \text{where} \quad C = \frac{\mu_0 N g^2 \mu_B^2 J(J+1)}{3k_B}$$

or

$$\boldsymbol{M}\left(1 - \frac{C\lambda}{T}\right) = \frac{Cb}{\mu_0 T} \tag{19.75}$$

and

$$\boldsymbol{M} = \frac{CB}{\mu_0 (T - T_c)}$$

where

$$T_c = C\lambda \quad \text{or} \quad \lambda = T_c / c \tag{19.76}$$

and susceptibility

$$\chi = \frac{\mu_0 \boldsymbol{M}}{B} = \frac{C}{T \; T_c} \tag{19.77}$$

which is Curie-Weiss Law that accounts for the temperature behaviour of the susceptibility of magnetic materials. It also gives excellent agreement with data for ferromagnetic substances above the Curie point with the small disagreement that the Curie point defined from the spontaneous magnetisation theory differs by a few degrees from the experimental value found for magnetic region. A plot of $\frac{1}{\chi}$ vs *T* is shown is Fig. (19.18) reveals these facts.

For Fe, Co, and Ni, T_c (the temperature below which these metals are ferromagnetic) and *T* (above which the paramagnetism sets in these metals) are listed below.

Table 19.4

	Fe	*Co*	*Ni*
T_C	1043	1393	631
T	1093	1428	650

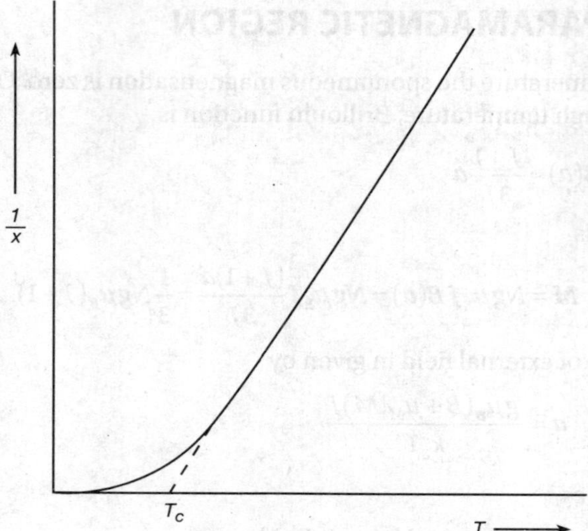

Fig. 19.18

These figures speak for themselves.

The Weiss theory contain features which agree with experiment, is still an emperical theory since it does not explain the origin of the Weiss field. The origin of the Weiss field was not known explanation until Heisenberg applied the quantum theory to the description of the ferromagnet.

19.13 THE NATURE OF FERROMAGNETIC CARRIERS

The fair agreement of the $\left[\dfrac{M(T)}{M_s(0)} \text{vs}\, T/T_c\right]$ curve with Brillouin function curve for $J = 1/2$ suggests that the ferromagnetic electrons must have $L = 0$, $J = s = 1/2$, $g = 2$. This in turn means that the entire magnetisation must essentially associates with the electron spin ($g = 2$) and not at all with the orbital motion ($g = 1$) of the electron. That this is indeed the case has been confirmed by gyro-magnetic experiments: these experiments consist in determining the value of g. For all the ferromagnetic metals it is found to very close to 2. Thus confirming that magnetisation is largely due to electron spin.

Now, the question arises, which electrons are involved. One can easily discard the argon core having electron distribution $1s^2\ 2s^2\ 2p^6\ 3s^2\ 3p^6$ as a source of ferromagnetism for Fe, Co and Ni. We know that 4s electrons are responsible for electrical conductivity, crystal binding and therefore only source left for ferromagnetism of metals are 3d electrons with unpaired spins. Now the question arises, what is the nature of ferromagnetic carriers.

Let us compare the theoretical and experimental values of saturation magnetisation. The theoretical value can be obtained simply by finding the number of electrons contributing towards magnetisation, i.e., unpaired 3d electrons because each electron contributes a magnetic moment $1\mu_B$ due to spin alone. The number of unpaired 3d electrons for the

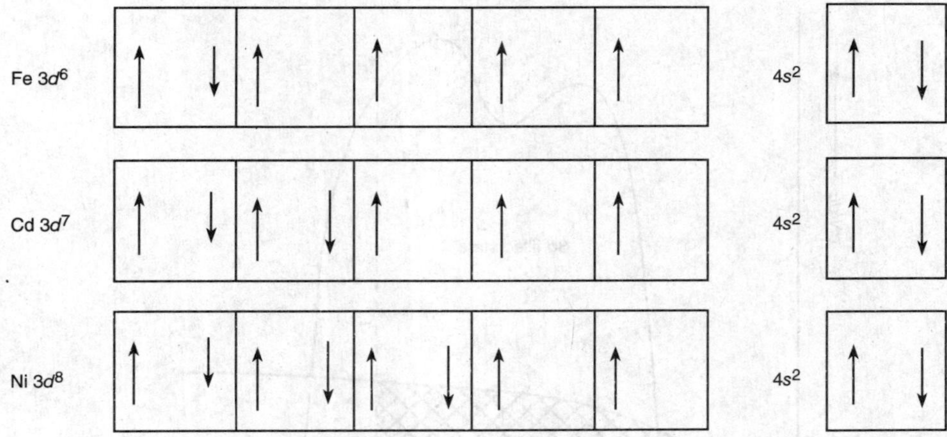

Fig. 19.19

single atom can be determined in accordance with the Hund's rules by the scheme shown in Fig. (19.19).

On this basis, one obtains. 4, 3 and 2 unpaired 3*d* electrons in Fe, Co and Ni respectively. The experimental value obtained form saturation magnetisation at $T = 0$ and the number of atoms per unit volume for Fe, Co and Ni are 2.22 μ_B, 1.72 μ_B and 0.54 μ_B respectively.

The comparison of these two results reveal that besides numerical disagreement in each case, the experimental values are non-integral. Therefore, one can conclude that the effective number of magnetic moment carriers per atom should be a non-integral, despite the fact that each atom has an integral number of electrons. It is worth while to mention that the discrepancy in not surprising as the above arguments apply to free atoms and the situation is essentially different when one deals with bound electrons in solids. Therefore one must take into account the band structure of theses metals.

Mott and Slater were the first to explain the non-integral values on the basis of wide 4*s* band over-lapping with a narrow 3*d* band Fig. (19.20). In general, there is on the average a certain fraction of the total number of 3*d* and 4*s* electrons in each band. The relative occupation of the two bands is determined by Fermi level (E_F). In the case of Nickel 10, 3*d* and 4*s* electrons are distributed in such a manner that on the average 9.4 electrons are in 3*d* band and 0.6 in the 4*s* band. Now applying Hund's rule, one finds that five electrons set their spin parallel to the field and the remaining 4.4 set anti parallel. Then a net parallel spin for nickel atom is $5 - 4.4 = 0.6$ and hence a magnetic moment of $0.6\mu_B$. This is in fair agreement with the experimental value of $0.54\mu_B$ for Ni. The accuracy of the model can also be checked by alloying Ni with Zn or Cu. The substitution of a Cu atom for Ni atom introduces one additional electron in the crystal, whereas Zn introduces two additional electrons in the crystal. These additional electrons will tend to fill the 3*d* states completely and hence the magnetisation should decrease. This is indeed the case observed from saturation magnetisation experiments. When 60% or 30% is added, the magnetisation of Ni virtually become zero.

Concluding, one can say that the number of magnetic moment carries in atoms of Fe, Co and Ni is definitely lower than the number of unpaired electrons in the 3*d* shell of the these metals and further these may have non-integral values.

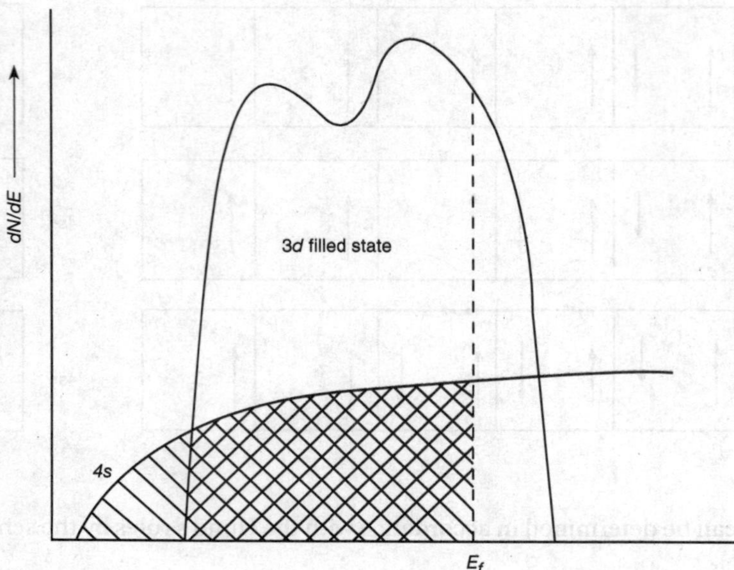

Fig. 19.20

19.14 NATURE AND ORIGIN OF WEISS MOLECULAR FIELD

We have seen that a substance is called Ferromagnetic if it possesses a spontaneous magnetic moment even in the absence of an applied magnetic field. The saturation or spontaneous magnetisation is defined as the spontaneous magnetic moment per unit volume. Weiss by introducing the concept of ferromagnetic domains (which were spontaneously magnetised and the spontaneous magnetisation of domain was assumed to be due to the molecular field) was able to explain the principal features of ferromagnetism. Let us have a rough estimate of the Weiss molecular field. Let us consider it as equivalent to an effective field B_m acting on the spins.

The interaction energy of a spin with the Weiss field B_m must be of the order of Curie temperature, i.e.,

$$\mu_B B_m = k_B T_C$$

For iron $T_C = 1000\,\text{K}$. This gives $B_m \cong 10^3 T$. From this one immediately concludes that the internal field is not due to simple dipole-dipole interaction between neighbours, because such fields would be of the order of $\frac{\mu_0 \mu_B}{4\pi a^3} \approx 10^{-1} T$ where a is lattice constant. We may also point out that in the ferromagnetic solids the molecular field constant $\lambda = \frac{B_m}{M} \simeq \frac{10^{-1}}{10^{-5}} = 10^4$, which is orders of magnitudes larger than the Lorentz factor $\frac{4\pi}{3}$. One might expect this for a simple model based on dipole-dipole interaction. Weiss did not touch upon the problem of nature and origin of the Weiss molecular field.

19.15 HEISENBERG'S EXCHANGE INTERACTION: QUANTUM THEORY OF FERROMAGNETISM

Heisenberg's explanation for the large value of the molecular field is based on the non-magnetic interaction called the exchange interaction between the atoms. The exchange interaction arises as a consequence of the Pauli exclusion principle, because of this principle we can not change the relative orientation of two spins without changing the special distribution of charge. This clearly indicates that the interaction exits between the two atoms. It is also clear that this interaction depends upon the relative orientation of the electron spins and not on the spin magnetic moments. The interaction energy between two moments may be written as

$$E_{ex} = -2J_0 \mathbf{S}_i . \mathbf{S}_j \tag{19.78}$$

where J_0 is called exchange constant (or exchange integral) and $\mathbf{S}_i$ and $\mathbf{S}_j$ are the total spins of the interacting atoms. E_{ex} is referred to as exchange energy. The vectors, $\mathbf{S}_i$ and $\mathbf{S}_j$ are related to the actual angular momenta by the relation $\mathbf{S}_i = S_i\hbar$ and $\mathbf{S}_j = S_j\hbar$. The constant J_0 has the dimensions of the energy.

The exchange constant J_0 is related to the overlap of the charge distribution of atoms i and j, i.e., on their interatomic separation. J_0 is positive for some what larger interatomic separation and negative for smaller ones. Expression Eq. (19.78) is called the Heisenberg model of exchange energy. It may be remarked that the exchange interaction has no classical analogue, it is purely electrostatic in origin. The spin entering into expression Eq. (19.78) is on account of Pauli's principle.

We observe from relation Eq. (19.78) that the exchange energy and hence the total energy of the two spin $\mathbf{S}_i$ and $\mathbf{S}_j$ is different when they are parallel than from that when they are anti parallel by an amount $2J_0$. Further if J_0 is positive, the energy of parallel configuration is lower than that of the anti-parallel one by this amount and thus the former configuration will be more stable favouring magnetisation to appear. Hence the parallel orientation of the spins of neighbouring atoms may too be advantageous from the energy point of view if their exchange constant is positive. This is necessary condition for ferromagnetism since the parallel arrangement of spins and consequently, of spin magnetic moments result in spontaneous magnetisation, which is characteristics of ferromagnetism. Thus the case $J_0 > 0$ leads to ferromagnetism and $J_0 < 0$ and to antiferromagnetism.

We now try to answer the question, why the exchange constant J_0 in general is positive only for the iron group atoms and negative for others: a negative value (corresponding to anti-parallel spin configuration and thus favouring non-magnetism) must be infact more probable, this is because the attraction between the nuclei and the electrons is greater than the repulsion between the nuclei and the electrons, and hence the smaller interatomic spacings. Following the qualitative analysis by Bethe J_0 is positive for the iron group atoms because there exists a certain critical ratio of the distance r_{ij} between neighbouring atoms i, j the these crystals to the radius of the unfilled $3d$-shell (r_{3d}). The behaviour of J_0 as the function of this ratio is plotted in Fig. (19.21) According to Slater, this ratio r_{ij}/r_{3d} is larger than 3 but not much larger. Some pertinent data in this respect are given in Table (19.5).

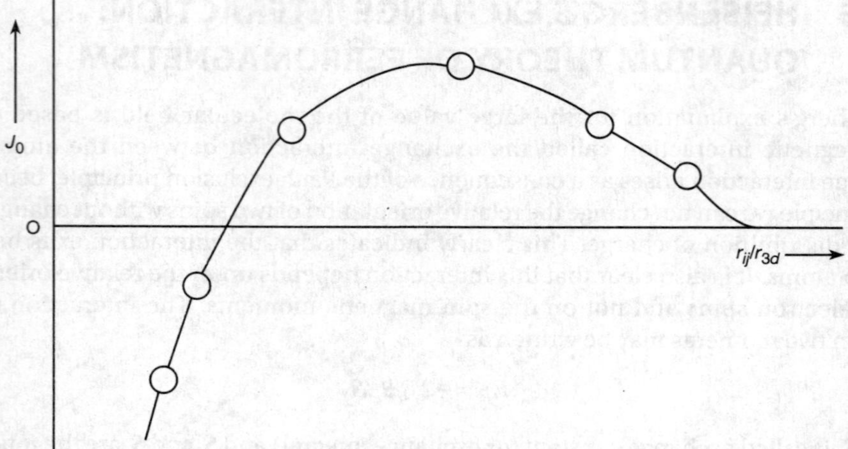

Fig. 19.21

Table 19.5

Element	***Fe***	***Co***	***Ni***	***Cr***	***Mn***	***Gd***
r_{ij}/r_{3d}	3.26	3.64	3.94	2.60	2.94	3.1
Ferromagnetic	Yes	Yes	Yes	No	No	Yes

Note that this explanation confirms the experimental observations on the magnetism of solids. One might raise the question whether an element with uncompensated spins, which itself is not ferromagnetic because the r_{ij}/r_{3d} value is not favourable, may be combined with another non-ferromagnetic element to form a compound for which the r_{ij}/r_{3d} value in suitable for ferromagnetism. That this seems indeed possible and is illustrated by the fact that MnAs and MnSb alloys are both ferromagnetic whereas pure, Mn, pure As and pure *sb* are not ferromagnetic. The lattice constants of these compounds are respectively 2.85 Å and 2.89 Å as compared to 2.58 Å for pure Mn. The ferromagnetism of the other alloys can be explained in a similar manner. Hence the necessary and adequate conditions for ferromagnetism are the existence of incomplete internal atomic shells and the positive sign of the exchange constant (integral) which causes the parallel orientation of the spins.

19.16 RELATION BETWEEN J_0 AND λ

Looking to the importance of the exchange integral (J_0) one would like to relate it to the Weiss constant (λ) and Curie temperature (T_C). Although it is very complicated problem but an approximate connection between J_0, λ an dT_C may be established by a simplified procedure suggested by Stoner.

Let us assume that the atom under consideration has z-nearest neighbours, each connected with the central atom by the interaction J_0. For more distant neighbours we take J_0 as Zero. Then using relation in Eq. (19.78) we may write for the exchange energy of a given atom i with its neighbors as

$$E=-2J_0\sum \boldsymbol{S}_i.\boldsymbol{S}_j \tag{19.79}$$

where the summation is over the nearest neighbours of atom i. The essential assumption of Stoner is that the instantaneous values of the neighbouring spins may be replaced by their time average. Thus if there are z-nearest neighbours, we have

$$E=-2Z\boldsymbol{J}_0[\boldsymbol{S}_{xi}.<\boldsymbol{S}_{xJ}>+\boldsymbol{S}_{yi}.<\boldsymbol{S}_{yj}>+\boldsymbol{S}_{zi}.<\boldsymbol{S}_{zJ}] \tag{19.80}$$

If magnetisation M is assumed to belong z-direction, one may write

$$<\boldsymbol{S}_{xj}>=<+\boldsymbol{S}_{yi}>=0 \text{ and } <\boldsymbol{S}_{zj}>=\frac{\boldsymbol{M}}{g\mu_B N}$$

Substituting these values, one obtains

$$E=-2ZJ_0\frac{\boldsymbol{S}_{zi}\boldsymbol{M}}{g\mu_B N} \tag{19.81}$$

Now, this exchange energy should be equal to the potential energy of spin in the Weiss field $\mu_0\lambda\,\boldsymbol{M}$, i.e.,

$$E=-g\boldsymbol{S}_{zi}\mu_B\lambda\boldsymbol{M}\mu_0 \tag{19.82}$$

Equating Eqs. (19.81) and (19.82), one obtains

$$\lambda=\frac{2Z\boldsymbol{J}_0}{\mu_0 Ng^2\mu_B^2} \tag{19.83}$$

and using Eq. (19.72), one finds

$$2Z\boldsymbol{J}_0=\mu_0\lambda Ng^2\mu_B^2=\frac{3k_BT_C}{\boldsymbol{J}(\boldsymbol{J}+1)} \tag{19.84}$$

and since ferromagnetism is due to exchange energies of spins and orbital motion has nothing to do with it. Thus $J=S$, we have

$$\therefore \qquad \boldsymbol{J}_0=\frac{3k_BT_C}{2Zs(s+1)} \tag{19.85}$$

for $\qquad s=1/2$

$$\frac{\boldsymbol{J}_0 Z}{k_BT_C}=2 \tag{19.86}$$

Eq. (19.85) is the relation as given by Weiss field theory between the exchange constant (integral) and Curie temperature T_C. More exact quantum statistics gives some what different result for zj_0/k_BT_c. According to R broke and Wood $k_BT_c/zj_0 = 0.280$, 0.325 and 0.346 respectively for the *sc*, BCC and FCC structures with $s = 1/2$. This may be compared with 0.5 from Eq. (19.86) for all three structures. These calculations enables us also to calculate the value of J_0 for BCC iron with $s = 1$, the observed Curie temperature (1093 K) corresponds to $\boldsymbol{J}_0 = 1.19 \times 10^{-2}$ eV.

19.17 FERROMAGNETIC DOMAINS

In a ferromagnetic material, the magnetisation curve (the *M-H* or the *B-H* curve) is known to show the phenomenon of **hysteresis**. On taking a ferromagnetic substance over a cycle of magnetisation, one finds that a fraction of the magnetic energy is converted to heat because of hysteresis, i.e., magnetisation lags behind the applied field. In order to give an explanation of this phenomenon, Weiss postulated that a ferromagnetic material consist of domains or blocks of macroscopic size which are spontaneously magnetised practically to saturation at a temperature below the Curie temperature, but whose magnetisation directions are oriented at random. As a result the ferromagnet does not show any magnetic property because the overall magnetisation then becomes Zero.

Under the influence of a magnetising field, the domain boundaries (called domain walls) become displaced and one domain of proper orientation grows in volume at the expense of the others. Fig. (19.22). This is supposed to be the mechanism of magnetisation in the initial protion of the *B-H* curve. Near saturation, there can also be a rotation of domain, i.e., force of magnetisation direction of other domains in the direction of the applied field. At large magnetisation the domain walls displacement is irreversible and so leads to a large remanent magnetisation. On the removal of the magnetising field alloying or precipitation of impurities also hinder reversible domain motion.

The existence of domains has been verified experimentally. On placing a drop of liquid containing a colloidal surpension of ferromagnetic particles on carefully polished surface of a ferromagnetic block, it is found that the colloidal particles collect along the domain boundaries, where strong local magnetic fields exist. Several types of complex domain configurations can then be seen under an optical microscope. The technique was first developed by Bitter and the domain structure seen on ferromagnetic surface some times called **Bitter patterns.**

Domain patterns are of intericate shape. These domain structures arise because the ferromagnet tries to reduce its magnetic energy to the minimum. In an effort to avoid free magnetic poles at the end closure domains are formed like caps on the bulk sample Fig. (19.23) and magnetic flux lines are closed within the ferromagnetic sample. In the domain configuration, in addition to the volume magnetisation energy boundaries between domains have anti-parallel field on either side. We know that the exchange interaction favours only parallel directions of spin alignment.

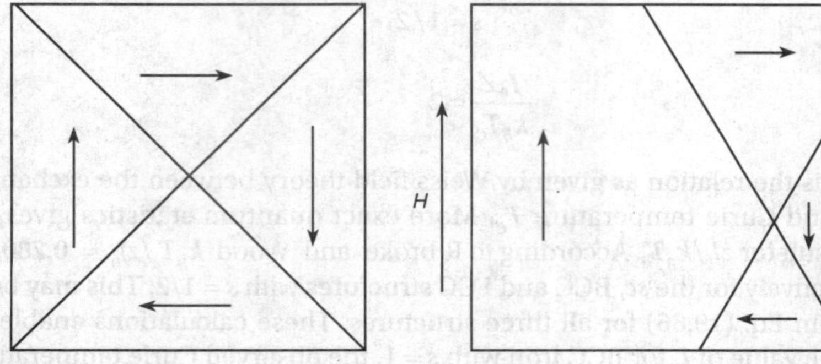

Fig. 19.22

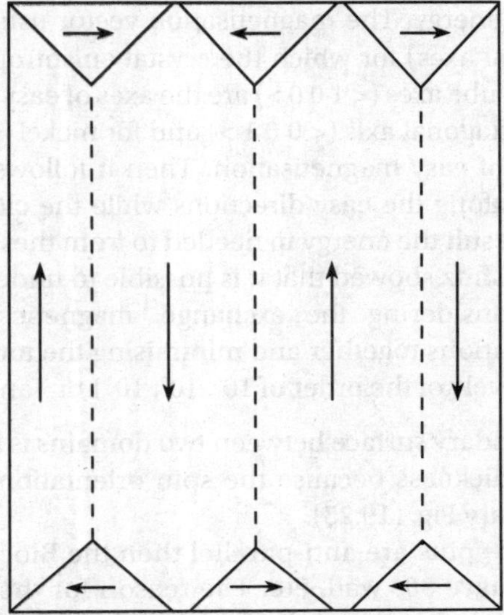

Fig. 19.23

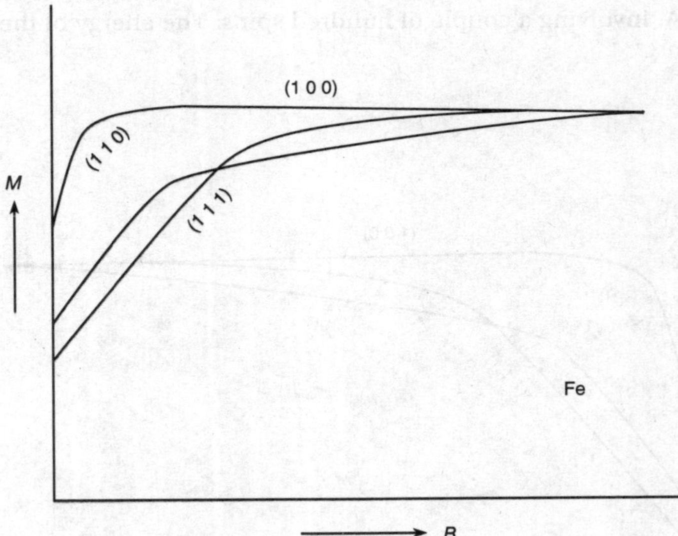

Fig. 19.24

This means that work has to be done in order to establish domain boundaries and the boundary layers become regions of high potential energy.

Further, it is found that in a ferromagnet there are crystallographic directions of easy magnetisation in which magnetisation in difficult and need an external field. This is known as **magnetic anisotropy**. A typical experimental curve is shown in Fig. (19.24). The energy needed to produce stable magnetisation in a certain direction is known as the crystal anisotropy energy. The magnetisation vector usually lies in the directions (with respect to the crystal axes) for which the crystal anisotropy energy is a minimum. For example, for iron the cube axes (< 1 0 0 >) are the axes of easy magnetisation, in Cobalt (hexagonal below) the hexagonal axis (< 0 0 1 >) and for nickel (FCC) the body diagonals (< 1 1 1 >) are direction of easy magnetisation. Then it follows that the basic domains Fig. (19.23) are oriented along the easy directions while the closure domains are along hard directions and as a result the energy in needed to from these closure domains.

In 1935 Landau and Lifshitz showed that it is possible to understand domain structure in a ferromagnet by considering the exchange, magnetic crystal anisotropy and boundary energy contributions together and minimising the total energy. The estimated contributions are respectively of the order of 10^5, 10^2, 10^3 J m^{-3} and 10^{-3} J m^{-2}.

The Bloch Wall: The boundary surface between two domains is known as the Bloch Wall. The Wall has a certain thickness because the spin orientation changes only gradually across the domain boundary Fig. (19.25).

If the adjacent domain spins are anti-parallel then the Bloch wall is called the 180º wall. Similarly, one can have 90º wall, etc. The reason for the gradual change of spin orientation can be shown to be due to the fact that the total energy across the wall is sum of the exchange and anisotropy energies ($E = E_e + E_a$) and at the equilibrium the energy is distributed over many spins. The thickness of the wall is estimated to be of the order 100 Å, involving a couple of hundred spins. The energy of the wall is of the order of 10^{-3} Jm^{-2}.

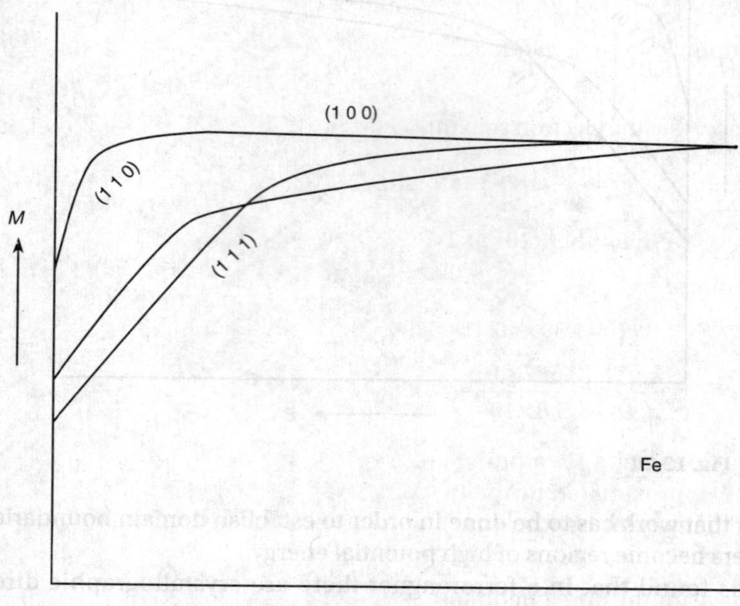

Fig. 19.25

19.18 MAGNETOSTRICTION

Magnetostriction is defined as the change in length of a ferromagnetic sample on magnetisation, there being no volume change. Polycrystalline nickel rods are found to contract uniformly on being magnetised, reaching a saturation, while cobalt first contracts up to a field of 0.3T and then the contraction slowly decreased to zero and elongation sets in. The behaviour of iron is opposite to that of cobalt, it first elongates, the elongation decreases to zero around 0.3T and there after iron contracts in length. Thus magnetostriction can be positive (elongation) or negative (contraction) depending on the material and can change from positive to negative in the same material. The fractional change in length in magnetostriction is very small of the order of 10^{-5}. The magnetostriction effect in nickel has been used to generate small uniform movements, e.g., in transducers and ultra-microtomes and also in oscillators. Investigation with single crystal show that magnetostriction is an isotropic and depends on the crystallogarphic direction in which the substance is magnetised, being positive in one direction and negative in another. The ansotropy gives rise to internal stresses inside a poly crystalline ferromagnetic material and affects their magnetic properties.

Mechanical straining of a ferromagnetic crystal can change the direction of domains magnetisation. For iron, if stretching is done in [1 1 1] direction, the magnetisation is decreased while a tension in the [1 0 0] direction increases it.

Example 19.1: A typical magnetic field achievable with an electromagnet with iron core is about 1 tesla. Compare the magnetic interaction energy of an electron spin magnetic dipole moment with $k_B T$ at room temperature and show that at ordinary temperature the approximation $\frac{k_B T}{\mu_B B} >> 1$ is valid.

Solution We know that the electron spin magnetic dipole moment is the Bohr magneton, i.e.,

$$\mu_B = \frac{e\hbar}{2m} = 9.3\times10^{-24}\,\text{J/Tesla}$$

$$\therefore \qquad \mu_B B = 9.3\times10^{-24}\times1 = 9.3\times10^{-24}\,\text{J} = 5.8\times10^{-5}\,\text{eV}$$

At room temperature

$$k_B T = 1.38\times10^{-23}\times300\,\text{J} = 2.587\times10^{-2}\,\text{eV}$$

$$\therefore \qquad \frac{k_B T}{\mu_B B} = \frac{2.587\times10^{-2}}{5.8\times10^{-5}} = 4.46\times10^{2} >> 1$$

Example 19.2: Consider a He atom in ground state (1s). The mean redius in Langevin formula may be approximated by the Bohr radius (= 0.53Å) Using $N = 27\times10^{23}$ per cm^3 for the atomic density of helium (gas); calculate the diamagnetic susceptibility of a helium atom.

Solution We have for the dimagnetic susceptibility of atom

$$x = -\frac{\mu_0 e^2}{6m} N Z r^2$$

Here, we have

$$\mu_0 = 4\pi\times10^{-7}\,\mathrm{Hm^{-1}}\; Z = 27\times10^{23}\,\mathrm{cm^{-3}} = 27\times10^{29}\,\mathrm{m^{-3}}$$
$$Z = 2 \text{ for He, } r = 0.53\,\text{Å} = 0.53\times10^{-10}\,\mathrm{m},\; m = 9.1\times10^{-31}\,\mathrm{kg}$$
$$e = 1.6\times10^{-19}\,\mathrm{C}$$

Substituting these values, we have

$$\chi = -\frac{4\pi\times10^{-7}\times(1.6\times10^{-19})^2\times27\times10^{29}\times2\times(0.53\times10^{-10})^2}{6\times9.1\times10^{-31}}$$
$$= -\frac{4\pi\times(1.6)^2\times27\times2\times(0.53)^2\times10^{-5}}{6\times9.1} = -9.03\times10^{-5}$$

Example 19.3: Estimate the order of magnitude of the exchange integral for a ferromagnet of Curie temperature 727°C. Also calculate the internal field. Given $\mu_B = 9.3\times10^{-24}$ J/T.

Solution The exchange integral is given by

$$J_0 = \frac{3}{2}\frac{k_B T_C}{Zs(s+1)}$$

where Z is the number of nearest neighbour. For $s = 1/2$

$$J_0 = \frac{3}{2}\frac{k_B T_C}{Z1/2(3/2)} = \frac{2k_B T_C}{Z}$$

and for cubic structure Z the nearest neighbours is 6.
Hence

$$J_0 = \frac{k_B T_C}{3} = \frac{1.38\times10^{-23}\times1000}{3} \qquad (727°\mathrm{C} = 1000\,\mathrm{K})$$
$$= 4.6\times10^{-21}\,\mathrm{J}$$

The internal field is given by

$$\mu_B B_m = k_B T_C$$

or

$$B_m = \frac{k_B T_C}{\mu_B} = \frac{1.38\times10^{-23}\times1000}{9.3\times10^{-24}} = 1.48\times10^{3}\,\mathrm{T}$$

Example 19.4: Compute the effective number of Bohr magnetons for a nickel ion. Assume first that orbital angular momentum is not quenched, then that Nickel has 8 electrons in the 3*d* shell.

Solution Since a *d* shell can hold up to 10 electrons, 5 electrons in nickel contribute +1/2 to spin and three contribute −1/2 to spin. As such net spin quantum number is 1 (taking the values from $s_1 - s_2$ to $s_1 + s_2$ differing by unity).

If quenching is ignored: 2 electrons contribute +2 to orbital magnetic quantum number, 2 contribute +1, 2 contribute zero, 1 contribute –1 and 1 contribute –2. Hence resultant value is 3. The shell is more than half filled so resultant $\boldsymbol{J} = L + \boldsymbol{S} = 4$ (Hunds rule) and Lande's g factor is

$$g = 1 + \frac{J(J+1) + S(S+1) - L(L+1)}{2J(J+1)}$$

$$= \frac{3}{2} + \frac{s(s+1) - L(L+1)}{2J(J+1)} = 1.5 + \frac{2 - 3\times4}{2\times4\times5} = 1.25$$

Hence effective number of Bohr magnetons is

$$P = g\sqrt{J(J+1)} = 1.25\sqrt{4\times5} = 5.59$$

If quenching is considered: Quenching means the orbital motion contributes nothing, i.e., $L = 0$, and $\boldsymbol{J} = 1$

$$\therefore \quad P = g\sqrt{J(J+1)} = 2\sqrt{2} = 2.83$$

Actually quenching is not quite complete in nickel. The experimental value is 3.2.

Example 19.5: Estimate the susceptibility of solid argon. Argon has atomic number 18 and at 4K its concentration is 2.66×10^{28} atoms m^{-3}. Take the root mean square distance of an electron from the nearest nucleus to be 0.6 Å. Also calculate the magnetisation of solid argon in a 2.0 T induction field.

Solution Susceptibility is given by

$$\chi = -\frac{\mu_0 \boldsymbol{M}}{\boldsymbol{B}} = -\frac{\mu_0 N Ze^2}{6m}\overline{r^2}$$

Substituting the values we have

$$\chi = -\frac{4\pi\times10^{-7}\times2.66\times10^{28}\times18(1.6\times10^{-19})^2\times(0.62\times10^{-10})^2}{6\times9.1\times10^{-31}}$$

$$= -\frac{4\pi\times2.66\times18(1.6)^2\times(0.62)^2}{6\times9.1}\times10^{-6}$$

or $$\chi = -1.09\times10^{-5}$$

The magnitude of magnetisation is given by

$$\boldsymbol{M} = \frac{|\chi|\boldsymbol{B}}{\mu_0} = \frac{1.09\times10^{-5}\times2.0}{4\pi\times10^{-7}}$$

The value justifies the approximation $\mu_0 \boldsymbol{M} << \boldsymbol{B}$.

Example 19.6: A sample of magnetic susceptibility χ in the form of a long cylinder is placed in a uniform applied field $\boldsymbol{B}$ along its axis. Find the magnitude of the magnetisation $\boldsymbol{M}$, the magnetic induction $\boldsymbol{B}$ and the magnetic field $\boldsymbol{H}$ at points inside it.

Solution Suppose the materials is paramagnetic so the magnetisation is in the same direction as the applied field. The magnetisation produces an induction field $\mu_0 \boldsymbol{M}$ in the same direction as applied field. So

$$\boldsymbol{B} = \boldsymbol{B}_a + \mu_0 \boldsymbol{M}$$

$$\boldsymbol{M} = \frac{\chi \boldsymbol{B}}{\mu_0(1+\chi)} = \frac{\chi(\boldsymbol{B}_a + \mu_0 \boldsymbol{M})}{\mu_0(1+\chi)}$$

or
$$\mu_0 \boldsymbol{M}(1+\chi) = \chi(\boldsymbol{B}_a + \mu_0 \boldsymbol{M})$$

or
$$\boldsymbol{M} = \frac{\chi \boldsymbol{B}_a}{\mu_0}$$

The induction field is $\boldsymbol{B} = \boldsymbol{B}_a + \mu_0 \boldsymbol{M} = \boldsymbol{B}_a + \chi \boldsymbol{B}_a = \boldsymbol{B}_a(1+\chi)$ and the magnetic field $\boldsymbol{H}$ is given by

$$H = \frac{\boldsymbol{B}}{\mu_0} - \boldsymbol{M} = \frac{B_a}{\mu_0}$$

The results for $\boldsymbol{M}$, $\boldsymbol{B}$ and $\boldsymbol{H}$ also hold if the material is diamagnetic. The susceptibility is then negative. $\boldsymbol{H}$ is the same as before but $\boldsymbol{B}$ is less then $\boldsymbol{B}_a$.

Example 19.7: Find Lande *g* factor for the ground state of a praseodymium ion, with 2*f* electrons and for the ground state of an erbium ion with 11*f* electrons.

Solution Both electrons in praseodymium have spin +1/2 so resultant S is = 1. The largest value of L is obtained if one electron contributes 3 and the other 2, so resultant $L = 5$. Note that they cannot both contribute 3 because their contributions to spin have same. Finally the shell is less than full, so

$$\boldsymbol{J} = /L - \boldsymbol{S}/ = 5 - 1 = 4$$

and
$$g = 1 + \frac{J(J+1) + S(S+1) - L(L+1)}{2J(J+1)}$$

$$= \frac{3}{2} + \frac{S(S+1) - L(L+1)}{2J(J+1)}$$

$$= \frac{3}{2} + \frac{1(1+1) - 5(5+1)}{2 \times 4(4+1)} = \frac{3}{2} - 0.7 = 0.8$$

Seven electrons in erbium contribute +1/2 to spin $\boldsymbol{S}$ while four contribute −1/2. Hence resultant $\boldsymbol{S} = \frac{7}{2} - \frac{4}{2} = \frac{3}{2}$.

Two electrons contribute 3 to L, two contribute 2, two contribute 1 and two contribute zero. The other three contribute −1, −2, −3 respectively. So result and is $L = 6$. $(2\times3 + 2\times2 + 1\times2 + 2\times0 - 1 - 2 - 3 = 6)$. The shell is more than half full, so $\boldsymbol{J} = \boldsymbol{L} + \boldsymbol{S} = \frac{15}{2}$.

Hence

$$g = \frac{3}{2} + \frac{S(S+1) - L(L+1)}{2J(J+1)}$$

$$= 1.5 + \frac{\frac{3}{2} \times \frac{5}{2} - 6 \times 7}{2 \times \frac{15}{2} \times \frac{17}{5}}$$

$$= 1.5 - 0.3 = 1.2$$

Example 19.8: Calculate the exchange co-efficient J_0 for iron using following datas. Curie temp = 1043 K, curie constant 1.77. The concentration of iron atoms is 8.5 × 10^{28} atoms m^{-3} and each atom has 12 nearest neighbours. Take $g = 2$.

Solution We have the exchange co-efficient is given by

$$J_0 = \frac{\mu_0 N g^2 \mu_B^2 \lambda}{2Z}$$

and $$\lambda = \frac{T_C}{C}$$

$\therefore$ $$J_0 = \frac{\mu_0 N g^2 \mu_B^2 T_C}{2ZC}$$

Substituting the values, we have

$$J_0 = \frac{4\pi\times10^{-7}\times 8.5\times10^{28}(2)^2(9.3\times10^{-24})^2\times 1043}{2\times12\times1.77}$$

$$= 9.07\times10^{-22}\ \text{J} = 5.6\times10^{-3}\ \text{eV} = 5.6\ \text{milli eV}$$

Example 19.9: A charge q is uniformly distributed over the surface of a spherical shell of radius r. If the sphere rotates with an angular velocity ω, find the magnetic dipole moment developed.

Solution The surface density of charge

$$\sigma = \frac{q}{4\pi r^2}$$

consider a ring between the planes AB and CD Fig. (19.26).

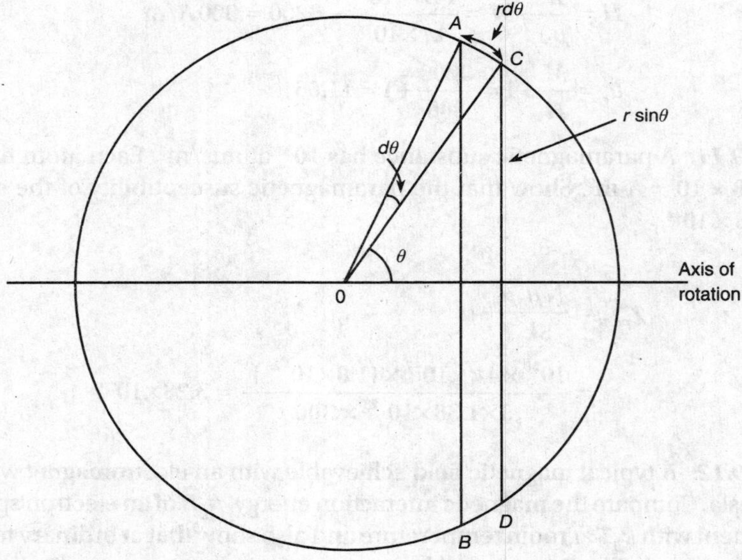

Fig. 19.26

Surface area of the ring = $2\pi r \sin\theta . r d\theta$ and charge on it = $2\pi\sigma r^2 \sin\theta . d\theta$

Since it is rotating with frequency $n = \frac{\omega}{2\pi}$, hence equivalent current

$$i = 2\pi\sigma r^2 \cdot \frac{\omega}{2\pi} \sin\theta . d\theta = \sigma r^2 \omega \sin\theta . d\theta$$

Area of the ring = $\pi(r\sin\theta)^2 = \pi r^2 \sin^2\theta$

$\therefore$ Equivalent dipole $d\mu$ = current × area = $\pi\sigma r^4 \omega \sin^3\theta . d\theta$

$$\therefore \qquad \mu = \int_0^\pi \pi\sigma r^4 \omega \sin^3\theta . d\theta = \int_0^\pi \pi\sigma r^4 \omega \sin\theta (1-\cos^2\theta)\, d\theta$$

Put $\cos\theta = x \quad \therefore \quad \sin d\theta = dx$

$$\therefore \qquad \mu = \int_1^{-1} \pi\sigma r^4 \omega (1-x^2)(-dx) = \pi\sigma r^4 \omega \int_1^{-1} (1-x^2) dx$$

$$= \pi\sigma r^4 \omega \left[x - \frac{x^3}{3} \right]_{-1}^{+1} = 2\pi\sigma r^4 \omega \frac{2}{3} = \frac{4\pi}{3} \sigma r^4 \omega$$

or

$$\mu = \frac{4\pi}{3} . \frac{q}{4\pi r^2} r^4 \omega = \frac{q r^2 \omega}{3}$$

Example 19.10: **A magnetic material has a magnetization of 3200 A/m and flux density of $14\pi \times 10^{-4}$ Wb/m². Show that the magnetising force and the relative permeability of the material are 300 A/m and relative respectively.**

Solution

$$H = \frac{B}{\mu_0} - M = \frac{14\pi \times 10^{-4}}{4\pi \times 10^{-7}} - 3200 = 300\,\text{A/m}$$

$$\mu_r = \frac{M}{H} + 1 = \frac{3200}{300} + 1 = 11.66$$

Example 19.11: **A paramagnetic substance has 10^{28} atoms/m³. Each atom has magnetic moment 1.8×10^{-23} A-m². Show that the paramagnetic susceptibility of the substance at 300 K is 3.28×10^{-4}.** **[Raj, Kerala]**

Solution

$$\chi_{para}^{M} = \frac{N\mu_0\mu_m^2}{3k_B T}$$

$$= \frac{10^{28} \times 4\pi \times 10^{-7} \times (1.8 \times 10^{-23})^2}{3 \times 1.38 \times 10^{-23} \times 300} = 3.28 \times 10^{-4}$$

Example 19.12: **A typical magnetic field achievable with an electromagnet with iron core is about 1 tesla. Compare the magnetic interaction energy, $\mu_B B$ of an electron spin magnetic dipole moment with $k_B T$ at room temperature and also show that at ordinary temperatures the approximation $k_B T/\mu_B B >> 1$ is valid.** **[Jodhpur, Mysore]**

Solution We have electron spin magnetic dipole in Bohr magneton as

$$\mu_B = \frac{eh}{4\pi m} = 9.27\times10^{-24}\ \text{A-m}^2$$

$$\therefore \quad \mu_B B = 9.27\times10^{-24}\ J, \quad \text{where} \quad B = 1\ \text{tesla}$$

Now, $$k_B T = 1.38\times10^{-23}\times300\,\text{J} = \frac{1.38\times10^{-23}\times300}{1.6\times10^{-19}}\text{eV} = 0.025\text{eV}$$

Thus $$\mu_B B / k_B T = \frac{5.8\times10^{-5}\ \text{eV}}{0.025\text{eV}} = 2.3\times10^{-3}$$

obviously $$\mu_B B / k_B T << 1$$

$$\therefore \quad k_B T / \mu_B >> 1$$

SUMMARY

1. **Magnetism in matter arises from the interactions of atomic-scale current loops. *Ferromagnetic materials* have strong interactions and exhibit the bulk magnetism associated with permanent magnets and with magnetic materials like iron. *Paramagnetism* and *diamagnetism* are weaker manifestation of magnetism in matter.**
2. **On the basis of the properties exhibited by magnetic materials, they can be placed under three classes: (i) diamagnet (ii) paramagnetic, (iii) ferromagnetic. The magnetic effects in these materials are due to atomic dipoles.**
3. **In vacuum, the magnetic induction $\boldsymbol{B}$ (or magnetic flux density) and magnetic field intensity $\boldsymbol{H}$ are related by**

$$\boldsymbol{B} = \mu_0 \boldsymbol{H}$$

where μ_0 is *permeability* of free space = $4\pi \times 10^{-7}\,\mu\text{m}^{-1}$ (or N/A^2).

4. **When a material is placed in a magnetic field, it gets magnetized, the *magnetization* $\boldsymbol{M}$ (magnetic moment per unit volume) of the sample and the magnetic induction inside the medium are related by**

$$\boldsymbol{B} = \mu_0(\boldsymbol{H} + \boldsymbol{M})$$

$\mu_0 \boldsymbol{H}$ is generated by external fields.

5. **$\boldsymbol{M} \propto \boldsymbol{H}$ or $\dfrac{\boldsymbol{M}}{\boldsymbol{H}} = \chi$ (magnetic susceptibility).**
6. **$\mu = \mu_0(1+\chi)$**

and $\mu_r = 1 + \dfrac{\boldsymbol{M}}{\boldsymbol{H}}$

where μ_r is relative permeability.

7. **Materials exhibit diamagnetism due to the fact that the effect of a magnetic field on the orbital motion of an electron is such as to produce a negative susceptibility.**
8. **$\omega_L = eB/2m$ is called the *Larmor precessional frequency.***

9. The precessional motion of the orbit produces a current given by

$$I = -\frac{e^2 B}{4\pi m}$$

10. $\chi_{\text{dia}} = \frac{\boldsymbol{M}}{\boldsymbol{H}} = -\frac{\mu_o Ze^2 N}{6m}\overline{r}^2$

where $\frac{2}{3}\overline{r}^2 = (\overline{x}^2 + \overline{y}^2)$ is called the mean square radius of the electrons from the field axis through the nucleus.

11. For paramagnetism from Langevin's theory

$$\chi = \frac{\boldsymbol{M}}{\boldsymbol{H}} = \frac{\mu_o N \mu^2}{3kT} = \frac{C}{T}$$

where $C = \frac{\mu_o N \mu^2}{3k}$ is called Curie constant.

12. According to classical theory of paramagnetism, all orientations of magnetic moment with respect to external field $\boldsymbol{B}$ is possible. However, quantum theory permits only certain finite set of orientations relative to $\boldsymbol{B}$. Paramagnetic susceptibility is given by

$$\chi_{\text{para}} = \frac{\boldsymbol{M}}{\boldsymbol{H}} = \frac{\mu_o N(g\mu_B)^2}{kT}$$

where g is the Lande's g-factor.

13. The existence of spontaneous magnetization in a magnetic material is called *ferromagnetism.*

14. Spontaneous magnetization is due to the alignment of appreciable number of electron spins in some favourable direction.

15. According to *Weiss theory,* ferromagnetic solids are made up of a number of small regions, called *domains.*

16. Ferromagnetic susceptibility

$$\chi = \frac{\boldsymbol{M}}{\boldsymbol{H}} = \frac{C}{T - T_C}, \quad T_C = C\lambda,$$

where C is called Curie temperature.

17. According to Weiss, the internal field in ferromagnetic materials is due to quantum *exchange interactions* between the electrons. These spin dependent interactions are electrostatic in nature and not magnetic.

18. Heisenberg showed that the energy of interaction of atoms i and j bearing spins S_i and S_j is of the form

$$U = -2J\boldsymbol{S}_i . \boldsymbol{S}_j$$

where U is exchange energy and J is the exchange integral.

19. *Antiferromagnetic substances* are those substances for which the dipoles have equal moments, but adjacent dipoles point in opposite directions.

20. There are certain substances for which the magnitude of magnetic moments of the two sublattices are not the same, due to differences in the number or in the nature of atoms in the sublattices. This results in a finite magnetic moment. This uncompensated antiferromagnetism is known as *ferrimagnetism* and related materials are called ferrites.

REVIEW QUESTIONS

1. Describe the origin to ferromagnetic domains and derive an expression for the thickness of the Block Wall. Enumerate various methods to study the domain structure and discuss one in detail.
2. Distinguish clearly between diamagnetism, paramagnetism and ferromagnetism.
3. Give an account of the quantum theory of paramagnetism. Discuss how the theory explains the behaviour of rare earth ions. What do you understand by quenching of orbital angular momentum?
4. Obtain an expression for paramagnetic susceptibility of free electrons on the basis of the classical laws. Discuss its inadequancy and show how Pauli modified it?
5. Give an account of Weiss theory of ferromagnetism. On the basis of this theory how will you explain hysteresis and Curie point. Explain clearly the basic difference between paramagnetic and ferromagnetic substances.
6. Describe the theory of ferromagnetism. Discuss the temperature variation of the saturation magnetization and point out the main conclusions you can derive from the corresponding experimental graph.
7. Draw temperature dependence of the susceptibility of all types of magnetic materials, comment. Explain the Heisenberg's exchange interaction in ferromagnetism.
8. Explain Heisenberg's exchange interaction. Show how it explains ferromagnetism?
9. Discuss Heisenberg's interpretation for the origin of Weiss molecular field. Relate the exchange integral to the Weiss constant and ferromagnetic Curie temperature. What are ferromagnetic domains?
10. A charge q is uniformly distributed over the surface of a spherical shell of radius r. If the sphere rotates with an angular velocity ω, find the magnetic dipole moment developed.
11. Give an account of the origin of atomic magnetism. Which source is important in the case of ferromagnetic materials?
12. Give an account of Langevin's theory for a paramagnetic gas like oxygen and explain why it fails when extended to the electron gas. Find an expression for the paramagnetic susceptibility of free electrons in metals.
13. Give an account of Weiss theory of ferromagnetism and show from the plot of the Langevin function, spontaneous magnetization exists below the Curie temperature and vanishes above the Curie temperature.
14. Give an account of the origin of atomic magnetism. Which source is important in the case of ferromagnetic materials?
15. Find the order of magnitude of the Weiss field. Why the field cannot be accounted for by the dipole interaction between the molecular magnets themselves.

16. Give an account of Heisenberg's theory for the origin of the Weiss field. Establish the criterion for ferromagnetism and point out the experimental evidence in its favour.

17. Explain the different contributions for the formation of domains in a ferromagnetic material and show how the hysteresis curve is explained on the basis of the domain theory.

PROBLEMS

1. A crystal contains n-atoms which posses spin and magnetic moment μ. Placed in a uniform magnetic field B; the atoms orient themselves in three directions parallel, perpendicular and antiparallel to the field. If the crystal is in thermal equilibrium at temperature T, prove that the mean magnetic moment is $\boldsymbol{M} = N\mu \dfrac{2\sinh x}{1+2\cosh x}$, where $x = \dfrac{\mu_B}{k_B T}$, if $x >> 1$ show that above equation assumes the form of the Curie law $\boldsymbol{M} = \dfrac{2N\mu_B^2}{3k_B T}$ and that for $x >> 1$, $\boldsymbol{M} = N\mu$.

2. For a paramagnetic gas n-atoms per unit volume and with $L = 0$, $S = 1/2$, calculate the number of atoms in the two levels at temperature T and the field H. Prove that the resultant magnetism is

$$\boldsymbol{M} = \frac{1}{2} Ng\mu_B \tanh\left(\frac{g\mu_B \boldsymbol{H}}{2k_B T}\right)$$

3. Consider a magnetic dipole $\boldsymbol{\mu}$ in a constant magnetic induction field B along the z-axis.

(a) Verify that $\mu_x = A\cos(\omega_L t + \phi)$ and $\mu_y = A\sin(\omega_L t + \phi)$

satisfy $\dfrac{d\mu}{dt} = -\left(\dfrac{e}{2m}\right)\mu \times \boldsymbol{B}$ if $\omega_L = \dfrac{eB}{2m}$

(b) What is the sense of rotation of μ?

(c) Show that μ has constant magnitude.

(d) Find expression for the componant of μ if $\boldsymbol{B}$ is in negative z-direction.

4. The diamagnetic core contribution to the molar susceptibility of sodium is -6.1×10^{-12} m^3/mole. Estimate the average distance of a core electron from the nearest nearest nucleus. Also estimate the dipole moment of a sodium core in a magnetic induction field of 0.5 T.

[**Hint:** $\chi = -\dfrac{\mu_0 n Z e^2}{6m}\langle r^2 \rangle$ where $n = \dfrac{N_0 d}{M_0}$, N_0 is Avogadro number, d density of sodium and M_0 molecular weight]

5. **(a)** Use Hund's rules to find L, S and J for the ground state of Vanadium ion, with three electrons in its 3d shell. Assume that orbital angular momentum is not quenched. What is value of Lande g factor? What is the magnitude of the magnetic dipole moment?

(b) If orbital angular momentum is completely quenched, what are L, $\boldsymbol{S}$, $\boldsymbol{J}$ and G? What is the magnitude of the dipole moment?

6. The saturation magnetization of iron is 1.75×10^6 A/m show that this corresponds to 2.22 Bohr magnetons per atom. The concentration of iron atoms is 8.50×10^{28} m^{-3}.

7. **(a)** Estimate the value of the effective field constant λ for nickel. Take $n = 9.14 \times 10^{28}$ m^{-3}, $g = 2$, $J = S = 0.3$.

(b) Estimate the curie constant C.

(c) Why does nickel have a lower Curie temperature than iron? Consider the concentration of atoms, effective fields and atomic dipole moments.

8. Show that the exchange co-efficient J_0 is given by

$$J_0 = \frac{3k_B T_C}{zS(S+1)}$$

for a ferromagnet with Curie temperature T_C. Each atom has Z indentical nearest neighbours and each has spin S. Orbital angular momentum is quenched. Use this expression to calculate J_0 for nickel. It has a face centered cubic structure and a Curie temperature of 631 K.

SHORT ANSWER-QUESTIONS

1. What is the cause of the macroscopic magnetic properties of materials?

Ans: These are the consequence of interactions between an external magnetic field and magnetic dipole moments of the constituent atoms. Associated with each individual electron are both orbital and spin magnetic moments. The net magnetic moment for an atom is just the sum of the contributions of each of its electrons, where in there will be spin and orbital moment cancellation of electron pairs.

2. How diamagnetism results?

Ans: Diamagnetism results from changes in electron orbital motion that are induced by an external field. All materials are diamagnetic.

3. What are paramagnetic materials?

Ans: These are the materials having permanent atomic dipoles, which are acted on individually and are aligned in the direction of an external field.

4. Why dia and paramagnetic materials are considered to be non magnetic?

Ans: The magnetizations for the types of materials are relatively small and persist only while an applied field is present.

5. How, a large and permanent magnetization is established in ferromagnetic materials?

Ans: Due to bonding forces, the spins of electrons of neighbouring atoms are permanently aligned in ferromagnetic materials. The common ferromagnetic elements are Fe, Co and Ni.

6. What are antiferromagnetic materials?

Ans: Antiparallel coupling of adjacent cation spin moment is found for some ionic materials. Those in which there is total cancellation of spin are termed anti-ferromagnetic.

7. What are ferrimagnetic materials?

Ans: These materials have an antiparallel arrangement of electron spins, but due to incomplete cancellation, there is a net magnetic moment that is comparable in order of magnitude to that of ferromagnetic materials.

8. How the net magnetization results for cubic ferrites?
Ans: The net magnetization for cubic ferrites results from the divalent ions (e.g., Fe^{2+}) that reside on octahedral lattice sites, the spin moments of which are mutually aligned.
9. How ferro and ferromagnetic materials behave with temperature?
Ans: With rising temperature, increased thermal vibrations tend to counteract the dipole coupling forces on ferro and ferrimagnetic materials. Consequently, the saturation magnetization gradually diminishes with temperature upto the Curie temperature (T_C), at which point it drops to zero, above T_C these materials are paramagnetic.
10. Explain domain character of ferro and ferrimagnetic materials.
Ans: Below Curie temperature (T_C) ferro- or ferrimagnetic materials are composed of domains-small volume regions where in all net dipole moments are mutually aligned and the magnetization is saturated. The total magnetization of the material is just the appropriately weighted vector sum of the magnetizations of all these domains. When an external magnetic field is applied, domains having magnetization vector oriented in the direction of the magnetic field grow at the expense of domains that have unfavourable magnetization orientations. At total saturation the entire solid material is a single domain and the magnetization is aligned with the field direction. The change in domain structure with increase of reversal of a magnetic field is accomplished by the motion of domain walls.
11. How will you explain both hysteresis and permanent magnetization in terms of domain walls?
Ans: Hysteresis (the log of ***B*** field behind the applied ***H*** field) and permanent magnetization (or remanence) result from the resistance to movement of domain walls.
12. What are soft magnetic materials?
Ans: Domain wall movement in these materials is easy during magnetization and demagnetization. Consequently they have small hysteresis loops and low energy losses i.e., low eddy current losses. We may note that by increasing the electrical resistivity of the magnetic medium eddy current losses may be reduced.
13. What are hard magnetic materials?
Ans: Because greater fields are required to demagnetize these materials, the magnetization is more peramanent.

OBJECTIVE QUESTIONS

1. Magnetic susceptibility has the dimension of
(a) Wb-m (b) dimensionless
(c) Wb/m² (d) Amp/m
2. 1 Bohr magneton is
(a) 9.27×10^{-13} amp-m² (b) 9.27×10^{24} amp-m²
(c) $9.27 \times 10–^{24}$ amp-m² (d) 9.27×10^{-6} amp-m²

3. The units of magnetic permeability are

(a) Wb-A^{-1}-m^{-1} (or H/m) (b) Wb/m^2

(c) A/m (d) None of these

4. SI units, tesla is expressed as

(a) kg/m^2 s^2 A (b) kg/s^2 A

(c) NA^{-1} m^2 (d) none of the above

5. Magnetic induction $\boldsymbol{B}$ and magnetic field intensity $\boldsymbol{H}$ are related by

(a) $\boldsymbol{B} = \mu_0 + \mu_0 \boldsymbol{H}$ (b) $\boldsymbol{B} = \mu_0 \mu_r \boldsymbol{H}$

(c) $\boldsymbol{B} = \mu_0 \boldsymbol{H}^2$ (d) $\boldsymbol{B} = \mu_0 + \boldsymbol{H}$

6. The magnetization ($\boldsymbol{M}$) of a solid, magnetic induction $\boldsymbol{B}$ and field strength $\boldsymbol{H}$ are related by

(a) $\boldsymbol{M} = (\boldsymbol{B}/\mu_0) = \boldsymbol{H}$ (b) $\boldsymbol{M} = 1 + \boldsymbol{B}/\boldsymbol{H}$

(c) $\boldsymbol{M} = 1 - \dfrac{\boldsymbol{B}}{\boldsymbol{H}}$ (d) $\boldsymbol{M} = \boldsymbol{B} - \mu_0 \boldsymbol{H}$

7. Which one of the following material does not have permanent magnetic dipoles?

(a) ferromagnetic (b) antiferromagnetic

(c) paramagnetic (d) diamagnetic

8. A suitable material for audio and TV-transformers is

(a) pure iron (b) Fe and 4% Si alloy

(c) Iron and 30% Ni alloy (d) Ferrite

9. The highest measured energy product in a permanent magnet is about

(a) 200 kJ/m^3 (b) 300 kJ/m^3

(c) 400 kJ/m^3 (d) 600 kJ/m^3

10. In a ferromagnetic material, susceptibility is

(a) very large and negative (b) very large and positive

(c) zero (d) negative

11. For Cu the resultant spin in Bohr magneton is

(a) 5 (b) 9

(c) 3 (d) 0

12. Below the ferromagnetic Curie temperature, the ferromagnetic material exhibit ***B-H*** curve whose from is

(a) straight line (b) exponential curve

(c) ***B-H*** loop (d) none of these

13. The temperature below which certain materials are ferromagnetic and above which they are paramagnetic is called.

(a) Neel temperature (b) Curie temperature
(c) Weiss temperature (d) None of these

14. At Curie temperature, the spontaneous magnetization for ferromagnetic materials is

(a) zero (b) infinite
(c) 10 (d) none of these

15. At Neel temperature

(a) susceptibility is maximum (b) susceptibility is minimum but not zero
(c) susceptibility is zero (d) none of these

16. The garnet crystal used in microprocessor has the formula

(a) $Y_3 Fe_5 O_{12}$ (b) $Y_2 Fe_3 O_5$
(c) $YFe_3 O_5$ (d) $YFe_5 O_{18}$

17. For paramagnetic materials, μ_r (relative permeability) at room temperature is

(a) 10 (b) 5
(c) 1 (d) zero

18. The susceptibility of a diamagnetic material is about

(a) -10^{-5} (b) 10^{-3}
(c) 10^5 (d) 10^7

ANSWERS

Question numbers	1	2	3	4	5	6	7	8	9	10
Answers	(b)	(c)	(a)	(b)	(b)	(a)	(d)	(d)	(c)	(b)
	11	12	13	14	15	16	17	18		
	(d)	(c)	(b)	(a)	(a)	(b)	(c)	(a)		

FILL IN THE BLANKS

1. Materials that lack permanent magnetic dipoles are called ______. **[diamagnetic]**
2. If the interaction between the atomic permanent dipole moments is zero or negligible, the material will be ______. **[paramagnetic]**
3. Diamagnetism is associated with ______ of the atoms, whereas paramagnetism is associated with ______ arising because of ______ electrons. **[neutral cores, moments, free]**
4. The ratio μ/μ_0 is called the ______ and denoted by ______. **[relative permeability, μ_r]**
5. Materials for which susceptibility (χ) < 0, i.e. magnetisation opposes the applied field, and μ_r < 1, are called ______ materials. **[diamagnetic]**

6. Materials for which χ is positive and has a small value and $\mu_r > 1$, are called ______ materials. The flux density inside the material is slightly ______ than outside. **[paramagnetic, greater]**
7. The relation between susceptibility (χ) and relative permeability is ______. **[$\chi = \mu_r - 1$]**
8. Ferromagnetic materials possess spontaneous magnetisation. The atomic magnetic ______ are arranged parallel to one another below a certain temperature called as the ferromagnetic ______. **[dipoles, Curie temperature]**
9. The atomic dioples in antiferromagnetic materials are arranged ______ to one another so that the net magnetization in the absence of a field is ______. Above the Neel temperature (T_N), dipoles becomes randomly oriented as in a ______ substances. **[antiparallel, zero, paramagnetic]**
10. The substances in which atomic dipoles are arranged ______ to one another but the moments in one direction have a larger magnitude so that the net magnetization is ______ are called ferrimagnetic. **[antiparallel, non-vanishing]**
11. The diamagnetic property is found in those substances in which the atoms have ______ number of electrons in their outermost orbit so that due to mutual opposite spins, the net magnetic moment of each atom is ______. **[even, zero]**
12. The susceptibility (χ) of a diamagnetic substance does not depend on the ______ and ______ of the applied external field. **[temperature, intensity]**
13. Above Curie temperature (T_c), all ferromagnets show normal ______ behaviour, and the Curie law takes the form $\chi =$ ______ (Curie-Weiss law). **[paramagnetic, $C/(T - T_C)$]**
14. Below the ferromagnetic Curie temperature, ferromagnets exhibit the well known ______ in ***B*** vs ***H*** curves. **[hysteresis]**
15. The boundary surface between two domains is known as ______. **[Block Wall]**
16. If the antiferromagnetic interaction in a solid is such that the spins of two neighbouring atoms are antiparallel but unequal in magnitude, the solid exhibits a net ______ magnetisation and is known as ______ material. **[spontaneous, ferrimagnetic]**
17. Ferrites form a class of ferrimagnetic materials which may be prepared from electrically ______ ceramic materials. **[non-conducting]**
18. Ferrite have a Curie temperature that varies from ______ °C to several ______ °C. **[100, hundred]**
19. Ferrites are ______ with very high resistivity (~10^{12} Ω-cm). **[insulators]**
20. A collective excitation associated with magnetic systems is called ______ (magnon). **[spin wave]**

20

Superconductivity

20.1 INTRODUCTION

The phenomenon of electrical conductivity in metals explain the finite resistance of these materials, which originates from the fact that a real crystal always exhibits deviations from perfect lattice periodicity, phonons and defects. An infinitely high electrical conductivity is unthinkable in this description, because (i) a crystal without a certain degree of disorder is inconceivable according to the second law of thermodynamics, and (ii) even in the absence of phonon and defect scattering electron-electron scattering will still cause resistance, mercury approaches an unmeasurably small value when it is cooled below 4.2 K. This phenomenon is called **"Superconductivity"** and in subsequent years many more materials were found to be superconducting when cooled to below a certain critical temperature (T_c).

Twenty-three of the chemical elements becomes supper-conducting at low temperatures. Metals in the left portions of transition series (Ti, V, Zr, Nb, Mo, Tc, Rh, Ru, La, Hf, Ta, W, Re, Os and Ir) belong to this group, as do elements just to the right (Al, Zn, Cd, Sn, Hg, Tl and Pb). Among the lanthanides and actinides Lu. Th and Pa are super-conducting. Other elements such as As, Ba, Ce, Ge, become superconducting when subjected to high pressure. In addition, a great many alloys and compounds, both stoichimetric and nonstoichiometric, are superconducting.

Over the years the highest transition temperature (T_c) had been gradually increased from the 4 K of Hg to 23 K in the compound Nb_3Ge, discovered by scientists at Westinghouse about 28 years ago. This remained the record until about 1986, when two scientists, K.A. Muller and J.G. Bednorz revealed that at temperature higher than 30 K the ceramic compounds of lanthanum, barium and copper oxides go into a superconducting state. In February, 1987 C.W. Chu et al. synthesized yttrium compound which go over into a superconducting state at temperatures T_c close to 90 K. Several months later new superconductors with $T_c > 30$ K were synthesized almost simultaneously at different laboratories in the world. The highest known T_c is 133 K for mercury cuprate system. These cuprate oxide conductors called as high-T_c Superconductors (HTSC). Besides these thigh T_c cuprate superconductors, recently several new superconductors e.g., alkali metal-doped

C_{60} or fullerenes A_XC_{60} have been discovered. C_{60} is made only of carbon atoms, and is similar to graphite. The observations for superconductivities of K_3C_{60} Rb_3C_{60}, and $C_3Rb_2C_{60}$ at 19.3 K, 30 K and 33 K respectively, have been reported only for polycrystalline samples.

Superconductors are nearly **perfect diamagnets:** The magnetic induction field vanishes in the interior of a bulk sample. A superconductor does not expel fields of arbitrary strength, however, fields with magnitudes greater than a certain value penetrate into the interior. For some superconductors, called type 1, penetration coincides with the destruction of superconductivity. The metal becomes normal and has a vanishing resistivity, For others called type II, the field strength must be increased to a second critical value before the resistivity ceases to vanish. In both cases a greater field strength is required at lower temperatures than at higher.

A microscopic explanation of superconductivity was not found for nearly a half century since its discovery in 1911. It remained to be very mysterious. It was not until shortly before 1960, that Bardeen, Copper and Schrieffer achieved the decisive breakthrough and proposed the theory now named after them (BCS theory). This theory employed a model hamiltonian based on the assumption that all electrons in a superconducting state near the Fermi surface are correlated in pair with equal and opposite momenta and spins. BCS theory has been found inadequate to explain the phenomenon of high-T_c superconductivity. Several non-phonon mechanisms of pairing in high-T_c superconductors have been proposed. A satisfactory theory of high-T_c superconductivity is still lacking. In the present chapter, a brief discussion of superconducting properties, BCS theory and high-T_c superconductors is presented.

Superconductivity is not only of theoretical interest, but is also of technological importance. The special properties of superconductor materials:

- Zero resistance to direct current
- High current density
- Low resistance at high frequency
- Low signal dispersion
- High sensitivity to magnetic field
- No penetration of externally applied magnetic field
- Rapid single flux quantum transfer
- Close to speed of light signal transfer

enable important applications, e.g. (i) Low resistance at high frequency and low signal dispersion play a fundamental role for microwave components and in communication technologies, (ii) High sensitivity to magnetic fields makes it possible to produce superconductive sensors with a sensing capability 1000 times higher than conventional devices, (iii) The capability of driving off external magnetic fields holds the potential for applications in magnetic levitation systems for transportation, (iv) Important applications in digital electronics and high speed computing derive from other peculiar superconductor properties such as the Josephson effect and the sharp transition to superconducting status.

Superconducting materials are the basis of extremely sensitive magnetic field and electromagnetic radiation detectors and, as such, have found widespread use in many scientific research fields and in medicine. Magnetic levitation is one of the most facinating

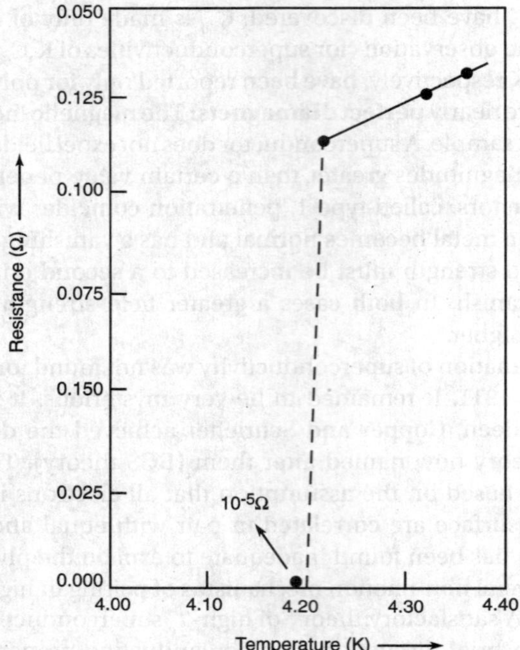

Fig. 20.1 The resistance of a mercury sample as a function of temperature. On cooling below 4.2 K, the electrical resistance falls below the experimental detection limit ($\approx 10^{-5}\Omega$)

demonstrations of superconductivity. With the recent discovery of high temperature superconductors, the number of technological applications of superconductivity is expected to increase enormously.

20.2 EXPERIMENTAL FEATURES OF SUPERCONDUCTIVITY

In the superconducting state several metals and alloys display a number of remarkable properties, e.g., zero resistance, Meissner effect, isotope effect etc., We must note that in the transition from the normal to the superconducting state a thermodynamic change of state or phase transition takes place. This is clearly manifested in other physical quantities also.

20.2.1 D.C. Electrical Resistance

The property that gives superconductivity its name was discovered by Onnes in mercury. Fig. (20.1) shows the classical experimental results on cooling below about 4.2K the electrical resistance of the mercury sample collapses. It was assumed at that time to be below 10^{-5} Ω.

In fact, it is fundamentally impossible to test experimentally whether the resistance is zero, one can only set an upper limit to the value of electrical resistance, and this of course still applies to today's much improved measurement methods.

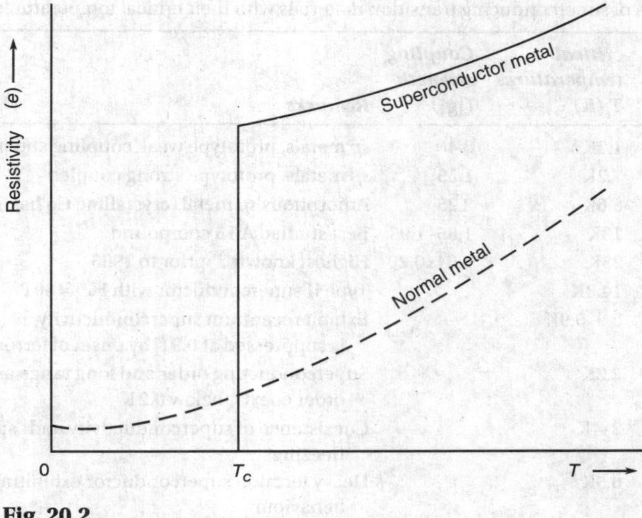

Fig. 20.2

4.2K roughly follows T^5 expected of normal metals ($\rho\ \alpha\ AT^5$). Below 4.2K it is superconducting. The transition occurs in a temperature range of less than 10^{-5} K. The behaviour and for normal as well as superconducting metals are shown in Fig. (20.2).

20.2.2 Critical Temperature (T_c)

The temperature at which the d.c. electrical resistance vanishes is called the critical temperature of the superconductor and denoted by T_c. About 23 metals and several alloys and compounds have been found to exhibit superconductivity, metals, which exhibit superconductivity are not particularly good conductors in the normal (above T_c) and the value of T_c vary from 0.4 K to 11.2 K: examples are: Pb (7.2K), Hg (4.2K), Sn (3.7K) al (1.2K). Some of the compounds which exhibit superconductivity are composed of elements which themselves are not supercomputing, e.g., CuS. The highest transition temperature is known so far is 133 K for layered mercury cuprate oxide compound. Table 20.1 gives values of T_c for various materials chosen to illustrate a wide range of properties.

We shall learn from B.C.S. theory that T_c is given by the relation.

$$k_B T_c = \hbar\omega_D \exp\left[1 - \frac{1}{|g|N(0)}\right]$$

where ω_D is the Debye phonon frequency, $N(0)$ is the electron density of states at the Fermi level and $|g|$ is the electron-phonon interaction strength. The search for a room temperature superconductor is centred on how to control the parameter $|g|$ in the fabrication of a superconducting alloy. Recent discovery of high temperature superconductivity in layered cuprates has raised the question, what is the possible mechanism of superconductivity in these systems?

Table 20.1 Examples of superconducing transition materials with their critical temperatures

Material	***critical temperatures $T_c(K)$***	***Coupling strength (|g|)***	***Remarks***
Al	1.2K	0.4	*sp* metals, prototype weak coupling superconductor
Pb	7.2K	1.55	*sp* metals, prototype strong coupler
Amorphous Ga	8.6K	2.25	Amorphous *sp* metal (crystalline Ga has $T_c = 1.1$K)
Nb_3Sn	18K	1.65–1.95	Best studied A 15 compound
Nb_3Ge (thin Film)	23K	1.64±0.20	Highest known T_c prior to 1986
$Gd_2PbMO_6S_8$	14.3K		Type II superconductor with $Hc_2 = 60T$
$ErRh_4B_4$	8.9–0.9K		Exhibit recentrant superconductivity, i.e., Superconductivity is suppressed at 0.9 K by onset of ferromagnetic order
$ErMO_6S_8$	2.2K		Superconducting order and long range antiferromagnetic order coexist below 0.2 K
$Gd_XTh_{1-X}Ru_2$	2–4K		Coexistence of superconductivity and "spin galss" magnetic freezing
$CeCu_2Si_2$	0.5K		Heavy fermion superconductor exhibiting unstable 4*f*-shell behaviour
$SrTiO_{3-X}$	0.05–0.5K	-	Doped semiconductor n<10^{20} cm^{-3}
$(TMTSF)_2CIO_4$	1.2–1.4K	-	Quasi 1-d "organic" material (atmospheric pressure)
$La_{2-x}Sr_xCuO_{4-\delta}$ (214)	36K		Layered cuprate high T_c superconductor
$YBa_2Cu_3O_{7-\delta}$ (123)	90K	-	"
$Bi_2Sr_2CuO_{6+\delta}$ (Bi-2201)	20K	-	"
$Bi_2Sr_2CaCuO_{8+\delta}$ (Bi-2212)	80K	-	"
$Bi_2Sr_2Ca_2Cu_3O_{10+\delta}$ (Bi-2223)	110K	-	"
$TI_2Ba_2CuO_{6+y}$ (TI-2201)	90K	-	"
$TI_2Ba_2CaCu_2O_{8+y}$ (TI-2212)	110K	-	"
$TI_2Ba_2Ca_2Cu_3O_{10+y}$ (TI-2223)	125K	-	"
$TIBa_2CaCu_2O_{7+y}$ (TI-1212)	100K	-	Layered cuprate high-T_c superconductor
$TIBa_2CaCu_3O_{9+y}$ (TI-1223)	120K	-	"
$Nd_{2-X}Ce_XCuO_{4+\delta}$	25K	-	Electron doped superconductor
$Ba_{1-X}K_XBiO_{3+y}$ (x <0.4)	30K	-	Non copper oxide high-T_c superconductor
K_3C_{60}	19.3K	-	Alkali metal doped high- T_c fullerence superconductor
Rb_3C_{60}	30K	-	"
$CsRb_2C_{60}$	33K	-	"
$Hg\ Ba_2\ Ca_2\ Cu_3\ O_9$	134K	-	Highest known layered cuprate high-T_c superconductor, under pressure T_c onset increases 164K
$Mg\ B_2$	40K	-	
$Sm_{0.95}\ La_{0.05}\ O_{0.85}\ F_{0.15}$ Fe As	57.3K	-	Highest Known Cu free HTSC superconductor

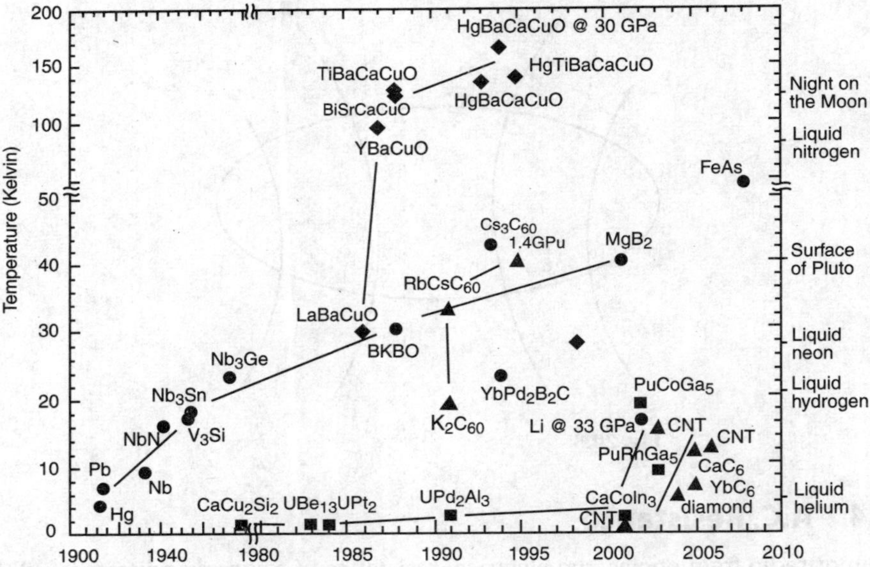

Fig. 20.2 (a) Superconducting Materials

Superconducting materials and their critical temperatures are shown in Fig. 20.2(a). Boosted by the growing demand for sustainable technologies and most efficient use of resources, the integration of superconductors in end-use technologies is one of the technology challenges of the 21st century. Superconductor technology can in fact provide cost-efficient and environmentally desirable solutions to many current and future needs. (Tables 20.4(a) and 20.4(b)).

20.2.3 Persistent Currents

One way to test for zero resistivity is to measure the current in a loop of superconducting wire in the absence of an electric field. If the resistivity is truly zero the current does not diminish with time. Current can be induced by directing a static magnetic field through a loop when it is in the normal state. The material is then cooled to below the critical temperature T_c, so that it becomes superconducting. Removal of the magnet alters the flux through the loop, thereby inducing a current. If the ring had a finite resistance R, the current would decrease according to

$$I(t) = I_0 \exp(-Rt / L) \tag{20.1}$$

where L is the self-inductance of the loop. It is found that the resistance of the material decreases by atleast 14 orders of magnitude when it enters the superconducting state, currents in superconducting loops have been observed to remain undiminished over longer than a year. Obviously, the resistivity is practically zero in the superconducting state. Such currents are termed as persistent currents. Persistent current in a superconducting ring is shown in Fig. (20.3). We must note that zero-resistivity and persistent current are one and the same property of the superconductors.

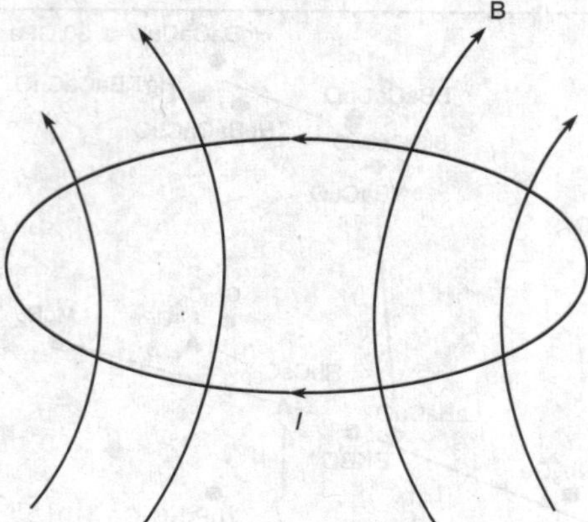

Fig. 20.3

20.2.4 A.C. Resistance

In audio or radio frequencies, the electrical resistance of a superconductor is still found to be zero. At very high frequencies, of the order of 10^{13} Hz, i.e., at infrared wave lengths ($\lambda = 10\mu$). However, the resistance becomes normal event at lowest temperature. (The resistance is measured in these ranges by measuring the reflectivity of the metal which gives the skin effect surface resistance).

The minimum photon energy needed to make the surface resistance equal to the normal resistance value at temperatures below T_c is found to be of the order of 4K T_c, where k is Boltzmann's constant. Since T_c is small, this energy is of the order of 10^{-3} eV only. This suggests that an energy gap E_g exists in the energy level distribution of the electrons in a metal in the superconducting state as shown in Fig. (20.4)

20.2.5 The Meissner Effect

A further distinctive property of a superconductor is its magnetic behaviour. The magnetic properties of a material that transforms to the superconducting state cannot simply be described by the Maxwell's equations with the additional requirement of a vanishing electrical resistance (R → O).

A superconductor placed in a magnetic field expels all flux lines from the interior (Meissner effect).

Fig. (20.5) shows exclusion of induction field lines in a superconducting sample in applied magnetic fields.

For the geometry shown, field lines that penetrate the sample skirt the edges of the superconducting sample, making the induction field higher at point near the surface then at analogous point near the normal sample. Flux exclusion can be confirmed by measuring the magnetic field in the vicinity of the sample.

Flux exclusion in not complete near a sample surface. Rather, the induction field decays exponantially with distance into the sample. The characterstic **penetration depth** is typically few hundred angstroms (10^{-7}m).

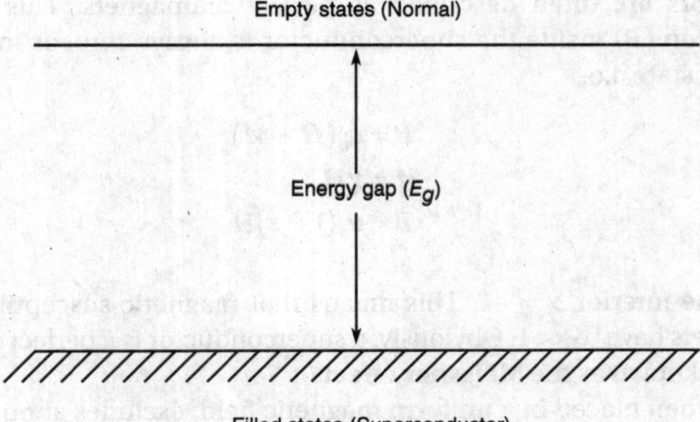

Fig. 20.4

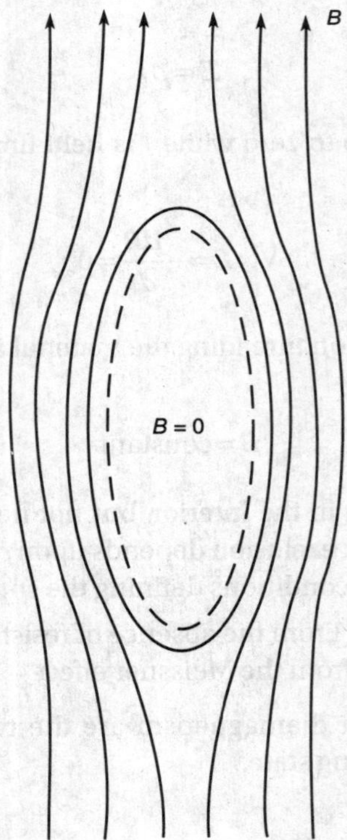

Fig. 20.5

Current exists at the surface of a supercurrent in a magnetic field, but not in its interior. According to the study state Maxwell equation $\nabla \times \boldsymbol{B} = \mu_0 \boldsymbol{J}$, so $\boldsymbol{J} = 0$ in the interior of field free region. Surface currents, on the other hand, create an induction field that cancels the applied field in the interior and augments in the region just outside the surface.

Superconductors are often described as perfect diamagnets. This means that the magnetic induction (B) inside the superconductor is always zero, as long as it is in the superconducting state, i.e.,

$$\boldsymbol{B} = \mu_0(\boldsymbol{H} + \boldsymbol{M})$$

where $$\boldsymbol{M} = \chi \boldsymbol{H}$$

$\therefore$ $$\boldsymbol{B} = \mu_0(1+\chi)\boldsymbol{H}$$

Since $B = 0$ in the interior, $\chi = -1$. This means that magnetic susceptibility is negative. Normal diamagnets have $|x| << 1$. Obviously, a superconductor is a perfect or 'ideal' diamagnet. This effect is known as the **Meissner effect**.

A diamagnet when placed in a uniform magnetic field, excludes about ten millionth of it, while a superconductor excludes it totally.

We must note that the disappearence of resistance (ρ) and the explusion of magnetic flux (B) are independent phenomena, the superconductor depends on first to maintain the second. From Ohm's

$$E = \rho \boldsymbol{J} \tag{20.2}$$

We see that if resistance ρ goes to zero while J is held finite, then $\boldsymbol{E}$ must be zero. From Maxwell's equation, we have

$$\nabla \times \boldsymbol{E} = -\frac{d\boldsymbol{B}}{dt} = 0 \tag{20.3}$$

where $\boldsymbol{B}$ is the magnetic induction threading the material and t is the time. The condition (ii) can be expressed as

$$\boldsymbol{B} = \text{constant} \tag{20.4}$$

Therefore $\boldsymbol{B}$ is constant in time in the interior, but not necessarily zero as occurs in the Meissner effect. Further, the flux exclusion depends upon $\rho = 0$ by means of the possibility of persistent currents. Thus the conditions defining the superconductivity are

$\boldsymbol{E} = 0$ (From the absence of resistance)

$\boldsymbol{B} = 0$ from the Meissner effect

i.e., zero resistivity and perfect diamagnetism are the two independent but essential properties of the superconducting state.

20.2.6 The Critical Field

Because of the Meissner effect, the magnetic state of a superconductor can be described as ideal diamagnetism. Persistent surface currents maintain a magnetization $\boldsymbol{M} = -\boldsymbol{H}_{ext}$ in the interior, and this magnetization is exactly opposite to the applied magnetic field $\boldsymbol{H}_{ext}$. If an external magnetic field is applied to a superconductor at a temperature $T < T_c$ then a certain amount of energy is consumed to induce the super currents which maintain the interior of the sample magnetically field free. If the magnetic field strength $\boldsymbol{H}_{ext}$ is increased, then

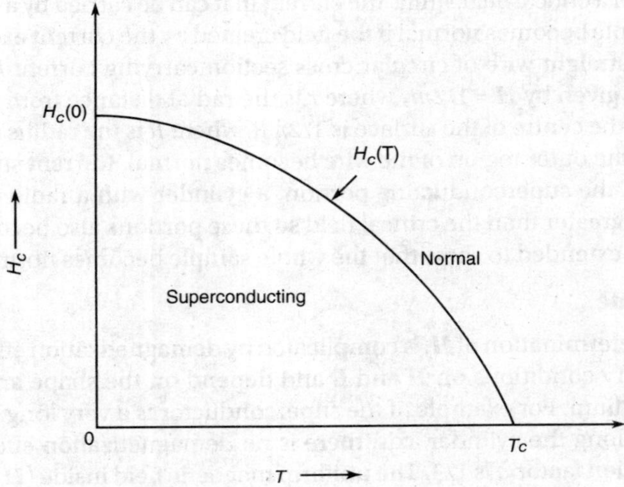

Fig. 20.6

at a critical field strength H_C, it is energetically more favourable for the material to convert to the normally conducting phase, in which the magnetic field penetrates the material.

This transition is admittedly associated with an increase of the free energy, but a strong increase of the field H_{ext} would lead to an even high energy expenditure in inducting the screening currents. There thus exists an upper critical magnetic field strength (H_c), above which the superconducting phase no longer exist, i.e., superconductivity disappears at H_c. This critical field is a function of temperature given approximately by **Tuyn's law,**

$$H_C = H_0\left[1-\left(\frac{T}{T_c}\right)^2\right] \tag{20.5}$$

There H_0, the critical field at 0° K has a specific value for each material. A qualitative phase diagram of a superconductor as a function of temperature T and external magnetic field is shown in Fig. (20.6).

Measurements of the H_C required to destroy the superconductivity near $T = 0°$ K shows that the condensation energy (energy difference between the normal and the superconducting states) is of the order $\frac{1}{2}\mu H_C^2(0)$. We have

$$\frac{H_C^2}{8\pi} = \frac{1}{2}N(0)\Delta^2 \tag{20.6}$$

where Δ is called the superconducting energy gap and is determined from the following relation

$$\Delta = 2\hbar\omega_D \exp\left(-\frac{1}{N(0)|g|}\right)$$

where ω_D is the Debye phonon frequency, $N(0)$ is the electron density of states at the Fermi level and $|g|$ is the electron-phonon interaction strength.

The existence of a critical field, limits the current that can be carried by a superconducting sample. The sample becomes normal if the field created by the current exceeds the critical field. For a long straight wire of circular cross section carrying current I, the field in the exterior region is given by $\boldsymbol{H} = 1/2\pi r$, where r is the radial distance from the centre of the wire. The field at the centre of the surface is $1/2\pi R$, where R is the radius of the wire and if this exceeds $\boldsymbol{H}c_1$ the outer region of the wire becomes normal. Current still exists but only on the surface of the superconducting portion, a cylinder with a radius smaller than R. There the field is greater than the critical field so these portions also become normal. The argument can be extended to show that the whole sample becomes normal.

Intermediate State

In practice, the determination of $\boldsymbol{H}_c$ is complicated by demagnetization effects. These arise from the boundary conditions on $\boldsymbol{H}$ and $\boldsymbol{B}$ and depend on the shape and orientation of the magnetic medium. For example, if the superconductor is a very long cylinder and the external field is along the cylinder axis, there is no demagnetization effect, for a sphere. the demagnetization factor n is 1/3. The uniform magnetic field inside ($\boldsymbol{H}_i$) is related to the externally imposed field ($\boldsymbol{H}_e$) by the relation

$$H_i = H_e - nM \tag{20.7}$$

where M is the magnetization. Now, for a superconductor, $\boldsymbol{M} = -\boldsymbol{H}_i$ so that $\boldsymbol{H}_i = \boldsymbol{H}_e/(1-n)$. Since $n = 1/3$ for a sphere, at $\boldsymbol{H}_e = 2/3\boldsymbol{H}_c$, $\boldsymbol{H}_i = \boldsymbol{H}_c$ and the sphere ought to go normal. But if it goes normal, $\boldsymbol{M} \cong 0$ so that $H_i = H_e = \frac{2}{3}H_c$, which is less than the critical fields: the sphere therefore ought to stay superconducting; obviously, for fields lying between $\frac{2}{3}H_c$, and $\boldsymbol{H}_c$, the sphere can be neither fully normal nor fully superconducting, but in an intermediate state, with a complicated distribution of superconducting and non superconducting regions. The phenomenon of intermediate state is similar to the coexistence of liquid and vapour for a certain regime of volume and temperature.

20.2.7 Coherence Length (ξ)

Pippard found experimentally that the penetration depth in impure tin is more than that in pure tin. Further the electrical resistance drops very sharply disappearing within about 1 millidegree of T_c in all superconductors. He concluded that these results indicate that there is a long range order among the electrons in the superconductor. The range of his ordering or the coherence length was estimated to be about 10^{-4} cm.

20.2.8 Types of Superconductors

[Type 1 (Soft) and Type II (Hard) Superconductors].

Superconductors may be classified in two and sometimes in three distinct groups, depending on their magnetization behaviour in an external magnetic field.

Type 1: (Soft) Superconductor

The Superconductor, in which the magnetic field is totally excluded from the interior of the Superconductor below a certain $\boldsymbol{H}_c$ and at $\boldsymbol{H}_c$ the material loses super-conductivity abruptly

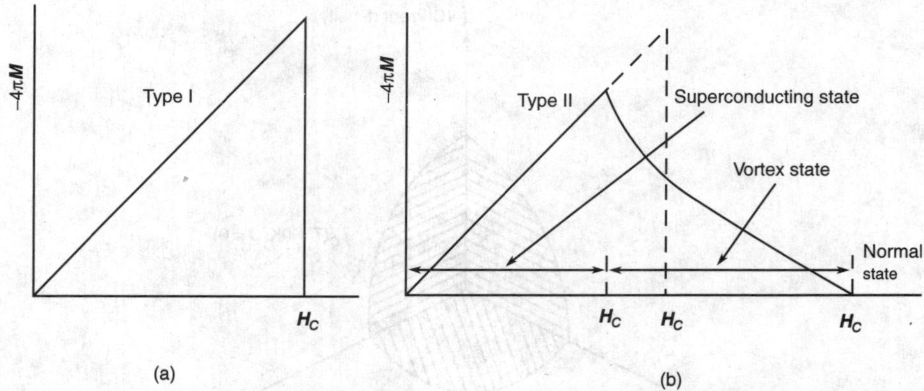

Fig. 20.7 (a) Magnetization curve for type I Superconductor (b) Magnetization curve for type II Superconductor

and the magnetic field penetrates fully are termed as type I or soft Superconductors. Clearly these conductors exhibit compete Meissner effect. The magnetization curve for type I super-conducting materials is shown in Fig. [20.7(a)]. The magnetization curve shows that transition at H_c, the material again acquires superconducting property and the field is expelled. Lead, Tin. Mercury fall into this category. The highest known critical fields for these materials are of the order of 1000 gauss (0.1 Tesla) making these materials unsuitable for use in high field super-conducting magnets. Type 1 materials are often called "soft" because of their tendency to give away to low magnetic fields.

Type II: (Hard) Superconductors

The materials which exhibit a magnetization curve of the form as shown in Fig. [20.7(a)] are known as type II or hard Superconductors. Alloys and transition metals with high values of the electrical resistivity in the normal state fall under this group of Superconductors. These Superconductors have two critical fields; Hc_1 (the lower critical field) and Hc_2 (the upper critical field). For this group of Superconductors, for applied field below Hc_1 the specimen is diamagnetic and hence the flux is completely excluded in range of magnetic field. At Hc_1 the flux begins to penetrate the specimen, and the penetration of the flux increases until the upper critical field, Hc_2 is reached. At Hc_2 the magnetization vanishes and the specimen returns to normal conducting state. In this group of Superconductors, the magnetization vanishes gradually as the field is increased, rather than suddenly as for the type I Superconductors. The value of the critical field for type II materials may be 100 times far more higher than the value of the critical field obtained for type I Superconductors. Critical fields, Hc_2 upto 30 Tesla have been observed. The materials which display type II behaviour are essentially inhomogeneous and include Nb-Zr, Nb-Ti alloys and Va-Ga and Nb-Sn inter-metallic compounds. Hard type II Superconductors are technically very useful materials in contrast to type I Superconductors.

For type II Superconductors, e.g., metals like niobium and vanadium and carefully homogenized solid solutions of indium with lead and indium with tin exhibit that transition is reversible, i.e., magnetization curves are reversible. These type II Superconductors,

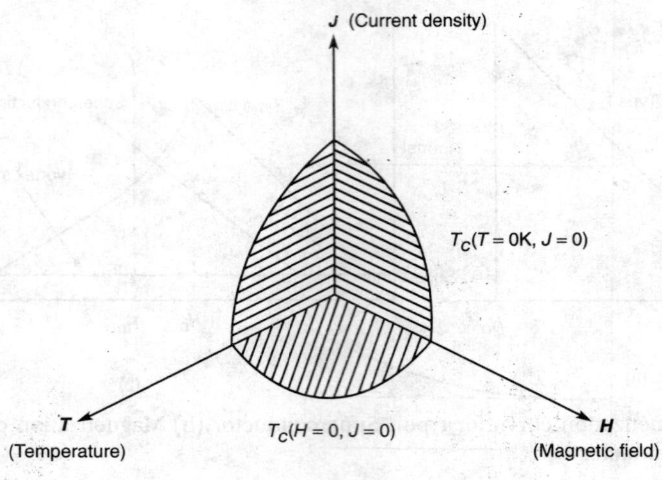

Fig. 20.8

like type I Superconductors are also ideal one. Many type II Superconductors exhibit irreversible magnetization behaviour i.e., hysteresis. The materials which display this behaviour are essentially inhomogeneous.

It is exciting to consider the possibility of power transmission in hard Superconductors. If the necessary refrigeration proved to be economically possible, then a Nb_3Sn conductor of 10^{-4} m^2 cross-section operating at say, 10^5 volts could carry, theoretical without loss, the enormous power of 10^{10} watts.

Abrikosov predicted that the upper critical field be proportional to the ratio of field penetration depth and coherence distance. The ratio increases for the impure materials, e.g., alloys. The hard Superconductors are characterized by three critical values, current density J_{c1}, H_{c2} and T_c which are independent of each other.

A distinguishing characteristic of type I and type II Superconductors is provided by a modification of the Meissner effect of type II Superconductors through flux penetration into the Superconductors. This is illustrated in Fig. [20.7(b)] in which it is clear that superconductivity is only partially destroyed in type II Superconductors for $Hc_1 << H << Hc_2$. This region is called the intermediate or vortex state. It is really a mixture of normal and super-conducting states. In general the vortex or intermediate state is unstable for type I (soft) Superconductors but stable for type II (hard) Superconductors.

Type-II Superconductors are preferred over type-I Superconductors for most practical applications by virtue of their higher critical temperatures and critical magnetic fields.

The three material parameters T_c, H_c and J_c are of extreme importance in the practical applications of super-conductivity. Fig. (20.8) shows schematically the boundary in temperature magnetic field current density space separating normal and super-conducting states. The position of this boundary will of course, depend on the material. For, temperature, magnetic field and current density values lying between the origin and this boundary, the material will be super-conductive, out side the boundary, conduction is normal.

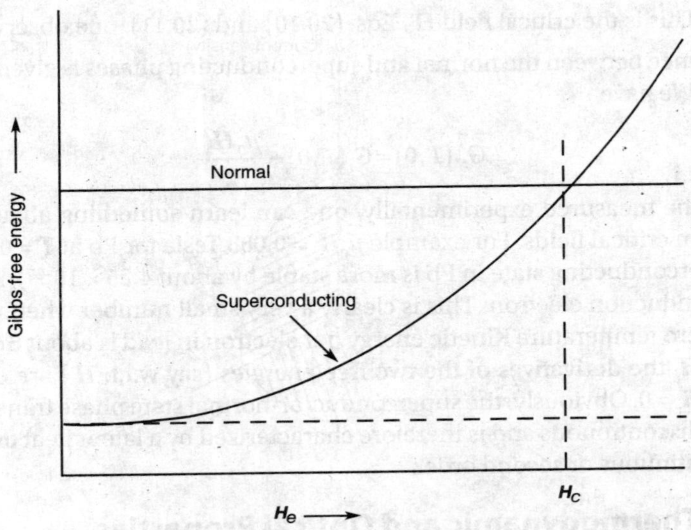

Fig. 20.9

20.2.9 Gibb's Free Energy

One can easily understand why the normal or the non-superconducting state becomes more stable at external magnetic fields higher than H_c. Let us consider the Gibb's free energy of the superconductor. If its magnetization is M in an external field H, the extra work done in bringing the superconductor into an external field H_e form far away (where $H_e = 0$) is given by

$$H = -\mu_0 \int_0^{H_e} M\,dH \tag{20.8}$$

Since $M = -H$ for a Superconductor, the extra work done is given by

$$W = \frac{\mu_0 H_e^2}{2} \tag{20.9}$$

This is extra Gibbs free energy of a superconductor. Diamagnets have a higher free energy in an external magnetic field tend to move to regions of lower field, this is a simple way of understanding the sign as well as the size, if we use ($x_{\text{dia}} = -1$). One therefore obtains Gibbs free energy G_s per unit volume as

$$G_s(T, H_e) = G_s(T, 0) + \frac{\mu_0 H_e^2}{2} \tag{20.10}$$

In the normal state, the magnetization is very small and the magnetic work term in Eq. (20.8) is therefore negligibly small. We therefore have

$$G_N(T, H_e) = G_N(T, 0) \tag{20.11}$$

The two Gibbs free energy curves are shown in Fig. (20.9).

One observe that since $G_s(T, H_e)$ increases quadratically with H_e, for some H_e the normal state has the same free energy as the superconductor, and beyond this the normal state is more stable. This is the critical field H_c, Eqs. (20.10) and (20.11), one observe that the free energy difference between the normal and superconducting phases is given by $\frac{\mu_0 H_c^2}{2}$ per unit volume. We have

$$G_N(T,0) = G_s(T,0) + \frac{\mu_0 H_e^2}{2} \tag{20.12}$$

Since H_c can be measured experimentally, one can learn something about the thermodynamics from critical fields. For example $\mu_0 H_c = 0.083$ Tesla for Pb at $T = 0$ means that at $T = 0$ the superconducting state in Pb is more stable by about 4.25×10^{-25} J/mole' or about 10^{-7} eV per conduction electron. This is clearly a very small number when we realize that the average zero temperature Kinetic energy per electron in lead is about 6 eV.

Finally, at H_c the derivatives of the two free energies (say w.r.t. H_c) are discontinuous, except when $H_c = 0$. Obviously, the superconductor-normal state phase transition is always first order or discontinuous and is therefore characterized by a latent heat except at $H_c = 0$, where it is continuous or second order.

20.2.10 Thermodynamic and Optical Properties

So far we have seen that free energy difference between the normal and the superconducting states can be related to the critical field. However, one can also directly measure the thermodynamic properties of a superconductor, e.g., specific heat, latent heat etc. What actually occurs in the transition from the normal to the superconducting state is a thermodynamic change of state, or phase transition, which is clearly manifest in other physical quantities.

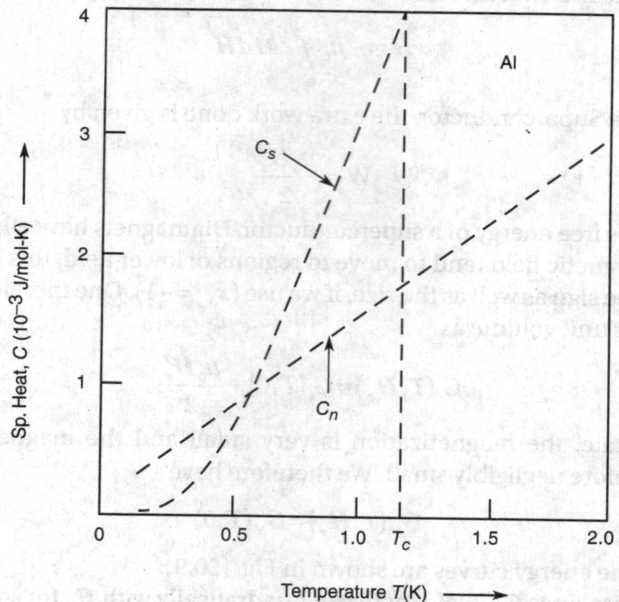

Fig. 20.10

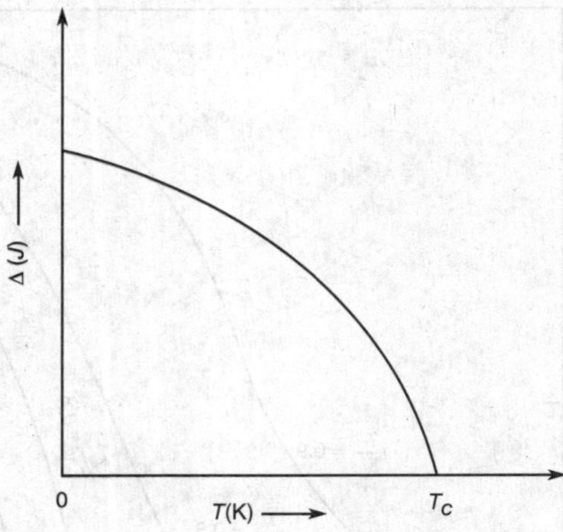

Fig. 20.11

(a) The specific heat

The specific heat plotted against temperature, for example, changes discontinuously at the transition temperature T_c. Fig. (20.10) shows the example of Al with a transition temperature of 1.19K.

The specific heat of a normal metal is seen to be of the form

$$C_c(T) = C_{ne} + C_{nl}$$
$$= \gamma T + \beta T^3 \tag{20.13}$$

where C_{ne} is the electronic part $(= \gamma T)$ and C_{nl} is a lattice-dynamical part $(= \beta T^3)$. In the normal conducting state $(n) C_n$ varies smoothly at low temperatures. At the transition to the superconducting state (at T_c) the specific heat jumps and then at very low temperatures sinks to below the value of normal phase. Normally one assumes that at the transition to the superconducting state the lattice dynamical part C_{nl} remains constant. A more exact analysis then shows that for a superconductor, the electronic part ($C_{nl} \sim T$ for a normal conductor) must be replaced by a component which, well below the critical temperature decreases exponentially

$$C_{es} = A \exp(-\Delta / k_B T) \tag{20.14}$$

An exponential dependence implies that it requires a finite energy Δ to excite an individual electron in a superconductor (in a normal metal, the energy required is zero). The superconductor has an energy gap against the excitation of electrons. The energy gap Δ is experimentally to be of the order $k_B T_c$. An energy gap (Δ) is a characteristic, but not universal feature of superconductors, i.e., there exists some superconductors in which energy gap is zero. The energy gap (Δ) is accounted for by BCS theory.

The existence of an energy gap in the superconducting electron levels has been confirmed by a number of experiments. It has been shown that the gap decrease from a value of about 3.5 $k_B T_c$ at 0K to zero at the transition temperature (T_c) for the normal or type I superconductors as shown in Fig. (20.11).

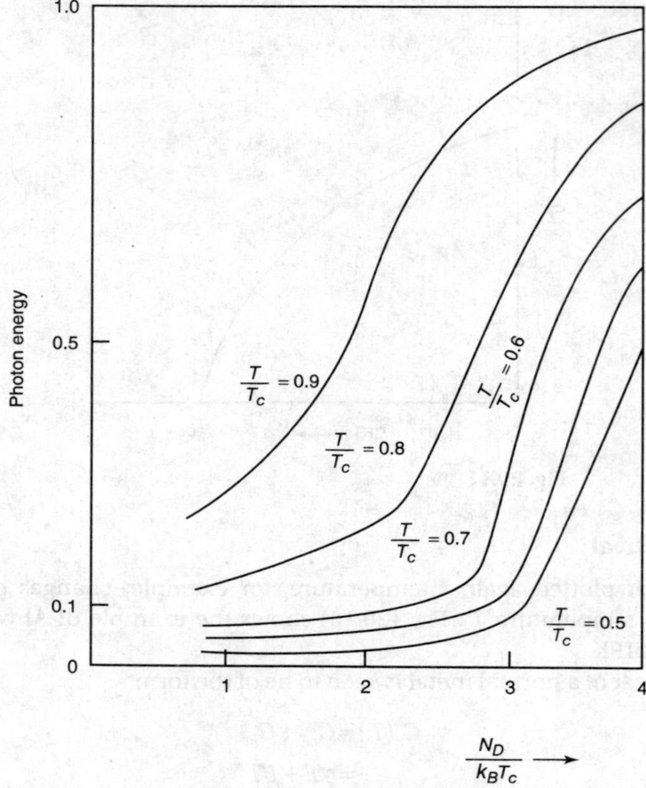

Fig. 20.12 Photon's energy vs surface resistance ratio curve for Al for various temperatures

A direct way of testing the presence of an energy gap in a material is by light absorption. If an energy Δ is needed to excite electrons, then a photon of energy $h\nu < \Delta$ should not be absorbed, and a superconductor ought to be more reflecting than the normal metal for light of frequencies $\nu < \Delta/h$.

Indeed superconductors have been found to be more reflecting than normal metals for frequencies $\nu < \Delta/h$.

(b) Thermal conductivity

Experimentally it was observed that for type I superconductors there is marked drop in the thermal conductivity when superconductivity sets in, suggesting that the superconducting electrons possibly playing no part in heat transfer.

20.2.11 Microwave and Infrared Properties

The existence of an energy gap in superconductors means that photons of energy ($h\nu$) less than the gap energy, i.e., $h\nu < E_g$ are not absorbed by superconductors. Nearly all the photons incident are reflected as for any metal because of the impedence mismatch at the boundary between vacuum and metal, but for a thin film of the order of ≈ 20Å, more photons are transmitted in the superconducting state than in the normal state. For photon energies less than the energy gap ($h\nu < Eg$) the resistivity (ρ) of a superconductor vanishes at absolute zero: At $T << T_c$ the resistance in the superconducting state has a sharp threshold at the gap energy. Photons of lower energy see a resistance that approaches that of the normal state because such photons cause transitions to unoccupied normal energy above the energy gap (E_g). As the temperature is increased not only does the gap energy decease, but resistivity for photons below the energy gap on longer vanishes, except at zero frequency. At zero frequency the super-conducting electrons short-circuit any normal electrons that have been thermally excited above the energy gap (E_g). At the finite frequencies the inertia of the superconducting electrons prevents them form completely screening the electric field, so that thermally excited normal electrons can absorb energy now. These results have been shown in Fig. (20.12) for Al for various temperatures.

20.3 THE ISOTOPE EFFECT AND ELECTRON-PHONON INTERACTION

There is another very important effect which suggests that superconductivity is related to lattice properties as well as electron properties. This is the isotope effect. In 1950 Reynolds, et al discovered that the transition temperature for superconductivity is proportional to $M^{-1/2}$, M being the isotopic mass of the ions of the superconducting metal. The experimental result within each series of isotopes may be fitted by a relation of the form

$$M^{\alpha}T_c = \text{constant} \tag{20.15}$$

where α is a number and for most of the metals its value is about 0.5 i.e.,

Although the observed isotope-shift exponent (α) is quite accurately +1/2 for a few classic superconductors, such as lead, recent experiments on other materials have shown that it can range all the away to zero or even change sign. Also it is found that $H_C \propto M^{1/2}$. Eq. (20.15) leads to

$$\frac{T_c}{\theta_D} = \text{constant} \tag{20.16}$$

where θ_D is Debye temperature. When one isotope is substituted for another the mass of the nuclei changes but the electronic properties of solid are left virtually unaltered. The existence of isotope effect shows that the superconducting transition must depend in someway on nuclear mass, i.e., on the mass of lattice ions (phonons). This discovery led Frohlich to propose the electron-phonon interaction as an important mechanism for superconducting phase transition. What is so strange is that nowhere in the theory of metals do the fine details of lattice vibrations (phonons) play an essential role.

20.4 THE EFFECT OF THE SUPERCONDUCTING TRANSITION ON THE PROPERTIES

Some properties change, and some do not when metal undergoes the superconducting transition. They are listed in Table 20.2.

Table 20.2

Property	*Affected or not Affected by the transition*	*Remarks*
Crystal Structure	No	
Shape and Size	No	
Electrical and optical property		
(a) Resistivity (ρ)	Yes	Vanishes
(b) Thermoelectricity	Yes	-
(c) Photoelectricity	No	-
(d) Reflectivity	No	-
Thermal Properties		
(a) Latent Heat	Yes	There is no latent heat at the transition unless there is a magnetic field
(b) Sp. Heat		A discontinuous change
(c) Thermal conductivity		Discontinuous change if in magnetic field
Elasticity	Yes	
Absorption of electron beams	No	

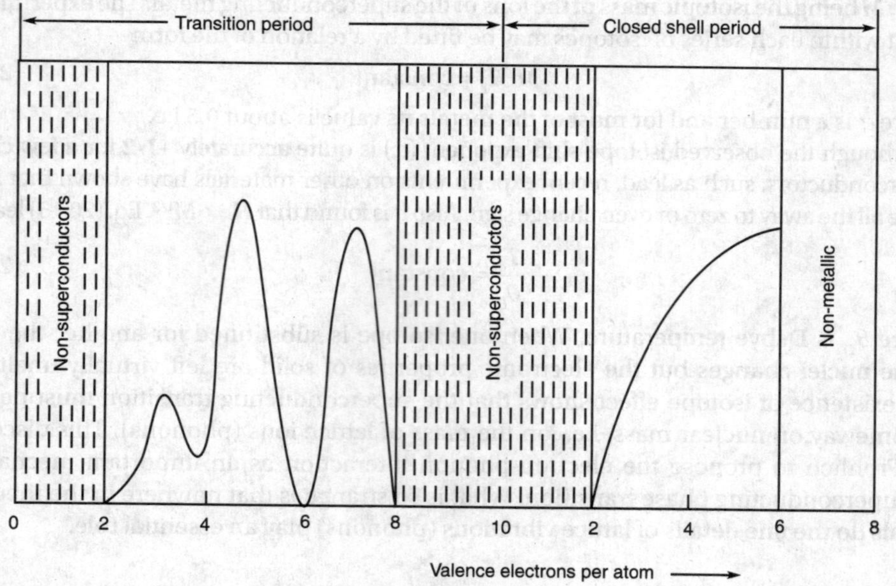

Fig. 20.13 Variation of T_c with number of valance electrons per atom

20.5 SPECIAL FEATURES OF SUPERCONDUCTORS

(i) Superconductors are, on the whole, poor conductors above the transition temperature and no element with less than two valence electrons or more than nine valance electrons per atom have yet been found to be superconductive Fig. (20.13). Essentially the same rule holds for compounds though exceptions with the average number of valence electrons per atom slightly lower than 2 or slightly greater than 8 are known.

(ii) Numerous alloys exhibit superconductive property, even though neither of its component may be a superconductor for e.g., CuS (copper sulphide), is a superconductor but copper and sulphur are not superconductors.

(iii) In transition metal elements the critical temperature varies in a periodic manner with the number of valence electrons per atom and exhibits maxima. In non-transition elements critical temperature increases as the number of valence electrons per atom increases.

20.6 THEORETICAL SURVEY

Theories of superconductivity developed along two lines. The phenomenological theories of London, Pippard, Ginzburg and Landau deal with the existence and order of magnitude of the penetration depth and coherence length and give a qualitative account of the electro-dynamic properties.

The microscopic theory of Frohlich introduced the concept of an interaction between superconducting electrons and phonons. In 1957 Bardeen, Cooper, and Schrieffer proposed a microscopic theory (BCS theory) which accounts for the observed properties of superconductors which we have described. The results of the phenomenological theories can be derived from the BCS theory, and superconductivity is now understood in terms of the electronic and atomic structures. BCS theory has also provided the basis for advance work. Josephson and Anderson discovered the importance of superconducting wave function.

London Equations

In 1935, F. London and H. London showed that the magnetic flux lines are not completely expelled from a superconductor but remain confined in a thin surface layer. So produced flux change induces a flow of super-current in a closed circuit within the thin layer. London brothers postulated a two fluid model with super fluid and normal densities and velocities n_s, n_n and v_s, v_n respectively. Let us consider the sum of the densities be $n_s + n_n = n_0$, where n_0 is the average number of electrons per unit volume due to the local charge neutrality condition. They postulated that the super and normal current densities satisfy the following relations

$$\frac{d\mathbf{J}_s}{dt} = \frac{n_s e^2}{m}\varepsilon \tag{20.17}$$

$$\mathbf{J}_n = \sigma n \varepsilon \tag{20.18}$$

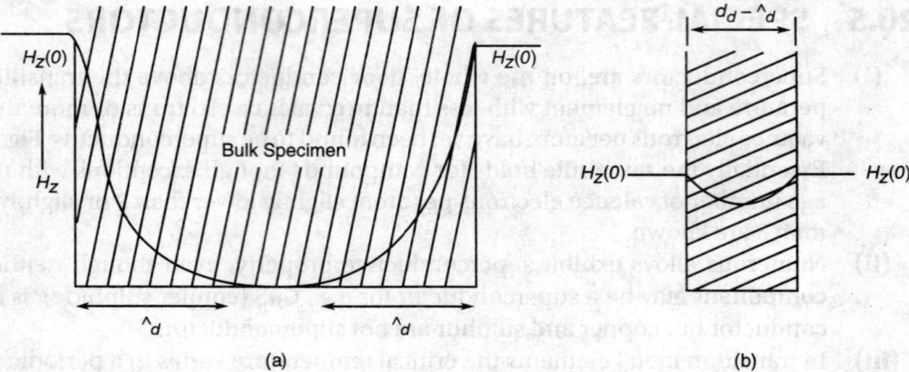

Fig. 20.14 Penetration of magnetic field in (a) bulk specimen, and (b) thin film specimen

We may note that Eq. (20.18) is the 'modified Ohm's law.' We obtain,

$$\nabla \times \boldsymbol{J}_s = -\frac{n_s e^2}{m\varepsilon_0 c^2}\boldsymbol{H} \tag{20.19}$$

Eq. (20.19) is the equation which explains the Meissner effect. We have the Maxwell equation which relates the magnetic field and the current density in the stationary case as

$$\nabla \times \boldsymbol{H} = \boldsymbol{J}_s \tag{20.20}$$

Since we are dealing with static Meissner effect and therefore we have neglected the displacement current and normal fluid density $\boldsymbol{J}_n$. Taking the curl of Eq. (20.20) and making use of Eq. (20.19), one obtains

$$\nabla \times \nabla \times \boldsymbol{H} = \nabla \times \boldsymbol{J}_s = -\frac{n_s e^2}{m\varepsilon_0 c^2}\boldsymbol{H} \tag{20.21}$$

Eq. (20.21) yield the London's penetration depth

$$\wedge_d = \left(\frac{m\varepsilon_0 c^2}{n_s e^2}\right)^{1/2} \tag{20.22}$$

If we apply magnetic field along z-direction on a semi-infinite slab extending in the x-direction, then Eq. (20.21) is expressed as

$$\frac{d^2 \boldsymbol{H}_z}{dx^2} = \frac{1}{\wedge_d^2}\boldsymbol{H}_z \tag{20.23}$$

The solution of Eq. (20.23) is

$$\boldsymbol{H}_z = \boldsymbol{H}_z(o)\exp(-x/\wedge_d) \tag{20.24}$$

Eq. (20.24) shows that magnetic field decays exponetially and $\wedge_d$ (penetration depth) is the effective depth upto which the magnetic field penetrates a superconductor. We may note that $\wedge_d$ ranges from 300 to 5000 Å depending on the material. Fig. (20.14) shows the penetration of magnetic field into a bulk specimen and thin film, whose thickness itself is

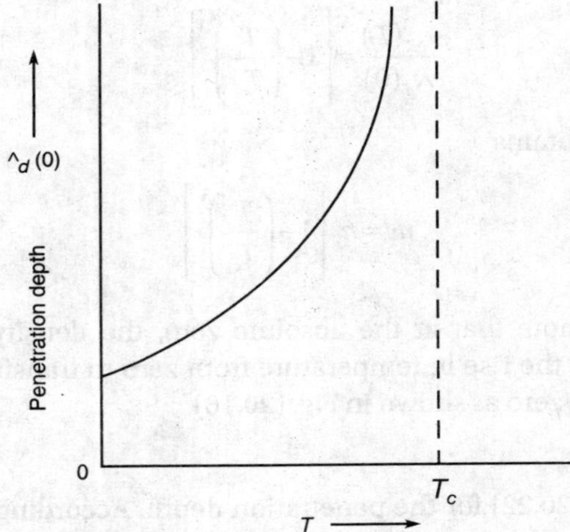

Fig. 20.15 Variation of penetration depth of a superconductor with temperature

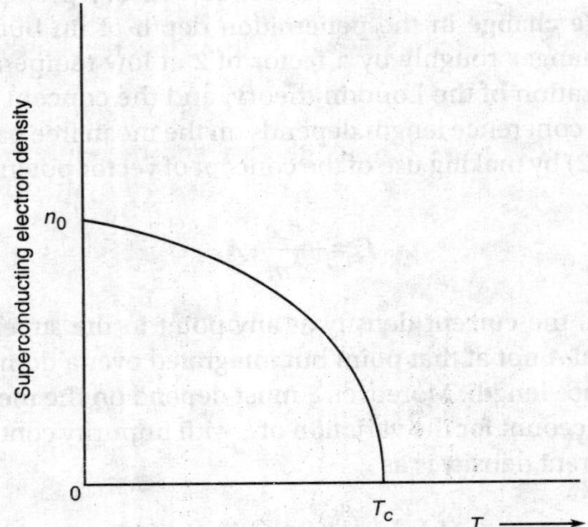

Fig. 20.16 Variation of superconducting electron density with temperature

of the order of penetration depth. However, in a bulk specimen, where the thickness of the specimen is much larger than penetration depth, the magnetic field does not penetrate a major portion of the specimen and obviously $\boldsymbol{M} = \boldsymbol{H}$ providing the Meissner effect. We may note that a thin film will not exhibit the same diamagnetism as a bulk specimen since it is at least partially penetrated by magnetic field.

We may note that penetration also depends on the temperature Fig. (20.15) and as the temperature approaches the transition temperature (T_c), the penetration depth becomes quite large. The following expression covered the said observation very well.

$$\frac{\wedge_d(T)}{\wedge_d(0)} = \left[1 - \left(\frac{T}{T_c}\right)^4\right]^{-1/2} \tag{20.25}$$

From Eq. (20.25), one obtains

$$n_s = n_0\left[1 - \left(\frac{T}{T_c}\right)^4\right] \tag{20.26}$$

From Eq. (20.26), we note that at the absolute zero, the density of superconducting electrons is n_0 and with the rise in temperature from zero to transition temperature (T_c), the density decreases to zero as shown in Fig. (20.16).

Pippard's Theory

Let us re-examine Eq. (20.22) for the penetration depth. According to this equation, the impurities should not change appreciably the penetration depth. However, Pippard performed an experiment by measuring the microwave surface impedance of tin diluted with small amounts of indium at about 3 cm wavelength and estimated the penetration depths from the surface impedance values. London theory predicts that there should not be an appreciable change in the penetration depth of tin but Pippard found that penetration depth changes roughly by a factor of 2 at low temperatures and this then lead to non-generalization of the London theory, and the concept of coherence length (ξ). We may note that coherence length depends on the mean free path.

Rewriting Eq. (20.22) by making use of the concept of vector potential as

$$\boldsymbol{J}_s = -\frac{n_s e^2}{m}\boldsymbol{A} \tag{20.27}$$

According to Pippard, the current density at any point in the superconductor depends on the vector potential $\boldsymbol{A}$ not at that point but integrated over a domain of radius ξ. Here ξ is known as coherence length. Moreover, ξ must depend on the mean free path $\wedge$ of the electrons in order to account for the variation of ξ with impurity content. The generalized expression for the current density is as

$$\boldsymbol{J}_s(\boldsymbol{r}) = \int K(\boldsymbol{r},\boldsymbol{r}')\boldsymbol{A}(\boldsymbol{r}')d\boldsymbol{r} \tag{20.28}$$

According to Pippard, the width of the Kernel function $K(\boldsymbol{r} - \boldsymbol{r}')$ defines the coherence length. Further, Pippard suggested the following form of Kernel as

$$K(\boldsymbol{r}-\boldsymbol{r}') = \exp[-(\boldsymbol{r}-\boldsymbol{r}')(1/\xi)] \tag{20.29}$$

where $\dfrac{1}{\xi} = \dfrac{1}{\xi_0} + \dfrac{1}{\alpha\wedge}$

Here ξ_0 represent the electromagnetic coherence length and α is an empirical constant (~1). We may note that the appearance of the electron mean free path reveals that the coherence length reduced by the presence of impurities in a superconductor.

20.7 FLUX QUANTIZATION

Based on some ideas about the nature of the superconducting state, London in 1950 speculated that the magnetic flux going through a superconducting ring should be quantized in multiples of h/e. The remarkable fact is that the flux of the magnetic field through a non-superconducting area is restricted to (nh/e) merely because it is surrounded completely by a superconductor. The flux quantum is rather small, being about 4×10^{-15} Weber. Deaver and Fairbank in USA and Doll and Nabaure in Germany practically simultaneously in 1961 showed from sensitive experiments that this is indeed true, but units of observed flux is quantized in units of $h/2e = 2.07 \times 10^{-15}$ Weber rather than h/e. This is consistent with the modern theory of superconductivity which is based on the idea that strongly correlated electron pairs are formed at the Fermi surface in the superconducting state.

20.8 BCS THEORY OF SUPERCONDUCTIVITY: COOPER PAIRS

Out of the various theories proposed on superconductivity, the most successful theory developed is by Bardeen, Cooper and Schrieffer (1957) known an BCS weak coupling theory is truly first successful microscopic theory. According to BCS theory, superconductivity occurs because of the existence of an attractive force between conduction electrons in a metal, leading to bond pairs of electrons, called Cooper pairs, which move without scattering (without resistance) with the lattice structure.

Although electrons repel each other due to Coulomb interaction, but in superconductors, this repulsion is dominated by an attractive force arising due to electron-lattice-electron interaction or electron-phonon-electron interaction (provided the thermal energy of the electrons is less than the energy associated with their interaction with the lattice vibration of phonon). There is a negatively charged electron which while moving through a lattice of positive ions attracts the positively charged ions. This distorts the lattice creating a wave of excess positive charge to which another electron of equal and opposite momentum and spin is attracted to form a pair known as Cooper pair. In order words, that the materials should become superconducting, the lattice of electrons at room temperature giving rise to electrical resistance, can at low temperatures (where there is not much thermal agitation) lead to the electron pairing. These paired electrons are scattered like normal electrons, because change in momentum in one of the electrons of a Cooper pair is counter balanced by an equal and opposite change in momentum of the other electron. Hence they can move through the material without resistance. A flow of such electron pairs constitutes a super-current. The accomplishments of the quantum theory of superconductors due to BCS include:

(i) An attractive interaction between the electrons through lattice phonons can lead to a ground state separated form excited states by an energy gap. The critical field, the thermal properties, and most of the electromagnetic properties of superconductors are consequence of the energy gap. In special circumstances, superconductivity may occur without an actual energy gap, called gapless superconductivity.

(ii) The electron-lattice-electron or electron-phonon-electron interaction leads to an energy gap of observed magnitude. The indirect interaction proceeds when

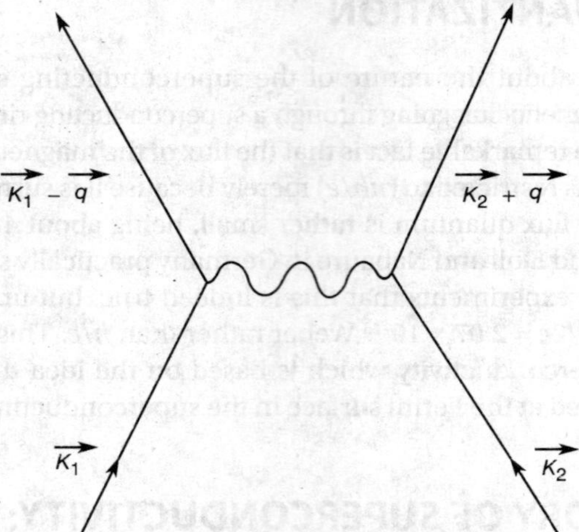

Fig. 20.17 Electron-Electron interaction through lattice phonons

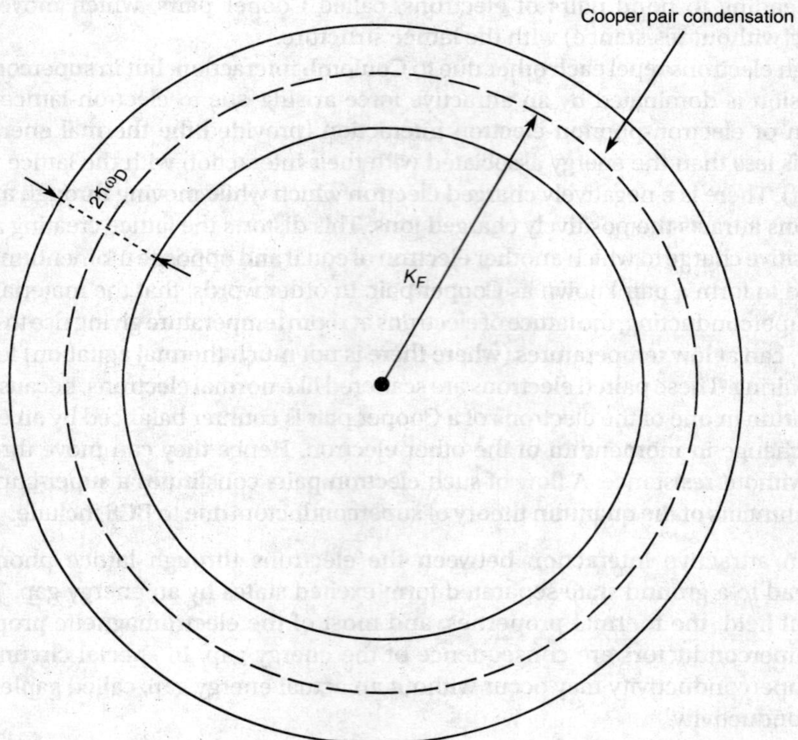

Fig. 20.18 Cooper pair condensation Fermi surface

one electron interacts with the lattice and deforms it, a second electron sees the deformed lattice and adjusts itself to take advantage of the deformation to lower its energy. Thus the second electron interacts with the first electron via the lattice deformation and forms an Cooper pair Figs. (20.17) and (20.18).

(iii) The penetration depth (λ) and the coherence length (ξ) emerge as natural consequences of the BCS theory. The London equations are obtained from the BCS theory for magnetic fields that vary slowly in space. Thus the Meissner effect, the central phenomenon in superconductivity is obtained form BCS theory in a natural way.

(iv) The superconducting transition temperature of an element or alloy is given by

$$k_B T_c = 1.14\hbar\omega_D \exp\left[-\frac{1}{|g|N(0)}\right]$$

since $\hbar\omega_D \approx k_B\theta_D$

and hence $T_c = 1.14\theta_D \exp\left[-\frac{1}{|g|N(0)}\right]$ (20.30)

where $|g|$ is the electron-phonon interaction, $N(0)$ is the electron density of orbitals or density of states at the Fermi level, and θ_D is Debye Temperature.

$|g|N(0)$ is always less than 1 and for some models even $gN(0) \le \frac{1}{2}$, the majority of known superconductors have $|g|N(0) \le \frac{1}{3}$ or $\frac{1}{4}$. The result for T_c is satisfied at least qualitatively by the experimental data.

(v) Magnetic flux through a superconducting ring is quantized and the effective unit of charge is $2e$ rather than e. As stated above, the BCS ground state involves pairs of electrons; thus flux quantization in terms of the pair charge $2e$ is a natural consequence of the theory.

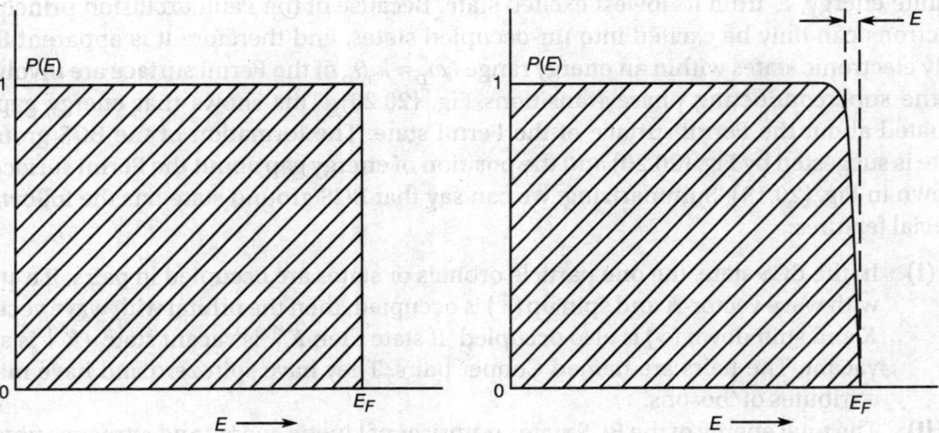

Fig. 20.19 Probability P of occupation of one electron states of energy (a) ground state of non-interacting Fermi gas, (b) BCS ground state. The latter differs from (a) only over a region of width E_g (energy gap) at the Fermi surface

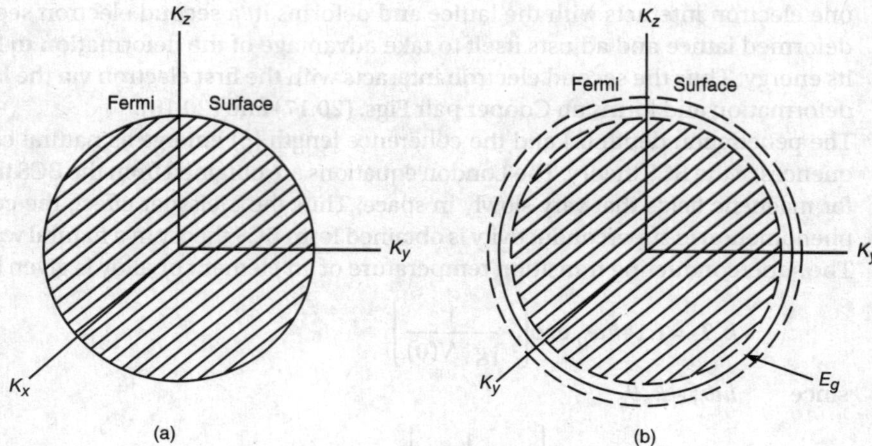

Fig. 20.20 Ground state of the Fermi gas of non-interacting electrons. Within the Fermi sphere, all one particle states are occupied whereas all other states outside the sphere are vacant. The lowest exited state is separated from the ground state by an arbitrarily low excitation energy. (b) BSC ground state of electrons. The states near Fermi surface are filled in accordance with Figs. [20.15(a)] and [20.15(b)]. Lowest excited state is separated from the ground state by an energy E_g

BCS Ground State

We know that the filled Fermi sea is the ground state of a Fermi gas of non-interacting electrons. We have read that it is just the Fermi surface. This state allows arbitrarily small excitations. One can form an excited state by taking an electron from the Fermi surface and raising it just above the Fermi surface. Taking an appropriate attractive interaction between electrons, BCS theory shows that superconducting ground state is separated by a finite energy, E_g from its lowest excited state. Because of the Pauli exclusion principle, electrons can only be excited into un-occupied states, and therefore it is apparent that only electronic states within an energy range $\hbar\omega_D \approx k_B\theta_D$ of the Fermi surface are involved in the superconducting phase transitions Fig. (20.20). This shows that energy gap is situated about the Fermi surface of the Fermi state. The formation of the BCS ground state is suggested by Fig. (20.20) and the position of energy gap about the Fermi surface is shown in Fig. (20.19). Summarizing, we can say that BCS ground state has the following special features:

(i) In the BCS state, the one particle orbitals or states are occupied in pairs: if a state with wave vector-$\boldsymbol{K}$ and spin up ($\uparrow$) is occupied, then the orbital with wave vector-$\boldsymbol{K}$ and spin down ($\downarrow$) is also occupied. If state then $\boldsymbol{K}\uparrow$ is vacant state $-\boldsymbol{K}\downarrow$ is also vacant. The pairs are named Cooper pairs. They have spin zero and have many attributes of 'bosons'.

(ii) The total energy of the BCS state comprises of kinetic energy and attractive potential. This energy is lower with respect of the Fermi state, because the Fermi state comprised simply of the kinetic energy. The attractive potential energy in the BCS state reduces the total energy of the state. As a consequence the BCS state becomes more stable than the Fermi state and Superconductivity persists.

The BCS Model

The model (BCS) Hamiltonian is

$$H_{BCS} = \sum_{K,\sigma} C^{+}_{K',\sigma} C_{K\sigma} - \sum_{K,K'} g_{KK'} b^{+}_{K'} b_K \tag{20.31}$$

The first term in Eq. (20.31) gives the unpertubed energy of the electrons forming the Cooper pairs. The latters have usual meanings. The second term in Eq. (20.31) is the pairing interaction in which a Cooper pair of electrons in $(\boldsymbol{K}\uparrow, -\boldsymbol{K}\downarrow)$ scatters to $(\boldsymbol{K}'\uparrow, -\boldsymbol{K}'\downarrow)$, $g_{KK'}$ is the strength of pairing interaction and generally it is taken independent of $\boldsymbol{K}$, $\boldsymbol{K}'$. $b^{+}_{K} = C^{+}_{K\uparrow} C_{-K\downarrow}$ and $b_K = C_{-K\downarrow} C_{K\uparrow}$ are respectively, creation and annihilation operators for a pair of electrons in $(\boldsymbol{K}\uparrow, \boldsymbol{K}'\downarrow)$. These operators obey the following commutation rules.

$$[C_{K\sigma}, C^{+}_{K'\sigma}] = \delta_{KK'} \delta_{\sigma\sigma'} \tag{20.32}$$

$$[C_{K\sigma}, C_{K'\sigma'}] = [C^{+}_{K\sigma}, C^{+}_{K'\sigma'}] = 0$$

and $C^{+}_{K\sigma} C_{K\sigma} = n_{K\sigma}$ is called the number operator.

The ground state of the BCS and its first excited states can conveniently be treated by introducing a linear combination of operators $C^{+}_{K\sigma} C_{K\sigma}$ via the so called **Bugoliubov transformations** as given by

$$\gamma^{+}_{K\uparrow} = u_K C^{+}_{K\uparrow} - v_K C_{-K\downarrow}$$

$$\gamma_{-K\downarrow} = v_K C^{+}_{K\uparrow} - u_K C_{-K\downarrow} \tag{20.33}$$

$$u_K^2 + v_K^2 = 1$$

Since, as is well known that spin $\frac{1}{2}$ particles can only be created or destroyed in pairs but never singly $\gamma^{+}_{K\uparrow}$ represents the simplest elementary excitation of an interacting spin $\frac{1}{2}$ system involving single particle creation and annihilation operator. We call $\gamma^{+}_{K\uparrow}$ and $\gamma_{K\downarrow}$ creation and annihilation operators for Bogoliubov quasiparticles called bugolons. From Eq. (20.33), we can express the C's in terms of γ's as

$$C^{+}_{K\uparrow} = u_K \gamma^{+}_{K\uparrow} + v_K \gamma_{-K\downarrow} \tag{20.34}$$

$$C_{-K\downarrow} = u_K \gamma_{-K\downarrow} - v_K \gamma^{+}_{K\uparrow}$$

Here u_K is even and v_K is odd.

i.e.,
$$u_K = u_{-K} \tag{20.35}$$

and
$$v_K = v_{-K} \tag{20.36}$$

γ's satisfy fermion commutation rules, if

$$u_K^2 + v_K^2 = 1$$

The variational calculations, yield the following results.

$$v_K^2 = \frac{1}{2}\left[1 - \frac{(\varepsilon_K - \mu)}{E_K}\right] \tag{20.37}$$

where $$E_K = [(\varepsilon_K - \mu)^2 + \Delta_K^2] \tag{20.38}$$

Here Δ_K called the energy gap, satisfying the so called gap equation

$$\Delta_K = -\sum_K \frac{g_{KK'}\Delta_{K'}}{2E_{K'}} \tag{20.39}$$

and μ has the physical significance of chemical potential (Fermi energy).

The form of E_K Eq. (20.38) shows that gap has opened up in the excitation spectrum. The states formerly in this gap are "pushed aside". Since the number of states remain the same, the superconducting density of states can be found by equating

$$N_s(E)dE = N_n(\varepsilon)d\varepsilon \approx N(0)d\varepsilon$$

$$\frac{N_s(E)}{N(0)} = \begin{cases} \dfrac{E}{(E^2 - \Delta^2)^{\frac{1}{2}}} & \neq E > 0 \\ 0 E < 0 \end{cases} \tag{20.40}$$

Further, from Eq. (20.40), we obtain the simple model for which

$$g_{KK'} = \begin{cases} g(\varepsilon_K - \mu) < \hbar\omega_D \quad \text{and} \quad |(\varepsilon_K^1 - \mu)| \hbar\omega_D \\ 0\,(\text{otherwise}) \end{cases} \tag{20.41}$$

the result that $\Delta K = \Delta$ (for $|\varepsilon - \mu| < \hbar\omega_D$) and zero otherwise where

$$\Delta = \hbar\omega_D \exp\left[-\frac{1}{N(0)|g|}\right] \tag{20.42}$$

In a bulk superconductor, all electrons are in the same quantum state characterized by a global phase ϕ. The charge carriers are condensed in a bound state of two electrons, the so called Cooper pair, which has a charge of-2e. The ground state wave function depends on the phase ϕ and can be written as

$$\Delta = \sqrt{n_s} \exp(i\phi)$$

where n_s is the total number of Cooper pairs.

Δ is the "superconducting order parameter" and specifies the choice for the quantum phase ϕ. There are no other states present in the interval $[\mu - \Delta, \mu + \Delta]$ because it takes an energy 2Δ to break a Cooper pair. An electron in this energy window cannot be injected in a superconductor alone, generating a "gap" in the transmission of charge carriers at a normal/superconductor interface.

Further, the condensation energy difference between the normal and superconducting states, is found to be

$$\Delta G = \frac{1}{2} N(0)\Delta^2 \tag{20.43}$$

Generalizing the gap equation to finite temperatures

$$\Delta(T) = -\frac{1}{2}\sum_K K \frac{\Delta_{K'}(T)}{E_{K'}(T)} \tan h\left(\frac{E_{K'}(T)}{k_B T}\right) \tag{20.44}$$

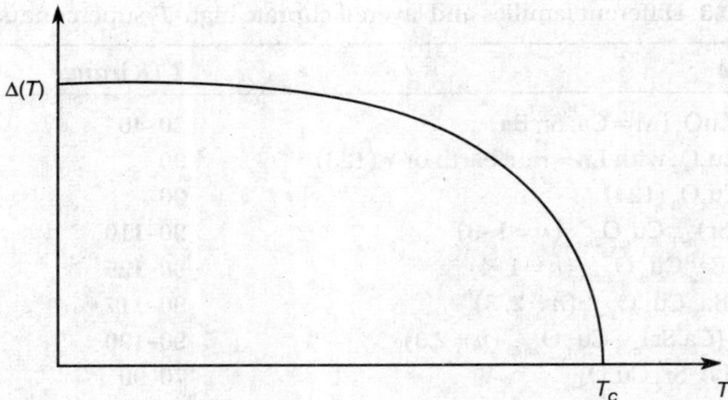

Fig. 20.21 Variation of Δ with *T*

If $g_{KK'}$ is approximated by Eq. (20.44), then Δ_K is again of the from Δ(independent of *K*) and $\Delta(\beta)\left(\beta = \dfrac{1}{k_B T}\right)$ satisfies

$$\frac{1}{N(0)|g|} = \int_0^{\hbar\omega_D} \frac{dx}{(\chi^2 + \Delta^2)^{1/2}} \tan h\left[\frac{1}{2}\beta(x^2 + \Delta^2)^{1/2}\right] \quad (20.45)$$

On solving numerically, we obtain the variation of Δ with *T* as shown in Fig. (20.21). The transition temperature is given by

$$\frac{1}{N(0)} = \int_0^{\hbar\omega_D} \frac{dx}{x} \tan h\left[\frac{x}{2k_B T_C}\right]$$

For the limit $gN(0) << 1$ (weak coupling limit),
we obtain,

$$k_B T_c = 1.14\hbar\omega_D \exp\left(-\frac{1}{|g|N(0)}\right)$$

But $\hbar\omega_D \approx k_B \theta_D$

$$T_C = 1.14\theta_D \exp\left(-\frac{1}{|g|N(0)}\right) \quad (20.46)$$

Also at $T = 0°\text{K}$

$$\Delta(0) = \frac{\hbar\omega_D}{\sin h[1/N(0)g]}$$ which in weak coupling limit, gives

$$\Delta(0) \approx 2\hbar\omega_D \exp\left(-\frac{1}{|g|N(0)}\right) \quad (20.47)$$

Table 20.3 Different families and layered cuprate high-T_c superconductors

Formula	T_c*(K)range*
$La_{2-x}M_xCuO_4$ (M = Ca, Sr, Ba)	20–40
Ln Ba_2 Cu_3O_7 with Ln = rare earth or Y (123)	90
Ln Ba_2 Cu_4O_8 (124)	90
Bi_2 (Ca, Sr)$_{n+1}$ Cu_nO_{2n+4} (n = 1–3)*	90–110
Tl_2 $Ca_{n-1}Ba_2$ Cu_n O_{2n+4} (n = 1–4)*	80–125
Tl Ca_{n-1} Ba_2 Cu_n O_{2n+3} (n = 2, 3)**	90–117
Tl_{1-x} Pb_x (Ca,Sr)$_{n+1}$ Cu_n O_{2n+3} (n = 2,3)	90–120
Tl Ca_{1-x} Ln_x Sr_2 Cu_2O_{6+x}	70–90
Tl $Sr_{n+1-x}Ln_xCu_nO_{2n+2}$ (n = 1, 2)	40–90
Pb_2Sr_2A Cu_3O_8	40–70
A (Ln or Ln + Sr or Ca)	

If follows that

$$\frac{2\Delta(0)}{k_B T_c} = 3.52 \tag{20.48}$$

is a universal constant. This ratio is in reasonably good agreement with experiment for weak coupling superconductors, like Al, but is too small for strong coupling superconductor, like Pb, high-T_c cuprates, etc.

20.9 HIGH TEMPERATURE SUPERCONDUCTIVITY

In the spring of 1986, K.A. Muller and J.G. Bednorz revealed that at temperatures higher than 30 K the ceramic compounds of lanthanum, barrium and copper oxides go into a superconducting state. In February, 1987, C.W. Chu, et al. synthesized Yttrium compounds which go over into a superconducting states at temperatures T_c close to 90 K. Several months later new superconductors with $T_c > 30$ K were synthesized almost simultaneously at different laboratories in the USA, Japan, the USSR and other countries. These superconductors were called as high-T_c superconductors (HTSC). In Table 20.3, we list a few families of layered cuprate high-T_c superconductors.

The previously known materials go into a superconducting state when they are cooled by liquid helium with a boiling temperature near 4 K. This is an extremely expensive cooler. Production of artificial materials going into a superconducting state at the liquid nitrogen temperature (77 K) or above offers wide use of superconductivity phenomena in science and technology.

The significance of the discovery of new high-T_c superconductors by Muller and Bednorz was emphasized by the fact that the authors were awarded the nobel prize in 1988-two years after discovery. We now present a brief survey of new high-T_c superconductors:

(i) $La_{2-x}M_xCuO_4$(M = Ba, Sr, Ca)

It has quasi two dimensional structure of k_2 NiF_4 Fig. (20.22). In this structure, the transition metal ion (Cu) can interact only in the *ab* plane. The parent compound

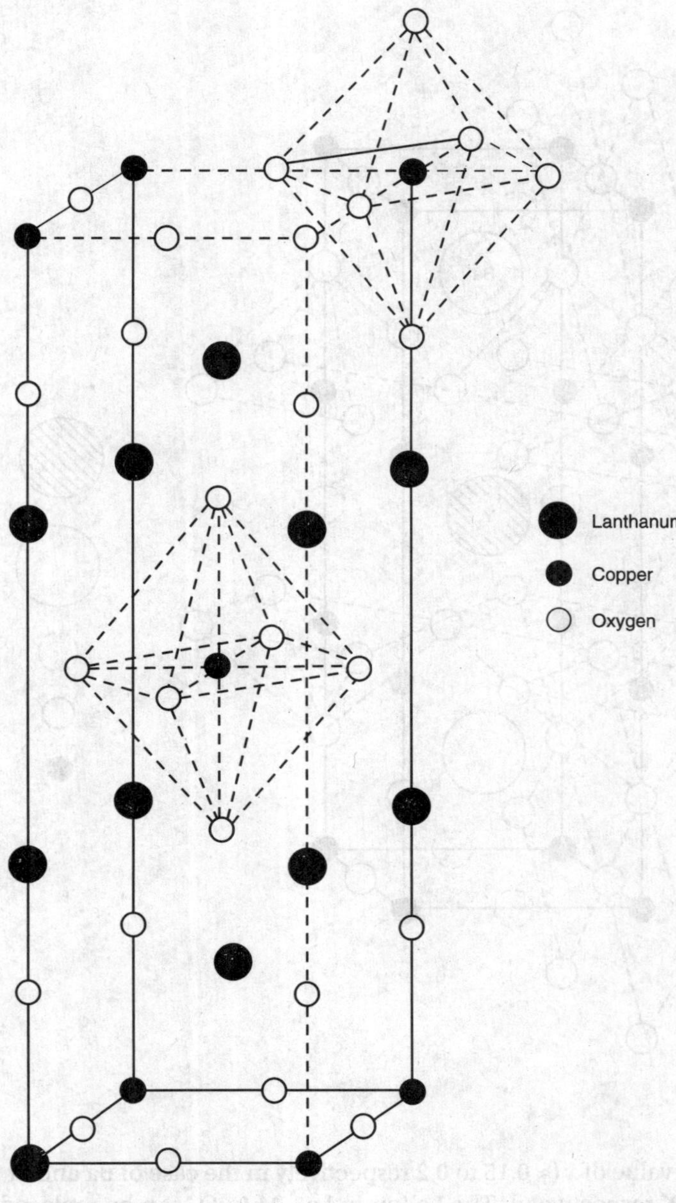

Fig. 20.22

La_2CuO_4 is orthorhombic at room temperature and becomes tetragonal around 500 K. It is insulating and antiferromagnetic with a Neel temperature (T_N) of 290 K. When there is some oxygen excess, La_2CuO_4 itself becomes superconducting around 35K. $La_{2-x}M_xCuO_4$(M = Sr, Ba) has a tetragonal structure at room temperature and becomes orthorhombic around 180 K. It shows the highest T_c around a

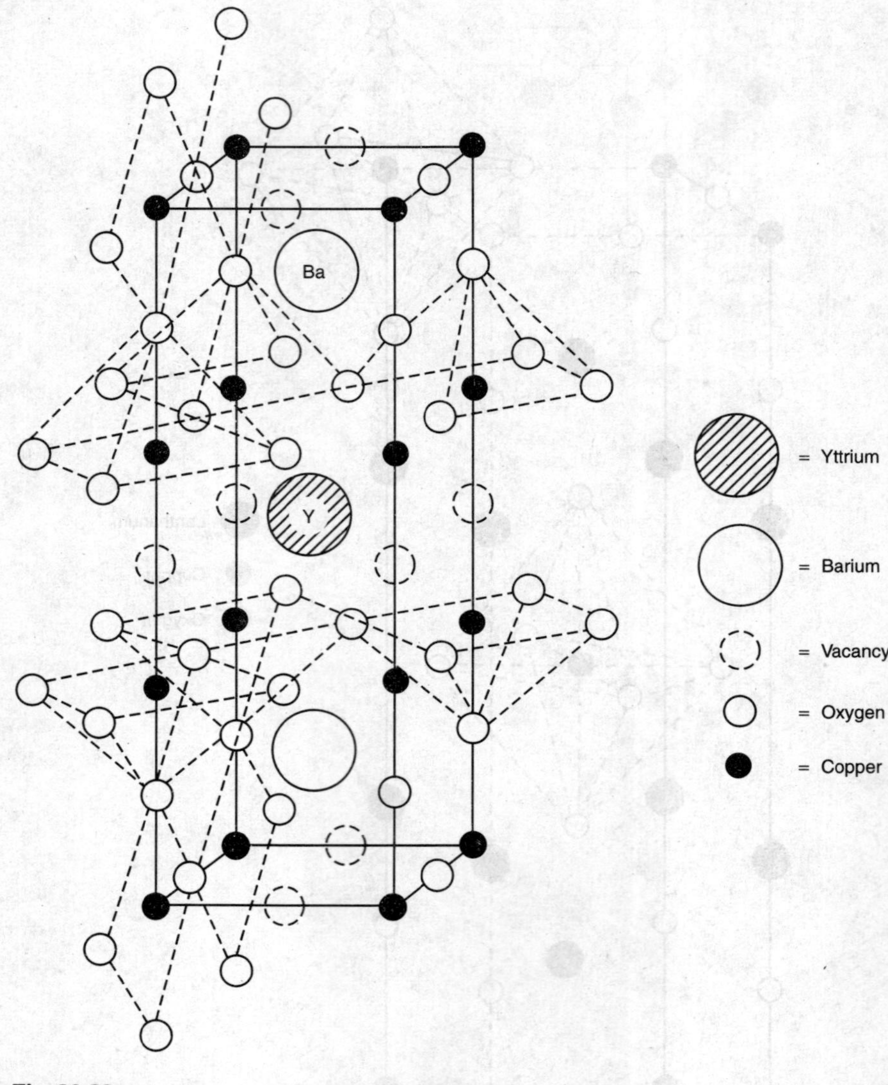

Fig. 20.23

critical value of x (≈ 0.15 to 0.2 respectively in the case of Ba and Sr with T_c's or 30 and 35K respectively). The La ion in $La_{2-x}M_xCuO_4$ can be replaced by Pr, Nd Gd and other rare earth ions up to a point without losing superconductivity.

(ii) Ln $Ba_2Cu_3O_7$ with Ln = rare earth or Y (1 2 3)

The discovery of superconductivity with T_c around 90K in '1 2 3' systems having general formula Ln Ba_2 Cu_3 $O_{6+\delta}$ (x ~0.95, Ln = Y, Nd, Sm, Eu, Gd, Dy, Ho, Er, Tm and Yb with orthorhombic structure Fig. (20.23).

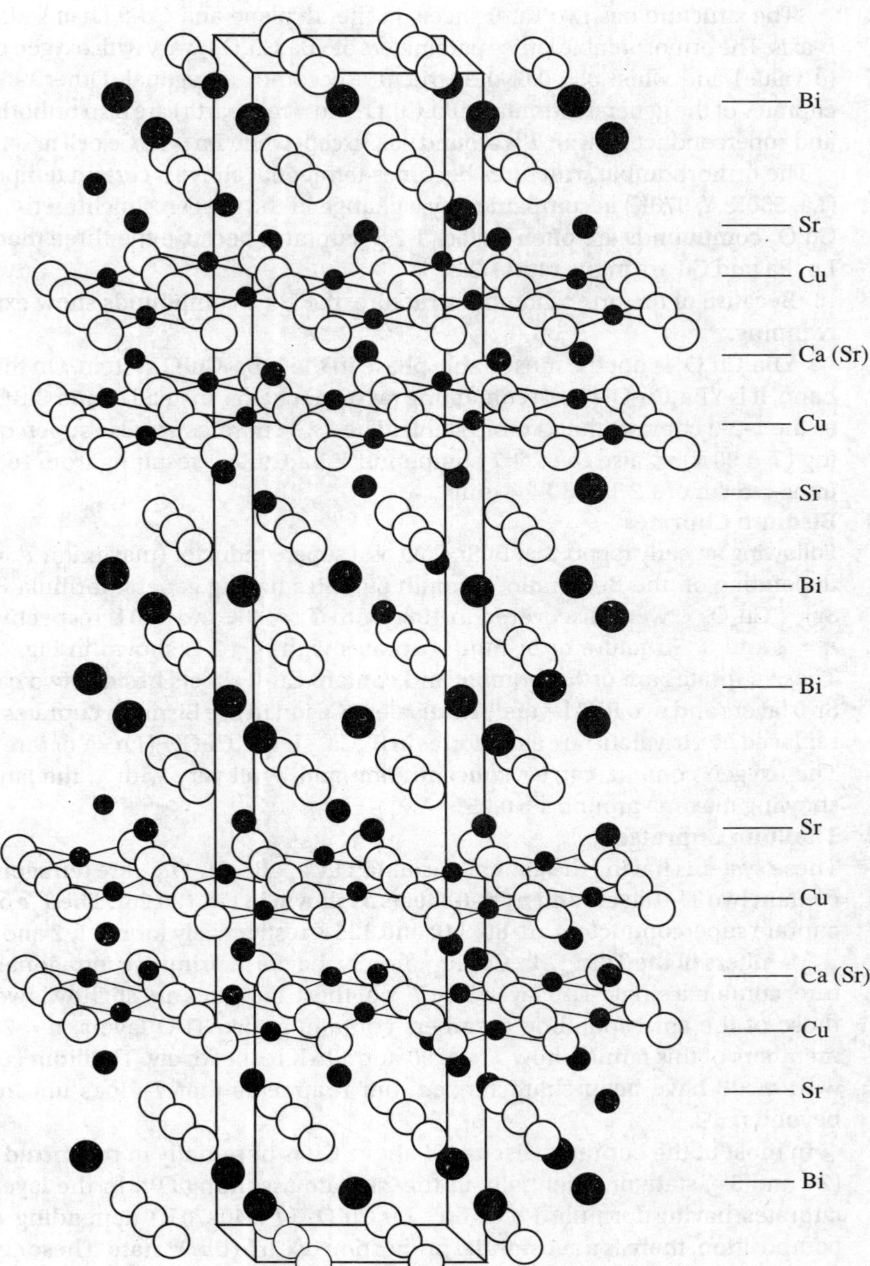

Fig. 20.24

The structure has two Cu-0 sheets in the *ab* plane and Cu-0 chains along the *b*-axis. The orthorhombic lattice parameters of $YBa_2Cu_3O_{7-\delta}$ vary with oxygen content (δ value) and when $\delta = 0.6$, the structure becomes tetragonal. Other rate earth cuprates of the general formula $LnBa_2Cu_3O_7$ (Ln = rare earth) are also orthorhombic and superconducting with T_c's around 90K except when Ln = Pr, Ce or Th.

The orthorhombic structures becomes tetragonal above a certain temperature (La, 590K; Y, 970K) accompanied by a change in the oxygen stoichimetry. Ln Ba_2 Cu_3O_7 compounds are often called '1 2 3' cuprates because the three metal ions Ln, Ba and Cu are in the ratio 1:2:3.

Because of the orthorhombic structure, the 1 2 3 compounds show extensive twinning.

$YBa_2Cu_3O_7$ is not the most stable phase in the Y-Ba-Cu-O system. On the other hand, it is $YBa_2Cu_4O_8$ (1 2 4) containing two CuO chains, instead of one Cu-O chain as the 1-2-3 cuprate, that is more stable. The 1 2 3 cuprates are also superconducting (T = 90 K) as also the 2 4 7 compound $Y_2Ba_4Cu_7O_{15}$, resulting from recurrent inter-growth of 1 2 3 and 1 2 4 units.

(iii) Bismuth Cuprates

Following an early report that $Bi_2Sr_2CuO_6$ is a superconductor (maximum $T_C \approx 22$ K), depending on the Bi/Sr ratio, bismuth cuprates having general formula Bi_2 (Ca, Sr)$_{n+1}$ Cu_nO_{2n+4} were discovered in 1988 with $T_c \approx 90$K and 100 K respectively for $n = 2$ and 3. Structure of bismuth cuprates with $n = 2$ is shown in Fig. (20.25). These cuprates are orthorhombic and contain Cu-0 sheets besides two calcium, Sr-0 layers and two Bi-0 layers. The divalent Ca ion in the Bismuth cuprates can be replaced by a trivalent rare earth ion as in $Bi_2Ca_{1-x}Ln_xSr_2CuO_{8+x}$ (Ln=Y or rare earth). The oxygen content, carrier concentration and T_c all vary with x, the latter two showing maxima around $x = 0.25$.

(iv) Thallium Cuprates

These systems having the general formula $Tl_2Ca_{n-1}Ba_2Cu_nO_{2n+4}$ are tetragonal and contain two Tl-0 layers and n Cu-0 sheets as shown in Fig. (20.26). The T_c's of these cuprate superconductors are 80, 110 and 125 K respectively for $n = 1$, 2 and 3.

Members of the $TlCa_{n-1}Ba_2Cu_nO_{2n+3}$ family, having a primitive tetragonal structure, contain a single Tl-0 layer. The T_c's of these cuprates are slightly lower than those of the corresponding members containing two Tl-O layers, $n = 2$ and 3 members of this family show T_c's of 90 and 115 K respectively. Thallium cuprates with $n > 3$ have been characterized, but it appears that T_c does not increase beyond $n = 3$.

In most of the cuprates discussed above Cu is nominally in the mixed valent (2+ and 3+) state or essentially in the 2+ state assuming O^{2-}. In the layered Pb cuprates having formula $Pb_2Sr_2Ca_{1-x}Ln_xCu_3O_8$ (T_c, 40–70 K) depending on the composition, there is a substantial proportion of Cu in the 1^+ state, These cuprates contain CuO_2 sheets stacked consecutively. The PbO layers joined by O-Cu (I)-O sticks.

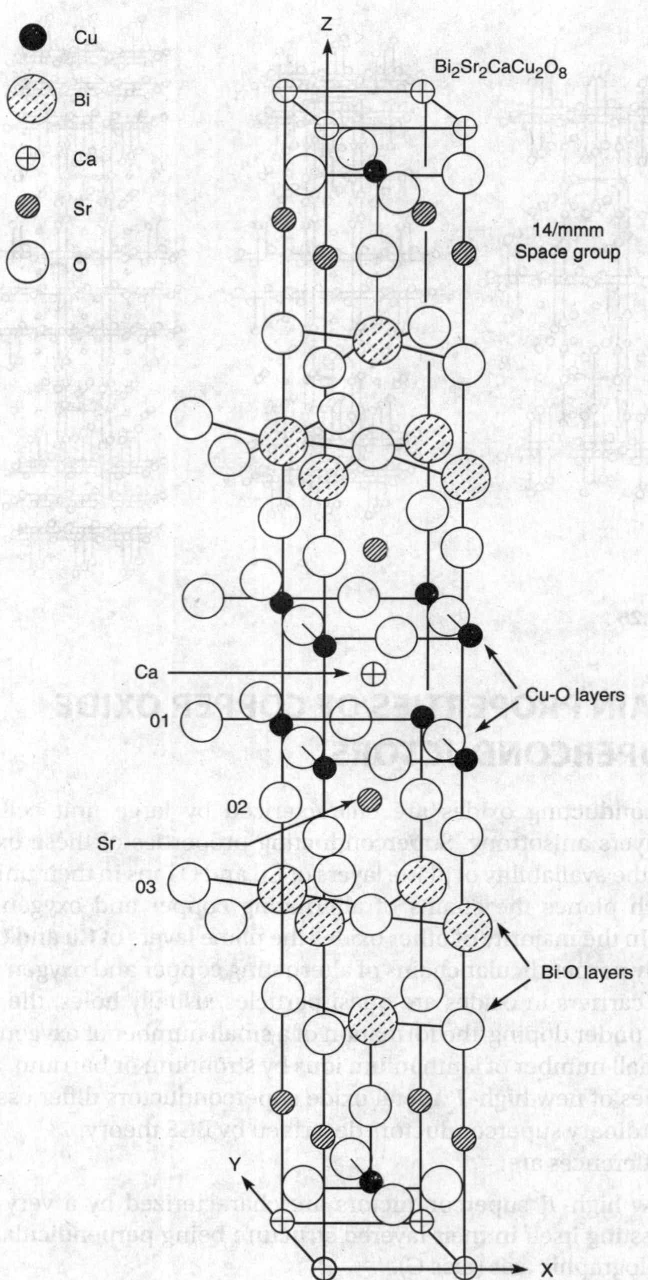
Cu
Bi
Ca
Sr
O
Z
$Bi_2Sr_2CaCu_2O_8$
14/mmm
Space group
Ca
Cu-O layers
01
02
Sr
03
Bi-O layers
Y
X

Fig. 20.25

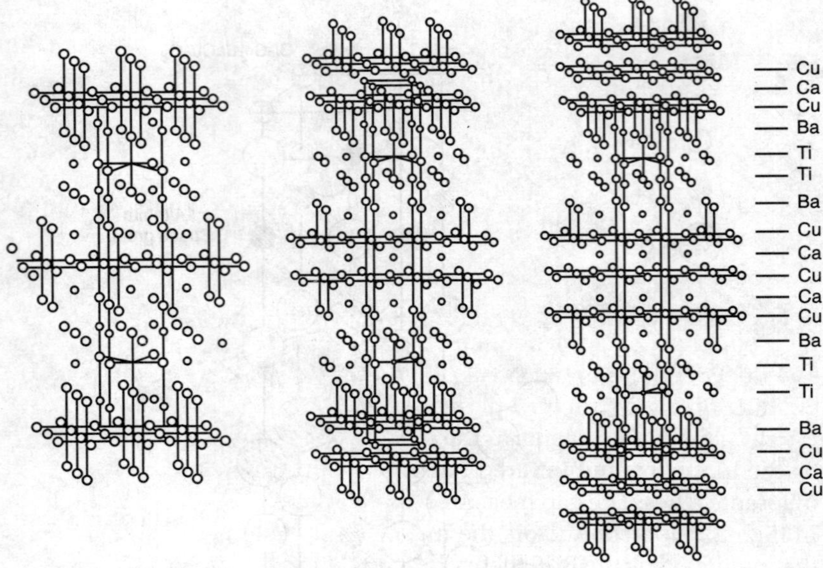

Fig. 20.26

20.10 MAIN PROPERTIES OF COPPER OXIDE SUPERCONDUCTORS

All the superconducting oxides are characterized by large unit cell volumes and a pronounced layers anisotropy. Superconducting properties of these oxides are caused, apparently, by the availability of plane layers of Cu and O ions in their unit cells. In yttrium crystals in such planes the chains of alternating copper and oxygen ions are clearly distinguished. In the majority of other oxides the plane layers of Cu and O ions are formed by two mutually perpendicular chains of alternating copper and oxygen ions.

The current carriers in oxides are quasi-particles, usually holes, the complex $[Cu\text{-}O]^+$ which appears under doping-the formation of a small number of oxygen-atom vacancies, or replacing small number of lanthanium ions by strontium or barrium, and so on.

The properties of new high-T_c metal oxide superconductors differ essentially from the properties of ordinary superconductors described by BCS theory.

The main differences are:

(i) All new high-T_c superconductors are characterized by a very large anisotropy manifesting itself in their layered structure being perpendicular to the principal crystallographic axis (axis C).

(ii) Large anisotropy becomes evident in the appreciable difference of coherence length for the direction along the c-axis and direction perpendicular to it. The largest anisotropy is found in the bismuth superconductor where the coherence length is ~2 Å along the c-axis, and ~40 Å in the transverse direction.

(iii) In ordinary superconductors the coherence lengths is ~10^{-4} cm. In oxide it is thousands of times smaller and equal to ~10^{-40} Å. The small value of the coherence length characterising the spatial stretching of the wave function of a Cooper pair points towards a strong coupling of quasi particles in the pair. Such a coupling results in high critical temperature.

(iv) The dependence of T_c on the concentration of charge carriers has a monotonic character. The maximum value of T_c is attained at a relative small density of charge carriers equal to $\sim 10^{15}\,m^{-3}$. In ordinary superconductors the temperature T_c rises monotonically with rising concentration. At small concentration of charge carriers their Fermi energy in the conduction band of the new superconductors has a small value about 0.4 eV.

(v) Due to the large coherence length ($\sim 10^{-6}$ m) in classical superconductors only a 10^{-4th} part of the electrons placed near the Fermi surface participate in forming Cooper pairs. In the new superconductors the space distribution of the wave function Cooper pairs is small, therefore many charge carriers involved in the conduction band structure, participate in their formation.

(vi) Energy gap measurements in the one particle spectrum of new superconductors by the tunneling and infrared radiation absorption and reflection methods give different values. The gap measured by the tunneling method is larger.

(vii) In high-T_c superconductors the magnitude of isotope effect is much lower than that predicted by the BCS theory based on the electron-phonon interaction.

(viii) In the BCS theory, as was shown by A.A. Abrikosov and L.P. Gorkov in 1960, the critical temperature T_c monotonically fast decreases with increasing concentration ν of magnetic impurities. In high-T_C superconductors the T_c is weakly dependent on ν, if ν is smaller than ν_c and suddenly vanishes at $\nu \geq \nu_c$.

(ix) high-T_c cuprates exhibit **pseudogap** in the normal state. The explanation of the origin of pseudogap is a most challenging problem today.

There are many unusual properties of high-T_c cuprate superconductors in the superconducting state as well as in the normal state ($T > T_c$) that are difficult to understand with the help of BCS theory.

20.11 ELECTRON SUPERCONDUCTORS

Most metals and alloys are generally electron superconductors. In 1989, it was reported that the cuprate $Nd_{2-x}Ce_x\,CuO_4$ is a superconductor ($T_C \sim 30\,K$) with electrons as charge carriers as distinct from the large number of copper oxide high-T_C superconductors described earlier, which are **hole superconductors**. It is not clear that oxygen holes are not present in these cuprates. Hall effect and other measurements have not been conclusive. Other similar systems discovered are

$$Nd_{2-x}\,TH_x\,CuO_4, Pr_{2-x}, TH_x\,CuO_4 \text{ and } Nd_2\,Cu_{4-x}\,F_x$$

20.12 COPPER FREE HIGH-T_C OXIDE SUPERCONDUCTORS

Copper free oxide superconductor $Ba_l - xk_x\,BiO_3$ has been found to be superconducting with a T_c reaching 35 K. This system displays a simple cubic perovskite crystal structure below T_c. This has a special significance from theoretical point of view as all the models for the high-T_c

superconductivity of cuprates make use of the *d*-electronic configuration of copper. Very recently MgB_2 ($T_c = 40$ K) has been reported as copper free highest T_c superconductor.

20.13 ALKALI METAL DOPED C_{60} FULLERENES SUPERCONDUCTORS

Recently high-T_c superconductivity has been observed in C_{60} doped with alkali metals: $T_c = 19.3$ K for K_3C_{60}, $T_c = 30$ K for Rb_3C_{60}, and $T_c = z33$ K for $CsRb_2C_{60}$. There is also a claim that doping with iodine give $T_c = 55$ K. Such high-T_c have not been observed in molecular organic systems before. In fact these molecular systems can join the ranks of inorganic cuprates as high-T_c superconductors.

20.14 IRON-PNICTIDE SUPERCONDUCTORS

Recently several series of new layered iron-pnictide superconductors with T_c about 40–57.3 K has attracted a lot of scientific interest. The discovery of T_c as high as 57.3 K in La-doped iron-based compound $Sm_{0.95}\ La_{0.05}\ O_{0.85}\ F_{0.15}$ Fe As is quite surprising since iron ions in many compounds have magnetic moments and they normally form an ordered magnetic state.

All iron-pnictide superconductors include a two-dimensional (2-D) Fe Pn (Pn: Pnictogen atom) layer with a tetragonal structure at room temperature. Therefore, at a glance their physical properties are considered to be highly two-dimensional, similar to those of cuprates.

20.15 NANOSCALE SUPERCONDUCTIVITY

Nanoscale superconductivity has been studied over the past four decades. It is important to know how the ground state properties change when one or more of the system dimensions are reduced below the characteristic length scales for a bulk superconductor: the coherence length (ξ_0) and the London penetration depth (λ_L). The superconducting properties for confined systems often show dramatic changes from those of the bulk. At much reduced dimensions, new phenomena not seen in bulk superconductors may also be observed. Nanoscale superconductivity has been studied in 2-D thin films, 1-D nanowires and 0-D nanoparticles.

20.16 COLOR SUPERCONDUCTIVITY

Theoretical studies suggest that baryon (proton and neutron are baryons) matter at sufficiently high density and sufficiently low temperature is a color superconductor. In nature, the highest densities of matter are reached in central regions of compact stars. There the density might be as large as $10\,\rho_0$ where $\rho_0 \approx 0.15\ \text{fm}^{-3}$ is the saturation density. It is possible that baryonic matter is deconfined under such condition and, perhaps, it is color superconducting.

20.17 MECHANISMS OF HIGH-T_C SUPERCONDUCTIVITY

The newly discovered high-T_c superconductors have many similar feathers to the low T_c conventional metallic superconductors, but they have also some important difference that weak coupling BCS theory alone cannot explain. On the basis of large number of experiments on these high-T_c systems which have been done in the past few years, one can conclude that the interaction responsible for high-T_c appears to be mediated by an excitation with an energy which is considerably larger than that of the highest energy phonons. The identification of these excitations is the most important and challenging problem today. Within a short time a large number of mechanisms have been suggested so far, some are half baked ideas, some are repetitions of the old models, and a few of doubtful parentage. The novel mechanisms that have been suggested for these high-T_c systems can be classified under two broad categories:

(a) One type of mechanism involves the exchange of electronic excitations, e.g., charge transfer, excitons, plasmons, charged bosons, biexcitons, etc., arising from the charge fluctuations in the system,

(b) The other type of mechanism is guided by the belief that superconductivity in cuprates is of magnetic origin. There are many variants of the magnetic mechanism. The detailed discussion of these mechanisms is beyond the scope of this book.

20.18 CONCLUSIONS

A brief account of superconductivity and recent advances in the field has been presented in this chapter. The currently very active research area of high-T_c superconductivity concentrates on following three major questions:

(i) The explanation of the mechanism by which the Fermi sea condenses into Cooper pairs at temperatures $T \geq 133\,\text{K}$.

(ii) The preparation of larger oxide single crystals and well defined epitaxial layers; and

(iii) A search for new materials with an even higher critical temperature, identically reaching 300 K, i.e., room temperature.

These problems are ofcourse related to one another, and a breakthrough in any may well provide solutions to the other problems.

20.19 TUNNEL EFFECT IN SUPERCONDUCTORS [JOSEPHSON EFFECTS]

Tunneling is a process by which electrons can travel from one metal, through a narrow vaccum or thin insulating barrier, into another metal. It is a quantum mechanical phenomenon. Classically a barrier if impassable if there is insufficient energy available. A ball will not get to the other side of a mountain unless it has sufficient kinetic energy, otherwise it rolls back. Quantum mechanics permits, if the mountain is sufficiently thin, the ball to get to the other side as if there were a tunnel bored through the mountain. Within the quantum world an electron can tunnel through a potential barrier.

Giaver in 1960 demonstrated experimentally the phenomenon of superconducting tunneling. In 1962, B Josephson, postulated that superconducting electron pairs could tunnel through a thin insulating layers sandwitched by two superconducting electrodes. Josephson effects are of two types:

(i) **dc Josephson effect:** This is an intrinsic quantum effect in superconductors. A supercurrent shows through an insulating barrier (or a weak link) without any externally applied voltage. The current is related to the phase difference. The current is related to the phase difference V of the order parameters in the two electrodes.

$$I = I_0 \sin\gamma$$

where I is the current flows through the junction and I_0 is the maximum current which can flow across the junction is I_0.

This surprising prediction (Josephson 1962) was rapidly experimentally confirmed leading to a new field in superconducting electronics. Until the mid 1990s, such superconducting "weak links" were either made of insulating barriers (the so-called Josephson tunnel junctions, at the origin of all the recent advances in *superconducting Qubits*), narrow diffusive metallic or semiconducting wires, or joint contact inserted between bulk electrodes. The first realization of ballistic weak links with a limited number of conduction channels and controlled at the atomic scale came in 1997, with the measurment transport through gold atomic contacts in close vicinity of superconducting aluminium contacts. In the last 5 years, it has been shown that carbon nanostructures can provide another class of weak links of good structural quality, for which strong quantum confinement offers gate tunability.

(ii) **ac Josephson effect:** When dc voltage is applied across the barrier, the supercurrent of Cooper pairs between two superconductors oscillates with a characteristic frequency

$$\omega_0 = \frac{2e}{\hbar} V_0$$

For a constant voltage V_0, the Josephson current is

$$J = J_0 \sin\omega t$$

For 1 μV voltage, the frequency ω_0 is 484 M Hz and the corresponding wavelength is 620. One can visualize the frequency-voltage relation by saying that the transfer of the Cooper pair from one side to the another requires an energy $2eV_0$ which appears as a photon of energy $\hbar\omega_0$.

The interesting point is that for some values of V_0 the Josephson current becomes a dc current, e.g., whenever the voltage obeys the equation

$$2eV_0 = n\hbar\omega$$

For such a voltage, in addition to the tunneling current, there is DC Josephson current.

Within a given circuit the Josephson function behaves as a resistively stunted function with capacitance.

The application of this phenomenon is in SQUIDS: Superconducting Quantum Interference Devices, the most sensitive detectors of magnetic flux.

Josephson junctions are sophisticated sandwich structures of superconducting films (usually of Nb) separated by extremely thin (~10Å) insulating oxide layers.

In a suitable submicroscopic circuit they act as the fastest switching elements available today.

Josephson junctions have also the lowest power consumption at their operating temperature, so they might provide the basis for the architecture of some of the fastest computers of the future. They are already used in the fastest commercially available oscilloscopes that operate at 10 GHz.

Example 20.1: Assuming the temperature dependence of H_c of the form

$$H_c(T)=H_c\left[1-\left(\frac{T}{T_C}\right)^2\right]$$

Calculate $\frac{C_n-C_s}{C_n}$ at T_c for aluminium from the following given data. H_o = 99 Gauss, T_c = 1.18 K, γ (sp. heat constant of the normal state = 1.35 × 10^4 ergs/mole-^{0}K)^2

Solution We have

$$C_n-C_s=\frac{T_c}{4\pi}\left[\left(\frac{dH_c}{dT}\right)^2+H_c\left(\frac{d^2H_c}{dT^2}\right)\right]_{T=T_c}$$

From the given relation

$$H_c(T)=H_c\left[1-\left(\frac{T}{T_c}\right)^2\right]$$

We obtain

$$\frac{dH_c}{dT}=-\frac{2H_0}{T_c^2}T$$

and

$$\frac{d^2H_c}{d^2T^2}=-\frac{2H_0}{T_c^2}$$

Substituting (ii) and (iii) in (i), one obtains

$$C_n-C_s=-\frac{H_0^2}{2\pi T_c}\left[\frac{T}{T_c}-\frac{3T^3}{T_c^3}\right]$$

At

$$T=T_c$$

$$C_n-C_s=H_0^2/\pi T_c$$

or

$$\frac{C_n-C_s}{C_n}=\frac{H_0^2/\pi T_c}{\gamma T_c}=\frac{H_0^2}{\pi T_c^2\gamma}$$

$$=\frac{(99)^2}{3.14\times(1.18)^2\times1.35\times10^4}=0.17.$$

Example 20.2: Calculate the maximum current that can be carried by a niobium wire with a radius of 0.050 mm, at the absolute zero of temperature.

Solution The maximum current is given by $J_{max}=2\pi RH_c$ where R is the wire radius and H_c is the critical field at T = 0 K. Since H_c = 1.58 × 10^5 A/M, we have

$$I_{max}=2\pi\times0.05\times10^{-3}\times1.58\times10^5=50\text{ A}.$$

Example 20.3: A magnetic field $\boldsymbol{H}_a$ is applied parallel to the axis of a long cylindrical superconducting sample. Find the surface density of supercurrent?

Solution Supercurrent flows around the circumference of the cylinder and produces a magnetic field (induction field) in the interior, if $\boldsymbol{K}$ is the supercurrent per unit length of cylinder, the field it produces is given by $\boldsymbol{B}_s = \mu_0\boldsymbol{K}$. $\boldsymbol{B}_s$ exactly cancels the applied induction field $\mu_0\boldsymbol{H}_a$ so $\boldsymbol{K} = \boldsymbol{H}_a$ when directions are taken into account the result can be written $\boldsymbol{K} = \boldsymbol{n} \times \boldsymbol{H}_a$, where $\boldsymbol{n}$ is the unit outward normal to the cylinder.

Example 20.4: Use the free electron medel to estimate the gap parameter Δ_0 for a superconducting sample with 5×10^{28} electrons/m³ and a critical field of 7×10^{28} A/m.

Solution

The Fermi energy,
$$E_F = \frac{\hbar^2}{2m}(3\pi^2 n)^{2/3}$$
$$= \frac{(1.5\times10^{-34})^2}{2\times9.11\times10^{-31}}\times(3\pi^2\times5\times10^{28})^{2/3}$$
$$= 7.86\times10^{-19}$$

Using $\rho = 3nT_c/4E_F$ and $H_c^2(0) = \dfrac{N(E_F)}{\mu_0 T_c}\Delta_0^2$

We obtain

$$\Delta_0 = \left(\frac{4\mu_0 E_F}{3n}\right)^{1/2} H_c\left(\frac{4\times4\pi\times10^{-7}\times7.86\times10^{-19}}{3\times5\times10^{28}}\right)\times7\times10^3$$
$$= 3.59\times10^{-23}\ J = 2.25\times10^{-4}\ \text{eV}.$$

Example 20.5: The critical fields for a superconducting are 1.4×10^5 and 4.2×10^5 A/m at 14 K and 13 K respectively. Show that the superconducting transition temperature and the critical field at 0 K are 14.47 K and 21.7 A/m respectively. [Kerala]

Solution We have

$$H_{c1} = H_c(0)\left[1-\left(\frac{T_1}{T_c}\right)^2\right]$$

and
$$H_{c2} = H_c(0)\left[1-\left(\frac{T_2}{T_c}\right)^2\right]$$

or
$$\frac{H_{c1}}{H_{c2}} = \frac{T_c^2 - Ti^2}{T_c^2 - T_1^2}$$

or
$$\frac{1.4\times10^5}{4.2\times10^5} = \frac{T_c^2-(14)^2}{T_c^2-(20.13)^2}$$

or
$$2T_c^2 = 3(14)^2-(20.13)^2$$

or
$$T_c = 14.47\ \text{K}$$

Now,
$$H_c(0) = \frac{H_c T_c^2}{T_c^2 - T_1^2} = \frac{1.4\times10^5\times(14.47)^2}{(14.47)^2-(14)^2} = 21.7\ \text{A/m}.$$

Example 20.6: **The London penetration depth for a superconductor at 3 K and 7.1 K are 39.6 nm and 173 nm respectively. Show that the superconducting transition temperature is 7.19 K.** [UJJAIN]

Solution We have

$$\frac{\wedge_d(T)}{\wedge_d(0)} = \left[1 - \left(\frac{T}{T_c}\right)^4\right]^{-1/2}$$

or $$\left(\frac{\wedge_d(T_1)}{\wedge_d(T_{c2})}\right)^2 = \frac{T_c^4 - T_2^4}{T_c^4 - T_1^4}$$

or $$\frac{(39.6)^2}{(173)^2} = \frac{T_c^4 - (7.1)^4}{T_c^4 - (20.6)^4}$$

or $$T_c = 7.19\,\text{K}.$$

Example 20.7: **The superconducti·g transition and Debye temperature for lead are 7.22 K and 96 K respectively. Using BCS equation show that $N(0)|g|$ and $\Delta(0)$ are 0.3678 and 10.97 × 10⁻⁴ eV respectively.** [Raj]

Solution We have

$$T_c = 1.14\theta_D \exp\left[-\frac{1}{N(0)|g|}\right]$$

or $$\frac{1}{N(0)|g|} = \log_e \frac{1.14\theta_D}{T_c} = \log_e \frac{1.14\times 96}{7.22} = 2.178$$

or $$N(0)|g| = 0.3678$$

Now, $$\Delta(0) = \frac{k_B\theta_D}{\sin h\left[\frac{1}{N(0)|g|}\right]} = \frac{1.38\times10^{-23}\times 96}{7.546\times1.6\times10^{-19}}\,\text{eV} = 10.97\times10^{-4}\,\text{eV}.$$

Example 20.8: **Estimate the energy gap E_g for niobium at T=0 and show that the minimum photon wavelength needed to break the cooper pair is 4.39 × 10⁻¹⁰ m.**

Solution We have

$$E_g(T=0) \approx 3.54\,k_B T_c$$
$$\approx 3.54\,(1.34\times10^{-23}\,\text{J/K})(9.25\text{K})$$
$$= 4.52\times10^{-22}\,\text{J}$$

Photon wavelength

$$\lambda = \frac{hc}{E_g} = 4.39\times10^{-4}\,\text{m}.$$

We can see the value of λ corresponds to far infra-red region of the electromagnetic spectrum. Photons with this wavelength or lower value have sufficient energy to break the Cooper pair in niobium. We may note that this estimate of E_g is close to the experimental value of 3.05 MeV.

20.20 APPLICATIONS OF SUPERCONDUCTIVITY

(i) Large scale applications

From the practical point of view the useful aspect of connentional type-II superconductors (predominantly Nb compounds) is their capacity to carry high transport currents with acceptably low energy dissipation. For e.g., one could construct a 10-Tesla superconducting magnet using copper wire but such a magnet would dissipate 2 MW of power. Therefore, one mostly uses Nb-compounds to construct commercial magnets. Wires made with these materials remain superconducting even at 4.2 K even in magnetic field considerably higher than the required performance peak of the magnet. Furthermore, for equal power and the uppermost field, the superconducting magnet is much smaller than its normal counterpart made out of copper wires.

Powerful superconducting magnets, super magnets are used in medicine for diagnostics, for e.g., in nuclear spin tomogaphs for energy storage, for magnetic levitation (trains), and in research laboratories, particularly in high-energy particle accelerators for fundamental physics research.

(ii) Electronics Applications

These are mainly thin film applications in ultra-fast micro-electronics or instrumentation.

Thin film applications in electronics together with superconducting magnets, are usually considered the most important area of superconducting-based technology. Most of these applications are based on Josephson effect, a quantum phenomenon that enables the construction of the fastest nanoscopic switches, Josephson junctions, and related device structures, SQUIDs; Superconducting Quantum Interference Devices.

Josephson junctions are sophisticated sandwich structures of superconducting films (usually of Nb) separated by extremely thin (~10Å) insulating oxide layers. In a suitable submicroscopic circuit they act as the fastest switching elements available today.

Josephson junctions have also the lowest power consumption at their operating temperature, so they might provide the basis for the architecture of some of the fastest computers of the future. They are already used in the fastest commercially available oscilloscopes that operate at 10 GHz.

As the name suggests, SQUIDs explore quantum interference effects: an analysis of superconducting loop shows that the magnetic flux that can thread the loop is quantized in quantum units of flux which have a value of 2×10^{-15} Weber. SQUIDs are suitably processed superconducting loops which detect minute changes in magnetic flux, i.e., they are high-sensitivity magnetic flux detectors that can be used in the finest precision instruments at the forefront of metrology.

There is a considerable research effort in superconductivity going on around the world; improved materials, technologies and applications are publicized every month. Thin films of HTSC oxides may have some impact in passive microwave devices already in the near future, large scale applications of HTSC ceramics, like cables or magnets, will most likely take longer to develop commercially. Presently, most efforts are focussed on three applications areas of HTSC oxides film:

(a) High quality epitaxial thin films of low surface impendance for passive microwave components like resonators, antennas, etc.

(b) Advanced epitaxial hetero-structures for novel superconductor-superconductor hybrid devices or chip interconnects.

(c) Superconductors insulator-superconductor (SIS) and superconductor-normal metal, superconductor (SNS) sandwich structures for high-T_c tunnel junctions and weak link devices.

In large scale technology, one should expect the continuation of the conventional Nb compound technology in wires, cables and magnets. HTSC oxides will most likely also gradually emerge in such applications.

The rapid progress in the field of superconductivity leads one to believe that applications of superconductors is limited only by one's imagination and time.

The Table 20.4 shows the potential applications of superconductors.

Table 20.4(a) Large-Scale Applications of Superconductivity

Application	***Major Technical Features***
Power Cables	higher current densities, smaller cable diameters, lower transmission losses
Current Limiters	highly non-linear super-normal conductor transition, self controlled current limitation
Transformers	higher current densities, smaller size, lower weight, lower losses
Motors/Generators	higher current densities, higher magnetic fields, smaller size, low weight & losses
Magnets for RTD, Magnetic Energy Storage, Magnetic Separation, NMR Spectroscopy, MRI, Magnetic Levitation	higher current densities, higher and ultra-higher magnetic fields, higher magnetic field gradients, smaller size, lower weight, lower losses, persistent currents, ultra-high temporal field stabilities, stronger levitation forces, larger air gaps
Cavities for Accelerators	lower surface resistance, higher quality factors, higher microwave-power handling
Magnetic Bearings (based on HTSC bulk materials)	higher current densities, lower losses, stronger levitation forces, self-controlled autostable levitation

Table 20.4(b) Electronics Applications of Superconductivity

Application	*Major Technical Features*
High Frequency Sensor Coils for NMR	lower resistive losses, higher quality factors, smaller size
Microwave Filters for Wireless Communication	lower surface resistance, smaller size, lower transmission losses, higher quality factors
Resonator for Oscillators & other passive microwave devices	lower resistive losses, higher quality factors, lower transmission losses, smaller size
Far-Infrared Bolometers	highly non-linear super-normal conductor transition, higher irradiation-mediated temperature sensitivities
Microwave Detectors	highly non-linear junction characteristic, higher conversion efficiency for frequency-mixing
X-Ray Detectors	lower particle excitation energies, higher photon energy resolution
SQUID Sensors for RTD, Medical Diagnosis and non Destructive Testing SQUID Amplifiers	persistent currents, quantum interference effects, ultra-high magnetic field sensitivities, low-noise low-signal amplification
Voltage Standards for Metrology & Industry	voltage steps in microwave irradiated junction arrays, quantum precision output voltages
Digital Circuit & Microprocessors	persistent current, single flux quantum signal levels, ultra-fast ultra-low power data transfer & processing

SUMMARY

1. Superconductors are the materials that exhibit the complete loss of d.c. resistance and expulsion of magnetic fields below their transition temperature (T_c). Above T_c these materials behave as normal metals.
2. BCS weak coupling microscopic theory is used to explain the behaviour of most superconductors.
3. The electronic properties of superconductors can be understood by assuming that an energy gap 2Δ exists in the electronic spectrum at the Fermi energy.
4. Superconducting metals are usually poor thermal conductors.
5. There is a critical field $H_c(T)$, which depends on temperature. The superconducting material is in the normal state when $H > H_c(T)$. In type-I superconductors, the magnetic induction B vanishes in the bulk of the superconductor for $H < H_c(T)$. In a type-II superconductor, the magnetic field starts penetrating the sample at an applied field H_{c_1} (lower critical field) lower than H_c. The Meissner effect is incomplete yet until at H_{c_2} (upper critical field). Between H_{c_2} and H_{c_1} flux penetrates the superconductor giving a *mixed state* consisting of superconductor penetrated by threads

of the material in its normal state or flux lines, Abrikosov showed that the mixed state consists of vertices each carrying a single flux $\phi = \dfrac{hc}{2e}$. These vertices are arranged in a regular two-dimensional array.

6. The specific heat shows a jump at T_c and decays exponentially with an energy Δ of the order of $k_B T_c$ as $\exp(-\Delta/k_B T_c)$ below T_c. There is a second order phase transition (constant entropy, constant volume, no latent heat) with discontinuity in the specific heat.
7. Different elements and compounds exhibiting superconductivity have different values of T_c. The highest known T_c is 134 K. The search for a room temperature superconductor has driven a great deal of research.
8. The search for room temperature or superconductors with higher T_c values continues today, with the aim of using them in a number of applications, e.g. low loss energy generation and transmission, magnetic levitation systems, superconducting electronic devices, high field magnets, etc. Superconductors are already used in research, industry, and medicine, and the extension of superconductors into other areas is an exciting possibility for the future.
9. When one superconductor is separated from another superconductor by a thin layer of insulator, the superconductor – insulator – superconductor junction is called a *Josephson junction* or *a weak link* and Cooper pair can tunnel across the thin insulator from one superconductor to the other superconductor. The effects of pair tunneling include: (i) DC Josephson effect, (ii) AC Josephson effect, and (iii) Macroscopic long range quantum interference.
10. A superconducting ring containing one or two Josephson junctions can be used a very sensitive flux meter, called as SQUID.

REVIEW QUESTIONS

1. Give an account of the experimental results which distinguish the superconducting state from the normal state of the matter.
2. Describe Meissner effect and distinguish between type I and type II superconductors.
3. Summarize the important experimental results for the phenomenon of superconductivity.
4. Explain the formation of a Cooper pair in a superconductivity.
5. Explain the following:
   (i) Persistent current (ii) Meissner effect
   (iii) Isotope effect (iv) Copper pairs
6. What are high temperature superconductors. Give few examples. How these high-T_c superconductors differ from normal superconductors.
7. Write short notes on the following:
   (a) Meissner Effect (b) Superconductivity
   (c) Cooper pair (d) High-T_c superconductors
   (e) London Equations (f) Pippard non-local theory
   (g) BCS theory of superconductivity (h) Josephson effect superconductors

PROBLEMS

1. Using the fact that the superconducting state displays perfect diamagnetism ($\boldsymbol{B}=0$) while the normal state has negligible diamagnetism ($\boldsymbol{M}=0$). Show that the entropy changes across the phase boundary by an amount

$$S_n - S_s = -\frac{v}{\mu_0}\boldsymbol{H}_c\frac{d\boldsymbol{H}_c}{dT}.$$

2. The heat capacity can be calculated from $C=-T\left(\frac{\partial^2 E}{\partial T^2}\right)$. Where E is the free energy,

(a) Derive an expression for the discontinuity in the heat capacity at $T=T_c$ in terms of the critical field. In particular show that

$$C_s - C_n = \mu_0 T_s T_c\left[\frac{\partial \boldsymbol{H}_c}{\partial T}\right]^2$$

at

$$T=T_c$$

[**Hint:** (a) $\boldsymbol{H}_c$ is linear in T for temperatures near T_c].

3. According to the BCS theory the temperature dependence of the gap Δ (T) parameter for a weak coupling superconductor can be found by solving

$$\frac{\Delta}{\Delta_0}=\tan h\left(\frac{T_c\Delta}{T\Delta_0}\right)$$

Let $\propto = T_c\Delta/T\Delta_0$ and $\tan h\,\alpha$ as a function of α. Use the plot to show that (a) $\Delta\to\Delta_0$ as $T\to 0\text{K}$ and (b) $\Delta=0$ for $T\geq T_c$ (c) use a expansion of $\tanh\alpha << 1$ to show that

$$\frac{\Delta}{\Delta_0}=\sqrt{3}\left[1-\frac{T}{T_c}\right]^{1/2}\left[\frac{T}{T_c}\right]$$

for T near T_c.

4. Using weak coupling BCS microscopic theory, calculate $N(0)g$ for Hg from

$$T_c = 4.16\,\text{K}, \theta = 70\,\text{K}.$$

5. Estimate the angle θ for Al at which B_c-T graph meets the t-axis. Given $T_c = 1.17\,\text{K}$, $B_0 = 0.99\,\text{Wb/m}^2$ **[Ans. $\theta = 149^{\circ}$]**

SHORT ANSWER-QUESTIONS

1. What are the two distinctive hallmarks of superconductivity?
Ans: (i) Zero resistivity, i.e., infinite conductivity below some critical temperature T_c and (ii) Meissner effect, i.e., $\boldsymbol{B}=0$ inside the superconductor.
2. When a superconducting state ceases to exist?
Ans: When temperature, magnetic field, or current density exceeds the critical value.
3. What are type-I and type-II superconductors?
Ans: Materials that always expel the flux before becoming normal are called type-I superconductors. All pure superconducting metals are type-I except for niobium and

vanadium. For type-II there are two critical fields: a lower one H_{c_1} and the upper one H_{c_2}. The flux is completely expelled only up to the field H_{c_1}. Between H_{c_1} and H_{c_2}, the superconductor is said to be in the mixed state. The Meissner effect is only partial. Above H_{c_2} the material returns to the normal state.

4. What are the characteristics of HTSC oxides are?

Ans: These are ionic metals with highly anisotropic, layered structures. These are quasi two dimensional insulators with an unusal normal state. Superconductivity with $T_c = 134\,K$ has been achieved. These are extreme-type-II superconductors with very short coherence lengths (ξ-10 Å) and large penetration depths ($\lambda \sim 3000$ Å). These allow inexpensive liquid nitrogen to be used as a coolant and therefore have a far reaching technological implications.

5. Mention some of the most important applications of superconductors.

Ans: The most important applications of superconductors are either in high current wires for powerful supermagnets or in thin film structures like Josephson Junctions for superconducting electronics. SQUIDs have been used to develop some of the most sensitive measurement instruments at the forefront of present technology.

6. What is the range of the critical current in Josephson Junction?

Ans: 100 μA-1 mA.

7. Does Pippard Coherence length depends on mean free path?

Ans. Yes.

8. Nb_3 Sn and Nb_3 Ge superconductors are Type-I or Type-II superconductors?

Ans: Type-II.

9. What is the attractive force that gives rise to a lowering of the ground state energy by the small amount Δ?

Ans. Phonon mediated electron-electron indirect interaction.

10. How it is possible to get a large number of electrons in to this new ground state (Q.9), given that electrons are fermions?

Ans: Formation of Cooper pairs. A Cooper pair acts as a single boson, even if the two electrons making up the pair may be many lattice sites away from each-other. We can then have a Bose-Einstein condensation, in which the Cooper pairs are the bosons. The Cooper pair all fall into same state described by a single wave function.

11. What is superconductivity?

Ans: Superconductivity is a macroscopic quantum effect that involves not only an absence of resistance, but also some special magnetic phenomena, including the existence of quantum of magnetic flux.

OBJECTIVE QUESTIONS

1. The temperature at which a metal becomes superconductor is called

(a) Curie temperature (b) Debye temperature
(c) Critical temperature (d) Neel temperature

2. The transition temperatures of most superconducting elements falls within the range

(a) Zero to 143 K (b) 0 to 10 K

(c) Zero to 23 K (d) 0 to 50 K

3. The relation between T_c and H_c for a superconductor is

(a) $H_c = H_0\left[1-\left(\frac{T}{T_c}\right)^2\right]$ (b) $H_c = H_0[1+T_c^2]$

(c) $H_c = H_0[(T-T_c)^2]$ (d) $H_c = \frac{H_0}{(1-T/T_c)^2}$

4. A type-I superconducting material when placed in a magnetic field will

(a) Expel all the magnetic lines of forces passing through it

(b) Attract the magnetic field toward its centre

(c) Not influence the magnetic field

(d) None of the above

5. A superconducting material on being subjected to critical magnetic field (H_c)

(a) Changes to normal state

(b) Remain unaffected

(c) Exhibit both normal as well as superconducting state properties

(d) None of the above

6. The width of the energy gap of a superconducting is maximum at

(a) T_c K (b) 0 K

(c) $(T_c/2)$ K (d) $(T_c/3)$ K

7. In superconductivity state

(a) Entropy decreases and thermal conductivity increases

(b) Entropy increases and thermal conductivity decreases

(c) Entropy and thermal conductivity both deceases

(d) Entropy and thermal conductivity both increases

8. A superconductor has

(a) a negative susceptibility

(b) a positive susceptibility

(c) Sometimes negative and sometimes positive susceptibility

(d) none of the above

9. Critical magnetic field of a superconductor

(a) increases with rise in temperature

(b) decreases with rise in temperature

(c) does not depend on superconducting transition temperature

(d) increases with decrease in temperature

10. Electron density of superconductor
(a) is infinite at absolute zero
(b) is zero at absolute zero
(c) is finite at absolute zero
(d) increases with the rise in temperature from absolute zero to transition temperature

11. The isotope effect coefficient for a superconductor
(a) is zero
(b) is generally greater than zero
(c) is generally in the range of 0.2 to 0.6
(d) is generally much greater than 1

12. Which of the following is a multiply connected superconductor?
(a) a ring
(b) a solid sphere
(c) a solid cylinder
(d) a solid rod

13. A cooper pair is a system of two electrons bound by exchange.
(a) of a photon between them
(b) of a phonon between them
(c) of a proton between them
(d) of a neutron between them

ANSWERS

Question numbers	1	2	3	4	5	6	7	8	9	10	11	12	13
Answers	(c)	(b)	(a)	(a)	(a)	(b)	(b)	(a)	(d)	(c)	(c)	(a)	(b)

FILL IN THE BLANKS

1. Superconductors are materials that behave as _____ metals at high temperatures ($T > T_c$ (critical temperature)) and below T_c their dc resistivity_____. **[Normal, vanishes]**
2. Superconductors are perfect _____; by this we mean that any magnetic field that is present in the bulk of the sample when $T > T_c$ is expelled when T is lowered through T_c (transition temperature). This is called the _____. **[diamagnet, Meissner effect]**
3. The electronic properties of superconductors can be understood by assuming that an _____ 2Δ exists in the electronic spectrum at _____ energy. **[energy gap, Fermi]**
4. Superconducting materials are usually_____ thermal conductors. **[poor]**
5. There is critical magnetic field (H_c) of a superconductor, which depends on temperature. When $H > H_c(T)$, the material is in the _____ state; when $H < H_c(T)$, the material is in_____ state. **[normal, superconducting]**
6. In a type-II superconductor, the magnetic field starts to penetrate the sample at an applied field_____ lower than H_c. The Meissner effect is _____ yet until at H_{c_2}. **[H_{c_1}, incomplete]**
7. Between H_{c_2} and H_{c_1} flux penetrates the superconductor giving a _____ state consisting of superconductor penetrated by threads of the material in its normal state or_____ lines. **[mixed, flux]**

8. Abrikosov showed that the mixed state consists of _____ each carrying a single flux $\phi = hc/2e$. These vortices are arranged in regular two-dimensional _____. **[vortices, array]**
9. The specific heat of superconductors shows a jump at _____ and decays _____ with an energy Δ of the order of k_BT_c as _____ below T_c. **[T_c, exponentially, $e^{-\Delta/k_BT_c}$]**
10. There is _____ order phase transition with _____ in the specific heat of super-conductors. **[second, discontinuity]**
11. The critical magnetic field (H_c) is found to be dependent on temperature and the variation is given by _____. **[$H_c(T) = H_c(0)\left[1-\left(\frac{T}{T_c}\right)^2\right]$]**
12. The specific heat for the superconducting state (C_s) at $T = T_c$ is _____ times the value of specific heat for the normal state (C_N). **[three]**
13. According to BCS theory two electrons interact via _____ vibrations. The electron-lattice interaction leads to _____ interaction. As a result electrons form _____ pairs in the superconducting state. **[lattice, electron-electron, bound]**
14. Below T_c, the energy gap parameter of a superconductor rises steeply from zero in accordance with _____. **[$2\Delta(T) \cong 3.52\, k_B\, T_c \left(1-\frac{T}{T_c}\right)^{1/2}$]**
15. The energy of pairs in a superconductor is lower than the total energy of free electrons. As a result, a _____ is formed near the Fermi energy, which is approximately given by _____. **[gap, $2\Delta = 3.52\, k_BT_c$]**
16. The expression for critical temperature of a superconductor is _____ **[$k_BT_c = 1.14\hbar\omega_D\left[-\left(\frac{1}{N(0)V}\right)\right]$]**

    (where $N(0) \rightarrow$ density of states and V is attractive interaction).
17. $\frac{2\Delta_0}{k_BT_c}$ is a _____ constant, and its value for weak coupling superconductors is _____. **[universal, 3.52]**
18. A DC current flows across the junction in the absence of any electric or magnetic field is called _____. **[DC Josephson effect]**
19. A DC voltage applied across the junction causes r.f. current oscillations across the junction is called _____. **[AC Josephson effect]**
20. A superconducting ring containing either one or two Josephson junctions can be used as a very sensitive flux meter. Such a device is known as _____. **[SQUID (Superconducting Quantum Interference Device)]**

21

Semiconductors and Semiconductor Devices

21.1 INTRODUCTION

The solid materials are classified into three groups mainly from the consideration of their current carrying capabilities: **conductors, semiconductors, and insulators**: conductors have an abundance of free electrons that act as charge carries, which means that they have high conductivity. An insulator, on the other hand, has hardly any free electrons and offers very low level of conductivity, i.e. a very poor conductor of electricity, A semiconductor is a material that has a conductivity level somewhere between that of a good conductor and that of an insulator. The term resistivity (inverse of conductivity) is often used when comparing the resistance levels of different materials. Typical resistivity values for three broad categories of materials are listed in Table (21.1).

It is observed that at 20°C, resistivity of semiconductors lies between that of metals and insulators. However, at a very low temperature (~ absolute zero), a semiconductor behaves as an insulator. We may note that semiconductors in very pure form (intrinsic) have negative temperature coefficient of resistance, i.e. their resistivity decreases (or conductivity increases) rapidly with the increase of temperature. This means that the number of charge carriers in pure semiconductors increases rapidly with temperature. On the contrary, conductors have positive temperature coefficient of resistance. We may further note that semiconductors have the special feature of having two types, of charge carriers: **electrons** and **holes**. "Holes" are absence of electrons in covalent bonds and act equivalently as positive charge carriers. Semiconductors also have an unique property, i.e., their conductivity increases by several orders of magnitude even by minute amount of doping with certain impurities. Such type of semiconducting materials are called **extrinsic semiconductors**. This property of semiconductors is exploited to develop electronic devices. Silicon (Si), Germanium (Ge) and Gallium Assenide (Ga As) semiconductors have received the broadest range of interest in the development of semiconductor devices. Moreover, the characteristics of semiconductors can be significantly altered through the application of light or heat. This has opened the applications of semiconducting materials as light and heat sensitive devices. Semiconductors constitute a large class of substances which have resistivities lying between those of insulators and conductors. The resistivity

Table 21.1 Typical Electrical resistivity values of few selected different materials at 20°C (Ω–m)

Conductor	***Semiconductor***	***Insulator***
Copper (Cu) ~ 1.7×10^{-8}	Silicon ~ 50×10^{1}	Mica ~ 7×10^{10}
Silver (Ag) ~ 1.6×10^{-8}	Germanium ~ 50×10^{-2}	Glass ~ $10^{6} - 10^{7}$
Aluminium (Al) ~ 2.8×10^{-8}	Indium-Antimonide ~ 200×10^{-2} (In Sb)	Diamond ~ 10^{10}
Iron (Fe) ~ 10×10^{-8}		
Nichrome (NiCr) ~ 100×10^{-6}		

of semiconductors varies within wide limits, i.e., 10^{-4} to 10^{4} Ω-m and is reduced to a very great extent with an increase in temperature (according to an exponential law) as shown in Fig. (21.1).

In Mendeleev's periodic table, semiconductors form the group IV of elements. The most typical and extensively employed semiconductors whose electrical properties have been well investigated, are Germanium (Ge), Silicon (Si) and Tellurium (Te). The study of their electrical properties reveals that

(i) Semiconductors have negative temperature Coefficient of resistance, i.e., the resistance of semiconductors decreases with increase in temperature and vice versa. For example, Germanium is actually an insulator at low temperatures but it becomes good conductor at high temperature.

(ii) The resistivity of semiconductors lies between that of a good insulator and of a metal conductor, i.e., lies within the range 10^{-4} to 10^{4} Ω-m.

(iii) The electrical conductivity of a semiconductors is very much affected when a suitable metallic impurity, e.g., Arsenic, Gallium, etc., is added to it. This property of semiconductors is most important from the point of view of applications.

Today no society is considered modern or developed unless it has sizable electronics industry. We know that there can be no electronics industry without the semiconductors and related technologies. Obviously, semiconductors form the backbone of electronics. We find that semiconductors affect all walks of life whether it is çommunications, computers, biomedical, power, aviation, defence, entertainment etc. The transistors, integrated circuits (IC), lasers and detectors, sensors and other semiconductor devices through the items of daily use affecting our daily life.

In recent years a number of devices based on semiconductors have been developed that are of great practical applications in electronics. Most important of these are semiconductor diodes, transistors and ICs. A transistor, which is a three terminal device having properties similar to those of a vacuum tube (triode), but requires no cathode power, and at the same time offers the hope of practically long life. In modem electronic systems, the complete electronic circuit, containing many transistors, resistors, diodes, capacitors etc., is fabricated on a single chip is called an Integrated circuit (IC). Recently, a new type of transistor which is called metal oxide semiconductor (MOS) transistor has become very useful in microelectronic circuits. This MOS transistor can be used also as a capacitor. In recent years, taking advantage of the silicon integrated circuit technology, a new field has emerged which attempts to combine the sensor, aculator and the control circuit on as one integrated unit. In this sense it emulates a biological system. These are known as smart sensors, micro system technology (MST) or micro-electromechanical systems (MEMS). It has been possible to fashion miniature mechanical devices such as gears, motors, springs etc. Their combination with sensing and actuating functions has given researchers and engineers the tools to build microsystems that could not be imagined earlier. This technology is doing the mechanical system what the invention of transistors and ICs did to the electronic systems. MEMS devices are small in size, light weight, low cost, reliable, with large batch fabrication technology. The MEMS technology involves a large number of materials. Silicon forms the backbone of this systems also due to its excellent mechanical properties as well as mature micro-fabrication technology including lithography, etching and bonding. As compared to electronic valves, semiconductor devices offer the following advantages: (i) low weight and small size (ii) no power for the filament (iii) long service life (thousands of hours) (iv) mechanical ruggedness (v) low power losses and (vi) low supply voltages. At the same time semiconductor devices suffer from a number of disadvantages: (i) marked spread in parameters between devices within the same type (ii) deterioration in performance with time (ageing); higher noise level than in electronic valves (iii) unsuitability of most transistors for use at frequencies over tens of megahertz, (iv) low input resistance as compared with vacuum triodes; (v) inability to handle large power (vi) deterioration in performance after exposure to radioactive emissions.

Continuing efforts in research and development, however, are eliminating or minimizing many of the demerits of semiconductor devices. There are semiconductor rectifiers for currents of thousands of amperes. Replacement of germanium with silicon makes crystal diodes and transistors suitable for operation at temperatures upto 125°C. There are transistors for operation at hundreds of megahertz and more, and also microwave devices such as gunn diode, tunnel diode, etc. The field of semiconductors is rapidly changing. This is expected to continue in the next decade with some of the changes that can be foreseen now. We expect organic semiconductors to play prominent role during this decade. Diamond as semiconductor will also be important. Optoelectronic devices will provide three-dimensional integration of circuits, and optical computing.

Semiconductor devices impose a very small drain on power sources and make it possible to miniaturize or even micro-miniaturize components and whole circuits. The minimum power for an electronic valve is 0.1 Watt, while for a transistor it may be one microwatt, or one hundred thousandth of the former.

Ofcourse, semiconductor devices will not replace electronic valves in each and all applications, for the valves are also being continually improved. Simply, crystal diodes and transistors may be advantageous in some uses and electronic valves in others.

Rapid progress in the fabrication of semiconductor structures has resulted into the reduction of three-dimensional bulk systems to two-dimensional, one-dimensional and finally to zero-dimensional systems. These reduced dimensional systems are used in future applications like improved semiconductor lasers and micro electronics.

Quantum Dots (QDs) represent ultimate reduction in the dimensionality of semiconductor devices. These are 3-dimensional semiconductor structures only nanometers in size confining electrons and holes. QDs operate at the level of a single electron which is certainly the ultimate limit for an electronic device and are used as gain material in lasers. QDs are used in quantum dot lasers, QD memory devices, QD photo-detectors and even quantum cryptography. The emission wavelength of a quantum clot is a function of its size. Obviously, by making quantum dots of different sizes, we can create light of different colours.

The semiconductor technology is based on the number of charges and their energy. The electronic devices such as transistors work due to flow of charge. The electron can be assumed as tiny rotating bar magnet with two possible orientations. spin-up or spin-down. An applied magnetic field can flip electrons from one state to another. In this way, spin can be measured and manipulated to represent the 0's I's of digital programming, analogous to the "current on and current off" states in a conventional silicon chip. The study of electron spin in materials is called **spintronics**. Spintronics is based on the direction of spin and spin coupling.

A revolutionary new class of semiconductor electronics based on the spin degree of freedom could be created. The performance of conventional devices is limited in speed and dissipation whereas spintronics devices are capable of much higher speed at very low power. Spintronics transistors may work at a faster speed, are also smaller in size and will consume less power.

Electron spins can be oriented in one direction or the other, called spin-up or spin-down. When electrons spins are aligned in one direction, these create a net magnetic moment as observed in magnetic materials like, iron and cobalt. Magnetism is an intrinsic physical property associated with the spins of electrons in a material. The electron spin may exist not only in the up or down state but also in infinitely many intermediate states because of its quantum nature depending on the energy of the system. This property may lead to highly parallel computation which could make a **quantum computer** work much faster for certain types of calculations. In quantum mechanics, an electron can be in both spin-up and spin-down states, at the same time. The mixed state could form the base of a computer, built around not binary bits but the quantum bits or qubit. It is any combination of a 1 or a 0. The simplest device using spin-dependent effect is a sandwich with two magnetic layers surrounding a non-magnetic metal or insulator. If the two magnetic layers are different, then the magnetization direction of one can be rotated with respect to the other. This leads to the utilization of these structures as sensor elements and for memory elements. Scientists and engineers are now trying to use the property of the electron-like spin rather than charge to develop new generation of microelectronic devices which may be more versatile and robust than silicon chips and circuit elements. Spins appears to be remarkably robust and move easily between semiconductors. In case of electron transport from one material to another, the spins do not lose its orientation or scatter from impurities or structural effects.

All spintronics devices work accordingly to simple principle: information is stored into the spins as a particular spin orientation (up or down); the spins being attached to mobile electrons, carry the information along the wire and the information is read at the terminal. Two recent discoveries:

(i) Optically induced long-lived coherent spin-state in semiconductors, and

(ii) Ferromagnetism semiconducting Ga Mn As will lead to revolutionary advances in **photonics** and electronics, such as, very fast, very dense memory and logic at extremely low power, spin quantum devices like Spin-LEDs, and Spin-RTDs and quantum computing in conventional semiconductors at room temperature. The emerging technology of spintronics may soon make it possible to store movies on a palmpilot or build a new computer.

To understand the operation of semiconductor devices, it is necessary to study the semiconductor materials in some detail.

21.2 ATOMIC STRUCTURE AND ENERGY LEVEL

To understand how semiconductors work, one must have a good knowledge of atomic structure. We know that matter is composed of compounds and elements. The elements are the basic materials found in nature. When elements are combined to form a new material, we have a compound. The smallest particle that an element can be reduced to and still retain its properties is called an atom.

Although the atoms of different elements have different properties, they all contain the same subatomic particles. There are a number of different subatomic particles, but only three of these are of interest in basic electronics-the proton, the neutron and the electron.

The protons and the neutrons are contained in the nucleus of the atom, and the electrons revolve around the nucleus along specific orbits. The electrons and the protons are the particles that have the electrical properties. Neutrons have no electrical charge. Usually, atoms have the same number of electrons and protons, and so they are electrically neutral. If an atom does have more electrons, it is called a negative ion. If it has more protons, it is called a positive ion.

Fig. (21.1) shows the representations of the atomic structures of H, B, Si, P and Ge atoms. Fig. [21.1(a)] represents the hydrogen atom. It contains one electron revolving around one proton which is the nucleus. The nucleus of H atom contains no neutrons.

Fig. [21.1(b)] represents the structure of a boron atom. Its nucleus contains 5 protons (P) and 5 neutrons (N). There are 5 electrons revolving around the nucleus in different orbits. There are two electrons in the first orbit, 2 electrons in the second orbit and only 1 electron in the outermost or valence orbit.

Fig. [21.1(c)] represents the structure of a silicon atom. It contains 14 protons and 14 neutrons in the nucleus. There are 14 electrons revolving around the nucleus in different orbits. There are 2 electrons in the first orbit, 8 electrons in the second orbit and 4 electrons in the valence orbit.

Figs. [21.1(d)] and [21.1(e)] represent the structure of phosphorus and germanium atoms respectively. We note that Phosphorus contains 5 electrons in the outermost orbit called valence electrons whereas Ge atom contains 4 electrons in the outermost orbit.

The electrons in the inner orbits of an atom do not normally leave the atom. But the electrons which are in the outermost orbit, so called valence electron do not always remain confined to the same atom. Some of these valence electrons in certain materials called metals move in a random manner and may travel from one atom to another in a crystal lattice. These electrons are called as free electrons. It is due to the presence of these free electrons in a material, conduction is possible and it constitutes the current. The electrons in the inner orbits of the atom remain bound to the nucleus and are, therefore, called bound electrons.

The tendency of an atom to give up its valence electrons depends on chemical stability. When an atom is stable, it resists giving up electrons, and when it is unstable, it tends to give up electrons. The level of stability is determined by the number of valence electrons, because the atom strives to have its outermost or valence shell completely field.

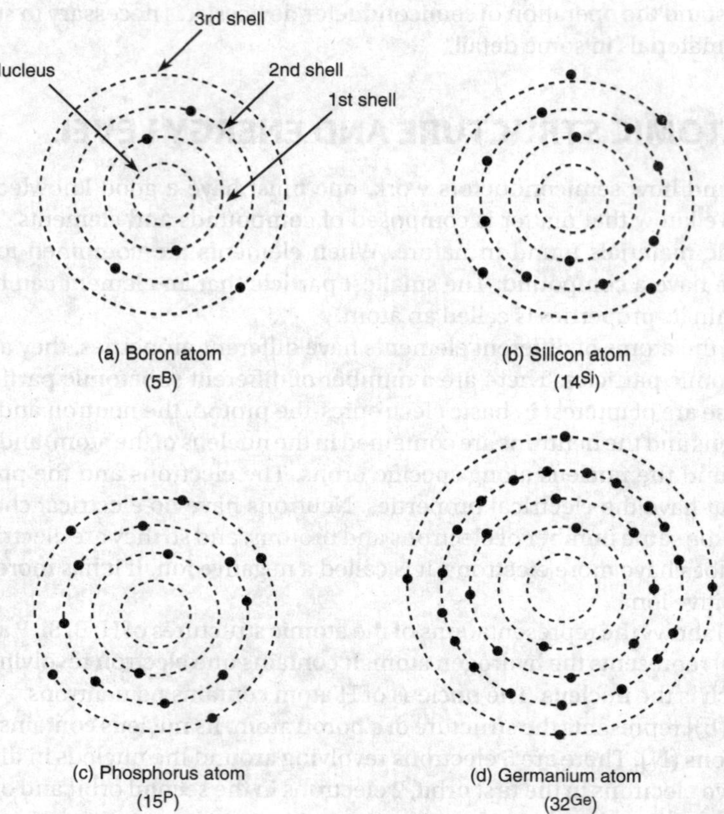

Fig. 21.1 Atomic structure of few atoms

If an atom's valence orbit is more than half filled, then atom tends to fill its orbit. So, since 8 is the maximum electrons that can be held in the valence orbit, elements with 5 or more valence electrons make good insulators, since they tend to accept rather than give up electrons. On the other band, atoms with less than 4 valence electrons tend to give up their electrons, thereby the valence shell is empty, this would allow the next shell, which is already filled, to be the outermost shell. These atoms make the best electrical conductors. The elements Si(14) and Ge(32) have 4 valence electrons, and are neither good conductors and nor good insulators. These are called semiconductors.

Most characteristics of semiconductors can be easily explained by means of an energy level diagram. We are familiar that each isolated atom has only a certain number of orbits available. These available orbits represent energy levels for the electrons in the atom. According to Bohr's theory of atomic structure only discrete values of electron energies are possible. An electron energy is usually expressed in electron volt ($1 \text{ eV} = 1.6 \times 10^{-19}$ J $= 1.6 \times 10^{-12}$ erg). An electron can have only certain permissible values, i.e., no electron can exist at an energy level other than a permissible one. Energy level diagram for hydrogen atom is shown in Fig. (21.3). The permissible energy levels for hydrogen atom are numbered $n = 1, 2, 3 \ldots$ in increasing order of energy.

In any atom, an electron orbiting very close to the nucleus in the first orbit is tightly bound to the nucleus and possesses only a small amount of energy. The greater the distance of an electron from the nucleus, the greater is its total energy. The total energy of an electron includes Kinetic and Potential energies. Obviously, an electron orbiting far from the nucleus would have a greater energy, and hence it can be easily knocked out of its orbit. This makes it clear that why the valence electrons having maximum energy part in chemical reactions and in bonding the atoms to form solids.

When radiations impinge on an atom, the energy of the electrons increases. As a result, electrons are excited to higher energy levels. The excited sate does not last long and very soon, the electron after emitting out energy in the form of heat light or other radiations, fall back to the original energy level.

21.3 ENERGY BANDS

The crystalline structure is common among most of the metals and semiconductors. We have read that any crystal is made up of a space array of molecules in regular repetition in three dimensions of some fundamental building block. For a gaseous element, the electronic energy levels are the same as for a single free atom because the individual atoms in a gas are well apart and has negligible influence on each other. However, in a crystal the individual atoms are so closely packed that the resulting energy levels are modified due to interaction between the atoms. When atoms form crystals, the energy levels of the inner-shell electrons are not appreciably affected, however, the levels of the outer-shell electrons are considerably altered as these electrons are shared by the adjacent atoms in the crystal. As a result of this interaction between the outer-shell electrons, the energy levels spread up to form a band of energy as shown in Fig. (21.2).

In order to understand the formation of energy band in a crystal, let us consider a silicon crystal made up of N-atoms. We shall also assume that the interatomic spacing can be varied without affecting the fundamental crystal structure. For very large interatomic spacing, say a in Fig. (21.2), the interaction between adjacent atoms is negligible, and energy levels same as that of isolated atoms. In silicon, the outermost sub shells, namely, $3s$ and $3p$ contain 2-electrons each. Hence, in a silicon crystal consisting of N-atoms, the outermost subshells $3s$ and $3p$ consist of 2N-electrons each. Thus, the $3s$ sub shell has 2N-electrons completely occupying the available 2N-states, and the $3p$ subshell has only 2N-electrons partially occupying the available 6N-states, all at the same energy level.

If the interatomic spacing is gradually decreased, i.e., moving from right to left in Fig. (21.2), there will be a gradual increase in the interaction between the neighbouring atoms. Due to this interaction, the atomic wave functions overlap, and the crystal becomes an electronic system which should obey the Pauli's exclusion principle. Hence, the 2N s-states spread out to form a band of energy. Actually, the separation between levels is small but since N is very large, of the order of 10^{23} cm^{-3} the total spread between the minimum and maximum energy levels becomes large. This spread will have several electron volts of energy and is referred to as energy band and is indicated by the lower shaded region in Fig. (21.2). The 2N-states in this band are completely filled with 2N-electrons.

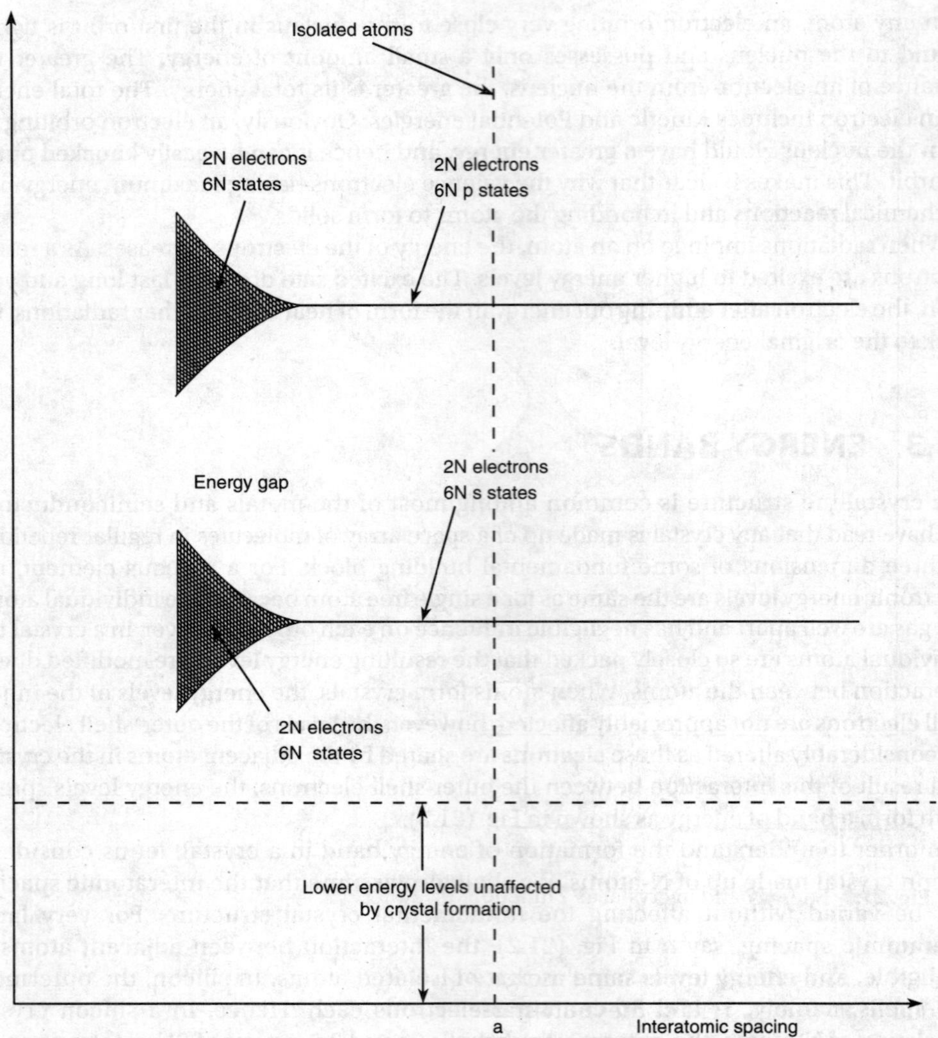

Fig. 21.2 Formation of energy bands in a crystal

Similarly, at the same value of interatomic spacing at *p*-level, 6N *p*-states spread up to form a band. This band is shown as the upper shaded region in Fig. (21.2). Though a total of 6N-states are available in this band, only 2N-states are occupied and 4N-states remain unoccupied.

An energy gap exists between the two energy bands. This energy gap is called as **forbidden energy gap,** as no electrons can occupy states in this gap. This forbidden energy gap decreases as the atomic spacing is decreased and becomes zero with further reduction in the interatomic spacing, say at *b*, as shown in Fig. (21.3). That is, the two energy bands will overlap when the interatomic spacing is small enough. Under such circumstances,

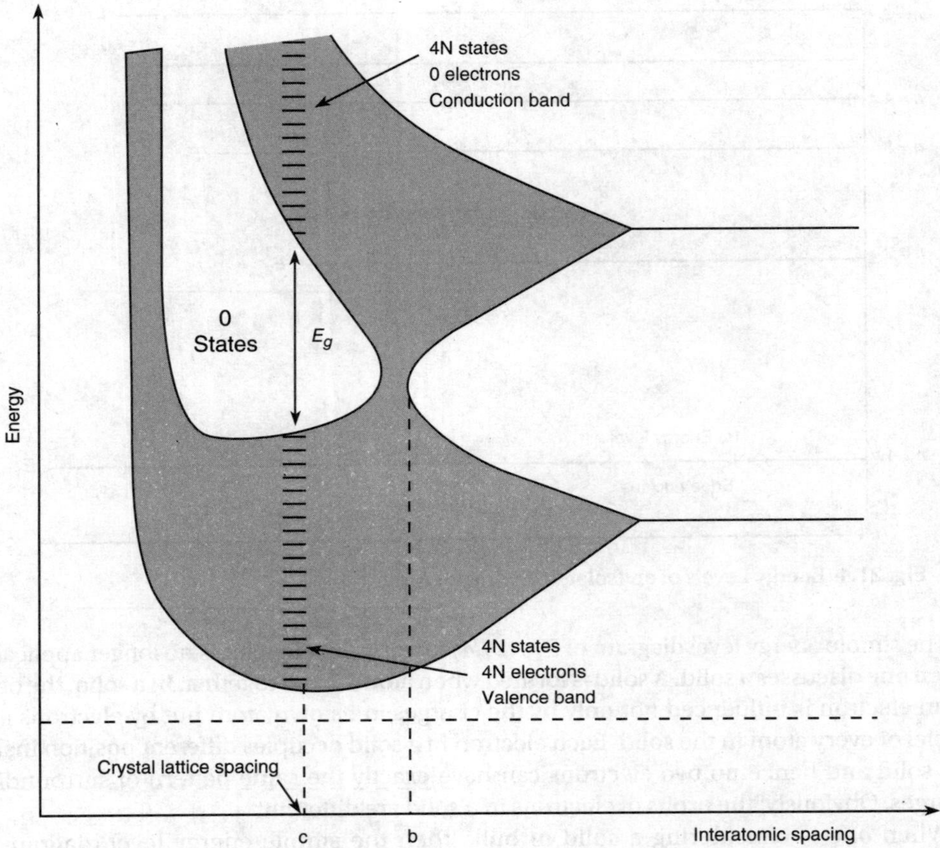

Fig. 21.3 Energy band in crystal as a function of interatomic spacing

the 6N *p*-states in the upper band merge with the 2N *s*-states in the lower band, giving a total of 8N-states. Half of these 8N-states are occupied by the 4N available electrons. These 4N-electrons now no longer belong to either *p*-subshell or *s* subshell but belong to the crystal as a whole. Thus, at this interatomic spacing, each atom in the crystal can contribute 4-electrons to the crystal. The band occupied by these contributed electrons is called the valence band.

If the interatomic spacing is reduced further, say *c* as shown' in Fig. (21.3), the interaction between the atoms becomes extremely large, and the energy band structure assumes the shape shown in Fig. (21.4). The exact energy band structure depends upon (i) the orientation of the atoms relative to one another in space and (ii) the atomic number of the atom, and may be obtain from solution of Shrodinger's wave number. For spacing *c*, 4 n-state in the valence band are completely filled by 4 n-electrons and this valence band is separated from the upper unfilled or empty band by a forbidden energy gap E_g. The forbidden energy gap contains no allowed states. The upper band has 4N unfilled states and is referred to as the conduction band.

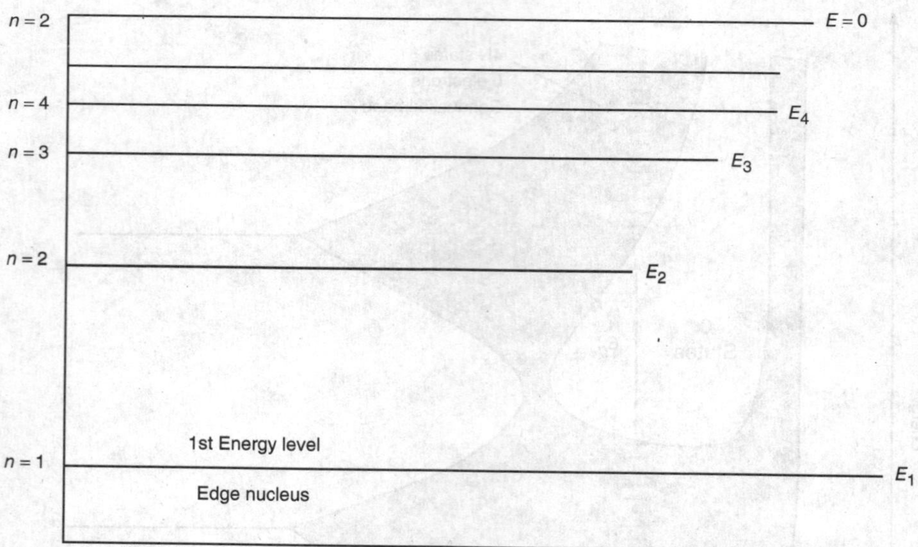

Fig. 21.4 Energy Levels of an Isolated Hydrogen Atom

The simple energy level diagram of Fig. (21.4) for electron energies is no longer applicable when one discusses a solid. A solid is formed when atoms bond together. In a solid, the orbit of an electron is influenced not only by the charges in its own atom but by electrons and nuclei of every atom in the solid. Each electron in a solid occupies different position inside the solid and hence no two electrons can have exactly the same pattern of surrounding charges. Obviously, the orbits of electrons in a solid are different.

When one is considering a solid in bulk, then the simple energy level diagram in Fig. (21.4) modifies to that shown in Fig. (21.5). All electrons belonging to the first orbits have slightly different energy levels because no two electrons see exactly the same charge environment. Since there are billions of first orbit electrons, the slightly different energy levels form a group or band. Similarly, the billions of second orbit electrons, all with slightly different energy levels, form the second energy band. And all third orbit electrons form the third band.

Silicon is a commonly used semiconductor having atomic number 14. Obviously, it has 4-electrons in its outermost or valence orbit. Clearly, the third band becomes the valence band Fig. (21.5). In Fig. (21.5), all these three bands are shown completely filled. Although the third shell of an isolated atom of silicon is not completely filled as it has only 4-electrons whereas it could accommodate a maximum of 8-electrons, the third energy band or valence band of a bulk silicon material behaves as if completely filled. It is so, because in solid silicon each atom positions itself between four other silicon atoms, and each of these neighbours share an electron with the central atom. In this manner, each atom now has 8-electrons associated with it thereby filling the valence band completely, i.e., all the permissible energy levels in the band are occupied by electrons. Obviously, no electron in a filled band can move and hence an electron in a completely filled band cannot contribute to electric current. At absolute zero temperature, electrons cannot move

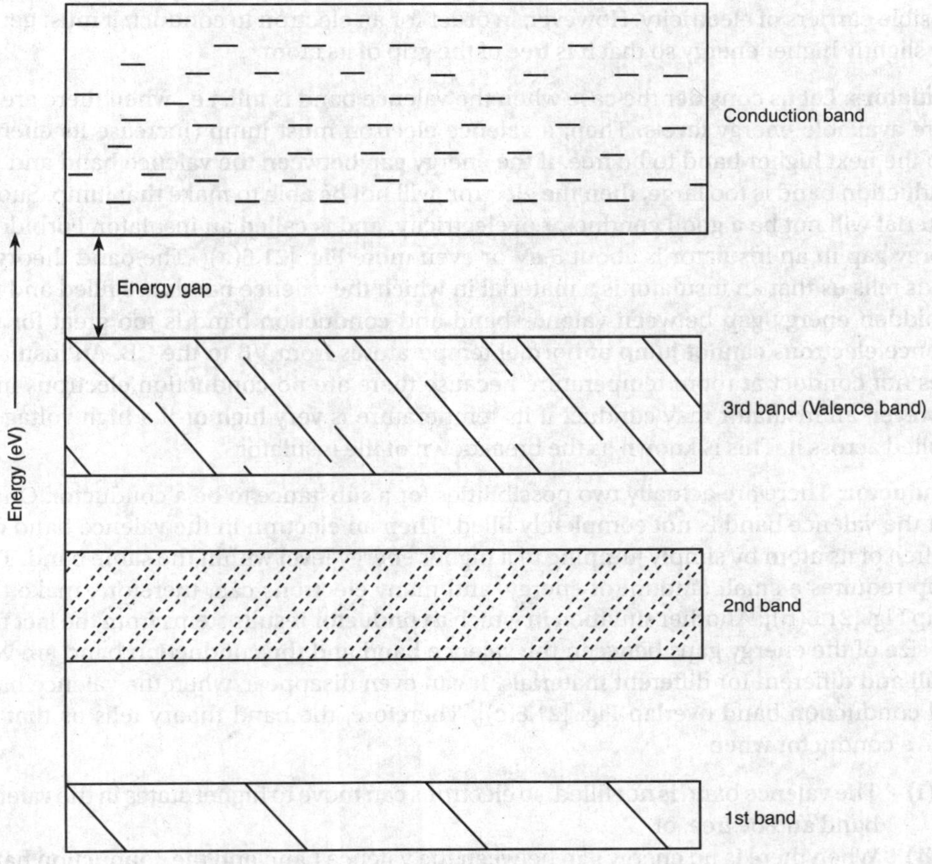

Fig. 21.5 Energy Bands of Silicon at Absolute Zero

through the solid Silicon material and hence it acts like a perfect insulator. Beyond the valence band there is a conduction band. At absolute zero temperature, the conduction band is empty, i.e., no electron has enough energy to go into a conduction band. Definite amount of energy is needed to shift the electron from the valence band to the conduction band. This amount differs from one substance to another substance and helps to classify them as conductors, insulators and semiconductors.

21.4 CONDUCTORS, SEMICONDUCTORS AND INSULATORS

The band structure in a solid determines whether the solid is an insulator or a conductor or a semiconductor. The bands are filled upto a certain level by the electrons within each atom. The highest band in which electrons are still predominantly attached to their atoms are found is called the valance band. This is the band in which the valence (outermost electrons from each atoms will be located. These are the electrons that are the possible carriers of electricity. However, in order for an electron to conduct, it must get up to a slightly higher energy so that it is free of the grip of its atom.

Insulators: Let us consider the case when the valence band is full, i.e., when there are no more available energy levels. Then, a valence electron must jump (increase its energy) into the next higher band to be free. If the energy gap between the valence band and the conduction band is too large, then the electron will not be able to make that jump. Such a material will not be a good conductor of electricity, and is called an insulator. Forbidden energy gap in an insulator is about 5 eV or even more Fig. [21.6(a)]. The band theory of solids tells us that an insulator is a material in which the valence bands are filled and the forbidden energy gap between valence band and conduction band is too great for the valence electrons cannot jump at normal temperatures from VB to the CB. An insulator does not conduct at room temperature because there are no conduction electrons in it. However, an insulator may conduct if its temperature is very high or if a high voltage is applied across it. This is known as the breakdown of the insulator.

Conductor: There are actually two possibilities for a substance to be a conductor. One is that the valence band is not completely filled. Then an electron in the valence band can get free of its atom by simply jumping to a higher energy level within the same band. This jump requires a small amount of energy, and many electrons can, therefore, make that jump Fig. [21.6(b)]. Another situation in which a conductor results stems from the fact that the size of the energy gaps between the valence band and the conduction band are very small and different for different materials. It can even disappear, when the valence band and conduction band overlap Fig. [21.6(c)]. Therefore, the band theory tells us that we have a conductor when

(i) The valence band is not filled, so electrons can move to higher states in the valence band and be free, or

(ii) When there is no energy gap between the valence band and the conduction band, so electrons can easily make the transitions from the valence to the conduction band, e.g., metals.

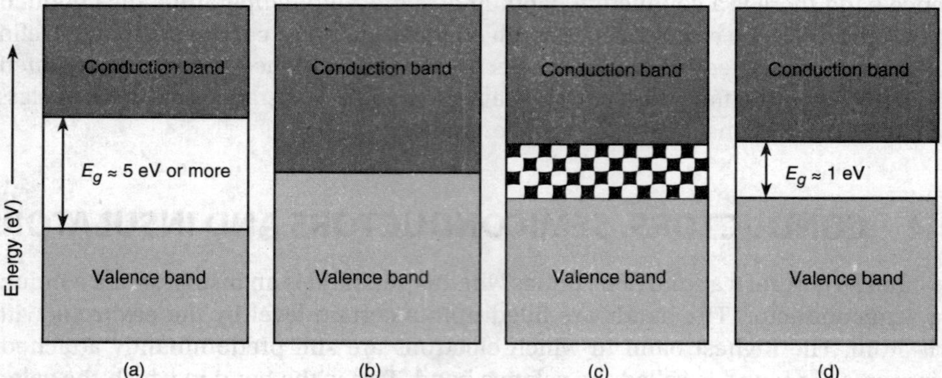

Fig. 21.6 Energy Band Diagram for (a) Insulators. (b) and (c) conductors, and (d) Semiconductors

Semiconductors: There is a case in between 'conductor' and 'insulator'. A material with intermediate properties is called a semiconductor. In a semiconductor,

(i) The valence band is filled, and
(ii) Although there is an energy gap between the valence band and the conduction band yet the energy gap is not very large.

The energy-band diagram for a semiconductor is shown in Fig. [21.6(d)]. In this case, the forbidden energy gap is of the order of IeV (for Silicon, $E_g = 1.12$ eV and for Ge, $E_g = 0.72$ eV). Silicon and Germanium are very good examples of semiconductors. The gaps between their valence and conduction bands are sufficiently small that the normal thermal energy of the solid at room temperature is enough to knock a few electrons into the conduction band. These electrons conduct a current, but. as the name 'Semiconductor' implies, a semiconductor does not conduct as best as a real conductor. The conductivity of these substances has a number of peculiarities:

(i) The dependence of conductivity on temperature is opposite to that of metals. The conductivity of semiconductors, in contradiction to that of conductors (metals), may decrease rapidly with temperature. At low temperatures, a semiconductor may become an insulator. Hence, the distinction between semiconductors and insulators is purely quantitative and, to a large extent, conventional. The resistance of most semiconductors is considerably more sensitive to changes in temperature than metals. Compact temperature meters of high sensitivity may be constructed using semiconducting thermal resistors (thermistors).
(ii) In a number of cases semiconductors may possess positive as well as negative temperature coefficient.

21.5 FERMI LEVEL

In the preceding article, we have seen that as far as solid-state theory is concerned only the upper energy bands (valence bands) are of interest, since electrons at lower levels practically do not take part in interactions among atoms. How can the behaviour of upper band electrons be described? Since we are dealing with a very large number of electrons, it is natural to use statistical physics methods and consider an aggregate of such electrons as a kind of gas, usually called as electron gas.

The state of each electron of such a gas may be represented by a point (p_x, p_y, p_z) in momentum space. The direction of motion of an electron is parallel to its radius vector P and the energy of an electron depends on its momentum. Let us consider as a crude approximation that the electrons in solid behave like free particles, i.e., we neglect the potential energy of the field in which the electrons move and the interaction between electrons.

If the electrons are free, the relationship between their energy and momentum is given by $E = (1/2m)P^2$. This means that in momentum space a surface of total energy is a sphere. Such a sphere is usually called a Fermi sphere. One may call the Fermi surface of maximum energy as the states of an electron gas are contained in a sphere in a sphere of radius

$$P_{\max} = \sqrt{2m E_{\max}}$$

It is most important to determine how the electrons may distribute themselves in a band among the energies from zero upto E_{max}. The number of electrons per unit volume then can be accommodated up to an energy E is given by

$$n = \frac{16\pi\left(2m_e^3\right)^{1/2}}{3h^3} E^{3/2} \tag{21.1}$$

If the metal is in its ground state, which occurs at absolute zero, all electrons occupy the lowest possible energy levels compatible which the Pauli exclusion (i.e., each can accommodate two electrons: one with spin up and one with spin down), as indicated in Fig. [21.7(b)]. If the total number of electrons per unit volume n_0 is less than total number of energy levels available in the band, the electrons will then occupy all energy states up to a maximum energy, designated by E_F, and called the Fermi energy. Thus the Fermi energy E_F is defined as the energy of the topmost filled level in the ground state of n electrons. If we set $E = E_F$ in Eq. (21.1), we must have $n = n_0$. Therefore for the Fermi energy, we obtain the value

$$E_F = \frac{h^2}{8m_e}\left(\frac{3n_0}{\pi}\right)^{\frac{2}{3}} \tag{21.2}$$

The energy distribution of electrons in the metal ground state corresponds to the shaded area in Fig. [21.7(a)]. When the Fermi energy is equal to the energy band width, the band is fully occupied.

When band is not completely filled, a small amount of energy is enough to excite the uppermost electrons to nearly energy levels as indicated in Fig. [21.7(c)]. However, only the uppermost electrons can be thermally about 0.025 eV, which is very small compared with E_F, and the exclusion principle make it impossible for the low energy electrons to be excited into nearby occupied states. The distribution of electrons among the energy level in a thermally excited state of the lattice corresponds to the shaded area in Fig. [21.7(d)]. The electrons which have been thermally excited are those with an energy greater than E_F. The states occupied by the excited electrons fall in an energy region of the order of 20 kT about E_F.

From the above discussion one can easily conclude that the concept of Fermi level serves the reference energy level from which all other energies are conveniently measured. The probability $F(E)$ of a state corresponding to energy E being occupied by an electron at temperature T K is given by

$$F(E) = \frac{1}{1+\exp\left(\dfrac{E-E_F}{kT}\right)} = \frac{1}{\exp\left[\beta\left(E-E_F\right)\right]+1} \tag{21.3}$$

Here k is Boltzmann constant. Three cases of interest are:

(i) At $T = 0°$K, if $E > E_F$ then $F(E) = 0$, i.e., energy state is empty.
(ii) At $T = 0°$K, if $E < E_F$ then $F(E) = 1$, i.e., energy state is occupied by an electron.
(iii) At $T \neq 0°$K, and $E = E_F$ then $F(E) = 1/2$, i.e., energy state is 50% occupied.

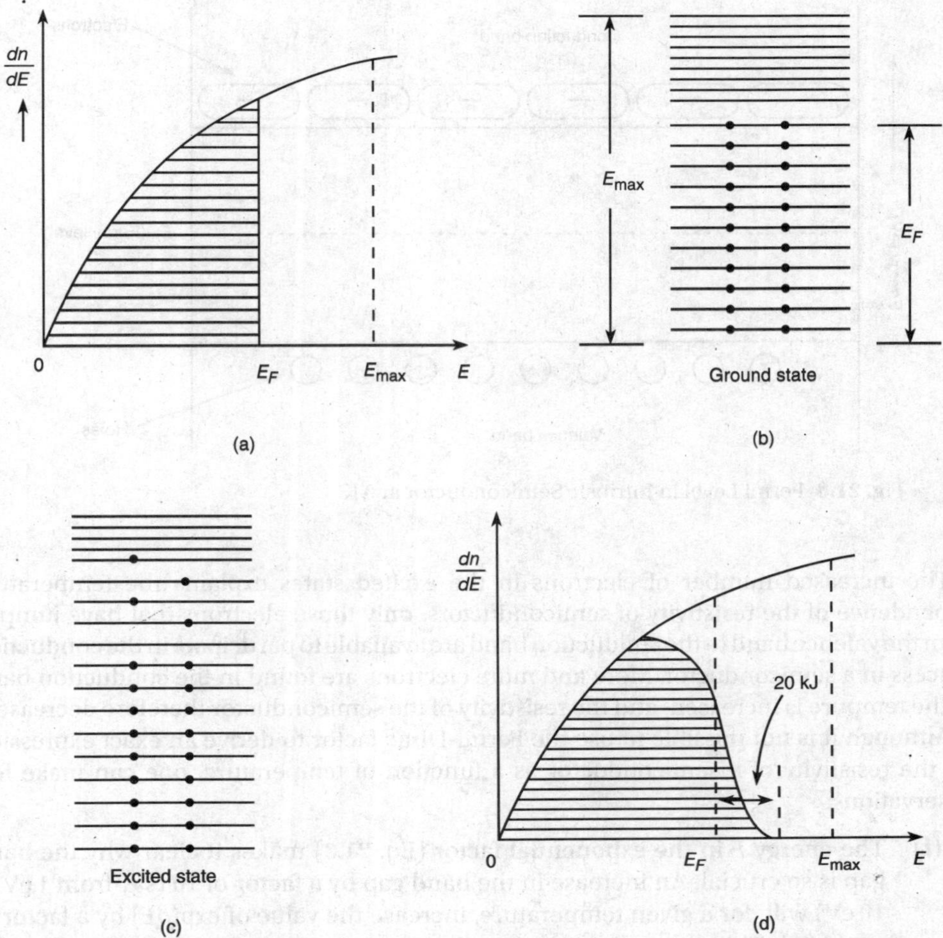

Fig. 21.7 (a) Density of energy states of free electrons in a solid (b) and (c). Distribution of free electrons among energy states in the conduction band. (d) Occupation of energy states at a temperature different from absolute zero

Obviously, one finds that at $T = 0$ K, all the energy states above E_F are empty, whereas all those below Fermi energy (E_F) are filled with electrons. With the rise in temperature, states above Fermi energy level (E_F), no more remain empty. They are then occupied by the electrons to some extent. Fig. (21.8) shows the position of Fermi level in energy band diagram in the case of pure or intrinsic semiconductors. It depicts that Fermi level for an intrinsic semiconductor is situated in the middle of forbidden gap, See Eq. (21.3), i.e., between conduction and valence band, and the position of Fermi level is independent of temperature. This reveals that when the temperature is raised there is a greater possibility of electrons being found above the Fermi level with an equal possibility of finding an electron vacancy so called 'hole' below Fermi level. We will discuss it in later sections.

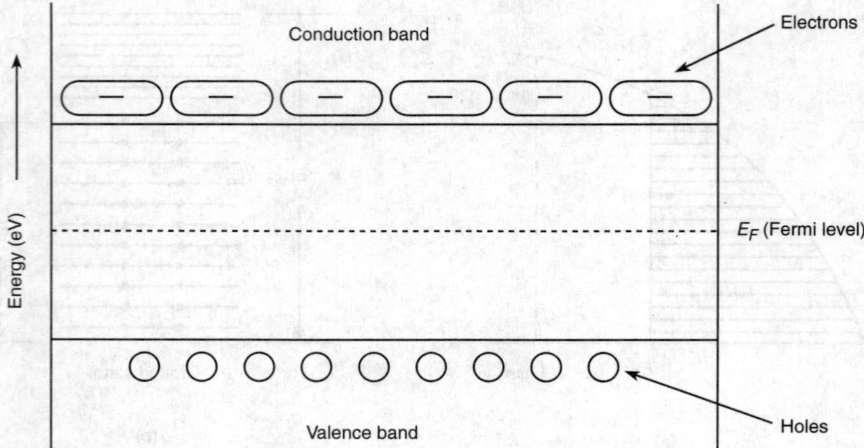

Fig. 21.8 Fermi Level in Intrinsic Semiconductor at 0 K

The increased number of electrons in the excited states explains the temperature dependence of the resistivity of semiconductors, only those electrons that have jumped from the valence band to the conduction band are available to participate in the conduction process in a semiconductor. More and more electrons are found in the conduction band as the tempure is increased, and the resistivity of the semiconductor therefore decreases.

Although it is not possible to use the Fermi-Dirac factor to derive an exact expression for the resistivity of a semiconductor as a function of temperature, one can make few observations:

(i) The energy E in the exponential factor (Eq. 21.3) makes it clear why the band gap is so crucial. An increase in the band gap by a factor of 10 (say from 1 eV to 10 eV) will, for a given temperature, increase the value of $\exp(\beta E)$ by a factor of $\exp(9\beta E)$. This generally makes the factor F_{FD} so small that the material has to be insulator.

(ii) Based on this analysis, one may expect the resistance of a semiconductor to decrease exponentially with increasing temperature. This is approximately true, because the function F(E) is not a simple exponential, and because the band gap does vary somewhat with temperature.

Clement and Quinnell developed the following empirical expression for the temperature variation of standard carbon resistor is

$$\text{Log}\, R + \frac{K}{\text{Log}\, R} = A + \frac{B}{T} \tag{21.3a}$$

where A, B and K are constants. The plot of Log $R + K/\text{Log}\, R$ versus $1/T$ is a straight line up to $1/T = 0.6$, so one can conclude that the Clement-Quinnell Eq. (21.3(a)) is good down to $T = 1\,K/0.6 = 1.7\,\text{K}$.

21.6 INTRINSIC SEMICONDUCTORS

Semiconductor devices, e.g., diodes and transistors, are made from a single crystal of semiconductors material, e.g., germanium or silicon. To make a semiconductor device, a sample of semiconductor must be in its purest form. A semiconductor in its purest form is called an intrinsic semiconductor. A semiconductor is not truly intrinsic unless its impurity content is less than 1 part impurity in 100 million parts of semiconductor. As stated earlier that at 0°K. temperature, the Fermi level (E_F) in intrinsic semiconductors, lies nearly mid way between the valence and conduction band.

As the temperature is raised, some electrons near the top of the valence band are thermally ionised into the lowest levels of the conduction band. The conduction band now contains a few electrons, and the valence band has a few missing electrons or holes[1]. Thermal ionisation, therefore, is said to create hole-electron pairs, since there is one hole for each electron. As both electrons and holes transport charge through the semiconductor, they are collectively termed charged carriers or simply carriers. In applied electric and magnetic fields a hole acts as if it is a positive electron. To understand the phenomenon of conduction of current in a semiconductor, it is essential to study its crystal structure.

21.6.1 Crystal Structure of Semiconductors

It is the nature of some solid materials to form themselves into bodies called crystals which have characteristic geometric shapes. These are the crystalline substances. Exact opposites are those other solids which form into shapeless masses and are said to be plastic, non crystalline, or amorphous. Quartz is a familiar example of crystalline substances where as a block of rubber is amorphous. Elements and compounds both can be crystalline. So can metals and non metals. Virtually all minerals are crystalline.

Externally, a crystal has several flat faces which are arranged symmetrically w.r.t. each other. Internally, it has a certain orderly arrangement of atoms in a repeating system called a lattice. Both externally and internally, a single crystal of a given true crystalline material looks like all other crystals of that material. A single crystal may be large to the point of hugeness or it may be so small as to be visible only with the aid of magnifier.

Inside the crystal lattice, certain loosely bound electrons (called valence electrons) in the outer orbit of one atom align themselves with similar electrons in adjacent atoms to form covalent bonds which hold the atoms together in the orderly structure of the lattice. Thus, in any covalent bond there are shared electrons, so called because they are shared by neighbouring atoms.

Germanium (Ge) and Silicon (Si) are two very typical semiconductors. In Germanium the atom has 32 electrons distributed in four orbits whereas in silicon there are 14 electrons distributed in three orbits. The outermost orbit called the valence orbit contain 4 electrons

---

[1]When an electron moves from the valence band it leaves behind a vacant energy state with a positive charge. Morcover as the electron moving from the valence band is located at the top edge of the band has a negative effective mass and its absence in the valence band is associated with a positive mass. The vacant energy state in the valence band therefore has a positive charge and a positive mass. The vacant energy state is called a 'hole'.

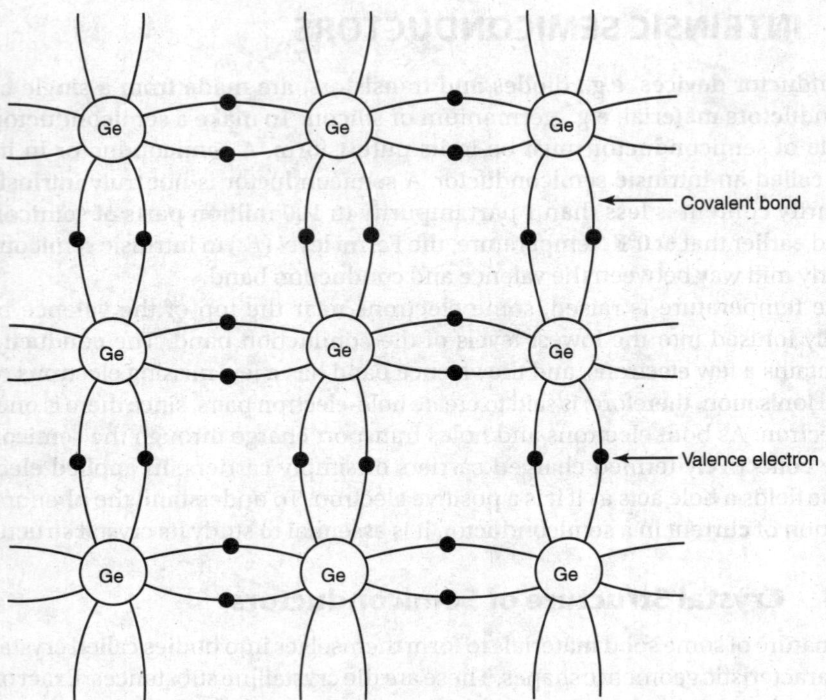

Fig. 21.9 A Simplified Representation of the Crystalline Structure of a Semiconductor (Ge) at Absolute zero

in each case Fig. (21.1). The crystals are tetrahedral in structure and each atom shares one of its electron with its neighbour. Such a sharing of its electrons between two neighbouring atoms is called a covalent bond. A simplified representation of the crystalline structure of a semiconductor (Ge) at absolute zero is shown in Fig. (21.9).

At very low temperature near absolute zero all the electrons in the atoms are tied up strongly by these bonds, but with the rise of temperature, a few electrons break away from some of the covalent bonds and get themselves freed creating vacant spaces, deficient of electrons known as holes. **A hole is equivalent to a net positive charge equal to that of electron**. Whenever a free electron is generated, a hole is created simultaneously, i.e., free electrons and holes are always generated in pairs. Obviously, the concentration of free electrons and holes will always be equal in an intrinsic semiconductor. This type of generation of free-electron hole pairs in semiconductors is called as thermal generation.

Owing to the thermal vibrations the free electrons in a semiconductor crystal jump from one bound position to another. This is equivalent to the motion of a hole relatively in opposite direction and thus gives an electric current in the direction of motion of the hole. The conductivity of semiconductor is, therefore, due to the motion of the holes and of the electrons Fig. [21.10(b)]. Such intrinsic semiconductors produce very weak electric current not adequate for useful work.

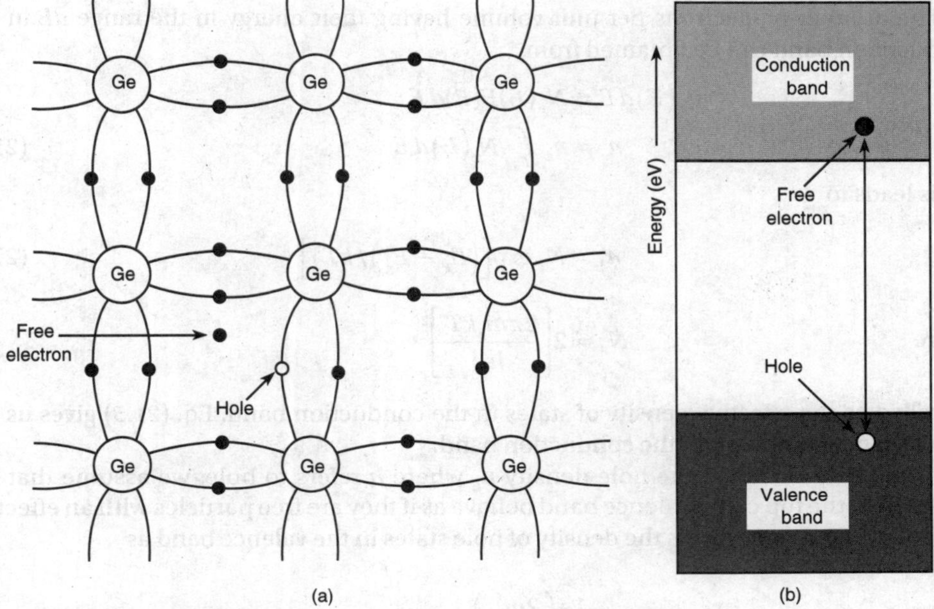

Fig. 21.10 (a) Crystalline structure of Ge at room temperature. (b) Generation of electron hole pair in an Intrinsic Semiconductor (Ge)

21.6.2 Carrier (or Electron and Hole) Concentrations in Intrinsic Semiconductors

In order to calculate the intrinsic carrier concentration, we first calculate the number of electrons excited into the conduction band at any temperature T K and which in turn are free to migrate in the crystal. In carrying out these calculations it will be assumed that the electrons in the conduction band behave as if they are free with effective mass m_e^* and energy will be measured from the top of the valence band. We further take the electron density of states in the conduction band equal to that for free electrons, i.e.,

$$N(E) = \frac{1}{(2\pi^2)}\left(\frac{2m_e^*}{h^2}\right)^{\frac{1}{2}}(E - E_g)^1/2 \tag{21.4}$$

In addition to the density of states function we also require the probability function for calculating the electron density in the conduction band. The appropriate probability function is the Fermi function $F(E)$ given by Eq. (21.4). As stated earlier that Fermi level E_F is in the middle of the band gap in the case of intrinsic semiconductor. If $(E - E_F) < 4kT$. i.e., the lower edge of the conduction band is about $4k_B$T above E_F, one can neglect 1 in the denominator of Eq. (21.1) and Fermi function $F(E)$ reduces to

$$F(E) = \exp(E_F = E)/kT \tag{21.5}$$

The number of electrons per unit volume having their energy in the range dE in the conduction band can be obtained from

$$n_e(E)dE = N_e(E)F(E)dE$$

or
$$n_e = n_e \int_{E_g}^{\infty} N_e(E)dE \tag{21.6}$$

This leads to

$$n_e = N_e \exp\left[\left(E_F - E_g\right)/kT)\right] \tag{21.7}$$

Here
$$N_e = 2\left[\frac{2\pi m_e^* kT}{h^2}\right]^{\frac{3}{2}}$$

is called as the effective density of states in the conduction band. Eq. (21.5) gives us the electron concentration in the conduction band.

In order to calculate the hole density n_h where h refers to hole, we assume that the holes near the top of the valence band behave as if they are free particles with an effective mass m_h^*. One can express the density of hole states in the valence band as

$$N_h(E)dE = \frac{1}{2\pi^2}\left(\frac{2m_h^*}{h^2}\right)^{\frac{3}{2}}(-E)^1/2 \text{per unit volume}$$

Here, one must remember that the energy is measured positive upwards from the top of the valence band, If $f_h(E)$ represents the probability of occupation of the holes in the valence band then it must be equal to the probability of the electron states being unoccupied in the valence band i.e.,

$$f_h(E) = 1 - f_e(E) = 1 - \frac{1}{\exp\left[\left(E - E_F\right)kT\right] + 1}$$

$$= \frac{\exp\left[\left(E - E_F\right)/kT\right]}{\exp[(E - E_F)/kT] + 1} \tag{21.8}$$

Since $E < E_F$, being in the valence band, so $\exp[(E - E_F)/kT] << 1$ and hence we can always neglect $\exp[(E - E_F)/kT]$ in comparison to 1 in the denominator and one obtains.

$$F_h(E) = \exp\left[\left(E - E_F\right)/kT\right] \tag{21.9}$$

Using the relation $n_h(E)dE = N_h(E)F_h(E)dE$, one obtains the hole concentration in the valence band as

$$n_h = N_v \exp[-E_F/kT] \tag{21.10}$$

where

$$N_v = 2\left[\frac{2\pi m_h^* kT}{h^2}\right]$$

is called the effective density of states in the valance band.

Since for intrinsic semiconductors, we have $n_e = n_h$ and, therefore, from Eqs. (21.8) and (21.11), one obtains

$$m_e^{*3/2} \exp\left[\left(E_F - E_g\right)/kT\right] = m_h^{*3/2} \exp[-E_F/kT]$$

or
$$E_F = \frac{E_g}{2} + \frac{3}{4}kT\log\left(\frac{m_h^*}{m_e^*}\right) \tag{21.11}$$

At $T = 0\,K$, Eq.(21.12) reduces to

$$E_F = \frac{E_g}{2} \tag{21.12}$$

21.6.3 Fermi level in Intrinsic Semiconductors

From Eq. (21.13) it follows that at $T = 0\,\text{K}$, $m_e^* = m_h^*$ and Fermi level in the intrinsic semiconductor lies exactly mid-way in the forbidden gap, i.e.,

$$E_F = E_g/2$$

Obviously, the Fermi level E_F which can be highest occupied energy level at $T = 0\,K$ lies near the middle of the band gap for an intrinsic semiconductor Fig. [21.10(c)]. The Fermi-Dirac (*FD*) distribution function $f(E)$ T at $0\,K$ is plotted in Fig. [21.10(d)].

In general $m_h^* > m_e^*$, i.e., m_h^* is slightly greater than m_e^* and so the Fermi level is raised slightly as the temperature T increases. But for all practical purposes the Fermi level in intrinsic semiconductors can be assumed to be constant over a wide range of temperatures. It is worthwhile to mention that Fermi level in semiconductors is not static level, but a dynamic one, since it changes appreciably with change of temperature and impurities. Variation of Fermi level with impurity concentration enables one, the operation of the various semiconductor devices.

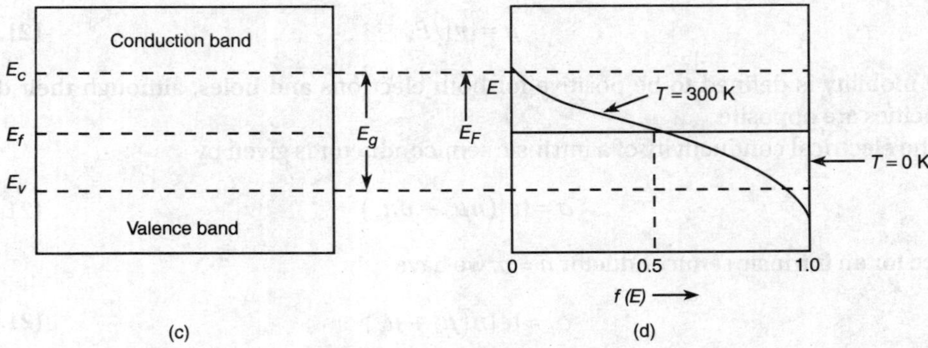

Fig. 21.10 (c) Energy band diagram. (d) *FD* distribution function $f(E)$ for an intrinsic semiconductor

21.6.4 Law of Mass Action

Multiplying Eqs. (21.8) and (21.11), one obtains

$$np = n_e n_h = 4\left(\frac{2\pi kT}{h^2}\right)^3 (m_e^* m_e^*)^{3/2} \exp\left(-\frac{E_g}{kT}\right)$$

where $n_e = n$ and $n_h = p$ and E_g is the band gap energy. Since band gap energy E_g is constant and hence the product np is also constant, i.e.,

$$np = n_i^2 = \text{const}\left[T^3 \exp(-E_g / kT)\right] \tag{21.13}$$

Here n_h is called the intrinsic density of either carriers. From Eq. (21.14) it is evident that n_i^2 or np is a constant depending on the temperature and the width of the forbidden gap. It does not depend on the impurities introduced as long as the impurities do not change the width of the forbidden energy gap. Eq. (21.14) is called the law of mass action. The result also holds good in the presence of impurities as well.

Because np is a constant independent of impurity concentration at a given temperature, the introduction of a small proportion of a suitable impurity to increase n, say must decrease p. This result is important in practice as one can reduce the total carrier concentration $(n + p)$ in an impurity crystal, sometimes enormously, by the controlled introduction of suitable impurities. Such a reduction is called compensation of one impurity type by adding another.

21.6.5 Electrical Conductivity

The electrical conductivity of intrinsic semiconductors called conductivity in the very low temperature (~0 K) is due to intrinsic charge carriers, i.e., due to electrons and holes. Such conductivity is sometimes termed as intrinsic conductivity.

Since there are two type of carriers in the intrinsic semiconductor, electrons and holes, its specific conductance is the sum of the conductivity $\sigma_e = |e|n\mu_e$ due to free electrons, with the concentration n and mobility μ_e, and of the conductivity $\sigma_h = |e|p\mu_h$ due to the presence of holes, with the concentration p and mobility μ_h. The mobility is defined as the magnitude of the drift velocity per unit electric field

$$\mu = |v| / E \tag{21.14}$$

The mobility is defined to be positive for both electrons and holes, although their drift velocities are opposite.

The electrical conductivity of a intrinsic semiconductor is given by

$$\sigma = |e|(n\mu_e + p\mu_h) \tag{21.15}$$

Since for an intrinsic semiconductor $n = p$, we have

$$\sigma_i = |e| n(\mu_e + \mu_h) \tag{21.16}$$

Here σ_h denotes the intrinsic conductivity. Substituting the value of *n*, one obtains

$$\sigma_i = 2|e|\left(\frac{2\pi kT}{h^2}\right)^{\frac{3}{2}}(m_e^{\cdot} m_h^{\cdot})^{3/4}\exp\left[-E_g/2kT\right](\mu_e + \mu_h) \qquad (21.17)$$

It is worthwhile to mention that the value of E_g is more in the case of silicon (E_g = 1.12 eV) than in the case of germanium (E_g = 0.72 eV). Obviously, less number of electron-hole pairs will be generated in silicon than in germanium at room temperature. This means that the conductivity of silicon will be less than that of germanium at room temperature.

When a battery is connected across a semiconductor Fig. (21.11), the electrons experience a force towards the positive terminal of the battery, and holes towards the negative terminal. The random motion of electrons and holes gets modified. Over and above the random motion, there also occurs a net movement, called drift. The random motion does not contribute to any electric current. The free electrons drift towards the positive terminal of the battery, whereas the holes towards the negative terminal. The electric current flows through the semiconductor in the same direction as in which the holes are moving. Since the electrons are negatively charged, the direction of conventional electric current is opposite to the direction of their motion as shown in Fig. (21.11). Although, the two types of charge carriers move in opposite directions, the two currents are in the same direction.

When the flow of charge carriers in a semiconductor is due to an applied voltage the resultant current is called a drift current. A second type of current called as diffusion current also exists in a semiconductor. The diffusion current flows as a result of a gradient of carrier concentration, i.e., the difference of carrier concentration from one region to another. The diffusion current is also due to the motion of both electrons and holes.

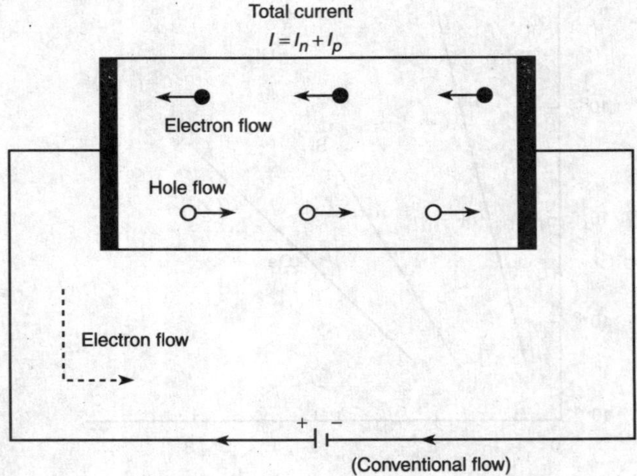

Fig. 21.11 Electric Current in an Intrinsic Semiconductor

Let us now investigate the effect of temperature on intrinsic semiconductor. From Eq. (21.18), it is evident that the exponential term $\exp[-E_g/2kT]$ dominates all other temperature dependence. Writing Eq. (21.18) as

$$\log\sigma_i = -\frac{E_g}{2kT} + \log\left[2|e|\left(\frac{2\pi kT}{h^2}\right)^{\frac{3}{2}}\left(m_e^* m_h^*\right)^{\frac{3}{4}}\left(\mu_e + \mu_h\right)\right]$$

or $$\log p = -\log\sigma_i$$

$$= -\frac{E_g}{2kT} - \log\left[2|e|\left(\frac{2\pi kT}{h^2}\right)^{\frac{3}{2}}\left(m_e^* m_h^*\right)^{\frac{3}{4}}\left(\mu_e + \mu_h\right)\right] \quad (21.18)$$

where p is the resistivity.

The intrinsic semiconductor has a small conductivity. In a sample of germanium at room, temperature the intrinsic carrier concentration is $2.5 \times 10^{19}/\text{m}^3$. With the rise in temperature, the conductivity increases, i.e., resistivity decreases. Fig. (21.12) shows plot of $\log p$ vs. I/T for some intrinsic semiconductors.

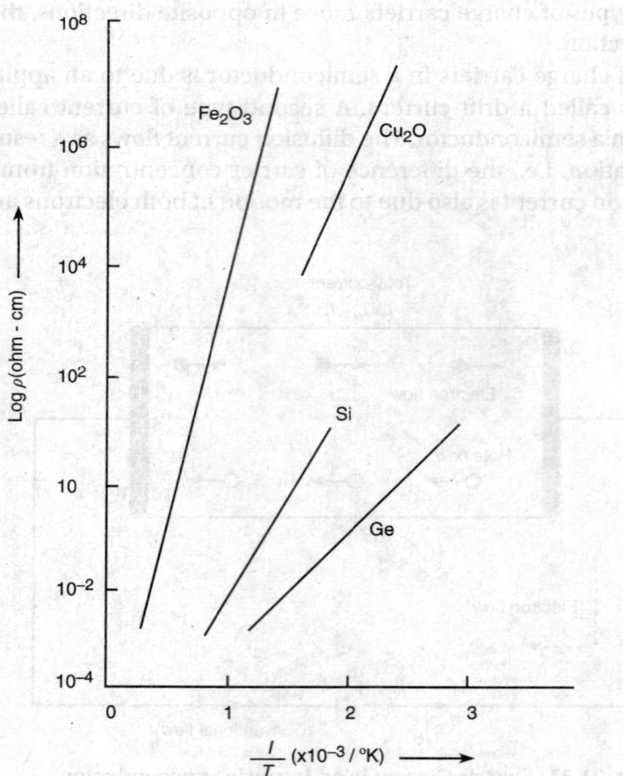

Fig. 21.12 Plot of $\log\rho$ vs. I/T for some Intrinsic Semiconductors

21.6.6 Thermoelectric Effect

The thermoelectric effect is often used to study the properties of semiconductors. When there is a temperature gradient in a thermoelectric material, an electric field appears. One can easily understand why this happens in a pure metal, in which case one can assume a gas of free electrons. As in an ideal gas, the density of free electrons is greater at the colder end of the wire, and therefore the electrical potential should be higher at the warmer end and lower at the colder end. No doubt the free electron model is not valid for semiconductors, and therefore one can reasonably believe that semiconductors should exhibit a thermoelectric effect.

In one dimension, the induced electric field E in a semiconductor is proportional to the temperature gradient, i.e.,

$$E = Q\frac{dT}{dx}$$

where Q is thermoelectric power. We may note that the direction of the induced field depends on whether the semiconductor is p-type or n-type, so the thermoelectric effect can be used to determine the extent to which n- or p-type carriers dominate in a complex system.

The thermoelectric effect is sometimes referred as *Seeback effect*. However, there are also other two types of of thermoelectric effect. In a normal conductor, heat is generated at the rate of I^2R. But a temperature gradient across the conductor causes additional heat to be generated. This is *Thomson effect*. The Thomson effect is entirely reversible. If the direction of the current is toward the higher temperature end of the conductor, heat is generated, but if the current flows toward the lower temperature end, heat is absorbed from the surroundings. The *Peltier* effect occurs when heat is generated at a junction between two conductors as current passes through the junction.

An important application of Seeback thermoelectric effect is in *thermometry*.

21.7 EXTRINSIC SEMICONDUCTORS

We have seen that the conductivity of intrinsic semiconductors is very small and hence they are not suitable for any useful work except as heat or light sensitive resistances. The conductivity can, however be enormously increased by addition of suitable impurity in a very small proportion, i.e., nearly 1 in 10^6 parts of the semiconductor. This process is called doping. Doping is done after the semiconductor material has been refined to high degree of purity. A doped or impurity semiconductor is know as an extrinsic semiconductor.

21.7.1 N-Type Semiconductors

Germanium and Silicon are tetravalent. The impurity atoms may be either pentavalent or trivalent, i.e., from group V and III of the periodic table. If a small quantity of a pentavalent impurity (having 5-electrons in the outermost orbit) like Arsenic (As), Antimony (Sb) or Phosphorus (P) is introduced in Germanium, it replaces equal number of Germanium atoms without changing the physical state of the crystal. Each of the four out of five

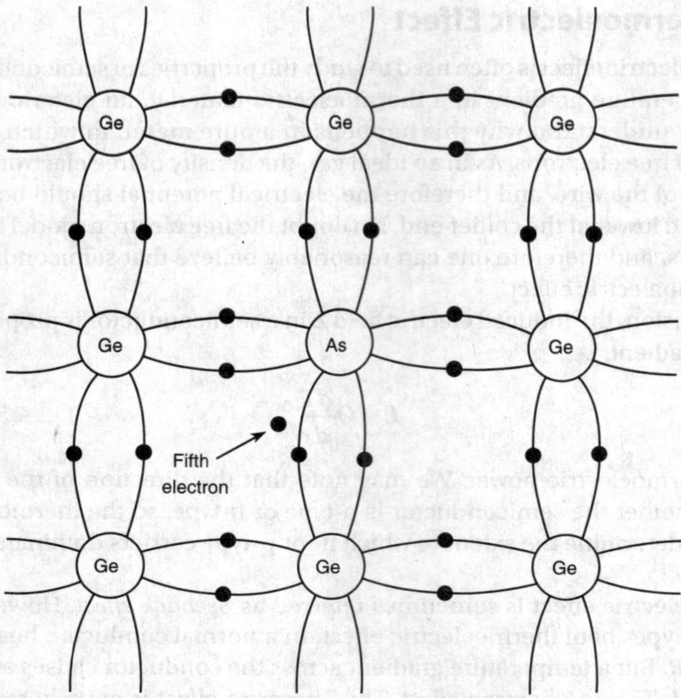

Fig. 21.13 N-type Semiconductor

valency electrons of impurity say of Arsenic enters into covalent bonds with Germanium, while the fifth valency electron is set free to move from one atom to the other as shown in Fig. (21.13). The impurity is called donor impurity as it donates electrons and the crystal is called N-type semiconductor. A small amount of Arsenic (impurity) injects billions of free electrons into Germanium thus increasing its conductivity enormously. In N-type semiconductors the majority carriers of charge are the electrons and holes are minority carriers. This is because when donor atoms are added to a semiconductor, the extra free electrons give the semiconductor a greater number of free electrons than it would normally have. And unlike, the electrons that are freed because of thermal agitation. donor electrons do not produce holes. As a result, the current carriers in a semiconductor doped with pentavalent impurities are primarily negative electrons. The impurity atom has five valence electrons. After donating one electron, it is left with +1 excess charge. It then becomes a positively charged immobile ion. It is immobile because it is held tightly in the crystal by the four covalent bonds as shown in Fig. (21.13).

It is important to understand that in N-type semiconductors, although electrons (negative charges) are the majority carriers, but the semiconductor doped with impurity remains electrically neutral. Free electrons and holes are generated in pairs due to thermal energy and negative charge of electrons donated by impurity atoms is exactly balanced by positive charge of the immobile ions. Representation of an N-type semiconductor in shown in Fig. (21.14), we have not shown Silicon or Germanium atoms in this figure. One

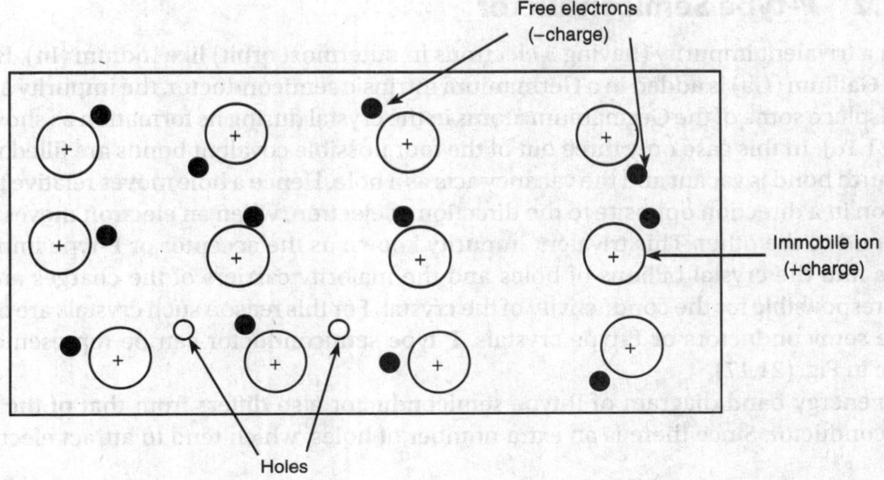

Fig. 21.14 Representation of an N-type Semiconductor

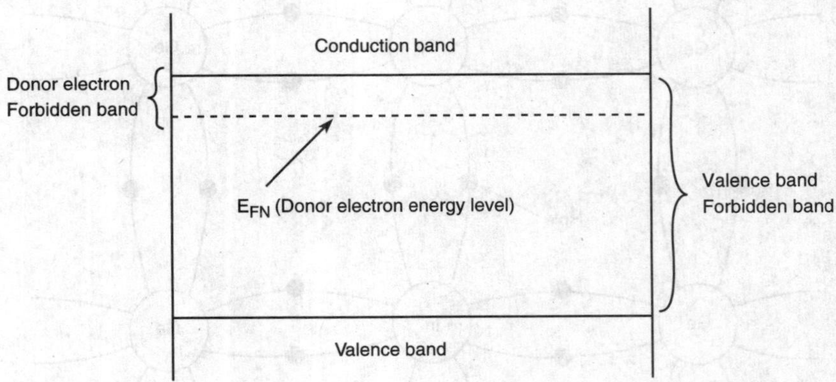

Fig. 21.15 Excess Free Electrons in N-type Semiconductors Produce a Donor Energy Level Close to the Conductor Band

should assume them as a continuous structure over the whole background. The fixed or immobile ions are regularly distributed in the crystal structure. The electrons and the holes, being free to move, are shown randomly distributed at any moment.

Since N-type semiconductors have extra free electrons, and pure semiconductors do not, the energy band diagram for a doped semiconductor is slightly different from that of a pure semiconductor. In effect, another energy level exists, a level for the donor electron, which is closer to the conduction band. The forbidden band for the donor electron is much narrower than the forbidden band for the valence electron, so one can see that it is much easier to cause electron flow in an N-type semiconductor Fig. (21.15).

21.7.2 P-type Semiconductor

When a trivalent impurity (having 3 electrons in outermost orbit) like Indium (In), Boron (B) or Gallium (Ga) is added in a Germanium intrinsic semiconductor, the impurity atoms will displace some of the Germanium atoms in the crystal during its formation as shown in Fig. (21.16). In this case only three out of the four possible covalent bonds are filled while the fourth bond is vacant and the vacancy acts as a hole. Hence a hole moves relative to the electron in a direction opposite to the direction of electron, when an electron moves from one bond to the other. This trivalent impurity known as the acceptor or P-type impurity injects into the crystal billions of holes and the majority carriers of the charges are the holes responsible for the conductivity of the crystal. For this reason such crystals are called P-type semiconductors or P-type crystals. P-type semiconductor can be represented as shown in Fig. (21.17).

The energy band diagram of P-type semiconductor also differs from that of the pure superconductor. Since there is an extra number of holes, which tend to attract electrons,

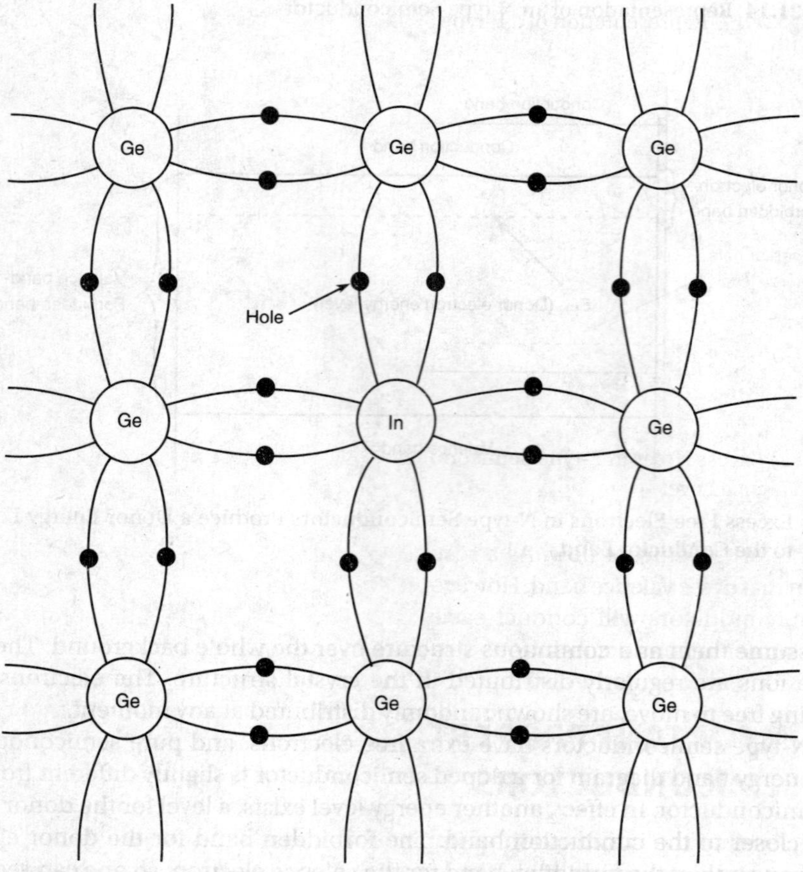

Fig. 21.16 P-type Semiconductor

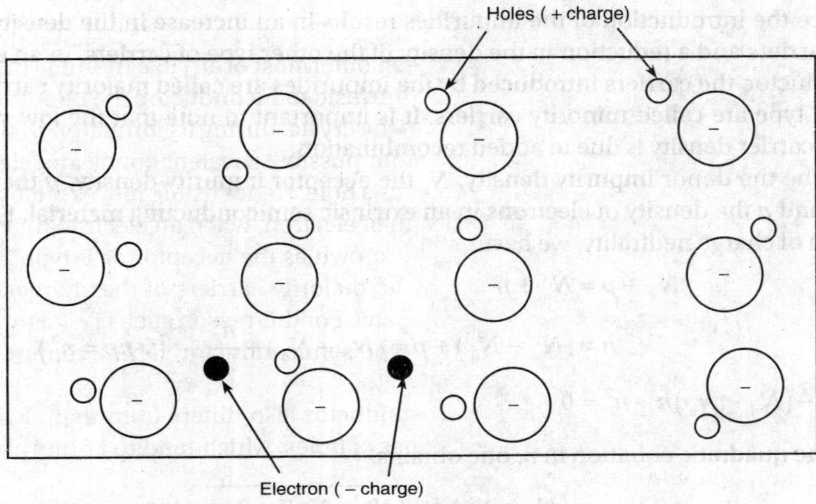

Fig. 21.17 Representation of a P-type Semiconductor

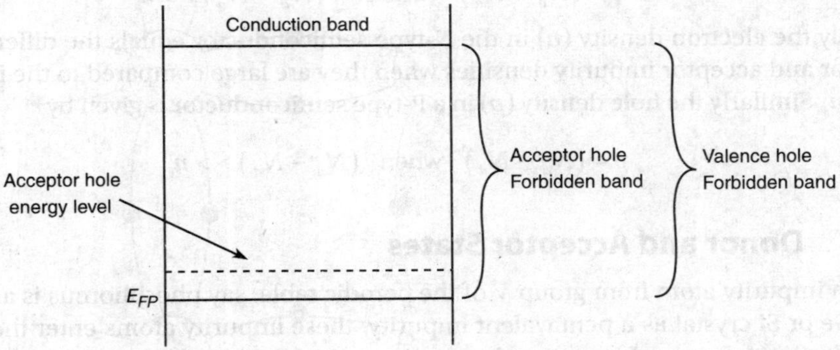

Fig. 21.18 **(a)** Excess Holes in P-type Semiconductors Introduce an Acceptor Energy Level Close to Valence Level

they aid in starting current flow. As a result. the acceptor energy level is also somewhat higher than that of the valence band. However, it is not as high as the donor level Fig. [21.18(a)]. P-type semiconductors will conduct easily than pure semiconductors, but not quite as easy as N-type semiconductors.

21.8 IMPORTANT PROPERTIES OF EXTRINSIC SEMICONDUCTORS

We have seen that the introduction of impurities in pure semiconductors increases the density of one type or another type of charge carriers. The product of holes and electrons in a semiconductor is constant depending on the width of energy gap and temperature and hence the introduction of the impurities results in an increase in the density of one type of carriers and a reduction in the density of the other type of carriers. In an extrinsic semiconductor, the carriers introduced by the impurities are called majority carriers and the other type are called minority carriers. It is important to note that the low value for minority carrier density is due to added recombination.

Let N_d be the donor impurity density, N_a the acceptor impurity density, p the density of holes and n the density of electrons in an extrinsic semiconducting material. From the condition of charge neutrality, we have

$$N_d + p = N_a + n$$

or $$n = (N_d - N_a) + p = (N_d - N_a) + \frac{n_i^2}{n} \quad (\because np = n_i^2)$$

or $$n^2 - (N_d - N_a)n - n_i^2 = 0 \tag{21.19}$$

Solving the quadratic equation in n, one obtains.

$$n = \frac{(N_d - N_a) \pm \sqrt{(N_d - N_a)^2 + 4n_i^2}}{2} \tag{21.20}$$

or $$n = (N_d - N_a) \quad \text{when} \quad (N_d - N_a) >> n_i$$

Obviously the electron density (n) in the N-type semiconductor equals the difference in the donor and acceptor impurity densities when they are large compared to the intrinsic density n_i, Similarly the hole density (p) in a P-type semiconductor is given by

$$p = (N_a - N_d) \quad \text{when} \quad (N_a - N_d) >> n_i$$

21.8.1 Donor and Acceptor States

When an impurity atom from group *V* of the perodic table, say phosphorous is added to a pure Ge or Si crystal as a pentavalent impurity, these impurity atoms enter the lattice by substitution for normal atoms, and not in interstitial positions. These impurity atoms contribute five electrons per atom to the valence band, i.e., we have an extra electron per impurity atom. These additional electrons (which cannot be accommodated in the valance band of the original lattice) occupy some discrete energy levels just below the conduction band, the separation may be a few tenths of an electron volt. These excess electrons are released by the impurity atoms and excited into the conduction band. The excited electrons then contribute to the electrical conductivity of the semiconductor. Conversely, the impurity may consist of atoms having fewer electrons than of a semiconductor Fig. [21.18(b)]. For the cases in which Si and Ge are the host substances, the impurity atoms could be boron or aluminium, each of which contributes only three electrons. In this situation the impurity introduces vacant discrete energy levels very close to the top of the valence band. Therefore it is easy to excite some of the more energetic electrons in the valance band into the impurity levels. This process produces vacant states or holes, in the valence band. These holes then act as positive electrons Fig. [21.18(c)]. We must note that to produce significant changes in the conductivity of a semiconductor, it is sufficient to have about one impurity atom per million semiconductor atoms.

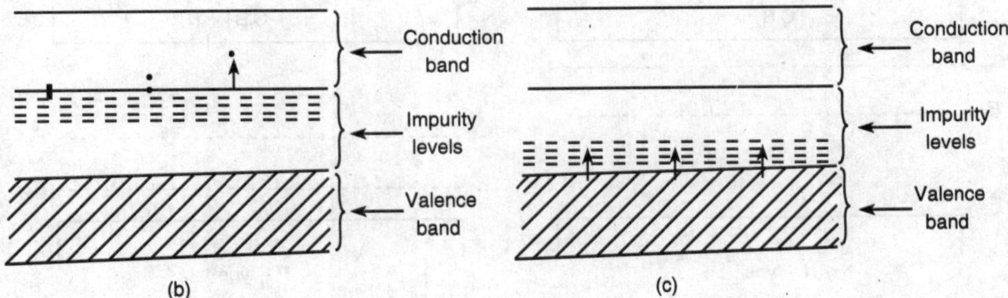

Fig. 21.18 Impurities in a semiconductor (b) Donors, or N-Type (c) Acceptors or P-Type

We have already seen that the crystal as a whole remains neutral because the electron remains in the crystal, We have further seen that the band diagram of N-type or P-type semiconductor differs from that of pure semiconductor. The band diagram for N-type or P-type semiconductor explains clearly why the conduction becomes possible in impure or extrinsic semiconductors at comparatively low temperatures.

21.8.2 Fermi Level in Extrinsic Semiconductor

We have read that Fermi level is situated in the middle of the band gap in an intrinsic semiconductor as the electron and hole densities are equal. When the intrinsic semiconductor is doped, the carrier densities change, consequently the position of the Fermi level also changes. The shift in the position of the Fermi level can easily be related to the majority carrier density in an extrinsic semiconductor, if it is assumed that the addition of impurities do not affect the densities of energy states in the conduction and valence bands.

Let N_c and N_v denote the density of states in the conduction band and density of states in the valence band respectively. We have for an intrinsic semiconductor

$$n = N_e \exp\left[\frac{E_{fi} - E_c}{kT}\right]$$

and

$$p = N_v \exp\left[\frac{E_c - E_\beta}{kT}\right] \tag{21.21}$$

Here E_β is the energy associated with the Fermi level in an intrinsic semiconductor. For an intrinsic semiconductor, we have $n = p$ and therefore from Eq. (21.22), we have

$$\frac{N_c}{N_v} = \exp\left[\frac{E_c - E_v - 2E_{fi}}{kT}\right]$$

Let E_{fn} be the energy associated with the Fermi level in an N-type semiconductor having an electron density n, we have

$$n = N_c \exp\left[\frac{E_{fn} - E_c}{kT}\right]$$

and

$$p = N_c \exp\left[\frac{E_v - E_{fn}}{kT}\right]$$

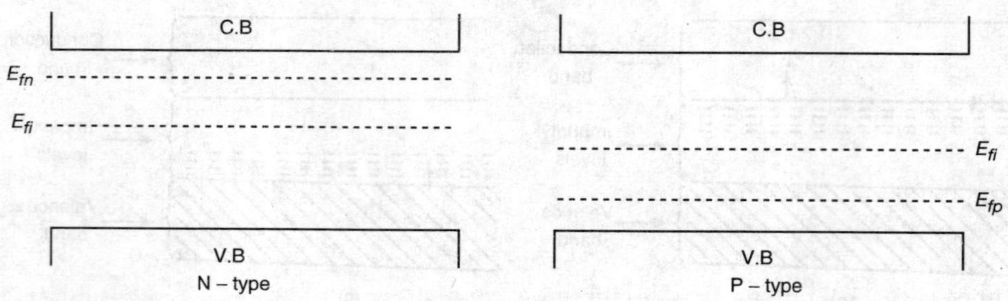

Fig. 21.19 Shift in the Fermi level in the N-and P-type of Semiconductors

$$\therefore \quad \frac{n}{p} = \frac{n^2}{n_i^2} = \frac{N_c}{N_v} = \exp\left[\frac{2E_{fn} - E_c - E_v}{kT}\right] = \exp\left[\frac{2(E_{fn} - E_{fi})}{kT}\right]$$

or
$$n = n_i \exp\left[\frac{E_{fn} - E_{fi}}{kT}\right] \tag{21.22}$$

Similarly, one obtains for a p-type semiconductor

$$p = n_i \exp\left[\frac{E_{fi} - E_{fp}}{kT}\right] \tag{21.23}$$

Thus the shift in the Fermi level in the n and p-type of semiconductor can be expressed as

$$\left.\begin{aligned} E_{fn} - E_{fi} &= kT\, In\frac{n}{n_i} \\ E_{fi} - E_{fp} &= kT\, In\frac{p}{p_i} \end{aligned}\right\}$$

Fig. (21.19) represents the shift the Fermi level in the N-and P-types of semiconductors.

21.8.3 Thermal Ionization or Extrinsic Semiconductors

When the temperature of an extrinsic semiconductor is raised above 0°K, the impurity atoms get ionized. Due to ionization the donor impurity atoms give rise to electrons in the conduction band and the acceptor impurity atoms give rise to holes in the valence band. These electrons or holes along with those generated by intrinsic action, then serve as the current carriers at a temperature. Two cases of interest are:

(i) Conduction electron concentration is equal to

$$n = n_0 \exp\left[(E_F - E_g)/kT\right] = N_d \tag{21.24}$$

This shows that under the present conditions the concentration of conduction electrons is approximately equal to the concentration of donors. This means that all the donors are ionized. Eq. (21.25) suggests that at room temperature the impurity concentration for Si and Ge upto 10^{14} to 10^{16} donors per cc suits this range, i.e., we dope an intrinsic semiconductor crystal with this concentration of donors, one can certainly predict that one will have ~ 10^{16} electrons/cc.

(ii) At higher temperature, the carrier concentration is proportional to $\sqrt{N_d}$.

21.8.4 Charge Densities in Extrinsic Semiconductors

The density of impurity atoms in N-and P-materials is so low compared to the density of semiconductor atoms that the rate of thermal pair generation is not affected appreciably the presence of impurity atoms. In the case of intrinsic semiconductors, we have seen that the concentration product

$$np = n_i^2 \tag{21.25}$$

was a constant at a given temperature.

Assuming all impurity atoms in extrinsic semiconductors to be ionized at the usual operating temperature, the free charge densities in impurity material can be based upon N_D, the donor atom density in N-material, or N_A the acceptor atom density in P-material.

The electrical neutrality of the material demands.

$$p + N_D = n + N_A \tag{21.26}$$

The L.H.S. of Eq. (21.27) gives the total positive charge as the sum of holes in the valence bond and the positive charge associated with the donor atoms have given up electrons and become positive ions. The R.H.S. of Eq. (21.27) sums the negative charge of free electron density n and the negative charge due to electrons held by the ionized acceptor atoms. Using Eq. (21.26), one can write

$$n = \frac{n_i^2}{p} \quad \text{and} \quad p = \frac{n_i^2}{n} \tag{21.27}$$

only donor impurities are introduced in N-type material, so $N_A = 0$. The donor density will be made much larger than the density of intrinsic holes, or

$$N_D >> p$$

and in N-type material the electron density is written from Eq. (21.27) as

$$n \cong N_D \tag{21.28}$$

on the similar reasoning for P-type material, we have $N_D = 0$ and $N_A >> n$. One obtains relations in P-type material as

$$n = \frac{n_i^2}{N_A} \tag{21.29}$$

From the above results we can conclude that the density of majority carriers approximated the impurity atom density at usual ambient temperatures and the density of minority carriers is reduced below the intrinsic level. This means that the increased electron density in N-type material raises the probability that an electron will meet and recombine with a hole, and so the hole density is decreased to maintain n^2 constant.

21.9 SEMICONDUCTOR DEVICES

A semiconductor device can be defined as a unit which consists, partly or wholly, of semiconducting materials and which can perform useful functions in electronic apparatus and solid state research. Examples of semiconductor devices are semiconductor diodes (P-N junction), transistors, intergrated circuits (1*c*) etc. Si, Ge and GaAs are most commonly used materials for the fabrication of semiconductor devices. The properties of these semiconductors are summarized in Table (21.2).

Almost all semiconductor devices are comprised of a single crystal semiconductor incorporating two or more semiconducting regions of different impurity density. The difference in the electric fields and carrier densities associated with differently doped

Table 21.2 Properties of Si, Ge, GaAs Semiconductors

Property	***Symbol***	***Unit***	***Value***		***GaAs***
			Germanium (Ge)	***Silicon (Si)***	
Atomic Numbers			32	14	
Atomic weight			72.6	28.08	144.63
Density		kg/m³	5.32×10^3	2.33×10^3	5.32×10^3
Atom Concentration		atoms/m³	4.4×10^{28}	5×10^{28}	2.21×10^{28}
Relative Dielectric Constant	ε_r		16	11.8	10.9
Band gap at 0 °K	E_{go}	eV	0.785	1.21	1.43
Band gap at 300 K	E_g	eV	0.72	1.1	1.32
Intrinsic Carrier (generation) at 300 K	n_i	Carriers m^3	2.5×10^{19}	1.5×10^{16}	9.0×10^{12}
Crystal Structure			Diamond	Diamond	Zincblende
Lattice Constant	a	Å	5.65748	5.43086	5.6534
Melting point		°C	936	1420	1250
Minority Carrier life time		second (s)	$\sim 10^{-3}$	$\sim 2.5 \times 10^{-3}$	$\sim 10^{-8}$
Breakdown Field		V/m	$\sim 10^{-7}$	$\sim 3 \times 10^7$	$\sim 4 \times 10^5$
Diffusion Constant	D_n (electrons)	m^2/s	0.009842	0.003367	0.001036
	D_n (holes)		0.004662	0.001295	0.000906
Effective Density of states in the conduction band	N_c	m^{-3}	1.04×10^{25}	2.8×10^{25}	4.7×10^{22}
Effective Density of states in the valence band	N_v	m^{-3}	6.4×10^{24}	1.02×10^{25}	7.0×10^{22}
Intrinsic conductivity	σ_i	Sm^{-1}	2.2428	0.4325×10^{-3}	1.2832×10^{-6}
Mobility (Drift) at 300 K	μ_n (electron)	$\dfrac{m^2}{V-s}$	0.38	0.13	0.85
	μ_p (holes)	$\dfrac{m^2}{V-s}$	0.18	0.05	0.04
work function	W	Volt (V)	4.4	4.8	4.7
Raman Phonon energy		eV	0.037	0.063	0.035

regions, called junctions, permit a wide range of essentially nonlinear conductivity effects in devices incorporating two, three or more distinct regions. Most semiconductor devices can be understood through the simplest of such junctions, called the P-N junction, which is a system of two semiconductors in physical contact, one with excess of electrons (N-type) and other with excess of holes (P-type).

21.10 P-N JUNCTION

When a P-type semiconductor is brought into contact with N-type semiconductor as the process of crystallization is taking place, the resulting combination is called a P-N junction. This junction has important properties and is, in effect the basis of modern semiconductor theory and practice. Most semiconductor devices contain one or more P-N junction. The most important characteristic of a P-N junction is its ability to conduct current in one direction only. In the reverse direction it offers very high resistance.

Formation of a P-N junction: Fig. (21.20) shows three types of such junctions. In Fig. [21.20(a)], P and N regions have been grown into the germanium block by mixing acceptor and donor impurities respectively, into the single crystal during its formation. This is known as a grown junction. It is worthwhile to mention that the grown type of P-N junction is not a sandwich made by attaching a P block to an N block, but actually consists of P and N layers in a single piece of germanium.

The diffused junction in Fig. [(21.20(b)] is made by placing a pellet of acceptor impurity, such as indium, on one face of a wafer of N-type germanium and then heating the combination to melt the impurity. Under proper conditions of temperature and time, a portion of the impurity metal will diffuse a short distance into the wafer. thereby creating a region of P-type germanium in intimate association with W-type bulk. This is also called an alloyed junction or fusion-alloy junction from the fact that a small amount of pellet material alloys with the germanium.

Fig. [(21.20(c)] shows a point contact type. Here a fine, pointed wire (catwhisker) makes pressure contact with the face of an N-type germanium wafer. After assembly, the device is electroformed by passing a high-current surge momentarily across the junction of Wafer and Whisker. The heat generated during the short interval drives a few electrons from the atoms in the region of the point contact, leaving holes and thus converting into p-type a

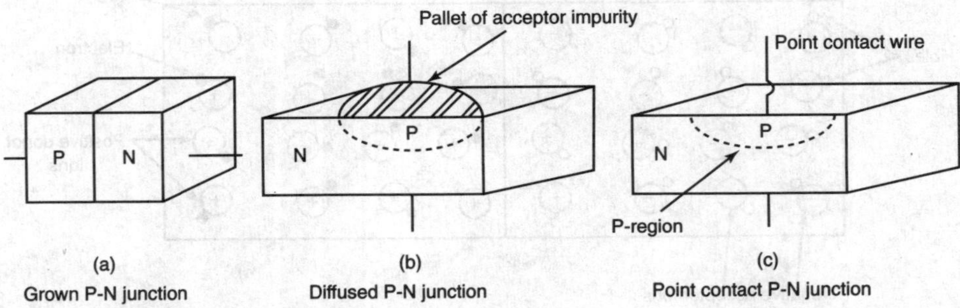

Fig. 21.20 Types of P-N Junction

small volume of germanium immediately under and around the point. Silicon P-N junction is produced in a similar manner. In most instances, the Silicon has been processed in such a way as to make it P-type. To create the junction, an N-type material is either inserted at the proper point in the crystal process or (in the diffused, junction process) an N-type material is later inserted into the body of P-type wafer. Like germanium diodes, silicon diodes also are produced both in junction and point contact types.

P-N junction with no external voltage: Fig. (21.21) shows a P-N junction just immediately after it is formed. There is no external voltage connected to the P-N junction. Since N-type material has a high concentration of free electrons while P-type material has a high concentration of holes, the following processes are initiated.

(i) At the junction, holes from the P-region diffuse into the N-region and free electrons from the N-region diffuse into the P-region. This process is called diffusion. Holes combine with the free electrons in the N-region whereas electrons combine with the holes in the P-region.

(ii) The diffusion of holes from P-region to N-region and electrons from N-region to P-region across the junction takes place because they move haphazardly due to thermal energy and also because there is a difference in their concentrations (The P-region has more holes whereas N-region has more free electrons) in the two regions.

(iii) As the free electrons move across the junction from N-type to P-type, positive donor ions are uncovered, i.e., they are robbed off free electrons. Hence a positive charge is built on the N-side of the junction. At the same time. the free electrons cross the junction and uncover the negative acceptor ions by filling in the holes. Therefore, a net negative charge is established on P-side of the junction. When a sufficient number of donor and acceptor ions is uncovered, further diffusion is stopped. It is because now, a barrier is set up against further movement of charge carriers. This is called potential barrier or junction barrier. The potential barrier is of the order of 0.1 to 0.3 Volt. Fig. (21.22) shows the electrostatic potential difference across the P-N junction. How this potential barrier is developed? When a sufficient number of donor and acceptor ions is uncovered, further diffusion

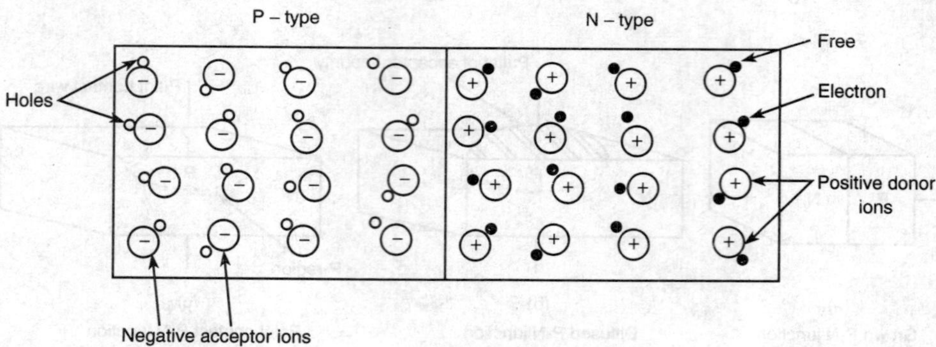

Fig. 21.21 A P-N Junction When Just formed

is prevented. It is because now positive charge (ions) on N-side repels holes to across from P-type to N-type and negative charge (ions) on N-side repels free electrons to enter from N-type to P-type. Because of this difference in potential exists between the two sections, which inhibits further electron-hole combinations at the junction, and the Fermi level of the two sides is in the same level as shown in Fig. [21.22(c)].

(iv) The region across the P-N junction in which the potential changes from positive to negative is called the depletioN-region. The width of this region is of the order of 6×10^{-8}m. Since this region has immobile (fixed) ions which are electrically charged, it is also called as the space-charge region. Outside this region on each side of the junction, the material is still neutral.

(v) The potential barrier for a silicon P-N junction is about 0.7V, whereas for a germanium P-N junction it is approximately 0.3V.

The potential barrier discourages the diffusion of majority carriers across the junction. However, the potential barrier helps minority carriers (few free electrons in the P-region and few holes in the N-region) to drift across the junction. The minority carriers are constantly generated due to thermal energy. But electric current cannot flow since no circuit has been connected to the P-N junction.

Width of Depletion Layer and Height of Potential Barrier: Let x_1 and x_2 be the width of depletion layer in P and N-sides respectively of a P-N diode junction, then total width of depletion layer

$$x = x_1 + x_2$$

Similarly if V_1 and V_2 are the potential barriers in P and N-sides of P-N junction, then the net junction potential barrier Fig. [21.22(b)]. The Poisson's equation in one dimension is

$$V_0 = V_1 + V_2$$

$$\frac{d^2V}{dx^2} = \frac{\rho_1}{\varepsilon} \quad (21.30)$$

where ε is the permittivity of medium. The charge density in P-side of depletion layer is given by

$$\rho = -eN_a \quad (21.31)$$

where N_a represent the density of acceptor ions.

Eq. (21.31) for P-side of depletion layer is given by

$$\frac{d^2V}{dx^2} = \frac{\rho N_a}{\varepsilon} \quad (21.32)$$

on integration, one obtains

$$\frac{dV}{dx} = \frac{eN_a}{\varepsilon}x + C_1 \quad (21.33)$$

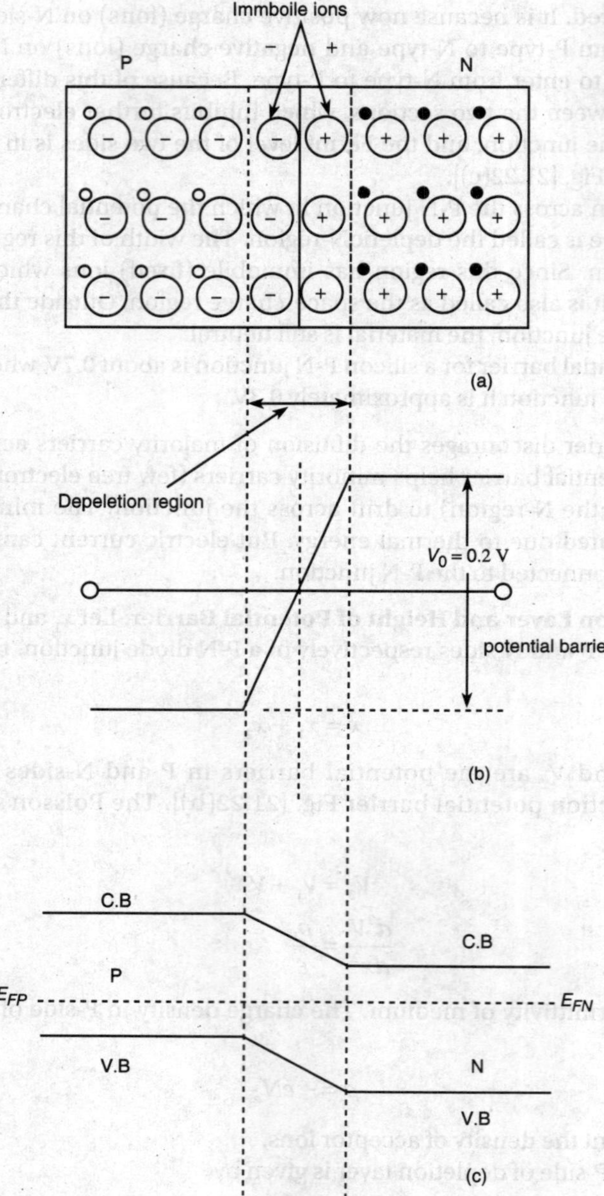

Fig. 21.22 Potential barrier across the P-N junction

where C_1 is constant of integration, using boundary condition, i.e., at

$$x = -x_1, \frac{dV}{dx} = 0$$

(since there is, no variation of potential in region excluding depletion layer). Eq. (21.34) gives

$$C_1 = \frac{eN_a}{\varepsilon} x_1$$

$$\therefore \quad \frac{dV}{dx} = \frac{eN_a}{\varepsilon} x_1 + \frac{eN_a}{\varepsilon} x_1 = \frac{eN_a}{\varepsilon}(x + x_1) \tag{21.34}$$

on integration, one obtains

$$V = \frac{eN_a}{\varepsilon}\left(\frac{x^2}{2} + x_1 x\right) + C_2$$

where C_2 is another constant of integration. Now at $x = 0$, $V = 0$ (at junction)

$$C_2 = 0$$

Hence

$$V = \frac{eN_a}{\varepsilon}\left(\frac{x^2}{2} + x_1 x\right) \tag{21.35}$$

At $\quad x = -x_1,\ V = V_1$

$$\therefore \quad V = \frac{eN_a}{\varepsilon}\left(\frac{x_1^2}{2} - x_1^2\right) = -\frac{eN_a x_1^2}{2\varepsilon}$$

$$\therefore \quad |V| = \frac{eN_a}{2\varepsilon} x_1^2$$

If N_d is density of donor ions in N-side of depletion layer, then Poission's equation for this becomes

$$\frac{d^2V}{dx^2} = -\frac{\rho}{\varepsilon} = -\frac{eN_d}{\varepsilon} \quad (\because \rho = eN_d) \tag{21.36}$$

Proceeding as above and applying boundary conditions that at $x = 0$,

$V = 0$ and at $x = x_2$, $\frac{dV}{dx} = 0, V = V_2$, one obtains

$$V_2 = \frac{eN_d x_2^2}{2\varepsilon} \tag{21.37}$$

$\therefore$ The height of the potential barrier across the junction is obtained as

$$V_0 |V_1| + |V_2| = e\left[\frac{N_a x_1^2 + N_d x_2^2}{2\varepsilon}\right] \tag{21.38}$$

Since the crystal as a whole is electrically neutral, the number of both sides of charge carriers must be equal, i.e.,

$$N_a x_1 A = N_d x_2 A$$

where A is the area of crystal

$$\therefore \qquad \frac{x_1}{x_2} = \frac{N_d}{N_a} \tag{21.39}$$

or

$$x_2 = \frac{N_a}{N_d} x_1$$

Substituting the value of x_2 in Eq. (21.39), one obtains

$$V_0 = \frac{\varepsilon N_a}{2\varepsilon} x_1^2 \left(1 + \frac{N_a}{N_d}\right)$$

This gives

$$x_1 \left[\frac{2\varepsilon V_0}{\varepsilon N_a \{1 + (N_a / N_d)\}}\right]^{\frac{1}{2}} = \left[\frac{2\varepsilon V_0}{e} \frac{(N_a / N_d)}{(N_a + N_d)}\right]^{\frac{1}{2}}$$

and

$$x_2 \left[\frac{2\varepsilon V_0}{eN_d \{1 + (N_d / N_a)\}}\right]^{\frac{1}{2}} = \left[\frac{2\varepsilon V_0}{e} \frac{(N_a / N_d)}{(N_a + N_d)}\right]^{\frac{1}{2}}$$

or

$$x = x_1 + x_2 = \left[\frac{2\varepsilon V_0}{e}\right]^{\frac{1}{2}} = \left[\left\{\frac{(N_d / N_a)}{(N_a + N_d)}\right\}^{\frac{1}{2}} + \left\{\frac{(N_a / N_d)}{(N_a + N_d)}\right\}^{\frac{1}{2}}\right]$$

or

$$x = \left(\frac{2\varepsilon V_0}{e}\right)^{\frac{1}{2}} \left[\left(\frac{N_a + N_d}{N_a N_d}\right)^{\frac{1}{2}}\right] \tag{21.40}$$

Width of the depletion region changes as square root of the voltage. In order to exemplify the order of thickness commonly met with, let us consider the example of Germanium PN diode in which $N_d = 10^{21}/\text{m}^3$ and $N_a = 10^{23}/\text{m}^3$. For these concentrations across the junction $V_0 = 0.31$ volt, and since for Germanium $\varepsilon_r = 16$ (Here $\varepsilon = \varepsilon_r \varepsilon_0$), one finds that the width of the unbiased junction is $x = 0.75 \times 10^{-6}$ m or 0.75 micron. We must note that in this numerical example $N_a >> N_d$. Obviously, Eq. (21.41) reduces to

$$x = \left(\frac{2\varepsilon V_0}{eN_d}\right)^{\frac{1}{2}} \tag{21.41}$$

This shows that heavier doping on the side of junction extends the depletion layer to the other side. Moreover, the depth of penetration or the width of the depletion region varies inversely as the square root of doping density on the opposite side.

Because of the presence of space charges of opposite kinds across the junction there is a voltage gradient confined within the depletion region. Obviously, the transition region thus acts like a parallel plate capacitor in which plates are separated by the distance x. The junction capacitance is thus

$$C_j = \frac{dQ}{dV} = \frac{\varepsilon}{x} A \quad \text{Farad} \tag{21.42}$$

where A is the area of the junction. Substituting Eq. (21.40) from Eq. (21.41), one obtains

$$\frac{C_J}{A} = \left[\frac{\varepsilon e}{2V_0}\frac{N_a N_d}{N_a + N_d}\right]^{\frac{1}{2}} \frac{\text{Farad}}{\text{m}^2} \tag{21.43}$$

For a special case, i.e., for $N_a >> N_d$, Eq.(21.44) reduces to

$$\frac{C_J}{A} = \left[\frac{\varepsilon e}{2V_0}N_d\right]^{\frac{1}{2}} \tag{21.44}$$

For the junction of 0.75μ in the above numerical example the junction capacitance from Eq. (21.43) or by substituting Eq. (21.42) in Eq. (21.43), one obtains.

$$\frac{C_J}{A} = \left[\frac{\varepsilon e}{2V_0}N_d\right]^{\frac{1}{2}} \tag{21.45}$$

For the junction width of 0.75 μ in the above numerical problem the junction capacitance from Eq. (21.43) works out as $C_J/A = 189\mu\text{F/m}^2$ or 18.9 mμF/cm². In a typical case if the junction area is $A = 10^{-3}$ cm², then we obtain its capacitance $C_J = 18$ pF.

Effect of Temperature on Barrier Voltage: The barrier voltage depends on doping density and temperature. For a given junction the doping density remains constant, therefore barrier voltage depends on temperature. With the rise in temperature, more minority charge carrier-s are produced, leading to their drift across the junction. Consequently the equilibrium occurs at a slightly lower barrier potential. Obviously, the barrier potential decreased with rise of temperature. It is found that for the both germanium and silicon semiconductors.

$$\Delta V_0 \simeq -2t\text{mV}$$

Here t °C denote the change in temperature, i.e., barrier potential decreases by about 2 mV per degree C.

21.11 FORWARD AND REVERSE BIASING

21.11.1 Forward Biasing

We have seen that the natural tendency of the majority carriers (free electrons in the N-section and holes in the P-section) was to combine at the junction. This is how the depletion region and potential barrier were formed. Actually, the combination of electrons and holes at the junction allows electrons to move in the same direction in both the P and N-sections. In the N-section, free electrons move toward the junction, in the P-section, for the holes to move toward the junction, valence electrons move away from the junction. Therefore, electron flow in both the sections is in the same direction. This, of course would be the basis of current flow.

With the P-N junction alone, the action stops because there is no external circuit and because of the potential barrier that build up. So, for current to flow, a battery can be connected to the diode to overcome the potential barrier. And the polarity of the battery should be such that the majority carriers in both sections are driven toward the junction. When the battery is connected in this way, it provides forward bias, causing forward or high current to flow, because it allows the majority carriers to provide the current flow.

To apply forward bias, positive terminal of the battery is connected to P-type and negative terminal to N-type as shown in Fig. (21.23). The applied forward potential establishes an electric field which acts against the potential barrier field. Obviously, the resultant field is weakened and the barrier height is reduced at the P-N junction, Fig. [21.23(b)] and Fig. [21.23(c)].

Since potential barrier height is very small (~ 0.2 V) and hence a small forward voltage is sufficient to completely eliminate the barrier. Obviously, at some forward voltage the

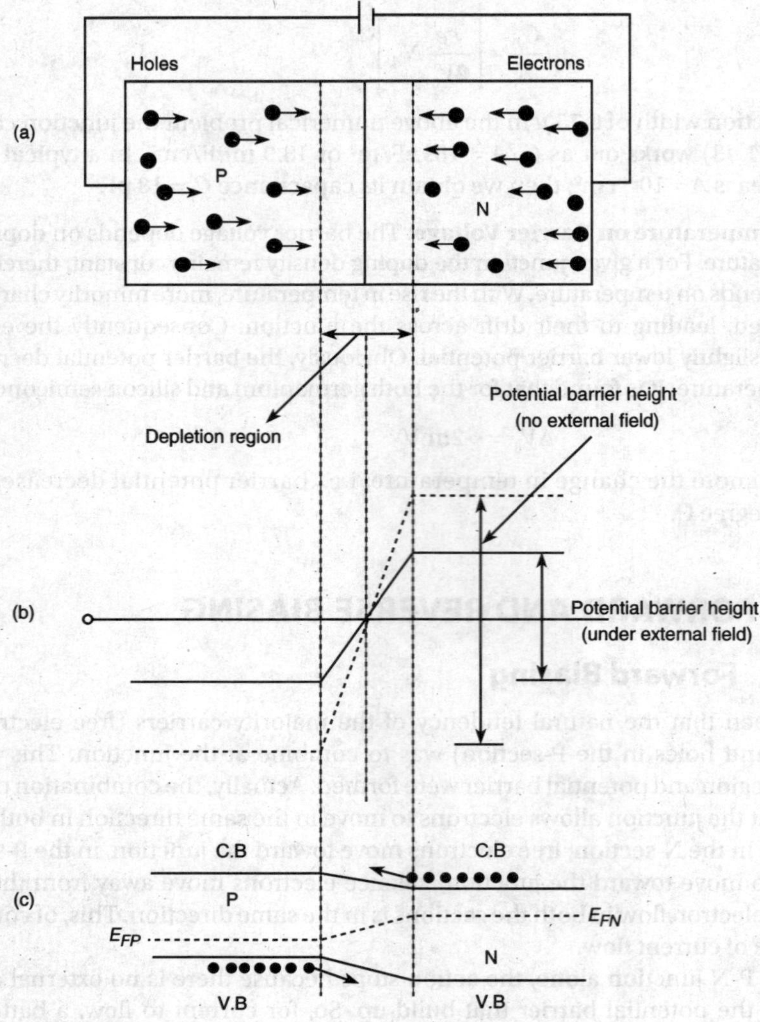

Fig. 21.23 P-N Junction Showing Forward Bias

potential barrier at the P-N junction can be eliminated altogether. Then the junction resistance will become almost zero and a low resistance path is established for the entire circuit. Thus a large current is generated in the circuit even for the small potential difference applied. Such a circuit is called forward bias circuit and the current is called forward current. The salient features of the forward bias circuit are summarized below:

(i) At some forward voltage, the potential barrier is eliminated altogether.
(ii) The P-N junction offers low forward resistance (r_f) to current flow.
(iii) The magnitude of current in the circuit due to the establishment of low resistance path depends upon the applied forward voltage, and it reduces as the voltage is increased there by forward current increases Fig. (21.28). It is given by

$$I = I_0\left(\left(e^{v/\eta V_T} - 1\right)\right)$$

where V = applied voltage (21.46)

$$V_T = \frac{T}{11600}, \text{ here } T \text{ is Temperature in Kelvin}$$

η = 1 for Ge and 2 for Si

The mechanism of current flow in a forward biased P-N junction is as follows:

(i) The free electrons from the negative battery terminal continue to arrive into the N-region while the free electrons from the negative battery terminal continue to arrive into the N-region while the free electrons in the N-region move towards the P-N junction.
(ii) The electrons travel through the N-region as free electrons. Obviously, current in N-region of the P-N junction is by free electrons.
(iii) When these free electrons reach the P-N junction, they combine with holes and become valence electrons. Since a hole is in the covalent bond and hence when a free electron combines with a hole, it becomes a valence electron.
(iv) Current in the P-region is by holes. The electrons travel through P-region as valence electrons.
(v) These valence electrons after leaving the crystal, flow into the positive terminal of the battery.

The current flow in a forward biased P-N junction is illustrated in Fig. (21.24).

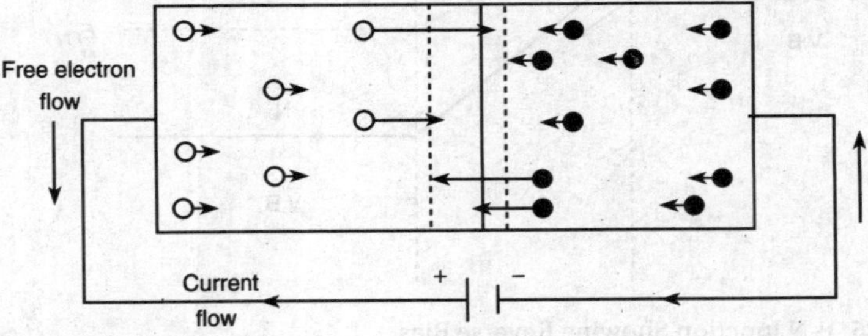

Fig. 21.24 Current Flow under Forward Bias

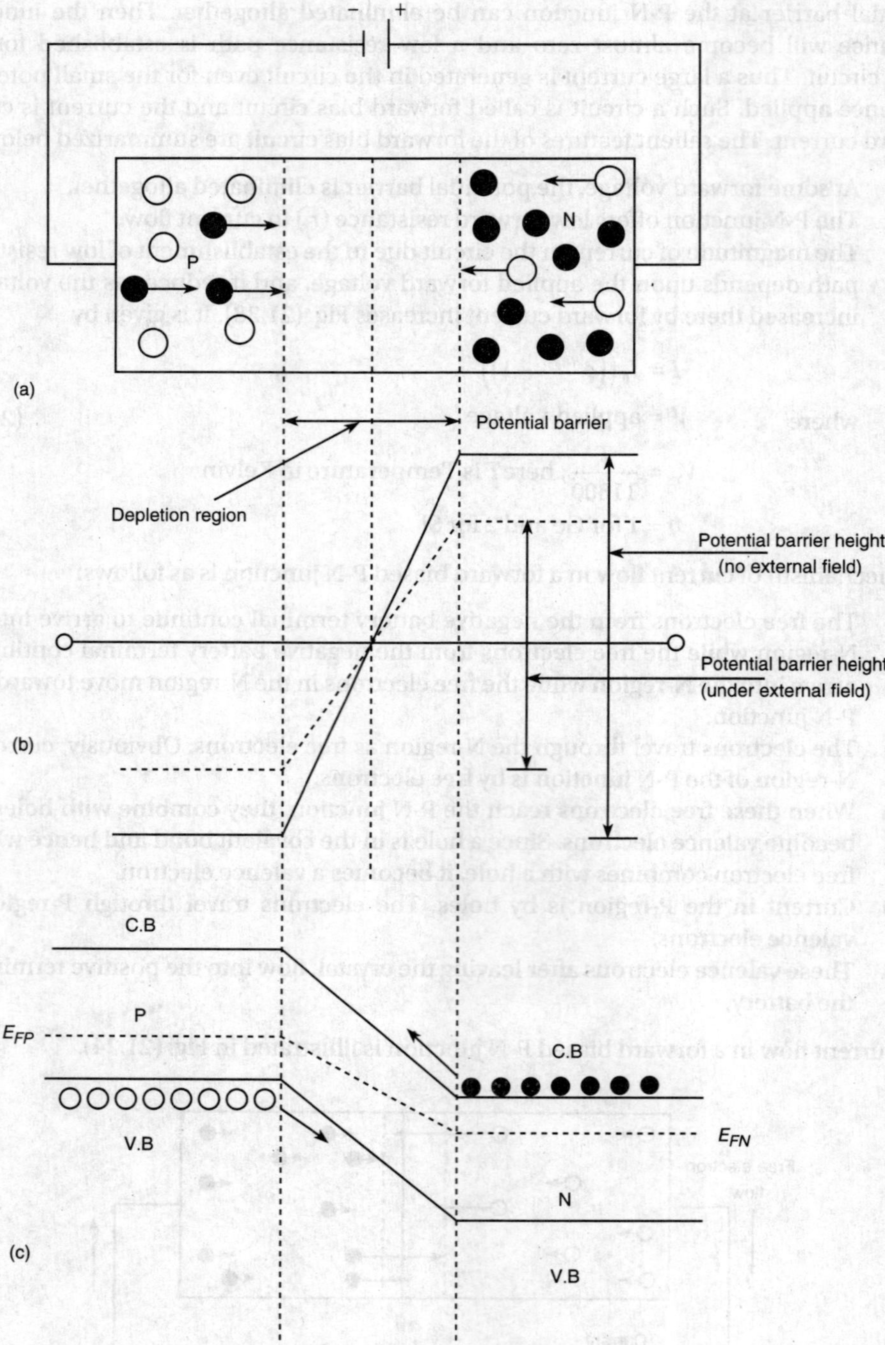

Fig. 21.25 P-N Junction Showing Reverse Bias

21.11.2 Reverse Biasing

We have seen that for forward current flow, the battery must be connected to drive the majority carriers towards the junction, where they combine to allow electrons to enter and leave the P-N junction. If the battery connections, are reversed, the potential at the N-side will draw the free electrons away from the junction, and the negative potential at the P-side will attract the holes away from the junction. With this battery connection then, the majority carriers cannot combine at the junction, and majority current cannot flow. For this reason, when a voltage is applied in this way, it is called reverse bias.

However, reverse bias can cause a reverse current to flow, because minority carriers are present in the semiconductor sections. Remember, that although the P-section is doped to have excess holes, yet some electrons are freed because of thermal agitation. Also, although the N-section is doped to have excess free electrons, some electrons are freed to produce holes in the N-section. The free electrons in the P-section, and the holes in the N-section are the minority carriers. Now, with reverse bias, one can see that the battery potentials repel the minority carriers toward the junction. As a result, these minority carriers cross the P-N junction in exactly the same way that the majority carriers did with forward bias. However, since there are much fewer minority carriers then there are majority carriers, this minority current, or reverse current as it is usually called, is much less and is of the order of μA, with the same voltage than majority or forward, current would be. Reverse bias P-N junction is shown in Fig. [21.25(a)]. The salient features of reverse biased P-N junction are following:

(i) The height of the potential barrier is increased and width of depletion region also increases, Fig. [21.25(b)] and Fig. [21.25(c)].

(ii) The reverse bias P-N junction offers very high resistance to current flow. This resistance is called **reverse resistance** (R_r).

(iii) Due to the establishment of high resistance path, very small current flows in the circuit. This current is usually called as the leakage current or reverse saturation current.

(iv) The current in the above situation is given by

$$I = I_0 \quad \left(e^{-V/\eta V_T} - 1\right) \tag{21.47}$$

Since applied voltage V is negative and as it increases the first term in R.H.S. reduces very fast and current from Eq. (21.48) reduces to $I = (-)I_0$ which is quite low and constant.

21.12 VOLT-AMPERE CHARACTERISTICS OF P-N JUNCTION

The P-N junction can be represented by a symbol of arrow and dash as shown in Fig. (21.26). The arrow head represents the P-section of the crystal and shows the direction of flow of holes or conventional current. Since the P-N junction diode's resistance changes according to the direction of current flow and hence it is called a nonlinear device. Basically, its nonlinearity is dependent on the polarity of the applied voltage. For current in the forward direction, it has a resistance of only a few hundred ohms. In the reverse direction, its resistance is often close to 100,000 ohms.

Fig. 21.26 Symbol of PN Junction

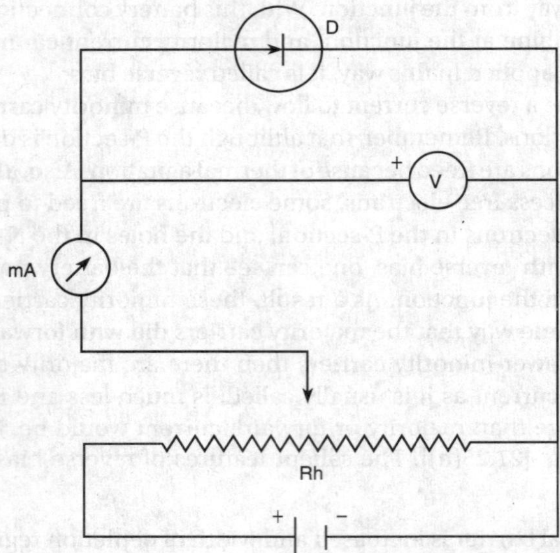

Fig. 21.27 Circuit Arrangement for Drawing V-I Characteristics of a P-N Junction

Fig. (21.27) shows the circuit arrangement for drawing the V-I characteristics of a P-N junction diode.

A graph between the potential difference across the P-N junction and the current through the Junction is called the V-I characteristic of the P-N junction diode and is shown in Fig. (21.28).

In the forward direction, though considerably more current flows and the current for the most part increases linearly as the bias voltage is increased. In the forward direction, then the P-N junction can be considered a linear device over a large portion of its operating curve. The small portion of the curve that is just above zero bias is nonlinear. This results because both majority and minority current actually comprise the overall current. Since the minority carriers are low energy carriers, majority current starts first, and then as the voltage is raised, minority current joins in, causing a nonlinear rise in current. But as the voltage is increased further, minority current becomes saturated since there are only few minority carriers. The curve then follows the majority current increase which is linear.

Because of the non-linearity of the curve, if a very small signal voltage is applied to the diode so that it only operates around the knee. the signal will be distorted. The signals must be large enough so that they operate mostly over the linear part of the curve.

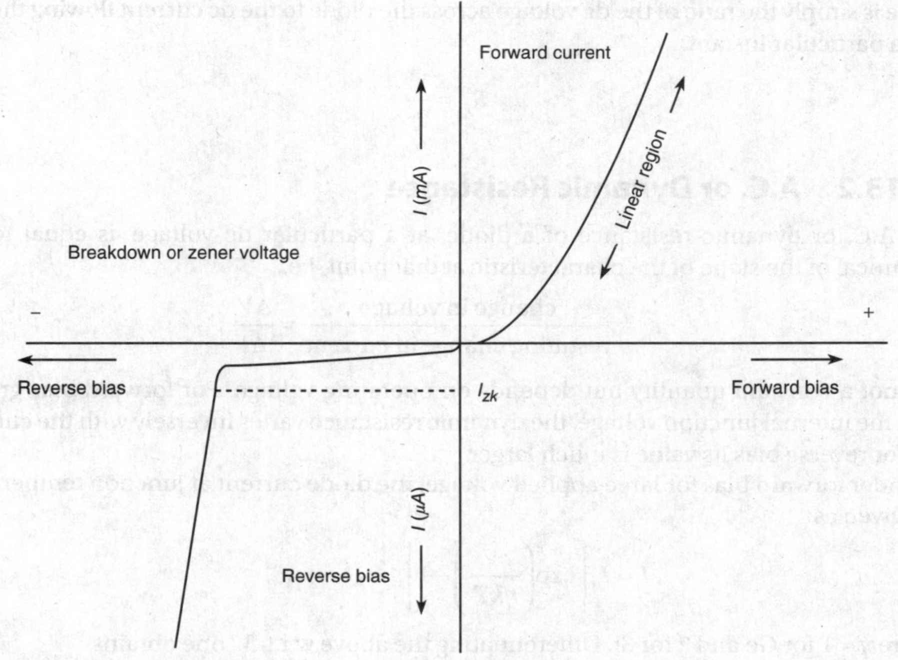

Fig. 21.28 V-I Characteristics of a P-N Junction Diode

When reverse bias is applied, a slight reverse current flows. This reverse current increases only negligibly as the bias voltage is increased a lot (20 to 25 volts). At this stage current suddenly rises in the reverse direction due to the breakdown of the crystal, i.e., covalent bonds of the crystal are broken in very large number. This breakdown reverse bias is called the Breakdown or zener voltage. Diodes are also designed to produce a useful wide range of Zener breakdown region, and are used in special voltage-regulating circuits. We will discussv Zener diodes later on.

21.13 STATIC AND DYNAMIC RESISTANCE OF A DIODE

One of the most important properties of the diode is its resistance in the forward and reverse biasing. An ideal diode must offer zero resistance in forward bias and infinitely large resistance in reverse bias. Truly speaking, no diode can act as an ideal diode, i.e., an actual diode does not behave as a perfect conductor when forward biased and as perfect insulator when reverse biased. We consider the two resistances of the diode

(i) D.C. or static resistance, (ii) A.C. or dynamic resistance

21.13.1 D.C. or Static Resistance

When P-N junction diode is forward biased, it offers a definite resistance in the circuit. This resistance is called as the dc or static resistance (R_F) of the diode, dc resistance of diode is simply the ratio of the dc voltage across the diode to the dc current flowing though it at a particular instant.

$$R_F = \frac{V}{I}$$

21.13.2 A.C. or Dynamic Resistance

The A.C. or dynamic resistance of a diode, at a particular dc voltage, is equal to the reciprocal of the slope of the characteristic at that point, i.e.,

$$r_f = \frac{\text{change in voltage}}{\text{resulting change in current}} = \frac{\Delta V}{\Delta I}$$

It is not a constant quantity but depends on operating voltage. For forward bias greater than the internal junction voltage, the dynamic resistance varies inversely with the current, but for reverse bias its value is much larger.

Under forward bias for large applied voltage, the diode current at junction temperature T is given as

$$I = I_0\left[\exp\left(\frac{eV}{\eta kT}\right) - 1\right]$$

where $\eta = 1$ for Ge and 2 for Si. Differentiating the above w.r.t. V, one obtains

$$\frac{dI}{dV} = I_0\left(\frac{e}{\eta kT}\right)e^{eV/\eta kT} = 1\left(\frac{e}{\eta kT}\right)$$

$$\therefore \qquad r_f = \frac{dV}{dI} = \frac{\eta kT}{Ie} = \frac{\eta kT}{Ie} = \eta\left(\frac{k}{e}\right)\frac{T}{I}$$

$$= \eta\left(\frac{1.38\times 10^{-23}}{1.6\times 10^{-19}}\right)\frac{T}{I}$$

or $$r_f = 8.625\times 10^{-5}\eta(T/I)$$

21.14 SPACE CHARGE (OR DEPLETION REGION) CAPACITANCE

When no external voltage is applied, the width (d) of the depletion region of a P-N junction diode comes out to be about 5×10^{-7} m and is mostly in the N-type material. The movement of majority carriers, i.e., holes and electrons across the junction causes opposite charge to be stored at this distance d apart. This is effectively a parallel plate capacitor whose capacitance C_T (often called space-charge capacitance) is approximately 20 pF with no external bias.

As forward bias is applied the depletion region decreases and the capacitance (C_T) increases. Under reverse-bias conditions, the depletion region increases and C_T decreases. This property of voltage variable capacitance is made use in varactors, varicaps or voltacaps.

Space charge or Transition Capacitance: The thickness of depletion layer changes with voltage applied across the junction. Reverse biasing of a P-N junction, causes majority carriers to move away from the junction, therefore uncovering more immobile charges. This increases the thickness of depletion layer. Similarly the forward biasing of a P-N junction decreases the amount of the uncovered charge and hence the thickness of depletion layer. This change in the uncovered charge with the applied voltage may be considered as a capacity effect. It gives rise to transition capacitance or junction capacitance C_T given by

$$C_T = \frac{dQ}{dV} = \frac{\varepsilon A}{x}$$

Here x is thickness of depletion layer and A is junction cross-sectional area. One can show that the width x of depletion layer is given by

$$x = \left[\frac{2\varepsilon V_0}{e}\frac{N_a + N_d}{N_a N_d}\right]^{\frac{1}{2}}$$

$$\therefore \qquad C_T = \left[\frac{\varepsilon e}{2V_0}\frac{N_a N_d}{N_a + N_d}\right]^{\frac{1}{2}} A$$

or $\qquad C_T \propto V_0^{-1/2}$

Obviously, the transition or junction capacitance is not constant but depends on applied voltage and hence it is also known as voltage-variance capacitance.

Diffusion of Storage Capacitance C_D: When forward biased the P-N junction develops a dominating diffusion capacitance C_D. This arises due to injected charge stored near the junction outside the transition region and is defined as the rate of change of injected charge with applied voltage, i.e.,

$$C_D = \frac{dQ}{dV}$$

C_D is found to be proportional to current (I) and is much large than the C_T C_D has larger value in forward direction and smaller value in reverse direction. The value of C_D in forward directions varies from 500 pF to 200 pF. Obviously, C_D predominates under forward bias and arises due to minority carriers, while C_T predominates under reverse bias and arises due to majority carriers.

21.15 EFFECT OF TEMPERATURE ON P-N JUNCTION DIODES

We know that temperature rise boosts the generation of electron-hole pairs in semiconductors and increases their conductivity. From a consideration of the energies of the carriers crossing the depletion region in a P-N junction diode, an involved calculation yields the following relation between diode current, voltage and temperature.

$$I = I_0 \exp[V/\eta V_T - 1] \tag{21.48}$$

Here

I = the diode current (forward if positive, reverse if negative).
I_0 = the diode reverse current, called the reverse saturation current, at temperature T.
V = the diode voltage, positive for forward bias, negative for reverse bias; in volts.
η = I for germanium and 2 for silicon.
V_T = $T/11600$, a quantity in volts, dependent upon the temperature and is known as volt temperature equivalent.
T = Temperature of the diode junction (°K).

On increasing the temperature the forward characteristic (Fig. (21.28) shifts to left, showing increase in current for same voltage and it shifts to right when temperature is decreased showing biased current.

21.16 ZENER DIODE

We have seen that in the breakdown region, large changes in diode current produce only small changes in diode voltage. So a semiconductor P-N diode designed to operate in the breakdown region may be employed as a constant voltage device. The diodes used in such a manner are called avalanche breakdown or zener diodes.

They are used as a voltage regulator in the manner shown in Fig. [21.29(b)]. The voltage source V and the resistance R are so selected that the diode operates in the breakdown region. The diode voltage in this region which is also the voltage across the load R_L is called zener voltage (V_Z) and the diode current is called zener current (I_Z). As the load current (I_L) or the supply voltage changes, the diode accommodates itself to these changes and maintains nearly constant load voltage (V_Z).

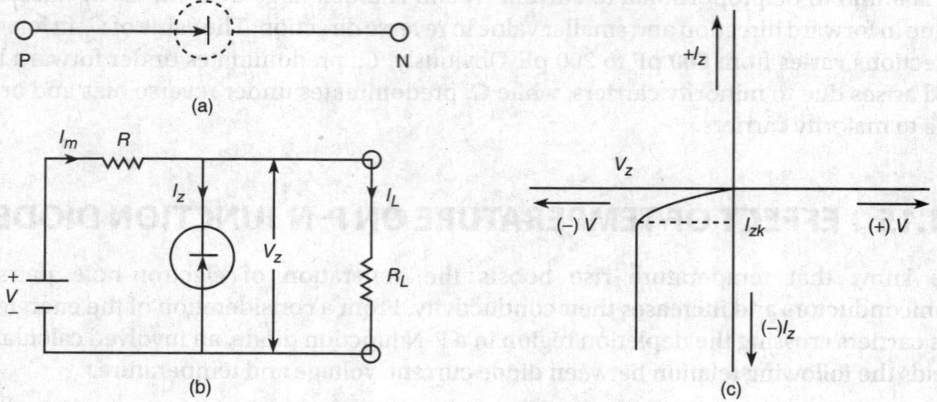

Fig. 21.29 (a) Symbol of Zener Diode. (b) Zener Diode as a Voltage Regulator

The diode will continue to regulate the voltage until the diode current falls to knee current I_{Zk} Fig. [21.29(c)]. Depending upon the nature of the semiconductor and its doping, the breakdown voltage in diode ranges from about 3 volt to several hundred volts. The breakdown phenomenon is reversible and harmless so long as the safe operating temperature is maintained. The mechanism of diode breakdown at reverse voltage is explained below:

(i) Avalanche Breakdown: In this mechanism, the minority charge carriers (electrons in P-type and holes in N-type) acquire sufficient energy from the applied reverse voltage to produce new charge carriers by removing valence electrons from the covalent bonds. The new carriers in turn produce additional charge carriers and the process multiplies to give large reverse currents. The diode is then said to be in the region of avalanche breakdown, usually, a junction with a broad depletion layer (therefore a low field intensity) break down by this mechanism. With the increase of temperature, the vibrations of the atoms in the crystal increase which increases the possibility of collisions of the charge carriers with the lattice atoms and reduce the possibility for the carriers to gain sufficient energy to start avalanche process. Thus at high voltage, the avalanche process is prominent and does take place to cause diode breakdown. The operating voltages in such diodes range from several volts to a few hundred volts.

(ii) Zener Breakdown: In this mechanism, the breakdown is initiated though a direct rupture of covalent bonds rather than avalanche process due to the existence of strong electric field across depletion layer. A junction having a narrow depletion layer $\left(\text{and hence high field intensity } E = \dfrac{V_r}{d}\right)$ will break down by this mechanism. An increase in temperature increases the energy of the valence electrons and makes it easier for these electrons to escape from these covalent bonds. Smaller applied voltage is, therefore required to pull these electrons from the crystal lattice. The zener effect plays an important role only in diodes with breakdown voltage below about 6 volts. The zener diode is always used in reverse biased condition Fig. [21.29(b)]. The range of voltages about the breakdown voltage in which zener diode conducts in reverse direction's called tolerance.

We must note that during manufacturing, it is very difficult to have exact doping for every zener diode of the same number, i.e., type. Obviously, breakdown voltages of zeners of same number also differ slightly. This range of breakdown voltages for the same type zener diodes is referred as tolerance. Let us consider, for example, a particular type of zeners marked 9V, 10% tolerance. These zener diodes may have breakdown voltage from 8.1V (9 – 0.9) to 9.9 V (9 + 0.9) instead of sharp 9V for all.

The breakdown voltage of zener diode depends upon operating temperature. It is found to decrease with increase in junction temperature. This is due to the increased reverse current, (i.e., increase in minority carriers) that flows with increasing temperature. The decrease in breakdown voltage is about 2 mV per degree centigrade rise in temperature. The maximum power which zener diode can dissipate (or handle) without damage is referred as its power rating and denoted by P_{ZM}. Zener diodes for the commercial purposes available in the market have power rating from 1/4 W to more than 50 W.

Power rating (P_{ZM}) is the product of maximum current I_{ZM} which a zener diode can handle and the rated or operating voltage of a zener diode (V_Z), i.e., $P_{ZM} = I_{ZM} V_Z$ A data sheet, sometimes includes the maximum current rating of a zener diode. The maximum value of current which a zener diode can handle at its rated voltage without damage is referred as its **maximum current rating** (I_{ZM}).

When a zener diode is operated in the breakdown region, at increase in current produces a slight increase in voltage. This means that a zener diode has a small a.c. resistance. This a.c. resistance is called **zener resistance** (often referred as **zener impedance** Z_{ZT}). Sometimes, we find the specification about this resistance on the data sheets supplied by the manufactures.

The opposition offered to the current flowing through the zener diode in the operating region is known as **zener resistance** (R_Z) **or Zener impedance** (Z_Z).

Zener diodes find wide commercial and industiral applications, e.g., voltage stablizer meter protection, wave shaping, etc.,

21.16.1 Constant Current Diode

These are the diodes that work exactly opposite to Zener diode. These diodes keep the current constant flowing through them when the voltage changes, i.e., instead of the holding the voltage constant, these diodes holds the current constant. The range of voltage over which a diode can keep the current constant is known as voltage compliance. These diodes are optimised for a particular voltage compliance.

21.17 TUNNEL DIODE

It is a device just like a P-N junction, which offers negative resistance under certain bias conditions. The negative resistance of the diode is due to tunneling and it is called a tunnel diode or Esaki diode. It is made very much like an ordinary P-N junction diode, except that both the P and N regions are heavily doped ($1:10^4$) thousand times more than ordinary diode. It is used as an active device in electronic circuits in the frequency range of few megahertz. The semiconductors with very high impurity concentration are referred to as **degenerate semiconductors**. Typical tunnel diode characteristic is shown in Fig. [21.29(a)].

As a result of high impurity levels, the contact potential is high, the depletion layer is very narrow and the Fermi levels lie in the conduction band for N-side and in the valence band of the P side Fig. [(21.29(b)]. Under these circumstances, a very abrupt transition from P to N-type material is achieved within the crystal. Since depletion region is very narrow, this gives rise to extremely large electric fields.

Tunnel effect: When a P-N semiconductor is heavily doped, (it has many majority carriers and ions), and under forward biased the hole and valence electron random drift is heavy. As a result, it is not uncommon for a large number of electrons to fill holes and release energy to only a few other valence electrons. These few valance electrons, then have their energy levels raised considerably so that they can cross from the N to

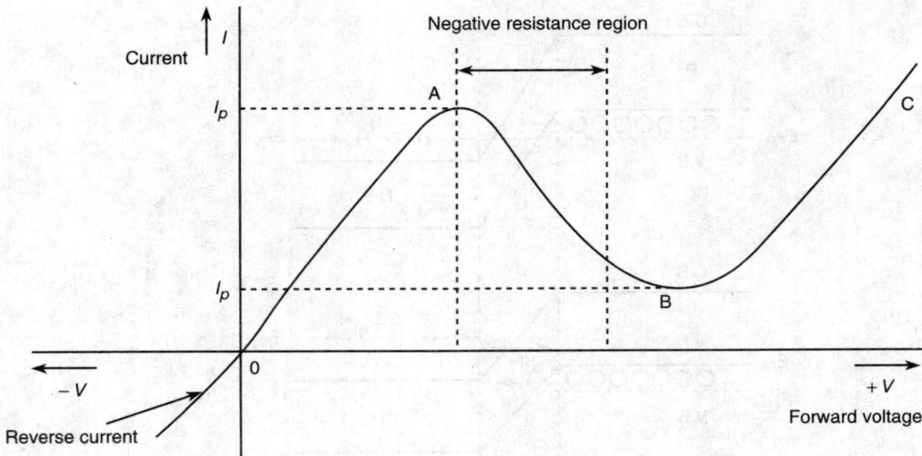

Fig. 21.30 (a)

the P section and current increases as shown by OA Fig. [21.30(a)], even with little or no applied voltage. This action which seems to allow a valence electron to cross a potential barrier without enough applied external energy is called the tunnel effect because it seems as though the valence electron tunnels' though the forbidden band as shown in Fig. [21.30(c)].

On further increasing voltage the barrier height decreases Fig. [21.30(d)] and Fermi level E_{FN} is much raised so that current decreases as shown by AB Fig. [21.30(a)]. This region is called **negative resistance region**. On further increasing voltage the conduction band of N-type is in level with CB of P-type and free electrons of N-type easily cross to P-type thereby current BC again increases like ordinary diode as shown by BC Fig [21.30(a)].

The negative resistance region AB Fig. (21.30) allows the diode to be used as an oscillator. It can also be used as an electronic switch since it has a good response in negative resistance region. By its nature, the tunnel diode has a rather high reverse current, but operation under this condition is not generally used.

The equivalent circuit of tunnel diode is as shown in Fig. (21.31). The series resistance R_s is the ohmic resistance. The series inductance Ls depends upon the lead length and geometry of the diode package. The junction capacitance C depends upon the bias and is usually measured at the valley point. The negative resistance $[-R_n]$ has a minimum at the point of inflection between peak current I_p and valley current I_V, Typical values for these parameters for a tunnel diode having peak current value $I_p = 10$ mA are $R_n = (-)30\Omega$, $R_s = 1\Omega$, $L_s = 5$ nH and $C = 20$ pF.

It is a voltage controllable device (like tetrode). Since between I_p and I_v, we have three values of voltage for a given current and on account of its multivalued feature it is used in pulse and digital circuiting.

Our main interest in the tunnel diode is its application as a very high speed switch. Since tunneling takes place at the speed of light, the transient response is limited only by total shunt capacitance (junction plus stray wiring capacitance) and peak driving current.

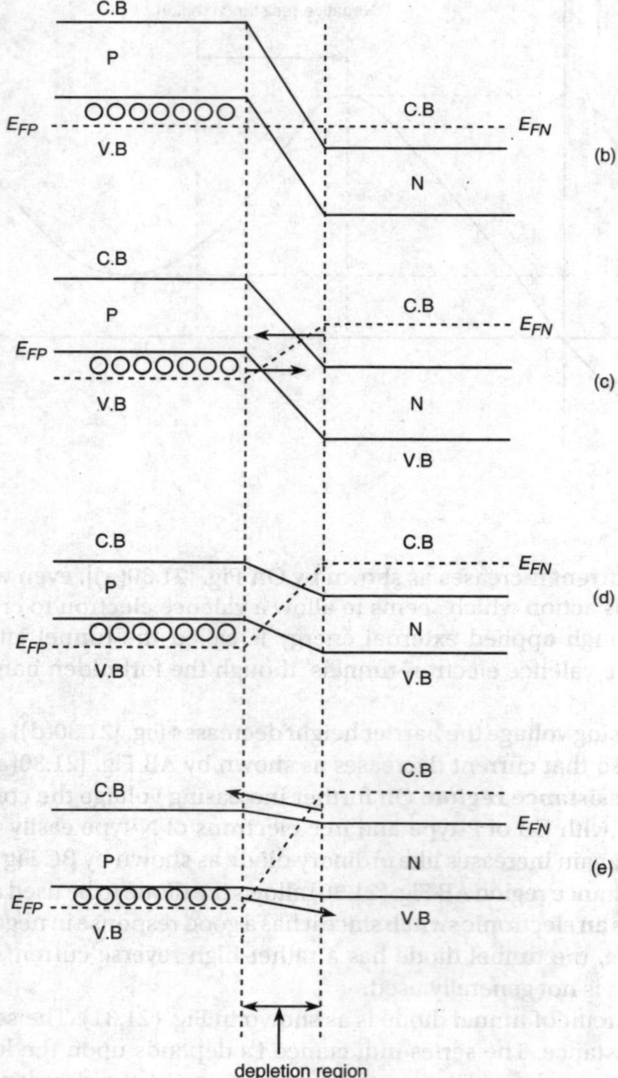

Fig. 21.30 (b–c) Tunnel Diode-Characteristics and Energy Levels

Switching times of the order of a nanosecond are reasonable and time as low as 50 p.secs have been obtained.

The advantages of a tunnel diode are (i) low cost, (ii) low noise, (in) high speed, (iv) environmental immunity and (v) low power.

The disadvantages of the diode are its low output voltage swing and the fact that it is a two terminal device unlike ordinary diode in which current flows only when forward biased. Because of this latter feature, there is no isolation between the input and output and it leads to serious circuit design problems.

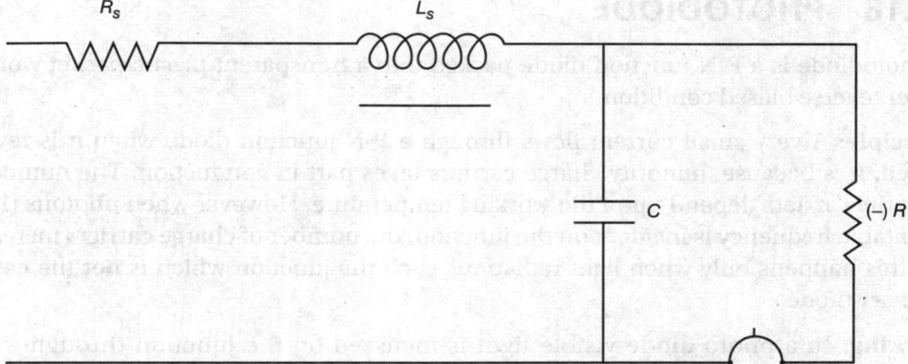

Fig. 21.31 Equivalent Circuit of a Tunnel Diode

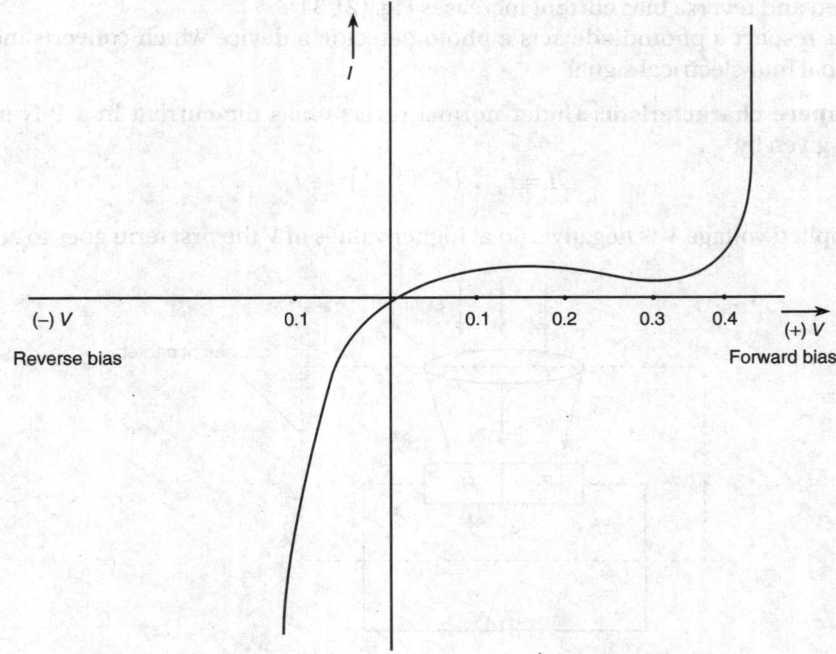

Fig. 21.32 V-I Characteristic of Backward Diode

However we may have a special type of tunnel diode whose peak value current is of the order of valley current. Such a diode is called Backward Diode. V.I Characteristic of a backward diode is shown in Fig. (21.32). We may use it under reverse biased condition as ordinary diode with the advantage that now the diode current is available just beyond zero volt whereas in ordinary diode, current is not available till applied voltage is more than (break over voltage) 0.7 volt in case of silicon and so in case a back diode it is conveniently used as rectifier for *V* small (mV) input signals.

21.18 PHOTODIODE

A photodiode is a P-N Junction diode packed into a transparent plastic packet working under reverse biased condition.

Principle: A very small current flows through a P-N junction diode when it is reverse biased. It is because, minority charge carriers takes part in conduction. The number of minority carriers depend upon the working temperature. However when photons (light) of suitable frequency is incident on the junction, the number of charge carriers increases. But this happens only when light radiations such the junction which is not the case in ordinary diode.

Working: In a photo diode visible light is focussed on the junction through a lens Fig. (21.33). Light on being incident produces free electrons and holes. Thus number of charge carriers increases and current increases. Generally a photodiode is optimized for its sensitivity to light. As the light intensity increases more and more, charge carriers are generated and reverse bias current increases Fig. (21.34).

In this respect a photodiode acts a photo detector, a device which converts incoming light signal into electrical signal.

Volt-Ampere characteristic: Under normal reverse bias the current in a P-N junction diode is given by

$$I = I_0 \quad \left(e^{-V/\eta kT - 1}\right) = -I_0$$

since applied voltage V is negative. So at higher values of V the first term goes to zero.

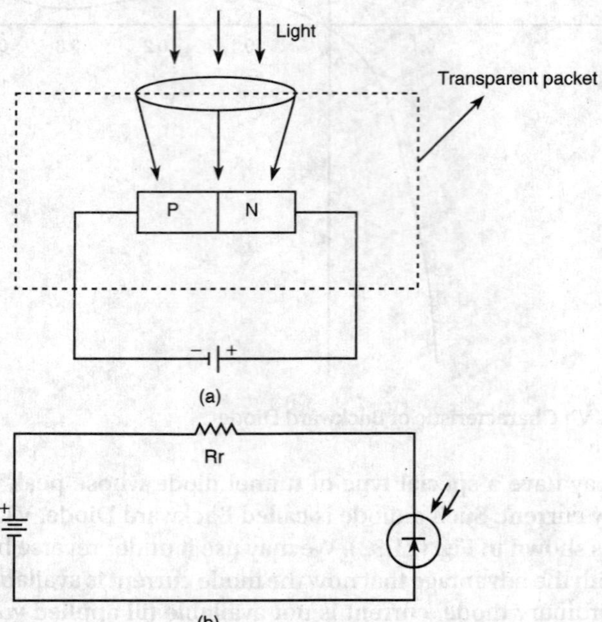

Fig. 21.33 Equivalent Circuit of a Photo-diode

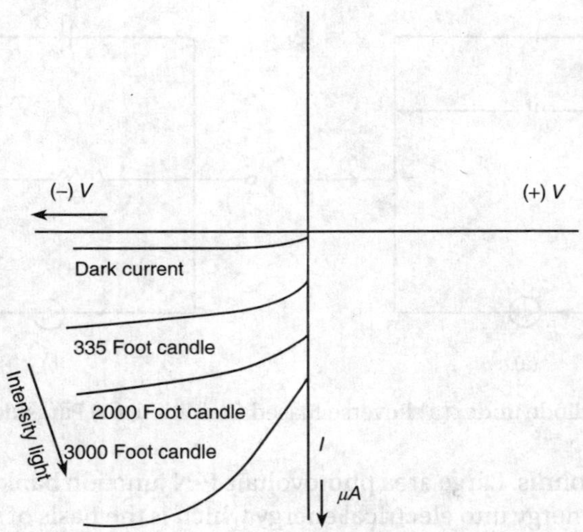

Fig. 21.34 V-I Characteristic of Photo-diode

On illuminating the diode under reverse biased conditions the number of hole-electron pairs increases, which is proportional to the intensity of incident light (number of photons). Hence the total current flowing through the diode,

$$I = I_0 + I_s$$

where I_s is the short circuit current generated on account of photoelectric effect. In general the volt-ampere characteristic is given by

$$I = I_s + I_0 \quad \left(e^{-V/\eta kT - 1} - 1\right) \tag{21.49}$$

This characteristic it drawn in Fig. (21.34).

Small Signal Model: Under ideal conditions a photo-diode under reverse biased condition is shown in Fig. [21.35(a)]. In Fig. [21.35(b)] parasitic elements *C, R* and *r* are taken into consideration, where *R* represents reverse biased resistance, *C* represents the transition capacitance and *r* represents the bulk ohmic resistance. The barrier capacitance *C* is of the order of 10 pF, $R = 50$ MΩ. In Fig. (21.34) the dark current I_0 corresponds to the reverse current due to thermally generated electron-hole pairs. From Eq. (21.50), we see that if diode terminals are open circuited then the total current *I* has to be zero. Under such conditions, a voltage known as photovoltaic V_{OC} will appear across diode. It is given by

$$V_{OC} = \eta V_T \, In\left(1 + \frac{I_s}{I_0}\right) \tag{21.50}$$

where V_0 is the volt equivalent of temperature and η is a constant equivalent to unity for Ge and 2 for Si.

Thus V_{OC} increases logarithmically with I_s or with intensity of incident light. Now if a load (R_L) is connected across the diode, the voltage drop across it will correspond to incident light intensity. The output voltage generated is of the order of few hundred milivolts

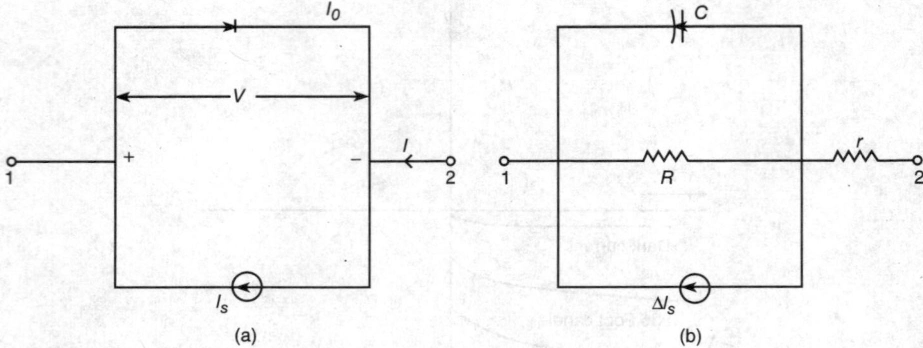

Fig. 21.35 Photo-diode under (a) Reverse Biased Condition, (b) Parasitic Elements

across R_L of few kilo-ohms. Large area photovoltaic P-N junction banks have been used for conversion of solar energy into electrical energy which is the basis of solar cells.

21.18.1 Solar Cells

Solar cells, which convert the sunlight directly into electricity, at present furnish the most important long duration power supply for satellites and space vehicles. These cells have also been successfully employed in small-scale terrestrial applications. Today, the solar cell is considered a major device for obtaining energy from the sun. Since it can covert sunlight directly to electricity with high conversion efficiency, and can provide nearly permanent power a low operating cost, and is virtually free of pollution. Recently, research and pollution development of low-cost, flat panel solar cells, thin film devices, concentrator systems, and many innovative concepts have increased. We can expect that in near future, the costs of small solar power modular unit and solar power plants will be economically feasible for large-scale use of solar energy.

The first solar cell were developed by Chapin et al. in 1954 using a diffused silicon P-N junction. Subsequently, the cadmium-sulphide solar cell was developed by Raynolds metal in 1954. To date, solar cells have been made of many other semiconductors, using device configuration and employing single-crystal, polycrystal, and amorphous thin film structures.

Photo-voltaic effect and solar cells

The photo-voltaic effect can be observed in nature in a variety of materials, but the materials that have shown the best performance in sunlight are the semiconductors. In photovoltaic conversion, the solar radiation falls on devices called solar cells which convert the sunlight directly into electricity. The principal advantages associated with solar cells are that they have no moving parts, require little maintenance, and work quite satisfactorily with beam or diffuse radiation.

A typical schematic representative of a conventional silicon solar cell is shown in Fig. [21.36(a)]. Silicon cells are thin wafers about 300 μm in thickness and 3 to 4 cm in diameter sliced from a single crystal of N-type or P-type doped silicon. A shallow P-N junction is formed at one end by diffusion. Metal electrodes made from a Ti-Ag

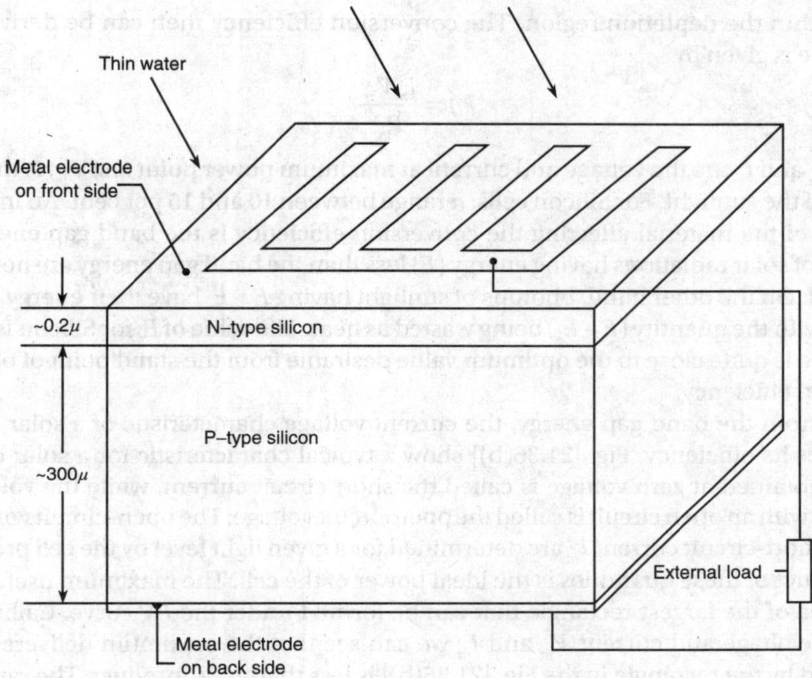

Fig. 21.36 (a) Schematic representation of a Silicon Solar Cell

solder are attached to the front and back side of the solar cell. On the front and back side of the cell are attached metal electrodes made from a Ti-Ag solder. On the front side, the electrode is in the from of a metal grid with fingers which permit the sunlight to go through, while on the backside, the electrode completely covers the surface. An antireflection contain of SiO having a thickness of about 0.1 μm is also put on the top surface of the cell.

When a monochromatic radiation of wavelength λ is incident on the front surface of a cell, it is absorbed and pairs of positive and negative charges, called electron-hole pairs are created. The generation rate of electron-hole pairs at a distance x from the semiconductor is given by

$$G(\lambda,x) = \alpha(\lambda)F(\lambda)\left[1 - R(\lambda)\right]\exp\left[-\alpha(\lambda)x\right] \qquad (21.51)$$

where $\alpha(\lambda)$ is the absorption coefficient, $F(\lambda)$ the number of incident photons/cm²/s per unit bandwidth, and $R(\lambda)$ the fraction of these photons reflected from the surface. The positive and negative charges are separated because of the P-N junction. The direct current thereby produced is collected by the metal electrodes and flows to the external load. The total photocurrent at a given wavelength is given by

$$J(\lambda) = J_p(\lambda) + N_n(\lambda) + J_{dr}(\lambda) \qquad (21.52)$$

where J_p and J_n are photo current densities due to holes and electrons respectively and J_{dr} is the photocurrent per unit band width, representing photo current generation taking place within the depletion region. The conversion efficiency then can be derived from the device is given by

$$\eta = \frac{I_m V_m}{P_{in}} \tag{21.53}$$

where V_m and I_m are the voltage and current at maximum power point and P_{in}, is the power density of the sunlight. For Silicon cells, n range between 10 and 15 per cent. An important property of the material affecting the conversion efficiency is the band gap energy (E_g). Photons of solar radiations having energy (E) less than the band gap energy are not usually absorbed. On the other hand, photons of sunlight having $E > E_g$ have their energy partially utilized with the quantity ($E - E_g$) being wasted as heat. The value of E_g for Silicon is 1.20 eV. This value is quite close to the optimum value desirable from the stand point of obtaining maximum efficiency.

Apart from the band gap energy, the current voltage characteristic of a solar cell also influences its efficiency. Fig. [21.36(b)] show a typical characteristic for a solar cell. The current obtained at zero voltage is called the short circuit current, while the voltage V_{OC} obtained with an open circuit is called the open circuit voltage. The open-circuit voltage V_{OC} and the short-circuit current I_{sc} are determined for a given light level by the cell properties. The product of these quantities is the ideal power of the cell. The maximum useful power is the area of the largest rectangle that can be formed under the *I-V* curve. Calling these values of voltage and current V_m and I_m we can see that the maximum delivered power illustrated by the rectangle in the Fig. [21.35(b)] is less than $V_{Oc} I_{sc}$ product. The ratio $I_m V_m / I_{sc} V_{oc}$ is called the fill factor, and is a Figure of merit for solar cell design. Typical values of the above quantities for a silicon cell are as follows: $V_{oc} = -450$ to 600 mV, $I_{sc} = -30$ to 50 mA/cm² and Fill factor = 0.65 to 0.80.

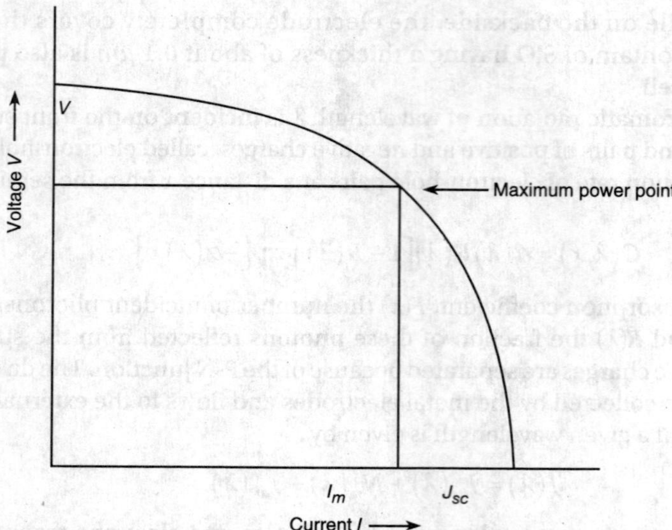

Fig. 21.36 (b) A typical current-voltage characteristic for a ideal solar cell

A well-made Si cell can have about 10 per cent efficiency for solar energy conversion, providing approximately 100 W/m^2 of electric power under full illumination. This is modest amount of power per unit solar cell area, considering the effort in fabricating a large area of Si cells. One approach to obtaining more power per cell is to focus considerable light onto the cell using mirrors. Although Si cells lose efficiency at the resulting high temperatures. Ga As and related compounds can be used at 100 °C or higher. In such solar concentrator system more effort and expense can be put into the solar cell fabrication, since fewer cells are required. For example, a Ga As–Al Ga As hetrofunction cell provides good conversion efficiency and operates at the elevated temperatures common in solar concentrator systems.

Applications: Some of the important applications of photodiodes are (i) Photo detection (both visible and invisible), (ii) Demodulation, (iii) Logic circuits, (iv) Switching, (v) Optical communication system.

21.19 LIGHT-EMITTING DIODE (LED)

LED is a solid state (P-N junction diode) light source which has replaced incandascent lamps in many day to day applications.

ED is just not an ordinary P-N junction diode where silicon is used. Here we use compounds having elements like. gallium, arsenic and phosphorus which are semi-transparent unlike silicon which is opaque (Gallium-arsenide gives infrared radiations and gallium-arsenide-phosphide gives visible light either red or yellow.)

Principle: Whenever a P-N junction is forward baised, free electrons and holes recombine at the junction. Here we may understand the situation by assuming that electrons fall into a hole, While jumping the electron releases (radiates) energy since electron jumps from a higher energy level (N-type) to a hole (P-type) in lower energy level (P-type). The energy radiated is in the form of heat and light Fig. [21.36(c)] but none of the light is visible since silicon is opaque and so no light can come out. Instead of silicon if we use a semitransparent material (like gallium, Arsenic compounds etc.,) then the light emitted due to recombination of free electrons and holes as the junction will escape to the surroundings and we shall perceive it. The colour of he radiations coming out will depend upon the material used. Mostly we find red, green or amber colour when the material used is gallium arsenide phosphide. We do get infrared radiations also when the compound used is gallium arsenide.

Working: An LED is forward biased with a small resistance in series Fig. (21.37). The construction of LED is such that they have typical voltage drop of 1.5 to 2.5 V for current between 10 and 50 mA when forward biased.

The exact voltage drop depends on the colour, tolerance, and other factors depending on the nature of construction.

Incidentally LED's have low reverse voltage ratings. A typical value is 3V for TIL 221 (red LED). So if an LED is reversed biased then voltage greater than 3V may destory it. One way to protect is to connect another ordinary diode in parallel with the given LED Fig. [21.37(b)].

Applications: LED's which produce visible radiations are used in instrument displays, calculators, digital clocks etc.,

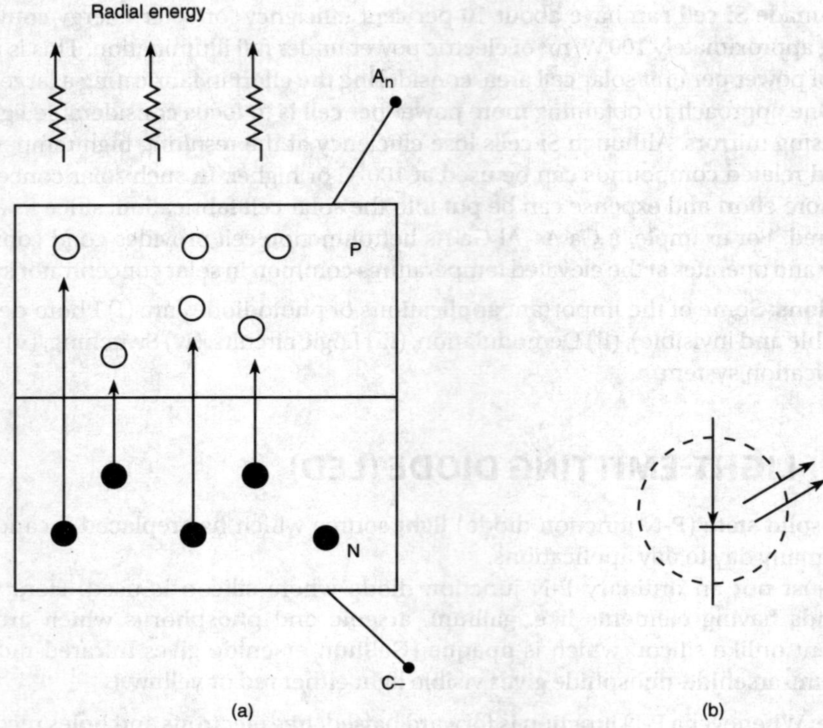

Fig. 21.36 (c) Schematic Construction of LED (b) symbol

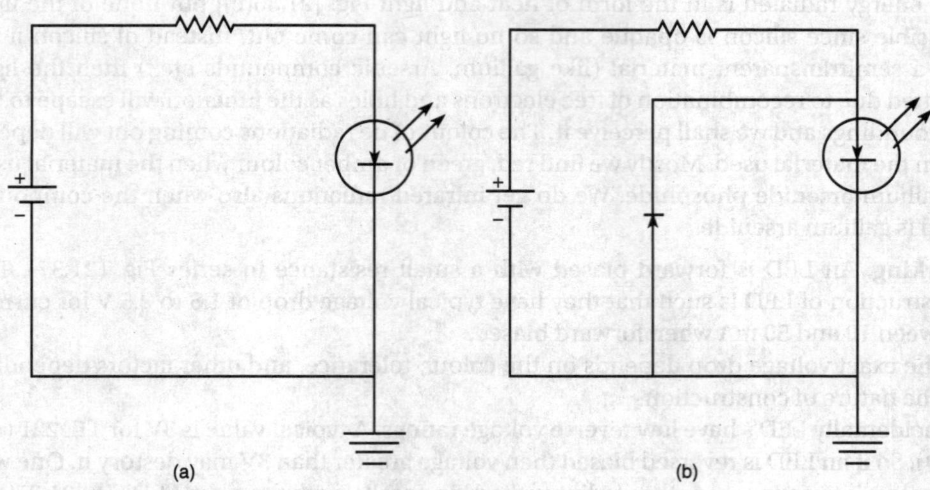

Fig. 21.37 LED Circuit

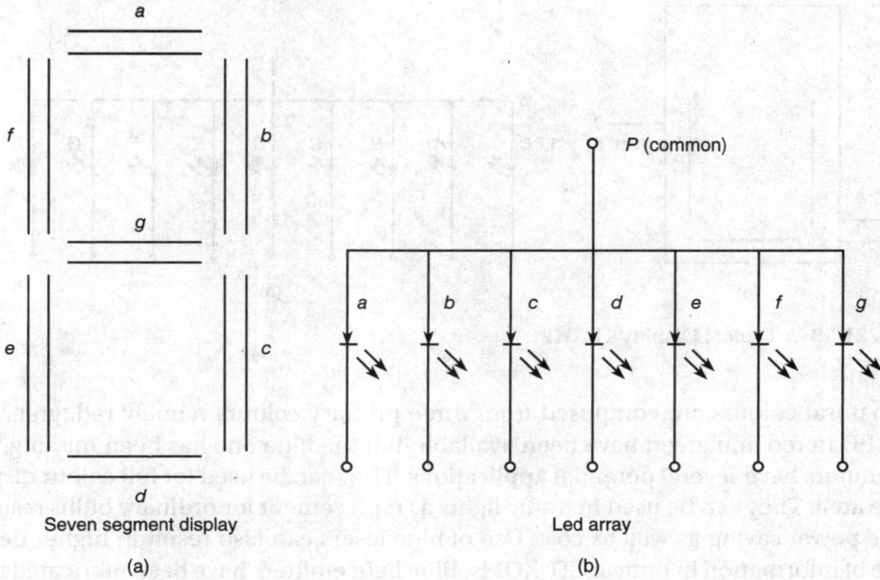

Fig. 21.38 LED Array

Here we have LED arrays. An LED array is a group of LED's that display numbers, letters or other symbols. The most common LED array is the seven segment display shown in Fig. [21.38(a)]. The display contains seven rectangular LED's (*A* to *G*). Each LED is called a segment because it forms part of the character being displayed. Fig. [21.38(b)] shows the schematic arrangement of the seven LED's constituting a display. Their anodes have been joined to common terminal 3. So a positive voltage when applied to it drives all anodes. By grounding one or more cathode, we can forward bias and of the diodes which in turn form any digit from 0 to 9. For instance by grounding the cathodes of *A, B* and *C*. We display the number '7' in Fig. [21.38(a)] or by grounding the cathodes of *A, B, C, D* and *G* we display the number '3'.

A seven segment array can also display capital *A, E, F,* plus lower case letters *b* and *d* which are commonly used in microwave ovens, stereo tuners and microprocessor trainer kits. A typical display TIL 312 shown in Fig. [21.39(b)] represents the schematic diagram of the same, showing left and right decimal points also. Fig. [21.39(a)] shows the pin-out of it. The advantages of an LED are:

(i) Low voltage of operation
(ii) Long life (more than 20 yrs)
(iii) Fast ON-OFF switching (10^{-9} sec)

The other uses of LED are:

(i) For indicating power ON/OFF (power level indicators)
(ii) Optical switching applications
(iii) Solid state video displays
(iv) Optical communication-Energy coupling circuits

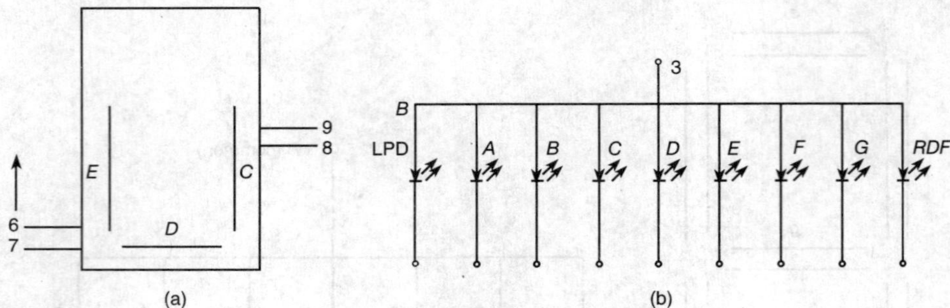

Fig. 21.39 A Typical Display TIL 312

All natural colours are composed from three primary colours namely red, green and blue. LED in red and green have been available-but the blue one has been missing, Blue light emitters have several potential applications. They can be used for full colour displays of large area. They can be used in traffic lights as replacement for ordinary bulbs resulting in huge power saving as well as cost. Use of blue lasers can also result in higher density storage of information in optical CD-ROMs. Blue light emitters have been fabricated using several materials including ZnO, ZnSe and SiC. These attempts have been reasonably successful. However maximum success in last few years has been achieved with GaN. This material has proved to be very useful as it is also capable of operating at high power density, high temperature and unfriendly environment. GaN having a band gap of 3.4 eV can give continuously varying band gap by combing with AIN to get upto 6.2 eV and with. In N to get a band gap down to 1.9 eV. The high thermal conductivity and superior stability of this material makes it ideal for several applications over other competing materials.

21.19.1 Liquid Crystal Display (LCD)

Just like LED we have another type of display that uses seven segment called Liquid Crystal Display also known as electro luminescent display. It consists of a thin layer of normally transparent liquid crystal material between two electrodes.

When an electric field is applied, the liquid crystal material between the two electrodes becomes turbulent, reflecting and scattering ambient light. It provides excellent brightness under high ambient light conditions and requires only 50μ W of power per segment, there by total power for one complete display of 7 segments is 350 μ W. This power is much less than that of an LED display, but the life expectancy is not as high as that for LED which is 10,000 hours minimum.

These displays are used in Watches, Pocket Calculators, Pocket televisions and portable instrument displays.

21.20 VARACTOR DIODE

An ordinary PN-junction diode under reverse biased condition causes the majority charge carriers to move away from the junction. thereby uncovering more immobile charge, i.e., ions.

Hence the thickness of the space charge layer at the junction increases with reverse voltage. This increase in uncovered charge with applied voltage may be considered a capacitive effect. We may define an incremental capacitance C_T by

$$C_T = \frac{dQ}{dV} \tag{21.54}$$

where dQ is the increase in covered charge caused by an increase dV in voltage. The quantity C_T is generally referred to as transition region or space charge or **Barrier Capacitance.**

This quantity is not constant but depends upon the magnitude of reverse voltage. A variation of C_T with applied reverse voltage is shown in Fig. (21.40).

The larger the reverse voltage, the larger is the space charge width and smaller the capacitance C_T. Similarly for increase in voltage in the forward biased condition the space charge width will decrease and C_T will increase.

Diodes which are based on voltage variable capacitance are called varactors. A varactor is also called varicap, voltacap, epicap or turning diode. A circuit for a varactor diode under reverse bias is shown in Fig. [21.41 (a)]. A schematic symbol of a varactor is shown in Fig. [21.41(b)].

The resistance R_s represents the body (ohmic) series resistance of the diode. Typical values of which is 8.5 Ω at 4V. The reverse diode resistance R_r = 1 MΩ and that's why it may be treated as open circuit across C_T.

The capacitance or a varactor is given by

$$C = \frac{\varepsilon A}{W} \tag{21.55}$$

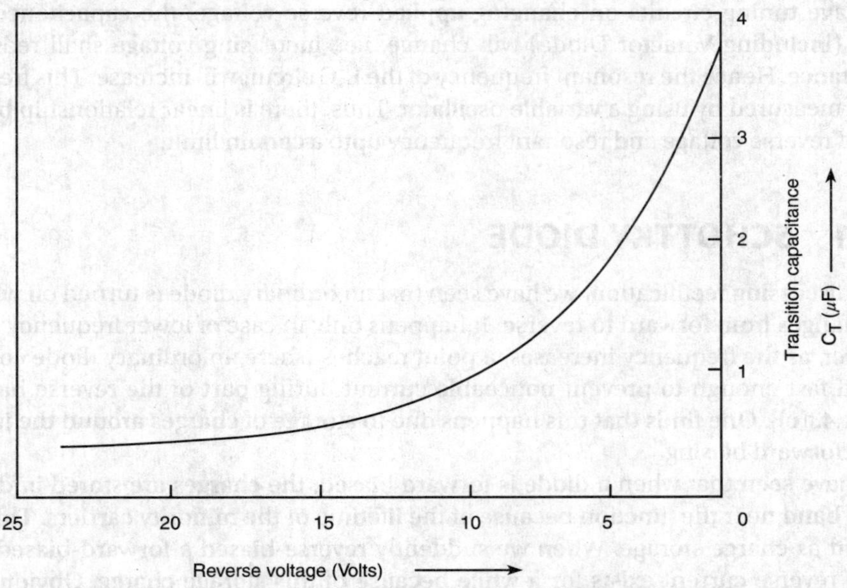

Fig. 21.40 Variation of C_T with reverse voltage for a typical silicon diode at 25 °C

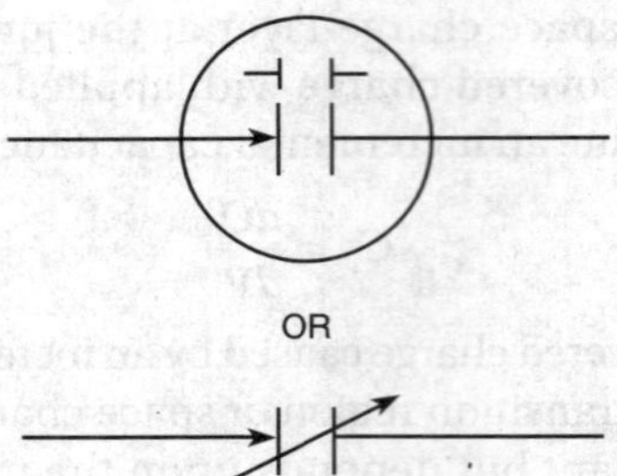

Fig. 21.41 (a) Varactor diode

where, ε is the permittivity of depleted region of semiconductor diode, A is the area of junction and W is the width of the depletion region. Since the width of the space charge (or depleted region) is approximately proportional to the square-root of the reverse bias voltage, V, whereas the area A and permittivity ε being constant; we have

$$C = \frac{K}{\sqrt{V}}, \quad \text{where} \quad K = \varepsilon A$$

Applications

Such diodes are useful in a number of circuits:

- **(i)** Voltage tuning of LC resonant circuit (Electronic tuning circuit in radio, television etc.,)
- **(ii)** Self balancing bridge circuits
- **(iii)** Parametric Amplifiers
- **(iv)** Automatic frequency control device

In voltage tuning circuits on changing applied reverse voltage, the capacitance of the circuit (including Varactor Diode) will change, i.e., increasing voltage shall reduce the capacitance. Hence the resonant frequency of the L-C circuit will increase. This frequency can be measured by using a variable oscillator. Thus, there is linear relationship between applied reverse voltage and resonant frequency upto a certain limit.

21.21 SCHOTTKY DIODE

While discussing rectification, we have seen that an ordinary diode is turned off when the bias changes from forward to reverse. It happens only in case of lower frequency signals. However, as the frequency increases, a point reaches where an ordinary diode could not turn off fast enough to prevent noticeable current during part of the reverse half cycle Fig. [21.43(c)]. One finds that this happens due to storage of charges around the junction during forward biasing.

We have seen that when a diode is forward biased, the charges are stored in different energy band near the junction because of the lifetime of the minority carriers. This effect is called as charge storage. When we suddenly reverse-biased a forward-biased diode, a large reverse current exists for a while because of this storage charge. Obviously, the charge storage prelongs the reverse recovery time (t_{rr}). The reverse recovery time is the

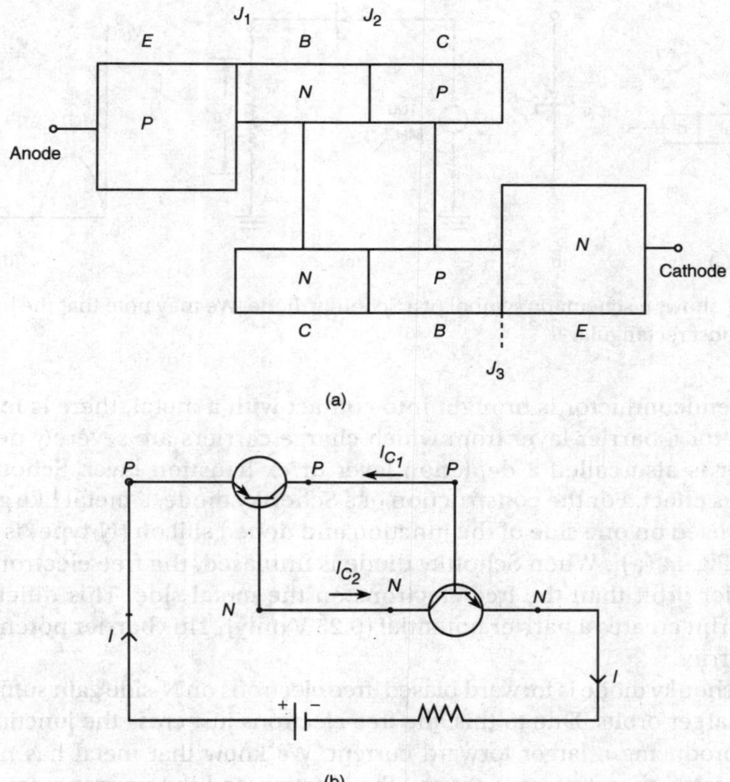

Fig. 21.42

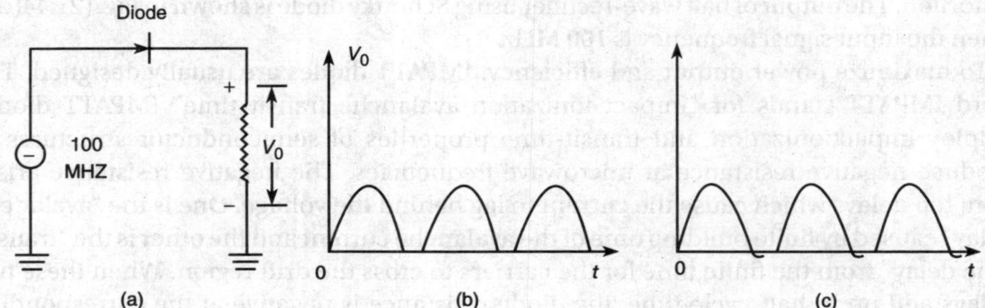

Fig. 21.43

time taken to turn off a forward biased time. More preciesly, it is time taken by a diode to drop the reverse current to 10% of forward current.

For ordinary diodes, t_{rr} is approximately 4ns. At low frequencies, i.e., blow 10 MHz, its effect is not even noticed Fig. [21.43(b)]. However, at higher frequencies, i.e., more than 10 MHz, its effect is noticeable, since the output signal begins to deviate from its normal shape Fig. [21.43(c)]. This problem is eliminated in Schottky diode.

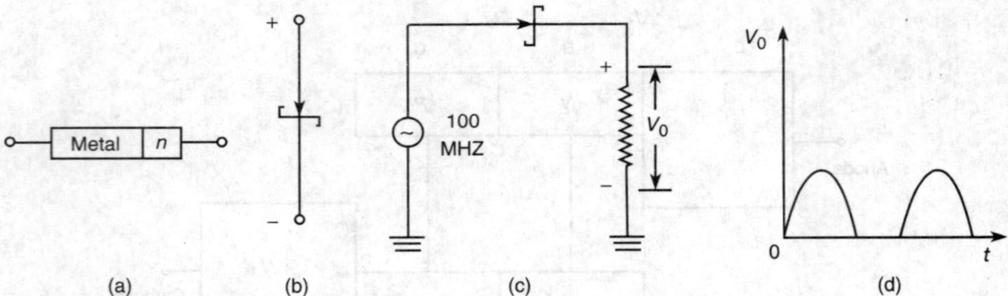

Fig. 21.44 (b) shows a schematic symbol of a Schottky diode. We may note that the bar (line) looks almost rectangular *S*

When a semiconductor is brought into contact with a metal, there is formed in the semiconductor a barrier layer from which charge carriers are severely depleted. The barrier layer is also called a depletion layer or ex-haustion layer. Schottky diode is based on this effect. For the construction of a Schottky diode, a metal like gold silver or platinum is used on one side of the junction and doped silicon (N-type) is used on the other side [Fig.44 (a)]. When Schottky diode is unbiased, the free electrons on N-side are in smaller orbit than the free electrons on the metal side. This difference in the size of the orbit creates a barrier potential (0.25 V only). This barrier potential is called Schottky barrier.

When a Schottky diode is forward biased, free electrons on N-side gain sufficient energy to travel to larger orbits. Due to this, the free electrons just cross the junction and enter the metal, producing a larger forward current. We know that metal has no holes and hence there is no charge storage around the junction and the reverse recovery time (t_{rr}) is zero. This means the Schottky diode is switched off at once when it is reverse-biased. This is why, Schottky diodes can easily rectify signals of frequency above 30 MHz without distortion. The output of half wave-rectifier using Schottky diode is shown in Fig. (21.44(d)). when the input signal frequency is 100 MHz.

To maximize power output and efficiency, IMPATT diodes are usually designed. The word IMPATT stands for "impact ionization avalanche transit time". IMPATT diodes employ impactionization and transit-time properties of semiconductor structures to produce negative resistance at microwave frequencies. The negative resistance arises from too delays which cause the current to lag behind the voltage. One is the "avalanche delay" caused by finite building time of the avalanche current and the other is the "transit-time delay" from the finite time for the carriers to cross the drift region. When these two delays add up to half-cycle time, the diode resistance is negative at the corresponding frequency.

Schottky diodes find several applications. We mention below few of them.

(i) We know that the speed of the digital computers depends upon the switching time of their diodes and transistors. Obviously, Schottky diodes find their best application in digital computers due to their fast switching operation.

(ii) Schottky diodes are best suited for low voltage rectification due to their small barrier potential (0.25V).

21.22 STEP RECOVERY DIODES OR CHARGE STORAGE DIODE OR SNAPBACK DIODE

In contrast to fast-recovery diodes, a charge storage diode is designed to store charge while conducting in the forward direction and upon switching to conduct for a short period in the reverse direction. A particularly interesting charge-storage diode is the step-recovery diode or snapback diode which conducts in a reverse direction for a short period then abruptly cuts off the current as the stored charges have been dispelled. This cutoff occurs in the range of picoseconds and results in a fast-rising wave front rich in harmonics.

The symbol of a step recovery diode is shown in Fig. [21.45(a)]. Step recovery diode has unusual doping profile in which the density of carriers decreased near the junction. This unusual distribution of carriers causes a phenomenon called reverse snap-off and that is why it is also called snapback diode.

When a high frequency (say 20 MHz) signal is applied in the circuit Fig. [21.45(b)], during positive half cycle, the diode conducts like an ordinary silicon diode. However, during the negative half cycle the reverse current exists for a while because of the stored charges, then suddenly the reverse current drops to zero Fig. [21.45(c)]. It is as if the diode is snapped off and hence it is also known as **snapback diode**.

We must note that output of this diode is non-sinusoidal periodic wave which contains a sharp spike in the negative half cycle. From the Fourier series theorem, we know that any non-sinusoidal periodic wave is equivalent to the superposition of sinusoidal components called harmonics. These harmonics have the frequencies f, $2f$, $3f$,...nf, where f is the fundamental frequency of the waveform.

The output of this diode Fig. [21.45(c)] is rich of harmonics and can be filtered to produce a sine wave of required higher frequency. Because of these characteristics, these diodes are used as harmonic generators and pulse formers most step-recovery diodes are made from Si with relatively long minority carrier life times ranging from 0.5 to 0.5 μs. We must note that the lifetimes are about 1000 times longer than for fast-recovery diodes.

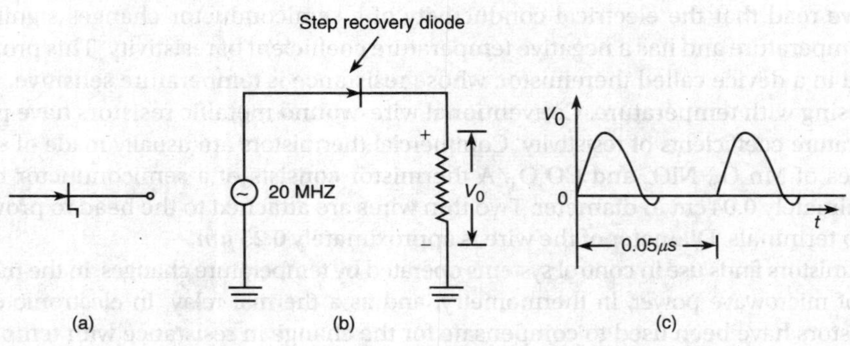

Fig. 21.45

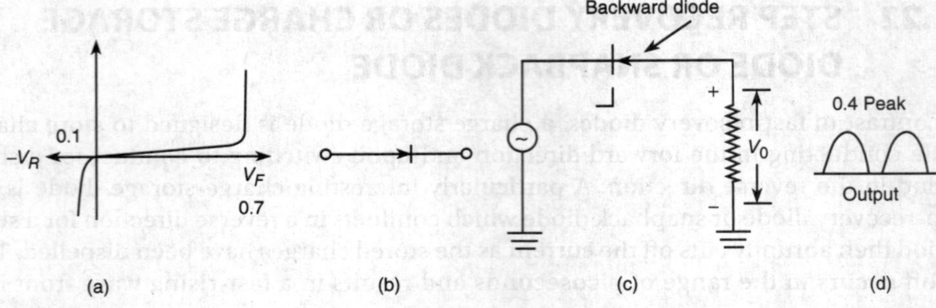

Fig. 21.46

21.23 BACKWARD DIODE

It is just a zener diode in which a doping level is increased to such an extent that its zener effect, i.e., reverse breakdown occurs near to zero.

Usually, zener diodes have breakdown voltages more than 2V. The backward diodes are prepared to have reverse breakdown voltage near to zero (–0.1V). Obviously, in backward diodes forward conduction still occurs around +0.7V but the reverse conduction starts approximately at –0.1V. This means that it conducts better in the reverse direction than in the forward direction.

Fig. [21.46(a)] shows V-1 characteristics of backward diodes. Its symbol is similar to zener diode and is shown in Fig. [21.46(b)].

The backward diode can be used for rectification of small signals, microwave detection, and mixing. Similar to the tunnel diode, the backward diode has a good frequency response because there is no minority carrier storage effect. In addition, the current-voltage characteristic is insensitive to temperature and to radiation effect, and the backward diode has very low I/*f* noise.

21.24 THERMISTORS AND BARRETTERS

We have read that the electrical conductivity of a semiconductor changes significantly with temperature and has a negative temperature coefficient of resistivity. This property is utilised in a device called theremistor, whose resistance is temperature sensitive, usually decreasing with temperature. Conventional wire-wound metallic resistors have positive temperature coefficients of resistivity. Commercial thermistors are usually made of sintered mixtures of Mn_2O_3, NiO_2 and CO_2O_3. A thermistor consists of a semiconductor bead of approximately 0.04 cm in diameter. Two thin wires are attached to the bead to provide for the two terminals. Diameter of the wire is approximately 0.25 μm.

Thermistors finds use in control systems operated by temperature changes, in the measurement of microwave power, in thermometry, and as a thermal relay. In electronic circuits, thermistors have been used to compensate for the change in resistance with temperature of ordinary component where variation of component values cannot be tolerated.

The 1-V characteristic of a thermistor has a negative slope. Devices that exhibit, in some region, negative slope in their 1-V characteristics are useful for making oscillators, amplifiers and switching circuits. However, thermistors are not suitable in these applications because their response characteristics are too slow. There are certain bulk semiconducting compounds which have negative resistance characteristics over a limited range of operating parameters utilizing mechanism unrelated to the temperature sensitivity of the resistivity. These materials have been used to obtain devices based on **Gunn effect**.

A heavily doped semiconductor shows metallic properties. It has a positive temperature coefficient of resistivity owing to the decrease of the carrier mobility with increasing temperature. Such a device is called **sensistor**. Thermistors also find extensive use as sensing elements in microwave power measuring equipments and as temperature sensors of electronic thermometers. Thermistors are capable of yielding power information over the power range of 10^{-5} mW to 20 mW with a typical burn out level 400 mW.

A barretter has a positive temperature coefficient of resistance, consists of an approximately mounted piece of "Wollaston" platinum wire having diameter of approximately $1.25\mu m$. Barretters are capable of yielding power information over the range of 10^{-5} mW to 20 mW. A typical burnout level is 20 mW. For low level *rf* power application below 10^3 MHz, 0.001 A, "Littlefuse" may be used as a barretter.

21.25 PHOTOCONDUCTOR

It consists simply of a slab of semiconductor (in bulk or thin film form) with ohmic contacts affixed to oppose ends as shown in Fig. (21.47). When light is incident on a semiconductor, electrons are excited from the valance bands into the conduction band. Obviously, electron-hole pairs in excess of those generated thermally are produced. The conductivity of the material increases owing to the increased number of carriers. Under the influence of light on the surface of a photo conductor, carriers are generated either by band-to-band transitions (intrinsic) or by transitions involving forbidden gap energy levels (extrinsic, resulting in an increase in conductivity.

For the intrinsic photoconductor, the conductivity is given by $\sigma = e(\mu_n + \mu_p P)$, and the increase of conductivity under illumination in mainly due to the increase in the number of carriers. The corresponding long-wavelength cut off of the photon of frequency v_c in this case is given by

$$\lambda_c = \frac{hc}{E_g} = \frac{1.24}{E_g(eV)}(\mu m) \tag{21.56}$$

where λ_c is the wavelength corresponding to the semiconductor band gap E_g, c is the speed of light. For wavelength (λ) shorter than λ_c, the incident radiation is absorbed by the semiconductor, and hole-electron pairs are generated. For the extrinsic case, photoexcitation may occur between a band edge and an energy level in the energy gap. Photoconductivity can take place by the absorption of photons of energy equal to or greater than the energy separation of the band gap levels and the conduction or valence band. In this case the long wavelength cut off is determined by the depth of the forbidden-gap energy level.

Fig. (21.48) shows the photo response of an intrinsic as well as extrinsic (a semiconductor with deep impurity level), e.g., thin film deposition on insulating substrate

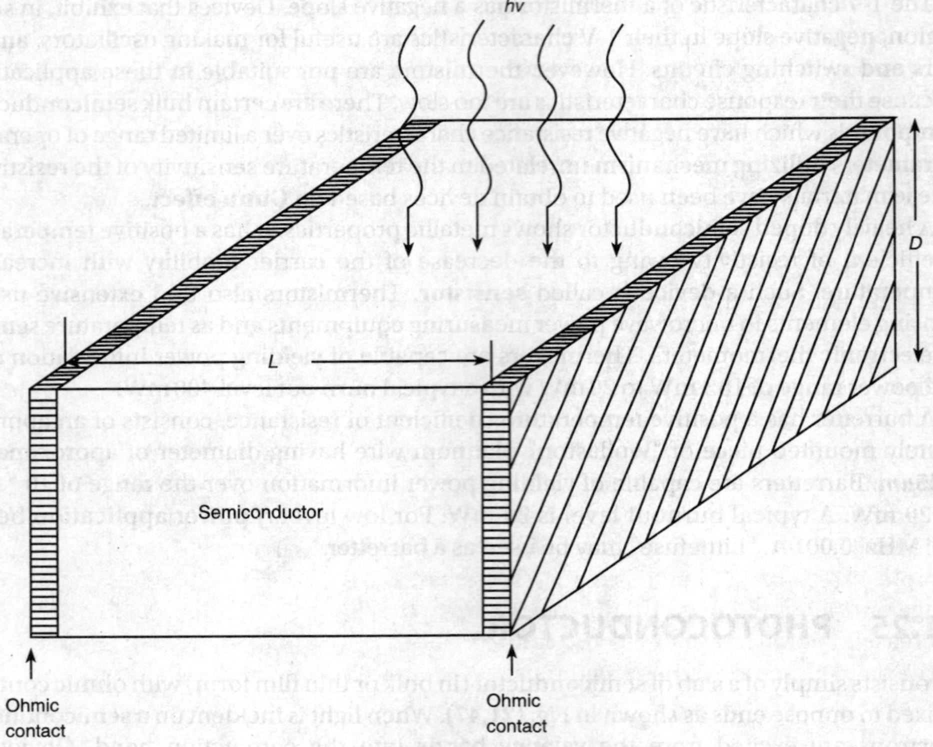

Fig. 21.47

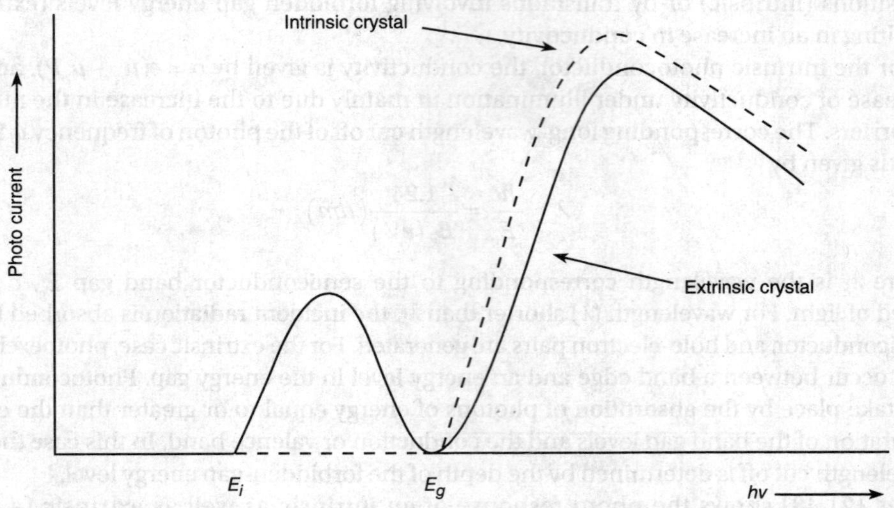

Fig. 21.48 Photo response of a photoconductor

which introduce many materials that find application as sensitive photoconductors are CdS and CdTe.

The performance of a photoconductor is measured in terms of (i) the photoconductivity gain and (ii) the response time of the detector. Typical gain and response time for a photoconductor are 10^5 and 10^{-3} seconds respectively. Photoconductivity devices are used as relays in digital or control circuits, as light meters, and to record a modulating light intensity such in a sound track.

21.26 GUNN EFFECT AND GUNN DIODE

Ridley-Watkins and independently Hilsum predicted that semiconductor materials under certain conditions can offer differential negative resistance. This differential negative resistance is a bulk effect and due to transfer electrons from one valley to another in the conduction band. Gunn while experimenting on a sample of N-type GaAs and some other III-V compounds, found that the current through the sample increased linearly with voltage till a certain threshold voltage. Beyond the threshold voltage a number of current pulses appeared with a time interval proportional to the length of the sample. This threshold field is high (~400 V/mm). The oscillations lie in the microwave range and set in the negative differential conductance (NDC) region where the current decreased with increase in electric firld E Fig. (21.49). This behaviour is a consequence of the band structure of these materials. In the region of velocity field, where v decreases

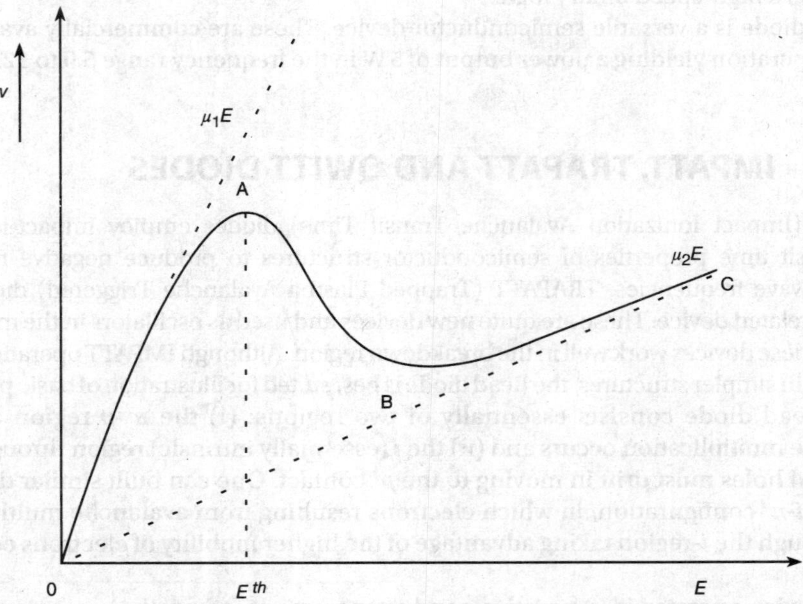

Fig. 21.49 A possible characteristic of electron drift velocity *vs* field for a semiconductor exhibiting the transferred electron mechanism

with increasing E, the differential mobility (dv/dE) of the electrons becomes negative. The reason is following.

n-GaAs has a direct energy band gap. The band gap is 1.4 eV. The free electrons in n-GaAs normally occupy the lowest energy states in the conduction band. The effective mass of electrons in this situation and their mobility μ_1 is high ($\approx 0.02\ m^2/v\text{-}s$). So as a result of the transfer of the electrons current begins to decrease with increase in field because $T = n_1 e\mu_1 E + n_2 e\mu_2 E$, where n_1 and n_2 are the concentration of electrons having the mobility μ_1 and μ_2 respectively. The average drift velocity of an electron is

$$v = \left(\frac{n_1\mu_1 + n_2\mu_2}{n_1 + n_2}\right)E \tag{21.57}$$

$(n_1 + n_2)$ being the total electron concentration.

If the transition of the electrons from the high-mobility state to the low-mobility state occurs rapidly over a range of field E, one observes that v diminished with field E beyond a certain threshold field E_{th}, shown by region AB in Fig. (21.49), when all the electrons move to the low-mobility state, the drift velocity is $v = \mu_2 E$. One finds that v continues to increase slowly with E. This builds up the successive oscillations.

Owing to the occurrance of the NDM in the velocity-field curve, Gunn diodes are used as sources of microwaves especially where high power is not a requirement. Gunn diodes are also used as local oscillators for mixers in microwave receivers over the frequency range 1 to 100 GHz. A Gunn device can carry just before the threshold voltage one of two possible currents, depending on the presence or absence of a domain. This can therefore, be used as a high-speed binary logic.

Gunn diode is a versatile semiconductor device. These are commercially available for pulsed operation yielding a power output of 5 W in the frequency range 5.0 to 12.0 GHz.

21.27 IMPATT, TRAPATT AND QWITT DIODES

IMPATT (Impact Ionization Avalanche Transit Time) diodes employ impact-ionization and transit time properties of semiconductor structures to produce negative resistance at microwave frequencies. TRAPATT (Trapped Plasma Avalanche Triggered) diode is an IMPATT related device. These are quite new devices and used as oscillators in the microwave region. These devices work well in the breakdown region. Although IMPATT operation can be obtained in simpler structures, the Read diode is best suited for illustration of basic principles.

The Read diode consists essentially of two regions: (i) the n^+-p region at which avalanche multiplication occurs and (ii) the i (essentially intrinsic) region through which generated holes must drift in moving to the p^+ contact. One can built similar devices in the p^+-n-i-n^+ configuration, in which electrons resulting from avalanche multiplication drift through the i-region taking advantage of the higher mobility of electrons compared to holes.

The device operates in a negative conductance mode, when the a.c. component of current is negative over a portion of cycle during which the A.C. voltage is positive, and vice-versa. The negative conductance occurs because two processes, causing the current

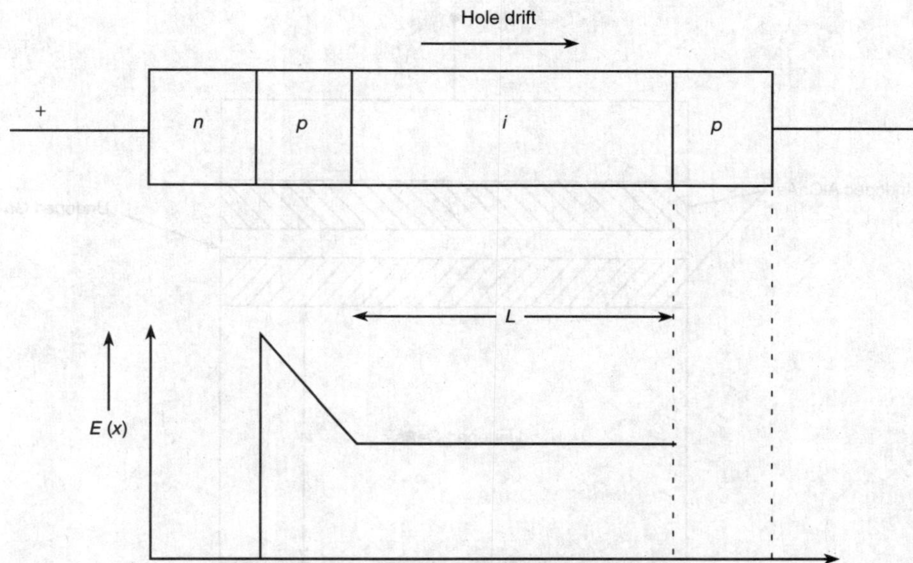

Fig. 21.50 The Read diode (a) The basic device configuration (b) electric field distribution in the device under reverse bias

to lag behind the voltage in time: (i) a delay due to avalanche process and (ii) a further delay due to the transit time of the carriers across the drift region. When these two delay times combine to produce a net phase-shift between 90° and 270°, the diode resistance is negative at the corresponding frequency. Consequently, negative conductance occurs and the device can be used for oscillation and amplification.

The structures that have been successfully employed for IMPATT devices are the PN, PIN, P^+NN^+, P^+NIN^+ and N^+PIP^+ (called read diodes). These devices are usually constructed from silicon and gallium arsenide. At present, IMPATT diode can generate the highest CW (Continuous Wave) power output at millimeter wave frequencies, i.e., microwave power above 30 GHz. A major drawback of the IMPATT diode for very high frequency operation is the fact that the avalanche process, which depends on random impact ionization events, is inherently noisy.

In TRAPATT diode, the TRAPATT mode of operation, i.e., the operating frequency is substantially lower than the transit time frequency and the efficiency is considerably higher. Under large-signal conditions the periodic avalanching of the diode begins at the high-field side and sweeps rapidly across the diode, leaving it substantially filled by a highly conducting plasma of holes and electrons whose space-charge depresses the voltage to very low values. Since the plasma cannot rapidly escape, this mode is called TRAPATT mode. This diode has been tried experimentally in pulsed transmitter of phased array radar systems.

A variety of approaches have been investigated to find alternative methods for injecting carriers into the drift region without relying on the avalanche mechanism. A particularly interesting device is QWITT (Quantum Well Injection Transit Time) diode,

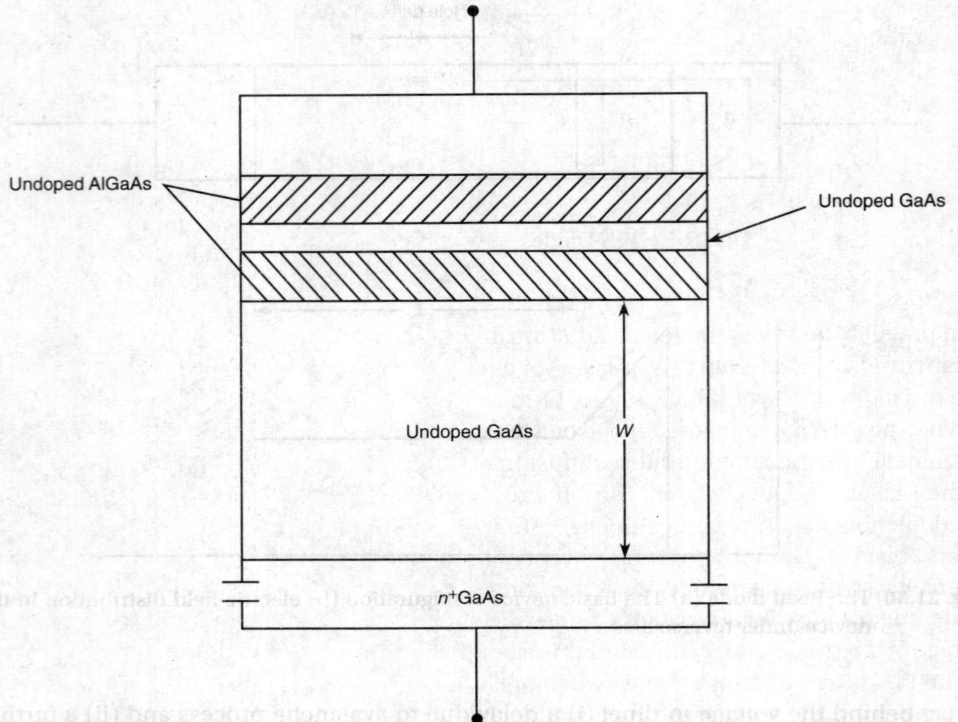

Fig. 21.51 Structure of a QWITT diode

which employes resonating tunneling through a quantum well to inject electrons into the drift region. The device consists of a single GaAs quantum well between two Al_xGa_{1-x} As barriers, in series with a drift region of undoped GaAs. This structure is then placed between two n^+-GaAs regions to form two contacts Fig. (21.51). In this device, one can achieve maximum resonant tunneling of electrons through the well if the dc bias is properly adjusted.

QWITT diode is a low-noise injection mechanism with superior high frequency characteristics.

The length of the transit time as well as the shape of the current pulse can be optimized to obtain the best power-frequency performance from the QWITT diode. This diode should extend the normal frequency limit associated with transit-time devices, while providing higher output power than simple quantum-well *RTD* oscillations.

21.28 PIN DIODES

This is a P-N junction with greatly improved switching times. Obviously this is a P-N junction with a doping profile tailored so that an intrinsic layer the '*I*-region" is sand witched between a player and an N-layer as shown in Fig. (21.52).

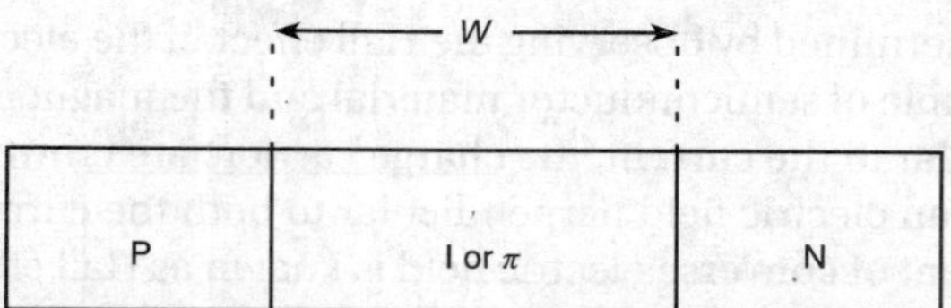

Fig. 21.52 PIN Diodes

In practice, however, the idealized I (intrinsic)-region is approximated by either a high resistivity P layer (referred to as π-layer) or a high resistivity N-layer (referred to as υ layer). The resistivity of I layer is typically 10^3 Ω-m.

When no external voltage is applied between the terminals of PIN diode, the concentration gradient across the function cause diffusion of carriers. The width of the depletion region in the I-layer is relatively large. Now, if a reverse bias is applied and increased gradually, the depletion region becomes thicker. At **swept out voltage** (a particular valued of the reverse bias), all the free carriers are swept out of the intrinsic layer. With further increase of reverse bias, the depletioN-region widens in the highly doped semiconductor regions. PIN diode offers a very high resistance under reverse bias. We must note that the breakdown voltage of the diode under reverse bias is also very large.

The PIN diode when used as a switch operates between the ON and OFF states. In ON state the diode is forward biased when minority carriers are infected into the high resistivity region between the highly doped regions. The increase in the densities of carriers reduces the resistance and the diode impedance is low. This causes a current flow. The forward resistance of the diode varies with the forward bias. In the OFF state the high resistivity region is completely swept out and the diode impedance is very high as the capacitance is very low.

The PIN diode has found wide applications in microwave circuits. It can be used as a microwave switch with essentially constant depletion layer capacitance and high power handling capability. When PIN diodes are used as microwave switches and when they are biased in the OFF condition the bias is usually beyond the swept-out voltage (usually-2V). PIN diodes can also be used as a **variolosser** (variable attenuator) by controlling the device resistance which varies approximately linearly with the forward current. PIN diodes can also modulate signals upto the GHz range. We must note that the forward characteristics of a thyristor in its ON state closely resemble those of PIN diode.

21.29 HALL EFFECT

An effect whereby a conductor carrying an electric current perpendicular to an applied strong magnetic field develops a voltage gradient which is transverse to both the current and the magnetic field. It was discovered by E.H. Hall in 1879. One can obtain the important information about the nature of conduction process in semiconductors and metals through analysis of this effect. Where a given sample of semiconductor material is *n* or P-type can be determined by observing the Hall effect. If the electric current is caused to flow through a sample of semiconductor material and the magnetic field is applied in a direction perpendicular to the current, the charge carriers are crowded to one side of the sample giving rise to an electric field perpendicular to both the current and the magnetic field. This development of converse electric field is known as Hall effect. The effect is also used to in the Hall probe for the measurements of magnetic fields, and in magnetically operated switching devices.

The strength of the electric field E_H produced is given by the relationship $E_H = R_H jB$, where j is the current density, B is the magnetic flux density, and R_H is a constant called the Hall coefficient. The value of R_H can be shown to be I/ne, where n is the number of charge carriers per unit volume and e is the electronic charge.

21.29.1 Hall Effect in Semiconductors

Hall effect in semiconductors originates from motion of both electrons and holes. This provides the best experimental evidence of the motion of holes, i.e., the production of a transverse electric field by the motion of charge carriers in a magnetic field. With an applied electrical field and a transverse component of magnetic field, both electrons and holes are deflected transversely toward the same side of the sample as shown in Fig. (21.53).

The drifting carriers experience Lorentz forces in the magnetic field H_z, and thereby suffer deflections. Both electrons and holes are deflected towards the same side of the sample as shown in Fig. (21.53). The situation in all sections is similar. The two types of carriers in semiconductors under this arrangement, so called Hall geometry, therefore, tend to combine and cancel each other at the front surface. However, this cancellation is incomplete, and thus there is a net charge which accumulates on the front surface

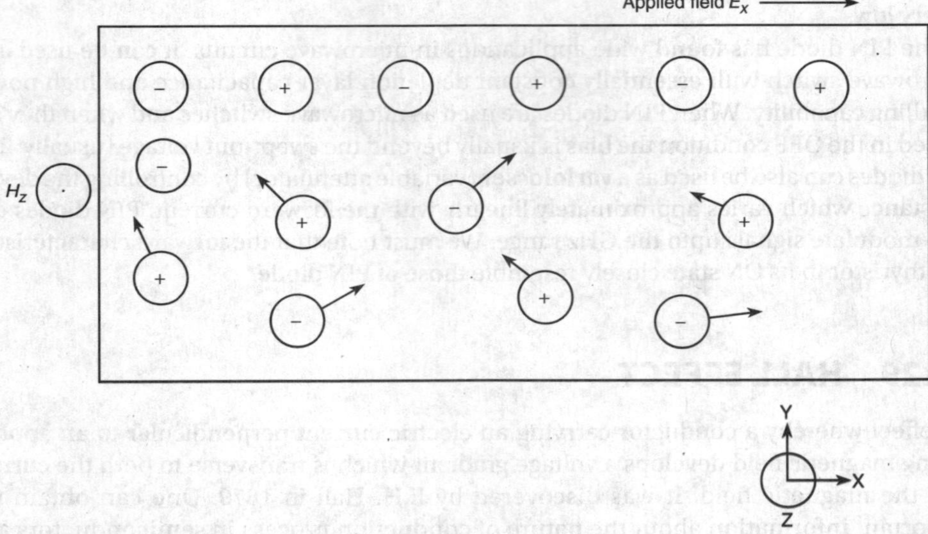

Fig. 21.53 Hall Effect in Semiconductors

of the sample. Because of these surface charges on the front and the rear surfaces of the sample, an electric field is produced in *y*-direction, known as Hall effect in semiconductors. ε_y may be calculated as follows,

The electron and hole current densities along *x*-axis can be expressed as

$$\left.\begin{aligned} J_{xe} &= n_e\mu_e\varepsilon_x \\ J_{xh} &= p_e\mu_h\varepsilon_x \end{aligned}\right\} \tag{21.58}$$

The total current density J_x is

$$J_x = J_{xe+}\,J_{xh=e}\,(n\mu_e = p\mu_h)\,\varepsilon_x \tag{21.59}$$

(Small changes in mobility due to magnetic resistance are neglected.)

Let us now obtain the electron current density (J_{ye}) along *y*-direction acting on an electron is $-e(v_e \times B) = eV_eB_z$, where v_e is one drift velocity of the electron. This Lorentz force is equivalent to an electric field $-v_eB_z$ since $v_e = -\mu_e\varepsilon_x$ and therefore this field may be expressed as $B_z\mu_e\varepsilon_\chi$. Using (21.59), one obtains,

$$J_{ye} = ne\mu_e\,(B_z\mu_e\varepsilon_x)$$

Replacing B_z by H_z we have

$$J_{ye} = ne\mu_e^2 H_z\varepsilon_\varepsilon$$

and hole current density is given by

$$J_{vh} = pe\mu_n^2 H_z\varepsilon_x$$

The total transverse current density is the sum of

$$J_y = J_{ye} + J_{yh} = (e\mu_e^2 - p\mu_h^2)H_z\varepsilon_x \tag{21.60}$$

A Hall field ε_y is required to made the net *Y*-current zero. The ratio $\varepsilon_y/\varepsilon_x$ is thus the same as J_y/J_x

$$e(n\mu_e + p\mu_h)\varepsilon_y = e(n\mu_e^2 - p\mu_h^2)H_z\varepsilon_x \tag{21.61}$$

$$\therefore \qquad \varepsilon_y = \frac{p\mu_h^2 - n\mu_e^2}{n\mu_e + p\mu_h}H_zE_x \tag{21.62}$$

The Hall coefficient R_H is given by

$$R_H = \frac{\varepsilon_y}{H_zJ_x} = \frac{p\mu_h^2 - n\mu_e^2}{e(n\mu_e + p\mu_h)^2} \tag{21.63}$$

Eq. (21.64) is like that for a metal if the semiconductor is extrinsic ($n >> p$ or $p >> n$). Measurements of a and R_H for strongly p-type and n-type extrinsic semiconductors permit determination of Hall mobilities from Eq. (21.5). Since μ_n, μ_p, n and p all are temperature dependent, Hall effect in an semiconductor exhibits wide temperature variations, corresponding to the increase of intrinsic carriers at higher temperatures.

From Eq. (21.64), it is obvious that the magnitude of Hall coefficient is inversely proportional to the carrier concentration. The coefficient of proportionality involves a factor which depends on the energy distribution of the carriers and the way in which the carriers are scattered in their motion. However, the value of this factor normally does not differ from unity by more than factor of two. The situation is more complicated when more than one type of carrier is important for the conduction. The Hall coefficient then depends on the concentration of the various types of carriers and their relative mobilities.

The product of the Hall coefficient and the conductivity is proportional to the mobility of the carriers when one type of carrier is dominant. The proportionality involves the same factor which is contained in the relationship between the Hall coefficient and the carrier concentration. The value obtained by taking this factor to be unity is referred to as the Hall mobility.

Experimental determination of the hall coefficient

To measure the carrier concentration directly, the most common method is the use of the Hall effect. Fig. (21.54) shows the basic set up, where a rectangular specimen of the semiconductor of width b and thickness t is placed between the pole pieces of an electromagnet such that a magnetic field B is applied along the z-direction. With the help of a battery, the current I is allowed to pass through the sample in the x-direction Fig. (21.54). One can measure the Hall voltage V_H with the help of two probes placed at the centres of the top and bottom surfaces of the sample. We have

$$V_H = E_H t \tag{21.64}$$

But $E_H = R_H J B$ and hence Eq. (21.65) takes the form

$$V_H = R_H \; Jt \tag{21.65}$$

We have the current density

$$J = \frac{I}{bt} \tag{21.66}$$

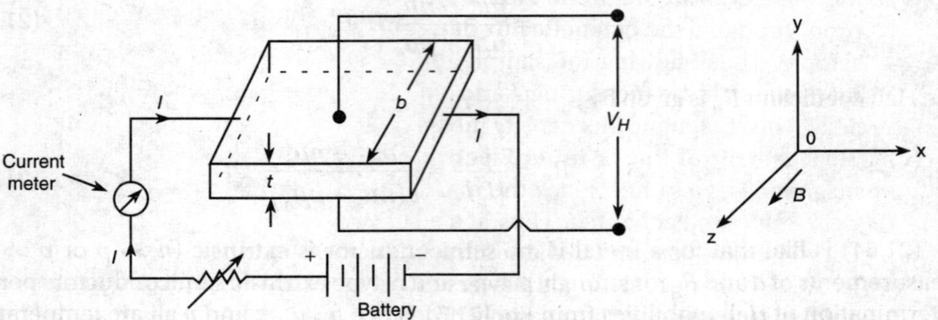

Fig. 21.54 Basic set up for the determination of Hall Coefficient

where bt is the cross-sectional area of the sample. Making use of Eq. (21.67), Eq. (21.66) becomes

$$V_H = \frac{R_H IB}{b}$$

or
$$R_H = \frac{V_H b}{IB} \tag{21.67}$$

one can measure the quantities appearing on the r.h.s. of Eq. (21.68) and hence R_H can be readily determined. In S.I. system, R_H is obtained in m³/Coulomb. We must note that the polarity of V_H is opposite for n-type and p-type of materials and hence R_H has opposite signs for the two type of semiconductors.

Uses of the Hall effect

(i) Determination of Semiconductor type: The Hall voltage V_H is measured by two probes at the centres of the top and bottom face of the sample. If the magnetic flux density is B Wb/m², then we have $n = 1/R_H^e$, where $R_H = 1/ne$.
Since R_H is positive for P-type and negative for N-type semiconductor, the sign of R_H is used to determine the type semiconductor specimen.

(ii) Determination of carrier concentration: We have $n = 1/|eR_H|$, where $R_H = 1/ne$. Obviously, by measuring R_H, one can find the electron concentration n and hole concentration p.

(iii) Determination of mobility of carriers: If the conduction is due to one type of carriers, e.g., electrons, we have

$$\sigma = ne\mu_n$$

$$\mu_n = \frac{\sigma}{ne} = \sigma R_H$$

i.e., $$\mu_n = \sigma\left(\frac{V_H b}{IB}\right) \quad \text{[using Eq. 21.9]}$$

Obviously, by knowing σ, one can determine μ_n. This mobility, i.e., given by $\sigma|R_H|$ is referred to as Hall mobility.

(iv) Measurement of Magnetic Flux Density: We have seen that Hall voltage V_H is proportional to the magnetic flux density B for a given current I passing through sample. Thus knowing the sample dimensions and R_H, one can determine the magnetic field by measuring I and V_H. One can also use it as the basic for the design of a magnetic flux density meter.

(v) Measurment of Power in an Electromagnetic wave: We know that in electromagnetic wave in free space the magnetic field H and the electric field E are at right angles to each other. Thus, if a semiconductor sample is placed parallel to E it will derive current I in the semiconductor. The semiconductor sample is subjected simultaneously to a transverse magnetic field H producing a Hall voltage across the semiconductor sample. The Hall voltage will be proportional to the product E and H across the sample, i.e., to the magnitude of the poynting vector of the electromagnetic wave. Obviously, the Hall effect can be used to determine the power flow in an electromagnetic wave.

(vi) **Hall Effect Multiple**: If the magnetic field B is produced by passing a current I through an air-core coil, B will be proportional to current I. The Hall voltage V_H is thus proportional to the product II. This forms the basis for the design of a multiplier.

21.30 TRANSISTOR

Transistors are three terminal (solid-state devices and are extremely important semiconductor devices in today's microelectronic circuitry). In a way they have revolutionised the entire field of electronics. They are capable of two primary types of functions. First, transistors can perform the same operation as their vacuum tube precursor, the triode valve; i.e., they can amplify an electrical signal (i.e., as amplifier). In addition they serve as switching devices in computers for the processing and storage of information. The major types of transistors are: the junction (or bimodal) transistor and the metal-oxide semiconductor field effect transistor (MOSFET).

21.30.1 Junction Transistor

This is composed of two P-N junctions arranged back to back in either the N-P-N or P-N-P configuration. The charge carriers in a junction transistor have positive as well as negative polarity and so the nomenclature Bipolar Junction Transistor is used.

In actual practice a junction transistor consists of a silicon (or germanium) crystal, in which a very thin layer of N-type Si (Ge) is sandwitched between two layers of P-type Si (Ge) and thereby we get PNP junction transistor. Alternatively, it may consist of a very thin layer of P-type between two layers of N-type material and we get a NPN transistor. The semiconductor sandwitched is extremely thin, and small and is hermatically sealed against moisture inside a metal or a plastic case. The size is quite small.

The three terminals taken from each section of a semiconductor are called (i) Emitter (ii) Base and Collector Fig. (21.55). The middle section, i.e., base is very thin in comparison with the other two. Metal leads *E, B* and *C* come out of the package for connection to the emitter, the base and the collector, respectively. In a junction transistor both the majority and the minority carriers are involved.

In a normal transistor operation, emitter-base junction is forward-biased and the collector-base junction is reverse-biased. The circuit symbols for both the types of transistors are

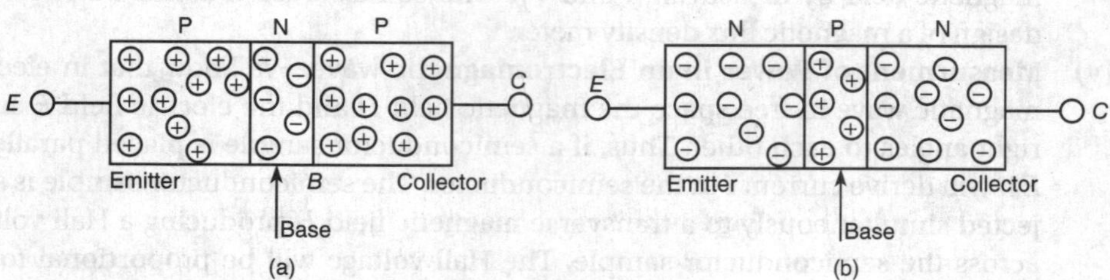

Fig. 21.55 Bipolar junction transistor

shown in Fig. (21.56). The arrow on the emitter shows the direction of the current when the emitter-base junction is forward-biased. Obviously, the current enters the transistor through the emitter terminal for a PNP transistor and leaves the transistor through the emitter terminal for a NPN transistor. In both the cases, the emitter (E), base (B) and collector (C) currents, I_E, I_B and I_C, respectively, are taken positive when the currents go into the transistor. The symbols V_{EB}, V_{CB} and V_{CE} represent respectively the emitter-base, collector-base and collector-emitter voltages. These are assumed positive when the terminal marked by the first subscript is positive with respect to the terminal marked by the second subscript.

As the emitter junction is usually forward-biased, I_E is negative in the case of an NPN transistor and positive for a PNP transistor. Since the collector junction is reverse biased, the voltage V_{CB} is negative for a PNP transistor and positive for an NPN transistor. We may note that signs are opposite for the two type of transistors. Transistors are generally fabricated with the help of the following four basic techniques: (i) Grown technique (ii) Alloy or fused technique (iii) Diffusion technique and (iv) Epitaxial technique.

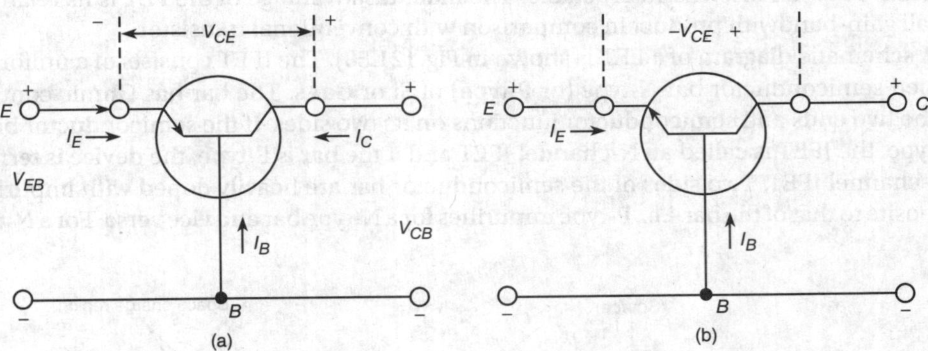

Fig. 21.56 Current symbols with reference to the current direction and voltage polarities for: (a) PNP transistor (b) NPN transistor

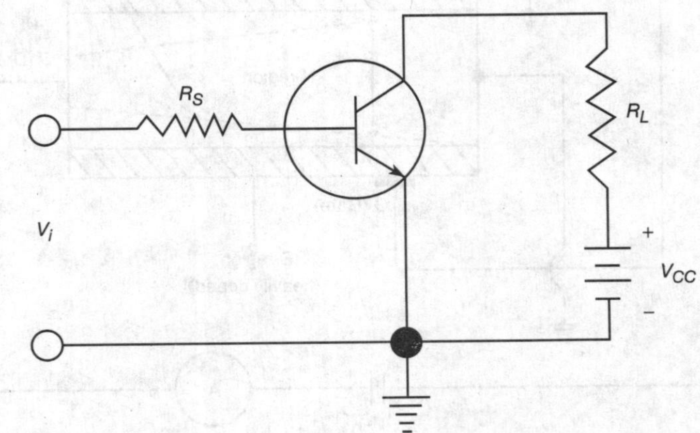

Fig. 21.57 Transistor as switching circuit

In many circuits such as those required in high speed counters and computer, a transistor is made to act as a switch. It is 'ON' when it is in saturation, and 'OFF' when it is in cutoff condition. By applying pulse voltages, the transistor is caused to switch over from one state to the other.

21.30.2 Field Effect Transistor (FET)

This is a semiconductor device with the output current controlled by an electric field. Since the current in FET is carried predominantly by one type of carriers, it is known as unipolar transistor. Obviously, FET is different from the bipolar transistor (BJT) which involves two types of carriers, i.e., both electrons and holes.

FET has several forms, e.g., Junction Field Effect Transistor (JFET), *Metal Oxide Semiconductor Field Effect (MOSFET) Transistor*, etc.,

The different classes of FETs are characterized by a high input impedance. These devices are used in controlled switching between conducting and non conducting states in digital circuits. FETs are also thermally stable. The main disadvantage of the FET is its relatively small gain-bandwith product in comparison with conventional transistor.

A schematic diagram of a FET is shown in Fig. (21.58). The JFET consists of a uniformly doped semiconductor bar N-type (or P-type) of Si or GaAs. The bar has Ohmic contacts at the two ends and semiconductor junctions on its two sides. If the semiconductor bar is N-type, the JFET is called an N-channel JFET and if the bar is P-type, the device is termed a P-channel JFET. Two sides of the semiconductor bar are heavily doped with impurities opposite to that of the bar, i.e., P-type impurities for a N-type bar and vice versa. For a N-type

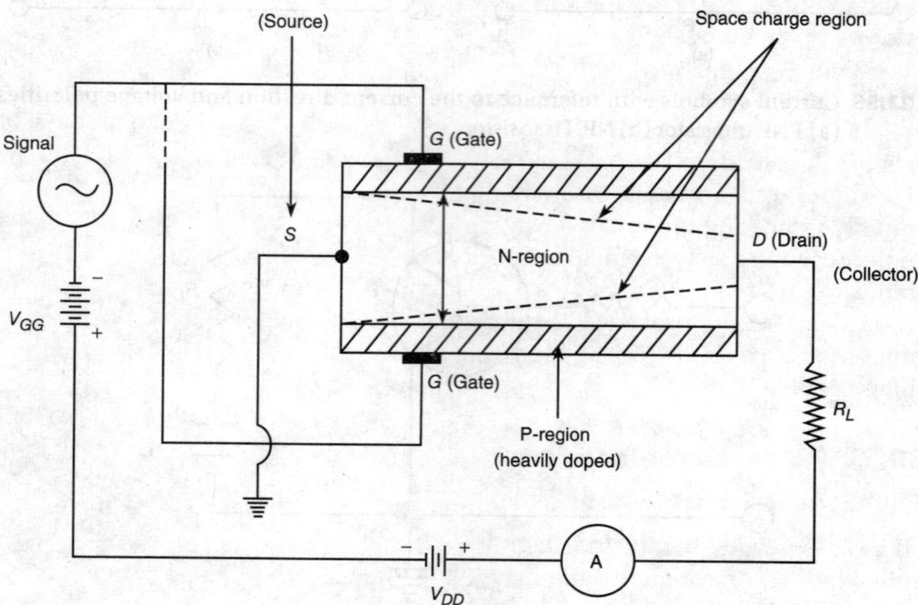

Fig. 21.58 Basic structure of an N-channel JFET

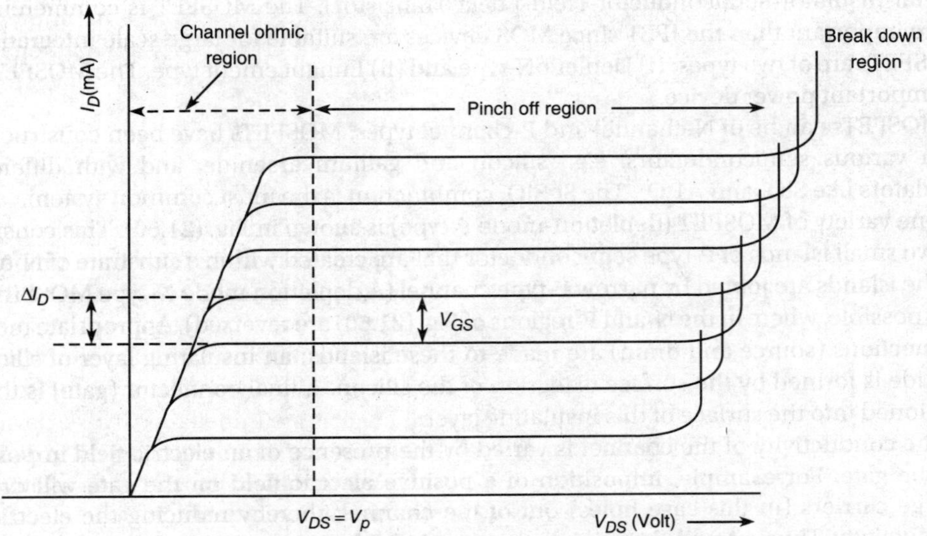

Fig. 21.59 $I_D - V_{DS}$ static characteristic of an N-channel JFET

JFET, we can denote the three regions by P^+, N and P^+, where P^+ denotes a very heavily doped P-region. By applying a voltage V_{DD} between the two ends of a semiconductor bar, a current is allowed to flow along the length of the bar. The central N-region through which the majority carriers (electrons) flow from the source (S) to the collector or drain (D) is called the channel. The P^+ regions form the gate. At each of the PN junctions there is a depletion layer extending into the N-region. These layers penetrate more and more as the V_{GG} (reverse bias) is increased, thus controlling the flow of electrons in the channel, i.e., the drain current I_D. The characteristic curves (l_D-V_{DD} curves) are shown in Fig. (21.59). When l_D is increased from zero at constant V_{GS}, l_D increases linearly at first confirming to the Ohm's law. With further increase in V_{DS}, the current attains saturation and this is called the **pinch-off** or **the saturatioN-region**. Beyond a certain value of V_{DS}, I_D increases very rapidly indicating the break-down region. The name field effect is used for the device because the transverse field produced by the gate gives the effect of controlling the drain current.

JFETs are less temperature sensitive and are not affected much by radioactive radiations. The input resistance of JFET is very high. One can achieve much lower switching time in switching devices.

21.30.3 Metal Oxide Semiconductor Field Effect Transistor (MOSFET)

Like JFET, the MOSFET is also a field effect transistor, whose drain current (I_D) is controlled by the voltage on the gate. MOSFET and JFET differ physically as well as in operation. The MOSFET is also referred to as the IGFET (Insulated-Gate Field-Effect Transistor), MOST (Metal-Oxide-Semiconductor field effect Transistor) and MISFET

(Metal-Insulator-Semiconductor-Field-Effect Transistor). The MOSFET is commercially more important than the JFET since MOS devices are suitable for large scale integration. MOSFETs are of two types: (i) DepletioN-type and (ii) Enhancement type. The MOSFET is an important power device.

MOSFETs can be of N-channel and P-channel types. MOSFETs have been constructed with various semiconductors, e.g., silicon and gallium arsenide, and with different insulators like SiO_2 and $A1_2O_3$. The Si-SiO_2 combination is the most common system.

One variety of MOSFET (depletion-mode P-type) is shown in Fig. (21.60). This consists of two small islands of P-type semiconductor that are created within a substrate of N-type Si, the islands are joined by narrow P-type channel (A depletion mode N-type MOSFET is also possible, where in the N-and P-regions of Fig. (21.60) are reversed). Appropriate metal connections (source and drain) are made to these islands; an insulating layer of silicon dioxide is formed by the surface oxidation of the silicon. A final connector (gate) is then fashioned into the surface of this insulating layer.

The conductivity of the channel is varied by the presence of an electric field imposed on the gate. For example, imposition of a positive electric field on the gate will drive charge carriers (in this case holes) out of the channel, thereby reducing the electrical conductivity. Thus, a small alteration in the electric field at the gate will produce a relatively large variation in current between the source and the drain. In some respects, then, the operation of a MOSFET is very similar to that described for the junction transistor. The primary difference is that the gate current is exceedingly small in comparison to the base current of a junction transistor. Obviously, MOSFETs are, therefore used where the signal sources to be amplified cannot sustain an appreciable current.

Another important difference between MOSFETs and junction transistors is that, although majority carriers dominate in the function of MOSFETs (i.e., holes for depletion-mode-P-type MOSFET Fig. (21.60), minority carriers do play a role with junction transistors (i.e., injected holes in the N-type base region).

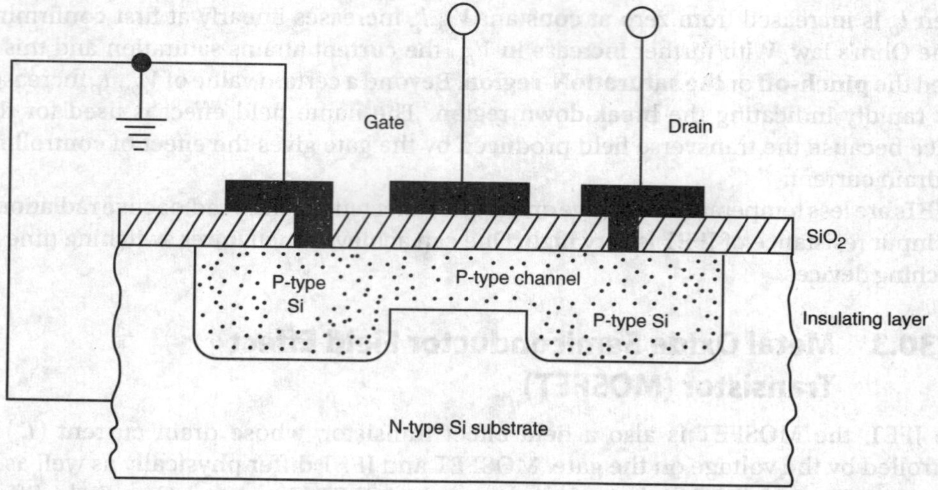

Fig. 21.60 Schematic cross-section view of a depletion-mode P-type MOSFET transistor

The MOSFETs can be used in any of the circuits wherever JFET is used, with the bulk (substrate) connected to source. We may note that handling of MOSFET is not so easy. It requires special precautions because of very fine layer of SiO_2 between gate and channel. It is very susceptible to high voltages and can get punctured. Even static electric charge can puncture it. This is why MOSFETs are protected by a shorting ring that is wrapped around all four terminals which must remain until the device is mounted and soldered properly.

21.31 SEMICONDUCTORS IN COMPUTERS

In addition to their ability to amplify an imposed electrical signal, semiconductor diodes and transistors may also act as **switching** devices, a feature utilized for arithmetic and logical operations, and also for storage of information in computers. Computer numbers and functions are expressed in terms of a binary code, i.e., numbers written to the **base 2**. Within this scheme or framework, numbers are represented by a series of two states (sometimes designated 0 and 1). Now, diodes and transistors within a digital circuit operate as switches that also have two states–ON and OFF, or conducting and non-conducting; 'OFF' corresponds to one binary state, and 'ON' to the other state. Obviously, a single number may be represented by a collection of circuit elements containing transistors that are appropriately switched.

21.32 INTEGRATED CIRCUITS (ICs)

The most important use of all these semiconductor devices today is not in discrete components, but rather in *integerated circuits*. It is possible to fabricate a silicon-based substrate (commonly called a *chip*) containing a hundred million or more components, including resistors, capacitors, transistors, and logic switches. This extreme miniaturization has done two things for electronic industry: (i) it has made it possible to put sophisticated ICs into even the smallest products, such as wrist watches and cameras, (ii) it has enabled those ICs to work much faster. Although there are still limits based on the speeds at which switches can operate, the transit time of an electronic signal has been reduced by packing the components so closely together. Those signals travel at essentially the speed of $-0.3\mu m/fs$, in units relevant to the distance/time scale of ICs.

ICs have had tremendous impact on the work of scientists and engineers by enhancing the computing power available to them. Today, personal computers (PCs), which fit on desktops and are plugged into a 110-volt outlet, can perform more than 10^9 operations/s, while the largest supercomputers are thousands of times faster. Presently supercomputers had achieved rates higher than 3×10^{13} operations/s.

One indication of the rapid progress in computing is the increase in the number of transistors that can fit on a single microchip. The rapid growth in microchip technology continues in the 21st century. The Pentium 4 chip, produced by Intel in 2002, contained 55 million transistors on a single chip. In order to achieve such high densities, it is necessary to build extremely small components. In the Pentium 4, for e.g; circuit elements are about 1.3×10^{-7} m across in each direction. We may note that this is less than the wavelength of visible light. For that reason chip manfacturers hope to use shorter ultraviolet lithographic techniques. Presently, manfacturers are trying to use shorter ultraviolet wavelengths to produce circuit elements less than 1×10^{-7} m across, thus producing even smaller and faster microprocessors.

An important factor in technology besides the computing itself is the storage and retrieval of information. Presently, there were two standard media: magnetic and optical. It is possible to store and reliably retrieve more than 10^{13} bits of information per m^2 of magnetic disk or tape. This is approaching a theoretical limit, because at this density as few as 100 magnetic particles are used to store each bit of information. At higher densities one begins to have problems with reliability, due to statistical fluctuations in the magnetizations of the particles and overlap of magnetic fields from one small bit to another.

21.33 MICRO ELECTROMECHANICAL SYSTEMS (MEMS)

Almost all applications in industry, defence, medicine and other fields require sensing and control of various parameters. In recent years, taking advantage of the silicon IC technology, a new field has emerged which attempts to combine **sensor, actuator** and **the control circuit** on as one integrated unit. In this sense it emulates a biological system Fig. (21.61). These are known as smart sensors, microsystems technology (MST) or micro-electromechanical systems (MEMS). It has been possible to fashion miniature mechanical devices such as gears, motors, springs etc., Their combination with sensing and actuating functions has given engineers and scientists the tools to build microsystems that could not be imagined earlier. Some examples of the smart structures and systems are given in Fig. (21.62).

MEMS devices are small in size, light weight, low cost, reliable, with large batch fabrication technology. They generally consist of sensors that gather environmental information such as pressure, temperature, acceleration, etc., integrated electronics to process the data collected and actuators to influence and control the environment in the desired manner.

The MEMS technology involves a large number of materials. Silicon forms the backbone of these systems due to its excellent mechanical properties as well as mature micro-fabrication technology including lithography, etching and bonding. Other materials having piezoelectric, piezoresistive, ferroelectric and other properties are widely used for sensing and actuating functions in conjunction with silicon. The field of MEMs is expected to touch all aspects of our lives during this decade with revolutions in aviation, automobiles, pollution, control and industrial processes.

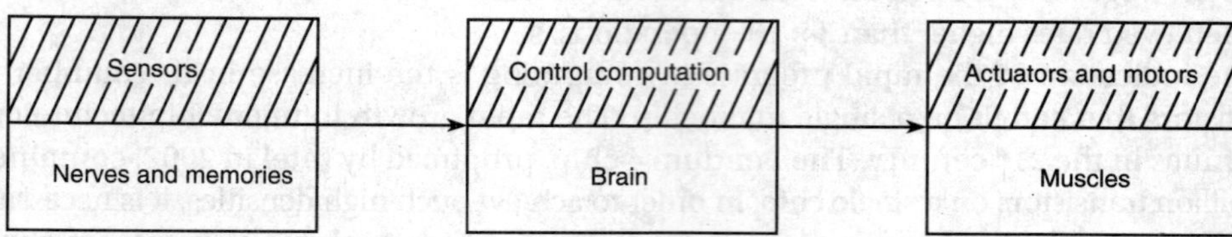

Fig. 21.61 Integrated sensor-actuator systems with controller are analogous to biological systems

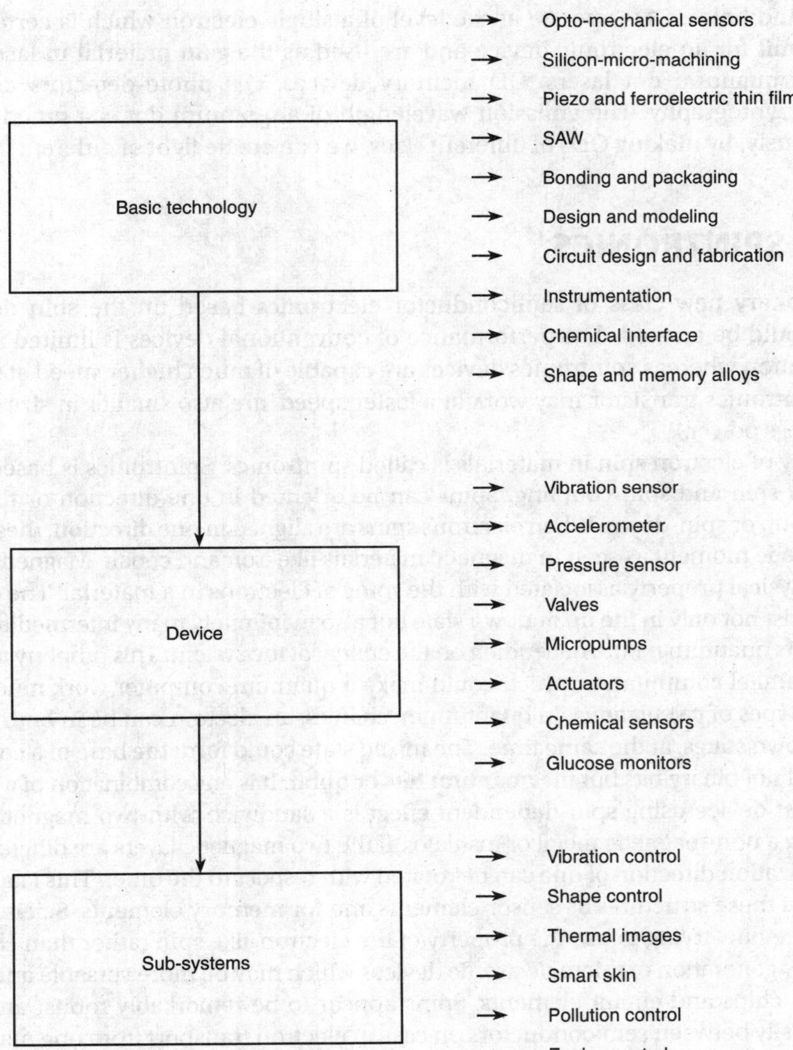

Fig. 21.62 Few examples of smart structures and systems and their applications

21.34 QUANTUM DOTS (QDs)

Rapid progress in the fabrication of semiconductor structures has resulted into the reduction of three dimensional systems to two-dimensional, one-dimensional, and finally to zero dimensional systems. These reduced dimensional systems are used in future applications like improved semiconductor lasers and microelectronics. Quantum dots represent the ultimate reduction in the dimensionality of semiconductor devices. These are three dimensional semiconductor structures only nanometers in size confining electrons and holes. QDs operate at the level of a single electron which is certainly the ultimate limit for an electronic device and are used as the gain material in lasers. QDs are use in quantum dot lasers, QD memory devices, QD photo-detectors and even quantum cryptography. The emission wavelength of a quantum dot is a function of its size. Obviously, by making QDs of different sizes, we can create light of different colours.

21.35 SPINTRONICS

A revolutionary new class of semiconductor electronics based on the spin degree of freedom could be created. The performance of conventional devices is limited in speed and dissipation whereas spintronics devices are capable of much higher speed at very low power. Spintronics transistor may work at a faster speed, are also smaller in size and will consume less power.

The study of electron spin in materials is called spintronics. Spintronics is based on the direction of spin and spin-coupling. Spins can be oriented in one direction or the other, called spin-up or spin-down. When electrons spins are aligned in one direction, these create a net magnetic moment as seen in magnetic materials like iron and cobalt. Magnetism is an intrinsic physical property associated with the spins of electrons in a material. The electron spin may exist not only in the up or down state but also in infinitely many intermediate states because of its quantum nature depending on the energy of the system. This property may lead to highly parallel computation which could make a quantum computer work much faster for certaiN-types of calculations. In quantum mechanics, an electron can be in both spin-up and spin-down states, at the same time. The mixed state could form the base of a computer, built around not binary bits but the **quantum bits** or **qubit**. It is any combination of a 1 or a 0. The simplest device using spin-dependent effect is a sandwich with two magnetic layers surrounding a non-magnetic metal or insulator. If the two magnetic layers are different, then the magnetization direction of one can be rotated with respect to the other. This leads to the utilization of these structures as sensor elements and for memory elements. Scientists and engineers are now trying to use the property of the electron-like spin rather than charge to develop new generation of microelectronic devices which may be more versatile and robust than silicon chips and circuit elements. Spins appear to be remarkably robust and move relatively easily between semiconductors. In case of electron transport from one material to another, the spins do not lose its orientation or scatter from impurities or structural defects.

Quantum Hall Effect

Klitzing et al. in 1980, observed that on the application of a large magnetic field on a silicon MOSFET at liquid helium temperature, the Hall voltage developed (V_H) is a function of the gate voltage (V_G) and there appear plateaus around those values for which the potential drop V_{pp} along the sample parallel to the current vanishes Fig. (21.63). Fig. (21.64) shows a plot of V_{pp} against V_G. Fig. (21.64) shows uniformly spaced zeros. We may note that the difference between the Hall voltages for two successive plateau is exactly 1/25813 Ω and serves as the definition of ohm in terms of Planck's constant (h) and electron charge (e).

In the channel of MOSFET, the electrons form a two dimensional electron gas and in a strong magnetic field. say along z-axis, the electron energies in the two dimensional

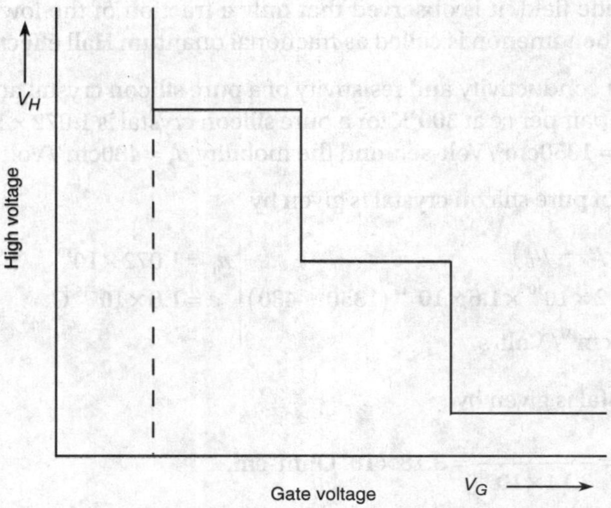

Fig. 21.63 Hall voltage (V_H) versus gate voltage (V_G) in a silicon MOSFET

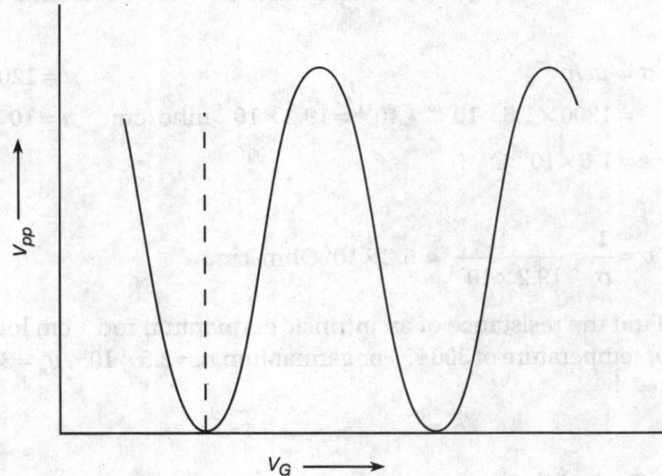

Fig. 21.64 In a silicon MOSFET the potential drop along the current parallel to the current V_s gate voltage (V_G)

electron gas are quantized into Landau Levels of energy $\left(n+\frac{1}{2}\right)\hbar\omega_c$, where ω_c is cyclotron frequency and n being an integer. When a Landau level of energy coincides with the Fermi level the potential drop V_{pp} tends to zero, because then electrons cannot be scattered to the already occupied states of energy. We may note that the gate voltage V_G controls the number of electrons per unit area of the channel and therefore the Fermi energy. This is why minima in V_{pp} and plateau in ϕ_H occur at intervals $\hbar\omega_c$. We may also note that with a sufficiently large magnetic field, it is observed that only a fraction of the lowest Landau level is occupied. This phenomenon is called as fractional quantum Hall effect.

Example 21.1: Find the conductivity and resistivity of a pure silicon crystal at 300 K. The density of electron-hole pair per cc at 300°K for a pure silicon crystal is 1.072×10^{10} and the mobility of electrons $\mu_n = 1350\text{cm}^2/\text{Volt-sec.}$ and the mobility $\mu_h = 480\text{cm}^2/\text{Volt-s}$.

Solution Conductivity of pure silicon crystal is given by

$$\sigma = n_i e(\mu_e + \mu_h) \qquad n_i = 1.072 \times 10^{10}$$

$$\sigma_i = 1.072 \times 10^{10} \times 1.6 \times 10^{-19}(1350 + 480) \qquad e = 1.6 \times 10^{-19}\,C$$

$$\mu_h = 480\text{cm}^2/\text{Volt-s}$$

Resistivity of silicon crystal is given by

$$\rho_i = \frac{1}{\sigma_i} = \frac{1}{3.14 \times 10^{-6}} = 3.18 \times 10^5 \text{ Ohm-cm.}$$

Example 21.2: A silicon wafer is doped with phosphorus of concentration 10^{13} atoms/cm³. If all the donor atoms active, what is its resistivity at room temperature? The electron mobility is 1200 cm²/Volt-s. Charge on the electron is 1.6×10^{-19} Coulomb.

Solution

$$\sigma = \mu e n \qquad \mu = 1200\text{cm/volt-s}$$

$$= 1200 \times 1.6 \times 10^{-10} \times 10^{13} = 19.2 \times 10^{-4}\text{ mho/cm} \qquad n = 10^{-13} = N_p$$

$$e = 1.6 \times 10^{19}\text{ C}$$

Resistivity

$$p = \frac{1}{\sigma} = \frac{1}{19.2 \times 10^{-4}} = 5.2 \times 10^2 \text{ Ohm-cm.}$$

Example 21.3: Find the resistance of an intrinsic germanium rod 1 cm long, 1 mm wide and 1 mm thick at temperature of 300 K. For germanium $n_i = 2.5 \times 10^{13}$, $\mu_n = 3900$ cm²/Volt-s at 300 K.

Solution

$$\sigma = n_i e(\mu_e + \mu_h)$$

$$= 25 \times 10^{13} \times 1.6 \times^{-19}(0.39 + 0.19)$$

$$= 2.32\text{mho/m}$$

$$\rho = \frac{1}{232}\text{Ohm} \times \text{m}$$

Now resistance $R = p_i\left[\frac{\text{length}}{\text{area of cross section}}\right] = 2.32 \times \frac{10^{-6}}{10^{-2}}$

$$R = \frac{1}{2.32 \times 10^{-4}}\text{ohm} = 4.31\text{k}\Omega.$$

Example 21.4: A sample of Germanium in made of *P* material by adding acceptor atoms at a rate of one atom per 4×10^8 Germanium atoms. The acceptor density is assumed to be zero and $n_i = 2.5 \times 10^{19}$ per m³ at 300°K. There are 4.4×10^{28} Germanium atoms/m³. The acceptor density is found to be 1.1×10^{20} atoms/m³.

Solution

$$n_i^2 = np = n_p N_a$$

$$n_p = \frac{n_i^2}{N_a} = \frac{6.25 \times 10^{38}}{1.1 \times 10^{20}} = 5.6 \times 10^{18}$$

$$\frac{n_p}{n_i} = \frac{5.6 \times 10^{18}}{2.5 \times 10^{19}} = 0.22.$$

Example 21.5: Find the value of the applied forward voltage for a P-N junction diode if $I_s = 50$ micro amp/cm, $I = 2$ amp/cm² and e/kT = 40 Volt.

Solution

$$I_s = V\left(e^{eV/kT} - 1\right)$$

$$e^{40V} = \frac{J}{J_s} - 1 \simeq \frac{2 \times 10^6}{30}$$

$$V = \frac{2303}{40} \log_{10} \frac{2 \times 10^5}{3} = 0.277 \text{ Volt.}$$

Example 21.6: The approximate value of P-N junction current under forward bias is given by $I = I_s e^{eV/kT}$. Show that the incremental resistance R_e defined by $\Delta V/\Delta I$ is equal to $1/40\,I$ at the room temperature (e/kT = 40/Volt).

Solution

$$I = I_s \exp[eV/kT] = I_s e^{40V}$$

$$R_e = \frac{\Delta V}{\Delta I} = \frac{1}{40I}$$

Example 21.7: Determine the number density of donor atoms which have to be added to an intrinsic germanium semi-conductor to produce an N-type semi-conductor of conductivity 5 mho/cm. The mobility of electrons in the N-type semi-conductor is 3850 cm²/Volt-sec.

Solution

$$\sigma_n = nq\mu_n = N_D q\mu_n \qquad \sigma_n = 5 \text{ mho/cm}$$

$$q = 1.6 \times 10^{-19} \text{ C}$$

$$\therefore \quad n = N_D = \sigma_n / q\mu_n = \frac{5}{1.6 \times 10^{-19} \times 3850} = 0.8 \times 10^{18} \text{ per cm}^3$$

$$\mu = 3850 \text{ cm}^2/\text{volt-s.}$$

Example 21.8: The saturation current density of a PN junction Germanium diode is 250 mA/m² at 300 °K. Find the voltage that would have to be applied across the junction to cause a forward current density of 10^5 A/m² flow.

$$I = I_s\left[\exp\left(\frac{eV}{kT}\right) - 1\right]$$

Solution Dividing throughout by volume, one obtains the equation in the form of current density as

$$J = J_s\left[\exp\left(\frac{eV}{kT}\right) - 1\right]$$

$$\therefore \quad \exp\left(\frac{eV}{kT}\right) - 1 = J/J_s$$

$$\therefore \quad \exp\left(\frac{eV}{kT}\right) - 1 = \frac{10^5}{250\times10^{-3}} = 4\times10^5 \qquad J_s = 250\,\text{mA/m}^2$$

$$J = 10^5\ \text{A/m}^2$$

or

$$\frac{eV}{kT} = \log_e\left(4\times10^5\right) = 129$$

$$V = \frac{12.9\times1.38\times10^{-23}\times300}{1.6\times10^{-19}} = 0.33\,\text{V}$$

Example 21.9: A PN junction in series with a resistance of 5×10^3 Ω is connected across a 50 and 1 mΩ respectively. Show that the forward and reverse bias currents in the circuit are 9.9 mA and 4.975×10^{-2} mA respectively.

Solution (i) Forward bias

$$\text{Current} = \frac{\text{Applied voltage}}{\text{Junction resistance + external resistance}} = \frac{50}{5000+50} = 9.9\,\text{mA}$$

$$\text{Current} = \frac{50}{5000\times10^6} = 4.975\times10^{-2}\,\text{mA}.$$

Example 21.10: Assume that the silicon diode in the following circuit requires a minimum current of 1 mA to be above the knee point (0.7 V) of its 0I-V characteristics. Also assume that the voltage across the diodes is independent of current above the knee point.

(i) If $V_B = 5$ V, what should be maximum value of R so that the voltage is above the knee point?

(ii) If $V_B = 5$ V, what should be the value of R to establish a current of 5 mA in the circuit?

(iii) What is the power dissipated in the resistance R and in the diode, when a current of mA flows in the circuit at $V_B = 6$ V.

(iv) If $R = 1$ kΩ, what is the minimum voltage V_B required to keep the diode above the knee point?

Solution

(i) The minimum voltage across the diode is = 0.7 V. Obviously, that is above the knee point of the characteristic curve. Thus the voltage across the resistance $R = v' = (5 - 0.7)\ V = 4.3$ V. The minimum current $i = 1$ mA $= 10^{-3}$ A.

$$R = \frac{V^1}{i} = \frac{4.3}{10^{-3}}\Omega = 4.3\text{k}\Omega$$

(ii) The current through the resistance $R = 5$ mA $= 5 \times 10^{-3}$ A and also the voltage drop across $R = (5 - 0.7)$ V $= 4.3$ V.

$$\therefore \qquad R = \frac{4.3}{5\times10^{-3}} = 0.86\times10^{-3}\,\Omega = 860\Omega$$

(iii) We have $V_B = 6$ V, and hence the voltage V^1 across the resistance is $V' = 6 - 0.7 = 5.3$ V and current $i = 5$ mA $= 5 \times 10^{-3}$ A.

$\therefore$ The power (P) dissipated through the resistance R is

$$P = iV' = 5\times10^{-3}\times5.3\,\text{W}$$

$$= 26.5\times10^{-3}\ \text{W} = 26.5\text{mW}$$

$\therefore$ The power P', dissipated in the diode is

$$p' = i\times0.7\,\text{W} = 5\times10^{-3}\times7\times10^{-1}\,\text{W}$$

$$= 3.5\times10^{-3}\,\text{W} = 3.5\text{mW}$$

(iv) To keep the diode above the knee point, the minimum current required is 1 mA.

$$V_R = \text{voltage drop across } R(=1\text{k}\Omega)$$

$$= 1\times10^{3}\times10^{-3} = \text{IV}$$

We have the minimum voltage drop across the diode = 0.7 V

$$V_B = (1 + 0.7)\text{V} = 1.7\text{V}$$

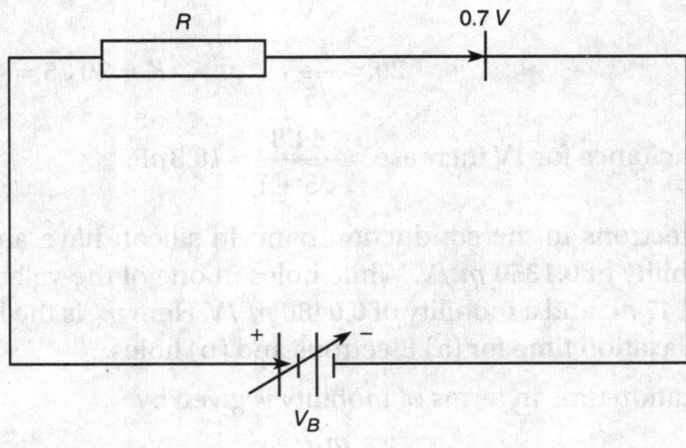

Fig. 21.65

Example 21.11: A Zener diode (BC-147) has a Zener voltage of 9V with 10% tolerance at 25 °C. Find the range of voltage ratings at which it will breakdown? If the temperature is raised to 75 °C, what will be the new breakdown voltage when decrease in Zener voltage is about 2mV per degree centigrade?

Solution Given, average breakdown voltage $V_z = 9\text{ V}$

$$\text{Tolerance} = 10\% = \frac{10}{100} \times 9 = 9\text{V}$$

∴ Range of breakdown voltage = (9 – 09) to (9 + 0.9) V = 8.2 V to 9.9 V.

In the second case, the increase in temperature = 75 – 25 = 50 ° C

Fall in breakdown voltage = 2 × 50 = 100mV = 0.1 V

∴ New breakdown voltage = 9 – 0.1 = 8.9 V.

Obviously, the range of new breakdown voltage

$$= (8.1 - 0.1)\text{to}(9.9 - 0.1)\text{V} = 8.0 \quad \text{to} \quad 9.8\text{V}$$

Example 21.12: Will Silicon and Germanium semiconductors at 1000°K ambient temperature still be semiconductors?

Solution We know the forbidden energy gap between conduction and valence bands for a semiconductor is nearly 1 eV. For Ge and Si the energy gap is 0.785 eV and 1.21 eV respectively at 0 °K. The energy gap decreases with increase in temperature which is represented by the expression

$$E_g(T) = 1.21 - 3.60 \times 10 - 4\text{T}(\text{for Si})$$

$$E_g(1000) = 1.21 - 3.60 \times 10^{-4}(1000) = 0.85\text{eV}$$

obviously, Si and Ge will remain semiconductors at 1000 K ambient temperature.

Example 21.13: An abrupt junction diode has transition temperature of 20 pF when reverse biased at 5 V. If the voltage is increased by 1 V, show that the capacitance will decrease by 1.7 pF.

Solution

$$C = \frac{K}{\sqrt{V}} \qquad (K = \text{eA})$$

$$20 = \frac{K}{\sqrt{5}} \qquad \therefore \qquad K = 20\sqrt{5} = 44.8$$

$$\therefore \qquad \text{Capacitance for 1V increase} = \frac{44.8}{\sqrt{5} + 1} = 18.3\text{pF}.$$

Example 21.14: Electrons in the conduction band in silicon have an effective mass of 0.259 m_0 and a mobility of 0.1350 m^2/V. While holes in one of the valance bands have an effective mass of 0.547 m_0 and a mobility of 0.0480 m^2/V. Here m_0 is the free electron mass. Find the average relaxation time for (a) Electrons and (b) holes.

Solution The relaxation time in terms of mobility is given by

$$\tau = \frac{m\mu}{e}$$

Taking effective mass of electron and holes in consideration, the relaxation time is given by

$$\tau = \frac{m^* \mu}{e}$$

(a) For electrons $m^* = 0.259 m_0$

$$\therefore \quad \tau_e = \frac{0.359 m_0 \mu}{e} = \frac{0.259 \times 9.1 \times 10^{-31} \times 0.1350}{1.6 \times 10^{-19}} = 1.99 \times 10^{-13}\ \text{S}$$

(b) For holes in the valance bands $m^* = 0.537 m_0$

$$\tau = \frac{0.537 \times 9.1 \times 10^{-31} \times 0.048}{1.6 \times 10^{-19}} = 1.47 \times 10^{-13}\ \text{S.}$$

Example 21.15: **Two Germanium diodes D_1 and D_2 are connected in series across 5V battery as shown in Fig. [21.66]. Find voltage across each diode assuming that breakdown voltage of diodes is greater than 5V. What is the effect of temperature? (b) Find current if $V_Z = 4.9$ V and $I_0 = 5 \mu A$.**

Solution (a) The two diodes are connected in series and hence the same current I flows in D_1 and D_2. Obviously, it is in forward direction through D_2 and in reverse direction through D_1. Since D_2 diode is forward biased, V_2 will be very small and hence V_1 (= 5 – V_2) will be very much large than V_T(= 0.026 V). This means the current will be equal to reverse saturation current I_0. Now we consider diode D_2. We have

$$I = I_0 \left[\exp\left(\frac{V}{V_T} \right) - 1 \right]$$

Putting $\quad I = I_0$ and $V = V_2$, we have

$$I_0 = I_0 \left[\exp\left(\frac{V_2}{V_T} \right) - 1 \right]$$

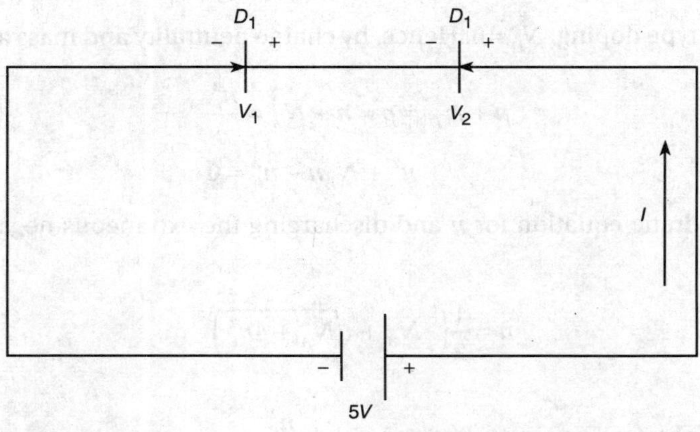

Fig. 21.66

or $$\exp\left(\frac{V_2}{V_T}\right)-1=1$$

or $$V_2 = V_T ln2 = 0.026 \times 0.693 = 0.018\,\text{V}$$

$\therefore$ $$V_1 = 5 - 0.018 = 4.892\,\text{V}$$

Effect of temperature: $V_2 = V_T \ln 2 = kT \ln 2$.
So V_2 will increase with temperature.
(b) If V_2 is 4.9 V then D_1 will breakdown. This means $V_1 = 4.9$ V.

$\therefore$ $$V_2 = 5 - 4.9 = 0.1\text{V}$$

Now using $I_0 = 5\mu A, V_2 = 0.1\text{V}$, one obtains

$$I = 5\left[\exp\left(\frac{0.1}{0.026}\right)-1\right] = 229\mu A.$$

Example 21.16: A LED has a greater forward voltage drop than does a common signal diode. A typical LED can be modeled as a constant forward voltage drop $V_D = 1.6$ V. Its luminous intensity I_V varies directly with forward current and is described by $I_V = 40\ i_D \approx$ millicandela (mcd). A series circuit consists of such an LED, a current-limiting resistance R, and a 5V *dc* source V_s. Find the value of LED current i_D such that the liminuous intensity is I mcd.

Solution We have

$$I_V = 40 i_D$$

$$i_D = \frac{I_V}{40} = \frac{1}{40} = 25\text{mA}.$$

Example 21.17: Light doping of P-type semiconductor material is defined as the case for which $p >> n$ is not valid ($n >> p$ is not valid for an N-type semiconductor). Derive a procedure to determine the number of mobile carriers for the case of light doping consider N_D and N_A are immobile donor and acceptor ions respectively.

Solution For P-type doping, $N_D = 0$. Hence, by charge neutrality and mass action law.

$$p + N_D = p = n + N_A = \frac{n_i^2}{n}$$

or $$n^2 + N_A n - n_i^2 = 0$$

Solving the quadratic equation for n and discharging the extraneous negative root, one obtains

$$n = \frac{1}{2}\left(-N_A + \sqrt{N_A^2 + 4n_i^2}\right)$$

Knowing n, one obtains from mass-action law $p = \dfrac{n_i^2}{n}$.

For N-type doping, $N_A = 0$. By analogous procedure,

$$p = \frac{1}{2}\left(-N_D + \sqrt{N_D^2 + 4N_i^2}\right); \qquad n = \frac{n_i^2}{p}$$

Example 21.18: A samples of Si is doped with 10^{17} phosphorus atoms/cm³. What would your expect to measure for its resistivity? What Hall voltage would you expect in a sample 100 μm thick if $I_x = 1$ mA and $B_z = 1$ KG (= 10^{-5} Wb/cm²)?

Solution Mobility for silicon 700 cm²/V-s. Obviously, the conductivity is

$$\sigma = e\mu_n n = 1.6\times10^{-19}\times700\times10^{17} = 11.2(\Omega\text{-cm})^{-1}$$

The resistivity is $\rho = \sigma^{-1} = 0.0893\Omega\text{-cm}$

The Hall coefficient is $R_H = -(en)^{-1} = -62.5\text{cm}^3/\text{C}$

The Hall voltage is $V_H = \frac{I_x B_z}{x} R_H = \frac{10^{-3}\times10^{-5}}{10^{-2}}\times(-62.5) = -62.5\,\mu\text{V}.$

Example 21.19: A Hall effect experiment is performed on a heavily doped n-type semi conductor sample with a length of 2.65 cm, a width of 1.70 cm, and a thickness of 0.052 cm, in a magnetic induction field of 0.5 T. The current in the sample along its length is 200 μA. The potential difference along the length of the sample in 195 mV and the potential difference across its which is 21.4 mV.

(a) What is the concentration of charge carriers.
(b) What is their mobility.

Solution (a) The Hall co-efficient is

$$R_H = \frac{E_H}{J_x B}$$

Here $E_H = \frac{V_H}{b}$, where V_H is Hall voltage = 21.4 m, $V = 21.4 \times 10^{-3}$ V and b is breadth = 1.7 cm = 1.7×10^{-2}m.

J_x the current density $= \frac{1}{bt}$

where I is current $= 200\mu A = 200\times10^{-6}$ amp

t is thickness $= 0.052\,\text{cm} = \text{cm} = 0.052\times10^{-2}$ m

B is magnetic field $= 0.5$T

$$\therefore \quad R_H = \frac{21.4\times10^{-3}}{1.7\times10^{-2}}\times\frac{1.7\times10^{-2}\times0.052\times10^{-2}}{200\times10^{-6}\times0.5} = -0.11128\text{m}^3/\text{C}$$

(b) n the electrons per unit volume

$$n = \frac{1}{eR_H} = \frac{1}{1.6\times10^{-19}\times0.11128} = 5.6\times10^{19}\ \text{electrons/m}^3$$

The resistance of the sample

$$R = \frac{V}{I} = \frac{195\times10^{-3}}{200\times10^{+6}} = 975\Omega$$

Since $\quad R = \frac{\ell}{A\sigma} = \frac{\ell}{Aen\mu}$

where σ is conductivity and μ mobility we have

$$\mu = \frac{\ell}{\text{AenR}}$$

$$= \frac{2.65\times10^{-2}}{1.7\times10^{-2}\times0.052\times10^{-2}\times1.6\times10^{-19}\times5.6\times10^{19}\times975} = 0.343\,\text{m}^3/\text{Vs}$$

Example 21.20: A CD stores 700 MB of data. The region of CD that stores data has an inner radius of 2.30 cm and outer radius of 5.80 cm, and the width of data track is 1.60 μm. Show that surface area allowed for each data bit is $1.6\times10^{-12}\,\text{m}^2$/bit and the approximate dimensions of each bit along the track is 1.0×10^{-6} m.

Solution We have 1 byte(B) = 8 bits. The usable surface area is the difference between the area of the disk of radius 5.80 cm and a disk of radius 2.30 cm effectively subtracting out the "hole" in the middle of the disk. Then the area per bit is that surface area divided by the number of bits.

(i) The usable area is

$$\text{A} = \pi[(0.058\,\text{m})^2 - (0.023\,\text{m})^2]$$

$$= 8.91\times10^{-3}\,\text{m}^2$$

The number of bits is 700 MB × 8 bits/B = 5.60×10^9 bits

$$\text{Area per bit} = \frac{\text{Area}}{\text{Bit}} = \frac{8.91\times10^{-3}\,\text{m}^2}{5.60\times10^9\ \text{bits}}$$

$$= 1.6\times10^{-12}\,\text{m}^2/\text{bit}$$

With a track width of 1.6 μm, the length of a bit (area divided by width) is

$$\text{Bit length} = \frac{1.6\times10^{-12}\,\text{m}^2}{1.6\times10^{-6}\,\text{m}} = 1.0\times10^{-6}\,\text{m}$$

obviously, the bit length is just 1 micron. Thus the length and width of the bit are comparable to *IR* laser wavelength (780 nm) use to scan it.

SUMMARY

1. Semiconductors have four electrons in the outer valence ring. The electrical conductivity of semiconductors goes up as temperature goes up.
2. In a very pure sample (intrinsic), the conductivity of a semiconductor is due to the excitation of electrons from valence to the conduction band by thermal fluctuations.

The resistivity at room temperature is between 10^{-2} Ω-cm and 10^{9} Ω-cm depending on the band gap of the material.

3. Si and Ge are prototypical bonded semiconductors.
4. The temperature dependence of conductivity of intrinsic semiconductor varies as $\exp\left(\frac{-E_g}{2k_BT}\right)$, where E_g is the minimum gap, the energy difference between the conduction band minimum and the valence band minimum.
5. In a semiconductor there exist a small band gap (about 1 eV) between the valence band and the conduction band. When a modest electric field is applied, sufficient numbers of electrons can overcome the band gap in order to conduct.
6. In their intrinsic form, semiconductors do not have enough free electrons to be useful. Therefore, they are modified by doping an impurity to help the conducting process.
7. After the intrinsic crystal is doped, it is called an *extrinsic* semiconductor. Doping material is of two types. The first has an excess of free electrons, and when it is added to a pure semiconductor crystal the resulting crystal is called an N-type semiconductor material. The second type of impurity has a deficiency of free electrons (an excess of free holes), and when it is added to a intrinsic semiconductor crystal the resulting crystal is called a P-type semiconductor.
8. Free carriers are created due to thermal excitations of electrons across the energy gap, or by ionization of donors and acceptors. In an intrinsic sample, only thermal excitation takes place, and the numbers of electrons and holes are equal, i.e., their concentration is

$$n = p = n_i$$

$$= 2(k_B \mathrm{T}/2\pi\hbar^2)^{3/2}\,(m_e m_h)^{3/4} \times e^{\frac{-E_g}{2k_BT}}$$

This concentration rises rapidly with temperature due to exponential factor. For an extrinsic semiconductor, in which cross-gap ionization is negligible as compared with the ionization from impurities, the carrier concentration are given apprimately by

$$n = N_d \quad \text{or} \quad p = N_a$$

for N and P-type samples, respectively.

9. *Conductivity* in semiconductors is usually expressed as

$$\sigma = ne\,\mu$$

where μ is the mobility. In intrinsic substances, conductivity increases rapidly with temperature due to the corresponding exponential increase in the carrier concentration n. Although μ also depends on temperature, this dependence is very weak compared with its dependence on the exponential factor.

10. When a high electric field is applied to a semiconductor, the carriers (electrons and holes) absorb appreciable energy from the field and their temperature rises above that of lattice, i.e., they become "hot". The effect of this is a decrease in their mobility.

    In certain semiconductors of appropriate band structure, e.g. GaAs, the heating of carriers results in the transfer of high-energy valleys of very low mobility. In such a case the application of electric field may cause a production of a region of *negative differential conductance.* Since such a situation is inherently unstable, the sample "breaks up" into coherent electrical oscillations, which is called *Gunn effect.*

11. *Optical absorption* in a semiconductor takes place when an electron in the valence band absorbs a photon from the optical beam, and transfers to the conduction band. Such a fundamental absorption is possible only if the photon energy is greater than the energy gap ($E_g < h\nu$). The shape of absorption curve is a function of the band structure, and therefore the absorption curve is often employed in the study of the lattice. Optical absorption may also happen due to *excitonic excitation,* or ionization of donors and/or acceptors.
12. The conductivity of a semiconductor is raised by shining a light beam on the sample. This is *photoconductivity.* The optical beam causes additional carriers to be excited across the energy gap, which causes a rise in conductivity.
13. *Luminescence* is the inverse of optical absorption. In some manner, electrons are first excited to higher states, and are subsequently allowed to fall to lower states. Photons are emitted in the process. These photons give rise to luminescence.
14. Diffusion coefficient (D) bears the following relation with the carrier mobility

$$D = -\mu k_B T / e$$

15. When P-type and N-types of materials are placed end to end to form, they form a PN junction, which is enormously useful. PN junction diode permits current to flow easily in one direction (forward bias) but hardly at all in opposite direction (reverse bias). During the manfacturing process of the *PN junction,* an interaction takes place between the two types of materials. The area affected by the combining holes and electrons is called the *depletion region or depletion.*
16. *Zener diode* is a semiconductor device designed to operate in the reverse breakdown area. This is used to regulate voltage in a circuit, even when there is a widely varying current.
17. There are several useful semiconductor devices, e.g. light emitting diodes, tunnel diodes, PIN diodes, photovoltaic cells, etc., Solar cells are currently the focus of an intense research effort.
18. *Bipolar Junction Transistor* (BJT) is the basic component of all modern electronics. Integrated circuits (ICs) are often nothing more than huge collections of BJTs all on the same piece of semiconductor material.
19. The transistor is very similar to a PN junction diode, but it has an additional PN junction. The output of a transistor provides amplification by taking the input signal and increasing its power.
20. *Junction Field Effect Transistors* (JFET) has three terminals similar to the BJT source, drain and gate and fourth JFET component is the channel. The channel is the material that connects the source to the drain. Unlike BJTs, which current-controlled devices, the JFET is a voltage-controlled device.
21. *Metal-oxide Semiconductor Field Effect* (MOSFETs) Transistor technology is used extensively in most calculators, watches, desktop computers, and other appliances. The main advantage of MOS circuits is their low current requirement. With less current, there is less heat dissipation. Another advantage is that MOS components can be made much smaller than BJT-components.

REVIEW QUESTIONS

1. What are semiconductors? How do they differ from conductors and insulators? Why an increase in temperature increases conductivity of a semiconductor?
2. Explain with suitable diagrams the conduction band, valence band and forbidden band and hence explain the behaviour of conductor, semiconductor and insulator. Explain electrons and holes contribution to electrical conduction.
3. Derive the expression for the conductivity of an intrinsic semiconductor.
4. What are intrinsic and extrinsic semiconductors? Discuss the locations of Fermi levels under suitable limiting conditions and give the necessary theory.
5. What is impurity conduction in semiconductors? Explain how the presence of a small impurity in a semiconductor modifies its conduction properties.
6. Explain the movement of electrons and holes in a semiconductor. In what respect N-type semiconductors differ from each other.
7. What is drift current and diffusion current in a semiconductor? Write expression for the total electron current density in a semiconductor.
8. How will you draw the characteristics of a semiconductor diode?
9. How an abrupt junction is formed?
10. What happens to the depletioN-region in a P-N diode under-forward bias and reverse bias conditions?
11. Explain the working of a P-N junction. Discuss forward and reverse biasing of P-N junction diode.
12. What happens to the depletioN-region in a P-N diode under forward bias and reverse bias conditions?
13. Explain with diagrams forward and reverse biasing of a P-N junction. What is meant by avalanche breakdown?
14. Explain with reference to Zener diode characteristic curve the following: (i) I_{ZK} (ii) I_{ZT} (iii) Z_Z, where symbols have usual meaning.
15. What do you understand by Zener breakdown? Draw a typical characteristic of a Zener diode. Identify the breakdown region and explain it.
16. A Zener diode is a P-N junction yet it is different from an ordinary P-N junction, explain.
17. Describe an experiment for drawing the characteristic of a zener diode.
18. Draw the (V-I) characteristics of a tunnel diode and discuss its applications.
19. Explain the mechanism of electrical conduction in a typical semiconductor like Germanium or Silicon. How the conductivity of a pure semiconductor is affected by adding traces of trivalent and pentavalent impurities?
20. What is mean by doping a semiconductor? What basic properties must these doping elements possess to make a pure semiconductor a P-type or N-type? Explain the physical principles involved.
21. Write short notes on the following.
(i) Energy bands in solids, (ii) Avalanche breakdown, (iii) Zener breakdown. (iv) P-type and N-type semiconductors, (v) Tunnel diode, (vi) Varactor diode, (vii) Light emitting diode, (viii) Solar cell. (ix) Schotty diode, (x) Backward diode.
22. Give the structure of light emitting diode. Explain its working and give some of its uses.

23. Explain, why Zener breakdown voltage has a negative temperature coefficient whereas the avalanche breakdown voltage has a positive temperature coefficient? why dynamic resistance of an ideal Zener diode is zero but the dc-resistance is not zero.
24. What is a breakdown diode? Explain the origin of breakdown of a junction.
25. Draw the volt-ampere characteristic of a tunnel diode. Explain the occurrence of the negative differential resistance in the characteristic of a tunnel diode. Write some of the uses of tunnel diode.
26. What is Hall effect? Briefly discuss physical origin of the Hall effect. Mention uses of Hall effect.
27. What is Hall coefficient? Show that for of a P-type semiconductor the Hall coefficient R_H is given by, $R_H = 1/eP$. Describe an experimental set up for the measurement of Hall Voltage, V_H.
28. For a semiconductor show that the Hall coefficient is given by

$$R_H = -\frac{1}{e}\frac{p\mu p^2 - n\mu_n^2}{\left(p\mu_p + n\mu_n\right)^2}$$

where μ_p and μ_N are the mobilities of holes and electrons respectively. Show that for a intrinsic semiconductor the above expression reduces to

$$R_H = \frac{1}{n_i e}\left(\frac{\mu_n - \mu_p}{\mu_n + \mu_p}\right)$$ **[Hint.** $n = p = n_i$ for an intrinsic semiconductor**]**

PROBLEMS

1. Mobilities of electrons and holes in a sample of intrinsic germanium at room temperature are 3600 cm²/volt-s, and 1700 cm²/volt-s, respectively. If the electron and hole densities are each equal to 2.5×10^{13} per cc, calculate its conductivity. **[Ans.** 2.12 mho/m**]**
2. The concentration of the acceptor atoms in P-type germanium crystal is 4×10^{15} per cc. Find the conductivity of the crystal at 300 K. Hole mobility in germanium at 300 K is 1900 cm² per volt-s. All the acceptor atoms are ionized at this temperature. **[Ans.** 121.6 mho/m**]**
3. Calculate the current produced in a small germanium plate of area 1 cm² and of thickness 0.3 mm. When a potential difference of 2 V is applied across the faces. Given concentration of free electrons in germanium is 2×10^{19} m³ and the mobility of electrons and holes are 0.36 m²/volt-s and 0.17 m²/volt-s respectively. **[Ans.** 1.13 amp**]**
4. A specimen of pure germanium at 300°K has a density of charge carriers of 2.5×10^{19} per m³. It is doped with a donor impurity atoms at the rate of one impurity atom for every 10^6 atoms of germanium. All impurity atoms may be supposed to be ionised. The density of germanium atom is 4.2×10^{28} atoms/m³. Find the resistivity of the doped germanium if electron mobility is 0.36 m²/volt-s. **[Ans.** $0.41 \times 10^{-3}\Omega$-m**]**

5. What concentration of donor atoms per cc is required in a germanium crystal so that the conductivity of the crystal is 2 mho/cm at 300 K. For germanium $\mu_n = 3900$. [**Ans.** 320.5×10^{19} m^3]

6. Find the conductivity of germanium which is doped with donor atoms of concentration $N_d = 10^{17}$ per cm^3 and compare it to that of intrinsic case. Given $n_i = 2.5 \times 10^{13}$/cm^3, $\mu_n = 3800$ cm^2/volt-s. [**Ans.** 6.25×10^{15} m^3]

7. Compare the number of electron hole pairs per cm^3 in a pure silicon crystal at the temperature of (i) 27°C and (ii) 57°C. For silicon $E_g = -1.13$ eV. [**Ans.** 8.39]

8. Find the value of P-N junction diode current in the forward bias condition. when the applied voltage across it is 0.05 volt and $I_s = 50\mu$V. Given e/kT = 40/volt. [**Ans.** 0.319 milli amp]

9. Determine the donor concentration in N-germanium having a resistivity of 0.015 Ω-m, $\mu_p = 0.36$ m^2/volt-s. Repeat for P-germanium of equal resistivity, $\mu_n = 0.1$ m^2/volt-s.

10. An N-type Si bar is 2 cm long and has a cross section of 2 mm × 2 mm. When a one volt battery is connected across it, a current or 8 mA flows. Determine: (i) doping level (ii) drift velocity. [**Ans.** (i) 192×10^{15}/cm^3 (ii) 650 cm/s]

11. **(a)** Show that the conductivity of a semiconductor is minimum when it is lightly doped with P-type impurity such that

$$P = n_i \sqrt{\frac{\mu_n}{\mu_p}}$$

(b) Show that the minimum conductivity is $2n_i \sqrt{\mu_n \mu_p q}$.

(c) With the help of the above result determine the minimum conductivity of silicon. [**Ans.** (c) $\sigma_{mn} = 3.87 \times 10^{-6}$ (Ohm-cm)^3]

12. A Silicon diode operates at a fixed bias of 0.4 V. Determine the factor by which the current will get multiplied when its temperature is raised from 25°C to 150°C. [**Ans.** 638]

13. When a diode is reverse biased with 8V it has junction capacitance of 15 pF. When the reverse biased is increased to 12 V, the capacitance drops to 13.05 pF. Find whether it is abrupt or graded junction? [**Ans.** graded junctioN-type]

SHORT ANSWER-QUESTIONS

1. What is the order of energy for forbidden band in (i) diamond (ii) silicon (iii) germanium, and (iv) aluminium?
Ans: (i) 9eV (ii) 1.2 eV (iii) 0.7 eV (vi) Zero.

2. What happens to the resistance and conductance of a semiconductor on heating?
Ans: Resistance decreases but conductance increases.

3. What is the order of resistively of a metal, insulator and semiconductor?
Ans: Resistively of (i) Metal $U \simeq 10^{-6}$Ω-cm, (ii) Insulator ≃ 10^{13} Ω-cm and (iii) semiconductor lies between 10^{13} Ω and 10^{-6} Ω-cm.

4. What are the important characteristics of a semiconductor?
Ans: (i) is has covalent bonding (ii) it is crystalline (iii) it has a negative temperature coefficient of resistance (iv) its conductivity increases with the addition of impurities.

5. What is mobility? Write its units?
Ans: The drift velocity per unit electric field is called mobility (μ). Its unit is $N^{-1} S^{-1}$ m C or N^{-1} mA.
6. Why metallic bodies are always opaque?
Ans: We know that metallic solids have partially filled conduction bands. The energy of photons in the visible light region varies between 1eV and 3eV. As light is incident to metallic solids, free electrons of the conduction band absorb the energy of the incident photons. Obviously, no photons are allowed to pass through, the metallic solids behave as though opaque.
7. A small portion of indium (In) is incorporated in germanium (Ge). Is the crystal N-type or P-type?
Ans: Obviously, the crystal is P-type as indium is trivalent and one of the covalent bonds will remain without electrons giving excess holes.
8. What are the charge carriers in N-type and P-type semiconductors?
Ans: The charge carriers in N-type semiconductors are the electrons while in P-type the charge carriers are holes.
9. Why a semiconductor is damaged by a strong current?
Ans: The safety limits of temperature for germanium and silicon are about 80°C and 200°C respectively. A strong current passing through a semiconductor heats it up beyond these temperatures. At these temperatures a large number of charge carriers are available for conduction and the specific resistance of the crystal become very low and thus they are damaged, i.e., loose the property of semiconductors.
10. What is the difference between hole-current and electron flow.
Ans: The hole current is in the direction of conventional current which is due to the flow of positive charges while the electron flow is opposite to the conventional current as it is due to the flow of negative charges.
11. What is the Germanium diode Zener voltage?
Ans: When a reverse bias of about 25 volts is applied to the crystal, the excessively high temperature destroys the covalent structure of germanium and the reverse current rise sharply. This breakdown voltage is called Zener voltage (V_z).
12. What do you understand by 'tolerance' of a Zener diode?
Ans: The range of voltages about the breakdown voltage in which a Zener diode conducts in reverse direction is called tolerance?
13. What do you understand by power rating of a Zener diode?
Ans: The maximum power which Zener diode can handle without damage is known as its power rating.
14. What do you understand by maximum current rating of a zener diode?
Ans: The maximum value of current which a zener diode can handle at its rated voltage without damage is known as its maximum current rating.
15. Which type of diode used for the rectification of very high frequency (MHz)?
Ans: Schottky diode.
16. Which diode is best suited for the rectification of very small peak voltage (~0.5 V)?
Ans: Backward diode.
17. Which diode has a negative resistance?
Ans: Tunnel diode.

18. What is a solar cell?
Ans: A Solar cell is a photodiode operated in its photovoltaic mode

19. What are optocouplers?
Ans: Optocouplers are devices that use light to optically couple a signal between two electrically isolated points. Optocouplers are also called Optoisolators.

20. What is phototransistor?
Ans: A phototransistor is a light detecting device that is also called a photo sensor. It is a transistor whose base current is supplied by the carriers generated due to the incident or striking light.

21. What is photodarlington?
Ans: A photodarlington is a phototransistor packed with another transistor connected in the Darlington configuration.

OBJECTIVE QUESTIONS

1. N-type germanium is obtained on doping intrinsic germanium by
(a) Phosphorous (b) Aluminium
(c) Boron (d) Gold

2. Depletion region is a zone which contains.
(a) holes only (b) electrons only
(c) both electrons and holes (d) neither electrons nor holes

3. In a semiconductor diode arrow represents.
(a) N-type material (b) P-type material
(c) both P and N type materials (d) none of the above

4. Zener diode is used for
(a) rectification (b) amplification
(c) stabilization (d) none of the above

5. For a tunnel diode a decrease in current causes.
(a) voltage constancy (b) decrease in voltage
(c) Increase in voltage (d) none of the above

6. PN-junction is formed when P-type semiconductors and N-type semiconductor are joined
(a) together (b) physically
(c) to get homogeneous material chemically
(d) in such a manner that electrons and holes diffuse to give depletion layer

7. The depletion region of a junction diode is formed.
(a) when forward bias is applied to it
(b) when the temperature of the junction is reduced

(c) under reverse bias

(d) during the manufacturing process

8. The width of the depletion layer of a junction

(a) is independent of applied voltage (b) is increased under reverse bias

(c) decreases with light doping (d) increases with heavy doping

9. The LED or the light emitting diode

(a) is made from one of the two basic semiconducting materials, silicon or germanium

(b) is made from the semiconducting compound gallium arsenide phosphide

(c) emits light when forward biased

(d) emits light when reverse biased

10. The P-side of a junction diode is earthed and the N-side is given a potential of 2V. The diode will

(a) not conduct (b) conduct partially

(c) breakdown (d) conduct fully

11. For detecting light intensity we use a/an

(a) photodiode in reverse bias (b) photodiode in forward bias

(c) LED in reverse bias (d) ED in forward bias

12. When a PN-junction diode is forward biased, the flow of current across the junction is mainly due to

(a) diffusion of charges (b) drift of charges

(c) depends on the nature of the material

(d) both drift and diffusion of charges

13. A PN-junction diode cannot be used

(a) as a rectifier

(b) for increasing the amplitude of an ac signal

(c) for getting light radiation

(d) for converting light energy into electrical energy

14. A strong electric field across a PN-junction that causes covalent bonds to break apart is called

(a) reverse breakdown (b) avalanche breakdown

(c) lever breakdown (d) low voltage breakdown

15. A light emitting diode produces light when

(a) forward biased (b) reverse biased

(c) unbiased (d) none of the above

16. A solar cell is an example of
(a) photo emissive cell (b) photo rediation cell
(c) photo rediation cell (d) photo conductive cell

17. When holes leave the P-material to fill electrons in the N-material the process is called.
(a) diffusion (b) depletion
(c) avalanche breakdown (d) Zener breakdown

18. A diode which has zero breakdown voltage is known as
(a) tunnel diode (b) Zener diode
(c) Schottky diode (d) backward diode

19. A Zener diode is used as
(a) a coupler (b) a rectifier
(c) an amplifier (d) a voltage regular

20. An empirical formula relating the resistance of many semiconductors to tempure is
(a) $\text{Log}\, R + \dfrac{K}{\log R} = A + \dfrac{B}{T}$ (b) $\text{Log}\, R + K \log R = \dfrac{B}{T}$
(c) $\text{Log}\, R + \dfrac{K}{\text{Log}\, R} = A$ (d) $\text{Log}\, R = A + \dfrac{B}{T}$
[*A*, *B* and *K* are constants]

ANSWERS

Question numbers	1	2	3	4	5	6	7	8	9	10
Answers	(a)	(d)	(b)	(c)	(c)	(d)	(d)	(b)	(b,c)	(d)
	11	12	13	14	15	16	17	18	19	20
	(a)	(a)	(b)	(c)	(b)	(c)	(a)	(a)	(d)	(a)

FILL IN THE BLANKS

1. Normal conductors have resistivities that increase with increasing temperature, but in semiconductors the resistivity ______ with increasing temperature. **[decreases]**
2. In a semiconductor there exists a small band gap, ______ between the valence band and the conduction band. **[1 eV]**
3. Some semiconductors are n-type, meaning that the majority carriers are ______ electrons. **[negative]**
4. Some semiconductors are p-type, meaning that the majority carrers are positive ______. **[holes]**
5. Materials of p-type and n-type can be placed end to end to form a ______ diode. **[PN junction]**

6. PN junction diode permits current to flow easily in ______ (forward bias) but hardly at all in the ______ (reverse bias). **[one direction, opposite direction]**
7. A Zener diode will operate in ______ bias once a sufficiently high voltage is applied. **[reverse]**
8. Semiconductor is a special case of the ______ crystal. **[insulating]**
9. In a semiconductor, the gap separating the filled and empty bands is ______ small, and at finite temperature some electrons from the filled valence band are thermally excited across the energy gap giving $n_e(T)$ electrons per unit volume in the conduction band and $n_h(T)$ holes per unit volume in the valence band, of course ______. **[very, $n_e = n_h$]**
10. Silicon and Germanium are the prototypical ______ bonded semiconductors. **[covalently]**
11. Semiconductors have ______ electrons in the outer valence ring. **[four]**
12. The electrical conductivity of semiconductors reacts to temperature in the same way it does in ______; i.e., conductivity goes up as temperature ______ **[insulators, goes up]**
13. After the intrinsic crystal is doped, is called an ______ semiconductor. **[extrinic]**
14. Doping material comes in ______ types. The first has an excess of ______ electrons, and when it is added to a pure semiconductor crystal the resulting crystal is called an ______ type semiconductor material. **[two, free, N]**
15. When doping material has a deficiency of ______ electrons (an excess of free holes), and when it is added to a pure semiconductor crystal the resulting crystal is called a ______ type semiconductor. **[free, P]**
16. When P and N types of materials are joined, they form the enormously useful ______ junction. This junction is vital to ______ and ______ operation **[PN, diode, semiconductor]**
17. The area affected by combining holes and electrons in a PN junction is called ______ or ______. **[depletion region, depletion layer]**
18. The negative and positive ions in the depletion region create an internal ______ voltage 0.5 V to 0.7 V (Si) and 0.25 V to 0.3 V (Ge) that prevents further current ______ from N-type to P-type material. **[barrier, flow]**
19. Barrier potential is present in all ______ junctions of diodes and transistors. **[PN]**
20. The behaviour of the PN junction can be changed by applying a ______ voltage. **[bias]**
21. The bias voltage changes the ______ of the PN junction's ______ region and therefore changes its resistance. **[width, depletion]**
22. A PN junction is ______ biased by making the P material positive with respect to the N material. **[forward]**
23. When the voltage is high enough, the PN junction's resistance drops rapidly, a very high current develops, and the PN junction is ______. The voltage at which this happens is called ______ voltage, ______ voltage, or ______ voltage. **[destroyed, reverse breakdown, peak inverse, avalanche]**
24. There are certain types of semiconductor devices designed to operate in the reverse breakdown area. The ______ diode is the most common of these devices. **[zener]**
25. The Zener diode is used to regulate voltage in a circuit, even when there is widely ______ current. **[varying]**

26. When forward biased, Zener diodes operate as ______ semiconductor (rectifier) diodes. **[normal]**

27. Tunnel diodes are normally operated in the ______ resistance region. They can also operate at ______ temperatures than Si or Ge diodes. **[negative, higher]**

28. Shottky or Shockley diodes are also known as ______ or ______ diodes, and they are used in ______ circuits and ______.
[hot carrier, surface barrier, switching, microwaves]

29. PIN diodes differs from the PN junction diode because it does not have a ______ junction. Instead it consists of ______ materials: P-type, N-type, and pure (intrinsic) silicon slice. **[PN, three]**

30. Varactor diodes are used primarily as ______ controlled capacitors in high frequency ______ circuits. **[voltage, tuned]**

31. NPN BJT transistor has a collector and emitter made of ______ type material and a base of ______ type material. **[P, N]**

32. A transistor can amplify ______, ______ or power, depending on the circuit values selected. This amplification is called ______. **[current, voltage, gain]**

33. There are ______ ways a transistor can be configured to produce different types of gain. **[three]**

34. The JFET has three terminals that are similar to BJT, i.e., source, drain and gate. A fourth JFET component is ______. **[channel]**

35. Metal oxide semiconductor (MOS) technology is used extensively in most calculators, watches, desktop ______ and other appliances. **[computer]**

36. There are two basic types of MOSFETs:
(i) the ______ MOSFET and
(ii) the ______ MOSFET. **[enhancement, depletion]**

37. The UJT has ______ leads but only ______ doped regions. The main function of UJT is ______. **[three, two, switching]**

22

Lasers

22.1 INTRODUCTION

We live in an age where laser light shows are common and lasers light up the sky. Within the past few years, lasers have come into wide spread use in science, medicine, manufacturing, and in many other fields. The LASER-the name is an acronym for '*Light Amplification by Stimulated Emission of Radiation*.' Laser is a device that produces a highly intense, concentrated and parallel beam of monochromatic and coherent light. In the words of C.H. Townes, inventor of Maser and laser, **'what laser does essentially is to give us electronic type control over light. It is a marriage of optics and electronics.**

Historically the laser is the outgrowth of 'maser' a device that microwaves instead of light waves. The world 'maser' is an acronym for Microwave (or Molecular) Amplification by Stimulated Emission of Radiation and is also known as **optical maser**. The principles on which lasers operate undoubtedly place them in the family of masers, but in structure, properties and applications, lasers are quite different from masers. The major difference between the two devices is that of their working frequency; the typical operating frequencies of masers are nearly 10^{10} Hz, while the frequencies in which lasers operate are of the order of 10^{15} Hz.

The basic idea behind the laser and maser is the process of stimulated emission that was first postulated by Einstein in his derivation of Planck's radiation law in 1917. If his considerations has been applied to non-equilibrium situations, the operating principle of the laser and maser could have been proposed at that time. In fact, it took about 35 years for others to come up with the basic idea and make this idea work.

Prior to 1960 the laser was just a twinkle in the eye of scientists. Today it is a brighter than the sun reality. The laser (it rhymes with razor) promises to be in measurement techniques, communications, military weaponry, and the conquest of space. It also figures in industrial processes, computers, medicine, chemistry and other fascinating possibilities. In brief the applications of laser light range from straight forward (such as surveying and welding) to the sophisticated (such as optical communications and holography) and it may be that more interesting applications are still to come.

22.2 IMPORTANT PROPERTIES OF LASER LIGHT

The salient features of a laser beam are: (i) directionality (ii) intensity (iii) monochromaticity and (iv) coherence. Let us discuss these features.

22.2.1 Laser Light is Highly Directional

A laser beam is highly directional, i.e., it spreads very little. A laser beam departs from strict parallelism only because of diffraction at the exit aperature of the laser. For example, a laser pulse used to measure the distance to the moon generates a spot on the moon's surface with a diameter of only a few meters. Light from an ordinary bulb can be made into an parallel beam by using a lens, but the beam divergence is much greater than for laser light and a parallel beam of light from a torch on the earth shall spread in an area of few square kilometers. Each point on a light bulb's filament forms its own separate beam, and the angular divergence of the overall composite beam is set by the size of the filament. The directionality of a beam can be measured by the *full angle beam divergence* which is twice the angle subtended by the outer edge of the beam with the beam axis. The outer edge of the beam is considered to be the line where the beam intensity decreases to 1/exp (I) times its value at the axis. Let us consider that an aperature of diameter d radiate a beam having a plane wave front. Let the beam travels as a parallel beam through the distance of *Rayleigh range,* which is of the order of d^2/λ, where λ is the radiated wavelength. Obviously, the beam begins to spread with distance owing to diffraction as shown in Fig. (22.1). In general, the aperature diameter (d) controls the angular spread $\Delta\theta$ radians of the far-field beam. The relationship between $\Delta\theta$ and d is given by

$$\Delta\theta = \frac{\lambda}{d} \tag{22.1}$$

Hence, the shorter the wavelength and larger the aperature diameter, the smaller the angular divergence of the beam. For a typical laser beam, the beam divergence is less than 10^{-5} radian, exhibiting that the beam spreading is less than 10^{-5} m for every meter.

22.2.2 Laser Light is Highly Intense, i.e., It can be Sharply Focussed

If two light beams transport the same amount of energy, the beam that can be focussed to the smaller spot will have the greater intensity at that spot. Lasers emit narrow beam of light, so that the energy is concentrated in a small region. The spatial and spectral confinement of energy in laser beam accounts for its high intensity. For laser light, the focused spot can be so small that an intensity of 10^{17} W/cm^2 is readily obtained. An oxyacetylene flame by contrast, has an intensity of only about 10^3 W/cm^2. Laser beams can generate power densities millions of times greater than those on the surface of the sun. It is found that when a laser beam is focused by a lens, the tremendous intensity at the focal point can produce a radiation pressure of the order of 10^6 kg per sq.cm.

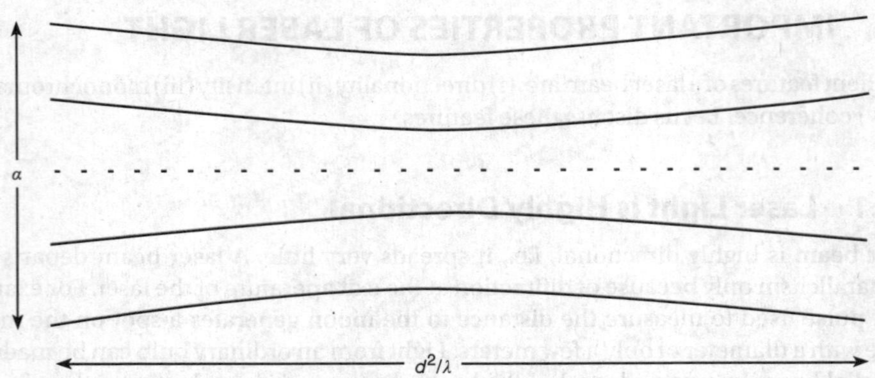

Fig. 22.1 Spreading of a beam beyond Rayleigh range

22.2.3 Laser Light is Highly Monochromatic

Light from an ordinary incondescent bulb is spread over a continuous range of wavelength and is certainly not monochromatic. The radiation from a fluorescent neon sign is monochromatic, to about 1 part in about 10^6. However, the sharpness of definition of laser light can be many times greater, as much as 1 part in 10^{15}.

22.2.4 Laser Light is Highly Coherent

Individual long wave (wave trains) for laser light can be reverses hundred kilometers long. When two separated beam that have traveled such distances over separate paths their common origin and are able to from a pattern of interference fringes. The corresponding coherence length for wave trains emitted by a light bulb is typically less than a meter, the wave from a laser is an orderly wave of one frequency when the whole beam is **spatially** in phase. There are two independent facts of coherence: namely (i) temporal coherence and (ii) spatial coherence.

(i) Temporal Coherence: This refers to the correlation between the radiation field at a point and the radiation field at the same point at a later time. i.e., the relation between $E(x, y, z, t_1)$ and $E(x, y, z, t_2)$ where these represent the radiation fields at the point (x, y, z) at times t_1 and t_2 respectively. If the phase difference between the two radiation field is constant during the period normally covered by observations (few microseconds), the wave is said to have temporal coherence. If the phase difference between two radiation fields changes many times and in an irregular way during the shortest period of observation (μs), the wave is said to be incoherent.

Time coherence comes about when each cycle of the wave takes exactly the same time to pass a point in space. This really means that the frequency of the wave is not varying and is given the name monochromatic.

(ii) Spatial Coherence: Two fields at different points on a wave front of given electromagnetic wave are said to be space coherent if they preserve a constant phase difference over any time t, i.e., space coherence requires that the waves not only are of same frequency, but they are in phase in space.

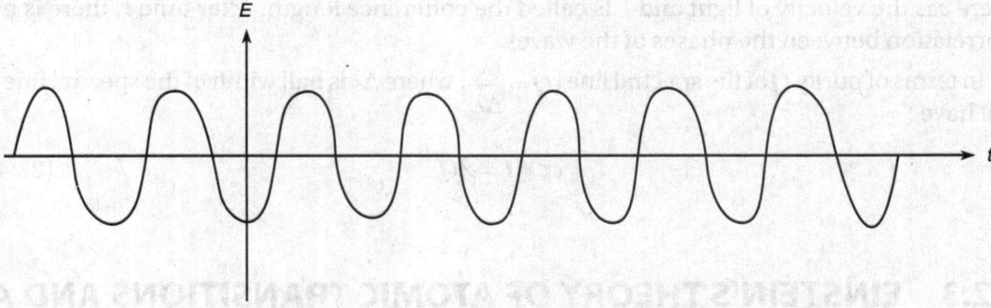

Fig. 22.2

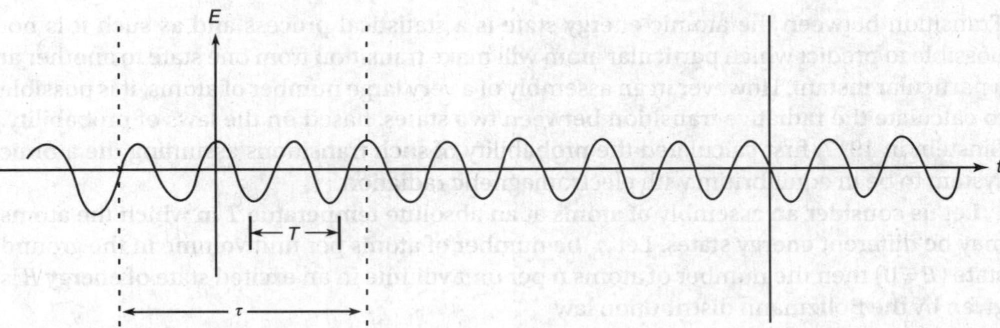

Fig. 22.3

Spatial coherence is possible even when the two beams are individually time incoherent, as long as any phase change in one of the beam is accompanied by a simultaneous equal phase change in the other beam. With the ordinary light sources, this is only possible if the two beams have been derived from the same part of the source.

Coherence Time and Coherence Length: Let us consider that the radiation field *E*, from a light source is an ideal sinusoidal function of time *t*, then at any given position one can write

$$E = a\cos(\omega_a t + \phi) \tag{22.2}$$

Here a is the magnitude of the field, ω_0 the angular frequency and ϕ, the phase, Fig. (22.2) depicts the field.

However, no emitted light produces a perfect sinusoidal variation of field *E* with time *t*, and for any actual light source the magnitude a and phase ϕ will vary with time. One can represent, a more realistic picture of the radiation from an emitting source of light as shown in Fig. (22.3).

In Fig. (22.3), τ represents the average time duration for which an ideal sinusoidal emission occurs, τ is called **coherence time or interval of coherence.** Let *T* denote the period of the oscillation, where $T = \frac{\omega_0}{2\pi}$. The spatial dimension, *L*, for which the light many be considered to be a perfect sinusoidal is given by

$$L = cT = \frac{\tau}{T}\lambda \tag{22.3}$$

Here c is the velocity of light and L is called the coherence length. After time τ, there is no correlation between the phases of the waves.

In terms of purity Q of the spectral line ($Q = \frac{\lambda}{\Delta\lambda}$, where $\Delta\lambda$ is half width of the spectral line), we have

$$c\tau = L = \lambda Q \tag{22.4}$$

22.3 EINSTEIN'S THEORY OF ATOMIC TRANSITIONS AND *A* AND *B* COEFFICIENTS

Transition between the atomic energy state is a statistical process and as such it is not possible to predict which particular atom will make transition from one state to another at a particular instant. However in an assembly of a very large number of atoms, it is possible to calculate the radiative transition between two states, based on the laws of probability. Einstein in 1917, first calculated the probability of such transitions assuming the atomic system to be in equilibrium with electromagnetic radiation.

Let us consider an assembly of atoms at an absolute temperatue T in which the atoms may be different energy states. Let n_0 be number of atoms per unit volume in the ground state ($E = 0$) then the number of atoms n per unit volume in an excited state of energy E is given by the Boltzmann distribution law

$$n = n_0 \exp(-E / k_B T) \tag{22.5}$$

where k_B(= 1.38 × 10^{-23} J/K) is the Boltzmann constant. The maximum number of atoms in an assembly of large number of atoms at an absolute temperature T is in ground state. Let n_1 be the number of atoms per unit volume in a lower energy state E_1 and n_2 be the same in a higher energy state E_2 Fig. (22.4), then from Eq. (22.5), one obtains

$$\frac{n_2}{n_1} = \exp[-(E_2 - E_1) / k_B T] \tag{22.6}$$

If $E_2 > E_1$, then $n_2 < n_1$.

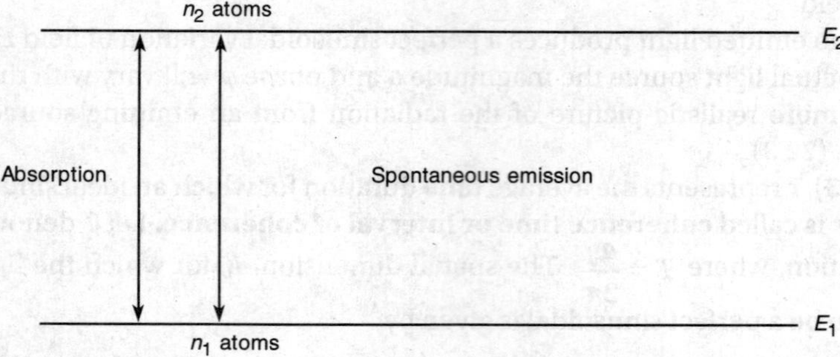

Fig. 22.4 Atomic transitions

The Atoms in the upper energy state E_2 makes spontaneous transitions to the lower energy state E_1 at a certain rate, determined by n_2 and the probability of spontaneous transition per unit time A_{21} from E_2 to E_1. Hence the number of spontaneous transition per unit volume per unit time is $A_{21}\, n_2$. These transitions lead to the emission of electromagnetic radiation of energy given by

$$h\nu = E_2 - E_1 \tag{22.7}$$

where h is Planck's constant and ν is frequency of the electromagnetic radiation. We must note that $1/A_{21}$ is a measure of the life time of the upper state against spontaneous decay to the lower state.

Now assume that the atoms are in equilibrium with e.m. radiation of frequency $\nu = (E_2 - E_1)/h$ having energy density u_ν. Due to absorption of energy from this radiation, some of the atoms in the lower energy state E_1 make upward transitions to the energy state E_2. The rate of such transitions is determined by the number n_1 and the energy density u_ν of the radiation and is given by $B_{12} n_1 u_\nu$ per unit volume per second. Obviously B_{12} is the probability of the **radiation induced transitions** per unit time from the lower state E_1 to the upper state E_2.

In addition to the above two types of transitions, there may also be induced transitions of atoms from E_2 to E_1 caused by interactions with the electromagnetic radiation. We call it as a stimulated or induced emission of radiation of frequency $\nu = (E_2 - E_1)/h$. The number of such induced or stimulated transitions per unit volume per unit time is $B_{12} n_2 u_\nu$ where B_{21} is the probability from E_2 to E_1 per unit time.

The coefficients A_{12}, B_{12} and B_{21} are known as Einstein's A and B coefficients. Under equilibrium, the time rate of transitions from E_2 to E_1 must be equal to that from E_1 to E_2. So one can write

$$A_{21} n_2 + B_{21} n_2 u_\nu = B_{12} n_1 u_\nu$$

or

$$u_\nu \frac{A_{21} n_2}{B_{12} n_1 - B_{21} n_2} = \frac{A_{21}}{B_{21}} \frac{1}{\left(\frac{B_{12}}{B_{21}}\right)\frac{n_1}{n_2} - 1} = \frac{A_{21}}{B_{21}} \frac{1}{\left(\frac{B_{12}}{B_{21}}\right)\exp\left(\frac{\hbar\nu}{k_B T}\right) - 1} \tag{22.8}$$

Eq. (22.8) must agree with Planck's radiation formula

$$u_\nu = \frac{8\pi h\nu^3}{c^3} \frac{1}{[\exp(h\nu/k_B T) - 1]} \tag{22.9}$$

Comparison of Eqs. (22.8) and (22.9) yields

$$B_{12} = B_{21} \tag{22.10}$$

and

$$\frac{A_{21}}{B_{21}} = \frac{8\pi h\nu^3}{c^3} \tag{22.11}$$

Eqs. (22.10) and (22.11) are called Einstein's relations obtained by Einstein in 1917. The relation (10) shows that the induced emission and absorption probabilities per unit time are equal. The calculation does not allow us to obtain the values of A_{21} and B_{12}, B_{21} they must be derived using quantum mechanical considerations. Using these relations, one obtains

$$\frac{A_{21}}{B_{21}\mu_v} = \exp\left(\frac{hv}{k_B T}\right) - 1 \tag{22.12}$$

which is the ratio of the number of spontaneous to stimulated transitions. Therefore if $hv >> k_B T$, spontaneous emission is much more probable than induced emission, which can be completely neglected. This holds true in the case of electrons transitions in atoms and molecules and in the case of radiative transitions in nuclei. But if $hv << k_B T$, as it is in the microwave region of the spectrum, induced or stimulated emission may become important.

Induced emission is the result of the action of the incoming radiation on the atoms (or molecules) of the substance. Hence the forced atomic oscillations bear a constant phase difference relative to the incoming radiation. This means that all atoms radiate in phase and **therefore induced emission is coherent**. On the other hand, spontaneous transitions occur at random, with no correlation between the times at which atoms undergo transitions. Therefore the phases of atomic radiations are distributed randomly. We than say spontaneous emission is incoherent.

22.4 PRINCIPLE OF LASER

The principle of laser exploits the phenomenon of stimulated transitions discussed in section 3. We know that light is emitted from a source when the atoms in the source make transitions from excited (E_2) state to a lower energy state (E_1) spontaneously. Normally, the mean life of excited atoms before spontaneous emission occurs is about 10^{-8}s. However, for some excited states, this mean life is perhaps as much as 10^5 times longer. We call such long-lived states metastable, they play an important role in laser operation. The energy of the emitted photon in the spontaneous transitions is

$$hv_{12} = E_2 - E_1 \tag{22.13}$$

on the other hand, if the atom is placed in an electromagnetic field that is alternating at frequency v, the atom can absorb an amount of energy hv_{12} (photom) from that field and move to the higher energy state E_2. This process is called absorption.

Besides absorption and spontaneous emission, a third type of transition from an upper energy state E_1 may be induced by an incident photon of energy $hv_{12} = E_2 - E_1$ giving rise to the stimulated emission of radiation. This process is called stimulated because the event is triggered by the external photon. The emitting photon in every way identical to the stimulating photon. Thus, the waves associated with photons have the same energy, phase, polarization, and direction of travel. All the above three of transition are shown in Fig. (22.5).

From Fig. [22.5(c)] it is obvious that we get two photons in this case. If these two photons are now incident on two other atoms in the excited state E_2, then it will result in the induced

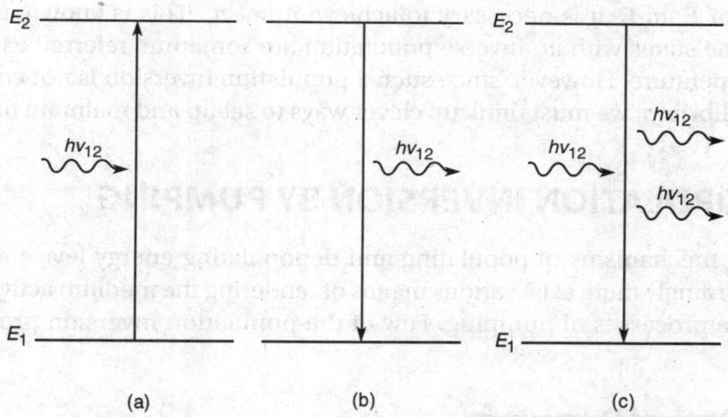

Fig. 22.5 (a) Spontaneous transition from an excited state E_2 to lower state E_1 with a emission of photon hv_{12}
(b) Transition from a lower state
(c) Induced or stimulated transition from an excited state E_2 to a lower state E_1. A passing light wave with photon energy hv_{12} causes the atom, to emit a photon of the same energy in creasing the energy of light wave

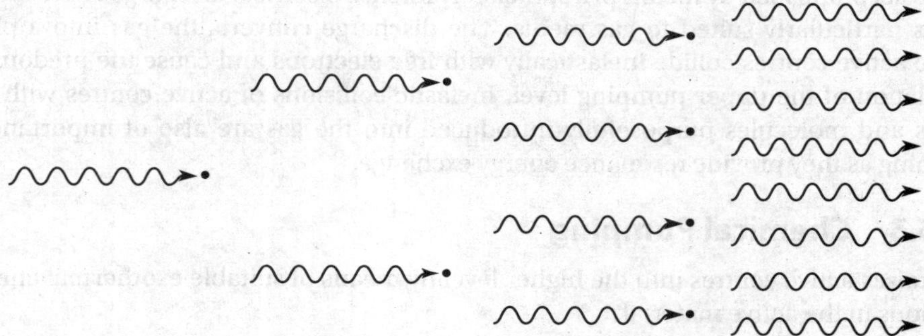

Fig. 22.6 Amplification due to induced emission

emission of two more photons so that there will be four coherent photons of the same energy result. The process goes on in a chain, yielding a large number of coherent photons of energy hv_{12}. This will increase the intensity of the coherent radiation enormously and amplification due to induced emission is achieved as shown in Fig. (22.6).

The necessary condition for the said amplification of the light intensity by the induced emission of radiation, is that we must have more photons emitted than absorbed, i.e., we must have a situation in which stimulated emission dominates. The direct way to achieve this is to start with more atoms in the excited state than in the ground state. Normally the number of atoms in the excited upper energy state (E_2) is much lower than in the lower energy state (E_1), i.e., $n_2 << n_1$ due to thermal equilibrium. As a result the rate of downward induced transition $B_{12}n_2u_v$ is much less than of the upward transitions $B_{12}n_2u_v$ between the same two energy levels due to absorption of photons. To increase the rate of induced emission from E_2 to E_1 it is necessary to achieve $n_2 >> n_1$. This is known as **population inversion**. The states with an inverse population are sometime referred as states with a negative temperature. However, since such a population inversion is not consistent with thermal equilibrium, we must think up clever ways to set up and maintain one.

22.5 POPULATION INVERSION BY PUMPING

The physical mechanisms of populating and depopulating energy levels are many and diverse. Accordingly there exist various means of rendering the medium active, collectively known as the processes of pumping. Few of the population inversion processes are as follows:

22.5.1 Optical Pumping

In this process, the predominant population of the upper level is achieved by means of light energy delivered from appropriately selected sources such as gaseous discharged flash tubes, or continuously burning tubes.

22.5.2 Electrical Pumping

This is accomplished by means of a sufficiently intense electrical discharge in the medium and is particularly suited to gas media. The discharge converts the gas into a plasma where active centres collide inelastically with free electrons and cause the predominant population of the upper pumping level. Inelastic collisions of active centres with other atoms and molecules purposefully introduced into the gas are also of importance for pumping as they provide resonance energy exchange.

22.5.3 Chemical Pumping

This raises active centres into the higher level by means of suitable exothermal chemical reactions in the active material.

22.5.4 Heat Pumping

In this scheme of pumping. the active material at first brought to a high temperature and then rapidly cooled down.

We may note that the situation of population inversion ($N_n > N_m$) corresponds to negative Kelvin temperature T in the equation

$$(N_n/N_m) = \exp[-(E_n - E_m)]/k_B T$$

22.6 OPTICAL RESONATORS

The selection of some photon states and the suppression of states can be realized by means of an optical resonator, a principal component of each laser. In its simplest version the optical cavity is a pair of mirrors set on an optic axis which defines the direction of the laser beam.

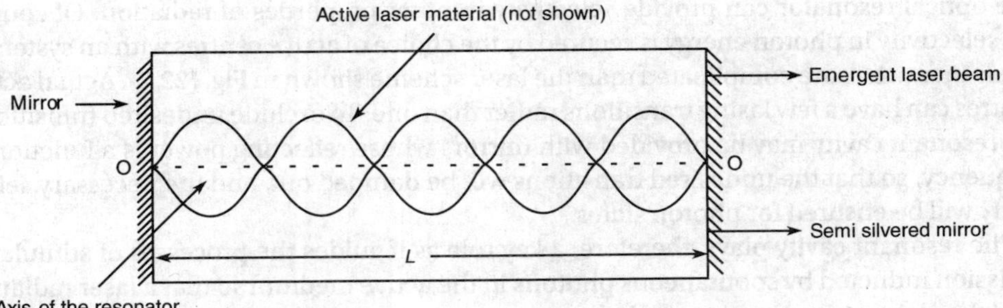

Fig. 22.7 The optical resonator determines the direction of laser action (along the resonator axis) and favours the process of stimulated emission in this direction

The active material is placed in between these mirrors. Solid active materials are often in the shape of a cylinder whose axis is aligned with the axis of the optic resonator, the length of the cylinder being about ten times its diameters. At least one of the mirrors of the resonator cavity is made semitransparent to serve as an output element passing the light out of the resonator.

Fig. (22.7) shows the schematic arrangement of a simple reasonator. The photons produced spontaneously in the OO direction or sufficiently close to it will travel within the active material a relatively long way, which is elongated by multiple reflections from the resonator's mirrors.

These photons interact with excited active centres to eventually initiate a powerful avalanche of stimulated photons constituting the laser beam. The photons emitted in other directions (and their stimulated avalanches) will traverse short path length in the material and die out soon.

Thus the optical resonant cavity provides the desired selectivity of photon states by primarily confining the possible direction of photon propagation. As a result, lasing action occurs in this direction.

Let L be length of the resonator cavity. The change in one round trip between the mirrors is given by

$$\delta = \frac{2\pi}{\lambda}2L + 2\gamma$$

where γ is the phase change at reflection off each mirror and λ is the wavelength of light. For resonance, i.e., reinforcement, δ must be a integral multiple of 2π. So, if m is an integer, one obtains

$$\delta = \frac{4\pi L}{\lambda} + 2\gamma = 2\pi m$$

or

$$\nu = \frac{c}{\lambda} = \frac{mc}{2L} = \frac{\gamma c}{2\pi L} \tag{22.14}$$

where c is the speed of light. Obviously, the light having frequencies that satisfy the condition (14) will resonate and set up stationary waves in the cavity as shown in Fig. (22.7). Each mode corresponds to one integral value of m. Since one of the mirrors of the resonator is partially transmitting, i.e., half silvered, some of the radiations leaks out, thus providing the collimated laser output.

The optical resonator can provide selectivity in other properties of radiation. Of course, the selectivity in photon energy is secured by the choice of active centres with an system of levels in much more complicated than the laser scheme shown in Fig. (22.7). Actual active centres can have a few lasing transitions rather than one. To exclude undesired transitions, the resonant cavity may be provided with mirrors whose reflecting power is a function of frequency, so that the undesired transitions will be damped out, and the necessary selectively will be ensured for photon states.

The resonant cavity plays, therefore, a key role as if guides the processes of stimulated emission inducted by spontaneous photons in the active medium so that a laser radiation of high coherence properties results.

22.7 GAIN CONSTANT

No energy level is sharp but has a **'finite energy spread'** due to Doppler effect and other factors. Thus all the atoms in an energy level cannot make transition due to the action of photons of a definite frequency ν. This means only a fraction. Say ΔN out of N, will undergo transition-upward or downward.

$\therefore$ Number of transitions/s per unit volume is given by

$$B_{12}u_\nu \Delta N_1 = B_{12}(I_\nu / c)\Delta N_1 \tag{22.15}$$

But the intensity I_ν is related to energy density u_ν given by $I_\nu = cu_\nu$, where c is the velocity of light.

Similarly, the number of transitions per second per unit volume is

$$B_{12}u_\nu \Delta N_2 = B_{21}(I_\nu / c)\Delta N_2 \tag{22.16}$$

Since quantity of energy $h\nu$ is (Photon) absorbed in each upward transition and in each downward induced transition an identical quantity is added to the beam. Thus the rate of change of energy density of the beam is

$$\frac{d}{dt}(u_\nu \Delta\nu) = \frac{h\nu}{c}(\Delta N_2 - \Delta N_1)B_{12}I_\nu \tag{22.17}$$

In obtaining Eq. (22.17), we have used $B_{12} = B_{21}$ and Eqs. (22.15) and (22.16).

Further, light moves over a distance dx in time dt, and hence $dt = dx/c$. Thus from Eq. (22.17), we have

$$\frac{d}{dt}(u_\nu \Delta\nu) = \frac{d}{dx}(u_\nu \Delta\nu) = \frac{d}{dx}(I_\nu \Delta\nu)$$

$$= \frac{h\nu}{c}(\Delta N_2 - \Delta N_1)B_{12}I_\nu$$

$$\therefore \quad \frac{dI_\nu}{dx} = \frac{h\nu}{c\Delta\nu}(\Delta N_2 - \Delta N_1)B_{12}I_\nu$$

$$\text{or} \quad \frac{dI_\nu}{I\nu} = \frac{h\nu}{c\Delta\nu}(\Delta N_2 - \Delta N_1)B_{12}dx - \beta_\nu dx$$

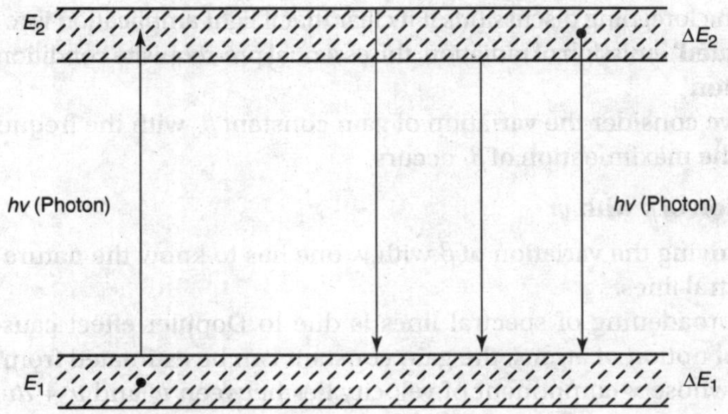

Fig. 22.8

where

$$\beta_\nu = \frac{h\nu}{c\Delta\nu}(\Delta N_2 - \Delta N_1)B_{12} \tag{22.18}$$

Here β_ν is called as the gain constant and is a function of the frequency ν.

Intensity

Integrating the relation Eq. (22.17), one obtains

$$I = I_{\nu o} e^{\beta_\nu x} \tag{22.19}$$

Since E_1 and E_2 have finite widths, the transitions between these states give rise to lines of different frequencies spread of spectral line due to transitions between two levels as shown in Fig. (22.8) Now, if $\Delta\nu$ be the spread about the mean about the mean frequency ν_0, then $\nu = \nu_0$, we have Fig. (22.8). Transitions between E_1 and E_2 levels with finite widths

$$\beta_\nu = \beta_{\nu_0} = \frac{h\nu_0}{c\Delta\nu}(N_2 - N_1)B_{12}$$

Now, assuming $\Delta N = N$ for each energy level,
we have

$$\beta_{\nu_0} = \frac{\lambda_0^2}{8\pi\Delta\nu}(N_2 - N_1)A_{12}$$

$$\left(\frac{A_{12}}{B_{12}} = \frac{8\pi h^3\nu^3}{c^3}\right)$$

Now, we examine the various situations:

(i) $N_2 > N_1$: As a result of population inversion, $\beta_{\nu_0} > 0$. Thus it follows from Eq. (22.18) that I_ν increases with increase of x. This clearly implies that as the beam of light progresses through the medium its intensity I_ν increases exponentially. Obviously, the beam intensity thus gets amplified.

(ii) $N_2 < N_1$: This is normally expected due to the Boltzman distribution, $\beta_{\nu_0} < 0$, i.e., negative and, from Eq. (22.18), I_ν decreases exponentially as x increases. We may note that there is no amplication in this case.

From the foregoing discussions it it clear that if light amplication is to be achieved by **stimulated'** emission of radiation, the essential pre-requisite condition is **'population inversion'**.

Now, we consider the variation of gain constant β_ν with the frequency ν and also when the maximiaation of β_ν occurs.

Variation of β with ν:

For studying the variation of β with ν, one has to know the nature of broadening of spectral lines.

If the broadening of spectral lines is due to Doppler effect caused by random thermal notion of atoms, the gain constant can be estimated from the fraction of atoms whose x-component of velocity lies between v_x and $v_x + dv_x$. We have this fraction as

$$\frac{dN}{N} = \left(\frac{m}{2\pi kT}\right)^{1/2} e^{-mv_x^2 kT} dv_x$$

Now, the Doppler shift in frequency is

$$\frac{\nu - \nu_0}{\nu_0} = \frac{v_x}{c} = v_x = \frac{c(\nu - \nu_0)}{\nu_0}$$

and $$dv_x = \frac{cd\nu}{\nu_0}$$

$\therefore$ $$dN = N\left(\frac{m}{2\pi kT}\right)^{1/2} \frac{c}{\nu_0} e^{-mv_x^2/2kT} d\nu$$

$$= N\left(\frac{m}{2\pi kT}\right)^{1/2} \frac{c}{\nu_0} e^{(\nu-\nu_0)^2 m/2kT\nu_0^2} B_{12}$$

Now, the gain constant β_ν,

$$\beta_\nu = h\left(\frac{m}{2\pi kT}\right)^{1/2} (N_2 - N_1) e^{-2e^2(\nu-\nu_0)^2/2kT\nu_0^2} B_{L2} \qquad (22.20)$$

Obviously, for a Gaussian broadening, the gain constant is also a Gaussian function of frequency ν.

At the centre, we have $\nu = \nu_0$. Thus, at the centre, β value is given by

$$\beta_{\nu_0} = h\left(\frac{m}{2\pi kT}\right)^{1/2} (N_2 - N_1) B_{12}$$

$$= \frac{A_{12}\lambda_0^3}{8\pi}\left(\frac{m}{2\pi kT}\right)^{1/2} (N_2 - N_1)$$

where we have substituted B_{12} in terms of A_{12}.

We may note that the above being the maximum value of β, the gain constant is highest at the centre of the spectral line.

The Quality Factor (Q)

The important characteristics of the optical resonator is its quality factor Q. In the standing wave pattern, the reflectors give two nodes. The wavelength of light λ being much smaller than the optical resonator cavity length L, the number of half waves between the two mirrors is very large. One obtains from Eq. (22.14) [section 22.6] for the difference in frequencies between two consecutive modes as

$$\Delta v = v_{m+1} - v_m = \frac{c}{2L}$$

Obviously, there is a spread in the frequency of the laser beam, which is controlled by the spectral character of the emission line contribution to the gain. The sharpness of the cavity resonance is determined by Q, a dimension less quantity. Q is used to show the ability of the cavity to store energy. The slower the energy of a radiation field falls off, i.e., the smaller the losses in the resonator, consequently, the higher the resonator's Q. Q of a cavity is defined

$$Q = 2\pi \frac{\text{Maximum energy stored per cycle in the mode}}{\text{Energy dissipated per cycle in the mode}}$$

In a passive resonant cavity, the light beam irradiance gradually declines with travel distance owing to various losses in the medium, namely, absorption, scattering, radiation through side surfaces etc. This decreases Q and reduce the sharpness of resonance. If $d\nu$ is the line width of the cavity mode, then one can easily show that $Q = \nu / d\nu$.

The smaller the line width, the higher the Q of the cavity, and the sharper the resonance. In the case of the He-Ne gas laser with a cavity of length 1 m, we have $\Delta v = c/2L = 1.5 \times 10^8$ Hz. Obviously, this is a small fraction of the mean frequency of the output radiation at 4.75×10^{14} Hz. This reveals the high degree of monochromaticity of the laser radiation.

22.8 AMPLIFICATION OF LIGHT AND THRESHOLD CONDITION

Let us consider a system of atoms through which a quasi-monochromatic light beam of frequency v propagates in the z-direction. Let S_1 and S_2 represents the two planes of area A perpendicular to the z-direction and located at z and $z + dz$, respectively, as shown in Fig. (22.9). The volume of the active laser material between the planes S_1 and S_2 = Adz. If n_1 is the concentration of the atoms in the lower state, u_v is the energy density of the radiation and B_{12} is the Einstein's B coefficient, then the time rate of stimulated absorption = $B_{12}\, n_1\, u_v Adz$, Obviously, the energy absorbed per unit time in the volume Adz is therefore equal to $B_{12} n_1 u_v hv Adz$. Similarly, one finds the energy gained per unit time in the volume Adz due to stimulated emission is $B_{21} n_2 u_v hv\ Adz$, where n_2 is the concentration of atoms in the upper state. Here we have ignored the radiation due to spontaneous emission because such radiations are directed at random and are lost. Now, the net energy gain per unit time in volume Adz in the frequency range v and $v + dv$ is

$$(B_{12} n_2 - B_{12} n_1) u_v hv\ Adz = B_{12}(n_2 - n_1) u_v hv\ Adz (B_{12} = B_{21})$$

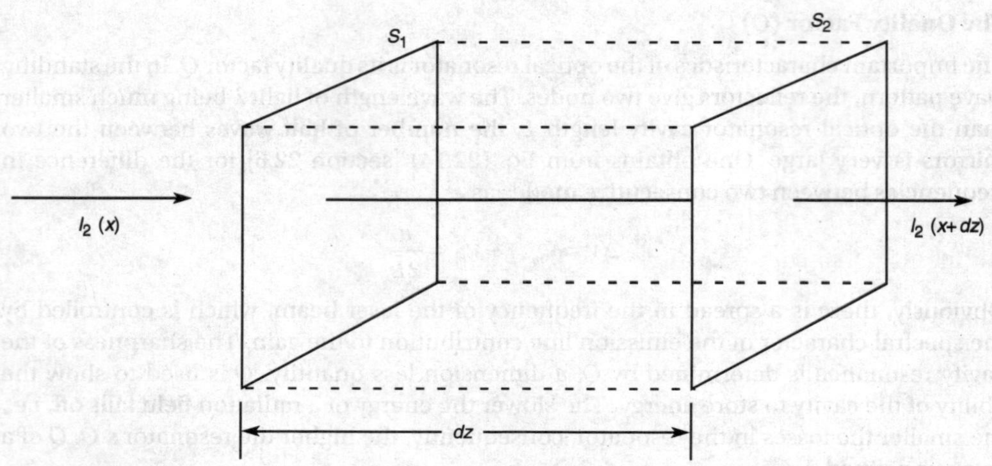

Fig. 22.9 (a) Amplification of a light beam passing though the plane S_1 and S_2 located at z and $z + dz$ respectively

Let us consider that I_ν be the intensity of the beam at S_1. Obviously, the energy entering the volume Adz per unit time through S_1 is $I_\nu\,(z)\,A$. Similarly, if $I_\nu\,(z + dz)$ be the intensity at S_2 then the energy moving out of the volume Adz per unit time through S_2 is

$$I_\nu(z+dz)A = \left(I_\nu(z) + \frac{\delta I_\nu}{\delta z}dz\right)A$$

obviously, net energy leaving the volume per unit time is $\frac{\delta I_\nu}{dz}Adz$. This must be equal to the energy to the energy gained by volume Adz due to stimulated transitions. Thus, one obtains

$$\frac{\delta I_\nu}{dz}Adz = B_{12}u_\nu h\nu(n_2 - n_1)Adz$$

$$\frac{\delta I_\nu}{dz} = B_{12}h\nu(n_2 - n_1)u_\nu \tag{22.21}$$

Also, we have

$$I_\nu = \upsilon u_\nu = \frac{c}{n}u_2 \tag{22.22}$$

where $\upsilon = c/n$ is the velocity of light in the medium having refractive index n. Using Eq. (22.22), Eq. (22.21) reduces to

$$\frac{I}{I_\nu}\frac{dI_\nu}{dz} = \alpha_\nu \tag{22.23}$$

where
$$\alpha_\nu = \frac{B_{12}h\nu n(n_2 - n_1)}{c} \tag{22.24}$$

On integration, Eq. (22.24) gives

$$I_\nu(z)=I_\nu(0)\exp(\alpha_\nu z) \tag{22.25}$$

Eq. (22.25) represents the growth of the intensity of the light as it traverses a distance z through the active medium of laser. Obviously, the growth result from the population inversion created by pumping and stimulated transitions. The quantity α_ν is called the gain constant having the dimension of reverse length. We can see that the gain constant α_ν increases with the degree of population inversion $n_2 > n_1$ and hence α_ν given by Eq. (22.24) is positive. This means that the intensity increases exponentially with distance z.

Let us consider that r_1 and r_2 be the reflection coefficients of two mirrors of the optical cavity. Obviously, in one complete round trip, the intensity of light decreases to $I_\nu(0)\ r_1 r_2$ where $I_\nu(0)$ is the initial intensity. Let us write $r_1 r_2 = \exp(-2\beta)$, where β is positive and can be considered to be a measure of the loss of intensity of light as it traverses from one mirror to the other. This means that on traversing a length L, the intensity drops to $I_\nu \exp(-\beta)$. We may assume that the reflection losses are the only losses in the cavity. Thus the intensity in a mirror to mirror passage becomes $I_\nu(0)\exp(\alpha_\nu L - \beta)$. Thus for a substained laser action the amplification must compensate for energy losses. Obviously, for the lasing action to build up, we must have

$$\exp(\alpha_\nu L-\beta)>1 \quad \text{or} \quad \alpha_\nu L-\beta$$

Consequently, one finds the threshold condition as

$$\alpha_\nu L=\beta \tag{22.26}$$

From Eq. (22.26) we can easily see that if $\alpha_\nu L = \beta$ the intensity dies out. At threshold, the population inversion is called the critical inversion. From Eqs. (22.24) and (22.26), one obtains the condition for critical inversion as

$$(n_2-n_1)_{\text{criti}}=\frac{\beta c}{L}(\mu\beta_{12}h\nu)^{-1} \tag{22.27}$$

22.8.1 Working of Laser

Fig. 22.9(b) shows a schematic block diagram of working of a laser. There are mainly three basic requirements of functioning of a laser; (i) an active medium, (ii) population inversion and (iii) some sort of feed back

In order to achieve the population inversion an active medium, i.e., a right group of atoms or molecules is required. This active medium can be a gas or mixture of gases or crystal or semiconductor, etc., With the help of pumping process, the population inversion is achieved. For pumping process, various techniques are used, e.g. optical, pressure-temperature cycle, physical separation, etc., To generate feed back, optical cavity is used. Optical cavity is formed by dielectric mirrors or polished ends of crystal rod which act as reflectors. The pumping process creates population inversion. If that particular energy state is short lived or unstable, atoms go to another lower state (usually metastable state) by way of fast decay. The metastable state is a long life state in comparison to other states. Usually, the life time of an energy state is 10^{-8} s, whereas life time for metastable (temporarily stable) state is 10^{-3} s or slightly more. This is the key to the functioning of a laser. We may note that in many atoms there can be one or more metastable states.

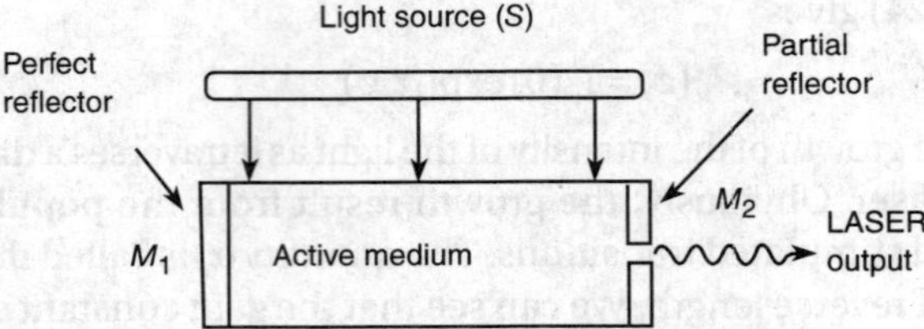

Fig. 22.9 (b) Schematic block diagram of functioning of a laser

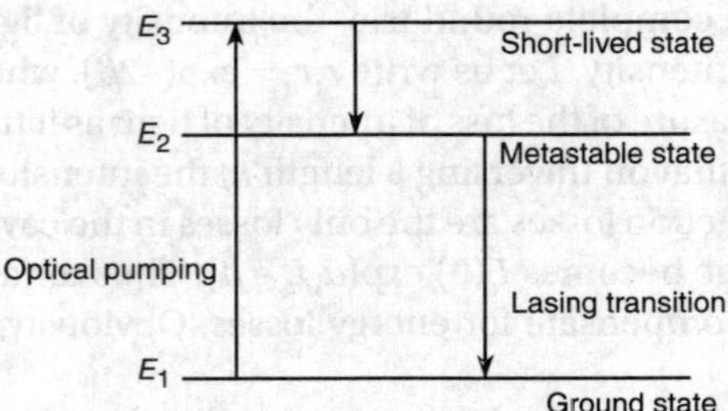

Fig. 22.9 (c) Transitions in a three level laser. The lasing transition takes the system from the metastable state to the ground state

After the population inversion, stimulated emission is triggered with the help of a photon of appropriate frequency. In some cases, the first atom, making the transition from metastable state to lower energy state generates triggering photon. Obviously, the stimulated emission produces more photons than the photons entering the system (Fig. 22.9(c)). In normal operation many more atoms will be in the excited (metastable) state than in the lower (ground) state. Obviously, this situation is, of course, contrary to what one finds under normal conditions. This is why it is known as *population inversion*.

There is potential difficulty with the three-level system as described. What happens to an atom after it has been returned to the ground state (E_1) from metastable state (E_2) by stimulated emission? We would like the external power source to return it immediately to E_3, but in practice this atom may take sometime for this to happen, during which it is possible that a photon with energy $E_2 - E_1$ (just the energy that a photon created by stimulated emission would have) can be absorbed. That would be undesirable, for that photon would then be unavailable for stimulating another transition and for passing through the partially silvered mirror, as we expect all of them to do eventually. The result will be much weaker laser beam, or perhaps none at all.

One can get over this hurdle by using a *four-level system* (Fig. 22.9(d)).

As before, atoms are pumped from the ground state to a higher state (now E_4), where they decay quickly to the metastable state (E_3). The stimulated emission takes atoms from E_3 to E_2. The spontaneous transition from E_2 to E_1 is not forbidden, so the state E_2 will not exist long enough for a photon to be used up in kicking the system from E_2 to E_3. Now, there is little chance that an atom in the ground state will absorb a photon with energy $E_3 - E_2$, so the lasing process can proceed efficiently.

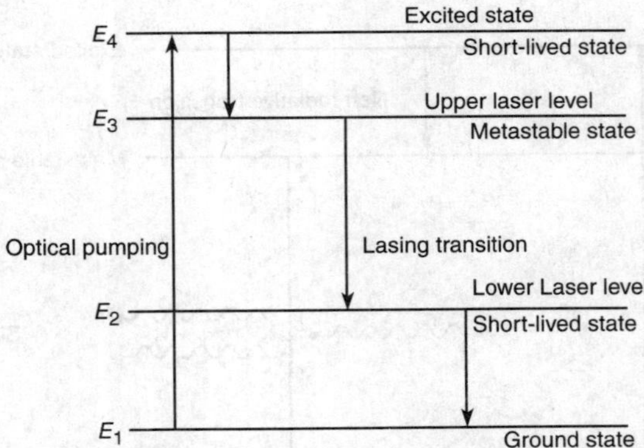

Fig. 22.9 (d) Four level laser system. The lasing transition takes the system from the metastable state (E_3) to another short-lived (lower laser level) state (E_2). The system then returns quickly to the ground state (E_1), so the photons cannot be reabsorbed in returning the system from E_2 to E_3

Now, it is clear that for ideal operation of laser, some conditions are to be satisfied by the system, e.g. the relaxation rates from excited state to upper laser level and from lower laser level to ground state should be high to maintain maximum population inversion between upper and lower laser levels. The pumping rate should be large enough to overcome the emission from upper to lower laser levels. The thermal equilibrium population of lower laser level should be as small as possible. Moreover, the decay of excited state to any level other than upper laser level should be as slow as possible. Further, the non-radiative decay of upper laser level should be slow.

The intense pattern is builtup between two mirrors (reflectors) M_1 and M_2 only if an exact number of waves fit between the mirrors. We may note that if only one wavelength that resonates, the laser gives out a single sharp line, if two or three wavelengths resonate, laser emits two or three lines.

22.9 VARIETIES OF LASERS

22.9.1 Solid-State Lasers

(i) Ruby Laser

The first successful operation of the laser was achieved by T. H. Maiman using ruby crystals. Ruby is aluminium oxide in which a few of the aluminium atoms have been replaced by chromium atoms (Al_2O_3: Cr^{3+}), the chromium atoms being the ones that take part in laser action. Maiman used 0.05% Cr^{3+} to get a pink colour, which is due to the fact that chromium absorbs, ultra-violet, green and yellow, and transmits only red and blue. This absorption raises chromium atoms to excited states from which two steps are required to carry them back to the ground energy level. In the first step chromium atoms give up some of their energy to lattice and fall in *metastable* state, characterized by a comparatively long life time. This transition is a non-radiative transition. In the second step, chromium atoms

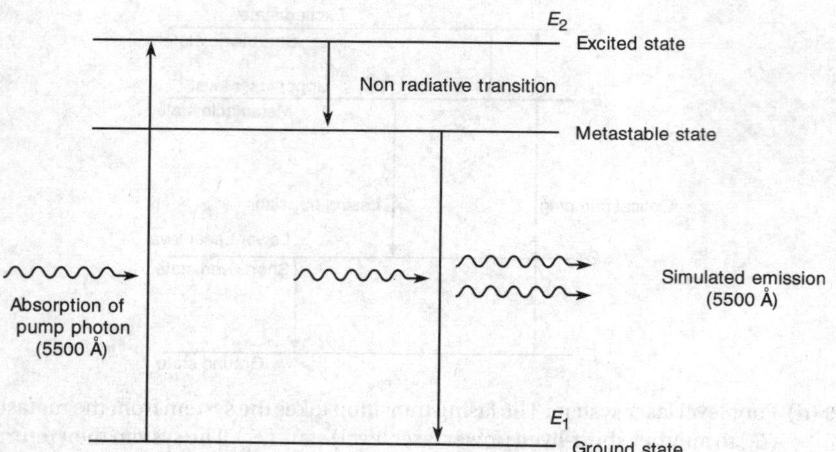

Fig. 22.10 Energy band scheme for laser action in ruby (Al_2O_3: Cr^{3+} ruby)

emit radiation of wave-length 6943 Å in the red region of the spectrum and fall to ground level. The metastable state has a life time of about 10^{-3} second compared to the extremely small life of 10^{-7} second of higher energy level. Thus the necessary time delay between the useless relaxation transition and the useful stimulated transition for the laser action is readily obtained. The energy levels in the ruby lasers are shown in Fig. (22.10).

Ruby optical laser apparatus is shown in Fig. (22.11). A ruby is machined into a rod typically 5 cm long and 0.5 cm in diameter with polished ends which are parallel and optically flat. This is placed near an electronic flash tube which provides the pumping energy. The ruby is cooled using liquid nitrogen. Upto a certain flash intensity, ordinary luminescent emission dominates, and powerful beam of red light flashes out from the ends of a ruby rod. This red light has the enormous power output of more than 10^4 Watts over a beam of cross-section 10^{-4} square meter. The emitted band is within a wave length interval of about 0.02 Å. The beam from the end face of the ruby rod has an angular spread of less than one degree. Typical output of a ruby laser is 10–20 Kilowatts in milli second pulses.

The stimulated emission predominates over spontaneous emission when the population of meta-stable exceeds considerably over the population predicted by Boltzmann distribution relation and in comparison with the ground energy state. The meta-stable consists of closely spaced double level R_2 and R_1. The larger contribution to laser action, however, comes from the lower level i.e., R_1. This is due to the fact that the spontaneous life for R_1 which is about 4×10^{-2} s exceeds the thermal relaxation time for energy transfer between R_2 and R_1. The minimum energy required to be pumped for laser action to start, is of the order of 50 to 100 Joules of absorbed energy.

It is important to note that this type of oscillator action has losses which are determined by the flatness and parallelism of the end mirror. The losses may be considerable, and may reduce the efficiency of the laser.

The maximum space coherence (directionality) obtainable is determined by the diffraction of the mirrors. The light-beam spread angle $\delta\theta$, is given by

$$\delta\theta \cong \frac{\lambda}{D} \qquad (22.28)$$

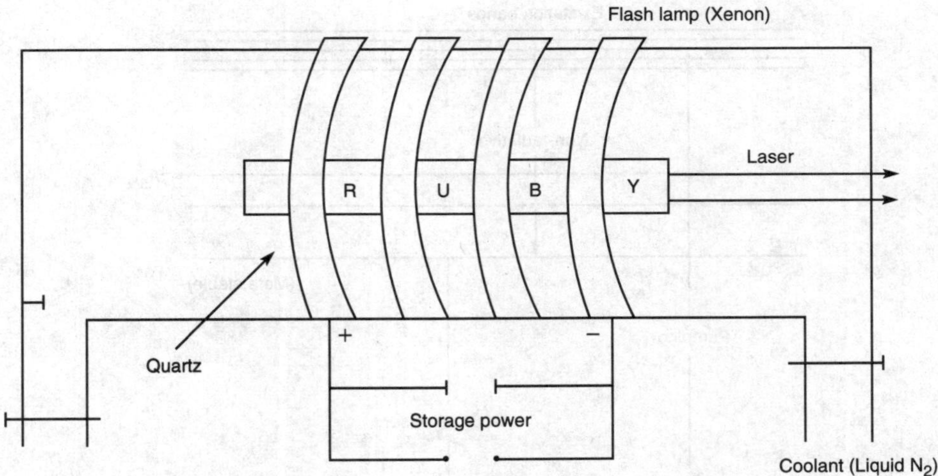

Fig. 22.11 Pulsed ruby optical Laser

where λ is the wavelength of the emitted radiation and D is the diameter of the source, which is the mirror in this case.

Pumping Power: One can roughly estimate the pumping power necessary for the population inversion as follows:

Let us consider that P be the power of the pumping light of frequency ν_p incident on a unit area of the active material. Let σ_p be the absorption cross-section of the pump light. Thus the pump power absorbed by the material = $\sigma_p p$. Obviously, this power is used in raising the atoms to the excited state. Let t be the life time of the upper (excited state), the power consumed in the transition to the lower state is $h\nu_p/t$. For population inversion to occur we have $\sigma_p P = h\nu_p/t$ or $P = h\nu_p/t\,\sigma_p$. In case of ruby laser, we have $h\nu_p = 3 \times 10^{-19}$ J, $\tau = 4$ ms and $\sigma_p = 10^{-23}$ square meter, one obtains $P = 7.5 \times 10^6$ W/s^2.

Apart from ruby, several other solid materials such as calcium-samarium, glass-neodymium etc. have been used as active materials for solid-state laser action. Basically, this class of lasers refers to the lasers based on ions in solid materials. These doped crystals gave pulsed output. The pulsed calcium fluoride lasers has four levels and was first developed by P.P. Sorokin and M.J. Stephenson during 1960–61. In general, most ordinary solid state laser are of the four level type as it is easier to establish population inversion in these systems. In a four-level laser system, excitation and radiation transfer proceed as in the case of a three level laser material, but there is now an additional energy level available above the ground level. Laser action, therefore, begins as soon as there is a significant population of the level above this additional level, which in turn increases efficiency.

Next to ruby, Nd^{3+} is the most commonly used solid material for lasers. It is capable of producing infrared radiations over a large range, the most useful wavelength being 1.064 μm. Rare earth materials used as solid laser give coherent radiation in the range 0.6 to 2.6 μm. Another common solid laser is the YAG laser which has four levels instead of the three levels in ruby. The material consists of yttrium aluminium garnet ($Y_3Al_5O_{12}$) doped with ND^{3+} ions in place of Y. On pumping optically by flash lamps, the ND^{3+} ions are excited to their characteristic bands. The excited ions decay non-radiatively into the meta

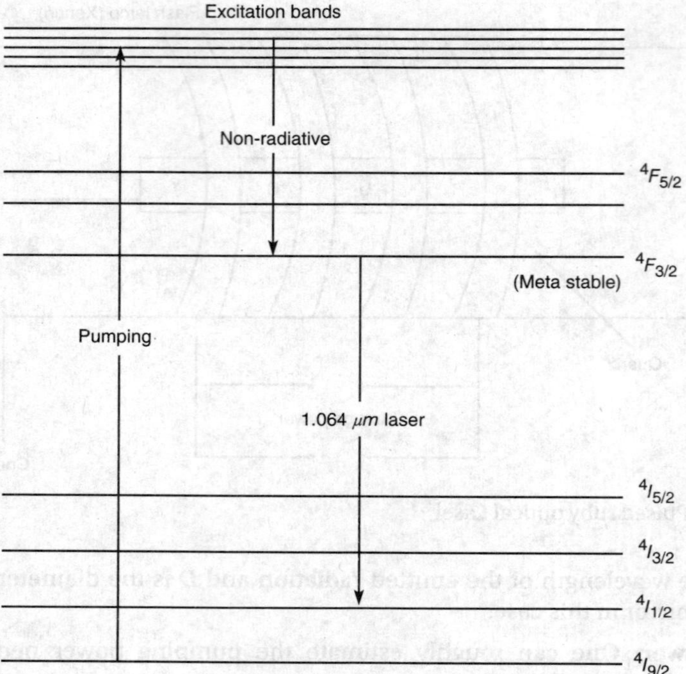

Fig. 22.12 Energy level diagram of Nd in YAG laser

stable state $4F_{3/2}$ which serves as the upper laser level and has a lifetime of about 0.25 milli second. The radiative (laser) transitions occur from this state predominantly to the $4I_{11/2}$ state below giving the 1.064 μm beam.

22.9.2 Semiconductor (PN Junction) Lasers

Coherent laser radiation ranging from the infrared, through the ultraviolet regions of the spectrum has been obtained from semiconductors and semiconductor junction diodes.

Laser radiation can be obtained by pumping a solid intrinsic semiconductor or a PN junction diode. The method of pumping varies with the type of semiconductor used. The most common type of semiconductor used is a III-V compound. Semiconductor lasers are different from other lasers in the following respects:

(i) Transition occurs between energy bands rather than between discrete energy levels, the emission being a result of electron transitions from the conduction to the valence bands.

(ii) The laser is very small in size, typically 50 $\mu m \times 250 \mu m \times 50 \mu m$.

(iii) The characteristics of the laser beam are strongly influenced by the properties of the junction material.

(iv) In PN junction lasers, population inversion occurs in the very narrow region about the junction. The pumping of the PN junction laser is accomplished by the application of a forward bias to the diode.

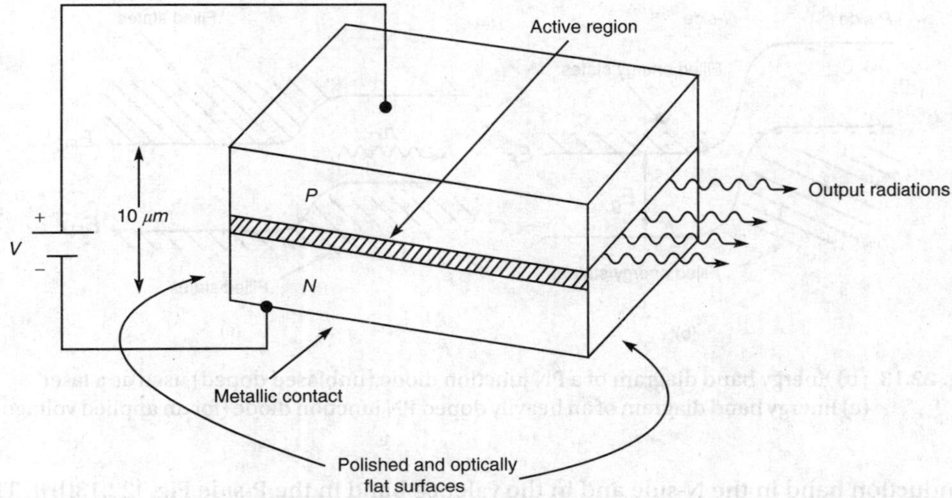

Fig. 22.13 (a) Semiconductor Laser

It is a property of semiconductors that the downward transitions of electrons recombining with holes may or not result in the emission of option of optical energy. Based on this property, semiconductors can be classified as *indirect and direct*. This classification is related to the energy band structure, silicon and germanium are indirect types, where as most of the III-V compounds are direct. In direct types, radiative transitions take place faster than non-radiative and impurity transitions. In indirect semiconductors, radiative recombination's occur faster than radiative recombinations, so that in essence photon emission is suppressed. In fact, some radiation in emitted by almost all semiconductor PN junctions, but in junctions using the indirect type of semiconductors, this radiation is very inefficient.

Fig. (22.13(a)) shows the basic structure of a PN junction laser. The diode laser is made of gallium arsenide (GaAs) by the standard diffusion method. The operating wavelength for this laser is usually in the range 8400 Å and 8500 Å. The index of refraction of semiconductors is usually so high that it is not necessary to polish the end faces to achieve reflection. Efficiencies as high as 70% have been observed in these lasers. A large number of materials are being used to get an operational range of wavelengths in the region of 0.33 to 30 μm. In these types of lasers due to band-to-band transitions involvement, the spectral purity is low and emission pattern is broad.

An important difference between a solid state laser and a semiconductor laser is that while in the former the bulk of the material is the host and the active material is as low as nearly one percent, whereas in the latter the entire material is participating in the laser action. It is important to mention that the semiconductor diode laser is essentially an electro-luminescent diode made from a degenerated directed band-gap semiconductor like GaAs, operating under a heavy forward bias.

Population inversion in a semiconductor laser

One can obtain the population inversion in a PN junction by heavily doping both the P and N regions and applying a forward bias. The heavy doping pushes the Fermi level E_F in the

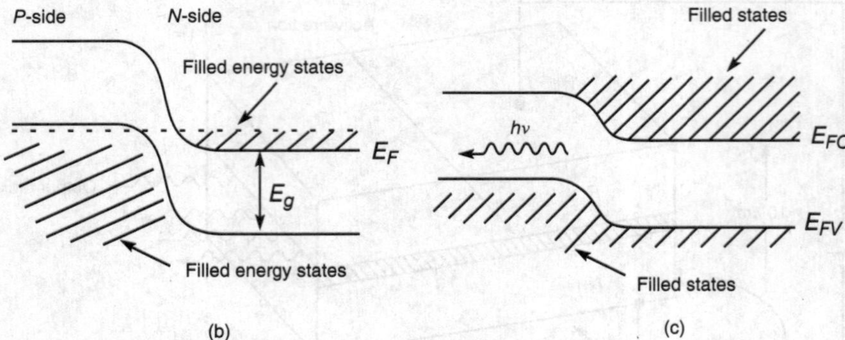

Fig. 22.13 (b) Energy band diagram of a PN junction diode (unbiased doped) used as a laser
(c) Energy band diagram of an heavily doped PN junction diode (for an applied voltage V)

conduction band in the N-side and in the valence band in the P-side Fig. [22.13(b)]. The system is in therodynamic equilibrium when there is no applied voltage and obviously there is no population inversion. To create population inversion, we will have to disturb the therodynamic equilibrium, i.e., a forward bias has to be applied. For an applied forward voltage V, the band diagram of the on junction is shown in Fig. [22.13(c)].

Let us assume that the charge carriers in the two bands are in thermal equilibrium with each other. This means that the population distributions in the conduction and in the valence bands are described by the quasi Fermi levels E_{FC} and E_{FV} respectively. One can express the occupation probability in the conduction band by

$$f_c(E_c) = \frac{1}{1 + exp(E_c - E_{Fc}) - k_B T} \tag{22.29}$$

and that in the valance band is expressed as

$$f_v(E_v) = \frac{1}{1 + exp(E_v - E_{Fv}) / k_B T} \tag{22.30}$$

For a light beam to be incident, one finds that the time rate of absorption of quanta R_a is proportional to (i) the probability per unit time of the transition from the valance band to the conduction band B (ii) the density of incident radiation $u(\omega)$, (iii) the probability of occupancy of the valance bands $f_c(E_v)$ and (iv) the probability of vacancy of the conduction $[1 - f_c(E_c)]$. Thus, one can write

$$R_a = CB_{vc} u(\omega) f_v(E_v)[1 - f_c(E_c)] \tag{22.31}$$

Here C (constant of proportionality) includes the density of states in the two bands. Similarly, one can write the expression for the time rate of emission of the quanta by stimulated emission R_e as

$$R_e = CB_{ev} u(\omega) f_c(E_c)[1 - f_v(E_v)] \tag{22.32}$$

For laser action, the emission must exceed absorption, we have $R_e > R_a$ i.e.,

$$f_c(E_c)[1 - f_v(E_v)] > f_v(E_v)[1 - f_c(E_c)] \tag{22.33}$$

One can easily see that Einstein's B coefficients B_{cv} and B_{ve} ae equal. Using Eqs. (22.29) and 24, Eq. (22.30) reduces to

$$E_{FC} - E_{FV} > F_c - E_v = h\nu \tag{22.34}$$

Since $E_{FC} - E_{FV} > eV$, where V is the applied forward bias, the inequality Eq. (22.34) reveals that the applied forward voltage must exceed $h\nu/e$, where ν is the frequency of the emitted light.

When PN junction is forward bias, the electrons flow from the N-side flow to the P-side, and holes from the P-side to the N-side. The injected electrons and holes appear in high concentrations in a part of the depletion region at the PN junction. The consequent large population initiates the lasing action when the PN diode current exceeds a threshold value.

We must remember that the electrons from the conduction band in this region jump the valence band to recombine with the holes, thus releasing the band energy gap in the form of coherent radiation. Morecover, the P and N regions of the diode absorbs the laser radiation which interacts with the lattice phonons. Obviously, the laser action is restricted to the very thin junction region of the diode.

22.9.3(a) Gas Lasers

Laser action can also be made to occur in some gases, such as CO_2 or a mixture of He and Ne. The credit for the discovery of first gas laser goes to A. Javan, W. Bennet and D.R. Herriot. They used the electric discharge through a gas for ionisation and pumping action, i.e., resonant transmission of excitation energy in a gas discharge. Sometimes excitation or pumping may be provided externally by a separate electrode. The cavity is about 20–50 cm long having a tube diameter of about 5 mm and two concave reflecting mirrors at the ends. A steady potnetial of a few thousand volts is used for getting a discharge. The main advantage of such lasers resides in that they operate continuously, though some gas lasers are also capable of pulsed working. These lasers display exceptionally high monochromatic, most pure spectrum and high stability of frequency. All these features make these lasers extremely useful in various branches of science and engineering. The tube used in these lasers in made up of glass or quartz. The ends of the tube terminate in flat glass windows inclined at Brewster angle to eliminate reflections.

He-Ne gas were the first to be developed and used. They operate in three distinct spectral ranges, in the red at 6328 Å, in the near infrared around 1.15 μm and in the infrared at 3.39 μm. The partial energy diagram of He-Ne is shown in Fig. [22.14(b)], which explains the origin of these lines. With the help of a dispersive prism these lines can be separated.

A gaseous mixture of He and Ne (about 10:1) is used as active medium. When the helium atoms in the mixture are excited to the meta stable state by means of a dc or radio-frequency source, they collide with the ground state neon atoms and resonant energy transfer takes place. One of the excited states of the neon has nearly the same excitation energy as He meta-stable state. Hence the Ne atoms are excited in the collision to a specific energy as the helium atom returns to the ground-level. This causes a population inversion in the energy states of neon and laser beam is obtained. He-Ne gas laser is shown in Fig. [22.14(a)].

22.9.3(b) Carbon Dioxide (CO_2) Laser

CO_2 laser is a molecular gas laser. The overall efficiency of most of the gas lasers is below 1%, but that of CO_2 laser is about 20%.

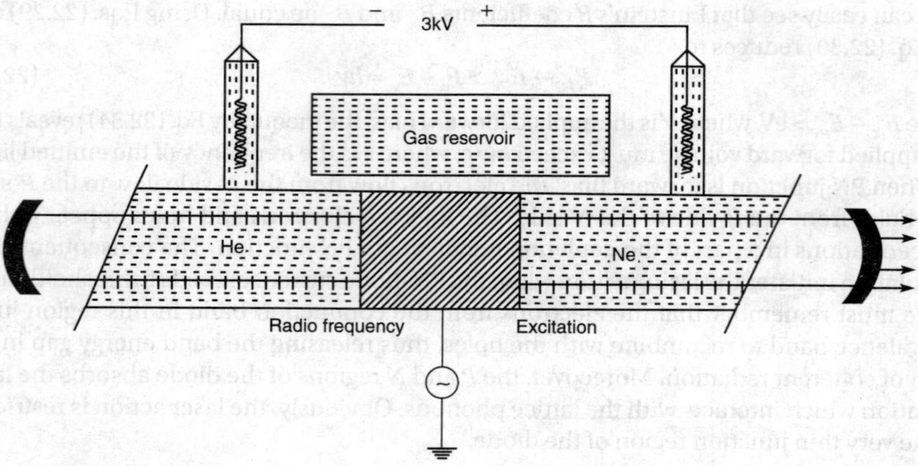

Fig. 22.14 (a) He-Ne Gas Laser

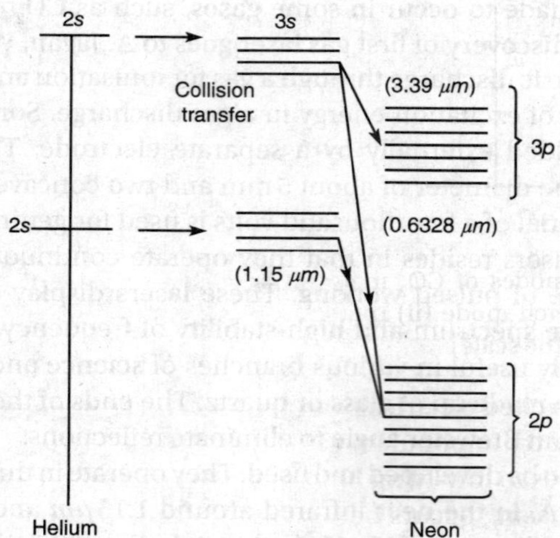

Fig. 22.14 (b) Partial Energy Levels of He-Ne Gas Laser

The construction of this laser is just similar to He-Ne laser. CO_2 laser is a discharge tube filled with a mixture of CO_2, N_2 and He gases in the ratio 1:4:5. However, the energy spectrum of molecular gas is more complex than the atomic gas. We know that the molecule has energy levels due to vibrations and rotations of the atoms along with the electronic energy levels of atoms. CO_2 molecule has three vibrational modes (Fig. 22.14(c)) and for each mode there are rotational modes.

The CO_2 molecule vibrates in three modes: (i) symmetric stretch mode (n 0 0), (ii) bending mode (0 n 0), and asymmetric strech mode (0 0 n), where n is an integer. Obviously, each mode is associated with a set of energy levels. The first excited asymmetric stretch state (0 0 1) is almost exactly equal to lowest vibrational level of nitrogen. We may note that the trick of energy transfer by resonant collision, used in He-Ne laser, is also used

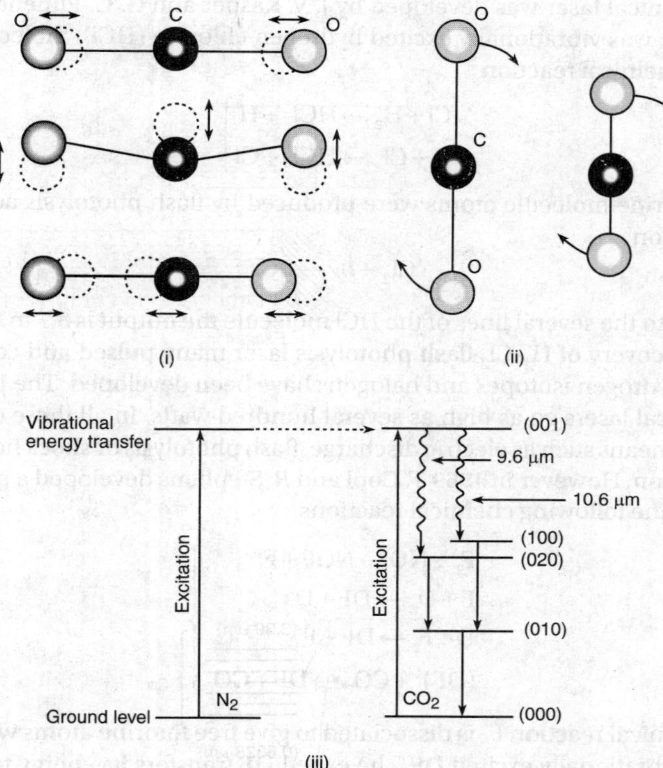

Fig. 22.14 (c) (i) Vibrational modes of CO_2 molecule: symmetric stretch mode, bending mode, asymmetric stretch mode (ii) rotational modes of CO_2 molecule (iii) energy level diagram (not to the scale)

in CO_2 laser. First, the N_2 is excited electrically and it delivers the energy to CO_2 (0 0 1) state through collisions. In case if try to populate (0 0 1) state of CO_2 directly by electron collisions, most of the energy goes to low-lying (0 1 0) and (0 2 0) states. This is why N_2 is used as vehicle. In CO_2 laser, the lasing occurs between (0 0 1) $\rightarrow$ (1 0 0) and (0 0 1) $\rightarrow$ (0 2 0), at $10.6\mu m$ and $9.6\mu m$ respectively. The vibrational-vibrational energy transfer pumping process results in high efficiency and enormous output power. This increases the utility of CO_2 laser. Helium is added in the mixture to improve the heat removal from the laser discharge. Moreover, addition of He in the mixture also helps in speeding up the transition from (1 0 0) to the ground state (0 0 0) to maintain population inversion.

22.9.4 Chemical Lasers

In chemical lasers, population inversion results from a chemical reaction. The high energy density that can be released in exothermic chemical reactions has led to the interest in converting this energy to coherent optical energy. If one could use the energy of such chemical reactions to obtain a population inversion directly, one could build compact and powerful lasers, Such lasers may also prove to be a powerful tool in the study of kinetics of chemical reactions.

The first chemical laser was developed by J. V. Kasper and G. C. Pimentel in 1965. The active molecule was vibrationally excited hydrogen chloride (HCl^*) molecule, formed by the following chemical reaction

$$\left.\begin{array}{l} Cl + H_2 \rightarrow HCl^* + H \\ H + Cl_2 \rightarrow HCl^* + Cl \end{array}\right\} \tag{22.35}$$

The initial chlorine molecule atoms were produced by flash photolysis according to the following reaction

$$Cl_2 + h\nu \rightarrow 2Cl \tag{22.36}$$

corresponding to the several lines of the HCl molecule the output is 3.7 to 3.84 μm.

Since the discovery of H_2-Cl_2 flash photolysis laser many pulsed and continuous laser based on the hydrogen isotopes and halogens have been developed. The powers of some of these chemical lasers go as high as several hundred watts. In all these chemical lasers some external means such as electric discharge, flash photolysis or shock heating starts the chemical reaction. However in 1969 *T.* Cool and *R.* Stephens developed a purely chemical laser based on the following chemical reactions

$$\left.\begin{array}{l} F_2 + NO \rightarrow NOF + F \\ F + D_2 \rightarrow DF + D \\ D + F_2 \rightarrow DF + F \\ (DF)^* + CO_2 \rightarrow DF + CO_2 \end{array}\right\} \tag{22.37}$$

In the first chemical reaction F_2 is dissociated to give free fluorine atoms which on mixing with D_2 yield vibrationally excited DF. The exited DF transfers its energy to CO_2 and laser action results at the wavelength 10.6 microns. The power output from this laser is about 500–600 watts.

22.9.5 Dye Lasers

P.P. Sorokin and J.R. Lankard in 1966 obtained laser action in organic molecules by their efficient fluorescence. They also produced laser action in many dye solutions which were pumped with micro-second long flashes from high intensity lamps. As shown in the energy level diagram Fig. (22.15), the optical excitation and emission in organic dyes take place between the rotational vibrational levels of the ground singlet S_0 and the excited singlet state S_1.

The active medium of an organic dye laser is an organic fluorescent material dissolved in a common solvent. These substances derive their colour from strong absorption in the visible region of spectrum, Sorokin obtained the laser action by using a solution containing a dye-chloro-aluminium pthalocynanine at 7650 Å. Dye laser can operate at room temperature and moreover they are tunable.

In a *tunable laser* the wavelength of the emitted radiation can be adjusted over a range as wide as 200 nm. Tunning is accomplished by changing the concentration of the dye and/or the length of dye cell. Semiconductors lasers are replacing the dye lasers for several applications.

Another kind of tunable laser is free electron laser (Fig. 22.15(b)).

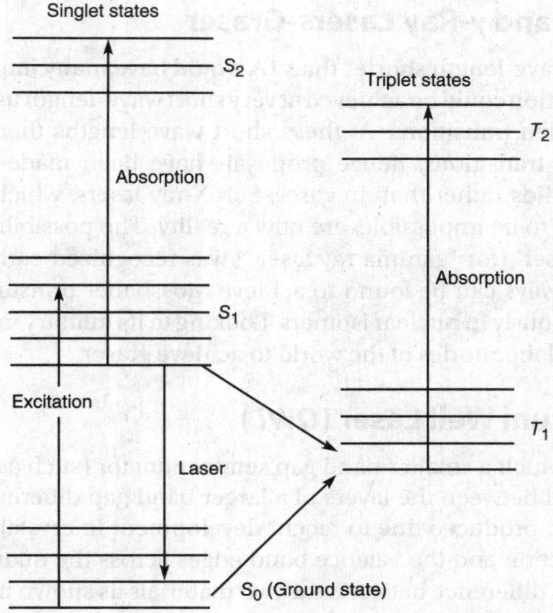

Fig. 22.15 Optical excitation and level emission between vibrational S_0 and S_1 in a dye laser

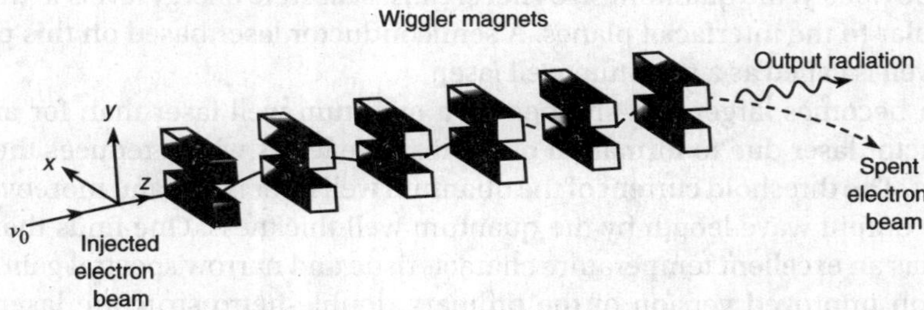

Fig. 22.15(b) Free-electron laser. The laser output is in the same direction (*z* in this figure) as the injected electron beam

The basic principle of this laser is that charged particles, (electrons in this case), emit electromagnetic radiation when accelerated. As shown in Fig. 22.15(b), a series of magnets called *Wigglers* is used to accelerate a beam of electrons transversely. The velocity of the electron is matched to the oscillations in the magnetic field in such a way that the emitted radiation is coherent as in a gas filled laser. Since the free electrons are not tied to atoms, they are not dependent upon atomic energy levels and can be tuned to wavelengths wall into the ultraviolet (UV) region of the spectrum. Free electron lasers produce wavelengths of 100 nm and shorter. *Electron*-beam lasers have been made to produce outputs in ultraviolet, visible and infrared parts of the spectrum. The exceptional fine-tunning that is possible with these devices, along with their relatively high efficiency, make these lasers and masers useful for a number of scientific applications, including accelerator technology, spectroscopy, radar, and fusion research.

22.9.6 X-Ray and γ-Ray Lasers-Graser

Lasers generating wave-length shorter than 1Å would have many important applications. In principle, laser action could be achieved at very short wave-length using radiation arising from nearly forbidden transitions. At these short wave-lengths there is no competition from non-radiative transitions, hence proposals have been made for achieving X-ray or γ-ray lasers in solids rather than in gases. Soft X-ray lasers, which a decade ago were generally conceded to be impossible, are now a reality. The possibility of a nuclear laser. often termed a 'graser' (for "gamma ray laser") was recognized early in 1960's. A graser may be possible if ways can be found to achieve Mossbauer transitions and population inversion simultaneously in nuclear isomers. Looking to its military importance efforts are going on in various laboratories of the world to achieve graser.

22.9.7 Quantum Well Laser (*QWL*)

Quantum wells in which a smaller band gap semiconductor (such as GaAs) some tens of Å thick, sandwiched between the layers of a larger band gap difference between the two materials have been produced due to recent development in crystal growth. There is an offset of the conduction and the valence band edges across the quantum well interfaces due to the band gap difference between the two materials as shown in Fig. (22.16).
When electrons are put into the quantum well, confinement in the well forces electrons to have discrete energy levels designated by $n = 1, 2,...$ in accordance with the quantum principle. Obviously, the quantum size effects causes discrete energy levels in the direction perpendicular to the interfacial planes. A semiconductor laser based on this principle of quantum well is called as a quantum well laser.

The gain becomes larger and sharper for a quantum well laser than for an ordinary semiconductor laser due to formation of the discrete levels which reduces the spread of the electron. The threshold current of the quantum well laser is low and moreover, one can control the output wave length by the quantum well thickness. One finds that quantum well laser has an excellent temperature characteristic and narrow spectral gain width.

QWL is an improved version of the ordinary double-hetro structure laser shown in Fig. (22.17). One can see that the thickness d of the active layer of Ga As contained between the outer layers of N-type (Ga, Al) As and P-type (Ga, Al) As and is about 200 nm. Electrons from the N-type (Ga, Al) As and holes from P-type (Ga, Al) As move into the conduction

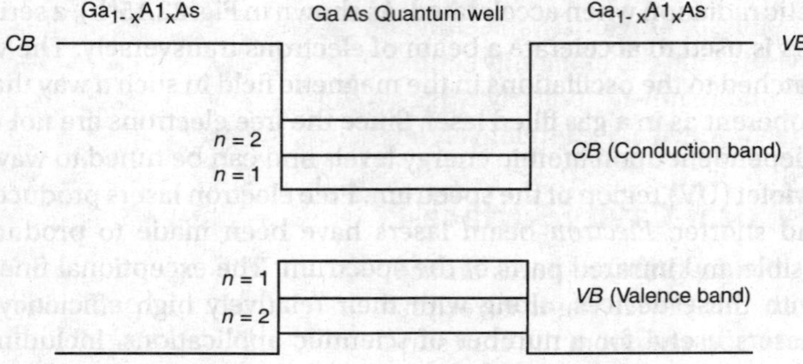

Fig. 22.16 Quantum well

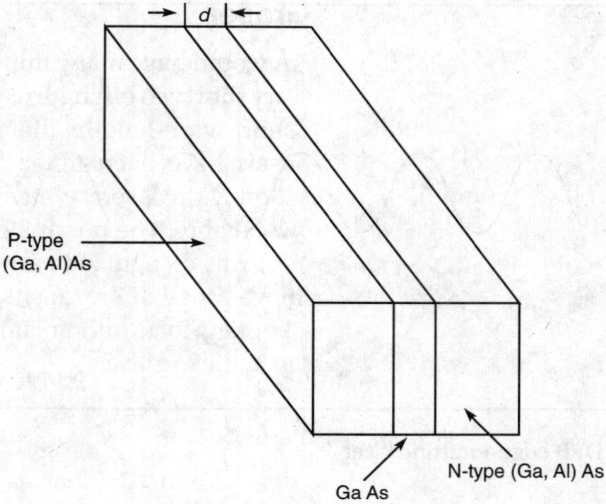

Fig. 22.17 A double structure laser

band and valence band of the active layer respectively. This causes a population inversion for lasing action. The light is guided between outer layers due to larger refractive index of the active layer. This offers a narrow beam. We can easily see that this is an additional feature over the confinement of the carriers. We can also see that d is greater in the ordinary double hetro structure laser than in *QWL*. This means that the energy level quantization in the valence and conduction bands in the double hetro structure laser is of no significance.

22.9.8 Distributed Feedback (DFB) Emitting Laser

The reflecting mirror facets in the conventional edge-emitting semiconductor laser are produced at either end of the optical cavity by cleaving or etching the material along two parallel crystal planes. In the DFB edge emitting laser, reflections are distributed along the cavity length by producing periodic index variations which individually reflects a small fraction of the photons as shown in Fig. (22.18).

We must note that collective reflection in DFB edge emitting laser is more or less the same as the reflection from the facets of a conventional laser. However, the feedback mechanism of the DFB laser is much more sensitive to the wave length in comparison to the conventional laser. Obviously, the emitted light of the DFB laser has a much narrower spectral line width. One can efficiently couple the coherent output beam to an optical fiber.

A thin, wide and relatively long cavity is required in either type of edge-emitting lasers. This means that the device size is relatively large and it requires some tens of mA to operate.

22.9.9 Vertical-Cavity Surface-Emitting Laser (VCSEL)

VCSEL developed in 1990s, requires an active light emitting region contained between two mirrors. In this device, light is recirculated vertically in an optical cavity so that emission occurs normally from the chip surface. In VCSEL, the length of the active region is considerably shorter than in other laser devices.

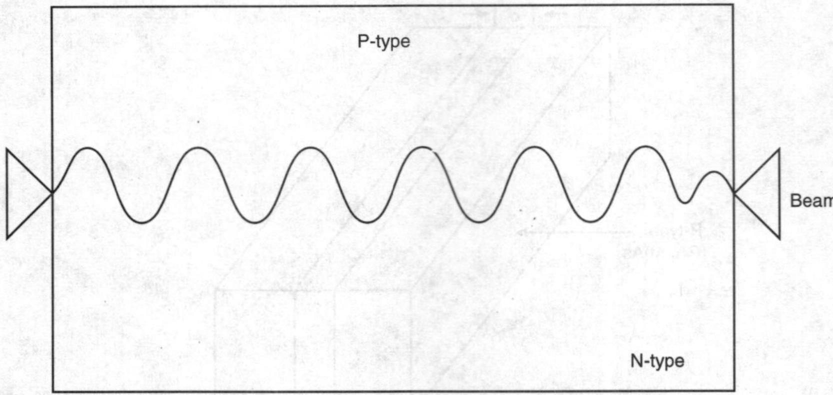

Fig. 22.18 A DFB edge-emitting laser

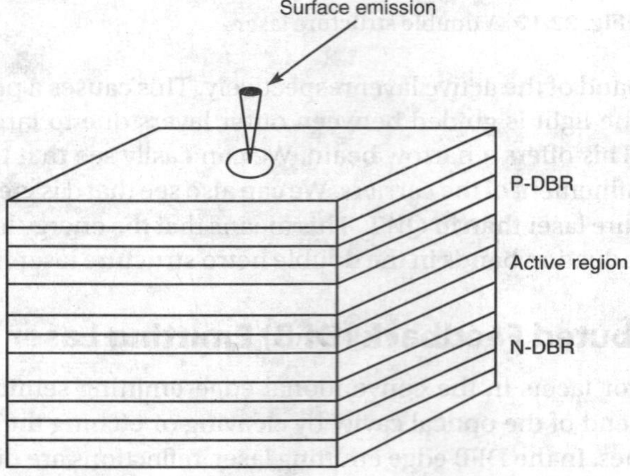

Fig. 22.19 A VCSEL laser

The shorter active length in VCSEL makes it essential to use many more highly reflecting mirrors formed by a stack layers of quarter-wavelength thick semiconductor material. These mirrors are called **distributed Bragg reflectors** (DBRs.) Since the mirrors are made from semiconductor materials and hence the current injected into the active region flows directly through these mirrors. One can form a PN junction with current flowing vertically through the device by doping one P-type and other DBR-N-type as shown in Fig. (22.19).

A VCSEL has compact size and the design flexibility of the out put aperature. The small size of VCSEL ensures that the current required for lasing action is quite small. The low capacitance of the device makes the operating speed inherently high. One can tailored the emission pattern to match the input of an optical fibre more effectively than possible with an edge emitting laser. We can see that VCSEL's form an ideal choice for high-speed optical data transmission. For the wavelength window of 780–980 nm. (Al, In, Ga) As system is used as material for VCSEL where as for the window of 1.3–1.55 μm, (In, Ga) (As, P) system is used.

22.9.10 Quantum Dot Lasers

Quantum dot lasers offer the potential to yield better properties as compared to quantum lasers. These emit light at wavelengths that are determined by the energy of dots. The research in nano-crystal quantum dots has opened the door for developing novel optical and optical and opto-electronic devices, such as tunable lasers, optical amplifiers and light emitting diodes from assemblies of these invisibly small particles. Quantum dot lasers work like any other semiconductor lasers such as those found in home audio compact disk players. The goal of a quantum dot laser is to manipulate the material into a high energy state. The result is the net release of energy which emerges as a photon. European group of scientists have launched a project called 'Big band' to develop ultra-wide band InP based quantum dot lasers amplifiers for telecom applications.

The primary goal BIG BAND is to produce QD-based semiconductor optical amplifiers (SOAs) operating over the entire telecommunication window. SOAs are seen as potential replacement for erbium doped fibre amplifiers in networks. Another goal is to push PnP device emission to longer wavelengths for gas sensing applications.

22.9.11 Quantum Cascade Laser (QCL)

This is a light source that fills the increasing need for compact and high performance mid-IR semiconductor lasers. It is a high power laser, radically different from double hetero structure lasers, and can be designed to emit essentially any wavelength in the mid-IR spectrum, extending into the far IR. QCL lasers in pulsed operation at room temperature generate peak optical powers of upto hundreds of mill watts concentrated in a single laser mode, and they can continuously tuned over a significant wavelength range. Recently, continuous wave (CW) operation at room temperature has been reported. within a decade's time after their invention QC lasers greatly out perform mid-IR diode lasers and are commercially available. They have been successfully used in trace gas analysis with a sensitivity of parts per billion in volume.

QC lasers may potentially be applied for optical wireless communication in the eye-safe atmospheric transmission windows at wavelengths of 3.5 nm and $8.13\mu m$.

22.9.12 Gyro Laser

A gyro laser detects slow rotational motions, down to μ rad/s and less as shown in Fig. (22.20). In one type, a hollow tube forms a ring of radius R mounted on an object rotating with angular speed ω. A beam splitter on the ring divides on incoming laser beam of wavelength λ into pieces, on travelling counter clockwise (CCW) and other clockwise (CW) around the ring. These beams recombine at the beam splitter after one circuit and are removed. Without rotation, the beam travel the same distance $2\pi R$ around the ring in a time $t = 2\pi R/c$ and are in phase when they recombine, however, there is rotation in the CW direction, then as the CW moving beam travels a circumference the beam splitter advances by a distance ωR. The beam must travel a total distance (the distance ωR is very small).

$$L_{CW} = 2\pi R + (\omega R)t = 2\pi R(1 + \omega R/c)$$

Similarly, the path length for the CCW-moving beam is decreases:

$$L_{CCW} = 2\pi R(1 - \omega R/c)$$

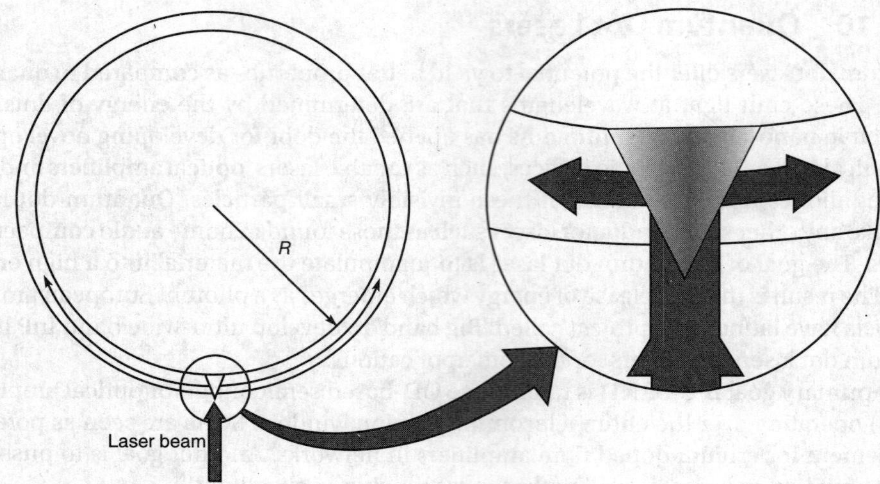

Fig. 22.20 The ideal gyro laser shown here has a cavity that forms a ring. Two laser beams enter and are directed in opposite directions

The two recombined beams are, in general, no longer in phase. Their phase difference is $\Delta\phi = 2\pi\,\Delta L/\lambda$, where ΔL is the path length difference,

$$\Delta L = L_{CW} - L_{CCW} = 2\pi\, R^2\Omega/c$$

Obviously, $\Delta\phi$ is proportional to the rotation speed. $\Delta\phi$ will manifest itself as a reduction in the intensity of the recombined beam. As ω changes, so does the interference.

In practical situations variants on this scheme are used. In one such scheme a laser medium that can send beams in both directions is put into the ring. We may note that in this case the frequencies supported in the cavity differ for the two directions. Let us suppose first that there is no rotation. Then the two beams will be maintained only if the travel distance $2\pi R$ is an integer N times a wavelength, i.e., $2\pi R = N\lambda = Nc/f_0$ or $f_0 = Nc/2\pi R$. (The subscript on f_0 signals "no motion"). This condition follows because without the integer requirement the beam will interfere destructively with itself when it makes a full circle.

With rotation, the length $2\pi R$ for which the beams "bite their tails" is changed to L_{CW} and L_{CCW} for the CW and CCW paths, respectively. Thus the maintained frequencies differ for the two directions, taking the values

$$f_{CW} = \frac{NC}{L_{CW}} = \frac{NC}{2\pi R\left(1+\dfrac{\omega R}{c}\right)} = \frac{f_0}{\left(1+\dfrac{\omega R}{c}\right)} \approx f_0\left(1-\frac{\omega R}{c}\right)$$

and

$$f_{CCW} = \frac{NC}{L_{CCW}} = \frac{NC}{2\pi R\left(1-\dfrac{\omega R}{c}\right)} = \frac{f_0}{\left(1-\dfrac{\omega R}{c}\right)} \approx f_0\left(1+\frac{\omega R}{c}\right)$$

When the two beams are extracted and combined their resultant contains easily detectable beats. The frequency of beat is

$$\Delta f = f_{CCW} - f_{CW} = 2 f_0 \omega \frac{R}{C} = \frac{2\omega R}{\lambda}$$

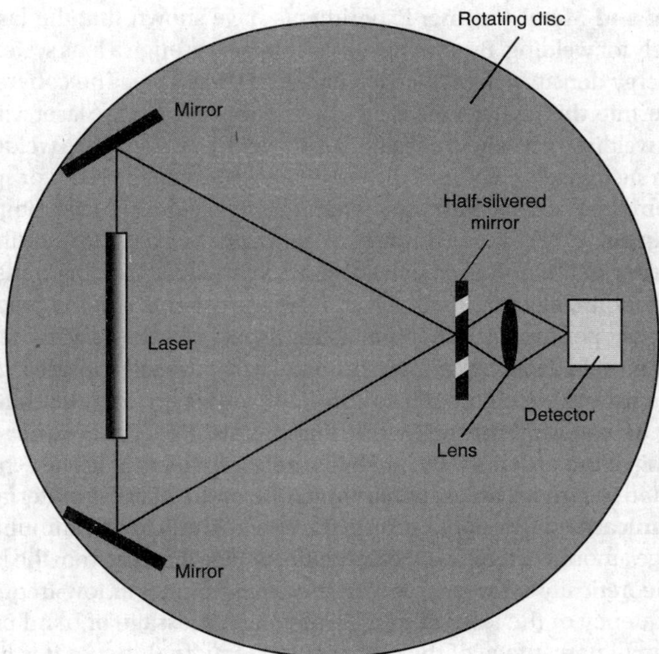

Fig. 22.21 In a real gyro with mirrors that can also be used to extract the beams and detect any beats that might to sult from the rotation of the platform

Measurement of the beat frequency thus translates into measurement of ω, and by this method one can measure rotational velocities too small to be detected by any other mans.

In practice the "ring" is not a circular cylinder, but rather a triangular path with mirrors at the corners Fig. (22.21); the triangle rests centered on the axis of the rotating object. We may note that the preceding formula for Δf remains the same if one replaces R by the ratio of twice the area enclosed by the path to the length of the path.

22.10 APPLICATIONS OF LASERS

With the discovery of laser, man's control of light has been and will continue to extend to an unpredictably large and diverse number of applications. The applications discussed follow from one or more of the special properties of the laser beam, namely; coherency, power or high intensity, directionality, and narrow frequency range. It would be practically impossible to list all the potential uses of the laser, but some of the important uses and applications of lasers are as follows.

(i) Surgery: The powerful, well collimated beam emitted by the laser shows promise in many medical applications. Exploration is being conducted in such fields as tooth drilling, eye surgery, cancer treatment, and cardiovascular surgery. In most of these cases the laser is like an optical knife which would be more accurate, less painful, and faster than the scalpel.

(ii) Welding and Metal Cutting: Experiments have shown that the laser can be used effectively for welding. By passing the laser beam through a lens system, an extremely high energy density can be focused in a very small area. The coherency of the light beam permits this high degree of focusing. It appears that the laser will be most useful for spot welding, with the electronic beam welder or resistance welder continuing to perform steam welds. Some of the advantages in using the laser for spot welding are: (a) no physical contact with the material being welded, (b) welding can be accompanied in air, (c) very little rise in temperature in the area surrounding the weld, and (d) a variety of metals have been successfully welded. Welding transistor functions, thermocouples, electronic micro-modules are examples of the precise work which the laser can perform. Another interesting aspect of welding with the laser is that it is possible to weld a joint even after the joint has been sealed inside a glass envelope.

With some minor changes in control, the same types of laser that perform welding can be used for cutting or burning metals. Experiments have succeeded in punching of the order of 0.005 inch diameter in sheets of metal. The laser has also succeeded in cutting holes in diamond, one of the hardest materials.

(iii) Communications: The application of the laser to the field of communication is receiving a large amount of technological attention. Since the light from the laser is coherent, it can theoretically carry messages in the same manner as low-frequency carriers. The frequency of the laser is so high that each message of band of frequencies is a very small percentage of the carrier frequency (in this case it is frequency of the laser). For instance, if a carrier is 1,10,000 Hz and we wave allowed to modulate this with messages up to 10 percent of the this frequency, then 1,00,000 Hz would be out from total allowable message band width. However at 1,00,000 Hz carrier frequency, the allowable 10 percent would have 1,00,00,000 Hz total message band width. Obviously, many more messages can be sent at higher carrier frequencies. Powerful though laser light may be, it is light and is therefore impeded to some extent by our atmosphere even under good conditions.

(iv) Lasers in Industry: The precision property of laser light has been of immense help in industry, particularly in testing the quality of optical components, e.g. prisms, lenses, etc., The use of lasers in industry has considerably increased the accuracy in the measurement of the size of the physical quantities.

(v) Lasers in Biology: The ability of laser beams to concentrate high power density of light at a focal point has opened a new area of micro Raman spectroscopic analysis. This enables one to extend such studies to biological and biomedical samples available only in very small quantities.

Following a suggestion of Arthur Ashkin, the radiation pressure of a laser beam has been made to act on viruses or bacteria as well as on atoms, but in a different way. As shown in Fig. (22.22), the electric dipole forces due to a split laser beam act on a neutral particle, e.g., a bacterium or a bead, in a way that pushes the particles into the brightest area. These forces act to stabilize the target and bring its centre towards the focus of the microscope lens through which the laser beam is directed. The forces are tiny, ~10^{-12}N (piconewtons), but they are adequate when acting on such small objects. Thus the bacterium or bead can be fixed and also pushed or pulled with a pair of laser beams, which thus act as **optical tweezers**.

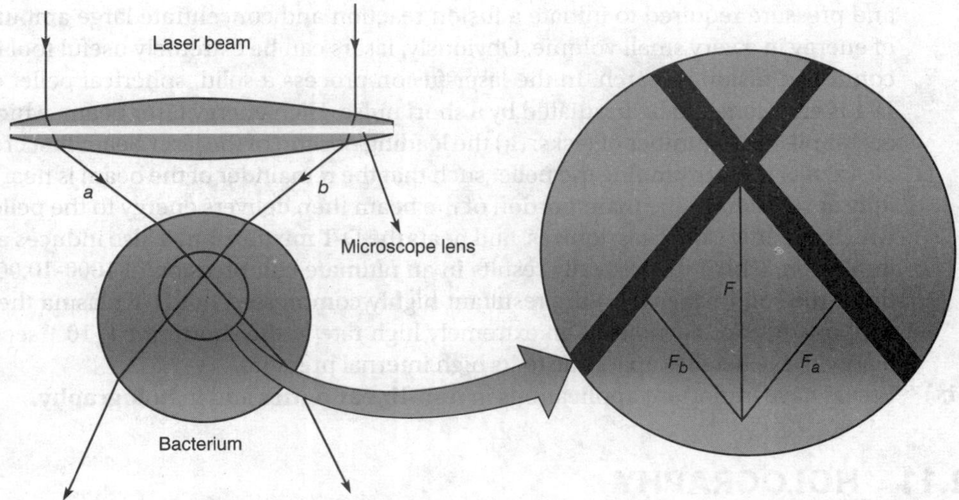

Fig. 22.22 The electric fields in a light beam can act on small objects. When the objects are neutral, the forces are due to dipoles that are inducted in them (by the separation of charges). By focusing laser beams through a lens, these forces act as restoring forces that bring the object to the focal point

Another laser beam operated with short powerful pulses can be used to cut the target. We may note that one of the interesting developments in this area is the study of the forces that curl up long molecules such as DNA.

(vi) Laser in Astronomy: Radio astronomers have found lasers extremely valuable for amplifying very faint radio signals from space. Radio telescopes have now an additional accessory-a laser (ruby), with the help of which the astronomers have been able to extend their range of observation.

(vii) Military Applications of Lasers: Prediction have been made that the laser beam could be used to disintegrate a striking missile is space. A tremendously large amount of energy would have to be produced to create a heat beam that would actually disintegrate its target. Along these same lines, it might be possible to create a truth death-ray gun to be used against humans.

A completely different use of the laser in military application involves a communication in secrecy. Since the laser beam is well collimated and because of this, communication link could be established that would be hard to detect unless one were in direct link between the points of communication.

(viii) Laser-Fusion: Fusion reactions occur when two light nuclei such as Deuterium (D), Tritium (T), or Helium (He) collide and rearrange themselves so as to form two other nuclei of smaller mass with a consequent release of energy. These reactions take place only at very high energy, i.e., high temperature (~10^8 K). At this extreme temperature the gas atoms are fully ionized and fully ionized gas is called **plasma.** For the release of a sufficient amount of thermonuclear energy and to sustain fusion reaction, it is necessary to confine the hot plasma, so that the fusion reaction continues for a long time. The laser has the potential to generate extreme temperatures and pressure required to initiate a fusion reaction and concentrate large amount of energy in a very small volume. Obviously, lasers can be extremely useful tool in controlled fusion research. In the laser-fusion process a solid, spherical pellet of D-T is envisioned to be irradiated by a short pulse. High energy laser beam, which accomplishes a number of tasks: (a) the leading portion of the laser beam first creates a plasma surrounding the pellet such that the remainder of the beam is nearly fully absorbed: (b) the main portion of the beam then delivers energy to the pellet which not only vaporizes, ionizes, and heats the D-T mixture, but it also induces an implosion, which theoretically results in an ultimate compression of 1000-10,000 times the solid density: (c) the resultant highly compressed hot D-T plasma then supports fusion reactions at an extremely high rate in the short time ($\sim 10^{-12}$ secs) before the pellet flies apart due to its high internal pressure.

(ix) Lasers have important applications in **non-linear optics** and in **holography.**

22.11 HOLOGRAPHY

Introduction

Holography is a way of recording and then reconstructing waves invented by Dennis Gabor in 1948. The waves may be of any kind-light, X-ray, corpuscular waves, etc.

The word 'holography' originates from the Greek 'holos' meaning 'the whole'. By using this word, the inventor of holography wanted to stress that it records complete information about a wave-both about its amplitude and its phase.

In conventional photography, only the distribution of the amplitude (more exactly, of its square) is recorded in two dimensional projection of an object onto the plane of the photograph. For this reason, when examining a photography from various directions, we do not obtain new angles of approach, and we cannot see, for instance, what is happening behind the objects in the foreground.

A hologram, on the contrary, regenerates not a two-dimensional, image of an object, but the field of the wave which it scatters. By changing our point of observation within the confines of this wave field, we see the object from different angles, sensing its three dimensional and realistic nature.

It is worthwhile to mention that Gabor did not have a laser when he formulated the idea of holography. He performed his first experiments with mercury arc lamp as source of light. The technique of preparing holograms became a practical proposition only after the advent of lasers. In 1963, E. Leith and J. Upatneiks prepared laser holograms for the first time. This opened the way to major advances in holography.

Principle

The Principle of holography can best be explained in two steps: (i) recording of hologram and (ii) reconstructing the image.

A schematic diagram showing how holograms are recorded is depicted in Fig. (22.23).

(i) **The object whose hologram is to be illuminated with a laser beam:** The scattered light wave impinges on a photographic plate. The reference beam-part of the light from the same laser reflected from a mirror-impinges on the same plate.

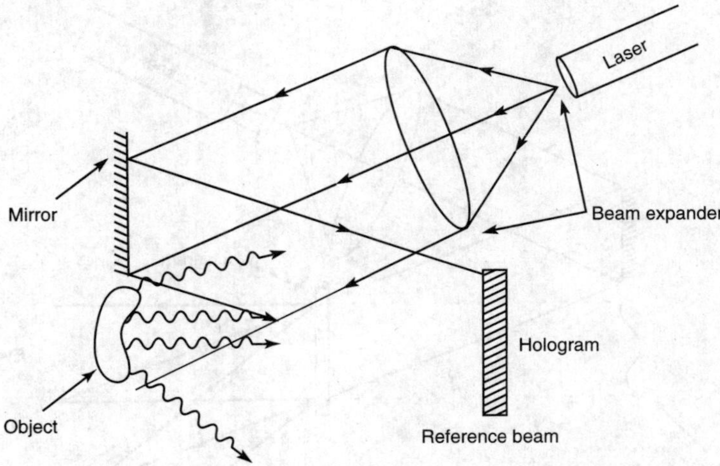

Fig. 22.23 Recording a hologram

The superposition of these two beams produces an interference pattern and this is recorded on that plate. The developed plate is called as the hologram. Since the interference pattern of a hologram is generally very fine, a developed hologram does not differ in appearance from a uniformly exposed plate. A hologram often contains rings and strips. They are due to the diffraction of light on particles of dust getting onto the mirrors and lenses, and have nothing in common with the interference microstructure containing the record of the light wave scattered by the object. The hologram contains, however, enough information to provide a complete reconstruction of the object.

(ii) Reconstruction of the image from a Hologram: Remove the object and put the hologram in the place where it was when formed and then switch on the laser Fig. (22.24). Now look though the hologram like through a window, you will see the object at its previous place, as if it were not removed, i.e., when light from a laser is directed at the hologram from the same direction as the reference beam, part of the light is 'bent' that it appears to come from the place once occupied by the object and the result is remarkably realistic 3-dimensional image. The visible object seems so real that we can detect parallax by changing the position of our head. When looking at the near and far parts of the object, we have to accommodate our eyes differently, and if we want to photograph it, we will have to choose such an aperture, as in ordinary photography, which will ensure an adequate field depth. If this is not done, some parts of the object will be sharp on the photograph, and other ones blurred. From Fig. (22.24) it is clear that two images are produced when the laser beam, which in now known as the readout wave, interacts with the interference pattern on the holographic plate and two images are produced by the diffracted waves. One of them appears at the original position occupied by the object called the virtual image and the other is called the real image which can be photographed directly without using a lens. Since, the real image reverses foreground and background and hence the interest of the observer lies in the virtual image.

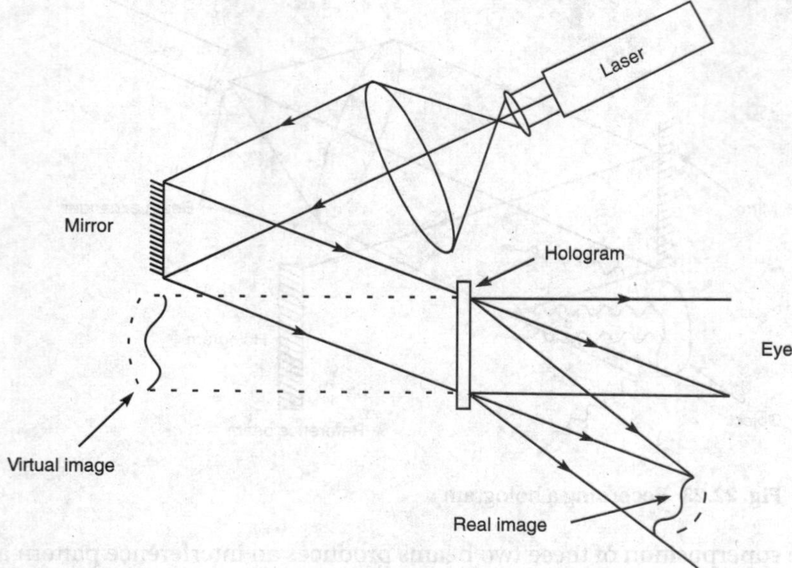

Fig. 22.24 Reconstruction of the image from a hologram

Theory: To understand the various characteristics of holograms it is essential to understand the theory on which recording of a hologram and reconstruction of the image is based.

Let us consider a small object *O*. Let it be illuminated by a source of light, but most of the light falls undistributed on a photographic plate Fig. (22.25). The light diffracted or scattered by the object also reaches the photographic plate *P*.

The direct beam, so called the reference beam, interferes with the light reaching from the object. Let us consider a point on the photographic plate (say *Q*). We are interested in finding the intensity at this point *Q*. One can write the field arriving at *Q* as the sum of the two fields (i) the field due to the reference beam, E_r and (ii) the field scattered from the object, E_s, i.e.,

$$E = E_r + E_s \tag{22.38}$$

It is worth-while to mention that the scattered field E_s is not simple, both amplitude and phase of the field vary greatly with position. The field of such a wave front can be represented by

$$E_s = \frac{A_0}{r_0} \exp[i(Kr_0 - \omega t)] \tag{22.39}$$

Here $r_0 = PQ$ (*Q* is not shown in Fig. (22.25)).

The reflected wave fronts are spherical and concentric around the point origin. One can represent the field E_r by the plane wave as follows

$$E_r = A_r \exp[i(K Z_0 - \omega t)] \tag{22.40}$$

where Z_0 is the distance of the photographic plate from *O*.

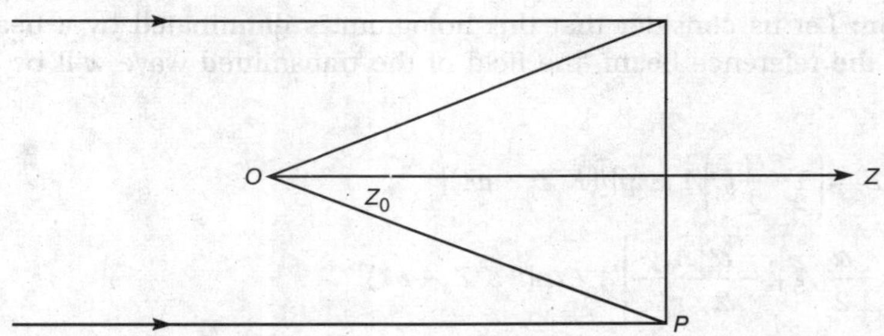

Fig. 22.25 Hologram of a point object O

The intensity at the point Q is given by

$$I = |E_r + E_s|^2$$

Substituting the values of E_s from Eqs.(22.39) and (22.40) respectively, one obtains

$$I = [A_r]^2 + \frac{[A_0]^2}{r_0^2} + \frac{A_0 A_r^*}{r_0}\exp[i\,K(r_o - Z_0)] + \frac{A_0^* A_r}{r_0}\exp[i\,K(Z_0 - r_0)] \tag{22.41}$$

Choosing the constants K and ϕ suitably, one can combine last two terms in Eq. (22.41) and obtain

$$I = [A_r]^2 + \frac{[A_0]^2}{r_0^2} + K\frac{\cos[(K(r_0 - Z_0) + \phi]}{r_0} \tag{22.42}$$

Eq. (22.42) shows that the blackening of the photographic plate depends on the three terms, last term being the most important as it gives the effect of the functions A_0 and ϕ. Obviously, the cosine terms shows that the total intensity I as function of r_0 shows a series of maxima and minima. The interference of the plane wave E_0 with the spherical wave E_s thus produces a set of circular interference fringes on the photographic plate, which when developed forms the hologram.

If one assumes that the response is proportional to the intensity, I, the power transmission of the plate represented by T^2 is given by

$$T^2 = 1 - I\alpha \tag{22.43}$$

where α is a constant. Eq. (22.43) can be approximated as

$$T \cong 1 - \frac{1}{2}I\alpha \tag{22.44}$$

Reconstruction: Let us consider that this hologram is illuminated by a beam which is identical to the reference beam, the field of the transmitted wave will be given the expression

$$E_1 = T\,E_r = \left(1 - \frac{\alpha}{2} I\right) A_r \exp[i(K\,Z_0 - \omega t)]$$

$$= \left[1 - \frac{\alpha}{2}[A_r]^2 - \frac{\alpha}{2}\frac{[A_0]^2}{r_0^2}\right] A_r \exp[i(K\,Z_0 - \omega t)]$$

$$- \frac{\alpha}{2}\frac{A_0 A_r^*}{r_0} \exp[i\,K(r_0 - Z_0)] A_r \exp[i(K\,Z_0 - \omega t)]$$

$$- \frac{\alpha}{2}\frac{A_0^* A_r}{r_0} \exp[i\,K(Z_0 - r_0)] A_r \exp[i(K\,Z_0 - \omega t)] \qquad (22.45)$$

$$= \left[1 - \frac{\alpha}{2}|A_r|^2 - \frac{\alpha}{2}\frac{|A_0|^2}{r_0^2}\right] A_r \exp[i(K\,Z_0 - \omega t)] - \frac{\alpha}{2}\frac{A_0^*[A_r]^2}{2r_0} \exp[i(Kr_0 - \omega t)]$$

$$- \frac{\alpha}{2}\frac{A_0^*|r|^2}{r_0} \exp[i(i\,K\,Z_0 - Kr_0 - \omega t)]$$

One can distinguish the following three terms on the R.H.S. of Eq. (22.45)

(i) The first term represents the attenuated incident wave.

(ii) The second term represents the wave surface identical to the wave surface emitted by the object except for a constant factor. Thus, this wave surface when projected back toward the illuminating source, seems to emanate from an apparent object located from a place where the original object located. We say that wave produce a virtual image, and an observer who looks through the plate would see the object in three dimensions Fig. (22.26). As stated earlier, one can take a photograph of the image by a camera.

(iii) The third term represents a wave surface which is also a replica of the original wave, except that they have conjugate, or reversed curvature, originally diverging spherical waves from an object are converted into converging spherical waves. It converges at point *O* and produces a real image at this point. One can photograph this real image without a lens. Thus a hologram produces a real image *O* and virtual image *O*.

Gas lasers have high coherence and continuous wave mode and hence they are often used for holography. Moreover, the emitting power of gas lasers is low and hence the time of exposure is large. Therefore, using gas lasers moving objects cannot be holographed. To prepare the holograph of moving objects and to record the development of a process in time, pulsed solid state lasers are often used. With the use of pulsed solid state lasers one can cut down the exposure time to about 10^{-3} sec.

Some important properties of a hologram

(i) The hologram acts somewhat like a simple periodic diffraction grating. It produces a beam direct and two first-order diffracted beams on either side of the

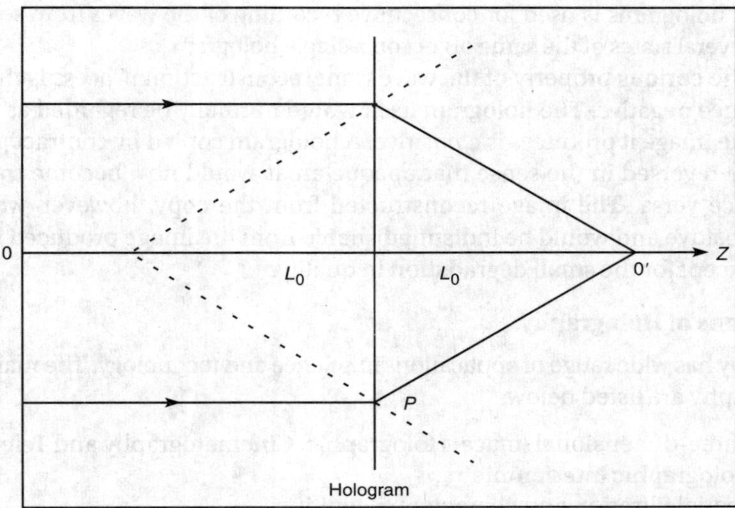

Fig. 22.26 Reconstruction of the image from a hologram

direct beam. The zero order of the diffracted light is the direct beam. while the two first orders on either side comprise the virtual and the real images.

(ii) The virtual image produced by a hologram appears in complete three-dimensional form. Each section of a hologram is capable of reconstructing an image of the entire object. As the pieces become smaller, the quality of the reconstructed image become poorer, small details will vanish, and a characteristic speckled structure will appear.

(iii) In holography the viewer always sees a positive image whether a positive or a negative photographic transparency is used for the hologram. The reason for this is that a negative hologram merely produces a wave field that is shifted 180° in phase w.r.t. that of a positive hologram. Since the eye is insensitive to this phase difference, the view seen by the observer is identical in the two cases.

(iv) A hologram is a reliable method of data storage. The information about a point object is recorded over the whole area of hologram and each such point of a real object is recorded on the whole hologram and not in one of its points as is the case in photographic plate.

(v) A hologram of a complex object can be considered as an interference (coherent) superposition of holograms from individual points or more complex portions of the object. The amplitudes of light waves are summated with account of the phase relations between them in such superposition.

We can also imagine a hologram that is an incoherent superposition of holograms of different objects or parts of the same object. In this case, the photolayer summates the illumination they create. If the number of such consecutive superpositions is not too great, then the hologram will simultaneously reconstruct several consecutively recorded light waves without any appreciable distortions. This property of holograms is used for consecutive recording of the waves from several object or several states of the same object on a single hologram.

(vi) The curious property of the wave front reconstruction process is that it does produce negative. The hologram itself would normally be regarded as a negative but the image it produces is a positive. A hologram copied by contract printing would be reversed in the sense that opaque areas would now become transparent and vice versa. The image reconstructed from the copy, however, would remain a positive and would be indistinguishable from the image produced by the original except for the small degradation in quality.

Applications of Holography

Holography has wide range of applications in science and technology. The main applications of holography are listed below:

(i) Three-dimensional image: Holographic, Cinematography and Television.
(ii) Holographic interferometry.
(iii) Spatial filtration and character recognition.
(iv) Spectroscopy.
(v) In production processes and optical engineering.

The conventional holography needs coherent light. This is a drawback. Recent advances in techniques, e.g. *sandwich holography* offer a new method of storing confidential data. The *rainbow* (white light transmission) holography is a significant achievement because it uses white light.

So far we have described a *transmission hologram,* i.e., one in which the reference beam is on the same side of the film as the object and the illuminating beam is on the opposite side. However, one can make a *reflecting hologram* by reversing the positions of the reference and illuminating beams, and if an overlapping series of holograms is made using several lasers of different colours, the result will be a *white light hologram,* in which the different colours contained in white light provide the colours observed in the image. This is what we commonly sees in magazine pictures and credit cards, because the casual observer donot have a laser handy for viewing a monochromatic hologram. With their complex interference patterns, it is difficult to counterfeit holograms, and are therefore popular security tools on instruments such as credit cards and checks.

Example 22.1: A laser beam has a power of 50 mW. It has an aperature of 5×10^{-3} m and it emits light of wavelength 7200 Å. The beam is focussed with a lens of focal length 0.1 m. Show that the areal spread and the intensity of the image are 2.074×10^{-10} m² and 2.4×10^{8} W/m² respectively.

Solution

$$d\theta = \frac{\lambda}{d} \qquad \lambda = 7200 \times 10^{-10}\text{ m}$$

or $$d\theta = \frac{7200 \times 10^{-10}}{5 \times 10^{-3}} = 1.44 \times 10^{-4}\text{ radian} \qquad d = 5 \times 10^{-3}\text{ m}$$

(i) $\quad$ Areal *l* spread $= (d\theta f)^2$

$$\text{Here } f = 0.1\text{m}$$

$$\therefore \quad \text{Areal spread} = (1.44\times10^{-4}\times0.1)^2 = 2.074\times10^{-10}\ \text{m}^2$$

(ii)
$$\text{Intensity} = \frac{\text{Power}}{\text{Area}} = \frac{\text{Power}}{\text{Area spread}} = \frac{50\times10^{-3}\ \text{W}}{2.074\times10^{-10}\ \text{m}^2}$$

$$= 2.411\times10^{8}\ \text{W/m}^2$$

Example 22.2: **A laser beam has a wavelength of 8 × 10⁻⁷ m and aperature 5 × 10⁻³ m. The laser beam is sent to moon. The distance of the moon is 4 × 10⁵ km. from the earth's surface. Determine (i) the angular spread of the beam and (ii) the axial spread when it reaches the moon's surface.**

Solution We have $d\theta = \lambda/d$

$$\text{Angular spread } d\theta = \frac{8\times10^{-7}}{5\times10^{-3}} = 1.6\times10^{-4}\ \text{radian} \qquad \lambda = 8\times10^{-7}\ \text{m}$$

$$\text{Areal spread} = (Dd\theta)^2 \text{ where } D = 4\times10^5\ \text{km} = 4\times10^8\ \text{m} \qquad d = 5\times10^{-3}\ \text{m}$$

$$= (4\times10^8\times1.6\times10^{-4})^2 = 4.096\times10^9\ \text{m}^2$$

Example 22.3: **A laser beam of wavelength 6000 Å on earth is focussed by a lens (or mirror) of dimater 2 m on to a crater on the moon's surface. The distance of the moon from earth is 4 × 10⁸ m. Neglecting the effect of earth's atmosphere, show that the angular spread of the spot on the moon will be 3 × 10⁻⁷ radian.**

Solution (i) Angular spread

$$d\theta = \frac{\lambda}{d} = \frac{6\times10^{-7}}{2} = 3\times10^{-7}\ \text{radian} \qquad |\ \lambda = 600\times10^{-10}\ \text{m}$$

$$d = 2\text{m}$$

$$D = 4\times10^8\ \text{m}$$

Example 22.4: **A laser beam has aperature d = 1.8 × 10⁻² m and emits radiation of wavelength λ = 5 × 10⁻⁷ m. Calculate**

(i) the semi angle of the cone of its beam (ii) the solid angle of this cone.

Solution The semi-angle of the conc. of the laser beam

$$\theta = \frac{\lambda}{d} = \frac{5\times10^{-7}\ \text{m}}{1.8\times10^{-2}\ \text{m}} = 2.72\times10^{5}\ \text{radian}$$

(ii) Solid angle of this cone

$$\phi = \pi\frac{D^2\theta^2}{D^2} = \pi\theta^2 = \frac{22}{7}\times(2.72\times10^{-5})^2 = 2.5\times10^{-9}\ \text{started}$$

Example 22.5: The coherence length for sodium light is 2.945×10^{-2} m. The wavelength of sodium light is 5890 Å. Determine (i) the number of oscillations corresponding to coherence length and (ii) the coherence time.

Solution (i) Number of oscillations in coherence length L.

$$n = \frac{L}{\lambda} = \frac{2.945 \times 10^{-2}}{5890 \times 10^{-10}} = 5 \times 10^4 \qquad \lambda = 5890 \times 10^{-10}\ \text{m}$$

$$c = 3 \times 10^{10}\ \text{m/s}$$

$$L = 2.945 \times 10^{-2}\ \text{m}$$

(ii) Coherence time $= L/c = \dfrac{2.945 \times 10^{-2}}{3 \times 10^8} = 9.8 \times 10^{-11}$ s

Example 22.6: The coherence length for sodium D_1 line is 2.5 cm. Determine (i) the spectral width of the line $\Delta\lambda$, (ii) the purity factor, Q and (iii) the coherence time τ. The wavelength of light is 6000 Å.

Solution (i) $\Delta\lambda = \dfrac{\lambda}{2L} = \dfrac{36 \times 10^{-10}}{5}\ \text{cm} = 7A$

(ii) $Q = \dfrac{\lambda}{\Delta\lambda} = \dfrac{6 \times 10^{-5}}{7 \times 10^{-8}} = 10^3$

(iii) $\tau = \dfrac{L}{C} = \dfrac{2.5}{3 \times 10^{-10}} \cong 0.8 \times 10^{-10}$ s

Example 22.7: Consider a He-Ne gas laser for which the laser radiation wavelength is 633 nm. (a) Calculate the n-value for a standing wave in a cylindrical cavity 80 cm long (b) given that the spread in frequencies, Δv, of neon atoms at room temperature is 1.7×10^9 Hz. What is the range of n-values covered by Δv? [Kerala]

Solution

(a) We have $n = \dfrac{2L}{\lambda_n} = \dfrac{1.6\text{m}}{633 \times 10^{-9}\ \text{m}} = 2.53 \times 10^6$

(b) $nl_n = nc/v_n = 2L =$ a constant, one finds that $n = \text{const} \times f_n$. This is turn implies that $(\Delta n)/n = (\Delta v_n)/v_n$. Thus, we have

$$\Delta n = \frac{n}{v_n}\Delta v_n = \frac{2L}{c}\Delta v_n = \frac{1.6\text{m}}{3 \times 10^8\ \text{m/s}}(1.7 \times 10^9\ \text{Hz}) = 9$$

Example 22.8: Consider a circular gyro laser of radius 1m that is to be used to detect a rotational velocity of 1 degree/day. What resolution of beat frequency is needed to achieve this level of detection if the wavelength of light used is 440 nm? [Mysore]

Solution Angular velocity, ω = 1 deg/day × (2π/360 rad/day) × [1/(24 × 3,600)day/s] we have $\Delta v = 2 \times (2 \times 10^{-7}$ rad/s) (1 m)/(440 × 10^{-9} m) = 0.9Hz.
we may not that this frequency is easily measurable.

Example 22.9: Calculate the gain constant (β) of a laser using following data: $N_2 - N_1 = 5 \times 10^{22}$ m^{-3} λ = 6500 Å, $\Delta\lambda$ (line width) = 15Å and the, life period for the spontaneous emission = 2 × 10^{-4}s. [Kerala]

Solution

$$\beta = \frac{\lambda^2}{8\pi\Delta v}(N_2 - N_1)A_{12}$$

$$\because \quad v = \frac{c}{\lambda} \quad \therefore \ \Delta v = \frac{c}{\lambda^2}\Delta\lambda$$

$$\text{or} \therefore \quad \Delta v = \frac{3\times10^8}{(65\times10^{-8})^2}\times15\times10^{-10}$$

$$\therefore \quad \beta = \frac{(65\times10^{-8})^2\times5\times10^{22}\times0.5\times10^4\times(65\times10^{-8})^2}{8\times3.14\times3\times10^8\times15\times10^{-10}} = 1.5\times10^4 s$$

$$= 3.94$$

$$\lambda = 6500\times10^{-10}\text{ m}$$

$$N_2 - N_1 = 5\times10^{22}\text{ m}^{-3}$$

$$A_{12} = 1/\text{life period}$$

$$= 1/(2\times10^{-4})$$

$$= 1.5\times10^4\text{ s}$$

Example 22.10: A ruby rod contains typically a total of 3 × 10^9 Cr^{3+} ions. It lases at 694.3 nm. Determine the (i) energy of one photon and (ii) total energy available per pulse assuming tahat there is total population inversion.

Solution

(i) Energy of Photon

$$E = \frac{1240}{\lambda(\text{nm})}\text{eV}$$

$$= \frac{1240}{694.3} = 1.79\text{eV}$$

(ii) Energy available per pulse

= (energy of one photon) (total no. of photons)
= (energy of one photon) (total no. of ions in the excited state)
= 1.79 × 3 × 1019 = 5.37 × 1019 eV = 8.6 J

SUMMARY

1. A *monochromatic source* contains only one wavelength (or colour). Theoretically and practically, one cannot produce single wavelength laser. When $\Delta\lambda$ is the wavelength range of main laser wavelength of λ, then Δv is the corresponding frequency range. The minimum frequency range for laser is 1 Hz.

2. The laser has a high degree of *directionality*; i.e. laser travels large distance without deviation. The reflecting surfaces or mirrors used at the ends of the active medium acts as collimating aperatures. The laser radiation undergoes multiple reflections with the region between the reflectors. The laser radiation travels long distance prior to coming out of the system, therefore curvature of wave front is negligible and radiation is a plane wave front.
3. *Coherence time* of laser is very short, $\Delta\tau = 10^{-14}$ s. ($\Delta\tau = 1/\Delta\nu$, sunlight has a bandwidth $\Delta\nu = 10^{14}$ Hz). In laser, there exists a finite bandwidth $\Delta\nu$, different frequencies (mean waves) get out of phase. When two waves differ in frequency by $\Delta\nu$ then in time $1/\Delta\nu$, they get out of phase by a full cycle. After this much time, the laser beam loses coherence.
4. The distance $\Delta L = c\Delta\tau$ is called *coherence* length of laser beam, where c is velocity of light. We may note that only portions of the same beam, separated by a distance less than coherence length, interfere constructively to produce, steady interference pattern.
5. The power flow per unit area, per unit bandwidth and per unit solid angle is called the *spectral brightness* of the source, i.e.,

$$B = \frac{\rho}{A\,\Delta\nu\,\Delta\Omega}$$

where A is surface area (source size), $\Delta\nu$ is bandwidth and $\Delta\Omega$ is the solid angle (beam divergence). B for sun is 10^{-12} W/cm². Hz. sr⁴, for He-Ne laser = 25 W/cm². Hz. sr⁴ and for mode-locked neodymium-glass laser of 10^4 MW = 10^8 W/cm². Hz. sr. However, tera-watt lasers have still much larger spectral brightness.
6. The basic mechanism in the operation of a laser is *stimulated emission*. A photon incident upon a molecule in an excited state can cause a stimulated emission.
7. Three-level and four-level systems can be used to control the output and efficiency of a laser.
8. The monochromatic, intense laser beam has a number of applications in research and industry, including biomedical uses and holography.
9. *Holography* is a process of recording the intensity and phase information of light reflected from an object to produce its three dimensional image.
10. The hologram is used to reconstruct the wave fronts with the same laser. The hologram acts as a diffraction gratting and produces three dimensional images of the object.

REVIEW QUESTIONS

1. Explain the terms absorption, spontaneous emission and stimulated emission of radiation. Obtain a relation between transition probabilities of spontaneous and stimulated emission.
2. What is a laser? Explain the salient features of a laser radiation. Mention some of its principal uses.
3. What are the important features of stimulated emission? Discuss the essential requirements for producing laser action.

4. What is difference between temporal and spatial coherence? Explain coherence time and coherence length.
5. Give an elementary theory of the laser.
6. Describe the construction and working of a ruby laser with necessary diagrams.
7. Describe a He-Ne gas laser. How is population inversion achieved in this type of laser?
8. What are Einstein's A and B coefficients. Obtain a relationship between them.
9. Explain the role of an optical resonator in a laser. How population inversion is achieved in a laser?
10. Explain, what do you understand by the modes of a cavity resonator. Explain quality factor.
11. Explain the principle and working of a semiconductor laser. Derive the condition for the predominance of the stimulated emission over absorption for this laser device.
12. Explain the principle of holography and discuss its salient features. Mention its uses.
13. Explain the basic principle of holography. What are salient features of holography. Mention some important uses of holography.
14. Write short notes on the following:

(a) Ruby laser (b) Gas laser
(c) Semiconductor laser (d) Einstein's A and B coefficients
(e) Optical resonator (f) Holography
(g) Quantum-well laser

PROBLEMS

1. Determine the minimum difference between the two arms of a Michelson interferometer that seeks to measure two closely spaced wavelengths using He-Ne laser for which the coherence length is 600 km. **[Ans.** 300 km.]

[**Hint.** $2d = n_1\lambda_1 = n_2\lambda_2$ or $2d\left(\frac{1}{\lambda_2} - \frac{1}{\lambda_1}\right) = n_2 - n_1 = m$ where n_1, n_2 and m are integers. Putting $\Delta\lambda = \lambda_1 - \lambda_2$, $m = \frac{2d\Delta\lambda}{\lambda^2} = \frac{2d\Delta\nu}{\nu\lambda} = \frac{2d}{L_c}$, where L_c is coherence length. Obviously, the minimum difference between the two atoms of the interferometer is $d = \frac{L_c}{2} = \frac{600}{2} = 300\,\text{km}$]

2. The output power of a gas laser is 1mW and the emitted wavelength is 630 nm. Show that the number of photons emitted per second is about 3×10^{15} per second.
3. The coherence length of Sodium D_2 line is 2.5 cm. If the velocity of light is 3×10^8 m/s, then show that the coherence time is 0.8×10^{-10} s.
4. Calculate the spot size of a He-Ne laser beam at the centre of confocal cavity if the length of cavity is 1.24m. The wavelength is 633 nm. **[Ans.** 0.5 mm]
5. Calculate the inversion density for He-Ne laser operating at 633 nm if the temperature of the discharge is 150°C and the gain constant is 3 percent per metre. The life time of the upper state for spontaneous transition is 10^{-7} s.

6. The bandwidth of a He-Ne laser is 500 Hz. Obtain the coherence time and the longitudinal coherence length. **[Ans.** (i) 2 ms. (ii) 600 km]
7. Calculate the gain constant β of a laser having the following parameters: wavelength = 650 nm, inversion density $(n_2 - n_1) = 5 \times 10^{22}/\text{m}^3$, life time for spontaneous emission 2×10^{-4}s and line width $\Delta\lambda = 15$ Å. **[Ans.** 3.95]
8. Show that for a normal optical source with temperature about 10^3K and wavelength 6000 Å, the emission is predominantly due to the spontaneous transitions.
9. A laser beam of wave length 6000 Å, power 10 mW, and angular spread 5×10^{-5} rad is focussed by a lens of focal length 10 cm. Calculate the radius and power density of the image. What is lateral coherence width?
   [Ans. (i) radius = 2.5×10^{-4} cm (ii) Power density = 5.1×10^4 W/cm^2 (iii) lateral coherence width = 1.2 cm.]
10. For sodium line, the wave lendth is 5890 Å and coherence time is 10^{-10} s. Show that the monochromaticity of the source is (5890 ± 0.0578) Å.

SHORT ANSWER-QUESTIONS

1. Write salient features of laser radiation.
2. What is the relation between the degree of nonmonochromaticity (ζ) and longitudinal coherence length (L_c) or simply coherence length?
   Ans: $\zeta = c/L_c V_a$
3. What is the relation between the coherence time (t) and frequency spread of the wave ($\Delta\nu$)?
   Ans: $t = 1/\Delta\nu$
4. In terms of quantum picture, when a light beam is said to be a perfectly coherent?
   Ans: When all the constituent photons of a light beam have the same energy, the same direction of momentum and identical polarization, the light beam is said to be perfectly coherent.
5. What is the relation between Einstein's A_{21} and B_{21} coefficient?
   Ans: $A_{21}/B_{21} = 8\pi h\nu^3/c^3$
6. What do you understand by a metastable state?
   Ans: Normally, the mean life of excited atoms before the spontaneous emission occurs is about 10^{-8} s. However, for some excited states, this mean life is perhaps as much as 10^5 times. We call such long-lived states metastable. They occupy an important role in action.
7. Mention any three methods of pumping for creating population inversion.
   Ans: (i) Photon excitation (ii) Electron excitation and (iii) Inelastic collision between atoms.

OBJECTIVE QUESTIONS

1. Laser light is considered to be coherent because it consists of
   (a) Many wavelengths (b) Uncoordinated wavelengths
   (c) coordinated wavelengths (d) divergent waves

2. The relation between Einstein's A_{21} and B_{21} coefficients is

(a) $\frac{A_{21}}{B_{21}} = 1:2$ (b) $\frac{A_{21}}{B_{21}} = \frac{8\pi h\nu^3}{c^3}$

(c) $\frac{A_{21}}{B_{21}} = 1:1$ (d) $\frac{A_{21}}{B_{21}} = \frac{\nu}{c}$

3. The ratio of the number of spontaneous to stimulated transitions is

(a) $\frac{A_{21}}{B_{21}u\nu} = 1:1$ (b) $\frac{A_{21}}{B_{21}u\nu} = 1:2$

(c) $\frac{A_{21}}{B_{21}u\nu} = \exp\left(\frac{h\nu}{k_B T}\right) - 1$ (d) $\frac{A_{21}}{B_{21}u\nu} = 8\pi\frac{h\nu^3}{c^3}$

4. When the atoms in the source go from an excited state (E_2) to a lower energy state (E_1), spontaneous emission of light takes place, the energy of the emitted photon being

(a) $h\nu_{12} = E_2 - E_1$ (b) $h\nu_{12} = E_2/E_1$

(c) $h\nu_{12} = E_1/E_2$ (d) $h\nu_{12} = 0$

5. The order of the pumping power necessary to achieve the population inversion in a ruby is

(a) 10^7 W/m^2 (b) 10^2 W/m^2

(c) 10^{10} W/m^2 (d) 10^{16} W/m^2

MARK WHICH OF THE FOLLOWING STATEMENTS AS TRUE OR FALSE

1. A LASER light is highly monochromatic and directional.
2. The concept of temporal coherence is intimately related with monochromaticity.
3. A spatially coherence beam is markedly directional.
4. In a laser, the atoms act together to produce light with several special characterstics.
5. Laser light is emitted when atoms make a transition from one quantum state to a quantum state of higher energy.
6. The two B coefficients of Einstein are not equal.
7. A laser can operate without population inversion.
8. The selectivity of photon states and the positive feedback necessary to off set the losses in a laser, can be acquired by enclosing the radiation in a resonator.

FILL IN THE BLANKS

1. A laser emit ______ waves.
2. The two B coefficients of Einstein are ______.

3. To active a markedly coherent radiation, a laser needs an optical ______.
4. Normally, the mean excited atoms before spontaneous emission occurs is about ______.
5. In the helium-neon laser, action occurs between two excited states of the ______ atoms.
6. Semiconductor lasers of very compact design consists of a forward bases ______ junction in a single crystal of a suitable impurity semiconductor.

ANSWERS

Section D	1 (c), 2 (b), 3 (c), 4 (a), 5 (a)
Section E	1 (True), 2 (True), 3 (True), 4 (False), 5 (False), 8 (True)
Section F	1. Coherent, 2. equal, 3. resonator, 4. 10^8 s, 5. neon, 6. PN

23

Optical Fibres

23.1 INTRODUCTION

Optical fibre is a wave guide through which light can be transmitted with very little leakage through the side walls. Optical fibres transport light signals from place to place as metallic conductors transport electrical signals. Optical fibres can guide light around bends and can carry light for long distances with very little attenuation (loss of light power). Telecommunications by modulating a beam of light (LED or Laser as a light source) and using optical waveguides has the inherent advantages of very large band width, small dimensions, low losses band width, and insensitivity to electrical interference with the discovery of optical fibers, new era of *photonics* has been ushered in.

An optical fibre is a cylindrical wave guide system through which the optical wave can propagate. The principle by which light wave travel through the fibre is **'total internal reflection'** without loss of incident intensity.

An important structure used in optical system is the layered structure or the **wave guide structure**. As the name implies, these structure are used to confine the optical waves in a well defined region and guide their propagation. One can make the layered structure from non-crystalline materials or from crystalline materials. For example glass is used to produce optical fibres used in optical communication whereas semiconductor wave guides are used in semiconductors laser.

An optical fibre is a very thin and flexible medium of cylindrical shape. It has three principal sections: **(i) the core (ii) the cladding and (iii) the jacket**. The inner most section of the fibre is referred as the cladding. We must note that the optical properties of the cladding are different from those of the core. The outermost section of the fibres is known as the jacket. Jacket is made of plastic or polymer and other materials and this protects the fibre structure from moisture, abrasion, mechanical shocks, and other environmental hazards.

The core forms the actual working structure of the optical fibre. The radiation (say light) entering the core, suffering a number of familiar phenomenon of total internal reflection at the core-cladding interface. Obviously, the interface between core and cladding acts as a mirror at which total internal reflection of the transmitted light takes place.

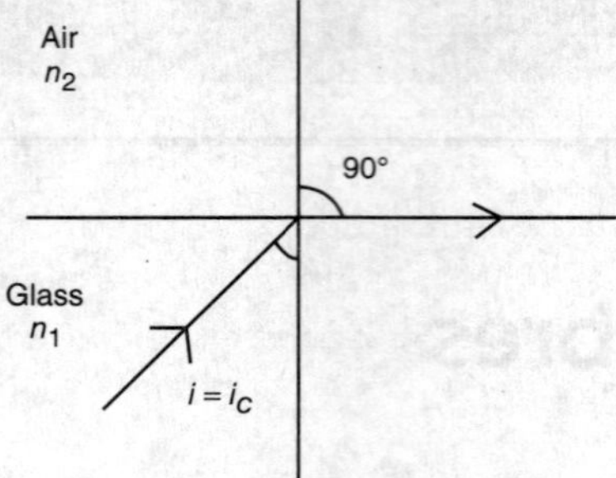

Fig. 23.1 Total internal reflection

Apart from the use as communication channel, optical fibres are widely used in many other areas, e.g., sensors for detecting electrical, mechanical and thermal energies, inexpensive and simple display systems, copying machines etc., Fibroscopes finds use in a variety of forms in medical diagnosia.

23.2 TOTAL INTERNAL REFLECTION

When a ray of light is incident at an angle i in denser medium (refractive index n_1), the refracted rays will bend away from the normal in rarer medium (refractive index n_2) and we have

$$\text{Snell's law,} \quad \frac{\sin i}{\sin r} = \frac{n_2}{n_1}$$

If the angle of incidence say i_c when the angle of refraction $r = 90°$, then

$$\sin i_c = \frac{n_2}{n_1} (\sin r = \sin 90° = 1) \tag{23.1}$$

The angle i_c is known as **critical angle.** The value of the critical angle (i_c) depends upon the refractive index of the denser medium (n_2). When the light in denser medium is incident at an angle i_c ($i > i_c$), then light will be reflected back in to the denser medium. This is known as **total internal reflection** and this phenomenon is used in optical fibre communication [Fig. (23.1)].

23.3 THE OPTICAL FIBRE

Fig. [23.2(a)] shows a glass fibre consisting of a (cylindrical) central core cladded by a material of slightly lower refractive index. One can represent the corresponding refractive index distribution (in the transverse direction given by)

$$\begin{aligned} n(r) &= n_1 \quad 0 < r < a \quad && \text{Core} \\ &= n_2 \quad r < a \quad && \text{Cladding} \end{aligned}$$

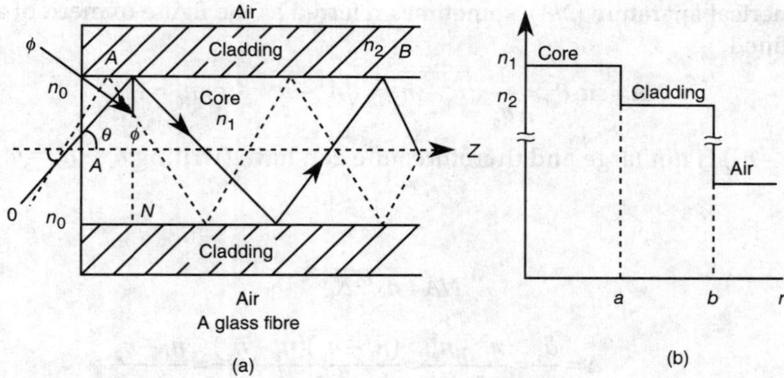

Fig. 23.2 (a) Refraction of a ray of light into core (b) Distribution of refractive index of a cladded optical fibre which consists of a cylindrical glass structure surrounded by a material of slightly lower refractive index

and has been shown in Fig. [23.2(b)]. Obviously n_1 is the refractive index of the core of a fibre, and n_2 that of the cladding ($n_1 > n_2$). A ray of light *OA*, incident on one end-face of the core, is refracted into the core along *AB*. If n_0 is the refractive index of air, we have from Snell's law

$$\frac{\sin i}{\sin \theta} = \frac{n_1}{n_0} \tag{23.2}$$

The ray *AB* is incident on the core-cladding interface at angle $\phi = (90 - \theta)$. We can see that as *i* decreases, θ also decreases so that ϕ increases. When this ray has to suffer total internal reflaction at the core-cladding interface

$$\sin\phi (= \cos\theta) > n_2 / n_1 \tag{23.3}$$

Thus

$$\sin\theta < \left[1 - \left(\frac{n_2}{n_1}\right)^2\right]^{\frac{1}{2}} \tag{23.4}$$

and one obtains

$$\sin i < \frac{n_1}{n_0}\left[1 - \left(\frac{n_2}{n_1}\right)^2\right]^{\frac{1}{2}} = [(n_1^2 - n_2^2)/n_0^2]^{1/2}$$

If $\left(n_1^2 - n_2^2\right) \geq n_0^2$ then for all values of *i*. Total internal reflaction will occur. Assuming $n_0 = 1$, one obtains the maximum value of sin *i* for a ray to be guided is

$$\begin{aligned} \sin i_m &= (n_1^2 - n_2^2)^{1/2} \quad \text{when} \quad n_1^2 < n_2^2 + 1 \\ &= 1 \quad \text{when} \quad n_1^2 > n_2^2 + 1 \end{aligned} \tag{23.5}$$

obviously, if a core of light is incident on one end of the fibre, it will be guided through it provided the semi angle of the core is less than i_m. This angle is a measure of the light gathering power of the fibre. The core of the semi vertical angle θ_m is termed as the acceptance core.

The numerical aperature (NA), sometimes referred as the figure of merit of an optical fibre, is defined

$$NA = \sin\theta_m = \frac{1}{n_0}\sqrt{n_1^2 - n_2^2} = \sqrt{n_1^2 - n_1^2} \quad (\because\ n_0 = 1) \tag{23.6}$$

Usually $(n_1 - n_2)$ is not large and therefore, one can have (writing $n_1 - n_2 = n_1\left(1 - \frac{n_2}{n_1}\right)$: $= n_1\sqrt{2\Delta}$).

$$NA = n\sqrt{2\Delta} \tag{23.7}$$

where
$$\Delta = \frac{\delta_n}{n_1} = \frac{n_1^2 - n_2^2}{2n_1^2} = \frac{(n_1 - n_2)(n_1 + n_2)}{2n_1^2} = \frac{n_1 - n_2}{n_1} \tag{23.8}$$

The quantity Δ is usually referred as the *fractional difference of the refractive indices of the core and the cladding*. Δ is also called *relative refractive index difference* when $n_1 = n_2$. For $n_1 = 1.48$ and $n_2 = 1.46$, one obtains $NA \approx 0.24$. this gives $i_m \approx 14°$. Obviously, the acceptance angle i_m is less than 30°, so that $NA < 0.5$. We can easily see that if NA is too small, it is difficult to launch power into the fibre.

To prevent energy losses via absorption and scattering, the cladding should be at least a few wavelengths thick. One can see that a typical value for n_2/n_1 is 0.99. For this value, one finds the critical angle is 0.142 rads or 8.11°. Thus rays travelling at an angle less than 8.11° relative to the reference axis will be totally internally reflected and guided by the fibre. Higher-angle rays will enter the cladding and be lost due to high levels of scattering and absorption.

We may note that the directric cladding on glass core reduces scattering loss, protects core from absorbing external optical disturbances and provides mechanical strength to main core glass fibre. Sometime there is buffer coating over cladding which adds further strength to main fibre and protects fibre from mechanical vibrations and impact.

23.4 THE COHERENT-BUNDLE

When a large number of optical fibres are put together then they from a bundle. When the optical fibres are not aligned, i.e., they are all jumbled up, then the bundle is said to be an incoherent bundle. However, when the optical fibres are aligned properly, i.e., if the relative position of the fibres in the input and output ends are the same, the bundle so formed is called a coherent bundle. When a fibre is illuminated at one of its ends, then there will be a bright spot at the other end of the fibre. Obviously, a coherent bundle will transmit the image from one end to another [Fig. (23.3)]. When the bundle is incoherent, then the output image will be scrambled. This property of the incoherence bundle can be used as a coder and transmitted image can be decoded by using a similar bundle at the output end. We must note that in a fibre bundle, since there can be hundred of thousands of fibres and hence decoding without the original bundle configuration is extremely difficult.

The important application of a coherent bundle is perhaps in a fibre optic endoscope where one can put it inside a human body and the interior of the body can be viewed from outside. The portion of the body can be viewed from outside. The portion of the body to be

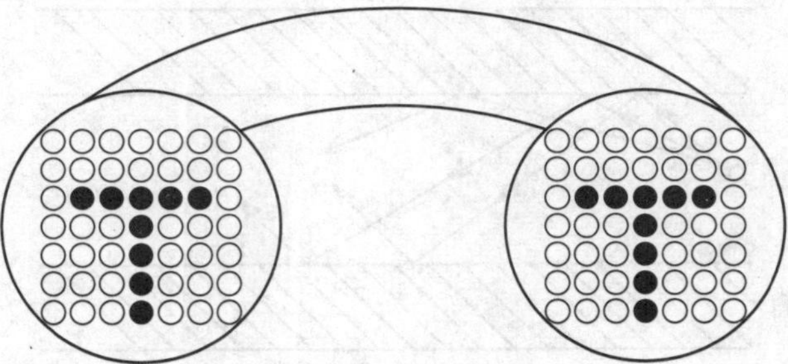

Fig. 23.3 A coherent, i.e., aligned bundle of fibres

seen is illuminated and for this purpose the bundle is enclosed in a sheath of fibres which can carry light from outside to the interior of the body.

23.5 CLASSIFICATION OF OPTICAL FIBRES

Usually the optical fibres are grouped under three categories, viz:

(i) Stepped index multimode fibre.
(ii) Stepped index monomode fibre.
(iii) Graded index multimode fibre.

(i) Stepped-index multimode fibre

Stepped-index is the simplest type of an optical fibre consists of a thin cylindrical structure of transparent glassy material of uniform refractive index n_1 surrounded by a cladding of another material of uniform but slightly lower refractive index n_2 [Fig. (23.3)]. These fibres are referred to as step index fibres due to the step discontinuity of the index profile at the core-cladding interface. The light rays incident on the end-face of the core at angles of incidence within the acceptance angle are guided along the core in different paths. It is observed that light rays of different colours of frequeencies are also separated as they travel inside the core. This means multimode propagation is possible in the fibre as shown in Fig. (23.4).

From Fig. (23.4), one can see that the time taken by the different light rays propagating along paths of different lengths upto the destination are different. This means the variations of the successive pulses of light may overlap, thus distorting the information being carried. Such distortions are usually called as the *transit-time dispersion*.

One can show that the time taken by a ray is function of the angle θ made by the ray with the *Z*-axis which leads to pulse dispersion and given by Fig. (23.3)

$$\tau = \frac{n_1 L}{c\cos\theta} \tag{23.9}$$

where $L = AB$. If we assume that all rays lying between θ and θ_c (critical angle) are present then the time taken by the rays corresponding to $\theta = \theta_c \approx \cos^{-1}(n_2/n_1)$ would be given by

$$\tau_{\min} = \frac{n_1 L}{c} \quad \text{and} \quad \tau_{\max} = \frac{n_1^2 L}{n_2 c} \tag{23.10}$$

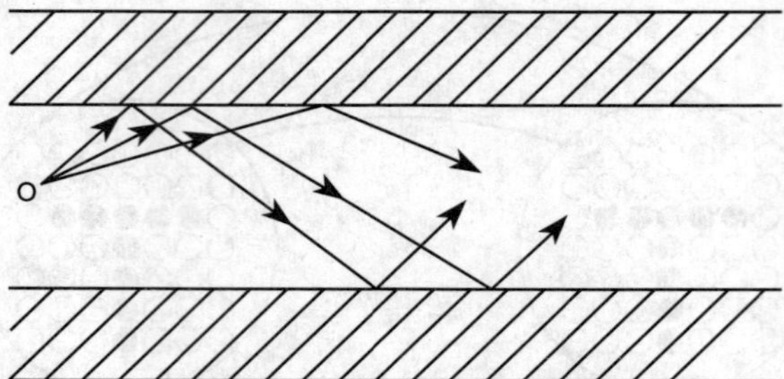

Fig. 23.4 Multimode propagation in a fibre

Thus, if all the input rays were excited simultaneously, the rays would occupy a time interval at the output end of a duration

$$\Delta\tau = \tau_{max} - \tau_{min} = \frac{n_1 L}{c}\left(\frac{n_1}{n_2} - 1\right) \tag{23.11}$$

For a typical fibre, one can assume $n_2 = 1.46$ (pure silica), $\Delta = \dfrac{n_1 - n_2}{n_1} \Rightarrow n = n_1 = 1.4746$

one obtains $\Delta\tau \cong 50 \times 10^{-9}$ s/km

$\cong 50$ ns/km

This shows that an impulse after traversing through the fibre of length 1 km broadens to a pulse dispersion. For this purpose two alternative solutions exist one involving the use of graded index fibres and the other involving single mode fibres.

(ii) Stepped Index (SI) Monomode Fibre

One can make core very thin, i.e., the core diameter is almost of the same order as the wave-lengh for the light to be propagated. This helps in minimizing the transit-time dispersion problems. Fig. (23.5) shows the propagation in a stepped index monomode optical fibre. Step index monomode fibre has low attenuation, small numerical aperature and a wide band widths.

An optical fibre is characterized by one more important *parameter*, known as *V-number*. *V*-number is generally called *normalized* frequency of the fibre. It is given by the relation

$$V = \frac{2\pi}{\lambda} a\sqrt{n_1^2 - N_2^2} \tag{23.12}$$

where a is the number of the core and λ is the free space wavelength.

One can show that when $V < 2.405$, only the LP_{01} mode can propagate, i.e., fibre becomes single mode fibre. For $V > 2.405$, the fibre can support many modes simultaneously, $V = 2.405$ is known as *cut off* wavelength of the fibre. Theoretically, Lp_{01} mode will propagate no matter how small the V value, i.e., no matter how small the core radius a.

$\because$ $$V = \frac{2\pi}{\lambda} a(NA)$$

or $\because$ $$a = \frac{V\lambda}{2\pi(NA)}$$

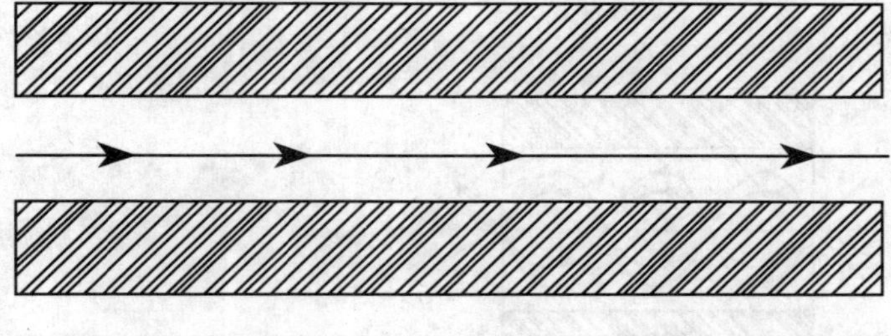

Fig. 23.5 Propagation in a stepped index monomode fibre

when $V < 2.405$, then

$$a < \frac{2.405\lambda}{2\pi(NA)}$$

or

$$a < \frac{2.405\lambda}{2\pi\sqrt{n_1^2 - n_2^2}}$$

for single mode fibre.

We may note that in general a single mode fibre core tend to be only a few microns in diameter, but it has an advantage from several points of view i.e., if core can be made as large as possible, one can easily note that this may be achieved by reducing the *NA* value, i.e., by making core and cladding *R.I.* very close to each other. One of the major advantages of using single mode fibre is the inter model dispersion no longer exists, and so they are capable of carrying very large bandwidth signal for long distance communication. For $V = 10$, the maximum number of modes (N_m) supported by a SI fibre is determined by $M_s \cong \frac{1}{2}V^2$. Thus for $V = 10$, $N_m = 50$.

(iii) Graded Index Multimode Optical Fibre

In comparison to stepped index mono-mode fibres, graded index multimode fibre is quite cost effective and offer a less expensive method of overcoming the problem of transit-time dispersion. It is observed that in such fibres, the refractive index of the core falls off gradually from the axis towards the core-cladding interface. This means that the individual light rays are gradually refracted in the core itself. Fig. (23.6) shows the wave propagation at different angle of incidence, travelling different distances from the central axis before suffering reflection to recross the central axis.

We may note that the number of guided modes is dependent upon the physical parameters, i.e., the refractive index difference between core and cladding $\left(\text{i.e., } NA = \sqrt{n_1^2 - n_2^2}\right)$ core radius of the fibre a and the wave length (λ) of the transmitted light, are all included in the normalized frequency V' for the fibre, where $V = 2\pi/\lambda\ a(N.A.)$ = Normalized frequency. As mention above the total number of guided modes for step index fibre is related to V by the following approximate expression

$$M_s = \frac{V^2}{2}$$

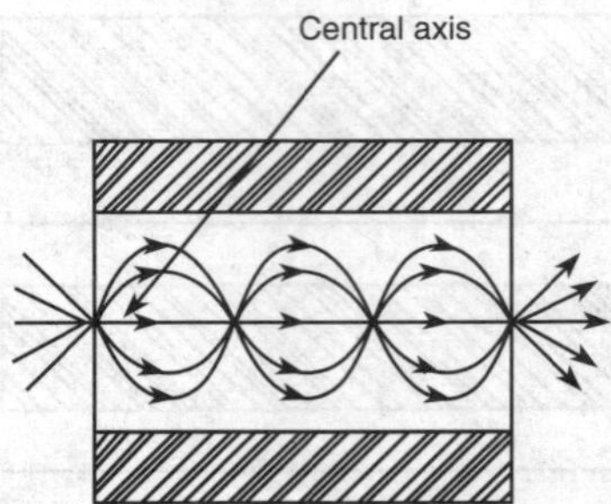

Fig. 23.6 Propagation in a graded index multimode optical fibre

From figure it is obvious that the light rays with large angles of incidence transverse greater paths than those with smaller angles of incidence. However, the velocity of propagation increases with a drop of the refractive index and thus all the rays recross the central axis at the nearly the same instant of time. This helps in reducing considerably the transit-time dispersion.

Fibre-optic systems use optical fibres to transmit information, in the form of coded pulses or fragmented images (using bundles of fibres). Over moderate distances they are used in telecommunications for which purpose they are becoming competitive with electric cables. In fibre communication, two bands are extensively used: 800 nm to 900 nm and 1200 nm to 1400 nm. The losses in the fibres for the said two bands are very low. Lasers and LED's (light emitting diodes) serve as light sources for fibres. Photodetectors are employed for processing the signal at the receiver end of the optical fibre link photodetectors convert the light signals into electrical wave forms. Optical fibres are also used in medical instruments (fibrescopes) to examine internal body cavities, such as stomach and bladder.

23.6 ADVANTAGES OF OPTICAL FIBRES

Fibres optic systems offer many advantages. Few of them are listed below:

(i) Optical fibres are light in weight and moreover occupy less space than coaxial (bundled twisted pair) cables.

(ii) Optical fibres are composed of dielectric materials and therefore they are totally immune to extraneous interfering electromagnetic signals.

(iii) There is virtually no signal leakage from optical fibres and hence cross-talks between neighbouring is not possible, where as it is quite frequent in conventional metallic systems.

(iv) The attenuation of signals in optical fibres is much smaller than in coaxial cables or twisted pair.

(v) Optical fibres are immune to electromagnetic interference and do not pick up line currents. Obviously, these can be safely used in high voltage environments, e.g. in a power station, and can be laid along side metallic power cables.

(vi) The basic raw material used in the fabrication of low loss fibres is silica, which is abundantly available in nature (although in an impure form), whereas copper constitutes the basic raw material in coaxial cables. Obviously, fibres are cost effective, i.e., more economical.

(vii) Optical fibres can withstand environmental hazards better and also have longer life compared to copper cables.

(viii) Optical fibres are non-conductive and non-inductive. This means there is no radiation and interference with other communication systems.

(ix) The speed of transmission through optical fibre is very fast because the signals are carried by light.

(x) The channel capacity of optical fibres is very large, compared to the audioband (5 to 10 Hz), the frequency of the carrier light wave is very high (~10^{14} Hz). Obviously, one can accommodate a large number of speech-signal channels in the fibres whose bandwidth is larger than that of a wire transmission line.

Disadvantages of optical fibre systems:

(i) Optical fibre systems are virtually useless themselves. To be practical, they must be connected to standard electronic facilities which often requires expensive interfaces.

(ii) Optical fibres by themselves have a significantly lower tensile strength than coaxial cable. This can be improved by coating the fibre with standard kevlar and a protective jacket of PVC.

(iii) Occasionally it is necessary to provide electrical power to remote interface or regenerating equipment. This can not be accomplished with the optical cable so additional metallic cables must be included in the cable assembly.

(iv) Optical fibre cable systems are relatively new and have not had sufficient time to prove their reliability.

(v) Optical fibres requires special tools to splice and repair cables and special test equipment to make routine measurements. Repairing fibre cables is also difficult and expensive and technicians working on optical fibre cables also require special skills and training.

In spite of all the attractive features that optical fibre communication system possess, a fully developed fibre optical communications system started becoming available in the commerical market only by 1975. Ever since the technology in this area is progressing well. The attainment of very low pass fibre has also stimulated forcefully efforts in the development of supporting technologies line long life semiconductor light sources, low loss optical fibre cable, fibre optic connectors, low pass splices, and highly sensitive fast response solid state photodetectors.

23.7 TYPES OF RAYS

The plane containing the central axis meridional rays remain in the meridian plane which can be divided in two categories: (a) bound rays that propatate along axis Figs. [23.7(a)], [23.7(b)] and (b) unbound rays that are not bound along axis and are refracted out of the core and absorbed in cladding or buffer.

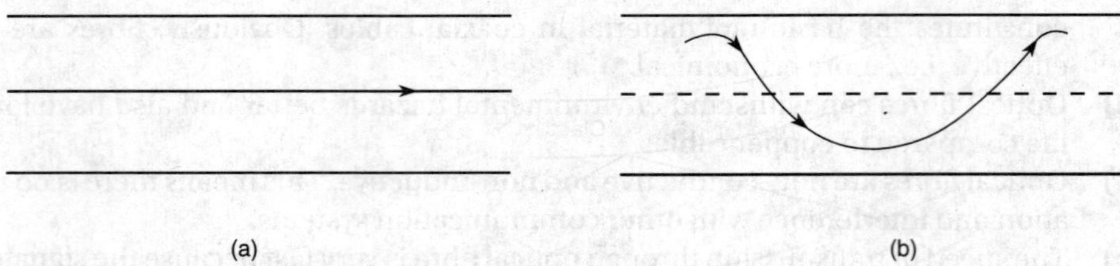

Fig. 23.7 Meridional rays propagation through optical fibre cable light

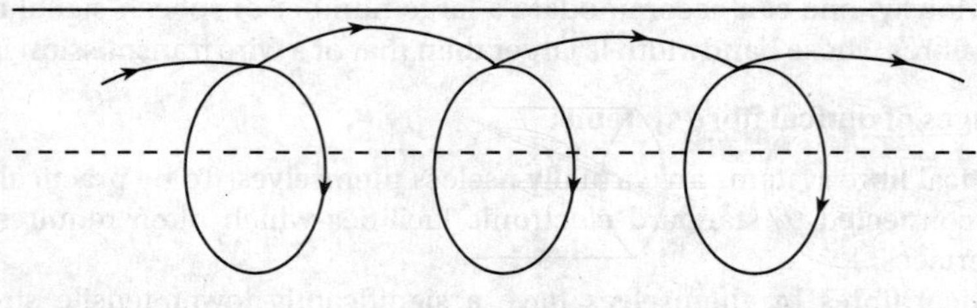

Fig. 23.8 Skewed ray propagation through optical fibre cable light

Skew Rays: These are the rays which enter the optical fibre obliquely and not confined to a plane and follow a helical path around the axis. These rays are soon scattered out of the core at bends and irregularities and thus don't contribute significantly in optical communication [Fig (23.8)].

23.8 MODES OF PROPAGATION IN OPTICAL FIBRE

Light propagation through straight and curve step index optical fibre with cladding is shown in Fig. (23.9).

Here we have shown only one ray and therefore this is called as ray diagram. However, in reality an infinite number of such rays Fig. [23.10 (a,b)] all slightly displaced from each other can be propagated at a time which are called different modes of propagation. At the time of propagation they will follow different paths in the optical fibre Fig. (23.11).

We have read that light can be described as an electromagnetic wave. It can be seen that the frequency spectrum of electromagnetic waves extends from the subsonic extends (a few Hz) to cosmic rays (10^{22} Hz). The light frequency spectrum can be divided into three general bands: infrared, visible and ultraviolet. Electromagnetic wave consists of a periodically varying electric field $\boldsymbol{E}$ and magnetic field $\boldsymbol{H}$ which are perpendicular to each other. The transverse modes are shown in Fig. (23.11).

These modes illustrates the case when the electric field $\boldsymbol{E}$ is perpendicular to the direction of propagation (z) and hence $\boldsymbol{E}_z = 0$, but the corresponding component of magnetic field $\boldsymbol{H}$ is in the direction of propagation. These modes are therefore called transverse electric (TE_m) modes. Similarly, when a component of electric field $\boldsymbol{E}$ is in the direction of propagation

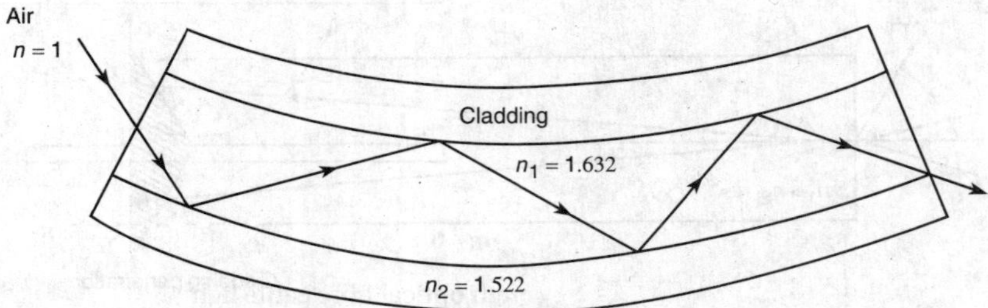

Fig. 23.9 Propagation of light through straight and curve. Step index optical fibre with cladding

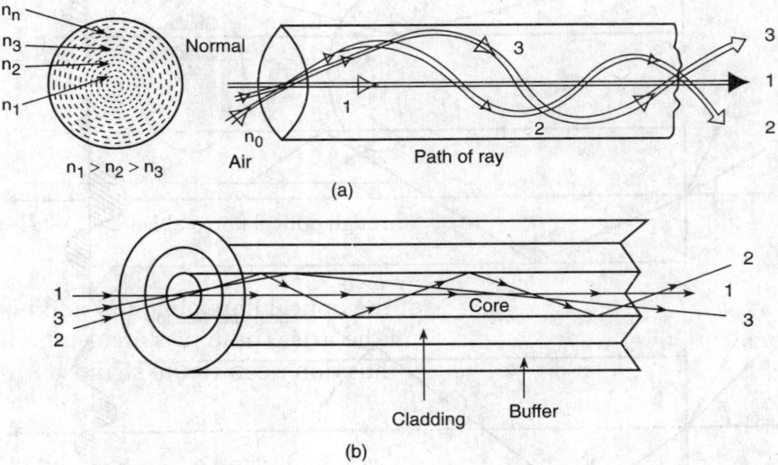

Fig. 23.10 Propagation of light in multimode graded index and step index fibre

but $\boldsymbol{H} = 0$, the modes so formed are called transverse magnetic (TM_m) modes. We have incorported the mode number m into this nomenclature by referring to TE_m or TM_m modes etc. We may note that each value of m is associated with a distinct wave pattern or mode as shown in Fig. (23.11), within the wave guide they are following different paths. We also find that for each value of m there is a value of $\gamma(\gamma_m)$, γ_m can take values in the range γ_c to $\pi/2$, which implies that the smaller the value of γ_m, i.e., γ_m closer to γ_c, higher the mode number and vice versa. We can see from Fig. (23.11) that the higher order modes, carry along with them a higher portion of mode energy within the cladding. This part of the wave which is carried in the cladding is generally referred as the "evanescent wave". We may note that (i) Fore free space communication, the total electromagnetic wave field lies in the transverse plane, where both $\boldsymbol{E} = 0$, and $\boldsymbol{H} = 0$. This is called transverse electromagnetic wave (TEM). However, TEM wave occur in metallic conductor like coaxial cable but not in optical wave guide (ii) for planar wave guide, generally TE ($E_z = 0$) and TM ($H_z = 0$) modes are obtained. (iii) As in optical fibre the cylindrical wave guide is bounded in two dimensions, rather than one as in plane wave guide. Obviously, two integers 1 and m are necessary in order

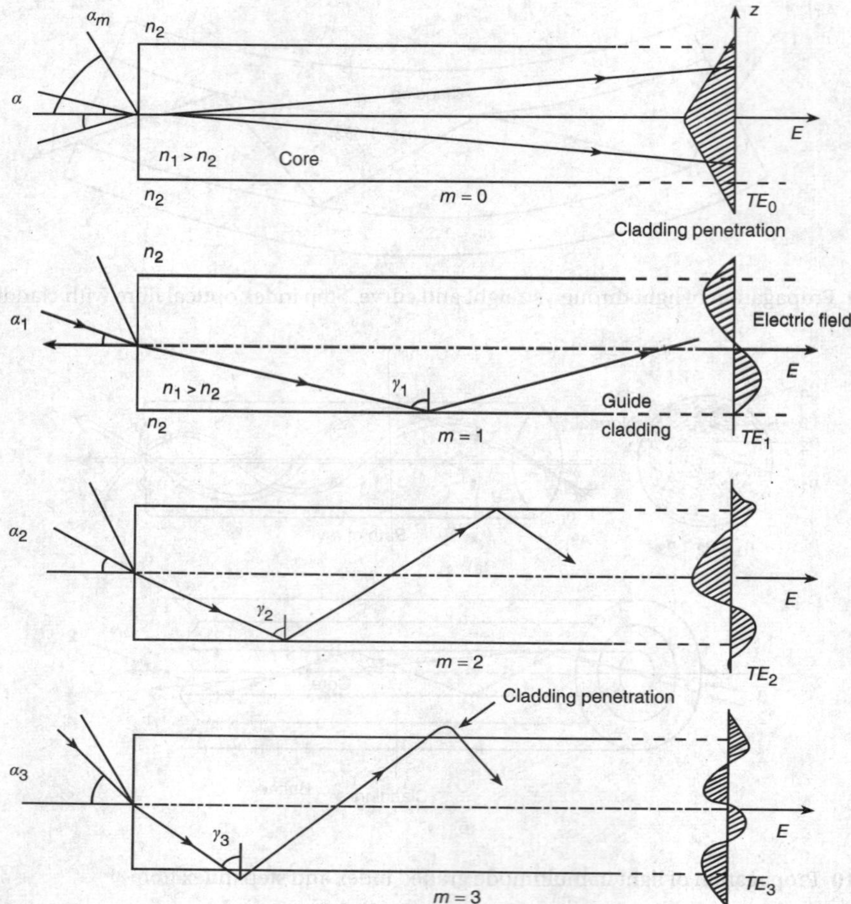

Fig. 23.11 Propagation of light in different modes through step index optical fibre, where α_m = acceptance angle r_c = critical angle, and $\pi/2 > r_1 > r_2 > r_3 ... > r_c$ where as $\alpha_0 < \alpha_1 < \alpha_2 < \alpha_3 ... < \alpha_m$

to specify the modes, in contrast to the single integer (m) required for planar guide. Thus, for cylindrical wave guide we refer TE_{1m} and TM_{1m} modes. The TE_{1m} and TM_{1m} modes corresponds to meridional rays [see Figs. 23.7(a) and 23.7(b)] which are travelling within the optical fibre. However, we may note that hybrid modes where $\boldsymbol{E}_z$ and $\boldsymbol{H}$ are non-zero also occur within the cylindrical wave guide. These modes which results from skew rays propagation as shown in Fig. (23.12), within the optical fibre are designated as HE_{1m} and EH_{1m} depending upon whether the components of $\boldsymbol{H}$ or $\boldsymbol{E}$ make larger contribution to the transverse field. These skew rays are known as each rays and corresponding modes are termed as leaky modes, not bound modes because the can propagate only appreciable distance prior being lost from the optical fibre.

In most practical wave guides, it is found that the refractive indices of core and cladding differ from each other by only a few percent and in that case the full set of modes

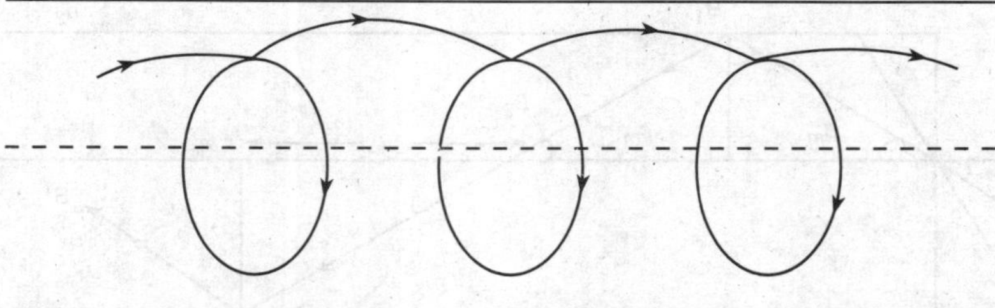

Fig. 23.12 Propagation of skewed rays through optical fibre cable light

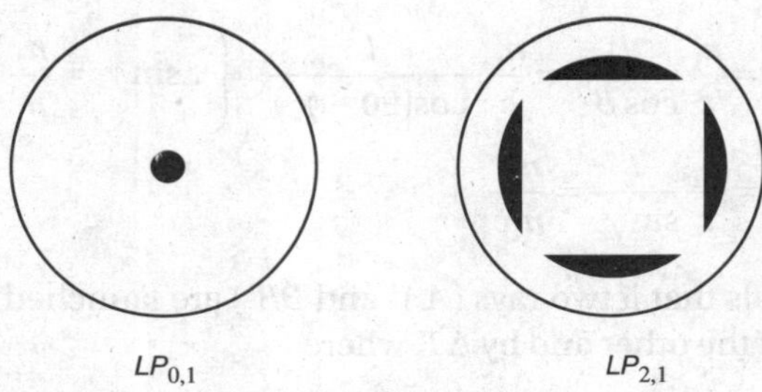

Fig. 23.13 Linearly polarised (LP_{1m}) modes

(i.e., *EH*, *HE* and *TE*) can be easily approximated by a single set called linearly polarized (LP_{1m}) modes. An (LP_{1m}) mode in general has *m* field maxima along a radius vector and 21 field maxima round a circumference Fig. (23.13).

23.9 DISPERSION-INTER MODEL DISPERSION

23.9.1 For Multimode Step Index Fibre

When the ray of light of different modes follow differnt paths, like some rays (shown by *AA′* in Fig. (23.14) which propagate along the optical fibre axis and some of which propagate along the trajectories of the oblique rays, for example *CC′* or *BB′* Fig. (23.14). Obviously, there will be difference in time taken for different modes to travel a distance *l*. We know that refractive index (*R.I*) of medium is simply a measure of velocity of propagation *v* of the light in medium. Thus we can write

$$R.I.(n) = \frac{V_a}{V_m} \quad \text{or} \quad V_m = \frac{V_a}{n} = \frac{c}{n}$$

where $c(= V_a)$ is the velocity of light in air and V_m is the velocity of light in medium.

The axial ray *AA′* travel an axial distance *l* along the optical fibre in time whereas the oblique ray *BB′* travel the same axial distance *l* along the path B_2B_3 in the time

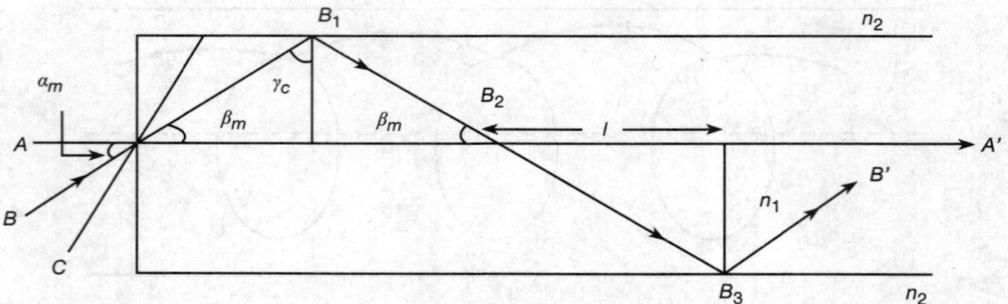

Fig. 23.14 Multipath time dispersion for multimode step index optical fibre

$$t_2 = \frac{n_1}{c}\frac{l}{\cos\beta_m} = \frac{n_1}{c}\cdot\frac{l}{\cos(90-\gamma_c)} \qquad \left(\because \sin\gamma_c = \frac{n_2}{n_1}\right)$$

$$= \frac{n_1}{c}\frac{l}{\sin\gamma_c} = \frac{n_1^2 l}{n_2 c}$$

The above result reveals that if two rays (AA' and BB') are launched together they will be reperated on arrival of the other and by ΔT, where

$$\Delta T = t_2 - t_1 = \frac{n_1 l}{c}\left(\frac{n_1}{n_2} - l\right)$$

$$= \frac{n_1 l}{n_2}\cdot\frac{1}{c}\Delta n \qquad (\text{where } \Delta n = n_1 - n_2)$$

Obviously, a pulse containning rays at all possible angles spreads out during propagation by an amount given by

$$\frac{\Delta T}{l} = \frac{n_1}{n_2}\frac{\Delta n}{c}$$

This is called as **multipath time dispersion** of the step index fibre and this causes pulse broadening.

23.9.2 For Graded Index Multimode Fibre

Different modes will follow different paths, sinusoidally. We may note that the rays making larger angles as shown in Fig. (23.15) traverse longer path but they do so in lower refractive index area and hence they travel with higher speed of propagation. However refractive index changes gradually in graded index optical fibre so there is a self focusing mechanism that leads to a smaller value of pulse dispersion than that of step index optical fibre. This shows that for long distance communication graded index optical fibre will be more useful than step index optical fibre, which causes less time dispersion.

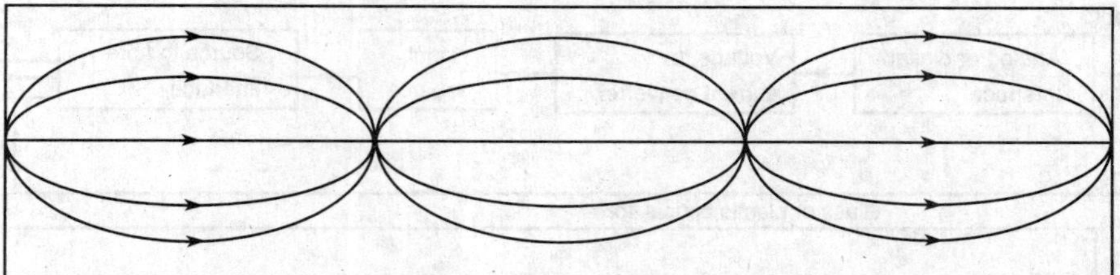

Fig. 23.15 Less multiple time dispersion for graded index optical fibre

23.9.3 Number of Guided Modes

We can define the normalized frequency as

$$f = \left(\frac{2\pi}{\lambda}\right) a\,(NA)$$

where a is the optical fibre core radius, NA is the numerical aperture and λ is the operating wave length. The number of guided modes in graded index fibre is $N_g = f^2/4$ and for multimode step-index fibre, $N_s = f^2/2$.

23.10 LOSSES IN OPTICAL FIBRE CABLE

The losses in optical fibre, i.e., attenuation are mainly caused by absorption and scattering of rays:

(i) **Absorption of rays:** Absorption of rays is basically a material property and it occurs at all wavelengths when electronic transitions takes place within material and are followed by non-radiative relaxation processes. This causes an increase of thermal energy in the material.

(ii) **Scattering:** Scattering is also partly a material property but in optical fibres it is caused by the imperfection in their gemoetries. Scattering occurs when the mode of propagation of light is changed, such that some of the optical energy leaves the optical fibre.

(iii) **Losses due to tight bending of fibres:** Tight bends of the optical fibre cause some of the light not to be internally reflected but to propagate into the cladding and be lost to propagate into the cladding.

(iv) **Loss due to imperfection in fibre:** The presence of imperfection in optical fibre material also causes power losses from the optical fibre cable.

(v) **Losses due to connectors and splices:** Power losses are also caused for long distance communication cables due to connectors and splices which required to connect two optical fibres temporarily.

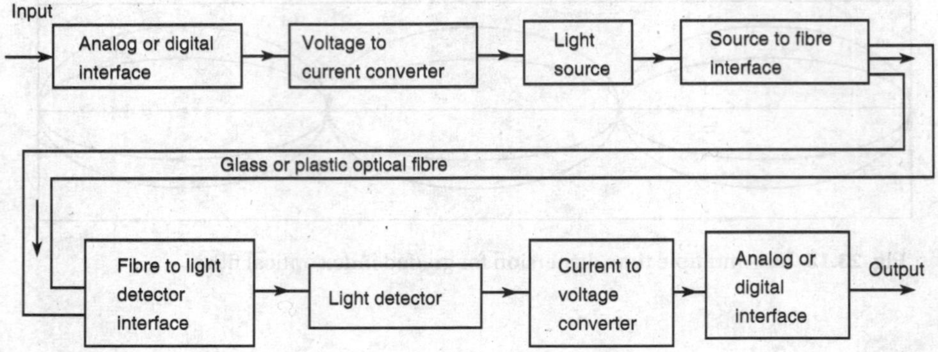

Fig. 23.16 Simplified fibre optic communication link

23.11 OPTICAL FIBRE COMMUNICATIONS SYSTEM

Fig. (23.16) shows a simplified block diagram of an optical fibre communications link. The three primary building blocks of the link are the *transmitter*, the *receiver*, and the *fibre guide*.

The transmitter consists of an analog or digital interface, a voltage to current converter, a light source, and a source to fibre light coupler. The fibre guide is either an ultrapure glass or plastic cable. The receiver includes a fibre to light detector coupling device, a photo detector, a current-to-voltage converter, an amplifier, and an analog or digital interface.

In an optical fibre transmitter, the light source can be modulated by a digital or an analog signal. For analog modulation, the input interface matches impedances and limits the input signal amplitude. For digital modulation, the original source may already be in digital form or, if in analog form, it must be converted to a digital pulse stream. For the latter case, an analog-to-digital converter nust be included in the interface.

The voltage-to-current converter serves as an electrical interface between the input circuitry and the light source. The light source is either a light-emitting diode (LED) or an injection laser diode (ILD). The amount of light emitted by either a LED or an ILD is proptional to the amount of drive current. Thus the voltage to current converter converts an input signal voltage to a current that is used to drive the light source.

The source-to-optical fibre coupler (such as a lens) is a mechanical interface. Its function is to couple the light emitted by the source into the optical fibre cable. The optical fibre consist of a glass or plastic fibre core, a cladding and a protective jacket. The fibre-to-light detector coupling device is also a mechanical coupler. Its function is to couple as much light as possible from the fibre cable into the light detector.

The light detector is very often either a PIN (p-type-intrinsic-n-type) diode or an APD (avalanche photodiode). Both the APD and the PIN diodes convert light energy to current. Consequently a current to voltage converter is required. The current to voltage converter transforms changes in detector current to changes in output signal voltage.

The analog to digital interface at the receiver output is also an electrical interface. If analog modulation is used, the interface matches impedances and signal levels to the output circuitry. If digital modulation is used, the interface must include a digital-to-analog converter.

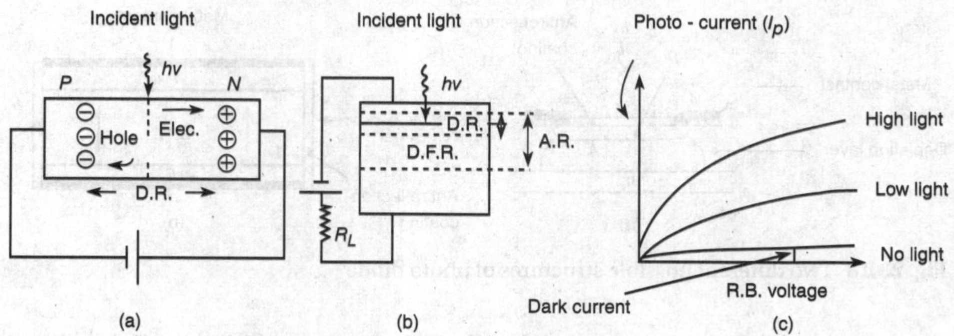

Fig. 23.17 Mechanism of P-N junction silicon phtodiode

An optical communication system is quite similar in concept to electronic (microwave) communication system (10^9Hz) only the frequency range is different (10^{15}Hz) and optical fibre instead of metallic fibre.

23.12 PHOTO DETECTOR

The most common semi-conductor photo detector is the photo diode. A photo detector detects the optical power incident upon it and converts the variation of this optical power into a correspondingly used system in optical fibre communication is IM/DD, where intensity of the optical carrier, modulated by biasing current in LED/diode laser (IM) and that variation of optical power directly detected in photo-detector (DD).

23.12.1 Photo-Diode Mechanism

The basic structure and extrisnic P-N junction silicon photo diode is shown in Fig. [23.17(a)]. In normal operation a sufficiently large reverse bias voltage is applied across the diode, so that depletion region as well as barrier potential will be large. Now the light incident on the reverse biased P-N junction diode, causes a photon ($h\nu$) to be absorbed. If the energy of the photon is more than the band gap energy (E_g), i.e., $h\nu > E_g$, then the absorbed photon excite an electron from valence band (VB) to conduction band (CB) and leaving a hole in VB. The electron-hole pair created due to incident light that is absorbed in the depletion region of the photo-diode is called photo-carriers.

Under reverse bias voltage, electron and hole flow in the depletion region (DR) in opposite direction, some of the electron-hole pairs will recombine and hence lost. The distance they will traverse prior to their recombination is called diffusion length (L_n and L_p for electron and holes respectively) and the time it takes for an electron and hole to recombine is known as carrier life time (T_n and T_p). The relation between the carrier life time and diffusion length can be expressed as

$$L_n = (D_n T_n)^{1/2}, L_p = (D_p T_p)^{1/2}$$

where D_n and D_p represents the electron and hole diffusion coefficients respectively, which is in cm^2/sec.

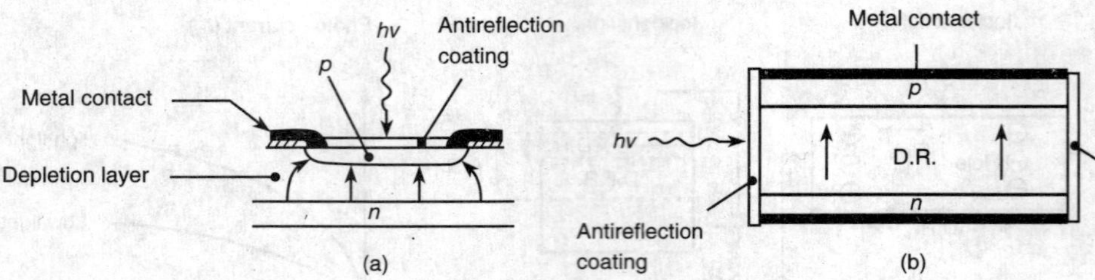

Fig. 23.18 Two different possible structures of photo diode

We may note that photons can be absorbed in depletion region (DR) as well as in diffusion region (DFR) too, which is usually indicated as absorption region (AR) [see Fig. 23.17(b)]. Obviously, this region becomes a void of free photo carriers. Now, within the depletion region those pairs of electron-hole cannot recombine because of the existence of main electrostatic field due to the depletion region voltage. Any photon induced electron-hole pair there in will be separated and drawn in opposite directions by the combined effect of equivalent depletion region voltage as well as the reverse bias voltage. We may note that the polarities of these two voltages are in the same direction. This effect leading to flow of photo-current in the external circuit [see Fig. 23.17(c)], as photo-carriers drift away across the depletion region. Fig. [23.17(c)] shows typical output characteristics for reverse bias (RB) PN photo diode. Different operating conditions are indicated from no light to high level light. When there is no incident light, then also revese current will be there. This is called as dark current.

If the electron-hole pairs are produced outside the absorption region (AR) then there will be higher probability to recombine and in that case no photo current will be available. Obviously this means that we have to make the top layer as thin as possible and effective depletion region as wide as possible in order to maximize the termed as quantum efficiency. One can increase the quantum efficiency of the photodiode if one goes from ordinary photo diode to PIN Photo Diode or Avalanche Photo Diode (APD) or Reach throught Avalanche Photo Diode (RAPD).

Usually, there are two different structures for photodiodes are possible as shown in Fig. (23.18). Fig. [23.18(a)] shows the photodiode with front illuminated operation at 0.8 to 0.9 μm range, where as Fig. [23.18(b)] shows the side illuminated photodiode structure operating at higher wavelength 1.09 μm range.

23.12.2 Quantum Efficiency (η) and Responsivity (R)

Quantum efficiency (η) and responsivity (R) or response speed (R) for a photo detector are two important characteristics. These two parameters, i.e., η and R depend on the material band gap, operating wavelength, percentage of doping and thickness of P and N regions. In order to achieve high quantum efficiency, the depletion layer must be thick enough to permit a large fraction of the incident light to be absorbed. This is why PIN or APD photo diode is better than ordinary photo diode. One can define quantum efficiency and responsivity as follows:

Quantum efficiency, $\eta = \dfrac{\text{Rate of generation of electron-hole pair (electron/s.)}}{\text{Rate of photons incident (Photons/s.)}}$

or $\eta = \dfrac{I_p/e}{P_o/h\nu}$

Responsivity, $R = I_p/P_o = \eta e/h\nu = \dfrac{he\lambda}{hc}$ A/watt

where $I_p \rightarrow$ photo current, $e \rightarrow$ electron charge, $p_o \rightarrow$ incident optical power, $h\nu \rightarrow$ photon energy and $c \rightarrow$ velocity of light.

23.12.3 Long Wave Length Cut-off

It is essential when considering intrinsic absorption process, that the energy of the incident photon ($h\nu$) should be greater than the band gap energy (E_g) of the material used as photo diode, otherwise electron-hole pair will not generate, i.e.,

$$h\nu \geq E_g \quad \text{or} \quad hc/\lambda \geq \lambda \leq \lambda_c \leq hc/E_g$$

Obviously, there will be threshold wave length for detection, which is commonly known as long wave length cut-off (λ_c). If $\lambda > \lambda_c$, i.e., the incident wave length of photon (λ) is more than λ_c then photo detector will not work.

23.13 OPTICAL FIBRE CABLE CONSTRUCTION

A single optical fibre can transmit light energy very effectively [Figs. (23.9) and (23.10)]. However, in order to transmit an image of an object one requires assembly or bundles of fine glass fibres or plastic fibres because each fibre transmits rays from a very small region of the object Fig. (23.19). The quality of the image depends upon the diameter of the optical fibre. It can be as small as 10^{-6}m. A bundle of the optical fibre may consist of thousands of individual optical fibres.

Essentially, there are three varieties of optical fibres that are available. All these three varieties are constructed of either glass, plastic, or a combination of glass and plastic. The three varieties are:

(i) Plastic core and cladding
(ii) Glass core with plastic cladding (often called PCS fibre, plastic-clad silica)
(iii) Glass core and glass cladding (often called SCS, silica-clad silica)

There is also a possiblity of using a fourth variety that uses a non-silicate substance, zinc chloride. Investigations have shown that fibres made of this substance will be as much as 1000 times as effective as glass their silica based counterpart.

Plastic fibres have several advantages over glass fibres. First plastic, fibres are more flexible and consequently, more rugged than glass. They are easy to install, can better withstand stress, are less expensive, and weigh approximately 60% less than glass.

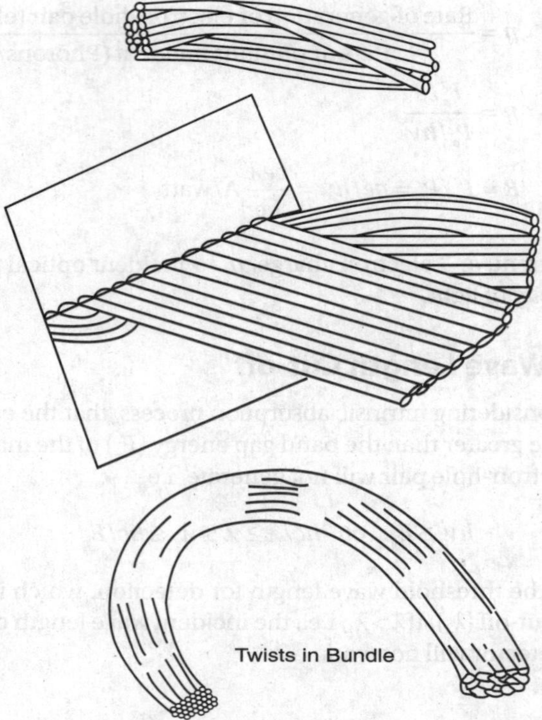

Fig. 23.19 Bundle of optical fibre

The disadvantage of plastic fibres is their high attenuation characteristic. They do not propagate light as efficiently as glass. Consequently plastic fibres are limited to relatively short runs, such as within a single building or a building complex.

Fibres with glass cores exhibit low attenuation characteristics, however, PCS fibres are slightly better than SCS fibres. PCS fibres also less affected by radiation and, therefore, are more attractive to military applications. SCS fibres have the best propagation characteristics and they are easier to terminate than PCS fibres. Unfortunately, SCS cables are the least suggested, and they are more susceptible to increase in attenuation when exposed to radiation.

The selection of fibre for a given application is a function of specific system requirements.

23.14 CABLE CONSTRUCTION

There are many different cable designs available today. Fig. (23.20) shows example of several fibre-optic cable configurations. Depending upon the configuration, the cable may include a **core**, a **cladding**, a **protective tube, buffers, strength numbers** and one or more **protective jackets.**

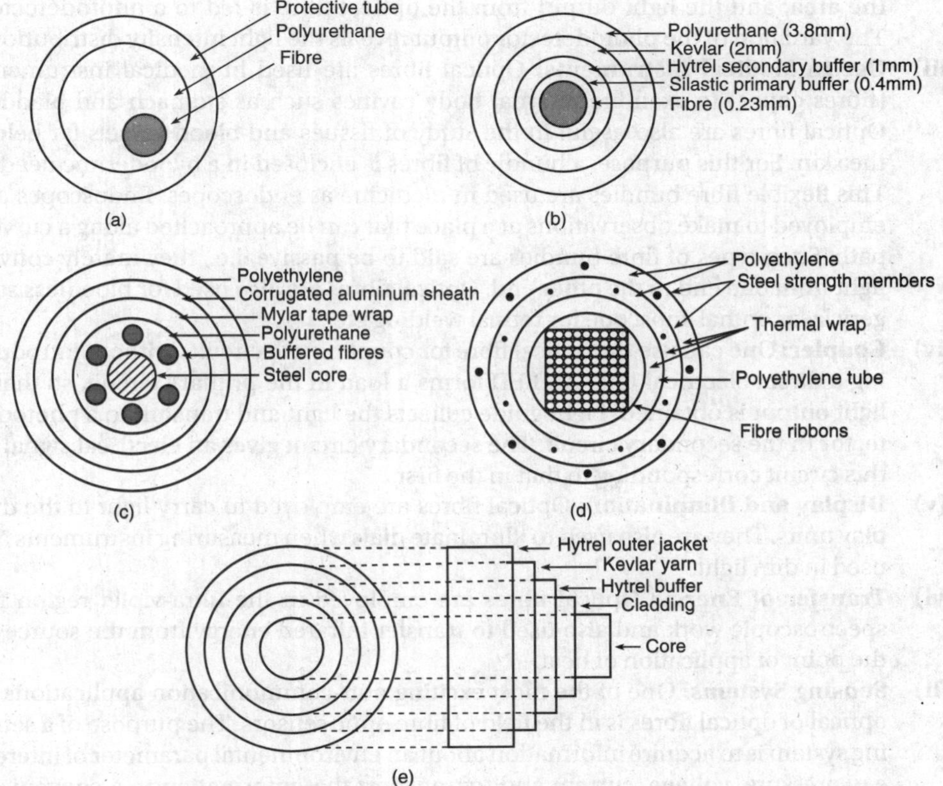

Fig. 23.20 Fibre optic cable configuration: (a) loose tube construction; (b) constrained fibre; (c) multiplestrands; (d) telephone cable; (e) plastic-clad silica cable

23.15 APPLICATIONS OF OPTICAL FIBRES

Fibre optics technology has matured as a powerful alternative for information transport and sensing. Few important applications of optical fibres are:

(i) Communication: Optical fibres are used to carry signals in optical communications due to their unique features, viz, large channel capacity, wide band width, good electrical isolation, no cross talks etc., Now a days telephone, television, satellite and computer links advanced communication system rely heavily on this technology.

The enormous bandwidth of optical fibre communication system finds their major use in long distance communication for transmission of speech, video and digital data signals. In the near future all the existing copper based trunk lines will be replaced by optical fibres.

(ii) Scanning: Optical fibres can be used to measure the distribution of light intensity over an illuminated area. For this purpose, the input end of the optical fibre scans the area, and the light output from the optical fibre is fed to a photodetector. The variation of the photodetector output reveals the light intensity distribution.

(iii) Use in Medical Instruments: Optical fibres are used in medical instruments (fibres copes) to examine internal body cavities such as stomach and bladder. Optical fibres are also useful in the study of tissues and blood vessels far below the skin. For this purpose, a bundle of fibres is enclosed in a hypodermic needle. This flexible fibre bundles are used in medicine as endoscopes. Endoscopes are employed to make observations at a place that can be approached along a curved path. Such types of fibre bundles are said to be passive, i.e., they merely convey light from one end to the other end. Optical fibres are also used for bloodless surgery laser opthal-moscope for retinal welding etc.,

(iv) Coupler: One can use the optical fibre for coupling two circuits without introducing a direct electrical link. An LED forms a load in the primary circuit, so that a light output is obtained. Light guide collects the light and transmit to a photodetector in the secondary circuit. The secondary circuit gives an electrical signal in this circuit corresponding to that in the first.

(v) Display and Illumination: Optical fibres are employed to carry light to the display units. They are also used to illuminate dials when measuring instruments are used in dim light.

(vi) Transfer of Energy: Optical fibres are employed in the ultra-violet region for spectroscopic work and also used to transfer infrared energy from the source to the point of application of heat.

(vii) Sensing Systems: One of the most exciting non-communication applications of optical of optical fibres is in the field of fibre optic sensors. The purpose of a sensing system is to acquire information about an environmental parameter of interest e.g. pressure, voltage, current and to transport that information to a convenient location for processing. One could build a sensing system where the acquisition function performed by conventional technology e.g. a piezoelectric pressure sensor and the transport is accomplished via an optical fibre link.

Optical sensing systems can be classified into two classes which work on considerable different principles, and which have substantially different areas of application. Incoherent sensing system are those which employ a sensor which has a variable transmission loss or reflection loss, which is read out by illuminating an optical signal into two parts which travel different optial paths, and which are then allowed to combine and interfere. One of the path contains a sensor which modulates the phase of the optical signal passing through it. The other path is a reference path.

(viii) Computers: Fibres are used because of their greater data handling capacities and higher memory densities. The fibres are used for mutual interfacing of central processor, linking them to peripheral devices, data transmission within the main frame. Hence, reduced bit errors and free environmental interference are assured.

(ix) LAN: In local area networks, fibres are used to link computers in ring, star or bus topology. They find application in office operations, private automatic branch exchanges etc.,

(x) Industrial Electronics: Fibres are used in power plants, rail road networks and metal industry for data acquisition control and signal processing, as transmission is not affected by high energy fields. In medical electronics by using fibres noise free control signals are possible. In automatic manufacturing fibres are used as sensors because of less weight.

Example 23.1: For a 3 μm diameter optical fibre with core and cladding index of refraction of 1.545 and 1.510 respectively; determine the cutoff wavelength.

Solution

$$\lambda_c = \frac{2\pi a n_1\sqrt{2\Delta}}{2.405}(\mu m)$$

where
$\lambda_c \rightarrow$ Cut off wavelength (μm)
$n_1 \rightarrow$ Core index of refraction (unitless)
$n_2 \rightarrow$ Cladding index of refraction (unitless)
$a \rightarrow$ Core radius (μm)

$$\Delta = \frac{n_1 - n_2}{n_1}$$

$$\Delta = \frac{1.545 - 1.510}{1.545} = 0.023$$

$$a = \frac{3\,\mu\text{m}}{2} = 1.5\,\mu m$$

$$\lambda_c = \frac{(2\pi)(1.5\mu m)(1.545)\sqrt{2(0.023)}}{2.405}$$

$$= 1.29\,\mu m.$$

Example 23.2: For a multimode index fibre $n_1 = 1.53$, $n_2 = 1.50$ and $\lambda = 1\mu m$. If the core radius is 50 μm then calculate the normalized frequency of the fibre (ν) and the number of guided mode.

Solution Normalised frequency, $\nu = \dfrac{2\pi a(NA)}{\lambda}$

$$= \frac{2\times 3.14\times 50\times 10^{-6}\times\sqrt{(1.53)^2-(1.50)^2}}{1\times 10^{-6}}$$

$$= 94.72$$

Total number of guided mode = $M_s = \dfrac{\nu^2}{2} = \dfrac{(94.72)^2}{2} = 4486.$

Example 23.3: For a multimode step index fibre with a glass core ($n_1 = 1.5$) and a fused quartz cladding ($n_2 = 1.46$), determine the crital angle (θ_c), acceptance angle (θ_m) and numerical aperture (NA). The source to fibre to medium is air.

Solution Critical angle, $\theta_c = \sin^{-1}\dfrac{n_2}{n_1} = \sin^{-1}\dfrac{1.46}{1.5}$

$$= 76.7^\circ$$

$$\theta_m = \sin^{-1}\frac{\sqrt{n_1^2 - n_2^2}}{n_0}$$

and

$$\theta_{m(\max)} = \sin^{-1}\sqrt{n_1^2 - n_2^2}$$

$$= \sin^{-1}\sqrt{(1.5)^2 - (1.46)^2}$$

$$= 20.2^\circ$$

$$NA = \sin\theta_m = 20.2 = 0.344$$

The maximum diameter a single-mode optical fibre can have is proportional to the wavelength of the light ray entering the cable and the numerical aperture of the fibre. Mathematically, the maximum radius of the core of the a single mode fibre is

$$r_{\max} = \frac{0.383\text{Å}}{NA} = \frac{0.383}{0.344} = 1.1$$

where $r_{\max}$ is maximum core radius in meters,
$NA \rightarrow$ numerical aperture (unitless) and
$\lambda \rightarrow$ light ray wavelength (meters).

Example 23.4: The numerical aperture (NA) of an optical fibre is 0.5 and core refractive index 1.54. Determine (i) refractive index (RI) of cladding (ii) change in core cladding refractive per unit RI of the core.

Solution

(i) $$NA = \sin\theta_m = \sqrt{n_1^2 - n_2^2}$$

$$\therefore (0.5)^2 = (1.54)^2 - n_2^2 \quad \text{or} \quad n_2 = 1.456$$

(ii) $$RI \text{ of the core} = \frac{n_1 - n_2}{n_1} = \frac{1.54 - 1.456}{1.54} = 0.0542$$

Example 23.5: A step index silica fibre consists of a core of refractive index 1.450 and diameter 90 μm. The numerical aperture (NA) of the fibre is 0.16. Calculate (a) the acceptance angle of the fibre (b) refractive index of cladding and (c) if the fibre is immersed in water (n_2 = 1.33), how does acceptance angle changes?

Solution

(a) We have read that the acceptance angle θ_i is the maximum angle of incidence at the entrance face of a step index fibre for which the rays propagate down the fibre:

we have $$\sin\theta_i = \frac{NA}{n_0} = \frac{0.16}{1} \quad (n_0 = 1 \text{ for air})$$

$$\therefore \quad \theta_i = 9.2^\circ$$

(b) $NA = \quad n_1^2 - n_2^2$

$\therefore \quad n_2 = n_1^2 - (NA)^2 = (1.45)^2 - (0.16)^2 = 0.0769$

$n_2^2 = 1.441$

We must note that the small but significant difference in refractive index of core ($n_1 = 1.450$) and cladding ($n_2 = 1.441$)

(c) When fibre is immersed in water, then $n_2 = 1.33$. Obviously, acceptance angle therefore decreases:

$$i_{\max} = \sin^{-1}\frac{0.16}{1.33} = 6.9°$$

Example 23.6: An optical fibre has the following data: n_1 (core) = 1.55, n_2 (cladding) = 1.50, core diameter = 50μm. Calculate the *NA* and the acceptance angle.

Solution

$$\theta_i = \sin^{-1}\sqrt{n_1^2 - n_2^2} = \sin^{-1}\sqrt{1.55^2 - 1.50^2}$$

$$NA = \sin\theta_i = \sin 230 = 0.3905$$

At the steepest angle, the angle of incidence on the core cladding interface just exceeds the critical angle θ_c. One obtains

$$\theta_c = \sin^{-1}\frac{1.50}{1.55} = 75.4°$$

Example 23.7: The refractive index of the core of W-step index fibre is 1.46 and the relative refractive index between the core is 1.46 and the refractive index between the core and the cladding of the fibre is 2%. Calculate (i) the numerical aperture (*NA*), (ii) the solid acceptance angle in the air for the fibre whose core index is 1.46 and (iii) the critical angle at the core-cladding interface within the fibre.

Solution

(i) $\Delta = \dfrac{n_1 - n_2}{n_1} = 0.02 \quad \text{and} \quad n_1 = 1.46$

$\therefore \; n_1 - n_2 = 0.02 \times 1.46 \quad \text{and}$

$n_2 = 1.46 - 0.02 \times 1.46 - 0.0292 = 1.43$

$NA = \left(n_1^2 - n_2^2\right)^{1/2} = (2.89 \times 0.03)^{1/2} = (0.0867)^{1/2} = 0.29$

(ii) For small angle, the solid acceptance angle air, ϕ, is given by $\phi = \pi\,(NA)^2 = \pi \times 0.0867$ = 0.272 radians.

(iii) $\sin\theta_c = \dfrac{n_2}{n_1} = \dfrac{1.43}{1.46} \quad \text{or} \quad \theta_c = \sin^{-1}\left(\dfrac{1.43}{1.46}\right)$

$= \sin^{-1} 0.98 = 78°$

Obviously, the critical angle at the core cladding interface is 78°.

Example 23.8: A waveguide is there with a core and cladding of refractive indices 1.46 and 1.45 respectively. Find the maximum number of modes with wavelength 900nm that travel if it is a planer waveguide of core thickness 10 μm.

Solution The maximum number of modes for a planer waveguide are given by

$$m = \frac{eV}{\pi}$$

where $$V = \frac{\pi d}{\lambda}\sqrt{n_1^2 - n_2^2} = 3.14\times 10^{-5}[(1.46)^2 - (1.45)^2]$$

$$= 5.95$$

$\therefore$ $$m = \frac{2\times 5.95}{3.14}2 \cong 3.8$$

where $n_1 = 1.46$

$n_2 = 1.45$

$d = 10\,\mu\text{m}$

$\lambda = 900\,nm.$

Example 23.9: The mean optical power launched into an optical fibre 150μW. The mean optical power at the fibre output is 10μW. Show that the signal attenuation per kilometer for the fibre is 2.352 dB/km.

Solution We have

$$\alpha_{AB} = 10\log_{10}\frac{p_i}{p_o}$$

$$= 10\log_{10}\frac{150\times 10^{-6}}{10\times 10^{-6}} = 11.76\text{dB}$$

$\therefore$ $$\alpha_{AB} \div L = \frac{11.76}{5} = 2.352\frac{\text{dB}}{\text{km}}.$$

Example 23.10: The core diameter of multimode step-index fibre is 46μm and numerical aperture is 0.3. Show that the number of guided modes at operating wavelength of 0.8μm is 1466.

Solution

$$f = \frac{2\pi}{\lambda}a(NA)$$

$$= \frac{2\pi}{0.8\times 10^{-6}}\times 23\times 10^{-6}\times 0.3$$

$$= 54.16$$

$\therefore$ $$N_g = \frac{f^2}{2} = \frac{(54.16)^2}{2} = 1466.$$

SUMMARY

1. Optical fibre is a waveguide through which light can be transmitted with very little leakage through the sidewalls.
2. Refraction and total internal reflection are the basis for optical fibres. Optical fibre consists of core of refractive index n_1 and cladding of refractive index n_2 ($n_1 > n_2$).
3. Numerical aperature (NA) for the optical fibre gives the relation between the acceptance angle and refractive index of three media involved, viz. core, cladding and air

$$NA = \sqrt{n_1^2 - n_2^2} = n_1\sqrt{2\Delta}$$

where Δ is relative refractive index between core and cladding,

$$\Delta = \frac{n_1 - n_2}{n_1}$$

4. A typical fibre consists of a glass core only $8\,\mu m$ in diammeter, surrounded by a so-called cladding consisting of glass with a lower refractive index.
5. Total internal reflection at the core-cladding interface guides light along the fibre.
6. The glass used in optical fibres is so clear that a 1 km thick slab would be as transparent as an ordinary window pane. Today's fibres carry light produced by semiconductor lasers at infrared wavelengths of 850, 1350, or 1550 nm.
7. Depending upon mode of transmission there are two types of fibres: (i) single mode fibres, and (ii) multimode fibres.
8. According to refractive index of core and cladding, there are two types of fibres: (i) step index, and (ii) graded index.
9. In the *step index fibre* a pure glass core, with a diameter between 6 and 250 micrometers, is surrounded by a coaxial or a plastic cladding of lower refractive index. The cladding usually between 10 and 150 micrometers thick. The interface between core and cladding acts as a cylindrical mirror at which total internal reflection of the transmitted light takes place. This structure enables light to travel through several kilometers of fibre.
10. In a *graded-index fibre*, each layer of glass, from the fibre axis to its outer wall, has a slightly lower refractive index than the layer inside it. This arrangement also prevents light from escaping through the fibre walls by a combination of refraction and total internal reflection, and can be made to give the same transit time for rays at different angles.
11. The main advantage of optical fibres over copper wires, or open-air electromagnetic-wave transmission is their huge rate of information flow, called *bandwidth.*
12. Optical fibre systems use optic fibres to transmit information, in the form of coded pulses or fragmented images from a source to receiver. Optical fibres provide the physical connectivity of the global internet and handle information ranging from telephone and television to light signals within medical, astronomical and industrial instruments.
13. Optical fibres used in *medical instruments* (*fibres copes*) to examine internal body cavities, such as stomach and bladder.

REVIEW QUESTIONS

1. Define an optical fibre system. What is numerical aperature of an optical fibre? What is its significance?
2. Obtain the expression for *NA* in terms of *RI* of the core and cladding of an optical fibre.
3. Outline the primary the building blocks of a fibre optic system.
4. Draw block diagram for optical communication through fibre optic cable. Mention the advantages of optical communication?
5. Explain the working of monomode and multimode optical fibres. For long distance communication which optical fibre is preferred and why?
6. Contrast glass and plastic fibre cables.
7. What is an optical fibre? What are its main sections? Explain the function of each section.
8. What is meant by propagation of modes in an optical fibre?
9. How light is guided by an optical fibre? Define acceptance angle and the numerical aperture (*NA*). How they are related to the refractive indices of the core and cladding?
10. Mention the different classes of optical fibres. What are the light sources for fibres? Write some advantages of optical fibres.
11. Discuss major applications of optical fibres.
12. What are the losses encountered in light propagation through optical fibres?
13. When an optical signal passes through a fibre, it is attenuated and distorted progressively. Mention the factors responsible for attenuation and distortion?

PROBLEMS

1. A multimode step index fibre has $n_1 = 1.53$, $n_2 = 1.50$ and $\lambda = 1\ \mu m$. In order to make this fibre single mode, what will be core radius?

$$\left[\textbf{Hint}: a \leq \frac{2.450\lambda}{2\pi(NA)},\quad NA = \sqrt{(1.53)^2 - (1.50)^2}\right.$$

$$\left. a \leq \frac{2.405\times 1\times 10^{-6}}{2\times 3.14\times\sqrt{(1.53)^2-(1.50)^2}} \quad \because a < 1.27\ \mu m\right]$$

2. Numerical aperture of a optical fibre is 0.5 and core refractive index is 1.48. Show that cladding refractive index is 1.393 and acceptance angle is 30°.

[**Hint** : $NA = \sqrt{n_1^2 - n_2^2} = 0.5$ *R.I.* of core = $n_1 = 1.48$ that is $n_2 = 1.393$ that is *R.I.* of cladding = $n_2 = 1.393$ Acceptance angle $\theta = \sin^{-1}(NA) = 30°$]

3. An optical fibre has a numerical aperature (*NA*) of 0.20 and cladding refractive index of 1.48. Find the acceptance angle in water which has a refractive index of 1.33. Also find the critical angle at the core cladding interface.
4. The refractive index of the core and that of the cladding of an optical fibre are 1.58 and 1.51 respectively. Calculate the acceptance angle and *N*. [**Ans.** 27.7°, 0.4651]

SHORT ANSWER–QUESTIONS

1. To prevent energy losses via absorption and scattering, what should be the order of cladding in an optical fibre?
Ans: Atleast a few wavelengths thick.

2. How numerical aperture (*NA*) of an optical fibre is defined?
Ans: $NA = \sqrt{\mu_1^2 - \mu_2^2}$, where μ_1 is the refractive index of the core of a fibre and μ_2 that of cladding ($\mu_1 > \mu_2$): One can define the numerical aperture of a fibre as the physical area that light must enter in order to propagate through the fibre. Similarly, sources have a numerical aperture for the light. The numerical apertures for light sources can be calculated in terms of angle involved with the output.

3. How will you define acceptance angle (θ_A) for an optical fibre?
Ans: $\theta_A = \sin^{-1}\sqrt{\mu_1^2 - \mu_2^2}$

4. What is major economic benefit offered by fibre optics in information technology?
Ans: Very high information transmission rate at a low cost per circuit-Km.

5. What type of reflaction is expected from an optical fibre?
Ans: Total internal reflection.

6. How an optical fibre is fabricated?
Ans: To fabricate fibre, one can use either different glasses with different refractive index for core and cladding or use fused silica (quartz glass) whose refractive index is modified by doping. The type of glass used will also with application, but the purity (lack of any foreign materials as well as lack of imperfections such as cracks or bubbles) must also be kept very high.

7. What is the benefit of having large core diameters and large NA for multimode step index fibres?
Ans: This makes it easier to couple light from an LED into the optical fibre.

8. Meridional rays are classified as either bound or unbound rays. While propagating through the fibre, the bound rays follow what property of light?
Ans: Total internal reflection.

9. What are the advantages of optical fibre over copper wire conductors?
Ans: Optical fibre transmission has following advantages over convention copper wire system:

(i) More data can be sent over the long distances through optical fibre due to its lowe transmission losses.
(ii) The light weight and the small thin hair-sized dimension of the optical fibres are quite advantageous, e.g. in aircraft.
(iii) Dielectric nature of optical fibre provides optical immunity to the electronic-magnetic interferences.
(iv) The principal material for the optical fibre, i.e. silica, whose source is sand, is abundantly available.

10. What is an optical fibre?
Ans: An optical fibre "pipes" light from one place to another by a series of total internal reflection, much as a bullet ricochets down a steel pipe. Light rays bounce along the inner walls, following the twists and turns of the fibre.

11. What is the main advantage of optical fibers over copper wires or open-air electromagnetic-wave transmission?
Ans: The main advantage of optical fibres over copper wires or open-air electromagnetic-wave transmission is their huge rate of information flow, called bandwidth.

12. Why optical fibres are less susceptible to electrical noise?
Ans: Optical fibres are made from insulators e.g. glass or plastic polymers, and hence they are less susceptible to electrical noise.

13. What is the order of the diameter of optical fibre?
Ans: The diameter of optical fibre is very small, of the order of μm.

FILL IN THE BLANKS

1. Optical fibre is ______ carrier and immune to ______ interference.
2. Optical fibre does not exit any ______ that might interfere with signals.
3. Light has a much ______ frequency than radio waves or modulated signals in copper wire.
4. Optical fibre does not rust or corrode and is much ______ and ______ than copper cable.
5. Scattering mechanism cause the transfer of ______ from one propagating mode into a different mode.
6. Major implementation of optical fibre transmission involves some from of digital ______.

ANSWERS

1. non-electrical, electromagnetic 2. noise 3. higher 4. smaller, larger 5. optical power 6. modulation

OBJECTIVE QUESTIONS

1. In fibre-optic communication, the propagation of light through the fibre is based on the principle of

(a) diffraction of light (b) interference of light
(c) total internal reflection (d) polarization of light **[Ans: (c)]**

2. The refractive index in a graded index fibre

(a) increase throughout the core
(b) remains constant throughout the core
(c) decreases parabolically throughout the core
(d) None of the above **[Ans: (c)]**

3. The acceptance angle or acceptance cone half angle for the optical fibre is

(a) $\theta_m = \sin^{-1}(n_1 - n_2)$ (b) $\theta_m = \sin^{-1}\sqrt{n_1^2 - n_2^2}$

(c) $\theta_m = \sqrt{n_1^2 - n_2^2}$ (d) $\theta_m = n_1^2 / n_2^2$ **[Ans: (b)]**

4. Numerical aperature of step index optical fibre is

(a) $NA = n_1\sqrt{2\Delta}$ (b) $NA = n_1^2\sqrt{2\Delta}$

(c) $NA = n_1^2 - n_2^2$ (d) $NA = n_1 2\Delta$ **[Ans: (b)]**

5. The jacket of an optical fibre enables

(a) to prevent from moisture trapping
(b) to prevent from mechanical abrasions
(c) to prevent from interaction with internal atmosphere
(d) all of the above **[Ans: (d)]**

6. When *V* parameter is less than 2.405, then optical fibre will support

(a) three modes (b) two modes
(c) one mode (d) none of the above **[Ans: (c)]**

7. The optical fibres are made of

(a) dielectric material (b) metallic material
(c) magnetic material (d) plastic doped with metallic impurities **[Ans: (a)]**

8. Pulse broadening can be minimized by making use of

(a) multimode graded index fibre (b) single mode index fibre
(c) multimode step index fibre (d) none of the above **[Ans: (a)]**

9. The total number of modes in a multimode fibre is given by

(a) $M = \frac{1}{2}\left[\frac{\pi d}{\lambda}NA\right]^2$ (b) $M = \left[\frac{2\pi d}{\lambda}NA\right]$

(c) $M = \frac{1}{2}\left[\frac{\pi d}{\lambda}NA\right]^{1/2}$ (d) $M = \frac{1}{2}\left[\frac{\pi d}{\lambda}NA\right]^3$ **[Ans: (a)]**

10. Attenuation is an optical fibre is measured by

(a) $\text{loss} = 10\log_{10}\left(P_i \big/ P_o^2\right)$ (b) $\text{loss} = -10\log_{10}\left(\frac{P_o}{P_i}\right)$

(c) $\text{loss} = 10\log\left(P_i^2 \big/ P_o\right)$ (d) $\text{loss} = -10\log\left(\sqrt{P_i \big/ P_o}\right)$ **[Ans: (b)]**

11. Information signals are transmitted from optical fibre in the form of

(a) analogue signals (b) digital signals
(c) binary and hexadecimal bits (d) none of the above **[Ans: (b)]**

12. Dispersion in optical fibres is due to

(a) chromatic dispersion (b) material dispersion
(c) intermodal dispersion (d) all of the above **[Ans: (d)]**

13. The propagation of light along the waveguide is decided by the

(a) interference (b) dispersion
(c) modes of the waveguides (d) none of the above **[Ans: (c)]**

14. In multimode propagation the light propagates the fibre in

(a) straight line (b) zig zag fasion
(c) circular path (d) elliptical path **[Ans: (b)]**

15. In order that mode remain guided, the propagation factor β must satisfy

(a) $n_2 < \beta < n_1$ (b) $n_2 \frac{2\pi}{\lambda} < \beta < n_1 \frac{2\pi}{\lambda}$

(c) $n_2 < \beta = n_1$ (d) $n_2 = \beta < n_1$ **[Ans: (b)]**

MARK THE FOLLOWING STATEMENTS AS TRUE OR FALSE

1. Total internal reflection of light is responsible for the propagation of light in an optical fibre.
2. The numerical aperture of an optical fibre can be greater than one.
3. The optical fibre is simply a cylindrical waveguide system operating at optical frequencies.
4. An optical fibre consists of a core at the centre and a cladding or sheath outside the core.
5. The core of an optical fibre is generally a cylindrical dielectric glass with refractive index n_1 and the cladding is the second dielectric sheath or cover usually of glass with a lower refractive index n_2.
6. A light source which may be a laser or a LED is capable of generating an signal at desired frequencies.
7. In a step index fibre, the refractive index n_1 of the core material is uniform throughout and undergoes an abrupt change in step at the core cladding boundary.
8. In the case of a grade-index fibre the refractive index of the core is made to vary as a function of radial distance from the centre of the fibre.
9. Usually in graded index fibres refractive index of core changes in a nearly parabolic manner according to relation: $n^2 = n_2^2(1 - a^2x^2)$, where x is radial distance from axis of core and a^2 is a constant; n_1 is refractive index at the axis, i.e., at $x = 0$. At the core cladding interface ($x = R$, R radius of core), $n = n_2$.

ANSWERS

Question Numbers	1	2	3	4	5	6	7	8	9
Answers	T	F	T	T	T	T	T	T	T

Table 23.1 Fundamental Constants

Constant	*Symbol*	*Value*
Velocity of light	c	$2.9979 \times 10^{8}\,ms^{-1}$
Elementary chare	e	$1.6021 \times 10^{-19}\,C$
Electron rest mass	m_e	$9.1091 \times 10^{-31}\,kg$
Proton rest mass	m_p	$1.6725 \times 10^{-27}\,kg$
Neutron rest mass	m_n	$1.6748 \times 10^{-27}\,kg$
Planck constant	h	$6.6256 \times 10^{-34}\,Js$
	$\hbar = h/2\pi$	$1.0545 \times 10^{-34}\,Js$
Charge-to-mass ratio for electron	e/m_e	$1.7588 \times 10^{11}\,kg^{-1}C$
Quantem charge ratio	h/e	$4.1356 \times 10^{-15}\,Js\,C^{-1}$
Bohr radius	a_0	$5.2917 \times 10^{-11}\,m$
crompton wavelength:		
of electron	$\lambda_{c.e}$	$2.4262 \times 10^{-12}\,m$
of proton	$\lambda_{c.p}$	$1.3214 \times 10^{15}\,m$
Rydberg constant	R	$1.0974 \times 10^{-7}\,m^{-1}$
Bohr magneton	μ_B	$9.2732 \times 10^{-24}\,JT^{-1}$
Avogardo constant	N_A	$6.0225 \times 10^{-23}\,mol^{-1}$
Boltzmann constant	k	$1.3805 \times 10^{-23}\,JK^{-1}$
Gas constant	R	$8.3143\,JK^{-1}\,mol^{-1}$
Ideal gas normal volume (STP)	V_0	$2.2414 \times 10^{-2}\,m^{3}\,mol^{-1}$
Faraday constant	F	$9.6487 \times 10^{-4}\,C\,mol^{-1}$
Coulomb constant	K	$8.9874 \times 10^{-9}\,Nm^{2}C^{-2}$
Vaccum permittivity	ε_0	$8.8544 \times 10^{-12}\,N^{-1}\,m^{-2}\,C^{2}$
Magnetic constant	K_m	$1.0000 \times 10^{-7}\,m\,kg\,C^{-2}$
Vacuum permeability	μ_0	$1.2566 \times 10^{-6}\,m\,kgC^{-2}$
Gravitational constant	γ	$6.670 \times 10^{-11}\,N\,m^{2}\,kg^{-2}$
Acceleration of gravity at sea level and at equator	g	$9.7805\,ms^{-2}$
Numerical constants	$\pi = 3.1429$; $e = 2.7183$;	$\sqrt{2} = 1.4142$; $\sqrt{3} = 1.7320$

Table 23.2 Units and Symbols

Quantity	*Symbol*	*Name of unit*	*Relations to fundamental units*	
			MKSC	***MKSA***
Length	l	meter	m	
Mass	m	kilogram	kg	
Time	t	second	s	
Velocity	v		m s^{-1}	
Acceleration	a		m s^{-2}	
Angular velocity	ω		s^{-1}	
Angular frequency	ω		s^{-1}	
frequency	ν	hertz (Hz)	s^{-1}	
momentum	p		m kg s^{-1}	
Force	F	newton (N)	m kg s^{-2}	
Angular momentum	L or J		m^2 kg s^{-1}	
Torque	τ		m^2 kg s^{-2}	
Work	W	Joules (J)	m^2 kg s^{-2}	
Power	P	watt (W)	m^2 kg s^{-3}	
Energy	E_k, E_p, U, E	joule (J)	m^2 kg s^{-2}	
Temperature	T	K	m^2 kg s^{-2}/particle	
Coefficient of diffusion	D		m^2 s^{-1}	
Coefficient of thermal conductivity	K		m kg $s^{-3}K^{-1}$	
Coefficient of viscosity	η		m^{-1} kg s^{-1}	
Young's modulus	Y		m^{-1} kg s^{-2}	
Bulk modulus	k		m^{-1} kg s^{-2}	
Shear modulus	G		m^{-1} kg s^{-2}	
Moment of inertia	I		m^2 kg	
Gravitational field	G		m s^{-2}	
Gravitational potential	V_g		m^2 s^{-2}	
Charge	q, Q	Coulomb	C	As
Electric current	I	ampere	s^{-1}C	A
Electric field	E		m kg s^{-2} C^{-1}	m kg s^{-3} A^{-1}
Electric potential	V	Volt (V)	m^2 kg s^{-2} C^{-1}	m^2 kg s^{-3} A^{-1}
Current density	j		m^2s^{-1}C	m^{-2}A

(continued)

Table 23.2 (*continued*)

Electric resistance	R	Ohm (Ω)	$m^2\ kg\ s^{-1}\ C^{-2}$	$m^2\ kg\ s^{-3}\ A^{-2}$
Inductance	L	henery (H)	$m^2\ kg\ C^{-2}$	$m^2\ kg\ s^{-2}\ A^{-2}$
Electric permittivity	ε_0		$m^{-3} kg^{-1}\ s^2\ C^2$	$m^{-3}\ kg^{-1}\ s^4\ A^2$
Polarization	P		$m^{-2}\ C$	$m^{-2}\ s\ A$
Dielectric displacement	D		$m^{-2}\ C$	$m^{-2} s\ A$
Magnetic field	B	tesla (T)	$kg\ s^{-1}\ C^{-1}$	$kg\ s^{-2}\ A^{-1}$
Magnetic permeability	μ_0		$m\ kg\ C^{-2}$	$m\ kg\ s^{-2}\ A^{-2}$
Magnetization	M		$m^{-1}\ s^{-1}\ C$	$m^{-1}\ A$
Magnetizing field	H		$m^{-1}\ s^{-1}\ C$	$m^{-1}\ A$
Magnetic flux	Φ_B	Weber (Wb)	$m^2\ kg\ s^{-2}\ C^{-1}$	$m^2\ kg\ s^{-2}\ A^{-1}$
Electric dipole moment	P		$m\ C$	$m\ s\ A$
Electric quadrupole moment	Q		$m^2\ C$	$m^2\ s\ A$
Magnetic dipole moment	M		$m^2\ s^{-1}\ C$	$m^2\ A$
Magnetic quadrupole moment	Q		$m^3\ s^{-1}\ C$	$m^3\ A$
Capacity	C	farad (F)	$m^2\ kg^{-1}\ s^2\ C^2$	$m^{-2}\ kg^{-1}\ s^4\ A^2$

Table 23.3 Electromagnetic Spectrum

Class	***Gamma rays***	***X rays***	***Ultraviolet radiation***	***Visible light***	***Infrared radiation***	***Microwaves***	***FM and TV broadcasting***	***Radio AM broadcasting***
Wavelength, *m*	10^{-12} 10^{-11} 0.01Å 0.1Å	10^{-10} 1Å	10^{-9} 10Å 1$m\mu$ 10^{-8} 100Å, 10$m\mu$	10^{-7} 1,0000Å 100$m\mu$ 10^{-6} 10,000Å 1μ	10^{-5} 10μ 10^{-4} 0.1 mm 100μ	10^{-3} 1 mm 10^{-2} 1 cm	10^{-1} 1 10 10^{2}	10^{3} 1km 10^{4} 10km 10^{5} 100km 10^{6} 1,000km 10^{7}
Frequency, c/s	10^{21}–10^{20}	10^{19}	10^{18}–10^{17}	10^{16}–10^{15}	10^{14}–10^{13}	10^{12}–10^{11}	10^{10},10^{9},10^{8},10^{7} 10^{4}mc/s 10^{3}mc/s 100mc/s 10mc/s	10^{6},$10^{5}$$10^{4}$,$10^{3}$,$10^{2}$ 1mc/s 100kc/s 10kc/s 1kc/s
Quantum energy $h\nu$ (eV)	10^{7} 10 MeV 10^{6} 1 MeV	10^{5} 100 keV	10^{4} 10keV 10^{3} 1keV	10^{3} 10	1 10^{-1}	10^{-2} 10^{-3}	10^{-4} 10^{-5} 10^{-6}	
Characteristic temperature $h\nu$/k, K	10^{10} 10^{9}	10^{8}	10^{7}–10^{6}	10^{5}–10^{4}	10^{3}–10^{2}	10–1	10^{-1}	
Type of transition	Nuclear	Inner electron	Outer electron	Molecular vibration	Molecular rotation	Election spin	Nuclear spin	
Artificial production	Betatron	X-ray tube	Gas discharge	Hot filaments		Megnetron, klystron, travelling-wave tube	Electrical circuit	AC generator
Detection	Geiger and scintillation counters, ionization chamber		Photoelectric and photomultiplier cells		Bolometer thermophile	Crystal	Electrical circuit	

24

Nanomaterials

24.1 INTRODUCTION

Nanoscience and Nanotechnology primarily study with the synthesis, characterization, exploration and exploitation of nanomaterials. Nanomaterials are characterized by at least one dimension in the nanometer (1 nm = 10^{-9}m) range. Nanostructures constitute a bridge between molecules and infinite bulk systems. Individual nanostructures include clusters, nanocrystals, nanowires, quantum dots, and nanotubes, whereas collections of nanostructures involve arrays, assemblies, and super lattices of individual nanostructures. The physical and chemical properties of nano structured materials can differ significantly of the atomic–molecular or bulk materials having same composition. The uniqueness of the structural characteristics, energetics, response, dynamics, and chemistry of nanostructured materials constitutes the basis of nanoscience. Suitable control of the properties and response of these materials can lead to new devices and technologies, i.e., nanotechnology.

Nanotechnology can be considered as a group of emerging technologies in which structure of matter is controlled at the nanometer scale, to produce new, interesting nanostructured materials and nano devices that have useful and unique properties. Nanotechnology imposes only limited control of structure at the nano-metre scale producing useful products.

The impact of nanotechnology is already being felt in the form of new computer memories that provide rapid access to stored data, that can hold more of this data than the minidrives used in iPODs and do not need any external power source to retain the data. It is being felt in the form of prototypes for photovoltaic cells that can literally be sprayed onto buildings or computers to provide cheap power sources. And it is being felt in the form of "nanoengineered" gels that spread the recovery of damaged nerve cells.

Nanoscience and nanotechnology have grown explosively in the past two decades due to increasing availability of methods of synthesis of these materials as well as tools of characterization and manipulation. Several innovative methods of synthesizing nanoparticles and nanotubes and their assemblies are now available. The size-dependent electrical, optical and magnetic properties of individual nanostructures, metals and other materials are being understood in a better manner. These size dependent properties of these materials can be utilized to tailor them as per the requirements of a specific application.

There is great vitality in the area of nanoscience and nanotechnology and immense opportunities. Truly speaking, nanoscience is a interdisciplinary area covering physics, chemistry, biology, materials and engineering, and technology.

Nanoscience and nanotechnology has tremendous scope for applications over a wide range. This is likely to benefit various industrial sectors, including chemical and electronic industries, as well as manufacturing. Health care, medical practice, space research and environmental protection will definitely benefit from nanoscience.

Studies are being conducted on the potential use of nanomaterials in diverse applications including hydrogen storage, ion-sensing and gas sensing, surface-modified nanoparticles for enhanced oil recovery adsorption of chemical and biological agent on to nanoparticles, active electrode materials for lithium-ion batteries, light emitting devices, dental composition, etc., Stronger and lighter metals could be used in space as well as implants in the human body. Drugs for immediate effect in a body can be produced and could be delivered to specific location in a very short time. Electronic devices of smaller size, e.g. transistors can be fabricated, helping computers operate faster. Nanotechnology also finds applications in producing quantum dot lasers or tiny lasers, high density storage devices, nanoelectromechanical systems (NEMS), etc., Nanoscience and nanotechnology has applications in biological sciences too. The list of applications is increasing day by day with the progress of research in this field. One of the major and difficult problems facing the design of nanostructure based systems in understanding how to interconnect and address them. Obviously, the success of this field will depend on the development of new device and manufacturing technologies.

One can also see nanotechnology at work in nature. There are some species living in the sea produce shells surrounding them by making use of nanoscale calcium carbonate particles. These shells are quite strong and do not yield to pressure easily. Moreover, cracks produced on the surface of these shells do not propagate inside and thus the animals remain safe.

The field of nanoscience and nanotechnology has matured so rapidly and so fast that it is probably hard to find a segment of any technical subject where the implications of nanomaterials have not been explored at least to a preliminary extent. There is every reason to believe the field will make much progress in the coming decade.

24.2 NANOSCALE MIRACLES

The key element of nanoscience and nanotechnology is 'carbon', single sixth chemical element denoted with the symbol 'C', occupying a unique position in the periodic table of elements. Sometimes nanotechnology is also termed as 'carbon nanotechnology'. Carbon is a fundamental element in the living world, i.e., it is the basis of existence of life on the earth. We know that this element, as compound forms is present in the human body, food items, atmospheres, plants, fuels, etc., Carbon has four electrons in its outermost orbit, i.e., its valency is 4. Thus, carbon can bind itself or other atoms and form different compounds. Carbon atoms can form complicated network of molecules, which is the basis of organic chemistry. Carbon is the only element in the periodic table that has allotropes from zero dimensions to three dimensions. **Diamond, graphite** and **amorphous carbon** are considered to be three main allotropic forms of carbon. Out of these three allotropic forms two well known solid forms of carbon, i.e., diamond and graphite exhibit entirely different properties. Each of them is quite useful in its own way. Diamond is a hard material and known as a precious gem, largely used in jewelry. Diamond owes its sparkle partly to its high refractive index. Diamond is a poor conductor of electricity. On the other hand, graphite is remarkably soft and fair electrical conductor used in pencils. The difference in their properties is due to variation in the hybridization of carbon atoms. Diamond is sp^3-bonded (hybridization) where as graphite is sp^2-bonded (hybridization) as shown in Fig. 24.1 and Fig. 24.2 respectively. In sp^2-bonding among the four valence electrons of carbon, three are

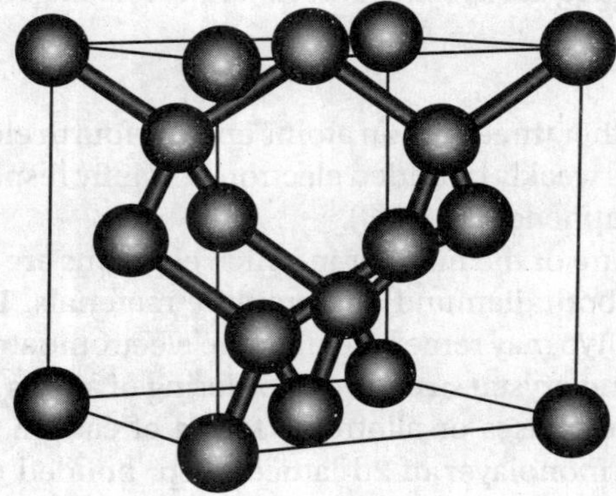

Fig. 24.1 Diamond structure

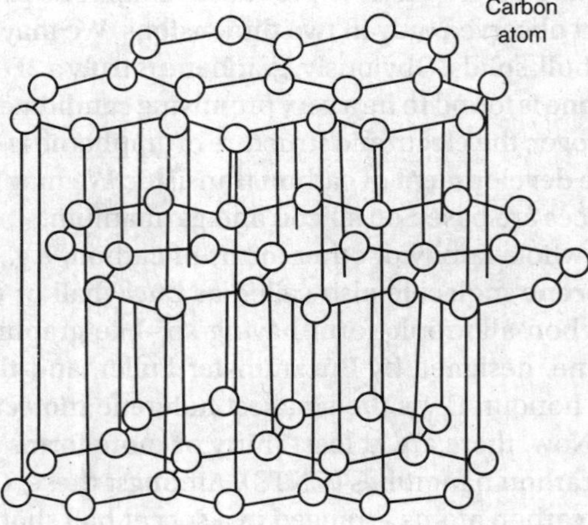

Fig. 24.2 The structure of graphite

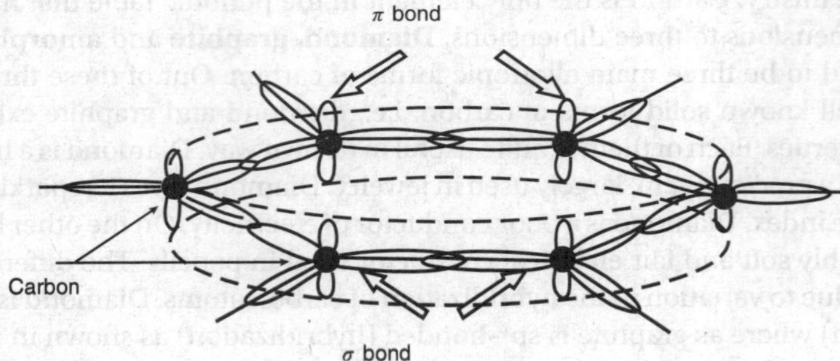

Fig. 24.3 Schematic of basic hexagonal bonding structure in grapheme sheet where carbons are denoted with circles, the σ-bonds are connecting carbon nuclei in plane and π-bonds are out of the plane

covalently bonded by other three carbon atoms and the fourth electron is weakly bounded to the carbon atom. This weakly bounded electron is mainly responsible for the interesting electron transport in graphene (Fig. 24.3).

The electronic structure or the manner in which electrons are placed according to their energy are different in both diamond and graphite materials. This is why their physical properties are different. We may remember that the electronic structure of a solid depends on its atomic number and crystal structure or ordering of atoms in the lattice.

There are other three phases or allotropic forms of carbon exist only in nano form. "Graphene", refers to a monolayer of 2d lattice of sp^2 bonded carbon atoms with bond length approximately 0.14 nm arranged in a honey comb lattice (with bonding similar to graphite). Graphene could be produced only recently in 2004 by micromechanical cleavage of graphite. We know that graphite is a layered structure with strong 2d bonds and very weak (van der Wall type) interlayer coupling. This property of graphite is used to cleave graphite crystal and 2d graphene is produced. Graphene exhibit ballstic transport and quantum hall effect observed only in two dimensions. We may note that these effects are not observed in 3D bulk solids. Obviously, graphene is truly a 2D crystal and is a gapless semiconductor. Graphene is found to be a very promising candidate to make **ballistic field effect transistor**. Moreover, the electronic structure of grapheme is quite peculiar and one can make its use for the development of carbon transistor. We may remember that almost all semiconductor devices are based on silicon and germanium.

To day, there exist a whole family of other forms of carbon, e.g., the hollow and **cage-like Buckminsterfullerene** molecule also called as buckyball or C_{60} fullerene. This is a zero "dimensional" carbon allotropic form having sp^2–like graphite bonding. The balls resemble geodesic dome, designed by Buckminster Fuller, and the buckballs or fullerenes are named in his honour. C_{60} is the smallest fullerene molecule in which two pentagons share an edge. Now, there are at least thirty or more forms of fullerenes, and also an extended family of carbon nanotubes (CNTS). Amongst these, C_{60} is the first spherical carbon molecule, with carbon atoms arranged in a soccer ball shape, i.e., the structure of C_{60} is a truncated icosahedrons, which resembles a soccer ball. Fig. 24. 4 is a schematic of C_{60} fullerene.

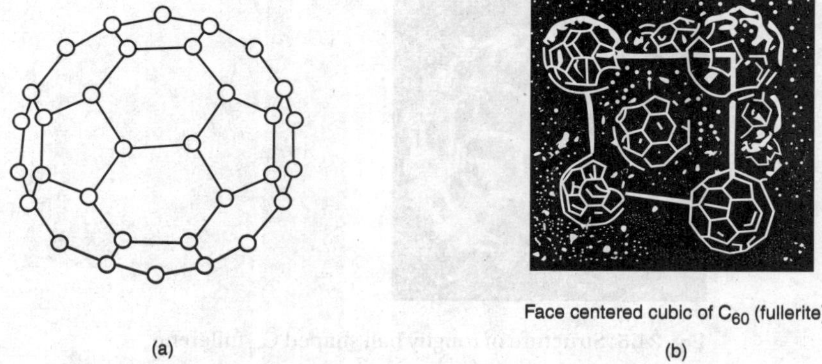

Fig. 24.4 (a) The structure of C_{60} molecule (b) Face centered cubic crystal of C_{60} fullerite

The structures of C_{60} molecule have 20 hexagons and 12 pentagons with a carbon atom at the vortices of each polygon and a bond along each polygon edge. The structure contains 60 carbon atoms in which five-membered rings is isolated by six-membered rings. Although the bonding in C_{60} is basically sp^2, but due to the curved surface of the icosahedron, some sp^3 hybridization is also present. C_{60} has a unique property of being able to carry something inside it. When doped with alkali metal atom, C_{60} fullerene molecule exhibit superconductivity. We may note that C_{60} molecule show wave-particle duality and hence it is a truly quantum mechanical system.

The main advantage gained upon functionalization of fullerenes is a substantial increase in their solubility. The existence of a great diversity of synthetic protocols combined with the high number of chemical reactions that have been mainly applied to C_{60} has led to the formation of a wide variety of functionalized fullerenes. Minute quantities of fullerenes in the form of C_{60}, C_{70}, C_{76} and C_{84} molecules are produced in nature. They have been observed in carbon root and also found to be formed by lightning discharges in the atmosphere.

Fig. 24.5 shows a slightly elongated and the spherical carbon molecule, has seventy carbon atoms and is known as C_{70}. The structure of C_{70} has extra six-membered carbon rings. Similarly there are also a good number of other potential structures containing the same number of carbon atoms.

However, their peculiar shape depend on whether five membered rings are isolated or not, or whether seven membered rings are present or not. So far no significant applications of fullerenes have been developed, yet, scientists are vey excited about the one new allotropes of carbon which is **carbon nanotube** (CNT). The current interest of researchers on CNTs is a direct consequence of the synthesis of **Buckminster-fullerene**, C_{60} and other fullerenes. They are molecules composed entirely of carbon, in the form of a tube (Fig. 24.6).

We have seen that carbon is the most amazing element in the periodic table. In the past 30 years, we have experienced the renaissance in three materials families: **graphite intercalation compounds** (GICs), **fullerene solids,** and **nanotubes**. GICs provided a wealth of detailed physical and chemical information, while immediate applications were frustrated by cost, lack of air stability, and most importantly, the lack of new or greatly improved applications and properties. Though the foundations of physics and chemistry remain

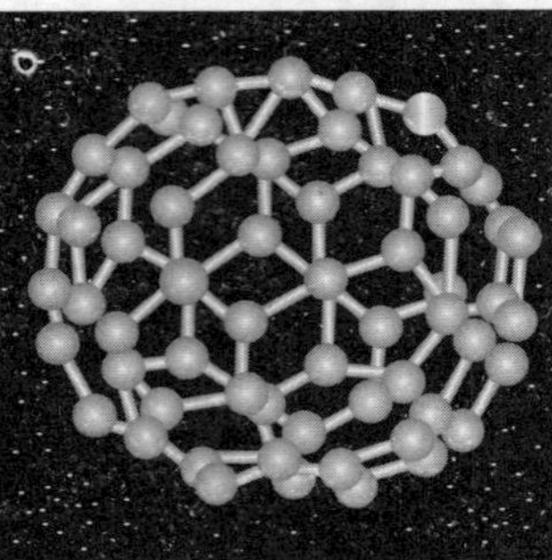

Fig. 24.5 Structure of roughy ball shaped C_{70} fullerene

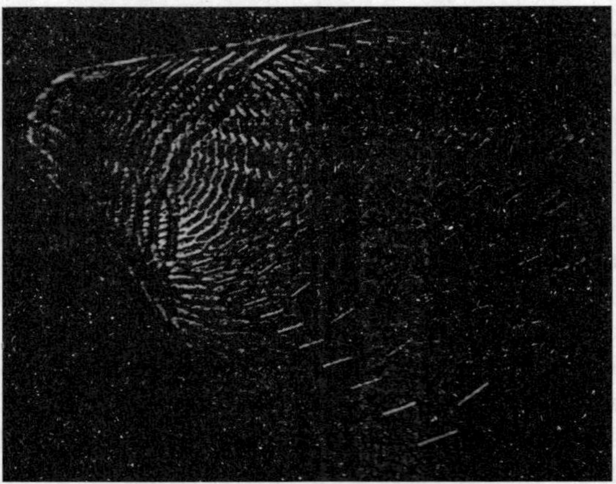

Fig. 24.6 Carbon nanotube structure

unshaken, the Li-ion battery industry has certainly benefited greatly from this knowledge. Fullerene solids fared better, but no large volume application has yet emerged. However, the elucidation of electronic properties in this highly correlated system has made invaluable contributions to fundamental science. Nanotubes may offer the best prospects yet. Much of the current research and commercialization of nanotechnology rely on tubes, balls and wires prepared from carbon atoms/or other materials. The carbon based tubes can be CNTs, bucky tubes and very long tubes, often referred to as **nanowires**. The combination of enhanced properties encourages the drive to multifunctional materials. Extension to other tubular and nano wire-like materials provide ample scope for new discoveries.

The various parameters of carbon atoms in different dimensional carbon materials are given in Table 24.1.

Nanomaterials have already shown a high potential that some of their applications in biology, medicine and drug delivery mechanisms have raised several ethical issues. We will discuss them in subsequent sections.

Table 24. 1 Various parameters of carbon atom in different dimensional carbon materials

Parameters	***Fullerene***	***Carbon Nanotubes***	***Graphite***	***Diamond***
Dimension	Zero	One	Two	Three
Hybridization	sp^2	sp^2	sp^3	sp^3
Crystals structure	FCC	FCC	Hexagonal	Cubic
Lattice constant (Å)	14.15 -		a=2.456 Å c=6.696 Å	3.513 Å -
Colour	Black solid	Black	Grey	Colorless
Density (g/cm^3)	1.72	1.2–2.0	2.26	3.51
Specific gravity	1.7–1.9	2	2.2	3.52
Bond length (Å)	1.4	1.44	1.42	1.54
Bond energy (eV/mole)	>25	>25	25	15
Melting point (°C)	>800 (sublimes)	3652	3652	3550
Thermal conductivity (Wm^{-1}K^{-1})	0.4		80–230	900–2320
Young's modulus (TPa)		1–1.2		
Tensile strength (GPa)		>60		
Band gap (eV)		0.5 eV (d=1.5 nm) 15 meV (d $\approx$ 1.5 nm)		5.47
Electrical conductivity (A/cm^2)		10^9		3.513

24.3 STUDY OF NANOSCALE MATERIALS

24.3.1 Transmission Electron Microscope (TEM)

Transmission electron microscope is very important tool in materials science for investigating the fine scale structure of materials.

TEM is a microscopy technique whereby a beam of electrons is transmitted through an ultra thin specimen, interacting with the specimen magnified and focused by an objective lens and appears on an imaging screen, a fluorescent screen in most TEMs, plus a monitor, or on a layer of photographic film, or to be detected by a sensor such as a CCD camera. The first practical TEM was built by Arebus and Hillier at the University of Toronto in 1938.

Figure 24.7 shows schematically a TEM illustrating the basic principles of its operation. An electron beam is produced by a heated tungsten filament at the top of an evacuated column and is accelerated down the column by a high voltage (usually from 75 to 120 kV). Electromagnetic coils are used to condense the electron beam, and then it is passed through a very thin *(i.e.,* about 100 nm thick or less) section of a thin metal specimen placed on the specimen stage (Fig. 24.7*)*. The specimen area examined must be very thin so that some of the electrons entering are able to pass through it. As the electrons pass through the specimen, some are absorbed and some of them are scattered so that they change direction. Differences in crystal atomic arrangement will cause electron scattering. After the electron beam has passed through the specimen, it is focused with the objective

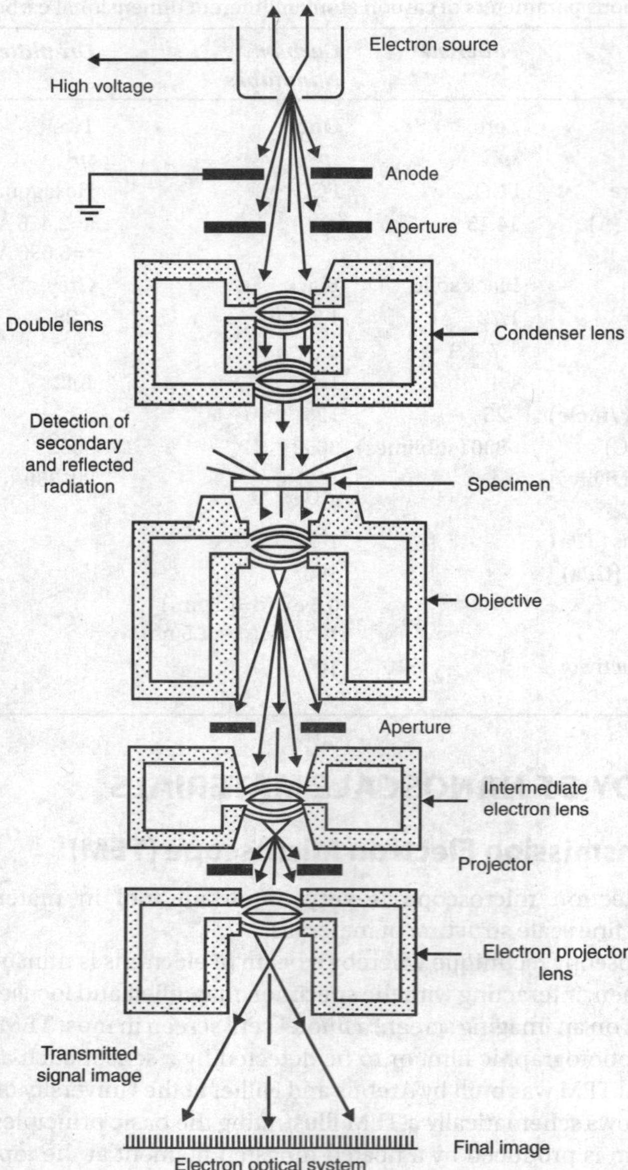

Fig. 24.7 **Schematic arrangement of electron lens system in a transmission electron microscope (TEM). All the lenses are enclosed in a column that is evacuated during operation. The path of the electron beam from the electron source to the final projected transmitted image is indicated by arrows. A specimen thin enough to allow an electron beam to be transmitted though it is placed between the condenser and objective lenses as indicated**

coil and then enlarged and projected on a fluorescent screen (Fig. 24.7). A region in a metal specimen that tends to scatter electrons to a high degree will appear dark on the viewing screen. Thus dislocations that have an irregular linear atomic arrangement will appear as dark lines on the electron microscope screen.

A TEM works much like a slide projector. A projector shines a beam of light through (transmits) the slide, as the light passes through, it is affected by the structures and objects on the slide. These effects result in only certain parts of the light beam being transmitted through certain parts of the slide. This transmitted beam is then projected into the viewing screen, forming an enlarged image of the slide.

TEMs work the same way except that they shine a beam of electrons (like the light) through the specimen (like the slide). Whatever part is transmitted is projected into a phosphor screen for the user to see. The working of typical TEMs is as follows:

The 'source' at the top represents the electron gun, producing a stream of monochromatic electrons.

- This stream of monochromatic electrons is focused to a small, thin, coherent beam by the use of condenser lenses. The first lens largely determines the 'spot' size, the general size range of the final spot that strikes the sample. The second lens (usually controlled by the "intensity of brightness knob") actually changes the size of the spot on the sample, changing it from a wide dispersed spot to a pinpoint beam.
- The beam is restricted by the condenser aperture (usually user selectable), knocking out high angle electrons (those for from the optic axis).
- The beam strikes the specimen and parts of it are transmitted.
- This transmitted portion of the beam is focused by the objective lens into an image.
- Optional objective and selected area metal apertures can restrict the beam, the objective aperture enhancing contrast by blocking out high-angle diffracted electrons, the selected area aperture enabling the user to examine the periodic diffraction of electrons by ordered arrangements of atoms in the sample.
- The image is passed down the column through the intermediate and projector lenses, being enlarged all the way.
- The image strikes the phosphor image screen and light is generated allowing the user to see the image. The darker areas of the image represent those areas of the sample that fewer electrons were transmitted through (they are thicker or denser). The lighter areas of the image represent those area of the sample that more electrons were transmitted through (they are thinner or less dense).

24.3.2 Applications of TEM

The TEM is used heavily in both material science/metallurgy and the biological sciences. In both cases the specimen must be very thin and able to withstand the high vacuum present inside the instrument.

For biological specimen, the maximum specimen thickness is roughly 1 micrometre. To withstand the instrument vacuum, biological specimen are typically held at liquid nitrogen temperatures after embedding in vitreous ice or fixated using a negative staining material such as uranyl acetate or by plastic embedding. Typical biological applications include tomographic reconstructions of small cells or thin sections of larger cells and 3-D reconstructions of individual molecules via single particle reconstruction.

24.3.3 Limitations

There are a number of drawbacks to the TEM technique. Many materials require extensive sample preparation to produce a sample thin enough to be electron transparent, which makes TEM analysis a relatively time consuming process with a low output of samples. The structure of the sample may also be changed during the preparation process. Also the field of view is relatively small, raising the possibility that the region analysed may not be characteristic of the whole sample. There is a possibility that the sample may be damaged by the electron beam, particularly in the case of biological materials.

24.3.4 Scanning Electron Microscope (SEM)

Figure 24.8 shows the schematic arrangement of SEM. Unlike the TEM, where electrons of the high voltage beam form the image of the specimen, the SEM produces images by

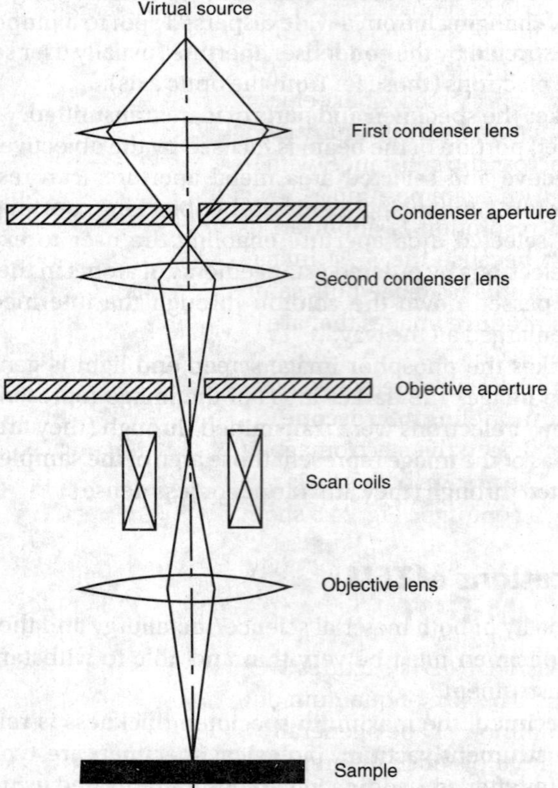

Fig. 24.8 Schematic diagram of SEM

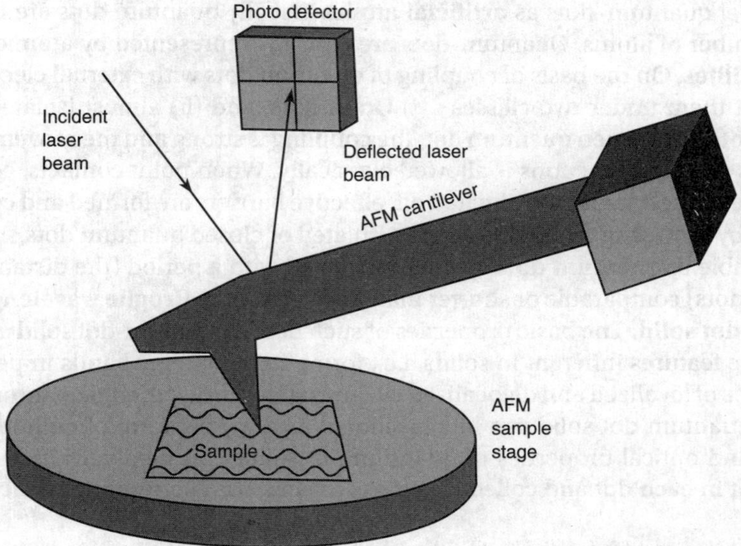

Fig. 24.9 Schematic diagram of AFM

detecting low-energy (<50 eV) secondary electrons that are emitted from the surface of the specimen due to excitation by the primary electron beam. In the SEM, the electron beam is allowed to move across the substance, with detectors generating an image by mapping the detected signals with beam position.

Usually, the TEM resolution is about an order of magnitude higher than the SEM resolution; however, because the SEM image relies on surface processes rather than transmission, it is possible to image bulk samples and also has a much greater depth of view. Thus SEM can produce images that are batter representation of the 3D structure of the sample.

Scanning probe microscopes are of two types: the (i) **atomic force microscope** (AFM) and the (ii) **Scanning tunneling microscope** (STM). In the AFM technique, a tiny tip moves up and down in response to the electromagnetic forces between the atoms of the surface and those on the tip. The motion of this tip is recorded and used to generate an image of the atomic surface using a computer. Fig. 24.9 shows a schematic arrangement of AFM.

24.4 QUANTUM-DOT

These are zero-dimensional (0D) structures 3–60 nm in size in which the carries are confined in all three directions. The size of quantum dots depends upon the application for which it is synthesized. Common shapes generally include pyramids, cylinders and spheres. Different synthesis route produce different kinds of quantum dots. The energy states of quantum dots are quantized in all directions and the density of states is represented by series of discrete, sharp peaks resembling that of an atom. On the basis of this comparison, one can label quantum–dots as **artificial atoms.** Usually quantum dots are formed by a definite number of atoms. Quantum-dots are typically represented by atomic clusters or **nanocrystallites.** On the basis of coupling of quantum dots with external electric circuits, one can put them under two classes : (i) **Open dots**, and (ii) almost isolated or **closed quantum dots**. In an open quantum dot, the coupling is strong and the movement of electrons across dot–lead junctions is allowed classically. When point contacts, connecting a dot with an external circuit, are pinched off, effective barriers are formed and conductance takes place by tunneling. These are almost–isolated or closed quantum dots.

An ensemble of **quantum dots** spatially arranged with a period (the distance between the nearest dots) comparable or shorter than the electron **de Broglie wavelength** is called a **quantum dot solid.** The basic properties of such dense quantum-dot solid are expected to reproduce features inherent in solids, i.e., formation of energy bands in perfect lattice and existence of localized and delocalized electron states in disordered quantum dot structures. The quantum dot solid can be considered as a special form of **colloidal crystals.** Electronic and optical properties of quantum dot solids can be determined by electron confinement in each dot and collective effects arising form periodic special organization of the dots.

Quantum dots can be fabricated with either a top-down technique or bottom-up technique. The top-down technique produces quantum dots with limited uniformity in diameters. Uniformity in diameters is crucial for the production of large arrays of dots that have consistent properties. Unfortunately, top-down approaches such as lithography are limited by the diffraction limit and cannot create dense arrays of quantum dots.

The most common way to produce a quantum dot is through the bottom-up approach. Bottom-up approach is a proper way to produce quantum dots in dense arrays that self-assemble in an orderly manner. The sites required for the nucleation of quantum dots seems to be provided by the high strain developed on the substrates. Such strain is developed at the interface between the substrate and the deposited layer. It is easier to produce quantum dots having consistent properties using approach.

A quantum dot is at least 10 times larger in size than an atom. A quantum dot emits only specific wavelength of electromagnetic radiation depending on its size. Quantum theory predicts that with decreasing size of quantum dots there will be a corresponding increase in energy of emitted light.

The ability to control the emission properties of a quantum dot has tremendous technological applications as well as of great academic interest to researchers. Unlike any other quantum structure, quantum dots have excellent confinement properties. This makes quantum dots extremely efficient in emitting light. Quantum dots have been the source of some of the world's most powerful lasers produced to date. **Quantum dot lasers** combines the length scales defining the confinemt of photons (100 nm) and the length scales defining the confinement of electrons and holes (10 nm). Moreover, quantum dot lasers exhibit reduced temperature resistivity. For fabrication, it is necessary to form high quality, uniform quantum dots in the active layer. The practicality of a **quantum dot laser** is still being improved. In medical studies, quantum dots are already in practice as **tags** that can be inserted into a patient's bodies. One can see these tags using most medical scanning technologies can help pinpoint the biological processes inside the body as they take place.

24.5 QUANTUM WIRE OR NANOWIRE

A nanowire or a quantum wire is a one dimensional (1D) structure that has a diameter of the order of a few nanometres. Such a wire is also known as a quantum wire. The length-to-width ratio, called the aspect ratio, for typical nanowires is about 1000 or more. As compared to quantum films, one more dimensions of the structure appears to be so small as to provide quantum confinement. Nanowires have many interesting properties that are not observed in bulk materials. This is because electrons in a nanowire occupy energy levels that are different from the energy levels or bands observed in bulk materials. Charge carriers are free to move only along the wire. Thus, only one kinetic component along with the confined energy contributes to the total energy. As a consequence, the density of states has $E^{1/2}$ dependence for each of the discrete pairs of states in the confined directions.

Quantum confinement exhibited by certain nanowires, e.g., carbon nanowires is reflected in the discrete values of the electrical conductivity. Such discrete values arise from a quantum mechanical restriction on the energy of electrons (and hence their number) travelling through nanowire. The difference in the successive values of conductivity is called *quantum conductivity*.

Nanowires have been grown out of metals, traditional semiconductors such as silicon and gallium, and a variety of polymers.

24.5.1 Conductivity of Nanowires

The conductivity of a nanowire is observed to be much less in comparison to the corresponding conductivity of the bulk material. This is due to a variety of reasons. One reason is the scattering of electrons from the boundaries of the wire. When the width of the wire is smaller than the mean free path of the free electrons in the bulk material the scattering of electrons from the boundaries increases. In copper, for example, the mean free path (at 300 K) is 40 nm. Copper nanowires less than 40 nm width will shorten the mean free path to the wire width, resulting in reduced conductivity. Silicon nanowires coated with nickel have shown improved conductivity.

Nanowire conductivity is found to be strongly influenced by the edge effects. The edge effects are due to atoms that are at the nanowire surface and are not fully bonded to neighbouring atoms as the atoms within the bulk of the nanowire. The unbounded atoms are often a source of defects within the nanowire and seems to cause the nanowire to conduct electricity more poorly in comparison to the bulk material. As a nanowire reduces in size, the surface atoms gets more numerous as compared to the atoms within the nanowire, and edge effects become more prominent.

The conductivity of quantum wire can undergo quantization. This can be explained on the basis of the quantization of energy of the electrons travelling through a nanowire. The conductivity of nanowire is thus the effect of the sum of the transport by separate *channels* of different quantized energy levels. The thinner the nanowire is, the smaller the number of channels available to the transport of electrons. One can verify this by measuring the conductivity of a nanowire while stretching it. Its conductivity decreases in a stepwise manner while its diameter shrinks.

24.5.2 Use of Nanowires

Applications of nanowires include sensors, interconnects, memory devices (potentially) logic devices and processors.

It is possible that semiconductor nanowire crossings may be important to the future of digital computing. No doubt, there are many other uses for nanowire, the only ones that actually take advantage of physics in the nanometre regime seems to be applications in fabrication of electronic devices.

Conducting nanowires offer the possibility of connecting molecular-scale devices in a **molecular computer**. Dispersions of conducing nanowires in different polymers are being investigated for use as transparent electrodes for flexible flat-screen displays.

Due to their high value of Young's moduli, application of nanowires in mechanically enhancing composites is also being investigated. Since nanowires appear in bundles, they may be used as tribological additives to improve friction characteristics and reliability of electronic transducers and actuators.

24.6 MODELLING OF QUANTUM SIZE EFFECT

The experimentally observed size effect can be explained with the help of following theoretical models.

1. ***Weak confinement region*** ($R >> a_B$)**:** The region wherein size of the cluster, R, is greater than the Bohr radius, a_B of the exciton in the bulk material is termed weak confinement region. In this region, due to size quantization of exciton, (exciton is a bound pair of electron and hole formed due to excitation of electron at an elevated energy level due to interaction with electromagnetic interaction) there is small increase, ΔE, in the exciton energy. This is given by $\Delta E = \hbar^2\pi^2/2MR^2$, where M is the mass of the exciton and is given by $M = m_e^* + m_h^*$, where m_e^* and m_h^* are effective masses of electron and hole respectively, and R is radius of the cluster.
2. ***Medium confinement region*** ($R \approx a_B$)**:** The region wherein size of the cluster is comparable to the Bohr radius of the exciton in the bulk material is termed medium confinement region. This is an usual situation observed in certain semiconductor compounds. In this region due to size quantisation of exciton there is increase in the exciton energy. This increase in energy is given by

$$\Delta E = \frac{\hbar^2\pi^2}{2m_e^* R^2} \quad \text{... (24.1)}$$

3. ***Strong confinement region*** ($R << a_B$)**:** The region wherein size of the cluster is smaller than the Bohr radius of the exciton in the bulk material is termed strong confinement region. There is an increase in the exciton energy due to size quantisation of exciton. This increase in energy is given by:

$$\Delta E = \frac{\hbar^2\pi^2}{2\mu R^2} \quad \text{... (24.2)}$$

where μ is the reduced effective mass of the electron.

The **EMA model** was further improved by Brus in 1984 and called as **Brus model** or **single-band effective-mass model**. In 1986, Kayanuma further improved this model and found to be in agreement with experimental results.

The **empirical tight binding model** (ETBM) proposed by Wang and Herron in 1990. was successful in explaining many phenomena observed in optical spectra.

Empirical pseudo-potential model (EPM) developed by Ram Krishna and Friesner in 1991 is well established and found to be a reliable approach. **Effective bond order model** (EBOM) developed by Einevoll in 1992 helped in calculating energy levels for particular phase of certain semiconductor compounds. However, it may be noted that all the models are not applicable for cluster sizes smaller than or equal to 1 nm.

24.7 SURFACE AND INTERFACE EFFECTS

Significant changes are reported in the properties of nanomaterials due to the spatial confinement of charge carriers in three dimensions. The surface properties of nanomaterials also play an important role in determining the properties of these materials. The larger surface-to-volume ratio observed at the nanoscale is one of the reasons for the typical properties of these materials. The number of surface atoms in a nanomaterial is appreciably larger than those in the bulk. This causes the defects to be located at the surface. These surface atoms and defects seems to playa major role in deciding the properties of these materials.

In a small cluster, a large fraction of atoms occupy surface positions, e.g., a 50-Å Cds cluster has about 15% of its atoms lying on its surface. The effect of relatively large proportion of atoms on or near the surface of clusters becomes prominent in the study of surface photons, absorbed species, and enhancement in the reactivity by using catalyst. The existence of this vast interface between the cluster and the surrounding medium seems to have a profound effect on the properties of the cluster. Whatever is the method of preparation, in most of the cases, the clusters have an imperfect surface. This leaves defects on the surface. These defects act as traps for the movement of the charge carriers after the photoexcitation. The presence of the trapped electrons or holes can modify the optical properties of these clusters and can give rise to **non-linear optical effects**. This can also alter the nature of the photochemical reactions which are of considerable interest in the field of photocatalysis. Controlling the cluster surface and in turn the cluster properties is a major challenge before researchers and provides a unique opportunity in this exciting and promising field.

The surface-to-volume ratio is important in the process of catalysis. The surface area of a catalyst is generally expressed in the units of square metre per gram and called as the **specific surface area** (S).

$$S = \frac{\text{area}}{\text{volume} \times \text{density}} \quad \text{... (20.3)}$$

The specific surface area per gram for geometrical shapes generally observed in nanomaterials are as follows:

$S = 6 \times 10^3/pd$ for a sphere with diameter d

$S = 6 \times 10^3/pa$ for a cube with side a

$S = 2 \times 10^3/pl$ for a thin disc whose length l is very small compared to its diameter d
$S = 4 \times 10^3/pd$ for a long cylinder with its diameter d very small compared to its length l.

24.8 PROPERTIES OF NANOMATERIALS

Research on Nanomaterials has been stimulated by their technological applications. These materials exhibit entirely new and special properties. Technologically useful properties of these materials are not limited to their structural, chemical or mechanical behaviours. For example, multilayers represent examples of materials in which one can modify or tune a property for specific application by sensitively controlling the individual layer thickness.

Very different physical laws dominate the nanoscale materails. Properties of material change. Some of these materials are better conductors of electricity or heat, some are mechanically stronger, some have different magnetic properties, and some reflect light better or change colours as their size changes. Nanoscale materials have a large surface area available for interactions compared to bulk materials of equal volume. We know that nanomaterials are formed by grouping a small number of molecules together. These are called **nanoclusters** or **simply clusters**. The smaller number of molecules in a cluster also gives rise to some properties that are observed only in the nanomaterials.

24.8.1 Optical Properties

Confining electrons to small geometries give rise to **'particle in a box'** energy levels. This quantum confinement creates new energy states and results in the optoelectronic properties of semiconductors. Because of their role in quantum dots, nano particles made of elements, which are normal constituents of semiconductors, have been the subject of much study, with particular emphasis on their electronic properties.

Optical properties such as the colour of a material depends its electronic structure. As the size of the material is reduced, the number of molecules that form the material gets smaller. *This* results in smaller size of the clusters that form the nanomaterial. The energy difference between two energy levels is inversely proportional to the size of the box, the size of a cluster (clusters (atomic) are large "molecules" containing typically from 10 to 2000 atoms. They have specific properties (mainly due to their large surface-to-volume ratio), which are size dependent and different from both those of atoms and the bulk materials) in this case. The energy levels of the electrons in the constituent molecules are no longer continuous as against the case of a solid. The energy levels get spread. *This* separation affects the energies required for the transitions of electrons to the excited states and the subsequent emission when the electron falls back to the ground state. This energy difference determines the colour of the light emitted. Clusters of different size therefore show different colours. The energy levels for electrons in clusters with different sizes are shown in Fig. 24.10.

For example, a piece of gold cut in smaller size is observed to change its colour in the nanosize regime. Depending on the size it can turn red, blue, yellow or some other colour. Materials having different size reflect and absorb light differently. Semiconductor cadmium selenide, in nanoscale size, can turn to any colour from red to blue depending on its size.

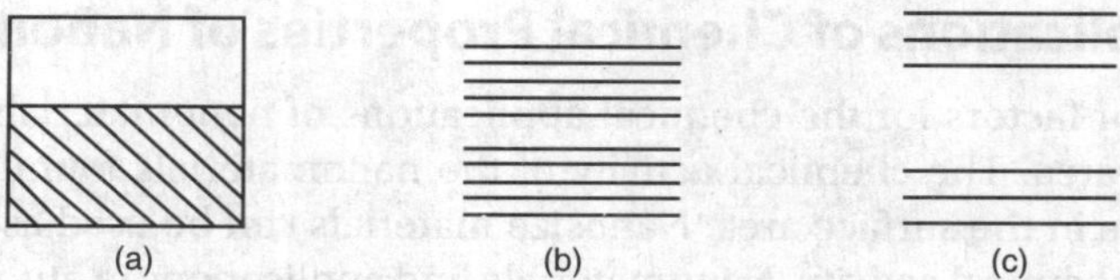

Fig. 24.10 Change in separation of energy level with size of cluster: (a) bulk metal, (b) large metal cluster, and (c) small metal cluster

Similar changes in colours have been reported in the nature also. For example, leaves of a tree or plant appear green because the atomic arrangement in them is such that it absorbs all colours except green which is reflected. When leaves die, chlorophyll, the major constituent responsible for their green colour, breaks down changing the atomic arrangement and they appear brown. Colours observed on a peacock feather are also due to the nanosize particles on it.

24.8.2 Electrical Properties

Nanomaterials can store considerably more energy than the conventional materials due to their large surface area. This property of nanomaterials finds very interesting and useful applications, e.g. high-energy-density batteries.

We known that conventional and rechargeable batteries are used in almost all applications that require electric power. The energy density (storage capacity) of these batteries is quite low and they require frequent recharging. Nanocrystalline materials are found to be good as separator plates in batteries as they can hold considerably more energy than the separator plates made form conventional bulk materials, e.g., nickel-metal hydride batteries made of nanocrystalline nickel and metal hydrides are likely to require far less frequent recharging and to last much longer.

24.8.3 Large Electrochromic Display Devices

The electrochromic (a colour change of matter caused by an electric current) device consists of materials whose optical properties can be changed by passing electric current through them or by applying an electric field. These devices are similar to *liquid-crystal displays* (LCD) commonly used in calculators and watches and are primarily used in public hoardings and ticker boards to display information. The resolution, brightness, and contrast of the display of these devices depend on the grain size of the material used in them. Nanomaterials, such as tungstic oxide gel (tungsten oxide), are being explored for this purpose.

24.8.4 Chemical Properties

The electronic structure of nanomaterials depends on their size and therefore their ability to react chemically depends on their size. The higher surface-to-volume ratio is also responsible for their higher chemical reactivity. For example, gold is fairly inert when in the bulk form, but at the nanoscale, it is highly reactive.

24.8.5 Applications of Chemical Properties of Nanomaterials

One of the major factors for the chemical applications of nanomaterials is the increase in their surface area. The chemical activity of the nanomaterials found to be increases with the increase in the surface area. Nanosize materials can be used as catalysts due to their enhanced chemical activity. Nanomaterials find applications in automobile catalytic converters and power generation equipment. Nanomaterials react with harmful and toxic gases e.g., carbon monoxide and nitrogen oxide to prevent environmental pollution arising from burning gasoline and coal. Fuel cell technology is another important application of the noble metal nanoparticles where they are used as catalysts. The present-day fuel cell catalysts are based on the platinum (Pt) group metals. Pt and Pt-Ru alloys are some of the most frequently used catalysts from this group. Obviously, the use of these metals is one major factor for higher costs of these cells. One possibility to produce economical catalysts is the use of bimetallic nanoparticles.

Gold nanoparticles are widely studied as they have several potential applications. For e.g., they have been used for the identification of the DNA sequences associated with disease, as well as viruses. New devices for use in computers can also be fabricated by using these materials. Obviously, these materials would therefore be excellent components in nanoelectronics, optics, and biology.

24.8.6 Mechanical Properties

Scientific challenges in nanoscience and nanotechnology include the development of nanomaterials with novel mechanical properties. The need for scratch, mar, and/or abrasion resistance is well established in various markets, including fingernail polishes, flooring, plastic glazing, headlamp covers and other automotive parts, transportation windows, and optical lenses, where clear scratch-resistant coatings are used. Many of the mechanical properties of materials are modified at the nanoscale. Hardness and elastic modulus, fracture toughness, scratch resistance, and fatigue strength are some properties that are modified. Energy dissipation, mechanical coupling within arrays of components, and mechanical non-linearities are influenced by structuring components at the nanometre scale.

A Number of important questions regarding the mechanical behaviour of nanocrystalline metals are yet to be answered; foundamentally the underlying mechanisms of deformation at both low and high temperatures are unknown. The major research challenge in this area currently lies in the areas of synthesis and processing. Until nanocrystalline materials can be produced which are dense, free of flaws, of high purity, and of sufficient quantity to carry out a variety of mechanical tests, measured values will be influenced by extrinsic rather than intrinsic properties. For high temperature applications, a great amount of research work needs to be performed in controlling grain sizes so that the correct balance of ductility, formability, and strength are achieved. Inspite of these challenging problems, nanocrystalline materials offer a wide variety of opportunities for fabricating new materials with precisely engineered microstructures where strength and ductility can be finely tuned. Moreover, the ability to structure materials on the nanoscale level makes it possible to test fundamental ideas about grain boundaries, dislocations, and deformations.

24.8.7 Mechanical Applications of Nanomaterials

Nanocrystalline materials possess unusual properties and their deformation behaviour has attracted intense activity in recent years.

Tougher and harder cutting tools can be developed using nanomaterials. Nanomaterials required for these purposes are made from tungsten carbide, tantalum carbide, and titanium carbide. Such tools are found harder, much more wear-resistant, erosion-resistant, and also last longer than their conventional (large-grained) counterparts. Moreover, for the miniaturisation of microelectronic circuits, the industry needs microdrills (drill bits with diameter less than the thickness of an average human hair or 100 μm) with enhanced edge retention and far better wear resistance. Nanocrystalline carbides are quite stronger, harder, and wear-resistant and therefore they are currently being used in these microdrills.

Nanomaterials can be used to manufacture automobiles with greater fuel efficiency Nanomaterials are quite stronger, harder, much more wear-resistant and hence they could be coated on spark plugs and other engine parts. Moreover, automobiles waste significant amounts of energy by losing the thermal energy generated by the engine. Nanocrystalline ceramics, e.g. zirconia and alumina retain heat much more efficiently. Engine cylinders coated with such materials are under consideration. This would result in complete and efficient combustion of the fuel.

One of the key properties required of aircraft components is their fatigue strength. Fatigue strength decreases with the age of the component. Moreover, the fatigue strength increases with a reduction in the grain size of the material. Nanomaterials provide such a significant reduction in the grain size over conventional materials that the fatigue life is increased by an average of 200-300%. In spacecraft, elevated-temperature strength of the material is crucial as the components (such as rocket engines, thrusters, and vectoring nozzles) operate at much higher temperatures than aircrafts and higher speeds. Nanomaterials are useful to enhance performance characteristics of aerospace components as well.

Ceramics are quite hard, brittle, and at same time difficult to machine even at high temperatures. However, with a reduction in grain size, the properties of ceramics change drastically. Ceramics made of nanocrystals are found to be ductile in nature. They can be pressed and sintered into various shapes at significantly lower temperatures. Zirconia in bulk form for example, is a hard, brittle ceramic. At nanoscale, it turns into a super plastic material, i.e., it can be deformed to a great extent (up to 300% of its original length). These ceramics have nanocrystalline grains. Ceramics based on silicon nitride (Si_3N_4) and silicon carbide (SiC) have been used in automotive applications as high-strength springs, ball bearings, and valve lifters. They have also been used as components in high-temperature furnaces as they possess good formability and machinabilty combined with excellent physical, chemical, and mechanical properties.

Nanocrystalline materials are found to provide better thermal insulation. Aerogels are porous and extremely light weight nanocrystalline materials. They can withstand 100 times their weight. They are currently being used for insulation in offices and homes. They are also being used as materials for "smart" windows, which darken when the sun is too bright and lighten otherwise.

24.8.8 Magnetic Properties

The magnetic properties of nanometer sized particles differ in several respects from the properties of their bulk counterparts. The large fraction of atoms located on surfaces or interfaces for example, whose local environments differ greatly, from those of the interior atoms leads to a blurring of the distinction between intrinsic and extrinsic properties since nano phase particles can be as much as 50% surface material, new magnetic properties characteristics of surfaces and interfaces become important from applications point of view.

Very interesting magnetic properties are observed in small **clusters**. The magnetic moment of each atom in cluster interacts with the moments of other atoms and forces all the other moments to orient in one direction with respect to some symmetry axis of the cluster. The cluster thus has a net magnetic moment and it is magnetised. As the cluster size decreases, it becomes easier for the atoms in a cluster to orient their magnetic moments in one direction and display ferromagnetic behaviour. In some cases, non-magnetic materials show a prominent increase in their magnetic moment when they contain less than 20 atoms.

The strength of a magnet is measured in terms of **coercivity** and **saturation magnetization**. It is observed that these values increase with a decrease in the grain size and increase in the specific surface area (surface area per unit volume) of the grains. Obviously, nanomaterials present magnetic properties that are not observed so far.

24.8.9 Magnetic Applications of Nanomaterials

Today information technologies ranging from personal computers to main frames use magnetic materials to store information on tapes, floppy diskettes, and hard disks. Our seemingly insatiable appetite for more computer memory will probably be met by a variety of magnetic recording technologies based on nanocrystalline thin-film media and magneto-optic materials. Personal computers and several of our consumer and industrial electronics components are now powered largely by lightweight switch mode supplies using new magnetic materials technology that was unavailable two decades ago. Magnetic materials touch many other aspects of our lives.

Magnetic properties of nanometer sized particles of both ferrimagnetic and ferromagnetic materials have attracted considerable attention in recent years. such interest has been generated for practical as well as theoretical regions. Some of the application areas being explored are in high-density magnetic recording media, as ferrofluids, giant magnetoresistive systems. Some notable applications of nanostructured magnetic materials are: Storage applications, giant magneto resistance (GMR) recording head, high power magnets.

Magnets made of nanocrystalline yttrium-samarium-cobalt grains possess very unusual magnetic properties due to their extremely large surface area. Such magnets are observed to possess high magnetic retention. Typical applications for these high-power rare-earth magnets include quieter submarines, automobile alternators land-based power generators, and motors for ships, ultra-sensitive analytical instruments, and magnetic resonance imaging (MRI) in medical diagnostics.

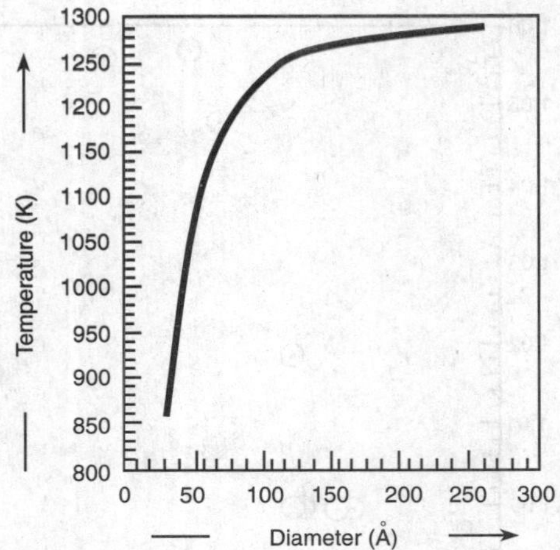

Fig. 24.11 Variation in the melting temperature of gold with reduction in the particle size

24.8.10 Melting Temperature

The large surface-to-volume ratio of nanomaterials exhibit a very unusual effect on their thermal properties. They show substantially lower melting temperatures as compared to the solids. Negative heat capacities are also reported in some nanomaterials. The theoretical analysis and calorimetric investigations shows that in the temperature range $10\ K \leq \theta_D$ (Debye temperature) the heat capacity of nanopowers is from 1.2 to 2 times higher than that of coarse grain bulk materials. One can easily explain these observations. When a large system e.g., a solid is heated, the energy added is completely converted into potential energy of the atoms/molecules. The solid is continuously converted to liquid form due to this added energy. The kinetic energy of the atoms/molecules and therefore the temperature of the solid thus remain constant. A small system like a nanomaterial on the other hand, tries to avoid partially molten states and converts some part of its kinetic energy into potential energy. Obviously, the kinetic energy of the nanomaterial decreases and it becomes cooler. Studies of the variation in the melting temperatures of gold reveal such behaviour. Variation in the melting temperature of gold with reduction in the particle size is shown in Fig. 24.11.

24.8.11 Structural Properties

One can determine the size of a nanocluster by the number of molecules that form it. Naturally, the arrangement of these molecules is quite different when the number of molecules forming the cluster changes.

The size dependence of the structural properties leads to changes in many other properties of nanomaterials. Structural properties of metallic nanocluster are studied using the XRD analysis.

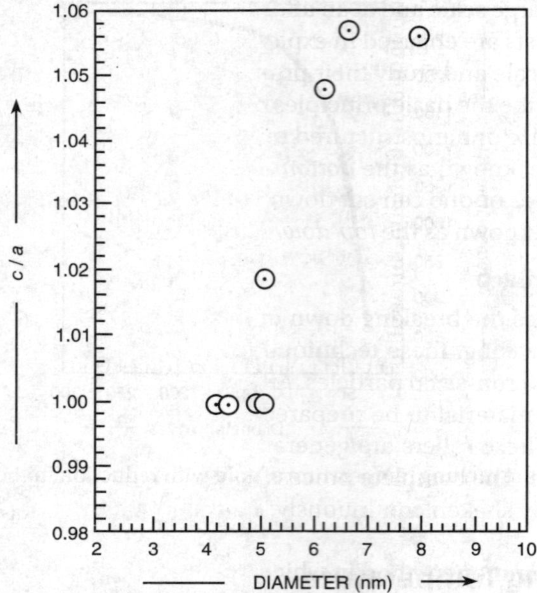

Fig. 24.12 Changes in the structure of aluminium

Aluminum is face-centered cubic in the bulk form. It is observed to be in the face-centered cubic hexagonal close packed, and the icosahedral structure at nanoscale. Similarly, indium is observed to undergo a transition from its tetragonal structure in the bulk form to the face centered. The changes in the structure of aluminium is shown in Fig. 24.12.

The lattice structure of nanometer-sized crystallites in nanocrystalline samples synthesized by various methods showed that the lattice structure of the nanometer-sized crystallites evidently deviates from the equilibrium state. One may classify the deviation as: (i) distorted lattice structures and stoichiometric line compounds, and (iii) formation of metastable phase below a critical crystalline size.

Structural changes in the nanomaterials also cause changes in the electronic structure of the clusters. The band levels and the valence energy change when the number of molecules in a cluster changes. The energy levels are in the form of a continuum when the material is in the bulk form. As the number of molecules is reduced, the energy levels become discrete and the material behaves in a different way. This change in the energy levels is shown in Fig. 24.12.

24.8.12 Preparation of Nanoscale Materials

Prior to the invention of transistor, metals were treated as the most important solids. Metals are obtained from earth crust and then purified. Usually, metals are used in the form of alloy. Similarly, elemental semiconductors such as silicon are also used in the form of either alloys or compounds. The compound semiconductors, ceramics, oxides, and insulators used in industry are prepared in the laboratory. Although naturally occurring nanoscale materials have been discovered, use of nanocrystals will be feasible only when they are prepared in large scale and at an affordable cost. As of today, research in this field is exploratory. Scientists are engaged in exploring various methods to produce nanoscale materials on a large scale and study their properties. A variety of methods are attempted. Here, we briefly describe the basic principles of preparation of nanoscale materials.

In principle, nanosize or nanostructured materials can be prepared by either arranging the atoms one by one, known as the *bottom-up approach,* using the techniques such as atom force microcopies or one can cut down the bulk or scrape the macroscopic solid to obtain nanosize solid known as the *top-down approach.*

(i) Top-Down Approach

This approach involves the breaking down of the bulk material into nanosize structures or particles. Truly speaking, these techniques are an extension of those that have been used for producing micron-sized particles. An example of such a technique is high energy wet ball milling. The material to be prepared is passed through a rolling mill consisting of two hard rollers. These rollers are generally made from hardened steel or are coated with tungsten carbide. The complete process is carried out in a sealed container, and the container is vigorously shaken continuously. This method produces the nanoparticles in powder form.

There is another common method in which the bulk material is cut to the required size and shape using an electron beam. The process requires a very high vacuum and involves sophisticated equipments for control. Other radiations such as lasers, X-rays or UV rays are also used in this technique. The size of the nanomaterials to be fabricated is governed by the wave length of the radiation used. Obviously, this puts some restrictions on the radiations that can be used for nanofabrication. Electron beams are generally used to fabricate nanoscale devices as the wave length of the beam can be controlled by changing its energy.

The drawback of the top-down approach is the imperfection of surface structure and significant crystallographic damage to the processed patterns. Further, it can also introduce internal stress in the material produced, which can cause changes in the properties of the material. This causes difficulties in the device design and fabrication. However, this approach is useful for the bulk production of nanomaterials.

(ii) Bottom-Up Approach

The bottom-up approach has the potential of creating less waste and hence more economical. This approach refers to the build up of a material from the bottom: atom-bottom molecule-by-molecule, or cluster-by-cluster. This is a method used for a long time to prepare small sized materials. There are several different methods involving the bottom-up approach. Wet chemical synthesis of nanomaterials or the sol-gel method is a very commonly used method. This method promises a better chance to obtain nanostructures with less defects, more homogeneous chemical composition.

(iii) Chemical Methods

Chemical methods of preparing nanomaterials are quite simpler in principle, versatile, and are inexpensive. Large number of materials such as metals, semiconductors, insulators, ceramics, and other complicated solids are grown in nanostructured form by employing these methods. The chemical reactions can be carried at low or at high temperatures. Low-temperature methods use reactions of aqueous solutions, while high temperature methods use different solvents. A salt containing cation and the one containing anion is mixed under controlled condition of concentration, rate of reaction, temperature, and pH of the solvents in presence of some organic molecules that work as binders. The large organic molecules bind to cations and/or anions and control the size of the precipitate. Depending on the concentration of cations, anions, and organic molecules nanocrystals of the required size can be formed in the solution. For e.g., one of the reactions for formation of zinc sulphide (ZnS) nanocrystals is: zinc acetate + sodium sulphide + thiophenol →zinc sulphide + sodium acetate + zinc-thiophenol

$$Zn(CH_3COO)_2 + Na_2S + C_6H_5SH \rightarrow ZnS + 2Na(CH_3COO) + Zn\text{-}SC_6H_5$$

Zinc sulphide is insoluble in organic solvents like methanol and hence precipitates out. However, sodium acetate is highly soluble in normal solvents and remains in the solution. Nanocrystal powder thus formed is separated out by using centrifugal machine and is washed several times so that the reaction byproducts and un-reacted solvents are removed. Different chemical reactions produce nanocrystals with different physical properties.

Nanocrystals can also be obtained by heating the powder of the material with glass, melting the mixture, and re-solidifying the same. Many old churches in Rome, old palaces in India have decorative coloured glasses made from nanocrystals grown in this fashion. We can say that knowledge of growing nanocrystals is very old while they are looked upon with new focus for the last three decades.

The sol-gel process, involves the formation of inorganic networks through the formation of a colloidal suspension *(sol)* and gelation of the sol to form a network in a continuous liquid phase *(gel)*. The precursors for synthesising these colloids consist usually of a metal or metalloid element surrounded by various reactive ligands. The starting material is processed to form a dispersible oxide and forms a sol in contact with water or dilute acid. Removal of the liquid from the sol yields the gel, and the sol/gel transition controls the particle size and shape. Calcination of the gel produces the oxide. Various steps involved in the sol-gel process can be summarized as follows:

(i) Formation of different stable solutions of the alkoxide or solvated metal precursor (the sol).

(ii) Gelation is the formation of an oxide- or alcohol-bridged network (the gel) by a polycondensation or polyesterification reaction that results in a dramatic increase in the viscosity of the solution.

(iii) Ageing of the gel (syneresis), during which the polycondensation reactions continue until the gel transforms into a solid mass, accompanied by contraction of the gel network and expulsion of solvent from gel pores. Some times the smaller particles merge with the larger particles and form coarse particles. This is called **as Ostwald ripening** or **coarsening.**

(iv) **Drying of the gel**, in this step water and other volatile liquids are removed from the gel network. However, this process is complicated due to fundamental changes in the structure of the gel.

(v) **Dehydration,** in this step the surface-bound M-OH groups are removed and the gel is stabilised against dehydration.

(vi) **Densification and decomposition** of the gels at high temperatures ($T > 800°C$). In this step, the pores of the gel network are collapsed and remaining organic species volatilised.

This synthesis method offers the possibility of synthesising non-metallic inorganic materials e.g., glasses, glass ceramics, or ceramic materials at very low temperatures compared to the high temperature process required by melting glass or firing ceramics.

It is very difficult to control the growth of particles and then stop the newly formed particles from mixing together. Moreover, we must ensure that the reactions are complete so that no undesired reactant is left on the product and completely removing any growth aids that may have been used in the process. However, the production rates of nanopowders are very low by this process. The main advantage of any bottom-up approach is that one can obtain same-sized particles.

(vii) Physical Methods

Most of the methods of preparing nanoscale materials discussed so far are physical methods of preparing nanoparticles. In addition to these methods, radio frequency (RF) plasma and molecular beam epitaxy (MBE) are the techniques also used to fabricate nanomaterials.

24.8.13 R.F. Plasma

Radio frequency plasma technique is used to synthesise metal nanoparticles. The metal is placed in a specially designed container and kept in a vacuum chamber. By using RF plasma it is evaporated with the help of RF coils surrounding the container. By passing helium gas high-temperature plasma is generated inside the RF coils. The metal vapours condense on to the helium gas atoms and diffuse towards a collector rod, i.e., at a lower temperature and form the nanoparticles. These are then passivated by allowing to pass some gas like oxygen through the chamber.

24.8.14 Molecular Beam Epitaxy

This is an advanced technique used to form nanoparticles. In this technique, a substance is evaporated under ultra-high vacuum (UHV $\approx 10^{-9}$ torr). The required material is evaporated at a very slow rate ($1\ \mu m/h$) and a thin film is deposited on a suitable substrate (supporting material) heated to the required temperature. Semiconductor heterojunctions can be grown in a controlled manner by this technique. The substrate material and the material to be deposited must be extremely pure so that any contamination in the device is avoided. This technique offers a very good control over the thickness of the material, it is reasonably reproducible, and the material prepared is also in its purest form. The UHV helps to make it possible to analyse and characterize the material deposited while it is being prepared. This method is highly advanced as well as expensive. However, nanodevices are fabricated by using this method.

24.8.15 Biological Nanomaterials Biomimicking

Biomaterials have received a considerable amount of attention over the last three decades as a means of treating deseases and easing suffering. Biomaterials have found application in about 8000 different kinds of medical devices, which have been used in repairing skeletal systems, returning cardiovascular functionality, replacing organs, and repairing and returning senses. Even though biomaterials have had a pronounced impact in medical treatment a need still exists to be able to design and develop better polymer, ceramic and metal systems.

Nanotechnology is playing an important role to improve biomaterial performance.

Most research focusing on biological micro-electro-mechanical systems (Bio MEMS) is for their use in diagnosis devices and for the detection of DNA, virsues, proteins, and other biologically derived molecules. Nanoscale Bio MEMS could allow for the real-time detection and analysis of signaling pathways, which would further our knowledge and understanding of the basic mechanisms and functions of the cell.

It is now believed that significant evidence exists that highlights the promise nanotechnology has for biological applications.

Many biological materials can be classified as biological nanomaterials e.g., proteins, viruses, and bacteria are some of the familiar examples. The building blocks of proteins are the amino acids. The size of each amino acid is about 0.6 nm. Out of more than 100 odd amino acids that occur naturally, only 20 are involved in the formation of proteins. A protein chain is formed by tying hundreds or some times thousands of amino acid molecules together. Such chains resemble nanowires. The basic building blocks for the DNA are four nucleotide molecules that bind together in the form of a double helix. Thus, the DNA in the form of a double helix is also an example of double nanowire.

(i) Microelectromechanical Systems (MEMSs)

Microelectromechanical systems (MEMSs) are small integrated devices or systems that combine electrical and mechanical components. To make MEMSs devices, lithographic techniques are combined with metal deposition processes. MEMSs involve a mechanical response to an applied electrical signal, or an electrical response resulting from a mechanical deformation. The size or MEMSs devices or systems vary in the sub-micrometre (or sub-micron) level to the millimetre level. A system can have any number of such devices, from a few to millions. The same technology used for fabrication of integrated circuits is extended to fabricate the MEMSs. The MEMSs has mechanical elements such as beams, gears, diaphragms, and springs to devices along with the electrical components on it.

The main advantage of MEMSs devices are miniaturization, multiplicity, and the ability to directly integrate these devices into micro electronics. **Multiplicity,** refers to the large number of devices and designs that can be manufactured rapidly, e.g., **miniaturization** has enabled the development of micrometer-sized accelerometers for activating airbags in cars.

MEMSs devices have applications which include inkjet-printer cartridges, accelerometers, miniature robots, microengines, locks, inertial sensors, microtransmissions, micromirrors, microactuators, optical scanners, fluid pumps, transducers, and chemical pressure and flow sensors. New applications or MEMSs devices and systems are emerging as the existing technology is applied to the miniaturisation and integration of conventional devices.

MEMSs systems can sense, control, and activate mechanical processes on the microscale and function individually or in arrays to generate effects on the macroscale. The microfabrication technology enables fabrication of large arrays of devices, which individually perform simple tasks, but in combination can accomplish complicated functions.

Future applications of MEMSs will be driven by processes enabling greater functionality through higher levels of electronic-mechanical integration and also greater numbers of mechanical components working alone or together to enable a complex action. Future MEMSs products will like to have higher levels of electrical mechanical integration and also more intimate interaction with the physical world. Advancing from their success as MEMSs sensors, MEMSs products will be embedded in larger non-MEMS systems, such as printers, automobiles, and biomedical diagnostic equipment, and will also enable new and improved systems. MEMSs devices are manufactured using batch fabrication techniques similar to those used for integrated circuits, unprecedented levels of functionality, reliability, and sophistication can be placed on a small silicon chip at a relatively low cost.

There are several devices and machines that have micrometer–sized elements.

(ii) Nanoelectromechanical Systems (NEMSs)

NEMSs are similar to MEMS but smaller in size and are related to nanotechnology and nanomechanics. NEMSs machines and devices are in the early states of development and many are still in conceptual stages. Numerous computer simulations of possibilities and ideas have been advanced. Because of their nanoscale size, NEMSs could be useful to measure small displacements and forces at a molecular scale. There are two standard methods of fabrication of NEMSs machine and devices: the top-down and the bottom-up approach. The top-down approach is accepted as the most standard approach to NEMSs. This approach may be summarised as "a set of tools designed to build a smaller set of tools," e.g., a millimetre-sized factory builds micrometre-sized factories, which in turn can build nanometre-sized devices and machines. The other approach can be thought of as putting together single atoms or molecules until a desired level of complexity and functionality is achieved in a device. Such an approach may utilise molecular self-assembly or mimic molecular biology systems.

The first very large scale integration NEMS device was demonstrated in 2000 by scientists at IBM. Its premise was an array of AFM tips that can heat/sense a deformable substrate in order to function as a memory device. A combination of these approaches may also be used, in which nanoscale molecules are integrated into a top-down framework, e.g., the carbon nanotube nanomotor.

More recently, a technique named **nanoimparint lithography** has been developed that may provide a low-cost, high-production rate manufacturing technology. This can produce patterns on a surface having 10 nm resolution at low cost and high rates because it does not require the use of a sophisticated radiation beam generating patterns for the production of each structure.

By exploiting nanoscale effects, NEMS offer a number of unique properties, which pave the way to applications such as force sensors, chemical sensors, and ultrahigh frequency resonators. The most interesting properties of NEMS arise from the behaviour of active parts, which is typically in the form of cantilevers or doubly damped beams with nanoscale dimensions. In NEMS, the charge controls the mechanical motion, and vice versa. The presence of mechanical motion decisions that the moving element to possess high-quality mechanical properties, including high Young's moduli and low density. While limitations in strength and flexibility compromise the performance of Si-based NEMs actuators. Carbon nano tubes (CNTs) are better candidate to realize the full potential of NEMS, in part due to their one-dimensional structure, high aspect ratio, perfect terminated surfaces, and exceptional electronic and mechanical properties. These properties, now complemented by significant advances in growth and manipulation techniques, make CNTs the most promising building blocks for next generation NEMS. The predicted behaviour of CNT based NEMS switches-basis of NEMS devices is favourable, and electronic properties have been shown to be reversible with mechanical deformation by a local probe.

24.9 CARBON NANOTUBES (CNTS)

Carbon nanotubes were discovered in 1991 by Sumio Iijima at the NEC laboratories. Although other nanotubes based on boron nitride and molybdenum have also been reported, currently NCTs are by far the most important group. CNTs have occupied a special identity in the field of nanotechnology. Since their discovery, theoretical and experimental studies in different fields, e.g., mechanics, optics, electronics etc., have focused on both the fundamental physical properties and on the potential applications of nanotubes. In all fields there has been substantial progress over the last two decades. Basically, CNTs are naturally self-organized nanostructure in the form of a tube composed of carbon atoms with completed bonds. The structure of CNTs consists of a single sheet of graphite, rolled into a tube, both ends of which are capped with C_{60} fullerene hemispheres. The tubes can be either open at their ends or capped at one or both ends with half a spheroidal fullerene. Figure 24.13 shows a schematic representation of a carbon nanotube. A nanotube is completely specified by what is referred to as a roll up vector, which identifies "its helical nature" and diameter. Depending on its roll-up vector, a nanotube can have either metallic (Fig. 24.13 (a)), semiconducting (Fig. 24.13 b, c) properties.

There are a vast number of ways in which a graphite sheet can be rolled up to form a seamless cylinder and hence a wide variety of nanotubes exist. Nanotubes are characterized by their *chiral* (or wrapping) vector, c such that $c = na_1 + ma_2$, where a_1 and a_2 are the basis vectors of the graphite (more correctly graphene) lattice and m, n are integers (Fig. 24.14). The chiral vector spans the circumference of the tube formed by joining the dotted lines shown in Fig. 24.14. Those tubes with chiral vectors of the form (n, o) are termed zig-zag tubes (Fig. 24.15 (a)), whereas $n = m$ a so called 'arm chair' tube results (Fig. 24.15 (b)). All other values of n and m produce a chiral tube (Fig. 24.15 (c)).

Closely following the synthesis of nanotubes came the remarkable theoretical prediction that their electronic properties could be changed between metallic and semiconducting simply by varying the tube diameter or its helicity, *i.e.*, by changing the values of n and m. A key theoretical result is that armchair nanotubes are metallic, whereas for all other SWNNTs, when n–m = 31 (1 an integer) the tubes are metallic; otherwise they are semiconducting.

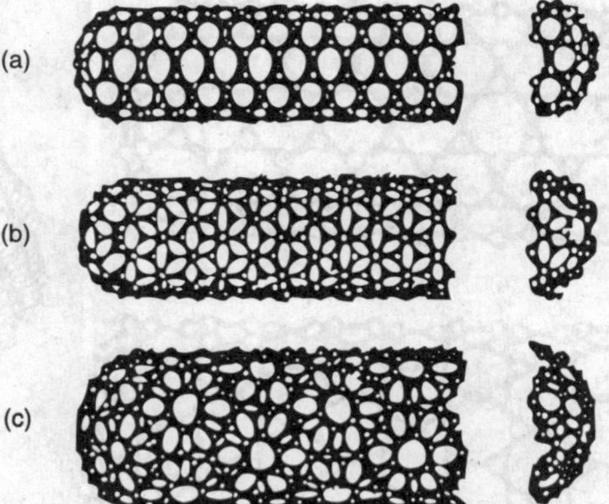

Fig. 24.13 The structure of carbon nanotubes (a) metallic (b) and (c) semiconducting

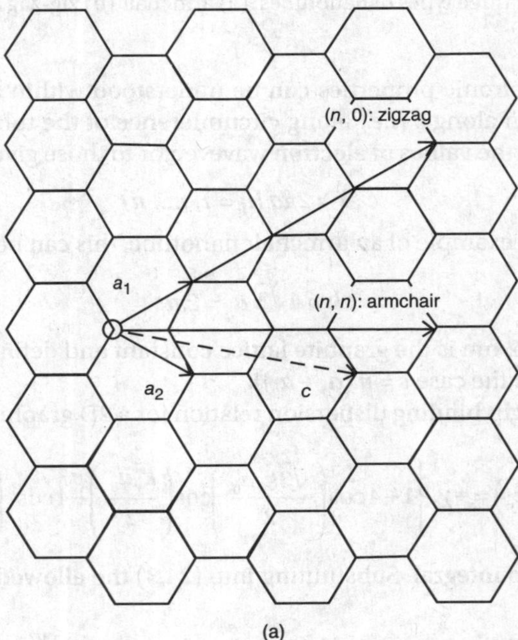

Fig. 24.14 The relationship between the graphite lattice basis vectors a_1, a_2 and the chiral vector $c = na_1 + ma_2$ is used to characterize carbon nanotubes. Two limiting cases are shown : $(n, 0)$ indices are associated with zig-zag tubes whereas (n, n) indices are associated with armchair tubes. All other tubes are chiral

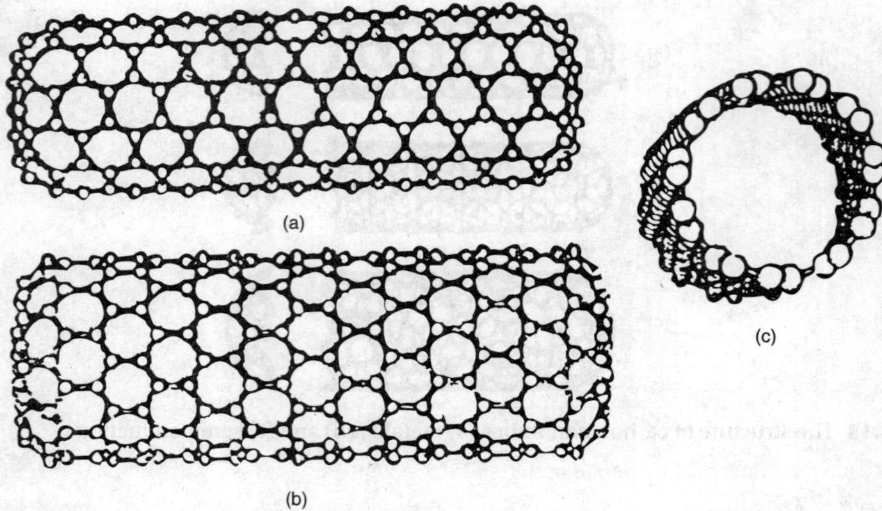

Fig. 24.15 Diagrams of three types of nanotubes: (a) armchair (b) zig-zag and (c) chiral

These striking electronic properties can be understood within a tight-binding framework. In the direction along *c (i.e.,* along circumference of the tube), periodic boundary conditions constrain the values of electron wavevector to those given by

$$c\,.\,k = 2\pi q\ (q = 1, ..., n) \qquad ...\ (24.1)$$

Taking the particular example of an armchair nanotube, this can be reduced to

$$k_x n\, 4\sqrt{3}\ a_0 = 2\pi q \qquad ...\ (24.2)$$

where $a_0 = \left(\sqrt{3}\right) 0.142$ nm is the graphite lattice constant and defines the chiral vector for the armchair tube (in the case $c = n\,(a_1 + a_2)$).

The well-known tight binding dispersion relation for a 2D graphite sheet is

$$E_{2D}\left(k_x, k_y\right) = +\gamma_0 \left\{1 + 4\cos\left(\frac{\sqrt{3}k_x a_0}{2}\right)\cos\left(\frac{k_y a_0}{2}\right) + 4\cos^2\left(\frac{k_y a_0}{2}\right)\right]^{1/2} \qquad ...\ (24.3)$$

γ_0 is the C–C overlap integral. Substituting into (24.3) the allowed values of k_x, given by (24.2) yields.

$$E_{10}^{A}\ (k_y) = \pm\ \gamma_0\ \left\{1 \pm 4\cos\left(\frac{q\pi}{n}\right)\cos\left(\frac{k_y a_0}{2}\right) + 4\cos^2\left(\frac{k_y a_0}{2}\right)\right]^{1/2}\ \left(-\pi < k_y a_0 < \pi\right) \ ...\ (24.4)$$

where the superscript A denotes that (24.4) is valid only for armchair tubes. It is assumed that the value of the overlap integral for graphite and nanotube are identical. The 1 D dispersion relations given by (24.4) are shown graphically in Fig. 24.16 for the case of (5, 5) armchair SWNT. There are six dispersion curves for both the valence and conduction bands. In each case, four bands are doubly degenerate (bold continuous lines) and

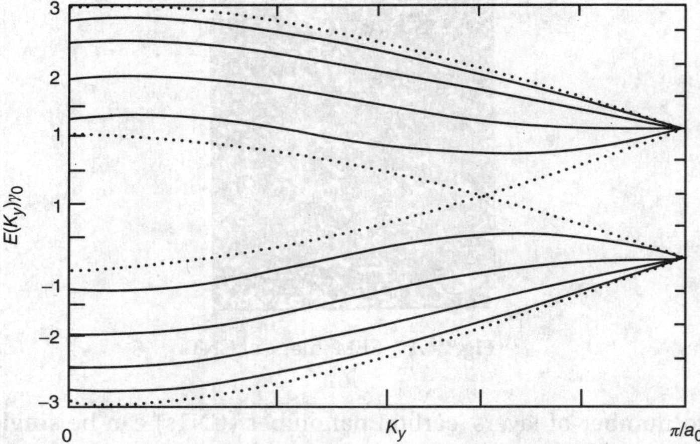

Fig. 24.16 One dimensional dispersion relation for an armchair (5, 5) nanotube

two bands are non-degenerate (thin, dashed lines). Thus ten electronic levels comprise the valence bands and ten comprise the conduction bands, as expected from the fact that there are ten hexagons around the circumference of a (5, 5) nanotube. There are two particularly noteworthy features of the dispersion relations shown in Fig. (24.15) (a), (b) and (c)). First there is a very high degree of degeneracy at the Brillouin zone boundary. One can readily understood this from the fact that at $k_y a_o = \pi$, Eqn. (24.4) reduces to

$$E_{1D}^{A} = \pm\gamma_0$$

The second and more important feature of Fig. 24.15 (a), (b) and (c) in terms of electronic properties of armchair nanotubes is that the highest valence band and lowest conduction band are degenerate at $k_y = \pm 2\pi/(3a_o)$ and each cross the Fermi energy ($E = 0$) at that point. This is true of all armchair nanotubes -thus all armchair nanotubes be metallic.

We must note that Eqns. (24.2) to (24.4) represent taking a 1D slice in a direction given by

$$E(\theta) = -KV\cos^2\theta$$

where K is the magnetic anisotropy constant, V the volume of the particle, θ the angle between the magnetization vector and an easy direction of magnetization and $E(\theta)$ is the magnetic anisotropy enegy for a uniaxial particle, through the 2D graphite band structure. The crossing at $E = 0$ in Fig. 24.16 occurs because the corresponding 2D graphite band cross at the K points (i.e., the corners) of the 2D Brillouin Zone. For armchair tubes, there is always an allowed K-vector that goes through K point of the 2D lattice. However, this alignment of the tube wave vector and K-point of the graphite Brillouin zone does not automatically occur for zig-zag or chiral tubes and hence these can have semiconducting character. Only for n–m = 31 there is an alignment of an allowed **k**-vector and a K-point of the 2D Brillouin Zone, giving rise to metallic behavior.

The *nano* in the nanotube prefix denotes that tube diameters are of the order of a nanometer (*i.e.*, 100 nm or less). Each nanotube is a single molecule composed of millions of atoms; the length of these molecules is much greater (on the order of the thousands of times greater) than its diameter.

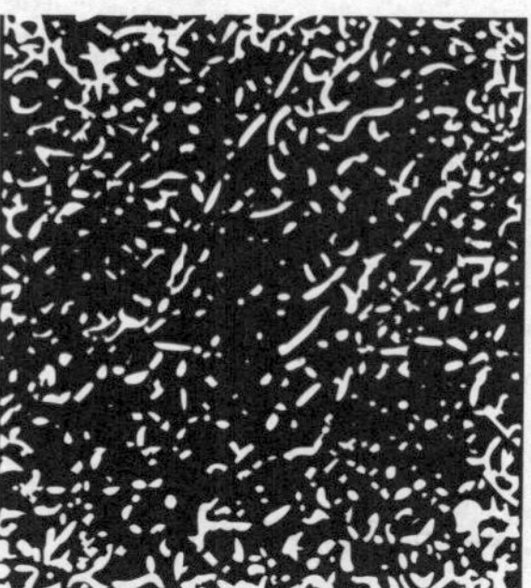

Fig. 24.17 SEM image of CNTs

According to number of layers, carbon nanotubes (CNTs) can be **single-walled** and **multi-walled** (MWNTs). In multi-walled nanotubes, more than one CNTs are coaxially arranged (Fig. 24.17). Iijima realized that graphite could be bent to form MWCNTs with different helicities or chiralities which refer to the way hexagonal rings are arranged with respect to the tube axis (Fig. 24.18).

It is found theoretically that the electronic properties of carbon nanotubes depended on the diameter and helicity: in particular, all of the so-called armchair type nanotubes are conductors (Fig. 24.18) and most zig-zag nanotubes are semiconductors (Fig. 24.18).

These nanotubes are extremely strong and stiff known fibres, and relatively ductile. According to indirect measurements, it has been found that carbon nanotubes are 100 times stronger than steel and six times lighter. For single walled nanotubes, tensile strength range between 50 and 200 GPa (approximately an order of magnitude greater than for carbon fibres); this is the strongest known material. Elastic modulus values are on the order of one tetrapascal [TPa : (1 TPa = 103 GPa)], with fracture strain between 5% and 20%. Furthermore, nanotubes have relatively low densities. The outstanding properties of the CNTs include extremely high tensile strength, high modulus, flexibility, light weight and high thermal conductivity. On the basis of unique properties, the carbon nanotube has been termed the 'ultimate fibre' and is extremely promising as a reinforcement in composite materials. The high stiffness, aspect ratio, and nanoscale tip make CNTs excellent probes for sensing, testing, manipulation, and fabrication on the nanoscale.

Carbon nanotubes also have unique structure sensitive electrical characteristics. Depending on the orientation of the hexagonal units in the graphite plane, i.e., tube well, with the tube axis, the nanotube may behave electrically as either a metal or a semiconductor. Semiconducting and metallic nanotubes recently are put into use in field effect transistors (FET) and single electron transistors. The flat-panel and full-colour displays, i.e., TV and computer monitors, have been fabricated using carbon nanotubes as field emitters. These displays should be cheaper to produce and will have lower power requirements than CRT and liquid crystal displays. The mechanical strength of CNT is amazingly 600 times tougher than steel, which finds applications in micro electro mechanical systems (MEMS) and in aerospace. Progress has been made in using CNTs as field emission devices e.g., diodes, transistors, etc., for high resolution display systems. Construction of nanoprobes and electrodes in biological and other applications are also being aimed.

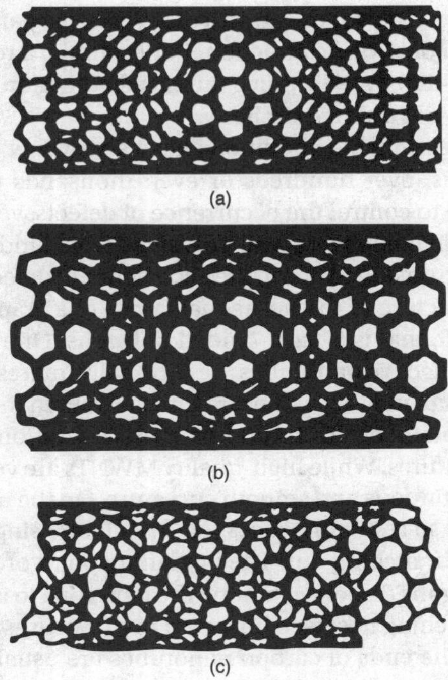

Fig. 24.18 Structures of carbon nanotubes (a) Zig-zag nanotube; (b) Armchair nanotube; and (c) Chiral or helical nanotube

24.9.1 Various Forms of Carbon Nanotubes

As already mentioned, the carbon nanotube is a generic term that includes both single and multiwalled nanotubes (SWNTs and MWNTs). The smallest nanotube reported to date has a diameter of only 4 Å. Since increasing the curvature of the graphite wall increases the strain energy, this is also predicted to be the smallest possible diameter a nanotube can have. Typical diameters for SWNTs are between 1 and 2 nm, corresponding to n and m numbers of around 1 0. In MWNTs, the distance between the adjacent layers is about 0.34 nm, which is much more than the atomic separation (1.42 Å) within the individual layers. While typical values for the outer diameter are 10-20 nm, the inner diameter is typically 3-5 nm. Thus an ordinary MWNT has 10-30 layers, each of which are SWNTs of different diameters. The shell structure of MWNTs has not been studied thoroughly yet, but from geometrical considerations it is concluded that adjacent layers will generally be incommensurate. Also, the larger the diameter of a MWNT is, the more likely the outer layers are disordered and noncircular.

The ends of nanotubes are generally closed in the growth process. While the wall of a carbon nanotube is made up of an exclusively hexagonal pattern, pentagons are needed to cap the tubes. More generally, pentagonal and heptagonal (5- and 7-corpered polygons, respectively) defects enable the graphitic sheet to take up more complicated structures than a simple tube, as we will discuss below.

Unless special setups are used for the growth of SWNTs, they are usually assembled into ropes by their mutual van der Waals attraction. For example, a rope of a typical diameter of about 10 nm contains ~100 SWNTs. Since multi walled nanotubes have a much higher bending stiffness, they do not arrange similarly.

Nanotubes grown under suitable conditions have a very low concentration of defects over μm distances, that is, over hundreds or even thousands of interatomic spacings. However, if one were able to control the occurrence of defects, very useful and interesting nanotube structures would emerge. We will next show what kind of modified SWNTs and MWNTs have already been observed due to the occurrence of defects within the nanotubes. A single, stable defect structure in the hexagonal lattice of a graphite sheet is made up of a pentagon-heptagon pair, that is 5- and 7-sided polygons. One such defect can cause a sharp bend in a SWNT. Defective nanotubes are especially interesting for electronic applications where the defect site may act as a tunnelling barrier. In the case of MWNTs, highly defective tubes may be obtained with certain synthesis conditions, for example with relatively low growth temperatures. While high-quality MWNTs are very straight and stiff, very defective ones have a continuous and smooth curvature. On the other hand, the curvature can be highly regular and so result in helices. An SWNT, or a single shell of a MWNT, can have a second nanotube branching out. The beginning point of the branch has similarly to the end sections pentagons dispersed among the hexagons to accommodate the necessary curvature. Such branching has been observed in a few cases to date.

As mentioned earlier, the ends of carbon nanotubes are usually closed. It is, however, possible to open nanotubes by a suitable chemical treatment. This is due to the stronger chemical reactivity of the end section with its larger curvature and pentagon structures. The opened nanotubes can be utilized for creating yet another side branch in the science of nanotubes, namely filled tubes. Both SWNTs and MWNTs have been filled with various materials, such as fullerenes, simple metals, and molecular compounds. Nanotubes with fullerenes inside are called peapods, and are presently intensively investigated. The study of the electronic properties of filled nanotubes is still in its infancy.

24.9.2 Chemistry of Carbon Nanotubes

The chemistry of carbon nanotubes, which encompasses the aspects of purification and functionalization, is a very important field for tailoring their structural and electronic properties and thus for the development of future applications. Purification aims at removing the nontubular material in raw nanotube samples. Several techniques, more or less complex, have been proposed, including solvent treatment followed by ultrafiltration, flocculalion using aqueous surfactants, oxidation and acid washing coupled with centrifugation, resuspension in surfactant solution, and cross-flow filtration steps. These methods however lead to a mixture of nanotubes of differing lengths and still containing potential contamination with nontubular material. Polymer suspensions and chromatographic purification are useful to remove nontubular material. Field-flow fractionation on purified, shortened nanotubes and size-exclusion chromatography on raw nanotubes have demonstrated some success at producing fractions separated by length. Capillary electrophoresis was reported to allow separation based on tube length.

Several methods for attachment of oxygen-containing groups to the nanotube surfaces have been developed. Carboxylic carbonyl and hydroxyl groups were found covering the nanotube walls. The concentration of acid groups on the surfaces of HNO_3 treated nanotubes was higher than that found for graphite treated under the same conditions. Sonication of nanotubes prior to the functionalization in acids increases the concentration of acid groups. Oxidation of nanotubes with a H_2SO_4–HNO_3 mixture also leads to a higher concentration of functional groups on the surface. Long-term acid treatment of nanotubes also induces modifications. Functional groups can be removed from the surface of nanotubes by healing. It should be noted that SWNTs obtained by different methods exhibit different behaviour in the course of acid treatment and subsequent oxidation. The sensitivity of SWNTs to chemical processing was investigated by TEM.

To enhance the possibilities of atomic force microscopes with nanotube probe tips, the tip ends were functionalized using both solutions and gases. Modification of tips using a discharge in N_2 results in the formation of nitrogen-containing heterocycles at the tip ends. High degree of saturation of nanotube tips with functional groups is achieved by treatment with a low-pressure ammonia plasma followed by oxidation with a $NaClO_3$ solution.

The reactions with silicon of individual nanotubes or nanotube bundles at 970°C in ultra-high vacuum resulted in the formation of Si-β-SiC (nanorod) nanotube heterojunctions.

24.9.3 Oriented or Localized Growth

The term "oriented" is used here to describe the formation of large quantities of CNTs aligned perpendicularly to the substrate surface, usually MWNTs of controlled dimensions (diameter 20-400 nm, length 0.1-60 μm). The key steps are substrate patterning, catalyst deposition, and the appropriate use of different CVD routes.

The term "localized" describes the self-assembled nanointerconnections made by CNTs between islands on prepatterned substrate. The directed growth of SWNTs parallel to the plane of a silicon substrate was reported. The SWNTs are suspended bridges grown from catalyst material placed on top of regularly patterned silicon tower structures. The key synthetic step in this approach involves developing a series of liquid-phase catalyst precursor materials, allowing for uniform film formation and large-scale catalyst patterning. Contact printing techniques were used to selectively deposit catalyst precursors on top of tac silicon tower arrays. Calcination led to the formation of catalyst particles confined on the tower tops and subsequent CVD (CH_3, 900° C, 20 min) growth yielded SWNTs emanating from the towers. Directed free-standing SWNT networks were formed by nanotubes growing to adjacent towers and becoming suspended above the surface. It is noted that a rational design of tower arrangements could lead to a variety directed SWNT architectures. The liquid phase catalyst precursors consist of three general components: inorganic chloride precursors, a removable triblock copolymer serving as the structure-directing agent for the chlorides, and an appropriate alcohol for dissolution of the inorganic and polymer compounds. The increase in yield was achieved through a so-called conditioning step that uses bulk amounts of catalysts placed near (upstream) the catalyst-patterned substrate in the CVD growth environment.

24.9.4 Electromechanical Properties

Along with their electronic properties, carbon nanotubes have been praised for their mechanical properties. The praise is based on high strength, flexibility, and large aspect ratio. A carbon nanotube is a freestanding conductor and its mechanical properties are highly relevant for the electronic properties and the potential technological applications of the tube. For example, nanotube-based mechanical oscillators may have applications in high-frequency electronic circuits, and carbon nanotubes are already used in scanning "probe microscopy both as STM and AFM tips. Thin strong and durable nanotubes come extremely close to an ideal scanning tip.

Mechanical properties of carbon nanotubes are closely related to the properties of a graphite sheet, but the tubular anisotropic form affects the mechanical behaviour. The basis is the graphite sp^2 bond, which is the strongest of chemical bonds. In nanotubes the overall density of defects can be extremely low depending on the synthesizing method and prevailing synthesizing parametres. This has led to predictions of a very high axial strength.

24.9.5 Basic Mechanical Properties

Theoretical calculations of Young's modulus for individual SWNTs center around 1 TPa or slightly higher, but values as high as 5.5 TPa have been presented. The spread is due to different interaction models and, also to differing values of nanotube wall thickness (if an isolated nanotube is not a well-defined quantity). Most of the theoretical attention has been on SWNTs because modelling the interlayer interaction in MWNTs is a complicated matter. Young's modulus values for MWNTs as well as SWNTs range from 0.97 TPa to 1.11 TPa with the value increasing slightly with the number of layers.

The small size of carbon nanotubes presents challenges also for experimental characterization. Nevertheless measurements have been performed. The first Young's modulus measurement related thermal vibration amplitudes of MWNTs to their Young's modulus and obtained an average value of 1.8 TPa with a large spread. After that, with AFM techniques, values such as 1.28 ± 0.59 TPa for arc-discharge-produced tubes have been obtained. The latest measuring technique is due to Poncharal group, who induced vibrations on MWNTs by alternating potential and have measured the vibration frequencies. Young's modulus values between 0.7 TPa and 1.3 TPa for tubes with a diameter less than 12 nm and between 0.1 TPa and 0.3 TPa for thicker tubes were reported. The drop is explained by an onset of a wave-like bending mode of the nanotube wall at the inner arc of the bending. Measuring SWNTs is more complicated than measuring MWNTs due to their small diameters and the tendency to bundle. Measurement of individual SWNTs using the thermal vibrations method provide an average value of 1.25 TPa. Although the current measurements suffer from inaccuracies due to vibration amplitude measurement and assumptions made on AFM tip characteristics, the current agreement is that defect free nanotubes, both SWNTs and MWNTs, have a Young's modulus value around slightly above 1 TPa, which is extremely high and sets nanotubes as the strongest known material albeit challenged by other nanotubular structures such as boron nitride tubes.

Theoretically, nanotube tensile strength is high, and this is supported by calculations in which SWNTs support as high as 30% of axial strain before brittle failure and by more recent kinetic activation based calculations that give a yield strain of 17% with a chirality

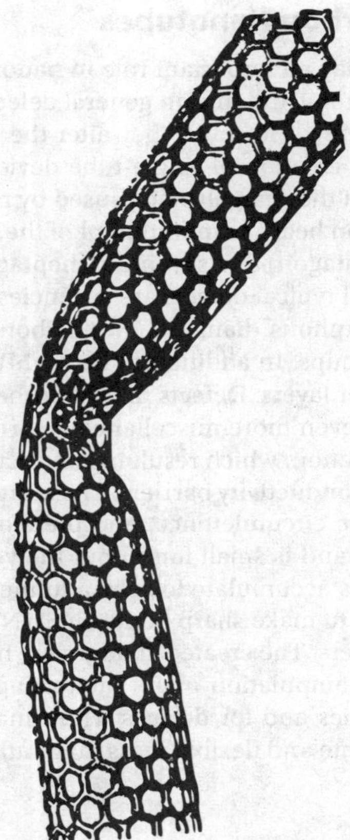

Fig. 24.19 A molecular dynamics simulations based example of nanotube behaviour under strain. The image shows portion of a (14, 0) nanotube of 2632 atoms (tube length is 20 nm) bent 36 degrees with constant bending rate in 300 K temperature and then, to enhance structural reconstruction, annealed in an elevated 3300 K temperature for a sequence of three 130 ps annealing and cooling simulation runs. In the bending process the tube buckled but remained hexagonally bonded. The defects visible in the image relax the strain and are a result of the third annealing

and temperature-dependent defect formation activation energy barriers. Tensile strength values ranging from 11 GPa to 63 GPa are reported. The role of the inner layers is not, however, clear, and the failure process may be more complicated. For individual SWNTs, the experimental value of tensile strength is still an open question, but for bundles of SWNTs tensile strength values ranging between a few GPa and several tens of GPa depending on the bundle an measurement characteristic.

As a result of the high tensile strength, nanotubes can bear large strain before plastic deformation or brittle failure. The tubes buckle, flatten, form ripples, and generally deform under strain but plastic deformation and relaxation occur only at elevated strain levels and temperatures. An example of simulated nanotube behaviour under strain is presented in Fig. 24.19.

24.9.6 Defects in Carbon Nanotubes

As in any material, defects play an important role in nanotube properties. Structurally, defects make the tube less strong and thus in general defects are not desirable from the purely mechanical point of view. However, they alter the electronic properties locally, which can be utilized in the creation of single-tube devices. Defects are generated in the synthesizing process, and they can also be caused by mechanical manipulation, or, for example, by ion or electron beam bombardment of the tube. The most typical structural defects are fivefold (pentagon) and sevenfold (heptagon) rings in the sixfold (hexagonal) lattice. Other types of typical defects are vacancies and miscellaneous bonding configurations such as amorphous diamond. Noncarbon based defects include substitutional atoms or atom groups. In addition to these, MWNTs exhibit diverse defects based on discontinuous inner layers. Defects may alter the tube form a straight tube to a bulging, kinked, spiral, or even more miscellaneous form. Defects and deformations induce scattering and localization, which result in conductivity drops, symmetry breaking based gap opening, and conductivity barriers. The scattering effects of isolated point defects are averaged over the circumference, and therefore their influence decreases with an increasing diameter and is small for all but the very thinnest tubes. However, if the defects or deformations accumulate locally, a tunnelling barrier may form. AFM manipulation has been used to make sharp bends in SWNTs, which has resulted in the formation of tunnelling barriers. The created changes are not, however, necessarily permanent. Nanotube defect manipulation offers interesting prospects both for devices with permanent characteristics and for devices with tunable properties, but a breakthrough requires more accurate and flexible mass manipulation of tube properties than is currently available.

24.9.7 Carbon Nanotubes as Mechanical Oscillators

Due to their small size, combined with a low mass density and high Young's modulus, nanotubes are predicted to have characteristic vibration frequencies in the gigahertz regime, which is extremely high for a mechanical vibrator. Such frequencies are not attainable with the current silicon-based micro-mechanical oscillator production techniques with which eigenfrequencies in the 10 MHz to 100 MHz regime can be attained. A first approximation for the fundamental modes of a nanotube vibrator can be obtained from elasticity theory and is for a thin cylindrical transversally vibrating rod that is clamped from one end

$$f_1 = 0.28\frac{d_1}{L^2}\sqrt{\frac{Y}{\rho}} \qquad \text{... (24.5)}$$

The values for tube diameter d_1 = 1 nm, Young's modulus Y = 1 TPa, and density ρ = 750 kg/m^3 are used. The mass density of a SWNT is not a well-defined quantity; the value presented here is a rough estimate based on the nanotube atomic spacial density and the volume encompassed by a tubular cylinder depicting the tube of 1 nm radius, to compare with graphite mass density 1900-2300 kg/m^3, a frequency of 10 MHz is obtained for a tube length of 1 μm and a frequency of 1 GHz for a length of 100 nm. For a rod clamped from both ends, the constant in front of the equation is altered according to the boundary conditions, but the overall dependency on d_1, L, Y, and ρ remains. Based on the discussion above, nanotubes, and especially SWNTs, are capable of significantly smaller dimensions and higher frequencies than conventional silicon-based vibrators.

Currently, only a few studies of excited mechanical vibrations in nanotubes exist. Pancharal group applied alternating potentials to MWNTs and measured Young's modulus and Roulet group have performed a basic study of acoustoelectric coupling in ropes of SWNTs, and they reported dependence on the excitation power, either electron heating or phase coherence breaking is observed.

There are many things in the mechanical properties that have not been measured, and the measurements that exist suffer from a degree of inaccuracy. Most theory concentrates on SWNTs, and although the SWNTs often show more clearly the nanotube characteristic and desirable electronic properties, the theoretical aspects due to multiple layers still have many questions to answer, as do defectous-structure related matters. However, in nanoscale science and technology, electronic and mechanical behaviour will become more intimately connected, and obviously, the electromechanical properties of carbon nanotubes will play an important role.

24.9.8 Electronic Structure and Properties

Tight Binding Model: The carbon atoms in nanotubes are sp^2-hybridized, thus each atom contributes one unpaired π-electron to the tube. Carbon nanotubes are electronically active materials due to these π-electrons. The basic electronic properties of the carbon nanotubes are calculated by extending the tight-binding approach of *2D* graphite to the geometrical configuration of the *1D* nanotube system. The crystal momentum vector perpendicular to the nanotube axis is quantized due to the periodic boundary condition imposed by the finite circumference of the nanotube. This relatively simple model has been proven, mainly by low temperature STM spectroscopy, to explain well the basic electronic structure of the carbon nanotube. According to this model, the chirality of the carbon nanotube determines its division into metallic and semiconducting tubes. In the model, the curvature of graphitic wall of the nanotube is ignored. This has lately been shown to bring about important modifications to the model. Still the model survives well enough to continue to serve as a basis for classification of the nanotubes into metallic and semiconducting ones.

A brief presentation of the basic tight binding model of the single shell carbon nanotube is as follows. Due to the tubular *1D* structure, the nanotube has *N* separate *1D* energy bands. In a general form these are derived from the graphite energy levels by the general formula

$$E_\mu(k) = E_{g2D}\left(k\frac{K_2}{|K_2|} + \mu K_1\right)\left(\mu = 0,\ldots N-1, \qquad -\frac{\pi}{T} < k < \frac{\pi}{T}\right) \quad \ldots (24.6)$$

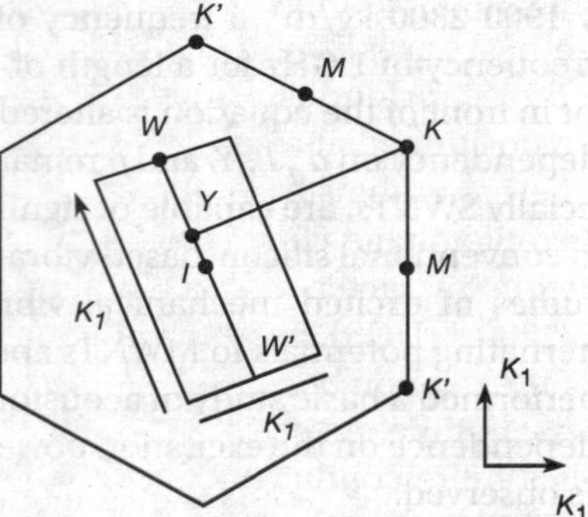

Fig. 24.20 Allowed K-vector or carbon nanotubes are formed by the vectors K_1 and K_2 in the reciprocal lattice space of graphite. The nanotube is metallic if the ratio of *YK* to K_1 is an integer

where E_{g2D} denotes the energy eigen values of a *2D* graphite sheet, K_1 and K_2 are reciprocal lattice vectors corresponding to C and *T*, respectively. Figure 24.20 shows how the carbon nanotube *k*-vectors are projected to the graphite momentum space lattice. In graphite the conduction and valence bands meet at the so called *K* points in the momentum space, which is responsible for its semimetallic behaviour. In the tight-binding picture, the metallicity of a carbon nanotube is determined by whether the 1D lines of allowed *K*-vectors run through these points.

In an energy diagram, whether the energy bands will cross the Fermi energy or leave an energy gap depends on the *(n, m)* that is the chirality, as follows: The armchair nanotubes, with indices *(n, n)*, have two bands that cross each other and the Fermi energy at the points $k = \pm 2\pi/3a$, they are thus always metallic, while only some of the zigzag tubes are so. The general condition for the metallicity of a nanotube is that the factor $2n + m$ is a multiple of 3, a condition that indeed is fulfilled for the armchair tubes. The theory also predicts that one-third of all nanotubes are metallic while two-thirds are semiconducting. These rates of occurrence have not been confirmed by the presently performed experiments, which however is not surprising considering doping effects and the fact that almost nothing is known about the possible preference in the growth processes for certain nanotube symmetries over others.

As a general result, in all metallic 1D systems with half-filled bands, the so-called Peierls distortion will, by an atomic displacement, change the unit cell so that the band becomes filled and a gap opens up at the Fermi energy. The phenomenon is well known among conducting polymers, charge transfer salts, etc., In carbon nanotubes, however, it has been shown theoretically that the effect very quickly diminishes as a function of nanotube radius.

24.9.9 Experimental Studies: STM Imaging and Spectroscopy

The electronic structure of the SWNT that is predicted by the tight-binding calculations described above was put to test at first primarily by scanning tunneling spectroscopy, done with low temperature STM. In such experiments, it is possible to image the nanotube with atomic resolution along the length of the tube and with a width of a few unit cells. The width of the imaging is of course limited by the curvature of the nanotube. This imaging reveals (in the best cases) the structure and diameter of the nanotube, enabling one to determine the *(n, m)* indices of the nanotube with some confidence. Furthermore, the STM tip be positioned at any point along the nanotube, separated by a tunneling distance from the nanotube surface, and spectroscopic measurements be performed by taking the current-voltage characteristics. The I-V characteristics reveal the metallic or semiconducting character of the tube since the measured current is proportional to its density of states (DOS). This experimental technique thus addresses the fundamental question of the relation between the structure and the basic electronic properties of the individual nanotube. As seen by STM spectroscopy, in the zero-bias region the metallic tube has a finite conductance while the semiconducting tube has zero, corresponding to the basic Fermi surface properties. At higher voltages asymmetric peaks in the conductance appear. These are the Van Hove peaks due to the 1D DOS. The low-temperature STM experiments were carried out with SWNTs, and thus they cover only a very limited range of diameters, between 1 and 2 nm. For this range, the measured E_g for semiconducting nanotubes-followed roughly the predicted $1/d_1$ dependence, being in the range 0.3-0.7 eV.

In recent scanning tunneling spectroscopy measurements on SWNTs, a curvature-induced "extra" gap was found in zig-zag type metallic nanotubes, but not in tubes of armchair type. This gap is also inversely dependent on the nanotube radius. The measured results were theoretically anticipated in refined calculations. The curvature-induced gap has a size of a few 10s of meV for subnanometer tube diameters. From this energy scale we see that for all but the very smallest of possible tubes, the effect is washed out at room temperature.

Other scanning tunneling spectroscopy experiments have been performed that explore other topics such as individual quantum states in short SWNTs and MWNTs doped nanotubes crossing nanotubes, defects and junctions in SWNTs and peapods (SWNTs filled with C_{60}).

24.10 EXPERIMENTAL STUDIES: RAMAN SPECTROSCOPY

Another important tool for nanotube characterization is Raman spectroscopy. Particularly, the strong, so-called radial breathing mode is derived from the tubular structure of the nanotube and is absent from, all other 2D graphitic forms of carbon. It is widely utilized to give a measure on the nanotube diameter. Due to the singularities in the DOS of the ID nanotubes and the electron-phonon coupling, very strong resonant Raman effects are observable in studies on carbon nanotubes. This enables the use of Raman spectroscopy to investigate the electronic structure of carbon nanotubes, particularly the occurrence of semiconducting and metallic properties. A possibly very important recent development is the ability to measure the Raman signal from individual nanotubes with the so-called micro-Raman technique. It was claimed to be possible to determine the *(n, m)* indices of the isolated nanotube with this technique. This, of course, opens up the possibility of transport measurements on a nanotube of known symmetry.

24.10.1 Optical Properties

The optical physics of nanotubes has not moved forward in pace with the other developments in the field. This is due to unability to make samples consisting of nanotubes of a single symmetry. Thus all optical signals are a mixture of signals from a very large variety of nanotubes. Furthermore, since nanotubes tend to aggregate together, especially SWNTs, any signal from semiconducting tubes is easily hidden by interaction with neighbouring metallic tubes. Unlike transport measurements, optical measurements on single nanotubes are experimentally less feasible. On the other hand, however, the 1D singularities in the optical response of nanotubes give strong signals, just as in the case of Raman spectroscopy. Also, if aligned assemblies, of nanotubes can be produced, the optical anisotropy of such a system is potentially huge. In a recent development, SWNTs were processed by sonication and surface-active chemicals to produce solutions of truly individually solvated tubes, coated with the surface-active hydrocarbon molecules. These SWNTs were rather short (due to the strong sonication), with an average length of only 130 nm. Perhaps for the first time, photoexcited luminescence from nanotube solutions were observed in this case. The luminescence spectrum was closely associated with the measured absorption spectrum, with only a minor red-shift. This work is intensively being pursued and is likely significantly widen research in carbon nanotube systems. Photoconductivity measurements on single nanotubes have so far not been successful.

The research on the optical properties of carbon nanotube may in the near future bring forward a number of potential applications within optoelectroates, thus adding to the list that already exists within other fields of nanotube-based nanotechnology.

24.10.2 Electronic Transport Properties

Within the realm of electronic transport physics, the carbon nanotube offers exciting possibilities both as a testing ground in basic research as well as in nanotechnology. In the case of the former, the near perfection in structure for lengths exceeding 1 μm makes the nanotube the best concrete example of a *1D* quantum wire. On the other hand, in nanotechnology the prospect of devices made from a single nanotube, that is, intrananotube devices, offer the possibility of truly molecular-level electronics. Even if molecular electronics is eventually pursued with smaller molecules, carbon nanotubes may be used as connecting wires between metal structures of ~100 nm size and molecules of ~1nm size.

Although the carbon nanotube is mainly "marketed" as a perfect *1D* object, this is not always the case. In fact, depending on the circumstances one may have to treat the nanotube as anything from a *0D* to a *3D* object. Obviously, if a SWNT is short enough, it becomes a *0D* object, or a quantum dot. If only the outer shell of a MWNT is considered, and the diameter of the tube is large enough with respect to some critical length scale of the particular situation, this outer shell becomes a *2D* object. Finally, if we again consider a larger MWNT (in diameter), and the charge carrier coupling between the layers is strong enough, then we may have at hand an anisotropic *3D* object. We can thus see that even without considering nanotubes of different chiralities and more elaborate topologies than the simple tubular one, the subject of electron transport in carbon nanotubes can be rather complicated.

As has been discussed, the energy gap in semiconducting SWNTs is ~0.5 eV, which is high enough with a good margin for the nanotube to function at room temperature as the active component e.g., in field effect transistors. Other effects, such as the energy level quantization in short nanotubes (L = 10–100 nm), require cryogenic temperatures for the effects to be observable. Roughly, it can be said that in metallic nanotubes the interesting transport phenomena occur at low temperatures, while in semiconducting tubes much of the work is carried out at room temperature.

24.10.3 Bulk Transport

In this section we review the work on the transport physics of single SWNTs, SWNT ropes, and MWNTs. This is clearly more significant than the subject of transport in macroscopic amounts of carbon nanotubes. However, the carbon nanotube may offer interesting applications as the conductive component in composites, when mixed together with an insulating host material. In order to have a composite conducting, the volume fraction of the conductive component has to exceed some critical value. Typically, tile conductive material consists μm-sized particles of a more or less rounded shape. For conduction to occur, the particles have to touch each other frequently enough so that the conductive channels are formed over macroscopic distances. As prescribed by percolation theory, this occurs at a volume fraction of 16% whereby the conductivity of the composite rises very sharply (as a function of filling percentage) with several orders of magnitude. However, it has already been demonstrated with conductive material consisting of 1D objects, for example fibrils of polyaniline a conducting polymer, that there is no apparent sharp threshold for the onset of conduction, and more importantly, the composite becomes conducting at a much lower volume fraction of the conductive component. Obviously, this should also be achievable with nanotubes as the conductive components. In fact nanotube-based composite systems are already being considered for commercial applications.

24.10.4 Carbon Nanocones

Theoretical studies predicted the formation of graphitic cones. Subsequently, isolated graphitic cones were produced by carbon condensation on a graphite substrate and by pyrolysis of heavy oils. More recently, single walled aggregates of conical graphitic structures have been prepared by laser ablation of graphite targets. In addition, conical structures consisting of other layered materials such as BN have also been prepared by reacting boron oxide vapours with multi-walled CNTs.

Recently, it is reported that pyrolysis of palladium precursors, always produce conical nanofibers. An important feature of these new nanostructures is that they are held together by van der Waals forces since the fibre is composed of an arrangement of stacked cones which can be opened (lamp-shade), or closed. It is believed that nanocones may be good electron field emitters. Calculations of the electronic properties of nanocones reveal that there is a charge accumulation towards the tip and that there are localized states near the Fermi level. These features make them suitable as field emitters.

24.11 HAECKELITES

The presence of defects such as pentagons and heptagons in fullerenes change the electronic properties. A new hypothetical type of graphene sheets has been proposed which admit pentagons, heptagons, and hexagons, observing that the number of heptagons and pentagons should be the same in order to compensate the negative curvature of the heptagons and the positive curvature of pentagons (Fig. 24.21). These arrangements have been named as haeckelites. They all exhibit a considerable signal at the Fermi level in the local density of states. Obviously, all these systems should possess a metallic behaviour. Thus, it is possible to roll up haeckelites to form nanotubes, which will be conductors independent of the diameter and the helical arrangement.

Another property of haeckelite tubes is that they keep the stiffness of classical graphite nanotubes composed with just hexagons; the Young's modulus of haeckelite tube is around 1.0 Tera Pa; of the same order as classical carbon nanotubes. Haeckelites also exhibit local rugosity due to the local bending produced by heptagons and pentagons. It might be possible, that electron irradiation experiments can generate heptagon-pentagon pairs (rugosity) in SWCNT.

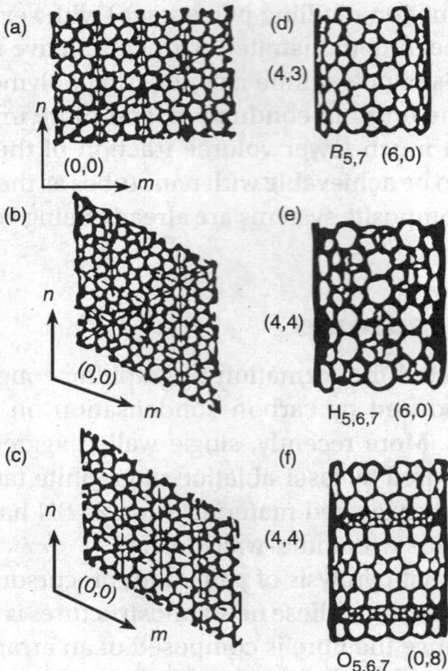

Fig. 24.21 Haeckelites (*a*) Rectangular haeckelite sheet (*R*); (*b*) Hexagonal haeckelite sheet (H); (c) Oblique haeckelite sheet (0); (*d*) to (*f*) are the corresponding haeckel tubes. All haeckelites are metallic (sheets and tubes)

24.11.1 Production Methods for CNTs

CNTs can be obtained by using four different methods: (i) arc-discharge (ii) pyrolysis (iii) laser vaporization and (iv) electrolysis.

The electric arc discharge setup is shown in Fig. 24.22. This method is very similar to the one used for obtaining fullerenes with two main differences: first, the pressure are higher, around 500 torr (for fullerenes the pressure is around 100 torr), and second, the nanotubes are found deposited in the cathode and not in the soot. This method produces well ordered MWCNT with diameters from 2 nm up to 30 nm. The separation between graphite layers is around 3.4 Å, which is very close to that of graphite 3.35 Å and corresponds to turbostratric graphite. The length of these nanotubes can be up to 30μ. The optimum conditions for generating nanotubes with this method consist of using currents of 150 amps, voltages of 25 volts and graphite electrodes between 6 mm and 8 mm in diameter with a separation between them of 1 mm under helium atmosphere. The electric arc reaction is too violent, so it is very difficult to control the formation reaction.

The pyrolysis method (Fig. 24.23) consists of heating a hydrocarbon or a precursor which contains carbon under the presence of a catalyst such as nickel, cobalt or iron. Prior to the discovery of C_{60}, it was well known that carbon fibres could be obtained by pyrolytic methods using hydrocarbons. We must note that carbon fibres are not nanotubes since the fibres are bigger in diameter and exhibit a great amount of defects and impurities. Nevertheless, the application of carbon fibres is widespread, showing a strength comparable to steel.

In 1994, Ajayan group found that carbon nanotubes could be aligned by embedding them in a polymer, thus forming a composite material. When the polymer was cut, they observed an aligned arrangement of nanotubes. In this context, a better alternative is to use laser ablation of cobalt thin films and then pyrolyse them with an organic precursor such as 2-amino 4, 6-dichloro-5-tryazine. The nanotubes obtained in this way are aligned in bundles which can grow up to 40μ and diameters between 30 nm and 50 nm.

The laser vaporization method involves firing a high power laser towards a graphite target inside a furnace at 1200° C. The condensation of material generated by the laser is

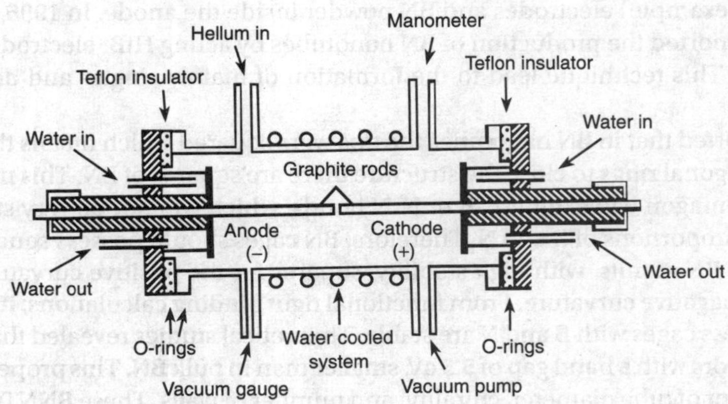

Fig. 24.22 Arc discharge set up

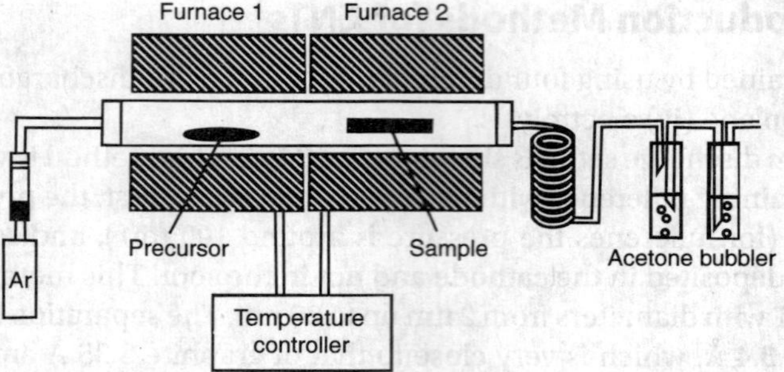

Fig. 24.23 Pyrolysis ste-up

responsible for the nanotube formation. If we add nickel or cobalt to the graphite target, we obtain single-walled carbon nanotubes (SWCNT). The SWCNT show diameters around 14 Å forming a two-dimensional crystal with a lattice constant of 17 Å. Arrangements of SWCNT can also be obtained by electric arc discharge with mixtures of carbon, nickel and yttrium.

The electrolysis method is based on using graphite electrodes inside a molten salt such as lithium chloride (700° C) under an argon atmosphere. Depending on conditions 20% to 30% consists of nanotubes. The depth of the cathode and the current (3-10 amps) play an important role in the formation of nanotubes.

24.11.2 Boron Nitride Curved Nanostructures (BNNTs)

Hexagonal boron nitride (h-BN) is another layered material with the structure very similar to graphite having hexagonal ring layers separated by 3.33 Å, in which every boron is connected to three nitrogen atoms and *vice-versa*; and B-N distance is 1.44 Å. Between the layers, every boron interacts with a nitrogen atom through a van der Waal's force. Regarding the electronic properties, h-BN is an insulator with a band gap of 5.8 eV, whereas graphite is a semi-metal (band overlap = 0.04 eV). The first BN nanotube were prepared using the electric arc discharge with a high melting point metal (tantalum and tungsten for example) electrodes and BN powder inside the anode. In 1996, Loiseau and coworkers reported the production of BN nanotubes by acting HfB_2 electrodes in an inert atmosphere. This technique lead to the formation of mainly single- and double-walled BN tubes.

It was reported that in BN nanotubes the tips were squared which means that instead of having pentagonal rings to close the structure there are squares of BN. This makes a sense because a pentagon will exhibit B-B or N-N bonds, which will not be very stable, changing also the proportions of B and N. Therefore, BN cages should possess squares. In terms of curvature, BN admits, with high stability, squares for the positive curvature and octagons for the negative curvature. From functional tight binding calculations, it is found that nitrogen excess cages with B and N are stable. Theoretical studies revealed that BNNTs are semiconductors with a band gap of 5.5 eV, smaller than in bulk BN. This property of BNNTs is independent of tube diameter, chirality, and number of walls. These BNNTs may also be extremely resistant to oxidation and should exhibit high Young's modulii. Therefore, the BNNTs may find important uses in ceramics.

BNNTs bundles can also be produced when carbon nanotubes are exposed to mixture of B_2O_3,V_2O_5, and MoO_3 in a nitrogen atmosphere at elevated temperatures. There are alternative approaches to generate BNNTs and other BN nanostructures: (a) thermolysis of borazine ($B_3N_6H_6$) over nickel boride catalyst particles at 1000-1100°C, (b) thermal annealing of powdered rhornbohedral boron and h-BN mixtures at 1200°C in lithium vapour, (c) plasma jets and (d) ball milling of h-BN, followed by thermal annealing at 1300°C.

At this stage, novel composites and nanoscale components need to be fabricated employing BNNTs, which appear to be extremely resistant to oxidation. Experimentally obtained values of Young's modulus of BNNTs (Y or $E = 1.22 \pm 0.24$ TPa) agrees well with the values obtained using theoretical approaches, and reveals that BN tubes are highly crystalline and probably, they may be the strongest insulating nanofibres.

Graphite and h-BN have opened new avenues for novel nanomaterials with different combinations of the type $B_xC_yN_z$. Pyrolysis experiments with melamine have been found very helpful in synthesizing C_xN_y nanofibres.

Boron doped carbon nanotubes have been synthesized by electric arc discharge with h-BN inside the graphite anode. During this experiment, it is observed that carbon nanotubes are highly graphitic and also very long (-20μ). It was found that B was located at the tips of the nanotubes and that the zig-zag chirality was preferred. It is reported by ab-initio calculations that boron was acting as surfactant avoiding the zig-zag tubes closure. In addition, boron doped nanotubes exhibit conducting properties.

It is reported that nitrogen doped carbon nanotubes can also exhibit interesting features such as pyridine like bonding in the graphitic mesh, which causes peak close to the Fermi level in the LDOS. All these nanotubes, arm chair and zig-zag are found to be conductors. Obviously, nitrogen is donating electrons to the nanotube system to produce conductivity.

24.12 KEY ISSUES IN NANOMANUFACTURING

Similar to large-scale fabrication, nanomanufacturing research issues revolve around precursor materials; fabrication processes and characterization techniques; instrumentation and equipment; theoretical modelling and control; and design and integration of structures into devices and systems.

24.13 NANOSTRUCTURES IN MOTION

Micromechanics, micromachines and moving nanostructures represent a new "micro-revolution". However, the concepts related to micromechanics and micromachining of silicon are not new. A number of commercial products including accelerometers, pressure transducers, thermal print heads and ink jet nozzle arrays have been fabricated for internal industrial use and for the open market.

Since 1982, the field of micromechanics has expanded considerably, adding micrometers, movable linkages, gears, micro-turbines, electrical drives and microweezers to an ever growing suite of micromechanical devices.

24.14 NANOMATERIAL ADVANTAGE

Nanocrystalline materials are three dimensional solids composed of nanometersized grains, or crystallites. Because of their unique structure, which is characterized by ultrafine grains and a rather high density of crystal lattice defect, these materials have extra ordinary fundamental properties that could be exploited to make **'next-generation' super-strong materials.**

Strength and ductility are the central mechanical properties of any material. They are determined by the physical nature of plastic deformation, which in conventional, coarse-grained metals is mainly carried by dislocations-line defects of regular crystal lattice-within individual grains. Study suggest, however, that the mechanism of plastic deformation in nanocrystalline materials may be different and so may lead to novel mechanical properties.

High strength, or hardness has already been observed in many nanomaterials. But in most cases such nanomaterials have very low ductility-they fail when their shape is changed. Some even become brittle when a force or deforming load is applied. Strength and ductility are usually opposing characteristics: the higher the strength, the lower the ductility, and vice versa. This correlation is associated with the nature of plasticity: the more difficult it is for dislocations to appear and to move, the stronger but brittle and less ductile is any crystalline material.

Very recently, Wang and coworkers report their success in engineering an unusual nanocrystalline material that combines two useful properties that are often mutually exelusive strength and ductility. They created a copper nanostructure by rolling the metal at temperature below 77 K and then heating it to around 450 K. The result was a 'bimodel' structure of micrometer-sized grains (at a volume fraction of around 25%) embedded in a matrix of nanocrystalline grains. The material showed extraordinary high ductility, but also retained its high strength. The reason for this behaviour seems to be that while the nanocrystalline grains provide strength, the embedded larger grains stabilize the tensile deformation of the material.

Processing nanomaterials to achieve both strength and ductility can enhance the material's resistance to fatigue. More than anything else, it is fatigue that at present limits the lifetime, and hence the range of applications, of many advanced materials. Ethics is the study of understanding what is good and what is bad. As nanotechnology steps into center stage and the bounty it promises begins to become reality, it will raise several issues of ethics, public policy, law and social responsibility. When we try to understand a new technology like the nanotechnology, it is very natural to think about the good and the bad effects of such technology. Any technology can be used for constructive as well as destructive use. This is also true for the nanotechnology. Nanotechnology as of now is still in its infant stage. A variety of applications are being proposed and explored. A majority of these are in the areas related to computer technology, defence, electronics, medicine and consumer products like cosmetics. These applications of nanomaterials can have different impacts on mankind and the world around. It is quite natural to think about the good and the bad effects of nanotechnology. When we think about such effects, we will be in a position to decide whether to take the nanotechnology in one direction or the other. We shall discuss the good and the bad about nanotechnology and the ethical issues it can raise.

24.15 APPLICATIONS OF NANOTECHNOLOGY

We would not like to focus exclusively on potential dangers of nanotechnology. Let us begin by considering some of the good things that the nanotechnology can provide. For example, nanotechnology helps to clean up the environment. Paints based on nanomaterials could convert sunlight into useable energy and remove CO_2 from the atmosphere. There are other materials which could remove other pollutants from the air. Lighter but stronger materials could be developed from designer molecules. Aircraft made of lighter materials with the strength of diamonds would be more fuel-efficient and safer. Clothing made of stronger materials would last longer.

In future, nanobots (nanoscale robots) would travel through blood vessels clearing away plaque and entering cancerous cells to destroy them. Nanotechnology is expected to manufacture food and clean water economically. Computer chips may be made at lower cost from chemical synthesis avoiding toxic byproducts. Obviously, nanotechnology is quite promising. Scientists have started understanding nanoscience and would like to exploit it to the benefit of mankind. Human life can be made healthier and safer by the use of nanoscience and nanotechnology. *The potential benefits of nanoscience and nanotechnology are immense but the potential hazards are equally serious.*

Today, nanotech is making its way into our lives, producing spill- and wrinkle-resistant clothing, lighter and stronger sports equipment, self-cleaning windows that break down dirt, etc., We will continue to see improvements in everyday products and advances in industries such as electronics. For instance, nanotechnology is helping to develop organic light-emitting diodes that can emit their own light, such as a glow worm. These would be brighter, faster, lighter, more energy efficient and would enable ultra-thin displays viewable from any angle without loss of quality. Nanotechnology will be useful in many other ways in the near and distant future. The following are some of the examples:

- *Transportation:* New materials will make vehicles lighter and stronger; these vehicles will use less fuel and can withstand more damage. Space travel will become inexpensive and no longer limited to the elite few.
- *Computers:* As processing chips become smaller, faster, and more powerful, computers will also shrink in size, allowing them to be truly ubiquitous and embedded in clothing and even in the human body.
- *Military:* Soldiers will wear an exoskeleton that can change its flexibility and become instant armor. Clothes will be able to store energy to deliver superhuman strength when needed, such as for jumping over a 20-ft wall. Weapons can be miniaturised, and smart bullets can be programmed to hit specific targets.
- *Energy:* More powerful and smaller batteries will enable devices to become smaller and operate longer. Efficient molecule-sized solar cells can be mixed into road

asphalt to continually harness the sun's energy. Fossil fuels and coal will be replaced with renewable energy sources, even turning garbage into fuel.

- *Environment:* One may be able to rebuild our thinning ozone layer and clean up the environment with nanobots that eat oil spills and other contaminants. More efficient manufacturing processes mean less pollution. We would not need to cut trees to produce paper anymore.
- *Medicine:* Surgeries will be performed with tools more than 1000 times precise than the sharpest scalpel today. Cosmetic surgery won't even require surgery-eyes can perhaps change colour and noses can change shape without cutting. One will able to use medical monitoring devices to detect diseases. New drugs can target specific cancer cells and viruses to destroy them. They can also fix defective genes or otherwise deliver precise treatment. To the extent that getting old is like any other physical disorder, repairing cells can slow, halt or reverse the ageing process.
- *Distant future:* Nanotechnology enables researchers to build materials atom by atom. Scientists therefore predict that they will be able to manufacture anything. Among other things, they can make food from dirt and water, eliminating famine worldwide.

24.16 LIMITATIONS OF NANOTECHNOLOGY

There is no doubt that there are many positive benefits from nanotechnology, but what about the bad effects of nanotechnology? Is nanotechnology ultimately worth the cost, or is the price too high. Let us examine some of the negative consequences predicted for nanotechnology:

- *Medical and Health:* Nanoparticles have been shown to be absorbed in the livers of research animals and even cause brain damage in fish exposed to them within 2 days. If nanoparticles find their way into living cells, they can enter the food chain through bacteria and pose a health threat, e.g., mercury in fish, pesticides in vegetables, or hormones in meat. Carbon nanotubes look very much like asbestos fibres, but what happens if they get released into air? Being carbon-based, they would not be detected and obviously they are repelled by the usual defence mechanism of our bodies.
- *Environment:* Since nanomaterials are as strong as diamonds, question arises, how decomposable will they be? Will they litter our environment further or pose other disposal problems like nuclear waste and nuclear hazards or space litter? In the distant future, will self-replicating nanobots-necessary to create the trillions of nanoassemblers needed to build any kind of product-run amok, spreading as quickly as virus?
- *Privacy:* With the shrink in size of products, eavesdropping devices too can become invisible to the naked eye and more mobile. This make it easier to invade our privacy. If nano devices are planted into our bodies, mind controlling nanodevices may be able to affect our thoughts by manipulating brain processes.
- *Terrorism:* Capabilities of terrorists also advances with advances in capabilities of military. Thus weapons become more powerful and portable. Obviously, these devices can also be turned against us. Nanotech may create new, unimaginable

forms of torture causing disintegration of a person at the molecular level. Radical groups or terrorists could let loose nanodevices targeted to kill a specific person.

- *Society:* With all these limitations, several expert advocate a strong system to regular and monitor nanotechnology developments. Nanotechnology is a multi disciplinary science. Nanotechnology could be used to restore environment to spread wealth, and to cure more illness. This depends on our action, working within the limits set by the real world.

Nanotechnology has become a very active and vital areas of research, which is fast developing in industrial sectors and spreading to almost every field of science and engineering. Control of matter on a molecular scale is the main aim of all researchers working in this field. Fundamental physical, mechanical, magnetic and biological properties of materials are improved remarkably as the size of their constituent grains decreases to a nanometer scale. Nanostructured materials and associated processing technologies have opened up exciting new possibilities for future applications in aerospace, automotives, electronics, coatings, nonvolatile memories, sensors, acuators, optoelectronics, drug delivery, etc., One of the difficult problems facing the design of nanostructures-based stems is understanding how to interconnect and address them. Nodoubt, the success of nanoscience will depend on the development of new device and manufacturing technologies.

SUMMARY

1. *Nanoscience* is, at its simplest, the study of fundamental principles of molecules and structures with at least one dimension roughly between 1 and 100 nanometers. These structures are known as nanostructures.
2. *Nanotechnology* is the application of nanostructures into useful nanodevices. Nanotechnology has applications in electronic devices including information storage retrieval, as well as life sciences.
3. *Molecular Nanotechnology* is the self assembly of the molecules into an ordered and functional structure.
4. *Carbon Nanotubes* (CNTS) are seamless graphene tubule structures with nanometer-size diameters and high aspect ratios. This new classes of one-dimensional materials is shown to have exceptional chemical, thermal and novel electronic properties. Nanotubes are categorized as single walled nanotubes (SWNTS) and multiwalled nanotubes (MWNTS). SWNTS has just the single shell of hexagons, whereas in MWNTs, multiple layers are nested like the rings in a tree trunk.
5. *Nanotubes* have very high electrical and thermal conductivities, along-with stability upto very high temperatures.
6. *The electronic* properties of nanotubes have been used to fabricate nanoscale transistors.
7. Another exciting nanotube application is in flat panel displays for televisions and computer monitors.
8. *Quantum dots* are nanometer-sized semiconductor crystals or electrostatically confined electrons. They are fabricated in semiconductor materials and have typical dimensions between nanometers to a few micron (10^{-6} m). These can be

easily connected to electrodes and are therefore excellent tools to study atomic like properties.

9. *Nanowires* are structures that have a thickness or diamater constrained to tens of nanometers or less and an unconstrained length. Many different types of nanowires exit, including metallic (e.g., Ni, Pt, Au), semi-conducting (e.g., Si, InP, GaN, etc.,), and insulating (e.g., SiO_2, TiO_2). Molecular nanowires are composed of repeating molecular units either organic (e.g., DNA) or inorganic (e.g., $Mo_6S_{9x}I_x$).

REVIEW QUESTIONS

1. What is Nanoscience and Nanotechnology? What is the difference between these two?
2. Explain the working of SEM and TEM. Compare these two.
3. Discuss the impact of nanoscience in various fields. Give examples of some natural phenomena where nanoscience plays an important role.
4. What are quantum dots ? How does the frequency of the light emitted from a quantum dot depend on the size of quantum dot.
5. What are nanowires ? In what respects the electrical properties of nanowires are different from conventional wires ?
6. Discuss the mechanical properties of nanomaterials. How these can be useful in various applications?
7. Described the sol-gel method of fabrication of nanomaterials.
8. How the nanotechnology will affect the life of a common man?
9. What are carbon nanotubes? Explain their electrical properties.
10. Mention important properties of nanoparticles.
11. How would one make and stabilize quantum dot ?
12. What are the different types of quantum dots investigated ? What makes quantum dots luminescence attractive?
13. What are the important concerns of materials scienctists in the nanoscience area.
14. How many types of nanotubes are there? Mention their characteristics.
15. What are the main allotropic forms of carbon? Compare their important properties.
16. What are the various techniques used to synthesize nanostructured materials/ CNTs? Discuss one of them briefly.
17. "From cars to medicine, nanotubes may be a miracle material". "They are stronger than steel and as flexible as plastic, conduct energy better than almost any material ever discovered and can be made from unexotic raw materials such as methane gas". Comment upon the above statement.

SHORT ANSWER-QUESTIONS

1. What are nanomaterials ?
   Ans: These are the materials containing nanocrystals, *i.e.,* their grain size is in the range of 1 to 100 nm. Nanomaterials may be metals, alloys, intermetallics or ceramics.
2. What do you understand by nanotechnology ?
   Ans: Nanotechnology deals with various structures of matter having dimensions of a billionth of a meter.

3. What is nanoscience and nanotechnology?
Ans: Nanoscience and nanotechnology primarily deal with the synthesis, characterization, exploration, and exploitation of nanostructured materials characterized by at least one dimension in the nanometer (1 nm = 10^{-9}m) range.
4. How nanoworld is uniquely different?
Ans: The Physical and chemical properties of nanostructures are distinctly different from those of a single atom (molecule) and bulk matter with the same chemical composition. These differences between nanomaterials and the molecular and condensed phase materials pertain to the spatial structures and shape, phase changes, energetic, electronic structure, chemical reactivity, and catalytic properties of large, finite systems, and their assemblies.
5. What are the most significant applications of nanomaterials?
Ans: Perhaps in nanodevices and nanoelectronics.
6. How nanoscience is interdisciplinary ?
Ans: Nanoscience is truly interdisciplinary area covering physics, chemistry, biology, materials and engineering.
7. What is a carbon nanotube?
Ans: Carbon nanotubes are long cylinders of 3-coordinated carbon, slightly pyramidalized by curvature from the pure sp^2 hybridization of graphene, toward the dimond like sp^3. Infinitely long in principle, a perfect tube is capped at both ends by hemi-fullerenes, leaving no dangling bonds. A single walled carbon nanotube (SWNT) is one such cylinder, while multi wall tubes (MWNT) consist of many nested cylinders whose successive radii differ by roughly the interlayer spacing of graphite.
8. What are the advantages of MWNTs over SWNTs?
Ans: Multiwall tubes have two advantages over their SWNTs cousins. The multishell structure is stiffer than the single wall one, especially in compression. Large-scale syntheses by enhanced chemical vapour deposition (CVD) processes are many, while for SWNTs, only the rice HiPco process appears to be scaleable.
9. What are the important concerns of material scientists in the nanoscience area?
Ans: (i) Nanoparticles or nanocrystals of metals and semiconductors, nanotubes, nanowires, and nanobiological systems.
(ii) Assemblies of nanostructures (e.g. nanocrystals and nanowires) and use of biological systems, such as DNA as molecular nanowires and templates for metallic or semiconducting nanostructures.
(iii) Theoretical and computational investigations that provide the conceptual framework for structure, dynamics, response and transport in nanostructures.
(iv) Applications of nanomaterials in biology, medicine, electronics, chemical processes, high strength material, etc.,
10. What are the present goals of the science and technology of nanomaterials?
Ans: To master the synthesis of nanostructures (nano-building units) and their assemblies of desired properties; to explore and establish nanodevices concepts, to generate new classes of high-performance nanomaterials, including biology inspired systems; and to improve techniques for the investigation of nanostructures.
11. What are nanocomposites?
Ans: Nanocomposites are the heterogeneous materials composed of at least one type of nanometer-sized particles.

12. What are van der Waals forces?

Ans: The van der Waals forces represent the weak (secondary) interaction between adjacent molecule or atoms, and arise from their electric dipoles. The vander Waals forces are long-range (even for distances larger than tens of nanometer), and may be attractive or repulsive depending on the distance between interacing atoms or molecules. In particles, the van der Waals forces between nanotubes become attractive when the distance is larger than the equilibrium distance, or repulsive when the distance is smaller than the equilibrium distance. The van der Waals forces between nearest nanotubes play a crucial role in physical properties of multiwall carbon nanotubes and single wall carbon nanotubes ropes.

13. How many kinds of CNTs other than SWCNT and MWCNT have been developed so far?

Ans: (i) **Carbon Nanocoil** having unique 3D structures.

(ii) **Carbon Nanofibre** (CNFS) : These are cylindrical or conical structures having length in the order of a few microns and diameter varying from 10 to 200 nm. The internal structure of CNFS varies and is comprised of different arrangements of modified grapheme sheets.

(iii) **Carbon Nanocone** (CNCON) : This has a prefect conical structure with symmetry different from other known carbon materials, including nanotubes, nanocoil and bucky balls.

(iv) **Carbon Nanohorn** (CNH) : This is made of the same graphite structure as CNT. CNH is a single wall carbon cone having structures similar to those of nanotube caps.

(v) **Carbon Nanoribbon** (CNR) : This is strip of a graphene sheet with morphology different from CNTs. It is highly graphitized and width of ribbon is in the range of 1 to 20 nm and length is in microscopic scale.

14. What is nanoelectronics ?

Ans: This is the field of science and engineering dealing with the fabrication, study and application of nanosize electronic devices for information technologies. Generally, quantum effects deermine the operational principles of these devices.

15. What is Nano-fish memory device ?

Ans: This is similar in structure to a MOSFET (MOS field effect transistor) except that is it is a three terminal device with two gate electrodes, one on the top of the other. The top electrode forms the control gate, below which a floating gate is capacitively coupled to the control gate. The memory cell operation involves putting charge on the floating gate or removing it, corresponding to two logic levels. These devices utilize single or multiple nanoparticles as the charge storage element.

16. What do you understand by nanophotonics ?

Ans: Nanophotonics is the field of science and engineering dealing with the fabrication, study and application of optical phenomena at the nanoscale. It deals with structures and process which are spatially localized in domains smaller than the wavelength of visible radiation. The field includes nanoscale confinement of radiation, nanoscale confinement of matter, nanoscale physical or photochemical transformations.

17. What are Quantum wires ?

Ans: These are 1D structures. As compared to quantum films, one more dimension of the structure appears to be so small as provide quantum confinement. Charge carriers are free to move only along the wire. Thus, only one kinetic component along with the confined energy contributes to the total energy. As a consequence, the density of states has $E^{1/2}$ dependence for each of the discrete pairs of states in the confined directions.

18. What are quantum dots?

Ans: These are 0D structures in which the carriers are confined in all three directions. Their energy state is quantized in all directions and the density of states is represented by series of discrete, sharp picks resembling that of an atom. Quantum dots are also called as **artificial atoms.** These are usually formed by a definite number of atoms. They are typically represented by atomic clusters or **nanocrystallites.** Considering the coupling of quantum dots with external electric circuits, they can be classified as open dots and almost isolated or closed quantum dots. In an open dot, the coupling is strong and the movement of electrons across dot-lead functions is classically allowed. When point contacts, connecting a dot with an external circuit, are pinched off, effective barriers are formed and conductance occurs only by tunneling. These are almost isolated or closed quantum dots.

19. What do you understand by bucky ball, C_{60} ?

Ans: This is a hollow cage carbon molecule consisting of 60 carbon atoms and has resemblance of its molecular structure to geodesic domes.

20. What is density matrix ?

Ans: A representation of mixed states generated by right-multiplying the state vector by its adjoint.

21. What is lithography ?

Ans: Literally, lithography means "stone writing". This refers generally to any technique for forming a patterned structure on a slid surface.

22. What is photolithography ?

Ans: This is a photochemical technique for etching specific patterns using projected light. It is the most widely used technique today.

23. What is a nanocomputer ?

Ans: A computer whose characteristic length scale is between $10^{-7.5}$ to $10^{-10.5}$ m (i.e., between ~ 32 and ~ 0.32 nm, i.e., closer to 1 nm than to 1 micrometer or 1 picometer on logarithmic scale).

24. What are smart materials ?

Ans: These are usually defined as those that can sense external stimuli and then respond to those stimuli by some means in real (or near real) time. Such materials are therefore a product that lies somewhere between a material and more complex nanodevice - a sensor, in fact. Some smart materials respond to light, others to pressure and so on. Some are self repairing. These materials find vide spread applications today. These materials should be recognized along with the familiar metals, plastics, ceramics, composites, powder metals, and specialty type and

multifunctional materials (shape metal alloys: (SMAs), micromechanical systems (MEMS), functionally graded materials (FGMS), and nanomaterials). By using these materials instead of adding mass, engineers can endow structures with built-in responses to a myriad of contingencies. In their various forms, these materials can perform as actuators, which can adapt to their environments by changing characteristics such as shape and stiffness, as sensors, which provide the actuators with information about structural and environmental changes, smart materials are beginning to play an important role in civil engineering drawings for dams, bridges, highways, and buildings. There are many more applications of these materials in process.

FILL IN THE BLANKS

1. The prefix nano in the word nanotechnology means a ________. (1×10^{-9}) **[billionth]**
2. Nanostructures constitute a bridge between molecules and infinite ________ systems. **[bulk]**
3. The structure of nanoparticles of Cds, CdSe and such materials is affected by ________. **[size]**
4. The synthesis of nanomaterials includes control of size, shape and ________. **[structures]**
5. The most significant applications of nanomaterials may be in nanodevices and ________. **[nanoelectronics]**
6. The skeleton of C_{60} consists of 20 hexagonal and ________ pentagonal rings fused all together. **[12]**
7. Composite materials reinforced with carbon or graphite are often used in ________ goods. **[sporting]**
8. The discovery and rapid evolution of carbon nanotubes has played a major role in triggering the explosive growth of ________ in nanotechnology. **[research and development]**
9. Graphene refers to a nanolayer of sp^2-bonded ________ atoms. **[carbon]**
10. Carbon nanotubes are hollow cylinders of ________ sheets. **[graphite]**
11. A carbon nanotube made of single graphite layer rolled up into a hollow cylinder is called a ________. **[SWNT]**
12. A tube comprising several concentrically arranged cylinders is referred to as ________. **[MWNT]**
13. Multiwall nanotubes have similar lengths to single walled tubes, but much lager ________. **[diameters]**
14. Nanotubes can behave like metals or ________. **[semiconductors]**
15. Graphene is a two-dimensional honeycomb net, made up of ________ bounded carbon atoms. **[sp^3]**
16. Armchair tubes always show ________ properties. **[metallic]**
17. The fundamental band gap in semiconducting nanotubes ranges from ________ eV to ________ eV being dependent on small variations of the diameter and bonding angle. **[0.4, 0.7]**

25

Atomic Nucleus

25.1 ATOMIC NUCLEUS (THE NUCLEUS AS A PART OF THE ATOM)

The atomic nucleus contains only two types of particles namely *proton* and *neutron.* It is true that further particles may be ejected from the nuclear assembly but they do no exist independently within the nucleus. Protons and neutrons are designated by the common name of the *nucleons.* The other particles of the modern physics are never found as independent entities within the nucleus. The atomic nucleus is not only the seat of the mass but also the origin of the energy of the atom. The atomic nuclei is occupying a small region at the centre of the atom with a diameter of the order of 10^{-14} m. The nuclear radius of the various nuclei can be estimated from the following relation.

$$R = r_0 A^{1/3} \qquad \text{... 25.1 (a)}$$

where r_0 is a constant (= 1.2×10^{-15} m), same for all nuclei. Various measurements for r_0 range between 1.0 and 1.5×10^{-15} m. Looking to the smallness of nucleus, we use the *femometer*, abbreviated fm, with 1 fm $=10^{-15}$ m. An alternative term for 10^{-15} m, *Fermi.* The term Fermi, has the same abbreviation, fm is more commonly used than femtometer. We may note that a nucleus has no clearly defined boundary. Obviously, the nuclear radius has only arbitrary meaning. Volume of an atomic nuclear radius, $V = 1.2 \times 10^{-45}\, A/\text{m}^3$. The mass of an atomic nucleus of mass number A is approximately.

$$M = 1.66 \times 10^{-27}\, A \text{ kg.}$$

Making use of these values, one obtains the average density of nuclear matter as

$$\rho = \frac{M}{V} = 1.49 \times 10^{18} \text{ kg/m}^3$$

which is independent of mass number A. This density of the nucleus is about 10^{15} times greater than the density of matter in bulk, and gives an idea of the degree of compactness of the nucleons in a nucleus. It also shows that matter in bulk is essentially empty, since most of the mass is concentrated in the nucleus.

Table 25.1

Particle	*Rest energy (MeV)*	*Charge*	*Atomic mass unit (u)*	*Spin, Magnetic moment*
Proton (P)	938.272	+e	1.0072765	½, 2.79 μ_N
Neutron (n)	939.566	o	1.0086649	½ , –1.91 μ_N
Electron (e)	0.51100	–e	5.4859×10^{-4}	½, –1.00116 μ_B

The primary constituents of nuclei are the proton and neutron and the nuclear mass is roughly the sum of its constituent proton and neutron masses (the slight difference being the binding energy of nucleus). The nuclear charge is +e times the number (Z) of protons.

The atomic (and nuclear) mass number A is the total integral number of protons and neutrons in a nucleus. We will designate an atomic nucleus by the symbol ${}^{A}_{Z}X_{N}$, where Z is atomic number (number of protons), A is mass number ($= Z + N$), N is neutron number (number of neutrons and X is symbol of chemical element. Each nuclear species with a given Z and A is called a **nuclide**. When we write the nuclide symbol for the better known elements, Z is often omitted, e.g. ^{16}O, ^{14}N, etc.

We may note that the chemical properties of an atom are determined by its electron configuration. Since the number of electrons and protons are equal in a neutral atom and hence the chemical properties are essentially determined by Z. Obviously, the dependence of the chemical properties on N is negligible.

Atomic masses are measured in mass units. Which are denoted by symbol u. Atomic mass units are defined in terms of the mass of the isotope ^{12}C, whose atomic mass is defined to be exactly 12 u.

$1u = 1.6605 \times 10^{-27}$ kg $= 931.49$ MeV/c^2. The masses of proton, neutron and electron are given in table 25.1.

25.2 ATOMIC NUMBER, MASS NUMBER AND ISOTOPES

A particular nucleus is identified by the number of protons it has, or its *atomic number Z and mass number A*. Thus the number of neutrons is $N = A – Z$. A nuclide is designated by the symbol of the chemical element to which it belong, according to the value of Z, with as superscript to the left indicating the value of the mass number A, such as ^{12}C, ^{16}O, ^{235}U. Sometimes it is convenient to write the atomic number Z as a subscript to the left, such as ${}^{12}_{6}$C, ${}^{16}_{8}$O, ${}^{235}_{92}$U. Due to great variety of nuclides, they are classified in three categories as mentioned below.

(i) Isotopes: These are the nuclides having the same atomic number Z but different neutron number N and different mass number A for e.g. ${}^{1}_{1}$H and ${}^{2}_{1}$H are isotopes of hydrogen, ${}^{238}_{92}$UI and ${}^{234}_{92}$UII are natural radioisotopes of ${}_{92}$U. Similarly ${}^{230}_{90}$Ionium , ${}^{232}_{90}$Th , ${}^{234}_{90}$UX$_1$ are natural isotopes. Let us now consider the radioactive disintegration.

$$ {}^{230}_{90}\text{Ionium} \xrightarrow{\alpha} {}^{226}_{88}\text{Ra} $$

Obviously ${}^{226}_{88}$Ra is the product of disintegration of ionium. It is found that the chemical properties of ${}^{230}_{90}$Ionium turned out to be similar to those of naturally occurring radioactive element. Thorium ($^{232}_{90}\text{Th}$), *i.e.,* when these two elements mixed could not be separated chemically even with extremely most sensitive methods. Also, Ionium and Thorium have been found to be spectroscopically identical. Clearly, $^{230}_{90}$ Ionium and $^{232}_{90}\text{Th}$ are isotopes. Thus, one can say that isotopes means the same position in the Periodic Table. In general, we can say that two species $^{A}_{Z}X$ and $^{A'}_{Z}Y$ which have different atomic masses (A and A') inspite of being different elements and have same atomic number (Z) and moreover identical chemical properties, then these different species are called *isotopes.*

(ii) Isobars: These are the nuclides having the same total number of nucleons (or A) but differ in Z and N. Examples of isobar are

$$^{14}_{6}\text{C and } ^{14}_{7}\text{N}, \ ^{234}_{90}\text{UX}_1, \ ^{234}_{91}\text{UX}_2 \ ^{234}_{92}\text{UII}$$

(iii) Isotones: These are the nuclides having the same number of neutrons N but *a* different Z and A. Example of isotones are $^{13}_{6}\text{C}$ and $^{14}_{7}\text{N}$. We may note that isotones having a given value of N obviously do not all correspond to the same chemical element. The analysis of the properties of **isotopes, isotones** and **isobars** helps us to disclose several features of nuclear structure. Such analysis helps us, e.g. to predict what will happen to the stability of a nucleus when an extra neutron (n) or proton (p) is added to it.

25.2.1 Size and Shapes of Nuclei

The nuclear radius may be approximated from a spherical charge distribution to be

$$R = r_0 A^{1/3}$$

where $r_0 = 1.2 \times 10^{-15}$m. Various measurements for r_0 range between 1.0 and 1.5×10^{-15}m. Since nucleus is so small, we use **femtometer** (*fm*), with 1 *fm* $=10^{-15}$m, Fermi (*fm*). The term Fermi, which has the same abbreviation, *fm*, is more commonly used than or 2.6 femtometers.

Hofstadter and coworkers at Stanford University in the 1950s performed the first precision electron scattering measurement of the nuclear charge distribution using electron energies from 100 to 500 MeV. In order to probe the actual shape of most nuclei one need a particle having a short wavelength. The de-Broglie Wavelength of a 500 MeV electron is about 2.5 fm, and by now, measurements have been made with much shorter wavelengths using higher energy electrons. These measurements are approximately described for all but the lightest nuclei by the Fermi distribution for the nuclear charge density $\rho(r)$ as

$$\rho(r) = \frac{\rho_0}{1 + \exp[(r - R)/a]} \quad \text{... 25.1(b)}$$

Fig. 25.1 (a) shows the shape of this distribution. Here ρ_0 is the central nuclear density, R is the distance at which the nuclear density has dropped to 50% of its central value.

25.2.2 Intrinsic Spin

The neutron and proton are fermions and spin quantum number of each is $f = ½$. The spin quantization rules for neutron and proton are same as for the electron.

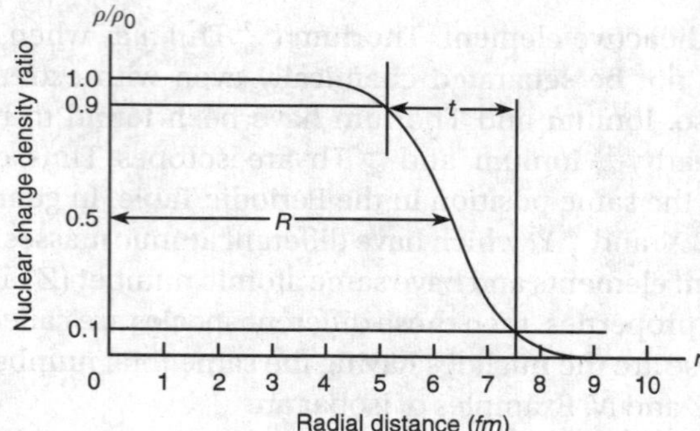

Fig. 25.1 The shape of the Fermi distribution for the nuclear charge density given by Eq. 25.1(b) where $\rho_0(r)$ is the central charge density, R is the distance at which the nuclear density has dropped to 50% of its central value

25.2.3 Intrinsic Magnetic Moment

Since the proton's charge is positive and hence the proton's intrinsic magnetic moment points in the same direction as its intrinsic spin angular momentum. We may note that this is in contrast with the electron, where the spin and magnetic moment point in the opposite direction. The unit of nuclear magnetic moment is nuclear magneton (μ_N). μ_N is defined by analogy to the Bohr magneton for electrons as follows :

$$\mu_N = \frac{e\hbar}{2m_p}$$

We may note that divisor in calculating μ_N is the proton mass m_p, which makes the nuclear magneton some 1800 times smaller than the Bohr magneton.

We have μ_p (proton magnetic moment) = 2.79 μ_N. We can see that this contrasts strongly with the magnetic moment of the electron μ_e= –1.00116 μ_s. We may note that neutron, which is electrically neutral, also has a magnetic moment, μ_n = –1.91 μ_N. Negative sign clearly indicates that the magnetic moment points opposite to the neutron spin. It is interesting to note the large deviation from unity of the proton's magnetic moment and the fact that the neutron even has a magnetic moment, clearly indicate that nucleons are more complicated structurally than electrons. The nonzero neutron magnetic moment clearly implies that the neutron has negative and positive internal charge components at different radii, and thus have a complex internal charge distribution.

25.3 NUCLEAR BINDING ENERGY

Mass spectroscopic studies of masses of the isotopes reveals that the mass number A of isotopes are all nearly whole numbers, the slight variation being only in the second place of decimal. This is called as **'Whole number rule.'** This rule lead to the general acceptance that atomic nuclei are made up of hydrogen nuclei (protons) and neutrons having almost the same mass. Although atomic masses are close to whole numbers but they differ from integers by small amounts, *i.e.*, they depart from the whole number rule systematically. The light nuclei have masses slightly higher than the nearest integer, e.g., ${}_1^1H = 1.100813$, ${}_2^4He = 4.00408$, ${}_3^6Li = 6.01614$. Large number of medium nuclei have masses somewhat lower, e.g. ${}^{20}Ne = 19.9988$, ${}_{17}^{35}Cl = 34.933$, ${}^{98}Mo = 97.944$. Very heavy nuclei have slightly higher mass again, e.g. ${}^{200}Hg = 200.028$, ${}^{205}Tl = 205.036$.

The difference between atomic mass and mass number of an atom is called **mass defect** (Δm), *i.e.*,

$$\Delta m = M - A \qquad \text{... 25.1(c)}$$

where M is the measured mass and A is mass number. Obviously, the mass defect represents the deviation of atomic mass from the whole number A. The deviations of the masses of nuclides from whole numbers was an important problem. Aston expressed these deviations in the form of a quantity called the **packing fraction** (f) defined by

$$f = \frac{\text{Actual atomic mass} - \text{Mass number}}{\text{Mass number}}$$

$$= \frac{M_{Z,A} - A}{A} = \frac{\Delta}{A} \qquad \text{... (25.2)}$$

The packing fractions is defined as the *mass defect per nucleon* in its nucleus. f is a very small quantity. Aston multiplied the values of f by 10^4 to obtain figures which are more convenient to record. Thus the usual quoted values of packing fractions are expressed as

$$f = \frac{\Delta}{A} \times 10^4 \qquad \text{... (25.3)}$$

Concluding, one may say that packing fraction (f) is the mean gain or loss per particle in atomic mass, in order to pack the nuclear elementary particle to form a stable nucleus. The variation of packing (f) with mass number for a large number of nuclides is shown in Fig. 25.2. The packing fraction, with the exception of those for ${}_2^4He$, ${}_8^{16}O$ and ${}_6^{12}C$ fall on or near the smooth curve. The values are high for elements of low mass number, apart from the nuclides mentioned. From the graph it is evident that the values of packing fractions at first fall rapidly with increasing mass number, pass through a relatively flat minimum and then rise slowly but steadily, for the heavy elements beyond $A = 180$. The knowledge of packing fraction was very useful in the study of isotopic masses, but they do not have a *precise theoretical significance* and *precise physical meaning.*

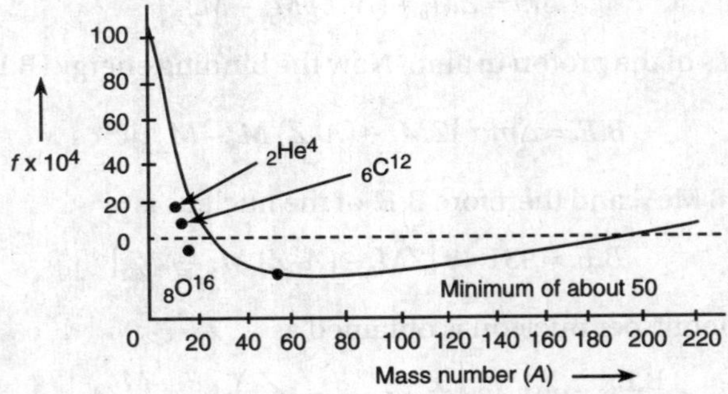

Fig. 25.2 Variation of atomic packing fraction ($f \times 10^4$) with mass number (A)

25.4 BINDING ENERGY

When a nuclide is formed from its constituent particles, it is found that sum of the mass of the nuclide is less than the sum of the masses of its constituent particles in the free state. For example deuterium atom ${}_1^2H(D)$ has a mass of 2.014102 u, while the rest mass of a free neutron and free electron together is equivalent to 2.016490 u, which is 0.002388 u greater. Obviously, when protons and neutrons combine to form a nucleus, a loss in mass results. This loss in mass is released in the form of equivalent energy when the nucleus is formed. If Δm is the decrease in mass when a number of protons, neutrons and electrons combine to form an atom, then the energy released ΔE will be in accordance with the Einstein's relation.

$$\Delta E = (\Delta m)c^2 \quad \text{... (25.4)}$$

where c is the velocity of light. We may note that the same amount of energy would be needed to break the atom into its constituent particles. In other words, one can say that if Δm is the increase in mass due to the combination of protons, neutrons and electrons to form the atom, then the energy $E = \Delta mc^2$ will be absorbed in this process and when the atom gets break, this amount of energy will be released. The loss in mass (Δm) is called the '*mass defect*' and its energy equivalent (ΔE) is called the '*binding energy*' of the nucleus. Obviously ΔE is also the energy that would be needed to break the nucleus into its constituent particles. Clearly, **the binding energy is a measure of the stability of the atomic nucleus**. Now, one can define the *nuclear energy as the energy equivalent of the mass defect* in *a nucleus,* one can also define it as the *difference between the rest mass energy of the constituent nucleons and the rest mass energy of the nucleus.*

Let us consider that the constituents of the atom be Z protons, Z electrons and $(A - Z)$ neutrons. The mass of Z protons and Z electrons are equivalent to mass of Z hydrogen atoms. Now, one can write the mass defect as

$$\Delta m = ZM_H + (A - Z)\, M_N - M_{Z,A}$$

where M_H is the mass of the hydrogen atom in u, M_N is the mass of neutron in amu and $M_{Z,A}$ is the atomic mass, the actual mass of the atom in amu.

We know that the mass of an electron is very negligible and therefore the main contribution to the mass of the atom is due to the nucleus, *i.e.,* due to the mass of nucleons (*i.e.,* mass of protons and mass of neutrons). Thus one can write the mass defect as

$$\Delta m = ZM_p + (A\text{–}Z)\, M_N - M_{Z,A}$$

where M_p is the mass of the proton in amu. Now the binding energy (B.E.) of the nucleus is

$$B.E. = \Delta mc^2\, [ZM_p + (A\text{–}Z)\, M_N - M_{Z,A}]c^2$$

since I amu = 931.48 MeV, and therefore B.E. of the nucleus is

$$B.E. = 931.48\, [ZM_p - (A\text{–}Z)\, M_N - M_{Z,A}] \quad \text{... (25.5)}$$

Now the binding energy, per nucleon is obtained as

$$\frac{\text{B.E.}}{A} = 931.48\left[\frac{Z}{A}M_P - \left(1 - \frac{Z}{A}\right)M_N - \frac{M_{Z,A}}{A}\right]$$

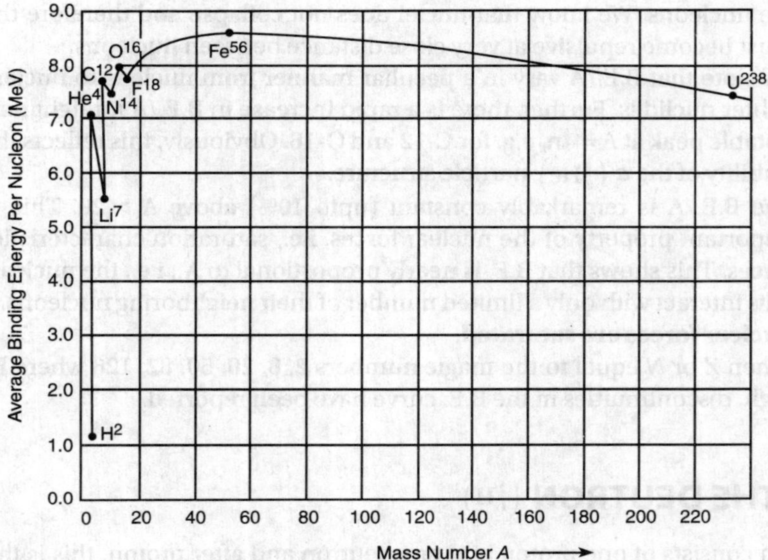

Fig. 25.3 Variation of binding energy per nucleon (BE/A) with mass number A

A graph of B.E./A (Binding energy per nucleon) as a function of the mass number is shown in Fig. 25.3. From the graph, we note that with the exception of ${}^{4}_{2}\text{He}$, ${}^{16}_{8}\text{O}$, and ${}^{12}_{6}\text{C}$ the values of B.E./A lie on or close to a single curve. We may note that B.E. is a measure of the stability of the nucleus. From the graph one can easily conclude as

(i) B.E./nucleon for very light nuclides e.g. H, He, etc. is very small.

(ii) B.E./nucleon rises sharply at first and then gradually until it reaches a maximum value of 8.79 MeV at A = 56 corresponding to ${}^{56}_{26}\text{Fe}$.

(iii) The maximum of B.E./nucleon curve is quite flat, and B.E./nucleon is still 8.4MeV at A = 140.

(iv) For higher mass numbers, the value of B.E./A drops slowly to about 7.6 MeV at the highest mass numbers, e.g. Uranium at A = 238. Looking at the pacing fraction curve (Fig. 25.2), we note that its two ends shows that packing fraction *f* is positive for light and heavy nuclides due to which B.E./A is also less for light and heavy nuclides. This fact immediately suggests that energy will be released if light nuclei can somehow be fused to form heavier ones.

This process of releasing nuclear energy by fusing lighter nuclei to form a medium size nucleus is known as *'fusion'* and is the source of energy in stars. Energy will also the released if heavy nuclei can somehow be split up into lighter ones. This method of releasing nuclear energy by breaking up of a heavy nucleus is known as *'fission'* and is the basis for the operation of nuclear power reactors.

(v) The B.E./A has a average value of about 8 MeV over a considerable range of mass numbers, where packing fraction *f* is negative.

(vi) The positive value of B.E./A for all values of A reveals the stability of the nuclei and at the same time also shows that there will be attractive nuclear forces between the nucleons. We know that nuclei does not collapse and therefore these forces must become repulsive at very close distance between nucleons.

(vii) We note that B.E./A vary in a peculiar manner from nucleus to nucleus for very lighter nuclides. Further, there is a rapid increase in B.E./A for light nuclei with a notable peak at A = 4n, e.g. for C-12 and O-16. Obviously, this reflects the peculiar stability of the α $\left({}_{2}^{4}\mathrm{He}\right)$ particle structure.

(viii) The B.E./A is remarkably constant (upto 10%) above A > 20. This reveals an important property of the nuclear forces, i.e., saturation characteristics of these forces. This shows that B.E. is nearly proportional to A., i.e., the nuclear constituents interact with only a limited number of their neighboring nucleons. Obviously **nuclear forces are saturated.**

(ix) When Z or N equal to the magic numbers 2, 8, 20, 50, 82, 126 where B.E. is very high, discontinuities in the B.E. curve have been reported.

25.5 THE DEUTRON $\left({}_{1}^{2}\mathrm{H}\right)$

A deuteron consists of one proton and one neutron and after proton, this is the simplest nucleus. The deutron mass is 2.013553u and the mass of a deuterium atom is 2.014102u. The difference in masses = 0.00549u, which is just the mass of an electron. The binding energy B_d of the deuteron can be calculated from

$$M\left({}_{1}^{2}H\right)=m_n+M\left({}_{1}^{1}H\right)-B_d/c^2 \qquad \text{... (i)}$$

Here $M\left({}_{1}^{2}H\right)=2.014102\ u$ (atomic deuteron mass), $M_n = 1.008665\ u$ (neutron mass)

$M\left({}_{1}^{1}H\right)=1.007825\ u$ (atomic hydrogen mass)

$$B_d/c^2 = m_n+M\left({}_{1}^{1}H\right)-M\left({}_{1}^{2}H\right) = 0.0023884 \qquad \text{... (ii)}$$

Taking $u = 931.5\ \text{MeV}/c^2$, we obtain

$$B_d/c^2 = 0.002388\ c^2 \,.\, u\left(\frac{931.5\ MeV}{c^2 u}\right) = 2.224\ \text{MeV}$$

Here, we have neglected atomic electron binding energy, 13.6 eV, as this is much smaller in comparison to 2.2 MeV.

Experimentally, by scattering gamma rays (photons) from deuterium gas

$$\gamma + d \rightarrow n + p \text{ (photo nuclear reaction)} \qquad \text{... (iii)}$$

binding energy of deuterium is obtained from

$$h\nu_{\min} = m_n c^2 + M\left({}_{1}^{1}\mathrm{H}\right) c^2 - M\left({}_{1}^{2}\mathrm{H}\right) c^2 = B_d$$

The above equation is not quite correct, as the momentum must also be conserved in the reaction (K_n and K_p can't both zero in mass - energy relation)

$$h\nu + M\left({}_{1}^{2}\mathrm{H}\right) c^2 = m_n c^2 + M\left({}_{1}^{1}\mathrm{H}\right) c^2 + K_n + K_p \qquad \text{... (iv)}$$

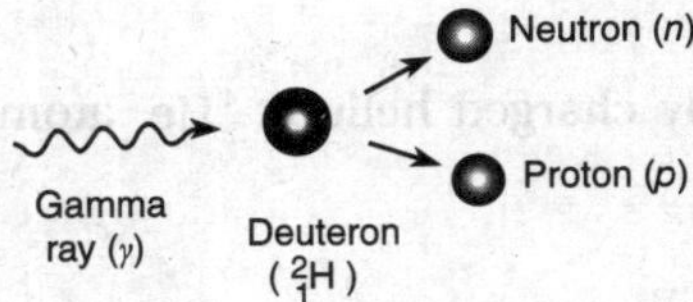

Fig. 25.4 A gamma ray photon of energy greater than 2.22 MeV is able to dissociate a deutron into a neutron and a proton. This experiment of photodisintegration confirms that the binding energy of deutron is 2.22 MeV

The precise relation can be written as

$$h\nu_{\text{min}} = B_d\left[1+\frac{B_d}{2M\left({}_1^2\text{H}\right)c^2}\right] \quad \text{... (v)}$$

The value $h\nu_{\text{min}}$ is almost exactly B_d, the deuteron binding energy (= 2.22 MeV), as the second term in (v) is so small. Experiments clearly reveals that a photon of energy less than 2.22 MeV cannot dissociate a deuteron.

Nuclear spin quantum number of deuteron is 1. This indicates that the spins of neutron and photon are aligned parallel to each other. We can see that **nuclear magnetic moment** of a deuteron is 0.86 μ_N, which is close to the sum of the values for the free proton and neutron: 2.79 μ_N - 1.91 μ_N = 0.88 μ_N. Obviously, this supports our hypothesis of parallel spins.

25.6 RADIOACTIVITY

25.6.1 Natural Radioactivity

Henry Becquerel in 1896 discovered the phenomenon of natural radioactivity. The atomic nuclei of certain substances, e.g. uranium and thorium minerals, radium, plutonium emit radiations which ionize gases and affect photographic plates. Such substances are said be *'radioactive'* substances and the phenomenon is known as *natural radioactivity*. All the naturally occurring radioactive elements lie in the range of atomic number Z = 81 to 92.

The radioactive behaviour of the substances remains unaffected by any physical and chemical changes. Radioactivity is basically a nuclear phenomenon. Uranium, radon, polonium, thorium, actinium, etc. are radioactive substances.

Radioactivity is a spontaneous and an irreversible self disintegrating activity. Moreover radioactivity is a drastic process because the element changes in kind due to this emission of radiation.

In general, radioactive substances are usually found to emit either α-rays or β-rays, sometimes α and β decay accompany γ-rays. It is not necessary that each substance should emit all the three, *i.e.*, α, β and γ-rays.

It is found that during, radioactive transformation due to either d-emission or β-emission, not all the nuclei undergo disintegration simultaneously Moreover, the rate at which the decay takes place depends on the number of parent atoms available at a given nuclei.

α-Rays

It is doubly ionized positively charged helium ${}_{2}^{4}\text{He}$ atom with its both the electrons removed.

β^{-}-Rays

This consists of electrons $\left({}_{-1}^{0}e\right)$ with negative charge and almost zero mass. They move with very high velocities.
We may note that β^{+}-ray denotes positron $\left({}_{+1}^{0}e\right)$ with positive charge, same in magnitude as an electron but with opposite charge and almost zero mass as an electron.

γ–Rays

They are electromagnetic waves (or photons) of very short wavelength (≈ 0.4 Å – 0.00 4 Å). Their frequency is higher than of X-rays. They do not carry any charge.

25.6.2 Radioactive Decay or Disintegration

Although nuclear forces are very strong but there are heavy elements like uranium, thorium, plutonium and radium, etc. are unstable, *i.e.,* these elements are constantly breaking up or disintegrating into fresh, radioactive atoms with the emission of α-, β- or γ-rays. from their nuclei. In other words, one can say that an unstable nucleus disintegrates to achieve a new configuration which is stable or one which leads to a stable one. We may note that in the process of disintegration, the parent (original) atom disappears and gives rise to daughter (new) atom. These new atoms are also, in general, radioactive and hence spontaneously disintegrates, in their turn, thereby leading to a chain of different radioactive elements forming a radioactive series, until an inactive element (usually lead) is reached. There are three well-known naturally occurring radioactive disintegration series. (i) **Uranium radium series,** (ii) **Thorium series** and (iii) **Actinium series**. This spontaneous breaking up of the radioactive nucleus is termed as *radioactive* disintegration.

We can represent the α and β disintegrations as

$${}_{Z}^{A}\text{X} \xrightarrow{\alpha({}_{2}^{4}\text{He})} {}_{Z-2}^{A-4}\text{Y} + {}_{2}^{4}\text{H}$$

where X is an element of mass number A and atomic number Z. This disintegrates into another element Y having mass number less than four, *i.e., A*-4 and atomic number less then 2, *i.e., Z* – 2. Obviously, in this process α-particle (${}_{2}^{4}\text{He}$ nucleus) is emitted. An example of above disintegration is decay of ${}_{92}^{238}\text{U}$ nuclei.

$${}_{92}^{238}\text{U} \xrightarrow[{}_{2}^{4}\text{He}]{\alpha} {}_{90}^{234}\text{Th} + {}_{2}^{4}\text{He}$$

Similarly, when a radioactive element disintegrating by the emission of β^{-}-rays, is turned into another new element whose mass number does not change but atomic number (number of proton in the nucleus) will increase by 1. In general, β^{-}-disintegration can be expressed by the following nuclear reaction.

$${}_{Z}^{A}X \xrightarrow[{}_{-1}^{0}e]{\beta^{-}} {}_{Z-1}^{A}Y + {}_{-1}^{0}e + \gamma\text{-rays}$$

An example of β disintegration is

$$^{214}_{82}\text{RaB} \xrightarrow[{}^{0}_{-1}e]{\beta^-} {}^{214}_{83}\text{RaC} + {}^{0}_{-1}e + \gamma\text{-}rays$$

25.6.3 Laws of Radioactive Disintegration

On the basis of experimental observations, Rutherford and Soddy in 1902 proposed the theory of radioactive disintegration. The theory accounts for the radioactive disintegration of uranium, radium and actinium. The important features of this theory are:

(i) The radioactive disintegration is a spontaneous phenomenon, producing new radioactive elements. The process continues till stable element is reached. Further the disintegration can neither be accelerated nor be retarded. The phenomenon is a *statistical one* and entirely dependent upon the *law of change.*

(ii) The atom of radioactive element constantly breaks into new atoms with the emission of α-rays, β-rays, which are usually but not necessarily accompanied by γ-radiation.

(iii) The radioactive atom has definite probability of disintegration. The number of atoms disintegrating in a certain time is directly proportional to the number of parent atoms remaining at that instant.

Let N be the number of atoms present in the radioactive substance at any instant t. Let dN be the number that disintegrates in a short interval dt. Then the rate of disintegration is $-dN/dt$, and is proportional to N. One can write

$$-\frac{dN}{dt} \propto N$$

or

$$\frac{dN}{dt} = -\lambda N = A/C \qquad \text{... (25.6)}$$

where the proportionality constant λ is a constant for a given substance and is known as *transformation constant* or *'decay constant'* or *'radioactive constant'* or *'disintegration constant'*.

The constant λ is independent of all external conditions like temperature, pressure. The negative sign is Eq. (25.6) indicates that there is a decrease in the original number of atoms in present element. In Eq. (25.6) A is called the **activity** and C is the **detection coefficient.** We may note that N and A are always interchangeable by the relation $A/C = -\lambda N$. Eq. (25.6) can be rewritten as

$$\frac{dN}{N} = -\lambda dt \qquad \text{... (25.7)}$$

If at $t = 0$, $N = N_0$ and at $t = t$, $N = N$, then by integrating Eq. (25.7) within the limits 0 to t, one obtains.

$$\int_{N_0}^{N} \frac{dN}{N} = -\lambda \int_{t_0}^{t} dt$$

or $$\left[\log e^{N}\right]_{N_0}^{N} = -[\lambda t]_0^t$$

or $$\log_e N/\log_e N_0 = -\lambda t$$

or $$N/N_0 = e^{-\lambda t}$$

or $$N = N_0 e^{-\lambda t} \quad \text{... (25.8)}$$

Eq. (25.8) shows that the number of atoms of a given radioactive substance decreases exponentially with time, provided no new atoms are added or introduced, *i.e.*, more rapidly at first and slowly afterwards (Fig. 25.5). From the graph, it is obvious that N is theoretically never zero but $N \to 0$ as $t \to \infty$. Eq. (25.8) is experimentally verified for the many radioactive elements. Figure 25.6 shows the decay of a radioactive element '*A*' and growth of other element '*B*'. Here the relative activity N/N_0 is plotted against decay time. The quantity dN/dt, *i.e.*, rate of decay is known activity or the strength of the radioactive substance. Using Eq. (25.8) and putting $t = 1/\lambda$, one obtains

$$N = N_0 e^{-1/\lambda \cdot \lambda} \quad \text{or} \quad \frac{N}{N_0} = 1/e \quad \text{... (25.9)}$$

Hence the **radioactive constant is the reciprocal of the time during which the original number of atoms of a radioactive substance reduce to 1/e of its value.**

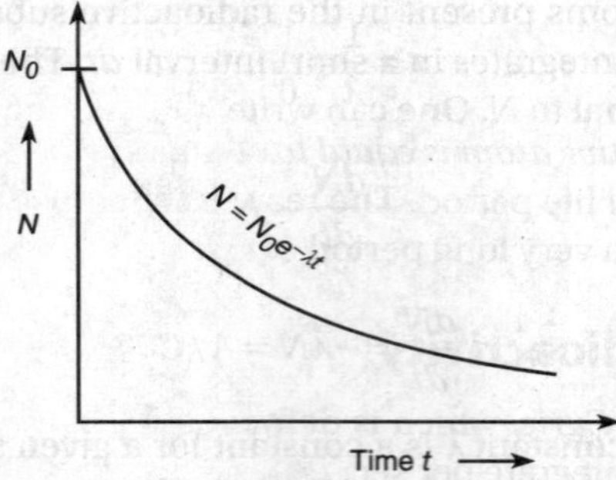

Fig. 25.5 Radioactive decay

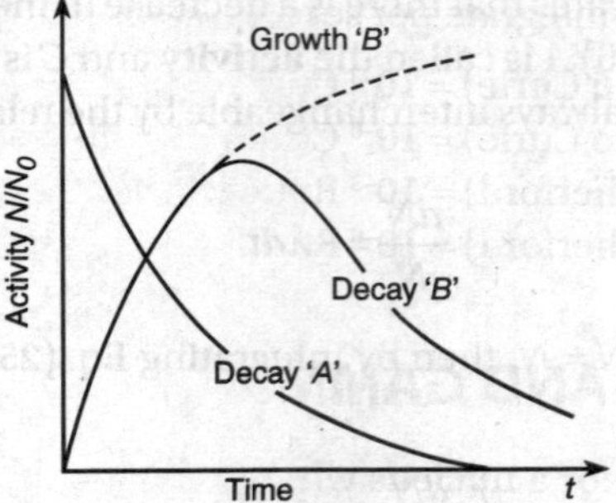

Fig. 25.6 Radioactive decay and growth

25.6.4 Half Life

The half-life period T of a radioactive substance is the time in which the amount of radioactive substance is reduced to half of its original value, *i.e.,*

$$N/N_0 = ½$$

In general $$N/N_0 = e^{-\lambda t}$$

When $$t = T \text{ then } N/N_0 = e^{-\lambda T} = ½$$

or $$\lambda T = \log 2 = \log_{10} 2 \times 2.3026$$
$$= 0.03010 \times 2.306 = 0.6931$$

or $$T = \frac{0.6931}{\lambda} \quad \text{... (25.10)}$$

This is the relation between half life and decay constant. The half life for radium is about 1600 years and that for uranium is 5000 million years.

25.6.5 Average Life (τ) or Mean Life or Radioactive Atom

The average life of a radioactive atom is equal to the sum of the lifetime of all the atoms divided by the total number of atoms.

Mathematically, average life of a radioactive substance is

$$\tau = \frac{1}{\lambda} = \frac{T}{0.693} \quad \text{... (25.11)}$$

i.e., mean life τ of a radioactive atom is equal to the reciprocal of its disintegration constant and it is greater than its half life period. The reason is that the last few atoms of a radioactive substance may last for a very long period of time.

25.6.6 Units of Radioactivity

The unit of radioactivity is *Curie,* which is defined as the activity of one gm of radium in which 3.7×10^{10} atoms disintegrate per sec, *i.e.,*

$$1 \text{ Curie} = 3.7 \times 10^{10} \text{ disintegrate/sec}$$

Radioactivity is also measured in **Rutherford** which is defined as the activity of 1 gm of radium in which 10^6 atoms disintegrate per second. The smaller units of radioactivity are

$$1 \text{ mC (milli Curie)} = 10^{-3} \text{ Curie} = 3.7 \times 10^7 \text{ dis/sec.}$$
$$1 \mu\text{C (micro Curie)} = 10^{-6} \text{ Curie} = 3.7 \times 10^4 \text{ dis/sec.}$$
$$1 \text{ mRd (milli Rutherford)} = 10^{-3} \text{ Rutherford (Rd)} = 10^3 \text{ dis/sec.}$$
$$1 \mu\text{Rd (micro Rutherford)} = 10^{-6} \text{ Rutherford (Rd)} = 1 \text{ dis/sec.}$$

25.7 ALPHA, BETA AND GAMMA DECAY

Radioactive decay may occur for a nucleus when some other combination of A nucleons has a lower mass. Consider the radioactive nucleus ${}^A_Z X$ be called the parent and have the mass $M({}^A_Z X)$. In the decay two or more products can be produced. In the case of two products let the mass of the lighter one be M_y and the mass of the heavier one (usually called the daughter) be M_D. According to principle of conservation of energy, we have

$$M({}_{Z}^{A}X) = M_D + M_y + Q/c^2 \quad \text{... (i)}$$

Where Q is the energy released and is equal to the total kinetic energy of the reaction products. In terms of mass, Q is given by

$$Q = [\,M({}_{Z}^{A}X) - M_D - M_y]\, c^2 \quad \text{... (ii)}$$

We may note that the **disintegration energy** Q is the negative of the binding energy (B). Interestengly, the binding energy normally refers to stable nuclei, where as Q is normally used with unstable nuclei. In case if $B > O$, a nuclide is bound and stable, if $Q > O$, a nuclide is unbound, unstable and decay. If we examine the naturally abundant radioactive nuclei, one finds that decays emitting nucleons do not occur, as the masses are such that $Q < O$. However, if nucleon decay were possible for these radioactive nuclei, it would have taken place too quickly for the nuclei to the naturally abundant.

25.7.1 Alpha Decay (α-decay)

Alpha (${}_{2}^{4}He$) nuclei is particularly stable. The binding energy of α nuclei is 28.3 MeV. In a α-nuclei, the combination of two neutrons and two protons is particularly strong because of the pairing effects. If the last two protons and two neutrons in a nucleus are bound by less then 28.3 MeV, than the emission of an α-decay, is energetically possible. We have

$${}_{Z}^{A}X \xrightarrow{\alpha} {}_{Z-2}^{A-4}Y + \alpha$$

$$Q = [\,M({}_{Z}^{A}X) - M)({}_{Z-2}^{A-4}Y) - M[{}_{2}^{4}He\,)]c^2 \quad \text{... (iii)}$$

If $Q > O$, then α-decay is possible. The familiar α-decay reaction is

$${}_{92}^{230}U \longrightarrow \alpha + {}_{90}^{226}Th$$

$$M({}_{92}^{230}U) = 230.033924$$

$$M({}_{2}^{4}He) = 4.0026034$$

$$M({}_{90}^{226}Th) = 226.0248914$$

$$\therefore Q = [230.033927u - 226.024891u - 4.002603u]\, c^2 \times \left(\frac{931.5\text{ MeV}}{c^2.u}\right)^{11(c)}$$
$$= 6.0\text{ MeV}$$

Obviously, $Q > O$ and hence α-decay is allowed. The mass of the products is less than the mass of the decaying nuclide. Many of the nuclei above $A = 150$ are found susceptible to α-decay. These heavy nuclei have increasingly stronger Coulomb repulsion as protons are added. The expulsion of two protons along with two neutrons in the form of α-particle may decrease this Coulomb energy and make the resulting nucleus more stable.

One may ask, why any nuclei with $A > 150$ exists ? We may note that nuclei are not necessarily made up of a collection of α-particles. For α-decay to occur, two neutrons and two protons group together within the nucleus prior to decay. Further, the α-particle, even when formed, has great difficulty in overcoming the nuclear attraction from the remaining nucleons to escape. Fig. 25.7 shows the potential energy diagram.

Normally, the barrier height V_B for α-particle is greater than 20 MeV. The kinetic energy of α-particles emitted from nuclei range from 4 MeV to 10 MeV. Classically, it is impossible for the α-particles to escape as the P.E. barrier is greater than the kinetic energy of α-particles. If we project 5 MeV α-particles onto a heavy nucleus, we find that α-particle is repelled by the Coulomb force as shown in Fig. 25.7 and does not get close enough to feel the attraction of the short range nuclear force. Classically, it is virtually impossible for an α-particle to reach the nucleus. Now the question arises, how then can the α-particle ever surmount the barrier if it trapped inside the potential barrier. However, quantum mechanically, the α-particles are able to tunnel through the barrier. Quantum mechanically, there is small, but finite change for the α-particle to appear on the other side to the barrier. We may note that the probability depends critically on the barrier height and width. Fig. 25.8 shows that a higher energy α-particle, E_2 has a much higher probability than does a lower energy α-particle, E_1, to tunnel through the barrier. We may note that the higher tunnelling probability corresponds to a shorter lifetime for the radioactive nuclide. A comparison of lifetimes of various α-emitters with the kinetic energies of the α-particles is shown in Fig. 25.9.

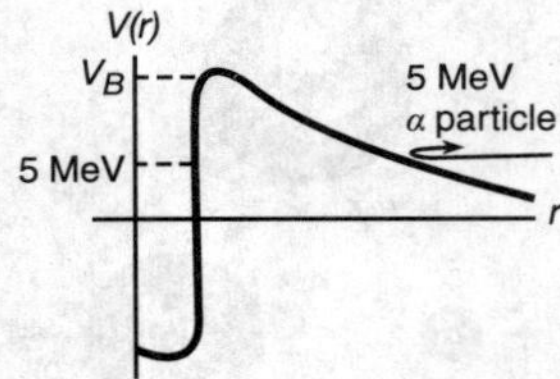

Fig. 25.7 The potential energy barrier for and α-particle. One can see that the Coulomb barrier V_B is much grater than typical α-particle energies produced by radioactive sources. We may note that classically a 5 MeV particle inside the nucleus or scattered from outside cannot penetrate the potential battier

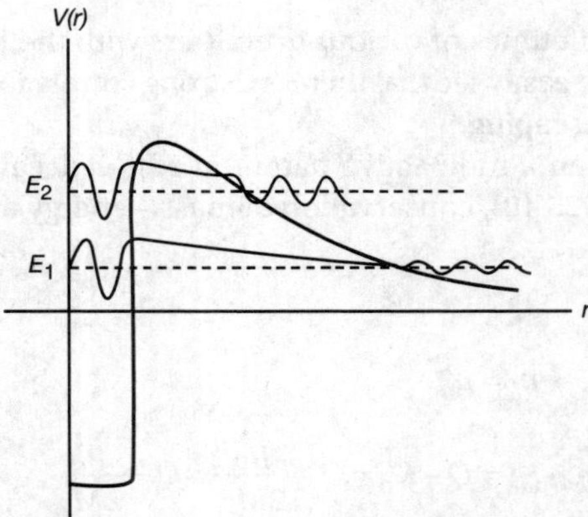

Fig. 25.8 Quantum mechanically, α-particles can tunnel through the barrier. A higher energy α-particle E_2 has a much higher probability (shorter life time) than a lower energy α-particle E_1. We may note the waves shown in figure are schematics only

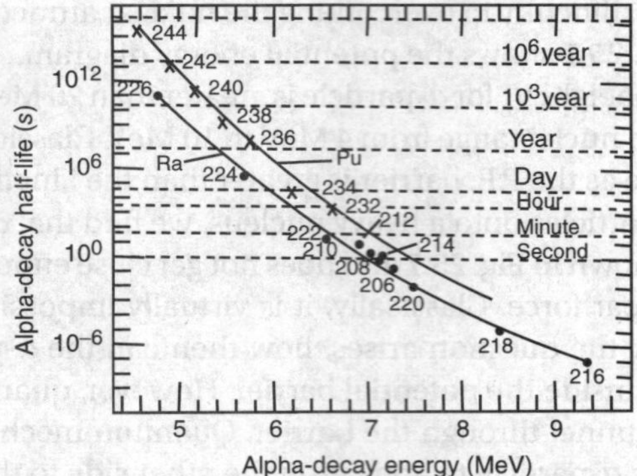

Fig. 25 9 A plot of half-lives for several radioactive α-emitters of radium and plutonium isotopes versus their α-energy. The two curves reveal that the higher energy α-particles result from nuclei having a much shorter lifetime, these α-particles have a higher probability of tunneling through the Coulomb barrier

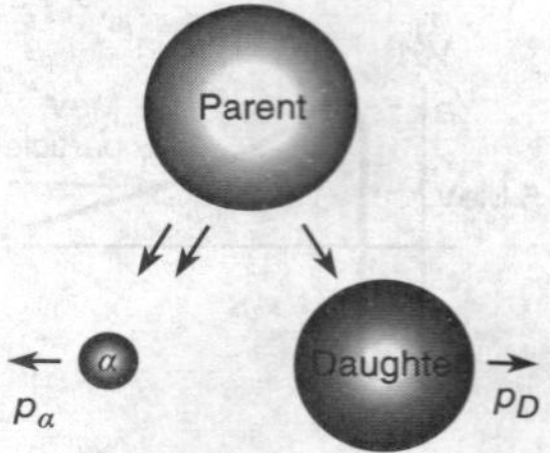

Fig. 25.10 α-decays of radioactive parent to an α-particle and a heavy daughter. Conservation of mass – energy and momentum still obeys. The momentum of α-particle and daughter are equal and opposite

A Comparison of the lifetimes of various α-emitters with the K.E.'s of the α-particles is shown in Fig. 25.9. We can easily see that there is a strong correlation between lower energies and greater difficulty of escaping.

We may note that when a radioactive parent at rest α-decays to an α-particle and a heavy daughter (D) (Fig. 25.10), conservation of mass – energy and momentum still must occur. We have

$$Q = K_\alpha + K_D \quad \text{... (iv)}$$

$$p_\alpha = p_D$$

$$K_\alpha = Q - K_D = Q - \frac{p_D^2}{2M_D} = Q - \frac{p_\alpha^2}{2M_D}$$

$$K_\alpha = Q - \frac{2M_\alpha K_\alpha}{2M_D} = Q - \frac{M_\alpha}{M_D} K_\alpha$$

or $$K_\alpha\left(1+\frac{M_\alpha}{M_D}\right)=Q$$

or $$K_\alpha=\frac{M_D}{M_D+M_\alpha}Q=\left(\frac{A-4}{A}\right)Q \quad \text{... (v)}$$

Since the parent mass A is normally over 150, α-particle takes most of the energy.

25.7.2 Beta (β) Decay

Some nuclides are not stable and therefore radioactive decay occurs. We have seen that in α-decay the parent nucleus reverts to daughter nucleus that is down 2 units in neutron number N and to the left 2 units in atomic number Z. In many cases, it is observed the α-decay leaves the daughter nucleus farther from the line of stability than the parent. Unstable nuclei may move closer to the line of stability by undergoing β^- decay, for example, the β-decay of neutron.

$$^1_0n \rightarrow {}^1_1p+\beta^-\left({}^{0}_{-1}e\right) \quad \text{... (vi)}$$

We know that electrons cannot exist within the nucleus, so when β^- decay occurs for nuclide, β^- particle is created at the time of decay. The β^- decay of ^{14}C to ^{14}N, a stable nucleus is

$$^{14}_{6}\text{C} \longrightarrow {}^{14}_{7}\text{N}+\beta^- \quad \text{... (vii)}$$

Like α-decay, this decay produces two products and one expect to measure a monoenergetic electron spectrum. However, the electron energy spectrum from the β^- decay of ^{14}C (Fig. 25.11) shows a continuous energy spectrum upto a maximum energy. The experimental result remains a major puzzle for several years. In addition to strange energy spectrum, there was no spin conservation.

We may note that in ^1_0n decay, the spin half ^1_0n cannot decay to two spin ½ particles, a proton and an electron. In case of ^{14}C decay, ^{14}C has spin 0, ^{14}N has spin 1 and the electron

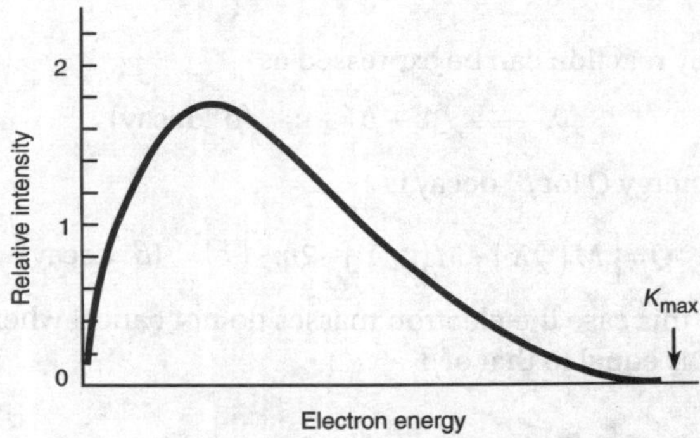

Fig. 25.11 β^- decay of ^{14}C, the relative intensity of electrons as a function of kinetic energy. If there were only two products in β^- decay, the electron energy would be monoenergetic and the energy equals to K_{max}

has spin ½. Clearly, in both the cases there is no conservation of spin, *i.e.,* we cannot combine spin ½ and 1 to obtain a spin of 0. Both the electron energy spectrum and the spin angular momentum posed major challenge before scientists to understand β^- decay.

W. Pauli in 1930 proposed correct explanation, who suggested that a third particle latter named as 'neutrons' (ν) must be produced in β^- decay. ν has spin quantum number ½, charge 0 and carries away the additional energy missing in fig. 25.11. In fig. 25.11 an occasional electron is detected with K.E. K_{max} required to conserve energy, but in the great majority of cases the electron's K.E. is less than K_{max}. This reveals that the neutrino has little or no mass, and its energy may be all kinetic, like the photon. We may note that the photon cannot be the massing particle as it has spin 1. Pauli's suggestion seemed to explain the difficulties faced in β^- decay spectrum, and all circum-stantial evidence supported the ν hypothesis. However, the detection of ν was difficult, and its existence was provided experimentally by Cowan and Reines in 1956.

We may note that neutrino interact so weakly with matter that they pass right through the earth with little chance to being absored. Neutrino has no charge and do not interact 'electromagnetically'. Neutrino are not affected by the strong force, *i.e.,* nuclear force. Recent experiment indicates that neutrino has very small mass. Now, it is believed that β^- decay is the reflection of a special kind of force, called simply the 'weak interaction'. Neutrinos are created (or absorbed) in weak processes, including β^- decay. We may note that the electromagnetic and weak forces are two manifestations of the **electroweak** force.

Now, we also know about antineutrinos $(\bar{\nu})$. The β^- decay of a free neutron and ^{14}C is how correctly written as

$$ {}_0^1n \longrightarrow {}_1^1p + \beta^- + \bar{\nu} \quad (\beta^-\text{-decay}) $$

25.7.3 β^+ decay

The nucleus ^{14}O is unstable and having an excess of protons. This decays by emitting a positron (β^+) to become a stable ^{14}N. The reaction is

$$ ^{14}O \longrightarrow {}^{14}N + \beta^+ + \nu \quad (\beta^+ \text{ decay}) $$

Nuclei near ^{14}N with an excess neutron (^{14}C) or proton (^{14}O) will decay to ^{14}N by the appropriate beta decay.

In general, β^+ decay reaction can be expressed as

$$ {}_Z^AX \longrightarrow {}_{Z-1}^{\;\;A}Y + \beta^+ + \nu \quad (\beta^+ \text{ decay}) \qquad \text{... (viii)} $$

The disintegration energy Q for β^+ decay is

$$ Q = \left[M\left({}_Z^AX\right) - M\left({}_{Z-1}^{\;\;A}Y\right) - 2m_e\right]c^2 \quad (\beta^+ \text{ decay}) \qquad \text{... (ix)} $$

We may note that in this case the electron masses do not cancel when atomic masses are used. The mass of β^+ is equal to that of β^-.

25.7.4 Electron Capture

For β-decay, there is one other important possibility. Since inner K-shell and L-shell electrons are tightly bound and their (classical) orbits are highly elliptical, these electrons spend a reasonable amount of time passing through the nucleus, thereby enhancing the possibility of atomic electron capture (EC). A proton in the nucleus absorbs the e^-, producing a neutron and a neutrino. One can write the reaction for a proton as

$$p = e^- \rightarrow n + \nu$$

The general reaction can be expressed as

$$^{A}_{Z}X + e^- \longrightarrow {}^{A}_{Z-1}Y + \nu \quad \text{(electron capture)} \qquad \text{... (x)}$$

When one of the inner atomic electrons is captured, another electron will take its place and thus producing a series of characteristic atomic X-ray spectra. This is a signature of electron capture as these X-rays are produced in the absence of any other kind of radiation. We may note that X-rays can also be produced in other kinds of nuclear decay, because the decay product may knock out electrons (e.g., an α-particle).

Interestingly, electron capture has the same effect as positron decay, a proton is converted to a neutron. We way note that electron capture occurs more frequently for higher - Z nuclides as the inner atomic electron shells are more tightly bound and there is greater probability of an electron being absorbed. One can write the disintegration energy Q for **electron capture** as

$$Q = \left[M\left(^{A}_{Z}X \right) - M\left(^{A}_{Z-1}Y \right) \right] c^2 \quad \text{(Electron capture)} \qquad \text{... (xi)}$$

Since $Q > 0$ for β^+ or electron capture (EC) to occur, here will be some cases where EC is possible, but not β^+ decay, due to the difference (viii) and (xi).

25.7.5 Gamma Decay

A nucleus in an excited state decays by emitting a photon, just like an atom. But the much higher energy associated with nuclear processes puts such photons in the gamma – ray region of the spectrum. Since the γ-ray photon is neutral and massless, it does not change the type of nucleus, therefore one write

$$^{A}_{Z}X^* \rightarrow {}^{A}_{Z}X + \gamma \text{ (gamma decay)} \qquad \text{... (xii)}$$

The γ-ray energy $h\nu$ is given by the difference of the higher energy state $E>$ (or E_2) and the lower energy $E<$ (or E_1)

$$h\nu = E> - E< = E_2 - E_1 \qquad \text{... (xiii)}$$

In order to conserve momentum, the nucleus normally must absorb some of this energy difference. However, for a nucleus initially at rest, eq. (xiii) is a very good approximation.

25.8 NUCLEAR FORCE

What hold nucleons in a nucleus together? The protons all carry a positive charge of electricity and they must repel one another. Yet the nucleons, both protons and neutrons are tightly bound in the nucleus. Obviously, there must also be a very strong force of attraction. This force cannot be electrical in nature since such a force would lead to repulsion. Further more, an electrical force could not involve the neutrons since they have no electrical charge.

There is evidently an entirely different type of force operating in a nucleus, and it must be so strong that it can then overcome the electrical repulsion of the closely packed protons. This special nuclear force is an aspect of what is called the strong interaction. It is the strongest type of force known. The nature of the nuclear force, holding nucleons in nuclei, has not yet fully been clarified. At the same time, much data have been obtained on the physical properties of the nuclei, as well as on the interaction of free nucleons in collisions over a wide range of kinetic energies, from 10^{-4} to 10^{-10} eV. The analysis of the observed phenomena allows the following conclusions to be drawn about the force between nucleons. We may note that the energy of the nuclear forces and of Coulomb interaction between the protons in the nucleus is equal to the *binding energy*.

(i) Nuclear Forces are Charge Independent Forces: The nuclear forces acting between two protons (p, p) or between two neutrons (n, n) are the same. This reveals that nuclear forces are of a non electrical nature.

(ii) Nuclear Forces are Forces of Attraction: Nuclear forces are extremely strong attractive forces over a distance of about nuclear radius (10^{-15} to 10^{-14} m); beyond this distance the force drops sharply almost to zero. The magnitude of nuclear forces is one million times greater than the force needed to separate an extra nuclear electron from the atom.

(iii) Nuclear Forces are of Short Range Action Forces: The nuclear forces only operate at distances between the nucleons that are comparable in order of magnitude with the size of the nucleons. These distances are called **range** R ($\approx 1.5 \times 10^{-15}$ m) of nuclear forces. Although the exact dependence on distance is not well established, there is good evidence that the force vanishes for all practical purposes at distances greater than a few times 10^{-15} m. According to Yukawa, the nuclear force between two nucleons is due to the interchange of particles called *mesons*. Yukawa represented the nuclear force field by a potential $V_p(r)$ given by

$$V_p(r) = -V_o r_o \frac{e^{-r/r_0}}{r} \quad \text{... (25.12)}$$

where V_o and r_o are empirical constants. Constant r_o is the range of the nuclear force and V_0 gives the strength of the interaction. However, there are some doubts about that the nuclear interaction can be described in terms of potential energy functions in the sameway that we have been able to explain the gravitational and electromagnetic interactions, which falls of rapidly at $r = 1/\alpha$.

According to this theory, the force between the two nucleons is the result of constant exchange of mesons between them.

(iv) Nuclear forces are spin dependent: Nuclear forces depend on the mutual orientation of spins of various nucleons and are different in parallel and antiparallel spins.

(v) Nuclear Forces are Strongest in Nature: The forces between the nucleons, are the strongest forces known in nature. The gravitational and the electromagnetic-interacting forces were known to man long before the knowledge of nuclear forces. The gravitational forces are the weakest forces ~10^{-40} of the strong interacting nuclear forces.

(vi) Nuclear forces get readily *saturated* by surrounding nucleons: Nuclear forces are the only ones known in nature that show saturation effect. The ability of nuclear forces to act upon other particles attain a point of saturation when a nucleons gets completely surrounded by other nucleons. However, the nucleons, those are located outside the surrounding nucleons do not 'feel' the interaction of the surrounded nucleons.

Summarising, we can say that nuclear forces between nucleons are attractive in nature when they are 0.5–25 F apart. These forces are short range forces having maximum value at about $\sim 2 \times 10^{-15}$ m and falls off sharply with distance. Obviously, they become negligible beyond this range. Nuclear forces are charge independent and therefore nuclear force between p-p, n-n and p-n are almost the same. Nuclear forces have the property of saturation, *i.e.,* a particular nucleon interacts with a limited number of nucleons around it and other surrounding ones remain unaffected. This is why nuclear forces become saturated over short distance. Nuclear forces are *spin dependent, i.e.,* depend on the mutual orientation of spins of various nucleons and are different in parallel and anti parallel spins.

In addition to *strong nuclear* short range *force* which is far stronger than Coulomb repulsion force, there is a **short range force** far more weaker than the nuclear force. It may be as small as 10^{-14} of strong nuclear force. This is termed as *weak interaction short range nuclear force* as indicated by *beta decay.* We may note that it is not of gravitational or Coulomb type. We may also note that weaker the force, the larger must be system in order that it might be of importance, e.g. strong interacting nuclear force hold the nucleons, the electromagnetic force holds the larger systems of atoms and molecules, while the gravitational force becomes important in astral bodies and systems. Now, we can summarises the chief forces of nature as: *strong nuclear* force, *weak interacting* force, *electromagnetic* force and *gravitational* force.

We may note that in accordance with Yukawa's theory of nuclear forces, protons and neutrons do not exist independently within a nucleus but constantly exchange charges by emission and absorption of π- *mesons* or *pions* (within the nucleus), in very short intervals $\sim 10^{-23}$ & to 10^{-24} s. As the exchange occurs in a very short time and hence no visible change in nucleonic mass would be observed in accordance with uncertainty principle. This gives rise to rapid exchange of meson or mesonic field between protons and neutrons in which meson acts as a quantum of nuclear force. We may also note that this process is analogous to exchange of photons between charge particles in electromagnetic interactions. Many observable properties of the nuclear force have been successfully explained by a model based on the exchange of pions.

25.9 STABLE NUCLEI OR NUCLEAR STABILITY

Not every combination of protons and neutrons will stick together indefinitely. Too many protons, and electrical repulsion wins out, sooner or later the nucleus decay by emitting a chunk of nuclear material. In larger nuclei most protons are far apart and therefore experience electrical repulsion more strongly than nuclear attraction (Fig 25.12).

To hold these nuclei together therefore requires more neutrons, which contribute attractive nuclear force but not electrical repulsion. So larger nuclei tend to have a higher ratio of neutrons to protons. Even this effect has its limits, though, and the result is that there are no stable nuclei for $Z > 83$.

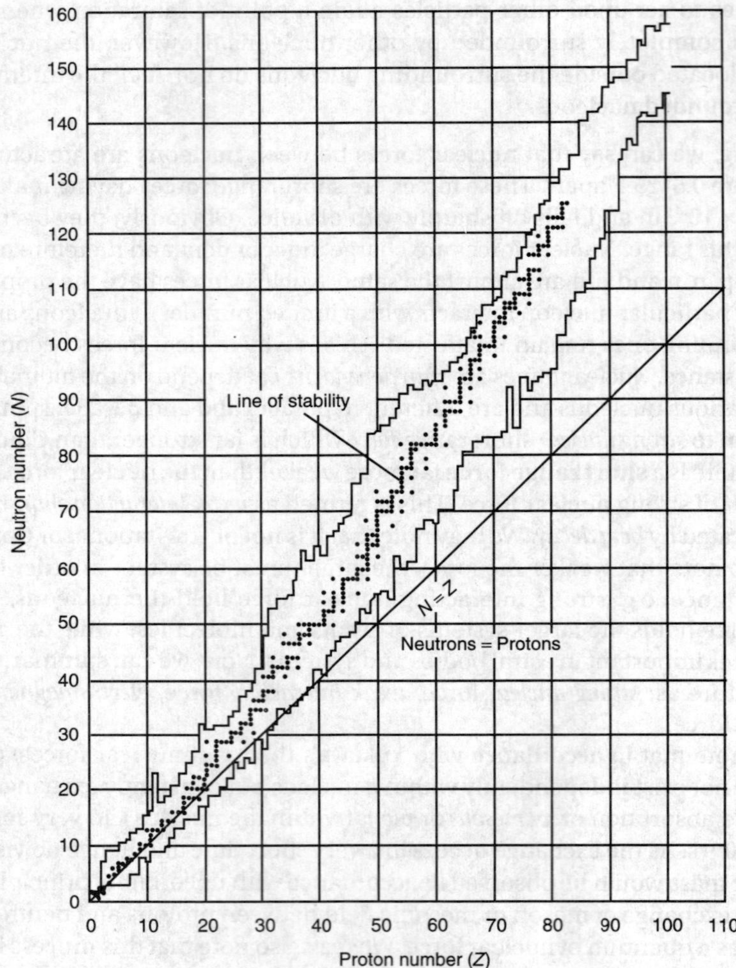

Fig. 25.12 A plot of the known nuclides with neutron number N versus proton number Z. The solid points represent stable nuclides, and the shaded area represents unstable nuclei. A smooth line through the solid points would represent the line of stability

A nucleus containing A nucleons is said to be **stable** if its mass is smaller than that of any other possible combination of A nucleons. The binding energy of a nucleus ${}_{Z}^{A}X$ against dissociation into any other possible combination of nucleons, for e.g. nuclei R and S, is given by

$$\text{B.E.} = [M(R) + M(S) - M({}_{Z}^{A}X)]c^2 \qquad \text{... (i)}$$

In particular, the energy required to remove one proton (or neutron) from a nuclide is called the proton (or neutron) **separation energy**, and Eq. (i) is quite useful for finding this energy. We may note that even if B.E. is negative for a particular dissociation, there may be other reasons why the nucleus is **stable**.

Fig. 25.12 shows known stable nuclei as well as many known unstable nuclei that are long - lived enough to be observed. We may note that Fig. 25.12 also includes nuclides that decay in a millisecond or less. The line representing the stable nuclides is called the **line of stability**. From the line of stability, one can extract several facts :

(i) It appears that for $A \leq 40$, nature prefers the number of protons and neutrons in the nucleus to be about the same, $Z = N$. However, for $A \geq 40$, there is a decided preference for $N > Z$. One can understand this difference as follows. The strength of the nuclear force is independent of whether the particles are *nn*, *np* or *pp*. Equal numbers of neutrons and protons may give the most attractive average inter nucleon nuclear force, but we must consider Coulomb force as well. We may note that with the number of protons increases, the Coulomb force between all the protons becomes stronger and stronger until it eventually affects the binding significantly.

(ii) One can calculate the **electrostatic energy** required to contain a charge Ze evenly spread throughout a sphere of radius R by determining the work required to bring the charge inside the sphere from infinity and comes out to be

$$\Delta E_{\text{coul}} = \frac{3}{5}\left(\frac{(\text{Ze})^2}{4\pi\,\varepsilon_0 R}\right) \quad \text{... (ii)}$$

In case of a single proton, Eq. (ii) gives for self energy

$$\Delta E_{\text{coul}} = \frac{3}{5}\left(\frac{e^2}{4\pi\,\varepsilon_0 R}\right) \qquad (\because Z = 1) \quad \text{... (iii)}$$

obviously, (iii) represents the work done to assemble the proton itself and we do not want to include it in the electrostatic self energy of a nucleus composed of Z - protons. This means that we must subtract Z such terms from the total given in Eq. (ii) to give us the total Coulomb repulsion energy in a nucleus:

$$\Delta E_{\text{coul}} = \frac{3}{5}\left(\frac{Z(Z-1)e^2}{4\pi\,\varepsilon_0 R}\right) \quad \text{... (iii)}$$

The delicate balance between neutrons and protons results in about 400 known stable nuclei, collectivity called nuclides. Fig. 25.12 reveals that there are no stable nuclei with $Z > 83$ because of the increasingly larger Coulomb force. The heaviest known stable nucleus is ${}^{209}_{83}\text{Bi}$. All nuclei with $Z > 83$ and $A > 209$ will eventually decay spontaneously into some combination of smaller masses. Adding one proton to a heavy nucleus adds a constant amount of nuclear B.E., but the repulsive Coulomb energy increases as $\Delta E_{\text{coul}} \approx (Z+1)^2 - Z^2 \approx 2Z$. Since the Coulomb force is long range, the proton interacts electromagnetically with the protons within the nucleus already there. And because this repulsive energy increases with Z, nuclei with higher Z eventually become unstable. However, the neutrons dilute the Coulomb repulsion slightly because they intersperse among the protons, causing the protons to be slightly father apart.

Fig. 25.12 reveals that most stable nuclides have both even Z and even N (called "even - even" nuclides), e.g. ${}^{2}_{1}\text{H}$, ${}^{6}_{3}\text{Li}$, ${}^{10}_{5}\text{B}$ and ${}^{14}_{7}\text{N}$. All the other stable nuclides are odd-even or even-odd i.e., with either an odd number of Z or N. Obviously, nature apparently prefers nuclei with even numbers of protons and neutrons.

Pauli exclusion principle helps to understand the above empirical observation. Neutrons and protons are distinguishable fermions, hence they separately obey the exclusion principle. Only two neutrons (or protons) may coexist in each spatial (quantum state), one with spin "up" and the other with spin "down". Obviously, each nuclear energy level is then able to hold two particles whose spins are paired to O. We may note that this configuration of opposite spins is particularly stable as placing the same number of particles in any other arrangement will produce a (less stable) state of higher energy. Thus, there in lies the preference for even N and Z.

25.10 NUCLEAR MODELS

We have seen how the right ratio of neutrons to protons is essential for stable nuclei and why that ratio increases for larger nuclei. Fig. 25.12 summarizes this information. But a closer look at Fig. 25.12 reveals that there are more stable nuclei for even values of Z, and some-those with the so-called magic numbers 2, 8, 20, 28, 50, 82 and 126 protons and neutrons - have many stable nuclei, why?

Answering this question and explaining the decay mechanisms and lifetimes of unstable nuclei require a theory of nuclear structure. There still is no complete nuclear theory analogous to atomic theory, that explains all aspects of all nuclei. Our still-imperfect knowledge of the nuclear force, and the tight packing of nucleons, render useless a simple two-particle model like the one for hydrogen. Instead, nuclear physicists resort to several models to explain different aspects of nuclear structure. Together, these models provide a good understanding of nucleus and accurately predict nuclear properties, although not with the precision available in atomic physics.

(i) Liquid Drop Model

Bohr and coworkers in 1930 were able to explain many nuclear phenomena by treating the nucleus as a collection of interacting particles in a liquid drop. This model of the nucleus is known as the liquid drop model provides a reasonable approximation for heavier nuclei, whose many nucleons behave somewhat like the molecules in a drop of liquid. Weizsacker in 1935 proposed the **semi-empirical mass formula** based on the liquid drop model, written in terms of the total binding energy is

$$B\left({}_{Z}^{A}X\right) = a_v A - a_A A^{2/3} - \frac{3}{5}\frac{Z(Z-1)e^2}{4\pi\varepsilon_0 r} - a_S\frac{(N-Z)^2}{A} + \delta \qquad \text{... (i)}$$

This is actually the binding energy. The first term in R.H.S. of (i) a_v indicates that the binding energy is approximately the sum of all the interactions between nucleons. Since the nuclear force is short range and each nucleon interacts only with its nearest neighbours this interaction is proportional to A, the total number of nucleons.

The second term, called the **surface effect**, is simply a correction to the first term (similar to surface tension), because the nucleons on the nuclear surface are not completely surrounded by other nucleons. The surface nucleons do not have saturated interactions, and one will have to apply a correction proportional to the liquid drop surface, $4\pi R^2$. Since $R \sim A^{1/3}$ the correction is proportional to $A^{2/3}$. The third term is Coulomb energy.

The fourth term is due to **symmetry energy**. In the absence of Coulomb forces, nucleus prefers to have $N = Z$. This term has a quantum-mechanical origin, depending on exclusion principle. We may note that the sign of this term is independent of the sign of N–Z.

The last term is due to the pairing energy and reflects the fact that the nucleus is more stable for even-even nuclides. One can determine this term empirically. One set of values given by Fermi for the parameters of Eq. (i) is

$$a_v = 14 \text{ MeV volume}, \; a_A = 13 \text{ MeV surface}, \; a_s = 19 \text{ MeV symmetry}$$

pairing $\delta = \begin{cases} +\Delta \text{ for even-even nuclei} \\ \text{o for odd A(even-odd, odd-even), nuclei} \\ -\Delta \text{ for odd-od nuclei} \end{cases}$

where $\Delta = 33 \text{ MeV. } A^{3/2}$

A liquid-drop nucleus can rotate, vibrate, and change shape as long as its volume does not change and the resulting quantized energy levels predict nuclear gamma-ray spectra that are in good agreement with observation. The liquid drop model also helps in explaining nuclear fission. But it cannot account for the dramatic effects of small changes in nucleon number, particularly, the role of magic numbers.

(ii) Nuclear Shell Model

Mayer and Jensen in 1940 advanced the idea that nucleus has a shell structure similar to that of atoms.

The shells occur because neutrons and protons obey the exclusion principle, and the magic numbers correspond to closed shell configurations analogues to the electronic structure of inert gases. Closed shell nucleons are tightly bound, making a magic nucleus particularly stable. Additional nucleons beyond a closed shell stay largely on the outskirts of the nucleus, where they are more readily excited to higher energy levels. Neutrons and protons behave independently in the shell model, and each has its own set of quantum numbers. Closed shell structure therefore occurs with magic numbers of either protons or neutrons. Some nuclei like ${}^{40}_{20}\text{Ca}$ ($Z = 20, N = 20$), are "doubly magic" and show exceptional stability.

(iii) Collective model

This model advanced by Aage Bohr, combines aspects of the liquid drop and shell models, emphasizing the collective quantum - mechanical behaviour of the nucleons. One remarkable prediction of this model is that large, non-magical nuclei may be more stable if they take non spherical shapes.

Active area of nuclear structure research involves the creation and exploration of exceptionally heavy or neutron rich nuclei. The creation of elements 115 and 116 in the early 2000 s suggests that physicists are approaching a region of longer - lived nuclei dubbed the "islands of stability", which may be associated with a new magic number of 184. Creation of silicon - 42 in 2005, whose relative stability implies that its atomic number $Z = 14$ becomes magic in this neutron - bloated ($N = 28$) species. Until we develop a complete nuclear theory, experiments like these will continue to challenge physicists with nuclear surprises.

25.11 NUCLEAR REACTIONS AND *Q* VALUE

A change in the structure of nuclei can be brought about by bombarding them with fast moving particles, e.g. neutron, proton, alpha particle, etc., In this process the characteristic or identify of incident nuclear particle undergoes a change which is called nuclear reaction. One can analyse a nuclear reaction quantitatively in terms of masses and energies of the nuclei and the particles involved. Consider a nuclear reaction represented by

$$X + x \rightarrow Y + y \qquad \text{... (25.13)}$$

where X is the target nucleus, x is the bombarding particle, Y is the product nucleus and y is the product particle. Assuming that the target nucleus X is at rest, the law of conservation of energy demands that

$$M_x c^2 + \left(E_x + m_x c^2\right) = \left(E_Y + M_Y c^2\right) + \left(E_y + m_y c^2\right) \qquad \text{... (25.14)}$$

where M_x, m_x, M_Y and m_y represent the masses of the target nucleus, incident particle, product nucleus and product particle respectively. The E's represent the kinetic energies. Rearranging the Eq. (25.14), one obtains

$$Q = E_Y + E_y - E_x = \left[M_x + m_x - M_y - m_y\right] c^2 \qquad \text{... (25.15)}$$

where $Q = E_Y + E_y - E_x$ is called the *'energy balance'* of the reaction or *Q*-value of the reaction and can be determined from the energy difference or from the mass difference. We may note that here *excited state of product nucleus, and the emission of more than one particle is not taken into account.*

If the *Q*-value for a reaction is positive, kinetic energy of the product is greater than that of the reactants, the reaction is then said to be *exothermic* (or *exoergic)* nuclear reaction. In other words, the total mass of the reactants is greater than that of the products. If the *Q*-value for a reaction is negative, it is called *endothermic* (or *endoergic)* reaction. Eq. (25.15) further suggests that whatever change occur in total kinetic energy of the system of particles, it must be balanced by equal change in rest mass energy.

25.12 ARTIFICIAL TRANSMUTATION OR ARTIFICIAL DISINTEGRATION

The general term' *transmutation'* is used for the conversion of one element into another. When by some artificial means, one type of atom is converted into another type of atom the process is known as *artificial transmutation.* We may note that this process is entirely different from the spontaneous transmutations in radioactive elements. The artificial transmutation is also called as artificial disintegration because the transmuted element disintegrates into a new element. Obviously, *the process of providing new stable nuclei from other stable nucleus is known as artificial transmutation or artificial disintegration of elements.*

25.12.1 Rutherford's Experiment for Artificial Disintegration

Rutherford in 1919 performed the earliest experiments on artificial nuclear disintegration. The disintegration was caused by α-particles from the "RaC" and resulted in the emission of a highly energetic proton from ordinary nitrogen nucleus.

When N_2 nucleus captures α-particles it is changed into unstable nucleus which immediately disintegrated into a stable oxygen nucleus by ejecting a proton. The disintegration is represented by the reaction

$$^{14}_{7}N + ^{4}_{2}He \rightarrow \left[^{18}_{9}F\right] \rightarrow ^{17}_{8}O + ^{1}_{1}H$$

Thus an atom of nitrogen is transformed into an atom of oxygen. This process is called transmutation of elements.

In the abbreviated notation, the above reaction can be expressed as $^{14}N\ (\alpha,p)\ ^{17}O$. This is read as: a nitrogen nucleus (^{14}N) (target nucleus), interacts with alpha particle (α) (incident or projectile), a proton (p) is ejected and an oxygen nucleus (^{17}O) (recoil nucleus) left behind.

25.13 ARTIFICIAL RADIOACTIVITY

In 1934 Curie-Jiliot bombarded aluminium and boron with α-particles. They found that these materials continued to emit radiation after the source of α-particles is removed. This shows that the product of disintegration must be radioactive and as usual it undergoes nuclear transformation. Moreover the intensity of this induced activity was found to decay exponentially. This phenomenon is called *artificial or induced radioactivity.* The determination of the deflections of the particles emitted in magnetic and electric field showed that these are positively charged particles called *positrons* (or *positive electrons*).

The reaction for aluminium can be represented follows:

$$^{27}_{13}Al + ^{4}_{2}He \rightarrow \left[^{30}_{15}P\right] + ^{1}_{0}n$$

Obviously, due to α-particle bombardment on aluminium, neutrons are emitted and an unstable isotope of phosphorous (half life 2.5 minutes) is formed. This isotope of phosphorus then decays spontaneously giving stable silicon and positron $\left(^{0}_{+1}e\right)$ represented by the following reaction

$$^{30}_{15}P \rightarrow ^{30}_{14}Si + ^{0}_{+1}e + \gamma$$

Some other artificial radioactive reactions are:

(i) $^{10}_{5}B + ^{4}_{2}He \rightarrow \left[^{14}_{7}N\right] \rightarrow ^{13}_{7}N^{*} + ^{1}_{0}n$ [* Unstable (Half life 10 min.)]

$$^{13}_{7}N^{*} \rightarrow ^{13}_{6}C + ^{0}_{+1}e + \gamma$$

(ii) $^{33}_{11}Na + ^{4}_{2}He \rightarrow ^{26}_{13}Al^{*} + ^{1}_{0}n$ [* Unstable]

$$^{26}_{13}Al \rightarrow ^{26}_{12}Mg + ^{0}_{+1}e + \gamma$$

(iii) $^{14}_{7}N + ^{4}_{2}He \rightarrow ^{17}_{9}F^{*} + ^{1}_{0}n$ [* Unstable]

$$^{17}_{9}F^{*} \rightarrow ^{17}_{8}O + ^{0}_{+1}e + \gamma$$

The elements with * mark in the above reactions are the *artificial radioactive isotopes* of that respective elements.

Practically by bombarding almost all the elements, with different particles, can be made radioactive, the different particles used as projectiles are α-particles, neutrons, protons, deutrons and γ-radiations. The α-particles have a small probability of colliding with the nuclei and hence they are very little employed for artificial radioactivity. Most useful projectiles are neutrons, obtained from nuclear reactor.

The study of artificial radioactivity helps to clarify the concepts of constitution and stability of nuclei. The artificial radioactive nuclides decay either by electron emission, or positron emission or orbital electron capture depends on the availability of disintegration energy. We may note that the nuclear reactions induced by protons, deutrons, neutrons and photons can also result in radioactive products. Moreover, there seems to be some correlation between the radioactivity of an artificial nuclide and the process and means of transformation by which it is produced. Electron emission is common for activities produced by (n, p), (n, α), (n, γ) and (d, p) nuclear reactions. These nuclear reactions decrease the positive charge to mass ratio of the nucleus, e.g.

$$^{115}_{49}\text{In}(n,\gamma)^{116}_{49}\text{In}^*$$

$$^{116}_{49}\text{In}^* \rightarrow {}^{116}_{50}\text{Sn} + {}^{0}_{-1}\text{e}$$

and $$^{14}_{7}\text{N}(n,p)^{14}_{6}\text{C}^*,\ ^{14}_{6}\text{C}^* \rightarrow {}^{14}_{7}\text{N} + {}^{0}_{-1}\text{e}$$

Positron emission is common for the activity produced by (p, n), (d, n) (α, n) and (γ, v) nuclear reactions, which increases the positive charge (+ve) charge to mass ratio of the nucleus, e.g.

$$^{13}_{6}\text{C}(p,\gamma)^{13}_{7}\text{N}^*,\ ^{13}_{7}\text{N}^* \rightarrow {}^{14}_{6}\text{C} + {}^{0}_{+1}\text{e}$$

and $$^{58}_{28}\text{Ni}(p,n)^{58}_{29}\text{Cu}^*,\ ^{58}_{29}\text{Cu}^* \rightarrow {}^{58}_{28}\text{Ni} + {}^{0}_{+1}\text{e}$$

We may note that usually (α, p), (p, α) and (d, α) reactions lead to stable products.

This discovery of artificial or induced radioactivity opened a new excited front of research and now hundreds of radioactive nuclides have been made by various methods, e.g. Iodine has only one stable species $^{127}_{53}\text{I}$ containing 53 p and 74 n. However, there are 17 artificially radioactive isotopes of iodine are possible, containing from 68 to 86 neutrons. We may note that the discovery of radiations from the natural and artificial radioactive nuclide has shown that analogous to atomic energy levels, **nucleus has also energy levels**. One can study these energy levels within the nucleus with the help of *nuclear spectroscopy*. Nuclear spectroscopy is an important source of information about the study of structure of the nucleus. Presently, more than 1200 radio nuclides have been identified.

25.14 RADIOISOTOPES

We have seen that in addition to the naturally occurring radioisotopes such as radium, thousands of others have been made artificially.

25.14.1 Radioisotopes of Transuranium Elements

All elements with $Z > 92$ are called transuranium elements because they all lie beyond $^{238}_{92}\text{U}$ in the Periodic Table. The first of these is neptunium. Neptunium was detected when a beam of slow neutron is incident on $^{238}_{92}\text{U}$ then the common radioisotope ^{239}U

is formed which gives β^- emission and the final product is neptunium ($^{239}_{93}Np$) having $Z = 93$. Neptunium changes into plutonium ($^{239}_{94}Pu$) with spontaneous emission of β^- particles. Plutonium is 94th element. In this way new members of Periodic Table are increasing and we have presently 109 elements.

An interesting feature of the transuranic elements in that they from a series beginning with actinium, referred to as the actinide series, which is analogous to the rate-earth series. The absence of the transuranic elements in nature, with the possible exceptions of neptunium and plutonium, may be attributed to their extremely short half life compared with the age of the earth.

25.14.2 Use of Radioisotopes

Radioisotopes have numerous applications in medicine, agriculture, industry and pure research. Many applications employ a special technique know as *'tracer technique.'* The important applications of radioactive isotopes mainly depends on the fact that the chemical properties of the radioisotopes of a given element are essentially identical. One can detect the radioisotope easily by its radioactivity and stable isotope by a mass spectrometer.

25.14.3 Tracer Technique

A small amount of a radioisotopes is introduced into the material to be studied and with the help of G.M. counter, its path is traced. For example a leakage in an underground water-pipe can be easily detected by this method. A small quantity of γ-emitter radio-sodium ^{24}Na is introduced into the pipe at its inlet. After the liquid has gone through the pipe, the ground around the leak will have larger γ-ray activity which can be detected by moving a G.M. counter on the ground. Similarly, in order to locate a blockage in an underground sewage pipe, one will have to introduce a rubber ball having ^{24}Na into the pipe. A G.M. counter above ground will give the position of the ball when it has come to rest.

One can also use radioisotopes in transporting different oils through underground pipe to distance place. When the type of the oil flowing through the pipe is changed, a small quantity of radioisotope is mixed exactly at the position where the change takes place. A Geiger counter is placed near the other end of the pipe line, which gives a signal when the radioisotope passes.

Tracer technique is also employed in the field of **medicine** in a number of ways. For example, a doctor can find out any obstruction in the circulation of the blood in the human body. Doctor injects radio phosphorous (^{32}P) into the blood of patient and examines the movement of the blood by detecting radiations emitted by ^{32}P with the help of G.M. counter. Doctor can thus locate clots of blood present in the body. In a similar way, one can also study the passage of a particular element in the body and the rate at which it accumulates in different organs. For, e.g. phosphorous accumulates in bones, and iodine in thyroid gland. When the thyroid gland suffers with some disease its rate of accumulations of iodine changes. In order to investigate it, ratio iodine (^{131}I) is given orally to the patient and the radiation emitted by his thyroid gland is measured externally by a G.M. counter at suitable intervals over the following 48 hours or so. Thus, one can diagonise an over active or under active (ailing) thyroid gland.

Tracer technique find use in **agriculture** to study the rate and direction of movement of an element in a plant. For this a radioisotope of that element is injected in the ground near the plant. After a few days the plant is laid on a photographic paper to produce an autoradiograph. The positions reached by the element are represented by the dark areas in the radiograph. This technique provided valuable information regarding the optimum season for fertilizing crops and for poisoning weeds.

Tracer technique finds use for testing the uniformity of mixtures in industry. For testing a chocolate mixture, a small quantity of short-lived radioisotope, such as ^{24}Na or ^{36}Mn is added to the primary ingredients. Several different sampled of the final product are then tested for radioactivity with the help of a G.M. counter. If each sample yield the same counting rate, then the mixing has been uniform. One can use this method in mixing processes occurring in the manufacture of chocolate, soap, cement, paint, fertilizers, cattle food and medical tablets.

The tracer technique is found to be extremely sensitive in testing the sealing process in making envelopes for radio valves. A sample valve is filled with radio-Krypton (Kr^{85}) and a G.M. counter is held outside the valve. The G.M. counter detects even an extremely poor leakage.

This technique also find use in research to study the exchange of atoms between various molecules, and to investigate the solubility and vaporization of materials.

Besides the uses of radioisotopes employing the tracer technique there are hundreds of other uses in various fields, as mentioned below:

25.14.4 Biological Uses

(i) A plant, when grows, absorbs phosphorus both from the soil and from fertilizer. To study the effect of fertilizer on the plant, one will have to know what properties of phosphorus comes from each other. If we are using radioisotope in fertilizer, we can easily find the exact proportion easily.

(ii) It is also helpful for blood transfusion, since one can detect whether the given blood is suitable for the patient or not.

25.14.5 Medical Uses

Radioisotopes have been widely used in locating and detecting the presence of *tumours* and particularly *brain tumours* which are quite difficult to detect.

To study the cases of *restricted circulation of blood,* radio sodium has been used. Using radio iodide or radio sodium, pumping action of the heart has been studied.

Radioisotopes have been used to study virtually every organ and tissue in the body. ^{3}H, ^{14}C and ^{32}P, all β-emitters are widely used in medical research. More than 1100 radioisotopes are available for clinical use. By far the most widely used is ^{99m}Tc. An isomer of technetium with a half life of 6h, produces a 140 keV γ-ray when it decays to the ground state of ^{99}Tc. A patient is given intravenous injection of ^{99m}Tc, e.g. TcO_4. The ^{99m}Tc is trapped by cancerous cells, e.g. in the thyroid or salivary glands, choroid plexus of the brain or the gastric mucosa. Few minutes after the injection, patient is scanned by an array of NaI γ-ray detectors. This helps to pinpoint the ^{99m}Tc activity. This is only one out of several applications for ^{99m}Tc.

The radioactive $^{67}_{31}$Ga is useful as a tumor - localizing agent for Hodgkin's disease.

^{153}Gd radioisotope is used for detection of bone mineral loss or osteoporosis in elderly people. ^{192}Ir is used for a large number of cancer treatments and also for industrial radiography for welds in steel, oil well rigs, and pipelines. ^{125}I reeds is implanted into the prostate to inhibit cancer. In addition to ^{125}I, which has a half life of 60 days, ^{103}Pd (half life 17 days) is also used. Both these nuclides emit useful low-energy γ-rays.

Radioisotopes are also used in **tomography**. This is a technique for displaying images or practically any part of the body to look for abnormal physical shapes or for functional characteristics of organs. Making use of the detectors together with computers, 3-dimentional images can be obtained. Tomography now includes single-photon emission computed tomography (SPECT), positron emission tomography (PET), magnetic resonance imaging (MRI), etc.

25.14.6 Industrial Uses

There are many different uses to which radioisotopes are put in industry.

(i) In industry, alloys are frequently subjected to different treatments such as annealing, quenching, etc., and radioisotopes have been used to study the effect of such treatment.

(ii) The movement of atoms in metal within the crystal lattice, *i.e., self diffusion* in metal has been studied using radioisotopes. During blending of different materials, uniformity of mixing has been achieved by labeling one of the constituents with radioactive tracer.

25.14.7 Agricultural Uses

(i) Radiations from certain radioisotopes are used for killing insects which damage the food grains. Certain seeds and canned food can be stored for longer periods by gently exposing them to radiations.

(ii) Better yield of milk from cows, and more eggs from hens have been obtained on the basis of information gained by mixing radioisotopes with their diet.

(iii) Radioisotopes are also used for determining the function of fertilizer in different plants. This increases the agricultural yield.

(iv) Certain seeds, when exposed to feeble radiations, develop into different varieties of plants, e.g. new and exciting colours have been given to some of the flowering plants.

25.14.8 Radiocarbon Dating

It our atmosphere, radioactive ^{14}C is produced by the bombardment of ^{14}N by neutrons produced by cosmic rays.

$$_0^1n + {}_7^{14}N \rightarrow {}_6^{14}C + {}_1^1P$$

In the atmosphere, a natural equilibrium of ^{14}C to ^{12}C exits. As shown in Fig. 25.13 all living organisms use or breathe CO_2 from the atmosphere.

However, after the death of living organisms their intake of ^{14}C ceases and the ratio $^{14}C/^{12}C$ (= R) decreased due to decay of ^{14}C. Measuring the ratio of ^{14}C to a stable ^{12}C in a sample of once – living matter and comparing with the ratio found in living material then provides the age. Archaeologists, art historians, geologists and other use radioactive decay

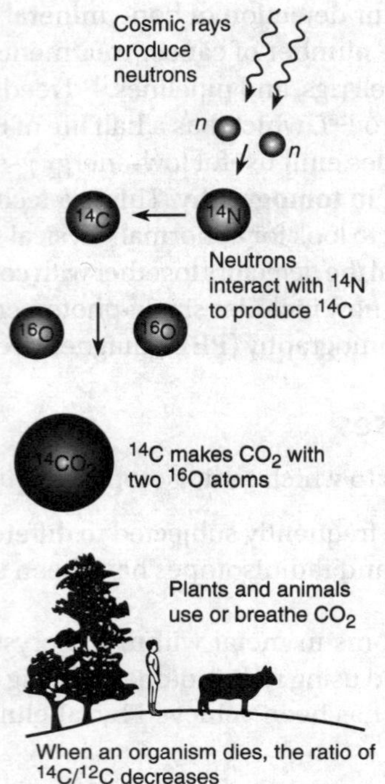

Fig. 25.13 Neutrons produced from cosmic rays react with ^{14}N in the atmosphere to produce ^{14}C. ^{14}C nuclei enter into living organisms in the form of CO_2. After the death of living organism, the ratio of $^{14}C/^{12}C$ in its remains decreases according to decay rate of ^{14}C

to date ancient objects. For ages upto a few tens of thousands of years, the 5730 year isotope C-14 is especially useful.

The cosmic ray flux at Earth varies with solar activity and so, therefore, does the $^{14}C/^{12}C$ ratio. Scientists correct for this effect with data from growth rings in ancient trees, which provide an independent measure of age. Measuring the actual radioactivity takes a fairly large sample, so today the most sophisticated dating is done instead by counting individual C-14 atoms separating them from ordinary C-12 using a mass spectrograph.

Radio carbon dating is quite accurate to about 20,000 years and can be used back about 50,000 years. For longer time spans, upto the billions of years characterizing the ages of rocks, ratios of longer lived isotopes provide age information. We may note that much knowledge of our own part, and our own planets and our solar system's comes from radioisotope dating.

Other particularly useful radioisotopes for dating purposes include ^{10}Be, ^{26}Al, ^{36}Cl and ^{130}I. The half-life of ^{10}Be is 1.5 million years and perhaps, it may be useful in studying the evolution both of humans and of the ice ages. The isotope ^{36}Cl is particularly of well suited for dating and tracing ground water movement and for determining the suitability radioactive waste depositories. The dating of ^{10}Be and ^{26}Al in marine sediments has confirmed their extraterrestrial origin, possibly from comets.

25.15 NUCLEAR REACTIONS

One can bring a change in the structure of the nuclei by bombarding them with fast moving particles. In this process the characteristics or identity of incident nuclear particle undergoes a change which is called a nuclear reaction. Usually, the following unclear particles are used α (${}^4_2\text{He}$), proton $p({}^1_1\text{H})$, deutron (${}^2_1\text{D}$), neutron (${}_0^1n$), high energy photons, *i.e.*, γ rays, etc., On the basis of the nature of projectile particle and outgoing particle, one can classify the nuclear reactions as follows:

25.15.1 Reactions Induced by α-Particles

(i) (α, p) reaction
One can write the general (α, p) reaction as

$$ {}^A_Z X + {}^4_2\text{He}(\alpha) \rightarrow \left[{}^{A+4}_{Z+2}\text{Cn}\right] \rightarrow {}^{A+3}_{Z+1}Y + {}^1_1\text{H} $$

(Cn → Compound nucleus)

Example: $ {}^{27}_{13}\text{Al} + {}^4_2\text{He} \rightarrow \left[{}^{31}_{15}\text{P}\right] \rightarrow {}^{30}_{14}\text{Si} + {}^1_1\text{H} $

(ii) (α, n) Reaction: General reaction

$$ {}^A_Z X + {}^4_2\text{He} \rightarrow \left[{}^{A+4}_{Z+2}\text{Cn}\right] \rightarrow {}^{A+3}_{Z+2}Y + {}^1_0 n $$

Examples: $ {}^9_4\text{Be} + {}^4_2\text{He} \rightarrow \left[{}^{13}_6\text{C}\right] \rightarrow {}^{12}_6\text{C} + {}^1_0 n $

$$ {}^{14}_7\text{N} + {}^4_2\text{He} \rightarrow \left[{}^{18}_9\text{F}\right] \rightarrow {}^{17}_9\text{F} + {}^1_0 n $$

We may note that (α, n) reaction is a well known reaction used for the detection of neutron by Chadwick in 1932.

25.15.2 Reactions Induced by Protons

(i) (p, α) reaction: General reaction

$$ {}^A_Z X + {}^1_1\text{H} \rightarrow \left[{}^{A+1}_{Z+1}\text{Cn}\right] \rightarrow {}^{A-3}_{Z-1}Y + {}^4_2\text{He} $$

Examples: $ {}^7_3\text{Li} + {}^1_1\text{H} \rightarrow \left[{}^8_4\text{Be}\right] \rightarrow {}^4_2\text{He} + {}^4_2\text{He} $

$$ {}^{27}_{13}\text{Al} + {}^1_1\text{H} \rightarrow \left[{}^{28}_{14}\text{Si}\right] \rightarrow {}^{24}_{12}\text{Mg} + {}^4_2\text{He} $$

(p, α) reaction provided one of the earliest quantitative proof of the famous Einstein's relation, $E=mc^2$.

(ii) (p, n) reaction: General reaction

$$ {}^A_Z X + {}^1_1\text{H} \rightarrow \left[{}^{A+1}_{Z+1}\text{Cn}\right] \rightarrow {}^{A}_{Z+1}Y + {}^1_0 n $$

Examples: $ {}^{11}_5\text{B} + {}^1_1\text{H} \rightarrow \left[{}^{12}_6\text{C}\right] \rightarrow {}^{11}_6\text{C} + {}^1_0 n $

$$ {}^{18}_8\text{O} + {}^1_1\text{H} \rightarrow \left[{}^{19}_9\text{F}\right] \rightarrow {}^{18}_9\text{F} + {}^1_0 n $$

We may note that in (*p*, *n*) reaction, the mass change is usually negative, so the reactions are endothermic.

(iii) (p, γ) reaction: General reaction

$$ {}_Z^AX + {}_1^1H \rightarrow \left[{}_{Z+1}^{A+1}Cn \right] \rightarrow {}_{Z+1}^{A+1}Y + \gamma $$

Example: ${}_{13}^{27}Al + {}_1^1H \rightarrow {}_{14}^{28}Si \rightarrow {}_{14}^{28}Si + \gamma$

25.15.3 Transmutation by Deutrons

(i) (d, α) reaction: General reaction

$$ {}_Z^AX + {}_1^2H \rightarrow \left[{}_{Z+1}^{A+2}Cn \right] \rightarrow {}_{Z-1}^{A-2}Y + {}_2^4He $$

Example: ${}_8^{16}O + {}_1^2H \rightarrow {}_9^{18}F \rightarrow {}_7^{14}N + {}_2^4He$

These reactions are in general **exothermic** as the mass change is usually positive.

(ii) (*d*, *p*) reaction: General reaction

$$ {}_Z^AX + {}_1^2H \rightarrow \left[{}_{Z+1}^{A+2}Cn \right] \rightarrow {}_{Z+1}^{A+1}X + {}_1^1H $$

Examples: ${}_6^{12}C + {}_1^2H \rightarrow {}_7^{14}N \rightarrow {}_6^{13}C + {}_1^1H$

${}_3^7Li + {}_1^2H \rightarrow [{}_4^9Be] \rightarrow {}_3^8Li + {}_1^1H$

We may note that here in (*d*, *p*) reaction isotopes will form. The *Q* value of these reactions is usually positive and obviously the reactions are in general ***exothermic***.

(iii) (*d*, *n*) reaction: General reaction

$$ {}_Z^AX + {}_1^2H \rightarrow \left[{}_{Z+1}^{A+2}Cn \right] \rightarrow {}_{Z+1}^{A+1}Y + {}_0^1n $$

Examples: ${}_3^7Li + {}_1^2H \rightarrow [{}_4^9Be] \rightarrow [{}_4^9Be] \rightarrow {}_4^8Be + {}_0^1n$

${}_6^{12}C + {}_1^2H \rightarrow [{}_7^{14}N] \rightarrow {}_7^{13}N + {}_0^1n$

25.15.4 Transmutation by Neutrons

(i) (n, α) reaction: General reaction

$$ {}_Z^AX + {}_0^1n \rightarrow \left[{}_Z^{A+1}Cn \right] \rightarrow {}_{Z-2}^{A-3}Y + {}_2^4He $$

Example: ${}_3^6Li + {}_0^1n \rightarrow [{}_3^7Li] \rightarrow {}_1^3H + {}_2^4He$

(ii) (*n*, *p*) reaction : General reaction

$$ {}_Z^AX + {}_0^1n \rightarrow \left[{}_Z^{A+1}Cn \right] \rightarrow {}_{Z-1}^AY + {}_1^1H $$

Examples: ${}_7^{14}N + {}_0^1n \rightarrow [{}_7^{15}N] \rightarrow {}_6^{14}C + {}_1^1H$

${}_{12}^{24}Mg + {}_0^1n \rightarrow [{}_{12}^{25}Mg] \rightarrow {}_{11}^{24}Na + {}_1^1H$

(iii) $(n, 2n)$ reaction: General reaction

$$^{A}_{Z}X + {}^{1}_{0}n \rightarrow \left[{}^{A+1}_{Z}\mathrm{Cn}\right] \rightarrow {}^{A-1}_{Z}X + {}^{1}_{0}n + {}^{1}_{0}n$$

Example: $^{27}_{13}\mathrm{Al} + {}^{1}_{0}n \rightarrow [{}^{28}_{13}\mathrm{Al}] \rightarrow [{}^{26}_{13}\mathrm{Al}] + 2\,{}^{1}_{0}n$

The reaction is in general **endothermic** as the mass change in these reactions is always negative.

(iv) (n, γ) reaction: General reaction

$$^{A}_{Z}X + {}^{1}_{0}n \rightarrow \left[{}^{A+1}_{Z}\mathrm{Cn}\right] \rightarrow {}^{A+1}_{Z}X + \gamma$$

Examples: $^{27}_{13}\mathrm{Al} + {}^{1}_{0}n \rightarrow [{}^{28}_{13}\mathrm{Al}] \rightarrow {}^{28}_{13}\mathrm{Al} + \gamma$

$$^{238}_{92}\mathrm{U} + {}^{1}_{0}n \rightarrow [{}^{239}_{92}\mathrm{U}] \rightarrow {}^{239}_{92}\mathrm{U} + \gamma$$

$$^{2}_{1}\mathrm{H} + {}^{1}_{0}n \rightarrow [{}^{3}_{1}\mathrm{H}] \rightarrow {}^{3}_{1}\mathrm{H} + \gamma$$

The Q value of these reactions is always positive and hence these reactions are *exothermic* with excess energy being carried away by γ-rays.

The slow neutrons, almost in all cases, induce (n, γ) reactions (radiative capture) and the product nuclei are often radioactive. These reactions are important source of artificial radioisotopes.

25.15.5 Transmutation by Photon (Photo Disintegration)

(i) (γ, n) reaction: General reaction

$$^{A}_{Z}X + \gamma \rightarrow \left[{}^{A}_{Z}\mathrm{Cn}\right] \rightarrow {}^{A-1}_{Z}X + {}^{1}_{0}n$$

Examples: $^{2}_{1}\mathrm{H} + \gamma \rightarrow [{}^{2}_{1}\mathrm{H}] \rightarrow {}^{1}_{1}\mathrm{H} + {}^{1}_{0}n$

$$^{9}_{4}\mathrm{Be} + \gamma \rightarrow [{}^{9}_{4}\mathrm{Be}] \rightarrow {}^{8}_{4}\mathrm{Be} + {}^{1}_{0}n$$

The reactions are **endothermic** as the Q value is negative.

(ii) (γ, p) reaction: General reaction

$$^{A}_{Z}X + \gamma \rightarrow \left[{}^{A}_{Z}\mathrm{Cn}\right] \rightarrow {}^{A-1}_{Z-1}Y + {}^{1}_{1}\mathrm{H}$$

Example: $^{25}_{13}\mathrm{Mg} + \gamma \rightarrow [{}^{25}_{13}\mathrm{Mg}] \rightarrow {}^{24}_{11}\mathrm{Na} + {}^{1}_{1}\mathrm{H}$

We may note that same nuclides can be transmuted in a number of different ways, depending upon the particle used to produce transmutation and upon the particle given off during the nuclear reaction.

25.16 NUCLEAR FISSION

Nuclear fission is a process in which a heavy nucleus after capturing a neutron splits up into two lighter nuclei of comparable masses. The product nuclei are called *fission fragments*. The process is accompanied by the release of a few fast neutrons and a huge amount of energy in the form of the kinetic energy of the fission fragments, and also as γ-rays. One of the typical fission reactions is

$$^{235}_{92}\text{U} + {}^{1}_{0}n \rightarrow \left[{}^{236}_{92}\text{U}\right] \rightarrow {}^{144}_{56}\text{Ba} + {}^{89}_{36}\text{Kr} + 3\,{}^{1}_{0}n + Q$$

Of course, many different fission reactions are possible, with many different fission products.

Both ^{144}Ba and ^{89}Kr are highly unstable and they reach stability by a series of β and γ emissions. The value of Q is found about to be 200 MeV per fission of ^{235}U atom. Recalling that 1 MeV = 1.6×10^{-13} J, the energy release per fission of ^{235}U is 3.2×10^{-13} J.

We may note that in a nucleus, there is a competition between the nuclear force, which holds the nucleus together, and the electrostatic repulsion of the nuclear protons, which tries to tear the nucleus apart. For most nuclei the nuclear force dominates this competition, but for heavy nuclei there is a delicate balance between the nuclear forces, a balance that is easily upset. We may further note that the energy necessary to cause fusion is typically about 6 MeV.

Later researches have shown that besides uranium, other nuclei are also fissionable. $^{232}_{90}$Th and $^{231}_{91}$Pa can be fissioned by fast neutrons, whereas transuranic element $^{239}_{94}$Pu and an artificial isotope of $^{233}_{92}$U can be fissioned by thermal neutrons.

Besides neutrons, accelerated protons, deutrons, and α-particles can also induce fission in the nuclei of thorium, uranium and transuranic elements. Even γ-rays can cause nuclear fission which is known as *'photo fission.'* Some lighter elements can also be fissioned by very high energy photons and deutrons.

'Spontaneous fission' (in which no bombarding particle is required) of thorium, uranium and transuranic elements has also been detected. It is a process (like natural radioactivity) in which a nucleus splits into two fragments of its own accord. However, the probability of this type of fission is very-very small.

There are three features of nuclear fission that make it useful as a means to generate electrical energy.

(i) Energy Dissipation

Most of the energy released in a nuclear fission process is released as kinetic energy of the fission fragments. These relatively heavy fragments do not travel very far through the reactor fuel element before they dissipate most of their kinetic energy in collisions with the atoms of the fuel element. One can extract the energy as heat and used to boil water; the resulting steam can then be used in conventional way to drive a turbine to generate electricity.

(ii) Neutron Multiplicity

The second important feature of the fission reaction that makes it useful is that the average number of neutrons produced is greater than one, making possible the *chain reaction.* How much greater than one it must be, in order to achieve a chain reaction, depends on the construction of the *nuclear reactor.*

(iii) Prompt and Delayed Neutrons

The third advantage of nuclear fission process is the one that enables an operator or the mechanical system to control the reaction and keep it from proceeding too rapidly. The two neutrons emitted in the fission process

$$^{233}_{92}\text{U} + n \rightarrow {}^{93}_{37}\text{Rb} + {}^{141}_{55}\text{Cs} + 2n$$

are called **promot neutrons** and constitute 99% of the total neutrons ejected in the process. About 1% of the neutrons in the fission process are **delayed neutrons** emitted following the decays of the heavy fragments while reaching the stable state. The best known example of this is

$$^{87}_{35}\text{Br} \xrightarrow{\beta^-} {}^{87}_{36}\text{Kr} \xrightarrow{n} {}^{87}_{36}\text{Kr}$$

the half life of which is 55.6 s. The delayed neutrons increase the mean life time of all fission neutrons are of importance in controlling the rate of reactions in a nuclear reactor.

25.16.1 Theory of Nuclear Fission

Bohr and Wheeler proposed that a heavy nucleus behaves like a liquid drop, and that the fission may be explained as a result of oscillations brought about by impinging neutrons. Some of the properties of nuclear forces (saturation, short range) are analogous to the properties of the nuclear forces which hold a liquid drop together. Attractive forces between nucleons, like the attractive forces between molecules, give rise to surface tension and the spherical state. Any disturbance of this state will require an external force, which distorts the sphere into an ellipsoid. If the force is large enough the ellipsoid narrows into a 'dumb-bell' shape and finally breaks at the neck into two major portions with some additional small drops. Fig. 25.14 explains the supposed mechanism of nuclear fission.

When an atomic nucleus undergoes fission, the incident neutron combines with the nucleus to form a compound nucleus which is highly energetic. The extra energy of the nucleus is partly the kinetic energy of the incident neutron but largely the added binding energy of the incident neutron. This excitation energy appears to initiate a series of rapid oscillations in the drop. When the excitation energy is low, the oscillations about the spherical shape are small, so that at its maximum deformation the nucleus adopts the ellipsoidal shape [Fig. 25.14 (b)]. The restoring force of the nucleus arises from

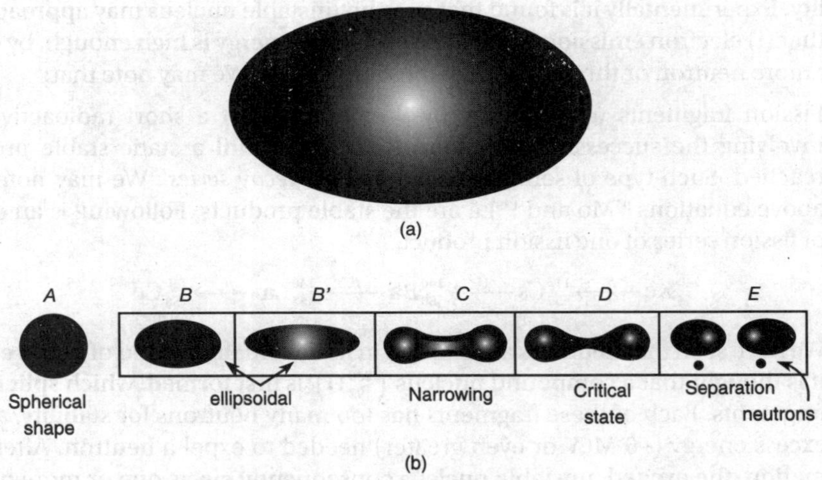

Fig. 25.14 (a) The elongated shape of heavy nucleus **(b)** Sequence of nuclear shapes in nuclear fission in liquid drop model of nucleus

the short-range inter nucleon forces. If the oscillations become so violent the stage D (Fig. 25.14 (b)) is reached, and as such 'half' is now positively charged, the final fission into stage E is inevitable. Clearly, there is a threshold energy or a critical energy required to produce stage D after which the nucleus cannot return to stage A, because of Coulomb repulsion between the two positive parts. During the fission a large amount of energy is released in fission process. The product nuclides must be in highly excited states at the instant of fission. Part of this energy is released almost immediately and carried away by the prompt neutrons, The remaining excess energy is released by neutrons or α-particles, or β-decays or γ-ray emission.

The value of critical deformation energy E_{crit} was first calculated by Bohr and Wheeler on liquid drop model and that they obtained the condition for spontaneous process as $Z^2/A \geq 45$. This condition shows that U^{238} requires more total excitation energy than U^{235} to initiate fission.

25.16.2 Fission products

In a given nucleus the fission may occur in a number of different ways. In general, the fissionable nucleus gives two fission fragments which thereafter decay by electron emission into stable end products. What particular fragments are produced by the given nucleus is a matter of chance, The uranium nucleus is capable of splitting in about 40 different ways. About 97% of the U^{235} nuclei undergoing nuclear fission yield products, which falls into two groups: (i) *'light' group* ($A = 85$ to 104) and a *'heavy'* group ($A = 130$ to 149). The following reactions makes it clear.

$$^{235}_{92}\text{U} + ^{1}_{0}n \rightarrow \left[^{236}_{92}\text{U}\right] \rightarrow \ ^{95}\text{Mo} + \ ^{139}\text{La} + 2^{1}_{0}n + Q$$

$$^{235}_{92}\text{U} + ^{1}_{0}n \rightarrow \left[^{236}_{92}\text{U}\right] \rightarrow \ ^{141}_{56}\text{Ba} + \ ^{92}_{36}\text{Kr} + 3^{1}_{0}n + Q$$

The products obtained from fission are in a highly excited state and have two many neutrons for stability. Experimentally it is found that such an unstable nucleus may approach stability by either (i) electron emission or (ii) if the excitation energy is high enough, by ejection of one or more neutron or through both of these processes. We may note that:

(i) Fission fragments which decay by β-emission start a short radioactive series involving the successive emission of electrons until a static stable product is reached. Such type of series is called *'fission decay series'*. We may note that in above equations ^{95}Mo and ^{139}La are the stable products. Following is an example of fission series of one fission product.

$$^{140}_{54}\text{Xe} \xrightarrow{\beta^-} \ ^{140}_{55}\text{Cs} \xrightarrow{\beta^-} \ ^{140}_{56}\text{Ba} \xrightarrow{\beta^-} \ ^{140}_{57}\text{La} \xrightarrow{\beta^-} \ \underset{\text{(Stable)}}{^{140}_{58}\text{Ce}}$$

(ii) We have seen that neutrons are emitted in fission. In the second of above equation it is thought that a compound nucleus $\left(^{236}_{92}\text{U}\right)$ is first formed, which split into two fragments. Each of these fragments has too many neutrons for stability, and also excess energy (~6 MeV or even greater) needed to expel a neutron. After its formation, the excited, unstable nucleus consequently ejects one or more neutrons with γ-rays within a very short time. This explanation seems to be consistent with experimental result.

25.16.3 Energy Released in Nuclear Fission

We have seen that when a heavy nucleus undergoes a fission process, a large amount of energy is released together with the emission of neutrons which further produced fission. This is the most striking properties of fission. The energy released per fission of ^{235}U atom is about 200 MeV as compared with several MeV for other nuclear reactions. A rough estimate of the energy released per fission can be made by using the binding energy curve. The B.E. curve shows that heavy nuclei whose mass numbers are above 240 have binding energies of about 7.6 MeV per nucleon; while fission fragments whose mass numbers lie in the range roughly from 70 to 160 have binding energies of about 8.5 MeV per nucleon. Obviously, the fission fragments have, on the average binding energy of about 0.9 MeV per nucleon greater than the (heavy) nucleus. The total amount of energy released per fission of a compound nucleus ^{236}U, which has 236 nucleons, should be roughly equal to the product of number of nucleons (236) multiplied by the excess B.E. /particle (0.9 MeV), *i.e.*, $236 \times 0.9 = 212$ MeV or roughly 200 MeV = 3.2×10^{-11} J. We can also calculate the total energy released per fission from the nuclear masses of ^{236}U and pair of fission products for the following reaction:

$$^{235}\text{U} + {}_0^1\text{n} \rightarrow [^{236}\text{U}] \rightarrow {}^{95}\text{Mo} + {}^{139}\text{La} + 2\,{}_0^1\text{n}$$

Now, Q value of the reaction

$$Q = (\text{Mass of the reactants - mass of products})c^2$$
$$= (236.133 - 235.919) \times 931 \text{ MeV}$$

The reaction is *exothermic* as the Q is positive and the amount of energy released per fission is obtained as

$$E = 0.124 \times 981 = 198 \text{ MeV} = \simeq 200 \text{ MeV} = 3.2 \times 10^{-11} \text{ J}$$

Although this energy seems quite small but we must remember that this is the energy released due to fission per nucleus of ^{235}U. There are N atoms in I gram atom of ^{235}U, where $N = 6.02 \times 10^{26}$ (Avogadro number). Thus 1 kg of ^{235}U has $\dfrac{6.02\times10^{26}}{235} = 2.56\times10^{24}$ atoms.

If all the nuclei of uranium atoms in 1 kg are fissioned, the released energy would be $2.56 \times 10^{24} \times 3.2 \times 10^{-11} = 8.2 \times 10^{10}$J. It would produce about 2.28×10^4 kWH of electrical energy. We may note that there are about 30 different ways in which the compound nucleus (^{236}U) can divide, but mass excess is approximately the same, for all these processes and in each case 200 MeV is the average amount of energy released per fission and in good agreement with experimental observations.

Table 25.2

Atomic mass of ^{235}U	= 235.124 u
Atomic mass of ${}_0^1$n	= 1.009 u
Total mass of reactants	= 236.133 u
Atomic mass of ^{139}La	= 138.955 u
Atomic mass of ^{95}Mo	= 94.946 u
Atomic mass of 2 ${}_0^1$n	= 2.018 u
Total mass of the products	= 2.35.919 u

25.16.4 Nuclear Fission as a Source of Energy

We have seen that large amount of energy is released in fission reaction and also with this more than one neutron is produced. This has made it possible to use the fission process as a source of energy. During each fission process, on the average 2.5 neutrons are emitted which attack other atoms causing fission. The number of neutrons goes on multiplying rapidly during' fission process till entire fissionable material is disintegrated. The reaction gets **self sustained** and is known as a *chain reaction.* The chain reaction may be of two types:

(i) Uncontrolled Chain Reaction

When more than one of the neutrons emitted in a particular fission process cause further fission, then amount of energy is liberated in each fission process and within a very short time the released energy takes a tremendous magnitude and is released as a violent explosion. This process takes place in a *nuclear bomb.* Such type of chain reaction is called an *'uncontrolled chain reaction.'*

(ii) Controlled Chain Reaction

If by some means, one can control the fission reaction in such a way that only one of the neutrons emitted in a fission causes another fission, then fission rate remains constant and the energy is released steadily such a chain reaction is called a *'controlled chain reaction'* and used in *nuclear reactor,* which can be used as a source of nuetrons or of nuclear power.

Energy Release: We now calculate the energy released by the complete fission of 1 kg of uranium. We have energy released/fission = 200 MeV.

$$1 \text{ MeV} = 1.6 \times 10^{-13} \text{ J} = 1.6 \times 10^{-13} \text{ Ws}$$

∴ Energy released per fission

$$= 200 \times 1.6 \times 10^{-13} \text{ J} = 3.2 \times 10^{-11} \text{ J}$$

$$= 3.2 \times 10^{11} \text{ Ws}$$

Avogadro's number = Number of atoms/ kg atoms

$$= 6.02 \times 10^{26} / \text{kg-atoms}$$

Now, the energy released by the fission of 1 kg atoms of ^{235}U or by 235 kg of ^{235}U is

$$= 3.2 \times 10^{-11} \times 6.02 \times 10^{26} = 1.93 \times 10^{16} \text{ J}$$

∴ Energy released per kg of ^{235}U

$$= 1.93 \times 10^{16}/235 = 8.2 \times 10^{13} \text{ J} = 8.2 \times 10^{13} \text{ Ws}$$

Power output $(P)/s = 8.2 \times 10^{13}$ W

Power output/day $$P = \frac{8.2 \times 10^{13}}{24 \times 60 \times 60}$$

$$= 9.49 \times 10^{8} \text{ W} = 10^{3} \text{ MW}$$

Obviously, the power output from 1 kg of ^{235}U is

$$= 10^{3} \text{MW}$$

Clearly, 1 kg ^{235}U on complete fission will supply 1000 MW of power/ day. We may note that this is equivalent to the output of a thermal power plant which consumes about 300 tonnes of coal/day.

Fission rate of ^{235}U for producing 1 W of electric power = $1/3.2 \times 10^{-11} = 3.1 \times 10^{10}$ fissions/s.

Time required for one fission = $1/3.1 \times 10^{10} = 3.2 \times 10^{-9}$ s.

This clearly reveals that fission rate is very large and large amount of fission energy is released in an extremely short interval of time, *i.e.* 10^{-9}s. Obviously, this much of production of energy leads to a very powerful explosion unless its rate of production is controlled.

We have already mentioned that controlled chain reactions are essential for building a nuclear reactor. In order to achieve a chain reaction with uranium, we will have to take care of following four competing processes:

(i) Fission of uranium nuclei, with the emission of more neutrons than are captured. On an average 2.5 neutrons per fission of uranium nuclei are released.

(ii) Non fission capture of uranium and other materials and escape of neutrons without being captured.

If the loss of neutrons by (ii) process is less than or equal to the surplus neutrons produced by (i) process, a nuclear chain reaction occurs, otherwise a chain reaction is not possible.

The *size of the sample* is one of the most crucial factor in making favourable condition for neutron balance in a reaction. We may note that if the size of the sample is less, then neutron will go out of the sample without producing fission, whereas if the size is too big then production rate or neutron will be more than escape, so explosion will occur. Obviously, there is a *critical size* of the sample, which is essential for the production of neutrons in an controlled chain reaction. Critical size calculation of the sample is one of the main problem in designing a nuclear reactor.

25.17 NUCLEAR REACTOR OR NUCLEAR PILE

A *nuclear reactor,* or a *nuclear pile,* is a device in which a self-sustaining controlled chain reaction is produced in a fissionable material. Obviously, nuclear reactor is a source of controlled energy.

The fissionable material, known *as fuel,* plays the key role in the operation of a nuclear reactor. Uranium, preferably enriched with the isotope ^{235}U is used as fuel. Natural uranium contains 98.28% abundance of ^{238}U and 0.72% abundance of ^{235}U. It will release about 2.5 fast neutrons with average energy 2 MeV, if some spontaneous fission occurs. For these fast neutrons, the fission cross section (probability) of two isotopes of uranium ^{238}U and ^{235}U are almost same. However, the abundance of ^{238}U is very large and hence the fission neutrons cause further fissions which will mainly occur in ^{238}U. We may note that these neutrons can suffer (i) fission (ii) radioactive capture and (iii) elastic or inelastic scattering. However, the last two processes are more probable and therefore chain reaction is not possible in natural uranium. It is reported that $^{239}_{94}$Pu will be produced in a non-fission capture of neutrons by ^{238}U (Fig. 25.15).

One can made possible the chain reaction in natural uranium by using **moderator**. A moderator slows down the neutrons to thermal energies by elastic collisions between

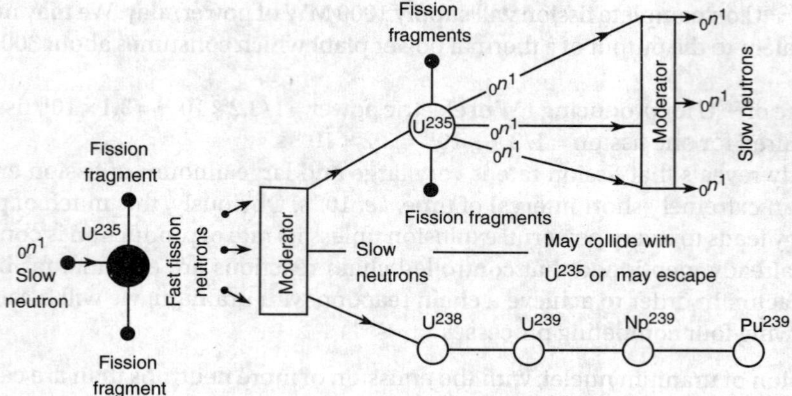

Fig. 25.15 A chain reaction exhibiting different steps occurring in fission process

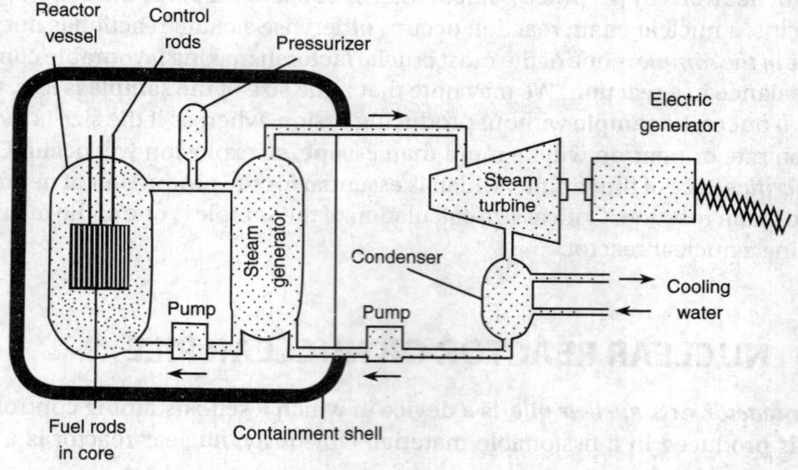

Fig. 25.16 A nuclear reactor (heterogeneous)

its nuclei and the fission neutrons. The thermal neutrons have a very high probability of producing fission in ^{235}U nuclei. Graphite and heavy water (D_2O) are most suitable moderators. Beryllium oxide, hydrides of metals and organic liquids are other materials used for moderators. We may note that the nuclei of these substances absorb neutrons only to a slight extent. When heavy water, D_2O is used as moderator, uranium can be dissolved in it. This type of nuclear reactor is known as *'homogeneous nuclear reactor.'*

Uranium is in the form of rods, when graphite is used as moderator and U-rods are distributed in a regular way throughout the graphite thereby forming a lattice. This type of reactor is called *heterogeneous reactor* (Fig. 25.16).

25.17.1 Critical Size for Nuclear Reactor

On average, if a chain reaction is to be sustained in a lump of uranium, at least one of the 2.5 neutrons born per fission must be preserved for further fission, *i.e.,* if all the neutrons in excess of one per fission are removed, the process will be self-sustaining. The *assembly of the fissionable material which meets, this criterion is called a critical assembly.* This is what we do actually in nuclear reactor. If the neutron from one fission are allowed to induce more than one secondary fissions, the assembly is called the *super critical assembly* (atom bomb). This we can express in some other way also. We can define a constant k (sometimes called the neutron multiplication factor), equal to the number of net effective fission neutrons born per fission as

$$k = \frac{\text{Number of neutrons present in one generation}}{\text{Number of neutrons present in the previous generation}}$$

If $k < 1$, the chain reaction will **slow down** and stop. If $k = 1$, the reaction will proceed at a **steady rate**. If $k > I$, however the reaction will grow to an **explosive rate**. These situations are called as *'sub critical,'* 'critical' and *'supercritical'* respectively.

The need for a favourable neutron balance sets certain conditions on the size of the reactor. If uranium fuel is distributed in a regular way throughout the assembly, the production of neutrons depends on the volume of the system, *i.e.,* proportional to r^3, whereas the probability of escape of neutron depends on surface area, *i.e.,* proportional to r^2. Now, if the size of the system is very small, then surface to volume ratio for the system $(= 1/r)$ is large. This means that the rate of non-fission capture and escape of neutron is more than production rate of neutron, in that case chain reaction cannot sustain. Clearly, it depends on the size of the system. Now, for larger size, greater will be its radius, and escape rate of-neutrons will be smaller as compared to the production rate, and hence chain reaction will sustain. However, if the size of the system is very big, then production rate of neutron will be much bigger than escape or absorption rate. This will result in an explosion. Concluding, we can say that for effective nuclear reactor we need a *critical size* of the system for which $k = 1$, *i. e.,* the production of neutrons by fission will be just equal to their loss by non-fission capture and escape.

Basically all types of nuclear reactor's consists of following essential basic parts:

(i) Fuel: The material that fissions and supplies neutrons for further fissions ^{235}U, ^{238}U or ^{239}Pu are generally used as fuels in nuclear reactors.

(ii) Reactor Core: Core is the main part of the reactor, where the reactor fuel like ^{235}U, ^{238}U ^{239}Pu, fissionable materials are used. Due to nuclear fission reaction, huge amount of heat is generated in the core. Usually the reactor core has a right circular cylinder with about a diameter of few meters.

Fuel elements are generally made up of uranium rods. Reactor core, in general has fuel elements, moderator, control rods, and cooling material. All these are housed in a pressure vessel as shown in Fig. 25.16.

(iii) Moderators: A major difficulty in trying to produce a chain reaction is the energy of the neutrons emitted in the fission process. Typically, the kinetic energies of these neutrons are a few MeV, such energetic neutrons have a relatively low probability of inducing new fissions, since the fission cross-section generally decreases rapidly with increasing neutron energy. We therefore must slow down, or moderate, these neutrons in order to increase their chances of initiating fission events. To slow down neutrons, they are just passed through moderators which are material rich in protons or in nuclei of small mass that would not capture them. Materials used to slow down the fast neutrons in a reactor are designated as moderators. These materials have high boiling point, large scattering cross section, small absorption coefficient and low atomic weight and nuclei of these substances absorb neutrons only to a slight extent. Commonly used materials as moderators are heavy water (D_2O), graphite, beryllium oxide, etc.

(iv) **Reflectors:** Reflectors surrounding the reactor core, containing the fissionable material and moderator are of the same substances as used for moderator. The efficiency of the reflector increases rapidly with its thickness and reaches its limiting value when this thickness is several times the mean free path of the neutrons in the substance. The reactor system including reflector, must be enclosed in a shield, usually of concrete. This reduces the intensity of the reduction of neutrons and γ-rays leaving the reactor.

(v) **Cooling System:** This is intended for removing the heat energy released in the reactor core. This heat is evolved from the kinetic energy of the fission fragments when they are slowed down in the fuel and moderator. The coolant or heat transfer agent (water, steam, He, CO_2, air and certain molten metals and alloys) is pumped through the reactor core. Then, through a heat exchanger the coolant transfers heat to the secondary thermal system of their reactor. The energy generated in a reactor is to be converted into electric power. For this purpose energy generated is transferred from the coolant to a working liquid, to produce steam or hot gas. The regulating vapour or gas is then used to generate electricity by means of turbines. Ordinary water and heavy water, both are found to be good coolants, as they can serve both as moderator and coolant. A coolant should have (i) high specific heat, (ii) high boiling point (ii) cheap, (iv) easy to pump and chemically stable.

(vi) **Controlled and Safety Systems:** The control and safety systems enable the chain reaction to be controlled, prevent it from spontaneously running away and also protect the space surrounding the reactor against intensive neutron flux and gamma rays existing in the reactor core. The first aim is achieved by pushing control rods of boron or cadmium into the reactor core. These rods have a large neutron absorption cross-section. The second aim is achieved by surrounding the reactor with massive layers of substances that strongly absorb neutrons and gamma rays and also by providing completely closed coolant circuits without any leakage. One can explain the control of the thermal reactor by a quantity K_e, known as **effective multiplication factor**, which controls the condition of reaction at steady state. How the neutron concentration would increase from generation to generation provided no neutrons leaked out from the system can be known from the multiplicative factor K_∞. To obtain a controlled chain reaction in a reactor, one will have to combine K_∞. in some way with the quantity which tells something about neutron leakage, and that is called *effective multiplication factor* (K_e) defined as

$$K_e = \frac{P}{A+L}$$

where P is production rate of neutrons, A is absorption rate of neutrons, L is leakage rate of neutrons, $K_e = 1$ denotes steady state, $K_e = 0$ critical condition, $K_e > 1$ is super critical and $K_e < 1$ is sub-critical condition.

We have mentioned earlier that production rate is proportional to r^3, the volume of the reacting system, whereas the escape or leakage rate is proportional to r^2, the surface area of the reacting system. Therefore

$$K_e \propto \frac{r^3}{r^2} \propto r$$

where r is radius of the reacting system. Clearly, the effective neutron multiplication factor (K_e) depends on r, *i.e.*, the size of the nuclear reacting system. For maintaining controlled chain reaction in a nuclear reactor, *i.e.*, to make effective multiplication factor $K_e = 1$, one needs a critical size of the system for which the production of neutrons by fission process will be just balanced by the non fission capture and escape. Now,

$$P\,(\text{Production rate of neutrons}) = NF$$

where N is average number of neutron emission per fission and F is rate of fission process.

$$K_e = \frac{NF}{A+L} = \frac{N(F/A)^{\cdot}}{1+(L/A)}$$

We may note that the ratio F/A depends upon the amount of fissionable and non-fissionable material and on their cross-sections of fission and neutron capture.

One can control the fission rate in a nuclear reactor by means of controlling rods. Controlling rods are generally made up of cadmium or boron steel. By adjusting the position of control rods in the core, one can adjust the rate of absorption of excess neutrons, *i.e.*, F/A can be controlled. Thus, one can maintain the value of $K_e = 1$, *i.e.*, obtain controlled fission. If the value $K_e < 1$, then one will have to add fresh fissionable material.

25.17.2 Thermal Power Nuclear Reactor

Primarily, the power reactors are intended for the production of electricity, *i.e.*, usable power. These reactors transform the heat generated from the fission into electricity. The heat so generated is allowed to pass to a fluid passing through the core of the reactor. The hot fluid is then allowed to pass through a heat exchanger where energy is transferred and utilized by proper means.

Figure 25.16 shows the basic principle of one type of power reactor. A definite quantity of pure enriched uranium forms the centre of heat energy source. By keeping the value of $K_e = 1$, the fission rate and hence the rate of production of heat energy is controlled, which is achieved with the help of controlling cadmium rods. One can decrease the temperature if desired, by pushing little further the cadmium rods so that they may absorb more neutrons thereby decreasing the fission rate (F/A). Cadmium rods are pulled out a little further for raising the temperature. When the reactor or pile becomes sub-critical, one will have to add fresh fissionable material.

As a primary coolant liquid sodium is used. Liquid sodium is not only a dense material with good thermal conductivity but it also has very low absorption cross-section for thermal neutrons. To lower the melting point, sodium is usually alloyed with potassium.

This primary coolant absorbs energy, which is transferred to water in a heat exchanger. The steam that is generated drives a turbine. The turbine is connected to an electric generator. The electric power so generated can be used for running factories, lighting cities or for other purposes. There are good numbers of power reactors operating in India. Besides producing energy, a nuclear reactor is also a powerful source of neutrons and of intense induced radioactivity.

25.17.3 Fast Breeder Reactors

The interest in fast breeder reactors for commercial power production lies in the prospect that they can serve as breeders. Such reactors could not only produce useful power but also regenerate more fossil material than is consumed. Ultimately if use is to be made of the relatively large amounts of ^{232}Th and ^{238}U, that are available in nature. Obviously, breeder reactors must assume a prominent role in the nuclear power program.

In a breeder reactor nuclei of one fissionable material are transformed as a result of nuclear reactions into nuclei of another fissionable material. Conversion of ^{232}Th and ^{238}U into ^{233}U and ^{239}Pu respectively as a result of neutron absorption is as follows:

$$^{232}\text{Th} + {}^1_0n \rightarrow {}^{233}\text{Th} + \gamma$$
$$\text{(fast) 22 min} \downarrow \beta^-$$
$$^{233}\text{Pa}$$
$$\downarrow 27\,d$$
$$^{233}\text{U} + \beta^-$$

and

$$^{238}\text{U} + {}^1_0n \rightarrow {}^{239}\text{U} + \gamma$$
$$\text{(fast) 24 min} \downarrow \beta^-$$
$$^{239}\text{Np}$$
$$2.3\,d \downarrow$$
$$^{239}\text{Pu} + \beta^-$$

Clearly, more fissionable material is produced than consumed. It is inevitable that some neutrons should be lost by leakage or by absorption, but so, long as more than two neutrons are produced in fission there is a possibility that more nuclear fuel may be produced than is consumed. This is the basic principle of the breeder reactor and the process is called *'Breeding.'* If one could made it work, a *breeder reactor* could make new fuel for itself and in addition, a stockpile of fissionable material could be built up for use as a fuel in other new reactors.

Breeder reactors are promise of low fuel costs and efficient utilization of fuel resources through breeding. Use of sodium as coolant permits achieving high temperatures at normal pressure; also sodium is a very efficient heat transfer medium. The handling of sodium introduces some design and operating complications. Experimental Breeder Reactor (EBR), has been built to study the feasibility of power production and breeding in such a reactor.

Research in the field of fast breeder reactors are in progress. The value of n for ^{233}U and neutrons is 2.28 ± 0.02, which makes the breeding of ^{233}U from ^{232}Th in thermal reactor system feasible. Research and developments are in progress in such systems. One design of such system involves the use of $^{233}U\ O_2SO_4$ in D_2O; whereas another design involves the use of a solution of ^{233}U Bi in liquid bismuth, with graphite as moderator. It is reported that ^{233}U-^{232}Th system could also be used in a fast breeder reactor, but the greater value of ν for ^{239}Pu favours the use of latter material in fast breeder reactors.

If we could achieve the conversion of ^{238}U and ^{232}Th into fissionable materials, the resulting increase in available nuclear fuels would be a step in the direction of fast breeder economic nuclear power reactor. We may note that much information is needed concerning the cross-sections for nuclear reactions of fast and intermediate energy neutrons with fissionable and fertile materials. Obviously, a lot of research work is to be done in the field of breeder reactors. Presently, several breeder reactors are operating. They have proved to be extremely expensive and have had operating problems.

25.17.4 Nuclear Reactor Problems

Today, nuclear power supplies about 18% of the world's total electrical energy. Nuclear power plant construction continues around the world. However, this is affected after recent Sunami's destruction in Japan. Problems certainly exist with nuclear power plants. Nevertheless, ongoing uncertainties about the risk of catastrophic nuclear accidents, long term waste storage and its disposal, terrorism, environmental problems, and weapon proliferation continue to haunt the nuclear power industry. Proliferation especially is a very real concern. Although nuclear power and weapons development are different enterprises, they share infrastructure and educated technological elite that can be put to either purpose. If nuclear power is to advance, it will need to do so under strict international guarantees against division of materials and expertise to weapons production. For large expansion of nuclear power scientists will have to overcome four critical problems lower costs, improved safety, better nuclear waster management, and lower proliferation risk.

25.18 NUCLEAR FUSION OR THERMO-NUCLEAR REACTIONS

When two or more very light nuclei moving at very high speeds are fused together to form a single nucleus then the process is called as *'nuclear fusion.'* The mass of the product nucleus is less than the sum of the masses of the nuclei which are fused. The lost mass is converted into energy which is released in the process according to relation $E = mc^2$. This property of the light nuclei is shown by the binding energy curve (Fig. 25.3), in which the average binding energy per nucleon rises rapidly with increase in mass number in the range of low mass number nuclei. Fusion reactions takes place at very high temperature (~10^8 K) so it is also called thermonuclear reaction. *Hydrogen bomb* is based on this principle and· also called as **nuclear fusion bomb.**

For example, two deutrons (heavy-hydrogen nuclei) can be fused to form a tritron (tritium nucleus) according to the following fusion reaction

$$ {}_1^2H + {}_1^2H \rightarrow {}_1^3H + {}_1^1H + 4\ \text{MeV} $$

The tritron so formed can further combine with a third deutron to form an α-particle (${}_2^4$He nucleus) according to following reaction

$$ {}_1^3\text{H} + {}_1^2\text{H} \rightarrow {}_2^4\text{He} + {}_0^1\text{n} + 17.6 \text{ MeV} $$

The net result of the two reactions is the burning of three deutrons and the formation of an α-particle, a ${}_0^1n$ and ${}_1^1$H (proton). The total energy released is 21.6 MeV. Obviously, the energy released per deutron burnt is 7.2 MeV. Most of the energy liberated in the fusion reaction appears as kinetic energies of neutron and proton. We may note that for fusion reaction to take place, the component nuclei must be brought to 10^{-14} m. In order to approach so closely, nuclei participating in reaction should be imparted high energies so that two positively charged deutrons may overcome the repulsive force and fused together. The only practical way of fusing the two nuclei of deuterons together is to raise their temperature of the order of 10^8 °C, typical of Sun's interior. At the required temperature, a fusion reactor's fuel will be in the form of *plasma,* which is fully ionized gas. Plasma is the *fourth state of matter.*

The energy output in a fusion reaction is much less than the energy released in fission reaction, which is about 200 MeV. However, this does not mean that fusion is a weaker source of energy in comparison to fission. In fact, the energy output per unit mass of the material consumed is much greater in the case of the fusion of the light nuclei then in the case of fission of heavy nuclei.

A successful fusion reactor has three basic requirements to fulfill as mentioned below:

(i) The plasma temperature must be high so that an adequate number of the ions have the speeds essential to come close enough together to react despite their mutual repulsion. Taking into account that many ions have speeds well above the average and that tunnelling through the potential barrier reduces the ion energy needed, the minimum temperature for igniting a D-T plasma is about 100 million Kelvin, which corresponds to an ion temperature of $kT \sim 10$ keV.

(ii) The plasma density (n ions/m^3) must be high to ensure that collisions between nuclei are frequent.

(iii) The plasma of reacting nuclei must remain together for a sufficiently long time τ. How long depends on the product $n\tau$, the confinement quality parameter. In the case of D-T plasma with $kT \sim 10$ keV, $n\tau$ must be greater than roughly 10^{20} s/m^3 for break even more than that for ignition.

The approach to the controlled release of fusion energy that has thus far shown the most promise uses a strong magnetic field to confine the reactive plasma. Laser beams have also received the most attention for inertial confinement but electron and proton beams have promise as well. However, magnetic confinement seems closer to the good of achieving a working fusion reactor.

A possible fusion reactor is shown in fig. 25.17.

25.18.1 The Source of Stellar Energy: Stellar Thermonuclear Reaction

Bethe in 1939 suggested that the production of stellar energy is by thermonuclear reactions in which protons are continuously transformed into helium nuclei. For comparatively

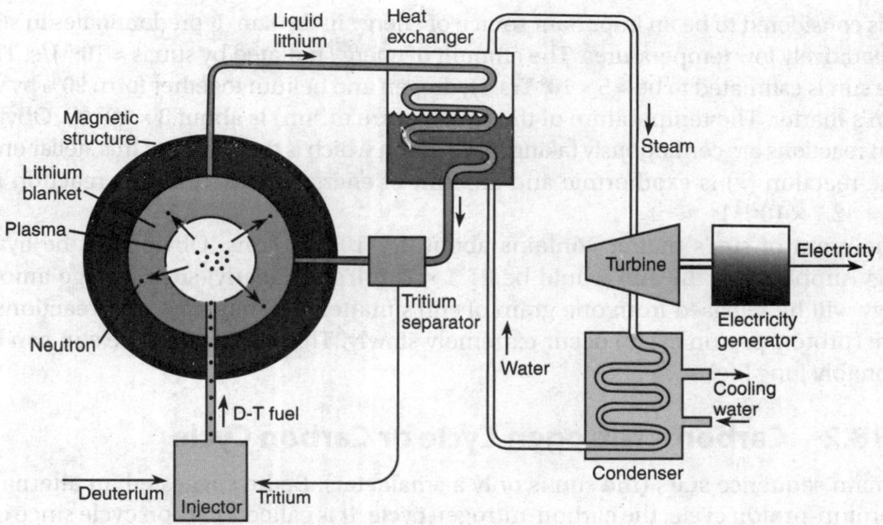

Fig. 25.17 Configuration of a possible fusion rector using magnetic confinement. The tritium is obtained from neutron reactions with lithium, and tritium must be separated prior to going back to fusion reactor as fuel in fission reactors, the heat produced by fusion is used to run a turbine, which in turn generates electric power

low stellar temperatures he proposed the proton-proton cycle represented by following chain reactions

$${}_1^1\text{H}+{}_1^1\text{H}\rightarrow {}_1^2\text{D}+{}_{+1}^{\;0}e(\beta^+)+\nu+0.42\text{ MeV} \quad \text{... (i)}$$

(Neutrino)

$${}_1^2\text{D}+{}_1^1\text{H}\rightarrow {}_2^3\text{He}+\gamma+5.5\text{ MeV} \quad \text{... (ii)}$$

$${}_2^3\text{He}+{}_1^1\text{H}\rightarrow {}_2^4\text{He}+{}_{+1}^{\;0}e(\beta^+)+\nu \quad \text{... (iii)}$$

(Neutrino)

$${}_2^3\text{He}+{}_2^3\text{He}\rightarrow {}_2^4\text{He}+2{}_1^1\text{H}+12.8\text{ MeV} \quad \text{... (iv)}$$

In this cycle of reactions, fusion reactions (i) and (ii) must occur twice to yield two ${}_2^3\text{He}$ nuclei in (iv). The net result of (i), (ii) and (iv) reactions is that four protons are fused to produce an α-particle, two positrons, two neutrinos and two γ-photons.

$$4\,{}_1^1\text{H}\rightarrow {}_2^4\text{He}+2\,{}_1^0\text{e}+2\nu+2\gamma$$

The energy released in the cycle is 24.6 MeV, for the mass difference is (4 × 1.0078 ~ 4.0026) or 0.0286 μ. Since the neutrinos escape, the energy left is about 24.3 MeV after the fusion of 4 ${}_1^1\text{p}$ into ${}_2^4\text{He}$ nucleus.

If we add (i), (ii) and (iii) equations, we obtain

$$4\,{}_1^1\text{H}\rightarrow {}_2^4\text{He}+2\,{}_1^0\text{e}(\beta^+)+2\nu+27\text{ MeV} \quad \text{... (v)}$$

This is considered to be an important source of energy in the sun. It predominates in states of comparatively low temperatures. The amount of energy radiated by sun is ≈ 10^{26} J/s. The age of the sun is estimated to be ≈ 5×10^9 Yrs. Hydrogen and helium together form 90% by weight of sun's matter. The temperature of the core (centre of sun) is about 3×10^7 °C. Obviously, fusion reactions are continuously taking place in sun which is the source of this stellar energy.

The reaction (ν) is exothermic and amount of energy released in this reaction is 26.7 MeV = 42.7×10^{-13} J.

One gram of sun's matter contains about 2×10^{23} protons. Obviously, the available energy supply from the sun would be 21.3×10^{10} J/gm. Clearly, such a huge amount of energy will be released from one gram of sun's matter. We may note that reactions cited above (proton-proton cycle) occur extremely slowly. This means that the sun can have a reasonably long life.

25.18.2 Carbon – Nitrogen Cycle or Carbon Cycle

For main sequence stars (the sun is only a small star), Bethe suggested an alternative to the proton-proton cycle, the carbon-nitrogen cycle. It is called a carbon cycle since carbon serves as a short of nuclear catalyst. The reactions of carbon cycle are

$$^{12}_{6}\text{C} + {}^{1}_{1}\text{H} \rightarrow {}^{13*}_{7}\text{N} + \gamma$$

$$^{13*}_{7}\text{N} \rightarrow {}^{13}_{6}\text{C} + {}^{0}_{+1}e(\beta^{+}) + \nu$$

$$^{13}_{6}\text{C} + {}^{1}_{1}\text{H} \rightarrow {}^{14}_{7}\text{N} + \gamma$$

$$^{14}_{7}\text{N} + {}^{1}_{1}\text{H} \rightarrow {}^{15*}_{8}\text{O} + \gamma$$

$$^{15}_{7}\text{N} + {}^{1}_{1}\text{H} \rightarrow {}^{12}_{6}\text{C} + {}^{4}_{2}\text{He}$$

On addition, we get

$$4\,{}^{1}_{1}\text{H} \rightarrow {}^{4}_{2}\text{He} + 2\,{}^{0}_{+1}e(\beta^{+}) + 3\gamma + 2\nu + 26\text{ MeV}$$

The initial ${}^{12}_{6}\text{C}$ acts as a kind of catalyst for the above process, since it reappears at the end.

25.19 COMPARISON BETWEEN FISSION AND FUSION

The table 25.3 depicts the comparison between fission and fusion.

Table 25.3 Comparison between fission and fusion

Fission	*Fusion*
1. Huge amount of energy is liberated.	1. Huge amount of energy is liberated.
2. A heavy nucleus is split up into two nuclei, e.g. ${}^{235}_{92}\text{U}$ into ${}^{95}\text{Mo}$ and ${}^{139}\text{La}$	2. Two lighter nuclei fused together at very high temperature, e.g. ${}^{2}_{1}\text{H}$, ${}^{3}_{1}\text{H}$
3. Fission process is possible even at room temperature.	3. Fusion is possible only at very high temperature (~ millions K).
4. The links of fission process are neutrons.	4. The links of this process are protons.

(*Continued*)

Table 25.3 (*Continued*)

5. The energy per nucleon in fission process is 0.85 MeV or 200 MeV per nuclide.	5. The energy per nucleon is 6.75 MeV.
6. Several isotopes are formed as product particles in this process.	6. In general isotopes are not formed in this process.
7. Availability of atoms used, e.g. ^{235}U, ^{238}U is limited and also costly.	7. Availability of atoms used, e.g., ${}_1^2H$, ${}_1^3H$ is unlimited and at the same time cheap also.
8. Disposal of radioactive waste is a challenging problem.	8. In comparison to fission, radioactive waste material produced is very small.
9. Nuclear fission power reactor is already in operation.	9. Fusion reactor development research is in progress.

25.20 NUCLEAR DETECTORS

Nuclear radiations (or particles) such as $\alpha^-, \beta^-, \beta^+$ and γ-rays cannot be observed directly. They are detected by the secondary effects which they produce in the materials through which they pass. There are such effects, e.g. ionization, photographic and fluorescence. Nuclear detecting instruments are based upon these effects. The instruments which are used for the detection of nuclear radiations are (i) Ionization chamber, (ii) Proportional counter (iii) Geiger – Muller counter and (iv) Cloud chamber, etc., Out of these four each of the first three detectors, used to detect radioactive radiation, is based on the production of ionization in a gas and the separation and collection of ions by means of an electrostatic field. One can explain the differences in these three systems with the help of Fig. 25.18. Figure 25.18 shows a cylindrical conducting chamber and insulated from it. The conducting chamber is filled with a gas at a pressure *of* about one atmosphere or even less then that. Between the wall and the central electrode through the resistance R shunted by capacitor C, a voltage V is applied as shown in Fig. 25.18. Relative to that of the central wall, the central electrode is at a positive potential.

When a charge particle enters the ionization chamber, ionization of gas takes place resulting in an ion pair formation. Due to the presence of voltage V, positive ions move to the chamber walls and negative ions (electrons) move to the central electrode. Now, for a

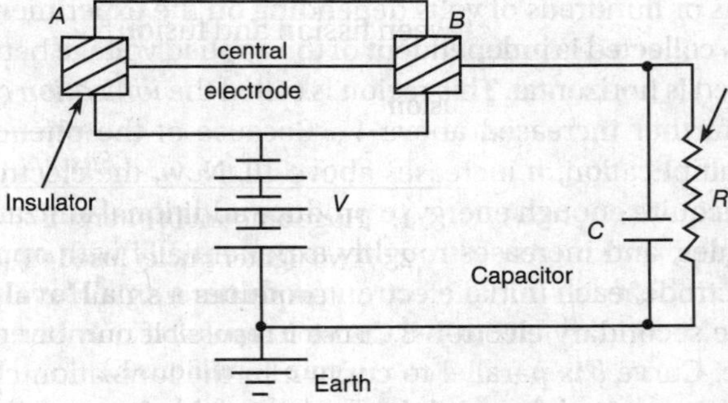

Fig. 25.18 Nuclear particle detector for detecting α and β particles

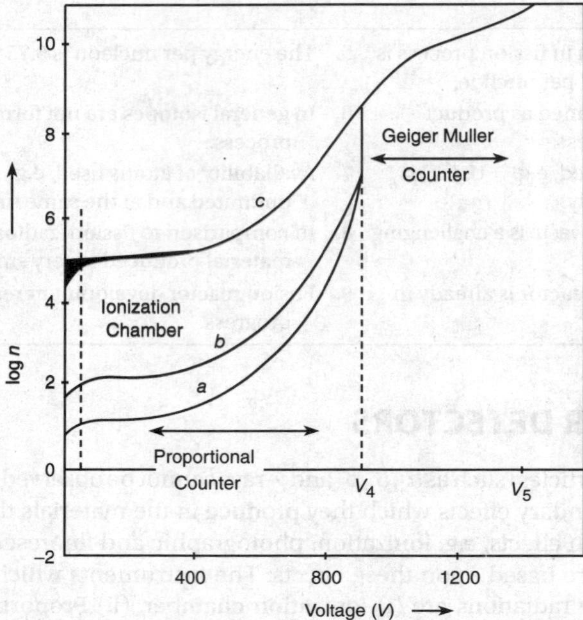

Fig. 25.19 Variation of total ion pair (log *n)* with applied voltage (*V*)

given initial ionization, how many ion pairs are collected or how many electrons reach the central electrode with the variation of the applied voltage, is shown in Fig. 25.19, where we have plotted the curves of total ion collection as functions of applied voltage. For convenience, we have taken the logarithm to the base 10 of the number (log *n)* of ion pairs as the ordinate. When there is no voltage across the electrodes, the ions will recombine and obviously, no charge will appear on the capacitor. With the increase in voltage say to a few volts, then there is competition between the loss of ion pairs by recombination and the removal of ions by collection on the electrodes. Some electrons will also reach the central electrode. Now, at a applied voltage V_1 (~10 V), it is found that the loss of ions by recombination becomes negligible and all electrons reach central electrode, log *n* attains the value unity. Now, on further increasing the voltage from V_1, *n* stays constant until a voltage V_2 is reached which may be some tens or hundreds of volts depending on the experimental conditions. The number of ion pairs collected is independent of the applied voltage between V_1 and V_2 region and curve obtained is horizontal. This region is called the *ionization chamber region.*

Now, the voltage is further increased above V_2. Because of the phenomenon of gas amplification or gas multiplication, *n* increases above 10. Now, the electrons released in the primary ionization acquire enough energy to produce additional ionization, when they collide with gas molecules, and increases roughly exponentially with applied voltage *V.* Close to the central electrode, each initial electron produces a small **'avalanche'** of electrons with most of these secondary electrons. Curve *b* results, if number of ion pairs are formed initially is more. Curve *b* is parallel to curve *a* in the ionization chamber region and we can see that it is separated from it by one unit on the *log n* scale. We may note that above applied voltage V_2, the behaviour of the two curves *a* and *b* is quite interesting.

However, for some range of voltages, upto V_3, each electron acts independently and gives its own avalanche, now they are being affected by the presence of other electrons. On the *log n* scale, the two curves *a* and *b* continue to be parallel with a difference between them of one unit. The number of ion pairs collected between V_2, and V_3 is then proportional to the initial ionization. This region is known as *'proportional counter operation' region.*

The gas multiplication effect continues to increase very rapidly above the applied voltage V_3, and as more electrons produce avalanches, the latter begin to interact with one another. The positive ion space charge of one avalanche inhibits the development of the next avalanche. In curve *b*, the discharge of initial electrons is affected with a less number of initial electrons which increases less rapidly than curve *a*. We note that curves *a* and *b* approach each other and eventually meet at V_4. The region between V_3 and V_4 is called the *region of limited proportionality.* The charge collected becomes independent of the ionization initiating at above V_4 and we can see that curves *a* and *b* becomes identical. We may note that gas multiplication increases the total number of ions to a value that is limited by the characteristics of the chamber and the external circuit. The region above the applied voltage V_4 is called the region of Geiger Mullar (GM) counter operation. The region of GM counter operation ends at a voltage V_5 where the discharge tends to propagate itself indefinitely. Obviously, the applied voltage V_5 marks the end of the useful voltage scale and the region above V_5 being that of continuous discharge.

Curve *c* is obtained if the initial ionization is very large. Curve *c* is similar to curves *a* and *b*, and parallel to them between V_2 and V_3, the proportional counter region ends sooner instead of at V_3. Obviously, the extent of that region depends on the initial ionization.

On the basis of the behaviour of the ions of the gas in the electrostatic field particles detection instruments have been developed as follows:

(i) Ionization Chamber

This is mainly used for detecting the α- and β-rays and to compare the activities of sources emitting these rays. This operates at voltages in the range V_1 and V_2 and moreover it is characterized by complete collection, without gas amplification, of all the electrons initially liberated by the passage of the particle. Subject to certain conditions, it will give a pulse proportional to the number of these particles, *i.e.,* electrons. Basically, it consists of a chamber fitted with a gas, like air or argon at normal pressure. This chamber could be used as an ionization chamber between 10 and 200 volts, approximately (Fig. 25.20). We may note that the numbers in Fig. 25.20 would be appropriate for a counter with an outer cylinder 0.01 m in radius, a wire central electrode of radius 10^{-4} m, and with gas to a pressure of a few centimeters of mercury, e.g. argon at about 6×10^{-2} m of Hg plus a little alcohol. The counter length is estimated to be about 0.1 m. We may note that the electrodes of an ionization chamber can also be parallel plates. Ionization chambers are usually of two types; (i) Non-integrating and (ii) Integrating. First type of ionization chamber consists of a cylindrical conducting chamber with a metallic electrode *AB* located on the axis of the chamber and insulated from it. The second type consists of a cylindrical chamber with two electrodes *A* and *B* placed at some distances from each other.

An ionization chamber is much less sensitive to β-particles (in comparison to α-particles) because β-particles produce fewer pairs of ions in their passage through the chamber. It is too intensive for γ-rays as these rays produce very little ionisation in the gas.

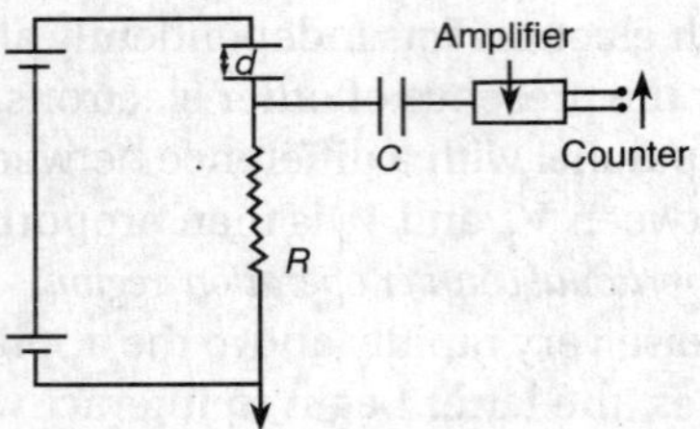

Fig. 25.20 Ionisation chamber

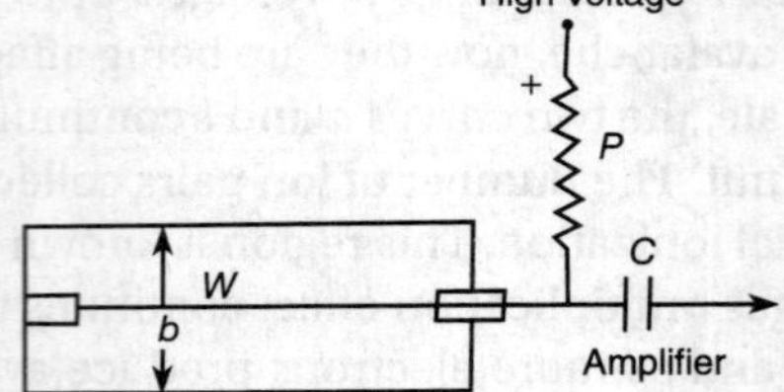

Fig 25.21 The proportional counter

For detecting γ-rays an ionization chamber of thick wall high atomic number (Pt, Bi) is employed. The γ-rays impinging on the walls of the chamber eject high-speed electrons which produce ionisation in the gas.

We may note that the ionisation chamber does not give a count of individual particles, but an average effect of a large number of particles.

25.20.1 Proportional Counter

This is used to detect and measure the energies of particles (α and β) and also of photons (γ- and X-rays). This operates in the region V_2 to V_3 and is characterised by a gas multiplication independent of the number of initial electrons. This counter functions satisfactorily at voltages 500 to 800 volts or above at pressure of about one atmosphere. Although gas multiplication is utilized in this counter, the pulse is always proportional to the initial ionization. The use of proportional counter permits both the counting and energy determination of particles which do not produce enough ions to yield a detectable pulse in the region of voltages V_1 to V_2. These pulses are displayed on an oscilloscope screen. The pulse height gives a measure of the energy of particle (or photon) entering the counter. Obviously, the proportional counter is thus able to distinguish directly (by virtue of the height of the pulse produced) between α- and β-particles and γ-rays since they differ so much in their inionizing powers. Clearly this counter offers advantages for pulse-type measurements.

25.20.2 Geiger-Mullar (GM) Counter or GM Tube

This is the most versatile, accurate and useful of the detecting instruments used for detecting and measuring the energies of α, β, γ and X-rays. It is sufficiently sensitive to detect individual α- and β-particles. It operates in the voltage region V_4 to V_5 and is characterized by the speed of the discharge throughout the entire length of the counter, resulting in a pulse independent of the initial ionization. GM counter is widely used is industry and medicine to locate radioisotopes, and in the laboratory for comparing the activity of radioactive specimen.

Usually a G. M. counter consists of a fine wire (usually tungsten) placed along the axis of a hollow metallic-cylinder electrode (cathode) enclosed in a glass tube (Fig. 25.22). The tube contains a suitable mixture of 90% argon at 10 cm pressure and 10% ethyl alcohol vapour at 1 cm pressure. We may note that different mixtures of gases at different pressure are used in different designs.

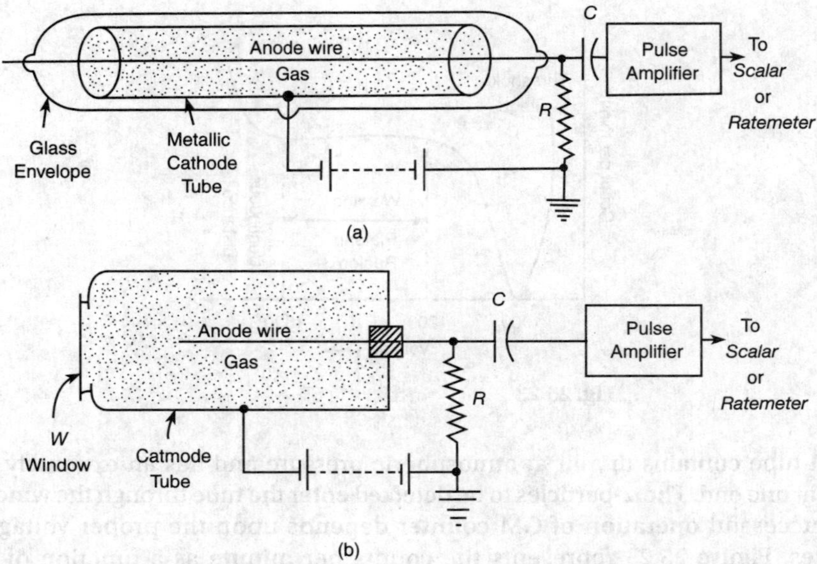

Fig. 25.22 Geiger – Muller counter

A potential difference between 800 to 2000 *V*, is usually applied between cylinder and wire which acts as anode. However, the value of this applied voltage is adjusted to be somewhat below the breakdown potential of the gaseous mixture.

When a charged particle passes through the GM counter tube, it ionizes the gas molecules, thereby releasing a number of ion-pairs. 'The central wire attracts the electrons (-ve ions) while the cylindrical electrode attracts the positive ions and the resulting pulse of a current through the tube produces a voltage pulse ($\approx 10\mu$V) across resistance *R*. The pulse is amplified by an electronic pulse amplifier to about 5 to 10 V. This amplified pulse is finally passed on either to a scalar or to a rate meter. The scalar records the arrival of each individual pulse separately, and so provides the exact number of particles entering the GM tube in a given time internal. The rate meter records the number of pulses in a given time, *i.e.*, it gives the average rate at which the particles enter the GM tube. Scalar is usually used in conjuction with the GM tube for very high pulse rates. The combination of GM tube with the recorder is called GM-counter.

The magnitude of the pulse produced does not depend upon the nature or energy of the entering particle and hence a GM tube does not distinguish between the arrival of an α-particle and a β-particle. We may note that a proportional counter does it. However, the particles can be distinguished by finding what absorber placed in their path will prevent them from entering the GM tube.

G.M. counter discussed so far is suitable for β- and γ-rays measurement and it is not suitable for α-particles, because the walls of glass tube are too thick and therefore a-particles cannot penetrate through the walls. For detection of α-particles, the *Geiger-Point counter* shown in Fig. 25.22(b) is generally used. It mainly consists of an axially mounted fine wire inside a tube which is maintained at a positive potential with respect to the GM tube.

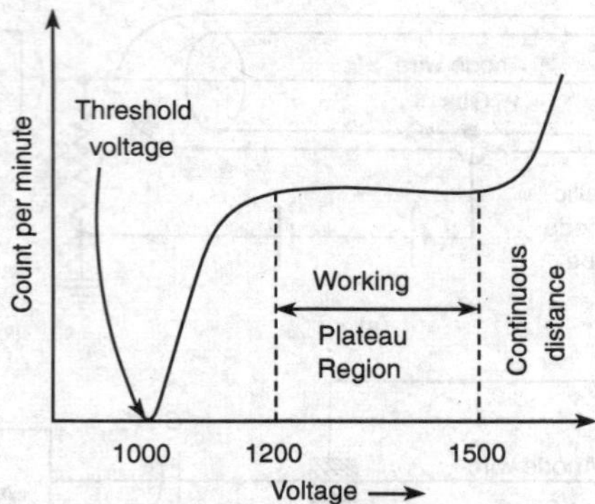

Fig. 25.23

This GM tube contains dry air at atmospheric pressure and has an extremely thin foil window at one end. The α-particles to be detected enter the tube through the window *(W)*.

The successful operation of GM counter depends upon the proper voltage to the electrodes. Figure 25.23 represents the counts per minute as a function of voltage. From figure 25.23, it is obvious that if the voltage is less than 1000 V, there is no discharge, *i.e.,* no secondary ionization. With the further increase in voltage, secondary ionization takes place. Now, the number of impulses increases almost linearly with applied voltage. This region is most suitable for **proportional counters**. On further increase of voltage to about 1200 V, the number of impulses remains constant over a certain region known as **'working plateau region**. In this region, the magnitude of impulses becomes independent of the amount of original ionization and is a function of potential, nature of gas, resistance *R and* geometrical conditions of apparatus. This is the most suitable region for GM counter. Beyond this voltage, *i.e.,* above this region, a continuous discharge will take place, which is undesirable and hence avoided.

25.21 PARTICLE ACCELERATORS

(i) Cyclotron

Cyclotron is first orbital type of accelerator in which charged atomic particles move under the action of a strong magnetic field in circular paths. This was invented by Lawrence in 1932. The important feature of the cyclotron is that, when the particles are to be accelerated, the phase of the accelerating field is always such as to speed them up. The machine is capable of giving energetic ions of larger intensity than that of linear accelerator. The principle of this accelerator is based upon resonance of electric and tmagnetic fields, and therefore it is called as **magnetic resonance accelerator**. However, the name cyclotron is much more popular as the acceleration of particles involves the cyclic motion.

To understand basic principle of a cyclotron, let us consider a charged particle of charge e and mass m, moving with uniform velocity v, in a transverse magnetic field B. This transverse magnetic field will exert a force on the particle perpendicular to the plane of v and B both. The magnitude of this force is evB. Under the influence of this force, the particle will move along a circular path; and will be subjected to centrifugal force given by $\frac{mv^2}{r}$, where r is radius of curvature of the path. Thus

$$\frac{mv^2}{r} = evB \quad \text{or} \quad v/r = \omega = \frac{e}{m} . B \qquad \text{... (25.16)}$$

where ω is angular velocity of rotation of particle.
The time T required by the particle in completing one revolution will be given by

$$T = \frac{2\pi r}{v} = \frac{2\pi m}{eB} \qquad \text{... (25.17)}$$

Equation (25.17) shows that irrespective of the velocity of the particle and for given value of B and e/m, the time of revolution remains constant.

The cyclotron consists of two short, hollow, semicircular cylinders D_1 and D_2, mounted rigidly along their diameter, separated from each other by a gap as shown in Fig. 25.24 (a). Because of their shapes, these semicircular cylinders are called *"Dees".* The dees are connected to a high frequency oscillator, which is really a short wave radio transmitter supplying energy to the dees. The entire arrangements of dees is placed inside a vacuum chamber within the pole pieces of a strong magnet. The complete cross-sectional diagram is shown in Fig. 25.24 (b). To obtain charged particles, say protons, at the centre of dees is placed a hot filament from which, by passing electric-current, thermions (electrons) can be obtained. Now, when a trace of hydrogen gas is admitted to the evacuated chamber, these electrons ionise some of the hydrogen atoms by collision, producing protons.

Consider protons in the neighbourhood of F, and the dees are connected to high frequency oscillator. At the instant when D_1 is negative and D_2 is positive, the protons will be accelerated towards D_1. As soon as the protons enter the space inside D_1, the electric field would not interact with its motion and under the influence of strong magnetic field their path would be circular. If after making half revolution, the potential difference between the dees D_1 and D_2 is reversed so that D_1 becomes positive and D_2 negative and the protons reach in the gap between dees D_1 and D_2, these would be accelerated towards D_2, (as D_2 negative) and repelled by D_1 causing the protons to increase in their speed. Hence inside the space of D_2 the particles will move in a circular path of larger radius. Now, again after

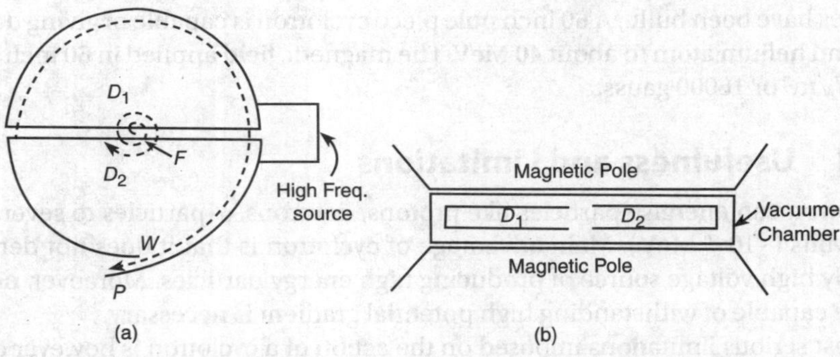

Fig. 25.24 (a) Position of dees of a cyclotron (b) Cross-sectional diagram of cyclotron

completing half revolution in D_2 as the protons reach in the gap, again the polarity of dees reverses and the protons would be accelerated towards D_1. In this way every time when the protons reach in the gap, these would be accelerated every time because of the electric field. When the radius of the path reaches to maximum, with the help of negatively charged plates, these protons can be made to pass through a narrow opening *W*. Due to negative charge, the path of the protons straightens and emerge from cyclotron in pulses because of alternating potential difference which is inevitable.

In the cyclotron alternating potential difference and magnetic field are so adjusted that by the time the proton moves along semicircular path inside the dees, their polarity reverses. Thus,

$$\text{frequency of electric field} = \text{cyclotron frequency} = \frac{eB}{2\pi m} \qquad \text{... (25.18)}$$

As the protons emerge from the cyclotron, their velocity and hence energy would be maximum. At the same time the radius of curvature of the path equals that of radius *R* of the dees. Thus

$$Bev_{\max} = \frac{mv_{\max}^2}{R} \quad \text{or} \quad v_{\max} = \frac{BeR}{m} \qquad \text{... (25.19)}$$

Hence maximum energy, $E_{\text{max is given by}}$

$$E_{\max} = \frac{1}{2}mv_{\max}^2 = \frac{1}{2}\frac{B^2e^2R^2}{m} \qquad \text{... (25.20)}$$

Thus we see that in a cyclotron the energy can be increased to any amount by increasing cyclotron radius. However, it is not possible. Because, as the velocity of the particles increases, the mass of the particles also increases according to the relativistic relation

$$m = \frac{m_0}{\sqrt{1-\frac{v^2}{c^2}}} \qquad \text{... (25.21)}$$

This relativistic mass increases with velocity limits the maximum energy obtainable from cyclotron. It is because of this reason electrons cannot be accelerated by cyclotron.

Lawrence first cyclotron with magnetic pole diameter 2.5 inches provided proton of energy 80 keV. At Berkeley, California U.S.A., so many cyclotrons of different diameters of pole pieces have been built., A 60 inch pole piece cyclotron is capable of giving deutrons at 24 MeV and helium atom to about 40 MeV. The magnetic field applied in 60 inch cyclotron was 1.6 W /m² or 16000 gauss.

25.21.1 Usefulness and Limitations

The cyclotron can energise particles like protons, deutrons, α-particles to several million electron volts (~10^{-50} MeV). Main advantage of cyclotron is that it does not demand any excessively high voltage source of producing high energy particles. Moreover, no accelerating tube capable of withstanding high potential gradient is necessary.

The most serious limitations imposed on the action of a cyclotron is however due to the relativistic increase in mass of the accelerated particle in accordance with the relation (25.21)

The frequency is given by

$$f = \left(\frac{q}{m}\right)\frac{B}{2\pi} = \left(\frac{q}{m}\right)\frac{Bc^2}{2\pi c^2}$$

$$= \frac{Bqc^2}{2\pi\left(m_0c^2 + W_K\right)} \quad \text{... (25.22)}$$

where W_k is kinetic energy of the accelerated particle at speed v.

From Eq. (25.22) it is obvious that as the speed of the particle increases, f decreases and W_K increases. Thus in a fixed frequency cyclotron, the particle as it spirals out from the source, will lag in phase behind the accelerating dee voltage. Obviously, it would thus cease to get accelerated. This means that one must suitably modify the frequency of oscillation to ensure resonance acceleration in the gaps. One cannot achieve this without upsetting the condition for low energy particles.

We may note that with electrons, the relativistic change in mass is more pronounced than with positive ions, e.g. proton, deutrons even for low energies.

Other limitations of the cyclotron that limit the maximum energy gain by the accelerating particle is set by the engineering difficulties and expense. The maximum kinetic energy of the particles when they leave the cyclotron is

$$W_m = \frac{1}{2}mv_{\max}^2 = \frac{1}{2}m\left(\frac{r_m qB}{m}\right)^2 = \frac{1}{2}\frac{q^2B^2}{m}r_m^2$$

obviously, $W_m \propto r_m^2$. This means that for a given particle the size of a cyclotron increases more rapidly than the corresponding increase in energy.

The compensation for relativistic mass gain has been achieved in *Synchrocyclotron.*

25.21.2 Betatron

It has been stated above that due to large relativistic increase in mass of the electrons at low energy, a cyclotron cannot be used to accelerate them. Kerst in 1941 invented another accelerator to accelerate β-particles or fast moving electrons, known as Betatron. The fundamental difference between a betatron and cyclotron is that, in a betatron a rapidly changing magnetic field is used to accelerate the electrons, and their path is an orbit of constant radius. The action of betatron depends upon the same principle as that of transformer, in which an alternating current applied to a primary coil, induces a current in the secondary coil. In the primary coil, oscillating magnetic field is produced due to alternating current which in turn produces oscillating potential in the secondary. In this way betatron is a transformer in which electron located inside an annular, doughnut-shaped vacuum chamber plays the role of secondary coil of one turn, and placed within the poles of laminated steel of electromagnet. The electrons gain energy by induction, because of the change of flux ϕ, with time, linking with the orbit.

Let us consider an electron moving in a circular orbit of radius r, where the total magnetic flux through the orbit is ϕ and the flux density at the orbit is B. This magnetic field B is perpendicular to orbital plane. The e.m.f. induced in the orbit is given by

$$\varepsilon = -\frac{d\phi}{dt} \quad \text{... (25.23)}$$

The work done on the electron of charge *e*, in one revolution is given by εe which equals $e\frac{d\phi}{dt}$. If F is tangential force on the orbit, the work can also be given by $F \,.\, 2\pi r$. Hence

$$F = \frac{e}{2\pi r} \cdot \frac{d\phi}{dt} \qquad \text{... (25.24)}$$

Due to influence of this applied force the energy of the electron will increase and hence it would try to move in an orbit of larger radius. If an orbit of constant radius is to be maintained, this increase in radius of curvature of the orbit be resisted. A radial force due to magnetic field and acting perpendicular to the direction of motion of the electron keeps it moving in the orbit of constant radius. This inward radial force is *Bev*, and must be equal to the centrifugal force mv^2/r.

Thus

$$Bev = \frac{mv^2}{r} \qquad \text{... (25.25)}$$

The tangential force on the electron is for the orbit of constant radius, is given by

$$F = \frac{d}{dt}(mv) = \frac{d}{dt}(Ber) = er\frac{dB}{dt} \qquad \text{... (25.26)}$$

Hence from Eqs. (25.24) and (25.26)

$$\frac{\text{e}}{2\pi r} \cdot \frac{d\phi}{dt} = er\frac{dB}{dt} \text{ or } d\phi = 2\pi r^2 dB \qquad \text{... (25.27)}$$

Integrating
$$\int_0^\phi d\,\phi = 2\pi r^2 \int_0^B dB \quad \text{or} \quad \phi = 2\pi r^2 B \qquad \text{... (25.28)}$$

This is known as **betatron condition**. The result shows that the total flux ϕ within the radius of the orbit *r*, must be twice the value which would be obtained if flux density *B* were uniform over whole area of the orbit and also this should be proportional to *B*. This condition holds for relativistic energies as well as non-relativistic energies. Such a flux distribution is obtained by specially shaped countered pole-piece faces, where the flux density at the centre of the orbit is greater than it is at the circumference of the orbit.

The construction of betatron is illustrated in Fig. 25.25 and Fig. 25.26 (a). The electrons are produced from a hot filament and are given primarily acceleration by the application of electric field of the order of a few thousands volts. These are then injected in the highly evacuated doughnut tube *DD* made of glass or ceramic at *G*. In the coils of electromagnet and alternating current from mains supply of normal frequency say 50 c/sec, is passed. Thus increasing flux in a given direction is obtained for quarter cycle only. This is shown in Fig. 25.26 (b). At any instant when magnetic field is just rising from zero in first quarter cycle, induces a potential in the doughnut tube, increasing the energy of the electrons. When the strength of the field passes its maximum and start decreasing in the next quarter cycle, the direction of induced e.m.f. reverses and the electrons slow down. To avoid this effect the electrons are removed from the stable orbit. This is done by discharging a capacitor through the primary coils or through auxiliary coils around primary, at the instant when electron receive the desired amount of energy. Due to this extra current momentarily the

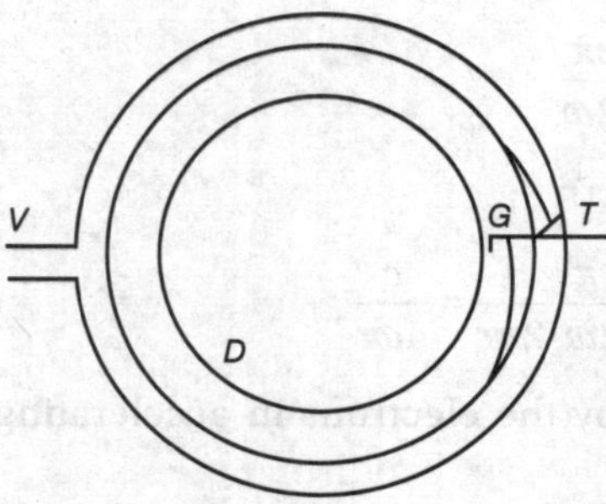

Fig. 25.25 Daughnut shaped annular chamber in a betatron

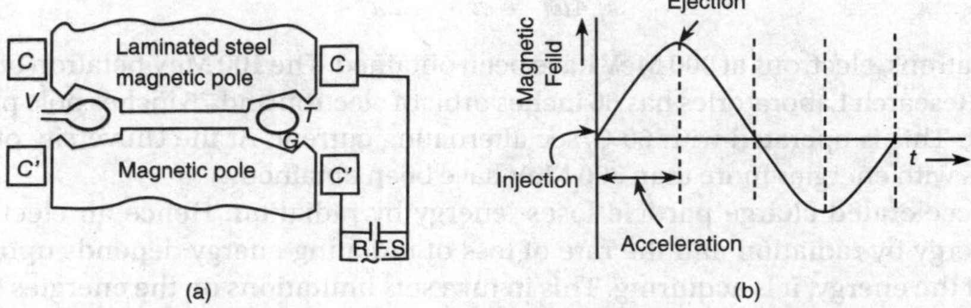

Fig. 25.26 (a) Position of D.T. in the pole pieces of electromagnet (b) Acceleration cycle in a betatron

total flux ϕ increases rapidly, but flux density B increases less rapidly, and causes orbital radius r to increase. Thus when the radius of final orbit increases the beam can be made to hit the target T, producing hard X-rays or γ-rays depending upon their energies in pulses.

The energy of the electron beam can be calculated from the average induced e.m.f. and the total number of revolutions made by the electron. Let the flux variation be given by

$$\phi = \phi_0 \sin \omega t$$

where ϕ_0 is amplitude and ω is angular frequency of changing flux. Hence the energy per revolution is given by

$$e\frac{d\phi}{dt} = \omega e \phi_0 \cos \omega t \quad \text{... (25.29)}$$

Hence the average energy per revolution

$$= \omega e \phi_0 \frac{1}{T/4} \int_0^{T/4} \cos \omega t \, dt = \frac{2\omega e \phi_0}{\pi} \quad \text{... (25.30)}$$

This represents energy per revolution. To calculate number of revolutions, we can approximately consider particle to travel with the velocity of light *c*, because electron of energy 1 MeV moves with velocity 0.94 *c*. Furthermore these accelerate only during quarter of alternating cycle, *i.e.,* for time

$$\frac{1}{4}T \; viz, \frac{1}{4} \cdot \frac{2\pi}{\omega} = \frac{\pi}{2\omega}$$

Distance traversed in this time $= \dfrac{c\pi}{2\omega}$

Distance traversed in one revolution $= 2\pi r$

$\therefore$ Number of revolution $= \dfrac{c\pi}{2\omega} \cdot \dfrac{1}{2\pi r} = \dfrac{c}{4\omega r}$ ... (25.31)

Hence total average energy acquired by the electrons in accelerating quarter cycle is given by

E = (No. of revolution energy per revolution)

$$= \frac{c}{4\omega r} \cdot \frac{2\omega e\phi_0}{\pi} = \frac{ec\phi}{2\pi r} \qquad \text{... (25.32)}$$

With betatrons, electrons at 300 MeV have been obtained. The 100 MeV betatron of General Electric Research Laboratories has 66 inches orbit of electron and 76 inches pole piece face diameter. This is operated with 60 C/sec alternating current. At the University of Illinois, electrons with energies more than 300 MeV have been obtained.

Any accelerated charge particle loses· energy by radiation. Hence an electron also loses energy by radiation and the rate of loss of radiating energy depends upon fourth power of the energy, it is acquiring. This in turn sets limitations on the energies to which electrons can be accelerated in a betatron. The maximum energy is reached when the energy lost per turn by radiation equals the maximum energy acquired by the electron per turn under practical conditions. Maximum electron energy can be reached to about 1000 MeV.

25.21.3 Uses and Limitations

Betatrons can energise electrons to the value of about 1000 MeV and are quite successful accelerators for electrons. Betatron is relatively small size accelerator and therefore quite suitable for laboratory work. Apart from its efficient use as an electron accelerator, it is quite useful and powerful source for X-ray generator. This is why betatron is used in hospitals for *therapeutic X-rays.* We may note that betatron can produce X-rays so short in wavelength as to surpass the γ-rays from radioactive sources.

Betatron suffers from the problem of *orbital stability.* One can achieve the stability by properly designing the machine. The main drawbacks of this accelerator are: (i) *huge magnet* (diameter exceeding ~2 m) is required for supplying the variable flux to accelerate the electrons (ii) an accelerated electron loses energy by radiation and the rate of loss of energy increases with the fourth power of energy so that a limit is set for the high energy electrons by *radiation loss,* and (iii) the maximum energy that a betatron under practical condition prevailing in the machine is about 1 BeV (=1000 MeV).

The disadvantage a betatron suffers from accelerating electrons have been eliminated to a great extent in *electron-synchrotron.* In fact, in electron synchrotron, one combines the accelerating system of the cyclotron with the ring-shaped pulsating guide field of a betatron.

A comparison of cyclotron and betatron is presented in Table 25.4.

Table 25.4 A comparison of Cyclotron and Betatron

Cyclotron	*Betatron*
(i) In a fixed frequency cyclotron alternating magnetic field is not employed, *i.e.,* magnetic field is constant.	(i) An oscillating field is used in betatron. This field is also called *induction field.*
(ii) In a cyclotron, particles are accelerated in helical orbits and the orbital radius goes on increasing and equals the radius of the dees in the limit.	(ii) In a betatron, the acceleration of the particles takes place in a circular orbit of almost constant radius.
(iii) For accelerating electrons a fixed-frequency cyclotron is not suitable.	(iii) Betatron is specially designed and meant for accelerating electrons.
(iv) A cyclotron suffers from *orbital stability* problem.	(iv) Betatron also suffers from the problem of orbital stability.
(v) Cyclotron cannot be used as an *X-ray* generator.	(v) Betatron is highly efficient generator for *X-rays.*

25.22 NUCLEAR STABILITY

Presently more than 2500 nuclides are known to us. Out of these only 300 nuclides are stable and rest are unstable. This shows that not all combinations of protons and neutrons form stable nuclei. For stable nuclei, a plot of neutron number $N = (A - Z)$ is plotted against the proton number Z (Fig. 25.27).

We may note from the graph that light stable nuclei fall on the line $N = Z$. Obviously, in such nuclei, the number of neutrons $(N = A - Z)$ is equal to the number of protons (Z). Heavy stable nuclei containing too many neutrons compared to the number of protons lie above the $N = Z$ line. One can understand this as follows: Light nuclei contain relatively small number of protons and therefore the disruptive Coulomb repulsive forces are less effective. This is why a relatively small number of neutrons is sufficient to provide attractive nuclear forces responsible for stability of these nuclei.

The heaviest known stable nucleus is ${}^{209}_{83}\text{Bi}$. Nuclei with $Z > 83$ or $A > 209$ are all unstable nuclei. The reason for the existence of this limit for stability of heavier nuclei lies in the fact that *nuclear forces are short range forces and also exhibit the property of saturation,* One immediate consequence of the saturation property of the nuclei is that the binding energy is proportional to A whereas the repulsive Coulomb forces are long range forces and moreover the electrostatic energy is proportional to the number of proton pairs, *i.e.,* equal to $\frac{1}{2}Z(Z-1)$ with the increase of number of nucleons, a situation, is reached where binding energy is not sufficient to overcome the repulsive Coulomb forces and therefore the nucleus becomes unstable. At $A = 210$ this critical situation reaches.

We may note that all the isotopes of a stable nucleus may not be stable, e.g. ^{206}Pb is stable but ^{210}Pb and ^{214}Pb are unstable. Interestingly, this happens not only with heavy nuclei but with all light nuclei too. Moreover, for maximum stability, there is an optimum value of neutron/proton ratio.

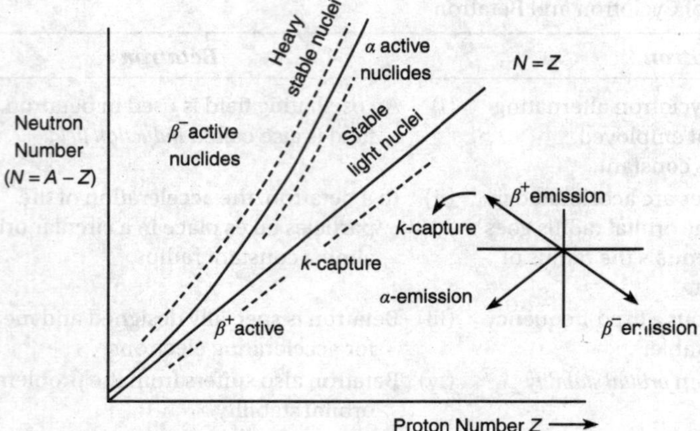

Fig. 25.27 A plot of neutron numbers $N = A - Z$, with proton number Z

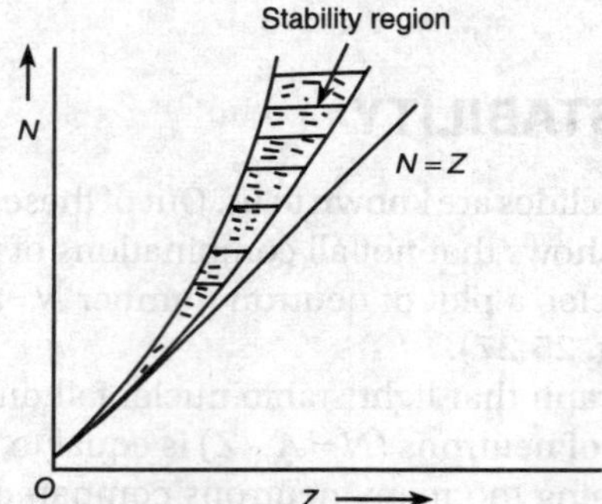

Fig. 25.28 Plot of number or neutrons required for maximum stability (shaded region)

We may note that nuclei which fall above and below the shaded region (Fig. 25.27) are unstable. Interestingly, artificial radioactive nuclei lie at the fringe of the region of stability shown. We may further note that an unstable nucleus emits particle and the new nucleus so formed lies closer to the region of stability.

Example 1: The half life of radioactive cobalt 60 is 5.26 years. Calculate the activity of one gram sample of cobalt. Also express the activity in Curie and Rutherford.

Solution

Activity $A = \lambda N$

$$\lambda = \frac{0.693}{T_{1/2}} = \frac{0.693}{5.26 \times 365 \times 24 \times 60 \times 60} \text{sec}^{-1} = 4.17 \times 10^{-9} \text{ sec}^{-1}$$

For cobalt, $N = \dfrac{6.02 \times 10^{23}}{60} = 1.002 \times 10^{22}$ dis./s

$$A = \lambda N = 4.17 \times 10^{-9} \times 1.002 \times 10^{22} \text{ dis/s} = 4.2 \times 10^{13} \text{ dis/s}$$

Activity (A) in Curie $= \dfrac{4.2\times10^{13}}{3.7\times10^{10}} = 1130$ Curie

Activity (A) in Rutherford $= \dfrac{4.2\times10^{13}}{10^{6}} = 4.2 \times 10^{7}$ rd.

Example 2: **A wooden piece of great antiquity weights 50 gm and shows ^{14}C activity of 320 dis./minute. Estimate the length of the time which has elapsed since this wood was part of a living tree. Assume that living tree show a ^{14}C activity of 12 dis./minute/gm. Given half life of ^{14}C is 5730 years.**

Solution Let us suppose that living tree had N_0 radioactive atoms just before it died. Its activity was

$$A_0 = \lambda N_0$$

When the tree died, the activity of the tree decreases exponentially. Thus, we have

$$A = -\frac{dN}{dt} = \lambda N = \lambda N_0 e^{-\lambda t} \quad \text{... (i)}$$

$$\therefore \quad \frac{A}{A_0} = e^{-\lambda t}$$

Given, A_0 = dis./min/gm.

$$A = \frac{320}{50} \text{ dis./min/gm}$$

and
$$\lambda = \frac{0.693}{5730} / \text{year}$$

From Eq. (i) we have

$$\log e\left(\frac{A}{A_0}\right) = -\lambda t$$

or
$$t = \frac{1}{\lambda} \times 2.303 \log_{10}\left(\frac{A_0}{A}\right)$$

$$= \frac{5730}{0.693} \times 2.303 \log_{10}\left(\frac{50\times12}{320}\right) \text{year} = 5170 \text{ years.}$$

Example 3: **Calculate the Q valve of the reaction $^{14}_{7}Na$ (α, p) $^{17}_{8}O$. Given the masses as follows:**

Mass of $^{14}_{7}Na$ = 14.00753 u

Mass of $^{4}_{2}He$ = 4.00386 u

Mass of $^{17}_{8}O$ = 17.00450 u

Mass of $^{1}_{1}H$ = 1.00813 u

Solution The given reaction is

$$^{14}_{7}\text{Na} + ^{4}_{2}\text{He} \rightarrow ^{17}_{8}\text{O} + ^{1}_{1}\text{H} + Q$$

Now, substituting the given values of masses, one obtains

$$14.00753 + 4.00386 = 17.00450 + 1.00813 + Q$$

or $$Q = 14.00753 + 4.00386 - 17.00450 - 1.00813$$

$$= -0.00124 \text{ u} = -0.00124 \times 931 \text{ MeV} = -1.154 \text{ MeV}.$$

Obviously, the reaction is **endoergic.**

Example 4: Assuming that energy released by the fission of a single uranium atom is 200 MeV, calculate the number of fissions per sec required to produce 1 Watt of power.

Solution Given energy released per fission

$$= 200 \text{ MeV} = 2 \times 10^8 \times 1.6 \times 10^{-19} \text{ J}$$

$$= 3.2 \times 10^{-11} \text{ J} = 3.2 \times 10^{-11} \text{ W}$$

Number of fissions required to produce 1 Watt of power

$$\frac{1}{3.2\times10^{-11}} = 3.125 \times 10^{10} \text{ fissions}$$

Example 5: Calculate the energy released in the fission of the following reaction:

$$^{235}_{92}\text{U} + ^{1}_{0}n \rightarrow ^{95}_{42}\text{Mo} + ^{139}_{57}\text{La} + 7\left(^{0}_{-1}e\right) + 2\,^{1}_{0}n + Q$$

Given: Mass of: $^{235}\text{U} = 235.0439$ u
Mass of neutron: $(^{1}_{0}n) = 1.0087$ u
Mass of: $^{139}\text{La} = 138.9061$ u

Solution
Total initial mass = 236.0526 u
Total final mass = 235.8332 u
Mass converted into the energy

$$\Delta m = 236.0526 - 235.8332 = 0.219 \text{ u}$$

∴ Energy equivalent

$$= 0.219 \times 931 = 204.3 \text{ MeV}$$

∴ $$Q = 204.3 \text{ MeV}$$

Example 6: A power reactor is developing energy at the rate of 30000 kW. How many atoms of ^{235}U undergo fission per second? How many kg of ^{235}U would be consumed in 1000 hours of operation. Assuming that on an average 200 MeV energy is released per fission?

Solution Given that the rate of development of energy by power reactor = 30000 kW $= 3 \times 10^7$ W

$$= 3 \times 10^7 \text{ J/s}$$

Energy released per fission = 200 MeV

$$= 200 \times 10^6 \times 1.6 \times 10^{-19} \text{ J} = 3.2 \times 10^{-11} \text{ J}$$

Number of atoms undergoing fission/s

$$= \frac{3\times10^7}{3.2\times10^{-11}} = 9.6 \times 10^{17}$$

Number of atoms whose fission takes place in 1000 hours

$= 9.6 \times 10^{17} \times 1000 \times 60 \times 60 = 3.384 \times 10^{23}$

$\therefore$ $= 6.02 \times 10^{23}$ atoms of ^{235}U weight 235×10^{-3} kg

$\therefore$ $= 3.384 \times 10^{23}$ atoms of ^{235}U weight will weight

$$= \frac{235\times10^{-3}}{6.02\times10^{23}}\times3.384\times10^{23} = 132.1 \times 10^{-3}\text{ kg} = 0.1321\text{ kg.}$$

Example 7: Calculate the energy liberated when a single helium nuclei is formed by the fusion of two deuterium nuclei.

Solution The desired equation is

$${}_1^2\text{H} + {}_1^2\text{H} \rightarrow {}_2^4\text{He} + Q$$

$$(\text{D}) \quad (\text{D}) \quad (\alpha)$$

Mass of 2D atoms $= 2 \times 2.01478 = 4.02956$ u

Mass of ${}_2^4$He atom $= 4.00388$ u

Mass defect $= (4.02956 - 4.00388)$ u

$= 0.02568$ u

Energy released, $Q = 0.02568 \times 931.8$ MeV $= 24$ MeV (appx.)

Example 8: Calculate the power output of a nuclear reactor which consumes 10 kg of ^{235}U per day. Given that the average energy released per ^{235}U fission is 200 MeV.

Solution Number of atoms in 1 kg of ^{235}U

$= 6.02 \times 10^{26}$, i.e., Avagadro number

$$= \frac{6.02\times10^{26}}{235}$$

or Number of atoms in 10 kg $= \dfrac{6.02\times10^{26}}{235} \times 10 = 2.56 \times 10^{25}$

Since energy released per fission = 200 MeV, therefore fission energy produced by these atoms $= 2.56 \times 10^{25} \times 200$ MeV

$= 2.56 \times 10^{25} \times 200 \times 1.6 \times 10^{-19}$ J

$= 819.2 \times 10^{12}$ J

Time taken to consume 10 kg uranium 235

$= 1$ day $= 24 \times 3600$ sec

$\therefore$ Power produced

$$= \frac{819.2\times10^{12}}{24\times3600} = 9.48 \times 10^{9}\text{ W}$$

Example 9: The magnetic flux density in a cyclotron is $2T$. What should be the frequency in order to accelerate protons or charge 1.6×10^{-19} C and mass 1.79×10^{-27} kg ?

Solution

$$n = \frac{Bq}{2\pi m} = \frac{2\times1.6\times10^{-19}}{2\times3.14\times1.79\times10^{-27}} = 2.845 \times 10^7 \text{ Hz} = 28.45 \text{ MHz}$$

Example 10: **Prior to emerging from a cyclotron deutrons describe a circle of radius 0.32 m. The frequency of the applied voltage is 10 MHz. Calculate the speed of the deutrons emerging from the cyclotron. You may neglect relativistic effect.**

Solution

$$f = \frac{Bq}{2\pi m} \quad \text{... (i)}$$

$$R = \frac{mv}{BQ} \quad \text{... (ii)}$$

Multiplying (i) and (ii), we have

$$Rf = \frac{v}{2\pi}$$

$$\therefore \quad v = 2\pi Rf = 2\pi \times 0.32 \times 10 \times 10^6 = 2.01 \times 10^7 \text{ m/s.}$$

Example 11: **A betatron has a magnetic current supply of frequency 60 Hz and the peak magnetic flux density at the orbit is 0.5 Wb /m². If the radius of the electron orbit is 0.75 m. Determine (i) The final energy of the electrons and (ii) The total time of flight of electrons.**

Solution We have the momentum of electron at the orbit

$$p = B\,er$$

We have to treat the electron relativistically. The final energy of the electrons is

$$E = mc^2 = pc = B\,erc$$

$$= 0.5 \times 1.6 \times 10^{-19} \times 0.75 \times 3 \times 10^8 \text{ J}$$

$$= \frac{0.5\times1.6\times10^{-19}\times0.75\times3\times10^8}{1.6\times10^{-19}} \text{ eV}$$

$$= 1.25 \times 10^8 \text{ eV} = 112.5 \text{ MeV}$$

The total number of revolutions n by the electron during acceleration time (one quarter cycle of magnetic flux) is

$$n = \frac{c}{4\omega r_0} = \frac{c}{8\pi v r_0} = \frac{3\times10^8}{8\times3.14\times60\times75} = 2.65\times10^5$$

Total time of flight of electron

$$= \frac{\text{Total distance covered}}{\text{velocity}}$$

$$= \frac{2\pi r_0 n}{c}$$

$$= \frac{2.65\times10^5\times3.14\times0.75}{3\times10^8} = 4.16 \times 10^{-3} \text{ s.}$$

Example 12: The maximum magnetic field at the orbit in a certain betatron was 0.4 T and operating at 50 c/s with a stable orbit of diameter 1.5 m. Calculate the average energy gained per revolution and the final energy of the electrons. You may assume that the electron moves nearly with the speed of light.

Solution The distance traversed by electrons during acceleration time in one quarter cycle is

$$d = \frac{1}{4}cT = \frac{c}{4f} \text{ where } f \text{ is the frequency of A.C.}$$

Let r be the radius of the stable orbit, then the number of revolutions made by electrons in one quarter cycle is given by

$$n = \frac{d}{2\pi r} = \frac{c}{2\pi rf} = \frac{3\times10^8}{8\times3.14\times0.75\times50} = 3.18 \times 10^5$$

Now, if E be the final energy acquired, the momentum p is given by,

$$p = \frac{E}{c} \text{ (relativistic)}$$

If B be the magnetic field at the *orbit* and v the final velocity acquired, we have

$$Bev = \frac{mv^2}{r}$$

or $$p = mv = Ber = E/c$$

∴ Final energy $$E = \text{B erc} = 0.4 \times 1.6 \times 10^{-19} \times 0.75 \times 37 \times 10^8$$

$$= 1.44 \times 10^{-11}\text{ J}$$

$$= \frac{1.44\times10^{-11}}{1.6\times10^{-13}}\text{ MeV} = 90\text{ MeV}$$

∴ Average energy/revolution

$$= \frac{90\times10^6}{3.18\times10^5}\text{ eV} = 283\text{ eV.}$$

Example 13: Find the number of ion pairs by 10 MeV proton. If in the proportionality region the amplification is 10^3, current pulse time is 10 μs and resistance between electrodes is 10^4, find voltage pulse height. The amount of energy required to produce one in pair is 34 eV.

Solution Number of ion pairs by 10 MeV protons

$$n = \frac{10\times10^6}{34} = 2.94 \times 10^5$$

Total ion pairs in proportionality region

$$= 2.94 \times 10^5 \times 10^3 = 2.94 \times 10^8$$

Hence charge on electrode due to these pair

$$q = 2.94 \times 10^8 \times 1.6 \times 10^{-19}\text{ Coulomb}$$

or $$q = 4.704 \times 10^{-11}\text{ C}$$

$\therefore$ $$\text{Current} = \frac{q}{t} = \frac{4.704\times10^{-11}}{10\times10^{-6}}$$

$$= 4.704 \times 10^{6} \text{ ampere}$$

$\therefore$ Voltage pulse height $= i \times R$

$$= 4.704 \times 10^{6} \times 10^{4} = 4.704 \times 10^{-2}$$

$$= 0.47 \text{ V}$$

Example 14: A halogen quenched G.M. counter works at 1 kV. Anode wire has radius 0.2 mm and radius of cathode is 20 mm. The guarantee period for the counter is 10^9 constant. Find maximum radial field. If the counter is operated 30 hrs a weak and count 3000 counts/min, find the life of the counter (take 1 year = 50 weeks).

Solution The field at an axial distance r is given by

$$E_r = \frac{V_0}{e\log_e(b/a)}$$

This would be maximum near the surface of wire

$\therefore$ $$(E_r)_{\max} = \frac{1\times10^{3}}{0.2\times\log_e 100}$$

Here $b = 20$ mm, $a = r = 0.2$ mm

or $$(E_r)_{\max} = \frac{1\times10^{3}}{0.2\times2.3026\times2}$$

$$= 1.087 \times 10^{3} \text{ V/mm}$$

If its life is x, we have

$$10^{9} = x \times 50 \times 30 \times 60 \times 3000$$

$$x = \frac{10^{3}}{5\times6\times3\times3} = \frac{1000}{270} \text{ yrs} = 3.7 \text{ yrs.}$$

SUMMARY

1. The primary constituents of nuclei are the photon and the neutron, and the nuclear mass is roughly sum of its constituent proton and neutron masses (the slight-different being the binding energy of the nucleus).
2. Each nuclear species with a given Z and A is called a **nuclide**.
3. A nuclear species can be represented by ${}_{Z}^{A}X$, where A is **mass number, i.e.,** the total number of nucleons, and Z is the **atomic number** i.e., the total number of protons. The total number of neutrons in a nucleus is the neutron number N, where $A = N + Z$.
4. Nuclides with the same Z but different A and N values are called **isotopes,** e.g. ${}_{1}^{2}\text{H}$, ${}_{1}^{3}\text{H}$ are isotope of ${}_{1}^{1}\text{H}$.
5. Nuclides with the same neutron number are called **isotones,** e.g. ${}_{6}^{14}\text{C}$, ${}_{7}^{15}\text{N}$, ${}_{8}^{16}\text{O}$ and ${}_{9}^{17}\text{F}$.
6. Nuclides with the same value of A are called isobars, e.g. ${}_{6}^{16}\text{C}$, ${}_{7}^{17}\text{N}$ and ${}_{8}^{16}\text{O}$.

7. Masses are measured in terms of atomic mass unit u.
8. Assuming that a nucleus is spherical, its radius is

$$r = r_0 A^{1/3}, \text{ where } r_0 = 1.2\, fm$$

9. Electron scattering is useful to measure the size and shape of nuclei.
10. The properties of nucleons are

Property	*Neutron*	*Proton*
Mass (u)	1.008665	1.007276
Change (e)	0	+1
Spin ($\hbar$)	½	½
Magnetic moment ($e\hbar/2m_p$)	- 1.91	+ 2.79

11. Nuclei are stable because of the **nuclear force** between nucleons. This short range (~ 3fm) force dominates the Coulomb repulsive force at distances of less than about 2 fm and is independent of charge and the study of the deutron and nucleon–nucleon scattering indicates that the nuclear force is attractive.
12. **Stable isotopes** require a delicate balance between protons and neutrons, with near equal numbers for ligher stable nuclei and more neutrons for heavier nuclei.
13. **Unstable** isotopes are radioactive and decay be shedding particles.
14. The curve of **binding energy** shows that energy can be released by either **fusion** of ligher nuclei or **fission** of heavier nuclei. The total binding energy for a nuclide is

$$B\left({}_Z^AX\right) = \left[Nm_n + Zm({}^1\text{H}) - M\left({}_Z^AX\right)\right]\ c^2$$

15. The von Weizsocker semi-empirical mass formula is useful in predicting the nuclear binding energy. There are no stable nuclei with $Z > 83$ or $A > 209$. Nuclei tend to be more stable with an even number of protons and/or neutrons. Nuclei near ^{56}Fe have the highest binding energy per nucleon, and the average binding energy per nucleon for most nuclei is about 8 MeV.
16. Radioactive processes include α-decay, β-decay and γ-decay. **Alpha** decay emits a ^{4}He nucleus:

$${}_Z^AX \rightarrow {}_{Z-2}^{A-4}Y + {}_2^4\text{He}$$

Beta decay emits an electron or a positron and antineutrino or a neutrino

$${}_Z^AX \rightarrow {}_{Z+1}^{A}Y + {}_{-1}^{0}e\ (\beta^-) + \overline{\nu}$$

17. In the **electron capture** process, the nucleus of an atom absorbs one of its own electrons (usually from the K shell) and emits a neutrino.
18. **Gamma ray** emits a high energy photon (gamma ray) when an excited nucleus drops to a lower energy state.

$$X^* \rightarrow X + \gamma$$

19. If a radioactive material contains N_0 radioactive nuclei at $t = 0$, the number N of nuclei containing at time t is

$$N = N_0\, e^{-\lambda t}$$

where λ is **decay constant**. The **decay rate** or **activity,** of a radioactive substrate is given by

$$R = \left|\frac{dN}{dt}\right| = N_0 \lambda e^{-\lambda t} = R_0 e^{-\lambda t}$$

Where $R_o = N_0\lambda$ is the activity at $t = 0$.

20. The half life $T_{1/2}$ is defined as the time interval required for half of a given number of radioactive nuclei to decay, where

$$T_{1/2} = \frac{\ln 2}{\lambda} = \frac{0.693}{\lambda}$$

21. Radioactivity is measured in **becquerels,** with 1 Bq equals to one decay per second. Sieverts (S_v) measure the biological effects of radiation.

22. There are only four radioactive series. For e.g., two of them begin with uranium isotopes, ^{235}U and ^{238}U.

23. Radioisotopes are useful to date objects like the age of the earth and ancient objects. Radiocarbon ^{14}C is one of the most useful.

24. Nuclear reactions can occur when the target nucleus X is bombarded by a particle 'a', resulting in a nucleus 'y' and an out going particle 'b' :

$$a + x \to y + b \quad \text{or} \quad X(a, b)\, y$$

The rest energy transformed to kinetic energy in such a reaction, called the **reaction energy** Q, is

$$Q = [M_a + M_x - M_y - M_b]\, c^2$$

A reaction for which Q is positive is called **exothermic**. A reaction for which Q is negative is called **endothermic**. The minimum kinetic energy of the incoming particle necessary for such a reaction to occur is called the **threshold energy**.

25. Different nuclear reaction mechanisms include compound nucleus, coulomb excitation, and direct reactions, among others. Nuclei have excited states, which may appear as resonances in a compound nucleus reaction. The lifetimes T of nuclear states are related to their width T by the uncertainty principle $T_T \approx \hbar/2$.

26. Heavy nuclei, e.g. $^{235}_{92}U$ and $^{239}_{94}Pu$ may fission into two nearly equal fission fragments when struck by low energy neutrons

$$^1_0n + {}^{235}_{92}U \to X + Y + b\,^1_0n$$

The extra neutrons produced in fission can sustain a **chain reaction** provided there is a **critical mass** of fissile material. Exponentially growing chain reactions power fission weapons, while controlled fission occurs in **nuclear reactors** used for power generation.

27. One may built nuclear reactions in different ways for special purposes. a breeder reactor produces more fissionable fuel than it consumes.

28. Nuclear fusion powers the sun and stars but has proved elusive on earth except in thermonuclear weapons. Isotopes of hydrogen, ^2_1H and ^3_1H, appear to be most useful. Although fusion is the source of sun's energy, it has yet been controlled on Earth. Inertial confinement or magnetic confinement fusion may one day provide us with nearly limitless energy.

29. Nuclear science have plentiful applications and include medicine, art, archaeology, crime detection, mining, oil, mineral studies and production, and small power systems.
30. The search for new elements continues in the quest for the island of stability for super heavy elements.

REVIEW QUESTIONS

1. What is radioactivity? What types radiations are emitted in radioactive disintegration?
2. What is radioactive decay? Explain the laws of radioactive disintegration.
3. What is half life? Obtain an expression of mean life time of radioactive element in terms of half life time.
4. What do you understand by activity of a radioactive substance ? Show how it varies?
5. Explain the terms decay constant, half life and mean life.
6. Define the Curie and Bacquerel.
7. Explain, how is the rate of disintegration affected when a solution of radioactive substance is heated?
8. Write notes on (i) activity of radioactive substances, (ii) natural radioactivity, (iii) radioactive dating (age of the earth).
9. What are radioactive isotopes? Give some of their uses.
10. Write notes on (i) Radioactive tracer technique and its uses (ii) uses of radioactive isotopes in medicine, agriculture and industry (iii) C-14 dating.
11. What are transuranic elements?
12. What are nuclear reactions? Give some examples of nuclear reactions produced by $\alpha, p, {}_1^2\text{H}, n$ and photon.
13. What do you understand by the Q-value of a nuclear reactions? Calculate Q value for the following reaction.

$${}_2^4\text{He} + {}_7^{14}\text{N} \rightarrow {}_8^{17}\text{O} + {}_1^1\text{H} + Q$$

Given: Mass of ${}_2^4\text{He}$ = 4.00388 u, ${}_7^{14}\text{N}$ = 14.00755 u, ${}_8^{17}\text{O}$ = 17.00453 u, ${}_1^1\text{H}$= 1.00815 u [–1.16 MeV]

14. What are exothermic (exoergic) and endothermic nuclear reactions?
15. Explain the terms: Mass defect, packing fraction, binding energy and bindind energy per particle. What is the significance of binding energy?
16. Explain the meaning of 1 u. What is the energy equivalent of 1 u in MeV. [931 MeV]
17. Draw a curve showing the variation of binding enegy per nucleon against the mass number, Use this curve to explain the release of energy in fusion of light nuclei and in fission of heavy nuclei.
18. The principle of releasing nuclear energy is the same in a nuclear reaction and in an atom bomb. Why then the control rods are used only in a nuclear reactor?
19. What is nuclear fission ? Explain it on the basis of liquid drop model of the nucleus. Why is it possible to product the fission of ^{235}U with slow neutrons whereas it is necessary to use fast neutróns to produce the fission of ^{238}U ?
20. What is a chain reaction? Explain uncontrolled and controlled chain reactions. Mention the condition for a self sustained chain reaction and how they are achieved?
21. Explain the function of a moderator to sustain a chain reaction?

22. Draw a labelled diagram of a nuclear reactor and explain its working and mention its important uses.
23. What is a nuclear fusion process? Give few examples ?
24. What is the difference between nuclear fission and fusion.
25. Explain the origin of solar energy.
26. Explain the principle of breeder reactor.
27. Write short note on: G.M. counter.
28. Explain construction and working of Geiger Muller counter.
29. Write short note on Scintillation counter.
30. Write short note on Photomultiplier tube.
31. (a) What do you mean by "dead time" in GM counter? Mention some of its applications. Draw a neat diagram of the GM counter
    (b) Describe the construction and working of a scintillation counter.
32. Through a schematic diagram, explain operation of a scintillation counter. How shall you select phosphor for protons, β-particles, slow neutrons and fast neutrons. Obtain an expression for output voltage pulse when radiation of energy E falls on the counting system.
33. Describe construction and working of Geiger-Muller counter. Explain the terms: dead time and quenching.
34. Draw curves to show qualitatively the variation of ionization current (assuming incident particle energy E in one case and $2E$ in another) versus voltage applied to an ionization chamber. Indicate operation regions of proportional and Geiger-Muller counters and give physical explanation to curves. What factors govern the counting rate of a GM counter?
35. Explain the principle of cyclotron. Give its construction, working and theory. Derive the relation for the energy of the particles emerging from a cyclotron.
36. Show that the kinetic energy of the particle in a cyclotron is independent of the voltage applied.
37. Wrtie note on:
    (i) Nuclear reactor (ii) Betatron
    (iii) Cyclotron (iv) Nuclear stability
    (v) Nuclear fission (vi) Nuclear fusion
    (vii) Fast breeder reactor (viii) Nuclear binding energy curve
    (ix) Geiger Muller counter (x) Nuclear models

PROBLEMS

1. ^{40}K has a half life of 4×10^8 year. Find its decay constant. [**Ans.** 1.75×10^{-9}/year]
2. The half-life of radium is 1600 years. How long it will take for 15/16 of a sample of radiation to decay? [**Ans.** 1600 year]
3. The half-life of radon in 3.8 days. Find when its 1/2th would remain. [**Ans.** 14.45 days]
4. Find decay constant of the radioisotope whose half-life is 5 hours. [**Ans.** 3.85×10^{-5}/s]
5. The half-life of tritium is 12.5 years. Find the fraction of the amount of a pure sample after 25 years. [**Ans.** 1/4th]
6. Calculate the B.E./nucleon for (a) ^{36}S (b) ^{42}Ca.

7. $_1^3$H and $_2^3$He have masses 3.016050 u and 3.016030 u respectively. Find their binding energies in MeV. What is the cause of the difference between their values? **[Ans.** 8.48 MeV, 8.65 MeV]
8. The mass of $_{17}^{35}$Cl is 34.9800 u. Calculate its binding energy and B.E. per nucleon. Given that mass of $_0^1n = 1.008665$ u and $_1^1\text{H} = 1.007825$ u. **[Ans.** 288 MeV, 8.22 MeV]
9. The binding energy of $_{17}^{35}$Cl is 298 MeV, calculate is mass in u. **[Ans.** 34.97 u]
10. Calculate the Q value of the following nuclear reaction:

$$_3^7\text{Li} + _1^1\text{H} \rightarrow _2^4\text{He} + _2^4\text{He} + Q$$

Mass of $_3^7\text{Li} = 7.016005$ u, $_1^1\text{H} = 1.007825$ u and $_2^4\text{He} = 4.00260$ u. **[Ans.** 17.34 MeV]
11. Calculate the Q valve of the reaction

$$_4^9\text{Be} + _2^4\text{He} \rightarrow {}_6^{12}\text{C} + _0^1n + Q$$

Mass of $_4^9\text{Be} = 9.015060$ u, and $_2^4\text{He} = 4.006674$ u, $_0^1n = 1.008986$ u, and $_6^{12}\text{C} = 12.003316$ u. **[Ans.** 10.83 MeV]
12. The fission of $_{92}^{235}$U releases approximately 200 MeV. What percentage of original mass $_{92}^{235}\text{U} + _0^1n$ disappears? **[Ans.** 0.1%]
13. A reactor is developing nuclear energy at the rate of 30,000 kW. How many kg of ^{235}U would be used in 1000 hours of operating that on an average energy of 200 MeV is released per fission. **[Ans.** 0.132 kg]
14. What is the energy released when 1 kg of nuclear fuel is consumed if the following fusion reaction is possible.

$$_1^2\text{H} + _1^2\text{H} \rightarrow _2^4\text{He}$$

[Ans. 5.74×10^{14} J]

SHORT ANSWER-QUESTIONS

1. What hold the nucleons together inside a nucleus?
Ans: Nuclear force.
2. What is the relation between the radius r of a nucleus and its nucleons A.
Ans: $r = r_0 A^{1/3}$
3. What is artificial transmutation of elements ?
Ans: An artificial disintegration of nuclei is a process in which a nucleus is transformed into a different species by its reaction with an energetic particle or photon.
4. What are transuranic elements?
Ans: Artificially produced elements beyond $Z = 92$ produced in the laboratory by the bombardment of certain heavier nuclei with appropriate particles are called transuranic elements.
5. Do the nuclear forces obey inverse square law ?
Ans: No
6. What is artificial transmutation ?
Ans: Artificial transmutation is the process of transformation of one element into another by artificial means of bombarding it with high energy particles.

7. Write five examples of isotopes.

Ans: Isotopes of Hydrogen $_1^1H$, $_1^2H$ and $_1^3H$

Isotopes of Oxygen $_8^{16}O$, $_8^{17}O$ and $_8^{18}O$

Isotopes of Chlorine $_{17}^{35}Cl$ and $_{17}^{37}Cl$

Isotopes of Potassium $_{19}^{39}K$ and $_{19}^{40}K$

Isotopes of Uranium $_{92}^{235}U$ and $_{92}^{238}U$

8. Write four examples of isobars.

Ans: $_6^{13}C$, $_7^{13}N$ $_{19}^{40}K$, $_{20}^{40}Ca$ $_{18}^{40}A$, $_{20}^{40}Ca$ $_{12}^{24}M$, $_{11}^{24}Na$

9. Give one example of isotone.

Ans: $_{14}^{31}Si$ and $_{15}^{32}P$

10. What do you understand by background radiations ?

Ans: Background radiations are those radio-active radiations that everyone is exposed to due to the presence of natural radioactive substances on the earth as well as the cosmic radiations, and the radioactive substances present in the body.

Internal background radiations are from the radioactive substances K-40, and C-14 present in the entire body. External background radiations are from cosmic rays and the radiations from the radioactive rocks on the earth.

11. What is a nuclear reactor?

Ans: It is a device in which a self sustaining controlled chain reaction is produced in a fissionable material. Obviously, a nuclear reactor or a nuclear pile is a source of controlled release of nuclear energy which is utilized for several useful purposes.

12. What are the essential units of a nuclear power plant ?

Ans: (i) Nuclear reactor (ii) Heat exchanger (iii) Steam turbine (iv) Electric generator or dynamo.

13. What is nuclear fusion?

Ans: A nuclear reaction in which lighter atoms combine together to form heavier nuclei resulting in the release of tremendous amount of energy.

14. The binding energy of $_3^7Li$ and $_2^4He$ are 39.2 MeV and 28.24 MeV respectively. Which one of the two nuclei is more stable ?

Ans: B.E. per particle of $_3^7Li = \frac{39.2}{7} = 5.6$ MeV

B.E. per particle of $_2^4He = \frac{28.24}{4} = 7.06$ MeV

Helium is more stable because its B.E. per particle is more.

15. What is hydrogen bomb?

Ans: Like atom bomb, hydrogen bomb is a weapon of mass destruction, employing thermonuclear reactions, which can release enormous energy within few minutes. The extremely high temperature required for the uncontrolled fusion reactions in a hydrogen bomb is produced by a fusion bomb containing ^{235}U or ^{239}Pu. Obviously, hydrogen bomb is a fission – fission bomb.

16. What is the order or the magnitude of temperature enabling a fusion reaction to take place?
Ans: About 10^6K.

17. What are fundamental pars of a nuclear reactor used for generation of electricity?
Ans: (i) Nuclear fuel, (ii) Moderator, (iii) Coolant (iv) Shield (v) Controlling rods.

18. What is the function of a moderator in a nuclear reactor?
Ans: Moderator is used to slow down the neutrons. Heavy water, graphite or beryllium oxide is used as moderator.

19. What is the function of controlling rods in a nuclear reactor?
Ans: Cadmium rods are used as controlling rods to control the rate of fission in the nuclear reactor. These rods absorb extra neutrons produced in the fission reaction.

20. What is a chain reaction?
Ans: A chain reaction is a series of nuclear fissions whereby the neutrons produced in each fission cause additional fissions releasing enormous amount of energy.

21. What do you understand by nuclear reaction?
Ans: Any reaction that tends to change the configuration of the nucleus of an atom can be termed as nuclear reaction.

22. How many types of nuclear reactions are there ?
Ans: (i) Nuclear fission reaction, and (ii) Nuclear fusion reactions.

23. What are fission reactions ?
Ans: Nuclear fission is a process in which the heavy nucleus of a radioactive substance like uranium is split into lighter nuclei by the bombardment of a low energy (slow moving) neutron, the reaction being accompanied by the release of energy and a two or three or more neutrons.

24. How U-235 isotope is used to determine the age of earth or rocks?
Ans: By determining the present ratio of U-235 to stable lead isotope ?

25. Give few examples of used of radioisotopes?
Ans: **(i)** $^{60}_{27}$Co is used to kill cancer cells in the blood of a human body.
(ii) $^{24}_{11}$Na is used to detect a clot in the blood in the human body
(iii) $^{131}_{53}$I is used to test the functioning of the thyroid gland.
(iv) $^{60}_{27}$Co like radio isotopes are used to detect the internal flow in a cast - materials.
(v) Radio-isotopes are used in agriculture in the selection of fertilizers.
(vi) Radio-isotopes are used by irradiating seeds to prevent unnecessary mutations of plants.
(vii) Perishable food items can be kept fresh for longer durations by exposing them to mild γ-radiations from a radio-isotope.
(viii) With the help of suitable detectors radioactive gases are injected into the atmosphere and study is made of the course of monsoon winds and the change they undergo.

26. What is carbon-dating ?
Ans: The process of estimating the age by measuring the disintegration of C-14 is known as carbon-dating.

27. Why are γ-rays emitted in nuclear processes and not in orbital electron transitions ?
Ans: The energy of γ-rays is of the order of MeV. Energies of this magnitude occur in nuclear processes and not in electron orbital transitions.

28. Why is the energy distribution of β-decay continuous ?
Ans: The phenomenon of β-decay arises due to the conversion of a neutron in the nucleus into a proton, electron and antineutrino. Because the energy –

$$_{0}^{1}n \rightarrow {}_{1}^{1}\text{H} + {}_{-1}^{0}e + \overline{\nu}$$

available in beta decay is shared by the electron and anti-neutrino in all possible ratios as they come out of the nucleus. This is why the β-ray spectrum is continuous.

FILL IN THE BLANKS

1. Nuclear forces are non ______ and non ______. **[electric, gravitational]**
2. $1\,\text{u} =$ ______ Mev. **[931]**
3. ${}_{17}^{35}\text{Cl} +$ ______ $= {}_{16}^{32}\text{S} + {}_{2}^{4}\text{He}$. **[${}_{0}^{1}n$]**
4. The total amount of energy released per fission of ^{235}U nucleus is roughly ______ MeV. **[200]**
5. The chain reaction is a process in which the nuclear ______ of an atom induces ______ in another atom which again induces fission in another atom and so on. **[fission, nuclear fission]**
6. Nuclear fusion is a process in which two ______ nuclei are ______ together to form a heavier and stable nucleus. **[lighter, fused]**
7. Source of steller energy is ______. **[nuclear fusion]**
8. In proton – proton cycle protons are continuously transformed into ______ nuclei. **[helium]**
9. In carbon-nitrogen cycle the synthesis of ______ nuclei with carbon (acting as a nuclear catalyst) takes place. **[hydrogen]**
10. A nuclear reaction is a device in which ______ is produced under a self sustaining controlled ______. **[nuclear fission, chain reaction]**
11. The penetrating power of α-particle is ______ than that of β-particle. **[less]**
12. Nuclei of heavy atoms contain more ______ than ______. **[protons, neutrons]**
13. The β-particle is emitted when a ______ inside the nucleus changes to ______. **[neutron, proton]**
14. Atoms of different elements whose nuclei have the same atomic weight but different atomic number are called ______. **[isobars]**
15. The radioactive isotope used in the treatment of cancer is a good source of ______ radiations. **[γ]**
16. Radioactive isotope of carbon is ______. **[^{14}C]**
17. ${}_{1}^{3}\text{H} +$ ______ $\rightarrow {}_{2}^{4}\text{He} + {}_{0}^{1}n + Q$. **[${}_{1}^{2}\text{H}$]**
18. ______ is responsible for keeping the nucleons together in a nucleus. **[nuclear force]**
19. The common product formed in the artificial transmutation by a proton is ______. **[helium]**
20. In plants and animals the ratio of ^{14}C to ^{12}C is ______. **[constant, i.e., 10^{-12}]**
21. When an α-particle is ejected, the atomic number of the atom decreases by ______. **[2]**

MULTIPLE CHOICE QUESTIONS

1. The nuclear radius may be approximated from a spherical charge distribution to be:
   (a) $R = r_o A^{2/3}$ (b) $R = r_o A^{1/3}$
   (c) $R = r_o A^{-2/3}$ (d) $R = r_o A^{-1/3}$
2. Nuclear Magneton is
   (a) $\mu_N = \frac{e\hbar}{2m_p}$ (b) $\mu_N = \frac{e\hbar}{2m_e}$
   (c) $\mu_N = \frac{e\hbar}{2}$ (d) $\mu_N = \frac{e\hbar}{2(m_p + m_e)}$
3. A deutron nucleus consists of
   (a) One proton and one electron (b) One neutron and one electron
   (c) One proton and one neutron (d) None of the above
4. In scattering neutrons from protons, a deutron is sometimes formed in the nuclear reaction
   (a) $n + p \rightarrow d$ (b) $n + p \rightarrow d + \gamma$
   (c) $n + p \rightarrow d + {}_{-1}^{0}e$ (d) $n + p \rightarrow d + {}_{+1}^{0}e$
5. The radioactive decay law is
   (a) $N(t) = N_o e^{\lambda t}$ (b) $N(t) = N_o e^{-\lambda t}$
   (c) $N(t) = \lambda N_o t$ (d) $N(t) = N_o e^{-\lambda t_{1/2}}$
6. The average (or mean) lifetime τ is calculated to be
   (a) $\tau = \frac{1}{\lambda}$ (b) $\tau = \lambda$
   (c) $\tau = e^{-\lambda}$ (d) $\tau = e^{\lambda}[1]$
7. Which one of the following is correct ?
   (a) $n \rightarrow p + \beta^- + \bar{\nu}$ (b) $n \rightarrow p + \beta^-$
   (c) $n \rightarrow p + \beta^- + \nu$ (d) $n \rightarrow p + \nu + \bar{\nu}$
8. Which one is a electron capture reaction ?
   (a) $p \rightarrow n + {}_{+1}^{0}e$ (b) $p + e^- \rightarrow n + \nu$
   (c) $p \rightarrow p + {}_{-1}^{0}e$ (d) ${}^{14}O \rightarrow {}^{14}N + \beta^+ + \nu$
9. Radioactive ^{14}C is produced in our atmosphere by the bombardment of ^{14}N by neutrons produced by cosmic rays. This can be expressed as
   (a) $n + {}^{14}N \rightarrow {}^{14}C + p$ (b) $n + {}^{14}N \rightarrow {}^{14}C$
   (c) $n + {}^{14}N \rightarrow {}^{14}C + {}_{+1}^{0}e$ (d) $n + {}^{14}N \rightarrow {}^{14}C + {}_{-1}^{0}e$
10. The total self energy of a sphere of radius R containing a charge Ze evenly spread throughout the sphere is given by
   (a) $\Delta E_{coul} = \frac{3}{5}\frac{(Ze)^2}{4\pi\varepsilon_0 R}$ (b) $\Delta E_{coul} = \frac{1}{2}\frac{(Ze)^2}{4\pi\varepsilon_0 R}$

(c) $\Delta E_{coul} = \dfrac{Ze}{4\pi \varepsilon_0 R^2}$ (d) $\Delta E_{coul} = \dfrac{1}{4}\dfrac{(Ze)^2}{4\pi \varepsilon_0 R}$

11. An example of induced fission is

(a) ${}^2_1H + {}^2_1H \rightarrow {}^1_0n + {}^3_1He + 3.3$ MeV (b) $n + {}^{235}_{92}U \rightarrow {}^{236}_{92}U^* \rightarrow {}^{90}_{40}Zr + {}^{134}_{52}Te + 3n$

(c) ${}^2H + {}^2H \rightarrow n + 4\,He + 17.6$ MeV (d) ${}^3He + {}^3He \rightarrow {}^4He + {}^1H + {}^1H + 12.86$ MeV

12. Which of the following radiations has the highest penetrating power?

(a) α (b) β (c) γ (d) X-rays

13. Which one of the following is an example of isobars –

(a) ${}^{14}_{13}Si$, ${}^{32}_{15}P$ (b) ${}^{14}_{7}N$, ${}^{15}_{7}N$

(c) ${}^{13}_{6}C$, ${}^{13}_{7}N$ (d) none of these

14. Higher value of decay constant indicates-

(a) slower decay (b) faster decay

(c) half life period (d) None of above

15. Which of the following is / are subatomic particles?

(a) positron (b) proton

(c) neutrino (d) all of the above

16. To convert ${}^{232}_{90}Th$ into ${}^{228}_{90}Th$, the least number of α and β particles that have to be emitted are –

(a) 2, 1 (b) 1, 1 (c) 1, 2 (d) 2, 2

17. ${}^{A}_{Z}X$ and ${}^{A}_{Z+1}Y$ are two radioactive elements. X emits one α-particle and Y-emits one β-particle. The daughter nuclei formed are –

(a) isotones (b) isotopes

(c) isotopes (d) none of the above

18. ${}^1_1H + {}^7_3Li \rightarrow$ - - - - - - + ${}^4_2He$

(a) ${}^1_1H$ (b) ${}^4_2He$ (d) ${}^9_4Be$ (d) ${}^8_4Be$

19. When ${}^{238}_{92}U$ disintegrates to form ${}^{206}_{82}Pb$, number of α and β particles emitted are –

(a) $8\alpha, 6\beta$ (b) $6\alpha, 8\beta$

(c) $8\alpha, 8\beta$ (d) $6\alpha, 6\beta$

ANSWERS

Question Numbers	1	2	3	4	5	6	7	8	9	10	11	12	13	14	15	16	17	18	19
Answers	(b)	(a)	(c)	(b)	(b)	(a)	(a)	(b)	(a)	(a)	(b)	(c)	(c)	(b)	(d)	(c)	(a)	(b)	(a)

26

Elementary Particles

26.1 INTRODUCTION

The discovery of electron by Thomson in 1898 and that of atomic nucleus and the proton by Rutherford in 1984 made it clear that the atoms themselves had an internal structure. The discovery of the neutron by Chadwick in 1932 in a way completed the picture of atom. By 1932 four "elementary" particles of matter were known: the electron, the proton, the neutron, and the neutrino. In addition, there were the positron, antiparticle to the electron, and the photon of electromagnetic radiation. Although, scientists thought of electrons, neutrons, and protons as elementary particles, because they thought that they are basic building blocks of matter. However, by 1960's with the advent of huge accelerating machines a plethora of particles, e.g. hyperons, kaons, leptons, and so on had been discovered. Today, several hundred particles known. Nowadays, the term elementary particle is used loosely to refer to hundreds of particles, most of which are unstable. We may note that adding energy to neutrons and protons can produce new elementary particles. Some of these particles are quite critical to man's understanding of matter and of the forces that hold matter together. There are four fundamental forces of nature: *gravity*, the *electromagnetic force*, the *nuclear force*, and the *weak force* that manifests itself in beta decay. The electromagnetic force is responsible for attracting the electrons to the nucleus, while the strong force (nuclear) is responsible for keeping neutrons and protons together within the nucleus.

Today, we know that neutrons and protons are made of even more fundamental particles called *quarks*. Quarks cannot be observed outside the nucleus, but it is believed that they must exist in order to explain observed experimental data.

We may not that there is an intertwined connection between cosmology with its grand dimensions, and particle physics, in which matter is studied in the smallest dimensions. Interestingly, the question of how the universe started and how universe may end is in fact lies in the mysteries of elementary particle physics. No doubt, the subject is complex, but of immense importance. The study of this field involves the meshing of relativity, quantum mechanics, electrodynamics, and gravitation. In this chapter we have tried to present some of the excitement and flavour of elementary particle physics.

26.2 FUNDAMENTAL INTERACTIONS OR FORCES IN NATURE

The number of elementary particles so far discovered is large, exceeding 200. The interactions amongst these particles may be considered as the direct source of the natural phenomena. Only four fundamental interactions known in the nature, e.g. strong, electromagnetic, weak and gravitational are apparently enough to account for all the physical processes and structures in the universe on all scale of size from hadrons to galaxies of stars. Table 26.1 summarized salient features of these basic interactions.

(i) Gravitational Interaction: Gravitation was the first interaction to be known and is usually discussed in terms of the forces between macroscopic objects or in terms of the general relativity. It operates quite certainly on elementary particle level, but it is so weak that there if any other interactions are present, it is over whelmed and (so far) is then undetectable. The strength of gravitational interaction

Table 26.1 Fundamental Interactions in Nature

Interaction	***Gravitational***	***Electromagnetic***	***Strong***	***Weak***
Relative magnitude	10^{-39}	10^{-3}	1	$10^{-13} - 10^{-11}$
Range	Infinite	Infinite	Short ~10^{-15} m	Short (~10^{-16} m)
Characteristic Time	10^{-16} sec	10^{-20} sec	10^{-23} sec	10^{-10} sec
Associated particles	Graviton	Photon	Pion, Kaon, Gluons	Intermediate boson, $W^{\pm}$, Z^0
Particles acted upon	Every particle having mass	Charged particles	Hadrons	Hadrons, Leptons
Examples	Astronomical forces	Atomic forces	Nuclear forces	Forces involved in β-decay
Strength (Natural units)	G (Newton) $= 6 \times 10^{-39}$	$g_e = \frac{1}{137}$	$g^2 = 1$	G Fermi $= 5 \times 10^{-17}$
Role in universe	Assembles matter into planets, stars and galaxies	Determines structure of atoms, molecules, solids and liquids. An important factor in astronomical universe	(i) Gluons holds quarks together to form nucleons (ii) Pions holds nucleons together to form atomic nuclei	Mediates transformation of quarks and leptons; helps determine composition of atomic nuclei
Conservation	–	Isospin violation	All	Flavour, CPT violation
Spin	2	1	1	1
Mass of field particle	0	0	0	$80.4, 80.4, 91.2 \frac{\text{GeV}}{c^2}$

is commonly stated in terms of gravitational constant $G = 6.7 \times 10^{-11}$ N-m²/kg².

The dimension less quantity $\frac{G m_e M_p}{\hbar c} \sim 3\times10^{-42}$ is usually taken as a constant, characterizing this interaction. The gravitational interaction between two bodies is believed to be mediated through the quantum of the this interaction called the *graviton*. Graviton is a hypothetical postulated particle to be its own antiparticle, to have zero charge and rest mass, and a spin of 2. Graviton is firmly predicted by theory but its direct observation is at present unlikely.

(ii) Electromagnetic Interaction: This is much stronger than gravitational interaction. This is the classical interaction of Maxwell-Lorentz field between electrically charged particles. The quantum of the field is photon. It is a long range square type interaction, whose strength is determined by Sommerfeld's fine structure constant $\alpha = \frac{e^2}{4\pi\varepsilon_0 \hbar} \sim \frac{1}{137}$. This electromagnetic interaction between a proton and an electron is about 10^{37} times stronger than gravitational force between them at the same distance.

This interaction is charge dependent. In terms of isobaric spin, this interaction depends on T and is governed by the isospin rule $\Delta T = 0, \pm1$. All other quantities, e.g. charge, baryon number, lepton number, hypercharge, parity, strangeness numbers are conserved.

The electromagnetic interaction is manifested in the chemical behaviors of atoms and molecules, Rutherford scattering and so on.

Photon capture can affect the production of mesons or hyperons by an electromagnetic interaction.

$$\gamma + p \begin{cases} \nearrow \pi^0 + p \text{ (photo-pion production)} \\ \searrow \pi^0 + K^+ \text{ (low strange particles)} \end{cases} \tag{26.1}$$

These decay processes illustrates particularly a important point which at first sight may appear paradoxical namely that uncharged particles can also be subjected to electromagnetic interaction, e.g.

$$\left.\begin{aligned} &\pi^0 \to \gamma + \gamma \\ &\eta^0 \to \pi^+ + \pi^- + \pi^+, \;\; \eta^0 \to \gamma + \gamma \\ &\Sigma^0 \to \Lambda^0 + \gamma \end{aligned}\right\} \tag{26.2}$$

In these processes no change of strangeness (S) is involved. The decay process such as $\Sigma^+ \to \pi^+ + \gamma$ are forbidden because the changes $\Delta S = 1$ is required. The paradox that the decay of neutral particle is by electromagnetic interaction is resolved by introducing an intermediate step in the overall nuclear reaction i.e., virtual production of nucleon-anti nucleon. We have

$$\pi^0 \xrightarrow[\text{Strong}]{} \underset{\text{Virtual}}{(N + \bar{N})} \xrightarrow[\text{electromagnetic}]{} \gamma + \gamma \tag{26.3}$$

An example of the electromagnetic interaction is the process of mutual annihilation of particles and antiparticles.

Unlike the gravitational force the electromagnetic interaction may be either attractive or repulsive.

(iii) Weak Interaction: This interaction is responsible for the nuclear β-decay and the weak decay of certain elementary particles, e.g. muons, pions, K-mesons and some hyperons. This interaction is about 10^{10} times weaker than the electromagnetic interaction and unlike gravitational or e.m. interaction, this is a very short range force. The range of weak interaction is given by $\dfrac{\hbar}{m_{\hat{w}}c} \sim 2.4 \times 10^{-18}$ m, where $m_{\hat{w}}$ is the mass of the W boson, mediating the weak force (half like ~10^{-17} s). The strength of the weak interaction may also be express in terms of the universal weak interaction coupling constant g_w, although this constant is not dimensionless as are others. Its value is 1.4×10^{-62} J-m^3 (from Fermi theory of β-decay). When expressed in dimensionless form, g_w is found to be small compared to the fine structural constant $\alpha \sim \dfrac{1}{137}$. Here $g_w = \left(\dfrac{g}{\hbar c}\right)(\lambda^2) = 3 \times 10^{-12}, \dfrac{\lambda_c}{2\pi}, \lambda_c$ being the Compton wavelength for the electron.

The weak interaction is mediated through the heavy vector bosons, known as $W^{\pm}$ and Z^0 bosons. They have the masses $m_w c^2 = 80.9$ GeV and $m_z c^2 = 93$ GeV. When one compares the characteristic times for the operation of the weak (~10^{-10} s) and strong (~10^{-23} s) interactions, one finds the ratio of the former to the latter ~10^{-13}.

Few examples of weak interaction are

$$\bar{\mu} + p \rightarrow n + \nu_{\mu}$$

$$\mu \rightarrow \bar{e} + \nu_{\mu} + \bar{\nu}_e \quad \text{(Muon decay)}$$

$$\bar{\nu}_e + p \rightarrow n + e^{+} \quad \text{(proton capture of anti neutrino)} \tag{26.1}$$

Other types of weak interactions which require coupling between other pairs of fermions,

$$\Lambda^0(S = -1) \rightarrow \pi(S = 0) + p(S = 0) \tag{26.2}$$

We must note that ν and $\bar{\nu}$ are associated exclusively with weak interactions just as photons are associated only with e.m. interaction.

An important characteristic of weak interaction which distinguishes it from other fundamental interactions, is that parity is violated (i.e., parity is not conserved) in weak interaction.

Recently, weak and electromagnetic interactions have been unified and the unified interaction is known as *electro-weak* interaction.

(iv) Strong Interaction: This is the dominant nuclear interaction responsible for most nuclear phenomena, nuclear energy levels, and so on. It is the interaction between protons and neutrons, having short range like the weak nuclear interaction, the range being $\dfrac{\hbar}{m_{\pi}c} \sim 10^{-15}$ m. Within this range it predominates over all the other fundamental interactions between the neutron and the proton with a characteristic strength parameter α ~0.3. Obviously, the value of α of strong interactions is much larger than the strength parameter $\alpha \sim \frac{1}{137}$ of e.m. interaction. The time for the operation of the strong interaction is about ~10^{-23} s, which is also the order of the characteristic nuclear time.

The quantum of the strong interaction is the *pi* meson or pion ($\pi^{\pm}$, π^0).

The nuclear strong interaction is responsible for the formation of the so called meson and baryon resonances, or excited states of the mesons and baryon particles. It is also the coupling that drives their decay, and for this reason their life times are short. A term usually used for the strongly interacting particles is hadron. The quantitative connection between the strong interactions and the structure of the nucleus is not yet fully clear.

Strong interaction is a short range force (~10^{-15} m) conserves baryon number (B), charge (Q), hypercharge (Y), Parity (π), isospin (T) and its component T_x. The following is an example of a strong interaction.

$$\pi^- + p \rightarrow \pi^0 + n \tag{26.3}$$

One can investigate the nature of the strong interaction by studying the properties of the *hypernuclei* in which a nucleon in a nucleus is replaced by a hyperon. The first hypernucleus discovered was a boron nucleus in which a neutron was replaced by a Λ^0-hyperon. So far several hypernuclei have been produced and their properties investigated. An example is Λ^0-nucleus of tritium. This decays according to $\Lambda^3\text{H} \rightarrow {}^3\text{He} + \pi^-$. The Q value for this is 41.5 MeV.

One can also obtain information about these systems through the study of the exotic atoms in which Σ^- hyperon is captures by the electric field of the nucleus. Sigma has the large mass and hence its orbit round the nucleus has a small radius. One obtain important information about the strong interaction by the measurement of the energies of the X-rays emitted during their transition from one orbit to another. This interaction between elementary particles is responsible for the total cross-section as a function of energy.

Recent studies reveal that protons and neutrons are made up of more fundamental entities known as quarks. It is believed that strong interaction is mediated through the exchange of massless quanta of spin $\hbar$ known as *gluons* between the quarks. Quarks and gluons have not been observed in the free state so far.

Examples of particles and their methods of interaction are shown in Table 26.2.

(v) Fifth Interaction There is an indication of the existence of fifth interaction besides four fundamental interactions discussed so far. It has been reported that the force between two masses be better represented by

$$F = G_{\infty}\frac{m_1 m_2}{r^2}\left[1 + \left(1 + \frac{r}{\lambda}\right)\alpha e^{-\frac{r}{\lambda}}\right] \tag{26.4}$$

with α = -0.007 and λ = 200 m. As α is negative, the second term in the square bracket represents a repulsive force. For $r >> 200$ m,

$$F = G\frac{m_1 m_2}{r^2} \tag{26.5}$$

Table 26.2 Elementary Particles and Corresponding Interaction Modes

Weak	*Electromagnetic*	*Strong*
e	e	π
μ	μ	K
ν	π	p
π	K	n
K	p	Λ
p	n	
n	Λ	
Λ		

Which is the force operative between two masses. For $r << 200$ m,

$$F = G_{\infty}\frac{m_1\, m_2(1+\alpha)}{r^2} = G\frac{m_1\, m_2}{r^2} \tag{26.6}$$

where $G' = G_{\infty}(1+\alpha)$

This is the force one measures in a Cavendish experiment. The value of G for small distances is about 1% less than the value of G for large distances.

26.3 DAWN OF PARTICLE PHYSICS

Prior to 1935, only six particles were known – the electron, proton, photon, positron and neutrino. At the time the most available source of high-energy particles was cosmic radiation – high-energy protons and other particles of extraterrestrial origin. In 1935, Hideki Yukawa proposed that the nuclear force should, like the electromagnetic force be mediated by exchange of a particle. Yukawa called his hypothetical particle a *meson*. This can be considered to be the beginning of particle physics. Keeping in mind the limited range of nuclear force, Yukawa argues that the proposed meson should have finite mass. One can understand the reason for this connection between mass and range from energy-time uncertainty relation. We have

$$\Delta E\, \Delta t \geq \hbar \tag{26.7}$$

The nuclear force has a finite range of about 1.5 fm. At close to the speed of light, the longest a particle mediating nuclear or strong force would need to exist is for a time Δt. One obtains,

$$\Delta t = \frac{\Delta x}{c} = \frac{1.5\times 10^{-15}\text{ m}}{3.0\times 10^{8}\text{ m/s}}$$
$$= 5\times 10^{-24}\text{ s}$$

Relation (26.1) gives

$$\Delta E \geq \frac{\hbar}{\Delta t} = \frac{1.05\times 10^{-34}\text{ J.s}}{5\times 10^{-24}\text{ s}}$$
$$= 2.1\times 10^{-11}\text{ J}$$
$$= 130\text{ MeV}$$

Yukawa therefore proposed a meson with mass 130 MeV/c^2, about 250 times the mass of the electron, i.e., the new particle would have a mass between that of the electron and that of the proton. Yukawa's prediction was eventually confirmed in 1937, but not before nature introduced yet another particle.

In 1937 Anderson and his colleagues identified in cosmic rays a particle with a mass 207 times (= 106 MeV/c^2) that of the electron. Subsequent experiments revealed that the newly discovered particle interacted very weakly with matter, however, and hence could not be the carrier of the nuclear or strong force. To resolve the puzzling situation, several theoretical physicists were inspired to propose that two mesons existed with slightly different masses. This idea was confirmed by Powell and Occhialini in 1947 by the discovery of *pi* (π) *meson*, or simply *pion*. The particle discovered by Anderson in 1937, the one thought to be Yukawa's predicted meson, is not really a meson, but a *muon* (μ). Muon takes part in the weak and electromagnetic interactions only. Two muons were found: the negatively charged muon (μ^-) and its antiparticle (μ^+).

The pion, Yukawa's carrier of the nuclear force, appear as three related particles, now called *pions*: positive π^+, negative π^-, and neutral π^0. The π^+ and π^- particles have masses of 139.6 MeV/c^2, and the π^0 particle has a mass of 135 MeV/c^2.

The new particles, i.e., pions and muons are all unstable, undergoing decays that ultimately result in well known stable particles. For example, the negative pion (π^-), which has a mean lifetime of 2.6 × 10^{-8} s (26 ns) first decays to a negative muon and an antineutrino:

$$\pi^- \to \mu^- + \overline{\nu}$$

The muon then decays with a 2.2 μs lifetime to an electron and a neutrino-antineutrino pair:

$$\mu^- \to e^- + \nu + \overline{\nu}$$

We may not that for chargeless particles (as well as some charged particles such as proton), a bar over the symbol represent an antiparticle of the particle, e.g. $\overline{\nu}$ is antiparticle of ν.

The interaction between two particles can be represented in space-time diagram (Fig. 26.1) called a Feynman diagram.

A *Feynman diagram* is a qualitative graph of time in the vertical direction versus space in the horizontal direction developed by American physicist R.P. Feynman. Feynman diagrams are qualitative in the sense that the actual values of time and space are not important, but the overall appearance of the graph reveals a representation of the process. One can approximate the time evolution of the process by starting at the bottom of the diagram and moving eyes upward.

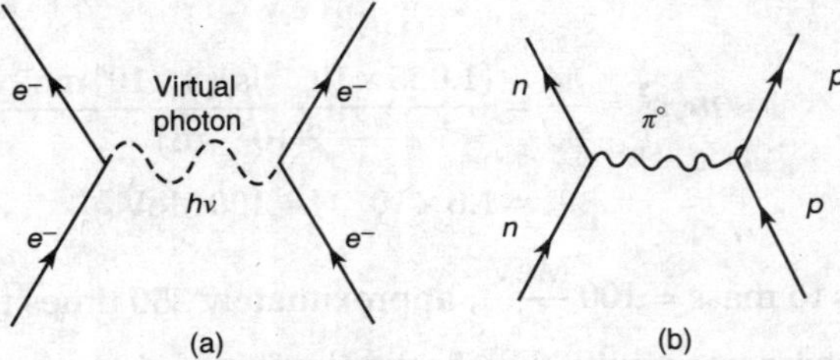

Fig. 26.1 Feynman diagram representing: (a) a photon mediating the electromagnetic interaction between two electrons, and (b) π-meson mediating between a proton and neutron

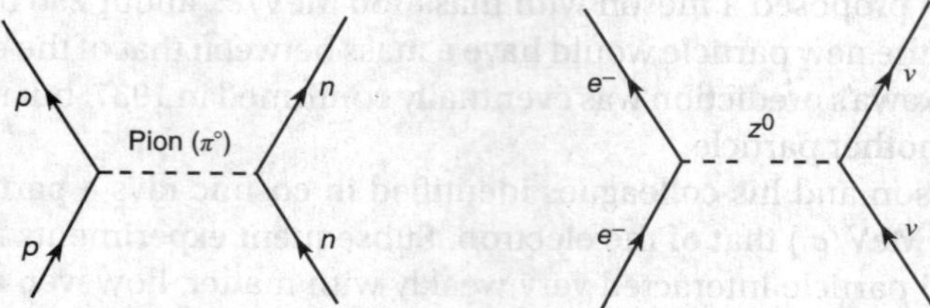

Fig. 26.2 Feynman diagram representing: (a) a proton and a neutron interacting via the nuclear force mediated by π^0 meson. However, this model is not the most fundamental model for nucleon interaction, (b) an electron and a neutrino interacting via the weak force mediated by Z^0 boson

In Fig. 26.1(a), in the electron-electron interaction, the photon which transfers energy and momentum from one electron to the other, is named a virtual photon, since it vanishes during the interaction without having been detected. We may note that this process does not violate the law of conservation of energy because the lifetime (Δt) of virtual photon is very short. This makes the uncertainty in the energy $\Delta E \approx \hbar/2t$ of the system consisting of two electrons and the photon greater than the photon energy. Obviously, within the constraints of uncertainty principle, the energy of the system is conserved.

Now, we consider a pion exchange between a proton and neutron according to Yukawa's theory of nuclear force (Fig. 26.2).

To create a pion of mass m_π, the energy required, $\Delta E_R = m_\pi c^2$. The very existence of pion would appear to violate the law of conservation of energy if the particle existed for a time greater than Δt, given by uncertainty relation $\Delta E_R\, \Delta t \approx \hbar/2$, i.e., $\Delta t \approx \hbar/2\Delta E_R$. Here Δt is the time interval required for the pion to transfer from one nucleon to another. Thus, one finds

$$\Delta t \approx \frac{\hbar}{2\Delta E_R} = \frac{\hbar}{2\, m_\pi c^2} \tag{26.8}$$

The maximum distance d, the pion can travel in a time interval Δt, $d = c\,\Delta t$ because pion cannot travel faster than the speed of light. Thus

$$d = c\Delta t = \frac{\hbar}{2m_\pi c} \tag{26.9}$$

Now, the range of the nuclear force is of the order of 10^{-15} fm. Using this value of d in Eq. (26.3), one obtains

$$m_\pi c^2 \approx \frac{\hbar c}{2d} = \frac{(1.055\times 10^{-34}\ \text{Js})(3\times 10^{8}\ \text{ms}^{-1})}{2(10^{-15}\ \text{m})}$$

$$= 1.6\times 10^{-11}\ \text{J} \approx 100\,\text{MeV}$$

i.e., this corresponds to mass $= 100\dfrac{\text{MeV}}{c^2}$, approximately 250 times the mass of electron. This is in reasonable agreement with the observed value of pion mass.

So far, we have dealt with the particles that mediate the nuclear force, pions, and the photons that mediate the electromagnetic force. The current studies indicate that the *nuclear force is more fundamentally describe as an average or residual effect of the force between quarks.* The graviton, the mediator of the gravitational force is yet to be observed. The $W^{\pm}$ and Z^0 particles that mediate the weak force were discovered in 1983 by C. Rubbia and coworkers.

26.3.1 π-Mesons

Using particle accelerators, pions are produced in the laboratory. The essential nuclear reactions are:

$$p+p \to p+p+\pi^{\circ} \qquad p+n \to p+p+\pi^{-}$$
$$p+p \to p+n+\pi^{+} \qquad p+p \to p+n+\pi^{0}$$

The incident proton from the accelerator must have sufficient energy to create the additional mass of the pion. The π^+ and π^- particles have masses of 139.6 MeV/c^2, and the π^0 particle has a mass of 135.0 MeV/c^2. These are very unstable particles. For example, the π^- with a meanlife of 2.6×10^{-8} s first decays to a muon and an antineutrino. The muon with a mean lifetime of 2.2 μs, then decays into an electron, a neutrino, and an antineutrino.

The recent theory of *quantum chromodynamics* (QCD) consider gluons in place of mesons as the mediator of strong interaction. We may note that strong interaction is more basic than the interaction in a nucleus.

[QCD is a quantum field theory, i.e., a *guage theory* of strong interactions based on the exchange of massless *gluons* between quarks and antiquarks. QCD is similar to quantum electrodynamics (QED), the quantum field theory of electromagnetic interactions, but with a gluon as the analogue of the photon and with a quantum number known as *colour* replacing that of electric charge.]

26.3.2 Muons, Kaons and Hyperons

μ meson has its mass 106 MeV/c^2 and does not interact strongly with matter. Muons are unstable and emit electrons:

$$\mu^{\pm} \to e^{\pm} + \nu + \overline{\nu} + 105\,\text{MeV}$$

The spin of muon is $\frac{1}{2}$. When ν and $\overline{\nu}$ are ejected together, at least in some cases the two together should annihilate giving two γ-rays:

$$\nu + \overline{\nu} \to 2\gamma$$

This was never observed. This means that ν and $\overline{\nu}$ are not antiparticles of each other. In other words, they could only be neutrinos of different species. Thus, two types of neutrinos associated with the muon decay should be in accordance with the following decay equations:

$$\mu^{+} \to e^{+} + \nu_e + \overline{\nu}_{\mu} + 105\,\text{MeV}$$
$$\mu^{-} \to e^{-} + \overline{\nu}_e + \nu_{\mu} + 105\,\text{MeV.}$$

where ν_{μ} and $\overline{\nu}_{\mu}$ are muon neutrino and muon antineutrino respectively. Obviously, muons behaves like an electron.

Kaons (K-Mesons)

Kaons exist as K^+, K^- and K^0. These particles have been observed in cosmic ray studies and also confirmed in accelerator experiments. Although these particles are similar to pions but have several more decay possibilities, e.g.

$$K^+ \to \pi^\pm + \pi^0 \qquad K^+ \to \mu^+ + \nu_\mu$$
$$K^+ \to \pi^+ + \pi^+ + \pi^- \qquad K^0 \to \pi^+ + \pi^-$$

All Kaon's have spin 0 and masses of K^+, K^-, K^0 are 966 m_e, 966 m_e and 975 m_e respectively.

Hyperons

There are three groups of hyperons, called lambda (Λ), sigma (Σ) and Xi (Ξ) are known today. These are unstable particles having masses greater than that of nucleon. These particles were discovered in 1947 in cosmic rays. Some of the decay modes of these particles are

$$\Lambda^0 \to \pi^+ + \pi^0 \qquad \Lambda^0 \to \nu^0 + \pi^0$$
$$\Sigma^+ \to \pi^+ + \pi^0 \qquad \Sigma^+ \to \nu^0 + \pi^+$$
$$\Sigma^- \to \nu^0 + \pi^0 \qquad \Sigma^0 \to \Lambda^0 + \gamma$$
$$\Xi^- \to \Lambda^0 + \pi^0 \qquad \Xi^0 \to \Lambda^0 + \pi^0$$

All hyperons have a spin ½ and can be produced in the laboratory by pion or kaon interaction.

26.4 PARTICLES AND ANTIPARTICLES

A subatomic particle that has the same mass and spin as another particle and equal but opposite values of other property or properties charge (Q), Baryon number (B), Strangeness (S), Charm (C) and isospin of equal magnitude but opposite sign, i.e., a conjugate particle. Examples of a particle and its antiparticle include the electron and positron, proton and antiproton, the positively and negatively charged pions, and the up quark and up antiquark. Positron has a positive charge equal in magnitude to the electron's negative charge. The neutron and the antineutron have magnetic moments opposite in sign relative to their spins. The antiparticle corresponding to a particle with a symbol a is usually denoted $\bar{a}$. When a particle and its antiparticle are identical, as with the photon and neutral pion, this is called a *self-conjugate* particle.

The existence of antiparticles is predicted by Dirac's relativistic quantum mechanics theory. According to Dirac, the energy spectrum of a free electron has two branches one starting at $+m_0c^2$ and extending to ∞ as the momentum $|p| \to \infty$ and the other branch starting at $-m_0c^2$ and extending to $-\infty$ as $|p| \to \infty$ (Fig. 26.3). We can see that two branches are separated by a forbidden gap $= 2m_0c^2$. Dirac's theory can be interpreted as a vacuum being filled with an infinite sea of electrons with negative energies. If enough energy is transferred to the sea, an electron can be ejected with positive energy. This leaves behind a hole, i.e., a *positron* (e^+). We may note the analogy with semiconductors where holes appear instead of positrons. Thus Dirac's relativistic theory of electron, along with refinements made by other physicists, opened the possibility of *antiparticle*, which have the same mass and lifetime as their associated particles, and the same magnitude, but the opposite sign for such physical quantities as electric charge and various quantum numbers as stated earlier.

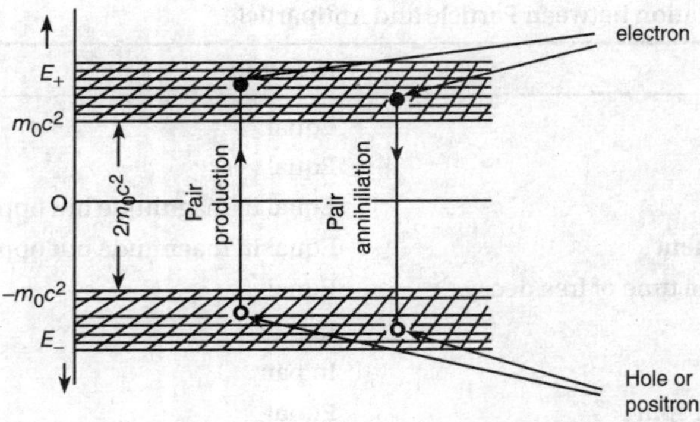

Fig. 26.3 Electron-positron pair creation and pair annihilation processes along with negative energy states in accordance with Dirac's relativistic theory

The energy required for the creation of *electron-positron* pair is equal to or greater than $2m_0c^2$. Dirac's interpretation was accepted with the discovery of positron in 1932 by Anderson. The positron (e^+) is usually referred as antiparticle of electron. The process of creation of an electron-positron pair is usually referred as *pair creation*. We may note that positron is a stable particle like electron. The ultimate fate of positrons is annihilation with a electron. After a positron slows down by allowing it to pass through matter, it is attracted by the Coulomb force to an electron, where its annihilation takes place through the reaction

$$e^+ + e^- \to 2\gamma$$

The process is called *electron-positron annihilation*. The main characteristic of positron annihilation is the emission of two oppositely directed photons having energy equal to the rest energy of the particle, i.e., 0.511 MeV. We may note that the kinetic energy of charged particles are usually very small when they meet and neglected.

It is believed that all particles including the neutral particles have antiparticles. For photons, neutral pion (π^0) and a few other particles, the particle and antiparticle are the same. Sometimes the particle and its antiparticle is written with the electric charge along with the symbol, e.g. e^- and e^+ or μ^- and μ^+. For other, bar is put over the symbol of antiparticle as ν and $\bar{\nu}$ or p and $\bar{p}$.

We may note that ordinary matter is composted of electrons, protons and neutrons. In the classification scheme one has to distinguish particles and antiparticles. According to existing theories of the creation of universe in the Big Bang *matter* and *antimatter* should have been created in exactly equal quantities. Antimatter consists of entirely of antiparticles.

For example, antihydrogen consists of an antiproton with an orbiting positron. Scientists had created artificially antihydrogen in the laboratory. The spectrum of antihydrogen should be identical to that of hydrogen. It appears that matter dominates over antimatter now in our universe, i.e., the universe consists overwhelmingly of normal matter, and explanations of the absence of large amounts of antimatter have been incorporated into cosmological models that involve the use of *grand unified theories* of elementary particles. Relation between particle antiparticle is summarized in Table 26.3.

Table 26.3 Relation between Particle and Antiparticle

Property	***Relation***
Mass	Equal
Spin	Equal
Charge	Equal in magnitude but opposite in nature
Magnetic Moment	Equal in magnitude but opposite in sign
Mean Life at the time of free decay	Equal
Production	In pair
Destruction	In pair
Total isotopical spin	Equal
Third component of Isotopic spin	Equal in magnitude but opposite in sign
Internal parity	Same for bosons Opposite for fermions
Strangeness Quantum number	Equal in magnitude but opposite in sign

26.5 CLASSIFICATION OF ELEMENTARY PARTICLES

On the basis of the property of elementary particles called "spin", one can place them into two broad categories, e.g. (i) **Fermions:** obeying FD statistics and (ii) **Bosons:** obeying BE statistics.

The most important difference between these two groups of particles is that there is no conservation law controlling the total number of bosons in universe, where as the total number of fermions is strictly conserved.

The fermions fall mainly into two categories:

(i) Leptons (Weakly interacting Fermions): The lightest family of fermions is known as the family of leptons and includes six lepton-antilepton pairs. The lightest of the four are the neutrinos, which have rest mass zero. The two kinds of neutrino known are *electron-neutrino* (ν_e) and muon neutrino (ν_μ). Each is associated with its corresponding *antiparticle.* A neutrino is said to have *negative helicity* equal to –1, and an antineutrino is said to have *positive helicity,* equal to +1 (Fig. 26.4). Both types of neutrino have spin ½, and neither has an electric charge. The two are differentiated from one another by their couplings, one to the electron field and other to the muon field. These couplings are through the weak interaction. When an electron is involved in a weak interaction, so is also the corresponding electron-neutrino. When a muon is involved in a weak interaction, the corresponding muon neutrino is also involved.

A typical reaction involving electrons and neutrinos is that of β-decay:

$$ {}_0^1n \rightarrow {}_1^1p + {}_{-1}^{0}e + \bar{\nu}_e \tag{26.10}$$

and

$$ {}_1^1p \rightarrow {}_0^1n + {}_{+1}^{0}e + \nu_e \tag{26.11}$$

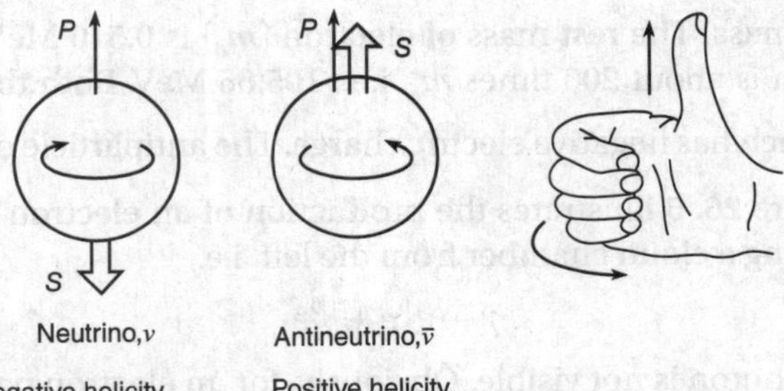

Fig. 26.4 The neutrino and antineutrino have opposite helicity

and the inverse of above are

$$\bar{\nu}_e + {}_1^1p \rightarrow {}_0^1n + {}_{+1}^{\ 0}e \tag{26.12}$$

and

$$\nu_e + {}_0n^1 \rightarrow {}_1^1p + {}_{-1}^{\ 0}e \tag{26.13}$$

Typical reactions involving muons and neutrinos are those of pion decay

$$\pi^- \rightarrow \mu^- + \bar{\nu}_\mu \tag{26.14}$$

and

$$\pi^+ \rightarrow \mu^+ + \nu_\mu \tag{26.15}$$

The muon production reaction is

$$\bar{\nu}_\mu + {}_1^1p \rightarrow {}_0^1n + \mu^+ \tag{26.16}$$

We note that in all these above reactions an electron is associated with its antineutrino when both appear on the same side of arrow, but associated with its neutrino when they appear on opposite sides. The same holds good for muon and its neutrino. This is a consequence of the law of conservation of particle number and in this case the law is the conservation of lepton number (L). Lepton number has a value $L = +1$ for leptons (${}_{-1}^{\ 0}e, \mu, \nu_e, \nu_\mu$), $L = -1$ for antileptons (${}_{+1}^{\ 0}e, \mu^+$, $\bar{\nu}_e$, $\bar{\nu}_\mu$) and $L = 0$ for all other Leptons.

Neutrinos carry only the weak interaction and so interact only very weakly with other particles. This is expressed through an exceedingly small interaction cross-section and correspondingly long mean free path through matter. A typical neutrino interaction cross-section is of the order of 10^{-49} m^2 or less, while the maximum known is about 10^{-42} m^2. A typical neutrino mean free path is thousands of light years long through condensed matter such as lead.

Neutrinos are the only known particles that appear to be absolutely polarized in spin with reference to their momentum vectors. Neutrinos spin counter clockwise (with negative helicity), and antineutrinos spin clockwise (with positive helicity) relative to their, directions of travel. Neutrinos are stable against decay into other particles.

We must note that the neutrinos involved in pion decays are not the same as those involved in β-decay. The existence of two classes of neutrinos was established in 1962.

Electrons and Muons: The two charged members of the lepton family are the electron (e) and muon (μ). These two particles act very much alike, although they differ in rest mass. The rest mass of electron (m_0) is 0.510 MeV whereas the rest mass of muon is about 200 times m_0, i.e., 105.66 MeV. Both these particles have spin $\frac{1}{2}$ and each has negative electric charge. The antiparticle of each is positively charged. Figure 26. 5 illustrates the production of an electron-positron pair by a photon entering a cloud chamber from the left, i.e.,

$$\gamma \rightarrow {}^{0}_{-1}e + {}^{0}_{+1}e \tag{26.17}$$

The track of photon is not visible. Obviously, for an electron pair to be produced, the photon must have an energy equal to at least $2m_e c^2 = 1.022$ MeV. In order for energy and momentum to be conserved in Equation (26.17), the process must occur near a nucleus.

The electron is stable against decay into other particles, while muon decays into an electron and two neutrinos:

$$\mu^- \rightarrow {}^{0}_{-1}e + \bar{\nu}_e + \nu_\mu \tag{26.18}$$

The positive muon (the antiparticle) decays into the charge conjugate states of the same three products:

$$\mu^+ \rightarrow {}^{0}_{+1}e + \nu_e + \bar{\nu}_\mu \tag{26.19}$$

Fig. 26.5 Electron-positron pair production (Courtesy: Brookhaven National Laboratory)

In these decays, the electron-neutrino is associated with the presence of an electron and the muon-neutrino with the muon as in the preceding reactions. The mean lifetime of the muon in these decays is 2.2×10^{-6} sec.

The electron was the first of the elementary particles to be found in nature by J.J. Thomson, while the muon was the first cosmic ray particle heavier than the electron was found. The great majority of cosmic ray secondary particles at sea level are muons. The muon lifetime is long enough for a negative muon sometimes to temporarily replace an atomic electron to form a muonic atom.

Salient features of leptons are:

(i) These are particles that don't experience the strong force.

(ii) They include the electron, the muon, massive particle tau, and three types of neutrinos, one associated with each of charged leptons.

(iii) Each of the leptons has an antiparticle as well.

(iv) There are six lepton-antilepton pairs, and experimental evidence strongly suggest that no other can exist.

(v) All leptons have spin ½ and are therefore *fermions,* which obey Pauli exclusion principle.

(vi) Leptons are believed to be true elementary point particles with zero size and no internal structure.

(ii) Hadrons: Hadrons are particles that do experience the strong force and fall into two subclasses: (i) mesons and (ii) baryons. *Mesons* have integer spins and are therefore bosons that don't obey Pauli exclusion principle. Mesons include pions and a host of other particles. All mesons are unstable. All mesons have antiparticles, but some neutral mesons are their own antiparticles. *Baryons* have ½ integer spins and are therefore fermions. Proton, neutron and some similar but more massive particles fall under this category. The baryons are grouped into pairs, triplets, and higher-multiple grouping of closely related particles. Each baryon has an antiparticle.

Baryons (Strongly interacting Particles): These are fermions of half integra spin and have masses equal to or greater than nucleon mass. These particles are subject to strong, weak and electromagnetic, i.e., all interactions. Particles belonging to this category and heavier than nucleons are called *hyperons*. The hyperons are usually found associated with the *K*-mesons and their precise role in the construction of matter is not fully understood. Members of this group are omega hyperon (Ω^-), Cascade hyperons ($\Xi^0\Xi^-$), Sigma hyperons (Σ^+, Σ^0, Σ^-), Lambda hyperon (Λ^0), and the nucleons (proton and neutron). The various hyperons may decay in a variety of ways, but the end result is always a proton and neutron, for example

$$\begin{aligned} \Omega^- &= \Xi^0 + \pi^- \\ &\downarrow \\ &\rightarrow \Lambda^0 + \pi^0 \\ &\downarrow \\ &\rightarrow p^+ + \pi^- \end{aligned} \tag{26.20}$$

$$\Lambda^0 \begin{cases} \nearrow p^+ + \pi^- \\ \nearrow n^0 + \pi^0 \\ \searrow p^- + \pi^+ \\ \searrow p^+ + e^- + \overline{\nu} \end{cases} \tag{26.21}$$

Baryons have an additively conserved internal quantum number called Baryon number B which has a value $B = +1$ for baryons, $B = -1$ for anti baryons and $B = 0$ for rest of the particles.

The proton, classed as a baryon, was long thought to be only stable hadron. A proton-antiproton pair may be produced in a high-energy proton-proton collision according to the scheme.

$$p^+ + p^+ \rightarrow p^+ + p^+ + p^+ + \overline{p}^- \tag{26.22}$$

(This is not the only process that may take place in a proton-proton collision). Proton-antiproton annihilation is a more complex process, involving the production of several particles, most of them pions. One such process is

$$p^+ + p^- \rightarrow 4\pi^+ + 4\pi^- + x\pi^0 \tag{26.23}$$

Proton-antiproton annihilation according to (26.23) took place in a bubble chamber is shown in Fig. 26.6. The experimental arrangement to observe an antiproton is shown in Fig. 26.7. Protons accelerated up to 6.2 *GeV*, hit a suitable target which produced several reactions involving *K's*, *π's* and *μ's* as well as *π*-particles. A deflecting magnet M_1 selected only negative particles, which were allowed to pass through an opening I the shielding. S_1, S_2 and S_3 are scintillation counters and C_1 and C_2 are Cerenkov detectors. These instruments are so designed that they are sensitive only to particles in a certain energy range. Between S_1 and S_2 there was placed a second deflecting magnet M_2, which acted as a momentum selector, since the radius of the path was fixed by the position of the detectors, and only particles with momentum $p = eBr$ were properly deflected toward S_2. Because it offered certain experimental advantages, the magnetic field was chosen to correspond to a momentum $p = 1.19$ GeV/c, which is slightly less than the momentum at which most protons should have produced (1.75 GeV/*c*).

Current theories of fundamental interactions require that protons should decay into leptons, but since the expected mean life is years, proton instability will not be easy to confirm.

Like fermions, bosons also fall into two categories:

(i) Massless Bosons (Subject to electromagnetic interaction only): This group consists of integral spin bosons, i.e., with spin 1, 2 or so and rest mass zero. Photon and graviton belongs to this group, the rest mass (m_0) of photon is zero and hence the general equation for the energy of a particle

$$E^2 = m_0^2c^4 + p^2c^2 = m^2c^4 \tag{26.24}$$

reduces to

$$E = pc = mc^2 \tag{26.25}$$

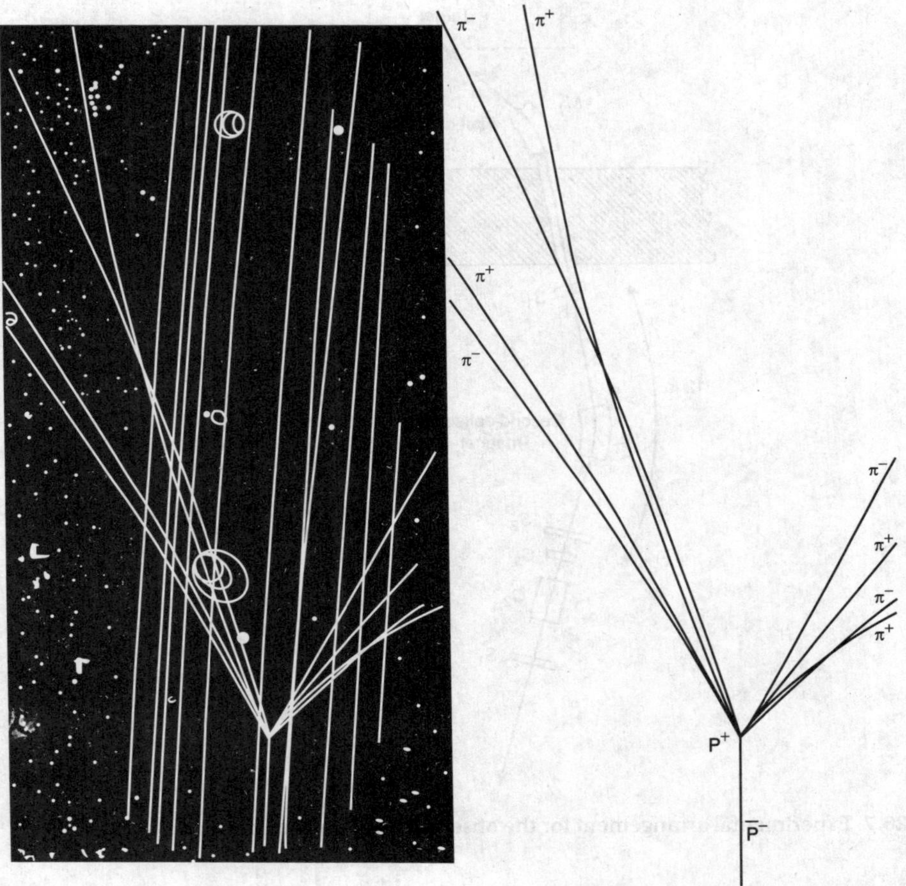

Fig. 26.6 Proton-antiproton annihilation (Courtesy: Brookhaven National Laboratory)

which yields a momentum for the photon as

$$p = \frac{E}{c} = mc \tag{26.26}$$

The rest mass of the photon is zero but its energy is finite. Therefore the velocity of photon is c (velocity of light in vacuum) and momentum $p = mc$. The energy of photon is also given by

$$E = h\nu \tag{26.27}$$

$$\therefore \qquad \text{Mass of the photon} = \frac{h\nu}{c^2} \tag{26.28}$$

The photon, given the symbol γ is the quantum of electromagnetic field. As such it is a *boson*. The photon has spin 1. The photon is itself uncharged, although it

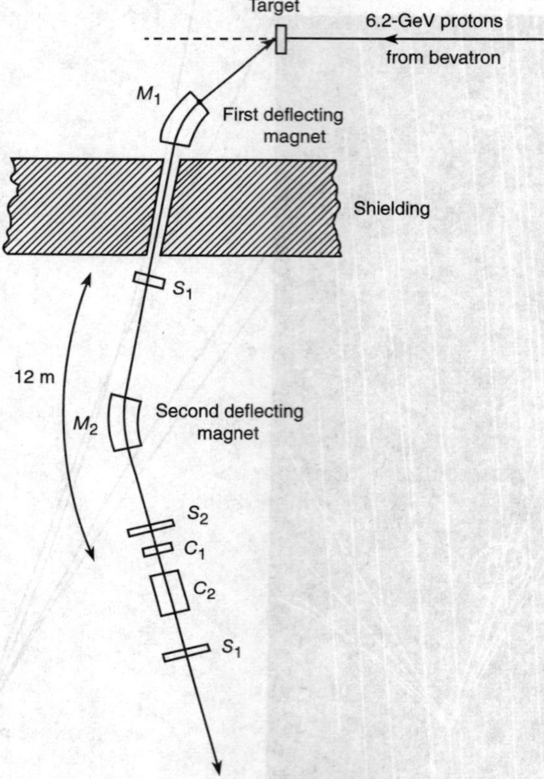

Fig. 26.7 Experimental arrangement for the observation of the antiproton

carries the Coulomb force field between electrical charges, either as 'virtual' particle in the static field or as a real particle in the radiation field.

The photon is stable and therefore, does not decay spontaneously, into any other particle. Obviously, the lifetime of photon is infinite so long as it does not interact with other particles, and so photons continuously reach the earth from the farthest distances of the universe. The photon may by thought of as the most commonly experienced elementary particle in its free state. The photon is its *own antiparticle*. Photon carries only the electromagnetic interaction, and so interacts with other particles only through this and through the gravitational field.

Another massless boson, so called *'graviton'* with a probable spin of 2 units, has been postulated as a field particle of gravity. Graviton has eluded detection so far.

These bosons, created by the electromagnetic field, are essentially of one kind.

(ii) Mesons (Strongly interacting particles): This group contains bosons of spin *O* and having masses intermediate between the leptons and the nucleons. Bosons of this group formed in the strong interactions are of two distinct kinds: (a) Pions or π mesons (π^+, π^- and π^0), (b) Kaons or K-mesons (K^+, K^- and K^0). The deay schemes for pion or kaon are as follows:

$$\left.\begin{aligned} \pi^+ &\rightarrow \mu^+ + \nu_\mu \\ \pi^- &\rightarrow \mu^- + \bar{\nu}_\mu \\ \pi^0 &\rightarrow \gamma + \gamma \end{aligned}\right\} \quad (26.29.a)$$

$$\left.\begin{aligned} K^+ &\rightarrow \mu^+ + \nu_\mu \\ &\rightarrow \pi^+ + \pi^0 \\ &\rightarrow \pi^+ + \pi^+ + \pi^- \\ &\rightarrow \pi^0 + e^+ + \nu_\varepsilon \\ &\rightarrow \pi^0 + \mu^+ + \nu_\mu \\ &\rightarrow \pi^+ + \pi^0 + \pi^0 \end{aligned}\right\} \quad (26.29.b)$$

Curiously enough, two distinct varieties of the neutral kaon are known. One K^0 has a mean lifetime of 8.8×10^{-11} sec and decays either into $\pi^+ + \pi^-$ or into $\pi^0 + \pi^0$. The lifetime of K^0 is 5.2×10^{-8} sec and it may decay into a pion, an electron, and a neutrino, a muon, and a mu-neutrino or into three pions.

Baryons and mesons together are called as *hadrons*. The name hadron means strongly interacting particle. Hadrons carry the electromagnetic, the weak, and the strong interactions. The decay modes of leptons, Baryons are summarized in Table 26.4.

(iii) Gauge Bosons or Field Particles: These are the quanta of different force fields and "carriers" of those forces. Field particles include the photon for the electromagnetic force; three particles called the W^+, W^-, and Z for the weak force; gluon; and a hypothetical graviton that would carry the gravitational force in a developing theory of quantum gravity. All these field particles are bosons, carrying spin 1 or spin 2 for graviton. We may note that Yukawa's meson is not appearing in this category in its role as carrier of the nuclear force. This indicates that the nuclear force is not really fundamental and gluon plays the more fundamental role.

Gluon is an elementary particle that mediates the strong interaction–between quarks (and antiquarks). Gluon is a gauge boson.

Fermions exert attractive or repulsive forces on each other by exchanging gauge bosons, which are force carriers.

One other boson that has bean predicted is necessary in quantum field theory to explain why the $W^\pm$ and Z have such large masses, yet the photon has no mass. This boson is named as the *Higgs boson* (or *Higgs particle*) after Peter Higgs, who first proposed it. We don't know whether the Higgs particle is an elementary boson or a composite particle. The standard model proposes that there is a field called the *Higgs field* that permeates space. By interacting with this field, particles acquire mass. Particles that interact strongly with the Higgs field have heavy mass; particles that interact weakly have small mass. The Higgs field has at least one particle associated with it, and that is the Higgs particle or Higgs boson.

The announcement of the discovery of the Higgs – like particle (God particle) with mass 125 GeV at large Hadron collider (LHC) at CERN, Geneva, on 4th July, 2012

Table 26.4 Particle Decay Modes

	Particle	*Decay mode*	*Relative probability, %*	*Half-life, s*
Leptons	Photon	Stable		
	Neutrino	Stable		1.52×10^{-6}
	Electron	Stable		
	Muon	$\mu^- \to e^- + \nu + \overline{\nu}$		
	Pion	$\pi^+ \to \mu^+ + \nu$	100	1.80×10^{-8}
		$\to e^+ + \nu$	$\sim 10^{-4}$	
		$\pi^0 \to \gamma + \gamma$	99	6×10^{-17}
		$\to \gamma + e^+ + e^-$	1	
Mesons	Kaon	$K^+ \to \mu^+ + \nu$	63	
		$\to \mu^+ + \pi^0$	21	8.56×10^{-9}
		$\to 2\pi^+ + \pi^-$	5.6	
		$\to \pi^0 + e^+ + \nu$	4.8	
		$\to \pi^0 + \mu^+ + \nu$	3.4	
		$\to \pi^+ + 2\pi^0$	1.7	
		$K^0 \to \pi^\pm + e^\mp + \nu$	18	
		$\to \pi^\pm + \mu^\mp + \nu$	14	4×10^{-8}
		$\to \pi^\pm + \pi^- + \pi^0$	6.3	
		$\to 3\pi^0$	11.3	
		$K^0 \to \pi^+ + \pi^-$	35	6.0×10^{-11}
		$\to 2\pi^0$	15	
	Eta	$\eta^0 \to \gamma + \gamma$	33	
		$\to \pi^+_{} + \gamma + \gamma$	20	$<10^{-6}$
		$\to 3\pi^0$	20	
		$\to \pi^+ + \pi^- + \gamma^0$	22	
		$\to \pi^+ + \pi^- + \gamma$	5	
Baryons	Proton	Stable		
	Neutron	$n^0 \to p^+ + e^- + \overline{\nu}$		7.0×10^2
	Lambda	$\Lambda^0 \to \pi^+ + \pi^-$	66	1.76×10^{-10}
		$\to n^0 + \pi^0$	34	
	Sigma	$\Sigma^+ \to p^+ + \pi^0$	53	5.6×10^{-11}
		$\to n^0 + \pi^+$	47	
		$\Sigma^0 \to \Lambda^0 + \gamma$		$<7 \times 10^{-15}$
		$\Sigma^- \to n^0 + \pi^-$		1.1×10^{-10}
	Xi	$\Xi^0 \to \Lambda^0 + \pi^0$		2.0×10^{-10}
		$\Xi^0 \to \Lambda^0 + \pi^-$		1.2×10^{-10}
	Omega	$\Omega \to \Lambda^0 + K^-$	50	10^{-10}
		$\to \Xi^0 + \pi^-$	50	

***To obtain the decay of antiparticles change all particles into antiparticles on both sides of the equations.**

is really a path breaking discovery since the beginning of 20th century. The collider with a circumference of 27 km is the biggest and most complex machine ever built by mankind.

Theoretically, the discovery of Higgs - like particle marks the completion of the standard model.

We may note that LHC has the potential not only to discover additional Higgs bosons, but can definitely throw light on many othe issues which are engaging the attention of high - energy scientists. For example, it could discover supersymmetry candidates for dark energy, extra dimensions, etc.

The classification scheme of the elementary particles is presented in Table 26.5. The particles in Table 26.5 all exist long enough to travel as distinct entities along paths of measurable length and their modes of decay can be observed in such devices as bubble chambers. A considerable number of experimental evidence also points to the existence of several hundred different hadrons whose lifetimes are only about 10^{-23} sec. These ultrashort-lived particles canot be detected by observing their creation and subsequent decay in a bubble chamber or other instruments because the distance they cover in 10^{-23} sec is only about $= 3 \times 10^{-15}$ m, even if they move at the speed of light - a length characteristic of hadron dimensions. Instead such particles appear as *resonant states* in the interactions of less remanescent (and hence more readily observable) particles. Resonant states occur in atoms as energy as revealed in Franck-Hertz experiment. We shall discuss resonance later in this chapter.

26.6 PARTICLE INSTABILITY

The creation and annihilation process, β decay etc., are manifestations of a more general property of elementary particles: their instability. In other words given the proper conditions, the fundamental particles can transform into the other particles a s result of their interaction. Particles which are unstable undergo spontaneous decay with a well defined half life. We must note that only four particles (and their antiparticles) are stable against spontaneous decay: the photon, the neutrino, the electron and the proton of all the unstable particles, the one having the longest half-life is the neutron. Mesons have a half life of the order of 10^{-8} s (π^0 and η^0 are exceptions), while the baryon half-lives are of the order of 10^{-10} s.

The necessary condition for spontaneous decay is the condition of conservation of energy, i.e., the rest mass of the parent-particle be larger than the sum of the rest masses of the daughter conserved. In addition to energy, momentum as well as angular momentum are supposedly conserved. Particle decay modes have been summarized in Table 26.5 Few examples are

(i) proton-antiproton annihilation

$$p^+ + \overline{p}^- \to \overline{\Xi}^+ + \Xi^-$$

$$\Xi^- \to \pi^- + \Lambda^0$$

$$\overline{\Xi}^+ \to \pi^+ + \pi^0$$

$$\pi^0 \to \pi^+ + p^- \qquad (26.30)$$

Table 26.5 Elementary Particles Stable Against Decay by the Strong Interaction Mesons are Jointly Considered Hadrons and Believed to be Composed of Quarks

Class	*Name*	*Particle*	*Anti-particle*	*Rest mass, MeV*	*Mean life, s*	*Spin*	L_e	L_m	L_t	*B*	*S*	*Y*	*I*	I_z
Massless Boson	Photon	γ	$(\overline{\gamma})$	0	Stable									
	e-neutrion	ν_e	$\overline{V}_e$	0	Stable	1/2	+1	0	0	0				
Lepton	μ-neutrion	ν_μ	$\overline{V}_\mu$	0	Stable	1/2	0	+1	0	0				
	τ-neutrion	ν_τ	$\overline{V}_\tau$	0	Stable	1/2	0	0	+1	0				
	Electron	*E*	e^+	0.51	Stable	1/2	+1	0	0	0				
	Muon	μ	μ^+	106	2.2×10^{-6}	1/2	0	+1	0	0				
	Tau	τ	τ^+	1784	5×10^{-13}	1/2	0	0	+1	0				
		π^+	π^-	140	2.6×10^{-8}									+1
Meson	Pion	π^0	(π^0)	135	8.3×10^{-17}	0	0	0	0	0	0	0	1	0
		π^-	π^+	140	2.6×10^{-8}									–1
		K^+	K^-	494	1.2×10^{-8}									+1/2
	Kaon	K^0	$\overline{K}_0$	498	9×10^{-13}	0	0	0	0	0	+1	+1	½	–1/2
					5×10^{-8}									
	Eta Meson	η^0	η^0	549	7×10^{-19}	0	0	0	0	0	0	0	0	0
	Proton	*p*	$\overline{p}$	938.3	Stable									+1/2
	Nucleon					1/2	0	0	0	+1	0	+1	½	
Baryon	Neutron	*n*	$\overline{n}$	939.6	932									–1/2
	Lambda hyperon	Λ^0	$\overline{\Lambda}^0$	1116	2.5×10^{-10}	1/2	0	0	0	+1	–1	0	0	0
	Sigma	Σ^+	$\overline{\Sigma}^-$	1189	8.0×10^{-11}									+1
	hyperon	Σ^0	$\overline{\Sigma}^0$	1192	10^{-14}	1/2	0	0	0	+1	–1	0	1	0
		Σ^-	$\overline{\Sigma}^+$	1197	105×10^{-10}									–1
		Ξ^0	$\overline{\Xi}^0$	1315	3.0×10^{-10}									+1/2
	Xi hyperon					1/2	0	0	0	+1	–2	–1	1/2	
		Ξ^-	$\overline{\Xi}^+$	1321	1.7×10^{-10}									–1/2
	Omega hyperon	Ω^-	Ω^+	1672	1.3×10^{-10}	3/2	0	0	0	+1	–3	–2	0	0

(ii) proton-proton collision

$$p^+ + p^+ \begin{matrix} \nearrow p^+ + p^+ + \pi^0 \\ \searrow p^+ + n + \pi^+ \end{matrix} \tag{26.31}$$

(iii) K^- and p^+ collision

$$K^- + p^+ \rightarrow \Xi + K^0 + \pi^+$$

$$K^0 \rightarrow \pi^+ + \pi^-$$

$$\Xi \rightarrow \pi^- + \Lambda^0$$

$$\Lambda^0 \rightarrow \pi^- + \pi^+ \tag{26.32}$$

Obviously, at high energies the collisions between particles are not elastic. The particles can no longer be considered as billiard balls, which is useful approximation at low energies.

26.7 CONSERVATION LAWS

Elementary particles follow a set of conservation rules that give clues as to their nature and which, incidentally, sum up much of the knowledge of physics.

(i) Conservation of energy: In common with all isolated physical phenomena, the reactions of fundamental particles conserve energy distributed in various forms including mass, e.g.

$$n \rightarrow p + e^- + \bar{\nu}_e + 0.782 \text{ MeV}$$

However the following process

$$K^0 \neq \pi^+ + \pi^- + \pi^e + \pi^0$$

is not observed, because the rest energy of the K^0 is not great enough to make four pions even if they all could be made at rest. However, the conservation of energy principle allows the following process

$$K^0 \Rightarrow \pi^+ + \pi^- + \pi^0$$

(ii) Conservation of linear momentum: Linear momentum is always conserved in all elementary particle reactions. It is related to the invariance of the physical laws under translation in space. The laws of interaction do not depend on the place of measurement. Obviously, the space is homogeneous. This demands that the annihilation of a slow positron and an electron cannot result in one photon, but must produce at least two, preceeding in opposite direction.

(iii) Conservation of angular momentum: This is also followed by all types of interactions. This is related to the invariance of the physical laws under rotation (isotropy of space). The conservation of angular momentum includes both types (orbital as well as spin) of angular momentum together. It should be noted that total angular momentum is conserved. The orbital and spin angular momentum may not be separately conserved. The orbital angular momentum is given by the motion of the object as a whole about any chosen external axis of rotation. The spin is the intrinsic angular momentum of each object about an axis through its centre of mass. Conservation of angular momentum places restrictions on the spins that the products of the reaction can have. Strongly interacting fermions have half integer spin ($S = \frac{1}{2}$ for $\Xi, \Sigma, \Lambda, n, p$ and $S = \frac{3}{2}$ for Ω), strongly interacting bosons (η, K, π) have $S = 0$; weakly interacting fermions, so called leptons (μ, e, ν_e, ν_μ) have $S = \frac{1}{2}$, massless bosons (photon) have $S = 1$ and gravitons have $S = 2$. The total spins of the reaction products may only differ by 0 or 1 from the total spins of the initial components. In other words, during a reaction $\Delta S_{\text{total}} = 0$ or 1. The following reaction exhibit the spin balance

$$n \rightarrow p + e^- + \bar{\nu}_e + Q$$

$$\left(\frac{1}{2}\right)\left(\frac{1}{2}\right)\left(\frac{1}{2}\right)\left(-\frac{1}{2}\right)$$

Thus $\Delta S = 0$. Spin angular momentum is given by $L = \sqrt{S(S+1)}\,\hbar$.

(iv) Conservation of Parity: Parity holds for strong nuclear and electromagnetic interactions, but is violated in weak interaction. Parity is related to the invariance of the physical laws under inversion of the space coordinates, i.e., x, y, z are to be replaced by $-x, -y, -z$. Obviously, this is equivalent to combined reflection and rotation. We must note that the physical laws do not depend on the left or right handedness of the coordinate system.

The spatial parity depends on the orbital angular momentum and is given by $P = (-1)^l$ where l is the azimutal quantum number. Obviously, the states of even l have parity ($P = +1$) while states of odd 1 have odd parity ($P = -1$). Since the two successive parity operations bring the system back to its original state and hence $P^2 = 1$. This means P have two possible eigen values ± 1:

Let $\psi(x, y, z) = \psi(r)$ represents the wave function of the state of a system, then we have

$$\psi(-r) = +\psi(r) \text{ for even parity states}$$
$$\psi(-r) = -\psi(r) \text{ for odd parity states}$$

Parity is a good quantum number as it is conserved in strong and electromagnetic interactions. We further note that an energy eigenstate is also an eigenstate of parity.

We must remember that parity operation symmetry represents discrete symmetry (reflection and rotation through 180°) unlike the symmetries under infinitesimal changes implied in conservation of energy, conservation of linear momentum and conservation of angular momentum.

Parity has a broader definition, when one is dealing with elementary particles. Every elementary particle with a non-zero mass has an *intrinsic* parity π. The states with $P_i = +1$ are called even intrinsic parity states and the states with $\pi = -1$ are said to be *odd intrinsic* parity states. Obviously, the total parity of a system of n particles is the product of their intrinsic parties and the orbital parity $(-1)^l$ and can be expressed as $\pi_1, \pi_2, \pi_3, \pi_4 \ldots \pi_n (-1)^l$.

The intrinsic parities of electron, proton and neutron are taken to be even. Then one finds that the intrinsic parities of other particles are fixed relative to the intrinsic parities of nucleons. We must note that the intrinsic parity of the pions ($\pi^{\pm}, \pi^0$) is odd.

Usually the parity of a particle is designated by the symbol (+) for even and (-) for odd and expressed as subscripts above the total angular momentum, i.e., (J^P). For nucleons one can write $(J^P) = \left(\frac{1}{2}+\right)$ and for pions ($J^P = 0$).

If space inversion symmetry exists, i.e., there is no distinction between the coordinate frame and the inverted coordinated frame, then

$$[H, P] = 0$$

i.e., P commutes with Hamiltonian H.

In a nuclear reaction the product of the parities of the initial components was always thought to be equal to the product of the parities of the final components. This conservation of parity law was shown to be invalid by the theoretical work of Tsung Lee and Chen Yang, and the experimental verification by Mrs. Chien-Shiung Wu at Columbia University in 1956. Lee and Yang questioned the validity of invariance under space reflection for processes due to the weak interaction. The Lee and Yang proposal was motivated by what at that time was called the $\tau - \theta$ puzzle. Each particle, described by its corresponding field or wave function is supposed to have a certain intrinsic parity. Mesons have negative parity, Kaons disintegrate in several ways: one resulting in three pions, another in two pions. The decay modes are $K^{\pm} \rightarrow 2\pi^{\pm} + \pi^{\mp}$ and $K^{\pm} \rightarrow \pi^{\pm} + \pi^0$. Now, the parity of two pions (two 'odds') is odd. If parity were always conserved, then the two pions would have come from a particle of even parity state (it was tentatively called the "theta" meson) while three pions would have come from a particle of odd parity state (it was called the "tau" meson). However, the "tau-theta puzzle" arose over the fact that except for parity, they were identical. Lee and Yang decided that they were identical, and that a single particle disintegrated in two ways, parity not being conserved. The theta and tau mesons are now realized to be the same thing–the kaon.

(v) Conservation of Charge: We note that without exception all of the elementary particles have either no electric charge or the precise charge $\pm e$. There is no variety, as there is of masses. The positive and negative electric charges are exactly equal. The net electric charge of a system is always conserved, e.g.

$$n \rightarrow p + {}_{-1}^{0}e + 0.782 \text{ MeV}$$

Nowadays, the word charge is used in a broader sense in fundamental particle physics. Apart from electric charge Q, one define the baryonic charge B for each baryon and leptonic charge l for each lepton. In addition, strangeness, charm and beauty have similar properties and are considered as charges. $B = +1$ for the baryons and –1 for the antibaryons. Similarly $l = \pm 1$ for the leptons and antileptons, while for baryons, mesons $l = 0$. We must note that there are actually three different types of leptonic charges l_e, l_m and l_t associated with the three different group of leptons, e.g. electron-positron (e^+e^-), $\mu^+\mu^-$ and $\tau^+\tau^-$ (τ-lepton or tauon). Each type of these leptons are associated with a neutrino, viz. the electron-neutrino (ν_e), the muon-neutrino (ν_μ) and the taun-neutrino (ν_τ). The association of the charged leptons with their neutrinos (neutral) is expected to arise from the empirical law of leptonic charge (lepton number) conservation. All the neutrinos associated with leptons are probably mass zero particles.

It is observed that leptons and baryons disappear in pairs. This means that the net number of baryons and each kind of leptons remains unchanged in the universe. The lepton conservation law is probably related to the gauge invariance of the weak nuclear field.

With the help of conservation laws, one can understand why some reactions do not take place, e.g.

$$n \rightarrow \pi^0 + \gamma$$
$$(B = 1 \quad 0 \quad 0)$$

does not occur, because baryonic charge B is not conserved. The another example is

$$\mu^+ \rightarrow e^+ + \gamma$$
$$(l_\mu = -1 \quad 0 \quad 0)$$
$$(l_e = 0 \quad -1 \quad 0)$$

The above reaction does not occur because of the non conservation of the leptonic charges. However, conservation of the baryonic and leptonic charges are probably not satisfied exactly. The two processes stated above may actually occur, though with very small probabilities. However, they have not yet been observed.

The violation of baryon number, if any, indicates that the lightest baryon, i.e., proton, may not be absolutely stable, but may show the decay $p \rightarrow e^+\pi^0$. This is the prediction of grand unification theory (GUT).

(vi) Conservation of Charge (c)-parity: This is related to the invariance under charge conjugation operation (c). This operation consists in replacing all particles by their antiparticles without changing any other physical property, such as momentum or spin. A system is invariant with regard to charge conjugation when, if a process is possible in the original system, the corresponding process is also possible in the charge conjugate system. Charge conjugation means reversal of the signs of all types of charge (electric, baryonic and leptonic) of the particles. We must note that the principle of charge conjugation applies not only to charged particles, but also to neutral particles. Obviously, the neutrino (electrically neutral), has an antiparticle (the antineutrino) which is different from the former, the leptonic charges of the two being equal and opposite. This reveals that they interact differently with matter.

There are self conjugate particles, e.g. π^0-meson, the photon, the η^0-meson, the $\frac{J}{\psi}$ particle and γ-meson for which all the different charges are zero ($Q=0, B=0, l=0$). These particles must be identical with their antiparticles. Obviously, these particles constitute the truly neutral particles.

Positronium is another example of a truly neutral system. In a positronium, an electron and a positron are bound together to form a hydrogen like structure. Q, B and l are all zero for it.

C parity is definite, i.e., even (+) or odd (–) for the really neutral particles. When all electrons are replaced by positrons, the e.m. field changes sign. The wave function of photon must change sign under charge conjugation. This means C is odd for photons. Similarly, one finds that the C-parities of the $\frac{J}{\psi}$ and γ-meson are odd, i.e.,

$$C|\gamma>=-|\gamma>, C|\nu>=-|\nu> \text{ and } C|\frac{J}{\psi}>=-|\frac{J}{\psi}>$$

However, C parity for π^0 must be even because it breaks up into two photons. For η^0 meson also C-parity is even:

$$C|\pi^0>=|\pi^0>, C|\eta^0>=|\eta^0>$$

The following system from the operation of C-parity also occurs in nature, $\pi^- \to \mu^- + \bar{\nu}$.

This shows that weak interactions are invariant under charge-parity operation. We must note that the electromagnetic and strong nuclear interactions are invariant under C, whereas the weak interaction is not invariant under C, which is evident from the asymmetry in the β-decay of ^{60}Co.

We must note that the quantum number C has no meaning for the charged particles like. Also for a state , the quantum number C has no meaning.

(vii) Conservation of *T*: The operation of T, which is related to the time-reversal invariance, means changing the time t to $-t$ in all equations of motion, i.e., reflection of the time axis as the origin of the time coordinate (in relativistic space-time continuum). Under time reversal the quantities velocity, momentum, and angular momentum are reversed and in a collision the initial and final states are exchanged. Obviously, T operation is also a discrete change like the parity operation.

$\frac{d\mathbf{p}}{dt}$ is invariant under time reversal, since it implies the replacements $p \to -p$ and $t \to -t$. Also the electric field e is invariant under time reversal. However, the magnetic field suffers the transformation $B \to -B$ because time reversal implies reversing the velocities and therefore also the currents. Then $\mathbf{V} \times \mathbf{B}$ is invariant under time reversal, because both factors have changed sign, and the equation of motion of charged particle, $\frac{d\mathbf{p}}{dt} = -q(\mathbf{E} + \boldsymbol{\nu} \times \mathbf{B})$, is invariant w.r.t. time reversal, T operation causes the reversal of sign of the angular momenta and transforms the wave function to its complex conjugate. One can easily verify that the time reversed wave function $\psi_{rev}(t) = \psi^*(-t)$ is a solution of the Schrodinger equation, which is obviously time-reversal invariant.

Strong and weak interactions are also considered invariant under time reversal, although lately the problem has been open to discussion. Recent (1966) experiments regarding a special form of decay of neutral Kaon, which occurs only in 0.3% of the cases, seem to indicate that weak interactions may violate the principle of invariance under time reversal.

(viii) CPT Theorem: All physical laws are considered to be invariant under the CPT operation, i.e., under the combined operations of time reversal T, Parity P and charge conjugation T. This theorem also known as Luders-Pauli theorem is true even in cases in which the laws may not be invariant under the individual operations. The orders of the three operations are immaterial. So far, all evidences point to the universal validity of this theorem. The CPT invariance implies that if an interaction is not invariant under any one operations C, P or T, its effect must be compensated by the combined effect of the remaining two operations. It is believed that probable CPT violation, if any can be observed in the properties of the neutral K^0 and $\bar{K}^0$ mesons.

(ix) Conservation of Isospin: If we consider that the neutron and the proton are the two charge states of the same entity known as nucleon then one finds the charge independence or isotopic invariance of nuclear force. This leads to the idea of isospin. One can take the isospin of the nucleons to be $I = \frac{1}{2}$ with the two components $I_3 = \frac{1}{2}$ for the proton and $I_3 = -\frac{1}{2}$ for the neutron. The charge being given by

$$Q = \frac{1}{2}(1 + 2I_3) \tag{26.33}$$

We must note that some of the books have used T for the isospin.

The isospin multiplicity is obtained by

$$M = 2I + 1 \tag{26.34}$$

In the presence of strong nuclear interaction, the two states of the nucleon with different I_3 are degenerate, i.e., symmetric. However, the symmetry is broken by the electromagnetic interaction. The electromagnetic interaction makes the energies of the two states, i.e., the proton and the neutron masses, slightly different. For both the states, the intrinsic spin and parity $\left(\frac{1^+}{2}\right)$ are the same.

We must note that the I is conserved in strong interaction, but not in electromagnetic interaction. One can see that $[H, I] = 0$, i.e., H commute with I. This implies that conservation of the isospin is related to the invariance of H under rotation of the axes in the hypothetical isospin space, which is assumed to be isotropic.

One can also introduce the concept of the isospin for the other strongly interacting particle multiplets having nearly same masses. Because of the symmetry breaking property of the electromagnetic interaction one finds that the energies, i.e., masses of the different charge states with different I_3 (same I) are different. Few example, isospin multiplets are the pions (π^+, π^0, π^-) with $I = 1$, $I_3 = +1, 0, -1$ respectively $(M = 3)$. Sigma hyperons $(\Sigma^+, \Sigma^0, \Sigma^-)$ with $I = 1$, $I_3 = +1, 0, -1$ respectively $(M = 3)$ and Kaons (K^+, K^0) with $I = \frac{1}{2}$ $I_3 = +\frac{1}{2}, -\frac{1}{2}$ respectively $(M = 2)$. Λ^0 hyperon and η^0 meson are examples of a particle multiplet having only one particle corresponding $I = 0$ and $I_3 = 0$.

The components of a particular isospin multiplet exhibit the similarity in observed properties, e.g. the mass spin, intrinsic parity and life time. This leads to the hypothesis of isotopic invariance in strong interactions involving the particles in the different cases states above. We shall see that the resonance particles (excided states of baryons and mesons) also form isospin multiplets. These particles have the same isospin multiplicity as the corresponding particles in their ground state. Δ-resonance is an example of an isospin quadruplet ($M = 4$) with $I = \frac{3}{2}$. This can exist in the following four charge states, e.g.

$$\Delta^{++}, \Delta^{+}, \Delta^{0}, \Delta^{-} \text{ with } I_3 = +\frac{3}{2} \quad +\frac{1}{2} \quad -\frac{1}{2} \text{ and } -\frac{3}{2} \text{ respectively.}$$

(x) Conservation of Strangeness: To describe the strange behaviour of certain groups of particles (strange particles) with respect to their production by strong interaction and decay by weak interaction a strangeness quantum number S is used. The law of conservation of strangeness requires that *the total strangeness S must remain the same in process due to strong or electromagnetic interactions.*

Particles and antiparticles have opposite strangeness. The following relationship holds between the change Q of a particle and its strangeness

$$Q = I_3 + \left(\frac{B + S}{2}\right) \tag{26.35}$$

Where B is the baryon number and I_3 is the third component of the isospin I.

One can see that the process $\pi^- + p^+ \to \eta^0 + \Lambda^0$, which complies with all the other conservation laws, is not observed because the total strangeness on the left is zero, while on the right it is –1. However, processes such as $\pi^+ + p^+ \to \Sigma^+ + K^+$ and $\pi^- + p^+ \to \Lambda^0 + K^0$ do occur. We can easily see that in both cases the total strangeness remains equal to zero. We have for the K^+ mesons $B = 0$, $S = 1$ and $Q = +1$. This gives $I_3 = +\frac{1}{2}$. On the other hand we have $Q = 0$, $I = -\frac{1}{2}$. Obviously, K^+ and K^0 form an isospin doublet with $I = \frac{1}{2}$. One finds that for the non-strange baryon, proton (or neutron) $B = 1$, $S = 0$ and hence $Q = I_3 + \frac{B}{2} = I_3 + \frac{1}{2}$. This requires one to put $I_3 = +\frac{1}{2}\left(\text{Or } -\frac{1}{2}\right)$ to make $Q = +1$ (or 0). Now, we define the hypercharge Y as

$$Y = B + S \tag{26.36}$$

This gives

$$Q = I_3 + \frac{Y}{2} \tag{26.37}$$

Since B and S are both conserved in strong interaction and hence Y is also conserved. We must note that the conservation of strangeness is not a rigorous law as strangeness is not conserved in weak interaction. Conservation of strangeness law can be violated by weak interactions which allow a change of strangeness of ±1. In most baryon decays, the conservation of strangeness is violated, indicating that the decays take place by means of weak interactions, e.g.

$$\Xi^- \to + \Lambda^0 + \pi^- \quad (\Delta S = +1)$$

$$\hookrightarrow \pi^- + \pi^+ \quad (\Delta S = +1)$$

Hyperons cannot decay into baryons through the strong interaction with conservation of strangeness. This explains why the hyperons decay so slowly (about 10^{-10} s) compared with the time involved in their production or annihilation in collisions

Table 26.6

Quantity	*Symbol*	*Weak*	*Electromagnetic*	*Strong*
Energy	E	Yes	Yes	Yes
Momentum	p	Yes	Yes	Yes
Parity	P	No	Yes	Yes
Isospin	I	No	No	Yes
Third Component of 1	I_3	No	Yes	Yes
Charge Conjugation	C	No	Yes	Yes
Time reversal	T	No	Yes	Yes
CP	–	No	Yes	Yes
CPT	–	Yes	Yes	Yes
Baryon Number	B	Yes	Yes	Yes
Lepton Number	l	Yes	Yes	Yes
Charge	Q	Yes	Yes	Yes
Strangeness	S	No	Yes	Yes
Hypercharge	Y	No	Yes	Yes
Charm	c	No	Yes	Yes

by means of strong interactions, which is about 10^{-23} s. The selection rule $\Delta S = 0, \pm 1$ is obeyed in weak decay of semi-stable strange particles. Few examples are

$$\Omega^- \rightarrow \Lambda^0 + K^- \quad (\Delta S = +1)$$
$$\Sigma^0 \rightarrow \Lambda^0 + \gamma \quad (\Delta S = 0)$$
$$\Sigma^+ \rightarrow p + \pi^0 \quad (\Delta S = +1)$$
$$\Xi^0 \rightarrow \Lambda^0 + \pi^0 \quad (\Delta S = +1)$$

The decay $\Xi^- \rightarrow n + p^-$ $(\Delta S = 2)$ is not observed.

Table 26.6 exhibit the behaviour of various quantities in different types of interactions.

26.8 RESONANCE PARTICLES

Extremely short-lived elementary particles that decay by strong interaction in about 10^{-24} s are called resonance particles. Obviously, they are hadrons. Resonances may be regarded as excited states of the more stable particles. For such short-lived particles, all one can do is infer their masses, their life times, and, indeed, their very existence from data on their decay products. If the energy of the colliding particles is slowly increased, a sudden peak in particle production is observed at the resonance energy; production drops off as with the further increase in energy. This reveals that a particle exists with an effective mass equal to the combined relativistic masses of the colliding particles. One cannot observe resonances directly, but resonances appear as spreads in the masses of more stable hadrons. Over one hundred baryon and meson resonances are known. Baryon resonances are normally denoted by the symbol of a similar baryon of lower mass, followed in brackets by the approximate resonance mass in MeV; e.g.

$$\text{N}(1450), \text{N}(1520), \text{N}(1535)$$
$$\Lambda(1405), \Sigma(1385), \Xi(1530)$$

where N denotes the symbol for nucleon. Most meson resonances have individual symbols, but we have shown mass in brackets, e.g.

$$\rho(775), \omega(784), K^*(892)$$
$$\eta'(958), \phi(1019), f(1260)$$

The spin of η' is 0 and that of f meson is 1.

26.9 QUARK MODEL

We have read that leptons appear to be truly elementary particles because their number is small, they have no measurable size or internal structure, and moreover they do not seem to breakdown to smaller units. On the other hand, hadrons are complex particles having size and structure. The discovery of resonance particles forced physicists for the new classifications for strongly interacting particles. In 1961, Gell-Mann noticed patterns in the then known particles. They called their patterns the *Eightfold way*, after Buddhist principles for right living.

According to Gell-Mann structural model, all hadrons are composite systems of two or three constituents called *quarks*. According to this model, three types of quarks exists, designated by the symbol, *u*, *d*, and *s*. They are arbitrary called as *up*, *down* and *strange*. For each there was a corresponding antiquark (initially called sideways) with charges $2e/3$, $-e/3$, and $-e/3$ respectively. The strangeness value of strange quark is –1 and those of other two quarks are zero. The fractional electronic charge is an unusual property of quarks. The various types of quarks are called *flavors*. All known hadrons at that time could be specified by some combinations of quarks and antiquarks. Baryons consists of threee quarks, and mesons consists of a quark and antiquark. Quark composition of two mesons and two baryons is shown in Fig. 26.8.

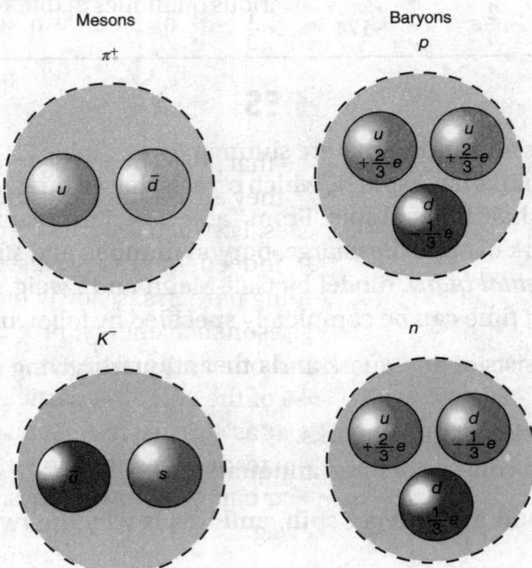

Fig. 26.8 Quark compositions as pictorial representation of few hadrons (two mesons and two baryons)

Table 26.7 Properties of quarks and antiquarks. All quarks have spin ½ and baryon number ⅓

Name	*Symbol*	*Spin*	*Charge*	*Mass (GeV/c^2)*	*Baryon Number*	*Strangeness (S)*	*Charm (C)*	*Bottomness (B)*	*Topness (T)*
Quarks									
Up	u	$\frac{1}{2}$	$+\frac{2}{3}e$	0.0015 – 0.005	$\frac{1}{3}$	0	0	0	0
Down	d	$\frac{1}{2}$	$+\frac{1}{3}e$	0.003 – 0.009	$\frac{1}{3}$	0	0	0	0
Strange	s	$\frac{1}{2}$	$+\frac{1}{3}e$	0.06 – 0.107	$\frac{1}{3}$	–1	0	0	0
Charmed	c	$\frac{1}{2}$	$+\frac{2}{3}e$	1.1 – 1.5	$\frac{1}{3}$	0	+1	0	0
Bottom	b	$\frac{1}{2}$	$+\frac{1}{3}e$	4.1 – 4.4	$\frac{1}{3}$	0	0	+1	0
Top	t	$\frac{1}{2}$	$+\frac{2}{3}e$	~174	$\frac{1}{3}$	0	0	0	+1
Antiquarks									
Anti-up	$\bar{u}$	$\frac{1}{2}$	$-\frac{2}{3}e$	0.0015 – 0.005	$-\frac{1}{3}$	0	0	0	0
Anti-down	$\bar{d}$	$\frac{1}{2}$	$+\frac{1}{3}e$	0.003 – 0.009	$-\frac{1}{3}$	0	0	0	0
Anti-strange	$\bar{s}$	$\frac{1}{2}$	$+\frac{1}{3}e$	0.06 – 0.107	$-\frac{1}{3}$	+1	0	0	0
Anti-charmed	$\bar{c}$	$\frac{1}{2}$	$-\frac{2}{3}e$	1.1 – 1.5	$-\frac{1}{3}$	0	–1	0	0
Anti-bottom	$\bar{b}$	$\frac{1}{2}$	$+\frac{1}{3}e$	4.1 – 4.4	$-\frac{1}{3}$	0	0	–1	0
Anti-top	$\bar{t}$	$\frac{1}{2}$	$-\frac{2}{3}e$	~174	$-\frac{1}{3}$	0	0	0	–1

Properties of quarks and antiquarks are summarized in Table 26.7.

We may note that quarks have spin ½, which reveals that all quarks are fermions, defined as any particle having half-integral spin. From Table 26.7, it is clear that with each quark is associated an antiquark of opposite charge, baryon number, and strangeness.

On the basis of *original quark* model by Gell-Mann and Zweig, the composition of all known hadrons at that time can be completely specified by following three rules:

(i) A meson consists of one quark and one antiquark, giving a baryon number of 0, as required.

(ii) A baryon consists of three quarks.

(iii) An antibaryon consists of three antiquarks.

The quarks all have odd half-integer spin, and this is why the two-quark mesons have integer spin.

Pauli exclusion principle precludes threee particles having the same quantum numbers, i.e., there must be an additional property distinguishing quarks. This is called *color*. This property is a kind of "charge" (not to be confused with electric charge) – that can take on any of three values whimsically, *red*, *green* and *blue*. The antiquarks have the colors *antired, antigreen*, and *antiblue*. The force binding quarks of different colors is the *strong force*. The quark theory is known as *quantum chromodynamics* (QCD). In QCD theory, *gluons* are field particles. Gluons play the role of photons in quantum electrodynamics (QED), binding particles subject to the strong force. Particles formed from quarks, i.e., the mesons and the baryons are always *colorless*. This is true of mesons because they contain a quark of one colour and another of its anticolour. In case of baryons, it is true because they contain three quarks of different colours, which combine to give the baryon zero net colour charge. We may note that the nuclear force, once thought to be fundamental, is actually a residual manifestation of the strong force, acting between the quarks in *colorless* particles – in much the same way that the van der Waals force act between neutral gas molecules is a "residue" of stronger electric force among the particles that make up the molecules.

We know that photons are uncharged but they mediate the electromagnetic force. On the other hand, gluons, like the quarks they bind, carry color charge. According to QCD, there are eight different gluons – all carrying two color charges, a color and an anticolor. Six carry combination likes red-antiblue ($R\overline{B}$), green-antired ($G\overline{R}$), and so on; the other two are colorless. When a quark emits or absorbs a gluon its color changes, e.g. a blue quark that emits a blue-antired ($B\overline{R}$) gluon becomes a red quark, and a red quark that absorbs this gluon becomes a blue quark.

A comparison of Yukawa's pion model and quark model is shown in Fig. 26.9. In Fig. 26.9(a), the charged pion mediates charge from one nucleon to the other, so the nucleons change their identities, i.e., p becomes n and n becomes p. Figure 26.9(b) presents the same interaction from the viewpoint of quark model. This depicts one possibility. A down quark in the neutron on the right emits a gluon. The energy of the gluon is then transferred to create a $u\overline{u}$ pair. We can see that the u quark stays within the nucleon (now changed to a proton), and the recoiling of quark and the $\overline{u}$ antiquark are transmitted to the proton

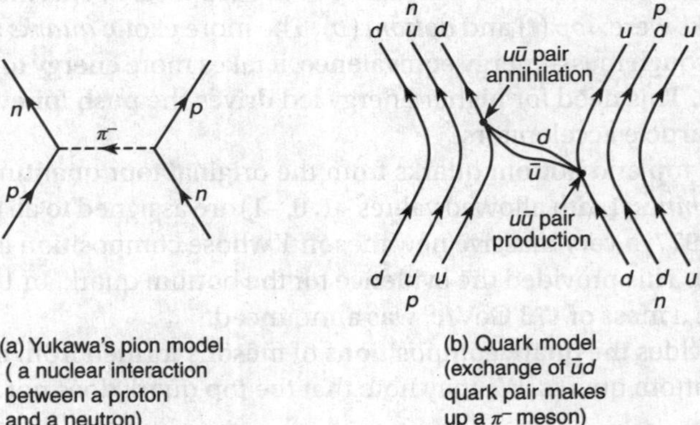

Fig. 26.9 Nuclear interaction between a proton and a neutron in view of (a) Yukawa's pion model and (b) Quark model

on the left hand side of figure. Obviously, $\overline{u}$ annihilates a u quark within the proton and the d is captured. Resulting effect is to change a u quark to a d quark, and the proton has changed to a neutron.

We can see that d quark and $\overline{u}$ antiquark in Fig. 26.9(b) transfer between nucleons, the d and $\overline{u}$ exchange gluons with each other. One can consider it to be bound to each other by means of strong force. This combination is a π^-, which is Yukawa's field particle, i.e., quark model is consistent with Yukawa's pion exchange model.

A very surprising aspect of quarks is that the strong force does not decrease with separation. This is why, it is impossible to isolate a single quark. This is the reason that we never see individual free particles with fractional electric charge. Current efforts are underway to form a *quark-gluon plasma*, a state of matter in which quarks are freed from neutrons and protons.

The up, down, and strange quarks soon found insufficient to account for all the observed properties of particles, i.e., discrepancies noticed between predictions of the model and certain experimental decay rates. In 1967, a forth quark, called the *charmed quark* designated by c, was proposed. A *charmed* quark has charge $+\frac{2}{3}e$, but its charm distinguishes it from the other three quarks. This addition of charm quark introduces a new quantum number C, representing charm. The new quark has $C = +1$, and its antiquark has $C = -1$ and all other quarks have $C = 0$ (Table 26.7). Like strangeness charm is conserved in strong and electromagnetic interactions, but not in weak interactions. Ten years later the discovery of a particle that implies the existence of the charmed quark was announced. The J/ψ (or simply ψ) has properties of a combination of the proposed charmed quark and its antiquark ($c\overline{c}$) and it does not fit into the u, d, s, i.e., three quark model. This charmed quark is much more massive than the other known mesons (~3 100 MeV/c^2), and its life time is quite longer than the life times of particles that decay via the strong force. Later, related mesons, corresponding to $\overline{c}d$ and $c\overline{d}$ quark combinations were discovered, all having large masses and long life times.

In 1975, tau lepton (t) with a mass of 1784 MeV/c^2 was reported. This is a fifth type of lepton and this led physicists to propose that more flavours of quarks may exist, based on symmetry arguments similar to those leading to the proposal of charmed quark. The predicted new quarks were: *top* (t) and *bottom* (b). The more exotic *quarks* are more massive, and therefore through mass-energy equivalence, it takes more energy to produce particles containing them. This need for higher energy led drives the push for ever more powerful and expensive particle accelerators.

To distinguish top and bottom quarks from the original four quantum numbers called *topless* and *bottomness* (with allowed values +1, 0, –1) are assigned to all quarks antiquarks (Table 26.7). In 1977, a very massive new meson Y whose composition is considered to be $b\overline{b}$ was reported. This provided the evidence for the bottom quark. In 1995, the discovery of top quark with a mass of 173 GeV/c^2 was announced.

Table 26.8 provides the quark compositions of mesons formed from up, down, strange, charmed, and bottom quarks. We may note that the top quark does not form mesons as it decays too quickly.

Table 26.8 Quark compositions of mesons formed from up, down, strange, charmed and bottom quarks

		Antiquarks				
		$\bar{b}$	$\bar{c}$	$\bar{s}$	$\bar{d}$	$\bar{u}$
	b	Υ $(\bar{b}b)$	B_c^- $(\bar{c}b)$	B_s^0 $(\bar{s}b)$	$\bar{B}^0$ $(\bar{d}b)$	B^- $(\bar{u}b)$
	c	B_c^+ $(\bar{b}c)$	J/ψ $(\bar{c}c)$	D_S^+ $(\bar{s}c)$	D^+ $(\bar{d}c)$	D^0 $(\bar{u}c)$
Quarks	**s**	B_S^0 $(\bar{b}s)$	D_S $(\bar{c}s)$	η, η' $(\bar{s}s)$	$\bar{K}^0$ $(\bar{d}s)$	K^- $(\bar{u}s)$
	d	B^0 $(\bar{b}d)$	D^- $(\bar{c}d)$	K^0 $(\bar{s}d)$	π^0, η, η' $(\bar{d}d)$	π^- $(\bar{u}d)$
	u	B^+ $(\bar{b}u)$	$\bar{D}^0$ $(\bar{c}u)$	K^+ $(\bar{s}u)$	π^+ $(\bar{d}u)$	π^0, η, η' $(\bar{u}u)$

Table 26.9 Quark Composition of some Baryons

Particle	*Quark Composition*
p	*uud*
n	*udd*
Λ^0	*uds*
Σ^+	*uus*
Σ^0	*uds*
Σ^-	*dds*
Δ^{++}	*uuu*
Δ^+	*uud*
Δ^0	*udd*
Δ^-	*ddd*
Ξ^0	*uss*
Ξ^-	*dss*
Ω^-	sss

Quark compositions for some baryons is shown in Table 26.9. We may note that some baryons have the same quark composition, such as the p and the Δ^+ and the n and Δ^0. In such cases, the Δ particles are considered to be excited states of the proton and the neutron.

26.10 THE STANDARD MODEL

Now, we have *six flavors* of quarks – up, down, strange, charmed, top and bottom. They seem to be truly elementary constituents of matter. We have seen that quarks join to form the hadrons, baryons and mesons. But other particles, i.e., leptons and field particles are not possible to be made from quarks. They, like quarks appear to be truly indivisible, elementary particles.

We have seen that particle physicists recognize three distinct "families": (i) the up and down quarks make the neutron and proton; together with the electron and its related neutrino, they account for the properties of ordinary matter; (ii) a family of the strange and charmed quarks, electron like muon, and the muon neutrino. The quarks of this family are more massive than the up and down quarks, and the muon is more massive than the electron; (iii) the third family consists of top and bottom quarks, the electron-like tau particle, and the tau neutrino. These are more massive than first two families. Table 26.7 provides details of these families.

Particle physicists now believe that there are three classifications of elementary particles: leptons, quarks, and field particles or gauge bosons. These three are further classified as either fermions or bosons. Quarks and leptons are fermions having spin ½, whereas the field particles are bosons having integral spin 1 or higher.

We have read that weak force is mediated by the W^+, W^-, and Z^0 bosons. These particles are said to have *weak charge* just as quarks have colour charge. Obviously, each elementary particle can have mass, electric charge, colour charge, and weak charge. Of course, one or more of these may be zero.

In 1979, Glashow, Salam and Weinberg unified the electromagnetic and weak interactions. The *electroweak* reveals that the weak and electromagnetic interactions have the same strength at very high particle energies. Electromagnetic and weak interactions are viewed as two different manifestations of single unifying electroweak theory. The photon and three bosons, $W^{\pm}$, Z^0 play a key role in this theory. According to this theory, the masses of W and Z particles should be about 82 GeV/c^2 and 93 GeV/c^2, respectively.

The combination of electroweak theory and QCD for strong interaction is called the *Standard model.* Currently, this describes the elementary particles and their interactions. This model includes the photons, the gluons, $W^{\pm}$ and Z particles, and an particle called Higgs boson. The discovery of higgs like particle with mass 125 GeV at Large Hadron Collider (LHC) at CERN (Geneva) was announced on 4th July, 2012. (Table 26.10).

We may note that the standard model does not include the gravitational force at present.

The standard model fails to answer, why, of the two mediators of electroweak interaction, the photon has no mass but the $W^{\pm}$ and Z bosons do. Summary of the interactions between particles described by the standard model is presented in Fig. 26.10.

Table 26.10 Field Particles and Forces (Standard Model)

Particle	*Mass (GeV/c^2)*	*Electric and color charges*	*Force mediated*	*Range*	*Approximate strength at 1fm (relative to strong force)*
Graviton	0	0, 0	Gravity	Infinite	10^{-38}
$W^{\pm}$	80.2	±1, 0	Weak	$<2.4 \times 10^{-18}$ m	10^{-13}
Z^0	91.2	0, 0	Weak	$<2.4 \times 10^{-18}$ m	10^{-13}
Photon, γ	0	0, 0	Electromagnetic	Infinite	10^{-2}
Gluon, γ (8 varieties)	0	0, 6 colour-anticolour combinations	Strong	Infinite	1
Higgs boson, H^0	40–1000?	0, 0	This particle is proposed to account for the masses of the other particles.		

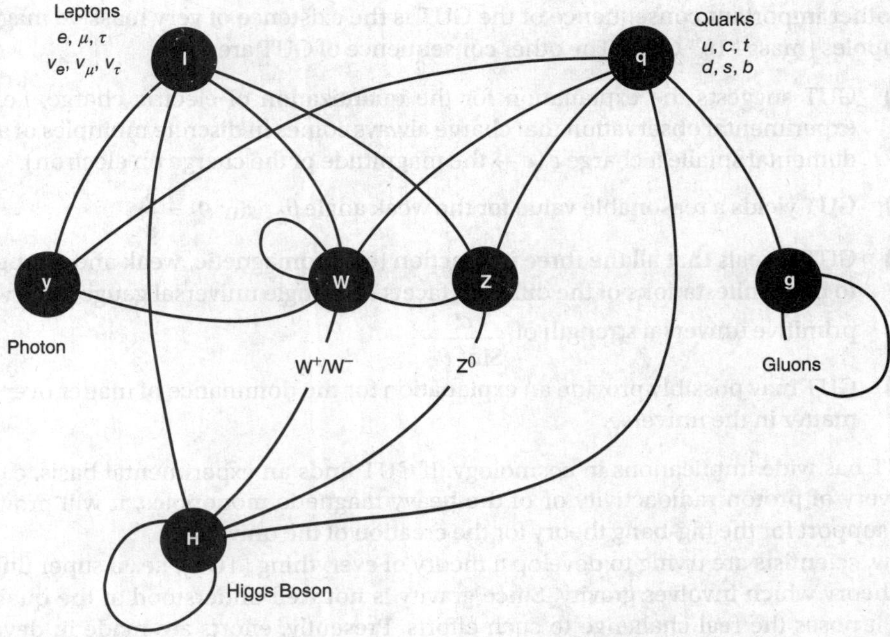

Fig. 26.10 Interactions between particles described by the Standard Model

26.11 GRAND UNIFIED THEORY

The theoreticians have further developed grand unification theory (GUT) of the strong, electromagnetic and weak forces. In such a GUT the strong, weak and electromagnetic interactions appear as different manifestations of one basic phenomenon, with leptons and quarks find natural places within the scheme. Obviously, GUT provides a super structure in which $SU(3)$ and $SU(2) \times U(1)$ theories can be embedded. The super structure would take the form of a larger symmetry in which quarks and leptons would be treated on equal footing. The search of large symmetry leads to a larger group $SU(5)$, the smallest simple group that can accommodate the constituent $SU(3)$ and $SU(2) \times U(1)$ symmetries. It is believed that $SU(5)$ to be a likely symmetry group-of nature.

A starling consequence of GUT is the disappearance of the distinction between quarks and leptons, i.e., a transition from a quark to a lepton or vice versa can take place. The interconversion of quarks and leptons is possible only in GUT. Obviously, the protons would then not be stable particles. The present theoretical estimated life of proton decay to be ~10^{30} years for the decay mode $\pi \to (e^+\pi^0)$. Proton decay requires that the baryon number violating interactions may be able to explain the asymmetry between matter and antimatter. There is as yet no experimental evidence in support of proton decay. Obviously, if proton can decay, the atom itself is unstable, and all matter is impermanent.

Another important consequence of the GUT is the existence of very massive magnetic monopoles (mass ~10^{16} GeV). The other consequence of GUT are

(i) GUT suggests the explanation for the quantization of electric charge, i.e., the experimental observation that charge always comes in discrete multiples of a fundamental smallest charge e ($e \rightarrow$ the magnitude of the charge on electron).

(ii) GUT yields a reasonable value for the weak angle $\theta_{\hat{w}}$, $\sin^2\theta_{\hat{w}} = \frac{3}{8}$.

(iii) GUT reveals that all the three interaction (electromagnetic, weak and strong) are to be manifestations of the different facets of a single universal gauge force with a primitive universal strength of $\dfrac{\alpha}{\text{Sin}^2\,\theta_w}$.

(iv) GUT may possibly provide an explanation for the dominance of matter over antimatter in the universe.

GUT has wide implications in cosmology. If GUT finds an experimental basis, e.g. the discovery of proton radioactivity or of the heavy magnetic monopoles, it will provide a good support for the big-bang theory for the creation of the universe.

Now, scientists are trying to develop a theory of everything (TOE), i.e., a super unification theory which involves gravity. Since gravity is not well understood at the quantum level, it poses the real challenge to such efforts. Presently, efforts are made in developing the so-called supergravity and *superstring field theories*. However, these efforts have important role to play in understanding the origin and history of early universe, matter generation and the inflation problems.

We now have a good understanding of the natural phenomena ranging from microcosm to macrocosm. Obviously, we have a unified view of the cosmos. However, the question about the ultimate structure of matter is still unanswered. It may have more surprises in its store. Having knowledge of the many elementary particles (6 leptons, 18 quarks and 13 quanta), some scientists have proposed *preon models*, i.e., next layer of matter which may be approach at the TeV energy scale. Obviously, there may be many more layers of the structure of the matter leading to the problem of infinite divisibility. This means ultimate constitutes of matter simply do not exist. On the contrary, there are scientists believe that there should be an end to the sequence of ever finer levels of the structure and the color confinement-provides an clear indication to that. We will have to wait for future experiments for deciding these exciting crucial issues.

SOLVED PROBLEMS

Example 26.1: If a pion decays from rest to give a muon of 4 MeV energy, what is the kinetic energy of the accompanying neutrino? What is the mass of the neutrino in this process?

Solution The pion decay is

$$\pi^+ \rightarrow \mu^+ + \nu_\mu + E$$

Here $E = M_\pi - M_\mu$ (neutrino has zero 'rest' mass)

$$= 273\, m_e - 207\, m_e = 66\, m_e$$

$$= 66 \times 0.51 \text{ MeV} (\because m_e = 0.51 \text{ MeV})$$
$$= 33.7 \text{ MeV}$$

Since the muon takes 4.0 MeV of energy, the energy shared by the neutrino is 33.7 – 4.0 = 29.7 MeV.

$$\text{Mass of this neutrino} = \frac{29.7}{0.51} m_e$$
$$= 58.2\, m_e$$

Example 26.2: Find the maximum kinetic energy of the electron emitted in the beta-decay of the free electron. The neutron-proton mass different is 1.30 MeV.

Solution The decay equation is

$$n^\circ \rightarrow p^+ + e^- + \nu_e + E$$

From this, we have

$$E = M_n - M_p - m_e$$
$$= (1.30 \text{ MeV}) - m_e$$
$$= 1.30 \text{ MeV} - 0.51 \text{ MeV}$$
$$= 0.79 \text{ MeV}$$

This energy release goes into the kinetic energy of the electron plus the kinetic energy of the antineutrino. The two particles can share the 0.97 MeV of energy in anyway they choose. Thus the electron may have a maximum kinetic energy of 0.79 MeV.

Example 26.3: Which of the following reactions are allowed and forbidden through the strong interactions?

(i) $\Sigma^+ + {}_0^1n \rightarrow \Sigma^- + {}_1^1p$

(ii) $\pi^+ + {}_0^1n \rightarrow K^+ + \Sigma^0$

(iii) $\Lambda^0 \rightarrow \Sigma^+ + \pi^+$

(iv) $\Lambda^0 + {}_0^1n \rightarrow \Sigma^- + {}_1^1p$

(v) $K^+ \rightarrow \pi^+ + \pi^\circ + \pi^0$

(vi) $K^- \rightarrow \pi^- + \pi^\circ$

(vii) $K^- + {}_0^1p \rightarrow K^\circ + {}_0^1n$

Solution

(i) Forbidden since charge is not conserved.
(ii) Allowed by strong interaction as strangeness is not conserved.
(iii) Forbidden as energy is not conserved.
(iv) Allowed, as strangeness is conserved.
(v) Forbidden as strangeness is not conserved.
(vi) Forbidden as strangeness is conserved.
(vii) Allowed, as strangeness is conserved.

Example 26.4: Classify the following processes as strong, electromagnetic, weak of totally forbidden and stable whether the following are conserved in each process: (i) parity (P), (ii) strangeness (S), (iii) Isospin (I), (iv) third component of isospin (I_3).

(i) $\pi^- + p \to \Lambda^0 + K^0$
(ii) $\pi^- + p \to n + \pi^\circ$
(iii) $\pi + \gamma \to p + \pi^\circ$
(iv) $\Sigma^0 \to \Lambda^0 + \gamma$
(v) $\Lambda^0 \to p + \pi^-$
(vi) $\Xi^- \to \Lambda^0 + \pi^-$
(vii) $K^0 \to \pi^+ + \pi^-$

Solution

(i) Strong: Parity, isospin, strangeness and I_3 are conserved.
(ii) Strong: Parity, isospin, strangeness and I_3 are conserved.
(iii) Electromagnetic: Parity, strangeness, I_3 are conserved but isospin not conserved.
(iv) Electromagnetic: Parity, strangeness, I_3 conserved but isospin not conserved.
(v) (vi) and (vii) Weak: Parity, strangeness, isospin and I_3 are not conserved.

Example 26.5: With the help of conservation laws determine which of the following reactions are allowed or forbidden?

(i) $\pi^+ + n \to \Lambda^0 + K^+$
(ii) $\pi^+ + n \to K^0 + K^+$
(iii) $\pi^0 + n \to K^0 + \Sigma^0$
(iv) $p \to e^+ + \gamma$
(v) $\pi^0 + \pi^- \to n + p$
(vi) $n \to p + e^+ + \overline{\nu}$
(vii) $\pi^+ + n \to \pi^- + p$

Solution

(i)

		π^+	+	n	$\to$	Λ^0	+	K^+	(Strong interaction)
Charge	q	1		0		0		1	
Baryon number	B	0		1		1		0	
Lepton number	L	0		0		0		0	
Isospin	I_3	+1		$-\frac{1}{2}$		0		$+\frac{1}{2}$	
Strangeness	S	0		0		-1		1	

Obviously, in the above strong interacting reactions all the conservation laws are obeyed and hence it is allowed.

(ii)

		π^+	+	n	$\to$	K^0	+	K^+	(Strong interaction)
Charge	q	+1		0		0		+1	
	B	0		1		0		0	
	L	0		0		0		0	

I_3	+1	$-\frac{1}{2}$	$-\frac{1}{2}$	$+\frac{1}{2}$
S	0	0	1	1

Since B, I_3 and S quantum numbers are not conserved and hence this reaction is not allowed.

(iii)

	π^+	+	n	$\rightarrow$	K^0	+	Σ^0	(Strong interaction)
q	0		0		0		0	
B	0		1		0		1	
L	0		0		0		0	
I_3	0		$-\frac{1}{2}$		$+\frac{1}{2}$		0	
S	0		0		-1		-1	

Since I_3 and S are not conserved and hence this reaction is not allowed.

(iv)

	p	$\rightarrow$	e^+	+	γ	(Weak interaction)
q	1		1		0	
B	1		0		0	
L	0		-1		0	

Since B and L are not conserved and hence this reaction is not allowed.

(v)

	π^+	+	π^-	$\rightarrow$	$\overline{n}$	+	p
q	0		-1		0		1
B	0		0		-1		+1
L	0		0		0		0

Obviously, q is not conserved and hence this is not allowed.

(vi)

	n	$\rightarrow$	p	+	e^+	+	n
q	0		1		1		0
B	1		1		0		0
L	0		0		-1		+1

Obviously, q is not conserved and hence this is not allowed.

(vii)

	π^+	+	n	$\rightarrow$	π^-	+	p
q	1		0		-1		1
B	0		1		0		1
L	0		0		0		0

Obviously, q is not conserved and hence this is not allowed.

Example 26.6: **If in the following reaction, the incident kaon has a kinetic energy of 1.63 GeV, calculate the total energy to be divided between the four recoiling particles.**

$$K^- + p^+ \rightarrow \Sigma^- + \pi^+ + \pi^- + \pi^+$$

The mass energy of p-mesons are 139.6 MeV, $\Sigma^- = 1197.3$ MeV, proton = 938.3 MeV and $K^- = 493.8$ MeV.

Solution Energy of particles taken together are as follows :

$$K^- = 0.4938 \text{ GeV} \qquad (\therefore 1 \text{ GeV} = 10^3 \text{ MeV})$$
$$p^+ = 0.9383 \text{ GeV}$$
$$E_{\bar{K}} = 1.63 \text{ GeV}$$

$\therefore$ Total energy, $E = 3.0621$ GeV

Energy of the four recoiling particles are:

$$\Sigma^- = 1.1973 \text{ GeV}$$
$$\pi^+ = 0.1396 \text{ GeV}$$
$$\pi^- = 0.1396 \text{ GeV}$$
$$\pi^+ = 0.1396 \text{ GeV}$$

Total energy, $= 1.6161$ GeV

$\therefore$ Excess energy $= 3.0621 - 1.6161 = 1.446$ GeV

$\therefore$ Average energy per particle $= \dfrac{1.446}{4} = 0.3615 \text{ GeV} = 361.5$ MeV

Example 26.7: If a pion decays from rest to give a muon of 4.0 MeV energy, what is the kinetic energy of the accompanying neutrino? What is the mass of the neutrino in the process?

Solution The decay mode of pion is given by

$$\pi^+ \rightarrow \mu^+ + \nu_\mu + E$$

$\therefore$ Energy, $E = m_\pi - m_\mu$ (neutrino has zero 'rest' mass)

$$= 273\, m_e - 207\, m_e = 66\, m_e$$
$$= 66 \times 0.51 \text{ MeV} \qquad (\because m_e = 0.51 \text{ MeV})$$
$$= 33.7 \text{ MeV}$$

Since muon takes 4.0 MeV of energy, the kinetic energy of the accompanying neutrino is

$$33.7 - 4.0 = 29.7 \text{ MeV}$$

$\therefore$ Mass of neutrino $= \left(\dfrac{29.7}{0.51}\right) m_e = 58.23\, m_e$

Example 26.8: Find the maximum kinetic energy of the electron emitted in the β-decay of the free neutron. The neutron-proton mass difference is 1.30 MeV.

Solution The decay scheme of a free neutron is

$$n^0 \rightarrow p^+ + e^- + \bar{\nu}_e + E$$

$\therefore$ Energy, $E = m_n - m_p - m_e$

$$= (1.30 \text{ MeV}) - m_e$$
$$= (1.30 - 0.51) \text{ MeV} \quad (\because m_e = 0.51 \text{ MeV})$$

This energy goes into the kinetic energy of the electron and antineutrino. The sharing of energy may be made in any manner so that the maximum kinetic energy of the electron is 0.70 MeV.

Example 26.9: A μ^- meson decays into an electron e^- and a pair of neutrons. Calculate the maximum available energy for the process and the average electron energy.

Solution The decay mode of μ^- meson is given by

$$\mu^- \rightarrow e^- + \nu_\mu + \nu_e$$

Taking $m_\mu = 207\ m_e$, the total available energy for the process is given by

$$207\ m_e \rightarrow m_e + 0 + 0 + E$$

($\because$ neutrino has zero 'rest' mass)

$$\therefore \quad E = (207 - 1)\ m_e = 206\ m_e$$

$$= 206 \times 0.51 \text{ MeV} \qquad (\because m_e = 0.51 \text{ MeV})$$

$$= 105 \text{ MeV}$$

If the above energy is equally shared by the particles, the average energy of the electron is

$$E_e = \frac{105}{3} = 35 \text{ MeV}$$

Example 26.10: A π^- is captured by a stable krypton atom, ${}^{84}_{36}\text{Kr}$. Calculate (a) the radius of the first and second Bohr orbits, (b) the energy radiated from transition $n = 2$ to $n = 1$, (c) the frequency of corresponding radiation and (d) the wavelength.

Solution We have $Z = 36$, $m = 273\ m_e$, $n = 1, 2$. We have also:

r_1 = radius of the first Bohr orbit for H-atom
$= 5.29 \times 10^{-11}$ m
h = Planck's constant $= 6.62 \times 10^{-34}$ Js.

(a) By direct substitution in the relation, we can find the r-values. For

$$n = 1, \quad r = r_1 \frac{n^2}{mZ} = 5.29 \times 10^{-11} \times \frac{1^2}{273 \times 36} = 5.38 \times 10^{-15} \text{ m}$$

$$n = 2, \quad r = r_1 \frac{2^2}{mZ} = 5.29 \times 10^{-11} \times \frac{4}{273 \times 36} = 21.52 \times 10^{-15} \text{ m}$$

(b) The energy value for $n = 1$ state is, also by direct substitution,

$$E_1 = R \frac{mZ^2}{n^2} = -2.18 \times 10^{-18} \times \frac{273 \times (36)^2}{1}$$

$$= -7.71 \times 10^{-13} \text{ J}$$

Similarly, for $n = 2$, $E_2 = \frac{1}{4} E_1 = -1.92 \times 10^{-13}$ J

$\therefore$ Energy radiated in transition is: $E = E_2 - E_1 = 5.79 \times 10^{-13}$ J.

(c) Frequency of radiation, n is given by

$$n = \frac{E}{h} = \frac{5.79 \times 10^{-13}\ \text{J}}{6.62 \times 10^{-34}\ \text{Js}}$$
$$= 8.74 \times 10^{20}\ \text{Hz}$$

Wavelength of radiation λ is given by

$$\lambda = \frac{c}{\nu} = \frac{3 \times 10^{8}\ \text{ms}^{-1}}{8.74 \times 10^{20}\ \text{s}^{-1}}$$
$$= 3.44 \times 10^{-13}\ \text{m} = 0.00344\ \text{A}$$

Example 26.11: Find the value of the third component of isotopic spin of Ξ^- in the following strong interaction:

$$\pi^+ + n \rightarrow \Xi^- + K^+ + K^+$$

Solution The interaction is $\pi^+ + n \rightarrow \Xi^- + K^+ + K^+$

$\therefore$ Isotopic spin, I: $\quad 1 + \frac{1}{2} \rightarrow \frac{1}{2} + \frac{1}{2} + \frac{1}{2}$

Third component of isospin, I_3: $\quad +1 - \frac{1}{2} \rightarrow I_3 + \frac{1}{2} + \frac{1}{2}$

$\therefore \quad I_3$ for $\Xi^- = -\frac{1}{2}$

Example 26.12: Identify the type of the following interaction from the conservation laws: $\Sigma^0 \rightarrow \Lambda^0 + \gamma$ (life time $\leq 10^{-14}$ s)

Solution We have: $\quad \Sigma^0 \rightarrow \Lambda^0 + \gamma$

Charge, Q:	$0 \rightarrow 0 + 0$	$\therefore \Delta Q = 0$
Baryon number, B:	$+1 \rightarrow +1 + 0$	$\therefore \Delta B = 0$
Lepton number, L:	$0 \rightarrow 0 + 0$	$\therefore \Delta L = 0$
Strangeness number, S:	$-1 \rightarrow -1 + 0$	$\therefore \Delta S = 0$
Hypercharge, Y:	$0 \rightarrow 0 + 0$	$\therefore \Delta Y = 0$

Since the strangeness number is conserved, the interaction is either a strong interaction or an electromagnetic one. Its half-life is $\leq 10^{-14}$ s which points to the fact that it cannot be a strong interaction, but is a weak decay. As S is conserved, it cannot be a weak interaction. So it is an electromagnetic interaction and a γ-photon is produced.

Example 26.13: Identify the unknown particle in the reactions given below, using the conservation laws.

(i) $\quad \mu^- + p \rightarrow {}^{1}_{0}n + \ldots$

(ii) $\quad \pi^- + p \rightarrow K^0 + \ldots$

Solution (i) The reaction is: $\mu^{-1} + p \to {}^{1}_{0}n + \ldots$

The unknown particle must be of zero charge and mass, spin $\frac{1}{2}$ and lepton number 1 so that the charge, mass and lepton numbers are conserved. Since the interacting particle is μ^- meson, the unknown particle is identified as mu-neutrino, ν_μ.

(ii) The reaction is $\pi^- + p \to K^0 + \ldots$

For charge conservation, Q: $-1 + 1 \to 0 + Q$ $\quad \therefore Q = 0$

Conservation of baryon number, B: $0 + 1 \to 0 + B$ $\quad \therefore B = +1$

Strangeness conservation, S: $0 + 0 \to -1 + S$ $\quad \therefore S = +1$

Third component of isospin, I_3: $-1 + \frac{1}{2} \to -\frac{1}{2} + I_3$ $\quad \therefore I_3 = 0$

Thus the unknown particle has charge zero, baryon number +1, strangeness number +1 and third component of isospin 0. So the particle could be Λ^0 or Σ^0.

We may note that Σ^0 has a rest mass greater than Λ^0. So Σ^0 will be produced only if the kinetic energy imparted to π^- is greater than that required for the production of Λ^0.

Example 26.14: Check if the following reactions are allowed or forbidden.

(i) $\pi^- + p \to \Lambda^0 + \pi^0$

(ii) $p + \bar{p} \to 2\pi^+ + 2\pi^- + 2\pi^0$

Solution (i) The reaction is: $\pi^- + p \to \Lambda^0 + \pi^0$

Q:	$-1 + 1 \to 0 + 0$	$\therefore \Delta Q = 0$	$Q \to$ conserved
B:	$0 + 1 \to +1 + 0$	$\therefore \Delta B = 0$	$B \to$ conserved
S:	$0 + 1 \to -1 + 0$	$\therefore \Delta S \neq 0$	$S \to$ not conserved

It is a strong interaction where the charge and baryon number are conserved. But since the strangeness number is not conserved, the reaction is forbidden.

(ii) The reaction is: $p + \bar{p} \to 2\pi^+ + 2\pi^- + 2\pi^0$

Q:	$+1 - 1 \to +2 - 2 + 0$	$\Delta Q = 0$
B:	$+1 - 1 \to 0 + 0 + 0$	$\Delta B = 0$
S:	$0 + 0 \to 0 + 0 + 0$	$\Delta S = 0$
Y:	$+1 - 1 \to 0 + 0 + 0$	$\Delta Y = 0$

Clearly, the above reaction is an allowed reaction.

Example 26.15: A μ^- meson decays at rest and emits an electron. Which other particle (s) will be emitted in the above decay? Use the conservation laws in support of your answer.

Solution Since μ^- decay from rest, the conservation of energy and linear momentum demands that two or more particles are emitted. Since one of the ejected particle is an electron, the charge conservation demands that the other particle (s) must be neutral. Since its rest mass must be less than m$^-$, it could be a neutrino. Now μ^- meson is a spin $\frac{1}{2}$ particle, and the vector sum of spins of $e^-\left(\frac{1}{2}\right)$ and that of $\nu\left(\frac{1}{2}\right)$ is either 0 or $\frac{1}{2}$. The conservation of spin requires that another neutral and spin $\frac{1}{2}$ particle should be emitted in the opposite direction. This could be an antineutrino, $\bar{\nu}$ and the decay scheme would be: $\mu^- \rightarrow e^- + \nu + \bar{\nu}$. However, if ν and $\bar{\nu}$ are simultaneously emitted they would annihilate to produce photons. Since photons are not observed, a particle and its antiparticle are not possibly emitted. So there should be two different neutrinos. Since m$^-$ and e^- are involved in the decay, the neutrinos could be ν_e and ν_μ. To conserve spin, one will be anti-muon-neutrino, $\bar{\nu}_\mu$. The other particle is electron neutrino, ν_e. We have

$$\mu^- \rightarrow e^- + \nu_e + \bar{\nu}_\mu$$

Example 26.16: An ultra-relativistic proton moves in a magnetic field. Can it radiate π^+, π^- and π^0 mesons, electrons and positrons?

Solution If the energy of the proton is sufficiently large, it can radiate π^0 and π^+ mesons, and also positrons. The reactions are:

$$p \rightarrow p + \pi^0;\ p \rightarrow n + \pi^+;\ p \rightarrow n + e^+ + \nu$$

But π^- mesons and electrons cannot be radiated.

Example 26.17: Which of the following reactions are permissible with regard to strangeness?

(a) $\pi^- + p \rightarrow \Lambda^0 + K^0$ (b) $\pi^- + p \rightarrow \Lambda^0 + \pi^0$
(c) $\Lambda \rightarrow p + \pi^-$ (d) $\Xi^- \rightarrow \Lambda + \pi^-$
(e) $\Xi^- \rightarrow 2\pi^- + p$ (f) $\pi^+ + p \rightarrow \Sigma^- + K^-$
(g) $\pi^+ + p \rightarrow \Sigma^- + \pi^-$ (h) $\pi^+ + n \rightarrow \Lambda^0 + K^+$
(i) $K^- + p \rightarrow \Sigma^- + \pi^-$ (j) $p + \Sigma^- \rightarrow K^+ + \pi^+$
(k) $p + \Sigma \rightarrow \pi^+ + \pi^+$ (l) $\pi^- + p \rightarrow \Xi + X^- + \pi$
(m) $\pi^- + p \rightarrow \Sigma^+ + K^-$ (n) $\pi^- + \bar{p} \rightarrow \Xi^- + K^+ + K^0$

Solution Reaction (a), (f), (h), (j), (l) and (n) gives $|\Delta S| = 0$ and they proceed through strong interaction. In reaction (b), (c), (d), (g), (i) and (k), $|\Delta S| = 1$. Hence disintegrations, (c), (d) have relatively small probabilities, while reactions (b), (g), (i) and (k) are not observed in practice. In (e) and (m), $|\Delta S| = 2$. They are not observed. Cascade hyperon decays in accordance with reaction (d) and afterward L decays according to (c).

SUMMARY

1. Elementary particles represent the most basic constituents of matter and are also the carriers of fundamental forces between particles, namely the gravitational forces, electromagnetic forces, strong nuclear forces and weak force that manifests itself in beta decay.
2. The innumerable particles, most of which are unstable, are crucial to our understanding of matter and the fundamental forces of nature that hold matter together.
3. There is an interwined connection between cosmology with its grand dimensions, and particle physics, in which matter is studied in its smallest dimensions.
4. The structure of ordinary matter and its interactions can be explained in terms of a handful of elementary particles: the *quarks,* which make up the proton and neutron, and a host of other *hadrons*: *leptons,* including electrons and the elusive neutrinos,

and the *gauge bosons*, which mediate the fundamental forces and include the photon for the electromagnetic force.

5. The known elementary particles can be grouped into three classes: *Leptons, Quarks,* and *Gauge bosons. Hadrons,* such strongly interacting particles as the *proton* and the *neutron,* which are bound states of quarks and/or antiquarks, are also sometimes called elementary particles.
6. *Leptons* are particles that does not experience the strong force. Six leptons are known, the electron, muon and tauons, plus three associated neutrinos: ν_e, ν_μ and ν_τ. The electron is a stable particle but the muon and tau leptons decay through the weak interactions with lifetimes of about 10^{-6} and 10^{-13} seconds. Neutrinos have small non-zero mass and that they "oscillate" among the three types. Neutrinos are stable neutral leptons, which interact only through the weak interactions. The leptons all have spin ½ and are therefore *fermions.* Obviously, leptons obey Pauli exclusion principle. Leptons are considered to be true elementary point particles with zero size and no internal structure.
7. *Hadrons* are the particles that do experience the strong nuclear force. They fall into two subclasses: (i) *mesons and* (ii) *baryons. Mesons* have integral spin and are therefore bosons. Bosons don't obey the Pauli exclusion principle. *Mesons* include Yukawa's pions and a host of others. All are unstable. *Baryons* have half-integer spins, i.e., they are fermions. Baryons include the familiar proton, neutron and other similar massive particles. Baryons are grouped into pairs, triplets, and higher-multiple groupings of closely related particles. Neutron and proton form a pair that differ in charge and slightly in mass. Each baryon has an antiparticle, as do most mesons, but some neutral mesons are their own particle.
8. *Quarks*: Corresponding to the leptons there are six quarks, namely the up (u), charm (c), and top (t) quarks with electric charge equal to +⅔ that of proton and the down (d), strange (s), and bottom (b) quarks of charge −⅓ the proton charge. Quarks have not been observed experimentally but reveal their presence indirectly in high-energy scattering experiments and through patterns observed in the properties of hadrons. They are believed to be permanently confined within hadrons, either in *baryons,* half integer spin hadron containing three quarks, or in *mesons,* integer spin hadrons containing a quark and antiquark.
9. *Gauge bosons or field particles*: Interactions between quarks and leptons are mediated by the exchange of gauge bosons, specifically the photon for electromagnetic interactions, the $W^{\pm}$ and Z° bosons for the weak interaction and eight massless gluons in the case of the strong interactions. There is a particle called gluon, and a *hypothetical graviton* that would carry the gravitational force in an as yet the incomplete theory of *quantum gravity.* All the field particles are bosons, having spin-1 or spin-2 for graviton.
10. Elementary particles are classified in different ways, e.g. fermions and bosons have half-integral and integer spins, respectively. Stable particles appear to be composed of constituent fermions. Leptons are considered to be truly elementary, point like particles: electrons, muons, taus, and their respective neutrinos. Hadrons consist of two classes: mesons and baryons. Mesons are unstable. Proton is the only stable free baryon, but some theories predict is also unstable. When bound within the nuclei, the neutrons and protons are stable. Baryons heavier than the nucleons are called *hyperons.*

11. The fundamental interactions are the gravitational, electroweak and the strong nuclear. For all practical purposed, i.e., except at very high energies, electroweak is two interactions, i.e., electromagnetic and weak. Photons are the mediators of the electromagnetic force, gluons mediate strong force, $W^{\pm}$ and Z bosons mediate the weak interaction, and the graviton (yet not observed particle), mediates the gravitational force.
12. Many new particles can be characterized by known properties such as mass, spin, and electronic charge. Of these, spin and charge are associated with conservation laws - conservation of angular momentum and conversation of electric charge. Allowed interactions among particles must conserve these quantities. Other particle properties appear to be conserve as well.
13. *Conservation of baryon number*: Associated with each baryon or antibaryon is its *baryon number*, assigned the value +1 for a baryon and –1 for an antibaryon. All experimental evidence to date points to conservation of baryon number; i.e., in all observed particle reactions, the sums of baryon numbers before and after the reactions are equal. Some theories suggest that baryon number is only approximately conserved. If it is so, then the proton is an unstable particle with a mean life time in excess of 10^{35} years.
14. *Conservation of Lepton number*: Lepton number seems to be conserved. The neutron and proton, being baryons, have lepton number zero, while the resulting electron and antineutrino have lepton numbers +1 and –1, respectively.
15. *Strangeness* is conserved in strong and electromagnetic interactions, but its value can change in weak interactions.
16. *Conservation of Parity*: The principle of parity invariange states that not fundamental distinction can be made between left and right; that the laws of physics are the same in a right handed system of coordinates as they are in left handed system. However, the parity of the total wave function describing a system of elementary particles is conserved in strong and electromagnetic interactions and not in weak interactions.
17. *Charge conjugation*: This is the property of elementary particles that determines the difference between a particle and its antiparticle. The property is not restricted to electrically charged particles, i.e., it applies to neutral particles also.
18. *CPT conservation*: A combination of charge conjugation (C), Parity reversal (P) and reversal of time coordinates (T), i.e., CPT conservation still holds. CPT makes a new physical process indistinguishable from the original. There may be a deep philosophical connection here with the direction of time, but the full implication of CPT symmetry and the failure of its individual components are still not fully understood.
19. No violation of CPT theorem is known experimentally. When C, P, and T (or any two of them) are violated, the principles of relativistic quantum field theory are not affected; however, violation of *CPT invariance* would drastically alter the fundamentals of relativistic quantum field theory. It is not known whether superstring theory obeys a version of the CPT theorem.
20. Quantum chromodynamics (QCD) theory establishes how quarks are combined to form various elementary particles. A property called *colour* is essential for understanding how quarks and antiquarks combine. In QCD, *gluons* are the field particles that play the role of photons in QCD, binding particles subject to strong force.

21. *Standard model*: The theory that currently describes elementary particles and their interactions is termed standard model. This model combines the electroweak and QCD theories and found quite successful in explaining elementary particle physics. This model includes the photons, the gluons, the *W* and *Z* particles, and an unexpected particle called the *Higgs boson*, believed responsible for other particle masses. The neutrino oscillation have solved a problem with the solar neutrino rate and a new field of neutrino physics is fast developing. The preponderance of matter over antimatter is getting proper attention, as is the search for Higgs boson.
22. *Grand Unification Theories* (GUTs): GUT attempts to combine the strong, weak, and electromagnetic interactions into a single '*gauge theory*' with a single symmetry group. Some versions of GUT predict that the proton should decay on the very long time scale of some 10^{36} years. Some also predict that the neutrino has nonzero mass. There is no evidence for proton decay at the moment. There is evidence that neutrinos have nonzero masses. It seems likely that GUTs need to be combined with supersymmetry.
23. *Superstring theory*: This is a unified theory of fundamental interactions involving supersymmetry, in which the basic objects are one-dimensional objects, i.e., *superstrings*. Superstrings are assumed to have a length scale of about 10^{-35} m and, since very short distances are associated with very high energies, they should have energy scales of the order of 10^{19} GeV, which is far beyond the energy of any particle accelerator that can envisaged. One of the salient feature of this theory is that it leads to spin-2 particles, which are identified as '*gravitons*'. Obviously, this theory automatically contains a quantum theory of the gravitational interaction.

REVIEW QUESTIONS

1. Classify the elementary particles. What do you know about leptons and mesons? What conservation laws are obeyed in the case of production and annihilation of particles?
2. What do you understand by antiparticles? Explain their characteristic properties taking examples of positron and antiproton.
3. Compare the properties of pions, muons and kaons.
4. Discuss and classify the different types of fundamental interactions known in nature. What attempts have been made to unify them?
5. Illustrate by taking examples the different conservation laws followed by the elementary particles.
6. Define Isospin, Strangeness and baryon number. Is there any relation connecting them? What conservation laws and selection rules are obeyed for the creation of strange particles?
7. Classify the elementary particles. How are these particles assigned: (i) isospin, (ii) baryon number, (iii) strangeness number and (iv) hypercharge. What conservation laws are obeyed in the case of annihilation of particles?
8. What are quarks? Give an account of quark model.
9. Mention salient features of GUT.
10. Describe the discovery of Ω^- Hyperon with experimental details.
11. What are resonance particles? Give their salient features.

12. Discuss the composition of hadrons based on quark model.
13. Give a brief account of quark structure of mesons and baryons.
14. Why free quarks could not be observed?
15. What is Quantum Chromodynamics? How it explains interaction of examples?
16. Mention salient features of Standard Model.
17. Write short notes on: (i) Baryon number, (ii) Strangeness, (iii) Isospin, Strangeness and Baryon number, (iv) Parity, (v) Quantum numbers and conservation laws of elementary particles, (vi) Quark model, (vii) QCD and (viii) GUT.
18. Mention the characteristics of following conservation laws: mass-energy, electric charge, linear momentum, baryon number, lepton number. Explain how these conservation laws are related to fundamental laws of nature.
19. Which families of particles seem to be truly elementary and why?
20. How, one can determine whether particles are mesons or baryons by observing their quark structure?
21. Does it appear that the total baryon number of the universe is zero? What does it convey?
22. Whether a baryon be produced when an antibaryon interacts with a meson? Explain.
23. Why recent discovery of Higgs – like particle with mass of 125 GeV at Large Hadron Collider at CERN is considered milestone in the field of elementary particles?

SHORT ANSWER-QUESTIONS

1. Write the names of fundamental interactions of nature with their relative strength, taking strength of strong interaction as 1.
   Ans: e.m. 10^{-1}, weak 10^{-13}, Grav. 10^{-37}
2. Write the exchange quanta in case of fundamental interactions.
   Ans: (i) Strong: Pions and in Quark model: gluons; (ii) e.m.: Photons; (iii) Weak: $W^{\pm}$, Z^0 bosons; and (iv) Gravitational: Graviton
3. Why the decay $n \rightarrow \pi^0 + \gamma$ is forbidden?
   Ans: B is not conserved, $B = 1$ for neutron, but $B = 0$ for both π^0 and γ
4. What is CPT theorem?
5. Write quark structure of the nucleons.
   Ans: $p = (uud)$, $n = (udd)$
6. Write the generalized expression for Q to include b and t.
   Ans: $Q = I_3 + \dfrac{B + S + C - b + t}{2}$
7. What is gluon?
   Ans: The particle that mediates the strong interaction between quarks is called a gluon. Gluon is massless and has spin1.
8. What are hadrons? Give examples.
   Ans: Particles that interact by the strong interactions are called hadrons, e.g. neutron, proton and mesons.
9. What are quarks? Why an isolated quark cannot be observed?
   Ans: Quark is a fundamental constituent of hadrons, i.e., of particles that take part in strong interaction. Quarks come in six flavors and three colors, join in threes to form *baryons*, including protons and neutrons, and in twos to form *mesons*. Quarks carry

fractional electric charges, but because the strong force between quarks does not decrease with distance, it appears impossible to have an isolated quark.

10. What are leptons?

Ans: Leptons are class of elementary particles that do not take part in strong interactions. They include the electron and its more massive cousins, e.g. the muon and tau, and the three types of neutrinos. Leptons have no substructure of quarks and are considered indivisible. They are all fermions.

11. What is quantum electrodynamics (QED)?

Ans: QED is a relativistic quantum-mechanical theory of electromagnetic interactions. QED is a gauge theory. The fundamental event in QED is the interaction of a photon with an electrically charged particle, which have been verified over a great range of distances and have led to highly accurate predictions.

12. What is quantum chromodynamics (QCD)?

Ans: QCD is a quantum field theory, which is a gauge theory of the strong interactions based on the exchange of massless gluons between quarks and antiquarks. QCD is similar to QED, but with the gluon as the analogue of the photon and with a quantum number known as *color* replacing that of electric charge. Gluons carry color and can interact with themselves.

13. What are gauge bosons?

Ans: These are quanta of the different force fields and "carriers" of those forces. These include the photon for the electromagnetic force; $W^{\pm}$ and Z for weak force; a particle called the gluon; and a hypothetical graviton that would carry the gravitational force in an yet incomplete theory of quantum gravity. All the field particles are bosons, carrying spin-1 and for the graviton, spin-2.

14. What is CPT theorem?

Ans: The theorem that combined operation of changing charge conjugation (C), parity (P), and time reversal (T), denoted CPT, is a fundamental symmetry operation of relativistic quantum field theory. No violation of this theorem reported so far. It is not known whether superstring theory obeys a version of this theorem.

15. What is standard model of elementary particles?

Ans: This model describes the elementary particles and their interactions. This model includes photons that mediate the electromagnetic interaction, the gluons of the strong force, the $W^{\pm}$ and Z particles that mediate the weak force, and an particle called the Higgs boson, believed responsible for other particle masses. This model can explain the phenomena of particle physics, but it leaves many fundamental questions unanswered, e.g. do the quarks and leptons have the particular masses they do? Why are only three families of elementary particles? Why are leptons and quarks distinct? Are these particles really elementary, or are there even smaller structures as hitherto unexplored scales?

16. What is electroweak unification?

Ans: This is a gauge theory, sometimes referred as *quantum flavour dynamics* (QFD), gives a unified description of the electromagnetic and weak interactions. This is known as Weinberg-Salam (WS) model or sometimes called as Glashow-Weinberg-Salam (GWS) model. The electroweak interaction is mediated by photons and by intermediate vector bosons, called the W-boson and the Z-boson. The theory successfully accounts for existing data for electroweak processes and also predicts the existence of a heavy particle with spin O, the Higgs boson. The discovery of Higgs – like particle (God particle), with mass of 125 GeV at Large Hadron Collider (LHC) at CERN (Geneva), on 4th July, 2012 was announced.

17. What do you understand by Grand Unification theories (GUTs)?

Ans: GUTs attempts to merge the electroweak and strong forces. Some versions of GUTs predict that the proton should decay on the very long time scale of about 10^{36} years. However, this could not be verified. The another GUT prediction – that neutrinos have mass – was verified.

18. What is string theory?

Ans: This theory pictures elementary particles as vibration modes on stringlike structures that may be as short as 10^{-35} m. This theory is set in a space-time with 10 or more dimensions. The extra dimensions other than 4 usual space-time dimensions are "compactified" in a way that makes them undetectable in normal interactions. Some researchers believe that string theories may fulfill the promise of a "theory of everything", explaining all our observations about the behaviour of the universe.

19. What is quantum gravity?

Ans: This is a theory that is consistent with quantum mechanics. In conventional quantum gravity theory, the gravitational force is mediated by a massless spin-2 particle, called the '*graviton*'. However, this work is quite speculative and the graviton has never been detected.

PROBLEMS

1. Which of the following reactions can occur? State the conservation laws followed and violated.

(i) $p + p \rightarrow n + p + \pi^+$ (ii) $p + p \rightarrow p + \Lambda^0 + e^+$

(iii) $e^+ + e \rightarrow \mu^+ + \pi^-$ (iv) $\Lambda^0 \rightarrow \pi^+ + \pi^-$

(v) $\pi^- + p \rightarrow n + \pi^0$

2. Discuss whether the following particle reactions are allowed or forbidden under conservation of charge *Q*, Baryon number *B* and strangeness *S*?

(i) $\pi^+ + n \rightarrow \Lambda^0 + K^+$ (ii) $\pi^+ + p \rightarrow \Lambda^0 + \pi^0$

3. Which of the following reaction can occur? State the conservation principles violated by the others.

(i) $\Lambda^0 \rightarrow \pi^+ + \pi^-$ (ii) $\pi^- + p \rightarrow n + \pi^0$

(iii) $\pi^+ + p \rightarrow \pi^+ + p + \pi^- + \pi^0$

[Ans. (i) *B* not conserved (ii) can occur (iii) charge not conserved**]**

4. Which of the following reactions can occur? State the conservation principles violated by the other.

(i) $p + p \rightarrow n + p + \pi^+$

(ii) $p + p \rightarrow p + \Lambda^0 + \Sigma^+$

(iii) $e^+ + e^+ \rightarrow \mu^+ + \pi^-$

5. Mention conservation laws and test the occurrence of following reactions.

(i) $\mu^+ \rightarrow e^+ + \bar{\nu} + \gamma$ (ii) $n \rightarrow p + e^- + \bar{\nu}$

(iii) $p \rightarrow n + e^- + \bar{\nu}$ (iv) $\gamma \rightarrow e^+ + e^-$

(v) $\pi^0 \rightarrow \pi^+ + \pi^-$ (vi) $\pi^+ \rightarrow \mu^+ + \nu + \bar{\nu}$

(vii) $\mu^+ \rightarrow e^+ + e^- + \nu$ (viii) $p + n \rightarrow n + \bar{p}$

6. Classify the following reactions with respective interactions and give reasons in support of your answer.

(i) $\pi^- \rightarrow p \rightarrow \pi^0 + n$ (ii) $\pi^- + p \rightarrow \Lambda^0 + K^0$

(iii) $\Sigma^0 \rightarrow \Lambda^0 + \gamma$ (iv) $p + \gamma \rightarrow + p + \pi^0$

(v) $\pi^0 \rightarrow \gamma + \gamma$ (vi) $\Lambda^0 \rightarrow p + \pi^-$

(vii) $K^0 \rightarrow \pi^+ + \pi^-$ (viii) $\Lambda^0 \rightarrow p + e^- + \bar{\nu}$

(ix) $\Xi^- \rightarrow \Lambda^0 + \pi^-$

[**Ans.** Short range weak: (vi), (vii), (viii), (ix);
Short range (i), (ii);
Electromagnetic: (iii), (iv), (v)]

7. Which of the following processes are allowed or forbidden? Give reasons in support of your answers.

(i) $\pi^- + p \rightarrow \Lambda^0 + \pi^0$ (ii) $\pi^- + n \rightarrow \Sigma^- + K^0$

(iii) $n \rightarrow \pi^0 + \nu$ (iv) $n \rightarrow p + \mu^- + \nu_\mu$

(v) $n \rightarrow p + e^-$ (vi) $p + p \rightarrow p + \bar{p}$

(vii) $p + p \rightarrow p + p + p + p$ (viii) $\pi^+ \rightarrow \mu^+ + \gamma$

(ix) $\pi^+ \rightarrow \mu^+ + \nu_\mu$

8. A negative pion may be captured into a stable orbit around the nucleus, constituting a mesic atom. Assume that one can compute the energy and the radius of the mesonic orbit by the same formulas used for electronic orbits in the hydrogen atom. (i) What energy is released when a free pion at rest is captured into the ground state around a proton? (ii) Compute the radius of the ground-state orbit of a pion moving around a pion. (iii) Estimate the nucleus for which the radius of the ground-state orbit of pion is equal to the radius of the nucleus.

[**Ans.** (i) 2.53 keV, (ii) 2.85×10^{-13} m (iii) ${}^{110}_{47}$Ag]

9. Consider the decay of a π^0 meson into two photons. Given that the photon energies are E_1 and E_2, show that they are related by

$$E_1 E_2 = \frac{1}{2} \frac{(m_\pi c^2)^2}{(1 - \text{Cos}\,\theta)}$$

where θ is the angle between photon directions as measure in the L-frame.

10. Analyse the process $K^- + p^+ \rightarrow \Omega^- + K^+ + K^0$ from the point of view of conservation laws. Calculate the threshold kinetic energy of the K-particle.

[**Ans.** I_3, I and S are conserved; 2710 MeV]

11. Find the energy of photon emitted in the decay $\Sigma^0 \rightarrow \Lambda^0 + \gamma$.

[**Ans.** 73.6 MeV]

12. A proton of kinetic energy K collides with a stationary proton, and a proton antiproton pair is produced. If the momentum of the bombarding proton is shared equally by the four particles that emerge form the collision, find the minimum value of K.

[**Ans.** $6\, m_p c^2$]

13. Why does a free neutron not decay into an electron and a positron but decay into a proton-antiproton pair?

14. According to the theory of the continuous creation of matter (which has turned out to be inconsistent with astronomical observations), the evolution of the universe can be traced to the spontaneous appearance of neutrons and antineutrons in free space. Which conservation law(s) would violate this process?

[**Ans.** Energy]

15. A μ^- meson collides with a proton, and a neutron plus another particle are created. What is the other particle?

[**Ans.** ν_μ (mu neutrino)]

16. All resonance particles have very short life times. Why does this suggest that they must be hadrons?

[**Ans.** Only the strong interaction can produce such rapid decays.]

17. Taking the range of nuclear force $R_N >> 1\, fm$ and using uncertainty principle, estimate the mass of the meson.

[**Hint:** $\Delta t \approx \hbar/2\Delta E = \hbar/2m_\pi c^2$ (i)

But $$m_\pi c^2 = \frac{\hbar}{2\Delta t} = \frac{\hbar c}{2R_N} \quad \text{(ii)}$$

$$= \frac{1.973 \times 10^2 \text{ eV.nm}}{2 \times 10^{-15} \text{ m}} \approx 100 \text{ MeV}$$

We may note that eq. (ii) is useful to relate the effective length of any force R and mass m that mediate the force R. One can write $R = \hbar/2mc = \dfrac{\hbar c}{2mc^2}$.]

18. Taking the mass of W^- particle 80.4 GeV/c^2, estimate the range of weak interaction responsible for the neutron β decay.

[**Hint:** $$R_w = \frac{\hbar c}{2m_{\hat{\omega}} c^2} = \frac{1.973 \times 10^2 \text{ eV.nm}}{2(80.4 \text{GeV}/c^2)(c^2)} = 1.2 \times 10^{-18} \text{ m}$$

We may note that the W^- travels near the speed of the light, because it is a very massive particle. In this case the violation of the conservation of energy is massive, so the violation must occur over a very short period of time. We have

$$\Delta t = \frac{\hbar}{2m_w c^2} \approx 4 \times 10^{-27} \text{ s}$$

Moreover, the life time of the neutron is much longer than the time it takes for the decay process itself.]

OBJECTIVE QUESTIONS

1. The suggested particle responsible for gravitational interaction is
(a) gluon (b) graviton
(c) photon (d) hadron
2. The spin of suggested particle 'graviton' is
(a) 0 (b) 1
(c) 2 (d) 1/2
3. Particles with integral spins are called
(a) fermions (b) bosons
(c) phonons (d) none of the above
4. Hadrons are the particles that interact through the
(a) strong force (b) electromagnetic force
(c) gravitational force (d) weak force
5. The antiproton was discovered in the process
(a) $n \rightarrow p + e + \overline{v}_e$ (b) $p + p \rightarrow p + p + p + \overline{p}$
(c) $K^- + p \rightarrow K^0 + K^+ + \Omega^-$ (d) $\pi^- + p \rightarrow K^0 + \Lambda^0$
6. The quanta of electromagnetic field is
(a) graviton (b) hadron
(c) meson (d) photon
7. Yukawa called his hypothetical particle as
(a) graviton (b) meson
(c) baryon (d) hadron
8. The particles that don't experience the strong force and have spin ½ are classified as
(a) leptons (b) hadrons
(c) baryons (d) gauge bosons
9. Strangeness is conserved in
(a) strong interactions only
(b) electromagnetic interactions only
(c) strong and electromagnetic interactions
(d) none of the above is correct
10. The force binding quarks of different colours is the
(a) weak force (b) electromagnetic force
(c) gravitational force (d) strong force
11. Electroweak unification theory combines
(a) gravitational and electromagnetic forces
(b) electromagnetic and strong forces

(c) electromagnetic and weak forces
(d) gravitational, electromagnetic and weak forces

12. Grand unification theories attempts to merge
(a) gravitational and electroweak forces
(b) gravitational and strong forces
(c) electroweak and strong forces
(d) gravitational, electroweak and strong forces

ANSWERS

Question Numbers	1	2	3	4	5	6	7	8	9	10	11	12
Answers	(b)	(c)	(b)	(a)	(b)	(d)	(b)	(a)	(c)	(d)	(c)	(c)

FILL IN THE BLANKS

1. The fundamental forces of nature are: gravity force, electromagnetic force, ______ and the ______. **[nuclear or strong force, weak force]**
2. The photon exchanged by two particles last only a short time, and therefore in accordance with uncertainty principle, its energy is ______. **[uncertain]**
3. Yukawa proposed that the nuclear force should, like the electromagnetic force, be mediated by exchange of a particle called ______. **[meson]**
4. The leptons all have spin ______ and are therefore ______, which obey the Pauli exclusion principle. **[½, fermions]**
5. Hadrons fall into two subclasses; ______ and ______. **[mesons, baryons]**
6. Baryons have ______ spins and are therefore ______. **[half integer, fermions]**
7. In all observed particle reactions baryon number is ______. **[conserved]**
8. Some theories suggest that baryon number is only approximately conserved. If that is so, then the proton itself is an ______ particle with a mean lifetime in excess of ______ years. **[unstable, 10^{35}]**
9. Strangeness is conserved in strong and electromagnetic interactions, but in weak interactions its value can ______. **[change]**
10. Parity is ______ if its value is unchanged in a particle interaction. **[conserved]**
11. The gluon is responsible for the ______ between the antiquark and quark that make up the pion, and the gluon is responsible for ______ between the quarks that make up the ______. **[attraction, attraction, nucleons]**
12. Standard model is not able to predict the particle ______. **[masses]**
13. The various GUT theories predict that the proton is unstable with a half life of ______ to ______ years. Current experimental measurements have shown the lifetime to be greater than ______ years. **[10^{29}, 10^{31}, 10^{32}]**
14. Neutrons and protons are stable when bound in ______. **[nuclei]**

15. Free baryon is the only______, but some theories predict it is also ______.
[stable, unstable]

16. Quarks have ______ electric charges and only exist as ______ of hadrons.
[fractional, constituents]

17. Quantum chromodynamics theory establishes how quarks are combined to form ______.
[elementary particles]

18. The standard model combines the electroweak and QCD theories and has been quire successful in explaining elementary particle ______. **[physics]**

19. A surprising aspect of quarks is that the strong force does'nt ______ with separation. For that reason it appears impossible to ______ single quark. **[decrease, isolate]**

20. Matter particle of standard model are ______. **[quarks]**

27

General Relativity

27.1 INTRODUCTION

The special theory of relativity is restricted to reference frames in uniform motion. Following special relativity, Einstein attempted to formulate a theory that would encompass observers in accelerated motion. However, he recognized that it is impossible to distinguish the effects of uniform acceleration from those of uniform gravitational field. Consequently, introduction of Einstein's general theory of relativity in 1916 became a theory of gravity. General relativity describes gravity as the geometrical curvature of four-dimensional spacetime. In this description, matter and energy curve spacetime in their vicinity, and objects moving through the curved spacetime follow the straighest possible paths- which are not the straightlines of Euclidean geometry. Fig. 27.1 shows a two-dimensional analogy for particles in curved surface.

General relativity's predictions differ substantially from those of Newton's theory of gravity only in regions of very strong gravitational fields - far stronger than those found anywhere in our solar system. For that reason general relativity has become a cornerstone of modern astrophysics, describing the physics of such bizarre objects as neutron stars, black holes, and so-called gravitational lenses that can produce multiple images of astrophysical objects. General relativity also addresses cosmological questions of the origin and ultimate fate of the universe. But we are not without terristrial uses for general relativity; the Global Positioning System (GPS) would be off by several kilometers if the effects of general relativity were not included in GPS calculations.

Einstein's conviction that the laws of nature should be expressed in the same form in every frame of reference accelerated as well as nonaccelerated, was the primary motivation that led him to the general theory of relativity. The two postulates of Einstein's general relativity are as follows:

(i) All the laws of nature have the same form for observers in any frame of reference, whether accelerated or not.

(ii) In the vicinity of any given point, a gravitational field is equivalent to an accelerated frame of reference in the absence of gravitational effects. (This postulate is known

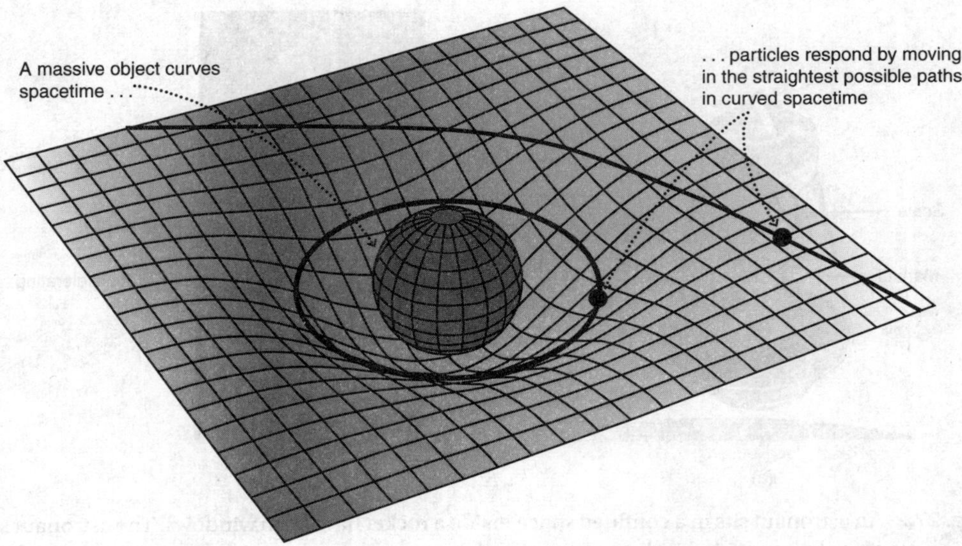

Fig. 27.1 A two-dimensional analog of curved spacetime in general relativity

as the principle of equivalence, i.e., observations made in an accelerated reference frame are indistinguishable from observations made in a Newtonian gravitational field). We discuss these tenets in next section.

27.2 PRINCIPLE OF EQUIVALENCE

The special relativity encompasses inertial frames of reference moving at uniform relative velocities. The special relativity can be used to explain the physics in any of the relative systems. But what about the systems moving in nonuniform motion w.r. to each other? Linearly accelerating and rotating systems are examples of such motion. One may ask, whether this kind of nonuniform motion also relative? Einstein believed so, and after several years of thinking about this problem, Einstein concluded that the laws of nature should be expressed in the same form in every frame of reference, accelerated as well as nonaccelerated and he presented his general theory of relativity.

Let us imagine *gedanken* experiment to help us to understand nonuniform motion. Let us consider an astronaut sitting in a confined space on a rocket placed on earth ready to blast off for a mission to Mars (Fig. 27.2(a)). The astronaut is strapped into a chair that is mounted on weighing scale. The weighing scale indicates mass M. The astronaut drops a safety manual which falls to the floor. Now, look at Fig. 27.2(b), which depicts the contrast situation. We can see that the rocket is now on its way to Mars and is far enough away from the Earth, that means the gravitational attraction is negligible. However, we may note that the rocket is still accelerating in order to gain speed for the trip- to Mars. Now, if the acceleration at this point has exactly the same magnitude as g on the earth,

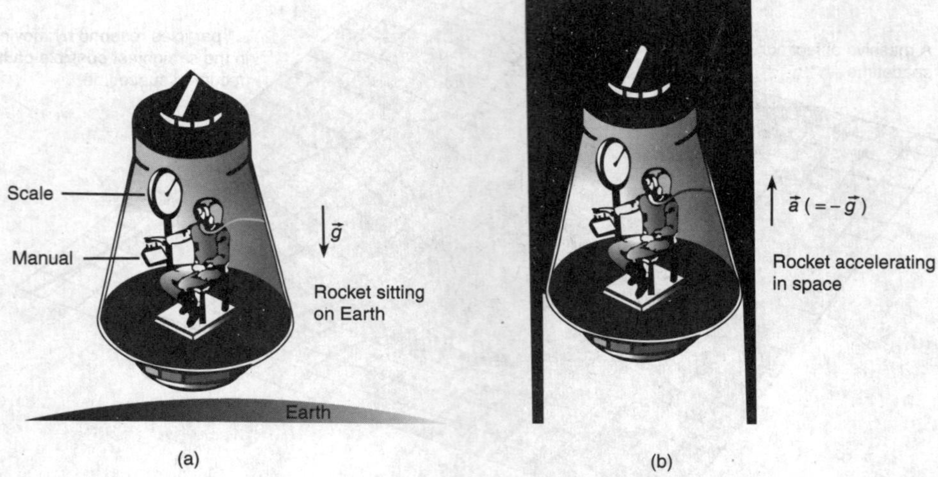

Fig. 27.2 An astronaut sits in a confined space inside a rocket having no windows. The astronaut sits on a chair provided with an ordinary weighing scale and drops a safety manual to the floor. (a) the rocket is at rest on earth's surface, and (b) the rocket is accelerating upward with acceleration **g**. If we forget about noise, the astronaut can do no experiment to distinguish between gravity and acceleration

then the weighing scale indicates precisely the same mass *M* that it indicated on earth, and the safety manual, when dropped, still falls with the same acceleration as measured by the astronaut. Now, the question arises, how can the astronaut tell whether the rocket is sitting on earth or accelerating in space? Einstein provided the answer in the *Principle of Equivalence*.

Astronauts inside rocket experience this equivalence to a certain degree when they are in free fall, rotating around the earth in space shuttle orbiter. Does the fact they feel '*weightless*' mean they are not attracted to the earth by gravity? No, in fact, they are "falling around the earth". We may note that inside of the orbiter is almost an intertial reference system, and the astronauts can perform experiments inside the orbiter on a small particle that obeys Newton's laws. However, according to an observer in an inertial frame of reference system fixed in the distant stars, the shuttle is in an accelerated frame of reference system. If the astronauts cannot see outside their capsule, can they tell whether they are somewhere in outer space or 'falling around the earth'? can astronauts understand precisely why they are weightless?

The astronauts with considerable experimentation, can figure out they are in gravitational field because weightlessness actually occurs only at the center of mass of the orbiter. If astronauts put a drop of water in a corner of orbiter, the drop does not have a precisely spherical shape because the gravitational field of the earth is not uniform. The drop of water becomes very slightly bulged on both sides closest and farthest from the earth due to the change in the earth's gravitational field with distance. Clearly, astronauts can tell that they are in gravitational field. The forces are termed '*tidal*' as the same forces are responsible for the ocean tides.

Now, we look at gravitational and inertial forces. According to Newton's second law of motion, a body accelerates in reaction to a force according to its inertial mass m_i, i.e.,

$$\boldsymbol{F} = m_i \boldsymbol{a} \tag{27.1}$$

where m_i is inertial mass, measures how strongly an object resists a change in its motion.

According to the law of gravitation, a gravitational mass, m_G reacts to the field **g** like that due to earth. We may note that the field **g** depends on the location of all the masses other than m_G. The force of gravitation measures how strongly an object m_G is attracted to other masses.

$$\boldsymbol{F} = m_G \boldsymbol{g} \tag{27.2}$$

For the same force, one can set the two preceding equations equal and obtains

$$\boldsymbol{a} = \left(\frac{m_G}{m_i}\right)\boldsymbol{g} \tag{27.3}$$

According to the principle of equivalence, the inertial and gravitational masses are equal.

The principle of equivalence is the key to Einstein's general relativity. It states *that observations made in an accelerated frame of reference* are *indistinguishable from observations made in a Newtonian gravitational field.* This equivalence would be interesting but not revolutionary if it could be applied only to mechanical phenomena, but Einstein went further and stated that principle of equivalence holds for all natural phenomena, it holds for optical and all electromagnetic phenomena as well.

The principle of equivalence follows from the experimental observation (to better than 1 part in 10^{12}) that the inertial mass of an object, which governs the object's acceleration when a force acts on it, is always equal to its gravitational mass, which governs the gravitational force another object exerts on it. (The two mass are actually proportional; the constant of proportionality is set equal to 1 by an appropriate choice of the constant of gravitation *G*). It was not until 1960s, aided by advances in technology like space probes, that experiments were finally conducted that confirmed Einstein's general relativity. General relativity has opened new fields of study in astrophysics and cosmology that include research in mathematics, particle physics, nuclear physics and astronomy.

27.2.1 Spacetime Curvature

We already learnt from special relativity that space and time are interrelated. Einstein thought of gravity not as a force but as a curvature of spacetime. Spacetime is flat in empty space, but spacetime becomes highly changeable near matter. It is like a flexible material that accomodates matter by curving in the vicinity of object. Fig. 27.1 shows a two dimensional analog of curved spacetime. Spacetime is curved by massive bodies, and we can imagine a small mass moving toward the massive body, not because of gravitational force, but because it rolls in on the spacetime curvature toward the massive body (Fig. 27.3).

The flow of time is determined by the magnitude of the gravitational field nearby. For example, light travels a longer path in strong gravitational fields, because the spacetime

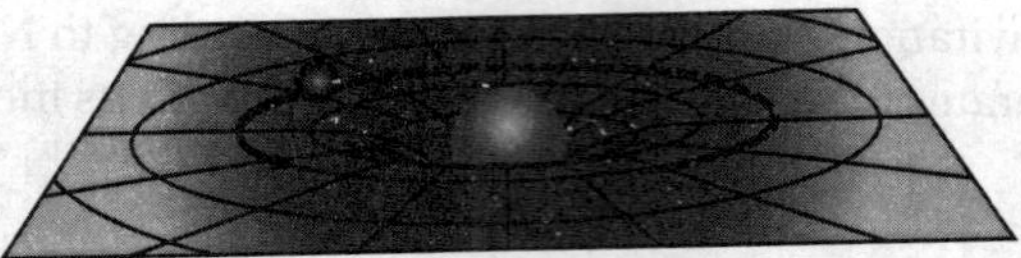

Fig. 27.3 One can imagine a smaller mass attracted to the larger mass by rolling along the spacetime curvature.

geometry has changed and expanded. One can simply put, mass-energy curves space and time. One can summarize as follows:

Mass-energy tells space-time how to curve. Spacetime curvature tells matter how to move.

27.3 TESTS OF GENERAL RELATIVITY

There are many tests of general relativity. Because of extreme experimental difficulties, only a few experimental tests of general relativity were done prior to 1960s. In recent years, a constant stream of experiments has been done. We will review few of them in this section.

(i) Bending of Light by Gravity

According to general relativity, light should be bent while passing through a strong gravitational field. Einstein predicted that starlight passing close to the Sun would be deflected (Fig. 27.4) by an angle of 1.75 seconds of arc, and the two measurements found 1.98 ± 0.16 and 1.61 ± 0.40 seconds.

Measuring this deflection has become a standard practice at every total eclipse since the first measurement were made during the total eclipse of 1919. Since 1919, many experiments, using both starlight and radiowaves from quasars, have confirmed Einstein's predictions about the bending of light with increasingly higher accuracy.

Spectacular evidence has now been reported for the gravitational bending of light. The effect is termed as '*gravitational lensing*'. When light from a distant object like a quasar passes a nearby galaxy on its way to earth, the light can suffer bent multiple times as it passes in different directions around the galaxy. The result can be like as shown in Fig. 27.5.

(ii) Gravitational Redshift

An interesting astronomical effect is suggested by the gravitational behaviour of light. If the frequency associated with a photon moving toward the earth increases, then the frequency of a photon moving away from the earth should decrease.

The earth's gravitational field is not particularly strong, however the fields of many stars are. The gravitation field of earth cannot slow down light, but it can do work on the light pulse to lower its energy. The energy of a light pulse depends on its frequency ν through Planck's relation, $E = h\nu$. As the light pulse travels up vertically, it loses its kinetic energy and therefore its frequency decreases. Its wavelength increases, and therefore the wavelengths of visible light are shifted toward the red end of the visible spectrum. One may say that the light is *redshifted*, and the general phenomenon is termed as the *gravitational redshift*.

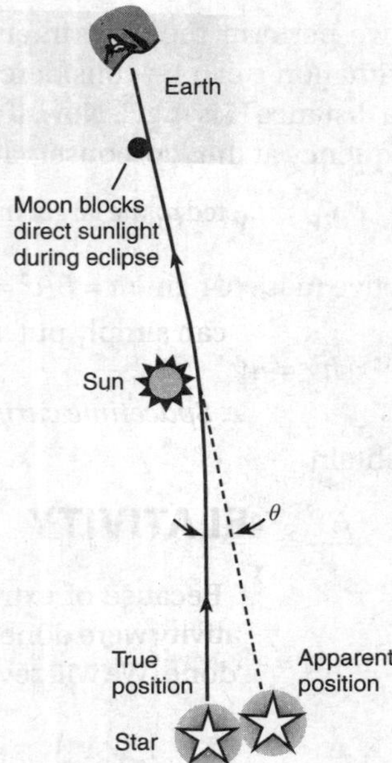

Fig. 27.4 Starlight bends as it grazes the sun due to gravitational attraction. The effect is that the apparent position of a star is not always its true position.

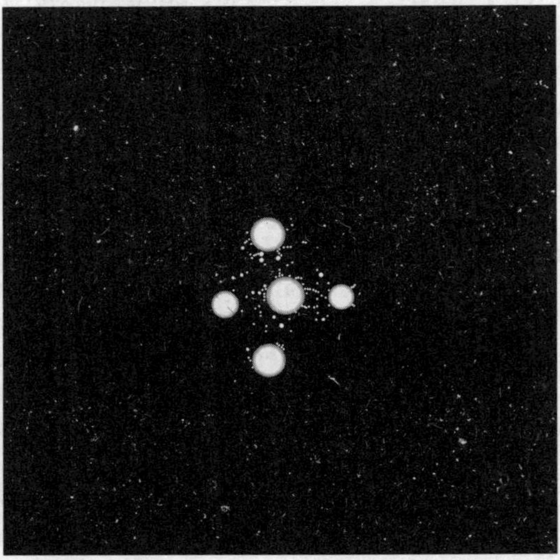

Fig. 27.5 Image of a gravitational lens effect by a Hubble space telescope. This image is called the "Einstein Cross". We can see that the light of a distant quasar has been bent around a nearby galaxy on its way on earth, forming the four outside images (Source: NASA).

Let us suppose that we perform the experiment close to the earth's surface, where to a good approximation g can be considered constant. The loss of energy while traveling upward a distance H is mgH. Now, if ν is the frequency of light at the bottom, and ν' is the frequency at the top, conservation of energy principle gives

$$h\nu = h\nu' + \text{mgH} \tag{27.4}$$

Now, substitute an effective mass of light $m = E/c^2 = h\nu/c^2$ in (27.4), one obtains

$$h\nu = h\nu' + h\nu\frac{\text{gH}}{c^2}$$

Writing $\Delta\nu = \nu - \nu'$, we obtain

$$\frac{\Delta\nu}{\nu} = \frac{\text{gH}}{c^2} \tag{27.5}$$

Since

$$\text{mgH} = -\text{GMm}\left(\frac{1}{r_1} - \frac{1}{r_2}\right)$$

$$\therefore \quad \frac{\Delta\nu}{\nu} = -\frac{\text{GM}}{c^2}\left(\frac{1}{r_1} - \frac{1}{r_2}\right) \tag{27.6}$$

Measurements comparing frequency differences are extremely sensitive, and such an experiment was performed by Pound and Rebka in 1960 using γ-rays from radioactive ^{57}Co ($^{57}Co + e^- \rightarrow {}^{57}Fe + \gamma$-ray). They allowed γ-rays down the tower, so the γ-rays gained energy, thereby increasing their frequency. In this case, a *blueshift* occurs. Both these scientists used the Mössbauer effect to obtain the needed sensitivity of $\Delta\nu/\nu \approx 10^{-15}$.

The general relationship for Eq. (27.5) does not depend on the gravitational potential energy mgH but rather on $\text{GMm}\left(\frac{1}{r_1} - \frac{1}{r_2}\right)$, where G is the universal gravitational constant. The ratio of the difference in frequency is given by (27.6), where clocks measuring the frequency are placed at distances r_1 and r_2 from the center of the gravitational field. When $r_1 < r_2$, $\Delta\nu$ is negative. Obviously, a signal leaving r_1 has a lower frequency when arriving at r_2. This clearly reveals that an atom emitting light signals in a strong gravitational field (r_1 is small near the center of mass) has its wavelength redshifted when the light arrives at r_2 far away from the center of mass.

It is difficult to measure gravitational redshifts from stars, because of the larger *Doppler redshift* due to the star moving away from the earth at high speed; a few stars are observed to be moving toward us. The existence of the gravitational redshift have been confirmed through several experiments using light from the sun and from white dwarfs. A very accurate experiment was performed by comparing the frequency of an atomic clock flown on a Scout D rocket at an altitude of 10,000 km with the frequency of a similar clock on the earth's surface. The measurement agreed with Einstein's general relativity to within 0.02%.

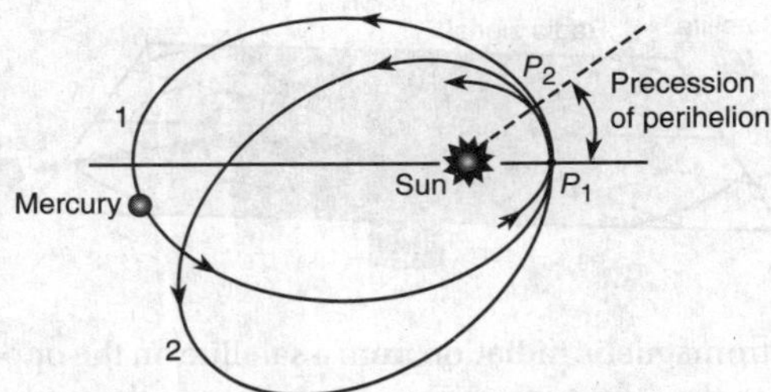

Fig. 27.6 Mercury's orbit slowly precesses about the Sun. For orbits 1 and 2, the points P_1 and P_2 are the perihelion. We have shown the effects exaggerated

(iii) Perihelion Shift of Mercury

Plants orbit are ellipses, and the point closest to the sun in a planetary orbit is termed the *perihelion*. The Mercury's orbit precesses about the sun as shown in Fig. 27.6. Although the point of perihelion precesses very slowly, but the rate of precession has been measured accurately. Most of the effect we observe is due to the rotation of the earth, but after that is substracted there is still a perihelion shift of 575 seconds of arc per century. Most of this shift is due to the gravitational perturbation caused by the other planets, but in 1859 U.J. Joseph Le verrier announced that after even these corrections were taken into account, there was still a remaining perihelion shift amounting to 43 seconds of arc per century. At that time, this was a great mystery to scientists, and the answer had to wait some 60 years, until Einstein's general relativity theory introduced in 1916. Einstein showed that general relativity theory also predicted a perihelion shift, and he showed that the shift was just the missing 43 seconds needed! Einstein was overjoyed by this calculation; together with the observed deflection of star light by the Sun in 1919.

(iv) Light Retardation

This is similar to the deflection of light (or any electromagnetic radiation) by gravitational fields. As light passes by a massive object, the path taken by the light is longer due to spacetime curvature (Fig. 27.7) with a radiowave.

The longer path causes a time delay for a light pulse traveling close to the Sun. The time difference between similar paths, with one being close to the Sun, can be compared with the prediction from general relativity.

I. Shapiro showed that such an effect could be measured by sending a radar wave to Venus, where it was reflected back to Earth. The position of venus should be in the 'superior conjunction' position on the other side of the Sun from the Earth (Fig. 27.8). In this path, the light signal passed close to the Sun and experienced a time delay. Shapiro's measurement in 1970, of a time delay of about 200 μs was in excellent agreement with general relativity theory (Fig. 27.9).

Several experiments have since been performed with spacecraft to measure the light retardation effect. The experiment to land part of the Viking spacecraft on Mars in 1976 produced agreement with the general relativity theory to with an experimental uncertainty of about 0.1%. Physicists in 2003 physicists reported on a measurement done with NASA cassini spacecraft. The spacecraft, on its way to its highly successful rendezvous with Saturn

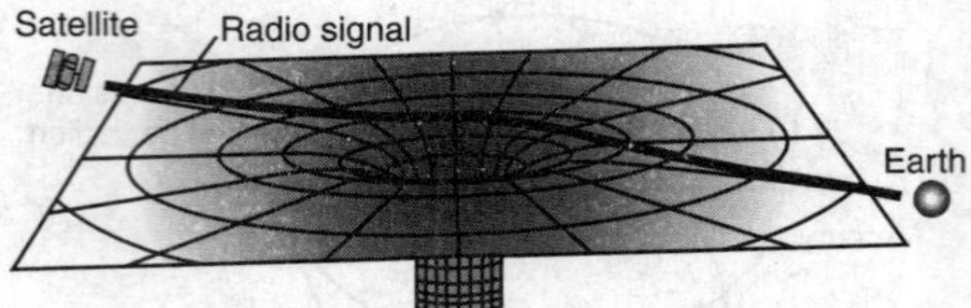

Fig. 27.7 Light or any electromagnetic radiation from a satellite on the opposite side of the Sun from Earth pass close to the Sun. As the light passes close to the Sun, the light is gravitationally attracted by the Sun and encounters a longer path due to spacetime curvature on its way to Earth. The effect is called light retardation

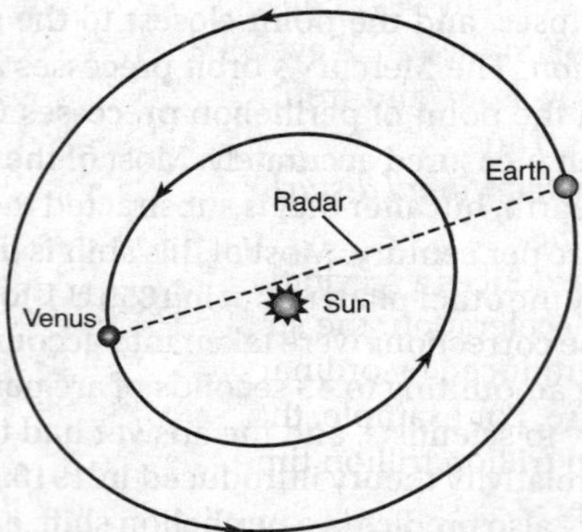

Fig. 27.8 The superior conjuction position of venus when it is on the other side of Sun from the Earth. The radar waves sent between two planets will be delayed slightly due to the gravitational attraction of the Sun

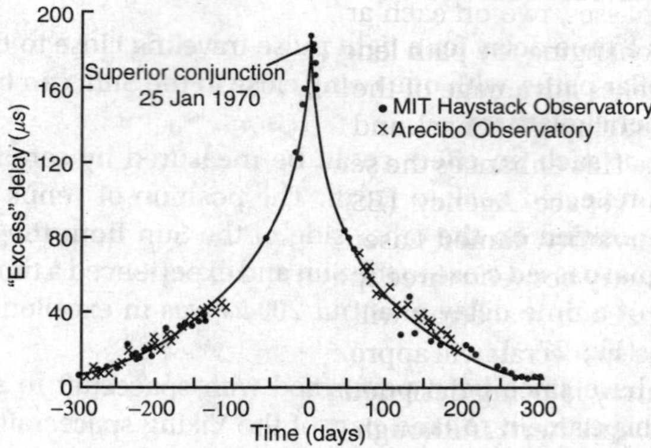

Fig. 27.9 The time delay measurements of the radar waves between Earth and Venus due to Shapiro in 1970. The solid line is due to general relativity prediction

in 2004, passed on the opposite side of the Sun from Earth in 2002. Scientists measured the frequency change of radio waves that passed close to the Sun while they were traveling to and fro through the spacecraft from Earth. The radiowaves passing close to the Sun traveled an extra distance due to the spacetime curvature. The results agreed with theory to 2 parts in 10^3.

27.4 GRAVITATIONAL WAVES

Every object has mass and therefore wraps the surrounding spacetime. When an object undergoes a change in motion, the surrounding wrap moves in order to readjust to the new position. These readjustments produce ripples in the overall geometry of spacetime. This is similar to moving a ball that rests on the surface of waterbed. A disturbance ripples across the waterbed surface in waves; if we move a more massive ball, then we get a greater disturbance and the production of even stronger waves. Similarly for spacetime in universe. Similar ripples travel outward from a gravitational source at the speed of light and are *gravitational waves*. Thus,

Gravitational wave is a gravitational disturbance generated by an accelerating mass that propagates through spacetime.

Any accelerating object produces a gravitational wave. In general, the more massive the object and the greater its acceleration, the stronger the resulting gravitational wave. But even the strongest waves produced by ordinary astronomical events are extremely weak, the weakest known in nature. For example, the gravitational waves emitted by a vibrating electric charge are a trillion trillion trillion times weaker than the electromagnetic waves emitted by the same charge.

Detecting gravitational waves is enormously difficult, and no confirmed detection has occurred to date. Recently completed wave detectors are expected to detect gravitational waves from supernovae, which may radiate away as much as 0.1% of their mass as gravitational waves, and perhaps from even more cataclysmic events, such as colliding black holes.

One possible way of detecting gravitational waves takes advantage of optical interference (Fig. 27.10). Four test masses, two on each arm of the apparatus, react to a gravitational wave passing through, changing the path lengths L_1 and L_2 slightly. This basic Michelson Interferometer device takes advantage of the Febry-Perot method by placing mirrors on the masses, so that the light bounces back and forth between masses (2 km or 4 km) several times before interfering. This enhances the sensitivity of the device greatly.

NASA and European space Agency (ESA) are jointly developing a space-based gravitational wave observatory named Laser Interferometer Space Antenna (LISA) that will use laser interferometry between three spacecraft located 5 million km apart in space in orbit around the Sun. Each spacecraft will carry two telescopes with associated lasers and optical systems. The spacecraft will approximate a triangle in the plane of Earth's orbit around the Sun. This array is being designed to be sensitive to gravitational waves with frequencies as low as 10 millihertz. Although the spacecraft will not be susceptible to the same kind of noise as the earth's surface based observatories, they will still have difficulties accounting for the solar wind and solar radiation pressure to achieve the sensitivities desired.

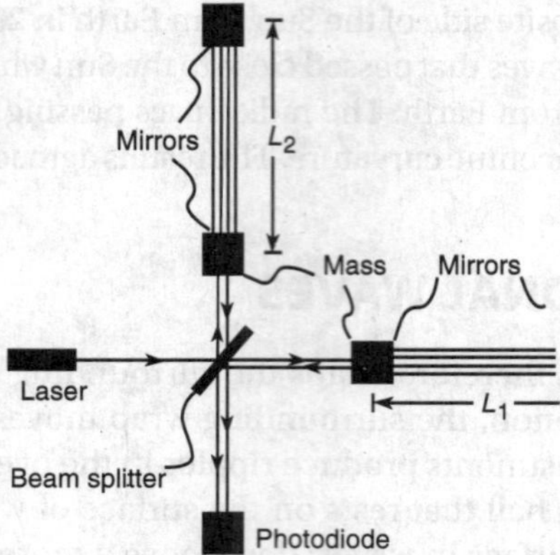

Fig. 27.10 LIGO device. A gravitational wave passing through large masses suspended on thin wires causes the distance L_1–L_2 to change. The interference pattern observed in photodiode may be capable of recognizing the gravitational wave

We may note that as weak as they are, gravitational waves are everywhere. Shake your hand back and forth; you have just produced a gravitational wave. It is not very strong, but it exists.

27.5 BLACK HOLES

Black hole is an object that has collapsed under its own gravitational forces to such an extent that its 'escape velocity' is equal to the speed of light. Black holes are believed to be formed in the gravitational collapse of very large stars at the ends of their lives.

At the end of a lifetime, a massive star uses up all its fuel and collapses under the influence of gravitational forces, causing a gigantic outburst of energy called a supernova explosion. The best-studied example of a remnant of a supernova explosion is the Crab Nebula, a chaotic, expanding mass of gas. The material that remains in the central core of such an object continues to collapse, and the core's ultimate fate depends on its mass. If the core has a mass less than 1.4 times the mass of our Sun, it gradually cools down and ends its life as a *white dwarf star*. If, however, the core's mass is greater than that, it may collapse further due to gravitational forces. What remains is a *neutron star*, in which the mass of a star is compressed to a radius of about 10 km.

An even more unusual star death may occur when the core has a mass greater than about three solar masses. The collapse may continue until the star becomes a very small object in space, commonly referred to as a *black hole*. In effect, black holes are the remains of stars that have collapsed under their own gravitational force. If an object such as a spacecraft comes close to a black hole, it experiences an extremely strong gravitational force and is trapped forever.

The escape speed from any spherical body depends on the mass and radius of the body. The escape speed for a black hole is very high because of the concentration of the star's mass into a sphere of very small radius. If the escape speed exceeds the speed of light(c),

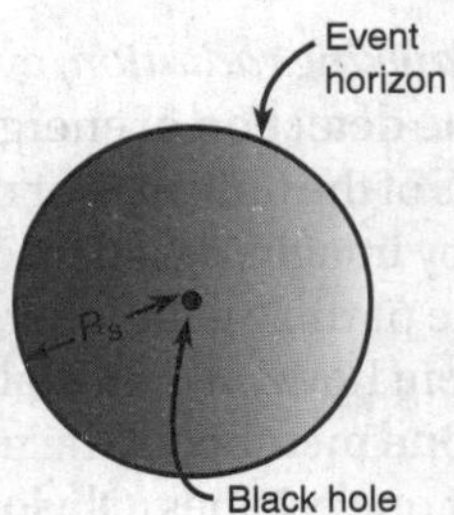

Fig. 27.11 A black hole. The distance R_s equals the Schwarzschild radius. Any event occurring within the boundary of radius R_s, called the event horizon, is invisible to an outside observer

radiation from the body (e.g. visible light) cannot escape and the body appears to be black, hence the origin of the term *black hole.* The critical radius R_s at which the escape speed is c is called the *Schwarzschild radius* (Fig. 27.11), is given by

$$R_s = \frac{2\,GM}{c^2} \tag{27.7}$$

$$= (1.5 \times 10^{-27} \text{ m/kg})M$$

This allows one to calculate the radius of a black hole for a given mass.

The imaginary surface of a sphere of this radius surrounding the black hole is called *event horizon*, which is the limit of how close you can approach the black hole and hope to be able to escape. The expression (7) is for a nonrotating spherical mass. The body continues contracting past the Schwarschild radius, but nothing happening inside the collapsing body can affect the spacetime region outside R_s. Any light ray or particle emitted inside the black hole is kept from going outside by the gravitational force.

An observer outside the Schwarzschild radius (R_s) cannot even tell when the radius passes through R_s. The light leaving the body right before the body crossed over to $R < R_s$ is so strongly attracted to the body that it takes progressively longer and longer to reach the observer. This is merely an optical effect, because physically the apparent luminosity of the body decreases rapidly with time, and the light is strongly redshifted.

The point at the center of a black hole is called the *singularity*. When we talk about the size of a black hole, we are referring to the size of the event horizon. Relation (27.7) allows us to calculate the radius of the black hole for a given mass.

Only very heavy stars end up as black holes. Lighter stars evolve into white dwarfs and neutron stars, which consists largely of neutrons. But as time goes on, the strong gravitational fields of both white dwarfs and neutron stars attract more and more cosmic dust and gas. When they have gathered up enough mass, they too will become black holes. If the universe lasts long enough, then we may expect everything in it may be in the form of black holes.

How can a black hole be detected if there is literally no-way to "see" it, i.e., light cannot leave a black hole? There are several indirect means to detect a black hole. As the star collapses during the formation of a black hole, the gravitational redshifts should be large. When the black hole is formed, the signals abruptly cease. However, observation of such signals would be purely a chance. Stephen Hawking, using quantum mechanics, outlined a method by which energy, called *Hawking radiation*, evaporates from the black hole much like a thermal energy spectrum. The detection of energy is very difficult, and the process is so slow that it would take many ages of the universe for the black hole to dissipate its energy. One can understand this process by imagining a particle being attracted from outside the black hole to the event horizon. The particle gains energy from the gravitational field of the black hole. As it approaches the event horizon, a virtual particle-antiparticle pair is created by utilising the large energy gain. One member of the newly created pair occasionally goes into the black hole, while the other one escapes. Obviously, the gravitational energy of the black hole has been used to create the particle-antiparticle pair.

Hawking gave the following relation for the calculation of the temperature of the blackbody radiating the energy

$$T = \frac{\hbar c^3}{8\pi k \mathrm{GM}} \tag{27.8}$$

The power radiated per unit area at temperature T can be obtained from Stefan-Boltzmann law, $R(T) = \varepsilon\sigma T^4$, where we let the emissivity $\varepsilon = 1$ for a blackbody and σ is the Stefan-Boltzmann constant. The radiated power $P(T)$ by a blackbody at temperature T becomes

$$P(T) = (4\pi r_s^2)\sigma T^4 \tag{27.9}$$

where r_s is the blackbody's Schwarzschild radius. Inserting T from (27.8) into (27.9), we obtain

$$P(T) = 4\pi\sigma r_s^2 \left(\frac{\hbar c^3}{8\pi k \mathrm{GM}}\right)^4 \tag{27.10}$$

Eq. (27.10) can be used to determine the rate of Hawking radiation emanating from a black hole.

One may detect black holes indirectly by their gravitational influence on their surroundings. Let us consider a binary system of a black hole and a companion star. The strong gravitational force of the black hole pulls gaseous matter off the star, but the gas does not go directly into the black hole. Because the companion star has some rotational motion, the gas moves into an orbit around the black hole (Fig. 27.12). This rotating mass collects into an *accretion disk* around the black hole. Friction among particles in the accretion disks results in transformation of mechanical energy into internal energy. As a result, the orbital height of the material above the event horizon decreases and the temperature rises. This high-temperature material emits a large amount of radiation, extending well into the X-ray region of the electromagnetic spectrum. These X-rays are characteristic of a black hole. Several possible candidates for black holes have been identified by observation of these X-rays.

Evidence also supports the existence of supermassive black holes at the centers of galaxies, with masses very much larger than the Sun. There is a strong evidence for a supermassive black hole of mass 2 to 3 million solar masses at the center of our galaxy. Theoretical models for these bizarre objects predict that jets of material should be evident along the rotation axis of the black hole.

Black holes seems to be of two sizes: (i) like Cygnus X-1 that have between 5 and 20 masses of the Sun (solar masses). These are called *stellar black holes*, (ii) *supermassive* black holes,

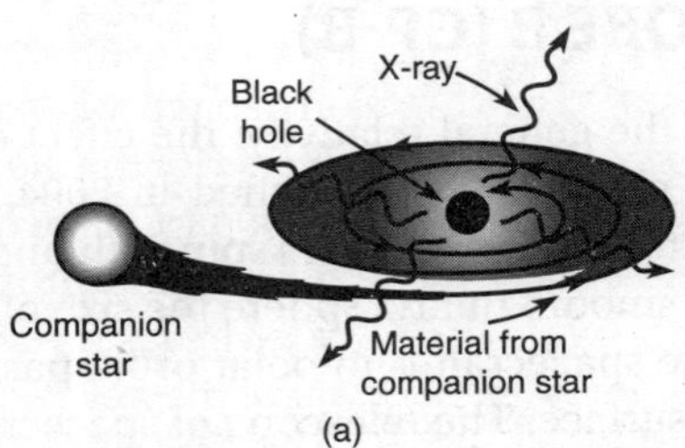

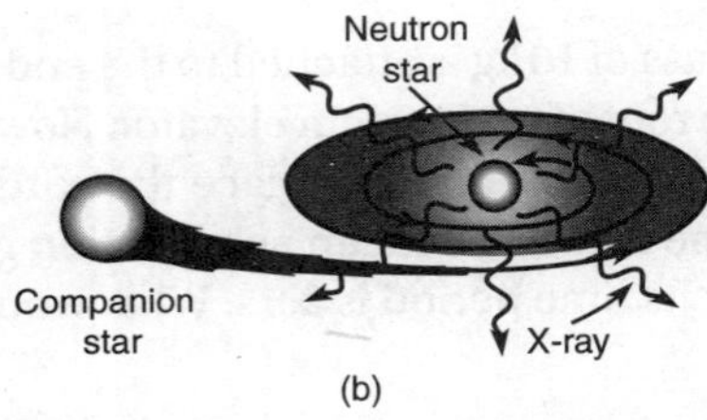

Fig. 27.12 Matter from a companion star is attracted by the strong gravitational pull of (a) a black hole, or (b) a neutron star. The matter spirals inward and form a disk before being pulled into the black hole or neutron star. Friction within the disk creates tremendous temperatures and cupious numbers of X-rays

which are thought to have masses of millions or billions of times that of our Sun. There is some controversy about black holes having intermediate sizes.

With the tremendous production of the Hubble Space Telescope and the output of new telescopes coming on line, and with better detection systems and computer analysis, one detects more and more black holes each year. We may note that scientific community generally accepts the existence of these strange phenomena that can only be inferred indirectly, as one can't see actually them.

Black holes are of considerable interest to those searching for *gravity waves* which are ripples in space-time caused by changes in a gravitational system. These ripples can be caused by a star collapsing into a black hole, a binary star consisting of a black hole and visible companion, and supermassive black holes at a galaxy center.

27.6 FRAME DRAGGING

Austrian physicists J. Lense and H. Thirring, using general relativity proposed in 1918 that a rotating body's gravitational force literally drag space and time around it as the body rotates. This effect, sometimes referred as *Lens-Thirring effect,* is usually called as *frame dragging.* All celestial bodies that rotate can modify the space-time curvature, and the larger the gravitational force, the greater the effect. We may note that it was one of the last predictions of general relativity to be confirmed.

27.7 GRAVITY PROBE B (GP-B)

To prove two predictions of the general relativity: the effect of spacetime curvature and frame dragging, the gravity probe B was launched in 2004. GP-B has four gyroscopes bathed in liquid helium that measure the Earth's spin with unprecedented accuracy. Each gyroscope has an incredibly smooth quartz sphere the size of a Ping-Pong ball that spins 10,000 times per minute. The spacecraft is in polar orbit, passing over each of the poles, about 650 km above Earth's surface. The telescope of spacecraft is locked onto the guide star IM Pegasus, about 300 light years away.

Gravity Probe B is designed to measure two effects: frame dragging and the geodetic effect. The geodetic effect is due to spacetime curvature.

Example 27.1: Suppose a mass of 10 kg is attached to the end of massless cord 1.5 m long whose other end is tied to the roof of an Einstein elevator. Now, you straighten the cord so that it is perpendicular to the elevator roof, displace the cord through a small angle-say less than 1°- and accelerate the elevator with an acceleration $g = 9.8\,\text{m/s}^2$. The system will act as a pendulum. Show that its time period is 2.5 s. What would be if the mass were 20 kg in place of 10 kg?

Solution In accordance with the principle of equivalence, one can replace this situation by one in which the system is in uniform gravitational field of sufficient strength to produce the acceleration, which is that at Earth's surface. Since the initial displacement is small, one can use the small-angle approximation to the period, i.e.,

$$T = 2\pi\sqrt{\frac{l}{g}} = 2\pi\sqrt{\frac{1.5\text{m}}{9.8\text{m/s}^2}} \cong 2.5\text{s}$$

where l is the length of the pendulum and g is the acceleration. We can see that this number is independent of the mass that is suspended. In fact, one can turn the argument around and say that this fact about pendulums is a a confirmation of the principle of equivalence, first noted by Galileo.

Example 27.2: The increase in energy of a fallen photon observed by Pound and Rebka in 1960 with H = 22.5 m. Show that the change in frequency of a photon of red light whose original frequency is 7.3×10^{14} Hz when it falls through 22.5 m is 1.8 Hz.

Solution

$$\begin{aligned} \nu' - \nu &= \left(\frac{gH}{c^2}\right)\nu \\ &= \frac{(9.8\,\text{m/s}^2)(22.5\,\text{m})}{(3\times10^8\ \text{m/s}^2)} \times 7.3\times10^{14}\ \text{Hz} \\ &= 1.8\,\text{Hz} \end{aligned}$$

We may note that Pound and Rebka used γ-rays of much higher frequency.

Example 27.3: Show that a black hole having 3 solar masses will take about 5.7×10^{68} y to radiate its energy. Compare it with lifetime of our universe.

Solution One may use mass energy Mc^2 to find that it loses energy at the rate

$$P(T)=\frac{-d\,(Mc^2)}{dt}=-c^2\frac{dM}{dt}$$

Now,
$$-c^2\frac{dM}{dt}=4\pi\sigma r_s^2\left(\frac{\hbar c^3}{8\pi k GM}\right)^4$$

$$=4\pi\sigma\left(\frac{2GM}{c^2}\right)^2\left(\frac{\hbar c^3}{8\pi k GM}\right)^4$$

$$=\frac{2\sigma\hbar^4 c^8}{(8\pi)^3 k^4 G^2}\frac{1}{M^2}$$

Thus the rate of mass loss is obtained as

$$\frac{dM}{dt}=-\frac{2\sigma\hbar^4 c^6}{(8\pi)^3 k^4 G^2}\frac{1}{M^2}=-\alpha\frac{1}{M^2} \qquad (1)$$

Where $\alpha=\dfrac{2\sigma\hbar^4 c^6}{(8\pi)^3 k^4 G^2}$ is a constant and its value comes out $= 3.69\times10^{15}\,kg^3\,s^{-1}$.

Rewriting Eq. (1) and integrating, one obtains

$$\int M^2\,dM=-\alpha\int dt$$

or
$$M^3=-3\alpha t+C=-3\alpha t+M_0^3$$

Where $C=M_0^3$ is a new constant and related to mass M_0 at $t=0$. One can find the time for all the mass to radiate by putting $M=0$ as

$$M_0^3=3\alpha t$$

or
$$\alpha=\frac{M_0^3}{3\alpha}$$

Time for the 3-solar mass black hole to evaporate

$$t=\frac{[3(1.99\times10^{30}\text{ kg})]^3}{3(3.96\times10^{15})}=1.8\times10^{76}\text{ s}$$

$$=5.7\times10^{68}\text{ years.}$$

The life time of our universe is 13.7 billion years. Obviously, t is many times the lifetime of our universe.

Example 27.4: Show that Schwarzschild radius for Sun and Earth are 3 km and 9 mm respectively.

Solution

$$R_s = (1.5 \times 10^{-27}\,\text{m/kg})M$$

$$\text{Mass of the Sun} = 2\times10^{30}\,\text{kg}$$

$$\therefore \quad R_s(\text{Sun}) = (1.5\times10^{-27}\,\text{m/kg})(2\times10^{30}\,\text{kg}) = 3\,\text{km}$$

$$\text{Mass of the Earth} = 6\times10^{24}\,\text{kg}$$

$$\therefore \quad R_s(\text{Earth}) = (1.5\times10^{-27}\,\text{m/kg})(6\times10^{24}\,\text{kg}) = 9\,\text{mm}.$$

Example 27.5: Find the Schwarzschild radius R_s for (a) a mass of 12 solar masses and (b) a 1-kg mass.

Solution

(a) $R_s(12M_s) = (1.5\times10^{-27}\,\text{m/kg})\times(12\times2\times10^{30}\,\text{kg})$
$= 36\,\text{km}$

(b) $R_s(M_{1\text{kg}}) = (1.5\times10^{-27}\,\text{m/kg})(1\text{kg})$
$= 1.5\times10^{-27}\,\text{m}.$

SUMMARY

1. Einstein's general relativity theory is a theory of gravity, which treats the effect of gravity on space and time. This theory replaces Newton's force laws by a system based on curvature of spacetime.
2. Einstein's principle of equivalence states that the behaviour of a system accelerating uniformly and behaviour of that system in a uniform gravitational field are indistinguishable. The principle of equivalence leads to the bending of light in a gravitational field. Spacetime curvature occurs near masses.
3. There are several tests of general relativity:

   (i) Light moving away from a mass is *redshifted;* light moving toward a mass is blueshifted. The gravitational red shift is given by

   $$\frac{\Delta\nu}{\nu} = \frac{gH}{c^2}$$

   Since this is also a shift in frequency, an observer will see a clock run more slowly when it moves closer to a mass and more rapidly when it moves away from a mass. Most experiments confirm the existence of gravitational redshifts.
4. Light is deflected in the presence of a mass. It's path is the straighest line possible in the distorted space whose curvature is produced by the presence of mass.

5. The general theory of relativity also predicts a noticeable effect on the perihelion of Mercury's orbit. Such a calculation accounted for an anomaly in astronomical measurements since 1859.
6. Non uniformly accelerating masses radiate energy in the form of gravitational radiation.
7. A gravitational disturbance generated by an accelerating mass that propogates through spacetime is termed as gravitational wave. However, gravitational waves has not yet been directly confirmed. These waves are currently of great interest.
8. The amount of redshift observed when light falls towards a mass depends on the value of the mass and the radius within which the mass sits. If the mass is large enough or the radius is small enough, the redshift goes all the way to infinite wavelength (zero frequency) and no light is observed at all. In that case one has a *black hole*. Black holes are objects that have collapsed under their own gravitational attraction. Nothing can escaple a black hole, not even light. The conditions for stars to evolve to black holes are entirely possible, as are the conditions for very massive black holes to form.
9. The size of a nonrotating spherical black hole is given by the Schwarzschild radius formula,

$$R_s = \frac{2GM}{c^2} = (1.5 \times 10^{-27} \text{ m/kg})M$$

Scientists believe that the existence of black holes has been confirmed, despite the fact that it is difficult to detect them.
10. The existence of frame dragging for black holes and perhaps for the earth has been confirmed recently.

REVIEW QUESTIONS

1. Explain the principal difference between the theory of special relativity and general relativity.
2. Explain the principle of equivalence in general relativity. Mention exactly what is equivalent in this principle.
3. Compare the bending of the paths of baseballs and photons by a gravitational field.
4. Why must the Sun be eclipsed to measure the deflection of starlight passing near the Sun?
5. Explain why true weightlessness occurs only at the center of mass of space shuttle orbiter as it rotates around the Earth?
6. What is the effect of strong gravitation on measurment of time?
7. Which runs slower, a clock at the top of the Sears Tower in Chicago or a clock on the shore of Lake Michigan?
8. How does the frequency of a particular spectral line observed in sunlight with the frequency of that line observed from a source on Earth?
9. Why is it difficult to experimentally verify the predictions of general relativity?
10. If we view events occuring on a star that is collapsing to become a black hole, do we see time speeded up or slowed down?

11. From equation

$$a = \left(\frac{m_G}{m_I}\right) g$$

We conclude that the inertial and gravitational masses are equal. Explain its significance.
12. Of all the planets, why is Mercury the best candidate for finding evidence of the relationship between gravitation and space.
13. What effect does mass have on spectrum?
14. What happens in the surrounding space when a massive object undergoes a change in its motion?
15. A star 10 light-years away explodes and produces gravitational waves. How long will it take these waves to reach the Earth?
16. Why are gravitational waves so difficult to detect?
17. Gravitational red shifts can be observed and measured during the collapse of a star into a black hole. When might the redshifts cease?

SHORT ANSWER-QUESTIONS

1. If we call the distance above Earth's surface h and if $h << R$ ($R \rightarrow$ Earth' radius), what will be Newton's law in this situation approximately?

   Ans: $m\dfrac{d^2h}{dt^2} = -\dfrac{GMm}{R^2}$

   where M is Earth's mass and m is the person's mass. Here we have ignored the fact that the gravitational force varies with the height above Earth.
2. Light moving towards Earth-or indeed any mass-acquires a blueshift. Express it mathematically to an increased frequency ν', for small values of ν/c.

   Ans: $\dfrac{\nu'}{\nu} = 1 + \dfrac{\nu}{c} = 1 + \dfrac{gh}{c^2}$

   where gh is the gravitational potential difference between the emitter and the detector.
3. What is gravitational redshift?

   Ans: Shift in wavelength toward the red end of the spectrum experienced by light leaving the surface of a massive object, as predicted by the theory of general relativity.
4. What is gravitational wave?

   Ans: Gravitational disturbance that propagates through spacetime made by a moving mass.
5. What is graviton?

   Ans: Graviton is quantum of gravity, similar in concept to the photon as a quantum of light.

6. What is gravitational lensing?
Ans: The effect, i.e., bending of light around masses is called gravitational lensing.
7. What is the rate at which energy is radiated (the Power) by two point masses M_1 and M_2 separated by a distance d and rotating about each other in circular orbits with an angular frequency ω?

Ans: $P = \frac{32}{5}\frac{G}{c^5}\omega^6 d^4 \mu^2$

where $\mu = \frac{M_1 M_2}{(M_1 + M_2)}$ is the reduced mass. This formula is an analogue to the Larmor formula for the power radiated by a rotating charge.
8. What stands for 'LIGO'?
Ans: LIGO stands for Laser Interferometer Gravitational-Wave observatory. These projects will open a whole new window to our universe.
9. What is the period for the star orbits the (nearby) stationary black hole?

Ans: $T = 2\pi\sqrt{\frac{R^3}{GM}}$

As long as the orbital plane of this system is not oriented perpendicular to us, the period can be determined by measurments of the Doppler shift of the luminous star-in one period it moves away from and then towards us.
10. It is believed that the star orbits the (nearly) stationary black hole. What will be its period?

Ans: $T = 2\pi\sqrt{\frac{R^2}{GM}}$

where M is mass of the black hole.

PROBLEMS

1. Radiation is emitted from the Sun over a wide range of wavelengths. Show that the gravitational redshift of light of wavelength 550 nm (the middle of the visible range) that is emitted from the Sun is 1.17×10^{-3} nm.
2. In an experiment of Pound and Rebka, a 14.4 keV gamma ray fell through a distance of 22.5 m near the Earth's surface. Calculate the change in frequency and the percentage change in frequency due to gravitational redshift?
[**Ans.** 8541 Hz; 2.45×10^{-13}%]
3. Show that the Schwarzschild radius for Jupitor ($m_{\text{Jupitor}} = 1.9 \times 10^{27}$ kg) is 2.82 m.
4. The Global Positioning System (GPS) satellites operate at an altitude of 20,200 km and use communication frequencies of 1575.42 MHz. Show that the gravitational frequency change with respect to Earth is 0.834 Hz.
5. One of the communication frequencies that the space shuttle orbiter uses is 294 MHz. Assuming g is constant, show that the gravitational frequency change with respect to Earth when the orbiter is at an altitude of 300 km is 9.6×10^{-3} Hz.

FILL IN THE BLANKS

1. All objects fall with the same ______ in the gravitational field at Earth's surface. **[acceleration]**
2. Light reaching us is redshifted by an amount______. **[$-\frac{GM}{Rc^2}$]**
3. The bending of light around masses has become a tool of modern astronomy through what are known as ______. **[gravitational lenses]**
4. The orbits of planets are ellipses, and the point closest to the Sun in a planetary orbit is called the ______. **[perihelion]**
5. Einstein's general theory of relativity is also a modern theory of gravitation, but does not include ______. **[quantum theory]**
6. The boundary region or surface of a black hole is called ______. **[surface horizon]**
7. The point at the center of a black hole is called the ______. **[singularity]**
8. The size of a nonrotating spherical black hole is given by______. **[Schwarzschild radius, $R_s = \frac{2GM}{c^2}$]**

28

Cosmology

28.1 INTRODUCTION

Most scientists and astronomers are quite certain that our universe evolved from a primordial event called the *Big Bang*. Some of that evidence comes from the Hubble space telescope (Fig. 28.1). Current evidence seems to indicate that our universe will expand forever.

28.2 EVIDENCE OF THE BIG BANG

Over the past half century, physicists and astronomers have proposed various theories about the origin of the universe. There were two rival theories of cosmology in the 1950s and 1960s, but data were insufficient to prove either of them. Both theories of the origin of the universe accounted for the known expansion of the universe, which was demonstrated conclusively by the redshift data of Hubble in 1929. The most reasonable and widely held explanation is that the redshift is caused by the Doppler effect. One theory, known as the *steady state theory,* held that matter is being continuously created; as the universe expands, the density of the universe remains constant. The theory known as the *Big Bang* theory, proposes that the distance galaxies are receding from us at speeds proportional to their distances. Based on additional evidence that we will describe shortly, scientists are quite certain that the universe began with such a *Big Bang*. Before presenting the Big Bang model, we mention the following three important pieces of evidence in favour of Big Bang:

(i) The observations by Hubble between 1929 and 1952 using the giant telescopes of Mount Wilson and Mount Palomar, that the galaxies of the universe are moving apart from each other at high speed. Obviously, the universe is apparently expanding outward from some primordial event.

(ii) In 1965 Arno Penzias and Robert Wilson at Bell laboratories found a faint "noise" of microwave radiation in a satellite communications antenna they were testing. Theorists at Princeton identified the "noise" as radiation dating to a much earlier

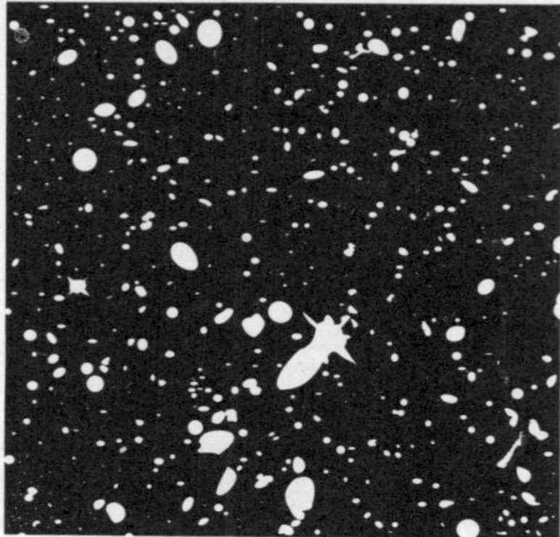

Fig. 28.1 A portion of the Hubble deep field, a Hubble space telescope image showing distant galaxies whose redshifts provide information about cosmic expansion

era in the universe. This *cosmic microwave background* radiation permeates the universe and this is the strongest evidence yet for the Big Bang.

(iii) There is good agreement between predictions of primordial nucleosynthesis of the elements and the known abundance of elements in the universe. This is applicable to the light elements that were produced in the early stages of the Big Bang.

We may note that there are still controversies concerning some of the predictions and results of Big Bang model of universe, however, its main ideas are widely accepted. Big Bang has been modified somewhat since it was first proposed. In the following sections we will discuss all these matters.

28.3 HUBBLE'S MEASUREMENTS

Hubble in the mid 1920's proved that some "nebulae" were indeed distant galaxies like our own Milky Way, each containing 10^9 to 10^{13} stars. Today cosmologists think of galaxies as individual "point particles" whose distribution traces the overall structure of the universe.

Hubble continued to study the galaxies throughout the 1920s, and by analyzing their spectra, he concluded that spectral lines from distant galaxies are shifted toward the end, with the amount of shift dependent on the distance to the galaxies. The most reasonable and widely held explanation is that the redshift is caused by the Doppler effect. Then the implication of Hubble's work is that the galaxies receding from us with speed v are related to the distance R from the earth by the relation known as *Hubble's law*:

$$V = HR \tag{28.1}$$

where V is the recession speed, R the distance and H is the *Hubble constant* whose value is now known to be very nearly 22 km/s/million light years of distance. Astronomers now use relation (28.1) to find the distances to remote galaxies, measuring their redshifts and to infer their distances.

Hubble parameter H is related to a scale factor a that is proportional to the distance between galaxies given by

$$H = \frac{1}{a}\frac{da}{dt} \tag{28.2}$$

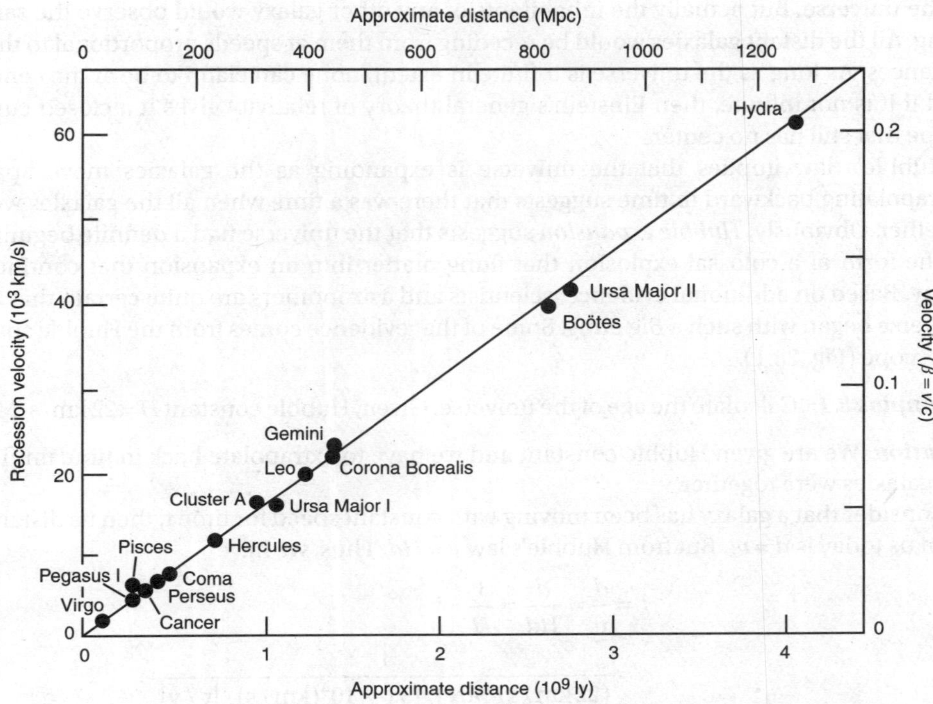

Fig. 28.2 The analysis of 15 clusters of galaxies for recession velocity as a function of distance. The solid line represents a Hubble constant of 15 km/s per million light years or 49 km/ s^{-1} / mpc^{-1}. Various analyses of the distance give different values of H

Because the universe has been expanding, Hubble's parameter H is not constant, but changes slowly over long periods of time. The value of H today is sometimes denoted by H_o (= 22 km/s/Mly). Hubble's law is now used to find the distances to remote galaxies, measuring their redshifts and using (28.1) to infer their distances.

To test, whether Hubble's law is valid, it is necessary to know the distance R to objects for which the redshift has been measured. Hubble developed sophisticated techniques that used the brightness of stars and galaxies to determine the distance R. Hubble was able to do this with some certainty for stars out to distances of 10 million lightyears and, with some more assumptions, for galaxies out to distances of 500 million lightyears.

By 1929, Hubble was able to show that 49 galaxies (which he called extragalactic nebulae) fit the velocity–distance relationship. Hubble's results required data of brightness and redshifts. By 1935 Hubble and his colleague had catalogued the redshifts of 100 additional galaxies; these data unequivocally proved that the galaxies farthest away from earth were moving at the highest speeds.

Hubble's equation (28.1) is valid. The linearity between v and R remains valid today (Fig. 28.2), although the distance measurements have been corrected over the intervening years as more data were available.

It may sound like Hubble's law puts us right in the center of things, in grotesque violation of modern science's view that Earth and its inhabitants don't occupy a favoured position in the universe. But actually the inhabitants of any other galaxy would observe the same thing. All the distant galaxies would be receding from them at speeds proportional to their distances. As long as the universe is infinite in extent, none can claim to be at the center. And if it is not infinite, then Einstein's general theory of relativity gives it a closed curve shape that still has no center.

Hubble's law implies that the universe is expanding as the galaxies move apart. Extrapolating backward in time suggests that there was a time when all the galaxies were together. Obviously, *Hubble expansion* suggests that the universe had a definite begining, in the form of a colossal explosion that flung matter into an expansion that continues today. Based on additional evidence, scientists and astronomers are quite certain that the universe began with such a *Big Bang*. Some of that evidence comes from the Hubble space Telescope (Fig. 28.1).

Example 28.1: Calculate the age of the universe. Given, Hubble constant $H = 22\,\text{km/s/Mly}$.

Solution We are given Hubble constant and we have to extrapolate back in time until all the galaxies were together.

Consider that a galaxy has been moving with constant speed for time t, then its distance from us today is $d = vt$. But from Hubble's law $v = Hd$. Thus, we have

$$t = \frac{d}{v} = \frac{d}{Hd} = \frac{1}{H}$$

$$= \frac{1}{(22\,\text{km/s/Mly})/[3.0\times10^5(\text{km/s})/\text{ly/y}]}$$

$$= 13.6\,\text{Gy}$$

obviously, universe is about 13.6 billion years old. This is based on the assumption that the expansion rate has not changed. We may note that this assumption is not quite correct, but nevertheless this result is a good measure of the age of the universe. We may also note that we used $c = 3.0 \times 10^5\,\text{km/s}$ in converting km/s to ly/y so our answer is in years.

Example 28.2: A star is one parsec away has a parallex of one second of arc relative to the earth's orbit about the sun. Taking one parsec is about 3.26 light years, determine Hubble's constant in km/s per Mpc.

Solution Astronomers use a measure of distance in parsec (pc). Parallex is the apparent displacement of a body due to its observation from two different positions.

We have $H = 22\,\text{km/s/Mly}$

$$= \frac{22\,\text{km/s}}{\text{Mly}} \cdot \frac{3.26\,\text{ly}}{\text{pc}}$$

$$= 72\,\text{km/s/Mpc}.$$

28.4 THE COSMIC MICROWAVE BACKGROUND RADIATION

Penzias and Wilson in 1964 at Bell Laboratories were studying radio emissions in the microwave wave length region. They kept picking up an annoying constant signal in their specially designed low-noise antenna. When repeated attempts to eliminate the source of the noise were unsuccessful, both of them were at a loss to understand their result. In a conversation with another colleague they learned of calculations by James and coworkers at Princeton that predicted a remnant background of radiation from the Big Bang. Dicke at Princeton was mounting to measure it. It was this remnant background radiation that Penzias and Wilson were observing. Theorists at Princeton identified "noise" as radiation dating to a much earlier era in the universe. This *cosmic microwave backward radiation* is the strongest evidence yet for the Big Bang.

The Big Bang theory suggests that the universe started very hot and then cooled as it expanded, doing work against its own gravitation. At first it was so hot that any atoms that formed would be dashed apart by collisions at the high thermal energy prevailing. Thus, in its early times the universe was populated by individual charged particles. These interacted readily with electromagnetic radiation, making the universe opaque. But by about 400,000 years the temperature had dropped to several thousand kelvins, and at that point atoms of hydrogen and helium could form. Since neutral atoms interact much more weakly with electromagnetic radiation, the universe became transparent, and photons emitted as the atoms formed could travel throughout the universe with little chance of being subsequently absorbed. Those photons became the cosmic background radiation, permeating the entire universe.

Expansion of the universe would cool not only matter but also radiation, the Big Bang scenario suggests that radiation from the time the atoms formed now be cooled to that of a blackbody at 2.7 K-radiation distribution that peaks in the microwave region of electromagnetic spectrum. Analysis of the microwave background shows that it has indeed precisely that blackbody character and temperature as shown in Fig. 28.3. Obviously, the observed cosmic microwave background is a direct reflection of conditions some 400,000 years after universe began.

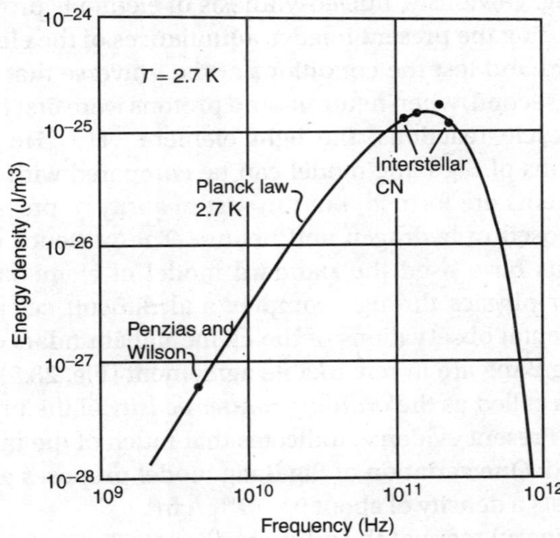

Fig. 28.3 Spectrum of the cosmic microwave background matches perfectly that of a blackbody at 2.726 K

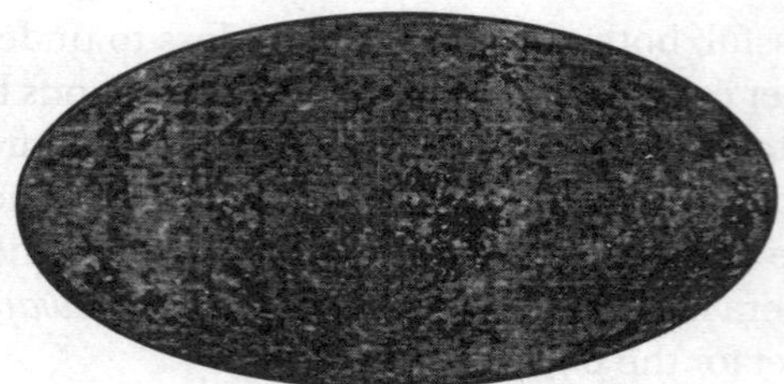

Fig. 28.4 All - sky map of the cosmic microwave background reveals minute spatial variations in the radiation intensity

We may note that the cosmic microwave background (CMB) is remarkably uniform, but not perfectly so. Today, one can record minute variations in the microwave background through satellite instruments, and analysis of these variations reveals a wealth of information about the early universe (Fig. 28.4). For example, the CMB variations are consistent with a universe whose overall geometry is flat, and the variations themselves represent the "seeds" of galaxies, galaxy clusters, and even larger structures that dominate the universe today. Analysis of CMB variations also provides our most accurate value for the Hubble constant, a value that independently corroborates galaxy redshift observations.

28.5 NUCLEOSYNTHESIS

The universe had cooled enough a few minutes after the Big Bang, so that protons and neutrons could undergo thermonuclear fusion and form light elements. The cosmic microwave background radiation provides a view back to about 400,000 years after universe came into existence. However, the formation of elements began after only a few minutes after the Big Bang. Obviously, nucleosynthesis of elements provides a stringent test of Big Bang. By measuring the present relative adundances of the elements, scientists are able to work backward and test the conditions of the universe that may have existed as early as a fraction of a second, when neutrons and protons were first joined to form nuclei.

In proton-proton cycle reactions, the light element ${}_{1}^{2}H$, ${}_{2}^{3}He$, ${}_{2}^{4}He$ and ${}_{3}^{7}Li$ are formed, and predictions of Big Bang model can be compared with recent observations. In stars, heavier elements are formed, but the vast majority of presently known mass in the universe is composed of hydrogen and helium. The presence of other elements is only minute. Scientists have used the standard model of elementary particle physics, together with nuclear physics through complex and difficult calculations to compare with current experimental observations of the elemental abundances. The observations and predictions of Big Bang are in remarkable agreement (Fig. 28.5), for a density about 3×10^{-31} g/cm³. This is called as the *ordinary matter density* of the universe, which is due primarily to baryons. Present evidence indicates that much of the matter in the universe has not been observed. One variation of Big Bang model proposes an inflatory universe, and that model suggests a density of about 9×10^{30} g/cm³.

The known (or ordinary) mass of the universe about 25% is composed of ${}^{4}He$ and the remainder being almost all free protons. The synthesis of ${}^{4}He$ depends critically on the

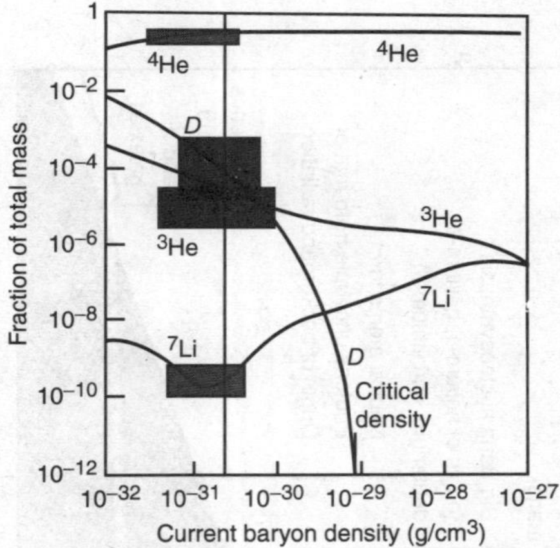

Fig. 28.5 Display of fractional mass abundances of several light elements against possible mass densities. The solid curves are calculations of standard model of the Big Bang and boxes represent experimental observations. The best agreement between observation and theory is for a density 3×10^{-31} g/cm^3. These data strongly support the standard Big Bang model, but do not agree with the critical density ~9×10^{-30} g/cm^3

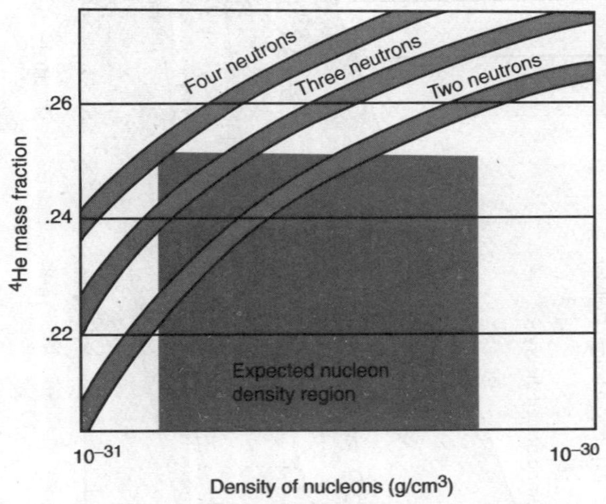

Fig. 28.6 The predicted ^{4}He mass fraction as a function of density of nucleons for three different possible groups of particles is shown. These results predict a maximum of four different families having neutrinos. Particle physics experiments indicates that there are only three families (three neutrinos), which is consistent with this figure

ratio of neutrons to protons and the density of matter during early stages of the universe. The ^{4}He abundance has been used to suggest that at most four families of elementary particles are possible. The suggestion and data seems to be in good agreement with the current belief from particle physics data that only three families exist (Fig. 28.6).

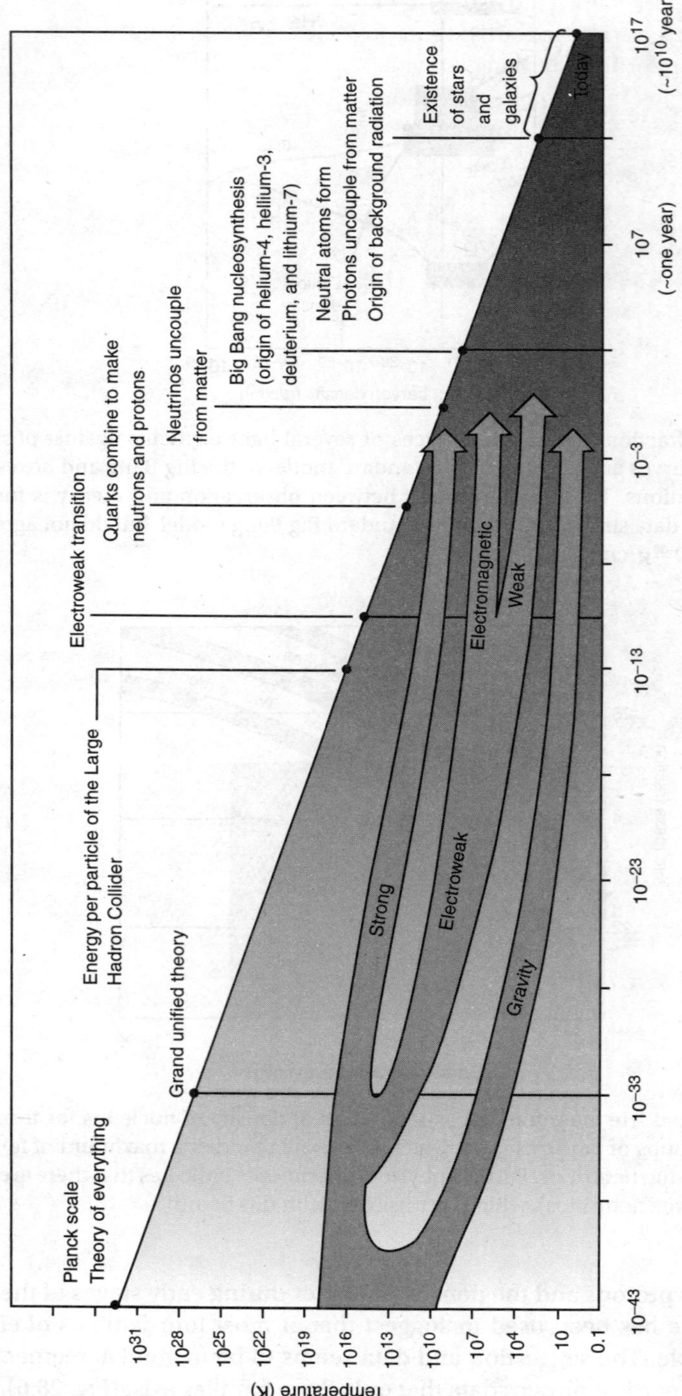

Fig. 28.7 Evolution of the universe from the earliest times to the present. We may note the highly logarithmic scales

In 2005, experiments using relativistic Heavy Ion Collider (RHIC) at Brookhaven National laboratory produced a so-called quark-gluon plasma similar to the state of the universe just microseconds after the Big Bang. Obviously, RHIC and other high energy machines accelerators can act as "time machines" allowing us to study conditions that prevailed in the very early universe. Fig. 28.7 summarizes one's understanding of universe's evolution, showing that the phenomena of particle physics and unification are inextricably tied with cosmic expansion.

28.6 THE BIG BANG

The Big Bang model rests on two following theoretical foundations:

(i) The general theory of relativity
(ii) *The cosmological principle,* which assumes that the universe looks roughly the same everywhere and in every disection. The universe is *isotropic* and *homogeneous.*

On the basis of these foundations, A. Friedmann in 1922 determined the evolution of the universe by his Friedmann equations. The solution of Friedmann suggest that the universe originated in a "Big Bang" explosion of a very hot, dense state and has been expanding in size ever since.

In 1927, G. Lemaitre independently arrived at results similar to Friedmann's. Then, Walker and Robertson in 1935 proved independently proved that the Robertson-Walker metric (A metric is a mathematical structure that provides information in space time about lengths and times. It is formally a symmetric nondegenerate tensor) is the simplest space time geometry consistent with an isotropic, homogeneous universe. A single parameter a that changes with time according to the Friedmann equation (28.3) describes the time evolution of space time. These universes go by various names, e.g. Friedmann, Lemaitre, Robertson and Walker (For e.g. FRW and FLRW universes). The idea is consistent with the model shown in Fig. 28.8 of the expanding ballon dotted with galaxies.

$$\left(\frac{1}{a}\frac{da}{dt}\right)^2 = \frac{8\pi G\rho_m}{3} - \frac{k}{a^2} + \frac{\Lambda}{3} \tag{28.3}$$

where a is the scale parameter introduced in eqn. (28.2), ρ_m is the average mass density of the universality, k is the curvature parameter of the universe, Λ is cosmological constant and G is the universal gravitational constant. Einstein introduced the constant Λ to form a static universe and a homogeneous and isotropic universe, because astronomers assured him that the universe was not in motion. $\Lambda/3$ term in (3) accounts for the energy of perfect vacuum in order to have the homogeneous and isotropic universe. After Hubble presented his evidence for the expanding universe, Einstein referred to Λ as his biggest blunder. Until a few years ago, cosmologists normally set $\Lambda = 0$, but now Λ is under active discussion for reasons we will discuss latter.

Using H of Eq. (28.2), rewritting (28.3) as

$$H^2 = \frac{8\pi G\rho_m}{3} - \frac{k}{a^2} + \frac{\Lambda}{3} \tag{28.4}$$

Fig. 28.8 How galaxies are receding with respect to each other? As the balloon is inflated, each dot is further away from the other dots. As the universe expands it seems to remain homogeneous

Eq. (28.4) is sometimes referred as *Friedmann equation*. This equation is extremely important in cosmology. Eq. (28.4) allow us to determine both the age and size of the universe if one can ascertain the unknown parameters in Eq. (28.4). Dividing both sides of Eq. (28.4) by H^2, we obtain

$$1 = \frac{8\pi G\rho_m}{3H^2} - \frac{k}{a^2 H^2} + \frac{\Lambda}{3H^2} \tag{28.5}$$

The terms on R.H.S. of (28.5) have special meaning in cosmology. The first term

$$\Omega_m = \frac{8\pi G\rho_m}{3H^2} \quad \text{Mass density,}$$

the second term

$$\Omega_k = -\frac{k}{a^2 H^2} \quad \text{Curvature}$$

and the third term

$$\Omega_\Lambda = \frac{\Lambda}{3H^2} \quad \text{Vacuum energy density}$$

The evidence that indicates the curvature term is zero will be discussed latter. The evolution of the universe is then determined by the two terms Ω_m and Ω_Λ, a striking and remarkable situation.

Since 1960s the Big Bang has generally been accepted as the event that began our universe. Using the knowledge of nuclear and particle physics, general relativity to understand the formation of our universe, or the Big Bang and describe the major steps in the intervening 13.7 billion years until today. The temperature of the universe is shown as a function of time in Fig. 28.9.

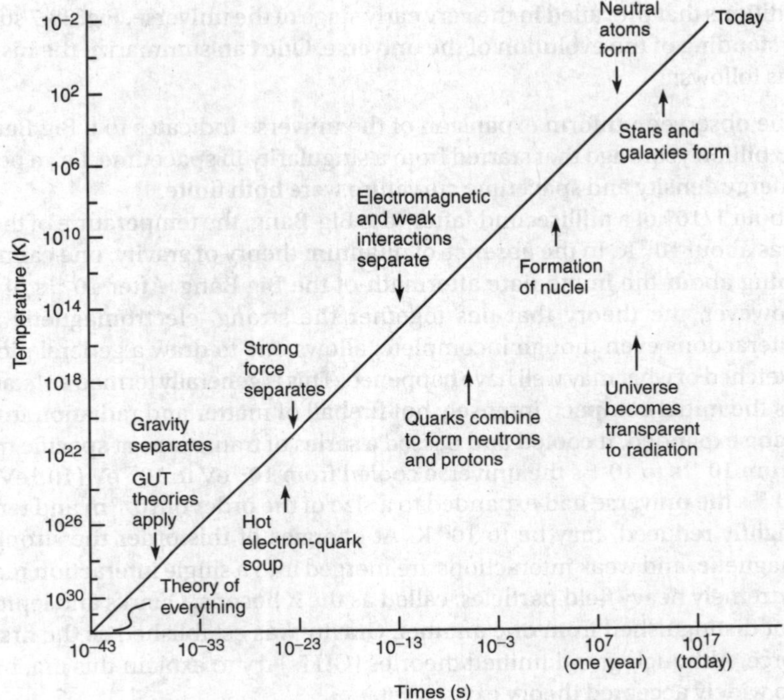

Fig. 28.9 Temperature of the universe as a function of time since the Big Bang. Both scales are logarithmic. The comments are placed vertically near the times at which the phenomenona occurred, except for the Theory of Everything which might be applicable for the time $<10^{-43}$ s

28.7 THE EARLY UNIVERSE

The cosmic microwave background reveals the universe as it was some 40,000 years after the Big bang. Nuclear physics allows us to look back even further, to the time from about 1 s to 30 minutes when the lightest nuclei were in the process of forming. The first composite nuclei were found the simplest deuterium, consisting of a proton and a neutron. We may note that the rate of formation of deuterium is critically sensitive to the expansion rate of the universe. Based on spectral lines from intersteller deuterium, the measurments of deuterium abundance provide direct evidence for conditions prevailing early in the Big Bang.

Particle physics provide evidence for still earlier times. Particle physics predicts the interactions and particle populations under the hot, high-energy conditions that existed a fraction of a second after the Big Bang. Experiments using the Relativistic Heavy Ion Collider (RHIC) at Brookhaven National Laboratory in 2005 produced a so-called quark-gluon plasma similar to the state of the universe just microseconds after the Big Bang. Obviously, high energy particle accelerators serve as "time machines" allowing one to study conditions that prevailed in the very early stage of the universe. Fig. 28.7 summarizes our understanding of the evolution of the universe. One can summarize the history of the universe as follows:

(i) The observed uniform expansion of the universe indicates to a Big Bang around 13 billion years ago that started from a singularity in spacetime, i.e., a point whose energy density and spacetime curvature were both finite.

(ii) About $1/10^{th}$ of a millisecond "after" the Big-Bang, the temperature of the universe was about 10^{12} K. In the absence of quantum theory of gravity, one cannot say any thing about the immediate aftermath of the Big Bang. After 10^{-43} s of Big Bang, however, the theory that ties together the strong, electromagnetic and weak interactions even though incomplete, allows one to draw a general picture to be sketched of what may well have happened. This is generally termed as "standard era".

(iii) As the initial compact, intensely hot fireball of matter and radiation from the Big Bang expanded, it cooled and passed a series of transitions at specific transitions. From 10^{-43} s to 10^{-35} s the universe cooled from 10^{28} eV to 10^{23} eV (10^{-4} eV ~ 1 K). By 10^{-35} s the universe had expanded to a size of the order of 10^{-30} m and temperature slightly reduced, may be to 10^{28} K. At energies of this order the strong, electromagnetic, and weak interactions are merged into a single interaction mediated by extremely heavy field particles, called as the X bosons. Quarks and leptons seems not distinguished from one another. Gravity was established as the first separate force. Although grand unified theories (GUTs) try to explain this era, but there is no widely accepted theory exist as yet.

(iv) From 10^{-35} s to 10^{-10} s the strong force might have broken off and become a separate force. The universe consisted of a dense soup of quarks and leptons whose behaviour was controlled by the strong, electroweak, and gravitational interactions. The fundamental particles, i.e., quarks and leptons had formed as well their antiparticles. One can describe the small universe as a hot, quark-electron soup. At 10^{-10} s the cooling had progressed to the point where the electroweak interaction became separated into the electromagnetic and weak interactions. No accelerator presently available has enough energy to explore what happened at these high energies.

(v) From 10^{-10} s to 10^{-3} s, quarks condensed into hadrons somewhere around 10^{-6} s. Only the two lightest quarks, the up and down, were very effective in the thermodynamics of the universe. At about 1 s energies of neutrino fell sufficiently for them to be unable to interact with the hadron–lepton soup, the "freezing out" of the weak interaction. At the end of this period the temperature cooled to about 10^{11} K. The four interactions of today had become distinct. The size of the universe at this time was out 1000 m and the universe mostly consisted of a soup of photons, electrons, neutrinos as well as various antiparticles, e.g., positrons and antiprotons.

(vi) From 10^{-3} s to 3 min, universe cooled to about 10^{9} K. This was the beginning of photosynthesis, the formation of nuclei Deutrons did not quickly fly apart when formed. The size of the universe at this time was about 10^{10} m.

(vii) From 3 min – 300,000 y, the main constituents of the universe after 5 min were ^{1}H and ^{4}He. After 5 min, the universe consisted of a plasma of hydrogen and helium nuclei and electrons in thermal equilibrium with radiation. Neutron decay occurred, so there were many more protons than neutrons. Temperature of the universe fell down to about 10^4 K at the end of this era. Now matter and radiation were decoupled and the universe became transparent. Atoms were not able to form because the intense electromagnetic radiation ionized these atoms as soon as they were formed. At this time the universe had expanded to about 10^{21} m and it continued to expand.

(viii) *From 300,000 y to present*: Universe had cooled enough that photons (electromagnetic radiations) decoupled from matter. During 300,000 to 700,000 years, the universe was dominated by radiations, i.e., most of the energy of universe was in the form of photons, which were continually absorbed and emitted by ions. The radiation left behind then continued to spread out with the rest of the universe, undergoing Doppler shifts to longer and longer wavelengths. Today, an observer would expect this remnant radiation to come equally strongly from all directions and exhibit a spectrum like that of a blackbody at 2.7 K. Interestingly, such radiations has actually been reported in microwave measurements undertaken from the earth as well as from satellites. These radiations are termed as *cosmic microwave background radiations*. In addition, the Bing Bang model predicts that there should be about 24% helium present in the universe as a left over from the primordial soup. Obviously, we have following three observations which support Big-Bang model of the cosmology:

(i) The uniform expansion of the universe.
(ii) The relative abundance of hydrogen and helium in the universe.
(iii) The cosmic background radiation.

Once matter and radiation were decoupled, gravitational interaction became the dominant influence on the evolution of the universe. Density fluctuations led to the formation of molecules, gas clouds, stars and eventually galaxies. Formation of the galaxies and stars adorn the night sky.

The Big Bang model has testable consequences and it is so successful that usually it is referred as the *standard model of cosmology*. However, this model presents at best an incomplete picture. There are following three unexplained difficulties that the simple Big Bang could not address:

(i) Why the Big Bang occurred in first place? There is no mechanism that explains it. The Big Bang itself is simply hypothesized.

(ii) *Why is the universe flat*? According to Einstein's theory of relativity space must bend due to the gravitational attraction. Depending on the amount of mass density (ρ) of universe (amount of matter per unit volume), space curves in on itself such that parallel lines converge above a certain density called the *critical density* (ρ_c). For the value of mass density (ρ) less than the critical density (ρ_c), i.e., $\rho < \rho_c$, parallel lines diverge, and the universe expands forever. This is termed as *open universe*. For the value of mass density (ρ) greater than the critical density (ρ_c), i.e., $\rho > \rho_c$, the expansion of the universe will eventually be halted and then collapse. This situation is termed as a *closed universe*. It appeared in 1980 that the universe may be just at that critical density where we have a *flat universe*, or nearly so. Obviously, such an occurrence would be quite an extraordinary circumstance.

(iii) *How homogeneous the universe is*? If we look closely enough homogeneity fails: Earth, the Sun, the Milky way galaxy all these are really inhomogeneities. On the reverse side, the universe is 13.7 billion years old, then the opposite sides of the universe are 27 billion light years apart. How can these regions have microwave radiation that is quite similar? In whatever direction we look universe, its temperature reflects the 3 K microwave radiation. How, the temperatures of two regions of the universe so similar and be 10^{19} m apart and not able to communicate? This question-marks about the laws of physics are same everywhere and were the same at that earlier epoch. This dilema is termed as the *horizon problem*.

(iv) *Magnetic monopoles* have not yet detected. The occurrence of magnetic monopoles brings symmetry to Maxwell's equations of electromagnetism. Moreover, occurrence of magnetic monopoles also satisfies other theories of physics.

28.8 THE INFLATIONARY UNIVERSE

The issues of inhomogeneity, the horizon problem and the flatness problem with the Big Bang model were relieved considerably in 1981 by a suggestion of Alan Guth of MIT, who proposed that the universe underwent a period of exponential expansion beginning at about 10^{-35} s and lasting until about 10^{-32} s (Fig. 28.10). This period is called *inflationary epoch*. The expansion was the result of a delay in the symmetry breaking that made the fundamental forces, i.e., nuclear and electroweak forces appear distinct. After the inflationary period, the universe resumed its evolution in accordance with the standard Big Bang model.

Inflationary idea may be hard to believe, but helps to solve the three problems of standard Big Bang listed earlier. This theory requires the mass density to be very close to critical density (ρ_c) of the universe. This resolves the flatness problem. According to this theory the universe was so tiny that it had already reached equilibrium prior the inflation occurred; this explains the homogenous universe. Due to tremendous expansion,

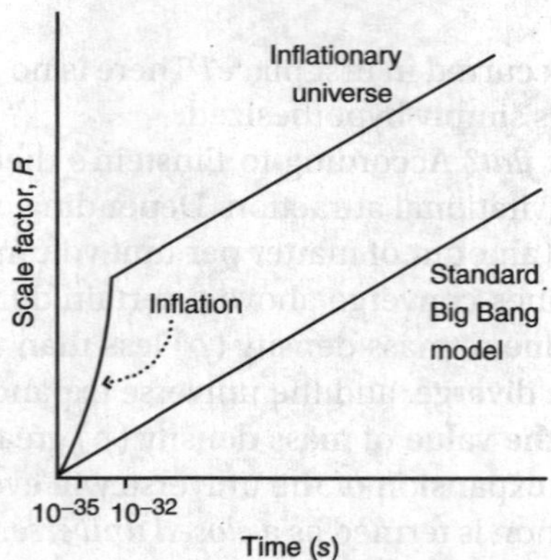

Fig. 28.10 Expansion of the universe in the standard and inflationary Big Bang models. The scale factor R measures the amount of expansion of the universe

now-distant locations would once have been close enough to reach the thermodynamic equilibrium that one observe today. Moreover, the inflationary expansion would "flatten out" any overall curvature, yielding the flat universe we see today. Several cosmologists have suggested that magnetic monopoles would have to occur along the boundaries or walls of different domains. These domains might be likened to different universes. However, one cannot observe the magnetic monopoles, if they exist and if they are at the edge of the universe. We may note that it is a neat, tidy package and moreover quite complicated. Interestingly, inflationary theory is crucially connected to theories of elementary particles.

28.9 DARK MATTER, DARK ENERGY AND FUTURE OF THE UNIVERSE

Will the universe continue to expand forever, or will the expansion eventually stop and reverse? The answer depends on how much matter the universe contains and on how fast the universe is expanding. One finds three possibilities:

(i) **Average density (ρ) of the universe smaller than a certain critical density (ρ_c), i.e., $\rho < \rho_c$:** The universe is open and the expansion will never stop (Fig. 28.11(a)). This implies that new galaxies and stars will cease to form and the existing one will end up as *black dwarfs*, neutron stars and black holes – an icy death.

(ii) **$\rho > \rho_c$:** The universe is *closed* and gravity will stop the expansion sooner or later. This means the universe will then begin to contract. Moreover, the progression of events will be the reverse of those that took place after the Big Bang, with an ultimate Big Crunch–a fiery death. Obviously, after this another Big Bang, i.e., the universe is cyclic, with no begining and no end.

(iii) **$\rho = \rho_c$:** This reveals that the expansion will continue at an ever-decreasing rate and the universe will not contract. The universe is said to be *flat* due to geometry

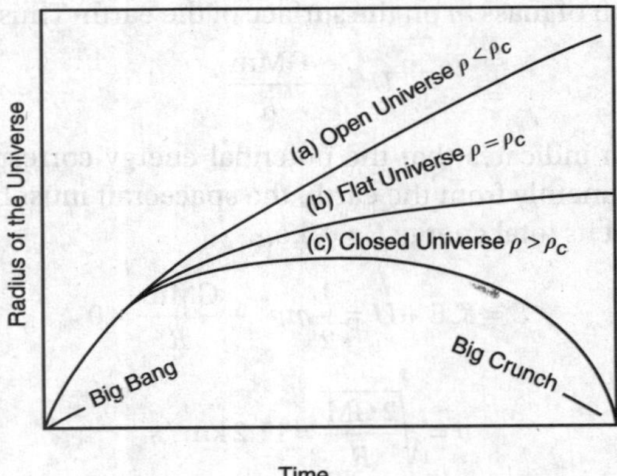

Fig. 28.11 Three models of the Universe that follow from equations of general relativity. ρ is the average density of the universe and ρ_c the critical density of the universe, is in the neighbourhood of 9×10^{-27} kg/m^3, equivalent to about 5 hydrogen atoms/cc

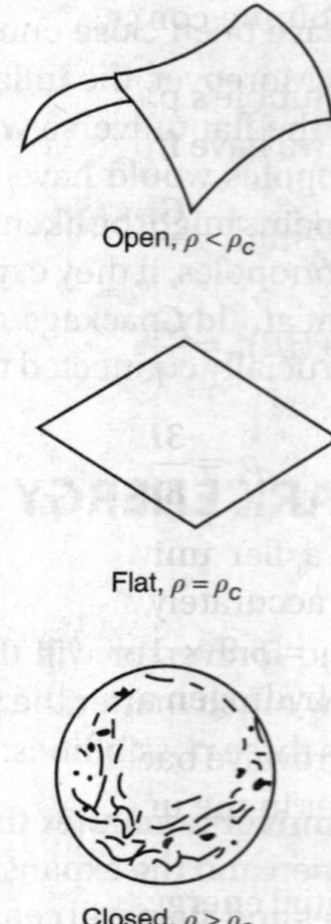

Fig. 28.12 Geometry of space in open, flat and closed universes, two dimensional analogies

of space in such a universe (Fig. 28.12). For $\rho < \rho_c$, space is negatively curved; a two dimensional analogy is a saddle. For $\rho > \rho_c$, space is positively curved; a two dimensional analogy is the surface of a sphere. We may note that in all cases, space time is curved.

Let us find the value of the critical density (ρ_c). For this, we first find the escape velocity from the earth. Let the mass of the earth is M and its radius is R. The gravitational potential energy U of a spacecraft of mass m on the surface of the earth. Thus,

$$U = -\frac{GMm}{R}$$

Here the negative sign indicates that the potential energy corresponds to an attractive force. To escape permanently from the earth, the spacecraft must have minimum kinetic energy $\frac{1}{2}mv^2$ such that its total energy $E = 0$, i.e.,

$$E = K.E + U = \frac{1}{2}mv^2 - \frac{GMm}{R} = 0 \tag{28.6}$$

or

$$v = \sqrt{\frac{2\,GM}{R}} = 11.2\,\text{km/s}$$

We now consider a spherical volume of the universe of radius R whose centre is the earth. If the distribution of matter in the universe is taken uniform, only the mass inside this volume affects the motion of a galaxy on the surface of the sphere. Let the density of matter inside this volume is ρ, the volume contains a total mass $M = \frac{4}{3}\pi R^3 \rho$. According to Hubble's law, $v = HR$, where H is Hubble's parameter. Taking the mass of the galaxy, m, if it has just enough speed to return, we have from Eq. (28.6)

$$\frac{1}{2}mv^2 = \frac{GMm}{R}$$

or
$$\frac{1}{2}m(HR)^2 = \frac{Gm}{R}\left(\frac{4}{3}\pi R^3 \rho_c\right) \tag{28.7}$$

or
$$\rho_c = \frac{3H^2}{8\pi G}$$

Obviously, the critical density for a flat universe depends only on Hubble constant (parameter) H, which is not known accurately. A reasonable value of H is taken 21 km/s per million light years. This gives $\rho_c = 8.9 \times 10^{-27}$ kg/m³. The mass of a hydrogen atom is 1.67×10^{-27} kg, so one finds $\rho_c \approx 5.3$ hydrogen atoms/m³.

Dark Matter: Analysis of cosmic microwave background fluctuations strongly suggest that actual density of the luminous matter in the universe is at the critical density, with $\Omega = 1$, where

$$\Omega = \frac{\text{Potential energy of the universel}}{\text{Kinetic energy of the universe}}$$
$$= \frac{8\pi G\rho}{3H^2} \tag{28.8}$$

However, when we total the visible matter in the universe, we obtain far less than the critical density. Adding in the mass equivalent equivalent of radiation ($E = mc^2$) in the universe increases the density only a little. Theories of elementary particles corroborate this observational result and suggest that there can be only enough ordinary matter–protons, neutrons, atoms and nuclei–to make up about 4% of the critical density. Moreover, the motions of stars in galaxies suggest the presence of the great deal more mass than is visible, i.e., large amount of *dark matter* is also present. About 90% of all matter in the universe is non-luminous. Further, the rotation speeds of the outer stars in spiral galaxies are unexpectedly high. This indicates that a spherical halo of invisible matter must surround each galaxy. Moreover, the motions of individual galaxies in the clusters of them imply gravitational fields about 10 times more powerful than the visible matter of galaxies provide. Still other observations support the idea of a preponderance of *missing mass*, i.e., dark matter in the universe. The composition of dark matter is unknown and which can't be the sort of matter – made mostly from quarks- with which one is familiar. Dark matter does not emit, scatter, or absorb light. The dark matter is also required to enable gravity to amplify the fluctuations cosmic microwave background radiation (CMB). Scientists are of the view that dark matter might be due to neutrinos, weakly interacting massive particles (WIMPs), or primordial black holes. One theory proposes that the problem is with Newton's laws of motion at large distances and not with missing mass. According to this *modified Newtonian dynamics* (MOND) theory, the force is smaller at large distances, thus eliminating the need for additional mass.

The source of the *cosmic acceleration* or the accelerating universe is *dark energy* Dark energy may be just another name for Einstein's cosmological constant or may be a different phenomenon with the same "antigravity" effect. Presently, we don't know, what exactly dark energy is? Preliminary analysis indicated that dark energy is fully 73% of the "stuff" that makes up our universe. Dark matter is about 23%. This means that only 4% of the universe is in the form of familiar matter made from baryons and thus ultimately from quarks. These data are from a confluence of recent observations of cosmic microwave wave background, distant supernovae, and surveys of distant galaxies; together, they tightly constrain the relative amounts of matter and dark energy (Fig. 28.13). Obviously, much of our universe consists of "stuff" about which we don't know much.

Quintessence: Astronomers have suggested a new form of energy with negative pressure called as quintessence. Although the cosmological constant

$$\Lambda = \frac{8\pi G}{c^4}\rho v \tag{29.9}$$

where ρ_v is the vacuum energy constant, is indeed constant, observations does not support that dark energy has always been constant. According to astronomers, dark energy might have become effective 5–10 billion years ago, but it had to be created no later than 10^{-35} s after the Big Bang. On the other hand, quintessence is a dynamic, time evolving spatially changing form of energy with negative pressure. There is a possibility that dark energy might be associated with inflation. Dark energy may be a problem with general relativity itself. So far, we have no clear idea about the dark energy.

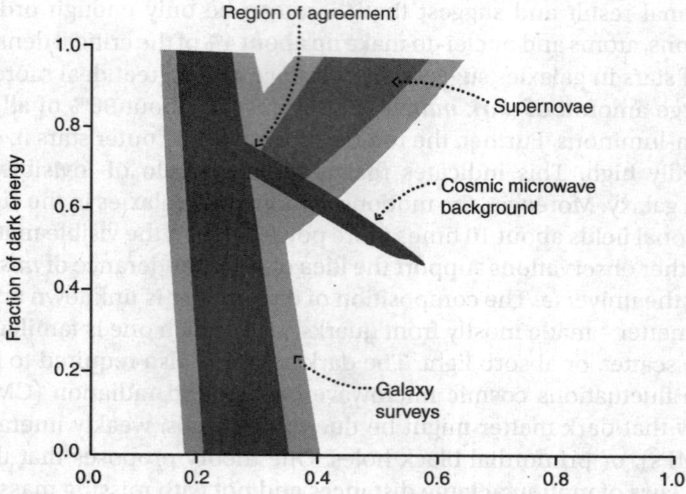

Fig. 28.13 Relative abundances of dark energy and of matter in the universe. Constraints on it from three independent sets of observations. All three together give 73% dark energy and 27% matter (23% dark matter + 4% ordinary matter)

28.10 UNDERSTANDING THE UNIVERSE

To understand the universe, one need to understand all aspects of universe from the largest to the smallest, i.e., all the forces, from gravity to the weak force; all the physical laws, from Newton's laws and Maxwell's laws to the laws of quantum physics; and the nature of elementary particles.

28.11 STELLER EVOLUTION AND ULTIMATE FATE

About 400,000 years after the Big Bang, matter in the universe in the form of protons, electrons, and ^{4}He drifted throughout the universe just like gas particles in a room. Thus, the temperature continued to decrease and gravitational forces brought the matter together into massive gaseous clouds. This provided the basis for evolution for stars. Protons were attracted together by their gravitational interaction. This helped to raise their kinetic energy. This process continued as the temperature of interior of these infant stars kept increasing. The temperature and density in the interior of a gaseous cloud was higher than the outside. The contraction of cloud continued until finally the temperature reached about 10^7 K and the nuclear fusion began. We may note that nuclear fusion is a characteristic of a star. A million or more years might have taken for the contraction of the cloud to be able to produce fusion for the star to born. Recent studies indicate it could have happened about 200,000 years ago. Fig. 28.14 shows the schematic formation of a star.

The end products of steller evolution, i.e., *White dwarfs, Neutron stars* and *Black holes* are quite fascinating objects by themselves.

Chandrasekhar showed that white dwarfs cannot have masses greater than $1.4 M_{sun}$ ($M_{sun} \rightarrow$ mass of the sun), known as Chandrasekhar limit. These stars do not collapse further because the degeneracy pressure of electrons is sufficient to hold back a gravitational

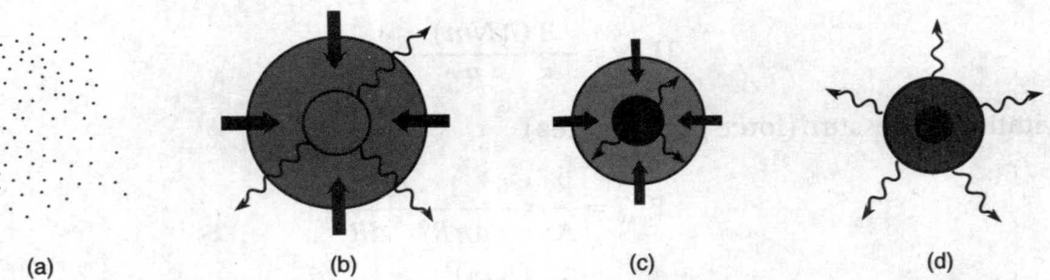

Fig. 28.14 Formation of a star: (a) Condensation of intersteller gas and clouds by gravitational attraction (b) With the contraction of matter, a core is formed that heats up and radiate energy (c) Outer region becomes so dense that the radiation from the hot inner core can no longer escape. The collapse gets slow, but the matter continues to heat and eventually a protostar having high density and temperature is born. For a star of about 1 solar mass, the contraction heat up enough to sustain nuclear fusion. (d) The radiation pressure due to nuclear fusion balances the gravitational contraction and star stabilizes into a main sequence star. A star is expected to burn for about 10^{10} billion years, converting its hydrogen into helium

collapse. We do not know enough about stars upto a mass of approximately 3.5 solar masses. Beyond this mass, nothing, nothing can hold back a gravitational collapse. That is not to say that all stars with an initial mass larger than 3.5 M_{sun} will form *black holes,* because in any collapse following the burning a lot of matter can be blown off, leaving *neutron stars.* Neutron star can be somewhat more massive than a white dwarf and still be stable. However, any compact object with a mass greater than 6 M_{sun} is generally thought to be a black hole. Huge black holes are thought to lie at the centres of galaxies, including our own galaxy. Mass of such black holes may be one to 100 billion suns. On a galactic scale 10^{11} M_{sun}, $R_{sun} = 3 \times 10^{14}$ m $\cong 3 \times 10^{-2}$ light yr is a very small size.

A neutron star is about 8 M_{sun} times heavy. The matter inside such a star is mostly in a *superfluid state.* This has posed fundamental problems for nuclear physics, low temperature physics and theory of elementary particles.

The large mass of a neutron star causes it to collapse abruptly when out of fuel, and then to explode violently. Due to explosion most of the star's mass flings in space. Such an event is called a *Supernova.* A supernova is a billions of times brighter than the original star ever was. There may be several such events during a century in the galaxy. Since these events are happening for about 10^{10} yr, there may be as many as 10^8 black holes of this type in our galaxy.

Not only heavy stars finally end up as black holes, but with the passage of time white dwarfs and neutron stars attract more and more cosmic dust and gas. After gathering up enough additional mass, they too will become black holes. If the universe lasts long, ultimately everything in the universe may be in the form of black holes.

Although black hole is a natural outcome of the general theory of relativity, but so far it remained as theoretical speculation, even though there is some evidence for its presence in X-ray binary Cygnus X-1.

The ultimate fate of a star depends on its mass. For stars somewhat more massive than the sun, the gravitational attraction of the mass continues till the density of the star is incredibly high. We now consider it in detail. Let there be N nucleons in the star. Mass of each nucleon is m. Total mass is Nm. The gravitational self-potential energy (U_{grow}) of a uniform sphere of radius R and mass Nm is

$$U_{grav} = -\frac{3}{5}\frac{G(Nm)^2}{R} \tag{28.10}$$

Gravitational pressure (force per unit area)

$$P_{grav} = \frac{F}{A} = -\frac{1}{4\pi R^2}\frac{dU_{grav}}{dR}$$

$$= \frac{3}{5}G\frac{(Nm)^2}{4\pi R^4} \tag{28.11}$$

Taking volume $V = \frac{4}{3}\pi R^3$, one can express the gravitational pressure in terms of the volume as

$$P_{grav} = 0.3G\frac{(Nm)^2}{V^{4/3}} \tag{28.12}$$

Outward electron pressure keeps the matter from total collapse due to Pauli exclusion principle. According to this principle two electrons cannot occupy the same state. However, for a sufficiently massive star, gravity force the electrons to interact with protons through the following reaction

$$e^- + p \rightarrow n + \nu_e \text{ (electron neutrino)} \tag{28.13}$$

The result, a *neutron star*, composed mostly of neutrons.

We may note that neutrons also obey Pauli exclusion principle. This means an outward pressure similar to that of the electrons will also result from neutrons. One can easily show that the pressure of an electron gas is

$$P_e = \frac{2f\pi^2}{3}\frac{\hbar^2}{2m_c}\left(\frac{N_e}{V}\right)^{5/3} \tag{28.14}$$

Where f denotes the fraction of the average energy between O and E_F (Fermi energy). Since the average energy for a Fermi gas is about $\frac{3}{5}E_F$ and therefore we can take $f = 3/5$. Using the same relation for neutrons simply by replacing N_e by N and m_e by m, we obtain neutron pressure as

$$P_n = \frac{3.9\hbar^2 (N/V)^{5/3}}{2m} \tag{28.15}$$

In equilibrium, the outward pressure of the neutrons due to Pauli exclusion principle is balanced by the gravitational pressure. Setting Eq. (28.13) and (28.15) equal, one obtains

$$0.3G\frac{(Nm)^2}{V^{4/3}} = \frac{3.9\hbar^2 (N/V)^{5/3}}{2m} \tag{28.16}$$

or

$$V^{1/3} = \frac{6.5\,\hbar^2}{N^{1/3}m^3G} \tag{28.17}$$

one can use (28.17) for the determination of the size of the neutron stars.

28.12 PLANCK'S TIME, LENGTH AND MASS

We are familiar with the three fundamental constants: gravitational constant G, Planck's constant h and speed of light c. We also know that the gravitational force depends on G, and several properties of a particle depend on c and h, e.g. energy, $E = mc^2$; wavelength as a function of particle momentum, $\lambda = h/p$ and the energy of photon, $E = h\nu$, where ν is the frequency of photon.

We now make dimensional analysis to determine some characteristic values of time, length, and mass making use of G, h and c. The values of these constants in SI system and their dimensions in terms of mass (M), length (L) and time (T) are

(i) $G = 6.6726 \times 10^{-11}$ Nm^2/kg^2
Dimensions of G : $M^{-1} L^3 T^{-2}$

(ii) $c = 2.9979 \times 10^8$ m/s
Dimensions of c : $M^0 L^1 T^{-1}$

(iii) $h = 6.6261 \times 10^{-34}$ Js
Dimensions of h: $M^1 L^2 T^{-1}$

Following dimensional analysis, we obtain mass (m) by using the combination $\sqrt{\frac{hc}{G}}$, we have

$$\text{Planck's mass}, m_P = \sqrt{\frac{hc}{G}} = 5.46 \times 10^{-8}\ \text{kg}$$

We may note that Planck's mass is of great significance. General relativity does not allow a black hole to be created from a mass smaller than Planck's mass, m_p.

Planck's length or characteristic length, λ_p.

Following dimensional analysis, we have

$$\lambda_P = \sqrt{\frac{Gh}{c^3}} = 4.05 \times 10^{-35}\ \text{m}$$

The time taken by light to travel across Planck's length is called *Planck's time* (t_p). We have

$$t_P = \sqrt{\frac{Gh}{c^5}} = \frac{\lambda_P}{c} = 1.35 \times 10^{-43}\ \text{s}$$

We can consider Planck's time as the creation of the universe, as the laws of physics seems to be not valid prior Planck's time of 10^{-43} s. Actually, physicists do not understand what happened prior to Planck's time in the creation of the universe.

Example 28.3: Assume that sun is a uniform sphere of mass $M = 1.99 \times 10^{30}$ kg and radius $R = 6.96 \times 10^8$ m and made entirely of protons, find the proton velocity.

Solution Self potential energy of the sun

$$\text{P.E.} = -\frac{3}{5}\frac{GM^2}{R}$$

$$= -\frac{3}{5} \times \frac{(6.67 \times 10^{-11}\ \text{Nm}^2/\text{kg}^2)(1.99 \times 10^{30}\ \text{kg})^2}{6.96 \times 10^8\ \text{m}}$$

$$= -2.28 \times 10^{41}\ \text{J}$$

obviously, the kinetic energy of the particles within the sun is 2.28×10^{41} J. Let N be the number of protons in the sun, then

$$N = \frac{M}{m_p}, \quad \text{M is the mass of the sun.}$$

Now, writing K. E. of sun $= \frac{1}{2} N m_p v^2$

or $$\frac{1}{2}\frac{M}{m_p} m_p v^2 = 2.28 \times 10^{41}\ \text{J}$$

or $$v^2 = \frac{2 \times (2.28 \times 10^{41}\ \text{J})}{1.99 \times 10^{30}\ \text{kg}}$$

or $$v = 4.8 \times 10^5\ \text{m/s}$$

Example 28.4: Calculate the mean temperature of the sun using the result of example 28.3.

Solution From kinetic theory of gases, we have the velocity of a gas particle at temperature T as

$$v = \sqrt{\frac{3kT}{m}}$$

or

$$T = \frac{mv^2}{3k}$$

$$= \frac{(1.67 \times 10^{-27}\,\text{kg}) \times (2.3 \times 10^{11}\,\text{m}^2/\text{s}^2)}{(3 \times 1.38 \times 10^{-23}\,\text{J/K})}$$

$$= 9 \times 10^6\,\text{K}$$

Example 28.5: Show that the radius of a neutron star with a mass of 2 solar masses is 11 km.

Solution

Number of neutrons N,

$$N = \frac{2\text{M}_{\text{sun}}}{\text{m}_{\text{neutron}}}$$

$$= \frac{2 \times 1.99 \times 10^{30}\,\text{kg}}{1.675 \times 10^{-27}\,\text{kg}} = 2.4 \times 10^{57}\ \text{neutrons}$$

we have for cube root of volume

$$V^{1/3} = \frac{6.5\,\hbar^2}{N^{1/3}\,m^3\,G}$$

$$= \frac{6.5 \times (1.06 \times 10^{-34}\,\text{Js})^2}{(2.4 \times 10^{57})^{1/3}(1.675 \times 10^{-27}\,\text{kg})^3 \times (6.67 \times 10^{-11}\,\text{J-m/(kg)}^2)}$$

$$= 1.7 \times 10^4\,\text{m}$$

But

$$V = \frac{4}{3}\pi R^3$$

$\therefore$

$$R = \left(\frac{3}{4\pi}\right)^{1/3} V^{1/3}$$

$$= \left(\frac{3}{4\pi}\right)^{1/3} (1.7 \times 10^4\,\text{m})$$

$$= 11\text{km}$$

SUMMARY

1. Our knowledge of elementary particles and their interactions together with general relativity and with observations of the *Hubble expansion* and *cosmic microwave background radiation* provide a picture of the origin of the structure of the universe.
2. The universe began some 14 billion years ago in a hot big bang.

3. The universe started as an extremely dense, hot fireball.
4. The simplest nuclei formed within the first half-hour, and the first atoms at about 400,000 years. At this point the universe became transparent and the cosmic microwave background formed.
5. By the time 10^{-13} s, quarks and leptons were formed, and by 10^{-3} s neutrons and protons were also formed. Nuclei formed by 3 min, but atoms came into existence only after 400,000 years.
6. Hubble's law relates the recessional velocity (v) and the distance from Earth (R) as

$$v = HR$$

7. The speed of light is finite means that the observations of the distant universe are observations of how things were long ago. Our observations about the universe reveals:
   (i) The universe is isotropic.
   (ii) The universe is on the average homogeneous.
   (iii) The universe is expanding in accordance with Hubble's law.
   (iv) Whether the curvature of the universe is flat, open or closed, is concerned with the average mass density of the universe and is still under investigation.
8. The basic equation due to Friedmann is

$$H^2 = \frac{8\pi G\rho_m}{3} - \frac{k}{a^2} + \frac{\Lambda}{3}$$

where H is Hubble's parameter related to a scale factor a that is the approximate separation distance between galaxies, i.e.,

$$H = \frac{1}{a}\frac{da}{dt}$$

9. The presence of a background blackbody radiation that, having evolved with the universe, is presently at a temperature of about 3 K. After the Big Bang, about 100,000 years this radiation decoupled from matter.
10. The fraction of helium present in the universe corresponds to how nuclear processes would have produced helium in the "oven" of the universe several minutes after.
11. When matter collected together due to gravitational force, stars were formed.
12. Nuclear fusion is the source of energy within the stars.
13. Stars emit energy and neutrinos due to nuclear fusion.
14. Solar neutrinos coming from the sun is due to oscillation of neutrinos.
15. When the thermonuclear fusion processes in a star are exhausted, massive stars undergo a gravitational contraction so strong that eventually the star consists mostly of neutrons. These *neutron stars* are incredibly dense.
16. Stars having mass less than about 1.4 m_s are able to exert enough pressure due to electrons to prevent gravitational collapse and they turn to be *white dwarfs*.
17. The expansion of the universe is accelerating under the influence of mysterious *dark energy*.
18. The mass density consists of 4% ordinary baryonic mass, 23% dark matter and 73% dark energy.

19. Dark energy consists of a repulse gravitational energy that is causing the expansion of the universe to accelerate. We don't know what the dark matter and dark energy is.
20. There are several astronomical objects, e.g. *galaxies,* a collections of many stars; *quasars,* a source of tremendous energy; *supernova,* occurred when a massive star underwent gravitational collapse after the thermonuclear fuel was exhausted; *gamma ray brusts,* originating from throughout the universe. Scientists believe that gamma ray brusts come from supernova.

REVIEW QUESTIONS

1. What is Hubble's parameter? How it is related to scale factor? Is Hubble parameter called Hubble's constant is not actually a constant?
2. How can Hubble's law hold without the universe having a centre?
3. What are cosmic microwave radiations? Describe their origin.
4. Explain how particle accelerators can help one to understand the early universe.
5. Why the universe cannot be older than the Hubble time? Explain.
6. What are the problems with Big Bang?
7. Explain Planck's time, length and mass and mention their significance.
8. How stars evolved? What is ultimate fate of stars?
9. Write Friedmann equation and explain how it is important in cosmology.
10. How did the universe begin? What is its overall structure? What will be its future?
11. Why Big Bang is referred as standard model of cosmology?
12. What is beyond the Big Bang?
13. Derive the expression for critical density (ρ_c). Explain $\rho < \rho_c$, $\rho = \rho_c$ and $\rho > \rho_c$ situations.
14. What is cosmological constant? Explain its significance.
15. If all the distant galaxies are moving away from us, explain why we are not at the centre of the universe?
16. How helium abundance support strongly Big Bang?
17. Explain the flatness problem and inflation.
18. Some scientists are of the view that some of the constants of nature may not in fact be constant–that they vary slightly with time. Explain it.
19. What are quasars? They are known to vary in brightness by just a few hours or day. What inference one can draw about the size of these quasars?
20. Write short notes on
    (a) Cosmic microwave radiation
    (b) Nucleosynthesis
    (c) Big Bang
    (d) Stellar evolution
    (e) Planck's time, length and mass
    (f) Quasars
    (g) Noval and Supernoval
    (h) Formation of stars and galaxies
    (i) Dark matter and dark energy
    (j) Accelerating universe

PROBLEMS

1. Suppose that the universe is flat, i.e., the density is the critical density. Find the value of the Hubble parameter (H) when the density of the universe was the density of ordinary matter. **[Ans.** 5.5×10^4/yr]
2. Determine the value of Planck's mass, Planck's length and Planck's time from dimensional analysis. [$m_p = 5.46 \times 10^{-8}$ kg, $\lambda_p = 4.05 \times 10^{-35}$ m and $t_p = 1.35 \times 10^{-43}$ s]
3. Calculate the temperature at which the universe was hot enough for photons to produce π^o. Also find approximately the time was this. [1.57×10^{12} K, 10^{-4} s]
4. Calculate the temperature of the universe when it has cooled enough that photons no longer dissociate the hydrogen atom. [1.58×10^5 K]
5. At what threshold temperature were the electron and muon first formed? Using these values calculate the gravitational pressure for the sun and the neutron star. [5.93×10^9K, 1.23×10^{12} K, 5×10^{13} N/m², 3.21×10^{33} N/m²]
6. Show that the critical density necessary for a closed universe for two extremes of the Hubble constant: 55 km.s⁻¹. MPc⁻¹ and 85 km·s⁻¹. MPc⁻¹ are 5.67×10^{-27} kg/m³ and 1.35×10^{-26} kg/m³ respectively.

SHORT ANSWER-QUESTIONS

1. What is main idea of Big Bang model?
   Ans. According to Big Bang model, the universe was created in a primeval fireball of incredible density and high temperature.
2. Why Big Bang model acceptable to most scientists as the most likely model of the universe?
   Ans. Big Bang model is widely accepted due to following reasons:
   (i) The observation that the galaxies of the universe are moving away from each other at high speeds. Obviously, the universe is apparently expanding outward from som primordial event.
   (ii) The observation that a cosmic microwave background radiation permeates the universe. This cosmic background radiation has been attributed to the Big Bang.
   (iii) There is quite good agreement between predictions of the primordial nucleosynthesis of the elements and the known abundance of elements in the universe. We may note that this applies to the light elements that were produced in the early stages of the Big Bang.
3. What is Hubble's law?
   Ans. The distant galaxies are receding from us at speeds proportional to their distance from earth. This result is known as Hubble's law

$$v = HR$$

where H is Hubble parameter, v is the recession speed. Hubble parameter H is related to a scale parameter a, i.e.,

$$H = \frac{1}{a}\frac{da}{dt}$$

i.e., H is proportional to distance between galaxies.

4. Can we use Hubble's law to find the distances to remote galaxies?
Ans. Yes. By measuring their red shift and using relation, $v = HR$.

5. What is the significance of Hubble's law?
Ans. Hubble's law puts us right in the centre of things, in grotesque violation of view of modern science that earth and its inhabitants don't occupy a favoured position in the universe. However, actually the inhabitants of any other galaxy would observe the same thing: all the distant galaxies would be receding from them at speeds proportional to their distances. Moreover, as long as the universe is infine in extent, no one can claim to be at the centre. However, if it is not infinite, then general theory of relativity gives universe a close shape that still has no centre.

6. What is the main implication of Hubble's law?
Ans. According to Hubble's law, universe is expanding as the galaxies moves apart. Extrapolation backward in time suggests that there was a time when the galaxies were together. This means, Hubble expansion suggests that universe had a definite beginning, in the form of colossal explosion that flung matter into an expansion that continues even today. Based on additional evidences, most scientists are quite certain that the universe began with such a Big Bang.

7. Calculate the age of the universe by using Hubble's law. Take $H = 22\,\text{km/Mly}$ and assume the expansion rate constant.
Ans. If the galaxy has been moving with a constant speed for time t, then the distance (d) from us today is

$$d = vt$$

But

$$v = Hd$$

$$\therefore \quad t = \frac{d}{v} = \frac{v}{H} \times \frac{1}{v} = \frac{1}{H}$$

$$= \frac{1}{(22\,\text{km/s/Mly})/[3\times 10^5\,\text{km/s/ly/y}]}$$

$$= 13.6\,\text{Gy}$$

These calculations shows that the universe is about 13.6 billion years old provided the expansion rate has not changed. However, this assumption is not quiet correct, but nevertheless it provides a good measure of the age of the universe.

8. How cosmic microwave background radiation (CMBR) provide the strongest evidence yet for the Big Bang?
Ans. The radiation present in the universe can be broadly divided into two categories: (i) those which originate from sources in a definite direction–like say, starlight or radiation from a quasar, and (ii) the radiation which forms a uniform background and not seems to be associated with any definite source in the sky. The first kind of radiation–though of great importance from the point of view of understanding the cosmic structures–is not of direct relevance to cosmology. On the other hand, the uniformly spread out, background radiation is of primary importance to cosmology.

A significant fraction of this background radiation is in the form of '***black body radiation***' with a temperature of about 2.7 K (i.e., the universe behaves as if it is enclosed in a box the walls of which are kept at 2.7 K). Since the peak intensity of blackbody radiation with 2.7 K occurs in the microwave region of electromagnetic spectrum, it is usually called CMBR or CMB, for short. This CMB is the strongest evidence yet for the Big Bang.

The Big Bang theory suggests that the universe started very hot and then cooled as it expanded, doing work against its own gravitation. At first universe was so hot that any atoms that formed would be dashed apart by collisions at the high thermal energy prevailing. Clearly, in its early times the universe was populated by individual charged particles. These interacted readily with electromagnetic radiation, making the universe opaque. But by about 400,000 years the temperature had dropped to several thousand Kelvins, and at that point atoms of hydrogen and helium could form. Since neutral atoms interact much more weakly with electromagnetic radiation, the universe became transparent, and photons emitted as the atoms formed could travel throughout the universe with little chance of being subsequently absorbed. Those photons became the CMBR, permeating the entire universe. There are about 500 photons/cc in the universe today. The total number of CMBR photons within the visible universe.

Expansion of the universe would cool not only matter but also radiation. The Big Bang suggests that radiation from the time the atoms formed should now be cooled to that of a blackbody at 2.7 K – a radiation distribution that peaks in the microwave region of electromagnetic spectrum. Analysis of microwave background shows that it has indeed precisely that blackbody character and temperature (Fig. 28.3). Clearly, the observed CMBR is a direct reflection of conditions some 400,000 year the universe began.

The CMBR is remarkably uniform, but not perfectly so. The temperature of CMBR remains the same in all directions of the sky and all times.

9. How the current abundance of deuterium support strongly to Big Bang?

Ans. The first composite nuclei were the simplest: deuterium consisting of a proton and a neutron. The rate of deuterium formation is critically sensitive to the expansion rate of the universe. Measurements of deuterium abundance, based on spectral lines from intersteller deuterium, therefore provide direct evidence for conditions early in the Big Bang.

10. How the current helium adundance favours strongly to Big Bang?

Ans. Source of energy generation in all stars including our own sun is nuclear fusion reactions. The temperature of the core of the Sun is about 14×10^6 K. Inside sun, through a sequence of reactions, four protons are converted into a ^{4}He nucleus with a net loss of mass. The difference in mass has been converted into radiant energy. The reactions are as follows:

$$p + p \rightarrow {}_1^2d + e^+ + \nu$$

The next two reactions involve the production of γ-rays that one eventually observe as sunlight. The first reaction is

$$d + p \rightarrow {}_2^3\text{He} + \gamma$$

The final reaction is

$${}_2^3\text{He} + {}_2^3\text{He} \rightarrow {}_2^4\text{He} + 2\text{p}$$

The overall reaction can be written as

$$4p \rightarrow {}_2^4He + {}_2e^+ + 2\nu + 26.7\,MeV$$

In a hotter stars there is a sequence that starts with carbon and a proton and ends up with carbon and helium. The end result is always to produce ${}_2^4He$. Further, the radiant energy produced in the cycle is known.

About 25% of the known (or ordinary) mass of the universe is composed of ^4He, the reminder being almost free protons. The synthesis of ^4He depended critically on the ratio of neutrons to protons and the density of matter during early stages of universe. The ^4He adundance has been used to suggest that at most four families of elementary particles are possible. Obviously, the suggestion and data are in good agreement with the current belief from elementary particle data that only three families exist.

11. What do you understand by natural time unit, i.e., Planck time (t_p)?

Ans. Natural time unit or Planck time (t_p) is obtained by constructing a quantity with the dimensions of time out of G, $\hbar$ and c, i.e.,

$$t_P = \sqrt{\frac{\hbar G}{c^5}} \cong 5.4 \times 10^{-44}\ s$$

In the above equation for t_p, G and $\hbar$ appear, tells us that this time has to do with quantum physics and gravity. For $t < t_p$, i.e., for times earlier when the universe packed into a very small space, we do not know to deal with physics, what happened before Planck's time in the creation of the universe. In other words, one can consider Planck's time to be the moment of creation of the universe.

12. What do understand by interstellar medium?

Ans. Intersteller medium contains highly complex molecules. Amino acids (biological molecules) are also expected to be present in space. This has raised the fundamental and intriguing question, whether life is a cosmic phenomenon, namely whether the life on earth had started with such molecules already present in solar nebular material, or these molecules formed on the earth itself in an independent way.

The synthesis of these molecules involve complex chemistry through a network of innumerable reactions involving ions, atoms, molecules, electrons, grains, radiation field, cosmic rays etc., has posed challenging problems. Even to start the network of reactions, the basic ingredient required is the H_2 molecule which is believed to have formed on grain surfaces and then released later. So far the surface properties and the physics of small grains are presently least understood and have let develop into a new area called solid-state astrophysics.

13. What is a galaxy?

Ans. A galaxy is a vast collection of stars, dust, and gas held together by the gravitational force of attraction between its components. Usually, galaxies are classified as elliptical, spiral or irregular in shape. *Elliptical* galaxies appear just like ellipsoidal clouds of stars, with very little internal structure apart from (in some cases) a denser nucleus. *Spiral* galaxies are disc-shaped collections of stars with prominent spiral arms. *Irregular* galaxies have no apparent structure or shape.

Our sun belongs to a spiral galaxy (G) or the *Milky way system*. There are about 10^{11} stars in this system, which is about 30000 parsecs across the maximum thickness at the centre of the galaxy. The galaxies are separated from each other by enormous distances, the nearest large galaxy to our own is *Andromeda galaxy*, being 6.7×10^5 parsecs away.

14. What is a galaxy cluster?

Ans. A group of galaxies containing several hundreds of members extending over a radius of upto a few megaparsecs. We may note that there also exist small group of galaxies, such as the 'Local Group', with a few tens of members. The richest and most regular clusters, e.g. the *Coma cluster*, with thousands of members, are gravitationally bound systems. There is no certainty, whether other less regular and less concentrated clusters are also bound. As well as galaxies, clusters contain hot *intra-cluster gas*, at temperatures between 10^7 and 10^8 K. One can detect this by its x-ray emission. There are also *superclusters* with extents of the order of a hundred megaparsecs, containing about a hundred galaxies. We don't know, whether superclusters are gravitationally bound.

15. What is a neutron star?

Ans. A neutron star is a compact steller object that is supported against collapse under self-gravity by the degeneracy pressure of the neutrons of which it is primarily composed. These stars are believed to be formed as the end products of the evolution of stars of mass greater than a few (4–10) solar masses. The core of the evolved star collapses and (assuming that its mass is greater than the 'Chandrasekhar limit' for a white dwarf), at a very high densities involved ($\sim 10^{14}$ kg/m^3), electrons react with protons in atomic nuclei to produce neutrons. The neutron – rich nuclei thus formed release free neutrons in a process known as *neutron drip*. The density increases to about 10^{17} kg/m^3, at which most of the electrons and protons have been converted to a degenerate gas of neutrons and the atomic nuclei have lost their independent identities. If the mass of the core exceeds the Oppenheimer–Volkoff limit, then further collapse will occur, leading to the formation of a 'black hole'.

16. What is a pulsar?

Ans. Pulsars are believed to be rapidly rotating magnetized neutron stars and several X-ray sources are thought to be neutron stars in binary systems with another star, from which material is drawn into an accretion disc.

17. What is a white hole?

Ans. White hole is a compact steller object that is supported against collapse under self gravity by the degeneracy pressure of electrons white dwarfs are formed as the end products of the evolution of stars of relatively low mass (about that of the sun). These consists of helium nuclei (and carbon and oxygen nuclei in the more massive cases) and a degenerate gas of electrons. The density of typical white dwarf is about 10^9 kg/m^3. Masses and radii of white dwarf are in the region of 0.7 m_{solar} and 10^3 km respectively. There is a maximum mass for white dwarfs, above which they are unstable to gravitational collapse. This is known as Chandrasekhar limit ($\sim 1.4\, m_{solar}$).

18. What is a white hole?

Ans. A region of space - time, that is the time reversal of a 'black hole', i.e., the matter will explode out of a white hole. Although white holes have sometimes been invoked to explain some violent events in the universe, there are strong theoretical arguments against their existence.

19. What are Novale and Supernovale?

Ans. Novale are simply stars that suddenly brighten, become visible, and then fade over some period. Supernova is an explosive brightening of a star in which the energy radiated by it increases by a factor of 10^{10}. It takes several years to fade and while it lasts dominates the whole galaxy in which it lies. It is believed that there are two types of supernovae: (i) type I showing no hydrogen spectral lines and (ii) type II that do. Type has been further put into 1a, 1b and 1c according to other criteria. It is estimated that there could be *supernovale explosion* in the Milky-way every 30-years, although only 6 have actually been observed in the last 1000 years. A supernovale explosion occurs when a star has burnt up all its nuclear fuel and the core collapses catastrophically to form a neutron star.

20. What is a cosmological constant?

Ans. $\Lambda = \frac{8\pi G}{c^4}\rho_v$

is cosmological constant, where ρ_v is the vacuum energy density.

21. What is the age of the universe?

Ans. Usually, the age of the universe is calculated by assuming that the expansion of the universe can be described by Big Bang. Age of the universe determined by the reciprocal of the value of the Hubble constant to be about 13.7 ± 0.2 billion years. There are several methods for the estimation of the age of the universe.

FILL IN THE BLANKS

1. Modern cosmology has its origin in Einstein's 1915 discovery of the ______. **[general theory of relativity and gravitation]**
2. Newton implicitly assumed that on the average the universe is ______ and ______. **[isotropic, homogeneous]**
3. Hubble found that to a good approximation the galaxies were receding from us with a speed ______. **[$v = HR$]**
4. Generally agreed value of Hubble's constant or parameter is very nearly ______. **[22 km/s/Mly]**
5. In accordance with the Hubble's law, the age of the universe is about ______. **[13.6 Gy]**
6. The cosmic microwave background radiation is the strongest evidence yet for the ______. **[Big Bang]**
7. The Big Bang scenario suggests that radiation from the time the atoms formed should now be cooled to that of a black body at ______. **[2.7 K]**
8. Guth's inflation theory of the universe holds that the universe underwent a period of exponential expansion beginning at about ______ and lasting until about ______. **[10^{-35} s, 10^{-32} s]**
9. Recent studies reveal that the universe is overall ______ and that its expansion is accelerating under the influence of mysterious ______. **[flat, dark energy]**
10. If critical density $\rho_c > \rho$, then the universe is ______, if $\rho < \rho_c$, then the universe is ______. **[closed, open]**

OBJECTIVE QUESTIONS

1. Hubble law can be expressed as

(a) $V = HR$ (b) $V^2 = HR/2$

(c) $V = H/R$ (d) $V = 1/HR$

2. Hubble parameter (H) is related to a scale factor a (proportional to the distance between galaxies) by

(a) $H = a\dfrac{d^2a}{dt^2}$ (b) $H = \dfrac{1}{a}\dfrac{da}{dt}$

(c) $H = a/t$ (d) $H = a^2t$

3. The Big Bang model rests on

(a) general theory of relativity only

(b) the cosmological principle, which assumes universe as both isotropic and homogeneous only

(c) (a) and (b) both

(d) none of the above

4. A neutron star is composed mostly of

(a) protons (b) neutrons

(c) electrons (d) none of the above

5. The time that it takes for light to travel across Planck's length is called

(a) Hubble parameter (b) Einstein's time

(c) Planck's time (d) Friedmann time

6. The critical density is defined by

(a) $\rho_c = \dfrac{3H^2}{8\pi G}$ (b) $\rho_c = \dfrac{3H}{8\pi G}$

(c) $\rho_c = H/R$ (d) $\rho_c = \dfrac{8\pi G}{3H^2}$

7. Chandrasekhar limit is

(a) about 1.4 solar masses

(b) about 14 solar masses

(c) about 144 solar masses

(d) none of the above

8. The value of Planck's time is

(a) 1.35×10^{-43} s (b) 1.35×10^{-14} s

(c) 1.35×10^{-8} s (d) 1.35×10^{-76} s

9. The presence of a background blackbody radiation that, having evolved along with the universe, is currently at a temperature around

(a) 1 K (b) 3 K

(c) 6 K (d) 273 K

10. According to Hubble's law, universe is

(a) contracting

(b) expanding

(c) static

(d) sometimes expanding and sometimes contracting

11. The cosmic background radiation decoupled from matter after the Big Bang about

(a) 10^5 yrs (b) 10^9 yrs

(c) 10^3 yrs (d) 10^{12} yrs

12. The cosmological constant can be represented by

(a) $\Lambda = \frac{8\pi G}{c}\rho_v$ (b) $\Lambda = \frac{8\pi G}{c^4}\rho_v$

(c) $\Lambda = \frac{8\pi G}{c^2}\rho_v$ (d) $\Lambda = \frac{8\pi G}{c^3}\rho_v$

ANSWERS

Question Numbers	1	2	3	4	5	6	7	8	9	10	11	12
Answers	(a)	(b)	(c)	(b)	(c)	(a)	(a)	(a)	(b)	(b)	(a)	(b)

SUGGESTED READINGS

1. Kenneth Krane, Modern Physics, John Wiley (2nd ed. 1996).
2. F. K. Richtmyer, E. H. Kennard and John N. Cooper, Introduction to Modern Physics, Mc Graw Hill (6th ed. 1956).
3. Arthur Beiser, Concepts of Modern Physics, Mc Graw Hill, (6th ed., 2003).
4. A. P. French, Special Relativity (Norton, Newyork, 1968).
5. R. Resnick, Introduction to Special Relativity (Wiley, Newyork, 1968).
6. R. Resnick and D. Nalliday, Basic Concepts in Relativity (Mac-Millan, Newyork, 1992).
7. S. L. Kakani, et al. Mechanics, Viva Books, New Delhi (2nd ed. 2011).
8. R. Feynam, et al., The Feynman Lectures on Physics (Addision-Wesely, Reading, 1965).
9. E. H. Wichmann, Quantum Physics, Vol. 4 of Berkeley Physics Course, McGraw Hill (1971).
10. S. L. Kakani, Quantum Physics, Sultan Chand, New Delhi (4th ed. 2004).
11. R. Eisbery and R. Resnic, Quantum Physics, John Wiley (2nd ed. 2002).
12. S. L. Kakani and C. Hemrajani, Solid State Physics (4th ed. 2005), Sultan Chand, New Delhi.
13. S. L. Kakani and Amit Kakani, Material Science, New Age, New Delhi (2004).
14. C. Kittel, Solid State Physics, John Wiley (7th ed., 1995).
15. M. N. Rudden and J. Wilson, Elements of Solid State Physics, John Wiley (2nd ed. 1993).
16. S. L. Kakani and K. C. Bhadari, Electronics, New age International, New Delhi (4th ed., 2011).
17. S. L. Kakani, Superconductivity: Key Problems, Bookman, Jaipur-4.
18. S. L. Kakani, Superconductivity: Current Topics Bookman, Jaipur-4.
19. K. P. Sinha and S. L. Kakani, High Temperature Superconductivity, Nova Science, Newyork (1995).
20. K. P. Sinha and S. L. Kakani, Magnetic Super-conductors, Nova Science, New York (1990).
21. S. L. Kakani, Superconductivity: New Age International (2nd ed. 2012).
22. R. J. Cava, Superconductors beyond 1-2-3, Scientific America 263, 42 (August 1990).
23. Mc Lachlam Jr., Statistical Mechanical Analogies, Prentice Hall, Englewood cliffs, (1968).
24. N. Ashby and S. C. Miller, Principles of Modern Physics, (2nd ed. 2011) San Francisco, Holden Day (1970).
25. C. Kittel, Thermal Physics, Wiley (1969).
26. S. L. Kakani, Thermodynamics and Statistical Physics, (2nd ed. 2011) Sultan Chand, New Delhi.
27. S. L. Kakani, and C. Hemrajani, Mathematical Physics, CBS Publishers, New Delhi-2 (2nd ed. 2003).
28. S. L. Kakani and C. Hemrajani, Electromagnetics, Electrodynamic and Plasma Physics, CBS Publishers, New Delhi-2 (2003).
29. R. D. Reed and R. R. Roy, Statistical Physics for student of Science and Engineering, Scranton, Intext (1971).
30. S. L. Kakani and Shubhra Kakani, Engineering Physics CBS Publishers, New Delhi-2 (2009).

31. A. Beiser, Perspectives of Modern Physics, Mc Graw Hill (1969).
32. E. R. Harrison, Cosmology, Cambridge University Press (1981).
33. J. V. Narlikar, Introduction to Cosmology, Cambridge University Press (1993).
34. W. J. Kaufaman, III, Relativity and Cosmology, Newyork (1977).
35. J. C. Polking horne, The Particle Play (Oxford, W. H. Freeman, 1981).
36. S. Weinberg, Dreams of Final Theory, Pantheon, Newyork (1992).
37. N. Semat and J. R. Albright, Introduction to Atomic and Nuclear Physics, Rinchart and Winston, Newyork, 1972.
38. B. L. Cohen, Concepts of Nuclear Physics, Mc Graw Hill, (1971).
39. R. D. Evans, The Atomic Nucleus, Mc Graw Hill, (1971).
40. I. Kaplan, Nuclear Physics, Addison-Wesley, 1962.
41. K. S. Karne, Introductory Nuclear Physics, Wiley, Newyork (1987).
42. J. W. Rohlf, Modern Physics from $\propto$ to Z^o, John Wiley and sons, Inc. (1994).
43. H. Haken and H.C. Wolf, Atomic and Quantum Physics, Springer-Verlog. World Publishing Corporation (1989).
44. G. Gamow and R. Stannard, The New World of Mr. Tompkins: Cambridge University Press (1999).
45. S. Gasiorowics, Quantum Physics, 3rd ed., John Wiley (2005).
46. D. Lu, University Physics, Higher Education Press, China. (2003).
47. C. Kittel, Solid State Physics (8th ed. 2005), John Wiley.
48. N. Zettili, Quantum Mechanics (2nd ed. 2009), John Wiley.

SUBJECT INDEX

E

F

G

H

I

J

K

L

R

S

T

U

V

W

X

Z